FLORA EUROPAEA

FLORA EUROPAEA

VOLUME 2

ROSACEAE TO UMBELLIFERAE

EDITED BY

T. G. TUTIN   V. H. HEYWOOD

N. A. BURGES   D. M. MOORE   D. H. VALENTINE

S. M. WALTERS   D. A. WEBB

WITH THE ASSISTANCE OF

P. W. BALL   A. O. CHATER   I. K. FERGUSON

CAMBRIDGE

AT THE UNIVERSITY PRESS

1968

Published by the Syndics of the Cambridge University Press
Bentley House, 200 Euston Road, London, N.W.1
American Branch: 32 East 57th Street, New York, N.Y.10022

Library of Congress Catalogue Card Number: 64–24315
Standard Book Number: 521 06662 X

Printed in Great Britain
at the University Printing House, Cambridge
(Brooke Crutchley, University Printer)

CONTENTS

THE FLORA EUROPAEA ORGANIZATION

vii

Denmark	A. HANSEN, København
Finland	J. JALAS, Helsinki
France	P. DUPONT, Nantes
	P. JOVET, Paris
	R. DE VILMORIN, Orsay
Germany	H. MERXMÜLLER, München
	K. WERNER, Halle
Greece	K. H. RECHINGER, Wien
Hungary	R. SOÓ, Budapest
	Z. E. KÁRPÁTI, Budapest
Iceland	E. EINARSSON, Reykjavik
Italy	G. MOGGI, Firenze
	R. E. G. PICHI-SERMOLLI, Genova
Jugoslavia	E. MAYER, Ljubljana
	V. BLEČIC, Beograd
Netherlands	S. J. VAN OOSTSTROOM, Leiden
Norway	R. NORDHAGEN, Oslo
Poland	B. PAWŁOWSKI, Kraków
Portugal	A. R. PINTO DA SILVA, Oeiras
	J. DO AMARAL FRANCO, Lisboa
	(assisted by
	MARIA DA LUZ DA ROCHA AFONSO)
Romania	A. BORZA, Cluj
	†E. I. NYÁRÁDY, Cluj
Spain	E. GUINEA, Madrid
	E. FERNÁNDEZ GALIANO, Sevilla
	O. DE BOLÓS, Barcelona
Sweden	N. HYLANDER, Uppsala
Switzerland	E. LANDOLT, Zürich
Turkey	P. H. DAVIS, Edinburgh
Russia	T. V. EGOROVA, Leningrad
	A. I. POJARKOVA, Leningrad
	V. N. TIKHOMIROV, Moskva

TECHNICAL CONSULTANT
Á. LÖVE, Boulder

GEOGRAPHICAL ADVISER
H. MEUSEL, Halle

The above lists refer to the Organization as it was constituted during the preparation of volume 2, which was completed in February 1967

LIST OF CONTRIBUTORS TO VOLUME 2

The following is a list of authors who have contributed accounts of genera or parts of them.

P. W. BALL, Liverpool
B. BAUM, Ottawa
K. BROWICZ, Kórnik, Poland
R. K. BRUMMITT, Kew
N. A. BURGES, Coleraine
M. S. CAMPBELL, Genève
J. F. M. CANNON, London
A. O. CHATER, Leicester
C. D. K. COOK, Liverpool
D. E. COOMBE, Cambridge
J. CULLEN, Liverpool
D. M. DALBY, London
J. DOSTÁL, Olomouc
T. T. ELKINGTON, Sheffield
I. K. FERGUSON, London
R. FERNANDES, Coimbra
J. DO AMARAL FRANCO, Lisboa
D. FRODIN, Cambridge
W. GAJEWSKI, Warszawa
P. E. GIBBS, St Andrews
E. GUINEA, Madrid
A. HANSEN, København
Y. HESLOP-HARRISON, Madison
V. H. HEYWOOD, Liverpool
R. B. IVIMEY-COOK, Exeter
Z. E. KÁRPÁTI, Budapest
I. KLÁSTERSKÝ, Průhonice u Prahy

P. LEINS, München
H. MERXMÜLLER, München
D. M. MOORE, Leicester
J. MCNEILL, Liverpool
G. NORDBORG, Lund
D. J. OCKENDON, Cambridge
B. PAWŁOWSKI, Kraków
A. R. PINTO DA SILVA, Oeiras
M. C. F. PROCTOR, Exeter
P. H. RAVEN, Stanford
N. K. B. ROBSON, London
A. SCHMIDT, Hamburg
V. SKALICKÝ, Praha
A. R. SMITH, Kew
A. TERPÓ, Budapest
C. C. TOWNSEND, Kew
T. G. TUTIN, Leicester
D. H. VALENTINE, Manchester
S. M. WALTERS, Cambridge
†E. F. WARBURG, Oxford
D. A. WEBB, Dublin
P. F. YEO, Cambridge
D. P. YOUNG, Sanderstead, United Kingdom
A. CHRTKOVÁ-ŽERTOVA, Průhonice u Prahy

PREFACE

The development of the Flora Europaea project was outlined in the Preface to Volume 1, and it is not necessary to recapitulate it here. It is sufficient to remind the reader of the successful publication of Volume 1, and the fulfilment of the promise to proceed as quickly as possible with Volume 2. That it has been possible to produce it in less than four years is gratifying to the Editorial Committee, and is a tribute to the unstinted collaboration of our advisers and friends in every part of Europe. We must express our gratitude here to our advisory editors and regional advisers, both old and new. The tradition of biennial Symposia has been maintained with meetings in Denmark in 1965, and in Spain in 1967.

Since the publication of Volume 1, the Editorial Committee has been strengthened by the addition of Dr D. M. Moore; and Dr I. K. Ferguson has been appointed as a third research assistant.

We again record our deep gratitude to the United Kingdom Science Research Council, whose continuing financial support has made it possible to maintain our organization and secretariat, and also to appoint from 1966 the additional Research Assistant. We have also been able to invite, for periods of three months, a number of visiting bursars, who have worked at British Universities and research institutes. Each has prepared an account of a particular genus or group of species for Volume 3. Visitors to date have been Dr A. Jasiewicz (Kraków), Dr S. Kožuharov (Sofija) and Dr J. Holub (Průhonice). The total amount of the Science Research Council grant for the period 1965–8 is £34,000.

In addition to this main grant, the project has received direct and indirect financial help from several countries, among which should be mentioned The Royal Society, London, the Danish Botanical Society and the Spanish Higher Council for Scientific Research. We gratefully acknowledge the continuing sponsorship of our project by the Linnean Society of London. A committee has been set up jointly by the Society and the Flora Europaea Organizing Committee to administer, in the interests of taxonomic research on the flora of Europe, a trust fund arising from the royalties from the Flora.

The British Museum (Natural History) has provided accommodation for Dr Ferguson; our special thanks are due to the Keeper of Botany and his staff for this and for many other favours. We are also grateful to the Director and Staff of the Herbarium and Library, Royal Botanic Gardens, Kew, for much help, willingly given; and to institutions abroad, notably the Naturhistorisches Museum, Wien and the Komarov Botanical Institute of the Academy of Sciences, Leningrad. We should like to mention specially Mr J. E. Dandy, one of our Advisory Editors, who has played an increasingly important part as adviser on nomenclature. In addition, many botanists, not formally associated with our organization, have helped us in various ways, notably H. Runemark (Lund), S. Kožuharov (Sofija) and the late N. Y. Sandwith (Kew). P. D. Sell (Cambridge) undertook the laborious task of preparing the index for the press.

As in Volume 1, the names of the authors primarily responsible for writing the accounts of families and genera are given in footnotes to the text. It should, however, be made clear that the Editorial Committee takes full responsibility for the form in which the text is published.

Acknowledgements are due to the Universities of Cambridge, Dublin, Durham,

Leicester, Liverpool and Manchester for their support in making facilities available to members of the Editorial Committee and their assistants. In particular, the University of Liverpool has continued to provide accommodation in the Hartley Botanical Laboratories for the Secretariat. We should in conclusion express our gratitude to the staff of the Secretariat, Mrs J. Beck, Mrs M. E. Donnelly and Mrs M. L. Pollard for their continuing efficiency and loyalty.

INTRODUCTION

The aim of the Flora is in general diagnostic, and the descriptions, while brief, are as far as possible comparable for related species. The Floras listed on pp. xvii–xix, and the monographs or revisions given when appropriate after the descriptions of families and genera, may assist the reader in obtaining more detailed information. Other references to published work are occasionally given in cases of special taxonomic difficulty.

All available evidence, morphological, geographical, ecological and cytogenetical, has been taken into consideration in delimiting species and subspecies, but they are in all cases definable in morphological terms. (Taxa below the rank of subspecies are not normally included.) The delimitation of genera is often controversial and the solution adopted in the Flora may be a somewhat arbitrary choice between conflicting opinions. We have endeavoured to weigh as fairly as possible the various opinions available, but there has been no consistent policy of 'lumping' or 'splitting' genera (or, for that matter, species). The order and circumscription of the families is that of Melchior in *Engler-Diels*, *Syllabus der Pflanzenfamilien* ed. 12 (1964). Since, however, this edition of the *Syllabus* did not appear until Volume 1 of the Flora had gone to press, there are some small discrepancies between the two with regard to the sequence of families. In particular, the Cactaceae and Guttiferae, which should have been in Volume 1, have been inserted in Volume 2.

All descriptions of taxa refer only to their representatives in Europe. In practice, we have relaxed this rule slightly for families and genera to avoid giving taxonomically misleading information, particularly in those cases where a large family or genus has only one or few, somewhat atypical, members in Europe. In such cases we have occasionally added 'in European members' or a similar phrase to emphasize the atypical representation. It should, however, never be assumed that the description is valid for all non-European taxa.

For the purpose of this Flora, we have tried as far as possible to interpret Europe in its traditional sense. The area covered is shown on the maps at the end of the volume.

Place-names used in the summaries of geographical distribution have been given in their English form when they refer to independent states (including the constituent republics of the U.S.S.R.) or to such geographical features of Europe as transcend national boundaries. All other place-names are given in the language of the country concerned. Thus we write *Sweden, Ukraine, Danube, Alps, Mediterranean* but *Corse, Kriti, Slovenija, Rodopi Planina, Ahvenanmaa*.

In *transliteration* from Cyrillic characters we have followed the ISO system recommended in the UNESCO *Bulletin for Libraries* **10**: 137 (1956) for place-names and titles of journals. With personal names, however, we have followed the list of transliterations given in the index-volume (1962) to *Not. Syst.* (*Leningrad*), and have transliterated personal names which do not occur in this list according to the conventions used there.

In transliterating place-names from Greek characters, we have, except for omitting the accents, followed *The Times Atlas of the World*, Mid-Century Edition, vol. 4 (London, 1956).

On pp. xvii–xix, we give a list of *Basic and Standard Floras*. The reason behind the choice of these Floras was not made clear in Volume 1. Basic Floras have been chosen as widely known Floras covering large or important parts of Europe. Standard Floras are considered to represent those Floras in current use and likely to be familiar to a large number of people in the particular country concerned; the list has been revised since the publication of Volume 1.

Synonyms, whether full or partial, are given in parentheses in the text only when they are used in one of the Basic Floras or when they are necessary to prevent confusion. (For primarily Iberian and Mediterranean species, synonyms used in the *Prodromus* of Willkomm & Lange, and the *Supplementum* by Willkomm (p. xix) are also included.) Synonyms (or the basionym) are also usually given in the text when the combination has not previously been used in a Flora or monograph, or when the nomenclature is otherwise unfamiliar or in need of explanation. Otherwise, synonyms are given in the Index only; but it is important to note that no attempt has been made to give a complete synonymy. Even at the binomial level, the number of names for European plants is four or five times the number of accepted species, and to include all these would be impracticable. Thus, in addition to the binomials in the text, the Index contains all synonyms at specific rank which are used in the Basic and Standard Floras, or in cited monographs, with an indication of the species in the text under which they have been relegated to synonymy. Some subspecific names also appear in the Index. In this way, we hope that users of any Basic or Standard Flora will be able to relate the names used in their own Floras to those in *Flora Europaea*. In cases where the name of a familiar species has been changed, an explanation of this is usually published as a Notula (see p. xvi).

Citations have been abbreviated, and the abbreviations used for authors and places of publication have been standardized; lists of these abbreviations are given in Appendices I, II and III. These lists apply only to the abbreviations used in Volume 2.

Species descriptions attempt to give, within the limits of length set by the Flora, both the diagnostic characters of the plant and a general idea of its appearance. Where dimensions are given, a measurement without qualification refers to length. Two measurements connected by × indicate length followed by width. Further measurements in parentheses indicate exceptional sizes outside the normal ranges. In order to save space and facilitate identification, descriptions may sometimes take the form of a comparison with another description. The conventional way of setting this out is, to give an example (p. 41):

16. Potentilla chamissonis Hultén...Like **15** but...

This implies that the description with which it is being compared (in this example **15. P. nivea** L.) applies to this taxon but for the differences noted. It does not necessarily mean that the two taxa are similar in general appearance. Additional descriptive information is sometimes also given, but in separate sentences.

The *diploid chromosome number* ($2n =$) is given where it has been possible to verify that the count was made on material of known wild European origin. For naturalized and cultivated species, the count is from material which is naturalized or is cultivated in the way which justifies its inclusion in the Flora. It is hoped to publish separately a list of references to the data on which the published numbers are based.

Ecological information is given sparingly, and only where the ecological characteristics of a species are clearly and concisely definable for its total European range. Sometimes a general statement, applicable to a whole genus or to a group of species, is made. There is an inevitable irregularity of treatment, as in a great many cases reliable ecological information is not available.

The description of each species is followed by an indication of its *distribution within*

Europe. This falls into two parts: (1) a summary in a short phrase; (2) a list of abbreviations of 'territories' in which the species occurs. The summary phrase makes use of everyday geographical phrases and concepts such as 'W. Europe', 'the Mediterranean region', 'the Balkan peninsula', etc. Maps IV and V and the legends accompanying them indicate the interpretation which is to be put on these phrases. We would emphasize that they are to be interpreted in a simple geographical sense, and do not attempt in any way to divide Europe phytogeographically.

Species believed to be endemic to Europe are distinguished by a symbol (●) before the summary of geographical distribution.

A more precise indication of distribution is given by the enumeration of the 'territories' (indicated by a two-letter abbreviation) in which the plant is believed to occur. The limits of these territories follow, with very few exceptions, existing political boundaries (see Map I). The territories, of course, vary greatly in size, and Ga, Hs or Ju gives very much less information than does Fa, Rs(K) or Tu. In all cases, however, the lists provide a guide to which national Floras should be searched for further detailed information, whether on taxonomy or on distribution. Occasionally, the list of territories is followed by a brief indication, in parentheses, of extra-European distribution. This is done only for plants of which the European range is but a small fraction of the total and for species not native in Europe.

In general the only infraspecific taxa described and keyed in the Flora are subspecies. Any formal treatment of variation below the level of subspecies would have been impossible in a Flora of this kind; the known variation of taxa is, however, covered in the descriptions. No 'experimental' categories, such as ecotypes, are used in the Flora in a formal systematic sense, though they are sometimes mentioned in notes.

Where it is difficult to distinguish between a number of closely similar species in a genus, an *ad hoc* 'group' has been made, and these groups, not the individual species, are keyed out in the main species key. They will serve for at least a partial identification. Following the description of a group in the text, a key to the component species is given, and they are then numbered and described, so that a more detailed study, or the availability of more adequate material, may enable the user to take the identification further. For example, in *Potentilla* there is the *P. argentea* group, which comprises the species *P. argentea* L., *P. calabra* Ten. and *P. neglecta* Baumg. Such groups have no nomenclatural status.

For inbreeding and apomictic groups, other *ad hoc* treatments have been devised. In Volume 2, the main problems have arisen in the Rosaceae; the methods used to overcome them are described in the notes following the descriptions of that family and the genera concerned.

Only those few *hybrids* which reproduce vegetatively and are frequent over a reasonably large area (e.g. *Circaea × intermedia*) are described and keyed as for species. Other common hybrids may be mentioned individually in notes (e.g. in *Viola*), or collectively for the whole genus (e.g. in *Epilobium*).

We have attempted to include the following categories of *alien species*:

(i) Aliens which are effectively naturalized. These include garden plants which have escaped to situations not immediately adjacent to those in which they are cultivated, as well as weeds and other plants which have been accidentally introduced; provided, in both cases, that the plant has been established in a single station for at least 25 years, or is reported as naturalized in a number of widely separated localities.

(ii) Trees or crop-plants which are planted or cultivated in continuous stands on a fairly extensive scale.

Casual aliens, i.e. those which do not persist without repeated re-introduction, are not included unless they have often been mistaken for a native or established species, or are for any other reason of special interest. In assessing the status of a species in any part of Europe we have, however, been dependent very largely on the information contained in the national Floras, and it is clear that the criteria used by different authors vary widely. All data on native, naturalized or casual status relating to synanthropic plants must, therefore, be regarded only as approximate.

It is the policy of the Committee not to publish new names in the Flora itself. To deal with the publication of much of this material, an arrangement has been made with the Editorial Board of *Feddes Repertorium*, by which taxonomic and nomenclatural notes are being published as part of a series entitled *Notulae Systematicae ad Floram Europaeam spectantes*; the first six parts of these have already appeared.

LISTS OF BASIC
AND STANDARD FLORAS

BASIC FLORAS

COSTE, H. *Flore descriptive et illustrée de la France, de la Corse et des Contrées limitrophes.* Vols. 1–3. Paris, 1900–1906.

HAYEK, A. VON. *Prodromus Florae Peninsulae balcanicae.* (In *Feddes Repert. (Beih.)* 30.) Vols. 1–3. Berlin-Dahlem, 1924–1933.

HEGI, G. *Illustrierte Flora von Mittel-Europa,* ed. 1. Vols. 1–7. München, 1906–1931. Ed. 2. Vols. 1– . München, 1936– .

HYLANDER, N. *Nordisk Kärlväxtflora.* Vols. 1– . Stockholm, 1953– .

KOMAROV, V. L. *et al.* (ed.). *Flora URSS.* Vols. 1–30. Leningrad & Moskva, 1934–1964.

STANDARD FLORAS

ARCANGELI, G. *Compendio della Flora italiana,* ed. 1. Torino, 1882.

BARCELÓ Y COMBIS, F. *Flora de las Islas Baleares.* Palma de Mallorca, 1879–1881.

BECK VON MANNAGETTA, G. *Flora Bosne, Hercegovine i Novipazarskog Sandžaka.* Vols. 1–4(1). Beograd & Sarajevo, 1903–1950.

BINZ, A. *Schul- und Exkursionsflora für die Schweiz,* ed. 11 by A. Becherer. Basel, 1964.

BINZ, A. & THOMMEN, E. *Flore de la Suisse,* ed. 2. Lausanne, 1953.

BOISSIER, E. *Flora orientalis.* Vols. 1–5. Genève, Bâle and Lyon, 1867–1884. *Supplementum.* 1888.

BORZA, A. *Conspectus Florae Romaniae.* Cluj, 1947–1949.

BRIQUET, J. *Prodrome de la Flore corse.* Vols. 1–3, Genève, Bâle, Lyon and Paris, 1910–1955.

CADEVALL I DIARS, J. *Flora de Catalunya.* Vols. 1–6. Barcelona, 1913–1937.

CLAPHAM, A. R., TUTIN, T. G. & WARBURG, E. F. *Flora of the British Isles,* ed. 2. Cambridge, 1962.

COUTINHO, A. X. PEREIRA. *Flora de Portugal,* ed. 2 by R. T. Palhinha. Lisboa, 1939.

DEGEN, A. VON. *Flora velebitica.* Vols. 1–4. Budapest, 1936–1938.

DIAPOULIS, K. A. *Ellenike Khloris.* Vols. 1–3. Athenai, 1939–1949.

DOMAC, R. *Flora za odredivanje i upoznavanje Bilja.* Zagreb, 1950.

DOSTÁL, J. *Květena ČSR.* Praha, 1948–1950.

FIORI, A. *Nuova Flora analitica d'Italia.* Vols. 1–2. Firenze, 1923–1929.

FIORI, A. & PAOLETTI, G. *Iconographia Florae italicae,* ed. 3. San Casciano, Val di Pesa, 1933.

FOMIN, A. V. *et al.* (ed.). *Flora RSS Ucr.,* ed. 1. Vols. 1–12. Kijiv, 1936–1965. Ed. 2. Vol. 1. Kijiv, 1938.

FOURNIER, P. *Les quatre Flores de la France, Corse comprise.* Poinsin-les-Grancey, 1934–1940. (Reprints with additions and corrections, Paris, 1946 and 1961.)

FRITSCH, K. *Exkursionsflora für Österreich und die ehemals österreichischen Nachbargebiete,* ed. 3. Wien and Leipzig, 1922.

GEIDEMAN, T. S. *Opredelitel' Rastenij Moldavskoj SSR.* Moskva and Leningrad, 1954.

GOFFART, J. *Nouveau Manuel de la Flore de Belgique et des Régions limitrophes,* ed. 3. Liège, 1945.

GORODKOV, B. N. & POJARKOVA, A. I. (ed.). *Flora Murmanskoj Oblasti.* Vols. 1– . Moskva & Leningrad, 1953– .

HALÁCSY, E. VON. *Conspectus Florae graecae.* Vols. 1–3. Leipzig, 1900–1904. *Supplementum* 1. Leipzig, 1908. *Supplementum* 2 (in *Magyar Bot. Lapok* 11). Budapest, 1912.

HEUKELS, H. *Flora van Nederland,* ed. 15 by S. J. van Ooststroom. Groningen, 1962.

HIITONEN, H. I. A. *Suomen Kasvio.* Helsinki, 1933.

HYLANDER, N. *Förteckning over Nordens Växter.* 1. *Kärlvaxter.* Lund, 1955. *Tillägg och Rättelser* (in *Bot. Not.* 112). Lund, 1959.

JANCHEN, E. *Catalogus Florae Austriae.* Vol. 1. Wien, 1956–1960. *Ergänzungsheft.* Wien, 1963. *Zweites Ergänzungsheft.* Wien, 1964. *Drittes Ergänzungsheft.* Wien, 1966.

JORDANOV, D. (ed.). *Flora na Narodna Republika Bălgarija.* Vol. 1– . Sofija, 1963– .

KNOCHE, H. *Flora balearica.* Vols. 1–4. Montpellier, 1921–1923.

LID, J. *Norsk og svensk Flora.* Oslo, 1963.
—— *The Flora of Jan Mayen.* Oslo, 1964.

LINDMAN, C. A. M. *Svensk Fanerogamflora,* ed. 2. Stockholm, 1926.

LÖVE, Á. *Íslenzkar Jurtir.* København, 1945.

MAEVSKIJ, P. F. *Flora srednej Polosy evropejskoj Časti SSSR,* ed. 9 by B. K. Schischkin. Leningrad, 1964.

MAYER, E. *Seznam praprotnic in Cvetnic Slovenskega Ozemlja* (in *Razpr. Mat.-Prir. Akad. Ljubljani. Dela* 5. *Inšt. Biol.* 3). Ljubljana, 1952.

MERINO Y ROMÁN, P. B. *Flora descriptiva é illustrada de Galicia.* Vols. 1–3. Santiago, 1905–1909.

NORDHAGEN, R. *Norsk Flora.* Oslo, 1940.

NYMAN, C. F. *Conspectus Florae europaeae.* Örebro, 1878–1882. *Supplementum* 1, 1883–1884. *Additamenta,* 1886. *Supplementum* 2, 1889–1890.

OSTENFELD, C. E. H. & GRÖNTVED, J. *The Flora of Iceland and the Faeroes.* København, 1934.

PALHINHA, R. T. *Catálogo das Plantas vasculares dos Açores.* Lisboa, 1966.

RACIBORSKI, M., SZAFER, W. & PAWŁOWSKI, B. (ed.). *Flora polska.* Vols. 1– . Kraków and Warszawa, 1919– .

RASMUSSEN, R. *Föroya Flora,* ed. 2. Tórshavn, 1952.

RAUNKIÆR, C. *Dansk Ekskursions-Flora,* ed. 7 by K. Wiinstedt. København, 1950.

RECHINGER, K. H. *Flora aegaea* (in *Denkschr. Akad. Wiss. Math.-Nat. Kl. (Wien)* 105(1)). Wien, 1943. *Supplementum* (in *Phyton (Austria)* 1). Horn, 1949.

ROBYNS, W. (ed.). *Flore Générale de Belgique. Spermatophytes.* Vols. 1– . Bruxelles, 1952– .

ROHLENA, J. *Conspectus Florae montenegrinae* (in *Preslia* 20 and 21). Praha, 1942.

RØNNING, O. I. *Svalbards Flora.* Oslo, 1964.

ROSTRUP, F. G. E. *Den danske Flora,* ed. 19 by C. A. Jørgensen. København, 1961.

ROTHMALER, W. *Exkursionsflora von Deutschland.* 2: *Gefässpflanzen.* Berlin, 1962. 4: *Kritischer Ergänzungsband: Gefässpflanzen.* Berlin, 1963.

ROUY, G. C. C. *Conspectus de la Flore de France.* Paris, 1927.

ROUY, G. C. C. *et al. Flore de France.* Vols. 1–14. Asnières, Paris and Rochefort, 1893–1913.

SAMPAIO, G. A. DE SILVA FERREIRA. *Flora portuguesa,* ed. 2 by A. Pires de Lima. Porto, 1947.

SĂVULESCU, T. (ed.). *Flora Republicii Populare Române.* Vols. 1– . Bucureşti, 1952– .

SCHMEIL, O. & FITSCHEN, J. *Flora von Deutschland,* ed. 67/68 by H. Voerkel & G. Müller. Jena, 1957.

SOÓ, R. DE & JÁVORKA, S. *A Magyar Növényvilág Kézikönyve.* Vols. 1–2. Budapest, 1951.

STANKOV, S. S. & TALIEV, V. I. *Opredelitel' vysših Rastenij Evropejskoj Časti SSSR*, ed. 2. Moskva, 1957.

STEFÁNSSON, S. *Flóra Íslands*, ed. 3 by S. Steindórsson. Akureyri, 1948.

STOJANOV, N., STEFANOV, B. & KITANOV, B. *Flora na Bălgarija*. Vols. 1–2, ed. 4. Sofija, 1966–67.

SZAFER, W., KULCZYŃSKI, S. & PAWŁOWSKI, B. *Rośliny polskie*. Warszawa, 1953.

TRELEASE, W. *Botanical Observations on the Azores* (in *Ann. Rep. Missouri Bot. Gard.* **8**). St Louis, 1897.

WEBB, D. A. *An Irish Flora*, ed. 5. Dundalk, 1967.

WEBB, D. A. *The Flora of European Turkey* (in *Proc. R. Irish Acad.* **65**B: 1–100). Dublin, 1966.

WEEVERS, T. *et al.* (ed.). *Flora neerlandica*. Vols. 1– . Amsterdam, 1948– .

WILLKOMM, H. M. *Supplementum Prodromi Florae hispanicae*. Stuttgart, 1893.

WILLKOMM, H. M. & LANGE, J. *Prodromus Florae hispanicae*. Vols. 1–3. Stuttgart, 1861–1880.

WULF, E. V. *Flora Kryma*. Vols. 1– . Yalta, Leningrad and Moskva, 1927– .

ZEROV, D. K. *et al.* (ed.). *Viznačnik Roslin Ukrajini*, ed. 2. Kijiv, 1965.

SYNOPSIS OF FAMILIES

KEY TO FAMILIES OF ANGIOSPERMAE

This key covers all the families of Angiospermum in volumes 1 and 2 and the great majority of those in volumes 3–4, though some introduced families and, doubtless, some anomalous genera, have been omitted. A comprehensive key will be included in volume 4

1 Plant free-floating on or below surface of water, not rooted in mud
  2 Plant with small bladders on leaves or on apparently leafless stems; leaves divided into filiform segments   **Lentibulariaceae**
  2 Not as above
    3 Plant without obvious differentiation into stems and leaves   **Lemnaceae**
    3 Plant with obvious stems and leaves
      4 Leaves with a cuneate basal part, 4–6 setaceous segments and a terminal orbicular lobe   **LXXI. Droseraceae**
      4 Leaves not as above
        5 Floating leaves sessile   **Hydrocharitaceae**
        5 Floating leaves long-petiolate
          6 Floating leaves orbicular, entire   **Hydrocharitaceae**
          6 Floating leaves rhombic, dentate in upper ⅔   **CXX. Trapaceae**
1 Land-plants or aquatics rooted in mud
  7 2- to 4-fid coloured staminodes present inside the sepals; leaves often fasciculate   **LIII. Molluginaceae**
  7 Not as above
    8 Perianth of 2 (rarely more) whorls differing markedly from each other in shape, size or colour
      9 Petals not all united into a tube at base, very rarely cohering at apex, or else flowers papilionate
        10 Ovary superior
          11 Carpels 2 or more, free, or united at the base only
            12 Sepals and petals 3
              13 Carpels more than 3
                14 Leaves lobed   **LXI. Ranunculaceae**
                14 Leaves entire   **Alismataceae**
              13 Carpels 3
                15 Leaves palmately divided; petioles spiny   **Palmae**
                15 Leaves simple, sessile   **LXXII. Crassulaceae**
            12 Sepals or petals more than 3
              16 Flowers zygomorphic; petals deeply divided   **LXIX. Resedaceae**
              16 Flowers actinomorphic; petals entire
                17 Stamens more than twice as many as petals
                  18 Shrubs or herbs with stipulate leaves; flowers perigynous   **LXXX. Rosaceae**
                  18 Herbs; stipules 0, though leaf-bases sometimes sheathing; flowers hypogynous
                    19 Fruit a head of achenes; sepals deciduous   **LXI. Ranunculaceae**
                    19 Fruit of 2–5 follicles; sepals persistent   **LXII. Paeoniaceae**
                17 Stamens not more than twice as many as petals
                  20 Leaves 3-foliolate   **LXXX. Rosaceae**
                  20 Leaves simple
                    21 Carpels spirally arranged on an elongated receptacle   **LXI. Ranunculaceae**
                    21 Carpels in 1 whorl
                      22 Trees with palmately lobed leaves; flowers in globose capitula   **LXXIX. Platanaceae**
                      22 Herbs or shrubs; leaves not palmately lobed; flowers not in globose capitula
                        23 Herbs or dwarf shrubs with terete stems; leaves ± succulent   **LXXII. Crassulaceae**
                        23 Shrubs with angular stems; leaves not succulent   **XCIII. Coriariaceae**
          11 Carpels obviously united for *c.* ½ their length or more, or carpel solitary
            24 Flowers actinomorphic
               25 Corona of long filaments present inside the petals   **CXI. Passifloraceae**

               25 Flowers without a corona
                26 Petals more than 10
                  27 Aquatic herbs with petiolate leaves
                    28 Leaves floating, usually with a deep basal sinus   **LVIII. Nymphaeaceae**
                    28 Leaves not floating, peltate   **LIX. Nelumbonaceae**
                  27 Terrestrial herbs or shrubs with sessile or subsessile leaves
                    29 Stamens 4–6   **LXIII. Berberidaceae**
                    29 Stamens numerous   **LII. Aizoaceae**
               26 Petals fewer than 10
                30 Stamens more than twice as many as petals
                  31 Stamens with their filaments united into a tube   **CVI. Malvaceae**
                  31 Stamens free or united into bundles
                    32 Perianth-segments persistent in fruit, 2 large and 2 small   **XLVII. Polygonaceae**
                    32 Perianth-segments not as above
                      33 Ovary on a long gynophore   **LXVII. Capparaceae**
                      33 Ovary sessile or nearly so
                        34 Ovary surrounded by a cup-shaped perigynous zone; ovule 1   **LXXX. Rosaceae**
                        34 No cup-shaped perigynous zone; ovules 2 or more
                          35 Leaves 2-pinnate or simple phyllodes present   **LXXXI. Leguminosae**
                          35 Leaves not as above
                            36 Carpel 1; leaves 2-ternate, lower leaflets stalked   **LXI. Ranunculaceae**
                            36 Carpels 2 or more; leaves not as above
                              37 Large trees; inflorescence with a conspicuous bract partly adnate to peduncle   **CV. Tiliaceae**
                              37 Not as above
                                38 Styles more than 1, free
                                  39 All or most leaves alternate; outer perianth-segments petaloid   **LXI. Ranunculaceae**
                                  39 All leaves opposite or verticillate; outer perianth-segments sepaloid   **CIX. Guttiferae**
                              38 Style 1 or 0
                                40 Petals 4   **LXVI. Papaveraceae**
                                40 Petals 5
                                    41 Ovary 1-locular or septate at base only; stamens numerous   **CXII. Cistaceae**
                                  41 Ovary 3-locular; stamens 15   **LXXXV. Zygophyllaceae**
          30 Stamens not more than twice as many as petals
            42 Trees, shrubs or woody climbers
                43 Flowers on tough leaf-like cladodes; leaves scale-like, brownish   **Liliaceae**
                43 Not as above
                  44 Leaves small, scale-like or ericoid
                    45 Perianth-segments in 2 whorls of 3; stamens 3   **Empetraceae**
                    45 Perianth-segments and stamens more than 3 in a whorl
                      46 Leaves opposite   **CXIV. Frankeniaceae**
                      46 Leaves alternate   **CXIII. Tamaricaceae**
                44 Leaves neither scale-like nor ericoid
                  47 Peduncles adnate to petioles; ovary on a short gynophore   **LXXXIX. Cneoraceae**

47　Not as above
　48　All leaves opposite
　　49　Leaves pinnate
　　　50　Shrubs; fruit a capsule　　　CI. **Staphyleaceae**
　　　50　Tree; fruit of 2 single-seeded samaras
　　　　　　　　　　　XCV. **Aceraceae**
　　49　Leaves entire or palmately lobed
　　　51　Fruit of 2 single-seeded samaras; leaves
　　　　　usually palmately lobed　XCV. **Aceraceae**
　　　51　Fruit a fleshy capsule; leaves not palmately
　　　　　lobed　　　　　　　C. **Celastraceae**
　48　At least some leaves alternate
　　52　Stamens 6　　　　　LXVIII. **Cruciferae**
　　52　Stamens 4, 5, 10 or 12
　　　53　Stamens 4 or 5
　　　　54　Stamens opposite petals
　　　　　55　Shrubs or small trees; petals shorter than
　　　　　　　sepals　　　　CIII. **Rhamnaceae**
　　　　　55　Woody climbers; petals longer than
　　　　　　　sepals　　　　CIV. **Vitaceae**
　　　　54　Stamens alternating with petals
　　　　　56　Bark resinous; ovule 1
　　　　　　　　　　XCIV. **Anacardiaceae**
　　　　　56　Bark not resinous; ovules several
　　　　　　　　　LXXVIII. **Pittosporaceae**
　　　53　Stamens 10 or 12
　　　　57　Leaves entire　　　　　**Ericaceae**
　　　　57　Leaves pinnate
　　　　　58　Spiny tree　　LXXXI. **Leguminosae**
　　　　　58　Unarmed shrubs or small trees
　　　　　　59　Stamens free　XCIV. **Anacardiaceae**
　　　　　　59　Stamens with connate filaments
　　　　　　　　　　　XCI. **Meliaceae**
42　Herbs, sometimes ± woody at base
　60　Sepals 2, petals 5
　　61　Stems erect or prostrate, not twining
　　　　　　　　LV. **Portulacaceae**
　　61　Stems twining　　LVI. **Basellaceae**
　60　Sepals as many as the petals
　　62　Leaves forming long pitchers; stigma very large,
　　　　peltate　　　　　LXX. **Sarraceniaceae**
　　62　Not as above
　　　63　Flowers strongly perigynous with a long tubu-
　　　　　lar or campanulate receptacle
　　　　　　　　　CXIX. **Lythraceae**
　　　63　Flowers hypogynous or perigynous with a flat
　　　　　or weakly concave receptacle
　　　　64　Cauline leaves opposite or whorled
　　　　　65　Leaves deeply divided, rarely only serrate
　　　　　　66　Petals 4　　LXVIII. **Cruciferae**
　　　　　　66　Petals 5
　　　　　　　67　Stamens without scales on the inner side
　　　　　　　　of the filaments　LXXXIII. **Geraniaceae**
　　　　　　　67　Stamens with scales on the inner side of
　　　　　　　　the filaments　LXXXV. **Zygophyllaceae**
　　　　　65　Leaves simple and entire
　　　　　　68　Leaves in 1 whorl; flower solitary, terminal
　　　　　　　　　　　　Trilliaceae
　　　　　　68　Leaves opposite or in more than 1 whorl
　　　　　　　69　Stipules present
　　　　　　　　70　Stipules scarious; land-plants
　　　　　　　　　　　LVII. **Caryophyllaceae**
　　　　　　　　70　Stipules not scarious; usually submerged
　　　　　　　　　aquatics　　CXV. **Elatinaceae**
　　　　　　　69　Stipules absent
　　　　　　　　71　Sepals united to more than half-way
　　　　　　　　　72　Styles connate; placentation parietal
　　　　　　　　　　　CXIV. **Frankeniaceae**
　　　　　　　　　72　Styles free; placentation free-central
　　　　　　　　　　　LVII. **Caryophyllaceae**
　　　　　　　　71　Sepals free or united at base only
　　　　　　　　　73　Ovary 1-celled; placentation free-central
　　　　　　　　　　　LVII. **Caryophyllaceae**

　　　　　　　　　73　Ovary 4- to 5-celled; placentation axile
　　　　　　　　　　　LXXXVI. **Linaceae**
　　　　64　Leaves alternate or all basal
　　　　　74　Leaves ternate　LXXXII. **Oxalidaceae**
　　　　　74　Leaves not ternate
　　　　　　75　Sepals and petals 2–3　XLVII. **Polygonaceae**
　　　　　　75　Sepals and petals 4–5
　　　　　　　76　Both whorls of perianth-segments green
　　　　　　　　　　　LXXX. **Rosaceae**
　　　　　　　76　Inner whorl of perianth-segments not green
　　　　　　　　77　Sepals and petals 4; stamens 4 or 6
　　　　　　　　　78　Stipules absent; stamens usually 6
　　　　　　　　　　LXVIII. **Cruciferae**
　　　　　　　　　78　Stipules present; stamens 4
　　　　　　　　　　LVII. **Caryophyllaceae**
　　　　　　　　77　Sepals and petals 5; stamens 5 or 10
　　　　　　　　　79　Leaves with conspicuous, red, viscid,
　　　　　　　　　　glandular hairs　LXXI. **Droseraceae**
　　　　　　　　　79　Not as above
　　　　　　　　　　80　Leaves with numerous pellucid glands,
　　　　　　　　　　　strongly scented when crushed
　　　　　　　　　　　LXXXVIII. **Rutaceae**
　　　　　　　　　　80　Leaves without pellucid glands
　　　　　　　　　　　81　Style 1; stigma entire or shallowly
　　　　　　　　　　　　lobed; anthers opening by pores
　　　　　　　　　　　　　　　Pyrolaceae
　　　　　　　　　　　81　Style or stigmas more than 1; anthers
　　　　　　　　　　　　opening by longitudinal slits
　　　　　　　　　　　　82　Stigmas 5
　　　　　　　　　　　　　83　Leaves lobed or pinnate
　　　　　　　　　　　　　　LXXXIII. **Geraniaceae**
　　　　　　　　　　　　　83　Leaves entire
　　　　　　　　　　　　　　84　Sepals united; leaves basal
　　　　　　　　　　　　　　　Plumbaginaceae
　　　　　　　　　　　　　　84　Sepals free; leaves cauline
　　　　　　　　　　　　　　　LXXXVI. **Linaceae**
　　　　　　　　　　　　82　Stigmas 2–4
　　　　　　　　　　　　　85　Flowers with conspicuous glandu-
　　　　　　　　　　　　　　lar-fimbriate staminodes
　　　　　　　　　　　　　　LXXIV. **Parnassiaceae**
　　　　　　　　　　　　　85　Glandular-fimbriate staminodes
　　　　　　　　　　　　　　absent
　　　　　　　　　　　　　　86　Stamens 5　LVII. **Caryophyllaceae**
　　　　　　　　　　　　　　86　Stamens 10　LXXIII. **Saxifragaceae**
24　Flowers zygomorphic
　87　Flowers saccate or spurred at base
　　88　Sepals 2, small　　　LXVI. **Papaveraceae**
　　88　Sepals 3 or 5
　　　89　Sepals 3, very unequal, 1 spurred; petals 3, not
　　　　　spurred　　　XCVIII. **Balsaminaceae**
　　　89　Sepals 5; petals 5
　　　　90　Leaves peltate　LXXXIV. **Tropaeolaceae**
　　　　90　Leaves not peltate
　　　　　91　Leaves alternate　　CX. **Violaceae**
　　　　　91　Leaves opposite　LXXXIII. **Geraniaceae**
　87　Flowers not saccate or spurred at base
　　92　All, or all but one, of the stamens united into a tube
　　　　　　　　LXXXI. **Leguminosae**
　　92　All stamens free
　　　93　Trees or shrubs
　　　　94　Leaves simple
　　　　　95　Ovary on a long gynophore　LXVII. **Capparaceae**
　　　　　95　Ovary sessile
　　　　　　96　Petals 4　　LXVIII. **Cruciferae**
　　　　　　96　Petals 5　　LXXXI. **Leguminosae**
　　　　94　Leaves compound
　　　　　97　Leaves trifoliolate or pinnate　LXXXI. **Leguminosae**
　　　　　97　Leaves palmate with more than 3 leaflets
　　　　　　　　XCVII. **Hippocastanaceae**
　　　93　Herbs
　　　　98　Ovary and fruit deeply 5-lobed
　　　　　99　Flowers in umbellate cymes; fruit with a long
　　　　　　beak　　　LXXXIII. **Geraniaceae**

99   Flowers in racemes; fruit not beaked **LXXXVIII. Rutaceae**

98 Ovary and fruit not deeply 5-lobed

100   Petals fimbriate or lobed   **LXIX. Resedaceae**

100   Petals entire or emarginate

101   Stamens 10   **LXXXI. Leguminosae**

101   Stamens not more than 6

102   Sepals inserted on a cup-like perigynous zone **LVII. Caryophyllaceae**

102   Sepals free

103   Ovary 2-locular; gynophore short or 0 **LXVIII. Cruciferae**

103   Ovary 1-locular; gynophore long **LXVII. Capparaceae**

10 Ovary inferior or partly so

104   Petals numerous

105   Aquatic plants; leaves not succulent **LVIII. Nymphaeaceae**

105   Land-plants; leaves succulent **LII. Aizoaceae**

104   Petals 5 or fewer

106   Petals and sepals 3

107   Flowers zygomorphic

108   Style and filaments obvious **Iridaceae**

108   Stigma and stamens sessile **Orchidaceae**

107   Flowers actinomorphic

109   Outer perianth-whorl sepaloid **Hydrocharitaceae**

109   Both perianth-whorls petaloid

110   Stamens 6 **Amaryllidaceae**

110   Stamens 3 **Iridaceae**

106   Petals and sepals 2, 4 or 5

111   Stamens numerous

112   Leaves opposite, with pellucid glands **CXXI. Myrtaceae**

112   Leaves alternate, without pellucid glands

113   Leaves entire; seeds covered with pulp **CXXII. Punicaceae**

113   Leaves serrulate; seeds dry

114   Styles free; fruit fleshy **LXXX. Rosaceae**

114   Styles united, except at the top; fruit a capsule **LXXV. Hydrangeaceae**

111   Stamens 10 or fewer

115   Aquatic; leaves pinnate, segments filiform; flowers in spikes **CXXIV. Haloragaceae**

115   Not as above

116   Trees, shrubs or woody climbers

117   Flowers in umbels

118   Climbers **CXXVIII. Araliaceae**

118   Erect shrubs

119   Evergreen; umbels flat **CXXIX. Umbelliferae**

119   Deciduous; umbels globose **CXXVII. Cornaceae**

117   Flowers not in umbels

120   Leaves palmately lobed **LXXVII. Grossulariaceae**

120   Leaves not lobed

121   Both perianth-whorls petaloid **CXXIII. Onagraceae**

121   Outer perianth-whorl sepaloid

122   Calyx-teeth very small; ovules 1 in each carpel; fruit a drupe **CXXVII. Cornaceae**

122   Calyx-teeth large; ovules numerous; fruit a capsule

123   Stamens 10 **LXXV. Hydrangeaceae**

123   Stamens 5 **LXXVI. Escalloniaceae**

116   Herbs

124   Both perianth-whorls sepaloid **LXXX. Rosaceae**

124   Inner perianth-whorl petaloid

125   Petals 5

126   Stamens 5 **CXXIX. Umbelliferae**

126   Stamens 10 **LXXIII. Saxifragaceae**

125   Petals 4 or 2

127   Flowers in umbels surrounded by 4 conspicuous white bracts **CXXVII. Cornaceae**

127   Flowers not in umbels; no conspicuous bracts **CXXIII. Onagraceae**

9 Petals all united at base into a longer or shorter tube

128 Ovary superior

---

129   Flowers papilionate

130   Sepals free; stamens 8 **XCII. Polygalaceae**

130   Sepals connate; stamens 10 **LXXXI. Leguminosae**

129   Flowers not papilionate

131   Stamens at least twice as many as corolla-lobes

132   Herbs with succulent leaves **LXXII. Crassulaceae**

132   Shrubs or trees

133   Flowers unisexual **Ebenaceae**

133   Flowers hermaphrodite

134   Anthers opening by pores; hairs simple or scale-like **Ericaceae**

134   Anthers opening by longitudinal slits; hairs stellate **Styracaceae**

131   Stamens as many as or fewer than corolla-lobes

135   Plant without chlorophyll; leaves scale-like

136   Flowers zygomorphic; stem stout, erect **Orobanchaceae**

136   Flowers actinomorphic; stem slender, twining **Convolvulaceae**

135   Green plants

137   Sepals 2; flowers actinomorphic

138   Petals 2; leaves in a rosette **Eriocaulaceae**

138   Petals 5; leaves not in a rosette **LV. Portulacaceae**

137   Sepals more than 2, or flowers zygomorphic

139   Ovary deeply 4-lobed with 1 ovule in each lobe

140   Leaves alternate **Boraginaceae**

140   Leaves opposite **Labiatae**

139   Ovary not 4-lobed

141   Flowers actinomorphic or nearly so

142   Carpels free

143   Leaves peltate; carpels 5 **LXXII. Crassulaceae**

143   Leaves not peltate; carpels 2

144   Corolla with a corona; styles 2, free but united by the stigma **Asclepiadaceae**

144   Corolla without a corona; styles 2, united except at the very base **Apocynaceae**

142   Carpels united

145   Stamens fewer than corolla-lobes

146   Herbs **Scrophulariaceae**

146   Shrubs or trees

147   Leaves opposite **Oleaceae**

147   Leaves alternate

148   Flowers yellow **Oleaceae**

148   Flowers not yellow **Scrophulariaceae**

145   Stamens as many as corolla-lobes

149   Stamens opposite the corolla-lobes

150   Styles or stigmas more than 1; ovule 1 **Plumbaginaceae**

150   Style 1; stigma 1; ovules numerous

151   Herbs **Primulaceae**

151   Shrubs **Myrsinaceae**

149   Stamens alternating with the corolla-lobes

152   Leaves opposite

153   Shrubs

154   Large, erect; leaves deciduous **Buddlejaceae**

154   Small, procumbent; leaves evergreen

155   Leaves elliptical or oblong; flowers pink **Ericaceae**

155   Leaves spathulate; flowers white **Diapensiaceae**

153   Herbs

156   Land-plants; leaves sessile **Gentianaceae**

156   Aquatic plants; leaves petiolate **Menyanthaceae**

152   Leaves alternate or all basal

157   Sepals, petals and stamens 4

158   Shrubs **XCIX. Aquifoliaceae**

158   Herbs

159   Corolla not violet-blue **Plantaginaceae**

159   Corolla violet-blue **Gesneriaceae**

157   Sepals, petals and stamens 5 (rarely sepals fewer)

160   Ovary 3-celled; stigmas 3 or 3-lobed

161   Leaves pinnate **Polemoniaceae**

161 Leaves simple **Diapensiaceae**
160 Ovary 2-celled; stigmas 2 or 1
162 Ovules 4 or fewer
163 Flowers numerous, in scorpioidal cymes; corolla-lobes distinct **Boraginaceae**
163 Flowers solitary or few, not in scorpioidal cymes; corolla not or scarcely lobed **Convolvulaceae**
162 Ovules numerous
164 Aquatic or bog-plants; corolla fimbriate **Menyanthaceae**
164 Land-plants; corolla not fimbriate
165 Leaves all basal **Gesneriaceae**
165 Some leaves cauline
166 Corolla-tube much shorter than lobes; stamens patent **Scrophulariaceae**
166 Corolla-tube long, or anthers connivent **Solanaceae**
141 Flowers strongly zygomorphic
167 Anthers opening by pores **Ericaceae**
167 Anthers opening by slits
168 Calyx with patent spines and erect, membranous, usually dark-spotted lobes **Primulaceae**
168 Calyx not as above
169 Flowers small, crowded in capitula **Globulariaceae**
169 Flowers not in capitula
170 Ovary 1-celled; carnivorous plants **Lentibulariaceae**
170 Ovary 2-celled; not carnivorous plants
171 Ovules numerous **Scrophulariaceae**
171 Ovules 4
172 Bracts shorter than calyx **Verbenaceae**
172 Bracts or bracteoles much longer than calyx **Acanthaceae**
128 Ovary inferior
173 Stamens 8–10, or 4–5 with filaments divided to base
174 Herb; anthers opening by slits; leaves ternate **Adoxaceae**
174 Woody; anthers opening by pores; leaves simple **Ericaceae**
173 Stamens 5 or fewer; filaments not divided
175 Leaves in whorls of 4 or more **Rubiaceae**
175 Leaves not in whorls
176 Stamens opposite corolla-lobes **Primulaceae**
176 Stamens alternating with corolla-lobes
177 Leaves opposite; stipules interpetiolar **Rubiaceae**
177 Leaves alternate, or stipules not interpetiolar
178 Flowers in capitula surrounded by an involucre of more than 2 bracts
179 Anthers coherent in a ring round the style
180 Ovule 1; calyx, if present, represented by hairs or scales **Compositae**
180 Ovules numerous; calyx-lobes conspicuous, green **Campanulaceae**
179 Anthers free
181 Ovules numerous; corolla-lobes longer than tube **Campanulaceae**
181 Ovule 1; corolla-lobes much shorter than tube **Dipsacaceae**
178 Flowers not in capitula, or bracts 2
182 Anthers coherent in a tube round the style **Lobeliaceae**
182 Anthers not cohering to one another
183 Anthers sessile; pollen-grains cohering in pollinia **Orchidaceae**
183 Stamens with filaments; pollen-grains free
184 Stamens 1–3 **Valerianaceae**
184 Stamens 4–5
185 Shrubs (sometimes small and creeping), or woody climbers **Caprifoliaceae**
185 Herbs
186 Tendrils present **CXVII. Cucurbitaceae**
186 Tendrils absent

187 Leaves pinnate **Caprifoliaceae**
187 Leaves not pinnate
188 Flowers hermaphrodite; fruit a capsule **Campanulaceae**
188 Flowers unisexual; fruit fleshy **CXVII. Cucurbitaceae**
8 Perianth not of 2 or more markedly different whorls
189 Perianth entirely petaloid
190 Parasites or saprophytes without chlorophyll
191 Flowers mostly unisexual; stamen 1 **XLVI. Balanophoraceae**
191 Flowers hermaphrodite; stamens 6–16
192 Filaments free **Monotropaceae**
192 Filaments united into a column **XLV. Rafflesiaceae**
190 Green plants
193 Perianth-segment 1, bract-like **Aponogetonaceae**
193 Perianth-segments more than 1, or perianth tubular
194 Stems succulent, leafless but with groups of spines **CXVIII. Cactaceae**
194 Not as above
195 Stamens more than 12
196 Herbs, or, rarely, woody climbers with pinnate leaves **LXI. Ranunculaceae**
196 Trees with simple leaves **LXIV. Magnoliaceae**
195 Stamens 12 or fewer
197 Flowers in ovoid capitula without an involucre **LXXX. Rosaceae**
197 Flowers not in capitula, or capitula with an involucre
198 Ovary superior
199 Perianth-segments 4
200 Flowers zygomorphic **XLI. Proteaceae**
200 Flowers actinomorphic
201 Perianth tubular below **CVII. Thymelaeaceae**
201 Perianth-segments free
202 Herbs **Liliaceae**
202 Shrubs **XLVII. Polygonaceae**
199 Perianth-segments more than 4
203 Carpels more than 1, free or nearly so
204 Leaves triquetrous, all basal **Butomaceae**
204 Leaves flat, cauline **LI. Phytolaccaceae**
203 Carpel 1, or carpels obviously united
205 Perianth-segments 6 **Liliaceae**
205 Perianth-segments 5
206 Stigmas 2–3; stipules sheathing, scarious **XLVII. Polygonaceae**
206 Stigma 1; stipules absent
207 Ovules numerous; perianth divided almost to base **Primulaceae**
207 Ovule 1; perianth with a long tube **L. Nyctaginaceae**
198 Ovary inferior, or flowers male
208 Leaves in whorls of 4 or more **Rubiaceae**
208 Leaves not in whorls
209 Flowers in capitula surrounded by an involucre
210 Anthers cohering in a tube round the style, or flowers unisexual **Compositae**
210 Anthers free; flowers hermaphrodite **Dipsacaceae**
209 Flowers not in capitula, though sometimes shortly pedicellate in compact umbels
211 Ovules numerous
212 Perianth-segments 3, or perianth tubular with a unilateral entire limb **XLIV. Aristolochiaceae**
212 Perianth-segments 6
213 Stamens 6 **Amaryllidaceae**
213 Stamens 3 **Iridaceae**
211 Ovules 1 or 2
214 Leaves opposite **Valerianaceae**
214 Leaves alternate
215 Flowers in simple cymes or solitary **XLII. Santalaceae**

215　Flowers in umbels or superposed whorls
　　　　　　　　　　CXXIX. Umbelliferae
189　Perianth not petaloid, often absent, if brightly coloured
　　　then dry and scarious
216　Trees or shrubs, sometimes small
217　Parasitic on branches of trees or shrubs
　　　　　　　　　　XLIII. Loranthaceae
217　Not parasitic
218　Stems creeping or climbing with adventitious roots;
　　　evergreen　　　　**CXXVIII. Araliaceae**
218　Not as above
219　Flowers borne on flattened evergreen cladodes;
　　　leaves small, brownish, scale-like　　**Liliaceae**
219　Not as above
220　Most leaves opposite or subopposite
221　Stems green and fleshy or leaves fleshy
　　　　　　　　　　XLVIII. Chenopodiaceae
221　Neither leaves nor stems fleshy
222　Styles 3　　　　　　**CII. Buxaceae**
222　Styles 4, or 1
223　Flowers in catkins　　**XXXI. Salicaceae**
223　Flowers not in catkins
224　Leaves pinnate; stamens 2　　**Oleaceae**
224　Leaves simple; stamens 4 or more
225　Stamens 5, alternating with sepals
　　　　　　　　　　CIII. Rhamnaceae
225　Stamens 8; sepals 5　　**XCV. Aceraceae**
220　Most leaves alternate
226　Leaves pinnate
227　Ovary inferior; styles 2; pith septate
　　　　　　　　　　XXXIII. Juglandaceae
227　Ovary superior; styles 3 or 1; pith not septate
228　Style 1; fruit a lomentum
　　　　　　　　　　LXXXI. Leguminosae
228　Styles 3; fruit a dry, 1-seeded drupe
　　　　　　　　　　XCIV. Anacardiaceae
226　Leaves simple
229　Leaves not more than 2 mm wide, oblong or
　　　linear
230　Stigma 1　　　　**CVII. Thymelaeaceae**
230　Stigmas 2–9
231　Stamens 3　　　　**Empetraceae**
231　Stamens 5　　**XLVIII. Chenopodiaceae**
229　Leaves more than 2 mm wide
232　Petiole with dilated base, enclosing the bud
　　　　　　　　　　LXXIX. Platanaceae
232　Petiole-base not enclosing the bud
233　Anthers opening by transverse valves
　　　　　　　　　　LXV. Lauraceae
233　Anthers opening by longitudinal slits
234　Flowers not in catkins or dense heads
235　Inflorescence of several male flowers, each
　　　of 1 stamen, and a female flower, appear-
　　　ing as a stalked ovary, all surrounded by
　　　4 or 5 conspicuous glands; latex present
　　　　　　　　　　LXXXVII. Euphorbiaceae
235　Inflorescence not as above; no latex
236　Flowers unisexual
237　Peltate scale-like silvery or ferruginous
　　　hairs present beneath the leaves and
　　　often elsewhere; ovary 1-locular; fruit
　　　fleshy　　　　**CVIII. Eleagnaceae**
237　No scale-like hairs; ovary 3-locular; fruit
　　　dry　　　**LXXXVII. Euphorbiaceae**
236　Flowers hermaphrodite
238　Trees; perianth-tube short, with stamens
　　　inserted near its base　**XXXVII. Ulmaceae**
238　Shrubs; perianth-tube long, with stamens
　　　inserted near its apex
　　　　　　　　　　CVII. Thymelaeaceae
234　Flowers in catkins or dense heads
239　Latex present; fruit or false fruit fleshy
　　　　　　　　　　XXXVIII. Moraceae

239　Latex absent; fruit dry
240　Dioecious; perianth absent
241　Bracts (catkin-scales) fimbriate or lobed
　　　at apex; flowers with a cup-like disk
　　　　　　　　　　XXXI. Salicaceae
241　Bracts (catkin-scales) entire; disk absent
242　Leaves without pellucid glands; sta-
　　　mens with long filaments; ovules
　　　numerous　　**XXXI. Salicaceae**
242　Leaves with pellucid glands; stamens
　　　with short filaments; ovule 1
　　　　　　　　　　XXXII. Myricaceae
240　Monoecious; perianth present in male or
　　　female flowers or both
243　Styles 3 or more; flowers of both sexes
　　　with perianth　　**XXXVI. Fagaceae**
243　Styles 2; perianth present in flowers of
　　　1 sex only
244　Male flowers 3 to each bract; perianth
　　　present　　**XXXIV. Betulaceae**
244　Male flowers 1 to each bract; perianth
　　　absent　　**XXXV. Corylaceae**
216　Herbs
245　Perianth absent or represented by scales or bristles,
　　　minute in flower; flowers in the axils of bracts, a
　　　number of which are usually closely imbricate on a
　　　rhachis, forming a spikelet; leaves usually linear,
　　　grass-like, sheathing below
246　Flowers usually with a bract above and below; sheaths
　　　usually open; stems usually with hollow internodes
　　　　　　　　　　Gramineae
246　Flowers with a bract below only; sheaths usually
　　　closed; stems usually with solid internodes
　　　　　　　　　　Cyperaceae
245　Perianth present, or flowers not arranged in spikelets
247　Aquatic plants; leaves submerged or floating; inflor-
　　　escence sometimes emergent
248　Leaves divided into numerous filiform segments
249　Leaves pinnately divided; flowers in a terminal
　　　spike　　　　**CXXIV. Haloragaceae**
249　Leaves dichotomously divided; flowers solitary,
　　　axillary　　　**LX. Ceratophyllaceae**
248　Leaves entire or dentate
250　Flowers in spikes
251　Rhizome densely covered with stiff fibres; spikes
　　　subtended by a group of leaf-like bracts (marine)
　　　　　　　　　　Posidoniaceae
251　Not as above
252　Flowers hermaphrodite, arranged all round or on
　　　2 sides of a terete rhachis (fresh or brackish
　　　water)　　　**Potamogetonaceae**
252　Flowers unisexual, arranged on one side of a flat
　　　rhachis (marine)　　**Zosteraceae**
250　Flowers not in spikes
253　Flowers solitary or few, sessile or shortly pedi-
　　　cellate, axillary
254　Leaves in whorls of 8 or more
　　　　　　　　　　CXXVI. Hippuridaceae
254　Leaves not in whorls of 8 or more
255　Carpels 2 or more, free
256　Carpels nearly or quite sessile in fruit
　　　　　　　　　　Zannichelliaceae
256　Carpels in fruit with stalks several times their
　　　own length　　　**Ruppiaceae**
255　Carpels united, or solitary
257　Female flowers with a very long filiform
　　　perianth-tube resembling a pedicel
　　　　　　　　　　Hydrocharitaceae
257　Perianth-tube short or 0
258　Perianth-segments 4–6; stamens 4 or more;
　　　leaves ovate to obovate
259　Perianth-segments 4; ovary inferior
　　　　　　　　　　CXXIII. Onagraceae

259 Perianth-segments 6; ovary superior **CXIX. Lythraceae**
258 Perianth-segments fewer than 4, or perianth absent; stamen 1; leaves linear to lanceolate
260 Leaves alternate (brackish) **Zannichelliaceae**
260 Leaves opposite (freshwater)
261 Leaves entire, without sheathing base; ovary compressed, deeply 4-lobed **Callitrichaceae**
261 Leaves spinulose-dentate, with sheathing base; ovary terete, not lobed **Najadaceae**
253 Flowers in heads on long peduncles or in compound inflorescences
262 Flowers hermaphrodite; heads few-flowered **Juncaceae**
262 Flowers unisexual; heads many-flowered
263 Leaves all basal; heads solitary on long scapes **Eriocaulaceae**
263 Some leaves cauline; inflorescence with female heads below and male heads above **Sparganiaceae**
247 Terrestrial plants or, if aquatic, with inflorescence and either stems or leaves emergent
264 Climbing plants with unisexual flowers
265 Leaves opposite; perianth-segments 5 **XXXIX. Cannabaceae**
265 Leaves alternate; perianth-segments 6 **Dioscoreaceae**
264 Not climbing, or rarely climbers with hermaphrodite flowers
266 Leaves linear
267 Flowers unisexual
268 Female flowers solitary; male flowers solitary or in short cymes **XLVIII. Chenopodiaceae**
268 Male and female flowers numerous, in dense heads or spikes
269 Male and female flowers in separate globose heads **Sparganiaceae**
269 Flowers in a dense cylindrical spike, male above, female below **Typhaceae**
267 Flowers hermaphrodite
270 Plant densely pubescent **XLVIII. Chenopodiaceae**
270 Plant glabrous or sparsely hairy
271 Flowers in dense spikes; spikes apparently lateral on a flattened leaf-like stem **Araceae**
271 Not as above
272 Carpel 1
273 Leaves not subverticillate, exstipulate **XLVIII. Chenopodiaceae**
273 Leaves subverticillate, with minute stipules **LVII. Caryophyllaceae**
272 Carpels more than 1
274 Carpels free (except at base); leaves with a conspicuous pore at apex **Scheuchzeriaceae**
274 Carpels ± completely united; leaves without a conspicuous pore at apex
275 Flowers in unbranched racemes; styles short or 0 **Juncaginaceae**
275 Flowers in cymes in a branched inflorescence; styles 3, distinct **Juncaceae**
266 Leaves lanceolate or wider, or sometimes small and scale-like, but never linear
276 Leaves compound
277 Flowers in compound umbels **CXXIX. Umbelliferae**
277 Flowers not in compound umbels
278 Flowers in capitula
279 Leaves simply pinnate; style 1 or 2 **LXXX. Rosaceae**
279 Leaves ternate; styles 3–5 **Adoxaceae**
278 Flowers not in capitula
280 Stamens numerous **LXI. Ranunculaceae**
280 Stamens 4 or 5(–10)
281 Epicalyx present **LXXX. Rosaceae**

281 Epicalyx absent **LXXXIII. Geraniaceae**
276 Leaves simple or apparently absent
282 Flowers numerous, small, crowded on an axis (spadix) subtended and often ± enclosed by a conspicuous bract (spathe) **Araceae**
282 Not as above
283 Inflorescence of several male flowers, each of 1 stamen, and a female flower, appearing as a stalked ovary, all surrounded by 4 or 5 conspicuous glands; latex present **LXXXVII. Euphorbiaceae**
283 Not as above
284 Leaves apparently absent; stem green and succulent **XLVIII. Chenopodiaceae**
284 Leaves obvious; stem not succulent
285 Lower leaves opposite, upper alternate; monoecious; male flowers with 2-partite perianth, female with tubular perianth **CXXV. Theligonaceae**
285 Not as above
286 Plant densely clothed with stellate hairs; ovary 3-locular with 1 ovule in each loculus **LXXXVII. Euphorbiaceae**
286 Not as above
287 Densely papillose annuals
288 Leaves oblong-lanceolate, never hastate; fruit opening by 5 valves **LII. Aizoaceae**
288 Leaves ovate-rhombic, often hastate; fruit indehiscent **LIV. Tetragoniaceae**
287 Not densely papillose annuals
289 Leaves whorled
290 Stigma 1; stems hollow **CXXVI. Hippuridaceae**
290 Stigmas 3; stems solid **LIII. Molluginaceae**
289 Leaves not in whorls
291 Leaves alternate or all basal (rarely the lower opposite)
292 Stamens numerous; carpels free except sometimes at base **LXI. Ranunculaceae**
292 Stamens 12 or fewer; carpels not free, or one only
293 Carpels attached to a central axis, otherwise free **LI. Phytolaccaceae**
293 Carpels united, or one only
294 Stamens 12 **XLIV. Aristolochiaceae**
294 Stamens 10 or fewer
295 Stipules united into a sheath **XLVII. Polygonaceae**
295 Stipules free or absent
296 Leaves very large, palmately lobed, all basal; inflorescence of dense many-flowered spikes much shorter than the leaves **CXXIV. Haloragaceae**
296 Not as above
297 Epicalyx present; stipules leaf-like **LXXX. Rosaceae**
297 Epicalyx 0; stipules small or 0
298 Ovary superior
299 Perianth tubular below
300 Ovule basal **XLVIII. Chenopodiaceae**
300 Ovule pendent **CVII. Thymelaeaceae**
299 Perianth-segments free or nearly so, rarely absent in female flowers
301 Perianth-segments 4
302 Flowers in ebracteate racemes **LXVIII. Cruciferae**
302 Flowers in axillary clusters **XL. Urticaceae**
301 Perianth-segments 5

303 Perianth herbaceous, rarely absent in female flowers
 XLVIII. Chenopodiaceae

303 Perianth scarious
 XLIX. Amaranthaceae

298 Ovary inferior

304 Leaves reniform, cordate
 LXXIII. Saxifragaceae

304 Leaves subulate to linear-lanceolate **XLII. Santalaceae**

291 Leaves opposite (rarely a few upper apparently alternate)

305 Leaves toothed or lobed

306 Flowers hermaphrodite

307 Ovary inferior; stigmas 2
 LXXIII. Saxifragaceae

307 Ovary superior; stigmas 5
 LXXXIII. Geraniaceae

306 Flowers unisexual

308 Perianth-segments 4 or 2; style 1
 XL. Urticaceae

308 Perianth-segments 3; styles 2
 LXXXVII. Euphorbiaceae

305 Leaves entire

309 Perianth 0; ovary compressed, 4-lobed
 Callitrichaceae

309 Perianth present; ovary not compressed and 4-lobed

310 Perianth-segments 3
 XLVII. Polygonaceae

310 Perianth-segments 4 or more

311 Ovary inferior **CXXIII. Onagraceae**

311 Ovary superior

312 Perianth-segments 6 or 12; style and stigma 1 **CXIX. Lythraceae**

312 Perianth-segments 4 or 5; styles or stigmas 2 or more

313 Leaves without a long spinose apex; fruit unwinged
 LVII. Caryophyllaceae

313 Leaves with a long spinose apex; fruit transversely winged
 XLVIII. Chenopodiaceae

EXPLANATORY NOTES ON THE TEXT

Signs and abbreviations

c.	*circa*, approximately
C.	central
cm	centimetre(s)
E.	eastern, east
incl.	including
loc. cit.	*loco citato*, on the same page in the work cited above
m	metre(s)
mm	millimetre(s)
N.	northern, north
2*n*	the somatic chromosome number
op. cit.	*opere citato*, in the work cited above
S.	southern, south
Sect.	Sectio
sp. ⎫ spp. ⎭	species
Subfam.	Subfamilia
Subgen.	Subgenus
Subsect.	Subsectio
subsp. ⎫ subspp. ⎭	subspecies
var.	varietas
W.	western, west
±	more or less
0	absent
●	endemic to Europe
[ ]	not native
*	status doubtful; possibly native
?	(before a two-letter geographical abbreviation) occurrence doubtful
†	extinct

Abbreviations of geographical territories

(For precise definitions of these territories, see map 1)

Al	Albania
Au	Austria
Az	Açores (Azores)
Be	Belgium and Luxembourg
Bl	Islas Baleares (Balearic Islands)
Br	Britain
Bu	Bulgaria
Co	Corse (Corsica)
Cr	Kriti (Crete)
Cz	Czechoslovakia
Da	Denmark
Fa	Færöer (Faroes)
Fe	Finland
Ga	France
Ge	Germany
Gr	Greece
Hb	Ireland
He	Switzerland
Ho	Netherlands
Hs	Spain
Hu	Hungary
Is	Iceland
It	Italy
Ju	Jugoslavia
Lu	Portugal
No	Norway
Po	Poland
Rm	Romania
Rs	U.S.S.R. (European part), subdivided thus:
	(N) Northern region
	(B) Baltic region
	(C) Central region
	(W) South-western region
	(K) Krym (Crimea)
	(E) South-eastern region
Sa	Sardegna (Sardinia)
Sb	Svalbard (Spitsbergen)
Si	Sicilia (Sicily)
Su	Sweden
Tu	Turkey (European part)

General notes

The sequence of families is that of Melchior in Engler-Diels, *Syllabus der Pflanzenfamilien* ed. 12 (1964), except that the Cactaceae and Guttiferae, which should have been in Volume 1, have been inserted in Volume 2.

Descriptions of taxa refer only to the European populations of the taxon in question. If extra-European representatives differ substantially, an explanatory note is sometimes added.

Groups of species have been used in some genera where the species are very difficult to separate. These groups have no formal nomenclatural status and are simply a device to enable a partial identification to be made.

Taxa below the rank of subspecies are neither keyed nor described, and varieties are mentioned only when there are special reasons.

Aliens are included only when they appear to be effectively naturalized or when planted in continuous stands on a fairly large scale.

Hybrids are mentioned only when they occur frequently.

A measurement given without qualification refers to length. Two measurements connected by × indicate length followed by width. Further measurements in parentheses indicate exceptional cases outside the normal range.

Synonyms given in the text are principally those names under which the species or subspecies is described in the Basic Floras listed on p. xvii. The index contains (in addition to these) names which occur in any of the Standard Floras (p. xvii) or in well-known monographs.

Chromosome numbers are given only when the editors are satisfied that the count has been made on correctly identified material known to be of wild European origin. For naturalized and cultivated species the count is from material which is naturalized or is cultivated in the way which justifies its inclusion in the Flora.

Ecological information is provided only when the habitat-preference of a species is sufficiently uniform over its European range to permit it to be summed up in a short phrase.

Geographical terms such as 'W. Europe', 'Mediterranean region', etc., are to be interpreted as shown on maps IV and V. The statement that a plant occurs in one or more of these regions does not necessarily imply that it occurs throughout the region.

Extra-European distribution is indicated only for those plants whose European range is small and whose range outside Europe is considerably greater, or for species which are not native in Europe.

SPERMATOPHYTA

ANGIOSPERMAE

DICOTYLEDONES

(continued)

ROSALES (*continued*)

LXXX. ROSACEAE[1]

Trees, shrubs or herbs. Leaves usually alternate and stipulate. Flowers regular, usually hermaphrodite, perigynous or epigynous. Hypanthium flat, concave or tubular. Sepals usually 5, sometimes with epicalyx. Petals usually 5, free, sometimes absent. Stamens usually 2, 3 or 4 times as many as the sepals, sometimes 1–5 or indefinite. Carpels 1 to numerous, free or connate, sometimes adnate to the hypanthium. Ovules usually 2, sometimes 1 or more, anatropous. Styles free, rarely united. Fruit of one or more achenes, drupes or follicles, or a pome, the hypanthium sometimes becoming coloured and fleshy. Endosperm usually absent.

It is convenient, in this family, to use the term 'hypanthium' to denote that part of the flower which bears the sepals, petals and stamens on its outer or upper margins, and on which the carpels are borne. The hypanthium is often, at least in part, receptacular in nature, but it is sometimes fused, to a variable extent, with the walls of the carpels, the exact line of demarcation being difficult to determine.

The family is notable for the large number of genera which are cultivated either for ornament or for food. Apomixis, either facultative or obligate, is a feature of the reproduction of several genera of the two largest European subfamilies, the Rosoideae and Maloideae; and in some of these, notably *Rubus*, *Alchemilla* and *Sorbus*, the number of taxa described at the level of species is very large. This situation has been met by describing and keying a number of species which represent the whole range of variation. The remaining species (at least those described in Standard Floras or important monographs) are then listed after the species which they most closely resemble, together with a note of the territories in which they occur.

```
1  Trees, shrubs or dwarf shrubs
2   Petals c. 8; procumbent dwarf shrub          16. Dryas
2   Petals 0, 4 or 5
3    Leaves pinnate or digitate
4     Flowers in dense capitula; petals 0
5      Stamens 2; fruit dry                      15. Acaena
5      Stamens numerous; fruit fleshy     14. Sarcopoterium
4     Flowers not in dense capitula; petals 4 or 5
6      Carpels and fruit exposed on the hypanthium
7       Prickly; fruit a head of drupelets        9. Rubus
7       Unarmed; fruit a head of achenes or follicles
8        Carpels 5, developing into follicles    1. Sorbaria
8        Carpels more than 5, developing into achenes
                                              19. Potentilla
6      Carpels enclosed in the hypanthium
9       Usually spiny shrubs; carpels numerous, free
                                                  10. Rosa
9       Unarmed trees; carpels 2–5, adnate to the hypanthium
                                                28. Sorbus
3    Leaves simple
10    Leaves opposite; sepals and petals 4     7. Rhodotypos
10    Leaves alternate; sepals and petals 5
11     Carpels not adnate to the hypanthium; fruit not a pome
12      Flowers yellow                            8. Kerria
12      Flowers white, pink, red or purple
13       Leaves lobed
14        Fruit of 1 or more drupelets            9. Rubus
14        Fruit of several follicles
15         Stipules absent; carpels free          3. Spiraea
```

```
15         Stipules caducous; carpels connate at base
                                              2. Physocarpus
13       Leaves not lobed
16        Carpel 1; fruit a drupe               35. Prunus
16        Carpels more than 1; fruit of several follicles
17         Carpels free                          3. Spiraea
17         Carpels connate at base               4. Sibiraea
11     Carpels enclosed in and adnate to the hypanthium; fruit
        a pome
18      Flowers solitary
19       Flowers less than 1 cm in diameter; fruit red or black
                                              31. Cotoneaster
19       Flowers more than 1 cm in diameter; fruit brown,
          green or yellow
20        Sepals shorter than petals, dentate   25. Cydonia
20        Sepals longer than petals, entire     33. Mespilus
18      Flowers in 2- to many-flowered inflorescences
21       Walls of carpels becoming stony in fruit
22        Leaves entire                        31. Cotoneaster
22        Leaves crenate-dentate or serrate or lobed
23         Evergreen; stipules caducous        32. Pyracantha
23         Deciduous; stipules persistent      34. Crataegus
21       Walls of carpels becoming cartilaginous in fruit
24        Flowers in compound corymbs or panicles
25         Evergreen; flowers in panicles      29. Eriobotrya
25         Deciduous; flowers in compound corymbs  28. Sorbus
24        Flowers in umbels, racemes or few-flowered clusters
26         Petals linear to oblong-ovate, not clawed
                                              30. Amelanchier
26         Petals obovate or orbicular, clawed
27          Styles free; flesh of fruit with stone-cells  26. Pyrus
27          Styles connate at base; flesh of fruit with few or no
             stone-cells                        27. Malus
1  Herbs
28  Petals 0
29   Leaves simple or digitately divided
30    Annual; stamens 1(–2)                     24. Aphanes
30    Perennial; stamens 4–10
31     Carpels 5–12                             20. Sibbaldia
31     Carpel 1                                 23. Alchemilla
29   Leaves pinnate
32    Stamens 2; hypanthium with 4 spines        15. Acaena
32    Stamens 4 or numerous; hypanthium without spines
                                              13. Sanguisorba
28  Petals 4 or more
33   Petals c. 8                                16. Dryas
33   Petals 4, 5 or 6
34    Sepals 4–6; epicalyx absent
35     Flowers yellow; fruit with hooked bristles  11. Agrimonia
35     Flowers white, cream, purple or red; fruit without bristles
36      Stipules absent; carpels 3              5. Aruncus
36      Stipules present; carpels 6 or more
37       Leaves pinnate, with small leaflets between the larger
          ones; fruit a head of achenes        6. Filipendula
37       Leaves undivided, digitate, or pinnate with equal leaf-
          lets; fruit a head of drupelets       9. Rubus
34    Sepals 4–5; epicalyx-segments 4–5
38     Carpels and achenes enclosed in the hypanthium
                                              12. Aremonia
38     Carpels and achenes exposed
39      Leaves pinnate or lyrate
40       Styles long, persistent               17. Geum
40       Styles short, deciduous               19. Potentilla
39      Leaves ternate, digitate or digitately lobed
41       Receptacle swollen, and fleshy or spongy in fruit
```

---

[1] Edit. D. H. Valentine and A. O. Chater.

42 Petals purple   **19. Potentilla**
42 Petals yellow or white
  43 Epicalyx-segments 3-toothed at apex; petals yellow
                                                      22. Duchesnea
  43 Epicalyx-segments not toothed at apex; petals white
                                                      21. Fragaria
41 Receptacle not swollen in fruit
44 Petals 1–2 mm; stamens 5(–10)   **20. Sibbaldia**
44 Petals more than 2 mm; stamens 10 or more
  45 Carpels more than 6   **19. Potentilla**
  45 Carpels 2–6
    46 Flowers 4-merous   **19. Potentilla**
    46 Flowers 5-merous   **18. Waldsteinia**

Subfam. **Spiraeoideae**

Stipules sometimes absent. Flowers 5-merous. Hypanthium flat, concave or campanulate, without carpophore; epicalyx absent; stamens 15 to numerous; carpels 1–5, whorled, free or connate at base, not sunk in hypanthium. Fruit of 1–5 follicles; seeds 2 or more. Basic chromosome number 8 or 9.

1. **Sorbaria** (Ser.) A. Braun[1]

Deciduous shrubs. Leaves pinnate, the leaflets serrate; stipules present. Inflorescence a terminal panicle. Petals ovate to orbicular, white; stamens 20–50; carpels 5, connate at base. Follicles dehiscent along the ventral suture. Seeds several.

Leaflets with less than 25 pairs of veins; stamens about twice as long as petals; pedicels erect in fruit   **1. sorbifolia**
Leaflets with more than 30 pairs of veins; stamens as long as petals; pedicels recurved in fruit   **2. tomentosa**

1. **S. sorbifolia** (L.) A. Braun in Ascherson, *Fl. Brandenb.* **1**: 177 (1864). Erect, up to 2 m, suckering. Leaflets 5–10 pairs, 4–10 × 1·5–3 cm, lanceolate to ovate-lanceolate, glabrous or puberulent beneath. Panicles 10–25 cm, with erect branches. Flowers 6–8 mm in diameter; stamens about twice as long as petals. Pedicels erect in fruit. *Cultivated for ornament in much of Europe and often locally naturalized.* [Da Fe Ga It No Su.] (*N. Asia.*)

2. **S. tomentosa** (Lindley) Rehder, *Jour. Arnold Arb.* **19**: 74 (1938). Up to 6 m. Leaflets 7–10 pairs, 5–10 × 1–2 cm, lanceolate to linear-lanceolate, long-acuminate, pubescent on the veins beneath when young. Panicles 20–30 cm, with patent branches. Flowers *c.* 6 mm in diameter; stamens as long as petals. Pedicels recurved in fruit. *Cultivated for ornament in S. & W. Europe; naturalized in France.* [Ga.] (*Himalaya.*)

2. **Physocarpus** (Camb.) Maxim.[1]

Deciduous shrubs. Leaves simple, usually lobed and serrate; stipules caducous. Inflorescence a terminal corymb. Petals suborbicular, white or pale pink, scarcely exceeding the sepals; stamens 20–40; carpels 1–5, connate at base. Follicles dehiscent along both sutures. Seeds 2–5.

1. **P. opulifolius** (L.) Maxim., *Acta Horti Petrop.* **6**(1): 220 (1879). Up to 3 m, glabrous or sparsely pubescent. Leaves 2–10 cm, ovate-orbicular, usually 3- to 5-lobed, the lobes crenate-dentate or crenate-serrate. Flowers *c.* 10 mm in diameter; carpels (3–)4–5. Follicles glabrous. *Cultivated for ornament in much of Europe and sometimes naturalized.* [Au Br Cz Ga Ho Ju No.] (*E. North America.*)

---

[1] By P. W. Ball.   [2] By J. Dostál.

3. **Spiraea** L.[2]

Deciduous shrubs. Leaves simple, rarely lobed, entire or serrate, usually shortly petiolate, exstipulate. Inflorescence paniculate, corymbose or umbellate. Petals white or pink; stamens 15 to numerous; carpels 5, free. Follicles dehiscent along the ventral suture. Seeds several.

In addition to the naturalized species described below, many species and hybrids are cultivated for ornament in gardens.

Literature: H. Zabel, *Die strauchigen Spiräen unserer Gärten.* Berlin. 1893. V. V. Shulgina, *Derev'ja i Kustarniki SSSR* **3**: 269–332 (1954). G. Krüssmann, *Handbuch der Laubholzkunde* **2**: 489–499. Berlin. 1962.

1 Inflorescence a panicle or compound corymb, terminating a long shoot
  2 Inflorescence a panicle, longer than wide
    3 Sepals erect in fruit; nectar-ring distinct
      4 Panicles cylindrical; flowers usually pink   **1. salicifolia**
      4 Panicles conical; flowers usually white   **2. alba**
    3 Sepals deflexed in fruit; nectar-ring absent
      5 Follicles glabrous   **3. douglasii**
      5 Follicles pubescent   **4. tomentosa**
  2 Inflorescence a compound corymb, at least as wide as long
    6 Stamens equalling or shorter than petals   **5. decumbens**
    6 Stamens at least 1·5 times as long as petals
      7 Leaves acute; petals pink   **6. japonica**
      7 Leaves obtuse; petals white or yellowish-white   **7. corymbosa**
1 Inflorescence a simple umbel or corymb, terminating a short shoot
  8 Inflorescence sessile or the lower very shortly pedunculate
                                    13. hypericifolia
  8 Inflorescence pedunculate
    9 Leaves with 3 longitudinal veins
      10 Veins conspicuous; petals shorter than stamens   **11. crenata**
      10 Veins inconspicuous; petals longer than stamens
                                12. × vanhouttei
    9 Leaves pinnately veined
      11 Branches angular; petals *c.* 6 mm   **8. chamaedryfolia**
      11 Branches terete; petals 2–3 mm
        12 Leaves of non-flowering shoots serrate towards apex, glabrous at maturity   **9. media**
        12 Leaves of non-flowering shoots entire, tomentose beneath   **10. cana**

Subgen. **Spiraea**. Inflorescence a terminal panicle or compound corymb, terminating a long shoot.

1. **S. salicifolia** L., *Sp. Pl.* 489 (1753). 1–2 m; branches erect, puberulent when young. Leaves 4–8 × 1–2 cm, elliptic-oblong, cuneate at base, acute or subobtuse, sharply and often doubly serrate, glabrous; petiole very short. Inflorescence 4–12 cm, paniculate; panicle cylindrical, dense, its lower branches usually ascending and not longer than the bracts; rhachis and pedicels pubescent. Flowers *c.* 8 mm in diameter; sepals triangular-ovate, puberulent, erect; petals pink, rarely white. Stamens about twice as long as petals; nectar secreted by a ring of tissue inside the stamens. Follicles erect, glabrous. *C. & E.C. Europe; rather local; widely cultivated and often naturalized elsewhere.* Au Bu Cz Hu Po Rm Rs (B, C, W) [Br ?Da Fe Ga Ge He Ho It Ju No Su].

2. **S. alba** Duroi, *Harbk. Baumz.* **2**: 430 (1772). Like 1 but leaves wider; inflorescence conical, wider at the base and often longer, its lower branches usually spreading and longer than their bracts; flowers white or rarely pink. *Locally naturalized in C. & W. Europe.* [Au Cz Da Fe Hu Ga Ge.] (*Eastern U.S.A.*)

4

3. S. douglasii Hooker, *Fl. Bor.-Amer.* **1**: 172 (1832). Up to 2·5 m, erect, compact. Leaves 3–10 cm, oblong, obtuse or acute, entire at the base, unequally serrate towards the apex; glabrous above, white-tomentose beneath. Inflorescence 10–20 cm, paniculate; panicle narrow or broad. Sepals white, tomentose, deflexed; petals deep pink; nectar-ring absent. Follicles glabrous, converging. *Locally naturalized from gardens in C. & N.W. Europe.* [Au Br Cz Ho Hu Po.] (*W. North America.*)

4. S. tomentosa L., *Sp. Pl.* 489 (1753). Like **3** but leaves ovate to ovate-oblong, acute, unequally and often doubly serrate, yellowish or greyish tomentose beneath; panicle narrow; follicles pubescent, usually diverging. *Locally naturalized from gardens in N. & C. Europe.* [Cz Da Ge.] (*E. North America.*)

5. S. decumbens Koch in Röhling, *Deutschl. Fl.* ed. 3, **3**: 433 (1831). Stems *c.* 0·25 m, procumbent and ascending. Leaves up to 4 cm, oblong-obovate, cuneate and entire at the base, serrate or dentate towards the apex. Inflorescence corymbose; corymbs 3–5 cm wide, wider than long, many-flowered. Incompletely dioecious; flowers 5–7 mm in diameter; petals white, as long as or rarely shorter than stamens. Follicles glabrous. *Calcareous rocks and screes.* ● *S.E. Alps.* Au It Ju.

(a) Subsp. **decumbens**: Leaves acutely serrate, pale green and glabrous beneath. Pedicels and rhachis of inflorescence glabrous. *460–800 m. Mainly in the eastern part of the range.*
(b) Subsp. **tomentosa** (Poech) Dostál, *Feddes Repert.* **79**: 34 (1968) (*S. decumbens* var. *tomentosa* Poech, *S. hacquetii* Fenzl & C. Koch): Leaves finely dentate, grey-tomentose beneath. Pedicels and rhachis of inflorescence tomentose. *600–1600 m. In the western part of the range.*

6. S. japonica L. fil., *Suppl.* 262 (1781). Up to 1·5 m., erect. Leaves up to 10 × 4 cm, ovate to oblong-lanceolate, cuneate at the base, acute, irregularly serrate, glaucous beneath and usually pubescent on the veins. Inflorescence corymbose; corymbs up to 12 cm wide; pedicels pubescent. Sepals deflexed; petals pink, shorter than stamens. Follicles glabrous. *Locally naturalized from gardens in C. Europe.* [Au Cz Ge Hu It.] (*Japan.*)

7. S. corymbosa Rafin., *Précis Découv. Somiol.* 36 (1814). Up to 1 m; branches erect or ascending, terete. Leaves *c.* 5 × 3 cm, ovate, rounded and entire at the base, obtuse or rounded and sharply serrate at the apex, glabrous. Inflorescence corymbose; corymbs 3–10 cm wide, somewhat convex. Sepals not deflexed; petals white or yellowish-white; stamens 3 times as long as petals. Follicles glabrous. *Locally naturalized from gardens.* [Cz.] (*Eastern U.S.A.*)

Subgen. **Nothospiraea** Zabel. Inflorescence a simple umbel or corymb, terminating a short shoot.

8. S. chamaedryfolia L., *Sp. Pl.* 489 (1753) (*S. ulmifolia* Scop.). Up to 2 m, densely branched; stems angular, brown, glabrous. Leaves up to 7 × 4 cm, ovate, ovate-lanceolate or rhombic-elliptical, acute, entire at the base, irregularly or doubly serrate towards the apex, glabrous. Inflorescence *c.* 4 cm, pedunculate, hemispherical, many-flowered; pedicels *c.* 10 mm. Sepals triangular-ovate, revolute; petals *c.* 6 mm, orbicular, shorter than stamens. Follicles glabrous, shining, with remains of style at apex. *Low woodland and scrub.* Carpathians, S.E. Alps, mountains of Balkan peninsula. Au Bu Cz It Ju Rm Rs (W) [Ga Ge He].

9. S. media Franz Schmidt, *Östr. Allgem. Baumz.* **1**: 53 (1792) (*S. oblongifolia* Waldst. & Kit., *S. chamaedryfolia* L. pro parte). Up to 1·5 m, erect; stems terete. Leaves up to 5 × 2 cm, broadly elliptical, rounded at apex, glabrous when mature, sparsely hairy or grey-tomentose beneath when young; those of the non-flowering shoots with 6–8 teeth near apex, those of the flowering shoots entire. Inflorescence up to 4 cm, pedunculate, almost spherical, many-flowered. Sepals *c.* 1 mm, half as long as hypanthium, revolute; petals *c.* 3 mm, orbicular, white or pale yellow, as long as stamens or shorter. Follicles glabrous or thinly hairy, with remains of style on dorsal side. $2n=10$. *Scrub on rocks. N. part of Balkan peninsula and S.C. Europe, extending eastwards to W. Ukraine; also in N. & E. Russia.* Au Bu Cz Hu Ju Po Rm Rs (N, C, W, E).

(a) Subsp. **media**: Inflorescence glabrous. Petals white, entire. *Throughout the range of the species.*
(b) Subsp. **polonica** (Błocki) Pawł., *Feddes Repert.* **79**: 34 (1968) (*S. polonica* Błocki): Inflorescence softly pubescent. Petals pale yellow, fimbriate. *Poland.*

The main area of distribution of this species is in E. Asia and Siberia, extending into the north and east parts of European Russia. It is separated by over 1300 km from the secondary area in C. Europe.

10. S. cana Waldst. & Kit., *Pl. Rar. Hung.* **3**: 252 (1807). Up to 1 m; stems terete, hairy when young. Leaves up to 3·5 × 1·5 cm, elliptical to broadly lanceolate, tapering abruptly at both ends, entire, or rarely with 2–3 teeth at the apex, dark green above, pale green and tomentose beneath. Inflorescence up to 2 cm wide, pedunculate. Hypanthium hairy; sepals revolute; petals *c.* 2 mm, orbicular, white or grey-white, shorter than stamens. Follicles hairy, with remains of style at apex. *Rocky places.* ● *N. Jugoslavia; one station in N.E. Italy.* It Ju.

11. S. crenata L., *Sp. Pl.* 489 (1753) (*S. crenifolia* C. A. Meyer). Up to 1 m; stems erect, finely hairy when young. Leaves up to 4 × 2 cm, lanceolate to obovate with a cuneate base, acute, entire or crenate-serrate towards the apex, with 3 very conspicuous veins running lengthwise from base to apex. Inflorescence *c.* 2 cm wide, pedunculate, the lowermost flowers usually in the axils of small leaves. Flowers *c.* 8 mm in diameter; hypanthium glabrous. Sepals erect in fruit; petals orbicular, white, shorter than stamens. Follicles subglabrous. *S.E. Europe, extending northwards to E. Czechoslovakia and to 55° N. in C. Russia.* Bu Cz Gr †Hu Ju Rm Rs (C, W, E) [Be He Hs].

12. S. × vanhouttei (Briot) Zabel, *Garten-Zeit.* (*Wittmack*) **3**: 496 (1884) (*S. cantoniensis × trilobata*). Up to 2 m; stems arcuate. Leaves 2–3·5 cm, rhombic or obovate, cuneate or rounded at the base, acute, incise-serrate and usually slightly 3- to 5-lobed. Inflorescence 2–5 cm wide, pedunculate. Pedicels 1 cm; petals white, twice as long as stamens. *Locally naturalized from gardens in C. Europe.* [Cz Ge Hu Rm.] (*Garden origin.*)

13. S. hypericifolia L., *Sp. Pl.* 489 (1753) (*S. flabellata* Bertol. ex Guss.). Shrub, up to 1·5 m. Stems terete or slightly angular, hairy or glabrous; the flowering stems arcuate, the non-flowering erect. Leaves 1–2·5(–3) cm, with 3–5 longitudinal veins, narrowly elliptical or obovate, cuneate, entire or (on flowering twigs) with 3–5 crenations at the apex, nearly or quite glabrous. Inflorescence sessile or the lower very shortly pedunculate; flowers numerous. Receptacle and sepals glabrous; petals white, longer than or equalling stamens. *S.W. & S.E. Europe.* Bu Ga Hs Lu Rs (C, W, E) [*Hu *It Rm].

(a) Subsp. **hypericifolia**: Leaves narrowly elliptical, acute, entire. Sepals shorter than hypanthium; petals ovate, rather longer than stamens. *Ukraine and S. Russia; one station in Bulgaria.*

(b) Subsp. **obovata** (Waldst. & Kit. ex Willd.) Dostál, *Feddes Repert.* **79**: 34 (1968). (*S.obovata* Waldst. & Kit. ex Willd.): Leaves narrowly obovate, obtuse, entire or with 3–5 crenations at apex. Sepals equalling hypanthium; petals *c.* 3 mm, obovate, equalling stamens. *S.W. Europe.*

4. Sibiraea Maxim.[1]

Like *Spiraea* but polygamo-dioecious, not hermaphrodite; inflorescence a terminal panicle; carpels connate at base; follicles dehiscent at the apex of the dorsal suture and along the ventral suture.

1. **S. altaiensis** (Laxm.) C. K. Schneider, *Ill. Handb. Laubholzk.* **1**: 485 (1905) (*S. laevigata* (L.) Maxim.). Procumbent shrub up to 1 m. Leaves 30–80 × 6–16 mm, oblong, cuneate at the base, obtuse, mucronate, entire, glabrous. Inflorescence *c.* 3 cm in flower, up to 7 cm in fruit. Sepals *c.* 1 mm, triangular; petals 2–2·5 mm, white; hypanthium tomentose. Fruit 3–4·5 mm. *Calcareous cliffs and rocks, 800–1600 m. W. Jugoslavia.* Ju. (*Mountains of C. Asia.*)

Known in Europe from 3 localities north of Mostar and 3 in the central part of the Velebit. These are separated by over 5000 km from the nearest Asiatic localities in E. Kazakhstan and E. Siberia. The European plant has been described as var. *croatica* (Degen) G. Beck, but when the range of variation of the Asiatic populations is considered, it is not possible to separate the European ones.

Sometimes cultivated for ornament. The plant in cultivation is usually erect and somewhat taller, with larger leaves and inflorescence. It is said to be naturalized in France.

5. Aruncus L.[2]

Polygamo-dioecious perennial herbs. Leaves compound, exstipulate. Inflorescence a panicle. Petals white or yellowish-white; stamens numerous; carpels 3, free. Follicles dehiscent along the ventral suture. Seeds several.

1. **A. dioicus** (Walter) Fernald, *Rhodora* **41**: 423 (1939). (incl. *A. sylvestris* Kostel., *A. vulgaris* Rafin., *Spiraea aruncus* L.). Rhizome stout, much branched; stems up to 2 m, simple, erect. Leaves up to 1 m, 2-pinnate; leaflets ovate, acute, cuneate to subcordate at base, acutely biserrate. Inflorescence large, pyramidal. Flowers *c.* 5 mm in diameter, subsessile, usually unisexual. Petals oblong- to obovate-cuneate. Fruit *c.* 3 mm, pendent. *Damp or shady places in mountain districts. From Belgium and the Pyrenees to S. Poland, C. Ukraine and N. Albania.* Al Au Be Cz Ga Ge He Hs Hu It Ju Po Rm Rs (W).

Subfam. Rosoideae

Stipules present, usually persistent. Flowers usually 4-, 5- or 6-merous. Hypanthium flat or concave, often with a central carpophore, sometimes campanulate or tubular; epicalyx sometimes present; stamens usually numerous, sometimes enclosed in or adnate to the hypanthium. Basic chromosome number 7, 8 or 9.

---

[1] By P. W. Ball.   [2] By T. G. Tutin.

6. Filipendula Miller[1]

Perennial, rhizomatous herbs. Leaves pinnate, usually with small leaflets between the larger ones. Inflorescence a cymose panicle. Flowers usually 5- or 6-merous; hypanthium flat or slightly concave; petals pale cream, sometimes purplish beneath; stamens 20–40; carpels 6–12, in one whorl. Fruit a head of achenes.

Basal leaves with at least 8 pairs of large leaflets; leaflets not more
    than 2 cm; petals 5–9 mm   **1. vulgaris**
Basal leaves with not more than 5 pairs of large leaflets; leaflets
    2 cm or more; petals 2–5 mm   **2. ulmaria**

1. **F. vulgaris** Moench, *Meth.* 663 (1794) (*F. hexapetala* Gilib., *Spiraea filipendula* L.). Subglabrous or sparsely pubescent; stems up to 80 cm, usually simple and with few leaves; roots bearing ovoid tubers. Basal leaves with 8–25 pairs of large leaflets; large leaflets 0·5–2 cm, oblong in outline, pinnatifid, the lobes often toothed. Inflorescence 3–10 cm, wider than long. Petals usually 6, 5–9 mm, purplish beneath; stamens about equalling petals. Achenes 3–4 mm, erect, pubescent. 2*n*=14, 16. *Dry grassland. Most of Europe, northwards to c. 64° N. in Norway.* Al Au Be Br Bu Cz Da Fe Ga Ge Gr Hb He Ho Hs Hu It Ju Lu No Po Rm Rs (N, B, C, W, K, E) Su Tu.

2. **F. ulmaria** (L.) Maxim., *Acta Horti Petrop.* **6**(1): 251 (1879) (*Spiraea ulmaria* L.). Pubescent to tomentose; stems 50–200 cm, simple or branched, leafy; roots not tuberous. Basal leaves with up to 5 pairs of large leaflets; large leaflets 2–8 cm, ovate-oblong to ovate-suborbicular, variously toothed or shallowly lobed. Inflorescence 5–25 cm, usually longer than wide. Petals 5(–6), 2–5 mm; stamens exceeding petals. Achenes *c.* 2 mm, spirally twisted. 2*n*=14, 16, 24. *Throughout Europe except some of the islands and much of the Mediterranean region.* All except Az Bl Co Cr Rs (K) Sa Sb Si Tu.

Variable in the indumentum and toothing of the leaflets. The following subspecies appear to be reasonably distinct.

1 Leaflets green with sparse, long, straight hairs beneath
                      (c) Subsp. denudata
1 Leaflets with crispate hairs and usually white-tomentose beneath
    2 Leaflets crenate-serrulate to shallowly biserrate; achenes glabrous
                    (a) Subsp. ulmaria
    2 Leaflets deeply biserrate or shallowly lobed with serrate lobes; achenes pubescent
                    (b) Subsp. picbaueri

(a) Subsp. **ulmaria**: Stem up to 200 cm, glabrous at least below. Leaflets ovate to ovate-suborbicular, plane, crenate-serrulate to shallowly biserrate, acute with a broadly triangular apex, sparsely to densely white-tomentose beneath. Inflorescence lax. Achenes glabrous. *Usually in damp or wet places. Throughout the range of the species.*

(b) Subsp. **picbaueri** (Podp.) Smejkal, *Preslia* **38**: 253 (1966) (*F. stepposa* Juz.): Stem not more than 100 cm, tomentose. Leaflets ovate to ovate-oblong, the margin crispate, deeply biserrate or shallowly lobed with serrate lobes, acuminate or acute with a long, narrowly triangular apex, always densely white-tomentose beneath. Inflorescence dense. Achenes pubescent. *Relatively dry grassland, steppes and scrub. From E. Austria and S. Czechoslovakia to Bulgaria; S.E. Russia.*

(c) Subsp. **denudata** (J. & C. Presl) Hayek, *Fl. Steierm.* **1**: 872 (1909) (*F. denudata* (J. & C. Presl) Fritsch): Stem up to 200 cm, glabrous or subglabrous. Leaflets ovate to ovate-oblong, plane, deeply biserrate or shallowly lobed with serrate lobes, acuminate or acute, with a long, narrowly triangular apex, green and sparsely hairy beneath with long straight hairs. Inflorescence lax. Achenes glabrous. *Usually in damp or wet places. E.C. & E. Europe.*

7. Rhodotypos Siebold & Zucc.[1]

Deciduous shrubs. Leaves simple, opposite. Flowers solitary, terminal, 4-merous; hypanthium flat; petals white; stamens numerous; carpels usually 4. Fruit a head of dry drupes.

1. **R. scandens** (Thunb.) Makino, *Bot. Mag. Tokyo* **27**: 126 (1913). Erect, up to 2 m or more. Leaves 4–8 cm, ovate or ovate-oblong, acuminate, biserrate, glabrous above, sericeous beneath when young. Flowers 3–5 cm in diameter; sepals dentate, persistent in fruit. Drupes 8–10 mm, black, shining. *Cultivated for ornament and sometimes naturalized.* [?Ga Rs (W).] (*Japan and C. China.*)

8. Kerria DC.[1]

Deciduous shrubs. Leaves simple. Flowers solitary, terminal on lateral branches, 5-merous; hypanthium flat; petals yellow; stamens numerous; carpels 5–8. Fruit a head of achenes.

1. **K. japonica** (L.) DC., *Trans. Linn. Soc. London* **12**: 157 (1818). Erect, up to 2·5 m; branches green. Leaves 2–10 cm, ovate, acuminate, biserrate, glabrous above, pubescent beneath. Flowers 3–5 cm in diameter. Achenes brownish-black. *Widely cultivated for ornament and sometimes naturalized.* [?Ga He ?Rm Rs (W).] (*C. & W. China.*)

Usually *flore pleno* in cultivation.

9. Rubus L.[2]

Perennial herbs or shrubs. Stems usually with prickles. Leaves usually pinnate, digitate or pedate, with 3–7 dentate leaflets. Flowers solitary or in racemose or paniculate inflorescences. Flowers usually 5-merous; hypanthium flat, with a large, usually convex receptacle; epicalyx absent; petals red, purple, pink or white; stamens numerous; carpels numerous; styles subterminal, usually deciduous; ovules 2. Fruit usually a coherent head of 1-seeded drupelets.

The European species of *Rubus*, native and naturalized, are placed in 5 subgenera. The first four contain 9 species, and present no taxonomic difficulties. In the remaining subgenus, *Rubus*, some 2000 species have been described. Almost all are agamospecies, segregated from *R. fruticosus* L. This name, which is based on a mixture of two species (**12**, *R. plicatus* and **38**, *R. ulmifolius*) belonging to different subsections, is now used only in an aggregate sense, and it covers the whole of the section *Rubus*, except for **75** and the *Corylifolii* (see note following **75**). Many of the species are tetraploid ($2n = 28$) but diploids, triploids, pentaploids and hexaploids are also known. It is likely that many of the species have arisen during the Pleistocene era as a result of hybridization and apomixis. The apomicts are all polyploid and pseudogamous; apomixis is often facultative so that new apomictic biotypes can arise at the present time by hybridization, thus increasing the number of potential taxa.

It is thus not profitable to treat fully and by conventional means the whole array of agamospecies, and no attempt at a detailed treatment is made here. The full spectrum of morphological variation has, however, been covered, and this has been done by recognizing 66 *circle-species* which are relatively wide-

spread and distinct. Round these the remaining species can be grouped. (This method was proposed by Gustafsson, *op. cit.*). The description of a circle-species applies only to the circle-species and does not cover the related species which are listed after the appropriate circle-species. All species recognised in Basic and Standard Floras are listed. The key should enable all European species to be run down to their respective subsections and sometimes series, but below this level only those species selected as 'circle-species' are keyed out.

The classificatory framework is based on that of Sudre (*op. cit.*).

In the descriptions the stem- and leaf-characters are those of the middle part of the first-year or non-flowering stem of a well-grown plant in a normal environment.

The armature may include large *prickles*, smaller *pricklets*, *acicles*, stalked *glands* and eglandular *hairs*. The acicles, which may be gland-tipped, are stouter and more rigid than the glands or hairs, but are always straight and never expanded at the base. Prickles which are straight but directed backwards along the stem are said to be *deflexed*; those which are curved as well as deflexed are said to be *falcate*. Sepals which are furnished with pricklets are *aculeolate*; sepals with the apex prolonged are *appendiculate*; descriptions of the indumentum of sepals refer to the outer surface.

Although the inflorescence is always determinate, it resembles a raceme, corymb or panicle in appearance and is called by these terms for convenience.

Endemic signs have not been used for the 'related species' because of lack of information, but it is probable that the great majority of those listed are endemic to Europe.

Literature: W. O. Focke, *Biblioth. Bot.* (*Stuttgart*) **72**(1–2): 1–223 (1910–11); **83** (1914). Å. Gustafsson, *Lunds Univ. Årsskr.* ser. 2, **39**(6) (1943). H. Sudre, *Rubi Europae.* Paris. 1908–1913. W. C. R. Watson, *Handbook of the Rubi of Great Britain and Ireland.* Cambridge. 1958. K. E. Weihe & C. G. Nees von Esenbeck, *Rubi Germanici.* Eberfeld. 1822–27.

Some of the more recent Floras treat *Rubus* very fully, e.g.: J. Legrain in W. Robyns, *Flore Générale de Belgique* **3**: 10–274 Bruxelles. 1958–59. E. I. Nyárády in T. Săvulescu, *Flora Republicii Populare Române* **4**: 276–580. Bucureşti. 1956. J. Dostál, *Květena ČSR* 572–630. Praha. 1950. Á. Kiss in R. Soó & S. Jávorka. *A Magyar Növényvilág Kézikönyve* 251–270. Budapest. 1951.

1 Stems herbaceous or nearly so; stipules free from petiole
  2 Leaves simple
    3 Dioecious; ripe fruit orange   **1. chamaemorus**
    3 Flowers hermaphrodite; ripe fruit red   **2. humulifolius**
  2 Leaves ternate
    4 All stems flower-bearing, without prickles; petals pink
        3. arcticus
    4 Some stems vegetative, with prickles; petals white   **4. saxatilis**
1 Stems woody; stipules united with petiole
  5 Leaves simple; receptacle flat   **5. odoratus**
  5 Leaves compound; receptacle convex
    6 Ripe fruit red or orange, separating from receptacle; leaves ternate or pinnate
      7 Leaves subglabrous beneath; flowers ±solitary; petals bright purple   **6. spectabilis**
      7 Leaves white-tomentose beneath; flowers in racemes; petals white or pink
        8 Stems densely covered with reddish bristles; petals pink, curved inwards   **9. phoenicolasius**
        8 Stems not densely covered with bristles; petals white, erect

---

[1] By P. W. Ball.   [2] By Y. Heslop-Harrison.

9 Inflorescence-axis glandular,with abundant acicles; leaves all ternate **8. sachalinensis**
9 Inflorescence-axis eglandular, with sparse acicles; leaves usually pinnate **7. idaeus**
6 Ripe fruit blackish, adhering to receptacle; leaves ternate, digitate or pedate
10 Stems pruinose; stipules lanceolate; leaflets 3; drupelets 2–20, pruinose (Subsect. *Caesii*) **75. caesius**
10 Stems usually not pruinose; stipules linear-lanceolate to filiform; leaflets 3–7; drupelets usually more than 20, rarely pruinose (*vide* also p. 25 *Corylifolii*)
11 Prickles nearly always very unequal, often scattered on the stem-faces; stem and inflorescence usually with stalked glands (Subsect. *Appendiculati*)
12 Stems with sparse glands; upper leaves usually covered with stellate hairs above; petals yellowish-white **44. canescens**
12 Stems usually with many glands; leaves without stellate hairs above; petals never yellowish
13 Stems weak and terete, often procumbent, pruinose; prickles weak, or broad-based and curved; leaves normally with 3–5 leaflets; inflorescence usually with weak prickles; sepals appressed to the young fruit; petals often small; stamens usually equalling or exceeding styles (Ser. *Glandulosi*)
14 Prickles compressed at base
15 Prickles on any one part of stem all similar; inflorescence with most glands shorter than the diameter of the axis; petals erect **70. scaber**
15 Prickles on stem distinctly unequal; inflorescence with most glands longer than the diameter of the axis; petals ± patent **71. schleicheri**
14 Prickles scarcely compressed at base
16 Terminal leaflet rounded and ± entire at base, shortly acuminate **72. glandulosus**
16 Terminal leaflet cordate or subcordate at base
17 Glands and acicles pale yellow **73. serpens**
17 Glands and acicles brown or purple **74. hirtus**
13 Plant not possessing this combination of characters
18 Inflorescence-glands often longer than the diameter of the axis; prickles very unequal and merging into numerous stalked glands, acicles, and pricklets of varying lengths
19 Stems robust, pruinose, with numerous, long, often glandular hairs and acicles; prickles straight or falcate; leaves pedate with 3–5 leaflets, grey-white tomentose beneath; inflorescence large, with long lower branches and pedicels **69. incanescens**
19 Plant without this combination of characters (Ser. *Histrices*)
20 Flowers usually less than 2 cm in diameter
21 Stem-prickles short, not confluent, not hairy; sepals without prickles; petals 5–7(–11) **66. rosaceus**
21 Stem-prickles large, confluent and hairy; sepals with prickles; petals 5 **64. pilocarpus**
20 Flowers usually more than 2 cm in diameter
22 Flowers white or pale pink **68. koehleri**
22 Flowers bright or deep pink
23 Stems glabrous or glabrescent **67. histrix**
23 Stems sparsely or densely hairy
24 Stems densely hairy; terminal leaflet ovate to suborbicular, acute, cordate at base **63. fuscater**
24 Stems sparsely hairy; terminal leaflet obovate, acuminate, subcuneate at base **65. lejeunei**
18 Inflorescence-glands shorter than the diameter of the axis; stem-prickles slightly or moderately unequal, but distinct from the pricklets
25 Stem-prickles only slightly unequal, mainly on the angles of the stem, with a few tuberculate pricklets on the faces and a few stalked glands (Ser. *Vestiti*)
26 Leaves green beneath
27 Terminal leaflet orbicular or obovate, with a ±

abrupt, short or long point; sepals patent, green, with white margins **48. mucronulatus**
27 Terminal leaflet ovate or ovate-oblong, gradually narrowed to a long point; sepals deflexed, tomentose **49. gremlii**
26 At least the upper leaves white-tomentose beneath
28 Petals suborbicular **45. vestitus**
28 Petals ovate or elliptical, longer than wide
29 Stems with sparse, tufted hairs and a few glands; sepals patent to erect **46. boraeanus**
29 Stems pubescent, with long and short hairs and sunken glands; sepals deflexed **7. adscitus**
25 Stem-prickles unequal, scattered over the stem surface making it rough and tuberculate, with many stalked glands
30 Inflorescence-axis glandular, tomentose or glabrous (Ser. *Rudes*)
31 Stems ± terete, becoming white-pruinose; leaflets usually 3 **62. vallisparsus**
31 Stems angled, not white-pruinose; leaflets usually 5
32 Cauline leaves cordate at base; inflorescence flexuous and lax **61. melanoxylon**
32 Cauline leaves subcuneate at base; inflorescence short, subcorymbiform **60. rudis**
30 Inflorescence-axis glandular and hairy, the glands mainly not exceeding the hairs (Ser. *Radulae*)
33 At least the upper leaves white-tomentose beneath
34 Inflorescence with few, or weak, slender prickles; stamens about equalling styles **52. apiculatus**
34 Inflorescence with numerous strong prickles; stamens exceeding styles
35 Sepals not appendiculate; carpels glabrous **50. radula**
35 Sepals appendiculate; carpels pubescent **51. genevieri**
33 Leaves all green beneath, though sometimes green-tomentose
36 Sepals usually deflexed after anthesis, or variable in the same plant
37 Stems densely hairy; inflorescence with many prickles; bracts long, linear-lanceolate **53. fuscus**
37 Stems sparsely hairy; inflorescence with very few or no prickles; bracts leaf-like **54. foliosus**
36 Sepals patent or becoming erect after anthesis
38 Stems with few or no hairs
39 Sepals green, with white margin, acute, without long apex; inflorescence short, few-flowered **55. infestus**
39 Sepals grey-tomentose, without white margin, with long, linear apex; inflorescence usually long, dense and tapered **56. thyrsiflorus**
38 Stems hairy
40 Flowers deep pink **58. obscurus**
40 Flowers white or pale pink
41 Inflorescence with very few prickles **57. pallidus**
41 Inflorescence with numerous prickles **59. menkei**
11 Prickles ± equal, mainly on the stem-angles; stem and inflorescence ± eglandular, or with sessile glands only
42 Stems glabrous, often suckering from base, rarely rooting apically; inflorescence a raceme or panicle (Subsect. *Suberecti*)
43 Stems high-arching; inflorescence a panicle
44 Leaflets 5–7, those of upper leaves persistently pubescent or almost tomentose beneath; receptacle densely pubescent **15. affinis**
44 Leaflets 5, often glabrous at maturity; receptacle glabrous **14. divaricatus**
43 Stems ± erect; inflorescence usually a raceme
45 Leaflets 5 or 7; prickles weak or short; ripe fruit dark red
46 Stems up to 300 cm; prickles short, conical, often purplish-black; leaflets plane, glabrescent beneath **10. nessensis**

8

46 Stems not more than 150 cm; prickles slender, subulate, yellowish; leaflets plicate, pubescent beneath **11. scissus**

45 Leaflets usually 5; prickles strong; fruit black
47 Stems with rather few, mainly patent, strong prickles; terminal leaflet rather long-acuminate, evenly serrate **13. sulcatus**
47 Stems with numerous, falcate or deflexed prickles; terminal leaflet shortly acuminate, unevenly serrate **12. plicatus**

42 Stems often hairy, usually rooting apically, not suckering; inflorescence a panicle
48 All leaves grey-white tomentose beneath; sepals grey-white tomentose externally, deflexed in fruit (Subsect. *Discolores*)
49 Stems pruinose
50 Leaves coriaceous, dark green above, white-tomentose beneath; stamens scarcely exceeding the styles **38. ulmifolius**
50 Leaves not coriaceous or dark green, tomentose and pubescent beneath; stamens greatly exceeding the styles **39. godronii**
49 Stems not or scarcely pruinose
51 Basal leaflets subsessile; inflorescence often with few prickles above; peduncles elongated **43. candicans**
51 Basal leaflets shortly stalked; inflorescence with many prickles; peduncles not distinctly elongated
52 Inflorescence-prickles patent, mostly straight. **40. bifrons**

52 Inflorescence-prickles falcate
53 Stems glabrescent, angled but not sulcate **41. discolor**
53 Stems hairy and sulcate **42. chloocladus**
48 Leaves green, or only the upper grey-white tomentose beneath; sepals green or grey-tomentose externally, their posture in fruit variable (Subsect. *Silvatici*)
54 Leaves laciniate **27. laciniatus**
54 Leaves not laciniate
55 At least some of the lower leaves grey- or white-tomentose beneath
56 Petals 3 cm or more (Açores) **36. hochstetterorum**
56 Petals less than 3 cm
57 Upper leaves white-tomentose beneath **35. rhamnifolius**
57 Upper leaves grey-tomentose beneath, or not tomentose
58 Panicle narrow, long; rhachis-prickles falcate; sepals white-tomentose **37. lindebergii**
58 Panicle not long and narrow; rhachis-prickles deflexed; sepals grey-white tomentose
59 Leaflets of cauline leaves sometimes 7, becoming distinctly convex; terminal leaflet cuspidate; sepals not long-pointed **34. polyanthemus**
59 Leaflets of cauline leaves always 5, not distinctly convex; terminal leaflet acuminate; sepals usually long-pointed **33. villicaulis**
55 At least the lower leaves green beneath
60 Petals fimbriate **18. pedatifolius**
60 Petals not fimbriate
61 Sepals erect or patent
62 Stems angled; leaflets usually 5; sepals never appendiculate; stamens usually exceeding styles
63 Inflorescence with stalked glands
64 Stem-prickles few, long-subulate **21. hypomalacus**
64 Stem-prickles numerous, some of them strong, broad-based, falcate **20. chaerophyllus**
63 Inflorescence with subsessile glands, or ± eglandular
65 Inflorescence nearly unarmed or with weak prickles **19. gratus**
65 Inflorescence with numerous strong prickles
66 Terminal leaflet elliptical or obovate, rounded

or subcordate at base, coarsely serrate; inflorescence-prickles falcate **17. vulgaris**
66 Terminal leaflet broadly ovate to elliptical, subcordate at base, finely serrate; inflorescence-prickles usually straight **16. lentiginosus**
62 Stems usually terete; leaflets usually 3; sepals usually appendiculate; stamens shorter than, or just equalling the styles
67 Stems angled, at least above; leaflets 5 **25. chlorothyrsos**
67 Stems terete or weakly angled; leaflets usually 3
68 Sepals green, with white margins **22. arrhenii**
68 Sepals grey-tomentose, without white margins
69 Stems and inflorescence-axis hairy; petals bright pink **23. sprengelii**
69 Stems and inflorescence-axis ± glabrous; petals white **24. myricae**
61 Sepals deflexed in fruit
70 Stems ± weak
71 Stems hairy; leaflets 5 **31. silvaticus**
71 Stems ± glabrous; leaflets usually 3 **32. egregius**
70 Stems strong, angled; leaflets usually 5
72 Leaves glabrescent beneath; petals emarginate **26. questieri**
72 Leaves pubescent or tomentose beneath; petals entire
73 Stems angled, with plane faces; terminal leaflet usually subcordate **29. pyramidalis**
73 Stems angled, with sulcate faces; terminal leaflet usually rounded or cuneate at base
74 Leaves green and pubescent, or slightly grey-tomentose beneath **30. macrophyllus**
74 Leaves white-tomentose beneath **28. rhombifolius**

Subgen. **Chamaemorus** (Hill) Focke. Dioecious. Stems annual, unarmed. Leaves simple. Fruit yellow to orange; receptacle convex.

1. **R. chamaemorus** L., *Sp. Pl.* 494 (1753). Stems 5–20 cm, arising from a creeping rhizome, those on male plants all flowering, glandular. Leaves reniform, rugose, with 5 obtuse, crenate-serrate lobes. Flowers solitary, terminal; pedicel and calyx shortly glandular. Sepals erecto-patent, ovate, acuminate; petals 5 or more, white, hairy, larger than the sepals. Drupelets about 20, large, edible. 2*n* = 56. *Mountain moors and bogs. N. Europe, extending southwards to N.W. Czechoslovakia.* Br Cz Da Fe Ge Hb No Po Rs (N, C) Sb Su.

Subgen. **Cyclactis** (Rafin.) Focke. Stems annual, armed or unarmed. Leaves simple or ternate. Flowers hermaphrodite. Fruit red to purplish, scarcely coherent; receptacle flat or convex.

2. **R. humulifolius** C. A. Meyer, *Beitr. Pfl. Russ. Reich.* **5**: 57 (1848). Stems 10–30 cm, erect or ascending, setose, pubescent or glabrous. Leaves 3- to 5-lobed, cordate, often wider than long, coarsely serrate or biserrate, pubescent above, glabrescent beneath except on the veins; stipules filiform, but often abortive in the upper leaves. Inflorescence of 1(–3) flowers. Sepals erecto-patent, lanceolate, puberulent; petals linear-lanceolate, acuminate, white, sometimes fugacious; stamens short, the outer filaments dilated, the inner filiform; carpels 5, glabrous; styles long. Drupelets 5–6 × 3 mm, often solitary, purplish-red, acid. *N. Russia.* †Fe Rs (N, C).

3. **R. arcticus** L., *Sp. Pl.* 494 (1753). Stems 10–30 cm, all flowering and unarmed. Leaves 3-lobed, or with 3(–5) ovate, unevenly serrate leaflets. Inflorescence of 1–3 long-pedicellate flowers 1·5–2·5 cm in diameter. Sepals and petals 5–7(–10); sepals glabrous; petals ovate, often toothed, pink; stamens purple, erect, incurved at apex, as long as the styles; carpels pubescent. Drupelets

numerous, dark red. $2n = 14$. *N. Europe.* †Br Fe No Rs (N, B, C) Su.

4. R. saxatilis L., *Sp. Pl.* 494 (1753). Stems 10–50 cm, the vegetative stems procumbent, terete, hairy, armed with small, straight prickles, often rooting at the apex, dying back nearly to their bases, from which arise the flowering-stems in the following year. Leaves ternate, ovate-elliptical, unevenly serrate, glabrescent above, slightly hairy beneath; stipules ovate. Inflorescence a 3- to 10-flowered corymb. Sepals lanceolate, acuminate, shortly pubescent; petals erect, narrow, small, white; stamens erect, white, exceeding the styles. Drupelets 2–6, red, shining. $2n = 28$. *Most of Europe, but rare in the south-west.* Al Au Be Br Bu Cz Da Fa Fe Ga Ge Gr Hb He Ho Hs Hu Is It Ju No Po Rm Rs (N, B, C, W, E) Su.

Subgen. **Anoplobatus** Focke. Stems biennial, woody, unarmed. Leaves simple, palmately lobed. Flowers hermaphrodite. Fruit red or orange; receptacle flat.

5. R. odoratus L., *Sp. Pl.* 494 (1753). Stems up to 300 cm, erect, glandular, hairy. Leaves up to 25 cm, 5-lobed, cordate at base, biserrate, hairy beneath. Inflorescence many-flowered; flowers 3–5 cm in diameter, fragrant. Petals purplish-pink. Drupelets small, red, hairy. *Cultivated for ornament and often more or less naturalized.* [Be Br Cz Fe Ga Hb Rm Rs.] (*E. North America.*)

Subgen. **Idaeobatus** Focke. Stems biennial, woody, armed. Leaves ternate or pinnate. Flowers hermaphrodite. Fruit red or orange, pubescent, separating from the convex receptacle when ripe.

6. R. spectabilis Pursh, *Fl. Amer. Sept.* **1**: 348 (1814). Stems 100–200 cm, erect, with numerous prickles. Leaves usually ternate; leaflets ovate, incise-serrate, thin, subglabrous beneath. Flowers 2·5 cm in diameter, usually solitary on lateral, leafy branches. Sepals triangular-ovate, pubescent; petals bright purple. Fruit large, orange, edible. *Cultivated for ornament and sometimes more or less naturalized.* [Br Ga Ge Ho.] (*W. North America.*)

7. R. idaeus L., *Sp. Pl.* 492 (1753). Suckering by adventitious buds from the roots; stems 100–150 cm, erect, terete, pruinose, often armed with numerous weak prickles. Leaves usually pinnate with 5–7 leaflets or ternate, glabrescent above, white-tomentose beneath; terminal leaflet ovate or oblong, sometimes slightly lobed, cordate, shortly acuminate; stipules filiform, ciliate. Infloresence of few-flowered, leafy, terminal and axillary racemes, the axis eglandular, with sparse acicles; flowers *c.* 1 cm in diameter, nodding. Sepals lanceolate, tomentose; petals narrow, erect, glabrous, white; stamens white, erect. Fruit red or orange. $2n = 14$. *Most of Europe, but only on mountains in the south.* All except Az Bl Cr Fa Is Lu Sb Tu; introduced in Sa.

Many variants, some unarmed or with simple leaves, are widely cultivated for their edible fruits (raspberry).

R. loganobaccus L. H. Bailey, *Gentes Herb.* **1**: 155 (1923) originated in 1881 in a Californian garden as a cross between *R. idaeus* subsp. *strigosus* (Michx) Focke and *R. ursinus* subsp. *vitifolius* Cham. & Schlecht., and is widely cultivated for its fruit (loganberry); it has robust, long-arching stems, large pinnate leaves with 5 leaflets, somewhat patent petals, large, purplish-red fruits which adhere to the receptacle, and $2n = 42$.

R. illecebrosus Focke, *Abh. Nat. Ver. Bremen* **16**: 278 (1899), from Japan, with short, erect, more or less herbaceous stems, pinnate leaves, white petals and red, ellipsoid-globose fruits, is cultivated for its fruit and is reported as locally naturalized in N. Europe.

8. R. sachalinensis Léveillé, *Feddes Repert.* **6**: 352 (1909) (*R. idaeus* subsp. *sachalinensis* (Léveillé) Focke). Like 7 but leaves all ternate; inflorescence-axis glandular, with abundant acicles; fruit rather dry ,velutinous. *N.E. Russia.* Rs (N, C). (*Siberia.*)

9. R. phoenicolasius Maxim., *Bull. Acad. Imp. Sci. Pétersb.* **17**: 160 (1872). Stems 200–300 cm, erect, densely covered with reddish, glandular bristles and sparse, slender prickles. Leaves usually ternate; leaflets broadly ovate, coarsely biserrate, slightly hairy above, white-tomentose beneath. Inflorescence a short, terminal raceme; flowers *c.* 1 cm in diameter. Sepals large, enclosing the young fruit, lanceolate, glandular-hairy; petals curved inwards, pink. Fruit *c.* 2 cm, ovoid, red, sweet. *Cultivated for ornament and for the edible fruit and occasionally naturalized.* [Au Br Cz Ge He.] (*E. Asia.*)

Subgen. **Rubus**. Stems biennial, woody. Leaves ternate, digitate or pedate, with 5–7 leaflets. Flowers hermaphrodite. Fruit black or red, more or less coherent and adherent to the convex receptacle.

Sect. RUBUS. The only section in Europe. Most species occur in open woodland, scrub, hedgebanks and neglected meadows.

Subsect. *Suberecti* P. J. Mueller. Stems usually suberect, glabrous, often suckering from the base and rarely rooting apically, often angled; prickles subequal; stalked glands usually absent. Leaves usually more or less green beneath. Inflorescence usually a raceme or corymb. Sepals green externally, often with white margin; petals often hairy. Fruit black or reddish-black. Flowering early.

10. R. nessensis W. Hall, *Trans. Roy. Soc. Edinb.* **3**: 21 (1794) (*R. suberectus* G. Anderson ex Sm.). Stems up to 300 cm, erect, glaucous, with rather short, conical, often purplish-black prickles. Leaflets 5(–7), plane, shining, thin, glabrescent; terminal leaflet ovate, acuminate, cordate, evenly and simply serrate; basal leaflets subsessile; petiole slightly sulcate above. Inflorescence subracemose, unarmed or bearing weak, falcate prickles; flowers large. Petals white, sometimes red-flushed externally, glabrous; stamens exceeding styles. Fruit dark red. $2n = 28$. *Heaths, mountains and upland woods. Most of Europe except the Mediterranean region and the extreme north.* Au Be Br Cz Da Ga Ge He Hb Ho Hu It Ju No Po Rm Rs (B, C, W) Su.

11. R. scissus W. C. R. Watson, *Jour. Bot.* (*London*) **75**: 162 (1937) (*R. fissus* auct. mult., non Lindley). Stems 50–150 cm, ascending, thinly pubescent, not pruinose, with numerous, scattered, subulate, yellowish prickles. Leaflets usually 7, imbricate, plicate, thick, densely pubescent beneath; terminal leaflet ovate-cordate, shortly acuminate, usually unevenly serrate; petiole distinctly sulcate above. Inflorescence short. Sepals patent to erect, green, with a white margin, often bearing a single pricklet; petals usually 10–15, narrowly obovate, pubescent, white, sometimes pink in bud; stamens shorter than styles; carpels and receptacle pubescent. Fruit dark red, partly abortive. $2n = 28$. *Heaths and upland woods. N. & N.C. Europe.* Be Br ?Cz Da Ga Hb Ho Hu No Po Rs (B) Su.

Related species include:

R. **graecensis** Maurer in Hegi, *Ill. Fl. Mitteleur.* ed. 2, 4(2): 315 (1965). Au Ju.

12. R. **plicatus** Weihe & Nees, *Rubi Germ.* 15 (1822). Stems erect, angled, glabrous, with falcate or deflexed, slender, yellow or crimson prickles. Leaflets 5–7, plicate, hairy on both surfaces, the lower surface sometimes almost grey-tomentose; terminal leaflet broadly ovate-cordate, rather shortly acuminate, coarsely and unevenly serrate; basal leaflets subsessile. Inflorescence racemose-corymbose, bearing only a few prickles. Sepals concave, cuspidate, patent, green, with a white margin; petals abruptly clawed, white or pink; receptacle densely hairy; carpels glabrous. Fruit black. $2n = 28$. *Heaths and upland woods. C. & N.W. Europe, extending to Italy and Bulgaria.* Au Be Br Bu Cz Da Ga Ge Hb He Ho Hu It Ju No Po Rm Rs (C, W) Su.

Related species include:

R. **ammobius** Focke, *Syn. Rub. Germ.* 118 (1877). Da Ge.
R. **bertramii** G. Braun ex Focke, *op. cit.* 117 (1877) (*R. biformis* Boulay). $2n = 28$. Be Br Da Ga Ge.
R. **opacus** Focke in Alpers, *Verz. Gefässpfl. Landdr. Stade* 25 (1875). $2n = 28$. Br Cz Ga Ge Ho.

13. R. **sulcatus** Vest ex Tratt., *Rosac. Monogr.* 3: 42 (1823). Stems up to 300 cm, slightly branched below, strongly sulcate nearly to the base, with rather few, strong, mainly patent prickles. Leaflets 5, large, pubescent beneath; terminal leaflet cordate-ovate, long-acuminate, evenly serrate; basal leaflets shortly stalked; petiole with falcate prickles. Inflorescence long, lax, subracemose, not or scarcely armed, hairy. Sepals somewhat deflexed, sometimes appendiculate, hairy and grey-tomentose, with a white margin; petals obovate, white, sometimes pink in bud; carpels glabrous; receptacle glabrous or subglabrous. Fruit black. $2n = 28$. *C. & N.W. Europe, extending to Italy.* Au Be Br Cz Da Ga Ge He Ho Hu It Ju No Po Rm Rs (W) Su.

Related species include:

R. **altissimus** Fritsch in Hayek, *Sched. Fl. Stir. Exsicc.* 5–6: 11 (1905). Au.

14. R. **divaricatus** P. J. Mueller, *Flora* (*Regensb.*) 41: 130 (1858) (*R. nitidus* Weihe & Nees, non Rafin.). Stems up to 100 cm, shining brown to violet, sparsely hairy or glabrous, suckering and also sometimes rooting at the apex, with numerous long, slender, straight prickles which are at first bright yellow. Leaflets 5, small, sparsely hairy above, pubescent beneath, often glabrous at maturity, with yellowish veins beneath; terminal leaflet ovate to obovate, shortly acuminate, subcordate or entire at base, unevenly serrate; petiole sulcate above, often with many hooked prickles. Inflorescence usually long and lax, broad, with 2- to 5-flowered branches, which are often branched near the base, often bearing numerous hooked prickles particularly near the base of the calyx. Sepals patent or slightly deflexed, green, with a white margin; petals hairy, white to deep pink; stamens white or pink, equalling or exceeding styles; carpels and receptacle glabrous. Fruit small, black. $2n = 21$. *Wet heaths and streamsides. W. & C. Europe, extending to Italy and Sweden.* Au Be Br Cz Da Ga Ge He Ho Hu It Lu Po Rm Su.

Related species include:

R. **contiguus** (Gelert) Raunk., *Dansk Ekskurs.-Fl.* ed. 3, 167 (1914). Da.

15. R. **affinis** Weihe & Nees, *Rubi Germ.* 18 (1822). Stems strong, arching, angled, with strong, straight, long-based prickles, rooting at apex, and occasionally also suckering from base. Leaflets 5–7, imbricate, sparsely hairy above and sometimes grey-tomentose beneath; terminal leaflet broad, ovate-cordate, acuminate, coarsely serrate, undulate. Inflorescence much-branched, leafy, and with long, strong, patent or falcate prickles; flowers large. Sepals somewhat deflexed, appendiculate, green-tomentose, with a whitish margin, covered with acicles; petals large, pink or white, ovate-orbicular; stamens exceeding styles; receptacle densely pubescent; carpels pubescent or glabrous. Fruit imperfect; drupelets large. $2n = 28$. *Moist heaths and grassland. N.W. & C. Europe, extending to Sweden.* Au Be Br Da Ga Ge He Ho ?Hs Hu Rm Su.

Related species include:

R. **fissus** Lindley, *Syn. Brit. Fl.* ed. 2, 92 (1835) (*R. rogersii* E. F. Linton). $2n = 28$. Br ?Cz Hb.
R. **senticosus** Koehler ex Weihe in Wimmer & Grab., *Fl. Siles.* 2(1): 51 (1829). Au Cz Ge Ho Hu Po. This species is placed here by Sudre, but in the opinion of most later authors it is in the separate Subsect. *Senticosi*.

Subsect. *Silvatici* P. J. Mueller. Stems arching, rooting apically in autumn; prickles mainly subequal and confined to the angles; stalked glands usually absent or few. Leaves green, or only the upper grey-tomentose beneath. Inflorescence often paniculate. Sepals green or grey-tomentose, their posture in fruit variable.

16. R. **lentiginosus** Lees in Steele, *Handb. Field Bot.* 60 (1847) (*R. carpinifolius* Weihe & Nees, non J. & C. Presl). Stems almost erect, robust, angled, sparsely hairy, with sessile glands and numerous, strong, broad-based, yellow to brick-red prickles. Leaflets 5–7, plicate, slightly hairy above and pubescent or grey-tomentose beneath; terminal leaflet broadly ovate to elliptical, acuminate, subcordate at base, finely and unevenly serrate. Inflorescence racemose or sometimes paniculate at base, leafy at base, with numerous, strong, usually straight prickles. Sepals patent, grey-green, sometimes hairy or aciculate, without white margins; petals ovate, white; stamens white, exceeding styles; carpels usually hairy; receptacle pubescent. $2n = 28$. *N.W. & N.C. Europe.* Be Br Cz Ga Ge Ho Hu.

17. R. **vulgaris** Weihe & Nees, *Rubi Germ.* 38 (1824). Stems robust, angled, often sulcate and black-purple, sometimes arching to touch the ground; prickles long, strong, nearly straight. Leaflets 5–7, plicate or undulate, pubescent to grey-tomentose beneath; terminal leaflet elliptical or obovate, rounded or subcordate at base, coarsely and unevenly serrate. Inflorescence leafy, subcorymbose at apex, pubescent, with subsessile glands and numerous strong, falcate prickles; bracts and sepals sometimes glandular. Sepals grey-tomentose, patent to somewhat deflexed; petals obovate, white or pink; stamens white, exceeding the green, red or yellowish styles; receptacle and carpels sometimes hairy. $2n = 21$. *N.W. & C. Europe.* Au Be Br Cz Ga Ge He Ho Hu Po Rm.

Related species include:

R. **selmeri** Lindeb. ex F. Aresch., *Bot. Not.* 1886: 76 (1886) (*R. nemoralis* sensu W. C. R. Watson, non P. J. Mueller). $2n = 28$. Br Da Ge Hb Ho No. Sudre treats *R. nemoralis* as a subspecies of **35**; the disagreement between Sudre and Watson may be due to the variable amount of leaf-tomentum.

R. incurvatus Bab., *Ann. Nat. Hist.* ser. 2, **2**: 36 (1848). $2n = 28$. Br Hb Hs. In Britain this species may have persistently tomentose leaves, and is placed by Watson near **23**.

18. R. pedatifolius Genev., *Mém. Soc. Acad. (Angers)* **8**: 93 (1860) (*R. clethraphilus* Genev.). Stems slender, angled, sulcate, brown and shining, glabrous, with rather few, subulate, yellowish prickles. Leaflets 3–5, glabrous above, shortly and densely pubescent beneath, those of the upper leaves sometimes grey-tomentose beneath; terminal leaflet elliptic-obovate, shortly acuminate, unevenly serrate. Inflorescence moderately long, with numerous, weak, straight prickles and a few glands and acicles; flowers *c.* 2·5 cm in diameter. Sepals somewhat deflexed or patent, appendiculate, grey-tomentose or pubescent; petals elliptical, fimbriate, pink; stamens white (sometimes drying pink), exceeding styles; carpels and receptacle pubescent. $2n = 28$. *W. Europe, extending to N. Italy.* Br ?Cz Ga Hs It.

19. R. gratus Focke in Alpers, *Verz. Gefässpfl. Landdr. Stade* 26 (1875). Stems robust, arching, sulcate, red, glabrescent, with a few patent or falcate prickles. Leaflets 5, large, green, glabrescent beneath; terminal leaflet ovate to obovate, acuminate, coarsely, unevenly and doubly dentate, usually entire at base. Inflorescence short and broad, pyramidal, leafy, pubescent, eglandular, almost unarmed, with weak subulate pricklets; peduncles long, ascending; flowers large, the petals, stamens and styles pink or pink at the base. Sepals appressed to fruit, greenish-grey, with white margin; filaments very long, investing the styles; anthers and receptacle usually hairy. Fruit large. $2n = 28$. *N.W. & C. Europe.* Be Br Cz Da Ga Ge Hb Ho Hu Po.

Related species include:

R. sciocharis (Sudre) W. C. R. Watson, *Jour. Ecol.* **33**: 339 (1946) (*R. sciaphilus* Lange, non P. J. Mueller & Lefèvre). $2n = 28$. Br Da Ge.

20. R. chaerophyllus Sagorski & W. Schultze, *Deutsche Bot. Monatsschr.* **12**: 1 (1894). Stems robust, angled, purple, slightly hairy, bearing mainly sessile glands and numerous unequal prickles, the largest broad-based and falcate. Leaflets 5, large, imbricate, pubescent beneath; terminal leaflet broad, orbicular-ovate to elliptical, shortly acuminate, more or less cordate at base, coarsely and unevenly serrate or biserrate. Inflorescence pyramidal, leafy, with long, patent, few-flowered branches, the axis hairy, with unequal prickles, numerous stalked glands and acicles; flowers up to 3 cm in diameter. Sepals patent or appressed to fruit, green-tomentose, with white margins, sometimes with acicles and glands; petals elliptic-oblanceolate, white or pinkish; stamens white, exceeding the greenish styles. Fruit rather large. *W. & C. Europe.* Be Br Cz Da Ga Ge Hb Ho Hs Po.

21. R. hypomalacus Focke, *Syn. Rub. Germ.* 274 (1877). Stems angled, sparsely pubescent, with a few long, subulate, yellow prickles and sometimes with a few pricklets, acicles and stalked glands. Leaflets 3–5, large, softly pubescent to glabrescent beneath; terminal leaflet elliptical or oblong, acuminate, subcordate, coarsely serrate. Inflorescence subracemose, rather few-flowered, leafy, often with long, ascending, paniculate lower branches, the axis with numerous hairs, sessile glands and acicles, and a few stalked glands; peduncles short, 2- to 3-flowered; flowers *c.* 2·5 cm in diameter. Sepals erecto-patent, slightly tomentose, and with short hairs, long acicles and a few glands; petals elliptical or obovate, white or pale pink; stamens usually white, equalling or exceeding the green styles; anthers sometimes slightly pubescent; carpels and receptacle pubescent. *N.W. & C. Europe.* Au Be Br Cz Da Ge Hb Ho.

Related species include:

R. bracteosus Weihe ex Lej. & Court., *Comp. Fl. Belg.* **2**: 162 (1831). Be Br Ge Rm.

22. R. arrhenii (Lange) Lange, *Haandb. Danske Fl.* ed. 2, 347 (1859). Stems rather weak, slender, terete, pubescent, usually eglandular; prickles numerous, small, yellow-based, deflexed. Leaflets 3–5, bright green, pubescent beneath; terminal leaflet elliptical, acuminate, finely and evenly serrate, subcordate to cuneate at base. Inflorescence very long, often pendent, lax, the branches erecto-patent, hairy or tomentose and with sparse glands and numerous fine, straight or falcate prickles. Sepals patent to erect, green, with white margins, hairy, appendiculate, sometimes aciculate and with sparse glands; petals ovate or orbicular, pink or white; stamens much shorter than styles; carpels glabrous or hairy. $2n = 28$. *N.W. & W.C. Europe.* Be Br Cz Da Ge Ho.

23. R. sprengelii Weihe, *Flora (Regensb.)* **2**: 17 (1819) (*R. borreri* Bell Salter). Stems terete, hairy, occasionally with a few pricklets and stalked glands; prickles slender, falcate or subuncinate. Leaflets 3–5, usually hairy beneath; terminal leaflet ovate, obovate or elliptical, acuminate, unevenly and sharply biserrate, entire at base. Inflorescence short, sub-corymbose, lax, with a hairy axis and patent branches and pedicels bearing small hooked prickles. Sepals more or less appressed to fruit, appendiculate, grey-tomentose and pubescent, sometimes slightly glandular or aciculate; petals crumpled, narrowly obovate-oblong, bright pink; stamens pink, about equalling the pink styles; carpels and receptacle hairy. Fruit rather small. $2n = 28$. *N.C. Europe, extending to S. Sweden and Ireland.* Be Br Cz Da Ga Ge Hb Ho Hu Po Su.

Related species include:

R. drejeri G. Jensen in Lange, *Icon. Pl. Fl. Dan.* **51**: 7 (1883). Da.
R. euchloos Focke in Ascherson & Graebner, *Syn. Mitteleur. Fl.* **6**(1): 470 (1902) (*R. orthoclados* A. Ley, non Boulay). ?Be Br Cz Hu. Watson equates this species with *R. bracteosus* Weihe ex Lej. & Court.
R. hemistemon P. J. Mueller ex Genev., *Mém. Soc. Acad. (Angers)* **24**: 314 (1868). Cz Ga Ge He.

24. R. myricae Focke in Alpers, *Verz. Gefässpfl. Landdr. Stade* 27 (1875). Stems procumbent to arcuate, slightly angled, nearly glabrous, with a few equal, subulate, prickles. Leaflets 3–5, hairy on both surfaces; terminal leaflet ovate-elliptical, acuminate, more or less cordate at base, unevenly dentate. Inflorescence with many ternate leaves, the lower branches many-flowered, the upper 1- to 3-flowered; axis only slightly hairy, with very few prickles; pedicels tomentose. Sepals appressed to young fruit, cuspidate, tomentose; petals orbicular or oblong, white; stamens white, much shorter than or just equalling the green styles; receptacle hairy; carpels glabrous. *N.C. & N.W. Europe; Romania.* Be Cz Ga Ge He Ho Rm.

Related species include:

R. cuiedensis E. I. Nyárády in Săvul., *Fl. Rep. Pop. Române* **4**: 900 (1956). Rm.
R. moldavicus E. I. Nyárády in Săvul., *op. cit.* 899 (1956). Rm.

25. R. chlorothyrsos Focke, *Abh. Nat. Ver. Bremen* **2**: 462 (1871). Stems low-arching or procumbent, angled at least above, densely hairy, sometimes with a few glands and acicles; prickles numerous, rather small, slender, subequal, hairy, deflexed or falcate. Leaflets 5, hairy or grey-tomentose beneath and with pectinately arranged hairs on the veins; terminal leaflet 4–5 times as long as its petiolule, usually elliptical, acuminate, rounded at base. Inflorescence long, leafy to the apex, the axis hairy, sparsely glandular and with numerous falcate or deflexed prickles; bracts leaf-like; flowers 1–2 cm in diameter. Sepals appressed to young fruit, grey- or white-tomentose and with longer hairs, appendiculate, sometimes with a few acicles and stalked glands; petals obovate, small, white; stamens white, slightly shorter than or equalling the greenish styles; receptacle hairy; carpels glabrous or pubescent. $2n = 28$. *N.W. & N.C. Europe, extending to Hungary.* Au Be Br Cz Da Ge Hb Ho Hu Po.

Related species include:

R. axillaris Lej. in Lej. & Court., *Comp. Fl. Belg.* **2**: 166 (1831) (*R. leyi* Focke, *R. scanicus* F. Aresch.). $2n = 28$. Be Br Da Ge Su.
R. carpinetorum Freyn, *Verh. Zool.-Bot. Ges. Wien* **31**: 373 (1882). Ju.
R. cimbricus Focke, *Abh. Nat. Ver. Bremen* **9**: 334 (1886). Cz Da Ge.
R. danicus Focke, *op. cit.* 322 (1886). Br Da Ge.
R. fictus Sudre, *Bat. Eur.* 70 (1907). Cz Ga Hu.
R. loretianus Sudre, *Rubi Eur.* 37 (1908). Ga Rm.
R. orthosepalus Halácsy, *Verh. Zool.-Bot. Ges. Wien* **35**: 664 (1886). Au.

26. R. questieri P. J. Mueller & Lefèvre, *Pollichia* **16–17**: 120 (1859). Stems robust, angled, with plane faces, glabrescent, armed with strong, brown, patent to falcate prickles and a few smaller gland-tipped pricklets; young shoots bronze-coloured. Leaflets 3–5, glabrescent on both surfaces; terminal leaflet elliptical to obovate, long-acuminate, rounded or subcordate at base, coarsely and unevenly serrate. Inflorescence elongate, narrow, leafy, the upper leaves grey-tomentose; pedicels shorter than the leaf-like bracts; axis grey-tomentose and with short hairs, a few glands and falcate prickles. Sepals somewhat deflexed at first, the apices sometimes rising as the fruit swells, long-pointed, tomentose, unarmed or with a few acicles; petals obovate, emarginate, pink; stamens white or pink-based, exceeding the yellowish or pinkish styles; carpels usually glabrous. $2n = 28$. *W. & C. Europe.* Au Be Br Cz Ga Hb Hu It Lu Rm.

Related species include:

R. castranus Samp. ex Coutinho, *Fl. Port.* 298 (1913). Lu.
R. maassii Focke ex Bertram, *Fl. Braunschw.* 75 (1876). Br Cz Ge Ho Rm.
R. mercicus Bagnall, *Jour. Bot. (London)* **30**: 372 (1892). $2n = 28$. Br Hs.
R. muenteri Marsson, *Fl. Neu-Vorpommern* 144 (1869). Be Br Ga Ge Lu Su.
R. nemoralis P. J. Mueller, *Flora (Regensb.)* **41**: 139 (1858). Br Da Ge No.
R. scheutzii Lindeb. ex F. Aresch., *Bot. Not.* **1886**: 38 (1886). $2n = 28$. Su.

27. R. laciniatus Willd., *Hort. Berol.* **2**(7): 82 (1806). Stems robust, sulcate, glabrescent, armed with numerous, equal, falcate prickles. Leaflets 5, long-stalked, divided into pairs of laciniate segments, glabrescent or hairy beneath. Inflorescence broad and leafy, with numerous short, falcate prickles. Sepals deflexed, often appendiculate, grey-tomentose; petals incised at the apex, white or pink. $2n = 28$. *Cultivated for ornament and widely naturalized in many areas.* [Be Br Cz Da Fe Ga Ge Ho Rm Su.] (*Origin unknown.*)

28. R. rhombifolius Weihe ex Boenn., *Prodr. Fl. Monast.* 151 (1824) (*R. argenteus* Weihe & Nees, non C. C. Gmelin). Stems angled and sulcate, deep red, slightly hairy, eglandular, with long, subulate or falcate prickles. Leaflets 5, large, subglabrous above, white-tomentose and pubescent beneath; terminal leaflet elliptical to ovate, gradually acuminate, *c.* 3 times as long as its petiolule, unevenly and finely serrate. Inflorescence large, pyramidal, with long-stalked, cymose branches, leafy; axis hairy, with subsessile, inconspicuous glands; prickles few to numerous, nearly straight; bracts not leaf-like, often glandular. Sepals deflexed after flowering, grey-green tomentose, unarmed or aciculate; petals suborbicular-ovate, entire, downy, pink; stamens white or pink, exceeding the pink styles; anthers pubescent; carpels glabrous or pubescent; receptacle hairy. $2n = 28$. *W. & C. Europe, extending to Bulgaria.* Au Be Br Bu Cz Da Ga Ge Hb He ?Ho ?Hs Hu Lu Po Rm.

Related species include:

R. albiflorus Boulay & Lucand ex Coste, *Fl. Fr.* **2**: 37 (1901). Ga Hu.
R. alterniflorus P. J. Mueller & Lefèvre, *Pollichia* **16–17**: 160 (1859). $2n = 28$. Be Br Cz Ga Ge Hu.
R. centronotus A. Kerner, *Ber. Naturw. Ver. Innsbruck* **2**: 411 (1871). Au.
R. cordifolius Weihe & Nees, *Rubi Germ.* 21 (1822), non J. & C. Presl. Cz Hu Po.
R. exornatus E. I. Nyárády in Săvul., *Fl. Rep. Pop. Române* **4**: 908 (1956). Rm.
R. lasiothyrsus Sudre, *Bull. Assoc. Fr. Bot.* **4**: 234 (1901). Be Br Cz Ga.
R. majusculus Sudre, *Bat. Eur.* 71 (1907). Br Ga Rm.
R. prolongatus Boulay & Letendre ex Coste, *Fl. Fr.* **2**: 41 (1901). Ga.
R. sampaianus Sudre in Samp., *Ann. Sci. Nat. (Porto)* **9**: 32 (1905). Ga Hs Lu.
R. schenkei Hruby, *Verh. Naturf. Ver. Brünn* **74** (Beih.) 80 (1944) (*R. cordifolius* × *candicans* ?). Cz.
R. silesiacus Weihe in Wimmer & Grab., *Fl. Siles.* **2**(1): 53 (1829). $2n = 28$. Br Cz Ge He Hu Po. Placed nearer to **43** by Focke and nearer to **49** by Watson.
R. subincertus Samp., *Ann. Sci. Nat. (Porto)* **9**: 31 (1905). Hs Lu.
R. sueviacus Sudre, *Bull. Soc. Bot. Fr.* **55**: 177 (1908). Au.
R. vallicola P. J. Mueller, *Pollichia* **16–17**: 188 (1859). Au Be Br Cz Hb. Placed near to **28** by Watson, but nearer to **34** by Sudre and Legrain.
R. wimmeranus Spribille, *Zeitschr. Naturw. Abt. Deutsch. Ges. Wiss. (Posen)* **9**: 117 (1902). Cz.

29. R. pyramidalis Kaltenb., *Fl. Aachen. Beck.* **2**: 275 (1844). Stems low-arching or procumbent, angled, with plane faces, becoming reddish, glabrous or slightly hairy and sometimes with a few stalked glands and acicles; prickles numerous, subequal, patent and straight or slightly falcate. Leaflets 5, yellow-green, glabrescent above, hairy and tomentose beneath, with shining, yellow hairs on the veins; terminal leaflet orbicular to elliptical, acuminate, usually subcordate at the base, coarsely and unevenly biserrate. Inflorescence dense, pyramidal during flowering, but the upper branches lengthening during fruiting, leafy at the base, hirsute and slightly glandular, with narrow-based, nearly straight prickles. Sepals deflexed at first but sometimes rising after anthesis, glandular, aciculate, slightly tomentose; petals obovate or ovate-elliptical, pale pink; stamens white, exceeding the green styles; carpels glabrous; receptacle hairy. $2n = 28$. *N.W. & C. Europe.* Au Be Br Cz Da Ga Ge Hb He Ho Hs Hu Po Su.

Related species include:

R. dumnoniensis Bab., *Jour. Bot.* (*London*) **28**: 338 (1890). $2n=28$. Be Br Ga Hb Hs.

30. **R. macrophyllus** Weihe & Nees, *Rubi Germ.* 35 (1824). Stems more or less distinctly angled and furrowed above, slightly glaucescent, hairy, sometimes with occasional glands; prickles short, subulate. Leaflets 5, large, usually green and rather pubescent beneath, but occasionally slightly greyish-tomentose, the veins not hairy; terminal leaflet very large, ovate-cordate, long-acuminate, twice as long as its petiolule, unevenly and finely dentate. Inflorescence rather short, subracemose, leafy at the base, the branches deeply divided; axis hairy, with a few stalked glands and weak prickles. Sepals deflexed after flowering, green-tomentose and unarmed; petals usually pale pink; stamens white, exceeding the green styles; receptacle very hairy; carpels glabrous. $2n=28$. *N.W. & C. Europe, extending to Bulgaria.* Au Be Br Bu Cz Da Ga Ge Hb He Ho Hu Ju Po Rm.

Related species include:

R. longebracteatus E. I. Nyárády in Săvul., *Fl. Rep. Pop. Române* **4**: 903 (1956). Rm.
R. montanus Libert ex Lej., *Fl. Spa* **2**: 311 (1813). Be Cz Ge Ho Hu.
R. neopyramidalis E. I. Nyárády in Săvul., *Fl. Rep. Pop. Române* **4**: 902 (1956). Rm.
R. orbifolius Boulay & Letendre ex Lefèvre, *Bull. Soc. Bot. Fr.* **24**: 224 (1877) in syn. Ga Hu.
R. piletostachys Gren. & Godron, *Fl. Fr.* **1**: 548 (1849). Au Ga Ju.
R. quadicus (Sabr.) G. Beck, *Fl. Nieder-Österr.* **2**(1): 726 (1892). Au.
R. schlechtendalii Weihe ex Link, *Enum. Hort. Berol. Alt.* **2**: 62 (1822). $2n=28$. Au Br Cz Ga Ge Hb.
R. slatinensis E. I. Nyárády in Săvul., *Fl. Rep. Pop. Române* **4**: 904 (1956). Rm.

31. **R. silvaticus** Weihe & Nees, *Rubi Germ.* 30 (1824). Stems rather weak, arching-procumbent, angled above, hairy, eglandular or with a few glands; prickles numerous and rather weak, subulate or subconical. Leaflets 5, green, hairy beneath; terminal leaflet elliptical, 4–5 times as long as its petiolule, acuminate, rounded at base, rather coarsely serrate or biserrate. Inflorescence pyramidal, dense, often leafy, the axis hairy, with numerous fine, deflexed prickles. Sepals deflexed, often appendiculate, pubescent, sometimes glandular and with acicles; petals obovate, white or pink; stamens white, much exceeding the green styles; anthers, carpels and receptacle hairy. $2n=28$. *C. Europe, extending to Ireland and Denmark.* Au Be Br Cz Da Ga Ge Hb He Ho ?Hs Hu Po Rm.

Related species include:

R. bicolorispinosus E. I. Nyárády in Săvul., *Fl. Rep. Pop. Române* **4**: 905 (1956). Rm.
R. nemorensis P. J. Mueller & Lefèvre, *Pollichia* **16–17**: 198 (1859). Be Br Cz Ga He Hu Po Rm.
R. splendidiflorus Sudre, *Bull. Soc. Étud. Sci. Angers* **31**: 69 (1902). Cz Ga Rm.

32. **R. egregius** Focke, *Abh. Nat. Ver. Bremen* **2**: 463 (1871). Stems rather weak, arcuate, procumbent or climbing, glabrescent, not pruinose, with few or no stalked glands or acicles; prickles short, subulate, yellow to orange. Leaves mostly ternate, often rather small, tomentose-pubescent beneath; terminal leaflet

suborbicular-obovate, acuminate, cordate at base, finely or coarsely serrate. Inflorescence a dense, long, tapering panicle, with very long, many-flowered lower branches, hairy, slightly glandular and with rather weak prickles; flowers rather large. Sepals patent, becoming deflexed; petals obovate, white; stamens exceeding the green styles; carpels and receptacle glabrous or glabrescent. Fruit of 15–20 rather large drupelets. $2n=28$. *N.W. & C. Europe.* Au Be Br Da Ga Ge Hb He Ho.

Varies greatly in leaf-form and -toothing.

Related species include:

R. ceticus Halácsy, *Verh. Zool.-Bot. Ges. Wien* **41**: 244 (1891). Au.
R. lespinassei Clavaud, *Act. Soc. Linn. Bordeaux* **37**: iii (1883) (*R. coutinhoi* Samp.). Ga Hs Lu.
R. lindleianus Lees, *Phytologist* (*Newman*) **3**: 361 (1848). $2n=28$. Be Br Ga Ge Hb Ho.

33. **R. villicaulis** Koehler ex Weihe & Nees, *Rubi Germ.* 30 (1824). Stems robust, arching or procumbent, angled, with dense, long, patent hairs and an occasional pricklet; prickles hairy, straight, mostly deflexed. Leaflets 5, pubescent or grey-tomentose beneath; terminal leaflet orbicular-ovate to rather narrowly elliptical, acuminate, cordate at base, unevenly or coarsely serrate; petioles with strong, falcate prickles. Inflorescence large, broad, with numerous strong, straight, deflexed prickles, occasionally with a few glands and pricklets; the lower axillary branches rather numerous, long and oblique; upper branches cymose; terminal flowers subsessile. Sepals usually long-pointed, aciculate, somewhat deflexed; petals rather broad, pink or white; stamens and styles pink or white; carpels glabrous or pubescent; receptacle hairy. $2n=28$. *N.W. & C. Europe.* Au Br Cz ?Ga Ge Hb He Ho ?Ju Po Rs (W).

Related species include:

R. atrocaulis P. J. Mueller, *Pollichia* **16–17**: 163 (1859). Br Ga Ge.
R. eduardii Borbás, *Österr. Bot. Zeitschr.* **40**: 247 (1890) (*R. villicaulis* var. *formanekianus* Borbás). Ju.
R. gelertii Frid., *Bot. Tidsskr.* **15**: 237 (1886). Br Da Ge.
R. insularis F. Aresch., *Bot. Not.* **1881**: 158 (1881). Da Su.
R. kelleri Halácsy, *Österr. Bot. Zeitschr.* **40**: 431 (1890). Au.
R. langei G. Jensen ex Frid. & Gelert, *Bot. Tidsskr.* **16**: 67 (1887) (*R. atrocaulis* auct. dan., non P. J. Mueller). Br Da Ge.
R. magurensis E. I. Nyárády in Săvul., *Fl. Rep. Pop. Române* **4**: 908 (1956). Rm.
R. ocnensis E. I. Nyárády in Săvul., *op. cit.* 907 (1956). Rm.
R. seciurensis E. I. Nyárády in Săvul., *op. cit.* 906 (1956). Rm.
R. septentrionalis W. C. R. Watson, *Jour. Ecol.* **33**: 338 (1946) (*R. confinis* Lindeb., non P. J. Mueller). Br No Su.
R. subvillicaulis E. I. Nyárády in Săvul., *Fl. Rep. Pop. Române* **4**: 906 (1956). Rm.

34. **R. polyanthemus** Lindeb., *Herb. Rub. Scand.* **1**: n. 16 (1882). Stems angled, moderately pubescent; prickles fairly strong, yellow or red. Leaflets 3–7, becoming convex, dull above, grey-tomentose or pubescent beneath; terminal leaflet suborbicular to broadly or narrowly obovate, cuspidate, almost simply serrate or serrate-dentate, rounded at base. Inflorescence long, slightly narrowed above, many-flowered; axis tomentose or pubescent, sometimes glandular and with few or numerous, strong, deflexed pricklets below. Sepals deflexed, grey-tomentose and pubescent, glandular and aculeolate; petals broad, obovate, pink; stamens pink or white, much exceeding the green- or red-

based styles; carpels and receptacle hairy. $2n=28$. *N.W. Europe, extending to S. Sweden and Germany.* Br Da Ge Hb Ho Su.

35. R. rhamnifolius Weihe & Nees, *Rubi Germ.* 22 (1822). Stems robust, angled, sulcate, often reddish, nearly glabrous, eglandular, armed with strong, equal, rather large-based, patent and straight or falcate prickles. Leaflets 5, coriaceous, glabrous above, pubescent beneath, those of the upper leaves white-tomentose beneath, unevenly and very finely serrate; terminal leaflet orbicular, 2–3 times as long as petiolule, shortly acuminate, rounded or subcordate at the base, with many hooked prickles. Inflorescence pyramidal, leafy at the base; axis hairy, with numerous strongly deflexed or falcate prickles. Sepals deflexed after flowering, aciculate, hairy or white-tomentose; petals orbicular and often rather large, white or pale pink; stamens white, exceeding the green styles; carpels and receptacle glabrescent; pollen imperfect. *C. Europe, extending to Belgium.* Be Ga Ge He Hu.

Related species include:

R. beirensis (Samp.) Samp., *Bol. Soc. Brot.* ser. 2, **10**: 111 (1935). Lu.
R. cardiophyllus P. J. Mueller & Lefèvre, *Pollichia* **16–17**: 86 (1859). $2n=28$. Br Da Ga Ge Hb Ho.
R. obtusangulus Gremli, *Excurs.-Fl. Schweiz* ed. 4, 144 (1881). He Lu.

36. R. hochstetterorum Seub., *Fl. Azor.* 48 (1844). Leaves of fertile branches with 3–5 large leaflets, pubescent above; terminal leaflet obovate, unevenly biserrate. Inflorescence large, long, the branches irregular, densely tomentose, eglandular; bracts lanceolate. Sepals deflexed, white-tomentose. Petals at least 3 cm, white, suborbicular; margins crenulate. ● *Açores.* Az.

A very imperfectly understood species, probably known only from the type. Most records are referable to **38**. Perhaps diploid.

37. R. lindebergii P. J. Mueller, *Pollichia* **16–17**: 292 (1859). Stems high-arching, then procumbent, branched, rather sulcate, sparsely hairy, glaucescent; prickles red-based, strong, patent or falcate. Leaflets 5, rather small, greyish above, softly pubescent to grey-tomentose beneath; terminal leaflet narrowly elliptic-obovate, shortly cuspidate-acuminate, rounded at base, finely serrate; petiole long, bearing many large, hooked prickles. Inflorescence long, narrow, leafy above; middle branches divided near base, 2–3 arising together from the same axis; prickles numerous, strong, falcate. Sepals deflexed, ovate, white-tomentose; petals narrowly obovate, white or pinkish; stamens longer than the styles; carpels glabrous. $2n=28$. *N.W. Europe.* Br Da No Su.

Related species include:

R. mercieri Genev., *Mém. Soc. Acad.* (*Angers*) **24**: 271 (1868). Br Ga He. Placed nearest to this species by Sudre, but nearer to **49** by Watson.

Subsect. *Discolores* P. J. Mueller. Stems arching, rooting apically in the autumn; prickles usually subequal and confined to the angles; stalked glands absent or few. All leaves grey-white tomentose beneath. Sepals grey-white tomentose, deflexed. Usually late-flowering.

38. R. ulmifolius Schott, *Isis* **1818**: 821 (1818) (*R. rusticanus* Merc., *R. discolor* sensu Syme, non Weihe & Nees, *R. amoenus* Portenschl., non Koehler). Stems robust, arching or procumbent,

angled, often sulcate, pruinose, glabrous to tomentose with semi-appressed hairs; prickles robust, broad-based, patent to falcate, hairy. Leaves pedate, often very small; leaflets 3–5, dark green and glabrous above, white-tomentose beneath, convex, variously toothed, coriaceous; terminal leaflet ovate or suborbicular to obovate. Inflorescence often long and narrow, sometimes pyramidal, leafy at the base, with patent branches and long pedicels; all axes with robust, broad-based, patent to falcate, hairy prickles. Sepals deflexed after flowering, sometimes slightly aciculate, white-tomentose; petals crumpled, orbicular or ovate, sometimes jagged at the apex, pink or occasionally white; stamens white or pink, equalling or just exceeding the green, pink or white styles; anthers glabrous; pollen completely fertile; carpels hairy, often tomentose. $2n=14$. *S., W. & C. Europe.* Al Au Az Be Bl Br Co Cr Cz Ga Ge Hb He Ho Hs It Ju Lu Sa Si Tu.

The only diploid, sexual species in Subgen. *Rubus* whose chromosome number has been verified from material of wild origin; it is very polymorphic in shape of leaves, branching of inflorescence, clothing of the axis, size of flowers and colour of petals, and up to 20 subspecies as well as 92 varieties have been described.

Related species include:

R. heteromorphus Ripart ex Genev., *Mém. Soc. Acad.* (*Angers*) **24**: 255 (1868) (*R. dalmatinus* Tratt. ex Focke). Al Ju.
R. portuensis Samp., *Ann. Sci. Nat.* (*Porto*) **8**: 10 (1904). Hs Lu.
R. sanguineus Friv., *Flora* (*Regensb.*) **18**: 334 (1835) (*R. sanctus* auct. plur., non Schreber, *R. anatolicus* Focke, *R. discolor* Boiss., non Weihe & Nees). Bu Rs (K).
R. thessalus Halácsy, *Consp. Fl. Graec.* **1**: 503 (1900) (*R. anatolicus* var. *cinereus* Hausskn.). Al Gr.

39. R. godronii Lecoq & Lamotte, *Cat. Pl. Centr. Fr.* 151 (1847) (*R. praecox* subsp. *godronii* (Lecoq & Lamotte) Hayek). Stems terete at base, slightly or strongly angled above, glaucescent, with somewhat unequal, straight or curved prickles with a subconical base. Leaflets 5, glabrous or hairy above, white-tomentose and hairy beneath; terminal leaflet ovate, elliptical or obovate, twice as long as its petiolule, rounded or slightly cordate at base, finely and almost simply dentate; stipules with sparse glands. Inflorescence densely or sparsely hairy, with falcate prickles; bracts with sparse, subsessile glands. Sepals deflexed after flowering, tomentose and pubescent, with no or few acicles; petals oblong, pale pink or sometimes white; stamens white, greatly exceeding the green styles; pollen sterile; carpels and receptacle hairy. *W. & C. Europe.* Be Br Cz Ga Ge Hb He Ho Lu Po.

Related species include:

R. caldasianus Samp., *Ann. Sci. Nat.* (*Porto*) **8**: 8 (1904). Lu.
R. winteri P. J. Mueller ex Focke, *Syn. Rub. Germ.* 196 (1877). $2n=28$. Au Br Ga Ge Hb He Ho Ju.

40. R. bifrons Vest ex Tratt., *Rosac. Monogr.* **3**: 28 (1823). Almost evergreen; stems much-branched, stout, red-brown, obtusely angled, with plane faces, not pruinose, usually glabrescent, occasionally with a few minute glands and acicles; prickles long, robust, subulate. Leaflets 3–5, dark green, glabrous or glabrescent above, white-tomentose beneath; terminal leaflet usually orbicular or shortly obovate, shortly acuminate or cuspidate, twice as long as its petiolule, truncate or subcordate at base; basal leaflets shortly petiolulate; stipules with sessile glands. Inflorescence elongate, the axis tomentose towards the base, variably hairy above and sometimes with some shortly stalked glands and a few acicles; prickles patent, mostly straight, subulate and strong. Sepals tomentose, pubescent, sometimes gland-dotted, aculeolate or unarmed; petals large, suborbicular, pale pink to red; stamens white or pale pink, exceeding the green or reddish styles; receptacle glabrescent; carpels sparsely pubescent. *W. & C. Europe*. Au Be Br Cz Ga Ge He Ho Hs Hu It Ju Lu Po Rm Rs (W).

Related species include:

R. banaticus E. I. Nyárády in Săvul., *Fl. Rep. Pop. Române* 4: 890 (1956). Rm.

R. cuspidifer P. J. Mueller & Lefèvre, *Pollichia* 16–17: 89 (1859). $2n = 28$. Be Br Ga He Lu Rm.

R. gillotii Boulay ex Coste, *Fl. Fr.* 2: 39 (1901). Ga.

R. hedycarpus Focke, *Syn. Rub. Germ.* 190 (1877) (*R. praecox* Bertol.). Bu Cz Ga.

R. istricus Pospichal, *Fl. Österr. Küstenl.* 2: 273 (1898). Ju.

R. margaritae Gáyer, *Magyar Bot. Lapok* 20: 13 (1921). Rm.

R. trifoliatus Pospichal, *Fl. Österr. Küstenl.* 2: 270 (1898). Ju.

41. R. discolor Weihe & Nees, *Rubi Germ.* 30 (1824) (*R. procerus* P. J. Mueller, *R. armeniacus* Focke, *R. karstianus* Borbás, *R. macrostemon* Focke, *R. praecox* subsp. *macrostemon* (Focke) Hayek). Stems tall, robust, light brown to purple, arching, angled, sparsely hairy at first, glabrescent, glaucescent; prickles sparse, strong, broad-based, straight or falcate. Leaflets 5, large, glabrescent above, white-tomentose or pubescent beneath; terminal leaflet ovate or suborbicular, shortly acuminate, twice as long as its petiolule, truncate at base; basal leaflets shortly petiolulate; stipules with sparse sessile glands. Inflorescence large, pyramidal-truncate, lax, floriferous, leafy, the axis villous, with numerous falcate or geniculate prickles; flowers *c.* 3 cm in diameter. Sepals deflexed after flowering, tomentose, hairy, unarmed; petals ovate to orbicular, pale pink or white; stamens white or pale pink, much exceeding the green styles; anthers pubescent; pollen sterile; carpels slightly pubescent, receptacle pubescent. Fruit very large. *S., W. & C. Europe*. Au Be Bu Cz Ga Ge He Ho Hu It Ju Lu Rm Tu [Br Da].

A vigorous variant (the so-called Himalayan blackberry) is often cultivated in gardens and more or less naturalized.

Related species include:

R. geniculatus Kaltenb., *Fl. Aachen. Beck.* 2: 267 (1844). Be Br Cz Ga Ge He Ho Rm.

R. hebetatus Sudre, *Bull. Soc. Étud. Sci. Angers* 31: 83 (1902). Au Ga Hs.

42. R. chloocladus W. C. R. Watson, *Watsonia* 3: 288 (1956) (*R. pubescens* Weihe ex Boenn., non Rafin.). Stems arching, somewhat sulcate, sparsely and minutely tomentose and with longer hairs, becoming purple, sometimes pruinose; prickles robust, deflexed or strongly curved. Leaflets 5, slightly hairy above, pubescent and grey-white tomentose beneath; terminal leaflet ovate to elliptical, 2–3 times as long as its petiolule; basal leaflets shortly petiolulate; stipules filiform, with subsessile glands. Inflorescence elongated and narrow, almost leafless, tomentose; prickles long-based and falcate. Sepals deflexed after flowering, white-tomentose, hairy, unarmed; petals ovate or obovate, white or pink; stamens white, exceeding the green styles; receptacle and carpels hairy. *W. & C. Europe*. Au Be Br Cz Ga Ge He Ho Hs Hu Ju Lu Po Rm.

Related species include:

R. evagatus Sudre, *Bull. Assoc. Pyr.* 12: 10 (1902). Au Ga.
R. vestii Focke, *Syn. Rub. Germ.* 155 (1877). Au Hu It Rm.

43. R. candicans Weihe ex Reichenb., *Fl. Germ. Excurs.* 601 (1832) (*R. thyrsoideus* Wimmer pro parte, *R. coarctatus* P. J. Mueller). Stems high-arching, not very woody, sulcate, not pruinose, glabrous or with a few hairs; prickles rather few but very long-based, some falcate. Leaves digitate; leaflets 5, glabrous above, white-tomentose and hairy beneath; terminal leaflet variable in shape, unevenly dentate; lower leaflets subsessile; stipules linear. Inflorescence broad and long, dense, leafy, often with few prickles above, with elongated, ascending peduncles. Sepals deflexed after flowering, tomentose, hairy; petals rather small, obovate to oblong, white to deep pink; stamens rather few, slightly exceeding the green styles; carpels glabrous to hairy; receptacle hairy; pollen and fruit sometimes largely sterile. $2n = 21$. *S., W. & C. Europe*. Au Be Bl Bu Cz Ga Ge Gr Hb He Ho Hs Hu It Ju Lu Po Rm Rs (K) Tu.

Various subspecies have been described, varying in hairiness of the stem, shape of leaves, vigour of flowering and the strength of the inflorescence-prickles.

Related species include:

R. aciodontus P. J. Mueller & Lefèvre, *Pollichia* 16–17: 83 (1859). Ga Hu Rm.

R. arduennensis Libert ex Lej., *Fl. Spa* 2: 317 (1813). Be Da Ga Ge Hu Rm.

R. cirlioarae E. I. Nyárády in Săvul., *Fl. Rep. Pop. Române* 4: 898 (1956). Rm.

R. constrictus P. J. Mueller & Lefèvre, *Pollichia* 16–17: 79 (1859). Cz Hu Ju Rm.

R. drautensis E. I. Nyárády in Săvul., *Fl. Rep. Pop. Române* 4: 893 (1956). Rm.

R. fragrans Focke, *Syn. Rub. Germ.* 172 (1877), non Salisb. Bu Ge ?Ju.

R. linkianus Ser. in DC., *Prodr.* 2: 560 (1825). [Ga Ju.] Included in *R. pubescens* by Focke, but placed by Sudre along with *R. arduennensis* in a special series of the *Discolores*, linking it with Ser. *Tomentosi* of the *Appendiculati*. *R. linkianus* is unknown in the wild state.

R. moestus Holuby, *Österr. Bot. Zeitschr.* 23: 375 (1873). Cz Hu Rm.

R. persicinus A. Kerner, *Ber. Naturw. Ver. Innsbruck* 2: 137 (1871). Cz Ju.

R. petnicensis E. I. Nyárády in Săvul., *Fl. Rep. Pop. Române* 4: 897 (1956). Rm.

R. phyllostachys P. J. Mueller, *Flora (Regensb.)* 41: 133 (1858). Au Be Cz Ga Ge He Hs Hu Ju Lu Po Rm.

R. saxosus E. I. Nyárády in Săvul., *Fl. Rep. Pop. Române* 4: 893 (1956). Rm.

R. severinensis E. I. Nyárády in Săvul., *op. cit.* 892 (1956). Rm.

R. subvillosus Sudre, *Bull. Assoc. Fr. Bot.* 5: 127 (1902). Co Ga Ge Hu Rm Si.

R. teregovensis E. I. Nyárády in Săvul., *Fl. Rep. Pop. Române* 4: 897 (1956). Rm.

R. thyrsanthus Focke, *Syn. Rub. Germ.* 168 (1877) (*R. villicaulis* auct. roman., non Koehler ex Weihe & Nees). Al Au Br Bu Cz Ga Ge Hb He Hu Ju No Po Rm Su.

R. tumidus Gremli, *Excurs.-Fl. Schweiz* ed. 2, 161 (1874). Ga He Rm.

Subsect. *Appendiculati* Genev. Stems arching, rooting at the apex; prickles nearly always unequal and scattered over the stem-faces; stems and inflorescence usually with stalked glands. Inflorescence often compound.

Series *Tomentosi* Wirtgen. Stems with sparse glands; leaves grey- or white-tomentose beneath; inflorescence usually eglandular, tomentose; petals yellowish-white.

44. R. canescens DC., _Cat. Pl. Hort. Monsp._ 139 (1813) (_R. tomentosus_ Borkh. pro parte). Stems arching, procumbent or nearly erect, usually glabrous, occasionally sparsely tomentose or pubescent, the hairs stellate, or occurring singly or in tufts, with a few glands and acicles; prickles weak, unequal, the largest curved. Leaves small, pedate; leaflets 3–5, usually greyish-tomentose above, grey-white-tomentose beneath; terminal leaflet rhombic, 4 times as long as its petiolule, subacuminate, entire or subcordate at base, very coarsely incise-serrate. Inflorescence rather long, nearly leafless, the axis tomentose, usually eglandular; peduncles ascending, slender, with small, yellowish, subulate prickles; bracts lanceolate, the lower trifid. Sepals deflexed after flowering, grey-white, sometimes aciculate; petals small, obovate-oblong, yellowish-white; stamens white, more or less equalling the green styles; pollen completely fertile; carpels glabrous. _S. & C. Europe, extending to Belgium._ Al Au Be Bu Co Cr Cz Ga Ge He Hs Hu It Ju Lu Po Rm Rs (W, K) Si Tu.

Very polymorphic. A sexual species, probably diploid.

Related species include:

R. **aipetriensis** Juz., _Not. Syst._ (_Leningrad_) 13: 104 (1950). Rs (K).
R. **almensis** Juz., _op. cit._ 93 (1950). Rs (K).
R. **collinus** DC., _Cat. Pl. Hort. Monsp._ 139 (1813) (?_R. canescens × ulmifolius_). Ga Hs.
R. **crimaeus** Juz., _Not. Syst._ (_Leningrad_) 13: 99 (1950). Rs (K).
R. **divergens** P. J. Mueller, _Flora_ (_Regensb._) 41: 182 (1858) (?_R. caesius × canescens_). Au Cz Ga Ge He Ju.
R. **eurythyrsiger** Juz., _Not. Syst._ (_Leningrad_) 13: 101 (1950). Rs (K).
R. **hrubyi** Rohlena, _Mém. Soc. Sci. Bohême_ (_Sci._) 1936 (22): 16 (1937). Ju.
R. **lloydianus** Genev., _Mém. Soc. Acad._ (_Angers_) 10: 26 (1861). Au Bu Cz Ga Hs It Hu Ju Rs (K).
R. **marschallianus** Juz., _Not. Syst._ (_Leningrad_) 13: 104 (1950). Rs (K).
R. **moestifrons** Juz., _op. cit._ 97 (1950). Rs (K).
R. **nanitauricus** Juz., _op. cit._ 91 (1950). Rs (K).
R. **oenoxylon** Juz., _op. cit._ 96 (1950). Rs (K).
R. **paratauricus** Juz., _op. cit._ 90 (1950). Rs (K).
R. **schultzii** Ripart ex P. J. Mueller, _Pollichia_ 16–17: 289 (1859) (?_R. canescens × vestitus_). Ga Ju.
R. **sericophyllus** P. J. Mueller & Wirtgen ex Focke, _Syn. Rub. Germ._ 240 (1877). Ge He ?Ju.
R. **stenophyllidium** Juz., _Not. Syst._ (_Leningrad_) 13: 92 (1950). Rs (K).
R. **stevenii** Juz., _op. cit._ 102 (1950). Rs (K).
R. **subtauricus** Juz., _op. cit._ 98 (1950). Rs (K).
R. **tauricus** Schlecht. ex Juz., _op. cit._ 88 (1950). Rs (K).
R. **tomentellus** Ripart ex Genev., _Mém. Soc. Acad._ (_Angers_) 24: 301 (1868) (?_R. canescens × ulmifolius_). Ga Ju.
R. **trachypus** Boulay & Gillot, _Ann. Soc. Bot. Lyon_ 8: 20 (1881) (?_R. canescens × ulmifolius_). Ga Ju.
R. **troitzkyi** Juz., _Not. Syst._ (_Leningrad_) 13: 105 (1950). Rs (K).
R. **undabundus** Juz., _op. cit._ 94 (1950). Rs (K).
R. **utshansuensis** Juz., _op. cit._ 94 (1950). Rs (K).

Series _Vestiti_ Focke. Stem-prickles only slightly unequal, mainly on the angles of the stems; glands rather sparse; stems and leaves often rather hairy.

45. R. vestitus Weihe & Nees in Bluff & Fingerh., _Comp. Fl. Germ._ 1: 684 (1825). Stems angled, hirsute and tomentose, not pruinose, red-brown, with few glands and pricklets; prickles deep purple, subulate. Leaflets (3–)5, dull green, slightly hairy above, white-tomentose beneath, with stout, pectinately arranged hairs on the veins; terminal leaflet orbicular or obovate, twice as long as its petiolule, shortly acuminate, subcordate, unevenly serrate-dentate, undulate. Inflorescence long; axis hairy, with some glands and pricklets; prickles long, slender, patent or deflexed; bracts mainly trifid; flowers 2·5–3 cm in diameter. Sepals deflexed after flowering, slightly glandular and aculeolate, tomentose; petals 5–6, suborbicular, villous, deep to pale pink; stamens numerous, exceeding the green or pink styles; anthers usually pubescent; filaments deep pink to white; receptacle hairy; carpels glabrous or slightly hairy. _W. & C. Europe, extending to Sweden._ Au Be Br Cz Da Ga Ge Hb He Ho Hu It Lu Po Rm Su.

Related species include:

R. **bakonyensis** Gáyer, _Feddes Repert._ 22: 190 (1925). Hu.
R. **conspicuus** P. J. Mueller, _Flora_ (_Regensb._) 42: 71 (1859). Br Ga Ge He Hu.
R. **holochloroides** Sudre & Sabr. in Sudre, _Rubi Eur._ 195 (1912). Au.
R. **holochloropsis** Sudre, _op. cit._ 219 (1913). Au.
R. **holochloros** (Sabr.) Fritsch, _Exkursionsfl. Österr._ ed. 3, 209 (1922). Au.
R. **leucanthemus** P. J. Mueller, _Pollichia_ 16–17: 122 (1859). Au Cz Ga Ge He Ho.
R. **leucotrichus** Sudre, _Bat. Eur._ 55 (1906). Au Ga.
R. **lipovensis** E. I. Nyárády in Săvul., _Fl. Rep. Pop. Române_ 4: 912 (1956). Rm.
R. **pilifer** Sudre, _Bull. Soc. Bot. Fr._ 46: 90 (1899). Cz Ga Ge Hu Rm.
R. **podophyllos** P. J. Mueller, _Bonplandia_ 9: 281 (1861). $2n = 28$. Au Br ?Cz Ga Ge He Hu. Placed near to 45 by Watson, but near to 47 by Sudre.
R. **saxigenus** Sudre, _Bull. Soc. Étud. Sci. Angers_ 35: 26 (1906). Au Ga.
R. **vaccarum** E. I. Nyárády in Săvul., _Fl. Rep. Pop. Române_ 4: 911 (1956). Rm.

46. R. boraeanus Genev., _Mém. Soc. Acad._ (_Angers_) 8: 87 (1860). Stems purple, with sparse, tufted hairs, unequal, scattered, stout pricklets, and a few acicles and short glandular hairs; prickles stout-based, straight or slightly falcate. Leaves pedate; leaflets 5, thick, grey-white tomentose beneath; terminal leaflet obovate or suborbicular-ovate, acuminate or cuspidate, cordate, coarsely toothed. Inflorescence broad, more or less cylindrical, with a few weak, deflexed or falcate prickles; branches long, erecto-patent; peduncles and pedicels often unarmed; flowers 1·5–2·5 cm in diameter. Sepals patent to erect; petals broadly obovate, usually pink, sometimes white; stamens pink or pink-based, equalling the pink-based styles; anthers, carpels and receptacle pubescent. $2n = 28$. _W. Europe, from Ireland to Switzerland._ Br Ga Hb He.

Related species include:

R. **augustus** Hormuzaki, _Mem. Secţ. Şti._ (_Acad. Română_) ser. 3, 2: 287 (1925). Au.
R. **breyninus** G. Beck, _Fl. Nieder-Österr._ 2(1): 729 (1892). Au.
R. **caflishii** Focke, _Syn. Rub. Germ._ 278 (1877). Au Cz Ga Ge He Hu Ju.
R. **calvarii** Hormuzaki, _Österr. Bot. Zeitschr._ 68: 225 (1919). Au.
R. **crucimontis** Hayek, _Verh. Zool.-Bot. Ges. Wien_ 66: 452 (1916). Au.
R. **epipsilos** Focke, _Syn. Rub. Germ._ 258 (1877). Au Ge.
R. **fimbrifolius** P. J. Mueller & Wirtgen ex Focke, _op. cit._ 256 (1877). Au Ga Ge.

R. graniticola Halácsy ex Topitz, *Österr. Bot. Zeitschr.* **42**: 202 (1892). Au Cz.

R. greinensis Halácsy ex Topitz, *loc. cit.* (1892). Au.

R. grossbaueri G. Beck, *Fl. Nieder-Österr.* **2**(1): 731 (1892). Au.

R. macrostachys P. J. Mueller, *Flora* (*Regensb.*) **41**: 150 (1858). Au Br Bu Cz Ga Ge He Hu Rm.

R. pseudotenellus Gilli, *Verh. Zool.-Bot. Ges. Wien* **80**: 68 (1931) (*R. tenellus* Hayek, non P. J. Mueller & Lefèvre). Au.

R. schlikkumii Wirtgen, *Flora* (*Regensb.*) **42**: 235 (1859) (*R. ocyriacus* Halácsy). Au Ge He Hu.

47. R. adscitus Genev., *Mém. Soc. Acad.* (*Angers*) **8**: 88 (1860) (*R. hypoleucus* P. J. Mueller & Lefèvre, non Vest). Stems angled, sulcate or the faces flat, densely pubescent with long and short hairs, acicles, and sunken glands; prickles subequal, all rather short, straight, yellowish; prickles on petiole straight or slightly falcate. Leaflets 3–5, softly hairy above, white-tomentose beneath; terminal leaflet oblong-obovate, shortly acuminate, rounded or sub-cordate at base, coarsely and unevenly serrate. Inflorescence pyramidal, with long, patent branches, many-flowered, leafy at the base; axis densely hairy, with many sunken glands and weak, deflexed prickles. Sepals deflexed after flowering, long-pointed, pubescent and tomentose, with a few glands; petals elliptic-ovate, pink; stamens white, or pink-based, equalling or exceeding the green styles; receptacle hirsute; carpels usually glabrous. *N.W. & C. Europe.* Au Be Br Cz Ga Hb Hu Rm.

Related species include:

R. dasyclados A. Kerner, *Ber. Naturw. Ver. Innsbruck* **2**: 155 (1871). Au Ge.

R. doftanensis E. I. Nyárády in Săvul., *Fl. Rep. Pop. Române* **4**: 913 (1956). Rm.

R. leucostachys Schleicher ex Sm., *Engl. Fl.* **2**: 403 (1824). $2n=28$. Br Da Ga.

R. macrothyrsus Lange, *Icon. Pl. Fl. Dan.* **48**: 6 (1871). $2n=28$. Br Da Ge.

R. tenuispinosus E. I. Nyárády in Săvul., *Fl. Rep. Pop. Române* **4**: 911 (1956). Rm.

48. R. mucronulatus Boreau, *Fl. Centre Fr.* ed. 3, **2**: 196 (1857) (*R. mucronatus* Bloxam, non Ser., *R. mucronifer* Sudre pro parte). Stems robust, subterete, striate, hairy at first and becoming deep purple, with a few slender prickles. Leaflets 3–5, imbricate, pubescent or greenish-tomentose beneath; terminal leaflet broad, orbicular or obovate, mucronate to long-cuspidate, subcordate to rounded at base, serrate to serrate-dentate. Inflorescence leafy, few-flowered, with long peduncles and pedicels; axis striate, tomentose, with prickles, unequal glands and acicles; flowers *c.* 3 cm in diameter. Sepals patent, pubescent, green, with a white margin; petals obovate, pink, rarely white; stamens pink, much longer than the pink styles; anthers and carpels hairy. *N.W. & C. Europe.* Au Be Br Cz Da Hb He Ho Ge Po.

Related species include:

R. fritschii Sabr. in Hayek, *Sched. Fl. Styr. Exsicc.* **7–8**: 15 (1906). Au.

R. henriquesii Samp., *Ann. Sci. Nat.* (*Porto*) **9**: 63 (1905) (*R. peratticus* Samp., *R. menkei* subsp. *henriquesii* (Samp.) Sudre). Ga Hs Lu.

R. mucronatoides A. Ley, *Jour. Bot.* (*London*) **45**: 446 (1907). Au Br.

49. R. gremlii Focke, *Syn. Rub. Germ.* 266 (1877) (*R. colemannii* subsp. *gremlii* (Focke) Sudre). Stems procumbent or climbing,

more or less hairy, with straight, yellow prickles. Leaves pedate; leaflets 3–5, glabrescent on both surfaces; terminal leaflet long-stalked, ovate or ovate-oblong, narrowed to a long point, rounded or cordate at base, unevenly serrate. Inflorescence rather long and narrow, often leafy; axis and pedicels densely pubescent with patent hairs and with small acicular prickles; glands various. Sepals deflexed, tomentose; petals narrowly obovate-cuneate, pubescent, pinkish or white; stamens often much exceeding the styles. *C. Europe; Britain.* Au Br Cz Ge He Hu Ju Rm.

Sudre breaks up the complex species **R. colemannii** Bloxam in Kirby, *Fl. Leicest.* ed. 2, 38 (1850) into several subspecies. Other species, listed below, link this complex with species in Ser. *Radulae*:

R. balatonicus Borbás, *Result. Wiss. Erforsch. Balaton* **2**(2): 146 (1907). Hu.

R. beckii Halácsy, *Verh. Zool.- Bot. Ges. Wien* **35**: 663 (1886). Au Cz.

R. chloroclados Sabr., *Österr. Bot. Zeitschr.* **41**: 413 (1891). Cz Rm.

R. clusii Borbás ex Sabr., *Erdész. Lapok* **1885**: 104 (1885). Au Cz Ju Rm. Sudre and Nyárády make this species synonymous with **49**.

R. condensatus P. J. Mueller, *Flora* (*Regensb.*) **41**: 167 (1858). Au Br Ga Ge He Hu.

R. diminutus Gáyer in Jáv., *Magyar Fl.* 495 (1924). Hu.

R. eriostachys P. J. Mueller & Lefèvre, *Pollichia* **16–17**: 225 (1859). Au Be Br Ga Hu.

R. ferox Vest ex Tratt., *Rosac. Monogr.* **3**: 40 (1823) (*R. apum* Fritsch, *R. lasiaxon* Borbás ex Waisb.). Au Ge Hu.

R. grandiflorus E. I. Nyárády in Săvul., *Fl. Rep. Pop. Române* **4**: 916 (1956). Rm.

R. gratiosus P. J. Mueller & Lefèvre, *Pollichia* **16–17**: 153 (1859). Be Cz Ga Hu.

R. gremblichii Halácsy, *Österr. Bot. Zeitschr.* **40**: 433 (1890). Au.

R. halacsyi Borbás ex Halácsy, *Verh. Zool.-Bot. Ges. Wien* **35**: 666 (1886). Au Cz Hu.

R. hebecaulis Sudre, *Bull. Assoc. Fr. Bot.* **3**: 101 (1900) (*R. hebeticaulis* auct.). Au Br Cz Ga Ge He Ho Hu Po Rm.

R. helveticus Gremli, *Beitr. Fl. Schweiz* 36 (1870). Au Br Ge He.

R. indotatus Gremli, *Österr. Bot. Zeitschr.* **21**: 128 (1871). Au He.

R. inopacatus P. J. Mueller & Lefèvre, *Pollichia* **16–17**: 117 (1859). Au Br Cz Ga Hu.

R. joannis G. Beck, *Fl. Nieder-Österr.* **2**(1): 736 (1892). Au.

R. laetecoloratus E. I. Nyárády in Săvul., *Fl. Rep. Pop. Române* **4**: 914 (1956). Rm.

R. morifolius P. J. Mueller, *Flora* (*Regensb.*) **41**: 164 (1858). Au Cz Ge Hu.

R. mulleri Lefèvre, *Pollichia* **16–17**: 180 (1859). Au Be Br Ga He Ho Hu Rm.

R. podophylloides Sudre, *Bull. Soc. Étud. Sci. Angers* **31**: 105 (1902). Au Cz Ga He Hu.

R. porphyrantherus Hormuzaki, *Mem. Secţ. Şti.* (*Acad. Română*) ser. 3, **2**: 290 (1925). Au.

R. rariglandulosus E. I. Nyárády in Săvul., *Fl. Rep. Pop. Române* **4**: 916 (1956). Rm.

R. salisburgensis Focke, *Syn. Rub. Germ.* 280 (1877). Au Br Cz Ga Ge He Hu.

R. schmidelyanus Sudre, *Bull. Soc. Bot. Fr.* **51**: 21 (1904). $2n=35$. Au Br Cz Ga Ge He Hu Po.

R. serratulifolius Sudre, *Compt. Rend. Congr. Soc. Sav.* (*Sci.*) **1908**: 206 (1909). Au Be Br Ga Ge.

R. subcoriaceus E. I. Nyárády in Săvul., *Fl. Rep. Pop. Române* **4**: 916 (1956). Rm.

R. taeniarum Lindeb., *Nov. Fl. Scand.* 5 (1858) (*R. infestus* auct., non Weihe ex Boenn.). Be Br Cz Da Ge ?Ho Su.

R. teretiusculus Kaltenb., *Fl. Aachen. Beck.* **2**: 282 (1844). Au Be Br Ge ?Ho Hu.

Series *Radulae* Focke. Stems rough and tuberculate; prickles scattered and very unequal, quite distinct from the smaller pricklets. Inflorescence-axis with glandular and eglandular hairs, the glandular hairs mostly not longer than the eglandular.

50. R. radula Weihe ex Boenn., *Prodr. Fl. Monast.* 152 (1824). Stems arching, robust, angled, glabrous to pubescent, glandular; prickles unequal, the smaller ones numerous, acicular. Leaves digitate; leaflets 5, rather large, deep green and nearly glabrous above, white-tomentose beneath; terminal leaflet 2–3 times as long as its petiolule, ovate or ovate-rhombic, gradually acuminate, rounded or slightly cordate at base, unevenly and finely serrate-dentate. Inflorescence pyramidal, often leafy to the apex, the axis hairy, with numerous unequal glands, acicles and prickles; the largest prickles long, robust, subulate, often nearly patent; flowers rather small. Sepals deflexed after flowering, not appendiculate, white-tomentose and hirsute, glandular and aculeolate; petals oblong-obovate, white or pinkish; stamens white, rarely pinkish, exceeding the green or pink-based styles; receptacle hairy; carpels glabrous. $2n=28$. *W. & C. Europe, extending to S. Norway.* Au Be Br Cz Da Ga Ge Hb He Ho Hu Lu No Po Rm Su.

Related species include:

R. gizellae Borbás, *Österr. Bot. Zeitschr.* **41**: 147 (1891). ?Cz Hu Ju.

R. uncinatus P. J. Mueller, *Flora (Regensb.)* **41**: 154 (1858). Au Ga Ge Hu.

51. R. genevieri Boreau, *Fl. Centre Fr.* ed. 3, **2**: 193 (1857). Stems slender, bluntly angled, with flat, striate faces, densely puberulent, with long, slender, strong prickles. Leaves rather small, pedate; leaflets (3–)5, glabrescent above, silky-tomentose beneath; terminal leaflet obovate-cuspidate, partly biserrate and partly dentate. Inflorescence very long and narrow, leafy, with flexuous, tomentose axis; pedicels long, prickly. Sepals appendiculate; petals obovate or spathulate, emarginate, tapering below, pale pink; stamens white or pink-based, much exceeding the reddish styles; receptacle and carpels pubescent. *W. & C. Europe.* Br ?Cz Ga Ge Hb He Ho Hu Lu.

Related species include:

R. brigantinus Samp., *Ann. Sci. Nat. (Porto)* **8**: 120 (1904). Lu.

R. echinatus Lindley, *Syn. Brit. Fl.* 94 (1829) (*R. discerptus* P. J. Mueller). $2n=28$. Br Cz Ga Hb He Lu.

R. maranensis Samp. ex Coutinho, *Fl. Port.* 301 (1913). Lu.

52. R. apiculatus Weihe & Nees in Bluff & Fingerh., *Comp. Fl. Germ.* **1**: 680 (1825). Stems robust, angled, striate, faces plane or slightly sulcate, glaucescent, slightly hairy, with scattered short glands and numerous acicles and pricklets, some gland-tipped; prickles unequal, slender, the largest patent or falcate, the more numerous smaller ones papilliform. Leaves large, pedate; leaflets (3–)5, glabrescent above, tomentose and rough beneath with short, patent hairs; terminal leaflet elliptic-obovate, shortly acuminate, often subcuneate and rounded at base. Inflorescence long, leafy, narrowed to the apex, with rather short branches; axis tomentose, with numerous glands and acicles and few, fine, sharply deflexed or patent yellowish prickles. Sepals somewhat

deflexed or patent after flowering, tomentose, glandular and with a few acicles; petals 5–6, ovate, pink; stamens white or pink, about equalling the cream-coloured styles; carpels glabrous or slightly hairy. $2n=28$. *W. & C. Europe.* Au Be Br Cz Ga Ge Hb He Ho Hu Lu Rm.

Related species include:

R. albicomus Gremli, *Beitr. Fl. Schweiz* 30 (1870). Au Cz Ga He Hu.

R. carinthiacus Halácsy, *Verh. Zool.-Bot. Ges. Wien* **41**: 254 (1891). Au.

R. dasycarpus (Sabr.) Fritsch, *Exkursionsfl. Österr.* ed. 3, 209 (1922). Au.

R. lumnitzeri (Sabr.) Fritsch, *op. cit.* 214 (1922). Au.

R. lusitanicus R. P. Murray, *Bol. Soc. Brot.* **5**: 189 (1887). Hs Lu.

R. micans Gren. & Godron, *Fl. Fr.* **1**: 546 (1849). Au Br Cz Ga Ge He Hu ?Po.

R. perdurus Holuby & Borbás ex Sabr., *Österr. Bot. Zeitschr.* **42**: 21 (1892). Cz.

R. roetensis Waisb., *Österr. Bot. Zeitschr.* **47**: 6 (1897). Au Cz Hu.

R. subcanus P. J. Mueller ex Genev., *Mém. Soc. Acad. (Angers)* **28**: 45 (1872). Au Cz Ga Ge Hu.

R. supinus Sabr., *Österr. Bot. Zeitschr.* **55**: 357 (1905). Au.

R. transmontanus Samp. ex Coutinho, *Fl. Port.* 302 (1913), non Focke. Lu.

R. verticalis Hormuzaki, *Mem. Secţ. Şti. (Acad. Română)* ser. 3, **2**: 292 (1925). Au.

53. R. fuscus Weihe & Nees in Bluff & Fingerh., *Comp. Fl. Germ.* **1**: 681 (1825). Stems robust, obtusely angled, reddish, densely hairy, slightly glandular; prickles unequal, the largest very long, straight or curved, the smallest acicular. Leaflets 5, deep green, slightly hairy above, softly hairy beneath; terminal leaflet twice as long as its petiolule, ovate, acuminate, cordate at the base, margins coarsely and partly biserrate. Inflorescence long and narrow or pyramidal, leafy, with hairy, glandular axis; prickles numerous, patent or falcate; bracts long, linear-lanceolate; peduncles nearly patent; pedicels long, tomentose. Sepals varying in posture, green-tomentose, glandular, aculeolate; petals large, not contiguous, ovate or obovate, white or pale pink; stamens white, exceeding the greenish styles; receptacle hairy; carpels usually glabrous. *W. & C. Europe, extending to S. Sweden.* Au Be Br Cz Da Ga Ge Hb He Ho Hu Rm Su.

Related species include:

R. granulatus P. J. Mueller & Lefèvre, *Pollichia* 16–17: 154 (1859). $2n=28$. Au Be Br Cz Ga Ge He Ho Hu Rm.

R. krasanii Sabr. in Hayek, *Fl. Steierm.* **1**: 771 (1909). Au.

R. timbal-lagravei P. J. Mueller ex Rouy & Camus, *Fl. Fr.* **6**: 97 (1900). Ga He Hs.

54. R. foliosus Weihe & Nees in Bluff & Fingerh., *Comp. Fl. Germ.* **1**: 682 (1825). Stems arching, then procumbent, angled, with plane faces, slightly hairy, glaucescent, with crowded, deep purple glands, acicles and pricklets; pricklets unequal, the largest very long, falcate. Leaflets 3–5, thick and tough, deep green, slightly hairy and shining above, green, tomentose beneath; terminal leaflet ovate to elliptical, acuminate, subcordate or rounded at base. Inflorescence large, many-flowered, variable; sometimes pyramidal, sometimes narrow, leafy, the branches either simple, branched near the base, or several arising from one axil; axis tomentose, with numerous dark purple glands and few, slender prickles; bracts leaf-like. Sepals varying in posture, appendiculate, green, tomentose, glandular; petals 5–7, fimbriate, white or pink; stamens white, slightly exceeding the pink-based styles; receptacle and carpels pubescent. *C. & N.W. Europe.* Au Be Br ?Cz Da Ga Ge Hb He Ho Hu.

Related species include:

R. **barbatus** (Sabr.) Fritsch, *Exkursionsfl. Österr.* ed. 3, 215 (1922). Au.

R. **brachystemon** Heimerl, *Österr. Bot. Zeitschr.* **32**: 109 (1882). Au Cz.

R. **corymbosus** P. J. Mueller, *Flora* (*Regensb.*) **41**: 151 (1858), non Weihe & Nees. Au Be Br Cz Ga Ge He.

R. **flexuosus** P. J. Mueller & Lefèvre, *Pollichia* **16–17**: 240 (1859). Au Be Br Cz Ga Ge He Ho Hu. According to Watson synonymous with **54**.

R. **insericatus** P. J. Mueller ex Wirtgen, *Flora* (*Regensb.*) **42**: 233 (1859). $2n=28$. Be Br Cz Ga Ge He Ho.

R. **microbelus** Sudre, *Bull. Soc. Bot. Fr.* **52**: 341 (1905). Au Be Ge.

R. **petri** Fritsch, *Österr. Bot. Zeitschr.* **60**: 310 (1910). Au.

R. **saltuum** Focke in Gremli, *Beitr. Fl. Schweiz* 30 (1870). Au He.

R. **truncifolius** P. J. Mueller & Lefèvre, *Pollichia* **16–17**: 139 (1859). Au Be Br Ga Ge.

55. R. infestus Weihe ex Boenn., *Prodr. Fl. Monast.* 153 (1824). Stems erect at first, then arching, becoming brownish-red, angled, with plane or sulcate faces, glabrous or sparsely hairy, not pruinose, with a few glands and acicles and often many unequal prickles, the largest prickles strong, curved and sometimes touching or coalescing at base. Leaflets 5, dark green and glabrescent above, pubescent to grey-green tomentose beneath; terminal leaflet ovate to broadly ovate, acuminate, subcordate, coarsely serrate. Inflorescence short, few-flowered, often leafy to the apex; axis sparsely hairy, with crowded, strong, unequal, straight or hooked prickles; pedicels long, tomentose. Sepals patent or appressed to the young fruit, green, hairy, glandular and aculeolate, with white margins; petals 5–7, orbicular or ovate, white or pale pink; stamens white at first, becoming red, and concealing the yellowish or green styles; receptacle pubescent; carpels glabrous. $2n=28$. *From Ireland and N. France to Poland.* Be Br Cz Da Ga Ge Hb Po.

Related species include:

R. **babingtonii** Bell Salter, *Ann. Nat. Hist.* **15**: 307 (1845). Br Da Ga.

R. **cunctator** Focke, *Syn. Rub. Germ.* 281 (1877). Au Lu.

56. R. thyrsiflorus Weihe & Nees in Bluff & Fingerh., *Comp. Fl. Germ.* **1**: 684 (1825). Stems procumbent, robust, terete, or angled with plane or slightly concave faces, glaucescent, glabrous or sparsely hairy, with a few glands and numerous, unequal, very short, deflexed and falcate prickles. Leaflets 3–5, large, imbricate, glabrescent above, shortly hairy beneath; terminal leaflet 4 times as long as its petiolule, suborbicular, acuminate, subcordate, serrate-dentate. Inflorescence long, dense-flowered and tapering, with several simple leaves and usually many, erecto-patent, 1- to 3-flowered branches in the upper part, the 2–3 lower branches with flowers in panicles; axis tomentose and hirsute, with acicles and sunken glands; prickles short. Sepals often 6, rising after anthesis, with linear apices, tomentose and glandular; petals often 6, white, elliptical, emarginate; stamens white, about equalling or slightly exceeding the green or reddish styles; receptacle pubescent; carpels usually glabrous. $2n=28$. *N.W. & C. Europe.* Au Be Br Cz Ga Ge He Hu Ju Rm.

Related species include:

R. **begoniifolius** Holuby, *Österr. Bot. Zeitschr.* **25**: 315 (1875). Al Cz.

R. **chloranthus** (Sabr.) Fritsch, *Exkursionsfl. Österr.* ed. 3, 213 (1922). Au Hu.

R. **prionatus** Sudre, *Compt. Rend. Congr. Soc. Sav.* (*Sci.*) **1908**: 214 (1909). Au Ga Ge.

R. **stylosus** Sabr. in Hayek, *Fl. Steierm.* **1**: 801 (1909) pro var. Au Ge.

57. R. pallidus Weihe & Nees in Bluff & Fingerh., *Comp. Fl. Germ.* **1**: 682 (1825). Stems arching, then procumbent, becoming purple, terete near base, angled above, with plane faces, densely hairy and with blackish glands; prickles very short, unequal, broad-based and mostly deflexed. Leaves pedate; leaflets 3–5, pale green and hairy above, green and glabrescent beneath; terminal leaflet ovate-elliptical, sometimes long-acuminate, cordate, serrate. Inflorescence broadly pyramidal-truncate, leafy below and sometimes to the apex, the middle branches long, patent; axis hairy, finely glandular, with few, weak, deflexed, pale prickles. Sepals narrow, deflexed after flowering but rising in fruit, purplish-greenish-grey tomentose, glandular and aculeolate; petals ovate or elliptical, white or sometimes pinkish; stamens white, variable in length; receptacle pubescent; carpels glabrous. $2n=28$. *N.W. & C. Europe.* Au Be Br Cz Da Ga Ge He Ho Hu Po.

Very variable; several subspecies and varieties have been described. Related species include:

R. **bloxamii** Lees in Steele, *Handb. Field Bot.* 55 (1847). $2n=28$. Be Br Da Ga Ge Hu. According to Watson, *R. multifidus* Boulay & Malbr. is synonymous with this species, but Sudre makes *R. multifidus* a subspecies of **59**.

R. **hirsutus** Wirtgen, *Prodr. Fl. Preuss. Rheinl.* 61 (1842), non Thunb. Au Be Br Cz Ge He.

R. **macrocalyx** Halácsy, *Österr. Bot. Zeitschr.* **40**: 433 (1890). Au.

R. **microstachys** Boulay, *Ronces Vosg.* 92 (1868). Au Ga Ge.

58. R. obscurus Kaltenb., *Fl. Aachen. Beck.* **2**: 281 (1844). Stems robust, with blunt angles, densely pubescent with long hairs, glandular, purple, glaucescent; prickles unequal, long, patent or falcate, bright red. Leaves pedate, more or less grey-white tomentose beneath; leaflets 3–5, glabrous above, becoming convex; terminal leaflet elliptic-obovate, acuminate, subcuneate and nearly entire at base; petioles short, with straight prickles. Inflorescence large and compound, pyramidal, the lower branches long and many-flowered; axis densely pubescent, with numerous sunken glands; prickles short, fine, straight, yellowish. Sepals erecto-patent in fruit, green, tomentose, glandular, aculeolate, appendiculate; petals narrow, deep pink; stamens white or pink, slightly exceeding the pink styles; receptacle and carpels pubescent. $2n=28$. *N.W. & C. Europe.* Au Be Br Cz Ga Ge He Ho Hu Po.

Related species include:

R. **castaneifolius** Sabr., *Mitt. Naturw. Ver. Steierm.* **52**: 289 (1916). Au.

R. **cruentatus** P. J. Mueller ex Focke, *Syn. Rub. Germ.* 312 (1877). Au Be Br Da Ga Ge.

R. **entomodontos** P. J. Mueller ex Focke, *loc. cit.* (1877). Br Cz Ga.

59. R. menkei Weihe & Nees in Bluff & Fingerh., *Comp. Fl. Germ.* **1**: 679 (1825). Stems procumbent, angled, with dense, patent hairs and unequal, yellowish, broad-based, deflexed prickles. Leaflets 3–5, slightly hairy above, hairy beneath, those of the upper leaves grey or grey-white tomentose; terminal leaflet elliptic-obovate, 2–3 times as long as its petiolule, cuspidate, rounded at base, unevenly serrate. Inflorescence more or less cylindrical, dense and subcorymbose at apex; axis hirsute, with numerous short glands and unequal acicles; prickles numerous strong, unequal, falcate. Sepals becoming appressed to fruit, green, tomentose, glandular and aculeolate; petals ovate-elliptical, white; stamens white, greatly exceeding the green or pink styles; receptacle hairy; carpels glabrous. *W. & C. Europe.* Au Be Br Cz Ga Ge He ?Ho Hu Ju Lu Po.

Related species include:

R. bregutiensis A. Kerner ex Focke, *Abh. Nat. Ver. Bremen* **13**: 152 (1894). Au Ga He Hu Ju.
R. suavifolius Gremli, *Beitr. Fl. Schweiz* 35 (1870). Au Ga Ge He Hu.

Series *Rudes* Sudre. Stems rough and tuberculate; prickles scattered, very unequal, quite distinct from the smaller pricklets. Inflorescence-axis tomentose to glabrous, always with some glandular hairs.

60. R. rudis Weihe & Nees in Bluff & Fingerh., *Comp. Fl. Germ.* **1**: 687 (1825). Small bush with deep purple, sulcate, glabrous not pruinose, very rough stems bearing numerous short glands and unequal, deflexed or falcate prickles. Leaves pedate; leaflets 3–5, shining and glabrous or glabrescent above, slightly tomentose and hairy beneath; terminal leaflet ovate to elliptical, acuminate, subcuneate at base, serrate-dentate, with coarse, angular, patent teeth. Inflorescence short, subcorymbose, with patent, cymose branches; axis tomentose (with short, unequal, purple glands longer than the tomentum on the pedicels) or glabrous; pedicels long; bracts lanceolate, the lower ones trifid; flowers 1·5–2 cm in diameter. Sepals appressed to fruit or imperfectly deflexed, triangular-attenuate, tomentose, glandular, slightly aculeolate; petals ovate-oblong, glabrescent, pink; stamens white, exceeding the green styles; receptacle and carpels glabrescent. *N.W. & C. Europe.* Au Be Br Bu Cz Ga Ge He Ho Hu Ju Po Rs (W).

Related species include:

R. amplus Fritsch ex Halácsy, *Verh. Zool.-Bot. Ges. Wien* **41**: 262 (1891). Au Ge Hu.
R. ctenodon (Sabr.) Fritsch, *Exkursionsfl. Österr.* ed. 3, 214 (1922). Au Ge.

61. R. melanoxylon P. J. Mueller & Wirtgen ex Genev., *Mém. Soc. Acad. (Angers)* **24**: 133 (1868). Stems purplish-black, angled, glabrescent, not pruinose, with sparse glands and unequal, strong, curved or straight prickles. Leaflets usually 5, glabrescent above, pubescent or slightly tomentose beneath; terminal leaflet ovate, shortly acuminate, subcordate, coarsely and unevenly serrate-dentate. Inflorescence flexuous, lax, rather leafy; axis glabrous or glabrescent, with numerous unequal, purple glands, and strong patent, deflexed or rather falcate prickles; peduncles patent or ascending, long, few-flowered. Sepals patent after anthesis, green, tomentose, glandular and aculeolate; petals ovate, bright pink; stamens white, exceeding the green styles; carpels hairy. *N.W. & C. Europe.* Au Be Br Cz Ga Ge He Hu Po Rm.

Related species include:

R. albicomiformis (Sabr.) Fritsch, *Exkursionsfl. Österr.* ed. 3, 209 (1922). Au.
R. amphistrophos (Focke) Fritsch, *op. cit.* 212 (1922). Au.
R. fagetanus E. I. Nyárády in Săvul., *Fl. Rep. Pop. Române* **4**: 919 (1956). Rm.
R. omalus Sudre, *Bull. Soc. Bot. Fr.* **52**: 324 (1905). Cz Ga Ge He Hu Rm.
R. rhodopsis Sabr. ex Fritsch, *Exkursionsfl. Österr.* ed. 3, 207 (1922). Au Ge.
R. rubristamineus E. I. Nyárády in Săvul., *Fl. Rep. Pop. Române* **4**: 918 (1956). Rm.
R. schummelii Weihe in Wimmer & Grab., *Fl. Siles.* **2**(1): 47 (1829). Cz Ga Ge.
R. thelybatos Focke, *Syn. Rub. Germ.* 279 (1877). Au Ga Ge Hu.

62. R. vallisparsus Sudre, *Bull. Soc. Étud. Sci. Angers* **35**: 42 (1906). Luxuriant; stems long, terete to obtusely angled, becoming white-pruinose, glabrescent, scabrid, slightly glandular, with unequal, rather short, deflexed or falcate prickles. Leaflets 3(–4), rather large, bright green and glabrescent above, green and glabrescent or slightly tomentose beneath; terminal leaflet ovate to elliptical, twice as long as its petiolule, acuminate, subcordate, variably serrate. Inflorescence pyramidal, truncate, leafy, the axis tomentose and sparsely pubescent, glaucescent below, with short, purple or deep red glands, and strong to medium patent or curved prickles. Sepals patent to erecto-patent in fruit, green, tomentose, acuminate, aculeolate, with short glands; petals ovate-elliptical, emarginate, not contiguous, pink; stamens white, becoming reddish, not much exceeding styles; styles yellowish, becoming red; receptacle glabrescent; carpels pubescent. $2n = 28$. *N.W. & C. Europe.* Au Be Br Ga Ge He Hu Rm.

Related species include:

R. alnicola Sudre, *Bull. Acad. Int. Géogr. Bot. (Le Mans)* **12**: 62 (1903). Ga. Treated by Sudre as a subspecies of *R. vallisparsus*, but according to Gustafsson perhaps a diploid species; it occurs in the Pyrenees.
R. glaucellus Sudre, *Bull. Assoc. Fr. Bot.* **1**: 91 (1898). Cz Ga Ge He Hu Po.
R. persimontis E. I. Nyárády in Săvul., *Fl. Rep. Pop. Române* **4**: 920 (1956). Rm.

Series *Histrices* Focke. Prickles very unequally distributed over the stem and showing a gradual transition into pricklets of varying lengths, acicles and stalked glands.

R. lasquiensis Spribille, *Verh. Bot. Ver. Brandenb.* **41**: 214 (1900), is a member of Ser. *Histrices*; it occurs in Poland, and its affinities are not known further.

63. R. fuscater Weihe & Nees in Bluff. & Fingerh., *Comp. Fl. Germ.* **1**: 681 (1825). Stems obtusely angled, deep purple-black, glaucous, densely hairy, with numerous unequal glands, pricklets and subulate prickles. Leaflets 3–5, deep green and glabrescent above, yellowish-green and pubescent beneath; terminal leaflet ovate or suborbicular, about twice as long as its petiolule, acute, cordate, unevenly and rather shallowly serrate-dentate. Inflorescence short, leafy at base; pedicels long; axis densely hairy, with many unequal purplish glands rarely longer than the diameter of the axis; prickles numerous, strong, straight or falcate; bracts lanceolate; flowers usually more than 2 cm in diameter. Sepals patent or almost appressed to fruit, shortly pointed, green, tomentose, white-margined, glandular and aculeolate; petals broad, ovate, fimbriate, deep pink or purple; stamens deep pink, rather short, remaining erect and coloured after petal-fall; styles greenish or deep pink; receptacle and carpels hairy. $2n = 28$. *N.W. & C. Europe.* Au Be Br Cz Ga Ge He Ho Hu.

Related species include:

R. adornatus P. J. Mueller ex Wirtgen, *Flora(Regensb.)* **42**: 234 (1859). $2n = 28$. Au Be Br Cz Ga Ge He Ho.
R. hartmanii Gand. ex Sudre, *Bull. Acad. Int. Géogr. Bot. (Le Mans)* **15**: 229 (1905) (*R. horridus* Hartman, non C. F. Schultz). $2n = 35$. Br Su.
R. oigocladus P. J. Mueller & Lefèvre, *Pollichia* 16–17: 134 (1859). Br Ga Ge He.

64. R. pilocarpus Gremli, *Beitr. Fl. Schweiz* 42 (1870) (*R. obtruncatus* subsp. *pilocarpus* (Gremli) Sudre). Stems becoming blackish-purple; some prickles large, hairy, strong, confluent, straight or curved, others smaller, stout and sometimes gland-tipped. Leaflets 3–5, imbricate, tough, tomentose beneath; terminal leaflet orbicular or broadly ovate-acuminate, coarsely biserrate. Inflorescence large, many-flowered with 1–2, long, compound, patent branches below and corymbose at apex; axis stout, tomentose, with prickles like those of the stem; flowers usually less than 2 cm in diameter. Sepals aculeolate; petals suborbicular, emarginate, fimbriate, pink or white; stamens white, about equalling the reddish or greenish styles; anthers sometimes pubescent; carpels densely pubescent. *N.W. & C. Europe.* Au Br Cz Ga Ge He Hu Ju.

Related species include:

R. glottocalyx G. Beck, *Fl. Nieder-Österr.* **2**(1): 739 (1892). Au.
R. obtruncatus P. J. Mueller, *Flora (Regensb.)* **41**: 152 (1858). Au Be ?Cz Ga Ge He Hu.

65. R. lejeunei Weihe & Nees in Bluff & Fingerh., *Comp. Fl. Germ.* **1**: 683 (1825). Stems robust, long, climbing, obtusely angled, not pruinose, sparsely hairy, with very short glands, and with tubercle-based acicles and pricklets; prickles unequal, long-based, tapering, straight or curved. Leaflets 3–5, large, glabrescent above, pubescent or glabrescent beneath; terminal leaflet rather long, obovate, acuminate, subcuneate at base, unevenly serrate or almost biserrate. Inflorescence pyramidal, with numerous paniculate lower branches and a subracemose apex; axis tomentose, with unequal red glands (some longer than the diameter of the rhachis), unequal acicles and rather short, subulate pricklets; flowers usually more than 2 cm in diameter. Sepals more or less deflexed after anthesis, tomentose, glandular, aculeolate, greenish, with a narrow, white margin; petals broadly elliptical, glabrous, bright pink; stamens red, pink or white, slightly exceeding the greenish, pink-based styles; receptacle pubescent; carpels glabrous or subglabrous. $2n = 35$. *W. & C. Europe.* ?Au Be Br Ga Ge He Hs Hu It Lu Po.

Related species include:

R. festivus P. J. Mueller & Wirtgen ex Focke, *Syn. Rub. Germ.* 314 (1877). Be Br Ge He.

66. R. rosaceus Weihe & Nees in Bluff & Fingerh., *Comp. Fl. Germ.* **1**: 685 (1825). Stems deep purple, obtusely angled, sulcate, glabrous or hairy, with numerous short glands and acicles; prickles unequal, short, patent, straight, some very small. Leaflets 3–5, deep green, glabrous or glabrescent above, pubescent beneath; terminal leaflet large, ovate or suborbicular, 3–4 times as long as its petiolule, acuminate, cordate, margins incise-serrate.

Inflorescence pyramidal, many-flowered, often pendent, racemose at the apex; axis hairy, with numerous long, purple glands, some longer than the diameter of the axis, and unequal, short, straight or falcate prickles; pedicels and middle and lower branches all more or less patent; pedicels short; flowers usually less than 2 cm in diameter. Sepals becoming appressed to fruit, tomentose, green, glandular, appendiculate; petals 5–7(–11), suborbicular-ovate or obovate, pink; stamens usually pink, exceeding the mostly red styles; receptacle slightly hairy; carpels glabrous or pubescent. $2n = 28$. *N.W. & C. Europe.* Au Be Br Cz Ga Ge He Ho Hu.

67. R. histrix Weihe & Nees in Bluff & Fingerh., *Comp. Fl. Germ.* **1**: 687 (1825). Stems angled, reddish, glaucescent, glabrous or glabrescent, with numerous glands, acicles and prickles; prickles very long, unequal, broad-based, straight or falcate. Leaflets 3–5, large, subimbricate, bright green, sparsely hairy above, pale green, tomentose beneath; terminal leaflet suborbicular-ovate to elliptical, emarginate, long-acuminate, coarsely and sharply serrate. Inflorescence pyramidal, lax, leafy; axis with glands longer than its diameter, and numerous long, very unequal purplish-brown, slender, often falcate prickles; upper bracts long, linear-lanceolate; flowers usually more than 2 cm in diameter. Sepals patent to erect after anthesis, long-acuminate, green, tomentose, glandular and aculeolate; petals 5–7, ovate, emarginate, glabrous or glabrescent, bright pink; stamens pink, exceeding the red or yellow-based styles; receptacle glabrescent; carpels glabrous. $2n = 28$. *N.W. & C. Europe.* Au Be Br ?Cz Ga Ge ?Ho Hu Po Rm.

Related species include:

R. abietinus Sudre, *Bull. Assoc. Fr. Bot.* **2**:3 (1899). Au Br Ga Ge.
R. pseudodoftanensis E. I. Nyárády in Săvul., *Fl. Rep. Pop. Române* **4**: 921 (1956). Rm.
R. rufescens P. J. Mueller & Lefèvre, *Pollichia* 16–17: 152 (1859). $2n = 28$. Be Br Ga Ge.

68. R. koehleri Weihe & Nees in Bluff & Fingerh., *Comp. Fl. Germ.* **1**: 681 (1825). Stems angled, glaucescent, hairy or glabrescent, with numerous glands, acicles (some gland-tipped) and pricklets; prickles very unequal and numerous, yellowish, some very long, patent and straight or slightly falcate. Leaflets 5, tough, glabrescent above, pubescent beneath; terminal leaflet elliptic-obovate, acuminate, subcordate to entire at base, coarsely and very unevenly dentate to finely serrate-dentate. Inflorescence rather long and narrow, more or less cylindrical, with or without some simple leaves; axis more or less hairy, with prickles like those of the stem, and some of the glands longer than the diameter of the axis; upper branches 1- to 3-flowered, patent, the lower more erect; flowers usually more than 2 cm in diameter. Sepals somewhat deflexed to patent after flowering, sometimes appressed to fruit in the terminal flowers, green, tomentose, glandular and aculeolate, appendiculate; petals ovate-elliptical, attenuate below, white or pale pink; stamens white, shorter than or much exceeding the greenish styles; receptacle and carpels glabrous, or the carpels slightly pubescent. $2n = 28$. *W. & C. Europe.* Au Be Br Cz Ga Ge Hb He Ho Hu It Ju Lu Po Rm.

Very variable.

Related species include:

R. apricus Wimmer, *Fl. Schles.* ed. 3, 626 (1857). $2n = 28$. Au Be Br Cz Ga Ge Hu Po Rs (W).
R. bavaricus (Focke) Sudre, *Bull. Acad. Int. Géogr. Bot. (Le Mans)* **15**: 229 (1905). Au Cz Ge He Hu.
R. caroli G. Beck, *Fl. Nieder-Österr.* **2**(1): 738 (1892). Au.

R. dasyphyllus (Rogers) Rogers, *Jour. Bot. (London)* **38**: 496 (1900). 2*n*=28. Br Da Ga Hb.

R. doranus Sudre, *Bull. Soc. Bot. Fr.* **51**: 23 (1904). Ga Hu.

R. gerezianus Samp. ex Coutinho, *Fl. Port.* 301 (1913). Lu.

R. hamatulus (Sabr.) Hayek, *Fl. Steierm.* **1**: 805 (1909). Au.

R. hebecarpos P. J. Mueller, *Bonplandia* **9**: 282 (1861) (*R. hebeticarpos* P. J. Mueller). Br Cz He Ho Hu Lu Po Rm.

R. impolitus Sudre, *Fl. Toulous.* 75 (1907). Au Cz Ga Ge.

R. perneggensis (Hayek) Fritsch, *Exkursionsfl. Österr.* ed. 3, 218 (1922). Au.

R. proximus Sudre, *Rubi Eur.* 186 (1912). Au.

R. pygmaeopsis Focke, *Syn. Rub. Germ.* 364 (1877). Au Be Br Cz Ge.

R. pygmaeus Weihe & Nees in Bluff & Fingerh., *Comp. Fl. Germ.* **1**: 687 (1825). Au Cz Ge.

R. spinulatus Boulay, *Ronces Vosg.* 101 (1868). Au Br Cz Ga He.

R. spissifolius Sudre, *Bull. Acad. Int. Géogr. Bot. (Le Mans)* **12**: 69 (1903). Ga Ge Hu.

R. subpygmaeopsis Spribille, *Verh. Bot. Ver. Brandenb.* **49**: 195 (1908). Au Ge Hu.

R. vagabundus Samp., *Ann. Sci. Nat. (Porto)* **9**: 69 (1905). Lu.

R. vastus (Sabr.) Hayek, *Fl. Steierm.* **1**: 796 (1909). Au.

R. vestitifolius Fritsch ex Halácsy, *Verh. Zool.-Bot. Ges. Wien* **41**: 252 (1891). Au.

69. R. incanescens Bertol., *Fl. Ital.* **5**: 223 (1844). Stems robust, obtusely angled, pruinose, with numerous, rather long, often glandular, hairs and acicles, otherwise glabrous; prickles straight or slightly curved, rather unequal, the largest very broad-based. Leaves pedate; leaflets 3–5, large, glabrous above, sparsely grey-white tomentose beneath; terminal leaflet obovate-acuminate, unevenly dentate. Inflorescence large, leafy at the base, with long, lax branches; axis densely tomentose, with numerous glandular hairs and acicles; prickles subulate; peduncles long. Sepals short, partially appressed to developing fruit; petals oblong, white or sometimes red; stamens exceeding the styles; carpels glabrous. *S.W. Europe.* Ga Hs It Lu.

The affinities of this species are uncertain, and it has some characteristics of the Ser. *Glandulosi*. It, pollen quality and size suggest it is diploid.

Series *Glandulosi* P. J. Mueller. Stems weak, terete, often procumbent, pruinose; prickles weak, or broad-based and curved. Inflorescence-axis usually with weak prickles; sepals usually appressed to the young fruit; petals often small; stamens usually equalling or exceeding styles.

R. merinoi Pau ex Merino, *Bol. Soc. Aragon. Ci. Nat.* **3**: 188 (1904) is a species in Ser. *Glandulosi* occurring in N.W. Spain; its affinities are not known.

70. R. scaber Weihe & Nees in Bluff & Fingerh., *Comp. Fl. Germ.* **1**: 683 (1825). Stems weak, terete or obtusely angled, glaucescent, more or less hairy, with short glands and weak prickles; prickles at base of stem large, weak and straight, those on upper part of the stem often falcate, with a stout base. Leaves pedate; leaflets 3–5, small, plicate, rugose, slightly hairy above and beneath; terminal leaflet ovate-elliptical or obovate, acuminate, cordate to rounded at base, finely or deeply and rather evenly serrate. Inflorescence pyramidal, dense at the apex, sometimes large and pendent, with patent branches; axis flexuous, with weak, yellowish, falcate prickles, shortly hairy, with pale glands shorter than the diameter of the axis, but longer than the tomentum and sometimes mixed with much longer glands;

flowers 1–2 cm in diameter. Sepals soon appressed to young fruit, or deflexed, triangular-lanceolate, acuminate, tomentose, finely glandular and aculeolate, green, with white margin; petals small, ovate or ovate-lanceolate, glabrous on the margin, erect, white or flesh-pink; stamens white, equalling or exceeding the greenish or reddish styles; receptacle hairy; carpels glabrous. 2*n*=28. *N.W. & C. Europe, extending to Bulgaria.* Au Be Br Bu Cz Ga Ge Hb He Hu Po Rm.

Related species include:

R. curtiglandulosus Sudre, *Bull. Acad. Int. Géogr. Bot. (Le Mans)* **12**: 87 (1903). Au Be Br Cz Ga Ge He Hu.

R. fragariiflorus P. J. Mueller, *Flora (Regensb.)* **41**: 173 (1858). Au Cz Ga Ge Hu.

R. miostilus Boulay, *Ronces Vosg.* 105 (1868) (*R. myostylus* auct.). Au Ga Ge Hs Lu.

R. tereticaulis P. J. Mueller, *Flora (Regensb.)* **41**: 173 (1858). Au Be Br Bu Cz Ga He Hu Rm.

71. R. schleicheri Weihe ex Tratt., *Rosac. Monogr.* **3**: 22 (1823). Stems trailing, usually terete, red, glaucescent, villous, with numerous yellowish-brown, unequal glands, acicles and prickles. Leaflets 3–5, rather small, hairy above and beneath, the midrib glandular beneath; terminal leaflet ovate or elliptical, long-acuminate, narrowed to a broad, truncate, often subcordate base; margins unevenly and deeply biserrate. Inflorescence nodding, racemose above, the middle branches crowded; bracts long, linear; axis usually tomentose and with long patent hairs, densely and unequally glandular, the glands mostly longer than the diameter of the axis; prickles yellowish, hooked, numerous on the pedicels; flowers 1–1·5 cm in diameter. Sepals patent to erecto-patent, rarely deflexed, tomentose, glandular, aculeolate, green; petals narrow, oblong-spathulate, sometimes emarginate, glabrous, more or less patent, white or pinkish; stamens white, usually exceeding the greenish styles; receptacle and carpels usually hairy. Fruit small, of *c.* 10 drupelets. 2*n*=28. *N.W. & C. Europe, extending to Bulgaria.* Au Be Br Bu Cz Ga Ge He ?Ho Hu Ju Po Rm.

Very variable in shape of leaves, clothing of axis of inflorescence, length of stamens and posture of sepals.

Related species include:

R. amplifrons Sudre, *Bull. Acad. Int. Géogr. Bot. (Le Mans)* **21**: 51 (1911). Au Cz Ga Ge Hu.

R. antonii Sabr., *Verh. Zool.-Bot. Ges. Wien* **58**: 83 (1908) (? non *R. koehleri* var. *antonii* Borbás). Au Cz Ge.

R. caeruleicaulis Sudre, *Bull. Soc. Étud. Sci. Angers* **35**: 47 (1906). Au Ga Ge Hu.

R. conterminus Sudre, *Bull. Assoc. Fr. Bot.* **4**: 6 (1901). Au Cz Ga Ge Hu.

R. eumorphus Kupcsok & Sabr., *Magyar Bot. Lapok* **9**: 225 (1910). Au.

R. fissurarum Sudre, *Bull. Acad. Int. Géogr. Bot. (Le Mans)* **12**: 75 (1903). Au Be Cz Ga Ge Hu.

R. furvus Sudre, *Bull. Assoc. Fr. Bot.* **4**: 3 (1901). Au Cz Ga Ge He Hu Po.

R. humifusus Weihe & Nees in Bluff & Fingerh., *Comp. Fl. Germ.* **1**: 685 (1825). Au Br Cz Ga Ge He Ho Hu Po Rs (W).

R. inaequabilus Sudre, *Bull. Acad. Int. Géogr. Bot. (Le Mans)* **12**: 78 (1903). Au Cz Ga Ge Hu.

R. laceratus P. J. Mueller, *Pollichia* **16–17**: 229 (1859). Au Cz Ga Ge Hu.

R. longicuspis P. J. Mueller ex Genev., *Mém. Soc. Acad. (Angers)* **24**: 120 (1868). Au Cz Ga Ge Hu.

R. **metschii** Focke, *Syn. Rub. Germ.* 359 (1877). Au Ju.

R. **mucronipetalus** P. J. Mueller, *Bonplandia* 9: 298 (1861). Au Be Cz Ga Ge Hu.

R. **opiparus** E. I. Nyárády in Săvul., *Fl. Rep. Pop. Romane* 4: 922 (1956). Rm.

R. **richteri** Halácsy, *Österr. Bot. Zeitschr.* 40: 434 (1890). Au.

R. **rosellus** Sudre, *Bull. Acad. Int. Géogr. Bot.* (*Le Mans*) 12: 72 (1903). Au Cz Ga Ge.

72. **R. glandulosus** Bellardi, *Mém. Acad. Sci.* (*Turin*) 5: 230 (1793) (*R. bellardii* Weihe & Nees). Stems robust, terete, reddish-brown, pruinose, glabrescent, with crowded, rather long glands and acicles; the numerous, unequal prickles weak and deflexed or occasionally falcate. Leaflets mostly 3, sparsely hairy on both surfaces; terminal leaflet elliptical or elliptic-obovate, shortly acuminate, more or less entire at base, usually finely and evenly serrate. Inflorescence broad, rather short, subcorymbose at apex, with 1–3 simple leaves, glandular on both surfaces; pedicels long; axis densely pubescent, densely glandular-aciculate, with many glands longer than the diameter of the axis; prickles often purple-black; flowers up to 2·5 cm in diameter. Sepals appressed to the developing fruit, triangular-ovate-attenuate, tomentose, green, glandular, aculeolate; petals oblanceolate, fimbriate at apex, white; stamens white, but drying pink, scarcely equalling the green- or red-based styles; receptacle pubescent; carpels glabrous. $2n=28$. *W. & C. Europe, extending to S. Sweden, Lithuania and Bulgaria.* Au Be Br Bu Cz Da Ga Ge He Ho Hs Hu It Ju Po Rm Rs (B) Su.

73. **R. serpens** Weihe ex Lej. & Court., *Comp. Fl. Belg.* 2: 172 (1831). Stems and leaves yellowish-green; stems procumbent, robust, terete, glaucescent, usually very hairy; glands and acicles pale yellow, the glands numerous, the acicles few; prickles setaceous, with a conical base. Leaflets 3–5, hairy on both surfaces; terminal leaflet ovate, acuminate, subcordate, minutely serrulate. Inflorescence short, leafy, arched; axis hairy, with numerous glands and weak prickles. Sepals appressed to the developing fruit, acuminate, tomentose, green, glandular, usually unarmed; petals small, oblong, white; stamens white, exceeding the greenish styles; carpels usually glabrous. $2n=28$. *W. & C. Europe, extending to Bulgaria.* Al Au Be Bu Cz Da Ga Ge He ?Ho Hs Hu Ju Po Rm Rs (W).

Related species include:

R. **aculeolatus** P. J. Mueller, *Pollichia* 16–17: 228 (1859). Au Be Ga He Hu.

R. **acutifolius** P. J. Mueller, *op. cit.* 211 (1859). Au Ge.

R. **analogus** P. J. Mueller & Lefèvre, *op. cit.* 232 (1859). Au Br Cz Ga Ge.

R. **angustifrons** Sudre, *Rubi Eur.* 217 (1913). $2n=28$. Au Be Br Cz Ga Ge Hu.

R. **angustisetus** Sudre, *Bull. Acad. Int. Géogr. Bot.* (*Le Mans*) 15: 231 (1905). Au Cz Ga Ge.

R. **asclepiadeus** Borbás, *Magyar Bot. Lapok* 2: 337 (1903). Au.

R. **bayeri** Focke, *Österr. Bot. Zeitschr.* 18: 99 (1868). Au Bu Cz Ju.

R. **biserratus** P. J. Mueller in Boulay, *Ronces Vosg.* 115 (1868). Au Be Ga Ge Hu.

R. **caliculatus** Kaltenb., *Fl. Aachen. Beck.* 2: 283 (1844) (*R. viridis* Kaltenb., non Presl ex Ortmann). Au Be Br Cz Ga Ge Hu.

R. **chlorostachys** P. J. Mueller, *Bonplandia* 9: 303 (1861). Au Cz Ga Hu.

R. **corylinus** P. J. Mueller, *Flora* (*Regensb.*) 41: 169 (1858). Au Cz Ga Ge.

R. **decurtatus** P. J. Mueller, *Pollichia* 16–17: 210 (1859). Au Ga Ge.

R. **divexiramus** P. J. Mueller ex Genev., *Monogr. Rubus* ed. 2, 88 (1880). Au Cz Ga.

R. **durotrigum** R. P. Murray, *Jour. Bot.* (*London*) 30: 15 (1892). Au Br Ge He Hu.

R. **hispidissimus** Sudre, *Bull. Soc. Étud. Sci. Angers* 35: 50 (1906). Au Cz Ga Ge Hu.

R. **horridulus** P. J. Mueller in Boulay, *Ronces Vosg.* 112 (1868). Au Cz Ga Ge Hu.

R. **incultus** Wirtgen ex Focke, *Syn. Rub. Germ.* 369 (1877). Au Be Br Cz Ga Ge He Hu.

R. **lamprophyllus** Gremli, *Österr. Bot. Zeitschr.* 21: 94 (1871) (*R. bayeri* subsp. *lamprophyllus* (Gremli) Focke). Au Bu Cz He Hu Ju.

R. **leptadenes** Sudre, *Bull. Acad. Int. Géogr. Bot.* (*Le Mans*) 15: 232 (1905). Au Be Br Cz Ga Ge He Hu.

R. **leptobelus** Sudre, *Bat. Eur.* 31 (1904). Au Be Cz Ga Ge Hu It.

R. **longisepalus** P. J. Mueller, *Bonplandia* 9: 297 (1861). Au Ga Ge ?Po.

R. **lusaticus** Rostock ex R. Wagner, *Ber. Deutsch. Bot. Ges.* 5: cliv (1887). Au Br Cz Ga Ge He Hu.

R. **napophiloides** Sudre, *Bat. Eur.* 32 (1904). Au Be Cz Ga Ge He Hu.

R. **niveoserpens** E. I. Nyárády in Săvul., *Fl. Rep. Pop. Romane* 4: 925 (1956). Rm.

R. **obrosus** P. J. Mueller, *Pollichia* 16–17: 234 (1859). Au Be Ge Hu.

R. **ochrosetus** Borbás, *Abauj-Torna Vármegye Fl.* 445 (1896). Al Gr Ju.

R. **oreades** P. J. Mueller & Wirtgen ex Genev., *Mém. Soc. Acad.* (*Angers*) 24: 89 (1868). Au Be Cz Ga Ge He Lu.

R. **pachychlamydeus** Sabr., *Mitt. Naturw. Ver. Steierm.* 52: 277 (1916). Au.

R. **parvulipetalus** Sudre, *Bull. Soc. Étud. Sci. Angers* 35: 49 (1906). Au Cz Ga Ge Hu.

R. **persericans** Sabr. ex Sudre, *Rubi Eur.* 219 (1913). Au.

R. **preissmannii** Halácsy, *Verh. Zool.-Bot. Ges. Wien* 41: 273 (1891). Au.

R. **renifrons** Sabr., *Österr. Bot. Zeitschr.* 42: 55 (1892). Au Cz.

R. **rivularis** Wirtgen & P. J. Mueller, *Flora* (*Regensb.*) 42: 237 (1859). Au Bu Cz Ho Hu Ju Rm Rs (W).

R. **subaculeatus** Borbás ex Fritsch, *Exkursionsfl. Österr.* ed. 3, 217 (1922) (*R. rivularis* var. *spinosulus* Sudre). Au Rm.

R. **subcaucasicus** Sabr. ex Sudre, *Rubi Eur.* 211 (1913). Au.

R. **vepallidus** Sudre, *Bull. Soc. Bot. Belg.* 47: 224 (1910). Au Be Cz Ga Ge He Hu.

R. **vindobonensis** Sabr., *Deutsche Bot. Monatsschr.* 7: 131 (1889). Au.

74. **R. hirtus** Waldst. & Kit., *Pl. Rar. Hung.* 2: 150 (1803–4). Stems robust, terete, more or less pruinose; glands and acicles brown or purple, the glands numerous, unequal, the acicles few; prickles unequal, long, subulate, fragile. Leaflets 3–5, hairy above, softly hairy beneath; terminal leaflet suborbicular-ovate, acuminate, cordate, unevenly serrate or sometimes finely mucronate-dentate. Inflorescence long, pyramidal, broad, lax, leafy, often pendent, with the lower and middle branches patent; axis hairy, with numerous purple glands mostly longer than the diameter of the axis, and a few rather acicular prickles; peduncles short, many-flowered; flowers 2–2·5 cm in diameter. Sepals appressed to the developing fruit, ovate, green, tomentose, densely glandular and aculeolate; petals obovate, glabrous, white; stamens white, slightly exceeding the greenish, rarely pinkish, styles; receptacle slender, cylindrical, hairy; carpels hairy. Fruit of rather small drupelets. $2n=28$. *W., C. & S.E.*

Europe. Al Au Be Br Bu Cz Ga Ge Gr Hb He Hs Hu Ju Po Rm Rs (W, K) Tu.

Many variants have been described varying in hairiness of stem, shape and indumentum of leaves, armature of inflorescence and length of stamens.

Related species include:

R. amoenus Koehler ex Weihe in Wimmer & Grab., *Fl. Siles.* 2(1): 54 (1829) (*R. purpuratus* Sudre). Au Be Br Cz Ga He Rm.

R. anisacanthoides Sudre, *Bull. Soc. Bot. Fr.* **52**: 328 (1905). Au Ga Ge Hu.

R. anoplocladus Sudre, *op. cit.* 337 (1905). Au Cz Ga Ge He Hu.

R. brumalis Sudre, *Bull. Assoc. Fr. Bot.* **4**: 5 (1901). Au Cz Ga Ge He.

R. capparidopsis Hormuzaki, *Mem. Secţ. Şti. (Acad. Română)* ser. 3, **2**: 299 (1925). Au.

R. carneus Sabr., *Mitt. Naturw. Ver. Steierm.* **52**: 270 (1916). Au Ge.

R. celtidifolius Focke ex Gremli, *Beitr. Fl. Schweiz* 33 (1870). Au He.

R. coriifrons (Sabr.) Hayek, *Fl. Steierm.* **1**: 822 (1909). Au.

R. crassus Holuby, *Österr. Bot. Zeitschr.* **23**: 381 (1873). Au Cz Hu Po.

R. declivis Sudre, *Compt. Rend. Congr. Soc. Sav. (Sci.)* **1908**: 233 (1909). Au Cz Ga Ge Hu.

R. elegantissimus Hayek, *Verh. Zool.-Bot. Ges. Wien* **66**: 459 (1916). Au.

R. erythrostachys (Sabr.) Sabr. ex Halácsy, *Verh. Zool.-Bot. Ges. Wien* **41**: 278 (1891). Au Bu Ju.

R. garrulimontis Hormuzaki, *Österr. Bot. Zeitschr.* **68**: 227 (1919). Au.

R. guentheri Weihe & Nees in Bluff & Fingerh., *Comp. Fl. Germ.* **1**: 679 (1825). Au Br Bu Cz Ga Ge He Hu Ju Po.

R. hercynicus G. Braun ex Focke, *Syn. Rub. Germ.* 370 (1877). Au Cz Ga Ge He Hu ?Ju ?Po.

R. hirtimimus Juz., *Not. Syst. (Leningrad)* **13**: 108 (1950). Rs (K).

R. interruptus Sudre, *Bull. Assoc. Fr. Bot.* **2**: 7 (1899). Au Cz Ga Ge He Hs Hu Po.

R. kaltenbachii Metsch, *Linnaea* **28**: 170 (1856). Au Be Cz Ga He Ho Hu ?Po ?Rs (W).

R. latifrons (Progel) Hayek, *Fl. Steierm.* **1**: 812 (1909). Ju.

R. minutidentatus Sudre, *Bull. Soc. Bot. Fr.* **52**: 323 (1905). Au Cz Ga Ge He Hs Hu Po.

R. minutiflorus P. J. Mueller, *Pollichia* **16–17**: 235 (1859). Au Cz Ho Hs Hu Ju ?Po.

R. nigricatus P. J. Mueller & Lefèvre, *op. cit.* 204 (1859). Au Be Cz Ga Ge Hb He Hu ?Rs (W).

R. offensus P. J. Mueller, *Bonplandia* **9**: 286 (1861). Au Be Cz Ga Ge He Hu.

R. pierratii Boulay, *Ronces Vosg.* 108 (1868). Au Ga Ge He Hu.

R. plusiacanthus Borbás ex Sudre, *Rubi Eur.* 200 (1912). Au Cz Hu Ju.

R. posoniensis Sabr., *Verh. Zool.-Bot. Ges. Wien* **36**: 90 (1886). Cz Hu.

R. praealpinus Hayek, *Fl. Steierm.* **1**: 822 (1909). Au.

R. praedatus Schmidely, *Bull. Herb. Boiss.* ser. 2, **3**: 79 (1903). Au Cz Ga Ge.

R. romanicus E. I. Nyárády in Săvul., *Fl. Rep. Pop. Romîne* **4**: 929 (1956). Rm.

R. rubiginosus P. J. Mueller, *Pollichia* **16–17**: 207 (1859). Au ?Cz Hu ?Po.

R. rubrisetoides Hormuzaki, *Mem. Secţ. Şti. (Acad. Română)* ser. 3, **2**: 305 (1925). Au.

---

[1] By I. Klášterský.

R. ruderalis Kupcsok, *Magyar Bot. Lapok* **9**: 236 (1910). Au Cz.

R. scenoreinus Juz., *Not. Syst. (Leningrad)* **13**: 106 (1950). Rs (K).

R. tenuidentatus Sudre, *Bull. Soc. Bot. Fr.* **52**: 344 (1905). Au Be Cz Ga Ge He Hs Hu Po.

R. topitzii Halácsy ex Topitz, *Österr. Bot. Zeitschr.* **42**: 203 (1892). Au.

R. trachyadenes Sudre, *Compt. Rend. Congr. Soc. Sav. (Sci.)* **1908**: 233 (1909). Au Cz Ga Ge Hu.

R. wittingii Halácsy, *Verh. Zool.-Bot. Ges. Wien* **41**: 271 (1891). Au.

Subsect. *Caesii* Focke. Stems usually terete. Leaves ternate or with 5 leaflets, the basal leaflets subsessile; stipules lanceolate. Inflorescence short and broad, or narrow with few-flowered branches; flowers often long-stalked. Petals often large, orbicular. Drupelets often large, partly abortive, usually pruinose.

75. R. caesius L., *Sp. Pl.* 493 (1753). Stems flagelliform, branched, terete, glabrous, pruinose, rarely with a few short glands; prickles few or many, weak, short, straight or falcate. Leaves ternate, slightly hairy above, more or less pubescent beneath; terminal leaflet ovate, sometimes more or less 3-lobed, shortly acuminate, subcordate, rather coarsely biserrate; lateral leaflets often bilobed; petiole sulcate; stipules ovate-lanceolate. Inflorescence short, consisting of a terminal and a few axillary 2- to 5-flowered corymbs; pedicels long, with short hairs, fine glands and sparse prickles; flowers 2–2·5 cm in diameter. Sepals appressed to the developing fruit, ovate-lanceolate, shortly acuminate, grey-green tomentose, with white margin and short glands; petals large, ovate, elliptical or suborbicular, white; stamens green, equalling the greenish styles. Drupelets 2–20, large, black, pruinose, loosely coherent. $2n=28$. *Somewhat calcicole. Most of Europe.* All except Az Co Cr Fa Is Sa Sb Tu.

Several varieties have been described, some with pink flowers.

Hybrids between *R. caesius* and plants belonging to other subsections of *Rubus* are termed **Rubi Corylifolii**. Many well-defined taxa have been described but they tend to have very restricted distributions and are not listed here; in Belgium for example, 38 *Corylifolii* have been described, in Britain 20, in Switzerland 50 and in Romania 28. *Corylifolii* are recognised by having at least some of the following characters; stems terete; stipules more or less lanceolate; leaves large; leaflets imbricate, the basal subsessile; inflorescence short and broad, or narrow and with few-flowered branches; flowers long-stalked; petals large, orbicular; drupelets large, partly abortive, rather pruinose. Some set good fruit consisting of large but few drupelets, whilst others tend to be somewhat infertile. In Great Britain and Scandinavia the chromosome numbers of many of the *Rubi Corylifolii* are higher than in the plants of other subsections (counts of 35 or 42 are relatively common), and this may well be true of the *Corylifolii* in other countries. Many are characteristic of ground which has been disturbed; they are particularly common at the edge of cultivated fields, in hedges, along ditch-banks, forest roads etc. The names *R. corylifolius* Sm. and *R. dumetorum* Weihe are no longer in use except as referring to rather broad aggregates.

10. Rosa L.[1]

Shrubs, usually deciduous. Stems usually with prickles. Leaves pinnate; stipules usually adnate to petiole. Flowers terminal, solitary or in corymbs, (4–)5-merous. Hypanthium urceolate, becoming coloured and fleshy in fruit; epicalyx absent; stamens and carpels numerous; styles protruding through the orifice of a disc, sometimes forming a short column; ovules 1. Fruit a pseudocarp of numerous achenes enclosed in the hypanthium.

Most species occur in scrub, woodland and hedges.

The description of leaves always refers to the best-developed leaves on the flowering stems.

The armature may include *prickles*, *acicles* (slender, needle-like structures), *setae*, stipitate *glands* and eglandular *hairs*.

The innumerable cultivars of *Rosa* to be found in European gardens (the great majority *flore pleno*) are mostly complex hybrids, of which the most important ancestors are **4**, **6** and **14**, described below, and also *R. chinensis* Jacq., *R. odorata* var. *gigantea* (Collett ex Crépin) Rehder & E. H. Wilson, *R. multiflora* Thunb., *R. wichuraiana* Crépin, all from E. Asia (though *R. chinensis* is not known in the wild state). Some modern garden roses, such as the 'hybrid polyanthas', include all seven of these species in their ancestry. For the origin of garden roses see C. C. Hurst, *Jour. Roy. Hort. Soc.* **66**: 73–82, 242–50, 282–9 (1941), and A. P. Wylie, *Jour. Roy. Hort. Soc.* **79**: 555–71 (1954); **80**: 8–24, 77–87 (1955).

In addition to these hybrids, and to the species described below, many species, mostly from E. Asia, are cultivated in gardens, and some are perhaps becoming naturalized locally. Two hybrids are cultivated on a field scale in Bulgaria and S. France for essential oil from their petals: these are **R.** × **bifera** (Poiret) Pers., *Syn. Pl.* **2**: 48 (1806) (*R. damascena* auct., non Miller), which is probably a hybrid between **4** and **14** and **R.** × **alba** L., *Sp. Pl.* 492 (1753), whose parentage is uncertain; it is perhaps a complex hybrid between **2**, **14** and a white-flowered member of Sect. *Caninae*.

Literature: F. Crépin, *Bull. Soc. Bot. Belg.* **8–21** (1869–82). H. Christ, *Rosen der Schweiz*. Basel. 1873. A. Déséglise, *Bull. Soc. Bot. Belg.* **15**: 176–405, 491–602 (1876). V. Borbás, *Primitiae Monographiae Rosarum Imperii Hungarici*. Budapest. 1880. J. Schwertschläger, *Die Rosen des südlichen and mittleren Frankenjura*. München. 1910. E. Willmott, *The genus* Rosa. London. 1910–14. G. Täckholm, *Acta Horti Berg.* **7**: 97–381 (1922). G. A. Boulenger, *Bull. Jard. Bot. Bruxelles* **10** (1924); **12** (1932). A. H. Wolley-Dod, *Jour. Bot.* (*London*) **68–69** (Suppl.): 1–111 (1930–1). R. Keller, *Synopsis Rosarum spontanearum Europae mediae*. Zürich. 1931. C. Vicioso, *Estudios sobre el Género 'Rosa' en España*. Ed. 2. Madrid. 1964.

1 Styles connate in a column
  2 Stylar column shorter than the inner stamens; styles sometimes becoming free in fruit   **15. stylosa**
  2 Stylar column at least as long as the inner stamens; styles not becoming free in fruit
    3 Inflorescence with 10–20 flowers
      4 Stylar column glabrous   **3. phoenicia**
      4 Stylar column hairy   **4. moschata**
    3 Flowers solitary or inflorescence with 2–7 flowers
      5 Leaves coriaceous, evergreen; stylar column usually hairy
        1. sempervirens
      5 Leaves herbaceous, deciduous; stylar column glabrous
        6 Erect shrub; stems arching; prickles very stout with very broad bases   **15. stylosa**
        6 Trailing shrub; stems weak; prickles ±slender   **2. arvensis**
1 Styles free
  7 Sepals ±entire
    8 Leaflets glabrous
      9 Leaflets simply serrate

        10 Stems with long prickles mixed with many short prickles and acicles   **5. pimpinellifolia**
        10 Stems without acicles
          11 Pedicels glandular-hispid   **13. virginiana**
          11 Pedicels glabrous   **9. glauca**
      9 Leaflets biserrate or compound-serrate
        12 Petals yellow; fruit *c*. 10 mm, globose   **6. foetida**
        12 Petals deep purplish-pink; fruit 15–25 mm, ovoid to elongate-pyriform, rarely globose   **12. pendulina**
    8 Leaflets pubescent, at least beneath
      13 Petals yellow   **6. foetida**
      13 Petals white, pink or purplish-pink
        14 Young stems, prickles and lower surface of the leaflets densely tomentose; leaves thick, rugose   **10. rugosa**
        14 Young stems and prickles glabrous, the lower surface of the leaflets ±sparsely hairy; leaves not rugose
          15 Pedicels glandular-hispid
            16 Leaflets biserrate or compound-serrate   **12. pendulina**
            16 Leaflets simply serrate
              17 Flowering stems usually densely covered with slender prickles or acicles; sepals erect and persistent after anthesis   **7. acicularis**
              17 Flowering stems usually without acicles; sepals patent and deciduous after anthesis   **13. virginiana**
          15 Pedicels glabrous
            18 Flowering stems usually densely covered with slender prickles or acicles; fruit ovoid, with a distinct neck below the disc   **7. acicularis**
            18 Flowering stems without acicles; fruit usually globose
              19 Stems usually with a pair of curved prickles at the nodes   **8. majalis**
              19 Stems without paired prickles at the nodes   **11. blanda**
  7 Outer 3 sepals distinctly pinnatifid or lobed
    20 Leaflets coriaceous; petals (25–)30–45 mm   **14. gallica**
    20 Leaflets not coriaceous; petals 8–25(–30) mm
      21 Leaflets glabrous or subglabrous, eglandular or very sparsely glandular
        22 Leaflets bluish-green or purplish; young stems pruinose
          23 Hypanthium and pedicels densely stipitate-glandular
            17. montana
          23 Hypanthium glabrous; pedicels rarely sparsely stipitate-glandular
            24 Sepals 2–3 mm wide; prickles sparse, rather slender, curved or straight, without stout bases   **9. glauca**
            24 Sepals 3–5 mm wide; prickles stout, curved or hooked, usually with stout bases   **(18–31). canina** group
        22 Leaflets green; young stems not pruinose
          25 Pedicels glabrous   **(18–31). canina** group
          25 Pedicels stipitate-glandular
            26 Sepals erect and persistent after anthesis   **17. montana**
            26 Sepals deflexed and deciduous after anthesis
              27 Leaflets not coriaceous; prickles usually hooked or curved   **(18–31). canina** group
              27 Leaflets subcoriaceous; prickles usually straight or slightly curved   **16. jundzillii**
      21 Leaflets distinctly hairy or glandular or both
        28 Leaflets coriaceous; petals (25–)30–45 mm   **14. gallica**
        28 Leaflets not coriaceous; petals 8–25(–30) mm
          29 Leaflets ±densely viscid-glandular beneath
            30 Leaflets glabrous or somewhat pubescent beneath, smelling of apples   **(39–47). rubiginosa** group
            30 Leaflets tomentose beneath, with a resinous smell
              (32–38). tomentosa group
          29 Leaflets eglandular beneath or with glands confined to main veins
            31 Prickles straight or slightly curved; leaflets always hairy and usually very tomentose, with a resinous smell
              (32–38). tomentosa group
            31 Prickles usually curved or hooked; leaflets glabrous or pubescent, very rarely tomentose and if tomentose then prickles distinctly curved or hooked and hypanthium glabrous; leaflets usually not scented
              (18–31). canina group

Sect. SYNSTYLAE DC. Trailing, climbing or creeping shrubs. Rhizome short. Prickles curved, all similar. Outer sepals usually pinnatifid, deflexed and deciduous after anthesis. Disc wide, with a narrow orifice. Carpels sessile. Styles connate in a column, at least as long as the inner stamens, not becoming free in fruit.

1. R. sempervirens L., *Sp. Pl.* 492 (1753). Evergreen, with long, creeping stems. Prickles sparse, curved, markedly decurrent at base. Leaflets. (3–)5–7, 30–60 × 10–20 mm, coriaceous, ovate-lanceolate, acuminate, serrate, glabrous, shining; stipules narrow. Inflorescence (1–)3- to 7-flowered, corymbose. Pedicels glandular-hispid, 2–4 times as long as the fruit. Sepals ovate, long-acuminate, usually entire, glandular with stalked glands on the margins and back, deflexed and deciduous after anthesis. Petals 10–20(–30) mm, white. Stylar column hairy or glabrous. Fruit *c.* 10 mm, globose or broadly ovoid, red. *Mediterranean region and S.W. Europe, northwards to 47° 30′ in W. France.* Al Bl Co Cr Ga Gr Hs It Ju Lu Sa Si Tu [Br].

2. R. arvensis Hudson, *Fl. Angl.* 192 (1762). Deciduous, with long, weak, trailing stems. Prickles sparse, hooked, the upper sometimes slender and almost straight. Leaflets 5–7, 15–40 × 10–20 mm, herbaceous, ovate to broadly elliptical, dull above, sparsely appressed-pubescent to subglabrous. Inflorescence 1- to 3(–5)-flowered. Pedicels stipitate-glandular, 2–3 times as long as fruit. Sepals eglandular on the back, deflexed after anthesis, the outer with long, narrow lobes. Petals 15–25 mm, white. Stylar column glabrous. Fruit 10–16 mm, globose to ovoid, red. *S., W. & C. Europe.* Al Au Be Bl Br Bu ?Co Cz Ga Ge Gr Hb He Ho Hs Hu It Ju ?Po Rm Rs (W) Si Tu.

3. R. phoenicia Boiss., *Diagn. Pl. Or. Nov.* 2(10): 4 (1849). Like 2 but stems climbing; prickles stout, hooked; leaflets 15–50 × 10–30 mm, usually densely hairy on both surfaces; inflorescence 10- to 20-flowered, corymbose; outer sepals with short and wide lobes; fruit *c.* 10 mm. *N.E. Greece (Thraki).* Gr. (*S.W. Asia.*)

4. R. moschata J. Herrmann, *Diss. Rosa* 15 (1762). Vigorous evergreen climber up to 12 m. Prickles sparse, stout, curved. Leaflets 5–7, 25–55 × 20–30 mm, ovate to elliptical, usually acute, serrate, usually pubescent, greyish-green beneath. Inflorescence many-flowered. Pedicels slightly pubescent and glandular. Sepals lanceolate, with a setiform apical appendage, more or less grey-pubescent, the outer with 2–4 lobes. Petals 10–15 mm, creamy-white. Stylar column hairy. Fruit 8–10 mm, ovoid. *Cultivated for ornament, mainly in S. & W. Europe; naturalized in the Mediterranean region.* [Cr Ga Gr Hs Si.] (*Himalaya, Iran.*)

Sect. PIMPINELLIFOLIAE DC. Deciduous shrubs. Rhizome long. Stems usually with straight prickles and acicles. Flowers solitary, bracteate. Sepals entire, erect and persistent after anthesis. Disc narrow, with a wide orifice. Carpels shortly stipitate. Styles free.

5. R. pimpinellifolia L., *Syst. Nat.* ed. 10, 2: 1062 (1759) (*R. spinosissima* L. pro parte, *R. myriacantha* DC.). Stems up to 1 m, erect, forming dense patches. Prickles abundant and long on the main stems, sparser and shorter on the flowering stems and mixed with acicles. Leaflets 5–11, 5–15(–20) × 4–9 mm, glabrous, suborbicular to broadly elliptical, usually simply serrate and eglandular. Flowers solitary. Pedicels longer than the fruit, eglandular or with stalked glands. Sepals narrowly lanceolate, acuminate, eglandular. Petals 10–20(–25) mm, white, rarely pink. Styles short, usually in a compact head, lanate. Fruit *c.* 6 mm, globose or depressed-globose, black. 2*n* = 28. *Europe except the north-east,* most of Fennoscandia, the extreme south-west and many of the islands. Al Au Be Br Bu Cz Da Ga Ge Gr Hb He Ho Hu Is It Ju No Po Rm Rs (C, W, K, E) †Su.

Many hybrids between **5** and members of Sect. *Caninae* have been described. The influence of **5** is usually recognizable in the presence of acicles and in the habit and leaflet-shape.

6. R. foetida J. Herrmann, *Diss. Rosa* 18 (1762). Stems up to 4 m. Prickles curved, more or less compressed and strongly decurrent at the base. Leaflets 5–7, 15–40 × 12–25 mm, ovate-elliptical, usually sparsely hairy and dark green above, sparsely glandular and slightly paler beneath, biserrate; teeth glandular. Flowers 1–3. Pedicels usually glabrous, sometimes hispid. Sepals attenuate at apex, glandular-hispid. Petals 20–30 mm, yellow. Styles long, lanate. Fruit *c.* 10 mm, globose, red. *Cultivated for ornament; locally naturalized in S. & C. Europe.* [Au Cz Ga Ge Gr Hs It Rm Tu.] (*S.W. Asia.*)

Sect. CASSIORHODON Dumort. (Sect. *Cinnamomeae* Crépin). Erect, deciduous shrubs. Rhizome long. Prickles at the nodes slender, straight, the others stout, hooked, or absent; acicles often present. Flowers in bracteate corymbs. Sepals usually entire, erect and persistent after anthesis. Disc narrow, with a wide orifice. Carpels lining the sides as well as the base of the hypanthium. Styles free.

7. R. acicularis Lindley, *Ros. Monogr.* 44 (1820). Stems up to 1 m, with numerous slender acicles and long straight prickles. Leaflets (3–)5–7, 15–55 × 15–28 mm, broadly elliptical to oblong, acute, usually rather coarsely serrate, usually glabrous above, pubescent beneath. Flowers solitary, scented. Bracts about equalling the pedicels, usually narrow. Pedicels usually glabrous, rarely glandular-hispid. Petals 18–24 mm, purplish-pink. Styles lanate. Fruit ovoid, with a distinct neck below the disc. *N.E. Europe, extending locally southwards to c. 52° N. in C. & E. Russia.* Fe Rs (N, C) Su [Au].

8. R. majalis J. Herrmann, *Diss. Rosa* 8 (1762) (*R. cinnamomea* sensu L. (1759), non L. (1753), *R. spinosissima* L., nom. ambig.). Stems up to 2 m, forming large patches; bark reddish-brown. Prickles slender, straight or slightly curved, in pairs at the nodes; flowering stems sometimes unarmed. Leaflets 5–7, 15–45 × 15–27 mm, elliptical to obovate, cuneate at base, serrate, pubescent, bluish-green above, pale bluish beneath. Flowers solitary. Bracts about equalling or a little longer than the pedicels, large. Pedicels glabrous. Petals 18–25(–30) mm, purplish-pink. Styles lanate. Fruit depressed-globose, rarely ovoid, glabrous, red. 2*n* = 14. *N. & C. Europe; U.S.S.R. except the south-west.* Au Cz Fe Ga Ge He It Ju No Po Rs (N, B, C, W, E) Su [Be Bu Da Ho].

9. R. glauca Pourret, *Mém. Acad. Toulouse* 3: 326 (1788) (*R. ferruginea* auct., non Vill., *R. rubrifolia* Vill.). Stems up to 3 m, erect bluish-green, pruinose when young, becoming brown. Prickles scattered, curved or straight, rather slender, without stout bases. Leaflets 5–9, 20–45 × 15–25 mm, elliptical to ovate, serrate, glabrous, bluish-green or purplish. Flowers 1–5. Pedicels glabrous, very rarely with stalked glands. Sepals 2–3 mm wide, sometimes almost entire but usually with a few linear lobes, glabrous. Petals 18–22 mm, narrow, deep pink. Disc flat. Styles white-lanate. Fruit *c.* 15 mm, globose, usually glabrous, brownish-red. *Mountains of C. Europe, extending southwards to the Pyrenees, C. Italy and N. Albania.* Al Au Bu Co Cz Ga Ge He Hs Hu It Ju Po Rm [Fe Su].

10. R. rugosa Thunb., *Fl. Jap.* 213 (1784). Stems up to 2·5 m, densely tomentose when young. Prickles hairy, at least at the base, straight, mixed with dense acicles. Leaflets 5–9, 20–50 × 18–25(–40) mm, ovate to elliptical, usually green and shiny above, thick, rugose. Flowers 1–3. Bracts large, enclosing the pedicels. Pedicels usually densely tomentose. Petals (25–)30–45(–50) mm, purplish-red, rarely white. Fruit depressed-globose, with a well-developed neck below the disc, glabrous, scarlet. 2*n* = 14. *Cultivated for ornament and for hedges; naturalized in parts of N., W. & C. Europe.* [Au Br Da Fe Ga Ge Hb Ho Hu No Po Rm Su.] (*E. Asia.*)

11. R. blanda Aiton, *Hort. Kew.* **2**: 202 (1789). Erect shrub up to 2 m. Stems unarmed, or with scattered acicles when young; bark brown. Leaflets 5–7(–9), 20–60 × 15–40 mm, elliptical to obovate-oblong, usually acute, coarsely serrate, dull and glabrous above, paler and usually pubescent beneath. Flowers 1–5. Bracts large, enclosing the pedicels. Pedicels glabrous. Petals 15–25 mm, pink. Fruit subglobose or ellipsoid, glabrous, red. *Cultivated for ornament; naturalized in C. Europe.* [Au Ge.] (*E. & C. North America.*)

12. R. pendulina L., *Sp. Pl.* 492 (1753) (*R. alpina* L.). Stems up to 2 m; bark green, yellowish-green or occasionally purplish. Prickles usually absent. Leaflets 7–11, 20–60 × 10–30 mm, oblong-ovate to -obovate, biserrate, glabrous or pubescent above, usually sparsely pubescent and sometimes glandular at least on the veins beneath, with glandular teeth. Flowers solitary. Bracts about as long as and enclosing the pedicels, soon deciduous. Pedicels glandular-hispid, recurved in fruit. Petals 15–25 mm, deep purplish-pink. Styles densely hairy. Fruit 15–25 mm, pendent, ovoid to elongate-pyriform, rarely globose, often glandular-hispid, red. *Mountains of C. & S. Europe.* Al Au Be Bu Cz Ga Ge Gr He Hs Hu It Ju Po Rm Rs (W).

Sect. CAROLINAE Crépin. Erect, deciduous shrubs. Stems slender, with straight, paired prickles and often acicles. Flowers few, in corymbs. Sepals patent and soon deciduous after anthesis. Pedicels and hypanthium glandular-hispid, rarely smooth. Carpels confined to the bottom of the hypanthium.

13. R. virginiana J. Herrmann, *Diss. Rosa* 19 (1762). Stems up to 2 m, with few or no suckers; bark bluish-green, becoming reddish-brown. Prickles hooked, curved or absent. Leaflets 5–9, 20–60 × 12–25 mm, elliptical to elliptic-obovate, often cuneate at base, acute, serrate, dull green above, glabrous or sparsely hairy beneath; teeth eglandular. Flowers 2–8. Bracts much shorter than the pedicels. Pedicels glandular-hispid. Sepals patent and deciduous after anthesis, glandular-hispid on the back. Petals 15–25(–30) mm, pink or white. Styles lanate. Fruit 10–15 mm, ovoid-globose to globose, glandular-hispid. *Cultivated for ornament and locally naturalized.* [Au Br Ga.] (*E. North America.*)

Sect. ROSA (Sect. *Gallicanae* DC.). Erect, usually low shrubs. Rhizome long. Stems usually with hooked prickles mixed with acicles. Outer sepals usually pinnatifid, deflexed and deciduous after anthesis. Disc wide, with a narrow orifice. Carpels sessile. Styles free.

14. R. gallica L., *Sp. Pl.* 492 (1753). Deciduous shrub 0·4–0·8 m, forming large patches. Stems with prickles and glandular setae. Leaflets 3–7, 20–60 × 18–30 mm, coriaceous, suborbicular or ovate to narrowly elliptical, rounded at the apex, usually compound-serrate, dull bluish-green and glabrous above, paler, pubescent and glandular beneath. Flowers solitary, rarely 2–4,

6–9 cm in diameter, strongly scented. Pedicels glandular-setose. Sepals glandular on the back. Petals (25–)30–45 mm, deep pink. Styles densely hairy, rarely glabrous. Fruit globose to fusiform, densely glandular-setose, bright red. *S. & C. Europe, extending to Belgium and C. France.* Al Au Be Bu Cz Ga Ge Gr He Hu It Ju Po Rm Rs (C, W, K) Tu [Co Hs Lu Sa Si].

Sect. CANINAE DC. Deciduous shrubs, with erect or arching stems. Rhizome short. Prickles usually numerous, hooked or straight; acicles usually absent. Flowers in bracteate corymbs. Outer sepals usually pinnatifid, deflexed, erect or patent, persistent or caducous after anthesis. Disc flat or conical, variable in size, narrow. Carpels long-stipitate. Styles free.

This section has long been recognized as critical, and very large numbers of taxa have been described. So far as is known, they are all polyploid, with 2*n* = 28, 35 or 42, and their reproduction is unusual. In the pentaploids, for example, 7 bivalents and 21 univalents are formed at meiosis. In the pollen the univalents are lost, so that most functional pollen-grains carry only 7 chromosomes; in the ovules, on the other hand, all the univalents go to one pole at the first meiotic division and the egg has 28 chromosomes. Thus, in sexual reproduction, most of the chromosomes of the offspring come from the seed-parent and have not paired at meiosis. Inheritance thus tends to be predominantly maternal, and this in turn means there is a tendency for the biotypes in Sect. *Caninae* to be relatively constant, though hybridization can produce new biotypes, some of which may survive and become stabilized. The situation is analogous to facultative apomixis, though *Rosa* is not apomictic.

In the account of Sect. *Caninae* which follows, an attempt has been made to cover the whole range of variation but to describe only a limited number of fairly well-defined species. It is recognized that many intermediates may occur. Because of the taxonomic difficulties, the geographical distribution of many of the species is imperfectly known, and can only be given in general terms.

15. R. stylosa Desv., *Jour. Bot. Rédigé* **2**: 317 (1809). Stems up to 3 m. Prickles hooked, some with very stout bases. Leaflets 5–7, 25–50 × 15–25 mm, elliptical to elliptic-oblong, usually acuminate, serrate, usually pubescent beneath at least on the veins, rarely pubescent above, eglandular. Flowers solitary to many. Pedicels long, usually glandular-hispid. Sepals deflexed and deciduous after anthesis. Petals 15 30 mm, usually white, sometimes pink. Disc conical. Styles connate in a column which is shorter than the inner stamens, sometimes becoming free in fruit. Fruit 10–15 mm, ovoid, rarely globose, glabrous, red. *From Ireland and W. Germany to S. Spain and Bulgaria; local.* Au Br Bu Ga Ge Hb He Hs Hu It Rm.

16. R. jundzillii Besser, *Cat. Horto Cremen.* 117 (1816) (*R. marginata* auct., non Wallr.). Stems up to 2 m, erect. Prickles slender, straight or slightly curved. Leaflets 5–7, 25–40 × 15–25 mm, subcoriaceous, elliptical to ovate, acute, biserrate, glabrous or rarely very sparsely pubescent, sparsely glandular on the margins and veins beneath; teeth glandular; rhachis sparsely glandular. Flowers solitary, slightly scented. Pedicels *c.* 25 mm, usually twice as long as the fruit, stipitate-glandular. Sepals glandular on the back, deflexed and deciduous after anthesis. Petals 15–25(–30) mm, pale to deep pink. Disc flat, wide. Styles ± densely pubescent. Fruit *c.* 12 mm, globose to ovoid, sparsely stipitate-glandular or glabrous, red. *C. & E. Europe, extending westwards to C. France and N.W. Italy.* Al Au Bu Cz Ga Ge ?Gr Hc Hu It Ju Po Rm Rs (C, W, K) ?Tu.

17. **R. montana** Chaix in Vill., *Hist. Pl. Dauph.* **1**: 346 (1786). Stems up to 3 m, erect. Prickles curved or nearly straight, abruptly dilated at base. Leaflets 7–9, 20–35 × 12–25 mm, broadly obovate, biserrate, glabrous, bluish-green, glandular beneath on the midrib and rarely also on the lateral veins; petiole and rhachis glandular. Flowers solitary or 2–3. Pedicels very densely stipitate-glandular. Sepals glandular on the back, erect and persistent after anthesis. Petals 14–22 mm, pale pink, becoming whitish. Disc as wide as or very little wider than the orifice. Styles lanate. Fruit 15–25 mm, ovoid to elongate-pyriform, stipitate-glandular, rarely glabrous. *S. Europe and parts of S.C. Europe.* Au Ga Gr He Hs It ?Ju Si.

(18–31). **R. canina** group. Stems up to 5 m, erect. Prickles usually curved or hooked, stout, usually all similar. Leaflets ovate, obovate, elliptical or suborbicular, glabrous or pubescent, eglandular or with a few glands on the main veins beneath, more rarely with numerous glands, the glands not strongly scented. Pedicels glabrous or stipitate-glandular. Flowers solitary or 2–5. Sepals 3–5 mm wide, usually deflexed after anthesis but sometimes erect or patent. Petals white or pink. Disc flat. Styles short or long, glabrous, villous or lanate. Fruit globose, ovoid or ellipsoid, glabrous or stipitate-glandular.

1 Sepals usually deflexed and deciduous in fruit; styles
   usually glabrous or villous, rarely lanate
2 Leaflets glabrous
3   Pedicels glabrous
4   Leaves eglandular
5   Disc narrow, the orifice more than 1 mm in diameter
                               24. subcanina
5   Disc wide, the orifice not more than 1 mm in diameter
                                   18. canina
4   Leaves glandular, at least on the rhachis, petioles and veins
6   Leaflets with glandular teeth   **21. nitidula**
6   Leaflets with eglandular teeth   **19. squarrosa**
3   Pedicels stipitate-glandular
7   Disc narrow, the orifice more than 1 mm in diameter
                                 24. subcanina
7   Disc wide, the orifice not more than 1 mm in diameter
8   Leaflets serrate, with short and rather wide teeth
                              20. andegavensis
8   Leaflets biserrate or compound-serrate, with long, acuminate teeth
9   Styles glabrous; leaves usually eglandular or rarely sparsely glandular on petiole and rhachis   **22. pouzinii**
9   Styles villous; leaves glandular on veins, petiole and rhachis   **21. nitidula**
2 Leaflets hairy
10 Pedicels glabrous
11   Disc narrow, the orifice more than 1 mm in diameter
                                 26. subcollina
11   Disc wide, the orifice not more than 1 mm in diameter
12   Leaves usually glandular beneath   **28. obtusifolia**
12   Leaves usually eglandular beneath   **29. corymbifera**
10 Pedicels ± stipitate-glandular
13   Leaves usually densely glandular beneath; styles usually long   **31. abietina**
13   Leaves with glands confined to rhachis and petiole; styles usually short   **30. deseglisei**
1 Sepals usually erect or patent in fruit, persistent; styles lanate
14 Acicles present   **27. rhaetica**
14 Acicles absent
15   Stems usually reddish; leaflets glabrous or rarely with a few scattered glands beneath   **23. vosagiaca**
15   Stems usually green; leaflets hairy and usually glandular
16   Sepals erect or patent in fruit; leaflets densely appressed-hairy beneath   **25. caesia**

16   Sepals deflexed or patent in fruit; leaflets usually sparsely hairy beneath   **26. subcollina**

18. **R. canina** L., *Sp. Pl.* 491 (1753). Stems green; internodes long. Prickles stout, curved or hooked; flowering stems rarely unarmed. Leaflets 5–7, 15–40 × 12–20 mm, ovate, obovate or elliptical, serrate or compound-serrate, glabrous and eglandular, dark to glaucous green, shining or dull above; petiole and rhachis often with acicles. Pedicels 10–20 mm, as long as or longer than the fruit, glabrous. Sepals deflexed and caducous after anthesis. Petals 15–25(–30) mm, pink to white. Disc wide, with the orifice less than 1 mm in diameter. Styles usually not long-exserted, densely villous to glabrous. Fruit 10–20 mm, globose, ovoid or ellipsoid, glabrous, red. 2*n* = 35. *Europe northwards to c. 62° N.* All except Az Fa Is Sb Tu.

19. **R. squarrosa** (Rau) Boreau, *Fl. Centre Fr.* ed. 3, **2**: 222 (1857). Like **18** but leaflets compound-serrate, glandular on the mid-veins and sometimes over the entire lower surface; stipules, petiole and rhachis densely glandular; styles long-exserted, villous. *Mainly in C. Europe; distribution not fully known.* Au Be Br Bu Cz Ga Ge Gr He Hu Po Rm.

20. **R. andegavensis** Bast., *Essai Fl. Maine Loire* 189 (1809). Like **18** but leaflets serrate; petiole and rhachis glandular and with acicles; pedicels stipitate-glandular; style usually long-exserted; fruit often stipitate-glandular. *W., S. & C. Europe.* Al Au Be Br Bu Cz Ga Ge Gr He Hs Hu It Ju Po Rm Rs (W) ?Sa Si.

21. **R. nitidula** Besser, *Cat. Pl. Jard. Krzemien. Suppl.* **4**: 20 (1815) (*R. blondaeana* Ripart ex Déséglise). Like **18** but leaflets with glandular teeth, glandular on the mid-veins and sometimes on the lateral veins; petioles and rhachis always glandular; pedicels sometimes stipitate-glandular; styles usually long-exserted, pubescent or lanate; fruit sometimes glandular-hispid. *From Britain and N. Portugal eastwards to S. Sweden, the Carpathians and Greece.* Au Be Br Bu Cz Ga Ge Gr He Hs Hu It Ju Lu Po Rm Rs (W).

22. **R. pouzinii** Tratt., *Rosac. Monogr.* **2**: 112 (1823). Like **18** but leaflets 15–25 × 10–18 mm, suborbicular to elliptical, with glandular teeth, sparsely glandular on rhachis and petiole; pedicels densely stipitate-glandular; styles long-exserted, glabrous; fruit occasionally glandular-hispid. *Mediterranean region, Portugal.* Bl Co Cr Ga Gr Hs It Lu Sa Si.

23. **R. vosagiaca** Desportes, *Ros. Gall.* 88 (1828) (*R. afzeliana* subsp. *vosagiaca* (Desportes) R. Keller & Gams, *R. glauca* subsp. *reuteri* (Godet) Hayek). Stems up to 2 m, pruinose when young; internodes long. Prickles crowded, rather short, curved or hooked. Leaflets 5–7, 20–40 × 15–25 mm, elliptical or ovate to obovate, serrate to compound-serrate, margins overlapping, glabrous, usually bluish-green and pruinose, or rarely green and not pruinose, rarely with scattered glands beneath. Pedicels 2–20 mm, rarely longer than fruit, glabrous. Sepals erect and persistent after anthesis. Petals 15–25 mm, bright pink. Disc narrow, with the orifice more than 1 mm in diameter. Styles lanate. Fruit 10–20 mm, globose, rarely ovoid, glabrous, deep red. *Most of Europe, northwards to Iceland and Fennoscandia, but rare in the south-west.* Al Au Bu Cz Da Fe Ga Ge Gr He Ho Hs Hu It Ju Lu No Po Rm Rs (N, B, C, W) Si Su.

24. **R. subcanina** (Christ) Dalla Torre & Sarnth., *Fl. Tirol* 6(2): 515 (1909) (*R. glauca* subsp. *subcanina* (Christ) Hayek). Like **23** but pedicels (10–)20–30 mm, often longer than the fruit,

rarely stipitate-glandular; sepals patent or deflexed. *Probably widespread in Europe, but rarer in the west.* Au Br Bu Cz Da Ga Ge He Hu It Ju No Po Rm Rs (B, C, W).

25. R. caesia Sm. in Sowerby, *Engl. Bot.* 33: t. 2367 (1812) (*R. coriifolia* Fries, *R. afzeliana* subsp. *coriifolia* (Fries) R. Keller & Gams). Like **23** but internodes short; leaflets pubescent to glabrous above, usually densely appressed-pubescent to almost tomentose beneath, rarely sparsely glandular; pedicels occasionally stipitate-glandular, sometimes pubescent when young; sepals patent or erect; fruits up to 25 mm, the central one often pyriform. 2n = 35. *Most of Europe eastwards to Estonia and C. Ukraine.* Au Be Br Bu Cz Da Fe Ga Ge Gr He Ho Hs Hu It Ju No Po Rm Rs (B, C, W) Si Su.

26. R. subcollina (Christ) Dalla Torre & Sarnth., *Fl. Tirol* 6(2): 516 (1909) (*R. coriifolia* subsp. *subcollina* (Christ) Hayek). Like **23** but leaves not bluish-green or pruinose, sometimes sparsely hairy above, hairy beneath at least on the veins and sometimes glandular on veins and rhachis; sepals patent or deflexed after anthesis and usually deciduous. *Probably widespread in Europe except the extreme north.* Au Be ?Br Bu Cz Da Ga Ge He Hs Hu ?It Ju No Po Rm Rs (W, E) Su.

27. R. rhaetica Gremli, *Excurs.-Fl. Schweiz* ed. 4, 164 (1881) (*R. afzeliana* subsp. *rhaetica* (Gremli) R. Keller & Gams). Like **23** but stout, hooked prickles and slender, straight acicles both present; leaflets not bluish-green or pruinose, somewhat glandular and more or less hairy on both surfaces; pedicels 5–10 mm, glabrous or stipitate-glandular. ● *Alps.* Au He It.

28. R. obtusifolia Desv., *Jour. Bot. Rédigé* 2: 317 (1809) (*R. tomentella* Léman, *R. klukii* Besser). Stems up to 2 m, green; prickles scattered, short, stout, compressed, strongly hooked. Leaflets 5–7, 15–35 × 14–25 mm, broadly ovate, simply to compound-serrate, softly appressed-pubescent on both surfaces, sometimes pubescent only on the veins beneath, usually glandular on the veins beneath; teeth glandular, short and wide; petiole and rhachis densely pubescent, more or less glandular, covered with minute acicles. Pedicels 5–15 mm, glabrous. Sepals eglandular, deflexed and caducous after anthesis. Petals 12–18(–24) mm, white or pale pink. Disc wide, with the orifice less than 1 mm wide. Styles long-exserted, villous or rarely glabrous. Fruit 10–20 mm, ovoid or globose, glabrous, red. *C., S. & N.W. Europe.* Al Au Be Br Bu Co Cz Da Ge Gr Hb He Hu It Ju Po Rm Rs (W, C) Sa Si Su.

29. R. corymbifera Borkh., *Vers. Forstbot. Beschr. Holzart.* 319 (1790) (*R. dumetorum* Thuill.). Like **28** but leaflets broadly elliptical to suborbicular, rarely narrower, simply serrate, sometimes glabrous above, usually eglandular. 2n = 35. *Probably widespread throughout Europe but rarer in the north & north-west.* Al Au Be Bl Br Bu Co Cr Cz Fe Ga Ge Gr Hc Ho Hs Hu It Ju Lu No Po Rm Rs (?B, C, W, K, E) Sa Si Su Tu.

30. R. deseglisei Boreau, *Fl. Centre Fr.* ed. 3, 2: 224 (1857). Like **28** but petiole and rhachis usually glandular; pedicels and sometimes the base of the hypanthium rather sparsely stipitate-glandular; styles usually short. *Mainly C. Europe; distribution not fully known.* Au Br Bu Cz Fe Ga Ge He Hu It Ju ?No Po Rs (?B, ?C, ?W) ?Su.

31. R. abietina Gren. ex Christ, *Ros. Schweiz* 132 (1873) (*R. obtusifolia* subsp. *abietina* (Gren. ex Christ) F. Hermann). Like **28** but prickles more abundant, curved; leaflets compound-serrate, densely glandular; pedicels stipitate-glandular; sepals

glandular on the back; fruit 10–25 mm, glabrous or densely stipitate-glandular. ● *Alps.* Au Ga Ge He It Ju.

(32–38). R. tomentosa group. Stems up to 3 m, erect. Prickles usually straight or slightly curved. Leaflets orbicular, ovate, oblong-ovate or elliptical, compound-serrate, very rarely simply serrate, usually densely tomentose, frequently glandular beneath, with a resinous smell. Pedicels usually glandular-hispid. Flowers solitary or 2–5. Sepals usually pinnatifid, usually erect after anthesis but sometimes patent or deflexed. Petals usually pink. Styles lanate or villous, very rarely glabrous. Fruit globose, ovoid or pyriform, glandular-hispid.

1 Sepals ± patent or deflexed after flowering
  2 Leaflets soft, densely tomentose or pubescent; sepals deciduous; styles usually villous or glabrous   **32. tomentosa**
  2 Leaflets rough, usually sparsely tomentose or pubescent; sepals persistent; styles densely villous to lanate   **33. scabriuscula**
1 Sepals erect and persistent after flowering
  3 Pedicels grey-puberulent and sometimes stipitate-glandular
    4 Young stems pruinose, glabrous, or ± pubescent; leaflets and pedicels eglandular   **37. heckeliana**
    4 Young stems not pruinose, densely pubescent; leaflets glandular, at least on the veins beneath   **38. orientalis**
  3 Pedicels stipitate-glandular but not grey-puberulent
    5 Prickles somewhat curved; base of sepals not swollen; sepals soon deciduous in fruit   **34. sherardii**
    5 Prickles straight; base of sepals swollen; sepals persistent in fruit
      6 Leaflets 30–50(–60) × 16–30 mm; young stems not pruinose   **35. villosa**
      6 Leaflets 12–35 × 8–18 mm; young stems usually pruinose   **36. mollis**

32. R. tomentosa Sm., *Fl. Brit.* 2: 539 (1800). Compact shrub. Stems up to 2 m. Internodes long; young stems and leaves pale green; prickles curved, or nearly straight on the flowering stems, somewhat slender. Leaflets 5–7, 20–40 × 12–20 mm, ovate, ovate-lanceolate or elliptical, rarely obovate, serrate or biserrate, usually densely pubescent or tomentose on both surfaces, glandular beneath and more or less sparsely glandular above, with a resinous smell. Pedicels *c.* 20 mm, longer than the fruit, glandular-hispid. Sepals densely glandular on the back, deflexed to patent, deciduous after anthesis. Petals 15–25 mm, pink or white. Disc 4–6 times the diameter of the orifice. Styles glabrous or villous. Fruit ovoid, globose or pyriform, stipitate-glandular, rarely glabrous, red. *Most of Europe except the extreme north.* Al Au Be Br Bu Co Cz Da Ga Ge Hb He Ho Hs Hu It Ju Lu No Po Rm Rs (B, C, W, K, ?E) Su.

33. R. scabriuscula Sm. in Sowerby, *Engl. Bot.* 27: t. 1896 (1808). Like **32** but more diffuse, with stems up to 3 m; leaflets sparsely tomentose or pubescent and often densely hairy only on the veins, rather rough to the touch, compound-serrate, teeth glandular; sepals persistent after anthesis; styles densely villous or lanate. *Probably widespread in most of Europe, except the northern and eastern margins.* Au Be ?Bl Br Bu Cz Da Ga Ge He Hs Hu It Ju No Po Rm Rs (W, C, ?E) Su.

34. R. sherardii Davies, *Welsh Botanol.* 1: 49 (1813) (*R. omissa* Déséglise). Like **32** but more compact; stems pruinose; internodes long; leaflets biserrate, with glandular teeth; pedicels 10–15(–20) mm, as long as the fruit; sepals erect and persistent after anthesis; petals pink; styles lanate, rarely villous. 2n = 28. *N., W. & C. Europe, eastwards to S.W. Finland and extending southwards to Bulgaria.* ?Be Br Bu Cz Da Fe Ga Ge Hb He Hs Ju Po Su.

35. R. villosa L., *Sp. Pl.* 491 (1753) (*R. pomifera* J. Herrmann). Compact shrub up to 1·5 m. Stems with short internodes, not pruinose. Prickles slender, long, straight, somewhat inflated at the base, scarcely decurrent. Leaflets 5–7, 30–50(–60) × 16–30 mm, oblong-ovate to broadly elliptical, biserrate, pubescent to tomentose on both surfaces, often densely glandular beneath, conspicuously bluish-green, with a resinous smell; teeth glandular. Pedicels 5–10(–15) mm, about as long as the fruit, stipitate- or setose-glandular. Sepals erect and persistent after anthesis. Petals 20–25 mm, pink. Disc narrow. Styles lanate. Fruit 10–20 mm, globose to pyriform, densely stipitate-glandular, dull red. *C. & S. Europe, extending northwards to the Netherlands.* Al Au Bu Ga Ge Gr He Ho Hu It Ju Po Rm Rs (C, W) [Cz Da No Rs (B) Su].

36. R. mollis Sm. in Sowerby, *Engl. Bot.* 35: t. 2459 (1812) (*R. villosa* auct., non L.). Like **35** but young stems usually pruinose; leaflets 12–35 × 8–18 mm, greyish-green; pedicels, sepals and sometimes fruit with sparse, slender, stipitate glands; fruit 10–15 mm, globose or broadly ovoid. $2n = 28$. *Mainly in N. & W. Europe, but extending locally to S.C. Russia.* Al Be Br Da Fa Fe Ga Ge Hb He Hs Ju Lu No Po Rs (B, C, W) Su.

37. R. heckeliana Tratt., *Rosac. Monogr.* **2**: 85 (1823) (*R. orphanides* Boiss. & Reuter). Stems up to 1 m, somewhat pruinose when young, more or less pubescent. Prickles sparse, curved or straight. Leaflets 5–7, 15–30 × 8–22 mm, orbicular to ovate, simply serrate or biserrate, densely pubescent above, grey-tomentose beneath, usually eglandular. Flowers solitary. Pedicels *c.* 5 mm, usually grey-puberulent, eglandular. Sepals more or less glandular-hispid on the back, erect and usually persistent after anthesis. Petals 12–15 mm, pink. Disc narrow. Styles lanate. Fruit 10–12 mm, globose to ovoid, glabrous or stipitate-glandular, red. *Mountains of E. Mediterranean region and Sicilia.* Al Cr Gr It Si.

38. R. orientalis Dupont ex Ser. in DC., *Prodr.* **2**: 607 (1825). Like **37** but usually not more than 0·5 m; young stems densely pubescent; prickles straight; leaflets usually somewhat glandular on the veins beneath; pedicels *c.* 10 mm, stipitate-glandular and tomentose; sepals more densely glandular-hispid on the back; fruit densely stipitate-glandular. *S. Jugoslavia, N. Albania, Greece.* Al Gr Ju.

(39–47). R. rubiginosa group. Erect shrubs up to 3·5 m. Prickles usually hooked or curved, sometimes mixed with acicles and glandular setae. Leaflets suborbicular, ovate, obovate or elliptical, rounded or cuneate at base, biserrate to compound-serrate, glabrous or somewhat pubescent, never tomentose, more or less densely glandular-viscid beneath, smelling of apples; teeth glandular. Pedicels glabrous or glandular-hispid. Flowers solitary or 2–3. Sepals pinnatifid, erect or deflexed after anthesis. Petals small, white or pink. Styles short or long, glabrous, villous or lanate. Fruit globose, ovoid or ellipsoid, glabrous or glandular-hispid.

1 Pedicels and hypanthium glabrous
  2 Styles lanate or densely villous
    3 Sepals deflexed and usually deciduous after anthesis; prickles usually mixed with acicles and glandular setae
      4 Leaflets 10–30 × 10–20 mm, cuneate at base; prickles curved or straight, mixed with setae   **43. caryophyllacea**
      4 Leaflets 8–12 × 6–10 mm, rounded at base; prickles curved or falcate, rarely mixed with setae   **47. serafinii**
    3 Sepals erect and persistent after anthesis; stems without acicles and glandular setae
      5 Prickles stout, curved or falcate; pedicels as long as or longer than fruit; stems up to 3·5 m   **40. elliptica**

      5 Prickles slender, nearly straight; pedicels *c.* ½ as long as fruit; stems not more than 0·5 m   **44. sicula**
  2 Styles glabrous or very sparsely villous
    6 Leaflets 8–12 × 6–10 mm, rounded at base   **47. serafinii**
    6 Leaflets 10–30 × 12–25 mm, cuneate at base   **41. agrestis**
1 Pedicels and hypanthium stipitate-glandular or glandular-pubescent
  7 Styles glabrous or subglabrous
    8 Leaflets cuneate at base   **41. agrestis**
    8 Leaflets rounded at base
      9 Prickles mixed with numerous setae and stipitate glands; stems not more than 0·5 m   **46. turcica**
      9 Prickles not mixed with setae and stipitate glands; stems up to 3·5 m   **42. micrantha**
  7 Styles villous or lanate
    10 Prickles slender, usually not mixed with glandular setae and acicles   **44. sicula**
    10 Prickles usually stout, mixed with glandular setae and acicles
      11 Leaflets usually pubescent and glandular on the upper surface; styles lanate
        12 Leaflets 7–15 × 5–15 mm, stems not more than 0·5 m   **45. glutinosa**
        12 Leaflets 10–25 × 8–15 mm; stems up to 3 m   **39. rubiginosa**
      11 Leaflets usually glabrous or subglabrous on the upper surface; styles villous
        13 Pedicels usually less than 5 mm; sepals deflexed and deciduous after anthesis   **46. turcica**
        13 Pedicels 10–15 mm; sepals erect and persistent after anthesis   **39. rubiginosa**

39. R. rubiginosa L., *Mantissa Alt.* 564 (1771) (*R. eglanteria* L., nom. ambig.). Up to 3 m. Prickles stout, curved or falcate, usually mixed with acicles and glandular setae especially on the flowering stems. Leaflets 5–7, 10–25 × 8–15 mm, suborbicular to broadly ovate to obovate, rounded at base, compound-serrate, glabrous or pubescent above, usually pubescent and more or less densely glandular beneath. Pedicels 10–15 mm, densely stipitate- or setose-glandular. Sepals glandular on the back, erect and persistent after anthesis. Petals 8–15 mm, deep pink. Disc narrow, with a wide orifice. Styles short, villous or lanate. Fruit 1–1·5 cm, subglobose, ovoid or ellipsoid, glabrous or glandular-hispid, bright red. $2n = 35$. *Most of Europe northwards to 61° N.* Al Au Be Bl Br Bu Cz Da Ga Ge Gr Hb He Ho Hu It Ju No Po Rm Rs (B, C, W, K) Su Tu.

40. R. elliptica Tausch, *Flora (Regensb.)* **2**: 465 (1819) (*R. graveolens* Gren. & Godron). Like **39** but stems without acicles or glandular setae; leaflets elliptical, cuneate at base, pubescent on both surfaces; pedicels glabrous; sepals eglandular on the back; fruit glabrous. *W. & C. Europe, extending south-eastwards to Albania and W. Ukraine.* Al Au Be Br Bu Cz Ga Ge Gr ?Hb He Hs Hu It Ju Po Rm Rs (W).

41. R. agrestis Savi, *Fl. Pis.* **1**: 475 (1798) (*R. sepium* Thuill., non Lam.). Up to 2 m. Internodes long; prickles curved, sometimes few or absent on the flowering shoots. Leaflets 5–7, 10–30(–50) × 12–25 mm, elliptical to oblong-obovate, acute, cuneate at base, biserrate to compound-serrate, glabrous or pubescent, always glandular, dull green; teeth glandular. Pedicels 10–20 mm, glabrous or sometimes sparsely stipitate-glandular. Sepals eglandular, glabrous, deflexed and deciduous after anthesis. Petals 10–20 mm, white. Disc wide, with a narrow orifice. Styles rather long, glabrous or slightly villous. Fruit 10–15 mm, subglobose, ovoid or ellipsoid, glabrous, red. *Most of Europe, rare in the north and east.* All except Az Cr Fa Fe Is No Rs (N, B, C, E) Sb.

42. R. micrantha Borrer ex Sm. in Sowerby, *Engl. Bot.* **35**: t. 2490 (1812). Like **41** but up to 3·5 m; leaflets usually rounded at base; pedicels stipitate-glandular; sepals glandular; fruit glabrous or glandular-hispid. *W., S. & C. Europe, extending to N. Ukraine.* Al Au Be Bl Br Bu Co Cz Ga Ge Gr Hb He Ho Hs Hu It Ju Lu Po Rm Rs (C, W, K) Si Tu.

43. R. caryophyllacea Besser, *Cat. Pl. Jard. Krzemien. Suppl.* **4**: 18 (1815). Up to 1 m. Prickles stout, broad, curved, often paired, nearly straight, mixed with setae on the flowering shoots. Leaflets 5–7, 10–30 × 10–20 mm, ovate to elliptical, compound-serrate, usually densely glandular on both surfaces, glabrous to pubescent. Pedicels glabrous. Sepals glandular or eglandular, deflexed after anthesis and tardily deciduous. Petals 12–20 mm, pale pink. Styles lanate. Fruit *c.* 10 mm, subglobose to ellipsoid, glabrous, bright red. *E.C. Europe, Balkan peninsula, Ukraine.* Au Bu Cz Gr Hu It Ju Po Rm Rs (C, W).

44. R. sicula Tratt., *Rosac. Monogr.* **2**: 86 (1823) (*R. thuretii* (Burnat & Gremli) Burnat & Gremli). Up to 0·5 m. Prickles usually sparse, slender, more or less curved, sometimes mixed with acicles and glandular setae. Leaflets 5–7, 7–15 × 5–15 mm, broadly ovate to suborbicular, compound-serrate, usually glabrous and sparsely glandular above, densely glandular and glabrous or sparsely pubescent beneath. Pedicels 3–5 mm, often stipitate-glandular, sometimes sparsely pubescent. Sepals glandular on the back, patent or erect after anthesis. Petals 10–15 mm, pink. Styles villous to lanate. Fruit *c.* 10 mm, globose, usually sparsely stipitate-glandular, red. *Mediterranean region.* Al Ga Gr Hs It ?Ju Si.

45. R. glutinosa Sibth. & Sm., *Fl. Graec. Prodr.* **1**: 348 (1809). Like **44** but prickles mixed with numerous stalked glands and setae; leaflets more densely glandular and usually densely pubescent to tomentose on the upper surface. *E. & C. Mediterranean region, Balkan peninsula.* Al Bu Cr Gr It Ju Si.

46. R. turcica Rouy, *Ill. Pl. Eur. Rar.* **6**: 45 (1896) (*R. ferox* Bieb., non Lawrance, *R. horrida* Fischer ex Crépin, non Sprengel). Like **44** but prickles abundant, stout, curved, mixed with numerous setae and stalked glands; sepals deflexed and deciduous after anthesis; styles glabrous or somewhat villous. *S.E. Europe.* Bu Gr Rm Rs (W, K) Tu.

47. R. serafinii Viv., *Fl. Lib.* 67 (1824). Like **44** but prickles hooked or falcate, rarely mixed with setae; leaflets 8–12 × 6–10 mm, shining and glabrous above; pedicels glabrous; sepals glandular or eglandular, deflexed and deciduous after anthesis; fruit glabrous. *C. Mediterranean region; Bulgaria and S. Jugoslavia.* Bu Co It ?Rm Sa Si.

11. Agrimonia L.[1]

Perennial, rhizomatous herbs. Stem erect, with glandular hairs. Leaves irregularly pinnate. Flowers 5-merous, in terminal, spike-like racemes; pedicels short, with 2 bracteoles. Hypanthium deeply concave, becoming hard in fruit; epicalyx absent; stamens 5–20; carpels 2; style terminal; ovules 1. Fruit of 1 to 2 achenes enclosed in the hypanthium, which is obconical, turbinate or cylindrical, and has hooked bristles at the upper end.

Literature: V. Skalický, *Acta Horti Bot. Prag.* **1962**: 87–108 (1962).

1 Petals pale yellow; mature hypanthium (including bristles)
  4–5 mm   **1. pilosa**

1 Petals golden yellow; mature hypanthium (including bristles)
  6–12 mm
2 Pedicels 4–10 mm; bracts often ovate and entire   **4. repens**
2 Pedicels 1–4 mm; lower bracts trilobed
3 Stem with both short and long eglandular hairs; mature
  hypanthium grooved for at least ¾ of its length   **2. eupatoria**
3 Stem with long eglandular hairs only; mature hypanthium
  grooved for half its length   **3. procera**

1. A. pilosa Ledeb., *Ind. Sem. Horti Dorpat., Suppl.* 1 (1823) (*A. dahurica* Willd. ex Ser.). Stem 50–150 cm, with long eglandular hairs. Leaves green on both surfaces, with glandular but very few eglandular hairs beneath; leaflets with 3–8 pairs of teeth, cuneate and entire at base. Raceme lax. Petals 2·5–4 mm, pale yellow, entire. Mature hypanthium (including bristles) 4–5 × 3–3·5 mm, deeply grooved throughout its length, more or less glabrous except for the bristles; inner bristles connivent and interwoven. *U.S.S.R. from c. 49° to c. 62° N., extending westwards to S. Finland, N.E. Poland and C. Romania.* Fe Po Rm Rs (N, B, C, W) [Cz].

2. A. eupatoria L., *Sp. Pl.* 448 (1753). Stem 15–150 cm, with both long patent, and short flexuous, eglandular hairs. Basal leaves often in a rosette and basal internodes short. Leaves with 3–6 pairs of main leaflets and 2–3 smaller pairs in between; leaflets serrate or crenulate almost to the base, dark green above, whitish- or greyish-tomentose beneath, with glandular hairs; stomata (21–)23–25(–27) μ. Pedicels at maturity 1–3 mm; lower bracts trilobed. Petals (3–)4–5(–6) mm, golden yellow, obovate, usually not emarginate. Mature hypanthium obconical to turbinate, deeply and narrowly grooved for at least ¾ of its length, and with many eglandular, appressed hairs; the inner bristles erect, the outer ascending, patent or deflexed. *Almost throughout Europe except the extreme north.* All except Cr Fa Is Sb.

(a) Subsp. **eupatoria**: Plants with rosette 15–40 cm, those without rosette up to 150 cm; not villous, except occasionally the tallest plants. Leaflets broadly obovate and coarsely crenulate to elliptical and serrate; glandular hairs concealed by the tomentum. Hypanthium (including bristles) 7–10 × 5–7 mm, grooved throughout its length; disc scarcely projecting; lowest bristles ascending or patent. $2n = 28$. *Throughout the range of the species, except Açores.*

(b) Subsp. **grandis** (Andrz. ex Ascherson & Graebner) Bornm., *Feddes Repert.* (*Beih.*) **89**: 244 (1940): Plants up to 150 cm, robust, villous, usually without rosette. Leaflets elliptical, coarsely serrate; glandular hairs concealed by the tomentum. Hypanthium (including bristles) 11 × 9 mm, grooved throughout its length; disc projecting by *c.* 0·4 mm; lowest bristles patent or slightly deflexed. *S., E. & E.C. Europe.*

(c) Subsp. **asiatica** (Juz.) Skalický, *Feddes Repert.* **79**: 35 (1968) (*A. asiatica* Juz.). Without rosette. Leaflets elliptical, coarsely and acutely dentate; glandular hairs not concealed by the tomentum. Hypanthium (including bristles) 7–8 × 7–8 mm, grooved for ¾ of its length; disc projecting by *c.* 0·5 mm; lowest bristles always distinctly deflexed. *S.E. Russia.* (C. & S.W. Asia.)

3. A. procera Wallr., *Erst. Beitr. Fl. Hercyn.* 203 (1840) (*A. odorata* auct., non Miller). Stem 50–120 cm, with long eglandular hairs; without basal rosette. Leaves green on both surfaces; stomata (27–)30(–33) μ; leaflets elliptical, coarsely and acutely serrate almost to the base. Pedicels at maturity (1·5–)2·5(–4) mm; lower bracts trilobed. Petals golden yellow, oblong to obovate, often emarginate. Mature hypanthium (including bristles) *c.* 11 × 11 mm, turbinate, with broad, shallow grooves for *c.* ½ its length, and with well-developed disc; lowest bristles distinctly deflexed.

---

[1] By V. Skalický.

$2n=56$. *W., C. & S. Europe, extending to S.W. Finland and E. Ukraine.* Au Be Br Bu Cz Da Fe Ga Ge Hb He Ho Hs Hu It Ju Lu No Po Rm Rs (B, C, W) Si Su.

4. A. repens L., *Syst. Nat.* ed. 10, **2**: 1046 (1759) (*A. odorata* Miller). Stem 50–100 cm, villous, with long and short eglandular hairs; without basal rosette. Leaves large, coriaceous, dark green above, greyish-green beneath; leaflets overlapping. Stipules large. Pedicels at maturity (4–)5(–10) mm; bracts often ovate and entire. Petals 5–7 mm, golden yellow, oblong, rounded at apex. Hypanthium (including bristles), $10–12 \times 12–14$ mm, cylindrical, with deep and narrow grooves for $\frac{3}{4}$ of its length; disc stout, projecting by *c.* 1·5 mm; lowest bristles distinctly deflexed. *Formerly cultivated in gardens and now naturalized in various parts of Europe.* [Au Be Cz Ge ?It Rm Tu.] (*Anatolia*.)

12. Aremonia Nestler[1]

Like *Agrimonia* but inflorescence a few-flowered cyme, each flower with an 8- to 12-lobed involucre; epicalyx present; stamens 5–10; hypanthium without bristles.

1. A. agrimonoides (L.) DC., *Prodr.* **2**: 588 (1825) (*Agrimonia agrimonoides* L.). Stem 5–35 cm, with basal leaf-rosette. Cauline leaves alternate, often 3-foliolate; leaflets obovate, crenate-dentate. Both open and cleistogamous flowers produced. Involucre up to 11 mm, concealing hypanthium at maturity. Epicalyx-segments *c.* 0·7 mm. Calyx-segments at maturity 2·5 mm on open flowers, minute and connivent on cleistogamous flowers. *Mountain woods. S. & C. Europe from Sicilia and S.W. Germany eastwards, and northwards to c. 49° N. in Czechoslovakia.* Al Au Bu Cz Ge Gr Hu It Ju Rm Si Tu [Br He].

(a) Subsp. **agrimonoides**: Stem *c.* 15 cm, slender. Leaves usually with 2–3 pairs of main leaflets, the 3 terminal leaflets much larger than the rest. Petals 4–5 mm. Fruit frequently formed from cleistogamous flowers. Hypanthium (with elaiosome) 5–6 mm in fruit, almost globose, brownish, abruptly narrowed to whitish, densely hairy elaiosome. *Throughout the range of the species, but rare in Greece.*

(b) Subsp. **pouzarii** Skalický, *Feddes Repert.* **79**: 36 (1968): Stem *c.* 25 cm, robust. Leaves usually with 3–4 pairs of main leaflets, the 5 or 7 terminal leaflets somewhat larger than the rest. Petals 6–8 mm. Fruits formed mainly from open flowers. Hypanthium (with elaiosome) 7–8 mm in fruit, broadly ellipsoid, covered with grey hairs, not narrowed to elaiosome. *Greece.*

Plants intermediate between (a) and (b) occur in S. Jugoslavia and S. Bulgaria.

13. Sanguisorba L.[2]

Perennial herbs. Leaves pinnate. Flowers hermaphrodite or polygamous, in dense, terminal capitula or spikes, with 2(–3) bracteoles below each flower. Hypanthium deeply concave; sepals 4; epicalyx and petals absent; stamens 4 or numerous, rarely 2; carpels 1–2(–3); style terminal. Fruit of 1(–2) achenes enclosed in the 4-angled hypanthium, which becomes dry and hard.

Literature: G. Nordborg, *Op. Bot.* (*Lund*) **11**(2): 1–103 (1966) and **16**: 1–166 (1967).

1 All flowers hermaphrodite
2 Sepals dull crimson   **1. officinalis**

2 Sepals green
  3 Fruiting hypanthium with narrow wings; stamens 4   **2. albanica**
  3 Fruiting hypanthium with broad wings; stamens 4–15
                                3. dodecandra
1 Upper flowers of capitulum female
  4 Plant viscid, with glandular hairs   **4. hybrida**
  4 Plant not viscid, without glandular hairs
    5 Rhizome clothed with the sheaths of old leaves   **5. ancistroides**
    5 Rhizome not clothed with the sheaths of old leaves
      6 Faces of the hypanthium reticulate, sculptured or irregularly ridged   **6. minor**
      6 Faces of the hypanthium with longitudinal ridges
                                 7. cretica

Subgen. **Sanguisorba**. Flowers hermaphrodite. Stamens (2–)4(–15). Stigmatic papillae short. Carpel 1.

1. S. officinalis L., *Sp. Pl.* 116 (1753) (*S. polygama* F. Nyl.). Glabrous. Stems 20–100 cm, erect, branched. Basal and lower cauline leaves with 3–7 pairs of leaflets; leaflets up to *c.* 5 cm, stalked, ovate or oblong-ovate, more or less cordate at base, glaucous beneath. Capitula 1–3 cm, subglobose or ellipsoid, erect. Sepals dull crimson. Stamens 4, equalling or slightly longer than the calyx; anthers dark crimson. Style 1, simple. Fruiting hypanthium with 4 narrow wings, smooth between the wings. $2n=28, 56$. *Damp, base-rich habitats. Most of Europe except most of the Mediterranean region and parts of the north.* Al Au Be Br Bu Cz *Da Ga Ge Gr Hb He Ho Hs Hu Is It Ju No Po Rm Rs (N, B, C, W, K, E) Su [Fe].

2. S. albanica Andrasovszky & Jáv., *Bot. Közl.* **19**: 23 (1921). Like **1** but leaves coriaceous, shining; capitula 2–8 cm, cylindrical, often interrupted at the base; sepals green; stamens longer than calyx; anthers yellow. *Scrub on serpentine, 450–600 m.* ● *N.E. Albania; ?Jugoslavia, c. 25 km S. of Peć.* Al ?Ju.

3. S. dodecandra Moretti, *Bibliot. Ital.* **70**: 436 (1833). Stems 40–100 cm, erect. Leaves elliptical, with 4–10 pairs of leaflets; leaflets stalked, linear-lanceolate to ovate, sharply serrate, light green beneath. Capitula 4–7 cm, cylindrical, long-stalked, greenish-yellow to whitish. Sepals green. Stamens 4–15, much longer than the calyx. Fruiting hypanthium with broad wings. $2n=56$. *Subalpine meadows and banks of streams.* ● *N. Italy* (*Prov. Sondrio*). It [Po].

Subgen. **Poterium** (L.) A. Braun & Bouché. Upper flowers of capitulum female, the middle and lower hermaphrodite. Stamens numerous. Stigmatic papillae long. Carpels 2.

4. S. hybrida (L.) Nordborg, *Op. Bot.* (*Lund*) **11**(2): 67 (1966) (*S. agrimonoides* Cesati). Erect, coarsely hairy, glandular-pubescent, viscid, with branched flowering stems up to 80 cm. Basal leaves with 3–5 pairs of oblong or elliptical leaflets, the terminal and subterminal leaflets the largest. Capitula rarely more than 1·5 cm, ovoid, compact, in a lax panicle with flexuous branches. Hypanthium 2·5–3 mm, more or less fusiform, glabrous, with longitudinal ridges, not winged. $2n=56$. *River-banks and wood-margins. W. half of Iberian peninsula.* Hs Lu.

5. S. ancistroides (Desf.) Cesati, *Icon. Stirp.* **2**: sub *Sang. dodecandra* (1841). Cushion-like, 10–20 cm; rhizome woody, above ground, clothed with the sheaths of old leaves. Leaves with 3–10 pairs of leaflets; leaflets 5–10 mm, mostly of equal size. Flowering stems usually leafless. Capitula *c.* 1 cm. Hypanthium *c.* 4 mm, fusiform, almost smooth or with faces slightly reticulate, sometimes with irregular longitudinal ridges, not winged. $2n=28$. *Limestone cliffs. S. half of Iberian peninsula.* Hs Lu.

[1] By V. Skalický.
[2] By M. C. F. Proctor and G. Nordborg.

6. S. minor Scop., *Fl. Carn.* ed. 2, **1**: 110 (1772) (*S. gaillardotii* (Boiss.) Hayek, *S. garganica* (Ten.) Bertol.). 10–90 cm, glabrous or hairy, with well-developed basal leaf-rosette; rhizome not clothed with the sheaths of old leaves. Flowering stems erect, leafy, rarely leafless. Leaves with 3–12 pairs of orbicular to elliptical leaflets; leaflets 0·5–2 cm, more or less stalked, crenate to incise-serrate, mostly of equal size. Capitula 1–3 cm, globose to ovoid. Hypanthium 3–8 mm, usually angled, ridged or winged and with faces reticulate or sculptured in various ways. *Dry grassland and rocky ground. S., W. & C. Europe, extending to S. Sweden and C. Russia; an occasional casual in the north-east.* All except Az Fa Is Rs (N, E) Sb, but only as a casual alien in Fe No Rs (B).

Extremely variable in size, habit and ornamentation of the hypanthium. Differences in the hypanthium are often clearly marked, but they are not always satisfactorily correlated with other characters. So far as is known, there are no sterility barriers between the subspecies which are here described.

1 Hypanthium not or scarcely angled, strongly verrucose
          (d) subsp. magnolii
1 Hypanthium ±4-angled, not verrucose
  2 Hypanthium distinctly hairy   **(b) subsp. lasiocarpa**
  2 Hypanthium glabrous or with very short hairs
    3 Hypanthium subglobose, the faces pitted and faintly reticulate
          (c) subsp. lateriflora
    3 Hypanthium ±elongated, the faces reticulate or ridged
      4 Angles of hypanthium ridged, the faces reticulate
          (a) subsp. minor
      4 Angles of hypanthium winged, the faces covered with irregular ridges
        5 Hypanthium up to 1·5 times as long as wide
          (e) subsp. muricata
        5 Hypanthium at least twice as long as wide
          (f) subsp. rupicola

(a) Subsp. **minor** (*Poterium sanguisorba* L., *S. dictyocarpa* (Spach) Franchet): 2*n*=28. *Throughout most of the range of the species.*

(b) Subsp. **lasiocarpa** (Boiss. & Hausskn.) Nordborg, *Op. Bot.* (*Lund*) **11**(2): 66 (1966) (*S. villosa* (Sibth. & Sm.) Dörfler): 2*n*=56. *Dry and rocky places. Turkey-in-Europe (near Istanbul).* (*Anatolia, Syria*).

(c) Subsp. **lateriflora** (Cosson) M. C. F. Proctor, *Feddes Repert.* **79**: 35 (1968): *Stony grassland. S. Spain.*

(d) Subsp. **magnolii** (Spach) Briq., *Prodr. Fl. Corse* **2**(1): 209 (1913) (*Poterium magnolii* Spach, *S. verrucosa* (Ehrenb.) A. Braun: 2*n*=28. *Mediterranean region.*

(e) Subsp. **muricata** Briq., *op. cit.* 210 (1913) (*Poterium polygamum* Waldst. & Kit., *S. muricata* (Spach) Gremli, *S. rhodopaea* (Velen.) Hayek): 2*n*=28, 56. *S. Europe; naturalized in C. and parts of N. Europe.*

(f) Subsp. **rupicola** (Boiss. & Reuter) Nordborg, *Op. Bot.* (*Lund*) **11**(2): 66 (1966): 2*n*=28, 56. *C. & S. Spain, Portugal, Sardegna, Sicilia.*

7. S. cretica Hayek, *Österr. Bot. Zeitschr.* **64**: 358 (1914). Like **6** but leaves with *c.* 6 pairs of larger leaflets 2·5–3 cm; capitula globose; hypanthium broadly winged, with longitudinal ridges on the faces. 2*n*=28. *Limestone cliffs. W. Kriti.* Cr.

14. Sarcopoterium Spach[1]

Like *Sanguisorba* but spiny shrubs; flowers unisexual; stamens numerous; carpels 2; fruit of 2 achenes enclosed in the hypanthium, which becomes red and fleshy.

1. S. spinosum (L.) Spach, *Ann. Sci. Nat.* ser. 3 (Bot.), **5**: 43 (1846) (*Poterium spinosum* L.). Much-branched shrub up to 60 cm. Shoots densely tomentose; lateral branches forming leafless spines. Leaves with 9–15 small, ovate, hairy leaflets. Capitula up to 3 cm, globose or oblong; the upper flowers female, the lower male. Calyx-teeth stellate-patent, caducous; stamens 10–30; hypanthium tubular-urceolate. 2*n*=28. *Dry places. Mediterranean region, from Sardegna eastwards.* Al Cr Gr It Ju Sa Si Tu.

15. Acaena Mutis ex L.[2]

Perennial herbs or dwarf shrubs. Leaves usually pinnate. Flowers hermaphrodite, in terminal capitula. Hypanthium deeply concave, contracted at the mouth; sepals 3–4; epicalyx and petals absent; stamens 1–10; carpels 2. Fruit of 1–2 achenes enclosed in the dry hypanthium which bears 4 to numerous spines.

1. A. anserinifolia (J. R. & G. Forster) Druce, *Rep. Bot. Exch. Club Brit. Is.* **4**: 484 (1917). Creeping, branched dwarf shrub with short, ascending, leafy, hairy stems up to 15 cm. Leaves 2–4 cm, pinnate with 3–5 pairs of leaflets; leaflets 4–12 mm, oblong, sessile, deeply crenate-serrate. Peduncles 4–7 cm in flower, up to 10 cm in fruit, bearing solitary capitula. Capitula 5–10 mm, globose, greenish, reaching 20 mm in fruit; each hypanthium with 4 long spines 5–6 mm, barbed at the apex and reddish-brown; stamens 2. *Naturalized from gardens in Britain and Ireland.* [Br Hb.] (*S.E. Australia, New Zealand.*)

16. Dryas L.[3]

Procumbent, branched dwarf shrubs. Leaves simple. Flowers hermaphrodite or polygamous, solitary. Hypanthium convex; sepals 7–10, epicalyx absent; petals (7–)8(–16); stamens and carpels numerous. Fruit a head of achenes with long, persistent, terminal, hairy styles.

1. D. octopetala L., *Sp. Pl.* 501 (1753). Stems up to 0·5 m. Leaves 5–40 × 2·5–20 mm, oblong to ovate, cordate or truncate at base, rugose, crenate, stipulate, petiolate; lower surface covered with a dense tomentum of white, simple hairs; veins usually with some large, brownish, branched hairs. Pedicels, calyx and hypanthium pubescent and usually with purple glands. Petals 7–17 mm, white, oblong. Styles 2–3 cm in fruit. 2*n*=18, 36. *On neutral and basic soils. Mountains of Europe, southwards to N. Spain, C. Italy and S. Bulgaria; also at low altitudes in the north.* Al Au Br Bu Cz Fe Ga Ge Hb He Hs Is It Ju No Po Rm Rs (N, C, W) Sb Su.

Plants which lack branched hairs on the leaves and which are found in a number of populations, have been described as **D. babingtoniana** A. E. Porsild, *Bull. Nat. Mus. Can.* **160**: 140 (1959). **D. punctata** Juz., *Bull. Jard. Bot. URSS* **28**: 320 (1929), which has large glands on the upper surface of the leaves, has been described from Arctic Russia. Neither of these taxa is worthy of more than varietal rank.

17. Geum L.[4]

Perennial herbs. Leaves pinnate or lyrate. Flowers solitary or in cymose, bracteate inflorescences, usually 5-merous. Hypanthium saucer-shaped, sometimes with a central, clavate carpophore; epicalyx present; petals more or less clawed, white, cream, yellow or red; stamens and carpels numerous; styles terminal. Fruit a head of achenes with long, persistent hairy styles, or with only the basal part of the style (rostrum) persistent.

---

[1] By M. C. F. Proctor.
[2] By D. H. Valentine.
[3] By T. T. Elkington.
[4] By W. Gajewski.

Several hybrids are recorded. The most frequent and widespread is **G. × intermedium** Ehrh., *Beitr. Naturk.* **6**: 143 (1791) (**5 × 9**), which is fertile; segregates approaching one or other of the parents, as well as intermediates, are often found. **G. × sudeticum** Tausch, *Hort. Canal.* **1**(1): t. [9] (1823) (*G. inclinatum* Schleicher ex Gaudin & Monnard) (**2 × 5**) occurs in scattered localities, with the parents, in the Alps and the Carpathians and in Jugoslavia.

Literature: F. Bolle, *Feddes Repert.* (*Beih.*) **72**: 1–119 (1933). W. Gajewski, *Monogr. Bot.* (*Warszawa*) **4**: 1–416 (1957).

1 Style persistent in its entirety
  2 Distal part of style covered with stiff, deflexed bristles; proximal part glabrous   **4. heterocarpum**
  2 Whole of style covered with long, ascending, soft hairs
    3 Plant with stolons; leaves pinnate   **1. reptans**
    3 Plant without stolons; leaves lyrate
      4 Flowers erect, golden yellow; inflorescence 1- to 3-flowered   **2. montanum**
      4 Flowers nodding, pale yellow or whitish; inflorescence 3- to 7-flowered   **3. bulgaricum**
1 Distal part of style deciduous; proximal part (rostrum) persistent, hooked
  5 Receptacle on a distinct carpophore 5–10 mm
    6 Petals clawed, cream to pink   **5. rivale**
    6 Petals not or scarcely clawed, yellow   **6. sylvaticum**
  5 Receptacle ± sessile
    7 Petals red   **7. coccineum**
    7 Petals yellow
      8 Petals usually not more than 8 mm
        9 Stipules not more than 1 cm   **13. hispidum**
        9 Stipules more than 1 cm
          10 Achenes c. 250, forming an ovoid head **12. macrophyllum**
          10 Achenes c. 70, forming a globose head **9. urbanum**
      8 Petals usually more than 8 mm
        11 Achenes 6–8 mm   **8. pyrenaicum**
        11 Achenes less than 6 mm
          12 Plant with a dense, soft pubescence   **11. molle**
          12 Plant not densely hairy, or hairs stiff   **10. aleppicum**

Subgen. **Oreogeum** (Ser.) F. Bolle. Style long, persistent in its entirety, with long, soft, ascending hairs throughout its length.

1. G. reptans L., *Sp. Pl.* 501 (1753) (*Sieversia reptans* (L.) Sprengel). Rhizome thick, ending in a leaf-rosette which produces several long stolons. Rosette-leaves pinnate, the segments deeply incised. Flowering stems 3–15 cm, usually 1-flowered, with few, small leaves. Flowers 25–40 mm in diameter, bright yellow. Achenes numerous; styles 20–25 mm. 2n=42. *Usually above 2000 m.* ● *Alps; Carpathians; mountains of N. Albania, Crna Gora and S.W. Bulgaria.* Al Au Bu Cz Ga Ge He It Ju Po Rm.

2. G. montanum L., *Sp. Pl.* 501 (1753) (*Sieversia montana* (L.) Sprengel). Rhizome thick, creeping; stolons lacking. Basal leaves lyrate; terminal leaflet c. 6 cm. Flowering stem 3–10(–30) cm, 1- to 3-flowered, with small leaves. Flowers 25–40 mm in diameter, golden yellow. Achenes numerous; styles 20 mm. 2n=28. ● *Mountains of C. & S. Europe.* Al Au Bu Co Cz Ga Ge Gr He Hs It Ju Po Rm Rs (W).

G. micropetalum Gasparr., *Not. Piante Lucan.* 11 (1833), from the S. Appennini, may possibly be a small-flowered variant of **2**, but the fruits are not known and further investigation is desirable.

3. G. bulgaricum Pančić, *Elem. Fl. Bulg.* 26 (1883). Rhizome thick; stem 30–50 cm, erect, with small leaves. Basal leaves large, lyrate; terminal leaflet 10–15 cm, cordate-reniform. Inflorescence 3- to 7-flowered; flowers c. 25 mm in diameter, nodding, campanulate. Calyx light green; petals triangular, emarginate, whitish to pale yellow. Achenes numerous; styles 10–15 mm. 2n=56. ● *Mountains of S. Jugoslavia, Albania and S.W. Bulgaria.* Al Bu Ju.

Subgen. **Orthostylus** (C. A. Meyer) F. Bolle. Style long, persistent in its entirety, the distal part with stiff, deflexed bristles.

4. G. heterocarpum Boiss., *Biblioth. Univ. Genève* ser. 2, **13**: 408 (1838) (*G. umbrosum* Boiss., non Dumort.). Plant softly hairy; stem 30–50 cm, branched. Basal leaves lyrate; terminal leaflet c. 6 cm, cordate, more or less lobed. Inflorescence 5- to 10-flowered; flowers c. 10 mm in diameter, campanulate. Petals elliptical or obovate, pale yellow; carpophore long. Achenes c. 15 mm, 5–15 in number, patent, the lowermost often deflexed; basal part of style glabrous; distal part long, straight. *Mountains of E. & S. Spain; isolated stations in S.E. France, C. Italy and Albania.* Al Ga Hs It.

Subgen. **Geum.** Style geniculate with hooked rostrum; distal part deciduous.

5. G. rivale L. *Sp. Pl.* 501 (1753). Rhizome short, thick; stem 20–30 cm, branched. Basal leaves pinnate, with 3–6 pairs of unequal leaflets; terminal leaflet 2–5 cm, suborbicular, incised or lobed; cauline leaves 3-partite; stipules c. 5 mm. Inflorescence 2- to 5-flowered; flowers nodding, campanulate. Calyx dark brownish-purple; petals 8–15 mm, erect, long-clawed, emarginate, cream to pink; carpophore 5–10 mm. Achenes 100–150, hairy, with long rostrum. 2n=42. *Most of Europe except the Mediterranean region.* Al Au Be Br Bu Cz Da Fa Fe Ga Ge Gr Hb He Ho Hs Is It Ju No Po Rm (N, B, C, W, E) Su.

6. G. sylvaticum Pourret, *Mém. Acad. Toulouse* **3**: 319 (1788). Whole plant hairy; rhizome usually short and thick; stem 15–40 cm, with small, simple cauline leaves. Basal leaves lyrate, with 1–2 pairs of lateral leaflets; terminal leaflet 3–5 cm, ovate, lobed; stipules small. Inflorescence 1- to 3-flowered; flowers c. 20 mm in diameter, erect. Petals patent, suborbicular, yellow. Carpophore 5 mm. Achenes 6–8 mm, 15–30 in number. 2n=42. *S.W. Europe, extending eastwards to Calabria.* Ga Hs Lu It.

7. G. coccineum Sibth. & Sm., *Fl. Graec. Prodr.* **1**: 354 (1809). Stem erect, branched, with small, lobed leaves. Basal leaves lyrate, with 2–3 pairs of lateral leaflets; terminal leaflet c. 8 cm, reniform. Inflorescence 2- to 4-flowered; flowers large, erect, long-stalked. Sepals deflexed after flowering; petals 10–18 mm, rounded, patent, red; carpophore absent. Achenes small, numerous, with bristly hairs; rostrum with a few long hairs at the base. 2n=42. *Mountains of Balkan peninsula.* Al Bu Gr Ju.

G. rhodopeum Stoj. & Stefanov, *Österr. Bot. Zeitschr.* **72**: 86 (1923), described from S. Bulgaria (Rodopi Planina), is like **7** but has 3–6 pairs of lateral pinnae, yellow petals 10–14 mm, shortly hairy achenes and hairy rostrum. Cultivation experiments indicate that it is a hybrid between **7** and some other, undetermined species of *Geum*. Further investigation is needed.

8. G. pyrenaicum Miller, *Gard. Dict.* ed. 8, no. 3 (1768). Rhizome short, thick; stem erect. Basal leaves lyrate, with 4–6 pairs of unequal leaflets; terminal leaflet c. 10 cm, rounded, cordate, lobed; cauline leaves very small. Inflorescence (1–)3(–5)-flowered; flowers large, erect. Petals 10–14 mm, rounded, patent, bright yellow; carpophore absent. Achenes 6–8 mm, numerous, with long rostrum. ● *Pyrenees.* Ga Hs.

G. gajew *lull* Pignatti, *Arch. Bot.* (*Forli*) 34: 12 (1958), from the C. Appennini, is like **8** but has smaller flowers. It has only been collected once, and the fruits are unknown; further investigation is desirable.

9. **G. urbanum** L., *Sp. Pl.* 501 (1753). Plant hairy. Rhizome short, thick; stem 20–60 cm, erect, branched. Basal leaves pinnate, with 1–5 pairs of unequal leaflets; terminal leaflets 2–10 cm, suborbicular and deeply lobed; cauline leaves 3- to 5-partite or 3-lobed, large; stipules 1–3 cm. Inflorescence (1–)2- to 5-flowered; flowers 10–15 mm in diameter, erect, long-stalked. Petals 4–7 mm, obovate or oblong, patent, rather pale yellow; carpophore absent. Achenes 3–6 mm, *c.* 70, hairy, forming a globose head; rostrum glabrous. $2n = 42$. *Most of Europe except the extreme north.* All except Az Bl Cr Fa Is Sb.

10. **G. aleppicum** Jacq., *Icon. Pl. Rar.* 1: 10 (1786). Plant hairy. Rhizome short, thick; stem 80–120 cm, erect, much-branched. Basal leaves pinnate, with 4–6 pairs of unequal leaflets; terminal leaflet *c.* 12 cm, deeply lobed, crenate and acutely dentate, cuneate at base; cauline leaves 3- to 5-partite, large; stipules 2–3 cm, deeply cut. Inflorescence (1–)3- to 6-flowered; flowers *c.* 20 mm in diameter, erect, in large cymes. Petals 8–10 mm, rounded, patent, yellow; carpophore absent. Achenes 2·5–5 mm, 200–250, hairy, forming an obovoid head; rostrum hairy at the base, otherwise glabrous. $2n = 42$. *E. & E.C. Europe.* ?Al Cz Hu Po Rm Rs (N, B, C, W, K) [Fe].

11. **G. molle** Vis. & Pančić, *Mem. Ist. Veneto* 10: 429 (1861). Plant with a dense, soft pubescence; stem 30–40 cm, erect. Basal leaves pinnate, with 2–3 pairs of small leaflets; terminal leaflet 5–6 cm, orbicular, subcordate, crenate; cauline leaves 3-partite. Stipules 1–2 cm, deeply cut. Inflorescence 3- to 5-flowered; flowers erect. Petals 10–12 mm, elliptical, patent, pale yellow; carpophore absent. Achenes less than 6 mm, numerous, with short, glandular rostrum. $2n = 42$. ● *Balkan peninsula; C. & S. Italy.* Al Bu Gr It Ju ?Sa.

12. **G. macrophyllum** Willd., *Enum. Pl. Hort. Berol.* 557 (1809). Stem *c.* 100 cm, erect, hairy. Basal leaves lyrate, very long stalked; lateral leaflets rather small and distant, the terminal large, 3- to 5-lobed, cordate-reniform; cauline leaves 3- to 8-partite, with lanceolate, incised stipules. Inflorescence 4- to 9-flowered; flowers *c.* 16 mm in diameter, yellow; carpophore absent. Achenes *c.* 250 pubescent, forming an ovoid head with glandular rostrum. *Cultivated in gardens; occasionally naturalized.* [Br Cz Ge No Rs (C).] (*E. Asia, North America.*)

13. **G. hispidum** Fries, *Fl. Halland.* 90 (1818). Plant hirsute and glandular; stem 30–40 cm, erect. Basal leaves pinnate, lanceolate in outline, with 3–5 pairs of leaflets; terminal leaflet *c.* 6 cm, elongated, deeply 3-lobed or incised; cauline leaves similar but smaller and with smaller terminal lobes; stipules not more than 1 cm. Inflorescence 3- to 5-flowered; flowers small, erect, long-stalked. Petals 5–8 mm, oblong or obovate, patent, pale yellow; carpophore absent. Achenes numerous, forming a globose head; rostrum rather short, glabrous. $2n = 42$. ● *S.E. Sweden; N.E. Spain.* Hs Su.

18. Waldsteinia Willd.[1]

Like *Geum* but leaves ternate or lobed; epicalyx small or absent; carpels 3–15; styles deciduous in fruit.

Basal leaves lobed; plant without stolons	**1. geoides**
Basal leaves ternate; plant with stolons	**2. ternata**

1. **W. geoides** Willd., *Ges. Naturf. Freunde Berlin Neue Schr.* 2: 106 (1799). Rhizome erect or shortly creeping. Leaves broadly cordate-reniform, with 5–7 lobes, coarsely serrate. Stems 15–25 cm, 3- to 7-flowered, with leaf-like bracts. Flowers 10–15 mm; petals auricled at the base. $2n = 14$. ● *E.C. Europe, extending to S. Bulgaria and W. Ukraine.* Bu Cz Hu Ju Rm Rs (W) [Ge].

2. **W. ternata** (Stephan) Fritsch, *Österr. Bot. Zeitschr.* 39: 449 (1889) (*W. trifolia* Rochel). Rhizome creeping, branched, with rooting stolons. Leaves ternate; leaflets sessile, cuneate at the base, shallowly lobed, serrate. Stem 10–15 cm, 3- to 7-flowered, with small bracts. Flowers *c.* 15 mm; petals not auricled at the base. *Carpathians, from c. 20° E. eastwards; a few stations in S.E. Austria and N.W. Jugoslavia.* Au Cz Ju Rm [Fe].

Known also from E. Asia (Japan to E. Siberia); the European plants appear to be identical although the two populations are separated by more than 5000 km. Further investigation is desirable.

19. Potentilla L.[2]

Perennial, rarely annual or biennial herbs, or small shrubs. Leaves digitate, pinnate or ternate. Flowers solitary or in cymes, (4–)5(–6)-merous. Hypanthium more or less flat, with a central, hemispherical, dry or spongy receptacle; epicalyx present; stamens 10–30; carpels (4–)10–80; style nearly basal, lateral or terminal. Fruit a head of achenes; styles usually not persistent.

Descriptions of leaves and leaflets refer only to the basal and lower cauline leaves.

Many species are cultivated in gardens for ornament. In addition to those native in Europe, the following species from the Himalayas are cultivated and are very locally naturalized: **P. atrosanguinea** Loddiges ex D. Don, *Prodr. Fl. Nepal.* 232 (1825), densely pubescent, with digitate leaves, 5 leaflets and red or purple petals longer than sepals; **P. nepalensis** Hooker, *Exot. Fl.* t. 88 (1824), sparsely hairy, with ternate leaves and orange-scarlet to red or purple petals longer than the sepals; **P. argyrophylla** Wallich ex Lehm., *Pugillus* 3: 36 (1831), with ternate, silvery-sericeous leaves and yellow petals 10–15 mm.

Interspecific hybridization and apomixis are of common occurrence in Subgen. *Potentilla* (**4–54**). Some species or groups of species are known to be complexes composed both of amphimictic, usually diploid, plants and apomictic polyploid plants. The experimental evidence at present available is somewhat fragmentary. **18–20**, **22**, **43** and **49** are known to be wholly or partially apomictic and morphological evidence suggests that **21**, **38–39** and **50** (and possibly others) are of hybrid origin and are therefore likely to be apomictic. Intermediates, presumably of hybrid origin, between any two of **15–51** may occur and be locally common. Some of these appear to reproduce apomictically, and now occur well outside the range of one of the putative parents, e.g. *P. subarenaria* Borbás ex Zimmeter (probably **49 × 51**) in Fennoscandia.

Because of hybridization, and apomixis it is often difficult to classify the material in clear-cut species. As a result of this, occasional individuals may not key out satisfactorily and may have characters which do not fully agree with the descriptions.

Literature: T. Wolf, *Biblioth. Bot.* (*Stuttgart*) 71: 1–715 (1908). B. Pawłowski, *Fragm. Fl. Geobot.* 11: 53–91 (1965) (Subgen. *Fragariastrum*). R. Czapik, *Acta Biol. Cracov.* (*Bot.*) 5: 43–61

---

[1] By W. Gajewski.
[2] By P. W. Ball, B. Pawłowski and S. M. Walters.

(1962). A. Müntzing, *Hereditas* **44**: 280–329 (1958), *Bot. Not.* **111**: 209–227 (1958). G. L. Smith, *New Phytol.* **62**: 264–300 (1963). The last 4 papers deal with apomixis in the genus.

1 Leaves pinnate
2 Petals white or purple
3 Petals purple, shorter than sepals **3. palustris**
3 Petals white, longer than sepals **5. rupestris**
2 Petals yellow
4 Flowers solitary in the axils of leaves or large bracts
5 Stoloniferous; petals longer than sepals **4. anserina**
5 Not stoloniferous; petals shorter than sepals **23. supina**
4 Flowers mostly in cymes; bracts small or 0
6 Shrub; achenes pubescent **1. fruticosa**
6 Herbs, sometimes woody at base; achenes glabrous
7 Leaflets entire or 2- to 3-fid at apex **2. bifurca**
7 Leaflets toothed or pinnatifid, at least some with more than 3 teeth or lobes
8 Leaflets toothed or pinnatifid about ½-way to the midrib
9 Petals 4–5 mm; calyx accrescent; sepals *c.* 10 mm in fruit **24. norvegica**
9 Petals 5 mm or more; calyx not accrescent; sepals less than 10 mm in fruit
10 Terminal leaflet of lower leaves 20–40 mm wide, larger than the lateral; sepals obtuse **6. geoides**
10 Terminal leaflet of lower leaves less than 20 mm wide, about the same size as the laterals; sepals acute
11 Stem and inflorescence tomentose or sericeous, usually eglandular
12 Leaflets 3–5; petals 5–8 mm **8. rubricaulis**
12 Leaflets (5–)7–19; petals 8–10 mm **11. pensylvanica**
11 Stem and inflorescence with short, glandular and eglandular hairs and a few setae
13 Leaflets 20–40 mm; petals scarcely longer than sepals **12. longifolia**
13 Leaflets 4–25 mm; petals 1½–2 times as long as sepals
14 Lower leaves with 7–12 pairs of leaflets; epicalyx-segments distinctly longer than sepals **13. pimpinelloides**
14 Lower leaves with 5–8 pairs of leaflets; epicalyx-segments as long as or slightly longer than sepals **14. visianii**
8 Leaflets pinnatisect, divided almost to the midrib into linear or oblong lobes
15 Leaflets 3–5; flowers 1–3(–6) on each stem
16 Petals 3–6 mm, only slightly longer than sepals; stems usually sericeous-villous **7. pulchella**
16 Petals 5–8 mm, distinctly longer than sepals; stems patent-pubescent **8. rubricaulis**
15 Leaflets 5 or more; flowers usually more than 5 on each stem
17 Sepals as long as or slightly longer than epicalyx-segments; stems and leaves tomentose **9. multifida**
17 Sepals about twice as long as epicalyx-segments; stems and leaves densely patent-hirsute **10. eversmanniana**
1 Leaves ternate or digitate
18 Petals white, cream or pink
19 Style hairy, at least in the lower half, persistent in fruit **75. saxifraga**
19 Style glabrous, or hairy at the extreme base, deciduous
20 Petals with a long claw
21 Leaflets crenate-dentate at least in the apical ⅔, tomentose and without straight hairs **66. speciosa**
21 Leaflets toothed only at apex, tomentose but with many straight hairs
22 Achenes pubescent **67. apennina**
22 Achenes glabrous

23 Petals purple; leaflets 4–8 mm **68. kionaea**
23 Petals white; leaflets 8–20 mm **69. deorum**
20 Petals without or with a short claw
24 Leaves ternate, rarely digitate
25 Petals 10–15 mm; stems silvery-grey sericeous **59. nitida**
25 Petals less than 10 mm; stems not sericeous
26 Stems 5- to many-flowered, exceeding the leaves; carpels sparsely hairy on the dorsal side, otherwise glabrous **65. grammopetala**
26 Stems 1- to 4-flowered, usually shorter than leaves; carpels pubescent on the ventral side, otherwise glabrous
27 Leaflets crenate-dentate only towards the apex; petals 6–9 mm **71. montana**
27 Leaflets serrate almost to the base; petals not more than 7 mm
28 Usually stoloniferous; filaments filiform, glabrous **72. sterilis**
28 Not stoloniferous; filaments broad and flat, densely ciliate at least up to the middle
29 Petals shorter than or as long as sepals; stems and pedicels eglandular **73. micrantha**
29 Petals longer than sepals; stems and pedicels with some pluricellular glandular hairs **74. carniolica**
24 Leaves digitate with 5 or more leaflets
30 Flowers stellate, the sepals and petals patent; petals usually longer than sepals
31 Stems and inflorescence silvery-sericeous
32 Leaves silvery-sericeous above; stems not more than 5 cm, 1- or 2-flowered **59. nitida**
32 Leaves glabrous above; stems 10–30 cm, usually many-flowered **60. alchimilloides**
31 Stems and inflorescence not sericeous
33 Leaflets silvery-sericeous beneath, glabrous or subglabrous and green above; achenes with a few long hairs at the point of attachment, otherwise glabrous **70. alba**
33 Leaflets green or somewhat sericeous, but the two surfaces always ± similar; achenes villous
34 Free part of stipules of the basal leaves ovate or ovate-lanceolate **58. crassinervia**
34 Free part of stipules of the basal leaves linear or linear-lanceolate
35 Petals 6–8 mm wide, much longer than sepals; filaments filiform, glabrous **57. clusiana**
35 Petals not more than 5 mm wide, slightly longer than sepals; filaments thickened and pubescent at base
36 Stems 5–30 cm, pubescent or villous; epicalyx-segments as long as or slightly longer than sepals **55. caulescens**
36 Stems 2–10 cm, densely villous; epicalyx-segments distinctly shorter than sepals **56. petrophila**
30 Flowers subcampanulate, the sepals and petals erecto-patent; petals shorter than or about equalling sepals
37 Achenes sparsely hairy on the dorsal side, otherwise glabrous; petals about equalling sepals **65. grammopetala**
37 Achenes villous; petals shorter than sepals
38 Leaflets grey- or silvery-tomentose or sericeous beneath
39 Stems and petioles with a very short, grey tomentum and with a few long, appressed or erecto-patent hairs **61. valderia**
39 Stems and petioles with dense, long, patent hairs **62. haynaldiana**
38 Leaflets pubescent and glandular beneath, green, sometimes sericeous when young
40 Leaflets crenate-serrate in the apical ½; epicalyx-segments about as long as sepals **63. doerfleri**
40 Leaflets with a few teeth in the apical ⅓; epicalyx-segments longer than sepals **64. nivalis**

18 Petals yellow
  41 Leaflets with at least some stellate or branched hairs beneath
    42 Leaflets 5–7, with relatively sparse stellate hairs beneath; stellate hairs with usually 5–10 rays   **50. pusilla**
    42 Leaflets 3–5, with a dense tomentum of stellate hairs beneath; stellate hairs with usually 15–30 rays   **51. cinerea**
  41 Stellate or branched hairs absent
    43 Sepals and petals 4, at least in some flowers
      44 Sepals and petals almost always 4; carpels 4–8(–20); most leaves with 3 leaflets (but with stipules resembling leaflets)   **52. erecta**
      44 Sepals and petals 4–5; carpels 20–50; some leaves with 4–5 leaflets   **53. anglica**
    43 Sepals and petals 5
      45 Flowering stems procumbent, rooting at the nodes; flowers solitary in the axils of leaves   **54. reptans**
      45 Flowering stems usually not rooting at the nodes; flowers usually in terminal cymes
        46 Leaflets densely tomentose, villous or sericeous beneath, the indumentum completely covering the surface of the leaflet
          47 Epicalyx-segments up to 2 times as long as sepals, about as wide as sepals at the base   **26. astracanica**
          47 Epicalyx-segments not or only slightly longer than sepals, narrower than sepals at the base
            48 Leaflets sericeous or villous beneath, the hairs all ±straight
              49 Petals (4–)5–7 mm; epicalyx-segments oblong or oblong-lanceolate, obtuse or subacute   **39. nevadensis**
              49 Petals 12–14 mm; epicalyx-segments linear-lanceolate or linear-triangular, long-acuminate   **27. detommasii**
            48 Leaflets tomentose, with most of the hairs crispate
              50 Leaflets entire in the basal ¼
                51 Style slightly clavate, but often somewhat distorted   **22. collina**
                51 Style conical, tapering towards apex
                  52 All hairs crispate   **18–20. argentea group**
                  52 Some hairs crispate, some long and simple   **21. inclinata**
              50 Leaflets toothed or lobed almost to the base
                53 Leaflets at least 3 times as long as wide   **21. inclinata**
                53 Leaflets less than 3 times as long as wide
                  54 All hairs on petiole crispate; leaflets 3   **15. nivea**
                  54 At least some hairs on petiole long and straight; leaflets 3–5
                    55 Leaflets with tufts of silky hairs at apex   **8. rubricaulis**
                    55 Leaflets without tufts of silky hairs at apex
                      56 Hairs on petiole all long and straight   **16. chamissonis**
                      56 Hairs on petiole both crispate and long and straight   **17. hookerana**
      46 Leaflets green or grey-green beneath, the indumentum not completely covering the surface
        57 Basal leaves ternate (rarely a few with 4–5 leaflets)
          58 Petals 10 mm or more
            59 Anthers less than 1 mm; leaflets with 2–4 pairs of teeth   **44. aurea**
            59 Anthers 1–1·5 mm; leaflets with 4–11 pairs of teeth
              60 Basal leaves pubescent between the main veins beneath   **31. grandiflora**
              60 Basal leaves pubescent only on the main veins beneath   **32. montenegrina**
          58 Petals less than 10 mm
            61 Annual or short-lived perennial without or with few non-flowering rosettes; epicalyx-segments longer than sepals in fruit   **24. norvegica**
            61 Perennial with numerous non-flowering rosettes; epicalyx-segments as long as or shorter than sepals in fruit
              62 Petals less than 6 mm

              63 Leaflets eglandular, glabrous or sparsely hairy beneath, glabrous above   **40. brauniana**
              63 Leaflets usually glandular, densely hairy beneath, sparsely to densely hairy above
                64 Petals 3–5 mm, as long as or only slightly longer than sepals; leaflets of basal leaves with elliptical or oblanceolate, very obtuse teeth   **41. frigida**
                64 Petals *c.* 5 mm, *c.* 1½ times as long as sepals; leaflets of basal leaves with triangular-oblong, acute or subobtuse teeth   **42. hyparctica**
            62 Petals 6 mm or more
              65 Petals cream or pale yellow, about as long as sepals; sepals linear-lanceolate or lanceolate   **65. grammopetala**
              65 Petals yellow, exceeding sepals; sepals ovate or triangular-ovate
                66 Terminal tooth of leaflet always smaller than the adjacent lateral; leaflets with appressed, sericeous hairs on the margin and main veins beneath   **44. aurea**
                66 Terminal tooth of leaflet subequal to the adjacent lateral; leaflets without sericeous hairs
                  67 Leaflets with triangular-oblong, acute to subobtuse teeth; petals not spotted; stems and leaflets usually glandular   **42. hyparctica**
                  67 Leaflets with oblong or elliptical, very obtuse teeth; petals often with an orange spot at the base; stems and leaflets usually eglandular   **43. crantzii**
  57 Basal leaves mostly digitate with 5 or more leaflets
    68 Flowering stems terminal; rosettes of leaves absent or few, lateral
      69 Petals 4–5 mm; sepals and epicalyx-segments accrescent, 15–20 mm in fruit   **25. intermedia**
      69 Petals 5–14 mm; sepals and epicalyx-segments not markedly accrescent
        70 Calyx and leaflets, at least on lower surface, with crispate hairs   **21. inclinata**
        70 Calyx and leaflets without crispate hairs
          71 Epicalyx-segments up to 2 times as long as the sepals, about as wide as sepals at the base   **26. astracanica**
          71 Epicalyx-segments shorter than to slightly longer than sepals, narrower than sepals at the base
            72 Leaflets with 3–7 teeth at apex; stems with long, simple hairs, without or with very few short hairs   **29. hirta**
             72 Leaflets coarsely toothed or pinnatifid to the base; stems with short glandular or eglandular hairs as well as with long, simple hairs
              73 Leaflets villous or sericeous beneath with many, crowded, short eglandular hairs   **27. detommasii**
              73 Leaflets with sparse, short eglandular hairs or with short glandular hairs beneath   **28. recta**
    68 Flowering stems lateral; stock with a terminal rosette of leaves, the plant often with numerous rosettes of leaves
      74 Stipules of cauline leaves adnate to the whole length of the petiole   **36. stipularis**
      74 Stipules of cauline leaves free or adnate only to the base of the petiole
        75 Lower surface of leaflets with crispate hairs, usually ±tomentose   **22. collina**
        75 Lower surface of leaflets with ±straight or slightly curved hairs, or glabrous
          76 Stipules of basal leaves linear to linear-triangular   **49. tabernaemontani**
          76 Stipules of basal leaves lanceolate to ovate
             77 Plant with sessile yellow glands; style papillose at base   **46. humifusa**
             77 Plant eglandular or with glandular hairs; style not papillose at base

78 At least some basal leaves with 6 or more leaflets
79 Epicalyx-segments narrowly linear **47. patula**
79 Epicalyx-segments oblong or linear-lanceolate
to broadly elliptical or lanceolate
80 Style conical-filiform
81 Leaflets glabrous above; petals *c.* 6 mm
37. longipes
81 Leaflets hairy above; petals usually more
than 6 mm **(38–39). chrysantha** group
80 Style conical at base, slightly clavate at apex
82 Hairs on stems and leaves with a minute
tubercle at base **45. heptaphylla**
82 Hairs on stems and leaves without a minute
tubercle at base **48. australis**
78 Basal leaves with not more than 5 leaflets
83 Terminal leaflet with a distinct petiolule
33. umbrosa
83 Terminal leaflet sessile or subsessile
84 Petals 10–15 mm
85 Terminal tooth of leaflet always smaller than
the adjacent lateral; style slightly clavate
44. aurea
85 Terminal tooth of leaflet subequal to the
adjacent lateral; style filiform-cylindrical
86 Cauline stipules acute or acuminate;
epicalyx-segments acute **34. delphinensis**
86 Cauline stipules obtuse or subacute;
epicalyx-segments subobtuse **35. pyrenaica**
84 Petals 4–10 mm
87 Hairs on stems and leaves with a minute
tubercle at base; hairs mostly patent
88 Petals (6–)7–10 mm; style conical-filiform
(38–39). chrysantha group
88 Petals 5–7 mm; style slightly clavate
45. heptaphylla
87 Hairs on stems and leaves without a minute
tubercle at base; hairs patent or appressed
89 Terminal tooth of leaflet always smaller
than adjacent lateral
90 Stems and leaflets variously hairy but
never sericeous; style conical-filiform
(38–39). chrysantha group
90 Upper part of stems and inflorescence
appressed-sericeous; leaflets with appressed sericeous hairs on the margin
and main veins beneath; style slightly
clavate **44. aurea**
89 Terminal tooth of leaflet subequal to the
adjacent lateral
91 Style slightly clavate; leaflets usually with
patent hairs on the margin **43. crantzii**
91 Style conical-filiform; leaflets usually with
appressed or semipatent hairs on the
margin
92 Petals 4–7 mm; leaflets sericeous or
sericeous-villous beneath (S. Spain)
30. nevadensis
92 Petals (6–)7–10 mm; leaflets usually not
sericeous beneath
93 Hairs on stem appressed or subappressed, never glandular
35. pyrenaica
93 Hairs on stem mostly patent or erecto-
patent, often glandular
(38–39). chrysantha group

Subgen. **Trichothalamus** (Lehm.) Reichenb. (*Dasiphora* Rafin.).
Shrubs. Leaves pinnate. Petals yellow or white. Receptacle hairy,
dry. Style sub-basal, clavate. Achenes densely pubescent.

1. P. fruticosa L., *Sp. Pl.* 495 (1753) (*Dasiphora fruticosa* (L.)
Rydb.). Much-branched, deciduous, more or less pubescent shrub
up to 1 m. Leaves pinnate; leaflets (3–)5(–7), 10–25 × 2–7 mm,
oblong-lanceolate or elliptical, entire. Flowers unisexual or
hermaphrodite, solitary or few in terminal cymes. Sepals
triangular-ovate; epicalyx-segments oblanceolate-linear, about
as long as sepals. Petals 8–12 mm, yellow, longer than sepals.
$2n = 14, 28$. *Britain and Ireland; Baltic region; Ural; Pyrenees;
Maritime Alps; Rodopi; widely cultivated for ornament and locally
naturalized, especially in Russia.* Br Bu Ga Hb Hs It Rs (N, B, C)
Su [No].

In N. Europe and Ural the plants are tetraploid with the
flowers usually unisexual, although the sterile carpels in male
flowers and the sterile stamens in female flowers are conspicuous.
In the Pyrenees the plants are diploid with hermaphrodite flowers.

P. glabrata Willd., *Ges. Naturf. Freunde Berlin Mag.* **7**: 285
(1816) (*P. davurica* Nestler), from E. Asia, is cultivated for ornament in much of Europe, and is recorded as an escape from
cultivation in France. It is very like **1** but has white petals.

Subgen. **Schistophyllidium** Juz. Perennial, woody at base.
Leaves pinnate. Petals yellow. Receptacle hairy, dry. Style sub-
basal, fusiform. Achenes pubescent near the point of attachment
when young, becoming glabrous at maturity.

2. P. bifurca L., *Sp. Pl.* 497 (1753) (*P. orientalis* Juz.). Rhizomatous perennial. Flowering stems up to 30 cm, subglabrous to
almost sericeous. Leaves pinnate; leaflets 5–15, 8–20 × 3–8 mm,
oblong-ovate, entire or 2- to 3-fid at apex. Flowers in a lax cyme.
Sepals oblong-ovate; epicalyx-segments linear-lanceolate, slightly
shorter than sepals. Petals 4–8 mm, yellow, longer than sepals.
*Steppes and dry sandy places. S.E. Europe, from E. Romania east-
wards, extending locally northwards to c. 53° N. in S.C. Russia.*
Rm Rs (C, W, E) [Su].

Subgen. **Comarum** (L.) Syme (*Comarum* L.). Herbs. Leaves
pinnate. Petals purple. Receptacle hairy, spongy. Style lateral,
filiform. Achenes glabrous.

3. P. palustris (L.) Scop., *Fl. Carn.* ed. 2, **1**: 359 (1772) (*Comarum palustre* L.). Plant with long, creeping, woody rhizome.
Flowering stems up to 45 cm, with scattered hairs. Leaves pinnate; leaflets (3–)5 or 7, 30–60 × 10–20 mm, oblong, coarsely
serrate, subglabrous beneath. Flowers in a lax terminal cyme.
Sepals 10–15 mm, ovate, acuminate, purplish, accrescent;
epicalyx-segments linear, much smaller than sepals. Petals about
half as long as sepals, deep purple, persistent. $2n = 28, 35, 42,
62–64$. *Marshes, bogs and acid fens. Europe from C. Spain, N. Italy
and S. Bulgaria northwards.* Au Be Br Bu Cz Da Fa Fe Ga Ge Hb
He Ho Hs Hu Is It Ju No Po Rm Rs (N, B, C, W, K, E) Su.

Subgen. **Potentilla.** Herbs. Leaves pinnate or digitate. Petals
yellow, rarely white, red or purple. Receptacle glabrous or
pubescent. Style usually subterminal. Achenes glabrous.

4. P. anserina L., *Sp. Pl.* 495 (1753). Perennial; stock short,
thick, with terminal rosette of leaves. Stems up to 80 cm, procumbent, stoloniferous. Leaves pinnate; leaflets 7–25, 10–40 ×
5–15 mm, oblong to ovate, serrate or crenate-serrate. Flowers
solitary, axillary. Sepals ovate or broadly elliptical; epicalyx-
segments triangular-lanceolate. Petals 7–10 mm, yellow, about
twice as long as sepals. Style lateral, filiform. *Most of Europe
except the extreme north-east and much of the south.* Au Be Br Bu
Cz Da Fa Fe Ga Ge Hb He Ho Hs Hu Is It Ju Lu No Po Rm Rs
(N, B, C, W, K, E) ?Si Su.

For an account of the variation of this species see A. Rousi,
Ann. Bot. Fenn. **2**: 47–112 (1965).

(a) Subsp. **anserina**: Leaflets 15–25, silvery-sericeous, at least beneath; hairs straight. Epicalyx-segments often toothed or lobed. Achenes grooved on the back. $2n = 28, 35, 42$. *Almost throughout the range of the species.*

(b) Subsp. **egedii** (Wormsk.) Hiitonen, *Suomen Kasvio* 449 (1933) (*P. egedii* Wormsk.): Leaflets 7–15, glabrous or sparsely pubescent beneath; hairs, except on the veins, crispate. Epicalyx-segments usually entire. Achenes not grooved. $2n = 28$. *Sea-shores. Coasts of N. Europe, southwards to c. 60° N. in Sweden and Finland.*

5. P. rupestris L., *Sp. Pl.* 496 (1753) (*P. corsica* Sieber ex Lehm.). Perennial. Flowering stems up to 60 cm, puberulent to densely pubescent, and glandular, at least above. Leaves pinnate; leaflets 5–7, 10–40 × 5–35 mm, ovate to suborbicular, irregularly crenate-dentate or doubly crenate-dentate. Flowers one to many. Sepals triangular, subacute to shortly acuminate; epicalyx-segments lanceolate, shorter than sepals. Petals 8–14 mm, white, longer than sepals. Style sub-basal, fusiform. $2n = 14$. *W. & C. Europe and Balkan peninsula, extending to S. Sweden, N. Italy and White Russia.* Al Au Be Br Bu Co Cz Ga Ge Gr He Hs Hu It Ju Lu No Po Rm Rs (W) Sa Su.

6. P. geoides Bieb., *Fl. Taur.-Cauc.* **1**: 404 (1808) (incl. *P. jailae* Juz.). Perennial. Flowering stems 15–50 cm, softly hirsute. Leaves pinnate; leaflets 7–9, 25–50 × 20–40 mm, suborbicular, doubly incise-serrate. Sepals ovate or triangular, obtuse or subacute, often 3-lobed at apex; epicalyx-segments linear-lanceolate, obtuse, often 2- to 3-lobed at apex, much shorter than sepals. Petals 10–12 mm, pale yellow, longer than sepals. Style sub-basal, fusiform. *N. Greece, Bulgaria; Krym.* Bu Gr ?Ju Rs (K).

Much confused with and perhaps not distinct from **5** in the Balkan peninsula. Plants described as *P. rupestris* var. *beniczkyi* (Friv.) T. Wolf, *P. rupestris* var. *mollis* (Pančić) Ascherson & Graebner and *P. rupestris* var. *strigosa* T. Wolf are either referable to this species or are intermediate between **5** and **6**.

7. P. pulchella R. Br. in Parry, *Jour. Voy. N.W. Pass.* (*Suppl. App.*) 277 (1824). Perennial. Flowering stems 2–15(–25) cm, usually sericeous-villous. Leaves pinnate; leaflets 3–5, 5–20 × 4–12 mm, obovate-cuneate in outline, pinnatisect, with 3–7 linear-lanceolate or oblong lobes, silvery-grey on both surfaces, or grey-green above. Flowers 1–3(–6). Sepals ovate; epicalyx-segments oblong, shorter than sepals. Petals 3–6 mm, yellow, oblong-obovate, slightly longer than sepals. Style conical-filiform, papillose at base, shorter than achene. $2n = 28$. *Spitsbergen and Vajgač.* Rs (N) Sb. (*Arctic Asia and America.*)

8. P. rubricaulis Lehm., *Pugillus* **2**: 11 (1830). Perennial. Flowering stems 3–20 cm, patent-pubescent, often reddish below. Leaves ternate or pinnate, the leaflets sometimes crowded; leaflets 3–5, 10–30 × 5–15 mm, obovate in outline, pinnatifid or pinnatisect with 3–7 oblong to ovate lobes, greyish-green above, white-tomentose beneath and sometimes sericeous. Flowers 1–3(–5). Sepals ovate-lanceolate; epicalyx-segments lanceolate or linear-oblong, as long as or slightly shorter than sepals. Petals 5–8 mm, yellow, broadly obovate, longer than sepals. Style conical-filiform, not or only slightly papillose. *Spitsbergen.* Sb. (*Arctic America.*)

9. P. multifida L., *Sp. Pl.* 496 (1753) (incl. *P. lapponica* (F. Nyl.) Juz.). Perennial. Flowering stems 10–40 cm, sparsely to densely tomentose. Leaves pinnate, the leaflets sometimes very crowded and so almost digitate; leaflets 5–9, 5–40 × 3–20 mm, deeply pinnatisect, with up to 5 linear lobes, green above, grey-green beneath. Flowers usually numerous. Sepals ovate-lanceolate;

epicalyx-segments oblong-linear, as long as or slightly shorter than sepals. Petals 5–7 mm, yellow, slightly longer than sepals. Style conical-filiform, much shorter than achene. *Fennoscandia; Ural; Pyrenees; Alps.* Ga He ?Hs It Rs (N, C, ?E) Sb Su [Fe].

10. P. eversmanniana Fischer ex Ledeb., *Fl. Ross.* **2**: 42 (1843). Like **9** but usually with dense, long, patent hairs; leaflets 11–19, the lobes often pinnatifid, grey-green or white above, white beneath; sepals twice as long as epicalyx-segments; petals about $1\frac{1}{2}$ times as long as sepals; style about as long as achene. ● *S.E. Russia.* Rs (?C, E).

11. P. pensylvanica L., *Mantissa* 76 (1767). Perennial. Flowering stems 15–80 cm, grey-pubescent. Leaves pinnate; leaflets (5–)7–19, 20–70 × 8–20 mm, oblong or lanceolate, coarsely toothed or pinnatifid, green or grey. Flowers numerous. Sepals oblong-lanceolate; epicalyx-segments linear-lanceolate, as long as sepals. Petals 8–12 mm, yellow, as long as or longer than sepals. Style conical-filiform. *C. & S. Spain; S.W. Alps; S. Ural; N.W. Russia.* Ga Hs It Rs (N, C, ?E).

This species extends to Asia and N. America and is very variable, especially in indumentum, leaflet-dissection and petal-size. The European populations can to some extent be distinguished from each other, but it is possible to match all of these with individuals from various parts of Asia and North America. It is therefore impossible to reach any satisfactory conclusion as to the status of the European populations without undertaking a complete revision of the whole complex. The taxa generally recognized in Europe are **P. pennsylvanica** L. *sensu stricto* (*P. strigosa* Pallas ex Ledeb., *P. sibirica* T. Wolf), from S.W. Alps, N.W. Russia and S. Ural, with the leaflets coarsely toothed, green and pubescent beneath and the petals about equalling the sepals; **P. conferta** Bunge in Ledeb., *Fl. Altaica* **2**: 240 (1830) (*P. sibirica* var. *pectinata* T. Wolf), from S. Ural, with the leaflets pinnatifid and grey-tomentose beneath and the petals about equalling the sepals; **P. hispanica** Zimmeter, *Gatt. Potent.* 7 (1884), from C. & S. Spain, with the leaflets usually grey-tomentose beneath and the petals at least $1\frac{1}{2}$ times as long as the sepals.

12. P. longifolia Willd., *Ges. Naturf. Freunde Berlin Mag.* **7**: 287 (1816) (*P. viscosa* Donn ex Lehm.). Perennial. Flowering stems 20–50 cm, densely hirsute and glandular. Leaves pinnate; leaflets 7–13, 20–40 × 7–12 mm, oblong-lanceolate, with 11–15 teeth, green. Flowers numerous, subcapitate, the pedicels less than 10 mm. Sepals ovate; epicalyx-segments oblong, about as long as sepals. Petals *c.* 8 mm, slightly longer than sepals. Style conical-filiform, papillose at base, about as long as achene. *C. Ural.* Rs (C).

13. P. pimpinelloides L., *Sp. Pl.* 497 (1753). Perennial. Flowering stems 15–40 cm, hirsute, glandular. Leaves pinnate; leaflets 15–25, 10–15 × 7–15 mm, orbicular-ovate to ovate, with 7–11 teeth, green. Flowers numerous, in a somewhat condensed terminal cyme. Sepals ovate-lanceolate; epicalyx-segments oblong-lanceolate, longer than sepals. Petals 7–8 mm, pale yellow, much longer than sepals. Style conical-filiform. *S.C. Russia (E. of Orel).* Rs (C). (*Caucasus.*)

14. P. visianii Pančić, *Mem. Ist. Veneto* **12**: 480 (1865). Perennial. Flowering stems up to 40 cm, pubescent with long patent hairs, glandular. Leaves pinnate; leaflets 11–17, 4–25 × 3·5–16 mm, cuneate-obovate, with 2–7 teeth. Flowers usually numerous in a lax terminal cyme; pedicels usually 10 mm or more. Sepals triangular-lanceolate; epicalyx-segments oblong to ovate-oblong, as long as or longer than sepals. Petals 8–10 mm, yellow, much longer than sepals. Style conical-filiform, shorter than achene. *Cliffs and dry grassland, usually on serpentine.* ● *N.W. part of Balkan peninsula.* Al Ju.

15. **P. nivea** L., *Sp. Pl.* 499 (1753). Perennial up to 20 cm. Flowering stems subglabrous to white-tomentose. Leaves ternate; leaflets 7–25 × 6–15 mm, ovate or obovate, crenate-serrate with 7–13 teeth, densely white tomentose beneath, very sparsely pubescent above; petiole tomentose with only crispate hairs. Flowers up to 12 in a terminal cyme. Sepals lanceolate or ovate; epicalyx-segments linear or linear-lanceolate, as long as or shorter than sepals. Petals 6–9 mm, yellow, slightly longer than sepals. Style conical-filiform. $2n = 56$. *Spitsbergen; N. & W. Fennoscandia; Ural; Alps and N. Appennini.* Au Ga Fe He It No Rs (N, C) Sb Su.

Plants described as subsp. **subquinata** (Lange) Hultén, *Bot. Not.* **1945**: 135 (1945) are apomictic, probably derived from hybridization between **15** and **16**. They are intermediate between these two species.

16. **P. chamissonis** Hultén, *Bot. Not.* **1945**: 140 (1945) (?*P. kuznetzowii* (Govoruchin) Juz. pro parte). Like **15** but flowering stems 15–30 cm; leaves sometimes digitate; leaflets pinnatifid with 5–9 oblong, obtuse or subacute lobes; petiole with only long, straight hairs; flowers often more than 12; petals 6–9 mm. $2n = 77$. *Arctic Europe.* Fe No Rs (N) Sb Su.

17. **P. hookerana** Lehm., *Ind. Sem. Hort. Bot. Hamburg. Add.* 10 (1849) (*P. kuznetzowii* (Govoruchin) Juz. pro parte). Like **15** but flowering stems up to 40 cm; leaves sometimes digitate; petiole with both crispate and long, straight hairs; flowers often more than 12. *Ural.* Rs (N, ?C).

(18–20). **P. argentea** group. Procumbent to erect perennial. Flowering stems terminal, densely tomentose. Leaves digitate; leaflets usually 5, (7–)10–30 × (4–)5–15 mm, cuneate-obovate, incise-dentate to pinnatifid, densely tomentose beneath, sparsely pubescent to densely white-tomentose above. Flowers numerous. Sepals ovate; epicalyx-segments linear, as long as sepals. Petals 4–7 mm, yellow, about as long as or longer than sepals. Style conical-filiform.

The distribution of the species recognized in this group is not known with any accuracy. The group as a whole occurs in the following territories: Al Au Be Br Bu Co Cz Da Fe Ga Ge Gr He Ho Hs Hu It Ju No Po Rm Rs (N, B, C, W, K, E) Si Su Tu.

The group consists of a polyploid complex, the members of which may be either amphi- or apomictic; some diploids and most polyploids are obligate apomicts. G. Marklund, *Mem. Soc. Fauna Fl. Fenn.* 9: 2–13 (1934), recognized 3 species and A. & G. Müntzing, *Bot. Not.* **1941**: 237–278 (1941), and A. Müntzing, *Bot. Not.* **111**: 209–227 (1958), have shown that one of Marklund's species is usually diploid and another usually hexaploid, though the correlation is not perfect. Tetraploid plants, intermediate between diploid and hexaploid parents, sometimes occur. The chromosome number of the 3rd species is not known.

1 Leaflets of the basal leaves mostly with 9–11 acute teeth or lobes; petals 5–7 mm   **19. neglecta**
1 Leaflets of the basal leaves with 2–7 obtuse or subobtuse teeth or lobes; petals 4–5 mm
2 Leaflets green above; pedicels and calyx relatively sparsely tomentose   **18. argentea**
2 Leaflets grey-green to white above; pedicels and calyx densely white-tomentose   **20. calabra**

18. **P. argentea** L., *Sp. Pl.* 497 (1753). Flowering stems up to 30 cm, usually procumbent or ascending. Leaflets green above,

with usually not more than 7 subobtuse teeth; middle cauline stipules entire or with 1 lobe. Pedicels and calyx relatively sparsely tomentose, grey-green. Petals 4–5 mm, not overlapping. Carpels pale yellow. $2n = 14$. *N. Europe, Alps; probably in the mountains elsewhere in C. & S. Europe.*

Usually apomictic.

19. **P. neglecta** Baumg., *Enum. Stirp. Transs.* 2: 63 (1816) (*P. impolita* auct., non Wahlenb.). Flowering stems up to 50 cm, usually erect. Leaflets grey-green above, usually with 9–11 acute teeth; middle cauline stipules with 1 or 2 lobes. Pedicels and calyx densely white-tomentose. Petals 5–7 mm, almost overlapping. Carpels dark yellow or orange-yellow. $2n = 42$. *Most of Europe but absent from many islands.*

Usually apomictic. A few plants with $2n = 35$ have been recorded. They are morphologically indistinguishable from **19**.

20. **P. calabra** Ten., *Fl. Nap.* 1, *Prodr.*: 68 (1811). Flowering stems up to 20 cm, usually procumbent. Leaflets grey-green to white above, flabellate or 2- to 3-lobed, the lobes obtuse; middle cauline stipules usually entire. Pedicels and calyx densely white-tomentose. Petals *c.* 5 mm. *Mountains of C. & S. Italy and Sicilia; W. part of Balkan peninsula.*

The plants from the Balkan peninsula are somewhat intermediate between **20** and **18** or **19**.

21. **P. inclinata** Vill., *Hist. Pl. Dauph.* 3: 567 (1788) (*P. canescens* Besser). Erect or ascending perennial. Flowering stems 15–50 cm, terminal, tomentose and with patent, long, simple hairs. Leaves digitate; leaflets 5–7, 15–40 × 5–15 mm, oblong-obovate, incise-dentate or -serrate, or pinnatifid with up to 12 pairs of teeth, tomentose beneath and with long simple hairs on the veins. Flowers numerous. Sepals ovate; epicalyx-segments linear, as long as sepals. Petals 5–7 mm, yellow, slightly longer than sepals. Style conical-filiform. $2n = 42$. *S. & C. Europe, extending northeastwards to C. Ukraine.* Al Au Bu Cz Ga Ge Gr He Hs Hu It Ju Po Rm Rs (W, K, E) Tu.

Very variable in the density of the indumentum and in the division of the leaflets. It is intermediate between the *P. argentea* group and **28** and may represent another apomictic hybrid complex.

P. pindicola (Nyman) Hausskn., *Mitt. Thür. Bot. Ver.* nov. ser., 5: 95 (1893) (*P. virescens* (Boiss.) Halácsy), from S.E. Europe, is like **21** but has epicalyx-segments linear-lanceolate, slightly longer than sepals, and petals sometimes up to 12 mm. It appears to be intermediate between the *P. argentea* group and **26** or **27**.

22. **P. collina** Wibel, *Prim. Fl. Werthem.* 267 (1799). Procumbent to erect perennial. Flowering stems up to 30(–40) cm, terminal or lateral, glabrous to white-tomentose. Leaves digitate; leaflets 5–7, oblong-obovate or oblanceolate, variously toothed or incised, sparsely pubescent to white-tomentose or sericeous beneath. Flowers few to numerous. Sepals ovate; epicalyx-segments linear or linear-lanceolate, usually as long as or shorter than sepals. Petals 4–7 mm, yellow, as long as or longer than sepals. Style conical at base, slightly clavate at apex, but often somewhat distorted. $2n = 42$. *C. Europe, extending to N. Italy, C. France, S. Sweden and the S. & W. parts of the U.S.S.R.* Au ?Be Bu Cz Da Ga Ge He Hu It ?Ju Po Rm Rs (B, C, W, K, E) Su.

As defined here this species is composed of a wide range of plants which are intermediate between the *P. argentea* group and **40–51**. Some are known to be apomictic and are probably of hybrid origin. It is sometimes very difficult to decide whether an individual should be referred to *P. collina* or to one of the putative parent species.

22 contains at least 12 taxa that have been described as species and are often recognized as such in Europe.

Those most closely resembling the *P. argentea* group are:

P. johanniniana Goiran, *Spec. Morph. Veg.* 45 (1875). It.
P. sordida Fries ex Aspegren, *Förs. Blek. Fl.* 38 (1823). Bu Cz Ga Ge.
P. collina *sensu stricto* (*P. wibeliana* T. Wolf). Au Cz Ga Ge Hu Rm Rs (C).

Those most closely resembling **40–51** are:

P. alpicola De la Soie ex Fauconnet, *Bull. Trav. Soc. Murith.* **5**: 18 (1876). He It.
P. opizii Domin, *Sitz.-Ber. Böhm. Ges. Wiss.* (*Math.-Nat. Kl.*) **1903** (25): 21 (1904). Cz.
P. rhenana P. J. Mueller ex Zimmeter, *Gatt. Potent.* 12 (1884). Ge.

The following are more or less intermediate:

P. argenteiformis Kauffm., *Mosk. Fl.* 159 (1866). Rs (E).
P. leucopolitana P. J. Mueller in Billot, *Annot.* 278 (1862). Au Cz Da Ga Ge Hu Po Rm Rs (C, W).
P. praecox F. W. Schultz, *Pollichia* 16–17: 5 (1859). Ge He.
P. silesiaca Uechtr., *Jahresb. Schles. Ges. Vaterl. Cult.* **44**: 82 (1867). ?Ge Po.
P. thyrsiflora Zimmeter in A. Kerner, *Sched. Fl. Exsicc. Austro-Hung.* **2**: 21 (1882). Cz Ge Hu It Po Rm Rs (C, W).
P. wiemanniana Günther & Schummel, *Sched. Cent. Siles. Exsicc.* **5** (1813). Au Cz Da Ga Ge He Hu It Ju Rs (B, C, W, E) Su.

23. P. supina L., *Sp. Pl.* 497 (1753). Annual or short-lived perennial. Flowering stems 10–40 cm, glabrous or setose at base, sparsely tomentose and with sessile glands above. Leaves pinnate; leaflets 5–11, 8–20 × 5–15 mm, oblong or obovate, incise-dentate. Flowers numerous. Sepals triangular-ovate; epicalyx-segments lanceolate to ovate, obtuse, longer than sepals. Petals 2·5–3 mm, yellow, shorter than sepals. Style conical-filiform. $2n = 28$. C., S. & E. Europe extending westwards to C. France and the Netherlands. Au Be Bu Cz Ga Ge Gr He Ho Hu It Ju Po Rm Rs (C, W, E) Tu.

24. P. norvegica L., *Sp. Pl.* 499 (1753). Annual or short-lived perennial. Flowering stems 10–70 cm, terminal, hirsute, sometimes with a few glands. Leaves ternate, rarely a few pinnate with 5 leaflets; leaflets 10–70 × 7–40 mm, obovate, elliptical or oblong, coarsely serrate or serrate-dentate or almost pinnatifid, green. Flowers numerous. Sepals *c.* 5 mm in flower, *c.* 10 mm in fruit, ovate, acute; epicalyx-segments oblong, subobtuse, longer than sepals in fruit. Petals 4–5 mm, yellow, shorter than or as long as sepals. Style conical-filiform. *N., C. & E. Europe; a frequent casual elsewhere and sometimes naturalized.* Au Cz Da Fe Ge It No Po Rm Rs (N, B, C, W, E) Su [Be Br Ga He Ho Ju].

This species appears to be spreading westwards; some authors believe it to be native only in E. & E.C. Europe.

25. P. intermedia L., *Mantissa* 76 (1767) (*P. heidenreichii* Zimmeter). Biennial or perennial. Flowering stems 20–50 cm, terminal, glabrous or setose at base, tomentose above. Leaves digitate; leaflets (3–)5, 10–40 × 5–18 mm, obovate or obovate-oblong, serrate-dentate to incise-serrate, green or sometimes grey-pubescent beneath. Flowers numerous. Sepals *c.* 5 mm in flower, 15–20 mm in fruit, ovate, acute; epicalyx-segments oblong-ovate, subobtuse, about as long as sepals. Petals 4–5 mm, yellow, as long as or slightly longer than sepals. Style conical-filiform. *N. & C. Russia; a common casual elsewhere, tending to become naturalized.* Rs (N, *B, C, *W) [Au Be Br Cz Da Fe Ga Ge He Ho It No Po Su].

26. P. astracanica Jacq., *Misc. Austr. Bot.* **2**: 349 (1781) (*P. taurica* sensu T. Wolf). Perennial. Flowering stems up to 30 cm, terminal, hirsute and often densely glandular. Leaves digitate; leaflets usually 5, 10–50 × 7–20 mm, oblong-oblanceolate to obovate, variously toothed. Flowers usually numerous, usually very crowded. Sepals triangular-ovate; epicalyx-segments triangular-lanceolate or -ovate, long-acuminate, usually 1½–2 times as long as sepals, always as wide as sepals at the base. Petals 12–15 mm, yellow, distinctly longer than calyx. Style conical-cylindrical. *S.E. Europe.* Bu Gr Ju Rm Rs (W, K, E).

Variable in the density and type of indumentum and in the shape and length of the calyx and epicalyx-segments. The following are often recognized as separate species: **P. astracanica** *sensu stricto* (*P. taurica* var. *genuina* T. Wolf), from S.E. Europe, with densely glandular stems, leaflets green or grey-green beneath and sepals and epicalyx-segments broadly triangular-ovate, subobtuse; **P. callieri** (T. Wolf) Juz. in Komarov, *Fl. URSS* **10**: 164 (1941), from Krym, like *P. astracanica* but with triangular-lanceolate, acute sepals and epicalyx-segments; **P. emilii-popii** E. I. Nyárády, *Bul. Grăd. Bot. Cluj* **8**: 87 (1928), from coasts of Romania and Bulgaria, with the stems eglandular, leaflets densely white sericeous-villous beneath and sepals and epicalyx-segments triangular-lanceolate, acute; **P. taurica** Willd., *Ges. Naturf. Freunde Berlin Mag.* **7**: 291 (1816), from Krym, with the stems eglandular or sparsely glandular, leaflets green or grey-green beneath and sepals and epicalyx-segments acute.

27. P. detommasii Ten., *Fl. Nap.* **1**, *Prodr.*: 61 (1811). Perennial. Flowering stems up to 30(–45) cm, terminal, hirsute and densely pubescent, eglandular. Leaves digitate; leaflets 5–7, 20–50 × 8–25 mm, obovate to obovate-oblong, crenate-serrate, sericeous-villous at least beneath. Flowers usually numerous; cymes crowded. Sepals triangular-ovate, acuminate; epicalyx-segments linear-lanceolate, long-acuminate, about as long as sepals. Petals 12–14 mm, yellow, as long as or longer than sepals. Style conical-cylindrical, shorter than achene. ● *Balkan peninsula, C. & S. Italy, Sicilia.* Al Bu Gr It Ju Si ?Tu.

28. P. recta L., *Sp. Pl.* 497 (1753) (incl. *P. adriatica* Murb., *P. hirta* auct. balcan., non L., *P. laciniosa* Kit. ex Nestler, *P. semilaciniosa* Borbás, *P. transcaspia* T. Wolf, ?*P. velenovskyi* Hayek). Perennial. Flowering stems 10–70 cm, terminal, densely pubescent, with long, patent or erecto-patent hairs and short, usually glandular hairs. Leaves digitate; leaflets 5–7, 15–100 × 5–35 mm, oblong to obovate, serrate to pinnatisect, green or grey. Flowers numerous; cymes lax. Sepals triangular-ovate; epicalyx-segments linear or linear-lanceolate, as long as or slightly longer than sepals. Petals 6–12 mm, yellow, as long as or longer than sepals. Style conical-cylindrical. $2n = 42$. C., E. & S. Europe. Al Au Bu Co Cz Ga Ge Gr He Hs Hu It Ju Po Rm Rs (C, W, K, E) Sa Si Tu [Be Br Fe Ho No Su].

A very variable species. It is not yet possible to make any satisfactory subdivision for the whole of Europe. In the western part of its range the achenes are narrowly but distinctly winged while in the eastern part they are not winged. The most distinct taxa are **P. recta** L. *sensu stricto*, from C., E. and parts of S. Europe, with serrate or incise-serrate, green or grey-green leaflets; **P. laciniosa** Kit. ex Nestler, *Monogr. Potent.* 45 (1816), from E.C. Europe, with pinnatifid or pinnatisect, green or grey-green leaflets; **P. pedata** Nestler, *op. cit.* 44 (1816), from S.E. Europe, with pinnatifid or pinnatisect, grey leaflets.

P. reuteri Boiss., *Diagn. Pl. Or. Nov.* 3(2): 51 (1856), from S. Spain, is probably of hybrid origin from **28** and **30**. It is like **28** but the basal leaves have not more than 7(–9) teeth or lobes and the achenes are without a narrow membranous margin.

29. P. hirta L., *Sp. Pl.* 497 (1753). Like **28** but stems and leaves with only long, patent, eglandular hairs; leaflets linear- to oblong-oblanceolate, rarely obovate, with 3–7 obtuse teeth or lobes at apex; petals longer than sepals. ● *W. Mediterranean region.* Co Ga Hs It.

30. P. nevadensis Boiss., *Elenchus* 40 (1838). Perennial. Flowering stems up to 30 cm, lateral, villous. Leaves digitate; leaflets 5, (4–)7–20 × (3–)5–15 mm, obovate or oblanceolate, crenate-serrate, sericeous-villous beneath. Flowers up to 4. Sepals ovate or ovate-lanceolate; epicalyx-segments oblong or oblong-lanceolate, slightly shorter than sepals. Petals (4–)5–7 mm, yellow, longer than sepals. Style conical-filiform. *Dry rocks and screes. S. Spain (Sierra Nevada).* Hs.

31. P. grandiflora L., *Sp. Pl.* 499 (1753). Perennial. Flowering stems 10–40 cm, lateral, densely hirsute with patent or sub-appressed hairs. Leaves ternate; leaflets 15–40 × 10–30 mm, obovate to almost suborbicular, coarsely dentate or crenate-serrate, sparsely to densely grey-tomentose beneath; terminal leaflet sessile or petiolulate. Flowers numerous. Sepals ovate, acute; epicalyx-segments oblong-lanceolate, acute, shorter than sepals. Petals 10–15 mm, yellow, about twice as long as sepals. Style conical-filiform. ● *Alps; C. & E. Pyrenees.* Au Ga He Hs It.

32. P. montenegrina Pant., *Österr. Bot. Zeitschr.* **23**: 5 (1873). Perennial. Flowering stems 30–80 cm, lateral, densely hirsute with subappressed hairs. Leaves ternate, very rarely a few digitate; leaflets 35–80 × 20–40 mm, broadly obovate to obovate-oblong, coarsely serrate or serrate-crenate, green, pubescent on the main veins beneath; terminal leaflet with distinct petiolule. Flowers numerous. Sepals triangular-lanceolate to ovate, acute; epicalyx-segments oblong, obtuse, or subacute, shorter than or as long as sepals. Petals 10–12 mm, yellow, about twice as long as sepals. Style conical-filiform. ● *N. part of Balkan peninsula.* Al Bu Ju.

33. P. umbrosa Steven ex Bieb., *Fl. Taur.-Cauc.* 3: 357 (1819). Like **32** but leaves digitate with 5 leaflets; sepals ovate, subobtuse; epicalyx-segments oblong-ovate. ● *Krym.* Rs (K)

34. P. delphinensis Gren. & Godron, *Fl. Fr.* 1: 530 (1849). Perennial. Flowering stems 30–50 cm, lateral, hirsute with sub-appressed hairs. Leaves digitate; leaflets 5, 25–60 × 15–30 mm, obovate, coarsely toothed, green; terminal leaflet sessile or sub-sessile. Flowers numerous. Sepals triangular-ovate, acute; epicalyx-segments lanceolate, acute, almost as long as sepals. Petals 10–12 mm, yellow, twice as long as sepals. Style conical-filiform. ● *S.W. Alps (Alpes Cottiennes, Massif du Pelrouse).* Ga.

35. P. pyrenaica Ramond ex DC. in Lam. & DC., *Fl. Fr.* ed. 3, 4: 459 (1805). Perennial. Flowering stems 10–40(–70) cm, lateral, sparsely to densely hirsute with appressed or subappressed hairs. Leaves digitate; leaflets 5, 12–60 × 7–30 mm, oblong or obovate-oblong, crenate-dentate, green; terminal leaflet sessile or sub-sessile. Flowers 5 or more. Sepals triangular-ovate, subacute; epicalyx-segments elliptic-linear, subobtuse, shorter than or as long as sepals. Petals (7–)10–15 mm, 1½–2 times as long as sepals. Style conical-filiform. ● *Pyrenees, N. & C. Spain.* Ga Hs.

Plants from Spain often have more coarsely and deeply toothed leaflets and sometimes the petals are less than 10 mm. They may be of hybrid origin from **28** or **29** and **35**.

36. P. stipularis L., *Sp. Pl.* 498 (1753). Perennial. Flowering stems 10–25(–35) cm, glabrous, or sometimes with pedicels, calyx and margin of leaflets sparsely hairy. Leaves digitate; leaflets 7–9, 6–20(–30) × 2–6 mm, oblong-obovate with 3–7 teeth at apex. Stipules 10–30 mm, adnate to the petiole throughout its length. Flowers up to 12. Sepals oblong-ovate; epicalyx-segments narrowly linear, slightly shorter than sepals. Petals 6–8 mm, yellow, 1½ times as long as sepals. Style conical at base, slightly clavate at apex. *N.E. Russia.* Rs (N). (*N. Siberia.*)

37. P. longipes Ledeb., *Fl. Ross.* 2: 50 (1843). Perennial. Flowering stems 15–50 cm, sparsely hairy. Leaves digitate; leaflets 7, 25–50 × 10–20 mm, oblong or oblong-obovate, coarsely toothed or almost pinnatifid, glabrous above, sparsely hairy beneath. Flowers usually numerous. Sepals ovate-lanceolate; epicalyx-segments linear-lanceolate, about as long as sepals. Petals *c.* 6 mm, yellow, longer than sepals. Style conical-filiform. *S.E. Russia.* Rs (C, E).

(**38–39**). **P. chrysantha** group. Perennial. Flowering stems 10–50(–70) cm, usually lateral, procumbent to erect, usually densely patent-pubescent or hirsute, and with short glandular and eglandular hairs. Leaves digitate; leaflets 5–9, 12–100 × 5–50 mm, oblong-lanceolate to obovate, crenate-serrate to coarsely serrate. Flowers in lax terminal cymes. Sepals ovate or oblong-ovate, acute; epicalyx-segments linear- or oblong-lanceolate, acute or subobtuse, about as long as sepals. Petals (6–)7–10 mm, yellow. Style more or less conical-filiform.

This group is morphologically intermediate between **28** and **43** and allied species and is probably of hybrid origin. There does not appear to be any definite evidence as to the breeding system in this group, but it is likely that it is, at least in part, apomictic.

Two species are briefly described below; both are frequently subdivided at infraspecific level.

38. P. chrysantha Trev., *Ind. Sem. Horto Wratisl.* 5 (1818). Leaflets usually 5. Petals 1½–2 times as long as sepals. *Balkan peninsula, Romania, Ural.* Bu Ju Rm Rs (C).

39. P. thuringiaca Bernh. ex Link, *Enum. Hort. Berol. Alt.* 2: 64 (1822) (*P. goldbachii* Rupr., *P. nestlerana* Tratt., *P. heptaphylla* sensu Coste, non L.). Leaflets 5–9. Petals 1–1½ times as long as sepals. *C. Europe, S.W. Alps, N. & C. Italy.* Cz Ga Ge He ?Hs It Rm [Fe No Rs (B) Su].

The records from N.E. Spain are probably referable to **35**.

40. P. brauniana Hoppe in Sturm, *Deutschl. Fl.* Abt. 1, Band 5, Heft 17 (1804) (*P. dubia* (Crantz) Zimmeter, non Moench, *P. minima* Haller fil.). Dwarf perennial. Flowering stems up to 5 cm, lateral, ascending, sparsely hairy and eglandular. Leaves ternate; leaflets 5–10(–15) × 3–7(–10) mm, oblong-obovate to obovate, shallowly dentate, glabrous above, sparsely hairy beneath. Stipules of basal leaves broadly ovate, obtuse. Flowers 1–3(–5).

Epicalyx-segments broadly elliptical, very obtuse, distinctly shorter than sepals. Petals 3–5 mm, yellow, as long as or up to $1\frac{1}{2}$ times as long as sepals. Style usually somewhat swollen at the base. $2n=14$. ● *Mountain grassland and rocks, usually above 2000 m; calcicole. Alps, Jura (Reculet), Pyrenees.* Au Ga Ge He Hs It Ju.

41. P. frigida Vill., *Hist. Pl. Dauph.* **3**: 563 (1788). Like **40** but flowering stems up to 10 cm; indumentum much thicker, with sessile glandular and long, ±patent, eglandular hairs; leaflets hairy above, more deeply incise-dentate; petals slightly shorter to slightly longer than sepals. $2n=28$. *Mountain grassland and rocks, usually above 2500 m; calcifuge.* ● *Pyrenees, Alps.* Au Ga He Hs It.

42. P. hyparctica Malte, *Rhodora* **36**: 177 (1934) (*P. emarginata* Pursh, non Desf.). Flowering stems up to 10 cm, lateral, ascending, sparsely hairy and sparsely to densely glandular. Leaves ternate; leaflets 5–15 mm, broadly obovate, with triangular-oblong, acute or subobtuse teeth, sparsely hairy above, sparsely to densely hairy and often glandular beneath. Stipules of basal leaves ovate. Flowers 1–3. Epicalyx-segments elliptical or oblong-elliptical, obtuse, a little shorter than the ovate sepals. Petals (4·5–)6–8 mm, yellow, about $1\frac{1}{2}$ times as long as sepals. Style conical at base, slightly clavate at apex. $2n=42$. *Arctic Europe.* Rs (N) Sb Su. (*Arctic America and W. Asia.*)

43. P. crantzii (Crantz) G. Beck ex Fritsch, *Excursionsfl. Österr.* 295 (1897) (*P., verna* L. nom. ambig., *P. alpestris* Haller fil., *P. salisburgensis* Haenke). Perennial with thick, woody stock; branches few, short, not or hardly rooting; lateral flowering stems up to 20 (–30) cm slender, ascending. Leaves digitate; leaflets (3–)5, 8–20(–40)×6–15(–20) mm, oblanceolate to obovate-cuneate or almost suborbicular, usually with broad, obtuse teeth, the terminal tooth subequal to the adjacent lateral, glabrous or sparsely hairy above, sparsely to densely hairy (mainly on the veins) beneath, the margin with patent hairs. Stipules of basal leaves ovate-lanceolate, subacute, often persistent in the withered state. Flowers 1–12. Epicalyx-segments oblong or elliptical, shorter than or almost as long as the triangular-ovate sepals. Petals 6–10 mm, yellow, often with an orange-coloured spot at base, longer than sepals. Style conical at base, slightly clavate at apex. $2n=28$, 42, 48, 64. *Open, rocky ground; usually calcicole. N. Europe; mountains of C. & S. Europe.* Al Au Br Bu Cz Fa Fe Ga Ge He Hs Is It Ju Lu No Po Rm Rs (N, B, C, W) Sb Su.

Very variable. Tetraploid plants from S. Poland (Tatra) are known to be sexual, but other plants in Scandinavia and Britain prove to be hexaploids or higher polyploids and apomictic. Plants from N. England and Sweden with characters intermediate between **43** and **49** are high polyploids and almost certainly the product of occasional hybridization.

P. serpentini Borbás ex Zimmeter, *Gatt. Potent.* 22 (1884), described from serpentine rocks in Austria (Burgenland), is a densely glandular-hairy variant; similar (but not identical) variants are known from serpentine in the Vosges and in Czechoslovakia and all probably merit varietal status only.

44. P. aurea L., *Cent. Pl.* **2**: 18 (1756). Like **43** but more mat-forming; upper part of stem and inflorescence covered with appressed silky hairs, never glandular; leaflets with appressed silky hairs on the margin and on the veins beneath, the terminal tooth always smaller than the adjacent lateral; epicalyx-segments linear-lanceolate; petals 6–11 mm. *Grassland and rocky places, usually between 1400 m and 2600 m; calcifuge. Mountains of S.*

and C. Europe, from N. Spain to the Carpathians and the Balkan peninsula. Al Au Bu Cz Ga Ge Gr He Hs It Ju Po Rm Rs (W).

(a) Subsp. **aurea**: Leaflets 5, with acute teeth. $2n=14$. ● *Throughout the range of the species except the S. & E. parts of the Balkan peninsula.*
Amphimictic and not very variable.

(b) Subsp. **chrysocraspeda** (Lehm.) Nyman, *Consp.* 225 (1878) (*P. ternata* C. Koch): Leaflets always 3, often with obtuse teeth. *Balkan peninsula, S. & E. Carpathians.*

45. P. heptaphylla L., *Cent. Pl.* **1**: 13 (1755) (*P. opaca* L., *P. rubens* (Crantz) Zimmeter, non Vill.). Perennial, with slender stock; branches short, ascending; lateral flowering stems up to 40 cm, slender, with soft patent hairs borne on minute tubercles, and shorter, thick pubescence, often with reddish multicellular glandular hairs. Leaves digitate; leaflets 5–7, 8–25×4–11 mm, ovate-lanceolate, dentate or crenate-dentate; stipules of basal leaves ovate-lanceolate, acute. Flowers 1–10. Sepals ovate-lanceolate; epicalyx-segments linear-lanceolate, as long as or shorter than sepals. Petals 5–7 mm, yellow. Achenes smooth. Style conical at base, slightly clavate at apex, smooth. $2n=14$ 28, 30, 35. *Dry, unshaded grassland; usually lowland and calcicole.* ● *C. Europe, extending to C. Jugoslavia and N. Ukraine, and northwards to 58° N. in S. Sweden.* Au ?Bu Cz Da Ga Ge He Hu It Ju Po Rm Rs (W) Su.

46. P. humifusa Willd., *Ges. Naturf. Freunde Berlin Mag.* **7**: 289 (1816) (*P. opaciformis* T. Wolf). Like **45** but with numerous more or less sessile, yellow, glandular hairs, and minutely papillose, conical-filiform style. *S. & E. parts of U.S.S.R.* Rs (C, W, K, E).

The style resembles that of **23–35**, but the otherwise close similarity to **45** makes it reasonable to retain this species here.

47. P. patula Waldst. & Kit., *Pl. Rar. Hung.* **2**: 218 (1805) (*P. schurii* Fuss ex Zimmeter). Stock relatively slender; lateral flowering stems up to 25(–70) cm, with long, rather stiff sub-appressed hairs. Leaves digitate; leaflets (5–)7–9, 5–15(–60)× 2–3(–15) mm, linear-oblanceolate, dentate with short linear lobes; stipules of basal leaves lanceolate, acute. Flowers 1–10. Sepals triangular-ovate; epicalyx-segments narrowly linear, more or less equalling sepals. Petals 6–7 mm, yellow. Achenes rugose, somewhat keeled; style conical at base, enlarged at apex. $2n=42$. *Dry, open grassland in the lowlands. E.C. & E. Europe, northwards to C. Czechoslovakia and to 54° N. in C. Russia.* Al Au Bu Cz Hu Ju Rm Rs (W, C, E).

48. P. australis Krašan, *Österr. Bot. Zeitschr.* **17**: 302 (1867). Like **47** but sometimes glandular; leaflets (5–)7, 7–20×3–6 mm, oblong-obovate to obovate; epicalyx-segments oblong to broadly elliptical; achenes smooth or only slightly rugose. *Dry grassland.* $2n=42$. ● *E.C. Europe, N.E. Italy, W. Jugoslavia; S.C. France.* Al Cz Ga Gr It Ju.

P. rigoana T. Wolf, *Biblioth. Bot. (Stuttgart)* **71**: 578 (1908), described from S. Italy (Monte Pollino), has the mat-forming and freely-rooting habit of **49** to **51**, but the hairiness and stipule shape of **48**. The petals are said to be concave and deeply emarginate, with an auriculate base. Its relationships are quite uncertain.

49. P. tabernaemontani Ascherson, *Verh. Bot. Ver. Brandenb.* **32**: 156 (1891) (*P. verna* auct., non L., *P. verna* subsp. *vulgaris* (Ser.) Gaudin). Mat-forming perennial, with numerous procumbent woody stems freely rooting at the nodes, and short, ascending, lateral flowering stems usually not more than 10 cm high;

hairiness variable, but consisting of simple hairs only. Leaves digitate; leaflets 5–7, 8–40 × 4–15 mm, cuneate-oblanceolate to -obovate, dentate or cuneate-dentate; stipules of basal leaves linear. Flowers 1–12. Sepals ovate; epicalyx-segments lanceolate, obtuse, usually much shorter than the sepals. Petals 6–10 mm, yellow, longer than sepals. Achenes rugose; style conical at base, slightly clavate at apex. $2n = 42, 49, 50, 56, c. 60, 63, 70, c. 84$. *Dry grassland and rocky places, usually lowland.* ● *N., W. & C. Europe, extending to the Baltic region, White Russia and Bulgaria.* Au Be Bu Br Co Cz Da *Fe Ga Ge He Ho Hs Hu It No Po Rm Rs (*N, B, C) Su.

Extremely variable. This polyploid complex is, so far as is known, wholly apomictic, and almost certainly arose by hybridization from related sexual species (**40–51**). Many taxa have been described, differing in habit, hairiness, leaflet-shape and -number, but no satisfactory taxonomic treatment is possible in the present state of knowledge.

The records of this species and **51** from Finland are almost all erroneous. They refer to **P. subarenaria** Borbás ex Zimmeter, *Gatt. Potent.* 21 (1884), an apomictic species probably of recent hybrid origin from **49** and **51**, which also occurs in C. Europe.

50. P. pusilla Host, *Fl. Austr.* **2**: 39 (1831) (*P. glandulifera* Krašan, *P. gaudinii* Gremli, *P. verna* subsp. *puberula* (Krašan) Hegi). Like **49** but whole plant rather sparingly covered with branched or stellate hairs interspersed with simple hairs, particularly on lower surface of leaflets; stellate hairs with 5–10 rays; petals 5–7 mm. *Mainly in subalpine grassland, up to 2000 m.* ● *C. Europe, extending to S.W. Alps.* Au Cz Ga Ge He Hu It Ju Po Rm.

Possibly of distant hybrid origin from 49×51, but with a wide distribution. Very variable.

51. P. cinerea Chaix ex Vill., *Prosp. Pl. Dauph.* 46 (1779) (incl. *P. arenaria* Borkh., *P. glaucescens* Willd., *P. incana* P. Gaertner, B. Meyer & Scherb., *P. tommasiniana* F. W. Schultz, *P. velutina* Lehm.). Mat-forming, with freely rooting, procumbent woody stems, and flowering stems up to 10 cm, the stem and leaves with a dense, usually continuous tomentum of stellate hairs with 15–30 rays, mixed with long, simple hairs and sometimes with glandular hairs. Leaves ternate or digitate; leaflets 3–5, 5–20 × 3·5–9 mm, oblong-obovate to broadly obovate, dentate or crenate-dentate, grey-green above, grey beneath; stipules of basal leaves linear. Flowers 1–6. Sepals ovate or ovate-lanceolate; epicalyx-segments lanceolate or elliptical, usually shorter than sepals. Petals 4–7 mm, yellow, longer than sepals. Achenes rugose; style conical at base, slightly clavate at apex. $2n = 14, 28$. *Dry places, up to 1600 m. C., E. & S. Europe from E. Spain eastwards; Baltic region.* Al Au Be Bu Cz Da Ga Ge Gr He Hs Hu It Ju Po Rm Rs (N, B, C, W, E) Su.

Plants from Poland are sexual with $2n = 28$; this number has also been recorded from Hungary and Czechoslovakia. The chromosome numbers $2n = 35, 42$ and 56 have been found in plants from Switzerland and S. France, but it is possible that they were hybrids between **49** and **51**.

52. P. erecta (L.) Räuschel, *Nomencl. Bot.* ed. 3, 152 (1797) (*P. tormentilla* Stokes). Perennial; stock stout, with a terminal rosette of leaves which are often dead at flowering time. Flowering stems (5–)10–30(–50) cm, procumbent to suberect, never rooting, appressed-pubescent. Leaves ternate or rarely digitate; leaflets 3, rarely 4 or 5, 5–30 × 2·5–10 mm, obovate-cuneate to oblong-lanceolate, dentate to incise-serrate at apex, glabrous or sparsely hairy above, sericeous-villous beneath. Flowers nearly

all 4-merous, usually many in a terminal cyme. Sepals ovate-lanceolate; epicalyx-segments linear-oblong. Petals (3–)4–6 mm, usually a little longer than sepals. Stamens 14–20. Carpels 4–8(–20). Style conical at base, slightly clavate at apex. $2n = 28$. *Almost throughout Europe, but rare in the Mediterranean region.* All except Bl Cr Gr Sb Si Tu.

53. P. anglica Laicharding, *Veg. Eur.* **1**: 475 (1790) (*P. procumbens* Sibth.). Like **52** but stock not so stout and with a persistent rosette of leaves; flowering stems 15–80 cm, procumbent, finally rooting at nodes; lower leaves often with 5 leaflets; flowers 4- to 5-merous, solitary, axillary or the upper forming a few-flowered cyme; petals 5–8 mm, up to twice as long as sepals; stamens 15–20; carpels 20–50. $2n = 56$. *W. & C. Europe, extending to S.W. Finland and the western borders of the U.S.S.R.* Az Be Br Co Cz Da Fe Ga Ge Hb Ho Hs It ?Ju Po Rm Rs (B, C, W) Su.

Hybrids between **52**, **53** and **54** are widely distributed in N. Europe and possibly elsewhere. The hybrid 52×54 closely resembles **53**, from which it is most readily distinguished by its high infertility.

54. P. reptans L., *Sp. Pl.* 499 (1753). Perennial; stock relatively slender with a persistent rosette of leaves; flowering stems 30–100 cm, procumbent, rooting at the nodes, glabrous or pubescent. Leaves digitate; leaflets 5(–7), 5–70 × 3–25 mm, obovate or oblong-obovate, dentate or serrate-dentate. Flowers 5-merous, all solitary, axillary. Sepals and epicalyx-segments variable. Petals (7–)8–12 mm, up to twice as long as sepals. Stamens *c.* 20. Carpels 60–120. Style conical at base, enlarged at apex. $2n = 28$. *Europe except the extreme north.* All except Fa Is Sb.

Subgen. **Fragariastrum** (Heister ex Fabr.) Reichenb. Herbs. Leaves ternate or digitate. Petals white or pink, rarely pale yellow. Receptacle densely hairy. Style subterminal, conical-filiform. Achenes usually pubescent.

Sect. FRAGARIASTRUM. Anthers oblong. Style deciduous, glabrous.

55. P. caulescens L., *Cent. Pl.* **2**: 19 (1756). Perennial. Flowering stems usually 5–30 cm, pubescent with erecto-patent (rarely subpatent) or subappressed hairs. Leaves digitate; leaflets 5–7, 10–30 mm, oblong or oblong-obovate, with few connivent teeth at apex. Flowers numerous, in lax cymes. Sepals lanceolate or ovate-lanceolate; epicalyx-segments as long as or slightly longer, but much narrower than sepals. Petals 6–10 × 2·5–5 mm, white, slightly longer than sepals, often slightly mucronulate, not or indistinctly emarginate. Filaments thickened towards the base, pubescent at least in the lower half. Style pale yellow. *Rock fissures; usually calcicole. Alps, and mountains of S. Europe from Spain to Jugoslavia.* Au Bl Ga Ge He Hs It Ju Si.

Very polymorphic, but none of the subordinate taxa seem to merit subspecific status.

56. P. petrophila Boiss., *Voy. Bot. Midi Esp.* **2**: 728 (1845). Like **55** but densely caespitose; whole plant densely silvery-villous; flowering stems 2–10 cm; leaflets 7–15 mm, 3- to 5(–7)-toothed at apex; cymes mostly dense; flowers mostly smaller; epicalyx-segments distinctly shorter than sepals; petals 5–7 × 3–5 mm. *Mountains of S. Spain.* Hs.

57. P. clusiana Jacq., *Fl. Austr.* **2**: 10 (1774). Perennial. Flowering stems 5–10 cm, pubescent with subappressed hairs. Leaves digitate; leaflets usually 5, 7–12 mm, obovate, truncate and 3- to 5-toothed at apex. Flowers usually 1–3. Sepals oblong-lanceolate; epicalyx-segments narrowly linear, slightly shorter than sepals. Petals 9–10 × 6–8 mm, white, broadly obovate, emarginate, much longer than sepals. Filaments filiform, glabrous. Style reddish. $2n = 42$. *E. Alps; mountains of W. Jugoslavia and Albania.* Al Au Ge It Ju.

58. P. crassinervia Viv., *Fl. Cors., App.* **1**: 2 (1825). Perennial. Flowering stems up to 20(–40) cm, densely pubescent with erecto-patent, long and short hairs, and glandular hairs. Leaves digitate; leaflets 5, 10–30 mm, obovate, crenate or crenate-dentate in the upper half; veins prominent beneath; stipules of basal leaves ovate or ovate-lanceolate. Flowers 5 or more. Sepals ovate; epicalyx-segments lanceolate or ovate-lanceolate, about as long as sepals. Petals *c.* 8 mm, white, entire, longer than sepals. Filaments glabrous or sparsely hairy at base. $2n = 14$. *Mountain rocks; calcifuge.* ● *Corse and Sardegna.* Co Sa.

59. P. nitida L., *Cent. Pl.* **2**: 18 (1756). Densely caespitose and silvery-grey, sericeous perennial. Flowering stems up to 5 cm. Leaves usually ternate; leaflets 5–10 mm, obovate or oblanceolate, with (0–)3(–7) teeth at apex; stipules lanceolate or ovate-lanceolate. Flowers 1–2. Sepals triangular-lanceolate; epicalyx-segments linear, shorter than sepals. Petals 10–12 × 7–10 mm, pink or white, emarginate, longer than sepals. $2n = 42$. *Calcareous rocks and screes.* ● *S.W. & S.E. Alps, N. Appennini.* Au Ga It Ju.

60. P. alchimilloides Lapeyr., *Mém. Acad. Toulouse* **1**: 212 (1782). Perennial. Flowering stems 10–30 cm, sericeous. Leaves digitate; leaflets 5–7, 10–25 mm, oblong-elliptical with usually 3 small teeth at apex, glabrous above, silvery-sericeous beneath and on margin; stipules lanceolate or linear-lanceolate. Flowers usually numerous. Sepals ovate-lanceolate; epicalyx-segments linear-lanceolate, as long as sepals. Petals 8–10 mm, white, emarginate, longer than sepals. *Mountain rocks and screes.* ● *Pyrenees.* Ga Hs.

61. P. valderia L., *Syst. Nat.* ed. 10, **2**: 1064 (1759). Perennial. Flowering stems up to 40 cm, grey-tomentose; tomentum of short hairs (0·2–0·5 mm) mixed with long erecto-patent hairs (1–2 mm) and with a few glandular hairs less than 0·1 mm. Leaves digitate; leaflets usually 7, 15–30 mm, narrowly obovate, dentate at least in the apical ½, green or grey above, grey-tomentose or almost sericeous beneath. Flowers usually numerous. Sepals triangular-ovate; epicalyx-segments linear-lanceolate, as long as or shorter than sepals. Petals 6–7 mm, white, shorter than sepals. Filaments villous. $2n = 14$. *Rocks and stony mountain pastures.* ● *Maritime Alps.* Ga It.

62. P. haynaldiana Janka, *Österr. Bot. Zeitschr.* **22**: 176 (1872). Like **61** but flowering stems without short hairs, with patent hairs 1·5–3 mm, and with numerous glandular hairs 0·1–0·3 mm; leaflets green above, silvery-sericeous-tomentose beneath; epicalyx-segments usually distinctly longer than sepals; filaments glabrous or pubescent. $2n = 14$. *Mountain rocks.* ● *Bulgaria and S. Carpathians.* Bu Rm.

63. P. doerfleri Wettst., *Biblioth. Bot. (Stuttgart)* **26**: 39 (1892) (*P. caulescens* var. *doerfleri* (Wettst.) T. Wolf). Perennial. Flowering stems 5–25 cm, densely pubescent with long patent hairs 1·5–2(–2·5) mm and with glandular hairs up to 0·15(–0·2) mm. Leaves digitate; leaflets 5, 15–40 mm, obovate, crenate-serrate in the apical half, pubescent, green or grey-green. Flowers numerous. Sepals triangular-lanceolate; epicalyx-segments linear-lanceolate, about as long as sepals. Petals 5–7 mm, white, shorter than sepals. Filaments densely pubescent in lower half. *Siliceous cliffs above* 2000 *m.* ● *S. Jugoslavia (Šar Planina).* Ju.

64. P. nivalis Lapeyr., *Mém. Acad. Toulouse* **1**: 210 (1782). Perennial. Flowering stems up to 30 cm, densely pubescent with long, patent hairs *c.* 1·5 mm and with short, glandular hairs. Leaves digitate; leaflets (5–)7(–9), 10–20 mm, obovate, with a few connivent teeth in the apical ⅓, pubescent, green or grey-green. Flowers numerous. Sepals triangular-lanceolate; epicalyx-segments linear-lanceolate, longer than sepals. Petals 6–7·5 mm, white, shorter than sepals. Filaments glabrous. ● *Mountains of N. & E. Spain, Pyrenees, S.W. Alps.* Ga Hs.

65. P. grammopetala Moretti, *Bot. Ital.* 4 (1826). Perennial. Flowering stems 10–30 cm, pubescent with long erecto-patent hairs (1–2 mm) and numerous glandular hairs (up to 0·8 mm). Leaves ternate or digitate; leaflets 3(–5), 15–25 mm, obovate, serrate-dentate in the apical half, pubescent, green. Flowers numerous. Sepals linear-lanceolate or lanceolate; epicalyx-segments linear or linear-lanceolate, as long as or slightly shorter than sepals. Petals 6–7·5 mm, cream or very pale yellow, about equalling sepals. *Mountain rocks; calcifuge.* ● *C. Alps.* He It.

66. P. speciosa Willd., *Sp. Pl.* **2**: 1110 (1800). Perennial. Flowering stems up to 30 cm, grey- or white-tomentose. Leaves ternate; leaflets 15–30 mm, broadly obovate to elliptic-obovate, crenate or crenate-dentate at least in the apical ⅔, white-tomentose beneath. Flowers numerous. Sepals broadly ovate; epicalyx-segments linear, as long as or longer than sepals. Petals *c.* 10 mm, white, slightly longer than sepals. *W. & S. parts of Balkan peninsula; Kriti.* Al Cr Gr Ju.

67. P. apennina Ten., *Fl. Nap.* 1, *Prodr.*: 30 (1811). Perennial. Flowering stems up to 20 cm, white-tomentose. Leaves ternate; leaflets 7–15 × 3–6 mm, silvery-sericeous, rarely glabrous above. Flowers 1–5(–7). Sepals ovate; epicalyx-segments linear, as long as or slightly shorter than sepals. Petals 8–12 mm, longer than sepals. Achenes pubescent. ● *Balkan peninsula; C. Appennini.* Al Bu It Ju.

(a) Subsp. **apennina**: Leaflets oblong-obovate, subentire or with 2–3 teeth at apex. Petals white. *C. Appennini, W. Jugoslavia, N. Albania.*

(b) Subsp. **stoianovii** Urum. & Jáv., *Magyar Bot. Lapok* **19**: 36 (1922): Leaflets broadly obovate, with 3–5 teeth at apex. Petals pale pink. *S.W. Bulgaria.*

68. P. kionaea Halácsy, *Verh. Zool.-Bot. Ges. Wien* **38**: 751 (1888) (*P. apennina* subsp. *kionaea* (Halácsy) Maire & Petit-mengin). Perennial. Flowering stems up to 10 cm, grey- or white-tomentose. Leaves ternate; leaflets 4–8 × 3–6 mm, broadly obovate, with 3–9 teeth at apex, silvery-sericeous, sometimes grey-green above. Flowers 1–3. Sepals ovate; epicalyx-segments linear, about as long as sepals. Petals *c.* 8 mm, purple, longer than sepals. Achenes glabrous. *Limestone cliffs and rocks above* 1800 *m.* ● *S.C. Greece (Giona).* Gr.

69. P. deorum Boiss. & Heldr. in Boiss., *Diagn. Pl. Or. Nov.* 3(2): 51 (1856). Perennial. Flowering stems up to 25 cm, grey- or white-tomentose. Leaves ternate; leaflets 8–20 × 4–9 mm, obovate to oblanceolate, with 3–9 teeth at apex, silvery-sericeous, sometimes grey-green above. Flowers 3–6, congested. Sepals ovate, purple on the inner surface; epicalyx-segments linear, about as long as sepals. Petals 10–12 mm, white, longer than sepals. Achenes glabrous. *Limestone cliffs and stony ground above* 2000 *m.* ● *C. Greece (Olimbos).* Gr.

70. P. alba L., *Sp. Pl.* 498 (1753). Perennial. Flowering stems up to 15 cm, pubescent with appressed or erecto-patent hairs. Leaves digitate; leaflets 5, 20–40 mm, oblong- to obovate-lanceolate with a few teeth at apex, glabrous and green above, silvery-sericeous beneath. Flowers several. Sepals lanceolate; epicalyx-segments linear, shorter than sepals. Petals 7–10 mm, white, longer than sepals. Filaments glabrous. Achenes smooth. $2n = 28$. ● *C. & E. Europe, extending southwards to N. Italy and Macedonia.* Au Bu Cz Ga Ge He Hu It Ju Po Rm Rs (C, W).

71. P. montana Brot., *Fl. Lusit.* **2**: 390 (1804) (*P. splendens* Ramond ex DC.). Perennial; usually with long stolons. Flowering stems 5–20 cm, pubescent with patent hairs. Leaves ternate or rarely digitate; leaflets 10–30 mm, oblong-obovate to obovate, crenate-dentate towards the apex, pubescent and green above, grey-sericeous beneath. Flowers 1–4. Sepals ovate; epicalyx-segments linear or lanceolate, as long as or slightly shorter than sepals. Petals 6–9 mm, white, longer than sepals. Achenes smooth. ● *N. part of Iberian peninsula, W. & C. France.* Ga Hs Lu.

72. P. sterilis (L.) Garcke, *Fl. Halle* **2**: 200 (1856) (*P. fragariastrum* Pers., *Fragaria sterilis* L.). Perennial; usually with long stolons. Flowering stems 5–15 cm, pubescent with patent hairs, eglandular. Leaves ternate; leaflets 10–25 mm, broadly obovate, crenate-serrate, sparsely pubescent and green above, grey-sericeous beneath. Flowers 1–3. Sepals ovate-lanceolate, yellow-green on inner side towards base; epicalyx-segments lanceolate, shorter than sepals. Petals *c.* 5 mm, white, slightly longer than sepals. Filaments filiform, glabrous. Achenes minutely rugulose. $2n = 28$. ● *W., C. & S. Europe, eastwards to Poland and Macedonia and extending northwards to S. Sweden.* ?Al Au Be Br Cz Da Ga Ge Hb He Ho Hs It Ju Lu Po ?Sa ?Si Su.

73. P. micrantha Ramond ex DC. in Lam. & DC., *Fl. Fr.* ed. 3, **4**: 468 (1805). Perennial. Flowering stems 1–5(–15) cm, densely pubescent with patent or subdeflexed hairs, eglandular. Leaves ternate; leaflets 10–50 mm, obovate, serrate or serrate-crenate, pubescent and green above, somewhat grey-sericeous beneath. Flowers 1–2(–3). Sepals and epicalyx-segments ovate-lanceolate, usually equal; sepals dark reddish on inner side towards base. Petals 3–5 mm, white, rarely pink, as long as or slightly shorter than sepals. Filaments densely ciliate, at least in lower half. Achenes minutely rugulose. $2n = 12, 14$. *S. & C. Europe, northwards to c. 50° 30′ N. in W. Germany.* Al Au Bu Co Cz Ga Ge Gr He Hs Hu It Ju Rm Rs (K) Sa Si.

74. P. carniolica A. Kerner, *Österr. Bot. Zeitschr.* **20**: 44 (1870) (*P. micrantha* var. *carniolica* (A. Kerner) T. Wolf). Like 73 but stems, pedicels and calyx with pluricellular glandular hairs as well as long, patent, simple hairs; flowers up to 4; sepals yellowish-green on inner side towards base; epicalyx-segments shorter than sepals; petals longer than sepals. ● *N.W. Jugoslavia.* Ju.

Sect. PLUMOSISTYLAE Pawł. Anthers globose. Style persistent in fruit, at least the lower half plumose.

75. P. saxifraga Ardoino ex De Not., *Ind. Sem. Horti Bot. Genuens.* **1848**: 25 (1848). Pulvinate perennial. Flowering stems up to 15(–20) cm, densely glandular and with appressed eglandular hairs. Leaves ternate or digitate; leaflets 3–5, 15–30 mm, linear to obovate-lanceolate, usually 3-toothed at apex, margin revolute, coriaceous, subglabrous and green above, silvery-sericeous beneath. Flowers (3–)5–12(–20). Sepals ovate or ovate-lanceolate; epicalyx-segments linear, shorter than sepals. Petals 4·5–6 mm, white, longer than sepals. $2n = 14$. *Limestone rock-fissures.* ● *Maritime Alps.* Ga It.

---

¹ By P. W. Ball.   ² By T. G. Tutin.

20. Sibbaldia L.¹

Like *Potentilla* but stamens usually 5 (rarely 4 or 10); carpels 5–12.

Sepals *c.* 3 × 1–1·5 mm; petal-veins not or scarcely anastomosing
　　　　　　　　　　　　　　　　　　　　　　　1. procumbens
Sepals *c.* 5 × 2 mm; petal-veins distinctly anastomosing　　**2. parviflora**

1. S. procumbens L., *Sp. Pl.* 284 (1753). Procumbent perennial herb with branched woody stock, each branch terminated by a rosette of leaves; flowering stems 1–5 cm, axillary, usually shorter than the leaves, pubescent. Leaves ternate; leaflets 5–20 mm, obovate-cuneate, 3-toothed or -lobed at the truncate apex, pubescent, green. Flowers few, in a dense cyme; sepals *c.* 3 × 1–1·5 mm; petals 1·5–2 mm, sometimes absent, yellow, oblanceolate, 3-veined, the veins not or scarcely anastomosing. Achenes 1–1·5 mm, more or less rugulose, shining. $2n = 14$. *N. Europe, and the mountains of C. & S. Europe.* Au Br Bu Co Cz Fa Fe Ga Ge He Hs Is Ju No Po Rs (N) Sb Su.

2. S. parviflora Willd., *Ges. Naturf. Freunde Berlin Neue Schr.* **2**: 125 (1799). Like **1** but more densely pubescent and greyish; sepals *c.* 5 × 2 mm; petals obovate, the veins distinctly anastomosing. *S. Jugoslavia (Galičica Planina, S.W. of Ohrid).* Ju. (*Mountains of W. & C. Asia.*)

21. Fragaria L.²

Like *Potentilla* but receptacle becoming fleshy and brightly coloured in fruit; achenes on the surface or sunk in pits.

1 Sepals patent or deflexed after flowering
2 Scape conspicuously longer than leaves; stolons 0 or few
　　　　　　　　　　　　　　　　　　　　　　　2. moschata
2 Scape not or scarcely longer than leaves; stolons usually numerous
3 Pedicels appressed-pubescent; achenes projecting, uniformly scattered over the receptacle　　　　　**1. vesca**
3 Pedicels patent-pubescent; achenes not projecting, confined to the upper part of the receptacle　　　　**3. viridis**
1 Sepals appressed after flowering
4 Achenes sunk in deep pits　　　　　　　**4. virginiana**
4 Achenes on the surface of the receptacle, projecting or not
5 Fruit *c.* 1 cm, achenes projecting; leaves pubescent above
　　　　　　　　　　　　　　　　　　　　　　　3. viridis
5 Fruit *c.* 3 cm, achenes not projecting; leaves ±glabrous above　　　　　　　　　　　　　**5. × ananassa**

1. F. vesca L., *Sp. Pl.* 494 (1753). Perennial herb, with long, epigeal, rooting stolons. Leaves 3-foliolate in a basal rosette; leaflets 1–6 cm, ovate or obovate to rhombic, coarsely serrate, bright green and sparsely hairy on upper surface. Scape 5–30 cm, little longer than the leaves, erect in fruit. Pedicels appressed-pubescent. Flowers *c.* 15 mm in diameter, white, usually hermaphrodite. Sepals patent or deflexed in fruit. Achenes uniformly scattered over and projecting from the usually red, glabrous receptacle. $2n = 14$. *Almost throughout Europe.* All except Bl Cr Fa Sb.

2. F. moschata Duchesne, *Hist. Nat. Frais.* 145 (1766) (*F. elatior* Ehrh.). Like **1** but up to 40 cm; stolons few or absent; scape longer than the leaves; pedicels patent-pubescent; flowers *c.* 20 mm in diameter, usually unisexual; receptacle without achenes near its base. *C. Europe, extending to N.W. France, C. Italy, Turkey and C. Russia; widely naturalized from gardens in N. Europe.* Al Au Be Bu Cz Ga Ge He Hs Hu It Ju Po Rm Rs (C, W) Tu [Br Da Fe Ho No Rs (N, B) Su].

3. F. viridis Duchesne, *op. cit.* 135 (1766) (*F. collina* Ehrh.). Like **1** but up to 20 cm, with short filiform stolons; leaflets appressed-pubescent, or glabrous on upper surface; flowers creamy-white, sometimes unisexual; sepals appressed or recurved after flowering; receptacle without achenes near its base. *Most of Europe, except the islands and the extreme north.* Al Au Be Bu Cz Da Fe Ga Ge Gr He Hs Hu It Ju No Po Rm Rs (B, C, W, K, E) Su.

(a) Subsp. **viridis**: Leaflets pubescent on upper surface. Pedicels with erecto-patent or appressed hairs; sepals appressed after flowering. Achenes projecting from the receptacle. $2n = 14$. *Throughout the range of the species except the south-east.*

(b) Subsp. **campestris** (Steven) Pawł., *Feddes Repert.* **79**: 35 (1968) (*F. campestris* Steven): Leaflets glabrous on upper surface. Pedicels with patent hairs; sepals recurved after flowering. Achenes in pits in the receptacle. *S.E. part of the range of the species.*

4. F. virginiana Duchesne, *Hist. Nat. Frais.* 204 (1766). Perennial herb; stolons numerous. Leaves 3-foliolate in a basal rosette, rather coriaceous, not or scarcely rugose, blue-green and nearly or quite glabrous above. Scapes up to 25 cm, decumbent in fruit. Flowers white, often unisexual, female much smaller than male. Sepals appressed after flowering. Receptacle *c.* 2 cm, glabrous, without achenes near its base; achenes deeply sunk in pits. *Cultivated and naturalized locally in E. Europe; exact distribution unknown through confusion with **5**.* (*E. North America.*)

5. F. × ananassa Duchesne, *op. cit.* 190 (1766) (*F. chiloensis* auct.). Like **4** but flowers hermaphrodite; receptacle *c.* 3 cm, in most cultivars covered all over with achenes which are on the surface or very slightly sunk. *Widely cultivated and naturalized throughout much of Europe.*

A hybrid between **4** and *F. chiloensis* (L.) Duchesne from W. America. Both the parents and the hybrid are hexaploid with $2n = 56$.

All species mentioned here, as well as many of their hybrids, are, or have been, cultivated for their edible fruits (strawberries). The common cultivated strawberry of Europe belongs to **5**, and the Alpine strawberry to **1**.

22. Duchesnea Sm.[1]

Like *Potentilla* but epicalyx-segments 3-toothed at apex; receptacle becoming fleshy and brightly coloured in fruit.

1. D. indica (Andrews) Focke in Engler & Prantl, *Natürl. Pflanzenfam.* 3(3): 33 (1888). Perennial herb, with epigeal, rooting stolons. Stems up to 50 cm. Leaves 3-foliolate, rather long-petiolate; leaflets obovate, crenate, with a cuneate base. Stipules lanceolate. Flowers solitary, yellow, not or only slightly exceeding the leaves; sepals *c.* 10 mm; epicalyx-segments broadly ovate, exceeding the sepals; petals *c.* 8 mm. Receptacle spongy, bright red, tasteless. *Naturalized in a few regions of W., C. & S. Europe.* [Au Az Ga He It.] (*Probably native in S. & E. Asia.*)

23. Alchemilla L.[2]

Perennial herbs with woody rhizome. Leaves palmate or palmately lobed. Inflorescence compound, cymose. Flowers small, green or yellowish, more or less aggregated into distinct clusters (glomeruli). Hypanthium urceolate; sepals 4(–5); epicalyx present; petals absent; stamens 4(–5), inserted on outer margin

---

[1] By D. H. Valentine.
[2] By S. M. Walters; Ser. *Elatae* by B. Pawłowski.

of disc; carpel 1; style basal. Fruit a single achene, wholly or partially enclosed in the thin, dry hypanthium.

It is known that many of the common European species of this genus reproduce apomictically; the pollen is largely or wholly abortive, and the seed develops precociously in the flower. Sexual reproduction is apparently confined to *A. pentaphyllea* and to a small group of species in Ser. *Hoppeanae*. The high chromosome numbers suggest complex, hybrid origins from parent sexual species which are now extinct.

More than 300 species have been described in Europe, mostly in Subsect. *Heliodrosium* Rothm. Some of these are widespread and easily recognized, and others, though very local, present no taxonomic difficulty. Many taxa, however, particularly in S.E. Europe, are still very inadequately known. The following account attempts to key and describe all the wide-ranging species, and some of the more easily recognizable local endemics, and to list as 'related species' other local taxa.

The key is designed for well-grown specimens with inflorescences and mature basal leaves. Late season's growth, or second growth after grazing or cutting, may differ considerably in leaf-shape and hairiness, and is often unidentifiable.

The term 'leaf', unless qualified, refers to mature, basal leaves; the term 'sinus' refers to the space or angle between the sides of the two basal lobes. The depth of division of the leaf is expressed by a fraction ($\frac{1}{2}$, $\frac{2}{3}$, etc.), which is the proportion of length of free lobe to the 'radius' of the leaf (measured from top of petiole to top of middle lobe). Between the lobes there may be developed a more or less obvious, toothless, V-shaped (rarely U-shaped) incision, so that each lobe has sub-parallel sides proximally. This incision is described as 'long' when it exceeds twice the length of the adjacent tooth. The number of teeth is given for one side of the middle lobe, and excludes the apical tooth.

In general, the large species are northern or montane hay-meadow or roadside plants, which have probably been introduced by human agency to part of their present geographical range; the medium-sized species occur in either semi-natural mountain grassland or grazed pasture, whilst many dwarf species are mountain plants. The extreme dwarf species characteristic of snow-patches on mountains, which Buser called 'sub-nival', belong to several different groups, and show striking parallelism in their relative hairlessness and in their deeply lobed leaves, usually with long incisions and large teeth (spp. **6, 12, 33, 62, 93–95, 117, 118**).

Literature: R. Buser, *Alchimilles Valaisannes*. Zürich. 1894 (also published in Jaccard, H., Catalogue de la Flore Valaisanne, *Neue Denkschr. Schweiz. Ges. Naturw.* **34**: 104–139 (1895). S. Juzepczuk in V. L. Komarov, *Flora URSS* **10**: 289–410. Mosqua & Leningrad. 1941. A. Maillefer, *Mém. Soc. Vaud. Sci. Nat.* **8**: 101–136 (1944). B. Pawłowski, *Flora Tatr* **1**: 442–503. Warsaw. 1956. W. Rothmaler, *Feddes Repert.* (*Beih.*) **100**: 59–93 (1938) (Subsect. *Calycanthum*). W. Rothmaler, *Feddes Repert.* **66**: 194–234 (1962) (C. Europe). G. Samuelsson, *Acta Phytogeogr. Suec.* **16** (1942) (Ser. *Pubescentes* and Ser. *Vulgares* in Fennoscandia).

1 Leaves palmately divided to more than $\frac{1}{2}$
  2 Plant glabrous, or with very few hairs mainly on the margins of the leaves
    3 Leaves almost compound, with (3–)5 segments; epicalyx-segments much smaller than sepals **1. pentaphyllea**
    3 Leaves lobed to $\frac{1}{2}-\frac{2}{3}$; epicalyx-segments at least as long as sepals **118. fissa**
  2 Plant hairy, at least on lower surface of leaf
    4 Leaf-segments 5–7, at least the middle segment completely free; pedical shorter than hypanthium (Ser. *Saxatiles*)

5 Teeth of leaf-segments 2–3 mm; leaves dull and only sparsely sericeous beneath **6. subsericea**

5 Teeth of leaf-segments usually *c.* 1 mm; leaves shiny and densely sericeous beneath

6 Leaf-segments always 5 **2. saxatilis**

6 Leaf-segments 6 or 7 on at least some leaves

7 Inflorescence with crowded glomeruli, not much exceeding leaves **4. alpina**

7 Inflorescence with more spaced glomeruli, much exceeding leaves

8 Leaf-segments ovate-lanceolate **3. transiens**

8 Leaf-segments broadly obovate **5. basaltica**

4 Leaf-segments (5–)7–9, usually joined at base; pedicel at least as long as hypanthium

9 Leaves lobed to more than $\frac{2}{3}$

10 Teeth of leaf-segments 2–3 mm; leaves dull and only sparsely sericeous beneath **12. grossidens**

10 Teeth of leaf-segments usually *c.* 1 mm; leaves shiny and densely sericeous beneath

11 At least the middle leaf-segment completely free **7. plicatula**

11 All leaf-segments joined at base

12 Leaves ± semicircular, with very wide sinus **9. anisiaca**

12 Leaves suborbicular, with narrow or closed sinus

13 Leaf-segments elliptical, acute **8. pallens**

13 Leaf-segments linear or oblong-linear, obtuse, sometimes subtruncate **10. hoppeana**

9 Leaves lobed to $\frac{2}{3}$ or less

14 Plant with patent or deflexed hairs

15 Hypanthium hairy **33. helvetica**

15 Hypanthium glabrous

16 Dwarf; epicalyx-segments less than $\frac{1}{2}$ as long as sepals **62. decumbens**

16 Medium-sized; epicalyx-segments at least $\frac{1}{2}$ as long as sepals **63. undulata**

14 Plant with appressed or subappressed hairs (present at least on distal half of veins on lower surface of leaf)

17 Leaves glabrous beneath, except on veins **117. incisa**

17 Leaves hairy beneath

18 Teeth on leaf-lobes indistinct, ± hidden by dense, sericeous indumentum **11. conjuncta**

18 Teeth on leaf-lobes obvious, not hidden by indumentum

19 Leaves glabrous or subglabrous above **13. faeroensis**

19 Leaves hairy above

20 Leaves rather sparsely hairy above and beneath **14. paicheana**

20 Leaves densely hairy above and beneath **72. buschii**

1 Leaves palmately lobed to $\frac{1}{2}$ or less

21 Stems and petioles with erecto-patent, patent or deflexed hairs

22 Hypanthium more or less densely hairy

23 Epicalyx-segments at least as long as sepals; hypanthium much shorter than mature achene

24 Stems and petioles with erecto-patent hairs **100. viridiflora**

24 Stems and petioles with patent hairs

25 Leaves glabrous or subglabrous above **101. albanica**

25 Leaves densely hairy above **102. phegophila**

23 Epicalyx-segments usually shorter than sepals; hypanthium as long as mature achene

26 Stems and petioles with at least some deflexed hairs

27 All pedicels hairy throughout their length

28 Leaf-lobes subtruncate, with long incisions **25. erythropoda**

28 Leaf-lobes rounded or subtriangular, with short incisions **26. lithophila**

27 At least some pedicels glabrous or with few hairs towards base only

29 Some epicalyx-segments ± equalling sepals (Krym) **38. hirsutissima**

29 Epicalyx-segments shorter than sepals

30 Dwarf mountain plants up to 15 cm; leaf-lobes with 4–5(–6) teeth

31 Leaves lobed to *c.* $\frac{1}{2}$; lobes with teeth 2 mm or more **33. helvetica**

31 Leaves lobed to *c.* $\frac{1}{4}$; lobes with teeth *c.* 1 mm **30. colorata**

30 Medium-sized or rather large plants up to 50 cm; leaf-lobes with (6–)7–9 teeth

32 Hairs on hypanthium with a small tubercle at base **36. gibberulosa**

32 Hairs on hypanthium without tubercle at base

33 Leaves ± reniform; lobes with distinct incisions **35. hebescens**

33 Leaves orbicular or suborbicular; lobes with short, indistinct incisions **37. bungei**

26 Stems and petioles with patent or erecto-patent hairs

34 All pedicels ± densely hairy

35 Leaf-lobes without incisions (or with very short, indistinct ones)

36 Base of petioles reddish; leaf-lobes with 6–9 teeth **57. filicaulis**

36 Base of petioles not reddish; leaf-lobes with 4–6 teeth **21. glaucescens**

35 Leaf-lobes with distinct incisions

37 Hairs on stems and petioles patent

38 Leaf-lobes truncate **23. flabellata**

38 Leaf-lobes rounded

39 Robust lowland plant up to 40 cm **22. hirsuticaulis**

39 Dwarf mountain plant not more than 15 cm **24. cinerea**

37 Hairs on stems and petioles erecto-patent or subappressed

40 Leaves lobed to $\frac{1}{4}$–$\frac{1}{3}$(–$\frac{2}{5}$); hairs on petioles never subappressed **23. flabellata**

40 Leaves lobed to more than $\frac{1}{3}$; hairs on petioles often subappressed

41 Leaves glabrous above, or hairy above only on folds **34. vetteri**

41 Leaves evenly hairy above **27. bulgarica**

34 At least some pedicels glabrous or subglabrous

42 Stems and petioles with patent hairs only

43 Dwarf plants up to 15 cm

44 Base of petioles reddish **57. filicaulis**

44 Base of petioles not reddish

45 Plant densely hairy **31. illyrica**

45 Plant rather sparsely hairy **58. minima**

43 Medium-sized plants up to 50 cm

46 Outer (spring) leaves glabrous above, inner (summer) leaves hairy above; flowers 2·5–3 mm wide **55. heterophylla**

46 All leaves ± hairy above; flowers at least 3 mm wide

47 Leaves lobed to $\frac{1}{2}$; lobes with distinct incisions **42. schistophylla**

47 Leaves lobed to $\frac{1}{3}$ or less; lobes with short or no incisions

48 Leaves ± orbicular, densely and evenly hairy above; lobes with equal teeth **41. monticola**

48 Leaves ± reniform, often sparsely or unevenly hairy above; lobes with unequal teeth **57. filicaulis**

42 Stems and petioles with erecto-patent hairs (sometimes mixed with patent hairs)

49 Leaves orbicular, with overlapping basal lobes

50 Leaf-lobes with no incisions **39. propinqua**

50 Leaf-lobes with rather long incisions **40. conglobata**

49 Leaves reniform or semicircular, with wide sinus

51 Leaves rather densely hairy, reniform; leaf-lobes with acute teeth **28. lapeyrousii**

51 Leaves rather sparsely hairy, almost semicircular; leaf-lobes with obtuse teeth **29. plicata**

22 Hypanthium glabrous or with sparse hairs

52 Epicalyx-segments at least as long as sepals; hypanthium much shorter than mature achene

53 Stems and inner (summer) petioles with erecto-patent
hairs, outer (spring) petioles glabrous   **106. aroanica**
53 Stems and all petioles with patent or slightly deflexed
hairs
54 Flowers 4·5–6·5(–7·2) mm wide; epicalyx-segments
mostly with 2–4 teeth   **96. achtarowii**
54 Flowers 3–5 mm wide; epicalyx-segments entire or with
1(–2) teeth
55 Leaves lobed to $\frac{2}{5}-\frac{1}{2}$; lobes with distinct incisions
105. peristerica
55 Leaves lobed to usually not more than $\frac{1}{3}$; lobes without
incisions
56 Leaves glabrous or sparsely hairy above
57 Leaves sparsely hairy beneath   **99. catachnoa**
57 Leaves densely hairy beneath
58 Epicalyx-segments often with 1 tooth; leaves
sparsely hairy above   **97. jumrukczalica**
58 Epicalyx-segments entire; leaves glabrous above
103. indivisa
56 Leaves densely hairy above
59 Upper cauline leaves glabrous above   **104. heterotricha**
59 All cauline leaves hairy above
60 Leaves lobed to $\frac{1}{5}(-\frac{1}{4})$ **98. mollis**
60 Leaves lobed to $(\frac{1}{4})\frac{1}{3}-\frac{2}{5}$ **107. zapalowiczii**
52 Epicalyx-segments usually shorter than sepals; hypan-
thium as long as mature achene
61 Petioles of outer (spring) leaves glabrous, those of inner
(summer) leaves hairy (Subser. *Heteropodae*)
62 Hairs on petioles of summer leaves erecto-patent
63 Leaf-lobes rounded; stems hairy up to the inflorescence
59. compta
63 Leaf-lobes truncate; stems glabrous at least in the
upper half
64 Robust, up to 50 cm; leaf-lobes with 6–12 teeth
60. rhododendrophila
64 Dwarf, up to 15 cm; leaf-lobes with 4–6 teeth
61. polonica
62 Hairs on petioles of summer leaves patent or deflexed
65 Leaves lobed to $\frac{2}{5}$ or more
66 Hairs on stems and petioles patent   **66. tatricola**
66 Hairs on stems and petioles deflexed
67 Dwarf; epicalyx-segments less than $\frac{1}{2}$ as long as
sepals   **62. decumbens**
67 Medium-sized; epicalyx-segments at least $\frac{1}{2}$ as long
as sepals   **63. undulata**
65 Leaves lobed to less than $\frac{2}{5}$
68 Stems glabrous, or only the 2 lowest internodes hairy
65. tirolensis
68 Stems hairy for at least part of their length above the
second internode
69 Leaves reniform, with very wide sinus   **64. rubristipula**
69 Leaves suborbicular or orbicular, with little or no
sinus
70 Cauline leaves lobed to not more than $\frac{1}{4}$; hypanthium rounded at base   **67. heteropoda**
70 Cauline leaves lobed to at least $\frac{1}{3}$; hypanthium
acute at base   **68. tenuis**
61 All petioles hairy (except sometimes the very first in the
season)
71 Very dwarf, rarely more than 5 cm
72 Leaves sericeous; leaf-lobes with 4–5 rather obtuse
teeth   **32. exigua**
72 Leaves not sericeous; leaf-lobes with 5–7 acute teeth
58. minima
71 Medium-sized or large, up to 70 cm
73 Leaves glabrous or subglabrous above, ± densely hairy
beneath
74 Flowers at least 3·5 mm wide   **56. curtiloba**
74 Flowers not more than 3 mm wide   **54. xanthochlora**
73 Leaves hairy above at least in folds, hairy or glabrous
beneath
75 Hairs on stems and petioles all patent or erecto-
patent

76 Hairs erecto-patent; hypanthium attenuate at base
53. gracilis
76 Hairs patent; hypanthium rounded at base
77 All leaves densely and evenly hairy on both surfaces
78 Leaves orbicular; leaf-lobes with equal teeth
41. monticola
78 Leaves ± reniform; leaf-lobes with unequal teeth
79 Inflorescence-branches ± densely hairy; flowers
usually at least 3 mm wide   **43. crinita**
79 Inflorescence-branches sparsely hairy; flowers
2–3 mm wide   **52. nemoralis**
77 Some leaves sparsely or unevenly hairy above
80 Leaf-lobes almost triangular; base of petioles not
reddish; hypanthium glabrous   **51. acutiloba**
80 Leaf-lobes rounded; base of petioles reddish;
hypanthium often somewhat hairy   **57. filicaulis**
75 At least some hairs on stems and petioles deflexed
81 Leaves rather sparsely hairy above, often only on
folds
82 Leaves flat, with basal lobes usually not touching
50. heptagona
82 Leaves undulate, with basal lobes touching
83 Leaf-lobes without incisions; flowers 3–4 mm wide
48. subcrenata
83 Leaf-lobes with short incisions; flowers 2·5–3 mm
wide   **49. cymatophylla**
81 Leaves densely and evenly hairy above
84 Leaves reniform, with wide sinus   **43. crinita**
84 Leaves orbicular or suborbicular, with little or no
sinus
85 Flowers 1·5–2·5 mm wide   **47. tytthantha**
85 Flowers at least 3 mm wide
86 Hypanthium attenuate at base; inflorescence
often glabrous or nearly so   **44. strigosula**
86 Hypanthium rounded at base; at least main inflorescence-branches hairy
87 Leaves orbicular, with basal lobes overlapping
45. subglobosa
87 Leaves suborbicular, with basal lobes not overlapping   **46. sarmatica**
21 Stems and petioles glabrous, or with appressed or subappressed hairs
88 Epicalyx-segments usually at least as long as sepals; hypanthium much shorter than mature achene
89 Whole plant glabrous (rarely a few hairs on summer
petioles)   **118. fissa**
89 At least the lowest internode and some of the petioles hairy
90 Plant ± densely subappressed-hairy throughout (Ural)
108. haraldii
90 At least pedicels and upper inflorescence-branches
glabrous
91 Leaf-lobes with long, almost U-shaped incisions
117. incisa
91 Leaf-lobes with V-shaped incisions or without incisions
92 Leaves lobed to $\frac{1}{3}-\frac{2}{5}(-\frac{1}{2})$; teeth long, equal   **116. pyrenaica**
92 Leaves lobed to less than $\frac{1}{3}$ (rarely to $\frac{1}{3}$); teeth medium
or short, often unequal
93 Flowers at least 4·5 mm wide; epicalyx-segments distinctly longer than sepals   **115. oculimarina**
93 Flowers usually less than 4·5 mm wide; epicalyx-
segments usually about as long as sepals, sometimes
slightly longer
94 Leaves very shallowly lobed to not more than $\frac{1}{5}$;
robust, up to 60 cm   **112. gorcensis**
94 Leaves lobed to more than $\frac{1}{5}$; usually medium-sized
or rather dwarf, not more than 30 cm
95 Epicalyx-segments slightly shorter than sepals
96 Leaf-lobes rounded   **109. fallax**
96 Leaf-lobes subtruncate   **110. sericoneura**
95 Epicalyx-segments at least as long as sepals
97 Rather robust; leaf-lobes with no incisions
111. flexicaulis
97 Dwarf; leaf-lobes with incisions

98 Flowers at least 4 mm wide; hairs on lowest 1–2(–3) internodes **113. cuspidens**

98 Flowers 3–4 mm wide; hairs on lowest internode only **114. othmarii**

88 Epicalyx-segments usually shorter than sepals; hypanthium as long as mature achene

99 Stems, up to and including at least the main inflorescence-branches, with some appressed or subappressed hairs

100 Whole plant, including pedicels, densely sericeous with subappressed hairs **27. bulgarica**

100 Plant glabrous or subglabrous on at least the upper surface of leaves or some pedicels or hypanthia

101 All hypanthia ±densely hairy

102 Leaves ±evenly hairy above

103 Leaves orbicular **19. fulgens**

103 Leaves reniform or semicircular

104 Leaf-lobes with short incisions **28. lapeyrousii**

104 Leaf-lobes with long incisions

105 Hairs subappressed; leaf-lobes with obtuse teeth **29. plicata**

105 Hairs appressed; leaf-lobes with acute teeth **70. camptopoda**

102 Leaves glabrous above or with hairs only along folds

106 Leaf-lobes 7 **34. vetteri**

106 Leaf-lobes 9–11

107 Leaf-lobes ±truncate, with distinct incisions **16. infravalesiaca**

107 Leaf-lobes not truncate, without incisions **17. schmidelyana**

101 At least some hypanthia glabrous or subglabrous

108 Leaf-lobes without incisions

109 Hypanthia attenuate at base **53. gracilis**

109 Hypanthia rounded at base

110 Leaf-lobes with wide, obtuse, unequal teeth; leaves often hairy above **69. glomerulans**

110 Leaf-lobes with narrow, acute, subequal teeth; leaves glabrous above **78. murbeckiana**

108 Leaf-lobes with incisions

111 Leaf-lobes 9–11; medium-sized, up to 30 cm **71. crebridens**

111 Leaf-lobes 5–7(–9); dwarf, not more than 15 cm

112 Pedicels mostly glabrous **72. buschii**

112 Pedicels hairy **20. kerneri**

99 Hairs on stems confined to lowest 2(–3) internodes

113 Leaves hairy at least on folds above

114 Petioles glabrous (rarely the latest summer petioles with some hairs)

115 Leaves orbicular; lobes with distinct incisions **84. glabricaulis**

115 Leaves reniform; lobes without incisions **85. versipila**

114 Petioles hairy

116 Hypanthium acute at base, attenuate into pedicel **74. wallischii**

116 Hypanthium rounded at base, clearly demarcated from pedicel

117 Leaves orbicular, with overlapping basal lobes **73. controversa**

117 Leaves reniform to suborbicular, with distinct sinus

118 Leaves lobed to *c.* $\frac{1}{2}$ **14. paicheana**

118 Leaves lobed to $\frac{1}{4}$–$\frac{1}{3}$ **75. connivens**

113 Leaves glabrous above

119 Petioles glabrous (rarely the latest summer petioles with some hairs)

120 Leaves lobed to not more than $\frac{1}{5}$–$\frac{1}{4}$

121 Outer leaves very shallowly lobed to *c.* $\frac{1}{7}$ **88. stanislaae**

121 All leaves lobed to $\frac{1}{5}$–$\frac{1}{4}$

122 Leaf-lobes with distinct incisions **86. coriacea**

122 Leaf-lobes without incisions

123 Leaves orbicular, without sinus **87. inconcinna**

123 Leaves reniform to suborbicular, with small or medium-sized sinus **83. glabra**

120 Leaves lobed to at least $\frac{1}{4}$

124 Leaf-lobes with distinct incisions

125 Leaves orbicular **93. demissa**

125 Leaves ±reniform

126 Leaves lobed to $\frac{1}{4}$–$\frac{1}{3}$ **89. trunciloba**

126 Leaves lobed to *c.* $\frac{2}{5}$

127 Leaf-lobes with equal teeth **90. sinuata**

127 Leaf-lobes with ±unequal teeth **94. pseudincisa**

124 Leaf-lobes without incisions

128 Medium-sized, with erect or ascending stems up to 40 cm **91. straminea**

128 Dwarf, with procumbent stems up to 25 cm

129 Leaves orbicular; lobes semicircular **92. aequidens**

129 Leaves reniform; lobes ovate **95. longana**

119 Petioles hairy

130 Epicalyx-segments less than $\frac{2}{3}$ as long as sepals

131 Leaves orbicular or suborbicular; leaf-lobes with distinct incisions **15. splendens**

131 Leaves reniform; leaf-lobes with small incisions **18. jaquetiana**

130 Some epicalyx-segments at least $\frac{2}{3}$ as long as sepals

132 Leaf-lobes with distinct incisions

133 Leaves thick, suborbicular, with narrow sinus; rather robust **76. baltica**

133 Leaves thin, orbicular, without sinus; usually rather slender **77. wichurae**

132 Leaf-lobes without incisions or with indistinct incisions

134 Leaf-lobes with narrow, acute, subequal teeth

135 Leaf-lobes with short incisions (often ±obscured by overlapping margins of lobes) **75. connivens**

135 Leaf-lobes without incisions

136 Leaves lobed to $\frac{2}{5}$–$\frac{1}{2}$ **79. acutidens**

136 Leaves lobed to $\frac{1}{4}$–$\frac{1}{3}$ **78. murbeckiana**

134 Leaf-lobes with wide, obtuse or subacute, usually unequal teeth

137 Sepals longer than wide; stems often glabrous above the second internode **83. glabra**

137 Sepals not or scarcely longer than wide; stems usually hairy in the lower part

138 Inflorescence narrow, usually much exceeding leaves **82. obtusa**

138 Inflorescence with divaricate branches, not much exceeding leaves

139 Flowers 4–5 mm wide **80. reniformis**

139 Flowers 3–3·5 mm wide **81. lineata**

Sect. PENTAPHYLLEAE Camus. Stoloniferous, rooting at the nodes. Leaves with almost separate segments, the middle segment obovate-cuneate. Fully sexual.

1. A. pentaphyllea L., *Sp. Pl.* 123 (1753). Dwarf, subglabrous. Flowering stems up to 15 cm long, procumbent or ascending. Leaves up to 3 × 3 cm; segments 3 or 5 (rarely 7), almost separate, obovate-cuneate, deeply incise-dentate in the distal half, often with a very few hairs mainly on the margins. Stipules lanceolate or ovate-lanceolate, conspicuous, especially on cauline leaves. Inflorescence few-flowered. Sepals *c.* 1·5 mm, 4 or 5, ovate, much larger than epicalyx-segments; anthers 4 or 5; pollen normal. Hypanthium *c.* 2 mm, glabrous. $2n = 64$. *North-facing slopes and snow-patches, usually above* 2000 *m; calcifuge.* ● *Alps, eastwards to c.* 11° *E. in Italy.* Au Ga He It.

Sect. ALCHEMILLA (*Brevicaulon* Rothm.). Stock woody, not stoloniferous. Leaves variably lobed, sometimes with free segments but the middle segment not cuneate. Almost wholly apomictic.

Subsect. *Chirophyllum* Rothm. (Sect. *Alpinae* Buser). Dwarf or medium-sized, rarely more than 30 cm, more or less densely sericeous. Leaves palmate or deeply palmately divided (always to at least ½), with dense (rarely sparse), appressed hairs beneath. Epicalyx-segments less than half as long as sepals. Mature achene wholly enclosed in hypanthium.

Series *Saxatiles* Buser. Dwarf, strongly rhizomatous. Leaves palmate with 5–7 leaflets, glabrous above, at least the middle leaflet practically free. Pedicels usually shorter than hypanthium. Sepals erect after flowering.

2. **A. saxatilis** Buser, *Not. Alchim.* 3 (1891). Stems erect, usually at least 3 times as long as the leaves. Leaflets 5, thick, dark green, broadly elliptical, with small and inconspicuous teeth on the rounded, distal margin. Inflorescence long, with distinct, separate glomeruli. *Siliceous rocks.* ● *Mountains of S. & C. Europe from Spain to Jugoslavia.* Co Ga Ge He Hs It Ju.

Related species include:

A. obovalis Buser in Dörfler, *Herb. Norm.* **47**: 198 (1906). *S.W. Alps (Col de Lautaret).* Ga.

3. **A. transiens** (Buser) Buser in Dörfler, *Herb. Norm.* **36**: 204 (1906). Like **2** but leaflets sometimes 6–7, ovate-lanceolate, and with more distinct, acute, distal teeth. *Siliceous rocks.* ● *Mountains of S. Europe, from Portugal to N. Italy.* Co Ga He Hs It Lu.

Intermediate between **2** and **4**. Related species include:

A. lucida Buser in Dörfler, *Herb. Norm.* **47**: 201 (1906). *S.W. Alps (Col de Lautaret).* Ga.
A. saxetana Buser in Dörfler, *op. cit.* 199 (1906). *W.C. Alps (Valais).* He.

4. **A. alpina** L., *Sp. Pl.* 123 (1753) (*A. alpina* subsp. *glomerata* (Tausch) Camus). Stems ascending, usually scarcely exceeding the leaves. Leaflets 5–7, green or yellow-green, lanceolate, with few but conspicuous teeth. Glomeruli dense. $2n = c. 120, c. 128, c. 140, c. 152.$ *Mainly calcifuge in the southern part of its range.* ● *N., W. & W.C. Europe, southwards to S. Spain, Appennini and E. Alps; only on mountains except in the extreme north.* Au Br Co Fa Fe Ga Ge Hb He Hs Is It No Rs (N) Sb Su.

Related species include:

A. argentidens Buser in Dörfler, *Herb. Norm.* **47**: 200 (1906). *W.C. Alps (Valais).* He.
A. brachyclada Buser in Dörfler, *op. cit.* 199 (1906). *S.W. Alps (Isère).* Ga.
A. viridicans Rothm., *Bol. Soc. Esp. Hist. Nat.* **34**: 150 (1934). *E. Pyrenees (Prov. Lérida).* Hs.

Other regional variation certainly occurs, but has not been treated taxonomically. In N. Europe the variation is less, but still detectable.

5. **A. basaltica** Buser, *Österr. Bot. Zeitschr.* **44**: 476 (1894). Like **4** but stems usually greatly exceeding leaves; leaflets broadly obovate, with truncate apex and rather long teeth. *Siliceous rocks.* ● *Mountains of S.W. & W.C. Europe, from Spain to E. Switzerland.* Ga He Hs.

Intermediate between **4** and **6**.

6. **A. subsericea** Reuter, *Compt. Rend. Soc. Hallér.* 2: 20 (1853–4). Stems not much exceeding leaves. Leaflets usually 5, lanceolate, grey-green and only sparsely sericeous beneath, with

conspicuous, straight teeth up to 3 mm. Glomeruli less dense than **4**. *Calcifuge; often in snow-patches.* ● *Mountains of S. Europe, from N. Spain to W. Austria.* Au Ga He Hs It.

Related species include:

A. amphibola Buser in Dörfler, *Herb. Norm.* **47**: 202 (1906). *W. Alps (Haute Savoie).* Ga. Exceptional in Ser. *Saxatiles* in growing on calcareous rock.
A. jucunda Buser ex Maillefer, *Mém. Soc. Vaud. Sci. Nat.* **8**: 116 (1944). *Alpes Maritimes.* Ga.
A. vaccariana Buser, *Bull. Soc. Bot. Ital.* **1906**: 61 (1906). *W.C. Alps (Valle d'Aosta).* It.

Maillefer suggests that these taxa may have arisen by hybridization between *A. pentaphyllea* and members of Ser. *Saxatiles* (though none of the latter are known to be sexual). See also **12**.

Series *Hoppeanae* Buser. Dwarf to medium-sized, with short rhizomes. Leaf-segments (5–)7–9, mostly slightly to distinctly connate. Pedicels usually at least as long as hypanthium. Sepals patent after flowering.

All species in this series are calcicole, and occur on mountains on rocks or in pastures.

7. **A. plicatula** Gand., *Rad Jug. Akad. Znan. Umj.* **66**: 34 (1883) (*A. hoppeana* subsp. *asterophylla* (Tausch) Gams). Leaves orbicular, densely sericeous beneath; segments 7(–9), lanceolate, with short, acute teeth, middle one usually completely free; segments in the young leaf usually distinctly folded along midrib. ● *Mountains of S. & S.C. Europe from the Alps and S. Carpathians to S. Spain and Albania.* Al Au Ga Ge He Hs It Ju Rm.

Variable. The most widespread taxon is var. *plicatula* (*A. visianii* Gand., *A. asterophylla* (Tausch) Buser, *A. alpigena* Buser), with the leaves glabrous above; in the south of the area is found var. *vestita* (Buser) Rothm. (*A. amphisericea* Buser), with the leaves more or less hairy above.

Related species with leaves glabrous above include:

A. buseri Maillefer, *Mém. Soc. Vaud. Sci. Nat.* **8**: 120 (1944). *S.W. & W.C. Alps (Isère; Vaud).* Ga He.
A. chirophylla Buser, *Bull. Soc. Nat. Ain* **13**: 24 (1903). *Jura (Ain).* Ga.
A. florulenta Buser in Briq., *Prodr. Fl. Corse* **2**(**1**): 205 (1913). *Jura (Ain).* Ga.
A. font-queri Rothm., *Bol. Soc. Esp. Hist. Nat.* **34**: 151 (1934). *S. Spain (Sierra Nevada).* Hs.
A. nitida Buser, *Bull. Soc. Nat. Ain* **13**: 33 (1903). *From S.C. France to N. Italy.* Ga He It.
A. scintillans Buser ex Jaquet, *Mém. Soc. Fribourg. Sci. Nat. (Bot.)* **2**: 62 (1907). *Jura; W.C. Alps (Vaud).* Ga He.

Related species with leaves hairy above include:

A. coruscans Buser in Dörfler, *Herb. Norm.* **47**: 205 (1906). *W.C. Alps (Isère).* Ga.
A. murisserica Maillefer, *Mém. Soc. Vaud. Sci. Nat.* **8**: 123 (1944). *W.C. Alps (Vaud).* He.
A. petraea Buser ex Maillefer, *loc. cit.* 122 (1944). *Jura (Ain).* Ga.

8. **A. pallens** Buser, *Not. Alchim.* 6 (1891). Leaves more or less orbicular, flat, not folded, glabrous above, only sparsely sericeous and pale blue-green beneath; segments 7(–9), elliptical, shortly but distinctly connate at the base, with the teeth shorter and

wider than in **7**. ● *Alps, eastwards to c. 12° 30′ E.; mountains of E. & S.C. France.* Au Ga Ge He It.

Related species include:

A. atrovirens Buser ex Jaquet, *Mém. Soc. Fribourg. Sci. Nat. (Bot.)* **2**: 4 (1905). *W.C. Alps (Fribourg).* He.

A. flavovirens Buser, *Bull. Soc. Nat. Ain* **13**: 33 (1903). *S.W. Alps (Col de Lautaret).* Ga.

A. longinodis (Buser) Maillefer, *Mém. Soc. Vaud. Sci. Nat.* **8**: 122 (1944). *W.C. Alps (Valais).* He.

9. A. anisiaca Wettst., *Biblioth. Bot. (Stuttgart)* **26**: 41 (1892) (*A. hoppeana* subsp. *anisiaca* (Wettst.) Gams). Leaves more or less semicircular, with very wide sinus, usually hairy above, densely sericeous beneath; segments narrowly lanceolate, shortly connate at base. ● *N.E. Alps.* Au Ge.

10. A. hoppeana (Reichenb.) Dalla Torre in Hartinger, *Atlas Alpenfl. (Text)* 94 (1882). Leaves orbicular, with more or less overlapping outer segments, glabrous above, sparsely appressed-hairy and greenish beneath; segments distinctly connate at base, linear to oblong-linear, with very obtuse apex and small, short apical teeth. ● *N. Alps, Jura, S.W. Germany.* Au Ge He.

11. A. conjuncta Bab., *Ann. Nat. Hist.* **10**: 25 (1842). Up to 40 cm. Leaves thick, orbicular, dull, blue-green and glabrous above, sericeous and shiny beneath; segments 7(–9), flat, elliptical, connate to $\frac{1}{3}(-\frac{1}{2})$, with the teeth very indistinct and almost hidden by sericeous marginal indumentum. ● *Jura and S.W. Alps; widely cultivated in gardens and naturalized in Britain.* Ga He [Br].

More robust than any other species in Subsect. *Chirophyllum.*

Related species include:

A. leptoclada Buser, *Alchim. Valais.* 4 (1894). *S.W. & W.C. Alps.* Ga He It.

12. A. grossidens Buser, *Not. Alchim.* 6 (1891) (*A. hoppeana* subsp. *grossidens* (Buser) Gams). Leaves suborbicular, grey-green and only sparsely sericeous beneath; segments 7, obovate, subtruncate, the middle segment almost free, with conspicuous straight teeth up to 3 mm. $2n=64$. ● *Alps, eastwards to c. 10° E. in Austria.* Au Ga Ge He It.

Related species include:

A. glacialis Buser, *Bull. Herb. Boiss.* ser. 2, **5**: 514 (1905). *S.E. Alps.* He It.

A. jugensis (Buser) Maillefer, *Mém. Soc. Vaud. Sci. Nat.* **8**: 120 (1944). *S.W. & W.C. Alps (Haute Savoie, Vaud).* Ga He.

A. grossidens and *A. glacialis* have been shown to be sexual species, and the latter seems to hybridize freely with *A. pentaphyllea* on the Gemmijoch. The whole group may well have originated by hybridization between *A. pentaphyllea* and members of Ser. *Hoppeanae.* See also **6**.

A. catalaunica Rothm., *Bol. Soc. Esp. Hist. Nat.* **34**: 150 (1934), described from the Spanish Pyrenees, and **A. petiolulans** Buser, *Bull. Soc. Nat. Ain* **13**: 24 (1903), described from Switzerland, have the prominent teeth of **12**, but the leaves have dense, sericeous hairs beneath.

Subsect. *Heliodrosium* Rothm. Very variable in habit and hairiness. Leaves rarely lobed to more than $\frac{1}{2}$, and never with separate segments. Epicalyx-segments at least half as long as sepals, but not longer and always narrower. Mature achene wholly enclosed in hypanthium.

Series *Splendentes* Buser. Like Subsect. *Chirophyllum* but leaves usually not lobed to more than $\frac{1}{2}$, and epicalyx-segments at least half as long as sepals.

A small group of species confined, except for **13**, to the Alps, Jura and Pyrenees, and possessing characters intermediate between Subsect. *Chirophyllum* and Subsect. *Heliodrosium.* They are presumably the products of ancient hybridization between members of these subsections. All have a compact, woody rhizome, and relatively short and usually slender stems not more than 30 cm. They are mostly calcicole.

13. A. faeroensis (Lange) Buser, *Ber. Schweiz. Bot. Ges.* **4**: 58 (1894). Dwarf but relatively robust, up to 15(–30) cm; sericeous throughout except for upper surface of leaves and some pedicels, which are usually glabrous. Leaves reniform, with very wide sinus, lobed always to more than $\frac{1}{2}$ and up to $\frac{2}{3}$; lobes usually 7, with long incisions and rather large, acute teeth extending more than halfway down side of lobe. Inflorescence compact. Epicalyx-segments not much more than $\frac{1}{2}$ as long as sepals. $2n=c.$ 220, *c.* 224. ● *Faeroes, E. Iceland.* Fa Is.

14. A. paicheana (Buser) Rothm., *Feddes Repert.* **66**: 226 (1962). Like **13** but more sparsely hairy; stems more slender; leaves lobed to $\frac{1}{2}$, sparsely appressed-hairy above, sparsely hairy and greenish beneath; epicalyx-segments distinctly longer, up to as long as sepals; inflorescence laxer. ● *W.C. Alps (Valais).* He.

15. A. splendens Christ ex Favrat, *Bull. Soc. Vaud. Sci. Nat.* **25**: 52 (1889). Stems up to 30 cm, slender, with few, small cauline leaves. Often glabrous except for sparse, appressed hairs on petioles, veins on lower surface of leaf and lowest internodes. Leaves orbicular or suborbicular, always glabrous above, lobed to $\frac{1}{4}-\frac{1}{3}$; lobes 9–11, subquadrangular, more or less truncate, with distinct incisions and 7–9 small, acute teeth. $2n=c.$ 163. ● *C. Alps; Jura.* Ga He.

16. A. infravalesiaca (Buser) Rothm., *Feddes Repert.* **66**: 226 (1962). Like **15** but with denser, appressed hairs often extending to the pedicels and flowers; and leaves lobed up to $\frac{1}{2}$. ● *W.C. Alps (Valais).* He.

17. A. schmidelyana Buser, *Not. Alchim.* 15 (1891). Stems up to 20 cm, more or less densely appressed-hairy up to and including the inflorescence-branches, with large cauline leaves. Leaves orbicular, lobed to $\frac{1}{3}-\frac{1}{2}$; lobes 9–11, subtriangular, without incisions and with *c.* 6 large, acute teeth. Leaves glabrous above, sparsely hairy beneath. ● *Jura; W. Alps (Savoie).* Ga He.

18. A. jaquetiana Buser, *Bull. Herb. Boiss.* ser. 2, **2**: 619 (1902). Like **17** but very sparsely appressed-hairy, the mature leaves and distal portions of stems subglabrous; leaves more or less reniform, with wide sinus and short incisions. ● *W.C. Alps (Fribourg).* He.

19. A. fulgens Buser, *Not. Alchim.* 15 (1891). Stems up to 30 cm, more or less densely sericeous throughout, but pedicels often subglabrous. Leaves orbicular, densely sericeous, often silvery beneath and sparsely appressed-hairy above, lobed to *c.* $\frac{1}{4}$; lobes 9–11, rounded, with short incisions and 7–9 equal, acute teeth. $2n=c.$ 140. ● *Pyrenees.* Ga Hs.

20. A. kerneri Rothm., *Feddes Repert.* **66**: 226 (1962). Stems up to 15 cm, densely sericeous throughout, including pedicels, but some flowers subglabrous. Leaves suborbicular with narrow sinus, glabrous above, sericeous beneath, lobed up to $\frac{2}{5}$; lobes 5–7(–9), subtruncate, with distinct incisions and 4–6 subobtuse teeth. ● *N.E. Alps, very local.* Au Ge.

Series *Pubescentes* Buser. Dwarf or medium-sized, usually with dense, soft, more or less patent hairs often covering all parts of the plant. Leaves less than 6 cm wide, usually shallowly palmately lobed to less than $\frac{1}{2}$; lobes with 4–5(–6) usually rather wide and obtuse teeth. Epicalyx-segments usually slightly shorter and narrower than sepals.

Characteristic of rather dry grassland or open, rocky habitats, in the lowlands in N. Europe and submontane in the south; not markedly synanthropic.

21. **A. glaucescens** Wallr., *Linnaea* 14: 134 (1840) (*A. hybrida* auct., non (L.) L., *A. minor* auct., *A. pubescens* auct., non Lam.). Rather small, up to 20 cm, usually compact, with neat, orbicular, densely hairy leaves, often sericeous; hairs patent or erecto-patent on all parts of the plant. Leaf-lobes (5–)7–9, rounded, with no incisions and 4–6 teeth. Stipules of cauline leaves dentate, not conspicuous; stipules, base of stem and petioles brownish, not reddish or purplish-red. Inflorescence with dense glomeruli. $2n = 103$–110. ● *Mainly in C. & N.E. Europe, but extending locally to Ireland, C. France, N. Italy, Bulgaria and Krym.* Au Be Br Bu Co Cz Da Fe Ga Ge Hb He Ho Hu It Ju No Po Rm Rs (N, B, C, W, K) Su.

22. **A. hirsuticaulis** H. Lindb., *Meddel. Soc. Fauna Fl. Fenn.* 30: 143 (1904). Robust, medium-sized, up to 40 cm, the leaves with dense hairs but not sericeous, the stems with patent hairs throughout. Leaf-lobes 5–7(–9), rounded, with long incisions and 4–6 teeth. Stipules of cauline leaves deeply laciniate, rather conspicuous. Inflorescence with very dense glomeruli. *N.E. Europe.* Fe Rs (N, B, C, E).

23. **A. flabellata** Buser, *Not. Alchim.* 12 (1891). Dwarf, up to 15 cm, with densely sericeous leaves and short, slender stems with erecto-patent or patent hairs throughout. Leaves lobed to $\frac{1}{4}$–$\frac{1}{3}$(–$\frac{2}{5}$); lobes 5–7, truncate, wider than long, with long incisions and 2–4(–6) teeth. Stipules, base of stems and petioles brownish. Inflorescence with dense glomeruli; pedicels more or less equalling hypanthium. *Usually calcifuge.* ● *Mountains of S. & C. Europe from the Pyrenees to the Carpathians and Krym.* Au Bu Cz Ga Ge He Hs It Ju Po Rm Rs (W, K).

24. **A. cinerea** Buser, *loc. cit.* (1891) (*A. lanuginosa* Rothm.). Like 23 but leaf-lobes rounded, with short incisions; hairs dense and patent throughout; pedicels often longer than hypanthium. ● *Alps, N. Appennini, mountains of N. half of Balkan peninsula.* Al Au Bu Ga It Ju.

Somewhat intermediate between 23 and 30.

Related species include:

A. pirinica Pawł., *Bull. Int. Acad. Sci. Cracovie* ser. B (1), 1951: 350 (1953). *S.W. Bulgaria (Pirin).* Bu.

25. **A. erythropoda** Juz. in Grossh., *Fl. Kavk.* 4: 323 (1934) (incl. *A. erythropodoides* Pawł., *A. jailae* Juz.). Rather robust, medium-sized, up to 30 cm. Stems and petioles with a mixture of patent and deflexed hairs, the latter particularly obvious on petioles of summer leaves and on the lower internodes. Leaves with dense, soft hairs; lobes usually 5–7, subtruncate, with long incisions and 3–6 obtuse to subacute teeth. Flowers up to 4 mm

wide, yellowish, in rather dense glomeruli. *W. Carpathians; mountains of Balkan peninsula; Krym.* Bu Cz Ju Rs (K). (*Caucasus.*)

Described by Juzepczuk, from the Caucasus, as developing a reddish-violet colour on the stems and mature petioles. Some, but not all, of the plants in the Balkan peninsula show this character.

26. **A. lithophila** Juz., *Not. Syst. (Leningrad)* 8: 12 (1938). Like 25 but leaf-lobes rounded or subtriangular, with very short incisions and 5–6 acute teeth. ● *Krym.* Rs (K).

Related species include:

A. exsanguis Juz., *Not. Syst. (Leningrad)* 8: 10 (1938) *Krym.* Rs (K).

27. **A. bulgarica** Rothm., *Feddes Repert.* 46: 125 (1939). Rather small, up to 20 cm, sericeous, with subappressed or erecto-patent hairs throughout. Leaves suborbicular, lobed up to $\frac{2}{5}$; lobes 5–7(–9), truncate or subtruncate, with long incisions and 4–6 rather narrow, acute teeth. Inflorescence small, with few, rather dense glomeruli. ● *Mountains of Balkan peninsula.* Bu Gr Ju.

Related species include:

A. consobrina Juz., *Not. Syst. (Leningrad)* 16: 143 (1954). *S. Ural.* Rs (C).
A. exul Juz., *Not. Syst. (Leningrad)* 14: 147 (1951). *S. Ural.* Rs (C).
A. helenae Juz., *Not. Syst. (Leningrad)* 16: 142 (1954). *S. Ural.* Rs (C).

These 4 species are closely related to **A. sericata** Reichenb., *Pl. Crit.* 1: 6 (1823) *sens. lat.* (incl. *A. rigida* Buser) from the Caucasus, which is commonly cultivated in gardens and has been recorded as an escape.

28. **A. lapeyrousii** Buser, *Bull. Herb. Boiss.* 1, **App. 2**: 18 (1893) (*A. hybrida* (L.) L., nom. ambig.). Medium-sized, up to 25 cm, with erecto-patent or subappressed hairs throughout, except on pedicels which are usually glabrous; in general less densely hairy than 21. Stems rather slender and diffuse. Leaves reniform, lobed to *c.* $\frac{1}{3}$; lobes 7–9, usually more or less triangular, with short incisions and 5–6 acute, somewhat connivent teeth. ● *Pyrenees and mountains of S.C. France.* Ga Hs.

29. **A. plicata** Buser, *op. cit.* 20 (1893). Like 28 but less densely hairy, so that neither leaf-surface is sericeous and shiny at maturity; leaf-lobes with rather long incisions (often obscured by folding) and obtuse teeth; only the lowest pedicels in the glomeruli hairy. ● *N.E. & C. Europe.* Au Cz Fe Ga Ge He Hu No Po Rm Rs (N, B, C, W) Su.

Related species include:

A. hungarica Soó, *Acta Bot. Acad. Sci. Hung.* 9: 424 (1963). *N. Hungary and S.E. Czechoslovakia.* Cz Hu.

30. **A. colorata** Buser, *Not. Alchim.* 10 (1891). Dwarf, up to 15 cm, more or less densely covered with soft hairs, some of which are slightly deflexed on the lower internodes and on mature petioles. Leaves reniform to suborbicular, somewhat undulate; lobes 7(–9), with short incisions. Stems quickly developing a reddish-purple colour. Pedicels glabrous, except for some lower flowers of glomeruli. *Usually calcicole.* ● *Mountains of S. & S.C. Europe from the Pyrenees to the Carpathians.* Al Au Cz Ga Ge He Hs It Ju Po Rm.

Very similar in geographical range to **23**, but usually rather clearly differentiated by soil preference.

Related species include:

A. **exilis** Juz., *Acta Horti Petrop.* 43: 537 (1931) (*A. egens* Juz.). *C. Russia (Ul'janovskaja Obl.).* Rs (C).

31. A. illyrica Rothm., *Feddes Repert.* **66**: 227 (1962). Like **30** but hairs patent, not deflexed; leaves more deeply lobed, the lobes more nearly truncate. ● *Mountains of N.W. Jugoslavia.* Ju.

32. A. exigua Buser ex Paulin, *Jahresb. Staatsgymn. Laibach* **1907**: 11 (1907) (*A. pusilla* Buser, non Pomel). Like **30** but very dwarf, up to 5(–10) cm; distal part of stems, including flowers, glabrous or subglabrous; hairs patent and sparser throughout. *Mountain grassland, often in damp hollows.* ● *Alps and mountains of Jugoslavia; E. Carpathians.* Au Ga Ge It Ju Rm.

33. A. helvetica Brügger, *Jahresb. Naturf. Ges. Graubündens* **23–24**: 64 (1880) (*A. intermedia* Haller fil.). Dwarf, with relatively sparse patent hairs except on the upper surface of the leaves and the pedicels, which are subglabrous. Leaves lobed to $\frac{1}{2}$–$\frac{2}{3}$; lobes 5–7, wide, with very long incisions and large distal teeth 2 mm or more. Inflorescence rather small. Flowers up to 4 mm wide. *Snow-patches.* ● *C. & E. Alps.* Au Ge He It.

A. **intermedia** subsp. **sooi** Palitz, *Feddes Repert.* **40**: 244 (1936), described from Romania (Transsilvania), seems to differ in having shallower leaf-lobes, and denser hairs on pedicels and upper leaf surface. It would probably be best treated as a separate species, but the information is insufficient.

34. A. vetteri Buser, *Bull. Herb. Boiss.* **2, App. 4**: 7 (1894). Medium-sized, up to 25 cm, sericeous and almost shiny, with subappressed hairs throughout, but mature summer leaves either glabrous above, or with a few hairs along the folds. Stems slender. Leaves lobed to c. $\frac{2}{5}$; lobes usually 7, rounded, with long incisions and narrow, acute teeth; cauline leaves small. ● *Mountains of S.W. Europe from Spain (Sierra de Gudar) to Maritime Alps.* Ga Hs It.

This species is intermediate between. Ser. *Pubescentes* and Ser. *Splendentes*, and might be included in either. It also resembles **27**, differing mainly in the leaves, which are subglabrous above.

Series *Vulgares* Buser. Very variable in habit and hairiness, but often robust and relatively glabrous. Leaves palmately lobed to less than $\frac{1}{2}$ (very rarely more); leaf-lobes with often 6 or more, variably shaped (but often acute) teeth.

A very heterogeneous collection of species which are usually grouped in subseries based on the degree of hairiness. This artificial grouping seems to be the only practicable one.

Subser. *Hirsutae* H. Lindb. Stem, at least on lower internodes and petioles, with more or less dense, erecto-patent, patent or deflexed hairs. Leaves with some erecto-patent or patent hairs.

35. A. hebescens Juz., *Acta Horti Petrop.* **43**: 537 (1931). Medium-sized plant up to 30 cm, with dense, deflexed hairs throughout except for pedicels, which are usually glabrous at least in their distal half. Leaves reniform, flat, shallowly lobed; lobes 9(–11), subtruncate, with distinct incisions and 6–8 rather small teeth. *C. Russia.* Rs (C). (*C. Asia.*)

Related species include:

A. **aemula** Juz., *Acta Inst. Bot. Acad. Sci. URSS* **1**: 120 (1933). *Krym.* Rs (K).
A. **pseudocalycina** Juz., *Not. Syst. (Leningrad)* **17**: 243 (1955). *S. Ural.* Rs (C).
A. **pycnantha** Juz., *Not. Syst. (Leningrad)* **8**: 15 (1938). *Krym.* Rs (K).

36. A. gibberulosa H. Lindb., *Acta Soc. Sci. Fenn.* **37(10)**: 4 (1909). Like **35** but leaves orbicular and strongly undulate; lobes rounded, with short or no incisions and rather long teeth; basal lobes often overlapping; hairs, especially on pedicels and hypanthia, with a small tubercle at base. ● *N.C. Russia.* Rs (N, C).

37. A. bungei Juz., *Animadv. Syst. Herb. Univ. Tomsk.* **1932 (5–6)**: 2 (1932) (*A. barbulata* Juz.). Like **35** but pedicels more often completely glabrous; leaves orbicular or reniform-orbicular; lobes 7–9, rounded, with short, indistinct incisions and 6–9 rather wide, acute teeth. *C. & S. Ural.* Rs (C). (*W. Siberia.*)

Related species include:

A. **argutiserrata** H. Lindb. ex Juz., *Animadv. Syst. Herb. Univ. Tomsk.* **1932(5–6)**: 4 (1932). *S. Ural.* Rs (C). (*W. Siberia.*)
A. **cheirochlora** Juz., *Not. Syst. (Leningrad)* **14**: 153 (1951). *E.C. Russia (Tatarskaja A.S.S.R.).* Rs (C).
A. **dasycrater** Juz., *op. cit.* 151 (1951). *E.C. Russia (Tatarskaja A.S.S.R.).* Rs (C).
A. **glyphodonta** Juz., *Not. Syst. (Leningrad)* **16**: 143 (1954). *C. Russia (Ivanovskaja Obl.).* Rs (C).
A. **macrescens** Juz., *Not. Syst. (Leningrad)* **14**: 148 (1951). *E.C. Russia (Tatarskaja A.S.S.R.).* Rs (C).
A. **oligantha** Juz., *Not. Syst. (Leningrad)* **17**: 245 (1955). *S. Ural.* Rs (C).
A. **trichocrater** Juz., *Sched. Herb. Fl. URSS* **14(82)**: 55 (1957). *E.C. Russia (Kirovskaja Obl.).* Rs (C).

38. A. hirsutissima Juz., *Not. Syst. (Leningrad)* **8**: 16 (1938). Like **35** but very densely hairy except for pedicels; leaves very shallowly lobed; epicalyx-segments often equalling sepals. ● *Krym.* Rs (K).

The large epicalyx gives this plant a characteristic appearance; it might be included in Subsect. *Calycanthum*, but the hypanthium is relatively too long.

39. A. propinqua H. Lindb. ex Juz., *Not. Syst. (Leningrad)* **4**: 184 (1923). Medium-sized, up to 35 cm, densely hairy throughout except for pedicels, which are usually completely glabrous. Hairs, particularly on petioles and upper part of inflorescence, usually erecto-patent. Leaves orbicular, shallowly lobed to c. $\frac{1}{4}$, with overlapping, basal lobes; lobes 7–9, more or less semicircular, with no incisions and 6–7 rather obtuse teeth. Flowers 3–4 mm wide; hypanthium distinctly hairy. ● *N. & C. Europe.* Cz Fe Ge No Po Rs (N, B, C) Su.

40. A. conglobata H. Lindb., *Acta Soc. Sci. Fenn.* **37(10)**: 36 (1909) (*A. juzepczukii* Alechin). Like **39** but leaf-lobes with rather long incisions and 6–9 narrower, often acute teeth; hypanthium usually subglabrous in upper part. *N. part of U.S.S.R.* Rs (N, B, C).

Related species include:

A. **confertula** Juz., *Not. Syst. (Leningrad)* **17**: 246 (1955). *S. Ural.* Rs (C).

A. **crassicaulis** Juz., *Not. Syst.* (*Leningrad*) **14**: 155 (1951). *C. Ural.* Rs (C).

A. **gortschakowskii** Juz., *Not. Syst.* (*Leningrad*) **17**: 249 (1955). *N. Ural.* Rs (N).

A. **languescens** Juz., *Not. Syst.* (*Leningrad*) **8**: 18 (1938). *Krym.* Rs (K).

A. **sibirica** Zamels, *Animadv. Syst. Herb. Univ. Tomsk.* **1931**(3): 3 (1931). *E. Russia.* Rs (C, ?E).

A. **stevenii** Buser, *Monit. Jard. Bot. Tiflis* **5**: 3 (1906). *Krym.* Rs (K).

41. A. **monticola** Opiz in Berchtold & Opiz, *Ökon.-Techn. Fl. Böhm.* **2**(1): 13 (1838) (*A. pastoralis* Buser). Medium-sized, up to 50 cm, rather robust; petioles, leaves and stems, up to and including inflorescence-branches, with dense patent hairs, but pedicels glabrous, and hypanthium often glabrous or nearly so. Leaves orbicular, sinus closed or nearly so; lobes 9–11, more or less semicircular, with short incisions and 7–9 rather regular, acute teeth. Glomeruli dense. Flowers c. 3 mm wide. $2n=c.$ 101, 103–109. *Most of Europe except the islands, but rare or local in the west, and only on mountains in the south.* Al Au Be Br Bu Cz Da Fe Ga Ge He Ho Hu It Ju No Po Rm Rs (N, B, C, W) Su.

Related species include:

A. **neostevenii** Juz., *Not. Syst.* (*Leningrad*) **8**: 19 (1938). *Krym.* Rs (K).

A. **prasina** Juz., *Not. Syst.* (*Leningrad*) **16**: 144 (1954). *C. Russia* (*Gor'kovskaja Obl.*). Rs (C).

42. A. **schistophylla** Juz., *Acta Inst. Bot. Acad. Sci. URSS* **1**: 121 (1933). Like **41** but leaves reniform, lobed up to ½; lobes 7–9, almost oblong, with very long incisions; hypanthium (and even some pedicels) usually rather densely hairy. ● *C. Russia, from Moscow region to Tatarskaja A.S.S.R.* Rs (C).

43. A. **crinita** Buser, *Scrin. Fl. Select.* (*Magnier*) **11**: 256 (1892). Like **41** but leaves reniform, very shallowly lobed, with wide sinus; lobes with no incisions and rather unequal, wide teeth; hairs on petioles and stems often somewhat deflexed; inflorescence rather lax; hypanthium almost always glabrous. *C. & S.E. Europe.* Bu Cz Ga Ge He Hu It Ju Po Rm.

Related species include:

A. **amicorum** Pawł., *Fragm. Fl. Geobot.* **1**(1): 61 (1954). *W. Carpathians.* Po.
A. **ladislai** Pawł., *op. cit.* 58 (1954). *W. Carpathians* (*Tatra*). Po.

44. A. **strigosula** Buser, *Bull. Herb. Boiss.* **1**, App. 2: 24 (1893). Medium-sized, up to 30 cm. Leaves orbicular, shallowly lobed; lobes c. 9, rounded, often overlapping at base, with short incisions usually more or less covered by folding, and 6–8 small, equal teeth. Leaves with patent hairs on both surfaces, rather rough; petioles and lower internodes with dense deflexed hairs. Inflorescence narrow, often almost glabrous. Flowers 3–4 mm wide, glabrous; hypanthium attenuate at base. ● *Mountains of S. and S.C. Europe from S.C. France to N. Jugoslavia.* Au Ga Ge He It Ju.

45. A. **subglobosa** C. G. Westerlund, *Redog. Allm. Lärov. Norr.-Söderköping* **1906–07**: 28 (1907). Like **44** but often up to 50 cm; more robust; inflorescence more spreading, and more or less hairy up to and including the smaller branches; pedicels glabrous; hypanthium rounded at base. $2n=c.$ 102, 108. ● *N. & N.C. Europe from Arctic Sweden to N.C. Germany.* Cz Ge No Rs (B) Su [Rs (N)].

46. A. **sarmatica** Juz., *Acta Inst. Bot. Acad. Sci. URSS* **3**: 202 (1936). Like **44** but leaves orbicular-reniform, often with narrow sinus; lobes usually 7–9; inflorescence more or less hairy up to and including the smaller branches; pedicels glabrous; hypanthium rounded at base. $2n=105$–106. ● *N.E. Europe, extending to W. Sweden.* Fe ?Po Rs (B, C) Su [Rs (N)].

Scandinavian authors distinguish **45** and **46** from **44**, principally on the degree of hairiness of the inflorescence, but plants occur in the Alps with sparse hairs on the inflorescence-branches and even on the hypanthium, and the distinction is not easy to maintain. Other species related to **44–46** include:

A. **breviloba** H. Lindb., *Acta Soc. Sci. Fenn.* **37**(10): 4 (1909). *C. Russia.* Rs (N, C).

A. **cyrtopleura** Juz. in Komarov, *Fl. URSS* **10**: 620 (1941). *S. Ural.* Rs (C).

A. **kornasiana** Pawł., *Fragm. Fl. Geobot.* **1**(1): 64 (1954). *W. Carpathians* (*Tatra*). Po.

A. **litwinowii** Juz., *Acta Inst. Bot. Acad. Sci. URSS* **1**: 122 (1933). *C. Russia and S. Ural.* Rs (C).

A. **substrigosa** Juz. in Majevski, *Fl. Sred. Ross.* ed. 7, 446 (1940). *C. & E. Russia.* Rs (C).

A. **tubulosa** Juz., *Not. Syst.* (*Leningrad*) **14**: 157 (1951). *C. & S. Ural.* Rs (C).

A. **walasii** Pawł., *Bull. Int. Acad. Sci. Cracovie* ser. B(1), **1951**: 345 (1953). *W. Carpathians* (*Tatra*). Cz Po.

47. A. **tytthantha** Juz., *Acta Inst. Bot. Acad. Sci. URSS* **1**: 123 (1933) (*A. multiflora* Buser ex Rothm.). Medium-sized, up to 50 cm. Leaves suborbicular, rather shallowly lobed, with narrow sinus, densely hairy on both surfaces, often somewhat sericeous beneath; lobes c. 9, rounded to subtriangular, with no incisions and 6–8 small, acute, subequal teeth. Cauline leaves rather large. Petioles and lower internodes densely hairy with at least some slightly deflexed hairs. Upper branches of inflorescence, including pedicels and hypanthium, glabrous. Flowers 1·5–2·5 mm wide. ● *Krym; naturalized in Scotland.* Rs (K) [Br].

Related species include:

A. **arcuatiloba** Juz. in Komarov, *Fl. URSS* **10**: 621 (1941). *Krym.* Rs (K).

A. **imberbis** Juz., *Not. Syst.* (*Leningrad*) **8**: 21 (1938). *Krym.* Rs (K).

48. A. **subcrenata** Buser, *Scrin. Fl. Select.* (*Magnier*) **12**: 285 (1893). Medium-sized, up to 50 cm. Leaves suborbicular, usually very undulate, usually rather sparsely hairy (often only on folds) above, more densely and evenly hairy beneath; lobes 7–9, rather wide and deep, the basal ones often touching and turned upwards, without incisions and with wide, coarse and unequal teeth. Petioles and lower internodes hairy; some hairs slightly deflexed. Inflorescence rather narrow and few-flowered, usually glabrous. Flowers 3–4 mm wide; hypanthium and pedicels always glabrous. $2n=96$, 104–110. *N. & C. Europe, extending southwards to the S.W. Alps and S.W. Bulgaria.* Au Br Bu Cz Da Fe Ga Ge He Is It Ju No Po Rm Rs (N, B, C, W) Su.

49. A. **cymatophylla** Juz., *Not. Syst.* (*Leningrad*) **3**: 41 (1922). Like **48** but leaf-lobes with short incisions and narrower, more equal teeth; leaves only sparsely hairy beneath; hairs on petioles and lower internodes strongly deflexed; flowers 2·5–3 mm wide. $2n=106$–107. *N.E. & N.C. Europe, extending to S. Poland and W. Sweden.* Cz Ge Po Rs (N, B, C) Su.

50. A. **heptagona** Juz., *op. cit.* 45 (1922). Like **48** but leaves more or less flat; lobes triangular, with narrow, acute teeth;

leaves often only sparsely hairy; hairs on petioles and lower internodes strongly deflexed; inflorescence of medium size. *Baltic region.* Rs (N, B, C) Su [Fe].

Species related to **48–50** include:

A. calvipes Juz., *Not. Syst.* (*Leningrad*) **16**: 164 (1954). *C. Russia* (*Gor'kovskaja Obl.*). Rs (C).

A. decalvans Juz., *Acta Horti Petrop.* **43**: 535 (1931). *E.C. Russia.* Rs (C).

A. devestiens Juz., *Not. Syst.* (*Leningrad*) **14**: 165 (1951). *E.C. Russia.* Rs (C).

A. hirtipes Buser, *Bull. Herb. Boiss.* ser. 2, **1**: 473 (1901). *E. Alps* (*Prov. Sondrio*). It.

A. homoeophylla Juz., *Not. Syst.* (*Leningrad*) **14**: 159 (1951). *E.C. Russia* (*Tatarskaja A.S.S.R.*). Rs (C).

A. hyperborea Juz., *op. cit.* 167 (1951). *C. Ural.* Rs (N, C).

A. obscura Buser, *Bull. Soc. Nat. Ain* **13**: 30 (1903). *W. & C. Alps, Jura.* Ga Ge He.

A. rhiphaea Juz., *Not. Syst.* (*Leningrad*) **14**: 169 (1951). *S. Ural.* Rs (C).

A. semilunaris Alechin, *Not. Syst.* (*Leningrad*) **3**: 132 (1922). *C. Russia and E. Baltic region.* Rs (N, B, C) [Fe].

A. stellaris Juz., *Acta Inst. Bot. Acad. Sci. URSS* **1**: 126 (1933). *N.C. Russia.* Rs (N, C).

A. stenantha Juz., *Not. Syst.* (*Leningrad*) **14**: 163 (1951). *S. Ural.* Rs (C).

A. submamillata Juz., *Not. Syst.* (*Leningrad*) **16**: 159 (1954). *C. Ural.* Rs (C).

51. A. acutiloba Opiz in Berchtold & Opiz, *Ökon.-Techn. Fl. Böhm.* **2**(1): 15 (1838) (*A. acutangula* Buser). Large, up to 65 cm, robust. Leaves usually more or less reniform, very variably hairy above with patent hairs, usually with hairs restricted to folds and distal portion of lobes, and with dense patent hairs beneath; lobes 9–11(–13), almost triangular, with straight sides and narrow, subtruncate apex (lobes of the late summer leaves of well-grown plants often longer than wide); teeth acute, very unequal, the largest in the middle. Petioles and lower half of stem with dense patent hairs; inflorescence glabrous. Flowers 3–4 mm wide; hypanthium rounded at base. $2n = c.$ 100, 105–109. *N., E. & C. Europe, extending to the S.W. Alps and Macedonia.* Au Be Br Bu Cz Da Fe Ga Ge Gr He Hu It Ju No Po Rm Rs (N, B, C, W, E) Su.

52. A. nemoralis Alechin, *Predv. Otčet Rabot. Nižegorod. Geobot. Eksped.* **1927**: 80 (1928). Like **51** but medium-sized, up to 50 cm; leaf-lobes short, ovate; leaves usually more or less hairy throughout on upper surface; flowers 2–3 mm. ● *C. Russia.* Rs (C).

Related species include:

A. brevidens Juz., *Not. Syst.* (*Leningrad*) **8**: 22 (1938). *Krym.* Rs (K).

A. denticulata Juz. in Komarov, *Fl. URSS* **10**: 622 (1941) (incl. *A. rubricaulis* Juz.). *C. Ural.* Rs (C).

A. lessingiana Juz., *Not. Syst.* (*Leningrad*) **14**: 161 (1951). *S. Ural.* Rs (C).

A. longipes Juz., *Not. Syst.* (*Leningrad*) **16**: 154 (1954). *S. Ural.* Rs (C).

A. rigescens Juz., *Animadv. Syst. Herb. Univ. Tomsk.* **1932** (5–6): 5 (1932). *C. & S. Ural.* Rs (C).

53. A. gracilis Opiz in Berchtold & Opiz, *Ökon.-Techn. Fl. Böhm.* **2**(1): 14 (1838) (*A. micans* Buser; incl. *A. opizii* Hadač). Medium-sized, up to 50 cm, usually rather slender. Leaves more or less reniform, subsericeous; lobes 9(–11), rounded, with no incisions and narrow, subequal teeth. Both surfaces of leaves, petioles and lower half of stem with more or less dense, erecto-patent or even almost subappressed, rather soft hairs. Inflorescence narrow, with diffuse glomeruli, its branches with sparse hairs or glabrous; pedicels glabrous. Hypanthium long, rather narrowly cuneate at base, glabrous. $2n = c.$ 93, 104–110. *N.E. & C. Europe, extending to S.W. Norway, E. France, N. Italy and C. Greece.* Au Be Bu Cz Da Fe Ga Ge Gr He Hu It Ju No Po Rm Rs (N, B, C, W) Su.

Easily distinguished (in well-grown specimens) from **41** and **51**, with which it often grows, by the subappressed hairiness of the upper leaf-surface, the narrow inflorescence and the elongated hypanthium.

Related species include:

A. hians Juz. in Komarov, *Fl. URSS* **10**: 621 (1941). *C. Ural.* Rs (C).

A. lindbergiana Juz., *Not. Syst.* (*Leningrad*) **4**: 181 (1923) (*A. atrifolia* Zamels). *C. & E. Russia.* Rs (?B, C).

A. malimontana Juz., *Not. Syst.* (*Leningrad*) **16**: 153 (1954). *S. Ural.* Rs (C).

54. A. xanthochlora Rothm., *Feddes Repert.* **42**: 167 (1937) (*A. pratensis* auct., vix Opiz, *A. vulgaris* auct., *A. sylvestris* auct.). Medium-sized, up to 50 cm, usually robust, often yellowish-green. Leaves reniform to orbicular-reniform, glabrous above or rarely with sparse hairs in the folds; lobes 9–11, rounded, with rather wide, acute, subequal teeth. Lower surface of leaf, petioles, and stems up to the inflorescence-branches with dense, patent or (especially on petioles and lower internodes) erecto-patent hairs. Pedicels and hypanthium usually glabrous or nearly so. Flowers 2·5–3 mm wide, with hypanthium *c.* 2 mm. $2n = c.$ 105. ● *W. & C. Europe, extending to S. Sweden, Latvia and C. Greece.* Au Be Br Bu Cz Da Ga Ge Gr Hb He Ho Hs Hu It Ju No Po Rm Rs (B, W) Su [Fe].

Easily distinguished from other widespread European species by its densely hairy stems and petioles, its leaves which are glabrous above, and its small flowers.

Related species include:

A. brevituba Juz., *Not. Syst.* (*Leningrad*) **14**: 181 (1951). *N.E. Russia* (*R. Unja*). Rs (N).

55. A. heterophylla Rothm., *Feddes Repert.* **46**: 128 (1939). Like **54** but up to 25 cm; later-developed leaves more or less hairy above; hairs on internodes patent or slightly deflexed; hypanthium usually with patent hairs. *Mountains of Balkan peninsula.* Al Bu Ju.

Related species include:

A. croatica Gand., *Rad Jug. Akad. Znam. Umj.* **66**: 33 (1883). *N. Jugoslavia.* Ju.

A. ivonis Pawł., *Acta Soc. Bot. Polon.* **22**: 251 (1953). *S.W. Bulgaria* (*Pirin*). Bu.

56. A. curtiloba Buser, *Mém. Soc. Fribourg. Sci. Nat.* (*Bot.*) **2**: 69 (1907). Like **54** but leaves lobed to not more than $\frac{1}{3}$; lobes with large teeth; flowers *c.* 3·5 mm wide. ● *W.C. Alps.* He.

Related species include:

A. flavicoma Buser ex Schroeter, *Ber. Schweiz. Bot. Ges.* **14**: 120 (1904). *C. Alps.* He.

A. leiophylla Juz., *Acta Inst. Bot. Acad. Sci. URSS* **1**: 127 (1933). *N. & C. Russia, extending to S. Ural.* Rs (C) [Fe].

A. multidens Buser, *Bull. Herb. Boiss.* **1**, App. **2**: 27 (1893). *S.W. Alps (Haute Savoie).* Ga.

57. A. filicaulis Buser, *Bull. Herb. Boiss.* **1**, App. **2**: 22 (1893). Small to medium-sized, up to 40 cm. Leaves reniform, with wide sinus, variably clothed with patent hairs on both surfaces, but often only on veins beneath and folds above; lobes 7(–9), rounded with no incisions and subacute, somewhat connivent teeth. Lower part of stem, and all petioles except those of earliest basal leaves, with patent hairs. Base of petioles and stipules purplish-red. Flowers 3·5–4 mm wide; hypanthium usually with patent hairs. ● *N., N.W. & N.C. Europe and on the principal mountains of W. & C. Europe from the Pyrenees to the Sudeten mountains.* Au Be Br Cz Da Fa Fe Ga Ge Hb He Ho Hs Is No Po Rm Rs (N, B, C) Su.

(a) Subsp. **filicaulis**: Upper part of stem, inflorescence-branches and pedicels glabrous. $2n = c.$ 96, 103–110, $c.$ 150. *Iceland, Fennoscandia and N. Russia; locally in the mountains of W. & C. Europe.*

(b) Subsp. **vestita** (Buser) M. E. Bradshaw, *Watsonia* **5**: 305 (1963) (*A. vestita* (Buser) Raunk., *A. minor* auct.): Hairy throughout. $2n = c.$ 96, 104–110. *Mainly in N.W. Europe, but extending to Arctic Norway, Finland, E. Austria and S. France.*

Related species include:

A. braun-blanquetii Pawł., *Bull. Inst. Acad. Sci. Cracovie* ser. B(1), **1951**: 343 (1953). *W. Carpathians (Tatra).* Po.

A. exuens Juz., *Not. Syst. (Leningrad)* **8**: 24 (1938). *Krym.* Rs (K).

A. fokinii Juz., *Not. Syst. (Leningrad)* **14**: 173 (1951). *C. Russia (Kirovskaja Obl.).* Rs (C).

A. macroclada Juz., *Not. Syst. (Leningrad)* **16**: 161 (1954). *S. Ural.* ?Rs (C). Doubtfully recorded for Europe.

A. strictissima Juz., *Not. Syst. (Leningrad)* **17**: 251 (1955). *S. Ural.* Rs (C).

58. A. minima Walters, *Watsonia* **1**: 10 (1949). Very dwarf, up to 5(–7) cm, sparsely hairy throughout, with subglabrous pedicels. Leaves reniform with wide sinus; lobes 5, with long incisions. Base of petioles and stipules brownish. Flowers $c.$ 2 mm, in small glomeruli. $2n = 103$–108. *Calcareous pastures.* ● *N. England.* Br.

Subser. *Heteropodae* Buser. Petioles of spring leaves glabrous; those of summer leaves with erecto-patent, patent or deflexed hairs. Stem (at least in lower half) and leaves usually somewhat hairy.

Members of this subseries can only be recognized by the contrast between the glabrous spring petioles and the hairy summer petioles on the same plant. Well-grown material is therefore essential for determination, and apparently well-developed specimens of some robust species (e.g. 65) may not always show the characteristically hairy summer petioles, even though the plant is flowering. Conversely, it should be noted that some species in Subser. *Hirsutae* have glabrous or subglabrous petioles to the earliest-formed spring leaves (e.g. 57).

The subseries is found in the main mountain ranges of Europe from the Pyrenees to the Carpathians and the west part of the Balkan peninsula, and also in Ural. Plants from the S. Carpathians have been referred to this subseries; they need further study, and have not been included here.

59. A. compta Buser, *Bull. Herb. Boiss.* ser. 2, **1**: 471 (1901). Medium-sized, up to 30 cm; stems erect, hairy up to the inflorescence, but the lowest internode glabrous. Leaves reniform, with sparse hairs on both surfaces, and appressed hairs on the veins beneath, lobed to $\frac{1}{4}$–$\frac{2}{5}$; lobes 9, rounded, with short incisions and 5–7 equal teeth. Petioles of summer leaves with erecto-patent hairs. Inflorescence not or scarcely exceeding leaves. Flowers 3–4 mm wide. ● *E. & C. Alps.* Au He It.

Related species include:

A. flaccida Buser, *Bull. Soc. Nat. Ain* **13**: 28 (1903). *S.W. & C. Alps, Jura.* Ga He.

A. kulczynskii Pawł., *Bull. Int. Acad. Sci. Cracovie* ser. B(1), **1951**: 338 (1953). *W. Carpathians (Tatra).* Po.

A. kvarkushensis Juz., *Not. Syst. (Leningrad)* **16**: 167 (1954). *C. Ural.* Rs (C).

A. szaferi Pawł., *Bull. Int. Acad. Sci. Cracovie* ser. B(1), **1951**: 341 (1953). *E. Carpathians (Stanislavskaja Obl.).* Rs (W).

60. A. rhododendrophila Buser, *Bull. Soc. Nat. Ain* **13**: 24 (1903). Robust, up to 50 cm; stems erect, glabrous except for the lowest internodes. Leaves reniform, glabrous on both surfaces or with sparse hairs beneath, lobed to $\frac{1}{4}$–$\frac{1}{3}$; lobes 9(–11), truncate, with short incisions and 6–12 unequal teeth. Petioles of summer leaves with erecto-patent hairs. Inflorescence large and divaricate, much exceeding the leaves, glabrous. Flowers $c.$ 3 mm wide. ● *S.W. & C. Alps, Jura, S.C. France.* Ga He.

61. A. polonica Pawł., *Fragm. Fl. Geobot.* **1**(1): 55 (1954). Dwarf, up to 15 cm; stems ascending, hairy in the lower half or glabrous. Leaves reniform, glabrous or hairy in the folds and at the margin above, sericeous on the veins and on the basal lobes beneath, lobed to $\frac{1}{4}$–$\frac{1}{3}$; lobes 7(–9), truncate, with distinct incisions and 5–6 subequal teeth. Petioles of summer leaves with erecto-patent hairs. Inflorescence somewhat exceeding leaves, glabrous, with erect branches. Flowers 3·5–4·5 mm wide. *Calcicole.* ● *W. Carpathians (Tatra).* Po.

62. A. decumbens Buser, *Bull. Herb. Boiss.* **2**: 44 (1894). Dwarf, with more or less procumbent stems up to 20 cm long, ascending only in inflorescence, subglabrous or hairy on lowest internode and at base of cauline stipules. Leaves reniform to suborbicular, hairy above only in the folds and beneath only on the veins, lobed to $c.$ $\frac{1}{2}$; lobes 7–9, rounded or quadrangular, usually with long incisions and with 4–7 long, narrow, acute teeth; cauline leaves with overlapping lobes. Petioles of summer leaves with somewhat deflexed hairs. Inflorescence usually greatly exceeding leaves, glabrous. Flowers 3–4 mm wide; epicalyx-segments less than $\frac{1}{2}$ as long as and narrower than the wide sepals. *Snow-patches and damp hollows, 1500–2600 m.* ● *Alps, Jura.* Au Ga Ge He It.

63. A. undulata Buser, *Bull. Herb. Boiss.* **1**, App. **2**: 26 (1893). Medium-sized, rather robust, with procumbent stems up to 40 cm long, often ascending in inflorescence, glabrous except for sparse hairs on the second internode; leaves orbicular, sparsely hairy on both surfaces, lobed to $c.$ $\frac{1}{2}$; lobes 9, rounded, with long incisions and 7–9 large, acute teeth; cauline leaves with divergent lobes. Petioles of summer leaves with somewhat deflexed hairs. Inflorescence exceeding leaves, usually glabrous. Flowers 2·5–3 mm; epicalyx-segments $\frac{1}{2}$–$\frac{2}{3}$ as long as sepals; hypanthium rather narrow. ● *Alps, Appennini.* Ga Ge He It.

64. A. rubristipula Buser in Dörfler, *Herb. Norm.* **36**: 217 (1898). Medium-sized, up to 30 cm; stems hairy up to at least the middle. Leaves sparsely hairy above, sericeous on the veins beneath, lobed to $\frac{1}{4}$–$\frac{1}{3}$; lobes 9(–11), rounded-triangular, with small incisions and 5–7 long, acute teeth. Petioles of summer leaves with patent hairs. Inflorescence only slightly exceeding leaves, glabrous or subglabrous. Flowers 2·5–3 mm. ● *C. Alps, Jura.* Ge He.

Related species include:

A. **amphisila** Juz., *Not. Syst.* (*Leningrad*) **17**: 252 (1955). *C. Ural.* Rs (C).
A. **iremelica** Juz., *Not. Syst.* (*Leningrad*) **14**: 174 (1951). *S. Ural.* Rs (C).

65. A. tirolensis Buser ex Dalla Torre & Sarnth., *Fl. Tirol* **6**(2): 536 (1909). Robust, up to 50 cm; stems ascending, glabrous or only the 2 lowest internodes hairy. Leaves reniform, sparsely hairy or subglabrous above, hairy beneath only on the veins, lobed to $\frac{1}{4}$–$\frac{1}{3}$; lobes 9(–11), rounded-triangular, overlapping, with short incisions and 8–12 unequal teeth. Petioles of summer leaves with patent hairs. Inflorescence exceeding leaves, glabrous. Flowers 3 mm. ● *C. & E. Alps, extending to W. Jugoslavia.* Au Ge He Ju.

Related species include:

A. **semispoliata** Juz., *Not. Syst.* (*Leningrad*) **17**: 253 (1955). *N. Ural.* Rs (N).

66. A. tatricola Pawł., *Bull. Int. Acad. Sci. Cracovie* ser. B(1), **1951**: 339 (1953). Dwarf, with more or less procumbent stems up to 15 cm long, hairy up to the first or second branch, but the lowest 1 or 2 internodes glabrous. Leaves orbicular or orbicular-reniform, sparsely hairy on both surfaces, sometimes only on the veins, folds and margins, lobed to $\frac{2}{5}$ or almost $\frac{1}{2}$; lobes 7(–9), rounded, the middle 3 often larger than the rest, with more or less short incisions and 5–8 usually unequal teeth. Petioles of summer leaves with patent hairs. Inflorescence usually exceeding leaves, usually glabrous. Flowers 3–4·5 mm. *Calcicole.* ● *W. Carpathians* (*Tatra*). Po.

67. A. heteropoda Buser, *Ber. Schweiz. Bot. Ges.* **4**: 73 (1894). Dwarf to medium-sized, up to 30 cm; stems hairy up to at least the middle; leaves suborbicular, sparsely hairy or glabrescent on both surfaces, lobed to $\frac{1}{4}$–$\frac{1}{3}$; lobes 7(–9), rounded-triangular, with wide teeth; cauline leaves lobed to not more than $\frac{1}{4}$; lobes truncate. Inflorescence narrow, exceeding the leaves. Flowers *c.* 4 mm; hypanthium rounded at base. ● *S.W. & C. Alps, Jura, Appennini.* Ga He It.

68. A. tenuis Buser, *op. cit.* 76 (1894). Like **67** but smaller and more slender; cauline leaves lobed to at least $\frac{1}{3}$; lobes ovate; flowers *c.* 3 mm; hypanthium acute at base. ● *Pyrenees, Alps, Appennini.* Au Ga Ge He It.

Related species include:

A. **sectilis** Rothm., *Feddes Repert.* **66**: 229 (1962). *E. Alps.* Au.

Subser. *Subglabrae* H. Lindb. Stems, at least on lowest internodes, and petioles with some appressed or subappressed hairs.

69. A. glomerulans Buser, *Bull. Herb. Boiss.* **1**, App. 2: 30 (1893). Medium-sized, up to 40 cm, often robust; stems not much exceeding leaves, with subappressed hairs up to the main inflorescence-branches. Leaves reniform to suborbicular, lobed to $\frac{1}{3}$–$\frac{1}{4}$, often undulate when fresh; lobes usually 9, semicircular, often overlapping when dried, with no incisions and 7–9 wide teeth. Petioles and usually both leaf-surfaces with sparse to rather dense, subappressed hairs. Flowers in dense glomeruli; hypanthium and pedicels glabrous. $2n=101$–109, *c.* 144. *Wet places, often in snow-patches. N. Europe, extending southwards in the mountains to the Pyrenees and C. Alps.* Br Da Fe Ga He ?Hs Is No Rs (N, B, C) Sb Su.

The indumentum of sub-appressed hairs, which extends into the inflorescence, distinguishes this species from all other widespread N. European species of Ser. *Vulgares*.

Related species include:

A. **borealis** Sam. ex Juz. in Pojark., *Fl. Murmansk.* **4**: 324 (1959). $2n=130$–152. *Arctic Fennoscandia.* Fe Rs (N) Su.
A. **kolaensis** Juz. in Pojark., *loc. cit.* (1959). *Arctic Russia* (*Murmanskaja Obl.*). Rs (N).
A. **obtusiformis** Alechin in Govoruchin, *Fl. Urala* 531 (1937). *N. Ural.* Rs (N).
A. **transpolaris** Juz., *Not. Syst.* (*Leningrad*) **16**: 179 (1954). *Arctic Russia* (*Murmanskaja Obl.*). Rs (N).

70. A. camptopoda Juz., *Not. Syst.* (*Leningrad*) **8**: 25 (1938). Medium-sized, up to 30 cm, with appressed hairs throughout, except on most pedicels; stems ascending, with few large cauline leaves. Leaves reniform, lobed to *c.* $\frac{1}{3}$; lobes 9, semicircular, with long incisions and 6–9 rather acute teeth. Inflorescence lax, few-flowered. Flowers *c.* 3 mm wide; hypanthium with dense erecto-patent hairs. ● *Krym.* Rs (K).

71. A. crebridens Juz., *op. cit.* 27 (1938). Like **70** but less densely hairy; leaves orbicular, lobed to *c.* $\frac{1}{4}$; lobes 9–11; hypanthium and pedicels glabrous or subglabrous. ● *Krym.* Rs (K).

72. A. buschii Juz., *Acta Inst. Bot. Acad. Sci. URSS* **1**: 128 (1933). Dwarf, with appressed hairs throughout except for pedicels and hypanthium; stems up to 10 cm long, slender, ascending. Leaves lobed to *c.* $\frac{1}{2}$; lobes 5, semicircular, with long incisions and 4–6 wide, obtuse teeth. Inflorescence narrow, lax, few-flowered. Flowers *c.* 4 mm wide; hypanthium and pedicels of lowest flowers sparsely hairy, the rest glabrous. ● *Krym.* Rs (K).

Species 70–72, distinguished by the almost complete indumentum of appressed hairs, seem to be related to Caucasian rather than to any other European species.

73. A. controversa Buser, *Bull. Soc. Nat. Ain* **13**: 25 (1903). Medium-sized, up to 30 cm; stems much exceeding leaves, with appressed hairs in lower half. Leaves orbicular, with more or less evenly distributed appressed hairs on both surfaces, lobed to $\frac{1}{3}$–$\frac{2}{5}$; lobes 9, overlapping, semicircular, with 6–8 equal teeth. Petioles with appressed hairs. Inflorescence glabrous. ● *Jura; W. Alps.* He.

Related species include:

A. **cleistophylla** Rothm. & O. Schwarz, *Feddes Repert.* **42**: 395 (1937). *N. Alps* (*Allgauer Alpen*). Ge.
A. **smytniensis** Pawł., *Fragm. Fl. Geobot.* **1**(1): 52 (1954). *W. Carpathians* (*Tatra*). Po.

74. A. wallischii Pawł., *Bull. Int. Acad. Sci. Cracovie* ser. B(1), **1951**: 333 (1953). Dwarf, with slender, procumbent or ascending stems up to 12 cm long, with subappressed hairs in lower half. Leaves reniform, with appressed hairs on folds and near margins above, and with appressed hairs mainly on the veins beneath, lobed up to *c.* $\frac{1}{3}$; lobes 7–9, broadly elliptical, somewhat overlapping, with long incisions, and 4–6 short, wide, obtuse teeth.

Petioles usually with rather dense, subappressed hairs. Inflorescence lax, few-flowered, glabrous. Flowers *c.* 4 mm wide; hypanthium narrow, attenuate into pedicel. *Wet granite rocks.* ● *W. Carpathians (Tatra).* Cz Po.

75. A. connivens Buser, *Bull. Herb. Boiss.* **2**: 107 (1894) (incl. *A. subconnivens* Pawł.). Medium-sized, up to 30 cm; stems with appressed hairs in lower half. Leaves reniform to suborbicular, usually somewhat sericeous above, with appressed hairs at least in folds and towards the margin, usually subglabrous beneath except for veins which have subappressed hairs, lobed to $\frac{1}{4}-\frac{1}{3}$; lobes 7–9, ovate-triangular, with short incisions (often more or less obscured by overlapping margins of lobes) and 7–10 narrow, acute, equal, connivent teeth. Inflorescence glabrous, much exceeding leaves. Flowers 3–4 mm wide. ● *Principal mountain ranges of S. & C. Europe from the Pyrenees to the Carpathians and S. Bulgaria.* Au Bu Cz Ga Ge He Hs It Ju Po Rm ?Rs (W).

Related species include:

A. czywczynensis Pawł., *Bull. Int. Acad. Sci. Cracovie* ser. B(1), **1951**: 334 (1953). *E. Carpathians (Čivčinskye Gory).* Rm Rs (W).
A. racemulosa Buser, *Bull. Herb. Boiss.* **1, App. 2**: 31 (1893). *S.W. Alps (Haute Savoie).* Ga.

76. A. baltica Sam. ex Juz. in Majevski, *Fl. Sred. Ross.* ed. 7, 449 (1940) (*A. nebulosa* Sam.). Medium-sized, rather robust, up to 40 cm. Leaves suborbicular, glabrous above, subglabrous beneath except for veins which have subappressed hairs throughout, lobed to *c.* $\frac{1}{4}$; lobes 9–11, ovate-triangular to semicircular, with short but distinct incisions and 7–9 acute, equal, connivent teeth. Petioles and lower half of stems usually with rather dense, appressed or subappressed hairs. Inflorescence lax, glabrous; flowers *c.* 4 mm wide. *N.E. Europe, extending to E. Poland.* Fe Po Rs (N, B, C) Su.

Related species include:

A. psiloneura Juz., *Acta Inst. Bot. Acad. Sci. URSS* **1**: 129 (1933). *C. Ural.* Rs (C).
A. stichotricha Juz., *Not. Syst. (Leningrad)* **14**: 176 (1951). *C. Russia (Tatarskaja A.S.S.R.).* Rs (C).

77. A. wichurae (Buser) Stefánsson, *Fl. Ísl.* 135 (1901). Like **76** but rather small, up to 30 cm; stems slender, usually glabrous above the second or third internode; leaves orbicular; lobes with deeper incisions. $2n=103$–106. ● *N. Europe, southwards to N. England and the Sudeten mountains.* Br Fa Fe Is No Po Rs (N, ?B) Su.

Related species include:

A. oxyodonta (Buser) C. G. Westerlund, *Redog. Allm. Lärov. Norr.-Söderköping* **1906–7**: 12 (1907). *C. Scandinavia.* No Su.

78. A. murbeckiana Buser, *Bot. Not.* **1906**: 142 (1906). Like **76** but stems often hairy up to the inflorescence-branches; leaves reniform, with wide sinus; lobes 9, with no incisions and rather wider teeth. $2n=102$–109. *Iceland; Fennoscandia; Ural.* Fe Is No Rs (N, C) Su.

Related species include:

A. turkulensis Pawł., *Bull. Int. Acad. Sci. Cracovie* ser. B(1), **1951**: 335 (1953). *E. Carpathians (Stanislavskaja Obl.).* Rs (W).

79. A. acutidens Buser, *Bull. Herb. Boiss.* **2**: 104 (1894). Like **76** but usually smaller, up to 25 cm; stems hairy only on lowest two internodes; leaves lobed to $\frac{2}{5}(-\frac{1}{2})$; lobes with no incisions and very narrow, acuminate teeth. ● *Alps.* Au Ga Ge He It.

Records of this species for the Carpathians are doubtful.

Related species include:

A. acuminatidens Buser, *Bull. Herb. Boiss.* ser. 2, **2**: 624 (1902). *C. Alps.* He.

76–79 have sometimes been treated as a single variable species, *A. acutidens*, and there is unusual taxonomic difficulty in the group. Nevertheless the regional variation reflected in the present treatment seems very significant, and well-grown specimens can usually be assigned to one of the four widespread species here adopted.

80. A. reniformis Buser, *Alchim. Valais.* 23 (1894). Robust; stems up to 40 cm, ascending, usually not greatly exceeding leaves, with subappressed hairs on lower half. Leaves reniform, with wide sinus, thick, glabrous above, with subappressed hairs beneath only on veins, very shallowly lobed to *c.* $\frac{1}{5}$; lobes 9, short, wide, obtuse, with no incisions and 7–9 rather wide, subequal teeth. Petioles with more or less dense, appressed hairs. Inflorescence diffuse, with divaricate branches. Flowers 4–5 mm wide; sepals up to 2 mm, not longer than wide, rounded. ● *Alps; Sudeten Mountains; W. Carpathians; mountains of Jugoslavia.* Au Cz Ga Ge It Ju Po.

Resembles species of Subser. *Glabrae* in robust habit and texture of leaves. Plants from the W. Carpathians previously referred to this species have very recently been distinguished as **A. obsoleta** Fröhner, *Preslia* **38**: 323 (1966). They have leaf-lobes with very unequal teeth and triangular, acute sepals longer than wide.

81. A. lineata Buser, *Alchim. Valais.* 27 (1894). Like **80** but leaves usually hairy on basal lobes beneath; leaf-lobes with small, equal teeth; flowers 3–3·5 mm wide. ● *Pyrenees; Vosges; Alps; mountains of W. part of Balkan peninsula.* Al Au Ga Ge He Hs It Ju.

Rothmaler gives this species for the Carpathians without more precise indication.

82. A. obtusa Buser, *op. cit.* 22 (1894). Like **80** but stems usually greatly exceeding leaves; leaves reniform to suborbicular, with rather narrow sinus, lobed to $(\frac{1}{5}-)\frac{1}{4}-\frac{1}{3}$; inflorescence narrow, with rather dense glomeruli; flowers 3–4 mm wide. $2n=c.$ 103. *N.E. Europe; mountains of C. & S. Europe extending southwards to the Maritime Alps.* Au Cz Fe Ga Ge He It Ju Po Rm Rs (N, B, C, W) Su.

Plants referable to **82** or the related species occur in the Balkan peninsula, but their exact identification is uncertain.

Related species include:

A. effusa Buser, *op. cit.* 24 (1894). *Cévennes; Jura; S. Alps; Appennini.* Ga He It.
A. inexpexa Buser, *op. cit.* 26 (1894). *Jura; W. & C. Alps; Appennini.* Au ?Ga Ge He It.

83. A. glabra Neygenf., *Enchirid. Bot. Siles.* 67 (1821) (*A. alpestris* auct.). Large, robust, up to 60 cm; stems much exceeding leaves, usually glabrous except on lowest 1(–2) internodes, which are usually sericeous with appressed hairs. Leaves reniform to suborbicular, glabrous except for appressed hairs on distal half of main veins beneath, lobed to *c.* $\frac{1}{4}$; lobes 9–11, triangular-ovate, with no incisions and 7–9 very unequal, rather wide teeth. Petioles with rather sparse, appressed hairs. Inflorescence rather narrow; flowers 3–4 mm wide; sepals triangular, acute, longer than wide. $2n=96$, c. 100, 102–110. *N. & C. Europe, extending southwards in the mountains to the Pyrenees and N. Balkan peninsula.* Au Be Br Bu Cz Da ?Fa Fe Ga Ge Hb He Ho Hs Hu Is It Ju No Po Rm Rs (N, B, C, W) Su.

Very variable, particularly in habit and leaf-shape, but a satisfactory taxonomic treatment is not yet possible.

Related species include:

A. boleslai Pawł., *Fragm. Fl. Geobot.* **1** (1): 49 (1954). *W. Carpathians (Tatra).* Cz Po.
A. cunctatrix Juz., *Not. Syst. (Leningrad)* **16**: 181 (1954). *N. Ural.* Rs (N, C).
A. glabriformis Juz., *Not. Syst. (Leningrad)* **14**: 178 (1951). *N. Ural.* Rs (N, C).
A. paeneglabra Juz., *op. cit.* 179 (1951). *C. Ural.* Rs (C).
A. ursina Fröhner, *Bot. Jahrb.* **83**: 389 (1965). *Mountains of E. Austria, W. Czechoslovakia and S. Germany.* Au Cz Ge.

Subser. *Glabrae* Rothm. Petioles and stems entirely glabrous (occasionally with a few hairs, especially on late summer petioles).

With the exception of **84**, which Juzepczuk included in Subser. *Hirsutae*, this subseries is confined to the main mountain ranges of S. and C. Europe from Spain to the Balkan peninsula.

84. **A. glabricaulis** H. Lindb., *Acta Soc. Sci. Fenn.* **37** (10): 3 (1909). Medium-sized, up to 35 cm; stems glabrous, usually only slightly exceeding leaves. Leaves suborbicular, with sparse patent hairs above, glabrous beneath except for sparse hairs distally on the main veins; lobes 7–9, short, broadly triangular, with short incisions and 5–8 short, wide teeth. Inflorescence narrow, completely glabrous, with rather few flowers 1·5–3 mm wide. *N.E. Europe.* Rs (N, B, C) [Fe].

A. parcipila Juz., *Not. Syst. (Leningrad)* **14**: 175 (1951), described from C. Ural (Osljanka), is said to differ from **84** in the sparse, patent hairs on the summer petioles, and was therefore included in Subser. *Heteropodae* by Juzepczuk. Some material determined by Juzepczuk as this species has entirely glabrous petioles, however, and cannot be distinguished from **84**.

85. **A. versipila** Buser, *Bull. Herb. Boiss.* **2**: 112 (1894). Like **84** but stems greatly exceeding leaves; leaves reniform; lobes without incisions; leaves sericeous with subappressed hairs above; flowers 3–4 mm wide. ● *Jura, Alps.* Au Ga Ge He.

Related species include:

A. aggregata Buser, *Alchim. Valais.* 17 (1894). *S.W. & C. Alps, Jura.* Ga He.
A. versipiloides Pawł., *Bull. Int. Acad. Sci. Cracovie* ser. B(1), **1951**: 330 (1953). *W. Carpathians (Tatra).* Po.

86. **A. coriacea** Buser, *Not. Alchim.* 19 (1891). Large, robust, glabrous, up to 50 cm, not turning reddish in late summer; stems usually greatly exceeding leaves. Leaves more or less orbicular, often very large (up to 15 × 17 cm), thick, lobed to c. $\frac{1}{3}(-\frac{1}{4})$; lobes 9–11, semicircular, with short incisions and 6–8 very wide, obtuse, unequal teeth. Inflorescence narrow, rather few-flowered. Flowers c. 4 mm wide; hypanthium c. 2 mm, truncate at base. ● *Mountains of S.W. and S.C. Europe, from Spain to W. Austria.* Au Ga Ge He Hs ?It.

87. **A. inconcinna** Buser, *Bull. Herb. Boiss.* **1**, App. **2**: 34 (1893). Like **86** but plant turning reddish-brown in late summer; leaf-

lobes without incisions; inflorescence spreading. ● *Mountains of S.W. and S.C. Europe from C. France to W. Austria.* Au Ga Ge He ?It.

Related species include:

A. subtatrica Pawł., *Bull. Int. Acad. Sci. Cracovie* ser. B(1), **1951**: 320 (1953). *W. Carpathians (Tatra).* Po.

88. **A. stanislaae** Pawł., *Bull. Int. Acad. Sci. Cracovie* ser. B(1), **1951**: 317 (1953). Like **86** but up to 30 cm, turning dull reddish-brown in late summer; outer leaves suborbicular, very shallowly lobed to c. $\frac{1}{7}$, inner leaves reniform, lobed to c. $\frac{1}{5}$; lobes with acute teeth; hypanthium of flower up to 2·7 mm, narrowing gradually to pedicel. ● *W. Carpathians (Tatra).* Cz Po.

89. **A. trunciloba** Buser, *Alchim. Valais.* 15 (1894). Medium-sized, glabrous, up to 35 cm, becoming dark reddish-brown in summer; stems slender, greatly exceeding leaves. Leaves reniform to suborbicular, thin, lobed to $\frac{1}{4}-\frac{1}{3}$; lobes 9(–11), semicircular, subtruncate, with long incisions and 6–8 short, wide, acute teeth. Inflorescence spreading. Flowers 3–3·5 mm wide. ● *Jura, Alps, Appennini.* Au Ga ?Ge He It.

90. **A. sinuata** Buser, *Bull. Herb. Boiss.* **2**: 102 (1894). Like **89** but leaves lobed to $\frac{1}{3}(-\frac{2}{5})$; lobes narrow, with very acute teeth; inflorescence narrow; flowers c. 4 mm wide. ● *Alps, Appennini.* Au Ga He It.

91. **A. straminea** Buser, *Alchim. Valais.* 13 (1894) (incl. *A. kotulae* Pawł.). Medium-sized, up to 40 cm, glabrous, rather slender, not or scarcely becoming reddish-brown in summer; stems greatly exceeding leaves. Leaves reniform, with wide sinus, thin, lobed to c. $\frac{1}{4}$; lobes 9–11, more or less triangular, with 7–9 narrow, acute, equal teeth. Inflorescence narrow. Flowers c. 3 mm wide. ● *Alps; W. Carpathians; mountains of N. part of Balkan peninsula.* Au Bu Cz Ga Ge He Hs It Ju Po.

Related species include:

A. longiuscula Buser, *Bull. Herb. Boiss.* **2**: 101 (1894). *W.C. Alps (Valais).* He.
A. squarrosula Buser, *Mém. Soc. Fribourg. Sci. Nat. (Bot.)* **1**: 126 (1902). *W. Alps.* He.

92. **A. aequidens** Pawł., *Fragm. Fl. Geobot.* **1** (1): 46 (1954). Dwarf, glabrous, with procumbent stems up to 15 cm long. Leaves orbicular, lobed to $\frac{1}{4}-\frac{1}{3}$; lobes 7–9, semicircular, with no incisions and (4–)5–8 subequal, obtuse or subacute teeth. Inflorescence narrow, many-flowered; flowers 3–4·5 mm wide. ● *W. Carpathians (Tatra).* Cz Po.

Related species (dwarf, with relatively short leaf-lobes) include:

A. sokolowskii Pawł., *Bull. Int. Acad. Sci. Cracovie* ser. B(1), **1951**: 328 (1953). *W. Carpathians (Tatra).* Cz Po.
A. zmudae Pawł., *loc. cit.* (1953). *W. Carpathians (Tatra).* Po.

93. **A. demissa** Buser, *Bull. Herb. Boiss.* **2**: 96 (1894). Dwarf, glabrous; stems procumbent or somewhat ascending, up to 25 cm long, not or scarcely exceeding leaves. Leaves suborbicular, lobed to c. $\frac{2}{5}$; lobes 7–9, obovate, more or less truncate, with distinct incisions, and 4–6 narrow, acute, equal, somewhat divergent teeth. Inflorescence narrow. Flowers 3·5–4 mm wide. *Wet places, often snow-patches.* ● *Pyrenees; Alps; Appennini.* Au Ga He ?Hs It.

94. A. pseudincisa Pawł., *Bull. Int. Acad. Sci. Cracovie* ser. B(1), **1951**: 326 (1953). Like **93** but leaves reniform to suborbicular; lobes oblong-elliptical, with acute, unequal and more or less connivent teeth; inflorescence spreading; flowers up to 5 mm wide. *Wet places on mountains, above 1500 m.* ● *W. Carpathians (Tatra).* Cz Po.

Species related to **93** and **94** include:

A. fissimima Buser, *Bull. Herb. Boiss.* **2**: 99 (1894). *W.C. Alps (Valais).* He.
A. semisecta Buser, *op. cit.* 94 (1894) (*A. vulgaris* subsp. *semisecta* (Buser) Gams). *Alps, Jura.* Au Ga He.

95. A. longana Buser, *Bull. Herb. Boiss.* ser. 2, **1**: 468 (1907) (incl. *A. marcailhouorum* Buser). Like **93** but leaves reniform; lobes ovate, with no incisions and short, rather wide, acute teeth. *Wet places, often snow-patches.* ● *Pyrenees; S. & E. Alps.* Au Ga He.

Related species (leaf-lobes long, without incisions) include:

A. frigens Buser ex Jaquet, *Mém. Soc. Fribourg. Sci. Nat. (Bot.)* **1**: 129 (1902) (*A. frigida* Buser, non Weddell). *W.C. Alps.* Ga He.

Subsect. *Calycanthum* Rothm. Very variable in habit and hairiness. Leaves rarely lobed to more than $\frac{1}{2}$. Epicalyx-segments at least as long as sepals, and usually almost as wide; both epicalyx-segments and sepals patent after anthesis, giving the appearance of an 8-pointed star. Hypanthium distinctly shorter than mature achene, and usually shorter than sepals. Flowers usually yellowish.

This subsection is confined to the mountains of S. & C. Europe and Ural.

Series *Elatae* Rothm. At least the lower internodes and some petioles with erecto-patent, patent, or deflexed hairs.

96. A. achtarowii Pawł., *Bull. Int. Acad. Sci. Cracovie* ser. B(1), **1951**: 301 (1953). Medium-sized, up to 45 cm; stems with patent hairs up to and including inflorescence-branches. Leaves suborbicular, with patent hairs, sparse above and dense beneath, lobed to $\frac{1}{5}-\frac{1}{4}$; lobes 9–11, semicircular, with no incisions and 8–10 acute teeth. Petioles with patent hairs. Pedicels glabrous or with sparse hairs. Flowers 4·5–6·5(–7·2) mm wide; epicalyx-segments with (1–)2–4(–5) teeth, much longer than sepals; hypanthium glabrous or subglabrous. ● *Bulgaria (Stara Planina).* Bu.

97. A. jumrukczalica Pawł., *op. cit.* 305 (1953). Like **96** but up to 20 cm; leaves lobed to *c.* $\frac{1}{3}$; lobes 7–9; flowers 3·5–4·5 mm wide; epicalyx-segments entire or with 1(–2) teeth, not or slightly longer than sepals. ● *Bulgaria (Stara Planina).* Bu.

98. A. mollis (Buser) Rothm., *Feddes Repert.* **33**: 347 (1934). Large, up to 80 cm, robust, usually with dense, patent hairs throughout, except on pedicels. Leaves often very large (up to 13 × 15 cm), lobed to $\frac{1}{8}-\frac{1}{4}$; lobes 9–11, semicircular, with 7–9 wide, ovate teeth. Pedicels glabrous or subglabrous. Flowers 3·5–5 mm wide; epicalyx-segments entire or with one tooth; hypanthium with patent hairs. *E. Carpathians.* Rm Rs (W).

Very widely cultivated in gardens in Europe.

99. A. catachnoa Rothm., *Feddes Repert.* (*Beih.*) **100**: 66 (1938). Rather large, up to 50 cm; stems with dense, patent hairs up to inflorescence. Leaves orbicular, glabrous above, with sparse hairs or subglabrous beneath, lobed to $\frac{1}{5}-\frac{1}{4}$; lobes 11–13, ovate-triangular, with no incisions and 8–12 ovate, obtuse and apiculate teeth. Petioles with dense patent hairs. Inflorescence compact, with dense glomeruli. Pedicels glabrous. Flowers 3·5–4 mm wide; epicalyx-segments entire, scarcely longer than sepals; hypanthium glabrous or with sparse hairs. ● *E. Albania; Bulgaria.* Al Bu.

100. A. viridiflora Rothm., *op. cit.* 73 (1938). Rather large, up to 50 cm; stems with dense erecto-patent hairs below, and patent hairs above and on the inflorescence-branches. Leaves suborbicular, with dense patent hairs on both surfaces, lobed to *c.* $\frac{1}{4}$; lobes 9–11, ovate-triangular, with no incisions and 6–9 large, subacute, unequal teeth. Petioles with dense, erecto-patent or subappressed hairs. Pedicels often glabrous. Glomeruli lax. Flowers 3·5–4 mm wide, greenish; hypanthium with dense, patent hairs. ● *C. Greece (Olimbos).* Gr.

101. A. albanica Rothm., *op. cit.* 76 (1938). Slender, up to 30 cm; stem, including the inflorescence-branches, with patent hairs. Leaves suborbicular, subglabrous or glabrous above and with rather sparse patent hairs beneath, lobed to $\frac{1}{5}-\frac{1}{4}$; lobes 9, truncate, with no incisions and 4–7 connivent teeth. Petioles with patent hairs. Pedicels often glabrous. Glomeruli dense. Flowers 3–4 mm wide; hypanthium with patent hairs. ● *Albania.* Al.

102. A. phegophila Juz., *Not. Syst.* (*Leningrad*) **16**: 182 (1954). Like **101** but with rather dense patent hairs throughout, including pedicels. ● *Krym.* Rs (K).

103. A. indivisa (Buser) Rothm., *Feddes Repert.* **33**: 346 (1934) (*A. acutiloba* Steven subsp. *indivisa* (Buser) Hayek). Medium-sized, up to 45 cm; stems with patent hairs up to and including main inflorescence-branches. Leaves suborbicular, glabrous above, with dense patent hairs beneath, lobed to $\frac{1}{4}-\frac{1}{3}$; lobes 11, ovate-elliptical, with no incisions and 8–10 wide teeth. Petioles with dense patent hairs. Ultimate inflorescence-branches and pedicels glabrous. Flowers 3·5–4 mm; hypanthium glabrous or with sparse hairs at base. ● *Mountains of Balkan peninsula.* Bu Gr Ju.

104. A. heterotricha Rothm., *Feddes Repert.* (*Beih.*) **100**: 75 (1938). Like **103** but up to 60 cm; basal leaves reniform, lobed to $\frac{1}{3}-\frac{1}{4}$, with dense, patent hairs on both surfaces; upper cauline leaves glabrous above; hypanthium hairy. ● *Mountains of N. Greece and S. Jugoslavia.* Gr Ju.

105. A. peristerica Pawł., *Bull. Int. Acad. Sci. Cracovie* ser. B(1), **1951**: 307 (1953). Medium-sized, up to 40 cm; stems with dense patent hairs up to main inflorescence-branches. Leaves glabrous above or with sparse hairs on margins and at base; with dense patent hairs beneath, lobed to $\frac{2}{5}-\frac{1}{2}$; lobes 9–11, ovate, with long incisions and 5–7 acute teeth. Pedicels glabrous. Flowers 3·5–4·5 mm wide; hypanthium with sparse hairs at base or glabrous. ● *S. Jugoslavia (Perister).* Ju.

106. A. aroanica (Buser) Rothm., *Feddes Repert.* **33**: 345 (1934). Medium-sized, up to 50 cm; stems with erecto-patent hairs in lower half. Leaves glabrous above, with appressed hairs beneath; lobes *c.* 9, ovate, with no incisions and 8 or 9 broadly ovate teeth. Petioles of outer leaves glabrous, of inner (summer) leaves with erecto-patent hairs. Inflorescence-branches and pedicels glabrous. Flowers 4·5–6·5 mm wide; hypanthium with sparse patent hairs. ● *S. Greece (Killini).* Gr.

107. A. zapalowiczii Pawł., *Bull. Int. Acad. Sci. Cracovie* ser. B(1), **1951**: 309 (1953). Medium-sized, up to 40 cm, with rather dense patent or slightly deflexed hairs, except for the ultimate inflorescence branches and pedicels which are glabrous. Leaves reniform to suborbicular, lobed to $\frac{1}{3}$–$\frac{2}{5}$; lobes 9(–11), more or less semicircular, with no incisions and 7–11 large, wide, obtuse teeth. Inflorescence narrow. Flowers 3·5–5 mm wide; epicalyx-segments usually slightly shorter and always much narrower than sepals, and both slightly longer than hypanthium; hypanthium glabrous or with sparse patent hairs. ● *E. Carpathians.* Rm Rs (W).

Related species include:

A. bandericensis Pawł., *Acta Soc. Bot. Polon.* **22**: 247 (1953). *S.W. Bulgaria* (*Pirin*). Bu.

107 and **108** are somewhat intermediate between Subsect. *Calycanthum* and Subsect. *Heliodrosium* in flower structure.

Series *Calycinae* Buser. Hairs, if present on stems and petioles, subappressed or appressed.

108. A. haraldii Juz., *Acta Inst. Bot. Acad. Sci. URSS* **1**: 130 (1933). Medium-sized, up to 30 cm, with appressed hairs throughout except on upper surface of some leaves, and pedicels and hypanthium of upper flowers, which are glabrous. Leaves reniform to suborbicular, lobed to about $\frac{1}{4}$; lobes 7–9, arcuate to semicircular, with short or no incisions, and 7–10 acute, somewhat connivent teeth. Inflorescence narrow, with lax glomeruli. Flowers 3–4 mm; epicalyx-segments slightly shorter and much narrower than sepals; hypanthium about as long as sepals. ● *S. Ural.* Rs (C).

Easily distinguished from all other European species of Subsect. *Calycanthum* by the appressed hairs in the inflorescence.

109. A. fallax Buser, *Ber. Schweiz. Bot. Ges.* **4**: 65 (1894). Medium-sized, with slender stems up to 30 cm. Leaves suborbicular, glabrous above, subglabrous or with sparse appressed hairs beneath but with dense appressed hairs, often sericeous, on the veins, lobed up to $\frac{1}{3}$; lobes 7–9, broadly triangular, with no incisions and 6–10 short teeth. Petioles and lower half of stems with more or less dense, appressed hairs. Inflorescence glabrous. Flowers up to 4 mm wide; sepals slightly larger than epicalyx-segments. *Usually calcicole.* ● *Pyrenees, S. Alps, Appennini, Balkan peninsula.* Au Ga Ge Gr He Hs It Ju.

110. A. sericoneura Buser, *op. cit.* 68 (1894). Like **109** but more robust; leaves more shallowly lobed; lobes 9–11, more truncate; flowers 4–5 mm wide. ● *C. & E. Alps; S. Carpathians.* Au He Rm.

Related species include:

A. babiogorensis Pawł., *Fragm. Fl. Geobot.* **3**(1): 34 (1957). *W. Carpathians* (*Tatra*). Cz Po.
A. giewontica Pawł., *op. cit.* 31 (1957). *W. Carpathians* (*Tatra*). Cz Po.
A. jasiewiczii Pawł., *op. cit.* 41 (1957). *W. Carpathians* (*Tatra*). Cz Po.
A. pycnoloba Juz., *Not. Syst.* (*Leningrad*) **14**: 183 (1951). *C. Ural.* Rs (C).
A. sericoneuroides Pawł., *Fragm. Fl. Geobot.* **3**(1): 37 (1957). *W. Carpathians* (*Tatra*). Cz Po.

111. A. flexicaulis Buser, *Bull. Herb. Boiss.* **1**, App. **2**: 32 (1893) (*A. glaberrima* subsp. *flexicaulis* (Buser) Gams). Large, robust, stems up to 60 cm. Leaves suborbicular, glabrous above, with very sparse, subappressed hairs beneath and not sericeous on the veins, lobed to not more than $\frac{1}{4}$; lobes 9–11, wide, rounded, with no incisions and 6–10 small teeth. Petioles and main in-

florescence-branches with appressed or subappressed hairs. Flowers *c.* 4 mm wide; sepals shorter than hypanthium. ● *Alps, Jura.* Ga Ge He.

112. A. gorcensis Pawł., *Bull. Int. Acad. Sci. Cracovie* ser. B(1), **1951**: 311 (1953). Like **111** but leaves reniform with very wide sinus, lobed to not more than $\frac{1}{5}$; lobes often triangular; flowers 4–5 mm wide; sepals about as long as hypanthium. ● *W. Carpathians; C. Jugoslavia* (*Bosna*). Cz Ju Po.

Related species include:

A. damianicensis Pawł., *Acta Soc. Bot. Polon.* **22**: 248 (1953). *S.W. Bulgaria* (*Pirin*). Bu.
A. vranicensis Pawł., *op. cit.* 250 (1953). *C. Jugoslavia* (*Bosna*). Ju.

113. A. cuspidens Buser, *Bull. Herb. Boiss.* **2**: 106 (1894). Small, with slender, ascending stems up to 25 cm, with appressed hairs on lowest 1–2(–3) internodes. Leaves reniform to suborbicular, glabrous above, with sometimes sparse subappressed hairs beneath, at least on basal lobes, and usually along whole length of veins, lobed up to $\frac{1}{3}$; lobes 7–9, semicircular or somewhat truncate, with short but distinct incisions and 5–7 acute, connivent teeth. Inflorescence small, glabrous. Flowers 4–6 mm wide. ● *E. & C. Alps.* Au Ge He.

Related species include:

A. asteroantha Rothm., *Feddes Repert.* (*Beih.*) **100**: 85 (1938). *Bulgaria* (*Stara Planina*). Bu.
A. eugenii Pawł., *Bull. Int. Acad. Sci. Cracovie* ser. B(1), **1951**: 315 (1953). *W. Carpathians* (*Tatra*). Cz Po.

114. A. othmarii Buser, *Bull. Herb. Boiss.* ser. 2, **1**: 464 (1901). Like **113** but dwarf; hairs restricted to lowest internode, petioles, and distal parts of veins on lower surface of leaf; flowers 3–4 mm wide. ● *C. Alps.* Au Ge He.

Related species include:

A. gracillima Rothm., *Feddes Repert.* (*Beih.*) **100**: 87 (1938) (?*A. riloensis* Ronniger). *S.E. Alps* (*Slovenija*); *Bulgaria* (*Rila Planina*). Bu Ju.
A. pseudothmarii Pawł., *Fragm. Fl. Geobot.* **1**(1): 44 (1954). *W. Carpathians* (*Tatra*). Cz Po.

115. A. oculimarina Pawł., *Fragm. Fl. Geobot.* **3**(1): 44 (1957). Medium-sized, with stems up to 50 cm long, procumbent below, usually with appressed hairs on only the 2 lowest internodes, sometimes glabrous. Leaves more or less reniform, glabrous above, usually with appressed hairs beneath only on the distal half of the veins, lobed to *c.* $\frac{1}{4}$; lobes 9(–11), semicircular, with short or no incisions and 6–8 rather large, subequal teeth. Petioles with appressed hairs or glabrous. Inflorescence spreading, with lax glomeruli. Flowers 4·5–6·5(–7·8) mm wide; sepals longer than hypanthium; epicalyx-segments distinctly longer than sepals, often with a single lateral tooth. *Wet granite rocks.* ● *W. Carpathians* (*Tatra*). Cz Po.

116. A. pyrenaica Dufour, *Ann. Gén. Sci. Phys.* (*Bruxelles*) **8**: 228 (1821) (*A. firma* Buser pro parte, *A. glaberrima* subsp. *firma* (Buser) Gams pro parte). Medium-sized, usually blue-green; stems up to 25 cm, ascending, glabrous except for lowest internode (rarely also the second) and distal half of veins on lower surface of leaf, which have some appressed hairs. Leaves suborbicular, lobed to $\frac{1}{3}$–$\frac{2}{5}$(–$\frac{1}{2}$); lobes 7–9, more or less semicircular, with short but distinct incisions and 5–8 long, acute, equal teeth. Inflorescence rather small. Flowers 3·5–4·5(–5·5) mm wide. ● *Pyrenees, Alps, Carpathians, mountains of N. part of Balkan peninsula.* Au Bu Cz Ga Ge He Hs It Ju Po Rm ?Rs (W).

Related species include:

A. venosula Buser, *Bull. Herb. Boiss.* ser. 2, **1**: 466 (1901). *E. Alps* (*Prov. Sondrio*). ?He It.

117. A. incisa Buser, *Scrin. Fl. Select.* (*Magnier*) **11**: 255 (1892). Dwarf, delicate; stems up to 15 cm, slender; like **116** in hairiness. Leaves reniform to suborbicular, lobed to $\frac{1}{3}$–$\frac{1}{2}$; lobes 7–9, long, narrow, with very long, almost U-shaped incisions, and (4–)5–7 long, narrow, connivent teeth, the longest up to 3 mm. Inflorescence small, lax. Flowers 3–4 mm wide. ● *Jura, Vosges, Alps, Carpathians.* Au Cz Ga Ge He Po Rs (W).

Related species include:

A. vallesiaca Rothm., *Feddes Repert.* **42**: 168 (1937) (*A. gracilis* Buser, non Opiz). *W.C. Alps* (*Valais*). He.

118. A. fissa Günther & Schummel, *Sched. Cent. Siles. Exsicc.* **9**· no. 2 (1819) (*A. glaberrima* auct., *A. glabra* Poiret). Delicate, glabrous, with procumbent or ascending stems up to 25 cm long. Leaves more or less orbicular, thin, lobed to $\frac{1}{2}$–$\frac{2}{3}$; lobes 5–7, long, subcuneate, with a wide, rounded or subtruncate apex, somewhat overlapping, with long, almost U-shaped incisions, and 4–6 very large teeth, the longest up to 4·5 mm. Inflorescence small. Flowers 3·5–5 mm wide. *In snow-patches and on wet rocks.* ● *Pyrenees, Alps, Sudeten mountains.* Au Cz Ga Ge He Hs Po.

Records of this species for the Carpathians are all, *fide* Pawłowski and Rothmaler, referable to **116** or **117**.

24. Aphanes L.[1]

Like *Alchemilla* but annual, with deeply dissected leaves and conspicuous, connate stipules; flowers in condensed, leaf-opposed cymes; stamen 1 (rarely 2), inserted on inner margin of disc.

Measurements of length of the fruiting hypanthium include the persistent sepals.

Literature: W. Rothmaler, *Feddes Repert.* **53**: 265–270 (1944).

1 Stems rather slender, ascending; all leaves petiolate; fruiting
   hypanthium less than 2·75 mm
2 Fruiting hypanthium more than 2 mm; lobes of stipules
   triangular **1. arvensis**
2 Fruiting hypanthium less than 2 mm; lobes of stipules oblong
   2. microcarpa
1 Stems rather robust, erect; at least the upper leaves sessile;
   fruiting hypanthium at least 2·75 mm
3 All leaves sessile; stipules densely imbricate **3. cornucopioides**
3 Lower leaves petiolate; stipules not or scarcely imbricate
   4. floribunda

1. A. arvensis L., *Sp. Pl.* 123 (1753) (*Alchemilla arvensis* (L.) Scop.). Rather slender, usually greyish-green, hairy. Stems up to 30 cm, ascending, usually much-branched from the base. Leaves deeply 3-fid, the segments divided at the apex into 3–5 oblong lobes; petiole short, adnate to the stipules. Lobes of stipules 5–7, triangular, about half as long as entire portion. Inflorescence more or less sessile; fruiting hypanthium 2·2–2·6 mm, usually slightly protruding beyond stipules; sepals somewhat spreading. $2n=48$. *Cultivated ground and other open habitats.* S., W. & C.

Europe, extending north-eastwards to S. Sweden, Latvia and N.E. Poland. All except ?Al ?Az Fa Fe Is No Rs (N, ?C, ?W, E) Sb.

Facultatively apomictic.

2. A. microcarpa (Boiss. & Reuter) Rothm., *Feddes Repert.* **42**: 172 (1937) (*A. arvensis* auct. pro parte, non L.). Like **1** but usually more slender and of a purer green; lobes of stipules oblong, usually almost as long as the entire portion; fruiting hypanthium (1–)1·4–1·9 mm, often not protruding beyond stipules; sepals connivent. $2n=16$. *Open habitats on sandy, usually acid soils. Europe northwards to Scotland and S. Sweden, and eastwards to N.E. Poland, the Carpathians and the Adriatic; a few isolated stations in the Balkan peninsula.* Al Au Az Be Br Bu Co Cz Da Ga Ge Gr Hb He Ho Hs Hu It Ju Lu Po Rm Sa Si Su Tu.

A sexual species, closely related to **1**.

Plants from Corse have been distinguished as var. *bonifaciensis* Buser; they differ in their smaller flowers and fruiting hypanthia and in their diffuse, branched habit. An even smaller-flowered variant is **Alchemilla minutiflora** Aznav., *Bull. Soc. Bot. Fr.* **46**: 141 (1899), described from Turkey-in-Europe. At present, it seems best to include all such plants from S. Europe in **2**.

3. A. cornucopioides Lag., *Gen. Sp. Nov.* 7 (1816) (*Alchemilla cornucopioides* (Lag.) Roemer & Schultes). Rather robust, very hairy. Stems up to 15 cm, erect, densely leafy. Leaves sessile, more or less 3-fid, each segment 2- to 3-lobed. Stipules densely imbricate, with triangular lobes, clearly adnate to base of leaf and forming an amplexicaul cup. Fruiting hypanthium 2·75–3·5 mm; sepals erecto-patent. *Dry, acid soils. Iberian peninsula.* Hs Lu.

4. A. floribunda (Murb.) Rothm., *Feddes Repert.* **42**: 172 (1937). Like **3** but stems not so densely leafy; lower leaves with short petioles; stipules not or scarcely imbricate. *Dry, acid soils. Mediterranean region, from Islas Baleares eastwards.* Bl Co Gr It Ju Sa Si Tu.

Subfam. Maloideae

Stipules present, usually caducous. Flowers 5-merous. Hypanthium tubular, not open at the apex and completely enclosing the 2–5 more or less connate carpels which are more or less adnate to the hypanthium; epicalyx absent; stamens numerous. Fruit a fleshy pome, developed from the carpels and hypanthium. Basic chromosome number 17.

25. Cydonia Miller[2]

Deciduous shrubs or trees. Leaves entire. Stipules caducous. Flowers solitary. Sepals shorter than petals, dentate, persistent; stamens 15–25; carpels 5, walls cartilaginous in fruit; ovules numerous; styles 5, free. Fruit many-seeded.

Chaenomeles speciosa (Sweet) Nakai, *Jap. Jour. Bot.* **4**: 331 (1929) (*Cydonia japonica* auct., non (Thunb.) Pers.) and **C. japonica** (Thunb.) Spach, *Hist. Vég.* (*Phan.*) **2**: 159 (1834), are widely cultivated for ornament, and may occur as escapes, though it is doubtful if they are naturalized. *Chaenomeles* is like *Cydonia* but has leaves serrate, sepals deciduous, stamens 40–60 and styles connate at the base.

1. C. oblonga Miller, *Gard. Dict.* ed. 8, no. 1 (1768) (*C. vulgaris* Pers.). Shrub or tree 1·5–6 m. Shoots at first villous, later glab-

---

[1] By S. M. Walters.   [2] By A. Terpó.

rous. Leaves 5–10 × 3·5–7·5 cm, ovate, entire. Flowers 4–4·5 cm in diameter; pedicels short, tomentose. Petals pink. Fruit 2·5–3·5 cm (5–12 cm in cultivation), globose or pyriform, fragrant, yellow, tomentose. 2*n* = 34. *Cultivated throughout a large part of Europe; naturalized in hedges and thickets in S. Europe and more locally in C. Europe.* [Al Au Bu Co Cz Ga Ge Gr He Hs Hu It Ju Lu Rm Rs (K) Sa Si Tu.] (*S.W. & C. Asia.*)

26. Pyrus L.[1]

Deciduous shrubs or trees, often spiny on lower branches or when young. Leaves simple, rarely lobed; stipules caducous. Flowers in corymbs. Petals clawed, white, rarely pinkish; stamens 15–30; anthers dehiscing centripetally, usually red; carpels 2–5, connate, walls cartilaginous in fruit; ovules 2; styles 2–5, free. Fruit pyriform, turbinate or globose; flesh containing stone-cells.

The leaves described are the middle leaves of the short shoots.

P. salicifolia Pallas, *Reise* 3: 734 (1776), a Caucasian species, has been recorded from Krym and European Turkey, but apparently in error for 11 and 9 respectively.

Literature: An. A. Federov in Sokolov, *Derev'ja i Kustarniki SSSR* 3: 151–306 (1954). A. Terpó, *Ann. Acad. Horti-Viticult. (Budapest)* **22**(6, 2): 1–258 (1960). J. do Amaral Franco & M. L. da Rocha Afonso, *Rev. Fac. Ci. (Lisboa)* ser. 2C, **13**(2): 175–213 (1965).

1 Fruit with deciduous calyx
  2 Leaves aristate-dentate; inflorescence and unfolding leaves glabrous   **2. magyarica**
  2 Leaves crenate-dentate, serrulate or entire; inflorescence and unfolding leaves pubescent or tomentose
    3 Leaves crenate-dentate or serrulate   **1. cordata**
    3 Leaves entire   **3. rossica**
1 Fruit with persistent calyx
  4 Fruit (5–)6–16 cm, fleshy, sweet-tasting   **13. communis**
  4 Fruit not more than 5·5 cm, hard, usually not sweet-tasting
    5 Leaves not more than 1½ times as long as wide
      6 Leaves cuspidate, entire   **5. caucasica**
      6 Leaves not cuspidate, crenulate or serrulate at least in part
        7 Petals 10–17 × 7–13 mm; leaves thin   **4. pyraster**
        7 Petals 8–10 × 5–7 mm; leaves thick   **6. bourgaeana**
    5 Leaves more than 1½ times as long as wide
      8 Leaves glabrous when unfolding   **7. syriaca**
      8 Leaves hairy when unfolding
        9 Mature leaves glabrous or papillose beneath
          10 Leaves crenulate, rounded at the base   **6. bourgaeana**
          10 Leaves entire to slightly serrulate, cuneate at the base   **8. amygdaliformis**
        9 Mature leaves pubescent or tomentose beneath
          11 Styles densely villous, at least at base
            12 Styles villous to the middle; fruit 2–3 cm   **10. elaeagrifolia**
            12 Styles villous only at base; fruit 3–5 cm   **11. nivalis**
          11 Styles ± glabrous
            13 Leaves usually less than 3·5 cm wide, entire   **9. salvifolia**
            13 Leaves usually more than 3·5 cm wide, serrulate-crenulate towards apex   **12. austriaca**

1. P. cordata Desv., *Obs. Pl. Env. Angers* 152 (1818). Shrub or small tree up to 8 m, with patent, usually spiny branches. Twigs purplish. Leaves 2·5–5·5 × 1·5–3·5 cm, ovate-lanceolate to ovate, crenate-dentate or serrulate, rounded to subcordate at base, acuminate to cuspidate at apex, pubescent when young; petiole 2–5 cm, slender. Corymbs and leaves densely tomentose when unfolding. Sepals 2–3 × 1–1·5 mm, triangular-subulate. Petals

[1] By A. Terpó and J. do Amaral Franco.

6–8 × 5–7 mm, obovate-elliptical. Fruit 0·8–1·5 (–1·8) cm, globose or obovoid, shiny, red, densely covered with lenticels; pedicel 1·2–2·5(–3·5) cm, slender; calyx deciduous. *Woods and hedges.* ● *Western margin of Europe, from C. Portugal to S.W. England.* Br Ga Hs Lu.

2. **P. magyarica** Terpó, *Ann. Acad. Horti-Viticult. (Budapest)* **22**(6, 2): 34 (1960). Like **1** but leaves aristate-dentate; inflorescence and leaves more or less glabrous; fruit 1·5–2 cm in diameter, not densely covered with lenticels. *In* Quercus *woodland*. ● *Hungary*. Hu.

3. **P. rossica** Danilov, *Not. Syst. (Leningrad)* **15**: 126 (1953). Tree 15–20 m, with spiny branches; bark peeling in wide, thin sheets. Leaves 3–7 × 2–6 cm, broadly ovate, entire, tomentose when unfolding; petiole 1½ times as long as lamina. Fruit 2–2·5 cm in diameter, globose, somewhat compressed, densely covered with lenticels; calyx deciduous. *In* Quercus *woodland.* ● *S.C. Russia (Kurskaja and Voronežskaja Oblasti)*. Rs (C).

4. **P. pyraster** Burgsd., *Anleit. Erzieh. Holzart.* **2**: 193 (1787) (*P. communis* auct., non L., *P. communis* var. *achras* Wallr.). Tree 8–20 m, with patent or ascending, usually spiny branches. Twigs grey to brown. Bud-scales 5–8. Leaves 2·5–7 × 2–5 cm, elliptical, ovate or orbicular, cuneate, rounded or cordate at base, acute or shortly acuminate, thin, crenulate-serrulate throughout or only at the apex, rarely entire, usually glabrous at maturity; petiole 2–7 cm, slender. Sepals 3–8 × 1–3·5 mm. Petals 10–17 × 7–13 mm, elliptical to orbicular. Fruit 1·3–3·5 × 1·8–3·5 cm, globose to turbinate, yellow, brown or black, the lenticels often conspicuous; pedicel 1–5·5 cm, slender; calyx persistent. *Thickets and open woods. S., W. & C. Europe and S. half of U.S.S.R.* Al Au Be ?Br Bu Cz ?Da Ga Ge Gr He Hs Hu It Ju Lu Po Rm Rs (C, W, E) Si.

This species, which comprises many of the variants of the wild and naturalized pears of Europe, has received a variety of taxonomic treatments. It has not yet been possible to subdivide it satisfactorily.

5. **P. caucasica** Fedorov in Grossh., *Fl. Kavk.* ed. 2, **5**: 421 (1952). Tree 15–25 m, with pyramidal crown and spiny lower branches. Twigs grey, glabrous. Leaves 3–5·5 × 2·5–4 cm, usually orbicular-ovate and caudate-cuspidate, chartaceous, entire, ciliate and sparsely pubescent while young, soon glabrous; petiole 2–5·5 cm, slender. Corymbs glabrous except the tomentose sepals. Sepals 4–5 × 1–1·5 mm, linear. Petals 11–13 × 9–11 mm, elliptical. Fruit 1·5–3·5 × 2–4 cm, globose-turbinate, dull, brown; pedicel 1–4 cm, stoutish; calyx persistent. *Thrace; Krym.* Gr Rs (K) ?Tu. (*Caucasus, N. Anatolia.*)

6. **P. bourgaeana** Decne, *Jard. Fruit.* **1**: t. 2 (1871) (*P. communis* var. *mariana* Willk., *P. communis* auct. iber., non L.). Tree up to 10 m, with open crown and spiny lower branches. Twigs grey, stoutish. Leaves 2–4 × 1·5–3·5 cm, ovate or ovate-cordate, rarely lanceolate, thick, crenulate; petiole 2–5 cm, slender. Corymbs and leaves tomentose when unfolding. Sepals 5–7 × 2–2·5 mm, oblong-lanceolate, mucronate. Petals 8–10 × 5–7 mm, obovate-cuneate. Fruit 1·7–2·5 cm in diameter, turbinate-globose, dull, yellow with brown spots to almost brown when mature; pedicel 1·5–4 cm, stout; calyx persistent. *Usually near seasonal streams. Portugal, W. Spain.* Hs Lu. (*Morocco.*)

7. **P. syriaca** Boiss., *Diagn. Pl. Or. Nov.* **2**(10): 1 (1849). Small, round-headed tree up to 10 m, with spiny branches. Twigs grey but reddish-brown when young, always glabrous. Leaves 3–9 × 1·5–3 cm, oblong-lanceolate, serrulate or crenulate, coriaceous,

shining, glabrous; petiole 2–5 cm, slender. Corymbs glabrous except for the sepals. Sepals 5 × 2·5 mm, lanceolate, tomentose; petals 10–12 mm, orbicular-elliptical. Fruit 3–3·5 cm in diameter, globose to turbinate; pedicel 2·5–4 cm, stout; calyx persistent. *Cultivated locally, and occasionally naturalized.* [Hu.] (*S.W. Asia.*)

8. P. amygdaliformis Vill., *Cat. Méth. Jard. Strasb.* 323 (1807) (*P. parviflora* Desf., *P. nivalis* sensu Lindley, non Jacq.). Shrub or small tree up to 6 m; branches sometimes spiny. Twigs grey, dull, tomentose while young. Leaves 2·5–8 × 1–3 cm, narrowly lanceolate to obovate, usually entire, rarely 3-lobed, with rounded or cuneate base, sparsely hairy when young, papillose beneath at maturity; petiole 2–5 cm. Sepals 5–6 × 1·5 mm, triangular, acuminate. Petals 7–8 × 5–6 mm, elliptical, usually emarginate at apex. Fruit 1·5–3 cm in diameter, usually globose, fulvous; pedicel stout, as long as or slightly longer than fruit; calyx persistent. *Dry, rocky places. Mediterranean region, Bulgaria.* Al Bu Co Cr Ga Gr Hs It Ju Sa Si Tu.

P. mecsekensis Terpó, *Ann. Acad. Horti-Viticult.* (*Budapest*) **22** (6, 2): 133 (1960), from Hungary, a tree 10–15 m, with ovate-lanceolate, crenulate leaves, is intermediate between **4** and **8** and is possibly of hybrid origin.

P. pyrainus Rafin., *Specch. Sci.* **2**: 173 (1814), from Sicilia, is doubtfully distinct from **8**.

9. P. salvifolia DC., *Prodr.* **2**: 634 (1825) (*P. communis* subsp. *salvifolia* (DC.) Gams). Small or medium-sized tree; branches usually spiny. Leaves 4–7 × 2–3·5 cm, lanceolate or elliptical, entire, glabrescent above, grey-tomentose beneath; petiole 2–5 cm. Styles more or less glabrous. Fruit turbinate or pyriform; calyx persistent. *Sunny slopes and dry, open woods.* ● *From Belgium to Greece and Krym.* Au Be Ga Gr Hu Ju Po Rm Rs (K).

Often cultivated, and perhaps only naturalized over part of its range.

10. P. elaeagrifolia Pallas, *Nova Acta Acad. Sci. Petrop.* **7**: 355 (1793). Shrub or small tree with stout, erect spiny branches. Twigs grey-tomentose. Leaves 3·5–8 × 2–3·5 cm, lanceolate to obovate-lanceolate, entire, or crenulate at apex, with a dense, grey-white tomentum; petiole shorter than lamina. Corymbs many-flowered, whitish-tomentose, subsessile. Sepals *c.* 5 × 1·5 mm, linear-triangular. Petals *c.* 10 × 7 mm, elliptical. Styles densely villous in lower half. Fruit 2–3 cm in diameter, pyriform to globose; pedicel 2–3 cm, stout; calyx persistent. *Dry places. S.E. Europe.* Al Bu Gr Rm Rs (K) Tu.

11. P. nivalis Jacq., *Fl. Austr.* **2**: 4 (1774) (*P. communis* subsp. *nivalis* (Jacq.) Gams). Tree 8–20 m, with stout, ascending, usually spineless branches. Twigs stout, white-tomentose when young, later blackish. Leaves 5–9 × 3–4 cm; lamina obovate, cuneate at base and decurrent, entire or slightly crenulate at apex, covered sparsely above and densely beneath with a whitish-grey pubescense; petiole 1–2 cm, tomentose. Corymbs tomentose-lanate. Sepals 6–8 × 3–4 mm, triangular-acuminate. Petals 14–16 × 12–14 mm, obovate-elliptical. Styles villous only at base. Fruit 3–5 cm in diameter, globose, yellowish-green with purple dots, becoming sweet when over-ripe; pedicel as long as or longer than fruit; calyx persistent. *Sunny slopes and dry, open woods.* ● *S. & S.C. Europe.* Au Bu Cz Ga He Hu It Ju Rm.

Several varieties with narrower leaves and smaller fruits are known, and many are grown as rootstocks.

---

¹ By A. Terpó.

12. P. austriaca A. Kerner, *Sched. Fl. Exsicc. Austro-Hung.* **7**: 15 (1896). Medium-sized or large tree, with black, spineless branches. Twigs stout, black, greyish-tomentose when young. Leaves 6–9 × 2·5–5 cm, lanceolate to obovate-lanceolate, serrulate-crenulate towards the acuminate apex, glabrescent above and with a yellow-grey tomentum beneath; petiole 1·5–6 cm. Corymbs tomentose. Sepals and petals similar to those of **11** but slightly smaller. Styles more or less glabrous. Fruit 2·5–5·5 × 2–4·5 cm, turbinate or pyriform; calyx persistent. ● *C. Europe.* Au Cz He Hu ?Rm.

Often cultivated, and perhaps only naturalized over part of its range.

13. P. communis L., *Sp. Pl.* 479, 1200 (1753). Tree up to 20 m; branches with or without spines, ascending on young and spreading on adult trees. Twigs stout, reddish-brown, soon becoming glabrous and shining. Leaves 5–8 × 3·5–5·5 cm, ovate and elliptical, more or less cuspidate at apex, crenulate-serrulate to subentire, usually glabrous at maturity; petiole equalling or shorter than lamina. Corymbs and leaves tomentose when unfolding. Sepals 6–8 × 3–4 mm, lanceolate-acuminate. Petals 12–14 × 10–12 mm, obovate. Fruit (5–)6–16 × 4–12 cm, oblong, pyriform, turbinate or subglobose, with a sweet taste; calyx persistent. *Cultivated on a field scale in most of Europe except the north and the drier regions of the south.*

More than a thousand cultivars of the garden pear are known. It is of hybrid origin, and many species are considered to be among its parents, including **4**, **7**, **9**, **11** and **12**.

27. Malus Miller[1]

Deciduous, rarely spiny shrubs or trees. Leaves simple, sometimes lobed. Flowers in umbels. Petals clawed, white, pink or red; stamens 15–50; anthers dehiscing centrifugally, yellow; carpels 3–5, connate, walls cartilaginous in fruit; ovules 2 or more; styles 2–5, connate at base. Fruit more or less globose; flesh usually without stone-cells.

Many species are grown for ornament.

Literature: W. Henning, *Züchter* **17–18**: 289–349 (1947).

1 Leaves lobed		
2 Petiole 2–7 cm; sepals 7–10 mm, persistent		**1. trilobata**
2 Petiole 0·5–2 cm; sepals 3–4 mm, deciduous		**2. florentina**
1 Leaves not lobed		
3 Mature leaves glabrous on both surfaces		**3. sylvestris**
3 Mature leaves tomentose, at least beneath		
4 Fruit more than 5 cm		**6. domestica**
4 Fruit less than 5 cm		
5 Leaves elliptical; styles glabrous		**4. dasyphylla**
5 Leaves ovate or obovate; styles sometimes villous in lower half		**5. praecox**

Sect. ERIOLOBUS (DC.) C. K. Schneider. Leaves conduplicate in bud, deeply lobed. Fruit with stone-cells.

1. M. trilobata (Labill.) C. K. Schneider, *Feddes Repert.* **3**: 179 (1906) (*Sorbus trilobata* (Labill.) Heynh.). Unarmed shrub up to 10 m; twigs densely pubescent at first, soon glabrous. Leaves 4–10 cm, often wider than long, serrulate, usually cordate at base, deeply 3-lobed, the main lobes usually 2- to 3-lobed, glabrous above, sparsely pubescent and becoming subglabrous beneath; petiole 2–7 cm. Flowers *c.* 3·5 cm in diameter, white. Sepals 7–10 mm, tomentose, persistent. Fruit 2–3 cm, obovoid, yellowish-green. *Evergreen scrub. N.E. Greece.* Gr [*Bu.] (*Syria, Lebanon, Israel.*)

2. M. florentina (Zuccagni) C. K. Schneider, *Ill. Handb. Laubholzk.* **1**: 724 (1906) ('*Sorbus torminalis* × *Malus pumila*' sensu Hayek). Unarmed small tree up to 4 m. Leaves 3–6 cm, broadly ovate, dentate, truncate or cordate at base, irregularly incise-lobed with several lobes on each side, white-tomentose beneath; petiole 0·5–2 cm. Flowers 1·5–2 cm in diameter, white. Sepals 3–4 mm, deciduous. Fruit *c.* 1 cm, ellipsoid or obovoid, red. ● *Italy; S. Jugoslavia and N. Greece; very local.* Al Gr It Ju.

Sect. MALUS. Leaves involute in bud, simple. Fruit without stone-cells.

3. M. sylvestris Miller, *Gard. Dict.* ed. 8, no. 1 (1768) (*M. communis* subsp. *sylvestris* (Miller) Gams, *M. acerba* Mérat). More or less spiny tree or shrub 2–10 m. Leaves 3–11 × 2·5–5·5 cm, ovate, elliptical or suborbicular, crenate or serrate, with rounded or cuneate base, shortly apiculate, glabrous when mature; petiole 1·5–3 cm. Flowers 3–4 cm in diameter, white or pink; sepals 3–7 mm, glabrous externally, tomentose internally; styles glabrous or sparsely villous at the base. Fruit 2·5–3 cm, subglabrous, yellowish-green. *Most of Europe, northwards to C. Fennoscandia, but rather local on mountains in the south; cultivated as a rootstock.* All except Az Bl Cr Fa Is Rs (N) Sa Sb.

4. M. dasyphylla Borkh., *Handb. Forstbot.* **2**: 1269 (1803) (*M. communis* subsp. *pumila* auct., non (Miller) Gams, *M. pumila* var. *paradisiaca* auct., non (L.) C. K. Schneider). Medium-sized, sparsely spiny tree; twigs tomentose, becoming glabrous. Leaves 3·5–11 × 2·5–5 cm, elliptical, acuminate, crenate, tomentose beneath; petiole 1–5 cm. Styles glabrous. Fruit 4 cm, yellowish (occasionally red along one side), acid; pedicel (1–)2·5(–5) cm. *Damp, lowland woods.* ● *Danube basin and N. part of Balkan peninsula.* Al Au Bu Gr Hu Ju Rm.

M. pumila Miller, *Gard. Dict.* ed. 8, no. 3 (1768) (*M. pumila* var. *paradisiaca* (L.) C. K. Schneider), a shrub up to 2 m, with abundant greenish, short-stalked, sweet fruits 3–5 cm in diameter, is widely cultivated as the paradise apple.

5. M. praecox (Pallas) Borkh., *Handb. Forstbot.* **2**: 1271 (1803) (*M. pumila* var. *praecox* (Pallas) C. K. Schneider). Small tree or shrub, sometimes spiny; twigs glabrous. Leaves 2–10 × 1–5 cm, ovate or obovate, abruptly acuminate, serrate, biserrate or crenate-serrate, tomentose beneath; petiole 0·5–5 cm. Flowers 4–5 cm in diameter; sepals tomentose both externally and internally; styles glabrous, or villous in lower half. Fruit 2–2·5 cm, equalling pedicel. *Deciduous woodland along rivers.* ● *U.S.S.R. northwards to c.* 55° N. *and eastwards to c.* 50° E. Rs (C, W, E).

6. M. domestica Borkh., *Handb. Forstbot.* **2**: 1272 (1803). Unarmed small to medium-sized tree; twigs tomentose. Leaves 4–13 × 3–7 cm, ovate-elliptical, serrate, with rounded, rarely cordate base, slightly tomentose above and densely tomentose beneath. Fruit more than 5 cm, varying in colour, sweet or acid, much longer than pedicel. *Cultivated for its fruit almost throughout Europe. Often escaping and occasionally naturalized.*

The apple is of hybrid origin, and has probably been derived from **3, 4, 5** and some Asiatic species. More than a thousand cultivars are grown.

28. Sorbus L.[1]

Deciduous trees or shrubs without spines. Leaves simple, lobed or pinnate. Flowers in compound corymbs. Petals white, rarely pink; stamens 15–25; carpels 2–5, partly free or connate, walls cartilaginous or membranous in fruit; ovules 2; styles free or connate at base.

The five most widespread European species (**1, 2, 3, 4** and **5**) are amphimictic, but some at least of the local species are polyploid and apomictic. Species **1** does not hybridize with the others in nature, but **2, 3, 4, 5** and their products form a hybrid complex. Some of the hybrid products are amphi- and some apomictic. It is likely that many of the apomictic species have arisen by hybridization, and their probable origin is given in the text. It is not practicable to describe all these species, partly because their taxonomy has not been elucidated for the whole of Europe. Accordingly, a number of representative species has been described (**6** to **18**), and after each of these have been listed the other species which seem to be most closely related to them. (This is analogous to the procedure in *Rubus*.)

Important characters used in the identification of the species are the depth and character of toothing and lobing of leaves of short shoots, and the colour of the fruit and the distribution of its lenticels. The leaves referred to in the descriptions are always those of the short shoots, unless otherwise specified.

A more detailed treatment is given in the monograph by T. Hedlund, *Kungl. Svenska Vet.-Akad. Handl.* nov. ser., **35**(1): 1–147 (1901). Accounts for particular countries include: E. F. Warburg in Clapham, Tutin and Warburg, *Flora of the British Isles* ed. 2, 423–437. Cambridge. 1962. Z. Kárpáti, *Feddes Repert.* **62**: 71–331 (1960) and *Bot. Közl.* **52**: 135–140 (1966). R. Düll, *Ber. Bayer. Bot. Ges.* **34**: 11–65 (1961). M. Kovanda, *Acta Dendrol. Čech.* **3**: 23–70 (1961). There is an account of cytology and the breeding system in two papers by A. Liljefors, *Acta Horti Berg.* **16**(10): 277–329 (1953) and **17**: 47–113 (1955).

1 Leaves pinnate, with the terminal leaflet about the same size as the others
  2 Bark shredding; styles 5; ripe fruit 20 mm or more, greenish or brownish **1. domestica**
  2 Bark smooth; styles 3–4; ripe fruit less than 14 mm, scarlet **2. aucuparia**
1 Leaves simple, lobed, or pinnatifid, or, if pinnate, with the terminal leaflet much larger than the others
  3 Leaves green beneath at maturity, more or less concolorous
    4 Tree; leaves lobed; fruit brown **3. torminalis**
    4 Shrub; leaves not lobed; fruit scarlet **4. chamaemespilus**
  3 Leaves white- or grey-tomentose beneath at maturity
    5 Leaves not or only very shallowly lobed
      6 Leaves grey-tomentose beneath
        7 Leaves not lobed, coarsely toothed **17. sudetica**
        7 Leaves very shallowly lobed, finely toothed **18. margittaiana**
      6 Leaves white-tomentose beneath
        8 Leaves usually widest at or below middle, rounded at base; fruit usually longer than wide **5. aria**
        8 Leaves usually widest above middle, cuneate at base; fruit not longer than wide
          9 Leaves with teeth symmetrical and patent; fruit usually less than 12 mm, with few lenticels **6. graeca**
          9 Leaves with teeth curved on outer edge; fruit usually more than 12 mm, with numerous lenticels **7. rupicola**
    5 Leaves distinctly lobed
      10 Leaves white-tomentose beneath, usually with less than 7 pairs of lateral veins **8. umbellata**
      10 Leaves with whitish-, yellowish- or grey-green tomentum beneath, usually with more than 7 pairs of lateral veins
        11 Leaves with at least 1 pair of free leaflets
          12 Leaves with 2 pairs of free leaflets **13. hybrida**
          12 Leaves with 4–5 pairs of free leaflets **14. meinichii**
        11 Leaves (at least those of the short shoots) without free leaflets
          13 Leaves deeply lobed, lobes extending to $\frac{2}{5}$ of the way to the midrib **12. dacica**

---

[1] By E. F. Warburg and Z. E. Kárpáti.

13 Leaves less deeply lobed
14 Leaves usually less than 8 cm; fruit 6–8 mm   **9. minima**
14 Leaves usually more than 8 cm; fruit more than 8 mm
   15 Leaf-lobes shallow, extending about $\frac{1}{8}$ of way to midrib   **16. latifolia**
   15 Leaf-lobes well-marked, extending $\frac{1}{4}$–$\frac{1}{3}$ of way to midrib
     16 Leaves $1\frac{1}{2}$–2 times as long as wide; fruit with few, small lenticels
       17 Leaves with whitish-grey tomentum beneath; fruit c. 10 mm, subglobose   **10. mougeotii**
       17 Leaves with yellowish-grey tomentum beneath; fruit 12–15 mm, much longer than wide
             15. intermedia
     16 Leaves c. $1\frac{1}{4}$ times as long as wide; fruit with many, large lenticels
       18 Fruit red   **11. austriaca**
       18 Fruit yellowish-brown   **16. latifolia**

1. S. domestica L., *Sp. Pl.* 477 (1753). Tree up to 20 m, with patent branches; bark shredding. Leaves pinnate, with 6–8 pairs of leaflets; leaflets 3–4·5 cm, oblong, serrate, pubescent beneath when young, finally glabrous. Flowers 16–18 mm in diameter, white; sepals triangular, longer than wide; styles 5. Fruit 20 mm or more, obovoid or pyriform, greenish or brownish, with numerous stone-cells. *S. Europe, extending northwards to C. Germany; planted for its fruit and for ornament in C. Europe and locally naturalized.* Al Bu Co Ga Ge Gr *He Hs Hu It Ju Rm Rs (K) Sa Si Tu [Au Cz].

2. S. aucuparia L., *loc. cit.* (1753). Tree up to 15(–20) m, usually with erecto-patent branches; bark smooth. Leaves pinnate, with 5–7 pairs of leaflets; leaflets 2·5–6(–9) cm, oblong, serrate. Flowers 8–10 mm in diameter, white; sepals deltate, sometimes rounded; styles 3–4. Fruit 6–9(–14) mm, subglobose, depressed-globose, or ovoid, scarlet, with few or no stone-cells. $2n = 34$. *Most of Europe.* All except Az Bl Cr Fa Sa Sb Tu.

1 Leaflets up to 9 cm; fruit depressed-globose   **(e) subsp. fenenkiana**
1 Leaflets 2·5–6 cm; fruit subglobose or ovoid
  2 Petiole not more than 2 cm   **(d) subsp. praemorsa**
  2 Petiole usually more than 2·5 cm
    3 Sepals glabrous   **(c) subsp. sibirica**
    3 Sepals hairy
      4 Inflorescence-axis ± hairy; fruit subglobose
          (a) subsp. aucuparia
      4 Inflorescence-axis glabrous; fruit longer than wide
          (b) subsp. glabrata

(a) Subsp. **aucuparia**: Buds, leaves beneath and inflorescence-axis more or less hairy. Petiole usually more than 2·5 cm. Leaflets firm, subobtuse or abruptly narrowed to an acute apex. Sepals deltate, hairy. Fruit subglobose. *Throughout most of the range of the species but rarer in the south.*

(b) Subsp. **glabrata** (Wimmer & Grab.) Cajander, *Suomen Kasvio* 360 (1906) (*S. glabrata* (Wimmer & Grab.) Hedl.): Less hairy than subsp. (a). Petiole usually more than 2·5 cm. Leaflets thin, gradually tapered to an acute apex, subglabrous or sparsely hairy on both surfaces. Inflorescence-axis glabrous or nearly so. Sepals rounded, hairy. Fruit longer than wide. *N. Europe and mountains of C. Europe.*

(c) Subsp. **sibirica** (Hedl.) Krylov, *Fl. Zap. Sibir.* 7: 1464 (1933) (*S. sibirica* Hedl.): Glabrous or nearly so. Petiole usually more than 2·5 cm. Leaflets gradually tapered to an acute apex, glabrous, or hairy only on midrib beneath. Inflorescence-axis glabrous. Sepals deltate, glabrous. *N.E. Russia.*

(d) Subsp. **praemorsa** (Guss.) Nyman, *Consp.* 241 (1878): Petiole not more than 2 cm. Leaflets $2\frac{1}{2}$ times as long as wide, subobtuse, bluntly serrate, hairy beneath. Fruit ovoid. *S. Italy, Sicilia, Corse.*

(e) Subsp. **fenenkiana** Georgiev & Stoj., *Bull. Soc. Bot. Bulg.* 5: 101 (1932): Leaflets up to 9 × 1–1·8 cm, linear-lanceolate, thin, sparsely hairy on midrib beneath. Inflorescence many- (up to 200-)flowered. Fruit 10–12 × 12–14 mm, depressed-globose. *Bulgaria.*

3. S. torminalis (L.) Crantz, *Stirp. Austr.* 2: 45 (1763). Tree up to 25 m. Leaves 5–9 cm, ovate, with 3–4 pairs of triangular-ovate to lanceolate lobes, decreasing in size towards the apex, serrate, green on both surfaces, glabrous above, pubescent beneath at least when young, the pubescence rarely persistent. Flowers 10–15 mm in diameter, white; styles 2. Fruit 12–18 mm, obovoid, rarely subglobose, brown, with numerous lenticels. $2n = 34$. *S., W. & C. Europe, extending to E. Denmark.* Al Au Be Br Bu Cz Da Ga Ge Gr He Hs Hu It Ju Lu Po Rm Rs (W, K) Sa Si Tu.

4. S. chamaemespilus (L.) Crantz, *op. cit.* 40 (1763). Shrub up to 1·5 m. Leaves 3–6 cm, elliptical to obovate, serrate, green on both surfaces, glabrous above, glabrous or sparsely pubescent beneath. Corymbs small and dense. Petals c. 5 mm, erect, pink; styles 2. Fruit 10–13 mm, subglobose or ovoid, scarlet, with few lenticels. $2n = 34$. ● *Mountains of C. & S. Europe, from the Vosges and the Carpathians to the Pyrenees, S. Italy and Bulgaria.* Al Au Bu Cz Ga Ge ?Gr He Hs It Ju Po Rm.

5. S. aria (L.) Crantz, *op. cit.* 46 (1763). Tree up to 25 m with wide crown, or a shrub. Leaves 5–12 cm, ovate or elliptical, usually widest at or below the middle and usually rounded at base, irregularly biserrate or shallowly lobed, the teeth curved on the outer edge and pointing towards the apex, densely and evenly white-tomentose beneath; veins (9–)10–14(–15) pairs. Flowers 10–15 mm in diameter, white. Fruit 8–15 mm, usually longer than wide, scarlet, usually with numerous small lenticels. $2n = 34$. *From Ireland and Spain eastwards to the Carpathians; southern limits uncertain because of confusion with allied species.* Al Au Be Bl Br Bu Co Cz Ga Ge Hb He Hs Hu It Ju Po Rm Rs (W) Sa Si.

(a) Subsp. **aria**: Leaves up to 12 cm, ovate or elliptical; veins distant; with greenish-white tomentum beneath. Amphimictic. *Throughout most of the range of the species.*

(b) Subsp. **lanifera** (A. Kerner) Jáv., *Magyar Fl.* 481 (1924): Leaves up to 7 × 5 cm, elliptical, somewhat coriaceous; veins relatively close; with pure white tomentum beneath. ● *Mountains of W. Jugoslavia.*

Related species include:

S. leptophylla E. F. Warburg, *Watsonia* 4: 44 (1957). Apomictic. $2n = 68$. Br.

6. S. graeca (Spach) Kotschy in Unger & Kotschy, *Ins. Cypern* 369 (1865) (*S. cretica* Lindley, *S. meridionalis* (Guss.) Fritsch). Shrub or small tree. Leaves 5–9 × 4–7 cm, obovate or suborbicular, not lobed, widest above the middle, broadly cuneate at the base, somewhat coriaceous, with 9–11 pairs of veins, usually with a thick, greenish-white tomentum beneath, biserrate; teeth symmetrical and patent. Fruit usually less than 12 mm, subglobose, crimson, with few, large lenticels. $2n = 34$. *S.E. & E.C. Europe, extending to Sicilia, Czechoslovakia and Germany.* Al Au Bu Cr Cz Ge Gr Hu It Ju Rm Rs (K) Si.

Amphimictic diploids and apomictic plants have both been reported in this species.

Related species include:

S. baldaccii (Degen & Fritsch ex C. K. Schneider) Zinserl. in Komarov, *Fl. URSS* 9: 398 (1939). Al Ju.

S. eminens E. F. Warburg, *Watsonia* **4**: 44 (1957). $2n=68$. Br.

S. hibernica E. F. Warburg, *op. cit.* 44 (1957). Hb.

S. lancastriensis E. F. Warburg, *op. cit.* 45 (1957). Br.

S. norvegica Hedl., *Nyt Mag. Naturvid.* (*Christiania*) **52**: 254 (1914). Apomictic. No Su.

S. pannonica Kárpáti, *Feddes Repert.* **62**: 182 (1960). Au Cz Ge Hu ?Rm.

S. porrigens Hedl., *Nyt Mag. Naturvid.* (*Christiania*) **52**: 255 (1914). Gr.

S. porrigentiformis E. F. Warburg, *Watsonia* **4**: 45 (1957). Apomictic. $2n=51$, 68. Br.

S. wilmottiana E. F. Warburg, *Watsonia* **6**: 296 (1967). Br.

S. eminens and *S. pannonica* are intermediate between **5** and **6**.

7. S. rupicola (Syme) Hedl., *Nyt Mag. Naturvid.* (*Christiania*) **52**: 256 (1914) (*S. salicifolia* (Hartman) Hedl.). Shrub *c.* 2 m, rarely a small tree. Leaves (6–)8–14·5 cm, *c.* 1½–2 times as long as wide, obovate or oblanceolate, widest above the middle, usually cuneate and entire at the base, with 7–9 pairs of veins, not lobed, rather thickly white-tomentose beneath, coarsely and unequally serrate, the teeth somewhat curved on the outer margin and directed towards the leaf-apex. Petals *c.* 7 mm. Fruit 12–15 mm, subglobose, wider than long, carmine, with numerous scattered lenticels. $2n=68$. *Apomictic.* ● *Britain and Ireland; Norway, S. Sweden and Estonia* (*Saaremaa*). Br Hb No Rs (B) Su.

Related species include:

S. vexans E. F. Warburg, *Watsonia* **4**: 46 (1957). Br.

8. S. umbellata (Desf.) Fritsch in A. Kerner, *Sched. Fl. Exsicc. Austro-Hung.* **7**: 18 (1896). Shrub or small tree. Leaves (3–)4–7(–10) × 3–6·5 cm, usually broadly obovate or suborbicular, with 4–7 pairs of veins, distally with shallow lobes and coarsely toothed, lobes acute or obtuse, densely and thickly white-tomentose beneath. Flowers *c.* 15 mm in diameter. Fruit globose, yellowish. *S.E. Europe.* Al Bu Gr Ju Rm Rs (K) ?Si.

1 Leaves deltate, longer than wide, acutely cuneate at base; lobes acuminate — (b) subsp. **banatica**
1 Leaves subrhombic, suborbicular or ovate-orbicular, ± as long as wide, obtusely cuneate at base; lobes obtuse
2 Lobes extending ⅓–½ of the way to midrib — (c) subsp. **flabellifolia**
2 Lobes extending ¼ of the way to midrib
3 Leaves up to 6 cm in diameter — (a) subsp. **umbellata**
3 Leaves up to 4 cm in diameter — (d) subsp. **koevessii**

(a) Subsp. **umbellata**: Leaves up to 6 cm in diameter, subrhombic-orbicular, with 5–6 pairs of veins; lobes obtuse, extending ¼ of the way to midrib. *Balkan peninsula, Romania.*

(b) Subsp. **banatica** (Jáv.) Kárpáti, *Feddes Repert.* **62**: 179 (1960): Leaves up to 10 × 6·5 cm, deltate, longer than wide, with 6–7 pairs of veins, acute, narrowly cuneate at base; lobes acuminate. *S.W. Romania,* ?*Jugoslavia.*

(c) Subsp. **flabellifolia** (Spach) Kárpáti, *op. cit.* 182 (1960): Leaves up to 6 cm in diameter, subrhombic-orbicular, with 5–6 pairs of veins; lobes obtuse, extending to ⅓–⅖(–½) of the way to midrib. *S. Jugoslavia; Greece; Krym.*

(d) Subsp. **koevessii** (Pénzes) Kárpáti, *Ann. Sect. Horti-Viticult. Univ. Sci. Agr.* (*Budapest*) **12**: 146 (1948): Leaves 3–4 cm in diameter, suborbicular, with 4–5 pairs of veins; lobes obtuse, extending ¼ of the way to midrib. *Mountains of Balkan peninsula.*

Related species include:

S. danubialis (Jáv.) Kárpáti, *Feddes Repert.* **62**: 185 (1960). Au Cz Ge Hu ?Ju Rm Rs (W).

S. taurica Zinserl. in Komarov, *Fl. URSS* **9**: 497 (1939). Rs (K).

S. turcica Zinserl. in Komarov, *loc. cit.* (1939). Rs (K).

These 3 species are intermediate between **6** and **8**; the following is intermediate between **5** and **8**.

S. subdanubialis (Soó) Kárpáti, *Feddes Repert.* **62**: 188 (1960). Cz Hu.

9. S. minima (A. Ley) Hedl., *Kungl. Svenska Vet.-Akad. Handl.* nov. ser., **35**(**1**): 61 (1901). Shrub up to *c.* 3 m. Twigs relatively slender. Leaves 6–8 cm, about twice as long as wide, with 8–9 pairs of veins, shallowly lobed, the deepest lobes extending ⅕–⅓(–½) of the way to the midrib, at maturity subglabrous above, evenly and thinly grey-tomentose beneath. Petals *c.* 4 mm. Fruit 6–8 mm, subglobose, scarlet, with few, small lenticels. *Limestone rocks.* ● *Wales.* Br.

S. minima is apomictic, as are probably many of the following related species. All have probably originated as a result of hybridization between **2** and **7**.

S. arranensis Hedl., *Kungl. Svenska Vet.-Akad. Handl.* nov. ser., **35**(**1**): 60 (1901). Br.

S. lancifolia Hedl., *Skr. Vid.-Selsk. Kristiania* (*Math.-Nat.*) **1911**(**6**): 166 (1912). No.

S. leyana Wilmott, *Proc. Linn. Soc. London* **146**: 78 (1934). Br.

S. neglecta Hedl. in Hyl., *Uppsala Univ. Årsskr.* **1945**(**7**): 221 (1945). No.

S. subpinnata Hedl., *Nyt Mag. Naturvid.* (*Christiania*) **49**: 198 (1911). No.

10. S. mougeotii Soyer-Willemet & Godron, *Bull. Soc. Bot. Fr.* **5**: 447 (1858). (*S. scandica* sensu Coste, non (L.) Hedl.). Shrub or tree up to 20 m. Leaves 7–10 × 3·5–5·5 cm, ovate or obovate, obtuse, cuneate at the base, with 8–10 pairs of veins, shallowly lobed, the larger lobes extending about ¼ of the way to the midrib, not overlapping, at maturity subglabrous above and with whitish-grey tomentum beneath. Petals 5–6 mm. Fruit *c.* 10 mm in diameter, subglobose, slightly longer than wide, red, with small and sparse lenticels. ● *Alps* (*mainly in the west*); *Pyrenees.* ?Au Ga He Hs ?It.

This species is apomictic so far as is known; it probably originated as a result of hybridization between **2** and **6**. It is related to **11**, which replaces it in the mountains of E.C. & S.E. Europe.

11. S. austriaca (G. Beck) Hedl., *Kungl. Svenska Vet.-Akad. Handl.* nov. ser., **35**(**1**): 65 (1901) (*S. mougeotii* subsp. *austriaca* (G. Beck) Hayek). Small tree. Leaves 8–13 × 7–8 cm, broadly ovate to ovate-elliptical, *c.* 1·3 times as long as wide, with 8–11 pairs of veins, shallowly lobed, the larger lobes extending *c.* ⅓ of the way to the midrib, usually somewhat overlapping, with whitish-grey tomentum beneath. Fruit up to 13 mm in diameter, subglobose, red, with rather large and numerous lenticels. ● *E. Alps, Carpathians, Balkan peninsula.* Au Bu Cz Hu Ju Rm Rs (W).

1 Leaves about as long as wide — (d) subsp. **serpentini**
1 Leaves 1·3–1·5 times as long as wide
2 Leaves not coriaceous — (a) subsp. **austriaca**
2 Leaves coriaceous
3 Leaves up to 13 cm — (b) subsp. **hazslinszkyana**
3 Leaves up to 8 cm — (c) subsp. **croatica**

(a) Subsp. **austriaca**: Leaves up to 12 × 8 cm, ovate, obtusely cuneate at base, with 9–12 pairs of veins. *Throughout most of the range of the species, but only in the mountains.*

(b) Subsp. **hazslinszkyana** (Soó) Kárpáti, *Feddes Repert.* **62**: 175 (1960): Leaves up to 13 × 10 cm, broadly ovate or ovate-elliptical, rounded or broadly cuneate at the base, with (8–)10–11 pairs of veins, coriaceous. *Hill-regions of N. Hungary and S.E. Czechoslovakia.*

(c) Subsp. **croatica** Kárpáti, *op. cit.* 177 (1960): Leaves up to 8 × 6 cm, elliptical or ovate-elliptical, obtusely cuneate at base, with 8–9 pairs of veins, with dense whitish tomentum beneath, coriaceous. *W. Jugoslavia* (*Velebit*).

(d) Subsp. **serpentini** Kárpáti, *op. cit.* 177 (1960): Leaves 10 cm, orbicular or suborbicular, with 8–9(–10) pairs of veins, not coriaceous. *E. Austria* (*Burgenland*).

Related species include:

S. anglica Hedl., *Nyt Mag. Naturvid.* (*Christiania*) 52: 258 (1914). $2n = 68$. Br Hb.
S. subsimilis Hedl., *op. cit.* 257 (1914). No.

Related species which are intermediate between **5, 6** and **11** include:

S. buekkensis Soó, *Acta. Biol. Acad. Sci. Hung.* 3: 224 (1952). Cz Hu.
S. carpatica Borbás in C. K. Schneider, *Ill. Handb. Laubholzk.* 1: 686 (1906). Au Cz Ju ?Rm.
S. hungarica (Bornm.) Kárpáti, *Feddes Repert.* 62: 199 (1960). Cz Ju ?Rm.
S. javorkae (Soó) Kárpáti, *op. cit.* 196 (1960). Cz Hu.
S. sooi (Máthé) Kárpáti, *op. cit.* 199 (1960). Cz Hu.
S. velebitica Kárpáti, *op. cit.* 190 (1960). Ju.

12. S. dacica Borbás, *Österr. Bot. Zeitschr.* 37: 404 (1887). Small tree. Leaves 10 × 7 cm, ovate, *c.* 1½ times as long as wide, with 8–9 pairs of veins, rather deeply lobed, the larger lobes extending ⅓–⅖ of the way to the midrib, rather coriaceous, with white or whitish-grey tomentum beneath. Fruit *c.* 10 mm. ● *W.C. Romania.* Rm.

Related species include:

S. pseudothuringiaca Düll, *Ber. Bayer. Bot. Ges.* 34: 55 (1961). Ge.

This species and **12** may have originated as a result of hybridization between **2** and **11**.

13. S. hybrida L., *Sp. Pl.* ed. 2, 684 (1762) (*S. fennica* (Kalm) Fries). Medium-sized tree. Leaves 7·5–10·5 cm, ovate, obtuse, slightly cordate at base, with 8–10 pairs of veins, coarsely serrate, lobed, with 2 pairs of free sessile leaflets proximally, separated from one another by 5–10 mm, and a third pair usually cut to the midrib at least on one side, rather coriaceous, with a greyish tomentum beneath. Fruit 10–12 mm, globose, red, with small and sparse lenticels. *S. & W. Fennoscandia, Bornholm.* Da Fe No Su.

Related species include:

S. borbasii Jáv., *Bot. Közl.* 14: 99 (1915). Rm.
S. pseudofennica E. F. Warburg, *Watsonia* 4: 43 (1957). Br.

14. S. meinichii (Lindeb.) Hedl., *Nyt Mag. Naturvid.* (*Christiania*) 52: 259 (1914). Small tree. Leaves obtuse, coarsely serrate, somewhat coriaceous, lobed, with 4–5 pairs of free, sessile leaflets proximally, subglabrous above, with greyish tomentum beneath; upper leaflets shortly decurrent on petiole, terminal leaflet larger than the lateral, ovate-rhombic, decreasingly lobed towards apex. Fruit up to 12 mm, subglobose, not or only slightly longer than wide, red. ● *S. & W. Norway.* No.

Related species include:

S. teodori Liljefors, *Acta Horti Berg.* 16: 283 (1953). $2n = 51$. Su.

13, 14 and their related species are all apomictic, so far as is known. They probably originated by hybridization between **2** and **7** or between **2** and **4**.

15. S. intermedia (Ehrh.) Pers., *Syn. Pl.* 2: 38 (1806) (*S. suecica* (L.) Krok & Almq., *S. scandica* (L.) Fries). Medium-sized tree. Leaves elliptical, 1·5–1·9 times as long as wide, with 7–9 pairs of lateral veins; lobes extending ¼–⅓(–½) of the way to the midrib, with yellowish-grey tomentum beneath. Leaves of vigorous long shoots often more deeply lobed. Fruit 12–15 × 8–10 mm, scarlet, with few and small lenticels. $2n = 68$. ● *S. Fennoscandia and Baltic region.* Da Fe Ge No Po Rs (B) Su [Cz].

Widely planted as an ornamental tree, this species is known to be apomictic; Liljefors considers that one of its parental species is **3**.

16. S. latifolia (Lam.) Pers., *loc. cit.* (1806). Medium-sized tree. Leaves broadly elliptical, 1–1¼ times as long as wide or somewhat longer, shortly acute, rounded or slightly cordate at base, with 7–9 pairs of veins, somewhat coriaceous, lobed (but never with free leaflets), glabrescent above, with grey-green tomentum beneath; lobes more or less triangular, the second pair usually the largest, acute or acuminate, rounded at the lower edge, biserrate or unequally serrate, the teeth terminating the main veins straight. Flowers *c.* 20 mm in diameter. Fruit 12–14 × 13–15·5 mm, subglobose, yellowish-brown, with rather many, large lenticels. ● *From E.C. Portugal to S.W. Germany.* Ga Ge Hs Lu [Su].

16 and the very numerous related species are, so far as is known, all apomictic, and probably originated by hybridization between **3** and one or other of the species **5, 6, 7** or **8**. The related species may be arranged in 2 series, as follows:

Series (*a*). Leaves with greenish-grey tomentum beneath. Fruit ovoid or subglobose, brown, reddish-brown or orange. Resembling **3**.

S. borosiana Kárpáti, *Ann. Sect. Horti-Viticult. Univ. Sci. Agr.* (*Budapest*) 12: 144 (1948). Hu.
S. decipiens (Bechst.) Irmisch in Petzold & Kirchner, *Arbor. Muscav.* 301 (1864). Ga Ge.
S. decipientiformis Kárpáti, *Ann. Sect. Horti-Viticult. Univ. Sci. Agr.* (*Budapest*) 14: 36 (1950). Hu.
S. degenii Jáv., *Magyar Bot. Lapok* 25: 85 (1926). Hu.
S. devoniensis E. F. Warburg, *Watsonia* 4: 46 (1957). Br Hb.
S. gayeriana Kárpáti, *Ann. Sect. Horti-Viticult. Univ. Sci. Agr.* (*Budapest*) 14: 35 (1950). Hu.
S. heilingensis Düll, *Ber. Bayer. Bot. Ges.* 34: 47 (1961). Ge.
S. joannis Kárpáti, *Bot. Közl.* 52: 140 (1966). Cz.
S. klasterskyana Kárpáti, *op. cit.* 136 (1966). Cz.
S. multicrenata Bornm. ex Düll, *Ber. Bayer. Bot. Ges.* 34: 49 (1961). Ge.
S. parumlobata Irmisch ex Düll, *op. cit.* 45 (1961). Ge.
S. pseudolatifolia Boros, *Mitt. Kgl. Ungar. Gartenb.-Lehranst.* 3: 51 (1937). Hu.
S. subcordata Bornm. ex Düll, *Ber. Bayer. Bot. Ges.* 34: 47 (1961). Ge.
S. zertovae Kárpáti, *Bot. Közl.* 52: 137 (1966). Cz.

Series (*b*). Leaves with whitish-grey tomentum beneath. Fruit subglobose, red or orange-red. Resembling **5–8**.

S. adamii Kárpáti, *Hung. Acta Biol.* 1: 112 (1949). Hu.
S. andreanszkyana Kárpáti, *Ann. Sect. Horti-Viticult. Univ. Sci. Agr.* (*Budapest*) 14: 35 (1950). Hu.
S. badensis Düll, *Ber. Bayer. Bot. Ges.* 34: 51 (1961). Ge.

S. **bakonyensis** (Jáv.) Kárpáti, *Hung. Acta Biol.* **1**: 116 (1949). Hu.

S. **balatonica** Kárpáti, *op. cit.* 121 (1949). Hu.

S. **barthae** Kárpáti, *Ann. Sect. Horti-Viticult. Univ. Sci. Agr.* (*Budapest*) **14**: 37 (1950). Hu.

S. **bohemica** Kovanda, *Acta Univ. Carol.* (*Biol.*) **1**: 77 (1961). Cz.

S. **bristoliensis** Wilmott, *Proc. Linn. Soc. London* **146**: 76 (1934). $2n = 51$. Br.

S. **dominii** Kárpáti, *Bot. Közl.* **52**: 140 (1966). Cz.

S. **eugenii-kelleri** Kárpáti, *Hung. Acta Biol.* **1**: 113 (1949). Hu.

S. **franconica** Bornm. ex Düll, *Ber. Bayer. Bot. Ges.* **34**: 49 (1961). Ge.

S. **futakiana** Kárpáti, *Bot. Közl.* **52**: 137 (1966). Cz.

S. **gerecseensis** Boros & Kárpáti, *Hung. Acta Biol.* **1**: 107 (1949). Hu.

S. **karpatii** Boros, *Ann. Sect. Horti-Viticult. Univ. Sci. Agr.* (*Budapest*) **13**: 153 (1949). Hu.

S. **kmetiana** Kárpáti, *Bot. Közl.* **52**: 137 (1966). Cz.

S. **latissima** Kárpáti, *Ann. Sect. Horti-Viticult. Univ. Sci. Agr.* (*Budapest*) **14**: 36 (1950). Hu.

S. **magocsyana** Kárpáti, *Bot. Közl.* **52**: 140 (1966). Cz.

S. **paxiana** Jáv., *Magyar Bot. Lapok* **25**: 89 (1926). Rm.

S. **pseudobakonyensis** Kárpáti, *Hung. Acta Biol.* **1**: 117 (1949). Hu.

S. **pseudosemiincisa** Boros, *Mitt. Kgl. Ungar. Gartenb.-Lehranst.* **3**: 53 (1937). Hu.

S. **pseudovertesensis** Boros, *op. cit.* 53 (1937). Hu.

S. **redliana** Kárpáti, *Hung. Acta Biol.* **1**: 118 (1949). Hu.

S. **semiincisa** Borbás, *Term.-Tud. Közl.* **11**: 34 (1879). $2n = 34$. Hu.

S. **simonkaiana** Kárpáti, *Ann. Sect. Horti-Viticult. Univ. Sci. Agr.* (*Budapest*) **14**: 38 (1950). Hu.

S. **slovenica** Kovanda, *Acta Univ. Carol.* (*Biol.*) **1**: 73 (1961). ?Au Cz.

S. **subcuneata** Wilmott, *Proc. Linn. Soc. London* **146**: 76 (1934). Br.

S. **vertesensis** Boros, *Mitt. Kgl. Ungar. Gartenb.-Lehranst.* **3**: 52 (1937). Hu.

17. **S. sudetica** (Tausch) Fritsch in A. Kerner, *Sched. Fl. Exsicc. Austro-Hung.* **7**: 20 (1896). Shrub. Leaves elongate- or obovate-elliptical, acute, *c.* 1¾ times as long as wide, with 8–9 pairs of veins, coarsely serrate, the teeth straight and not pointing forward, not lobed, glabrous above, with a dense grey tomentum beneath. Petals *c.* 5 mm. Fruit red. ● *N.W. Czechoslovakia* (*Krkonoše*). Cz.

18. **S. margittaiana** (Jáv.) Kárpáti, *Feddes Repert.* **62**: 304 (1960). Like 17 but leaves obtuse, finely serrate, very slightly lobed, with a less dense tomentum beneath. ● *E. Czechoslovakia* (*Fatra*). Cz.

17 and 18 are probably apomictic species which have arisen by hybridization between 4 and one or other of the species 5, 6, 7 or 8.

29. Eriobotrya Lindley[1]

Evergreen shrubs or trees. Leaves simple. Inflorescence a terminal panicle. Sepals persistent in fruit; petals ovate or suborbicular, clawed, white; stamens 20; carpels 2–5, walls thin in fruit; ovules 2; styles connate only at base. Fruit with 1 or a few large seeds.

1. **E. japonica** (Thunb.) Lindley, *Trans. Linn. Soc. London* **13**: 102 (1821). Small tree up to 10 m. Leaves 12–25 cm, obovate to elliptic-oblong, dentate, reddish-brown-tomentose beneath. Inflorescence densely reddish-brown-tomentose. Flowers *c.* 1 cm in diameter. Fruit 3–6 cm, pyriform or ellipsoid, yellow. Seeds 1–1·5 cm. *Cultivated in S. Europe for the edible fruit, and elsewhere for ornament. More or less naturalized in Açores, Portugal and Kriti.* [Az Cr Hs It Lu.] (*C. China.*)

30. Amelanchier Medicus[2]

Deciduous shrubs or small trees, without spines. Leaves simple, serrate. Stipules caducous. Flowers in terminal racemes, rarely solitary. Petals linear to oblong-obovate, not clawed, white, rarely pink; stamens 10–20; carpels 5, connate or partly free, with walls cartilaginous in fruit; ovules 2; styles 2–5, free or connate at base. Fruit small, 4- to 10-celled, bluish- or purplish-black, usually juicy and sweet.

This genus includes ornamental species (chiefly from North America) which flower profusely and have yellow or red leaves in the autumn. Some of these, and a few hybrids, are occasionally grown in European gardens, and their nomenclature has been very confused.

1 Styles free; leaves coarsely serrate (3–5 teeth per cm)   **1. ovalis**
1 Styles connate; leaves finely serrulate (6–12 teeth per cm)
  2 Young leaves whitish; racemes erect; petals 4–10 mm **2. spicata**
  2 Young leaves purplish; racemes nodding; petals 15–18 mm
      3. grandiflora

1. **A. ovalis** Medicus, *Gesch. Bot.* 79 (1793) (*A. vulgaris* Moench, *A. rotundifolia* Dum.-Courset). Erect or spreading shrub up to 3 m; bark blackish; young twigs lanate. Leaves 2·5–5 cm, ovate to obovate, rounded or emarginate and mucronate at apex, rather coarsely serrate, lanate beneath when young. Racemes 3- to 8-flowered, erect, lanate. Sepals at first lanate, soon glabrous; petals 10–13 mm; styles 5, free. Fruit bluish-black; pedicels 5–10 mm. *Rocky places and open woods, mainly on limestone and in mountain areas. S. & C. Europe northwards to Luxembourg and S. Poland.* Al Au Be Bl Bu Co Cr Cz Ga Ge Gr He Hs Hu It Ju Lu Po Rm Rs (K) Sa Si Tu.

The plant from Kriti and S. Greece, usually a smaller shrub with more persistent tomentum on young twigs, leaves, pedicels and calyces, is regarded by some authors as a distinct variety, var. *cretica* (Willd.) Fiori (*A. cretica* (Willd.) DC.) and may deserve higher status.

2. **A. spicata** (Lam.) C. Koch, *Dendrologie* **1**: 182 (1869). Like 1 but up to 4 m; leaves finely serrulate, densely white-tomentose when young; petals 4–10 mm; styles connate at base; apex of ovary lanate. $2n = 68$. *Cultivated for ornament and sometimes naturalized in N. Europe.* [Fe Ga Ge No Rs (C) Su.]

Formerly considered to be a garden hybrid between 1 and *A. canadensis* (L.) Medicus, but now considered to be conspecific with *A. humilis* Wieg., from N.E. North America.

A. canadensis (L.) Medicus, *Gesch. Bot.* 79 (1793) (*A. oblongifolia* (Torrey & A. Gray) M. J. Roemer), native of eastern U.S.A., is closely related to 2, but differs mainly in having the leaves usually oblong, and the apex of the ovary more or less glabrous. It is sometimes cultivated in gardens in N. and C. Europe, but its name has been frequently used for other species.

3. **A. grandiflora** Rehder, *Jour. Arnold Arb.* **2**: 45 (1920) (*A. confusa* Hyl., *A. laevis* auct. eur., non Wieg.). Shrub or small tree up to 9 m; young twigs hairy. Leaves 3–7 cm, broadly elliptical, cordate, finely serrulate, purplish and floccose-tomentose when young but soon green and glabrous. Racemes many-flowered,

---

[1] By P. W. Ball.   [2] By J. do Amaral Franco.

nodding, slightly villous. Petals 15–18 mm; styles 5, connate at base. Fruit dark purple; pedicels 20–22 mm. *Of garden origin; now commonly cultivated for ornament, and sometimes naturalized in W. Europe.* [Be Br Ga Ge Ho.]

This taxon has been much confused in gardens both with *A. arborea* (Michx fil.) Fernald (*A. canadensis* sensu W. Darlington, non (L.) Medicus) and *A. laevis* Wieg. The former, from eastern U.S.A., is a taller tree up to 20 m, with young leaves tomentose-lanate on both surfaces, racemes silky-tomentose, and fruit brown; the latter, from N.E. North America, has young twigs nearly glabrous, leaves always glabrous, and fruit smaller, on longer pedicels 30–50 mm.

31. Cotoneaster Medicus[1]

Shrubs, rarely small trees, without spines. Leaves entire. Stipules caducous. Flowers small, in cymes or corymbs, or solitary. Sepals persistent; petals white or pink; stamens *c.* 20; carpels 2–5, free on the ventral side, with walls stony in fruit; ovules 2; styles 2–5, free. Fruit red or black, with mealy flesh; pyrenes 2–5.

Some of the species are very similar to one another, and difficult to distinguish in the vegetative state. They are also variable, particularly in the indumentum and the shape of the leaf-apex. Interspecific hybrids are fairly common. Many species are cultivated as ornamental shrubs, and some are naturalized by bird-dispersal, often in places remote from gardens.

Literature: G. Klotz, *Wiss. Zeitschr. Univ. Halle* (*Math.-Nat.*) **6**(6): 945–982 (1957).

1 Petals patent, white (rarely pink)
  2 Evergreen shrub not more than 1 m   **11. microphyllus**
  2 Deciduous shrub usually more than 1 m
    3 Leaves pubescent beneath at first, soon glabrous; inflorescence lax   **8. granatensis**
    3 Leaves persistently tomentose beneath; inflorescence dense
      4 Hypanthium and sepals tomentose   **9. nummularius**
      4 Hypanthium and sepals slightly pubescent   **10. tauricus**
1 Petals erect, ±tinged with red or pink
  5 Fruit black   **7. niger**
  5 Fruit red or purplish
    6 Flowers solitary or in pairs; shrubs not more than 0·7 m, usually procumbent
      7 Leaves not more than 1·2 cm, shining above   **1. horizontalis**
      7 Leaves more than 2 cm, dull above   **4. cinnabarinus**
    6 Flowers in clusters of 2–5 or more; shrubs usually more than 1 m, erect
      8 Twigs strigose-pubescent   **2. simonsii**
      8 Twigs appressed-villous or almost glabrous
        9 Calyx glabrous, or pubescent at margin   **3. integerrimus**
        9 Calyx densely pubescent or tomentose
          10 Leaves obtuse or rarely subacute, tomentose beneath   **5. nebrodensis**
          10 Leaves acuminate or acute, usually glabrescent   **6. acuminatus**

Sect. COTONEASTER. Petals erect, reddish, pinkish or white with red markings.

1. C. horizontalis Decne, *Fl. Serres Jard. Eur.* **22**: 168 (1879). Deciduous or semi-evergreen shrub *c.* 0·5 m, with horizontally patent, much-branched stems. Leaves up to 1·2 cm, suborbicular or broadly elliptical, acute or mucronate at apex, dark green and shining above, glabrous or with a few hairs beneath. Petioles 1–2 mm, strigose-pubescent. Flowers 1–2, subsessile. Petals reddish or whitish. Fruit 5–6 mm, subglobose, bright red;

pyrenes usually 3. *Frequently cultivated and rarely naturalized.* [Au Br.] (*W. China.*)

2. C. simonsii Baker in Saunders, *Refug. Bot.* **1**: t. 55 (1869). Erect shrub up to 4 m, deciduous to semi-evergreen. Twigs with strigose pubescence persisting for 2–3 years. Leaves 1–3 cm, orbicular-obovate, broadly cuneate at the base, acute or slightly acuminate, dark green above and pubescent while young, paler green beneath and sparsely strigose, chiefly on the veins; petioles 2–4 mm. Flowers 2–4 in short cymes. Petals white with red markings. Fruit 8 mm, shortly ellipsoid or obovoid, scarlet; pyrenes 3–4. *Frequently cultivated, and occasionally naturalized in N.W. Europe.* [Br Ga Hb No.] (*E. India.*)

3. C. integerrimus Medicus, *Gesch. Bot.* 85 (1793) (*C. vulgaris* Lindley). Erect, branched, deciduous shrub up to 2 m. Young twigs tomentose, soon glabrous. Leaves 2–5 × 0·5–3 cm, suborbicular to ovate, obtuse or, especially on the long shoots, acute, usually mucronate, rounded at the base, glabrous above, greyish-tomentose beneath, petiolate. Flowers 2–3(–4) in short, glabrous cymes. Calyx glabrous, or pubescent at margin. Petals pink. Fruit 6–8 mm, subglobose, red; pyrenes (2–)3(–4). *Dry, stony places, mainly in the mountains; somewhat calcicole. Much of Europe, but absent from most of the U.S.S.R., the extreme north, and much of the Mediterranean region.* Al Au Be Br Bu Cz Da Fe Ga Ge Gr He Hs Hu It Ju No Po Rm Rs (B, W, K) Su.

4. C. cinnabarinus Juz., *Not. Syst.* (*Leningrad*) **13**: 32 (1950). Like **3** but rarely more than 0·7 m, with decumbent branches; leaves broadly ovate or orbicular, with obtuse apex; flowers solitary or rarely in pairs; fruit reddish-orange. ● *N.W. Russia* (*Kol'skij Poluostrov.*) Rs (N).

C. × antoninae Juz., *Not. Syst.* (*Leningrad*) **13**: 33 (1950), from Kol'skij Poluostrov (N. Russia) and from Finland, is possibly a hybrid between **4** and **7**. Further investigation is needed.

C. uniflorus Bunge in Ledeb., *Fl. Altaica* **2**: 220 (1830), from C. Asia and W. China, has been recorded from Kol'skij Poluostrov and from Finland, but the records are all referable to **4**.

5. C. nebrodensis (Guss.) C. Koch, *Hort. Dendrol.* 179 (1853) (*C. tomentosus* Lindley). Deciduous shrub up to 3 m. Young twigs tomentose. Leaves 3–6 cm, suborbicular, broadly ovate or elliptical, obtuse or rarely subacute, with white pubescence above at first, later glabrescent, and with whitish or greyish tomentum beneath; petiole 3–6 mm, tomentose. Axis of inflorescence, peduncles and calyx tomentose; flowers 3–12, in nodding cymes. Petals usually reddish on the outer side. Fruit 7–8 mm, subglobose, red, with a whitish tomentum; pyrenes 3–5. *Dry, stony places, mainly in the mountains; somewhat calcicole. S. & S.C. Europe, northwards to the Vosges and S. Poland.* Al Au Bu Cz Ga Ge Gr He Hs Hu It Ju Po Rm Si [Da No Su].

C. × intermedius Coste, *Bull. Soc. Bot. Fr.* **40**: cxxii (1893), a hybrid between **3** and **5**, occurs sporadically throughout the area of the parents and is sometimes treated as a species.

6. C. acuminatus Lindley, *Trans. Linn. Soc. London* **13**: 101 (1821). Erect, deciduous shrub up to 4 m, often with pendent branches. Young twigs densely pubescent. Leaves 3–6 cm, twice as long as wide, elliptic-ovate to ovate-lanceolate, acuminate or acute, broadly cuneate at the base, more or less pubescent on both surfaces when young, glabrous above and subglabrous beneath at maturity. Flowers 2–5 in short, pubescent cymes. Calyx pubescent; petals pink or whitish. Fruit 9–10 mm wide, ellipsoid, bright red, slightly hairy near apex; pyrenes 2. *Frequently cultivated and occasionally naturalized.* [Ga Ge.] (*Himalaya.*)

---

[1] By K. Browicz.

7. C. niger (Thunb.) Fries, *Summa Veg. Scand.* 175 (1846) (*C. orientalis* A. Kerner, *C. melanocarpus* Loddiges ex C. K. Schneider). Deciduous shrub up to 2–2·5 m. Young twigs more or less pubescent, the older glabrous, shining, reddish-brown. Leaves up to 5 cm, broadly ovate to ovate-oblong, obtuse or sometimes subacute and mucronulate, especially on the long shoots, dark green and sparsely pubescent above at first, whitish-tomentose beneath; petiole 1–5 mm. Flowers 3–8(–15) in nodding, pubescent or almost glabrous cymes. Calyx glabrous or slightly pubescent; petals reddish or reddish-white. Fruit 6–9 mm, subglobose, black, pruinose; pyrenes 2. *N., E. & E.C. Europe, southwards to Macedonia.* Bu Cz Da Hu Ju No Po Rm Rs (N, B, C, W, K, E) Su.

C. matrensis Domokos, *Mitt. Kgl. Ungar. Gartenb.-Lehranst.* **7**: 50 (1941), from Hungary and Czechoslovakia, and **C. alaunicus** Golitsin, *Nov. Syst. Pl. Vasc.* (*Leningrad*) **1964**: 145 (1964), from C. Russia, are probably hybrids of 3 and 7 or variants of 7. They need further study.

Sect. CHAENOPETALUM Koehne. Petals patent, white (rarely pink).

8. C. granatensis Boiss., *Elenchus* 41 (1838) (*C. multiflorus* var. *granatensis* (Boiss.) Wenzig). Deciduous shrub or sometimes small tree up to 4·5 m, with slender, arching branches. Twigs pubescent at first, soon glabrous. Leaves 2–5 cm, elliptical or orbicular, obtuse or mucronulate, rounded or broadly cuneate at base, glabrous above, with scattered hairs beneath. Inflorescence many-flowered, branched, lax, pubescent. Calyx somewhat hairy. Fruit 6–9 mm, pyriform, red; pyrenes 2. ● *S. Spain (Sierra Nevada).* Hs.

9. C. nummularia Fischer & C. A. Meyer, *Ind. Sem. Horti Petrop.* **2**: 34 (1835). Erect, deciduous shrub up to 1·5 m, but sometimes dwarfed by exposure; young twigs grey-tomentose. Leaves 1–2·5 × 0·9–2·2 cm, broadly elliptical, obovate, or suborbicular, obtuse, mucronate, sometimes emarginate, sparsely hairy above when young, whitish- or greyish-tomentose beneath; petiole 1·5–2·5 mm, tomentose. Inflorescence 3- to 7-flowered, dense. Hypanthium and calyx tomentose. Fruit *c.* 8 mm, subglobose, red, whitish-pubescent when young, becoming glabrous; pyrenes 2. *Kriti.* Cr. (*C. & S.W. Asia, N. Africa.*)

The plant from Kriti has pink petals, though over the rest of its range the species has white petals.

10. C. tauricus Pojark., *Not. Syst.* (*Leningrad*) **8**: 138 (1938). Like 9 but leaves oblong-elliptical or oblong-ovate; hypanthium and calyx slightly pubescent. *Stony mountain slopes. Krym.* Rs (K).

11. C. microphyllus Wallich ex Lindley, *Bot. Reg.* **13**: t. 1114 (1827). Evergreen shrub up to 1 m, with stiff, patent branches. Twigs strigose-pubescent. Leaves 0·5–0·8 cm, obovate to obovate-oblong, obtuse or emarginate, rarely subacute, cuneate at base, dark green, glabrous and shining above, glaucous and pubescent beneath; petiole 1–2 mm. Flowers solitary, or sometimes 2–3. Calyx pubescent. Fruit 6 mm, globose, scarlet; pyrenes 2. *Cultivated, and naturalized in Britain and Ireland.* [Br Hb.] (*Himalaya and S.W. China.*)

32. Pyracantha M. J. Roemer[1]

Evergreen, usually spiny shrubs. Leaves simple. Stipules minute, caducous. Flowers in compound corymbs. Sepals persistent; petals suborbicular, white; stamens 20; anthers yellow; carpels 5, free on the ventral side, with walls stony in fruit; ovules 2. Fruit red, orange or yellow, with 5 pyrenes.

1. P. coccinea M. J. Roemer, *Syn. Monogr.* **3**: 219 (1847) (*Cotoneaster pyracantha* (L.) Spach). Up to 2 m (up to 6 m in cultivation). Leaves 2–4 cm, elliptical to obovate-elliptical, glabrous or sparsely pubescent beneath when young, crenate-dentate. Flowers 7–8 mm in diameter. Fruit 5–7 mm, bright red, rarely orange or yellow. *S. Europe, westwards to N.E. Spain.* Al Bu Ga Gr Hs It Ju Rs (K) Tu [Br Lu].

This and several closely related species are widely cultivated for the ornamental flowers and fruit.

33. Mespilus L.[1]

Deciduous trees or shrubs, sometimes spiny. Leaves simple. Stipules deciduous. Flowers solitary. Sepals longer than petals, entire, persistent; stamens 30–40; anthers red; carpels 5, connate, with walls stony in fruit; ovules 2; styles 5, free. Fruit with foliaceous sepals.

1. M. germanica L., *Sp. Pl.* 478 (1753). Shrub or small tree up to 6 m. Leaves 5–12 cm, lanceolate or oblanceolate to obovate, pubescent, but sometimes glabrous above, entire or serrulate towards the apex. Flowers 3–4 cm in diameter. Sepals 10–16 mm, linear-triangular. Petals white. Fruit 2–3 cm, brown, pyriform to depressed-globose. *S.E. Europe, extending to Sardegna and Sicilia; cultivated and naturalized in C. & W. Europe.* Bu Gr *It Rs (K) *Sa *Si [Au Be Br Cz Ga Ge He Ho Hs Ju Rm].

Cultivated for the fruit, which after incipient decay becomes soft and edible.

34. Crataegus L.[2]

Deciduous, usually spiny shrubs or small trees. Leaves simple, lobed or pinnatifid, serrate. Stipules persistent. Flowers in corymbs. Petals obovate, white, rarely pink; stamens 5–25; carpels 1–5, free on the ventral side, walls stony in fruit; styles 1–5; ovules 2. Fruit red, yellow or black, usually with mealy flesh; pyrenes 1–5.

In many species the leaves and stipules are much larger and more deeply cut on non-flowering shoots than on flowering shoots. The characters given in the keys and descriptions are to be interpreted as referring only to the leaves of the flowering shoots.

In regions where the areas of two taxa overlap, hybrids are commonly found. Notes about some of these hybrids are given in the text.

1 Lateral veins of leaves ending only in the apices of the teeth or lobes
  2 Leaves obovate to oblong-obovate, simply and acutely serrate above, entire at base; spines 70–100 mm   **2. crus-galli**
  2 Leaves ovate or elliptic-ovate, biserrate; spines 10–30 mm
    3 Sepals glandular-serrate; stamens 10   **1. intricata**
    3 Sepals entire; stamens 20
      4 Leaves lobed to less than ½ way to the midrib; petioles 8–15 mm   **3. sanguinea**
      4 Leaves lobed to more than ½ way to the midrib; petioles 15–35 mm   **4. altaica**
1 Lateral veins of leaves ending both in the apices of the teeth or lobes and in the sinuses
  5 Young twigs, leaves, pedicels and hypanthia glabrous or with straight, patent hairs
    6 Leaf-lobes serrulate; stipules serrate

¹ By P. W. Ball.   ² By J. do Amaral Franco.

7 Fruit light or bright red, crowned by erect or erecto-patent
   sepals
  8 Leaf-lobes strongly acuminate; fruit oblong-cylindrical
                          9. calycina
  8 Leaf-lobes subobtuse; fruit subglobose   **10. microphylla**
7 Fruit dark red to blackish, crowned by deflexed or patent
   sepals
  9 Style 1; fruit with 1 pyrene   **9. calycina**
  9 Styles 2–3; fruit with 2 pyrenes
   10 Corymb 5- to 10-flowered, glabrous; twigs glabrescent
    11 Leaves lobed to less than ⅓ way to midrib, the lobes
       obtuse; fruit 8–12 mm, without protuberances at the
       base   **5. laevigata**
    11 Leaves lobed ½ way to the midrib, the lobes acute; fruit
       12–15 mm, usually with 5 protuberances at the base
                          6. macrocarpa
   10 Corymb 12- to 20-flowered, villous; twigs villous
    12 Leaves with 3–5 obtuse or acute lobes; lobes serrulate
       in upper half only; sepals triangular, acute   **7. taurica**
    12 Leaves with 5–7 acuminate lobes; lobes serrulate almost
       to the base; sepals lanceolate, long-attenuate **8. ucrainica**
6 Leaf-lobes entire or with a few acute teeth; stipules entire
  13 Leaf-lobes acute to subobtuse; ripe fruit 6–10 mm, dark
     or bright red   **14. monogyna**
  13 Leaf-lobes acuminate; ripe fruit 10–14 mm, dark purple
     to blackish
    14 Leaf-lobes with only 1 or 2 teeth at apex; leaves 40–
       80 mm; main veins curving towards base of leaf
                         13. ambigua
    14 Leaf-lobes unevenly serrate in their apical half; leaves
       30–35 mm; main veins straight
     15 Corymb villous   **11. pallasii**
     15 Corymb glabrous   **12. karadaghensis**
5 Young twigs, leaves, pedicels and hypanthia tomentose, lanate
  or sericeous (rarely twigs lanate and leaves glabrous)
  16 Petiole 10–30 mm; leaf-lobes serrate; flowers 10–15 mm in
    diameter
   17 Ripe fruit red and pruinose, with 1 pyrene; twigs pruinose
                        15. sphaenophylla
   17 Ripe fruit blackish, with 4–5 pyrenes; twigs not pruinose
    18 Leaves with 3–7 lobes; indumentum arachnoid-lanate;
       corymb compound, many-flowered   **16. pentagyna**
    18 Leaves with 7–11 lobes; indumentum tomentose; corymb
       simple, few-flowered   **17. nigra**
  16 Petiole 2–10 mm; leaf-lobes entire or with 1–3 teeth at apex;
    flowers 15–20 mm in diameter
   19 Leaves not more than 30 mm; fruit 7–10 mm in diameter
    20 Indumentum lanate; styles 1–3   **19. heldreichii**
    20 Indumentum sericeous; styles 5   **22. pycnoloba**
   19 Leaves 30–50 mm; fruit 15–25 mm in diameter
    21 Twigs and leaves lanate but soon glabrescent; leaf-lobes
       broad; fruit dark red   **18. schraderana**
    21 Twigs and leaves persistently hairy; leaf-lobes narrow;
       fruit orange-red to yellow
     22 Leaves with 3–7 acute, sparsely incise-dentate lobes;
       styles 3–5   **20. laciniata**
     22 Leaves with 3(–5) subobtuse, entire lobes; styles 1–2
                          21. azarolus

1. **C. intricata** Lange, *Bot. Tidsskr.* **19**: 264 (1895) (*C. coccinea*
auct. plur., non L.). Shrub up to 3 m; twigs glabrous, purplish-
brown; spines up to 30 mm. Leaves 35–70 × 20–60 mm, elliptic-
ovate, acute, bright green, glabrous above and nearly so beneath;
lobes 7–11, short, acute, serrate; petiole 15–25 mm. Corymb
slightly villous. Sepals glandular-serrate; stamens 10; anthers
yellow. Fruit 10–12 mm, subglobose, reddish-brown; pyrenes 3–4.
Cultivated for ornament in C. Europe and rarely naturalized. [Rm.]
(*E. North America.*)

C. mollis (Torrey & A. Gray) Scheele, *Linnaea* **21**: 569 (1848),
with leaves densely pubescent beneath when young (but later
usually only on the veins), and usually pyriform, scarlet fruit,
from C. North America, and **C. submollis** Sarg., *Bot. Gaz.*
31: 7 (1901), with leaves softly pubescent beneath when young
(but later only puberulent on the veins), and larger, bright
orange-red fruit, from E. North America, are both widely planted
in gardens and for hedges in N.W. Europe; they have both
passed for a long time under the name *C. coccinea* L.

2. **C. crus-galli** L., *Sp. Pl.* 476 (1753). Like **1** but a larger shrub
or a small tree; twigs light or greyish-brown; spines 70–100 mm;
leaves 20–80 × 5–20 mm, obovate to oblong-obovate, not lobed,
acutely serrate above the entire base, glabrous; petiole 5–15 mm;
sepals entire; fruit red; pyrenes 2. *Cultivated for ornament and
locally naturalized.* [?Co Cz ?Ga.] (*E. & C. North America.*)

3. **C. sanguinea** Pallas, *Fl. Ross.* **1**(1): 25 (1784). Shrub up to
4 m; twigs soon glabrous, purple; spines absent or small. Leaves
50–80 × 45–65 mm, ovate, acute, cuneate, slightly pubescent on
both surfaces; lobes 5–9, short, acute, acutely serrate; petiole
8–15 mm. Corymb glabrous; flowers 10–15 mm in diameter.
Sepals entire; stamens 20; anthers pink or purple; styles 2–5.
Fruit 8–12 mm, globose, bright red (var. *sanguinea*) or yellow
(var. *chlorocarpa* (C. Koch) C. K. Schneider); pyrenes 5. *C. & E.
Russia, northwards to c.* 60° *N; W. Kazakhstan.* Rs (C, E) [Au
?Ga.] (*Siberia.*)

4. **C. altaica** (Loudon) Lange, *Rev. Crat.* 42 (1897). Small tree;
twigs glabrous, dark brown; spines 6–20 mm. Leaves 50–90 ×
40–80 mm, ovate, acute, truncate to broadly cuneate, glabrous,
lobed; lobes 5–9, extending more than ½ way to the midrib,
acuminate, acutely and unevenly serrate; petioles 15–35 mm;
stipules c. 20 × 15 mm, incise-serrate. Corymb glabrous; flowers
8–12 mm in diameter. Sepals entire, short; stamens 20; anthers
whitish; styles 4–5. Fruit 8–10 mm, yellow; pyrenes 4–5. *On the
borders of S.E. Russia and Kazakhstan* (*near Ural'sk*). Rs (E).
(*S.C. Asia.*)

5. **C. laevigata** (Poiret) DC., *Prodr.* **2**: 630 (1825) (*C. oxyacan-
thoides* Thuill., *C. oxyacantha* auct.). Shrub; twigs glabrescent,
brown; spines 6–15 mm. Leaves obovate, slightly coriaceous,
light green beneath; lobes 3–5, short, broad, obtuse, serrulate;
petiole 6–18 mm; stipules 5–10 × 1–3 mm, acuminate, unequally
incise-serrate. Corymb 5- to 10-flowered, glabrous; flowers
15–18 mm in diameter. Sepals entire; anthers red; styles 2–3.
Fruit globose or ellipsoid, deep red; pyrenes 2. *Woods.* ● *N.W.,
N.C. & C. Europe, from England, C. Sweden and Latvia to the W.
Pyrenees, and N. Italy.* Au Be Br Cz Da Ga Ge He Ho Hs Hu It
Po ?Rm ?Rs (B, W) Su [No].

Nearly all the glabrous or slightly pubescent species in Europe
have been called *C. oxyacantha* by regional authors, so that references in the literature are very confused.

(a) Subsp. **laevigata**: Leaves 15–35 mm, sparsely pubescent or
almost glabrous, without hairs in vein-axils beneath; sepals
broadly triangular, about as long as wide; hypanthium glabrous;
fruit 8–10 mm, globose. *Commoner in the western part of the
range, and in the lowlands.*

(b) Subsp. **palmstruchii** (Lindman) Franco, *Feddes Repert.* **74**:
25 (1967) (*C. palmstruchii* Lindman): Leaves 30–50 mm, more
densely pubescent on the veins beneath and with tufts of hair in
the vein-axils beneath; sepals nearly twice as long as wide;
hypanthium villous; fruit 10–12 mm, ellipsoid. *Commoner in the
eastern part of the range, and in the mountains.*

C. × media Bechst., *Diana* **1**: 88 (1797) (*C. monogyna ×
laevigata*) occurs in N.W. and C. Europe. Plants derived from
C. monogyna subsp. *nordica* and *C. laevigata* subsp. *laevigata* have
cuneate leaves, with 3–5 subacute, subentire lobes extending ¾ of

the way to the midrib, almost straight main veins, flowers 10–14 mm in diameter, triangular-acute sepals 1·5–2 × 2 mm, villous hypanthium and 1–2 styles. Plants derived from *C. monogyna* subsp. *monogyna* and *C. laevigata* subsp. *laevigata* differ in having leaves with the lobes extending not more than ½ way to the midrib, serrulate in their apical half, flowers 12–16 mm in diameter, broadly triangular sepals and glabrous hypanthium.

C. × schumacheri Raunk., *Bot. Tidsskr.* **42**: 247 (1933) (*C. calycina × laevigata*) also occurs in N.W. and C. Europe. It has orbicular-ovate leaves, rounded-cuneate at base, 3–5 ovate, acuminate, serrulate lobes with almost straight main veins, triangular-subulate sepals longer than wide, glabrous hypanthium, oblong fruit 8–10 mm, crowned by erect, but later recurving sepals.

6. C. macrocarpa Hegetschw., *Fl. Schweiz* 464 (1840). Like **5** but usually procumbent shrub; with leaves 30–50 × 20–40 mm, lobes extending ⅓ way to the midrib, ovate, acute; fruit 12–15 × 10–12 mm, ellipsoid, usually with 5 protuberances at the base, and crowned by sepals which are longer than wide. ● *E. Alps; Czechoslovakia.* Au Cz Ga Ge He It.

7. C. taurica Pojark. in Komarov *Fl. URSS* **9**: 501 (1939). Shrub; twigs villous; spines *c.* 10 mm. Leaves ovate-rhombic to obovate, dark green and sparsely villous above, lighter green and densely and softly hairy beneath; lobes 3–5, broad, obtuse or acute, serrulate to the middle or only near the apex, extending ⅔–¾ of the way to the midrib; petiole 8–20 mm. Corymb 12- to 20-flowered, villous. Hypanthium and pedicels strongly white-villous. Sepals triangular, acute; styles (1–)2(–3). Fruit 12–15 × 10–14 mm, subglobose, usually distinctly 5-angled below, deep red, becoming sparsely villous to subglabrous when ripe, crowned by deflexed sepals; pyrenes usually 2. *Scrub in rocky places. E. Krym.* Rs (K).

8. C. ucrainica Pojark. in Komarov, *Fl. URSS* **9**: 502 (1939). Like **7** but twigs and leaves less villous; spines 12–20 mm; leaf-lobes 5–7, ovate-acuminate, serrulate almost to the base, extending ⅓–½ of the way to the midrib; hypanthium and pedicels less villous; sepals lanceolate, long-attenuate; fruit 10–14 × 9–12 mm, globose-ovoid, obtusely 5-angled. *Margins of woods.* ● *N., W. & C. Ukraine.* Rs (C, W).

9. C. calycina Peterm., *Deutschl. Fl.* 176 (1849). Shrub or small tree; twigs glabrous, purplish- or cinnamon-brown; spines up to 13 mm, or absent. Leaves 30–60(–85) × 30–50(–65) mm, obovate-rhombic to broadly ovate, cuneate at base, thin, light green beneath, glabrescent; lobes 3–5(–9), broadly ovate, acute to acuminate, serrulate almost to the base, extending ⅓–⅔ of the way to the midrib; petiole 10–35 mm; stipules 10–15(–25) × 4–8 mm, falcate-incurved or oblong, subulate at apex, incise-serrate. Corymb glabrous; flowers 15–20 mm in diameter. Sepals 3–4 × 1–2 mm, triangular-subulate to long-acuminate, entire; style 1. Fruit 6–13 × 5–10 mm, oblong-cylindrical to subglobose, light to dark red; pyrene 1. ● *N.W. & C. Europe, extending to S. Russia and W. Bulgaria.* Au Be Bu Cz Da Fe Ga Ge Hu Ju No Po Rm Rs (B, C, W, E) Su.

(a) Subsp. **calycina**: Fruit oblong-cylindrical, light red, crowned by more or less erect sepals. *N.W. & C. Europe, extending to E. Romania.*

(b) Subsp. **curvisepala** (Lindman) Franco, *Feddes Repert.* **79**: 39 (1968) (*C. curvisepala* Lindman): Fruit ellipsoid to subglobose, dark red, crowned by deflexed sepals. *Woods. E.C. Europe, extending to S. Russia and W. Bulgaria; S. Sweden and S. Finland.*

C. × kyrtostyla Fingerh., *Linnaea* **4**: 372 (1829) (*C. calycina × monogyna, C. heterodonta* Pojark.), found in the lowlands of N.W. Europe, is like **9(a)** but differs mainly in the cuneate leaves with oblong, acute lobes with only a few teeth at the apex, main veins curving downwards and fruits crowned by deflexed sepals.

C. plagiosepala Pojark., *Nov. Syst. Pl. Vasc.* (*Leningrad*) **1965**: 135 (1965), from S.E. Poland, is very like **9(b)** but has long-cuneate, usually 3-lobed leaves and patent or erecto-patent sepals 4·5–5·5 mm; its status is uncertain.

10. C. microphylla C. Koch, *Verh. Ver. Beförd. Gartenb. Preuss.* nov. ser., **1**: 288 (1853). Slender shrub; twigs glabrescent, purplish-brown; spines 5–12 mm, slender. Leaves 15–30 × 10–25 mm, ovate, light green beneath, glabrescent; lobes 3–5, wide, subobtuse, serrulate; petiole 4–16 mm; stipules falcate-incurved, incise-serrate. Corymb glabrous or puberulent; flowers 8–13 mm in diameter. Sepals short, triangular-ovate; style 1. Fruit 9–12 mm, subglobose, bright red, crowned by erect or erecto-patent sepals; pyrene 1. *Krym.* Rs (K). (*Caucasus, N. Iran.*)

11. C. pallasii Griseb., *Spicil. Fl. Rumel.* **1**: 89 (1843) (*C. beckerana* Pojark., *C. stevenii* Pojark.). Shrub; twigs glabrescent, reddish; spines 10–20 mm or absent. Leaves 30–35 × 30–35 mm, ovate, acute, light green beneath, glabrescent, main veins straight; lobes 5–7, acuminate, unevenly serrate in the apical half, extending ¾ of the way to the midrib; stipules entire, subulate. Corymb villous; flowers *c.* 15 mm in diameter. Styles (1–)2. Fruit 10–12 mm, ellipsoid-globose, first yellow, later red, blackish when ripe, crowned by deflexed sepals; pyrenes (1–)2. *Mountain slopes. S.E. Russia; Krym.* Rs (K, E).

12. C. karadaghensis Pojark., *Not. Syst.* (*Leningrad*) **37**: 167 (1963). Like **11** but leaf-lobes always 5; corymbs glabrous; styles and pyrenes 1(–2). ● *Krym.* Rs (K).

13. C. ambigua C. A. Meyer ex A. Becker, *Bull. Soc. Nat. Moscou* **31**(1): 12, 34 (1858) (*C. volgensis* Pojark.). Shrub or small tree up to 8 m; twigs glabrescent, dark brown; spines 8–15 mm or absent. Leaves 40–80 × 40–80 mm, ovate, broadly cuneate, thin, with hair-tufts in axils of veins beneath, main veins curving towards base of leaf; lobes 5–7(–9), oblong, acuminate, acutely serrate in the apical half, extending more than ¾ of the way to the midrib; stipules entire, subulate. Corymb and hypanthium villous, later glabrescent; flowers *c.* 15 mm in diameter. Sepals 2–3 mm; styles (1–)2. Fruit 11–14 mm, subglobose, dark red, crowned by patent sepals; pyrenes (1–)2. *Wooded mountain slopes. S. Russia, E. Ukraine.* Rs (C, W, E).

14. C. monogyna Jacq., *Fl. Austr.* **3**: 50 (1775) (*C. oxyacantha* L., nom. ambig.). Shrub or small tree up to 10 m; spines 7–20 mm. Leaves obovate to rhombic, cuneate, discolorous; lobes 3–7, oblong, acute or subobtuse, entire or sparsely toothed near the apex, extending ¾ of the way to the midrib, the sinuses usually open and deep; stipules entire, lanceolate-subulate. Flowers 8–15 mm in diameter. Style 1. Fruit 6–10 mm, dark or bright red, crowned by deflexed sepals which are usually slightly longer than wide; pyrene 1. *Hedges and thickets. Almost throughout Europe except the northern and eastern margins.* All except Az Fa Is Rs (N, B, E) Sb.

A variable species. The taxa here included in *C. monogyna* have been variously treated by different authors.

1 Twigs and leaves densely pubescent   **(f) subsp. azarella**
1 Twigs and leaves glabrous or sparsely pubescent
  2 Leaves coriaceous, ± glaucous or glaucescent beneath
    3 Leaves 30–50 mm; petiole 15–30 mm; fruit brownish-red
          (c) subsp. leiomonogyna

3 Leaves 10–30 mm; petiole 3–15 mm; fruit bright red, slightly
   pruinose while young   **(d)** subsp. **brevispina**
2 Leaves slightly coriaceous, greyish or light green beneath
  4 Leaves up to 20 mm; flowers 8–10 mm in diameter
                             (e) subsp. **aegeica**
  4 Leaves more than 20 mm; flowers 10–15 mm in diameter
    5 Leaves (3–)5- to 7-lobed, large; petiole 20–25 mm; hypan-
      thium villous   **(a)** subsp. **nordica**
    5 Leaves deeply 3(–5)-lobed, small; petiole 5–15 mm; hypan-
      thium glabrous   **(b)** subsp. **monogyna**

(a) Subsp. **nordica** Franco, *Feddes Repert.* **79**: 37 (1968):
Leaves 30–50 × 18–40 mm, ovate-cuneate, (3–)5- to 7-lobed,
slightly coriaceous, greyish beneath, glabrescent; petiole 20–
25 mm; hypanthium villous; fruit 8–10 mm in diameter, deep
red, subglobose. 2*n* = 34. *Lowlands.* ● *N. & C. Europe.*

(b) Subsp. **monogyna**: Leaves 25–35 × 25–35 mm, obovate,
deeply 3(–5)-lobed, slightly coriaceous, light green beneath,
glabrescent; petiole 5–15 mm; hypanthium glabrous (rarely with
a few scattered hairs); fruit 6–9 × 5–7 mm, dark red, globose-
urceolate. *Submontane. From France to S. Ukraine.*

C. × **degenii** Zsák, *Bot. Közl.* **32**: 191 (1935) (*C. monogyna ×
nigra*), found in E.C. Europe, is like **14**(b) but has sparsely lanate
twigs with few spines, leaves sparsely pubescent above and more
or less lanate beneath, at least near the mid-vein, 5–9 unevenly
serrate lobes, usually 2–3 styles and a subglobose, brownish
fruit.

(c) Subsp. **leiomonogyna** (Klokov) Franco, *Feddes Repert.* **79**:
37 (1968) (*C. leiomonogyna* Klokov): Leaves 30–50 × 15–35 mm,
obovate, 3- to 5-lobed, coriaceous, glaucescent beneath, glabrous
or nearly so; petiole 15–30 mm; hypanthium glabrous; fruit
7–9 × 5–7 mm, shortly ellipsoid, brownish-red. *Forest-steppe.*
● *C. & S. Ukraine.*

(d) Subsp. **brevispina** (G. Kunze) Franco, *Collect. Bot.*
(*Barcelona*) 7: 463 (1968) (*C. brevispina* G. Kunze): Leaves
10–30 × 10–30 mm, obovate, 3(–5)-lobed, coriaceous, glaucous
beneath, glabrescent; petiole 3–15 mm; hypanthium glabrous
(very rarely villous); fruit 7–10 mm, globose, bright red, but
slightly pruinose while young. *Hedges and thickets, usually near
streams. Iberian peninsula and Islas Baleares.*

(e) Subsp. **aegeica** (Pojark.) Franco, *op. cit.* **79**: 37 (1968)
(*C. aegeica* Pojark.): Leaves 10–20 × 7–15 mm, obovate-cuneate
to almost flabellate, 3(–5)-lobed, chartaceous, light green beneath, glabrous; petiole 3–6 mm; flowers 8–10 mm in diameter;
hypanthium glabrous; fruit 7–8 mm, subglobose. *E. Aegean
region, from Karpathos to Thasos.*

(f) Subsp. **azarella** (Griseb.) Franco, *Collect. Bot.* (*Barcelona*)
7: 471 (1968) (*C. azarella* Griseb.): Twigs and young leaves densely
pubescent, usually retaining some hairs; leaves 15–30 × 7–30 mm,
flabellate, deeply 3- to 5(–7)-lobed, subcoriaceous, light green
beneath; petiole 4–13 mm; hypanthium villous; fruit 7–10 mm,
subglobose, brownish-red. *Dry mountain thickets.* ● *S.E.
Europe, Sicilia, S. & E. Italy and S. & E. Spain.*

C. × **polyacantha** Jan, *Elench. Hort. Parm.* 8 (1827) (*C. laciniata
× monogyna,* ?*C. oxyacantha* auct. balcan.) recorded originally
from Sicilia, and perhaps found also in the Balkan peninsula, is
like **14**(f) but has blackish twigs; more slender spines; leaves
15–20 × 15–20 mm, retaining the hairs on the veins beneath;
stipules 6–9 × 3–5 mm, semisagittate, serrate; corymb few-
flowered; styles (1–)2; fruit 4–7 × 4–7 mm, deep red, globose-
urceolate but flattened sideways, verruculose; pyrenes usually 2.

15. C. sphaenophylla Pojark. in Komarov, *Fl. URSS* 9: 502
(1939). Unarmed shrub; twigs reddish-brown, villous, becoming
glabrous and pruinose. Leaves 30–50 × 20–30 mm, coriaceous,

villous but glabrescent, obovate-cuneate, the upper part with
3 lobes; lobes extending $\frac{1}{3}$ to $\frac{1}{2}$ way to the midrib, the middle lobe
wide, incise-serrate, the laterals narrower, acute, serrate; petiole
10–25 mm; stipules falcate-lanceolate, entire. Hypanthium and
pedicels densely tomentose; flowers 10–15 mm in diameter.
Style 1. Fruit 10–14 × 9–12 mm, red but pruinose, crowned by
patent sepals; pyrenes 1. *Scrub on hills.* ● *Krym.* Rs (K).

C. **dipyrena** Pojark., *op. cit.* 508 (1939), from Krym, is probably
a hybrid between **15** and **16**. It differs from **15** by the less pubescent, 5- to 7-lobed, ovate or rhombic leaves, less tomentose
hypanthium and pedicels, and deep red or purplish-black fruit
with usually 2 pyrenes.

16. C. pentagyna Waldst. & Kit. ex Willd., *Sp. Pl.* **2**: 1006
(1800). Shrub or small tree; twigs sparsely arachnoid-lanate,
later glabrescent, grey-brown; spines *c.* 10 mm. Leaves 20–60 ×
20–40 mm, subrhombic-ovate to obovate, coriaceous, dark green
and nearly glabrous above, lighter green and sparsely arachnoid-
lanate beneath, finally nearly glabrous; lobes 3–7, irregularly
serrate, wide, suboctuse, extending $\frac{2}{3}$ of the way to the midrib,
with acute sinuses; petiole 15–30 mm; stipules narrow, falcate
long-acuminate, remotely dentate to entire. Corymb compound,
many-flowered; hypanthium and pedicels tomentose; flowers
12–15 mm in diameter. Sepals 1 × 2 mm, broadly triangular,
soon deciduous; styles 3–5; top of ovary tomentose. Fruit 10–15 mm,
globose-ellipsoid, blackish-purple, dull; pyrenes 4–5. *Margins of
woods. E.C. Europe and N. part of Balkan peninsula, extending
eastwards to S. Ukraine.* Al Bu Cz Hu Ju ?Rm Rs (W, K, E) ?Tu.

C. **klokovii** Ivaschin, *Ukr. Bot. Žur.* 21(6): 61 (1964), from the
E. Ukraine, a tree with the leaves less hairy beneath, flowers
10–12 mm in diameter, and usually 3 styles and pyrenes, is perhaps a subspecies of **16**.

17. C. nigra Waldst. & Kit., *Pl. Rar. Hung.* **1**: 62 (1801). Like
16 but leaves 40–80 × 25–55 mm, ovate or triangular, acute,
thinner, tomentose on both surfaces; lobes 7–11, sharply and
irregularly serrate, acute, extending $\frac{1}{2}$ way to the midrib; petiole
10–20 mm; stipules wide, ovate-falcate; corymb simple, few-
flowered; top of ovary glabrous; fruit black, lustrous. *Woods.*
● *E.C. Europe, extending to Albania and C. Jugoslavia.* Al Cz
Hu Ju ?Rm.

18. C. schraderana Ledeb., *Fl. Ross.* **2**: 91 (1843). Shrub; twigs
thinly lanate, soon glabrescent, purplish; spines up to 10 mm, or
absent. Leaves 30–50 × 30–50 mm, obovate to rhombic, coriaceous, sparsely lanate beneath but finally nearly glabrous; lobes
3–5, wide, suboctuse, extending $\frac{2}{3}$–$\frac{3}{4}$ of the way to the midrib,
with narrow sinuses, entire or with 1–3 coarse teeth near the apex;
petiole 5–8 mm; stipules large, semicordate, serrate. Hypanthium
and pedicels tomentose; flowers 15–18 mm in diameter. Sepals
short, triangular; styles 2–4. Fruit 12–14 × 15–16 mm, depressed-
globose, dark red, crowned by subpatent sepals; pyrenes 2–4.
Mountains. N. Greece; Krym. Gr Rs (K).

19. C. heldreichii Boiss., *Diagn. Pl. Or. Nov.* **3**(2): 47 (1856)
(*Mespilus heldreichii* (Boiss.) Ascherson & Graebner). Shrub;
twigs lanate; spines small. Leaves 15–30 × 12–30 mm, broadly
ovate, coriaceous, lanate on both surfaces; lobes 3–5, wide, acute,
sparsely serrate in the apical half, extending $\frac{2}{3}$ of the way to the
midrib; petiole 3–10 mm; stipules semiorbicular-falcate, entire
or with a few teeth near the base. Hypanthium and pedicels
lanate; flowers 15–18 mm in diameter. Styles 1–3. Fruit *c.* 7 mm,
globose, red, crowned by erect, later recurved, sepals; pyrenes 1–3.
Mountains. ● *S. Albania, C. & S. Greece, Kriti.* Al Cr Gr.

20. C. laciniata Ucria, *Nuovo Racc. Opusc. Aut. Sic.* **6**: 251 (1793) (*C. orientalis* Pallas ex Bieb.). Shrub or small tree up to 10 m; twigs lanate, blackish after the fall of the hairs; spines few. Leaves 30–50 × 25–40 mm, rhombic to obovate-oblong, cuneate, coriaceous, lanate on both surfaces; lobes 3–7, narrow, oblong, acute, extending $\frac{7}{8}$ of the way to the midrib, incise-dentate with 1–3 teeth at apex; petiole 3–8 mm; stipules semilunate, serrate, caducous. Hypanthium and pedicels lanate; flowers 15–20 mm in diameter. Sepals long-acuminate; styles 3–5, rarely connate. Fruit 15–20 mm, globose or pyriform, brick-red to yellowish-orange, lanate while young, crowned by deflexed sepals; pyrenes 3–5. *Mountain thickets and rocky slopes. S.E. Europe; Sicilia; S.E. Spain.* Al Bu Cr Gr Hs Ju Rs (W, K) Si [Ga].

(a) Subsp. **laciniata**: Leaves usually less than 30 mm wide; fruit globose, brick- to orange-red. *Throughout the range of the species.*

(b) Subsp. **pojarkovae** (Kossych) Franco, *Feddes Repert.* **79**: 37 (1968) (*C. pojarkovae* Kossych): Leaves usually more than 30 mm wide; fruit pyriform, yellowish-orange. *Mountain slopes.* ● *Krym.*

21. C. azarolus L., *Sp. Pl.* 477 (1753). Like **20** but twigs tomentose; leaves tomentose on both surfaces; lobes 3(–5), subobtuse, entire, the sinuses less deep; stipules falcate, coarsely serrate; hypanthium and pedicels densely tomentose; sepals shortly acuminate; styles 1–2(–3); fruit 20–25 mm in diameter, subglobose, orange-red or yellow; pyrenes 1–3. *Dry hillsides and mountains. Kriti; cultivated for its edible fruits in S. Europe and locally naturalized.* Cr [Ga Hs It ?Ju Si].

The wild plant from Kriti is var. *aronia* L.

C. × ruscinonensis Gren. & Blanc, *Billotia* **1**: 71 (1866) (*C. azarolus × monogyna*), is found occasionally with the parents in the W. Mediterranean region. It is like **21** but has thinner, subglabrous or glabrous leaves glossy green above; petioles up to 12 mm; corymb laxer, glabrescent; flowers 13–16 mm in diameter; sepals more acute; fruit 10–15 mm in diameter, less tomentose while young.

22. C. pycnoloba Boiss. & Heldr. in Boiss., *Diagn. Pl. Or. Nov.* **3**(2): 46 (1856). Shrub; twigs silvery-sericeous; spines strong but short. Leaves 15–25 × 15–25 mm, obovate-cuneate, densely silvery-sericeous on both surfaces; lobes 5–7, narrow, acute, entire, extending $\frac{2}{3}$ of the way to the midrib, with narrow sinuses; petiole 2–4 mm; stipules obovate, acute, entire, caducous. Hypanthium and pedicels densely silvery-sericeous; flowers *c.* 15 mm in diameter. Sepals triangular-lanceolate, acute, erecto-patent; styles 5. Fruit 8–10 mm, globose, brick-red, sericeous while young, crowned by deflexed sepals; pyrenes 3–5. *Mountain woods at 1000–1800 m.* ● *S. Greece (Peloponnisos).* Gr.

Subfam. **Prunoideae**

Stipules present, often small and caducous. Flowers 5-merous. Hypanthium concave or tubular, with a single free carpel at the base; epicalyx absent; stamens numerous. Fruit a 1-seeded drupe. Basic chromosome number 8.

¹ By D. A. Webb.

35. **Prunus** L.¹

Shrubs or trees. Leaves simple, usually crenate or serrate, petiolate. Stipules free, narrow, more or less scarious, often deciduous. Flowers 5-merous, solitary or in clusters, umbels, corymbs or racemes. Petals pink or white.

The delimitation of the genus and its subgenera here adopted is that of Rehder. Many authors elevate the subgenera or sections to the rank of genus.

Literature: A. Rehder, *Bibliography of Cultivated Trees and Shrubs*, 318–350. Jamaica Plain. 1949.

1 Ovary and fruit hairy
  2 Leaves broadly ovate to suborbicular, convolute in bud
    5. armeniaca
  2 Leaves at least twice as long as wide, longitudinally folded in bud
    3 Small shrub without spines; hypanthium about twice as long as wide
      4. tenella
    3 Tree or spiny shrub; hypanthium about as long as wide
      4 Leaves 6–9 mm wide; fruit 20–25 mm **3. webbii**
      4 Leaves 12–40 mm wide; fruit at least 35 mm
        5 Petals white or pale pink when expanded; mesocarp coriaceous; endocarp compressed, pitted **2. dulcis**
        5 Petals usually deep pink throughout anthesis; mesocarp succulent; endocarp ± globose, deeply sulcate **1. persica**
1 ˙Ovary and fruit glabrous
  6 Flowers in racemes or corymbs
    7 Flowers in short corymbs of 3–10 **16. mahaleb**
    7 Flowers in elongated racemes of 12–100
      8 Leaves evergreen, coriaceous; fruit ovoid-conical, ± acuminate
        9 Racemes usually considerably exceeding the subtending leaf; petioles and young twigs dark red **20. lusitanica**
        9 Racemes equalling or slightly exceeding the subtending leaf; petioles and young twigs pale green **21. laurocerasus**
      8 Leaves deciduous, thin and soft; fruit ± globose
        10 Petals 6–9 mm **17. padus**
        10 Petals 2·5–4 mm
          11 Bark aromatic; leaves with at least 15 pairs of inconspicuous lateral veins; fruit purplish-black **18. serotina**
          11 Bark not aromatic; leaves with 8–11 pairs of very distinct lateral veins; fruit dark red **19. virginiana**
  6 Flowers solitary, or in clusters or umbels
    12 Pedicel at least twice as long as ripe fruit
      13 Shrub seldom more than 1 m; petals 5–7 mm **13. fruticosa**
      13 Tree or large shrub; petals 9–15 mm
        14 Tree with well-defined trunk; leaves dull above; hypanthium urceolate **14. avium**
        14 Shrub, or small tree without well-defined trunk; leaves glossy above; hypanthium broadly campanulate **15. cerasus**
    12 Pedicel shorter than ripe fruit, or only slightly longer
      15 Petals pink
        16 Small tree; leaves 5–15 cm; hypanthium campanulate to pelviform **1. persica**
        16 Low shrub; leaves 1–3 cm; hypanthium tubular **12. prostrata**
      15 Petals white, rarely veined with red
        17 Leaves entirely glabrous
          18 Branches not spiny; fruit yellow **11. cocomilia**
          18 Branches spiny; fruit bluish-black
            19 Bark blackish; leaves dull; flowers mostly solitary **8. spinosa**
            19 Bark silvery-grey; leaves glossy above; flowers mostly in clusters of 2–3 **9. ramburii**
        17 Leaves hairy, at least on the veins beneath
          20 Young twigs dull, usually hairy
            21 Bark blackish; fruit 10–15 mm, ± erect **8. spinosa**
            21 Bark brown; fruit 20 mm or more, pendent **10. domestica**

20 Young twigs glossy, glabrous
22 Leaves boldly and irregularly serrate   **6. brigantina**
22 Leaves regularly crenate, or serrate with inconspicuous
    teeth directed strongly towards apex   **7. cerasifera**

Subgen. **Amygdalus** (L.) Focke. Deciduous; leaves longitudinally folded in bud. Shoots with terminal bud. Flowers subsessile, in clusters of 1–3, appearing before the leaves on shoots of the previous year's growth, each flower-bud flanked by 2 leaf-buds. Fruit usually pubescent or tomentose; endocarp sulcate or pitted.

1. **P. persica** (L.) Batsch, *Beytr. Entw. Pragm. Gesch. Nat.-Reiche* 30 (1801) (*Persica vulgaris* Miller). Tree up to 6 m with straight, glabrous, reddish, angular twigs. Leaves 5–15 × 2–4 cm, oblong-lanceolate, acute to acuminate, serrulate, glabrescent. Flowers subsessile, mostly solitary; hypanthium about as wide as long; sepals tomentose; petals 10–20 mm, deep (rarely pale) pink. Fruit 40–80 mm, globose, velutinous (glabrous in var. *nucipersica* (Borkh.) C. K. Schneider, the nectarine), yellow or pale green, tinged with red; mesocarp succulent, pale green or orange; endocarp deeply sulcate. *Extensively cultivated for its fruits (peaches) as a field crop in S. & S.C. Europe, and on a small scale in gardens further north; occasionally escaping and locally naturalized.* [Al Au Bl Bu Co Cr Cz Ga Ge Gr He Hs Hu It Ju Lu Rm Rs (W, K, E) Sa Si Tu.] (*China.*)

2. **P. dulcis** (Miller) D. A. Webb, *Feddes Repert.* **74**: 24 (1967) (*Amygdalus communis* L.; *A. dulcis* Miller, *P. communis* (L.) Arcangeli, non Hudson, *P. amygdalus* Batsch). Shrub or tree up to 8 m, in wild plants spiny and intricately branched, in cultivated plants with straight, spineless branches. Leaves 4–12 × 1·2–3 cm, oblong-lanceolate, crenate-serrate, glabrous. Flowers mostly in pairs; hypanthium broadly campanulate; sepals tomentose at least on the margin; petals *c.* 20 mm, bright pink in bud, fading to pale pink or almost white. Fruit 35–60 mm, ovoid-oblong, compressed, tomentose, grey-green; mesocarp coriaceous, eventually splitting and separating away from the finely pitted, keeled endocarp. *Extensively cultivated for its edible seeds (almonds) as a field crop in S. & S.C. Europe, and in gardens for ornament further north; frequently naturalized in the Mediterranean region.* [Al Au Bl Bu Co Cr Cz Ga Ge Gr He Hu It Ju Lu Rm Rs (W, K, E) Sa Si Tu.] (*C. & S.W. Asia, N. Africa.*)

3. **P. webbii** (Spach) Vierh., *Österr. Bot. Zeitschr.* **65**: 21 (1915). Like wild plants of **2** but branches strongly divaricate; leaves 3·5 × 0·9 cm or less; petals *c.* 10 mm, deep pink; fruits 20–25 mm, less densely tomentose, scarcely compressed, with endocarp scarcely keeled and only slightly pitted. *S. part of Balkan peninsula; Kriti; S. Italy.* Al Bu Cr Gr It Ju.

4. **P. tenella** Batsch, *Beytr. Entw. Pragm. Gesch. Nat.-Reiche* 29 (1801) (*P. nana* (L.) Stokes, non Duroi, *Amygdalus nana* L.). Shrub up to 1·5 m, spreading by suckers, glabrous except for bud-scales and fruit; branches suberect, grey. Leaves up to 5 × 2 cm, but often smaller, lanceolate to oblong-elliptical, serrate. Flowers mostly solitary; hypanthium tubular, about twice as long as wide; petals 10–15 mm, bright pink (rarely white). Fruit 12–20 mm, subglobose, densely villous with yellowish hairs; mesocarp coriaceous; endocarp with a reticulum of shallow furrows. *Dry grassland. E. & E.C. Europe, from S. Bulgaria to c. 55° N. in C. Russia; cultivated for ornament elsewhere and occasionally naturalized.* Au Bu Cz Hu Ju Rm Rs (C, W, K, E) [Ga].

Subgen. **Prunus.** Deciduous; leaves convolute in bud. Shoots without terminal bud. Flowers subsessile or shortly pedicellate,

solitary or in small, axillary clusters; flower-bud without accompanying leaf-buds. Hypanthium broadly campanulate. Fruit usually glabrous, often pruinose; mesocarp succulent; endocarp smooth or somewhat rugose.

5. **P. armeniaca** L., *Sp. Pl.* 474 (1753) (*Armeniaca vulgaris* Lam.). Shrub or small tree 3–6(–10) m, glabrous except for the flower and fruit; young twigs and young leaves reddish. Leaves 5–10 × 5–8 cm, broadly ovate to suborbicular, acuminate to cuspidate, serrate, truncate or subcordate at the base; petiole 2–4 cm. Flowers subsessile, solitary or in pairs, appearing before the leaves; hypanthium and calyx hairy; petals 10–15 mm, white or very pale pink. Fruit 4–8 cm, subglobose, velutinous, reddish-orange to yellow; mesocarp orange-yellow; endocarp lenticular, smooth, with 3 narrow ridges along one margin. *Cultivated for its fruits (apricots) as a field crop in S. & S.C. Europe, and in gardens further north; locally naturalized.* [Al Au Az Bl Bu Co Cr Cz Ga Ge Gr He Hs Hu It Ju Lu Rm Rs (W, K, E) Sa Si Tu.] (*C. Asia and China.*)

6. **P. brigantina** Vill. in L., *Syst. Pl. Eur.* **1**, *Fl. Delph.*: 49 (1785) (*P. brigantiaca* Vill.). Shrub or small tree 2–6 m, with spreading branches; young twigs glabrous and glossy. Leaves 5–8 × 2·5–5 cm, ovate to elliptical, acuminate, truncate or subcordate at the base, boldly and irregularly serrate, glabrous and glossy above, pubescent on the veins beneath; petiole 1–2 cm, pubescent. Flowers very shortly pedicellate, in clusters of 2–5, appearing before the leaves; petals *c.* 8 mm, white. Fruit *c.* 25 mm, subglobose, slightly apiculate, glabrous, yellow; endocarp lenticular, smooth. *Dry, stony slopes, 1200–1800 m.* ● *S.W. Alps, northwards to 45° N.* Ga It.

7. **P. cerasifera** Ehrh., *Beitr. Naturk.* **4**: 17 (1789) (incl. *P. divaricata* Ledeb. Shrub or tree up to 8 m, with numerous intricate, fine, sometimes spiny branches; young twigs glabrous and glossy. Leaves 4–7 × 2–3·5 cm, oblong-obovate, cuneate at the base, regularly crenate or appressed-serrate, glabrous and glossy above, pubescent on the veins beneath. Flowers mostly solitary, appearing with or slightly before the leaves; pedicel *c.* 15 mm, glabrous; petals 8–10 mm, usually white. Fruit 20–30 mm, globose, glabrous, red or yellow; endocarp subglobose, keeled, smooth. *Balkan peninsula; Krym; planted elsewhere for its fruit or for hedges and locally naturalized.* Al Bu Gr Ju Rs (K) Tu [Au Br Da Ga Ge Hu It Rm].

Var. *pissardii* (Carrière) L. H. Bailey, with dark red leaves and flowers tinged with reddish-pink, is often grown in gardens for ornament.

8. **P. spinosa** L., *Sp. Pl.* 475 (1753). Dense shrub up to 4 m, spreading by suckers and with numerous, divaricate, intricate, spiny branches; bark blackish; young twigs usually pubescent. Leaves 2–4 cm, obovate to oblanceolate, finely crenate or serrate, cuneate at the base, dull green and glabrous above, usually pubescent on the veins beneath. Flowers mostly solitary, appearing before the leaves, very numerous; pedicel *c.* 5 mm, glabrous; petals 5–8 mm, white. Fruit 10–15 mm, globose, erect, bluish-black, very pruinose, acid and astringent; endocarp subglobose, smooth or slightly rugose. $2n=32$. *Europe, except the north-east and extreme north.* All except Az Cr Fa Is Rs (N) Sb.

Natural hybrids between **8** and **10b**, with $2n=40$, have been recorded, and at least some of the plants that have been named **P. fruticans** Weihe, *Flora (Regensb.)* **9**: 748 (1826), are hybrids of this parentage; but as the two parents are variable and rather similar, the morphological delimitation of the hybrid is almost impossible.

9. P. ramburii Boiss., *Elenchus* 39 (1838). Like **8** but completely glabrous; bark of older branches silvery-grey; leaves 1·5–2·5 cm, elliptical to linear-oblong, glossy; flowers mostly in clusters of 2–3, appearing with the leaves; fruit smaller. *Dry, calcareous slopes.* ● *S. Spain (Sierra Nevada, Sierra de Gádor).* Hs.

10. P. domestica L., *Sp. Pl.* 475 (1753). Shrub or tree up to 10 m, in cultivated plants with straight, spineless branches, but in wild plants often somewhat spiny, often spreading by suckers; bark dull brown; young twigs dull, usually pubescent. Leaves 3–8 × 1·8–5 cm, obovate to elliptical, crenate-serrate, glabrous and dull green above, densely pubescent to subglabrous beneath. Flowers usually in clusters of 2–3, appearing with the leaves; pedicel 5–20 mm; petals 7–12 mm, white. Fruit 20–75 mm, globose to oblong, usually pendent, purple, red, yellow or green, sweet or acid, not astringent; endocarp somewhat rugose. *Cultivated for its fruits as a field crop in most of Europe except the north-east and extreme north, and widely naturalized.* Probably all except Az Cr Fa Is Rs (N) Sb. *(Caucasus.)*

Cultivated plants referable to this species are hexaploid, with $2n = 48$; they are usually interpreted as allopolyploids derived from **7** (diploid) and **8** (tetraploid); triploid and hexaploid hybrids of this parentage have been found in the Caucasus.

The numerous cultivars and the very variable naturalized plants are best arranged in 2 subspecies; (a) comprises the plums and (b) the damsons and greengages.

(a) Subsp. **domestica** (subsp. *oeconomica* (Borkh.) C. K. Schneider): Tree, without spines. Young twigs and pedicels often subglabrous. Petals greenish-white. Fruit 40–75 mm, longer than wide, usually bluish-black, purple or red. Endocarp compressed, keeled, often separating from mesocarp. *Naturalized mainly near houses.*

(b) Subsp. **insititia** (L.) C. K. Schneider, *Ill. Handb. Laubholzk.* **1**: 630 (1906) (*P. insititia* L.; incl. *P. domestica* subsp. *italica* (Borkh.) Hegi): Shrub or tree, sometimes spiny. Young twigs and pedicels densely pubescent. Petals pure white. Fruit 20–50 mm, subglobose, purple, red, yellow or green. Endocarp subglobose, scarcely keeled, adherent to mesocarp. *Widely naturalized in hedges and woods.*

11. P. cocomilia Ten., *Fl. Nap.* **1**, *Prodr.*: 68 (1811) (*P. pseudarmeniaca* Heldr. & Sart. ex Boiss.). Small tree or shrub (at high altitudes dwarfed and procumbent), completely glabrous. Leaves 2·5–4 × 1·2–2·5 cm, elliptic-obovate, crenulate. Flowers in clusters of 2–4, appearing with the leaves; pedicel 2–4 mm. Petals *c.* 6 mm, white. Fruit 12–40 × 8–25 mm, oblong, subacute, yellow flushed with purple-red, glabrous. *Mountains of S. part of Balkan peninsula, S. Italy and Sicilia.* Al Gr It Ju Si.

Subgen. **Cerasus** (Miller) Focke. Deciduous; leaves longitudinally folded in bud. Shoots with terminal bud. Flowers usually in umbels, sometimes in corymbs or solitary. Fruit glabrous, not pruinose; mesocarp succulent; endocarp subglobose, smooth or slightly sulcate.

12. P. prostrata Labill., *Icon. Pl. Syr.* **1**: 15 (1791). Low, spreading shrub up to 1 m, with deflexed-arcuate branches. Young twigs puberulent. Leaves 9–12(–27) × 3–6(–9) mm, linear-oblong to broadly elliptical-ovate or -obovate, conspicuously incise-serrate at least in apical half, glabrous above, grey-tomentose to glabrous beneath. Flowers mostly solitary, subsessile, appearing before the leaves; hypanthium tubular; petals *c.* 7 mm, bright pink. Fruit *c.* 8 mm, broadly ovoid, red; endocarp slightly sulcate. *Mountains of Mediterranean region; local.* Al Co Cr Gr Hs Ju Sa.

13. P. fruticosa Pallas, *Fl. Ross.* **1**(1): 19 (1784) (*Cerasus fruticosa* (Pallas) Woronow). Low, spreading shrub 0·3–1·5 m (rarely more), completely glabrous. Leaves of long shoots 3–5·5 × 1·5–2·7 cm, elliptic-oblanceolate, finely crenate-serrate, dark, shining green; those of short shoots 1·5–2·5 cm, obovate, otherwise similar. Flowers in subsessile umbels of 2–5, with leaf-like bud-scales at the base; pedicels 2–3 cm. Hypanthium campanulate; petals 5–7 mm, white. Fruit 7–10 mm, globose, dark red; endocarp smooth. *Thickets and dry grassland. E., S.E. & C. Europe, northwards to c. 57° in E. Russia, southwards to Bulgaria and Crna Gora, and westwards to W. Germany.* Au Bu Cz Ge Hu It Po Rm Rs (C, W, E).

The hybrid between this species and **15** is locally common in C. Europe.

14. P. avium L., *Fl. Suec.* ed. 2, 165 (1755) (*Cerasus avium* (L.) Moench). Tree 10–20(–30) m, with well-defined trunk and horizontal branches; young twigs glabrous; bark reddish-brown, peeling off in paper-like strips. Leaves 8–15 × 4–7 cm, obovate-oblong, acuminate, crenate-serrate with deep but obtuse teeth, glabrous but dull above, usually with some persistent pubescence beneath, drooping when young; petiole 2–5 cm, with 2 conspicuous glands at the top. Flowers in sessile umbels of 2–6, with mainly scarious bud-scales at the base; pedicels 2–5 cm; hypanthium urceolate (constricted at mouth). Petals 9–15 mm, white. Fruit 9–12 mm, globose, dark red (also creamy-yellow, bright red or black in cultivars), sweet or bitter; endocarp subglobose, smooth. $2n = 16$. *Most of Europe, except the extreme north and east, but rare as a native in the Mediterranean region. Widely cultivated for its fruit, and often naturalized.* Al Au Be Br Bu ?Co Cz Da Ga Ge Gr Hb He Ho Hs Hu It Ju Lu No Po Rm Rs (C, W, K) Sa Su Tu [Bl].

From this species are derived most of the sweet cherries. There seems to be little evidence for the assertion made by many authors that this species is native only to W. Asia; its native status in N.W. & C. Europe is attested by archaeological and subfossil evidence.

15. P. cerasus L., *Sp. Pl.* 474 (1753) (*Cerasus vulgaris* Miller; incl. *C. austera* (L.) Borkh., *C. collina* Lej. & Court.). Like **14** but usually a shrub (sometimes a small tree up to 8 m, but with ill-defined trunk and many suckers); leaves somewhat smaller, glossy above, glabrescent beneath, firmer and not drooping; petiole 1–3 cm, often without glands; many bud-scales at base of umbels with leaf-like tip; hypanthium broadly campanulate; fruit bright red, acid. *Cultivated for its fruit, and sometimes for hedges, and widely naturalized.* [Al Au Br Bu Cz Da Fe Ga Ge Gr Hb He Ho Hs Hu It Ju Lu No Po Rm Rs (B, C, W, K) Su.] *(S.W. Asia.)*

Cultivars form the sour or Morello cherries used for preserving. Hybrids between **14** and **15** (*P.* × *gondouinii* (Poiteau & Turpin) Rehder) are cultivated in W. Europe as 'Duke' cherries.

16. P. mahaleb L., *Sp. Pl.* 474 (1753) (*Cerasus mahaleb* (L.) Miller). Shrub, or rarely small tree, up to 10 m; young twigs glandular-puberulent. Leaves 4–7 cm, broadly ovate, cuspidate, rounded to subcordate at the base, crenate-serrate with conspicuous marginal glands, glabrous, or slightly pubescent beneath. Flowers fragrant, in short, corymbose racemes of 3–10, which terminate short, lateral, leafy shoots; pedicels *c.* 10 mm; hypanthium campanulate. Petals 5–8 mm, white. Fruit 8–10 mm, ovoid, black; mesocarp thin, bitter; endocarp smooth. *Dry hill-*

sides, thickets and open woods. C. & S. Europe, extending to Ukraine and S. Belgium. Al Au Be Bu Co Cz Ga Ge Gr He Hs Hu It Ju Lu Rm Rs (W, K) Si [No Su].

Subgen. **Padus** (Miller) Focke. Like Subgen. *Cerasus* but flowers numerous, in elongated racemes which terminate short, leafy shoots.

17. P. padus L., *Sp. Pl.* 473 (1753) (*Cerasus padus* (L.) Delarbre). Tree or shrub with ascending branches; bark foetid. Leaves 6–10 cm, obovate to elliptic-oblong, acuminate, acutely and finely serrate, dull green above, paler beneath. Racemes with 15–35 flowers. Petals 6–9 mm, often erose-denticulate. Fruit 6–8 mm, subglobose, shining black, bitter and astringent; endocarp sulcate. $2n = 32$. *Most of Europe except the Mediterranean region, Balkan peninsula and S.E. Russia.* Au Be Br Bu Cz Da Fe Ga Ge Hb He Ho Hs It Ju Lu No Po Rm Rs (N, B, C, W, K, E) Su.

(a) Subsp. **padus**: Tree up to 17 m; young shoots glabrescent; leaves glabrous beneath, or with hairs only in axils of veins; racemes more or less pendent; flowers with heavy scent. *Throughout the range of the species.*

(b) Subsp. **borealis** Cajander, *Suomen Kasvio* 353 (1906) (subsp. *petraea* (Tausch) Domin): Shrub, seldom more than 3 m; young shoots pubescent; leaves pubescent beneath, with prominent veins; racemes horizontal or ascending; flowers scarcely scented. $2n = 32$. *N. & W. Fennoscandia; mountains of C. Europe from the Vosges to the Carpathians and S.E. Alps.*

18. P. serotina Ehrh., *Beitr. Naturk.* 3: 20 (1788). Tree up to 20 m; bark aromatic. Leaves obovate to elliptic-oblong, acuminate, finely serrate with flattened, forwardly directed teeth, dark, shining green above, paler and slightly pubescent beneath; lateral veins at least 15, not very conspicuous. Racemes 6–15 cm, with *c*. 30 flowers. Calyx persistent in fruit; petals 3–4, denticulate, creamy white. Fruit 8 mm, depressed-globose, purplish-black; endocarp smooth. *Planted, mainly in C. Europe, for timber and elsewhere for ornament, and occasionally naturalized.* [Au Cz Da Ga Ge Ho Hu Ju Po Rm Rs Su.] (*E. North America.*)

19. P. virginiana L., *Sp. Pl.* 473 (1753) (*Padus virginiana* (L.) M. J. Roemer). Like **18** but not more than 5 m, and usually a shrub; bark not aromatic; leaves dull, with very acute and more patent teeth; lateral veins 8–11, conspicuous and distinct; calyx deciduous; fruit dark red. *Planted on a small scale and occasionally naturalized.* [Cz Ga Rs.] (*E. North America.*)

Subgen. **Laurocerasus** (Duh.) Rehder. Leaves evergreen, coriaceous, glabrous and glossy, longitudinally folded in bud. Shoots with terminal bud. Flowers in elongated, axillary racemes with leafless peduncle. Fruit glabrous.

20. P. lusitanica L., *Sp. Pl.* 473 (1753) (*Cerasus lusitanica* (L.) Loisel.). Shrub or tree, 3–8(–20) m; young twigs and petioles dark red, glabrous. Leaves 8–13 × 2·5–7 cm, elliptic-ovate to oblong-lanceolate, acuminate, regularly crenate or dentate, without glands, very dark green above. Racemes, with peduncle, 10–28 cm, suberect. Petals 4–7 mm, dull white. Fruit 8–13 mm, ovoid to subglobose, somewhat acuminate, purplish-black. $2n = 64$. *Iberian peninsula, just extending to S.W. France; Açores. Often planted for ornament in W. Europe.* Az Ga Hs Lu.

(a) Subsp. **lusitanica**: Up to 17 m; leaves 2·5–5 cm wide, oblong-lanceolate. Racemes 15–28 cm, with 50–100 flowers. Fruit about as long as its pedicel. *Iberian peninsula and S.W. France.*

(b) Subsp. **azorica** (Mouillefert) Franco, *Bol. Soc. Port. Ci. Nat.* ser. 2, **10**: 82 (1964): Not more than 4 m; leaves 4·5–6·5 cm wide, ovate-elliptical. Racemes 10–17 cm, with 20–30 flowers. Fruit much longer than its pedicel. *Açores.*

21. P. laurocerasus L., *Sp. Pl.* 474 (1753) (*Cerasus laurocerasus* (L.) Loisel., *Laurocerasus officinalis* M. J. Roemer). Like **20** but seldom more than 8 m; young twigs and petioles pale green; leaves lighter green, more rigid, sometimes subentire, usually with 1 or more sessile, circular glands on lower side near the petiole; racemes 8–13 cm, scarcely exceeding the subtending leaf; fruit *c*. 12 mm. *E. part of Balkan peninsula; extensively planted in parks and gardens in S. & W. Europe and locally naturalized.* Bu Ju Tu [Br Co Ga Hb Lu].

LXXXI. LEGUMINOSAE[1]

Trees, shrubs or herbs. Leaves alternate, rarely opposite, simple to 2-pinnate, stipulate. Flowers usually hermaphrodite, usually 5-merous. Sepals usually united. Petals free or somewhat connate. Stamens usually 10, sometimes less than 10 or numerous. Ovary a single unilocular carpel; style 1. Fruit a dehiscent, 2-valved or indehiscent, occasionally lomentaceous legume. Seeds usually without endosperm.

A large number of species, both native and introduced, are cultivated for food, for fodder and for ornament. Those most frequently utilized as food are to be found in *Apios*, *Arachis*, *Cicer*, *Glycine*, *Glycyrrhiza*, *Lens*, *Phaseolus*, *Pisum*, *Vicia* and *Vigna*. The edible part is usually the seed or legume, or both. Species of these genera and many others, particularly in *Anthyllis*, *Ceratonia*, *Coronilla*, *Galega*, *Lathyrus*, *Lotus*, *Lupinus*, *Medicago*, *Melilotus*, *Trifolium* and *Trigonella* are cultivated for fodder on a large scale or are planted to improve pasture.

Cultivated ornamental plants are to be found in the majority of European genera. In addition to some of those mentioned above,

the following are of particular importance in this respect: *Acacia*, *Amorpha*, *Caragana*, *Cercis*, *Colutea*, *Cytisus* and *Genista* and related genera, *Laburnum*, *Robinia*, *Sophora*, *Spartium* and *Wisteria*.

1   Stamens numerous, free, longer than the corolla   **4. Acacia**
1   Stamens not more than 10, the filaments often partly or completely united, usually shorter than the corolla
   2   Corolla absent; stamens 5, free   **2. Ceratonia**
   2   Corolla present; stamens more than 5 or the filaments united
     3   At least some leaves 2-pinnate; corolla scarcely zygomorphic
       3. Gleditsia
     3   Leaves simple, 3-foliolate or 1-pinnate; corolla strongly zygomorphic
       4   Flowers with 1 petal   **42. Amorpha**
       4   Flowers with 5 petals, 2 or more sometimes connate
         5   Leaves consisting of a tendril only (but with large leaf-like stipules)   **51. Lathyrus**
         5   Leaves not consisting of a tendril only
           6   At least some leaves paripinnate, imparipinnate or digitate; leaflets 2, 4 or more
             7   Leaves paripinnate; rhachis often ending in a spine or tendril

1 Edit. V. H. Heywood and P. W. Ball.

8 Shrubs or small trees; rhachis often ending in a spine
  9 Pedicels 5 mm or more, articulated   **36. Caragana**
  9 Pedicels usually less than 5 mm, not articulated
    10 Stipules forming spines   **35. Halimodendron**
    10 Stipules not forming spines   **38. Astragalus**
8 Herbs; rhachis not ending in a spine
  11 Stipules adnate to the petiole; calyx bilabiate, the upper lip with 4 teeth, the lower with 1 tooth   **74. Arachis**
  11 Stipules not adnate to the petiole; calyx actinomorphic or if bilabiate, the upper lip with 2 teeth, the lower with 3 teeth
    12 Stem and leaves glandular-pubescent   **48. Cicer**
    12 Stem and leaves not glandular-pubescent
      13 Stem winged   **51. Lathyrus**
      13 Stem not winged
        14 Leaflets parallel-veined   **51. Lathyrus**
        14 Leaflets pinnately veined
          15 Calyx-teeth all equal and at least twice as long as the tube   **50. Lens**
          15 At least 2 calyx-teeth less than twice as long as the tube
            16 Calyx-teeth ± leaf-like; stipules up to 10 cm   **52. Pisum**
            16 Calyx-teeth not leaf-like; stipules not more than 2 cm
              17 Style pubescent all round or on the lower side, or glabrous   **49. Vicia**
              17 Style pubescent on the upper side only   **51. Lathyrus**
7 Leaves imparipinnate or digitate
  18 Principal lateral veins of the leaflets terminating at the margin, often in a tooth
    19 Glabrous or glabrescent, eglandular   **57. Trifolium**
    19 Variously hairy and glandular, sometimes sparsely so
      20 Calyx gibbous at base; stipules free from petiole   **48. Cicer**
      20 Calyx not gibbous at base; stipules adnate to petiole   **53. Ononis**
  18 Lateral veins of the leaflets anastomosing and not reaching the margin
    21 At least some flowers in terminal or apparently terminal inflorescences
      22 Leaves digitate   **28. Lupinus**
      22 Leaves imparipinnate
        23 Climbing shrubs; flowers in pendent racemes; legume velutinous   **31. Wisteria**
        23 Not climbing; flowers in erect inflorescences; legume not velutinous
          24 Trees; flowers in large panicles   **5. Sophora**
          24 Herbs or small shrubs; flowers in heads or racemes
            25 Stamens free; flowers in racemes 5 cm or more   **5. Sophora**
            25 Stamens connate; flowers in heads   **63. Anthyllis**
    21 All flowers axillary or in axillary inflorescences
      26 Plant glandular, at least in part
        27 Racemes pendent; legume 5–10 cm   **30. Robinia**
        27 Racemes erect; legume not more than 3 cm
          28 Dwarf shrubs; corolla 20–25 mm   **37. Calophaca**
          28 Herbs; corolla less than 20 mm   **41. Glycyrrhiza**
      26 Plant eglandular
        29 Flowers in umbels or clusters, the pedicels arising ± from the same point
          30 Legume lomentaceous
            31 Keel obtuse; legume strongly reticulate-veined   **64. Ornithopus**
            31 Keel acute; legume not or only faintly reticulate-veined
              32 Segments of the legume lunate or horseshoe-shaped to rectangular with a semicircular to orbicular sinus which has a curved protuberance at its base   **66. Hippocrepis**

              32 Segments of the legume linear or oblong, straight or slightly curved
                33 Stamens diadelphous; legume glabrous   **65. Coronilla**
                33 Stamens monadelphous; legume pubescent   **67. Hammatolobium**
          30 Legume dehiscent or indehiscent, not lomentaceous
            34 Keel beaked
              35 Leaves with 4–7 pairs of leaflets   **62. Securigera**
              35 Leaves with 2–3 pairs of leaflets or simple
                36 Lower leaves simple; upper leaves with 2–3 pairs of leaflets; legume spirally twisted and flattened so that it is circular in outline   **61. Hymenocarpus**
                36 All leaves with 2 pairs of leaflets; legume linear or oblong, straight or curved   **59. Lotus**
            34 Keel not beaked
              37 Keel very dark red or black   **58. Dorycnium**
              37 Keel not dark red or black
                38 Leaves imparipinnate
                  39 Umbels without an involucre   **38. Astragalus**
                  39 Umbels with an involucre of scarious bracts   **72. Ebenus**
                38 Leaves digitate or apparently so
                40 Pedicels 15–20 mm; leaflets with a spinescent apex   **36. Caragana**
                40 Pedicels not more than 5 mm; leaflets without a spinescent apex   **27. Lotononis**
        29 Flowers in racemes or condensed panicles or solitary
          41 Leaflets distinctly parallel-veined   **51. Lathyrus**
          41 Leaflets pinnately veined or the lateral veins obscure
            42 Leaflets with a spinescent apex   **36. Caragana**
            42 Leaflets without a spinescent apex
              43 Spiny shrubs   **69. Eversmannia**
              43 Unarmed herbs, shrubs or trees
                44 Legume lomentaceous
                  45 Racemes (2–)4- to many-flowered   **70. Hedysarum**
                  45 Flowers solitary, axillary
                    46 Corolla 4–7 mm; segments of the legume flat and rectangular with a suborbicular sinus   **66. Hippocrepis**
                    46 Corolla 10–14 mm; segments of the legume ovoid-oblong, terete   **67. Hammatolobium**
                44 Legume not lomentaceous
                  47 Legume indehiscent, usually toothed or spiny
                    48 Legume oblong, dorsiventrally compressed, the valves sinuate-dentate on the back; corolla blue, or yellow with a blue apex   **40. Biserrula**
                    48 Legume ± orbicular, the margin usually toothed, the sides reticulate-veined or foveolate and the veins often toothed; corolla white, pink or purple   **71. Onobrychis**
                  47 Legume usually dehiscent, not toothed or spiny
                    49 Racemes 10 cm or more, pendent; stipules usually forming spines; leaflets stipellate   **30. Robinia**
                    49 Racemes usually less than 10 cm, erect; stipules not forming spines
                      50 Leaflets stipellate   **44. Apios**
                      50 Leaflets not stipellate
                        51 Legume strongly inflated, membranous
                          52 Shrubs up to 2 m or more   **33. Colutea**
                        52 Acaulescent herbs   **38. Astragalus**
                        51 Legume not or only slightly inflated, not membranous

53 Keel beaked   **59. Lotus**
53 Keel not beaked but sometimes mucronate
  54 Keel mucronate at apex
    55 Mucro on the adaxial side of the keel   **38. Astragalus**
    55 Mucro on the abaxial side of the keel   **39. Oxytropis**
  54 Keel not mucronate at apex
    56 Stamens monadelphous
      57 Rhachis very short so that the leaves are almost digitate; leaflets 1–2 pairs; corolla bright pink   **72. Ebenus**
      57 Rhachis long; leaflets 4–10 pairs; corolla yellow or white to bluish-violet
        58 Corolla 10–15 mm; perennial   **32. Galega**
        58 Corolla *c.* 3 mm; annual   **38. Astragalus**
    56 Stamens diadelphous
      59 Style glabrous   **38. Astragalus**
      59 Style pubescent on the lower side   **49. Vicia**
6 Leaves simple, 1-foliolate or 3-foliolate, sometimes very small
  60 Leaves 7–12 cm, simple, suborbicular, cordate; adaxial petal innermost   **1. Cercis**
  60 Leaves 3-foliolate, or simple, but never suborbicular and cordate; adaxial petal outermost
    61 Principal lateral veins of the leaflets terminating at the margin; leaflets often toothed
      62 Plant glandular-pubescent, at least above
        63 Stamens monadelphous; legume straight or very slightly curved   **53. Ononis**
        63 Stamens diadelphous; legume falcate to spirally coiled, rarely almost straight   **56. Medicago**
      62 Plant not glandular-pubescent
        64 At least some petal-claws adnate to the staminal tube; corolla usually persistent in fruit   **57. Trifolium**
        64 Petal-claws free from the staminal tube; corolla deciduous
          65 Filaments of at least 5 stamens dilated at the apex   **57. Trifolium**
          65 Filaments all filiform
            66 Legume coiled in 1 or more turns of a spiral   **56. Medicago**
            66 Legume straight or curved
              67 Perennial
                68 Legume obovate or ovate to subglobose   **54. Melilotus**
                68 Legume oblong, oblong-falcate, oblong-reniform, reniform or variously curved   **56. Medicago**
              67 Annual or biennial
                69 Corolla blue   **55. Trigonella**
                69 Corolla white or yellow
                  70 Legume linear or oblong, at least 3 times as long as wide   **55. Trigonella**
                  70 Legume ovate or obovate to subglobose or reniform, less than 3 times as long as wide
                    71 Legume reniform   **56. Medicago**
                    71 Legume ovate or obovate to subglobose
                      72 Legume without or with a very short beak and without a membranous wing   **54. Melilotus**
                      72 Legume with a long, curved beak or with a broad membranous wing on the margin   **55. Trigonella**
    61 Principal lateral veins of the leaflets anastomosing and not reaching the margin, sometimes obscure; leaflets not toothed (leaves sometimes caducous or reduced to a spine-tipped phyllode)
      73 Plant spiny
        74 Corolla pink, red, purple or violet-blue
          75 Leaves *c.* 5 mm; calyx bilabiate   **23. Erinacea**
          75 Leaves 10–20 mm; calyx actinomorphic   **73. Alhagi**
        74 Corolla yellow
          76 Leaves of adult plants reduced to persistent spine-tipped phyllodes
            77 Leaves and branches mostly alternate; legume scarcely exserted from the calyx   **24. Ulex**
            77 Leaves and branches mostly opposite; legume conspicuously exserted from the calyx   **25. Stauracanthus**
          76 Leaves not spine-tipped, often caducous
            78 Calyx tubular, with 5 short teeth, the upper portion breaking away at anthesis to leave a cup-like remnant   **10. Calicotome**
            78 Upper part of calyx not breaking away at anthesis
              79 Calyx with 5 ± equal teeth, not or only slightly bilabiate   **63. Anthyllis**
              79 Calyx ± distinctly bilabiate
                80 Upper lip of calyx with 2 short teeth; leaves 3-foliolate   **13. Chamaecytisus**
                80 Upper lip of calyx deeply 2-fid; leaves often 1-foliolate
                  81 Calyx not more than 7 mm   **16. Genista**
                  81 Calyx 7 mm or more
                    82 Leaves and branches mostly alternate; calyx not inflated   **16. Genista**
                    82 Leaves and branches mostly opposite; calyx somewhat inflated   **18. Echinospartum**
      73 Plant not spiny
        83 Young stems broadly winged   **17. Chamaespartium**
        83 Young stems not broadly winged
          84 Leaflets stipellate; leaves 3-foliolate
            85 Corolla not more than 7 mm; plant with reddish-brown hairs   **47. Glycine**
            85 Corolla 10 mm or more; plant glabrous or with whitish hairs
              86 Beak of the keel forming 1⅓–2 turns of a spiral   **45. Phaseolus**
              86 Beak of the keel recurved   **46. Vigna**
          84 Leaflets not stipellate; leaves simple or 3-foliolate
            87 Legume with prominent glandular tubercles   **26. Adenocarpus**
            87 Legume without glandular tubercles (sometimes with glandular hairs)
              88 Leaves simple or 1-foliolate, sometimes very small
                89 Annual herbs
                  90 Leaves linear, grass-like; legume dehiscent   **51. Lathyrus**
                  90 Leaves obovate or elliptical, not grass-like; legume indehiscent   **68. Scorpiurus**
                89 Shrubs or perennial herbs, woody at base
                  91 Corolla violet; leaves 1–3 mm, scarious   **34. Eremosparton**
                  91 Corolla white or yellow; leaves usually larger, herbaceous, sometimes caducous
                    92 Calyx caducous after anthesis   **20. Lygos**
                    92 Calyx not caducous
                      93 Calyx split to the base adaxially   **21. Spartium**
                      93 Calyx not split to the base
                        94 Calyx ± tubular; legume ± included in the persistent calyx   **63. Anthyllis**
                        94 Calyx campanulate; legume exserted, or the calyx not persistent
                          95 Upper lip of calyx with short teeth   **12. Cytisus**
                          95 Upper lip of calyx deeply 2-fid or deeply toothed

96 Legume ovoid, oblong or falcate, dehiscent, not inflated **16. Genista**
96 Legume globose-inflated, not or tardily dehiscent **20. Lygos**
88 At least some leaves 3-foliolate
97 Leaflets conspicuously glandular-punctate **43. Psoralea**
97 Leaflets not or very minutely glandular-punctate
98 Legume lomentaceous
99 Annual; corolla 4–8 mm **65. Coronilla**
99 Perennial; corolla 10–14 mm **67. Hammatolobium**
98 Legume not lomentaceous
100 Calyx usually bilabiate, the upper lip with 4 teeth, the lower with 1 tooth, somewhat shorter than the upper **27. Lotononis**
100 Calyx actinomorphic or bilabiate, but never with the upper lip with 4 teeth and the lower with 1 tooth
101 Annual or perennial herbs, sometimes with a woody stock
102 Stamens free; flowers in clusters of 3 arranged in a terminal leafy raceme **6. Thermopsis**
102 Stamens connate; flowers not in clusters of 3 arranged in a terminal leafy raceme
103 Calyx inflated, 4·5–6 mm wide in flower, up to 12 mm wide in fruit and enclosing the legume **63. Anthyllis**
103 Calyx less than 4·5 mm wide, not inflated
104 Keel very dark red or black **58. Dorycnium**
104 Keel not very dark red or black
105 Stamens monadelphous; stipules completely free from the petiole **29. Argyrolobium**
105 Stamens diadelphous; stipules inserted on or adnate to the base of the petiole
106 Stipules inserted at the base of the petiole; legume not longitudinally winged **59. Lotus**
106 Stipules inserted on the stem, and adnate to the base of the petiole; legume with 2 or 4 longitudinal wings **60. Tetragonolobus**
101 Shrubs or trees
107 Stamens free; stipules connate, conspicuous **7. Anagyris**
107 Stamens variously connate; stipules free, often minute or absent
108 Legume ±included in the persistent calyx; calyx with 5 ±equal teeth
109 Flowers solitary or in clusters of 2–3; calyx-teeth shorter than tube; corolla yellow **63. Anthyllis**
109 Flowers in dense axillary racemes; calyx-teeth longer than tube; corolla bright pink **72. Ebenus**
108 Legume exserted or the calyx deciduous; calyx bilabiate
110 Flowers in pendent racemes **8. Laburnum**
110 Flowers in erect inflorescences
111 Legume broadly winged **9. Podocytisus**
111 Legume not winged
112 Upper lip of calyx deeply 2-fid
113 Calyx-tube distinctly shorter than lips **29. Argyrolobium**

113 Calyx-tube as long as or longer than lips
114 Petiole 15–50 mm; legume 35–50 mm **22. Petteria**
114 Petiole not more than 15 mm; legume not more than 25 mm
115 Standard 4–5·5 mm; calyx campanulate **19. Gonocytisus**
115 Standard 7–20 mm; calyx tubular-campanulate
116 Pedicel 5–10 mm; legume glabrous **12. Cytisus**
116 Pedicel 1–3 mm; legume hairy
117 Flowers in umbellate heads **16. Genista**
117 Flowers axillary or in axillary clusters
118 Standard distinctly shorter than keel; seeds estrophiolate **16. Genista**
118 Standard longer than keel; seeds strophiolate **15. Teline**
112 Upper lip of calyx with 2 short teeth
119 Calyx tubular **13. Chamaecytisus**
119 Calyx campanulate
120 Flowers axillary, arranged in leafy racemes **12. Cytisus**
120 Flowers in leafless, terminal heads or racemes
121 Flowers in usually 2- or 4-flowered heads **14. Chronanthus**
121 Flowers in long racemes
122 Twigs hairy **11. Lembotropis**
122 Twigs glabrous **12. Cytisus**

Subfam. Caesalpinioideae

Flowers more or less zygomorphic. Sepals and petals imbricate; the adaxial petal innermost and so overlapped by the lateral petals. Stamens usually not more than 10, free (in European spp.).

1. Cercis L.[1]

Deciduous trees or shrubs. Leaves simple, digitately veined; stipules small, caducous. Flowers fasciculate (in European sp.), hermaphrodite. Calyx campanulate, with 5 equal teeth; corolla strongly zygomorphic, the 3 upper petals much smaller than the lower 2; stamens 10. Legume linear-oblong, compressed, narrowly winged on the ventral suture, more or less dehiscent. Seeds usually numerous.

1. **C. siliquastrum** L., *Sp. Pl.* 374 (1753). Tree up to 10 m, cauliflorous. Leaves 7–12 cm, suborbicular, obtuse or emarginate, cordate, glabrous, long-petiolate. Corolla 15–20 mm, pinkish-purple. Legume 6–10 × 1·5–2 cm, brown, glabrous. *Mediterranean region, extending to E. Bulgaria; cultivated elsewhere for ornament and sometimes naturalized.* Al Bu Cr Ga Gr It Ju Si Tu [Hs Lu Rs (K)].

2. Ceratonia L.[1]

Polygamous or dioecious. Evergreen trees or shrubs. Leaves paripinnate; stipules minute, caducous. Flowers in short axillary racemes. Calyx with 5 short, caducous teeth; petals 0; stamens 5. Legume linear-oblong, compressed, indehiscent, with a sugary pulp between the seeds. Seeds numerous.

1. **C. siliqua** L., *Sp. Pl.* 1026 (1753). Tree or shrub up to 10 m. Leaflets 2–5 pairs, 30–50 × 30–40 mm, elliptical or obovate to suborbicular, coriaceous, dark green and shining above, pale

[1] By P. W. Ball.

green beneath. Flowers green. Legume 10–20 × 1·5–2 cm, brownish-violet, pendent. *Native to the Mediterranean region, but also extensively cultivated there (and in Portugal) for fodder and widely naturalized, so that the limits as a native are hard to determine.* Al Bl Cr Ga Gr Hs It Ju Sa Si [Co Lu].

3. Gleditsia L.[1]

Polygamous. Deciduous trees, usually with stout, simple or branched spines on the trunk and branches. Leaves paripinnate and 2-pinnate; stipules minute. Flowers usually in axillary racemes. Calyx 3- to 5-lobed; corolla scarcely zygomorphic, with 3–5 petals; stamens 6–10. Legume compressed, indehiscent or tardily dehiscent. Seeds numerous (in European sp.).

1. **G. triacanthos** L., *Sp. Pl.* 1056 (1753). Tree up to 45 m. Pinnate leaves with 10–15 pairs of leaflets; 2-pinnate with 8–14 pinnae; leaflets 20–35 × 7–12 mm (pinnate leaves) or 8–20 × 3–8 mm (2-pinnate leaves), oblong-lanceolate, remotely crenate-serrate. Flowers 2·5–3 mm, greenish-white. Legume 30–45 × 2–3 cm, falcate and often twisted. *Planted for ornament and in hedges in C. & S. Europe; occasionally naturalized.* [Au Bu Cz Ga Ge Hs It Lu Rm.] (*C. & E. North America.*)

Subfam. Mimosoideae

Flowers actinomorphic. Sepals and petals valvate. Stamens numerous, free.

4. Acacia Miller[2]

Trees or shrubs. Leaves 2-pinnate and dorsiventral in the juvenile state and remaining so in the adult state or soon reduced to simple phyllodes; stipules rudimentary or becoming large spines. Flowers small, yellow to white, in cylindrical spikes or in globose capitula arranged in racemes or panicles; calyx and corolla 4- to 5-merous; stamens numerous, free, long and conspicuous. Legume usually dehiscent. Seeds with a filiform funicle ending in a cupuliform strophiole.

1 Leaves all 2-pinnate
  2 Leaves deciduous, with 2–8 pairs of pinnae; stipules spinescent
    3 Leaflets 3–5 mm; stipular spines 25 mm on older branches; legume cylindrical-fusiform, inflated   **1. farnesiana**
    3 Leaflets 6–10 mm; stipular spines up to 100 mm on older branches; legume linear-falcate, flattened   **2. karoo**
  2 Leaves evergreen, with 8–20 pairs of pinnae; stipules rudimentary
    4 Twigs and young leaves whitish-tomentose; leaflets 3–4 mm; legume 10–12 mm wide, not or scarcely constricted between the seeds   **3. dealbata**
    4 Twigs and young leaves yellowish-villous; leaflets 2 mm; legume 5–7 mm wide, distinctly constricted between the seeds   **4. mearnsii**
1 Adult leaves reduced to flattened phyllodes, occasionally mixed with some 2-pinnate leaves
  5 Flowers in axillary spikes; legume terete   **5. longifolia**
  5 Flowers in capitula usually arranged in racemes; legume compressed
    6 Phyllodes with 2–6 longitudinal veins; legume twisted or annular
      7 Procumbent shrub; phyllodes 3–8 cm; flowers bright yellow   **6. cyclops**
      7 Tree; phyllodes 6–13 cm; flowers creamy-white   **7. melanoxylon**
    6 Phyllodes with a single longitudinal vein; legume almost straight
      8 Phyllodes strongly falcate, oblong-lanceolate to obovate; 10–20 capitula in each raceme   **8. pycnantha**

      8 Phyllodes not or scarcely falcate, linear to lanceolate or oblanceolate; 2–10 capitula in each raceme
        9 Capitula 10–15 mm in diameter; legume distinctly constricted between seeds; funicle short, whitish   **9. cyanophylla**
        9 Capitula 4–6 mm in diameter; legume not or scarcely constricted between seeds; funicle encircling the seed, scarlet   **10. retinodes**

1. **A. farnesiana** (L.) Willd., *Sp. Pl.* **4**: 1083 (1806). Shrub up to 4 m. Leaves deciduous, 2-pinnate, bright green, glabrous, with 3–8 pairs of pinnae; leaflets 10–25 pairs, 3–5 × 1–1·5 mm, linear-oblong; stipular spines on older branches 25 mm, straight. Capitula 10–12 mm in diameter, 2–3 together in axils of older leaves, rarely solitary; peduncles 1–2 cm, pubescent. Flowers bright yellow, fragrant. Legume 50–90 × 10–15 mm, cylindrical-fusiform, thick, dark brown. *Cultivated in S.W. Europe for ornament and for perfumery.* [Ga Hs It Si.] (*Dominican Republic.*)

2. **A. karoo** Hayne, *Darst. Beschr. Arzn. Gewächse* **10**: t. 33 (1827) (*A. horrida* auct., non Willd.). Like **1** but leaves with 2–7 pairs of pinnae; leaflets 5–14 pairs, 6–10 × 2–4 mm, oblong; stipular spines on older branches 50–100 mm; capitula in fascicles of 4–6 in the axils of the upper leaves; peduncles up to 2·5 cm, glabrous; flowers slightly fragrant; legume 80–130 × 6–8 mm, linear-falcate, flattened, slightly constricted between the seeds, greyish-brown when mature. *Cultivated in S.W. Europe for ornament and for hedges, and locally naturalized.* [Co Hs Lu Si.] (*South Africa.*)

3. **A. dealbata** Link, *Enum. Hort. Berol. Alt.* **2**: 445 (1822). Tree up to 30 m; bark smooth, grey. Twigs and young leaves whitish-tomentose. Leaves 2-pinnate, glaucous-green, with 8–20 pairs of pinnae; leaflets 30–50 pairs, 3–4 mm, linear; leaf-rhachis with glands only at the insertion of the pinnae. Capitula 5–6 mm in diameter, in profuse panicles which are longer than leaves. Flowers pale yellow. Legume 40–100 × 10–12 mm, compressed, not or scarcely constricted between the seeds, brown, pruinose. *Planted for ornament, for timber and for soil-stabilization; widely naturalized in S. Europe.* [Az Ga Hs Lu It Ju Rm Sa.] (*S.E. Australia, Tasmania.*)

4. **A. mearnsii** De Wild., *Pl. Bequaert.* **3**: 61 (1925) (*A. mollisima* auct.). Like **3** but not more than 15 m; softly yellow-villous; leaves dark green with 8–14 pairs of pinnae; leaflets 25–40 pairs, 2 mm; rhachis with glands between the pinnae; legume 5–7 mm wide, distinctly constricted between the seeds, blackish-brown. *Planted for ornament and for tanning in the Iberian peninsula and Italy; locally naturalized.* [Co Hs Lu It.] (*S.E. Australia, Tasmania.*)

5. **A. longifolia** (Andrews) Willd., *Sp. Pl.* **4**: 1052 (1806). Shrub or small tree up to 8 m; bark smooth, grey. Twigs stiff, glabrous. Phyllodes 7–15 × 0·8–3 cm, oblong to oblong-lanceolate, straight, subobtuse, light green, 2- to 4-veined. Spikes 25–50 × 7–9 mm, axillary, cylindrical, subsessile. Flowers bright yellow, strong-smelling. Legume 70–150 × 4–5 mm, linear, terete, almost straight, rostrate, constricted between the seeds, brown when mature. Funicle very short, whitish. *Widely planted in S.W. Europe for ornament and for stabilizing coastal dunes.* [?Ga Hs It Lu.] (*New South Wales.*)

6. **A. cyclops** A. Cunn. ex G. Don fil., *Gen. Syst.* **2**: 404 (1832). Procumbent, caespitose shrub up to 3 m; bark brownish, fissured. Twigs glabrous. Phyllodes 3–8 × 0·4–1·8 cm, oblong-linear to oblanceolate, obtuse and mucronate, 3- to 6-veined. Capitula 4–6 mm in diameter, solitary or in groups of 2–3, shortly pedun-

---

     [1] By P. W. Ball.   [2] By J. do Amaral Franco.

culate. Flowers yellow. Legume 40–80 × 10–13 mm, compressed, undulate or twisted, scarcely constricted between the seeds, reddish-brown. Funicle encircling the seed in a double fold, scarlet. *Naturalized on maritime cliffs in C. Portugal.* [Lu.] (*Western Australia.*)

7. A. melanoxylon R. Br. in Aiton, *Hort. Kew.* ed. 2, **5**: 462 (1813). Tree up to 40 m; stem straight, erect; bark dark brown, deeply furrowed. Young twigs hirsute. Phyllodes 6–13 × 0·7–2 cm (larger on suckers), lanceolate to oblanceolate, slightly falcate, obtuse to acute, dull, dark green, 3- to 5-veined; 2-pinnate and transitional leaves often occurring here and there on young trees, the pinnae with 14–20 oblong leaflets 5–7 mm. Capitula 10 mm in diameter. Flowers creamy white. Legume 70–120 × 8–10 mm, compressed, twisted, reddish-brown. Funicle encircling the seed in a double fold, scarlet. *Planted for timber in S.W. Europe and locally naturalized.* [Az Br Ga Hs It Lu.] (*S.E. Australia, Tasmania.*)

8. A. pycnantha Bentham, *London Jour. Bot.* (*Hooker*) **1**: 351 (1842). Tree up to 12 m; branches curved upwards; bark smooth, grey. Twigs glaucous. Phyllodes 8–20 × 1–3·5 cm (up to 10 cm wide on suckers and then oblique, obovate), falcate, acute, bright green, 1-veined. Capitula 8–10 mm in diameter, in long racemes of 10–20. Flowers deep yellow. Legume 80–130 × 5–6 mm, almost straight, compressed, dark brown or almost black. Funicle short, whitish. *Planted for tanning, and sometimes for ornament; locally naturalized in S. Portugal and C. & S. Italy.* [It Lu Sa.] (*South Australia, Victoria.*)

9. A. cyanophylla Lindley, *Bot. Reg.* **25** (*Misc.*): 45 (1839) (*A. saligna* auct., non (Labill.) Wendl. fil.). Tree up to 10 m; bark smooth, grey, later greyish-brown and fissured. Twigs glaucous, pendent. Phyllodes 10–20(–35) × 0·6–2(–3) cm (up to 8 cm wide, ovate, undulate, on suckers), linear to lanceolate, not or scarcely falcate, subacute, more or less glaucous, 1-veined. Capitula 10–15 mm in diameter, in short racemes of 2–6. Flowers bright yellow. Legume 60–120 × 4–8 mm, compressed, distinctly constricted between the seeds, glaucous when young, later brownish. Funicle short, whitish. *Planted for stabilizing coastal dunes, and also for ornament; naturalized in S. Europe.* [Co Ga Gr Hs It Lu Sa Si.] (*Western Australia.*)

Commonly mistaken for *A. saligna* (Labill.) H. L. Wendl., which has smaller capitula.

10. A. retinodes Schlecht., *Linnaea* **20**: 664 (1847) (*A. floribunda* auct., non Willd.). Like **9** but twigs usually brown, not pendent; phyllodes 6–15 × 0·4–1·8 cm, acute to obtuse, light green; capitula 4–6 mm in diameter, in racemes of 5–10; racemes sometimes paniculate; flowers pale yellow; legume not or very slightly constricted between the seeds; funicle encircling seed and bent back upon itself in a double fold, scarlet. *Widely planted for ornament in S. Europe and locally naturalized.* [Az Br Ga Hs It ?Ju Lu Rm.] (*South Australia.*)

Subfam. **Lotoideae**

Flowers zygomorphic. Sepals and petals imbricate; the adaxial petal (standard) outermost, the 2 lateral petals (wings) free, the 2 lower petals innermost and usually partially adhering to each other by means of interlocking hairs on the margin to form the keel. Stamens 10, rarely 5, free, or more usually all or 9 of the filaments united.

---

[1] By P. W. Ball.

5. Sophora L.[1]

(Incl. *Goebelia* Bunge ex Boiss.)

Deciduous or evergreen trees or shrubs, or stout perennial herbs. Leaves imparipinnate; stipules small, usually linear, scarious. Flowers in terminal racemes or panicles. Calyx tubular or tubular-campanulate, slightly bilabiate; stamens free. Legume stipitate, constricted between the seeds and often moniliform, indehiscent or tardily dehiscent. Seeds few or many, rarely 1.

1 Trees; flowers in panicles   **1. japonica**
1 Herbs; flowers in racemes
  2 Leaflets densely pubescent or villous above; corolla white
                  2. alopecuroides
  2 Leaflets glabrous or sparsely pubescent above; corolla pale yellow   **3. jaubertii**

1. S. japonica L., *Mantissa* 68 (1767). Tree up to 25 m. Leaflets 3–8 pairs, 25–50 × 12–20 mm, ovate or ovate-lanceolate, dark green and shining above, glaucous or pubescent beneath. Flowers in large terminal panicles. Corolla 10–15 mm, creamy white or pale pink. Legume 50–80 mm, glabrous. *Frequently planted for ornament and locally naturalized.* [Cz Ga Rm.] (*E. Asia.*)

2. S. alopecuroides L., *Sp. Pl.* 373 (1753) (*Goebelia alopecuroides* (L.) Bunge ex Boiss.). Erect perennial 40–100 cm. Leaflets 5–13 pairs, 15–40 × 5–15 mm, oblong, elliptical or lanceolate, very densely grey- or white-tomentose or -villous. Flowers in terminal racemes 5–15 cm. Corolla 17–22 mm, white, usually sparsely pubescent at the base of the standard. Legume 5–12 cm, pubescent. *S.E. Russia* (*just north of the Terek river*); *Krym.* Rs (K, E). (*W. Asia.*)

3. S. jaubertii Spach in Jaub. & Spach, *Ill. Pl. Or.* **4**: 45 (1851) (*S. alopecuroides* sensu Hayek, non L., *S. reticulata* var. *buxbaumii* (Aznav.) Hayek, *S. prodanii* E. Anderson, *Goebelia prodanii* (E. Anderson) Grossh.). Like **2** but leaflets glabrous or sparsely pubescent; racemes 12–26 cm; corolla pale yellow. *Turkey-in-Europe; E. Romania; Krym.* Rm Rs (K) Tu. (*Anatolia, W. Caucasus.*)

6. Thermopsis R. Br.[1]

Stout perennial herbs. Leaves digitately 3-foliolate; stipules herbaceous, free. Flowers in leafy terminal racemes. Calyx campanulate with 5 subequal teeth (in European sp.); stamens free. Legume oblong, compressed, dehiscent. Seeds usually numerous.

1. T. lanceolata R. Br. in Aiton, *Hort. Kew.* ed. 2, **3**: 3 (1811). Stems 10–30(–40) cm, pubescent. Leaflets 30–60(–70) × 5–12 mm, oblong-elliptical, glabrous or sparsely pubescent above, densely pubescent beneath; stipules 15–20 × 3–5 mm, lanceolate or ovate-lanceolate. Flowers in fascicles of 3 in the axil of each bract. Corolla 25–28 mm, yellow. Legume 50–60 × 8–10 mm, oblong. Seeds 12–18. *S.E. Russia, W. Kazakhstan.* Rs (E). (*Temperate Asia.*)

7. Anagyris L.[1]

Shrubs. Leaves 3-foliolate; stipules connate, leaf-opposed. Flowers in short axillary racemes. Calyx campanulate, subbilabiate; stamens free. Legume stipitate, constricted and septate between the seeds, compressed, dehiscent. Seeds few.

1. A. foetida L., *Sp. Pl.* 374 (1753). Foetid shrub up to 4 m. Leaflets 30–70 × 10–30 mm, elliptical or lanceolate-elliptical, subobtuse, mucronulate; stipules *c.* 5 mm. Corolla 18–25 mm, yellow, the standard much shorter than the wings and keel, often with a black spot. Legume 100–200 × 15–20 mm. Seeds violet or yellow. *Mediterranean region and S. Portugal.* Al Bl Co Cr Ga Gr Hs It Ju Lu Sa Si Tu.

8. Laburnum Fabr.[1]

Trees or shrubs. Leaves 3-foliolate. Flowers in simple, axillary or pseudoterminal, leafless racemes, pendent at anthesis. Calyx campanulate, slightly bilabiate, the lips undivided or shortly toothed; corolla yellow; stamens monadelphous. Legume slightly constricted between the seeds, dehiscent. Seeds numerous, estrophiolate, compressed.

Twigs appressed-pubescent, greyish-green; legume appressed-
  pubescent, later ± glabrous   **1. anagyroides**
Twigs glabrescent, green; legume glabrous   **2. alpinum**

1. L. anagyroides Medicus, *Vorl. Churpf. Phys.-Ökon. Ges.* **2**: 363 (1787) (*L. vulgare* J. Presl, *Cytisus laburnum* L.). Shrub or small tree up to 7 m. Twigs greyish-green, appressed-pubescent. Leaflets 3–8 cm, elliptical to elliptic-obovate, usually obtuse and shortly mucronate, greyish-green, appressed-pubescent beneath when young. Racemes 10–30 cm, lax. Corolla *c.* 2 cm, golden-yellow. Legume 4–6 cm, appressed-pubescent when young, subglabrous at maturity; upper suture unwinged. Seeds black. *Woods and scrub. Mountains of S.C. Europe and Italy, extending to E. France and W. Jugoslavia. Frequently cultivated for ornament and sometimes naturalized.* Au Cz Ga Ge He Hu It Ju Rm [Br Co Hb].

Often divided on the basis of small differences in the calyx, corolla and inflorescence into geographical variants treated as subspecies, but many transitions occur between them.

2. L. alpinum (Miller) Berchtold & J. Presl, *Rostlinář* **3**: 99 (1835) (*Cytisus alpinus* Miller). Shrub or tree up to 5 m (up to 10 m in cultivation). Twigs glabrous (sometimes hairy when young), green. Leaflets 3–8 cm, light green beneath. Racemes 15–40 cm, rather dense. Corolla 1·5(–2) cm, yellow. Legume 4–5 cm, glabrous; upper suture with wing 1–2 mm wide. Seeds brown. *Mountains of S.C. Europe, Italy and W. part of Balkan peninsula.* Al Au Cz Ga He It Ju.

9. Podocytisus Boiss. & Heldr.[1]

Unarmed shrubs. Leaves 3-foliolate. Flowers in terminal, erect racemes. Calyx shortly campanulate, bilabiate, the teeth equal; corolla yellow; stamens monadelphous. Style involute; stigma capitate. Legume compressed, not or tardily dehiscent, broadly winged. Seeds 3–6, estrophiolate.

1. P. caramanicus Boiss. & Heldr. in Boiss., *Diagn. Pl. Or. Nov.* **2**(9): 7 (1849). Erect, up to 2 m. Branches virgate, glaucous, terete when young. Leaflets 5–15 mm, obovate, mucronulate; petioles 3–10 mm. Racemes 5–15 cm, often pyramidal. Pedicels 5–8 mm, with bracteoles at the middle. Standard *c.* 15 mm, orbicular; keel as long as standard. Legume 5–7 cm, ovate to oblong, stipitate; the valves thin, papery; upper suture broadly winged. *Balkan peninsula, southwards from 42° N.* Al Gr Ju.

10. Calicotome Link[2]

Spiny shrubs; branches alternate. Leaves 3-foliolate, petiolate. Flowers axillary, solitary or in umbellate fascicles or ebracteate racemes. Calyx tubular, with 5 short teeth, but with the apical portion breaking away as the flower expands, leaving a cup-like

remnant. Corolla yellow. Legume narrowly oblong, the sutures usually somewhat thickened. Seeds several, estrophiolate.

Literature: W. Rothmaler, *Bot. Jahrb.* **74**: 276–287 (1949).

Flowers mostly solitary; legume ± glabrous   **1. spinosa**
Flowers mostly in fascicles of 2–15, or in ebracteate racemes; legumes villous or sericeous, sometimes sparsely so   **2. villosa**

1. C. spinosa (L.) Link, *Enum. Hort. Berol. Alt.* **2**: 225 (1822). Erect shrub up to 3 m; branches with stout lateral spines; young twigs sparsely sericeous. Leaves 3-foliolate; leaflets 5–15 mm, obovate, with sericeous hairs beneath, glabrous above. Flowers mostly solitary, but occasionally in fascicles; pedicels 4–8 mm; bracteoles entire or somewhat 3-fid, sparsely pubescent, borne just below the calyx. Calyx sparsely pubescent. Corolla 12–18 mm, glabrous. Legume *c.* 30 mm, narrowly oblong, glabrous (occasionally with sparse hairs), the sutures not or scarcely thickened. *Evergreen scrub and dry rocky ground. W. Mediterranean region.* Bl Co Ga Hs It Sa ?Si.

Plants from N.W. Italy with distinctly 3-fid bracteoles have been distinguished as subsp. **ligustica** Burnat, *Fl. Alp. Marit.* **2**: 57 (1896). There are many intermediates, however, and it is doubtful if it merits even varietal status.

2. C. villosa (Poiret) Link in Schrader, *Neues Jour. Bot.* **2**(2): 51 (1808) (incl. *C. infesta* (C. Presl) Guss.). Like **1** but with slender spines; young twigs, lower surface of the leaves, bracteoles and calyx densely sericeous or villous; flowers mostly in umbellate fascicles of 2–15 or ebracteate racemes; legume usually densely sericeous or villous and with the sutures distinctly thickened. 2*n*=48. *Evergreen scrub and dry rocky ground. Mediterranean region and S. Portugal.* Al Co Cr Gr Hs It Ju Lu Sa Si Tu.

11. Lembotropis Griseb.[3]

Unarmed shrubs. Leaves 3-foliolate. Flowers in terminal, leafless racemes. Calyx campanulate, bilabiate; upper lip with 2 teeth, lower with 3 teeth; corolla yellow, becoming black when dry; stamens monadelphous; stigma capitate. Legume dehiscent. Seeds with rudimentary strophiole.

1. L. nigricans (L.) Griseb., *Spicil. Fl. Rumel.* **1**: 10 (1843) (*Cytisus nigricans* L.). Erect, up to 1(–1·5) m; twigs up to 45 cm or more, 1–2 mm in diameter, terete, hairy, flowering in first year. Leaflets (6–)10–30 × (2–)5–10(–16) mm, obovate to elliptical or linear, dark green above, lighter green beneath, appressed-hairy on both surfaces when young, the upper glabrescent. Pedicels 4–8 mm, appressed-hairy, with 1 linear, long-persistent bract. Standard 7–10 × 6 mm; wings shorter than the rostrate keel. Legume 20–35 × 5–6(–7) mm, linear-oblong, appressed-pubescent. *C. & S.E. Europe, extending north-eastwards to C. Russia.* Al Au Bu Cz Ge Gr He Hu It Ju Po Rm Rs (C, W) Sa.

(a) Subsp. **nigricans**: Twigs appressed-pubescent or sericeous; inflorescence elongate, many-flowered. *Throughout the range of the species.*

(b) Subsp. **australis** (Freyn ex Wohlf.) J. Holub, *Preslia* **36**: 253 (1964) (*Cytisus nigricans* var. *australis* (Freyn ex Wohlf.) Hayek): Twigs more or less densely covered with white, patent hairs; inflorescence short, few-flowered. *Italy, W. part of Balkan peninsula; ?C. Romania.*

12. Cytisus L.[1]

Unarmed shrubs. Leaves 1- or 3-foliolate, alternate, sometimes crowded on older branches. Flowers axillary, forming leafy,

[1] By D. G. Frodin and V. H. Heywood.
[2] By P. E. Gibbs.
[3] By V. H. Heywood.

terminal racemes. Calyx bilabiate, upper lip with 2 teeth, lower with 3 teeth, the teeth usually much shorter than the lips; corolla yellow to white; stamens monadelphous; stigma capitate or introrse. Legume linear or oblong, dehiscent. Seeds usually numerous, usually strophiolate.

Literature: J. Briquet, *Études sur les Cytises des Alpes Maritimes.* Genève & Bâle. 1894. W. Rothmaler, *Feddes Repert.* **53**: 137–150 (1944). A. Skalická, *Preslia* **39**: 10–29 (1967). C. Vicioso, *Genísteas Españolas* **2**: 179–223. (*Bol. Inst. For. Inv. Exper.* No. 72). Madrid. 1955.

1 Flowers in leafless racemes   **1. sessilifolius**
1 Flowers in leafy racemes
  2 Twigs 5- to 10-angled, the angles wing-like, ±T-shaped in transverse section
    3 Leaves all 1-foliolate
      4 Twigs and leaves densely lanate; standard 8–10 mm   **11. agnipilus**
      4 Twigs and leaves villous; standard 12–15 mm   **9. procumbens**
    3 Leaves at least on the lower branches 3-foliolate
      5 Corolla white; pedicels 10 mm   **6. multiflorus**
      5 Corolla yellow; pedicels 4–7 mm
        6 Twigs 2–3 mm in diameter; upper leaves 1-foliolate, caducous   **5. purgans**
        6 Twigs 1 mm in diameter; all leaves 3-foliolate, ± persistent during growing season   **7. ardoini**
  2 Twigs terete or angled but the angles not T-shaped in transverse section
    7 Style convolute after anthesis
      8 Leaves 1-foliolate, except sometimes on new growth
        9 Branches incurved to form a rounded bush; calyx glabrous; legume glabrous   **21. reverchonii**
        9 Branches erect, not incurved; calyx sericeous; legume hirsute   **13. commutatus**
      8 Leaves mostly 3-foliolate except on new growth
        10 All leaves petiolate; keel straight on the upper side, not beaked; style glabrous
          11 Branches and twigs ± terete, slightly striate; legume sparsely pubescent or glabrescent   **15. malacitanus**
          11 Branches and twigs 7- to 8-angled; legume densely villous or lanate   **16. baeticus**
        10 All leaves or at least the 1-foliolate leaves sessile; keel strongly curved on the upper side, ± beaked; style ciliate below
          12 Legume ± inflated, the valves densely hirsute; twigs usually 8- to 10-angled   **17. striatus**
          12 Legume strongly compressed, the valves glabrous or with appressed hairs; twigs usually 5-angled
            13 All leaves sessile, those on young twigs 1-foliolate   **18. grandiflorus**
            13 1-foliolate leaves sessile; 3-foliolate leaves petiolate
              14 Twigs 5- to 8-angled, the angles somewhat rounded; legume densely hairy   **19. cantabricus**
              14 Twigs almost always 5-angled, the angles flattened and ridge-like, sometimes slightly winged laterally; legume glabrous   **20. scoparius**
    7 Style straight or arcuate after anthesis
      15 Leaves mainly 1-foliolate
        16 Erect, up to 200 cm; standard 20–23 × 20 mm
          17 Leaflets 10–15 mm; corolla greenish-yellow   **13. commutatus**
          17 Leaflets 20–30 mm; standard cream; wings and keel yellow   **14. ingramii**
        16 Procumbent or ascending, 10–40 cm; standard 10–14 mm
          18 Twigs patent-villous; calyx hirsute or sparsely pubescent   **10. decumbens**
          18 Twigs appressed-pubescent or glabrous; calyx glabrous   **12. pseudoprocumbens**
      15 Leaves mainly 3-foliolate
        19 Mature legume at least 9 mm wide; standard 18–22 mm   **2. aeolicus**
        19 Mature legume less than 9 mm wide; standard 8–18 mm
          20 Leaves subsessile or very shortly petiolate; bracteoles ovate   **23. tribracteolatus**
          20 Leaves distinctly petiolate; bracteoles narrowly elliptical or linear, often caducous
            21 Plant 100–250 cm; standard 15–20 mm
              22 Leaflets obovate, sparsely pubescent beneath; legume 12–15 mm, glabrous   **22. patens**
              22 Leaflets oblong to elliptical, villous beneath; legume hirsute or villous, glabrescent   **3. villosus**
            21 Plant 20–60(–100) cm; standard 8–12 mm
              23 Petiole 10–25 mm; legume glabrous   **4. emeriflorus**
              23 Petiole less than 10 mm; legume villous, at least on the margins   **8. sauzeanus**

Subgen. **Cytisus**. Leaves 3-foliolate. Flowers in leafless racemes.

1. C. sessilifolius L., *Sp. Pl.* 739 (1753). Erect, up to 200 cm, glabrous; branches terete. Leaves petiolate on flowering branches, more or less sessile on non-flowering branches; leaflets 8–20 mm, obovate to broadly elliptical. Flowers 3–12, pedicellate, with bracts and bracteoles persistent at anthesis. Corolla yellow; standard 11 mm, suborbicular; keel incurved at apex, beaked. Legume 25–40 × 10 mm, oblong to linear. *Scrub and mountain woods.* ● *E. Spain, S. France, Italy.* Ga Hs It.

Subgen. **Sarothamnus**. Leaves 1- or 3-foliolate. Flowers in leafy racemes.

Sect. TRIANTHOCYTISUS Griseb. Twigs terete or angular. Calyx-lips with straight margins. Leaves 3-foliolate, petiolate. Style straight or arcuate. Strophiole small or absent.

2. C. aeolicus Guss. ex Lindley, *Bot. Reg.* **22**: t. 1902 (1836). Erect, 100–200 cm; branches rigid, terete, appressed-tomentose or glabrous. Petioles 10–20 mm; leaflets 20–50 × 7–20 mm, elliptical or lanceolate, subcoriaceous, glabrous above, sericeous beneath. Flowers in groups of 3; pedicels 6–12 mm, subtomentose. Corolla yellow; standard 18–22 mm, broadly obovate. Legume 40–50 × 9–12 mm, oblong to linear, more or less curved at maturity, glabrous. ● *Isole Eólie.* Si.

3. C. villosus Pourret, *Mém. Acad. Toulouse* **3**: 317 (1788) (*C. triflorus* L'Hér.). Erect, 100–200 cm; branches ascending, rigid, 5-angled, pubescent towards the apex. Petioles 2–8(–10) mm; leaflets (10–)15–30 × 5–15 mm, oblong to elliptical, the central longer than the lateral, glabrescent above, villous beneath. Flowers solitary or in groups of 2–3; pedicels 5–10 mm, subtomentose. Corolla yellow; standard 15–18 mm, dark-red-striate at the base. Legume 20–45 × 4–7 mm, hirsute or villous, glabrescent. *Woods and scrub; somewhat calcifuge. S. Europe.* Al Co Ga Gr Hs It Ju Sa Si.

4. C. emeriflorus Reichenb., *Fl. Germ. Excurs.* 524 (1832) (*Genista glabrescens* Briq.). Procumbent to erect, 30–60(–100) cm; branches rigid, angular; young twigs appressed-villous. Petioles 10–25 mm; leaflets 10–20 mm, obovate to lanceolate, glabrescent above, sericeous-pubescent beneath. Racemes terminal, densely leafy; flowers in clusters of 1–4; pedicels 12 mm, sparsely hairy. Corolla yellow; standard 10–12 mm. Legume 25–35 × 6–7 mm, glabrous. ● *S. Alps (around Como and Lugano).* He It.

Sect. ALBURNOIDES DC. Twigs usually with wing-like angles which are T-shaped in transverse section. Leaves 1- or 3-foliolate. Calyx-lips with convex margins. Style curved. Stigma capitate. Strophiole large.

5. C. purgans (L.) Boiss., *Voy. Bot. Midi Esp.* **2**: 134 (1839) (*Sarothamnus purgans* (L.) Godron, *Genista purgans* (L.) DC.).

Erect or ascending, (20–)30–100 cm, much-branched, fastigiate; branches rigid, striate, terete; twigs virgate, almost leafless, erect, pubescent or sericeous when young, later glabrous. Leaves 1-foliolate on flowering twigs, 3-foliolate on lower branches, sessile, caducous; leaflets 6–12 mm, oblanceolate, subspathulate to linear-lanceolate, sericeous-pubescent beneath, subglabrous above. Flowers at ends of twigs, solitary or paired, smelling of vanilla; pedicels 5 mm, with 2 small bracteoles at apex, pubescent. Calyx pubescent. Corolla deep yellow; standard 10–12 mm. Legume (12–)15–30 × 5–7(–8) mm, oblong, straight or slightly incurved, appressed-villous, black when mature. *Dry stony slopes; calcifuge. S.W. Europe.* Ga Hs Lu.

6. **C. multiflorus** (L'Hér.) Sweet, *Hort. Brit.* 112 (1826) (*Cytisus albus* (Lam.) Link, non Hacq., *C. lusitanicus* Willk., *Genista alba* Lam.). Erect, much-branched, 100–300 cm; branches flexible, 5-angled; twigs striate, sericeous when young, glabrescent at maturity. Leaves 3-foliolate on lower branches, 1-foliolate on upper branches; sessile or very shortly petiolate. Leaflets up to 10 mm, linear-lanceolate or oblong, silvery-sericeous. Flowers in profuse fascicles of 1–3; pedicels 10 mm. Calyx 5 mm, sericeo-puberulent. Corolla white; standard 9–12 mm. Legume 15–25 × 5–9 mm, oblong, strongly compressed, appressed-pubescent or hirsute. *Woods, heaths and river-banks; calcifuge.* ● *N.W. & C. Spain, N. & C. Portugal.* Hs Lu.

7. **C. ardoini** E. Fourn., *Bull. Soc. Bot. Fr.* 13: 389 (1866). Low-growing, 20–60 cm; young twigs villous, strongly ridged, with 8–10 wings. Leaves 3-foliolate; petioles *c.* 6 mm; leaflets 4–10 × 1–3 mm, narrowly oblong to obovate, appressed-pubescent above, sparsely so beneath. Flowers 1–3 on short axillary twigs; pedicels 4–7 mm. Calyx sparsely villous. Corolla yellow; standard 8–12 mm, suborbicular. Legume 20–25 mm, villous or hirsute. *Calcareous rocks.* ● *S.W. Alps (Alpes Maritimes).* Ga.

8. **C. sauzeanus** Burnat & Briq. in Briq., *Cytises Alp. Marit.* 157 (1894). Like 7 but young twigs 5-angled, without wings; leaflets wider; legume with long hairs on margins, sometimes glabrous on the sides. ● *Dry slopes and screes. S.E. France (mountains between Grenoble and Sisteron).* Ga.

Possibly the result of hybridization between 7 and 3.

Sect. COROTHAMNUS (Koch) Nyman. Twigs simply angular, or with wing-like angles T-shaped in transverse section. Leaves 1-foliolate. Style curved. Stigma introrse. Strophiole large.

9. **C. procumbens** (Waldst. & Kit. ex Willd.) Sprengel, *Syst. Veg.* 3: 224 (1826) (*Genista procumbens* Waldst. & Kit. ex Willd., *G. pedunculata* subsp. *procumbens* (Waldst. & Kit. ex Willd.) Gams). Decumbent or ascending, 20–40(–80) cm; branches terete, leafy, appressed-villous; young twigs strongly ridged, with 8–10 wings which are T-shaped in transverse section. Leaves in fascicles of 3–6 on lower twigs; leaflets 15–20 × 4 mm, oblong-lanceolate, acute, glabrous above, appressed-villous beneath. Flowers solitary or in groups of 2–3; pedicels 1½–2 times as long as calyx. Calyx 5 mm, subglabrous or sparsely appressed-villous. Corolla golden-yellow; standard 12–15 mm, ovate. Legume 30–32 × 5 mm, compressed, appressed-villous. *Dry grassland and open woods.* ● *E.C. Europe and Balkan peninsula.* ?Al Au Bu Cz Gr Hu Ju Rm Rs (W).

10. **C. decumbens** (Durande) Spach, *Ann. Sci. Nat.* ser. 3 (Bot.), 3: 156 (1845) (*Genista prostrata* Lam., *G. pedunculata* subsp. *decumbens* (Durande) Gams, *Spartium decumbens* Durande). Procumbent or ascending, 10–30 cm; branches leafy; twigs with 5 unwinged angles, patent-villous. Leaves shortly petiolate in fascicles on lower branches; leaflets 8–20 mm, obovate, oblong or lanceolate, pubescent beneath, slightly so or glabrous above, the margins hairy. Flowers in fascicles of 1–3; pedicels 2–3(–4) times as long as calyx, slender, hirsute. Calyx 5 mm, hirsute (rarely sparsely pubescent). Corolla yellow; standard 10–14 mm, obovate. Legume 20–32 × 6 mm, patent-pubescent (rarely glabrous), becoming black at maturity. ● *From C. France southwards to S. Italy and Albania.* Al Ga He It Ju.

Variable in the indumentum of the leaves, calyx and legume.

11. **C. agnipilus** Velen., *Fl. Bulg.* 643 (1891). Like 10 but erect; branches densely crispate-lanate, with 5 wing-like angles; leaflets oblong-lanceolate, with yellowish crispate hairs on both surfaces; pedicels not more than twice as long as calyx; calyx with crispate hairs; standard 8–10 mm; legume villous. ● *Albania; S. & E. Bulgaria.* Al Bu ?Gr.

12. **C. pseudoprocumbens** Markgraf, *Ber. Deutsch. Bot. Ges.* 44: 213 (1926) (*C. diffusus* Vis. pro parte, *Genista diffusa* auct., non Willd., *G. pedunculata* subsp. *diffusa* Gams pro parte). Procumbent or ascending, 20–40 cm; branches leafy, 5-angled, glabrous at maturity, the angles unwinged. Leaves 1-foliolate; leaflets 10–50 mm, oblong-lanceolate, sparsely ciliate when young. Flowers solitary or in groups of 2–3; pedicels 1½–3(–4) times as long as calyx. Calyx glabrous. Corolla golden-yellow, glabrous; standard 10–12 mm, about as long as keel. Legume 15 mm, usually glabrous, black. *Dry, rocky places.* ● *N.W. part of Balkan peninsula, N. Italy.* Al It Ju.

Sect. SAROTHAMNUS (Wimmer) Bentham. Leaves 1- or 3-foliolate. Style usually convolute after anthesis. Stigma capitate. Strophiole large.

13. **C. commutatus** (Willk.) Briq., *Cytises Alp. Marit.* 151 (1894) (*Sarothamnus commutatus* Willk.). Erect, up to 100 cm; branches strongly 5-angled, shallowly grooved between the angles; young twigs sericeous-villous. Leaves 1-foliolate, sessile; leaflets on young twigs solitary, lanceolate, acute to obtuse; those on older branches in clusters, oblanceolate, spathulate or obovate, mucronulate. Flowers solitary, usually in the centre of leaf-clusters; pedicels about as long as calyx, villous, with 2 bracteoles. Calyx sericeous. Corolla greenish-yellow; standard *c.* 20 × 20 mm. Style more or less convolute. Legume 30–50 × 10 mm, arcuate, patent, hirsute, becoming black at maturity. *Mountain scrub.* ● *N. Spain (Cordillera Cantábrica).* Hs.

14. **C. ingramii** Blakelock, *Bot. Mag.* 169: t. 211 (1953). Erect, up to 200 cm; branches angular when young, sparsely hairy. Leaves 1-foliolate, sessile (some leaves on lower branches and new growth 3-foliolate); leaflets 20–30 × 12–20 mm, glabrous above, silvery-appressed-hairy beneath. Flowers solitary; pedicels longer than the calyx, appressed-hairy, with 3 bracteoles. Calyx densely appressed-sericeous. Standard 22–23 × 20 mm, cream; wings and keel yellow. Style arcuate. Legume 30–35 × 8–9 mm, straight, covered with long hairs. ● *N.W. Spain.* Hs.

15. **C. malacitanus** Boiss., *Elenchus* 32 (1838) (*Sarothamnus malacitanus* (Boiss.) Boiss.; incl. *Sarothamnus rotundatus* Pau). Erect, much-branched, 100–500 cm. Branches and twigs terete, slender, slightly striate, attenuate and recurved at the apex; young twigs pubescent. Leaves all 3-foliolate or those on the upper branches and young branchlets 1-foliolate; all leaves petiolate, the petiole longer than the leaflets; leaflets oblong-linear to ovate-elliptical, glabrous, or pubescent and later glab-

rous above, sericeous beneath. Flowers solitary or in clusters of 2–3; pedicels longer than calyx, sericeous or pubescent, with 2 or 3 bracteoles. Calyx sericeous to glabrous. Corolla yellow; standard 14–16 mm. Legume 25–40 mm, ovate-oblong to linear, compressed, straight or arcuate, laxly sericeous or with sparse, long, white hairs when young, often glabrescent. *Spain, S. France, S. Portugal.* Ga Hs Lu.

(a) Subsp. **malacitanus**: Leaves on upper branches and young branchlets 1-foliolate; leaflets oblong-linear; pedicels with 2 bracteoles. *Dry hillsides. S. Spain, S. Portugal.*

(b) Subsp. **catalaunicus** (Webb) Heywood, *Feddes Repert.* **79**: 22 (1968) (*Sarothamnus catalaunicus* Webb): All leaves 3-foliolate; leaflets ovate-elliptical to oblong-obovate; pedicels with 3 bracteoles. *N.E. Spain, S. France.*

16. C. baeticus (Webb) Steudel, *Nomencl. Bot.* ed. 2, **1**: 477 (1840) (*Sarothamnus baeticus* Webb). Erect shrub, sometimes arborescent, 200–700 cm. Branches 7- to 8-angled, nodular, striate; young twigs more or less puberulent, becoming glabrous. Leaves 3-foliolate, petiolate (the upper often subsessile); leaflets obovate, very obtuse, glabrous above, sparsely sericeous beneath. Flowers solitary or in fascicles of 2–7; pedicels 3–4 times as long as calyx, pubescent, with 3 bracteoles. Calyx pubescent or glabrous. Corolla yellow; standard 15–20 mm. Legume 20–40 × 5–8 mm, straight, not or only slightly constricted between seeds, densely lanate-villous, with patent hairs. *S.W. Spain, S. Portugal.* Hs Lu.

Sometimes regarded along with **15**(a) and (b) as subspecies of *C. arboreus* (Desf.) DC. from N.W. Africa.

17. C. striatus (Hill) Rothm., *Feddes Repert.* **53**: 149 (1944) (*C. pendulinus* L. fil., *Genista striata* Hill, *Sarothamnus patens* Webb, quoad descr.; incl. *Sarothamnus eriocarpus* Boiss. & Reuter, *S. welwitschii* Boiss. & Reuter). Erect, much-branched, 100–300 cm. Branches and twigs cylindrical, striate, usually 8- to 10-angled when young, often drying black; young branches sericeous or villous, later glabrescent and becoming leafless. Leaves solitary or sometimes fasciculate, those on lower branches 3-foliolate, petiolate, those on the middle and upper branches 3- or 1-foliolate, sessile; leaflets 4–16 × 1–6 mm, ovate, elliptical to linear-lanceolate, glabrous and glaucous above, sericeous or villous beneath. Flowers solitary or in pairs, rarely in clusters of 3; pedicels as long as or up to twice as long as calyx. Calyx sericeous. Corolla yellow; standard 10–25 mm. Legume 18–35(–40) × 8–12 mm, straight or slightly curved, oblong-ovate to -elliptical, more or less inflated, densely hirsute, erect or semi-patent. *Woods, hedges and scrub.* ● *Portugal, W. & C. Spain.* Hs Lu [Ga].

Extremely variable in dimensions of leaves and flowers, habit and indumentum.

18. C. grandiflorus DC., *Prodr.* **2**: 154 (1825) (*Sarothamnus grandiflorus* (DC.) Webb, *S. virgatus* Webb). Erect or ascending, 200–300 cm. Older branches terete, striate; twigs virgate, 5-angled, glabrous. Leaves 1-foliolate on young twigs, 3-foliolate on older twigs, all sessile, glabrous, caducous; leaflets obovate to elliptic-lanceolate, obtuse or acute. Inflorescence leafy; flowers solitary or in pairs; pedicels slightly longer than calyx, with 3 minute bracteoles. Calyx glabrous. Corolla golden-yellow; standard 18–25 mm. Legume 20–45 × 8–12 mm, strongly compressed, straight or slightly curved, densely clothed with long white hairs, becoming black at maturity. *Scrub. S. Spain, C. & S. Portugal.* Hs Lu.

19. C. cantabricus (Willk.) Reichenb. fil. in Reichenb. & Reichenb. fil., *Icon. Fl. Germ.* **22**: 15 (?1869) (*Sarothamnus cantabricus*

Willk.). Erect, 100–200 cm. Branches and twigs 5- to 8-angled and furrowed, glabrous. Leaves 1-foliolate, and sessile on young twigs; 3-foliolate and petiolate on older twigs. Leaflets obovate to lanceolate, the lateral shorter than the central, sparsely sericeous-villous or glabrous above, densely so beneath. Flowers solitary; pedicel twice as long as calyx. Calyx glabrous. Corolla yellow; standard 15–18 mm. Legume 30–50 × 6–8 mm, arcuate, compressed, covered with long white hairs, becoming black at maturity. *Scrub and woods.* ● *N. Spain, just extending to S.W. France.* Ga Hs.

20. C. scoparius (L.) Link, *Enum. Hort. Berol. Alt.* **2**: 241 (1822) (*Sarothamnus scoparius* (L.) Wimmer ex Koch). Much-branched shrub, up to 200(–250) cm; branches erect, ascending or procumbent, green, usually leafy, 5-angled, glabrous, or sericeous when young. Leaves usually 3-foliolate, petiolate to subsessile, but 1-foliolate and sessile on young twigs; leaflets 6–20 × 1·5–9 mm, elliptic-oblong to obovate, with appressed hairs or glabrous. Flowers axillary, solitary or in pairs; pedicels twice as long as calyx. Calyx glabrous. Corolla golden-yellow; standard 16–18(–20) mm. Legume 25–40(–70) × 8–10(–13) mm, oblong, strongly compressed, with brown or white hairs on the margin, otherwise glabrous, black at maturity. *W., S. & C. Europe, extending northwards to S. Sweden and eastwards to W.C. Ukraine.* Au Be Br Co Cz Da Ga Ge Hb He Ho Hs Hu It Ju Lu No Po Rm Rs (B, C, W) Sa Si Su [Az].

(a) Subsp. **scoparius**: Branches erect or ascending, 150–200(–250) cm; leaves and young twigs glabrous or sparsely sericeous. 2n=46. *Woods, heaths, dunes and mountain slopes; usually calcifuge. Throughout the range of the species.*

(b) Subsp. **maritimus** (Rouy) Heywood, *Proc. Bot. Soc. Brit. Is.* **3**: 176 (1959) (*Genista scoparia* var. *maritima* Rouy): Branches procumbent with occasional erect or ascending branches from the centre of the plant up to 40 cm; leaves and young twigs densely sericeous. 2n=46. *Maritime cliffs. Coasts of N.W. Europe.*

Plants of this subspecies from Britain, Ireland and the Channel Islands retain their characteristics in cultivation. Records from W. France, Germany and Corse require confirmation.

A distinctive variant occurs in Denmark (Jylland) with ascending branches 50–60 cm, and sericeous, 1-foliolate leaves on young twigs. It remains constant in cultivation and may deserve recognition as a further subspecies. Intermediates between it and subsp. *scoparius* occur in Jylland due to hybridization following the introduction of the latter subspecies.

21. C. reverchonii (Degen & Hervier) Bean, *Kew Bull.* **1934**: 224 (1934) (*Sarothamnus reverchonii* Degen & Hervier). Like **20** but branches incurved to form a rounded bush; leaves 1-foliolate except initially on new growth; flowers more shortly pedicellate. 2n=48. *Mountain woods.* ● *S.E. Spain.* Hs.

22. C. patens L., *Syst. Veg.* ed. 13, 555 (1774) (*Genista patens* DC.). Stems 100–250 cm, erect. Leaves 3-foliolate, petiolate; leaflets 10–15 mm, obovate, shortly mucronate, sparsely puberulent beneath; petioles 4–15 mm. Flowers in axillary clusters; pedicel 5–10 mm, with 3 linear bracteoles. Calyx *c.* 10 mm, sericeous. Standard 15–20 mm, broadly ovate, sericeous; keel and wing glabrous. Legume 12–15 mm, glabrous. Seeds 2–3 mm. *Scrub; calcicole.* ● *E. Spain.* Hs.

23. C. tribracteolatus Webb, *Iter Hisp.* 51 (1838) (*Genista tribracteolata* (Webb) Pau). Stems decumbent or ascending, branches rather flexible, 5-angled. Leaves 3-foliolate, subsessile or very shortly petiolate; leaflets 8–10 × 3–9 mm, ovate to obovate, sericeous, especially beneath. Flowers in axillary clusters. Pedicels with 3 ovate bracteoles. Calyx 5–6 mm, sericeous. Corolla yellow; standard 8–12 mm, broadly ovate, glabrous. Legume 15–30 × 5–8 mm, oblong, sericeous when immature, finally glabrous. *Dry hillsides. S.W. Spain* (*prov. Cádiz*).　● Hs.

13. Chamaecytisus Link[1]

Unarmed shrubs or the branches sometimes spinose. Leaves 3-foliolate. Flowers in leafy racemes or in terminal heads subtended by a leafy involucre. Calyx tubular, bilabiate; upper lip with 2 teeth, lower with 3 teeth, the teeth much smaller than the lips; corolla yellow, rarely white or purple; stigma capitate or extrorse. Legume dehiscent, somewhat compressed, black. Seeds strophiolate.

A taxonomically difficult genus in which interspecific hybridization appears to be of frequent occurrence.

Literature: as for *Cytisus* and the following: A. Holubová-Klásková, *Acta Univ. Carol.* (*Biol.*) **1964** (*Suppl.* **2**): 1–24 (1964).

1　Flowers in leafy racemes
2　Older branches becoming spinose
3　Legume pubescent only along the sutures; standard 25–35 mm; leaflets appressed-pubescent above　**3. spinescens**
3　Legume pubescent all round; standard 12–25 mm; leaflets glabrous or subglabrous above
4　Young branches glabrous; legume white-villous　**1. creticus**
4　Young branches appressed-pubescent; legume appressed-grey-tomentose　**2. subidaeus**
2　Older branches not becoming spinose
5　Corolla purple, white or pinkish-white
6　Corolla purple　**4. purpureus**
6　Corolla white or pinkish-white
7　Leaflets glabrous above; legume glabrous　**17. graniticus**
7　Leaflets appressed-pubescent on both surfaces; legume appressed-pubescent　**18. skrobiszewskii**
5　Corolla yellow
8　Legume glabrous, ciliate or sparsely hairy **(5–10). hirsutus** group
8　Legume uniformly pubescent
9　Standard pubescent outside
10　Standard with spots near the base
11　Leaves appressed-hairy above; calyx appressed-hairy　**11. borysthenicus**
11　Leaves glabrous above; calyx patent-hairy　**(5–10). hirsutus** group
10　Standard without spots
12　Branches patent-pubescent or -hirsute; calyx patent-pubescent or -hirsute　**(5–10). hirsutus** group
12　Branches glabrous below, densely ±appressed-pubescent above; calyx densely villous-tomentose　**12. lindemannii**
9　Standard glabrous outside
13　Standard without spots at the base
14　Branches appressed-villous; leaflets sparsely appressed-pubescent above　**16. ruthenicus**
14　Branches patent-pubescent or -hirsute; leaflets densely pubescent above　**(5–10). hirsutus** group
13　Standard with spots at the base
15　Standard with orange-red spots; inflorescence 1-sided, some branches procumbent; leaves glabrous when mature　**13. ratisbonensis**
15　Standard with violet or brown spots; inflorescence not 1-sided; branches all erect or ascending; leaves remaining pubescent at least beneath
16　Flowers 1–2 in a fascicle; spots violet　**14. zingeri**
16　Flowers 2–5 in a fascicle; spots brown
17　Indumentum golden; leaves glabrous above　**15. paczoskii**
17　Indumentum white; leaves pubescent on both surfaces　**16. ruthenicus**

1　Flowers in heads, subtended by a leafy involucre
18　Corolla white
19　Plant procumbent; indumentum of crispate and straight hairs　**34. kovacevii**
19　Plant erect; indumentum of straight hairs
20　3 upper teeth of calyx triangular-lanceolate; stem with both appressed and patent hairs　**30. albus**
20　3 upper teeth of calyx linear-subulate; stem with all hairs semi-patent　**35. nejceffii**
18　Corolla yellow or pale yellow
21　Branches with at least some patent or semi-patent hairs
22　Leaflets linear, green on both surfaces　**27. dorycnioides**
22　Leaflets elliptical to lanceolate, greyish or whitish at least beneath
23　Legume appressed-hairy
24　Calyx patent-hairy
25　Leaflets 7–10 mm wide, sparsely patent-hairy beneath　**33. podolicus**
25　Leaflets 3–6 mm wide, densely whitish-sericeous beneath　**21. austriacus**
24　Calyx appressed-hairy
26　Hairs on branches mainly appressed; flowers in heads of 5–8　**31. banaticus**
26　Hairs on branches all patent; flowers in heads of 12–18　**32. rochelii**
23　Legume patent-hairy
27　Leaflets (10–)15–35 mm, sparsely pubescent to subglabrous above　**19. supinus**
27　Leaflets 10–17 mm, densely appressed-sericeous on both surfaces　**20. eriocarpus**
21　Branches appressed-hairy
28　Leaflets glabrous or sparsely pubescent above
29　Calyx appressed-hairy
30　Leaflets 5–8 mm; some branches procumbent　**24. pygmaeus**
30　Leaflets 12–30 mm; branches erect or ascending
31　Corolla pale yellow; legume 30–40 mm, shortly appressed-hairy　**31. banaticus**
31　Corolla deep yellow; legume 25–28 mm, densely sericeous　**23. heuffelii**
29　Calyx patent-hairy
32　Leaflets 15–20 mm, green on both surfaces
33　Leaflets obovate; legume villous　**22. tommasinii**
33　Leaflets linear; legume lanate　**27. dorycnioides**
32　Leaflets 20–35 mm, greyish-hairy beneath
34　Branches erect, with appressed hairs only; flowers in heads of 6–8　**28. litwinowii**
34　Branches ascending, with both patent and appressed hairs; flowers in heads of 10–12　**32. podolicus**
28　Leaflets densely appressed-hairy on both surfaces
35　Calyx appressed-hairy
36　Leaflets 10 × 2–2·5 mm, linear to linear-spathulate　**25. jankae**
36　Leaflets 12–30 × 4–8 mm, lanceolate to oblong-obovate
37　Branches with appressed hairs only; leaflets 12–22 × 4–8 mm　**29. blockianus**
37　Branches with both patent and appressed hairs; leaflets 20–30 × 5–6 mm　**31. banaticus**
35　Calyx patent-hairy
38　Legume sericeous　**21. austriacus**
38　Legume lanate
39　Leaflets elliptical, lanceolate or obovate; branches with long patent hairs　**20. eriocarpus**
39　Leaflets linear; branches sericeous　**26. danubialis**

(i) Inflorescence racemose.

1. C. creticus (Boiss. & Heldr.) Rothm., *Feddes Repert.* **53**: 143 (1944) (*Cytisus creticus* Boiss. & Heldr.). Up to 30 cm, much-branched; branches spreading, glabrous, with short internodes, later becoming spinose. Leaflets 3–7 × 1·5–2 mm, obovate, acute, apiculate or obtuse, subglabrous above, appressed-pubescent beneath. Flowers solitary. Calyx appressed-pubescent. Corolla

---

[1] By V. H. Heywood and D. G. Frodin.

yellow; standard 10–12 mm. Legume white-villous. *Dry, rocky places.* ● *Kriti.* Cr.

2. **C. subidaeus** (Gand.) Rothm., *loc. cit.* (1944) (*Cytisus subidaeus* Gand.). Up to 150 cm, much-branched; branches appressed-pubescent when young, later becoming spinose. Leaflets cuneate-ovate, glabrous above, grey-pubescent beneath. Flowers solitary. Calyx appressed-pubescent. Corolla yellow; standard 12–23 mm. Legume appressed-tomentose. *Rocky scrub.* ● *Kriti.* Cr.

3. **C. spinescens** (C. Presl) Rothm., *loc. cit.* (1944) (*Cytisus spinescens* C. Presl, *C. subspinescens* Briq.). Up to 20 cm, much-branched; branches appressed-white-pubescent when young, later becoming spinose. Leaflets 3–11 × 1–3 mm, obovate-cuneate, obtuse or acute, appressed-pubescent above, sericeous beneath. Flowers 1 or 2 in a fascicle. Calyx appressed-pubescent. Corolla pale yellow; standard 23–33 mm. Legume pubescent along the sutures only. ● *C. & S. Italy; W. Jugoslavia.* ?Gr It Ju.

Plants with leaves silvery-pubescent on both surfaces and the legume wholly sericeous occur in various parts of the range.

4. **C. purpureus** (Scop.) Link, *Handb.* **2**: 154 (1831) (*Cytisus purpureus* Scop.). Up to 30 cm; branches subglabrous, unarmed. Leaflets obovate, mucronate, subglaucous, glabrous. Flowers in fascicles of 2–3 forming a leafy raceme. Calyx with sparse, patent hairs. Corolla lilac-pink to purplish, the standard 15–25 mm, with a darker patch in the centre. Legume 15–25 × 4–5 mm, glabrous. *Bushy and rocky places; calcicole.* ● *S. & S.E. Alps, N. Albania, N. Jugoslavia.* Al Au It Ju [Ge].

(5–10). **C. hirsutus** group. 30–200 cm; branches hairy, at least when young. Leaflets glabrous to hirsute above, hairy beneath at least when young. Flowers in fascicles of 1–4. Calyx glabrous or hairy. Corolla yellow or pale yellow; standard 20–25 mm. Legume glabrous, ciliate or villous.

The vernal state of **19**, *C. supinus*, closely resembles **6**, *C. polytrichus*, but can be distinguished from it by its smaller flowers.

1　Leaflets glabrous on both surfaces at maturity; calyx glabrous
　　or sparsely appressed-pubescent　　**10. leiocarpus**
1　Leaflets pubescent beneath at maturity; calyx with patent hairs
　　or with both patent and appressed hairs
　2　Calyx with both patent and appressed hairs　　**9. glaber**
　2　Calyx with only patent hairs
　　3　Hairs of branches and leaves appressed　　**7. wulfii**
　　3　Hairs of branches and leaves patent
　　　4　Branches procumbent; legume villous　　**6. polytrichus**
　　　4　Branches usually erect or ascending; legume sparsely to
　　　　densely hairy
　　　　5　Leaflets (4–)6–20(–30) × (2–)4–10(–18) mm　　**5. hirsutus**
　　　　5　Leaflets (10–)20–30 × (6–)10–15 mm　　**8. ciliatus**

5. **C. hirsutus** (L.) Link, *Handb.* **2**: 155 (1831) (*Cytisus hirsutus* L., *C. pumilus* De Not., *C. leucotrichus* Schur, *C. hirsutus* subsp. *leucotrichus* (Schur) Ascherson & Graebner). 20–100(–200) cm; branches erect or ascending, rarely procumbent, patent-pubescent or hirsute. Leaflets (4–)6–20(–30) × (2–)4–10(–18) mm, obovate to elliptical, glabrous to hirsute above, pubescent to hirsute beneath. Calyx patent-pubescent or hirsute. Corolla yellow or pinkish-yellow; standard with or without brown spots. Legume 25–40 × 5–8 mm, linear, hirsute all round or only at margins. *Somewhat calcifuge.* *C. & E. Europe.* Al Au Bu Cz Ga Gr He Hu It Ju ?Po Rm Tu.

Extremely variable in habit and indumentum. Numerous variants have been described at subspecific or specific rank but are not clearly distinguishable.

6. **C. polytrichus** (Bieb.) Rothm., *Feddes Repert.* **53**: 144 (1944) (*Cytisus hirsutus* subsp. *polytrichus* (Bieb.) Hayek, *C. polytrichus* Bieb.). 10–25 cm, branches procumbent, with long patent hairs when young. Leaflets (5–)10–15 × (1–)4–6 mm, elliptical, patent-hirsute on both surfaces, rarely glabrous above. Calyx patent-hirsute. Corolla yellow; standard with brown spots; wings rounded, entire. Legume 20–35 × 5–6 mm, villous. ● *Mountains of S. Europe, from the Maritime Alps to Krym, and extending northwards to C. Romania.* Bu Ga Gr It Ju Rm Rs (K).

7. **C. wulfii** (V. Krecz.) A. Kláskova, *Preslia* **30**: 214 (1958) (*Cytisus wulfii* V. Krecz.). Like **6** but branches and leaves appressed-pubescent; corolla with emarginate wings; legume 30 × 7 mm, with long, silvery, appressed hairs. *Krym.* Rs (K).

8. **C. ciliatus** (Wahlenb.) Rothm., *Feddes Repert.* **53**: 144 (1944) (*Cytisus hirsutus* subsp. *ciliatus* (Wahlenb.) Ascherson & Graebner). Up to 100 cm; branches erect or rarely procumbent, patent-pubescent or -villous. Leaflets (10–)20–30 × (6–)10–15 mm, obovate to elliptical, glabrous to pubescent above, patent-pubescent beneath. Calyx patent-pubescent or -villous. Corolla yellow; standard with or without brown spots. Legume 20–40 mm, ciliate or hirsute at the sutures, otherwise glabrous or rarely sparsely to densely pubescent. ● *E.C. Europe and Balkan peninsula.* Al Au Bu Cz Gr Hu Ju Rm.

9. **C. glaber** (L. fil.) Rothm., *Feddes Repert.* **53**: 143 (1944) (*Cytisus elongatus* Waldst. & Kit.). Up to 150 cm; branches erect or arching, densely appressed-pubescent when young, later glabrous. Leaflets 20–25(–30) × 8–12 mm, obovate or oblong, appressed-pubescent on both surfaces, becoming glabrous above, ciliate. Calyx densely or sparsely pubescent with both appressed and patent hairs. Corolla yellow; standard with brown spots. Legume 20–25 mm, densely appressed-sericeous. *Calcicole.* ● *W. & C. Romania, N.E. part of Balkan peninsula.* Bu Ju Rm.

10. **C. leiocarpus** (A. Kerner) Rothm., *op. cit.* 144 (1944). Up to 30 cm; branches procumbent, sericeous when young, later glabrous. Leaflets 25–40 × 10–20 mm, sparsely pubescent beneath, glabrous when mature. Calyx glabrous or sparsely appressed-pubescent. Corolla yellow; standard with brown spots. Legume 30–35 × 6–8 mm, glabrous or sparsely hairy at the sutures and on the faces. ● *W. & C. Romania, N. part of Balkan peninsula.* Al Bu Ju Rm.

11. **C. borysthenicus** (Gruner) A. Kláskova, *Preslia* **30**: 214 (1958) (*Cytisus borysthenicus* Gruner). Up to 120 cm; branches erect, densely sericeous. Leaflets 25–35(–60) × 4–6 mm, obovate, densely sericeous above. Calyx shortly appressed-pubescent. Corolla yellow; standard (20–)25–30 mm, pubescent outside, with orange spots. Legume 20–25 × 7–8 mm, broadly linear, densely silvery-pubescent. *Sand-flats, river-banks and dunes.* S. *Russia, Ukraine, W. Kazakhstan.* Rs (C, W, E).

12. **C. lindemannii** (V. Krecz.) A. Kláskova, *loc. cit.* (1958) (*Cytisus lindemannii* V. Krecz.). Like **11** but branches ascending, glabrous in the lower parts, densely pubescent in the upper parts, the hairs greenish, somewhat appressed; calyx densely villous-tomentose; standard without orange spots; legume 30 × 6 mm, linear, densely villous-tomentose. *Woods and scrub.* *Ukraine and S. Russia.* Rs (W, E).

13. **C. ratisbonensis** (Schaeffer) Rothm., *Feddes Repert.* **53**: 144 (1944) (*Cytisus ratisbonensis* Schaeffer, *C. biflorus* L'Hér.). (10–)30–45 cm; branches procumbent or ascending, sericeous, glabrous later. Leaflets 10–15 × 4–6 mm, obovate to obovate-lanceolate, glabrous above, sericeous beneath, glabrous when

mature. Inflorescence unilateral, with flowers in fascicles of 1–2(–3). Calyx appressed-sericeous, becoming yellow. Corolla yellow; standard 16–22 mm, with orange-red spots. Legume 20–30 × 4–5 mm, appressed-pubescent all round. ● *C. Europe, extending locally eastwards to S.E. Ukraine.* Au Bu Cz Ge Hu Po Rm Rs (W, C).

Plants from S.E. Ukraine, with stems 20–50 cm, inflorescence not unilateral and calyx grey-villous, have been described as **Cytisus kreczetoviczii** Wissjul. in Fomin, *Fl. RSS Ucr.* **6**: 588 (1954), but their status is not yet clear.

14. C. zingeri (Nenukow ex Litv.) A. Klásková, *Preslia* **30**: 214 (1958) (*Cytisus zingeri* (Nenukow ex Litv.) V. Krecz.). 40–120(–150) cm; vegetative branches shortly appressed-pubescent, the hairs yellow; flowering branches glabrous. Leaflets (15–)20–25(–30) × (8–)10–12(–15) mm, obovate. Flowers solitary or in pairs. Corolla yellow; standard 16–23 mm, usually with violet spots. Legume 25–30 × 2–5 mm. *U.S.S.R., from c. 50° N. to c. 60° N.* Rs (N, C, W, E).

15. C. paczoskii (V. Krecz.) A. Klásková, *loc. cit.* (1958) (*Cytisus paczoskii* V. Krecz.). 50–60 cm; branches ascending, villous, the hairs silvery or golden. Leaflets 15–20 × 5–6 mm, elliptical or narrowly obovate, glabrous above, appressed-villous beneath. Flowers in fascicles of 2–4. Corolla yellow; standard 18–25 mm, with brown spots; keel pubescent. Legume 30 mm, villous-tomentose. ● *W. Ukraine.* Rs (W).

16. C. ruthenicus (Fischer ex Wołoszczak) A. Klásková, *Preslia* **30**: 214 (1958) (*Cytisus ruthenicus* Fischer ex Wołoszczak). 25–150(–200) cm; branches erect or arcuate, grey-white-appressed-villous. Leaflets 10–17 × 4–8 mm, elliptical or lanceolate, sparsely appressed-pubescent above, densely so beneath. Flowers in fascicles of (2–)3–5. Calyx densely appressed-pubescent. Corolla yellow; standard (20–)25–30 mm, with or without spots; keel crispate-hairy. Legume 30–35 × 6–8 mm, densely appressed-villous. *U.S.S.R. southwards from c. 59° N., E. Poland.* Po Rs (C, W, K, E).

17. C. graniticus (Rehmann) Rothm., *Feddes Repert.* **53**: 144 (1944) (*Cytisus graniticus* Rehmann). Up to 30 cm; branches ascending, sericeous. Leaflets 10–15 × 3–5·5 mm, obovate-elliptical, glabrous or subglabrous above, sericeous beneath. Flowers in pairs. Calyx sericeous. Corolla white; standard *c.* 20 mm. Immature legume glabrous. *Granite cliffs.* ● *Krym.* Rs (K).

18. C. skrobiszewskii (Pacz.) A. Klásková, *Preslia* **30**: 214 (1958) (*Cytisus skrobiszewskii* Pacz.). 15–30 cm; branches ascending, shortly appressed-pubescent. Leaflets 10–18 × 5–6 mm, elliptical or obovate, appressed-pubescent on both surfaces. Flowers in fascicles of 2–4. Calyx sparsely appressed-pubescent. Corolla white, rarely pale pink; standard 18–22 mm. Legume 30–45 × 5–6 mm, densely appressed-pubescent. ● *S.W. Ukraine.* Rs (W).

(ii) Inflorescence capitate.

19. C. supinus (L.) Link, *Handb.* **2**: 155 (1831) (*Cytisus supinus* L.). 20–60(–120) cm; branches erect or ascending, rarely procumbent, patent-pubescent to villous or hirsute. Leaflets (10–)15–35 × (5–)7–14 mm, elliptical, oblong or obovate, sparsely patent-pubescent to subglabrous above, patent-pubescent beneath. Flowers in heads of 2–8(–10) (occasionally in racemes, in the vernal state, and then closely resembling members of the *C. hirsutus* group). Calyx patent-pubescent to hirsute. Corolla

yellow; standard 20–25 mm, (17–21 mm in vernal state) usually brown-spotted. Legume 20–35 × 5–6 mm, oblong, patent-villous or lanate. *C. & S. Europe, extending northwards to C. France, S. Germany and W. Ukraine.* Al Au Bu Cz Ga Ge Gr He Hs Hu It Ju Po Rm Rs (W) Tu.

Plants from Ukraine, Romania, Bulgaria, Hungary and Poland have been distinguished as **Cytisus aggregatus** Schur, *Enum. Pl. Transs.* 149 (1866), differing in their many-flowered capitula and in other floral features, but similar variants occur occasionally throughout the range of the species and a separation does not seem possible.

20. C. eriocarpus (Boiss.) Rothm., *Feddes Repert.* **53**: 144 (1944) (*Cytisus eriocarpus* Boiss.; incl. *C. absinthoides* Janka). 40–80 cm; branches erect, with long, patent and, sometimes, short, appressed hairs. Leaflets 10–17 mm, elliptical, lanceolate or obovate, densely sericeous, appressed-villous, or lanate on both surfaces. Calyx with patent hairs, rarely mixed with some appressed hairs. Corolla yellow. Legume densely patent-lanate. ● *Balkan peninsula.* Bu Gr Ju.

21. C. austriacus (L.) Link, *Handb.* **2**: 155 (1831) (*Cytisus austriacus* L., *C. smyrnaeus* auct., non Boiss.; incl. *C. pindicola* Hálácsy). 15–70 cm; branches erect or procumbent, more or less densely appressed-hairy. Leaflets (10–)15–25 × 3–6 mm, oblong, obovate or lanceolate, more or less densely appressed hairy above, densely whitish-sericeous beneath, sometimes whitish-sericeous on both surfaces. Flowers in heads, usually numerous. Calyx with patent hairs, rarely mixed with some appressed hairs. Corolla deep yellow; standard 15–22 mm, sericeous outside. Legume 20–30 × 5 mm, sericeous. *E.C. & S.E. Europe, extending to S.C. Russia.* Al Bu Cz Gr Hu Ju Rm Rs (W, K, E) ?Tu.

Variable in habit, leaf-shape, size and indumentum. Several variants with restricted distribution have been recognized as separate species, but they are linked by numerous intermediates and do not form a satisfactory subspecies pattern.

22. C. tommasinii (Vis.) Rothm., *Feddes Repert.* **53**: 144 (1944) (*Cytisus tommasinii* Vis.). Like **21** but branches with short, appressed hairs; leaflets up to 20 mm, glabrous above, sparsely appressed-hairy beneath, green on both surfaces; legume patent-villous. ● *Mountains of W. Jugoslavia and N. Albania.* Al Ju.

Variable in number of flowers per head and in corolla-size, and not always clearly separable from **21** and **23**.

This species has also been recorded from N.W. Greece, probably in error for relatively sparsely hairy variants of **23**.

23. C. heuffelii (Wierzb.) Rothm., *loc. cit.* (1944) (*Cytisus heuffelii* Wierzb.). 20–50 cm; branches erect, appressed-hairy. Leaflets 12–30 × 2–9 mm, oblong, obovate or lanceolate, glabrous or sparsely appressed-hairy above, usually densely so beneath. Flowers in heads of 2–5(–8). Calyx appressed-hairy. Corolla yellow; standard 15–25 mm, sparsely hairy outside. Legume 25–28 × 4–5 mm, densely silvery-sericeous. ● *Balkan peninsula, Romania and Hungary.* Al Bu Gr Hu Ju Rm.

24. C. pygmaeus (Willd.) Rothm., *loc. cit.* (1944) (*Cytisus pygmaeus* Willd.). Dwarf shrub 5–15 cm; branches procumbent and ascending, sparsely appressed-sericeous. Leaflets 5–8 × 3–4 mm, obovate, elliptical or linear, subglabrous or sericeous along the midrib and at the margins. Flowers in heads of 1–5. Calyx appressed-hairy. Corolla yellow; keel glabrous. Legume 15–20 × 6 mm, appressed-hairy. *E. part of Balkan peninsula.* Bu Ju Tu.

25. C. jankae (Velen.) Rothm., *loc. cit.* (1944) (*Cytisus jankae* Velen.). Dwarf shrub; branches procumbent, appressed-sericeous; stems 8–10 cm, erect, slender. Leaflets 10 × 2–2·5 mm, linear to linear-spathulate, densely silvery-sericeous. Flowers in heads of 2–4. Calyx sericeous. Corolla pale yellow; standard *c.* 15 mm; keel densely sericeous outside. Legume 15 × 5 mm, sericeous. ● *Balkan peninsula.* Al Bu Ju.

26. C. danubialis (Velen.) Rothm., *loc. cit.* (1944) (*Cytisus danubialis* Velen.). 40–70 cm, much-branched; branches appressed silvery-sericeous. Leaflets 20 × 2–3 mm, linear, silvery-sericeous on both surfaces. Flowers in heads of 3–7. Calyx patent-villous. Corolla pale yellow; standard *c.* 20 mm, minutely hairy outside. Legume 20 × 4 mm, densely lanate. ● *Bulgaria, Romania.* Bu Rm.

27. C. dorycnioides (Davidov) Frodin & Heywood, *Feddes Repert.* **79**: 21 (1968) (*Cytisus dorycnioides* Davidov). Like **26** but branches appressed- or patent-hirsute; leaflets 15–20 × 3–5 mm, sparsely appressed-hairy and green on both surfaces; flowers in heads of 5–12. ● *N.E. Greece* (*near Xanthi*). Gr.

28. C. litwinowii (V. Krecz.) A. Klásková, *Preslia* **30**: 214 (1958) (*Cytisus litwinowii* V. Krecz.). 20–50 cm; branches erect, appressed-pubescent. Leaflets 20–35 × 4–8 mm, subglabrous above, appressed-hairy beneath. Flowers in heads of 6–8. Corolla golden-yellow; standard 17–20 mm. Legume not known. ● *S.C. Russia; Ukraine.* Rs (C).

29. C. blockianus (Pawł.) A. Klásková, *loc. cit.* (1958) (*Cytisus blockianus* Pawł., *C. blockii* V. Krecz.). 20–50 cm, branches ascending, sparsely appressed-hairy. Leaflets 12–22 × 4–8 mm, oblong-obovate to lanceolate, appressed-hairy on both surfaces. Flowers in heads of (2–)5–10. Calyx sparsely appressed-hairy. Corolla pale yellow; standard 18–25 mm. Legume 20–30 × 4–5 mm, densely villous. ● *Moldavia, W. Ukraine.* Rs (W).

30. C. albus (Hacq.) Rothm., *Feddes Repert.* **53**: 144 (1944) (*Cytisus albus* Hacq., *C. leucanthus* Waldst. & Kit., *C. leucanthus* subsp. *albus* (Hacq.) Hayek). 30–80 cm; branches erect or ascending, with both patent and appressed hairs. Leaflets (10–)20–30 × 7–12 mm, oblong-obovate, appressed-hairy on both surfaces or only beneath. Flowers in heads of (2–)5–8. Calyx with semi-patent whitish hairs, the 3 upper teeth triangular-lanceolate. Corolla white; standard 16–20 mm. Legume 20–30 × 5–6 mm, appressed-villous. ● *C. & S.E. Europe, northwards to S. Poland.* Al Bu Cz Gr Hu Ju Po Rm Rs (W) Tu.

31. C. banaticus (Griseb. & Schenk) Rothm., *loc. cit.* (1944) (*Cytisus leucanthus* subsp. *pallidus* (Schrader) Hayek). Like **30** but hairs on branches mainly appressed; leaflets 20–30 × 5–6 mm, obovate to lanceolate, always appressed-hairy on both surfaces; calyx appressed-hairy; corolla pale yellow; legume 30–40 mm, appressed-hairy. ● *E.C. Europe and N.E. part of Balkan peninsula.* Bu Cz Ju Hu Rm.

32. C. rochelii (Wierzb.) Rothm., *loc. cit.* (1944) (*Cytisus rochelii* Wierzb., *C. leucanthus* subsp. *obscurus* (Rochel) Hayek). 50–100 cm; branches erect, with patent or semi-patent hairs. Leaflets 20–25 × 6–8 mm, densely appressed-hairy. Flowers in heads of 12–18. Calyx appressed-hairy. Corolla pale yellow; standard 18–20 mm. Legume *c.* 30 mm, appressed-villous. ● *From Bulgaria and E. Jugoslavia to W. Ukraine.* Bu Ju Rm Rs (W).

33. C. podolicus (Błocki) A. Klásková, *Preslia* **30**: 214 (1958) (*Cytisus podolicus* Błocki). 30–50 cm; branches ascending, densely covered with both patent and appressed hairs when young. Leaflets 25–30 × 7–10 mm, elliptical to lanceolate, subglabrous above, with sparse, long, patent hairs beneath. Flowers in heads of 10–12. Calyx densely patent-hairy. Corolla pale yellow; standard 23–25 mm. Legume 25–30 mm, densely appressed-villous. ● *Moldavia, W. Ukraine.* Rs (W).

34. C. kovacevii (Velen.) Rothm., *Feddes Repert.* **53**: 144 (1944) (*Cytisus kovacevii* Velen.). Small shrub; branches slender, with both patent and crispate hairs. Leaflets elliptical, densely sericeous on both surfaces. Calyx crispate-hairy. Corolla white; standard *c.* 10 mm. Legume not known. ● *N. Bulgaria.* Bu.

35. C. nejceffii (Urum.) Rothm., *loc. cit.* (1944) (*Cytisus nejceffii* Urum.). 30–40 cm; branches erect, with dense semi-patent hairs. Leaflets (10–)20–25(–30) × 3–5 mm, linear-elliptical long-acuminate, sparsely appressed-hairy above, densely so beneath. Flowers in heads of (3–)5–8(–10). Calyx with the 3 upper teeth linear-subulate, with numerous semi-patent hairs. Corolla white; standard 15–22 mm. Legume not known. ● *N. Bulgaria.* Bu.

14. Chronanthus (DC.) C. Koch[1]

Unarmed shrubs. Leaves 3-foliolate. Flowers in small heads, rarely solitary. Calyx campanulate, bilabiate; upper lip with 2 teeth, not divided to the base, lower with 3 teeth; corolla yellow, persistent and enclosing the legume; stamens monadelphous; stigma introrse. Legume compressed, dehiscent. Seeds 2–3, or 1 by abortion, with a small strophiole.

1. C. biflorus (Desf.) Frodin & Heywood, *Feddes Repert.* **79**: 21 (1968) (*Spartium biflorum* Desf., *Cytisus fontanesii* Spach ex Ball). Erect or ascending shrub 20–50 cm. Branches 5- to 10-angled, glabrous, flowering in first year. Leaflets 4–9 × 0·5–1 mm, linear to lanceolate, with short, appressed hairs. Flowers in groups of 2 or 4 (rarely 1, 3 or 5). Standard 8–12 mm, cordate; keel nearly as long as standard, with rounded beak. Legume 10–15 mm, the valves translucent, glabrous. *Scrub and woods.* E. & S. Spain; Islas Baleares. Bl Hs.

15. Teline Medicus[2]

Unarmed shrubs. Leaves 3-foliolate. Flowers in axillary or terminal racemes. Calyx tubular-campanulate, bilabiate; upper lip deeply 2-fid, the lower with 3 distinct teeth; corolla yellow; standard broadly ovate, somewhat exceeding the wings and keel. Legume narrowly oblong, compressed, dehiscent. Seeds 2–6, strophiolate.

Literature: C. Vicioso, *Genísteas Españolas* 1: 125–135 (*Bol. Inst. For. Inv. Exper. Madrid* No. 67). Madrid. 1953. (sub *Genista* L.).

Standard glabrous; leaves petiolate; leaflets obovate
1. monspessulana

Standard sericeous; leaves subsessile; leaflets linear-oblanceolate
2. linifolia

1. T. monspessulana (L.) C. Koch, *Dendrologie* **1**: 30 (1869) (*Cytisus monspessulanus* L., *C. candicans* (L.) DC., *Genista candicans* L.). Stems 100–300 cm, erect. Leaves 8–20 mm, petiolate; leaflets obovate, with sparse to dense patent hairs on both surfaces, especially the lower, emarginate to shortly mucronate.

---

[1] By D. G. Frodin and V. H. Heywood. [2] By P. E. Gibbs.

Flowers in axillary clusters. Pedicel 1–2 mm, with 3 linear bracteoles. Calyx 5–6 mm, densely sericeous or with patent hairs, the lower lip longer than the upper. Standard 10–12 mm, broadly ovate, glabrous; keel sparsely sericeous; wings glabrous. Legume *c.* 20 mm, narrowly oblong, densely sericeous or with patent hairs. Seeds 3–6. *Scrub and open woodland. Mediterranean region, Portugal, Açores.* Az Co Ga Gr Hs It Ju Lu Sa Si ?Tu.

2. T. linifolia (L.) Webb & Berth., *Phyt. Canar.* **2**: 41 (1842) (*Genista linifolia* L., *Cytisus linifolius* (L.) Lam.). Stems 50–150 cm, erect. Leaves subsessile; leaflets 10–15 mm, linear-oblanceolate, glabrescent above, appressed-sericeous beneath. Flowers in congested terminal racemes. Pedicels 2–3 mm, with three linear bracteoles. Calyx 7–8 mm, sericeous. Standard 10–18 mm, broadly ovate, sericeous; keel sericeous; wings glabrous. Legume 15–20 mm, narrowly oblong, densely pubescent. Seeds 2–3. *Woodland and scrub; calcifuge. W. Mediterranean region.* Bl Ga Hs.

16. Genista L.[1]

Spiny or unarmed shrubs. Leaves 1- or 3-foliolate, often caducous; stipules absent or small, often represented by a subglobose swelling (pulvinus). Flowers in heads or racemes or in axillary clusters, rarely solitary. Calyx bilabiate; upper lip deeply 2-fid, lower 3-toothed; corolla yellow; stamens monadelphous. Legume dehiscent or indehiscent, ovoid to linear-oblong. Seeds 1–many, estrophiolate.

Literature: P. E. Gibbs, *Notes Roy. Bot. Gard. Edinb.* **27**: 11–99 (1966). C. Vicioso, *Genísteas Españolas* 1 (*Bol. Inst. For. Inv. Exper. Madrid* No. 67). Madrid. 1953.

Most species grow on dry heaths or stony hillsides, or in scrub or dry woodland.

```
 1  Plant ±spiny
  2  Legume narrowly oblong and compressed; seeds 2–12; stan-
       dard broadly ovate, as long as the wings and keel
    3  Plant with weak spines; bracteoles absent        14. pulchella
    3  Plant with stout spines; bracteoles present
      4  Spines axillary and recurved; standard glabrous
        5  Leaves 3-foliolate                           27. morisii
        5  Leaves simple
          6  Pedicels c. 1 mm; stems and branches somewhat winged;
               legume sericeous                          26. carpetana
          6  Pedicels 2–5 mm; stems and branches terete; legume
               glabrous
            7  Flowering branches subtending the axillary spines; lips
                 of the calyx as long as the tube        25. corsica
            7  Flowering branches or flowers borne directly on the
                 axillary spines; lips of the calyx shorter than the tube
                                                         24. scorpius
      4  Spines terminating the main branches, not axillary; stan-
           dard ±sericeous
        8  Most leaves 3-foliolate                       23. aspalathoides
        8  All leaves simple
          9  Most flowers borne singly in the axil of each bract
                                                         (17–19). lobelii group
          9  Most flowers in pairs or clusters in the axil of each bract
            10  Calyx 2·5–5 mm, the lips as long as or shorter than the
                  tube; flowers in long racemes
              11  Standard glabrous to sparsely sericeous  21. hystrix
              11  Standard densely sericeous             22. polyanthos
            10  Calyx 4–7 mm, the lips longer than the tube; flowers
                  in short clusters
              12  No pulvini with spinose stipules; standard uniformly
                    sericeous                            20. salzmannii
              12  Some pulvini with spinose stipules; standard with a
                    median ridge of sericeous hairs      21. hystrix
  2  Legume ovoid-acuminate, usually 1-seeded, or rhomboid-
       falcate and inflated, 2- to 12-seeded; standard triangular or
       ovate with a rounded or acute apex, usually shorter than the
       keel
    13  Plant with spiny branches; axillary spines absent
      14  Branches alternate; leaves simple             17. lobelii
      14  Branches opposite; leaves 3-foliolate          54. acanthoclada
    13  Plant with recurved axillary spines
      15  Flowers or flowering branches borne directly on the
            axillary spines                             55. fasselata
      15  Flowers or flowering branches not borne directly on the
            axillary spines
        16  Most leaves 3-foliolate
          17  Calyx and leaves with sparse, patent hairs; flowering
                branches sometimes terminated by a spine  43. cupanii
          17  Calyx and leaves subglabrous; flowering branches never
                terminated by a spine
            18  Leaves without spinose stipules          41. triacanthos
            18  Leaves with spinose stipules             42. tridens
        16  All leaves simple
          19  Leaves with spinose stipules
            20  Young stems and calyx with patent hairs; legume
                  5–15 mm, more or less falcate, 4- to 6-seeded
                                                         30. berberidea
            20  Young stems and calyx with appressed hairs or glab-
                  rous; legume 6–7 mm, ovoid-acuminate, 1- to
                  2-seeded
              21  Standard c. 10 mm, cordate at the base  39. lucida
              21  Standard c. 6 mm, truncate at the base  42. tridens
          19  Leaves without spinose stipules
            22  Bracts, at least of the lowermost flowers, 1 mm or less,
                  or absent; bracteoles minute
              23  Calyx glabrous or very sparsely sericeous; legume
                    falcate; seeds 2–12                  29. falcata
              23  Calyx densely hairy; legume ovoid-acuminate; seeds
                    1–2
                24  Flowers congested in heads; standard about as long
                      as the keel                        35. hispanica
                24  Flowers in lax racemes; standard ½–⅔ as long as the
                      keel                               36. germanica
            22  Bracts, at least of the lowermost flowers, more than
                  1 mm; bracteoles usually conspicuous
              25  Leaves glabrous; legume falcate; seeds 2–12
                26  Calyx 3–4 mm, glabrous; legume glabrous  28. anglica
                26  Calyx 5–6 mm, patent-pubescent; sutures of the
                      legume with patent hairs           30. berberidea
              25  Leaves pubescent beneath; legume ovoid-acuminate;
                    seeds 1–2
                27  Flowering branches terminated by a spine
                                                         40. anatolica
                27  Flowering branches not terminated by a spine
                  28  Plant decumbent, less than 30 cm; inflorescence
                        lax
                    29  Base of the standard truncate to cuneate, the claw
                          2 mm or less                   33. sylvestris
                    29  Base of the standard subcordate, the claw more
                          than 2 mm                      34. aristata
                  28  Plant erect, usually more than 30 cm; inflorescence
                        congested
                    30  Apex of the standard emarginate; upper teeth of
                          the calyx c. ½ as long as the lower teeth
                                                         38. tournefortii
                    30  Apex of the standard acute; upper teeth of the
                          calyx about as long as the lower teeth  37. hirsuta
 1  Plant not spiny
  31  Leaves simple
    32  Flowers in heads
      33  All branches alternate; upper surface of the leaves glab-
            rous                                         13. subcapitata
```

[1] By P. E. Gibbs.

33 Some branches opposite; upper surface of the leaves sericeous   **56. umbellata**
32 Flowers in racemes or axillary clusters
34 Legume ovoid-acuminate; seeds 1–2; standard triangular or rhombic, usually shorter than the keel
35 All branches alternate; standard 4·5–8 mm, triangular
36 Calyx with sericeous hairs   **33. sylvestris**
36 Calyx glabrous
37 Leaves narrowly elliptical; keel exceeding the standard by *c.* 1 mm   **31. micrantha**
37 Leaves linear-oblong; keel exceeding the standard by *c.* 2 mm   **32. carinalis**
35 At least some branches opposite; standard 7–12 mm, ovate or rhombic
38 Calyx subglabrous; upper teeth obtuse, *c.* ⅓ as long as the lip   **51. aetnensis**
38 Calyx sericeous; upper teeth acute, at least as long as the lip
39 Standard 7–9 mm, longer than wide; claw less than 1 mm wide   **52. spartioides**
39 Standard 10–12 mm, as long as wide; claw more than 1 mm wide   **53. haenseleri**
34 Legume narrowly oblong; seeds 2 or more; standard broadly ovate, as long as the keel
40 Keel and standard glabrous
41 Stems usually 3-winged; leaves with a narrow, hyaline, obscurely denticulate margin   **2. januensis**
41 Stems terete; leaves without a hyaline, denticulate margin
42 Leaves on the main stem (9–)12–50 mm, ovate, lanceolate, elliptical, oblong or oblanceolate, pubescent or glabrous and ciliate on the margin and midrib beneath   **1. tinctoria**
42 Leaves on the main stem 3–10 mm, linear-oblanceolate, subglabrous   **3. lydia**
40 Keel, and usually the standard, sericeous
43 Standard subglabrous, or with a narrow, median ridge of sericeous hairs
44 Most flowers in pairs in the axil of each bract; bracts fasciculate   **5. cinerea**
44 All flowers borne singly in the axil of each bract; bracts not fasciculate
45 Leaves usually more than 8 mm, shortly petiolate   **6. florida**
45 Leaves usually less than 6 mm, sessile   **7. valentina**
43 Standard uniformly sericeous, often densely so
46 Plant erect, sparingly branched, with long, flexuous branches
47 Bracts solitary; most flowers borne in fascicles on short lateral branches 5–15 mm   **4. ramosissima**
47 Bracts fasciculate; most flowers paired, borne directly on the main branches   **5. cinerea**
46 Plant decumbent, much-branched
48 Upper surface of the leaves glabrous or subglabrous
49 Bracteoles present (often small)
50 Flowers in lax racemes; leaves usually less than 10 × 3 mm   **9. pseudopilosa**
50 Flowers in dense, terminal racemes; leaves usually more than 10 × 3 mm   **12. sericea**
49 Bracteoles absent
51 Standard 12–14 mm; flowers 1–2, subterminal   **8. obtusiramea**
51 Standard not more than 12 mm; flowers axillary or in racemes
52 Flowers in clusters near the apices of the main branches; leaves sessile   **15. albida**
52 Flowers usually in long racemes on ascending branches; leaves shortly petiolate or subsessile   **16. pilosa**
48 Upper surface of the leaves pubescent
53 Bracteoles borne just below the calyx
54 Bracteoles less than 1 mm   **10. teretifolia**
54 Bracteoles 2–3 mm   **11. sakellariadis**

53 Bracteoles halfway along the pedicel or absent
55 Flowers borne singly in the axil of each bract; pedicels 2–5 mm; calyx *c.* 4 mm; standard 7–10 mm   **14. pulchella**
55 Flowers sometimes in pairs in the axil of each bract; pedicels 1–3 mm; calyx 5–7 mm; standard 9–12 mm   **15. albida**
31 Leaves 3-foliolate
56 Calyx 3 mm or less; lower teeth *c.* 0·5 mm (Ibiza)   **47. dorycnifolia**
56 Calyx at least 3 mm; lower teeth 1 mm or more
57 Flowers in 2- to 12-flowered terminal heads; standard slightly shorter to slightly longer than the keel
58 Lowest bracts simple; standard glabrous or with sparse hairs   **44. radiata**
58 Lowest bracts 3-foliolate; standard with dense sericeous hairs
59 Bracteoles 2–3 mm, as long as the calyx-tube   **45. holopetala**
59 Bracteoles 1 mm or less, shorter than the calyx-tube   **46. hassertiana**
57 Flowers in often long and lax racemes; standard ½–¾ as long as the keel
60 Stem, leaves and calyx with dense, long, patent hairs; leaflets of the main cauline leaves 2–6 mm wide   **49. nissana**
60 Stem, leaves and calyx with appressed hairs; leaflets 1–2 mm wide
61 Most bracts 3-foliolate, persistent; standard deltate or triangular   **48. sessilifolia**
61 Bracts (except the lowermost) simple, fugacious or absent; standard broadly ovate or rhombic   **50. ephedroides**

Sect. GENISTA. Unarmed, with simple leaves. Corolla glabrous, the calyx and leaves usually so. Standard broadly ovate, equalling the wings and keel. Legume narrowly oblong. Seeds 3–10.

1. G. tinctoria L., *Sp. Pl.* 710 (1753) (incl. *G. depressa* Bieb., *G. hungarica* A. Kerner, *G. marginata* Besser, *G. mayeri* Janka, *G. ovata* Waldst. & Kit., *G. tanaitica* Smirnov, *G. tetragona* Besser, *G. patula* Bieb.). Procumbent to erect shrub, 10–200 cm. Leaves 9–50 × 2·5–15 mm, simple, very variable in shape; leaves, calyx and legume glabrous to densely sericeous. Flowers borne singly in the axil of each bract in short racemes towards ends of branches, or in long, simple or compound racemes. Bracts foliaceous; bracteoles *c.* 1 mm; pedicel 1–2 mm. Calyx 3–7 mm; corolla glabrous; standard 8–15 mm, broadly ovate. 2*n* = 48. *Most of Europe from S. Scotland and Estonia southwards, but absent from many of the islands.* Al Au Be Br Bu Cz Da Ga Ge Gr He Ho Hs Hu It Ju No Po Rm Rs (B, C, W, K, E) Su Tu.

Very variable in habit, leaf-shape and degree of hairiness. Populations occur showing different combinations of these characters and these have been variously referred to as distinct species or subspecies. Much of the variation is continuous and there is little correlation between the different characters. It is not at present possible to give a comprehensive account of the variation in this species, but the taxa that are generally recognized can be arranged in the following 4 groups:

(i) (*G. anxantica* Ten., *G. campestris* Janka, *G. elata* Wenderoth, *G. tenuifolia* Loisel., *G. tinctoria* L. sensu stricto, *G. virgata* Willd.): Plant 20–200 cm, erect or ascending; leaves (10–)15–35 × 2·5–6(–9) mm, oblong or lanceolate, glabrous with ciliate margins and midrib and conspicuous lateral veins; flowers numerous, in simple or branched racemes; calyx and legume usually glabrous.

(ii) (*G. alpestris* Bertol., *G. tinctoria* subsp. *littoralis* (Corb.) Rothm.): Plant not more than 20 cm, procumbent; leaves 9–12 ×

3–4 mm, ovate-elliptical to elliptic-oblong, glabrous with ciliate margins and midrib, usually with conspicuous lateral veins; flowers few; calyx and legume usually glabrous.

(iii) (*G. hungarica* A. Kerner, *G. lasiocarpa* Spach, *G. mantica* Pollini, *G. mayeri* Janka, *G. ovata* Waldst. & Kit., *G. perreymondii* Loisel.): Plant 20–200 cm, erect or ascending; leaves 20–50 × 6–15 mm, ovate or elliptical, usually pubescent, with conspicuous lateral veins; flowers numerous, in simple or branched racemes; calyx and legume usually pubescent.

(iv) (*G. csikii* Kümmerle & Jáv., *G. depressa* Bieb., *G. friwaldskyi* Boiss., *G. tetragona* Besser): Plant usually not more than 20 cm, procumbent; leaves 10–20 × 3–5 mm, lanceolate, elliptical, oblong or oblanceolate, usually pubescent, without conspicuous lateral veins; flowers few; calyx and legume glabrous or pubescent.

The precise distribution of these variants is not clear; (i) and (ii) probably occur almost throughout the range of the species, (iii) in C. & S. Europe and possibly in E. Europe, (iv) in S.E. Europe.

2. **G. januensis** Viv., *Elench. Pl. Horti Bot.* 19 (1802) (*G. triangularis* Willd., *G. lydia* var. *spathulata* (Spach) Hayek). Procumbent to erect shrub 10–50 cm; stems and branches usually 3-winged. Leaves of the flowering-branches 5–12 × 2–4 mm, elliptical to obovate; leaves of the non-flowering branches 5–40 × 3–7 mm, elliptical to lanceolate; all leaves glabrous, with a narrow, hyaline, obscurely denticulate margin. Flowers in short racemes on ascending lateral branches. Calyx 3·5–4 mm, subglabrous, the lips shorter than the tube. Standard 9–10 mm, broadly ovate. *Calcicole. Balkan peninsula and Italy, extending northwards to Slovenija and W. Romania.* Al Bu Gr It Ju Rm.

3. **G. lydia** Boiss., *Diagn. Pl. Or. Nov.* 1(2): 8 (1843) (*G. rumelica* Velen., *G. rhodopea* Velen.). Procumbent or erect shrub up to 100 cm; stems not winged. Leaves 3–10 × 1–3 mm, linear-oblanceolate or linear-oblong, subglabrous, entire and without a hyaline margin. Flowers in short racemes on lateral branches. Calyx 3·5–5 mm, glabrous, the lips almost as long as the tube. Standard 10–12 mm, broadly ovate. *E. part of Balkan peninsula.* Bu Gr Ju Tu.

Sect. SPARTIOIDES Spach. Unarmed, with simple leaves. Standard broadly ovate, equalling the wings and keel, usually sericeous. Keel and legume sericeous. Legume narrowly oblong with appressed to semi-patent hairs. Seeds 2 or more.

4. **G. ramosissima** (Desf.) Poiret in Lam., *Encycl. Méth. Bot.*, *Suppl.* 2: 715 (1812). Erect shrub with lax, flexuous branches. Leaves 5–10 × 2–3 mm, elliptical to obovate, sericeous beneath, glabrous above, sessile. Inflorescence an irregular raceme, the flowers solitary, or in fascicles of 2–5, on main branches or on short lateral branches. Bracts solitary. Calyx 5–7 mm. Standard 10–12 mm, ovate, with dense, semi-patent hairs. *S.E. Spain.* Hs.

5. **G. cinerea** (Vill.) DC. in Lam. & DC., *Fl. Fr.* ed. 3, 4: 494 (1805). Like 4 but flowers mostly paired and borne directly on the main branches; bracts fasciculate; standard glabrous or with a median ridge of hairs (rarely uniformly sericeous). *S.W. Europe.* Bl Ga Hs It Lu.

(a) Subsp. **cinerea**: Branches sericeous; adult leaves usually more than 5 × 2 mm. *Throughout the range of the species except Mallorca.*

(b) Subsp. **leptoclada** (Willk.) O. Bolós & Molinier, *Collect.*

Bot. (*Barcelona*) 5: 807 (1958): Branches with dense, white, pseudo-farinose, appressed hairs; mature leaves usually less than 5 × 2 mm. *S.E. Spain* (*Prov. Murcia*), *Mallorca.*

6. **G. florida** L., *Syst. Nat.* ed. 10, 2: 1157 (1759) (*G. polygaliphylla* Brot., *G. leptoclada* Gay ex Spach). Like 4 but leaves 5–25 × 2–5 mm, oblanceolate, sometimes sparsely hairy above, shortly petiolate to subsessile; flowers borne singly in the axil of each bract in long, lax racemes; lowermost bracts leaf-like, simple, the upper ones reduced in size; calyx 4–6 mm; standard broadly ovate, subglabrous. *Spain, N. Portugal.* Hs Lu.

7. **G. valentina** (Willd. ex Sprengel) Steudel, *Nomencl. Bot.* ed. 2, 1: 671 (1840) (*G. oretana* Webb ex Willk.). Erect shrub, with lax, flexuous branches. Leaves 2·5–5 × 0·8–1·5 mm, simple, narrowly elliptical to obovate, sericeous beneath, glabrous above, sessile. Flowers borne singly in the axil of each bract in lax, elongate racemes. Bracts leaf-like, simple. Calyx 2·5–5 mm, sparsely sericeous. Standard 8–12 mm, ovate, with short appressed hairs. *Mountains of E. & S.E. Spain.* Hs.

8. **G. obtusiramea** Gay ex Spach, *Ann. Sci. Nat.* ser. 3 (Bot.), 2: 116 (1845). Procumbent to erect, much branched shrub; branches with internodes 5–10 mm and prominent pulvini. Leaves 2–8 × 1–3 mm, elliptical to obovate, pubescent beneath, glabrous above, sessile. Flowers solitary or paired, borne near the apex of each branch. Calyx 5–6 mm. Standard 12–14 mm, broadly ovate, densely sericeous. *Mountain heaths.* ● *N.W. Spain, C. Portugal.* Hs Lu.

9. **G. pseudopilosa** Cosson, *Not. Pl. Crit.* 102 (1851). Decumbent shrub, with flexuous, ascending branches. Leaves 4–12 × 1–4 mm, elliptical to oblanceolate, appressed-sericeous beneath, glabrous above, sessile, involute. Flowers borne singly in the axil of each bract in lax terminal racemes. Bracts almost leaf-like; bracteoles less than 1 mm, borne at about the middle of the pedicel; pedicels 1–3 mm. Calyx 5–6 mm. Standard 8–12 mm, broadly ovate, appressed-sericeous, the base usually truncate. *S. & S.E. Spain.* Hs.

10. **G. teretifolia** Willk., *Flora* (*Regensb.*) 34: 617 (1851). Like 9 but leaves sericeous on both surfaces; bracteoles borne near the apex of the pedicel. *Dry pastures.* ● *N. Spain* (*near Pamplona*). Hs.

11. **G. sakellariadis** Boiss. & Orph. in Boiss., *Diagn. Pl. Or. Nov.* 3(6): 42 (1859). Like 9 but leaves sericeous on both surfaces; bracteoles 2–3 mm, borne near the apex of the pedicel; pedicels 3–5 mm. ● *C. Greece* (*Olimbos*). Gr.

12. **G. sericea** Wulfen in Jacq., *Collect. Bot.* 2: 167 (1789). Much-branched shrub. Leaves 5–25 × 2–5 mm, subsessile, narrowly elliptical, oblanceolate or obovate, shortly mucronate, sericeous beneath, glabrous or subglabrous above, the margins involute. Flowers borne singly in the axil of each bract, in terminal clusters of 2–5; flowering branches often slender and flexuous. Bracteoles c. 1 mm, borne at the middle of the pedicel; pedicels 2–3 mm. Standard 10–14 mm, broadly ovate, sericeous, base cuneate. ● *Mountains of W. part of Balkan peninsula and N.E. Italy.* Al ?Gr It Ju.

13. **G. subcapitata** Pančić, *Fl. Princ. Serb.* 224 (1874) (*G. involucrata* auct. pro parte, non Spach). Like 12 but flowers sessile in heads with an involucre of leaf-like bracts; bracteoles 2–3 mm, sub-foliaceous. ● *C. part of Balkan peninsula.* Al Bu Ju.

14. **G. pulchella** Vis., *Flora (Regensb.)* **13**: 51 (1830) (*G. villarsii* G. C. Clementi). Spreading, unarmed or sometimes weakly spiny shrub. Young branches and both surfaces of the leaves with dense, long, sericeous appressed or patent hairs. Leaves 2–9 × 1·5–3 mm, sessile, narrowly elliptical. Flowers borne singly in the axil of each bract in congested racemes. Bracteoles absent; pedicels 2–5 mm. Calyx *c.* 4 mm. Standard 7–10 mm, ovate, with dense sericeous hairs. *Mountains of S.E. France, W. Jugoslavia and Albania.* Al Ga Ju.

15. **G. albida** Willd., *Sp. Pl.* **3**: 942 (1802) (*G. scythica* Pacz., *G. involucrata* auct. pro parte, non Spach). Like **14** but never spinose; young branches and leaves with short hairs; leaves 3–10 × 1·5–4 mm, elliptical to obovate, sometimes glabrous above; flowers borne singly or in pairs in the axil of each bract; bracteoles sometimes present; pedicels 1–3 mm; calyx 5–7 mm; standard 9–12 mm. *E. part of Balkan peninsula, S. Ukraine.* Bu Gr ?Ju Rm Rs (W, K).

G. halacsyi Heldr., *Sched. Herb. Graec. Norm.* no. 1526 (1899), from S. Greece, and **G. millii** Heldr. ex Boiss., *Fl. Or., Suppl.* 160 (1888), from S.E. Greece, are perhaps conspecific with **15**.

16. **G. pilosa** L., *Sp. Pl.* 710 (1753). Procumbent to suberect shrub up to 150 cm. Leaves 5–12 mm, usually oblanceolate, shortly petiolate to subsessile, appressed-sericeous beneath, glabrous above. Flowers borne singly or in pairs in the axil of each bract, in lax racemes on ascending branches. Bracteoles absent. Calyx 4–5 mm. Standard 8–10 mm, broadly ovate with sparse, appressed-sericeous hairs. *W. & C. Europe, extending to S. Sweden, C. Italy and Macedonia.* Al Au Be Br Bu Cz Da Ga Ge He Hs Ho Hu It Ju Po Rm ?Rs (W).

Sect. ERINACOIDES Spach. Branches spiny. Leaves usually simple. Standard broadly ovate, equalling keel; standard and keel usually sericeous. Legume narrowly oblong, sericeous. Seeds 1 to many.

(17–19). **G. lobelii group.** Spreading, much-branched, spiny shrubs. Leaves simple. Most flowers borne singly in the axil of each bract.

The species of this group show an overall similarity but the differential characters appear to be constant even where the taxa are sympatric, e.g. **17** and **18** in the Sierra de Segura.

1 Some pulvini with spinose stipules   **19. baetica**
1 No pulvini with spinose stipules
  2 Most pedicels 4–9 mm, slender; flowers usually only 1 or 2 on each branch   **17. lobelii**
  2 Most pedicels 3–4 mm, stout; flowers usually in short racemes   **18. pumila**

17. **G. lobelii** DC. in Lam. & DC., *Fl. Fr.* ed. 3, **4**: 499 (1805). Branches sometimes flexuous. Leaves 2–5 × 0·5–2 mm, elliptical to obovate, sericeous beneath, subglabrous above, caducous. Flowers borne singly in the axil of each bract, usually only one or two on each branch. Bracteoles less than 1 mm, borne at the middle of the pedicel. Pedicels 4–9(–11) mm, slender. Calyx 4–6 mm, with short appressed hairs. Standard 8–12 mm, ovate, densely sericeous. *Calcicole.* ● *Mountains of S.E. France, S. & S.E. Spain.* Ga Hs.

Sometimes divided into two subspecies: subsp. **lobelii**, from S.E. France, with the pulvini not bidentate and the legume 13–15 mm, lanceolate-oblong, 3- to 4-seeded; and subsp. **longipes** (Pau) Heywood, *Collect. Bot. (Barcelona)* **5**: 519 (1957), from S. & S.E. Spain, with the pulvini bidentate and the legume

9–10 mm, ovate-oblong, 1- to 2-seeded. The pulvinus is not a reliable character and recent collections from at least one locality in Spain have the longer, 3- to 4-seeded legume of subsp. *lobelii*. Occasional plants from France also have a short 1- to 2-seeded legume.

18. **G. pumila** (Debeaux & Reverchon ex Hervier) Vierh., *Verh. Zool.-Bot. Ges. Wien* **69**: 181 (1919). Like **17** but branches stout, rigid; pedicels (1–)3–4(–5) mm, stout; flowers usually in short racemes. *Calcicole.* ● *Mountains of S., S.E. & E.C. Spain.* Hs.

19. **G. baetica** Spach, *Ann. Sci. Nat.* ser. 3 (Bot.), **2**: 113 (1845). Like **17** but branches stout; some pulvini with spinose stipules; pedicels 2–4 mm; standard 11–13 mm. *Calcicole.* ● *S. Spain (Sierra Nevada).* Hs.

20. **G. salzmannii** DC., *Prodr.* **2**: 147 (1825). Much-branched, spiny shrub. Leaves 3–8 × 1–3 mm, simple, the lower surface sericeous, the upper glabrous or with sparse sericeous hairs; pulvini without spinose stipules. Bracteoles 1 mm or less, borne at the middle of the pedicels. Pedicels 1–4 mm. Flowers usually in pairs in the axils of each bract. Standard *c.* 10 mm, broadly ovate, sparsely to densely hairy. ● *N. Italy, Elba, Sardegna, Corse.* Co It Sa.

G. parnassica Halácsy, *Magyar Bot. Lapok* **11**: 136 (1912), from S.C. Greece (Parnassos), and possibly also Samothraki, is perhaps conspecific with **20**. It has the leaves silvery-sericeous on both surfaces and the standard 9–13 mm, densely sericeous.

21. **G. hystrix** Lange, *Descr. Icon. Ill.* 2 (1864). Spiny shrub. Leaves 3–5 × 1–3 mm, simple, appressed-hairy beneath, glabrescent above. Flowers mostly 2 or more in the axil of each bract, in lax racemes 4 cm or more. Bracteoles 1 mm or less, borne at the middle of the pedicel. Pedicels 2–5 mm. Calyx 2·5–7 mm. Standard *c.* 10 mm, glabrous or with a median ridge of sericeous hairs or sparsely sericeous. ● *N. Spain, N. Portugal.* Hs Lu.

(a) Subsp. **hystrix**: Erect, up to 150 cm; pulvini without spinose stipules. Lips of calyx as long as or shorter than tube. *N.W. Spain, N. Portugal.*
(b) Subsp. **legionensis** (Pau) P. Gibbs, *Notes Roy. Bot. Gard. Edinb.* **27**: 57 (1966): Spreading, up to 30 cm; pulvini with spinose stipules. Lips of calyx longer than tube. *N. Spain (Picos de Europa).*

Flowering from late May to July.

22. **G. polyanthos** R. de Roemer ex Willk., *Linnaea* **25**: 20 (1852). Like **21**(a) but up to 200 cm; pulvini with spinose stipules; standard always uniformly, densely sericeous. ● *S.W. Spain, S. & E.C. Portugal.* Hs Lu.

Flowering from March to early May.

The distinction between **21** and **22** is not always clear and it might perhaps be better to regard **21** as a subspecies of **22**.

23. **G. aspalathoides** Lam., *Encycl. Méth. Bot.* **2**: 620 (1788). Erect, spiny shrub. Leaves mostly 3-foliolate; leaflets 3–12 × 1–3 mm, narrowly oblanceolate, with involute margins and short grey hairs on both surfaces. Flowers 1 or more in the axil of each bract, in lax racemes; bracteoles *c.* 2 mm, usually 2–3, the final pair just below the calyx; pedicels 2–4 mm. Calyx 5–6 mm, sericeous, the lips longer than the tube. Standard 10–12 mm, sparsely to densely sericeous. *Sicilia.* Si.

Sect. SCORPIOIDES Spach. Shrubs with axillary spines and alternate branching. Leaves simple or 3-foliolate. Corolla usually glabrous; standard equal to the wings and keel. Legume oblong. Seeds 2–8.

24. G. scorpius (L.) DC. in Lam. & DC., *Fl. Fr.* ed. 3, **4**: 498 (1805). Erect, rarely spreading, intricately branched shrub, with stout axillary spines. Leaves 3–11 × 1·5–2 mm, simple, sparsely hairy beneath, subglabrous above. Flowers borne on short branches arising from the spines, or directly on the spines. Pedicels 2–5 mm. Calyx 3–5 mm, glabrous or nearly so, the lips shorter than the tube. Standard 7–12 mm. Legume 15–40 mm, glabrous. 2*n* = 40, 48. *Spain and S. France.* Ga Hs.

G. melia Boiss., *Diagn. Pl. Or. Nov.* **2**(9): 2 (1849), described from the Aegean region (Milos), has been collected once only. It is said to differ from **24** in the appressed (not crispate) hairs; the flowers in many-flowered clusters; and the lower lip of the calyx with a long median tooth.

25. G. corsica (Loisel.) DC. in Lam. & DC., *Fl. Fr.* ed. 3, **5**: 548 (1815). Like **24** but the flowers or flowering branches never borne directly on the spines; lips of the calyx as long as the tube; legume 12–20 mm. ● *Corse and Sardegna.* Co Sa.

26. G. carpetana Leresche ex Lange, *Vid. Meddel. Dansk Naturh. Foren. Kjøbenhavn* **1877–78**: 237 (1878). Spreading shrub, the branches somewhat winged and with axillary spines. Leaves 4–6 × 2 mm, simple, sericeous; stipules weakly spinose. Flowers borne singly or in clusters on the branches or spines; pedicels *c.* 1 mm. Calyx *c.* 4 mm, sparsely sericeous or with subpatent hairs. Standard 8–11 mm. Legume 12–15 mm, sericeous. *N. & W.C. Spain.* Hs.

27. G. morisii Colla, *Herb. Pedem.* **2**: 65 (1834). Spreading shrub with slender, weak, axillary spines. Leaves 3-foliolate; leaflets 3–9 × 2 mm, both surfaces with sparse, long, subpatent hairs. Bracteoles linear-lanceolate, equalling or exceeding the calyx-tube; pedicels 1–2 mm. Calyx 5–8 mm, with sparse, subpatent hairs. Standard *c.* 10 mm. Legume *c.* 20 mm, with dense patent hairs. ● *Sardegna.* Sa.

Sect. PHYLLOSPARTIUM Willk. Shrubs with axillary spines and simple leaves. Standard usually glabrous, ovate with an acute apex, usually shorter than the keel. Legume falcate, inflated. Seeds 4–12.

28. G. anglica L., *Sp. Pl.* 710 (1753). Decumbent to erect shrub with glabrous or hairy branches and axillary spines. Leaves 4–10 × 2–3 mm, lanceolate or elliptical, glabrous. Stipules not spinose. Flowers in short racemes. Bracts leaf-like, fasciculate; bracteoles less than 1 mm. Calyx 3–4 mm, glabrous, the lips longer than the tube. Standard 6–8 mm. Legume glabrous. 2*n* = 42. *W. Europe, extending eastwards to S. Sweden, N. Germany and S.W. Italy.* Be Br Da Ga Ge Ho Hs It Lu Su.

29. G. falcata Brot., *Phyt. Lusit.* 52 (1800). Like **28** but young branches and lower surface of leaves sparsely sericeous; leave ovate to oblong-lanceolate, obtuse or acute; bracts and bracteoles minute or absent; calyx *c.* 5 mm, glabrous to sparsely pubescent, the lips as long as the tube; standard *c.* 9 mm. *Calcifuge.* ● *W. half of the Iberian peninsula.* Hs Lu.

30. G. berberidea Lange, *Descr. Icon. Ill.* 1 (1864). Like **28** but young branches with dense patent hairs; stipules somewhat spinose; bracteoles *c.* 1 mm; calyx 5–6 mm, with dense patent hairs, the lips twice as long as the tube; standard 8–10 mm;

legume with sparse patent hairs along the sutures. *Damp meadows and bogs.* ● *N.W. Spain, N. Portugal.* Hs Lu.

Sect. VOGLERA (P. Gaertner, B. Meyer & Scherb.) Spach. Leaves simple or 3-foliolate. Standard usually triangular, or ovate with an acute apex, usually shorter than the keel. Legume ovoid-acuminate. Seeds 1–2.

31. G. micrantha Ortega, *Hort. Matrit. Descr.* 68 (1798). Unarmed, spreading shrub. Leaves *c.* 12 × 2 mm, simple, narrowly elliptical, glabrous. Flowers in terminal racemes. Bracts narrowly elliptical; bracteoles *c.* 2 mm, borne just below the calyx; pedicels *c.* 1 mm. Calyx 2·5–5 mm, glabrous. Standard 5–7 mm, triangular, glabrous; base truncate to obtuse; keel with sparse sericeous hairs, *c.* 1 mm longer than the standard. ● *N. & E. Spain, N. Portugal.* Hs Lu.

32. G. carinalis Griseb., *Spicil. Fl. Rumel.* **1**: 3 (1843). Like **31** but leaves linear-oblong; base of the standard cordate-truncate; keel *c.* 2 mm longer than the standard. *E. part of Balkan peninsula.* Bu Gr Ju Tu.

33. G. sylvestris Scop., *Fl. Carn.* ed. 2, **2**: 53 (1772). Decumbent shrub with weak axillary spines; young branches with sericeous hairs. Leaves 10–20 × 1–3 mm, simple, narrowly oblong or elliptical, the lower surface with sparse hairs. Flowers in long, lax terminal racemes. Calyx 5–7 mm, with sparse sericeous hairs, the upper lip about as long as the lower, the lower teeth equal. Standard 7–8 mm, triangular, the base subcordate, the claw less than 2 mm. ● *Albania and Jugoslavia; C. & S. Italy.* Al It Ju.

34. G. aristata C. Presl in J. & C. Presl, *Del. Prag.* 34 (1822). Like **33** but branches with stouter spines; leaves and calyx with patent hairs; calyx *c.* 5 mm, the upper lip *c.* ½ as long as the lower, the median lower tooth usually exceeding the lateral by 1 mm or more; standard 6–8 mm, triangular, the base truncate, the claw 2 mm or more. ● *Sicilia.* Si.

35. G. hispanica L., *Sp. Pl.* 711 (1753). Decumbent to erect shrub 10–50 cm, with axillary spines. Leaves 6–10 × 3–5 mm, simple, sessile, lanceolate to oblanceolate, the lower surface with dense appressed or patent hairs. Flowers in dense, terminal, subcapitate racemes. Bracts *c.* 1 mm; bracteoles absent. Standard glabrous, broadly ovate, slightly emarginate, about equalling the wings and keel. ● *N. Spain and S. France.* Ga Hs.

(a) Subsp. **hispanica**: Branches and leaves with patent hairs; standard 6–8 mm. *S. France, westwards to E. Pyrenees, and E. Spain.*

(b) Subsp. **occidentalis** Rouy, *Fl. Fr.* **4**: 226 (1897) (*G. occidentalis* (Rouy) Coste): Branches and leaves with appressed hairs; standard 8–11 mm. *W. Pyrenees and N. Spain.*

36. G. germanica L., *Sp. Pl.* 710 (1753). Erect shrub, with axillary spines, rarely unarmed. Leaves 8–20 × 4–5 mm, simple, elliptical or lanceolate, the lower surface with long, subpatent hairs. Flowers in lax racemes. Bracts *c.* 1 mm; bracteoles absent. Calyx *c.* 5 mm, sericeous. Standard *c.* 8 mm, ovate with an acute apex, about ⅔ as long as keel. 2*n* = *c.* 46. ● *From S.W. France to C. Russia and from S. Sweden to C. Italy and Bulgaria.* Au Be Bu Co Cz Da Ga Ge He Ho Hu It Ju Po Rm Rs (B, C, W) Su.

37. G. hirsuta Vahl, *Symb. Bot.* **1**: 51 (1790). Erect shrub with stout axillary usually unbranched spines. Leaves 6–15 × 3–5 mm, simple, lanceolate, the lower surface and margins with sparse, long, subpatent hairs, the upper glabrous. Flowers in congested terminal racemes. Bracts leaf-like, borne just below the bracteoles; bracteoles 3–5 mm, borne just below the calyx. Calyx 9–12 mm, the lower lip longer than the upper. Standard ovate, acute, glabrous to uniformly sericeous. ● *W. Spain, S. Portugal, Islas Baleares*. Bl Hs Lu.

38. G. tournefortii Spach, *Ann. Sci. Nat.* ser. 3 (Bot.), **2**: 269 (1844). Like **37** but spines much-branched; bracts borne at the base of the pedicel; upper lip of the calyx $\frac{1}{2}$ as long as the lower lip; standard emarginate, and always sparsely sericeous. *S. & C. Spain, Portugal*. Hs Lu.

39. G. lucida Camb., *Mém. Mus. Hist. Nat. (Paris)* **14**: 231 (1827). Erect shrub with stout axillary spines. Leaves 3–8 × 3 mm, simple, narrowly elliptical, the lower surface sericeous; stipules spinose. Flowers in terminal racemes. Bracts leaf-like; bracteoles minute, at the middle of the pedicel. Calyx *c.* 5 mm, sericeous. Standard *c.* 10 mm, ovate, subacute, glabrous, shorter than the keel. ● *Islas Baleares* (*Mallorca*). Bl.

40. G. anatolica Boiss., *Diagn. Pl. Or. Nov.* **1**(2): 8 (1843). Spreading or erect shrub, the young branches patent-pubescent. Leaves 5–10 × 1–2·5 mm, simple, narrowly elliptical, the lower surface sericeous or pubescent; without spinose stipules. Flowers in short terminal racemes terminated by a spine. Bracts 2–3 mm, leaf-like; bracteoles *c.* 7 mm, borne at the apex of the pedicel. Calyx *c.* 6 mm, sericeous or pubescent. Standard *c.* 8 mm, broadly ovate, emarginate, glabrous, shorter than keel. *S. Bulgaria, Turkey-in-Europe*. Bu Tu. (*W. Anatolia*.)

41. G. triacanthos Brot., *Phyt. Lusit.* 54 (1800). Erect shrub, with axillary spines. Leaves 3-foliolate; leaflets 3–8 × 1–2 mm, oblanceolate, subglabrous; without spinose stipules. Flowers in lax terminal or sometimes intercalary racemes. Lowest bracts *c.* 2 mm, simple, the uppermost much reduced. Calyx 2·5–4 mm, glabrous. Standard *c.* 6 mm, triangular, glabrous, shorter than the keel. *W. part of Iberian peninsula*. Hs Lu.

42. G. tridens (Cav.) DC., *Prodr.* **2**: 148 (1825) (*G. gibraltarica* DC.). Like **41** but some leaves simple and the stipules spinose. *S.W. Spain*. Hs.

43. G. cupanii Guss., *Cat. Pl. Boccad. 77* (1821). Spreading shrub with axillary spines; young branches with dense patent hairs. Leaves 3-foliolate; leaflets 4–10 × 0·8–1·5 mm, narrowly elliptical, with sparse patent hairs. Flowers in lax terminal racemes, the flowering branches sometimes terminated by a spine. Bracts *c.* 2 mm, linear. Calyx 4–5 mm, with sparse hairs. Standard *c.* 8 mm, triangular, glabrous, shorter than the keel. ● *Sicilia* (*Madonie*). Si.

Sect. ASTEROSPARTUM Spach. Unarmed shrubs with simple or 3-foliolate leaves. Branches and leaves usually opposite or subopposite. Standard broadly or angular-ovate, equalling or shorter than the keel. Legume ovoid-acuminate to falcate. Seeds 1–2.

44. G. radiata (L.) Scop., *Fl. Carn.* ed. 2, **2**: 51 (1772) (*Cytisanthus radiatus* (L.) O.F. Lang). Erect shrub with opposite branches at almost every node. Leaves 3-foliolate, opposite; leaflets 5–20 × 2–4 mm, oblanceolate, sericeous beneath, subglabrous above. Flowers subopposite and subsessile in terminal clusters of 4–12. Lowermost flowers with simple, usually shortly 3-fid, scarious bracts much shorter than the flowers; bracteoles 1–3 mm. Calyx 4–6 mm, the lips about as long as the tube. Standard 8–14 mm, broadly ovate, as long as or slightly shorter than the keel, glabrous or with a median ridge of sericeous hairs. *Calcicole*. ● *S. Alps extending to E. Switzerland, C. Italy and W. Jugoslavia, and*

very locally to S.W. Romania and C. Greece. Al Au Ga Gr He It Ju Rm.

45. G. holopetala (Fleischm. ex Koch) Bald., *Mem. Accad. Ist. Bologna* ser. 5, **9**: 238 (1902). Like **44** but with 2–4 flowers on each branch, the lowermost pair with leaf-like, 3-foliolate bracts; upper flowers (if any) with rudimentary bracts; bracteoles 2–3 mm; standard 8–10 mm, with dense sericeous hairs. ● *N.W. Jugoslavia*. Ju.

46. G. hassertiana (Bald.) Bald. ex Buchegger, *Österr. Bot. Zeitschr.* **62**: 416 (1912). Like **44** but leaflets 4–10 × 1–2 mm, markedly inrolled and sub-linear; flowers in terminal clusters of 2–4; lowermost bracts leaf-like, 3-foliolate; bracteoles 1 mm or less; standard *c.* 8 mm, with dense sericeous hairs. ● *Albania*. Al.

47. G. dorycnifolia Font Quer, *Butll. Inst. Catalana Hist. Nat.* **20**: 46 (1920). Like **44** but some branches and leaves alternate; bracteoles 0·5 mm; calyx 2·5–3 mm, the lips $\frac{1}{3}$ as long as the tube; standard *c.* 8 mm, slightly shorter than the keel. $2n = 48$. ● *Islas Baleares* (*Ibiza*). Bl.

48. G. sessilifolia DC., *Prodr.* **2**: 146 (1825) (*G. trifoliolata* Janka). Erect shrub with few subopposite, flexuous branches, mainly from near the base. Leaves 3-foliolate, sessile or shortly petiolate; leaflets 5–25 × 1–2 mm, linear, often inrolled, sericeous beneath, glabrous above. Flowers alternate or subopposite, in long lax racemes. Bracts leaf-like, 3-foliolate, at least the lowest equalling or exceeding the flowers. Calyx 4–5 mm, with short, appressed, sericeous hairs; lips and upper teeth as long as the tube. Standard 7–10 mm, deltate or triangular, $\frac{1}{2}$–$\frac{2}{3}$ as long as the keel, sericeous. *Calcicole. S. Jugoslavia, S. & E. Bulgaria*. Bu Ju.

49. G. nissana Petrović, *Add. Fl. Agri Nyss.* 51 (1885). Like **48** but branches, both surfaces of the leaves, calyx and standard with dense, long, patent hairs; leaflets of the main cauline leaves 2–6 mm wide, linear-oblanceolate to elliptical, not inrolled. ● *S. Jugoslavia*. Ju.

50. G. ephedroides DC., *Mém. Lég.* t. 36 (1825). Shrub 50–100 cm. Leaves 3-foliolate, caducous; leaflets 4–15 × 2–3 mm, sericeous on both surfaces. Flowers alternate or subopposite in lax racemes. Lowermost bracts 3-foliolate, uppermost simple. Calyx 3–6 mm, sericeous; lips and upper teeth about as long as the tube. Standard 7–10 mm, broadly ovate or rhombic, *c.* $\frac{3}{4}$ as long as the keel, sparsely sericeous. *Sardegna, Sicilia, S. Italy* (*Ponza*). ?Co It Sa Si.

51. G. aetnensis (Biv.) DC., *Prodr.* **2**: 150 (1825). Like **50** but up to 500 cm; leaves simple, fugacious; bracts leaf-like, fugacious; calyx *c.* 3 mm, subglabrous; upper teeth obtuse and *c.* $\frac{1}{4}$ as long as the tube; lower teeth minute; standard subglabrous. ● *Sardegna, Sicilia*. Sa Si.

52. G. spartioides Spach, *Ann. Sci. Nat.* ser. 3 (Bot.), **2**: 243 (1844) (*G. retamoides* Spach ex Cosson). Erect shrub 30–100 cm, with alternate and opposite branches. Leaves 3–8 × 2–3 mm, simple, elliptical, sericeous on the lower surface. Flowers usually clustered in interrupted racemes. Calyx 2·5–4 mm, sericeous, the lips about as long as the tube. Standard 7–9 mm, rhombic, glabrous or sparsely sericeous. *S. Spain*. Hs. (*N.W. Africa*.)

53. G. haenseleri Boiss., *Elenchus* 31 (1838). Like **52** but 100–200 cm; flowers solitary or in clusters; calyx *c.* 6 mm, somewhat inflated; standard 10–12 mm, with a median sericeous band. ● *S. Spain*. Hs.

Sect. ACANTHOSPARTUM Spach. Shrubs with opposite branches, the branches terminated by a spine. Leaves 3-foliolate, opposite or alternate. Standard rhombic, shorter than or exceeding the keel. Legume ovoid-acuminate. Seeds 1–2.

54. G. acanthoclada DC., *Prodr.* **2**: 146 (1825). Erect shrub; older branches with prominent pulvini. Leaflets 5–10 × 1–3 mm, narrowly oblanceolate. Flowers borne singly in the axil of each bract, subopposite, towards the ends of branches. Bracts leaf-like, the uppermost simple. Calyx 2·5–5 mm, sparsely sericeous, the lips almost as long as tube. Standard 6–10 mm, sericeous, shorter than the keel. *Greece and Aegean region.* Cr Gr.

Sect. FASSELOSPARTUM P. Gibbs. Shrubs with opposite and alternate branches and axillary spines. Leaves simple or 3-foliolate. Standard rhombic, shorter than the keel. Legume ovoid, acuminate. Seeds 1–2.

55. G. fasselata Decne, *Ann. Sci. Nat.* ser. 2 (Bot.), **4**: 360 (1835) (*G. sphacelata* Spach). Shrub up to 150 cm; pulvini obscure and scale-like. Leaflets 3–15 × 1–3 mm, narrowly oblanceolate, sericeous. Flowers borne singly or in lax clusters in the axil of each bract, on spines or unarmed branches. Bracts leaf-like; bracteoles minute. Calyx 4–5 mm, glabrous, the lips about ⅓ as long as tube. Standard 6–7 mm, glabrous. *S. Aegean region (Karpathos, Kasos).* Cr. (*E. Mediterranean region.*)

Sect. CEPHALOSPARTUM Spach. Unarmed shrubs with opposite and alternate branches. Leaves simple or 3-foliolate. Standard ovate, equalling or shorter than the keel. Legume oblong or ovoid-acuminate.

56. G. umbellata (L'Hér.) Poiret in Lam., *Encycl. Méth. Bot.,* *Suppl.* **2**: 715 (1812). Caespitose shrub 20–60 cm; branches mostly opposite. Leaves 5–15 × 2–3 mm, simple, narrowly elliptical, sericeous. Flowers in heads of 4–16. Bracts leaf-like. Calyx *c.* 5 mm, with dense patent hairs. Standard 8–12 mm, broadly ovate, densely sericeous or with sub-patent hairs. Legume narrowly oblong. Seeds 2–5. *S. & S.E. Spain.* Hs.

17. Chamaespartium Adanson[1]

(*Genistella* Ortega, *Pterospartum* (Spach) Willk.)

Unarmed dwarf shrubs, the young stems distinctly winged and flattened (cladodes). Leaves simple or absent. Flowers in congested, terminal racemes. Calyx tubular, bilabiate; upper lip deeply 2-fid, lower with 3 distinct teeth; corolla yellow, the standard broadly ovate, equalling the wings and keel. Legume narrowly oblong, dehiscent. Seeds 2–5, with or without a strophiole.

Literature: C. Vicioso, *Genísteas Españolas* **1**: 136–146 (*Publ. Inst. For. Inv. Exper. Madrid* No. 67). Madrid. 1953. (Sub *Genistella.*)

Plant with simple, elliptical leaves; wings of the stem entire at the
nodes **1. sagittale**
Plant without leaves; wings of the stem 3-toothed or 3-lobed at the
nodes **2. tridentatum**

1. C. sagittale (L.) P. Gibbs, *Feddes Repert.* **79**: 54 (1968) (*Genista sagittalis* L., *Genistella sagittalis* (L.) Gams, *Pterospartum sagittale* (L.) Willk.). Plant with procumbent, woody, mat-forming stems, and usually erect, herbaceous, simple or little-branched flowering stems 10–50 cm; wings constricted at the nodes, but entire, without teeth or lobes. Leaves 5–20 × 4–7 mm, elliptical, glabrous or subglabrous above, pubescent beneath.

Calyx 5–8 mm, sericeous; corolla 10–12 mm, the standard usually glabrous. Legume 14–20 × 4–5 mm, pubescent. Seeds estrophiolate. ● *C. Europe, extending to S.E. Belgium and southwards locally in the mountains to S. Spain, Calabria and Greece.* Al Au Be Bu Cz Ga Ge Gr He Hs Hu It Ju Rm Rs (W) [Po].

Plants from S. France, with more or less procumbent, sericeous flowering stems and smaller flowers with a sericeous standard, are sometimes treated as a separate subspecies (*Genista delphinensis* Verlot).

2. C. tridentatum (L.) P. Gibbs, *Feddes Repert.* **79**: 54 (1968) (*Genistella tridentata* (L.) Samp., *Pterospartum tridentatum* (L.) Willk.; incl. *P. cantabricum* (Spach) Willk., *P. lasianthum* (Spach) Willk., *P. stenopterum* (Spach) Willk.). Erect or procumbent, much-branched shrub 30–70 cm; wings undulate, rather coriaceous, contracted at each node to form 3 teeth or lobes. Leaves absent. Calyx 4–7 mm, sericeous; corolla 8–12 mm, the standard glabrous to densely sericeous. Legume 10–12 × *c.* 4 mm, pubescent. Seeds strophiolate. *Heaths and scrub on acid soils.* ● *W. part of the Iberian peninsula.* Hs Lu.

18. Echinospartum (Spach) Rothm.[1]

Small shrubs with opposite spiny branches. Leaves 3-foliolate, shortly petiolate or sessile. Calyx inflated, campanulate, bilabiate; upper lip deeply 2-fid; lower with 3 prominent teeth; all teeth as long as or longer than the tube; corolla yellow; stamens monadelphous. Legume ovoid-acuminate, dehiscent, villous. Seeds 1–3.

Literature: W. Rothmaler, *Bot. Jahrb.* **72**: 79–84 (1941).

1 Flowers in terminal clusters of 3–9 **3. lusitanicum**
1 Flowers 2 on each branch, sometimes with a further 2 lower
down
2 Flowering branch without a terminal spine; standard glabrescent or sparsely sericeous **1. horridum**
2 Flowering branch terminated by a small spine; standard
densely sericeous **2. boissieri**

1. E. horridum (Vahl) Rothm., *Bot. Jahrb.* **72**: 80 (1941) (*Genista horrida* (Vahl) DC.). Up to 40 cm, spiny. Leaves 4–9 mm, narrowly oblanceolate, sericeous beneath, glabrescent above; pulvini prominent. Flowers usually 2 on each branch, opposite. Calyx 7–12 mm, sparsely sericeous; teeth acute. Standard 12–16 mm, glabrescent to sparsely sericeous. Legume 9–14 × 4–4·5 mm. 2*n* = 44. *Exposed mountain rocks and slopes; calcicole.* ● *Pyrenees and S.C. France.* Ga Hs.

2. E. boissieri (Spach) Rothm., *op. cit.* 83 (1941) (*Genista boissieri* Spach). Like **1** but flowering branches terminated by a small spine; flowers sometimes 4 on each branch with the lower 2 remote; calyx with sparse to dense hairs and teeth with a filiform apex; standard densely sericeous. 2*n* = 44. *Exposed mountain slopes; calcicole.* ● *S.E. Spain.* Hs.

3. E. lusitanicum (L.) Rothm., *op. cit.* 82 (1941) (*Genista lusitanica* L.). Like **1** but up to 200 cm; flowers in terminal clusters of 3–9; calyx 10–18 mm; standard 12–20 mm; legume 15–20 × 4–7 mm. *Exposed mountain rocks and slopes; calcifuge.* ● *W. part of Iberian peninsula.* Hs Lu.

(a) Subsp. **lusitanicum**: Calyx and standard densely sericeous or with patent hairs. 2*n* = *c.* 52. *W.C. Spain, C. Portugal.*

(b) Subsp. **barnadesii** (Graells) C. Vicioso, *Bol. Inst. Estud. Astur.* (*Supl. Ci.*) ser. C, **5**: 41 (1962) (*Genista barnadesii* Graells): Calyx sparsely to densely sericeous; standard glabrous or sparsely sericeous. 2*n* = *c.* 52. *W. Spain.*

¹ By P. E. Gibbs.

19. Gonocytisus Spach[1]

Unarmed shrubs. Leaves 3-foliolate. Flowers in terminal, leafless racemes. Calyx campanulate, bilabiate, membranous, the upper lip divided to the base, with asymmetrical teeth; corolla yellow, not persistent in fruit; stamens monadelphous; stigma extrorse, subcapitate. Legume compressed, dehiscent, unwinged. Seeds estrophiolate.

1. **G. angulatus** (L.) Spach, *Ann. Sci. Nat.* ser. 3 (Bot.), **3**: 153 (1845) (*Genista angulata* (L.) Lam.). Erect, up to 5 m. Branches terete; twigs distinctly ridged, becoming terete, sparsely appressed-puberulent. Leaflets 4–22 × 1–6 mm, narrowly elliptical to ovate-oblong, acute or obtuse, apiculate, appressed-hairy on both surfaces, the upper glabrescent; petioles usually absent or rudimentary, up to 1 mm in larger leaves. Pedicels 1–2 mm, with 2 caducous bracteoles. Standard 4–5·5 mm, oblong; keel longer than standard. Legume 10–15 mm, ovoid to rhomboid, mucronulate, appressed-pubescent, the margins slightly thickened. *Cliffs and dry hillsides. Turkey-in-Europe (S. half of Gelibolu peninsula).* Tu. (*W. & S. Anatolia.*)

20. Lygos Adanson[1]

(*Retama* Boiss.)

Unarmed shrubs. Leaves simple, soon deciduous. Flowers in racemes. Calyx urceolate, campanulate or turbinate, bilabiate; corolla white to yellow; stamens monadelphous; style filiform, incurved. Legume ovoid to globose, indehiscent or finally incompletely dehiscent along ventral suture. Seeds 1(–2).

Literature: M. Zohary, *Bull. Res. Counc. Israel* Sect. D, Bot. **7D**: 1–12 (1959) (sub *Retama*).

1 Corolla 5–8 mm, yellow   **1. sphaerocarpa**
1 Corolla 10–17 mm, white, sometimes becoming cream-coloured on drying
 2 Keel cuspidate; legume obovoid, with a short mucro
         2. monosperma
 2 Keel obtuse or acute, rarely acuminate; legume obovoid-ellipsoid, attenuate into a beak   **3. raetam**

1. **L. sphaerocarpa** (L.) Heywood, *Feddes Repert.* **79**: 53 (1968) (*Retama sphaerocarpa* (L.) Boiss.). Up to 2 m, much-branched. Branches erect or ascending, glabrous. Leaves linear to linear-lanceolate, sericeous-pubescent, deciduous. Racemes dense. Calyx 3 mm, glabrous or pubescent, persistent in fruit; upper lip 2-lobed, the lower with 3 long teeth. Corolla 5–8 mm, yellow; standard suborbicular, glabrous or sparsely hairy; wings lanceolate, shorter than the obtuse keel. Legume 7–9 mm, ovoid to globose, muticous or shortly apiculate, smooth. *Dry places, mainly on sandy soils. E. Portugal, C. & S. Spain.* Hs Lu.

2. **L. monosperma** (L.) Heywood, *Feddes Repert.* **79**: 53 (1968) (*Retama monosperma* (L.) Boiss.). Stems up to 3 m, erect, divaricately-branched. Branches pendent, sericeous when young. Leaves linear-lanceolate, the later ones linear-subspathulate, all sericeous-pubescent, deciduous. Racemes lax. Calyx 3·5 mm, urceolate or campanulate, glabrous, circumscissilely caducous after anthesis; the upper lip with 2 triangular teeth, the lower with 3 linear-subulate teeth, the teeth often ciliate. Corolla 10–12 mm, white; standard rhombic-ovate, hairy; wings oblong, obtuse, as long as or shorter than the cuspidate-acuminate keel. Legume 12–16 mm, obovoid, with a short mucro directed towards the

ventral suture, rugose when mature. *Maritime sands. S.W. Spain, S. Portugal.* Hs Lu ?Sa.

3. **L. raetam** (Forskål) Heywood, *Feddes Repert.* **79**: 53 (1968) (*Retama raetam* (Forskål) Webb & Berth.). Stems up to 2 m, erect, much-branched. Branches deflexed, pubescent. Leaves linear, sericeous, deciduous. Racemes dense. Calyx urceolate-campanulate, circumscissilely caducous after anthesis; the upper lip with 2 broadly triangular teeth, the lower with 3 short lanceolate teeth. Corolla 15–17 mm, white, becoming cream-coloured on drying; standard ovate-oblong; wings oblong, longer than the obtuse keel. Legume 10–20 mm, obovate-ellipsoid, attenuate into a beak. *Maritime sands. S. Sicilia.* Si. (*N. Africa, S.W. Asia.*)

The above description refers to subsp. **gussonei** (Webb) Heywood, *Feddes Repert.* **79**: 53 (1968) (*Retama gussonei* Webb), the European representative of this species. It is distinguished from the other subspecies by its larger corolla and its short keel.

21. Spartium L.[1]

Unarmed shrubs. Leaves 1-foliolate. Flowers in lax, terminal, leafless, many-flowered racemes. Calyx spathe-like, split above, irregularly unilabiate (rarely bilabiate), with 5 short teeth; corolla yellow; stamens monadelphous. Legume linear-oblong, dehiscent, subseptate between the seeds. Seeds numerous, estrophiolate.

1. **S. junceum** L., *Sp. Pl.* 708 (1753). Up to 3 m (or more in cultivation). Branches cylindrical, striate, medullated, flexible, glaucous-green, glabrous. Leaves 10–30 × 2–5 mm, sparse, oblong-linear to lanceolate, glabrous above, appressed-sericeous beneath, subsessile, caducous. Flowers showy, sweet-scented, borne singly; pedicels with a small caducous bract at the base and two bracteoles at the apex. Corolla 20–25 mm. Legume flat, sericeous, becoming glabrous. Seeds 10–18. *Mediterranean region and S.W. Europe.* Al Az Bl Co Cr Ga Gr Hs It Ju Lu Sa Si Tu [Bu Rm Rs (K)].

22. Petteria C. Presl[1]

Unarmed shrubs. Leaves 3-foliolate. Flowers in terminal, erect, leafless racemes. Calyx campanulate-tubular, bilabiate; upper lip divided to about two thirds, lower 3-toothed; corolla yellow; stamens monadelphous. Legume linear-oblong, dehiscent, straight or slightly curved, somewhat inflated. Seeds 5–9, estrophiolate.

1. **P. ramentacea** (Sieber) C. Presl, *Abh. Böhm. Ges. Wiss.* ser. 5, **3**: 570 (1845) Erect, up to 2 m. Branches 2–3 mm wide, terete or obscurely angled, glabrous; twigs loosely appressed-hairy when young, later subglabrous. Leaves petiolate; petioles 15–50 mm; leaflets 20–70 × 12–30 mm, elliptical to obovate, rounded, occasionally slightly emarginate, dull green on both surfaces, glabrous above, with appressed hairs along the mid-vein and margins beneath. Inflorescence 4–7 cm, 10- to 20-flowered. Calyx appressed-hairy. Standard 16–20 × 14–15 mm, pentagonal, emarginate. Legume 35–50 × 8–10 mm, stipitate or sessile, light brown, glabrous, the margins slightly thickened, the apex mucronate or mucronulate. Seeds orange-brown. *Mountain scrub.* ● *Jugoslavia and N. Albania.* Al Ju.

23. Erinacea Adanson[2]

Spiny shrubs with opposite or alternate branches. Leaves simple, sometimes 3-foliolate, shortly petiolate. Flowers 1–3, in axillary or subterminal clusters. Calyx inflated-campanulate, bilabiate;

---

[1] By V. H. Heywood.   [2] By P. E. Gibbs.

upper lip with 2 teeth; lower with 3 teeth, the teeth ⅓ as long as the tube; corolla blue-violet; stamens monadelphous. Legume narrowly oblong, dehiscent. Seeds (1–)4–6, estrophiolate.

1. **E. anthyllis** Link, *Handb.* **2**: 156 (1831) (*E. pungens* Boiss.). Hummock-forming, 10–30 cm; branches with stout spines. Leaves *c.* 5 mm, narrowly oblanceolate. Corolla 16–18 mm. Legume 12–20 mm, glandular-villous. *Stony mountain slopes; usually calcicole. Spain, mainly in the south and east, just extending to France in E. Pyrenees.* Ga Hs.

24. Ulex L.[1]

Very spiny shrubs. Leaves usually alternate, exstipulate, 3-foliolate on seedlings, but on mature plants reduced to scale-like or narrow, usually spine-like phyllodes. Flowers axillary, solitary or in small clusters, sometimes aggregated into racemes or umbel-like inflorescences; bracteoles 2, immediately below the flower. Calyx more or less yellow, persistent, divided to the base into 2 lips; upper lip with 2 teeth, the lower with 3 teeth; corolla yellow, persistent; stamens monadelphous. Legume broadly ovoid to linear-oblong, hairy, dehiscent, scarcely exserted from calyx. Seeds 1–6, strophiolate.

Most of the branches may be classified as long or short shoots, though a few are intermediate; the short shoots have usually 2–8 nodes. *Primary phyllodes* are those borne on long shoots; *secondary phyllodes* are those borne on short shoots. The part of a short shoot beyond the last node is here called the *terminal spine*; the lateral branches of the shoot, which may or may not bear nodes, are the *lateral spines*.

In spite of recent monographic work, the taxonomy of this genus is very imperfectly understood. It seems possible that hydridization is more frequent than has generally been supposed.

Literature: G. Sampaio, *Brotéria* (*Bot.*) **21**: 142–168 (1924). W. Rothmaler, *Bot. Jahrb.* **72**: 69–116 (1941). C. Vicioso, *Revisión del Género 'Ulex' en España.* Madrid. 1962.

1　Bracteoles at least 2 mm wide; calyx with ±patent hairs
　　　　　　　　　　　　　　　　　　　　1. europaeus
1　Bracteoles not more than 1·5 mm wide; calyx usually with
　　appressed hairs or glabrescent (rarely villous)
　2　Most of the secondary phyllodes less than half as long as the
　　　lateral spines which they subtend; primary phyllodes not
　　　more than 5 mm, narrow, rigid, often glabrous
　　3　Short shoots and spines with persistent, appressed hairs
　　　　　　　　　　　　　　　　　　　　7. argenteus
　　3　Short shoots and spines glabrous, or with crispate or patent
　　　hairs
　　　4　All petals exceeding the calyx; calyx 5–6·5 mm　**6. micranthus**
　　　4　Standard and keel about equalling the calyx; wings shorter;
　　　　calyx at least 6·5 mm　　　　　　　**5. parviflorus**
　2　Most of the secondary phyllodes more than half as long as the
　　　lateral spines which they subtend; primary phyllodes at least
　　　5 mm, ±leaf-like, usually villous
　　5　Plant pale green; short shoots erecto-patent; phyllodes soft
　　　and flexible (W.C. Portugal)　　　　　　**4. densus**
　　5　Plant dark green; short shoots diverging at right angles when
　　　mature; phyllodes ±rigid
　　　6　Calyx usually less than 10 mm; standard exceeding calyx by
　　　　less than 2 mm; petals clear yellow　　　**2. minor**
　　　6　Calyx usually at least 10 mm; standard exceeding calyx by
　　　　2·5–3·5 mm; petals deep golden-yellow　　**3. gallii**

1. **U. europaeus** L., *Sp. Pl.* 741 (1753). 60–200 cm, with main stems erect or ascending, densely branched in younger parts but eventually bare at the base; young twigs and spines somewhat glaucous. Twigs hirsute to tomentose, with grey to reddish-brown hairs. Primary phyllodes *c.* 8 mm. Terminal spines 12–25(–30) mm, stout, straight, glabrous. Bracteoles 2–7 mm wide. Calyx 12–16(–20) mm, with more or less patent hairs. Petals 15–20 mm, clear yellow; wings straight, longer than keel. Legume 11–20 mm, densely villous. Flowering season variable, but chiefly in spring or late winter. *Usually on well-drained, neutral or moderately acid soils. W. Europe, extending eastwards to Italy; cultivated elsewhere for fodder, bedding and hedges, and widely naturalized.* Br Co Ga Ge Hb He Ho Hs It Lu [Au Az Be Cz Da No Su].

(a) Subsp. **europaeus**: Bracteoles 2–4 mm wide, ovate, subacute. 2n=96. *Throughout the range of the species.*
(b) Subsp. **latebracteatus** (Mariz) Rothm., *Bot. Jahrb.* **72**: 115 (1941): Bracteoles 4–7 mm wide, suborbicular, obtuse. 2n=64. *N.W. Spain and N. & C. Portugal, mainly near the coast.*

2. **U. minor** Roth, *Catalecta Bot.* **1**: 83 (1797) (*U. nanus* T. F. Forster ex Symons). 10–100(–150) cm, with main stems often procumbent; young twigs and spines not glaucous. Twigs hirsute with reddish-brown hairs. Primary phyllodes *c.* 5 mm, villous, scarcely pungent; secondary phyllodes more than half as long as lateral spines. Spines usually slender and rather weak, straight or slightly curved, villous at base, the terminal mostly 8–15 mm. Bracteoles *c.* 0·5 mm wide. Calyx 6–9·5(–11) mm, appressed-pubescent. Petals clear yellow; wings and keel equal, about equalling calyx; standard 1–2 mm longer than calyx. Legume *c.* 8 mm, villous. Seeds 2–6. Flowers in late summer and autumn. 2n=32. ● *W. Europe, northwards to England.* Br Ga Hs Lu [Az].

3. **U. gallii** Planchon, *Ann. Sci. Nat.* ser. 3 (Bot.), **11**: 213 (1849). Like **2** but usually taller and more robust and often with stouter and slightly longer spines; bracteoles *c.* 0·75 mm wide; calyx (9–)10–13(–14) mm; corolla deep golden-yellow; wings and keel usually slightly exceeding calyx, and standard exceeding it by 2·5–3·5 mm. 2n=80. *Calcifuge.* ● *W. Europe, from Scotland to N.W. Spain.* Br Ga Hb Hs.

4. **U. densus** Welw. ex Webb, *Ann. Sci. Nat.* ser. 3 (Bot.), **17**: 291 (1852). 20–50 cm, with numerous erect stems, forming a compact, dense bush. Young twigs, phyllodes and spines hirsute to villous with white or pale brown hairs. Primary phyllodes 5–8 mm, narrowly triangular, leaf-like except for a short spine at the apex; secondary phyllodes more than half as long as lateral spines. Spines slender and rather weak, straight, the terminal 8–20 mm. Flowers mostly in terminal, subumbellate racemes; bracteoles ovate-elliptical, *c.* 0·5 mm wide. Calyx 11–16 mm, appressed-pubescent or subglabrous; standard equalling or slightly exceeding calyx; wings *c.* 2 mm shorter than calyx and slightly shorter than keel. Seeds 1–2. 2n=64. *Dry calcareous soils.* ● *W.C. Portugal.* Lu.

5. **U. parviflorus** Pourret, *Mém. Acad. Toulouse* **3**: 334 (1788). Up to 150 cm but often less; habit very variable. Long shoots villous, crispate-pubescent or glabrescent; short shoots and spines crispate-pubescent to glabrous. Primary phyllodes 2–5(–6) mm, deltate-acuminate to narrowly triangular, pubescent or glabrous; secondary phyllodes mostly less than half as long as lateral spines. Spines straight or recurved, the terminal 4–30 mm. Bracteoles up to 1·5 mm, narrowly ovate to suborbicular. Calyx 6·5–12(–14) mm, appressed-pubescent when young, often more or less glabrescent, rarely with long, patent hairs. Standard and keel about equalling calyx; wings shorter. Seeds 1–3. 2n=32, 64, 96. *S.W. Europe, from Portugal to S. France.* Bl Ga Hs Lu.

---

[1] By E. Guinea and D. A. Webb.

Very variable, especially in S. Spain and S. Portugal; more than 15 variants have been given binomials. The most conspicuous differences between individuals (habit, length and shape of spines, arrangement of flowers) are often found within a single population; other characters (size of flowers, indumentum of various parts) show a confusing amount of reticulation. The treatment below must be regarded as only a provisional attempt to chart the pattern of variation.

1 Calyx with long, patent hairs, at least in bud   (c) subsp. **funkii**
1 Calyx appressed-pubescent
   2 Short shoots and spines densely crispate-pubescent
                                     (d) subsp. **eriocladus**
   2 Short shoots and spines ±glabrous
     3 Calyx usually less than 10 mm; long shoots glabrous or crispate-pubescent but not villous   (a) subsp. **parviflorus**
     3 Calyx at least 10 mm; long shoots often villous throughout their first year   (b) subsp. **jussiaei**

(a) Subsp. **parviflorus**: Short shoots and spines more or less glabrous; long shoots crispate-pubescent to subglabrous. Calyx 6·5–9·5(–11·5) mm, appressed-pubescent when young. *Almost throughout the range of the species.*

(b) Subsp. **jussiaei** (Webb) D. A. Webb, *Feddes Repert.* **74**: 5 (1967) (*U. jussiaei* Webb): Like subsp. (a) but long shoots often villous as well as crispate-pubescent; calyx 10–12(–14) mm. ● *Portugal and W. Spain; local.*

(c) Subsp. **funkii** (Webb) Guinea, *Feddes Repert.* **74**: 5 (1967) (*U. willkommii* var. *funkii* Webb): Like subsp. (a) but calyx 9–11 mm, with long, patent, white hairs, at least when young. *S. Spain. (N.W. Africa.)*

(d) Subsp. **eriocladus** (C. Vicioso) D. A. Webb, *Feddes Repert.* **74**: 5 (1967) (*U. eriocladus* C. Vicioso, *U. ianthocladus* Webb, excl. var. *calycotomoides* Webb): Long shoots, short shoots and spines grey-pubescent with crispate hairs. Calyx 6·5–12·5 mm, appressed-pubescent with blackish hairs when young. ● *S.E. Portugal, S.W. Spain.*

6. U. micranthus Lange, *Vid. Meddel. Dansk Naturh. Foren. Kjøbenhavn* 1877–78:235 (1878). 20–50(–80) cm; main stems erecto-patent or arcuate, rather sparingly branched, dark green, forming a rather open bush. Young twigs puberulent with very short, tightly crispate hairs; phyllodes and spines glabrous. Primary phyllodes 2–3·5 mm, with broad, deltate base and long, spinose apex; secondary phyllodes less than half as long as the lateral spines. Spines recurved, rather stout, the terminal 7–12 mm. Flowers distributed along the branches; bracteoles c. 0·75 mm wide, broadly ovate. Calyx 5–6·5 mm, appressed-pubescent. Petals 6·5–9 mm, all exceeding the calyx. Seeds 1–2. 2n=32. *Heaths on acid soils.* ● *N.W. Portugal, just extending to N.W. Spain.* Hs Lu.

7. U. argenteus Welw. ex Webb, *Ann. Sci. Nat.* ser. 3 (Bot.), **17**: 291 (1852). (10–)25–120 cm; habit variable. Young twigs and spines covered with short, persistent, whitish, appressed hairs. Primary phyllodes 2–4·5 mm, linear-oblong; secondary phyllodes less than half as long as lateral spines. Bracteoles 0·6–2 mm, ovate. Calyx 7–14 mm, appressed-pubescent or sericeous. Standard about equalling calyx; wings 1·5–2·5 mm shorter. Seeds usually 2. 2n=64, ?96. ● *S. Portugal; one locality in S.E. Spain.* Hs Lu.

Three taxa, within the limits of the species as here defined, are usually recognized, and are treated as subspecies below. There are, however, many plants, especially in the neighbourhood of

Faro (S. Portugal), which do not fit well into any of the subspecies and require further investigation.

1 Calyx 8·5–14 mm; bracteoles 1·2–2 mm; plant silvery-glaucous
                                   (c) subsp. **erinaceus**
1 Calyx 6–12 mm; bracteoles 0·5–1 mm
   2 Calyx 6–9 mm; plant somewhat glaucous   (a) subsp. **argenteus**
   2 Calyx 10–12 mm; plant not glaucous   (b) subsp. **subsericeus**

(a) Subsp. **argenteus**: Low-growing; more or less glaucous. Terminal spines 5–9 mm. Bracteoles 0·5–1 mm, narrower than pedicel. Calyx (6–)7–8(–9) mm, with short and often sparse hairs. *Quercus-woods and scrub. S. Portugal (Algarve and S.W. Alentejo).*

(b) Subsp. **subsericeus** (Coutinho) Rothm., *Bot. Jahrb.* **72**: 96 (1941): Like subsp. (a) but taller and not glaucous; calyx 10–12 mm. *Sandy soil near the sea. S. Portugal (around Faro).*

(c) Subsp. **erinaceus** (Welw. ex Webb) D. A. Webb, *Feddes Repert.* **74**: 6 (1967) (*U. erinaceus* Welw. ex Webb): Plant silvery-glaucous. Terminal spines 10–20(–35) mm. Bracteoles 1·2–2 mm, as wide as the pedicel. Calyx (8·5–)9·5–12(–14) mm, sometimes with longer and more abundant hairs than in the other subspecies. *Rocky headlands. S.E. Spain (Cabo de Gata), S.W. Portugal (Cabo de São Vicente).*

25. Stauracanthus Link[1]

(Incl. *Nepa* Webb)

Like *Ulex* but with leaves often opposite or subopposite; calyx tubular at the base, at least in bud, and with the upper lip sometimes deeply 2-fid; legume conspicuously exserted from calyx.

Literature: As for *Ulex*.

Phyllodes spine-tipped; standard much longer than calyx; legume 8–12 mm   **1. boivinii**
Phyllodes scale-like, not spine-tipped; standard about equalling calyx; legume 15–25 mm   **2. genistoides**

1. S. boivinii (Webb) Samp., *Lista Esp. Herb. Port., Ap.* **3**: 8 (1914) (*Ulex boivinii* Webb, *Nepa boivinii* (Webb) Webb). 20–50 cm, densely and intricately branched. Twigs and spines glabrous or appressed-pubescent with brownish hairs. Short shoots somewhat zig-zag, freely branched. Primary phyllodes 2–3 mm, lanceolate-acuminate, spine-tipped. Spines straight, rather slender, the terminal 3–4 mm, the lateral often branched. Flowers borne mainly on short shoots; bracteoles c. 1 mm, ovate-lanceolate. Calyx 5–7(–9) mm, densely sericeous with brown hairs, tubular in basal half up to anthesis but later cleft nearly to the base; upper lip shortly 2-toothed. Standard and keel densely sericeous, much longer than calyx. Legume 8–12 mm, ovate-oblong. Seeds 2. 2n=c. 128. *Sandy or gravelly places near the coast. S.W. Spain, S. Portugal.* Hs Lu.

2. S. genistoides (Brot.) Samp., *Ann. Sci. Acad. Polyt. Porto* **7**: 53 (1912) (*Ulex genistoides* Brot.). 30–100 cm, laxly and diffusely branched. Branches and spines opposite or subopposite, sericeous with silvery hairs. Primary phyllodes 1–1·5 mm, scale-like, not spine-tipped. Spines straight, the terminal 5–17 mm, the lateral often branched. Flowers in relatively spine-free racemes or small panicles, on long or short shoots. Calyx 9–16 mm, densely sericeous with golden-brown hairs, shortly tubular at the base; upper lip deeply 2-fid. Standard and keel about as long as calyx. Legume 15–25 mm, linear-oblong. Seeds 3–6. 2n=48. *Sandy soils. S. & W. Portugal, S.W. Spain.* Hs Lu.

(a) Subsp. **genistoides**: Spines slender, often flexuous. Bracteoles 1–2 mm wide, linear-lanceolate to broadly ovate. Calyx

9–13 mm. Standard subglabrous to moderately hairy.
● *Throughout the range of the species in Europe.*

(b) Subsp. **spectabilis** (Webb) Rothm., *Bot. Jahrb.* **72**: 88 (1941). Spines rather stout, rigid. Bracteoles 2–4·5 mm wide, more or less orbicular. Calyx 10–16 mm. Standard very hairy. *S.W. Portugal (S.W. part of Baixo Alentejo). (W. Morocco.)*

26. Adenocarpus DC.[1]

Unarmed shrubs with alternate branches. Leaves 3-foliolate, sometimes fasciculate. Flowers in terminal racemes or clusters. Calyx tubular, bilabiate, sometimes with glandular tubercules; upper and lower lips prominent, the upper deeply bifid, the lower with three distinct teeth; corolla orange-yellow. Legume oblong, dehiscent, with glandular tubercles. Seeds numerous, estrophiolate.

Literature: C. Vicioso, *Genisteas Españolas* 2 (*Bol. Inst. For. Inv. Exper. Madrid* No. 72). Madrid. 1955.

1 Leaflets 1–1·5 mm wide, very narrowly elliptical, markedly
   involute and appearing linear   **4. decorticans**
1 Leaflets usually more than 3 mm wide, obovate, oblanceolate
   or ovate, the margin not, or only slightly involute
2 Pedicel 7–15 mm; standard 15–23 mm   **3. hispanicus**
2 Pedicel not more than 5 mm; standard 10–16 mm
3 Bracteoles 2–4 mm, ovate or lanceolate; calyx 8–11 mm, the
   lower teeth ⅓ to ½ the total length of the lip   **2. telonensis**
3 Bracteoles 1 mm or less, linear; calyx 5–8 mm, the lower
   teeth not more than ⅓ the total length of the lip   **1. complicatus**

1. **A. complicatus** (L.) Gay, *Ann. Sci. Nat.* ser. 2 (Bot.), **6**: 125 (1836) (incl. *A. commutatus* Guss., *A. intermedius* DC., *A. villosus* Boiss.). Erect, up to 4 m; twigs and leaves sparsely to densely sericeous or with patent hairs. Leaflets 5–25 × 2–7 mm, oblanceolate. Bracteoles *c.* 1 mm, linear (sometimes absent); pedicels 3–5 mm. Calyx 5–7 mm, with or without glandular tubercules, subglabrous or sericeous, lower teeth not more than ⅓ the total length of the lip. Standard 10–15 mm, sericeous. Legume 15–45 × 4–6 mm, narrowly oblong, glandular-tuberculate. *Woods and scrub. S.W. Europe and Mediterranean region, extending to N.W. France.* Ga Gr Hs It Lu Si.

Very variable in degree of pubescence, size of calyx and leaves, and presence or absence of glandular tubercules on the calyx. The following subspecies may be recognized although their distributions do not form a very satisfactory geographical pattern and there is much morphological intergradation.

1 Calyx densely glandular-pubescent; twigs villous, glabrescent
          (a) subsp. complicatus
1 Calyx eglandular, very rarely with a few glands
2 Twigs and calyx densely villous   **(b) subsp. aureus**
2 Twigs and calyx shortly pubescent, sometimes glabrescent
          (c) subsp. commutatus

(a) Subsp. **complicatus** (*A. divaricatus* var. *graecus* (Griseb.) Boiss., *A. intermedius* DC.): *N.W. & C. Portugal, N. & C. Spain, W. France, S. Italy, Sicilia, E. Greece.*

(b) Subsp. **aureus** (Cav.) C. Vicioso, *Anal. Inst. Bot. Cavanilles* **6**(2): 43 (1946) (*A. villosus* Boiss.): *C. Portugal, C. Spain, C. Italy.*

(c) Subsp. **commutatus** (Guss.) Coutinho, *Fl. Port.* 320 (1913) (*A. commutatus* Guss.): *Mediterranean region, E. Portugal.*

2. **A. telonensis** (Loisel.) DC. in Lam. & DC., *Fl. Fr.* ed. 3, **5**: 550 (1815) (*A. grandiflorus* Boiss.). Erect, up to 1 m. Leaflets

3–8 × 1–4 mm, densely clustered along the branches, sparsely sericeous beneath, glabrous above. Flowers in clusters of 2–7; bracteoles 2–4 mm, ovate or lanceolate; pedicels 2–3 mm. Calyx 8–11 mm, densely sericeous; lower teeth ⅓–½ the total length of the lip. Standard 13–16 mm, broadly ovate, sericeous. Legume 15–30 × 4–5 mm, narrowly oblong, sericeous and glandular. *Scrub. S. France, C. & S. Spain, S. Portugal.* Ga Hs Lu.

3. **A. hispanicus** (Lam.) DC. in Lam. & DC., *op. cit.* 549 (1815). Erect, up to 2 m. Leaflets 15–30 × 3–8 mm, oblanceolate, acuminate, sericeous on both surfaces. Flowers in congested racemes; bracteoles absent; pedicels 7–15 mm. Calyx 8–12 mm, with semi-patent hairs and glandular papillae, especially on the teeth. Standard 15–23 mm, broadly ovate, sericeous. Legume 20–50 × 8–10 mm, oblong, densely glandular-tuberculate. *C. & S.W. Spain, S. Portugal.* Hs Lu.

(a) Subsp. **hispanicus**: Leaves sparsely hairy above; calyx with numerous glandular papillae, especially on the teeth; standard sparsely hairy. *W.C. Spain.*

(b) Subsp. **argyrophyllus** Rivas Goday, *Ann. Inst. Bot. Cavanilles* **12**(2): 307 (1954): Leaves densely silvery-white-hairy above; calyx with few or no glandular papillae; standard densely pubescent. *S.W. Spain, S. Portugal.*

4. **A. decorticans** Boiss., *Biblioth. Univ. Genève* ser. 2, **13**: 407 (1838). Erect, up to 3 m. Leaflets 9–18 × 1–1·5 mm, very narrowly elliptical and markedly involute, sericeous on both surfaces. Flowers in racemes; bracteoles absent; pedicels 4–8 mm. Calyx *c.* 8 mm, densely sericeous, not glandular. Standard *c.* 15 mm, broadly ovate, sericeous. Legume 20–60 × 8–10 mm, oblong, densely glandular-tuberculate. *Mountains of S. Spain.* Hs.

27. Lotononis (DC.) Ecklon & Zeyher[2]

(*Amphinomia* DC.)

Unarmed shrubs. Leaves 3-foliolate or 5-foliolate, digitate, alternate or subopposite. Flowers solitary or in terminal or rarely in axillary inflorescences. Calyx weakly bilabiate, the upper lip deeply 4-toothed, the lower 1-toothed; corolla yellow; stamens monadelphous, the 5 shorter with dorsifixed anthers, the 5 longer with basifixed anthers. Legume oblong, compressed, dehiscent. Seeds numerous.

A large genus occurring mainly in Africa and S.W. Asia.

Leaves 5-foliolate; calyx 8–10 mm   **1. lupinifolia**
Leaves 3-foliolate; calyx 6–7 mm   **2. genistoides**

1. **L. lupinifolia** (Boiss.) Bentham, *London Jour. Bot.* (*Hooker*) **2**: 607 (1843). Grey-pubescent to silvery-sericeous, caespitose dwarf shrub up to 20 cm. Leaflets 5, 5–10 × 1–4 mm, linear to obovate-lanceolate, silvery-sericeous beneath. Flowers in sessile to shortly pedunculate, 2- to 4-flowered, usually axillary clusters. Calyx 8–10 mm; corolla *c.* 12 mm, sericeous; standard about as long as keel. Legume 12–15 × 4–5 mm, up to twice as long as persistent calyx. *Dry places. S. Spain.* Hs.

2. **L. genistoides** (Fenzl) Bentham, *loc. cit.* (1843) (*Amphinomia genistoides* (Fenzl) Hayek). Like **1** but leaflets 3, 5–8 × 1–4 mm, oblong-oblanceolate to obovate; flowers in 1- to 3-flowered clusters; calyx 6–7 mm; corolla 8–9 mm; standard shorter than keel; legume only slightly longer than the persistent calyx. *Dry scrub. S.E. Bulgaria, N.E. Greece.* Bu Gr. (*Anatolia.*)

---

28. Lupinus L.[1]

Annual or perennial herbs, rarely shrubs. Leaves usually digitate, petiolate; stipules adnate to the base of the petiole. Flowers in terminal racemes. Calyx bilabiate, divided almost to the base; corolla variously coloured; wings connate at apex; keel beaked; stamens monadelphous. Legume dehiscent, compressed, usually constricted between the seeds. Seeds 3–12, with a sunken hilum.

1 Annuals; seeds usually more than 5 mm, dull
2 Upper lip of the calyx shallowly bidentate; seeds 8–14 mm
    5. albus
2 Upper lip of the calyx 2-partite; seeds not more than 10 mm
  3 Corolla yellow, cream, pale pink or lilac; flowers regularly arranged in distant whorls
    4 Corolla bright yellow; leaflets sparsely appressed-villous above   **1. luteus**
    4 Corolla at first cream becoming pale pink or lilac; leaflets glabrous above except along the margins **2. hispanicus**
  3 Corolla blue or bluish; flowers alternate or irregularly whorled
    5 Corolla 15–17 mm; legume 13–20 mm wide **6. varius**
    5 Corolla 10–14 mm; legume not more than 12 mm wide
      6 Leaflets 2–5 mm wide, linear; lower lip of calyx 6–7 mm
        3. angustifolius
      6 Leaflets 5–15 mm wide, obovate; lower lip of calyx 10–12 mm   **4. micranthus**
1 Perennial herbs, rarely shrubs; seeds not more than 5 mm, shining
  7 Shrub 100–300 cm; leaflets 5–10 mm wide; corolla usually yellow   **10. arboreus**
  7 Perennial herbs up to 150 cm; leaflets 10–30 mm wide; corolla never yellow
    8 Leaflets 9–17, 70–150 × 15–30 mm; upper lip of calyx entire   **9. polyphyllus**
    8 Leaflets 6–11, 15–60 × 8–15 mm; upper lip of calyx emarginate
      9 Stems glabrous or pubescent; upper lip of calyx 4–6 mm   **7. perennis**
      9 Stems villous or hirsute; upper lip of calyx *c.* 8 mm   **8. nootkatensis**

1. L. luteus L., *Sp. Pl.* 722 (1753). Hairy annual 25–80 cm. Leaflets 40–60 × 8–12 mm, obovate-oblong, mucronate, sparsely villous; stipules dimorphic, those of the lower leaves 8 mm, subulate, those of the upper leaves 22–30 × 2–4 mm, linear-obovate. Racemes 5–16 cm; flowers regularly verticillate, scented; peduncle 4–12 cm. Upper lip of calyx 6–7 mm, 2-partite, the lower 10 mm, shallowly 3-dentate. Corolla 13–16 mm, bright yellow. Legume 40–50 × 10 mm, densely villous, black. Seeds 4–6, 6–8 × 4·5–6·5 mm, orbicular-quadrangular, compressed, smooth and dull, black marbled with white, with a white curved line on each side. *Light, acid soils. Iberian peninsula, Italy and islands of W. Mediterranean. Widely cultivated elsewhere for fodder and green manure, and sometimes naturalized.* Co Hs It Lu Sa Si [Az Be Bu Cz Ga Ge Ho Hu Ju Po Rm Rs (C, W)].

2. L. hispanicus Boiss. & Reuter, *Diagn. Pl. Nov. Hisp.* 10 (1842). Like **1** but leaflets not mucronate, glabrous above except along the margins; stipules of the upper leaves often falcate; flowers not scented; corolla at first cream, becoming pale pink or lilac; seeds 4·5–6 mm, light red-brown with darker spots, with a dark, curved line on each side. $2n = 52$. *Light, acid soils.* ● *N. & C. Portugal, W. Spain.* ?Gr Hs Lu.

3. L. angustifolius L., *Sp. Pl.* 721 (1753). Shortly hairy annual 20–80 cm. Leaflets 10–50 × 2–5 mm, linear to linear-spathulate, glabrous above and sparsely villous beneath; stipules linear-subulate. Racemes 10–20 cm; flowers alternate; peduncle 1–3 cm. Upper lip of calyx *c.* 4 mm, 2-partite, the lower 6–7 mm, irregularly 3-dentate to subentire. Corolla 11–13 mm, blue. Legume shortly hirsute, yellow to black. Seeds 4–6, ellipsoid, smooth and dull, yellow-brown, dark brown or grey with yellow spots. *Light, acid soils. S. Europe.* Bu Co Cr Ga Gr Hs It ?Ju Lu Sa Si Tu [Au Az Cz Ge He Hu Po Rm Rs (C, W)].

(a) Subsp. **angustifolius** (incl. *L. leucospermus* Boiss.): Plant 50–80 cm; leaflets 30–40 × 4–5 mm, flat, linear-spathulate; legume 40–60 × 8–13 mm; seeds 6–8 × 4–7 mm. *Inland loamy soils. Throughout the range of the species, except Sardegna.*

(b) Subsp. **reticulatus** (Desv.) Coutinho, *Fl. Port.* 315 (1913) (*L. reticulatus* Desv.): Plant 20–40 cm; leaflets 10–20 × 2 mm, conduplicate, linear; legume 35–45 × 6–8 mm; seeds 4·5–5 × 3–3·5 mm. *Maritime sands, rarely inland. S.W. Europe and W. Mediterranean region; naturalized in Jugoslavia.*

4. L. micranthus Guss., *Fl. Sic. Prodr.* **2**: 400 (1828) (*L. hirsutus* sensu L. 1763 pro parte, et auct., non L. 1753). Brown-hirsute annual 10–40 cm. Leaflets 15–70 × 5–15 mm, obovate-cuneate to obovate-oblong, mucronate, sparsely hirsute; stipules linear-subulate. Racemes up to 12 cm; lower flowers alternate, the upper irregularly verticillate; peduncle up to 1·5 cm. Upper lip of calyx 6 mm, 2-partite, the lower 10–12 mm, deeply 3-dentate. Corolla 10–14 mm, blue, the standard white in the middle and the keel blackish-violet at apex. Legume 30–50 × 10–12 mm, hirsute, red-brown. Seeds 3–4, 5–8 mm, orbicular-quadrangular, compressed, smooth and dull, pinkish-grey to brown, with dark veins and dots. *Acid soils. Mediterranean region, C. & S. Portugal.* Al Bl Co Cr Ga Gr Hs It Ju Lu Sa Si Tu.

5. L. albus L., *Sp. Pl.* 721 (1753). Shortly hairy annual up to 120 cm. Leaflets of the lower leaves 25–35 × 14–18 mm, obovate, of the upper leaves 40–50 × 10–15 mm, obovate-cuneate, all mucronulate, nearly glabrous above, sparsely villous beneath; stipules setaceous. Racemes 5–10 cm, sessile, flowers alternate. Calyx 8–9 mm, both lips shallowly dentate. Corolla 15–16 mm, white to blue. Legume 60–100 × 11–20 mm, becoming longitudinally rugulose, shortly-villous, glabrescent, yellow. Seeds 4–6, 8–14 mm, orbicular-quadrangular, compressed or depressed, smooth and dull, light yellow, sometimes with dark variegation. *Acid soils. S. part of Balkan peninsula and Aegean region; widely cultivated elsewhere.* Al Bu Cr Gr Ju Tu [Au Az Co Cz Ga Ge He Hs Hu It Lu Rm Sa Si].

(a) Subsp. **albus** (*L. termis* Forskål): Corolla white, the keel pale blue at apex; legume 80–100 × 17–20 mm; seeds 12–14 mm, unspotted. *Cultivated in C. & S. Europe for its edible seeds and for fodder.*

(b) Subsp. **graecus** (Boiss. & Spruner) Franco & P. Silva, *Feddes Repert.* **79**: 52 (1968) (*L. graecus* Boiss. & Spruner): Corolla deep blue; legume 60–70 × 11–13 mm; seeds 8–9 mm, dark variegated. ● *Balkan peninsula and Aegean region.*

6. L. varius L., *Sp. Pl.* 721 (1753). Sericeous or hirsute annual up to 50 cm. Leaflets 25–35 × 6–9 mm, oblong-obovate, mucronulate, sericeous to softly villous; stipules linear-subulate. Racemes up to 10 cm; flowers irregularly verticillate; peduncle 1–4 cm. Upper lip of calyx *c.* 8 mm, 2-partite, the lower 10–12 mm, shallowly 3-dentate to subentire. Corolla 15–17 mm, blue, the standard with a white and yellow or pale purple blotch. Legume 40–50 × 13–20 mm, softly hirsute, dark red-brown. Seeds 3–4, orbicular, compressed, tuberculate-scabrid, iridescent, brown- and black-marbled on cream and purple, with a black, curved

---
[1] By J. do Amaral Franco and A. R. Pinto da Silva.

line on each side. *Light, acid soils. Mediterranean region, Portugal.* Al Co Cr Gr Hs It Ju Lu Si [Au ?Ge Rm].

(a) Subsp. **varius**: Stems and leaflets sericeous and calyx densely villous, the hairs red-brown; racemes up to 6 cm; legume 13–15 mm wide; seeds 7–9 mm. $2n=32$. *Western part of the range of the species.*

(b) Subsp. **orientalis** Franco & P. Silva, *Feddes Repert.* **79**: 52 (1968) (*L. digitatus* Forskål, *L. pilosus* L.): Stem and calyx hirsute and leaflets softly villous, the hairs white; racemes 5–10 cm; legume 18–20 mm wide; seeds 10–12 mm. *Greece and Aegean region. (Syria to Egypt.)*

7. **L. perennis** L., *Sp. Pl.* 721 (1753). Stout perennial 20–70 cm; stems glabrous or pubescent. Leaflets 7–11, $15-50 \times 8-12$ mm, oblanceolate, glabrous above, sparsely pubescent beneath; stipules setaceous. Racemes up to 30 cm, lax; flowers alternate or verticillate; peduncle up to 10 cm. Upper lip of calyx 4–6 mm, emarginate, lower lip *c.* 8 mm, entire. Corolla 12–16 mm, purple-blue, pink or white or parti-coloured; keel ciliate. Legume 30–50 mm, pubescent. Seeds 4–6. *Cultivated for ornament and for fodder, mostly in C. Europe, and sometimes naturalized.* [Au Cz Ga ?Ju No Rm.] (*E. North America.*)

8. **L. nootkatensis** Donn ex Sims, *Bot. Mag.* **32**: t. 1311 (1810). Like **7** but usually stouter, with villous or hirsute stems; leaflets 6–8, $20-60 \times 10-15$ mm; upper lip of calyx *c.* 8 mm, lower lip 8–10 mm. $2n=48$. *Naturalized in Norway and Scotland.* [Br No.] (*N.W. North America, N.E. Asia.*)

Possibly only a subspecies of **7**.

9. **L. polyphyllus** Lindley, *Bot. Reg.* **13**: t. 1096 (1827). Minutely pubescent, stout, usually unbranched perennial 50–150 cm. Leaflets 9–17, $70-150 \times 15-30$ mm, obovate-lanceolate, mostly glabrous above, sparingly sericeous beneath; stipules subulate. Racemes 15–60 cm, rather dense; flowers verticillate; peduncle 3–8 cm. Lips of calyx entire. Corolla 12–14 mm, blue, purple, pink or white; keel glabrous. Legume 25–40 mm, sparsely hairy, brown. Seeds 5–9, *c.* 4 mm, variously spotted. *Cultivated for ornament and for fodder in a large part of Europe, and widely naturalized.* [Au Cz Da Fe Ga Ge He Ho Hu Ju No Po Rm Rs (C) Su.] (*W. North America.*)

The hybrid **9 × 10** (*L. × regalis* Bergmans) is frequently cultivated for ornament in C. & N. Europe and often occurs as a casual. It may be locally naturalized.

10. **L. arboreus** Sims, *Bot. Mag.* **18**: t. 682 (1803). Much-branched shrub 100–300 cm. Leaflets 5–12, $20-60 \times 5-10$ mm, obovate-oblong, mucronate, strigose beneath, strigose or glabrous above; stipules subulate. Racemes 10–30 cm, lax; flowers alternate or subverticillate, scented; peduncle 4–10 cm. Upper lip of calyx emarginate, the lower entire. Corolla 14–17 mm, usually yellow, sometimes white, purple to blue or variegated. Legume 40–80 mm, strigose, brown. Seeds 8–12, 4–5 mm, ellipsoid, dark brown, more or less mottled, with a pair of spots near micropyle. *Naturalized near the sea in Britain and Ireland.* [Br Hb.] (*California.*)

29. Argyrolobium Ecklon & Zeyher[1]

Herbs or small shrubs. Leaves 3-foliolate; stipules free from the petiole. Flowers in terminal fascicles or short racemes. Calyx deeply bilabiate, the lips longer than the tube; upper lip deeply 2-fid; lower lip 3-toothed; corolla yellow; wings free; keel obtuse;

stamens monadelphous. Legume linear or oblong, compressed, dehiscent. Seeds 4–10 (in European spp.).

Stem and lower surface of leaflets densely appressed silvery-sericeous; leaflets elliptical to lanceolate   **1. zanonii**
Stem and leaflets hirsute, green; leaflets ovate or obovate
    2. biebersteinii

1. **A. zanonii** (Turra) P. W. Ball, *Feddes Repert.* **79**: 41 (1968) (*A. linnaeanum* Walpers, *A. argenteum* (L.) Willk., non (Jacq.) Ecklon & Zeyher, *Cytisus argenteus* L., *C. zanonii* Turra). Stems up to 25 cm, procumbent, woody at base, silvery-sericeous with dense appressed hairs. Leaflets $5-20 \times 3-6(-8)$ mm, those of the lower leaves elliptical, of the upper lanceolate, glabrous or sparsely hairy above, densely sericeous beneath. Flowers solitary or up to 3 in small terminal fascicles. Corolla 9–12 mm. Legume $15-35 \times 4\cdot5-5\cdot5$ mm, somewhat torulose, sericeous-villous. $2n=48$. *Dry open places. S. Europe from Albania westwards, extending northwards to 46° N. in France.* Al Bl Co Ga Hs It Ju Lu Sa ?Si.

A. dalmaticum (Vis.) Ascherson & Graebner, *Syn. Mitteleur. Fl.* **6**(2): 234 (1907) (*Chamaecytisus dalmaticus* Vis.), described from W. Jugoslavia, is a dubious species probably known only from the original collection. It has diadelphous stamens and is considered by some authors to be an abnormal plant of **1**.

2. **A. biebersteinii** P. W. Ball, *Feddes Repert.* **79**: 41 (1968) (*A. calycinum* Jaub. & Spach, nom. illegit., *A. pauciflorum* (Bieb. ex Willd.) Hayek, non Ecklon & Zeyher, *Cytisus pauciflorus* Bieb. ex Willd.). Like **1** but diffuse shrub 30–50 cm, hirsute, green; leaflets $15-25 \times 6-17$ mm, ovate or obovate; flowers 2–10 in fascicles or short racemes; legume hirsute. *Krym.* ?Ju Rs (K). (*S.W. Asia.*)

30. Robinia L.[1]

Deciduous trees or shrubs. Leaves imparipinnate; stipels often present; spinose stipules usually present. Flowers in pendent axillary racemes. Calyx campanulate, slightly bilabiate; corolla white, pink or purple; stamens diadelphous. Legume linear or oblong, compressed, dehiscent. Seeds 3–10.

Tree up to 25 m; racemes many-flowered; flowers 15–20 mm
    1. pseudacacia
Stoloniferous shrub up to 1 m; racemes 3- to 5-flowered; flowers *c.* 25 mm   **2. hispida**

1. **R. pseudacacia** L., *Sp. Pl.* 722 (1753). Tree up to 25 m. Leaflets 3–10 pairs, $25-45 \times 12-25$ mm, elliptical or ovate, glabrous or subglabrous; usually with stipular spines. Racemes 10–20 cm, many-flowered. Corolla 15–20 mm, white, the base of the standard yellow. Legume $5-10 \times c.$ 1 cm, glabrous. *Commonly planted for ornament and for stabilizing dry soil. Extensively naturalized in S. & W. Europe and more locally in C. & E. Europe.* [Al Au Be Br Bu Cz Ga Ge Gr He Ho Hs Hu It Ju Rm Sa Si Tu.] (*C. & E. North America.*)

R. viscosa Vent., *Descr. Pl. Jard. Cels* t. 4 (1800), from S.E. North America, is often planted in C. & S. Europe and may be locally naturalized. It is like **1** but with glandular-viscid twigs, petioles and peduncles; racemes 5–8 cm, 6- to 18-flowered; corolla purple, and legume sparsely glandular-hispid.

2. **R. hispida** L., *Mantissa* 101 (1767). Stoloniferous, usually hispid shrub up to 1 m. Leaflets 3–6 pairs, $20-35 \times 15-30$ mm, ovate-oblong to suborbicular. Racemes 3- to 5-flowered. Corolla *c.* 25 mm, pink or purple. Legume 5–8 cm, glandular-hispid. *Planted for ornament and hedges and locally naturalized in S. Europe.* [Ga.] (*S.E. North America.*)

---

[1] By P. W. Ball.

31. Wisteria Nutt.[1]

Climbing, deciduous shrubs. Leaves imparipinnate; stipels often present; stipules caducous. Flowers in pendent, terminal racemes. Calyx campanulate, slightly bilabiate; corolla violet, rarely white; stamens diadelphous. Legume linear or oblong, compressed, tardily dehiscent. Seeds 1–8.

1. **W. sinensis** (Sims) Sweet, *Hort. Brit.* 121 (1826). Climbing up to 10 m. Leaflets 3–6 pairs, 5–8 × 2–3 cm, ovate-oblong, acuminate, glabrescent. Racemes 15–30 cm; pedicels densely pubescent. Corolla *c.* 25 mm. Legume 10–15 cm, velutinous. *Commonly cultivated for ornament and locally naturalized.* [Ga.] (*E. Asia.*)

W. floribunda (Willd.) DC., *Prodr.* **2**: 390 (1825) (*W. multijuga* Van Houtte), from Japan, with 6–9 pairs of leaflets, racemes 20–50 cm and pedicels sparsely pubescent, is also commonly cultivated. It is often confused with **1** and may also be naturalized.

32. Galega L.[1]

Perennial herbs. Leaves imparipinnate; stipules small, free. Flowers in axillary racemes. Calyx campanulate, with 5 subequal teeth; corolla white to blue-violet; keel subobtuse; stamens monadelphous. Legume cylindrical, torulose, dehiscent. Seeds numerous.

Stipules ½-sagittate; corolla white to pale purplish-blue; legume
patent or erecto-patent   **1. officinalis**
Stipules ovate or ovate-orbicular; corolla blue-violet; legume
deflexed   **2. orientalis**

1. **G. officinalis** L., *Sp. Pl.* 714 (1753) (incl. *G. patula* Steven). Stems 40–150 cm, glabrous or sparsely pubescent. Leaflets 4–8 pairs, 15–50 × 4–15 mm, oblong, elliptical or lanceolate, acute or obtuse, mucronate, glabrous or pubescent beneath; stipules ½-sagittate. Calyx glabrous or sparsely pubescent, the teeth about as long as tube; corolla 10–15 mm, white to pale purplish-blue. Legume 20–50 × 2–3 mm, patent or erecto-patent. *E., C. & S. Europe; cultivated for fodder and for ornament and naturalized elsewhere.* Al Au Bu Cz Ga Ge Gr Hs Hu It Ju Po Rm Rs (W, K, E) Tu [Be Br He Lu].

2. **G. orientalis** Lam., *Encycl. Méth. Bot.* **2**: 596 (1788). Like **1** but leaflets 30–60 × 10–25 mm, acuminate; stipules ovate or ovate-orbicular; calyx pubescent, the teeth shorter than tube; corolla blue-violet; legume deflexed. *Cultivated for fodder and locally naturalized.* [Au Ga.] (*Caucasus.*)

33. Colutea L.[2]

Deciduous shrubs. Leaves imparipinnate; leaflets entire. Stipules small. Flowers in axillary racemes. Calyx campanulate, slightly bilabiate; corolla yellow or orange-red; stamens diadelphous; stigma large, inserted obliquely on the inner edge of the style and surrounded by hairs. Legume very inflated, with papery walls, indehiscent or dehiscent near the apex. Seeds numerous, reniform smooth.

Literature: K. Browicz, *Monogr. Bot.* (*Warszawa*) **14**: 1–136 (1963).

1 Keel beaked; legume curved upwards at the apex; corolla
orange-red   **4. orientalis**

1 Keel without beak; legume not curved upwards at the apex;
corolla yellow
2 Wings longer than the keel, with a distinct spur on the lower
edge   **3. cilicica**
2 Wings as long as or shorter than the keel, rarely slightly longer
but then rounded on the lower edge, without a spur
3 Ovary glabrous, or if pubescent then the surface visible;
leaflets variable, elliptical, obovate or ovate   **1. arborescens**
3 Ovary tomentose, the indumentum completely covering the
surface; leaflets always elliptical   **2. atlantica**

1. **C. arborescens** L., *Sp. Pl.* 723 (1753). Much-branched shrub up to 6 m. Young shoots puberulent or subglabrous, later occasionally glabrous. Leaflets (3–)4–5(–6) pairs, up to 30 × 20 mm, broadly elliptical, more rarely obovate or ovate. Inflorescence 3- to 8-flowered, puberulent when young, glabrescent. Corolla 16–20 mm, yellow. Legume 5–7 × 3 cm. Seeds up to 4 × 3·5 mm. *Dry slopes and open woods; somewhat calcicole. S. & S.C. Europe, extending to N.C. France; frequently cultivated in gardens and naturalized in N.W. & N.C. Europe.* Al Au Bu Co Cz Ga Ge Gr He Hs Hu It Ju Rm Sa Si ?Tu [Be Br].

(a) Subsp. **arborescens** (incl. *C. melanocalyx* sensu Hayek pro parte): Ovary completely glabrous, or very slightly pubescent along the ventral suture. *From E. France and Italy eastwards.*

(b) Subsp. **gallica** Browicz, *Monogr. Bot.* (*Warszawa*) **14**: 128 (1963): Ovary sparsely pubescent. *From Austria and Jugoslavia westwards.*

The hybrid **1** × **4** (*C.* × *media* Willd.) is commonly cultivated and is occasionally naturalized. It is variable and may resemble either parent in certain characters.

2. **C. atlantica** Browicz, *Monogr. Bot.* (*Warszawa*) **14**: 127 (1963) (*C. arborescens* auct. hisp. pro parte). Like **1**(b) but young shoots tomentose, the hairs persisting in the second year, and the ovary silvery-tomentose; leaflets 10–15(–18) × 7–9(–10) mm, always elliptical; inflorescence 1- to 3(–4)-flowered. *C. & S. Spain.* Hs.

Plants intermediate between **1**(b) and **2** occur in C. Spain.

3. **C. cilicica** Boiss. & Balansa in Boiss., *Diagn. Pl. Or. Nov.* **3**(5): 83 (1856) (*C. melanocalyx* sensu Hayek pro parte). Like **1**(a) but corolla up to 22 mm; wings distinctly longer than keel and with a distinct spur on the lower edge; ovary completely glabrous. *Sunny slopes and open oakwoods. N.E. Greece; Krym; Turkey-in-Europe.* Gr Rs (K) Tu. (*Caucasus, Anatolia.*)

Plants intermediate between **1** and **3** occur in C. Greece.

4. **C. orientalis** Miller, *Gard. Dict.* ed. 8, no. 3 (1768) (*C. cruenta* Aiton). Shrub up to 3 m. Young shoots slender, completely glabrous. Leaflets 3–4 pairs, up to 18 × 15 mm, broadly obovate or orbicular, bluish-green. Corolla 11–13 mm, orange-red; keel with a small beak, darker at the apex; wings falcate, shorter than keel. Ovary glabrous. Legume up to 4 × 2 cm, narrowing and curved upwards towards the apex, dehiscing at the apex. Seeds up to 2·5 mm. *Often cultivated in gardens, and occasionally naturalized in S. Europe.* [Ju Rs (K).] (*Caucasian region.*)

34. Eremosparton Fischer & C. A. Meyer[1]

Junciform shrubs with green stems and minute, scarious leaves. Flowers solitary, axillary. Calyx campanulate, with 5 subequal teeth; corolla violet; stamens diadelphous. Legume broadly ovate or suborbicular, compressed, dehiscent. Seeds 1–2.

---

[1] By P. W. Ball.   [2] By K. Browicz.

1. E. aphyllum (Pallas) Fischer & C. A. Meyer, *Enum. Pl. Nov.* **1**: 76 (1841). Stems 50–100 cm, erect, glabrous or pubescent. Leaves 1–3 mm, linear-oblong, obtuse. Pedicels 4–6 mm. Calyx pubescent, the teeth broadly triangular; corolla 5–7 mm. Legume 6–10 × 5–7 mm, tomentose. *Sandy deserts. S.E. Russia, W. Kazakhstan.* Rs (E). (*W.C. Asia.*)

35. Halimodendron Fischer ex DC.[1]

Shrubs. Leaves paripinnate, with a persistent, usually spine-tipped, rhachis; leaflets 1 or 2 pairs; stipules subulate, spinose. Flowers in axillary racemes. Calyx pelviform, with 5 short teeth; corolla pale purple; stamens diadelphous. Legume oblong to obovoid, inflated, shortly beaked, dehiscent. Seeds few.

1. H. halodendron (Pallas) Voss in Voss & Siebert, *Vilmorin's Blumeng.* ed. 3, **1**: index [35] (1896). Grey or bluish spinose shrub up to 2 m. Leaflets 15–35 × 5–10 mm, oblong-obovate or oblanceolate, glabrous or sericeous. Inflorescence 1- to 3-flowered; peduncles 3–4 cm. Calyx persistent; corolla 14–18 mm. Legume 10–30 × 3–5 mm, yellowish. *Steppes and maritime sands. S.E. Ukraine, S.E. Russia.* Rs (W, E). (*C. & S.W. Asia.*)

36. Caragana Fabr.[1]

Trees or shrubs. Leaves paripinnate, often with spine-tipped rhachis; stipules small, deciduous or sometimes persistent and spinose. Flowers solitary or fasciculate; pedicels articulate. Calyx tubular or campanulate, with 5 subequal teeth; corolla usually yellow; stamens diadelphous. Legume linear, sometimes inflated, usually acute, dehiscent. Seeds usually numerous.

1 Leaflets 4–6 pairs; rhachis 30–70 mm **1. arborescens**
1 Leaflets 2 pairs; rhachis not more than 15 mm
  2 Flowers often fasciculate; calyx campanulate, about as long
     as wide, slightly gibbous at base **2. frutex**
  2 Flowers solitary; calyx tubular, much longer than wide,
     strongly gibbous at base **3. grandiflora**

1. C. arborescens Lam., *Encycl. Méth. Bot.* **1**: 615 (1785). Shrub or small tree up to 7 m. Leaflets 4–6 pairs, 10–35 × 5–13 mm, obovate or elliptic-oblong, glabrescent; rhachis 30–70 mm, deciduous. Flowers usually in fascicles of 2–5; pedicels 15–60 mm. Calyx campanulate, slightly gibbous at base; teeth 1–1·5 mm, broadly triangular; corolla 15–22 mm. Legume 30–60 × 3–5 mm, acute. *Planted for ornament and occasionally naturalized.* [Ga.] (*N. Asia.*)

2. C. frutex (L.) C. Koch, *Dendrologie* **1**: 48 (1869) (incl. *C. mollis* (DC.) Besser). Glabrous or pubescent dwarf shrub up to 1 m (–3 m in cultivation). Leaflets 2 pairs, 5–25 × 2–15 mm, obovate with spinescent apex; rhachis up to 15 mm, persistent or not. Flowers solitary or in fascicles of 2–3; pedicels 15–20 mm. Calyx campanulate, slightly gibbous at base; teeth 1–1·5 mm, broadly triangular; corolla 15–25 mm. Legume 25–45 × 3–4 mm, acute. *E. Europe, from E. Bulgaria to C. Russia and W. Kazakhstan.* Bu Rm Rs (C, W, K, E).

3. C. grandiflora (Bieb.) DC., *Prodr.* **2**: 268 (1825) (incl. *C. scythica* (Komarov) Pojark.). Pubescent dwarf shrub up to 0·5(–1) m. Leaflets 2 pairs, 4–7 × 1–2 mm, linear-oblanceolate with spinescent apex; rhachis up to 5 mm, persistent, spinescent. Flowers solitary; pedicels 5–8 mm. Calyx tubular, much longer than wide, strongly gibbous at base; teeth 1·5–2 mm, triangular-

lanceolate; corolla 18–25 mm. Legume 20–25 × *c.* 2·5 mm, acuminate. *S. Ukraine, Moldavia.* Rs (W, E).

The plant described above is var. *scythica* Komarov, endemic to Europe. Var. *grandiflora*, from the Caucasus and S.C. Asia, differs only in being larger in all its parts.

37. Calophaca Fischer[1]

Dwarf shrubs. Leaves imparipinnate; stipules scarious or herbaceous, adnate to the petiole. Flowers in axillary racemes. Calyx tubular, with 5 subequal teeth; corolla yellow; stamens diadelphous. Legume oblong, terete, dehiscent, stipitate-glandular when young. Seeds usually 1–2.

1. C. wolgarica (L. fil.) Fischer, *Cat. Jard. Gorenk.* ed. 2, 68 (1812). Dwarf shrub up to 1 m. Leaflets 6–8 pairs, 3–15 × 3–10 mm, ovate-elliptical to suborbicular, grey-pubescent beneath; stipules scarious. Racemes 4- to 6-flowered, glandular-hairy. Calyx glandular; corolla 20–25 mm. Legume 20–30 × *c.* 5 mm. *Dry places.* ● *S.E. Russia, S. & E. Ukraine.* Rs (W, K, E).

38. Astragalus L.[2]

Annual or perennial herbs or small shrubs. Leaves imparipinnate or paripinnate, sometimes terminating in a spine; leaflets entire. Flowers in racemes or axillary clusters, sessile or pedicellate. Calyx infundibuliform, tubular or campanulate, sometimes inflated in fruit, with distinct, equal or unequal teeth; keel not mucronate at apex (very rarely adaxially mucronate); stamens 10, diadelphous (very rarely 5, monadelphous); stigma and style glabrous. Legume usually dehiscent, very varied in shape and texture, glabrous or hairy, unilocular to bilocular. Seeds 1–many.

In this account descriptions of peduncles and racemes refer to their appearance at anthesis; when the length of the peduncle is compared with the length of the leaf, the leaf subtending it is intended.

1 Leaves with the rhachis ending in a spine, paripinnate or the
    terminal leaflet caducous
  2 Hairs on leaves and stems medifixed
    3 Standard 11–12 mm; leaflets 3–5 pairs **78. balearicus**
    3 Standard more than 12 mm; leaflets more than 5 pairs
      4 Hairs on calyx and legume ascending or almost patent
         80. sirinicus
      4 Hairs on calyx and legume ± appressed
        5 Leaflets narrowly elliptical to linear; calyx-teeth $\frac{1}{4}$–$\frac{1}{2}$ as
          long as tube **77. angustifolius**
        5 Leaflets oblong or elliptical; calyx-teeth $\frac{1}{5}$–$\frac{1}{4}$ as long as
          tube **79. massiliensis**
  2 Hairs on leaves and stems simple
    6 Flowers in shortly pedunculate racemes; calyx not hidden by
      its indumentum, not splitting to the base in fruit
      7 Calyx strongly inflated in fruit, more than 10 mm wide
         74. clusii
      7 Calyx scarcely inflated in fruit, less than 10 mm wide
        8 Calyx-teeth equalling tube **75. sempervirens**
        8 Calyx-teeth twice as long as tube **76. giennensis**
    6 Flowers sessile or subsessile in axils of leaves; calyx hidden
      by its dense, villous indumentum, splitting to the base in
      fruit
    9 Bracteoles absent, or much shorter than calyx and less
      hairy; calyx usually less than 10 mm
      10 Standard with acute auricles at base of limb **61. granatensis**
      10 Standard with rounded auricles at base of limb
        11 Terminal spine of leaf much longer than terminal pair of
          leaflets **60. creticus**

---

[1] By P. W. Ball.   [2] By A. O. Chater.

11 Terminal spine of leaf equalling or shorter than terminal pair of leaflets
  12 Leaflets sparsely hairy or glabrescent **60. creticus**
  12 Leaflets densely hairy **62. arnacantha**
9 Bracteoles almost as long as calyx and equally hairy; calyx usually 10 mm or more
  13 Stipules and the stems beneath them almost glabrous
    14 Calyx less than 15 mm **63. parnassi**
    14 Calyx more than 15 mm **64. thracicus**
  13 Stipules and the stems beneath them densely tomentose
    15 Bracts ovate, navicular **65. trojanus**
    15 Bracts linear-oblanceolate, ±plane **64. thracicus**
1 Leaves with the rhachis not ending in a spine, imparipinnate
16 Annual with slender stock; sterile shoots absent at time of flowering
  17 Hairs on leaves or stems medifixed, with one arm much shorter than the other, usually straight and appressed
    18 Standard less than 5 mm; legume triangular-ovate **21. epiglottis**
    18 Standard more than 5 mm; legume not triangular-ovate
      19 Valves of legume sharply keeled; leaflets obtuse or sub-acute **22. edulis**
      19 Valves of legume not keeled; leaflets truncate or emarginate
        20 Legume 10–20 × 5–7 mm, inflated **23. cymbicarpos**
        20 Legume 20–50 × 2–3 mm, not inflated **24. hamosus**
  17 Most hairs on leaves or stems simple, basifixed
    21 Legume curved to form a ±complete ring, flattened, blackish **1. contortuplicatus**
    21 Legume not forming a complete ring, not blackish
      22 Valves of legume with a distinct keel
        23 Leaflets 10–15 pairs; legume usually more than 6 mm wide **2. boeticus**
        23 Leaflets less than 10 pairs; legume less than 6 mm wide
          24 Legume c. 5 mm wide; standard striped; keel exceeding wings **3. striatellus**
          24 Legume 2–3 mm wide; standard not striped; keel shorter than wings
            25 Legume with very strong, raised, reticulate venation **5. reticulatus**
            25 Legume with faint, reticulate venation or smooth **19. oxyglottis**
      22 Valves of legume not or obscurely keeled
        26 Calyx at least 8 mm; standard at least 15 mm
          27 Calyx 8–9 mm; standard 15–18 mm
            28 Legume more than 35 mm **10. peregrinus**
            28 Legume less than 35 mm **11. pamphylicus**
          27 Calyx 10–13 mm; standard 20–25 mm; legume less than 35 mm
            29 Legume straight, weakly rugose-tuberculate, with a straight beak **12. haarbachii**
            29 Legume curved, strongly tuberculate, with a hooked beak **13. suberosus**
        26 Calyx less than 8 mm; standard less than 15 mm
          30 Leaflets 1–4 pairs **4. arpilobus**
          30 Leaflets more than 4 pairs
            31 Legume covered with contorted lamellae **20. echinatus**
            31 Legume not covered with lamellae
              32 Legume linear, not widened at base, curved
                33 Peduncles not more than 0·3 cm **8. scorpioides**
                33 Peduncles more than 0·3 cm
                  34 Standard with an abrupt, long, acute apex **6. ankylotus**
                  34 Standard rounded or emarginate
                    35 Standard 12–14 mm; calyx-teeth at least as long as tube **9. longidentatus**
                    35 Standard 9–10 mm; calyx-teeth shorter than tube **7. stalinskyi**
              32 Legume lanceolate or linear-lanceolate, widened at base, ±straight
                36 Most peduncles more than 1 cm
                  37 Legumes stellately patent at maturity; standard 5–6 mm **16. polyactinus**

37 Legume erect or erecto-patent at maturity; standard 9–11 mm **17. stella**
36 Most peduncles less than 1 cm
  38 Legume erect at maturity; flowers 5–10 **18. sesameus**
  38 Legume patent at maturity; flowers 4–6
    39 Legume with short, appressed hairs only **15. tribuloides**
    39 Legume with both short, appressed and long, patent or ascending hairs **14. sinaicus**
16 Perennial, with stout stock; sterile shoots present at time of flowering
  40 Most hairs simple, basifixed
    41 Acaulescent or almost so, the peduncles or racemes arising from a rosette of leaves
      42 Calyx strongly inflated in fruit and c. 15 mm wide; standard 35–45 mm **66. physocalyx**
      42 Calyx not strongly inflated in fruit and less than 15 mm wide; standard less than 35 mm
        43 Plant with usually appressed hairs which are white when dry
          44 Standard 15–20 mm; peduncles longer than leaves **32. austraegaeus**
          44 Standard 10–14 mm; peduncles not longer than leaves **39. depressus**
        43 Plant with usually patent hairs which are pale brown when dry
          45 Standard hairy on back
            46 Calyx 12–25 mm; standard (19–)23–30 mm **(45–47). dasyanthus** group
            46 Calyx 9–10 mm; standard 15–22 mm **(54–57). nummularius** group
          45 Standard glabrous
            47 Most leaflets more than 12 pairs
              48 Most peduncles more than ½ as long as leaves; legume less than 15 mm **33. turolensis**
              48 Most peduncles less than ½ as long as leaves; legume at least 15 mm **(48–53). exscapus** group
            47 Most leaflets less than 12 pairs
              49 Legume sparsely hairy or glabrous **(48–53). exscapus** group
              49 Legume densely hairy **(54–57). nummularius** group
    41 Caulescent, with leaves separated by well-developed internodes
      50 Calyx becoming inflated in fruit; corolla persistent
        51 Peduncles more than 2 cm **73. ajubensis**
        51 Peduncles not more than 2 cm
          52 Calyx-teeth less than ½ as long as tube **67. ponticus**
          52 Calyx-teeth more than ½ as long as tube
            53 Bracts longer than calyx; calyx-teeth longer than tube **70. alopecuroides**
            53 Bracts not longer than calyx; calyx-teeth equalling or shorter than tube
              54 Leaflets 17–27 pairs; standard 5–9 mm wide
                55 Peduncles 1(–2) cm **69. alopecurus**
                55 Racemes sessile **72. centralpinus**
              54 Leaflets 11–15 pairs; standard 9–11 mm wide
                56 Leaflets less than 10 mm **71. grossii**
                56 Most leaflets more than 10 mm **68. vulpinus**
      50 Calyx not becoming inflated in fruit; corolla not persistent
        57 Corolla uniformly yellow, or greenish- or whitish-yellow
          58 Standard at least 17 mm
            59 Standard hairy on back **(45–47). dasyanthus** group
            59 Standard glabrous
              60 Legume grooved beneath
                61 Legume curved, strongly tuberculate, with a hooked beak **13. suberosus**
                61 Legume straight, weakly rugose-tuberculate, with a straight beak **12. haarbachii**
              60 Legume not grooved beneath
                62 Standard 18–20 mm; legume lanate, with patent hairs **33. turolensis**

62 Standard more than 20 mm; legume glabrous or tomentose
   63 Leaflets ovate or cordate-orbicular; legume tomentose   **58. graecus**
   63 Leaflets linear-lanceolate; legume glabrous   **59. drupaceus**
58 Standard less than 17 mm
  64 Legume ovoid-globose; upper stipules connate   **25. cicer**
  64 Legume not ovoid-globose; upper stipules free
   65 Legume less than 18 mm   **34. galegiformis**
   65 Legume more than 18 mm
    66 Leaflets less than 7 mm wide; legume usually more than 10 mm wide   **36. penduliflorus**
    66 Leaflets at least 7 mm wide; legume usually less than 10 mm wide
     67 Legume slightly inflated; calyx-teeth *c.* 0·5 mm, broadly triangular   **35. frigidus**
     67 Legume not inflated; calyx-teeth more than 1 mm, linear
      68 Stipules 15–20 mm; calyx 5–6 mm, glabrous or with hairs only on teeth   **43. glycyphyllos**
      68 Stipules 10–15 mm; calyx 6–8 mm, densely hairy   **44. glycyphylloides**
57 Corolla not uniformly yellow, but violet or purple at least in part (rarely white)
  69 Stipules connate around the stem for at least ⅓ of their length
   70 Standard not or scarcely exceeding calyx; leaflets densely hairy above   **31. setosulus**
   70 Standard distinctly exceeding calyx; leaflets sparsely hairy or glabrous above
    71 Standard 9–12 mm
     72 Legume more than 9 mm, at least twice as long as calyx   **29. bourgaeanus**
     72 Legume less than 9 mm, scarcely longer than calyx   **26. glaux**
    71 Standard more than 12 mm
     73 Stems caespitose, woody at base; calyx 12–13 mm   **32. austraegaeus**
     73 Stems ascending, not woody at base; calyx 6–10 mm
      74 Legume 7–8 mm; upper stipules 2·5–5 mm   **27. danicus**
      74 Legume 10–15 mm; upper stipules 5–10 mm
       75 Keel mucronate at apex   **30. pseudopurpureus**
       75 Keel not mucronate at apex   **28. purpureus**
  69 Stipules free from each other (rarely connate around stem for less than ⅓ of their length)
   76 Standard 20–35 mm; legume 50–70 mm   **42. lusitanicus**
   76 Standard less than 20 mm; legume less than 50 mm
    77 Young legume glabrous   **41. australis**
    77 Young legume hairy
     78 Calyx 7–9 mm; standard 16–18 mm   **37. umbellatus**
     78 Calyx less than 7 mm; standard less than 16 mm
      79 Internodes shorter than stipules   **39. depressus**
      79 Internodes longer than stipules
       80 Leaflets usually 6–7 pairs; calyx-teeth triangular, obtuse   **40. norvegicus**
       80 Leaflets usually 7–12 pairs; calyx-teeth lanceolate, ± acute   **38. alpinus**
40 Most hairs on leaves and stems medifixed (sometimes with one arm very short) or forked from the base
81 Hairs on leaves flexuous and erecto-patent
  82 Calyx-teeth as long as or longer than tube
   83 Calyx more than 9 mm; standard more than 15 mm
    84 Leaflets 4–7 pairs; bracts *c.* 12 mm   **104. lacteus**
    84 Leaflets 8–20 pairs; bracts *c.* 5 mm   **103. dolichophyllus**
   83 Calyx less than 9 mm; standard less than 15 mm
    85 Legume included in calyx; corolla red   **99. autranii**
    85 Legume exserted from calyx; corolla yellowish, tinged with blue   **100. agraniotii**
  82 Calyx-teeth much shorter than tube
   86 Calyx less than 10 mm; standard *c.* 15 mm   **98. idaeus**
   86 Calyx at least 10 mm; standard at least 18 mm

87 Legume less than 6 mm wide, oblong or linear; peduncles usually equalling leaves
  88 Leaflets less than 7 pairs   **101. arcuatus**
  88 Leaflets more than 7 pairs   **102. reduncus**
87 Legume more than 6 mm wide, ovoid; racemes usually sessile
  89 Surface of leaves hidden by dense, erecto-patent hairs
   90 Standard constricted in middle   **107. testiculatus**
   90 Standard not constricted in middle   **108. rupifragus**
  89 Surface of leaves not hidden by indumentum, which consists of ± parallel hairs
   91 Legume glabrous; calyx-teeth *c.* ½ as long as tube   **106. wilmottianus**
   91 Legume densely hairy; calyx-teeth *c.* ¼ as long as tube   **105. baldaccii**
81 Hairs on leaves ± straight, appressed
92 Flowers pendent at anthesis
  93 Leaflets 15–20 pairs; stems 40–80 cm   **83. falcatus**
  93 Leaflets 6–14 pairs; stems up to 30 cm
   94 Stipules free from each other; calyx-teeth ⅔ as long as tube   **39. depressus**
   94 Stipules connate around stem; calyx-teeth less than ½ as long as tube
    95 Legume 4–5 mm wide, semi-lunate   **82. algarbiensis**
    95 Legume 2–3 mm wide, oblong-lanceolate   **81. odoratus**
92 Flowers patent to erect
  96 Acaulescent or almost so, with branched stock, the racemes arising from a rosette of leaves
   97 Calyx-teeth as long as or longer than tube
    98 Legume included in calyx; corolla red   **99. autranii**
    98 Legume exserted from calyx; corolla yellowish, tinged with blue   **100. agraniotii**
   97 Calyx-teeth much shorter than tube
    99 Racemes sessile
     100 Legume glabrous; calyx-teeth *c.* ½ as long as tube   **106. wilmottianus**
     100 Legume densely hairy; calyx-teeth *c.* ¼ as long as tube   **105. baldaccii**
    99 Racemes distinctly pedunculate
     101 Corolla yellow, sometimes tinged with violet
      102 Leaves not more than 3 cm; calyx *c.* 7 mm   **98. idaeus**
      102 Most leaves more than 3 cm; calyx at least 8 mm
       103 Most leaflets more than 3 mm wide; standard at least 20 mm   **97. helmii**
       103 Leaflets 1·5–3 mm wide; standard 16–19 mm   **133. fialae**
     101 Corolla white, purplish, violet or reddish, but never yellow
      104 Legume globose, strongly inflated and membranous   **109. physodes**
      104 Legume not inflated and membranous
       105 Most leaves with at least 10 pairs of leaflets   **110. monspessulanus**
       105 Most leaves with less than 10 pairs of leaflets
        106 Leaves 2–3 cm; stipules 1–2 mm   **113. sericophyllus**
        106 Leaves more than 3 cm; stipules more than 2 mm
         107 Standard 13–18 mm; legume 8–10 × 3–4 mm   **95. leontinus**
         107 Standard 19 mm or more; legume 10–25 × 4–8 mm
          108 Calyx 12–13 mm   **111. spruneri**
          108 Calyx 7–11 mm   **112. incanus**
96 Caulescent, with leaves separated by well-developed internodes
  109 Stipules connate around the stem for at least part of their length
   110 Ovary and legume glabrous; leaflets orbicular to ovate, very remote   **96. amarus**
   110 Ovary and legume (at least when young) hairy; leaflets not as above

111 Calyx 2·5–3 mm   **89. tenuifolius**
111 Calyx more than 3 mm
112 Leaflets more than 10 mm wide   **85. roemeri**
112 Leaflets less than 10 mm wide
113 Calyx-teeth more than $\frac{1}{2}$ as long as tube   **94. onobrychis**
113 Calyx-teeth not more than $\frac{1}{2}$ as long as tube
114 Legume not more than 2 mm wide, usually more than 20 mm long   **116. subuliformis**
114 Legume more than 2 mm wide, usually less than 20 mm long
115 Leaflets 2–6 × 0·5–1 (–1·5) mm (Atlantic coasts)   **91. baionensis**
115 Leaflets larger
116 Peduncles shorter than leaves
117 Legume 2·5–3 mm wide; calyx 7–9 mm, the teeth $\frac{1}{3}-\frac{2}{3}$ as long as tube   **92. tenuifoliosus**
117 Legume 3·5–4 mm wide; calyx 4–5 mm, the teeth *c*. $\frac{1}{4}$ as long as tube   **90. arenarius**
116 Peduncles equalling or longer than leaves
118 Keel mucronate at apex   **30. pseudopurpureus**
118 Keel not mucronate
119 Standard with ovate limb   **95. leontinus**
119 Standard with linear-oblong limb
120 Racemes with less than 10 flowers   **93. mesopterus**
120 Racemes with more than 10 flowers   **94. onobrychis**
109 Stipules free from each other, except sometimes those at base of stem
121 Calyx 2–4 mm; standard not more than 12 mm
122 Calyx-teeth triangular-ovate, up to $\frac{1}{4}$ as long as tube   **86. austriacus**
122 Calyx-teeth linear-lanceolate, $\frac{1}{2}-\frac{3}{4}$ as long as tube
123 Standard more than 9 mm; leaflets 2·5–6 mm wide; branches erecto-patent   **88. clerceanus**
123 Standard less than 9 mm; leaflets 0·5–1(–2·5) mm wide; branches strict   **87. sulcatus**
121 Calyx at least 5 mm; standard more than 12 mm
124 Stems herbaceous throughout
125 Corolla yellow   **84. asper**
125 Corolla violet, purple or whitish
126 Leaflets less than 10 mm wide   **95. leontinus**
126 Leaflets more than 10 mm wide   **85. roemeri**
124 Stems woody at least at base
127 Legume 50–70 mm   **119. pugionifer**
127 Legume less than 50 mm
128 Leaflets of longer leaves 0·5–1 mm wide, setaceous
129 Racemes subumbellate   **117. corniculatus**
129 Racemes oblong, lax
130 Corolla lilac or purplish; leaflets hairy above   **118. muelleri**
130 Corolla usually yellow; leaflets glabrous above   **116. subuliformis**
128 Leaflets of longer leaves more than 1 mm wide, not setaceous
131 Legume at least 10 times as long as wide
132 Leaves 5–10 cm; calyx 7–9 mm; bracts 0·5–1 mm (Ural)   **115. karelinianus**
132 Leaves 1–5 cm; calyx 9–11 mm; bracts 1–2 mm
133 Leaflets 4–6 pairs, silvery; peduncles 4–5 times as long as leaves   **113. sericophyllus**
133 Leaflets 5–7 pairs, green; peduncles up to 3 times as long as leaves   **114. apollineus**
131 Legume less than 10 times as long as wide
134 Leaflets 2–4 pairs
135 Calyx 9–12 mm, not becoming inflated; corolla white
136 Calyx-teeth 1·5–2·5 mm, $\frac{1}{5}-\frac{1}{4}$ as long as tube   **126. zingeri**
136 Calyx-teeth 3–4 mm, $\frac{1}{3}-\frac{1}{2}$ as long as tube   **125. glaucus**

135 Calyx 12–15 mm, becoming ±inflated in fruit; corolla yellow
137 Bracts ovate-acuminate; standard 22–27 mm   **132. medius**
137 Bracts linear-lanceolate; standard 18–23 mm   **131. albicaulis**
134 Leaflets mostly more than 4 pairs
138 Legume with very closely appressed hairs (Spain)
139 Legume curved; leaflets convolute   **130. hegelmaieri**
139 Legume straight; leaflets plane   **129. hispanicus**
138 Legume with ascending to patent hairs
140 Legume equalling the calyx (or only the beak exceeding the calyx), with almost patent hairs   **127. vesicarius**
140 Legume exceeding the calyx, with ascending hairs
141 Racemes dense, ovoid, usually less than 5 cm
142 Peduncles not exceeding leaves   **120. cornutus**
142 Peduncles at least 1$\frac{1}{2}$ times as long as leaves
143 Standard emarginate at apex; leaflets linear to linear-lanceolate   **128. peterfii**
143 Standard rounded at apex; leaflets lanceolate to oblong-lanceolate
144 Calyx-teeth 1·5–2·5 mm, $\frac{1}{5}-\frac{1}{4}$ as long as tube   **126. zingeri**
144 Calyx-teeth 3–4 mm, $\frac{1}{3}-\frac{1}{2}$ as long as tube   **125. glaucus**
141 Racemes lax, elongate, usually more than 5 cm
145 Corolla white or yellowish; calyx-teeth (2–)3–4 mm
146 Corolla white; calyx not inflated after anthesis   **124. pallescens**
146 Corolla yellowish; calyx slightly inflated after anthesis   **128. peterfii**
145 Corolla violet; calyx-teeth usually less than 3 mm
147 Peduncles 2–3 times as long as leaves   **123. macropus**
147 Peduncles less than twice as long as leaves
148 Calyx 10–15 mm; legume 3·5–4 mm wide; standard emarginate at apex   **121. brachylobus**
148 Calyx 8–10 mm; legume usually 2–3 mm wide; standard rounded at apex   **122. varius**

Subgen. **Trimeniaeus** Bunge. Usually annuals; hairs usually simple. Leaves imparipinnate.

1. **A. contortuplicatus** L., *Sp. Pl.* 758 (1753). Stems up to 50 cm, ascending or erect. Leaves 2–10 cm; leaflets 6–10 pairs, ovate or obovate, emarginate, densely hairy beneath, sparsely so above. Peduncles $\frac{1}{2}$ as long as leaves; racemes dense, with 5–12 flowers. Calyx *c*. 5 mm, the teeth almost twice as long as tube. Corolla yellow; standard *c*. 6 mm; keel longer than wings. Legume *c*. 7 mm in diameter, curved to form a usually complete ring, membranous, blackish, hairy. *E.C. Europe, extending to N. Bulgaria and S.E. Russia.* Bu ?Cz Hu It Ju Rm Rs (W, K, E).

2. **A. boeticus** L., *Sp. Pl.* 758 (1753). Stems up to 60 cm, erect. Leaves 5–20 cm; leaflets 10–15 pairs, narrowly oblong to oblong-obovate, truncate or emarginate, sparsely hairy beneath, glabrous above, rarely some hairs medifixed. Peduncles $\frac{1}{2}$ as long as leaves; racemes dense, with 5–15 flowers. Calyx 5–7 mm, the teeth as long as tube. Corolla yellow; standard 12–14 mm; wings longer than keel. Legume 20–40 × (5–)7–8 mm, oblong, triangular in

transverse section, grooved beneath; beak hooked; valves keeled, with short, appressed hairs. *Sandy places. Mediterranean region, S. Portugal.* Bl Co Cr Ga Gr Hs It Ju Lu Sa Si.

3. **A. striatellus** Pallas ex Bieb., *Fl. Taur.-Cauc.* **2**: 189 (1808). Stems up to 20 cm, ascending. Leaves 3–7 cm; leaflets 6–8 pairs, oblong-cuneate, truncate or emarginate, glabrous. Peduncles $\frac{1}{2}$ as long to as long as leaves; racemes lax, with 2–5 flowers. Calyx 3·5–4 mm, the teeth almost as long as tube. Corolla whitish; standard 7–11 mm, purple-striped; keel longer than wings. Legume 15–30 × 5 mm, oblong-lanceolate, curved to as much as a semicircle, ovate-triangular in transverse section, grooved beneath; beak straight; valves sharply keeled, glabrous. *Krym, and adjacent coast of Ukraine.* Rs (W, K). (*W. Asia.*)

4. **A. arpilobus** Kar. & Kir., *Bull. Soc. Nat. Moscou* **15**: 336 (1842). Stems up to 15 cm, procumbent. Leaves 2–7 cm; leaflets 1–3(–4) pairs, suborbicular to oblong, emarginate, hairy on both surfaces. Peduncles absent or up to almost twice as long as leaves; racemes lax, with 2–5(–8) flowers. Calyx 2·5–3·4 mm, the teeth $\frac{2}{3}$ as long as tube. Corolla usually whitish; standard *c.* 7 mm, often violet; wings equalling keel. Legume 15–25 × 3 mm, linear, curved to as much as a semicircle, oblong in transverse section, grooved beneath; beak incurved; valves very obtusely keeled, almost smooth, with short, dense, appressed hairs. *S.E. Russia (Oz. Baskunčak).* Rs (E). (*W.C. Asia.*)

5. **A. reticulatus** Bieb., *Fl. Taur.-Cauc.* **3**: 491 (1819). Stems up to 10(–20) cm, ascending or erect. Leaves 3–7 cm; leaflets 5–7 pairs, cuneate, emarginate, sparsely hairy or subglabrous beneath, glabrous above. Peduncles $\frac{1}{2}$ as long as leaves; racemes dense, with 3–5 flowers. Calyx 2–3 mm, the teeth almost as long or as long as tube. Corolla whitish; standard 5–6 mm; wings longer than keel. Legume 10–20 × 2–3 mm, linear-subulate, slightly curved, quadrangular in transverse section, slightly grooved beneath; beak straight; valves distinctly but obtusely keeled, with very strong, raised, reticulate venation, glabrous. *S.E. Russia (Krasnoarmejsk, near Volgograd).* Rs(E). (*N. Kazakhstan.*)

6. **A. ankylotus** Fischer & C. A. Meyer, *Ind. Sem. Horti Petrop.* **2** (*Animadv.*): 27 (1835). Acaulescent, or stems up to 1(–5) cm. Leaves 3–8 cm; leaflets 5–7 pairs, narrowly oblong, obtuse, rarely truncate, hairy on both surfaces. Peduncles $\frac{1}{2}$ as long as or longer than leaves; racemes fairly dense, with 3–6 flowers. Calyx *c.* 6 mm, the teeth $\frac{1}{3}$ as long as tube. Corolla whitish; standard *c.* 9 mm, with an abrupt, long, acute point; wings longer than keel. Legume 20–30 × 3–4 mm, linear, slightly curved, cordate in transverse section; beak 3–4 mm, hooked; valves not keeled, faintly reticulately veined, with short appressed and longer ascending hairs. *S.E. Russia, W. Kazakhstan.* Rs (E).

7. **A. stalinskyi** Širj., *Feddes Repert.* **53**: 75 (1944) (*A. brachymorphus* Nikif.). Like **6** but stems often up to 5(–20) cm, procumbent; leaflets subacute to emarginate; racemes sometimes subsessile; calyx-teeth $\frac{2}{3}$ as long as tube; standard lingulate, obtuse or emarginate; legume 15–25 × 3–4 mm; beak 2–3 mm. *Lower Volga.* Rs (E). (*W.C. Asia.*)

8. **A. scorpioides** Pourret ex Willd., *Sp. Pl.* **3**: 1280 (1802). Stems 5–30 cm, ascending or erect. Leaves 3–8 cm; leaflets 7–8 pairs, oblong or ovate-oblong, emarginate or truncate, sparsely hairy on both surfaces or glabrous above. Peduncles up to 0·3 cm; flowers solitary or 2–3 together. Calyx 6–7 mm, the teeth $\frac{1}{2}$–$\frac{2}{3}$ as long as tube. Corolla bluish or yellowish; standard 9–11 mm; wings longer than keel. Legume 30–50 × 2·5–4 mm,

linear, curved slightly or to as much as a semicircle, cordate in transverse section; beak 1–2 mm, hooked; valves not keeled, slightly rugose, with short appressed hairs. *Corse; S.C. Spain.* Co Hs †It.

9. **A. longidentatus** Chater, *Feddes Repert.* **79**: 47 (1968) (*A. mauritanicus* Cosson, non Steven). Stems up to 30 cm, ascending. Leaves 5–10 cm; leaflets (4–)6–8 pairs, oblong-obovate, truncate or emarginate, sparsely hairy beneath, glabrous above. Peduncles slightly shorter than leaves; racemes with 5–12 flowers. Calyx 5–6·5 mm, the teeth as long as or longer than tube. Corolla pink or pale purple; standard 12–14 mm; wings longer than keel. Legume 25–35 × 5 mm, linear-oblong and acute, slightly curved, narrowly cordate in section; valves not keeled, densely hairy with long, patent hairs. *S.E. Spain (Sierra Alhamilla).* Hs. (*N.W. Africa.*)

10. **A. peregrinus** Vahl, *Symb. Bot.* **1**: 57 (1790). Stems up to 50 cm, robust, procumbent. Leaves 3–10 cm; leaflets 8–10 pairs, oblong or obovate, emarginate or truncate, hairy beneath, usually glabrous above. Peduncles equalling or slightly shorter than leaves; racemes lax, with 3–5 flowers. Calyx 8–9 mm, the teeth $\frac{2}{3}$–$\frac{3}{4}$ as long as tube. Corolla white; standard 15–18 mm; wings longer than keel. Legume 40–60 × 7 mm, linear-lanceolate, straight or scarcely curved, circular in transverse section, grooved beneath; beak straight; valves not keeled, rugose-tuberculate, with sparse, appressed hairs. *Kriti (Koufonisi).* Cr. (*N. Africa.*)

11. **A. pamphylicus** Boiss., *Fl. Or.* **2**: 239 (1872). Stems 5–20 cm, procumbent or ascending. Leaves 2–8 cm; leaflets 8–12 pairs, oblong-cuneate, obtuse to emarginate, hairy beneath, glabrous above. Peduncles almost as long as leaves; racemes dense, with 2–6 flowers. Calyx 8–9 mm, the teeth $\frac{1}{2}$ as long as tube. Standard *c.* 18 mm. Legume 20–30 × 5–6 mm, oblong-linear, slightly curved, circular in transverse section, deeply grooved beneath; beak short; valves not keeled, weakly tuberculate-reticulate, sparsely hairy. *S. Greece (E. Peloponnisos).* Gr. (*S. Anatolia.*)

12. **A. haarbachii** Spruner ex Boiss., *Diagn. Pl. Or. Nov.* **1(2)**: 50 (1843). Annual to perennial; stems 5–50 cm, robust, procumbent or ascending. Leaves 4–10 cm; leaflets 8–12 pairs, oblong to obovate, emarginate, hairy beneath, subglabrous above. Peduncles equalling or longer than leaves; racemes dense, with 7–18 flowers. Calyx 10–13 mm, the teeth $\frac{2}{3}$ as long as tube. Corolla yellowish; standard 20–25 mm, lanceolate, acuminate. Legume 25–30 × 10–15 mm, ovate-oblong, straight, dorsiventrally compressed, with a deep, rounded groove beneath; beak short, straight; valves not keeled, weakly rugose-tuberculate, sparsely to densely lanate. *E. Bulgaria and N.E. Greece; Kriti.* Bu Cr Gr.

13. **A. suberosus** Banks & Solander in A. Russell, *Nat. Hist. Aleppo* ed. 2, **2**: 260 (1794). Like **12** but leaflets usually truncate; legume 20–25 × 8 mm, obovate, slightly curved, with a shallow, angled groove beneath; beak long, hooked; valves strongly tuberculate. *Sardegna.* Sa. (*S.W. Asia.*)

A. verrucosus Moris, *Stirp. Sard.* **1**: 12 (1827), is like **12** but has the legume 20–25 mm, with a shallow groove beneath, and the valves strongly tuberculate; it differs from **13** chiefly in the wider, more abruptly beaked legume and the membranous stipules connate for more than $\frac{1}{2}$ their length (not foliaceous and connate only at the base as in **13**). **A. maritimus** Moris, *Fl. Sard.* **1**: 523 (1837), is like **12** but has the legume *c.* 12 mm, slightly curved, and more strongly tuberculate; it differs from **13** chiefly in the stouter stems and smaller legumes. Both seem to be known only from the original gatherings on Sardegna and satisfactory material has not been traced. The specimens named *A. verrucosus* that have been seen are all referable to **2** or **13**.

14. A. sinaicus Boiss., *Diagn. Pl. Or. Nov.* **2**(9): 57 (1849). Stems up to 10 cm, procumbent or ascending. Leaves 3–8 cm; leaflets 7–10 pairs, oblong or elliptical, obtuse, rarely emarginate, densely hairy on both surfaces. Peduncles very short; racemes dense, with 4–6 flowers. Calyx 4–6 mm, the teeth ⅔ as long to as long as tube. Corolla violet, rarely pale yellow; standard 7–8 mm. Legumes 8–15 × 2·5–4 mm, stellately patent, lanceolate-oblong, almost straight, laterally compressed, but slightly dorsiventrally compressed at base; beak straight or curved; valves not keeled, densely hairy with both short, appressed hairs and long, patent or ascending hairs. *Greece and Aegean region; S. Jugoslavia; Krym.* Cr Gr Ju Rs (K).

15. A. tribuloides Delile, *Descr. Égypte, Hist. Nat.* **2**: 70 (1813). Like **14** but stems up to 20 cm; leaflets 5–10 pairs; peduncles 0·1–0·3(–1) cm; corolla whitish or violet-tinged; legume 5–10 × 2–4 mm, more strongly dorsiventrally compressed at base; valves with only short, appressed hairs. *S.E. Russia (near Astrakhan).* Rs (E). (*W. & C. Asia; India; Egypt.*)

16. A. polyactinus Boiss., *Fl. Or.* **2**: 226 (1872) (*A. cruciatus* auct. hisp., non Link). Stems 2–12 cm, procumbent or ascending. Leaves 2·5–4 cm; leaflets 6–11 pairs, linear to linear-elliptical, densely hairy beneath, sparsely so above. Peduncles (0·2–)1–2 cm, the lower up to half as long as leaves, the upper much shorter; racemes dense, with 6–14 flowers. Calyx 3·5–5 mm, the teeth ⅔ as long as tube. Corolla yellow; standard 5–6 mm. Legumes 10–12 × 3–4 mm, stellately patent even when mature, linear-lanceolate, slightly curved, laterally compressed, expanded and somewhat dorsiventrally flattened at base; beak straight or curved; valves densely hairy with both short, appressed and long, ascending hairs. *S.E. & E. Spain.* Hs. (*N.W. Africa.*)

17. A. stella Gouan, *Obs. Bot.* 50 (1773). Stems up to 20 cm, procumbent or ascending. Leaves 2–8 cm; leaflets 9–11 pairs, oblong to obovate, obtuse or emarginate, sparsely or densely hairy on both surfaces. Peduncles ½ as long as to slightly longer than leaves; racemes dense, with 7–12 flowers. Calyx 5–6 mm, the teeth slightly shorter than tube. Corolla yellowish; standard 9–11 mm. Legumes (7–)10–15 × 3–4 mm, erect or suberect, lanceolate, almost straight, laterally compressed, scarcely dorsiventrally compressed at base; beak *c.* 2 mm, curved; valves densely hairy with appressed or ascending hairs. *W. Mediterranean region; one station in S. Portugal; twice recorded from Greece.* Ga Gr Hs Lu.

18. A. sesameus L., *Sp. Pl.* 759 (1753). Like **17** but stems up to 30 cm, more robust; leaflets usually only sparsely hairy; racemes subsessile, with 5–10 flowers; calyx-teeth as long as tube; standard 8–9 mm, sometimes bluish; legumes strictly erect, with less dense, appressed hairs. *C. & W. Mediterranean region, extending to S. Portugal and S. Bulgaria.* Bu Ga Gr Hs It Ju Lu Sa Si.

19. A. oxyglottis Steven ex Bieb., *Fl. Taur.-Cauc.* **2**: 192 (1808). Stems up to 15 cm, procumbent or ascending. Leaves 2–7 cm; leaflets 5–8 pairs, cuneate, emarginate, sparsely hairy beneath, usually glabrous above. Peduncles shorter than leaves, sometimes only 0·2 cm; racemes dense, with 4–8 flowers. Calyx *c.* 2·5 mm, the teeth ½ as long as tube. Corolla pink; standard 5–8 mm. Legumes 10–15 × 3 mm, stellately patent, linear-lanceolate and acute, not expanded or flattened at base, quadrangular in section with the sides concave; valves sharply keeled, faintly reticulate or almost smooth, glabrous or with appressed hairs. *S.E. Russia; Krym.* Rs (K, E). (*S.W. & C. Asia.*)

20. A. echinatus Murray, *Prodr. Stirp. Gotting.* 222 (1770) (*A. pentaglottis* L.). Stems up to 60 cm, ascending. Leaves 4–8 cm; leaflets 6–9 pairs, obovate or oblong, emarginate or truncate, sparsely (rarely densely) hairy beneath, subglabrous or glabrous above. Peduncles equalling or exceeding leaves; racemes very dense, with 10–15 flowers. Calyx 6–7 mm, the teeth as long as tube. Corolla purplish; standard *c.* 9 mm. Legume 12–15 × 6 mm, ovate, laterally compressed; beak usually hooked; valves not keeled, covered with contorted lamellae and hairs. *Mediterranean region, S. Portugal.* Cr Ga Gr Hs It Lu Si.

Subgen. **Epiglottis** (Bunge) Willk. Annuals; hairs medifixed, straight, appressed, with one arm much shorter than the other. Leaves imparipinnate.

21. A. epiglottis L., *Sp. Pl.* 759 (1753). Stems 5–25(–50) cm, ascending. Leaves 2–4 cm; leaflets 5–10 pairs, hairy on both surfaces. Racemes dense, with 7–12 flowers. Calyx 2·5–3 mm. Corolla yellow; standard 3 mm; stamens 5. Legume 7–9 mm, triangular-ovate, dorsiventrally flattened and broadly cordate at base, laterally compressed near apex, densely hairy. *Mediterranean region, C. & S. Portugal.* Cr Gr Hs It Lu Sa Si [†Ga].

(a) Subsp. **epiglottis**: Leaflets 5–7(–9) pairs, narrowly elliptic-ovate. Peduncles less than 1 cm, shorter than leaves and shorter than the globose infructescence. *Throughout the range of the species.*

(b) Subsp. **asperulus** (Dufour) Nyman, *Consp.* 196 (1878) (*A. asperulus* Dufour): Leaflets up to 10 pairs, narrowly elliptical. Peduncles equalling leaves and at least twice as long as the oblong infructescence. *S. Spain.*

22. A. edulis Durieu ex Bunge, *Astrag. Geront.* **1**: 9 (1868). Stems up to 40 cm, ascending or erect. Leaves 5–8 cm; leaflets 6–8 pairs, elliptical, obtuse, hairy beneath, glabrous or subglabrous above. Peduncles equalling leaves; racemes lax, with 5–10 flowers. Calyx 4–5 mm. Corolla usually yellow; standard *c.* 9 mm; stamens 10. Legume 15–20 × 5–7 mm, oblong, triangular in transverse section with convex sides and concave base; valves very sharply keeled, strongly reticulately veined, subglabrous. *S.E. Spain (near Almería).* Hs. (*N.W. Africa.*)

23. A. cymbicarpos Brot., *Phyt. Lusit.* 63 (1800). Stems up to 25 cm, procumbent. Leaves 5–10 cm; leaflets 7–10 pairs, cuneate or oblong-obovate, emarginate, hairy beneath, glabrous above. Peduncles absent or up to as long as leaves; racemes lax, with 2–5 flowers. Calyx *c.* 5 mm. Corolla white; standard *c.* 7 mm; stamens 10. Legume 10–20 × 5–7 mm, elliptical, slightly curved, inflated; beak long, curved; valves not keeled, rugose, with short appressed hairs. *Damp sandy pastures. Portugal; C. Spain.* Hs Lu.

24. A. hamosus L., *Sp. Pl.* 758 (1753). Stems up to 60 cm, leaves 5–10(–15) cm; leaflets 9–11 pairs, oblong-obovate, emarginate or truncate, hairy beneath, glabrous or subglabrous above. Peduncles ½ as long as leaves; racemes fairly dense, with 5–14 flowers. Calyx 5–6 mm. Corolla yellow; standard 7–8 mm; stamens 10. Legume 20–50 × 2–3 mm, linear, acuminate at apex, curved for about a semicircle, laterally compressed; beak short; valves not keeled, almost smooth, with short appressed hairs. *S. Europe.* Al Bl Bu Co Cr Ga Gr Hs It Ju Lu Rm Rs (K, E) Sa Si Tu.

Subgen. **Hypoglottis** Bunge. Perennials; hairs simple (rarely a few hairs medifixed.) Leaves imparipinnate; stipules connate for ⅓–½ of their length, forming a sheath round the stem opposite the petiole. Flowers subsessile, in dense heads. Calyx not inflated in fruit, the mouth not oblique.

25. A. cicer L., *Sp. Pl.* 757 (1753). Stems (5–)25–60(–100) cm, robust, ascending or suberect. Leaves (6–)9–13 cm; leaflets (8–)10–15 pairs, lanceolate to ovate-lanceolate, with appressed, short, often sparse hairs on both surfaces, rarely subglabrous above. Peduncles ½–⅔ as long as leaves. Calyx 7–10 mm, the teeth almost ½ as long as tube. Corolla yellow; standard 14–16 mm. Legume 10–15 mm, ovoid-globose, inflated, membranous, with short, black and white hairs. *From Belgium and N.C. Russia southwards to N. Spain, Bulgaria and Krym; occasionally naturalized further north.* Au Be Bu Cz Ga Ge He Hs Hu It Ju Po Rm Rs (B, C, W, K, E).

26. A. glaux L., *Sp. Pl.* 759 (1753) (incl. *A. granatensis* Lange, non Lam.). Stems 5–30 cm, robust, ascending. Leaves 3–5 cm; leaflets 12–15 pairs, linear- to obovate-oblong, hairy beneath, glabrous or subglabrous above. Peduncles ½ as long as or up to a little longer than leaves, sparsely to densely hairy with appressed or patent hairs; racemes globose, with many flowers. Calyx *c.* 5 mm, the teeth as long as tube. Corolla purplish; standard 10–12 mm. Legume 5–8 mm, ovoid-trigonous, sulcate beneath, hard, white-villous; beak short, hooked. *Dry pastures. S.W. Europe.* Ga Hs Lu.

27. A. danicus Retz., *Obs. Bot.* 3: 41 (1783) (*A. hypoglottis* auct., ?non L.). Stems 8–30 cm, slender, ascending. Leaves 4–10 cm; leaflets 6–13 pairs, oblong-ovate or oblong, obtuse or emarginate, sparsely hairy on both surfaces; upper stipules 2·5–5 mm. Peduncles 1½–2 times as long as leaves; racemes globose to ovoid-oblong, with many flowers. Calyx 6–8 mm, the teeth ⅖–½ as long as tube. Corolla purplish or bluish-violet; standard 15–18 mm. Legume 7–8 × 5 mm, ovoid, inflated, white-villous. $2n = 16$. *From Ireland and subarctic Russia southwards to the S.W. Alps, Austria and C. Ukraine; rather local.* Au Br Cz Da Ga Ge Hb It Po Rs (N, B, C, W, E) Su.

28. A. purpureus Lam., *Encycl. Méth. Bot.* 1: 314 (1783) (incl. *A. gremlii* Burnat). Stems 10–40 cm, slender, ascending, sometimes with some hairs medifixed. Leaves 4–8 cm; leaflets 7–15 pairs, elliptic-oblong, emarginate, hairy beneath, subglabrous above; upper stipules 5–10 mm. Peduncles 1–2 times as long as leaves; racemes globose, with many flowers. Calyx 8–10 mm, the teeth ⅔ as long as tube. Corolla purplish, rarely whitish; standard *c.* 18 mm, deeply emarginate. Legume 10–15 mm, ovoid, inflated, white-villous. *S. & W. Europe, from W. France and E. Spain to the west part of the Balkan peninsula; mainly in mountains.* Al Ga Hs It Ju.

29. A. bourgaeanus Cosson, *Not. Pl. Crit.* 160 (1852). Like **28** but caespitose, stems 5(–15) cm; peduncles up to 1 cm, much shorter than leaves; calyx *c.* 4 mm; standard 9–10 mm, scarcely emarginate; legume 10–12 mm, cylindric-ovate, subtrigonous, with more closely appressed hairs. ● *E. Spain.* Hs.

30. A. pseudopurpureus Guşuleac, *Bul. Fac. Şti. Cernăuţi* 6: 291 (1933). Like **28** but corolla bluish-violet; standard weakly emarginate; keel mucronate at apex on adaxial side. *Limestone rocks.* ● *E. Carpathians (near Târgu Mureş and Bacău).* Rm.

31. A. setosulus Gontsch., *Not. Syst. (Leningrad)* 10: 33 (1947). Stems 3–10 cm, suberect. Leaves 2·5–4 cm; leaflets 6–8 pairs, lanceolate or oblanceolate, subobtuse, densely hairy on both surfaces. Peduncles ½ as long as to equalling leaves; racemes ovoid, with many flowers. Calyx 12–16 mm, the teeth 1½–2 times as long as tube. Corolla purplish; standard 14–16 mm, obtuse. Legume 9–10 mm, oblong, with dense, ascending hairs. ● *Krym (mountains above Alušta).* Rs (K).

32. A. austraegaeus Rech. fil., *Phyton (Austria)* 1: 202 (1949) (*A. tauricola* auct., non Boiss.). Acaulescent, with branched stock, or stems up to 5 cm. Leaves 4–12 cm; leaflets 12–16 pairs, oblong-ovate, acute, densely hairy beneath with appressed or ascending hairs, sparsely so above. Peduncles 1–1½ times as long as leaves; racemes oblong. Calyx 12–13 mm, the teeth *c.* ¾ as long as tube. Corolla yellowish, violet at apex; standard 15–22 mm, attenuate and emarginate at apex. Legume 8–10 mm, ovoid-trigonous, with dense, semi-appressed, white hairs; beak curved. *S. Aegean region (Karpathos, Kasos).* Cr. (*Rodhos.*)

33. A. turolensis Pau, *Not. Bot. Fl. Esp.* 1: 20 (1887) (*A. aragonensis* Freyn). Acaulescent, with branched stock, or stems up to 20 cm. Leaves 4–8 cm; leaflets 12–17 pairs, elliptic-oblong, obtuse, densely hairy beneath with ascending hairs, sparsely hairy above. Peduncles ½–1½ times as long as leaves; racemes ovate-oblong. Calyx 12–13 mm, the teeth ⅔ as long as tube. Corolla yellow; standard 18–20 mm, slightly attenuate and emarginate. Legume *c.* 10 mm, ovoid-trigonous, brown-lanate. *E. Spain (Prov. Teruel and Cuenca).* Hs. (*N. Africa.*)

Subgen. **Phaca** (L.) Bunge. Perennials, usually sparsely hairy or subglabrous; hairs simple, white, usually appressed. Leaves imparipinnate; stipules usually free from each other, but sometimes partly adnate to petiole. Flowers pedicellate, usually more or less pendent at anthesis. Calyx campanulate, not inflated in fruit, the mouth oblique.

34. A. galegiformis L., *Sp. Pl.* 756 (1753). Stems 40–100 cm, stout, erect. Leaves 10–20 cm; leaflets 11–16 pairs, elliptic- or lanceolate-oblong, obtuse or acute, sparsely hairy beneath, glabrous above; stipules 7–10 mm. Peduncles ½ as long as leaves; racemes 10–20 cm. Calyx 5–6 mm, the teeth ⅓ as long as tube. Corolla yellow; standard 14–15 mm, weakly emarginate. Legume 10–16 mm, linear-lanceolate, slightly curved, triangular in transverse section, glabrous, stipitate. *W. Ukraine; Romania.* Rm Rs (W). (*Caucasus.*)

35. A. frigidus (L.) A. Gray, *Proc. Amer. Acad. Arts Sci.* 6: 219 (1864) (*Phaca frigida* L.). Stems 10–35 cm, stout, erect, usually unbranched, glabrous. Leaves 5–15 cm; leaflets 3–8 pairs, 7–15 mm wide, ovate or broadly elliptical, glabrous above; stipules 10–20 × 5–10 mm. Peduncles 1–1½ times as long as leaves; racemes 2–5 cm, with 5–20 flowers. Calyx 5–6 mm, the teeth *c.* 1/10 as long as tube, broadly triangular. Corolla yellowish-white; standard 12–14 mm, emarginate; wings and keel both 11–13 mm. Legume 20–30 × 6 mm, ellipsoid, flattened above, slightly inflated beneath, densely black- or white-hairy at first but glabrescent. *N. Europe; mountains of C. Europe.* Au Cz Fa Ga Ge He It No Po Rm Rs (N) Su.

(a) Subsp. **frigidus**: Stems 10–35 cm; leaflets 4–8 pairs, sparsely hairy or glabrous beneath; calyx often reddish, sparsely white-hairy on tube, densely black-hairy on teeth. $2n = 16$. *Throughout the range of the species.*

(b) Subsp. **grigorjewii** (B. Fedtsch.) Chater, *Feddes Repert.* 79: 47 (1968) (*A. grigorjewii* B. Fedtsch.): Stems 10–15 cm; leaflets 3–4 pairs, sparsely or densely hairy beneath; calyx green, sparsely white-hairy on tube, densely white-hairy on teeth. ● *N. Russia (Poluostrov Kanin).*

36. A. penduliflorus Lam., *Fl. Fr.* 2: 636 (1778) (*Phaca alpina* L. pro parte). Stems 20–50 cm, stout, erect, usually branched, hairy at least below. Leaves 5–10 cm; leaflets 7–15 pairs, 3–6 mm wide, elliptical to oblong-lanceolate, sparsely hairy on both surfaces or glabrous above; stipules 7–10 × 3 mm. Peduncles 1–1½ times as long as leaves; racemes 2–4 cm, with 5–20 flowers. Calyx 5–7 mm, the teeth ⅕–⅓ as long as tube, triangular-lanceolate or linear. Corolla yellow; standard 10–12 mm, obtuse; wings 9–10 mm; keel 8–9 mm. Legume 20–30 × 10–15 mm, ovoid, flattened above, very strongly inflated beneath, densely blackish-hairy at first but glabrescent. *E. & C. Pyrenees, Alps, Carpathians; C. Sweden.* Au Cz Ga Ge He Hs It Ju Po Rm Su.

37. A. umbellatus Bunge, *Astrag. Geront.* 1: 24 (1868). Stems 2–5 cm, stout, erect, hairy. Leaves 3–6 cm; leaflets 3–4(–5) pairs, ovate, obtuse, usually densely hairy beneath, glabrous above; stipules 5–10 × 3–7 mm. Peduncles slightly longer than leaves; racemes umbellate, with 5–10 flowers. Calyx 7–9 mm, the teeth $\frac{1}{3}$–$\frac{1}{4}$ as long as tube, triangular. Corolla white; standard 16–18 mm; wings and keel both 13–17 mm. Legume 12–18 × 7–10 mm, ovoid-ellipsoid, inflated, densely black-hairy to maturity. *Arctic Russia.* Rs (N). (*Arctic Asia and America.*)

38. A. alpinus L., *Sp. Pl.* 760 (1753) (*Phaca alpina* L. pro parte, *P. astragalina* L.). Stems (4–)8–30 cm, procumbent or ascending, slender. Leaves 4–8(–12) cm; leaflets 7–12 pairs, elliptical, acute or obtuse, sparsely hairy on both surfaces or glabrous above; stipules 3–5 mm, sometimes shortly connate around stem. Peduncles 1–2 times as long as leaves; racemes 2–3 cm, lax, with 5–15 flowers. Calyx 4–5 mm, the teeth lanceolate. Standard 10–14 mm. Legume 8–15 × 3–4 mm, scarcely inflated, blackish-hairy but later glabrescent. *N. Europe, southwards on the mountains to 56° 30′ N. in Scotland; Pyrenees; Alps; Carpathians.* Au Br Cz Fe Ga Ge He Hs It Ju No Po Rm Rs (N) Su.

(a) Subsp. **alpinus**: Calyx-teeth almost as long as tube. Corolla whitish; keel bluish-violet, almost as long as standard and *c.* 1 mm longer than wings. Legume 10–15 mm, oblong, almost completely unilocular. $2n = 16$. *Throughout the range of the species except for for much of arctic Europe.*

(b) Subsp. **arcticus** Lindman, *Svensk Fanerogamfl.* ed. 2, 384 (1926) (*A. subpolaris* Boriss. & Schischkin): Calyx-teeth $\frac{1}{2}$ as long as tube. Corolla purplish-violet; keel *c.* 1 mm shorter than both standard and wings. Legume 8–11 mm, ovoid, often with septum reaching $\frac{1}{2}$ way across. $2n = 16$. *Arctic Europe.*

The differences between the subspecies are much more distinct in E. European Russia than in Fennoscandia, where there is less satisfactory correlation of characters.

39. A. depressus L., *Cent. Pl.* 2: 29 (1756). Acaulescent with branched stock, or stems up to 10 cm, procumbent; hairs medifixed with one arm very short, or almost simple. Leaves 2–30 cm; leaflets 6–14 pairs, obovate to obcordate, appressed-hairy beneath, glabrous above; stipules 4–13 mm, not connate. Peduncles 1–6 cm, never exceeding leaves; racemes oblong, with 6–14 flowers. Calyx (3–)4–6 mm, the teeth $\frac{2}{3}$ as long as tube, lanceolate. Corolla whitish or bluish-purple; standard 10–12(–14) mm, emarginate. Legume (6–)15–22 × 3–4 mm, linear-lanceolate, valves slightly keeled, appressed-hairy but glabrescent, faintly reticulately veined, bilocular. $2n = 16$. *Mountains of S. Europe, extending northwards to N.E. Switzerland.* Al Bu Cr Ga Gr He Hs It Ju Rm Si.

One of the most variable species, particularly in size of parts; no useful subdivision has yet been made.

40. A. norvegicus Weber, *Pl. Min. Cogn. Dec.* 13 (1784) (*A. oroboides* Hornem.). Stems (5–)20–40 cm, stout, erect, glabrous, or sparsely hairy below. Leaves 5–10 cm; leaflets (5–)6–7(–8) pairs, oblong-ovate, usually emarginate, glabrous, but hairy beneath when young; stipules 5–8 mm, ovate-lanceolate. Peduncles $1\frac{1}{2}$–2 times as long as leaves; racemes oblong, dense, with 10–30 flowers. Calyx 4–6 mm, the teeth *c.* $\frac{1}{4}$ as long as tube, triangular, obtuse. Corolla pale violet; standard 10–12 mm, weakly emarginate; wings and keel both 8–10 mm. Legume *c.* 10 × 4 mm, ovoid, compressed, blackish-hairy when young. *Arctic Europe, extending to mountains of Fennoscandia and S. Ural; Carpathians; E. Alps.* Au Cz No ?Rm Rs (N, C) Su.

41. A. australis (L.) Lam., *Fl. Fr.* 2: 637 (1778) (*Phaca australis* L.; incl. *A. krajinae* Domin). Stems 10–30 cm, ascending. Leaves 2–10 cm; leaflets 4–6(–9) pairs, narrowly elliptical to ovate-lanceolate, acute, subglabrous to densely appressed-hairy on both surfaces; stipules 4–10 mm, ovate to lanceolate. Peduncles $1\frac{1}{2}$–2 times as long as leaves; racemes elongate, with 8–15 flowers. Calyx 4–7 mm, the teeth $\frac{1}{2}$ as long as tube. Corolla yellowish-white or white; standard 10–15 mm, sometimes dark violet at apex; wings 8–12 mm, entire or 2-lobed at apex; keel 7–8 mm, dark violet at apex. Legume 10–30 × 5–9 mm, oblong-ovoid, inflated, glabrous. *Mountains of Europe, from the Carpathians and C. Ural southwards to the Pyrenees, C. Appennini and W. & C. Bulgaria; one lowland station in subarctic Russia.* $2n = 32, 48$. Au Bu Cz Ga Ge He Hs It Ju Po Rm Rs (N, C, W).

42. A. lusitanicus Lam., *Encycl. Méth. Bot.* 1: 312 (1783). Stems 30–70 cm, stout, erect. Leaves 8–12(–18) cm; leaflets 8–10 pairs, oblong-lanceolate to elliptical, apiculate, appressed-tomentose-sericeous beneath. Peduncles $\frac{1}{4}$–$\frac{3}{4}$ as long as leaves; racemes oblong, dense, with many flowers. Calyx 10–15 mm, the teeth $\frac{1}{4}$–$\frac{1}{2}$ as long as tube. Corolla white; standard 20–35 mm. Legume 50–70 × 10–20 mm, oblong, slightly inflated, reddish-brown or blackish, tomentose or puberulent. *Portugal and S.W. Spain; S. Greece (Peloponnisos).* Hs Gr Lu.

(a) Subsp. **lusitanicus**: Leaflets glabrous above; calyx-teeth $\frac{1}{3}$–$\frac{1}{2}$ as long as tube; legume usually more than 60 mm. *Portugal and S.W. Spain.* (*N.W. Africa.*)

(b) Subsp. **orientalis** Chater & Meikle, *Feddes Repert.* 79: 48 (1968): Leaflets appressed-tomentose-sericeous above; calyx-teeth $\frac{1}{4}$–$\frac{1}{3}$ as long as tube; legume usually less than 60 mm. *Peloponnisos.* (*S.W. Asia.*)

43. A. glycyphyllos L., *Sp. Pl.* 758 (1753). Stems 30–100 (–150) cm, procumbent. Leaves 10–20 cm; leaflets (3–)4–6(–7) pairs, ovate or broadly elliptical, obtuse, often apiculate, sparsely appressed-hairy beneath, glabrous above; stipules 15–20 mm, lanceolate. Peduncles usually less than $\frac{1}{2}$ as long as leaves; racemes oblong, dense, with many, suberect flowers. Calyx 5–6 mm, glabrous or with black hairs on the teeth; teeth $\frac{1}{3}$–$\frac{1}{2}$ as long as tube. Corolla pale cream; standard 11–15 mm, emarginate. Legume 30–40 × 4–5 mm, linear-oblong, slightly curved, slightly compressed laterally, glabrous. $2n = 16$. *Most of Europe except the extreme north, but mainly on mountains in the south.* Al Au Be Br Bu Co Cz Da *Fe Ga Ge Gr Hb He Ho Hs Hu It Ju Lu No Po Rm Rs (B, C, W, K, E) Su Tu.

44. A. glycyphylloides DC., *Prodr.* 2: 292 (1825). Like **43** but stems erect or ascending; leaflets 5–8 pairs; stipules 10–15 mm, ovate-lanceolate; calyx 6–8 mm, densely blackish-hairy throughout; corolla yellow or whitish-yellow; standard truncate; legume 20–25 × 4–5 mm, oblong-lanceolate. *Balkan peninsula; Krym.* Bu Gr Ju Rs (K).

Plants from Srbija with leaflets patent-hairy beneath, very narrow, membranous (not herbaceous) stipules, longer calyx-teeth and almost straight legumes have been called **A. serbicus** Pančić ex G. Beck, *Fl. Bosn. Herceg.* 3: 278 (1927). It is not clear how constant these characters are and whether they are always correlated.

Subgen. **Astragalus** (Subgen. *Caprinus* Bunge). Perennials, usually densely hairy; hairs simple, usually flexuous or patent and pale brown when dry. Leaves imparipinnate; stipules usually free from each other, sometimes partly adnate to petiole. Flowers pedicellate, suberect. Calyx tubular, not inflated in fruit, the mouth not oblique. Legume semi-bilocular to bilocular.

This subgenus contains many species of narrow geographical range, differing from each other in a combination of minor, often overlapping characters. Most of them have here been aggregated into 3 groups.

(45–47). **A. dasyanthus** group. Leaves 15–35 cm; leaflets 8–20 pairs, 10–30 × 5–12 mm, ovate-oblong to elliptic-lanceolate, sparsely to densely hairy beneath, sparsely hairy above; stipules 12–25 mm, lanceolate to ovate. Peduncles up to ½ as long as leaves. Calyx 12–15(–25) mm, the teeth about equalling tube. Corolla yellow; standard 17–30 mm, hairy. Legume 13–20 × c. 8 mm, ovoid, trigonous, villous, semi-bilocular.

1	Caulescent; standard 17–21 mm	**45. dasyanthus**
1	Acaulescent with branched stock; standard more than 21 mm	
2	Racemes with 4–10 flowers; pedicels 3–6 mm	**46. pubiflorus**
2	Racemes with 10–30 flowers; pedicels c. 2·5 mm	**47. tanaiticus**

45. A. dasyanthus Pallas, *Reise* 3: 749 (1776). Stems (3–) 20–45 cm, ascending or erect. Leaflets 11–20 pairs. Peduncles 10–20 cm; racemes dense, with 10–30 flowers; bracts equalling calyx; pedicels c. 1 mm. Standard 17–21 mm. Legume c. 20 mm. ● *S.E. Europe, extending northwards to Hungary and to c. 53° N. in C. Russia.* Bu ?Cz Hu Ju Rm Rs (C, W, E).

46. A. pubiflorus DC., *Astrag.* 216 (1802) (*A. exscapus* auct. ross., non L.). Acaulescent, with branched stock. Leaflets 8–14 pairs. Peduncles c. 1 cm; racemes dense, with 4–10 flowers; bracts equalling calyx; pedicels 3–6 mm. Standard 23–30 mm. Legume c. 15 mm. ● *Ukraine and adjacent regions of S.C. Russia; outlying stations in E. Romania and Bulgaria.* Bu Rm Rs (W, K, E).

47. A. tanaiticus C. Koch, *Linnaea* 24: 94 (1851). Acaulescent, with branched stock. Leaflets 12–18 pairs. Peduncles 5–10 cm; racemes lax, with 10–30 flowers; bracts ⅔ as long as calyx; pedicels c. 2·5 mm. Standard 23–30 mm. Legume c. 15 mm. ● *S. & E. Ukraine and S. Russia.* Rs (W, E).

(48–53). **A. exscapus** group. Acaulescent or almost so, with branched stock. Leaves 7–30 cm; leaflets 6–30 pairs, narrowly oblong to orbicular-ovate; stipules 10–20 mm, ovate- to linear-lanceolate. Peduncles up to ½ as long as leaves, or racemes subsessile; pedicels 3–7 mm; bracts ⅔ as long as or equalling calyx. Calyx-teeth ⅓–⅔ as long as tube. Corolla yellow; standard glabrous. Legume trigonous, semi-bilocular.

1	Standard 30–35 mm; calyx 15–20 mm	**48. longipetalus**
1	Standard less than 30 mm; calyx not more than 15 mm	
2	Leaflets ovate to ovate-elliptical	
3	Legume 15–25 mm, densely villous	**49. exscapus**
3	Legume c. 30 mm, sparsely hairy or glabrous	**50. huetii**
2	Leaflets oblong-ovate or narrower	
4	Legume glabrous	**53. wolgensis**
4	Legume hairy	
5	Leaflets 10–15 pairs; legume c. 15 mm wide, ovoid	**51. utriger**
5	Leaflets 15–30 pairs; legume 5–10 mm wide, oblong-ovoid	**52. henningii**

48. A. longipetalus Chater, *Feddes Repert.* **79**: 48 (1968) (*A. longiflorus* Pallas pro parte). Leaflets 8–16 pairs, 5–20 × 3–8 mm, ovate or orbicular-ovate, hairy beneath, glabrous above. Ped-

uncles 2–15 cm; racemes with 6–10 flowers. Calyx 15–20 mm; tube glabrous, rarely sparsely hairy; teeth often hairy. Standard 30–35 mm; wings 25–30 mm; keel 20–25 mm. Legume 20–25 × 15 mm, oblong to broadly ovoid, glabrous; beak short, abrupt. *S.E. Russia and W. Kazakhstan.* Rs (E).

49. A. exscapus L., *Mantissa Alt.* 275 (1771). Acaulescent, with branched stock or rarely with short, poorly developed stems. Leaflets 12–19 pairs, 10–25 × 5–10 mm, elliptic-ovate, sparsely or densely hairy on both surfaces. Peduncles very short, rarely up to ½ as long as leaves; racemes with 3–10 flowers. Calyx 12–15 mm, densely hairy. Standard 20–30 mm; wings 15–20 mm; keel 12–15 mm. Legume 15–25 × 7–10 mm, oblong, densely villous; beak very short. *C. Europe, extending to S. Alps; S.E. Spain; Albania; Bulgaria and N.E. Greece.* Al Au Bu Cz Ge Gr He Hs Hu It Rm.

A. ictericus Dingler, *Flora* (*Regensb.*) 64: 381 (1881), and **A. maroniensis** Dingler, *op. cit.* 382 (1881), both described from N.E. Greece, are perhaps variants of **49**, but more information is required; the type specimens have been destroyed.

50. A. huetii Bunge, *Astrag. Geront.* 1: 37 (1868). Leaflets 6–13 pairs, 8–20 × 3·5–10 mm, ovate, subglabrous beneath, glabrous above. Peduncles very short or up to 5 cm. Calyx 11–15 mm, sparsely hairy. Standard 20–30 mm; wings 17–22 mm; keel 15–18 mm. Legume c. 30 × 10 mm, oblong, acuminate, glabrous or sparsely hairy; beak long. ● *Sicilia.* Si.

51. A. utriger Pallas, *Spec. Astrag.* 75 (1800). Leaflets 10–15 pairs, 5–20 × 4–10 mm, ovate- to lanceolate-oblong, sparsely hairy beneath, glabrous above. Peduncles 1–5 cm; racemes with 3–7 flowers. Calyx 12–15 mm, densely hairy. Standard 20–25 mm; wings 17–20 mm; keel 13–17 mm. Legume 15–30 × 15 mm, ovoid, acuminate, densely villous. *Krym.* Rs (K). (*N.W. Caucasus.*)

52. A. henningii (Steven) Boriss. in Komarov, *Fl. URSS* 12: 199 (1946). Leaflets 15–30 pairs, 8–20 × 1·5–6 mm, linear-lanceolate to oblong, hairy on both surfaces or glabrous above. Peduncles 1–5(–15) cm; racemes with 2–7 flowers. Calyx 10–15 mm, densely hairy. Standard 18–23 mm; wings 15–19 mm; keel 13–18 mm. Legume 15–30 × 5–10 mm, oblong-ovoid, acuminate, sparsely hairy. ● *S.E. Russia and S. Ukraine.* Rs (W, E).

53. A. wolgensis Bunge, *Astrag. Geront.* 1: 36 (1868). Leaflets 10–20 pairs, 5–15 × 3–7 mm, lanceolate- to ovate-oblong, subglabrous or sparsely hairy on both surfaces. Peduncles up to 15 cm; racemes with 3–7 flowers. Calyx 10–15 mm, sparsely hairy. Standard 20–25 mm; wings 17–22 mm; keel 15–20 mm. Legume 15–25 × 7–15 mm, ovoid or oblong-ovoid, acute, glabrous; beak short. *E.C. & S.C. Russia.* Rs (C, E).

(54–57). **A. nummularius** group. Acaulescent, with branched stock. Leaves 3–25 cm; leaflets 8–12 pairs, ovate-elliptical to suborbicular; stipules 7–20 mm, lanceolate. Peduncles up to 5 cm; racemes with 3–10 flowers; pedicels 3–7 mm; bracts ½–⅔ as long as tube. Calyx villous, the teeth ⅓ as long as or up to slightly longer than tube. Corolla yellow; standard glabrous or sparsely hairy. Legume laterally compressed or trigonous, semi-bilocular.

1	Calyx-teeth shorter than tube	
2	Legume keeled on back; leaflets truncate or emarginate	**54. nummularius**
2	Legume grooved on back; leaflets obtuse	**55. hellenicus**
1	Calyx-teeth as long as or longer than tube	
3	Leaves 10–25 cm; standard 20–23 mm	**56. anatolicus**
3	Leaves 3–6 cm; standard c. 15 mm	**57. tremolsianus**

54. A. nummularius Lam., *Encycl. Méth. Bot.* 1: 317 (1783). Leaves 5–10 cm; leaflets 4–12 × 3–8 mm, ovate or suborbicular, truncate or emarginate, lanate on both surfaces. Racemes subsessile. Calyx 9–10(–13) mm, the teeth $\frac{1}{3}$–$\frac{1}{2}$ as long as tube. Standard 15–22 mm, glabrous or sparsely hairy; wings 10–18 mm; keel 7–15 mm. Legume 12–20 mm, oblong-ovate, laterally compressed, keeled on back, sparsely hairy. ● *E. Kriti*. Cr.

55. A. hellenicus Boiss., *Fl. Or.* 2: 292 (1872). Leaves 5–10 cm; leaflets 7–20 × 5–15 mm, ovate to ovate-elliptical, obtuse, densely sericeous on both surfaces. Peduncles very short. Calyx 10–15 mm, the teeth $\frac{1}{3}$–$\frac{1}{2}$ as long as tube. Standard 20–27 mm, glabrous; wings 15–20 mm; keel 10–17 mm. Legume 15–22 mm, ovate, compressed-trigonous, with a dorsal groove, densely hairy. *S. Greece*. Gr. (*Anatolia*.)

56. A. anatolicus Boiss., *Diagn. Pl. Or. Nov.* 1(2): 77 (1843). Leaves 10–25 cm; leaflets 10–30 × 7–15 mm, ovate, obtuse, sparsely hairy or subglabrous on both surfaces. Peduncles 1–4 cm. Calyx 10–12 mm, the teeth as long as or slightly longer than tube. Standard 20–23 mm, glabrous; wings 15–18 mm; keel 13–16 mm. Legume 20–25 mm, oblong-ovoid, trigonous, without a dorsal groove, sparsely hairy. *Turkey-in-Europe (Gelibolu peninsula)*. Tu. (*W. Anatolia*.)

57. A. tremolsianus Pau, *Mem. Mus. Ci. Nat. Barcelona (Bot.)* 1(3): 17 (1925). Leaves 3–6 cm; leaflets 5–10 × 2–8 mm, oblong to suborbicular, obtuse, glabrous above, sparsely hairy beneath. Peduncles very short. Calyx *c.* 10 mm, the teeth slightly longer than tube. Standard *c.* 15 mm, glabrous. Legume *c.* 10 mm, oblong-ovoid, trigonous, densely hairy. ● *S.E. Spain (Sierra de Gádor, near Almería)*. Hs.

58. A. graecus Boiss. & Spruner in Boiss., *Diagn. Pl. Or. Nov.* 1(2): 57 (1843). Stems 30–40 cm, stout, erect, with patent hairs. Leaves 15–30 cm; leaflets 20–35 pairs, 5–15 × 5–10 mm, ovate or cordate-orbicular, lanate beneath, glabrous above; stipules 15–30 mm, linear-lanceolate. Peduncles up to 2 cm; racemes confined to middle part of stem, with 5–10 flowers. Calyx 13–20 mm, lanate, the teeth *c.* $\frac{1}{3}$ as long as tube. Corolla yellow; standard 30–40 mm, glabrous. Legume 25–30 mm, oblong-ovate, slightly laterally compressed, tomentose. *C. & S. Greece*. Gr.

59. A. drupaceus Orph. ex Boiss., *Diagn. Pl. Or. Nov.* 3(2): 32 (1856). Like **58** but smaller in all its parts; leaves 8–17 cm; leaflets 8–25 pairs, 5–15 × 2–5 mm, linear-lanceolate; stipules 10–15 mm; peduncles very short; racemes with 3–5 flowers; calyx 11–15 mm, the teeth almost as long as tube; standard 20–30 mm; legume 10–17 mm, glabrous. ● *S. Greece (Peloponnisos)*. Gr.

Subgen. **Tragacantha** Bunge. Perennials, woody at the base; hairs simple. Leaves paripinnate, the rhachis ending in a sharp spine; stipules adnate to petiole for at least $\frac{1}{2}$ their length, but free from each other. Flowers in dense, sessile racemes in the axils of leaves, partly concealed by the stipules, usually confined to the middle of the stem. Calyx not inflated in fruit, hidden by its dense, villous indumentum, the teeth splitting to the base in fruit. Legume *c.* 5 mm, ovoid-ellipsoid, villous.

60. A. creticus Lam., *Encycl. Méth. Bot.* 1: 321 (1783). Forming large, hemispherical tussocks. Leaves 2–5 cm; leaflets (5–)6–7 pairs, 5–12 × 0·5–2 mm, linear-lanceolate; stipules glabrous. Flowers in pairs in axils of leaves; bracts navicular, lanceolate; bracteoles absent. Calyx 6–10 mm, the teeth slightly shorter than or almost as long as tube. Standard 10–15 mm, the limb oblong, sometimes slightly constricted in the middle, with rounded auricles at base. *Mountains of S. part of Balkan peninsula; Kriti*. Al Cr ?Ju Gr.

(a) Subsp. **creticus**: Leaflets densely hairy on both surfaces, the uppermost pair much shorter than the terminal spine of the leaf. Corolla yellowish; standard 10–12 mm; wings 8–11 mm; keel 7–11 mm. *Kriti*.

(b) Subsp. **rumelicus** (Bunge) Maire & Petitmengin, *Mat. Étude Fl. Géogr. Bot. Or.* 2: 15 (1907): Leaflets sparsely hairy or glabrescent, the uppermost pair longer than the terminal spine of the leaf. Corolla pinkish-lilac, rarely yellowish. Standard 13–15 mm; wings 12–13 mm; keel 10–12 mm. ● *Balkan peninsula*.

61. A. granatensis Lam., *Encycl. Méth. Bot.* 1: 321 (1783). Forming large, hemispherical tussocks. Leaves 2–4 cm; leaflets 4–9 pairs, 5–10 × 0·5–2 mm, linear-lanceolate; stipules glabrous or tomentose in the centre. Flowers 4–7, in axils of leaves; bracts navicular, linear-lanceolate; bracteoles shorter than calyx, linear-oblanceolate, and much less hairy than calyx, or absent. Calyx 5–9 mm, the teeth slightly shorter than tube. Standard 12–15 mm, the limb oblong, with acute auricles at base. *Sicilia; C. & S. Spain*. Hs Si.

(a) Subsp. **granatensis** (*A. siculus* subsp. *plumosus* Arcangeli; *A. boissieri* Fischer): Leaflets densely grey-tomentose on both surfaces. Calyx-teeth linear-lanceolate, more or less abruptly aristate at apex. Corolla whitish-yellow, the standard with pinkish veins. *N. Sicilia (Nebrodi); C. & S. Spain*.

(b) Subsp. **siculus** (Biv.) Franco & P. Silva, *Feddes Repert.* 79: 49 (1968) (*A. siculus* Biv.): Leaflets glabrescent, or appressed-hairy at maturity. Calyx-teeth linear-subulate, gradually narrowed to apex. Corolla pinkish, the standard with darker veins. ● *Sicilia (Etna)*.

62. A. arnacantha Bieb., *Fl. Taur.-Cauc.* 2: 205 (1808). Like **61**(a) but leaflets 4–5(–6) pairs; flowers in pairs in axils of leaves; corolla whitish-yellow or pink; standard 14–18 mm, the limb with rounded auricles at base. ● *Krym; E. Bulgaria*. Bu Rs (K).

63. A. parnassi Boiss., *Diagn. Pl. Or. Nov.* 2(9): 80 (1849). Forming large hemispherical tussocks. Leaves 1·5–7 cm; leaflets 5–8(–15) pairs, 5–10 × 1–3 mm, linear-lanceolate to elliptical; stipules glabrous or subglabrous. Bracts linear-oblanceolate; bracteoles almost as long as calyx, linear-subulate, villous like the calyx. Calyx 10–12 mm, the teeth about as long as the tube. Corolla purplish or pink, rarely yellowish; standard with ovate limb. *S. part of Balkan peninsula; Calabria*. Al Gr It Ju Tu.

1 Leaflets moderately hairy on both surfaces, the terminal pair
    shorter than terminal spine of leaf   **(c) subsp. cylleneus**
1 Leaflets sparsely hairy beneath, glabrous above, the terminal
    pair longer than terminal spine of leaf
   2 Standard 14–17 mm   **(a) subsp. parnassi**
   2 Standard 18–24 mm   **(b) subsp. calabrus**

(a) Subsp. **parnassi**: Leaflets sparsely hairy beneath, glabrous above, the terminal pair longer than the terminal spine of the leaf. Racemes with 3–4 flowers. Standard 14–17 mm, with acute auricles at base. ● *From N.W. Macedonia and Thrace to S.C. Greece*.

(b) Subsp. **calabrus** (Fiori) Chater, *Feddes Repert.* 79: 49 1968) (*A. calabrus* Fiori): Like subsp. (a) but racemes with up to 6 flowers; standard 18–24 mm. ● *Calabria*.

(c) Subsp. **cylleneus** (Boiss. & Heldr. ex Fischer) Hayek, *Prodr. Fl. Penins. Balcan.* 1: 781 (1926): Leaflets moderately hairy on both surfaces, the terminal pair usually slightly shorter than the terminal spine of the leaf. Standard 14–20 mm, with subacute or obtuse auricles at base. ● *S. Greece (mountains of N. Peloponnisos)*.

64. A. thracicus Griseb., *Spicil. Fl. Rumel.* **1**: 55 (1843). Like **63(a)** but not forming tussocks; leaves up to 9 cm; stipules tomentose, at least in the centre; racemes with up to 7 flowers; corolla white; standard 17–20 mm. ● *S.E. part of Balkan peninsula.* Bu Gr Ju Tu.

A. monachorum Širj., *Feddes Repert.* **47**: 242 (1939), from Athos, appears to be intermediate between **63(a)** and **64**; it is like **64** but has glabrous stipules, calyx 20–23 mm and standard 20–24 mm. **A. jankae** Degen & Bornm., *Magyar Bot. Lapok* **18**: 17 (1919), from S. Bulgaria, appears to be intermediate between **64** and **65**; it is like **64** but has the leaflets hairy above, and the bracts ovate. The relationship between these two taxa and **63(a)**, **64** and **65** is obscure.

65. A. trojanus Steven ex Fischer, *Syn. Astrag. Trag.* 88 (1853). Like **63(a)** but not forming tussocks; stipules usually tomentose; bracts ovate, navicular; calyx 11–14 mm; corolla white; standard 18–22 mm. *Turkey-in-Europe* (*S. part of Gelibolu peninsula.*) Tu. (*W. Anatolia.*)

Subgen. **Calycophysa** Bunge. Perennials; hairs simple. Leaves usually imparipinnate; stipules adnate to petiole for up to ½ their length, free or almost free from each other. Flowers suberect, often in dense, subcapitate racemes. Calyx more or less inflated and not splitting to the base in fruit. Corolla persistent. Legume included in calyx.

66. A. physocalyx Fischer, *Bull. Sci. Acad. Imp. Sci. Pétersb.* **2**: 74 (1837). Caespitose, acaulescent, with branched stock. Leaves 15–20 cm, imparipinnate; leaflets 15–22 pairs, ovate or elliptical, acute, sparsely hairy. Peduncle very short; racemes with 1–5 flowers; bracts 5–8 mm, ovate. Calyx 15–17 mm at anthesis, 25 × 15 mm and vesicular-inflated in fruit. Corolla yellow; standard 35–45 mm. Legume 12–15 mm, oblong-lanceolate, glabrous. $2n = 16$. ● *S.W. Bulgaria* (*Kulata*). *Extinct in its only other known locality* (*Plovdiv*). Bu.

67. A. ponticus Pallas, *Spec. Astrag.* 14 (1800). Stems 50–100 cm, stout, erect, sparsely hairy. Leaves 10–30 cm, imparipinnate; leaflets 15–25 pairs, oblong-elliptical, glabrous except on mid-vein beneath. Peduncles up to 1 cm; racemes ovoid; bracts as long as calyx. Calyx 10–15 mm, pubescent with short hairs, slightly inflated in fruit, the teeth less than ½ as long as tube. Corolla yellow. Standard 15–22 mm; wings slightly longer than keel. Legume ovoid-compressed, sparsely hairy. *S.E. Europe, from Bulgaria to S.E. Russia.* Bu Rm Rs (W, K, E).

68. A. vulpinus Willd., *Sp. Pl.* **3**: 1259 (1802). Stems 15–40 cm, stout, erect, sparsely hairy. Leaves 7–20(–25) cm, imparipinnate; leaflets 12–15 pairs, broadly ovate or obcordate, with short, dense hairs beneath, subglabrous above; stipules glabrous. Peduncles up to 1 cm; racemes ovoid or oblong; bracts ⅔ as long as calyx. Calyx 15–25 mm, villous with long, dense hairs, slightly inflated in fruit, the teeth about as long as tube. Corolla yellowish; standard 22–28 mm, the limb 9–11 mm wide; wings not or scarcely longer than keel. Legume ovoid, with long, dense hairs. *S.E. Russia and W. Kazakhstan.* Rs (E).

69. A. alopecurus Pallas, *Spec. Astrag.* 11 (1800). Stems 50–100 cm, stout, erect, densely hairy. Leaves 20–30 cm, imparipinnate; leaflets 17–27 pairs, lanceolate to ovate-lanceolate, glabrous above, sparsely hairy beneath; stipules hairy. Peduncles up to 1(–2) cm; racemes ovoid or oblong; bracts as long as calyx. Calyx 12–17 mm, villous with long, dense hairs, slightly inflated in fruit, the teeth ⅔–¾ as long as tube. Corolla pale yellow; standard 18–23 mm, the limb 5–7 mm wide; wings slightly longer

than keel. Legume ovoid-globose, with long, dense hairs. *S. Ural.* Rs (C). (*N. & W. Asia.*)

70. A. alopecuroides L., *Sp. Pl.* 755 (*A. narbonensis* Gouan). Stems 15–75 cm, stout, erect, densely hairy to subglabrous. Leaves 15–20 cm, imparipinnate; leaflets 12–15 pairs, oblong-elliptical to oblong, densely hairy to subglabrous beneath, glabrous above; stipules usually hairy. Peduncles up to 2 cm; racemes globose; bracts slightly longer than calyx. Calyx 12–17 mm, villous with long, dense hairs, slightly inflated in fruit, the teeth slightly longer than tube. Corolla pale yellow; standard 22–27 mm; wings slightly shorter than keel. Legume triangular-ovate, strongly laterally compressed, with long, dense hairs. ● *E., C. & S. Spain, extending northwards into S. France* (*Hérault, Aude*). Ga Hs.

71. A. grossii Pau, *Mem. Mus. Ci. Nat. Barcelona* (*Bot.*) **1**(3): 16 (1925). Stems 20–30 cm, stout, erect, sparsely hairy. Leaves 10–20 cm, imparipinnate; leaflets 11–12 pairs, 5–7 mm, oblong, subglabrous; stipules glabrous. Peduncles 1–2 cm; racemes oblong; bracts *c.* ½ as long as calyx. Calyx *c.* 15 mm, villous, slightly inflated in fruit, the teeth slightly shorter than tube. Corolla pale yellow; standard *c.* 25 mm. Legume unknown. ● *S.E. Spain* (*Sierra de Gádor, near Almería*). Hs.

The status of this plant is uncertain; it may be only a variant of **70**.

72. A. centralpinus Br.-Bl., *Feddes Repert.* **79**: 49 (1968) (*A. alopecuroides* auct., non L.). Stems 50–100 cm, stout, erect, densely hairy. Leaves 20–30 cm, imparipinnate; leaflets 20–30 pairs, elliptic- to ovate-lanceolate, sparsely hairy beneath, glabrous above; stipules glabrous or hairy. Racemes ovoid, sessile; bracts equalling calyx. Calyx 14–20 mm, villous with long, dense hairs, slightly inflated in fruit, the teeth *c.* ⅔ as long as tube. Corolla yellow; standard 15–20 mm; wings equalling keel. Legume ovoid-compressed, with long, dense hairs. ● *S.W. Alps; S. Bulgaria* (*W. Rodopi*). Bu Ga It.

73. A. ajubensis Bunge, *Astrag. Geront.* **1**: 61 (1868) (*A. durhamii* Turrill). Stems 50–100 cm, stout, erect, glabrous. Leaves 10–20 cm, imparipinnate; leaflets 10–15 pairs, oblong-lanceolate, glabrous above, subglabrous beneath; stipules glabrous. Peduncles 3–8 cm; racemes globose; bracts *c.* ½ as long as calyx. Calyx 17–19 mm, villous with long, dense hairs, slightly inflated in fruit, the teeth as long as tube. Corolla yellow; standard 20–23 mm; wings about equalling keel. Legume obovoid, with long, dense hairs. *Turkey-in-Europe* (*Gelibolu peninsula*). Tu. (*S.W. Asia.*)

74. A. clusii Boiss., *Diagn. Pl. Or. Nov.* **2**(9): 101 (1849) (*A. tumidus* Willd. pro parte). Caespitose, spiny; stems 10–15 cm, woody. Leaves 3–5 cm, paripinnate, the rhachis ending in a spine; leaflets 4–8 pairs, linear-lanceolate to obovate, densely hairy on both surfaces; stipules densely hairy. Peduncles short; racemes with 2–3 flowers; bracts ovate, acuminate; bracteoles present. Calyx *c.* 10 mm at anthesis, densely appressed-hairy, *c.* 20 × 15 mm and vesicular-inflated in fruit. Corolla whitish; standard 15–20 mm. Legume obovoid, densely hairy. ● *S. & E. Spain.* Hs.

75. A. sempervirens Lam., *Encycl. Méth. Bot.* **1**: 320 (1783). Sometimes caespitose, spiny; stems 5–40 cm, woody at base, procumbent or ascending. Leaves 2–7 cm, paripinnate, the rhachis ending in a spine; leaflets 4–10 pairs, linear-oblanceolate; stipules adnate to petiole for *c.* ½ their length. Peduncles short; racemes with (3–)4–8 flowers; bracts ovate- to linear-lanceolate, acuminate;

bracteoles absent. Calyx 7–15 mm, villous, slightly inflated in fruit, the teeth as long as tube. Corolla white to purple, rarely yellow; standard longer than wings. Legume ovoid, densely hairy. *Mountains of S. Europe, extending northwards to C. Switzerland.* Ga Gr He Hs It.

1 Standard *c.* 1½ times as long as calyx
 2 Not caespitose; standard 16–20 mm (a) subsp. **sempervirens**
 2 Caespitose; standard 10–12 mm (b) subsp. **cephalonicus**
1 Standard equalling calyx
 3 Corolla pale purplish or yellow; bracts with weak lateral veins (c) subsp. **muticus**
 3 Corolla purple; bracts without lateral veins (d) subsp. **nevadensis**

(a) Subsp. **sempervirens** (*A. aristatus* L'Hér.): Not caespitose. Leaflets 6–10 pairs, sparsely to densely hairy beneath, sparsely hairy or subglabrous above. Bracts ovate-lanceolate, with prominent, branched or anastomosing lateral veins. Corolla whitish to pale purplish; standard *c.* 1½ times as long as calyx. ● *From the Alps and Appennini to N.E. Spain; ?C. part of Balkan peninsula.*

(b) Subsp. **cephalonicus** (C. Presl) Ascherson & Graebner, *Syn. Mitteleur. Fl.* **6(2)**: 779 (1909): More or less caespitose. Leaflets 6–7 pairs, usually densely hairy on both surfaces. Bracts ovate-lanceolate, with prominent, anastomosing lateral veins. Corolla whitish, rarely pale purplish; standard *c.* 1½ times as long as calyx. *Greece.*

(c) Subsp. **muticus** (Pau) Rivas Goday & Borja, *Anal. Inst. Bot. Cavanilles* **19**: 406 (1961): Caespitose. Leaflets 6–7 pairs, usually obtuse, densely appressed-hairy on both surfaces. Bracts lanceolate, with weak lateral veins. Corolla pale purplish or yellow; standard about equalling calyx. ● *E.C. Spain.*

(d) Subsp. **nevadensis** (Boiss.) P. Monts., *Collect. Bot.* (*Barcelona*) **2**: 266 (1949): Caespitose. Leaflets 6–7 pairs, acute, densely appressed-hairy on both surfaces. Bracts linear-lanceolate, with a single, central vein. Corolla purple; standard about equalling calyx. ● *S. Spain.*

76. A. giennensis Heywood, *Feddes Repert.* **79**: 50 (1968). Like **75(d)** but more robust, with stronger spines; leaflets 6–9 pairs, less densely hairy; peduncles *c.* 1 cm; bracts lanceolate or linear-lanceolate, with 1–2 distinct lateral veins; calyx-teeth about twice as long as tube; corolla pale purplish. ● *S.E. Spain* (*Prov. Jaén*). Hs.

Subgen. **Cercidothrix** Bunge. Perennials, sometimes suffruticose; hairs usually medifixed. Leaves imparipinnate; rhachis sometimes spine-like and then terminal leaflet caducous; stipules free, adnate to the petiole, or connate to each other around the stem. Flowers erecto-patent to erect, or pendent, usually subsessile. Calyx tubular, not inflated in fruit.

77. A. angustifolius Lam., *Encycl. Méth. Bot.* **1**: 321 (1783). Caespitose, forming dense tussocks, spiny; stems 5–20 cm, woody at base. Leaves 2·5–6 cm, the spine-like rhachis slender; leaflets 6–10 pairs, 2–7 × 0·75–2 mm, narrowly elliptical to linear, acute or subobtuse, usually densely appressed-hairy on both surfaces; stipules glabrous or hairy. Peduncles very short or up to slightly longer than leaves; racemes ovoid, with 3–12 flowers; bracts 4–8 mm, linear-lanceolate. Calyx 5–9 mm, appressed-hairy, the teeth ¼–½ as long as tube. Corolla white; standard 13–23 mm; wings 10–20 mm; keel 8–15 mm, sometimes purplish. Legume 10–15 mm, exceeding calyx, oblong-lanceolate, triquetrous, with short, dense, appressed hairs. *Mountain rocks. Balkan peninsula, Thasos, Kriti.* Al Bu Cr Gr Ju.

Two subspecies may be recognized. There is also clinal variation in certain characters, particularly in the length of the calyx-teeth, which decreases towards the south.

(a) Subsp. **angustifolius**: Leaves 2·5–4 cm, the rhachis not strongly pungent. Peduncles equalling or shorter than leaves; racemes with 3–8 flowers. *Throughout the range of the species.*

(b) Subsp. **pungens** (Willd.) Hayek, *Prodr. Fl. Penins. Balcan.* **1**: 790 (1926): Leaves 4–6 cm, the rhachis strongly pungent. Peduncles longer than leaves; racemes with 6–12 flowers. *N. Greece.*

78. A. balearicus Chater, *Feddes Repert.* **79**: 51 (1968) (*A. poterium* auct., non Vahl). Like **77** but smaller in all its parts; leaves 1–3 cm; leaflets 3–5 pairs, usually sparsely hairy on both surfaces; stipules glabrous; bracts 1–4 mm; racemes with 1–5 flowers; calyx 4–5 mm; standard 11–12 mm; legume 7–9 mm, lanceolate, with sparse, appressed hairs. ● *Islas Baleares.* Bl.

79. A. massiliensis (Miller) Lam., *Encycl. Méth. Bot.* **1**: 320 (1783) (*A. tragacantha* L. pro parte). Usually laxly caespitose and not forming dense tussocks, spiny; stems 10–30 cm, woody at base. Leaves 2–7 cm, the spine-like rhachis stout; leaflets 6–12 pairs, 4–6 × 1·5–2·5 mm, oblong or elliptical, obtuse or truncate and mucronate, densely hairy beneath, densely or sparsely so above; stipules densely appressed-hairy. Peduncles up to 3 cm; racemes with 3–8 flowers; bracts 3–4 mm, ovate-lanceolate. Calyx 5–7 mm, appressed-hairy, the teeth ⅕–¼ as long as tube. Corolla white; standard 13–17 mm; wings 11–15 mm; keel 9–13 mm. Legume 9–10 mm, oblong, acute, with short, dense, more or less appressed hairs; beak absent or less than 2 mm. ● *S.W. Europe.* Co Ga Hs ?It Lu Sa ?Si.

80. A. sirinicus Ten., *Fl. Neap. Prodr. App. Quinta* 23 (1826). Like **79** but more densely caespitose; leaflets often subacute; stipules glabrous; racemes with up to 15 flowers; bracts 5–8 mm; calyx 6–10 mm, patent-hairy, with teeth up to ½ as long as tube; corolla yellowish, tinged with violet; standard 14–19 mm; legume 10–13 mm, with dense erecto-patent hairs or glabrescent; beak up to 3 mm. ● *C. Mediterranean region and C. & S.W. part of Balkan peninsula.* Al Co Ga Gr It Ju Sa.

(a) Subsp. **sirinicus** (incl. *A. angustifolius* subsp. *tymphresteus* (Boiss. & Spruner) Hayek): Leaflets obtuse to subacute. Racemes with 8–15 flowers; peduncles 1–3 cm, *c.* ½ as long as leaves; bracts 6–8 mm. Calyx-teeth *c.* ½ as long as tube. Standard 14–17 mm. Legume with dense erecto-patent hairs persistent to maturity; beak 2–3 mm. *Appennini, Balkan peninsula.*

Often confused with **77(a)** in the Balkan peninsula; they can best be distinguished by the pubescence on the legume and calyx.

(b) Subsp. **genargenteus** (Moris) Arcangeli, *Comp. Fl. Ital.* 187 (1882): Leaflets obtuse, mucronate. Peduncles very short; racemes with 3–5 flowers; bracts *c.* 5 mm. Calyx-teeth ¼–⅓ as long as tube. Standard 17–19 mm. Legume glabrescent; beak less than 2 mm. *Corse, Sardegna.*

81. A. odoratus Lam., *Encycl. Méth. Bot.* **1**: 311 (1783). Stems up to 30 cm, ascending or erect. Leaves 4–12 cm; leaflets 9–14 pairs, 10–17 mm, lanceolate to elliptic-oblong, obtuse or acute, slightly hairy beneath, glabrous above; stipules connate. Peduncles about equalling leaves; racemes dense; flowers pendent. Calyx 4–5 mm, the teeth *c.* ¼ as long as tube. Corolla whitish or yellow; standard 9–12 mm. Legume 8–10 × 2–3 mm, oblong-lanceolate, laterally compressed, smooth, with sparse appressed hairs, glabrescent. *N.E. Greece and S.E. Jugoslavia.* Gr Ju. (*S.W. Asia.*)

Occasionally recorded elsewhere as a casual, and perhaps naturalized in C. Italy.

82. A. algarbiensis Cosson ex Bunge, *Astrag. Geront.* **1**: 9 (1868). Like **81** but leaflets oblong-cuneate, truncate or emarginate; corolla yellow; calyx-teeth $\frac{1}{3}-\frac{1}{2}$ as long as tube; legume 8–12 × 4–5 mm, semi-lunate, rugose, glabrous. ● *S. Spain and S. Portugal.* Hs Lu.

Placed by Bunge in Subgen. *Epiglottis*, but clearly a perennial and closely related to **81**.

83. A. falcatus Lam., *Encycl. Méth. Bot.* **1**: 310 (1783). Like **81** but stems 40–80 cm, erect; leaves 8–17 cm; leaflets 15–20 pairs, 15–25 mm, oblong, densely hairy beneath; only the lowest stipules connate; corolla yellow; legume 15–25 mm, oblong and acute, slightly curved, laterally compressed, with persistent, appressed hairs. *C. & S.E. Russia.* Rs (C, E) [*Rm].

84. A. asper Jacq., *Misc. Austr. Bot.* **2**: 335 (1781). Stems 20–60 cm, stout, herbaceous, erect; branches strict. Leaves 5–10 cm; leaflets 8–15 pairs, 15–30 mm, linear-lanceolate, sparsely appressed-hairy on both surfaces or subglabrous; stipules not connate. Peduncles 1–2 times as long as leaves; racemes long, dense; bracts 4–8 mm, linear-lanceolate. Calyx 8–14 mm, the teeth $\frac{1}{3}-\frac{1}{2}$ as long as tube. Corolla yellow; standard 15–20 mm. Legume 12–25 × 3 mm, linear-lanceolate and acuminate, straight or curved at apex, with appressed hairs. *E.C. Europe and south part of U.S.S.R., extending southwards to N. Bulgaria.* Au Bu Cz Hu Rm Rs (C, W, K, E).

85. A. roemeri Simonkai, *Term.-Tud. Közl.* (*Pótfüz.*) **19**: 138 (1892). Stems 50–70 cm, stout, herbaceous, erect. Leaves 7–12 cm; leaflets 4–6 pairs, 25–60 mm, elliptical or lanceolate, sparsely appressed-hairy beneath, subglabrous above; stipules not or only shortly connate round stem. Peduncles about twice as long as leaves; racemes ovoid-oblong; bracts *c.* 3 mm, linear-lanceolate. Calyx 8–10 mm, the teeth almost as long as tube; corolla whitish-lilac; standard 20–25 mm. Legume 15–20 × 5–8 mm, oblong-lanceolate, straight, densely appressed-hairy; beak short, curved. *Calcareous rocks and screes.* ● *E. Carpathians.* Rm.

86. A. austriacus Jacq., *Enum. Stirp. Vindob.* 263 (1762). Often some hairs simple. Stems 10–60 cm, slender, erect or ascending; branches erecto-patent. Leaves 3–7 cm; leaflets 5–10 pairs, 5–15 × 0·75–2·5 mm, usually linear, subglabrous; stipules connate only near base of stem. Peduncles equalling or slightly longer than leaves; racemes long, lax, with 7–25 flowers; bracts 0·5–1 mm, ovate; calyx 2–3 mm, the teeth *c.* $\frac{1}{4}$ as long as tube, triangular-ovate. Corolla blue and violet; standard 5–8 mm. Legume 5–12 × 1·5–2·5 mm, linear-oblong and acute, appressed-hairy. *E.C. & E. Europe, from Austria to S. Ural, extending locally southwards to N. Bulgaria and Krym; S.W. Alps; N.E. Spain.* Au Bu Cz Ga Hs Hu It Ju Rm Rs (C, W, K, E).

87. A. sulcatus L., *Sp. Pl.* 756 (1753). Stems 25–80 cm, erect; branches strict. Leaves 5–7 cm; leaflets 6–11 pairs, 10–30 × 0·5–1 mm, linear to linear-oblong, sparsely hairy beneath; stipules not connate. Peduncles 1–1½ times as long as leaves; racemes long, lax, with 5–20 flowers; bracts 1–2·5 mm, triangular-lanceolate. Calyx 2·5–4 mm, the teeth $\frac{1}{3}-\frac{1}{4}$ as long as tube, linear-lanceolate. Corolla pale lilac; standard 7–8 mm. Legume 8–12 × 2 mm, narrowly oblong and acute, straight, sparsely appressed-hairy, unilocular, rarely semi-bilocular. *From E. Austria eastwards to S. Ural and W. Kazakhstan.* Au Cz Hu Rm Rs (C, W, E).

88. A. clerceanus Iljin & Krasch., *Not. Syst.* (*Leningrad*) **5**: 113 (1924). Like **87** but stems suberect; branches erecto-patent; leaflets 5–8 pairs, 2·5–6 mm wide, ovate-oblong; standard 10–12 mm; legume 10–20 × 3 mm, bilocular or almost so. ● *S. Ural* (*near Sterlitamak*). Rs (E).

89. A. tenuifolius L., *Sp. Pl.* ed. 2, 1065 (1763) (*A. scopiformis* Ledeb., *A. tauricus* Pallas pro parte). Subcaespitose, woody at base; stems 1–7 cm, many. Leaves 3–8 cm; leaflets 5–7 pairs, 5–15 × 0·5–1(–2·5) mm, linear-setaceous, with short, white, appressed hairs on both surfaces; stipules connate. Peduncles 1–2 times as long as leaves; racemes oblong, lax, with 10–25 flowers; bracts *c.* 1 mm, triangular-ovate. Calyx 2·5–3 mm, the teeth *c.* $\frac{1}{3}$ as long as tube, triangular-lanceolate. Corolla whitish or pale lilac; standard 6–7 mm. Legume oblong-conical and acuminate, with appressed hairs. *S.E. part of U.S.S.R., from Krym to Baškirskaja A.S.S.R.* Rs (C, K, E).

90. A. arenarius L., *Sp. Pl.* 759 (1753). Stems 15–30 cm, slender, procumbent or ascending, woody at base. Leaves 3–5 cm; leaflets 2–9 pairs, 10–20(–30) × 2–4 mm, lanceolate or linear, with appressed hairs on both surfaces; stipules 3–6 mm, connate. Peduncles *c.* $\frac{2}{3}$ as long as leaves; racemes oblong, lax, with 3–8 flowers; bracts 1–2 mm, linear-lanceolate. Calyx 4–5 mm, the teeth *c.* $\frac{1}{4}$ as long as tube, triangular. Corolla purplish or lilac, rarely whitish or yellowish; standard 13–17 mm. Legume 12–20 × 3·5–4 mm, oblong and acute or truncate, with short appressed hairs, sometimes glabrescent; beak short. ● *U.S.S.R., from c. 49° to 61° N., N.C. Europe eastward to C. Germany and extending to S. Sweden; locally naturalized elsewhere.* Cz Ge Po Rs (N, B, C, W, E) Su [Fe Ge].

91. A. baionensis Loisel., *Fl. Gall.* 474 (1807). Like **90** but leaflets 2–6 × 0·5–1(–1·5) mm, oblong-linear; peduncles usually equalling leaves; corolla pale blue; standard 12–14 mm; legume 8–10 × 4 mm, truncate. *Maritime sands.* ● *Coasts of W. France and N. Spain.* Ga Hs.

92. A. tenuifoliosus Maire, *Bull. Soc. Hist. Nat. Afr. Nord* **39**: 134 (1949) (*A. tenuifolius* Desf., non L.). Like **90** but leaflets 5–10 × 1–2 mm; racemes with up to 12 flowers; calyx 7–9 mm, the teeth $\frac{1}{3}-\frac{2}{3}$ as long as tube, linear to lanceolate; standard up to 20 mm; legume 2·5–3 mm wide, linear, long-acute. *S.E. Spain* (*Murcia*). Hs. (*N. Africa.*)

93. A. mesopterus Griseb., *Spicil. Fl. Rumel.* **1**: 49 (1843). Like **90** but leaflets 10–12 pairs; peduncles twice as long as leaves; corolla purplish; calyx *c.* 6 mm. *Stony places.* ● *N.E. Greece and Turkey-in-Europe.* Gr Tu.

94. A. onobrychis L., *Sp. Pl.* 760 (1753) (incl. *A. skorpilii* Velen., *A. sofianus* Velen., *A. pancicii* Heuffel, *A. borysthenicus* Klokov, *A. circassicus* Grossh.). Stems 10–60 cm, procumbent and ascending, woody at base. Leaves 3–10 cm; leaflets 8–15 pairs, 4–15 × 1–3(–5) mm, elliptic-lanceolate, acute or subobtuse, appressed-hairy on both surfaces; stipules 1·5–12 mm, connate. Peduncles 1–3 times as long as leaves; racemes ovoid or oblong, with usually more than 10 flowers; bracts 2–4 mm, lanceolate. Calyx 6–8 mm, the teeth $\frac{1}{4}-\frac{3}{4}$ as long as tube, linear to lanceolate. Corolla pale or dark violet, rarely white or yellowish; standard 15–30 mm, the limb linear-oblong. Legume 7–15 mm, ovoid-lanceolate to oblong, acute, compressed, densely appressed-hairy; beak distinct. *Europe, northwards to S. France, Austria and C. Ural.* Al Au Bu Cz Ga †Ge Gr He Hs Hu It Ju Po Rm Rs (C, W, K, E) Tu.

Extremely variable throughout almost the whole of its range. *A. onobrychioides* Bieb., *Tabl. Prov. Casp.* 117 (1798) (*A. cephalotes* Pallas) from the Caucasus, is like **94** but has suberect stems 3–8 cm, linear-lanceolate or subulate bracts 7–10 mm, and calyx 9–12 mm, with the teeth almost as long as tube. It appears to be quite distinct; it has been doubtfully reported from the Lower Don.

95. A. leontinus Wulfen in Jacq., *Misc. Austr. Bot.* **2**: 59 (1781). Stems up to 20 cm, slender; sometimes acaulescent. Leaves 5–12 cm; leaflets 5–10 pairs, 5–15 × 2–5 mm, ovate to narrowly elliptical, obtuse or truncate, usually densely appressed-hairy beneath, subglabrous above; stipules 3–8 mm, connate or free. Peduncles $1\frac{1}{2}$–2 times as long as leaves; racemes ovoid, with 10–20 flowers; bracts 3–5 mm, linear-lanceolate. Calyx 5–8 mm, the teeth $\frac{1}{4}$–$\frac{1}{3}$ as long as tube. Corolla violet to pale purplish, rarely whitish; standard 13–18 mm, the limb ovate. Legume 8–10 × 3–4 mm, ovoid-oblong, scarcely compressed, densely or sparsely appressed-hairy. $2n=32$. ● *Alps and mountains of N.W. Jugoslavia.* Au Ga He It Ju.

96. A. amarus Pallas, *Spec. Astrag.* 8 (1800). Stems 10–40 cm, erect, woody at base. Leaves 5–15 cm; leaflets 3–5(–7) pairs, 5–12 × 3–12 mm, very remote, orbicular (rarely ovate), subglabrous; stipules connate. Peduncles about as long as leaves; racemes very lax, with 7–20 flowers; bracts 3–5 mm, lanceolate. Calyx 10–12 mm, glabrous, the teeth *c.* $\frac{1}{6}$ as long as tube. Standard 20–30 mm. Legume 10–16 × 4–5 mm, oblong, acute, slightly curved, glabrous; beak straight. *S.E. Russia & W. Kazakhstan.* Rs (E).

97. A. helmii Fischer ex DC., *Prodr.* **2**: 301 (1825). Acaulescent, with branched stock. Leaves 3–8 cm; leaflets 4–7 pairs, 7–15 × 3–8 mm, elliptical, acute, densely appressed-hairy on both surfaces. Peduncles equalling leaves; racemes capitate, with 5–10 flowers; bracts 2–4 mm, linear-lanceolate. Calyx 8–10 mm, densely hairy, the teeth $\frac{2}{5}$–$\frac{1}{2}$ as long as tube, linear-subulate. Corolla yellowish; standard 20–25 mm. Legume 10–15 × *c.* 5 mm, ovoid-oblong, densely patent-hairy. *Stony places. E.C. Russia.* Rs (C, E).

98. A. idaeus Bunge, *Astrag. Geront.* **1**: 107 (1868). Acaulescent, with branched stock. Leaves up to 3 cm; leaflets 6–12 pairs, 3–6 × 1–2 mm, oblong-elliptical, densely sericeous with subappressed hairs on both surfaces. Peduncles shorter than or equalling leaves; racemes capitate. Calyx *c.* 7 mm, densely hairy, the teeth *c.* $\frac{1}{2}$ as long as tube. Corolla yellow; standard *c.* 15 mm. Legume unknown. *Mountain rocks. Kriti.* Cr.

99. A. autranii Bald., *Bull. Herb. Boiss.* **3**: 196 (1895). Like **98** but leaflets 4–9 pairs, elliptic-lanceolate; peduncles longer than leaves; calyx-teeth slightly longer than tube; corolla red; legume included in calyx, villous. ● *Albania.* Al.

100. A. agraniotii Orph. ex Boiss., *Diagn. Pl. Or. Nov.* **3(2)**: 29 (1856). Dwarf, densely caespitose, acaulescent, with branched stock. Leaves 1–1·5 cm; leaflets 4–8 pairs, *c.* 3 × 1·5 mm, elliptical, densely sericeous with subappressed hairs on both surfaces. Peduncles shorter than or slightly longer than leaves; racemes capitate. Calyx 5–6 mm, densely hairy, the teeth longer than the tube. Corolla yellowish, tinged with blue; standard *c.* 10 mm. Legume 6–7 mm, exceeding calyx, ovoid, densely villous. *Mountain rocks.* ● *S. Greece (Malevo).* Gr.

101. A. arcuatus Kar. & Kir., *Bull. Soc. Nat. Moscou* **14**: 407 (1841). Stems up to 7 mm, procumbent, woody at base. Leaves 1·5–2 cm; leaflets 2–4 pairs, 5–10 × 1·5–3 mm, linear-lanceolate to oblanceolate, with flexuous erecto-patent hairs, dense beneath and sparser above; stipules 3–4 mm. Peduncles from slightly shorter than to twice as long as leaves; racemes subglobose, with 5–7 flowers; bracts 1–2 mm, ovate-lanceolate. Calyx 8–13 mm, the teeth *c.* $\frac{1}{3}$ as long as tube. Corolla purplish; standard 17–25 mm. Legume 15–30 × *c.* 3 mm, linear, curved, with dense, erecto-patent hairs. *W. Kazakhstan.* Rs (?C, E).

102. A. reduncus Pallas, *Spec. Astrag.* 109 (1800). Subacaulescent, with branched stock. Leaves 3–20 cm; leaflets 5–10(–20) pairs, 3–12 × 2–5 mm, oblong-ovate, with flexuous erecto-patent hairs, dense beneath and sparser above; stipules 4–7 mm. Peduncles equalling or slightly exceeding leaves; racemes ovoid, with 5–20 flowers; bracts 3–5 mm, linear-lanceolate. Calyx 11–13 mm, densely hairy, the teeth $\frac{1}{4}$–$\frac{1}{2}$ as long as tube. Corolla yellowish, rarely violet; standard 17–27 mm. Legume 10–20(–25) × *c.* 4 mm, oblong or oblong-lanceolate, acute, curved, with dense, erecto-patent hairs. ● *S. Ukraine and S.E. Russia.* Rs (W, K, E).

103. A. dolichophyllus Pallas, *Spec. Astrag.* 84 (1800). Acaulescent, with branched stock, or stems up to 2 cm. Leaves 5–15 cm; leaflets 8–15(–20) pairs, 5–15 × 1·5–5 mm, oblong or lanceolate, with dense, flexuous, erecto-patent hairs on both surfaces. Peduncles very short or absent; racemes with 5–20 flowers; bracts *c.* 5 mm, linear. Calyx 10–13 mm, villous, the teeth slightly longer than tube. Corolla pale yellowish; standard 20–30 mm, emarginate at apex. Legume 6–12 × 4–5 mm, ovoid-oblong, obtuse, with dense, erecto-patent, short hairs; beak *c.* 3 mm. *S. part of U.S.S.R., northwards to c. 51° N., and just extending into E. Romania.* Rm Rs (C, W, K, E).

104. A. lacteus Heldr. & Sart. ex Boiss., *Diagn. Pl. Or. Nov.* **3(2)**: 31 (1856). Like **103** but leaflets 4–7 pairs, more remote; racemes with 3–5 flowers; bracts *c.* 12 mm, lanceolate; calyx-teeth equalling tube; corolla white; standard *c.* 20 mm, truncate at apex. ● *S. Greece (Parnon Oros).* Gr.

105. A. baldaccii Degen, *Österr. Bot. Zeitschr.* **46**: 415 (1896). Acaulescent, with branched stock. Leaves 4–7 cm; leaflets 4–9 pairs, 6–9 × 1–3 mm, ovate-elliptical to lanceolate, densely hairy with more or less appressed hairs on both surfaces. Peduncles absent. Calyx 10–14 mm, densely appressed-hairy, the teeth *c.* $\frac{1}{4}$ as long as tube. Corolla whitish, tinged with lilac; standard 23–25 mm. Legume 12–15 × 5–6 mm, ovoid-oblong, densely hairy. ● *S. Albania, C. & N. Greece.* Al Gr.

106. A. wilmottianus Stoj., *Bull. Soc. Bot. Bulg.* **1**: 73 (1926). Like **105** but calyx-teeth *c.* $\frac{1}{2}$ as long as tube; corolla purplish; legume ovoid, glabrous. ● *W. Bulgaria.* Bu.

107. A. testiculatus Pallas, *Spec. Astrag.* 82 (1800). Acaulescent, with branched stock, or rarely with stems up to 6 cm. Leaves 3–10 cm; leaflets 7–12 pairs, 3–15 × 1–5 mm, oblong or elliptical, with very dense, rather long, erecto-patent hairs. Peduncle very short or absent; racemes with 3–10 flowers. Calyx 10–12 mm, with dense patent hairs, the teeth $\frac{1}{4}$–$\frac{1}{2}$ as long as tube. Corolla whitish, pale violet or purplish; standard 18–27 mm, the limb oblong-obovate, constricted in the middle, emarginate. Legume 10–20 × 7–10 mm, ovate, compressed, with dense patent hairs, usually pendent; peduncle elongating up to *c.* 3 cm in fruit. *S.E. part of U.S.S.R., from Krym to C. Ural.* Rs (C, K, E).

108. A. rupifragus Pallas, *Spec. Astrag.* 86 (1800). Like **107** but more often caulescent, with stems up to 5(–20) cm; calyx 10–16 mm; limb of standard oblong-elliptical, not constricted in the middle; legume usually sessile, erect. *Stony places. S.E. Russia, Krym.* Rs (C, K, E).

109. A. physodes L., *Sp. Pl.* 760 (1753) (incl. *A. suprapilosus* Gontsch.). Acaulescent, with branched stock. Leaves 5–15 cm; leaflets 8–15 pairs, 5–12 × 2–4 mm, lanceolate or oblong-elliptical, sparsely appressed-hairy beneath, sparsely hairy to glabrous above. Peduncles *c.* $\frac{2}{3}$ as long as leaves; racemes ovoid, with many flowers. Calyx 7–10 mm, with short, erecto-patent hairs, the teeth *c.* $\frac{1}{6}$ as long as tube. Corolla violet; standard 15–20 mm. Legume 10–25 mm, globose to ovoid, membranous and strongly inflated, glabrous. *S.E. Russia and W. Kazakhstan; Krym.* Rs (K, E).

110. A. monspessulanus L., *Sp. Pl.* 761 (1753). Acaulescent, with branched stock. Leaves (3–)7–20 cm; leaflets (7–)10–20 pairs, 5–10 × 1·5–5 mm, suborbicular to oblong, obtuse, sparsely appressed-hairy beneath, glabrous above. Peduncles 1–2 times as long as leaves; racemes ovoid or oblong, with 7–30 flowers; bracts 3–10 mm, linear-lanceolate. Calyx 9–16 mm, the teeth $\frac{1}{4}$–$\frac{3}{4}$ as long as tube. Corolla purplish or red, rarely whitish; standard 20–30 mm, entire to 2-lobed at apex. Legume 25–45 × 3–6 mm, linear, cylindrical, acute, slightly curved (rarely for as much as a semicircle), scarcely rugose, sparsely appressed-hairy, glabrescent. *S. Europe, extending to W. Ukraine.* Al Bu Ga Gr He Hs It Ju Rm Rs (W) Si ?Tu.

Both the following subspecies are extremely variable.

(a) Subsp. **monspessulanus** (incl. var. *atticus* (Nyman) Hayek): Leaflets ovate to oblong. Corolla purplish-violet, rarely whitish. Legume 4–5 times as long as calyx and usually *c.* 10 times as long as wide. $2n = 16$. *Throughout the range of the species.*

A. teresianus Sennen & Elias, *Bol. Soc. Iber. Ci. Nat.* **26**: 911 (1927), a dwarf, caespitose plant from N.C. Spain, is probably not more than varietally distinct from **110(a)**.

(b) Subsp. **illyricus** (Bernh.) Chater, *Feddes Repert.* **79**: 51 (1968) (*A. illyricus* Bernh.): Leaflets orbicular to ovate. Corolla red or flesh-coloured. Legume 2–3 times as long as calyx and usually *c.* 5 times as long as wide. *Balkan peninsula, extending to Trieste.*

111. A. spruneri Boiss., *Diagn. Pl. Or. Nov.* **1(2)**: 79 (1843). Like **110(b)** but leaflets 5–8(–12) pairs; calyx 12–18 mm; corolla white, pale purplish or violet; legume 12–20 × 5–7 mm, *c.* 3 times as long as wide, obovoid or oblong-obovoid, curved, strongly rugose. *Balkan peninsula, mainly in the south and east; S.E. Romania; Kikhlades.* Al Bu Gr Ju Rm Tu.

112. A. incanus L., *Syst. Nat.* ed. 10, **2**: 1175 (1759). Acaulescent, with branched stock. Leaves 4–10 cm; leaflets 6–10(–14) pairs, 5–9 × 1·5–5 mm, elliptical or oblong to suborbicular, more or less densely appressed-hairy and silvery beneath, usually less densely so and green above. Peduncles about equalling or slightly longer than leaves; racemes subglobose to ovoid, with 6–20 flowers; bracts 2–7 mm, linear-lanceolate. Calyx 7–11 mm, appressed-hairy, the teeth *c.* $\frac{1}{4}$ as long as tube. Corolla whitish to purple; standard 19–24 mm. Legume 10–25 × 4–8 mm, inflated-oblong to oblong-cylindrical, straight or slightly curved, appressed-hairy. $2n = 16$. *S.W. Europe.* ?Bl Ga Hs.

1 Legume less than 5 mm wide, cylindrical
2 Legume 15–20 mm, not spotted; leaflets acute ... (a) subsp. **incanus**
2 Legume 20–25 mm, purple-spotted; leaflets obtuse and mucronate ... (b) subsp. **incurvus**
1 Legume more than 5 mm wide, inflated-oblong
3 Legume 15–20 mm; leaflets obtuse or emarginate ... (c) subsp. **nummularioides**
3 Legume 10–15 mm; leaflets mucronate ... (d) subsp. **macrorhizus**

(a) Subsp. **incanus**: Leaflets oblong or obovate, acute. Standard emarginate. Legume 15–20 × 4 mm, not spotted. ● *S. France, E., C. & S. Spain.*

(b) Subsp. **incurvus** (Desf.) Chater, *Feddes Repert.* **79**: 51 (1968) (*A. incurvus* Desf.): Leaflets suborbicular, obtuse and mucronate. Standard emarginate. Legume 20–25 × 4 mm, purple-spotted. *C. & S.E. Spain.*

(c) Subsp. **nummularioides** (Desf. ex DC.) Maire in Jahandiez & Maire, *Cat. Pl. Maroc* **2**: 414 (1932) (*A. nummularioides* Desf. ex DC.): Leaflets elliptical to suborbicular, obtuse or emarginate. Standard obtuse or subacute. Legume 15–20 × 6–8 mm, not spotted. *Mountains of S.E. Spain.*

(d) Subsp. **macrorhizus** (Cav.) Chater, *Feddes Repert.* **79**: 52 (1968) (*A. macrorhizus* Cav.): Leaflets elliptical to suborbicular, obtuse and mucronate. Standard emarginate. Legume 10–15 × 7–8 mm, not spotted. *E., C. & S. Spain.*

113. A. sericophyllus Griseb., *Spicil. Fl. Rumel.* **1**: 52 (1843). Stems 1–2 cm, woody at base. Leaves 2–3 cm; leaflets 4–6 pairs, 5–8 × 1–2·5 mm, narrowly elliptical, subacute, densely appressed-hairy, silvery; stipules 1–2 mm. Peduncles 4–5 times as long as leaves; racemes lax, with 8–15 flowers. Calyx *c.* 10 mm, with appressed black and patent white hairs, the teeth *c.* $\frac{1}{4}$ as long as tube, linear. Corolla reddish-purple and whitish; standard 18–20 mm. Legume *c.* 25 × 2 mm, linear, straight, acute, with dense, appressed or ascending hairs. ● *S. part of Balkan peninsula.* Al Gr Ju.

114. A. apollineus Boiss. & Heldr. in Boiss., *Diagn. Pl. Or. Nov.* **3(2)**: 27 (1856). Like **113** but stems up to 5 cm; leaflets 5–7 pairs, elliptical to oblong, obtuse or truncate, less densely hairy and not silvery; peduncles 2–3 times as long as leaves; corolla violet; legume 30–40 mm. ● *S.C. Greece (Parnassos).* Gr.

115. A. karelinianus M. Popov in Komarov, *Fl. URSS* **12**: 695 (1946). Stems 10–20 cm, erect, woody at base. Leaves 5–10 cm; leaflets 5–7 pairs, 5–10 × 1–2 mm, linear-lanceolate, appressed-hairy on both surfaces; stipules 3–4 mm, not connate. Peduncles 1$\frac{1}{2}$–2 times as long as leaves; racemes dense, with 5–10 flowers; bracts 0·5–1 mm, ovate. Calyx 7–9 mm, the teeth $\frac{1}{5}$–$\frac{1}{4}$ as long as tube, linear-subulate. Corolla whitish or violet, rarely yellowish; standard *c.* 20 mm, the limb *c.* 6 times as long as claw. Legume 20–30 × 2–3 mm, linear-subulate, almost straight, with ascending hairs. ● *C. & S. Ural.* Rs (C).

116. A. subuliformis DC., *Astrag.* 134 (1802) (*A. subulatus* Pallas, non Desf.; incl. *A. ucrainicus* M. Popov & Klokov, *A. pseudotataricus* Boriss.). Stems up to 5(–10) cm, erect, woody at base. Leaves 2–6 cm; leaflets 2–7 pairs, 4–15 × 0·5–1 mm, narrowly linear to oblong, appressed-hairy beneath, glabrous above; stipules 1–2 mm, free or connate only near the base of stem. Peduncles 1–1$\frac{1}{2}$ times as long as leaves; racemes very lax, with 3–10 flowers; bracts 0·5–1 mm, ovate. Calyx 8–12 mm, the teeth $\frac{1}{6}$–$\frac{1}{5}$ as long as tube, linear-subulate. Corolla yellowish (rarely whitish or violet); standard 17–23 mm, the limb *c.* 4 times as long as claw. Legume 15–35 × 1·5–2 mm, linear-subulate, straight, with dense, appressed hairs. *S.E. Europe, from Macedonia to S.E. Russia, where it extends northwards to c.* 53° *N.* Ju Rm Rs (C, W, K, E) Tu.

117. A. corniculatus Bieb., *Cent. Pl.* **1**: t. 45 (1810). Like **116** but leaflets densely hairy on both surfaces; stipules 2–3 mm; racemes dense, umbelliform; corolla violet; calyx teeth $\frac{1}{4}$–$\frac{1}{3}$ as long as tube, lanceolate; legume often slightly curved. ● *S. Ukraine, Moldavia, S.E. Romania.* Rm Rs (W, K).

118. A. muelleri Steudel & Hochst., *Flora (Regensb.)* **10**: 72 (1827) (*A. vegliensis* Sadler ex Ascherson & Graebner). Like **116** but leaflets densely hairy on both surfaces; peduncles 2–3 times as long as leaves; corolla lilac or purplish. ● *Coasts of Jugoslavia; C. Italy.* It Ju.

119. A. pugionifer Fischer ex Bunge, *Astrag. Geront.* **1**: 125 (1868). Stems 25–40 cm, erect, woody at base. Leaves 5–7 cm;

leaflets 3–4 pairs, 10–20 × 2–3 mm, linear, appressed-hairy on both surfaces; stipules 2–3 mm, not connate. Peduncles about twice as long as leaves; racemes lax, with 5–10 flowers; bracts 3–4 mm, linear. Calyx 14–18 mm, the teeth $\frac{1}{4}$–$\frac{1}{3}$ as long as tube, linear-subulate. Corolla white, tinged with pale purple; standard *c.* 25 mm. Legume 50–70 × 2 mm, linear, straight, appressed-hairy. ● *N. Macedonia and Thrace.* Bu Ju Tu.

120. A. cornutus Pallas, *Reise* **1**: 499 (1771). Stems up to 40(–100) cm, erect, woody at base. Leaves (3–)5–9 cm; leaflets 4–9 pairs, 10–35 × 1–7 mm, broadly obovate to linear-oblong, appressed-hairy on both surfaces; stipules 3–9 mm, not connate. Peduncles about as long as leaves; racemes 2–4(–5) cm, ovoid, dense, with 10–40 flowers; bracts 3–5 mm, linear. Calyx 7–12 mm, the teeth $\frac{1}{4}$–$\frac{1}{3}$ as long as tube, linear-subulate. Corolla violet; standard 18–20 mm, emarginate. Legume 10–18 × 3–4 mm, oblong, straight, with dense, ascending hairs. *S.E. Europe, from N.E. Bulgaria to S. Ural and W. Kazakhstan.* Bu Rm Rs (C, W, E).

121. A. brachylobus DC., *Prodr.* **2**: 285 (1825). Like **120** but peduncles *c.* $1\frac{1}{2}$ times as long as leaves; racemes 6–15 cm, elongate, lax; calyx 10–15 mm, the teeth $\frac{1}{5}$–$\frac{1}{4}$ as long as tube; standard 18–26 mm. *Krym, S.E. Russia, W. Kazakhstan.* Rs (K, E).

122. A. varius S. G. Gmelin, *Reise Russl.* **1**: 116 (1770) (*A. virgatus* Pallas). Stems 10–50 cm, erect, woody at base. Leaves 4–8 cm; leaflets 5–11 pairs, 10–25 × 1·5–5 mm, linear to oblong-lanceolate, appressed-hairy on both surfaces; stipules 2–5 mm, not connate. Peduncles 1–$1\frac{1}{2}$ times as long as leaves; racemes 8–20 cm, lax, with 15–20 flowers; bracts 2–3 mm, linear. Calyx 8–10 mm, the teeth $\frac{1}{5}$–$\frac{1}{4}$ as long as tube, linear-subulate. Corolla violet; standard 15–20 mm, rounded at apex. Legume 12–20 × 2–3 mm, linear-oblong, straight, with dense, usually ascending hairs. *S. part of U.S.S.R., extending to E. Hungary and E. Bulgaria.* Bu Hu Rm Rs (C, W, E).

123. A. macropus Bunge, *Astrag. Geront.* **1**: 125 (1868). Like **122** but peduncles 2–3 times as long as leaves; racemes 5–8 cm, denser, with 8–15 flowers; corolla paler violet; standard 20–25 mm. *S.C. & S.E. Russia, W. Kazakhstan.* Rs (C, E).

124. A. pallescens Bieb., *Fl. Taur.-Cauc.* **3**: 489 (1819). Like **122** but peduncles 2–3 times as long as leaves; racemes 3–8 cm, lax, with 5–12 flowers; calyx 11–15 mm, the teeth $\frac{1}{3}$(–$\frac{1}{2}$) as long as tube; corolla white; standard 18–23 mm. ● *S. & E. Ukraine and adjacent parts of S.E. Russia.* Rs (W, E).

125. A. glaucus Bieb., *Fl. Taur.-Cauc.* **2**: 186 (1808) (*A. dealbatus* Pallas pro parte). Stems up to 10 cm, erect, woody at base, with sparse, appressed, white hairs or subglabrous. Leaves 3–6 cm; leaflets 3–6 pairs, 10–25 × 3–6 mm, lanceolate to oblong-lanceolate, appressed-hairy on both surfaces; stipules 3–5 mm, not connate. Peduncles about twice as long as leaves; racemes 2–5 cm, ovoid, dense, with 8–20 flowers; bracts 3–5 mm, linear-lanceolate. Calyx 9–12 mm, the teeth 3–4 mm, $\frac{1}{3}$–$\frac{1}{2}$ as long as tube, subulate. Corolla whitish; standard 18–25 mm, rounded at apex. Legume 10–15 × 3–4 mm, oblong, straight, with dense, ascending hairs. ● *E. Romania (Dobruja); Krym.* Rm Rs (K).

126. A. zingeri Korsh., *Acta Horti Petrop.* **11**: 297 (1890). Like **125** but stems up to 25 cm, with denser, appressed hairs; stipules 1–2 mm; leaflets often glabrous above; calyx teeth 1·5–2·5 mm, $\frac{1}{5}$–$\frac{1}{4}$ as long as tube, triangular-lanceolate; standard emarginate; legume up to 20 mm. *S.E. Russia, W. Kazakhstan.* Rs (C, E).

Subgen. **Calycocystis** Bunge. Perennials; hairs medifixed. Leaves imparipinnate. Calyx more or less inflated in fruit; corolla not persistent.

A largely Asiatic subgenus, characterized by the strong inflation of the calyx in fruit, but the European representatives show this to only a slight degree. Species **127–133** are extremely close to each other, and may also easily be confused with **125** and **126**.

127. A. vesicarius L., *Sp. Pl.* 760 (1753). Stems up to 25 cm, erect, woody at base, sparsely to densely hairy with appressed, white hairs. Leaves 4–8 cm; leaflets (4–)5–10 pairs, linear-lanceolate to oblong, rarely linear, more or less densely hairy with appressed, white hairs on both surfaces; stipules 2–4 mm, not connate. Peduncles 2–3 times as long as leaves; racemes 2–5 cm, with usually 10–20 flowers; bracts 3–5 mm, lanceolate. Calyx 8–14 mm, with dense, ascending or patent, white and black hairs, the teeth $\frac{1}{6}$–$\frac{1}{2}$ as long as tube, triangular-lanceolate to linear. Standard 17–23 mm. Legume 8–15 × 3–5 mm, equalling (or only beak exceeding) the calyx, oblong, acute, densely hairy with almost patent hairs. *S.E. & E.C. Europe, extending from N.W. Greece and Bulgaria to E. Austria and Ukraine; S.W. Alps; N. & C. Italy; S. Spain.* Al Au Bu Cz Ga Gr Hs Hu It Ju Rm Rs (?C, W, K).

In need of further study; three subspecies are recognized here, but others probably occur in E. Europe.

1 Flowers concolorous, yellow (c) subsp. **pastellianus**
1 Flowers bicolorous
2 Standard entire at apex; bracts linear (b) subsp. **carniolicus**
2 Standard emarginate at apex; bracts lanceolate (a) subsp. **vesicarius**

(a) Subsp. **vesicarius** (incl. *A. albidus* Waldst. & Kit.): Stems up to 5(–10) cm. Leaflets oblong to lanceolate, densely hairy. Calyx 8–14 mm, with patent or ascending white hairs, and short or long, patent or appressed black hairs. Corolla bicolorous, the standard purplish or violet, the wings and keel paler or whitish; standard emarginate at apex. $2n=16$. *Throughout the range of the species except most of the Balkan peninsula.*

Plants from Albania and N.W. Greece with linear leaves perhaps represent another subspecies.

(b) Subsp. **carniolicus** (A. Kerner) Chater, *Feddes Repert.* **79**: 52 (1968) (*A. carniolicus* A. Kerner): Stems up to 15 cm. Leaflets oblong to lanceolate, sparsely hairy. Calyx 8–10 mm, with sparse, ascending, white hairs and more numerous short, appressed, black hairs. Corolla bicolorous, the standard purplish or violet, the wings and keel paler or whitish; standard entire at apex. $2n=32$. ● *W. part of the Balkan peninsula, extending to Trieste.*

(c) Subsp. **pastellianus** (Pollini) Arcangeli, *Comp. Fl. Ital.* 186 (1882): Stems up to 25 cm. Leaflets oblong-elliptical, intermediate in pubescence between subspp. (a) and (b). Calyx 10–11 mm, with very sparse, ascending, white hairs and numerous short, appressed, black hairs. Corolla uniformly yellowish; standard entire or rarely weakly emarginate at apex. ● *Valleys of E. Italian Alps; ?Bulgaria.*

A. tarchankuticus Boriss., *Not. Syst. (Leningrad)* **14**: 222 (1951), from Krym, and **A. pseudoglaucus** Klokov, *Not. Syst. (Leningrad)* **15**: 50 (1953), from S. Ukraine, are perhaps to be included in this subspecies; the status of the plants from Bulgaria is also uncertain.

128. A. peterfii Jáv., *Sched. Fl. Hung. Exsicc.* **4**: 38 (1916). Like **127**(a) but stems up to 20 cm; leaflets linear or linear-lanceolate; racemes sometimes up to 15 cm; corolla yellowish; legume 15–25 mm, $1\frac{1}{2}$–2 times as long as calyx, with dense ascending hairs. *Steep, grassy slopes.* ● *C. Romania (Suatu).* Rm.

129. A. hispanicus Cosson ex Bunge, *Astrag. Geront.* **1**: 135 (1868). Like **127**(a) but stems up to 20 cm; leaflets linear or linear-lanceolate, plane; standard 23–30 mm; legume 15–30 mm, 1½–2 times as long as calyx, with dense, closely appressed hairs. ● *S.E. Spain.* Hs.

130. A. hegelmaieri Willk., *Suppl. Prodr. Fl. Hisp.* 235 (1893). Like **127**(a) but more suffruticose at base; leaflets linear, convolute; standard 20–25 mm; legume 15–25 mm, slightly curved, with dense, closely appressed hairs. *Grassy slopes.* ● *S.E. Spain* (*near Crevillente*). Hs.

Further information is required about this plant, and no material of it has been traced; it is perhaps conspecific with **129**.

131. A. albicaulis DC., *Astrag.* 166 (1802). Stems 5–20 cm, erect, woody at base, densely hairy with appressed, white hairs. Leaves 2–6 cm; leaflets 3–4 pairs, oblong or ovate, with appressed hairs on both surfaces; stipules 2–3 mm, not connate. Peduncles 2–3 times as long as leaves; racemes 2–7 cm, lax, with 10–20 flowers; bracts 3–5 mm, linear-lanceolate. Calyx 12–15 mm, with ascending hairs, the teeth ¼–⅓ as long as tube, subulate. Corolla yellowish; standard 18–23 mm, emarginate at apex. Legume 10–15 × 3–4 mm, equalling calyx, oblong, with dense almost patent hairs. *S.E. Russia and W. Kazakhstan.* Rs (C, E).

132. A. medius Schrenk, *Bull. Phys.-Math. Acad. Pétersb.* **2**:196 (1843). Like **131** but leaflets 2–4 pairs; bracts ovate, acuminate; calyx-teeth lanceolate; standard 22–27 mm. *S. Ural.* Rs (C).

133. A. fialae Degen, *Österr. Bot. Zeitschr.* **50**: 242 (1900). Acaulescent, with branched stock, woody at base. Leaves 1·5–5(–10) cm; leaflets 8–12 pairs, 4–12 × 1·5–3 mm, linear-oblong to elliptic-lanceolate, densely appressed-hairy on both surfaces or less so above; stipules 4–7 mm, not connate. Peduncles ½ as long or up to as long as leaves; racemes dense; bracts 3–10 mm, ovate to linear. Calyx 9–10 mm, with dense, almost patent hairs, the teeth ¼–½ as long as tube. Corolla yellowish, tinged with violet; standard 16–19 mm. Legume 10–12 mm, oblong, straight, with dense, ascending hairs. ● *S. Jugoslavia and N. Albania.* Al Ju.

39. Oxytropis DC.[1]

Perennial, acaulescent or caulescent herbs. Leaves imparipinnate (in European spp.); stipules adnate to petiole, free or connate. Flowers in axillary racemes. Calyx tubular, with subequal teeth; corolla violet, purple, white or pale yellow; keel with a tooth at apex on abaxial side; stamens diadelphous. Legume oblong to ovoid, dehiscent, stipitate or sessile, unilocular, semi-bilocular or almost completely bilocular. Seeds several.

Literature: W. Gutermann & H. Merxmüller, *Mitt. Bot. Staatssamm.* (*München*) **4**: 199–275 (1961). P. Leins & H. Merxmüller, *op. cit.* **6**: 19–31 (1966).

```
1  Leaflets 1–2 pairs                              9. mertensiana
1  Leaflets more than 2 pairs
 2  Legume unilocular, on a carpophore at least ½ as long as calyx-
      tube, usually not splitting the calyx
  3  Caulescent
   4  Stipules connate for at least half their length; legume
        pendent                                      1. lapponica
   4  Stipules ± free from each other
    5  Legume pendent; corolla whitish               2. deflexa
    5  Legume ± erect; corolla purplish-violet       3. jacquinii
  3  Acaulescent
   6  Carpophore equalling or longer than calyx-tube
    7  Scape with short, appressed hairs; legume 20–25 × 8 mm,
         with hairs 0·2–0·25 mm long                 4. carpatica
```

```
    7  Scape with long, ± patent hairs; legume 15–20 × 5–7 mm,
         with hairs c. 1 mm long                     7. pyrenaica
   6  Carpophore c. ½ as long as calyx-tube
    8  Corolla pale blue to pale purple
     9  Scape 0·6 mm in diameter; leaflets mostly 10–12 pairs,
          lanceolate; legume narrowly oblong          5. gaudinii
     9  Scape c. 1 mm in diameter; leaflets mostly 13–20 pairs,
          elliptic-lanceolate; legume shortly ovoid   6. amethystea
    8  Corolla deep purplish-violet
    10  Scape stout, with patent hairs, usually with more than
          7 flowers                                   7. pyrenaica
    10  Scape slender, with appressed hairs, with 3–5 flowers
                                                      8. triflora
 2  Legume at least semi-bilocular, with a septum developing in-
      wards from the ventral suture, ± sessile and usually ± split-
      ting the calyx
  11  Legume with both ventral and dorsal septum, almost com-
        pletely bilocular
   12  Both septa ± equally wide, or the dorsal somewhat
         narrower; inflorescence elongating after flowering
                                                      18. halleri
   12  Dorsal septum wider than ventral; inflorescence not elon-
          gating                                      19. uralensis
  11  Legume with only a ventral septum, semi-bilocular
   13  Whole plant, including legume, viscid, covered with sessile,
          aromatic glands                             24. fetida
   13  Plant eglandular, or with a few glands only on margins of
          stipules or at the base of the bracts
    14  Caulescent; calyx-teeth equalling or longer than tube
     15  Corolla purple
      16  Inflorescence oblong, many-flowered; standard
            8–10 mm; stipules ovate, more or less connate with
            each other                                23. floribunda
      16  Inflorescence capitate, few-flowered; standard c.
            14 mm; stipules lanceolate to ovate-lanceolate, free
                                                      22. purpurea
     15  Corolla light yellow; stipules narrowly triangular, free
      17  Standard 12–14 mm; legume 15–20 mm           20. pilosa
      17  Standard 16–18 mm; legume 20–30 mm           21. pallasii
    14  Acaulescent; calyx-teeth distinctly shorter than tube
     18  Limb of standard obtusely 2-lobed; limb of wings
           orbicular                                  17. ambigua
     18  Limb of standard subentire to emarginate; limb of
           wings oblong or obovate
      19  Leaflets 17–25 pairs; raceme 5–10 cm
       20  Corolla pale yellowish; bracts c. 5 mm      10. hippolyti
       20  Corolla purplish; bracts 7–12 mm           16. songorica
      19  Leaflets mostly less than 17 pairs; racemes usually less
             than 5 cm
       21  Stipules adnate to petiole for at least ⅓ of their length
        22  Stipules many-veined; leaflets 10–15 pairs; tooth of
              keel 1–1·5(–2) mm                       11. campestris
        22  Stipules 1-veined; leaflets 6–8 pairs; tooth of keel
              c. 0·5 mm                               12. prenja
       21  Stipules adnate to petiole for not more than ¼ of their
             length
        23  Tooth of keel 1 mm; legume oblong-ovoid, with
              short, appressed hairs                  15. spicata
        23  Tooth of keel c. 0·5 mm; legume with patent hairs
              c. 3 mm long
         24  Corolla light yellow, becoming reddish near apex;
               legume ovoid                           13. urumovii
         24  Corolla lilac; legume narrowly ellipsoid
                                                      14. foucaudii
```

(A) Legume unilocular, without septa.

1. O. lapponica (Wahlenb.) Gay, *Flora* (*Regensb.*) **10**: 30 (1827) (*Astragalus lapponicus* (Wahlenb.) Burnat). Stems up to 10 cm. Leaflets 8–14 pairs, lanceolate or oblong-lanceolate, appressed-hairy on both surfaces; stipules connate with each other for at least half their length and quite shortly adnate to petiole. Racemes subglobose, scarcely elongating after flowering;

flowers pendent after anthesis. Corolla violet-blue; standard 8–12 mm. Legume 8–15 mm, pendent, narrowly oblong, with short appressed hairs. $2n=16$. *Mountains of Fennoscandia; Pyrenees; Alps; one station in E. Albania.* Al Au Fe Ga He It Ju No Su.

2. O. deflexa (Pallas) DC., *Astrag.* 96 (1802). Like **1** but leaflets 11–16 pairs; stipules free; racemes oblong; corolla whitish. $2n=16$. *Arctic Norway* (S.W. Finnmark.). No.

In Europe only as subsp. **norvegica** Nordh., *Svensk Bot. Tidskr.* **58**: 159 (1964). Subsp. **deflexa**, from N. Asia, has 15–25 pairs of leaflets and bluish or purplish corolla.

3. O. jacquinii Bunge, *Arb. Naturf.-Ver. Riga* **1**: 226 (1847) (*O. montana* subsp. *jacquinii* (Bunge) Hayek, *Astragalus montanus* L. pro parte). Stems 5–40 cm, stout. Leaflets 14–20 pairs, ovate-lanceolate to lanceolate, sparsely hairy; stipules connate at the base, and adnate to petiole for up to $\frac{1}{3}$ of their length. Racemes subglobose, slightly elongating after flowering. Calyx-teeth c. $\frac{1}{4}$ as long as tube. Corolla purplish-violet; standard 10–13 mm. Legume 20–30 × 8 mm, lanceolate-ovoid, acuminate, suberect, with hairs 0·35–0·65 mm; carpophore equalling or longer than calyx-tube. ● *Alps; French Jura.* Au Ga Ge He It.

4. O. carpatica Uechtr., *Österr. Bot. Zeitschr.* **14**: 218 (1864). Acaulescent. Leaflets mostly more than 12 pairs, lanceolate, subglabrous; stipules free. Scapes 10–20 cm; racemes subglobose. Calyx-teeth $\frac{1}{3}$–$\frac{1}{2}$ as long as tube. Corolla bright blue; standard 10–16 mm. Legume 20–25 × 8 mm, narrowly ovoid, with sparse hairs 0·2–0·25 mm; carpophore equalling or longer than calyx-tube. ● *Carpathians.* Cz Po Rm Rs (W).

5. O. gaudinii Bunge, *Arb. Naturf.-Ver. Riga* **1**: 226 (1847) (*Astragalus triflorus* var. *gaudinii* (Bunge) Gams). Acaulescent with slender scape, grey-sericeous. Leaflets 10–12 pairs, lanceolate; stipules free from each other, but shortly adnate to petiole. Scapes 3–10 cm; racemes ovoid. Calyx-teeth c. $\frac{1}{2}$ as long as tube. Corolla pale lilac-blue; standard 10–15 mm. Legume c. 25 × 5 mm, narrowly oblong, patent or pendant, hairy; carpophore c. $\frac{1}{2}$ as long as calyx-tube. ● *S.W. & W.C. Alps.* Ga He It.

6. O. amethystea Arvet-Touvet, *Ess. Pl. Dauph.* 24 (1871) (*Astragalus montanus* L. pro parte). Acaulescent with stout scape, densely lanate. Leaflets 13–20 pairs, elliptic-lanceolate; stipules free from each other but shortly adnate to petiole. Racemes ovoid. Calyx-teeth at least $\frac{1}{2}$ as long as tube. Corolla pale purplish, becoming dull grey-lilac; standard 10–14 mm. Legume 18 × 7–8 mm, shortly ovoid, apiculate, densely hairy; carpophore c. $\frac{1}{2}$ as long as calyx-tube. $2n=16$. ● *S.W. Alps; E. Pyrenees* (Sierra de Cadi). Ga Hs ?It.

Hybrid swarms with **3** occur in the Alps.

7. O. pyrenaica Godron & Gren. in Gren. & Godron, *Fl. Fr.* **1**: 449 (1849) (*O. montana* subsp. *samnitica* (Arcangeli) Hayek). Acaulescent with stout scape, shortly tomentose. Leaflets 12–20 pairs, oblong-elliptic or lanceolate; stipules free from each other but shortly adnate to petiole. Scapes 3–20 cm; racemes ovoid-globose, with 8–20 flowers. Calyx-teeth usually more than $\frac{1}{2}$ as long as tube. Corolla purplish- or bluish-violet; standard 10–12 mm. Legume 15–20 × 5–7 mm, narrowly ovoid, acuminate, sparsely hairy; carpophore usually c. $\frac{1}{2}$ as long as calyx-tube. $2n=16$. Calcicole. ● *Mountains of S. & S.C. Europe.* ?Al Au Ga ?Gr He Hs It Ju Rm.

Plants of hybrid origin intermediate between **3** and **7** are frequently found.

8. O. triflora Hoppe in Sturm, *Deutschl. Fl.* Abt. 1, Band 12, Heft 49 (1827) (*Astragalus triflorus* (Hoppe) Gams, non A. Gray). Like **7** but more slender, with more appressed indumentum; leaflets usually less than 12 pairs, ovate-lanceolate; racemes with 3–5 flowers; calyx-teeth c. $\frac{1}{2}$ as long as tube; legume 15–17 × 5 mm, lanceolate; carpophore c. $\frac{1}{2}$ as long as calyx-tube. ● *E. Alps.* Au ?It.

9. O. mertensiana Turcz., *Bull. Soc. Nat. Moscou* **13**: 68 (1840). More or less acaulescent. Leaflets 1–2 pairs, elliptic- or linear-lanceolate, subglabrous; stipules almost free from each other but adnate to petiole for at least half their length. Scapes 2·5–7 cm; racemes ovoid, with 2–4 flowers. Calyx-teeth c. $\frac{1}{2}$ as long as tube; corolla reddish- or whitish-violet; standard 12–15 mm. Legume 15–20 × 5–6 mm, oblong, acuminate, hairy, usually unilocular; carpophore c. $\frac{1}{2}$ as long as calyx-tube. *N.E. Russia.* Rs (N).

(B) Legume with a septum arising from the ventral suture; rarely also with a septum from the dorsal suture.

10. O. hippolyti Boriss., *Sovetsk. Bot.* **1936** (4): 121 (1936). Acaulescent. Leaflets 17–25 pairs, oblong-lanceolate, sparsely hairy beneath, subglabrous above; stipules free from each other but adnate to petiole. Scapes 25–40 cm; racemes 5–10 cm, with 10–25 erect flowers. Calyx-teeth much shorter than tube. Corolla pale yellowish; standard 18–20 mm, subentire to emarginate at apex. Legume 15–25 × 4–5 mm, narrowly ovoid. ● *E. Russia* (near Belebej, Baškirskaja A.S.S.R.). Rs (E).

11. O. campestris (L.) DC., *Astrag.* 74 (1802) (*Astragalus campestris* L.). Acaulescent, with appressed or erecto-patent hairs. Leaflets 10–15 pairs, elliptical or lanceolate; stipules many-veined, connate for $\frac{1}{4}$–$\frac{3}{4}$ of their length and adnate to petiole for $\frac{1}{3}$–$\frac{1}{2}$ of their length. Scapes 5–20 cm; racemes ovoid, with 5–15 flowers. Corolla light yellow to whitish or light violet; standard 15–20 mm; keel often violet or blackish-violet at apex with tooth 1–1·5(–2) mm. Legume 14–18 × 6–8 mm, ovoid to oblong-cylindrical, erect, with appressed or semi-patent hairs up to 1 mm. $2n=48$. *N. Europe, and mountains of C. & S. Europe.* Au Br Bu Cz Fe Ga He Hs It Ju No Po Rm Rs (B, C, N, W) Su.

1 Limb of standard narrowly elliptical, more than twice as long as wide; calyx-teeth (0·8–)1·5(–2) mm **(b) subsp. tiroliensis**
1 Limb of standard elliptical, broadly elliptical or obovate, not more than twice as long as wide
  2 Calyx-teeth (1·5–)2(–3) mm; legume ovoid **(a) subsp. campestris**
  2 Calyx-teeth (2–)3(–4) mm; legume oblong-cylindrical **(c) subsp. sordida**

(a) Subsp. **campestris**: Calyx 7–10 mm, the teeth (1·5–)2(–3) mm. Corolla usually yellowish; standard 1$\frac{1}{4}$–1$\frac{1}{2}$ times as long as wings; limb of standard emarginate, elliptical, broadly elliptical or obovate, usually less than twice as long as wide. Legume ovoid. $2n=48$. *Throughout most of the range of the species, but northwards only to S.E. Sweden.* Au Br Bu Cz Ga He Hs It Ju Po Rm Rs (B, W) Su.

(b) Subsp. **tiroliensis** (Sieber ex Fritsch) Leins & Merxm., *Mitt. Bot. Staatssamm.* (München) **6**: 27 (1966) (*O. tiroliensis* Sieber ex Fritsch): Calyx 6–8 mm, the teeth (0·8–)1·5(–2) mm. Corolla usually light violet or whitish; standard 1$\frac{1}{4}$–1$\frac{1}{2}$ times as long as wings; limb of standard narrowly ovate, more than twice as long as wide. Legume ovoid. ● *E. & C. Alps.* Au He It.

(c) Subsp. **sordida** (Willd.) Hartman fil. in Hartman, *Handb. Skand. Fl.* ed. 11, 305 (1879) (*O. sordida* (Willd.) Pers.): Calyx 7–13 mm, the teeth (2–)3(–4) mm. Corolla yellowish or light violet; standard 1$\frac{1}{4}$–1$\frac{1}{2}$ times as long as wings; limb of standard elliptical, broadly elliptical or obovate, usually less than twice as long as wide. Legume oblong-cylindrical, somewhat curved. $2n=48$. *N. & E. Fennoscandia and arctic Russia.* Fe No Rs (N, C).

O. nuriae Sennen, *Bol. Soc. Ibér. Ci. Nat.* **26**: 120 (1926) (*O. halleri* var. *ochroleuca* Costa), from the E. Pyrenees, is like **11**(a) but has 8–10 pairs of leaflets and 3–6 flowers only. It may represent a further subspecies of **11**.

O. gmelinii Fischer ex Boriss., *Sovetsk. Bot.* **1936**(4): 120(1936), from C. Ural, is like **11**(a) but has the limb of the standard rounded at apex. It is probably not a distinct species.

12. O. prenja (G. Beck) G. Beck in Reichenb. & Reichenb. fil., *Icon. Fl. Germ.* **22**: 124 (1903). Acaulescent; scape and petioles with appressed to semi-patent hairs. Leaflets 6–8 pairs, oblong-ovate, appressed-hairy; stipules 1-veined, connate at the base, adnate to petiole for at least ⅓ of their length. Scapes up to 10 cm; racemes globose. Corolla purple; standard 15–19 mm, emarginate at apex; tooth of keel *c.* 0·5 mm. Legume *c.* 15 mm, ovoid. ● *Mountains of S.W. Jugoslavia and N. Albania.* Al Ju.

13. O. urumovii Jáv., *Magyar Bot. Lapok* **19**: 34 (1922) (*O. campestris* subsp. *dinarica* Murb.). Acaulescent; scape and petioles with patent hairs. Leaflets 8–15 pairs, ovate-lanceolate, densely villous-sericeous on both surfaces; stipules many-veined, connate, and adnate to petiole for less than ¼ of their length. Scapes 5–20 cm; racemes ovoid, with (4–)6–10(–15) flowers. Corolla light yellow, soon becoming reddish at the apex; standard 10–14 mm, the limb elliptical or broadly elliptical; tooth of keel *c.* 0·5 mm. Legume 10–15 × 4 mm, ovoid, with dense, white, erecto-patent hairs up to 3 mm long. ● *W. Jugoslavia and Albania; S.W. Bulgaria.* Al Bu Ju.

14. O. foucaudii Gillot, *Bull. Soc. Bot. Fr.* **42**: 517 (1895) (*O. lazica* auct., non Boiss.). Like **13** but 3–10 cm; leaflets 12–16 pairs; corolla lilac; legume 15–18 × 4 mm, narrowly ellipsoid. 2*n* = 32. ● *Pyrenees.* Ga Hs.

15. O. spicata (Pallas) O. & B. Fedtsch., *Beih. Bot. Centr.* **22**(2): 218 (1907). Acaulescent. Leaflets 12–17 pairs, oblong-lanceolate, sparsely hairy above, more densely so beneath; stipules adnate to petiole for less than ¼ of their length. Scapes 25–35 cm; racemes oblong, dense, with many, erect flowers. Calyx 5–8 mm; corolla purple; standard 15–18 mm, the limb orbicular, abruptly contracted into claw; tooth of keel 1 mm. Legume (10–)15–20 mm, oblong-ovoid, with short, dense, appressed, black and white hairs. *S.E. Russia.* Rs (E).

16. O. songorica (Pallas) DC., *Astrag.* 73 (1802). Like **15** but leaflets 18–24 pairs; racemes lax, with patent flowers; calyx 4–12 mm; standard with limb oblong-obovate, gradually narrowed into claw. *S.E. Russia.* Rs (E).

17. O. ambigua (Pallas) DC., *Astrag.* 70 (1802). Acaulescent. Leaflets 12–16 pairs, oblong-ovate to lanceolate, appressed-hairy on both surfaces; stipules many-veined, adnate to petiole in lower part. Scapes 15–30 cm; racemes ovoid, with many flowers. Corolla purple; standard 17–21 mm, the limb obovate, bluntly 2-lobed; limb of wings orbicular. Legume 15–20 × 5 mm, oblong-ovoid, with appressed black and white hairs. *S.E. Russia.* Rs (E). (*N. Asia.*)

18. O. halleri Bunge ex Koch, *Syn. Fl. Germ.* ed. 2, 200 (1843) (*O. sericea* (DC.) Simonkai, non Nutt., *Astragalus sericeus* Lam. pro parte). Acaulescent, appressed- to patent-hairy. Leaflets (8–)10–14(–18) pairs, ovate-lanceolate to lanceolate; stipules free or somewhat connate, shortly adnate to petiole. Scapes 4–30 cm; racemes ovoid, elongating after flowering, with 5–15 flowers. Corolla blue to purple, rarely paler; standard 15–20 mm; tooth of keel 1–1·5 mm. Legume 15–20 × 5–6 mm, ovoid to narrowly ellipsoid, with

short, dense appressed hairs; ventral and dorsal septa equal in width, or the ventral wider. ● *Pyrenees, Alps, Carpathians; Scotland; one station in E. Albania.* Al Au Br Cz Ga He Hs It Po Rm.

(a) Subsp. **halleri**: Scapes usually 1–2 mm in diameter; scapes and petioles with appressed or erecto-patent hairs, sometimes only sparsely hairy. Corolla bluish-purple. 2*n* = 32. *Pyrenees, Alps, Carpathians; Scotland.*

(b) Subsp. **velutina** (Sieber) O. Schwarz, *Mitt. Thür. Bot. Ges.* **1**: 107 (1949): Scapes usually 2–3 mm in diameter; scapes and petioles villous with dense, patent hairs. Corolla pale purplish. 2*n* = 16. *C. Alps.*

Plants from E. Albania (Korab), with the stipules 1- or 2-veined and the legume with an extremely narrow dorsal septum, have been described as **O. sericea** subsp. **korabensis** Kümmerle & Jáv., *Bot. Közl.* **19**: 24 (1920). Further investigation is required as these plants may not be correctly placed in this species.

19. O. uralensis (L.) DC., *Astrag.* 63 (1802). Like **18** but racemes not elongating after flowering; dorsal septum of legume wider than the ventral. *Ural.* Rs (C).

20. O. pilosa (L.) DC., *Astrag.* 91 (1802) (*Astragalus pilosus* L.). Stems 20–50 cm; stock and petioles with patent hairs. Leaflets (7–)9–13(–15) pairs, oblong or linear-oblong, with appressed hairs; stipules narrowly triangular, free from each other, very shortly adnate to petiole. Racemes ovoid to oblong, with many flowers. Corolla light yellow; standard 12–14 mm, the limb broadly ovate. Legume 15–20 mm, narrowly ovoid to narrowly cylindrical, with long, dense, erecto-patent hairs, with a ventral septum only. 2*n* = 16. *C. & E. Europe, northwards to Estonia and extending to S. Sweden, the S.W. Alps, C. Italy and C. Jugoslavia.* Au Bu Cz Ga Ge He Hu It Ju Po Rm Rs (B, C, W, K, E) Su.

Var. **pygmaea** G. Beck in Reichenb. & Reichenb. fil., *Icon. Fl. Germ.* **22**: 120 (1903), is a dwarf variant of **20**.

21. O. pallasii Pers., *Syn. Pl.* **2**: 334 (1807). Like **20** but standard 16–18 mm; legume 20–30 mm. *Krym.* Rs (?W, K).

22. O. floribunda (Pallas) DC., *Astrag.* 94 (1802). Grey-hairy; stems 2–15 cm. Leaflets (5–)8–12 pairs, lanceolate or oblong; stipules ovate, more or less connate, very shortly adnate to petiole. Racemes oblong, dense, with many flowers. Corolla purple; standard 8–10 mm, the limb broadly ovate. Legume 15–20(–25) × 2·5–4 mm, narrowly cylindrical, with erecto-patent black and white hairs, with a narrow, ventral septum only. *S.E. Russia, W. Kazakhstan.* Rs (E).

O. cretacea Basil. in Komarov, *Fl. URSS* **13**: 544 (1948), from W. Kazakhstan (near Ural'sk), is like **22** but has stems 15–25 cm, dimorphic leaves and ovoid racemes. The legume is unknown so its relationships are uncertain.

23. O. purpurea (Bald.) Markgraf, *Feddes Repert.* (*Beih.*) **45**: 130, 192 (1927). Like **22** but stipules lanceolate to ovate-lanceolate, free; racemes capitate; standard *c.* 14 mm. ● *Albania, N. Greece.* Al Gr.

24. O. fetida (Vill.) DC., *Astrag.* 60 (1802) (*Astragalus fetidus* Vill.). Acaulescent, viscid with aromatic, sessile glands. Leaflets 15–25 pairs, thick, lanceolate or oblong-lanceolate; stipules adnate to petiole for ½ their length. Scapes 3–15 cm; racemes ovoid, with 3–7 flowers. Corolla yellowish; standard 12–22 mm, the limb elliptical. Legume 18–22 × 5–6 mm, oblong-cylindrical, often somewhat curved, glandular, with a ventral septum only. 2*n* = 16. *S.W. & W.C. Alps.* Ga He It.

40. Biserrula L.[1]

Annual herbs. Leaves imparipinnate; stipules small, free. Flowers in axillary racemes. Calyx campanulate, with 5 subequal teeth; keel obtuse; stamens diadelphous, only 5 fertile. Legume oblong, indehiscent, dorsiventrally compressed, each valve sinuate-dentate on the back. Seeds numerous.

1. **B. pelecinus** L., *Sp. Pl.* 762 (1753). Stems 10–40 cm, shortly pubescent. Leaflets 7–15 pairs, 5–10 × 1–5 mm, linear-oblong to obovate-orbicular, emarginate. Corolla 4–6 mm, blue or pale yellow with blue tip. Legume 10–40 × 4–8 mm, brown. *Mediterranean region, extending to S. Bulgaria and S. Portugal.* Al Bl Bu Co Cr Ga Gr Hs It Lu Sa Si.

41. Glycyrrhiza L.[2]

Perennial herbs, with some or all parts of the plant glandular and often viscid. Leaves imparipinnate. Stipules membranous, caducous. Calyx weakly bilabiate; corolla whitish-violet, rarely pale yellow; stamens diadelphous or monadelphous. Legume compressed or constricted between the seeds, indehiscent or tardily dehiscent, usually brown. Seeds 1–8.

1 Corolla 4–6 mm; raceme (excluding peduncle) not more than 20 mm at anthesis; capitate at anthesis and in fruit **5. echinata**
1 Corolla 7 mm or more; raceme (excluding peduncle) usually more than 20 mm; rarely capitate at anthesis, not capitate in fruit
  2 Plant not more than 25 cm, scabrid; legume strongly curved, terete, constricted between the seeds, glabrous **4. aspera**
  2 Plant usually more than 25 cm, rarely scabrid; legume straight or nearly so, compressed, sometimes setose
    3 Standard pale yellow; legume *c.* 14 mm, fusiform, densely setose **1. foetida**
    3 Standard pale violet or whitish; legume usually more than 15 mm, linear-oblong, straight or slightly curved, setose to glabrous
      4 Leaflets 9–17, 20–40(–55) mm, elliptical to oblong-ovate, obtuse, sometimes mucronate; legume straight, with straight sutures **2. glabra**
      4 Leaflets 5–11, 10–20(–30) mm, broadly elliptical to obovate, acute or obtuse; legume slightly curved, with undulate sutures **3. korshinskyi**

1. **G. foetida** Desf., *Fl. Atl.* 2: 170 (1799). Stem 25–50 cm. Leaflets 9–11, 10–35 mm, obovate, elliptical or ovate-lanceolate, apex acute or obtuse, mucronate. Racemes about equalling the leaves, dense or lax. Corolla 7–10 mm; standard pale yellow. Legume 14 mm, fusiform, densely covered with glandular bristles and with sessile and short-stalked glands. Seeds 2–3. *Sandy soil. S. Spain.* Hs. *(N.W. Africa.)*

2. **G. glabra** L., *Sp. Pl.* 742 (1753) (*G. glandulifera* Waldst. & Kit.). Stem 50–100 cm. Stem and petioles pubescent, sometimes scabrid. Leaflets 9–17, 20–40(–55) mm, elliptical, ovate or oblong, obtuse, sometimes mucronate, often viscid. Racemes exceeded by their subtending leaves at least at anthesis, lax, elongate. Corolla 8–12 mm. Legume up to 30 mm, linear-oblong, compressed, straight, glabrous or glandular-setose, the sutures straight. Seeds (2–)3–5. *Dry open habitats. S. & E. Europe, but doubtfully native in the south-west. Cultivated as a source of liquorice and frequently naturalized.* Al Bu Cr *Ga Gr *Hs It Ju Rm Rs (C, W, K, E) Sa Si Tu [Au Cz He Hu Lu].

The name **G. glandulifera** Waldst. & Kit., *Pl. Rar. Hung.* 1: 20 (1800), has been given to variants with glandular-setose legumes.

3. **G. korshinskyi** Grigoriev, *Bull. Jard. Bot. URSS* 29: 94 (1930). Erect or ascending, up to 70 cm. Leaflets 5–11, 10–20(–30) mm, broadly elliptical to obovate, acute or obtuse. Racemes about equalling the leaves, rather lax. Corolla 10–13 mm. Legume (10–)15–25(–30) mm, linear-oblong, compressed, slightly curved, covered with sessile and short-stalked glands, the sutures undulate. Seeds usually 4–7. *Saline steppes and meadows. S.E. Russia.* Rs (C, E). *(W.C. Asia.)*

4. **G. aspera** Pallas, *Reise* 1: 499 (1771). Plant scabrid. Stem 10–25 cm, flexuous, ascending. Leaflets 7–9, 5–30 mm, obovate or elliptical, mucronate, the terminal much the largest. Racemes about equalling the leaves. Corolla 10–18 mm. Legume 30–35 mm, strongly curved, terete, constricted between the seeds, glabrous. Seeds up to 8. *Steppes. S.E. Russia, W. Kazakhstan.* Rs (E). *(C. & S.W. Asia.)*

5. **G. echinata** L., *Sp. Pl.* 741 (1753) (incl. *G. macedonica* Boiss. & Orph., *G. inermis* Boros). Stem up to 130 cm. Leaflets 5–13, 5–45 mm, obovate to lanceolate, mucronate. Racemes (excluding the peduncles) at anthesis 12–20 mm, much exceeded by their subtending leaves, capitate. Corolla 4–6 mm. Legume 12–16 mm, elliptical, compressed, more or less densely glandular-setose or sometimes glabrous. Seeds 1–3. *Cultivated for liquorice. S.E. Europe, extending to Hungary and S. Italy.* Bu Cr Gr Hu It Ju Rm Rs (W, K, E) Tu.

42. Amorpha L.[1]

Deciduous shrubs, rarely herbs. Leaves imparipinnate; stipules small, caducous. Flowers in a terminal spike or a cluster of spikes. Calyx campanulate, with 5 teeth; standard blue, whitish, or purple; wings and keel absent; stamens diadelphous. Legume short, indehiscent. Seeds 1(–2).

1. **A. fruticosa** L., *Sp. Pl.* 713 (1753). Shrub up to 6 m. Leaflets 5–12 pairs, 15–40 × 8–20 mm, ovate or elliptical, pubescent or subglabrous, glandular-punctate. Inflorescence 7–15 cm. Standard *c.* 6 mm, blue or purplish. Legume 7–9 mm, glandular-punctate. *Planted for ornament and naturalized locally in C. & S. Europe.* [Al Au Bu Cz Ga He It Ju Rm.] *(C. & E. North America.)*

43. Psoralea L.[1]

Perennial herbs or shrubs. Leaves 3-foliolate (in European spp.), glandular-punctate; stipules small, free. Flowers in axillary heads or racemes, with a pair of 3-fid bracts at the base. Calyx campanulate, with 5 unequal teeth; corolla blue-violet to white; keel obtuse; stamens monadelphous. Legume indehiscent. Seeds 1.

Leaflets entire; flowers in heads **1. bituminosa**
Leaflets dentate; flowers in racemes **2. americana**

1. **P. bituminosa** L., *Sp. Pl.* 763 (1753). Stems 20–100 cm, sparsely to densely pubescent, smelling of bitumen. Leaflets 10–60 × 3–20(–30) mm, linear-lanceolate to ovate-orbicular, entire. Flowers in heads; peduncles usually longer than the leaves. Corolla 15–20 mm, blue-violet. Legume ovoid, compressed, with a falcate beak up to 15 mm long. *S. Europe.* Al Bl Bu Co Cr Ga Gr Hs It Ju Lu Rm Rs (K) Sa Si Tu.

2. **P. americana** L., *Sp. Pl.* 763 (1753) (*P. dentata* DC.). Like **1** but leaflets 8–50 × 6–35 mm, rhombic-orbicular to ovate, dentate; flowers in racemes; peduncles about equalling the leaves; corolla *c.* 8 mm, white, the keel often with a violet tip; legume not beaked. *S. Italy, Sicilia; S.W. part of the Iberian peninsula.* Hs It Lu Si.

[1] By P. W. Ball.   [2] By P. F. Yeo.

44. Apios Fabr.[1]

Twining perennial herbs with tuberous roots. Leaves imparipinnate; leaflets 3–9, stipellate; stipules small. Flowers in axillary racemes. Calyx campanulate, bilabiate; corolla purple, the keel coiled or strongly recurved; stamens diadelphous; style glabrous; stigma terminal. Legume linear, compressed, dehiscent. Seeds numerous.

1. A. americana Medicus, *Vorl. Churpf. Phys.-Ökon. Ges.* **2**: 355 (1787) (*A. tuberosa* Moench). Stem 30–120 cm, glabrous or pubescent. Leaflets 3–10 cm, ovate or ovate-lanceolate. Racemes usually with more than 20 flowers. Calyx 3–5 mm; corolla 10–12 mm, brownish-purple or pale purple. Legume 6–12 cm. Seeds *c.* 6 mm, dark brown. *Cultivated locally in S. Europe for its edible tubers and sometimes elsewhere as a curiosity or for ornament, and occasionally naturalized.* [Ga Ge It.] (*C., S. & E. North America.*)

45. Phaseolus L.[1]

Annual or perennial, usually climbing herbs. Leaves 3-foliolate; stipels present; stipules small. Flowers in axillary racemes. Calyx campanulate, bilabiate; corolla variously coloured; keel with a spirally coiled beak; stamens diadelphous; style hairy on the inside; stigma oblique. Legume linear-oblong, dehiscent. Seeds usually numerous.

Racemes shorter than leaves, with not more than 6 flowers   **1. vulgaris**
Racemes longer than leaves, many-flowered   **2. coccineus**

1. P. vulgaris L., *Sp. Pl.* 723 (1753). Annual up to 4 m. Leaflets 5–10 × 4–6 cm, ovate or ovate-orbicular, acuminate. Racemes shorter than leaves, up to 6-flowered. Corolla 10–18 mm, white, pink or purple; beak of keel forming 2 turns of a spiral. Legume up to 50 × 2·5 cm, brown. *Cultivated for the edible legume* (French bean) *and seeds throughout Europe; often occurring as a relict of cultivation.* (*South America.*)

2. P. coccineus L., *Sp. Pl.* 724 (1753). Like **1** but perennial; racemes many-flowered, longer than leaves; corolla 15–30 mm, scarlet, sometimes with white wings and keel; beak of keel forming 1–1½ turns of a spiral. *Cultivated for the edible legume* (Runner bean) *and seeds, and for ornament, throughout Europe; often occurring as a relict of cultivation.* (*Tropical America.*)

46. Vigna Savi[1]

Like *Phaseolus* but the keel with a recurved, not spirally coiled, beak.

1. V. unguiculata (L.) Walpers, *Repert. Bot. Syst.* **1**: 779 (1842) (*V. sinensis* (L.) Savi ex Hassk.). Glabrous or subglabrous annual 30–200 cm, usually not twining. Leaflets 5–16 cm, ovate-lanceolate. Racemes 2- to 12-flowered. Calyx 7–8 mm; corolla 20–25 mm, white or pale yellow with pink base, or pink or pale red. Legume 15–30 × 0·5–1 cm, pendent. Seeds 10–15, 10–15 mm. *Cultivated in S. Europe for the edible seed and for fodder.* (*Tropical Africa.*)

V. cylindrica (L.) Skeels, *U.S. Dept. Agric. Bur. Pl. Ind. Bull.* **282**: 32 (1913), with the legume 6–15 cm, erect or ascending, and seeds *c.* 4 mm, is also cultivated in S.W. Europe for the edible seed.

47. Glycine Willd.[1]

Erect or twining herbs. Leaves 3-foliolate; stipels present; stipules small. Flowers in axillary racemes. Calyx campanulate or tubular-campanulate, somewhat bilabiate; corolla usually purple, not or only slightly exceeding calyx, the keel not coiled at apex; stamens diadelphous or monadelphous; style glabrous. Legume linear or oblong, constricted between the seeds and septate, dehiscent. Seeds 2–4.

Literature: F. J. Hermann, *U.S. Dept. Agric. Techn. Bull.* **1268**: 1–82 (1962).

1. G. max (L.) Merr., *Interpr. Rumph. Herb. Amb.* 274 (1917) (*G. hispida* (Moench) Maxim.). Erect annual 30–200 cm, hispid with reddish-brown hairs. Leaflets 3–15 cm, ovate-elliptical. Racemes 5- to 8-flowered. Calyx (4–)5–7 mm; corolla (4·5–)6–7 mm, violet, pink or white. Legume 25–80 × 8–15 mm, pendent. Seeds 2–4, 6–11 mm. *Cultivated in S.E. Europe for the extraction of oil, for the edible seeds, and for fodder.*

The origin of this species (the soya-bean of commerce) is not known. It has possibly been derived from **G. soja** Siebold & Zucc., *Abh. Akad. Wiss.* (*München*) **4(2)**: 119 (1846) (*G. ussuriensis* Regel & Maack), a native of E. Asia.

48. Cicer L.[1]

Annual or perennial herbs with glandular hairs. Leaves usually imparipinnate, rarely paripinnate and terminated by a tendril; stipules herbaceous. Flowers solitary or in axillary racemes. Calyx gibbous at base, bilabiate, but sometimes the teeth subequal; corolla white or violet; stamens diadelphous; style glabrous. Legume ovate or oblong, dehiscent. Seeds 1–4.

1   Corolla 25–28 mm; calyx-teeth distinctly unequal   **3. montbretii**
1   Corolla 10–22 mm; calyx-teeth subequal
   2   Upper leaves terminated by a tendril; corolla 20–22 mm; calyx strongly gibbous at base   **4. graecum**
   2   Upper leaves terminated by a leaflet; corolla 7–12 mm; calyx slightly gibbous at base
     3   Peduncles shorter than leaves; corolla 10–12 mm; calyx-teeth at least twice as long as tube   **1. arietinum**
     3   Peduncles at least twice as long as leaves; corolla 7–10 mm; calyx-teeth only slightly longer than tube   **2. incisum**

1. C. arietinum L., *Sp. Pl.* 738 (1753). Erect, pubescent annual 20–50(–100) cm. Leaves imparipinnate; leaflets 3–8 pairs, 8–18 × 3–10 mm, ovate or elliptical, deeply toothed. Peduncles shorter than leaves, 1-flowered. Calyx slightly gibbous at base, the teeth subequal, at least twice as long as tube; corolla 10–12 mm, pale purple or white. Legume 20–30 × 10–15 mm. Seeds 1–2. *Widely cultivated for the edible seed and persisting as an escape from cultivation in S. Europe.* [Ga It.] (? *S.W. Asia.*)

2. C. incisum (Willd.) K. Malý in Ascherson & Graebner, *Syn. Mitteleur. Fl.* **6(2)**: 900 (1909). Procumbent or ascending, sparsely pubescent perennial 10–30 cm. Leaves imparipinnate; leaflets 1–3 pairs, 2–10 × 1–5 mm, 3- to 5-fid. Peduncles at least twice as long as leaves, 1- to 2-flowered. Calyx slightly gibbous at base, the teeth subequal, slightly longer than the tube; corolla 7–10 mm, violet. Legume *c.* 10 mm. Seeds 1–2. *Screes. S. Greece, Kriti.* Cr Gr. (*S.W. Asia.*)

3. C. montbretii Jaub. & Spach, *Ann. Sci. Nat.* ser. 2 (Bot.), **18**: 229 (1842). Erect, pubescent perennial 25–40 cm. Leaves imparipinnate; leaflets 6–8 pairs, 12–25 × 5–10 mm, oblong,

[1] By P. W. Ball.

128

dentate or serrate. Peduncles shorter than or equalling leaves, (1–)2- to 5-flowered. Calyx strongly gibbous at base, the teeth distinctly unequal; corolla 25–28 mm, white with violet spot on standard. Legume *c.* 25 × *c.* 12 mm. Seeds 3–4. *S. Albania; S.E. Bulgaria and Turkey-in-Europe.* Al Bu Tu. (*Anatolia.*)

4. C. graecum Orph. ex Boiss., *Diagn. Pl. Or. Nov.* **3**(2): 43 (1856). Erect, pubescent perennial 20–50 cm. Lower leaves imparipinnate, upper leaves paripinnate, terminated by a tendril; leaflets 3–8 pairs, 8–18 × 4–10 mm, oblong, elliptical or obovate, deeply toothed. Peduncles slightly shorter than leaves, 1- to 4-flowered. Calyx strongly gibbous at base, the teeth subequal, longer than tube; corolla 20–22 mm, white. Legume 20–30 × 7–9 mm. ● *S. Greece.* Gr.

49. Vicia L.[1]

Annual or perennial herbs, often climbing by means of tendrils. Leaves paripinnate, usually with a tendril, very rarely imparipinnate; stipules usually small, herbaceous. Flowers solitary, axillary or in axillary fascicles or racemes. Calyx actinomorphic to bilabiate; keel obtuse; stamens diadelphous; style pubescent all round or on the lower side, or glabrous. Legume more or less oblong, compressed, dehiscent. Seeds usually 2 or more.

Several species in this genus are important fodder plants. The most frequently utilized are **21, 27, 28, 45, 46, 54** and **55**, but many other species are cultivated locally, especially in S. Europe.

V. dennesiana H. C. Watson in Godman, *Nat. Hist. Azores* 155 (1870), was a pubescent perennial with 8–12 pairs of leaflets, 10- to 25-flowered, pedunculate racemes, brown corolla (purple in bud) *c.* 25 mm, standard shorter than wings and keel (the wings recurved at the apex), glabrous legume *c.* 50 × 10 mm and the hilum comprising *c.* ⅓ of the circumference of the seed. It was found only once, on damp earthy cliffs on Açores (São Miguel), and is known to have become extinct soon after its discovery, though it persisted in cultivation for some years. *Vide* J. D. Hooker, *Bot. Mag.* **113**: t. 6967 (1887).

1 Inflorescence sessile or with the peduncle shorter than flowers
2 Standard pubescent on the back
  3 Calyx-teeth subequal; corolla 14–22 mm   **45. pannonica**
  3 Calyx-teeth unequal; corolla 18–30 mm   **52. hybrida**
2 Standard glabrous on the back
  4 All leaves without a tendril, the rhachis terminated by a short mucro
    5 Corolla white with black wings, rarely purple; legume 80 mm or more, pubescent   **55. faba**
    5 Corolla yellow; legume not more than 50 mm, glabrous
      6 Leaflets 1–4 pairs, at least 15 mm wide   **39. oroboides**
      6 Leaflets 7–13 pairs, not more than 10 mm wide
   40. truncatula
  4 At least the upper leaves with a tendril
    7 Mouth of the calyx-tube oblique or the calyx-teeth unequal, the lowest tooth much longer than the upper teeth
      8 Standard yellow sometimes suffused with purple; wings yellow sometimes with black tip
        9 Legume glabrous except for the pubescent margin; wings greenish-yellow with black tip; keel purple   **50. melanops**
        9 Legume densely hairy, rarely completely glabrous; wings and keel yellow sometimes with a purple tinge   **51. lutea**
      8 Standard and wings purple
        10 Leaflets not more than 3 mm wide, linear, usually with 3 acute points at the apex   **49. peregrina**
        10 Leaflets at least 4 mm wide, ovate, elliptical or lanceolate, obtuse or emarginate, mucronate

---

1 By P. W. Ball.

11 Perennial; leaflets 3–9 pairs; legume 5–8 mm wide, glabrous   **42. sepium**
11 Annual; leaflets 1–3 pairs; legume 10–15 mm wide, pubescent on the margin   **54. narbonensis**
7 Mouth of the calyx-tube not oblique, the calyx-teeth equal or subequal
  12 Standard and keel yellow sometimes tinged with purple
    13 Corolla 23–35 mm; wings yellow sometimes with black tip; legume pubescent   **43. grandiflora**
    13 Corolla 18–22 mm; wings blue; legume glandular
   44. barbazitae
  12 Corolla purple sometimes with whitish wings and keel, never yellowish
    14 Corolla not more than 10 mm; seeds tuberculate
      15 Legume 15–30 × 3–4 mm, with a short curved beak; leaflets of upper leaves shortly mucronate
   47. lathyroides
      15 Legume 30–40 × 4–5 mm, with a long straight beak; leaflets of upper leaves long-mucronate   **48. cuspidata**
    14 Corolla usually more than 10 mm; seeds smooth
      16 Leaflets 1–3 pairs; wings and keel whitish   **53. bithynica**
      16 Leaflets of the upper leaves 3–many pairs; wings and keel purple
        17 Perennial; upper leaves usually with a simple tendril; hilum ⅕–¼ of the circumference of the seed
   41. pyrenaica
        17 Annual; upper leaves with a branched tendril; hilum ⅙–⅕ of the circumference of the seed   **46. sativa**
1 Inflorescence pedunculate, the peduncle much longer than the flowers
18 Each pair of stipules dimorphic, one linear, the other palmatifid, with subulate segments   **27. articulata**
18 Stipules of each pair ± identical
  19 Calyx-teeth equal, all equalling or longer than tube
    20 Leaves without tendril; legume torulose   **28. ervilia**
    20 At least the upper leaves with a tendril; legume not torulose
      21 Calyx-teeth about twice as long as tube; legume long-stipitate   **30. vicioides**
      21 Calyx-teeth not more than 1½ times as long as tube; legume not or only shortly stipitate
        22 Corolla 6–10 mm; seeds 3–6
          23 Perennial; leaflets 3–6 pairs; stipules entire   **23. glauca**
          23 Annual; leaflets 5–10 pairs; stipules dentate   **29. leucantha**
        22 Corolla 2–4(–5) mm; seeds usually 2
          24 Lower stipules linear-lanceolate, usually dentate; legume usually pubescent   **31. hirsuta**
          24 Lower stipules linear-setaceous, entire; legume glabrous   **32. meyeri**
  19 Calyx-teeth unequal, at least the upper shorter than tube
    25 All leaves without a tendril
      26 Corolla purple, blue or white
        27 Leaflets 1–3 pairs   **9. sicula**
        27 Leaflets 6 or more pairs
          28 Villous-sericeous; corolla 18–25 mm; legume villous
   7. argentea
          28 Glabrous or pubescent; corolla 12–15(–20) mm; legume glabrous
            29 Leaflets entire, mucronate   **1. orobus**
            29 Leaflets minutely denticulate, obtuse   **2. montenegrina**
      26 Corolla yellow sometimes with reddish or purplish tinge
        30 Leaflets 1–4 pairs, 15–45 mm wide   **39. oroboides**
        30 Leaflets 6 or more pairs, not more than 10 mm wide
          31 Racemes 20- to 30-flowered; corolla 7–13 mm; plant glabrous   **5. ochroleuca**
          31 Racemes not more than 20-flowered; corolla 12–25 mm; plant sparsely pubescent
            32 Leaflets obtuse, mucronate; racemes 3- to 8-flowered
   40. truncatula
            32 Leaflets acute; racemes 6- to 20-flowered
              33 Lower calyx-teeth shorter than tube; legume glabrous   **3. sparsiflora**
              33 Lower calyx-teeth about equalling tube; legume sparsely pubescent   **4. pinetorum**

25 At least the upper leaves with a long, usually branched
    tendril
34 Corolla 4–9 mm, whitish or pale purple; racemes 1- to
    8-flowered
35 Leaflets 1 pair   **38. bifoliolata**
35 Leaflets 2 or more pairs
36 Seeds usually 2; racemes shorter than leaves **34. disperma**
36 Seeds 3–6; racemes equalling or longer than leaves
37 Stem densely pubescent; legume 20–40 × 8–14 mm
       33. durandii
37 Stem glabrous or sparsely pubescent; legume not
    more than 17 × 5 mm
38 Lower calyx-teeth as long as or longer than tube;
    leaflets 3–5 mm wide   **37. pubescens**
38 Lower calyx-teeth shorter than tube; leaflets
    0·5–3 mm wide
39 Racemes (1–)2- to 5-flowered, longer than leaves;
    hilum $\frac{1}{12}$–$\frac{1}{8}$ of the circumference of the seed
       35. tenuissima
39 Racemes 1- or 2-flowered, about equalling leaves;
    hilum $\frac{1}{5}$ of the circumference of the seed
       36. tetrasperma
34 Corolla 9 mm or more; rarely smaller and then violet,
    with 10- to 40-flowered racemes
40 Corolla yellow; leaflets 15–40 mm wide   **6. pisiformis**
40 Corolla white, purple, violet or blue, rarely with yellow-
    ish wings; leaflets usually less than 15 mm wide
41 Stipules denticulate to serrate, or bipartite
42 Stipules bipartite, the lobes entire; racemes shorter
    than leaves   **26. monantha**
42 Stipules denticulate to serrate, not bipartite; racemes
    usually longer than leaves
43 Leaflets 2–3 pairs; racemes 1- to 3-flowered
       53. bithynica
43 Leaflets 3 or more pairs; racemes (2–)4- to 20-
    flowered
44 Calyx-tube strongly gibbous at base; limb of
    standard about $\frac{1}{2}$ as long as claw   **22. benghalensis**
44 Calyx-tube not or slightly gibbous at base; limb of
    standard about as long as claw
45 Stipules lunate; hilum $\frac{1}{2}$–$\frac{2}{3}$ of the circumference of
    the seeds
46 Leaflets 5–12 pairs; corolla white with purple
    veins; legume black   **16. sylvatica**
46 Leaflets 3–5 pairs; corolla blue or purple;
    legume brown   **20. dumetorum**
45 Stipules $\frac{1}{2}$-sagittate; hilum not more than $\frac{1}{3}$ of the
    circumference of the seed
47 Corolla 10–15 mm   **24. biennis**
47 Corolla 15–24 mm
48 Leaflets 1–4 mm wide; corolla violet with pale
    keel; hilum $\frac{1}{4}$–$\frac{1}{3}$ of the circumference of the
    seed   **18. onobrychoides**
48 Leaflets 4–9 mm wide; corolla white with bluish
    veins on the standard; hilum $\frac{1}{10}$ of the cir-
    cumference of the seed   **19. altissima**
41 Stipules entire
49 Calyx strongly gibbous at base; limb of the standard
    about $\frac{1}{2}$ as long as claw
50 Legume not or only shortly stipitate, the stipe not
    exceeding the calyx; leaflets 1–6 pairs   **25. cretica**
50 Legume stipitate, the stipe exceeding the calyx; leaf-
    lets 4–12 pairs
51 Corolla reddish-purple, usually black at tip;
    racemes shorter than or equalling leaves
       22. benghalensis
51 Corolla violet or purple, sometimes with white or
    yellow wings; racemes usually longer than leaves
       21. villosa
49 Calyx only slightly gibbous at base; limb of the stan-
    dard usually as long as or longer than claw
52 Corolla (17–)18 mm or more; racemes not more
    than 12-flowered

53 Plant densely hirsute; legume hirsute   **8. serinica**
53 Plant glabrous or pubescent; legume glabrous or
    sparsely pubescent
54 Corolla violet; leaflets 4–11 pairs, 10–35 mm
       18. onobrychioides
54 Corolla white or yellowish with purple tip; leaflets
    1–6 pairs, 4–10 mm   **25. cretica**
52 Corolla not more than 18 mm; racemes up to 40-
    flowered
55 Racemes 1- to 6-flowered; leaflets 1–6 pairs
       25. cretica
55 Racemes 4- to 40-flowered; leaflets 4 or more pairs
56 Plant villous; legume villous   **14. sibthorpii**
56 Plant glabrous or pubescent; legume glabrous or
    sparsely pubescent
57 Racemes shorter than or equalling leaves
58 Corolla white with violet tip and veins; leaflets
    mostly 3–6 pairs; seeds 4–6   **24. biennis**
58 Corolla purple, violet or blue; leaflets 5 or more
    pairs
59 Legume 4–6 mm wide; seeds 4–8; lower tooth
    of calyx ± equalling tube **(10–13). cracca** group
59 Legume 6–8 mm wide; seeds 1–3; lower tooth
    of calyx distinctly shorter than tube
       15. cassubica
57 Racemes exceeding leaves
60 Corolla 8–12 mm; limb of the standard about
    as long as claw; legume 4–6 mm wide
       (10–13). cracca group
60 Corolla (10–)12–18 mm; legume 6–8 mm wide
61 Limb of the standard longer than claw; leaf-
    lets 6–20 pairs; tendrils usually branched
       (10–13). cracca group
61 Limb of the standard shorter than claw; leaf-
    lets 4–8 pairs; tendrils usually simple
       17. multicaulis

Sect. CRACCA S. F. Gray. Leaflets usually numerous (more than 5 pairs); flowers usually numerous in long-pedunculate racemes; calyx bilabiate, somewhat gibbous at base; corolla usually large (more than 10 mm); style equally pubescent all round.

1. V. orobus DC. in Lam. & DC., *Fl. Fr.* ed. 3, **5**: 577 (1815). Pubescent perennial up to 60 cm. Leaves shortly mucronate at apex without a tendril; leaflets 6–15 pairs, 8–23 × 3–8 mm, oblong to elliptical, obtuse, mucronate. Racemes 6- to 20-flowered. Calyx-teeth unequal, shorter than tube; corolla 12–15(–20) mm, white with purple veins. Legume 20–30 × 4–7 mm, yellow, glabrous. Seeds 4–5; hilum $\frac{1}{3}$–$\frac{1}{2}$ of the circumference. *W. Europe.* Be Br Da Ga Ge Hb He Hs Lu No.

2. V. montenegrina Rohlena, *Feddes Repert.* **3**: 146 (1907) (incl. *V. orbelica* Stoj. & Stefanov). Glabrous or sparsely pubescent perennial 40–60 cm. Leaves without a tendril, often with a terminal leaflet; leaflets 8–17 pairs, 10–25 × 4–8 mm, oblong, obtuse or emarginate, minutely denticulate. Racemes 10- to 30-flowered. Calyx-teeth unequal, shorter than tube; corolla *c.* 14 mm, violet-blue or white with violet tinge. Legume *c.* 20 × 5–6 mm, glabrous. Seeds 4–5. ● *W. Jugoslavia, S.W. Bulgaria.* Bu Ju.

Possibly not specifically distinct from **V. abbreviata** Fischer ex Sprengel, *Pugillus* **1**: 50 (1813), from the Caucasus.

3. V. sparsiflora Ten., *Fl. Nap.* **5**: 110 (1836). Sparsely pubescent perennial up to 100 cm. Leaves without a tendril; leaflets 6–14 pairs, 8–25 × 4–10 mm, oblong to elliptical, acute. Racemes 6- to 20-flowered. Calyx-teeth unequal, shorter than tube; corolla 15–25 mm, pale yellow. Legume 20–40 × 4–9 mm, pale yellow, glabrous. Seeds *c.* 4. $2n=12$. *From S.E. Czechoslovakia to S. Italy and S.E. Bulgaria; local.* Bu Cz Hu It Ju Rm.

4. V. pinetorum Boiss. & Spruner in Boiss., *Diagn. Pl. Or. Nov.* **1**(**2**): 104 (1843). Sparsely pubescent perennial 30–50 cm. Leaflets 10–16 pairs, 10–20 × 3–7 mm, elliptic-lanceolate, acute. Racemes 8- to 20-flowered. Calyx-teeth unequal, the lower about as long as tube; corolla 12–19 mm, yellow. Legume 15–18 × 3·5–4 mm, dark brown, sparsely pubescent. Seeds *c.* 6; hilum $\frac{1}{5}$–$\frac{1}{6}$ of the circumference. ● *S. Aegean region.* Cr Gr.

5. V. ochroleuca Ten., *Fl. Nap.* **1**, *Prodr.* 42 (1811). Glabrous perennial 30–60 cm. Leaflets 7–12 pairs, linear to oblong or elliptical. Racemes 20- to 30-flowered. Calyx-teeth unequal, shorter than tube; corolla 7–13 mm, pale yellow or yellow-green. Legume 15–30 × 5·5–7·5 mm, brown, glabrous. Seeds *c.* 6; hilum $\frac{1}{9}$–$\frac{1}{8}$ of the circumference. *N.W. part of the Balkan peninsula, Italy, Sicilia.* Al Ju It Si.

(a) Subsp. **ochroleuca**: Leaflets (15–)20–45 × 3–9 mm, linear or oblong; calyx 2·7–4 mm. $2n=12$. *Italy, Sicilia.*

(b) Subsp. **dinara** (Borbás) K. Malý ex Rohlena, *Sitz.-Ber. Böhm. Ges. Wiss.* (*Math.-Nat. Kl.*) **1912**(**1**): 36 (1913): Leaflets 10–20 × 2–5 mm, usually elliptical; calyx 2·4–3 mm. *W. Jugoslavia, N. Albania.*

6. V. pisiformis L., *Sp. Pl.* 734 (1753). Glabrous perennial 100–200 cm. Leaflets 3–5 pairs, 15–40(–60) × 15–40 mm, ovate or ovate-orbicular, obtuse, mucronate. Racemes 8- to 30-flowered. Calyx-teeth unequal, shorter than tube; corolla 13–20 mm, yellow. Legume 25–40 × 6–10 mm, pale brown, glabrous. Seeds 6–7; hilum $\frac{1}{2}$ of the circumference. *C. & E. Europe, extending to S.E. Norway, E. France, N. Italy and C. Jugoslavia.* Al Au Bu Cz Ga Ge He Hu It Ju No Po Rm Rs (C, W, K, E) Su.

7. V. argentea Lapeyr., *Hist. Abr. Pyr.* 417 (1813). Villous-sericeous perennial 10–40 cm. Leaves without a tendril, usually with a terminal leaflet; leaflets 7–9 pairs, 8–16 × 1–4 mm, linear, obtuse; stipules entire. Racemes 3- to 7-flowered. Calyx-teeth unequal, the lower longer than the tube; corolla 18–25 mm, white with violet veins. Legume 22–30 × 7·5–10 mm, brown, villous. Seeds 4–7; hilum *c.* $\frac{1}{10}$ of the circumference. ● *Pyrenees.* Ga Hs.

8. V. serinica Uechtr. & Huter, *Österr. Bot. Zeitschr.* **55**: 81 (1905) (*V. argentea* auct. ital., non Lapeyr). Like **7** but grey-hirsute; leaves with a simple or branched tendril; leaflets 8–11 pairs, linear-lanceolate, mucronate; racemes 4- to 10-flowered; legume hirsute. ● *S. Italy* (*Monte del Papa, near Lagonegro*). It.

9. V. sicula (Rafin.) Guss., *Fl. Sic. Syn.* **2**: 292 (1844). Glabrous perennial 40–80 cm. Leaves without a tendril; leaflets 1–3 pairs, 20–50 × 1–3 mm, linear, acute. Racemes 15- to 25-flowered. Calyx-teeth unequal, shorter than tube; corolla 15–20 mm, pale purple. Legume 25–50 × 4–10 mm, glabrous, yellow. Seeds *c.* 6. *S. Italy, Sicilia.* ?Gr It Si.

10–13. V. cracca group. Perennial up to 200 cm. Leaflets 5–20 pairs; stipules entire. Racemes 8- to 40-flowered. Calyx-teeth unequal; corolla purple, violet or blue; limb of the standard equalling or longer than claw. Legume brown, glabrous. Seeds 4–8.

A critical group of species not yet fully understood and frequently misidentified. Flower-colour may be of taxonomic importance in the group, but it is not clear from the information available to what extent the colour varies in each species.

1 Corolla 8–12(–13) mm; limb of the standard about as long as claw
2 Stems glabrous or pubescent with appressed hairs; lower calyx-teeth about as long as tube **10. cracca**
2 Stems densely pubescent with patent hairs; lower calyx-teeth *c.* $1\frac{1}{2}$ times as long as tube **11. incana**
1 Corolla (10–)12–18 mm; limb of the standard longer than claw
3 Leaflets 2–6 mm wide, linear-lanceolate **12. tenuifolia**
3 Leaflets 1–2 mm wide, linear to setaceous **13. dalmatica**

10. V. cracca L., *Sp. Pl.* 735 (1753). Stems glabrous or pubescent with appressed hairs. Leaflets 6–15 pairs, 5–30 × 1–6 mm, linear to ovate-oblong. Racemes 10- to 30-flowered, dense. Lower calyx-teeth almost equalling tube; corolla 8–12 mm, bluish-violet; limb of the standard about equalling claw. Legume 10–25 × 4–6 mm, with stipe shorter than calyx. Seeds with hilum $\frac{1}{4}$–$\frac{1}{3}$ of the circumference. $2n=14, 27, 28, 30$. *Almost throughout Europe.* All except Az Bl ?Lu Sb.

V. oreophila Žertová, *Nov. Bot. Horti Bot. Univ. Carol. Prag.* **1962**: 51 (1962), described from mountains in Czechoslovakia, and probably occurring elsewhere in the mountains of C. & S. Europe and in Scandinavia, is probably best treated as a subspecies of **10**. It is smaller, with the stems 5–30 cm, the leaflets 6–10 pairs, lanceolate, the racemes shorter than the leaves and the corolla 10–13 mm. It has $2n=28$.

11. V. incana Gouan, *Fl. Monsp.* 189 (1765) (*V. cracca* subsp. *gerardii* Gaudin, *V. gerardii* All.). Like **10** but stems densely pubescent with patent hairs; leaflets 10–22 pairs; racemes 20- to 40-flowered; lower calyx-teeth *c.* $1\frac{1}{2}$ times as long as tube; legume with stipe equalling or longer than calyx. $2n=12$. *Mountains of C. & S. Europe.* Al Au Bu Co Cz Ga Gr He Hs It Ju Lu Si.

12. V. tenuifolia Roth, *Tent. Fl. Germ.* **1**: 309 (1788) (incl. *V. boissieri* Freyn, *V. elegans* Guss.). Stems glabrous or appressed-pubescent. Leaflets 5–13 pairs, 10–30(–40) × 2–6 mm, linear or linear-oblong. Racemes 15- to 30-flowered, usually dense. Lower calyx-teeth shorter than or sub-equal to the tube; corolla (10–)12–18 mm, purple, pale lilac or bluish-lilac; limb of the standard longer than claw. Legume 20–35 × 5–8 mm. Seeds with hilum $\frac{1}{5}$–$\frac{1}{4}$ of the circumference. $2n=24$. *C., S. & E. Europe, extending to 60° N. in Sweden; often occurring as a casual in W. & N. Europe and perhaps locally naturalized.* Al Au Be Bu Co ?Cr Cz Da Ga Ge ?Gr He Hs Hu It Lu Ju Po Rm Rs (N, B, C, W, K, E) Sa Si Su Tu [Br Ho].

Many of the records from S.E. Europe are probably referable to **13**.

13. V. dalmatica A. Kerner, *Sched. Fl. Exsicc. Austro-Hung.* **4**: 2 (1886) (*V. tenuifolia* subsp. *stenophylla* Velen.). Like **12** but leaflets 1–2(–2·5) mm wide, linear or setaceous, and racemes 8- to 20-flowered, lax. $2n=12$. *S.E. Europe, extending northwards to Hungary.* Al Bu Cr Gr Hu It Ju Rm Rs (K).

14. V. sibthorpii Boiss., *Diagn. Pl. Or. Nov.* **2**(**9**): 122 (1849). Densely villous perennial 30–100 cm. Leaflets 6–12 pairs, 5–20 × 2–6 mm, elliptical, oblong-elliptical or linear-elliptical; stipules entire. Racemes 8- to 25-flowered. Calyx-teeth unequal, the lower about equalling the tube; corolla 12–15 mm, bluish-purple, sometimes with white wings; limb of standard equalling or longer than claw. Legume 12–20 × 5–6 mm, brown, villous. Seeds 4–8; hilum $\frac{1}{6}$ of the circumference. *Greece and Aegean region.* Cr Gr ?Ju.

15. V. cassubica L., *Sp. Pl.* 735 (1753). Pubescent or subglabrous perennial 30–60(–100) cm. Leaflets 5–16 pairs, 7–30 × 3–10 mm, linear-lanceolate to oblong or elliptical; stipules entire. Racemes 4- to 15-flowered. Calyx-teeth unequal, shorter than tube; corolla 10–13 mm, purple or blue, wings and keel whitish. Legume 15–30 × 6–8 mm, oblong-rhombic, yellow, glabrous. Seeds 1–3; hilum $\frac{1}{3}$ of the circumference. $2n = 12$. *S.C. & E. Europe, extending westwards to France and northwards to 60° N. in Fennoscandia; absent from most islands.* Al Au Bu Cz Da *Fe Ga Ge Gr Hu It Ju No Po Rm Rs (N, B, C, W, K, E) Si Su Tu.

16. V. sylvatica L., *Sp. Pl.* 734 (1753). Usually glabrous perennial 60–200 cm. Leaflets 5–12 pairs, 6–20 × 3–10 mm, oblong to ovate-oblong; stipules dentate. Racemes 5- to 20-flowered. Calyx-teeth unequal, shorter than tube; corolla 12–20 mm, white with purple veins. Legume 25–30 × 5–10 mm, black, glabrous. Seeds 4–5; hilum $\frac{2}{3}$ of the circumference. $2n = 14$. *N., C. & E. Europe, extending southwards to Italy and Crna Gora.* Al Au Br Co Cz Da Fe Ga Ge Hb He ?Hs Hu It Ju No Po Rm Rs (N, B, C, W, E) Su.

17. V. multicaulis Ledeb., *Icon. Pl. Fl. Ross.* 1: 12 (1829). Pubescent perennial 20–45 cm. Leaves usually with a simple tendril; leaflets 4–8 pairs, 15–22 × 1·5–2·5 mm, linear-oblong or elliptic-oblong; stipules entire. Racemes 10- to 20-flowered. Calyx-teeth unequal, the lower almost equalling tube; corolla 15–18 mm, violet. Legume 20–30 × 6–7 mm, brown, glabrous. Seeds *c.* 5. *Rocks and stony slopes. C. Ural.* Rs (C).

18. V. onobrychioides L., *Sp. Pl.* 735 (1753). Glabrous or pubescent perennial 30–120 cm. Leaflets 4–11 pairs, 10–35 × 1–4 mm, linear or oblong-lanceolate; stipules entire or with few teeth. Racemes 4- to 12-flowered. Calyx-teeth unequal, the lower equalling the tube; corolla 17–24 mm, violet with pale keel. Legume 25–40 × 5–7 mm, reddish-brown, glabrous, stipitate. Seeds 5–10; hilum $\frac{1}{4}$–$\frac{1}{3}$ of the circumference. *S. Europe.* Al Bu Ga Gr He Hs It Ju Lu Tu.

19. V. altissima Desf., *Fl. Atl.*, 2: 163 (1799). Glabrous perennial 60–200 cm. Leaflets 5–9 pairs, 10–25 × 4–9 mm, oblong, sometimes denticulate; stipules dentate. Racemes 5- to 15-flowered. Calyx-teeth unequal, shorter than tube; corolla 15–19 mm, white with bluish veins on the standard. Legume 40–50 × 5–7 mm, brown, subglabrous. Seeds 6–10; hilum $\frac{1}{10}$ of the circumference. *W. Mediterranean region, westwards to S. France.* Co Ga It Sa Si.

20. V. dumetorum L., *Sp. Pl.* 734 (1753). Subglabrous perennial (30–)80–150(–200) cm. Leaflets 3–5 pairs, 12–40 × 6–20 mm, ovate; stipules dentate. Racemes 2- to 14-flowered. Calyx-teeth unequal, shorter than tube; corolla 12–20 mm, blue or purple. Legume 25–60 × 6–10 mm, brown, glabrous. Seeds 6–10; hilum *c.* $\frac{1}{2}$ of the circumference. *From S.C. Sweden and White Russia southwards to C. Italy and Greece, and westwards to E. France.* Au Bu Cz Da Ga Ge Gr He Hu It Ju Po Rm Rs (B, C, W, E) Su.

21. V. villosa Roth, *Tent. Fl. Germ.* 2(2): 182 (1793). Annual 30–200 cm. Leaflets 4–12 pairs, linear to elliptical; stipules entire. Calyx strongly gibbous at the base; calyx-teeth unequal; corolla 10–20 mm, violet, purple or blue, sometimes with white or yellow wings; limb of the standard *c.* $\frac{1}{2}$ as long as claw. Legume 20–40 × (4–)6–12 mm, brown, stipitate. Seeds 2–8; hilum $\frac{1}{12}$–$\frac{1}{8}$ of the circumference. *Europe, southwards from N. France and White Russia; widely cultivated for fodder and naturalized further north.* Al Au Bl Bu Co Cr Cz Ga Ge Gr He Hs Hu It Ju Lu Po Rm Rs (C, W, K, E) Sa Si Tu [Be Br Da Fe Ho No Rs (N, B) Su].

1 Stems villous; lower calyx-teeth longer than tube   **(a) subsp. villosa**
1 Stems glabrous or appressed-pubescent; lower calyx-teeth
   shorter than tube
  2 Legume pubescent

3 Racemes 5- to 20-flowered   **(c) subsp. eriocarpa**
3 Racemes 2- to 6- flowered   **(d) subsp. microphylla**
2 Legume glabrous or glabrescent
  4 Racemes 10- to 30-flowered   **(b) subsp. varia**
  4 Racemes 2- to 10-flowered
    5 Wings usually yellow; legume glabrous **(e) subsp. pseudocracca**
    5 Wings purple, violet or white; legume usually sparsely
      pubescent when young   **(d) subsp. microphylla**

(a) Subsp. **villosa**: Villous. Leaflets (8–)10–35 × 2–8 mm. Racemes 10- to 30-flowered. Calyx-teeth plumose, the lower as long as or longer than tube; corolla 10–20 mm; wings variously coloured. Legume glabrous. *Almost throughout the range of the species; widely naturalized.*

(b) Subsp. **varia** (Host) Corb., *Nouv. Fl. Normand.* 181 (1893) (*V. dasycarpa* auct., ?an Ten., *V. varia* Host): Glabrous or appressed-pubescent. Leaflets (8–)10–30 × 2–8 mm. Racemes 10- to 30-flowered. Calyx-teeth glabrous or appressed-pubescent, all shorter than tube; corolla 10–16(–18) mm; wings violet, purple, blue or white. Legume glabrous. *Almost throughout the range of the species; widely naturalized.*

(c) Subsp. **eriocarpa** (Hausskn.) P. W. Ball, *Feddes Repert.* **79**: 45 (1968) (*V. eriocarpa* (Hausskn.) Halácsy): Like subsp. **(b)** but leaflets 5–15(–20) × 1–5 mm; racemes 5- to 20-flowered; legume pubescent at least when young. *Greece and Aegean region; Sicilia.*

(d) Subsp. **microphylla** (D'Urv.) P. W. Ball, *Feddes Repert.* **79**: 45 (1968) (*V. microphylla* D'Urv.): Like subsp. **(b)** but leaflets 3–10 × 1–4 mm; racemes 2- to 6-flowered; legume glabrous or pubescent. *S. Greece and Aegean region.*

(e) Subsp. **pseudocracca** (Bertol.) P. W. Ball, *Feddes Repert.* **79**: 45 (1968) (*V. pseudocracca* Bertol.; incl. *V. elegantissima* R. J. Shuttlew): Like subsp. **(b)** but leaflets 5–20 × 1–5 mm; racemes 3- to 10-flowered; wings usually yellow. $2n = 14$. *S.W. Europe.*

22. V. benghalensis L., *Sp. Pl.* 736 (1753) (*V. atropurpurea* Desf.). Villous annual or short-lived perennial 20–80 cm. Leaflets 5–9 pairs, 10–25 × 1·5–6 mm, linear, oblong or elliptical; stipules entire or dentate. Racemes 2- to 12-flowered. Calyx strongly gibbous at base; calyx-teeth unequal, the lower longer than tube; corolla 10–18 mm, reddish-purple, usually black at tip; limb of the standard about $\frac{1}{2}$ as long as claw. Legume 25–40 × 8–11 mm, brown, pubescent, at least on the suture, shortly stipitate. Seeds 3–5; hilum $\frac{1}{3}$ of the circumference. $2n = 14$. *Mediterranean region, Portugal, Açores.* Az Bl Co Ga Gr Hs It Lu Sa Si.

23. V. glauca C. Presl in J. & C. Presl, *Del. Prag.* 37 (1822). Pubescent perennial 10–40 cm. Leaves with a simple tendril; leaflets 3–6 pairs, 5–12 × 1–3 mm, oblong to ovate; stipules entire. Racemes 4- to 8-flowered. Calyx-teeth equal, about equalling tube; corolla 8–10 mm, pale purple. Legume 15–25 × 5–9 mm, dark red-brown, pubescent, at least on the suture. Seeds *c.* 4; hilum $\frac{1}{8}$–$\frac{1}{7}$ of the circumference. *Sardegna, Sicilia.* Sa Si. (*N.W. Africa.*)

24. V. biennis L., *Sp. Pl.* 736 (1753) (*V. picta* Fischer & C. A. Meyer). Glabrous or sparsely pubescent annual 30–150 cm. Leaflets 3–6(–8) pairs, 10–40 × 2–10 mm, oblong; stipules entire or dentate. Racemes 5- to 20-flowered. Calyx-teeth unequal, shorter than or equalling tube; corolla 10–15 mm, white with violet at the tip and violet veins. Legume 25–35 × 6–7 mm, brown, glabrous. Seeds 4–6; hilum $\frac{1}{4}$–$\frac{1}{3}$ of the circumference. $2n = 14$. *From Hungary to W. Kazakhstan; local.* Hu Rm Rs (C, W, E).

25. V. cretica Boiss. & Heldr. in Boiss., *Diagn. Pl. Or. Nov.* 2(9): 118 (1849). Sparsely pubescent annual 10–30 cm. Leaflets 1–6 pairs, 4–10 × 1–4 mm, linear to elliptical; stipules entire.

Racemes 1- to 6-flowered. Calyx somewhat gibbous at base; calyx-teeth unequal, shorter or the lower longer than tube; corolla 9–20 mm, white or yellowish, purple at tip. Legume 20–30 × 5·5–8 mm, brown, glabrous or sparsely pubescent. Seeds 4–5; hilum $\frac{1}{15}$ of the circumference. *S. Aegean region.* Cr Gr.

(a) Subsp. **cretica**: Racemes usually equalling or shorter than leaves; calyx 3·5–5(–6) mm; corolla 9–16 mm. *Throughout the range of the species except for most of the Kikladhes.*

(b) Subsp. **aegaea** (Halácsy) P. W. Ball, *Feddes Repert.* **79**: 42 (1968): Racemes longer than leaves; calyx 5·5–8 mm; corolla 18–20 mm. *Kikladhes.*

26. **V. monantha** Retz., *Obs. Bot.* **3**: 39 (1783) (*V. calcarata* Desf.). Subglabrous annual 30–60 cm. Leaflets 5–8 pairs, 10–25 × 1–6 mm, linear-oblong; stipules bipartite, lobes entire. Calyx-teeth unequal, shorter than tube; corolla pale purple. Legume 20–50 × 6–12 mm, yellow, glabrous. Seeds 3–7; hilum $\frac{1}{6}$ of the circumference. *Mediterranean region.* Bl Gr Hs It Sa Si [Ga].

(a) Subsp. **monantha**: Racemes 1- to 2-flowered; corolla 10–15 mm; legume 20–35 × 6–8·5 mm; seeds less than 3·5 mm, brown. *Throughout the range of the species.*

(b) Subsp. **triflora** (Ten.) B. L. Burtt & P. Lewis, *Kew Bull.* **1949**: 510 (1950): Racemes 2- to 4-flowered; corolla 14–20 mm; legume 30–50 × 8·5–12 mm; seeds more than 3·5 mm, blackish. *S. Italy; Lampedusa; Greece.*

27. **V. articulata** Hornem., *Enum. Pl. Hort. Haun.* 41 (1807) (*V. monanthos* (L.) Desf., non Retz.; incl. *V. smyrnaea* Boiss.). Glabrous annual 20–70 cm. Leaflets 5–9 pairs, 10–25 × 1–4 mm, linear-oblong; each pair of stipules dimorphic, one simple, linear, the other palmatifid with 8–17 linear-subulate segments. Racemes 1- or 2-flowered. Calyx-teeth slightly unequal, longer than tube; corolla 8–17 mm, white or pale blue, sometimes with black tip. Legume 15–35 × 6–10 mm, yellow, glabrous. Seeds 2–4; hilum $\frac{1}{10}$ of the circumference. *S. Europe.* Bu Gr Hs It Ju Lu Sa Si [Au Cz Ga Ge Po Rm].

Sect. ERVUM (L.) S. F. Gray. Leaflets usually numerous (more than 4 pairs); flowers few, in long-pedunculate racemes; calyx not gibbous at base; corolla usually less than 10 mm; style glabrous or equally pubescent all round.

28. **V. ervilia** (L.) Willd., *Sp. Pl.* **3**: 1103 (1802). Glabrous or pubescent annual 15–50(–70) cm. Leaves without tendril; leaflets 8–15(–20) pairs, 5–15 × 1–4 mm, oblong or linear; stipules entire or palmatifid. Racemes 1- to 4-flowered. Calyx-teeth equal, longer than tube; corolla 6–9(–12) mm, white tinged with red or purple. Legume 10–30 × 4–6 mm, yellow, glabrous, torulose. Seeds 2–4; hilum $\frac{1}{12}$ of the circumference. *S. Europe.* Al Bu Cr Ga Gr It Ju Lu Tu [Au Cz Ge He].

29. **V. leucantha** Biv., *Stirp. Rar. Sic. Descr.* **1**: 9 (1813). Glabrous or pubescent annual 30–50 cm. Leaflets 5–10 pairs, 5–15 × 1–5 mm, oblong or linear; stipules dentate. Racemes 2- to 12-flowered. Calyx-teeth equal, longer than the tube; corolla 6–10 mm, pale purple, white at base. Legume 15–30 × 5–10 mm, pale brown, pubescent or subglabrous, beaked. Seeds 3–6. *C. Mediterranean region.* ?Bl It Ju Sa Si. (*N.W. Africa.*)

30. **V. vicioides** (Desf.) Coutinho, *Fl. Port.* 363 (1913) (*V. erviformis* Boiss.). Pubescent annual 15–50 cm. Racemes 5- to 20-flowered. Calyx-teeth subequal, twice as long as tube; corolla 4–8 mm, pale purple. Legume 13–16 × 5–10 mm, yellowish, glabrous or sparsely sericeous, long-stipitate. Seeds 2. *S. Spain, S. Portugal.* Hs Lu. (*N.W. Africa.*)

31. **V. hirsuta** (L.) S. F. Gray, *Nat. Arr. Brit. Pl.* **2**: 614 (1821). Pubescent annual 20–70 cm. Leaflets 4–10 pairs, 5–20 × 1–3(–5) mm, linear- or ovate-oblong; stipules entire, the lower linear-lanceolate, often with 2–4 setaceous teeth. Racemes 1- to 8-flowered, almost equalling leaves. Calyx-teeth equal, longer than tube; corolla 2–4(–5) mm, dirty white with purplish tinge. Legume 6–11 × 3–5 mm, black, usually pubescent. Seeds usually 2; hilum $\frac{1}{3}$ of the circumference. $2n = 14$. *Almost throughout Europe.* All except Cr Fa Sb; introduced in Is.

32. **V. meyeri** Boiss., *Fl. Or.* **2**: 595 (1872). Like **31** but subglabrous; leaflets 0·5–1·5 mm wide; stipules all linear-setaceous, ciliate; racemes shorter than leaves; legume glabrous. *Krym.* Rs (K).

33. **V. durandii** Boiss., *Diagn. Pl. Or. Nov.* **2(9)**: 116 (1849) (*V. baetica* Lange). Densely pubescent annual 30–60 cm. Leaflets 6–10 pairs, 7–15 × 2–3 mm, lanceolate to elliptic-ovate; stipules entire. Racemes 5- to 8-flowered, longer than leaves. Calyx-teeth unequal, the lower longer than tube; corolla *c.* 8 mm, pale purple. Legume 20–40 × 8–14 mm, glabrous. Seeds 3–5. *S.W. Spain.* Hs. (*N.W. Africa.*)

34. **V. disperma** DC., *Cat. Pl. Horti Monsp.* 154 (1813). Sparsely pubescent annual 10–50 cm. Leaflets 5–10 pairs, 8–12 × 1·5–4·5 mm, linear to elliptical; stipules entire. Racemes (1–)2- to 6-flowered, shorter than leaves. Calyx-teeth unequal, the lower slightly longer than tube; corolla 4–5 mm, pale purple. Legume 12–20 × 5–8 mm, brown, glabrous. Seeds usually 2; hilum $\frac{1}{7}-\frac{1}{6}$ of the circumference. *S.W. Europe.* Az Bl Co Ga Hs It Lu Sa Si.

35. **V. tenuissima** (Bieb.) Schinz & Thell., *Viert. Naturf. Ges. Zürich* **58**: 70 (1913) (*V. gracilis* Loisel., Banks & non Solander). Subglabrous annual 15–60 cm. Leaflets 2–5 pairs, (6–)10–25 × 1–3 mm, linear; stipules entire. Racemes (1–)2- to 5-flowered, longer than leaves. Calyx-teeth unequal, shorter than tube; corolla (5–)6–9 mm, pale purple. Legume 12–17 × 3–4 mm, brown, glabrous or pubescent. Seeds 4–6; hilum $\frac{1}{12}-\frac{1}{8}$ of the circumference. *S. & W. Europe northwards to C. England.* Al Az Be Bl Br Ga Gr He Ho Hs It Ju Lu Rm Rs (K) Sa Si Tu.

36. **V. tetrasperma** (L.) Schreber, *Spicil. Fl. Lips.* 26 (1771). Subglabrous annual 10–60 cm. Leaflets 3–6(–8) pairs, 5–20 × 0·5–3 mm, linear or linear-oblong; stipules entire. Racemes 1- or 2-flowered about equalling leaves. Calyx-teeth unequal, shorter than tube; corolla 4–8 mm, pale purple. Legume 9–16 × 3–5 mm, brown, usually glabrous. Seeds 3–5; hilum $\frac{1}{4}$ of the circumference. $2n = 14$. *Throughout Europe northwards to c. 62° N. in Fennoscandia and Russia.* All except Az Co Cr Fa Is Sb.

37. **V. pubescens** (DC.) Link, *Handb.* **2**: 190 (1831). Like **36** but sparsely pubescent; leaflets 3–5 pairs, 10–20 × 3–5 mm, elliptical to ovate-oblong; racemes up to 6-flowered, sometimes longer than leaves; lower calyx-teeth equalling or longer than tube; legume 12–16 × 3–4 mm, usually pubescent; hilum $\frac{1}{12}-\frac{1}{10}$ of the circumference. *S. Europe.* Bl Bu Co Cr Gr Hs It Ju Lu Rs (K) Sa Si.

38. **V. bifoliolata** Rodr., *Bull. Soc. Bot. Fr.* **25**: 239 (1878). Glabrous annual 20–80 cm. Leaflets 1 pair, 10–14 × 0·7–1·5 mm, linear or linear-oblong; stipules entire. Racemes (1–)2-flowered, longer than leaves. Calyx-teeth unequal, shorter than tube; corolla 4–8 mm, purple. Legume 10–20 × 4–5 mm, brown, glabrous, stipe longer than calyx. Seeds 1–6; hilum $\frac{1}{6}$ of the circumference. ● *Islas Baleares (Menorca).* Bl.

Sect. VICIA. Leaflets usually more than 3 pairs; flowers solitary, axillary or in few-flowered, sessile or shortly pedunculate racemes; corolla usually large (more than 10 mm); style pubescent on the lower side beneath the stigma.

39. V. oroboides Wulfen in Jacq., *Collect. Bot.* **4**: 323 (1791). Glabrescent or sparsely pubescent perennial 25–50 cm. Leaves without a tendril; leaflets 1–4 pairs, 40–80 × 15–45 mm, ovate, acute. Racemes 2- to 12-flowered, subsessile, or shortly pedunculate. Calyx-teeth subequal, about equalling tube; corolla 14–19 mm, pale yellow. Legume 20–40 × 6–9 mm, black, glabrous. Seeds *c.* 15; hilum $\frac{3}{4}$ of the circumference. $2n=14$. *Meadows and mountain woods; calcicole.* ● *E. Alps, extending to W. Hungary and C. Jugoslavia.* Au Hu It Ju.

40. V. truncatula Fischer ex Bieb., *Fl. Taur.-Cauc.* **3**: 473 (1819). Sparsely pubescent perennial 30-50 cm. Leaves without a tendril; leaflets 7–13 pairs, 10–35 × 3–10 mm, elliptic- or linear-oblong, obtuse, mucronate. Racemes 3- to 8-flowered, subsessile or shortly pedunculate. Calyx-teeth unequal, shorter than tube; corolla 17–20 mm, pale yellow with reddish tinge. Legume 25–30 × *c.* 7 mm, dull brown, glabrous. Seeds 2–5; hilum $\frac{1}{2}$ of the circumference. *Lower Danube basin.* Bu Ju Rm. (*Caucasus.*)

41. V. pyrenaica Pourret, *Mém. Acad. Toulouse* **3**: 333 (1788). Glabrous or subglabrous, procumbent, stoloniferous perennial 5–30 cm. Leaflets 3–6 pairs, 4–12 × 2–6 mm, oblong to suborbicular, truncate or emarginate, mucronate; tendrils usually unbranched; stipules entire. Flowers solitary. Calyx-teeth equal, shorter than tube; corolla 16–25 mm, bright violet-purple. Legume 25–50 × 4–6 mm, black, glabrous. Seeds 6–12; hilum $\frac{1}{5}$–$\frac{1}{4}$ of the circumference. *Alpine pastures and screes.* ● *Mountains of Spain and S. France.* Ga Hs.

42. V. sepium L., *Sp. Pl.* 737 (1753). Usually pubescent perennial 30–100 cm. Leaflets 3–9 pairs, 7–30 × 4–14 mm, ovate to ovate-oblong, obtuse or emarginate, mucronate; stipules more or less entire, spotted. Flowers 2–6 together, sometimes shortly pedunculate. Calyx-teeth unequal, shorter than tube; corolla 12–15 mm, dull bluish-purple. Legume 20–35 × 5–8 mm, black, glabrous. Seeds 3–7; hilum $\frac{1}{2}$–$\frac{3}{4}$ of the circumference. $2n=14$. *Almost throughout Europe.* All except Al Az Bl Co Cr Fa Sb Tu.

43. V. grandiflora Scop., *Fl. Carn.* ed. 2, **2**: 65 (1772). Pubescent annual 30–60 cm. Leaflets 3–7 pairs, 10–20 × 2–8 mm, linear to suborbicular; stipules toothed at base. Flowers 1–2(–4) together, very shortly pedunculate. Calyx-teeth equal, shorter than tube; corolla 23–35 mm, yellow, sometimes with purple tinge, wings sometimes black at tip. Legume 30–50 × 6–8 mm, black, pubescent. Seeds *c.* 15; hilum $\frac{2}{3}$–$\frac{3}{4}$ of the circumference. *C. & S.E. Europe, extending westwards to Italy and Sicilia.* Al Au Bu ?Cr Cz Gr Hu It Ju Rm Rs (W, K, E) Si Tu [Ge Po].

44. V. barbazitae Ten. & Guss., *Ind. Sem. Horti Neap.* **1839**: 12 (1839). Like **43** but calyx-teeth about as long as tube; corolla 18–22 mm, yellow with blue wings; legume brown, glandular, glabrous at ends; hilum $\frac{1}{6}$ of the circumference. *Balkan peninsula and C. Mediterranean region; local.* Bu Co Ga Gr It Ju Si.

45. V. pannonica Crantz, *Stirp. Austr.* ed. 2, **2**: 393 (1769). Pubescent annual 10–60 cm. Leaflets 4–10 pairs, 8–30 × 2–7 mm, oblong or linear-oblong, obtuse or truncate, mucronate; stipules entire, spotted. Flowers (1–)2–4 together. Calyx-teeth subequal, shorter than tube; corolla 14–22 mm, purple or yellow; standard pubescent on back. Legume 20–35 × 7–11 mm, yellow, pubescent. Seeds 2–8; hilum $\frac{1}{4}$–$\frac{1}{6}$ of the circumference. *Europe, extending northwards to C. France, Czechoslovakia and N. Ukraine.* Al Au Bu Cz Ga Gr Hs Hu It Ju Rm Rs (?C, W, K) Tu [Ge He Ho].

(a) Subsp. **pannonica**: Corolla pale yellow; limb of standard shorter than claw; seeds black. *From Ukraine to Czechoslovakia, N. Italy and Greece.*

(b) Subsp. **striata** (Bieb.) Nyman, *Consp.* 209 (1878) (subsp. *purpurascens* (DC.) Arcangeli, *V. purpurascens* DC.): Corolla dirty purple; limb of standard about as long as claw; seeds black marbled with brown. *Throughout the range of the species except the north-eastern part.*

46. V. sativa L., *Sp. Pl.* 736 (1753). Pubescent annual up to 80 cm. Leaflets 3–8 pairs, 6–20(–30) × 1–6 mm, linear to obcordate, acute to emarginate, mucronate; stipules dentate, usually with a dark spot. Flowers 1–2(–4) together. Calyx-teeth equal, longer or shorter than tube; corolla (8–)10–30 mm, purple. Legume 25–70 × (3–)4–10 mm, yellow-brown to black, glabrous or pubescent, breaking the calyx when mature. Seeds 6–12; hilum $\frac{1}{6}$–$\frac{1}{5}$ of the circumference. *Throughout Europe to 69° N. in Russia.* All territories, but only as an alien in Fa Is Sb.

Frequently cultivated for fodder.

A very variable species often divided into a number of species or subspecies. The most recent work by Mettin & Hanelt, *Kulturpfl.* **12**: 163–225 (1964), and Yamamoto, *Mem. Fac. Agric. Kagawa Univ.* **21**: 1–104 (1966), shows that the species has considerable variation in the basic chromosome number. The different chromosome numbers are largely correlated with morphology, and artificially raised hybrids show a fairly high degree of sterility, although this varies somewhat according to the parent plants used. Despite the sterility, evidence was obtained to show that gene exchange between plants of different chromosome number could occur with reasonable facility and this is supported by the almost continuous range of variation found in herbarium specimens. It is therefore proposed to treat this variation at subspecific rather than specific rank.

1 At least some leaflets toothed or incised (d) subsp. **incisa**
1 All leaflets entire, or at most crenate-dentate at apex
 2 Plant with underground stems bearing apetalous flowers and white, 1- to 2-seeded legumes. (b) subsp. **amphicarpa**
 2 Plant without underground stems
  3 Corolla (8–)10–18 mm; calyx-teeth shorter than tube; legume black or very dark brown (a) subsp. **nigra**
  3 Corolla 18–30 mm; calyx-teeth as long as or longer than tube; legume yellow-brown to dark brown, rarely almost black
   4 Legume 4·5–6 mm wide (c) subsp. **cordata**
   4 Legume 6–11 mm wide
    5 Legume contracted between the seeds, brown or yellow-brown; seeds 3·5–6·5 mm (e) subsp. **sativa**
    5 Legume not contracted between the seeds, dark brown or almost black; seeds 5·5–8 mm (f) subsp. **macrocarpa**

(a) Subsp. **nigra** (L.) Ehrh., *Hannover. Mag.* **1780(15)**: 229 (1780) (*V. angustifolia* L., *V. cuneata* Guss., *V. heterophylla* C. Presl, *V. pilosa* Bieb.): Leaflets linear to oblong-cuneate, acute, obtuse or truncate. Calyx-teeth shorter than tube; corolla (8–)10–18 mm; standard light reddish-purple, the wings similar or somewhat darker. Legume 25–55 × (2·5–)3–6 mm, not contracted between the seeds, black or brownish-black, usually glabrous. Seeds 2–4 mm. $2n=12$, 14. *Throughout the range of the species.*

(b) Subsp. **amphicarpa** (Dorthes) Ascherson & Graebner, *Syn. Mitteleur. Fl.* **6(2)**: 974 (1909) (*V. amphicarpa* Dorthes): Plant with underground stems bearing minute apetalous flowers and white, irregularly ovate 1- or 2-seeded legumes *c.* 15 mm. Leaflets linear, acute to obcordate. Calyx-teeth shorter than tube; corolla 20–25 mm; standard dark reddish-purple, the wings much darker. Legume 25–35 × 4–6 mm, not contracted between the seeds, dark brown, usually glabrous. Seeds 4·5–5 mm. *S. Europe.*

The chromosome number $2n = 10$ has been reported by a number of authors for this subspecies. According to Mettin & Hanelt, *loc. cit.* (1964), these plants should be referred to subsp. (c). They record $2n = 14$ for a non-European collection of subsp. *amphicarpa*.

(c) Subsp. **cordata** (Wulfen ex Hoppe) Ascherson & Graebner, *op. cit.* 968 (1909) (*V. cordata* Wulfen ex Hoppe): Leaflets oblong- or obovate-cuneate, truncate to emarginate. Calyx-teeth longer than tube; corolla 18–22 mm; standard reddish-purple, the wings dark red. Legume 30–50 × 4·5–6 mm, not contracted between the seeds, dark brown or almost black, usually glabrous. Seeds 3–4·5 mm. $2n = 10$. *S. Europe.*

(d) Subsp. **incisa** (Bieb.) Arcangeli, *Comp. Fl. Ital.* 201 (1882) (*V. incisa* Bieb.): Most leaflets toothed or incised, obovate in outline, truncate or emarginate. Calyx-teeth about as long as tube; corolla *c.* 20 mm; standard pale blue to violet, the wings usually darker. Legume *c.* 40 × 5–6 mm, not contracted between the seeds, glabrous. Seeds *c.* 4 mm. *Krym, Bulgaria, N.E. Greece, ?Italy.*

(e) Subsp. **sativa**: Leaflets oblong-cuneate to obcordate, truncate or emarginate. Calyx-teeth as long as or longer than tube; corolla 18–30 mm; standard pink to dark reddish-purple, the wings darker. Legume 35–70 × 6–11 mm, contracted between the seeds, brown or yellow-brown, usually hairy. Seeds 3·5–6·5 mm. $2n = 12$. *Almost throughout the range of the species, but introduced in the northern half.*

(f) Subsp. **macrocarpa** (Moris) Arcangeli, *Comp. Fl. Ital.* 201 (1882): Like subsp. (e) but legume 8–10(–12) mm wide, reticulate-veined, not contracted between the seeds, dark brown to black, more or less glabrous; seeds 5·5–8 mm. *Mediterranean region, S. Bulgaria.*

47. V. lathyroides L., *Sp. Pl.* 736 (1753) (incl. *V. olbiensis* Reuter). Pubescent annual up to 20 cm. Leaflets 2–4 pairs, 4–14(–20) × 0·5–4 mm, obovate-elliptical to linear, very shortly mucronate; tendrils simple; stipules entire, not spotted. Flowers solitary. Calyx-teeth equal, about as long as tube; corolla 5–8 mm, purple. Legume 15–30 × 3–4 mm, black, glabrous, with a short curved beak, not breaking the calyx when mature. Seeds 6–12, cubic, tuberculate. $2n = 12$. *Most of Europe northwards to S.W. Finland (Ahvenanmaa).* All except Az Bl Fa Is Rs (N, E) Sb.

48. V. cuspidata Boiss., *Diagn. Pl. Or. Nov.* 1(2): 104 (1843). Like **47** but leaflets 2–6 pairs, those of the upper leaves acuminate and with a long mucro; corolla 10–14 mm; legume 30–40 × 4–5 mm, with a long straight beak. *N.E. Greece, Turkey-in-Europe.* Gr Tu. (*E. Mediterranean region.*)

49. V. peregrina L., *Sp. Pl.* 737 (1753) (incl. *V. megalosperma* Bieb.). Sparsely pubescent annual up to 100 cm. Leaflets 3–7 pairs, 8–30 × 0·5–2(–3) mm, linear or oblong, mucronate and emarginate with acute lobes, so that the apex appears to be 3-lobed; stipules entire. Flowers solitary or 2 together. Calyx-teeth unequal, the lowest equalling tube; corolla 10–16 mm, purple. Legume 30–40 × 8–12 mm, brown, pubescent. Seeds 4–6; hilum $\frac{1}{12}-\frac{1}{10}$ of the circumference. *S. Europe.* Al Bl Bu Cr Ga Gr Hs It Ju Lu Rm Rs (K, E) Sa Si Tu [He Hu].

50. V. melanops Sibth. & Sm., *Fl. Graec. Prodr.* 2: 72 (1813) (*V. pichleri* Huter). Pubescent annual 15–80 cm. Leaflets 5–10 pairs, 5–20 × 2–8 mm, oblong or ovate, obtuse or emarginate; stipules entire. Flowers 1–4 together. Calyx-teeth unequal, the lower equalling tube; corolla 15–22 mm, greenish-yellow, wings black-tipped, keel purple. Legume 20–50(–80) × 6–12 mm, brown,

glabrous, margin tuberculate, pubescent. Seeds 4–7; hilum $\frac{1}{8}-\frac{1}{4}$ of the circumference. *Balkan peninsula, Italy, Sicilia, S. France.* Al Bu Ga Gr It Ju Si Tu [Cz].

51. V. lutea L., *Sp. Pl.* 736 (1753). Subglabrous to villous annual up to 60 cm. Leaflets 3–10 pairs, 10–25 × 1–5 mm, linear or oblong; stipules entire or dentate. Flowers 1–3 together. Calyx-teeth unequal, the lower longer than tube; corolla (15–)20–35 mm, pale yellow often purple-tinged. Legume 20–40 × 8–14 mm, yellowish-brown to black, pubescent, the hairs tuberculate at base, rarely glabrous. Seeds 3–9; hilum $\frac{1}{4}-\frac{1}{2}$ of the circumference. *S. & W. Europe, extending northwards to England, Switzerland, Hungary and Moldavia; a frequent casual elsewhere and locally naturalized.* Al *Az Bl *Br Bu Cr Ga Gr He Hs Hu It Ju Lu Rm Rs (W) Sa Si Tu [?Au Cz Ge].

(a) Subsp. **lutea**: Subglabrous to densely pubescent; leaflets of upper leaves subobtuse, mucronate; legume with white hairs with small tubercle at base; seeds 4–10. $2n = 14$. *Throughout the range of the species.*

(b) Subsp. **vestita** (Boiss.) Rouy, *Fl. Fr.* 5: 219 (1899) (*V. vestita* Boiss.): Villous; leaflets of upper leaves acuminate; legume with red-brown to red hairs with a large tubercle at base; seeds 3–4. *S.W. Europe.*

52. V. hybrida L., *Sp. Pl.* 737 (1753). Subglabrous or pubescent annual 20–60 cm. Leaflets 3–8 pairs, 6–15 × 1·5–7 mm, oblong or obovate-elliptical, emarginate to obtuse, mucronate; stipules entire. Flowers solitary. Calyx-teeth unequal, the lower longer than tube; corolla 18–30 mm, pale yellow or purplish, standard pubescent on back. Legume 25–40 × 8–10 mm, brown, pubescent. Seeds 5–6; hilum $\frac{1}{10}-\frac{1}{8}$ of the circumference. *S. Europe.* Al Bl Bu Co Cr Ga Gr Hs It Ju Rm Rs (K) Sa Si Tu [He].

Sect. FABA (Miller) S. F. Gray. Leaflets 1–3 pairs; flowers solitary, axillary or in few-flowered, sessile or shortly pedunculate racemes; corolla large (more than 10 mm); style pubescent on the lower side beneath the stigma.

53. V. bithynica (L.) L., *Syst. Nat.* ed. 10, **2**: 1166 (1759). Glabrous or pubescent annual 20–60 cm. Leaflets 2–3 pairs, 20–50 × 2–20 mm, oblong-lanceolate to ovate; stipules dentate. Flowers 1–3 together. Calyx-teeth unequal, longer than tube; corolla 16–20 mm; standard purple, wings and keel white. Legume 25–50 × 7–10 mm, brown or yellow, pubescent. Seeds 4–7; hilum $\frac{1}{7}-\frac{1}{6}$ of the circumference. $2n = 14$. *S. & W. Europe northwards to England.* Al Az Bl Br Bu Cr Ga Gr Hs It Ju Lu Rs (K) Sa Si Tu.

54. V. narbonensis L., *Sp. Pl.* 737 (1753) (incl. *V. serratifolia* Jacq.). Pubescent, erect annual 20–60 cm. Lower leaves without tendril; leaflets 1–3 pairs, 20–50 × 10–40 mm, ovate or elliptical, obtuse or emarginate, entire or serrate; stipules *c.* 10 mm, entire or dentate. Flowers 1–6 together. Calyx-teeth unequal, the lower longer than tube; corolla 10–30 mm, dark purple. Legume 30–70 × 10–15 mm, black or brown, glabrous with tuberculate-dentate pubescent margin. Seeds 4–8, 4–6 mm; hilum $\frac{1}{8}$ of the circumference. *S. Europe extending northwards to Hungary.* Al Bu Co Ga Gr Hs Hu It Ju Lu Rm Rs (K) Sa Si Tu [Au Cz Ge].

55. V. faba L., *Sp. Pl.* 737 (1753). Like **54** but more robust; leaves without tendril; leaflets 40–80(–100) × 10–20(–40) mm; corolla usually white with black wings; legume 80–200 × 10–20 mm, densely pubescent but becoming sparsely pubescent when mature; seeds 20–30 mm, ovoid-oblong, compressed. *Cultivated throughout Europe since prehistoric times for the edible* *seeds and immature legume and as fodder; often occurring as a relict of cultivation and perhaps locally naturalized.*

The origin of this species is not known. Some authorities consider it to be native to S.W. Asia and others to N. Africa, but no undoubtedly wild plants are known from these areas. An alternative theory is that it has been developed under cultivation from **54**, which it closely resembles in many characters.

50. Lens Miller[1]

Like *Vicia* but calyx-teeth equal and at least twice as long as the tube; style pubescent on the upper side; legume strongly compressed; seeds flat, orbicular.

1 Stipules semi-hastate or dentate
  2 Legume glabrous; peduncle usually aristate   **1. nigricans**
  2 Legume pubescent; peduncle not aristate   **4. ervoides**
1 Stipules oblong-lanceolate, entire
  3 Legume 12–16 × 6–12 mm; racemes about equalling leaves
        2. culinaris
  3 Legume 7–11 × 4·5–6·5 mm; racemes slightly longer than leaves   **3. orientalis**

1. L. nigricans (Bieb.) Godron, *Fl. Lorr.* **1**: 173 (1843) (*Ervum nigricans* Bieb., *L. culinaris* subsp. *nigricans* (Bieb.) Thell., *Vicia nigricans* (Bieb.) Cosson & Germ.). Annual 10–30 cm, patent-pubescent. Leaves sometimes without tendril; leaflets 2–5 pairs, 5–10 × 1·5–2 mm, linear to oblong; stipules semi-hastate or dentate. Racemes 1- to 3-flowered, longer than leaves; peduncle with an articulation near the apex and aristate. Calyx-teeth 2–4 times as long as tube; corolla 4–7 mm, pale blue or lilac. Legume 9–12 × 4–6 mm, yellowish, glabrous. Seeds 1–3. *S. Europe.* ?Al ?Bl Bu Co Cr Ga Gr Hs It Ju Lu Rs (K, E) Sa Si.

2. L. culinaris Medicus, *Vorl. Churpf. Phys.-Ökon. Ges.* **2**: 361 (1787) (*L. esculenta* Moench, *Ervum lens* L., *Vicia lens* (L.) Cosson & Germ.). Like **1** but usually more robust; stems up to 40(–50) cm; leaflets 3–8 pairs, up to 20 × 8 mm, oblong or elliptical; stipules oblong-lanceolate, entire; racemes about equalling leaves; calyx-teeth up to 6 times as long as tube; legume 12–16 × 6–12 mm. *Widely cultivated in C., S. & E. Europe for its edible seeds* (*lentils*) *and sometimes naturalized.* [Al Au Az Bu Co Cr Cz Ga Ge Gr He Hs Hu It Ju Lu Rm Rs (B, C, W, K, E) Sa Si.] (*Origin not known.*)

3. L. orientalis (Boiss.) M. Popov, *Bull. Univ. Asie Centr.* **15** (Suppl.): 22 (1927). Like **1** but stipules oblong-lanceolate, entire; peduncles not or shortly aristate; legume 7–11 × 4·5–6·5 mm. *S.E. Greece; Krym.* Gr Rs (K). (*S.W. Asia.*)

Only recently detected in Europe and perhaps more widespread.

4. L. ervoides (Brign.) Grande, *Bull. Orto Bot. Napoli* **5**: 58 (1918) (*L. lenticula* (Schreber) Alef., *Ervum ervoides* (Brign.) Hayek). Annual up to 30 cm, appressed-pubescent. Leaves without tendril or uppermost with a short tendril; leaflets 2–4 pairs, 8–18 × 1–3(–4) mm, linear to elliptical; stipules semi-hastate or dentate. Racemes 1- to 2-flowered, longer than leaves; peduncle with an articulation near the apex, not aristate. Calyx-teeth 2–4 times as long as tube; corolla 4–6 mm, pale blue. Legume 9–11 × 4–5 mm, yellowish, pubescent. Seeds 1–3. *S. Europe, but very local in the west.* Al Bl Bu Cr Gr It Hs Ju Rs (K) Sa Si.

[1] By P. W. Ball.

51. Lathyrus L.[1]

(Incl. *Orobus* L.)

Annual or perennial herbs, often climbing by means of tendrils. Leaves usually paripinnate and terminated by a tendril, rarely reduced to a tendril or a grass-like phyllode; leaflets usually distinctly parallel-veined; stipules usually herbaceous. Flowers in axillary racemes, or solitary, axillary. Calyx actinomorphic to bilabiate; keel usually obtuse; stamens diadelphous; style pubescent on the upper side, rarely glabrous. Legume usually oblong, compressed, dehiscent. Seeds 2 or more.

No clear distinction exists between this genus and *Vicia*. The majority of the species can readily be separated from *Vicia* by the often winged stem, the parallel-veined, and often fewer leaflets and the style-pubescence. On the other hand species **1–10** can be separated only by style-pubescence, and it must be doubtful whether this character alone can justify the assignment of these species and, for example *Vicia orobus* or *V. oroboides*, to separate genera.

In a number of species, particularly **12–14** and **39–45**, the colour of the corolla changes markedly on drying, becoming pale purple or blue instead of a bright red or reddish-purple. Because of this some of the descriptions of corolla-colour may not be accurate.

Most species occur in dry grassland or scrub or as ruderals.

Literature: M. Bässler, *Feddes Repert.* **72**: 69–97 (1966). K. Brunsberg, *Bot. Not.* **118**: 377–402 (1965). Z. Czefranova, *Nov. Syst. Pl. Vasc.* (*Leningrad*) **1965**: 152–167 (1965).

1 Leaves without leaflets
  2 Rhachis forming a grass-like phyllode; stipules minute; corolla crimson   **53. nissolia**
  2 Rhachis forming a tendril; stipules 6–30(–50) mm, ovate, hastate; corolla yellow   **54. aphaca**
1 At least the upper leaves with 1 or more pairs of leaflets
  3 Stem winged, at least in the upper part
    4 Lower leaves without leaflets, the rhachis broadly winged and resembling a leaf
      5 Upper leaves with 1–2 pairs of leaflets; corolla yellow; dorsal suture of legume with 2 wings   **52. ochrus**
      5 Upper leaves with 2–4 pairs of leaflets; corolla purple with violet, lilac, white or pink wings; dorsal suture of legume not winged
        6 Wings violet or lilac; standard emarginate; dorsal suture of legume channelled   **50. clymenum**
        6 Wings white or pink; standard mucronate; dorsal suture of legume not channelled   **51. articulatus**
    4 All leaves with 1 or more pairs of leaflets
      7 At least some leaves with 2 or more pairs of leaflets; racemes 2- to many-flowered
        8 Leaves without a tendril; rhachis mucronate
          9 Corolla cream, sometimes with a pink or purple tinge; hilum *c.* $\frac{1}{8}$ of the circumference of the seed   **15. pannonicus**
          9 Corolla red-purple or crimson; hilum *c.* $\frac{1}{4}$ of the circumference of the seed
            10 Calyx pubescent   **18. alpestris**
            10 Calyx glabrous, sometimes with ciliate teeth **19. montanus**
        8 Leaves with a tendril
          11 Rhachis at least 4 mm wide, broadly winged   **34. heterophyllus**
          11 Rhachis not more than 3 mm wide
            12 Stipules usually at least 25 mm, about as large as leaflets   **9. pisiformis**
            12 Stipules less than 25 mm, distinctly smaller than leaflets
              13 Lowest tooth of the calyx about as long as tube   **23. palustris**

13 Lowest tooth of the calyx distinctly shorter than tube
  14 Leaflets (3–)4–5 pairs; legume 30–35 × 5–5·5 mm; seeds smooth   **11. incurvus**
  14 Leaflets 2–3 pairs; legume 40–70 × 7–11 mm; seeds tuberculate   **26. cirrhosus**
7 All leaves with only 1 pair of leaflets; rarely some with 2 pairs and then the flowers solitary
  15 Racemes (3–)5- to many-flowered
    16 Stipules less than ½ as wide as stem   **32. sylvestris**
    16 Stipules at least ½ as wide as stem
      17 Margin of leaflets undulate; calyx-teeth only slightly unequal, the lowest tooth not more than 1½ times as long as the upper 2   **31. undulatus**
      17 Margin of leaflets not undulate; calyx-teeth very unequal, the lowest tooth at least twice as long as the upper 2
        18 Leaflets not more than twice as long as wide, the lateral veins not extending more than about ½-way to the apex   **30. rotundifolius**
        18 Leaflets more than twice as long as wide, with 3 or more parallel veins ± reaching the apex   **33. latifolius**
  15 Racemes 1- to 3(–4)-flowered
    19 Corolla yellow, sometimes with a pinkish tinge
      20 Corolla 7–12 mm; legume 5–7 mm wide; seeds rugulose   **47. hierosolymitanus**
      20 Corolla 12 mm or more; legume 7–12 mm wide
        21 Calyx-teeth equalling or slightly longer than tube; corolla 12–18 mm; seeds tuberculate or papillose   **46. annuus**
        21 Calyx-teeth 2–3 times as long as tube; corolla 18–25 mm; seeds smooth   **48. gorgoni**
    19 Corolla variously coloured, but never yellow
      22 Corolla 20 mm or more
        23 Peduncles not more than 60 mm; legume with 2 wings on the dorsal suture   **44. sativus**
        23 Peduncles more than 70 mm; legume not winged
          24 Pubescent, at least on the calyx and legume   **37. odoratus**
          24 Glabrous
            25 Leaflets 1–4 mm wide, linear-lanceolate; calyx-teeth longer than tube   **35. tremolsianus**
            25 Leaflets 4–18 mm wide, lanceolate to ovate; calyx-teeth shorter than tube   **36. tingitanus**
      22 Corolla less than 20 mm
        26 Calyx-teeth not or only slightly longer than tube
          27 Leaflets 5–12 mm; corolla blue; seeds smooth   **25. neurolobus**
          27 Leaflets 15–90 mm; corolla crimson with blue wings, or orange-red; seeds tuberculate or reticulate-rugose
            28 Sparsely pubescent; corolla crimson with pale blue wings; legume tuberculate and densely pubescent   **49. hirsutus**
            28 Glabrous; corolla orange-red; legume glabrescent but persistently pubescent on the suture   **42. setifolius**
        26 Calyx-teeth 1½–3 times as long as tube
          29 Corolla red; legume with 2 wings on the dorsal and ventral sutures   **45. amphicarpos**
          29 Corolla white, pink or purple
            30 Legume with 2 keels on the dorsal suture; peduncles 10–30 mm   **43. cicera**
            30 Legume with 2 wings on the dorsal suture; peduncles 30–60 mm   **44. sativus**
3 Stem not winged
  31 At least the upper leaves with a tendril
    32 Corolla 25–30 mm; legume 60–90 mm   **29. grandiflorus**
    32 Corolla less than 25 mm; legume less than 60 mm
      33 Racemes 2- to many-flowered; calyx-teeth ± distinctly unequal
        34 Corolla yellow   **(20–22). pratensis group**
        34 Corolla purple or bluish
          35 Leaflets 1 pair   **27. tuberosus**
          35 Leaflets 2–5 pairs

            36 Leaflets linear, oblong or lanceolate, distinctly parallel-veined, the lateral veins extending ± to the apex   **23. palustris**
            36 Leaflets lanceolate-elliptical to orbicular-elliptical, pinnately veined
              37 Stipules 10–20 mm wide, triangular-hastate; legume 6–10 mm wide   **4. japonicus**
              37 Stipules 2–10 mm wide, linear-lanceolate to orbicular-ovate, semi-sagittate; legume 4·5–5·5 mm wide   **10. humilis**
  33 Flowers solitary, very rarely 2 together; calyx-teeth usually equal
    38 Peduncles 2–5 mm; seeds smooth, with hilum less than $\frac{1}{20}$ of the circumference   **41. inconspicuus**
    38 Peduncles 5–70 mm; seeds with hilum more than $\frac{1}{20}$ of the circumference
      39 Legume 7–11 mm wide, pubescent when immature; seeds 2–3, coarsely reticulate-rugose; peduncles not aristate   **42. setifolius**
      39 Legume 3–7 mm wide, glabrous; seeds 8–15; peduncles usually aristate
        40 Peduncles 5–20 mm; legume 4–7 mm wide, with prominent longitudinal veins; seeds smooth or slightly rugose   **39. sphaericus**
        40 Peduncles 20–70 mm; legume 3–4 mm wide, with indistinct reticulate venation; seeds tuberculate   **40. angulatus**
31 All leaves without a tendril
  41 Flowers solitary, with peduncles 2–10 mm; calyx-teeth ± equal
    42 Pubescent; leaflets of upper leaves 7–20 mm; legume 5–7 mm wide, glabrous   **38. saxatilis**
    42 Glabrous; leaflets of upper leaves 25–40 mm; legume 2–5 mm wide, densely pubescent when young   **41. inconspicuus**
  41 Flowers in racemes, rarely solitary and then with peduncles more than 10 mm and calyx-teeth unequal
    43 Corolla pale cream, yellow or orange-yellow, sometimes with a red or purple tinge
      44 Leaflets pinnately veined; stipules 2–12 mm wide, lanceolate to ovate   **(5–8). laevigatus group**
      44 Leaflets distinctly parallel-veined; stipules 0·5–2(–4) mm wide, linear or linear-lanceolate, semi-sagittate, semi-hastate or sagittate
        45 Corolla yellow; leaflets 1 pair, 8–20 mm; legume black   **(20–22). pratensis group**
        45 Corolla cream or pale yellow sometimes with a red or purple tinge; leaflets 1–5 pairs, usually more than 20 mm; legume pale brown
          46 Calyx glabrous or ciliate on the margin; style filiform   **15. pannonicus**
          46 Calyx pubescent or villous; style dilated at apex
            47 Pubescent; leaflets 1–3 pairs; legume glabrous   **16. pallescens**
            47 Villous; leaflets 3–5 pairs; legume villous   **17. pancicii**
    43 Corolla red-purple, purple or bluish
      48 Stipules 8–20 mm wide, usually hastate or sagittate, almost as large as the leaflets
        49 Leaflets 2–5 pairs; calyx-teeth less than twice as long as tube; legume 6–10 mm wide   **4. japonicus**
        49 Leaflets 1 pair; calyx-teeth 2–3 times as long as tube; legume 3–5 mm wide   **24. laxiflorus**
      48 Stipules 0·5–8 mm wide
        50 Leaflets pinnately veined or very feebly parallel-veined, the lateral veins much weaker than midrib
          51 Legume covered with brown glands; racemes 6- to 30-flowered   **2. venetus**
          51 Legume eglandular; racemes 1- to 10-flowered
            52 Leaflets acuminate; stipules 10–25 mm   **1. vernus**
            52 Leaflets obtuse or subacute; stipules 4–10 mm
              53 Leaflets 3–6(–11) pairs; legume black   **3. niger**
              53 Leaflets 1 pair; legume pale reddish-brown   **28. roseus**
        50 Leaflets parallel-veined, the lateral veins reaching ± to the apex of the leaflets

54 Roots fusiform, fleshy; leaflets 1–2 pairs; calyx-teeth only slightly unequal; legume 5–9 mm wide **14. digitatus**
54 Roots slender; leaflets 2–4 pairs; calyx-teeth usually very unequal; legume 4–6 mm wide
55 Leaflets 5–11 mm wide, usually less than 10 times as long as wide, obtuse or subacute, mucronate **18. alpestris**
55 Leaflets 2–6 mm wide, always at least 10 times as long as wide, acuminate
56 Keel more or less winged at apex; style dilated at apex **12. filiformis**
56 Keel acute, not winged at apex; style ±filiform, not dilated at apex **13. bauhinii**

1. L. vernus (L.) Bernh., *Syst. Verz. Erfurt* 247 (1800). Glabrous or sparsely pubescent perennial; stem 20–40(–60) cm, not winged. Leaves without tendril; leaflets (1–)2–4 pairs, 30–70(–100) × (1–)10–30 mm, ovate or lanceolate, rarely linear, acuminate, feebly parallel-veined; stipules 10–25 × 2–8 mm, ovate-lanceolate, rarely linear, semi-sagittate. Racemes 3- to 10-flowered. Calyx-teeth unequal; corolla 13–20 mm, reddish-purple, becoming blue. Legume 40–60 × 5–8 mm, brown, glabrous. Seeds 8–14, smooth; hilum ¼ of the circumference. $2n=14$. *Most of Europe except the islands and parts of the south and west.* Al Au Bu Cz Da Fe Ga Ge Gr He Hs Hu It Ju No Po Rm Rs (N, B, C, W, K, E) Su [Be Ho].

2. L. venetus (Miller) Wohlf. in Koch, *Syn. Deutsch. Fl.* ed. 3, 714 (1892) (*L. variegatus* (Ten.) Gren. & Godron). Like **1** but leaflets ovate-orbicular, acute; stipules ovate-orbicular; racemes 6- to 30-flowered; corolla 10–15 mm; legume covered with brown glands. $2n=14$. *S.E. & E.C. Europe, extending westwards to Corse.* Al Au Bu Co Cz Gr He Hu It Ju Rm Rs (C, W, E) Si.

3. L. niger (L.) Bernh., *Syst. Verz. Erfurt* 248 (1800). Glabrous or sparsely pubescent perennial; stem 15–90 cm, not winged. Leaves without tendril; leaflets 3–6(–11) pairs, lanceolate to elliptical, obtuse, mucronate, more or less pinnately veined; stipules 4–10 × 1–2 mm, linear, semi-sagittate. Racemes 2- to 10-flowered. Calyx-teeth unequal; corolla 10–15 mm, purple becoming blue. Legume 35–60 × 4–6 mm, black, glabrous. Seeds 6–10, smooth; hilum ¼ of the circumference. *Europe except for most of the north-east, the extreme south and many islands.* Al Au *Be *Br Bu ?Co Cz Da Fe Ga Ge Gr He *Ho Hs Hu It Ju Lu No Po Rm Rs (B, C, W, K, E) ?Si Su Tu.

(a) Subsp. **niger**: Stem 30–90 cm; roots usually not tuberous; leaflets 10–40 × 5–16 mm, lanceolate to elliptical. $2n=14$. *Throughout the range of the species.*

(b) Subsp. **jordanii** (Ten.) Arcangeli, *Comp. Fl. Ital.* 198 (1882): Stem 15–25 cm; roots fusiform-tuberous; leaflets 8–25 × 3–7 mm, oblong-lanceolate. ● *S. Italy.*

4. L. japonicus Willd., *Sp. Pl.* 3: 1092 (1802). Somewhat glaucous, glabrous or pubescent perennial; stem up to 90 cm, not winged. Leaves sometimes without tendril; leaflets 2–5 pairs, (10–)20–40 × (5–)10–20 mm, elliptical, pinnately veined; stipules 10–25 × 10–20 mm, triangular-hastate. Racemes 2- to 12-flowered. Calyx-teeth more or less unequal; corolla 14–22(–25) mm, purple becoming blue. Legume 30–50 × 6–10 mm, brown, glabrescent. Seeds 4–11, smooth; hilum ¼ of the circumference. *Maritime sands and shingle; rarely on shores of large lakes. Coasts of W. & N. Europe; inland in N.W. Russia and N. Norway.* Br Da Fe †Ga Ge Hb ?Hs Is No Po Rs (N, B, C) Su.

(a) Subsp. **japonicus**: Racemes 2- to 7-flowered; calyx pubescent; corolla 18–22(–25) mm. $2n=14$. *Arctic Europe.*

The European plant may be distinct from typical *L. japonicus* and may represent another subspecies; cf. E. G. Pobedimova, *Not. Syst.* (*Leningrad*) **19**: 20–39 (1959).

(b) Subsp. **maritimus** (L.) P. W. Ball, *Feddes Repert.* **79**: 45 (1968) (*L. maritimus* Bigelow, *Pisum maritimum* L.): Racemes 5- to 12-flowered; calyx usually glabrous; corolla 14–18(–20) mm. $2n=14$. *W. Europe; Baltic region; subarctic Russia.*

(5–8). L. laevigatus group. Perennial; stem 20–60 cm, not winged. Leaves without tendril; leaflets 2–6 pairs, 30–100 × 5–50 mm, pinnately veined; stipules 5–30 × 2–12 mm, lanceolate to ovate. Racemes 2- to 20-flowered. Calyx-teeth unequal; corolla yellow or orange-yellow. Legume 50–75 × 5–8 mm, brown. Seeds 6–12, smooth; hilum ⅕–⅓ of the circumference.

1 Legume densely glandular when young; leaflets with brownish glands beneath; corolla brown- or orange-yellow **8. aureus**
1 Legume eglandular; leaflets eglandular; corolla yellow
2 Corolla 25–30 mm **7. gmelinii**
2 Corolla 15–25 mm
3 Calyx-teeth less than ⅓ as long as tube **5. laevigatus**
3 Calyx-teeth more than ⅓ as long as tube
4 Sparsely pubescent; leaflets 5–30 mm wide **5. laevigatus**
4 Glabrous; leaflets 20–50 mm wide **6. transsilvanicus**

5. L. laevigatus (Waldst. & Kit.) Gren., *Mém. Soc. Émul. Doubs* ser. 3, **10**: 193 (1865). Glabrous or sparsely pubescent. Leaflets 2–6 pairs, 5–40 mm wide, oblong or elliptical to ovate. Corolla 15–25 mm, yellow. Legume glabrous. ● *C. Europe, extending to N. Spain, N. part of Balkan peninsula and W. part of U.S.S.R.* Au Cz Ga Ge He Hs Hu It Ju Po Rm Rs (B, C, W).

(a) Subsp. **laevigatus**: Usually glabrous; leaflets 20–40 mm wide, elliptical to ovate; lower calyx-teeth less than ⅓ as long as tube. $2n=14$. *E.C. & E. Europe.*

(b) Subsp. **occidentalis** (Fischer & C. A. Meyer) Breistr., *Bull. Soc. Bot. Fr.* **87**: 53 (1940): Sparsely pubescent; leaflets 5–30 mm wide, oblong, elliptical or ovate-elliptical; lower calyx-teeth ⅓ as long to as long as the tube, the lateral teeth triangular. $2n=14$. *C. & S.W. Alps, Pyrenees, N. Spain.*

The populations in C. Europe are often intermediate between the two subspecies.

6. L. transsilvanicus (Sprengel) Fritsch, *Sitz-ber. Akad. Wiss. Wien* (*Math.-Nat.*) **104**: 517 (1895). Glabrous. Leaflets 2–4 pairs, 20–50 mm wide, elliptical to ovate. Lower calyx-teeth as long as or slightly shorter than tube, the lateral oblong- or triangular-ovate; corolla 20–25 mm, yellow. Legume glabrous. $2n=14$. ● *Carpathians, N. Romania.* Cz Hu Rm Rs (W).

7. L. gmelinii Fritsch, *op. cit.* 516 (1895) (*L. luteus* (L.) Peterm., non Moench). Glabrous or sparsely pubescent. Leaflets 2–4 pairs, 20–50 mm wide, elliptical to ovate. Lower calyx-teeth ¼–½ as long as tube; corolla 25–30 mm, yellow. Legume glabrous. *C. & S. Ural.* Rs (C). (*Mountains of C. Asia.*)

8. L. aureus (Steven) Brandza, *Prodr. Fl. Române* 546 (1883). Sparsely pubescent; leaflets 3–6 pairs, 25–50 mm wide, elliptical to ovate, with brownish glands beneath. Lower calyx-teeth about as long as tube, triangular or ovate-triangular; corolla 17–22 mm, brownish- or orange-yellow. Legume densely glandular when young. *Black Sea region, from Bulgaria to Krym.* Bu ?Gr Rm Rs (K).

9. L. pisiformis L., *Sp. Pl.* 734 (1753). Glabrous perennial; stem 50–100 cm, winged. Leaflets 3–5 pairs, 25–60 × (7–)10–30 mm, ovate or elliptical, pinnately or feebly parallel-veined; stipules 20–50 × 10–20 mm, ovate or elliptical. Racemes 8- to

15(–20)-flowered. Calyx-teeth unequal; corolla 10–15(–20) mm, reddish-purple. Legume 40–50 × 4–5 mm, dark brown, glabrous. Seeds 10–20, smooth; hilum $\frac{1}{8}$–$\frac{1}{6}$ of the circumference. $2n=14$. *E.C. Europe and U.S.S.R.* Cz Hu Po Rs (N, B, C, W, E).

10. L. humilis (Ser.) Sprengel, *Syst. Veg.* **3**: 263 (1826). Glabrous or sparsely pubescent perennial; stem 20–50 cm, not winged. Leaflets 3–5 pairs, 15–40 × 7–20 mm, lanceolate-elliptical to orbicular-elliptical, pinnately veined; stipules 6–18 × 2–10 mm, linear-lanceolate to orbicular-ovate, semi-sagittate. Racemes (1–)2- to 4-flowered. Calyx-teeth unequal; corolla 16–20 mm, purple. Legume 30–50 × 4–5·5 mm, dark brown, glabrous. Seeds 4–5, smooth, black. *C. Ural.* Rs (C). *(N. & C. Asia.)*

11. L. incurvus (Roth) Willd., *Sp. Pl.* **3**: 1091 (1802). Sparsely pubescent perennial; stem 40–80 cm, narrowly winged. Leaflets (3–)4–5 pairs, 15–50 × 6–17 mm, oblong-lanceolate or oblong-elliptical, feebly parallel-veined; stipules 5–25 × 1·5–3 mm, linear-lanceolate. Racemes 5- to 12-flowered. Calyx-teeth slightly unequal; corolla 13–15 mm, bluish-purple. Legume 30–35 × 5–5·5 mm, brown, glabrous. Seeds 6–11, smooth; hilum $\frac{1}{4}$ of the circumference. $2n=14$. *S.E. Russia, S. & E. Ukraine, W. Kazakhstan.* Rs (C, W, K, E) [Su].

12. L. filiformis (Lam.) Gay, *Ann. Sci. Nat.* ser. 4 (Bot.), **8**: 315 (1857) (*L. canescens* (L. fil.) Gren. & Godron). Glabrous or sparsely pubescent perennial; stem 15–50 cm, winged; roots slender. Leaves without tendril; leaflets 2–4 pairs, 30–60 × 2–6 mm, linear-lanceolate, acuminate; stipules 9–12 × 0·5–1·5 mm, linear. Racemes 4- to 10-flowered. Calyx-teeth unequal; corolla 14–22 mm, bright reddish-purple; keel more or less winged at apex; style dilated at apex. Legume 45–70 × 4–6 mm, brown, glabrous. Seeds *c.* 10, smooth; hilum $\frac{1}{8}$ of the circumference. *Mountain rocks; calcicole. E. Spain, S. France, N. Italy.* Ga Hs It.

13. L. bauhinii Genty, *Bull. Soc. Dauph. Éch. Pl.* ser. 2, **3**: 90 (1892) (*L. filiformis* (Lam.) Gay var. *ensifolius* (Lapeyr.) Hayek). Like **12** but corolla 20–27 mm; keel acute, not winged at apex; style not dilated at apex; hilum $\frac{1}{4}$–$\frac{1}{3}$ of the circumference of seed. ● *Pyrenees; Jura; Alps; N.W. part of Balkan peninsula.* Al Ga Ge He Hs Ju.

14. L. digitatus (Bieb.) Fiori in Fiori & Paol., *Fl. Anal. Ital.* **2**: 105 (1900) (incl. *L. sessilifolius* (Sibth. & Sm.) Ten., *L. tempskyanus* (Freyn & Sint.) K. Malý). Glabrous perennial; stem 10–40 cm, not winged; roots fusiform, fleshy. Leaves usually almost digitate, without tendril; leaflets 1–2 pairs, 15–80 × 2–8 mm, linear; stipules 6–8 × 0·5–1 mm, lanceolate, semi-sagittate. Racemes 4- to 10-flowered. Calyx-teeth slightly unequal; corolla 15–30 mm, bright reddish-purple. Legume 40–70 × 5–9 mm, brown, glabrous. Seeds 5–7, smooth; hilum $\frac{1}{5}$ of the circumference. $2n=14$. *S.E. Europe, S. Italy.* Al Bu Gr It Ju Rs (K) Tu.

15. L. pannonicus (Jacq.) Garcke, *Fl. Nord-Mittel-Deutschl.* ed. 6, 112 (1863) (*L. albus* (L. fil.) Kittel). Glabrous or sparsely pubescent perennial; stem 15–50 cm, not or very narrowly winged; roots tuberous, fleshy. Leaves without tendril; leaflets 1–4 pairs, 15–75 × 2–5(–8) mm, linear to oblong-lanceolate or oblong-elliptical; stipules 10–20 × 1–2·5 mm, linear-lanceolate, semi-sagittate. Racemes 3- to 9-flowered. Calyx glabrous or ciliate on the margin, the teeth unequal; corolla 12–20 mm, pale cream with reddish or purplish tinge; style filiform. Legume 30–65 × 3–8 mm, pale brown, glabrous. Seeds 12–20, smooth; hilum $\frac{1}{8}$ of the circumference. $2n=14$. *From C. Spain, S. Italy and S.E. Russia northwards to N.W. France and to 54° N. in C. Russia.* Al Au Bu Cz Ga Ge Hs Hu It Ju Po Rm Rs (C, W, K, E).

A variable species that has been variously divided into varieties and subspecies or even separate species. There is no general agreement as to the number of taxa that can be recognized. The fullest account is given by Širjaev, *Bull. Assoc. Russe Sci. Prague (Sci. Nat. Math.)* **5**: 239–261 (1937), who recognized 6 varieties in Europe. Bässler, *Feddes Repert.* **72**: 89 (1966), gives a list of 6 subspecies from Europe, without any explanation, which differs somewhat from Širjaev's treatment, a modification of which is given here.

1 Lowest calyx-tooth about as long as or longer than tube; stem usually very narrowly winged near the apex **(e)** subsp. **varius**
1 Lowest calyx-tooth shorter than tube (sometimes only slightly shorter); stem not winged
2 Leaflets 15–30 × 4–7 mm, oblong-lanceolate or oblong-elliptical **(d)** subsp. **hispanicus**
2 Leaflets 35–75 × 2–6(–7) mm, linear or linear-lanceolate or linear-elliptical
3 Peduncles shorter than or about as long as the subtending leaf; calyx-teeth usually glabrous, the lowest $\frac{1}{2}$–$\frac{4}{5}$ as long as the tube **(b)** subsp. **collinus**
3 Peduncles usually exceeding the subtending leaf; calyx-teeth usually ciliate, the lowest usually not more than $\frac{1}{2}$ as long as the tube
4 Tubers 2–5 cm, oblong-ovoid **(a)** subsp. **pannonicus**
4 Tubers mostly 10 cm or more, slender or fusiform **(c)** subsp. **asphodeloides**

(a) Subsp. **pannonicus**: Stems up to 50 cm, not winged; tubers 2–5 cm, oblong-ovoid. Leaflets 35–60 × 2–4 mm, linear or linear-lanceolate. Peduncles 6–9 cm, usually exceeding the subtending leaf. Calyx-teeth ciliate, the lowest *c.* $\frac{1}{2}$ as long as tube. *Damp pastures. E.C. Europe and N.W. part of Balkan peninsula.*

(b) Subsp. **collinus** (Ortmann) Soó, *Scripta Bot. Mus. Transs.* **1**: 46 (1942) (*L. lacteus* (Bieb.) Wissjul., *L. pannonicus* var. *versicolor* auct. pro parte): Stems up to 50 cm, not winged; tubers usually more than 10 cm, slender or fusiform. Leaflets 35–70 × 2–6(–7) mm, linear, linear-lanceolate or linear-elliptical. Peduncles 3–7(–8) cm, usually shorter than the subtending leaf. Calyx-teeth usually glabrous, the lowest $\frac{1}{2}$–$\frac{4}{5}$ as long as tube. *Dry grassland and scrub. E.C. & E. Europe, N. part of Balkan peninsula.*

(c) Subsp. **asphodeloides** (Gouan) Bässler, *Feddes Repert.* **72**: 89 (1966): Like subsp. **(b)** but peduncles 5–9 cm, usually exceeding the subtending leaf; calyx-teeth usually ciliate, the lowest *c.* $\frac{1}{2}$ as long as tube. *Dry grassland and scrub. From W. France to the Alps and S. Italy.*

(d) Subsp. **hispanicus** (Lacaita) Bässler, *loc. cit.* (1966): Stems 15–30 cm, not winged; tubers usually more than 10 cm, slender or fusiform. Leaflets 15–30 × 4–7 mm, oblong-lanceolate or oblong-elliptical. Peduncles 4–8 cm, usually exceeding the subtending leaf. Calyx-teeth glabrous or sparsely ciliate, the lowest *c.* $\frac{1}{2}$ as long as tube. *Dry grassland and scrub.* ● *C. & E. Spain.*

Intermediates between subspp. **(c)** and **(d)** occur in the Pyrenees and in N.E. Spain. This subspecies has also been recorded from Krym but all the specimens seen from there are referable to subsp. **(b)**.

(e) Subsp. **varius** (C. Koch) P. W. Ball, *Feddes Repert.* **79**: 47 (1968) (*L. pannonicus* var. *versicolor* auct. pro parte): Stems up to 50 cm, narrowly winged, at least near the apex; tubers 3 cm or more, oblong-ovoid to fusiform. Leaflets 35–75 × 3–7 mm, linear-lanceolate or linear-elliptical. Peduncles 5–11 cm, as long as to much longer than the subtending leaf. Calyx-teeth usually ciliate, the lowest as long as or longer than tube. *Dry grassland and scrub.* ● *N. Italy, N.W. Jugoslavia.*

Very variable in the size of the tubers and the length of the peduncle. Plants with short tubers and short peduncles are possibly intermediate between subspp. **(a)** and **(e)**.

16. L. pallescens (Bieb.) C. Koch, *Linnaea* **15**: 723 (1841). Like **15** but pubescent; roots slender, not fleshy; leaflets 1–3 pairs; stipules often semi-hastate; calyx pubescent; corolla 17–22 mm; style dilated at apex; hilum $\frac{1}{4}$ of the circumference of the seed. *S.E. Europe, extending to Hungary and to c.* 53° *N. in E. Russia.* Bu Hu Ju Rm Rs (C, W, K, E).

17. L. pancicii (Juriš.) Adamović, *Prosv. Glasn.* **22**: 1246 (1901). Villous perennial; stem 30–70 cm, not winged. Leaves without tendril; leaflets 3–5 pairs, 40–75 × 5–8 mm, linear or linear-lanceolate; stipules 7–9 mm, linear, semi-sagittate. Racemes 4- to 10-flowered. Calyx-teeth slightly unequal; corolla 15–19 mm, pale yellow; style dilated at apex. Legume 50–65 × 6–7 mm, pale brown, villous. Seeds *c.* 10, smooth; hilum $\frac{1}{3}$ of the circumference. ● *S.E. Jugoslavia, S.W. Bulgaria.* Bu Ju.

18. L. alpestris (Waldst. & Kit.) Kit. ex Čelak., *Österr. Bot. Zeitschr.* **38**: 86 (1888) (incl. *L. friedrichsthalii* (Griseb.) K. Malý). Glabrous or pubescent perennial; stem 15–60 cm, sometimes narrowly winged at apex. Leaves without tendril; leaflets 2–3 pairs, 20–50 × 5–11 mm, linear to lanceolate-oblong or elliptical; stipules 12–20 × 2–4 mm, lanceolate, semi-sagittate. Racemes 3- to 6-flowered. Calyx-teeth unequal; corolla 12–16 mm, red-purple. Legume 30–40 × 5–6 mm, brown, glabrous. Seeds 10–14; hilum $\frac{1}{4}$ of the circumference. *Mountain woods. Balkan peninsula from c.* 40° *to* 42° 30′ *N.* Al Bu Gr Ju.

19. L. montanus Bernh., *Syst. Verz. Erfurt* 247 (1800) (*L. macrorrhizus* Wimmer). Glabrous or subglabrous perennial; stem 15–50 cm, winged. Leaves without tendril; leaflets (1–)2–4 pairs, 10–50(–100) × 1–12(–16) mm, linear to elliptical; stipules 5–25 × 2–8 mm, linear or lanceolate, semi-sagittate. Racemes 2- to 6-flowered. Calyx-teeth unequal; corolla 10–16 mm, crimson, becoming bluish. Legume 25–45 × 4–5 mm, red-brown, glabrous. Seeds 4–10, smooth; hilum $\frac{1}{5}$–$\frac{1}{4}$ of the circumference. $2n=14$. *S., W. & C. Europe, extending to the Baltic region and White Russia.* Al Au Be Br Co Cz Da Fe Ga Ge Hb He Ho Hs Hu It Ju Lu No Po Rs (B, C) Su.

20–22. L. pratensis group. Glabrous or pubescent perennial; stem not winged. Leaflets usually 1 pair; stipules sagittate. Racemes 2- to 12-flowered. Calyx-teeth unequal; corolla 10–20 mm, yellow. Legume 20–40 × 4–7 mm, black, glabrous or pubescent. Seeds 5–12, smooth; hilum $\frac{1}{7}$–$\frac{1}{5}$ of the circumference.

1 Leaves mucronate or with a very short simple tendril **21. binatus**
1 At least the upper leaves with a distinct tendril
2 Lower calyx-teeth slightly shorter than tube **20. pratensis**
2 Lower calyx-teeth as long as or longer than tube
3 Leaflets 2–9 mm wide, linear-lanceolate to lanceolate; stipules smaller than leaflets **20. pratensis**
3 Leaflets 8–15 mm wide, ovate; stipules as large as or larger than leaflets **22. hallersteinii**

20. L. pratensis L., *Sp. Pl.* 733 (1753). Stem 30–120 cm. Leaves with a tendril; leaflets 10–30(–40) × 2–9 mm, linear-lanceolate to lanceolate or elliptical; stipules (5–)10–30 × 3–6(–12) mm, linear to lanceolate, rarely ovate, smaller than or about as long as the leaflets. Racemes (2–)5- to 12-flowered. Lower calyx-teeth shorter to longer than tube. Corolla 10–16(–18) mm. $2n=7$, 8, 9, 12, 14, 16, 21, 28, 42. *Usually in grassland or scrub. Almost throughout Europe.* All except ?Az Bl Cr Sb.

Very variable and possibly containing a number of subspecies. There does not seem to be any correlation between morphology and chromosome number.

21. L. binatus Pančić, *Fl. Princ. Serb.* 256 (1874). Stem up to 30 cm. Leaves without or with a very short simple tendril; leaflets 8–20 × 2·5–5 mm, lanceolate; stipules 6–12 × 1–4 mm, linear-lanceolate to lanceolate, smaller than the leaflets. Racemes 1- to 4-flowered. Lower calyx-teeth as long as or slightly longer than tube; corolla 15–20 mm. $2n=14$. *Screes.* ● *C. Jugoslavia.* Ju.

Plants intermediate between **20** and **21** sometimes occur in C. Jugoslavia.

22. L. hallersteinii Baumg., *Enum. Stirp. Transs.* **2**: 333 (1816). Stem 30–120 cm. Leaves with a tendril; leaflets 25–40 × 8–15 mm, ovate; stipules 20–40 × 6–12 mm, ovate, equalling or larger than leaflets. Racemes 2- to 6-flowered. Lower calyx-teeth longer than tube; corolla 14–20 mm. $2n=14$. *Woods or scrub.* ● *Romania and C. part of Balkan peninsula.* Gr Ju Rm.

Plants from the central part of the Balkan peninsula are sometimes intermediate between **20** and **22**.

23. L. palustris L., *Sp. Pl.* 733 (1753) (*L. pilosus* Cham.). Sparsely pubescent perennial 20–120 cm. Leaflets 2–5 pairs, 25–80 × 3–12(–16) mm, linear to oblong or lanceolate; stipules 10–20 × 3–8 mm, lanceolate or ovate, semi-sagittate. Racemes 2- to 8-flowered. Calyx-teeth unequal; corolla 12–20 mm. Legume 25–60 × (5–)7–9 mm, brown, glabrous. Seeds 3–12(–20), smooth; hilum $\frac{1}{4}$ of the circumference. *Wet places. Most of Europe, but very rare in the Mediterranean region.* Al Au Be Br Bu Co Cz Da Fe Ga Ge Hb He Ho Hs Hu Is It Ju Lu No Po Rm Rs (N, B, C, W, E) Su Tu.

(a) Subsp. **palustris**: Stem narrowly winged; corolla purplish-blue; seeds mottled. $2n=42$. *Throughout the range of the species except the Iberian peninsula.*
(b) Subsp. **nudicaulis** (Willk.) P. W. Ball, *Feddes Repert.* **79**: 47 (1968): Stem not winged; corolla bright red-purple; seeds black, not mottled. ● *N. & W. parts of Iberian peninsula.*

24. L. laxiflorus (Desf.) O. Kuntze, *Acta Horti Petrop.* **10**: 185 (1887) (*L. inermis* Rochel ex Friv.). Pubescent or subglabrous perennial; stem 20–50 cm, not winged. Leaves without tendril; leaflets 1 pair, 20–40 × 10–20 mm, lanceolate to suborbicular; stipules 10–30 × 8–15 mm, lanceolate to ovate-orbicular, sagittate or semi-sagittate. Racemes 2- to 6-flowered. Calyx-teeth subequal, pubescent, 2–3 times as long as tube; corolla 15–20 mm, blue-violet. Legume 30–40 × 3–5 mm, pubescent. Seeds *c.* 6; hilum $\frac{1}{10}$–$\frac{1}{6}$ of the circumference. $2n=14$. *Mountain woods. S.E. Europe, S. Italy.* Al Bu Cr Gr It Ju Rs (K) Tu.

25. L. neurolobus Boiss. & Heldr. in Boiss., *Diagn. Pl. Or. Nov.* **2(9)**: 125 (1849). Glabrous perennial; stem 15–50 cm, winged. Lower leaves without tendril, upper with simple tendril; leaflets 1 pair, 5–12 × 1–4 mm, oblong to ovate-oblong; stipules 2–3 mm, linear. Racemes 1- to 2-flowered. Calyx-teeth equal; corolla 6–10 mm, blue. Legume 20–30 × 3–4 mm, brown, glabrous, with very prominent longitudinal veins. Seeds 3–6, smooth; hilum $\frac{1}{6}$ of the circumference. $2n=14$. *Woodland streams.* ● *W. Kriti.* Cr.

26. L. cirrhosus Ser. in DC., *Prodr.* **2**: 374 (1825). Glabrous perennial; stem 90–120 cm, winged. Leaflets 2–3 pairs, 15–40 × 5–12 mm, elliptical or oblong-lanceolate; stipules 6–20 × 1–5 mm, linear or lanceolate, semi-sagittate. Racemes 4- to 10-flowered. Calyx-teeth unequal; corolla 12–17 mm, pink. Legume 40–70 × 7–11 mm, pale brown, glabrous, the dorsal suture with 3 keels. Seeds 10–15, tuberculate; hilum $\frac{1}{6}$ of the circumference. ● *Pyrenees; Cevennes.* Ga Hs.

27. L. tuberosus L., *Sp. Pl.* 732 (1753). Glabrous or subglabrous perennial; stem 30–120 cm, not winged. Leaflets 1 pair, 15–45 ×

5–15 mm, elliptical to oblong, feebly parallel-veined; stipules 5–20 × 1–4 mm, linear to lanceolate, semi-sagittate. Racemes 2- to 7-flowered. Calyx-teeth slightly unequal; corolla 12–20 mm, bright red-purple. Legume 20–40 × 4–7 mm, brown, glabrous. Seeds 3–6, smooth or slightly tuberculate; hilum $\frac{1}{10}$–$\frac{1}{8}$ of the circumference. $2n = 14$. *Most of Europe except the north and extreme south.* Al Au Be Bu Cz Ga Ge Gr He Ho Hs Hu It Ju Po Rm Rs (B, C, W, K, E) [Br Da Su].

28. L. roseus Steven, *Mém. Soc. Nat. Moscou* **4**: 92 (1813). Glabrous perennial; stem 40–150 cm, not winged. Leaves without tendril; leaflets 1 pair, 25–50 × 15–35 mm, ovate-elliptical to obovate-orbicular, pinnately veined; stipules 4–15 mm, linear-triangular. Racemes 1- to 5-flowered. Calyx-teeth unequal; corolla 12–20 mm, pink. Legume 30–55 × 5–9 mm, pale brown, glabrous. Seeds up to 10, smooth; hilum $\frac{1}{8}$ of the circumference. *Krym.* Rs (K). (*Anatolia and Caucasus.*)

29. L. grandiflorus Sibth. & Sm., *Fl. Graec. Prodr.* **2**: 67 (1813). Pubescent perennial; stem 30–150 cm, not winged. Leaflets 1(–3) pairs, 25–50 × 10–35 mm, ovate, feebly parallel-veined; stipules 2–10 × 0·5–1·5 mm, linear, sometimes sagittate. Racemes 1- to 4-flowered. Calyx-teeth equal, shorter than tube; corolla 25–30 mm, the standard violet, wings purple, keel pink. Legume 60–90 × 6–7 mm, brown, glabrous. Seeds 15–20, smooth; hilum $\frac{1}{6}$ of the circumference. *Shady places in the mountains. S. half of the Balkan peninsula; S. Italy, Sicilia.* Al Bu Cr Gr It Ju Si.

30. L. rotundifolius Willd., *Sp. Pl.* **3**: 1088 (1802) (incl. *L. litvinovii* Iljin). Glabrous perennial; stem 40–80 cm, winged. Leaflets 1 pair, 25–60 × 20–40(–45) mm, elliptical to suborbicular or ovate-orbicular, usually not more than twice as long as wide, feebly parallel-veined, the margin plane; stipules 10–25 × 3–6 mm, oblong or lanceolate, hastate. Racemes 3- to 8-flowered. Calyx-teeth unequal, the lowest at least twice as long as upper 2; corolla 15–22 mm, purple-pink. Legume 40–70 × 4–10 mm, brown, glabrous. Seeds 8–10, reticulate-rugose; hilum $\frac{1}{5}$ of the circumference. $2n = 14$. *E. Russia; Krym.* Rs (C, K). (*W. Asia.*)

31. L. undulatus Boiss., *Diagn. Pl. Or. Nov.* **3**(2): 41 (1856) (*L. rotundifolius* auct., non Willd.). Like **30** but leaflets 30–70 × 16–35 mm, elliptical to ovate, up to 4 times as long as wide, the margin undulate-crispate; racemes 5- to 10-flowered; lowest calyx-tooth *c.* 1½ times as long as upper 2; legume 60–80 × 7–11 mm. *Turkey-in-Europe (eastern half).* Tu. (*Anatolia.*)

32. L. sylvestris L., *Sp. Pl.* 733 (1753). Glabrous or pubescent perennial; stem 60–200 cm, winged. Leaflets 1 pair, (20–)50–150 × 5–20(–40) mm, linear to lanceolate; stipules 10–30 × 2–5 mm, linear to lanceolate, semi-sagittate, less than ½ as wide as stem. Racemes 3- to 12-flowered. Calyx-teeth unequal; corolla 13–20 mm, purple-pink. Legume 40–70 × 5–13 mm, brown, glabrous. Seeds 10–15, reticulate-rugose; hilum $\frac{1}{3}$–$\frac{1}{2}$ of the circumference. $2n = 14$. *Most of Europe except the extreme north and extreme south.* Al Au Be Br Bu Co Cz Da Fe Ga Ge Gr He Ho Hs Hu It Ju Lu No Po Rm Rs (N, B, C, W, E) Sa Su.

33. L. latifolius L., *Sp. Pl.* 733 (1753) (*L. megalanthus* Steudel; incl. *L. membranaceus* C. Presl). Glabrous or pubescent perennial; stem 60–300 cm, winged. Leaflets 1 pair, (30–)40–150 × 3–50 mm, linear to ovate- or elliptic-orbicular; stipules (20–)30–60 × 2–11 mm, lanceolate to ovate, semi-hastate, more than ½ as wide as stem. Racemes 5- to 15-flowered. Calyx-teeth unequal; corolla (15–)20–30 mm, purple-pink. Legume 50–110 × 6–10 mm, brown, glabrous. Seeds 10–15, reticulate-rugose; hilum $\frac{1}{3}$–$\frac{1}{2}$ of the circumference. $2n = 14$. *C. & S. Europe, extending to N. France;*

cultivated for ornament and sometimes naturalized. Al Au Bl Bu Co Cz Ga Gr He Hs Hu It Ju Lu Po Rm Rs (W, K) Sa Si [Be Br Ge].

Very variable in leaflet-shape, and sometimes divided, on this basis, into two species or subspecies, but there is little correlation between leaflet-shape and other characters.

34. L. heterophyllus L., *Sp. Pl.* 733 (1753). Like **33** but leaflets usually 2–3 pairs on upper leaves, and corolla 12–22 mm. $2n = 14$. *S.W. & C. Europe, extending to C. Sweden.* Au Cz Ga Ge He Hs It Lu Po Sa Su.

The status of this species is very uncertain. It is possibly only a variant of **33**.

35. L. tremolsianus Pau, *Not. Bot. Fl. Esp.* **4**: 29 (1891) (*L. elegans* Porta & Rigo, non Vogler, *L. latifolius* var. *angustifolius* sensu Willk. pro parte). Glabrous annual; stem up to 100 cm, winged. Leaflets 1 pair, 30–100 × 1–4 mm, linear-lanceolate; stipules 10–30 × 2–4 mm, linear-lanceolate, semi-sagittate. Racemes 1- to 3-flowered. Calyx-teeth subequal, at least the lower longer than tube; corolla 20–30 mm, pink with blue wings. Legume *c.* 70 × 6 mm, glabrous. ● *S.E. Spain.* Hs.

36. L. tingitanus L., *Sp. Pl.* 732 (1753). Glabrous annual; stem 60–120 cm, winged. Leaflets 1 pair, 20–80 × 4–18 mm, linear-lanceolate to ovate; stipules 12–25 × 3–12 mm, lanceolate to ovate, semi-sagittate or -hastate. Racemes 1- to 3-flowered. Calyx-teeth subequal, shorter than tube; corolla 20–30 mm, bright purple. Legume 60–100 × 8–10 mm, brown, glabrous. Seeds 6–8, smooth; hilum $\frac{1}{6}$ of the circumference. *S. & E. part of Iberian peninsula; Sardegna; Açores.* Az Hs Lu Sa.

37. L. odoratus L., *Sp. Pl.* 732 (1753). More or less pubescent annual; stem 50–200 cm, winged. Leaflets 1 pair, 20–60 × 7–30 mm, ovate-oblong or elliptical, sometimes feebly parallel-veined; stipules 15–25 × 3–4 mm, lanceolate, semi-sagittate. Racemes 1- to 3-flowered. Calyx-teeth subequal, longer than tube; corolla 20–35 mm, purple. Legume 50–70 × 10–12 mm, brown, pubescent. Seeds *c.* 8, smooth; hilum $\frac{1}{4}$ of the circumference. $2n = 14$. ● *S. Italy and Sicilia; widely cultivated for ornament and perhaps naturalized in some parts of C. & S. Europe.* It Si [Au Ga Lu].

In cultivation the corolla may be white, pink, purple, violet or blue or some combination of these; some of these colour variants may occasionally occur as casuals.

38. L. saxatilis (Vent.) Vis., *Fl. Dalm.* **3**: 330 (1852) (*L. ciliatus* Guss., *Vicia saxatilis* (Vent.) Tropea). Pubescent annual; stem 10–30 cm, not winged. Leaves without tendril; leaflets 1–3 pairs, those of the lower leaves 3–8 × 1·5–3 mm, obcordate, with 3 teeth at apex, those of the upper leaves 7–20 × 0·5–2 mm, linear; stipules 2–3 × 0·5 mm, linear, semi-hastate, the lobe often irregularly toothed. Flowers solitary; peduncles 2–10 mm, articulated near the middle or apex. Calyx-teeth subequal, slightly shorter than tube; corolla 6–9 mm, pale blue or yellowish. Legume 15–30 × 5–7 mm, brown, glabrous. Seeds 3–8, smooth; hilum $\frac{1}{12}$ of the circumference. *Mediterranean region; Krym.* Bl Co Cr Ga Gr Hs It Ju Rs (K).

39. L. sphaericus Retz., *Obs. Bot.* **3**: 39 (1783). Glabrous or pubescent annual; stem 10–50 cm, not winged. Leaflets 1 pair, 20–60(–100) × 1–7 mm, linear to linear-lanceolate; stipules 6–15 × 0·5–1 mm, linear, semi-sagittate. Flowers solitary; peduncles 5–20 mm, aristate, articulated near the middle or apex. Calyx-teeth equal, as long as or slightly longer than tube; corolla

6–13(–16) mm, orange-red. Legume 30–70 × 4–7 mm, brown, glabrous, with prominent longitudinal veins. Seeds 8–15, smooth or slightly rugose; hilum $\frac{1}{15}$–$\frac{1}{11}$ of the circumference. $2n = 14$. *S. Europe, extending northwards to N.W. France and S. Hungary; Denmark and S. Sweden.* Al Bl Bu Co Cr Da Ga Gr He Hu Hs It Ju Lu Rm Rs (K) Sa Si Su Tu.

40. L. angulatus L., *Sp. Pl.* 731 (1753). Like **39** but stipules 7–14 × 1·5–2 mm; peduncles 20–70 mm, articulated near the apex and rarely 2-flowered; corolla purple or pale blue; legume 25–50 × 3–4 mm, with rather indistinct reticulate venation; seeds rugose to finely tuberculate; hilum $\frac{1}{18}$–$\frac{1}{14}$ of the circumference. *Sandy soils. Mediterranean region and S.W. Europe, extending to N.W. France.* Co Cr Ga Gr Hs It Ju Lu Sa.

41. L. inconspicuus L., *Sp. Pl.* 730 (1753). Glabrous annual; stem 10–30 cm, not winged. Leaves without or with a usually simple tendril; leaflets 1 pair, 25–40 × 1–4 mm, linear-lanceolate; stipules 7–10 × 0·5–2 mm, linear or lanceolate, semi-sagittate. Flowers solitary; peduncles 2–5 mm, articulated near the base or middle. Calyx-teeth equal, as long as tube; corolla 4–9 mm, pale purple. Legume 30–60 × 2–5 mm, pale brown, densely pubescent when young, glabrescent. Seeds 5–14, smooth; hilum $\frac{1}{25}$–$\frac{1}{20}$ of the circumference. *Mediterranean region.* Al ?Bl Bu Ga Gr Hs It Ju Tu.

42. L. setifolius L., *Sp. Pl.* 731 (1753). Glabrous annual; stem 10–60 cm, narrowly winged. Leaflets 20–90 × 0·5–3(–4) mm, linear; stipules 2–15 × 0·2–2 mm, linear, semi-sagittate. Flowers solitary; peduncles 10–40 mm, articulated near apex. Calyx-teeth slightly unequal, as long as or slightly longer than tube; corolla 8–11(–18) mm, orange-red. Legume 15–30 × 7–11 mm, pale brown, glabrescent, but persistently pubescent on suture. Seeds 2–3, finely papillose; hilum $\frac{1}{9}$–$\frac{1}{8}$ of the circumference. *S. Europe.* Al Bl Bu Co Cr Ga Gr Hs It Ju Lu Rm Rs (K) Si.

43. L. cicera L., *Sp. Pl.* 730 (1753) (? incl. *L. aegaeus* Davidov). Glabrous annual; stem 20–100 cm, winged. Leaflets 1(–2) pairs, 10–90(–110) × 1–6(–15) mm, linear to lanceolate; stipules 10–20 × 2–5 mm, lanceolate, semi-sagittate. Flowers solitary; peduncles 10–30 mm, articulated near the middle or apex. Calyx-teeth equal, 2–3 times as long as tube; corolla (5–)10–14(–20) mm, reddish-purple. Legume 20–40 × 5–10 mm, brown, glabrous, with 2 keels on dorsal suture. Seeds 2–6, smooth; hilum $\frac{1}{13}$ of the circumference. $2n = 14$. *Grassland and cultivated ground. S. Europe; often cultivated elsewhere for fodder and sometimes persisting.* Al Bl Bu Co Cr Ga Gr He Hs It Ju Lu Rm Rs (K, E) Sa Si Tu [Au].

L. stenophyllus Boiss. & Heldr. in Boiss., *Diagn. Pl. Or. Nov.* 2(9): 126 (1849), from Anatolia, has been once recorded from Kriti, but its presence there requires confirmation. It is like **43** but has calyx-teeth as long as the tube and the legume with 3 keels on the dorsal suture.

44. L. sativus L., *Sp. Pl.* 730 (1753). Like **43** but leaflets 25–150 × 3–7(–9) mm; peduncles 30–60 mm; corolla 12–24 mm, white, pink or blue; legume 10–18 mm wide, with 2 wings on dorsal suture; hilum $\frac{1}{16}$–$\frac{1}{15}$ of the circumference of seed. $2n = 14$. *Cultivated for fodder in C., S. & E. Europe and widely naturalized.* [Al Au Az Be Bu Cr Ga Gr He Hs Hu Ju Lu Rm Rs (C, W, K, E) Tu.] (*Origin not known.*)

45. L. amphicarpos L. *Sp. Pl.* 729 (1753) (*L. quadrimarginatus* Bory & Chaub.). Glabrous annual; stem 5–25 cm, narrowly winged. Leaves without or with a simple tendril; leaflets 1 pair, 5–30 × 1–4 mm, linear to ovate-oblong; stipules 5–10 × 2–4 mm, lanceolate, semi-sagittate. Flowers solitary; peduncles 20–40 mm,

articulated near apex. Calyx-teeth equal, *c.* 1½ times as long as tube; corolla 10–12 mm, red. Legume 20–25 × 9–11 mm, brown, glabrous, with 2 wings on both dorsal and ventral sutures. Seeds 2–4, smooth; hilum $\frac{1}{15}$–$\frac{1}{13}$ of the circumference. *S. Greece; Sicilia; Islas Baleares; S.W. part of Iberian peninsula.* Bl ?Cr Gr Hs Lu Si.

46. L. annuus L., *Demonstr. Pl.* 20 (1753). Glabrous annual; stem 40–150 cm, winged. Leaflets 1 pair, 50–150 × (1–)4–18 mm, linear or linear-lanceolate; stipules 10–25 × 0·3–0·8 mm, linear, semi-sagittate. Racemes 1- to 3-flowered. Calyx-teeth equal, as long as or slightly longer than tube; corolla 12–18 mm, yellow or orange-yellow. Legume 30–80 × 7–12 mm, pale brown, glandular when young, glabrescent. Seeds 7–8, tuberculate or papillose; hilum $\frac{1}{10}$–$\frac{1}{9}$ of the circumference. *Cultivated ground. Mediterranean region, Portugal; sometimes cultivated for fodder.* Al Bl Co Cr Ga Gr Hs It Ju Lu Sa Si Tu [Az].

47. L. hierosolymitanus Boiss., *Diagn. Pl. Or. Nov.* 2(9): 127 (1849). Like **46** but corolla 7–12 mm, pinkish-yellow; legume 5–7 mm wide; seeds rugulose. *Kriti, Kikladhes, N.E. Greece.* Cr Gr. (*S.W. Asia.*)

48. L. gorgoni Parl., *Gior. Sci. Sic.* 62: 3 (1838). Glabrous annual; stem 20–60 cm, winged. Leaflets 1 pair, 30–60 × 3–9 mm, linear to linear-elliptical; stipules 25–45 × 3–5 mm, lanceolate, semi-sagittate. Flowers solitary; peduncles 30–45 mm, articulated near apex. Calyx-teeth equal, 2–3 times as long as tube; corolla 18–25 mm, reddish-yellow. Legume 25–50 × 7–10 mm, brown, glabrous, with 3 keels on dorsal suture. Seeds 5–8, smooth; hilum $\frac{1}{11}$–$\frac{1}{9}$ of the circumference. *Sardegna; Sicilia; Turkey-in-Europe.* Sa Si Tu. (*S.W. Asia.*)

49. L. hirsutus L., *Sp. Pl.* 732 (1753). Sparsely pubescent annual; stem 20–120 cm, winged. Leaflets 1 pair, 15–80 × 3–20 mm, linear or oblong; stipules 10–18 × 1–2 mm, linear, semi-sagittate. Racemes 1- to 3(–4)-flowered. Calyx-teeth equal, as long as or slightly longer than tube; corolla 7–15(–20) mm, red with pale blue wings. Legume 20–50 × 5–10 mm, brown, pubescent, tuberculate. Seeds 5–10, tuberculate; hilum $\frac{1}{6}$–$\frac{1}{5}$ of the circumference. $2n = 14$. *C. & S. Europe.* Al Au Be Bu Co Cz Ga Ge Gr He Hs Hu It Ju Lu Po Rm Rs (K, E) Sa Si Tu [Br].

50. L. clymenum L., *Sp. Pl.* 732 (1753) (incl. *L. tuntasii* Heldr. ex Halácsy). Glabrous annual; stem 30–100 cm, winged. Leaves with broad leaf-like petiole and rhachis, the lower linear-lanceolate, without leaflets, the upper with 2–4(–5) pairs of leaflets; leaflets 20–60(–80) × (3–)6–11(–20) mm, linear to elliptical or lanceolate; stipules 9–18 × 2–6 mm, linear to ovate, semi-hastate. Racemes 1- to 5-flowered. Calyx-teeth equal, shorter than tube; corolla 15–20 mm, crimson with violet or lilac wings, very rarely pale yellow; style aristate. Legume 30–70 × 5–12 mm, brown, glabrous, channelled on the dorsal suture, not torulose. Seeds 5–7, smooth; hilum $\frac{1}{7}$–$\frac{1}{6}$ of the circumference. *Mediterranean region.* Al Bl Co Cr Ga Gr Hs It Ju Lu Si Tu [Az].

51. L. articulatus L., *Sp. Pl.* 731 (1753). Like **50** but leaflets 0·5–5(–11) mm wide; corolla with white or pink wings; style obtuse; legume 5–8 mm wide, not channelled on the dorsal suture, somewhat torulose. *Mediterranean region, Portugal.* Bl Co Cr Ga Gr Hs It Ju Lu Sa Si.

52. L. ochrus (L.) DC. in Lam. & DC., *Fl. Fr.* ed. 3, 4: 578 (1805). Like **50** but the lower leaves ovate-oblong; upper leaves with 1–2 pairs of leaflets; leaflets 15–35 × 6–20 mm, ovate; stipules 6–12 mm; racemes 1- to 2-flowered; calyx-teeth slightly unequal, as long as tube; corolla pale yellow; legume 40–60 ×

10–12 mm, with 2 wings on the dorsal suture. 2n = 14. *S. Europe.* Al Bl Co Cr Ga Gr Hs It Ju Lu Sa Si Tu.

53. L. nissolia L., *Sp. Pl.* 729 (1753). Glabrous or sparsely pubescent annual; stem 10–90 cm, not winged. Petiole and rhachis up to 130 mm, forming a linear, grass-like phyllode, without leaflets or tendril; stipules 0·5–2 mm, filiform. Racemes 1- to 2-flowered. Calyx-teeth equal, much shorter than tube; corolla (6–)8–18 mm, crimson. Legume 30–60 × 2–4 mm, pale brown, glabrous or pubescent. Seeds 12–20, hilum ⅓–⅛ of the circumference. 2n = 14. *Grassy places. W., C. & S. Europe.* Al Au Be Br Bu Co Cz Ga Ge Gr He Ho Hs Hu It Ju Lu Po Rm Rs (W, K, E) Sa Si Tu.

54. L. aphaca L., *Sp. Pl.* 729 (1753). Glabrous annual; stem up to 100 cm, angled. Seedling leaves with 1 pair of small leaflets; mature leaves without leaflets but with a tendril; stipules 6–50 × 5–40 mm, ovate, hastate. Flowers usually solitary; peduncles 20–50 mm. Calyx-teeth equal, 2–3 times as long as tube; corolla 6–18 mm, yellow. Legume 20–35 × 3–8 mm, brown, glabrous. Seeds 6–8; hilum 1/10 of the circumference. 2n = 14. *W., C. & S. Europe, but only as an alien in much of the northern part of its range.* Al *Au *Az Bl Bu Co Cr Ga Gr Hs *Hu It Ju Lu *Rm Rs (W, K, E) Sa Si Tu [Be Br Ge He Ho].

52. Pisum L.[1]

Like *Lathyrus* but stems terete; calyx-teeth large and more or less leaf-like; wings adnate to keel; style dilated at apex, longitudinally grooved with recurved margins.

1. P. sativum L., *Sp. Pl.* 727 (1753). Glabrous annual up to 200 cm. Leaflets 1–3 pairs, 2–7 × 1–4 cm, suborbicular to elliptical or oblong; stipules up to 10 × 6 cm, ovate to elliptical, semicordate at base. Racemes 1- to 3-flowered. Corolla 15–35 mm, white to purple. Legume 30–120 × 10–25 mm, yellow or brownish, reticulate-veined. Seeds up to 10. *S. Europe; cultivated almost throughout Europe since prehistoric times for the edible seed and for fodder, and often occurring as an escape from cultivation.* Al Bu Co Ga Gr Hs Ju It Lu Rm Rs (W, K) Sa Si Tu [Au Rs (C)].

(a) Subsp. **sativum** (incl. *P. arvense* L.): Racemes shorter than or only slightly exceeding the leaves; corolla white to purple; seeds globose or somewhat angular, smooth or rugose. *Widely cultivated and sometimes naturalized, particularly in S. Europe.*

(b) Subsp. **elatius** (Bieb.) Ascherson & Graebner, *Syn. Mitteleur. Fl.* 6(2): 1064 (1910) (*P. elatius* Bieb.): Racemes exceeding the leaves; corolla lilac or purple; seeds globose, granular. *S. Europe.*

53. Ononis L.[2]

Annual or perennial herbs or dwarf shrubs, usually glandular-hairy. Leaves 3-foliolate, sometimes simple or imparipinnate, leaflets usually toothed; stipules adnate to the petiole. Flowers in panicles, spikes, or racemes. Calyx campanulate or tubular; corolla yellow, pink or purple, sometimes nearly white; keel more or less beaked; stamens monadelphous. Legume oblong or ovate, dehiscent. Seeds 1-many.

Literature: G. Širjaev, *Beih. Bot. Centr.* 49(2): 381–665 (1932).

1 Flowers in panicles, sometimes condensed and with the primary branches 1-flowered, the pedicels then distinctly articulated more than 1·5 mm from the base; legume linear or oblong, usually deflexed
2 Perennial; stems woody, at least at the base
3 Corolla pink or purple, occasionally whitish

4 Terminal leaflet with a long petiolule — **1. rotundifolia**
4 Terminal leaflet subsessile
  5 Stems and peduncles densely white tomentose; leaflets with few rather blunt teeth — **2. tridentata**
  5 Stem and peduncles sparsely to moderately hairy with straight hairs; leaflets serrate or serrulate with usually numerous teeth
    6 Stems 25–100 cm, erect, usually many-flowered; all or most bracts without leaflets — **3. fruticose**
    6 Stems 5–35 cm, procumbent, 1- to 6-flowered; all bracts with 1 or 3 leaflets — **4. cristata**
3 Corolla yellow, sometimes with violet or red veins
  7 Filament of the adaxial stamen connate with the rest only in the lower ½ (Bulgaria) — **5. adenotricha**
  7 Filament of the adaxial stamen connate with the rest in the lower ⅔–¾
    8 Margin of leaflets not undulate; inflorescence lax — **6. natrix**
    8 Margin of leaflets undulate; inflorescence dense — **7. crispa**
2 Annual; stem herbaceous
  9 Legume torulose — **8. ornithopodioides**
  9 Legume not torulose
    10 Primary branches of the inflorescence 2-flowered
      11 Standard yellow, sometimes with pink veins, glabrous — **9. biflora**
      11 Standard pink or purple, glandular-hairy — **10. maweana**
    10 Primary branches of the inflorescence 1-flowered
      12 Wings with tooth on inner margin
        13 Calyx-teeth 5-veined; seeds smooth — **17. pubescens**
        13 Calyx-teeth 3-veined; seeds tuberculate
          14 Legume not or slightly inflated — **18. viscosa**
          14 Legume much inflated — **19. crotalarioides**
      12 Wings without tooth on inner margin
        15 Peduncles shortly aristate; corolla yellow — **11. sicula**
        15 Peduncles muticous; corolla pink or purple
          16 Leaves all 1-foliolate — **16. verae**
          16 At least the lower leaves 3-foliolate
            17 Calyx-teeth with glandular hairs only — **15. laxiflora**
            17 Calyx-teeth with at least some of the hairs eglandular
              18 Corolla more than 10 mm, exceeding the calyx; leaflets 10–15 mm — **14. pendula**
              18 Corolla less than 8 mm, equalling or shorter than the calyx; leaflets 5–8 mm
                19 Calyx-teeth entire — **12. reclinata**
                19 Calyx-teeth dilated at apex and 3-toothed — **13. dentata**
1 Flowers in spikes or racemes, sometimes with several flowers at each node; pedicels not articulated or with an articulation not more than 1·5 mm from the base; legume ovate or rhombic
20 Legume deflexed; pedicels 5–15 mm — **20. cintrana**
20 Legume erect or patent; pedicels less than 5 mm or absent
21 Perennial
  22 Pedicels articulated at base; calyx-teeth 1½–2½ times as long as tube
    23 Leaflets 15–23 mm; flowers in dense spike-like inflorescences — **21. speciosa**
    23 Leaflets not more than 10 mm; flowers in lax inflorescences
      24 Leaflets 4–10 mm; corolla 12–18 mm — **22. aragonensis**
      24 Leaflets 2–4 mm; corolla not more than 10 mm — **23. reuteri**
  22 Pedicels not articulated; calyx-teeth 2½–4 times as long as tube
    25 Lower leaves pinnate; corolla pink
      26 Corolla 10–14 mm, equalling the calyx; calyx-teeth 5-veined — **25. leucotricha**
      26 Corolla 15–23 mm, exceeding the calyx; calyx-teeth 7-veined — **24. pinnata**
    25 Leaves 1- to 3-foliolate; corolla pink, purple or yellow
      27 Corolla yellow; standard glabrous
        28 Corolla shorter than or equalling the calyx
          29 Calyx-teeth glabrous or with short glandular hairs — **28. minutissima**
          29 Calyx-teeth with long eglandular and glandular hairs
            30 Inflorescence without conspicuous leaf-like bracts; stipules less than 6 mm — **30. cephalotes**

---

[1] By P. W. Ball.   [2] By R. B. Ivimey-Cook.

30 Inflorescence with conspicuous leaf-like bracts; stipules more than 6 mm   **26. pusilla**
28 Corolla exceeding the calyx
31 Racemes 2- to 4-flowered or flowers solitary; leaves not glandular-hairy   **29. striata**
31 Racemes many-flowered; leaves glandular-hairy
32 Racemes with numerous leaf-like bracts equalling or exceeding the flowers   **27. saxicola**
32 Racemes with scarious bracts (except sometimes at the base) shorter than the flowers   **30. cephalotes**
27 Corolla pink or purple; standard hairy
33 Unarmed shrub; stems hispid; leaves with subsessile glands beneath   **31. hispida**
33 More or less herbaceous, sometimes spiny; stem variably hairy; leaves without subsessile glands beneath
34 Inflorescence dense; middle and upper bracts without leaflets   **35. masquillieri**
34 Inflorescence lax; bracts 1-foliolate
35 Stems usually procumbent, usually rooting; leaflets obtuse or emarginate   **33. repens**
35 Stems usually erect or ascending, not rooting; leaflets obtuse or acute
36 Plant usually spiny; flowers usually borne singly at each node of the raceme   **32. spinosa**
36 Plant unarmed; flowers usually in pairs at each node of the raceme   **34. arvensis**
21 Annual
37 Calyx campanulate
38 Standard hairy
39 Corolla pink
40 Seeds smooth   **42. cossoniana**
40 Seeds tuberculate
41 Leaves 3-foliolate; flowers in short, dense, terminal racemes   **36. filicaulis**
41 Leaves 1-foliolate; flowers in long lax racemes
42 Stems erect, glabrous; inflorescence not elongating after anthesis   **37. alba**
42 Stems procumbent or ascending, villous; inflorescence elongating after anthesis   **38. oligophylla**
39 Corolla yellow
43 Stems procumbent or ascending; seeds smooth   **39. variegata**
43 Stems erect; seeds tuberculate   **40. euphrasiifolia**
38 Standard glabrous
44 Lowest bracts 1-foliolate or without leaflets; corolla pink   **46. subspicata**
44 Lowest bracts 3-foliolate; corolla variously coloured
45 Seeds smooth   **41. hirta**
45 Seeds tuberculate
46 Calyx equalling or slightly exceeding corolla; calyx-teeth about as long as tube   **45. tournefortii**
46 Calyx shorter than corolla; calyx-teeth distinctly longer than tube
47 Middle cauline leaves with leaflets 10–20 mm, denticulate, with usually 10–16 teeth; legume with 1–3 seeds   **43. diffusa**
47 Middle cauline leaves with leaflets 6–10 mm, dentate, with 4–6 teeth; legume with 2–5 seeds   **44. serrata**
37 Calyx tubular
48 Leaves and lower bracts 1-foliolate   **48. alopecuroides**
48 Upper leaves and lower bracts 3-foliolate
49 Corolla 10–12 mm, exceeding the calyx   **47. mitissima**
49 Corolla more than 12 mm, about equalling the calyx   **49. baetica**

Sect. NATRIX Griseb. Flowers in panicles, the primary branches 1- to 3-flowered; fruiting pedicels usually more or less deflexed. Legume oblong.

1. O. rotundifolia L., *Sp. Pl.* 719 (1753). Erect, branched dwarf shrub 35–50 cm; stems villous and glandular. Leaves 3-foliolate; leaflets *c.* 25 mm, elliptical to orbicular, obtuse, coarsely toothed, sparingly glandular; terminal leaflet with long petiolule. Primary branches of the inflorescence *c.* 30 mm, up to 60 mm in fruit, muticous; pedicels 3–6 mm. Corolla 16–20 mm, pink or whitish. Legume 20–30 mm. Seeds 10–20, *c.* 3 mm, minutely tuberculate. *Usually calcicole.* ● *From S.E. Spain to E. Austria & C. Italy, mainly in the mountains.* Au Ga He Hs It.

2. O. tridentata L., *Sp. Pl.* 718 (1753). Erect or procumbent dwarf shrub 15–40 cm; stem tomentose. Leaves mostly 3-foliolate; leaflets 12–14(–40) mm, linear to obovate, somewhat fleshy, entire to coarsely toothed. Primary branches of the inflorescence up to 10 mm, usually aristate; arista 4–5 mm; bracts *c.* 5 mm. Corolla 10–17 mm, pink. Legume 13–20 mm, compressed. Seeds 2–3, *c.* 2·75 mm, brown and mealy. $2n = 30$. *Gypsaceous soils. E., C. & S. Spain.* Hs.

3. O. fruticosa L., *Sp. Pl.* 718 (1753). Erect dwarf shrub 25–100 cm; young stems shortly pubescent. Leaves mostly 3-foliolate, subsessile; leaflets 7–25 mm, oblong-lanceolate, subcoriaceous, glabrous. Primary branches of the inflorescence 10–30 mm, in the axils of scarious bracts; pedicels 2–5 mm. Corolla 10–20 mm, pink. Legume 18–22 mm. Seeds *c.* 4, 2·5 mm, minutely tuberculate. *Dry rocky places, mainly on mountains. C. & E. Spain and C. Pyrenees; S.E. France.* Ga Hs.

4. O. cristata Miller, *Gard. Dict.* ed. 8, no. 9 (1768) (*O. cenisia* L.). Procumbent, rhizomatous perennial 5–35 cm; stems shortly hairy and glandular. Leaves 3-foliolate; leaflets 5–10 mm, oblong or oblanceolate, subcoriaceous. Primary branches of the inflorescence 10–30 mm, shortly or minutely aristate. Corolla 10–14 mm, pink. Legume 9–12 mm. Seeds *c.* 5, 2·5 mm, tuberculate. *S.W. Alps; C. Appennini; E. Pyrenees and mountains of E. Spain.* Ga Hs It.

5. O. adenotricha Boiss., *Diagn. Pl. Or. Nov.* **1**(2): 14 (1843). Rhizomatous perennial 5–30 cm; stems several, glandular-hairy, arising from a woody base. Lower leaves pinnate, becoming 1- to 3-foliolate towards the apex of the stem; leaflets 6–7 mm, ovate, emarginate. Primary branches of the inflorescence *c.* 7 mm, muticous or aristate. Corolla 7–9 mm, yellow. Legume 10–12 mm. Seeds 2–4, *c.* 2 mm, tuberculate. *Dry stony places; calcicole. Mountains of S. Bulgaria.* Bu. (*Anatolia, Lebanon.*)

6. O. natrix L., *Sp. Pl.* 717 (1753). Erect, much-branched dwarf shrub; stems 20–60 cm, densely glandular-hairy. Leaves 3-foliolate, the lower rarely pinnate; leaflets variable, ovate to linear. Flowers in lax leafy panicles, the primary branches 1-flowered. Corolla 6–20 mm, yellow, frequently with red or violet veins. Legume 10–25 mm. Seeds 4–10, *c.* 2 mm, smooth or minutely tuberculate. *S. & W. Europe, northwards to N. France, but very rare in the Balkan peninsula.* Bl Co Cr Ga Ge He Hs It Ju Lu Sa Si.

1 Leaflets 2–5(–8) mm; corolla 6–12 mm; seeds greyish-brown, smooth   (c) subsp. **hispanica**
1 Leaflets (5–)8 mm or more; corolla (10–)12–20 mm; seeds dark brown, minutely tuberculate
2 Stem with dense, long (0·5–2 mm) glandular and eglandular hairs; calyx teeth 2·5–4 times as long as tube   (a) subsp. **natrix**
2 Stem with short (0·2–0·5 mm) glandular hairs, sometimes with some longer hairs; lower leaves 1- to 3-foliolate; calyx-teeth 1·5–2·5 times as long as the tube   (b) subsp. **ramosissima**

(a) Subsp. **natrix**: Stems erect, with dense, long (0·5–2 mm) eglandular and glandular hairs and some short glandular hairs. Lower leaves 3-foliolate, sometimes pinnate; leaflets 12–20(–30) mm. Calyx-teeth 2·5–4 times as long as tube; corolla 12–20 mm, usually with red or violet veins. Legume 12–25 mm. Seeds dark brown, minutely tuberculate. *Almost throughout the range of the species, from S.E. Spain eastwards.* Very variable, especially in the western part of its range.

(b) Subsp. **ramosissima** (Desf.) Batt. in Batt. & Trabut, *Fl. Algér.* (*Dicot.*) 213 (1889) (*O. ramosissima* Desf.): Stems erect, often much-branched with glandular hairs 0·2–0·5 mm and sometimes with longer eglandular hairs. Lower leaves 1- to 3-foliolate; leaflets (5–)8–15(–25) mm. Calyx-teeth 1·5–2·5 times as long as tube; corolla (10–)12–15 mm. Legume 10–15 mm. Seeds dark brown, minutely tuberculate. *W. Mediterranean region & Portugal, usually near the coast.*

(c) Subsp. **hispanica** (L. fil.) Coutinho, *Fl. Port.* 331 (1913) (*O. virgata* G. Kunze, *O. hispanica* L. fil.): Like subsp. (b) but stems often procumbent and with very dense glandular hairs; leaflets 2–5(–8) mm; corolla 6–12 mm; legume 10–12 mm; seeds greyish-brown, smooth. *Coasts of Spain and Portugal; Islas Baleares.*

7. **O. crispa** L., *Sp. Pl.* ed. 2, 1010 (1763). Erect, dwarf shrub 15–50 cm; stem densely glandular-hairy, leafy. Lower leaves often 5-foliolate, upper 3-foliolate; leaflets 7–9 mm, orbicular, the margins undulate. Inflorescence dense, the primary branches 1- or 2-flowered, muticous or shortly aristate. Corolla 15–20 mm, yellow, the standard often with red veins. Legume 15–20 mm. Seeds 12, *c.* 2 mm, minutely tuberculate. ● *S.E. Spain, Islas Baleares.* Bl Hs ?Lu.

8. **O. ornithopodioides** L., *Sp. Pl.* 718 (1753). Erect annual up to 30 cm; stems glandular-hairy. Leaves 3-foliolate; leaflets 10–15 mm, obovate to oblanceolate, the terminal long-petiolate. Primary branches of the inflorescence 1- or 2-flowered, aristate, as long as the leaflets. Corolla 6–8 mm, yellow, shorter than the calyx. Legume 12–20 mm, subfalcate, torulose. Seeds 6–10, *c.* 1·5 mm, acutely tuberculate. *Mediterranean region.* Al Bl Co Gr Hs It Ju Sa Si.

9. **O. biflora** Desf., *Fl. Atl.* 2: 143 (1798) (*O. geminiflora* Lag.). Erect annual 10–50 cm; stems sparsely glandular-hairy. Leaves 3-foliolate; leaflets 15–20 mm, elliptical, fleshy. Primary branches of the inflorescence 15–40 mm, 1- to 2-flowered, aristate. Corolla 12–16 mm, yellow, sometimes with pink veins; standard glabrous. Legume *c.* 20 mm. Seeds 12–14, *c.* 3 mm, acutely tuberculate. *Mediterranean region, Portugal.* Gr Hs It Lu Sa Si.

10. **O. maweana** Ball, *Jour. Bot.* (*London*) 11: 304 (1873). Erect annual 8–35 cm; stems glandular hairy. Lower leaves 1- to 5-foliolate; leaflets variable, obovate to linear. Primary branches of the inflorescence less than 5 mm, 2-flowered, muticous; pedicels up to 5 mm. Corolla 9–13 mm; standard pink to purple, glandular-hairy; wings and keel yellow. Legume 6–12 mm. Seeds 12, *c.* 1 mm, minutely tuberculate. *Sandy soils. S.W. Portugal.* Lu. (*N.W. Morocco.*)

11. **O. sicula** Guss., *Cat. Pl. Boccad.* 78 (1821). Slender, erect annual 10–35 cm; stems villous and glandular-hairy. Leaves 3-foliolate, the upper 1-foliolate; leaflets 10–15 mm, oblong, finely dentate. Primary branches of the inflorescence 10–20 mm, aristate, 1-flowered. Corolla 5–9 mm, yellow, usually shorter than the calyx. Legume 10–17 mm. Seeds many, *c.* 1 mm, tuberculate. *W. Mediterranean region.* Hs It ?Sa Si.

12. **O. reclinata** L., *Sp. Pl.* ed. 2, 1011 (1763). Procumbent annual 2–15 cm; stems villous and glandular-hairy. Leaves 3-foliolate; leaflets 5–8 mm, oblanceolate to obovate-orbicular. Primary branches of the inflorescence up to 10 mm, muticous, 1-flowered. Corolla 5–10 mm, pink or purple, about equalling the calyx. Legume 8–14 mm. Seeds up to 20, 0·5–1 mm, acutely tuberculate. *Dry grassy places. S. & W. Europe, northwards to S. England.* Al Bl Br Bu Co Cr Ga Gr Hs It Ju Lu Sa Si Tu.

13. **O. dentata** Solander ex Lowe, *Trans. Camb. Philos. Soc.* 4: 34 (1831). Like **12** but the 4 upper calyx-teeth dilated at the apex and 3-toothed or lobed; seeds up to 1·75 mm. *Sicilia, S. Spain, S. Portugal.* Hs Lu Si.

Possibly not specifically distinct from **12**.

14. **O. pendula** Desf., *Fl. Atl.* 2: 147 (1798). Erect annual 20–40 cm; stems variably glandular-hairy. Leaves 3-foliolate, upper 1-foliolate; leaflets 10–18 mm, obovate to suborbicular, apex rounded. Primary branches of the inflorescence 5–7 mm, muticous, 1-flowered. Corolla 12–18 mm, pink. Legume 10–12 mm. Seeds 8–12, *c.* 1 mm, acutely tuberculate. *S. Spain, Sicilia.* Hs Si.

15. **O. laxiflora** Desf., *op. cit.* 146 (1798). Ascending annual 8–30 cm; stems branched, pubescent and glandular-hairy. Leaves 3-foliolate; leaflets 10–15 mm, ovate. Primary branches of the inflorescence up to 20 mm, muticous, 1-flowered. Corolla 8–10 mm, pink or whitish, scarcely exceeding the calyx. Legume 16–22 mm, not deflexed. Seeds many, *c.* 1 mm, acutely tuberculate. *S. Spain; one station in E.C. Portugal.* Hs Lu.

16. **O. verae** Širj., *Beih. Bot. Centr.* 49(2): 517 (1932). Ascending or procumbent annual 6–12 cm; stems hairy and glandular. Leaves all 1-foliolate, glandular and sparsely hairy; leaflets of cauline leaves orbicular, those of the uppermost leaves linear. Primary branches of the inflorescence 6–8 mm, muticous, 1-flowered. Corolla 10–12 mm, pink. Legume 6–8 mm. Seeds 8, *c.* 1 mm, acutely tuberculate. ● *Kriti.* Cr.

17. **O. pubescens** L., *Mantissa Alt.* 267 (1771). Erect annual 15–35 cm; stems viscid and villous. Upper and lower leaves 1-foliolate, middle 3-foliolate; leaflets 10–25 mm, elliptical. Inflorescence dense, the primary branches 5–10 mm, muticous, 1-flowered. Corolla *c.* 15 mm, yellow. Legume 8–10 mm, acuminate. Seeds 2–3, *c.* 3 mm, dark brown, smooth. *Mediterranean region; frequent in the east and west but very rare in the centre.* ? Bl Cr Ga Gr Hs ?It Lu ?Si Tu.

18. **O. viscosa** L., *Sp. Pl.* 718 (1753). Erect annual 10–80 cm; stems densely and softly hairy and glandular. Leaves all 1-foliolate or the middle cauline leaves 3(–5)-foliolate; leaflets 10–20 mm, variable, usually elliptical to obovate, obtuse. Primary branches of the inflorescence 10–20 mm, aristate, 1-flowered. Corolla up to 12 mm, yellow, the standard frequently with red veins. Legume up to 20 mm, not or only slightly inflated. Seeds many, 1–1·5 mm, yellow-brown, acutely tuberculate. $2n = 32$. *Mediterranean region, Portugal.* Al Bl Cr Ga Gr Hs It Ju Lu Sa Si Tu.

Very variable especially in the western part of its range, where several subspecies and numerous varieties have been recognized.

1 Legume 12–20 mm, exceeding the calyx
2 Primary branches of the inflorescence muticous or with arista less than 3 mm **(d) subsp. sieberi**
2 Primary branches of the inflorescence with arista 6–15 mm
3 Corolla not exceeding calyx **(c) subsp. breviflora**
3 Corolla exceeding calyx
4 Arista with simple hairs **(a) subsp. viscosa**
4 Arista without simple hairs **(e) subsp. subcordata**
1 Legume 6–10 mm, about equalling calyx
5 Primary branches of the inflorescence not exceeding the leaves, muticous or with short arista **(d) subsp. sieberi**
5 Primary branches of the inflorescence exceeding leaves
6 Leaves mostly 1-foliolate; corolla yellow, rarely with pink veins **(b) subsp. brachycarpa**
6 Middle cauline leaves 3-foliolate; corolla pinkish-purple **(f) subsp. foetida**

(a) Subsp. **viscosa**: Leaves 1- to 3-foliolate. Primary branches of inflorescence longer than leaves; arista 6–15 mm. Corolla exceeding the calyx, yellow. Legume exceeding calyx. Seeds numerous. *C. & W. Mediterranean region*; *S. & N.E. Portugal*.

(b) Subsp. **brachycarpa** (DC.) Batt. in Batt. & Trabut, *Fl. Algér.* (*Dicot.*) 212 (1889): Leaves 1- to 5-foliolate. Primary branches of the inflorescence 14–18 mm; arista 2–6 mm. Corolla 9–11 mm, slightly exceeding the calyx, yellow, rarely with pink veins. Legume *c.* 8 mm, about equalling the calyx. *Spain and Portugal*.

(c) Subsp. **breviflora** (DC.) Nyman, *Consp.* 161 (1878) (*O. breviflora* DC.): Middle cauline leaves 3-foliolate, rarely all leaves 1-foliolate; apex of leaflets often emarginate. Primary branches of the inflorescence equalling or shorter than leaves; arista 6–15 mm. Corolla shorter than calyx. Legume 12–18 mm. Seeds numerous. *Almost throughout the range of the species.*

(d) Subsp. **sieberi** (Besser ex DC.) Širj., *Beih. Bot. Centr.* 49 (2): 526 (1932) (*O. sieberi* Besser ex DC.): Leaves 1- to 3-foliolate. Primary branches of the inflorescence shorter than leaves, often nearly muticous; flowers aggregated at apices of branches. Corolla exceeding calyx, standard often pink. Legume rather longer than calyx. Seeds 4–6. ● *From Sardegna to Kriti.*

(e) Subsp. **subcordata** (Cav.) Širj., *op. cit.* 527 (1932): Stems 15–25 cm, glabrous below. Leaves all 1-foliolate; leaflets subcordate. Primary branches of the inflorescence up to twice as long as leaves; arista 6–15 mm. Corolla yellow. Legume 2–3 times as long as calyx. *S. Spain.*

(f) Subsp. **foetida** (Schousboe ex DC.) Širj., *op. cit.* 527 (1932): Stems up to 70 cm, rather slender, glabrous below, variably hairy above. Middle leaves 3-foliolate, leaflets acute. Primary branches of the inflorescence 3–4 times as long as leaves. Corolla pinkish-purple, exceeding calyx. Legume about equalling calyx. Seeds 3–4. *S. Spain.*

19. O. crotalarioides Cosson, *Not. Pl. Crit.* 155 (1852). Like **18** but with leaves all 1-foliolate; leaflets up to 28 mm, elliptical, obovate or ovate-oblong; primary branches of the inflorescence *c.* 20 mm; corolla yellow, shorter than the calyx; legume 20–25 × 8–9 mm, much inflated; seeds up to 2 mm. *S. Spain.* Hs.

Sect. ONONIS (Sect. *Bugrana* Griseb.). Flowers in racemes or very condensed panicles with the primary branches not more than 1·5 mm; legume usually erect or patent, ovate or rhombic; seeds few.

20. O. cintrana Brot., *Phyt. Lusit.* ed. 3, **1**: 138 (1816). Erect annual 10–35 cm; stems densely hairy and glandular. Leaves 1-foliolate, the middle cauline 3-foliolate. Flowers in dense terminal panicles, the primary branches 1- to 2-flowered; pedicels 5–15 mm. Corolla 12–14 mm, yellow with red veins. Legume *c.* 5 mm. Seeds *c.* 4, *c.* 1 mm, brown, acutely tuberculate. *Dry places. C. & S. Portugal, S.W. Spain.* Hs Lu.

21. O. speciosa Lag., *Gen. Sp. Nov.* 22 (1816). Erect or ascending dwarf shrub up to 1 m; stems with very short, dense glandular hairs; inflorescence with long hairs, some glandular. Leaves 3-foliolate; leaflets 15–23 mm, elliptical to suborbicular, subcoriaceous, glabrous, viscid with sessile glands. Flowers in dense, oblong panicles, the primary branches 1- to 3-flowered; pedicels 2–6 mm; bracts subscarious, caducous. Corolla 15–20 mm, golden-yellow. Legume *c.* 6 mm. Seeds 3 mm, black, smooth. *S. Spain.* Hs.

22. O. aragonensis Asso, *Syn. Stirp. Arag.* 96 (1779). Dwarf shrub 15–30 cm; stems often much contorted, densely hairy above, sometimes with scattered short glandular hairs inter-mixed. Leaves 3-foliolate; leaflets 4–10 mm, elliptical or suborbicular, obtuse, or emarginate, coriaceous. Flowers in long, lax, terminal panicles, the primary branches 1- to 2-flowered; pedicels 2–4 mm. Corolla 12–18 mm, yellow. Legume 7–8 mm. Seeds 1–2, 3–4 mm, dark greenish-brown, smooth. *Pyrenees, E. & S. Spain.* Ga Hs.

23. O. reuteri Boiss. in Boiss. & Reuter, *Pugillus* 30 (1852). Like **22** but leaflets 2–4 mm and corolla not more than 10 mm. *S.W. Spain.* Hs.

24. O. pinnata Brot., *Fl. Lusit.* **2**: 99 (1804) (*O. rosifolia* DC.). Dwarf shrub 45–80 cm; stems numerous, strict, villous, sparsely glandular. Leaves pinnate; leaflets 5–9, 10–15 mm, ovate, rounded or emarginate at apex, base cuneate. Flowers borne singly at each node in dense racemes, which elongate later; pedicels 4 mm. Corolla 15–23 mm, pink. Legume 6–7 mm. Seeds 3, *c.* 1·5 mm, brown, smooth. *Scrub. S. half of Iberian peninsula.* Hs Lu.

25. O. leucotricha Cosson, *Not. Pl. Crit.* 34 (1849). Dwarf shrub 15–40 cm, villous and with numerous, short glandular hairs. Leaves pinnate; leaflets 5–7, 10–30 mm, ovate or elliptical, caducous, but the petioles long-persistent. Flowers subsessile, borne singly at each node, in dense, terminal racemes, which elongate after anthesis. Corolla 10–14 mm, pinkish-purple. Legume 6–8 mm. Seeds 4–6, 1·5–2 mm, brown, smooth. *S.W. Spain* (*near Cádiz*). Hs.

26. O. pusilla L., *Syst. Nat.* ed. 10, **2**: 1159 (1759) (*O. columnae* All.). Perennial up to 25 cm, somewhat woody at the base; stems variably hairy. Leaves 3-foliolate, long-petiolate; leaflets 5–13 mm, elliptical to suborbicular, sometimes emarginate. Flowers sessile in lax spikes, with leaf-like bracts (longer than the flowers) to the apex. Corolla 5–12 mm, yellow, about equalling the calyx. Legume 6–8 mm. Seeds *c.* 6, *c.* 2 mm, yellow-brown, minutely tuberculate. *S. Europe, extending northwards to N. France and Czechoslovakia.* Al Au Bu Co Cz Ga Gr He Hs Hu It Ju Lu Rm Rs (K) Sa Si Tu [Be].

27. O. saxicola Boiss. & Reuter, *Pugillus* 32 (1852). Like **26** but up to 20 cm, procumbent and slender; corolla *c.* 12 mm, exceeding the calyx. ● *S.W. Spain* (*Serrania de Ronda*). Hs.

28. O. minutissima L., *Sp. Pl.* 717 (1753). Dwarf shrub 5–30 cm Stems often procumbent and rooting, nearly glabrous. Leaves 3-foliolate; leaflets 3–6 mm, sessile, oblong-oblanceolate to obovate, caducous. Flowers pedicellate, borne singly at each node, in dense terminal racemes. Calyx-tube whitish, with long, subulate teeth; corolla 8–10 mm, yellow, not exceeding the calyx. Legume 6–7 mm. Seeds 3–6, 1·5–2 mm, brown, smooth. $2n=30$. *W. Mediterranean region, extending eastwards to Jugoslavia.* Bl Co Ga Hs It Ju Sa Si.

29. O. striata Gouan, *Obs. Bot.* 47 (1773). Somewhat woody, rhizomatous up to 20 cm; flowering stems ascending, somewhat tomentose. Leaves 3-foliolate; leaflets 3–6 mm, oblanceolate to orbicular-obovate, often emarginate, the veins very prominent when dry. Flowers borne singly at each node, in few-flowered terminal racemes; pedicels short. Corolla (5–)10–13 mm, yellow, exceeding the calyx. Legume 6–7 mm. Seeds 1–3, *c.* 2 mm, dark greenish-brown, smooth. *Mountains; calcicole.* ● *S.W. Europe.* Ga Hs It.

30. O. cephalotes Boiss., *Elenchus* 33 (1838) (*O. montana* Cosson). Perennial with woody rhizome 8–20 cm; stems densely glandular-hairy and villous. Leaves 3-foliolate; leaflets 3–7 mm,

suborbicular to broadly elliptical or obovate, obtuse or emarginate, densely glandular-hairy. Flowers subsessile, borne singly at each node, in short, dense, terminal racemes without conspicuous leaf-like bracts. Corolla *c.* 10 mm, yellow, equalling or exceeding the calyx. Legume *c.* 6 mm. Seeds 1–2, *c.* 3 mm, brown, minutely tuberculate. *S.E. Spain.* Hs.

31. O. hispida Desf., *Fl. Atl.* **2**: 146 (1798). Shrub 30–150 cm; stems hispid. Leaves 3-foliolate; leaflets 5–15 mm, elliptical or ovate, obtuse; lower bracts 3-foliolate, the upper without leaflets. Flowers shortly pedicellate, borne singly at each node, in lax racemes. Corolla 13–20 mm, pink. Legume 6–7 mm. Seeds *c.* 2, *c.* 1·5 mm, black, tuberculate. *Sicilia.* Si.

32. O. spinosa L., *Sp. Pl.* 716 (1753). Dwarf shrub 10–80 cm; stems variously hairy and sparsely glandular, usually erect or ascending, usually spiny. Leaves mostly 3-foliolate; leaflets very varied in shape. Bracts usually 1-foliolate. Flowers borne singly, rarely in pairs, at each node, in lax racemes. Calyx glandular-pubescent, and shortly hirsute at the mouth. Corolla 6–20 mm, pink or purple, usually much exceeding the calyx. Legume 6–10 mm. Seeds 1 or few, *c.* 2 mm, brown or blackish, tuberculate, rarely smooth. $2n=30$. *W., C. & S. Europe, extending to S. Norway and N.W. Ukraine.* All except Az Fa Fe Hb Is Rs (N, B, E) Sb.

One pair of chromosomes possesses a long constriction so that it frequently appears to be two pairs; this probably accounts for the records of $2n=32$ for this species and for records of $2n=32$ and 64 for **33**.

1 Plant with weak spines; legume shorter than calyx; corolla
   15–20 mm   **(d) subsp. austriaca**
1 Plant with robust spines; legume equalling or exceeding calyx
  2 Corolla 10–20 mm; legume 2- to 4-seeded   **(a) subsp. spinosa**
  2 Corolla 6–10 mm, little longer than calyx; legume 1-seeded
    3 Seeds tuberculate; leaflets 6–10 mm   **(b) subsp. antiquorum**
    3 Seeds smooth; leaflets 10–20 mm   **(c) subsp. leiosperma**

(a) Subsp. **spinosa** (*O. campestris* Koch & Ziz): Stems erect, usually spiny and with two opposite rows of hairs when young. Leaflets ovate-oblong, acute, more than 3 times as long as wide. Flowers borne singly at each node; corolla 10–20 mm. *In the northern part of the range of the species, southwards to S. Italy and N.E. Portugal.*

(b) Subsp. **antiquorum** (L.) Arcangeli, *Comp. Fl. Ital.* 157 (1882) (*O. antiquorum* L.; incl. *O. decipiens* Aznav., *O. diacantha* Sieber ex Reichenb.): Stems erect, irregularly hairy, usually, and often exceedingly, spiny. Leaflets 6–10 mm, sometimes less. Corolla 6–10 mm. Seeds tuberculate. *S. Europe.*

(c) Subsp. **leiosperma** (Boiss.) Širj., *Beih. Bot. Centr.* **49**(2): 590 (1932) (*O. leiosperma* Boiss.): Like subsp. (b) but leaflets larger and seeds smooth. *S. & E. parts of Balkan peninsula, Aegean region; Krym.*

(d) Subsp. **austriaca** (G. Beck) Gams in Hegi, *Ill. Fl. Mitteleur.* **4**(3): 1224 (1923): Stems simple, irregularly hairy, unarmed or slightly spiny; somewhat foetid. Leaflets up to 20 mm. Flowers borne singly at each node, in long lax racemes. Corolla 15–20 mm. Seeds tuberculate. *C. Europe.*

33. O. repens L., *Sp. Pl.* 717 (1753) (*O. spinosa* subsp. *procurrens* (Wallr.) Briq.). Shrubby perennial 40–70 cm; stems procumbent or ascending, often rooting, unarmed or with usually soft spines, variably hairy with long, eglandular and short, glandular hairs, not in two distinct rows. Leaves 1- to 3-foliolate; leaflets usually ovate, obtuse or emarginate, less than 3 times as long as wide. Flowers borne singly, rarely in pairs, at each node, in lax leafy racemes. Calyx densely hirsute; corolla (7–)15–20 mm, pink or purple, usually much exceeding the calyx. Legume 5–7 mm. Seeds 1–2, *c.* 2·5 mm, brown or blackish, tuberculate. $2n=30, 60$. *W. & C. Europe, extending to c. 66° N. in E. Sweden, to Estonia, and to the N. half of the Balkan peninsula.* Au Be Br Bu Cz Da Ga Ge Hb He Ho Hs It Ju Lu No Po Rm Rs (B, W, K) Su Tu [Fe].

34. O. arvensis L., *Syst. Nat.* ed. 10, **2**: 1159 (1759) (*O. hircina* Jacq., *O. intermedia* auct.). Shrubby perennial 50–100 cm; stems erect, variably hairy. Leaves mostly 3-foliolate; leaflets 10–25 mm, elliptical to ovate. Flowers pedicellate, borne in pairs at each node, in dense terminal racemes; bracts 1- to 3-foliolate. Corolla 10–20 mm, pink. Legume 6–9 mm, about equalling the calyx. Seeds 1–3, *c.* 2·5 mm, dark brown, tuberculate. $2n=30$. *Europe from c. 65° N. in Norway, E. Germany and Albania eastwards, but absent from N. Russia.* Al Au Bu Cz Da Fe Ge Gr Hu It Ju No Po Rm Rs (B, C, W, K, E) Su.

A variable species, possibly divisible into a number of subspecies. One of the more distinct is **O. spinosiformis** Simkovics, *Österr. Bot. Zeitschr.* **27**: 158 (1877) (*O. semihircina* Simkovics), from E.C. Europe, with 2 rows of hairs on the upper part of the stem and the corolla 10–15 mm.

35. O. masquillierii Bertol., *Hort. Bot. Bon.* **2**: 11 (1839). Dwarf shrub up to 40 cm; stems ascending or decumbent, with unilateral long hairs but few glandular hairs. Leaves 3-foliolate; leaflets 10–18 mm, oblong, ovate or elliptical, acute or obtuse, the terminal long-petiolate. Flowers pedicellate, borne singly at each node, in usually dense racemes; bracts without leaflets. Corolla 11–15 mm, pink, exceeding the calyx. Legume 3–6 mm, shorter than the calyx. Seeds *c.* 2, *c.* 2 mm, dark brown, tuberculate. $2n=30$. *Dry clay soils. N. & C. Italy.* It.

36. O. filicaulis Salzm. ex Boiss., *Voy. Bot. Midi Esp.* **2**: 153 (1839). Procumbent or ascending annual 10–50 cm; stems variably hairy and glandular. Leaves 3-foliolate; leaflets 4–10 mm, obovate, obtuse; lowest bracts 3-foliolate, upper 1-foliolate. Flowers shortly pedicellate, borne singly at each node, forming a short, dense terminal raceme. Corolla *c.* 12 mm, pink, equalling or slightly exceeding the calyx. Legume 5 mm. Seeds *c.* 3, 1–1·5 mm, brown, minutely tuberculate. *S.W. Spain.* Hs.

37. O. alba Poiret, *Voy. Barb.* **2**: 210 (1789). Erect annual 15–45 cm; stems branching from the base, strict, nearly glabrous. Leaves 1-foliolate; leaflets 10–45 mm, usually lanceolate or oblong. Flowers borne singly at each node, in lax terminal racemes; pedicels 1–2 mm. Corolla 12–16 mm, pink or whitish, exceeding the calyx. Legume 5 mm. Seeds *c.* 4, *c.* 1·5 mm, tuberculate. *Pastures and cultivated fields. Italy and Sardegna.* It Sa.

38. O. oligophylla Ten., *Fl. Nap.* 1, *Prodr.*: 70 (1811). Like **37** but stems procumbent or ascending, villous and with scattered glandular hairs; pedicels *c.* 5 mm, up to 15 mm after anthesis; corolla 10–13 mm; seeds minutely tuberculate. ● *C. & S. Italy, Sicilia.* It Si.

39. O. variegata L., *Sp. Pl.* 717 (1753). Procumbent or ascending annual 10–30 cm; stems glandular-tomentose and villous. Leaves 1-foliolate; leaflets 5–10 mm, obovate. Flowers borne singly at each node in a lax, terminal, often branched raceme; pedicels up to 4 mm. Corolla 12–14 mm, yellow, much exceeding the calyx. Legume *c.* 8 mm. Seeds 10–14, *c.* 1·5 mm, reddish-brown, smooth. *Maritime sands. Mediterranean region, S. Portugal.* Al Co Cr Gr Hs It Lu Sa Si.

40. O. euphrasiifolia Desf., *Fl. Atl.* **2**: 141 (1798). Erect annual; stems 7–20 cm, hairy and glandular. Leaves 1-foliolate; leaflets 10–25 mm, lower ovate, upper linear. Flowers shortly pedicellate, borne singly at each node in congested racemes which elongate after anthesis. Corolla 12–15 mm, yellow, exceeding the calyx. Legume *c.* 8 mm. Seeds *c.* 12, 0·5–1 mm, reddish-brown, tuberculate. *Maritime sands. S.E. Spain (Cabo de Gata).* Hs. (*N.W. Africa.*)

41. O. hirta Poiret in Lam., *Encycl. Méth. Bot., Suppl.* **1**: 741 (1811) (*O. baetica* auct., non Clemente, *O. ellipticifolia* Willk.). Procumbent annual 10–30 cm; stems sometimes simple and ascending, tomentose and with scattered glandular hairs. Leaves 3-foliolate, lower sometimes 1-foliolate; leaflets 6–10 mm, ovate to ovate-orbicular. Flowers borne singly at each node in a dense raceme which becomes lax after anthesis; pedicels up to 2 mm. Corolla 8–12 mm, purple. Legume 5 mm. Seeds 2–3, *c.* 2 mm, smooth. *Grassy scrub. S.W. Spain; W.C. Portugal.* Hs Lu.

42. O. cossoniana Boiss. & Reuter, *Pugillus* 33 (1852). Procumbent or ascending annual 10–35 cm, often somewhat woody at the base; stems glandular-hairy. Leaves 3-foliolate; leaflets 7–15 mm, obovate or elliptical; lowest bracts 3-foliolate, upper without leaflets. Flowers shortly pedicellate, borne singly at each node, in a fairly dense raceme. Corolla 13–15 mm, pink, exceeding the calyx. Legume 5–7 mm, ovate. Seeds *c.* 4, 2·5–3 mm, smooth. *Maritime sands. S.W. Spain; one station in S.W. Portugal.* Hs Lu.

43. O. diffusa Ten., *Fl. Nap.* **1**, *Prodr.*: 41 (1811). Procumbent or ascending annual 10–40 cm; stems glandular-hairy, viscid. Leaves 3-foliolate, but the basal sometimes pinnate; leaflets 10–20 mm, oblanceolate to suborbicular, denticulate, with usually 10–16 more or less appressed teeth; lowest bracts 3-foliolate, upper without leaflets. Flowers very shortly pedicellate, borne singly at each node in a dense terminal raceme, elongating considerably after anthesis. Corolla 9–11 mm, pink, equalling or exceeding the calyx. Legume 5–8 mm. Seeds 1–3, *c.* 2 mm, reddish-brown, tuberculate. *Coasts of the Mediterranean region and of Portugal and N.W. Spain.* Co Cr Gr Hs It Lu Sa Si.

44. O. serrata Forskål, *Fl. Aegypt.* 130 (1775). Diffuse, procumbent annual 5–30 cm; viscid. Leaves 3-foliolate; leaflets 6–10 mm, oblong or oblong-linear, dentate, with 4–6 patent or somewhat recurved teeth; lowest bracts 3-foliolate, the upper without leaflets. Flowers shortly pedicellate, borne singly at each node in a dense terminal raceme which elongates after anthesis. Corolla *c.* 8 mm, white or pale pink, equalling or slightly exceeding the calyx. Legume 6–7 mm. Seeds 2–5, *c.* 1 mm, brown, minutely tuberculate. *Karpathos.* Cr. (*S.W. Asia, N. Africa.*)

45. O. tournefortii Cosson, *Not. Pl. Crit.* 34 (1849). Branched, ascending annual 10–40 cm; stems densely hairy below, glandular-hairy above. Leaves 3-foliolate; leaflets 5–10 mm, obovate, emarginate, somewhat fleshy, with prominent veins; lower bracts 3-foliolate, upper without leaflets. Flowers shortly pedicellate, borne singly at each node in a dense raceme which elongates after anthesis. Corolla 6–8 mm, white to pale yellow, variously veined with purple, equalling or slightly shorter than the calyx. Legume 7–8 mm, ovoid-oblong. Seeds 4–6, *c.* 1·5 mm, brown, minutely tuberculate. *S.W. Spain.* Hs. (*Morocco.*)

46. O. subspicata Lag., *Period. Soc. Med. Cádiz* **4**: 1 (1824) (*O. picardii* Boiss.). Erect or ascending annual 5–30 cm; stems

variably glandular-hairy. Leaves 3-foliolate, the lower with long petioles up to 3 cm; leaflets 6–18 mm, linear-oblanceolate to ovate, serrate; bracts mostly without leaflets. Flowers shortly pedicellate, borne singly at each node, in a short dense raceme which elongates after anthesis. Corolla 8–14 mm, pink, exceeding the calyx. Legume *c.* 7 mm, ovoid-oblong. Seeds 4–6, *c.* 1 mm, brown, tuberculate. *Maritime sands. W. Spain, Portugal.* Hs Lu.

47. O. mitissima L., *Sp. Pl.* 717 (1753). Erect or procumbent annual; stems 15–60 cm, somewhat tomentose or nearly glabrous. Leaves 3-foliolate; leaflets 10–20 mm, obovate or elliptical, sometimes caducous; lower bracts 3-foliolate, upper without leaflets, concave and membranous. Flowers shortly pedicellate, borne singly at each node, in a dense terminal raceme. Calyx-tube glabrous, with very prominent white veins, margins of teeth glandular-ciliate. Corolla 10–12 mm, pink, exceeding the calyx. Legume 5–6 mm, ovate. Seeds 2–3, 1·5–2 mm, dark brown, spinulose. *Mediterranean region, Portugal.* Bl Co Cr Ga Gr Hs It Ju Lu Sa Si Tu.

48. O. alopecuroides L., *Sp. Pl.* 717 (1753). Robust annual 10–65 cm, variable in habit; stems nearly glabrous below, densely hairy and glandular above. Leaves 1-foliolate; leaflets 20–50 cm, elliptical or elliptic-orbicular, glabrous. Flowers subsessile, borne singly at each node, in dense terminal racemes, elongating slightly after anthesis. Corolla 13–16 mm, pink, exceeding the calyx. Legume 8–10 mm. Seeds 2–3, 2–3 mm, orange-brown, smooth and shining. *Cultivated ground and waste places. W. Mediterranean region; casual elsewhere in S. Europe.* *Co Hs It Si.

49. O. baetica Clemente, *Ens. Vid* 291 (1807) (*O. salzmanniana* Boiss. & Reuter). Like **48** but less robust; upper leaves and lower bracts 3-foliolate; corolla about equalling the calyx. *Sandy places. S.W. Spain, S. Portugal.* Hs Lu. (*N.W. Africa.*)

54. Melilotus Miller[1]

Annual, biennial or short-lived perennial herbs. Leaves 3-foliolate; leaflets usually toothed. Flowers in axillary racemes. Calyx-teeth subequal; corolla yellow or white, rarely tinged with blue or violet, deciduous, free from the staminal tube; stamens diadelphous, the filaments not dilated. Legume globose to obovoid, rarely lanceolate-rhomboid, straight, indehiscent or very tardily dehiscent. Seeds 1–2, rarely more.

Several species are cultivated locally as fodder, and occur frequently as casuals outside the geographical limits given below.

Many species smell strongly of coumarin, especially when dry.

Literature: O.E. Schulz, *Bot. Jahrb.* **29**: 660–735 (1901).

1 Corolla white, tinged with blue or violet (Turkey)   **16. physocarpa**
1 Corolla white or yellow, without a blue or violet tinge
  2 Stipules of the middle leaves entire or minutely denticulate
   3 Corolla white
    4 Ovary and young legume pubescent; legume with conspicuous transverse veins   **7. taurica**
    4 Ovary and young legume glabrous; legume reticulate-veined
     5 Pedicels 1–1·5 mm   **3. alba**
     5 Pedicels 2–4 mm   **4. wolgica**
   3 Corolla yellow
    6 Corolla 2–3 mm; legume 1·5–3 mm   **10. indica**
    6 Corolla 3–9 mm; legume 3–8 mm
     7 Racemes with not more than 10 flowers
      8 Racemes 4–6 cm; legume 7–8 mm, lanceolate-rhomboid   **6. polonica**

---

[1] By A. Hansen.

8 Racemes *c.* 1 cm; legume 3–3·5 mm, globose with a short
   beak **9. neapolitana**
7 Racemes with at least 10 flowers
9 Ovary and young legume glabrous; legume with transverse veins
10 Biennial; legume with indistinct transverse veins **5. officinalis**
10 Annual; legume with very prominent transverse veins **11. elegans**
9 Ovary and young legume pubescent; legume reticulate-veined
11 Biennial or perennial; legume flattened **2. altissima**
11 Annual; legume globose **9. neapolitana**
2 Stipules of the middle leaves toothed
12 Legume reticulate-veined
13 Corolla 3–3·5 mm; legume faintly veined, compressed **1. dentata**
13 Corolla 6–9 mm; legume foveolate-rugose, globose **8. italica**
12 Legume concentric-striate
14 Racemes much shorter than subtending leaf; legume acute **15. messanensis**
14 Racemes in fruit at least as long as subtending leaf; legume rounded at apex
15 Standard as long as or longer than keel; legume blackish-brown **12. infesta**
15 Standard shorter than keel; legume yellowish-brown
16 Corolla 3–4 mm; legume broadly sessile; leaflets oblong-cuneate **13. sulcata**
16 Corolla 4–8 mm; legume stipitate; leaflets obovate-cuneate **14. segetalis**

1. M. dentata (Waldst. & Kit.) Pers., *Syn. Pl.* **2**: 348 (1807). Erect or ascending, branched biennial 20–150 cm. Leaflets oblong-elliptical or lanceolate-ovate, serrate. Stipules subulate, dentate at base. Racemes many-flowered. Corolla 3–3·5 mm, bright yellow; wings shorter than standard, longer than keel. Legume 4·5–5·5 mm, obovoid, slightly reticulate-veined, glabrous, blackish-brown when ripe. 2*n*=16. *Salt steppes, saline meadows and river-banks. E. & C. Europe, extending northwards to S. Sweden.* Au Cz Da Ge Hu Ju Po Rm Rs (C, W, E) Su.

2. M. altissima Thuill., *Fl. Paris* ed. 2, 378 (1799). Erect, branched biennial or short-lived perennial 60–150 cm. Leaflets oblong-ovate or cuneate, obtuse, serrate. Stipules subulate-setaceous, entire. Racemes 2–5 cm, many-flowered, elongating in fruit. Corolla 5–7 mm, yellow; wings, standard and keel equal. Legume 5–6 mm, obovoid, acute, reticulate-veined, pubescent, black when ripe, usually 2-seeded; style long and persistent. 2*n*=16. *Damp or saline habitats, and as a ruderal. Throughout a large part of Europe, but rare in the east and absent as a native from the islands.* Al Au Be Cz Da Ga Ge Gr He Ho Hs Hu It Ju No Po Rm Rs (C, W, E) Su [*Br Hb].

3. M. alba Medicus, *Vorl. Churpf. Phys.-Ökon. Ges.* **2**: 382 (1787). Erect, branched annual or biennial 30–150 cm. Leaflets narrowly oblong-obovate to suborbicular, serrate. Stipules setaceous, entire. Racemes lax and slender, many-flowered. Corolla 4–5 mm, white; wings and keel nearly equal, shorter than standard. Legume 3–5 mm, obovoid, mucronate, reticulate-veined, glabrous, greyish-brown when ripe. 2*n*=16. *Open habitats, often as a weed or ruderal. Almost throughout Europe, except for most of the islands, but doubtfully native, especially in the north.* Al Au Bu Cz *Da *Fe Ga Ge Gr He *Ho Hs Hu It Ju Lu *No Po Rm Rs (*N, *B, C, W, K, E) *Su Tu [Be Br].

4. M. wolgica Poiret in Lam., *Encycl. Méth. Bot., Suppl.* **3**: 648 (1814). Erect, branched biennial 40–120 cm. Lower leaflets rhombic-ovate, serrate, the upper oblong-lanceolate to linear, usually entire. Stipules linear, setaceous, entire. Racemes 5–10 cm, lax and slender, many-flowered, elongating in fruit; pedicels

2–4 mm. Corolla 3–3·5 mm, white; wings and standard longer than keel. Legume 4–5 mm, obovoid, acute, distinctly reticulate-veined, glabrous, brownish-yellow when ripe, usually 1-seeded. *Usually on saline soils. S.E. Russia, S. & E. Ukraine, W. Kazakhstan; casual in N. Europe.* Rs (W, E).

M. arenaria Grec., *Consp. Fl. Roman., Supl.* 198 (1909), from sand dunes in Romania (S. of Constanţa) and Ukraine (Danube delta), with racemes grouped 3–5 together, corolla 5–6 mm and legume 6–7 mm with 2 or more seeds, may be separable from **4**, but further information is needed.

5. M. officinalis (L.) Pallas, *Reise* **3**: 537 (1776) (*M. arvensis* Wallr.). Decumbent or erect, branched biennial 40–250 cm. Leaflets of lower leaves obovate to ovate, the upper ovate-lanceolate, all serrate. Racemes lax and slender, many-flowered. Corolla 4–7 mm, yellow; wings and standard equal, longer than keel. Legume 3–5 mm, transversely rugose, mucronate, glabrous, brown when ripe, usually 1-seeded; style often deciduous. 2*n*=16. *Cultivated ground, often on clay or saline soils. Most of Europe except the extreme south, but only as an alien in much of the north.* Al Au *Be Bl Bu Cz Ga Ge Gr He *Ho Hs Hu It Ju Po Rm Rs (*N, *B, C, W, K, E) Sa Tu [Br Da Fe Hb No Su].

6. M. polonica (L.) Pallas, *Reise* **3**: 537 (1776). Erect or ascending, branched biennial 35–150 cm. Lower leaflets obovate, acute, dentate, the upper spathulate, subentire. Racemes 4–6 cm, lax, 4- to 9-flowered; pedicels 4–5 mm. Corolla 5·5–6·5 mm, bright yellow; wings and standard a little longer than keel. Legume 7–8 mm, lanceolate or oblong-rhomboid, acute, reticulate-veined, yellow or pale brown when ripe. *Sandy soils. S. Ukraine, S.E. Russia, W. Kazakhstan.* Rs (W, E). (*W.C. Asia.*)

7. M. taurica (Bieb.) Ser. in DC., *Prodr.* **2**: 188 (1825). Erect, branched biennial 30–80 cm. Lower leaflets rhombic-obovate or suborbicular-cuneate, the upper oblong, obtuse or truncate, all serrate. Stipules linear-subulate, entire. Racemes 5–9 cm, lax, 40- to 60-flowered, elongating in fruit. Corolla *c.* 6 mm, white; wings, standard and keel subequal; ovary and young legume pubescent. Legume 4–5 mm, obovoid, transversely striate, with prominent veins, pale brown when ripe, usually 1-seeded. *Dry hillsides and cultivated ground. Krym.* ?Rm Rs (K). (*N. Anatolia.*)

8. M. italica (L.) Lam., *Fl. Fr.* **2**: 594 (1778). Erect, branched annual 20–60 cm. Lower leaflets orbicular-obovate, the upper narrower, all obtuse or truncate, serrate above middle. Stipules incise-dentate. Racemes 1·5–3 cm, lax, many-flowered, elongating in fruit. Corolla 6–9 mm, yellow; standard longer than wings and keel. Legume 5–6 mm, globose, obtuse with an apiculus, strongly reticulate-veined, yellowish or greyish-brown when ripe. *Dry, open habitats. Mediterranean region.* Al Bl Co Cr Ga Gr Hs It Ju Sa Si [Lu].

9. M. neapolitana Ten., *Fl. Nap.* 1, *Prodr.*: 62 (1811). Erect, branched annual 15–50 cm; stem pubescent above. Lower leaflets obovate-orbicular, the upper oblong-linear, all obtuse, serrate. Stipules lanceolate, entire. Racemes *c.* 1 cm, lax, 8- to 20-flowered, elongating in fruit. Corolla 4–6 mm, bright yellow; standard, wings and keel equal; ovary pubescent. Legume 3–3·5 mm, globose, reticulate-veined, narrowing to a conical beak 0·5–1 mm, becoming glabrous, light brown when ripe. 2*n*=16. *Dry, open habitats. S. Europe.* Al Bl Bu Co Cr Ga Gr Hs It Ju Lu Rs (K) Sa Si Tu.

10. M. indica (L.) All., *Fl. Pedem.* **1**: 308 (1785) (*M. parviflora* Desf.). Erect or ascending annual 15–50 cm. Leaflets lanceolate-oblong, serrate. Stipules subentire. Racemes dense, (10–)many-flowered. Corolla 2–3 mm, pale yellow; wings and keel equal, shorter than standard. Legume 1·5–3 mm, subglobose, strongly reticulate-veined, glabrous, whitish-grey when young. *Mediterranean region and S.W. Europe; naturalized in C. & N.W. Europe.* Al *Az Bl Co Cr Ga Gr Hs It Ju Lu Sa Si Tu [Au Be Br Cz Ge He Ho].

11. M. elegans Salzm. ex Ser. in DC., *Prodr.* **2**: 188 (1825). Erect annual 20–150 cm; stem pubescent above. Lower leaflets obovate-orbicular, the upper oblong, all obtuse or truncate, serrate. Lower stipules triangular-lanceolate, the upper linear-setaceous, entire. Racemes 1·5–2 cm, lax, 15- to 30-flowered. Corolla 4–5 mm, yellow; standard and wings equal, shorter than keel; ovary glabrous. Legume 3·5–4 mm, obovoid, compressed, with transverse or sigmoid veins, brownish-yellow when ripe. *Grassland, usually near the sea. W. & C. Mediterranean region, Portugal.* Al Bl Co Ga ?Gr Hs It Ju Lu Sa Si.

12. M. infesta Guss., *Fl. Sic. Prodr.* **2**: 486 (1828). Erect or ascending, branched annual 30–50 cm. Lower leaflets triangular or cuneate-obovate, the upper oblong-cuneate. Stipules semi-ovate or sagittate, dentate. Racemes 2–3 cm, lax, 15- to 50-flowered, elongating in fruit. Corolla 6–7·5 mm, yellow, standard and keel subequal, shorter than wings; ovary glabrous. Legume 4–5 mm, subglobose or obovoid, concentric-striate, blackish-brown when ripe. *W. Mediterranean region.* Bl Co It Si.

13. M. sulcata Desf., *Fl. Atl.* **2**: 193 (1799). Erect, branched or simple annual 10–40 cm. Leaflets oblong-cuneate, obtuse, serrate. Stipules dentate. Racemes 1–1·5 cm, 8- to 25-flowered, elongating in fruit and then as long as or longer than the leaves. Corolla 3–4 mm, yellow. Legume 3–4 mm, globose, concentric-striate, pale yellow or yellowish-brown when ripe. *Cultivated ground and other open habitats. Mediterranean region, S. Portugal.* Al Bl Co Cr Ga Gr Hs It Ju Lu Sa Si Tu.

14. M. segetalis (Brot.) Ser. in DC., *Prodr.* **2**: 187 (1825). Erect annual 40–60 cm. Leaflets obovate-cuneate, obtuse, serrate. Lowest stipules entire, the upper dentate. Racemes *c.* 3 cm, dense, 30- to 50-flowered, *c.* 3 times as long as their subtending leaf. Corolla 4–8 mm, yellow. Legume 2·5–5·5 mm, oblong-globose, concentric-striate, yellow when ripe. *Damp places. Mediterranean region, Portugal.* Bl Co Cr Ga Gr Hs It Ju Lu Sa Si.

This species is composed of two taxa which may merit specific rank. Typical *M. segetalis*, from the W. Mediterranean region and C. & S. Portugal, has the legume 4·5–5·5 mm with 12–15 concentric striations. The second taxon, from the Mediterranean region and Portugal, has the legume 2·5–3 mm with up to 8 concentric striations. The correct name for the latter is uncertain.

15. M. messanensis (L.) All., *Fl. Pedem.* **1**: 309 (1785) (*M. sicula* (Turra) B. D. Jackson). Erect or ascending, branched annual 20–40 cm. Leaflets obovate-orbicular or lanceolate-cuneate, serrate. Lower stipules triangular-lanceolate, denticulate, the upper lanceolate, entire but denticulate at base. Racemes 0·7–1 cm, 3- to 10-flowered, shorter than their subtending leaves. Corolla 4–5 mm, yellow; standard and keel subequal, longer than wings. Legume 5–8 mm, oblique-ovoid, acute, concentric-striate, yellowish-brown when ripe. *Cultivated ground and damp places, especially near the coast. Mediterranean region, Portugal.* Bl Co Ga Hs It Ju Lu Sa Si Tu.

16. M. physocarpa Stefanov, *Bull. Soc. Bot. Bulg.* **3**: 79 (1929). Erect annual 10–15 cm, diffusely branched from base; branches crispate-pubescent. Leaflets cuneate-obovate, denticulate, slightly fleshy. Stipules lanceolate, entire. Racemes short, few-flowered,

longer than subtending leaf. Corolla *c.* 5 mm, white, tinged with blue or violet. Legume 4–4·5 mm, ovoid-globose, abruptly mucronate, pendent, with few obscure longitudinal veins, sparsely crispate-pubescent. *Dry, rocky places.* ● *Turkey-in-Europe (Kumbag, near Tekirdag).* Tu.

This species is apparently known only from the original collection, the location of which is uncertain. It is very similar to and may not be distinct from **M. bicolor** Boiss. & Balansa in Boiss., *Diagn. Pl. Or. Nov.* **3**(6): 46 (1859), from W. Anatolia.

55. Trigonella L.[1]

Annual. Leaves pinnately 3-foliolate; leaflets usually toothed. Flowers solitary or in sessile or pedunculate axillary heads or short racemes. Calyx-teeth equal or unequal; corolla yellow, blue or purplish, free from the staminal tube, deciduous; stamens diadelphous or monadelphous; filaments not dilated. Legume usually linear or oblong, straight or curved, indehiscent or dehiscing along one suture. Seeds 1–many.

Literature: G. Širjaev, *Publ. Fac. Sci. Univ. Masaryk* **102**: 1–57 (1928); **110**: 1–37 (1929); **128**: 1–31 (1930); **136**: 1–33 (1931); **148**: 1–43 (1932); **170**: 1–37 (1933). I. T. Vassilczenko, *Acta Inst. Bot. Acad. Sci. URSS* **10**: 124–269 (1953).

1 Calyx 5–12 mm, tubular
  2 Legume 1–2 mm wide, not beaked; seeds smooth; pedicels *c.* 2 mm   **7. grandiflora**
  2 Legume 2·5–7 mm wide, beaked; seeds tuberculate; pedicels not more than 0·5 mm
    3 Flowers (3–)10–15 in pedunculate heads; beak of legume 3·5–5 mm   **20. coerulescens**
    3 Flowers solitary or paired and not pedunculate; beak of legume 10–40 mm
      4 Leaflets 20–50 × 10–15 mm; legume (excluding beak) 60–110 mm; seeds quadrangular   **23. foenum-graecum**
      4 Leaflets 5–12 × 3–8 mm; legume (excluding beak) 15–40 mm; seeds ovoid
        5 Corolla 8–10 mm; legume straight; beak 10–20 mm   **21. gladiata**
        5 Corolla *c.* 18 mm; legume curved; beak 30–40 mm   **22. cariensis**
1 Calyx 2–5 mm, usually campanulate
  6 Legume flat, membranous, with a broad membranous wing on the suture
    7 Calyx *c.* 3 mm; corolla 7–10 mm; legume 12–20 × 10–15 mm   **1. graeca**
    7 Calyx *c.* 2 mm; corolla *c.* 5 mm; legume 11–14 × 8–10 mm   **2. cretica**
  6 Legume not membranous and not winged
    8 Corolla blue, rarely white; legume 4–5 mm
      9 Racemes globose, scarcely elongating after anthesis; legume abruptly contracted into a beak   **18. caerulea**
      9 Racemes subglobose, elongating after anthesis; legume gradually attenuate into a beak   **19. procumbens**
    8 Corolla yellow, sometimes tinged with purple; legume 5 mm or more
      10 Racemes sessile or subsessile with peduncles less than 0·5 cm
        11 Legume glabrous
          12 Legume erect or patent   **13. arcuata**
          12 Legume pendent
            13 Stems sparsely hairy; calyx-teeth shorter than tube   **10. spinosa**
            13 Stems densely appressed-pubescent; calyx-teeth longer than tube   **17. monspeliaca**
        11 Legume pubescent at least when young
          14 Legume 7–17 mm, pendent   **17. monspeliaca**
          14 Legume (10–)20–50 mm, erect or patent
            15 Legume 1·5–2 mm wide, with reticulate veins   **15. polyceratia**

---

[1] By R. B. Ivimey-Cook.

15 Legume 1–1·5 mm wide, with oblique anastomosing
    veins   **16. orthoceras**
10 Racemes pedunculate
 16 Legume glabrous
  17 Calyx-teeth equal
   18 Legume pendent; calyx-teeth ½ as long as tube **3. maritima**
   18 Legume erect or patent; calyx-teeth slightly longer than
      tube   **12. striata**
  17 Calyx-teeth unequal
   19 Legume 5–8 mm, ovate   **11. spicata**
   19 Legume 10–16 mm, linear or oblong
    20 Peduncles c. 1·5 cm, about as long as leaves
                 6. rechingeri
    20 Peduncles 2–6 cm, at least twice as long as leaves
     21 Wings shorter than keel; legume acuminate
                 4. corniculata
     21 Wings as long as keel; legume subacute **5. balansae**
 16 Legume hairy
  22 Legume pendent; seeds smooth
   23 Stems glabrous; legume with thick, oblique veins
                 3. maritima
   23 Stems villous; legume with indistinct, reticulate veins
                 8. sprunerana
  22 Legume erect or patent; seeds tuberculate or tuberculate-
    rugose, rarely smooth
   24 Legume 5–8 mm; flowers in ±elongated racemes
    25 Legume contracted between the seeds, pubescent;
      seeds smooth   **9. smyrnaea**
    25 Legume not contracted between the seeds, glabres-
      cent; seeds finely tuberculate **11. spicata**
   24 Legume 10 mm or more; flowers in subumbellate
    racemes
    26 Calyx-teeth shorter than tube; legume 1·5–2 mm wide
                 15. polyceratia
    26 Calyx-teeth about as long as tube; legume 1–1·5 mm
      wide
     27 Corolla 5–7 mm; legume with transverse veins
                 14. fischerana
     27 Corolla 4–5 mm; legume with oblique veins
                 16. orthoceras

Subgen. **Trigonella.** Calyx usually campanulate. Legume not inflated.

1. T. graeca (Boiss. & Spruner) Boiss., *Fl. Or.* **2**: 91 (1872). Stems 10–30 cm, usually ascending, glabrous. Leaflets 10–15 × 7–12 mm, obovate to suborbicular, truncate, somewhat fleshy. Racemes subcapitate, many-flowered; peduncles 4–6 cm; pedicels 3–4 mm. Calyx c. 3 mm, the teeth ½ as long as tube; corolla 7–10 mm, yellow. Legume 12–20 × 10–15 mm, ovate-orbicular, flat, membranous, with transverse anastomosing veins, and with a membranous wing on the upper suture. Seeds 2–3, c. 4 mm, ovoid, brown, tuberculate. *Stony places.* ● *S. & W. Greece.* Gr.

2. T. cretica (L.) Boiss., *loc. cit.* (1872) (*Pocockia cretica* (L.) Ser.). Like **1** but calyx c. 2 mm; corolla c. 5 mm; legume 11–14 × 8–10 mm with 1(–2) seeds. *Calcareous screes. Kriti.* ?cr. (*W. Anatolia.*)

The records from Europe are probably all erroneous.

3. T. maritima Delile ex Poiret in Lam., *Encycl. Méth. Bot., Suppl.* **5**: 361 (1817). Stems 5–40 cm, procumbent, glabrous. Leaflets 5–10 × 5–8 mm, ovate, truncate or emarginate, denticulate, glabrous or sparsely hairy beneath. Racemes subumbellate, (3–)5- to 10-flowered; peduncles 1–2 cm; pedicels 1–1·5 mm. Calyx 2–2·5 mm, the teeth ½ as long as tube; corolla 6–7 mm, yellow. Legume 10–16 × 2–3 mm, pendent, linear, somewhat curved and deflexed, subglabrous or sparsely hairy, with thick, oblique veins. Seeds c. 1 mm, ovoid, brown, smooth. *Dry places. C. Mediterranean region.* It Sa Si. (*N. Africa, S.W. Asia.*)

4. T. corniculata (L.) L., *Syst. Nat.* ed. 10, **2**: 1180 (1759). Stems 10–55 cm, procumbent to erect, glabrous or subglabrous. Leaflets 10–40 × 7–35 mm, linear-lanceolate to obovate, obtuse, sometimes emarginate. Racemes ovate-oblong, 8- to 15-flowered; peduncles up to 6 cm; pedicels c. 3 mm. Calyx 3–4 mm, the teeth unequal, as long as or shorter than tube; corolla 6–7 mm, yellow; wings shorter than keel. Legume 10–16 × (1·5–)2–3 mm, pendent, linear, acuminate, compressed, somewhat curved, glabrous, with thin transverse veins. Seeds 1–1·5 mm, oblong, tuberculate. 2n = 16. *Mediterranean region.* Al Bu ?Cr Ga Gr Hs It Ju Si.

5. T. balansae Boiss. & Reuter in Boiss., *Diagn. Pl. Or. Nov.* **3(5)**: 79 (1856). Like **4** but racemes globose at anthesis; wings equalling the keel; legume 2–4 mm wide, subacute. *S. Greece and Aegean region.* Cr Gr.

Perhaps only a subspecies of **4**.

6. T. rechingeri Širj., *Österr. Bot. Zeitschr.* **85**: 58 (1936). Stems 5–20 cm, sparsely pubescent or subglabrous. Leaflets 7–10 × 7–9 mm, obovate, denticulate, glabrous or subglabrous. Racemes capitate, 6- to 10-flowered; peduncles c. 1·5 cm; pedicels 1·5–2 mm. Calyx 3–4 mm, the teeth unequal, the longest about equalling tube; corolla c. 7 mm, yellow. Legume 10–14 × 3–4 mm, pendent, oblong, glabrous, with transverse veins. Seeds 1·5–2 mm, ovoid, yellow, finely tuberculate. *Coastal rocks; calcicole.* ● *S. Aegean region.* Cr Gr.

7. T. grandiflora Bunge, *Arb. Naturf.-Ver. Riga* **1**: 218 (1847). Stems 5–25(–35) cm, glabrous or subglabrous. Leaflets 8–15 × 5–10 mm, obovate-cuneate, truncate, dentate, sparsely pubescent on the veins beneath. Racemes subumbellate, (1–)2- to 3(–5)-flowered; peduncles c. 1 cm; pedicels c. 2 mm. Calyx 5–7 mm, the teeth ⅓–½ as long as tube; corolla 13–16 mm, yellow. Legume 20–75 × 1–2 mm, erect or patent, linear, curved, glabrous, with oblique veins, the apex hooked. Seeds c. 3 mm, cylindrical, yellow with a few red spots. *Dry slopes and cultivated ground. S.E. Russia.* Rs (E). (*C. Asia, Iran.*)

8. T. sprunerana Boiss., *Diagn. Pl. Or. Nov.* **1(2)**: 17 (1843). Stems up to 15 cm, villous. Leaflets 6–13 × 3–9 mm, obovate, villous. Racemes subcapitate, 5- to 10-flowered; peduncles 0·5–1·5 cm; pedicels c. 2 mm. Calyx 3–3·5 mm, the teeth slightly unequal, the longest as long as or slightly longer than tube; corolla 5–7 mm, yellow. Legume 15–20 × c. 2 mm, pendent, linear-oblong, curved, villous, with indistinct, reticulate veins. Seeds c. 2 mm, ovoid, smooth. *Aegean region.* Gr Tu. (*S.W. Asia.*)

9. T. smyrnaea Boiss., *op. cit.* 19 (1843). Stems 10–15 cm, subglabrous. Leaflets 6–7 × 3–4 mm, obovate, emarginate, puberulent beneath. Racemes ovate, 5- to 7-flowered; peduncles 1·5–3·5 cm; pedicels c. 1 mm. Calyx c. 4 mm, the teeth unequal, the longest about as long as tube; corolla c. 8 mm, yellow. Legume 6–7 × 2 mm, erect or patent, shortly cylindrical, curved, pubescent, contracted between the seeds, with indistinct, longitudinal veins. Seeds c. 2 mm, ovoid, smooth. *Krym.* Rs (K). (*Anatolia.*)

10. T. spinosa L., *Sp. Pl.* 777 (1753). Stems 15–30 cm, procumbent or ascending, sparsely hairy. Leaflets 5–8(–12) × 4–5(–8) mm, obovate, denticulate or subentire, glabrous above, sparsely hairy beneath. Racemes subumbellate, (1–)4- to 6-flowered, sessile or subsessile; pedicels c. 0·5 mm. Calyx c. 2·5 mm, the teeth shorter than tube; corolla c. 4 mm, yellow. Legume (15–)20–35 × 1–2 mm, pendent, linear, compressed, curved, glabrous, with transverse veins. Seeds *c.* 2·5 mm, linear, tuberculate. *S. Aegean region* (*Gavdhos*, ?*Kriti*). Cr. (*E. Mediterranean region.*)

11. T. spicata Sibth. & Sm., *Fl. Graec. Prodr.* **2**: 108 (1813). Stems 10–40 cm, ascending, glabrous. Leaflets 6–14 × 3–6 mm, obovate to elliptical, denticulate, glabrous above, subglabrous beneath. Racemes capitate, many-flowered; peduncles 2–4 cm; pedicels *c.* 2 mm. Calyx *c.* 4 mm, the teeth unequal, the longest about as long as tube; corolla 6–7 mm, yellow. Legume 5–8 × 3 mm, erect or patent, ovate, compressed, with long, patent hairs when young, glabrescent, with reticulate veins and with a long, curved beak. Seeds 1, *c.* 2·5 mm, ovoid, finely tuberculate. *S.E. Europe.* Bu Gr Rs (K) Tu.

12. T. striata L. fil., *Suppl.* 340 (1781) (*T. tenuis* Fischer ex Bieb.). Stems up to 30 cm, usually erect, puberulent. Leaflets 5–10 × 4–7 mm, elliptical to obovate, denticulate, glabrous or puberulent on the veins beneath. Racemes subumbellate, usually 4- to 5-flowered; peduncles 1–4 cm; pedicels *c.* 1 mm. Calyx 3–4 mm, the teeth slightly longer than tube; corolla 4–5 mm, yellow. Legume 15–20 × 1·3–2 mm, erect or patent, linear, curved, glabrous, with transverse anastomosing veins. Seeds 1·5–2 mm, oblong, tuberculate-rugose. *S.E. Europe.* Bu Gr Ju Rs (K, E).

13. T. arcuata C.A. Meyer, *Verz. Pfl. Cauc.* 136 (1831) (?*T. cancellata* Pers.). Like **12** but leaflets more or less triangular; racemes 4- to 8-flowered, sessile; legume *c.* 15 mm and more strongly curved. *S.E. Russia, W. Kazakhstan.* Rs (E). (*S.W. & C. Asia.*)

14. T. fischerana Ser. in DC., *Prodr.* **2**: 183 (1825). Stems 5–30 cm, usually densely pubescent. Leaflets 3–7 × 3–6 mm, usually obovate, denticulate, puberulent. Racemes subumbellate, 4- to 10-flowered; peduncles 0·5–2 cm; pedicels *c.* 1 mm. Calyx 3–5 mm, the teeth about as long as tube; corolla 5–7 mm, yellow. Legume 10–25 × 1–1·5 mm, erect or patent, linear, curved, pubescent, with transverse anastomosing veins. Seeds *c.* 2 mm, oblong, tuberculate-rugose. *Krym.* Rs (K, ?E). (*S.W. Asia.*)

15. T. polyceratia L., *Sp. Pl.* 777 (1753). Stems 20–45 cm, somewhat hairy. Leaflets 5–13 × 5–10 mm, ovate, obovate or suborbicular, entire to pinnatifid, usually sparsely hairy. Racemes subumbellate, 1- to 6(–8)-flowered, subsessile, rarely with peduncle up to 3·5 cm; pedicels *c.* 0·5 mm. Calyx 3–4 mm, the teeth shorter than tube; corolla 4–6(–9) mm, yellow. Legume (10–)20–50 × 1·5–2 mm, erect or patent, straight or slightly curved, pubescent, with reticulate veins. Seeds 1·5–2 mm, oblong, yellow, tuberculate. *Spain and N. Portugal, just extending to S. France.* Ga Hs Lu.

16. T. orthoceras Kar. & Kir., *Bull. Soc. Nat. Moscou* **14**: 399 (1841). Like **15** but stems usually glabrous; calyx-teeth as long as tube; corolla 4–5 mm; legume 1–1·5 mm wide, with oblique anastomosing veins. *S.E. Russia, W. Kazakhstan.* Rs (E). (*S.W. & C. Asia.*)

17. T. monspeliaca L., *Sp. Pl.* 777 (1753). Stems up to 35 cm, densely appressed-pubescent. Leaflets 4–10 × 3–7 mm, obovate-cuneate, entire to incise-serrate. Racemes subumbellate, 4- to 14-flowered, subsessile; pedicels up to 1 mm. Calyx *c.* 3 mm, the teeth slightly longer than tube; corolla *c.* 4 mm, yellow. Legume 7–17 × 1–1·5 mm, pendent, linear, slightly curved upwards, usually pubescent, with thick oblique veins. Seeds *c.* 1·5 mm, brown, finely tuberculate. *C. & S. Europe, extending to Belgium and W. France.* Al Au Be Bl Bu Cr Co Cz Ga Gr He Hs Hu It Ju Lu Rm Rs (W, K) Sa Si Tu.

Subgen. **Trifoliastrum** (Moench) G. Beck. Calyx campanulate. Legume inflated.

18. T. caerulea (L.) Ser. in DC., *Prodr.* **2**: 181 (1825). Stems 20–60(–100) cm, erect, sparsely hairy, hollow. Leaflets 20–50 × 5–20 mm, ovate to oblong, emarginate, denticulate. Racemes globose, dense, many-flowered; peduncles 2–5 cm; pedicels *c.* 1 mm. Calyx *c.* 3 mm, the teeth about equalling tube; corolla *c.* 6 mm, blue or white. Legume 4–5 × 3 mm, erect or patent, rhomboid-obovate, abruptly contracted to a beak *c.* 2 mm. Seeds *c.* 2 mm, ovoid, brown, finely tuberculate. *Cultivated for fodder throughout much of Europe, and widely naturalized or casual as a weed or ruderal. Apparently nowhere indigenous; probably derived from* **19.**

19. T. procumbens (Besser) Reichenb., *Pl. Crit.* **4**: 35 (1826) (*T. besserana* Ser., *T. caerulea* subsp. *procumbens* (Besser) Thell.). Like **18** but stems solid; leaflets 12–28 × 3–8 mm, linear-lanceolate; racemes subglobose, becoming oblong, lax; legume ovoid, gradually attenuate into a beak. *E.C. & S.E. Europe.* Au Bu ?Cz Gr Hu Ju Rm Rs (?C, W, K, E) Tu [Ga].

Subgen. **Foenum-graecum** Širj. Calyx tubular. Legume not inflated.

20. T. coerulescens (Bieb.) Halácsy, *Consp. Fl. Graec.* **1**: 351 (1900). Stems up to 40 cm, densely hairy. Leaflets 8–20 × 7–12 mm, ovate-triangular or obovate, denticulate, villous. Racemes ovate, (3–)10- to 15-flowered; peduncles 1–2(–5) cm (rarely shorter); pedicels *c.* 0·5 mm. Calyx 5–8 mm, villous, the teeth as long as tube; corolla 11–13(–16) mm, blue. Legume (excluding beak) 10–15 × 2·5–3 mm, erect or patent, lanceolate, compressed, more or less straight, villous, with oblique anastomosing veins; beak 3·5–5 mm. Seeds *c.* 2·5 mm, ovate, finely tuberculate. *Aegean region; Krym.* Gr Rs (K).

21. T. gladiata Steven ex Bieb., *Fl. Taur.-Cauc.* **2**: 222 (1808). Stems 5–25 cm, densely pubescent. Leaflets 5–12 × 3–6 mm, obovate or oblanceolate, emarginate, denticulate, usually sparsely pubescent. Flowers usually solitary; pedicels *c.* 0·5 mm. Calyx *c.* 6 mm, the teeth ⅓–½ as long as tube, pubescent with subpatent hairs; corolla 8–10 mm, pale yellow, sometimes tinged with purple. Legume (excluding beak) 15–40 × 3–7 mm, erect or patent, linear-oblong, compressed, more or less straight, pubescent, with longitudinal anastomosing veins; beak 10–20 mm. Seeds *c.* 3·5 mm, ovoid, tuberculate. *S. Europe.* Bu Cr Ga Gr Hs Hu It Ju Rm Rs (K) Sa Si Tu.

22. T. cariensis Boiss., *Diagn. Pl. Or. Nov.* **1**(2): 21 (1843). Like **21** but stems up to 35 cm; leaflets 6–10 mm wide, broadly obovate-cuneate, glabrous above, sparsely pubescent beneath; calyx *c.* 11 mm; corolla *c.* 18 mm; legume curved; beak 30–40 mm; seeds finely tuberculate. *S.E. Greece.* Gr. (*S. & W. Anatolia, Cyprus.*)

23. T. foenum-graecum L., *Sp. Pl.* 777 (1753). Stems 10–50 cm, sparsely pubescent. Leaflets 20–50 × 10–15 mm, obovate to oblong-oblanceolate, denticulate. Flowers solitary or paired, subsessile. Calyx 6–8 mm, the teeth about as long as tube; Corolla 12–18 mm, yellowish-white tinged with violet at the base. Legume (excluding beak) 60–110 × 4–6 mm, erect or patent, linear, somewhat curved, glabrous or glabrescent, with longitudinal veins; beak (10–)20–30 mm. Seeds *c.* 5 × 3 mm, quadrangular, somewhat compressed, yellow or pale brown, finely tuberculate. $2n=16$. *Cultivated for fodder, mainly in C. & S. Europe, and widely naturalized.* [Al Au Be Bu Cr Cz Ga Ge Gr He Hs Hu It Ju Lu Rm Rs Si Tu.] (?*S.W. Asia.*)

56. Medicago L.[1]

Annual or perennial herbs or small shrubs. Leaves 3-foliolate, stipulate. Flowers in axillary pedunculate racemes. Calyx campanulate, with 5 nearly equal teeth; corolla caducous; stamens diadelphous; filaments filiform. Legume longer than the calyx, nearly always indehiscent, usually spirally coiled, sometimes falcate, reniform or almost straight, often spiny. Seeds 1-several.

All the annual species grow in more or less open habitats, such as sea-shores, roadsides, cultivated fields or grassy places.

Literature: R. Nègre, *Trav. Inst. Sci. Chérif.* ser. bot., **5**: 1–119 (1956); *Bull. Soc. Hist. Nat. Afr. Nord* **50**: 267–314 (1959). S. J. van Ooststroom & Th. Reichgelt, *Acta Bot. Neerl.* **7**: 90–123 (1958). I. Urban, *Verh. Bot. Ver. Brandenb.* **15**: 1–85 (1873). C. C. Heyn, *Scripta Acad. Hierosol.* **12** (1963).

1 Margin of legume with one longitudinal vein, with which the transverse veins join, sometimes running into spines, without a strong submarginal vein close to the marginal one
 2 Legume ±falcate or reniform
  3 Legume up to 3 mm, reniform; seed solitary
   4 Racemes 10- to 50-flowered; transverse veins of legume slightly anastomosing, forming an elongated network **1. lupulina**
   4 Racemes 3- to 10-flowered; transverse veins of legume freely anastomosing, forming a ±isodiametric network **2. secundiflora**
  3 Legume more than 3 mm, ±falcate; seeds usually several
   5 Pedicels erect in fruit **5. sativa**
   5 Pedicels deflexed in fruit
    6 Stipules dentate; legume with transverse veins anastomosing freely **3. hybrida**
    6 Stipules entire; legume with transverse veins scarcely anastomosing **4. cretacea**
 2 Legume spirally coiled
  7 Perennials with a stout, woody stock
   8 Shrub 100–400 cm **11. arborea**
   8 Herbs less than 100 cm
    9 Pedicels erect in fruit
     10 Legume with long, slender spines **13. carstiensis**
     10 Legume not spiny
      11 Leaflets obovate to almost linear, long-cuneate at base; corolla (5–)6–11 mm **5. sativa**
      11 Leaflets broadly obovate, shortly cuneate at base; corolla 3–6 mm **10. suffruticosa**
    9 Pedicels recurved or deflexed in fruit
     12 Transverse veins of legume thick, anastomosing to form a conspicuous intermediate vein parallel to the marginal
      13 Leaflets narrowly linear; legume convex at apex **8. saxatilis**
      13 Leaflets obovate-cuneate to oblong-cuneate; legume flat at apex **9. cancellata**
     12 Transverse veins of legume thin, sometimes indistinct, not forming an intermediate vein parallel to the marginal
      14 Pedicels 2–2½ times as long as calyx in fruit; transverse veins of legume anastomosing freely, strongly curved **6. prostrata**
      14 Pedicels about as long as calyx in fruit; transverse veins of legume scarcely anastomosing, nearly straight **7. rupestris**
 7 Annuals
  15 Margin of legume with distinct (usually long) spines
   16 Legume pubescent and glandular **15. ciliaris**
   16 Legume glabrous (except sometimes for the spines) and eglandular **14. intertexta**

  15 Margin of legume without spines, occasionally with small rounded projections
   17 Legume strongly pelviform **16. scutellata**
   17 Legume flat or convex at the ends
    18 Transverse veins of legume anastomosing abundantly **19. soleirolii**
    18 Transverse veins of legume scarcely anastomosing
     19 Transverse veins of legume not becoming thickened towards the thin margin **12. orbicularis**
     19 Transverse veins of legume becoming thickened towards the thick marginal vein **18. rugosa**
1 Legume with a strong submarginal vein, or with a wide veinless border
 20 White-tomentose perennial; legume with a distinct hole through the centre **21. marina**
 20 Annuals or nearly glabrous perennials; legume without a hole through the centre
  21 Submarginal vein of legume connected to the marginal vein by cross-veins which form a regular subquadrate network **17. blancheana**
  21 Legume without cross-veins connecting the submarginal and marginal veins
   22 Legume in a lax spiral, the young legume projecting from the calyx as soon as the petals have fallen
    23 Legume without apparent transverse veins or with a veinless border ⅛–⅓ as wide as the radius of the spiral
     24 Legume discoid with no apparent transverse veins; upper turn of spiral without spines, much smaller than the other spiny ones **35. disciformis**
     24 Legume shortly cylindrical, with a wide veinless border; upper turn of spiral spiny, about the same size as the others **36. tenoreana**
    23 Legume with prominent transverse veins and no wide veinless border
     25 Margin of legume wide, flat, with 2 rows of spines, one upward- and one downward-pointing, parallel to the legume **34. coronata**
     25 Margin of legume keeled or rounded; spines usually patent
      26 Leaflets nearly always with a dark spot; marginal vein of legume sulcate and margin therefore with 3 conspicuous grooves **30. arabica**
      26 Leaflets never with a dark spot; marginal vein of legume not sulcate
       27 Groove between the submarginal and marginal veins of the legume very wide, visible when the legume is viewed from the edge
        28 Legume discoid to shortly cylindrical, usually glabrous or nearly so; transverse veins curved but not sigmoid, anastomosing freely **31. polymorpha**
        28 Legume subglobose, sparsely villous and often glandular; transverse veins sigmoid, not anastomosing **37. minima**
       27 Groove between the submarginal and marginal veins narrow, not visible when the legume is viewed from the edge
        29 Leaflets usually incise-dentate or almost pinnatifid; transverse veins of legume sigmoid, sparingly branched, not anastomosing **33. laciniata**
        29 Leaflets dentate near the apex; transverse veins of legume strongly curved but not sigmoid, anastomosing freely near the submarginal vein **32. praecox**
   22 Legume in a very close spiral, the young legume concealed within the calyx when the petals fall
    30 Legume with a veinless border ¼–⅓ as wide as the radius of the spiral
     31 Legume usually in a left-handed spiral; margin with 1 keel; spines usually short, obtuse **28. turbinata**
     31 Legume always in a right-handed spiral; margin with 3 keels; spines usually long, acute **29. murex**
    30 Legume without a wide veinless border
     32 Legume densely glandular-pubescent

[1] By T. G. Tutin.

33 Nearly glabrous perennial; calyx-teeth glabrous
24. pironae
33 Densely pubescent annuals; calyx-teeth pubescent
34 Legume discoid to cylindrical; transverse veins scarcely anastomosing **23. rigidula**
34 Legume globose to ellipsoid; transverse veins anastomosing freely near the submarginal vein **26. aculeata**
32 Legume glabrous or with sparse hairs
35 Marginal and submarginal veins of legume confluent at maturity and forming a single acute or convex keel
36 Transverse veins of legume nearly straight; spines not sulcate at base **25. littoralis**
36 Transverse veins of legume strongly curved; spines sulcate at base **27. globosa**
35 Marginal and submarginal veins of legume separated by a distinct groove at maturity and so forming 3 keels
37 Racemes usually with more than 3 flowers; legume glabrous, usually without spines; spines, if present, usually short, conical, straight or uncinate
20. tornata
37 Racemes usually with 1–3 flowers; legume nearly always sparsely villous, always spiny; spines usually long, curved **22. truncatula**

Subgen. **Medicago**. Legume without a submarginal vein parallel with and close to the marginal vein; margin of legume usually thin.

1. M. lupulina L., *Sp. Pl.* 779 (1753). More or less pubescent annual or short-lived perennial 5–60 cm. Leaflets orbicular, obovate, or rhombic- or oblong-cuneate, rounded to emarginate, usually apiculate; stipules lanceolate to ovate, serrate or entire. Racemes 10- to 50-flowered. Corolla 2–3 mm. Legume 1·5–3 mm, reniform, black when ripe; transverse veins strongly curved, slightly anastomosing and forming an elongated network. Seed 1, black. $2n=16$. *Throughout Europe, except the extreme north; sometimes cultivated for forage.* All except Sb; probably introduced in Fa Is.

2. M. secundiflora Durieu in Duchartre, *Rev. Bot.* 1: 365 (1845). Like **1** but whitish-pubescent annual 2–20 cm; stipules lanceolate, entire or with 2 teeth at base; racemes 3- to 10-flowered, secund; legume 3·5–4 mm; transverse veins freely anastomosing forming a more or less isodiametric network. *S. France (Aude), N.E. Spain, Islas Baleares; doubtfully native in Italy.* Bl Ga Hs *It.

3. M. hybrida (Pourret) Trautv., *Bull. Sci. Acad. Imp. Sci. Pétersb.* 8: 271 (1841) (*M. pourretii* Noulet). Glabrous perennial 10–40 cm. Leaflets broadly obovate to suborbicular, crenulate-serrate; stipules ½-ovate, sagittate, dentate, acute. Racemes 2- to 5-flowered. Legume oblong-falcate, acuminate, dehiscent; transverse veins forming a strong, more or less isodiametric network. Seeds 2–3. *Woods and mountain slopes; calcicole.* ● *Corbières and Pyrenees.* Ga.

4. M. cretacea Bieb., *Fl. Taur.-Cauc.* 2: 223 (1808) (*Trigonella cretacea* (Bieb.) Grossh.). Like **3** but stipules entire; legume 1-seeded, indehiscent, with prominent transverse veins which are almost parallel and scarcely anastomose. *Krym.* Rs (K). (*W. Transcaucasus.*)

5. M. sativa L., *Sp. Pl.* 778 (1753). More or less pubescent perennial up to 80 cm. Leaflets obovate to almost linear, long-cuneate, dentate at apex; stipules lanceolate to linear-subulate, entire or dentate at base. Racemes 5- to 40-flowered. Pedicels short, stout, erect in fruit. Corolla (5–)6–11 mm. Legume nearly

straight, falcate or in a spiral of 1–3 turns with a hole through the centre, glabrous, pubescent or glandular, not spiny; transverse veins anastomosing and forming a transversely or radially elongated network. *Throughout most of Europe, except the north and some islands; often naturalized.* All except Az Fa Is Sb. Introduced in Fe.

Very variable, particularly in S. and S.E. Europe, where a number of distinct species have been recognized. In view of the overall variation and the known frequency of hybridization between *M. sativa* and *M. falcata*, two of the most distinct taxa, it seems best to treat them all as subspecies. Hybrids between subspecies with bluish and yellow corollas often have green or almost black corollas. The best-known of these is subsp. *falcata* × subsp. *sativa* (*M.* × *varia* Martyn). What appear to be hybrids between subsp. *glomerata* and subspp. (a)–(c) have been called *M. polychroa* Grossh.

1 Corolla blue to purple
2 Corolla 7–11 mm **(a) Subsp. sativa**
2 Corolla 5–6(–7) mm
3 Legume falcate **(b) Subsp. ambigua**
3 Legume in a spiral of 2–3 turns **(c) Subsp. caerulea**
1 Corolla yellow
4 Legume nearly straight or falcate **(d) Subsp. falcata**
4 Legume in a spiral of 1½–3 turns **(e) Subsp. glomerata**

(a) Subsp. **sativa**: Corolla 7–11 mm, blue to violet; legume 4–6 mm in diameter, in a spiral of 1½–3½ turns. $2n=32$. *Cultivated as forage and naturalized throughout the range of the species. Origin uncertain.*

(b) Subsp. **ambigua** (Trautv.) Tutin, *Feddes Repert.* 79: 53 (1968) (*M. falcata* var. *ambigua* Trautv., *M. trautvetteri* Sumnev.): Corolla 5–6(–7) mm, violet-blue; legume falcate. *S.E. Russia (Orenburg).* Rs (E).

(c) Subsp. **caerulea** (Less. ex Ledeb.) Schmalh., *Fl. Sred. Juž. Ross.* 1: 226 (1895) (*M. caerulea* Less. ex Ledeb.): Corolla 5–6 mm, violet, rarely whitish; legume 2–3(–5) mm in diameter, in a spiral of 2–3 turns. $2n=16$. *S.E. Russia, W. Kazakhstan and Krym.* Rs (K, E).

(d) Subsp. **falcata** (L.) Arcangeli, *Comp. Fl. Ital.* 160 (1882) (*M. borealis* Grossh., *M. falcata* L., *M. romanica* Prodan): Corolla 5–8 mm, yellow; legume almost straight to falcate. $2n=16, 32$. *Throughout the range of the species, except for a few islands.*

(e) Subsp. **glomerata** (Balbis) Tutin, *Feddes Repert.* 79: 53 (1968) (*M. glomerata* Balbis, *M. glutinosa* Bieb., *M. polychroa* Grossh. pro parte): Corolla 6–10 mm, yellow; legume 3–5 mm in diameter, in a spiral of 1½–3 turns. *S. Europe.*

6. M. prostrata Jacq., *Hort. Vindob.* 1: 39 (1770). Slender, nearly or quite glabrous perennial 10–40 cm. Leaflets oblong- or linear-cuneate, dentate at apex; stipules ovate-lanceolate, the lower deeply dentate. Racemes usually 1- to 6-flowered. Corolla *c.* 4 mm. Pedicels 2–2½ times as long as calyx, slender, recurved in fruit. Legume 3–4 mm in diameter, in a rather lax spiral of 2–3 turns, pendent, glabrous or glandular, not spiny; transverse veins slender, anastomosing freely, but not forming an intermediate vein parallel to the marginal vein. $2n=32$. *Dry grassland.* ● *From E. Austria and Italy to the Black Sea.* Al Au Cz Hu It Ju Rm.

7. M. rupestris Bieb., *Fl. Taur.-Cauc.* 2: 225 (1808). Appressed-pubescent perennial *c.* 20 cm, with a woody stock. Leaflets narrowly linear, dentate at apex; stipules subulate, almost entire. Racemes 2- to 8-flowered. Corolla 5–7 mm. Pedicels about as long as calyx, slender, recurved in fruit. Legume 3–4 mm in diameter, in a spiral of about 1 turn, pubescent, not spiny;

transverse veins scarcely anastomosing, nearly straight. *Rocky places. Krym.* Rs (K).

8. M. saxatilis Bieb., *Fl. Taur.-Cauc.* **2**: 225 (1808) (*M. rhodopea* Velen.). Like **7** but legume in a spiral of 3–4 turns, convex at apex, transverse veins thick, conspicuous, anastomosing to form a prominent intermediate vein parallel to the margin; margin with a few, small straight spines. *Rocky places. Bulgaria and Krym.* Bu Rs (K).

9. M. cancellata Bieb., *Fl. Taur.-Cauc.* **2**: 226 (1808). Like **7** but leaflets obovate-cuneate; legume in a spiral of 2–3 turns, flat at apex, pendent, not spiny; transverse veins forming a regular network and a prominent intermediate vein parallel to the margin. *Steppe. S.E. Russia.* Rs (E).

10. M. suffruticosa Ramond ex DC. in Lam. & DC., *Fl. Fr.* ed. 3, **4**: 541 (1805). Perennial 5–35 cm. Leaflets obovate or obcordate, cuneate at base, denticulate; stipules semi-ovate-lanceolate, dentate. Racemes 3- to 8-flowered. Corolla 3–6 mm. Legume 4–6 mm in diameter, in a spiral of 2–4 turns, not spiny, with a small hole in the middle. *Rocky places. S. France and E. Spain.* Ga Hs.

(a) Subsp. **suffruticosa**: More or less densely pubescent; stipules acuminate; peduncle equalling or slightly longer than the subtending leaf; corolla *c.* 6 mm; legume with a thin margin. ● *Corbières, Pyrenees.*

(b) Subsp. **leiocarpa** (Bentham) P. Fourn., *Quatre Fl. Fr.* 544 (1936): Glabrous or nearly so; stipules not acuminate; peduncle longer than the subtending leaf; corolla *c.* 3 mm; legume with a thickened margin. ● *Languedoc, E. Pyrenees to Valencia.*

11. M. arborea L., *Sp. Pl.* 778 (1753). Sericeous shrub 100–400 cm. Leaflets obovate, cuneate at base, entire or denticulate at apex; stipules lanceolate, entire. Racemes very short, almost capitate, 4- to 8-flowered. Corolla 12–15 mm. Legume 12–15 mm in diameter, in a spiral of 1–1½ turns, not spiny, reticulately veined, with a hole through the centre. *Rocky places. S. Mediterranean region.* Al Bl Cr Gr Hs It Sa Si [Ga Lu].

12. M. orbicularis (L.) Bartal., *Cat. Piante Siena* 60 (1776). Glabrous or sparsely hairy, procumbent annual 20–90 cm. Leaflets obovate-cuneate, dentate at the apex or in the upper ⅔; stipules laciniate. Racemes 1- to 5-flowered. Corolla 2–5 mm. Legume 10–17(–20) mm in diameter, in a spiral of 4–6 turns, lenticular, convex on both faces, glabrous or somewhat glandular-pubescent, not spiny; transverse veins with a few, usually weak anastomosing branches. 2*n*=16. *S. Europe.* Al Bl Bu Co Cr Ga Gr Hs It Ju Lu Rm Rs (K) Sa Si [Hu].

13. M. carstiensis Jacq., *Collect. Bot.* **1**: 86 (1787). Nearly or quite glabrous perennial up to 60 cm. Leaflets obovate or elliptical, denticulate; stipules ovate-lanceolate or lanceolate, laciniate or dentate. Racemes (1–)5- to 12(–20)-flowered. Corolla 6–8 mm. Legume 6–8 mm in diameter, in a spiral of 5–8 turns, shortly cylindrical, flat at both ends, glabrous; transverse veins few, strong, scarcely anastomosing; margin thickened, with nearly straight, rather slender spines, half as long to as long as the diameter of the legume. *Bushy places.* ● *From N.E. Italy and S.E. Austria to Bulgaria.* Al Au Bu It Ju.

14. M. intertexta (L.) Miller, *Gard. Dict.* ed. 8, no. 4 (1768). Nearly glabrous procumbent or ascending annual up to 50 cm. Leaflets obovate, cuneate, denticulate, sometimes with a dark spot; stipules ovate to ovate-lanceolate, incise-dentate. Racemes 1- to 7(–10)-flowered. Corolla 6–9 mm. Legume 12–15 mm in

diameter, in a spiral of (3–)6–10 turns, ovoid, cylindrical or rarely discoid, convex at both ends, glabrous except often for the spines; transverse veins strong, anastomosing and forming a tangentially elongated network; spines usually 3–4 mm, curved and appressed to the legume. *Mediterranean region and Portugal.* Gr Hs It Lu Sa Si [Ga].

M. granadensis Willd., *Enum. Pl. Hort. Berol.* 803 (1809), from S. Spain (near Málaga), appears to be a variant of **14** with discoid fruits.

15. M. ciliaris (L.) All., *Fl. Pedem.* **1**: 315 (1785). Like **14** but legume villous, with glandular hairs; spines usually shorter, less appressed, straight except for the more or less curved apices. *Mediterranean region extending to Portugal.* Bl Co Ga Gr Hs It Lu Sa Si.

16. M. scutellata (L.) Miller, *Gard. Dict.* ed. 8, no. 2 (1768). More or less densely glandular-pubescent annual 20–60 cm. Leaflets obovate to elliptical, cuneate, dentate in the upper part; stipules ovate-lanceolate to lanceolate, incise-dentate. Racemes 1- to 3-flowered. Corolla 6–7 mm. Legume 9–18 mm in diameter, in a spiral of 4–8 pelviform, imbricate turns, glandular-pubescent, not spiny; transverse veins numerous, conspicuous, freely anastomosing and joining the strong marginal vein. *S. Europe.* Bl Co Cr Ga Gr Hs It Ju Lu Rs (W, K) Sa Si [Au].

Subgen. **Cymatium** (Pospichal) Gams. Legume with a submarginal vein parallel with and close to the marginal vein; margin of legume usually thick.

17. M. blancheana Boiss., *Diagn. Pl. Or. Nov.* 3(5): 75 (1856). Pubescent annual up to 60 cm. Leaflets obovate or elliptical, cuneate, dentate to deeply incise-dentate at base. Racemes 1- to 3-flowered. Corolla 7–9 mm. Legume 8–12 mm in diameter, in a spiral of 4–6 turns, lenticular to globose, glabrous to more or less pubescent, not spiny; transverse veins strong, with abundant, weaker anastomosing branches forming a strong subquadrate network joining the submarginal and marginal veins. *Naturalized in Italy and C. & S. Portugal, a casual elsewhere in Europe.* [It Lu.] (*Asia Minor.*)

18. M. rugosa Desr. in Lam., *Encycl. Méth. Bot.* **3**: 632 (1792) (*M. elegans* Jacq. ex Willd.). Glandular-pubescent annual 10–50 cm. Leaflets obovate, cuneate, dentate in the upper part; stipules lanceolate, incise-dentate. Racemes 1- to 5-flowered. Corolla 2–4 mm. Legume 6–10 mm in diameter, in a spiral of 2–3 turns, discoid, glabrescent, not spiny; transverse veins scarcely anastomosing, becoming strongly thickened towards the strong marginal vein. *C. & E. Mediterranean region, local.* Cr Gr It Sa Si [?Co Ga ?Hs Lu].

19. M. soleirolii Duby, *Bot. Gall.* **1**: 124 (1828). Like **18** but stipules laciniate; corolla 8–9 mm; legume 5–7 mm in diameter; transverse veins with abundant anastomosing branches, not becoming thicker towards the marginal vein. *Krym; probably introduced in Italy, S. France and Corse.* *Co *Ga *It Rs (K).

20. M. tornata (L.) Miller, *Gard. Dict.* ed. 8, no. 3 (1768). More or less patent-pubescent annual up to 60 cm. Leaflets obovate, cuneate, dentate near the apex; stipules lanceolate, dentate to laciniate near the base. Racemes 1- to 10-flowered. Corolla 5–7 mm. Legume 5–8 mm in diameter, in a spiral of 1¼–8 turns, lenticular to cylindrical, flat at both ends, glabrous, spiny or not; transverse veins scarcely anastomosing except near the marginal vein, where they form a submarginal vein from which the spines, if present, arise; spines conical, usually short, but up to $\frac{1}{2}$ as long as the diameter of the legume, patent, straight or uncinate. *W. Mediterranean region extending to Portugal.* Co Hs It Lu Sa Si.

21. M. marina L., *Sp. Pl.* 779 (1753). Procumbent, white-tomentose, densely leafy perennial 20–50 cm. Leaflets obovate, cuneate at base, denticulate at apex; stipules ovate, acuminate, entire or toothed. Racemes almost capitate, 5- to 12-flowered. Corolla 6–8 mm. Legume 5–7 mm, in a spiral of 2–3 turns with a small hole through the middle, cylindrical, densely white-tomentose; submarginal and marginal veins thick, with two rows of short, conical spines. *Maritime sands. Shores of the Mediterranean, Black Sea and Atlantic to c. 48° N.* Al Bl Bu Co Cr Ga Gr Hs It Ju Lu Rm Rs (K) Sa Si ?Tu.

22. M. truncatula Gaertner, *Fruct. Sem. Pl.* 2: 350 (1791) (*M. tribuloides* Desr.). Sparsely villous annual up to 50 cm. Leaflets obovate or obcordate, cuneate, denticulate near the apex; stipules ovate-lanceolate to narrowly lanceolate, incise-dentate to laciniate in the lower half. Racemes 1- to 3(–5)-flowered. Corolla 5–6 mm. Legume 5–8 mm in diameter, in a spiral of 3–6 turns, cylindrical, nearly always sparsely villous, spiny; transverse veins scarcely anastomosing, joining a very thick submarginal vein separated from the slender but distinct marginal vein by the groove; spines up to more than half as long as the diameter of the legume, curved or uncinate, each arising partly from the marginal and partly from the submarginal vein. *Mediterranean region extending to Portugal and W. France.* Bl Co Cr Ga Gr Hs It Ju Lu Sa Si ?Tu.

23. M. rigidula (L.) All., *Fl. Pedem.* 1: 316 (1785) (*M. gerardii* Waldst. & Kit. ex Willd., *M. agrestis* Ten. ex DC.). Like **22** but corolla 6–7 mm; legume in a spiral of 4–7 turns, discoid to cylindrical, nearly always densely glandular-pubescent, rarely glabrescent, nearly always spiny; transverse veins strongly curved, scarcely anastomosing; submarginal veins at first separated from the marginal vein by a shallow groove, becoming confluent with it and forming a convex margin when fully ripe; spines usually about half as long as the diameter of the legume, somewhat curved, uncinate. $2n = 14, 16$. *S. Europe.* Al Bl Bu Co Cr Ga Gr Hs It Ju Lu Rm Rs (W, E) Sa Si ?Tu [Cz Hu].

24. M. pironae Vis., *Linnaea* 28: 365 (1857). Nearly glabrous perennial 20–40 cm. Leaflets obovate to obcordate, cuneate, denticulate near apex; stipules ovate, entire or incise-dentate. Racemes 2- to 6-flowered. Corolla 7–8 mm. Legume 5–7 mm in diameter, in a spiral of 3–4 turns, shortly cylindrical, glandular-pubescent, spiny; transverse veins anastomosing and forming an irregular network; marginal and submarginal veins forming a convex margin; spines short, flattened and sulcate at base. $2n = 16$. *Rocky places.* ● *N.E. Italy.* It.

25. M. littoralis Rohde ex Loisel., *Not. Pl. Fr.* 118 (1810). Sparsely villous annual up to 40(–110) cm. Leaflets obovate or obcordate, cuneate, dentate towards the apex; stipules lanceolate, incise-dentate. Racemes 1- to 6-flowered. Corolla 5–6 mm. Legume 4–6 mm in diameter, in a spiral of 3–6 turns, discoid to cylindrical, glabrous, spiny or not; transverse veins nearly straight, scarcely anastomosing except near the submarginal vein; submarginal vein at first separated from the marginal vein by a shallow groove, becoming confluent with it and forming a keeled margin when fully ripe; spines varying from short and conical to half as long as the diameter of the legume, arising from the submarginal vein. *Mediterranean region extending to Portugal and W. France.* Al Bl Bu Co Cr Ga Gr Hs It Ju Lu Sa Si.

26. M. aculeata Gaertner, *Fruct. Sem. Pl.* 2: 349 (1791) (*M. turbinata* Willd., non (L.) All.). Like **25** but stipules ovate-lanceolate; legume 7–10 mm in diameter, in a spiral of 5–7 turns, globose to ellipsoid, with dense, short, sometimes glandular hairs, sometimes glabrescent, spiny or not; transverse veins somewhat curved; spines usually conical, straight or curved. *Mediterranean region, extending to Portugal and Bulgaria.* Al ?Bl Bu Co Cr Ga Gr Hs It Ju Lu Sa Si.

27. M. globosa C. Presl in J. & C. Presl, *Del. Prag.* 45 (1822). Like **25** but transverse veins strongly curved, joining the submarginal vein at a very acute angle; spines sulcate at base. *Aegean region and perhaps Sicilia.* Bu Cr Gr ?Si.

28. M. turbinata (L.) All., *Fl. Pedem.* 1: 315 (1785) (*M. tuberculata* (Retz.) Willd.). Sparsely villous annual up to 50 cm. Leafle obovate, cuneate, dentate near the apex; stipules lanceolate to ovate-lanceolate, dentate to incise-dentate. Racemes 1- to 8-flowered. Corolla 5–6 mm. Legume 5–7 mm in diameter, in a spiral of 5–6 turns, cylindrical or somewhat conical, glabrous, spiny; transverse veins slender, curved, not or very rarely anastomosing, ending at a veinless border $\frac{1}{4}-\frac{1}{3}$ as wide as the radius of the spiral; marginal vein forming a keel; spines usually short, broad, obtuse, rarely acute and up to $\frac{1}{2}$ as long as the diameter of the legume. *Mediterranean region, extending to Portugal and Bulgaria.* Bl Bu Co Cr Ga Gr Hs It Ju Lu Sa Si.

29. M. murex Willd., *Sp. Pl.* 3: 1410 (1802) (*M. sphaerocarpos* Bertol.). Like **28** but leaflets obcordate or obtriangular; corolla 4–5 mm; legume 5–9 mm in diameter, in a spiral of 5–9 turns, sometimes not spiny; submarginal and marginal veins distinct, forming 3 keels; spines usually longer and less conical. *Mediterranean region extending to Portugal.* Bl Co Cr Ga Gr Hs It Ju Lu Sa Si ?Tu.

30. M. arabica (L.) Hudson, *Fl. Angl.* 288 (1762) (*M. maculata* Sibth.). Sparsely pubescent to glabrous annual up to 50 cm. Leaflets usually obcordate, cuneate, dentate near the apex, usually with a dark spot. Racemes 1- to 4(–6)-flowered. Corolla 5–7 mm. Legume (4–)5–6 mm in diameter, in a lax spiral of 4–7 turns, subglobose to shortly ellipsoid, somewhat flattened at both ends, glabrous, usually spiny; transverse veins curved, anastomosing and forming a tangentially elongated network near the margin; margin with 3 grooves, the lateral deeper and wider than the central; spines usually $\frac{1}{2}-\frac{3}{4}$ as long as the diameter of the legume, deeply sulcate. *S. Europe extending north-westwards to Britain and the Netherlands and eastwards to Krym; a frequent casual further north.* Al Be Bl Br Bu Co Ga Ge Gr Ho Hs Hu It Ju Lu Rm Rs (K) Sa Si ?Tu [Cz Hb He Su].

31. M. polymorpha L., *Sp. Pl.* 779 (1753) (*M. denticulata* Willd., *M. hispida* Gaertner, *M. lappacea* Desr., *M. nigra* (L.) Krocker *M. polycarpa* Willd.). Glabrous or pubescent annual up to 40 cm. Leaflets obovate to obcordate, cuneate, dentate near the apex; stipules lanceolate to ovate-lanceolate, laciniate. Racemes 1- to 5(–8)-flowered. Corolla 3–4·5 mm. Legume 4–8(–10) mm in diameter, in a lax spiral of $1\frac{1}{2}$–6 turns, usually glabrous and spiny; transverse veins strong, anastomosing freely, at least near the submarginal vein; submarginal vein conspicuous, separated from the marginal vein by a deep groove, the margin consequently with 3 keels separated by 2 grooves; spines absent or up to more than the diameter of the legume in length, each arising from both the submarginal and marginal veins and therefore deeply sulcate. $2n = 14$. *S. Europe, extending northwards to Britain and eastwards to Krym; a frequent casual further north.* Al Au Az Bl Br Bu Co Cr Ga Ge Gr Hs It Ju Lu Rm Rs (K) Sa Si ?Tu [Be Cz He Ho Hu].

32. M. praecox DC., *Cat. Pl. Horti Monsp.* 123 (1813). Somewhat pubescent annual up to 30 cm. Leaflets obcordate, cuneate, dentate near the apex; stipules lanceolate to ovate-lanceolate, incise-dentate to pectinate. Racemes 1- to 2-flowered. Corolla 2–3 mm. Legume 3–4 mm in diameter, in a lax spiral of 4–5 turns, subglobose to cylindrical, sparsely puberulent, spiny, transverse vein strongly curved, anastomosing near the submarginal vein and forming a tangentially elongated network; submarginal vein separated from the marginal vein by a narrow groove not visible when the legume is viewed from the edge; marginal vein broad and rounded; spines up to as long as the diameter of the legume, usually somewhat curved and uncinate, deeply sulcate. *Mediterranean region extending to Krym.* Bl Bu Co Ga Gr Hs It Rs (K) Sa.

33. M. laciniata (L.) Miller, *Gard. Dict.* ed. 8, no. 5 (1768). Somewhat puberulent annual up to 40 cm. Leaflets obcordate, cuneate, often incise-dentate or almost pinnatifid; stipules ovate to ovate-lanceolate in outline, dentate to pectinate. Racemes 1- to 2-flowered. Corolla *c.* 5 mm. Legume 2·5–5 mm in diameter, in a lax spiral of 3–7 turns, globose to ellipsoid, glabrous or sometimes pubescent, spiny; transverse veins sigmoid, sparingly branched; submarginal vein broad, separated from the flat or convex marginal vein by a narrow groove; spines up to as long as the diameter of the legume, nearly straight but uncinate, deeply sulcate at base. *Naturalized locally in the Mediterranean region, extending to the Ukraine.* [Co Ga Hs It Rs (W).] (*N. Africa, Asia Minor.*)

(a) Subsp. **laciniata**: Stipules divided more than halfway to the middle; legume 4–5 mm in diameter, with 10–16 transverse veins per turn of the spiral. *Throughout the range of the species.*

(b) Subsp. **schimperana** P. Fourn., *Quatre Fl. Fr.* 545 (1936): Stipules divided less than halfway to the middle; legume 2·5–4 mm in diameter, with 7–10(–12) transverse veins per turn of the spiral. *France, perhaps elsewhere.*

34. M. coronata (L.) Bartal., *Cat. Piante Siena* 61 (1776) Pubescent annual 10–30 cm. Leaflets obovate, cuneate, dentate near the apex. Racemes 3- to 8-flowered. Corolla 2·5–3 mm. Legume 2–4 mm in diameter, in a lax spiral of 2 turns, shortly cylindrical, usually pubescent, spiny; transverse veins strongly curved, scarcely anastomosing; submarginal vein separated from the marginal by a wide groove; marginal vein expanded into a flat border from which arise 2 rows of spines, one upward- and one downward-pointing, parallel to the legume; spines short, straight, deeply sulcate. $2n = 16$. *Mediterranean region extending to Bulgaria.* Al Bu Co Cr Ga Gr Hs It Ju [Lu].

35. M. disciformis DC., *Cat. Pl. Horti Monsp.* 124 (1813). Softly pubescent annual 10–30 cm. Leaflets obovate, cuneate, dentate near the apex; stipules lanceolate, dentate near the base. Racemes 1- to 4-flowered. Corolla 4–5 mm. Legume *c.* 6 mm in diameter, in a lax spiral of 5 turns, discoid, glabrous, spiny; transverse veins apparently absent; upper turn of the spiral much smaller than the others and without spines; spines slender, somewhat curved, about half as long as the diameter of the legume. *Mediterranean region extending to Bulgaria.* Bu Cr Ga Gr Hs It.

36. M. tenoreana Ser. in DC., *Prodr.* 2: 180 (1825). Like **35** but legume *c.* 5 mm in diameter, in a lax spiral of 4–5 turns, shortly cylindrical, flat at both ends; transverse veins anastomosing to form a network, with a wide veinless border; turns of the spiral all subequal and spiny; spines in 2 rows, one upward- and

[1] By D. E. Coombe.

one downward-pointing parallel to the legume, nearly straight. *W. Mediterranean region, local.* Ga Hs It Sa Si.

37. M. minima (L.) Bartal., *Cat. Piante Siena* 61 (1776). Villous annual up to 40 cm. Leaflets obovate or obcordate, dentate near the apex; stipules lanceolate to ovate-lanceolate, sometimes shallowly dentate near the base. Racemes 1- to 6(–8)-flowered. Corolla 4–4·5 mm. Legume 3–5 mm in diameter, in a lax spiral of 3–5 turns, subglobose, sparsely villous and somewhat glandular, nearly always spiny; transverse veins slender, curved, not anastomosing; submarginal vein wide, separated from the narrow, flat or convex marginal vein by a wide groove; spines very short to longer than the diameter of the legume, patent, usually uncinate, deeply sulcate at base. $2n = 16$. *Most of Europe, except the north and some islands.* Al Au Be Bl Br Bu Co Cr Cz Da Ga Ge Gr He Ho Hs Hu It Ju Lu Po Rm Rs (W, K, E) Sa Si Su ?Tu.

57. Trifolium L.[1]

Annual, biennial or perennial herbs, rarely somewhat woody. Leaves 3-foliolate, very rarely digitate with 5(–8) leaflets; leaflets usually toothed. Flowers in heads or short spikes, very rarely solitary. Calyx-teeth equal or unequal; petals persistent or deciduous, adnate to each other and to the staminal tube; stamens diadelphous; all or 5 of the filaments dilated at the apex. Legume included in the calyx or shortly exserted, rarely much exceeding the calyx, indehiscent or dehiscent by a ventral suture or by an indurated lid. Seeds 1–4(–10).

Literature: E. G. Bobrov, *Fl. Syst. Pl. Vasc.* 6: 164–344 (1947). C. Vicioso, *Anal. Inst. Bot. Cavanilles* 10: 347–398 (1952), 11: 289–383 (1953). R. Hendrych, *Preslia* 28: 403–412 (1956). M. Hossain, *Notes Roy. Bot. Gard. Edinb.* 23: 387–481 (1961).

A large genus extensively cultivated for fodder. The most important species are 10, 14, 29, 30, 62, 63, 88 and 97, but others may be used locally.

1 Some leaves with 5 or more leaflets   **2. lupinaster**
1 All adult leaves 3-foliolate
 2 Calyx with 5(–6) veins; leaves often pinnately 3-foliolate; corolla always persistent and scarious in fruit
  3 Calyx-teeth subequal or the 2 upper longer than the rest; corolla before anthesis white or cream; flowers ±umbellate, subtended by membranous bracts; legume 2- to 4-seeded
   4 Calyx-teeth subequal, subulate, separated by broad obtuse sinuses   **14. hybridum**
   4 Calyx-teeth unequal, the 2 upper longer than the rest, narrowly lanceolate, separated by narrow acute sinuses   **10. repens**
  3 Calyx-teeth unequal, the 2 upper shorter than the rest; corolla before anthesis yellow, orange, lilac or violet; flowers ±spicate; bracts represented by a few short, red, evanescent glandular hairs; legume 1(–2)-seeded
   5 Corolla violet or reddish-violet before anthesis
    6 Fruiting heads ovoid, lax, with peduncles exceeding the leaves; corolla 8–10 mm   **34. speciosum**
    6 Fruiting heads subglobose, dense, with peduncles equalling or shorter than the leaves; corolla 4–6 mm   **36. lagrangei**
   5 Corolla yellow, rarely orange or lilac before anthesis
    7 Uppermost leaves subopposite; heads solitary, paired or few, pseudoterminal
     8 Perennial; corolla 7–9 mm, bright chestnut-brown after anthesis   **32. badium**
     8 Annual or biennial; corolla *c.* 6 mm, very dark brown after anthesis   **33. spadiceum**
    7 All leaves alternate; heads usually numerous, lateral, axillary

9 All leaflets of the upper leaves subsessile, their petiolules subequal
10 Fruiting pedicels 2–4 times as long as the upper limb of the calyx-tube **45. sebastiani**
10 Fruiting pedicels not more than 1½ times as long as the upper limb of the calyx-tube, usually much shorter
11 Fruiting pedicels 1–1½ times as long as the upper limb of the calyx-tube; corolla 2–3(–4) mm **47. micranthum**
11 Fruiting pedicels distinctly shorter than the upper limb of the calyx-tube; corolla 5–8 mm
12 Stipules of the upper leaves lanceolate-ovate, not dilated at the base **42. aureum**
12 Stipules of the upper leaves semicordate-ovate, often auriculate
13 Leaflets oblong-ovate or rhombic, widest near the middle **43. velenovskyi**
13 Leaflets obovate-cuneate or narrowly elliptic-obovate, widest near the apex
14 Stems 20–50 cm; leaflets up to 18 mm, narrowly elliptic-obovate; corolla 5–7 mm **41. patens**
14 Stems 5–20 cm; leaflets 8–10 mm, obovate-cuneate; corolla 7–8 mm **37. brutium**
9 Terminal leaflet of the upper leaves distinctly petiolulate, its petiolule longer than that of the lateral leaflets
15 Corolla 3–3·5 mm, scarcely sulcate **46. dubium**
15 Corolla 4–10 mm (if less, then markedly sulcate)
16 Corolla (3–)4–5(–6) mm; legume 3–6 times as long as the style **44. campestre**
16 Corolla (5–)6–10 mm; legume scarcely exceeding the style
17 Peduncles shorter than or equalling the leaves; pedicels 1–2 mm, longer than the upper limb of the calyx-tube **35. boissieri**
17 Peduncles exceeding the leaves; at least the lower pedicels not longer than the upper limb of the calyx-tube
18 Lower pedicels about as long as the upper limb of the calyx-tube, upper pedicels somewhat longer **40. dolopium**
18 All pedicels much shorter than the upper limb of the calyx-tube
19 Corolla 5–7 mm; stems 20–50 cm **41. patens**
19 Corolla 7–9 mm; stems 5–20 cm
20 Corolla 8–9 mm; orange before anthesis **39. aurantiacum**
20 Corolla 7–8 mm, yellow before anthesis
21 Stipules oblong-lanceolate, not semicordate-auriculate **38. mesogitanum**
21 Stipules semicordate-ovate, often auriculate **37. brutium**
2 Calyx with more than 5(–6) veins, usually 10, 20 or more; leaves digitately 3-foliolate; petals deciduous or marcescent, sometimes scarious
22 Flowers subtended by small, sometimes connate, bracts; throat of calyx not closed by a ring of hairs or by an annular or bilabiate callosity; legume usually 2- to 8-seeded, included or exserted
23 Calyx-tube not inflated in fruit
24 Heads 1- to 6-flowered
25 Corolla 6–8 mm; legume greatly exserted **1. ornithopodioides**
25 Corolla 12–25 mm; legume not or slightly exserted
26 Peduncles less than 15 mm, usually concealed by the stipules; bracts free; calyx-tube cylindrical **24. uniflorum**
26 Peduncles 15–150 mm, evident; bracts connate, forming 1 or 2 minute involucres; calyx-tube campanulate
27 Peduncles 50–150 mm; corolla 18–25 mm **3. alpinum**
27 Peduncles 15–25 mm; corolla 12–14 mm **4. pilczii**
24 Heads 7- to many-flowered
28 All heads sessile

29 Internodes 10–80 mm; heads ±remote; corolla longer than the calyx **22. glomeratum**
29 Internodes usually less than 5 mm; heads congested or confluent; corolla shorter than the calyx **23. suffocatum**
28 At least some heads with peduncles 5 mm or more
30 Stipules denticulate or fimbriate
31 Veins of the leaflets and stipules ending in glandular teeth; standard slightly exceeding the calyx **5. strictum**
31 Veins of the leaflets and stipules not ending in glandular teeth; standard about twice as long as the calyx **6. nervulosum**
30 Stipules not denticulate or fimbriate
32 Stems creeping and rooting at the nodes
33 Upper calyx-teeth narrowly lanceolate; leaflets with translucent lateral veins and usually light or dark markings **10. repens**
33 Upper calyx-teeth ovate-lanceolate or triangular; leaflets with opaque lateral veins and usually unmarked **11. occidentale**
32 Stems not creeping or rooting at the nodes
34 Corolla 18–25 mm; bracts connate, forming 1 or 2 minute involucres **3. alpinum**
34 Corolla 5–16 mm; bracts not connate
35 Fruiting pedicels as long as or longer than the calyx-tube
36 Perennial
37 Calyx-teeth subulate, separated by broad obtuse sinuses **14. hybridum**
37 Calyx-teeth lanceolate or narrowly triangular, with narrow acute sinuses
38 Pedicels 4–5×0·5 mm, strongly thickened in fruit; standard straight **15. bivonae**
38 Pedicels 1–4×0·1–0·2 mm, scarcely thickened in fruit; standard ±recurved
39 Flowers not or scarcely deflexed after anthesis; calyx-tube about as wide as long **13. thalii**
39 Flowers strongly deflexed after anthesis; calyx-tube longer than wide
40 Standard ovate-lanceolate **10. repens**
40 Standard broadly ovate or elliptical **12. pallescens**
36 Annual
41 All calyx-teeth 2–4 times as long as the tube
42 Heads 20–25 mm wide; corolla 8–11 mm; seeds 2 mm **18. michelianum**
42 Heads 10–15 mm wide; corolla 6–8 mm; seeds 1 mm **19. angulatum**
41 Upper 2 calyx-teeth equalling or only slightly exceeding the tube
43 Heads 10–20 mm wide; corolla 6–9 mm **17. nigrescens**
43 Heads 8–12 mm wide; corolla 4–5 mm **21. cernuum**
35 Fruiting pedicels shorter than the calyx-tube
44 Annual
45 Heads 8–11 mm wide, globose; corolla 4–5 mm **20. retusum**
45 Heads 15–25 mm wide, hemispherical or cylindrical; corolla 9–12 mm **16. isthmocarpum**
44 Perennial
46 Calyx sparsely pubescent, at least at the base of the teeth
47 Heads globose or ovate; corolla 7–9 mm **7. montanum**
47 Heads oblong; corolla 10–15 mm **8. ambiguum**
46 Calyx glabrous
48 Peduncles 10–30 mm; corolla 5–8 mm; calyx-teeth ovate-triangular (Greece) **9. parnassi**
48 Peduncles (10–)50–150 mm; corolla 7–12 mm; calyx-teeth lanceolate to subulate
49 Calyx-tube as wide as long, somewhat inflated in fruit; corolla 6–10 mm **13. thalii**

49 Calyx-tube longer than wide, cylindrical; corolla 10–12 mm **15. bivonae**
23 Calyx-tube slightly to conspicuously inflated or gibbous in fruit
50 Perennial
51 Calyx-tube in fruit glabrous, ± regular, only the 10 longitudinal veins conspicuous **13. thalii**
51 Calyx-tube in fruit pubescent and strongly gibbous above, with numerous longitudinal and transverse veins forming a reticulum
52 Bracts 0·5–1 mm, free **28. physodes**
52 Bracts 3–4 mm, ± united below into an irregular involucre **29. fragiferum**
50 Annual
53 Fruiting calyx pubescent, tomentose or lanate (sometimes finally glabrescent), adaxially gibbous; bracts inconspicuous and ± concealed; heads lateral
54 Fruiting heads ± pedunculate; calyx pyriform, pubescent to tomentose, finally glabrescent, its 2 upper teeth evident, divergent **30. resupinatum**
54 Fruiting heads subsessile; calyx ± globose, lanate, its 2 upper teeth ± concealed **31. tomentosum**
53 Fruiting calyx glabrous, inflated ± equally on all sides; bracts prominent, glumaceous, striate; heads pseudo-terminal
55 Leaflets broadly obovate-cuneate; corolla slightly exceeding the calyx; bracts shorter than the fruiting calyx-tube **25. spumosum**
55 Leaflets (at least the upper) oblong, elliptical or lanceolate, rarely obovate-elliptical or suborbicular-cuneate; corolla much exceeding the calyx; bracts about as long as the fruiting calyx-tube
56 Calyx-tube usually much inflated in fruit with 24–35 prominent longitudinal and ± prominent transverse veins **26. vesiculosum**
56 Calyx-tube scarcely inflated in fruit without or with c. 24 weak longitudinal veins and without transverse veins **27. mutabile**
22 Flowers ebracteate but heads sometimes involucrate; throat of the calyx usually ± closed by a ring of hairs or an annular or bilabiate callosity at maturity; legume 1(–2)-seeded, almost always included in the calyx-tube
57 Fertile flowers 2–12; inner flowers consisting only of sterile calyces developing either at or after anthesis
58 Sterile flowers developing after anthesis from a central nodule; fruiting heads appressed to the ground or subterranean **97. subterraneum**
58 Sterile flowers developing simultaneously with the fertile ones; fruiting heads aerial
59 Fertile flowers 10–15, in 2 rows; mature heads 20–25 mm in diameter **98. globosum**
59 Fertile flowers 4–6, in 1 row; mature heads 8–15 mm in diameter **99. pauciflorum**
57 Fertile flowers usually numerous; sterile flowers absent
60 All or at least some calyces evidently 20-veined, or the 20 veins completely obscured by dense sericeous hairs
61 Perennial
62 Apex of stipules pubescent, subulate or narrowly linear, scarious **79. alpestre**
62 Apex of stipules glabrous or glabrescent, lanceolate or narrowly triangular, herbaceous
63 Stipules entire, adnate by less than ½ their length to the petiole; leaflets obscurely denticulate **74. medium**
63 Stipules often serrate above, adnate by more than ½ their length to the petiole; leaflets spinose-denticulate **80. rubens**
61 Annual
64 Stems less than 4 cm; heads crowded
65 Leaflets deeply emarginate; corolla not exceeding the calyx **70. congestum**
65 Leaflets obtuse; corolla about twice as long as the calyx **71. barbeyi**
64 Stems 5–40 cm; heads solitary

66 Calyx-tube glabrous or glabrescent; fruiting heads shortly pedunculate, not involucrate **69. lappaceum**
66 Calyx-tube hairy; heads sessile, with an involucre formed by the upper stipules
67 Free part of the stipules (except of the uppermost leaves) long, linear-lanceolate, straight; corolla exceeding the calyx **72. hirtum**
67 Free part of the stipules (except of the uppermost leaves) short, ovate-lanceolate, often recurved; corolla not exceeding the calyx **73. cherleri**
60 Calyx 10-veined, or some calyces with up to 14 veins
68 Perennial
69 Stems 1–5 cm; leaflets 2–8 mm **68. ottonis**
69 Stems 5–100 cm; leaflets 10–60 mm
70 Lowest calyx-tooth about 2–3 times as long as the other 4, linear
71 Corolla 20–25 mm; peduncles 40–80 mm **86. pannonicum**
71 Corolla 15–20 mm; peduncles not more than 25 mm **85. ochroleucon**
70 Calyx-teeth subequal or the lowest not more than 1½ times as long as the other four, setaceous, filiform or linear
72 Upper internodes with patent hairs
73 Calyx-tube glabrous
74 Corolla reddish-purple; upper stipules lanceolate **74. medium**
74 Corolla cream; upper stipules ovate-lanceolate **78. pignantii**
73 Calyx-tube hairy
75 Stipules of the middle cauline leaves abruptly contracted into a setaceous arista; heads sessile, involucrate **63. pratense**
75 Stipules of the middle cauline leaves with a triangular-lanceolate, acuminate apex; heads often shortly pedunculate **66. noricum**
72 Upper internodes with appressed hairs, or glabrous
76 Stems 5–8 cm; stipules with an oblong, obtuse apex **67. wettsteinii**
76 Stems usually more than 10 cm; stipules acute
77 Stipules of the middle cauline leaves abruptly contracted into a setaceous arista **63. pratense**
77 Stipules of the middle cauline leaves with a linear, lanceolate or ovate, ± herbaceous apex
78 Petioles united along their length to the lower part of the stipules
79 Leaflets linear-oblong; calyx-tube densely pubescent **76. patulum**
79 Leaflets obovate-cuneate; calyx-tube glabrous **77. velebiticum**
78 Petioles united along only part of their length to the stipules
80 Calyx-tube glabrous or glabrescent **74. medium**
80 Calyx-tube persistently hairy
81 Calyx-teeth filiform, all of them longer than the tube; corolla pink **75. heldreichianum**
81 Calyx-teeth lanceolate-subulate, ± herbaceous, the upper 4 teeth equalling or shorter than the calyx-tube; corolla yellowish-white, rarely pink
82 Leaflets not emarginate; lowest calyx-tooth usually distinctly longer than the tube **85. ochroleucon**
82 Leaflets deeply emarginate; lowest calyx-tooth scarcely longer than the tube **87. canescens**
68 Annual
83 Heads sessile, axillary or terminal, involucrate
84 Heads few-flowered, scarcely exceeding the subtending stipules; lowest calyx-tooth shorter than the tube **51. saxatile**
84 Heads many-flowered, much exceeding the subtending

stipules; lowest calyx-tooth as long as or longer than the tube

85   All leaves alternate

86   Lateral veins of the leaflets ± straight

87   Corolla not or scarcely exceeding the lowest calyx-tooth

88   Calyx-teeth subequal, all longer than the tube, divergent in fruit   **56. gemellum**

88   Calyx-teeth unequal, only the lowest equalling or slightly exceeding the tube, connivent or somewhat divergent in fruit

89   Fruiting calyx readily abscissing, with ± inflated tube and erecto-patent teeth   **48. striatum**

89   Fruiting calyx not readily abscissing, tube not inflated, teeth straight or connivent   **52. bocconei**

87   Corolla exceeding the lowest calyx-tooth

90   Corolla twice as long as the calyx; upper leaflets linear   **53. tenuifolium**

90   Corolla 1½ times as long as the calyx; upper leaflets narrowly oblong to obcordate-cuneate, deeply emarginate   **54. trichopterum**

86   Lateral veins of the leaflets recurved, often ± thickened towards the margins

91   Corolla 4–5 mm, equalling or slightly exceeding the calyx; axillary heads numerous   **58. scabrum**

91   Corolla 8–10 mm, twice as long as the calyx; axillary heads few   **59. dalmaticum**

85   At least the two uppermost leaves opposite   **96. clypeatum**

92   Calyx-teeth triangular-lanceolate (-acuminate) or subulate from an expanded triangular base, spreading or recurved in fruit

93   Corolla much shorter than the calyx; calyx-teeth subequal, up to 4 times as long as the tube   **61. dasyurum**

93   Corolla nearly equalling or exceeding the calyx; calyx-teeth unequal or subequal, not more than twice as long as the tube and usually less

94   Throat of the calyx closed with a ring of hairs

95   Calyx-teeth triangular-lanceolate, dilated at the base, as long as or shorter than the tube   **55. phleoides**

95   Calyx-teeth lanceolate-setaceous, scarcely dilated at the base, longer than the tube   **56. gemellum**

94   Throat of the calyx closed by a bilabiate callosity, leaving only a narrow vertical slit

96   Calyx-tube campanulate, glabrous or sparsely hairy above; leaflets 10–20 mm   **93. squamosum**

96   Calyx-tube ovoid, densely hairy; leaflets 20–40(–70) mm   **94. squarrosum**

92   Calyx-teeth subulate, setaceous or filiform, ± straight and erect in fruit

97   Corolla whitish or pink, much exceeding the calyx; calyx-teeth 1–1½ times as long as the tube   **64. pallidum**

97   Corolla reddish-purple, not or scarcely exceeding the calyx; calyx-teeth twice as long as the tube   **65. diffusum**

83   Heads pedunculate, terminal or axillary

98   Upper leaves alternate

99   Leaflets of upper leaves lanceolate, linear or linear-oblong

100   Heads 10–25 mm, usually numerous; calyx-throat not closed by a bilabiate callosity

101   Calyx-teeth triangular-lanceolate, dilated at the base, sparsely ciliate or glabrescent   **55. phleoides**

101   Calyx-teeth setaceous, usually densely pubescent or villous

102   Corolla much shorter than the calyx   **49. arvense**

102   Corolla equalling or longer than the calyx   **50. affine**

100   Heads 20–110 mm, one or few, usually terminal;

calyx-throat at maturity narrowed to a vertical slit by a bilabiate callosity

103   Corolla 10–12 mm, not or scarcely exceeding the calyx   **81. angustifolium**

103   Corolla 13–25 mm, much exceeding the calyx

104   Corolla 16–25 mm; stem robust, little branched   **82. purpureum**

104   Corolla 13–15 mm; stems weak and diffusely branched   **83. desvauxii**

99   Leaflets of upper leaves obovate-cuneate or obcordate

105   Heads capitate, ± globose in fruit; stipules denticulate   **60. stellatum**

105   Heads spicate, oblong, cylindrical or conical in fruit; stipules entire or obscurely dentate

106   Corolla much shorter than the calyx; calyx-tube obconical or campanulate; stipules lanceolate, entire   **57. ligusticum**

106   Corolla equalling or exceeding the calyx; calyx-tube ovoid or globose; stipules ovate, at least at the apex, entire or obscurely dentate or angled

107   Calyx-teeth unequal, the 4 upper ones shorter than the tube, all with long patent hairs   **84. smyrnaeum**

107   Calyx-teeth subequal, all as long as or longer than the tube, with erecto-patent hairs   **62. incarnatum**

98   At least the two uppermost leaves opposite

108   Corolla 20–25 mm; calyx-teeth broadly ovate-triangular, with many veins.

108   Corolla less than 20 mm; calyx-teeth linear-subulate to ovate-lanceolate, with 1–3 veins

109   Corolla much shorter than the calyx; calyx-teeth subequal, up to 4 times as long as the tube   **61. dasyurum**

109   Corolla equalling or exceeding the calyx; calyx-teeth unequal or subequal, the lowest one not more than twice as long as the tube

110   Leaflets up to 60 × 4·5 mm, linear-oblong, acute   **91. latinum**

110   Leaflets 8–40(–70) × 4–15 mm, relatively shorter and broader, obtuse

111   Legume exserted slightly from the mouth of the calyx-tube

112   Calyx-teeth spinescent in fruit, the lowest one 3-veined, at least at the base   **88. alexandrinum**

112   Calyx-teeth scarcely spinescent in fruit, the lowest one 1-veined   **89. apertum**

111   Legume not exserted, concealed by the closed bilabiate callosity at the mouth of the calyx-tube

113   Calyx-teeth subequal

114   Calyx-teeth ovate-lanceolate, each with 3–5 veins; heads 20–35 mm, ovate, shortly pedunculate   **95. obscurum**

114   Calyx-teeth lanceolate, each with 3 veins; heads 10–15 mm, globose; peduncles 30–120 mm   **92. leucanthum**

113   Calyx-teeth unequal

115   Calyx-teeth 1-veined, or 3-veined only at the base   **90. echinatum**

115   Calyx-teeth distinctly 3-veined to the middle or above

116   Calyx-tube campanulate, glabrous or sparsely hairy above; leaflets 10–20 mm   **93. squamosum**

116   Calyx-tube ovoid, densely hairy; leaflets 20–40(–70) mm   **94. squarrosum**

Subgen. **Falcatula** (Brot.) D. E. Coombe. Flowers bracteate. Calyx-throat open, without a ring of hairs or a callosity. Legume oblong, slightly curved, exceeding the calyx, dehiscent. Seeds 5–9.

1. T. ornithopodioides L., *Sp. Pl.* 766 (1753) (*Trigonella ornithopodioides* (L.) DC.) Stems 5–10(–20) cm, procumbent, glabrous. Leaflets 4–10(–14) mm, obovate or obcordate, cuneate, truncate, mucronate and serrate, shortly petiolulate. Petioles 20–40(–50) mm, longer than the leaves, stipules and heads. Stipules 7–10 mm, lanceolate, acuminate. Heads (1–)2- to 4(–5)-flowered. Peduncles up to 8 mm. Corolla 6–8 mm, white or pink. Calyx-teeth subequal, longer than the tube. Standard narrowly oblong. Legume 6–8 mm, exserted. $2n = 16$. *Open habitats, moist or wet in winter. W. Europe northwards to Ireland and the Netherlands, and extending eastwards to Italy; S.E. part of C. Europe.* ?Az Bl Br Co Ga Ge Hb Ho Hs Hu It ?Ju Lu Rm Sa.

Plants growing in winter in shallow water have floating leaves with petioles up to 10 cm.

Subgen. **Lotoidea** Pers. Flowers subtended by free or united bracts (or short glandular hairs). Calyx-throat open, without a ring of hairs or a callosity. Legume included in the calyx or exserted. Seeds (1–)2–4(–10).

Sect. LUPINASTER (Fabr.) Ser. Perennial. Flowers often in 2 superposed whorls each subtended by a minute involucre of connate bracts. Calyx ± regular. Corolla persistent and becoming scarious. Legume shortly stalked.

2. T. lupinaster L., *Sp. Pl.* 766 (1753). Stems 15–50 cm, erect or ascending, glabrous or glabrescent. Upper leaves with 5(–8) lanceolate or linear leaflets, their petioles up to 10 mm, shorter than and largely united to the stipules. Heads lax, 10- to 20-flowered; peduncles 10–30 mm. Corolla 15–20 mm, red or white. Legume 1- to 9-seeded. *E. Europe.* Cz Po Rm Rs (N, C, W, E).

The variability in habit, leaflet-shape, corolla-colour and chromosome number require further study. Plants from European Russia regarded as identical with *T. lupinaster* (originally described from Siberia) are said to be rhizomatous, with dark-green, lanceolate leaflets 30–50 × 5–20 mm, red (or white) corolla and light green seeds; $2n = 32$ ($2n = 40$ in some Asiatic plants from S. Sayan). **T. ciswolgense** Sprygin ex Iljin & Truchaleva, *Dokl. Akad. Nauk SSSR* **132**: 219 (1960) (*T. lupinaster* var. *albiflorum* Ser.), is said to be non-rhizomatous, taller, more erect, leaves light green, corolla white, seeds violet; it has $2n = 16$ and occurs mainly in the S. and W. parts of the range of the species. Plants with linear leaflets, 8–10 times as long as broad, corolla red or white and $2n = 32$, occurring in the W. and S.W. parts of the range of the species have been described as **T. litwinowii** Iljin in Iljin & Truchaleva, *loc. cit.* (1960) (*T. lupinaster* subsp. *angustifolium* (Litv.) Bobrov).

3. T. alpinum L., *Sp. Pl.* 767 (1753). Glabrous, densely caespitose perennial with massive tap-root; stems very short, hidden by dead leaf-bases. Leaves 3-foliolate; leaflets 10–40(–70) mm, lanceolate or linear. Petioles 20–50(–120) mm. Stipules up to 40(–90) mm, concealing the stems, largely adnate to the petioles. Heads 3- to 12-flowered. Peduncles 50–150 mm. Corolla 18–25 mm, pink, purple or rarely cream, strongly scented. Legume 1- to 2-seeded. $2n = 16$. *Meadows and pastures, mainly between 1700 and 2500 m; calcifuge.* ● *Alps, N. & C. Appennini, mountains of S. France and N. Spain.* Au Ga He Hs It.

Some plants from west of the Rhône have short, obtuse, elliptical leaflets with strongly curved veins; they need further study.

4. T. pilczii Adamović, *Denkschr. Akad. Wiss. Math.-Nat. Kl. (Wien)* **74**: 130 (1904). Like **3** but smaller, less robust; leaflets 8–12 mm, oblong to obovate; petioles up to 20 mm; stipules up to 7 mm; heads 2- to 8-flowered; peduncles 15–25 mm;

corolla 12–14 mm, purplish. ● *S. Jugoslavia and Albania.* Al Ju.

Sect. PARAMESUS (C. Presl) Godron. Annual. Inflorescence of 1–4 closely superposed whorls, each subtended by a minute involucre of ± connate bracts. Legume (1–)2-seeded, with a swollen and indurated wall, indehiscent, exceeding the calyx-tube.

5. T. strictum L., *Cent. Pl.* **1**: 24 (1755) (*T. laevigatum* Poiret). Erect or ascending, 3–15(–25) cm. Leaflets 8–20 mm, linear to oblong-elliptical, the upper often lanceolate; veins ending in stalked glands. Stipules conspicuous, ovate or rhombic, glandular-denticulate. Heads 7–10 mm, axillary or pseudoterminal; peduncles 1–2 times as long as the leaves. Involucral bracts exceeding the pedicels. Corolla 5–6 mm, pink, the standard slightly exceeding the calyx. Legume nearly orbicular, dorsally gibbous, 2-seeded. *Grassland; calcifuge. W. & S. Europe, northwards to Britain.* Br Bu Co Cz Ga Gr Hs Hu It Lu Rm Sa Si Tu [Ge].

6. T. nervulosum Boiss. & Heldr. in Boiss., *Diagn. Pl. Or. Nov.* **2**(9): 25 (1849). Procumbent or ascending, 4–15 cm. Leaflets narrowly oblanceolate; marginal glands absent or obscure. Stipules ovate, laciniate-dentate; glands absent or obscure. Heads 10–15 mm, solitary, pseudoterminal; peduncles more than twice as long as the leaves. Involucral bracts equalling the pedicels, or obscure. Corolla 6–7 mm, pink; standard 2 or more times as long as the calyx. Legume oblong, dorsally gibbous. *N. Albania; N.E. Greece.* Al Gr.

Sect. LOTOIDEA. Annual or perennial. Flowers umbellate, rarely spicate, numerous, pedicellate, subtended by lanceolate membranous bracts. Calyx-teeth unequal. Legume (1–)2- to 4(–5)-seeded.

7. T. montanum L., *Sp. Pl.* 770 (1753). Perennial; stock woody, with several erect, almost unbranched, more or less lanate stems 15–60 cm. Leaflets of basal leaves 30–70 mm, ovate, lanceolate or elliptical. Leaflets of cauline leaves 15–40(–60) mm, elliptical or oblong. Leaves glabrous above, sericeous or glabrescent beneath. Heads 15–30 mm, often in pairs, dense and many-flowered, globose or ovoid; peduncles 10–70 mm, densely hairy, ± erect. Calyx more or less hairy; teeth subulate. Corolla 7–9 mm, white or yellowish, rarely pink, yellowish-brown after anthesis; standard recurved. Legume usually 1-seeded. $2n = 16$. *Dry grassy places. Europe, but absent from much of the west, the Mediterranean region and the north.* Au Be Bu Cz Fe Ga Ge He Hs Hu It Ju No Po Rm Rs (N, B, C, W, K, E) Su.

Var. *gayanum* Godron, from S.W. France and Spain, may merit subspecific rank. It has the leaflets of the basal leaves broadly elliptical or sub-orbicular; heads solitary; peduncles 100–150 mm, sparsely hairy; pedicels 2 mm, as long as the calyx-tube and exceeding the bracts, deflexed in fruit; and the calyx-tube glabrescent, with oblique mouth.

T. balbisianum Ser. in DC., *Prodr.* **2**: 201 (1825), from mountains of S. France and N.W. Italy, may also merit subspecific rank. It has the corolla *c.* 10 mm, the calyx-teeth subequal, longer than the tube, and $2n = 16$.

8. T. ambiguum Bieb., *Fl. Taur.-Cauc.* **2**: 208 (1808). Robust perennial; stems 8–60 cm, procumbent or ascending, sparsely hairy or glabrescent. Leaflets 10–50(–70) mm, ovate-lanceolate or ovate-elliptical, glabrous or glabrescent. Heads 25–40 mm, hemispherical becoming oblong-ovate, usually solitary; peduncles 50–100(–200) mm. Bracts much longer than the pedicels. Calyx-tube glabrous or slightly hairy above. Corolla 10–15 mm, white, becoming reddish. Legume 1- to 2-seeded. *S. part of U.S.S.R.* ?Rm Rs (W, K, E).

9. T. parnassi Boiss. & Spruner in Boiss., *Diagn. Pl. Or. Nov.*
1(2): 30 (1843). Suffruticose perennial, glabrous, caespitose,
stems 1–3 cm. Leaflets 3–6 mm, obovate or obcordate. Stipules
ovate, acuminate. Heads up to 12-flowered, not paired; peduncles
10–30 mm, equalling or longer than the leaves. Bracts longer
than the pedicels. Calyx glabrous; teeth ovate-triangular,
imbricate, ½ as long as the tube. Corolla 5–8 mm, pink.
● *Mountains of Greece.* Gr.

10. T. repens L., *Sp. Pl.* 767 (1753). Glabrous or glabrescent
perennial, usually with extensively creeping stems rooting at the
nodes. Leaflets usually bright green with either light or dark
marks along the veins, or both; lateral veins translucent in the
living plant. Stipules large, membranous, sheathing, contracted
into a subulate apex. Heads usually globose. Flowers scented;
calyx-teeth narrowly lanceolate, the 2 upper longer than the
rest, separated by narrow acute sinuses; corolla 8–13 mm,
becoming light brown and strongly deflexed after anthesis;
standard ovate-lanceolate. Legume linear, compressed, constricted between the 3–4 seeds. *Grassy places, mainly on well-
drained soils. Throughout Europe to c. 71° N.* All except Sb.

Extensively cultivated for fodder and many cultivars have been
selected and are grown for this purpose. These often persist or
become more or less naturalized.

There is, in addition, a considerable amount of variation in
wild plants. A comprehensive treatment of this variation is not
yet possible, but the following are among the more distinct subspecies that may be recognized.

1 Heads 25–30 mm wide; corolla yellow; standard 3–4 times as
  long as the calyx   **(d) subsp. ochranthum**
1 Heads not more than 25 mm wide; corolla white or pink;
  standard not more than 3 times as long as the calyx
 2 Peduncles 10–20 mm, scarcely exceeding the leaves
            (e) subsp. orphanideum
 2 Peduncles usually exceeding the leaves
  3 Calyx with only 6 distinct veins   **(c) subsp. orbelicum**
  3 Calyx 10-veined
   4 Petioles densely hairy   **(f) subsp. prostratum**
   4 Petioles glabrous
    5 Leaflets 10 mm or more; heads 15–25 mm wide
            (a) subsp. repens
    5 Leaflets less than 10 mm; heads less than 20 mm wide
            (b) subsp. nevadense

(a) Subsp. repens: Stems rooting at the nodes. Leaflets
10–25(–40) mm, obovate or elliptical; petioles 20–200 mm.
Heads usually 40- to 80-flowered; peduncles 50–300 mm. Upper
calyx-teeth longer than the rest; corolla 7–10 mm, white or pale
pink, rarely deep red; standard 2–2½ times as long as the calyx.
$2n = 32$. *Almost throughout the range of the species.*
Plants from the high mountains of Europe are usually more
dwarf with short creeping stems, smaller leaves, peduncles
10–30(–60) mm and heads with fewer flowers.
(b) Subsp. nevadense (Boiss.) D.E. Coombe, *Feddes Repert.*
79: 54 (1968): Stems rooting at the nodes. Leaflets minute, cuneate
obcordate; petioles relatively long; peduncles exceeding the
leaves. Upper calyx-teeth a little longer than the others; corolla
white; standard 3 times as long as the calyx. ● *S. Spain
(Sierra Nevada).*
The identity of similar white-flowered plants from the Alps is
uncertain.
(c) Subsp. orbelicum (Velen.) Pawł., *Zapiski Fl. Tatr* **4**:
9 (1949): Stems often short, but sometimes elongate and rooting
at the nodes. Leaflets 6–8 mm, obovate-cuneate. Heads dense;
peduncles up to 120 mm. Calyx-tube 6- to 8-veined with only
6 veins distinct; upper calyx-teeth 1½ times as long as the tube;

corolla 8–10 mm, cream; standard 2–2½ times as long as the
calyx.   ● *Carpathians and mountains of Balkan peninsula.*
  (d) Subsp. **ochranthum** E.I. Nyárády, *Bul. Grăd. Bot. Cluj* **20**:
45 (1940): Stems long, rooting. Leaflets up to 13 mm, broadly
obovate or nearly orbicular, weakly denticulate; petioles 30–60
mm. Heads 25–30 mm wide. Corolla 10–12 mm, light yellow
or greenish-yellow; standard 3–4 times as long as the calyx.
● *Bosna; Romania.*
  (e) Subsp. **orphanideum** (Boiss.) D.E. Coombe, *Feddes Repert.*
79: 54 (1968): Stems 1–5 cm, procumbent and rooting or subcaespitose and non-rooting. Leaflets 5–7 mm, broadly obcordate
or obovate; petioles 10–20 mm. Heads 8- to 12-flowered;
peduncles 10–20 mm. Corolla pale pink; standard 2–2½ times as
long as the calyx.   ● *Greece, Kriti.*
  (f) Subsp. **prostratum** Nyman, *Consp.* 178 (1878) (*T. biasolettii*
Steudel & Hochst.): Stems usually rooting at the nodes. Leaflets
5–10(–15) mm, broadly obcordate; petioles densely hairy. Heads
14–18 mm wide. Corolla *c.* 9 mm, pale pink.   ● *C. part of the
Mediterranean region from S. France and Corse to Albania.*

11. T. occidentale D.E. Coombe, *Watsonia* **5**: 70 (1961).
Like **10** but leaflets 6–10 mm, thicker, almost orbicular, obtuse or
emarginate, glaucous, without light or dark markings; lateral
veins not translucent in the living plant; petioles with sparse but
persistent hairs; stipules vinous-red; heads 20–24 mm wide,
20(–40)-flowered; flowers scentless; upper calyx-teeth ovate-
lanceolate or triangular, often with 1–2 teeth on the upper
margin; corolla 8–9 mm. $2n = 16$. *Sand-dunes and dry grassy
places near the sea.*   ● *Coasts of S.W. England and N.W.
France.* Br Ga ?Hs.

12. T. pallescens Schreber in Sturm, *Deutschl. Fl.* Abt. 1,
Band 4, Heft 15 (1804) (incl. *T. glareosum* (Ser.) Schleicher ex
Boiss., non Dumort., *T. arvernense* Lamotte). Glabrous, caespitose perennial; tap-root often massive; stems 5–10(–20) cm,
numerous, procumbent or ascending, with 1–3 non-rooting
nodes. Leaflets 6–20 mm, elliptical or obovate, bright-green.
Stipules ovate-lanceolate, membranous, with acute apex. Heads
15–25 mm wide, pseudoterminal, at first globose; peduncles
20–90 mm, stout; pedicels 1·5–4 mm, longer than the bracts and
the calyx-tube. Flowers sweetly scented, deflexed after anthesis.
Corolla 6–10 mm, yellowish-white to pink, becoming dark
brown; standard broadly ovate or elliptical, 2–3 times as long
as the calyx. $2n = 16$. *Damp screes and pastures above* 1800 *m;
calcifuge.*   ● *Mountains of C. & S. Europe.* Al Au Bu Ga Gr
He Hs It Ju Rm.

13. T. thalii Vill., *Prosp. Pl. Dauph.* 43 (1779). Like **12** but
stems shorter; leaflets obovate, dull green; peduncles (10–)50–
120 mm; pedicels 1–1·5 mm, shorter than or rarely equalling the
calyx-tube; flowers not or scarcely deflexed after anthesis;
calyx-tube in fruit about as wide as long, somewhat inflated,
strongly ribbed, with somewhat patent teeth; standard twice as
long as calyx. *Meadows and pastures; calcicole.*   ● *Alps,
Appennini, Pyrenees and mountains of N. Spain.* Au Ga Ge He
Hs It.

Often confused with **12** or with mountain plants of **10** but
especially distinctive in fruit.

14. T. hybridum L., *Sp. Pl.* 766 (1753) (*T. fistulosum* Gilib.).
Perennial, glabrous or glabrescent; stems (5–)20–40(–90) cm,
erect or ascending, and lax, or rarely densely caespitose and procumbent, not rooting at the nodes. Leaflets 10–20(–30) × 10–15
(–20) mm, obovate or obcordate; petioles up to 10 cm; stipules
partly herbaceous, ovate to ovate-lanceolate, gradually contracted
into a subulate apex. Heads globose, pseudoterminal and

axillary; peduncles longer than the leaves; pedicels slender, the upper ones 4–5 mm, up to twice as long as the calyx-tube, deflexed after anthesis. Calyx-tube 1–1·5 mm, with 5 veins distinct, the other 5 often obscure; calyx-teeth 2–3 mm, longer than the tube, the two upper slightly longer than the others, subulate, separated by broad obtuse sinuses; corolla (5–)7–10 mm, purple or white at first, pink later, becoming brown. Legume 2- to 4-seeded. *Meadows and pastures. Most of Europe; widely cultivated as a forage plant and native distribution uncertain.* *Au Bu Cr *Cz Ga Gr *He Hs *Hu It Ju *Rm Rs (*W, K, *E) Tu [Be Br Da Fe Ge Hb Ho No Po Rs (N, B, C) Su].

1 Stems 5–10(–15) cm, densely caespitose, procumbent or
   ascending; heads 12–15 mm wide; corolla purple
                      (c) subsp. anatolicum
1 Stems more than 15 cm; heads more than 15 mm wide; corolla
   white and pink
2 Stems sparingly branched, erect, fistulose; heads *c.* 25 mm wide
                      (a) subsp. hybridum
2 Stems much-branched, procumbent or ascending, scarcely
   fistulose; heads 16–19 mm wide   **(b) subsp. elegans**

(a) Subsp. **hybridum**: $2n=16$. *Native distribution uncertain; widely cultivated in C., N. & E. Europe and frequently naturalized.*

(b) Subsp. **elegans** (Savi) Ascherson & Graebner, *Syn. Mitteleur. Fl.* 6(2): 496 (1907): *Scattered in the southern part of the range of the species, rarer in the north.*

Probably the wild progenitor of subsp. (a).

(c) Subsp. **anatolicum** (Boiss.) Hossain, *Notes Roy. Bot. Gard. Edinb.* 23: 466 (1961): *Mountains of Bulgaria and Greece.* (*Anatolia.*)

15. **T. bivonae** Guss., *Fl. Sic. Prodr.* 2: 512 (1828). Perennial; stems 10–20 cm, erect or ascending, numerous, simple, from a stout stock. Leaflets up to 20 × 10 mm, broadly obovate, elliptical or ovate, obtuse, with numerous curved lateral veins, prominent near the margin. Peduncles 80–150 mm; pedicels 4–5 × 0·5 mm, thick, often longer than the calyx-tube. Flowers strongly deflexed and imbricate after anthesis. Calyx glabrous; teeth lanceolate, acuminate, straight, the two upper slightly longer than the others and about as long as the cylindrical tube; corolla 10–12 mm, pink; standard straight. Legume usually 1-seeded. *Mountain grassland.* ● *Sicilia.* Si.

Very similar before anthesis to **28**, but readily distinguished by the floral characters after anthesis.

16. **T. isthmocarpum** Brot., *Phyt. Lusit.* ed. 3, 1: 148 (1816). Glabrous annual; stems 5–60 cm, branching from the base, procumbent or ascending, often fistulose. Leaflets 10–25 mm, obovate, broadly elliptical or obtriangular; petioles up to 10 cm. Stipules membranous, abruptly contracted into a subulate apex. Heads 15–25 mm wide, hemispherical, or cylindrical, dense; peduncles 20–100 mm, longer than the leaves. Flowering pedicels not longer than the calyx-tube, weakly deflexed in fruit. Calyx-teeth subequal, triangular-acute to lanceolate-subulate, shorter than or as long as the calyx-tube, straight or recurved; corolla 9–12 mm, pink; standard 1½–2 times as long as the calyx. Legume oblong, constricted between the two seeds. *Moist grassland on sandy soil. W. Mediterranean region and Portugal.* Co Hs It Lu Si [Ga *Tu].

(a) Subsp. **isthmocarpum**: Heads hemispherical. Calyx-teeth narrowly triangular, acute, straight or recurved in fruit. Corolla pink; standard twice as long as the calyx. *Spain and Portugal.*

(b) Subsp. **jaminianum** (Boiss.) Murb., *Lunds Univ. Årsskr.* 33(12): 67 (1897): Heads cylindrical. Calyx-teeth linear-subulate,

patent or recurved in fruit. Corolla white or pale pink; standard 1½ times as long as the calyx. *Italy, Sicilia, Corse.*

Intermediates between subspp. (a) and (b) occur in S.W. Spain.

17. **T. nigrescens** Viv., *Fl. Ital. Fragm.* 12 (1808). Glabrous or glabrescent annual; stems 5–40 cm, often numerous, procumbent, erect or ascending. Leaflets 8–15(–25) mm, obovate or obcordate. Stipules triangular-lanceolate, acuminate. Heads 10–20 mm wide, globose, lax; peduncles longer than the leaves; pedicels equalling or longer than the calyx-tube, deflexed in fruit. Calyx-teeth lanceolate or linear, without wide sinuses between, the upper teeth usually slightly longer than the others and equalling or exceeding the tube. Corolla 6–9 mm, white, cream or pink, becoming brown. Legume 1- to 5-seeded. $2n=16$. *Grassland and waste places. S. Europe.* Al Az Bl Bu Co Cr Ga Gr Hs It Ju Lu Sa Si Tu.

(a) Subsp. **nigrescens**: Ovules (3–)4(–6); legume shallowly constricted between the seeds. *Almost throughout the range of the species.*

(b) Subsp. **petrisavii** (G. C. Clementi) Holmboe, *Stud. Veg. Cyprus* 106 (1914): Ovules 2; legume deeply constricted between the seeds. *Damp grassland. Balkan peninsula and Sicilia.*

Robust plants with 1-seeded legumes from damp places in the E. part of the Balkan peninsula have been named **T. meneghinianum** G. C. Clementi, *Sert. Or.* 31 (1855). **T. macropodum** Guss., *Fl. Sic. Syn.* 2: 388 (1844), from Sicilia, with short, densely caespitose stems, sulcate peduncles, subequal calyx-teeth and 2-seeded legumes, may be another subspecies. Both require further study.

18. **T. michelianum** Savi, *Fl. Pis.* 2: 159 (1798). Annual; stems erect, up to 65 cm, 2–6 mm thick, fistulose, striate, branching, often constricted at the nodes. Leaflets 10–30 mm, oblong or obovate, dentate; petioles up to 70 mm. Stipules ovate, acuminate. Heads 20–25 mm wide, globose, many-flowered, lax; peduncles equalling or exceeding the leaves; pedicels 3–6 mm, 5–10 times as long as the calyx-tube. Flowers deflexed after anthesis. Calyx *c.* 5 mm; teeth subequal, linear-subulate, 3–4 times as long as the tube; corolla 8–11 mm, pink. Legume obovate or orbicular, stipitate, 2-seeded, thinly pubescent; seed 2 mm. *In wet meadows and by standing water, S. Europe, extending northwards to N. France.* Bu Co ?Cr Ga Gr Hs It Ju Lu Rm Sa Si.

Less robust plants with solid stems, shorter pedicels, and calyx-teeth only twice as long as the tube are recorded from the Balkan peninsula as **T. balansae** Boiss., *Diagn. Pl. Or. Nov.* 3(5): 81 (1856); they may represent a distinct subspecies.

19. **T. angulatum** Waldst. & Kit., *Pl. Rar. Hung.* 1: 26 (1800). Like **18** but with heads 10–15 mm wide; peduncles as long as or shorter than the leaves; calyx 3–3·5 mm; teeth about twice as long as the tube; corolla 6–8 mm, reddish; legume oblong, glabrous, 3- to 5-seeded; seed 1 mm. *Damp saline places. E.C. Europe, extending southwards to Macedonia; S.E. Russia (near Temrjuk).* Cz Hu Ju Rm Rs (E) [Ga].

20. **T. retusum** L., *Demonstr. Pl.* 21 (1753) (*T. parviflorum* Ehrh.). Glabrous or glabrescent annual; stems 10–20(–40) cm, numerous, branching, procumbent or ascending. Leaflets 8–18 mm, the lower obovate-lanceolate, the upper oblong, mucronate, denticulate; veins curved, prominent; petioles 10–70 mm. Stipules triangular, acuminate, membranous. Heads 8–11 mm wide, globose, dense, the upper nearly sessile, the lower with peduncles up to 30 mm. Pedicels *c.* 1 mm or less, much shorter than the bracts and calyx-tube, not or slightly deflexed in fruit.

Calyx exceeding the petals. Calyx-teeth very unequal, finally recurved, the upper longer than the tube; corolla 4–5 mm, white or pink; standard ovate, not emarginate. Legume 2-seeded. *Dry, grassy places. C. & S.E. Europe, extending locally westwards to N.W. Spain and E. Portugal.* Au Bu Cz Ga *Ge Hs Hu Ju Lu Rm Rs (W, K, E) Tu.

21. **T. cernuum** Brot., *Phyt. Lusit.* ed. 3, **1**: 150 (1816). Like **20** but leaflets truncate or emarginate; pedicels 2 mm, about as long as the calyx-tube and longer than the bracts, strongly deflexed in fruit; calyx shorter than the corolla with teeth subequal, scarcely as long as the tube; petals pink; standard deeply emarginate. *Dry, grassy places. S.W. Europe.* Az Co Ga Hs Lu.

22. **T. glomeratum** L., *Sp. Pl.* 770 (1753). Glabrous annual; stems (2–)10–20(–35) cm, numerous, procumbent or ascending. Internodes 10–80 mm. Leaflets 5–10(–20) mm, obovate, mucronate; petioles 10–20(–70) mm; stipules ovate, acuminate. Heads 8–12 mm wide, globose, dense, sessile or subsessile, mostly remote. Flowers sessile. Calyx-tube glabrous with 10(–12) distinct veins, a little longer than the teeth; teeth subequal, triangular-ovate, auriculate, acuminate, deflexed; corolla 4–5 mm, pink, a little longer than the calyx. Legume (1–)2-seeded. *Dry places. S. & W. Europe, northwards to c. 52° 30′ in England.* Al Az Bl Br Bu Co Cr Ga Gr Hb Hs It Ju Lu Sa Si Tu.

23. **T. suffocatum** L., *Mantissa Alt.* 276 (1771). Glabrescent, caespitose annual; stems 1–3(–5) cm, procumbent. Internodes rarely reaching 5 mm. Leaflets 3–8 mm, obovate-cuneate, emarginate; petioles 10–60 mm; stipules ovate, acuminate. Heads 5–6 mm, sessile, numerous, usually confluent, rarely somewhat separated, then ovate. Calyx sparsely pubescent at first, glabrescent; teeth all as long as the tube, lanceolate or subulate, recurved; corolla 3–4 mm, white; standard a little shorter than the calyx. Legume 2-seeded. *Dry places. S. & W. Europe, northwards to c. 53° 30′ in England.* Al Az Bl Br Co Cr Ga Gr Hs It Ju Lu Rm Sa Si Tu.

Sect. CRYPTOSCIADIUM Čelak. Perennial. Heads axillary, 1- to 3(–5)-flowered; peduncles usually very short and covered by the imbricate stipules. Pedicels evident, curved in fruit. Calyx cylindrical, 10-veined. Legume 3- to 10-seeded.

24. **T. uniflorum** L., *Sp. Pl.* 771 (1753). Taproot woody; stems 1–3(–6) cm, caespitose, procumbent. Internodes very short. Leaflets 4–10 mm, orbicular, obovate or rhombic, acute or obtuse, apiculate, strongly veined with cusped teeth, often appressed-pubescent beneath. Petioles 10–30(–70) mm, glabrous or appressed-pubescent. Stipules broadly triangular, long-acuminate, membranous, imbricate. Pedicels 1–7 mm, usually shorter than the calyx-tube, curved or deflexed and sometimes much thickened in fruit. Calyx-tube 6–7 mm, glabrous or pubescent; teeth subequal, narrowly lanceolate, straight, usually much shorter than the tube; corolla (12–)15–20(–27) mm, white, cream, purple or parti-coloured; standard strongly recurved. Legume linear, acute, pubescent above. *Dry pastures and stony places. E. Mediterranean region, extending to Sicilia.* Cr Gr It Si Tu [*Ga].

Very variable in indumentum, in size and shape of the leaflets, length of peduncles, length, thickness and curvature of the fruiting pedicels, length of calyx-teeth (which may equal the tube), and length and colour of the corolla. **T. savianum** Guss., *Fl. Sic. Prodr.* **2**: 488 (1828), from Sicilia and Calabria, has strongly recurved and thickened fruiting pedicels and is more pubescent than plants from other areas. It may merit subspecific rank.

Sect. MISTYLLUS (C. Presl) Godron. Glabrous annuals. Heads pseudoterminal with prominent, glumaceous, striate, free bracts. Calyx 20- to 35-veined, inflated more or less equally on all sides in fruit; teeth setaceous, recurved. Legume included, 1- to 4-seeded.

25. **T. spumosum** L., *Sp. Pl.* 771 (1753). Stems 10–30(–50) cm, procumbent or ascending. Leaflets 10–20(–30) mm, broadly obovate-cuneate, thin, not strongly veined, denticulate. Heads globose to ovate; peduncles 10–40(–100) mm. Bracts conspicuous, shorter than the mature calyx-tube. Calyx-tube much inflated, pyriform, with transverse as well as longitudinal striations; corolla pink, slightly exceeding the calyx. Legume 3- to 4-seeded. *Dry, grassy places and disturbed ground. Mediterranean region, Portugal.* Bl Co Cr Ga Gr Hs It Lu Sa Si Tu.

26. **T. vesiculosum** Savi, *Fl. Pis.* **2**: 165 (1798). Stems (5–)15–50(–70) cm, rigid. Leaflets (5–)15–30(–60) mm, almost coriaceous, those of the lower leaves obovate, of the upper usually oblong, elliptical or lanceolate, rarely obovate-elliptical or suborbicular-cuneate, long-apiculate, spinulose-denticulate; stipules with long setaceous apices. Heads 20–60 × 20–35 mm, globose, ovoid or oblong; peduncles 10–50 mm; bracts about as long as the calyx-tube. Calyx-tube turbinate and contracted at the mouth, or broadly cylindrical to ovoid, with 24–35 prominent longitudinal veins connected by numerous more or less evident transverse veins; teeth as long as or a little shorter than the tube. Corolla white, becoming pink; standard 1½–2 times as long as the calyx. Legume 2- to 3-seeded. *Dry, grassy places. S. Europe from Corse eastwards, extending northwards to Hungary.* Al Bu Co Cr Gr Hu It Ju Rm Rs (W, K, E) Tu [Ga].

Very variable in habit, size of the parts, shape of the calyx-tube and degree of development of the transverse veins of the calyx-tube. Some plants from the Balkan peninsula, Calabria and Sicilia, with cylindrical or ovoid calyx-tube with indistinct transverse veins have been separated as **T. multistriatum** Koch, *Syn. Fl. Germ.* ed. 2, 190 (1843) (*T. setiferum* Boiss., *T. rumelicum* (Griseb.) Halácsy), but they do not appear to be specifically distinct.

27. **T. mutabile** Portenschl., *Enum. Pl. Dalmat.* 16 (1824) (*T. leiocalycinum* Boiss. & Spruner). Like **26** but more robust; calyx-tube broadly cylindrical or ovoid, the longitudinal veins slender, faint, not prominent and the transverse veins absent. *Dry grassy places.* ● *S. Italy, Sicilia; Greece, Albania, islands of W. Jugoslavia.* Al Gr It Ju Si.

Sect. VESICASTRUM Ser. Bracts free or united into a small involucre. Flowers subsessile. Calyx inflated in fruit, upper lip externally densely hairy (rarely glabrous), scarious and reticulately veined, its two teeth often setaceous.

28. **T. physodes** Steven ex Bieb., *Fl. Taur.-Cauc.* **2**: 217 (1808). Glabrous perennial; stems 5–25 cm, procumbent or ascending, not rooting at the nodes. Leaflets 10–20(–25) mm, ovate or elliptical to obovate-orbicular; stipules lanceolate, aristate. Heads 15–20 mm wide, globose or ovoid; peduncles 10–80 mm. Bracts 0·5–1 mm, free, shorter than or equalling the pedicels. Upper lip of the calyx hairy, its teeth lanceolate, porrect; lower calyx-teeth subulate, straight, somewhat longer than the tube; corolla 8–14 mm, pink. *Balkan peninsula, S. Italy, Sicilia; Portugal.* Al Cr Gr It Ju Lu Si Tu.

T. rechingeri Rothm., *Bot. Jahrb.* **73**: 438 (1944) (*T. physodes* var. *sericocalyx* Gibelli & Belli) and **T. sclerorrhizum** Boiss., *Diagn. Pl. Or. Nov.* **2**(9): 28 (1849) (*T. physodes* var. *psilocalyx*

Boiss.), both from Kriti, may merit subspecific rank. The former has obovate leaflets; hairy stems, petioles, stipules and bracts; smaller heads and calyx entirely densely villous-lanate; the latter has the calyx completely glabrous or with a few scattered hairs.

29. T. fragiferum L., *Sp. Pl.* 772 (1753). Usually more or less hairy perennial; stems (2–)10–30(–40) cm, several, procumbent, often rooting at the nodes, rarely caespitose with stems short and not rooting. Leaflets (3–)8–20 mm; ovate, elliptical or obcordate. Stipules lanceolate-subulate, membranous. Heads 10–14 mm wide, hemispherical in flower, 10–22(–35) mm, globose, ellipsoid or irregularly cylindrical in fruit; peduncles up to 200 mm, often hairy, exceeding the leaves. Bracts 3–4 mm, whorled, the lowest ones united below, forming a deeply dissected, irregular involucre. Upper lip of calyx-tube greatly inflated in fruit; corolla 6–7 mm, pale pink. Legume included, 1- to 2-seeded. *Almost throughout Europe northwards to 60° 30′.* All except Az Fa Is Rs (N) Sb.

(a) Subsp. **fragiferum**: Heads 10–22 mm in fruit, globose. Calyx 4–4·5 mm in flower, with teeth longer than the tube, 8–10 mm in fruit, concealing all or most of the persistent corolla. 2n = 16. *Probably throughout the range of the species, but possibly absent from much of S. Europe.*
(b) Subsp. **bonannii** (C. Presl) Soják, *Nov. Bot. Horti Bot. Univ. Carol. Prag.* **1963**: 50 (1963) (*T. neglectum* C. A. Meyer): Heads (10–)15–25(–35) mm in fruit, subglobose to irregularly cylindrical; calyx 3·5–4 mm in flower, the teeth not longer than the tube, 4–6 mm in fruit, the corolla exserted by 2–2·5 mm. *Mainly in the south but extending northwards to Poland and S. England.*

Much of the variation in habit, size and indumentum is phenotypic. Plants from dry places in the Mediterranean region approach **28** in habit but are readily distinguished by their characteristic involucral bracts.

30. T. resupinatum L., *Sp. Pl.* 771 (1753). Glabrous annual; stems 10–30(–60) cm, procumbent, ascending or erect. Leaflets 7–20 mm, obovate-cuneate. Bracts minute, united at the base. Flowers resupinate, scented or scentless. Heads in fruit 8–20(–25) mm, globose, stellate; peduncles shorter than to twice as long as the leaves. Calyx 5–10 mm in fruit, pyriform, sparsely pubescent to tomentose, glabrescent, crowned by the two divergent upper calyx-teeth; corolla 2–8 mm, pink, rarely reddish-purple. *Grassy places or disturbed, usually damp ground; sometimes cultivated. Doubtfully native in S. Europe; frequently introduced in W. & C. Europe.* *Az *Al *Be *Bl Bu *Co *Cr *Ga *Gr *Hs *It *Ju *Lu *Rm *Rs (K) *Sa *Si *Tu [Au Be Br Cz Ge He Ho Hu].

Very variable in habit and the size of its parts. Var. *majus* Boiss. (*T. suaveolens* Willd.), with tall, fistulose stems; peduncles twice as long as the leaves; flowers strongly scented; fruiting heads 20 mm or more and corolla 7–8 mm is anciently cultivated for fodder. It is naturalized in Portugal and probably also in the Mediterranean region.

31. T. tomentosum L., *Sp. Pl.* 771 (1753). Like **30** but caespitose; stems not more than 15 cm, usually procumbent; fruiting heads 7–11(–14) mm, subsessile; upper lip of calyx almost spherical in fruit, lanate, its two teeth short and usually concealed. *Dry places. Mediterranean region, Portugal, Açores.* Al Az Bl Co Cr Ga Gr Hs It Ju Lu Sa Si Tu.

Sect. CHRONOSEMIUM Ser. Leaves often pinnately 3-foliolate. Bracts represented by a few, short, red glandular hairs. Calyx 5-veined, upper teeth shorter than the lower. Corolla eventually darkening, persistent and scarious. Legume stalked, slightly exceeding calyx, 1(–2)-seeded.

32. T. badium Schreber in Sturm, *Deutschl. Fl.* Abt. 1, Band 4, Heft 16 (1804). Perennial with a massive tap-root. Stems 10–25 cm, many, ascending, appressed-hairy or glabrescent. Uppermost leaves more or less opposite; leaflets 10–20 mm, sessile or with short petiolules, elliptical, rhombic or deltate. Stipules 10–15 mm, ovate-lanceolate. Heads up to 25 mm in fruit; pedicels about as long as the calyx-tube. Corolla 7–9 mm, golden-yellow, becoming bright chestnut-brown after anthesis; standard cochleate, sulcate. Legume less than twice as long as the style. *Usually calcicole.* ● *Mountains of C. & S. Europe.* Al Au Bu Cz Ga Ge He Hs It Ju Po Rm.

33. T. spadiceum L., *Fl. Suec.* ed. 2, 261 (1755). Annual or biennial; stems 20–40 cm, slender, erect, little-branched, glabrescent. Uppermost leaves more or less opposite; leaflets up to 20 mm, all sessile, those of the upper leaves oblong; petioles exceeding the oblong-lanceolate stipules. Heads up to 20 mm in fruit, dense, cylindrical, pseudoterminal, often in pairs; peduncles erect; pedicels much shorter than the calyx-tube. Corolla *c.* 6 mm, golden-yellow, becoming very dark brown after anthesis; standard sulcate. Legume about 4 times as long as the style. *Grassy places; calcifuge. N., C. & E. Europe, extending westwards to N. Spain.* Au Bu Cz Fe Ga Ge He Hs It Ju No Po Rm Rs (N, B, C, W) Su.

34. T. speciosum Willd., *Sp. Pl.* 3: 1382 (1802) (incl. *T. violaceum* Davidov). Annual; stems 10–30 cm, with appressed or patent hairs, erect. Leaflets 10–18(–24) mm, oblong-elliptical, glabrous or hairy, the terminal petiolulate. Stipules semi-ovate or oblong. Heads up to 30 mm in fruit, ovoid, lax; peduncles 2–3 times as long as the leaves; pedicels *c.* 1 mm, about as long as the upper limb of the calyx-tube. Upper calyx-teeth equalling or shorter than the upper limb of the calyx-tube; lower calyx-teeth 2–3 times longer than the upper. Corolla 8–10 mm, violet. *Mediterranean region from Sicilia eastwards.* Al Bu Cr Gr It Ju Si Tu.

35. T. boissieri Guss. ex Boiss., *Fl. Or.* 2: 152 (1872). Like **34** but hairs patent; peduncles shorter than or equalling the leaves; pedicels 1–2 mm, longer than the upper limb of the calyx-tube; corolla pale yellow; standard narrower, more or less folded longitudinally over the legume. *Greece and Aegean region.* Cr Gr.

36. T. lagrangei Boiss., *Fl. Or.* 2: 154 (1872). Annual; stems erect or ascending, appressed-hairy. Leaflets 7–14 mm, obovate or ovate, the terminal petiolulate. Heads 8–15 mm, subglobose, dense; peduncles equalling or shorter than the leaves; pedicels about ½ as long as the calyx-tube. Upper calyx-teeth shorter than the tube, the lower ones subulate, twice as long as the tube; corolla 4–6 mm, violet-red; standard 3 times as long as the calyx. *Rocky places. Aegean region.* Cr Gr.

37. T. brutium Ten., *Viagg. Calabr.* 126 (1827). Annual; stems 5–20 cm, hairy, ascending. Leaflets 8–10 mm, obovate-cuneate, truncate or emarginate, the terminal leaflet of the upper leaves with a short petiolule (up to 1·5 mm) or subsessile. Stipules semicordate-ovate, often auriculate. Heads 15–20 mm in fruit, subglobose to ovoid; peduncles 30–50 mm, 2–4 times as long as the leaves; pedicels up to 0·5 mm, shorter than the upper limb of the calyx-tube. Upper calyx-teeth shorter than the upper limb of the calyx-tube; lower calyx-teeth more or less equalling the lower limb of the calyx-tube. Corolla 7–8 mm, yellow before anthesis, limb of standard broadly obovate, sulcate. Legume scarcely longer than the style. ● *S. Italy.* It.

38. T. mesogitanum Boiss., *Diagn. Pl. Or. Nov.* **1**(2): 34 (1843). Like **37** but terminal leaflet of the upper leaves with a longer petiolule (more than 1·5 mm); stipules oblong-lanceolate; limb of standard oblong. *Stony meadows. Turkey-in-Europe (near Kesan).* Tu. (*W. Anatolia.*)

39. T. aurantiacum Boiss. & Spruner in Boiss., *Diagn. Pl. Or. Nov.* **1**(2): 33 (1843). Like **37** but terminal leaflet of the upper leaves with a longer petiolule (more than 1·5 mm); stipules oblong-lanceolate, the upper sometimes cordate-auriculate; heads 20–25 mm in fruit, ovoid; corolla 8–9 mm, orange before anthesis; limb of standard obovate-oblong. ● *Greece and Kriti.* Cr Gr.

40. T. dolopium Heldr. & Hausskn. ex Gibelli & Belli, *Malpighia* **3**: 228 (1889). Annual; stems erect, appressed-hairy. Leaflets 6–14 mm; the lower oblong-ovate, cuneate; the upper oblong to oblong-lanceolate, acute. Heads 8–10 mm, ovoid or globose, lax; peduncles 20–40 mm, 2–3 times as long as the leaves; pedicels of the upper flowers longer than the calyx-tube. Upper calyx-teeth shorter than the upper limb of the calyx-tube; lower calyx-teeth 1–2 times as long as the lower limb of the calyx-tube. Corolla *c.* 8 mm, golden-yellow; limb of standard obovate-oblong. Legume scarcely longer than the style. ● *N. Greece.* Gr.

Intermediate between **39** and **41**.

41. T. patens Schreber in Sturm, *Deutschl. Fl.* Abt. 1, Band 4, Heft 16 (1804). Annual, sparsely hairy; stems 20–50 cm, flexuous, erect or ascending. Leaflets up to 18 mm, narrowly elliptic-obovate, the terminal subsessile or with a petiolule up to 2 mm. Stipules ovate, dilated and rounded at the base. Heads 10–15 mm in fruit; peduncles 20–50 mm, slender, usually much longer than the leaves. Flowers almost sessile. Upper calyx-teeth shorter than the upper limb of the calyx-tube; lower calyx-teeth equal to or longer than the lower limb of the calyx-tube. Corolla 5–7 mm, yellow; limb of standard oblong-ovate, sulcate. *Moist grassland. C. & S. Europe.* Al Au Bu Co Cr Cz Ga Gr He Hs Hu It Ju Po Rm Tu.

42. T. aureum Pollich, *Hist. Pl. Palat.* **2**: 344 (1777) (*T. agrarium* L., nom. ambig., *T. strepens* Crantz, nom. illeg.). Robust biennial; stems 15–30(–40) cm, many, erect and branched, usually appressed-hairy. Leaflets up to 15 mm, oblong-ovate or rhombic, widest near the middle, the terminal one nearly sessile. Stipules lanceolate-ovate, not dilated below. Peduncles up to 50 mm, stout, equalling or exceeding the leaves. Heads up to 16 mm, dense, many-flowered. Flowers nearly sessile. Upper calyx-teeth shorter than the upper limb of the calyx-tube; lower calyx-teeth 1–2 times as long as the lower limb of the calyx-tube. Corolla 6–7 mm, golden-yellow; limb of standard obovate, sulcate. $2n=16$. *Thickets, margins of woods and clearings. Much of Europe, but absent from most of the extreme north, most of the west and the Mediterranean region.* Al Au Be Bu Cz Da Fe Ga Ge Gr He Ho Hs Hu It Ju No Po Rm Rs (N, B, C, W, K, E) Su [Br].

43. T. velenovskyi Vandas, *Sitz.-Ber. Böhm. Ges. Wiss. (Math.-Nat. Kl.)* **1888**: 441 (1889). Like **42** but stipules dilated and rounded below; heads laxer and fewer flowered (30–40); peduncles longer; pedicels ½ as long as the calyx-tube; lower calyx-teeth about 3 times as long as the lower limb of the calyx-tube; corolla not becoming darker after anthesis. ● *N. part of the Balkan peninsula.* Al Bu Ju.

44. T. campestre Schreber in Sturm, *Deutschl. Fl.* Abt. 1, Band 4, Heft 16 (1804) (*T. procumbens* L., nom. ambig.). Annual; stems up to 30(–50) cm, hairy, erect or ascending. Leaflets 8–10 mm, obovate, the terminal one petiolulate. Stipules semiovate, dilated and rounded at the base. Heads up to 15 mm, dense, 20- to 30-flowered; pedicels ½ as long as the calyx-tube. Upper calyx-teeth as long as or shorter than the upper limb of the calyx-tube; lower calyx-teeth 1–2 times as long as the lower limb of calyx-tube. Corolla (3–)4–5(–6) mm, yellow; limb of standard broadly cochleate, sulcate. Legume 3–6 times as long as the style. $2n=14$. *Dry, grassy places. Throughout Europe except the extreme north and east.* All except Is Rs (N) Sb; introduced in Fa Fe.

Variable in habit and size. **T. pumilum** Hossain, *Notes Roy. Bot. Gard. Edinb.* **23**: 479 (1961), from the Aegean region (Amorgos), described as being similar to **46**, appears to be a small variant of **44**.

45. T. sebastianii Savi, *Diar. Med. Flajani* 2 (1815). Annual; 5–30 cm, sparsely hairy, erect or ascending. Leaflets up to 18 mm, obovate-lanceolate or oblong, with 15–25 pairs of lateral veins; petiolules subequal, very short. Heads 8–10 mm, 8- to 20-flowered; pedicels deflexed, 2–4 times as long as the calyx-tube. Upper calyx-teeth 2–3 times as long as upper limb of calyx-tube; lower calyx-teeth 3–4 times as long as the lower limb of the calyx-tube. Corolla 4 mm, yellow, becoming reddish-brown after anthesis; limb of standard broadly obovate or orbicular, sulcate. *C. & S. Italy, Sicilia; Balkan peninsula, very local.* Bu Gr It Ju Si Tu.

46. T. dubium Sibth., *Fl. Oxon.* 231 (1794) (*T. minus* Sm., *T. filiforme* auct.). Annual; stems up to 25(–50) cm, usually hairy, procumbent or ascending. Leaflets up to 11 mm, obcordate or obovate with 4–9 pairs of lateral veins; terminal leaflet petiolulate. Stipules 4–5 mm, equalling or exceeding the petioles of the upper leaves, broadly ovate. Heads 8–9 mm, 3- to 15(–25)-flowered; peduncles not capillary; pedicels stout, shorter than the calyx-tube. Calyx-teeth unequal, the lower about as long as the lower limb of the calyx-tube. Corolla 3–3·5 mm, yellow, becoming yellowish-brown after anthesis; standard narrowly oblong, nearly smooth. $2n=28, 32$. *Dry grassy places. Most of Europe except the extreme north.* All except Bl Cr Fa Is Rs (B, K) Sa Sb Tu [Fe Rs (N)].

47. T. micranthum Viv., *Fl. Lib.* 45 (1824) (*T. filiforme* L., nom. ambig.) Annual; stems 2–10(–20) cm, glabrescent, procumbent or ascending. Leaflets up to 5(–8) mm, obcordate or obovate with 4–9 pairs of lateral veins; terminal leaflet subsessile. Stipules oblong or ovate. Heads *c.* 4 mm, 1- to 6-flowered; peduncles capillary; pedicels capillary, as long as or longer than the upper limb of the calyx-tube. Calyx-teeth unequal, the lower longer than the lower limb of the calyx-tube. Corolla 2–3(–4) mm, yellow, becoming yellowish-brown after anthesis; standard oblong, nearly smooth. $2n=16$. *Grassy places. W. & S. Europe, extending northwards to Hungary.* Al Az Be Bl Br Bu Co Cr Da Ga Ge Gr Hb Ho Hs Hu It Ju Lu No Rm Sa Si.

Subgen. **Trifolium.** Flowers ebracteate. Calyx-throat usually more or less closed with a ring of hairs or an annular or 2-lobed callosity. Legume nearly always included in the calyx-tube, 1- to 2-seeded.

Sect. TRIFOLIUM. Heads usually spicate, rarely capitate. Flowers usually sessile, all fertile.

48. T. striatum L., *Sp. Pl.* 770 (1753). Annual, softly hairy; stems 4–30(–50) cm, spreading or ascending. Leaflets 6–16 mm, obovate-cuneate; lateral veins almost straight; stipules ovate, their apex abruptly setaceous. Heads 10–15 mm, ovoid or oblong, sessile, axillary and pseudoterminal, usually not paired. Calyx-tube 2·5–3 mm, ovoid, hairy, somewhat inflated and readily abscissing at maturity; calyx-teeth subulate, subequal, straight, erect or somewhat patent. Corolla 4–5 mm, pink; standard free, usually equalling or exceeding the upper calyx-teeth. $2n=14$. *Dry places. S. W. & C. Europe, extending northwards to S. Sweden.* All except Cr Fa Fe Is No Rs (N, B, C, W, E) Sb.

Variable in size, habit and the relative lengths of calyx-tube, calyx-teeth and petals. Robust plants from S. Europe, with stems up to 50 cm, heads cylindrical and corolla not exceeding the upper calyx-teeth, have been described as **T. tenuiflorum** Ten., *Fl. Nap.* **1**, *Prodr.*: 44 (1811) (*T. incanum* C. Presl). This taxon may merit subspecific rank.

49. **T. arvense** L., *Sp. Pl.* 769 (1753). Annual or biennial; stems 4–40 cm, erect or diffusely branched, whitish- or reddish-pubescent, rarely glabrescent. Upper leaves sessile, their leaflets 5–20 mm, linear-oblong. Lower stipules lanceolate-subulate, the upper subulate from an ovate base. Heads up to 20 mm, numerous, ovoid or oblong, pedunculate, with numerous, densely sericeous flowers. Calyx 3·5–7(–9) mm, the tube globose in fruit, often covered with dense hairs, rarely glabrescent; teeth 1–3(–5) times as long as the tube, reddish, subequal, setaceous, with long hairs. Corolla *c.* 4 mm, whitish or pink, much shorter than the calyx. $2n=14$. *Dry places; somewhat calcifuge. Most of Europe except the extreme north.* All except Fa Is Sb.

Very variable in habit, indumentum, pigmentation, size and shape of the leaflets and length of the calyx-teeth. Several taxa are recorded over most of Europe. Plants with smaller and less hairy calyx (3·5–4·5 mm) are often called T. gracile Thuill., *Fl. Paris* ed. 2, 383 (1799); they seem most frequent in W. Europe but occur sporadically elsewhere. **T. longisetum** Boiss. & Balansa in Boiss., *Diagn. Pl. Or. Nov.* 3(6): 47 (1859), with calyx-teeth 4–5 times as long as the tube may deserve recognition as a subspecies of the Mediterranean region.

50. **T. affine** C. Presl, *Symb. Bot.* **1**: 54 (1832) (*T. preslianum* Boiss.). Like 49 but standard about as long as the calyx and wings hairy outside. *Turkey-in-Europe, S. Bulgaria.* Bu Tu. (Anatolia.)

51. **T. saxatile** All., *Mélang. Philos. Math. Soc. Roy. Turin (Misc. Taur.)* 5: 77 (1774). Annual, greyish-pubescent; stems 5–15 cm, often numerous, procumbent or ascending. Leaflets 3–6 mm, narrowly obovate-cuneate, emarginate; stipules ovate or lanceolate, acute, the upper ones dilated, reddish with darker veins. Heads 6–10 mm wide, depressed-globose, few-flowered, sessile and sheathed at the base by the stipules. Calyx readily abscissing at maturity, its tube ovoid, densely hairy; teeth straight or incurved, unequal, all shorter than the tube. Corolla 3–4 mm, whitish or pinkish, not exceeding the calyx. *Dry gravel and moraines.* ● *Alps.* Au Ga He It.

52. **T. bocconei** Savi, *Atti Accad. Ital. (Firenze)* **1**: 191 (1808). Annual; stems (2–)5–25(–30) cm, densely pubescent, erect or ascending, sparingly branched. Leaflets of upper leaves 7–23 mm, narrowly cuneate-oblong, glabrescent, denticulate, veins nearly straight. Stipules lanceolate, abruptly contracted above, the upper ones not dilated. Heads 9–15 mm, dense, cylindrical or conical, the terminal ones often paired but unequal. Calyx-tube cylindrical, pubescent; teeth subulate, erect or connivent, unequal, the lowest one equalling the tube. Corolla 4–5 mm, pinkish, equalling the calyx, persistent in fruit. *Dry places. S. & W. Europe, northwards to S.W. England.* Bl Br Bu Co Cr Ga Gr Hs It Ju Lu Sa Si Tu [Ge Hs].

53. **T. tenuifolium** Ten., *Fl. Nap.* **1**, *Prodr.*: 44 (1811). Like 52 but stems more branched and spreading; upper leaflets linear or oblong; calyx-teeth not connivent in fruit; corolla twice as long as the calyx, pink or cream. ● *S. Italy, Balkan peninsula, Aegean region.* Al Bu Cr Gr It Ju Tu.

54. **T. trichopterum** Pančić, *Verh. Zool.-Bot. Ges. Wien* **6**: 480 (1856). Like 52 but leaflets 5–10 mm, obcordate-cuneate, deeply emarginate, usually persistently hairy on the lower surface; calyx with long dense, white or ultimately brown, plumose hairs, the teeth erect; corolla 5–6 mm, about 1½ times as long as the calyx. *Dry stony slopes.* ● *Balkan peninsula.* Al Bu Gr Ju.

55. **T. phleoides** Pourret ex Willd., *Sp. Pl.* **3**: 1377 (1802). Annual; stems 10–35 cm, erect, sparsely appressed-hairy. Upper leaves subsessile; leaflets 10–25 mm, linear or oblanceolate, denticulate. Stipules oblong, with a subulate apex. Heads up to 25 mm, dense, oblong, ovoid or conical, solitary or paired; peduncles 50–60 mm. Calyx-tube ovoid, with 10 distinct veins and a ring of hairs at the throat; teeth 1·5–2 mm, subequal, as long as or a little shorter than the tube, triangular-lanceolate, dilated at the base, ultimately divergent. Corolla 4–5 mm, a little shorter than the calyx. *Dry, stony grassland, mainly in the mountains. S. Europe.* Al Bu Co Gr It Hs Lu Sa Si [Ga Ge].

56. **T. gemellum** Pourret ex Willd., *Sp. Pl.* **3**: 1376 (1802). Like 55 but leaflets 7–15 mm, elliptic-obovate; heads sessile or very shortly pedunculate; throat of the calyx with an annular, hairy callosity; calyx-teeth 3–4 mm, lanceolate-setaceous, scarcely dilated at the base, longer than the tube. *Dry, sandy and stony places. Spain and Portugal.* Hs Lu.

57. **T. ligusticum** Balbis ex Loisel., *Fl. Gall.* 731 (1807). Annual; stems 10–40(–60) cm, dark green, with sparse patent hairs, ascending or diffuse. Leaves all petiolate; leaflets 10–20 × 5–13 mm, broadly obovate. Stipules ovate or oblong with setaceous apex. Heads 6–15 mm, ovoid or oblong, often paired, then one axillary and long-pedunculate, the other terminal but laterally displaced and shortly pedunculate. Calyx 4–6 mm; mouth of the calyx-tube closed with a callosity; calyx-teeth ciliate, setaceous, ultimately divergent, subequal, 1–2 times as long as the tube. Corolla 3–4 mm, much shorter than the calyx. *Dry places; calcifuge. S.W. Europe.* Az Bl ?Bu Co Ga Hs It Lu Sa Si ?Tu [Ge].

58. **T. scabrum** L., *Sp. Pl.* 770 (1753). Annual; stems 5–25 cm, rigid, flexuous, numerous, procumbent or ascending. Leaflets 5–10 mm, obovate-cuneate, coriaceous, denticulate; lateral veins recurved and prominent at the margins. Stipules ovate or oblong with setaceous apex, entire. Heads 5–12 mm, numerous, mostly axillary, sessile, globose or ovoid, attenuate and scarcely clasped at the base by the stipules. Calyx persistent in fruit; teeth rigid, spinose, slightly recurved in fruit, the lowest one longer than the tube. Corolla 4–5 mm, whitish, rarely pink, usually shorter than the calyx. $2n=10$. *Dry places. S. & W. Europe, northwards to Scotland.* Al Az Be Bl Br Bu Co Cr Ga Ge Gr Hb He Ho Hs It Ju Lu Rm Rs (K) Sa Si Tu.

Plants from the Mediterranean region with cylindrical heads, more recurved calyx-teeth and pink corolla, slightly exceeding the calyx, have been described as **T. lucanicum** Gasparr. ex Guss., *Fl. Sic. Prodr.* **2**: 494 (1828) and **T. scabrum** subsp. **turcicum** Velen., *Sitz.-Ber. Böhm. Ges. Wiss. (Math.-Nat. Kl.)* **1893**(37): 23 (1894); they may merit subspecific rank. Plants from Istria and Greece identified with **T. compactum** Post, *Fl. Syr. Pal. Sin.* 239 (1896), perhaps should be included in 58.

59. T. dalmaticum Vis., *Flora (Regensb.)* **12** (Ergänz. 1): 21 (1829). Like **58** but stems ascending, scarcely flexuous; leaflets 8–15 mm; heads globose becoming oblong, involucrate, mostly terminal and solitary, rarely paired; all calyx-teeth longer than the tube, strongly recurved in fruit; corolla 8–10 mm, pink, twice as long as the calyx. *Dry, rocky and grassy places.* ● *Balkan peninsula.* Al Bu Gr Ju Tu.

Sometimes confused with **58** but appears to be quite distinct.

T. filicaule Boiss. & Heldr. in Boiss., *Diagn. Pl. Or. Nov.* 2(9): 24 (1849), recorded from S. Greece (Lakonia), probably belongs with *T. dalmaticum;* it differs principally in the corolla being less than twice as long as the calyx.

60. T. stellatum L., *Sp. Pl.* 769 (1753). Annual; stems (2–)8–20(–35) cm, erect, simple or branching from the base, with dense patent hairs. Leaflets 8–12 mm, obcordate, denticulate towards the apex; stipules ovate, obtuse, acutely denticulate, the margin and veins bright green. Heads 15–25 mm, globose or ovoid; peduncles (5–)30–100 mm, with appressed or patent hairs. Calyx-teeth twice as long as the tube, patent in fruit, 3-veined, triangular-lanceolate with a subulate-acuminate apex. Corolla 8–12 mm, pink, rarely purple or yellow, equalling the calyx, rarely much longer. $2n=14$. *Fields, roadsides and stony slopes. Mediterranean region, Portugal.* Al Bl Co Cr Ga Gr Hs It Ju Lu Sa Si Tu [Br].

61. T. dasyurum C. Presl, *Symb. Bot.* **1**: 53 (1832) (*T. formosum* D'Urv., non Savi). Like **60** but usually more robust, branching above, with the 2 uppermost leaves subopposite; leaflets 20–25 mm, oblong-elliptical or lanceolate, acute, entire; stipules lanceolate-acuminate, entire; heads 20–35 mm, often paired; calyx-teeth up to 4 times as long as the tube; corolla *c.* 16 mm, shorter than the calyx. *Greece and Aegean region.* Cr Gr.

Variable in stature and length of peduncle.

62. T. incarnatum L., *Sp. Pl.* 769 (1753). Annual; stems (10–)20–50 cm, simple or branching only from the base, erect or ascending; hairs usually patent below and appressed above. Leaflets 8–25 mm, obovate-cuneate to suborbicular, denticulate towards the apex. Stipules ovate, blunt, often herbaceous, sometimes pigmented, angled or obscurely dentate. Heads 10–40 mm, solitary, oblong-ovoid to cylindrical; peduncles long. Calyx-teeth as long as or longer than the tube, linear, acute, patent in fruit. Corolla 10–12 mm, equalling or exceeding the calyx, blood-red, pink, cream or white. *S. & W. Europe; cultivated also in a large part of Europe and widespread as an escape except in the extreme north.* Al Au Be Br Bu Co Cr Cz Da Fe Ga Ge Gr He Hs Ho Hu It Ju Lu No Po Rm Rs (W, K, E) Sa Si Su Tu.

(a) Subsp. **incarnatum**: Stems robust, erect, often unbranched, not very hairy. Heads dense. Corolla blood-red, rarely pure white, equalling or slightly exceeding the calyx. *Cultivated and widely naturalized.*

(b) Subsp. **molinerii** (Balbis ex Hornem.) Syme in Sowerby, *Engl. Bot.* ed. 3, 3: 45 (1864): Stems usually several, ascending, less robust, densely hairy. Heads less dense. Corolla usually yellowish-white, rarely pink, much exceeding the calyx. *Certainly native on cliff-tops exposed to sea-spray; also inland in parts of the Mediterranean region. S. & W. Europe, northwards to S.W. England.*

The subspecies are very distinct in N.W. Europe but are connected by intermediates in the south. In some regions the status of the plants is uncertain.

63. T. pratense L., *Sp. Pl.* 768 (1753) (incl. *T. borysthenicum* Gruner). Perennial, caespitose, more or less hairy; stems 5–100 cm. Leaflets obovate, or oblong-lanceolate to nearly orbicular, hairy below, often glabrescent above. Stipules triangular above, abruptly contracted into a setaceous, usually ciliate point; upper stipules very wide. Heads 20–40 mm, globose or ovoid, solitary or paired, usually sessile and involucrate. Calyx-tube 10-veined, usually appressed-hairy; teeth triangular with filiform apex, straight, ciliate, separated by broad sinuses, the lowest one about twice as long as the tube. Corolla 12–15 mm, usually reddish-purple or pink, rarely cream or white, (1–)2 times as long as the lowest calyx-tooth. Legume ovate, with a thickened apex. $2n=14$. *Meadows and pastures on fertile and moist but well-drained soils from sea-level to 3150 m; extensively cultivated as a forage crop. Throughout Europe except for parts of the extreme north and parts of the extreme south; introduced in Iceland and the Faroes.* All except Bl Cr Sb.

Extremely variable both in the wild and cultivated state in habit, stature, indumentum, size and shape of leaflets and size and colour of flowers. The non-rhizomatous perennial habit, the stipules and the calyx afford the best means of identification. Many ecologically specialized wild populations (for example, of high mountains and coastal habitats) are locally distinct, but it is impossible at present to bring the numerous local taxa into a comprehensive scheme for the whole of Europe. The following indicates some of the variation within the species.

Var. *pratense*. Long-lived perennial. Stems usually 20–40 cm, solid, procumbent or ascending, appressed-hairy, or some rarely with patent hairs above. Heads often solitary. In natural or semi-natural habitats throughout the range of the species. Includes the so-called var. *parviflorum* Bab. with heads shortly pedunculate, flowers often pedicellate, sometimes bracteate, and corolla not exceeding the calyx, a widely-occurring monstrosity often confused with unrelated species.

Var. *sativum* Sturm (*T. sativum* (Sturm) Crome). Short-lived perennial. Stems 40–70(–100) cm, hollow, more or less erect, glabrescent or glabrous. Leaflets up to 50 mm or more. Heads large, often paired; corolla usually pink. Includes most of the important cultivars. **T. baeticum** Boiss., *Voy. Bot. Midi Esp.* **2**: 726 (1845), from S.W. Spain and Sicilia, is similar but has yellow flowers.

Var. *americanum* C. O. Harz. Stems with stiff, patent hairs. Leaflets never emarginate, the upper ones lanceolate. Calyx with patent, villous hairs. Corolla deep red. Native in parts of S.E. Europe; introduced into cultivation in C. Europe by way of N. America in 1883, but declining in cultivation since 1910. Often identified with **T. expansum** Waldst. & Kit., *Pl. Rar. Hung.* 3: 237 (1807).

Var. *maritimum* Zabel (var. *villosum* Wahlberg). Stems slender, ascending; hairs dense, patent below, often appressed above. Leaflets and heads relatively small. Corolla pink, white or cream. Maritime pastures and dune-slacks, mainly on the S. coast of the Baltic.

Var. *frigidum* Gaudin. Stems 5–30 cm, stout, procumbent or ascending, densely hairy, at least above. Stipules often hairy over the whole outer surface. Heads large, 30 mm or more wide. Corolla dirty white, often yellowish or pinkish, 1½ times as long as the calyx. $2n=14$. Alps; often treated as a subspecies. Somewhat similar plants from the mountains of E. Portugal, the Pyrenees, Carpathians and the mountains of the Balkan peninsula differ in minor characters, especially flower colour; they are closer to var. *pratense* in size of heads and indumentum.

Other mountain plants of S. Europe have small, relatively few-flowered heads.

64. T. pallidum Waldst. & Kit., *Pl. Rar. Hung.* **1**: 35 (1800–1). Annual or biennial; stems 15–40(–60) cm, usually much-branched, hairy. Leaflets 15–25 mm, obovate-cuneate to elliptical. Stipules abruptly contracted into a filiform arista. Heads 15–25 mm, globose to ovoid, involucrate, sessile. Calyx-tube 10-veined; veins evident; teeth setaceous, triangular and 5-veined at the base, 1½ times as long as the tube. Corolla *c.* 12 mm, whitish or pale pink, rarely bright orange-pink, 1½–3 times as long as the calyx. *Dry grassy places, rocks and screes. S. Europe, westwards to Corse and extending northwards to Hungary.* Al Bu Co Cr Gr Hu It Ju Rm Si Tu [Cz].

65. T. diffusum Ehrh., *Beitr. Naturk.* **7**: 165 (1792). Like **64** but leaflets narrower; apex of stipules linear, herbaceous; heads denser; calyx-teeth subulate-filiform, 3-veined at the base, twice as long as the tube; corolla reddish-purple, not or scarcely exceeding the calyx. *Damp or shady, grassy places. S., S.E. & E.C. Europe.* Al Bl Bu Co Cz Ga Gr Hs Hu It Ju Lu Rm Rs (W, K, E) Sa Si.

Sometimes confused with **72** but readily distinguished by the 10-veined calyx-tube.

66. T. noricum Wulfen, *Arch. Bot. (Roemer)* **3**: 387 (1805) (incl. *T. praetutianum* Guss. ex Ser.). Caespitose perennial; stems 8–20 cm, simple, with dense, patent hairs. Cauline leaves 2 or 3; leaflets 15–25 mm, oblong, elliptical or obovate; stipules whitish, gradually narrowed into a long, acute apex. Heads 25–40 mm wide, globose, often nodding. Calyx-tube campanulate-cylindrical with dense, patent hairs; all teeth subequal, equalling or a little longer than the tube; corolla *c.* 15 mm, cream; standard about 1½ times as long as the calyx. Legume dehiscing ventrally. *Alpine pastures, screes and stony slopes; calcicole; from about 1600 to 2600 m.* ● *Appennini, E. Alps, mountains of W. part of Balkan peninsula.* Al Au Gr Ju It.

67. T. wettsteinii Dörfler & Hayek, *Österr. Bot. Zeitschr.* **70**: 16 (1921). Like **66** but stems 5–8 cm, appressed hairy; leaflets 5–10 mm, obovate; stipules with obtuse apex; calyx glabrous at the base; corolla pink. *Mountain rocks.* ● *N.E. Albania.* Al ?Ju.

68. T. ottonis Spruner ex Boiss., *Diagn. Pl. Or. Nov.* **1**(2): 28 (1843). Like **66** but densely caespitose; stems 1–5 cm; leaflets 2–8 mm, ovate; calyx-teeth usually shorter than the tube; heads 15–20 mm; corolla 12–15 mm, deep purple; standard 1½–2 times as long as the calyx. *Rocky places on mountains.* ● *S. Greece.* Gr.

69. T. lappaceum L., *Sp. Pl.* 768 (1753). Annual, bright green, with numerous usually branched glabrescent, erect or ascending stems 5–40 cm. Leaflets 5–20 mm, obovate-cuneate, obtuse, hairy. Stipules oblong, conspicuously veined; apex long, lanceolate or subulate, herbaceous, hairy. Heads 12–20 mm wide, globose, rarely ovoid; peduncles up to 35 mm in fruit. Calyx-tube with 20 conspicuous veins, glabrous or glabrescent, rarely hairy; teeth 3·5–6 mm, longer than the tube, prominently 5-veined and triangular below, filiform and hairy above. Corolla 7–8 mm, pink, equalling the calyx at anthesis, much shorter than the calyx in fruit. Legume ovate, with a thickened apex. *S. Europe.* Al Az Bl Bu Co Ga Gr Hs It Ju Lu Rs (K) Sa Si Tu [Cz].

Varies considerably in habit and size. The only taxon possibly deserving more than varietal rank is subsp. **adrianopolitanum** Velen., *Fl. Bulg., Suppl.* 80 (1898), from S. Bulgaria, with upper stipules broadly ovate and shortly acuminate, and heads elliptical, with rather long peduncles.

70. T. congestum Guss., *Cat. Pl. Boccad.* 81 (1821). Annual; stems 1–3 cm, numerous, stout, woody, procumbent, glabrous. Leaflets 5–9 mm, narrowly cuneate, deeply emarginate, obscurely denticulate, densely appressed-pubescent above and beneath; petioles up to 30 mm; stipules membranous, broadly ovate, abruptly contracted into a short point. Heads 8–10 mm, pseudo-terminal and axillary, sessile, involucrate, more or less congested. Flowers mostly erect. Calyx-tube 20-veined, densely but shortly pubescent; teeth nearly twice as long as the tube, subulate from a narrowly triangular base, glabrous and 5-veined below, with sparse patent hairs above the middle. Corolla 6–7 mm, white, not longer than the calyx. *Fields and roadsides on clay soils. Sicilia, S. Italy, Malta.* It Si.

71. T. barbeyi Gibelli & Belli, *Atti Accad. Sci. Torino* **22**: 610 (1887). Like **70** but stems densely pubescent; leaflets oblong-obovate, cuneate, obtuse; stipules linear; calyx-teeth shorter than the tube, densely hairy, broadly triangular at the base, subulate above; corolla pink, about twice as long as the calyx. *Grassy places. Karpathos.* Cr. (*Rodhos.*)

72. T. hirtum All., *Auct. Fl. Pedem.* 20 (1789). Annual; stems up to 35 cm, often with patent branches; hairs patent. Leaflets 8–20 mm, obovate-cuneate, denticulate above, rarely emarginate. Stipules lanceolate, abruptly contracted into a long setaceous apex with spreading hairs. Heads 15–20(–25) mm wide, persistent in fruit, densely hairy, globose, solitary, sessile, with an involucre formed of dilated stipules and one or sometimes two 3-foliolate leaves. Calyx 20-veined, the veins obscured by dense hairs; teeth twice as long as the tube. Corolla 12–15 mm, purple, longer than the calyx. *Dry places. S. Europe.* Al Bu Co Cr Ga Hs It Ju Lu ?Rm Rs (K) Tu.

73. T. cherleri L., *Demonstr. Pl.* 21 (1753). Like **72** but stems 5–15(–30) cm, rarely branching above; leaflets obcordate-cuneate, almost entire; stipules with a short, ovate-lanceolate, herbaceous and often recurved apex; heads hemispherical, readily abscissing below the involucre in fruit; calyx-teeth more or less equalling the tube; corolla pinkish-white, equalling or shorter than the calyx. *Dry places, mainly lowland. S. Europe.* Al Bl Bu Co Cr Ga Gr Hs It Ju Lu Sa Si Tu.

74. T. medium L., *Amoen. Acad.* **4**: 105 (1759). Rhizomatous. Stems (10–)30–45(–65) cm, ascending, more or less flexuous, often branched, sparsely appressed-hairy or glabrescent, rarely with spreading hairs above. Leaflets 20–60 × (5–)9–20(–35) mm, ovate, obovate or elliptical, almost entire, petioles usually exceeding the stipules. Stipules lanceolate, herbaceous and ciliate above, usually adnate by less than half their length to the petioles. Heads 25–35 mm, globose or ovoid, usually solitary, ultimately shortly pedunculate. Calyx-tube glabrous or glabrescent, rarely with sparse spreading hairs above, 10(–20)-veined; teeth filiform. Corolla 12–20 mm, light purple-red. Legume dehiscing longitudinally. $2n = c.$ 70, 78–80, 84. *In open woodland, scrub and poor pastures. Throughout Europe except the extreme north and south.* All except Az Bl Cr Fa Is Sa Sb Si.

Rather variable in C. & S. Europe; intermediates occur between the subspecies which do not accommodate some plants from S. Europe.

1 Calyx-tube 13- to 20-veined   **(c) subsp. sarosiense**
1 Calyx-tube 10(–14)-veined
  2 Stems with spreading hairs above   **(d) subsp. balcanicum**

2 Stems appressed-hairy or glabrescent above
  3 Upper 4 calyx-teeth not longer than the tube
                               (a) subsp. **medium**
  3 Upper 4 calyx-teeth longer than the tube   (b) subsp. **banaticum**

(a) Subsp. **medium**: Stems (10–)30–45 cm, appressed-hairy or glabrescent above; leaflets 20–50 × 9–20 mm; calyx-tube glabrous or glabrescent, 10-veined; upper 4 calyx-teeth not longer than the tube. *Throughout the range of the species, possibly excepting Greece.*

(b) Subsp. **banaticum** (Heuffel) Hendrych, *Preslia* **28**: 405 (1956): Stems up to 50 cm, appressed-hairy or glabrescent above; leaflets 25–45 × 15–30 mm; calyx-tube glabrous or glabrescent, 10(–14)-veined; upper 4 calyx-teeth longer than the tube. ● *Czechoslovakia, Hungary, Romania.*

(c) Subsp. **sarosiense** (Hazsl.) Simonkai, *Enum. Fl. Transs.* 180 (1887): Stems 20–65 cm, appressed-hairy or glabrescent above; leaflets up to 20–60 × 15–35 mm; calyx-tube glabrous or shortly hairy, 13- to 20-veined; upper 4 calyx-teeth up to twice as long as the tube. ● *Foothills of the Carpathians.*

(d) Subsp. **balcanicum** Velen., *Fl. Bulg.* 135 (1891): Stems 10–30 cm, with spreading hairs above; leaflets 25–30 × 12–13 mm; calyx-tube glabrous or with a few sparse spreading hairs above; upper 4 calyx-teeth longer than the tube. ● *Balkan peninsula.*

75. T. heldreichianum Hausskn., *Mitt. Thür. Bot. Ver.* nov. ser., **5**: 72 (1893). Like **74** but stems 15–30 cm, slender, shortly branched throughout; leaflets 12–20 × 8–10 mm, coriaceous and with prominent veins beneath, finely denticulate; stipules membranous, appressed-hairy; heads *c.* 20 mm; calyx-tube appressed-pubescent, 10-veined; calyx-teeth glabrous and with a transparent membranous margin below, the 4 upper *c.* 1½ times, the lowest about twice as long as the tube; corolla pink. *In woods.* ● *Crna Gora, Bulgaria, N. Greece.* Bu Gr Ju.

76. T. patulum Tausch, *Syll. Pl. Nov. Ratisbon. (Königl. Baier. Bot. Ges.)* **2**: 245 (1828). Like **74** but slender and delicate; stems 20–60 cm, appressed-hairy, much-branched; leaflets 15–50 × 4–5 mm, linear-oblong; petioles short, united along their length with the lanceolate, subulate stipules; heads up to 40 mm, ovoid or oblong, lax; calyx-tube densely pubescent; teeth longer than the tube, eventually curved and patent. *Stony woodlands and scrub.* ● *S. Italy, W. part of Balkan peninsula.* Al Ju It Gr.

77. T. velebiticum Degen, *Magyar Bot. Lapok* **10**: 113 (1911). Like **74** but leaflets 10–15 × 5 mm, obovate-cuneate; petioles usually entirely adnate to the stipules; free apex of the stipules ovate-lanceolate; heads globose; calyx-tube glabrous; upper 4 calyx-teeth subulate, glabrous below, sparsely hairy or glabrous above, ½ as long as the tube. *Rocky places.* ● *N.W. Jugoslavia.* Ju.

78. T. pignantii Fauché & Chaub. in Bory, *Expéd. Sci. Morée* **3**(2): 219 (1832). Rhizomatous; stems 15–45 cm, ascending, with patent hairs and axillary, leafy short shoots. Leaflets 10–30 × 10–15 mm, broadly obovate or elliptical, obtuse or emarginate. Upper stipules ovate-lanceolate, herbaceous, the greater part free from the petiole. Heads 20–30 mm, globose; flowers shortly pedicellate. Calyx-tube glabrous; teeth filiform, with long patent hairs, subequal, equalling or longer than the tube, ultimately patent or recurved; corolla 15–18 mm, yellowish-white. *Woods and thickets, mainly in the mountains.* ● *Balkan peninsula.* Al Bu Gr Ju.

79. T. alpestre L., *Sp. Pl.* ed. 2, 1082 (1763). Rhizomatous; stems (5–)15–40 cm, erect or ascending, usually simple and hairy. Leaflets 20–50(–60) × 5–13 mm, lanceolate or narrowly elliptical,

hairy or glabrous above; veins very numerous, curved. Stipules adnate by more than ½ their length to the petioles, often ciliate and pubescent, the free part linear or subulate, usually scarious. Heads 15–25 mm, paired or solitary, globose or ovoid, subsessile or with peduncles up to 1 cm. Calyx-teeth subulate or filiform, straight, hairy, the lowest one much longer than the others. Corolla *c.* 15 mm, purple, rarely pink or white. Legume ovoid, dehiscing longitudinally. *Dry open woods, scrub and pastures.* C., E. & S. Europe, *northwards to Denmark and Estonia.* Al Au Be Bu Cz Da Ga Ge Gr He ?Hs Hu It Ju Po Rm Rs (B, C, W, K, E).

Very variable. The widespread plant has appressed hairs on the stem and a 20-veined calyx; plants from Italy and the Balkan peninsula frequently have dense, patent hairs; var. *durmitoreum* Rohlena from Crna Gora, has a usually 10-veined calyx, slender stems and smaller flowers.

80. T. rubens L., *Sp. Pl.* 768 (1753). Rhizomatous, usually glabrous. Stems (20–)30–60 cm, erect, usually simple and glabrous. Leaflets up to 70 × 10 mm, oblong-lanceolate, rarely elliptical, spinulose-denticulate; veins very numerous, curved. Stipules ovate or lanceolate, adnate by more than ½ their length to the petioles, usually glabrous, the upper ones often forming an obconical sheath, the free part herbaceous, often serrate. Heads up to 80 × 25 mm, cylindrical, solitary or paired; peduncles up to 4 cm. Calyx-tube 20-veined, glabrous, rarely hairy; teeth straight, subulate, hairy. Corolla *c.* 15 mm, purple, rarely white. Legume ovoid, dehiscing longitudinally. *Dry open woodland and scrub.* ● C. Europe, *extending locally to N. Spain, C. Italy, Crna Gora and Ukraine region.* Al Au Cz Ga Ge He Hs Hu It Ju Po Rm Rs (W, C).

81. T. angustifolium L., *Sp. Pl.* 769 (1753). Annual; stems 10–50 cm, appressed-hairy, few, one often taller and stiffly erect, the others shorter and ascending, branching at the base. Leaflets (10–)20–80 × (1–)2–4 mm, linear-lanceolate, acute; stipules lanceolate-subulate. Heads (15–)20–80 mm, solitary, ovoid or conical-cylindrical; peduncles (10–)20–40(–60) mm. Calyx-teeth subequal, linear or subulate, finally patent, ciliate, the apex glabrous or with a few short hairs. Corolla 10–12 mm, pink, shorter than or equalling the calyx-teeth. *Dry places; calcifuge.* S. Europe. Al Az Bl Bu Co Cr Cz Ga Gr It Ju Lu Rm Rs (K) Sa Si Tu.

The plant known under the illegitimate name T. **intermedium** Guss., *Cat. Pl. Boccad.* 82 (1821), non Lapeyr., from the Mediterranean region and Bulgaria, is like **81** but has stems 10–20 cm; leaflets 10–20 mm, narrowly elliptical or lanceolate; calyx-teeth subequal, more or less evenly ciliate from the base to the apex. It may merit subspecific rank.

82. T. purpureum Loisel., *Fl. Gall.* 484 (1807). Robust annual; stems up to 60 cm, often branching above; hairs somewhat patent. Leaflets 20–50 × 2·5–5 mm, linear to elliptic-oblong. Heads 20–110 mm; flowers opening gradually from below upwards. Calyx-teeth very unequal, the lowest one twice as long as the others; corolla 16–25 mm, bright reddish-purple, much exceeding the calyx-teeth. *Dry, often disturbed ground. S. France, Sicilia, Balkan peninsula; doubtfully elsewhere in the Mediterranean region.* Al Bu ?Co Ga Gr Ju Rm Si Tu.

83. T. desvauxii Boiss. & Blanche in Boiss., *Diagn. Pl. Or. Nov.* **3**(2): 12 (1856). Like **82** but stems weak and diffusely branched; heads smaller; corolla 13–15 mm. *Bulgaria, Thrace.* Bu Gr. (*E. Mediterranean region.*)

Often treated as a variety of **82** but very distinct in cultivation.

84. T. smyrnaeum Boiss., *Diagn. Pl. Or. Nov.* 1(2): 25 (1843) (*T. lagopus* Pourret ex Willd., non Gouan, *T. sylvaticum* Gérard sec. C. Vicioso, *T. hervieri* Freyn). Annual, softly grey-hairy; stems 5–20 cm, often stout, often spreading, with divaricate branches; hairs patent. Leaflets 5–10 mm, obovate-cuneate, often emarginate; stipules broadly ovate, sometimes denticulate above. Heads ovoid or cylindrical, dense, often more or less paired. Fruiting calyx with a subglobose tube; calyx-teeth subulate, ciliate, unequal, the lowest a little longer than the tube. Corolla 7–8 mm, pale pink, equalling or exceeding the calyx. *Dry places; usually calcifuge. S. Europe.* Bu Co Cr Ga Gr Hs It Ju Lu Tu [Ge].

85. T. ochroleucon Hudson, *Fl. Angl.* 283 (1762) (incl. *T. caucasicum* Tausch, *T. pallidulum* Jordan). Shortly rhizomatous or caespitose perennial; stems 20–50 cm, ascending; upper internodes long. Leaflets 15–30(–50) × 5–8 mm, oblong-elliptical or lanceolate. Stipules with a linear-lanceolate, herbaceous apex. Heads 20–40 mm, globose or oblong, shortly pedunculate or subsessile. The 4 upper calyx-teeth equalling or shorter than the tube, the lowest usually longer and deflexed in fruit, with one distinct central vein, rarely with two lateral veins. Corolla 15–20 mm, exceeding the calyx, at first yellowish-white, rarely pink, ultimately falling. Legume thickened at the apex. *Mainly in shady or somewhat damp habitats. W., C. & S. Europe.* Al Au Be Br Bu Co Cz Ga Ge Gr He Hs Hu It Ju Lu Po Rm Rs (W) Sa Si Tu.

Variable in indumentum, petal-colour and length of the calyx-teeth, especially the lowest. Plants from S. Europe (var. *roseum* (C. Presl) Guss.) often have pink petals and the lowest calyx-tooth 2-3 times as long as the tube, sometimes with two indistinct lateral veins; some plants from Ukraine and S. Russia have yellowish-white petals and 2 distinct lateral veins on the lowest calyx-tooth, the latter twice as long as the tube (*T. caucasicum* Tausch).

86. T. pannonicum Jacq., *Obs. Bot.* 2: 21 (1767). Shortly rhizomatous, hairy perennial; stems 20–50(–100) cm, erect. Upper leaflets 30–60 × 8–18 mm, oblong-lanceolate or elliptical; stipules with a long, linear, herbaceous apex. Heads up to 50(–80) mm, ovoid or cylindrical; peduncles up to 80 mm. Calyx-teeth linear-subulate, the lowest twice as long as the others. Corolla 20–25 mm, yellowish-white. *Meadows, meadow-steppes and open scrub.* ● *E.C. & S.E. Europe extending westwards to the S.W. Alps, occasionally cultivated elsewhere.* Al Bu Cz Ga Gr Hu It Ju Po Rm Rs (W) ?Tu.

87. T. canescens Willd., *Sp. Pl.* 3: 1369 (1802). Caespitose perennial; stems 10–30 cm, ascending, simple, clothed with appressed, white hairs. Stipules 20–30 mm, lanceolate, membranous below, herbaceous above. Leaflets of lower leaves 15–30 mm, thin, broadly oblong, deeply emarginate. Heads terminal, solitary, ovoid, lax. Calyx appressed-hairy; teeth 1-veined, the lowest only a little longer than the others and scarcely longer than the tube. Corolla *c.* 20 mm, creamy-white; standard twice as long as the calyx, emarginate. *Mountain grassland. Krym.* Rs (K).

88. T. alexandrinum L., *Cent. Pl.* 1: 25 (1755). Annual; stems 40–70 cm, erect, branching, sparsely appressed-hairy. Leaflets 15–25 × 5–10 mm, oblong or lanceolate; free part of the stipules subulate, marginal hairs of the upper ones dilated at the base. Heads 15–20(–25) mm, ovoid or oblong-conical; peduncles up to 30 mm. Calyx hairy; tube obconical; teeth unequal, triangular-subulate, spinescent, the lowest 3-veined, at least at the base, about as long as the tube, others shorter, 1-veined. Corolla 8–10 mm, cream, about twice as long as the calyx. Apex of the mature legume slightly exserted, not concealed by the ring of hairs in the calyx-throat, which is devoid of a bilabiate callosity. *Widely cultivated in the warmer parts of Europe and locally naturalized.* [Au Bu He Hs Hu Lu.]

Long cultivated in N.E. Africa, possibly native in the E. Mediterranean region. Plants from S.E. Europe appear to belong to *T. constantinopolitanum* (*vide* **90**) which is sometimes identified with *T. alexandrinum* var. *phleoides* (Boiss.) Boiss.

89. T. apertum Bobrov in Komarov, *Fl. URSS* 11: 391 (1941). Like **88** but calyx narrower and less hairy; veins less prominent; teeth 1-veined, narrower, less spinescent, subpatent in fruit; corolla larger. *S.E. Russia.* Rs (E). (*N.W. Caucasus, N.E. Anatolia.*)

90. T. echinatum Bieb., *Fl. Taur.-Cauc.* 2: 216 (1808) (*T. supinum* Savi). Annual; stems 10–60 cm, procumbent or ascending, sparsely to densely hairy or glabrous. Leaflets 8–25 × 4–12 mm, obovate or oblanceolate; stipules short. Heads 8–15 mm in flower, ovoid or globose; peduncles 20–50(–80) mm. Fruiting calyx with tube 1·5–2 mm, campanulate or obconical, sparsely hairy or glabrous, the furrows between the 10 veins not reaching the widened mouth of the tube; teeth linear-subulate, 1-veined or 3-veined only at the base, stiff and spinose, the lowest much longer than the others and twice as long as the tube. Corolla 8–12 mm, pink or cream, at least twice as long as the calyx. Legume included. *Grassy, often damp places. Balkan peninsula extending to Italy and S. Romania.* Al Bu Gr It Ju Rm Si Tu.

Very variable in habit, height, indumentum and length of the calyx-teeth.

T. constantinopolitanum Ser. in DC., *Prodr.* 2: 193 (1825), is a plant of doubtful identity and affinity, variously treated as identical with or as a variety of **88**, or as a subspecies of **90**. Plants from Bulgaria, N. Greece and Turkey-in-Europe so treated are said to have stems with patent hairs, obovate leaflets, calyx-tube somewhat constricted at the apex, the lowest calyx-tooth scarcely longer than the tube, and corolla pale yellow. The legume is included in the calyx-tube as in **90** and not exserted as in **88**.

91. T. latinum Sebastiani, *Rom. Pl.* 1: 7 (1813). Like **90** but peduncles 60–80 mm; leaflets up to 60 × 4·5 mm, linear-lanceolate, acute; calyx-tube pubescent. *S.E. part of the Balkan peninsula; C. Italy.* Bu Gr It Tu.

Often regarded as a hybrid between **90** and **92**.

92. T. leucanthum Bieb., *Fl. Taur.-Cauc.* 2: 214 (1808). Annual; stems (5–)15–30 cm, with dense patent hairs. Leaflets 10–20 mm, cuneate-oblong. Heads 10–15 mm, nearly globose, often paired; peduncles 30–120 mm, appressed hairy. Fruiting calyx urceolate, densely pubescent, the veins reaching the base of the subequal, lanceolate, 3-veined, patent teeth. Corolla 6–8 mm, white or pink, a little longer than the calyx. Legume included. *Dry, stony places. S. Europe.* Al Bu Co Cr Gr Hs It Ju Lu Rs (K) Sa Si Tu [Ga].

93. T. squamosum L., *Amoen. Acad.* 4: 105 (1759) (*T. maritimum* Hudson). Annual; stems 10–40 cm, procumbent or erect, hairy or glabrescent. Upper leaflets 10–20 × 6–8 mm, narrowly obovate-cuneate or oblong, often apiculate. Stipules linear, the free herbaceous part longer than the rest. Fruiting heads 10–20 mm, ovoid; peduncles up to 20 mm. Fruiting calyx tough, with a campanulate tube, glabrous or thinly hairy above, the 10 veins and furrows disappearing below the dilated mouth; teeth lanceolate-acuminate, spreading, herbaceous, the lowest distinctly 3-veined and equalling the tube, the other four 1- or 3-veined, about ½ as long as the tube. Corolla 5–7 mm, pale pink, exceeding the calyx. Legume included. *Damp, grassy places especially near the sea. Mediterranean region and W. Europe, northwards to England.* Al Az Bl Br Co Cr Ga Gr Hs It Ju Lu Sa Si Tu.

Moderately variable in habit, size of the parts and indumentum. **T. xatardii** DC. in Lam. & DC., *Fl. Fr.* ed. 3, **5**: 558 (1815), from S.W. France, has the calyx-tube sparsely hairy to near the base and the lowest calyx-tooth sometimes barely longer than the others, but is probably not clearly distinct from 93. **T. cinctum** DC., *Cat. Pl. Horti Monsp.* 152 (1813), from the coasts of Jugoslavia and Albania, has heads subtended by a 6- or 7-fid bracteiform involucre; it may merit subspecific rank.

94. T. squarrosum L., *Sp. Pl.* 768 (1753) (*T. panormitanum* C. Presl). Robust annual; stems 20–80 cm, erect or ascending, appressed-hairy or glabrous. Leaflets 20–40(–70) × 8–15 mm, oblong. Stipules 20–50 mm, with a herbaceous, linear, 3-veined, free apex; marginal hairs dilated at the base. Heads 15–30 mm, ovoid; peduncles (0–)10–60 mm. Calyx-tube ovoid, contracted at the mouth, densely hairy; teeth triangular-lanceolate, herbaceous, 3-veined, the lowest wider, twice as long as the others, deflexed in fruit; marginal hairs dilated at the base so that the calyx-teeth appear finely denticulate. Corolla 9–12 mm, pale pink or white, equalling or slightly exceeding the lowest calyx-tooth. Legume included. *Grassy places. S. Europe.* Al Co Ga Gr Hs It Ju Lu Rs (W) Sa Si Tu.

95. T. obscurum Savi, *Obs. Trif.* 31 (1810). Annual; stems 20–60 cm, little-branched, often flexuous, glabrous below, appressed-hairy above. Leaflets 15–20 × 7–12 mm, obovate to oblong-lanceolate. Free part of the stipules lanceolate or linear, herbaceous. Heads 20–35 mm, shortly pedunculate, globose, oblong-ovoid or conical-cylindrical and rather lax in fruit. Fruiting calyx with an ovoid, glabrescent or hairy tube, usually somewhat constricted above; teeth almost equal, about as long as the tube, ovate-lanceolate, all with 3 or 5 veins, sometimes with anastomoses, herbaceous, patent. Corolla 4–5 mm, whitish or pale pink, not exceeding the calyx. Legume included. *Moist, sandy fields. C. & S. Italy; S. Spain.* Hs It. (*N.W. Africa.*)

(a) Subsp. **obscurum**: Upper leaflets obovate, 2–2½ times as long as wide; free apex of stipules 6–10 mm, lanceolate. Heads not more than 20 mm, globose or ovoid; peduncles longer than the upper stipules. ● *C. & S. Italy.*

(b) Subsp. **aequidentatum** (Pérez Lara) C. Vicioso, *Anal. Inst. Bot. Cavanilles* **11**(2): 344 (1953) (*T. isodon* Murb.): Upper leaflets oblong-lanceolate, often apiculate, 3–3½ times as long as wide; free apex of stipules 12–20 mm, linear. Heads up to 35 mm, conical-cylindrical or ovoid; peduncles scarcely exceeding the upper stipules. *S. Spain.*

96. T. clypeatum L., *Sp. Pl.* 769 (1753). Annual; stems 8–35 cm; hairs deflexed. Leaflets 10–20 mm, broadly obovate-cuneate, often apiculate. Stipules ovate, acute. Heads solitary; peduncles 10–60 mm, hairs often appressed. Calyx-teeth patent, ovate-triangular, foliaceous, with numerous prominent veins. Corolla 20–25 mm, pale pink. *Karpathos.* Cr. (*Mediterranean region.*)

Sect. TRICHOCEPHALUM Koch. Heads capitate; flowers sessile, ebracteate, the outer fertile, the inner consisting only of sterile calyces.

97. T. subterraneum L., *Sp. Pl.* 767 (1753). Annual; stems up to 20(–30) cm, numerous, procumbent. Leaflets broadly obcordate; petioles usually long; stipules semi-ovate. Fruiting-heads globose, appressed to or buried in the soil by the long, deflexed peduncle. Fertile flowers 2–5(–7); corolla 8–14 mm, whitish. Sterile flowers numerous, developing after anthesis, becoming strongly deflexed over the fruiting calyces. Legume 1-seeded, ovoid, somewhat exserted. $2n = 16$. *Dry grassy places, often on sandy soils. S. & W. Europe, northwards to England, the Netherlands and S.E. Hungary.* Al Az Be Bl Br Bu Co Cr Ga Gr Hb Ho Hs Hu It Ju Lu Rm Rs (W, K) Sa Si Tu.

Very variable in the size of its vegetative parts.

98. T. globosum L., *Sp. Pl.* 767 (1753) (*T. radiosum* Wahlenb., *T. nidificum* Griseb.). Annual with patent hairs on stems and petioles. Leaflets obovate-cuneate or obcordate; stipules ovate. Peduncles longer than the subtending leaves. Fertile flowers 10–15, in two rows. Sterile calyces densely hairy, developing simultaneously with the fertile ones. Fruiting heads 20–25 mm, globose, becoming detached at maturity. *Dry grassy places. S.E. part of Balkan peninsula.* Bu Gr Tu.

European plants are often identified with **T. radiosum** Wahlenb. in Jakob Berggren, *Res. Eur. Österländ.* **2** App.: 43 (1827), and treated as specifically distinct from *T. globosum* L.

99. T. pauciflorum D'Urv., *Mem. Soc. Linn. Paris* **1**: 350 (1822) (*T. olivieranum* Ser., *T. globosum* auct., non L.). Like **98** but stems and peduncles with more or less appressed hairs; peduncles more or less equalling the subtending leaves; heads *c.* 15 mm, with 4–6, uniseriate, fertile flowers; sterile calyces with more slender lobes. *Dry places. Aegean region.* Gr Tu.

58. Dorycnium Miller[1]

Perennial herbs or small shrubs. Leaves 5-foliolate, the lowest pair simulating stipules; stipules minute, free. Flowers in axillary heads. Calyx campanulate, with 5 equal or unequal teeth; corolla white or pink, with an obtuse, dark red or black keel; stamens diadelphous. Legume oblong to ovoid-globose, dehiscent. Seeds 1–many.

Literature: M. Rikli, *Bot. Jahrb.* **31**: 314–404 (1901).

1 Corolla 10–20 mm   **1. hirsutum**
1 Corolla 3–7 mm
   2 At least the lower and middle cauline leaves with a rhachis at least 5 mm long   **2. rectum**
   2 Leaves without or with a very short rhachis
     3 Calyx-teeth equal; legume 5–7 mm, cylindrical   **3. graecum**
     3 Calyx-teeth ± unequal, the lowest distinctly longer than the upper 2; legume 3–5 mm, ovoid-globose   **4. pentaphyllum**

1. D. hirsutum (L.) Ser. in DC., *Prodr.* **2**: 208 (1825) (*Bonjeanea hirsuta* (L.) Reichenb.). Perennial herb or small shrub 20–50 cm, usually villous. Leaves without or with a very short rhachis; leaflets 7–25 × 3–8 mm, oblong-obovate. Heads 4- to 10-flowered. Calyx-teeth unequal; corolla 10–20 mm, standard and wings white or pink. Legume 6–12 mm, oblong-ovoid, the valves not contorted at maturity. $2n = 14$. *Mediterranean region and S. Portugal.* Al Bl Co Cr Ga Gr Hs It Ju Lu Sa Si Tu.

2. D. rectum (L.) Ser. in DC., *loc. cit.* (1825) (*Bonjeanea recta* (L.) Reichenb.). Perennial herb or small shrub 30–150 cm, appressed-pubescent. Leaves with rhachis 5–10 mm; lowest pair of leaflets 7–25 × 6–18 mm, ovate to subreniform, upper 3 15–35 × 8–20 mm, obovate or obovate-oblong. Heads 20- to 40-flowered. Calyx-teeth equal; corolla 5–6 mm, standard and

---

[1] By P. W. Ball.

wings white or pink. Legume 10–20 mm, linear-oblong, the valves contorted at maturity. *Mediterranean region, C. & S. Portugal.* Al Bl Co Cr Ga Gr Hs It Lu Sa Si Tu.

3. D. graecum (L.) Ser. in DC., *loc. cit.* (1825). Perennial herb or small shrub 20–80 cm, appressed-pubescent. Leaves without or with a very short rhachis; leaflets 10–25 × 3–10 mm, obovate-oblong. Heads 10- to 25-flowered. Calyx-teeth equal; corolla 6–7 mm, standard and wings white. Legume 5–7 mm, oblong or oblong-ovoid, the valves not contorted at maturity. *E. part of Balkan peninsula; Krym.* Bu Gr Rs (K) Tu.

4. D. pentaphyllum Scop., *Fl. Carn.* ed. 2, **2**: 87 (1772). Perennial herb or small shrub 10–80 cm. Leaves without rhachis; leaflets linear to obovate-oblong. Calyx-teeth unequal; corolla 3–6(–7) mm, standard and wings white. Legume 3–5 mm, ovoid-globose. *C. & S. Europe.* Al Au Bl Bu Co Cz Ga Ge Gr He Hs Hu Ju It Lu Po Rm Rs (K) Sa Si Tu.

The following subspecies are treated by many authors as species, but subspp. (a), (b) and (c) are often difficult to separate on purely morphological grounds. Subsp. (d) is generally well-defined and few intermediates occur except in S.E. France.

1 Heads 12- to 25-flowered; pedicels as long as or longer than calyx-tube
  2 Calyx-teeth about as long as tube   **(c) subsp. gracile**
  2 Calyx-teeth much shorter than tube   **(d) subsp. herbaceum**
1 Heads 5- to 15-flowered; pedicels shorter than or as long as calyx-tube
  3 Leaflets of the upper leaves 6–12 × 2–3 mm; standard apiculate
    (a) subsp. pentaphyllum
  3 Leaflets of the upper leaves (8–)10–20 × 2–4 mm; standard not or only slightly apiculate   **(b) subsp. germanicum**

(a) Subsp. **pentaphyllum** (*D. suffruticosum* Vill.): Stems 10–50 cm, appressed-pubescent. Leaflets of upper leaves 6–12 × 2–3 mm, linear-oblanceolate. Heads 5- to 15-flowered; pedicels usually shorter than calyx-tube. Calyx-teeth shorter than tube; corolla 4–6(–7) mm, standard apiculate. *S.W. Europe, extending to S. Italy.*

Plants from N. Portugal and N. Spain are like subsp. (a) but have the leaflets 12–18 × 2–3(–4) mm and the heads with up to 25 flowers. They may represent another subspecies.

(b) Subsp. **germanicum** (Gremli) Gams in Hegi, *Ill. Fl. Mitteleur.* **4**(3): 1380 (1923) (*D. germanicum* (Gremli) Rikli): Like subsp. (a) but leaflets of upper leaves (8–)10–20 × 2–4 mm, oblong-obovate; standard not or only slightly apiculate. ● *C. Europe and Balkan peninsula.*

(c) Subsp. **gracile** (Jordan) Rouy, *Fl. Fr.* **5**: 137 (1899) (*D. jordanii* Loret & Barr.): Like subsp. (a) but stems 30–80 cm; leaflets of upper leaves 10–20 × 2–4 mm; heads 12- to 20-flowered; pedicels usually longer than calyx-tube; calyx-teeth about equalling tube; corolla 3–5 mm. *Mediterranean coasts of France and Spain.*

(d) Subsp. **herbaceum** (Vill.) Rouy, *op. cit.* 135 (1899) (*D. herbaceum* Vill.): Stems 20–65 cm, usually patent-pubescent. Leaflets of upper leaves 5–20 × 2–6 mm, oblong-obovate. Heads 12- to 25-flowered; pedicels usually as long as calyx. Calyx-teeth not more than ⅓ as long as tube; corolla 3–5 mm; standard not apiculate. *C. & S.E. Europe, extending to S. Italy and Sicilia.*

59. Lotus L.[1]

Annual or perennial herbs, often woody at base. Leaves imparipinnate; leaflets 5, the lowest pair resembling stipules, rarely one of the lowest pair absent; stipules minute. Flowers solitary or in heads. Calyx campanulate or tubular-campanulate, actinomorphic or bilabiate; keel beaked; stamens diadelphous. Legume cylindrical, sometimes compressed, dehiscent. Seeds numerous.

Literature: A. Brand, *Bot. Jahrb.* **25**: 166–232 (1898). In addition a large number of papers have been published in recent years dealing with the genetics and cytotaxonomy of *Lotus*. The following are some of the most recent, and references to earlier papers may be found in them: W. F. Grant, *Canad. Jour. Genet. Cytol.* **7**: 457–471 (1965). K. Larsen & A. Žertová, *Bot. Tidsskr.* **59**: 177–194 (1963). N. Miniaev, *Not. Syst.* (*Leningrad*) **18**: 119–141 (1957). J. Ujhelyi, *Ann. Hist.-Nat. Mus. Hung.* **52**: 185–200 (1960).

1 Leaves with 4 leaflets, the upper 3 broadly obcordate, mucronate   **30. tetraphyllus**
1 Leaves with 5 leaflets, the upper 3 never obcordate
  2 Calyx ±bilabiate, the lateral teeth usually shorter than the rest, the upper teeth usually curved upwards
   3 Leaves with rhachis less than ⅓ as long as the lowest pair of leaflets
    4 Perennial; corolla 8 mm or more, usually about twice as long as calyx   **20–23. creticus** group
    4 Annual; corolla 5–8(–9) mm, not more than 1½ times as long as calyx   **25. halophilus**
   3 Leaves with rhachis more than ⅓ as long as the lowest pair of leaflets
    5 Peduncles not more than 1½(–2) times as long as the subtending leaves in fruit
     6 Bracts shorter than calyx; corolla purple or purple and yellow   **28. azoricus**
     6 Bracts usually longer than the calyx; corolla yellow
      7 Peduncles mostly longer than the subtending leaves in fruit; lowest pair of leaflets triangular or rhombic-orbicular   **26. ornithopodioides**
      7 Peduncles shorter than the subtending leaves in fruit; lowest pair of leaflets ovate or elliptic-ovate   **27. peregrinus**
    5 Peduncles at least twice as long as the subtending leaves in fruit
     8 Annual; standard apiculate   **29. arenarius**
     8 Perennial, woody at base; standard not apiculate
      9 Calyx campanulate; lateral calyx-teeth not more than 0·5 mm wide at middle, filiform or linear-triangular with a triangular base   **1–12. corniculatus** group
      9 Calyx tubular-campanulate; lateral calyx-teeth at least 0·5 mm wide at middle, oblong or triangular-oblong   **20–23. creticus** group
  2 Calyx not bilabiate, the teeth all ±equal, but sometimes curved
   10 Calyx-teeth all shorter than tube
    11 Perennial with woody stock; corolla bright yellow, orange or red   **1–12. corniculatus** group
    11 Annual; corolla white or pale yellow.   **24. strictus**
   10 Calyx-teeth longer than tube
    12 Corolla purple (Thasos)   **19. aduncus**
    12 Corolla white, yellow, orange or reddish
     13 Corolla 18–25 mm; standard exceeding keel by at least 2 mm   **13. aegaeus**
     13 Corolla not more than 18 mm; standard not or only slightly exceeding keel
      14 Calyx-teeth at least 3 times as long as tube **14. parviflorus**
      14 Calyx-teeth less than 3 times as long as tube
       15 Legume 4–8 mm in diameter, sulcate on back   **17. edulis**
       15 Legume 1–3 mm in diameter, not sulcate on back
        16 Perennial, woody at base; legume usually 2–3 mm in diameter   **1–12. corniculatus** group
        16 Annual; legume 1–2 mm in diameter
         17 Legume not more than 3 times as long as calyx; seeds 8–12; keel obtusely angled on the lower edge   **15. subbiflorus**
         17 Legume at least 3 times as long as calyx; seeds more than 12; keel with a right-angle on the lower edge

---

[1] By P. W. Ball (*L. corniculatus* group in collaboration with A. Chrtková-Žertová).

18 Legume straight or slightly curved at apex; peduncles 1- to 3-flowered, shorter or longer than leaves; corolla yellow **16. angustissimus**

18 Legume strongly curved; peduncles 1-flowered, shorter than leaves; corolla white or pink **18. conimbricensis**

Sect. LOTUS. Calyx usually actinomorphic; corolla usually yellow; style not toothed. Legume not inflated.

1–12. L. corniculatus group. Perennial herbs usually with woody stock. Calyx-teeth usually more or less equal, but sometimes curved; corolla 8–18 mm, yellow, sometimes orange or red and yellow. Legume cylindrical, straight. Seeds many.

A widespread, variable group containing diploid and tetraploid species. It is sometimes treated as a single species with a number of subspecies and varieties. Recent work has shown that there are a number of relatively local diploid taxa throughout C. & S. Europe and Asia (in addition to **1** and **9**), together with more widespread tetraploids. The data available at the present are insufficient to produce a comprehensive account of the group, and it is likely that some of the species recognized here are heterogeneous, while others may not be distinct species.

1 Leaflets of upper leaves linear or linear-lanceolate, at least (3–)4 times as long as wide
  2 Calyx-teeth usually shorter than tube; corolla often becoming greenish on drying **1. tenuis**
  2 Calyx-teeth c. 1½ times as long as tube; corolla becoming reddish on drying **2. krylovii**
1 Leaflets lanceolate to obovate, usually not more than 3 times as long as wide
  3 Calyx-teeth at least 1½ times as long as tube
    4 Corolla 10–15 mm, about twice as long as calyx **11. preslii**
    4 Corolla 6–10 mm, not more than 1½ times as long as calyx **12. palustris**
  3 Calyx-teeth not more than 1½ times as long as tube
  5 Stem hollow
    6 Leaflets obovate; upper 2 calyx-teeth separated by an acute sinus in bud **9. uliginosus**
    6 Leaflets rhombic; upper 2 calyx-teeth separated by an obtuse sinus in bud **10. pedunculatus**
  5 Stem solid, sometimes with a narrow hollow at base
    7 Calyx-teeth shorter than or almost equalling tube
      8 Flowering stems usually more than 10 cm; heads mostly 3- to 6-flowered; leaflets 5–15 mm **7. corniculatus**
      8 Dwarf plants with stout stock and procumbent or ascending flowering stems 2–10 cm; heads 1- to 3(–5)-flowered; leaflets 2–6 mm **8. alpinus**
    7 At least the lower calyx-teeth equalling or longer than tube
    9 Corolla not more than 10(–11) mm
      10 Calyx-teeth all ± straight; calyx and leaves glabrous or subglabrous **5. stenodon**
      10 Upper and lateral calyx-teeth strongly curved; calyx and leaves pubescent with silvery appressed hairs or with patent hairs **6. glareosus**
    9 Corolla 10–18 mm
      11 Stems procumbent or ascending
        12 Calyx ± zygomorphic, the upper and lateral teeth curved **4. delortii**
        12 Calyx actinomorphic, the teeth all straight **7. corniculatus**
      11 Stems ± erect
        13 Leaflets elliptic-lanceolate **3. borbasii**
        13 Leaflets elliptic-oblanceolate or obovate **5. stenodon**

1. L. tenuis Waldst. & Kit. ex Willd., *Enum. Pl. Hort. Berol.* 797 (1809) (*L. tenuifolius* (L.) Reichenb., non Burm. fil.). Stems 20–90 cm, glabrous or sparsely pubescent. Leaflets 5–15 × 1–4 mm, linear or linear-lanceolate. Heads 1- to 4(–6)-flowered.

Calyx-teeth equal, usually shorter than tube; corolla 6–12 mm, yellow; wings obovate-oblong. Legume 15–30 × 2–2·5 mm. $2n = 12$. *Most of Europe except the north-east and extreme north.* Al Au Be Bl Br Bu Co Cr Cz Da Ga Ge Gr He Ho Hs Hu It Ju Lu Po Rm Rs (W, K, E) Sa Si Su Tu [Fe No].

2. L. krylovii Schischkin & Serg., *Animadv. Syst. Herb. Univ. Tomsk.* 1932 (7–8): 5 (1932) (*L. frondosus* (Freyn) Kuprian.). Stems 10–35 cm, glabrous or sparsely pubescent. Leaflets 7–13 × 1–4(–6) mm, linear-lanceolate to oblong-obovate. Heads 1- to 2(–4)-flowered. Calyx-teeth equal, longer than tube; corolla 7–10 mm, yellow, with standard red on back and red-veined. Legume 20–35 × 2–3 mm. *Saline steppes. S.E. Russia, W. Kazakhstan.* Rs (E). (*C. & S.W. Asia.*)

3. L. borbasii Ujhelyi, *Ann. Hist.-Nat. Mus. Hung.* 52: 187 (1960) (*L. corniculatus* subsp. *major* auct. pro parte). Stems 15–30(–50) cm, more or less erect, glabrous or villous. Leaflets 5–18(–20) × 1·5–8 mm, elliptic-lanceolate. Heads 2- to 6-flowered. Calyx-teeth slightly unequal, longer than tube, linear-triangular; corolla 13–18 mm; wings rhombic-ovate. Legume 10–30 × 2–4 mm. $2n = 12$. ● *E.C. Europe, extending southwards to Hercegovina.* Au Cz Hu Ju.

4. L. delortii Timb.-Lagr. ex F. W. Schultz, *Arch. Fl. Fr. Allem.* 201 (1852) (incl. *L. pilosus* Jordan). Stems 10–20 cm, procumbent, villous. Leaflets 4–9 × 1·5–3 mm, usually obovate-oblong. Heads 2- to 4-flowered. Calyx-teeth slightly unequal, longer than tube, linear-triangular, the upper and lateral teeth often curved; corolla 12–15 mm; wings obovate. Legume 20–35 × 2·5–3 mm. ● *N. Italy, S. France E. Spain.* Ga Hs It.

5. L. stenodon (Boiss. & Heldr.) Heldr., *Sched. Herb. Graec. Norm.* no. 1419 (1897) (*L. orphanidis* Ujhelyi; incl. *L. preslii* var. *rostellatus* (Heldr.) Hayek). Stems up to 20 cm, procumbent or erect, glabrous or sparsely hairy. Leaflets 3–12 × 1·5–7 mm, elliptic-oblanceolate to obovate. Heads 2- to 5-flowered; peduncle 1–6 cm; pedicels 0·5–2 mm. Calyx-teeth unequal, longer than tube; corolla 8–17 mm. Legume 10–30 × 1·2–2 mm. ◐ *W. part of Balkan peninsula.* Al Gr Ju.

6. L. glareosus Boiss. & Reuter, *Pugillus* 36 (1852). Stems up to 20 cm, procumbent or ascending, sparsely to densely villous, or with dense, silvery, appressed hairs. Leaflets 2–10 × 1·5–5 mm, obovate or elliptic-obovate. Heads 1- to 6-flowered, calyx-teeth ± unequal, longer than tube, linear-triangular or linear with triangular base, the lateral and upper teeth curved; pedicels 0·5–1·5 mm. Calyx-teeth unequal, as long as or longer than tube; corolla 8–10 mm, usually reddish. Legume 15–25 × 2–2·5 mm. ● *Mountains of S. Spain, C. Portugal.* Hs Lu.

Variable and possibly containing two species. One relatively dwarf with dense, silvery, appressed hairs and small leaflets, peduncles and pedicels, the other larger, with patent hairs and larger leaflets, peduncles and pedicels.

7. L. corniculatus L., *Sp. Pl.* 775 (1753) (incl. *L. ambiguus* Besser ex Sprengel, *L. caucasicus* Kuprian.). Stems 5–35 cm, procumbent or ascending, glabrous to villous. Leaflets 4–18 × 1–10 mm, lanceolate or oblanceolate to suborbicular. Heads (1–)2- to 7-flowered; pedicels 1–2·5 mm. Calyx-teeth equal, shorter or slightly longer than the tube, triangular to filiform with triangular base; corolla 10–16 mm, usually yellow. Legume 15–30 × 2–2·5 mm. $2n = 24$. *Almost throughout Europe.* All except Sb; introduced in Is.

As defined here this species is very variable. It may eventually be possible to recognize a number of subspecies, but the native distribution of this and some related species is very confused, owing to their widespread use as a forage crop.

The main variants which occur in Europe are as follows: (i) sparsely to densely pubescent; calyx-teeth shorter than tube (*N. part of the range of the species, locally in the south*); (ii) glabrous or sparsely pubescent; leaflets small, fleshy; calyx-teeth about ½ as long as tube (*coasts of W. & N. Europe*); (iii) villous or densely pubescent; calyx-teeth slightly longer than tube (*C. & S. Europe*).

Dwarf plants resembling **8** in many characters also occur in the mountains.

8. L. alpinus (DC.) Schleicher ex Ramond, *Mém. Mus. Hist. Nat.* (*Paris*) **13**: 275 (1825). Like **7** but stems usually not more than 10 cm; leaflets 2–6 × 1·5–4 mm; heads 1- to 3(–5)-flowered; corolla 12–18 mm. 2n=12. *Pyrenees, Alps, ?Balkan peninsula.* ?Al Au ?Bu Ga Ge He Hs It Ju.

9. L. uliginosus Schkuhr, *Handb.* **2**: 412 (1796) (*L. pedunculatus* auct., non Cav., *L. corniculatus* subsp. *major* auct. pro parte). Stems 30–100 cm, erect or ascending, subglabrous to villous, hollow. Leaflets 8–25 × 3–15 mm, obovate, obtuse, often mucronate, glaucous beneath. Heads 5- to 12(–15)-flowered; pedicels 1–2 mm. Calyx-teeth about as long as tube, the upper pair separated by an acute sinus in bud; corolla 10–18 mm. Legume 15–35 × 2–2·5 mm. 2n=12. *Marshes and wet grassland. W., C. & S. Europe, extending northwards to 60° N. in Fennoscandia and eastwards to c. 25° E. in Ukraine; often occurring as a casual elsewhere in Europe.* Al Au Az Be Br Bu Co Cr Cz Da Ga Ge Gr Hb He Ho Hs It Ju Lu Po Rm Rs (B, W) Sa Su [Fa Fe Hu No].

10. L. pedunculatus Cav., *Icon. Descr.* **2**: 52 (1793). Like **9** but leaflets rhombic, acute; heads 3- to 8(–10)-flowered; calyx-teeth longer than tube, the upper 2 separated by an obtuse sinus in bud. Stems 40–120 cm, pubescent; leaflets 15–35 × 5–12 mm; legume 15–40 × 2–3 mm. *W. & C. Spain, E.C. Portugal.* Hs Lu.

L. granadensis Žertová, *Folia Geobot. Phytotax.* (*Praha*) **1**: 79 (1966), from S. Spain (Sierra Nevada), is probably a subspecies of **10**. It has broadly obovate to orbicular, obtuse or subacute, rather densely hairy, bright green leaflets.

11. L. preslii Ten., *Fl. Nap.* **5**: 160 (1836). Stems 15–80 cm, procumbent, subglabrous or pubescent. Leaflets 6–15 × 3–8 mm, obovate. Heads 1- to 6-flowered; pedicels c. 1·5 mm. Calyx-teeth linear, with triangular base, at least 1½ times as long as tube, the upper pair separated by an acute sinus in bud; corolla 10–15 mm. Legume 20–30 × c. 2 mm. *Wet places. Mediterranean region; local.* Al Bl Ga Gr Hs ?It Ju Si.

12. L. palustris Willd., *Sp. Pl.* **3**: 1394 (1802). Stems 50–100 cm, procumbent or ascending, pubescent or villous. Leaflets 6–20 × 4–8 mm, obovate or oblanceolate. Heads 2- to 4-flowered; pedicels c. 1 mm. Calyx-teeth about twice as long as tube, linear-triangular, curved; corolla 6–10 mm, yellow or whitish, only slightly exceeding calyx. Legume 12–30 × c. 2 mm. *Wet places. S. Albania, Greece and Aegean region.* Al Cr Gr.

13. L. aegaeus (Griseb.) Boiss., *Fl. Or.* **2**: 167 (1872). Villous perennial 20–60 cm. Leaflets 10–20 × 5–10 mm, obovate. Heads 1- to 5-flowered. Calyx-teeth slightly longer than tube; corolla 18–25 mm, bright, pale yellow; standard much longer than keel. Legume 30–50 × 3–4 mm, slightly torulose. *C. & E. parts of Balkan peninsula.* Bu Gr Ju.

14. L. parviflorus Desf., *Fl. Atl.* **2**: 206 (1799). Villous annual up to 40 cm. Leaflets 7–15 × 2·5–7 mm, obovate to oblong-lanceolate. Heads 3- to 7-flowered; peduncles longer than leaves; becoming recurved in fruit. Calyx-teeth 3–4 times as long as tube; corolla 5–10 mm, yellow; keel with a right-angle on the lower edge and a long beak. Legume 4–6 × c. 1·5 mm, not or only slightly longer than calyx; valves not contorting on dehiscence. 2n=12. *S. Europe.* Az Bl Co Cr Ga Gr Hs It Ju Lu Sa Si.

15. L. subbiflorus Lag., *Varied. Ci. Lit. Artes* (*Madrid*) **2**(4): 213 (1805) (*L. hispidus* Desf. ex DC. 1815, non 1805). Annual up to 50(–100) cm. Leaflets 5–20 × 1–8 mm, oblong, lanceolate or obovate-oblong. Heads (1–)2- to 4-flowered; peduncles longer than leaves. Calyx-teeth longer than tube, corolla 5–10 mm, yellow; keel obtusely angled on the lower edge, with a long beak. Legume (4–)6–16 × 1·5–2 mm, straight, up to 3 times as long as calyx; valves contorting on dehiscence. *W. Europe, northwards to c. 52° N. in Britain, and extending eastwards to Sicilia.* Az Bl Br Co Ga Hb Hs It Lu Sa Si.

(a) Subsp. **subbiflorus**: Stems villous or almost hirsute at least at the apex; calyx-teeth 1½–3 times as long as tube. 2n=24. *Almost throughout the range of the species.*

(b) Subsp. **castellanus** (Boiss. & Reuter) P. W. Ball, *Feddes Repert.* **79**: 41 (1968) (*L. castellanus* Boiss. & Reuter): Stems glabrous or pubescent; calyx-teeth c. 1½ times as long as tube. ● *W. & C. Spain, C. & S. Portugal.*

16. L. angustissimus L., *Sp. Pl.* 774 (1753) (incl. *L. praetermissus* Kuprian., *L. thessalus* Hayek). Like **15** but keel with a right-angle on the lower edge and a short beak; legume (10–)15–30 × 1–1·5 mm, usually at least 4 times as long as calyx. Heads 1- to 3-flowered; peduncles shorter or longer than leaves; corolla 5–12 mm, sometimes with purple veins. 2n=12. *S. Europe, extending northwards to S. England and N. Ukraine.* Al Az Bl Br Bu Co Cr Ga Gr Hs Hu It Ju Lu Rm Rs (C, W K, E) Sa Si Tu.

Sect. KROKERIA (Moench) Willk. Like Sect. *Lotus* but the legume very inflated and sulcate on the back.

17. L. edulis L., *Sp. Pl.* 774 (1753). Sparsely pubescent annual 10–50 cm. Leaflets 5–16 × 3·5–10 mm, obovate to obovate-oblong. Heads 1- to 2-flowered; peduncles longer than leaves. Calyx-teeth subequal, longer than tube; corolla 10–16 mm, yellow. Legume 20–40 × 4–8 mm, curved. 2n=14. *Mediterranean region, S. Portugal.* Al Bl Co Cr Ga Gr Hs It Ju Lu Sa Si Tu.

Sect. ERYTHROLOTUS Brand. Calyx actinomorphic; corolla white, pink or purple; style not toothed.

18. L. conimbricensis Brot., *Phyt. Lusit.* 59 (1800) (*L. coimbrensis* Brot. ex Willd.). Sparsely pubescent or glabrous annual 5–30 cm. Leaflets 4–10 × 2–5 mm, obovate or rhombic. Peduncles shorter than leaves, 1-flowered. Calyx-teeth longer than tube; corolla 5–8 mm, white or pale pink, with violet keel. Legume 20–60 × 1–2 mm, curved upwards. 2n=12. *Mediterranean region, S. Portugal.* Co Cr Ga Gr Hs It Lu Sa Si.

19. L. aduncus (Griseb.) Nyman, *Syll.* 298 (1855). Villous perennial 10–15 cm. Leaflets 5–14 × 2–6 mm, obovate or oblong-obovate. Heads 2- to 5-flowered; peduncles as long as or longer than leaves. Calyx-teeth 1½–2 times as long as tube; corolla 15–20 mm, purple. Legume not known. *Limestone rocks above 1000 m.* ● *Aegean region* (*Thasos*). Gr.

Sect. LOTEA (Medicus) Willk. Calyx bilabiate; corolla yellow; style not toothed.

20–23. L. creticus group. Perennials up to 50 cm, glabrous to densely sericeous. Leaflets obovate to oblong-oblanceolate. Heads (1–)2- to 6-flowered. Upper 2 calyx-teeth curved upwards, lateral 2 shorter than the lower. Legume 20–50 × 1·5–2 mm.

1 Lateral calyx-teeth acute, only slightly shorter than the upper; keel with a long straight beak
2 Leaflets densely sericeous; keel with a purple beak   **22. creticus**
2 Leaflets subglabrous to pubescent; keel with a yellow or brownish-yellow beak   **23. collinus**
1 Lateral calyx-teeth obtuse, much shorter than the upper; keel with a short, curved beak
3 Legume straight or slightly curved   **20. cytisoides**
3 Legume curved in a semi-circle or circle   **21. drepanocarpus**

20. L. cytisoides L., *Sp. Pl.* 776 (1753) (*L. creticus* auct., non L.). Leaflets 4–14 × 2–8 mm. Calyx 6–7·5 mm, the lateral teeth obtuse, much shorter than the upper. Corolla 8–14 mm; standard emarginate; wings slightly longer than keel; keel with a short, curved, purple beak (about as long as the rest of the keel). Legume straight or slightly curved. 2*n* = 14. *Mediterranean region.* Bl Co Cr Ga Gr Hs It Ju Sa Si.

21. L. drepanocarpus Durieu in Duchartre, *Rev. Bot.* **2**: 438 (1847). Like **20** but legume curved in a semi-circle or circle. *Naturalized in S. France and Jugoslavia.* [Ga Ju.] (*N. Africa.*)

22. L. creticus L., *Sp. Pl.* 775 (1753) (*L. commutatus* Guss.). Leaflets 7–18 × 4–9 mm, densely sericeous. Calyx 7·5–9 mm, the lateral teeth acute, almost as long as the upper. Corolla 12–18 mm; standard entire; wings much longer than keel; keel with a long, straight, purple beak (up to twice as long as the rest of the keel). Legume straight or slightly curved. 2*n* = 28. *Maritime. Mediterranean region, Portugal.* Gr Hs Ju Lu Si.

23. L. collinus (Boiss.) Heldr., *Sched. Herb. Graec. Norm.* no. 1320 (1897) (*L. creticus* subsp. *collinus* (Boiss.) Briq.). Like **22** but leaflets subglabrous to pubescent; keel with a yellow or brownish-yellow beak. *Dry places inland. Greece and S. Spain, local.* Gr Hs. (*N. Africa, S.W. Asia.*)

L. longisiliquosus R. de Roemer, *Linnaea* **25**: 22 (1852), from S. Spain, is probably identical with **23**, although it was originally placed in the *L. corniculatus* group. Records of *L. longisiliquosus* from Islas Baleares, however, probably refer to **7**.

24. L. strictus Fischer & C. A. Meyer, *Ind. Sem. Horti Petrop.* **1**: 32 (1835). Glabrous or subglabrous annual up to 100 cm. Leaves shortly petiolate; leaflets 6–15 × 4–8 mm, oblong-obovate. Heads 2- to 10-flowered; peduncles longer than leaves. Calyx-teeth subequal, the upper teeth curved upwards; corolla 15–20 mm, white or pale yellow. Legume 25–30 × 3–4 mm, straight or curved at apex. *E. Bulgaria, N.E. Greece.* Bu Gr. (*S.W. Asia.*)

25. L. halophilus Boiss. & Spruner in Boiss., *Diagn. Pl. Or. Nov.* **1**(2): 37 (1843) (*L. villosus* Forskål, non Burm. fil., *L. pusillus* Viv., non Medicus). Pubescent annual 10–30 cm. Leaves sessile or shortly petiolate; leaflets 3–7 × 1·5–3 mm, oblong-obovate. Heads 1- to 9-flowered; peduncles longer than leaves. Lateral calyx-teeth slightly shorter than upper; corolla 5–8(–9) mm. Legume 20–30 × 1·5–2 mm, slightly torulose; curved at apex. *Maritime sands. E. Mediterranean region, extending to Sicilia.* Cr Gr It Si.

---

[1] By P. W. Ball.

26. L. ornithopodioides L., *Sp. Pl.* 775 (1753). Pubescent annual 10–50 cm. Leaves petiolate; leaflets 8–30 × 4–16 mm, the upper 3 obovate to rhombic, the lower 2 ovate-rhombic, cordate or cuneate at base. Heads 2- to 5-flowered; peduncles equalling or slightly longer than leaves. Lateral calyx-teeth very short, obtuse; corolla 7–10 mm. Legume 20–50 × 2–3 mm, torulose, curved. 2*n* = 14. *S. Europe.* Al Bl Co Cr Ga Gr Hs It Ju Lu Rs (W, K) Sa Si Tu.

27. L. peregrinus L., *Sp. Pl.* 774 (1753). Like **26** but leaflets 5–15 × 3–8 mm, the lower 2 ovate, always cuneate; peduncles shorter than or equalling leaves; corolla 6–9 mm; legume not or only slightly torulose, straight. *Greece and Aegean region; Linosa.* Cr Gr Si.

Sect. PEDROSIA (Lowe) Brand. Calyx more or less bilabiate; corolla yellow or purple; style toothed.

28. L. azoricus P. W. Ball, *Feddes Repert.* **79**: 40 (1968) (*L. macranthus* auct. azor., non Lowe). Diffuse annual or perennial 20–40 cm, with dense, appressed, silvery indumentum. Upper 3 leaflets 5–10 × 2–6 mm, obovate, basal 2·5–8 × 2·5–6 mm, suborbicular or ovate-orbicular. Peduncles 1-flowered, shorter than leaves, with a trifoliolate bract inserted near the middle or towards the apex. Calyx-teeth slightly unequal, the upper 2 longer than the lower 3 and curved upwards, all longer than tube. Corolla 20–25 mm, purple or yellow and purple; standard shorter than the long-beaked keel. Legume 40–55 × 3–4 mm, straight. ● *Açores (Santa Maria and São Miguel).* Az.

29. L. arenarius Brot., *Fl. Lusit.* **2**: 120 (1804). Pubescent or subglabrous annual up to 50 cm. Leaves shortly petiolate; leaflets 5–15 × 2·5–8 mm, obovate. Heads 2- to 6-flowered; peduncles longer than leaves. Calyx-teeth more or less unequal, the lateral 2 usually shorter than the others. Corolla 10–15 mm, yellow with red or purple striations. Legume 25–50 × 2–2·5 mm, straight. *S. Spain, S. & C. Portugal.* Hs Lu.

Sect. QUADRIFOLIUM Brand. Leaflets 4; calyx-teeth slightly unequal; corolla yellow, with red or purple striations; style not toothed.

30. L. tetraphyllus L., *Syst. Veg.* ed. 13, 575 (1774). Sparsely pubescent perennial up to 15(–30) cm. Upper 3 leaflets 3–8 × 2–5 mm, obcordate, mucronate, lowest leaflet oblong or oblong-obcordate. Peduncles 1-flowered, longer than leaves. Upper 2 calyx-teeth curved upwards, lateral 2 slightly shorter than the others; corolla 7–10 mm. Legume 20–25 × *c.* 2 mm, straight. ● *Islas Baleares.* Bl.

60. Tetragonolobus Scop.[1]

Like *Lotus* but leaves 3-foliolate; stipules herbaceous; flowers solitary or paired; calyx-teeth equal; legume almost square in transverse section, with the angles winged or keeled.

1 Peduncles at least twice as long as the leaves; calyx-teeth shorter than tube; corolla yellow or orange
2 Flowers solitary; corolla 25–30 mm; legume 30–60 × 3–5 mm, glabrous or subglabrous   **1. maritimus**
2 Flowers 1–4 on each peduncle; corolla 17–25 mm; legume 20–40 × 4–6 mm, pubescent   **2. biflorus**
1 Peduncles shorter than or equalling leaves; calyx-teeth longer than tube; corolla pink, red or purple
3 Legume with 4 wings at least 2 mm wide; calyx-teeth not more than twice as long as tube; style not membranous-winged at the tip   **3. purpureus**

3 Legume with wings less than 2 mm wide, often only keeled; calyx-teeth 2–3 times as long as tube; style with a unilateral membranous wing at the tip
4 Legume 4-winged or -keeled; seeds 3–5 mm in diameter
4. conjugatus
4 Legume with 2 wings on the upper side but neither winged nor keeled beneath; seeds 2–2·5 mm in diameter **5. requienii**

1. T. maritimus (L.) Roth, *Tent. Fl. Germ.* **1**: 323 (1788) (*Lotus siliquosus* L.). Glabrous or pubescent perennial 10–40 cm. Leaflets up to 30 × 15 mm, oblanceolate to obovate; stipules ovate, acute or subobtuse. Peduncles much longer than leaves. Flowers solitary. Calyx-teeth shorter than tube; corolla 25–30 mm, pale yellow; style with a unilateral membranous wing at the tip. Legume 30–60 × 3–5 mm, glabrous or subglabrous; wings *c*. 1 mm wide. $2n=14$. *C. & S. Europe, extending northwards to S. Sweden and eastwards to E. Ukraine, but absent from most of the Mediterranean region.* Au Bu Co Cz Da Ga Ge He Hs Hu It Ju Po Rm Rs (W, K) Sa Su [Br].

2. T. biflorus (Desr.) Ser. in DC., *Prodr.* **2**: 215 (1825) (*Lotus biflorus* Desr.). Like **1** but annual; stipules ovate-orbicular, obtuse; flowers 1–4 on each peduncle; corolla 17–25 mm, deep bright orange; legume 20–40 × 4–6 mm, pubescent. *S. Italy, Sicilia; N.W. Greece.* Gr It Si.

3. T. purpureus Moench, *Meth.* 164 (1794) (*Lotus tetragonolobus* L.). Pubescent annual 10–40 cm. Leaflets up to 40 × 25 mm, obovate or obovate-rhombic; stipules ovate, more or less acute. Peduncles shorter than or equalling leaves. Flowers solitary or paired. Calyx-teeth 1–2 times as long as tube; corolla 15–22 mm, crimson; style not winged at the tip. Legume 30–90 × 6–8 mm, glabrous; wings 2–4 mm wide. *S. Europe, extending northwards to c. 47°N. in Ukraine.* Bl Cr *Ga Gr Hs It Rs (W, K) Sa Si [Cz Lu].

T. wiedemannii Boiss., *Fl. Or.* **2**: 176 (1872) (*Lotus wiedemannii* (Boiss.) Nyman) is an imperfectly known species from the Aegean region (Paros). It is like **3** but with leaflets *c*. 4 mm, emarginate (not acute to obtuse); stipules very shortly stalked (not sessile); corolla *c*. 10 mm; and calyx-teeth obtuse (not acute), shorter than the tube.

4. T. conjugatus (L.) Link, *Enum. Hort. Berol. Alt.* **2**: 264 (1822). Pubescent annual 10–30 cm. Leaflets up to 25 × 15 mm, obovate or obovate-rhombic, acute or mucronate; stipules ovate, acute. Peduncles shorter than or equalling leaves. Flowers solitary or paired; calyx-teeth 2–3 times as long as tube; corolla 14–20 mm, bright red; style with a unilateral membranous wing at the tip. Legume 20–40 × 3–6 mm, glabrous, with 4 wings or keels up to 2 mm wide. Seeds 3–5 mm in diameter. *Sicilia.* Si.

5. T. requienii (Mauri ex Sanguinetti) Sanguinetti, *Fl. Rom. Prodr. Alt.* 581 (1864) (*T. conjugatus* auct., non (L.) Link, *Lotus requienii* Mauri ex Sanguinetti). Like **4** but corolla 13–15 mm; legume 25–65 × 3·5–4·5 mm with 2 wings on the upper side and neither winged nor keeled beneath; seeds 2–2·5 mm in diameter. *Mediterranean region, S. Portugal.* Bl ?Co Gr Hs It Lu.

61. Hymenocarpos Savi[1]

(*Circinnus* Medicus)

Annual. Leaves simple to imparipinnate; stipules minute, membranous. Flowers in heads. Calyx campanulate, with 5 equal teeth; corolla yellow; keel beaked; stamens diadelphous.

Legume indehiscent, spirally twisted and flattened so that it is suborbicular in outline; outer margin membranous-winged.

1. H. circinnatus (L.) Savi, *Fl. Pis.* **2**: 205 (1798). Stems up to 30 cm, densely patent-pubescent. Lower leaves simple, obovate-oblong, upper with 2–3 pairs of leaflets, the terminal much larger than the lateral. Calyx-teeth filiform, much longer than tube; corolla 5–7 mm. Legume 10–15 mm in diameter, the outer margin often toothed. *S. Europe, westwards to S.E. France.* Al Bu Co Cr Ga Gr It Ju Sa Si Tu.

62. Securigera DC.[1]

(*Bonaveria* Scop.)

Annual. Leaves imparipinnate; stipules small. Flowers in axillary heads. Calyx campanulate, bilabiate; corolla yellow; keel beaked; stamens monadelphous. Legume linear, compressed, with thickened margins and a long beak, tardily dehiscent.

1. S. securidaca (L.) Degen & Dörfler, *Denkschr. Akad. Wiss. Math.-Nat. Kl.* (*Wien*) **64**: 718 (1897) (*Bonaveria securidaca* (L.) Reichenb.). Stems 10–50 cm, glabrous. Leaflets 4–7 pairs, oblong-obovate, truncate or emarginate, mucronate. Heads 4- to 8-flowered. Corolla 8–12 mm. Legume 5–10 cm (including beak), the beak 1·5–3 cm, recurved at tip. $2n=12$. *S. Europe westwards to S.E. France.* Al Bu Co Cr Ga Gr It Ju Rs (K) Si Tu.

63. Anthyllis L.[2]

(Incl. *Physanthyllis* Boiss., *Cornicina* Boiss. (DC.) and *Dorycnopsis* Boiss.)

Shrubs or herbs. Leaves usually imparipinnate, rarely simple or 3-foliolate. Stipules small, caducous. Flowers usually in dense heads, rarely in fascicles or borne singly in the bract-axils. Calyx tubular, campanulate or constricted near the apex, with equal or unequal teeth; corolla variously coloured; stamens monadelphous, or the upper stamen free for up to ½ its length. Legume sessile or stipitate, frequently indehiscent or tardily dehiscent, usually completely included within the persistent calyx. Seeds 1–many.

Measurements of the calyx given in the keys and descriptions are at anthesis and include the teeth unless these are specifically excluded.

1 Shrubs or undershrubs with woody branches
2 Flowers in terminal heads; bracts subtending the heads palmatisect
3 Leaflets 3–5, the terminal leaflets much larger than the lateral
6. henoniana
3 Leaflets very numerous, all ± the same size
4 Leaflets narrowly elliptical; calyx 4–6 mm **4. barba-jovis**
4 Leaflets very narrowly elliptical to linear; calyx 6·5–9 mm
5. aegaea
2 Flowers in axillary fascicles or solitary; bracts simple
5 Plant spiny; branches tortuous **3. hermanniae**
5 Plant not spiny; branches straight or flexuous
6 At least the upper leaves 3-foliolate; calyx 4·5–7 mm
1. cytisoides
6 All leaves simple; calyx 3–4·5 mm **2. terniflora**
1 Herbs, sometimes slightly woody at the base, but never with woody branches
7 Calyx inflated at anthesis, gibbous, constricted at the apex
8 Calyx unequally 5-toothed, the mouth oblique; legume 1-seeded, or if 2-seeded then not constricted between the seeds **15. vulneraria**

8 Calyx ±equally 5-toothed, the mouth straight; legume 2-seeded, constricted between the seeds   **16. tetraphylla**
7 Calyx not inflated at anthesis (rarely slightly inflated in fruit), neither gibbous nor constricted at the apex
  9 Annual
    10 Legume straight; heads 4- to 8-flowered   **17. lotoides**
    10 Legume curved; heads more than 8-flowered
      11 Legume not winged, falcate, beaked, the beak exserted from the calyx; calyx 7–8 mm   **19. hamosa**
      11 Legume winged, circinate or annular, not beaked, completely included within the calyx; calyx 5–7 mm   **18. cornicina**
  9 Perennial, woody at the base
    12 Calyx 1·5–6 mm; middle cauline leaves with 5–11 leaflets
      13 Calyx 4–6 mm, the teeth linear, plumose   **12. ramburii**
      13 Calyx 1·5–3 mm, the teeth broadly triangular, not plumose
        14 Leaves glaucous, mostly near the base of the stem; corolla yellow   **13. onobrychoides**
        14 Leaves green, evenly distributed along the stem; corolla pink   **14. gerardi**
    12 Calyx (4–)6·5–12 mm; middle cauline leaves with 13–41 leaflets
      15 Leaves confined to lower ⅓ of the stem; petiole dilated, sheathing and striate
        16 Corolla red to purple; calyx-teeth plumose, ± as long as the tube   **7. montana**
        16 Corolla yellow; calyx-teeth not plumose, much shorter than the tube   **8. aurea**
      15 Leaves evenly distributed along the stem; petiole not dilated, sheathing or striate
        17 Stems procumbent or decumbent; leaves densely tomentose   **10. tejedensis**
        17 Stems erect or ascending; leaves villous to hirsute
          18 Calyx 6–9 mm, with teeth as long as the tube; standard obovate-orbicular   **9. polycephala**
          18 Calyx 10–12 mm, with teeth shorter than the tube; standard ovate   **11. rupestris**

1. A. cytisoides L., *Sp. Pl.* 720 (1753). Shrub up to 60 cm; branches woody, erect, straight to flexuous, whitish- or greyish-tomentose-puberulent. Lower leaves simple; upper leaves 3-foliolate; the terminal leaflet narrowly elliptical, much larger than the laterals. Flowers solitary or in fascicles of 2–3 in the axils of simple, elliptical to ovate bracts, forming a spiciform inflorescence. Calyx 4·5–7 mm, villous to hirsute, with 5 equal teeth, the teeth shorter than the tube. Corolla yellow. Seed 1. *S. & E. Spain, Islas Baleares, S. France.* Bl Ga Hs.

2. A. terniflora (Lag.) Pau, *Bull. Acad. Int. Géogr. Bot. (Le Mans)* **16** (Mém.): 75 (1906) (*A. genistae* Dufour ex DC.). Like **1** but less robust; indumentum finely sericeous; leaves all simple, oblong to narrowly elliptical; calyx 3–4·5 mm. *S. & S.E. Spain.* Hs.

3. A. hermanniae L., *Sp. Pl.* 720 (1753). Shrub up to 50 cm, with tortuous, woody branches, the ends of the branches becoming spiny. Leaves simple or 3-foliolate; leaflets oblong-spathulate or oblong-obovate, sericeous above, strongly so beneath. Flowers solitary or in axillary fascicles of up to 3 flowers, forming an interrupted raceme. Calyx 3–5 mm, tubular, the teeth triangular, much shorter than the sparsely sericeous tube. Corolla yellow. Seed 1. *Mediterranean region.* Al Bl Co Cr Gr It Ju Sa Si Tu.

4. A. barba-jovis L., *Sp. Pl.* 720 (1753). Shrub up to 90 cm, with woody branches. Leaves imparipinnate with (9–)13–19 subequal, narrowly elliptical to narrowly obovate leaflets, sparsely green-sericeous above, densely silver-sericeous beneath. Flowers in terminal heads, subtended by digitate bracts close beneath the flowers. Heads usually more than 10-flowered. Calyx

4–6 mm, tubular-campanulate, the teeth long-triangular, shorter than the whitish-hairy tube. Corolla yellow. Seed 1. ● *Mediterranean region from E. Spain to N. Jugoslavia. Often cultivated.* ?Al Co Cr Ga Gr Hs It Ju Sa Si.

5. A. aegaea Turrill, *Kew Bull.* **1939**: 189 (1939). Like **4**, but leaflets very narrowly elliptical to linear, often revolute; heads with fewer (5–9) flowers; calyx 6·5–9 mm. ● *Kriti and Kikladhes (Pholegandros, Amorgos).* Cr Gr.

6. A. henoniana Cosson ex Batt. in Batt. & Trabut, *Fl. Algér. (Dicot.)* 250 (1889) (*A. sericea* Lag., non Willd.). Shrub up to 60 cm, with woody branches. Leaves with 3–5 leaflets, sericeous above and beneath; terminal leaflet larger than the laterals; rhachis winged and somewhat sheathing. Heads terminal, 5- to 8-flowered, subtended by digitate bracts just beneath the flowers. Calyx 5–6·5 mm, tubular-campanulate, hirsute, the teeth triangular-subulate, *c.* ½ as long as the tube. Corolla yellow. Seed 1. *S. & E. Spain.* Hs.

7. A. montana L., *Sp. Pl.* 719 (1753). Perennial with woody stems, often forming large clumps. Leaves mostly near the base of the flowering stems, rarely up to the middle or above, imparipinnate; leaflets 17–41, narrowly elliptical to narrowly obovate-oblong, subequal, pubescent on both surfaces; leaf-bases dilated, sheathing, striate. Flowering stems erect, sericeous to hirsute. Flowers in dense heads, subtended by palmatisect bracts borne just beneath the flowers. Calyx tubular, with subulate, plumose teeth about equalling the tube. Corolla red to purple, the standard much exceeding the other petals. $2n = 12$. *Alps and mountains of S. Europe.* Al Au Bu Ga Gr He Hs It Ju Rm.

Polymorphic; three rather indistinct subspecies may be recognised.

1 Calyx-tube and calyx-teeth both *c.* 2 mm; terminal leaflets rounded, differing somewhat in shape from the laterals   **(c) subsp. hispanica**
1 Calyx-tube and calyx-teeth both 3 mm or more; terminal leaflets similar in shape to the laterals, rarely larger or smaller
  2 Calyx-tube and calyx-teeth both 3–4 mm; corolla pink; bracts usually exceeding the flowers   **(b) subsp. jacquinii**
  2 Calyx-tube and calyx-teeth both 4–5 mm; corolla purple; bracts usually not exceeding the flowers   **(a) subsp. montana**

(a) Subsp. montana: *Alps from France to Austria; Appennini.*
(b) Subsp. jacquinii (A. Kerner) Hayek, *Prodr. Fl. Penins. Balcan.* **1**: 885 (1926) (*A. jacquinii* A. Kerner): *E. Alps and mountains of the Balkan peninsula.*
Intermediates between this and subsp. **(a)** occur in Austria and N. Jugoslavia.
(c) Subsp. hispanica (Degen & Hervier) Cullen, *Watsonia* **6**: 389 (1968) (*A. montana* var. *hispanica* Degen & Hervier): *S. Spain.*

Variants similar to subsp. **(c)** occur in the Pyrenees. Their status is very doubtful and the whole complex is in need of revision.

8. A. aurea Welden in Host, *Fl. Austr.* **2**: 319 (1831). Slightly woody perennial. Leaves mostly near the base, or in the lower ⅓ of the stem, rarely a small leaf near the head, imparipinnate; leaflets 13–19, subequal, narrowly ovate to elliptical, sericeous above and beneath; leaf-bases dilated, sheathing, striate. Heads dense. Calyx 6·5–8 mm, tubular, ridged, sericeous-hirsute, with triangular-subulate teeth much shorter than the tube. Corolla yellow. ● *Balkan peninsula.* Al Bu Gr Ju.

9. A. polycephala Desf., *Fl. Atl.* **2**: 150 (1798) (*A. podocephala* Boiss.). Tall, more or less erect perennial up to 60 cm, woody below. Leaves evenly distributed along the stem, imparipinnate; leaflets 13–15 (often fewer in the uppermost leaves), narrowly elliptical, subequal, villous to hirsute above and beneath, without dilated sheathing leaf-bases. Heads arranged in a raceme, the lower pedunculate, the upper subsessile, dense. Calyx 6–9 mm, tubular-campanulate, hirsute, with long, plumose, awn-like teeth, the lowermost more or less equalling the tube. Corolla yellow; standard obovate-orbicular. *S. Spain.* Hs.

10. A. tejedensis Boiss., *Biblioth. Univ. Genève* ser. 2, **13**: 408 (1838). Like **9** but more or less procumbent, forming dense tufts; indumentum very dense, brownish; capitula not in a racemose inflorescence but aggregated together; calyx 8·5–10 mm, all teeth about as long as the tube, the lowermost slightly longer; corolla yellow to orange, sometimes flushed brownish-violet. *S. Spain.* Hs.

11. A. rupestris Cosson, *Not. Pl. Crit.* 155 (1852). Like **9** but leaflets 10–15, oblong-lanceolate; heads 2 or 3, not in a raceme; calyx 10–12 mm, the teeth shorter than the tube; standard ovate. *Crevices of limestone rocks.* ● *S.E. Spain (Sierra de Segura, S. de Cazorla and neighbouring mountains).* Hs.

12. A. ramburii Boiss., *Elenchus* 35 (1838). Perennial up to 55 cm, woody below. Stems erect to ascending. Leaves mostly in the lower ½ of the stem, imparipinnate; leaflets 7–11, subequal, narrowly elliptical to oblong, sparsely sericeous above and beneath. Heads small, subtended by digitate bracts borne just beneath the flowers. Calyx 4–6·5 mm, more or less tubular, hirsute; teeth triangular, slightly less than ½ as long as the tube. Corolla yellow, becoming reddish later. ● *S. & E. Spain.* Hs.

13. A. onobrychoides Cav., *Icon. Descr.* **2**: 40 (1793). Perennial up to 50 cm, very woody below. Stems erect to ascending. Leaves mostly in the lower half of the stem, imparipinnate; leaflets 5–11, equal, very narrowly elliptical or very narrowly obovate to linear, glaucous, very sparsely pubescent above, more densely so beneath. Heads small, ebracteate. Calyx 2–3 mm, tubular-campanulate, sericeous, the teeth triangular, less than half as long as the tube. Corolla yellow. ● *S. & E. Spain.* Hs.

14. A. gerardi L., *Mantissa* 100 (1767) (*Dorycnopsis gerardii* (L.) Boiss.). Perennial up to 60 cm, woody below. Stems ascending, diffuse, branched. Leaves remote, evenly distributed along the stem, imparipinnate; leaflets 5–11, equal, linear, oblanceolate or very narrowly elliptical, glabrous above, sparsely pubescent beneath. Heads small, ebracteate. Calyx 1·5–2·5 mm, tubular-campanulate, very sparsely hairy, the teeth triangular, up to half as long as the tube. Corolla pink. *S.W. Europe eastwards to Capraia.* Co Ga Hs It Lu Sa.

15. A. vulneraria L., *Sp. Pl.* 719 (1753). Annual, biennial or perennial. Lowermost leaves reduced to a terminal leaflet, or imparipinnate with a much larger terminal leaflet; upper leaves imparipinnate, equifoliolate or not. Heads many-flowered, subtended by 2 palmatisect bracts borne close beneath the flowers. Calyx inflated at anthesis, constricted at the apex, with 5 unequal teeth and the mouth oblique. Corolla yellow, red, purple, orange, whitish or parti-coloured. Legume 1(–2)-seeded. *Throughout Europe.* All except Az Sb.

A very polymorphic species divisible into about 30 infraspecific taxa (many of them frequently recognized as species),

between which intermediates occur, often over a large area. 24 of these taxa occur in Europe, and they are recognized as subspecies below. They fall into 2 fairly distinct groups ((a)–(r) and (s)–(x) below). The occurrence of intermediates often makes identification difficult and many specimens can be identified only by comparison with herbarium specimens.

The term bract refers only to the large outer bracts of each inflorescence. Leaves on the vegetative stems or rosettes are referred to as the lowest leaves.

1 Calyx 2–4(–5) mm wide, the lateral teeth small, appressed to the upper teeth (often only visible when fresh); bract-lobes narrowly deltate, acute; upper cauline leaves equifoliolate
  2 Indumentum of stems composed entirely of patent hairs
    3 Calyx 7–10 mm; plant delicate; stems 6–15 cm
      (k) subsp. **vulnerarioides**
    3 Calyx (10–)11–13 mm; plant robust; stems 20–25 cm
     4 Leaves confined to the lower part of the stem, fleshy, glabrous above   **(r)** subsp. **corbierei**
     4 Leaves evenly distributed along the stem, thin, sparsely hairy above   **(q)** subsp. **hispidissima**
  2 Indumentum of stems with at least some appressed hairs
    5 Indumentum of calyx evenly appressed, shining
     6 Calyx 14–17 mm; leaves ±evenly distributed along the stem   **(p)** subsp. **maura**
     6 Calyx 10–13·5(–14) mm; leaves confined to the lower part of the stem
      7 Lower cauline leaves inequifoliolate, with 1–3 leaflets; calyx (11–)12–13·5(–14) mm **(n)** subsp. **praepropera**
      7 Lower cauline leaves subequifoliolate, with 7–11 leaflets; calyx 10–12 mm **(o)** subsp. **weldeniana**
    5 Indumentum of calyx patent or semipatent, never shining
     8 Bracts divided for more than ½ of their total length, usually as long as or longer than the calyces
      9 Leaves evenly distributed along the stem
       10 Calyx red at apex, weakly semipatent-hairy; stems thin, not woody at the base; lower cauline leaves subequifoliolate **(a)** subsp. **vulneraria**
       10 Calyx concolorous, strongly patent-hairy; stems usually woody at the base; lower cauline leaves markedly inequifoliolate
        11 Hairs on the lower part of the stem patent; stems ascending to erect **(c)** subsp. **polyphylla**
        11 All cauline hairs appressed; stems decumbent **(b)** subsp. **maritima**
      9 Leaves confined to the lower part of the stem
       12 Calyx 7–9 mm; plants small, delicate, with stems 10–15(–20) cm
        13 Lower cauline leaves subequifoliolate with 5–9(–11) leaflets **(k)** subsp. **vulnerarioides**
        13 Lower cauline leaves inequifoliolate with 1–5 leaflets **(f)** subsp. **pulchella**
       12 Calyx 10–13 mm; plants tall, robust, with stems 20–40 cm
        14 Calyx 10–11 mm; lower cauline leaves inequifoliolate
         15 Bracts exceeding the flowers; calyx concolorous **(d)** subsp. **bulgarica**
         15 Bracts shorter than the flowers; calyx red at apex **(e)** subsp. **boissieri**
        14 Calyx 11–17 mm; lower cauline leaves equifoliolate
         16 Calyx 11–13 mm; red at apex **(l)** subsp. **forondae**
         16 Calyx 14–17 mm, concolorous **(m)** subsp. **pindicola**
     8 Bracts divided to ½ or less, usually shorter than the calyces
      17 Leaflets of rosette and lower cauline leaves broadly elliptical to orbicular, very densely silvery-sericeous above, obscuring the green colour of the leaf **(h)** subsp. **argyrophylla**
      17 Leaflets of rosette and lower cauline leaves narrower, glabrous above, or if hairy, then the hairs grey but not obscuring the green colour of the leaf

18 Leaves confined to the lower part of the stem; stems stiff, ascending to erect, robust **(i) subsp. reuteri**

18 Leaves mostly at the base of the thin, flexuous stems; plants decumbent, delicate

19 Calyx 5–7(–8) mm; lower cauline leaves obviously inequifoliolate; rosette-leaves usually reduced to the terminal leaflet **(g) subsp. arundana**

19 Calyx 8–10 mm; lower cauline leaves subequifoliolate; rosette-leaves usually with 3–5 leaflets **(j) subsp. atlantis**

1 Calyx (4·5–)5–7 mm wide, the lateral teeth obvious, not appressed to the upper; bract-lobes parallel-sided, obtuse; upper cauline leaves inequifoliolate

20 Calyx appressed-sericeous

21 Plants much-branched, with branches from most axils; calyx-indumentum dense; stem- and leaf-indumentum canescent **(v) subsp. iberica**

21 Plants without axillary branches; calyx-indumentum sparse; stem- and leaf-indumentum inconspicuous, not canescent

22 Calyx red at apex; corolla red to pink; plant decumbent **(u) subsp. pyrenaica**

22 Calyx concolorous, white; corolla yellow; plant ascending to erect **(s) subsp. carpatica**

20 Calyx ±patent-villous to -hirsute

23 Calyx (12–)13–15(–18) mm, with smoke-grey indumentum **(t) subsp. alpestris**

23 Calyx 9–12 mm, with white indumentum

24 Stems 10–15 cm; sinus between upper calyx-teeth wide, obvious **(x) subsp. borealis**

24 Stems 15–40 cm; sinus between upper calyx-teeth narrow, obscure **(w) subsp. lapponica**

(a) Subsp. **vulneraria** (*A. linnaei* Juz.): Stems 5–55 cm, decumbent to ascending, sometimes branched, sericeous. Lowest leaves inequifoliolate with 5–7 leaflets; uppermost leaves equifoliolate with 9–15 leaflets. Calyx usually with a red apex; corolla mostly yellow. ● *N. Europe from Ireland to Finland and Latvia.* Be Br Da Fe Ga Ge Hb Ho No Rs (B) Su.

Red-flowered variants are often recognized as var. *coccinea* L. Plants from the coasts of Britain and France to Denmark are intermediate with subsp. (**v**) and have been recognized as var. *langei* Jalas. For intermediates with subsp. (**s**) see that subspecies. Plants intermediate between subsp. (**a**) and subsp. (**c**) have been described as *A. colorata* Juz.

(b) Subsp. **maritima** (Schweigger) Corb., *Nouv. Fl. Norm.* 148 (1894) (*A. maritima* Schweigger): Stems 20–60 cm, usually decumbent, white-sericeous, branched. Lowest leaves inequifoliolate with 1–3 leaflets; uppermost leaves equifoliolate with 7–11 leaflets. Calyx concolorous; corolla yellow. ● *S. part of the Baltic region, near the coast.* Da Ge Po Rs (B) Su.

(c) Subsp. **polyphylla** (DC.) Nyman, *Consp.* 164 (1878) (*A. arenaria* (Rupr.) Juz., *A. polyphylla* (DC.) Kit. ex G. Don fil., *A. schiwereckii* (DC.) Błocki): Stems 30–90 cm, stout, ±erect, branched, usually hirsute below, rarely glabrous. Leaves numerous; lowest inequifoliolate with 3–7 leaflets, uppermost equifoliolate with 11–15 leaflets. Calyx concolorous; corolla yellow, rarely reddish. 2*n*=12. *C. & E. Europe, extending to C. Jugoslavia; rarely introduced elsewhere.* Au Cz Da Hu It Ju Po Rm Rs (N, B, C, W, K, E) [Be Ge].

For intermediates to subsp. (**s**) and subsp. (**o**) see those subspecies; intermediates to subsp. (**d**) occur in the Balkan peninsula. *A. schiwereckii* (DC.) Błocki is a rather rare, subglabrous variant. The subspecies has been cultivated as a fodder plant in various areas of Europe.

(d) Subsp. **bulgarica** (Sagorski) Cullen, *Watsonia* 6: 389 (1968): Stems 25–35 cm, ascending, branched, hirsute below. Lowest leaves inequifoliolate with 1–3 leaflets; uppermost leaves equifoliolate with 9–11 leaflets. Calyx concolorous, usually much exceeded by the bract-lobes; corolla pale yellow to almost whitish. ● *Balkan peninsula.* Al Bu Gr Ju.

(e) Subsp. **boissieri** (Sagorski) Bornm., *Feddes Repert.* 50: 135 (1941) (*A. boissieri* Sagorski, *A. taurica* Juz.): Like subsp. (**d**) but with shorter bract-lobes and calyx red at apex; corolla sometimes reddish. *Krym.* Rs (K).

(f) Subsp. **pulchella** (Vis.) Bornm., *Bot. Jahrb.* 59: 483 (1925) (*A. albana* Wettst., *A. scardica* Wettst., *A. biebersteiniana* (Taliev) Popl. ex Juz.): Stems 6–20 cm, decumbent, usually sericeous, more or less unbranched. Lowest leaves inequifoliolate with 1–3 leaflets; uppermost leaves equifoliolate with 7–11 leaflets. Calyx red at apex; corolla yellow or reddish. *Balkan peninsula, Krym.* Al Bu Gr Ju Rs (K).

Intermediates to subspp. (**t**) and (**d**) occur in N. Greece.

(g) Subsp. **arundana** (Boiss. & Reuter) Vasc., *Anais Inst. Vinho Porto* 1: 73 (1941) (*A. arundana* Boiss. & Reuter): Stems 5–20 cm, sericeous, thin and flexuous. Lowest leaves equifoliolate with 1–3 leaflets; uppermost leaves equifoliolate with 9–11 leaflets. Calyx very small, purple at apex; corolla purplish red. ● *S. Spain.* Hs.

This subspecies grows at higher altitudes than subsp. (**h**). Intermediates between them occur in the middle regions of the mountains of S. Spain.

(h) Subsp. **argyrophylla** (Rothm.) Cullen, *Watsonia.* 6: 389 (1968) (*A. argyrophylla* Rothm., *A. webbiana* auct. mult., non Hooker): Stems 10–20 cm, ascending, hirsute near the base. Lowest leaves inequifoliolate with 1–3 leaflets; uppermost equifoliolate with 7–11 leaflets, most of the terminal leaflets broadly elliptical to orbicular with dense silvery indumentum. Calyx purple at apex; corolla red to purple. ● *S. Spain.* Hs.

A. webbiana Hooker, *Bot. Mag.* 60: t. 3284 (1833), the name most frequently applied to this taxon, was described from Madeira.

(i) Subsp. **reuteri** Cullen, *Watsonia* 6: 389 (1968) (*A. hispida* Boiss. & Reuter, non *A. vulneraria* var. *hispida* Boiss.): Stems 15–30 cm, hirsute-villous below. Lowest leaves inequifoliolate with 1–3 leaflets. Calyx purple at apex; corolla red. ● *S. & E. Spain, at low altitudes.* Hs.

Some specimens resemble subsp. (**n**), which is absent from mainland Spain.

(j) Subsp. **atlantis** Emberger & Maire, *Bull. Soc. Hist. Nat. Afr. Nord* 24: 209 (1933) (*A. nivalis* (Willk.) G. Beck): Stems 8–20 cm, decumbent, hirsute below. Lowest leaves subequifoliolate with 5–9 leaflets; uppermost equifoliolate with 7–13 leaflets. Calyx purple at apex; corolla purplish. *High mountains of S. & E. Spain.* Hs.

(k) Subsp. **vulnerarioides** (All.) Arcangeli, *Comp. Fl. Ital.* ed. 2, 502 (1894) (*A. vulnerarioides* (All.) Bonjean ex Reichenb., *A. bonjeanii* G. Beck): Stems 6–15 cm, ascending, completely hirsute, or hirsute only below. Lowest leaves inequifoliolate with 1–5 leaflets; uppermost equifoliolate with 9–13 leaflets. Calyx red at apex; corolla yellow. ● *Pyrenees, S.W. Alps, C. Appennini.* Ga Hs It.

Specimens from C. Appennini (Monte Majella) approach subsp. (**f**). *A. bonjeanii* G. Beck is the name applied to variants with stems appressed-hairy in the upper part.

(l) Subsp. **forondae** (Sennen) Cullen, *Watsonia* 6: 389 (1968) (*A. forondae* Sennen): Stems 20–40 cm, ascending to erect, hirsute below. Lowest leaves equifoliolate with 9–13 leaflets, most of the leaflets broadly elliptical or almost orbicular. Calyx red at apex; corolla usually yellow. ● *Mountains of N.E. Spain, Pyrenees, S.W. Alps.* Ga Hs It.

More or less sympatric with subsp. (**k**) but at much lower altitudes. Intermediates between this and subsp. (**u**) (*A. sampaiana* Rothm.), are found in N.E. Portugal and N.W. Spain.

(m) Subsp. **pindicola** Cullen, *Watsonia* 6: 389 (1968).

Like subsp. (l), but calyx longer and concolorous and the whole plant usually very sparsely hairy. ● *N.W. Greece.* Gr.

(n) Subsp. **praepropera** (A. Kerner) Bornm., *Bot. Jahrb.* **59**: 483 (1925) (*A. praepropera* (A. Kerner) G. Beck, *A. spruneri* (Boiss.) G. Beck, *A. dillenii* auct. mult., non Schultes ex G. Don fil., nec sensu Rothm.): Stems 10–35 cm, ascending to erect, hirsute below. Lowest leaves usually with only 1 leaflet; uppermost leaves equifoliolate with 7–13 leaflets. Calyx purple at apex; corolla red to purplish. *Mediterranean region, but absent from Spain and much of Italy.* Al Bl Co Cr Ga Gr It Ju Sa Tu.

Occasional robust variants approach subsp. (p).

(o) Subsp. **weldeniana** (Reichenb.) Cullen, *Watsonia* 6: 389 (1968) (*A. weldeniana* Reichenb.): Like subsp. (n) but lowest leaves with 5–9 broadly elliptical leaflets. ● *Italy, N. & W. Jugoslavia.* It Ju.

Intermediates to subsp. (n) are common in the coastal part of Jugoslavia. For intermediates to subsp. (s) see that subspecies. Intermediates to subsp. (c) (*A. tricolor* Vuk.) occur inland in N. Jugoslavia.

(p) Subsp. **maura** (G. Beck) Lindb., *Acta Soc. Sci. Fenn.* nov. ser., B, **1**(2): 77 (1932) (*A. maura* G. Beck): Stems 30–60 cm, hirsute below; lowest leaves with 1–3 leaflets, the terminal leaflet often very large (5–10 cm); uppermost leaves equifoliolate with 9–13 leaflets. Calyx red at apex; corolla usually red. *W. & S. Portugal, S. & E. Spain, S. Italy, Sicilia.* Hs It Lu Si. (*N. Africa.*)

Variants from Spain (except the extreme south) differ from the typical plant in being less robust, and with the leaves not so evenly distributed along the stem; they appear to be intermediate to either subsp. (n) or subsp. (i), and have been given a variety of names, the most common being *A. gandogeri* Sagorski and *A. font-queri* Rothm. Very fleshy-leaved plants (*A. pachyphylla* Rothm.) occur in coastal regions of C. Portugal. They are apparently an ecological modification.

(q) Subsp. **hispidissima** (Sagorski) Cullen, *Watsonia* 6: 389 (1968) (*A. hispidissima* Sagorski): Stems 20–25 cm, ascending, hirsute over their whole length; lowermost leaves with 1–3 leaflets; uppermost leaves equifoliolate with 9–11 leaflets. Calyx concolorous; corolla yellow when dry. *Macedonia.* Gr Ju. (*Anatolia.*)

(r) Subsp. **corbierei** (Salmon & Travis) Cullen, *Watsonia.* 6: 295 (1967) (*A. maritima* var. *corbierei* Salmon & Travis): Like subsp. (q) but the leaves fleshy; indumentum of the whole plant less dense. ● *W. Britain (Anglesey, Cornwall); Channel Islands (Sark).* Br Ga.

Probably occurs also on the N. & W. coasts of France.

(s) Subsp. **carpatica** (Pant.) Nyman, *Consp., Suppl.* **2**(1): 87 (1889) (*A. vulgaris* A. Kerner pro parte, *A. carpatica* Pant.): Stems 10–30 cm, ascending, sparsely sericeous; all leaves inequifoliolate with 1–7(–9) leaflets. Calyx sparsely sericeous, usually concolorous; corolla pale yellow, rarely deep yellow or reddish. ● *N.W. & C. Europe. Cultivated for fodder elsewhere and often more or less naturalized.* Be Br Cz Da Ga Ge Hb He It Ju Po.

Intermediates occur with almost every other subspecies with which it comes into geographical contact. Intermediates to subsp. (t), with which it is largely sympatric, occur throughout the Alps; the two subspecies appear to be ecologically partially isolated, subsp. (t) occurring at high altitudes. Intermediates to subsp. (c) (*A. affinis* Brittinger ex A. Kerner) are widespread in the lowlands of Europe where they have been cultivated as fodder plants, this perhaps explaining their rather surprising occurrence in S. Belgium. Intermediates to subsp. (a) (*A. pseudovulneraria* Sagorski) are common in N.W. Europe. Intermediates to subsp. (o) (var. *versicolor* Sagorski) occur in N. Italy and Slovenija.

(t) Subsp. **alpestris** Ascherson & Graebner, *Syn. Mitteleur. Fl.* 6(2): 626 (1908) (*A. alpestris* Hegetschw., non Reichenb.): Like subsp. (s) but calyx longer, hirsute with greyish-black hairs. $2n = 12$. ● *Cordillera Cantábrica; Alps; Carpathians; mountains of the Balkan peninsula.* Au Bu Cz Ga Ge Gr He Hs Hu It Ju Po Rm.

Replaced in the Pyrenees by subsp. (u). A small, red-flowered variant (*A. valesiaca* G. Beck) occurs in Switzerland (Valais).

(u) Subsp. **pyrenaica** (G. Beck) Cullen, *Feddes Repert.* **79**: 52 (1968) (*A. coccinea* f. *pyrenaica* G. Beck): Plant sparsely hairy, stem 10–25 cm. Leaves all inequifoliolate with 1–7 leaflets. Calyx red at apex; corolla pink to red. ● *Cordillera Cantábrica; Pyrenees.* Ga Hs.

(v) Subsp. **iberica** (W. Becker) Jalas, *Bull. Jard. Bot. Bruxelles* 27: 409 (1957) (*A. asturiae* W. Becker): Stems 20–40 cm, mostly decumbent, sericeous, usually very branched; leaves inequifoliolate with 3–9 leaflets. Calyx red at apex; corolla red. ● *W. coast of Europe.* Be Ga Hs Lu.

(w) Subsp. **lapponica** (Hyl.) Jalas, *Ann. Bot. Soc. Zool.-Bot. Fenn. Vanamo* 24(1): 38 (1950) (*A. kuzenevae* Juz.): Stems 15–40 cm, ascending to erect, sericeous. Leaves all inequifoliolate with 1–9 leaflets. Calyx concolorous or red at apex; corolla yellow. $2n = 12$. ● *Fennoscandia, N. Britain, Chibiny mountains of the Kol'skij Poluostrov.* Br Fe No Rs (N) Su.

Specimens from S. Finland (*A. vulneraria* subsp. *fennica* Jalas) have laxer, fewer-flowered heads and narrower leaflets, and show differences in ecology.

(x) Subsp. **borealis** (Rouy) Jalas, *op. cit.* 40 (1950): Very like subsp. (w) but indumentum denser, stems up to 15 cm. ● *Iceland.* Is.

16. A. tetraphylla L., *Sp. Pl.* 719 (1753) (*Physanthyllis tetraphylla* (L.) Boiss.). Procumbent annual. Stems villous to hirsute. Leaves imparipinnate, with (1–)5 leaflets, the large, more or less obovate terminal leaflet much exceeding the small, lateral leaflets, hairy on both surfaces, much less above. Flowers in axillary fascicles of 1–7. Calyx $12–15 \times 4\cdot5–6$ mm, inflated at anthesis, later up to 12 mm wide, gibbous, especially so in fruit, densely sericeous, frequently reddish near the apex, the teeth subequal, the mouth of the calyx straight. Corolla yellow, the keel often red at apex. Legume usually 2-seeded, constricted between the seeds. *Mediterranean region, S. Portugal.* Al Bl Co Cr Ga Gr Hs It Ju Lu Sa Si.

17. A. lotoides L., *Sp. Pl.* 720 (1753) (*Cornicina lotoides* (L.) Boiss.). Erect annual. Stems villous to hirsute. Lowermost leaves often reduced to the terminal leaflet, the upper irregularly imparipinnate with (3–)5–7 subequal, narrowly lanceolate to oblanceolate leaflets, hairy above and beneath, strongly so on the margin. Heads 4- to 8-flowered, subtended by digitate bracts borne just beneath the flowers. Calyx 9–11 mm, tubular, with 5 unequal triangular teeth and an oblique mouth, long-pubescent. Corolla yellow to orange. Legume straight, 6- to 10-seeded, erect, the infructescence forming a compact long, narrow head. *Spain and Portugal.* Hs Lu.

18. A. cornicina L., *Sp. Pl.* 719 (1753) (*Cornicina loeflingii* Boiss.). Annual. Stems ascending to erect, villous to hirsute. Lowermost leaves reduced to the terminal leaflet, the upper with up to 9 subequal, very narrowly elliptical leaflets, sericeous above and beneath. Heads 9- or more-flowered, subtended by digitate bracts. Calyx 5–7 mm, not inflated or gibbous at anthesis, but inflated in fruit, with 5 unequal teeth and an oblique mouth. Corolla yellow-orange. Legume winged, curved into an almost complete circle, completely included within the calyx. Seeds 3 or more. *C. & S. Spain, Portugal.* Hs Lu.

19. A. hamosa Desf., *Fl. Atl.* **2**: 151 (1798) (*Cornicina hamosa* (Desf.) Boiss.). Like **18** but upper leaves often with up to 11 leaflets; calyx 7–8 mm, more or less tubular, curved; legume deflexed and upcurved, many-seeded, beaked, the beak exserted from the calyx. *S. Spain, Portugal*. Hs Lu.

64. Ornithopus L.[1]

Annual. Leaves imparipinnate; stipules small, free, linear. Flowers in axillary heads. Calyx tubular or campanulate, with 5 equal teeth; keel obtuse; stamens diadelphous. Legume lomentose, terete or compressed, usually constricted between the segments, strongly reticulate.

All species occur in dry, often acid, sandy places, or as ruderals.

1 Heads ebracteate or with minute scarious bracts **4. pinnatus**
1 Heads subtended by a pinnate leafy bract
  2 Corolla yellow; legume not or only slightly contracted between the segments **1. compressus**
  2 Corolla pink or white; legume strongly contracted between the segments
    3 Corolla 6 mm or more; bracts shorter than the flowers, usually about equalling the calyx **2. sativus**
    3 Corolla not more than 5 mm; bracts much longer than the flowers **3. perpusillus**

1. O. compressus L., *Sp. Pl.* 744 (1753). Pubescent, stems 10–50 cm. Leaflets 7–18 pairs, oblong, elliptical or oblong-lanceolate. Heads 3- to 5-flowered; bracts with 7–9 leaflets. Calyx-teeth at least $\frac{1}{2}$ as long as the tube; corolla 5–8 mm, yellow. Legume 20–50 mm, curved, more or less compressed, not or only slightly contracted between the segments; segments 5–8, oblong; beak 7 mm or more, curved. *S. Europe*. Al Az Bl Bu Co Cr Ga Gr Hs It Ju Lu Sa Si Tu.

2. O. sativus Brot., *Fl. Lusit.* **2**: 160 (1804). Pubescent, stems 20–70 cm. Leaflets 9–18 pairs, lanceolate or elliptical to ovate. Heads 2- to 5-flowered; bracts with 5–9 leaflets, shorter than the flowers. Calyx-teeth slightly shorter than to about equalling tube; corolla 6–9 mm, white or pink. Legume 12–40 × 2–2·5 mm, compressed, contracted between the segments; segments 3–7, elliptic-oblong. *S.W. Europe; cultivated as a fodder plant in most of Europe and locally naturalized.* Az Ga Hs Lu [Ge Po Rs (C, W)].

(a) Subsp. **sativus** (*O. roseus* Dufour): Legume 12–25 mm, straight; beak usually not more than 5 mm, straight, sometimes hooked at the tip. *Native in S.W. France, N. half of Iberian peninsula and Açores, cultivated elsewhere.*

(b) Subsp. **isthmocarpus** (Cosson) Dostál, *Květena ČSR* 788 (1948): Legume 20–40 mm, curved, usually with a long, narrow, cylindrical constriction between the segments; beak 10 mm or more, curved. *S.W. part of Iberian peninsula.*

The two subspecies are apparently linked by intermediates (var. *macrorrhynchus* Willk.) in C. Portugal and W.C. Spain.

3. O. perpusillus L., *Sp. Pl.* 743 (1753). Pubescent, stems up to 30 cm. Leaflets 7–13 pairs, elliptical or oblong. Heads 3- to 8-flowered; bracts with 5–9 leaflets, longer than the flowers. Calyx-teeth not more than $\frac{1}{2}$ as long as tube; corolla 3–5 mm, white or pink. Legume 10–18(–25) × 1·5–2 mm, straight, compressed, contracted between the segments; segments 4–9, elliptic-oblong, beak not more than 3 mm, straight, often hooked at the tip. $2n=14$. *W. & W.C. Europe, extending eastwards to Italy, Poland and S. Sweden; rarely naturalized further east.* Az

---

[1] By P. W. Ball.

Be Br Co Da Ga Ge Hb He Ho Hs It Lu Po Rm Sa Su [Au Cz Rs (B, C, W)].

Hybrids between **2** and **3** occur frequently.

4. O. pinnatus (Miller) Druce, *Jour. Bot.* (*London*) **45**: 420 (1907) (*O. ebracteatus* Brot.). Glabrous or sparsely pubescent, 10–50 cm. Leaflets 3–7 pairs, linear to oblanceolate. Heads 1- to 5-flowered, ebracteate or with minute scarious bracts. Calyx-teeth not more than $\frac{1}{2}$ as long as tube; corolla 6–8 mm, yellow. Legume 20–35 mm, curved, terete, not contracted between the segments; segments (6–)8–12, cylindrical; beak not more than 5 mm. $2n=14$. *W. Mediterranean region and W. Europe, northwards to S.W. England* (*Isles of Scilly*). Az Bl Br Co Ga Gr Hs It Lu Sa Si.

65. Coronilla L.[1]

Annual or perennial herbs or dwarf shrubs. Leaves imparipinnate, rarely simple or 3-foliolate; stipules various, free or connate. Flowers in axillary heads. Calyx campanulate, more or less bilabiate; keel acute; stamens diadelphous. Legume lomentaceous, terete or longitudinally ridged or angled, not constricted between the segments.

Literature: A. Uhrová, *Beih. Bot. Centr.* **53 B**: 1–174 (1935).

1 Lower leaves simple or 3-foliolate, the terminal leaflet much larger than the lateral
  2 Upper leaves simple or 3-foliolate **12. scorpioides**
  2 Upper leaves imparipinnate **13. repanda**
1 Lower leaves imparipinnate, the leaflets more or less equal
  3 Claw of the standard 2–3 times as long as the calyx **1. emerus**
  3 Claw of the standard equalling or slightly longer than the calyx
    4 Corolla yellow
      5 Leaflets with a narrow scarious margin
        6 Herb 30–70 cm; heads 12- to 20-flowered; pedicels 4–6 mm **6. coronata**
        6 Small shrubs not more than 50 cm; heads usually not more than 10-flowered; pedicels 2–4 mm
          7 Leaflets shortly petiolulate; stipules 3–8(–10) mm, herbaceous with membranous tip, deciduous **3. vaginalis**
          7 Leaflets sessile; stipules *c.* 1 mm; membranous, persistent **4. minima**
      5 Leaflets without a scarious margin
        8 Annual; legume strongly curved **11. rostrata**
        8 Small shrubs; legume ± straight
          9 Stem not junciform; leaves persistent; leaflets elliptical to obovate-orbicular, not fleshy **2. valentina**
          9 Stem junciform with long internodes; leaves caducous; leaflets linear or oblong, fleshy **5. juncea**
    4 Corolla white, pink or purple
      10 Annual; heads 3- to 6(–9)-flowered
        11 Corolla 4–7 mm; legume ± straight, with a curved beak **10. cretica**
        11 Corolla 7–11 mm; legume strongly curved, with a straight beak **11. rostrata**
      10 Perennial; heads (5–)10- to 40-flowered
        12 Heads up to 40-flowered; segments of the legume 9–10 mm (Kriti) **9. globosa**
        12 Heads not more than 20-flowered; segments of the legume 4–6 mm
          13 Leaflets of the lower leaves (5–)7–12 pairs, not more than 12 mm wide **7. varia**
          13 Leaflets 3–5(–6) pairs, 10–20 mm wide **8. elegans**

1. C. emerus L., *Sp. Pl.* 742 (1753). Small shrub up to 100(–200) cm. Leaflets 2–4 pairs, 10–20 mm, obovate, mucronate, glaucous; stipules 1–2 mm, free, membranous. Corolla 14–20 mm, pale yellow; claw of the standard 2–3 times as long as the calyx. Legume 50–110 mm; segments 3–12, 8–10 mm. *C. & S.E. Europe, extending locally to S. Norway, the Pyrenees and E. Spain.* Al Au Bu Co Cr Cz Ga Ge Gr He Hs Hu It Ju No Rm Rs (K) Si Su Tu [Da].

(a) Subsp. **emerus**: Peduncles about equalling the leaves; heads 1- to 5-flowered; segments of the legume obtusely angled. $2n=14$. ● *Throughout the range of the species except the south-east.*

(b) Subsp. **emeroides** (Boiss. & Spruner) Hayek, *Prodr. Fl. Penins. Balcan.* **1**: 917 (1926) (*C. emeroides* Boiss. & Spruner): Peduncles longer than the leaves; heads up to 8-flowered; segments of the legume subterete. *S.E. Europe, E. & S. Italy.*

Plants from Italy and N. Jugoslavia are often intermediate between the two subspecies.

2. **C. valentina** L., *Sp. Pl.* 742 (1753). Small shrub up to 100 cm. Leaflets 2–6 pairs, up to 20 mm, obovate, emarginate; stipules free, deciduous. Heads 4- to 12-flowered. Corolla 7–12 mm, yellow. Legume 10–50 mm; segments 1–10, 5–7 mm, fusiform, subcompressed, with two obtuse angles. *Scrub and cliffs. Mediterranean region, S. Portugal.* Al Bl Co Cr Ga Gr Hs It Ju Lu Sa Si [Br].

(a) Subsp. **valentina**: Leaflets 3–6 pairs; stipules 5–10 mm, herbaceous. Legume with 3–7 segments. *C. part of Mediterranean region (S.E. France to Albania).*

(b) Subsp. **glauca** (L.) Batt. in Batt. & Trabut, *Fl. Algér. (Dicot.)* 285 (1889) (*C. glauca* L.): Leaflets 2–3 pairs; stipules 2–6 mm, ovate or lanceolate, membranous. Legume with 1–4(–10) segments. *Almost throughout the range of the species.*

Subsp. **pentaphylla** (Desf.) Batt., *loc. cit.* (1889) (*C. pentaphylla* Desf.) occurs in S. Spain. It is intermediate between subspp. (a) and (b) with 2–3 pairs of leaflets; stipules 5–10 mm, herbaceous, and the legume with 1–3 segments. Similar intermediate plants occur in S. France and in Italy.

3. **C. vaginalis** Lam., *Encycl. Méth. Bot.* **2**: 121 (1786). Small shrub up to 50 cm. Leaflets 2–6 pairs, 4–10 mm, oblong-ovate or obovate to suborbicular, shortly petiolate, margin scarious; stipules 3–8(–10) mm, connate, herbaceous, with membranous tip, deciduous. Heads 4- to 10-flowered; pedicels 2–4 mm. Corolla 6–10 mm, yellow. Legume 15–35 mm; segments 3–8, 4·5–5 mm, ovoid, obtusely 6-angled, 4 of the angles winged. $2n=12$. *Dry grassland, scrub and open woods; calcicole.* ● *Mountain regions of C. Europe, Italy, Jugoslavia and Albania.* Al Au Cz Ga Ge He Hu It Ju ?Rm.

4. **C. minima** L., *Cent. Pl.* **2**: 28 (1756). Small shrub up to 30(–45) cm. Leaflets 2–6 pairs, 2–15 mm, elliptical or obovate to suborbicular, sessile, margin scarious; stipules *c.* 1 mm, connate, membranous, persistent. Heads up to 10(–15)-flowered; pedicels 2–4 mm. Corolla 5–8(–12) mm, yellow. Legume 10–35 mm; segments 1–7, 4·5–5·5 mm, oblong, 4-angled. $2n=24$. *Dry, open habitats. S.W. Europe extending to N.W. France, S.W. Switzerland and E. Italy.* Ga He Hs It Lu.

5. **C. juncea** L., *Sp. Pl.* 742 (1753). Small junciform shrub 20–100 cm, with long internodes. Leaves caducous; leaflets (1–)2–3 pairs, 5–25 mm, linear or oblong, fleshy; stipules 1–3(–5) mm, free, membranous. Heads 5- to 12- flowered. Corolla 6–12 mm, yellow. Legume 10–50 mm; segments 2–11, 4–5 mm, obtusely 4-angled. *Dry open habitats. W. Mediterranean region, extending to W. Jugoslavia and S. Portugal.* Bl Ga Hs It Ju Lu.

6. **C. coronata** L., *Syst. Nat.* ed. 10, **2**: 1168 (1759). Perennial herb 30–70 cm. Leaflets 3–6(–7) pairs, 15–30(–40) mm, elliptical or obovate, petiolate, margin narrowly scarious; stipules 3–5 mm, connate, membranous, deciduous. Heads 12- to 20-flowered; pedicels 4–6 mm. Corolla 7–11 mm, yellow. Legume 15–30 mm; segments 1–5(–9), 6–7·5 mm, ovoid-oblong, obtusely 4-angled. $2n=10$. *Dry woods, scrub and grassland; calcicole. C. Europe and W. part of Balkan peninsula, extending to C. France, N. Italy and Krym.* Al Au Cz Ga Ge Gr He Hu It Ju ?Rm Rs (W, K).

Recorded from E. Spain, probably in error for robust plants of **4**.

7. **C. varia** L., *Sp. Pl.* 743 (1753). Perennial herb 20–120 cm. Leaflets (5–)7–12 pairs, 6–20 × 3–12 mm, oblong or elliptical, margin narrowly scarious; stipules 1–6 mm, free, membranous. Heads (5–)10- to 20-flowered. Corolla (8–)10–15 mm, white, pink or purple. Legume 20–60(–80) mm; segments 3–8(–12), 4–6 mm, oblong, 4-angled. $2n=24$. *C. & S. Europe, extending to C. Russia, but native limits not clear; often cultivated for fodder and naturalized in W. & N. Europe.* Al Au Bu Cr Cz Ga Ge Gr He *Ho Hu Hs It Ju Po Rm Rs (*B, C, W, K, E) Tu [Be Br Da No Rs (N) Su].

8. **C. elegans** Pančić, *Fl. Princ. Serb.* 262 (1874) (*C. latifolia* (Hazsl.) Jáv.). Like **7** but leaflets 3–5(–6) pairs, (15–)20–50 × (7–)10–20 mm, pruinose beneath; heads 6- to 18-flowered; corolla 8–10 mm; legume 50–80 mm; segments 5–10. $2n=12$. *Woods and scrub.* ● *From Albania and Czechoslovakia eastwards to Ukraine.* Al Au Cz Gr Hu Ju Rm Rs (W).

9. **C. globosa** Lam., *Encycl. Méth. Bot.* **2**: 122 (1786). Like **7** but leaflets 15–30 × 5–13 mm; heads 15- to 40-flowered; corolla 9–11 mm, usually white; legume 30–70 mm; segments 2–5, 9–10 mm. *Cliffs.* ● *Kriti.* Cr.

10. **C. cretica** L., *Sp. Pl.* 743 (1753). Annual up to 90 cm. Leaflets 3–8 pairs, 5–20 mm, obovate-oblong, obtuse or truncate; stipules 1–3 mm, free, linear, membranous. Heads 3- to 6(–9)-flowered. Corolla 4–7 mm, white or pink. Legume 30–80 mm, straight, with curved beak; segments 5–9, 6–7 mm, linear-oblong, 4-angled. *Grassy places and as a ruderal. S.E. Europe, extending to S. & E. Italy.* Al Bu Cr Gr It Ju Rs (W, K) Tu.

11. **C. rostrata** Boiss. & Spruner in Boiss., *Diagn. Pl. Or. Nov.* **1** (2): 100 (1843) (*C. parviflora* Willd., non Moench). Like **10** but leaflets emarginate; stipules ovate; corolla 7–11 mm, pink, white or pale yellow; legume strongly curved, with straight beak; segments obtusely 2-angled. *S. Albania, Greece and Aegean region; Krym.* Al Cr Gr Rs (K) Tu.

12. **C. scorpioides** (L.) Koch, *Syn. Fl. Germ.* 188 (1835). Annual up to 40 cm. Leaves simple or 3-foliolate; terminal leaflet up to 40 mm, elliptical or suborbicular, much larger than the reniform-orbicular lateral leaflets; stipules 1–2 mm, connate, membranous. Heads 2- to 5-flowered. Corolla 4–8 mm, yellow. Legume 20–60 mm, curved; segments 2–11, oblong, more or less straight, obtusely 4- to 6-angled. $2n=12$. *Dry open habitats, often as a weed or ruderal. S. Europe; often casual elsewhere.* Al Bl Bu Co Cr Ga Gr Hs It Ju Lu Rm Rs (K) Sa Si Tu.

13. **C. repanda** (Poiret) Guss., *Fl. Sic. Syn.* **2**: 302 (1844). Like **12** but upper leaves imparipinnate; leaflets 2–4 pairs, subequal, 4–15 mm, oblanceolate to obovate, truncate or emarginate; segments of the legume distinctly curved. *S. half of Iberian peninsula; Islas Baleares; Italy and Sicilia.* Bl Hs It Lu Si.

(a) Subsp. **repanda**: Leaflets 8–15 mm, oblanceolate, truncate or obtuse; corolla (6·5–)8–9 mm; legume 50–70 × 1·4–2 mm. *Sandy soil; usually maritime. Throughout the range of the species, except S.E. Spain.*

(b) Subsp. **dura** (Cav.) Coutinho, *Fl. Port.* 356 (1913) (*Ornithopus durus* Cav.): Leaflets 4–10(–15) mm, obcordate, emarginate; corolla 5–6·5 mm; legume 30–40 × *c.* 1 mm. *Inland. C. & S. Portugal, S. Spain.*

66. Hippocrepis L.[1]

Annual or perennial herbs. Leaves imparipinnate; stipules small, linear or lanceolate, free. Flowers in axillary heads, rarely solitary. Calyx tubular-campanulate with 5 subequal teeth; corolla yellow; keel acute; stamens diadelphous. Legume lomentaceous, laterally compressed; segments lunate to horseshoe-shaped, or flat and rectangular with a semicircular to orbicular sinus, which has a curved protuberance at its base enclosing the seed.

Literature: F. Bellot, *Anal. Inst. Bot. Cavanilles* **7**: 197–334 (1947). A. Hrabětová-Uhrová, *Acta Acad. Sci. Nat. Mor.-Sil.* **21**(4) (1949); **22**: 99–158, 219–250, 331–356 (1950).

A taxonomically difficult genus in which there is considerable difference of opinion as to the status and affinities of many of the taxa, particularly in species **1–4**.

All species occur in dry, usually sunny, situations.

Width of the legume refers to the width at the articulation.

1 Peduncles not more than 5 mm, usually 1-flowered
    10. unisiliquosa
1 Peduncles more than 5 mm, with 2 or more flowers
2 Annual, slender and herbaceous at the base
  3 Peduncles longer than the leaves in fruit; corolla 12–15 mm
    9. salzmannii
  3 Peduncles shorter than or equalling the leaves; corolla 3–8 mm
    4 Corolla 3–5 mm; legume with long papillae on the seed-protuberance **7. ciliata**
    4 Corolla 5–8 mm; legume glabrous or with very small papillae **8. multisiliquosa**
2 Perennial, more or less woody and much branched at the base, or caespitose
  5 Seed-protuberance, or sometimes the whole legume, covered with white papillae at least 0·5 mm long **4. squamata**
  5 Legume glabrous or with much shorter papillae
    6 Legume with red or brown papillae **3. comosa**
    6 Legume glabrous, or with white papillae
      7 Legume without broad, flattened regions between the seed-protuberances; segments lunate **1. glauca**
      7 Legume with broad, flattened regions between the seed-protuberances; segments with a semicircular to orbicular sinus
        8 Legume more or less densely papillose, particularly on the seed-protuberance **2. scabra**
        8 Legume glabrous, rarely very sparsely papillose
          9 Leaflets 3–6 pairs, the largest *c.* 5 mm wide, obovate; claw of the standard about as long as the calyx **5. valentina**
          9 Leaflets 5–10 pairs, the largest usually not more than 2 mm wide, linear or oblong; claw of the standard twice as long as the calyx **6. balearica**

1. H. glauca Ten., *Fl. Nap.* 1, *Prodr.*: 43 (1811) (*H. comosa* subsp. *glauca* (Ten.) Rouy). Perennial up to 40 cm, woody at base. Leaflets 4–7 pairs, 2–10 × 0·5–3 mm, linear to obovate, densely white pubescent beneath. Heads 4- to 8-flowered; peduncles 2–3 or more times as long as the leaves. Corolla 6–12 mm; claw of the standard about as long as the calyx. Legume (20–)30–40 × 1–2(–3) mm, without broad, flattened regions between the seed-protuberances; segments lunate, with white papillae. *Mediterranean region.* ● Al Ga Gr Hs It Ju Si.

Not always clearly separable from **2** and **3**.

2. H. scabra DC., *Prodr.* **2**: 312 (1825). Perennial up to 40 cm, woody at base. Leaflets (2–)3–8 pairs, 3–12 × 0·5–5 mm, oblong to obovate, usually pubescent beneath. Heads 2- to 8-flowered; peduncles 2–5 times as long as the leaves. Corolla 6–12 mm; claw of the standard up to twice as long as the calyx. Legume 12–25 × (2–)3–6 mm, with broad flat regions between the seed-protuberances; segments with horseshoe-shaped to orbicular sinuses, with white papillae. *Spain.* Hs.

This species, as defined here, is composed of three taxa, which are more or less separable geographically: *H. scabra* DC., *sensu stricto*, from S. Spain, with corolla 8–12 mm and the legume with orbicular sinuses; **H. commutata** Pau, *Bol. Soc. Aragon. Ci. Nat.* **2**: 274 (1903), from C. & N. Spain, with corolla 6–8 mm and the legume with orbicular sinuses; **H. bourgaei** (Nyman) Hervier, *Bull. Acad. Int. Géogr. Bot.* (*Le Mans*) **17**: 37 (1907), from S.E. Spain, with corolla 5–8 mm and the legume with semicircular sinuses. The characters by which these taxa are separated are not very satisfactory, and there is considerable variation within a single population. Intermediates between **1** and **2** occur in N.E. Spain, while *H. bourgaei* has sometimes been included in **1** and sometimes in **3**.

3. H. comosa L., *Sp. Pl.* 744 (1753). Perennial up to 40(–60) cm, woody at base. Leaflets 3–8 pairs, (2–)5–15 × (1–)2–4 mm, obovate to linear, subglabrous to densely pubescent beneath. Heads (2–)5- to 12-flowered; peduncles up to 4 times as long as the leaves. Corolla 6–10(–14) mm; claw of the standard usually distinctly longer than the calyx. Legume 15–30 × 2–3 mm; segments horseshoe-shaped or with semicircular sinuses, with red-brown papillae. 2*n*=28. *W., C. & S. Europe, northwards to N. England and N.C. Germany.* Al Au Be Br Bu Cz Ga Ge Gr He Ho Hs Hu It Ju Rm Rs (W) Sa.

An extremely variable species which can be distinguished from all the other perennial species of *Hippocrepis* by the red-brown papillae on the legume. Plants in S.E. Europe, otherwise closely resembling this species, sometimes have white papillae, and it is not clear whether these are variants of **3** or of **1** or **2**.

4. H. squamata (Cav.) Cosson, *Not. Pl. Crit.* 105 (1851). Caespitose perennial 10–40 cm, woody at base. Leaflets 3–7 pairs, 2·5–8 × 1·5–5 mm, lanceolate to suborbicular, usually very densely pubescent, sericeous or silvery. Heads 2- to 8-flowered; peduncles 2–4 times as long as the leaves. Corolla 6–12 mm; claw of the standard about as long as the calyx. Legume 10–25 × 2·5–4 mm; segments lunate or with a semicircular sinus; at least the seed-protuberance covered with white papillae 0·5 mm or more long. ● *C., S. & E. Spain.* Hs.

(a) Subsp. **squamata**: Leaflets up to 7 pairs, lanceolate to obovate; legume with papillae covering the seed-protuberance and sometimes the sinuses. *C., E. & S.E. Spain.*

(b) Subsp. **eriocarpa** (Boiss.) Nyman, *Consp.* 187 (1878): Leaflets up to 5 pairs, suborbicular; legume completely covered by long papillae. *S. Spain.*

5. H. valentina Boiss., *Elenchus* 38 (1838). Perennial 20–50 cm, woody at base. Leaflets 3–6 pairs, 9–12 × 3–6 mm, obovate or obovate-elliptical, obtuse, mucronate. Heads 2- to 10- flowered; peduncles up to 1½ times as long as the leaves. Corolla 9–12 mm; claw of the standard about as long as the calyx. Legume 10–30 × 3–5 mm with broad flat regions between the seed-protuberances; segments with semicircular sinus, glabrous. *Calcareous rocks.* ● *S.E. Spain (Alicante prov.).* Hs.

6. H. balearica Jacq., *Misc. Austr. Bot.* 2: 305 (1781). Like **5** but leaflets 5–10 pairs, 4–12 × 1–2(–5) mm, linear or oblong, acute; peduncles up to 3 times as long as the leaves; corolla 10–15 mm; claw of the standard about twice as long as the calyx; legume 15–45 mm; segments sometimes with orbicular sinus. *Calcareous rocks.* ● *Islas Baleares.* Bl.

7. H. ciliata Willd., *Ges. Naturf. Freunde Berlin Mag.* 2: 173 (1808). Slender annual up to 30 cm. Leaflets 3–6(–7) pairs, 5–15 × 0·5–3 mm, linear or oblong. Heads 2- to 6-flowered; peduncle about equalling the leaves. Corolla 3–5 mm. Legume 15–25 × 2·5–4 mm, curved so that the sinuses open on the concave edge; segments with orbicular sinuses and with long papillae on the seed-protuberance. *S. Europe.* Al Bl Bu Co Cr Ga Gr Hs It Ju Lu Rs (K) Sa Si Tu.

8. H. multisiliquosa L., *Sp. Pl.* 744 (1753). Slender annual 10–60 cm. Leaflets 3–8 pairs, 5–15 × 2–5 mm, obovate-oblong. Heads 2- to 6-flowered; peduncle about equalling leaves. Corolla 5–8 mm. Legume 20–40(–60) × 3–5 mm, curved so that the sinuses open on the convex edge; segments with orbicular sinuses, glabrous or very sparsely papillose. 2*n*=14. *W. Mediterranean region and S. Portugal; Greece.* Bl Co Gr Hs Lu Sa Si.

9. H. salzmanii Boiss. & Reuter in Boiss., *Diagn. Pl. Or. Nov.* 1(2): 101 (1843). Like **8** but the peduncle distinctly longer than the leaves in fruit; corolla 12–15 mm; legume 25–50 × 4–7 mm. *Maritime sands and dry, stony ground. S.W. Spain (near Cádiz).* Hs. (*N.W. Morocco.*)

10. H. unisiliquosa L., *Sp. Pl.* 744 (1753) (incl. *H. biflora* auct. eur., non Sprengel). Slender annual up to 40 cm. Leaflets 3–7 pairs, 2–12 × 1–5 mm, linear to obovate. Flowers usually solitary, axillary, rarely 2 or 3 together and shortly pedunculate. Corolla 4–7 mm. Legume 15–40 × 4–5 mm; segments with an orbicular sinus, glabrous or sparsely papillose. *S. Europe.* Al Bl Bu Co Cr Ga Gr Hs It Ju Lu Rs (K) Sa Si Tu.

67. Hammatolobium Fenzl[1]

Perennial herbs. Leaves imparipinnate or 3-foliolate, sometimes almost digitate; stipules small, free, linear. Flowers solitary or in axillary heads, subtended by a simple or 3-foliolate bract. Calyx tubular-campanulate, with 5 subequal teeth; keel acute; stamens monadelphous. Legume lomentaceous; segments terete, finely reticulate-veined.

1. H. lotoides Fenzl, *Pugillus* 3 (1842). Much branched and woody at base with procumbent villous stems 10–15 cm. Leaflets 5–15 × 4–10 mm, obovate. Heads 1- to 5-flowered. Corolla 10–14 mm, yellow or purple. Legume 10–25 mm, oblong, pubescent, beaked; segments 2–8, ovoid-oblong. *Mountain cliffs. S. Greece.* Gr.

[1] By P. W. Ball. [2] By A. Chrtková-Žertová.

68. Scorpiurus L.[1]

Annual. Leaves simple, with 3–5 parallel veins; stipules free, linear. Flowers solitary or in axillary heads. Calyx campanulate, with 5 equal teeth; corolla yellow or purplish; keel acute; stamens diadelphous. Legume lomentaceous or indehiscent, curved or variously contorted, longitudinally ridged, usually with spines or tubercles on the outer ridges.

Literature: C. C. Heyn & V. Raviv, *Bull. Torrey Bot. Club* 93: 259–267 (1966).

Fruit smooth or with tubercles or spines on the outer ridges; flowers usually 2–5 in a head **1. muricatus**
Fruit with capitate tubercles on the outer ridges; flowers solitary, rarely 2 together on a peduncle **2. vermiculatus**

1. S. muricatus L., *Sp. Pl.* 745 (1753) (incl. *S. subvillosus* L., *S. sulcatus* L.). Stems up to 80 cm, glabrous or pubescent with appressed or patent hairs. Heads (1–)2- to 5-flowered. Corolla 5–10(–12) mm. Legume with the ridges smooth or the outer 4–8 tuberculate or spinose. Seeds lunate, attenuate at the ends. 2*n*=28. *S. Europe; a rare casual elsewhere.* Al Bl Bu Co Cr Ga Gr Hs It Ju Lu Rs (K) Sa Si Tu.

Often divided into 3 taxa, of varying rank, on the type of indumentum, the length of the peduncle in relation to the leaves, the length of the calyx-teeth in relation to the tube and the degree of spininess and the contortion of the legume. The correlation between these characters is inconstant, and there does not seem to be any satisfactory geographical separation.

2. S. vermiculatus L., *Sp. Pl.* 744 (1753). Stems up to 70 cm, hirsute. Flowers solitary or rarely 2 together on a peduncle. Corolla 10–20 mm. Legume with stout capitate tubercles on the outer ridges. Seeds elliptical or oblong. *S.W. Europe.* *Co *Ga ?Gr Hs It Lu Sa.

69. Eversmannia Bunge[1]

Spiny shrubs, the spines formed from axillary branches. Leaves imparipinnate; stipules connate. Flowers in axillary racemes. Calyx tubular-campanulate, the teeth more or less unequal; corolla pink; stamens monadelphous or diadelphous. Legume lomentaceous, disarticulating tardily, oblong, compressed, not constricted between the segments.

1. E. subspinosa (Fischer ex DC.) B. Fedtsch., *Acta Horti Petrop.* 24: 173 (1905). Grey-pubescent shrub 12–60 cm, with tortuous branches. Leaflets 3–7 pairs, 5–10 × 3–4 mm, elliptical or obovate, obtuse. Corolla 12–17 mm. Legume 30–50 × 4–5 mm, glabrous, flexuous, with the dorsal edge persisting after the disarticulation of the legume. *S.E. Russia (by the lower Volga).* Rs (E). (*Kazakhstan.*)

70. Hedysarum L.[2]

Annual or perennial herbs. Leaves imparipinnate; stipules free or connate. Flowers in axillary racemes. Calyx campanulate, with 5 subequal teeth; corolla pink, purple or violet, rarely white or yellow; stamens diadelphous. Legume lomentaceous, more or less compressed, with up to 8 segments. Seeds 1 in each segment.

1 Annual
2 Leaflets 1–3(–5) pairs, the terminal leaflet 15–45 × 9–40 mm **2. flexuosum**
2 Leaflets (2–)4–8 pairs, the terminal leaflet 5–15 × 2–5 mm

3 Corolla 8–11 mm, 1½–2 times as long as the calyx
3. spinosissimum

3 Corolla 14–20 mm, 2½–5 times as long as the calyx
4. glomeratum

1 Perennial
  4 Acaulescent, with all the leaves in a basal rosette
    5 Calyx about as long as the corolla; legume unarmed
15. candidum
    5 Calyx shorter than the corolla; legume with short spines or setae
      6 Corolla yellow; calyx about as long as the wings
16. grandiflorum
      6 Corolla purple-violet; calyx exceeding the wings
17. biebersteinii
  4 Caulescent; plants with leafy and usually branched stems
    7 Leaflets glabrous or sparsely hairy beneath; legume glabrous or sparsely pubescent, unarmed
      8 Leaflets emarginate; bracts 1–1·5 mm; corolla cream or white, sometimes with bluish veins   **7. boutignyanum**
      8 Leaflets obtuse or acute; corolla reddish-violet
        9 Bracts 6–15 mm, longer than the pedicel; segments of legume with a wide membranous margin   **5. hedysaroides**
        9 Bracts 1·5–3 mm, shorter than the pedicel; segments of legume with a very narrow membranous margin
6. alpinum
    7 Leaflets pubescent or sericeous beneath; legume pubescent or with spines or setae
     10 Legume unarmed
      11 Leaflets lanceolate or linear; corolla pale pink or lilac
9. razoumowianum
      11 Leaflets oblong, elliptical or obovate; corolla purple or violet
        12 Calyx-teeth shorter than tube   **10. cretaceum**
        12 Calyx-teeth as long as or longer than tube   **12. tauricum**
     10 Legume with spines or setae
      13 Calyx-teeth about as long as tube; legume glabrous
1. coronarium
      13 Calyx-teeth longer than tube; legume pubescent
        14 Corolla yellow   **14. varium**
        14 Corolla purple to violet
          15 Corolla 15–24 mm   **8. gmelinii**
          15 Corolla 9–14 mm
            16 Leaflets elliptical to obovate; calyx 3–3·5 mm
18. macedonicum
            16 Leaflets linear to oblong-elliptical; calyx 4–9 mm
              17 Spines on legume more than 1 mm   **11. ucrainicum**
              17 Spines on legume not more than 1 mm   **13. humile**

Sect. HEDYSARUM. Caulescent. Stipules free. Segments of the legume spinulose.

1. H. coronarium L., *Sp. Pl.* 750 (1753). Perennial 30–100 cm, sparsely appressed-pubescent. Leaflets 3–5 pairs, 15–35 × 12–18 mm, elliptical to obovate-orbicular, glabrous or subglabrous above, pubescent beneath. Racemes 10- to 35-flowered, dense. Calyx sparsely to densely pubescent, the teeth about as long as the tube; corolla 12–15 mm, bright reddish-purple. Legume with 2–4 spinulose, but otherwise glabrous segments. $2n=16$. *C. & W. part of Mediterranean region; cultivated for fodder and naturalized elsewhere in S. Europe.* Hs It Sa Si [Bl Ga Gr Ju Lu].

2. H. flexuosum L., *Sp. Pl.* 750 (1753). Annual 20–60 cm, glabrous or sparsely pubescent. Leaflets 1–3(–5) pairs, 15–45 × 9–40 mm, pubescent, oblong or obovate, the terminal larger than the lateral. Racemes 15- to 40-flowered. Calyx sparsely pubescent, the teeth longer than the tube; corolla 8–12 mm, purple or pink. Legume with 2–8 setose and spinulose segments. *Sandy places near the sea. S.W. Spain, S.W. Portugal.* Hs Lu [Rs (K)]. (*N. Africa.*)

3. H. spinosissimum L., *Sp. Pl.* 750 (1753). Annual 15–35 cm, appressed-pubescent. Leaflets (2–)4–8 pairs, 5–12 × 2–5 mm,

elliptical or oblong, subglabrous or pubescent. Racemes 2- to 10-flowered. Calyx 4–6 mm, sparsely pubescent, the teeth as long as or longer than the tube; corolla 8–11 mm, white to pale pinkish-purple, 1½–2 times as long as calyx. Legume with 2–4 spinulose and pubescent segments. *Mediterranean region.* Bl Co Cr Ga Gr Hs ?It Sa Si.

4. H. glomeratum F. G. Dietrich, *Vollst. Lexic.* **4**: 534 (1804) (*H. capitatum* Desf., non Burm. fil., *H. spinosissimum* sensu Coste pro parte). Like **3** but leaflets sometimes obovate; corolla 14–20 mm, pinkish-purple, 2½–5 times as long as the calyx. *Mediterranean region, Portugal.* Co Ga Gr Hs It Ju Lu Sa Si.

Sect. GAMOTION Basiner (Sect. Obscura B. Fedtsch.) Caulescent. Stipules usually united. Segments of the legume unarmed, reticulate-veined.

5. H. hedysaroides (L.) Schinz & Thell., *Viert. Naturf. Ges. Zürich* **58**: 70 (1913) (*H. obscurum* L.). Perennial 10–40 cm, glabrous or sparsely pubescent. Leaflets 3–10 pairs, 10–25 × 5–12 mm, obtuse. Racemes 3–8 cm, 15- to 35(–48)-flowered. Bracts 6–15 mm. Calyx-teeth shorter or longer than the tube; corolla 13–25 mm, reddish-violet, rarely white. Legume with 2–5 segments, with a membranous margin. ● *Mountains of S.C. Europe; Arctic Russia, N. & C. Ural.* Au Cz Ga Ge He ?Hs It Ju Po Rm Rs (N, C, W).

1 Leaflets 6–10 pairs; longest calyx-teeth 3–4·5 mm
(b) subsp. exaltatum
1 Leaflets usually 4–6 pairs; longest calyx-teeth 1·5–3 mm
  2 Leaflets ovate or elliptical   **(a) subsp. hedysaroides**
  2 Leaflets oblong-elliptical   **(b) subsp. arcticum**

(a) Subsp. **hedysaroides**: Stems 10–15 cm. Leaflets 4–6 pairs, ovate or elliptical. Calyx-teeth triangular, the longest 1·5–3 mm. $2n=14$. *Almost throughout the range of the species.*

This subspecies contains two variants, one with calyx-teeth longer than the tube occurring in the Alps, the other with calyx-teeth shorter than the tube occurring in the mountains of E.C. Europe and the Alps.

(b) Subsp. **exaltatum** (A. Kerner) Žertová, *Feddes Repert.* **79**: 47 (1968) (*H. exaltatum* A. Kerner): Stems 30–40 cm. Leaflets 6–10 pairs. Calyx-teeth lanceolate or subulate, the longest 3–4·5 mm. *S. Alps.*

(c) Subsp. **arcticum** (B. Fedtsch.) P. W. Ball, *Feddes Repert.* **79**: 47 (1968) (*H. arcticum* B. Fedtsch.): Stems 20–30 cm. Leaflets 4–6 pairs, oblong-elliptical. Calyx-teeth shorter than the tube, the longest *c*. 1·5 mm. *Arctic Russia, N. & C. Ural.*

6. H. alpinum L., *Sp. Pl.* 750 (1753). Perennial 30–100 cm, glabrous or sparsely pubescent. Leaflets 6–14 pairs, 10–30 × 4–10 mm, oblong or lanceolate, obtuse or acute. Racemes 6–15 cm, 20- to 30(–60)-flowered. Bracts 1·5–3 mm. Calyx-teeth shorter than the tube; corolla 12–15 mm, reddish-violet. Legume with 2–6 sparsely pubescent segments, with a very narrow membranous margin. *N. & E. Russia.* Rs (N, C, E). (*N. Asia.*)

7. H. boutignyanum Alleiz., *Bull. Soc. Bot. Fr.* **75**: 38 (1928). Glabrous perennial 50–60 cm. Leaflets 4–8 pairs, 10–30 × 6–18 mm, elliptical or obovate, emarginate; petioles glandular. Racemes 5–20 cm. Bracts 1–1·5 mm. Calyx-teeth shorter than the tube; corolla 12–16(–20) mm, cream or white, sometimes with bluish veins. Legume with 2–5 segments, glabrous except for hairy margin. ● *S.W. Alps.* Ga.

Sect. MULTICAULIA (Boiss.) B. Fedtsch. Caulescent. Stipules united. Segments of the legume unarmed or spinose, rugose.

8. **H. gmelinii** Ledeb., *Mém. Acad. Sci. Pétersb.* 5: 551 (1815). Pubescent perennial 20–70 cm. Leaflets (3–) 4–11 pairs, 7–30 × 3–14 mm, oblong or elliptical, pubescent. Racemes 15- to 45-flowered. Calyx appressed-pubescent or subglabrous, the teeth longer than the tube; corolla 15–24 mm, purple. Legume with 3–6 pubescent, spinose, rugose segments. *S.E. Russia, W. Kazakhstan.* Rs (E). (*N.C. Asia.*)

9. **H. razoumowianum** Helm & Fischer ex DC., *Prodr.* 2: 342 (1825). Perennial up to 50 cm, appressed-pubescent. Leaflets 4–8 pairs, 10–30 × 1·5–4·5 mm, linear or lanceolate, glabrous or sparsely pubescent above, pubescent beneath. Racemes 8- to 20-flowered. Calyx appressed-pubescent, the teeth as long as or longer than the tube; corolla 15–20 mm, pale pink or lilac. Legume with 2–7 pubescent, spinose segments, sulcate on the margin. *S.E. Russia.* Rs (E). (*N. Kazakhstan.*)

10. **H. cretaceum** Fischer ex DC., *loc. cit.* (1825). Sparsely pubescent perennial 20–50 cm. Leaflets 5–12 pairs, 5–12 × 1·5–3 mm, oblong or elliptical, subglabrous above, appressed-pubescent beneath. Racemes 12- to 20-flowered. Calyx 3–4 mm, sparsely appressed-pubescent, the teeth shorter than the tube; corolla 11–17 mm, deep violet. Legume with 2–4 pubescent, reticulate-veined segments. ● *S.E. Russia.* Rs (E).

11. **H. ucrainicum** B. Kaschm., *Bull. Jard. Bot. Pétersb.* 5: 59 (1905). Pubescent perennial 7–20(–35) cm. Leaflets 5–10 pairs. 8–14(–20) × 1·5–3 mm, oblong-elliptical, subglabrous above, densely white-pubescent beneath. Racemes 8- to 25-flowered. Calyx 4–5·5 mm, pubescent, the teeth 2–3 times as long as the tube; corolla 10–12 mm, purple. Legume with 2–4 sparsely pubescent, spinose segments. ● *E. Ukraine.* Rs (E).

12. **H. tauricum** Pallas ex Willd., *Sp. Pl.* 3: 1208 (1802). Sparsely pubescent perennial 15–20 cm. Leaflets 6–8 pairs, 7–12 × 2–3·5 mm, oblong or elliptical to obovate, pubescent. Racemes 8- to 15-flowered. Calyx sparsely pubescent, the teeth as long as or longer than the tube; corolla 11–18 mm, purple. Legume with 2–5 pubescent segments, sulcate on the margin. *N.E. Bulgaria, Krym.* Bu Rs (K).

Two distinct variants occur: one with corolla 11–14(–15) mm, from Bulgaria and Krym, the other with corolla 14–18 mm, from Krym.

13. **H. humile** L., *Syst. Nat.* ed. 10, 2: 1171 (1759). Pubescent perennial 20–50 cm. Leaflets 6–16 pairs, 4–10 × 1–3 mm, linear or oblong, glabrous above, pubescent beneath. Racemes 15- to 25-flowered. Calyx 4–9 mm, with long white hairs, the teeth longer than the tube; corolla 9–14 mm, purple-violet. Legume with (1–)2–3 pubescent, shortly spinose segments, sometimes with spines only on the margin. *W. Mediterranean.* Ga Hs.

14. **H. varium** Willd., *Sp. Pl.* 3: 1206 (1802). Pubescent perennial 30–60 cm. Leaflets 7–8 pairs, 5–12 mm, oblong or elliptical, pubescent. Racemes 15- to 30-flowered. Calyx 5–8 mm, pubescent, the teeth about 1½ times as long as the tube; corolla 14–22 mm, yellow, the keel sometimes purple at apex. Legume with 2–4 pubescent segments. *Turkey-in-Europe (S.W. of Gelibolu);* ?*Macedonia.* Tu. (*N. & C. Anatolia.*)

H. formosum Fischer & C. A. Meyer ex Basiner, *Mém. Sav. Étr. Pétersb.* 6: 69 (1846), from W. Asia, with ovate leaflets and yellow corolla, has been recorded from Bulgaria. The specimen on which this record is based has purple flowers, and cannot be determined with certainty.

¹ By P. W. Ball.

Sect. SUBACAULIA (Boiss.) B. Fedtsch. Acaulescent. Stipules united. Segments of the legume unarmed or spinulose.

15. **H. candidum** Bieb., *Fl. Taur.-Cauc.* 2: 176 (1808). Perennial with peduncles 20–40 cm, patent-pubescent. Leaflets 2–6 pairs, 10–35 × 8–25 mm, oblong or elliptical, silver-white-pubescent above, sericeous beneath. Racemes 15- to 50-flowered. Calyx about as long as corolla, the teeth much longer than the tube; corolla 15–22 mm, yellow, pink or purple; standard as long as or slightly shorter than keel. Legume with 2–4 unarmed, tomentose segments. *Krym.* Rs (K). (*W. Caucasus.*)

16. **H. grandiflorum** Pallas, *Reise* 2: 743 (1773). Like **15** but sometimes subglabrous; leaflets 2–5 pairs, 15–40 × 8–30 mm, elliptical or ovate; calyx distinctly shorter than corolla; corolla 18–25 mm, yellow; standard exceeding keel; segments spinulose. *S.E. Europe, from Bulgaria to c. 55°N. in E. Russia.* Bu Rm Rs (W, E).

17. **H. biebersteinii** Žertová, *Feddes Repert.* 79: 47 (1968) (*H. argenteum* Bieb., non L.). Like **15** but peduncles with appressed hairs; leaflets 3–7 pairs, oblong or oblong-ovate, sericeous; corolla purple-violet; standard exceeding keel; segments shortly spinulose. *S.E. Russia.* Rs (E).

Sect. CRINIFERA (Boiss.) B. Fedtsch. Caulescent. Stipules united. Segments of the legume setose.

18. **H. macedonicum** Bornm., *Mitt. Thür. Bot. Ver.* nov. ser., 36: 43 (1925). Perennial 20–50 cm, appressed-pubescent. Leaflets 8–12(–14) pairs, 3–5(–8) × 2–3(–5) mm, elliptical or obovate, glabrous above, white pubescent beneath. Racemes 15- to 30-flowered. Calyx 3–3·5 mm, the teeth slightly longer than the tube; corolla 10–14 mm, purple-violet. Legume with 2–3 segments. $2n = 16$. ● *S. Jugoslavia.* Ju.

71. Onobrychis Miller¹

Annual or perennial herbs. Leaves imparipinnate; stipules free or connate. Flowers in axillary racemes. Calyx campanulate, with 5 equal teeth; corolla white, pink or purple, rarely yellow; stamens diadelphous. Legume indehiscent, more or less orbicular and compressed, with a distinct, usually toothed, margin and foveolate to reticulate-veined sides which often have teeth on the veins or ridges. Seeds 1–3.

Literature: H. Handel-Mazzetti, *Österr. Bot. Zeitschr.* 59: 369–378, 424–430, 479–488 (1909); 60: 5–12, 64–71 (1910). G. Širjaev, *Publ. Fac. Sci. Univ. Masaryk* 56: 1–195 (1925); 76: 1–165 (1926); 242: 1–14 (1937); *Bull. Soc. Bot. Bulg.* 4: 7–24 (1931).

1 Annual; racemes not more than 8-flowered
  2 Corolla 7–8 mm; legume with linear-triangular to subulate teeth on the margin   **22. caput-galli**
  2 Corolla 10–14 mm; legume with broadly triangular teeth on the margin   **23. aequidentata**
1 Perennial; racemes usually at least 10-flowered
  3 Standard pubescent on the back; legume with 2 rows of teeth on the margin, rarely unarmed; stipules usually free
    4 Leaflets sparsely pubescent and green beneath; margin of the legume with teeth 0·5–1·5 mm   **3. radiata**
    4 Leaflets densely grey-tomentose beneath; margin of the legume with teeth not more than 0·5 mm, or unarmed
      5 Sides of the legume unarmed or tuberculate; margin unarmed or with minute teeth   **1. hypargyrea**
      5 Sides of the legume toothed; margin with teeth up to 0·5 mm   **2. pallasii**

3 Standard glabrous; legume with 1 row of teeth on the margin
   or unarmed; stipules usually connate
6 Standard at least 1·2 times as long as the keel
  7 Calyx-teeth 2–3 times as long as tube
    8 Calyx densely hirsute; legume with long hairs (0·5 mm or
      more) **6. ebenoides**
    8 Calyx sparsely pubescent; legume with short hairs (less
      than 0·5 mm) **7. supina**
  7 Calyx-teeth 1–2 times as long as tube
    9 Calyx glabrous except for the pubescent margin; margin
      of the legume unarmed or with teeth not more than
      3 mm **8. gracilis**
    9 Calyx pubescent; margin of the legume with some teeth at
      least 3 mm **9. pindicola**
6 Standard shorter than to slightly longer than keel
  10 Standard at least 5 mm shorter than keel **10. stenorhiza**
  10 Standard not more than 2 mm shorter than keel
    11 Wings 7–8 mm, distinctly exceeding calyx
      12 Margin of the legume unarmed or tuberculate; stems
        usually very short, rarely up to 40 cm **4. saxatilis**
      12 Margin of the legume with 6–7 teeth; stems 30–60 cm
        5. petraea
    11 Wings not more than 6(–7) mm, shorter than calyx
      13 Legume densely pubescent or villous, the hairs at least
        0·5 mm
        14 Standard emarginate; margin of the legume with teeth
          3–6 mm **14. peduncularis**
        14 Standard entire or truncate; margin of the legume with
          teeth not more than 3 mm
          15 Leaflets 12–16 pairs; corolla purple; margin of legume
            with 8–10 teeth **13. sphaciotica**
          15 Leaflets 6–12 pairs; corolla white or pink; margin of
            legume with 3–6 teeth
            16 Stems pubescent, the hairs less than 1 mm; legume
              4–7 mm **11. alba**
            16 Stems villous, the hairs 1 mm or more; legume
              8–10 mm **12. degenii**
      13 Legume with hairs less than 0·5 mm
        17 Calyx longer than keel **16. reuteri**
        17 Calyx shorter than keel
          18 Standard 1–2 mm shorter than keel **17. montana**
          18 Standard less than 1 mm shorter than to slightly longer
            than keel
            19 Corolla (9–)10–14 mm
              20 Corolla usually white with pink veins; margin of the
                legume with teeth at least 3 mm (Spain)
                14. peduncularis
              20 Corolla pink or purple; margin of the legume with
                teeth not more than 3(–4) mm
                21 Leaflets densely silvery pubescent beneath
                  15. argentea
                21 Leaflets green or greyish beneath
                  22 Legume (6–)7–12 mm, the teeth on the margin
                    usually 2–4 mm **17. montana**
                22 Legume 4–8 mm, the teeth on the margin not
                  more than 2 mm
                  23 Calyx-teeth with long ± patent hairs **21. viciifolia**
                  23 Calyx-teeth with appressed or ± erecto-patent
                    hairs
                    24 Plant with non-flowering rosettes of leaves at
                      the base; margin of the legume with teeth
                      not more than 1 mm **15. argentea**
                    24 Plant without non-flowering rosettes of leaves;
                      margin of the legume with teeth 0·5–2 mm
                      18. arenaria
             19 Corolla 5–10 mm
              25 Margin of the legume with teeth 2–4 mm **18. arenaria**
              25 Margin of the legume with teeth not more than
                2 mm
                26 Leaflets ± densely grey-pubescent, at least be-
                  neath; calyx-teeth with ± patent hairs
                    19. oxyodonta
                26 Leaflets green or grey-green, glabrous to pubes-

cent; calyx-teeth with appressed or erecto-patent
    hairs, or glabrous
  27 Legume ± toothed **18. arenaria**
  27 Legume unarmed, rarely slightly tuberculate
    20. inermis

Sect. HYMENOBRYCHIS DC. Usually perennial; racemes many-flowered; standard pubescent on back; ovary usually curved; margin of the legume with 2 rows of teeth, rarely unarmed.

A large section, centred in Asia and N. Africa, with a few marginal European representatives.

1. O. hypargyrea Boiss., *Diagn. Pl. Or. Nov.* 1(2): 91 (1843). Villous perennial up to 100 cm or more. Leaflets 4–7 pairs, 25–50 × 6–20 mm, ovate-oblong or oblong, acute, glabrous above, densely grey-tomentose beneath; stipules not connate, but adnate to petiole. Calyx villous, the teeth slightly longer than tube; corolla 15–20 mm, pale yellow with pink veins. Legume 14–18 mm, villous; sides unarmed or tuberculate; margin entire or shortly denticulate. *Macedonia.* Gr Ju. (*C. & W. Anatolia.*)

2. O. pallasii (Willd.) Bieb., *Cent. Pl.* 1: t. 35 (1810). Like **1** but leaflets 5–8 pairs, 15–30(–40) × 10–15 mm, ovate or ovate-oblong; calyx-teeth 1·5–2 times as long as tube; legume with toothed sides; margin denticulate, the teeth up to 0·5 mm. ● *Krym.* Rs (K).

3. O. radiata (Desf.) Bieb., *Cent. Pl.* 2: t. 55 (1832). Villous perennial 40–60 cm. Leaflets 5–9 pairs, 10–20(–30) × (6–)8–12 mm, elliptical or ovate-oblong, obtuse or subacute, glabrous above, sparsely pubescent beneath; stipules free or connate. Calyx villous, the teeth 2–3 times as long as tube; corolla 15–20 mm, pale yellow with red veins. Legume 11–16(–17) mm, glabrous or pubescent; sides toothed; margin denticulate, the teeth 0·5–1·5 mm. *S.E. Russia.* Rs (E). (*Caucasus.*)

Sect. ONOBRYCHIS. Perennial; stipules usually connate; racemes many-flowered; standard glabrous; ovary straight; margin of the legume with 1 row of teeth, rarely unarmed.

4. O. saxatilis (L.) Lam. *Fl. Fr.* 2: 653 (1778). Caespitose, subacaulescent perennial, rarely with stems up to 40 cm, pubescent with appressed hairs. Leaflets 6–15 pairs, 7–25 × 1–3(–4) mm, linear, glabrous above. Calyx sparsely pubescent, the teeth 1·5–2 times as long as tube; corolla 9–14 mm, pale yellow with pink veins; wings *c.* 8 mm, longer than calyx. Legume 5–8 mm, unarmed, but the margin sometimes tuberculate. *W. Mediterranean region.* Ga Hs It.

5. O. petraea (Bieb. ex Willd.) Fischer, *Cat. Jard. Gorenki* ed. 2, 73 (1812). Like **4** but with stems 30–60 cm, leaflets 8–25 × 2·5–5·5 mm, elliptical or elliptic-oblong; calyx-teeth 1·5–2·5 times as long as tube; corolla pale purple; margin of the legume with 6–7 teeth *c.* 0·5 mm. *Krym.* Rs (K). (*Caucasus.*)

6. O. ebenoides Boiss. & Spruner in Boiss., *Diagn. Pl. Or. Nov.* 1(2): 97 (1843). Greyish pubescent perennial up to 45 cm, with appressed hairs. Leaflets 4–7 pairs, 4–15(–20) × 1·5–2·5(–5) mm, linear-elliptical or elliptical. Calyx densely hirsute, the teeth 2–3 times as long as tube; corolla 7–10 mm, pink with darker veins, rarely white; standard distinctly longer than keel. Legume 5–7 mm, densely tomentose; sides toothed; margin with 4–5 teeth. ?● *C. & S. Greece.* Gr.

7. O. supina (Chaix) DC. in Lam. & DC., *Fl. Fr.* ed. 3, 4: 612 (1805). Sparsely pubescent perennial up to 40(–60) cm. Leaflets 5–12(–20) pairs, 6–15(–25) × 1·5–3(–4) mm, oblong or elliptical. Calyx usually sparsely pubescent, the teeth 2–3(–4) times as long as tube; corolla 7–10 mm, pink, or white with pink veins, the standard distinctly longer than keel. Legume 4–5 mm, villous; sides toothed; margin with 3–4 teeth up to 2 mm. $2n=14$. *S.W. Europe from E. Spain to N.W. Italy; Calabria.* Ga Hs It.

8. O. gracilis Besser, *Enum. Pl. Volhyn.* 74 (1822) (incl. *O. longeaculeata* Pacz.). Glabrous or sparsely pubescent perennial 25–75 cm. Leaflets 3–8 pairs, 12–25 × 1–3(–5) mm, usually linear. Calyx glabrous except for the pubescent margin, the teeth 1–2 times as long as tube, ciliate on the margin; corolla 5–7 mm, pink, the standard distinctly longer than keel. Legume 2·5–5(–7) mm, pubescent with short appressed hairs; sides toothed; margin unarmed or with 4–6 teeth up to 3 mm. *S.E. Europe, from Macedonia to Krym.* Bu Gr Ju Rm Rs (W, K) Tu.

9. O. pindicola Hausskn., *Mitt. Thür. Bot. Ver.* nov. ser., **5**: 87 (1893). Like **8** but leaflets 7–11(–16) × 2–5(–6) mm, oblong or elliptical; calyx-tube and teeth usually shortly pubescent; corolla 6–9 mm; legume 4–7 mm, with teeth up to 3·5(–5) mm. *Greece and S. Jugoslavia.* ?Bu Gr Ju.

10. O. stenorhiza DC., *Prodr.* **2**: 346 (1825). Densely grey-pubescent perennial up to 40 cm. Leaflets of the lower leaves 5–13 pairs, 5–12 × 2–3·5 mm, elliptical or oblong-elliptical; of the upper leaves 3–8 pairs, 10–13 × 1–2 mm, linear or oblong. Calyx appressed-pubescent, the teeth 1½–2 times as long as tube; corolla 10–13 mm, pink with darker veins, the standard not more than ½ as long as keel. Legume 4–5 mm, pubescent; sides toothed; margin with 5–7 teeth up to 4 mm. ● *S.E. Spain.* Hs.

11. O. alba (Waldst. & Kit.) Desv., *Jour. Bot. Appl.* **3**: 83 (1814). Pubescent perennial up to 60 cm. Leaflets of the lower leaves 6–10 pairs, 5–35 × 1·5–4(–6) mm, linear to elliptical. Calyx pubescent with patent hairs or villous, rarely with the tube glabrous, the teeth at least twice as long as tube; corolla 8–14 mm, white or pink often with pink veins, the standard about as long as keel. Legume 4–7 mm, villous, the hairs at least 0·5 mm; sides toothed; margin with 2–6 teeth. *Balkan peninsula, S. Romania; C. & S. Italy.* Al Bu Gr It Ju Rm.

1 Corolla 8–10 mm; calyx-teeth 3–4 times as long as tube
                            (d) subsp. calcarea
1 Corolla 10–14 mm; calyx-teeth 2–3 times as long as tube
  2 Corolla 12–14 mm; calyx-tube villous   **(c) subsp. echinata**
  2 Corolla 10–12 mm; calyx-tube usually sparsely pubescent
    3 Leaflets of lower leaves 8–35 mm; corolla usually white; calyx with white hairs   **(a) subsp. alba**
    3 Leaflets of lower leaves 5–12 mm; corolla usually pinkish; calyx with brownish hairs   **(b) subsp. laconica**

(a) Subsp. alba: Leaflets of lower leaves 8–35 × 1·5–4 mm; calyx-tube subglabrous to pubescent with white hairs, the teeth 2–3 times as long as the tube; corolla 10–12 mm, white sometimes with pinkish veins; legume 5–7 mm, the margin with (3–)4–6 teeth up to 2(–3) mm. ● *S. Romania, N. & C. part of Balkan peninsula, C. & S. Italy.*

(b) Subsp. laconica (Orph. ex Boiss.) Hayek, *Prodr. Fl. Penins. Balcan.* **1**: 928 (1926): Leaflets of the lower leaves 5–12 × 1·5–2·5 mm; calyx-tube pubescent with brownish hairs, the teeth 2–3 times as long as the tube; corolla 10–12 mm, pink, or white and pink; legume 4–6 mm, the margin with 3–4 teeth up to 1·5 mm. ● *S. & W. parts of Balkan peninsula; S. Italy.*

(c) Subsp. echinata (G. Don fil.) P. W. Ball, *Feddes Repert.* **79**: 42 (1968) (*O. echinata* G. Don fil.): Leaflets of lower leaves

8–25 × 3–6 mm; calyx-tube villous, the teeth 2–3 times as long as tube; corolla 12–14 mm, pink; legume 5–6 mm, the margin with 3–5 teeth up to 3 mm. *S. Italy.*

(d) Subsp. calcarea (Vandas) P. W. Ball, *Feddes Repert.* **79**: 41 (1968)(*O. calcarea* Vandas): Leaflets of lower leaves 4–15 × 1–4 mm; calyx-tube more or less pubescent, the teeth 3–4 times as long as tube; corolla 8–10 mm, white; legume *c.* 5 mm, the margin with 2–4(–5) teeth up to 2 mm. *C. part of Balkan peninsula.*

O. bertiscea Širj. & Rech. fil., *Feddes Repert.* **38**: 325 (1935), from W. Jugoslavia (near Gusinje, Crna Gora), is possibly another subspecies of **11**. It has the leaflets of the lower leaves 6–11 × 1·5–2·5 mm; the calyx-tube sparsely hairy, with teeth twice as long as the tube; the corolla 8–9(–10) mm, purple with darker veins, and the legume 5–6·5 mm, the margin with 3–4 teeth up to 0·75 mm.

12. O. degenii Dörfler, *Denkschr. Akad. Wiss. Math.-Nat. Kl. (Wien)* **64**: 718 (1897). Villous perennial 30–50 cm, the hairs at least 1 mm. Leaflets of the lower leaves 9–12 pairs, 12–20 × 2–4·5 mm, elliptical or oblong-elliptical, very densely hairy with appressed hairs; stipules free or united up to the middle. Calyx villous, the teeth 3–4 times as long as tube; corolla 10–14 mm, white, the standard about as long as keel. Legume 8–10 mm, villous with the hairs at least 0·5 mm; sides with long teeth; margin with 3–4 teeth up to *c.* 3 mm. ● *C. Macedonia (Alšar, S.E. of Prilep).* Ju.

13. O. sphaciotica W. Greuter, *Candollea* **20**: 213 (1965). Densely silvery-sericeous perennial up to 60 cm. Leaflets of the lower leaves 12–16 pairs, up to 30 × 5 mm, oblong-elliptical or broadly linear. Calyx villous, the teeth 3 times as long as tube; corolla 11–12 mm, purple, the standard as long as or slightly shorter than keel. Legume *c.* 10 mm, sericeous-pubescent, with the hairs at least 0·5 mm; sides toothed; margin with 8–10 teeth, 1–3 mm. *Limestone cliffs, 1300–1750 m.* ● *W. Kriti.* Cr.

14. O. peduncularis (Cav.) DC., *Prodr.* **2**: 346 (1825). Pubescent perennial 10–60 cm. Leaflets of the lower leaves 4–14 pairs, 4–20 × 1–6 mm, linear to elliptical, glabrous above, pubescent beneath. Calyx glabrous or pubescent, the teeth twice as long as tube; corolla 10–15 mm, white to purplish, often with purple veins. Legume 5–15 mm; sides toothed; margin with 4–10 teeth 3–6 mm. *C. & S. Spain, S. Portugal.* Hs Lu.

(a) Subsp. peduncularis (*O. eriophora* Desv.): Leaflets 7–14 pairs; standard usually shorter than keel; legume 8–15 mm, densely pubescent with long hairs (at least 1·5 mm), rarely glabrous. *Almost throughout the range of the species.*

(b) Subsp. matritensis (Boiss. & Reuter) Maire, *Bull. Soc. Hist. Nat. Afr. Nord* **19**: 84 (1928) (*O. longeaculeata* (Boiss.) Pau, *O. matritensis* Boiss. & Reuter): Leaflets 3–8 pairs; standard equalling the keel; legume 5–8 mm, pubescent with short hairs (less than 0·5 mm). ● *E.C. & S.E. Spain.*

15. O. argentea Boiss., *Voy. Bot. Midi Esp.* **2**: 188 (1839). Pubescent perennial 10–35 cm. Leaflets 5–8 pairs, 4–10 × 2–3 mm, elliptical to linear-oblong. Calyx pubescent, the teeth 2–3 times as long as tube; corolla 10–14 mm, pink, often with darker veins. Legume 4·5–8 mm, pubescent; sides toothed; margin with 4–7 teeth up to 1(–4) mm. *S. & E. Spain, Pyrenees.* Ga Hs.

(a) Subsp. argentea: Leaflets densely silver-pubescent beneath; racemes comose before anthesis; calyx-teeth with dense patent hairs; standard slightly shorter than keel. *S. Spain.*

(b) Subsp. hispanica (Širj.) P. W. Ball, *Feddes Repert.* **79**: 42 (1968) (*O. hispanica* Širj.): Leaflets pubescent but not silvery beneath; racemes not comose; calyx-teeth with erecto-patent hairs on the margin, glabrous or subglabrous on the surfaces; standard equalling or slightly exceeding keel. ● *From S.E. Spain to the Pyrenees.*

O. pyrenaica (Sennen) Sennen ex Širj., *Publ. Fac. Sci. Univ. Masaryk* **56**: 135 (1925), from the Pyrenees, is a rare taxon of somewhat uncertain affinities related to **15b** and **17**. It is a dwarf (up to 12 cm) sparsely pubescent perennial, with leaflets 2–4 × 1·5–2·5 mm; calyx-teeth about as long as tube, with a few short hairs on the margin; corolla 8–10 mm; legume with 4–6 teeth up to 1·5 mm on the margin.

16. O. reuteri Leresche in Leresche & Levier, *Deux Excurs. Bot.* 73 (1880). Pubescent perennial up to 30 cm, with appressed hairs. Leaflets 7–11 pairs, 3–7 × 1·5–2·5 mm, linear to obovate. Calyx pubescent, exceeding the keel, the teeth 1·5–2·5 times as long as tube; corolla 7–9 mm, Legume 6–7 mm, pubescent; sides toothed; margin with 6–7 teeth up to 2 mm. ● *N. Spain.* Hs.

17. O. montana DC. in Lam. & DC., *Fl. Fr.* ed. 3, **4**: 611 (1805). Subglabrous or sparsely pubescent perennial up to 50 cm. Leaflets 5–8 pairs, elliptical or ovate to oblong. Calyx sparsely pubescent, the hairs appressed or erecto-patent; corolla (8–)10–14 mm, pink, usually with purple veins. Legume pubescent; sides toothed or unarmed; margin with 4–8 teeth. *Mountains of C. Europe, Italy and the Balkan peninsula; doubtfully in the Pyrenees.* Al Au Bu Cz Ga Ge Gr He It Ju Po Rm.

1 Standard equalling or slightly shorter than keel; legume (6–)7–12 mm, the margin with teeth up to 4 mm
        (c) subsp. cadmea
1 Standard *c.* 2 mm shorter than keel; legume 6–8 mm, the margin with teeth not more than 2 mm
  2 Leaflets (5–)10–20 mm; calyx-teeth not more than 3 times as long as tube   **(a) subsp. montana**
  2 Leaflets 4–7(–10) mm; calyx-teeth 3–4 times as long as tube
        (b) subsp. scardica

(a) Subsp. **montana**: Leaflets (5–)10–20 × (2–)3–5 mm, usually glabrous above, pubescent beneath; racemes 3–7 cm before anthesis; calyx-teeth 1·5–3 times as long as tube; standard *c.* 2 mm shorter than keel; legume 6–8 mm, the teeth on the margin 0·5–2 mm. 2*n*=28. ● *Alps, N. & C. Apennini, Carpathians, W. Jugoslavia.*

(b) Subsp. **scardica** (Griseb.) P. W. Ball, *Feddes Repert.* **79**: 42 (1968) (*O. sativa* var. *scardica* Griseb.): Leaflets 4–7(–10) × 2–4 mm, glabrous above, sparsely pubescent beneath; racemes 1–3 cm before anthesis; calyx-teeth 3–4 times as long as tube; standard *c.* 2 mm shorter than keel; legume 6–7 mm, the teeth on the margin *c.* 0·5 mm. *Balkan peninsula.*

(c) Subsp. **cadmea** (Boiss.) P. W. Ball, *Feddes Repert.* **79**: 42 (1968) (*O. cadmea* Boiss.): Leaflets 3–11 × 2–6 mm, pubescent; racemes 2–3 cm before anthesis; calyx-teeth 1·5–3 times as long as tube; standard equalling or slightly shorter than keel; legume (6–)7–12 mm, the teeth on the margin up to 4 mm. *C. & S. Greece.*

18. O. arenaria (Kit.) DC., *Prodr.* **2**: 345 (1825). Subglabrous or pubescent perennial 10–80 cm. Leaflets usually 3–12 pairs, linear-oblong to elliptical. Calyx glabrous or pubescent, the teeth with appressed or erecto-patent hairs on the margin; corolla pink with purple veins. Legume 4–6(–7) mm, pubescent; sides toothed; margin with 3–8 teeth, rarely unarmed. *C., E. & S.E. Europe, extending to C. France and C. Italy.* Al Au Bu Cz Ga Ge Gr He Hu It Ju Po Rm Rs (B, C, W, E) Tu.

1 Margin of the legume with teeth 2–4 mm long; leaflets usually 3–7 pairs

2 Calyx-teeth 2·5–4 times as long as tube; racemes not more than 7 cm in flower   **(f) subsp. lasiostachya**
2 Calyx-teeth about twice as long as tube; racemes up to 12 cm in flower   **(g) subsp. cana**
1 Margin of the legume with teeth 0·5–2 mm long; leaflets 5–12 pairs
  3 Standard *c.* 1 mm shorter than keel   **(c) subsp. sibirica**
  3 Standard ± equalling keel
    4 Calyx-teeth 1·5–2·5 times as long as tube
      5 Corolla 8–10 mm; standard pink on the back
        (a) subsp. arenaria
      5 Corolla (9–)10–12 mm; standard very pale pink on the back   **(b) subsp. taurerica**
    4 Calyx-teeth (2–)2·5–4 times as long as tube
      6 Calyx 5–7(–8) mm; corolla 7·5–9 mm   **(d) subsp. miniata**
      6 Calyx usually 7–10 mm; corolla 9–12·5 mm
        (e) subsp. tommasinii

(a) Subsp. **arenaria** (*O. sativa* auct. gall. pro max. parte, non Lam. *O. borysthenica* (Širj.) Klokov, *O. tanaitica* Sprengel): Leaflets 5–12 pairs, 10–30 × 1·5–4(–8) mm; calyx 4–7 mm, the teeth 1·5–2·5 times as long as tube, sparsely pubescent; corolla 8–10 mm; legume 4–6 mm, the margin with 4–6 teeth (0·5–2 mm). 2*n*=14, 28. *C. & E. Europe, extending westwards to C. France and southwards to C. Italy.*

(b) Subsp. **taurerica** Hand.-Mazz., *Feddes Repert.* (*Beih.*) **100**: 53 (1937): Like subsp. (a) but corolla (9–)10–12 mm; standard very pale pink on the back. ● *S.E. Alps.*

(c) Subsp. **sibirica** (Turcz. ex Besser) P. W. Ball, *Feddes Repert.* **79**: 42 (1968) (*O. sibirica* Turcz. ex Besser): Like subsp. (a) but leaflets 15–30 × 4–6 mm, and standard *c.* 1 mm shorter than keel (not equalling). *E. Russia.* (*N. Asia.*)

(d) Subsp. **miniata** (Steven) P. W. Ball, *Feddes Repert.* **79**: 42 (1968) (*O. miniata* Steven): Like subsp. (a) but leaflets 6–15 × 2·5–3(–5) mm; 2·5–4 times as long as tube with numerous erecto-patent hairs on the margin; corolla 7–9(–11) mm. *Krym.* (*Caucasus.*)

(e) Subsp. **tommasinii** (Jordan) Ascherson & Graebner, *Syn. Mitteleur. Fl.* **6**(2): 882 (1909) (*O. tommasinii* Jordan, *O. ocellata* G. Beck): Leaflets 7–14 pairs, (5–)10–25 × 2–5 mm; calyx (5–)7–10 mm, the teeth (2–)2·5–3·5 times as long as tube, ciliate on the margin; corolla 9–12·5 mm; legume 4–7 mm, the margin with 3–7 teeth (0·5–2 mm), or rarely unarmed. 2*n*=14. ● *W. Jugoslavia, N. Albania, Italy.*

(f) Subsp. **lasiostachya** (Boiss.) Hayek, *Prodr. Fl. Penins. Balcan.* **1**: 928 (1926): Leaflets 4–7 pairs, 5–20 × 2–3 mm; racemes not more than 7 cm in flower; calyx 5–8 mm, the teeth 2·5–4 times as long as tube, with long erecto-patent hairs on the margin; corolla 5–9 mm; legume 4–5 mm, the margin with 3–5 teeth (2–4 mm). *Balkan peninsula, except the north-west.*

(g) Subsp. **cana** (Boiss.) Hayek, *op. cit.* 929 (1926): Like subsp. (f) but leaflets 5–10(–13) × 1·5–3·5 mm; racemes up to 12 cm in flower; calyx 5–7 mm, the teeth about twice as long as tube; margin of the legume with 2–4 teeth. *Turkey-in-Europe.* (*W. & C. Anatolia.*)

Two taxa of uncertain status closely related to **18, 19, 20** and **21**, occur in S.E. Russia, just north of the Terek river. These are **O. novopokrovskii** Vassilcz., *Bull. Jard. Bot. URSS* **29**: 624 (1930), with linear leaflets, calyx-teeth 3–4 times as long as tube, subglabrous except for patent hairs on the margin, corolla 8–10 mm and legume 5–6 mm with 5–6 teeth, 1–2 mm, on the margin, and **O. cyri** Grossh., *Sci. Pap. Appl. Sect. Tiflis Bot. Gard.* **6**: 123 (1929), with calyx-teeth subglabrous except for patent hairs on the margin, corolla (6–)8–12(–13) mm and legume 5–6 mm with 4–5 teeth on the margin.

19. O. oxyodonta Boiss., *Diagn. Pl. Or. Nov.* **1**(2): 98 (1843). Like **18** subsp. (a) but more or less grey-pubescent; leaflets 4–8 pairs, grey-pubescent, sometimes subglabrous above; racemes not more than 3(–4) cm in flower; calyx-teeth densely pubescent with more or less patent hairs; corolla 7–9 mm. *Balkan peninsula.* Al Gr Ju.

Possibly another subspecies of **18**.

20. O. inermis Steven, *Bull. Soc. Nat. Moscou* **29**(2): 165 (1856). Subglabrous perennial 30–70 cm. Leaflets 6–8 pairs, 8–24 × 3–5 mm, linear to elliptical. Calyx *c.* 4 mm, sparsely pubescent, the teeth 1·5–3·5 times as long as tube, with erecto-patent hairs on the margin; corolla 7–9 mm, pink with purple veins. Legume 4–6 mm, pubescent; sides unarmed or with a few tubercles; margin unarmed or rarely with 2–3 tubercles. *Krym and S.E. Russia.* Rs (K, E).

21. O. viciifolia Scop., *Fl. Carn.* ed. 2, **2**: 76 (1772) (*O. sativa* Lam.). Subglabrous to pubescent perennial 10–80 cm. Leaflets 6–14 pairs, 10–35 × 4–7 mm, ovate to oblong, rarely linear. Racemes up to 9 cm in flower. Calyx 5–8 mm, pubescent, the teeth 2–3 times as long as tube; corolla (8–)10–14 mm, pink with purple veins. Legume 5–8 mm, pubescent; sides toothed; margin usually with 6–8 teeth up to 1 mm. $2n = 28$. *Possibly native in C. Europe; widely cultivated for fodder and naturalized.* *Al *Au *Cz *Hu *Ju *Rm [Be Br Da Ga He Jo It Lu Po Rs (C, W) Su].

Sect. LOPHOBRYCHIS Hand.-Mazz. Like Sect. *Onobrychis* but annual and racemes 2- to 8-flowered.

22. O. caput-galli (L.) Lam., *Fl. Fr.* **2**: 651 (1778). Glabrous or sparsely pubescent annual up to 90 cm. Leaflets 4–7 pairs, 4–20 × 2–6 mm, obovate to linear. Peduncles about as long as leaves in flower. Calyx-teeth 2–4 times as long as tube; corolla 7–8 mm, reddish-purple. Legume 6–10 mm; sides with long teeth; margin with 4–9 linear-triangular to subulate teeth 3–5 mm. $2n = 14$. *Mediterranean region extending to C. Bulgaria.* Al Bu Cr Ga Gr Hs It Ju Si Tu.

23. O. aequidentata (Sibth. & Sm.) D'Urv., *Mém. Soc. Linn. Paris* **1**: 346 (1822). Pubescent annual up to 40 cm. Leaflets 5–8 pairs, 4–25 × 2–5 mm, elliptical to linear. Peduncles much longer than leaves in flower. Calyx-teeth 4–5 times as long as tube; corolla 10–14 mm, reddish-purple. Legume 6–12 mm; sides toothed; margin with 4–7 broadly triangular teeth 2–5 mm. *Mediterranean region (rare in the west), S. & W. parts of the Balkan peninsula.* Al Bu Cr Ga Gr Hs It Ju Si Tu.

O. crista-galli (L.) Lam., *Fl. Fr.* **2**: 652 (1778), from N. Africa and S.W. Asia, with peduncles equalling or shorter than leaves; calyx-teeth 4–5 times as long as tube; corolla 7–8 mm; and legume 8–14 mm with 3–5 broadly triangular toothed teeth 4–6 mm, has been recorded, probably in error for **23**, from Greece and Kriti. It has also been recorded as a casual from S. France.

72. Ebenus L.[1]

Perennial herbs or small shrubs. Leaves 3-foliolate or imparipinnate; stipules connate, scarious, divided at the apex. Flowers in axillary heads or racemes, subtended by scarious bracts. Calyx tubular-campanulate, with 5 equal teeth; corolla pink or purple; keel obtuse, obliquely truncate; stamens usually monadelphous. Legume included in the calyx, indehiscent, compressed. Seeds 1–2.

Small shrub; stipules bifid at apex; flowers in dense racemes   **1. cretica**
Herb, woody at base; stipules 3- to 4-fid at apex; flowers in heads
                                      2. sibthorpii

1. E. cretica L., *Sp. Pl.* 764 (1753). Small shrub up to 50 cm. Leaves 3-foliolate or imparipinnate with 2 pairs of leaflets on a short rhachis so that the leaves are almost digitate; leaflets 15–30 mm, elliptic-oblong, sericeous; stipules bifid at apex. Flowers in dense racemes, the bracts not forming an involucre at the base; peduncles scarcely exceeding the leaves. Corolla 10–15 mm, bright pink; standard and keel subequal. *Cliffs and rocks.* ● *Kriti.* Cr.

2. E. sibthorpii DC., *Prodr.* **2**: 351 (1825). Herb, woody at the base, with flowering stems 15–30 cm. Leaves with 2–4 pairs of leaflets; rhachis elongated; leaflets 8–20 mm, elliptic-oblong to obovate; stipules 3- to 4-fid at apex. Flowers in heads, with scarious bracts forming an involucre at the base; peduncles at least as long as the leaves. Corolla 10–15 mm, reddish-purple; standard shorter than keel. *Cliffs and rocks. S.E. Greece.* Gr.

73. Alhagi Gagnebin[1]

Spiny shrubs, the spines formed from short, axillary branches which bear scale leaves and the flowers. Leaves simple; stipules minute, free. Flowers solitary or paired. Calyx campanulate, with 5 equal teeth; corolla pink, red or purple; keel curved upwards at the apex, obtuse; stamens diadelphous. Legume linear, indehiscent, strongly contracted between the 1–5 seeds.

Leaves glabrous; calyx-teeth not more than 0·5 mm, very shortly
   triangular, much wider than long   **1. pseudalhagi**
Leaves densely pubescent; calyx-teeth 0·5 mm or more, triangular-ovate, about as long as wide   **2. graecorum**

1. A. pseudalhagi (Bieb.) Desv., *Jour. Bot. Appl.* **1**: 120 (1813). Much-branched shrub up to 60 cm, glabrous. Leaves 10–20 mm, oblong, lanceolate or ovate. Flowers 3–8 on each spine; calyx-teeth not more than 0·5 mm, very shortly triangular, wider than long, the sinus between the teeth obtusely angled; corolla 7–10 mm, red or pink. Legume 8–30 mm, blackish-brown. *Steppes. S.E. Russia, W. Kazakhstan.* Rs (E).

2. A. graecorum Boiss., *Diagn. Pl. Or. Nov.* **2**(9): 114 (1849). Like **1** but stems and leaves appressed-pubescent; calyx-teeth at least 0·5 mm, triangular-ovate, about as long as wide, the sinus between the teeth acutely angled; corolla purple. *Maritime sands. S.E. Greece, Kikladhes.* Gr.

74. Arachis L.[1]

Annual. Leaves usually paripinnate with 2 pairs of leaflets; stipules linear, adnate to the petiole. Flowers sessile, in axillary racemes and with a long, tubular hypanthium. Calyx bilabiate, the upper lip (3–)4-toothed, the lower lip entire. Corolla yellow, adnate to the base of the stamen-tube; stamens monadelphous, the 2 adaxial stamens sterile. Legume cylindrical, torulose, indehiscent, underground. Seeds 1–3.

After fertilization the base of the ovary elongates rapidly and buries the ovary in the ground. Subsequent development of the legume takes place underground.

Literature: B. W. Smith, *Amer. Jour. Bot.* **37**: 802–815 (1950).

1. A. hypogaea L. *Sp. Pl.* 741 (1753). Stems up to 50 cm, much-branched, pubescent. Leaflets 25–60 × 15–30 mm, elliptical to obovate. Corolla 15–20 mm. Legume 20–40 × 10–15 mm, glabrous, reticulate. *Cultivated for the edible seed (ground-nut) in S. Europe. (Tropical and subtropical South America.)*

---

[1] By P. W. Ball.

GERANIALES

LXXXII. OXALIDACEAE[1]

Herbs, rarely small shrubs. Leaves usually compound. Flowers 5-merous, hermaphrodite, actinomorphic. Ovary superior, 5-locular; placentation axile. Fruit a capsule. Seeds with endosperm.

1. Oxalis L.[2]

Perennial herbs, sometimes bulbous. Leaves palmately 3(–8)-foliolate, with or without stipules; leaflets usually indented at the apex, otherwise entire, showing sleep-movements. Inflorescence axillary, cymose, sometimes a cymose umbel or a single flower. Flowers often heterostylous. Sepals 5, free; petals 5, free or weakly united. Stamens 10, obdiplostemonous. Ovules numerous. Styles 5, free. Fruit a loculicidal capsule. Seeds projectile, with an elastic integument.

Several bulbous species, originally introduced by gardeners, have become established to the extent of being serious weeds; these species often develop tuberous roots. They multiply vegetatively by fragmentation into bulbils, but are virtually seed-sterile. The resultant propagation of various clones has obscured the taxonomy of the genus, which is in need of revision.

Literature: A. Chevalier, *Rev. Bot. Appl.* **20**: 651–694 (1940). R. Knuth in Engler, *Pflanzenreich* (**IV, 130**): 1–481 (1930). D. P. Young, *Watsonia* **4**: 51–69 (1958).

1 Petals yellow
  2 Aerial stem absent; bulbils present at base of plant
                                     10. pes-caprae
  2 Aerial stem present; bulbils absent
    3 Stem rooting at nodes; leaves alternate; stipules auriculate
      4 Leaflets 5–18 × 8–23 mm; capsule 10–25 mm **1. corniculata**
      4 Leaflets 3–5 × 3–6 mm; capsule 5–7 mm **2. exilis**
    3 Stem not rooting at nodes; leaves mostly subopposite; stipules not auriculate
      5 Inflorescence umbellate; fruiting pedicels deflexed; stipules oblong; stem with non-septate hairs **3. stricta**
      5 Inflorescence not umbellate; fruiting pedicels not deflexed; stipules absent; stem with septate hairs **4. europaea**
1 Petals white, red, violet or purple
  6 Stem rhizomatous; bulb absent (although rhizome may be swollen)
    7 Rhizome 1–2 cm thick at apex; inflorescence corymbose **5. articulata**
    7 Rhizome less than 1 cm thick; flowers solitary **6. acetosella**
  6 Stem erect or absent; bulb present
    8 Aerial stem erect, leafy; petals pale lilac **12. incarnata**
    8 Aerial stem absent; petals pink, red or violet
      9 Leaflets not emarginate; petals 25–35 mm **11. purpurea**
      9 Leaflets emarginate; petals 15–20 mm
        10 Leaves 4-foliolate **9. tetraphylla**
        10 Leaves 3-foliolate
          11 Leaflets widest at or below middle, pubescent, and punctate beneath near the margin **7. corymbosa**
          11 Leaflets widest near apex, subglabrous, not punctate **8. latifolia**

1. O. corniculata L., *Sp. Pl.* 435 (1753) (*O. repens* Thunb.). Creeping, pubescent perennial, but flowering soon after germination. Stems up to 50 cm, procumbent, rooting at the nodes. Leaves alternate; petioles 2–8 cm, with small auriculate stipules; leaflets 5–18 × 8–23 mm, obcordate, deeply emarginate. Inflorescence umbellate, of 1–7 flowers; fruiting pedicels deflexed. Sepals lanceolate; petals 4–7 mm, yellow. Capsule 10–25 mm, cylindrical, hoary. Seeds transversely ridged, brown. 2*n*=24. *Dry open habitats, especially cultivated ground. S. Europe, extending locally northwards to N. France, Hungary and C. Ukraine; often naturalized or casual further north.* Al Az Bl Bu Co Cr Ga Gr He Hs Hu It Ju Lu Rs (W, K) Sa Si Tu [Au Be Br Cz Fe Ge Hb Ho No Po Rm Su].

Var. *repens* (Thunb.) Zucc., with leaflets 5–9 × 6–12 mm, inflorescence 1- to 2-flowered, and capsule 10–15 mm, from S. Africa, is frequently found as a garden escape, as are also purple-suffused cultivars.

2. O. exilis A. Cunn., *Ann. Nat. Hist.* 3: 316 (1839). Like **1** but very dwarf with creeping, filiform stems; leaflets 3–5 × 3–6 mm; inflorescence 1-flowered; capsule 5–7 mm. *Naturalized from gardens in Britain and Channel Islands.* [Br Ga.] (*Australasia.*)

3. O. stricta L., *Sp. Pl.* 435 (1753) (*O. dillenii* Jacq., *O. navieri* Jordan). Like **1** but caespitose, short-lived; stems up to 20 cm, ascending, with non-septate hairs, not rooting at the nodes; leaves subopposite or in groups; stipules oblong, inconspicuous; ridges on seed often with white markings. 2*n*= c. 24. *Locally naturalized in S., W. & C. Europe as a weed of cultivated ground.* [Al Au Br Cz Da Ga Ge It Ju Sa.] (*E. & C. North America.*)

4. O. europaea Jordan in F. W. Schultz, *Arch. Fl. Fr. Allem.* **1**: 309 (1854) (*O. stricta* auct. plur., non L.). Like **1** but stems up to 40 cm, erect, not rooting, emitting filiform, underground stolons; leaves subopposite or subverticillate; stipules absent; stems and petioles with crisped, septate hairs; inflorescence cymose, with pedicels not deflexed in fruit; capsule 8–12 mm, not hoary. 2*n*=24. *Naturalized as a weed of cultivated ground in most of Europe except the extreme north and south.* [Al Au Be Br Bu Co Cz Da Fe Ga Ge Hb He Ho Hs Hu It Ju No Po Rm Rs (B, C, W, K) Su.] (*North America and E. Asia.*)

O. valdiviensis Barn. in C. Gay, *Hist. Chil. Bot.* 1: 446 (1846), from Chile, a glabrous, caespitose plant with dichasia of purple-veined, yellow flowers and ovoid capsule, is cultivated for ornament and occasionally becomes naturalized.

5. O. articulata Savigny in Lam., *Encycl. Méth. Bot.* 4: 686 (1798). Pubescent, caespitose perennial up to 35 cm. Rhizome with swollen oblong segments up to 2 cm in diameter, bearing the scarious remains of leaf-bases. Leaves in terminal rosettes. Petioles 5–30 cm; leaflets 12–40 × 15–50 mm, obcordate, deeply emarginate, covered with orange or brown tubercles. Inflorescence a corymbose cyme; sepals lanceolate, with 2 apical tubercles; petals 12–20 mm, pink. Capsule 10 mm, cylindrical-ovoid. *Cultivated in gardens and naturalized in waste places in parts of W. Europe.* [Az Br Ga Hb Hs Lu.] (*E. temperate South America.*)

Somewhat variable in size and morphology.

6. O. acetosella L., *Sp. Pl.* 433 (1753). Sparsely pubescent, creeping perennial. Rhizome above ground, slender, bearing the fleshy, toothlike remains of leaf-bases; leaves scattered. Petioles 5–10 cm; leaflets 10–27 × 15–30 mm, obcordate, emarginate. Flowers solitary, campanulate; peduncles 5–10 cm; sepals oblong-lanceolate; petals 8–15 mm, white with lilac veins to pale purple or violet. Late flowers apetalous, cleistogamous. Capsule 3–4 mm, angular-ovoid; seeds light brown, longitudinally ridged. 2*n*=22. *Woods and shady places. Most of Europe but rarer in the South.* All except Az Bl Cr Rs (K, E) Sa Si Sb.

---

[1] Edit. D. H. Valentine.   [2] By D. P. Young.

7. O. corymbosa DC., *Prodr.* **1**: 696 (1824) (*O. martiana* Zucc.). Pubescent, acaulescent perennial with subterranean bulb 15–30 mm, which soon develops into a mass of sessile bulbils 3–6 mm. Petioles 5–15(–30) cm, flexuous; leaflets 20–45 × 20–55 mm, obcordate or orbicular, with a narrow indentation at the apex, punctate beneath, especially near the margin. Inflorescence a corymbose cyme; flowers infundibuliform; petals 15–20 mm, purplish-pink. Sterile in Europe. *Naturalized as a weed and on disturbed ground in parts of W. Europe.* [Az Br Ga He Hs Lu.] (*South America; widely naturalized in subtropical countries.*)

O. debilis Kunth in Humb., Bonpl. & Kunth, *Nov. Gen. Sp.* **5**: 236 (1822), doubtfully distinct from **7**, from which it apparently differs only in its smaller (2–3 mm) bulbils and brick-red petals, is occasionally naturalized in England, and perhaps elsewhere.

8. O. latifolia Kunth in Humb., Bonpl. & Kunth, *Nov. Gen. Sp.* **5**: 237 (1822). Like **7** but almost glabrous; bulbils emitted from the base of the bulb, usually on horizontal stolons 1–2(–30) cm, less often on short, erect stalks; leaflets up to 60 × 100 mm, obdeltate, rather shallowly emarginate, not punctate; inflorescence umbellate; petals pale to deep pink. *Naturalized in cultivated fields and waste places in parts of W. Europe.* [Az Br Ga Hb Hs Lu.] (*Tropical South America.*)

9. O. tetraphylla Cav., *Icon. Descr.* **3**: 19 (1795). Like **7** but bulb up to 4 cm; bulbils on stolons up to 10 cm; leaves 4-foliolate; leaflets up to 35 × 40 mm, obdeltate, shallowly emarginate, sometimes with a purple zone; inflorescence umbellate; petals bright red. *Naturalized in cultivated ground in a few regions.* [Au Ga Ju.] (*Mexico.*)

O. deppei Loddiges ex Sweet, *Brit. Flower Gard.* ser. 2, **1**: t. 96 (1831), from Mexico, differing from **9** by its sessile bulbils and truncate (scarcely emarginate), leaflets, is cultivated and occasionally becomes established.

O. lasiandra Zucc. in Otto & A. Dietr., *Allgem. Gartenz.* **2**: 245 (1834), of doubtful origin, is a similar plant, but with 7–8 narrowly oblanceolate leaflets. It is cultivated and rarely naturalized.

10. O. pes-caprae L., *Sp. Pl.* 434 (1753) (*O. cernua* Thunb.). Sparsely pubescent, caespitose perennial, with a deeply-buried bulb, which emits an annual, ascending, subterranean stem bearing bulbils and a rosette of leaves at soil level. Petioles up to 20 cm; leaflets 8–20 × 12–30 mm, obcordate, deeply emarginate. Flowers infundibuliform (sometimes *flore pleno*), in umbellate cymes; petals 20–25 mm, yellow. Capsule short, rarely formed. *Cultivated ground and other open habitats. Extensively naturalized in the Mediterranean region and W. Europe.* [Bl Br Co Cr Ga Gr Hs It Lu Sa Si.] (*South Africa.*)

11. O. purpurea L., *Sp. Pl.* 433 (1753). Pubescent or villous, caespitose perennial, with a bulb which emits an ascending subterranean stem bearing a rosette of leaves at soil level. Petioles 3–10 cm; leaflets 8–23 × 9–30 mm, rhombic, obtuse, punctate and often purple beneath. Flowers solitary, infundibuliform; peduncles 1–10 cm; sepals lanceolate; petals 25–35 mm, purplish-pink, white at base. Capsule *c.* 5 mm. *Open habitats. Naturalized from gardens in S.W. Europe.* [Az Co Hs Lu Si.] (*South Africa.*)

12. O. incarnata L., *Sp Pl.* 433 (1753). Glabrous, erect perennial, with a subterranean bulb. Stem up to 20 cm, bearing sessile, axillary bulbils. Leaves subopposite, mostly crowded; petioles 2–6 cm; leaflets 5–18 × 8–20 mm, delicate, obcordate, deeply emarginate. Flowers infundibuliform, solitary; peduncles 3–7 cm; sepals with 2 apical tubercles; petals 12–20 mm, pale lilac with darker veins. Sterile. *Walls and roadsides. Naturalized from gardens in S.W. England and the Channel Islands.* [Br Ga.] (*South Africa.*)

LXXXIII. GERANIACEAE[1]

Herbs, or rarely small shrubs with soft stems. Leaves stipulate, usually lobed or divided, and sometimes more or less compound. Flowers in cymes, umbels or spikes, 5-merous, actinomorphic or somewhat zygomorphic. Sepals 5, free; petals 5, free; stamens obdiplostemonous in two whorls of 5, some of them sometimes reduced to staminodes. Ovary superior, of 5 united carpels, separating in fruit into 5 1-seeded mericarps; styles united in flower, sometimes separating in fruit.

Literature: R. Knuth in Engler, *Pflanzenreich* **53** (IV, 129): 1–640 (1912).

Several species and hybrids of *Pelargonium* from S. Africa are widely cultivated for ornament or (*P. radula* and hybrids) for essential oils. They are reported as occasional escapes, or as persisting among native vegetation in abandoned fields, and are sometimes planted in roadside hedges, but none appears to be truly naturalized. The plants likely to be encountered include **P. peltatum** (L.) L'Hér. in Aiton, *Hort. Kew.* **2**: 427 (1789), with fleshy, peltate, 5-lobed leaves and procumbent stems; **P. radula** (Cav.) L'Hér. in Aiton, *op. cit.* 423 (1789) and its hybrids, with fragrant, deeply pinnatisect leaves and with a dark patch on the upper 2 petals; and **P. × hybridum** (L.) L'Hér. in Aiton, *op. cit.*

424 (1789) (*P. inquinans × zonale*), with bright scarlet flowers and leaves usually marked with a dark ring.

1 Flowers in spikes; stigmas united, capitate **3. Biebersteinia**
1 Flowers in cymes or umbels; stigmas free, linear
  2 Leaves palmately (rarely ternately) divided or lobed; beak of mericarp straight or curved in a simple arc, sometimes absent **1. Geranium**
  2 Leaves pinnately divided or lobed; beak of mericarp spirally twisted at maturity **2. Erodium**

1. Geranium L.[2]

Herbs. Leaves more or less orbicular in outline, palmately (rarely ternately) lobed or divided, many or all of them basal; cauline leaves, if present, usually opposite near base of stem, but often alternate in inflorescence. Inflorescence cymose; ultimate peduncles usually 2-flowered. Flowers actinomorphic. Stamens all fertile, or rarely 3–5 reduced to staminodes. Stigmas 5, filiform. Mericarps usually dehiscent, separating from the base upwards, usually retaining outer part of style in the form of a long beak, of which the apex remains for a while attached to the central axis formed by the still coherent inner parts of all 5 styles.

In all European species, except where the contrary is implied

---

[1] Edit. D. A. Webb.
[2] By D. A. Webb and I. K. Ferguson.

in the description, the basal leaves (to which the descriptions refer) have long petioles; the cauline leaves, if present, have progressively shorter petioles and lamina often with fewer lobes; all leaves bear rather short, appressed hairs on both surfaces; and the sepals are hairy, obtuse to subacute, and mucronate or aristate.

In the description below *lobe* is used to indicate a primary division of the leaf, *segment* a division of a lobe. Measurements of sepals refer to the fruiting condition, and include the arista.

1 Annual or biennial
  2 Petals with conspicuous claw; sepals erect during flowering
    3 Sepals keeled; leaves not deeply divided   **36. lucidum**
    3 Sepals not keeled; leaves very deeply divided, so as to appear compound
      4 Petals 9–13 mm; pollen orange; mericarps with rather few ridges   **37. robertianum**
      4 Petals 5–9 mm; pollen yellow; mericarps with numerous ridges   **38. purpureum**
  2 Petals without distinct claw; sepals ±patent during flowering
    5 Petals entire   **30. rotundifolium**
    5 Petals 2-lobed, emarginate or crenulate at apex
      6 Mericarps (excluding style) glabrous
        7 Uppermost leaves opposite; arista of sepal at least 2 mm   **33. columbinum**
        7 Uppermost leaves alternate; sepals with short mucro (*c.* 0·5 mm)
          8 Plant greyish-green, densely pubescent; lowest leaf of inflorescence considerably exceeding the subtended peduncle, and with petiole considerably longer than lamina   **31. molle**
          8 Plant green, sparsely or moderately hairy; lowest leaf of inflorescence shorter than or slightly exceeding the subtended peduncle, and with petiole scarcely longer than lamina   **32. brutium**
      6 Mericarps (excluding style) hairy
        9 Mericarps transversely ridged, separating without a stylar beak   **25. divaricatum**
        9 Mericarps without transverse ridges, and with long stylar beak
          10 Sepals shortly mucronate; 3–5 stamens lacking anthers   **33. pusillum**
          10 Most of the sepals with long arista; all 10 stamens with anthers
            11 Leaves divided for *c.* 95 % of the radius, with ±linear segments
              12 Peduncles shorter than subtending leaf   **35. dissectum**
              12 Peduncles longer than subtending leaf   **34. columbinum**
            11 Leaves divided for *c.* 75 % of the radius, with ovate-oblong segments
              13 Seeds yellowish-grey, with irregular, dark brown patches, almost smooth; cotyledons with deep lateral notches   **26. bohemicum**
              13 Seeds uniformly brown, distinctly foveolate; cotyledons without lateral notches   **27. lanuginosum**
1 Perennial, often with conspicuous rhizome, tuber or woody stock
  14 Petals with a distinct claw at least ⅓ as long as the limb
    15 Stamens 8–10 mm, only slightly exceeding the sepals
      16 Leaves all ±basal; petioles *c.* 5 mm   **4. humbertii**
      16 Several cauline leaves present; petioles of basal leaves 100–200 mm   **39. cataractarum**
    15 Stamens 14–22 mm, more than twice as long as the sepals
      17 Lamina of basal leaves 4–10 cm wide, divided for 75–80 % of the radius; lobes lanceolate, irregularly pinnatifid or incise-dentate   **1. macrorrhizum**
      17 Lamina of basal leaves 2·5–4 cm wide, divided for 90 % of the radius; lobes cuneate, entire except for shortly 3-toothed apex   **2. dalmaticum**
  14 Petals with very short claw or none
    18 Petals entire, apiculate, erose or very slightly emarginate

19 Petals spreading horizontally or deflexed; mericarps with 2–4 conspicuous ridges below the base of the style
  20 Arista of sepal 4–6 mm   **17. aristatum**
  20 Arista of sepal *c.* 1 mm
    21 Petals 6–10 mm wide, patent or slightly deflexed   **15. phaeum**
    21 Petals 2·5–4 mm wide, sharply deflexed   **16. reflexum**
19 Petals curving upwards, giving a ± cup-shaped flower; mericarps smooth, or with a single, indistinct ridge
  22 Inflorescence compact, somewhat corymbose, with usually more than 10 flowers
    23 Pedicels deflexed during maturation of the fruit; sepals 11–15 mm   **7. pratense**
    23 Pedicels erect during maturation of the fruit; sepals 6–12 mm   **8. sylvaticum**
  22 Infloresence diffuse, not corymbose, with usually less than 10 flowers
    24 Petals *c.* 6 mm   **28. sibiricum**
    24 Petals at least 12 mm
      25 Rhizome long, slender, horizontal   **9. endressii**
      25 Rhizome short
        26 Roots fleshy, fusiform, arising in a cluster from a very short rhizome; arista of sepals not more than 1·25 mm   **22. asphodeloides**
        26 Roots fibrous, not clustered; arista of sepals usually more than 1·5 mm
          27 Sepals hairy only on the veins   **23. palustre**
          27 Sepals hairy both on and between the veins   **24. collinum**
18 Petals 2- or 3-lobed, or distinctly emarginate
  28 Stock small and inconspicuous; petals not more than 10(–12) mm
    29 Mericarps smooth, hairy
      30 Leaf-lobes cuneate, truncate; hairs on mericarp appressed   **29. pyrenaicum**
      30 Leaf-lobes lanceolate, acute; hairs on mericarp patent   **28. sibiricum**
    29 Mericarps rugose, glabrous
      31 Plant greyish-green, densely pubescent; lowest leaf of inflorescence considerably exceeding the subtended peduncle, and with petiole considerably longer than lamina   **31. molle**
      31 Plant green, sparsely or moderately hairy; lowest leaf of inflorescence shorter than or only slightly exceeding the subtended peduncle and with petiole scarcely longer than lamina   **32. brutium**
  28 Plant with conspicuous tuber, rhizome or woody stock; petals usually more than 10 mm
    32 Peduncles 1-flowered   **6. sanguineum**
    32 Peduncles 2-flowered
      33 Pedicels with long glandular hairs
        34 Leaves divided to the base   **18–21. tuberosum** group
        34 Leaves divided for 65–80 % of the radius
          35 Leaf-segments acute; sepals 5–7 mm; petals usually white   **12. albiflorum**
          35 Leaf-segments obtuse; sepals *c.* 10 mm; petals bluish-violet   **13. peloponesiacum**
      33 Pedicels eglandular, or with subsessile glands (sometimes also with much longer eglandular hairs)
        36 Leaves not more than 4 cm wide, usually all basal
          37 Leaves silvery-sericeous on both sides, with linear-oblong segments   **5. argenteum**
          37 Leaves green, at least above, with broadly oblong to suborbicular segments   **3. cinereum**
        36 Larger leaves at least 5 cm wide, some of them cauline
          38 Pedicels with subsessile glands
            39 Petals *c.* 16 mm, pale lilac with darker veins   **10. versicolor**
            39 Petals *c.* 25 mm, deep purple   **14. ibericum**
          38 Pedicels eglandular
            40 Rhizome long, slender; leaves divided for *c.* 70 % of the radius   **11. nodosum**
            40 Rhizome short, often tuberous; leaves divided to the base   **18–21. tuberosum** group

1. **G. macrorrhizum** L., *Sp. Pl.* 680 (1753). Perennial, with stout, horizontal rhizome. Stem (6–)20–50 cm, erect, with 0–1 pairs of cauline leaves as well as the bracts. Leaves 4–10 cm wide, fragrant, divided for 75–80 % of the radius into 5–7 obovate, pinnatifid lobes; segments 3–4 on each side, obtuse but conspicuously mucronate. Inflorescence with 2–3 peduncles subtended by 2 bracts, which are subsessile and 3- to 5-lobed but otherwise like the basal leaves; lateral peduncles each with 4–9 nodding flowers in a dense corymb or umbel. Sepals erect, reddish; petals *c.* 15 mm, with claw at least half as long as the obovate, entire, patent or deflexed, dull purplish-red limb. Stamens 18–22 mm; filaments curved. Style up to 40 mm, of which the terminal half is not thickened and drops off before the fruit ripens. Mericarps glabrous, without ridges. 2*n*=46. *Shady places, usually among mountains; calcicole.* ● *Balkan peninsula, S. & E. Carpathians, S. Alps, Appennini; cultivated elsewhere for ornament and often naturalized.* Al Au Bu Ga Gr It Ju Rm *Rs (W) [Be Br Ge Rs (K)].

Dwarf mountain plants from Greece are often nearly glabrous and with very small, scarious bracts; they approach closely to **2** but differ in the shape of the leaf-lobes.

2. **G. dalmaticum** (G. Beck) Rech. fil., *Magyar Bot. Lapok* 33: 28 (1934) (*G. macrorrhizum* subsp. *microrhizon* Freyn). Like **1** but smaller and more delicate; glabrous except for pedicels and sepals; lamina 2·5–4 cm wide, divided nearly to the base into 5 cuneate lobes with straight, entire sides and 3 triangular teeth at the apex; cauline leaves always absent; bracts always very small and scarious; petals *c.* 13 mm; stamens 14–18 mm. *Rocky places.* ● *S.W. Jugoslavia and N. Albania.* Al Ju.

3. **G. cinereum** Cav., *Monad. Class. Diss. Dec.* 204 (1787). Perennial, with very stout, vertical rhizome. Leaves all basal, 2–3 cm wide, pubescent on both sides, sometimes greyish-sericeous beneath, divided for 80 % of the radius (but often apparently less from overlapping of the lobes) into 5–7 obovate-cuneate or obdeltate, usually almost contiguous lobes, each with 3 obtuse or mucronate segments or teeth at apex. Peduncles 5–10 cm, 2-flowered; bracts usually small, scarious. Sepals aristate; petals *c.* 15 mm, obovate, emarginate, with very short claw. Mericarps sericeous, with 1–3 ridges below the style. *Rocky or grassy places in mountains. S. & W. part of Balkan peninsula; C. & S. Italy; Pyrenees.* Al Ga Gr Hs It Ju.

Two rather ill-defined subspecies may be recognized:

(a) Subsp. **cinereum**: Leaf-segments *c.* 2 mm; sepals pubescent but with long hairs very few or only on margin; petals pale lilac with darker veins. ● *Pyrenees.*

(b) Subsp. **subcaulescens** (L'Hér. ex DC.) Hayek, *Prodr. Fl. Penins. Balcan.* 1: 572 (1925) (*G. subcaulescens* L'Hér.): Leaf-segments up to 5 mm; usually at least some of the sepals with plentiful long, white hairs; petals usually deep reddish-purple, but sometimes pale. *S. & W. part of Balkan peninsula; C. & S. Italy.*

Plants which resemble **3** in many features but differ in others have been described from calcareous mountains in several regions of Spain. They occur in disjunct populations, and it is difficult to give them satisfactory taxonomic treatment; they should probably be regarded as variants or subspecies of **3**. **G. subargenteum** Lange in Willk. & Lange, *Prodr. Fl. Hisp.* 3: 525 (1878), from the Cordillera Cantábrica, has the leaves very densely appressed-pubescent (grey-green above, silvery beneath); sepals without long hairs; and petals deep purplish-pink. **G. dolomiticum** Rothm., *Bol. Soc. Esp. Hist. Nat.* 34: 151 (1934), from N.W. Spain (S. of Ponferrada), is robust, with petioles up

to 15 cm; leaves pale green beneath; stem sometimes branched and with cauline leaves near the base; sepals acute to acuminate, with long marginal hairs; and petals apparently fairly deep purple. **G. cazorlense** Heywood, *Bull. Brit. Mus. (Bot.)* 1: 112 (1954), from S.E. Spain (Sierra de Cazorla), is small and densely caespitose, with closely contiguous leaf-lobes bearing short, rounded segments; sepals small, acute or shortly mucronate; and petals 12 mm, white with violet veins.

4. **G. humbertii** Beauverd, *Bull. Soc. Bot. Genève* ser. 2, 31: 447 (1940). Like **3**, but stem said to be branched, and perhaps with some cauline leaves near the base; petals *c.* 10 mm, with well-developed claw *c.* 5 mm long, white with pink veins. Fruit unknown. *Mountain rocks.* ● *N. Greece (Kaimakchalan).* Gr.

Only once collected and rather inadequately described, this plant needs further investigation. It may be a variety of **3**, but the clawed petals appear to be very distinctive.

5. **G. argenteum** L., *Cent. Pl.* 2: 25 (1756). Like **3** but leaves densely silvery-sericeous on both sides, divided for 95 % of the radius; lobes deeply divided, each with usually 3 linear-oblong segments; petals pale pink. *Calcareous rocks and screes.* ● *Mountains of N. & C. Italy, extending to S.E. France and N.W. Jugoslavia.* Ga It Ju.

6. **G. sanguineum** L., *Sp. Pl.* 683 (1753). Perennial, with stout, horizontal rhizome. Stems diffuse, branched, erect to decumbent, with long, white, patent hairs and sessile glands. Leaves mostly cauline, 3–5(–8) cm wide, divided for 85 % of the radius into 5–7 pinnatisect lobes, each with 1–3 linear-oblong, acute segments on each side. Peduncles (with pedicel) 7–15 cm, 1-flowered (very rarely 2-flowered), with 2 small bracts at a node near the middle, representing the junction with the pedicel. Sepals 8–13 mm; petals 15–20 mm, obovate, emarginate, bright reddish-purple (rarely pink). Mericarps somewhat hairy, without ridges. 2*n*=84 (52–56). *Rocky or sandy ground or on well-drained soils. Most of Europe southwards from c. 60° N.* All except Az Bl Cr Fa Ho Is Rs (N) Sa Sb.

7. **G. pratense** L., *Sp. Pl.* 681 (1753). Perennial, with stout, oblique rhizome. Stems 30–80 cm, erect, with deflexed hairs below and patent, glandular hairs above. Leaves *c.* 10 cm wide, divided almost to the base into 5–7 ovate, deeply pinnatifid lobes; segments oblong, acute. Inflorescence compact, with suberect branches; pedicels deflexed after flowering, becoming erect again when fruit is ripe. Sepals 11–15 mm; petals 15–20 mm, obovate, entire, bright violet-blue. Mericarps hirsute, without ridges. 2*n*=28. *Throughout a large part of Europe, but rare in the Mediterranean region and much of the north; frequently cultivated for ornament and widely naturalized.* Au Be Br Bu Cz Fe Ga Ge Gr Hb He Ho Hs Hu It Ju No Po Rm Rs (N, B, C, W, E) Su [Da].

8. **G. sylvaticum** L., *Sp. Pl.* 681 (1753). Perennial, with stout, oblique rhizome. Stems 20–60 cm, erect, with variable indumentum. Leaves 5–12 cm wide, divided for 80–95 % of the radius into 5–7 ovate, dentate to pinnatisect lobes. Inflorescence more or less compact, with suberect branches; pedicels remaining erect after flowering. Sepals 6–12 mm; petals 6–18 mm, obovate, entire, variable in colour. Mericarps hirsute, without ridges. *Most of Europe, but only on mountains in the south, and absent from many islands.* Al Au Be Br Bu Cz Da Fa Fe Ga Ge Gr Hb He Hs Hu Is It Ju No Po Rm Rs (N, B, C, W, E) Su [Ho].

The taxa recognized below as subspecies are often very distinct in particular regions, but in view of the great variability of (a) their satisfactory diagnosis, on a continental scale, as separate species, is not possible on the information available.

1 Petals white with red veins; pedicels without glandular hairs
                                      (b) subsp. **rivulare**
1 Petals blue, purple or lilac, rarely white, not red-veined
2 Leaves divided for less than 90 % of the radius; lobes dentate or shortly pinnatifid; petals usually reddish-purple
                                      (a) subsp. **sylvaticum**
2 Leaves divided for more than 90 % of the radius; lobes deeply pinnatifid or pinnatisect; petals blue, violet or lilac
3 Petals 13–18 mm, bright violet-blue; pedicels without glandular hairs
                                      (c) subsp. **caeruleatum**
3 Petals 6–15 mm, pale blue to lilac; pedicels with or without glandular hairs   **(d)** subsp. **pseudosibiricum**

(a) Subsp. **sylvaticum**: Leaves less deeply divided than in other subspecies. Pedicels and mericarps often with patent, glandular hairs. Petals *c.* 15 mm, usually reddish-purple, but sometimes white, pink, lilac or blue. 2*n*=28. *Throughout the range of the species.*

(b) Subsp. **rivulare** (Vill.) Rouy, *Fl. Fr.* **4**: 81 (1897) (*G. rivulare* Vill.): Stem seldom more than 30 cm. Pedicels and mericarps with appressed, eglandular hairs only. Petals 11–15 mm, white with red veins. 2*n*=28. ● *Alps, mainly in the West; calcifuge.* Ga He It.

(c) Subsp. **caeruleatum** (Schur) D. A. Webb & I. K. Ferguson, *Feddes Repert.* **74**: 25 (1967) (*G. caeruleatum* Schur): Like subsp. (b) but petals 13–18 mm, bright violet-blue. ● *S. Carpathians and mountains of Balkan peninsula.* Al Bu Ju Rm.

(d) Subsp. **pseudosibiricum** (J. Mayer) D. A. Webb & I. K. Ferguson, *Feddes Repert.* **74**: 26 (1967) (*G. pseudosibiricum* J. Mayer): Pedicels with or without glandular hairs. Petals 6–15 mm, pale blue to lilac. *C. & S. Ural and adjacent parts of E. Russia.* Rs (C, E). (*N. Asia.*)

In subsp. *sylvaticum*, and perhaps in others, some plants are often found with only female flowers, which are considerably smaller than the normal ones.

9. **G. endressii** Gay, *Ann. Sci. Nat.* **26**: 228 (1832). Perennial, with long, slender, creeping rhizome. Stems 30–80 cm, erect, hirsute. Leaves 5–8 cm wide, divided for 80 % of the radius into 5 ovate-rhombic, nearly contiguous, irregularly incise-dentate lobes. Inflorescence diffuse, with rather few flowers on long peduncles; pedicels densely hirsute with patent, eglandular hairs, and also bearing numerous subsessile glands. Sepals 9–10 mm; petals *c.* 16 mm, obovate, entire or very slightly emarginate, pink, without darker veins. Mericarps pubescent, without ridges. *Wet places.* ● *S.W. France (W. half of Basses-Pyrénées), just extending into Spain; cultivated for ornament in W. Europe and locally naturalized.* Ga Hs [Be Br].

10. **G. versicolor** L., *Cent. Pl.* **1**: 21 (1755) (*G. striatum* L.). Like 9 but leaf-lobes sometimes more widely separated; pedicels with the long, eglandular hairs much sparser and sometimes absent; petals deeply emarginate, white or pale lilac with violet veins. *Mountain woods.* ● *S. part of Balkan peninsula, C. & S. Italy, Sicilia; naturalized from gardens in N.W. Europe.* Al Gr Ju It Si [Br Ga Hb].

G. endressii × *versicolor* is cultivated in gardens, and is naturalized in Britain and Ireland, in the absence of the parent species. It has emarginate, pale pink petals with darker veins.

11. **G. nodosum** L., *Sp. Pl.* 681 (1753). Perennial, with creeping, fairly slender rhizome. Stems 20–50 cm, erect, subglabrous or with short, deflexed hairs. Leaves 6–10 cm wide, divided for 65–80 % of the radius into 3–5 ovate-elliptical, toothed lobes;

teeth obtuse, shortly mucronate. Inflorescence diffuse, few-flowered; pedicels densely pubescent with mainly deflexed hairs. Sepals 8–9 mm, appressed-pubescent; petals 12–17 mm, oblanceolate-cuneate, deeply emarginate, bright pink or violet with darker veins. Mericarps hairy, with a transverse ridge, sometimes rather faint, below the style. *Mountain woods.* ● *From C. France to the Pyrenees, C. Italy and C. Jugoslavia; occasionally naturalized from gardens elsewhere.* Co Ga He Hs It Ju [Be Br Ge Ho].

12. **G. albiflorum** Ledeb., *Icon. Pl. Fl. Ross.* **1**: 6 (1829). Perennial, with rather short, vertical rhizome. Stems 40–60 cm, erect, glabrous at least below. Leaves 10–18 cm wide, hairy above, subglabrous beneath, divided for 75–80 % of the radius into 5–7 rhombic, pinnatifid or dentate lobes; segments acute. Flowers numerous, in a fairly compact, corymbose inflorescence; pedicels with long, glandular hairs. Sepals 5–7 mm, reddish, sparsely hairy; petals 9–15 mm, obovate, emarginate, suberect, white or pale lilac. Mericarps hairy. *Woods and stream-sides. N.E. Russia (N. & C. Ural, and lower Pečora basin).* Rs (N). (*N. & C. Asia.*)

13. **G. peloponesiacum** Boiss., *Diagn. Pl. Or. Nov.* **3**(1): 110 (1853). Perennial. Stems erect, branched from near the base, with eglandular hairs below and glandular above. Leaves 8–12 cm wide, divided for 65 % of the radius into 3–5 ovate-rhombic, irregularly pinnatifid lobes; segments obtuse. Inflorescence corymbose; pedicels densely pubescent with short, eglandular and longer, glandular hairs. Sepals *c.* 10 mm, glandular-hairy; petals *c.* 15 mm, obovate, emarginate, bluish-violet. Mericarps hirsute. *Shady places.* ● *Albania; S. Greece.* Al Gr.

14. **G. ibericum** Cav., *Monad. Class. Diss. Dec.* 209 (1787). Perennial, with stout, oblique rhizome. Stems erect, branched only in upper half, with patent, eglandular hairs. Leaves 7–11 cm wide, divided for 75–80 % of the radius into 5–7 rhombic, pinnatifid lobes; segments acute. Inflorescence few-flowered, compact; pedicels with subsessile glands and very long, whitish, eglandular hairs. Sepals 12–15 mm; petals *c.* 25 mm, deep purple, emarginate, sometimes with a tooth in the apical notch. Mericarps hirsute. *Naturalized from gardens in N.W. France.* [Ga.] (*Caucasus.*)

15. **G. phaeum** L., *Sp. Pl.* 681 (1753). Perennial, with short, stout, oblique rhizome. Stems 40–70 cm, erect, with short glandular and long, patent eglandular hairs. Leaves 6–15 cm wide, divided for *c.* 70 % of the radius into (5–)7 broadly oblong, weakly pinnatifid lobes. Inflorescence axillary or leaf-opposed, and also terminal, rather lax, many-flowered; pedicels remaining erect after flowering. Sepals 8–9 mm, with arista *c.* 1 mm, densely glandular-puberulent and also with long eglandular hairs. Petals 8–10 × 6–10 mm, obovate-orbicular, entire, erose or apiculate, patent, blackish- or brownish-purple or (var. *lividum* (L'Hér.) DC.) dull lilac. Mericarps hirsute, with 2–4 strong transverse ridges below the style. 2*n*=28. *Damp or shady places. C. Europe, extending to the Pyrenees, C. Italy, Bulgaria and the W. borders of U.S.S.R. Frequent also as an escape from gardens both within this range and further north.* Al Au Bu Cz Ga Ge He Hs Hu It Ju Po Rm Rs (C, W) [Be Br Da Hb Ho Su].

16. **G. reflexum** L., *Mantissa Alt.* 257 (1771). Like 15 but arista of sepals often obsolete; petals 7–9 × 2·5–4 mm, oblong, sharply deflexed in flower, dull lilac or purple; pedicels deflexed during the maturation of the fruit. *Mountain woods and meadows.* ● *S. part of Balkan peninsula; C. Italy.* Al Bu Gr It Ju [Ge].

17. G. aristatum Freyn & Sint., *Bull. Herb. Boiss.* **5**: 587 (1897). Like **15** but with deflexed hairs on lower part of stem; pedicels glandular-hirsute, deflexed during the maturation of the fruit; sepals with arista 4–6 mm, and with very long (2–3 mm) eglandular hairs; petals 13 × 17 mm, oblong-elliptical, apiculate, somewhat deflexed, pale lilac with darker veins; mericarps sericeous towards the base, subglabrous above, with ridges as in **15**. ● *Mountains of S. Albania, S. Jugoslavia (Makedonija) and N.W. Greece*. Al Gr Ju.

(18–21) G. tuberosum group. Perennials with short, thick, often tuber-like rhizome. Stem single, 20–60 cm, erect, pubescent, with 0–2 pairs of cauline leaves in addition to the pair of large, leaf-like bracts at the base of the inflorescence, which consists of 2 main branches and sometimes a smaller terminal branch between them, each branch terminating in a corymbose cyme. Basal leaves divided for at least 95 % of the radius into 5–9 lobes, which are variously toothed, lobed or pinnatisect. Cauline leaves (if present) and bracts also deeply lobed, but the lobes narrower and less deeply segmented. Petals broadly obovate, deeply emarginate, purplish-pink. Mericarps hairy, without ridges.

The four species included in this group appear to differ constantly in well-correlated characters, despite a considerable similarity in general appearance.

1 Hairs on pedicels glandular   **19. macrostylum**
1 Hairs on pedicels eglandular
  2 Petals 17–22 mm   **21. malviflorum**
  2 Petals 8–15 mm
    3 Cauline leaves other than bracts absent; lobes of basal leaves with 3–6 segments on each side   **18. tuberosum**
    3 At least 1 cauline leaf present; lobes of basal leaves with 0–1(–2) segments on each side   **20. linearilobum**

18. G. tuberosum L., *Sp. Pl.* 680 (1753). Rhizome a small, globose tuber. Basal leaves 6–8 cm wide; lobes pinnatisect, with 3–6 linear-oblong segments on each side. Cauline leaves absent. Lobes of lower bracts toothed or pinnatifid. Hairs on pedicels eglandular. Sepals 4–7 mm; petals 8–13 mm. Styles 18–20 mm in fruit, stout and hairy right up to the stigmas. *Usually in cultivated ground. S. Europe, from S.E. France to the Aegean region and Krym*. Al Bu Cr ?Co *Ga Gr It Ju Rs (K) Sa Si Tu.

19. G. macrostylum Boiss., *Diagn. Pl. Or. Nov.* **1**(1): 58 (1843). Rhizome tuberous, but somewhat irregular and lobed. Basal leaves as in **18**. A pair of cauline leaves present; lobes of these and of the lower bracts deeply pinnatifid or pinnatisect. Hairs on pedicels glandular. Sepals 8–11 mm; petals 12–17 mm. Styles 16–18 mm in fruit, the apical 2 mm glabrous and more slender than the rest. *Greece and Albania*. Al Gr.

20. G. linearilobum DC. in Lam. & DC., *Fl. Fr.* ed. 3, **5**: 629 (1815). Rhizome a small, globose tuber. Basal leaves 4–7 cm wide; lobes with usually only 1 linear segment diverging on each side. 1–2 pairs of cauline leaves present; lobes of these and of bracts mostly entire. Hairs on pedicels eglandular. Sepals 5–9 mm; petals 9–15 mm. Styles *c.* 22 mm in fruit, the apical 2 mm glabrous and more slender than the rest. *Dry places. S.E. Ukraine and S. Russia*. Rs (W, K, E).

21. G. malviflorum Boiss. & Reuter, *Pugillus* 27 (1852). Rhizome ovoid or oblong, sometimes contorted, scarcely tuber-like. Basal leaves *c.* 7 cm wide, with usually only 5 deeply pinnatisect lobes, each with 3–4 linear-oblong segments on each side. A pair of cauline leaves present, or sometimes a single leaf. Hairs on pedicels eglandular. Sepals 8–10 mm; petals 17–22 mm. Styles 25 mm in fruit, glabrous below the stigmas but not markedly narrowed. *Rocky hillsides. S. Spain (Grazalema to Sierra Nevada and northwards to Jaén)*. Hs.

22. G. asphodeloides Burm. fil., *Spec. Bot. Geran.* 28 (1759) (incl. *G. tauricum* Rupr.). Perennial, with fleshy, fusiform roots arising from a very short rhizome. Stems 30–75 cm, erect, with deflexed, sometimes glandular hairs. Leaves 4–6 cm wide, divided for 80–85 % of the radius into 5–7 obovate-cuneate, apically 3-fid lobes; segments entire to incise-dentate. Inflorescence diffuse, few-flowered; pedicels slender, glandular-pubescent. Sepals 10 mm, with arista 0·5–1·25 mm; petals *c.* 15 mm, entire or slightly emarginate, pinkish-lilac with darker veins. Mericarps pubescent, without ridges. *S. Europe, from Sicilia eastwards*. Al Bu Gr It Rm Rs (K) Si Tu.

23. G. palustre L., *Cent. Pl.* **2**: 25 (1756). Perennial, with thin, fibrous roots, arising from a short rhizome. Stems 20–60 cm, erect or spreading, with patent or somewhat deflexed hairs. Leaves 5–10 cm wide, divided for 80–85 % of the radius into 5–7 obovate-cuneate, incise-dentate lobes. Inflorescence diffuse, few-flowered; fruiting pedicels deflexed, 2–4 times as long as the sepals, densely pubescent. Sepals 10–13 mm, with arista usually *c.* 2 mm, glabrous except for short, appressed hairs on the veins; petals 12–18 mm, entire or very slightly emarginate, purple or lilac, hairy on inner surface at extreme base. Mericarps hairy, without ridges. $2n = 28$. *Throughout a large part of Europe, but absent from the islands, the Mediterranean region and much of the north*. Au Be Bu Cz Da Fe Ga Ge He ?Hs Hu It Ju Po Rm Rs (N, B, C, W, E) Su.

24. G. collinum Stephan ex Willd., *Sp. Pl.* **3**: 705 (1800). Like **23** but leaf-segments usually more acute; hairs on stem and petioles appressed; sepals densely appressed-pubescent all over and with slightly shorter arista; petals lilac-pink, hairy only on margins at base, not on inner surface. *Wet places. S. part of U.S.S.R.; Romania*. Rm Rs (C, W, K, E).

G. acutilobum Coincy, *Jour. Bot. (Paris)* **12**: 56 (1898), known from two places in N. & E. Spain, does not appear to be separable from **24**; it requires further investigation.

25. G. divaricatum Ehrh., *Beitr. Naturk.* **7**: 164 (1792). Annual; stems 30–60 cm, erect or ascending, divaricately branched, with long eglandular and short glandular hairs. Leaves 4–9 cm wide, divided for 75–85 % of the radius into 3–5 shortly pinnatifid lobes with obtuse segments; often asymmetrical with the central lobe not the largest. Peduncles shorter than the subtending leaves; fruiting pedicels deflexed, much longer than the calyx. Sepals shortly mucronate; petals 5–7 mm, equalling the sepals, emarginate, pink. Filaments pubescent at the base. Mericarps pubescent, transversely ridged, separating without a stylar beak. *C. & S. Europe; S. & W. parts of U.S.S.R.* Al Bu Cz Ga Ge Gr He Hs Hu It Ju Po Rm Rs (B, C, W, K, E) [Au].

26. G. bohemicum L., *Cent. Pl.* **2**: 25 (1756). Annual or biennial; stems 26–60 cm, erect, hirsute. Leaves 2–6 cm wide, divided for 75 % of the radius into 5–7 obovate or rhombic, irregularly pinnatifid lobes; segments ovate-oblong, subacute. Peduncles longer than the subtending leaves; fruiting pedicels erect, about equalling the calyx. Sepals 9–12 mm, aristate; petals 8–9 mm, emarginate, bright violet-blue with darker veins. Filaments ciliate at the base. Mericarps hairy, without ridges; seeds yellowish-grey, with dark brown patches, almost smooth. Cotyledons with deep lateral notches. $2n = 28$. *E. & C. Europe, extending to S. Norway, E. France, N. Italy and Albania*. Al Bu Cz Fe Ga Ge Gr He Hu It Ju No Po Rm Rs (N, B, C, W, K, E) Su.

27. G. lanuginosum Lam., *Encycl. Méth. Bot.* **2**: 655 (1788). Like **26**, but leaf-lobes more obtuse; petals 7·5 mm, with a white base; seeds uniformly brown, distinctly foveolate; cotyledons without lateral notches. $2n=48$. *Mediterranean region; Sweden.* Al Bu Co Ga Gr Hs It Sa Si *Su.

28. G. sibiricum L., *Sp. Pl.* 683 (1753). Perennial, with short, inconspicuous rhizome. Stems 30–60 cm, procumbent or ascending, with long, deflexed hairs. Leaves 4–7 cm wide, divided for 85 % of the radius into 5–7 lanceolate, deeply pinnatifid lobes; segments ovate-lanceolate, acute. Inflorescence diffuse; peduncles sometimes 1-flowered; pedicels with numerous patent or deflexed hairs. Sepals 6–7 mm, aristate; petals *c.* 6 mm, more or less emarginate, lilac with darker veins. Mericarps with patent hairs, without ridges. *From S. Ural across C. Russia and Ukraine to E. Romania; widely naturalized in C. Europe and still spreading westwards.* Rm Rs (B, C, W, E) [Au Cz Ga Ge He Hu Po].

29. G. pyrenaicum Burm. fil., *Spec. Bot. Geran.* 27 (1759). Perennial, with short, inconspicuous rhizome. Stems 25–70 cm, erect, with short glandular and long, patent eglandular hairs. Leaves 2–5 cm wide, divided for 65 % of the radius into 5–7 contiguous, cuneate, truncate lobes, each with straight, entire sides and a broad, crenate apex. Fruiting pedicels deflexed, densely pubescent. Sepals 4–5 mm, shortly mucronate; petals 7–10 mm, deeply emarginate, purple to lilac. Mericarps with appressed hairs, without ridges. $2n=26$. *S. & W. Europe; widely naturalized elsewhere.* Al Br Bu Co Ga Gr Hb He Hs Hu It Ju Lu Rm Si Tu [*Au Be Cz Da Fe *Ge Ho Hu No Po Rs (C, W, K, E) Su].

30. G. rotundifolium L., *Sp. Pl.* 683 (1753). Annual; stems 10–40 cm, erect or ascending, with long and short hairs, both glandular and eglandular mixed. Basal leaves 3–7·5 cm wide, divided for 25–40 % of the radius into 5–7 contiguous, cuneate lobes, which are deeply crenate or divided apically into short, obtuse segments; upper leaves more deeply divided, with more acute segments. Peduncles usually shorter than subtending leaves. Sepals 5–6 mm, mucronate; petals 5–7 mm, entire or very slightly emarginate, pink. Mericarps hairy, without ridges. $2n=26$. *Most of Europe except the north.* Al Au Az Be Bl Br Bu Co Cr Cz Ga Ge Gr Hb He Hs Hu It Ju Lu Rm Rs (B, C, W, K, E) Sa Si Tu.

31. G. molle L., *Sp. Pl.* 682 (1753). Densely pubescent, usually greyish-green annual, perhaps rarely perennial; stems 10–40 cm, decumbent or ascending, branched from the base, with long, soft white hairs and shorter, often glandular hairs. Basal leaves 1·5–4 cm wide, divided for 70 % of the radius into 5–7 obovate-cuneate, shortly 3-fid lobes; uppermost leaves alternate, sessile. Lowest leaves of inflorescence considerably exceeding subtended peduncle, and with petiole considerably longer than lamina. Sepals 4–5 mm, shortly mucronate; petals 3–7 mm, deeply emarginate, pinkish-purple. Filaments glabrous. Mericarps glabrous, usually with transverse ridges. $2n=26$. *Europe, except the extreme north.* All except Is Rs (N) Sb, but only as a naturalized alien in Fa.

32. G. brutium Gasparr., *Rendic. Accad. Sci. (Napoli)* **1**: 49 (1842) (*G. villosum* Ten., non Miller). Like **31** but probably more often perennial; green, and usually rather sparsely pubescent; stems up to 70 cm, usually erect; lowest leaves of inflorescence shorter than or scarcely exceeding subtended peduncle, and with petiole usually shorter than lamina; petals 6–11 mm, bright reddish-purple. *Balkan peninsula; S. Italy and Sicilia.* Al Bu Gr It Ju Si Tu.

31 and **32** are very distinct when well-grown, but in dwarfed plants the distinctive characters are often obscured. **32** is often confused with **29**, but is easily distinguished by the glabrous, rugose mericarps, as well as by the brighter and redder petals.

33. G. pusillum L., *Syst. Nat.* ed. 10, **2**: 1144 (1759). Like **31** but with only short hairs on stem; sepals *c.* 4 mm; petals 2–4 mm, pale lilac; 3–5 of the stamens reduced to staminodes; and mericarps hairy, without transverse ridges. $2n=36$, 26. *Most of Europe except the extreme north.* All except Az Bl Fa Is Sb Si; probably not native in Hb.

34. G. columbinum L., *Sp. Pl.* 682 (1753) (incl. *G. schrenkianum* Trautv. ex Krylov). Annual; stems 10–60 cm, ascending or erect, usually with short, deflexed hairs. Leaves 2–5 cm wide, divided almost to the base into 5–7 contiguous, rhombic, deeply pinnatifid lobes; segments linear-oblong. Uppermost leaves opposite, distinctly petiolate. Peduncles longer than the subtending leaves; pedicels with short, deflexed eglandular hairs. Sepals 9–13 mm, with appressed eglandular hairs, mainly on the veins; arista at least 2 mm. Petals 7–10 mm, purplish-pink. Mericarps glabrous or sparsely hairy, without ridges. $2n=18$. *Europe, except the extreme north.* All except Az Fa Is Rs (N) Sb.

35. G. dissectum L., *Cent. Pl.* **1**: 21 (1755). Like **34** but basal leaves sometimes larger; peduncles shorter than the subtending leaves; pedicels and sepals densely pubescent, with some of the hairs glandular; sepals 5–6 mm, with shorter arista; petals *c.* 5 mm; mericarps hairy. $2n=22$. *Most of Europe except the extreme north, but doubtfully native in much of Fennoscandia and U.S.S.R.* All except Fa Is Sb Rs (N).

36. G. lucidum L., *Sp. Pl.* 682 (1753). Annual, shining green, often tinged with red, usually sparsely hairy. Stems 10–40 cm, erect or ascending. Leaves 2–6 cm wide, divided for 60–70 % of the radius into 5 obovate-cuneate lobes, which are crenate or shortly 3-fid at the apex, with broad, obtuse, mucronate segments. Sepals 5–7 mm, aristate, erect and connivent during flowering, strongly keeled and with transverse ridges. Petals 8–10 mm, with a well-marked claw longer than the obovate, entire, pink limb. Mericarps usually separating without a stylar beak, laterally compressed, with 5 strong, longitudinal ridges at apex, irregularly reticulate-rugose below, pubescent on upper and inner sides. $2n=20$. *Most of Europe except the northeast.* Al Au Be Bl Br Bu Co Cr Cz Da Fe Ga Ge Gr Hb He Hs Hu It Ju Lu No Rm Rs (B, W, K) Sa Si Su Tu [Po ?Rs (E)].

37. G. robertianum L., *Sp. Pl.* 681 (1753). Annual or biennial, often turning red, more or less hairy. Stems 10–50 cm, procumbent to ascending. Leaves 3–8 cm wide, very deeply divided, so as to appear compound, with 3(–5) principal divisions, which are 2-pinnatisect with oblong, mucronate or apiculate segments. Pedicels with long, patent glandular and deflexed eglandular hairs. Sepals 7–9 mm, erect, lanceolate, aristate, hirsute. Petals 9–13(–15) mm; limb 6–9 mm, obovate-cuneate, entire, bright pink, abruptly contracted to a narrow claw. Pollen orange. Mericarps separating without stylar beak, but remaining attached to the axis by a strand of delicate fibres; with 1 or 2 strong, transverse ridges at the apex and a few low, irregular ridges forming an open reticulum elsewhere; ridges usually hairy. $2n=32$, 64. *Europe except the extreme north.* All except ?Bl Fa Is Sb.

38. G. purpureum Vill. in L., *Syst. Pl. Eur.* **1**, *Fl. Delph.*: 72 (1785). Like **37** but sepals ovate, mucronate or shortly aristate; petals 5–9 mm, purplish-pink, with limb 3–5 mm, elliptic-oblong, longer than broad, and contracted less abruptly to a relatively broader claw; pollen yellow; mericarps usually glabrous, with about 4 strong, transverse ridges near the apex, and covered elsewhere by a close reticulum of lower but conspicuous ridges. $2n=32$. *S. & W. Europe, northwards to 52° N. in Britain.* Al Az Bl Br Bu Co Cr Ga Gr Hb He Hs It Ju Lu ?Rm Rs (K) Sa Si Tu.

In parts of Europe this species is difficult to distinguish satisfactorily from **37**, and is often treated as a subspecies.

39. G. cataractarum Cosson, *Not. Pl. Crit.* 99 (1851). Perennial with woody stock, sparsely hirsute to villous. Stems up to 40 cm, pseudodichotomous. Leaves 4–7 cm wide, suborbicular in outline, divided to the base into 5 obovate-cuneate, irregularly pinnatifid lobes; petiole 5–25 cm. Sepals 6–7 mm, erect, ellipticoblong, mucronate or shortly aristate. Petals *c.* 15 mm, bright pinkish-purple; limb about twice as long as claw. Stamens 8–10 mm; pollen yellow. Mericarps glabrous, separating without stylar beak, and without suspensory fibres, reticulate-rugose. *Damp or shady limestone rocks. S.E. Spain (Sierra de Segura and adjacent ranges).* Hs.

2. Erodium L'Hér.[1]

Annual to perennial herbs usually with hermaphrodite flowers, rarely dioecious. Leaves mostly opposite, usually longer than wide, pinnatifid to pinnate, or rarely undivided, usually with appressed hairs. Inflorescence an umbel (rarely reduced to a single flower), subtended by 2 or more usually scarious bracts. Flowers actinomorphic or slightly zygomorphic. Stamens 5, antesepalous, with a nectary at the base of the filament, alternating with 5 scale-like staminodes. Mericarps indehiscent, separating from the base upwards, retaining during dispersal the outer part of the style as a long beak, which in most species becomes twisted into a spiral at maturity, the pitch of the spiral varying with the humidity. Stigmas 5, filiform.

In some species the principal leaflets of the leaf alternate irregularly with very much smaller lobes or leaflets. The latter are referred to as *intercalary lobes* or *leaflets*.

The sepals of most species are accrescent. The measurements below refer to the fruiting condition, and include the terminal mucro or arista.

At the top of the mericarp, near the base of the style, are two flattened areas or depressions, referred to below as *pits*, which in some species are divided into two or more sections by one or more ridges. As the ridges lie near the lower margin of the pit, the impression created is of a pit with one or more *furrows* below it, and this terminology is used in the descriptions.

Literature: R. Knuth in Engler, *Pflanzenreich* 53 (IV. 129): 221–290 (1912). F. Vierhapper, *Verh. Zool.-Bot. Ges. Wien* 69: 112–155 (1919).

1 Leaves undivided, pinnatifid or pinnatisect, sometimes compound at the base, but if so with only 1(–2) pairs of distinct leaflets
  2 Beak of fruit not more than 17 mm
    3 Leaves at least 3 cm wide
      4 Apical pits of mericarp with a furrow at the base
          6. malacoides
      4 Apical pits of mericarp without a furrow at the base
          7. alnifolium
    3 Leaves less than 3 cm wide
      5 Annual; apical pits of mericarp with a furrow at the base
        6 Petals twice as long as sepals; flowers in umbels of 2–5
          9. sanguis-christi

      6 Petals rarely exceeding sepals, often absent; flowers solitary, rarely in pairs   **8. maritimum**
     5 Perennial; apical pits of mericarp without a furrow at the base
      7 Leaves grey-green, villous or densely pubescent; pedicels with numerous patent hairs   **10. corsicum**
      7 Leaves green, sparsely hairy; pedicels with usually few appressed hairs   **11. reichardii**
  2 Beak of fruit more than 17 mm
    8 Plant acaulescent (S. Spain)   **3. boissieri**
    8 Plant normally caulescent
      9 Perennial
        10 Beak of fruit not more than 40 mm
          11 Sepals 5–7 mm; mericarps less than 5 mm   **4. chium**
          11 Sepals 10–12 mm; mericarps *c.* 8 mm   **12. ruthenicum**
        10 Beak of fruit more than 40 mm
          12 Roots tuberous; beak of mericarp with long, yellowish hairs on inner face (Kriti)   **34. hirtum**
          12 Roots not tuberous; beak of mericarp with inner face glabrous, but usually with hairs on basal part of outer face
            13 Leaves divided to midrib, at least near the base; lobes oblong, pinnatifid or incise-dentate
              14 Sepals *c.* 12 mm, with arista 2–3 mm   **12. ruthenicum**
              14 Sepals *c.* 8 mm, with arista 1 mm   **21. alpinum**
            13 Leaves divided for not more than $\frac{2}{3}$ of the distance to midrib; lobes ovate to orbicular, usually crenate
              15 Bracts suborbicular, glabrous; mericarps *c.* 5 mm, with beak 40–60 mm   **1. gussonii**
              15 Bracts triangular-lanceolate, hairy; mericarps *c.* 9 mm, with beak 65–100 mm   **2. guttatum**
      9 Annual or biennial
        16 Bracts at base of umbel 2, suborbicular to reniform
          5. laciniatum
        16 Bracts at base of umbel at least 3, ovate to lanceolate
          17 Beak of fruit less than 45 mm
            18 Apical pits of mericarp with a furrow at the base
              6. malacoides
            18 Apical pits of mericarp without a furrow at the base
              19 Beak of fruit more than 25 mm; leaves lobed   **4. chium**
              19 Beak of fruit less than 25 mm; leaves not lobed   **7. alnifolium**
          17 Beak of fruit more than 45 mm
            20 Apical pits of mericarp with two conspicuous furrows at the base   **14. botrys**
            20 Apical pits of mericarp without furrows, or with a single shallow one
              21 Mericarps 6–7 mm; sepals 7–10 mm   **13. hoefftianum**
              21 Mericarps 9–14 mm; sepals 12–20 mm
                22 Leaf with intercalary lobes or leaflets; apical pits of mericarp glandular-hairy   **16. ciconium**
                22 Lobes of leaf diminishing regularly from base to apex; apical pits of mericarp smooth or foveolate, not hairy   **15. gruinum**
1 Leaves pinnate for most of their length (sometimes pinnatisect towards the apex)
  23 Intercalary leaflets, much smaller than the principal ones, present
    24 Annual or biennial   **16. ciconium**
    24 Perennial, with stout, woody rhizome
      25 Petals yellow   **18. chrysanthum**
      25 Petals pink, purple or white
        26 Plant acaulescent (France and Spain)
          27 Leaves densely white-sericeous above, green and ± glabrous beneath   **22. rupestre**
          27 Leaves without a conspicuous difference in indumentum between the two surfaces
            28 Leaves glabrous to sparsely strigulose, with linear segments; petals 15–20 mm   **24. rodiei**
            28 Leaves usually densely hairy, with ovate-oblong to lanceolate segments; petals not more than 13 mm   **23. petraeum**
        26 Plant caulescent (S.E. Europe and Italy)

---

[1] By D. A. Webb and A. O. Chater.

29 Leaves silvery-sericeous on both surfaces **17. guicciardii**
29 Leaves ± hairy, but not silvery
  30 Leaflets pinnatifid or somewhat pinnatisect, with
    oblong-lanceolate segments; flowers hermaphrodite
                         21. alpinum
  30 Leaflets pinnate or deeply pinnatisect, with linear
    segments; dioecious
    31 Leaves not more than 5 cm, green, with usually rather
      few glandular hairs; beak of fruit usually more than
      40 mm **19. absinthoides**
    31 Leaves up to 10 cm, greyish, with numerous eglandular
      hairs; beak of fruit not more than 40 mm
                         20. beketowii
23 Intercalary leaflets absent
  32 Plant caulescent
    33 Most of the leaflets divided less than half-way to the mid-
      rib; apical pits of mericarp glandular **26. moschatum**
    33 Most of the leaflets divided more than half-way to the
      midrib; apical pits of mericarp eglandular **25. cicutarium**
  32 Plant acaulescent
    34 Bracts herbaceous, united to form a cupule **33. manescavi**
    34 Bracts scarious, free or united only at base
    35 Annual or biennial **25. cicutarium**
    35 Perennial
      36 Many of the leaflets, at least of the later leaves of the
        season's growth, entire **31. astragaloides**
      36 Leaflets of well-grown plants all pinnatifid or pinnati-
        sect
      37 Mericarps 7·5–9 mm **32. daucoides**
      37 Mericarps 5–7 mm **27–30. acaule** group

1. E. gussonii Ten., *Fl. Nap.* 1, *Prodr.*: 39 (1811). Perennial; stems 15–30 cm, with long, white hairs. Leaves up to 4 cm, broadly ovate, cordate, usually 3-lobed, crenate or obtusely dentate. Stipules large, dark brown. Umbels with 4–10 flowers; bracts suborbicular, glabrous, brown. Sepals 5–7 mm; petals *c.* 12 mm, purplish. Mericarps *c.* 6 mm, with short, appressed, white hairs; apical pits shallow, with a few glands, with or without a very shallow furrow at the base. Beak 40–60 mm. $2n=20$. *Grassland.* ● *S. Italy.* It.

2. E. guttatum (Desf.) Willd., *Sp. Pl.* 3: 636 (1800). Perennial; stems 3–30 cm, with short, deflexed hairs. Leaves 1–2·5 cm, triangular to ovate, cordate or truncate at base, the lower undivided, the upper rather deeply 3-lobed. Umbels with 2–3 flowers; bracts triangular-lanceolate, acute, hairy. Sepals 11–13 mm; petals *c.* 10 mm, deep violet with a black basal spot. Mericarps *c.* 9 mm, with brownish, ascending hairs; apical pits with a few small glands and with a shallow furrow at the base. Beak 65–100 mm. *Sandy or rocky places. S. Spain (Málaga prov.).* Hs. (*N. Africa.*)

3. E. boissieri Cosson, *Bull. Soc. Bot. Fr.* 20: 244 (1873) (*E. asplenioides* auct. hisp., vix (Desf.) Willd.). Acaulescent perennial. Leaves 2–5 cm, ovate, variously lobed or pinnatifid in apical half, pinnatisect towards base and usually with a pair of distinct leaflets at base, covered with short, white, appressed hairs. Umbels with 1–5 flowers; bracts ovate, brown or greenish; pedicels densely glandular-hairy. Sepals *c.* 9 mm, with arista *c.* 1·5 mm; petals 12–15 mm, lilac with purple veins. Mericarps 7–9 mm, with somewhat appressed, white hairs; apical pits shallow, usually eglandular, without a furrow at the base. Beak 50–75 mm. $2n=20$. *Calcareous screes.* ● *S. Spain (Sierra Nevada).* Hs.

4. E. chium (L.) Willd., *Phytogr.* 1: 10 (1794). Stems 5–50 cm, with deflexed hairs at least near the base. Leaves ovate, very variably dissected. Umbels with 2–8 flowers; bracts 3 or more, ovate, acute, brown. Sepals 5–7 mm; petals 5–9 mm, purplish. Mericarps 3·5–4·5 mm, with short, appressed, whitish

hairs; apical pits small but rather deep, covered with minute glands, without a furrow at the base. Beak 30–40 mm. $2n=20$. *Dry, mainly open habitats. Mediterranean region, Portugal.* Bl Co Cr Ga Gr Hs It Lu Sa Si Tu.

Very variable; the following subspecies may be recognized, but correlation of characters is rather poor. and intermediates are fairly common, especially in Corse.

(a) Subsp. **chium**: Robust annual or biennial. Leaves divided for $\frac{1}{3}$–$\frac{2}{3}$ of distance to midrib, with obtusely dentate lobes. Pedicels often eglandular. Staminodes ciliate. *Throughout the range of the species.*

(b) Subsp. **littoreum** (Léman) Ball, *Jour. Linn. Soc. London* (*Bot.*) 16: 387 (1878) (*E. littoreum* Léman): Slender perennial. Leaves pinnatisect, with strongly dentate or pinnatifid lobes. Pedicels usually glandular. Staminodes glabrous. *S.E. Spain, Islas Baleares, Corse, S. France.*

5. E. laciniatum (Cav.) Willd., *Sp. Pl.* 3: 633 (1800). Annual or biennial; stems 7–50 cm, with deflexed hairs at least near the base. Leaves 2–7 cm, oblong to broadly ovate, very variously dissected: undivided and irregularly serrate, or with 3 pinnatifid lobes, or almost bipinnatisect with linear-lanceolate segments. Umbels with 4–9 flowers; bracts 2, suborbicular to reniform, glabrous, brown; hairs on pedicels and sepals eglandular. Sepals *c.* 7 mm, distinctly mucronate; petals 7–10 mm, purplish. Mericarps 4·5–6·5 mm, with short, whitish hairs; apical pits shallow, eglandular, without a furrow at the base. Beak 35–90 mm. $2n=20$. *Maritime sands and other dry places. Mediterranean region, S. Portugal.* Al Co Cr Gr Hs It Lu Sa Si Tu [Ga].

Plants from S.W. Spain (near Cádiz) show some approach to **6** in the structure of their fruit, and may perhaps be hybrids.
Some plants from C. & E. Spain and the Aegean region differ in some or all of the following characters: bracts 3 or more, acute, white; calyx and pedicels glandular-pubescent; sepals scarcely mucronate; beak 30–40 mm. In S.W. Asia there are plants in which these characters are fairly well correlated, and they have been distinguished as subsp. **pulverulentum** (Cav.) B. L. Burtt & P. Lewis, *Kew Bull.* 1954: 405 (1954) (*E. cavanillesii* Willk.). In Europe, however, the correlation is poor, and no such distinction seems practicable.

6. E. malacoides (L.) L'Hér. in Aiton, *Hort. Kew.* 2: 415 (1789) (incl. *E. subtrilobum* Jordan, *E. aragonense* Loscos). Annual or biennial; stems (3–)10–60 cm, with deflexed hairs, often glandular. Leaves 2–10 × 1–5 cm, ovate to oblong, cordate, dentate, sometimes pinnatifid or 3-lobed. Umbels with 3–7 flowers; bracts several, ovate-orbicular, often hairy, whitish; hairs on pedicels and sepals usually glandular. Sepals 5–7 mm; petals 5–9 mm, purplish. Mericarps 5 mm, with white or brownish hairs; apical pits deep, usually glandular, with a wide, deep furrow at the base. Beak 18–35 mm. $2n=40$. *Dry, open habitats. S. Europe, extending northwards in W. France to 49°N.* Al Az Bl Co Cr Ga Gr Hs It Ju Lu Rs (K) Sa Si Tu.

7. E. alnifolium Guss., *Fl. Sic. Prodr.* 2: 307 (1828). Annual; stems up to 50 cm, hispid with deflexed hairs. Leaves 5–10 cm, the lower cordate-suborbicular, undivided or slightly lobed, crenate, the upper ovate, dentate or serrate. Umbels with 4–7 flowers; bracts ovate, subacute, glabrous or ciliate, brown or whitish; pedicels and sepals glandular-hairy. Sepals 6–7 mm; petals *c.* 6 mm, pale pink. Mericarps *c.* 4 mm, with brownish hairs; apical pits deep, eglandular, without a furrow at the base. Beak 10–20 mm. $2n=20$. *Dry grassland. C. & S. Italy, Sicilia, Sardegna.* It Sa Si.

8. E. maritimum (L.) L'Hér. in Aiton, *Hort. Kew.* **2**: 416 (1789). Annual or biennial; stems up to 20 cm, hairy. Leaves 0·5–2·5 cm, mostly basal, broadly ovate, obtusely dentate or pinnatifid. Flowers solitary, rarely in pairs; bracts ovate, brown; pedicels with appressed, often glandular hairs. Sepals 3–4·5 mm; petals 3 mm or less, pink or white, very often absent. Mericarps 3 mm, with short, brownish hairs and a few longer hairs near the apex; apical pits deep, usually eglandular, with a pronounced furrow at the base, delimited by a high, narrow ridge. Beak *c.* 10 mm. 2*n*=20. *Dry places, usually near the sea. N.W. Europe, from N.W. France to S.W. Scotland; W.C. Mediterranean region; one station in N.W. Spain.* Br Co Ga Hb Hs It Sa Si.

Outside Europe known only from Tenerife and a small island off Tunisia.

9. E. sanguis-christi Sennen, *Ann. Soc. Linn. Lyon* nov. ser., **72**: 12 (1926). Annual, at first acaulescent, but sometimes with procumbent, leafy flowering stems later. Leaves 1–1·5 cm, oblong-lanceolate, pinnatifid to pinnatisect with obtuse, suborbicular lobes, usually with one pair of free leaflets at the base, greyish with appressed hairs on both surfaces and with sessile, vesicular glands beneath. Flowers in umbels of 2–5; bracts 2, suborbicular, whitish; pedicels glandular-pubescent. Sepals 3·5–4 mm, with scarious border, scarcely mucronate; petals 6 mm, deep red. Mericarps 4 mm, with yellowish hairs of equal length; apical pits eglandular, with a rather narrow furrow at the base. Beak 12–15 mm. 2*n*=20. *Dry limestone rocks near the sea.* ● *E. Spain.* Hs.

10. E. corsicum Léman in Lam. & DC., *Fl. Fr.* ed. 3, **4**: 842 (1805). Perennial; stems up to 25 cm, hairy. Leaves 10–25 mm, suborbicular to oblong-ovate, crenate-dentate, the upper sometimes pinnatifid, grey-green, villous or densely pubescent. Flowers solitary or in umbels of 2–3; bracts 2, ovate, hairy, pale brown; pedicels with patent, eglandular hairs. Sepals 3–7 mm, without mucro; petals 5–10 mm, white or pink. Mericarps 3·5 mm, with numerous, long, white hairs; apical pits deep, glandular, without a furrow at the base. Beak 10–15 mm. 2*n*=20 (?18). *Rocks near the sea.* ● *Corse, Sardegna.* Co Sa.

11. E. reichardii (Murray) DC., *Prodr.* **1**: 649 (1824) (*E. chamaedryoides* L'Hér.). Like **10** but acaulescent; leaves 5–15 mm, green, sparsely hairy; flowers always solitary; bracts glabrous; hairs of pedicels appressed; petals white with purple veins; mericarps with appressed hairs and shallower apical pits. 2*n*=20. *Damp rocks.* ● *Islas Baleares.* Bl ?Co.

12. E. ruthenicum Bieb., *Cent. Pl.* **1**: t. 48 (1810) (*E. serotinum* Steven). Perennial; stems 15–50 cm, usually with patent hairs. Leaves up to 8 cm, triangular-ovate, with one pair of free pinnae at the base, the upper part pinnatifid; lobes dentate or pinnatifid. Umbels with 3–13 flowers; bracts several, linear-lanceolate, hairy, brown. Sepals 10–12 mm; petals *c.* 12 mm, violet. Mericarps *c.* 8 mm, densely hairy; apical pits glandular, without a furrow at the base. Beak 30–70 mm. *Dry open habitats. Ukraine and Moldavia; once reported from Dobrogea.* Rm Rs (W, ?E).

13. E. hoefftianum C. A. Meyer, *Mém. Acad. Sci. Pétersb.* ser. 6 (Sci. Nat.), **7** (Bot.): 3 (1855) (incl. *E. neilreichii* Janka). Like **12** but annual or biennial; leaves not more than 6 cm, occasionally without free pinnae at base; umbels with 1–8 flowers; sepals 7–10 mm; petals *c.* 8 mm; mericarps 6–7 mm, with apical pits more or less eglandular; beak 50–75 mm. *S.E. & E.C. Europe.* Bu Cz Gr Hu Ju Rm Rs (W, ?K, E) Tu.

14. E. botrys (Cav.) Bertol., *Amoen.* 35 (1819). Caulescent annual; stems (5–)10–40 cm, with long, patent or deflexed hairs. Leaves up to 5 cm, usually appressed-setose, oblong or ovate, at least the upper deeply pinnatifid or pinnatisect; lobes pinnatifid or dentate. Umbels with 1–4 flowers; bracts ovate-lanceolate, acute, subglabrous, brown. Sepals 10–13 mm; petals *c.* 15 mm, violet. Mericarps 8–11 mm, with ascending, whitish hairs; apical pits deep, eglandular, with two furrows at the base, the upper larger. Beak 50–110 mm. 2*n*=40. *Dry places. S. Europe; sometimes naturalized or casual further north.* Bl Bu Co Cr Ga Gr Hs It Ju Lu Sa Si Tu.

15. E. gruinum (L.) L'Hér. in Aiton, *Hort. Kew.* **2**: 415 (1789). Caulescent annual or biennial; stems 15–50 cm, with patent or deflexed hairs. Leaves up to 10 cm, ovate or ovate-lanceolate, deeply pinnatifid or pinnatisect, sometimes with a pair of free leaflets at the base; lobes and leaflets irregularly dentate or serrate. Umbels with 2–6 flowers; bracts lanceolate, acute, glabrous, whitish. Sepals 15–20 mm, usually with few eglandular hairs; petals 20–25 mm, violet. Mericarps *c.* 14 mm, with numerous ascending, whitish hairs; apical pits deep, foveolate, with a wide furrow at the base. Beak 60–110 mm. *Dry grassland and maritime sands. Aegean region; Sicilia.* Cr Gr ?Hs Si [Ga].

16. E. ciconium (L.) L'Hér. in Aiton, *Hort. Kew.* **2**: 415 (1789). Annual or biennial; stems 10–70 cm, with short, usually deflexed and glandular hairs. Leaves up to 9 cm, pinnate at least near the base; leaflets pinnatisect, the ultimate segments dentate or pinnatifid; intercalary lobes present. Umbels with 3–10 flowers; bracts ovate-lanceolate, densely hairy. Sepals 12–15 mm, glandular-hairy; petals *c.* 8 mm, bluish or lilac, with darker veins. Mericarps 9–11 mm, with numerous whitish hairs; apical pits deep, densely glandular, without a furrow at the base. Beak 60–100 mm. 2*n*=18. *Dry, sandy or disturbed ground. S. Europe, extending northwards to 48° N. in E.C. Europe.* Al Bl Bu Co Cr Cz Ga Gr Hs Hu It Ju Rm Rs (W, K) Sa Si Tu.

17. E. guicciardii Heldr. ex Boiss., *Diagn. Pl. Or. Nov.* **3**(6): 40 (1859). Dioecious perennial; stems up to 20 cm, with short, dense, white, appressed hairs. Leaves up to 5 cm, bipinnate, or pinnate with pinnatisect leaflets, silvery-sericeous on both surfaces; ultimate segments linear-lanceolate or -oblanceolate; intercalary leaflets present. Umbels with 2–7 flowers; bracts several, ovate-lanceolate, acute, brown, with eglandular hairs. Sepals 8–12 mm, with appressed, eglandular hairs and arista 2–3 mm; petals 8–10 mm, pink. Mericarps *c.* 9 mm, with very dense, ascending, white hairs; apical pits glandular-hairy, without a furrow at the base. Beak 45–65 mm. ● *High mountains of S. Albania and C. & N.W. Greece.* Al Gr.

18. E. chrysanthum L'Hér. ex DC., *Prodr.* **1**: 645 (1824). Like **17** but usually smaller, with leaves seldom more than 3 cm; bracts often obtuse, with glandular hairs; sepals with glandular hairs and mucro less than 2 mm; petals bright yellow; mericarps 6–7 mm. *Open stony ground; calcicole.* ● *Mountains of C. & S. Greece.* Gr.

19. E. absinthoides Willd., *Sp. Pl.* **3**: 627 (1800). Dioecious perennial; stems up to 20 cm, glandular-pubescent at least above. Leaves up to 5 cm, bipinnate, or pinnate with pinnatisect leaflets, green, with relatively sparse, appressed hairs, some of them glandular; ultimate segments linear to linear-lanceolate; intercalary leaflets present. Umbels with 2–8 flowers. Sepals 10–13 mm, with patent, glandular hairs; petals *c.* 10 mm, violet. Mericarps as in **17**. *Rocks and stony ground. Mountains of Macedonia.* Bu Gr Ju.

Described from Anatolia. The European plant is distinguished by its narrower leaf-segments as subsp. **elatum** (Form.) P. H. Davis & J. Roberts, *Notes Roy. Bot. Gard. Edinb.* **22**: 18 (1955).

20. E. beketowii Schmalh., *Fl. Sred. Juž. Ross.* **1**: 198 (1895). Perennial; stems 15–30 cm. Leaves *c.* 10 cm, bipinnate, with narrowly linear segments, greyish with numerous appressed, glandular hairs; intercalary leaflets present. Umbel with 5–15 flowers. Sepals *c.* 9 mm; petals *c.* 8 mm, lilac. Mericarps *c.* 7 mm; beak 30–40 mm. *Rocky and gravelly places.* ● *S.E. Ukraine (Kal'mius and Kal'čik valleys, near Ždanov).* Rs (W).

Apparently distinct, but needing further investigation.

21. E. alpinum L'Hér., *Geraniologia* t. 3 (1792). Perennial; stems up to 20 cm, shortly and sparsely hairy. Leaves up to 7 cm, more or less appressed-hairy but not silvery, oblong-lanceolate, pinnate, with pinnatifid, rarely pinnatisect leaflets; ultimate segments oblong-lanceolate, acute; intercalary leaflets present. Umbels with 2–9 flowers; bracts 3 or more, ovate-lanceolate, acute, subglabrous, brown. Sepals 8–10 mm, glandular-hairy; petals 10–14 mm, violet. Mericarps *c.* 10 mm, with ascending, white hairs; apical pits small, shallow, glandular-hairy, without a furrow at the base. Beak 40–60 mm. $2n = 18$. ● *Mountains of C. Italy.* It.

22. E. rupestre (Pourret ex Cav.) Guittonneau, *Bull. Soc. Bot. Fr.* **110**: 244 (1963) (*E. supracanum* L'Hér.). Acaulescent perennial. Leaves 1–3 cm, whitish-silvery on upper surface with dense, straight, short, appressed hairs, green and subglabrous beneath, pinnate, with intercalary leaflets; leaflets pinnatisect, with oblong-lanceolate, usually entire segments. Umbels with 1–3 flowers; bracts lanceolate, appressed-hairy. Sepals 5–7 mm; petals *c.* 10 mm, pale pink with darker veins. Mericarps *c.* 5 mm, with ascending, white hairs; apical pits with conspicuous, sessile glands, without a furrow at the base. Beak 12–15 mm. $2n = 20$. *Dry, calcareous rocks.* ● *N.E. Spain.* Hs.

23. E. petraeum (Gouan) Willd., *Sp. Pl.* **3**: 626 (1800). Acaulescent perennial. Leaves 1–7 cm, ovate-oblong, subglabrous to densely hairy, sometimes canescent or somewhat silvery, without a conspicuous difference in indumentum between the surfaces, pinnate, with intercalary leaflets; leaflets pinnatisect, with oblong- to linear-lanceolate, usually pinnatifid segments. Umbels with 1–5 flowers; bracts triangular-ovate to linear-lanceolate, appressed-hairy. Sepals 6–10 mm; petals 8–13 mm, equal or unequal, violet, pink or white, the 2 upper sometimes with a black basal patch. Mericarps 5–6 mm, with ascending, white hairs; apical pits without a furrow at the base. Beak 18–33 mm. $2n = 20$. *Rocky places, mainly in the mountains. Spain and S.W. France.* Ga Hs.

Very variable, especially in stature, size of leaves, nature and density of indumentum, width of bracts, size and colour of petals and length of beak of fruit. There is, however, much reticulation of these characters; on the information at present available it seems best to recognize 5 subspecies in Europe, with probably a sixth in Morocco.

1 Apical pits of mericarp eglandular, or with very few, usually minute and sessile glands
2 Leaves subglabrous and odourless; petals white with red veins
(b) subsp. **lucidum**
2 Leaves ± hairy, fetid; petals bright pink (a) subsp. **petraeum**
1 Apical pits of mericarp with conspicuous stalked glands
3 Petals equal, without a dark patch at the base
(e) subsp. **valentinum**
3 Upper 2 petals larger than lower 3, and with a dark patch at the base

4 Petals violet to purple; hairs on leaves often glandular
(c) subsp. **glandulosum**
4 Petals white to pale pink or lilac, with red or purple veins; hairs on leaves usually eglandular (d) subsp. **crispum**

(a) Subsp. **petraeum**: Leaves somewhat hairy, fetid; petals usually larger than in other subspecies, bright pink, equal, without a dark patch at the base; apical pits of mericarps more or less glandular; beak 18–24 mm. *Calcicole.* ● *S. France (C. Pyrenees to Aude).*

(b) Subsp. **lucidum** (Lapeyr.) D. A. Webb & Chater, *Feddes Repert.* **74**: 17 (1967) (*E. lucidum* Lapeyr.): Like subsp. (a) but leaves almost glabrous, not fetid, with ultimate segments usually broader; petals smaller, white, veined with pink. *Usually calcifuge.* ● *E. & C. Pyrenees.*

(c) Subsp. **glandulosum** (Cav.) Bonnier, *Fl. Compl. Fr.* **2**: 88 (1913) (*E. macradenum* L'Hér.): Leaves usually rather densely glandular-pubescent, fetid; petals violet to purple, the upper 2 with a dark patch at the base; apical pits of mericarp with conspicuous stalked glands; beak 25–32 mm. ● *Pyrenees and N. Spain.*

(d) Subsp. **crispum** (Lapeyr.) Rouy, *Fl. Fr.* **4**: 101 (1897) (*E. cheilanthifolium* Boiss.): Leaves usually more or less densely white-pubescent, with rather short segments, often with revolute margins; petals white, pale pink or lilac, with red or purple veins, the upper 2 with a dark patch at the base; apical pits of mericarp with conspicuous stalked glands; beak 20–35 mm. *Calcicole. E. Pyrenees and Corbières; E. & S. Spain.*

(e) Subsp. **valentinum** (Lange) D. A. Webb & Chater, *Feddes Repert.* **74**: 17 (1967) (*E. petraeum* var. *valentinum* Lange): Like subsp. (d) but upper petals without a dark patch at the base. *Calcicole.* ● *S.E. Spain (S. Valencia to N.E. Granada).*

24. E. rodiei (Br.-Bl.) Poirion, *Feddes Repert.* **74**: 14 (1967) (*E. petraeum* subsp. *rodiei* Br.-Bl.). Acaulescent perennial. Leaves 3–5 cm, ovate, long-petiolate, subglabrous, pinnate, with intercalary leaflets; leaflets deeply pinnatisect; ultimate segments linear, entire. Peduncles 8–20 cm. Umbels with 2–8(–10) flowers; bracts long-acuminate. Sepals 9–11 mm; petals 15–20 mm, bright pink. Mericarps 7–8 mm, with ascending, pale brown hairs; apical pits eglandular, without a furrow at the base. Beak 30–35 mm. *Fissures in dolomite rock.* ● *S.E. France (N.W. of Grasse).* Ga.

Some variants of **23**(a) from shaded situations resemble **24**, but their petals are always smaller and their leaves hairier.

25. E. cicutarium (L.) L'Hér. in Aiton, *Hort. Kew.* **2**: 414 (1789). Usually caulescent and annual, often somewhat fetid; stem up to 60(–100) cm. Leaves up to 15 cm, pinnate, without intercalary leaflets, with variable indumentum; leaflets pinnatifid to pinnate, but always divided for more than half-way to midrib. Umbels with up to 12 flowers; bracts brownish. Sepals 5–7 mm; petals 4–11 mm, purplish-pink, lilac or white. Mericarps 4–7 mm, with ascending hairs; apical pits eglandular. Beak 10–70 mm. *Cultivated or disturbed ground, sandy places and dry grassland. Throughout Europe, but probably introduced in much of the centre, north and east.* All except Fa Is Sb.

A very variable and difficult complex. Within the compass of the species as here delimited over 30 binomials have been proposed for European plants (omitting those of A. Jordan); but the delimitation of these supposed species has been attempted, if at all, with reference only to immediately adjacent populations. The only recent attempt to survey the field more widely is that of Litardière in Briquet, *Prodr. Fl. Corse* **2**(2): 28–35 (1936), and we follow here his taxonomic scheme, subject to nomenclatural revisions suggested by Guittonneau, *Bull. Soc. Bot. Fr.*

110: 43–48, 241–244 (1963). Each of the 3 subspecies here proposed has a fairly distinctive facies, but intermediates are common, whose cytological status is often unknown. Plants intermediate between subsp. (a) and (b) are especially common in N.W. Europe, and have been named **E. cicutarium** subsp. **dunense** Andreas, *Nederl. Kruidk. Arch.* **54**: 198 (1947) (*E. glutinosum* subsp. *dunense* (Andreas) Rothm.). They have mostly $2n=40$; but one variant has been found with $2n=60$, which has been interpreted as an amphidiploid hybrid between subspp. (a) and (b) and named **E. danicum** K. Larsen, *Biol. Meddel. Kong. Danske Vid. Selsk.* **23**(6): 14 (1958).

1 Mericarp without or with a very faint furrow below apical pit
    (b) subsp. bipinnatum
1 Mericarp with a distinct furrow below apical pit
2 Hairs on mericarp arising from blackish tubercles; beak 40–70 mm   **(c) subsp. jacquinianum**
2 Mericarp without blackish tubercles; beak 10–40 mm
    (a) subsp. cicutarium

(a) Subsp. **cicutarium** (incl. *E. salzmannii* Delile, *E. primulaceum* (Lange) Welw. ex Lange): Often robust, with stems up to 100 cm; sparsely or densely hairy, eglandular or somewhat glandular. Leaflets pinnatifid or somewhat pinnatisect. Bracts several, ovate. Petals usually purplish-pink, the upper 2 larger and often with a blackish basal patch. Mericarps 5–7 mm, without blackish tubercles; apical pits large, with a furrow at the base. Beak 10–40 mm. $2n=20, 36, 40, 48, 54$. *Throughout the range of the species.*

(b) Subsp. **bipinnatum** Tourlet, *Cat. Pl. Indre Loire* 103 (1908) (*E. bipinnatum* Willd., *E. staphylinum* Bertol., *E. sabulicola* (Lange) Lange): Usually less robust than subspp. (a) and (c); often densely glandular-hairy. Leaflets deeply pinnatisect or almost pinnate. Bracts several, ovate. Umbels with 3–7 flowers. Petals usually lilac or white, equal, without black patch. Mericarps 4–5 mm, without black tubercles; apical pits small, without a furrow at the base, or with a very faint one. Beak 10–40 mm. $2n=20, 40$. *Usually on maritime sands. W. Europe, northwards to the Netherlands, and eastwards to Sardegna.*

(c) Subsp. **jacquinianum** (Fischer, C. A. Meyer & Avé-Lall.) Briq. in Engler, *Pflanzenreich* **53**(IV.**129**): 281 (1912) (*E. aethiopicum* auct., non *Geranium aethiopicum* Lam.): Usually robust; often densely glandular-hairy. Leaflets deeply pinnatisect or almost pinnate. Bracts 2–3, suborbicular. Petals usually equal and without black patch. Mericarps 5–7 mm, with hairs arising from blackish tubercles; apical pits large, with a conspicuous furrow at the base. Beak 40–70 mm. $2n=20$. *Sandy places. S. Spain, S. Portugal, Sardegna, Elba. (N. Africa.)*

26. **E. moschatum** (L.) L'Hér. in Aiton, *Hort. Kew.* **2**: 414 (1789). Annual or biennial, smelling of musk. Stems 10–50 cm, hispid with usually deflexed hairs, dense above, sparse below. Leaves up to 20 cm, oblong-lanceolate, pinnate almost throughout their length, without intercalary leaflets; leaflets ovate, dentate, serrate or somewhat pinnatifid, the lower ones remote. Umbels with 5–12 flowers; bracts several, broadly ovate, subacute, subglabrous, pale brown. Sepals 6–9 mm; petals c. 15 mm, violet or purple. Mericarps 5–6 mm, with patent, brown or white hairs; apical pits very wide, glandular, with a wide, deep furrow at the base. Beak 20–45 mm. $2n=20$. *Cultivated ground and waste places. S. & W. Europe; frequently naturalized or casual elsewhere.* Al Az Bl *Br Co Cr Ga Gr Ho Hs It Ju Lu Sa Si Tu [Au Be Cz Ge *Hb He Hu].

27–30. E. acaule group. Acaulescent perennials. Leaves up to 15 cm, oblong to ovate-lanceolate, pinnate, without intercalary leaflets; leaflets ovate-elliptical to ovate-lanceolate, pinnatifid

or pinnatisect. Umbels with 3–10 flowers; bracts several, ovate-lanceolate, subglabrous. Mucro of sepals small or absent. Mericarps 6–7 mm, with ascending, white hairs. Beak 25–50 mm.

1 Petiole almost as long as lamina   **30. rupicola**
1 Petiole not more than half as long as lamina
2 Petals without a dark patch; hairs on sepals appressed, eglandular   **27. acaule**
2 Upper two petals with a black basal patch; hairs on sepals ± patent, glandular
3 Bracts and stipules brown; hairs on leaves mostly eglandular; apical pits of mericarp eglandular   **28. carvifolium**
3 Bracts and stipules whitish; hairs on leaves mostly glandular; apical pits of mericarp densely glandular   **29. paui**

27. **E. acaule** (L.) Becherer & Thell., *Feddes Repert.* **25**: 215 (1928) (*E. romanum* (Burm. fil.) L'Hér.). Ultimate segments of leaves elliptic-lanceolate; petiole very short, sparsely eglandular-pubescent. Peduncles eglandular below; bracts brown. Sepals 5–8 mm, with appressed, eglandular hairs. Petals 7–12 mm, lilac, equal, without black patch. Apical pits of mericarp eglandular, with a fairly distinct furrow at the base. $2n=40$. *Dry places. Mediterranean region, S. Portugal.* Co Cr Ga Gr Hs It Lu Sa Si Tu.

Sometimes regarded as a subspecies of **25**, but the combination of acaulescent with perennial and usually robust habit is distinctive. Acaulescent variants of **25** are small and annual.

28. **E. carvifolium** Boiss. & Reuter, *Diagn. Pl. Nov. Hisp.* 9 (1842). Hairs on leaves mostly eglandular; ultimate segments linear-lanceolate; petiole short. Peduncles eglandular below; bracts brown. Sepals 5–8 mm, usually with patent, glandular hairs, sometimes also with appressed, eglandular hairs. Petals 7–12 mm, purple, the upper 2 larger and with a blackish, basal patch. Apical pits of mericarp eglandular, without a furrow at the base or with a very faint one. $2n=20, 40$. *Pinewoods and mountain pastures.* ● *N.C. & W.C. Spain.* Hs.

29. **E. paui** Sennen, *Bol. Soc. Ibér. Ci. Nat.* **26**: 83 (1927). Leaves 1·5–3 cm, somewhat greyish on both surfaces from a dense covering of mostly glandular hairs; ultimate segments lanceolate or linear-lanceolate; petiole densely hairy, short. Bracts and stipules whitish. Sepals c. 10 mm, densely hairy, glandular or eglandular. Petals c. 12 mm, violet, the upper 2 larger and with a blackish basal patch. Apical pits of mericarp densely glandular, without a furrow at the base. Beak 25–30 mm. ● *N.C. Spain (Pico de Urbión).* Hs.

30. **E. rupicola** Boiss., *Voy. Bot. Midi Esp.* **2**: 724 (1845). Leaves with mostly patent hairs; leaflets pinnatifid; ultimate segments oblong-lanceolate; petiole densely glandular-hairy, about as long as lamina. Bracts and stipules brownish. Sepals with long, patent, eglandular, and shorter, glandular hairs. Petals as in **29**. Apical pits of mericarp eglandular, without a furrow at the base. $2n=20$. *Rock-crevices.* ● *S. Spain (Sierra Nevada).* Hs.

31. **E. astragaloides** Boiss. & Reuter, *Pugillus* 130 (1852). Acaulescent perennial. Leaves up to 4 cm, whitish-tomentose on both surfaces, oblong-lanceolate, pinnate, without intercalary leaflets; leaflets ovate, often entire, but some of those of the older leaves usually dentate or pinnatifid. Stipules reddish-brown. Umbels with 2–8 flowers; bracts lanceolate, whitish-tomentose. Sepals 8–10 mm, densely glandular-villous; mucro small or absent. Petals 10–12 mm, purplish. Mericarps 6–8 mm, with ascending, white hairs; apical pits with conspicuous glandular hairs, without a furrow at the base. Beak 25–45 mm. *Sandy soil.* ● *S. Spain (Sierra Nevada).* Hs.

32. E. daucoides Boiss., *Elenchus* 28 (1838). Acaulescent perennial. Leaves up to 8 cm, eglandular-villous to glandular-pubescent, lanceolate or linear-lanceolate, pinnate, without intercalary leaflets; leaflets ovate to ovate-lanceolate, pinnatifid or dentate. Stipules whitish or pale brown. Peduncles 4–13 cm; umbels with 1–7 flowers; bracts lanceolate to ovate, white, somewhat hairy. Sepals 8–12 mm. Petals *c.* 10 mm, pale lilac or purplish, the upper 2 with a dark basal patch. Mericarps 7·5–9 mm, with ascending, white hairs; apical pits shallow, with conspicuous glandular hairs, without a furrow at the base. Beak 24–45 mm. $2n = 40, 60, 80$. *Calcareous mountain rocks. Spain.* Hs.

Plants from S.E. Spain (Sierra de Cazorla), with eglandular leaves, broad and distinct leaflets, peduncles 14–18 cm, and up to 9 flowers in the umbel, but otherwise identical with **32**, have been distinguished as **E. cazorlanum** Heywood, *Bull. Brit. Mus.* (*Bot.*) **1**: 116 (1954).

33. E. manescavi Cosson, *Ann. Sci. Nat.* ser. 3 (Bot.), **7**: 205 (1847). Acaulescent perennial. Leaves up to 30 cm, lanceolate, pinnate, without intercalary leaflets; leaflets ovate, deeply pinnatifid, with acutely dentate segments. Peduncles up to 50 cm; umbels with 5–20 flowers; bracts suborbicular, herbaceous, united to form a cupule. Sepals 12–14 mm; petals 15–20 mm, purple. Mericarps 10–12 mm, with patent, brown or white hairs; apical pits eglandular, without or with a very slight furrow at the base. Beak 40–70 mm. $2n = 40$. *Meadows and pastures.* ● *W. & C. Pyrenees, westwards from c.* 0° 20′ *W.* Ga ?Hs.

34. E. hirtum (Forskål) Willd., *Sp. Pl.* **3**: 632 (1800). Perennial; roots with globose tubers up to 1·5 cm. Stems 10–30 cm, ascending to erect, with patent, white hairs. Leaves up to 6 cm, deeply pinnatisect; lobes incise-dentate or pinnatifid; petiole at least as long as lamina in basal leaves, very short in upper cauline leaves. Umbels with 3–5 flowers; bracts broadly ovate, pale, scarious, hairy; pedicels 10–20 mm, stiffly deflexed in fruit. Sepals 6–7 mm, with short mucro; petals *c.* 8 mm, pale pink. Mericarps *c.* 5 mm; apical pits small, with a conspicuous furrow at the base. Beak (50–)80–100 mm, not spirally twisted, with numerous long, brown or yellow hairs on the inner side. *Sandy hillsides. S.E. Kriti* (*near Ierapetra*); *probably introduced.* [*Cr.] (*N. Africa, E. Mediterranean region.*)

The only representative in Europe of the section *Plumosa* Boiss., distinguished by the persistently straight beak to the mericarps, furnished with long, yellowish hairs on its inner side.

3. Biebersteinia Stephan[1]

Perennial herbs with pinnate leaves. Flowers actinomorphic, in a terminal spike. Stamens all fertile; filaments connate at the base. Style arising from the inner side of the carpel, near the base; stigmas united, capitate. Mericarps without beak.

1. B. orphanidis Boiss., *Diagn. Pl. Or. Nov.* 3(1): 113 (1853). Plant glandular-pubescent throughout. Stock stout, woody. Stem 35–50 cm, erect, stout. Leaves up to 35 cm, mostly basal but some cauline, alternate; all oblanceolate in outline, shortly petiolate, pinnate; leaflets deeply and irregularly pinnatisect, with toothed or lobed segments. Stipules large, scarious, brown. Spike 5–10 cm, compact. Sepals accrescent, unequal, the larger (outer) 15–20 mm in fruit, ovate-deltate, imbricate, connivent round the fruit. Petals shorter than sepals, pink, with long claw and obovate, fimbriate limb. Mericarps 6×4 mm, rugose, dark brown. *Once collected in S. Greece* (*Killini Oros*); *perhaps extinct in Europe.* †Gr. (*E. Anatolia.*)

LXXXIV. TROPAEOLACEAE[2]

Flowers solitary, axillary, hypogynous, zygomorphic, hermaphrodite. Sepals 5, the dorsal produced into a spur; petals 5, clawed; stamens 8, free, unequal; ovary 3-locular, each loculus with 1 pendent ovule; placentation axile; style 1, apical; stigmas 3, linear. Fruit of 3 indehiscent, 1-seeded carpels, which separate from the central axis when mature.

1. Tropaeolum L.[3]

Somewhat succulent herbs, procumbent or climbing by coiling petioles; leaves alternate; stipules usually absent.

1. T. majus L., *Sp. Pl.* 345, [1231] (1753). Glabrous annual or perennial. Leaves 4–15 cm, peltate, orbicular, subentire to somewhat angular or sinuate. Flowers 3–6 cm in diameter, orange or red to yellow, or parti-coloured; spur 2–4 cm, straight, cylindrical; limb of petal orbicular, more or less equalling claw; 3 lower petals ciliate at base. *Widely cultivated in gardens; frequently escaping and locally naturalized.* [Au Bl Ga Hs.] (*Peru to Colombia.*)

LXXXV. ZYGOPHYLLACEAE[4]

Herbs or shrubs. Leaves stipulate, usually pinnate. Flowers hermaphrodite, usually actinomorphic, (4–)5-merous. Disc present. Stamens usually twice as many as petals. Ovary superior, usually angled or winged. Fruit dry or fleshy.

1 Spiny shrub   **5. Nitraria**
1 Herbs, sometimes woody at base
  2 At least the basal leaves alternate; sepals persistent
    3 Perennial; not succulent; flowers with long pedicels
                       1. Peganum
    3 Annual; succulent; flowers subsessile   **6. Tetradiclis**
  2 Leaves all opposite; sepals deciduous
    4 Stipules spinose   **2. Fagonia**
    4 Stipules not spinose
      5 Flowers yellow; fruit spiny on the back   **4. Tribulus**
      5 Flowers white; fruit not spiny   **3. Zygophyllum**

[1] By D. A. Webb.   [2] Edit. N. A. Burges.
[3] By D. M. Moore.   [4] Edit. T. G. Tutin.

1. Peganum L.[1]

Herbs. Sepals 4–5, persistent; petals 4–5, neither cucullate nor clawed. Disc annular; stamens 12–15. Ovary globose. Fruit a capsule; seeds with endosperm.

1. **P. harmala** L., *Sp. Pl.* 444 (1753). Glabrous perennial 30–60 cm. Stems terete below, angled above, much-branched. Leaves alternate, deeply and irregularly pinnatisect, somewhat fleshy; lobes linear-lanceolate, acute; stipules small, linear, acuminate. Flowers 10–20 mm, solitary, pedicellate. Sepals linear, sometimes toothed near the base, persistent; petals greenish-white. Fruit 7–10 mm, stipitate, globose. $2n = 24$. *Steppes and dry waste places. Mediterranean region and S.E. Europe.* Bu ?Cr Gr Hs It Ju Rm Rs (W, K, E) Sa Tu [Ga Hu].

2. Fagonia L.[1]

Herbs, often with spinose stipules. Sepals 5, deciduous; petals 5, clawed, caducous. Disc inconspicuous; stamens 10. Ovary 5-angled. Fruit a capsule; seeds with endosperm.

1. **F. cretica** L., *Sp. Pl.* 386 (1753). Almost glabrous, procumbent perennial 10–40 cm. Stems branched, angled and striate. Leaves opposite, 3-foliolate, petiolate; leaflets 5–15 mm, lanceolate or linear-lanceolate, asymmetrical, rather coriaceous; stipules shorter than petioles, spinose. Flowers *c.* 10 mm, solitary, axillary. Sepals acuminate; petals purplish. Fruit 8–10 mm (including the persistent style), the 5 loculi very sharply angled and ciliate on the angles. *Dry, stony places. S. part of Mediterranean region.* Bl Cr Gr Hs Si.

3. Zygophyllum L.[1]

Herbs, sometimes woody at base. Leaves opposite. Sepals 4–5, deciduous; petals 4–5, clawed. Disc fleshy, angled; stamens 8–10. Ovary and style 4- to 5-angled. Fruit a capsule; seeds with endosperm.

1	Leaves with 4–5 pairs of leaflets	1. macropterum
1	Leaves with 1 pair of leaflets	
2	Plant arachnoid-tomentose	4. album
2	Plant glabrous	
3	Leaflets obovate-orbicular to elliptical	2. fabago
3	Leaflets linear-oblong	3. ovigerum

1. **Z. macropterum** C. A. Meyer in Ledeb., *Fl. Altaica* 2: 102 (1830). Scabrid-pubescent perennial with a woody stock. Stems *c.* 10 cm, procumbent; rhizome fleshy. Leaves pinnate; leaflets 6–8 mm, 4–5 pairs, elliptical; rhachis rather wide, ending in a mucro; stipules membranous. Flowers solitary in the dichotomies of the stem. Sepals obovate-oblong; petals spathulate, orange. Fruit 15–25 mm, very broadly winged on the angles. *Saline places. S. Ural.* Rs (C, E). (*W. & C. Asia.*)

2. **Z. fabago** L., *Sp. Pl.* 385 (1753). Glabrous perennial. Stems 60–100 cm, erect. Leaves with 1 pair of obovate-orbicular to elliptical, asymmetrical, rather fleshy leaflets; rhachis forming a short projection between the leaflets; stipules herbaceous. Flowers solitary, axillary. Sepals oblong-ovate; petals oblong-ovate, obtuse or weakly emarginate, cream in upper half, orange below. Fruit 20–35 mm, oblong-cylindrical, at least 3 times as long as wide. *Dry places. S.E. Europe, from c.* 28° *E. eastwards; naturalized locally in the W. Mediterranean region.* Rm Rs (W, K, E) [Ga Hs Sa].

3. **Z. ovigerum** Fischer & C. A. Meyer ex Bunge, *Arb. Naturf.-Ver. Riga* 1: 200 (1847). Like **2** but leaflets linear-oblong; fruit 10–15 mm, about as long as wide. *Dry, saline soils. S.E. Russia, W. Kazakhstan.* Rs (E). (*W.C. Asia.*)

4. **Z. album** L. fil., *Dec. Prim. Pl. Rar. Hort. Upsal.* 11 (1762). Greyish, arachnoid-tomentose small shrub. Stems *c.* 40 cm, spreading, the smaller ones herbaceous and rather succulent. Leaves with 1 pair of elliptical or obovate, fleshy leaflets and a fleshy, oblong petiole; stipules scarious. Flowers solitary, axillary. Sepals elliptical; petals obovate, white, clawed. Fruit 5–10 × 4–7 mm, sharply 5-angled. *Sea-shores. Kasos and islets around Kriti; N.E. Spain.* Cr Hs. (*W. Asia, N. Africa.*)

4. Tribulus L.[1]

Herbs. Sepals 5, deciduous; petals 5, fugacious. Disc annular, 10-lobed; stamens 10. Ovary 5-lobed. Fruit splitting into 5 indehiscent portions; seeds without endosperm.

1. **T. terrestris** L., *Sp. Pl.* 387 (1753). Pubescent, procumbent annual 10–60 cm. Stems simple or freely branched. Leaves opposite, often unequal, paripinnate; pinnae 5–8 pairs, elliptical or oblong-lanceolate. Flowers 4–5 mm; petals yellow. Fruit of 5 stellately arranged, hard, rugose carpels which are keeled and tuberculate on the back, and with 2 or more stout spines on the sides. *Dry open habitats, often as a weed. S. Europe, extending locally northwards to N.W. France, S.E. Czechoslovakia and E.C. Russia.* Al Au Bl Bu Co Cr Cz Ga Gr Hs Hu It Ju Lu Rm Rs (C, W, K, E) Sa Si Tu.

Varies from green and rather sparsely appressed-pubescent to almost silvery-tomentose.

Two subspecies, based on the degree of hairiness and development of spines on the fruit, are sometimes recognized. These do not appear to have any discrete patterns of geographical distribution and are therefore best regarded as varieties.

5. Nitraria L.[1]

Much-branched, spiny shrub. Calyx 5-fid, fleshy, persistent; petals 5, cucullate, not clawed. Disc inconspicuous; stamens 15. Ovary oblong-pyramidal. Fruit fleshy; seeds without endosperm.

1. **N. schoberi** L., *Syst. Nat.* ed. 10, 2: 1044 (1759). Glabrous. Stems 50–200 cm. Leaves alternate or fasciculate, simple, oblanceolate, obtuse, fleshy, sessile; stipules membranous, caducous. Flowers *c.* 4 mm, shortly pedicellate, in dichasia. Calyx-lobes triangular; petals greenish-white. Fruit 10–12 mm, ovoid-conical. *Saline soils. S.E. Europe, from S.E. Romania to W. Kazakhstan; local.* Rm Rs (K, E). (*W. & C. Asia.*)

6. Tetradiclis Steven ex Bieb.[1]

Herb. Sepals 4, persistent; petals 4, very shortly clawed. Disc annular, inconspicuous; stamens 4. Ovary 4-angled, depressed in the middle. Fruit a capsule; seeds with little endosperm.

1. **T. tenella** (Ehrenb.) Litv., *Trav. Mus. Bot. Acad. Pétersb.* 3: 122 (1907) (*T. salsa* C. A. Meyer). Glabrous annual up to 10 cm. Leaves succulent, pinnatisect or laciniate, the lowest opposite, the others alternate. Flowers 0·5–1 mm, subsessile in the axils of the leaf-like bracts, forming a scorpioid, spicate inflorescence. Fruit *c.* 3 mm in diameter. *Saline places. S. Ukraine; S.E. Russia (Volga delta).* Rs (W, K, E). (*C. & S.W. Asia.*)

[1] By T. G. Tutin.

LXXXVI. LINACEAE[1]

Herbs or small shrubs. Leaves exstipulate, simple, entire. Inflorescence cymose. Flowers 4- or 5-merous, actinomorphic. Sepals free; petals usually free (sometimes joined at base); fertile stamens in one whorl, sometimes with a whorl of staminodes. Ovary superior, usually 8- or 10-celled. Styles usually free. Fruit a loculicidal capsule; seeds usually 1 in each loculus.

Flowers 5-merous; seeds flat	**1. Linum**
Flowers 4-merous; seeds ovoid	**2. Radiola**

1. Linum L.[2]

Herbs or small shrubs. Leaves sessile, usually narrow, 1-veined or parallel-veined. Flowers 5-merous. Sepals entire. Petals clawed, longer than the sepals. Stamens 5, alternating with 5 tooth-like staminodes; filaments united at base. Capsule dehiscing with 10 valves, often with a short beak. Seeds flat.

Measurements of the capsule exclude the beak. Pedicel characters refer to the mature fruiting stage. The inflorescence is basically cymose, but in many species (Sect. *Dasylinum*, *Linastrum* and *Linum*) well-developed inflorescences have branches which are pseudoracemose (cincinni). The difference between an irregular dichasium and an inflorescence composed of cincinni seems to be at least in part a question of individual development, and is not therefore of great taxonomic importance.

Unless otherwise stated, the species occur in rather open habitats on rocks or well-drained, calcareous or sandy soils.

1 Leaves all opposite; petals less than 7 mm, white (Sect. *Cathartolinum*)   **36. catharticum**
1 At least upper cauline leaves alternate; petals usually coloured, if white then more than 7 mm
 2 At least upper leaves with a pair of glands at the base; stem with narrow wings decurrent from leaf-bases (Sect. *Syllinum*)
  3 Annual; leaf-margins rough, finely serrulate   **12. nodiflorum**
  3 Perennial; leaf-margins smooth, entire
   4 Leaves hairy
    5 Petals white   **10. leucanthum**
    5 Petals yellow   **11. pallasianum**
   4 Leaves glabrous
    6 Shrub up to 1 m, with thick, persistent leaves often in dense clusters at the ends of woody branches   **1. arboreum**
    6 Plants with variably developed woody stock, sometimes with basal leaf-rosettes; flowering stems not more than 60 cm, annual
     7 Inflorescence subcapitate   **2. capitatum**
     7 Inflorescence ± laxly cymose, or flowers solitary
         (3–9). flavum group
 2 Leaves without glands at the base; stem terete or striate
  8 Plant pubescent, with hairs more than 1 mm; uppermost leaves usually with glandular margins (Sect. *Dasylinum*)
   9 Annual, usually less than 20 cm   **28. pubescens**
   9 Perennial, usually more than 20 cm
    10 Plant woody at base; flowering stems slender; short non-flowering shoots present at time of flowering
         27. spathulatum
    10 Plant not woody at base; flowering stems stout; non-flowering shoots absent at time of flowering
     11 Middle cauline leaves usually glandular-ciliate; petals pink (blue when dry)   **25. viscosum**
     11 Middle cauline leaves not glandular-ciliate; petals blue
         26. hirsutum

  8 Plant glabrous or with hairs less than 1 mm; uppermost leaves eglandular
   12 Capsule less than 3·5 mm; petals yellow, pink or white (Sect. *Linastrum*)
    13 Petals pink or white; capsule more than 2·7 mm
     14 Homostylous; petals 2–2½ times as long as sepals
         32. tenuifolium
     14 Heterostylous; petals 3–4 times as long as sepals
         33. suffruticosum
    13 Petals yellow; capsule less than 2·7 mm
     15 Perennial; lower leaves opposite; sepals scarcely exceeding capsule   **29. maritimum**
     15 Annual; lower leaves alternate; sepals much exceeding capsule
      16 Leaf-margins smooth, entire
       17 Petals 4–6 mm; stems less than 30 cm   **30. trigynum**
       17 Petals 8–18 mm; stems more than 30 cm   **31. tenue**
      16 Leaf-margins rough, finely serrulate (but sometimes revolute)
       18 Leaves narrowly lanceolate, more than 1 mm wide
         34. strictum
       18 Leaves linear or setaceous, often with revolute margins, less than 1 mm wide   **35. setaceum**
   12 Capsule more than 3·5 mm; petals blue (rarely red or white) (Sect. *Linum*)
    19 Leaf-margins scabrid
     20 Homostylous; petals red or pink   **22. decumbens**
     20 Heterostylous; petals blue
      21 Annual; stigmas capitate   **17. virgultorum**
      21 Perennial; stigmas clavate
       22 Leaves with 3–5 veins; petals 20–24 mm   **14. nervosum**
       22 Leaves with 1 vein; petals 12–18 mm   **15. aroanium**
    19 Leaf-margins smooth
     23 Bracts with scarious margins; sepals 10–14 mm; petals 25–40 mm   **13. narbonense**
     23 Bracts without scarious margins; sepals 3·5–9 mm; petals 10–25 mm
      24 Styles united almost to the apex   **16. hologynum**
      24 Styles free almost to the base
       25 Stigmas capitate; usually heterostylous
         (18–21). perenne group
       25 Stigmas linear or clavate; homostylous
        26 Sepals long-acuminate; petals red or pink
         22. decumbens
        26 Sepals shortly acuminate; petals blue
         27 Usually biennial or perennial; stems several; capsule 4–6 mm   **23. bienne**
         27 Annual; stem usually solitary; capsule 6–9 mm
          24. usitatissimum

SECT. SYLLINUM Griseb. Stem with narrow wings decurrent from leaf-bases. Leaves alternate, with a pair of glands at base. Sepals sometimes glandular-ciliate. Petals slightly joined at base of claw, usually yellow. Stigmas usually linear or oblong-linear. Homostylous or heterostylous.

1. L. arboreum L., *Sp. Pl.* 279 (1753). Glabrous shrub up to 1 m. Leaves (5–)10–20 × 3–10 mm, spathulate, thick, persistent, 1-veined, with cartilaginous margins, often crowded in more or less dense rosettes. Inflorescence usually few-flowered, rather compact. Sepals 5–8 mm, lanceolate, acuminate, not ciliate. Petals 12–18 mm, yellow. Capsule (5–)6–8 × 5–6 mm, beaked, about equalling sepals. *Limestone rocks. S. Aegean region.* Cr Gr. (*S.W. Anatolia and Rodhos.*)

L. caespitosum Sibth. & Sm., *Fl. Graec. Prodr.* **1**: 216 (1806), is smaller in all its parts, particularly the sepals (4·5 mm), but

---

[1] Edit. S. M. Walters.   [2] By D. J. Ockendon and S. M. Walters.

otherwise does not differ from **1**. It occurs in Kriti, mainly on conglomerate rocks. **L. doerfleri** Rech. fil., *Österr. Bot. Zeitschr.* **84**: 147 (1935), also described from Kriti, differs from **1** in its minutely serrulate (not entire) sepals and somewhat smaller capsule. The status of both these variants is doubtful.

2. L. capitatum Kit. ex Schultes, *Östreichs Fl.* ed. 2, **1**: 528 (1814). Stock woody, with well-developed rhizomes often terminating in leaf-rosettes. Flowering stems 10–40 cm, robust, green, angular. Rosette-leaves oblong-spathulate, obtuse; cauline linear-lanceolate, acute. Inflorescence a subcapitate cyme, 5- to 10(–15)-flowered. Sepals 5–6 mm, oblong-lanceolate, acuminate. Petals 15–20 mm, yellow, with obovate, obtuse limb. Capsule *c*. 5 mm; beak *c*. 1 mm. Heterostylous. *Rocky slopes on mountains.* ● *Balkan peninsula; C. & S. Italy.* Al Bu Gr It Ju.

(3–9). **L. flavum** group. Glabrous perennials with a variably developed woody stock and erect or ascending flowering stems 5–60 cm. Sepals lanceolate, usually glandular-ciliate. Petals 10–35 mm, yellow. Capsule globose; beak 1–2·5 mm. Heterostylous.

A difficult group, occurring in S., C. and E. Europe, from Spain and S. Germany eastwards. Although the extreme taxa differ strikingly in habit, much of the variation is not clearly discontinuous, and it is not clear how much variation is genetically based.

1 Petals (22–)25–35 mm, gradually narrowed into claw; beak of capsule *c*. 2 mm   **8. campanulatum**
1 Petals 10–25(–30) mm, usually ± abruptly narrowed into claw; beak of capsule usually *c*. 1 mm
  2 Inflorescence with 1–9 flowers; flowering stems not more than 20 cm   **9. elegans**
  2 Inflorescence with more than 10 flowers; flowering stems up to 60 cm
    3 Inflorescence with (20–)25–40 flowers; stock erect or ascending, little-branched
      4 Sepals (5–)6–8 mm, scarcely exceeding capsule   **3. flavum**
      4 Sepals 8–10 mm, up to twice as long as capsule   **4. thracicum**
    3 Inflorescence usually with 10–20 flowers; stock much-branched, often with ± slender rhizomes
      5 Sepals 4–6·5 mm, shortly acuminate, usually not exceeding capsule   **7. ucranicum**
      5 Sepals 5–9 mm, narrowly acuminate, clearly exceeding capsule
        6 Leaves of non-flowering stems obtuse; petals 20–25 (–30) mm   **6. uninerve**
        6 Leaves of non-flowering stems acute or subacute; petals usually less than 20 mm   **5. tauricum**

3. L. flavum L., *Sp. Pl.* 279 (1753). Robust, with erect flowering stems up to 60 cm from a compact stock, and few or no non-flowering rosettes. Leaves 20–35 × 3–12 mm, 3(–5)-veined, the lower spathulate, the upper lanceolate. Inflorescence branched, usually with 25–40 flowers. Sepals (5–)6–8 mm, lanceolate, acuminate. Petals *c*. 20 mm, with obovate limb and relatively short claw. Capsule 5–6 mm, beak *c*. 1 mm. *C. & S.E. Europe, extending to N.E. Italy and northwards to c. 55°N. in C. Russia.* Al Au Bu Cz Ge Hu It Ju Po Rm Rs (C, W, K, E).

L. basarabicum (Săvul. & Rayss) Klokov ex Juz. in Komarov, *Fl. URSS* **14**: 133 (1949), from Moldavia and W. Ukraine, is said to differ mainly in its oblong-elliptical, 5-veined lower leaves and inflorescence with 2–20(–25) flowers. It could be placed here or with **5**.

4. L. thracicum Degen, *Österr. Bot. Zeitschr.* **43**: 55 (1893). Like **3** but stems usually 20–40 cm, more numerous, slender, ascending, and sepals 8–10 mm, narrowly acuminate, 1½–2 times as long as capsule. ● *Balkan peninsula.* Bu Gr Ju Tu.

L. rhodopeum Velen., *Sitz.-Ber. Böhm. Ges. Wiss.* (*Math.-Nat. Kl.*) **1895**(37): 3 (1896), from S. Bulgaria and Samothraki, seems to differ only in its few stems and petals 16–18 mm. **L. turcicum** Podp., *Verh. Zool.-Bot. Ges. Wien* **52**: 637 (1902), from Greece (Olimbos), is a robust plant with ovate, 5-veined leaves, narrowly-acuminate calyx and beak of capsule *c*. 2·5 mm. The status of these two variants is doubtful.

5. L. tauricum Willd., *Enum. Pl. Hort. Berol.* 339 (1809) (*L. serbicum* Podp.; incl. *L. bulgaricum* Podp., *L. pseudelegans* Podp., *L. orientale* sensu Hayek). Stock much-branched, often woody, with numerous non-flowering rosettes and flowering stems up to 40 cm. Lower leaves narrowly spathulate, more or less acute, usually 3-veined; upper cauline leaves often lanceolate, 1(–3)-veined. Inflorescence usually 10- to 20-flowered. Sepals 6–8 mm, narrowly acuminate, much exceeding capsule. Beak of capsule *c*. 1 mm. *S.E. Europe.* Al Bu Gr Ju Rm Rs (W, K) Tu.

The plant from Krym described by Willdenow is said by Russian authors to differ from other material from the U.S.S.R. in its lanceolate cauline leaves and persistent basal leaves. The more widespread plant in W. Ukraine, Moldavia and S.E. Romania (and also in Krym) has been distinguished as **L. linearifolium** (Lindem.) Jáv., *Magyar Bot. Lapok* **9**: 156 (1910), with linear-spathulate basal and linear or linear-lanceolate cauline leaves, the former often withered at time of flowering.

L. euboeum Bornm., *Bot. Jahrb.* **59**: 443 (1925), described from S.E. Greece (Evvoia), has 5-veined leaves and short, triangular-lanceolate sepals.

6. L. uninerve (Rochel) Jáv., *Magyar Bot. Lapok* **9**: 156 (1910). Like **5** but with dense central stock and thin, creeping, woody rhizomes, spathulate basal leaves *c*. 10 mm wide, and petals 20–25(–30) mm. ● *C. & S. Romania.* Rm.

7. L. ucranicum Czern., *Consp. Pl. Charc.* 12 (1859) (incl. *L. uralense* Juz.). Like **5** but sepals not more than 6·5 mm, shortly acuminate and not exceeding capsule. *From E. Ukraine across S.C. Russia to S. Ural.* Rs (C, W, E).

8. L. campanulatum L., *Sp. Pl.* 280 (1753). Stock woody, usually somewhat branched, and often with slender rhizomes and non-flowering rosettes. Flowering stems up to 25 cm, slender, erect or ascending. Lower leaves spathulate, the upper cauline usually oblanceolate, all, or at least the cauline, 1-veined. Inflorescence usually 3- to 5-flowered, subcorymbose. Sepals narrowly acuminate, enlarging in fruit and much exceeding capsule. Petals (22–)25–35 mm, with long claw, giving the appearance of a tubular corolla. Stigmas oblong-linear. Beak of capsule *c*. 2 mm. ● *W. Mediterranean region, from E. Spain to C. Italy.* Ga Hs It.

9. L. elegans Spruner ex Boiss., *Diagn. Pl. Or. Nov.* **3**(1): 99 (1853). Dwarf plant with woody, branched stock and compact basal leaf-rosettes. Flowering stems usually less than 15 cm. Lower leaves obovate to spathulate, 3-veined, thick and with conspicuous hyaline margin. Inflorescence with (1–)3–7 flowers. Sepals 7–8(–10) mm, narrowly lanceolate, acuminate. Petals 15–20 mm, obovate, with relatively short claw. Capsule much shorter than sepals. *Rocky places on mountains.* ● *Balkan peninsula.* Al ?Bu Gr Ju.

Perhaps conspecific with **L. boissieri** Ascherson & Sint. ex Boiss., *Fl. Or., Suppl.* 137 (1888), from N.W. Anatolia.

L. goulimyi Rech. fil., *Anzeig. Akad. Wiss. (Wien)* **93**: 96 (1956), from Greece (Evvoia), has stems up to 30 cm, and inflorescence with up to 10 flowers. Similar plants have been called *L. elegans* var. *elatius* Halácsy, and are not clearly distinguishable from few-flowered variants of **5**.

L. dolomiticum Borbás, *Term.-Tud. Közl.* **29**: 208 (1897), restricted to dolomitic rocks in Hungary (near Pilisszentiván), differs principally in its shorter sepals (6–7 mm) and petals (10–16 mm). It has $2n = 28$.

10. L. leucanthum Boiss. & Spruner in Boiss., *Diagn. Pl. Or. Nov.* **1**(1): 55 (1843) (incl. *L. gyaricum* Vierh.). Like **9** but leaves more or less densely covered with very short, stiff hairs, and petals white. Heterostylous. ● *S.E. Greece; Aegean region (Yioura).* Gr.

The original description mentions the remarkable indumentum, but most gatherings from Imittos (and some elsewhere) contain a proportion of subglabrous plants. It is possible that such plants are the result of hybridization with other species of this section (particularly **9**), but there is insufficient evidence to support this view, and later authors have amended the description to include both glabrous and hairy variants.

11. L. pallasianum Schultes in Roemer & Schultes, *Syst. Veg.* **6**: 758 (1820) (incl. *L. borzeanum* E. I. Nyárády, *L. czerniaevii* Klokov). Caespitose, with branched, woody stock and flowering stems up to 30 cm. Basal leaves linear to linear-spathulate, subacute; cauline linear. Stems, at least in lower part, and leaves more or less densely grey-pubescent. Inflorescence usually with 2–4 flowers. Sepals (4–)6–9 mm, lanceolate, acuminate, subglabrous, with hyaline, glandular-denticulate margin. Petals 15–20 mm, pale yellow. Capsule 4–6 mm. Heterostylous. $2n = 28$. *S. & E. Ukraine, S.E. Romania.* Rm Rs (?C, W, K, E).

12. L. nodiflorum L., *Sp. Pl.* 280 (1753) (incl. *L. luteolum* Bieb.). Glabrous annual; stems up to 40 cm, solitary or few, somewhat branched, winged. Lower leaves spathulate, upper linear, 1(–3)-veined, all with small brown glands at base and with rough, finely serrulate margin. Inflorescence very lax, the branches pseudoracemose, with subsessile terminal and axillary flowers. Sepals 8–13 mm, linear-subulate, acuminate. Petals *c.* 20 mm, yellow, with long claw. Capsule 5–6 mm. Homostylous. *Mediterranean region, eastwards from S.E. France; Balkan peninsula; Krym.* Al Bu Cr Ga Gr It Ju Rs (K) Tu.

Sect. LINUM. Leaves alternate, glabrous, without basal glands. Sepals eglandular. Petals free, blue, purple or pink. Stigmas capitate, clavate or linear. Homostylous or heterostylous.

13. L. narbonense L., *Sp. Pl.* 278 (1753). Glabrous perennial; stems up to 50 cm, erect or ascending. Leaves 1–5(–10) mm wide, linear or lanceolate, long-acuminate, 1- to 3(–5)-veined. Sepals 10–14 mm, lanceolate, long-acuminate, with minutely serrulate-ciliate, scarious margins. Petals 2½–3 times as long as sepals, bright blue; stigmas linear. Capsule 7–9 mm, subglobose, with a narrow beak 2–2·5 mm. Heterostylous. $2n = 30$. *W. & C. Mediterranean region, N. Spain, N.E. Portugal.* Bl Co Ga Hs It Ju Lu Si [He].

14. L. nervosum Waldst. & Kit., *Pl. Rar. Hung.* **2**: 109 (1802–3) (incl. *L. jailicola* Juz.). Glabrous or puberulent perennial; stems up to 60 cm, erect. Leaves 3–6(–10) mm wide, lanceolate-acuminate, 3- to 5-veined, with scabrid margins. Sepals 7–11 mm, ovate-lanceolate, narrowly acuminate, with a setaceous apex and narrow, scarious, ciliate margins. Petals 2–2½ times as long as sepals, blue. Stigmas clavate. Capsule 6–10 mm, subglobose. Heterostylous. $2n = 18$. *S.E. Europe, extending northwards to 53°N. in S.C. Russia.* Bu Ju Rm Rs (C, W, K).

15. L. aroanium Boiss. & Orph. in Boiss., *Diagn. Pl. Or. Nov.* **3**(1): 96 (1853). Glabrous perennial; stems up to 40 cm, rather thin and flexuous, decumbent or ascending. Leaves 1–3(–4) mm wide, lanceolate-acuminate, 1-veined, with scabrid margins. Sepals 5–7 mm, lanceolate, narrowly acuminate, with narrow, scarious, ciliate margins. Petals 2½–3 times as long as sepals, blue. Stigmas clavate. Capsule 5–7 mm, subglobose. Heterostylous. *Mountains of S. & C. Greece.* Gr.

16. L. hologynum Reichenb., *Fl. Germ. Excurs.* 833 (1832). Glabrous perennial; stems up to 40 cm, ascending or erect. Leaves 0·5–1·5 mm wide, linear or filiform, 1-veined, with smooth or sometimes scabrid margins. Sepals 5–8 mm, lanceolate, long-acuminate, with scarious, not ciliate, margins; lower part of mid-vein of sepals very prominent and more or less keel-like. Petals 2½ times as long as sepals, blue, violet, or pinkish-purple, Styles united almost to the apex; stigmas linear. Capsule 4–6 mm, subglobose. Homostylous. ● *Balkan peninsula, just extending into S.W. Romania.* Al Bu Gr Ju Rm.

Distinguished from all other European species of *Linum* by its united styles.

17. L. virgultorum Boiss. & Heldr. ex Planchon, *London Jour. Bot. (Hooker)* **7**: 172 (1848). Glabrous annual; stems 5–20 cm, erect or ascending below. Leaves up to 2 mm wide, linear-lanceolate, 1-veined, with scabrid margins. Sepals 6–8 mm, lanceolate-acuminate, with scabrid margins. Petals 2–2½ times as long as sepals, blue. Stigmas capitate. Capsule 4·5–5 mm, subglobose. Heterostylous. *S. Greece (Athinai).* Gr. (*Anatolia.*)

(18–21). L. perenne group. Glabrous perennials, sometimes caespitose and woody at the base; stems up to 60 cm, erect, ascending or decumbent. Leaves 1–3(–4) mm wide, linear or linear-lanceolate, entire, 1- to 3-veined. Sepals 3·5–6 mm, unequal, the outer narrower than the inner and with an acute or acuminate apex, the inner with a rounded or mucronate apex and entire, scarious margins. Petals 3–4 times as long as sepals, blue. Stigmas capitate. Capsule 3·5–7(–8) mm, subglobose, beak very short or absent. Usually heterostylous.

A difficult group in need of further study. There is considerable variation in habit, width of leaves, number of flowers in the inflorescence, shape and size of petals, sepals and capsules, and posture of fruiting pedicels. At least some of this variation can be environmentally induced. The variation of most characters is almost continuous and intermediates between taxa can be found; most of these taxa have therefore been treated here as subspecies, since the incomplete morphological separation is often accompanied by geographical isolation.

1 Homostylous, with anthers and stigmas at about the same
   height   **21. leonii**
1 Heterostylous, with anthers and stigmas at different heights
 2 Leaves very crowded, 6 mm or less   **20. punctatum**
 2 Leaves moderately crowded, the longest at least 10 mm
  3 Pedicels deflexed or flexuous   **19. austriacum**
  3 Pedicels erect, ± straight   **18. perenne**

18. L. perenne L., *Sp. Pl.* 277 (1753). Stems 10–60 cm, decumbent, ascending or erect. Middle cauline leaves 1- to 3-veined. Inflorescence usually many-flowered. Inner sepals acute or obtuse. Pedicels erect. Capsules 5–8 mm. Heterostylous. *C. & E. Europe, extending locally westwards to Britain and the*

Pyrenees. Al Au Br Bu Cz Ga Ge Gr He Hs Hu It Ju Po Rm Rs (N, C, W, E).

1 Upper cauline leaves 3-veined; inner sepals twice as long as wide   **(e) subsp. extraaxillare**
1 Upper cauline leaves 1-veined; inner sepals less than twice as long as wide
  2 Inner sepals very obtuse, longer than the outer (lowlands)
    3 Stems erect from an ascending base; pollen grains with furrows   **(a) subsp. perenne**
    3 Stems usually decumbent or ascending; pollen grains with pores   **(b) subsp. anglicum**
  2 Inner sepals acute or obtuse, equalling the outer (mountains)
    4 Pollen grains with furrows   **(c) subsp. alpinum**
    4 Pollen grains with pores   **(d) subsp. montanum**

(a) Subsp. **perenne**: Stems (20–)30–60 cm, erect from an ascending base. Middle cauline leaves 1–2·5 mm wide, 1-veined or obscurely 3-veined. Inner sepals 4·5–5·5 mm, very obtuse, exceeding the outer sepals by 0·5–1 mm. Capsules 5–7 mm. Pollen grains with furrows. $2n = 18$. *C. & E. Europe.*

(b) Subsp. **anglicum** (Miller) Ockendon, *Feddes Repert.* **74**: 20 (1967) (*L. anglicum* Miller): Like subsp. (a) but stems decumbent or ascending, occasionally suberect; pollen grains with pores. $2n = 36$. ● *Britain.*

(c) Subsp. **alpinum** (Jacq.) Ockendon, *Feddes Repert.* **74**: 20 (1967) (*L. alpinum* Jacq.; incl. *L. boreale* Juz.): Stems 5–30 cm, decumbent or ascending. Middle cauline leaves 1–3 mm wide. Inner sepals 4·5–6 mm, equalling the outer sepals. Pollen grains with furrows. $2n = 18$. *Pyrenees, Alps, Appennini, Rodopi; N. Ural.*
Plants from the Pyrenees are particularly variable. Some have the wide leaves and long inner sepals characteristic of subsp. (e), while others have suberect or flexuous pedicels.

(d) Subsp. **montanum** (DC.) Ockendon, *Feddes Repert.* **74**: 20 (1967) (*L. montanum* Schleicher ex DC.): Stems 20–40 cm, ascending or erect. Middle cauline leaves 1–3 mm wide. Inner sepals 5·6–6·5 mm, acute, equalling the outer sepals. Pollen grains with pores. $2n = 36$. ● *Jura and N. Alps.*

(e) Subsp. **extraaxillare** (Kit.) Nyman, *Consp., Suppl.* **2**(1): 71 (1889) (*L. extraaxillare* Kit.): Like subsp. (d) but upper cauline leaves 2–4 mm wide, 3-veined; pollen grains with furrows. $2n = 18$. ● *Carpathians; mountains of Balkan peninsula.*

19. **L. austriacum** L., *Sp. Pl.* 278 (1753). Stems (6–)10–60 cm, erect or ascending. Middle cauline leaves 1- or obscurely 3-veined. Inflorescence many-flowered. Inner sepals acute or obtuse. Pedicels deflexed or flexuous. Capsules 3·5–7·5 mm. Heterostylous. *C. & S. Europe.* Al Au Bu Cz Ga Ge Gr He Hs Hu It Ju Po Rm Rs (C, W, K, E) Si Tu [Da].

1 Capsules 5–7·5 mm   **(b) subsp. collinum**
1 Capsules 3·5–5 mm
  2 Leaves less than 1 mm wide; pedicels flexuous
    (c) subsp. euxinum
  2 Leaves more than 1 mm wide; pedicels deflexed
    (a) subsp. austriacum

(a) Subsp. **austriacum**: Stems (20–)30–60 cm. Middle cauline leaves 1–3 mm wide. Inner sepals 3·5–5·5 mm, obtuse. Pedicels deflexed. Capsule 3·5–5 mm. $2n = 18$. *Throughout most of the range of the species but absent from the south-west.*
(b) Subsp. **collinum** Nyman, *Consp.* 125 (1878) (*L. collinum* Guss., nom. provis.): Stems 6–30(–40) cm. Middle cauline leaves 1–3 mm wide. Inner sepals 4·5–5·5 mm, acute or obtuse. Pedicels deflexed. Capsules 5–7·5 mm. $2n = 18$. *Mediterranean region, C. France.*
A heterogenous subspecies comprising a few very disjunct populations.

(c) Subsp. **euxinum** (Juz.) Ockendon, *Feddes Repert.* **74**: 21 (1967) (*L. euxinum* Juz.; incl. *L. marschallianum* Juz.): Stems 12–36 cm, densely leafy. Leaves 0·1–1 mm wide, revolute. Inner sepals 3–4 mm, very obtuse. Pedicels flexuous or squarrose. Capsules 4–5·5 mm. *Krym.*

20. **L. punctatum** C. Presl in J. & C. Presl, *Del. Prag.* 58 (1822) (incl. *L. pycnophyllum* Boiss. & Heldr.). Stems up to 15 cm, very crowded, very densely leafy throughout. Leaves 3–5(–7) mm, narrowly ovate, 1-veined. Flowers solitary or in terminal groups of 2 or 3. Inner sepals 5–6 mm, obtuse. Pedicels erect or somewhat flexuous. Capsules 5·5–7 mm. Heterostylous. *Mountains of Sicilia and Greece.* Gr Si.

21. **L. leonii** F. W. Schultz, *Flora* (*Regensb.*) **21**: 644 (1838). Stems 7–30(–40) cm, erect when young, later decumbent. Middle cauline leaves 1-veined. Inflorescence 1- to 6(–12)-flowered. Inner sepals 3·5–6 mm, acute or obtuse. Pedicels deflexed, flexuous or suberect. Capsules 5–7(–8) mm. Homostylous. $2n = 18$. ● *France, W. Germany.* Ga Ge.

22. **L. decumbens** Desf., *Fl. Atl.* **1**: 278 (1798). Glabrous annual; stems up to 40 cm, decumbent, ascending or suberect. Leaves 1–3(–5) mm wide, linear or linear-lanceolate, acute, 1- to 3(–5)-veined, usually with scabrid margins. Sepals 7–9 mm, ovate, long-acuminate, with wide, scarious margins below and narrow, ciliate margins above. Petals 2–2½ times as long as sepals, pink or red. Stigmas linear. Capsule 4·5–6·5 mm, subglobose; beak 0·5–1 mm, acuminate. Homostylous. *S. Italy, Sardegna, Sicilia.* It Sa Si.

23. **L. bienne** Miller, *Gard. Dict.* ed. 8, no. 8 (1768) (*L. angustifolium* Hudson). Biennial or perennial (rarely annual); stems 6–60 cm, usually branched, slender, ascending or erect. Leaves 0·5–1·5 mm wide, linear or linear-lanceolate, acuminate, 1- to 3-veined. Sepals 4–5·5 mm, subequal, ovate-acuminate, with a conspicuous midvein; margin of inner sepals scarious and ciliate, margin of outer sepals entire. Petals 2–3 times as long as sepals, blue. Stigmas linear. Capsule 4–6 mm, subglobose; beak *c.* 1 mm, acuminate. Homostylous. $2n = 30$. *W. & S. Europe, northwards to 54° N. in Britain.* Al Bl Br Bu Co Cr Ga Gr Hb Hs It Ju Lu Rs (K) Sa Si Tu.

24. **L. usitatissimum** L., *Sp. Pl.* 277 (1753) (incl. *L. crepitans* (Boenn.) Dumort., *L. humile* Miller). Like **23** but more robust; always annual; stem usually single; leaves 1·5–3 mm wide, 3-veined; sepals 6–9 mm; capsule 6–9 mm; beak *c.* 1 mm. *Not known wild. Of uncertain origin, perhaps derived from* **23**. *Formerly cultivated throughout most of Europe for the fibre (flax) and oil from the seed (linseed); now much less commonly grown, especially in the west, but still recorded as a casual throughout Europe.*

Sect. DASYLINUM (Planchon) Juz. Leaves alternate, pubescent, without basal glands. Sepals glandular-ciliate. Petals free, blue or pink. Stigmas linear. Heterostylous.

25. **L. viscosum** L., *Sp. Pl.* ed. 2, 398 (1762). Pubescent perennial; stems up to 60 cm, erect or ascending below. Leaves 3–8(–11) mm wide, lanceolate or ovate-lanceolate, 3- to 5-veined with glandular margins (margins of middle and lower leaves occasionally eglandular). Sepals 6–9 mm, lanceolate. Petals 2½–3 times as long as sepals, pink (blue when dry). Capsule 3·5–4·5 mm, subglobose; beak *c.* 0·5 mm, shortly acuminate. ● *S. & S.C. Europe, from N. Spain to S. Germany and N. Jugoslavia; rather local and mainly in the mountains.* Au Ga Ge Hs It Ju.

26. L. hirsutum L., *Sp. Pl.* 277 (1753). Pubescent perennial; stems up to 45 cm, robust, erect, many-flowered. Leaves 10–45 mm, up to 10 mm wide, ovate or oblong, acute or obtuse, 3- to 5-veined, eglandular, except the uppermost. Sepals 8–12 mm, ovate-lanceolate. Petals about 2½ times as long as sepals, blue. Capsule 3·5–5 mm, subglobose; beak *c.* 0·5 mm, shortly acuminate. $2n = 16$. *E.C. & E. Europe, northwards to Czechoslovakia and S.C. Russia.* Al Au Bu Cz Gr Hu Ju Po Rm Rs (C, W, K) Tu.

(a) Subsp. **hirsutum** (incl. *L. lanuginosum* Juz.): Stems pubescent above; middle cauline leaves oblong, usually with 5 veins. *Throughout the range of the species.*

(b) Subsp. **glabrescens** (Rochel) Soó, *Magyar Biol. Int. Munkái* 6: 132 (1933) (*L. pannonicum* A. Kerner): Stem almost glabrous above; middle cauline leaves linear-oblong, usually with 3 veins. *Czechoslovakia, Hungary, Jugoslavia.*

27. L. spathulatum (Halácsy & Bald.) Halácsy, *Consp. Fl. Graec.* 1: 258 (1900) (*L. hirsutum* subsp. *spathulatum* (Halácsy & Bald.) Hayek). Like **26** but woody at the base; stems up to 25 cm, slender, decumbent, few-flowered. ● *Mountains of N. Greece and S. Albania.* Al Gr.

28. L. pubescens Banks & Solander in A. Russell, *Nat. Hist. Aleppo* ed. 2, **2**: 268 (1794). Pubescent annual with 1–2 stems on each plant; stems 7–20(–35) cm, slender, erect, few-flowered. Leaves up to 20 × 5 mm, lanceolate. Sepals 8–10 mm, narrowly lanceolate. Petals twice as long as sepals, pink. Capsule 3·5–4·5 mm, subglobose. *Albania, Greece, Kriti.* Al Cr Gr.

Sect. LINASTRUM (Planchon) Bentham. Leaves mostly alternate, glabrous, without basal glands. Sepals glandular-ciliate. Petals free, yellow, pink or white. Stigmas linear or capitate, rarely clavate. Homostylous or heterostylous.

29. L. maritimum L., *Sp. Pl.* 280 (1753) (incl. *L. mulleri* Moris). Glabrous or sometimes hairy perennial; stems up to 80 cm, erect or ascending. Leaves 2–4(–5) mm wide, lanceolate or narrowly elliptical; lower leaves opposite, 3-veined; middle and upper leaves alternate, 1-veined. Sepals 3 mm, ovate, acute, inconspicuously ciliate. Petals 8–15 mm, yellow. Stigmas clavate. Capsule 2–3 mm, wider than long, subglobose; beak minute or absent. Heterostylous. $2n = 20$. *Damp, usually saline soils. Mediterranean region and Portugal; E. Austria.* Au Bl Co Ga Gr Hs It Ju Lu Sa.

30. L. trigynum L., *Sp. Pl.* 279 (1753) (*L. gallicum* L.). Glabrous annual; stems 10–30 cm, erect or ascending. Leaves 1–2(–3) mm wide, linear-lanceolate to narrowly elliptical, with smooth margins. Sepals 3–4 mm, shortly acuminate with glandular-ciliate margins and setaceous apex. Petals 4–6 mm, yellow. Stigmas linear. Capsule *c.* 2 mm, subglobose; beak *c.* 0·3 mm. Homostylous. *S. Europe, extending northwards to C. France and E. Czechoslovakia.* Al Bl Bu Co Cr Cz Ga Gr Hs Hu It Ju Lu Rm Rs (W) Sa Si Tu.

31. L. tenue Desf., *Fl. Atl.* 1: 280 (1798). Like **30** but more robust; stems up to 70 cm; leaves 1–4(–5) mm wide, minutely serrulate; sepals 3–5 mm; petals 8–18 mm. Heterostylous. *S. Spain, S. & C. Portugal.* Hs Lu.

32. L. tenuifolium L., *Sp. Pl.* 278 (1753). Glabrous (or sometimes hairy) perennial with few short non-flowering shoots; stems (10–)20–45 cm, erect, ascending or decumbent, slightly branched below. Leaves 0·5–1(–2) mm wide, linear, 1-veined; leaf-margins rough, minutely serrulate, flat or slightly inrolled. Sepals 5–8 mm, lanceolate-acuminate, 1-veined, with glandular-

ciliate, minutely serrulate margins. Petals 2–2½ times as long as sepals, pink or almost white. Styles patent; stigmas capitate. Capsule 2·7–3·5(–4) mm, subglobose; beak *c.* 0·7 mm, acuminate. Homostylous. $2n = 16, 18$. *C. & S. Europe, extending northwards to Belgium and C. Ukraine.* Al Au Be Bu Co Cz Ga Ge Gr He Hs Hu It Ju Po Rm Rs (W, K, E) Sa Si Tu.

33. L. suffruticosum L., *Sp. Pl.* 279 (1753). Like **32** but sometimes woody at base, with many short non-flowering shoots; stems 5–40(–50) cm, procumbent with ascending lateral shoots, much branched below; leaves 0·2–1 mm wide, linear or setaceous with rough, minutely serrulate, strongly inrolled margins; sepals 4–6 mm, ovate-acuminate, 3-veined; petals white with a violet or pink claw, 3–4 times as long as sepals; styles erect; heterostylous. $2n = 72$. *S.W. Europe, from C. Spain to N.W. Italy and extending northwards to N. C. France.* Ga Hs It.

(a) Subsp. **suffruticosum**: Woody at base; sometimes densely puberulent-scabrid; stems 20–50 cm; leaves linear, rough; petals 20–30 mm. *Spain.*

(b) Subsp. **salsoloides** (Lam.) Rouy, *Fl. Fr.* **4**: 71 (1897) (*L. salsoloides* Lam.): Scarcely woody at base; stems 5–25 cm; leaves filiform, minutely serrulate; petals 10–20 mm. *Throughout the range of the species.*

Dwarf plants with very densely leafy stems and small imbricate leaves have been called **L. ortegae** Planchon, *London Jour. Bot. (Hooker)* 7: 184 (1848), and may deserve subspecific or specific rank.

34. L. strictum L., *Sp. Pl.* 279 (1753). Annual; stems 10–45 cm, erect, inconspicuously hairy below. Leaves 1·5–3(–5) mm wide; margins minutely serrulate, very rough, often inrolled. Sepals 4–6 mm, ovate-lanceolate, long-acuminate, minutely serrulate and glandular-ciliate. Petals 6–12 mm, yellow. Stigmas capitate. Capsule 2–2·5 mm, subglobose; beak *c.* 0·3 mm. Homostylous. $2n = 18$. *S. Europe.* Al Bl Bu Co Cr Ga Gr Hs It Ju Lu Rs (K) Sa Si Tu.

(a) Subsp. **strictum**: Stems robust, seldom branched below; inflorescence a dense, spike-like cyme or a corymb; flowers sessile, or subsessile with thick pedicels rarely longer than the calyx. *Throughout the range of the species.*

(b) Subsp. **corymbulosum** (Reichenb.) Rouy, *Fl. Fr.* **4**: 60 (1897) (*L. corymbulosum* Reichenb., *L. liburnicum* Scop.): Stems slender, often branched below; inflorescence lax, spreading; pedicels slender, equalling or longer than the calyx. *Throughout the range of the species.*

The form of the inflorescence is very variable in both subspecies.

35. L. setaceum Brot., *Phyt. Lusit.* 43 (1800). Like **34** but leaves 0·5 mm wide, setaceous with inrolled margins, densely crowded in middle of stem; inflorescence lax, much-branched. $2n = 18$. *S. Spain, C. & S. Portugal.* Hs Lu.

Sect. CATHARTOLINUM (Reichenb.) Griseb. Leaves opposite, without basal glands. Petals small, free, white. Stigmas capitate. Homostylous.

36. L. catharticum L., *Sp. Pl.* 281 (1753). Usually annual, very slender, glabrous, rarely more than 15 cm. Leaves 1-veined, lower obovate-lanceolate, upper lanceolate. Inflorescence a lax dichasium; flowers nodding in bud, on long slender pedicels. Sepals 2–3 mm, lanceolate, glandular-ciliate. Petals obovate, entire or shallowly emarginate, about twice as long as sepals, white, with a yellow claw. Capsule globose; seeds compressed.

$2n=16$. *Europe, northwards to $69°$ N. in Fennoscandia; mainly on mountains in the south.* All except Az Bl Cr Sa Sb Si.

In Fennoscandia, the Alps and the Carpathians more or less perennial variants occur, which may have a slightly woody stock and much-branched stems. It does not seem possible to distinguish such plants sufficiently clearly from the widespread annual plant for them to merit subspecific rank.

2. Radiola Hill[1]

Like *Linum* but flowers 4-merous; sepals toothed at apex; petals more or less equalling sepals; capsule dehiscing with 8 valves; seeds ovoid.

1. R. linoides Roth, *Tent. Fl. Germ.* **1**: 71 (1788). Glabrous annual, rarely more than 10 cm; stems filiform, usually with free dichotomous branching. Leaves up to 3 mm, opposite, more or less elliptical, 1-veined. Flowers numerous, in dichasia; pedicels short; sepals *c.* 1 mm, (2–)3(–4)-toothed at apex; petals *c.* 1 mm, obtuse or emarginate, white. Capsule *c.* 1 mm, globose. *Seasonally damp, bare, sandy or peaty ground; calcifuge.* $2n=18$. *Most of Europe except the north-east and extreme north, but rare and local in much of the centre & south-east.* Al Be Bl Br Bu Co Cz Da Ga Ge Gr Hb †He Ho Hs Hu It Ju Lu No Po Rm Rs (B, C, W, E) Sa Si Su Tu.

LXXXVII. EUPHORBIACEAE[2]

Dioecious or monoecious herbs or shrubs, often with latex. Leaves usually alternate, simple; usually stipulate. Flowers usually actinomorphic, often apetalous and sometimes without sepals. Male flowers with one to many stamens with free or connate filaments. Female flowers with a usually 3-locular superior ovary and 3 styles; disc usually present, annular, pulvinate or cyathiform; ovules 1–2 in each loculus. Fruit a capsule, often dehiscing explosively; seeds often carunculate.

1	Inflorescence usually umbellate; flowers without perianth, in small groups surrounded by an involucre; latex present	7. Euphorbia
1	Inflorescence not umbellate; perianth present; flowers not surrounded by an involucre; latex absent	
2	Plant with stellate hairs	3. Chrozophora
2	Plant glabrous or with simple hairs	
3	Shrubs	
4	Spiny; leaves up to 1·5 cm, entire	2. Securinega
4	Unarmed; leaves up to 60 cm, palmately lobed	6. Ricinus
3	Herbs, sometimes woody at base	
5	Leaves palmately lobed	6. Ricinus
5	Leaves entire, serrate or crenate	
6	Leaves opposite	4. Mercurialis
6	Leaves alternate	
7	Perennial; leaves entire	1. Andrachne
7	Annual, leaves crenate	5. Acalypha

1. Andrachne L.[3]

Monoecious. Small shrubs. Flowers solitary or in small fascicles in the leaf-axils. Sepals 5–6, free or shortly connate; petals 5–6, very small in male flowers. Male flowers with 5–6 stamens, free or connate round the rudimentary ovary; glands between the petals and stamens free or connate. Female flowers with 3-locular ovary and 3 free or shortly connate styles.

1. A. telephioides L., *Sp. Pl.* 1014 (1753). Stems up to 30 cm, green, simple or little-branched, arising from a brown, much-branched stock. Leaves up to 10 mm, obovate to elliptical, acute or subacute, entire, glaucous, closely and evenly spaced; petioles 0·5–2 mm; stipules silvery, often red at base. Petals yellowish. Styles deeply 2-fid. Capsule 2–3 mm, subglobose, glabrous; seeds triquetrous, with a convex, punctulate back. *Dry places. Mediterranean region; Krym.* Bu Cr Gr Hs It Ju Rs (K) Si [Ga].

2. Securinega Commerson[3]

Dioecious. Shrubs or small trees, often spiny. Leaves distichous. Flowers axillary, solitary or fasciculate. Calyx 5- to 6-partite; petals absent. Male flowers with 5–6 stamens; glands alternating with the calyx-lobes, outside the stamens. Female flowers with 3, free or shortly connate styles.

1. S. tinctoria (L.) Rothm., *Feddes Repert.* **49**: 276 (1940) (*S. buxifolia* auct., non (Poiret) Müller Arg.). Spiny shrub up to 150 cm. Leaves 8–15 × 2–4 mm, oblong-obovate, obtuse or emarginate, mucronulate, distichous on the long shoots; petiole *c.* 1 mm; stipules very small, subulate. Styles 2-fid. Capsule *c.* 3 mm, bluntly trigonous, with 3 shallow grooves. *Sandy river-banks.* ● *C. & S.W. Spain; E. Portugal.* Hs Lu.

3. Chrozophora Λ. Juss.[3]

Monoecious. Annual herbs, covered with stellate hairs. Male flowers in terminal spike-like racemes or in axillary fascicles; calyx 5-fid; petals 5; stamens 5–10, monadelphous; rudimentary ovary absent. Female flowers solitary at the base of the male; calyx 10-partite; petals absent or very small; ovary 3-locular; styles 3, 2-fid.

Literature: D. Prain, *Kew Bull.* **1918**: 103 (1918).

Plant green or grey-green; leaves cuneate at base; stamens 9–11	1. tinctoria
Plant whitish; leaves truncate to subcordate at base; stamens 4–5(–7)	2. obliqua

1. C. tinctoria (L.) A. Juss., *Euphorb. Tent.* 84 (1824). Plant green or grey-green, rather thinly stellate-tomentose. Stems up to 50 cm, more or less branched. Leaves ovate to rhombic, entire or sinuate-dentate, subobtuse, cuneate at base; petiole as long as to twice as long as lamina. Male flowers with 9–11 stamens. Capsule mucronate, covered with peltate hairs; seeds *c.* 4 mm, rough. *Dry places near the coast. S. Europe.* Al Bl Bu Co Cr Ga Gr Hs It Ju Lu Rs (K) Sa Si Tu.

2. C. obliqua (Vahl) A. Juss. ex Sprengel, *Syst. Veg.* 3: 850 (1826) (*C. verbascifolia* (Willd.) A. Juss. ex Sprengel). Like **1** but plant whitish, very densely stellate-tomentose; leaves truncate to subcordate at base; male flowers with 4–5(–7) stamens; capsule sparsely and shortly muricate. *Dry places. Mediterranean region.* Cr Gr Hs It ?Rs (K).

[1] By S. M. Walters.　　[2] Edit. T. G. Tutin.　　[3] By T. G. Tutin.

4. Mercurialis L.[1]

Usually dioecious. Herbs with watery sap. Leaves opposite; stipules small. Male flowers usually in clusters on long axillary spikes; calyx-lobes 3; stamens 8–15. Female flowers solitary or few, axillary, subsessile or pedunculate; calyx-lobes 3; styles 2. Fruit with 2 cells, each with 1 seed.

1 Rhizomatous; aerial stem simple
  2 Lower leaves usually scale-like; petiole (3–)5–10(–18) mm
                      6. perennis
  2 Lower leaves like the upper, but smaller; petiole 1–2 mm
                      7. ovata
1 Not rhizomatous; stems branched
  3 Densely tomentose   **5. tomentosa**
  3 Glabrous or sparsely hairy
    4 Annual, without a thick woody stock   **1. annua**
    4 Perennial, with a thick woody stock
      5 Sparsely hairy; leaves incise-dentate **4. reverchonii**
      5 Glabrous; leaves crenate-dentate or shallowly sinuate-dentate
        6 Leaves crenate-dentate; fruit 3–4 × 5–6 mm; seed smooth
                    2. elliptica
        6 Leaves shallowly sinuate-dentate; fruit *c.* 2 × 3 mm; seed rugulose   **3. corsica**

1. **M. annua** L., *Sp. Pl.* 1035 (1753). Glabrous or sparsely hairy annual 10–50 cm. Stem branched, often from the base. Leaves 1·5–5 cm, ovate to elliptic-lanceolate, crenate-serrate; petiole 2–15 mm. Rarely monoecious. Female flowers axillary, few, subsessile; calyx-lobes triangular-ovate, acute. Fruit 2–3 × (2–)3–4 mm, hispid, rarely nearly glabrous; seed *c.* 2 mm, ovoid, rugulose. 2*n*=16, 48, 64, 80, 96, 112. *Cultivated ground and waste places. Most of Europe, but introduced in much of the north and west.* Al Au *Az Be Bl *Br Bu Co Cr Cz *Da Ga Ge Gr He Ho Hs Hu It Ju Lu Po Rm Rs (C, W, K) Sa Si Tu [Fe Hb No Su].

B. Durand, *Ann. Sci. Nat.* ser. 12 (Bot.), **4**: 579–736 (1963), has shown that diploid plants are nearly always dioecious, are widespread, and very variable. Most are completely interfertile, but some saxicolous populations from S.E. France form sterile hybrids with other diploid populations, though some introgression seems to occur. These local diploid populations have been described as **M. huetii** Hanry, *Billotia* **1**: 21 (1864), and are characterized by small size, long, patent branches, and fruit up to 2 mm wide, ciliate, but not hispid on the surface.

The polyploids are characterized by being usually monoecious. Plants with 2*n*=48 occur in the coastal regions of Portugal and S. and E. Spain, and also in isolated populations on the eastern coast of Corse, and probably correspond to **M. ambigua** L. fil., *Dec. Prim. Pl. Rar. Hort. Upsal.* 15 (1762). The remaining polyploids (2*n*=64, 80, 96, 112) are restricted, as far as is known, to Corse and Sardegna.

In the absence of clear-cut morphological characters it does not seem possible to give specific rank to any of the taxa in the complex, though they appear to be distinguishable on the average characters of population-samples.

2. **M. elliptica** Lam., *Encycl. Méth. Bot.* **4**: 119 (1797). Glabrous perennial up to 60 cm. Stems branched; stock stout and woody. Leaves (1–)2–4 cm, elliptical or ovate, obtuse or subacute, crenate-dentate; petiole 2–8 mm. Female flowers axillary, 1 or few; petioles up to 3 mm; calyx-lobes orbicular-ovate, obtuse. Fruit 3–4 × 5–6 mm, glabrous; seed *c.* 2 mm, globose or ovoid,

smooth or nearly so. *Sandy waste places. C. & S. Portugal, S. Spain.* Hs Lu.

3. **M. corsica** Cosson, *Not. Pl. Crit.* 63 (1850). Like **2** but leaves remotely and shallowly sinuate-dentate; fruit *c.* 2 × 3 mm, often hispidulous; seeds *c.* 1·5 mm, ovoid, rugulose. *Rocky places and disturbed ground.* ● *Corse and Sardegna.* Co Sa.

4. **M. reverchonii** Rouy, *Naturaliste (Paris)* **9**: 199 (1887). Like **2** but sparsely pubescent; leaves 1·5–6 cm, the lower suborbicular, the others ovate, all incise-dentate, subobtuse, truncate to subcordate at base; petioles 5–10 mm; female flowers 3–5, in a usually branched, shortly pedunculate axillary inflorescence; fruit hispid; seeds reticulate-veined. *S.W. Spain.* Hs. (*Morocco.*)

5. **M. tomentosa** L., *Sp. Pl.* 1035 (1753). Densely tomentose perennial up to 60 cm. Stems freely branched. Leaves 1–5 cm, oblong-lanceolate to oblong-obovate, acute or obtuse and mucronate, entire or weakly serrate; petiole 1–3 mm. Female flowers 1 or few, axillary, subsessile; calyx-lobes ovate-lanceolate, acute. Fruit 4 × 6 mm, densely tomentose; seed *c.* 3 mm, ovoid, rugulose. *Rocky slopes.* ● *S.W. Europe.* Bl Ga Hs Lu.

Sterile hybrids between **2** and **5** occur with the parents.

6. **M. perennis** L., *Sp. Pl.* 1035 (1753). More or less pubescent rhizomatous perennial up to 40 cm, becoming blackish when dried. Stems simple. Leaves 2–8 cm but usually scale-like in the lower part of the stem, elliptic-lanceolate to elliptic-ovate, crenate-serrate, crowded in the upper part of the stem; petiole (3–)5–10(–18) mm. Female flowers 1–3, in axillary inflorescences up to 7 cm in fruit. Fruit *c.* 4 × 7 mm, pubescent; seed 3–3·5 mm, globose, rugulose. 2*n*=42, 64 (♀), 66 (♂) *c.* 80, 84. *Shady places, usually in* Quercus- *or* Fagus-*woods. Most of Europe, northwards to 66°N. in Norway.* Al Au Be Br Bu Co Cz Da Fe Ga Ge Gr *Hb He Ho Hs Hu It Ju Lu No Po Rm Rs (B, C, W, K, E) Si Su.

7. **M. ovata** Sternb. & Hoppe, *Denkschr. Bayer. Bot. Ges. Regensb.* **1**: 170 (1815). Like **6** but plant yellowish-green when dried: stems with foliage leaves throughout, not crowded towards the top, ovate to suborbicular; petiole usually 1–2 mm. 2*n*=32. *C. & S.E. Europe.* Al Au Bu Cz Ge Gr He Hu It Po Rm Rs (W, K, E) Tu.

A sterile hybrid, **M.×paxii** Graebner in Ascherson & Graebner, *Syn. Mitteleur. Fl.* **7**: 408 (1916) (**6 × 7**), occurs locally.

5. Acalypha L.[1]

Monoecious. Annual herbs. Flowers in small axillary spikes. Male flowers with 4-partite perianth; stamens 8–16; filaments united at base. Female flowers with 3- to 5-partite perianth; ovary 3-locular; styles 3. Seeds 3, carunculate.

1. **A. virginica** L., *Sp. Pl.* 1003 (1753). Pubescent annual 10–50 cm. Leaves 2–10 cm, lanceolate to ovate, crenate; petiole ⅓–½ as long as lamina. Inflorescence with male flowers above and female at base; bract of female flowers with 7–9 lanceolate lobes. Fruit pubescent; seed ellipsoid, smooth; caruncle small. *Naturalized locally in Italy, S. Switzerland and S. Austria.* [Au He It.] (*North America.*)

6. Ricinus L.[1]

Monoecious shrubs or large herbs. Leaves alternate, deeply palmately lobed. Flowers paniculate, male above, female below.

---

[1] By T. G. Tutin.

Male flowers with membranous perianth; stamens numerous; filaments repeatedly branched. Female flowers with caducous, membranous perianth; ovary 3-locular. Seeds 3, carunculate.

1. **R. communis** L., *Sp. Pl.* 1007 (1753). Annual herb or shrub up to 4 m. Leaves up to 60 cm, peltate, palmately 5- to 9-fid; lobes lanceolate to ovate-lanceolate, acuminate, irregularly dentate. Panicle erect. Fruit 10–20 mm, with long conical projections or smooth. Seed 9–17 mm, smooth, shiny, reddish-brown to blackish, marked in various shades of white, grey or brown; caruncle large. *Cultivated for the oil obtained from the seeds and for ornament. Naturalized in S. & S.C. Europe.* [Al Az Bl Bu Co Cr Ga Gr Hs It Ju Lu Rm Sa Si.] (*Tropics.*)

7. Euphorbia L.[1]

Monoecious. Herbs or small shrubs, with latex. Flowers in small groups surrounded by a more or less deeply lobed involucre with 4–5 (rarely more) glands at the top and usually pubescent within, the whole forming a *cyathium*. 'Inflorescence' of cyathia usually umbellate, often with axillary rays below the umbel. Perianth absent. Male flowers of a single stamen jointed to the pedicel. Female flowers solitary, pedicellate, surrounded by several male flowers; ovary 3-locular, with 3, usually free, styles; ovules solitary in each loculus. Seeds usually carunculate.

In the following account the inflorescence is referred to as an umbel, though it is not strictly one. The leaves subtending the primary branches (rays) of the umbel are called ray-leaves and those subtending the ultimate branches raylet-leaves. The bracts, when present, subtend the individual male flowers in the cyathium.

In addition to the species described below several others are reported as casual aliens in S. Europe, and some of them may be in process of naturalization. These include **E. indica** Lam., *Encycl. Méth. Bot.* 2: 423 (1788), **E. missurica** Rafin., *Atl. Journ.* 1: 146 (1832), **E. engelmannii** Boiss., *Cent. Euphorb.* 15 (1860) and **E. rhytisperma** (Klotzsch & Garcke) Engelm. ex Boiss. in DC., *Prodr.* 15(2): 43 (1862).

E. marginata Pursh, *Fl. Amer. Sept.* 2: 607 (1814), from North America, is cultivated for ornament and frequently escapes in S.E. Europe. It is an annual with the ray- and raylet-leaves with wide, petal-like, white margins and the glands with conspicuous white appendages.

Literature: E. Boissier in DC., *Prodr.* 15(2): 7–187 (1862).

1 Stipules present; leaves usually asymmetrical at the base, opposite, but not decussate
  2 Ripe seeds smooth
    3 Leaves linear-oblong   **3. polygonifolia**
    3 Leaves ovate or falcate-oblong
      4 Seeds 3 mm   **2. peplis**
      4 Seeds less than 1·5 mm   **4. humifusa**
  2 Ripe seeds rugulose
    5 Capsule hairy
      6 Capsule hairy on the keels only   **7. prostrata**
      6 Capsule hairy all over
        7 Hairs on capsule closely appressed   **6. maculata**
        7 Hairs on capsule patent   **5. chamaesyce**
    5 Capsule glabrous
      8 Leaves 10–30(–36) mm, serrate   **1. nutans**
      8 Leaves (1–)3–7(–11) mm, entire or obscurely serrulate   **5. chamaesyce**
1 Stipules absent; leaves symmetrical at the base, usually alternate

---

[1] By A. R. Smith and T. G. Tutin.

  9 Leaves opposite and decussate; capsule spongy   **60. lathyris**
  9 Leaves alternate, rarely opposite but then not decussate; capsule indurate
    10 Glands suborbicular or ovate, sometimes irregularly lobed, but neither horned nor with the outer margin truncate or emarginate
      11 Shrubs with stout branches; glands suborbicular, entire or irregularly lobed
        12 Leaves 25–65 mm, glabrous; capsule smooth or nearly so   **9. dendroides**
        12 Leaves *c.* 150 mm, sparsely hairy beneath; capsule verrucose   **8. stygiana**
      11 Herbs, or slender, wiry dwarf shrubs; glands ovate, entire
        13 Annual or biennial; stems usually solitary
          14 Capsule winged   **54. pterococca**
          14 Capsule not winged
            15 Capsule tuberculate
              16 Tubercles in 2 rows on each valve of the capsule; seeds verruculose   **53. cuneifolia**
              16 Tubercles not in 2 rows on each valve of the capsule; seeds smooth
                17 Tubercles ending in a long bristle; capsule usually indehiscent   **17. akenocarpa**
                17 Tubercles not ending in a bristle; capsule dehiscent
                  18 Lowest raylet-leaves like the ray-leaves; capsule not more than 2·5 mm; seeds reddish   **51. serrulata**
                  18 Lowest raylet-leaves markedly different from the ray-leaves; capsule 2–3 mm; seeds olive-brown   **50. platyphyllos**
            15 Capsule not tuberculate
              19 Seeds reticulate-rugose or transversely sulcate
                20 Leaves entire; seeds with 3 transverse depressions on each face   **56. phymatosperma**
                20 Leaves serrate in the upper half; seeds reticulate-rugose   **55. helioscopia**
              19 Seeds smooth
                21 Capsule glabrous, even when young
                  22 Capsule indehiscent, scarcely sulcate   **17. akenocarpa**
                  22 Capsule dehiscent, usually distinctly sulcate
                    23 Leaves entire or weakly sinuate; umbel with 3 rays; capsule 5–7 mm   **14. lagascae**
                    23 Leaves sharply serrate; umbel with 4–5 rays; capsule *c.* 3 mm   **15. arguta**
                21 Capsule hairy, at least when young
                  24 Leaves rounded and mucronulate at apex; capsule 3–4 mm, persistently sericeous or with scattered bristles   **17. akenocarpa**
                  24 Leaves, except the lowest, acute; capsule *c.* 2·5 mm, pubescent when young, becoming glabrous   **16. microsphaera**
        13 Perennial herbs or small shrubs; stems usually numerous
          25 Capsule smooth or nearly so
            26 Stems slender, not scaly below   **13. corallioides**
            26 Stems stout, scaly below
              27 Rhizome without tubers; raylet-leaves longer than wide; seeds not more than 3·2 mm   **12. villosa**
              27 Rhizome with large tubers; raylet-leaves wider than long; seeds *c.* 4 mm   **11. isatidifolia**
          25 Capsule tuberculate or cristate
            28 Seeds minutely tuberculate   **52. pubescens**
            28 Seeds quite smooth, rarely shallowly vermiculate-rugose
              29 Procumbent or ascending; inflorescence usually reduced to a single cyathium, rarely with 2–3 short rays
                30 Stems not scaly at base; glands 8   **49. capitulata**
                30 Stems scaly at base; glands 4   **48. chamaebuxus**
              29 Erect; inflorescence always of several cyathia
                31 Plant with a large, subterranean, napiform tuber; stems usually less than 20 cm   **31. apios**
                31 Plant without a large, subterranean tuber; stems more than 20 cm
                  32 Capsule with 2 crests on the back of each carpel   **23. gregersenii**

32 Capsule not cristate
33 Shrubs with slender stems
  34 Rays of the umbel and dead twigs persistent
    35 Rays of the umbel spinescent; cauline leaves and ray-leaves similar   **47. acanthothamnos**
    35 Rays of the umbel not spinescent; ray-leaves much wider than the cauline   **45. spinosa**
  34 Rays of the umbel and dead twigs not persistent
    36 Leaves mucronate; ray-leaves rhombic- or orbicular-ovate, mucronate   **44. squamigera**
    36 Leaves not mucronate; raylet-leaves ovate or obovate, not mucronate
      37 Capsule with long, slender tubercles   **46. glabriflora**
      37 Capsule with low, broad tubercles   **43. bivonae**
33 Herbs, sometimes woody at base
  38 Tubercles on capsule ± hemispherical
    39 Umbel with more than 5 rays   **18. palustris**
    39 Umbel with 1–5 rays
      40 Stems scaly at base
        41 Raylet-leaves ovate-lanceolate or elliptical; cyathium long-pedunculate; capsule with small, hemispherical tubercles   **36. carniolica**
        41 Raylet-leaves broadly rhombic; cyathium ± sessile; capsule with large, irregular tubercles   **37. duvalii**
      40 Stems not scaly at base
        42 Plant nearly or quite glabrous
          43 Leaves acuminate
            44 Axillary rays not overtopping the umbel; raylet-leaves 10–17 mm; capsule deeply sulcate   **19. velenovskyi**
            44 Axillary rays overtopping the umbel; raylet-leaves 4–10 mm; capsule scarcely sulcate   **20. soongarica**
          43 Leaves obtuse, rarely acute
            45 At least the lower leaves widest above the middle
              46 Ray-leaves much shorter than the rays; rays long, patent   **42. clementei**
              46 Ray-leaves about as long as the rays, at least at flowering time; rays short, crowded   **38. brittingeri**
            45 All leaves widest at or below the middle
              47 Middle cauline leaves linear-lanceolate, acute; capsule 5–7 mm   **39. ruscinonensis**
              47 Middle cauline leaves broadly ovate to ovate-lanceolate, obtuse or subobtuse; capsule 3·5–4 mm
                48 Leaves 10–30 mm; ray-leaves broadly ovate to suborbicular   **40. welwitschii**
                48 Leaves 50–80 mm; ray-leaves narrowly rhombic   **41. monchiquensis**
        42 Plant distinctly pubescent
          49 Capsule sparsely tuberculate   **30. oblongata**
          49 Capsule densely tuberculate   **38. brittingeri**
  38 Some of the tubercles on capsule at least twice as long as wide, often filiform
    50 Stems scaly at base
      51 Rhizome swollen and jointed; stems *c.* 2 mm in diameter; raylet-leaves triangular
        52 Stems terete   **34. dulcis**
        52 Stems finely ribbed in upper part   **35. angulata**
      51 Rhizome cylindrical; stems *c.* 5 mm in diameter; raylet-leaves ovate   **12. villosa**
    50 Stems not scaly at base
      53 Leaves distinctly serrate or serrulate
        54 Stems stout; ray-leaves orbicular   **(25–29). epithymoides** group
        54 Stems very slender; ray-leaves ovate to obovate
          55 Leaves linear-oblong; ray-leaves obovate-cuneate   **33. uliginosa**

          55 Leaves obovate to obovate-oblong; ray-leaves ovate
            56 Upper leaves acute or subacute; all leaves about equal in size   **38. brittingeri**
            56 Upper leaves rounded at apex; lower leaves much smaller than upper   **32. polygalifolia**
      53 Leaves entire or almost entire
        57 Plant glabrous
          58 Ray-leaves obovate or suborbicular   **(25–29). epithymoides** group
          58 Ray-leaves ovate
            59 Stems 70–150 cm, stout, erect; axillary rays numerous   **21. ceratocarpa**
            59 Stems 5–30 cm, slender, decumbent; axillary rays absent   **32. polygalifolia**
        57 Plant pubescent
          60 Leaves glabrous above; tubercles on capsule usually mixed long and short, the long not tapering   **22. hyberna**
          60 Leaves ± hairy above; tubercles on capsule all long, tapering, purple-tipped
            61 Upper leaves acute; raylet-leaves reniform-deltate   **24. squamosa**
            61 All leaves obtuse; raylet-leaves elliptical to ovate   **(25–29). epithymoides** group
10 Glands horned or with a truncate or emarginate outer margin
  62 Horns of glands short, often dilated and minutely lobed at apex; bracts between the male flowers absent
    63 Leaves obovate to suborbicular   **57. myrsinites**
    63 Leaves lanceolate
      64 Leaves lanceolate; seeds smooth   **58. rigida**
      64 Leaves linear-lanceolate; seeds rugose-vermiculate   **59. broteri**
  62 Horns of glands usually slender, or outer margin of glands truncate or emarginate; bracts between the male and female flowers present, hirsute or plumose (except **86**)
    65 Annual; seeds not smooth
      66 Seeds tuberculate or rugulose
        67 Lower leaves narrowly linear, very numerous and usually closely imbricate   **61. aleppica**
        67 Lower leaves not narrowly linear and closely imbricate
          68 Seeds vermiculate-rugulose
            69 Seeds oblong, black, sometimes with white ridges   **62. medicaginea**
            69 Seeds ovoid, brown or grey
              70 Leaves linear to linear-oblanceolate   **64. exigua**
              70 Leaves obovate   **67. peplus**
           68 Seeds tuberculate
            71 Capsule 2·5 × 2·8 mm   **63. dracunculoides**
            71 Capsule 1·6–2 × 1·6 mm   **64. exigua**
      66 Seeds pitted or sulcate or both
        72 Seeds sulcate, sometimes also pitted
          73 Seeds sulcate and pitted; capsule with 2 ridges on each valve   **67. peplus**
          73 Seeds sulcate, not pitted; capsule not ridged
            74 Seeds transversely sulcate   **65. falcata**
            74 Seeds longitudinally sulcate   **66. sulcata**
        72 Seeds pitted, but not sulcate
          75 Raylet-leaves linear   **68. ledebourii**
          75 Raylet-leaves rhombic-trullate to transversely ovate
            76 Raylet-leaves obliquely rhombic-trullate   **69. taurinensis**
            76 Raylet-leaves rhombic-deltate to transversely ovate   **70. segetalis**
    65 Perennial; seeds smooth or not
      77 Seeds pitted, rugulose or tuberculate
        78 Procumbent or ascending; umbel not developed or with not more than 3 rays
          79 Capsule not winged   **90. maresii**
          79 Capsule winged   **91. herniariifolia**
        78 Erect; umbel usually with more than 3 rays
          80 Seeds tuberculate   **93. pithyusa**

80 Seeds rugulose or pitted
81 Seeds rugulose; axillary rays aggregated to form a
 whorl below the umbel **77. biumbellata**
81 Seeds pitted; axillary rays not aggregated into a
 whorl below the umbel
82 At least some leaves serrate; teeth slender
 10. serrata
82 Leaves entire, or serrulate towards the apex
83 Capsule at least 4 mm wide; raylet-leaves lanceolate to ovate
84 Ray-leaves little shorter than cauline leaves;
 cyathial lobes very densely ciliate
 78. boetica
84 Ray-leaves distinctly shorter than cauline leaves;
 cyathial lobes sparsely ciliate **79. bupleuroides**
83 Capsule not more than 3·5 mm wide; raylet-leaves
 reniform, deltate, rhombic, transversely ovate
 or suborbicular
85 Glands truncate **80. nicaeensis**
85 Glands with (1–)2 horns
86 Glands with short horns; seeds irregularly or
 shallowly pitted
87 Seeds evenly and shallowly pitted **74. petrophila**
87 Seeds irregularly pitted **75. transtagana**
86 Glands with long horns; seeds evenly and fairly
 deeply pitted
88 Leaves linear-lanceolate, acute **71. pinea**
88 Leaves obovate to oblanceolate, usually obtuse
 and mucronate
89 Seeds less than 2 mm **72. portlandica**
89 Seeds more than 2 mm **73. deflexa**
77 Seeds smooth
90 Raylet-leaves usually connate in pairs at base; capsule
 glabrous or pubescent; stems usually flowering in
 the second year
91 Capsule densely villous **105. characias**
91 Capsule glabrous
92 Axillary rays arranged in 2–5 whorls below the umbel
 104. heldreichii
92 Axillary rays not in whorls **103. amygdaloides**
90 Raylet-leaves not connate; capsule glabrous; stems
 usually flowering in the first year
93 Capsule 4–6 mm wide
94 Up to 15 cm, with horizontal rhizome (Greece)
 82. orphanidis
94 Up to 70 cm, with vertical stock (widespread)
95 At least some leaves sharply serrate **10. serrata**
95 All leaves entire or minutely serrulate
96 Glands with long horns; caruncle prominent,
 boat-shaped; leaves imbricate and succulent
 102. terracina
96 Glands emarginate, with short horns; caruncle
 minute, conical; leaves imbricate, ± succulent
97 Leaves obovate-oblong to ovate; rays 3–6
 94. paralias
97 Leaves linear-oblong to linear-oblanceolate;
 rays 8–12 **79. bupleuroides**
93 Capsule not more than 4·5 mm wide
98 Leaves palmately veined, somewhat coriaceous, often
 glaucous
99 Leaves obviously irregularly and sharply serrulate
 near the apex **93. pithyusa**
99 Leaves nearly or quite entire
100 Umbel with (5–)7–20(–30) rays
101 Umbel with not more than 8 rays
102 Leaves (10–)40–75 mm **80. nicaeensis**
102 Leaves not more than 20 mm **89. gayi**
101 Umbel with up to 30 rays
103 Leaves usually acute; seeds 1·5–2 mm
 92. seguierana
103 Leaves usually obtuse; seeds 2–2·5 mm
 80. nicaeensis
100 Umbel with (1–)2–5(–9) rays

104 Leaves 17–27 mm wide, prominently 9- to 11-
 veined **81. bessarabica**
104 Leaves not more than 17 mm wide, with fewer
 than 9 weak veins
105 Ray-leaves longer than the cauline leaves
 88. variabilis
105 Ray-leaves not longer than the cauline leaves
106 Plant with basal rosette of leaves
107 Rosette-leaves oblanceolate, usually obtuse;
 axillary rays 0–4(–7) **84. triflora**
107 Rosette-leaves linear-oblanceolate, usually
 emarginate; axillary rays 0(–1) **85. saxatilis**
106 Plant not forming rosettes
108 Glands 2-horned
109 Horns shorter than the gland
110 All leaves less than 6 mm wide
 76. matritensis
110 Some leaves more than 6 mm wide
 83. barrelieri
109 Horns at least as long as the gland
111 Leaves entire **84. triflora**
111 Leaves finely serrulate **83. barrelieri**
108 Glands truncate or crescentic
112 Leaves usually more than 10 mm wide
 80. nicaeensis
112 Leaves less than 10 mm wide
113 Cyathial cup glabrous within **86. valliniana**
113 Cyathial cup pubescent within
114 Capsule 3–3·5 mm **87. minuta**
114 Capsule 2·5 mm **89. gayi**
98 Leaves pinnately veined, usually membranous and
 not glaucous
115 Leaves cordate at base
116 Stems up to 25 cm; leaves not more than 30 mm
 95. nevadensis
116 Stems up to 90 cm; leaves up to 80 mm
 96. agraria
115 Leaves rarely cordate at base
117 Plant puberulent or pubescent
118 Leaves lanceolate to ovate-lanceolate; umbel
 with up to 16 rays **98. salicifolia**
118 Leaves linear to linear-lanceolate; umbel with
 up to 6 rays **99. esula**
117 Plant glabrous
119 Leaves usually emarginate or 3-cuspidate
120 Leaves linear **99. esula**
120 Leaves oblanceolate to elliptic-obovate
121 Not rhizomatous; leaf-margins flat **99. esula**
121 Rhizomatous; leaf-margins undulate
 100. undulata
119 Leaves not emarginate or 3-cuspidate
122 Upper surface of leaves shiny **97. lucida**
122 Upper surface of leaves dull
123 Leaves linear, crowded, especially on the
 lateral shoots **101. cyparissias**
123 Leaves lanceolate to broadly ovate, not
 crowded **99. esula**

Subgen. **Chamaesyce** Rafin. Usually procumbent annuals.
Leaves stipulate, opposite, distichous, usually asymmetrical
at base, petiolate. Cyathia axillary or clustered, not in umbels.
Glands often with petaloid appendages. Seeds without a
caruncle.

1. E. nutans Lag., *Gen. Sp. Nov.* 17 (1816) (*E. preslii* Guss.).
Procumbent to ascending annual up to 60 cm. Stems pubescent
above when young, otherwise nearly glabrous. Leaves 10–30(–36)
× 5–10(–14) mm, elliptic-oblong, obtuse, subacute or acute,
asymmetrical at base, serrate, occasionally sparsely pubescent
above, glabrous beneath; petiole 1–2 mm. Stipules 0·5 mm,
triangular, connate or free. Glands transversely ovate, yellow,
with small, pale pink appendages. Capsule 1·8–2 × 2 mm, rather

deeply sulcate, smooth, glabrous. Seeds 1·1 mm, ovoid-quadrangular, irregularly transversely rugulose, blackish. *Disturbed ground. Locally naturalized in S. & S.C. Europe, and casual elsewhere.* [Au Az Bu Ga He Hs Hu It ?Ju Lu Rm Si.] (*North America.*)

2. **E. peplis** L., *Sp. Pl.* 455 (1753). Procumbent, somewhat fleshy, glabrous annual, usually with 4 branches from the base; branches up to 40 cm. Leaves (4–)5–11(–16) × 2·5–5(–10) mm, falcate-oblong, obtuse or emarginate, entire or almost so; base obliquely truncate; petiole 2–3 mm. Stipules 1·5 mm, subulate. Glands semicircular, reddish-brown, with small, paler appendages. Capsule (3–)3·5–4(–4·5) × 4–5 mm, rather deeply sulcate, nearly smooth, purplish. Seeds 3 mm, ovoid-pyriform, smooth, pale grey, occasionally brown-mottled. *Sandy sea-shores, rarely inland. Coasts of S. & W. Europe, northwards to S.W. England.* Al Az Bl Br Bu Co Cr Ga Gr †Hb Hs It Ju Lu Rm Rs (W, K, E) Sa Si Tu.

3. **E. polygonifolia** L., *Sp. Pl.* 455 (1753). Procumbent, somewhat fleshy, glabrous annual; branches up to 18 cm. Leaves (4–)9–12·5 × (1–)2–3 mm, linear to linear-oblong, obtuse, minutely apiculate, somewhat asymmetrical at the base, entire; petiole 1·5 mm. Stipules 1 mm, triangular. Glands suborbicular, concave. Capsule 3 × 3 mm, shallowly sulcate, nearly smooth. Seeds 2 mm, ovoid-pyriform-quadrangular, smooth, pinkish-grey. *Sea-shores. Naturalized in S. France and N. Spain.* [Ga Hs.] (*E. North America.*)

4. **E. humifusa** Willd., *Enum. Pl. Hort. Berol., Suppl.* 27 (1813). Procumbent, glabrous, more or less glaucous annual, with 4 branches from the base; branches up to 13 cm. Leaves (2–)5–8(–9·5) × (1–)2–4(–5) mm, ovate, obtuse, asymmetrical at base, serrulate, especially in the upper half; petiole 0·5–1 mm. Stipules c. 1 mm, subulate-filiform. Glands transversely ovate to suborbicular, concave, stipitate. Capsule 1·5 × 1·5–2 mm, transversely rugulose, nearly smooth. Seeds 1·2 mm, ovoid, smooth, mottled grey-brown. *Stony or disturbed ground. S. Ukraine and S.E. Russia; naturalized elsewhere in parts of S. & C. Europe as a weed and ruderal.* Rs (W, K, E) [Au Co ?Cz Ga Ge He Hu It ?Ju Po Rm Sa Si.] (*W. & C. Asia.*)

E. serpens Kunth in Humb., Bonpl. & Kunth, *Nov. Gen. Sp.* 2: 52 (1817), has smaller, entire leaves, stipules often connate, and smaller, more or less quadrangular seeds. It is native of America, and occurs frequently as a casual in S. France and Spain, where it is perhaps locally naturalized.

5. **E. chamaesyce** L., *Sp. Pl.* 455 (1753). Procumbent, glabrous or villous annual with branches up to 30 cm. Leaves (1–)3–7(–11) × (1–)2·5–4·5(–6) mm, asymmetrically ovate-suborbicular to oblong, obtuse or emarginate, oblique at the base, nearly entire or obscurely serrulate; petiole c. 1 mm. Stipules up to 1 mm, triangular. Glands suborbicular, with small, whitish appendages. Capsule 2 × 2 mm, rather deeply sulcate, smooth, glabrous to densely patent-pubescent. Seeds 1·2 mm, ovoid-quadrangular, irregularly tuberculate-rugulose, greyish. *Open habitats. S. Europe, extending northwards to E.C. Russia.* Al Bl Bu Co Cr Ga Gr Hs It Ju Lu Rm Rs (C, W, K, E) Sa Si Tu.

(a) Subsp. **chamaesyce**: Glabrous or pubescent. Leaves less than 10 mm, suborbicular-ovate, usually emarginate, often entire. Appendages not more than twice as wide as the glands, usually entire. *Throughout the range of the species.*

(b) Subsp. **massiliensis** (DC.) Thell. in Ascherson & Graebner, *Syn. Mitteleur. Fl.* 7: 457 (1917): Villous. Leaves usually up to 10 mm, ovate-oblong to oblong, obtuse, serrulate. Appendages

more than twice as wide as the glands, often 3-lobed. *S.E. Europe and C. Mediterranean region.* Bu Cr Gr It Ju Rm Rs (W, K) Sa Si.

6. **E. maculata** L., *Sp. Pl.* 455 (1753). Procumbent annual, with branches up to 20 cm. Stems pubescent. Leaves (2–)4–7(–13) × (0·5–)1–2(–4) mm, ovate-oblong to oblong, slightly curved, obtuse or subacute, obliquely truncate at the base, serrulate near the apex, sparingly pubescent above, more densely so beneath, usually with a purple blotch on the midrib; petiole 0·5–1 mm. Stipules c. 1 mm, triangular-subulate. Glands transversely ovate, with small, purplish appendages. Capsule 1–1·3 × 1·2–1·5 mm, shallowly sulcate, smooth, sparsely covered with closely appressed hairs. Seeds 0·8 mm, ovoid-quadrangular, with 3–4 transverse furrows on each face, brownish. *Naturalized as a weed and ruderal. S. & S.C. Europe.* [Au Az Bu Ga ?Ge He Hs Hu It Ju Lu Rm Sa Si.] (*North America.*)

7. **E. prostrata** Aiton, *Hort. Kew.* 2: 139 (1789). Procumbent annual, with branches up to 20 cm. Stems usually glabrous below, pubescent above. Leaves (2–)8–10(–15) × (1–)4–6(–8) mm, ovate, obtuse, asymmetrical at the base, serrulate to subentire, sparsely pubescent to glabrescent on both surfaces; petiole c. 1 mm. Stipules c. 1 mm, triangular, the upper free, the lower often connate. Glands transversely ovate, with small appendages. Capsule 1·5 × 1·5 mm, shallowly sulcate, sharply keeled, smooth, glabrous except for the ciliate keels. Seeds 1 mm, ovoid-quadrangular, deeply transversely furrowed, greyish. *Naturalized as a weed and ruderal. Mediterranean region and Portugal.* [Gr Hs It Lu Si.] (*North America.*)

Subgen. **Esula** Pers. Usually erect herbs or sometimes shrubs. Leaves exstipulate, usually alternate, symmetrical at base, sessile or subsessile. Cyathia almost always in umbels. Glands without petaloid appendages. Seeds usually with a caruncle.

Sect. PACHYCLADAE (Boiss.) Tutin. Shrubs with stout branches. Leaves alternate, entire, present only on the current year's growth. Glands suborbicular, entire or irregularly lobed. Capsule with indurated pericarp. Seeds smooth.

8. **E. stygiana** H. C. Watson, *London Jour. Bot. (Hooker)* 3: 605 (1844). Branches erect. Leaves c. 150 × 40 mm, oblong, mucronate, glaucous and sparsely pubescent beneath, dark green above. Ray-leaves oblong, villous. Rays c. 4, repeatedly branched. Glands suborbicular. Capsule c. 6 mm, shallowly sulcate, rounded, verrucose. *Rocky, bushy places, particularly in small volcanic craters; 500–800 m.* ● *Açores.* Az.

9. **E. dendroides** L., *Sp. Pl.* 462 (1753). Stems up to 200 cm, apparently dichotomously branched. Leaves 25–65 × 3–8 mm, oblong-lanceolate, obtuse, mucronulate. Ray-leaves like the cauline but rather shorter and wider; raylet-leaves broadly rhombic, yellowish. Rays 5–8, dichotomous. Glands suborbicular, irregularly lobed. Capsule 5–6 mm, the valves laterally compressed, smooth or nearly so. Seeds 3 mm, laterally compressed, grey. 2n=18. *Rocky places near the sea. Mediterranean region.* Al Bl Co Cr Ga Gr Hs It Ju Sa Si.

Sect. CARUNCULARES (Boiss.) Tutin. Herbs. Leaves sharply serrate. Glands transversely ovate, truncate or obscurely crescentic. Capsule with indurated pericarp. Seeds smooth or minutely punctate.

10. **E. serrata** L., *Sp. Pl.* 459 (1753). Glabrous, glaucous perennial 20–50 cm. Stock slender, woody. Leaves linear-oblong

to ovate-lanceolate, acute or obtuse, with fine, patent teeth; the upper broadly ovate at base. Ray-leaves lanceolate-acuminate to suborbicular; raylet-leaves ovate to suborbicular, yellow. Rays 3–5, once to several times dichotomous. Capsule 5–6 mm. Seeds *c.* 3 mm, smooth or shallowly punctate, grey. *S.W. Europe, extending northwards to 46° N. in W. France and eastwards to Pantellaria.* Bl Ga Hs It Lu Sa Si.

Sect. HELIOSCOPIA Dumort. Herbs or shrubs. Glands transversely ovate, not truncate, emarginate or with horns. Bracts present between the male flowers. Capsule with indurated pericarp.

(A) Perennial; capsule smooth or more or less tuberculate or rugulose; seeds smooth, rarely weakly reticulate.

11. E. isatidifolia Lam., *Encycl. Méth. Bot.* **2**: 430 (1788). Robust glabrous perennial 30–45 cm, with a stout rhizome bearing pendent, pyriform tubers the size of a hen's egg; latex yellow. Stems scaly below, very stout, sometimes with axillary rays. Leaves oblong, obtuse, entire, crowded. Ray-leaves cordate at base; raylet-leaves deltate-cordate, wider than long. Rays 5, dichotomous. Capsule 7–8 mm, rugulose. Seeds *c.* 4 mm, pale brown, weakly reticulate. *Dry, calcareous pasture and scrub.* ● *E. Spain.* Hs.

12. E. villosa Waldst. & Kit. ex Willd., *Sp. Pl.* **2**: 909 (1800) (*E. pilosa* auct. eur., non L.; incl. *E. austriaca* A. Kerner, *E. carpatica* Wołoszczak, *E. semivillosa* Prokh., *E. tauricola* Prokh.). Stout, glabrous or pubescent, rhizomatous perennial 30–120 cm. Stems numerous, often with non-flowering branches as well as axillary umbels, scaly below. Leaves oblong, oblong-lanceolate to oblong-ovate or elliptical, 2–6 times as long as wide, obtuse to acute, often mucronate, entire, or serrulate near apex. Ray-leaves ovate, obtuse, mucronate; raylet-leaves smaller and relatively wider, yellowish. Rays (4)5 or more, trichotomous and then dichotomous. Capsule 3–6 mm, smooth, minutely tuberculate or with tubercles longer than broad (*E. carpatica*), glabrous to densely villous. Seeds 2·5–3·2 mm, smooth, brown. *Damp meadows, open woods and river-banks. S.E., S. & E.C. Europe, extending northwards to C. Russia and N.W. France.* Al Au †Br Bu Cz Ga Ge Gr Hs Hu It Ju Po Rm Rs (C, W, K, E).

Variable in leaf-shape, indumentum and development of tubercles on the capsule. Several variants occupying limited areas have been described as species, e.g. **E. austriaca** A. Kerner, *Sched. Fl. Exsicc. Austro-Hung.* **3**: 62 (1884) and **E. carpatica** Wołoszczak, *Spraw. Kom. Fizyogr. Krakow.* **27**: 153 (1892), but intermediates between them occur. It may be possible to recognize subspecies, but a thorough investigation of the whole complex is needed. *E. pilosa* L. is an Asiatic species.

13. E. corallioides L., *Sp. Pl.* 460 (1753). Caespitose, villous perennial 40–60 cm. Stems few, rather slender, with few axillary rays below the terminal umbel, not scaly below. Leaves oblong to oblanceolate. Ray-leaves like the cauline but wider; raylet-leaves green or red-tinged. Capsule 3–4 mm, usually densely pubescent, finely granulate. Seeds *c.* 2·5 mm, reddish-brown. *Woods.* ● *C. & S. Italy, Sicilia.* It Si [Br].

(B) Annual; capsule smooth or nearly so; seeds smooth.

14. E. lagascae Sprengel, *Neue Entdeck.* **2**: 115 (1820). Glabrous annual 30–45 cm. Lower leaves ovate, upper oblong-lanceolate, obtuse, entire or weakly sinuate. Ray-leaves ovate-lanceolate or triangular-ovate, subcordate at base, obtuse and mucronate; raylet-leaves ovate-rhombic. Rays 3. Capsule

5–7 mm, ovoid, acutely keeled, not sulcate, weakly reticulate-veined. Seeds 3·6–4·2 mm, grey or brownish with darker spots. *Cultivated ground.* ● *C. & S. Spain; Sardegna.* Hs Sa ?Si.

15. E. arguta Banks & Solander in A. Russell, *Nat. Hist. Aleppo* ed. 2, **2**: 253 (1794). Softly pubescent annual 10–40 cm. Stem with axillary rays below the terminal umbel. Leaves oblanceolate or narrowly obovate, sharply and deeply serrate, acute. Ray-leaves like the cauline but rather wider; raylet-leaves rhombic to triangular, serrate. Rays 4–5, stout. Capsule *c.* 3 mm, finely punctate or reticulate, shallowly sulcate. Seeds grey. *Cultivated ground. S. Greece (Peloponnisos).* Gr. (*E. Mediterranean region.*)

(C) Annual; capsule often setose; seeds smooth.

16. E. microsphaera Boiss., *Diagn. Pl. Or. Nov.* **1**(7): 87 (1846). Rather stout, subglabrous annual 15–30 cm. Leaves oblong, serrulate towards the apex, the lowest obtuse, the rest acute. Ray-leaves ovate, obtuse; raylet-leaves orbicular-triangular obtuse. Rays 5. Capsule *c.* 2·5 mm, subglobose, pubescent when young, becoming glabrous, smooth, not sulcate, tardily dehiscent; seeds ovoid, laterally compressed, dark brown. *Arable land. Turkey-in-Europe (Marmaraereğlisi).* *Tu. (*S.W. Asia.*)

17. E. akenocarpa Guss., *Cat. Pl. Boccad.* 75 (1821) (incl. *E. cybirensis* Boiss., *E. zahnii* Heldr. ex Halácsy). Glabrous or pubescent annual 15–45 cm. Leaves elliptical to obovate-cuneate, serrulate near the acute or rounded, mucronulate apex. Ray-leaves elliptic-ovate. Rays 2–5, stout, usually much longer than the ray-leaves in fruit. Capsule 3–4 mm, subsessile, woody, usually indehiscent, persistent, scarcely sulcate, usually with tubercles ending in a long bristle, sometimes glabrous or sericeous. Seeds 2·6 mm, blackish. *Disturbed ground. Mediterranean region.* Cr Gr Hs It Si Ju.

(D) Perennial; capsule with elongated or hemispherical tubercles; seeds smooth or nearly so.

18. E. palustris L., *Sp. Pl.* 462 (1753). A very robust, caespitose, glabrous, glaucous perennial with creeping rhizome. Stems 50–150 cm, with numerous non-flowering branches and some axillary rays below the terminal umbel. Leaves 20–60(–80) × 3–15 mm, lanceolate or oblong-lanceolate, turning purplish-red in autumn. Ray-leaves ovate, somewhat shorter than rays; raylet-leaves orbicular-ovate, yellowish. Rays more than 5. Capsule 4·5–6 mm, covered with many short tubercles. Seeds 3·2–3·7 mm, brown. 2*n*=20. *Damp places, especially by rivers, in swampy woods or near the sea. Most of Europe from S. Finland, S. Norway and N. France southwards, but rare in the Mediterranean region.* Al Au Bu Co Cz Fe Ga Ge Gr He Ho Hs Hu It Ju No Po Rm Rs (B, C, W, K, E) Su [Be].

19. E. velenovskyi Bornm., *Bot. Jahrb.* **66**: 117 (1933) (*E. soongarica* sensu Hayek, non Boiss.). Robust, glabrous perennial *c.* 60 cm, with axillary rays which do not overtop the terminal umbel. Leaves 15–30(–90) × 4–6(–18) mm, linear-lanceolate, acuminate; margin cartilaginous, sharply serrate in upper half. Ray-leaves 15–40 × 10–18 mm, ovate, scarcely serrate, yellowish in flower; raylet-leaves 10–17 × 10–15 mm, rhombic-ovate. Rays 5–10, 50–80 mm, with 4 cyathia. Capsule 4·2–4·4 × 5 mm, deeply sulcate, with few, small tubercles. Seeds dark brown. ● *Bulgaria, Makedonija, N.E. Greece.* Bu Gr Ju.

20. E. soongarica Boiss., *Cent. Euphorb.* 32 (1860). Like **19** but axillary rays overtopping the terminal umbel; leaves 20–110 × 5–22 mm, usually wider than in **19**; ray-leaves 10–30 ×

2–8 mm; raylet-leaves 4–10 × 2–8 mm; rays 20–35 mm, with 3 cyathia; capsule 4–5 × 4–5 mm, scarcely sulcate, with sparse, hemispherical tubercles. *E.C. Russia* (*Kujbyševskaja Obl.*). Rs (E). (*Temperate Asia.*)

21. E. ceratocarpa Ten., *Fl. Nap.* **1**, *Prodr.* 28 (1811). Glabrous perennial 70–150 cm. Stems with numerous axillary rays. Leaves lanceolate, acute, margin weakly undulate, entire. Ray-leaves ovate-lanceolate; raylet-leaves ovate, narrowed at base. Rays 5–6, first 3- to 5-chotomous, then 2- to 3-chotomous. Capsule 4–5 mm, glabrous, with long flattened-conical tubercles, sulcate. Seeds *c.* 3 mm, dark grey. *Dry places.* ● *S. Italy, Sicilia.* It Si.

22. E. hyberna L., *Sp. Pl.* 462 (1753). Perennial with a stout rhizome. Stems 30–60 cm, usually with axillary rays. Leaves oblong to oblanceolate-oblong, obtuse or emarginate, entire, glabrous above, sparsely villous beneath, turning pinkish-red. Ray-leaves like the cauline leaves. Rays (4–)5(–6). Capsule 5–6 mm, usually with short and long, slender tubercles. Seeds 3·4–3·8 mm. 2*n* = 36. *Damp or shady places, mainly on mountains in the south.* ● *W. & S. Europe, eastwards to N. Italy and northwards to Ireland.* Br Co Ga Hb Hs It Lu Sa.

1 Capsule distinctly pedicellate **(a) subsp. hyberna**
1 Capsule nearly sessile
2 Glands with thickened, rugose margins; ripe seed smooth
(b) subsp. insularis
2 Glands with thin, flat margins; ripe seed rugulose
(c) subsp. canuti

(a) Subsp. **hyberna**: Axillary rays usually 0–5; glands after flowering with thin, flat margins; capsule distinctly pedicellate; seeds nearly smooth, pale brownish-grey. *Throughout most of the range of the species.*

(b) Subsp. **insularis** (Boiss.) Briq., *Prodr. Fl. Corse* **2**(2): 77 (1935) (*E. insularis* Boiss.): Axillary rays usually numerous; glands after flowering with thickened, rugose margins; capsule nearly sessile; seeds nearly smooth, pale brownish-grey. *Corse, Sardegna, N.W. Italy.*

(c) Subsp. **canuti** (Parl.) Tutin, *Feddes Repert.* **79**: 55 (1968) (*E. canuti* Parl.): Axillary rays 0–1; glands after flowering with thin, flat margins; capsule subsessile; seeds rugulose, reddish. *Maritime Alps.*

E. gibelliana Peola, *Malpighia* **6**: 249 (1892), is like subsp. (c) but with undulate glands. It occurs on serpentine in N.W. Italy and probably represents an ecotype of subsp. (c).

23. E. gregersenii K. Malý ex G. Beck, *Glasn. Muz. Bosni Herceg.* **32**: 90 (1920). A softly pubescent perennial. Stems with non-flowering axillary branches. Leaves oblong, obtuse or emarginate, entire, glabrous above, pilose beneath. Raylet-leaves broadly elliptical. Rays 4–5. Glands brown. Capsule glabrous, with 2 crests formed of elongated, confluent tubercles on the back of each valve. *Woods and meadows on serpentine.* ● *C. Jugoslavia.* Ju.

24. E. squamosa Willd., *Sp. Pl.* **2**: 918 (1800). Pubescent perennial 45–60 cm. Rhizome nodose-thickened. Stems usually pubescent, with axillary rays. Leaves oblong to oblong-elliptical, entire or nearly so, appressed-pubescent, particularly beneath, very shortly petiolate, the lower obtuse, the upper acute. Ray-leaves rhombic-elliptical; raylet-leaves reniform-deltate, obtuse, glabrous. Rays 5–8, usually dichotomous, slender. Capsule 5 mm, covered with cylindrical-filiform tubercles and often pubescent. Seeds 3 mm, smooth. *Woods. S.E. Russia* (*Rostovskaja Obl.*). Rs (E). (*Caucasian region.*)

(25–29). **E. epithymoides** group. Perennial, with a stout stock. Axillary rays few or none. Leaves lanceolate to elliptical or obovate, obtuse. Rays 4–5. Capsule with long, slender, often purple-tipped tubercles.

A group of closely related taxa centred in the Balkan peninsula; some may be best regarded as subspecies, but further investigation is required.

1 Leaves serrate **29. montenegrina**
1 Leaves entire or serrulate near apex
2 Plant glabrous **27. gasparrinii**
2 Plant pubescent
3 Leaves 3–4 times as long as wide; capsule *c.* 6 mm
26. lingulata
3 Leaves 2–3 times as long as wide; capsule 3–5 mm
4 Leaves usually 30–50 mm; raylet-leaves elliptical; capsule 3–4 mm **25. epithymoides**
4 Leaves usually 10–20 mm; raylet-leaves broadly ovate; capsule 4–5 mm **28. fragifera**

25. E. epithymoides L., *Sp. Pl.* ed. 2, 656 (1762) (*E. polychroma* A. Kerner). Softly and rather densely pubescent. Stems 20–40 cm, robust, not woody below. Leaves usually 30–50 × 11–26 mm, 2–3 times as long as wide, obovate-oblong or ellipticoblong, rounded at the base, entire or obscurely serrulate. Ray-leaves like the cauline but yellow, sometimes purple-tinged in flower; raylet-leaves elliptical. Rays about as long as ray-leaves; lobes of the cyathium as long as the cup; glands small. Capsule 3–4 mm. Seeds 2·5–2·9 mm, brown or yellowish-grey, with a brown, raised reticulum. 2*n* = 16. *Somewhat calcicole. C. & S.E. Europe, from S.E. Germany to C. Ukraine and S. Bulgaria.* Al Bu Cz Ge Gr Hu It Ju Po Rm Rs (W).

E. jacquinii Fenzl ex Boiss. in DC., *Prodr.* **15**(2): 136 (1862), of unknown origin, but perhaps from N. Jugoslavia, is like **25** but has rugose seeds. It may be a hybrid or an abnormality.

26. E. lingulata Heuffel, *Verh. Zool.-Bot. Ges. Wien* **8**: 192 (1858). Like **25** but cauline leaves 3–4 times as long as wide; rays much longer than ray-leaves; capsule *c.* 6 mm. *Shady places, mainly in the mountains.* ● *Albania, C. & S. Jugoslavia, S. Romania, N. Greece.* Al Gr Ju Rm.

27. E. gasparrinii Boiss. in DC., *Prodr.* **15**(2): 125 (1862). Like **25** but glabrous; stems woody below; leaves ovate-oblong or elliptical; ray-leaves obovate or suborbicular; capsule *c.* 4 mm; seeds minutely punctate or smooth. *Damp places in the mountains.* ● *C. Italy, Sicilia.* It Si.

28. E. fragifera Jan, *Cat. Pl. Phaen.* 76 (1818). Softly pubescent. Stems 10–30 cm, woody below. Leaves usually 10–20 mm, lanceolate to obovate, rounded at the base, entire. Ray-leaves ovate; raylet-leaves broadly ovate. Rays about as long as ray-leaves. Capsule 4–5 mm, densely covered with filiform papillae, which become dark red on drying. Seeds 3·4–3·9 mm, brownish- or bluish-grey, with a paler, raised reticulum. *Rocky places.* ● *W. Jugoslavia and Albania, just extending to N.E. Italy.* Al It Ju.

29. E. montenegrina (Bald.) K. Malý ex Rohlena, *Sitz.-Ber. Böhm. Ges. Wiss.* (*Math.-Nat. Kl.*) **1912**(1): 110 (1913). Like **28** but leaves elliptical or elliptic-lanceolate, serrate, narrowed at the base; ray-leaves suborbicular, but attenuate and subpetiolate at the base; seeds 2·5 mm, punctate-scabrid. *Mountain rocks.* ● *S. Jugoslavia.* Ju.

30. E. oblongata Griseb., *Spicil. Fl. Rumel.* **1**: 136 (1843). Stout, caespitose, densely pubescent perennial up to 80 cm. Stems robust, often with axillary rays below the terminal umbel.

Leaves narrowly obovate, obtuse, serrulate. Ray-leaves ovate, narrowed, or rounded at base. Rays 5, about equalling or somewhat exceeding the ray-leaves. Glands often only 2–3. Capsule 3·8–4·3 mm, sparsely covered with short tubercles, glabrous. Seeds 2·5 mm, brown. *Shady places. S. part of Balkan peninsula and Aegean region.* Al Bu Cr Gr Ju Tu.

31. E. apios L., *Sp. Pl.* 457 (1753). Decumbent to erect, pubescent perennial with a subterranean, napiform tuber 2–7 cm. Stems 5–20 cm, slender, scaly at base, with 0–1(–3) axillary rays. Leaves linear-lanceolate to oblong or obovate-cuneate, obtuse to subacute, serrulate. Ray-leaves like the cauline; raylet-leaves suborbicular, cuneate. Umbel with 3–5 rays. Capsule *c.* 3 mm, with short, conical tubercles. Seeds *c.* 2 mm, smooth, dull, dark brown. 2*n* = 12. *Dry, rocky or bushy places. S.E. part of Balkan peninsula; Aegean region; S.E. Italy.* Bu Cr Gr It.

E. dimorphocaulon P. H. Davis, *Phyton* (*Austria*) **1**: 196 (1949), from Kriti, is probably a seasonal form of **32**, flowering in autumn instead of spring, at the time when non-flowering shoots are present.

32. E. polygalifolia Boiss. & Reuter in Boiss., *Cent. Euphorb.* 34 (1860). Glabrous or rarely pubescent perennial with a woody stock and numerous slender, decumbent or ascending stems 5–30 cm. Lower leaves obovate, much smaller than the upper; upper obovate-oblong; all obtuse, entire or serrulate. Ray-leaves ovate; raylet-leaves rhombic-ovate or reniform, mucronulate. Rays (3–)4–6. Capsule *c.* 3·5 mm, depressed-globose, with numerous stout, cylindrical tubercles. Seeds smooth, grey, shiny. *Dry grassland and thickets.* ● *N. & E. Spain.* Hs.

E. mariolensis Rouy, *Bull. Soc. Bot. Fr.* **29**: 127 (1882), may be subspecifically distinct, but requires further investigation. It is pubescent, has narrower leaves and larger, less densely tuberculate capsules. It is recorded from calcareous hillsides in N.E. Spain and S. France.

33. E. uliginosa Welw. ex Boiss. in DC., *Prodr.* **15**(2): 127 (1862). Glabrous or somewhat pubescent perennial 20–60 cm, with a stout, woody stock. Stems very slender, woody at base. Leaves 5–20 × 1–3 mm, linear-oblong, serrulate, obtuse, coriaceous. Ray-leaves linear-lanceolate to obovate-cuneate, shorter than umbel-rays; raylet-leaves broadly obtriangular. Rays 2–5. Capsule 2·5–3 mm, densely covered with short, clavate tubercles. Seeds 2 mm, smooth, dark brown. *Temporary pools and wet heaths.* ● *W. Portugal, N.W. Spain.* Lu Hs.

34. E. dulcis L., *Sp. Pl.* 457 (1753). More or less pubescent perennial 20–50 cm. Rhizome long, thicker than the stems, fleshy, swollen and jointed. Stems slender, scaly at base, terete, with (0–)4–8 axillary rays. Leaves 25–70 mm, elliptical to oblong. Ray-leaves shorter than rays, like the cauline but wider; raylet-leaves triangular-subcordate, serrulate. Rays (3–)5(–8), slender. Glands dark purple after flowering. Capsule 3–4 mm, deeply sulcate, glabrous or pubescent, irregularly and sometimes sparsely covered with cylindrical and hemispherical tubercles. Seeds 2·3–2·6 mm, smooth, dark brown. 2*n* = 12, 24. *Damp or shady places.* ● *W. & C. Europe, extending locally southwards to C. Italy and Macedonia.* Au Be Bu Co Cz Ga Ge He Ho Hs Hu It Ju Lu Po Rm Rs (W) [Br Da].

E. deseglisei Boreau ex Boiss. in DC., *Prodr.* **15**(2): 128 (1862), resembles **34** in its leaves, but is said to differ from it in having the rays shorter than the ray-leaves. It occurs sporadically in France and may be an abnormality or a hybrid, but requires further investigation.

35. E. angulata Jacq., *Collect. Bot.* **2**: 309 (1789). Like **34** but glabrous; stems finely ribbed in the upper part, the ribs with sharp angles; leaves 10–25 mm; glands yellowish-red after flowering; capsule 2·5 mm. ● *S. & E.C. Europe.* Au Cz Ga Hs Hu It Ju Lu Po Rm Rs (W).

36. E. carniolica Jacq., *Fl. Austr.* **5**: 34 (1778). Glabrous or slightly pubescent perennial 20–55 cm, with a stout, woody stock and long, creeping rhizome. Stems in small tufts, scaly at base, usually with axillary rays below the terminal umbel. Leaves (20–)40–70 mm, obovate-oblong, cuneate, obtuse or cuspidate, entire. Ray-leaves like the cauline; raylet-leaves ovate-lanceolate, narrowed at base, entire. Rays 3–5. Cyathium long-pedunculate. Glands brownish-yellow. Capsule 5 mm; valves keeled, covered with hemispherical tubercles. Seeds 3·2–3·7 mm, smooth, brown. *Mountain woods and bushy slopes.* ● *E. Alps, N. part of Balkan peninsula, E. Carpathians.* Au He It Ju Rm Rs (W).

37. E. duvalii Lecoq & Lamotte, *Cat. Pl. Centr. Fr.* 327 (1847). Glabrous perennial 20–40 cm, with a stout woody stock. Stems tufted, scaly at base, often with axillary rays below the terminal umbel. Leaves obovate to lanceolate, obtuse, subcordate at base, usually serrulate. Ray-leaves broadly elliptical to almost reniform, shorter than rays; raylet-leaves rhombic to suborbicular. Rays (3–)5. Cyathium more or less sessile. Capsule *c.* 4 mm, covered with large, irregular tubercles. Seeds *c.* 3 mm, smooth, brown. *Rocky pastures and thickets.* ● *S. France.* Ga.

38. E. brittingeri Opiz ex Samp., *Lista Esp. Herb. Port., Ap.* **2**: 5 (1914) (*E. verrucosa* L. 1759, non L. 1753). More or less pubescent perennial 20–45 cm, with a woody stock and numerous slender, herbaceous stems without scales at the base, and usually with several axillary non-flowering shoots. Leaves 20–35 mm, oblong-elliptical to obovate, serrulate. Ray-leaves ovate to broadly elliptical; raylet-leaves yellowish at flowering time, green or purplish later. Rays (4–)5. Capsule 3–4 mm, weakly sulcate, with crowded tubercles. Seeds 2–2·5 mm, dark brown with paler, raised markings when quite ripe. *Woods and grassy places.* ● *W. & C. Europe, N. & C. Italy, N. part of Balkan peninsula.* Al Au Be Cz Ga Ge Gr He Hs Hu It Ju Rm.

E. flavicoma DC., *Cat. Pl. Horti Monsp.* 110 (1813), from Spain, S. France and N. Italy, is probably not specifically distinct from **38**. It is a smaller plant with rather coriaceous leaves and 1–5 rays, which are usually shorter than the ray-leaves; the seeds are said to be rather larger than in **38**.

39. E. ruscinonensis Boiss., *Cent. Euphorb.* 33 (1860). Nearly glabrous perennial 10–20 cm. Stems woody below, without scales at base, with non-flowering branches and axillary rays. Leaves serrulate, the lower elliptical, obtuse, sparsely hairy beneath, the middle linear-lanceolate, acute, glabrous; upper cauline and ray-leaves ovate or subcordate, obtuse. Rays 5, short. Capsule 5–7 mm, scarcely sulcate, covered with small, hemispherical-conical tubercles. *Calcareous slopes.* ● *S. France* (*Corbières*). Ga.

40. E. welwitschii Boiss. & Reuter, *Pugillus* 108 (1852). Nearly or quite glabrous perennial with a napiform tuber. Stems 30–60 cm, stout, woody at base, often with axillary rays. Leaves 10–30 × 10–15 mm, broadly ovate to ovate-lanceolate, obtuse or subobtuse, serrulate or almost entire. Ray-leaves broadly ovate to suborbicular, obtuse or subobtuse, shorter than rays; raylet-leaves orbicular-cordate. Rays 5(–6). Capsule 3·5–4 mm, covered with hemispherical tubercles, sometimes

EUPHORBIACEAE

sparsely hairy. Seeds 2·8–3 mm, ovoid, smooth, dark brown. *Grassy slopes and margins of fields. C. & S. Portugal.* ?Hs Lu. (*Morocco.*)

41. E. monchiquensis Franco & P. Silva, *Feddes Repert.* **79**: 56 (1968) (*E. rupicola* var. *major* Boiss.). Like **40** but up to 100 cm; leaves 50–80 × 12–20 mm, lanceolate, entire, glabrous or very sparsely villous; ray-leaves narrowly rhombic; capsule rather sparsely tuberculate; seeds ellipsoid. ● *S.W. Portugal.* Lu.

42. E. clementei Boiss., *Elenchus* 82 (1838). More or less glabrous perennial 30–70 cm, with a napiform tuber. Stems terete, ascending, woody at base, sometimes with numerous slender axillary rays, usually bearing a single cyathium. Leaves obovate-oblong, obtuse, narrowed to a very short petiole, serrulate. Ray-leaves broadly ovate or rhombic, much shorter than rays; raylet-leaves suborbicular, entire. Rays 5. Capsule 4–5 mm, with a few hemispherical tubercles. Seeds 2·5–3·5 mm, smooth, reddish-brown. *Scrub and calcareous rocky places. S. Portugal, S. Spain.* Hs Lu. (*N.W. Africa.*)

43. E. bivonae Steudel, *Nomencl. Bot.* ed. 2, **1**: 610 (1840). Glabrous shrub up to 150 cm. Stems leafless below, densely leafy above, not persistent and spiny when dead. Leaves linear-lanceolate, to ovate-lanceolate, acute or acuminate, rarely obtuse, entire. Ray-leaves ovate, as long as or longer than rays; raylet-leaves broadly obovate. Rays 5, short, dichotomous. Capsule 3·5–4·8 mm, glabrous, sulcate, with low, broad tubercles. Seeds 3 mm, smooth, dark brown, shiny. *Calcareous rocks near the sea. Sicilia (W. coast), Malta.* Si. (*N. Africa.*)

44. E. squamigera Loisel., *Fl. Gall.* 729 (1807) (*E. rupicola* Boiss.). Shrub 60–120 cm. Stems branched, bare below, densely leafy above. Leaves 25–50 mm, linear-lanceolate to narrowly elliptical, mucronate, glabrous or pubescent beneath, entire or obscurely serrulate. Ray-leaves elliptic-ovate; raylet-leaves rhombic- or orbicular-ovate, obtuse, mucronate, yellowish at flowering time. Rays 5. Capsule 4–5 mm, sulcate, covered with short, cylindrical tubercles. Seeds 2·8–3 mm, smooth, brown. *Calcareous rocks. S. & E. Spain.* Hs. (*N. Africa.*)

E. carthaginensis Porta & Rigo ex Willk., *Ill. Fl. Hisp.* **2**: 154 (1892), from S.E. Spain, is doubtfully distinct from **44**. It is said to be pruinose and to have a capsule *c.* 3 mm, with broad, scale-like tubercles and slightly smaller seeds.

45. E. spinosa L., *Sp. Pl.* 457 (1753). Glabrous, freely branched shrub 10–30 cm. Dead branches and umbel-rays more or less persistent but not pungent. Leaves 5–15(–20) mm, lanceolate or linear-lanceolate, entire. Ray-leaves obovate, about as long as rays, yellowish. Rays 1–5, very short, each usually with 1 cyathium. Capsule 3–4 mm, weakly sulcate, usually with long (rarely short) tubercles. Seeds 2–3 mm, smooth, brown. 2*n* = 14. *Dry, rocky places.* ● *Mediterranean region, from France to Albania.* Al Co Ga It Ju Sa Si.

46. E. glabriflora Vis., *Mem. Ist. Veneto* **12**: 477 (1864). Small, glabrous shrub with a thick, woody stock. Stems 10–20 cm, woody below, with slender, annual flowering branches. Leaves linear-lanceolate to elliptic-oblong, acute or obtuse, glaucous, punctate. Ray-leaves rather wider than the cauline, equalling or exceeding rays; raylet-leaves ovate or obovate, yellow. Rays (1–)3–5. Capsule *c.* 4 mm, with long, slender tubercles. Seeds smooth. *Stony slopes on mountains.* ● *Balkan peninsula.* Al Gr Ju.

47. E. acanthothamnos Heldr. & Sart. ex Boiss., *Diagn. Pl. Or. Nov.* 3(4): 86 (1859). Glabrous, intricately branched shrub 10–30 cm, with the branches usually terminating in paired spines formed from the indurated, forked rays of the umbels. Leaves elliptical to elliptic-obovate, obtuse or acute, entire. Ray-leaves like the cauline; raylet-leaves cuneate-obovate, yellow. Rays 3(–4). Capsule 3–4 mm, sulcate, with short, conical tubercles. Seeds 2 mm, smooth, brown. *Rocky places; lowland. Greece and Aegean region.* Cr Gr ?Ju.

48. E. chamaebuxus Bernard ex Gren. & Godron, *Fl. Fr. Prosp.* 8 (1846). Glabrous perennial 5–15 cm, with a slender rhizome. Stems scaly at base. Leaves obovate to elliptical, obtuse or acute, mucronulate, entire. Ray-leaves like the cauline, nearly equalling the rays. Rays 2–3, or the umbel often reduced to a single cyathium. Glands 4, reddish. Capsule *c.* 5 mm, covered with obtuse, flattened, flap-like tubercles. Seeds 1·8–2 mm smooth, brown. *Mountain rocks and screes.* ● *W. Pyrenees, Cordillera Cantábrica, Sierra Nevada.* Ga Hs.

49. E. capitulata Reichenb., *Fl. Germ. Excurs.* 873 (1832). Glabrous perennial 1–10 cm, with numerous stems from a slender rhizome. Leaves less than 10 mm, obovate, obtuse, entire, usually imbricate. Ray-leaves like the cauline but shorter. Cyathium solitary. Glands 8, purple, Capsule sulcate, with hemispherical or cylindrical, inflated tubercles. Seeds smooth. *Mountain rocks and screes.* ● *Balkan peninsula.* Al Gr Ju.

(E) Annual; capsule tuberculate; seeds smooth.

50. E. platyphyllos L., *Sp. Pl.* 460 (1753). Glabrous or pubescent annual 15–80(–110) cm. Stems with numerous axillary rays. Leaves obovate- to oblong-lanceolate, serrulate, acute, deeply cordate at base. Ray-leaves elliptic-oblong; raylet-leaves deltate, the lowest markedly different from the ray-leaves and similar to those subtending the cyathia. Rays usually 5. Capsule 2–3 mm, covered with hemispherical tubercles, shallowly sulcate. Seeds 1·8–2·2 mm, olive-brown. *S., W. & C. Europe.* Al Au Be Bl *Br Bu Co Cr Cz Ga Ge Gr He *Ho Hs Hu It Ju Po Rm Rs (W, K) Si Tu.

E. gaditana Cosson, *Not. Pl. Crit.* 46 (1849), was collected once only in cultivated ground in S.W. Spain (Sanlucar de Barrameda, N. of Cádiz). It is probably an abnormal form of **50** with the leaves narrowed at the base (or the lower shortly petiolate) and with smaller seeds (*c.* 1 mm).

51. E. serrulata Thuill., *Fl. Paris* ed. 2, 237 (1799) (*E. stricta* L., nom. illegit.). Like **50** but usually more slender and smaller, always glabrous; umbel with (2–)4–5 rays; raylet-leaves becoming narrower downwards and passing gradually into the ray-leaves; capsule 2·5 mm or less, deeply sulcate, covered with cylindrical tubercles which are longer than wide; seeds 1·2–1·5 mm, red-brown. *S., C. & W. Europe.* Al Au Be Br Bu Co Cz Ga Ge Gr He Ho Hs Hu It Ju Po Rm Rs (W, K) Tu.

(F) Perennial; capsule tuberculate; seeds minutely tuberculate.

52. E. pubescens Vahl, *Symb. Bot.* **2**: 55 (1791). Densely pubescent to subglabrous perennial up to 100 cm. Stem stout, sometimes with axillary rays. Leaves oblong-lanceolate to linear, acute to subobtuse, cordate at base, denticulate or nearly entire. Ray-leaves elliptical to obovate; raylet-leaves rhombic-ovate, subcordate at base. Rays 5–6. Capsule (2–)3–4 mm, deeply sulcate, villous or subglabrous, with oblong tubercles. Seeds (1·5–)2–2·5 mm, dark brown, with paler, small, irregular tubercles. 2*n* = 14. *Damp meadows and river-banks. S. Europe.* Al Bl Co Cr Ga Gr Hs It Ju Lu Sa Si Tu.

(G) Annual; capsule with 2 rows of tubercles on each valve; seeds verruculose.

53. E. cuneifolia Guss., *Pl. Rar.* 190 (1826). Glabrous annual 5–25 cm. Stem slender, simple or sparingly branched. Leaves broadly obovate, long-cuneate, denticulate near the rounded apex. Ray-leaves like the cauline; raylet-leaves suborbicular. Rays 5, trichotomous then dichotomous. Capsule 1·5 mm, scarcely sulcate. Seeds 1–1·3 mm. *Damp grassy places. C. Mediterranean region.* Co It Sa Si.

(H) Annual; capsule winged; seeds alveolate-reticulate.

54. E. pterococca Brot., *Fl. Lusit.* **2**: 312 (1804). Glabrous annual 10–30 cm, often with axillary rays below the terminal umbel. Leaves obovate or spathulate, obtuse, serrulate, the lower shortly petiolate. Ray-leaves like the cauline; raylet-leaves rhombic-ovate. Rays 5 (rarely more), trichotomous then dichotomous. slender. Capsule *c.* 1·5 mm, smooth, with 2 undulate wings on each valve. Seeds *c.* 1·3 mm, dark brown. *Grassy places. C. Portugal to S.E. Greece, mainly in the islands.* Bl Co Gr Hs It Lu Sa Si.

(I) Annual; capsule smooth, unwinged; seeds reticulate-rugose or tranversely sulcate.

55. E. helioscopia L., *Sp. Pl.* 459 (1753). Erect, glabrescent annual, usually with a single stem 10–50 cm. Leaves obovate-spathulate, obtuse, serrate in the upper half. Ray- and raylet-leaves like the cauline but smaller. Rays 5, trichotomous then dichotomous. Capsule 2·5–3·5 mm. Seeds 2 mm, reticulate-rugose. 2*n*=42. *Disturbed ground. Almost throughout Europe, but only as a casual in the extreme north.* All except Az Fa Is Sb.

E. helioscopioides Loscos & Pardo, *Ser. Pl. Arag.* 93 (1863) and **E. dominii** Rohlena, *Sitz.-Ber. Böhm. Ges. Wiss.* (*Math.-Nat. Kl.*) 1904(38): 83 (1905) seem to be identical and are probably dwarf variants of **55**. They have several decumbent stems 3–10 cm; crowded, imbricate leaves; rays 2–3, short, and seeds *c.* 1 mm.

56. E. phymatosperma Boiss. & Gaill. in Boiss., *Diagn. Pl. Or. Nov.* 3(4): 83 (1859). Glabrous, glaucous annual 8–12 cm. Leaves *c.* 6 mm, obovate, obtuse, entire. Ray-leaves elliptical; raylet-leaves ovate-rhombic, mucronulate. Rays 3(–5). Capsule 3–4 mm. Seeds 2–2·5 mm, with 3 transverse depressions on each face. *S. Italy, N. Greece.* Gr It. (*N. Africa, S.W. Asia.*)

Represented in Europe and N. Africa by subsp. **cernua** (Cosson & Durieu) Vindt, *Trav. Inst. Sci. Chérif.* ser. bot., **2**: 82 (1953).

Sect. MYRSINITEAE (Boiss.) Tutin. Herbs. Glands crescentic, with short horns usually dilated and minutely lobed at apex. Bracts between male flowers absent.

57. E. myrsinites L., *Sp. Pl.* 461 (1753). Glabrous, glaucous perennial up to 40 cm. Stems decumbent or ascending, numerous, stout, simple, densely leafy, with 0–3 axillary rays. Leaves obovate to suborbicular, cuspidate or mucronate, rather thick and fleshy. Ray-leaves obovate-spathulate to orbicular, mucronate. Raylet-leaves suborbicular to broadly cordate, mucronate. Rays (1–)5–12, once or twice dichotomous, variable in length. Glands with dilated, often weakly lobed horns. Capsule (4–)5–7 mm, glabrous, smooth or minutely tuberculate. Seeds (2–)3–4 mm, vermiculate-rugose or rarely smooth, greyish-brown. *Rocky places. S. Europe, from Islas Baleares to Krym.* Al Bl Bu Co Cr Gr It Ju Rm Rs (K) Si Tu [Cz].

Variable in size, length of rays, and degree of rugosity of seed. Populations on some of the Mediterranean islands are nearly homogeneous for these features and have been described as species: **E. fontqueriana** W. Greuter, *Candollea* 20: 170 (1965), in Islas Baleares, **E. corsica** Req., *Ann. Sci. Nat.* 5: 384 (1825), in Corse, and **E. rechingeri** W. Greuter, *Candollea* 20: 172 (1965), in Kriti. They may merit subspecific rank, but further investigation of the species throughout its range is desirable.

58. E. rigida Bieb., *Fl. Taur.-Cauc.* **1**: 375 (1808) (*E. biglandulosa* Desf.). Glabrous, intensely glaucous perennial with a thick woody stock. Stems 30–50 cm, stout, erect or ascending, densely leafy. Leaves lanceolate, acuminate, thick and fleshy, the lower patent, the upper almost imbricate. Ray-leaves obovate; raylet-leaves suborbicular, often mucronate. Rays 6–12, short, once or twice dichotomous. Glands with capitate, minutely lobed horns. Capsule 5–8 mm, strongly trigonous, sparsely papillose when dry. Seeds smooth, whitish when ripe. 2*n*=20. *Dry, rocky places. S. Europe; local.* Al ?Cr Gr It Lu Rs (K) Si Tu.

59. E. broteri Daveau, *Bol. Soc. Brot.* **3**: 33 (1885). Like **58** but less glaucous; leaves linear-lanceolate, acute to subobtuse; capsule minutely hyaline-punctate and whitish-granulate; seeds shallowly and irregularly rugose-vermiculate. *On acid, sandy soils.* ● *E. Portugal, Spain.* Hs Lu.

Sect. LATHYRIS Dumort. Biennial; cauline leaves decussate; capsule with spongy pericarp.

60. E. lathyris L., *Sp. Pl.* 457 (1753). Glabrous, glaucous biennial up to 150 cm, with numerous axillary shoots. Leaves 30–150 × 5–25 mm, linear to oblong-lanceolate, entire. Ray-leaves ovate-lanceolate. Raylet-leaves triangular-ovate, acute, paler green than the cauline and ray-leaves. Rays 2–4, up to 8 times dichotomous. Glands with two clavate horns. Capsule 9–13 × 13–17 mm, shallowly sulcate, more or less smooth. Seeds 5 mm, barrel-shaped, rugulose, brown or grey. *A ruderal and weed of cultivated ground. S., W. & C. Europe, but probably native only in E. & C. Mediterranean region.* Co *Ga Gr It *Ju Sa [Au Az Be Bl Br Bu Cz Ge He Ho Hs Lu Rm].

Sect. CYMATOSPERMUM (Prokh.) Prokh. Annuals; cauline leaves opposite or alternate; capsule with indurated mesocarp; seeds ornamented.

(A) Seeds tuberculate or rugulose.

61. E. aleppica L., *Sp. Pl.* 458 (1753). Glabrous or minutely papillose, somewhat glaucous annual up to 40 cm, sometimes with up to 30 basal branches, and 1–9(–12) axillary rays. Leaves 10–25(–50) × 0·2–3(–5) mm, very dense and closely imbricate, linear-subulate to linear-oblanceolate, entire. Ray-leaves like the upper cauline. Raylet-leaves ovate-rhombic to trullate or falcate, entire or irregularly toothed. Rays 2–4(–6) up to 5 times dichotomous. Glands with 2 horns, the horns paler than the glands. Capsule 2 mm, shallowly sulcate, more or less smooth. Seeds 1·5 mm, ovoid-tetragonous, grey with white tubercles. *Cultivated or stony ground. E. & C. Mediterranean region; Krym.* Bu ?Co ?Cr *Ga Gr It Ju Rs (K) Si Tu.

62. E. medicaginea Boiss., *Elenchus* 82 (1838). Glabrous annual up to 40 cm, with (0–)2–8 axillary rays. Leaves 20–35 × 6–12 mm, obovate-cuneate to elliptic-oblong, subobtuse or emarginate, minutely serrulate. Ray-leaves like the upper cauline. Raylet-leaves up to 13 × 19 mm, rhombic-deltate, upper ones often yellowish. Rays (3–)5, up to 5 times dichotomous. Glands with 2 horns, often longer than the glands. Capsule 2·5 × 2·5–3 mm, deeply sulcate, finely granulate on the keels. Seeds 1·5–1·75 mm, oblong-quadrangular, irregularly and closely white vermiculate-rugulose on a blackish ground. *Pastures. S. Portugal, S. Spain, Islas Baleares.* Bl Hs Lu.

63. E. dracunculoides Lam., *Encycl. Méth. Bot.* **2**: 428 (1788). Glabrous annual up to 10 cm, often much-branched from the base, and with 0–6 axillary rays. Leaves 4–15 × 1–4·5 mm, obovate to linear-spathulate, often 2- to 6-toothed at the apex. Ray- and raylet-leaves smaller than the cauline leaves, variable in shape. Rays (2–)3, up to 4 times dichtomous. Capsule 2·5 × 2·8 mm, shallowly sulcate, smooth. Seeds 1·5 mm, ovoid, grey with white conical tubercles. *Stony ground. S.E. Spain (Cabo de Gata, Prov. Almería).* Hs.

This species is represented in Europe only by subsp. **inconspicua** (Ball) Maire, *Bull. Soc. Hist. Nat. Afr. Nord* **20**: 202 (1929) (*E. glebulosa* var. *almeriensis* Lange). The typical subspecies occurs throughout most of the drier tropics and subtropics of the Old World.

64. E. exigua L., *Sp. Pl.* 456 (1753). Glabrous annual up to 35 cm, often much-branched from the base, with 0–3 (rarely more) axillary rays. Leaves 3–25 × 1–2 mm, linear to oblong-cuneate, entire. Ray-leaves like the upper cauline. Raylet-leaves obliquely triangular-ovate-lanceolate, rarely 1- to 2-toothed on one side near the base. Rays 3–5, up to 7 times dichotomous. Glands with 2 horns, rarely the horns much reduced. Capsule 1·6–2 × 1·6 mm, shallowly sulcate, smooth but granulate on the keels. Seeds 1·2 mm, ovoid-quadrangular, vermiculate-rugose, grey. $2n=24$. *Cultivated ground. Most of Europe northwards to c. 65° N., but absent from much of the east.* All except Fa Fe Is Rs (N, K, E) Sb.

(B) Seeds transversely sulcate.

65. E. falcata L., *Sp. Pl.* 456 (1753) (incl. *E. acuminata* Lam.). Glabrous annual up to 40 cm, simple or with 2–3(–9) branches from the base, with (0–)8(–16) axillary rays. Leaves 5–30 × 3–5 mm, obovate-spathulate to linear-oblong, cuneate, mucronate, entire. Ray-leaves like the cauline. Raylet-leaves up to 21 × 10 mm, asymmetrically suborbicular or elliptic-ovate, acuminate to aristate, subentire. Rays 4–5, up to 5 times dichotomous. Glands broad, with 2 horns. Capsule 1·5–2 × 1–2·5 mm, shallowly sulcate, smooth. Seeds 1·2 mm, flattened-ovoid-quadrangular, pale grey or brown. $2n=16, 36$. *Disturbed ground and as a weed and ruderal. Europe from N.C. France and C. Russia southwards.* Al Au Bl Bu Co Cr Cz Ga Ge Gr He Hs Hu It Ju Lu Po Rm Rs (C, W, K, E) Sa Si Tu.

(C) Seeds longitudinally sulcate.

66. E. sulcata De Lens ex Loisel., *Fl. Gall.* ed. 2, **1**: 339 (1828). Glabrous annual up to 10 cm, often becoming much-branched from the base, with 0–1 axillary rays. Leaves 4–7 × 1–1·5 mm, linear to linear-oblanceolate, entire. Ray-leaves slightly larger than the upper cauline. Raylet-leaves like the ray-leaves but somewhat wider. Rays (2–)3–4(–5), often many times dichotomous. Glands with 2 horns. Capsule 1·7 × 1·7 mm, deeply sulcate, smooth. Seeds 1·25–1·5 mm, ovoid-hexagonal, with a longitudinal furrow on each face, pale grey, often darker in the furrows. *Dry, open ground. S.W. Europe.* Bl Ga Hs It.

(D) Seeds pitted, or sulcate and pitted.

67. E. peplus L., *Sp. Pl.* 456 (1753). Glabrous annual up to 40 cm, with 2 or more branches from the base and with 0–3 axillary rays. Leaves 5–25 × 3–15 mm, with petioles up to 8 mm,

ovate, suborbicular or obovate, entire. Ray-leaves like the cauline, but with shorter petioles. Raylet-leaves smaller, slightly obliquely ovate. Rays 3, up to 5 times dichotomous. Glands with 2 filiform horns. Capsule 2 × 2 mm, shallowly sulcate, smooth; each valve with two dorsal ridges. Seeds 1·1–1·4 mm, ovoid-hexagonal, sulcate ventrally and pitted dorsally, pale grey, darker in the depressions. $2n=16$. *Weed of cultivated ground. Most of Europe northwards to c. 65° N.* All except Fa Is Rs (N, ?K) Sb.

E. peploides Gouan, *Fl. Monsp.* 174 (1765), appears to be merely a dwarf variant of **67**, with a poorly developed umbel and smaller seeds with fewer pits. It occurs in dry places in the Mediterranean region and Portugal.

E. calabrica Huter, Porta & Rigo, *Österr. Bot. Zeitschr.* **57**: 436 (1907), from Calabria, is like **67** but the capsule is without ridges and the seeds are vermiculate-rugose. It is doubtfully distinct from **67** and is in need of further investigation.

68. E. ledebourii Boiss., *Cent. Euphorb.* 35 (1860). Glabrous annual up to 20 cm, simple or branched from the base, with 0–2 axillary rays. Leaves 10–20 × 0·5–2 mm, subopposite or opposite, linear, entire. Ray- and raylet-leaves like the upper cauline but somewhat larger. Rays (2–)5, often much-branched. Glands with 2 horns. Capsule *c.* 2·7 × 2·7 mm, shallowly sulcate, finely granulate on the keels. Seeds 2 mm, ovoid-cylindrical, pitted, the pits circular on the dorsal faces and elongate and irregularly confluent on the ventral faces, pale grey, often darker in the depressions. *Cliffs and stony ground. Krym (near Sudak).* Rs (K). (*Caucasian region.*)

69. E. taurinensis All., *Fl. Pedem.* **1**: 287 (1785) (incl. *E. graeca* Boiss. & Spruner). Glabrous annual up to 15 cm, simple or with 2 branches from the base, with up to 6 axillary rays. Leaves 20–30 × 3·5–5 mm, shortly petiolate, linear-oblanceolate, entire. Ray-leaves like the upper cauline. Raylet-leaves somewhat obliquely rhombic-trullate, acute, subentire. Rays (3–)4(–5), often much-branched. Glands yellow, with 2 pink horns. Capsule 3 × 3·5 mm, shallowly sulcate, finely granulate on the keel. Seeds 1·8–2 mm, ovoid, rather deeply pitted, greyish-white, dark grey in the pits. *Disturbed ground. S. Europe; a frequent casual elsewhere.* Al Bl Bu Cr Ga Gr Hs It Ju Rs (K) Tu [Au Hu].

Sect. PARALIAS Dumort. Annual or perennial; leaves alternate, palmately veined; capsule with indurated mesocarp; seeds ornamented or smooth.

(A) Seeds distinctly pitted.

70. E. segetalis L., *Sp. Pl.* 458 (1753) (*E. tetraceras* Lange). Glabrous annual up to 35 cm, rarely perennating, simple or branched from the base, with (0–)4–5 axillary rays. Leaves 10–30(–60) × 1–3 mm, linear to linear-lanceolate, entire. Ray-leaves elliptic-oblong. Raylet-leaves deltate-rhombic, obtuse; base cuneate to subcordate. Rays 5(–6), up to 5 times dichotomous. Glands emarginate or with 2, rarely 4, horns. Capsule 2·5–3 × 3–3·5 mm, deeply sulcate, granulate-rugulose on the keels. Seeds 1·5–2 mm, ovoid, pale grey. $2n=16$. *Open, sandy ground, often near the sea. S.W. Europe and Mediterranean region, eastwards to N.W. Jugoslavia; locally naturalized in C. Europe as a weed.* Bl Co ?Cr Ga Hs It Ju Lu [Cz Ge He Hu Rm].

71. E. pinea L., *Syst. Nat.* ed. 12, 333 (1767) (*E. segetalis* subsp. *pinea* (L.) Hayek, *E. segetalis* var. *pinea* (L.) Willk.). Like **70** but always perennial, with usually densely leafy stems up to 50 cm, often much-branched from the base and usually with more axillary rays. _Sandy seashores. W. & C. Mediterranean region, S.W. Europe._ Al Az Bl Co Ga Hs It Ju Lu Sa Si.

72. E. portlandica L., _Sp. Pl._ 458 (1753) (_E. segetalis_ var. _littoralis_ Lange). Glabrous perennial up to 40 cm, usually much-branched from the base; pattern and number of axillary branches very variable. Leaves 5–25 × 2–6 mm, obovate to narrowly oblanceolate, usually obtuse, entire, rarely denticulate near the apex. Ray-leaves usually like the upper cauline, rarely ovate-suborbicular. Raylet-leaves deltate-rhombic, obtuse, mucronate. Rays (3–)4–5(–6), up to 4 times dichotomous. Glands yellowish, with 2 horns. Capsule 2·8–3 × 3–3·2 mm, deeply sulcate, granulate on the keels. Seeds 1·5–1·8 mm, ovoid, pale grey, darker in the pits. _Maritime sands. W. Europe, from Gibraltar to S.W. Scotland._ Br Ga Hb Hs Lu.

73. E. deflexa Sibth. & Sm., _Fl. Graec. Prodr._ **1**: 328 (1809). Glabrous, glaucous, ascending, caespitose perennial up to 35 cm, branched at the base, with 0–4(–7) axillary rays. Leaves 3–15 × 2–6 mm, orbicular to obovate or oblong, entire, shortly petiolate. Ray-leaves ovate-deltate or rhombic. Raylet-leaves rhombic, transversely ovate or more or less reniform. Rays (3–)5(–9), up to 3 times dichotomous. Glands with 2 long horns. Capsule 3 × 3 mm, deeply sulcate, granulate on the keels. Seeds 2·5 mm, ovoid or ovoid-cylindrical, pale grey, darker in the pits. _Mountain rocks and screes._ ● _Aegean region and C. Greece._ Cr Gr.

74. E. petrophila C. A. Meyer, _Mém. Acad. Sci. Pétersb._ ser. 6 (Sci. Nat.), **7** (Bot.): 9 (1855). Glabrous or minutely papillose, caespitose perennial up to 25 cm, branched at the base, with 0–2(–3) axillary rays. Leaves 10–16 × 3–5 mm, oblong to suborbicular-ovate, more or less entire. Ray-leaves suborbicular to rhombic-trullate. Raylet-leaves suborbicular to deltate or reniform. Rays (2–)4–5, up to 3 times dichotomous. Glands with 2 horns, or emarginate to transversely oblong with 1 or 2 small horns. Capsule 2·5–3 × 2·5–3 mm, shallowly sulcate, nearly smooth. Seeds 2 mm, evenly and shallowly pitted, ovoid, with a dorsal ridge. _Rocks and stony slopes; calcicole. Krym & E. Ukraine._ ?Rm Rs (K, E).

75. E. transtagana Boiss., _Diagn. Pl. Or. Nov._ **3**(4): 88 (1859). Glabrous or minutely papillose, glaucous perennial 4–15(–22) cm, with a long, slender rhizome and 0–5 axillary rays. Leaves 4–12 × 1–5 mm, the lowest small, oblong, the upper larger, linear-lanceolate, serrulate near apex. Ray-leaves broadly ovate to rhombic. Raylet-leaves rhombic-deltate or reniform, obtuse, mucronate, entire or slightly serrulate. Rays (2–)3–5, twice dichotomous. Glands suborbicular, pitted, with 2 short horns. Capsule 3·5 × 3·5 mm, shallowly sulcate, smooth. Seeds 2·3 mm, irregularly pitted, ovoid-pyriform, whitish to light brown, grey in the pits. _Quercus-scrub._ ● _C. & S. Portugal._ Lu.

(B) Seeds smooth, rugulose, tuberculate or indistinctly pitted.

76. E. matritensis Boiss., _Cent. Euphorb._ 35 (1860). Glabrous or minutely papillose, slightly glaucous perennial up to 40 cm, simple or branched at the base from a stout, woody stock, with 0–5(–10) axillary rays. Leaves 8–25(–35) × 2–6 mm, linear-oblanceolate to elliptic, entire. Ray-leaves somewhat wider than the upper cauline. Raylet-leaves suborbicular to rhombic. Rays 5, up to 3 times dichotomous. Glands with 2 short horns. Capsule 3·5–4 × 3·5 mm, shallowly sulcate, finely granulate on the keels. Seeds 2·5 mm, ovoid, indistinctly pitted, pale grey. _Dry, open habitats._ ● _C. & S. Spain, E. Portugal._ Hs Lu.

77. E. biumbellata Poiret, _Voy. Barb._ **2**: 174 (1789). Glabrous perennial up to 65 cm, occasionally branched from the base,

with (6–)8–20(–27) axillary rays usually clustered together to form a whorl below the umbel, and with 0–10(–19) more widely spaced axillary rays. Leaves 20–55 × 2–12 mm, linear to linear-lanceolate, entire. Ray-leaves lanceolate to ovate-deltate. Raylet-leaves reniform to deltate-rhombic. Rays 8–21, up to 4 times dichotomous. Glands with somewhat clavate horns. Capsule 3–3·8 × 3·5–4 mm, shallowly sulcate, granulate on the keels. Seeds 2·3 mm, ovoid-cylindrical, irregularly and shallowly rugulose, pale grey, darker in the depressions. _Rocky or sandy ground near the coast. W. Mediterranean region; local._ Bl Co Ga Hs It Si.

E. megalatlantica subsp. **briquetii** (Emberger & Maire) Losa & Vindt, _Trav. Inst. Sci. Chérif._ ser. bot., **19**: 456 (1960), occurs between Lorca and Puerto de Lumbreras in S.E. Spain. It is like **77** but has not more than 10 axillary rays or 5 umbel-rays and a more prominently tuberculate capsule; it may be more closely related to _E. biumbellata_ than to _E. megalatlantica._

78. E. boetica Boiss., _Cent. Euphorb._ 36 (1860). Glabrous, somewhat glaucous, caespitose perennial up to 40 cm, with a far-creeping rhizome; stems simple or branched at the base, with 0–6 axillary rays. Leaves 10–30 × 1–5 mm, linear to linear-lanceolate, entire, obsoletely 3(–5)-veined. Ray-leaves linear-lanceolate. Raylet-leaves lanceolate to ovate. Rays 4–6, up to 3 times dichotomous, occasionally proliferating. Glands variable in shape. Cyathial lobes usually very densely ciliate. Capsule (3·5–)5 × 4(–5) mm, shallowly sulcate, weakly punctate on the keels. Seeds 2·5 mm, ovoid, shallowly pitted, grey. _Scrub on dry, sandy, acid soil, usually near the sea._ ● _S.W. Spain, S. Portugal._ Hs Lu.

79. E. bupleuroides Desf., _Fl. Atl._ **1**: 387 (1798). Glabrous, somewhat glaucous perennial up to 60 cm, with 0–7 axillary rays. Leaves 20–40 × 3–8 mm, linear-oblong to linear-oblanceolate, entire, coriaceous. Ray-leaves shorter than the cauline, often acute. Raylet-leaves elliptic-lanceolate to ovate-oblong, not much shorter than the ray-leaves, obtuse. Rays 8–12, once or twice dichotomous. Glands with 2 short horns. Cyathial lobes sparsely ciliate. Capsule 4·3 × 4·3 mm, shallowly sulcate, nearly smooth. Seeds 2·9 mm, ovoid, smooth or with numerous shallow depressions, greyish. _Calcareous grassland, c._ 1300 m. _S.E. Spain_ (_La Sagra_). Hs. (_N.W. Africa._)

Represented in Europe only by subsp. **luteola** (Cosson & Durieu ex Boiss.) Maire, _Bull. Soc. Hist. Afr. Nord_ **30**: 363 (1939) (_E. luteola_ Cosson & Durieu ex Boiss.). The typical subspecies occurs in Algeria.

80. E. nicaeensis All., _Fl. Pedem._ **1**: 285 (1785) (_E. goldei_ Prokh., _E. pannonica_ Host, _E. stepposa_ Zoz ex Prokh., _E. volgensis_ Krysht.). Glabrous or minutely papillose, glaucous, often reddish-suffused perennial up to 80 cm, with 0–10(–20) axillary rays. Leaves 10–75 × 3–18 mm, lanceolate to oblong or occasionally ovate, nearly entire, obtuse, coriaceous, 3(–7)-veined. Ray-leaves elliptic-ovate to suborbicular. Raylet-leaves transversely ovate or reniform, often yellowish. Rays (3–)5–18, once or twice dichotomous. Glands truncate to emarginate, or sometimes with 2 short horns. Capsule 3–4·5 × 3–4 mm, shallowly sulcate, rugulose, sometimes pubescent. Seeds 2–2·5 mm, ovoid, nearly smooth, rarely indistinctly pitted, pale grey. _Dry, open ground. S., E. & E.C. Europe, northwards to c._ 53° N. _in E. Russia._ Al Au Bu Cz Ga Gr Hs Hu It Ju Lu Rm Rs (C, W, K, E) Tu.

Variable in size and shape of leaves, number of rays and degree of indentation of the glands. A number of more or less local populations can be recognized and have often been given specific rank; the distinctions between them are of a minor nature and do not seem to be clear-cut, but 2 subspp. can be recognized.

(a) Subsp. **nicaeensis**: Rays (3–)9–18: capsule 3·5–4·5 mm. *S. Europe.*

(b) Subsp. **glareosa** (Pallas ex Bieb.) A.R. Sm., *Feddes Repert.* **79**: 55 (1968) (*E. glareosa* Pallas ex Bieb.): Rays (3–)7–8; capsule 3 mm. *E. & E.C. Europe.*

81. E. bessarabica Klokov in Fomin, *Fl. RSS Ucr.* **7**: 629 (1955). Like **80** but with broadly ovate, 9- to 11-veined cauline and ray-leaves 17–27 mm wide; umbel with 3–4 rays. *S.W. Ukraine, ?Moldavia.* Rs (W).

82. E. orphanidis Boiss., *Diagn. Pl. Or. Nov.* **3**(4): 89 (1859). Glabrous, glaucous, prostrate to ascending perennial with an extensive, branched, fleshy, articulated rhizome and stems up to 15 cm, with 0–3 axillary rays. Leaves 6–18 × 2–7 mm, usually obovate, entire, petiolate. Ray-leaves oblong-oblanceolate to ovate-oblong. Raylet-leaves rhombic-deltate to reniform, obtuse, occasionally emarginate. Rays 3–5, up to 4 times dichotomous. Glands with 2 rather long, occasionally bifid horns. Capsule 4·5–5 mm, deeply sulcate, smooth. Seeds 3 mm, broadly ovoid, dark grey. *Rocks and screes above 1500 m. Greece (Parnassos).* Gr.

83. E. barrelieri Savi, *Bot. Etrusc.* **1**: 143 (1808) (*E. baselicis* Ten.). Glabrous or minutely papillose, somewhat glaucous perennial up to 40 cm, with 0–2(–4) axillary rays. Leaves 4–20 × 2–11(–16) mm, linear-oblanceolate to ovate-deltate, entire or serrulate in the upper half. Ray-leaves ovate to reniform. Raylet-leaves ovate-rhombic to transversely ovate, mucronate or aristate, often purple-tinged. Rays (1–)3–5(–12), once or twice dichotomous. Glands with 2 short or long, occasionally bifid horns. Capsule 3–4 × 2·5–3 mm, not or weakly sulcate, smooth. Seeds 1·7–2 mm, ovoid, smooth, pale grey. *Shady and rocky places. Balkan peninsula, Italy, S. France.* Bu Ga Gr It Ju Tu.

Plants from Italy differ from those from the Balkan peninsula in having entire leaves and usually more rays.

84. E. triflora Schott, Nyman & Kotschy, *Analect. Bot.* 63 (1854). Glabrous, somewhat glaucous perennial up to 15 cm, with a far-creeping rhizome and no axillary rays. Leaves 5–15 × 2–8 mm, entire, oblong to ovate-deltate. Ray-leaves like the upper cauline leaves; raylet-leaves rhombic or transversely ovate. Rays (3–)5, once or twice dichotomous. Glands with 2, 3-fid horns. Capsule 3 × 3 mm, shallowly sulcate, smooth. Seeds 2 mm, ovoid, smooth, pale grey. *Alpine meadows and open pine-woods.* ● *W. Jugoslavia (Velebit).* Ju.

E. kerneri Huter in A. Kerner, *Sched. Fl. Exsicc. Austro-Hung.* **2**: 48 (1882), from the S.E. Alps, is like **84** and may be conspecific. It is up to 35 cm and has 0–4(–7) axillary rays. The leaves are rather larger and the longest are clustered in a rosette.

85. E. saxatilis Jacq., *Fl. Austr.* **4**: 23 (1776). Glabrous, somewhat glaucous, ascending perennial up to 20 cm, with a far-creeping branched rhizome and 0(–1) axillary rays. Leaves 2–25 × 2–7 mm, entire, the longest ones clustered in a rosette; rosette-leaves linear-oblanceolate; cauline leaves oblong to suborbicular-rhombic. Ray-leaves ovate or suborbicular, obtuse, often emarginate, cordate. Raylet-leaves rhombic, transversely ovate or reniform. Rays (4–)5, never more than once dichotomous. Glands with 2 horns. Capsule 3–4 × 3 mm, shallowly sulcate, smooth. Seeds 2·3 mm, ovoid, smooth, whitish. $2n = 18$. *Rocky places, mainly in the mountains; calcicole.* ● *N.W. Jugoslavia, E. Austria.* Au ?Cz Ju.

86. E. valliniana Belli, *Ann. Bot. (Roma)* **1**: 9 (1903). Glabrous, glaucous perennial up to 16 cm, with a far-creeping rhizome, and with 0–4 axillary rays. Leaves 3–19 × 2–9 mm, suborbicular to ovate-lanceolate or oblanceolate, entire. Ray-leaves like the upper cauline. Raylet-leaves reniform, deltate or rhombic. Rays 5–7, up to 3 times dichotomous. Glands shallowly emarginate, pitted. Cyathial cup completely glabrous within; bracts between the male flowers absent. Capsule 3–3·5 × 3·5–4 mm, shallowly sulcate, slightly granulate on the keels. Seeds 2·5–2·8 mm, ovoid, smooth, somewhat shiny, dark brown. *Dry, open, stony, calcareous slopes.* ● *S.W. Alps.* Ga It.

87. E. minuta Loscos & Pardo, *Ser. Pl. Arag.* 96 (1863) (*E. pauciflora* Dufour, non Hill). Glabrous, somewhat glaucous perennial with a branched rhizome; stem up to 13 cm, simple or branched from the base, with 0(–2) axillary rays. Leaves 2–12(–18) × 1–4·5(–9) mm, usually oblanceolate to elliptical, nearly or quite entire, slightly coriaceous. Ray- and raylet-leaves suborbicular to rhombic-deltate. Rays (2–)3–5, once or twice dichotomous. Glands emarginate or truncate. Capsule 3–3·5 × 3 mm, shallowly sulcate, nearly smooth. Seeds 2·1–2·3 mm, ovoid, irregularly and shallowly rugulose or almost smooth, whitish. *Dry, stony ground.* ● *N. & E. Spain.* Hs.

88. E. variabilis Cesati, *Bibliot. Ital.* **91**: 348 (1838). Glabrous, somewhat glaucous, rhizomatous perennial 35(–60) cm, with (0–)1–4(–7) axillary rays. Leaves 4–35 × 1–6(–15) mm, elliptic-ovate to lanceolate, entire. Ray-leaves like the cauline but longer. Raylet-leaves deltate. Rays (3–)4–5, once dichotomous. Glands with 2 horns. Capsule 2·5–3 × *c.* 3·5 mm, shallowly sulcate, smooth or finely granulate on the keels. Seeds 2·5 mm, ovoid, smooth, grey. *Scrub on calcareous slopes.* ● *N. Italy, just extending into S.E. France.* Ga It.

89. E. gayi Salis, *Flora (Regensb.)* **17**(2): Beibl. 6 (1834). Glabrous, somewhat glaucous, rather weak perennial up to 17(–26) cm, with a far-creeping, slender rhizome; stems simple or branched from the base, and with 0–1(–6) axillary rays. Leaves (1·5–)4–12(–20) × 1·5–6·5 mm, obovate to oblanceolate-oblong, usually obtuse, entire. Ray-leaves like the upper cauline. Raylet-leaves linear-lanceolate to elliptic-ovate or rarely deltate. Rays 2–3(–8), often unbranched. Glands emarginate, pitted. Capsule 2·5 × 2·5 mm, deeply sulcate, smooth or finely granulate on the keels. Seeds 2 mm, ovoid, smooth, greyish. *Rocky and bushy places.* ● *Corse; C. Spain.* Co Hs.

Spanish plants have been distinguished as **E. sennenii** Pau, *Bol. Soc. Aragon. Ci. Nat.* **6**: 29 (1907). They have wider raylet-leaves and more numerous axillary rays and umbel-rays, but are otherwise indistinguishable from plants from Corse.

90. E. maresii Knoche, *Fl. Balear.* **2**: 161 (1922). Glabrous, much-branched perennial with slender, procumbent-ascending stems up to 40 cm, with no axillary rays and with a far-creeping, much-branched rhizome. Leaves 1·5–12 × 1–4 mm, obovate to linear-oblanceolate, entire, often becoming purplish; petiole up to 1 mm. Ray-leaves like the upper cauline. Raylet-leaves orbicular to elliptic-lanceolate. Umbel not developed, or with 2–3 rays. Rays up to twice dichotomous, occasionally proliferating. Glands truncate or with 2 horns, rugulose. Capsule 2·5 × 2·5 mm, shallowly sulcate, granulate-rugose on the keels. Seeds 1·5 mm, ovoid-cylindrical, shallowly pitted, whitish, grey in the pits. *Shady crevices in limestone rocks.* ● *Islas Baleares.* Bl.

91. E. herniariifolia Willd., *Sp. Pl.* **2**: 902 (1800). Glabrous or tomentose, somewhat glaucous perennial with numerous procumbent to ascending stems up to 20 cm, forming dense mats, much-branched but with no axillary rays. Leaves 0·4–10 × 0·4–5 cm, orbicular to obovate-elliptical, subacute or obtuse, entire; petiole up to 1·5 mm. Ray- and raylet-leaves very like the cauline. Rays 2–3, not or once dichotomous, occasionally proliferating. Glands with 2 horns. Capsule 3 × 3·5 mm, shallowly sulcate, smooth, with 2 wings on each keel, glabrous or tomentose. Seeds 2 mm, cylindrical, irregularly and shallowly pitted, pale grey. *Mountain rocks; calcicole. Albania, Greece and Kriti.* Al Cr Gr.

92. E. seguierana Necker, *Acta Akad. Theod.-Pal.* **2**: 493 (1770) (*E. gerardiana* Jacq.). Glabrous, glaucous, somewhat caespitose perennial up to 60 cm, branched or not from the base and with up to 30 axillary rays. Leaves 10–35 × 2–8 mm, linear to elliptic-oblong, entire, somewhat coriaceous. Ray-leaves ovate-lanceolate to ovate. Raylet-leaves rhombic-deltate to reniform. Rays 5–30, up to 4 times (usually twice) dichotomous. Glands transversely ovate, truncate. Capsule 2–3 × 2–3 mm, shallowly sulcate, weakly granulate on the keels, elsewhere smooth. Seeds 1·5–2 mm, ovoid-fusiform, smooth, pale grey. *Dry places. Most of Europe except the north and extreme south.* Al Au Be Bu Co Cz Ga Ge Gr He Ho Hs Hu It Ju Rm Rs (C, W, K, E) Tu.

(a) Subsp. **seguierana**: Axillary rays fewer than 10. Leaves usually erect. Umbel usually with fewer than 15 rays. Capsule 2·5–3 × 2·5–3 mm. *Throughout the range of the species, but rare in the Balkan peninsula.*

(b) Subsp. **niciciana** (Borbás ex Novák) Rech. fil., *Ann. Naturh. Mus.* (*Wien*) **56**: 212 (1948) (*E. niciciana* Borbás ex Novák): Axillary rays up to 30. Leaves often patent. Umbel with (15–)20–30 rays. Capsule 2–2·5 × 2–2·5 mm. *Balkan peninsula.* Al Bu Gr Ju Tu.

93. E. pithyusa L., *Sp. Pl.* 458 (1753). Minutely papillose, glaucous, often suffruticose perennial up to 55 cm, much-branched at the base and with 0–20(–30) axillary rays, occasionally forming a whorl. Leaves 5–28(–45) × 1–12 mm, deflexed and closely imbricate at the base of the stem, linear-lanceolate to ovate-lanceolate. Ray-leaves ovate, obtuse, mucronate, entire or irregularly serrulate. Raylet-leaves transversely ovate to suborbicular. Rays 5–8, up to 4 times dichotomous, occasionally proliferating. Glands variable in shape. Capsule 2·3–3·2 × 2·5–3·5 mm, shallowly sulcate, granulate on the keels. Seeds 1·5–1·7(–2) mm, ovoid, rugulose, tuberculate or almost smooth, dark grey and whitish. 2*n* = 36. *W. Mediterranean region.* Bl Co Ga It Sa Si.

(a) Subsp. **pithyusa**: Non-flowering branches up to 40; axillary rays not more than 20, never whorled. Leaves 5–28 mm. Glands without horns. Capsule 2·3–2·5 × 2·5–2·7 mm; seeds up to 1·7 mm. *Sandy and rocky shores. Throughout the range of the species, except Sicilia.*

(b) Subsp. **cupanii** (Guss. ex Bertol.) A.R. Sm., *Feddes Repert.* **79**: 66 (1968) (*E. cupanii* Guss. ex Bertol.): Non-flowering branches absent; axillary rays up to 30, sometimes whorled. Leaves up to 45 mm. Glands with 2, multifid horns. Capsule 3–3·2 × 3–3·5 mm; seeds 2 mm. *Bushy and rocky places inland. Islands of W. Mediterranean region.*

94. E. paralias L., *Sp. Pl.* 458 (1753). Glabrous, glaucous, somewhat fleshy, caespitose perennial up to 70 cm, branched from the base, and with 0–9 axillary rays. Leaves 3–30 × 2–15 mm, lowest ones obovate-oblong, middle ones elliptic-oblong and upper ones ovate; all entire, adaxially concave, imbricate. Ray-leaves like the upper cauline. Raylet-leaves suborbicular-rhombic to reniform, strongly adaxially concave. Rays 3–6, up to 3 times

dichotomous. Glands emarginate. Capsule 3–5 × 4·5–6 mm, deeply sulcate, granulate on the keels. Seeds 2·5–3·5 mm, broadly ovoid, smooth, pale grey. 2*n* = 16. *Sandy sea-shores. Coasts of W. & S. Europe, from Ireland and the Netherlands to Romania.* Al Be Bl Br Bu Co Cr Ga Gr Hb Ho Hs It Ju Lu Rm Sa Si Tu.

Sect. ESULA. Perennials; cauline leaves alternate, pinnately veined; glands truncate or with 2 horns; capsules with indurated pericarp; seeds smooth.

(A) Raylet-leaves not connate; axillary non-flowering branches usually present.

95. E. nevadensis Boiss. & Reuter, *Pugillus* 110 (1852). Glabrous, somewhat glaucous perennial up to 25 cm, simple or branched from the base and with up to 4 axillary non-flowering shoots and 3–12 axillary rays. Leaves 5–20(–30) × 2–15 mm, lanceolate to broadly ovate, rounded or cordate at the base, almost entire. Ray-leaves like the cauline but shorter. Raylet-leaves reniform, deltate-rhombic or suborbicular. Rays 5–9, not or once dichotomous. Glands with 2 horns. Capsule (2–)3 × 3·5–4 mm, deeply sulcate, granulate on the keels. Seeds 2·5 mm, ovoid, grey, shiny. ● *Mountains of S. & E. Spain.* Hs.

96. E. agraria Bieb., *Fl. Taur.-Cauc.* **1**: 375 (1808). Glabrous perennial up to 90 cm, unbranched at the base and with 0–3 axillary non-flowering branches and (2–)6–24 axillary rays. Leaves (15–)25–45(–80) × 7–27 mm, triangular-ovate to oblong or (var. *subhastata* (Vis. & Pančić) Griseb.) lingulate-panduriform or (var. *euboea* (Halácsy) Hayek) linear-lanceolate, broadly cordate-auriculate at the base, amplexicaul, more or less entire. Ray-leaves like the cauline. Raylet-leaves triangular-subreniform. Rays (6–)8–14, up to three times dichotomous. Glands emarginate or with 2 short horns. Capsule 2·5–3 × 3–3·5 mm, deeply sulcate, granulate-rugulose on the keels. Seeds 2 mm, ovoid, grey. *S.E. Europe.* Al Bu Gr Ju Rm Rs (W, K) Tu.

97. E. lucida Waldst. & Kit., *Pl. Rar. Hung.* **1**: 54 (1801). Robust, glabrous, perennial up to 140 cm, usually unbranched at the base but with (2–)4–6(–10) axillary non-flowering branches and 6–20 axillary rays. Leaves (25–)50–110(–130) × (7–)15–23(–35) mm, lanceolate to ovate-lanceolate, entire, somewhat shiny. Ray-leaves ovate to ovate-oblong. Raylet-leaves suborbicular-deltate. Rays 7–11, twice dichotomous. Glands with 2 horns. Capsule 3·5–4 × 4 mm, deeply sulcate, smooth. Seeds up to 3 mm, oblong-ovoid, pale grey. 2*n* = 36. *Marshes and riverbanks. C. & S.E. Europe, extending to White Russia and C. Ukraine.* Au Bu Cz Ge Gr Hu Ju Po Rm Rs (C, W) Tu.

98. E. salicifolia Host, *Syn. Pl. Austr.* 267 (1797). More or less glandular-pubescent perennial up to 80 cm, unbranched at the base, with 0–3 axillary non-flowering branches and 6–18 axillary rays. Leaves 25–100 × 7–35 mm, lanceolate to ovate-lanceolate, entire. Ray-leaves ovate. Raylet-leaves orbicular-deltate to reniform. Rays 9–16, up to 3 times dichotomous. Glands with 2 horns. Capsule 3 × 3·5 mm, deeply sulcate, granulate. Seeds 2 mm, ovoid, shiny, brownish. 2*n* = 36. *Lowland meadows.* ● *C. & S.E. Europe.* Al Au Bu Cz Ge Gr Hu Ju Rm Rs (W) ?Tu.

99. E. esula L., *Sp. Pl.* 461 (1753) (incl. *E. gmelinii* Steudel, *E. subtilis* Prokh., *E. zhiguliensis* Prokh.). Glabrous or pubescent perennial up to 120 cm. Stems usually unbranched at the base but with up to 11 axillary non-flowering branches and 0–20(–30) axillary rays. Leaves 15–85 × 0·5–15(–22) mm, linear to broadly ovate or obovate, entire. Ray-leaves shorter and often wider than cauline. Raylet-leaves rhombic, deltate or reniform. Rays (4–)5–17, once or twice dichotomous. Glands emarginate or with 2 horns. Capsule 2·5–3 × 3·5 mm, deeply sulcate, granulate on the keels. Seeds 2 mm, ovoid, grey or brownish. *Throughout a large part of continental Europe, but only as an alien in the north.* Al Au Be Bu Cz Ga Ge Gr Ho Hs Hu It Ju Lu Po Rm Rs (*B, C, W, K, E) Tu [Br Da Fe He No Rs (N) Su].

(a) Subsp. **esula** (incl. *E. borodinii* Sambuk, *E. filicina* Portenschl., *E. imperfoliata* Vis., *E. pancicii* G. Beck, *E. pseudagraria* Smirnov): Axillary rays 8–20; leaves oblanceolate to broadly ovate or obovate, obtuse or slightly emarginate; umbel usually with 8–17 rays. 2*n* = 60, 64. *Throughout most of the range the species.*

(b) Subsp. **tommasiniana** (Bertol.) Nyman, *Consp.* 652 (1881) (*E. virgata* Waldst. & Kit., non Desf.; incl. *E. subcordata* Ledeb., *E. tenuifolia* Lam., *E. uralensis* Fischer ex Link). Axillary rays usually 2–12; leaves linear to lanceolate, sometimes widened and rounded at base, acute or subacute; umbel usually with 5–9 rays. 2*n* = 20. *S., E. & E.C. Europe; the hybrid with subsp. esula often naturalized elsewhere.*

E. **leptocaula** Boiss. in DC., *Prodr,* 15(2): 159 (1862) (*E. astrachanica* C. A. Meyer ex Prokh., *E. borszczowii* Prokh.), from E. Romania and U.S.S.R., is like subsp. *tommasiniana* but usually smaller. It is often puberulent and usually has narrowly linear leaves up to 85 mm but less than 4 mm wide. *E.* **sareptana** A. Becker ex Boiss. in DC., *Prodr.* 15(2): 159 (1862) (*E. tanaitica* Pacz.), from S.E. Russia, is similar but has leaves up to 11 mm wide, oblanceolate to obovate or elliptic-obovate and strongly emarginate. They may merit subspecific rank.

100. E. undulata Bieb., *Fl. Taur.-Cauc.* 1: 371 (1808). Glabrous, glaucous ascending perennial up to 15 cm, simple or branched from the base, with 1–6 axillary rays and with a slender, extensive rhizome. Leaves 10–20 × 3–12 mm, oblanceolate to elliptic-obovate; margins usually undulate. Ray-leaves like the cauline. Raylet-leaves ovate-trullate. Rays 2–5, once dichotomous. Capsule 3·5–4 mm, deeply sulcate, smooth. Seeds 2 mm, ovoid. *Steppes and semi-deserts. S.E. Russia, W. Kazakhstan.* Rs (C, E).

101. E. cyparissias L., *Sp. Pl.* 461 (1753). Glabrous perennial up to 50 cm, rhizomatous, usually unbranched at the base but with up to 16 axillary non-flowering branches and 0–7 axillary rays. Leaves 5–40 × 0·5–3 mm, linear, entire. Ray-leaves linear to oblong. Raylet-leaves reniform, rhombic or suborbicular. Rays (5–)9–18(–22), once or twice dichotomous. Glands with 2 horns. Capsule 3 × 3·5 mm, deeply sulcate, granulate on the keels. Seeds 1·75 mm, ovoid, somewhat shiny, grey. 2*n* = 20, 40. ● *Most of Europe except the extreme north and the extreme south, but only as an alien in most of the north.* Al Au Be ?Bl Bu Cz Ga Ge Gr He Ho Hs Hu It Ju Po Rm Rs (C, W, E) Tu [Br Da Fe No Rs (N, *B, K) Su].

102. E. terracina L., *Sp. Pl.* ed. 2, 654 (1762). Glabrous perennial; stem up to 70 cm, simple or branched from the base and with 0–5 axillary rays. Leaves 15–40(–55) × 4–7(–11) mm, linear-lanceolate to elliptic-oblong, rarely (var. *angustifolia* Lange) 1·5–2 mm wide, linear, minutely serrulate. Ray-leaves resembling the upper cauline. Raylet-leaves deltate-rhombic, sometimes slightly asymmetrical, occasionally coarsely serrulate. Rays 4–5, up to 5 times dichotomous. Glands with 2 long, slender

horns. Capsule 3–5 × 4–5 mm, deeply sulcate, smooth. Seeds 2–2·5 mm, ovoid, pale grey. *Dry sandy ground, often near the sea. Mediterranean region, extending to N.W. Spain and Portugal.* Al Bl Co Cr Ga Gr Hs It Ju Lu Sa Si.

(B) Raylet-leaves connate; axillary non-flowering shorts absent.

103. E. amygdaloides L., *Sp. Pl.* 463 (1753). Pubescent, caespitose perennial; stem up to 90 cm, with up to 30 axillary rays. Leaves (10–)25–70(–130) × (5–)10–20(–30) mm, oblanceolate to obovate or spathulate, entire. Ray-leaves broadly ovate. Raylet-leaves partially or wholly connate, rarely more or less free (var. *ligulata* (Chaub.) Bory). Rays (3–)5–11, up to 4 times dichotomous. Glands with 2 horns. Capsule 3–4 × 2·5–4 mm, deeply sulcate, punctate. Seeds 2–2·5 mm, ovoid, blackish. 2*n* = 18. *Woods. C., S. & N.W. Europe, extending eastwards to C. Ukraine.* Al Au Be Br Bu Co Cz Ga Ge Gr *Hb He Ho Hs Hu It Ju Lu Po Rm Rs (W, K) Sa Si Tu.

(a) Subsp. **amygdaloides**: Stems flowering in the second year. Leaves of the first year's growth larger than those of the second and crowded at the top of the stem. Horns of the glands usually convergent. *Throughout the range of the species, except Corse and Sardegna.*

(b) Subsp. **semiperfoliata** (Viv.) A.R. Sm., *Feddes Repert.* 79: 65 (1968) (*E. semiperfoliata* Viv.): Stems usually flowering in the first year. Leaves more or less uniform in size and evenly spaced. Horns of the glands usually parallel. ● *Corse and Sardegna.*

104. E. heldreichii Orph. ex Boiss., *Diagn. Pl. Or. Nov.* 3(4): 90 (1859) (*E. semiverticillata* Halácsy). Pubescent to glabrescent perennial with biennial stem up to 100 cm, with numerous axillary rays arranged in 2–5 whorls of (3–)5–9 rays. Leaves (15–)40–80(–110) × (3–)10–20(–30) mm, oblong, oblanceolate or obovate, entire; those of the first year's growth usually larger than those of the second. Ray-leaves obovate to ovate. Raylet-leaves suborbicular-deltate to semicircular, not or partially connate at the base. Rays 5–8, up to three times dichotomous. Glands with 2 long horns. Capsule 3–3·5 × 3·5 mm, deeply sulcate, smooth. Seeds 2·2 mm, more or less ovoid, pale grey. *Dry mountain slopes.* ● *Greece and S. Albania.* Al Gr.

105. E. characias L., *Sp. Pl.* 463 (1753) (*E. melapetala* Gasparr.). Densely tomentose, rarely glabrescent, glaucous, caespitose perennial, sometimes with biennial stems up to 180 cm, with 13–30(–40) axillary rays. Leaves (14–)30–130 × 4–10(–17) mm, linear to oblanceolate or occasionally obovate, entire; those of the first year's growth usually larger than those of the second. Ray-leaves like the upper cauline. Raylet-leaves suborbicular-deltate, usually connate in pairs at the base. Rays 10–20, usually twice, but up to 4 times dichotomous. Glands variable. Capsule 4–7 × 5–6 mm, deeply sulcate, smooth, densely villous. Seeds 2·5–3·8 mm, ovoid, silver-grey. *Dry, fairly open ground. Mediterranean region, Portugal.* Al Bl Co Cr Ga Gr Hs It Ju Lu Sa Si.

(a) Subsp. **characias**: Stems usually up to 80 cm; glands dark reddish-brown, rarely yellow, with short horns or emarginate. 2*n* = 20. *W. Mediterranean region, Portugal.*

(b) Subsp. **wulfenii** (Hoppe ex Koch) A.R. Sm., *Feddes Repert.* 79: 55 (1968) (*E. wulfenii* Hoppe ex Koch, *E. veneta* sensu Hayek): Stems up to 180 cm; glands yellowish, with long horns. *E. Mediterranean region.*

RUTALES

LXXXVIII. RUTACEAE[1]

Herbs, shrubs or trees. Leaves alternate or opposite, simple or compound, dotted with translucent glands, exstipulate, sometimes reduced to spines. Flowers usually hermaphrodite and actinomorphic. Sepals 4–5, free or connate below. Petals 4–5. Disc present. Stamens as many or twice as many as the petals, rarely more, free or rarely monadelphous. Ovary superior, usually syncarpous and often 4- to 5-locular, but carpels occasionally united at the base only or rarely free. Styles as many as the carpels, free or connate.

1 Trees or large shrubs; fruit either fleshy or a samara
2 Leaves simple   **4. Citrus**
2 Leaves compound
3 Leaflets 5–13; fruit a black, fleshy drupe   **5. Phellodendron**
3 Leaflets 3(–5); fruit a flat, broadly winged samara   **6. Ptelea**
1 Herbs or dwarf shrubs; fruit a capsule
4 Leaves imparipinnate; leaflets 2·5–7·5 cm, distinctly serrulate; flowers zygomorphic, never yellow   **3. Dictamnus**
4 Leaves simple, 3(–5)-sect or 2- to 3-pinnatisect; lobes less than 2·5 cm, entire or obscurely crenate-serrate; flowers actinomorphic, yellow
5 Leaves 2- to 3-pinnatisect; petals 4, except in the central flower, dentate or fimbriate, rarely entire; filaments glabrous below   **1. Ruta**
5 Leaves simple or 3-sect, very rarely 5-sect; petals 5, entire; filaments hairy below on inner surface   **2. Haplophyllum**

Subfam. Rutoideae

Fruit a capsule, usually 4- to 5-valved. Seeds with endosperm.

1. Ruta L.[2]

Perennial herbs, more or less woody below. Leaves alternate, 2- to 3-pinnatisect, ultimate segments linear to obovate. Inflorescence cymose, bracteate. Sepals and petals 4, or frequently 5 in the central flower. Petals cucullate, yellow, dentate or ciliate or more rarely entire. Stamens twice as many as the petals; filaments glabrous, attenuate. Capsule 4- to 5-lobed, dehiscent. Styles connate.

All species grow in dry, usually rocky situations.

1 Leaf-segments linear; pedicels shorter than the capsule; petals not denticulate or ciliate   **1. montana**
1 Leaf-segments oblanceolate to oblong-obovate; pedicels as long as or longer than the capsule; petals denticulate or ciliate
2 Petals fringed with long cilia
3 Bracts not or scarcely wider than the branches which they subtend; plant glandular-puberulent above   **2. angustifolia**
3 Lower bracts much wider than the branches which they subtend; plant glabrous throughout   **3. chalepensis**
2 Petals denticulate, without long cilia
4 Sepals lanceolate, acute; pedicels slightly longer than the capsule   **4. graveolens**
4 Sepals deltate-ovate, obtuse; pedicels at least twice as long as the capsule   **5. corsica**

1. R. montana (L.) L., *Amoen. Acad.* **3**: 52 (1756). Stem 15–70 cm, glabrous below the inflorescence. Lower leaves petiolate, the upper sessile; ultimate segments up to 1 mm wide, rather thick, linear, those of the upper leaves up to 12 mm. Inflorescence dense, the ultimate branches often subracemose; pedicels shorter

than capsule; inflorescence-branches, pedicels, bracts and sepals more or less densely glandular. Sepals lanceolate, acuminate. Petals oblong, undulate but scarcely denticulate. Capsule glabrous; segments obtuse and rounded at apex. *S.W. Europe; very locally also in C. & E. Mediterranean region.* Bl Ga Gr Hs It Lu Tu.

2. R. angustifolia Pers., *Syn. Pl.* **1**: 464 (1805). Stem 25–75 cm, glabrous below the inflorescence. Lower leaves shortly petiolate; ultimate segments 1·25–3·5 mm wide, obovate-lanceolate to narrowly oblong. Inflorescence rather lax; pedicels as long as or longer than the capsule; bracts lanceolate, not or scarcely wider than the subtended branch; inflorescence branches, pedicels, bracts and sepals glandular-puberulent. Sepals deltate-ovate, subacute. Petals oblong, fringed with cilia frequently as long as the width of the petal. Capsule glabrous; segments acuminate. *W. Mediterranean region, extending to N.W. Jugoslavia.* Bl Co Ga Hs It Ju Lu Sa Si.

3. R. chalepensis L., *Mantissa* 69 (1767) (*R. bracteosa* DC.). Stem 20–60 cm, glabrous. Lower leaves more or less long-petiolate; ultimate segments 1·5–6 mm wide, narrowly oblong-lanceolate or obovate. Inflorescence lax; pedicels as long as or longer than the capsule; branches and pedicels glabrous, rarely a very few minute glands above; bracts wider than the subtended branch, the lower several times so, cordate-ovate. Sepals glabrous, deltate-ovate. Petals oblong, fringed with cilia not as long as the width of the petal. Capsule glabrous; segments acuminate. *S. Europe.* Al Az Bl Co Cr Ga Gr Hs It Ju Lu Sa Si.

R. fumariifolia Boiss. & Heldr. in Boiss., *Diagn. Pl. Or. Nov.* **2**(8): 125 (1849), from Kriti, appears to be only a depauperate form or small variety of **3**.

4. R. graveolens L., *Sp. Pl.* 383 (1753). Glabrous throughout; stem 14–45 cm. Lower leaves more or less long-petiolate, the uppermost subsessile; ultimate segments 2–9 mm wide, lanceolate to narrowly oblong or obovate. Inflorescence rather lax; pedicels as long as or longer than the capsule; bracts lanceolate, leaf-like. Sepals lanceolate, acute. Petals oblong-ovate, denticulate, undulate. Capsule glabrous; segments somewhat narrowed above to an obtuse apex. *Balkan peninsula and Krym; perhaps elsewhere in Mediterranean region; widely naturalized from gardens in S. & S.C. Europe.* Al *Bl Bu Ga *Hs Ju Rs (K) Tu [Au Co Cz Ga Ge Gr He Hu It Rm].

5. R. corsica DC., *Prodr.* **1**: 710 (1824). Glabrous throughout; stem 18–45 cm, much-branched from the base upwards. Lower leaves with petioles up to 8 cm, the uppermost subsessile; ultimate segments 1·5–7 mm wide, obovate to cuneate-orbicular. Inflorescence lax, virgate; pedicels ascending, the lower up to 7 times as long as the capsule, the upper twice as long. Sepals deltate-ovate, obtuse. Petals broadly ovate, pale yellow, denticulate, undulate. Capsule-segments acuminate. ● *Mountains of Corse and Sardegna.* Co Sa.

2. Haplophyllum A. Juss.[2]

Perennial herbs, sometimes woody below. Leaves alternate, entire, lanceolate to elliptical or linear, rarely broadly ovate, or 3-sect with lanceolate or linear segments, rarely some leaves

---
[1] Edit. T. G. Tutin.   [2] By C. C. Townsend.

with up to 5 segments. Inflorescence cymose, bracteate. Sepals and petals 5. Petals concave, yellow, entire. Stamens 10; filaments free, more or less expanded below and pubescent on the inner surface. Capsule 5-lobed, dehiscent. Styles connate.

1 Ovules 4 in each loculus of the ovary
  2 Leaves from the middle of the stem usually 3-sect, the upper and lower entire; each capsule-segment crowned with a large, ± denticulate appendage   **4. coronatum**
  2 Leaves from the middle of the stem usually entire; each capsule-segment crowned with a blunt, tuberculiform appendage (at times dissected into shortly cylindrical tubercles) or with no obvious appendage
    3 Leaves not reaching the inflorescence, a length of stem below the inflorescence naked
      4 Lower bracts small, not leaf-like   **1. linifolium**
      4 Lower bracts large and leaf-like   **5. balcanicum**
    3 Leaves reaching the inflorescence, the upper often subverticillate
      5 Sepals suborbicular to obtusely deltate, distinctly erose-denticulate   **3. thesioides**
      5 Sepals linear-lanceolate to lanceolate, acute
        6 Leaves linear-lanceolate; plant glabrous throughout (except the filaments)   **5. balcanicum**
        6 Leaves lanceolate to oblong-lanceolate; plant crispate-pubescent, to sublanate at least in the inflorescence   **2. suaveolens**
1 Ovules 2 in each loculus
  7 Loculi of the ovary, even when very young, quite rounded at the apex, with no conspicuous corniculate or tuberculiform appendage   **8. buxbaumii**
  7 Loculi of the ovary each with a prominent corniculate or tuberculiform appendage at the apex, at least when young
    8 Most leaves 3-sect to the base   **6. patavinum**
    8 All leaves simple   **7. boissieranum**

1. H. linifolium (L.) G. Don fil., *Gen. Syst.* **1**: 780 (1831) (*H. hispanicum* Spach). Stems 15–40 cm, woody below, glabrous or crispate-pubescent. Leaves 10–35 × 0·75–7 mm, sessile, lanceolate-ovate to linear, glabrous or crispate-pubescent, becoming smaller upwards and usually ceasing some distance below the inflorescence. Inflorescence more or less compact, flat-topped. Sepals suborbicular to deltate-lanceolate, glabrous to white-lanate. Capsule glabrous to densely pubescent, especially around the apical appendages, which are blunt and tuberculiform, often dissected into numerous shortly cylindrical tubercles. *Dry, sunny slopes. C., E. & S. Spain.* Hs [Ga].

2. H. suaveolens (DC.) G. Don fil., *loc. cit.* (1831) (*H. ciliatum* Griseb.). Stems 15–30(–50) cm, crispate-pubescent, sometimes more or less lanate in the inflorescence. Leaves sessile, entire, rarely 3-sect, lanceolate to oblong, acute, shortly hairy especially along the margins, the middle 7–30 × 2–11 mm, not becoming appreciably smaller upwards, reaching the inflorescence, the uppermost often forming an involucre-like whorl. Inflorescence compact. Sepals lanceolate, acute, glabrous to sparsely lanate. Capsule tuberculate, usually glabrous, sometimes pubescent on the inner surface at the top; tubercles prominent and increasingly so above. Apical tubercle of young ovary-segments erect, conical. *S.E. Europe, from Macedonia to E. Ukraine.* Bu Ju Rm Rs (W, E) Tu.

Very like **1**, but with a characteristic facies. In view of the small specific distinctions obtaining in the genus as a whole, it seems best to maintain the two as distinct.

3. H. thesioides (Fischer ex DC.) G. Don fil., *loc. cit.* (1831) (*H. tauricum* Spach, *H. suaveolens* sensu Vved., non (DC.) G. Don fil.). Stems 10–35 cm, subglabrous or more or less crispate-pubescent, sometimes more or less lanate in the inflorescence.

Leaves sessile, entire, oblanceolate to oblong-ovate, subacute, glabrous or thinly hairy on the margins and lower surface, the middle 11–30 × 3–13 mm, not becoming appreciably smaller upwards, sometimes forming an involucre-like whorl. Inflorescence compact. Sepals suborbicular to obtusely deltate, distinctly erose-denticulate, glabrous or ciliate. Capsule glabrous, tuberculate-glandular. Apical tubercle of young ovary-segments flattened, smooth, large. *Bulgaria; Krym; Don valley.* Bu Rs (K, E).

Distinguished from **2** chiefly in the shape of the sepals and the broad, flat, apical tubercles of the young ovary-segments.

4. H. coronatum Griseb., *Spicil. Fl. Rumel.* **1**: 129 (1843). Stems 12–35 cm, shortly crispate-pubescent, white-lanate in the inflorescence. Leaves variable; middle leaves or central leaflets 8–25 × 1–4 mm. Inflorescence compact. Sepals linear-lanceolate. Capsule pubescent on the inner surface towards the top, tuberculate-glandular, each segment crowned with a large, somewhat laterally compressed, more or less denticulate appendage. ● *Balkan peninsula, from N. Albania to C. Greece.* Al Gr Ju.

5. H. balcanicum Vandas, *Magyar Bot. Lapok* **4**: 264 (1905). Stems 15–25 cm, plant quite glabrous except for filaments. Leaves 10–15 × 2·5–3·5 mm, lanceolate to oblanceolate, acute, attenuate below, not becoming appreciably smaller upwards, reaching the inflorescence or almost so, passing into the large, leaf-like lower bracts. Inflorescence compact. Sepals narrowly deltate to deltate-ovate. Capsule smooth to almost pitted below, the glands flat or slightly immersed, each segment crowned with a smooth, feebly delimited, obtuse apical tubercle. ● *N.E. Greece, S.W. Bulgaria.* Bu Gr.

6. H. patavinum (L.) G. Don fil., *Gen. Syst.* **1**: 780 (1831) (*Ruta patavina* L.). Stems 10–30 cm, finely crispate-pubescent. Leaves crowded; basal leaves simple, lanceolate-oblong to oblanceolate; middle leaves 3-sect to the base, the middle segment 15–33 × 1·5–5 mm, somewhat longer and wider than lateral ones, pale beneath; uppermost leaves linear, simple or 3-sect. Inflorescence dense. Sepals deltate-lanceolate to linear-lanceolate. Capsule with scattered, sharply conical-tuberculate glands, with a small, conical, tuberculiform appendage on the outer surface near the top of each segment. ● *W. part of Balkan peninsula, just extending to N.E. Italy and S.W. Romania.* Al Gr It Ju Rm.

7. H. boissieranum Vis. & Pančić, *Mem. Ist. Veneto* **15**: 14 (1870) (*H. albanicum* (Bald.) Bornm.). Stems 6–25 cm, crispate-pubescent, more or less lanate in the inflorescence. Leaves 10–40 × 1·5–3 mm, lanceolate to oblanceolate, long-attenuate below, sparingly hairy or glabrous. Inflorescence more or less compact. Sepals linear-lanceolate. Capsule pubescent on the inner surface near the top, with scattered more or less prominently tuberculate glands below, each segment with a conical apical appendage which is sometimes dissected into smaller tubercles. ● *Jugoslavia and Albania.* Al Ju.

Very like **4** but distinguished by the much smaller apical appendages of the capsule-segments, the invariably simple leaves which are more or less glabrous when mature, and by having 2 ovules in each loculus.

8. H. buxbaumii (Poiret) G. Don fil., *Gen. Syst.* **1**: 780 (1831). Stems 15–50 cm, single or rather few, sparsely to densely crispate-pubescent. Leaves 25–50 × 2·5–13 mm, oblanceolate, or linear-lanceolate, entire, or 3-sect almost or quite to the base, with the middle segment larger than the lateral ones, more or less long-attenuate and subpetiolate below, crispate-pubescent.

Inflorescence wide and lax, often wider than high; branching often beginning about half-way up the stem. Sepals deltate-ovate, glabrous or ciliate. Capsule glabrous, densely furnished with pale-margined glands, the apex of each segment rounded even in the very young ovary. *Turkey-in-Europe; Kriti.* Cr Tu [Ga]. (*S.W. Asia.*)

3. Dictamnus L.[1]

Perennial herbs with alternate, pellucid-punctate, imparipinnate leaves. Flowers large and showy, in terminal, bracteate racemes. Sepals 5. Petals 5, narrow, the lowest declinate. Stamens 10; filaments declinate and upwardly curving. Styles connate. Capsule hard, deeply 5-lobed.

1. **D. albus** L., *Sp. Pl.* 383 (1753). Bushy, densely leafy perennial 40–80 cm, more or less woody below; stems and branches glandular-punctate, thinly hairy. Leaflets 3–6 pairs, lanceolate to ovate, pubescent at least on the principal veins beneath; inflorescence-axis, pedicels, bracts and sepals densely furnished with short, patent hairs and dark, stipitate glands. Sepals lanceolate. Petals 2–2·5 cm, elliptic-lanceolate, white to pink or bluish, with purple dots or streaks. Filaments about as long as the petals, glandular above. *S. & S.C. Europe, extending northwards to c. 54°N. in E. Russia.* Al Au Bu Cz Ge Gr He Hs Hu It Ju Po Rm Rs (W, K, E).

The following variants are frequently maintained as separate species: **D. caucasicus** Fischer ex Grossh., *Fl. Kavk.* 3: 20 (1932), from S. Russia; **D. gymnostylis** Steven, *Bull. Soc. Nat. Moscou* 29(2): 333 (1856), from Ukraine and S. Russia; and **D. hispanicus** Webb ex Willk., *Suppl. Prodr. Fl. Hisp.* 263 (1893), from E. & S.E. Spain. These are chiefly founded on indumentum of the vegetative parts, ovary and style, leaf-characters (size, shape and number of leaflets, development of wing of petiole), habit, and development of the ovary-appendages. All these characters, however, show much variation and cannot be convincingly correlated with distribution. Several are also found to be of little taxonomic value in closely allied genera. Accordingly, *D. albus* is treated here as a single polymorphic species.

Subfam. Aurantioideae

Fruit a berry with a thick, coriaceous rind or a harder shell, and a juicy pulp. Seeds without endosperm.

4. Citrus L.[1]

Small trees. Young twigs with single spines in the leaf-axils, but older branches often unarmed. Leaves alternate, simple, coriaceous, thin; lateral veins few. Petioles often more or less winged or margined and articulated with the lamina. Flowers white, solitary and axillary or in short axillary racemes. Sepals 4–5; petals (4–)5–(8). Stamens 4–10 times as many as the petals. Ovary usually 10- to 14-locular; ovules in 2 rows. Seeds surrounded by stipitate, fusiform pulp-vesicles.

Literature: W. T. Swingle, *The Botany of* Citrus *and its wild relatives*, in H. J. Webber & L. D. Batchelor, *The* Citrus Industry 1: 129–474. Berkeley and Los Angeles. 1943. T. Tanaka, *The species problem in* Citrus (*Revisio Aurantiacearum* IX). Tokyo. 1954.

A genus of considerable taxonomic difficulty. The present account is based on the work of the above two authorities. The species described are those most commonly cultivated for their

fruit and essential oils in the Mediterranean region. In addition, **C. bergamia** Risso & Poiteau, *Hist. Nat. Orang.* 111 (1818), is cultivated in Calabria for the essential oil yielded by its rind. It is a small tree with winged petioles and oblong-ovate leaves, and has a pale yellow, pyriform fruit 7·5–10 cm in diameter. All the cultivated species are probably derived from plants which are native in tropical and subtropical parts of S.E. Asia.

1 Petiole terete or carinate-margined but not winged; fruit 15–25 cm in diameter   **1. medica**
1 Petiole distinctly winged; fruit not more than 15 cm in diameter
   2 Stamens generally more than 4 times as many as the petals; flowers of two sorts, hermaphrodite and functionally male; ripe fruit light yellow, with a mammilliform process at apex
    3 Fruit acid; flowers tinged or streaked with purple   **2. limon**
    3 Fruit insipidly sweet; flowers pure white   **3. limetta**
   2 Stamens about 4 times as many as the petals; flowers usually all hermaphrodite; fruit yellow or orange without a mamilliform process at apex
    4 Leaves narrowly elliptical; fruit rarely more than 6·5 cm, depressed; rind very easily detached from the segments   **4. deliciosa**
    4 Leaves broadly elliptical; fruit 7–15 cm, spherical or broadly ovoid, not or only slightly flattened above and below; rind adhering to the segments
     5 Fruit 10–25 cm in diameter; rind yellow; petiole usually broadly winged
      6 Twigs and underside of midrib glabrous   **5. paradisi**
      6 Twigs and underside of midrib sparsely hairy   **6. grandis**
     5 Fruit usually 7–9 cm; rind orange or orange-yellow; petiole rather narrowly winged, obovate to oblanceolate, usually at least twice as long as wide
      7 Fruit with a rough rind and bitter sour pulp; petioles obovate in outline   **7. aurantium**
      7 Fruit with a sweet taste and nearly smooth rind; petioles oblanceolate in outline   **8. sinensis**

1. **C. medica** L., *Sp. Pl.* 782 (1753) (Citron). Small tree. Twigs angular when young, soon terete, glabrous, with short, stout axillary spines. Leaves glabrous, elliptic-ovate to ovate-lanceolate, crenate to serrate; veins prominent on both surfaces. Petiole terete or narrowly margined. Flowers in short, few-flowered racemes, hermaphrodite or functionally male; petals often pink or purplish on the outer surface. Stamens very numerous, coherent in groups of four or more. Fruit 15–25 cm 10- to 13-locular; rind very thick, often rough and warty, yellow when ripe; pulp pale green or yellow, acid or sweetish.

2. **C. limon** (L.) Burm. fil., *Fl. Ind.* 173 (1768) (Lemon). Small tree. Twigs angled when young, soon rounded, glabrous, with stout axillary spines. Leaves broadly elliptical, acute, serrate or crenate. Petiole with narrow wing or merely margined, distinctly articulated with the lamina. Flowers solitary or in short, few-flowered racemes, hermaphrodite or functionally male; petals purplish-suffused on the outer surface. Stamens 25–40, coherent in groups. Fruit 6·5–12·5 cm, 8- to 10-locular, yellow when ripe, oblong or ovoid, with a broad, low, mamilliform process at apex; rind somewhat rough to almost smooth; pulp acid.

3. **C. limetta** Risso, *Ann. Mus. Hist. Nat.* (*Paris*) 20: 195 (1813) (Sweet Lime). Like 2 but flowers pure white; fruit shorter, sweet.

According to Tanaka, a mutant of **2**.

4. **C. deliciosa** Ten., *Ind. Sem. Horti Neap.* 9 (1840) (Tangerine). Small, spreading tree. Twigs spiny, slender. Leaves narrowly elliptical. Flowers solitary or in small axillary clusters. Fruit 5–7·5 cm in diameter, depressed-globose; rind thin, easily separated from the pulp, bright orange when ripe; pulp sweet.

---

[1] By C. C. Townsend.

5. C. paradisi Macfadyen in Hooker, *Bot. Misc.* **1**: 304 (1830) (Grapefruit). Spiny tree with rounded crown. Twigs angular, glabrous. Leaves 10–15 cm, broadly elliptical, rounded or sometimes cordate at the base, subacute at the apex; midrib glabrous. Petiole very broadly winged, frequently up to 15 mm wide near the top, obcordate in outline and tapering below. Flowers in axillary clusters or terminal racemes. Stamens 20–25. Fruit 10–15 cm in diameter, depressed-globose or subpyriform; rind thick, pale yellow when ripe; pulp with coarse vesicles.

6. C. grandis (L.) Osbeck, *Dagb. Ostind. Resa* 98 (1757) (Shaddock, Pomelo). Like **5** but a large tree with few spines; twigs and midrib pubescent; petiole less broadly winged; fruit up to 25 cm in diameter; pulp with slender vesicles.

7. C. aurantium L., *Sp. Pl.* 782 (1753) (Seville Orange). Tree with a rounded crown. Twigs angular when young, soon terete, with slender axillary spines. Leaves 7·5–10 cm, broadly elliptical, subacute at the apex, cuneate or rounded below. Petioles rather broadly winged above, tapering to a wingless base. Flowers solitary or few in the axils, very fragrant. Fruit *c.* 7·5 cm in diameter, subglobose, slightly flattened at both ends, 10- to 12-locular; rind thick, rough, orange when ripe; pulp acid; core hollow when ripe.

8. C. sinensis (L.) Osbeck, *Dagb. Ostind. Resa* 41 (1757) (Orange). Tree with rounded crown. Twigs angular when young, soon terete, with few slender, rather flexible axillary spines Leaves acute, rounded below. Petioles narrowly winged. Flowers in short, lax racemes or solitary, fragrant. Fruit depressed-globose to shortly ovoid, 10- to 13-locular; rind thin to rather thick, nearly smooth, orange to orange-yellow when ripe; pulp sweet; core remaining solid when ripe.

Subfam. **Toddalioideae**

Fruit a drupe or a samara. Seeds with or without endosperm.

5. **Phellodendron** Rupr.[1]

Dioecious. Deciduous trees. Leaves opposite, imparipinnate. Flowers small, in terminal panicles or corymbs, green. Sepals 5–8. Male flowers with 5–6 stamens, longer than the petals; ovary rudimentary. Female flowers with connate styles; stigma 5-lobed; stamens reduced to 5–6 small staminodes. Fruit a black, subglobose drupe.

1. P. amurense Rupr., *Bull. Phys.-Math. Acad. Pétersb.* **15**: 353 (1857). Tree up to 15 m, with deeply fissured, light grey, corky bark; branches forming a wide canopy. Leaflets 5–10 × 1·6–4·5 cm, 5–13, narrowly ovate to ovate-lanceolate, long-acuminate, shortly attenuate or rounded at base, shallowly crenulate, dark green and shining above, paler and almost glabrous beneath. Flowers in terminal panicles; sepals *c.* 1 mm, acute; petals 4–6 mm, oblong, cucullate, densely pubescent within. Fruit *c.* 8 mm in diameter. *Cultivated for ornament and occasionally for timber in S.E. Europe.* [Bu Rm.] (*N.E. Asia.*)

6. **Ptelea** L.[1]

Deciduous polygamous shrubs or small trees. Leaves alternate, 3(–5)-foliolate. Flowers small, greenish-white, in terminal corymbs. Sepals and petals 4–5. Stamens 4–5; filaments villous below; female flowers with 4–5 small staminodes. Styles connate, short. Fruit a compressed, broadly winged samara.

1. P. trifoliata L., *Sp. Pl.* 118 (1753). Shrub or small tree up to 8 m, with rounded crown. Leaflets 6–12 × 2–3·5 cm, 3(–5), ovate to elliptic-oblong, narrowed at each end or shortly acuminate at the apex, entire or obscurely crenulate, dark green and shining above, paler beneath. Sepals 1 mm, free, acute. Petals 4 mm, densely hairy within, oblong. Samara 1·5–2·5 cm, suborbicular, emarginate, straw-coloured, reticulate-veined. *Cultivated in gardens and locally naturalized.* [Ga Ge Hu Rm.] (*North America.*)

LXXXIX. CNEORACEAE[2]

Flowers 3- to 4-merous, hermaphrodite, actinomorphic, in small axillary cymes. Sepals and petals 3–4. Receptacle elongated and forming a gynophore in fruit. Ovary superior with 3–4 loculi, each with 2 pendent ovules. Fruit of usually 3 drupe-like cocci, attached to the central gynophore. Seed with curved embryo and fleshy endosperm.

1. **Cneorum** L.[3]

Small shrubs. Leaves entire, coriaceous. Flowers yellow.

1. C. tricoccon L., *Sp. Pl.* 34 (1753). Nearly glabrous, ever-green shrub 30–100 cm. Leaves 10–30 × 3–7 mm, oblong, obtuse, mucronate, narrowed at base but sessile. Calyx-lobes *c.* 1 mm, ovate, persistent. Petals *c.* 5 mm. Fruit of usually 3 cocci *c.* 5 mm. *Rocky slopes, usually calcareous.* ● *W. Mediterranean region.* Bl Ga Hs It Sa.

XC. SIMAROUBACEAE[2]

Trees or shrubs with bitter bark. Leaves pinnate, alternate. Flowers actinomorphic, 5-merous. Petals free. Disc 10-lobed. Stamens 10 in male flowers, 2–3 in hermaphrodite flowers; filaments free. Ovary superior; carpels 5–6, more or less connate, unilocular, each with 1 ovule attached to the inner angle; styles 2–5, connate. Fruit drupe-like, a berry or a group of samaras.

1. **Ailanthus** Desf.[3]

Deciduous trees. Flowers polygamous, in large terminal panicles. Fruit a group of samaras.

---

[1] By C. C. Townsend.   [2] Edit. T. G. Tutin.
[3] By T. G. Tutin.

1. A. altissima (Miller) Swingle, *Jour. Washington Acad. Sci.* 6: 490 (1916). Up to 20 m, freely suckering. Bark smooth, grey. Leaves 45–60 cm, glabrous or nearly so; leaflets 7–12 cm, 13–25, lanceolate-ovate, long-acuminate, ciliate, with 2–4 teeth near the base, each with a large gland beneath. Panicles 10–20 cm.

Flowers 7–8 mm in diameter, greenish. Samaras 3–4 cm, reddish when young. *Planted for ornament, shade and soil-conservation; extensively naturalized in C., S. & W. Europe.* [Al Au Az Be Br Bu Co Cz Ga Ge Gr He Hs Hu It Ju Lu Rm Rs (W, K) Sa Si.] (*China.*)

XCI. MELIACEAE[1]

Trees or shrubs. Leaves pinnate, alternate. Flowers actinomorphic, usually hermaphrodite, usually 5-merous. Petals free or connate at base. Stamens twice as many as petals, monadelphous. Ovary superior, with 2–8 loculi; placentation axile; style 1. Fruit a capsule, drupe or berry.

1. Melia L.[2]

Deciduous trees or shrubs. Flowers in large, axillary panicles. Sepals 5–6; petals 5–6, free. Ovary with 5–8 loculi; ovules 2 in each loculus. Fruit a drupe.

1. M. azedarach L., *Sp. Pl.* 384 (1753). Up to 15 m. Bark furrowed. Leaves up to 90 cm, 2-pinnate; leaflets 2·5–5 cm, ovate-lanceolate to elliptical, acute, serrate or lobed. Panicles 10–20 cm. Flowers lilac, fragrant; petals *c.* 18 mm. Fruit 6–18 mm, subglobose, yellow. *Widely planted in S. Europe for ornament and shade and locally naturalized.* [Cr Ga Ju.] (*S. & E. Asia.*)

XCII. POLYGALACEAE[3]

Leaves simple, exstipulate. Flowers hermaphrodite, zygomorphic, in spikes or racemes. Sepals 5, free; petals 3–5, more or less united; stamens 8, with filaments partly united. Ovary 2-locular, with 1 seed in each loculus.

1. Polygala L.[4]

Small, perennial herbs, rarely annuals or small shrubs. Leaves usually alternate, entire or subentire. Flowers usually in terminal racemes, rarely in axillary racemes or solitary in leaf-axils. Sepals unequal, the 2 inner (*wings*) much larger than the 3 outer. Petals 3, united proximally into a corolla-tube, free distally, the lower (*keel*) of different form from the 2 upper, and usually bearing a fimbriate crest. Filaments partly or wholly united into a tube, which is partly adnate to the corolla-tube. Gynophore often present, usually elongating in fruit. Stigma 2-lobed, only the posterior lobe receptive. Capsule compressed, usually with a marginal wing. Seeds hairy, with a usually 3-lobed strophiole.

Descriptions and measurements of the petals refer to the free portion and exclude the corolla-tube. Terminal racemes are sometimes displaced by the growth of an axillary branch immediately below them; they are then termed *pseudolateral*.

Literature: R. Chodat, *Arch. Sci. Phys. Nat.* (*Genève*) ser. 3, 18: 281–299 (1887); *Bull. Soc. Bot. Genève* 5: 123–185 (1889); *Mém. Soc. Phys. Hist. Nat. Genève* Suppl. Centen. No. 7 (1891); 31(2), No. 2 (1893). P. Graebner in Ascherson & Graebner, *Syn. Mitteleur. Fl.* 7: 309–387 (1915–16). B. Pawłowski, *Fragm. Fl. Geobot.* 3: 35–68 (1958).

1　Leaves coriaceous; wings patent, not enclosing corolla, deciduous
2　Crest of keel inconspicuous, with 2–6 short lobes; capsule 6–8 mm　　　　　　　　　　　　　　**2. chamaebuxus**
2　Crest of keel conspicuous, with 5–9 long, narrow lobes; capsule 9–13 mm　　　　　　　　　　　**3. vayredae**

1　Leaves scarcely coriaceous; wings enclosing at least a part of the corolla, persistent in fruit
3　Shrub 100–250 cm; wings 15–20 mm　　　　　**4. myrtifolia**
3　Herb, or shrub less than 60 cm; wings 3–15 mm
4　Keel without crest; lower cauline leaves often caducous　　　　　　　　　　　　　　　　　**1. microphylla**
4　Keel with usually prominent, fimbriate crest; leaves usually persistent at least to anthesis
5　Annual; corolla much shorter than wings
6　Wings 3–3·5 mm; corolla purple　　　　　**6. exilis**
6　Wings 6–8 mm; corolla whitish　　　　**10. monspeliaca**
5　Perennial; corolla longer than wings, or only slightly shorter
7　At least some racemes subterminal or pseudolateral
8　Lower leaves ± opposite　　　　　　**26. serpyllifolia**
8　All leaves alternate
9　Wings 4–4·5 mm　　　　　　　　　　**33. alpina**
9　Wings 5–8 mm
10　Wings conspicuously asymmetrical, curving upwards; outer sepals subequal, not inflated　**7. sibirica**
10　Wings ±symmetrical; upper outer sepal inflated and longer than lower pair
11　Wings uniformly membranous, with 3–5 branched veins　　　　　　　　　　　　**9. supina**
11　Wings herbaceous, with a membranous margin, and with 1 vein
12　Leaves linear to oblong, apiculate　**5. rupestris**
12　Leaves obovate, obtuse　　**8. subuniflora**
7　Racemes all terminal
13　Corolla-tube at least ⅔ as long as wings; keel clearly exserted
14　Wings ciliate　　　　　　　　**17. lusitanica**
14　Wings glabrous
15　Gynophore longer than ovary in flower and more than 2 mm in fruit
16　Gynophore less than 3 mm in fruit; wings 7–10 mm in flower　　　　　　　**12. anatolica**
16　Gynophore more than 3 mm in fruit; wings 9–13 mm in flower
17　Racemes with (10–)30–60 flowers; bracts 3–6 mm　　　　　　　　　　　　**11. major**
17　Racemes with 5–20 flowers; bracts 1·5–2 mm　　　　　　　　　　　　　**13. boissieri**

[1] Edit. N. A. Burges.　　　[2] By T. G. Tutin.
[3] Edit. D. H. Valentine and D. A. Webb.
[4] By J. McNeill.

15 Gynophore much shorter than ovary in flower and less than 1 mm in fruit
18 Wings less than 7 mm **16. sardoa**
18 Wings 7·5–11 mm
 19 Leaves entire; wings obtuse **14. venulosa**
 19 Leaves slightly serrulate; wings acute **15. preslii**
13 Corolla-tube less than ⅔ as long as wings; keel not exserted
20 Corolla and wings yellow in flower
21 Wings 7·5–9 mm **20. flavescens**
21 Wings 4–7 mm **23. comosa**
20 Corolla and wings purple, blue, pink or white (rarely wings green) in flower
22 Leaf-rosettes present, with leaves much longer than cauline leaves
23 Stems with a decumbent, usually leafless portion below the leaf-rosette **30. calcarea**
23 Leaf-rosettes basal
24 Leaves not bitter; flowering stems arising laterally from rosette **33. alpina**
24 Leaves bitter; flowering stems arising from centre of rosette
25 Wings 4·5–8 mm in fruit, elliptical **31. amara**
25 Wings 2–4·5(–5) mm in fruit, oblong to obovate **32. amarella**
22 Leaf-rosettes absent; basal leaves not longer than most of the cauline leaves
26 Wings 7·5–11 mm in flower and more than 8 mm in fruit
27 Racemes ± ovoid, dense, not elongating in fruit **19. doerfleri**
27 Racemes conical or cylindrical, lax, elongating in fruit
28 Stems filiform; corolla-tube curved; flowers blue **18. baetica**
28 Stems stout; corolla-tube straight; flowers usually pink, sometimes blue or white
29 Upper petals scarcely longer than keel **22. apiculata**
29 Upper petals much longer than keel **21. nicaeensis**
26 Wings 3–7 mm in flower and not more than 8 mm in fruit
30 Lower leaves ± opposite
31 Cauline leaves increasing in size upwards **29. edmundii**
31 Cauline leaves ± equal **26. serpyllifolia**
30 All leaves alternate
32 Racemes grouped in a corymbose panicle **28. cristagalli**
32 Racemes solitary
33 Upper cauline leaves crowded, larger than those below; peduncle 5 mm or less **25. alpestris**
33 Upper cauline leaves not crowded, not larger than those below; peduncle usually more than 10 mm
34 Crest of keel 4- to 6-lobed; wings greenish in flower **27. carueliana**
34 Crest of keel 8- to 40-lobed; wings pink, blue, purple or white
35 Bracts not exceeding pedicels at anthesis **24. vulgaris**
35 Bracts exceeding pedicels at anthesis
36 Wings (6–)8–11 mm, with 3–5 veins; bracts scarcely projecting beyond apex of developing raceme **21. nicaeensis**
36 Wings 4–6(–8) mm, with 1–3 veins; bracts projecting beyond apex of developing raceme **23. comosa**

Subgen. **Brachytropis** (DC.) Chodat. Flowers in axillary racemes. Wings persistent, petaloid; keel without a crest, enclosed by upper petals; filaments united almost to the apex; anthers opening by two large pores on the inner surface. Seeds with little endosperm.

1. P. microphylla L., *Sp. Pl.* ed. 2, 989 (1763) (*Brachytropis microphylla* (L.) Willk.). Dwarf shrub 10–30 cm, with the habit of an *Ephedra* and erect or ascending, glabrous, sulcate stems. Leaves 5–15 × 1–2 mm, linear to linear-lanceolate, acute, deciduous, the lower usually falling before anthesis. Racemes with 3–8 flowers. Wings 8–10 mm, broadly ovate to suborbicular. Corolla 7–10 mm, blue; upper petals asymmetrical, wider than long; keel *c.* 2·5 mm. *Dry, rocky places.* ● *W. Spain, N. & C. Portugal.* Hs Lu.

Subgen. **Chamaebuxus** (DC.) Duchartre. Flowers axillary. Wings deciduous, petaloid; keel with a small crest; filaments united only at base; anthers opening by a valve on the inner surface. Seeds with much endosperm.

2. P. chamaebuxus L., *Sp. Pl.* 704 (1753). Decumbent dwarf shrub 5–15 cm. Leaves 15–30 × (3–)5–10 mm, coriaceous, ovate to linear-lanceolate. Flowers solitary or in pairs in leaf-axils. Outer sepals unequal, the upper larger; wings patent, white to yellow, sometimes pinkish-purple. Corolla 10–14 mm; upper petals shorter than keel; keel with very small, 2- to 6-lobed crest; tube and upper petals similar in colour to wings; keel bright yellow, becoming purple or brownish-red. Capsule 6–8 mm, sessile, surrounded by a wing less than 1 mm wide. $2n = 38$, *c.* 46. *Woods, pastures and rocky slopes, mainly in the mountains.* ● *Alps and W.C. Europe, northwards to c. 51°N. in Germany, and extending southwards to S. Italy and W. Jugoslavia.* Au Cz Ga Ge He ?Hu It Ju Rm.

3. P. vayredae Costa, *Introd. Fl. Cataluña* ed. 2, *Supl.* 10 (1877) (*Chamaebuxus vayredae* (Costa) Willk.). Like **2** but leaves linear-lanceolate to linear; wings, corolla-tube and upper petals pinkish-purple; keel with a small, fimbriate crest with 5–9 narrow lobes; capsule 9–13 mm, with wing 2–2·5 mm wide. ● *E. Pyrenees.* Hs.

Subgen. **Polygala**. Flowers in terminal or pseudolateral racemes. Wings persistent, often petaloid; keel with a prominent, fimbriate crest; filaments united for at least half their length; anthers opening by a large, subapical pore. Seeds with much endosperm.

4. P. myrtifolia L., *Sp. Pl.* 703 (1753). Erect shrub 100–250 cm. Leaves 2·5–5 cm, oblong to obovate, obtuse. Racemes short, few-flowered, pseudolateral; bracts persistent. Wings 15–20 mm, violet-purple. Corolla 13–18 mm, lilac, shading to deep violet at apex of keel; upper petals short, 2-lobed, the upper lobe reflexed. Capsule elliptical-orbicular, emarginate, narrowly winged. *Cultivated for ornament and locally naturalized in the W. Mediterranean region.* [Co Ga Si.] (*South Africa.*)

5. P. rupestris Pourret, *Mém. Acad. Toulouse* 3: 325 (1788). Stems puberulent, arising from a woody stock. Leaves slightly coriaceous, linear to oblong, apiculate; margins revolute. Racemes with 1–3(–8) flowers, pseudolateral. Upper outer sepal *c.* 4 mm; lower pair *c.* 3 mm; wings 6–8 mm, obovate, greenish, with membranous margin and with 3 indistinct veins appearing as a single midrib. Corolla *c.* 5·5 mm, white, tipped with purple; upper petals *c.* 3·5 mm, narrowly oblong, slightly exceeding the keel; keel with a large crest. Filaments free for the upper ½–⅔ of their length, the united part ciliate throughout. Capsule suborbicular, narrowly winged. *W. Mediterranean region.* Bl Ga Hs.

6. P. exilis DC., *Cat. Pl. Horti Monsp.* 133 (1813). Low, branching annual 5–20 cm with terminal and pseudolateral racemes. Leaves 10–25 mm, narrowly oblong to linear, obtuse. Upper outer sepal 1·5–1·75 mm, narrowly obovate; lower pair less than 1 mm, linear; wings 3–3·5 mm, oblanceolate, obtuse, whitish, with 1 main vein and obscure branches. Corolla *c.* 2·5 mm, purple; upper petals narrowly oblong, shorter than keel; keel with a small crest. Filaments united for ⅔ of their length, ciliate at the base. Capsule 2·5–3 × 1·5–2 mm, obcordate, wider than wings. ● *W. Mediterranean region, from N. Italy to E. Spain.* Ga Hs It.

7. P. sibirica L., *Sp. Pl.* 702 (1753). Erect or ascending perennial 10–20 cm; stems numerous, sparsely crispate-pubescent. Leaves ovate to lanceolate, acute, somewhat hairy. Racemes pseudoterminal, lax, with 5–10 flowers. Outer sepals subequal; wings 6–7 × 2–3 mm, curving upwards so as to expose the keel, obliquely ovate-lanceolate, finely ciliate, greenish. Corolla about as long as wings, lilac or blue; upper petals narrowly spathulate; keel exserted; crest conspicuous, with very fine lobes. Filaments mostly free in the upper ⅕ of their length, ciliate at the base. Style bent at right angles. Capsule *c.* 5 mm, suborbicular. *Dry, calcareous slopes. S. & C. Russia, Ukraine, C. & E. Romania.* Rm Rs (C, W, E).

8. P. subuniflora Boiss. & Heldr. in Boiss., *Diagn. Pl. Or. Nov.* **3(1)**: 59 (1853). Small, decumbent perennial 5–8 cm. Leaves numerous, closely imbricate, obovate, obtuse. Racemes pseudolateral, with 1 or 2 blue flowers. Wings oblong-obovate, exceeding the corolla. Upper petals spathulate, exceeding the keel; crest of keel small, 5- to 7-lobed. Filaments united except for a small apical portion. Style short, erect; lower lobe of stigma acute, about as long as style. *Mountain rocks.* ● *S. Greece (Aroania Oros).* Gr.

9. P. supina Schreber, *Icon. Descr. Pl.* 19 (1766). Procumbent to suberect perennial with woody stock; stems 8–30 cm, numerous, puberulent. Lower leaves obovate to orbicular, obtuse or mucronate. Flowers blue. Wings slightly oblique, suborbicular to lanceolate, slightly exceeding corolla. Filaments united throughout their length. Style more than twice as long as ovary, slightly curved near apex; lower lobe of stigma small. Capsule obovate to suborbicular, emarginate. *Stony slopes and mountain rocks. S.E. Europe.* Al Bu Gr Ju Rm Rs (K) Tu.

1 Wings acute; upper outer sepal scarcely inflated
　　　　　　　　　　　　　　　　　　(b) subsp. hospita
1 Wings obtuse or mucronate; upper outer sepal distinctly inflated
　2 Wings less than half as wide as capsule; gynophore less than
　　　0·3 mm in fruit　　　　　　　**(c) subsp. rhodopea**
　2 Wings almost as wide as capsule; gynophore 0·5–1 mm in
　　　fruit　　　　　　　　　　　　**(a) subsp. supina**

(a) Subsp. **supina** (incl. *P. andrachnoides* Willd.): Stems procumbent to ascending. Upper leaves orbicular, ovate or elliptical. Racemes often subterminal, with 1–12 flowers. Wings suborbicular to ovate-elliptical. *C. part of Balkan peninsula; Krym.*

(b) Subsp. **hospita** (Heuffel) McNeill, *Feddes Repert.* **79**: 30 (1968): Stems ascending. Upper leaves lanceolate. Racemes with 1–3 flowers. Wings oblanceolate to lanceolate, slightly longer than capsule and almost as wide. Capsule 7–8 mm. ● *N. part of Balkan peninsula, S.W. Romania.*

(c) Subsp. **rhodopea** (Velen.) McNeill, *Feddes Repert.* **79**: 31 (1968) (*P. hohenackeriana* subsp. *rhodopea* (Velen.) Hayek): Stems ascending to suberect. Upper leaves narrowly elliptical to oblanceolate. Racemes distinctly pseudolateral, with 6–9 flowers. Wings oblong to oblanceolate. ● *S. Bulgaria, N.E. Greece.*

10. P. monspeliaca L., *Sp. Pl.* 702 (1753). Erect annual. Leaves 10–25 mm, lanceolate to linear-lanceolate, acute. Racemes

terminal. Outer sepals *c.* 3 mm, linear-lanceolate, subequal; wings 6–8 mm, narrowly elliptical, acute, greenish-white, with 3 main veins and numerous lateral branches, not anastomosing. Corolla *c.* 4 mm, whitish; crest of keel large. Filaments united for most of their length, ciliate above. Lower lobe of stigma large. Capsule sessile, obcordate. $2n = c.$ 38. *Mediterranean region, Portugal, Bulgaria.* Al Bl Bu Co Cr Ga Gr Hs It Ju Lu Sa Si Tu.

11. P. major Jacq., *Fl. Austr.* **5**: 6 (1778) (incl. *P. moldavica* Kotov). Stems 15–60 cm, suberect or ascending from a woody stock, sparsely puberulent. Leaves glabrous or sparsely puberulent, the basal narrowly obovate to oblanceolate, the others lanceolate to linear. Racemes with (10–)30–60 flowers; bracts 3–6 mm, caducous. Flowers reddish-purple to violet-blue, rarely milky white. Wings 9–13 mm (10–15 mm in fruit), ovate-orbicular to ovate. Corolla-tube 9–14 mm, bent upwards. Gynophore 3–4 mm in fruit. Capsule 5–6 × 4·5–5 mm, oblong, narrowly winged. Strophiole with lateral lobes shorter than central. $2n = 32$. *Meadows. S.E. Europe, extending westwards to Italy and S. Czechoslovakia.* Al Au Bu Cz Gr Hu It Ju Rm Rs (W, K) Tu.

Variable, especially in size and colour of flowers and shape of wings.

12. P. anatolica Boiss. & Heldr. in Boiss., *Diagn. Pl. Or. Nov.* **3(1)**: 57 (1853). Like **11** but stems densely caespitose, not more than 40 cm; flowers usually lilac-pink; wings 7–10 mm (up to 12 mm in fruit); gynophore less than 3 mm in fruit; strophiole more or less equally 3-lobed. *S.E. part of Balkan peninsula.* Bu Gr Tu.

13. P. boissieri Cosson, *Not. Pl. Crit.* 100 (1851). Like **11** but stems not more than 40 cm, less densely leafy; basal leaves broader; racemes lax, with 5–20 flowers more than 1 cm apart; bracts 1·5–2 mm; flowers purple to pink, rarely white; corolla-tube straight; capsule broadly winged. *Mountains of S. Spain.* Hs.

14. P. venulosa Sibth. & Sm., *Fl. Graec. Prodr.* **2**: 52 (1813). Stems 5–30 cm, ascending from a woody stock, pubescent. Leaves entire, puberulent, the basal spathulate to obovate, the others elliptical to linear-lanceolate. Bracts caducous, equalling or slightly exceeding pedicels. Wings 7·5–9·5 × 2·5–3·5 mm, lanceolate to narrowly elliptical, obtuse, white or lilac with green veins, glabrous. Corolla bluish; tube longer than wings; upper petals greatly exceeding keel. Style 3–4 times as long as stigma. Capsule sessile, shorter and wider than wings. *Rocky hillsides. S. Greece and Aegean region.* ?Bu Cr Gr.

15. P. preslii Sprengel, *Syst. Veg.* **5**: 531 (1828). Like **14** but less hairy, larger and more erect; bracts slightly shorter, persisting throughout anthesis; flowers pink or white; wings larger, acute; corolla with tube scarcely as long as wings; style 2–3 times as long as stigma; capsule subsessile, scarcely as wide as wings. ● *Sicilia.* ?It Si.

16. P. sardoa Chodat, *Bull. Soc. Bot. Genève* ser. 2, **5**: 109 (1913). Like **14** but glabrous, or slightly pubescent above; stems not more than 15 cm; leaves all linear-lanceolate; bracts slightly shorter; wings 7 × 3 mm, oblong-elliptical, acute, whitish; corolla-tube slightly shorter than wings; style less than twice as long as stigma. ● *Sardegna.* Sa.

17. P. lusitanica Welw. ex Chodat, *Mém. Soc. Phys. Hist. Nat. Genève* **31(2)**, No. 2: 441 (1893). Stems 20–40 cm, erect, sparsely puberulent. Leaves more or less glabrous, the basal oblanceolate to lanceolate, the others linear. Racemes lax;

bracts half as long as pedicels; pedicels crispate-puberulent. Flowers blue. Wings 8·5–9 × 3·5–4 mm, oblong-elliptical to subspathulate, with a long claw, ciliate. Corolla-tube narrow, $\frac{2}{3}$ as long as wings. Style twice as long as stigma. ● *N.W. Spain, N. Portugal*. Hs Lu.

18. **P. baetica** Willk. in Willk. & Lange, *Prodr. Fl. Hisp.* **3**: 559 (1878). Stems filiform, spreading or climbing. Leaves glabrous, lanceolate to linear-lanceolate. Racemes lax, with 5–20 flowers; bracts lanceolate, shorter than pedicels in flower. Flowers blue. Wings *c.* 9 mm in flower (*c.* 12 mm in fruit), broadly ovate, with a short, slender claw and with green, strongly anastomosing veins. Corolla scarcely longer than wings; tube curved. Gynophore *c.* 1·5 mm in fruit. Style about 1½ times as long as stigma. Capsule broadly ovate, winged. *Shady places. W. Spain*. Hs.

19. **P. doerfleri** Hayek, *Denkschr. Akad. Wiss. Math.-Nat. Kl. (Wien)* **94**: 159 (1918). Stems *c.* 40 cm, numerous, erect, unbranched. Leaves glabrous, linear-lanceolate, acute. Racemes ovoid-oblong, dense, even in fruit; bracts slightly exceeding pedicels. Flowers pink. Wings 7–9 mm (11 mm in fruit), broadly obovate, with anastomosing veins. Corolla-tube *c.* 6 mm, about equalling the upper petals. Gynophore very short. Capsule shorter and narrower than wings. Lobes of strophiole not more than $\frac{1}{4}$ as long as seed. *Mountain grassland*. ● *N.E. Albania*. Al ?Ju.

20. **P. flavescens** DC., *Cat. Pl. Horti Monsp.* 134 (1813). Stems 15–40 cm, ascending to erect. Lower leaves obovate, the others lanceolate to linear-lanceolate. Racemes with 12–25 flowers; bracts 2·5–3·5 mm, exceeding pedicels in flower. Wings 7·5–9 mm, lanceolate to elliptical, acute to shortly acuminate, yellow in flower, greenish-yellow in fruit, with anastomosing veins. Corolla yellow; tube *c.* 3·5 mm; upper petals much longer than keel and about half as long as wings. Gynophore less than 1 mm in fruit. Capsule 6–7 × 4·5–5 mm, oblong-obovate, shorter and slightly wider than wings. Seeds oblong-ovoid; lateral lobes of strophiole at least half as long as seed. ● *Italy*. ?Co It.

Plants from sandy grassland near the coast of E.C. Italy, with obtuse wings and shorter lobes to the strophiole, have been distinguished as **P. pisaurensis** Caldesi, *Nuovo Gior. Bot. Ital.* **11**: 189 (1879).

21. **P. nicaeensis** Risso ex Koch in Röhling, *Deutschl. Fl.* ed. 3, **5**: 68 (1839). Stems (10–)15–40 cm, usually ascending from a decumbent, woody base. Lower leaves spathulate to linear-lanceolate, the others lanceolate to linear. Racemes 3–15 cm, with 8–40 flowers, usually very lax; bracts exceeding pedicels and slightly exceeding flower-buds, caducous. Flowers pink, less often blue or white. Wings (5–)8–11 mm, 3- to 5-veined. Corolla-tube about half as long as wings, shorter than upper petals. Gynophore up to 1·5 mm in fruit. Capsule shorter than wings. Lateral lobes of strophiole about half as long as seed. *Dry, grassy or stony places. S. & S.C. Europe, extending to S.E. Russia*. Al Au Bu Co Ga Ge Gr He Hs Hu It Ju ?Rm Rs (C, W, E).

1 Outer sepals at least half as long as wings
2 Outer sepals scarcely longer than corolla-tube; flowers usually blue   (g) subsp. **carniolica**
2 Outer sepals almost as long as keel (excluding crest); flowers usually pink   (h) subsp. **forojulensis**
1 Outer sepals less than half as long as wings
3 Upper petals scarcely longer than wings   (e) subsp. **caesalpinii**
3 Upper petals longer than wings, usually clearly exserted
4 Wings 6–8 mm, lanceolate   (f) subsp. **gariodiana**
4 Wings (7–)8–11 mm, usually elliptical to suborbicular
5 Stem and leaves ± tomentose   (c) subsp. **tomentella**
5 Stem and leaves glabrous to pubescent
6 Basal leaves obovate to spathulate, persistent to anthesis   (a) subsp. **nicaeensis**
6 Leaves all linear-lanceolate to lanceolate (the basal sometimes deciduous before anthesis)
7 Seeds oblong; lateral lobes of strophiole straight   (b) subsp. **mediterranea**
7 Seeds ovoid to pyriform; lateral lobes of strophiole crescentic   (d) subsp. **corsica**

(a) Subsp. **nicaeensis**: Leaves pubescent, the basal obovate to spathulate, the others linear-lanceolate to linear. Flowers usually blue. Outer sepals 3–4 mm; wings 7–8 mm. Upper petals exceeding wings. Capsule broadly obovate. ● *S.E. France, N.W. Italy*.

(b) Subsp. **mediterranea** Chodat, *Bull. Soc. Bot. Genève* **5**: 179 (1889): Glabrous or minutely puberulent. Leaves all linear to linear-lanceolate. Flowers usually pink. Outer sepals 3–4 mm; wings 7·5–10 mm. Upper petals exceeding wings. Capsule oblong-obovate to obcordate. Seeds oblong; lateral lobes of strophiole straight. *S.E. Europe, extending westwards to N. Italy, and northeastwards to S. Russia*.

The most widespread and variable of the subspecies, which should, perhaps, be divided. The plants from Ukraine and S. Russia included here are somewhat isolated geographically. They have been distinguished as **P. cretacea** Kotov, *Žur. Inst. Bot. URSR* **21–22**: 238 (1939), and may, perhaps, be referable to **P. hybrida** DC., *Prodr.* **1**: 325 (1824).

(c) Subsp. **tomentella** (Boiss.) Chodat, *Bull. Soc. Bot. Genève* **5**: 179 (1889) (subsp. *graeca* Chodat): Like subsp (b) but stems and leaves densely hairy, often tomentose. Wings lanceolate, acute. ● *Greece and Aegean region*.

(d) Subsp. **corsica** (Boreau) Graebner in Ascherson & Graebner, *Syn. Mitteleur. Fl.* **7**: 337 (1916): Leaves glabrous, the lower linear-lanceolate, the upper narrowly linear. Wings *c.* 10 × 6 mm, soon becoming colourless and membranous after flowering. Capsule *c.* 7 × 4·5 mm, obcordate. Seeds ovoid to pyriform; lateral lobes of strophiole crescentic. $2n = 34$. ● *Corse, N. Italy*.

(e) Subsp. **caesalpinii** (Bubani) McNeill, *Feddes Repert.* **79**: 32 (1968) (*P. pedemontana* sensu Chodat, non Perr. & B. Verlot, *P. rosea* sensu Willk. pro parte, non Desf., *P. vulgaris* subsp. *pedemontana* auct. hisp.): Leaves glabrescent, the lower obovate to oblanceolate, the upper linear to linear-lanceolate, usually shortly apiculate. Flowers usually pink. Outer sepals *c.* 3 mm; wings 6·5–8 mm. Upper petals about as long as wings. Capsule obcordate. ● *N.E. Spain, S. France*.

(f) Subsp. **gariodiana** (Jordan & Fourr.) Chodat, *Bull. Soc. Bot. Genève* **5**: 180 (1889): Leaves linear to linear-lanceolate, rather thick. Wings 6–8·5 mm, lanceolate. Upper petals exceeding the wings. Capsule narrowly obovate, cuneate, subsessile. ● *S.E. France, N.W. Italy*.

(g) Subsp. **carniolica** (A. Kerner) Graebner in Ascherson & Graebner, *Syn. Mitteleur. Fl.* **7**: 339 (1916) (*P. carniolica* A. Kerner): Lower leaves oblanceolate to obovate, the upper linear to linear-lanceolate. Flowers usually blue. Outer sepals 4–5 mm, scarcely exceeding corolla-tube; wings 6–8 × 3·5–5 mm. Upper petals considerably exceeding wings. Gynophore in fruit 1–1·5 mm. Capsule obcordate, enclosed by wings. ● *E.C. Europe and W. part of Balkan peninsula*.

(h) Subsp. **forojulensis** (A. Kerner) Graebner, *op. cit.* 338 (1916): Like subsp. (g) but flowers usually pink; outer sepals 4·5–5 mm, clearly exceeding corolla-tube; wings 7–9 mm; upper petals scarcely exceeding wings; capsule subsessile. ● *S. Alps*.

22. P. apiculata Porta, *Nuovo Gior. Bot. Ital.* **11**: 238 (1879). Stock woody, branched; stems 20–40 cm, erect. Lower leaves elliptical, obtuse; upper leaves lanceolate to linear, acute to acuminate. Racemes conical, rather lax; bracts longer than pedicels. Wings 8–9 × 4–5 mm; veins scarcely anastomosing. Corolla about as long as wings; upper petals scarcely longer than keel. Capsule oblong, emarginate, subsessile. Central lobe of strophiole short, appressed; lateral lobes flattened, erect, forming a crest about half as long as the seed.   ● *S. Italy.* It.

23. P. comosa Schkuhr, *Handb.* **2**: 324 (1796). Stems 7–20(–40) cm, erect or ascending. Lower leaves narrowly spathulate to obovate, obtuse, usually falling before anthesis; upper leaves linear to linear-lanceolate, acute. Racemes with 15–50 flowers, conical to cylindrical, dense; bracts 2–5 mm, linear, acuminate, exceeding the flower-buds, often persisting throughout anthesis. Flowers usually lilac-pink. Wings 4–6(–8) mm. Corolla about as long as wings. Capsule obcordate-cuneate, shorter than wings and about as wide, narrowly winged. Strophiole hairy; lateral lobes about ⅓ as long as seed. 2n=28–32, 34. *Mainly in C. & E. Europe, but extending to S. Sweden, Belgium, N. Spain and N. Italy.* Al Au Be Bu Cz Fe Ga Ge Gr He Ho Hs Hu It Ju Po Rm Rs (N, B, C, W, K) Su Tu.

Rather variable, and not always clearly distinguishable from **21** and **24**; its eastern and southern limits are, therefore, somewhat uncertain.

24. P. vulgaris L., *Sp. Pl.* 702 (1753) (incl. *P. oxyptera* Reichenb.). Stock woody, branched. Stems 7–35 cm, ascending to erect, glabrous or sparsely hairy. Leaves alternate, not bitter, the lower obovate to elliptical, the upper longer, linear-lanceolate. Racemes with 10–40 flowers, rather dense, conical at first, elongating in fruit; bracts membranous except for the midrib, scarcely exceeding pedicels in flower and shorter than flower-buds, caducous. Flowers blue, pink or white. Wings (3–)4–7(–8) mm, with 3 anastomosing veins. Corolla-tube usually longer than upper petals. Style usually 1–1½ times as long as stigma. Capsule about as long as wings. Seeds oblong-ellipsoid; lobes of strophiole about ⅓ as long as seed. 2n=28, 32, 48, c. 56, 68, c. 70. *Most of Europe westwards from N.W. Russia and C. Ukraine.* All except ?Bl Cr Is Rs (K, E) Sb ?Si.

Extremely variable; of the more widespread variants **P. oxyptera** Reichenb., *Pl. Crit.* **1**: 25 (1823), seems to have the strongest claim for recognition at specific rank; it is distinguished by the lanceolate to lanceolate-elliptical, acute wings, never wider than the capsule. It appears, however, that although in some regions (e.g. Poland) it is constant and readily distinguishable, in others (e.g. Britain) it is not.

P. carniolica var. *stojanovii* (Stefanov) Stoj. & Stefanov, from S.W. Bulgaria, appears to be indistinguishable from **24** except for the presence of axillary racemes below the terminal one.

25. P. alpestris Reichenb., *Pl. Crit.* **1**: 25 (1823). Stems 7–15 cm, few, decumbent or ascending. Leaves increasing in size upwards, the upper broadly lanceolate. Racemes 1·5–3·5 cm, with 5–20 flowers, dense; bracts shorter than pedicels at anthesis, caducous. Flowers blue or white. Wings 4–6·5 mm, ovate to narrowly obovate, 1- to 3-veined. Corolla about as long as wings, distinctly articulated between tube and keel, and between keel and crest. Capsule wider than wings. Seeds ellipsoid; dorsal lobe of strophiole appressed. *Mountain pastures and meadows.* ● *S. & S.C. Europe, from the Alps to the Pyrenees and Greece.* Al Au Ga Ge Gr He Hs It Ju.

(a) Subsp. **alpestris**: Wings 4–4·5(–5) mm, narrower than capsule; veins scarcely anastomosing. Capsule sessile. Lateral lobes of strophiole c. ⅓ as long as seed. 2n=34. *Pyrenees, Jura, Alps, Appennini.*

(b) Subsp. **croatica** (Chodat) Hayek, *Prodr. Fl. Penins. Balcan.* **1**: 597 (1925) (*P. calcarea* subsp. *croatica* (Chodat) Graebner): Wings 5–6·5 mm, wider than capsule; veins usually distinctly anastomosing. Capsule on short gynophore. Lateral lobes of strophiole ⅓ as long as seed. *W. part of Balkan peninsula, S. Italy.*

26. P. serpyllifolia J. A. C. Hose, *Ann. Bot.* (*Usteri*) **21**: 39 (1797) (*P. serpyllacea* Weihe). Stems 6–25 cm, slender, decumbent to ascending, not woody at base. Leaves 3–15 mm, the lower elliptical to obovate, opposite or subopposite, the upper lanceolate to linear-lanceolate, alternate or opposite. Racemes 1–3(–4) cm, with 3–10 flowers, terminal or pseudolateral; bracts shorter than pedicels at anthesis. Flowers usually blue. Outer sepals 1·5–2·5 mm; wings 4·5–5·5 mm, oblanceolate to elliptical; veins anastomosing. Upper petals usually longer than wings; crest of keel 10- to 25-lobed. Capsule shorter and wider than wings. Seeds ovoid; lateral lobes of strophiole c. ⅓ as long as seed. 2n=32, 34, c. 68. *Calcifuge. W. & C. Europe, eastwards to E. Germany and N.W. Jugoslavia.* Au ?Az Be Br Co Cz Da Fa Ga Ge Hb He Ho Hs It Ju Lu No.

27. P. carueliana (A. W. Benn.) Burnat ex Caruel in Parl., *Fl. Ital.* **9**: 117 (1890). Like **26** but leaves alternate, remote, linear to subspathulate, obtuse; racemes all terminal; outer sepals 2·5–3·5 mm; wings 6 mm, greenish, sometimes with a purple tinge, slightly falcate, with veins scarcely anastomosing; corolla brownish-purple; crest of keel inconspicuous, 4- to 6-lobed. ● *N.W. Italy (Alpi Apuane).* It.

28. P. cristagalli Chodat, *Bull. Soc. Bot. Genève* ser. 2, **5**: 110 (1913). Stems 15–20 cm, flexuous. Leaves glabrescent, acute, the lower broadly lanceolate, the upper linear-lanceolate. Racemes 1–2 cm, terminal on main axis and on upper axillary branches, forming a corymbose panicle. Outer sepals 3 mm, subulate; wings 5 × 1·5 mm, lanceolate, apiculate; veins not anastomosing. Crest of keel with long, narrow lobes. Capsule ¾ as long as wings. Seeds oblong-ellipsoid; lobes of strophiole flattened, erect, forming an apical crest.   ● *S. Greece.* Gr.

29. P. edmundii Chodat, *Bull. Herb. Boiss.* **4**: 911 (1896). Stems c. 5 cm, numerous, slender, woody at the base. Leaves glabrous, increasing in size upwards, the lower 2–4 mm, spathulate, obtuse, opposite, the upper up to 10 mm, elliptical, obtuse, alternate, forming a small rosette below the raceme. Racemes terminal, sessile, very short, corymbose. Wings c. 5 mm, ovate, shortly clawed; veins anastomosing. Corolla-tube short; upper petals oblong, obtuse.   ● *N.W. Spain (Picos de Europa).* Hs.

30. P. calcarea F. W. Schultz, *Flora* (*Regensb.*) **20**: 752 (1837). Stems 10–20 cm, with decumbent, usually leafless stolons terminating in leaf-rosettes, from which arise a number of almost erect flowering stems; non-flowering shoots also present, arising from the stock or from the rosette. Leaves glabrous or sparsely hairy, not bitter. Rosette-leaves spathulate to obovate; leaves of flowering stems smaller, linear-lanceolate, obtuse. Racemes with 6–20 flowers; bracts linear-lanceolate. Flowers usually blue or white. Wings c. 5 mm, obovate to oblong-elliptical; veins anastomosing. Corolla exceeding wings. Capsule 4–6 mm. Seeds ovoid-oblong; lateral lobes of strophiole about ½ as long as seed. 2n=34. *Calcicole.* ● *W. Europe, northwards to S. England.* Be Br Ga Ge He Hs.

235

31. P. amara L., *Syst. Nat.* ed. 10, **2**: 1154 (1759). Stems 5–20 cm, numerous, arising from the centre of a basal rosette. Leaves glabrous, bitter, the basal 15–35 × 6–10 mm, elliptical to obovate, the upper lanceolate to oblong, widest near the middle, acute. Racemes with 8–25 blue, violet, pink or white flowers. Wings 4·5–8 mm in fruit, elliptical. Corolla 3·5–6·5 mm, distinctly articulated between tube and keel and between keel and crest. Capsule 3·5–5·5 mm; seeds 2·3–2·8 mm. ● *Mountains of E.C. Europe and N. Jugoslavia.* Au ?Bu Cz Ge Hu Ju Po Rm Rs (W).

(a) Subsp. **amara**: Wings 6–8 mm in fruit. Corolla 4·5–6·5 mm; crest of keel with 15–30 lobes; capsule much shorter than wings and only slightly wider. 2n=28. *E. Alps, W. Carpathians, Hungary, Jugoslavia.*

(b) Subsp. **brachyptera** (Chodat) Hayek, *Sched. Fl. Stir. Exsicc.* 9–10: 21 (1906) (*P. subamara* Fritsch): Wings 4·5–6·5 mm in fruit. Corolla 3·5–5·5 mm; crest of keel with 10–20 lobes; capsule slightly shorter and much wider than wings. 2n=28. *Carpathians and S. Poland; a few localities in S.E. Alps.*

32. P. amarella Crantz, *Stirp. Austr.* ed. 2, **2**: 438 (1769) (*P. amara* subsp. *amarella* (Crantz) Chodat, *P. austriaca* Crantz). Like 31 but cauline leaves obtuse, widest near apex; wings 2–5 mm in fruit, oblong to obovate; corolla 2–4 mm, scarcely articulated; crest of keel with 5–15 lobes; capsule 3–4 mm, usually much wider than wings; seeds 1·5–2·3 mm. 2n=34. ● *Much of Europe, but absent from most of the south.* Au Be Br ?Bu Cz Da Fe Ga Ge He Hu It Ju No Po Rm Rs (N, B, C, W) Su.

33. P. alpina (Poiret) Steudel, *Nomencl. Bot.* 642 (1821). Stems decumbent, bearing leaf-rosettes from which arise lateral flowering stems 1–5 cm. Leaves not bitter; those of rosettes obovate to oblong, the others much smaller, oblong to linear-oblong. Bracts caducous. Flowers bright blue. Wings 4–4·5 mm, with sparingly branched veins, not anastomosing. Capsule nearly as long as wings and about twice as wide. Lateral lobes of strophiole *c.* ⅓ as long as seed. 2n=c. 34. *Alpine pastures; usually calcicole.* ● *Pyrenees; Alps, eastwards to c. 11° 30′ E.* Ga It He Hs.

SAPINDALES

XCIII. CORIARIACEAE[1]

Shrubs with opposite or whorled, simple, exstipulate leaves. Flowers hermaphrodite or unisexual, 5-merous, in axillary or terminal racemes. Stamens 10. Carpels superior, free, 1-seeded, in a single whorl.

1. Coriaria L.[2]

Stems angular. Sepals free. Petals free, green and shorter than the sepals in flower, enlarging, darkening and becoming succulent in fruit. Styles free, filiform, stigmatic all over. Fruit a collection of achenes, enclosed by the corolla until ripe.

1. C. myrtifolia L., *Sp. Pl.* 1037 (1753). Glabrous; stems 1–3 m, arcuate, with 4-angled, suberect branches. Leaves 3–6 cm, ovate-lanceolate, acute or cuspidate, sessile, opposite, or rarely in whorls of 3–4. Racemes 2–5 cm in flower, longer in fruit, axillary or terminating short, lateral branches. Sepals ovate, acute, persistent. Petals dark reddish-brown in fruit, strongly keeled on inner side, eventually separating to reveal the fruit. Flowers male, female and hermaphrodite, with fairly conspicuous rudiments of stamens in female and carpels in male flowers. Fertile stamens exserted, pendent; sterile stamens included, erect. Fertile styles long-exserted. Achenes 4 mm, ridged, shining black. *Dry woods, hedges and rocky places. S.W. Europe, from S. Spain to N.W. Italy.* Bl Ga ?Gr Hs It [Lu].

XCIV. ANACARDIACEAE[3]

Trees or shrubs. Leaves alternate, usually pinnate or digitate. Calyx usually 5-partite; petals 5, rarely absent, free or more or less connate; ovary superior, 1-locular, with one ovule; placentation basal or apical. Fruit a drupe.

1	Leaves simple	**2. Cotinus**
1	Leaves pinnate or digitate	
2	Leaves digitate	**1. Rhus**
2	Leaves pinnate	
3	Mature leaves and young twigs hairy	**1. Rhus**
3	Mature leaves and young twigs glabrous, except sometimes the petioles	
4	Leaves up to 15 cm; petals 0	**3. Pistacia**
4	Leaves 25 cm or more; petals 5	**4. Schinus**

1. Rhus L.[4]

Polygamous or dioecious shrubs or small trees, often with resinous bark. Buds naked. Leaves digitate or pinnate. Flowers small, in axillary or terminal panicles. Petals 5. Stamens 5. Placentation basal. Fruit with short, non-plumose pedicels; styles terminal.

1	Leaves pinnate	
2	Rhachis of leaves winged, at least between the distal leaflets	**1. coriaria**
2	Rhachis of leaves completely unwinged	**2. typhina**
1	Leaves digitate	
3	Bark of twigs grey; at least some leaves with 5 leaflets; leaflets entire or 3-dentate at apex	**3. pentaphylla**
3	Bark of twigs brown; leaves with 3 leaflets; leaflets entire or with 2–3 teeth on each side	**4. tripartita**

1. R. coriaria L., *Sp. Pl.* 265 (1753). Almost evergreen shrub or small tree up to 3 m. Young twigs and petioles densely hispid. Leaves imparipinnate; leaflets 1–5 cm, 7–21, ovate to oblong, coarsely crenate-serrate, sometimes with 1–2 small lobes at base; rhachis hispid, winged, at least between the distal leaflets. Inflorescence *c.* 10 cm, the branches more or less concealed by the flowers. Sepals ovate, greenish; petals oblong, white, longer

---

[1] Edit. D. A. Webb. [2] By D. A. Webb.
[3] Edit. T. G. Tutin. [4] By T. G. Tutin.

than the sepals. Drupe shortly hispid, brownish-purple. *Rocky places and scrub at low altitudes. S. Europe.* Al Az Bu Cr Ga Gr Hs It Ju Lu Rs (K) Si Tu.

2. R. typhina L., *Cent. Pl.* **2**: 14 (1756) (*R. hirta* (L.) Sudworth). Like **1** but deciduous, up to 10 m; leaflets 5–12 cm, oblong-lanceolate; rhachis not winged; inflorescence 10–20 cm; drupe crimson. *Cultivated for ornament, and locally naturalized in S. Europe.* [Bu Cz Ga He Ju It Rm.] (*E. North America.*)

3. R. pentaphylla (Jacq.) Desf., *Fl. Atl.* **1**: 267 (1798). Thorny tree up to 7 m. Bark of twigs grey, glabrous or nearly so. Leaves digitate; leaflets 3–5, up to 2 cm, oblanceolate to obtriangular, entire to shallowly lobed, most often 3-dentate at apex; petioles winged in upper half. Inflorescence axillary, slender, little-branched, almost raceme-like, shorter than the leaves. Sepals ovate; petals ovate, pale yellow. Drupe with 3 tubercles at apex, red. *Dry calcareous places. Sicilia.* Si. (*N. Africa.*)

4. R. tripartita (Ucria) Grande, *Bull. Orto Bot. Napoli* **5**: 62 (1918). Like **3** but bark of twigs brown; all leaves with 3 leaflets; leaflets entire or with 2–3 teeth on each side; inflorescence obviously paniculate; petals greenish. *Dry places. Sicilia.* Si. (*N. Africa, S.W. Asia.*)

2. Cotinus Miller[1]

Like *Rhus* but leaves simple and entire; buds with several imbricate scales; pedicels in fruit long, slender, with long, patent hairs; styles lateral.

1. C. coggygria Scop., *Fl. Carn.* ed. 2, **1**: 220 (1772) (*Rhus cotinus* L.). Rounded, glabrous shrub up to 5 m. Leaves 3–8 cm, ovate or obovate, glaucous; petioles not winged. Inflorescence 15–20 cm, terminal, with long, slender branches. Pedicels numerous, many without fruits, all plumose. Drupe 3–4 mm, reniform. *Dry rocky slopes. S. Europe, from S.E. France eastwards, and extending northwards to S.E. Czechoslovakia and C. Ukraine.* Al Au Bu Cr Cz Ga Gr He Hu It Ju Rm Rs (W, K, E) Tu [Ge Hs].

3. Pistacia L.[1]

Dioecious trees or shrubs with resinous bark. Buds with several scales. Leaves pinnate, occasionally some reduced to 1 leaflet. Flowers in lateral panicles. Petals absent. Stamens 3–5. Placentation basal.

1 Petioles glabrous
2 Leaves imparipinnate; panicles with long branches **1. terebinthus**
2 Leaves paripinnate; panicle spike-like **4. lentiscus**
1 Petioles pubescent or puberulent
3 Leaves coriaceous; drupe *c.* 5 mm **2. atlantica**
3 Leaves thin; drupe *c.* 25 mm **3. vera**

1. P. terebinthus L., *Sp. Pl.* 1025 (1753). Small deciduous tree or shrub up to 5 m. Leaves imparipinnate; leaflets usually

2–8·5–1 × 3·5 cm, 3–9, ovate to obovate or oblong, mucronate, coriaceous; rhachis not winged; petioles glabrous. Inflorescence with long branches. Flowers brownish. Drupe 5–7 × 4–6 mm, obovoid, compressed, apiculate, at first reddish, becoming brown. *Dry, open woods and rocky, usually calcareous slopes. Mediterranean region, Portugal.* Al Bl Bu Co Cr Ga Gr Hs It Ju Lu Sa Si Tu.

2. P. atlantica Desf., *Fl. Atl.* **2**: 364 (1799) (*P. mutica* Fischer & C. A. Meyer). Like **1** but leaflets lanceolate, obtuse, not mucronate; petioles puberulent; rhachis narrowly winged. *N.E. Greece; Turkey-in-Europe; Krym.* Gr Rs (K) Tu.

3. P. vera L., *Sp. Pl.* 1025 (1753). Like **1** but leaflets 1–3, thin, puberulent when young; petioles pubescent; rhachis scarcely winged; drupe *c.* 25 mm. *Cultivated for its edible seeds in S. Europe and perhaps locally naturalized.* [?Ga ?Gr ?Hs ?Si.] (*Temperate Asia.*)

4. P. lentiscus L., *Sp. Pl.* 1026 (1753). Small evergreen tree or shrub 1–8 m. Leaves paripinnate; leaflets 1–5 × 0·5–1·5 cm, (4–)8–12, lanceolate to obovate-lanceolate, mucronate, coriaceous; rhachis broadly winged; petioles glabrous. Inflorescence compact, spike-like. Flowers yellowish or purplish. Drupe *c.* 4 mm, globose, apiculate, red becoming black. *Dry open woods and scrub. Mediterranean region, extending to Portugal.* Al Bl Co Cr Ga Gr Hs It Ju Lu Sa Si.

P. x raportae Burnat, *Fl. Alp. Marit.* **2**: 54 (1896) (*P. lentiscus × terebinthus*) occurs locally in France, Italy, Sardegna and, perhaps, Portugal.

4. Schinus L.[1]

Polygamous or dioecious trees or shrubs. Buds with several scales. Leaves pinnate. Flowers in lateral and terminal panicles. Sepals 5. Petals 5. Stamens 10. Placentation apical.

Rhachis unwinged; drupe 6–7 mm, pink **1. molle**
Rhachis winged in upper part; drupe 4–5 mm, bright red **2. terebinthifolia**

1. S. molle L., *Sp. Pl.* 388 (1753). Evergreen tree or shrub usually up to 8 m. Branches slender, pendent. Leaflets 2·5–6 × 3–8 mm, 7–13 pairs linear-lanceolate, often serrate, pubescent when young; rhachis unwinged. Inflorescence much-branched, lax. Flowers white. Drupe 6–7 mm, globose, pink. *Planted for ornament in S. Europe and more or less naturalized.* [Gr Hs It Lu Sa Si.] (*Mountains of C. & S. America, from Mexico to N. Chile and N. Argentina.*)

2. S. terebinthifolia Raddi, *Mem. Mat. Fis. Soc. Ital. Sci.* **18** (Fis.): 399 (1820). Like **1** but branches not pendent; leaflets 10–20 mm wide, 2–7 pairs; rhachis winged in upper part; inflorescence dense; drupe 4–5 mm, bright red. *Planted for ornament in S.W. Europe and locally naturalized.* [Hs Lu.] (*S.W. Brazil, Paraguay.*)

XCV. ACERACEAE[2]

Trees or shrubs with opposite, exstipulate leaves. Flowers actinomorphic, sometimes perigynous, in racemes, panicles or corymbs. Sepals 5, free; petals 5, free, rarely absent; stamens usually 8, inserted on the usually well-developed disc. Ovary superior, of 2 carpels, each with 2 ovules; styles 2. Fruit of 2 winged, single-seeded mericarps (samarae).

---

[1] By T. G. Tutin.   [2] Edit. S. M. Walters.

1. Acer L.[1]

Trees or shrubs, usually deciduous. Leaves long-petiolate, usually palmately lobed. Flowers greenish or yellowish, often unisexual, sometimes apetalous. Samarae winged on outer side only.

Literature: F. Pax in Engler, *Pflanzenreich* **8** (**IV. 163**): 6–80 (1902). A. I. Pojarkova, *Acta Inst. Bot. Acad. Sci. URSS* **1**: 225–374 (1933).

Two North American species, **A. saccharophorum** C. Koch, *Hort. Dendrol.* 80 (1853) (*A. saccharum* auct.), and **A. saccharinum** L., *Sp. Pl.* 1055 (1753) (*A. dasycarpum* Ehrh.) are planted locally for timber in C. Europe. *A. saccharophorum* has leaves rather like those of **1**, but paler beneath and without latex. *A. saccharinum* has leaves more deeply lobed and silvery-white beneath, also without latex.

```
1  Leaves pinnately 3- to 7-foliolate              15. negundo
1  Leaves simple, sometimes deeply palmately lobed
  2  Leaves less than 8 cm, ± coriaceous
    3  Wings of fruit horizontal; leaves ciliate    3. campestre
    3  Wings of fruit subparallel or diverging at an acute angle;
         leaves not ciliate
      4  Leaves glabrous and green beneath, evergreen  14. sempervirens
      4  Leaves pubescent or subglaucous beneath, deciduous
        5  Leaves 3-lobed; lobes usually entire     13. monspessulanum
        5  Leaves (3–)5-lobed; lobes dentate       (8–12). opalus group
  2  Leaves up to 15 cm, not coriaceous
    6  Leaves undivided (rarely slightly 3-lobed)   4. tataricum
    6  Leaves distinctly 3- to 7-lobed
      7  Middle lobe of leaf separated nearly to base  6. heldreichii
      7  Leaf lobed to not more than ⅔ of distance to base
        8  Inflorescence paniculate; petioles without latex
          9  Inflorescence erect, broadly pyramidal; leaves irregularly
               dentate                             7. trautvetteri
          9  Inflorescence pendent, narrow; leaves serrate
                                                   5. pseudoplatanus
        8  Inflorescence corymbose; petioles with or without latex
          10  Petioles without latex; wings of fruit diverging at an
                acute angle                        (8–12). opalus group
          10  Petioles with latex; wings of fruit diverging at an obtuse
                angle, sometimes nearly horizontal
            11  Leaves usually less than 7 cm, ciliate  3. campestre
            11  Leaves (5–)7–15 cm, not ciliate
              12  Leaf-lobes almost entire         2. lobelii
              12  Leaf-lobes sinuate-dentate       1. platanoides
```

Sect. PLATANOIDEA Pax. Usually monoecious; leaves (3–)5- to 7-lobed; latex present; inflorescence corymbose; stamens inserted on middle of disc.

1. A. platanoides L., *Sp. Pl.* 1055 (1753). Spreading tree up to 30 m. Leaves (5–)10–15 cm, 5- to 7-lobed; lobes acuminate, with few large, acuminate teeth. Flowers in erect, glabrous corymbs appearing before the leaves. Fruit with widely divergent to subhorizontal wings. $2n = 26$. *Most of Europe except the extreme north, the extreme west and the islands; only on mountains in the south. Planted for ornament and occasionally naturalized.* Al Au Be Bu Cz Fe Ga Ge Gr He Hs Hu It Ju No Po Rm Rs (N, B, C, W, E) Su Tu [Br Ho].

2. A. lobelii Ten., *Cat. Pl. Horti Neap.*, *App.* ed. 2, 69 (1819). Like **1** but with a narrow, columnar habit; leaf-lobes almost entire; calyx hairy. *Mountain woods.* ● *C. & S. Italy.* It.

3. A. campestre L., *Sp. Pl.* 1055 (1753). Shrub or small tree up to 20(–25) m. Leaves 4–7 cm, obtusely 3- to 5-lobed, ciliate, often thick in texture, sometimes subcoriaceous. Flowers few,

greenish, in erect, pubescent corymbs, opening with the leaves. Fruit usually pubescent, sometimes glabrous (var. *leiocarpum* (Opiz) Wallr.), with horizontal wings. *Most of Europe from N. England, S. Sweden and C. Russia southwards, but rare in the Mediterranean region; planted for hedges or ornament and occasionally naturalized.* Al Au Be Br Bu Co Cz Da Ga Ge Gr He Ho Hs Hu It Ju Po Rm Rs (C, W, K) Sa Si Su Tu [Hb].

Variable, especially in S.E. Europe, where some of the variation may result from hybridization with **13** and **14**. Such plants, with 3 triangular, acute, entire leaf-lobes, have been recorded from Jugoslavia, Hungary and Romania, as subsp. **marsicum** (Guss.) Hayek, *Prodr. Fl. Penins. Balcan.* **1**: 606 (1925).

Sect. ACER. Usually monoecious; leaves 3- to 5(–7)-lobed, more rarely undivided; no latex; inflorescence paniculate or corymbose; stamens inserted on inner margin of disc.

4. A. tataricum L., *Sp. Pl.* 1054 (1753). Shrub or small tree up to 10 m. Leaves 6–10 cm, oblong, acute, cordate at base, usually undivided, more rarely shallowly 3-lobed; margin irregularly incise-biserrate. Flowers greenish-white, in suberect panicles. Fruit glabrescent, with straight, subparallel wings. *S.E. Europe, extending westwards to 16° E. in Czechoslovakia and northwards to c. 55° in C. Russia.* Al Au Bu Cz Gr Hu Ju Rm Rs (C, W, E).

5. A. pseudoplatanus L., *Sp. Pl.* 1054 (1753). Spreading tree up to 30 m. Leaves (7–)10–15 cm, 5-lobed to about ½ way; lobes acute, coarsely serrate. Flowers numerous, greenish, in narrow, pendent panicles, usually appearing with the leaves. Fruit glabrous, with acute wings usually diverging at about a right angle. $2n = 52$. *C. & S. Europe, mainly in and around the mountains, from Belgium and N. Poland to C. Portugal, Sicilia and C. Greece. Widely planted elsewhere for shelter and ornament and frequently naturalized.* Al Au Be Bu Co Cz Ga Ge Gr He Ho Hs Hu It Ju Lu Po Rm Rs (W, C) Si Tu [Br Da Hb Su].

6. A. heldreichii Orph. ex Boiss., *Diagn. Pl. Or. Nov.* **3**(**5**): 71 (1856). Tree up to 25 m. Leaves 5–14 cm, deeply 5-lobed with middle lobe free nearly to base; lobes acute, with 2 or 3 large teeth on each side. Flowers rather few, subglabrous, yellowish, in suberect panicles, opening with leaves. Fruit glabrous, with arcuate wings usually diverging at an obtuse angle. ● *Mountains of Balkan peninsula.* Al Bu Gr Ju.

(a) Subsp. **heldreichii**: Leaves 5–8 cm, glaucous beneath; samarae 2–3 cm. *Throughout most of the range of the species.*

(b) Subsp. **visianii** K. Malý, *Magyar Bot. Lapok* **7**: 219 (1908) (*A. macropterum* Vis.): Leaves up to 14 cm, scarcely glaucous beneath; samarae 4–5 cm. *C. & S. Jugoslavia, W. Bulgaria.*

7. A. trautvetteri Medv., *Izv. Kavk. Obšč. Ljub. Est.* **2**: 8 (1880). Tree up to 15 m. Leaves 10–15 cm, deeply 5-lobed; lobes acute, with 3–4 large teeth on each side. Flowers rather few, subglabrous, pale green, in erect, pyramidal panicles, appearing after the leaves. Fruit glabrescent with subparallel wings. *Near Istanbul (forest of Belgrad).* Tu. (*N. Anatolia, W. Caucasus.*)

Sect. GONIOCARPA Pojark. Usually monoecious; leaves 3- to 5-lobed (more rarely undivided), subcoriaceous, sometimes evergreen; no latex; inflorescence corymbose; stamens inserted on inner margin of disc.

(8–12). A. opalus group. Small tree or shrub up to 15 m. Leaves very variable in size, shape and texture, usually 5-lobed. Flowers rather few, yellowish, in subsessile corymbs with slender

---

[1] By S. M. Walters.

pedicels, opening before the leaves. Fruit glabrous, with straight wings diverging at an acute angle.

1 Leaf-lobes parallel-sided, with deep sinuses often reaching
　half-way to the base
2 Lower surface of mature leaf pubescent **10. granatense**
2 Lower surface of mature leaf glabrous or subglabrous (except
　for veins)
3 Leaves glaucous, those on flowering shoots with deep sinuses
　often reaching more than half-way to the base **12. stevenii**
3 Leaves usually green above, with sinuses rarely reaching more
　than half-way to the base **11. hyrcanum**
1 Leaf-lobes ± triangular-ovate, with shallow sinuses never reach-
　ing half-way to the base
4 Leaf-lobes acute, lower surface often glabrescent (except for
　veins) **8. opalus**
4 Leaf-lobes obtuse, lower surface usually rather densely pubes-
　cent or tomentose **9. obtusatum**

8. A. opalus Miller, *Gard. Dict.* ed. 8, no. 8 (1768) (*A. opuli-folium* Chaix). Leaves up to 10 cm, but usually less than 8 cm, with 5 wide, short, acute or subacute lobes. Lower surface of leaf often glabrescent except for veins and vein-axils. Peduncles glabrous. ● *S.W. Europe, extending locally northwards to c. 50° 30′ in W. Germany.* Co Ga Ge He Hs It.

9. A. obtusatum Waldst. & Kit. ex Willd., *Sp. Pl.* **4**(2): 984 (1806) (incl. *A. aetnense* Tineo ex Strobl). Leaves up to 12 cm, with (3) 5 short, wide, obtuse lobes. Lower surface of leaf more or less densely and persistently hairy, often tomentose. Peduncles hairy. *Balkan peninsula, C. & S. Italy, Sicilia, Corse.* Al Co Gr It Ju Si.

10. A. granatense Boiss., *Elenchus* 25 (1838). Leaves up to 7 cm, with 3 long, parallel-sided main lobes and 2 subsidiary basal ones. Lower surface of leaf, young petioles and young branches usually more or less densely hairy (glabrous in var. *nevadense* (Boiss. ex Pax.) Font-Quer & Rothm.). *S. Spain; Mallorca.* Bl Hs. (*N. Africa.*)

11. A. hyrcanum Fischer & C. A. Meyer, *Ind. Sem. Horti Petrop.* **4**: 31 (1837) (incl. *A. intermedium* Pančić). Leaves up to 10 cm, with 5 long, narrow, parallel-sided lobes. Lower surface of leaf glabrous or slightly hairy. *Balkan peninsula.* Al Bu Gr Ju.

A. reginae-amaliae Orph. ex Boiss., *Diagn. Pl. Or. Nov.* **3**(1): 109 (1853), described from the mountains of C. Greece, has 3(–5)-lobed, glabrous leaves only 2–4 cm. It is closely related to **11** and perhaps best treated as a variety.

12. A. stevenii Pojark., *Acta Inst. Bot. Akad. Sci. URSS* **1**: 150 (1933). Like **11** but with somewhat glaucous leaves, and narrower and longer leaf-lobes (most leaves on flowering shoots with deep sinuses reaching more than half-way to base). ● *Krym.* Rs (K).

13. A. monspessulanum L., *Sp. Pl.* 1056 (1753). Shrub or small tree up to 12 m. Leaves 3–8 cm, 3-lobed, coriaceous, shiny above, somewhat glaucous beneath, long-petiolate. Flowers greenish-yellow, in corymbs, erect at first, somewhat pendent later; pedicels long, slender. Fruit glabrescent, with subparallel wings. *S. Europe, extending locally northwards to c. 50° N. in W. Germany.* Al Bu Co Ga Ge Gr Hs It Ju Lu Rm Sa Si ?Tu [Au].

A. martinii Jordan, *Pug. Pl. Nov.* 52 (1852), described from France (near Lyon), differs from **13** in its larger leaves more cordate at the base, with 3–5 dentate lobes. Other taxa with leaves somewhat intermediate in shape and texture between **13** and **8** have been described from S. France, Italy and Spain. The status of these taxa is uncertain.

14. A. sempervirens L., *Mantissa* 128 (1767) (*A. orientale* auct. non L., *A. creticum* auct. non L.). Evergreen shrub up to 5(–12) m. Leaves 2–5 cm, 3-lobed to undivided, coriaceous, green beneath, shortly petiolate. Flowers few, greenish-yellow, in erect, glabrous corymbs. Fruit with wings subparallel or diverging at an acute angle. *Greece and Aegean region.* Cr Gr.

Sect. NEGUNDO (Boehmer) Pax. Dioecious; leaves pinnate; no latex; floral disc absent.

15. A. negundo L., *Sp. Pl.* 1056 (1753). Tree up to 20 m. Leaves 5–10 cm, imparipinnate, with 3 or 5 (7) ovate-acuminate leaflets. Flowers apetalous, greenish, opening before the leaves; male inflorescence a corymb, female a lax, pendent raceme. Fruit glabrous, with arcuate wings diverging at an acute angle. *Widely planted for ornament and occasionally naturalized.* [Au Bu Cz Ga Ge He Hs Hu Rs (C, W).] (*E. North America.*)

XCVI. SAPINDACEAE[1]

Trees, shrubs, woody climbers or rarely (as in the only European representative) herbs, with alternate, usually compound leaves. Flowers hermaphrodite, hypogynous, usually small. Sepals and petals 3–5, free; stamens 8, deflexed. Ovary superior, 3-locular, with one seed in each loculus.

Koelreuteria paniculata Laxm., *Novi Comment. Acad. Sci. Petrop.* **16**: 561 (1772), a graceful tree with pinnate leaves, yellow flowers and an inflated capsule, native to E. Asia, is widely planted in S. Europe in parks and by roadsides, and is perhaps locally naturalized in E. Romania and W. Ukraine.

1. Cardiospermum L.[2]

Herbs, sometimes woody at the base. Leaves stipulate, ternately divided. Sepals and petals 4. Fruit an inflated, membranous capsule.

Literature: L. Radlkofer in Engler, *Pflanzenreich* **98a**(IV.165): 370–413 (1931).

1. C. halicacabum L., *Sp. Pl.* 366 (1753). Annual, somewhat woody at the base, sparsely hairy throughout, climbing by means of branched, axillary tendrils. Stem up to 2 m but often much less, strongly ridged. Leaves more or less deltate in outline, ternate; leaflets deeply ternatisect, with a large, rhombic-lanceolate, irregularly incise-dentate terminal lobe and 2 small lateral lobes. Flowers in small, long-pedunculate, axillary cymes, often with tendrils intermixed with the flowers. Sepals 1·5 mm, ovate-orbicular. Petals 4 mm, white. Capsule up to 3 × 3 cm, trigonous or subglobose, papery; seeds 5 mm in diameter, black with a conspicuous, white, heart-shaped hilum. *Cultivated as a curiosity in S. Europe and locally naturalized.* [Gr Hs Ju.] (*Widespread in the warmer regions of both hemispheres.*)

---

[1] Edit. D. A. Webb.　　　[2] By D. A. Webb.

XCVII. HIPPOCASTANACEAE[1]

Trees or large shrubs. Leaves opposite, digitate, exstipulate. Flowers male and hermaphrodite, somewhat zygomorphic. Sepals 5, connate; petals 5, free; stamens 5–9, free, hypogynous; ovary superior, 3-locular, with 2 ovules in each loculus; style and stigma 1. Fruit a large, 1- to 2-seeded, loculicidal capsule, opening by 3 valves.

Represented in Europe by one species. The family is otherwise confined to America and S. & E. Asia.

1. Aesculus L.[2]

Deciduous. Flowers in large, terminal, erect panicles. Calyx tubular or campanulate, 5-toothed.

1. **A. hippocastanum** L., *Sp. Pl.* 344 (1753). Tree up to 25 m. Buds up to 3·5 cm, resinous, viscid. Leaflets 5–7, 8–25 cm, obovate, cuneate, usually acuminate, irregularly crenate-serrate, glabrous above, tomentose or glabrescent beneath. Panicle 15–30 cm, cylindrical. Petals *c.* 1 cm, white with yellow to pink spot at base. Fruit *c.* 6 cm in diameter, spiny. Seeds 2–4 cm, brown, with a large white hilum. *Mountain woods.* ● *C. part of the Balkan peninsula; one station in E. Bulgaria. Extensively planted for ornament and as a shade tree in most of Europe except the extreme north, and locally for timber; locally naturalized in thickets and hedges in W. & C. Europe.* Al Bu Gr Ju [Au Br Cz Ga Ge Hb He].

A. carnea Hayne in Guimpel, Otto & Hayne, *Abbild. Fremd. Holzart.* 25 (1825) (*A. rubicunda* Loisel.), is also often planted. It is like **1** but is usually smaller in all its parts, with the buds not viscid, the petals pink or red and the fruit almost smooth. It is an allopolyploid of garden origin derived from *A. hippocastanum* and *A. pavia* L., a native of E. North America.

XCVIII. BALSAMINACEAE[3]

Herbs. Leaves simple, exstipulate. Flowers solitary or in racemes, hermaphrodite, strongly zygomorphic. Sepals usually 3, free. Petals 5, the 4 lower connate in 2 lateral pairs. Stamens 5, alternate with petals; anthers connate. Ovary superior, 5-celled; ovules anatopous, axile, numerous, uniseriate in each cell; stigma sessile, 5-toothed. Fruit a loculicidal capsule; valves 5, dehiscing elastically and coiling. Seeds without endosperm.

1. Impatiens L.[4]

Sepals 3, the lowest large, petaloid, saccate, usually spurred, the lateral ones small, ovate, usually green. Upper petal largest, each lateral pair connate except for 2 apical lobes.

1 Leaves opposite or verticillate   **4. glandulifera**
1 Leaves alternate
  2 Flowers yellow or orange
   3 Flowers, including spur, not more than 1·8 cm; sepal-sac wider than long; upper leaves usually largest, usually with 20 or more teeth on each side   **3. parviflora**
   3 Flowers, including spur, usually 2 cm or more; sepal-sac longer than wide; upper leaves smaller than lower, usually with 16 or fewer teeth on each side
    4 Flowers yellow; sepal-sac gradually contracted to spur; spur usually curved through less than 90°   **1. noli-tangere**
    4 Flowers orange; sepal-sac abruptly contracted to spur; spur bent through 180°   **2. capensis**
  2 Flowers purplish-pink to reddish, occasionally white
   5 Partial inflorescences exceeding the subtending leaf; capsule 2–4 cm, glabrous   **5. balfourii**
   5 Partial inflorescences shorter than the subtending leaf; capsule not more than 1·3 cm, pubescent   **6. balsamina**

1. **I. noli-tangere** L., *Sp. Pl.* 938 (1753). Glabrous annual 20–180 cm; stems simple or branched. Leaves 1·5–10 × 1·5 cm, alternate, ovate-elliptical to ovate-lanceolate or oblong; base cuneate to subcordate; apex obtuse to acute, mucronate; margin serrate to crenate, often glandular near base; teeth 7–16(–20) on each side, usually mucronate. Flowers (2–)3–6 in axillary racemes, the early ones often cleistogamous, the others (1·5–)2–3·5 cm, yellow with small brownish spots; sepal-sac (8–)10–20 × 7–13 mm, longer than wide, gradually contracted to spur; spur 6–12 mm, curved, rarely bent through 90° or more. Capsule *c.* 1·5 cm, linear, glabrous. 2*n* = 20, 40. *Damp, shady places. Most of Europe, but absent from the extreme north and parts of the south.* Au Be Br Bu Cz Da Fe Ga Ge Gr He Ho Hs Hu It Ju No Po Rm Rs (N, B, C, W, E) Su.

2. **I. capensis** Meerb., *Afbeeld. Zelds. Gewass.* t. 10 (1775) (*I. biflora* Walter). Like **1** but leaves with 5–12(–14) teeth on each side, often undulate; flowers orange with large reddish-brown blotches; sepal-sac abruptly contracted to spur; spur 5–9 mm, bent through 180° to lie parallel with sac. *Naturalized by rivers and canals in Britain and France.* [Br Ga.] (*North America.*)

3. **I. parviflora** DC., *Prodr.* 1: 687 (1824). Glabrous annual 10–100 cm; stems simple or sometimes branched. Leaves 4–20 × 2–9 cm, alternate, elliptical to ovate-elliptical, the uppermost usually the largest; base cuneate, decurrent on petiole; apex acuminate; margin serrate or crenate-serrate, often glandular near base; teeth (13–)20–35 on each side, mucronate. Flowers 3–10 in axillary racemes, the early ones often cleistogamous, the others 0·6–1·8 cm, pale yellow; sepal-sac 3–5 × 4–6 mm, wider than long, gradually contracted to spur; spur 1–7 mm, straight or slightly curved. Capsule 1–2·5 cm, clavate or linear, glabrous. 2*n* = 24, 26. *Naturalized in woods, on river-banks and on disturbed ground in a large part of Europe.* [Au Be Br Cz Da Fe Ga Ge He Hu It ?No Po Rm Rs (C, W) Su.] (*C. Asia.*)

4. **I. glandulifera** Royle, *Ill. Bot. Himal. Mount.* 151 (1835) (*I. roylei* Walpers). Glabrous annual 100–200 cm; stems stout, simple or sometimes branched. Leaves 5–18 × 2·5–7 cm, opposite or in whorls of 3, lanceolate to elliptical; base cuneate, shortly decurrent on petiole; apex acuminate; margin serrate, glandular near base; teeth (18–)25–50 on each side, mucronate. Flowers (3–)5–12 in axillary racemes, 2·5–4 cm, purplish-pink, rarely white; sepal-sac 12–20 × 9–17 mm, longer than wide, abruptly contracted to spur; spur 2–5(–7) mm, straight. Capsule 1·5–3 cm,

---

[1] Edit. N. A. Burges.   [2] By P. W. Ball.
[3] Edit. S. M. Walters.   [4] By D. M. Moore.

clavate, glabrous. 2*n*=18, 20. *Naturalized on river-banks and in waste places in a large part of Europe.* [Au Be Br Cz Da Fe Ga Ge Hb He Ho Hu It Ju No Po Rm Rs (B, C, W) Su.] (*Himalaya.*)

5. I. balfourii Hooker fil., *Bot. Mag.* **124**: t. 7878 (1903). Glabrous annual 40–80 cm; stems simple or branched. Leaves 2–13 × 1·5–7 cm, ovate-lanceolate; base cuneate, shortly decurrent on petiole; apex long-acuminate; margin serrate, glandular near base; teeth 20–40 on each side, mucronate. Flowers 3–8 in axillary racemes, 2·5–4 cm, pinkish-purple; sepal-sac 8–9 × 6–8 mm, longer than wide, gradually contracted to spur; spur 12–18 mm, straight or slightly curved. Capsule 2–4 cm, linear to sub-

clavate, glabrous. *Locally naturalized on disturbed ground and at wood-margins. C. & S. Europe.* [Ga He Hu It.] (*Himalaya.*)

6. I. balsamina L., *Sp. Pl.* 938 (1753). Pubescent or glabrous annual 10–60 cm; stems simple. Leaves 5–12 × 1–2·5 cm, alternate, elliptical to lanceolate-ovate; base cuneate, decurrent on petiole; apex acute to acuminate; margin serrate; teeth 15–20 on each side. Flowers 1(–3) in leaf-axils, 1–2·5 cm, pinkish to purplish or white; pedicels 1–2 cm; sepal-sac 3–5 × 8–12 mm, wider than long, abruptly contracted to spur; spur 4–10 mm, curved, often absent. Capsule 0·8–1·3 cm, ellipsoid, pubescent. *Widely cultivated in gardens, and occasionally naturalized.* [Au Cz Ga ?Ju.] (*S. & E. Asia.*)

XCIX. AQUIFOLIACEAE[1]

Trees or shrubs with simple, alternate leaves. Stipules inconspicuous. Flowers actinomorphic, usually unisexual, in axillary cymes. Sepals united; petals free or slightly united at base; stamens equal in number to the petals. Ovary superior, syncarpous; fruit a drupe with several pyrenes.

Literature: T. Loesener, *Nova Acta Acad. Leop.-Carol.* **78**: 1–598 (1901), **89**: 1–313 (1908).

1. Ilex L.[2]

Evergreen shrubs or small trees; dioecious, but with conspicuous vestiges of gynoecium in male and of stamens in female flowers. Flowers 4-merous. Ovary 4-locular; stigma 4-lobed, sessile; drupe containing (2–)4 pyrenes with stony endocarp.

1 All leaves entire, or with 1–3 very small and forwardly-directed teeth **3. perado**
1 Many of the leaves boldly spinose-dentate or serrate, with at least 4 ± patent, spinose teeth
2 Petiole with a wide, shallow groove; leaves yellowish or greenish when dried, usually strongly undulate **1. aquifolium**
2 Petiole with a narrow, deep groove; leaves blackish when dried, only slightly undulate **2. colchica**

1. I. aquifolium L., *Sp. Pl.* 125 (1753) (incl. *I. balearica* Desf.). Shrub or small tree 2–10 m (up to 24 m in cultivation), glabrous except for puberulent young shoots and inflorescences; bark pale grey. Leaves 5–12 cm, ovate, spinose-acuminate or -cuspidate, 1·5–3 times as long as wide, dark green and very glossy above, paler and duller beneath, mostly with strongly spinose, undulate margin, but sometimes (commonly on upper branches of old trees, and rarely over most of the plant) flat and entire; petiole short, with a wide, shallow groove. Flowers 8 mm in diameter, in crowded cymes. Vestigial ovary in male flowers small; vestigial

stamens in female flowers with full-sized filaments but small anthers; functionally hermaphrodite flowers have been recorded. Fruit 8–10 mm, globose, bright red, usually longer than its pedicel. 2*n*=40. *S. & W. Europe, extending north-eastwards to N. Germany and Austria.* Al Au Be Bl Br Bu Co Da Ga Ge Gr Hb He Ho Hs It Ju Lu No *Rm Sa Si †Su.

Widely cultivated for ornament in all but the coldest parts of Europe; numerous cultivars and hybrids are found in gardens, some of them approaching **2** or **3** in leaf-shape. Wild plants with a large proportion of their leaves entire seem to predominate in S. & E. Spain and the Islas Baleares; they have been mistaken for **3**, or distinguished as **I. balearica** Desf., *Hist. Arb.* **2**: 362 (1809). They all, however, possess some leaves with at least a few strong, patent, marginal spines, and in all other characters agree with **1**.

2. I. colchica Pojark., *Ref. Nauč.-Issled. Rabot. Akad. Nauk SSSR(Biol.)***1945**:9 (1947). Like **1** but always a shrub 1–3 m; leaves oblong, *c.* 2·5 times as long as wide, all spinose-serrate and only slightly undulate, turning black on drying; petiole with a narrower and deeper groove. *Turkey-in-Europe.* Tu. (*Caucasus, N. Anatolia.*)

3. I. perado Aiton, *Hort. Kew.* **1**: 169 (1789). Like **1** but leaves 2·5–6 cm, elliptic-oblong to suborbicular, 1·1–1·8 times as long as wide, shortly mucronate or emarginate, entire or with 1–3 fine marginal spines directed strongly towards the apex; petiole somewhat winged; young shoots glabrous; corolla pinkish; fruit 7–9 mm, usually shorter than its pedicel. *Açores.* Az. (*Madeira, Canarias.*)

The above description applies to subsp. **azorica** Tutin, *Jour. Bot.* (*London*) **71**: 100 (1933). Other subspecies have longer leaves, often with spinose-undulate margins, and larger fruits.

CELASTRALES

C. CELASTRACEAE[3]

Trees, shrubs or climbers. Leaves simple; stipules small or absent. Inflorescence usually cymose. Flowers actinomorphic, usually hermaphrodite. Calyx 4- to 5-lobed; petals 4–5, rarely absent, usually small and greenish; stamens 4–5, opposite the calyx-

lobes; disk usually present; ovary superior, 1- to 5-locular; ovules 1–2(–many) in each loculus.

Unarmed; ovary and capsule 4- to 5-locular **1. Euonymus**
Densely spiny; ovary and capsule 2-locular **2. Maytenus**

---

[1] Edit. D. A. Webb.   [2] By D. A. Webb.   [3] Edit. T. G. Tutin.

1. Euonymus L.[1]

Unarmed shrubs or small trees. Leaves usually opposite. Flowers 4- to 5-merous. Capsule 4- to 5-locular, dehiscent; seeds partly or completely covered by a bright orange, fleshy aril.

1 Stems creeping and rooting, with ascending branches; leaves linear to linear-oblong; cymes all 1-flowered   **4. nanus**
1 Stems never creeping and rooting; leaves lanceolate to broadly obovate; cymes (1–)2- to 12-flowered
2 Evergreen   **5. japonicus**
2 Deciduous
3 Twigs subterete, covered with dark brown tubercles   **3. verrucosus**
3 Twigs ± quadrangular, without tubercles
4 Leaves 3–8(–10) cm; buds 2–4 mm, ovoid, acute; flowers usually 4-merous   **1. europaeus**
4 Leaves (5–)8–16 cm; buds 7–12 mm, fusiform, acuminate; flowers usually 5-merous   **2. latifolius**

1. E. europaeus L., *Sp. Pl.* 197 (1753) (*E. vulgaris* Miller). Much-branched, glabrous, deciduous shrub or small tree 2–6 m. Twigs green, quadrangular, without brown tubercles. Buds 2–4 mm, ovoid, acute. Leaves up to 10 × 3·5 cm, opposite, ovate-lanceolate to elliptical, acute or acuminate, crenate-serrulate. Cymes 3- to 8-flowered. Flowers usually 4-merous. Capsule 10–15 mm wide, angled, pink; seeds covered by the aril. $2n = 64$. *Most of Europe, except the extreme north and much of the Mediterranean region.* All except Az Bl Cr Fa Fe Is Rs (N) Sb.

2. E. latifolius (L.) Miller, *Gard. Dict.* ed. 8, no. 2 (1768) (*Kalonymus latifolia* (L.) Prokh.). Like **1** but twigs less distinctly quadrangular; buds 7–12 mm, fusiform, acuminate; leaves up to 16 × 7 cm, oblong-elliptical to obovate, acuminate, serrulate; cymes 4- to 12-flowered; flowers usually 5-merous; capsule 15–20 mm wide, narrowly winged on the angles. *S.C. & S.E. Europe, extending to C. Italy and S. France.* Al Au Bu ?Cz Ga Ge Gr He It Ju Rm Rs (K) Tu.

3. E. verrucosus Scop., *Fl. Carn.* ed. 2, **1**: 166 (1772). Much-branched, nearly glabrous shrub 1–3 m. Twigs slender, green, subterete, covered with dark brown tubercles. Leaves up to 6 × 3·5 cm, opposite, elliptic-oblong to ovate, acute or acuminate, crenate-serrulate, often puberulent on the veins beneath. Cymes

1- to 3-flowered. Flowers 4-merous. Capsule *c.* 10 mm wide, with rounded angles; seeds black, partly covered by the aril. *E.C. & E. Europe northwards to c. 57°N., extending to N. Italy and Albania.* Al Au Bu Cz Gr Hu It Ju Po Rm Rs (N, B, C, W, K, E) Tu.

4. E. nanus Bieb., *Fl. Taur.-Cauc.* **3**: 160 (1819). Procumbent or ascending, glabrous, more or less evergreen shrub 0·2–2 m. Twigs quadrangular. Leaves up to 3·5 × 0·7 cm, alternate or sometimes opposite or verticillate, linear to linear-oblong, obtuse or subacute, entire or remotely denticulate. Cymes 1-flowered. Flowers 4-merous. Capsule *c.* 10 mm wide, sharply angled; seeds brown, partly covered by the aril. *N.E. Romania, Moldavia, W. & C. Ukraine.* Rm Rs (W).

A species of very disjunct distribution, being recorded outside Europe only from Mongolia, Tibet and the north flank of the Caucasus.

5. E. japonicus L. fil., *Suppl.* 154 (1781). Erect, glabrous, evergreen shrub or small tree up to 6 m. Twigs weakly angled, grey. Leaves up to 7 × 3·5 cm, opposite, elliptical to obovate, acute or obtuse, crenate-serrate. Flowers 4-merous. Capsule *c.* 8 mm wide, with rounded angles; seeds covered by the aril. *Commonly planted for ornament and locally naturalized in S. Europe.* [Bu Ga Hs It Ju.] (*Japan.*)

2. Maytenus Molina[1]

Usually spiny shrubs. Leaves alternate. Calyx-lobes, petals and stamens (4–)5. Capsule (1–)2(–3)-locular, dehiscent; seeds with a fleshy aril round the base.

1. M. senegalensis (Lam.) Exell, *Bol. Soc. Brot.* ser. 2, **26**: 223 (1952) (*Catha europaea* (Boiss.) Boiss.). Intricately branched, very spiny, evergreen shrub 1–2 m. Leaves 1–3 × 0·3–0·6 cm, ovate-oblong to obovate-rhombic, entire, long-cuneate at base, somewhat glaucous. Calyx-lobes, petals and stamens 5. Capsule 5–7 mm wide, globose, 2-locular or 1-locular through abortion; seeds reddish-brown, shiny. *Rocky places. S. Spain (between Málaga and Almería).* Hs. (*Tropical Asia and Africa, extending to N.W. Africa.*)

CI. STAPHYLEACEAE[2]

Trees or shrubs. Leaves pinnate, stipulate. Flowers hermaphrodite, actinomorphic. Sepals and petals 5, free; stamens 5, free; ovary superior, 2- to 3-locular, with numerous ovules in each loculus; styles 2–3; stigma capitate. Fruit a few-seeded capsule.

1. Staphylea L.[3]

Leaves opposite; lateral leaflets sessile; stipules deciduous. Flowers in terminal panicles. Fruit inflated, dehiscent, 2- to 3-lobed. Seeds without aril.

1. S. pinnata L., *Sp. Pl.* 270 (1753). Shrub up to 5 m. Leaflets 5–7, 5–10 cm, ovate-oblong, acuminate, serrulate, glabrous. Panicles 5–12 cm, oblong, pendent. Sepals ovate, whitish, about as long as petals; petals 6–10 mm, oblong, whitish. Fruit 2·5–4 cm, subglobose. Seeds *c.* 1 cm, yellowish-brown. *C. Europe, extending to S. Italy, Bulgaria and W. Ukraine.* Au Bu Cz Ga Ge He Hu It Ju Po Rm Rs (W) [Br].

---

[1] By T. G. Tutin.   [2] Edit. N. A. Burges.   [3] By P. W. Ball.

CII. BUXACEAE[1]

Leaves simple, exstipulate. Flowers small and inconspicuous, actinomorphic, unisexual, in axillary, bracteolate clusters. Corolla absent. Ovary superior, 3-locular, with 1–2 pendent ovules in each loculus.

1. Buxus L.[2]

Evergreen shrubs or small trees with entire, coriaceous, opposite leaves. Monoecious, each axillary cluster containing a terminal female flower with some male flowers below it. Male flowers with 4 sepals, 4 stamens and vestigial ovary. Female flowers without clearly defined perianth but subtended by spirally arranged bracteoles; styles stout, persistent; stigmas 2-lobed. Fruit a loculicidal capsule, with 2 seeds in each loculus. Seeds with caruncle.

Leaves 15–30 mm; inflorescence *c.* 5 mm in diameter; styles less
than half as long as capsule   **1. sempervirens**
Leaves 25–40 mm; inflorescence *c.* 10 mm in diameter; styles nearly
as long as capsule   **2. balearica**

1. B. sempervirens L., *Sp. Pl.* 983 (1753). Shrub or small tree 2–5(–8) m, mainly glabrous but with persistent, whitish pubescence on proximal part of leaf and usually on the 4-angled young shoots. Leaves 15–30 × 7–15 mm, dark, glossy green above, paler beneath, ovate, oblong or elliptical, usually emarginate, shortly petiolate; margin somewhat revolute. Inflorescence *c.* 5 mm in diameter, with ovate, acute bracteoles; male flowers sessile. Anthers 1–2 mm. Capsule *c.* 7 mm, broadly oblong; styles *c.* 2·5 mm in fruit, patent, straight. Seeds 5–6 mm, black, glossy. *Usually on dry, base-rich soils. S.W. & W.C. Europe; gregarious and locally abundant, but absent from wide areas.* Al Au Be Br Co Ga Ge Gr He Hs It Ju Lu Sa *Tu [Az Rm].

Several cultivars are known in gardens, the commonest being a dwarf one used for edging of flower-beds.

2. B. balearica Lam., *Encycl. Méth. Bot.* **1**: 511 (1785). Like **1** but glabrous or soon glabrescent; shoots stouter and more stiffly erect; leaves 25–40 × 9–18 mm, less glossy and paler green above; inflorescence *c.* 10 mm in diameter, with suborbicular, obtuse bracteoles; male flowers pedicellate; anthers *c.* 2·5 mm; styles of mature capsule 5–7 mm, arcuate.   ● *Sardegna; Islas Baleares; a few localities on the coast of S. & E. Spain.* Bl Hs Sa.

B. longifolia Boiss., *Diagn. Pl. Or. Nov.* **2**(12): 107 (1853), known only from the S.E. corner of Asiatic Turkey, is extremely similar, and should, perhaps, be considered conspecific.

RHAMNALES

CIII. RHAMNACEAE[3]

Trees or shrubs. Leaves simple, usually stipulate. Inflorescence cymose. Flowers perigynous. Calyx 4- to 5-lobed, lobes valvate in bud. Petals 4–5, often small and sometimes absent, inserted at mouth of the hypanthium and often hooded over the stamens. Stamens 4–5, alternating with the calyx-lobes; anthers versatile. Ovary superior, 2- to 4-locular; ovules solitary. Fruit often fleshy.

1 Stipules spinescent, persistent
2 Young twigs puberulent; fruit dry, winged   **1. Paliurus**
2 Young twigs glabrous; fruit fleshy, unwinged   **2. Ziziphus**
1 Stipules soft, often caducous
3 Winter buds with scales; flowers usually 4-merous, often
unisexual   **3. Rhamnus**
3 Winter buds naked; flowers usually 5-merous, hermaphrodite
                                                     4. Frangula

1. Paliurus Miller[4]

Stipules spinescent. Flowers 5-merous, hermaphrodite. Styles 2–3. Fruit dry, hemispherical, with a wide membranous wing round the top.

1. P. spina-christi Miller, *Gard. Dict.* ed. 8 (1768) (*P. australis* Gaertner). Nearly glabrous, much-branched shrub up to *c.* 3 m. Twigs flexuous, puberulent when young. Leaves 2–4 cm, alternate and distichous, ovate, crenate-serrate, shortly petiolate. Flowers in small, axillary, shortly pedunculate cymes. Fruit 18–30 mm in diameter; wing undulate. *Dry slopes; also often used for hedges.*

Mediterranean region (except the islands), Balkan peninsula and Black Sea coast. Al Bu Ga Gr Hs It Ju Rm Rs (K) Tu [Co Hu].

P. microcarpus Wilmott, *Jour. Bot.* (*London*) **56**: 145 (1917), is a variant from N. Greece (Makedhonia) with the fruit 9–10 mm in diameter and very narrowly winged.

2. Ziziphus Miller[4]

Stipules spinescent. Flowers 5-merous, hermaphrodite. Styles 2–3. Fruit a fleshy drupe.

Twigs green; leaves denticulate; drupe ovoid-oblong   **1. jujuba**
Twigs grey; leaves very shallowly crenate; drupe subglobose **2. lotus**

1. Z. jujuba Miller, *Gard. Dict.* ed. 8, no. 1 (1768). Shrub or small tree up to 8 m. Twigs flexuous, glabrous when young, green. Leaves 2–5·5 cm, alternate, oblong, obtuse, glandular-denticulate, shortly petiolate. Flowers few, in a small, axillary cyme which is longer than its peduncle. Drupe 1·5–3 cm, ovoid-oblong, dark reddish or almost black, edible. *Cultivated in S. Europe for its edible fruits and frequently naturalized in the Mediterranean region and on the Black Sea coast.* [Al Bu Cr Ga Gr Hs It Ju Rm Si Tu.] (*Temperate Asia.*)

2. Z. lotus (L.) Lam., *Encycl. Méth. Bot.* **3**: 317 (1789). Like **1** but always a shrub; twigs grey; leaves very shallowly glandular-crenate; cyme shorter than its peduncle; drupe subglobose, deep yellow. *Dry places. Spain; Sicilia; Greece.* Gr Hs Si. (*N. Africa and Arabia.*)

---

[1] Edit. D. A. Webb.   [2] By D. A. Webb.
[3] Edit. T. G. Tutin.   [4] By T. G. Tutin.

3. Rhamnus L.[1]

Germination epigeal. Winter buds with scales. Stipules subulate, caducous. Flowers 4-merous or sometimes 5-merous, usually unisexual but monoecious. Styles 3–4; stigmas small. Fruit a drupe with 2–4 pyrenes.

Literature: S. Rivas Martinez, *Anal. R. Acad. Farm.* (*Madrid*) **28**: 363–397 (1962). W. Vent, *Feddes Repert.* **65**: 3–132 (1962) (Sect. *Rhamnastrum*).

1 Leaves evergreen
  2 Spines present; flowers usually 4-merous, in cymose fascicles or rarely solitary   **3. lycioides**
  2 Spines 0; flowers 5-merous, in racemes, or rarely solitary
    3 Leaves entire or remotely denticulate   **1. alaternus**
    3 Leaves strongly and closely spinose-denticulate
        2. ludovici-salvatoris

1 Leaves deciduous
  4 Leaves all alternate; spines 0
    5 Leaves tomentose on both surfaces   **13. sibthorpianus**
    5 Leaves glabrous or somewhat hairy but not tomentose on both surfaces
      6 Usually erect, 100–400 cm; leaves with (5–)7–20 pairs of lateral veins   **11. alpinus**
      6 Procumbent, 5–20 cm; leaves with 4–9(–13) pairs of lateral veins   **12. pumilus**
  4 Lower leaves mostly opposite or subopposite; lateral twigs usually spinescent
    7 Lamina of leaves on short shoots 1–2½ times as long as petiole; petiole 2–3 times as long as stipules
      8 Lamina of leaves on short shoots 2–2½ times as long as petiole, ovate to elliptical   **8. catharticus**
      8 Lamina of leaves on short shoots about as long as petiole, usually orbicular   **9. orbiculatus**
    7 Lamina of leaves on short shoots 3–6 times as long as petiole; petiole not or scarcely longer than stipules
      9 Leaves mostly less than 1 cm; lateral veins inconspicuous
        10 Leaves oblong or ovate   **7. prunifolius**
        10 Leaves obovate   **3. lycioides**
      9 Leaves 1–6 cm; lateral veins conspicuous
        11 Mature leaves pubescent beneath
          12 Leaves 3½–4½ times as long as wide, glabrous above   **10. persicifolius**
          12 Leaves rarely more than 3 times as long as wide, pubescent above   **5. rhodopeus**
        11 Mature leaves glabrous or with hairs on veins and petiole only
          13 Leaves ± orbicular   **6. intermedius**
          13 Leaves lanceolate, ovate or obovate
            14 Leaves crenate-serrate   **4. saxatilis**
            14 Leaves entire   **3. lycioides**

Sect. ALATERNUS (Miller) DC. Not spiny. Leaves alternate, evergreen. Flowers 4- to 5-merous, in racemose inflorescences, rarely solitary.

1. R. alaternus L., *Sp. Pl.* 193 (1753). Nearly glabrous shrub up to 5 m, very variable in habit. Leaves (1–)2–6 cm, lanceolate to ovate, acute to obtuse, often mucronate, entire or remotely denticulate, coriaceous; petiole 1–8 mm. Inflorescence dense, more or less pubescent; bracteoles ciliolate, usually caducous. Calyx-lobes lanceolate, acute, yellow; petals absent. Drupe 4–6 mm, not fleshy, obovoid, reddish becoming black; pyrenes 3. *Mediterranean region, extending to Portugal.* Al Bl Co Ga Gr Hs It Ju Lu Sa Si [Rs (K)].

R. myrtifolius Willk., *Linnaea* **25**: 18 (1852), from Spain, was distinguished on account of its procumbent habit, small, oblong-

lanceolate leaves, solitary flowers and smaller drupe. There appears, however, to be intergradation between this and typical *R. alaternus.*

2. R. ludovici-salvatoris Chodat, *Bull. Soc. Bot. Genève* ser. 2, **1**: 242 (1909) (*R. balearicus* (DC.) Willk., non Link). Nearly glabrous shrub up to 2 m. Leaves 1–2·5 cm, elliptical to suborbicular, strongly and closely spinose-denticulate. Inflorescence dense; bracteoles glandular-denticulate, usually persistent. Calyx-lobes ovate-lanceolate, yellow; petals absent. Drupe obovoid-globose. ● *Islas Baleares; E. Spain* (*near Valencia*). Bl Hs.

The bracteoles are said to be persistent in **2** and caducous in **1**, but this is by no means always true.

Sect. RHAMNUS. Spiny. Leaves often opposite and fascicled, usually deciduous. Flowers 4-merous, in cymose fascicles, rarely solitary.

3. R. lycioides L., *Sp. Pl.* ed. 2, 279 (1762). Much-branched, glabrous or puberulent shrub up to 1 m. Leaves 0·5–2 cm, usually coriaceous, evergreen or deciduous, obovate to linear, obtuse or emarginate, sometimes mucronate, entire, rarely crenulate; petiole 2–3 mm; stipules caducous. Calyx-lobes lanceolate, acute, yellowish; petals absent or very small. Drupe 4–6 mm, obovoid, compressed laterally, yellowish or black when ripe; pyrenes 2. *Mediterranean region, extending to Portugal.* Bl Cr Ga Gr Hs Lu Sa Si.

1 Leaves linear to linear-spathulate; lateral veins invisible on upper surface; drupe black when ripe
  2 Leaves and young twigs nearly or quite glabrous   (a) subsp. **lycioides**
  2 Leaves and young twigs densely puberulent   (b) subsp. **velutinus**
1 Leaves usually obovate, rarely linear-obovate; lateral veins distinctly visible on upper surface; drupe usually yellowish
  2 Leaves evergreen; veins conspicuous   (c) subsp. **oleoides**
  2 Leaves deciduous; veins inconspicuous   (d) subsp. **graecus**

(a) Subsp. **lycioides**: Young twigs and leaves nearly or quite glabrous; leaves narrowly linear, rarely linear-spathulate; midrib slender; lateral veins invisible on upper surface; flowers usually hermaphrodite; drupe black when ripe. *Spain and Islas Baleares.*

(b) Subsp. **velutinus** (Boiss.) Tutin, *Feddes Repert.* **79**: 56 (1968): Like subsp. (a) but young twigs and leaves densely puberulent; midrib very wide, occupying most or all of the space between the recurved margins of the leaf. *S.E. Spain.*

(c) Subsp. **oleoides** (L.) Jahandiez & Maire, *Cat. Pl. Maroc* **2**: 476 (1932) (*R. oleoides* L.): Leaves 1–4 × 0·3–1 cm, usually obovate, coriaceous; lateral veins distinctly visible on upper surface; flowers unisexual; drupe yellowish or sometimes blackish when ripe. *Almost throughout the range of the species.*

(d) Subsp. **graecus** (Boiss. & Reuter) Tutin, *Feddes Repert.* **74**: 26 (1967) (*R. graecus* Boiss. & Reuter): Like subsp. (c) but leaves 0·6–1·8 × 0·4–0·8 cm, deciduous, not coriaceous. *S. Greece and Aegean region.*

4. R. saxatilis Jacq., *Enum. Stirp. Vindob.* 39, 212 (1762). Much-branched, procumbent to erect shrub up to 2 m. Leaves 1–5 cm, lanceolate, ovate or obovate, acute, crenate-serrate, glabrous above when mature, deciduous; lateral veins conspicuous; petiole not or little longer than stipules. Drupe 5–8 mm, black when ripe. *Calcicole.* ● *S. & S.C. Europe.* Al Au Bu Cz Ga Ge Gr He Hs Hu It Ju Rm Si.

(a) Subsp. **saxatilis** (incl. *R. infectorius* L.): Procumbent; young twigs glabrous; leaves 1–3 cm, thin, glabrous beneath when mature. *Almost throughout the range of the species.*

(b) Subsp. **tinctorius** (Waldst. & Kit.) Nyman, *Consp.* 146

(1878) (*R. tinctorius* Waldst. & Kit.): Usually erect; young twigs pubescent; leaves 2–5 cm, thick, pubescent on the veins beneath when mature. *E.C. & S.E. Europe.*

5. R. rhodopeus Velen., *Fl. Bulg.* 119 (1891). Like **4** but twigs puberulent; leaves densely pubescent, the lateral veins soon branching into a reticulum. ● *S.E. part of Balkan peninsula.* Bu Gr Ju Tu.

6. R. intermedius Steudel & Hochst., *Flora* (*Regensb.*) **10**: 74 (1827). Like **4** but leaves 1–1·5 cm, broadly ovate to orbicular, cuspidate. ● *W. Jugoslavia and Albania.* Al Ju.

Perhaps 4 × 9.

7. R. prunifolius Sibth. & Sm., *Fl. Graec. Prodr.* **1**: 157 (1806). Small, procumbent, very spiny shrub. Leaves 0·5–1·2 cm, oblong or ovate, crenate, glabrous, deciduous; lateral veins inconspicuous; petiole not or little longer than stipules. Drupe black when ripe. *Rocky places. Greece and Kriti.* Cr Gr.

8. R. catharticus L., *Sp. Pl.* 193 (1753). Shrub or small tree 4–6 m. Leaves 3–7 cm, ovate to elliptical, obtuse or cuspidate, glabrous or somewhat pubescent, deciduous; lateral veins 2–4 pairs, conspicuous; petiole much longer than stipules; lamina of leaves on short shoots 2–2½ times as long as petiole. Drupe 6–8 mm, black when ripe. $2n=24$. *Usually calcicole. Most of Europe, northwards to 61° 45′ N. in Sweden; absent from the extreme south.* Al Au Be Br Bu Cz Da Fe Ga Ge ?Gr Hb He Ho Hs Hu It Ju Lu No Po Rm Rs (N, B, C, W, K, E) Si Su.

9. R. orbiculatus Bornm., *Österr. Bot. Zeitschr.* **37**: 225 (1887). Like **8** but small shrub; leaves 1–3 cm, orbicular or rarely ovate or elliptical; lamina of leaves on short shoots almost as long as petiole; drupe dehiscent when ripe. *S.W. Jugoslavia, Albania.* Al Ju.

10. R. persicifolius Moris, *Stirp. Sard.* **2**: 2 (1827). Like **8** but leaves elliptic-lanceolate, crenate-serrate, pubescent beneath; petiole ¼–⅓ as long as lamina; drupe reddish when ripe. ● *Sardegna.* Sa.

Sect. RHAMNASTRUM Rouy. Unarmed. Leaves all alternate, deciduous. Flowers 4-merous, in cymose fascicles, rarely solitary. (*Oreoherzogia* W. Vent.)

11. R. alpinus L., *Sp. Pl.* 193 (1753). Usually erect shrub up to 4 m. Leaves (1·5–)4–10(–15) cm, broadly elliptical, obtuse or cuspidate; lateral veins (5–)7–20 pairs, prominent beneath; petiole up to 20 mm; stipules caducous. Drupe 4–6 mm, black. *Usually calcicole. Mountains of S. & S.C. Europe.* Al Au Bu Co Ga Gr He Hs It Ju Sa.

1 Twigs pubescent; bud-scales pubescent **(a)** subsp. **alpinus**
1 Twigs glabrous; bud-scales glabrous or ciliate
  2 Erect; leaves green on both surfaces; lateral veins straight **(b)** subsp. **fallax**
  2 Procumbent; leaves glaucous above, grey beneath; lateral veins curved **(c)** subsp. **glaucophyllus**

(a) Subsp. **alpinus**: Erect. Twigs pubescent; bud-scales pubescent. Leaves green on both surfaces; lateral veins (5–)7–12 pairs. *S.W. Europe; C. Alps; Italy.*

(b) Subsp. **fallax** (Boiss.) Maire & Petitmengin, *Mat. Étude Fl. Géogr. Bot. Or.* **4**: 60 (1908) (*R. fallax* Boiss.): Erect. Twigs glabrous; bud-scales ciliate but otherwise glabrous. Leaves green

on both surfaces; lateral veins 10–20 pairs. *E. Alps; Balkan peninsula.*

(c) Subsp. **glaucophyllus** (Sommier) Tutin, *Feddes Repert.* **74**: 26 (1967) (*R. glaucophyllus* Sommier): Procumbent. Twigs glabrous; bud-scales rarely ciliate. Leaves glaucous above, grey beneath; lateral veins 6–8 pairs. ● *N.W. Italy* (*Alpi Apuane*).

12. R. pumilus Turra, *Gior. Ital. Sci. Nat. Agric. Arti Commerc.* **1**: 120 (1764). Usually dioecious. Usually procumbent, up to 20 cm high. Leaves 1·5–6(–7·5) cm, usually obovate or elliptical, acute, acuminate or sometimes obtuse; lateral veins 4–9(–13) pairs, somewhat arcuate, prominent beneath; petiole up to 10(–15) mm; stipules caducous. Flowers rarely hermaphrodite. Drupe 6–8 mm, black. *Mountain rocks.* ● *Alps, and mountains of S. Europe from Spain to Albania.* Al Au Ga Ge He Hs It Ju Sa.

13. R. sibthorpianus Roemer & Schultes, *Syst. Veg.* **5**: 286 (1819) (incl. *R. guicciardii* (Boiss.) Heldr. & Sart. ex Halácsy). Dioecious or polygamous. Erect shrub. Leaves 1–9 cm, ovate to suborbicular, acuminate to obtuse, entire or serrulate, tomentose; lateral veins 6–12 pairs, prominent beneath; petiole up to 10 mm, usually shorter than the caducous stipules. Drupe *c.* 4 mm, subglobose. *Mountain rocks.* ● *S. Greece.* Gr.

4. Frangula Miller[1]

Like *Rhamnus* but germination hypogeal; winter buds naked; flowers usually 5-merous and hermaphrodite; style 1.

1 Shrub up to 80 cm; flowers in umbellate cymes with a distinct peduncle **3. rupestris**
1 Tall shrub or small tree; flowers solitary, or fascicled in sessile cymes
  2 Leaves 2–7 cm, glabrous or sparsely pubescent beneath when mature; calyx glabrous **1. alnus**
  2 Leaves 10–18 cm, persistently pubescent beneath; calyx pubescent **2. azorica**

1. F. alnus Miller, *Gard. Dict.* ed. 8, no. 1 (1768) (*Rhamnus frangula* L.). Shrub or small tree usually 4–5 m. Leaves 2–7 cm, obovate, cuspidate, entire, pubescent beneath when young, later nearly or quite glabrous, petiolate; lateral veins 7–9 pairs. Flowers axillary, solitary or fascicled; pedicels and calyx glabrous. Drupe 6–10 mm in diameter, glabrous, red becoming black. $2n=20, 26$. *Most of Europe, except the extreme north and much of the Mediterranean region.* Al Au Be Br Bu Cz Da Fe Ga Ge Gr Hb Hc Ho Hs Hu It Ju Lu No Po Rm Rs (N, B, C, W, K, E) Su Tu.

2. F. azorica Tutin in Palhinha, *Cat. Pl. Vasc. Açores* 72 (1966) (*Rhamnus latifolius* L'Hér., non *Frangula latifolia* Miller). Like **1** but up to 10 m; leaves 10–18 cm, persistently pubescent beneath; lateral veins 10–13 pairs; pedicels and calyx pubescent; drupe 10–13 mm in diameter, puberulent. *In Laurus-Myrica woods. Açores.* Az. (*Madeira.*)

3. F. rupestris (Scop.) Schur, *Enum. Pl. Transs.* 142 (1866) (*Rhamnus rupestris* Scop.). Ascending or procumbent shrub up to 80 cm. Leaves 2–5 cm, elliptical to suborbicular, obtuse or acute, denticulate or sometimes entire, pubescent on the veins beneath, petiolate; lateral veins 5–8 pairs. Flowers in small, axillary, pedunculate, umbellate cymes; pedicels pubescent. Drupe *c.* 6 mm in diameter, glabrous, red becoming black. *Rocky places, usually on mountains.* ● *W. part of Balkan peninsula, extending to N.E. Italy.* Al Gr It Ju.

---

[1] By T. G. Tutin.

CIV. VITACEAE[1]

Shrubs, usually diffuse and climbing by means of leaf-opposed tendrils. Leaves alternate, stipulate, usually palmately lobed or divided. Flowers small, actinomorphic, in terminal or leaf-opposed cymes or panicles. Sepals 5, united; petals 5, free or united distally; stamens 5, antepetalous. Ovary superior, 2-locular; style and stigma 1; fruit a berry with 2–4 seeds.

Literature: K. Süssenguth in Engler & Prantl, *Die natürlichen Pflanzenfamilien*, ed. 2, **20 d**: 174–371. Leipzig & Berlin. 1953.

Leaves simple; bark usually peeling off in shreds; disc distinct from ovary; petals united distally, falling at anthesis   **1. Vitis**
Leaves simple or digitate; bark not peeling off in shreds; disc merged in base of ovary; petals free, patent, persisting for at least a short time after anthesis   **2. Parthenocissus**

1. Vitis L.[2]

Bark usually peeling from old stems in long shreds. Leaves simple, usually palmately lobed. Petals cohering at the apex, falling at anthesis without separating. A 5-lobed, glandular disc present at base of ovary.

Literature: Hegi, *Illustrierte Flora von Mittel-Europa* **5**: 359–425. München. 1925. L. H. Bailey, *Gentes Herb.* **3**: 151–244 (1934). W. Scherz & J. Zimmermann in Engler & Prantl, *Die natürlichen Pflanzenfamilien*, ed. 2, **20 d**: 334–371. Leipzig & Berlin. 1953.

1. **V. vinifera** L. *Sp. Pl.* 202 (1753). Stems up to 35 m, climbing over trees, but in cultivation usually reduced by annual pruning to 1–3 m. Leaves 5–15 cm, orbicular in outline, cordate, usually palmately 5- to 7-lobed, irregularly toothed, glabrescent above, often with persistent tomentum beneath. Tendrils branched, normally occurring opposite 2 leaves out of every 3. Flowers numerous, in rather dense panicles, which replace the tendrils in the upper part of the stem. Calyx very shortly 5-lobed. Petals *c.* 5 mm, pale green. Seeds 2–3. *S. & S.C. Europe, extending northwards in large-scale cultivation to c. 52°N in S.W. Poland.* Al Au Bu Co Cz Ga Ge Gr He Hu It Ju Rm Rs (W, K) Sa Si Tu [Az Be Bl Cr Hs Lu Po Rs (E)].

(a) Subsp. **sylvestris** (C. C. Gmelin) Hegi, *Ill. Fl. Mitteleur.* **5**: 364 (1925) (*V. sylvestris* C. C. Gmelin): Dioecious, with dimorphic foliage, the leaves of male plants being more deeply lobed. Fruit *c.* 6 mm, ellipsoid, bluish-black, acid; seeds usually 3, subglobose, with short, truncate beak. River-banks and damp woods. *S.E. & S.C. Europe, extending locally to Corse and W. Germany.*

(b) Subsp. **vinifera** (subsp. *sativa* Hegi): Flowers hermaphrodite. Fruit 6–22 mm, ellipsoid to globose, green, yellow, red or purplish-black, sweet; seeds 0–2, pyriform, with a rather long beak. $2n = 38$. *Cultivated for wine-making and for the edible fruit in S. Europe and much of C. Europe, and widely naturalized.*

The cultivated vine, here treated as a subspecies, is, like most plants long established in cultivation, impossible to accommodate satisfactorily in any orthodox taxonomic scheme. The vines of Europe are certainly derived, at least in part, by selection from subsp. *sylvestris*, which is native in a large part of C. & S.E. Europe, though the practice of cultivation (and therefore the cultivated clones) probably originated in S.W. Asia. Confusion of the native subspecies with naturalized plants of subsp. *vinifera* has made difficult the precise determination of the original

geographical limits, and some wild vines are probably recent hybrids between the two subspecies. Some authors, moreover, have suggested that other species of *Vitis* from the E. Mediterranean region, now extinct in the wild state, have contributed to the cultivated vine, but the evidence for this theory is slender. The situation has been further complicated in the past century by the introduction to Europe of many species of *Vitis* from North America. These are more or less resistant to the attacks of *Viteus vitifolii* ('phylloxera'), a parasitic aphid which did immense damage to European vines from 1867 onwards. American vines are now used almost exclusively as stocks; the scions grafted on these are either cultivars of *V. vinifera* subsp. *vinifera*, or hybrids between it and American species, or of purely American species or hybrids. These American vines are locally naturalized, especially around neglected or abandoned vineyards. The species planted on a large scale in Europe include **V. aestivalis** Michx, *Fl. Bor. Amer.* **2**: 230 (1803), **V. berlandieri** Planchon, *Compt. Rend. Acad. (Paris)* **91**: 425 (1880), **V. cordifolia** Lam., *Tabl. Encycl. Méth. Bot.* **2**: 134 (1797), **V. labrusca** L., *Sp. Pl.* 203 (1753), **V. rotundifolia** Michx, *Fl. Bor. Amer.* **2**: 231 (1803) (*V. vulpina* auct., non L., *Muscadinia rotundifolia* (Michx) Small), **V. rupestris** Scheele, *Linnaea* **21**: 159 (1848) and **V. vulpina** L., *Sp. Pl.* 203 (1753) (*V. riparia* Michx). The following key may help in their identification, but it does not attempt to deal with hybrids.

1 Bark not shredding from old stems; cavity in young stems continuous across nodes; tendrils simple   **rotundifolia**
1 Bark shredding from old stems; cavity in young stems interrupted by a diaphragm at each node; tendrils branched
  2 A tendril or inflorescence present at every node; leaves covered beneath with a continuous, red-brown tomentum   **labrusca**
  2 Some nodes (usually 1 in 3) without tendril or inflorescence; leaves glabrous to floccose beneath, but without continuous tomentum
    3 Leaves floccose with ferruginous hairs on veins beneath   **aestivalis**
    3 Leaves without ferruginous hairs beneath
      4 Leaves cordate, with conspicuous and fairly deep basal sinus
        5 Young shoots angled, floccose; leaves mostly as wide as long   **berlandieri**
        5 Young shoots terete, not floccose; leaves mostly longer than wide   **cordifolia**
      4 Leaves truncate at base, or subcordate with wide, shallow sinus
        6 Stems long, diffuse; tendrils well-developed   **vulpina**
        6 Compact bush; tendrils few or none   **rupestris**

In addition to fruiting vines, several species of *Vitis* from E. Asia are cultivated in gardens for their ornamental foliage; of these **V. coignetiae** Pulliat ex Planchon, *Vigne Amér. Viticult. Eur.* **7**: 186 (1883), and **V. thunbergii** Siebold & Zucc., *Abh. Akad. Wiss. (München)* **4(2)**: 198 (1846), are reported as locally naturalized. Both have a reddish-brown tomentum on the leaves, at least when young, as in *V. labrusca*, but have every third node without a tendril. *V. coignetiae* has terete young shoots and leaves scarcely lobed; *V. thunbergii* has angled young shoots and leaves deeply lobed.

2. Parthenocissus Planchon[2]

Bark not peeling in long shreds from old stems. Leaves digitate or simple. A tendril or inflorescence present at every node. Petals

---

[1] Edit. D. A. Webb.   [2] By D. A. Webb.

free, patent or deflexed, persistent for at least a short time after anthesis. Disc adnate to base of ovary and not visible as a distinct structure. Berry bluish-black, usually pruinose.

1 Most of the leaves simple and 3-lobed   **3. tricuspidata**
1 Leaves all digitate, mostly 5-foliolate
  2 Tendrils with (3–)5–8(–12) branches, ending in adhesive discs
                                              1. quinquefolia
  2 Tendrils with 3–5 branches, often swollen at the apex, but without adhesive discs   **2. inserta**

1. P. quinquefolia (L.) Planchon in A. & C. DC., *Monogr. Phan.* **5**: 448 (1887). Stems up to 30 m, climbing and trailing. Leaves digitate; leaflets (3–)5(–7), 5–10 cm, obovate-elliptical, coarsely and obtusely serrate, dull and somewhat glaucous beneath. Tendrils mostly with 5–8 branches, each of which develops, on touching a solid support, a terminal, adhesive disc. Flowers in terminal and leaf-opposed, more or less thyrsoid panicles with several lateral branches. Petals *c.* 3 mm, deflexed, green. Fruit *c.* 6 mm, globose, with 2–3 seeds. *Formerly widely cultivated for ornament; now largely replaced in some countries by* **2** *and* **3**, *but naturalized locally in C. Europe and Britain.* [Au Br Ge He Ho.] (*E. part of U.S.A.*)

2. P. inserta (A. Kerner) Fritsch, *Exkursionsfl. Österr.* ed. 3, 321, 789 (1922) (*P. inserens* Hayek, *P. vitacea* (Knerr) A. S. Hitchc., *P. quinquefolia* auct. eur. med., non (L.) Planchon). Like **1** but leaflets more acutely serrate and shining green beneath; tendrils longer, with 3–5 branches, clinging by coiling or by swelling of the apex inside a crevice, but without adhesive disc; inflorescences all leaf-opposed, smaller, and with conspicuously dichotomous branching. *Widely cultivated for ornament and frequently naturalized.* [Au Cz Da Ga Ge Ho Hu It Ju Po Rm.] (*S. Canada, N. & W. parts of U.S.A.*)

Hybrids between **1** and **2** have been reported, and it is possible that some of the naturalized plants are hybrids.

3. P. tricuspidata (Siebold & Zucc.) Planchon in A. & C. DC., *Monogr. Phan.* **5**: 452 (1887). Stems up to 20 m, climbing. Leaves on old plants mostly suborbicular-deltate in outline, with 3 acuminate lobes, but on young shoots some 3-foliolate and some simple, unlobed leaves may occur. Tendrils short, much-branched, clinging by preformed adhesive discs. Inflorescences as in **1**, but mostly on short shoots. Fruit with 1–2 seeds. *Widely cultivated for ornament on walls; naturalized in Jugoslavia and perhaps elsewhere.* [Ju.] (*China and Japan.*)

MALVALES

CV. TILIACEAE[1]

Leaves simple, alternate; stipules usually caducous. Flowers hermaphrodite, actinomorphic, 5-merous. Sepals and petals free. Stamens numerous, sometimes united at the base into fascicles. Ovary superior, of 5 carpels; fruit a capsule or nut.

Corchorus olitorius L., *Sp. Pl.* 529 (1753), extensively cultivated in warmer countries for its fibres and as a vegetable, is occasionally cultivated in the Aegean region and has been reported as locally naturalized or casual. It is an erect, glabrous annual with ovate, serrate leaves, small, yellow, leaf-opposed flowers and a narrow, many-seeded capsule.

1. Tilia L.[2]

Deciduous trees with large, obtuse buds. Leaves distichous, petiolate, cordate or truncate at the base, serrate or denticulate. Stipules caducous. Flowers fragrant, yellowish or whitish, in cymes; peduncle adnate in its basal half to a large, membranous bract. Stamens up to 80, free or in 5 antepetalous fascicles; epipetalous staminodes sometimes present. Ovary 5-locular; stigma 5-lobed. Fruit a unilocular nut with 1–3 seeds.

All European species are interfertile, and natural hybrids are common, especially in S.E. & E.C. Europe. They have often been described as species, varieties or forms, and many of them are extensively planted in parks and gardens. The existence of these hybrids makes difficult the accurate delimitation of the geographical range of some of the species.

Literature: V. Engler, *Monographie der Gattung* Tilia. Breslau. 1909. J. Wagner, *Mitt. Deutsch. Dendrol. Ges.* **44**: 316–345 (1932); **45**: 5–60 (1933). J. Wagner, *Mitt. Kgl. Ungar. Gartenb.-Lehranst.* **7–10** (1940–44).

1 Leaves white-tomentose beneath with stellate hairs; staminodes present   **1. tomentosa**

1 Leaves glabrous beneath, or pubescent with simple hairs; staminodes absent
  2 Fruit strongly ribbed
    3 Style pubescent   **2. dasystyla**
    3 Style glabrous
      4 Leaves serrate, but without aristate teeth   **3. platyphyllos**
      4 Leaves with acuminate-aristate teeth   **4. rubra**
  2 Fruit smooth or slightly ribbed
    5 Cymes obliquely erect; tertiary veins of leaves not prominent   **5. cordata**
    5 Cymes pendent; tertiary veins of leaves prominent   **6. × vulgaris**

1. T. tomentosa Moench, *Verz. Ausl. Bäume Weissenst.* 136 (1785) (*T. argentea* DC.). Up to 30 m, broadly pyramidal. Young twigs tomentose. Leaves 8–10 cm, suborbicular-cordate, serrate, biserrate or slightly lobed, dark green and glabrescent above, white-tomentose with stellate hairs beneath; petiole usually less than half as long as lamina. Bract lanceolate to oblong, subsessile. Flowers dull white, in pendent cymes of 6–10. Staminodes present. Fruit 6–8 mm, usually ovoid, minutely verrucose. *Balkan peninsula, extending northwards to N. Hungary and W. Ukraine; often planted elsewhere for ornament.* Al Bu Gr Hu Ju Rm Rs (W) Tu.

Variable, especially in habit (ascending or somewhat pendent branches) and in length of petiole. An extreme variant, perhaps best treated as a cultivar, with conspicuously pendent branches, long and slender petioles, spathulate bracts and a depressed-globose fruit (nearly always sterile), is frequently planted for ornament, usually under the name of **T. petiolaris** auct., ? an DC.

2. T. dasystyla Steven, *Bull. Soc. Nat. Moscou* **4**: 260 (1832). Up to 20 m. Twigs pubescent at first, later glabrous. Leaves 7–11 × 5–8 cm, firm, orbicular-ovate, subcordate, serrate with

---

[1] Edit. D. A. Webb.   [2] By K. Browicz.

aristate teeth, dark, shining green and glabrous above, light green beneath, with tufts of yellowish hairs in the vein-axils. Flowers 3–7, usually 5. Style pubescent. Fruit *c.* 10 mm; pericarp woody, pubescent, strongly ribbed. Quercus-*woods; rare.* ● *Krym.* Rs (K).

3. **T. platyphyllos** Scop., *Fl. Carn.* ed. 2, **1**: 373 (1772) (*T. officinarum* Crantz *pro parte*). Up to 40 m; branches spreading. Leaves 6–9(–12) cm, soft, broadly ovate, abruptly acuminate, symmetrically or obliquely cordate at the base, regularly serrate with acute but not aristate teeth, pubescent or almost glabrous. Cymes pendent; flowers 2–5, usually 3. Fruit 8–10 mm, subglobose to pyriform; pericarp woody, tomentose, strongly 5-ribbed. *C. & S. Europe, extending eastwards to W. Ukraine and locally northwards to N. France and S.W. Sweden. Often planted in parks and gardens.* Al Au Be Bu Co Cz Da Ga Gr He Ho Hs Hu It Ju Po Rm Rs (W) ?Si Su ?Tu [*Br].

Very variable, particularly in the amount of pubescence on leaves and young shoots. Three fairly distinct subspecies may be recognized.

1 Leaves pubescent on both surfaces; young twigs pubescent
　　　　　　　　　　　　　　　　　　(b) subsp. **cordifolia**
1 Leaves glabrous, at least above
　2 Leaves distinctly pubescent on midrib beneath, sometimes also on secondary veins; young twigs glabrous or pubescent
　　　　　　　　　　　　　　　　　(a) subsp. **platyphyllos**
　2 Leaves glabrous beneath, or very slightly pubescent on midrib; young twigs glabrous　　　　　　(c) subsp. **pseudorubra**

(a) Subsp. **platyphyllos** (*T. officinarum* subsp. *platyphyllos* (Scop.) Hayek): *Mainly in the C. & S. parts of the range of the species.*

(b) Subsp. **cordifolia** (Besser) C. K. Schneider, *Ill. Handb. Laubholzk.* **2**: 376 (1909) (*T. cordifolia* Besser; incl. subsp. *eugrandifolia* C. K. Schneider, *T. officinarum* subsp. *grandifolia* (Ehrh. ex Hoffm.) Hayek): *In the N., E. & C. parts of the range of the species.*

(c) Subsp. **pseudorubra** C. K. Schneider, *op. cit.* 378 (1909) (*T. officinarum* subsp. *flava* (Wolny) Hayek): *Mainly in the S. part of the range of the species.*

4. **T. rubra** DC., *Cat. Pl. Horti Monsp.* 150 (1813) (*T. officinarum* subsp. *rubra* (Weston) Hayek & subsp. *corinthiaca* (V. Engler) Hayek). Like **3** but leaves firmer, glabrous or sub-

glabrous, more obliquely cordate or truncate, and with markedly acuminate-aristate teeth. Fruit variable in shape. *S.E. & E.C. Europe; distribution uncertain, because of confusion with* **3**(c). ?Al Bu Gr Hu ?Ju ?Rm Rs (K).

(a) Subsp. **rubra**: Young twigs pubescent. Leaves sparingly pubescent on the veins beneath. *Throughout the range of the species, except Krym.*

(b) Subsp. **caucasica** (Rupr.) V. Engler, *Monogr. Gatt. Tilia* 107 (1909) (*T. caucasica* Rupr.): Young twigs glabrous; leaves glabrous except sometimes for small tufts of hairs in the vein-axils. *Krym.* (*Caucasus.*)

5. **T. cordata** Miller, *Gard. Dict.* ed. 8, no. 1 (1768) (*T. parvifolia* Ehrh. ex Hoffm.). Up to 30 m, with large, spreading crown. Young twigs glabrous or subglabrous. Leaves 3–9 cm, suborbicular, abruptly acuminate, acutely and finely serrate, cordate at the base, glabrous except for some tufts of reddish-brown hairs in the vein-axils beneath; tertiary veins not prominent. Cymes obliquely erect, with 4–15 flowers; bracts petiolate. Fruit *c.* 6 mm, globose; pericarp membranous, smooth or slightly ribbed. *Throughout Europe except the extreme north, the extreme south and some islands.* Al Au Br Bu Co Cz Da Fe Ga Ge ?Gr He Ho Hs Hu It Ju No Po Rm Rs (N, B, C, W, K, E) Su Tu.

6. **T. × vulgaris** Hayne, *Darst. Beschr. Arzn. Gewächse* **3**: t. 47 (1813) (*T. europaea* L. pro parte, *T. intermedia* DC.) (*T. cordata × platyphyllos*). Young twigs usually glabrous. Leaves 6–10 cm, broadly ovate, shortly acuminate, obliquely cordate or nearly truncate at the base, dull green above, paler beneath, glabrous except for whitish hairs in the vein-axils beneath; tertiary veins prominent. Cymes pendent, with 5–10 flowers. Fruit *c.* 8 mm, subglobose or broadly ovoid, rounded at the ends; pericarp woody, slightly ribbed. *Occasional as a natural hybrid in most regions of Europe where the parent species grow together; also planted in parks and gardens, especially in N.W. Europe, and sometimes naturalized.* Au Cz Ga Ge Gr He Hs Hu It Po Rm Rs (W) Si Su [Br].

T. × euchlora C. Koch, *Wochenschr. Gärtn. Pflanzenk.* **9**: 284 (1866) (probably *T. cordata × dasystyla*) is often planted for ornament, especially in C. Europe. It differs from **6** in the leaves, which are dark, shining green above, and with reddish-brown hairs in the vein-axils beneath, the fewer (3–7) flowers, and the fruit tapered at the ends.

CVI. MALVACEAE[1]

Herbs, shrubs or small trees with simple (but sometimes deeply palmatisect), alternate, stipulate leaves. Flowers regular, hermaphrodite. Bracteoles usually present immediately below the calyx, forming an epicalyx. Sepals 5, united at the base; petals 5, free or united slightly at the base. Stamens numerous, the filaments united for most of their length to form a tube which surrounds the ovary and styles. Ovary superior, of 4 or more (usually numerous) carpels; styles usually free. Fruit a schizocarp or a loculicidal capsule.

1 Epicalyx absent
　2 Annual, with suborbicular leaves; fruit hairy, each carpel with several seeds　　　　　　　　　　　　**9. Abutilon**
　2 Woody perennial, with oblong-lanceolate leaves; fruit glabrous, each carpel 1-seeded　　　　　　　　**3. Sida**

1 Epicalyx present
　3 Carpels 3–5; fruit a capsule
　　4 Epicalyx-segments broadly ovate to deltate　　**11. Gossypium**
　　4 Epicalyx-segments linear
　　　5 Fruit depressed, 5-angled, with 1 seed in each loculus
　　　　　　　　　　　　　　　　　　14. Kosteletzkya
　　　5 Fruit at least as long as wide, rounded, with several seeds in each loculus
　　　　6 Fruit not more than twice as long as wide; calyx persistent in fruit　　　　　　　　　　　**12. Hibiscus**
　　　　6 Fruit more than twice as long as wide; calyx deciduous
　　　　　　　　　　　　　　　　　　13. Abelmoschus
　3 Carpels at least 6; fruit a schizocarp
　　7 Epicalyx-segments 5 or more
　　　8 Mericarps in several superimposed planes, forming a ± globose head　　　　　　　　　　**2. Kitaibela**
　　　8 Mericarps in a single whorl, forming a circular disc
　　　　9 Staminal tube terete; flowers not more than 30 mm in

---

[1] Edit. D. A. Webb.

248

diameter, at least some of them on conspicuous pedicels
or peduncles **7. Althaea**
9 Staminal tube 5-angled; flowers at least 30 mm in diameter,
subsessile in a spike-like inflorescence **8. Alcea**
7 Epicalyx-segments 2–3
10 Epicalyx-segments united at the base, at least in bud
6. Lavatera
10 Epicalyx-segments free, even in bud
11 Mericarps forming a ± globose head; epicalyx-segments
wider than sepals **1. Malope**
11 Mericarps in a single whorl, forming a circular disc;
epicalyx-segments narrower than sepals
12 Stigmas terminal, capitate; mericarps 2-seeded **10. Modiola**
12 Stigmas lateral, filiform; mericarps 1-seeded
13 Mericarps inflated; petals rounded at apex **4. Malvella**
13 Mericarps not inflated; petals emarginate or 2-lobed
5. Malva

1. Malope L.[1]

Herbs. Flowers long-pedicellate, solitary in the leaf-axils.
Epicalyx-segments 3, free, ovate to orbicular, cordate, wider
than the sepals. Petals not emarginate. Stigmas lateral, filiform.
Mericarps numerous, 1-seeded, indehiscent, glabrous, rugulose,
irregularly arranged in a globose head.

M. multiflora Cav., *Monad. Class. Diss. Dec.* 85 (1786),
from S. Spain, has never been seen since its original discovery,
and may belong to some other genus. It was said to have 3–4
small, white flowers in each leaf-axil.

Leaves all longer than wide; petals 20–40 mm **1. malacoides**
Many of the leaves at least as wide as long; petals 35–60 mm **2. trifida**

1. **M. malacoides** L., *Sp. Pl.* 692 (1753) (incl. *M. stipulacea*
Cav.). Perennial, or perhaps sometimes annual, hispid at least
above. Stems 20–50 cm, several, ascending, usually simple.
Lower leaves 20–50 × 12–35 mm, oblong-lanceolate to ovate,
crenate; the upper similar or 3-lobed. Epicalyx-segments 8–12 ×
6–8 mm in flower, strongly accrescent, cordate-orbicular,
acuminate, wider and shorter than the lanceolate, acuminate
sepals. Petals 20–40 mm, deep pink or purple. *Mediterranean
region.* Al Co Cr Ga Gr Hs It Sa Si Tu [Rs (W)].

2. **M. trifida** Cav., *Monad. Class. Diss. Dec.* 85 (1786).
Annual, glabrous except for marginal cilia on stipules, bracts and
sepals. Stem up to 150 cm, single, erect, stout. Leaves up to
8 × 10 cm, but often much less, long-petiolate, crenate or obtusely
dentate, the lower suborbicular, the upper with 3–5 broad,
triangular lobes. Epicalyx-segments orbicular-cordate, 12–15 mm
in flower, increasing to 30 mm in fruit. Sepals broadly lanceolate
to ovate. Petals 35–60 × 20–40 mm, deep purplish-red. *S.W.
Spain and S.C. Portugal (very local); cultivated for ornament,
and casual, or perhaps naturalized, elsewhere.* Hs *Lu [Cz Ga
Rs]. (*N.W. Africa.*)

2. Kitaibela Willd.[1]

Perennial herbs. Flowers in axillary cymes. Epicalyx-segments
6–9, slightly connate at the base. Stigmas lateral, filiform. Fruit
a schizocarp; mericarps numerous, 1-seeded, eventually dehiscent,
arranged in about 5 superposed whorls so as to form a depressed-
globose head.

1. **K. vitifolia** Willd., *Ges. Naturf. Freunde Berlin Neue Schr.*
2: 107 (1799). Stems up to 3 m, robust, sparingly branched;
stems, petioles and inflorescence hispid with white hairs. Leaves
long-petiolate; lamina up to 18 cm, rhombic to suborbicular-

cordate, with 5–7 triangular, dentate lobes, glabrescent. Cymes
with 1–4 flowers. Epicalyx-segments ovate, acuminate, slightly
longer and wider than the sepals. Petals 25 × 20 mm, obdeltate-
cuneate, entire or slightly retuse, white. Mericarps dark brown,
hairy. *Damp thickets and grassland.* ● *Jugoslavia, from S.E.
Hrvatska and Vojvodina to N. Makedonija.* Ju [Hu Rm].

3. Sida L.[1]

Herbs or small shrubs. Epicalyx absent. Stigmas terminal,
capitate. Fruit a schizocarp; mericarps fairly numerous, indehiscent, in a single whorl, each with a single, pendent ovule;
pericarp not inflated.

1. **S. rhombifolia** L., *Sp. Pl.* 684 (1753). Low shrub, puberulent,
especially on lower surface of leaves. Stems 30–80 cm, slender,
stiff, straight. Leaves 3–6 × 1–3 cm, rhombic-ovate to narrowly
oblong, crenate-serrate at least in apical half; petiole short.
Flowers axillary, solitary; pedicels 2–4 cm. Sepals rhombic,
apiculate; petals 10–12 mm, obovate-cuneate, dull yellow.
Mericarps 7–12, pale brown, reticulate, with 1 or 2 prominent
stylar beaks. *Cultivated as a medicinal plant, and naturalized on
roadside and waste ground in the Açores and Portugal.* [*Az Lu.]
(*Africa, America, S. & E. Asia.*)

4. Malvella Jaub. & Spach[1]

Perennial herbs. Flowers solitary in leaf-axils. Epicalyx-segments
3, free. Petals not emarginate. Stigmas lateral, filiform. Fruit
a schizocarp; mericarps 9–12, in a single whorl, 1-seeded, with
pendent ovule and inflated, membranous pericarp.

1. **M. sherardiana** (L.) Jaub. & Spach, *Ill. Pl. Or.* 5: 47 (1855).
Villous with white, simple, and yellowish, stellate hairs. Stems
20–45 cm, procumbent, woody at the base. Leaves 13–35 ×
20–35 mm, orbicular to reniform, crenate, shortly petiolate.
Epicalyx-segments linear, much smaller than the ovate-triangular
sepals. Petals *c.* 10 mm, deep pink. Mericarps pale brown, hairy,
finely reticulate. *Cultivated fields and waste places. C. Spain;
N. & C. Greece and S. Bulgaria; Krym.* Bu Gr Hs Rs (K).

5. Malva L.[2]

Herbs. Epicalyx-segments 2–3, free. Petals emarginate or 2-lobed,
purple, pink or white. Stigmas lateral, filiform. Fruit a schizocarp; mericarps numerous, 1-seeded, indehiscent, dark brown or
black, arranged in a single whorl around the short, conical or
flattened apex of the receptacle.

Many species are variable in indumentum and leaf-shape.
In most species the basal leaves are subentire. Sepal-characters
refer to the fully open flower, unless the contrary is stated; the
measurements exclude the basal part of the sepal which is fused
to others.

Three hybrids have been reported among European species:
5 × 6; 8 × 12; 11 × 12. It is possible that the apparent variability of some species is due, at least in part, to hybridization.

All European species except 6 are found principally in dry,
open habitats; they are also nitrophilous, and usually occur,
therefore, as ruderals or weeds of cultivated ground. It is almost
impossible, for this reason, to determine accurately the geographical limits as natives of many of the species.

1 Sepals linear to narrowly triangular, more than 3 times as long
as wide
2 Staminal tube glabrous **4. cretica**
2 Staminal tube pubescent **3. stipulacea**

---

1 Sepals ovate or triangular, not more than 3 times as long as wide
  3 Epicalyx-segments ovate or ovate-deltate, not more than 3 times as long as wide
    4 Ripe mericarps smooth or faintly ribbed; lower flowers solitary in leaf-axils **5. alcea**
    4 Ripe mericarps distinctly reticulate; lower flowers 2 or more in each leaf-axil
      5 Petals 12–30 mm, 3–4 times as long as sepals; lower surface of sepals with numerous small, stellate hairs **8. sylvestris**
      5 Petals 10–12 mm, not more than 2·5 times as long as sepals; lower surface of sepals with few stellate hairs or none **9. nicaeensis**
  3 Epicalyx-segments linear to narrowly ovate, at least 3 times as long as wide
    6 Lower flowers solitary in leaf-axils, or all flowers in a terminal cluster
      7 Petals about equalling sepals **2. aegyptia**
      7 Petals at least twice as long as sepals
        8 Middle cauline leaves lobed for at most $\frac{1}{10}$ of the radius; staminal tube glabrous **1. hispanica**
        8 Middle cauline leaves lobed for at least $\frac{1}{5}$ of the radius; staminal tube hairy
          9 Epicalyx-segments 2, linear; mericarps with flat dorsal face and sharp angles **3. stipulacea**
          9 Epicalyx-segments 3, narrowly oblong; mericarps with rounded dorsal face and angles
            10 Pedicels with stellate hairs; mericarps ± glabrous **7. tournefortiana**
            10 Pedicels without stellate hairs; mericarps with numerous white hairs **6. moschata**
    6 Lower flowers in groups in each leaf-axil
      11 Dorsal face of ripe mericarps smooth, or only faintly ridged
        12 Petals at least twice as long as sepals; calyx not accrescent **12. neglecta**
        12 Petals less than twice as long as sepals; calyx strongly accrescent **13. verticillata**
      11 Dorsal face of ripe mericarps distinctly reticulate
        13 Petals at least 12 mm, usually bright purple or pink **8. sylvestris**
        13 Petals less than 12 mm, pale pink or lilac
          14 Calyx strongly accrescent; fruiting pedicels usually less than 10 mm; angles of mericarps winged **10. parviflora**
          14 Calyx only slightly accrescent; fruiting pedicels usually more than 10 mm; angles of mericarps not winged
            15 Staminal tube ± glabrous; epicalyx-segments about 6 times as long as wide **11. pusilla**
            15 Staminal tube with numerous hairs; epicalyx-segments not more than 3 times as long as wide **9. nicaeensis**

Sect. BISMALVA (Medicus) Dumort. Flowers solitary in leaf-axils, or in a congested, terminal raceme.

1. M. hispanica L., *Sp. Pl.* 689 (1753). Erect annual up to 70 cm, stellate-pubescent and also villous with long, patent hairs. Leaves 2–3 cm wide, semicircular, crenate-serrate or slightly lobed. Flowers mostly axillary. Epicalyx-segments 2, linear to triangular-lanceolate; sepals 6–9 mm, rhombic-ovate; petals *c.* 20 mm, pale pink; staminal tube glabrous. Mericarps glabrous, without ridges; dorsal face rounded. *Spain and Portugal.* Hs Lu.

2. M. aegyptia L., *Sp. Pl.* 690 (1753). Erect to decumbent annual; stems up to 25 cm, strigose with branched hairs. Leaves suborbicular in outline, deeply dissected into narrow segments. Flowers mostly in terminal clusters. Epicalyx-segments 2(–3), linear; sepals 7–11 mm, broadly triangular-ovate, somewhat acuminate; petals about equalling sepals, lilac, glabrous; staminal tube pubescent. Mericarps usually glabrous; lateral faces conspicuously ridged; dorsal face flat, transversely ridged. *Spain; Aegean region.* Cr Gr Hs. (*N. Africa, S.W. Asia.*)

3. M. stipulacea Cav., *Monad. Class. Diss. Dec.* 62 (1786) (*M. trifida* Cav.). Like **2** but sepals at least twice as long as wide; petals twice as long as sepals, bearded; mericarps densely pubescent, with lateral faces less strongly ridged. ● *S., E. & C. Spain.* Hs.

4. M. cretica Cav., *op. cit.* 67 (1786). Erect annual up to 40 cm, more or less hispid with long, patent hairs. Lower leaves suborbicular, crenate or slightly lobed; upper leaves usually deeply divided into 3(–5) oblong, dentate lobes. Flowers axillary, on long pedicels. Epicalyx-segments 3, linear to narrowly triangular; sepals 7–10 mm, similar in shape; staminal tube glabrous. Mericarps glabrous; dorsal face flat, with numerous small, transverse ridges; angles slightly winged. *Mediterranean region.* Co Cr Gr Hs It Sa Si [Ga].

(a) Subsp. **cretica**: Pedicels usually with small, stellate, as well as simple hairs; petals 1–1·5 times as long as sepals, bluish-lilac to pink. *Throughout the range of the species except Spain.*

(b) Subsp. **althaeoides** (Cav.) Dalby, *Feddes Repert.* **74**: 26 (1967) (*M. althaeoides* Cav.): Pedicels with simple hairs only; petals twice as long as sepals, pale pink. *S. & E. Spain.*

Plants intermediate between the two subspecies are found in S. Italy and Malta.

5. M. alcea L., *Sp. Pl.* 689 (1753). Erect perennial 30–125 cm; stems sparsely hirsute below, stellate-pubescent above. Upper leaves usually palmatisect, with (3–)5 obtusely dentate or pinnatifid lobes; lower leaves cordate-orbicular, less deeply lobed. Pedicels with stellate hairs only. Epicalyx-segments 3, ovate-deltate, densely pubescent; petals 20–35 mm, bright pink; staminal tube with long hairs. Mericarps glabrous or pubescent; dorsal face and angles rounded, smooth or with faint ridges. $2n=84$. ● *Most of Europe, northwards to S. Sweden, but rare in the Mediterranean region.* Au Be Bu Co Cz Da Ga Ge He Ho Hs Hu It Ju Po Rm Rs (B, C, W, E) Sa Su [Fe No].

Very variable in leaf-shape, indumentum, and shape and size of petals. Numerous variants have been described as species, but the correlation between the variable characters does not seem to be good enough to permit of taxonomic recognition. **M. excisa** Reichenb. in Reichenb. & Reichenb. fil., *Icon. Fl. Germ.* **5**: 18 (1841), represents the most extreme variant. It has deeply bifid petals and deeply divided upper leaves with narrow, almost simple lobes.

6. M. moschata L., *Sp. Pl.* 690 (1753). Like **5** but leaves with 5–7 acutely 2-pinnatifid lobes; hairs on pedicels all simple; epicalyx-segments linear to narrowly ovate, glabrous or sparsely hirsute; mericarps with long, white hairs. $2n=42$. *Most of Europe from England and Poland southwards; naturalized from gardens further north.* Al Au Be Br Bu Co Cz Ga Ge Gr *Hb He Ho Hs It Ju Po Rm Rs (C, W, E) Si [Da Fe Hu No Rs (B) Su].

7. M. tournefortiana L., *Cent. Pl.* **1**: 21 (1755). Like **5** but less robust; leaves often deeply dissected into narrow, linear segments; epicalyx-segments linear to narrowly ovate, glabrous or sparsely hirsute; petals 15–25 mm, pale pink; mericarps with dorsal face rounded or flat. ● *S.W. Europe.* Ga Hs Lu.

Sect. MALVA. Flowers 2 or more in each leaf-axil.

8. M. sylvestris L., *Sp. Pl.* 689 (1753) (incl. *M. ambigua* Guss., *M. erecta* C. Presl, *M. mauritiana* L.). Biennial or perennial, with simple and stellate hairs. Stems up to 150 cm, erect to decumbent, woody at the base. Leaves very variable in size, reniform to suborbicular-cordate, more or less palmatifid, with 3–7 semicircular to oblong, crenate lobes. Epicalyx-segments oblong-lanceolate to elliptical; sepals stellate-pubescent beneath; petals 12–30 mm, pink to purple, with darker veins, bearded. Mericarps glabrous or pubescent, strongly reticulate; dorsal face flat; angles sharp, but not winged. $2n=42$. *Almost throughout Europe except the extreme north.* All except Fa Is Sb, but only as an alien in much of the north.

Very variable in habit, indumentum, leaf-shape, corolla and length of pedicels. The variants that have been recognized represent local races, and the correlation of characters which they exhibit is not maintained on a continental scale.

9. M. nicaeensis All., *Fl. Pedem.* 2:40 (1785) (*M. montana* auct., vix Forskål). Like **8** but annual or biennial; leaves semicircular in outline, scarcely cordate; sepals usually glabrous beneath; petals 10–12 mm, pale lilac without darker veins, glabrous or nearly so. *S. Europe; occasionally as a casual elsewhere.* Az Bl Co Cr Ga Gr Hs It Ju Lu Rs (K) Sa Si Tu.

10. M. parviflora L., *Demonstr. Pl.* 18 (1753). Glabrous or pubescent annual; stem 20–50 cm, erect, with ascending branches. Leaves long-petiolate, suborbicular-cordate, mostly with 5–7 deltate, crenate lobes. Flowers in groups of 2–4; fruiting pedicels mostly less than 10 mm. Epicalyx-segments linear to lanceolate; sepals orbicular-deltate, with short cilia or none, strongly accrescent, patent and scarious in fruit; petals 4–5 mm, slightly exceeding the sepals, pale lilac-blue, glabrous; staminal tube subglabrous. Mericarps glabrous or pubescent; dorsal face strongly reticulate; angles slightly winged. *Mediterranean region and S.W. Europe; a frequent casual elsewhere and occasionally naturalized.* Al Az Bl Co Cr Ga Gr Hs It Ju Lu Sa Si [Br].

11. M. pusilla Sm. in Sowerby, *Engl. Bot.* 4: t. 241 (1795) (*M. rotundifolia* L.). Like **10** but stem usually decumbent; flowers in clusters of up to 10; fruiting pedicels often more than 10 mm; sepals conspicuously long-ciliate, scarcely accrescent; petals pale pink, glabrous; angles of mericarps sharp but not winged. *N., E. & C. Europe, extending to Belgium and N. Italy.* Au Az Be Bu Cz Da Ge Gr Ho Hu It Ju No Po Rm Rs (N, B, C, W, K, E) Su Tu [Br Fe Ga].

12. M. neglecta Wallr., *Syll. Pl. Nov. Ratisbon.* (*Königl. Baier. Bot. Ges.*) 1: 140 (1824) (*M. rotundifolia* auct. plur., non L.). Annual, usually densely stellate-pubescent; stems 16–60 cm, ascending or decumbent. Leaves reniform to orbicular-cordate, with 5–7 crenate-dentate lobes. Flowers in clusters of 3–6; fruiting pedicels mostly more than 10 mm. Epicalyx-segments linear to ovate-oblong, shorter than the sepals; sepals triangular-ovate, ciliate; petals 9–13 mm, at least twice as long as the sepals, pale lilac to whitish, bearded; staminal tube pubescent. Mericarps pubescent, scarcely ridged; dorsal face rounded. $2n=42$. *Most of Europe, except the extreme north and extreme south.* All except Az Bl Cr Fa Is Si Sb; only as an alien in Fe.

13. M. verticillata L., *Sp. Pl.* 689 (1753) (incl. *M. crispa* (L.) L., *M. meluca* Graebner ex Medv., *M. mohileviensis* Downar). Like **12** but calyx usually strongly accrescent; petals c. 7 mm, not more than twice as long as sepals, sometimes glabrous; staminal tube subglabrous; fruiting pedicels less than 10 mm; mericarps often weakly ridged on dorsal face. *Cultivated as a salad plant; widely naturalized in S. & C. Europe, and casual or locally naturalized further north.* [Au Br Cz Ga Ge Gr Ho It Ju Po Rm Rs (N, B, C, W, E).] (*E. Asia.*)

¹ By R. Fernandes.

6. Lavatera L.¹

Herbs or soft-wooded shrubs, usually stellate-pubescent. Flowers solitary or in clusters in the leaf-axils. Epicalyx-segments 3, more or less united at the base, at least in bud. Petals emarginate. Stigmas lateral, filiform. Fruit a schizocarp; mericarps numerous, 1-seeded, usually indehiscent, arranged in a single whorl.

The separation of this genus from *Malva* on the basis of its epicalyx-segments united at the base is very unsatisfactory, as in at least two species, traditionally assigned to *Lavatera* by their general facies, the epicalyx-segments in the fully expanded flower are virtually free. No better taxonomic treatment has, however, been so far proposed.

The compound hairs in this genus are described as *stellate* when they have numerous branches radiating in all directions, and as *fasciculate* when the branches, which are usually fewer and longer, are only slightly divergent.

1 Flowers in clusters
  2 Indumentum including simple, glandular hairs; stipules broad, sometimes amplexicaul **11. triloba**
  2 Indumentum without simple, glandular hairs; stipules narrow
    3 Epicalyx-segments longer than sepals, at least in fruit; stems woody in lower part **3. arborea**
    3 Epicalyx-segments shorter than sepals; stems herbaceous
      4 Sepals strongly accrescent; mericarps ridged, with flat dorsal face and sharp, ± denticulate angles **2. mauritanica**
      4 Sepals only slightly accrescent; mericarps smooth or slightly ridged; dorsal face and angles rounded **1. cretica**
1 Flowers solitary (rarely in pairs) in leaf-axils
  5 Annual; stems not densely tomentose
    6 Central axis of fruit expanded above to a disc which covers and conceals the ripe mericarps; stem hispid with simple or few-rayed, deflexed hairs **10. trimestris**
    6 Central axis of fruit not expanded above mericarps; stem sparsely and minutely stellate-pubescent **9. punctata**
  5 Perennial; stems densely tomentose, at least when young
    7 Leaves ovate-lanceolate or -oblong, about twice as long as wide, not lobed **5. oblongifolia**
    7 Lower leaves nearly as long as wide; upper leaves usually lobed
      8 All leaves suborbicular, scarcely lobed; mericarps with concave dorsal face and very sharp angles **4. maritima**
      8 At least the middle leaves distinctly 3- to 5-lobed; mericarps with rounded dorsal face and angles
        9 Herb; pedicels usually more than 1 cm at anthesis and more than 1·3 cm in fruit **8. thuringiaca**
        9 Shrub; pedicels not more than 1 cm at anthesis and 1·3 cm in fruit
          10 Epicalyx invaginated at insertion of pedicel, its segments shorter than the sepals; mericarps glabrous **7. bryoniifolia**
          10 Epicalyx not invaginated at insertion of pedicel, its segments ± equalling the sepals; mericarps hispid or tomentose **6. olbia**

1. L. cretica L., *Sp. Pl.* 691 (1753). Annual or biennial, stellate-pubescent to slightly hispid. Leaves up to 20 cm, suborbicular-cordate, with 5–7 short lobes. Flowers in clusters of 2–8; pedicels unequal, shorter than the subtending petiole. Epicalyx-segments c. 6 mm, free nearly to the base, ovate; sepals c. 8 mm, triangular-ovate, acuminate; petals 10–20 mm, lilac. Mericarps 7–9(–11), smooth or slightly ridged; angles rounded. *Waste places. Mediterranean region and W. Europe, northwards to S.W. England.* Al Az Bl Br Co Cr Ga Gr Hs It Ju Lu Sa Si Tu.

Like *Malva sylvestris* in general appearance, and sometimes confused with it. *L. cretica* is best distinguished by the broader epicalyx-segments, mericarps with rounded angles, and absence of patent, simple hairs on the peduncles.

2. **L. mauritanica** Durieu in Duchartre, *Rev. Bot.* 2: 436 (1847). Erect, stellate-tomentose annual. Leaves suborbicular-cordate, shortly 5- to 7-lobed. Flowers in clusters; pedicels shorter than the subtending petiole. Epicalyx-segments ovate to oblong, free nearly to the base, shorter than the sepals; sepals broadly triangular-ovate, strongly accrescent, completely covering the ripe fruit; petals 8–15 mm, purple. Dorsal face of mericarps flat, reticulate; lateral faces strongly ridged; angles sharp, denticulate. *Maritime rocks. C. & S. Portugal; ?E. Spain.* ?Hs Lu. (*N.W. Africa.*)

The European plants belong to subsp. **davaei** (Coutinho) Coutinho, *Fl. Port.* 402 (1913). Subsp. *mauritanica* has rather larger mericarps, epicalyx-segments virtually free and sepals less accrescent.

3. **L. arborea** L., *Sp. Pl.* 690 (1753). Biennial up to 3 m, woody at the base; younger parts stellate-tomentose. Leaves up to 20 cm, orbicular, shortly 5- to 7-lobed. Flowers in clusters of 2–7; pedicels 1–2·5 cm, shorter than the subtending petiole. Epicalyx-segments 8–10 mm, suborbicular to ovate-oblong, obtuse, longer than the sepals, strongly accrescent, patent in fruit; sepals *c.* 4 mm, triangular, acute, connivent in fruit; petals 15–20 mm, lilac, with purple veins and base. Mericarps 6–8, glabrous or tomentose; dorsal and lateral faces ridged; angles sharp. *Rocky places, especially near the sea; also in hedges and waste places, but there usually as an escape from gardens. Mediterranean region and coasts of W. Europe, northwards to 55°N. in Ireland.* Al Bl Br Co Cr Ga Gr Hb Hs It Ju Lu Sa Si [Az].

4. **L. maritima** Gouan, *Obs. Bot.* 46 (1773) (*L. africana* Cav.). Shrub 30–120 cm; older branches bare, grey, glabrous, rugose; younger parts densely whitish-tomentose with minute stellate hairs. Leaves up to 7 × 8 cm but usually smaller, suborbicular, usually slightly 5-lobed; petiole 6–30 mm. Flowers solitary or in pairs; pedicels longer than the subtending petiole. Epicalyx-segments 3–8 mm, shorter than the sepals, elliptic-lanceolate to ovate, nearly free; sepals triangular-ovate, acuminate, accrescent, connivent in fruit; petals 1·5–3 cm, pale pink with purple base. Mericarps 9–13, strongly ridged; dorsal face concave; angles very sharp, denticulate. *Dry, rocky places. W. Mediterranean region.* Bl Co Ga Hs It Sa.

5. **L. oblongifolia** Boiss., *Biblioth. Univ. Genève* ser. 2, **13**: 407 (1838). Perennial 60–150 cm, woody at the base, covered with a dense, floccose tomentum of yellowish stellate hairs. Leaves up to 7·5 × 4 cm, ovate-lanceolate to oblong, cordate, crenate-dentate; petiole up to 15 mm. Flowers solitary in the leaf-axils; pedicels up to 15 mm. Epicalyx-segments 6–8 mm, deltate, obtuse, about half as long as the sepals; sepals lanceolate, acute, erect; petals 1·5–2·5 cm, pink with purplish base. Mericarps smooth, usually glabrous; dorsal face and angles rounded. *Rocky and bushy places; calcicole.* ● *S. Spain.* Hs.

6. **L. olbia** L., *Sp. Pl.* 690 (1753). Shrub 60–200 cm; stems hispid; young parts of plant stellate-tomentose. Leaves up to 15 × 13 cm, the lower 3- to 5-lobed, the upper oblong-ovate to lanceolate, often slightly 3-lobed. Flowers solitary in leaf-axils; pedicels 2–7 mm. Epicalyx-segments 7–13 mm, ovate, shortly acuminate; sepals 10–14 mm, ovate, acuminate; petals 1·5–3 cm, purple. Mericarps *c.* 18, tomentose or hispid, not ridged; dorsal face and angles rounded. *W. Mediterranean region, C. & S. Portugal.* Bl Co Ga Hs It Lu Sa Si.

7. **L. bryoniifolia** Miller, *Gard. Dict.* ed. 8, no. 11 (1768) (*L. unguiculata* Desf., *L. sphaciotica* Gand.). Like **6** but with floral leaves hastately 3-lobed, the basal lobes nearly as wide as the central; flowers more distant; epicalyx invaginated at insertion of pedicel, with segments shorter than the sepals and very abruptly acuminate; mericarps glabrous. *Greece and Aegean region; Sicilia.* Cr Gr Si.

8. **L. thuringiaca** L., *Sp. Pl.* 691 (1753). Tomentose perennial; stems 60–200 cm, erect, herbaceous. Leaves up to 9 cm, orbicular-cordate, (3–)5-lobed. Flowers solitary in the leaf-axils. Epicalyx invaginated at insertion of pedicel; segments united to about half-way, broadly ovate, shortly acuminate, shorter than the sepals. Sepals *c.* 12 mm, triangular, acuminate, accrescent; petals (1·5–)2–4·5 cm, purplish-pink. Mericarps *c.* 20; dorsal face smooth but keeled; lateral faces smooth or slightly ridged; angles rounded. *C. & S.E. Europe, extending to C. Italy and N.C. Russia; rarely naturalized or casual elsewhere.* Al Au Bu Cz Ge Gr Hu It Ju Po Rm Rs (B, C, W, K, E) Tu [Fe Ga Rs (N) Su].

(a) Subsp. **thuringiaca**: Upper leaves with obtuse or subacute lobes. Inflorescence lax. Pedicels up to 8 cm in fruit. Petals up to 4·5 cm. *Throughout the range of the species except Greece and much of Italy.*

(b) Subsp. **ambigua** (DC.) Nyman, *Consp.* 128 (1878) (*L. ambigua* DC.): Upper leaves with acute to cuspidate lobes. Inflorescence rather dense. Pedicels not more than 3·5 cm in fruit. Petals not more than 3·5 cm. *S. & W. parts of Balkan peninsula; Italy.*

9. **L. punctata** All., *Auct. Fl. Pedem.* 26 (1789). Annual; stem 20–90 cm, erect, branched, usually flushed with purple-red and sparsely dotted with minute, white stellate hairs. Lower leaves up to 4·5–5 cm, reniform or semicircular, shortly 5-lobed; upper leaves hastate, with a long central lobe and shorter, patent lateral lobes. Flowers solitary in the leaf-axils; pedicels up to 15 cm in fruit. Epicalyx-segments 6–8 mm, broadly ovate, sometimes slightly 3-lobed, acuminate; sepals 8–9 mm, triangular, acuminate, accrescent, connivent in fruit; petals 1·5–3 cm, lilac-pink. Mericarps 14–17, glabrous, ridged; angles blunt. *Mediterranean region.* Al Bl Co Cr Ga Gr ?Hs It Si Tu.

10. **L. trimestris** L., *Sp. Pl.* 692 (1753). Annual; stems up to 120 cm, erect or ascending, more or less strigose with simple or few-rayed, deflexed hairs. Leaves 3–6 × 2·5–7 cm, suborbicular-cordate, the upper somewhat 3- to 7-lobed. Flowers solitary in the leaf-axils; pedicel usually exceeding the subtending petiole. Epicalyx-segments shorter than the sepals, accrescent, united for most of their length, the free part broadly ovate, cuspidate; sepals 9–14 mm, oblong-lanceolate, acute, connivent in fruit; petals 2–4·5 cm, bright pink. Mericarps *c.* 12, glabrous, ridged, covered by a disc-like expansion of the central axis; dorsal face and angles rounded. *Mediterranean region, Portugal; cultivated elsewhere in gardens and locally naturalized.* Bl Co Ga Gr Hs It ?Ju Lu Sa Si [Rs (B, C, W, K)].

11. **L. triloba** L., *Sp. Pl.* 691 (1753). Musk-scented perennial, woody at the base, with simple, glandular, as well as fasciculate or stellate hairs. Leaves cordate-orbicular, slightly 3-lobed; petiole up to 9 cm; stipules up to 2·5 × 1·5 cm, sometimes amplexicaul. Flowers in clusters of 3–7; pedicels shorter than subtending petiole. Epicalyx-segments 8–15 mm, lanceolate to broadly ovate; sepals ovate, acuminate, accrescent, somewhat connivent in fruit; petals 1·5–3 cm. Mericarps 12–16, glabrous or glandular-ciliate, smooth; dorsal face and angles rounded. *Often*

in damp or saline habitats. *W. Mediterranean region, S. Portugal.* Bl Hs It Lu Sa Si.

1 Stem, petioles and inflorescence softly tomentose with slender fasciculate hairs **(c) subsp. agrigentina**
1 Stem, petioles and inflorescences more or less hispid and with rigid stellate hairs
2 Epicalyx-segments united in their lower third, at least along two of the lines of junction; leaves flat or slightly undulate **(a) subsp. triloba**
2 Epicalyx-segments mostly free or very nearly so at anthesis; leaves strongly undulate **(b) subsp. pallescens**

(a) Subsp. **triloba** (*L. rotundata* Láz.-Ibiza & Tubilla): Stems, petioles and pedicels floccose-tomentose to hispid; rigid stellate hairs present at least on pedicels and epicalyx. Leaves up to 8 cm, crenate, flat or slightly undulate. Epicalyx-segments usually united in their lower third; petals purple, sometimes suffused with yellow, or pure yellow. *C., S. & E. Spain, S. Portugal, Sardegna.*

(b) Subsp. **pallescens** (Moris) Nyman, *Consp.* 128 (1878) (*L. minoricensis* Camb.): Indumentum as in (a). Leaves smaller, crispate-undulate. Epicalyx-segments usually free or very nearly so; petals purple or pale pink, only slightly exceeding the sepals. ● *Islas Baleares (Menorca); Sardegna (island of San Pietro).*

(c) Subsp. **agrigentina** (Tineo) R. Fernandes, *Feddes Repert.* 74: 20 (1967) (*L. agrigentina* Tineo): Softly tomentose all over, with numerous long, slender fasciculate hairs and without rigid stellate hairs. Petals pure yellow. ● *Calabria, Sicilia.*

7. Althaea L.[1]

Herbs. Flowers rather small, usually distinctly pedunculate or pedicellate, in racemes or panicles. Epicalyx-segments 6–9, united at the base. Petals obovate, entire or emarginate. Staminal tube terete, hairy. Stigmas lateral, filiform. Mericarps indehiscent, arranged in a single whorl, unilocular, 1-seeded.

Literature: E. G. Baker, *Jour. Bot.* (*London*) 28: 140–141 (1890).

1 Annual; indumentum of simple and stellate hairs; anthers yellow
2 Stipules entire; petals scarcely exceeding sepals **1. hirsuta**
2 Stipules deeply divided into narrow lobes; petals twice as long as sepals **2. longiflora**
1 Perennial; indumentum of stellate hairs only; anthers purplish-red
3 Sepals erect in fruit; mericarps glabrous **3. cannabina**
3 Sepals curved over the fruit; mericarps stellate-pubescent
4 Leaves entire or lobed to about halfway to base; peduncle of axillary inflorescences shorter than the subtending leaf **4. officinalis**
4 Leaves lobed nearly or quite to base; peduncle of axillary inflorescences at least as long as the subtending leaf **5. armeniaca**

1. A. hirsuta L., *Sp. Pl.* 687 (1753). Annual up to 60 cm, with numerous stiff, simple and some stellate hairs. Leaves suborbicular, cordate, crenate or dentate towards base of stem, becoming progressively more deeply lobed upwards; upper cauline leaves palmately 3- to 5-lobed; lobes linear, dentate or serrate. Stipules entire. Flowers solitary; pedicels longer than the subtending leaves. Epicalyx-segments lanceolate, acuminate, nearly as long as the calyx. Sepals lanceolate or ovate, long-acuminate, erect in fruit. Petals *c.* 15 mm, scarcely exceeding the sepals, pinkish-lilac. Anthers yellow. Mericarps glabrous, transversely rugose, with a fine longitudinal rib on the dorsal

face. *Dry places, often as a weed of cultivated ground; somewhat calcicole. S. & S.E. Europe, extending to S.E. Czechoslovakia.* Al Bl Bu Co Cr Cz Ga *Ge Gr *He Hs Hu It Ju Lu Rm Rs (W, K E) Sa Si Tu [Be].

2. A. longiflora Boiss. & Reuter, *Diagn. Pl. Nov. Hisp.* 9 (1842). Like **1** but stipules deeply divided into narrow lobes; petals *c.* 25 mm, twice as long as sepals; mericarps keeled on the dorsal face. *E., C. & S. Spain and S.E. Portugal.* Hs Lu [Ga].

3. A. cannabina L., *Sp. Pl.* 686 (1753) (incl. *A. kotschyi* Boiss.). Pubescent perennial up to 180 cm; all hairs stellate. Leaves palmately lobed, often to the base, with 3–5 lanceolate or linear, irregularly dentate and sometimes lobed segments. Flowers solitary or in clusters, on long, often branched, axillary peduncles, and often also forming a terminal panicle. Epicalyx-segments linear-lanceolate to ovate-acuminate. Sepals ovate, acuminate, erect in fruit. Petals 15–30 mm, pink. Anthers purplish-red. Mericarps glabrous, transversely rugose on the dorsal face; angles rounded. *S. & E.C. Europe.* Al Bu Cz Ga Gr Hs Hu It Ju Lu Rm Rs (W, K) Sa Si Tu.

A. narbonensis Pourret ex Cav., *Monad. Class. Diss. Dec.* 94 (1786), described from S. France, does not appear to merit more than varietal rank. The taxon to which this name is applied by Iljin in Komarov, *Fl. U.R.S.S.* 15: 142 (1949) does not seem to be the same, and may represent a subspecies found in S.E. Europe.

4. A. officinalis L., *Sp. Pl.* 686 (1753) (incl. *A. kragujevacensis* Pančić, *A. taurinensis* DC.). Densely grey-pubescent perennial up to 200 cm; all hairs stellate. Leaves triangular-ovate, acute, crenate-serrate, undivided or palmately lobed, often somewhat plicate. Flowers solitary or clustered in axillary and terminal inflorescences; peduncle of axillary inflorescences shorter than the subtending leaf. Epicalyx-segments linear-lanceolate. Sepals ovate, acute, curved over the fruit. Petals 15–20 mm, very pale lilac-pink, rarely deeper pink. Anthers purplish-red. Mericarps more or less densely covered with stellate hairs, smooth. $2n=42$. *Damp places. Most of Europe, from England, Denmark and C. Russia southwards.* Al Au Be Br Bu Co Cz Da Ga Ge Gr *Hb Ho Hs Hu It Ju Lu Po Rm Rs (C, W, K, E) Sa Si Tu [He].

5. A. armeniaca Ten., *Ind. Sem. Horti Neap.* 1 (1837). Like **4** but sparsely pubescent; lower leaves palmatipartite, upper 3-partite; peduncle of axillary inflorescences at least as long as the subtending leaf; mericarps rugose and slightly sulcate on the dorsal face. *River-banks. S.E. Russia; rarely casual elsewhere.* Rs (E). (*C. & S.W. Asia.*)

A. broussonetiifolia Iljin in Komarov, *Fl. URSS* 15: 678 (1949), from S.E. Russia (banks of the lower Volga) and W.C. Asia, is doubtfully distinct from **5**. The sinuses between the lobes of the leaves are dilated above the base and the leaf-base is often broadly cuneate, instead of being truncate or subcordate.

8. Alcea L.[2]

Tall perennials with erect stems. Flowers large, subsessile in terminal, spike-like racemes. Epicalyx-segments 6(–7), united at the base, smaller than the sepals or about equalling them. Petals emarginate. Staminal tube 5-angled, glabrous. Stigmas lateral, filiform. Carpels 18–40. Fruit a schizocarp; mericarps indehiscent, arranged in a single whorl, hairy at least in centre of dorsal face, each divided by an internal septum into an upper, empty cell and a lower one with a single seed.

The genus is centred in S.W. & C. Asia, and is very difficult taxonomically on account of reticulation of characters and

---

[1] By T. G. Tutin.　　[2] By D. A. Webb.

(probably) hybridization. Until a complete revision has been undertaken this treatment of European plants must be regarded as provisional.

Literature: M. Zohary, *Bull. Res. Counc. Israel* **11D**: 210–229 (1963); *Israel Jour. Bot.* **12**: 1–26 (1963).

1 Petals entirely yellow **6. rugosa**
1 Petals white, pink, purple or violet in apical part (sometimes yellow at the base)
 2 Upper leaves palmatisect almost to the base **5. lavateriflora**
 2 Upper leaves not divided for more than ¾ of the radius
  3 Stem sparsely setose or ± glabrous at maturity
   4 Furrow on dorsal face of mericarp rather shallow, largely obscured by hairs; angles not winged **1. setosa**
   4 Furrow on dorsal face of mericarp deep; angles produced into membranous wings **2. rosea**
  3 Stem with persistent, dense tomentum of stellate hairs
   5 Epicalyx not more than half as long as calyx **4. heldreichii**
   5 Epicalyx at least ⅔ as long as calyx **3. pallida**

1. A. setosa (Boiss.) Alef., *Österr. Bot. Zeitschr.* **12**: 255 (1862) (*Althaea pontica* (Janka) Baker fil., *Althaea rosea* auct. balcan. pro parte, non (L.) Cav.). Stem sometimes purple-spotted, setose with rather distant groups of deflexed hairs, tomentose-pubescent only on youngest parts. Leaves cordate-orbicular to deltate, the upper usually divided for ½–⅔ of the radius into 3–5 oblong or deltate, obtuse, crenate lobes. Epicalyx-segments triangular, acute; sepals similar but rather larger, without prominent veins. Petals 35–50 × 40–55 mm, almost contiguous, violet, usually with yellow base. Mericarps 6 mm; dorsal face densely hairy, with a shallow furrow; angles rugose, not winged; lateral faces appressed-setose. *Turkey-in-Europe; Kriti; cultivated elsewhere for ornament and occasionally naturalized.* Cr Tu [It Ju]. (*S.W. Asia, Egypt.*)

2. A. rosea L., *Sp. Pl.* 687 (1753) (*Althaea rosea* (L.) Cav.). Stem glabrescent or sparsely setose with deflexed hairs, tomentose-pubescent only on youngest parts. Leaves cordate-orbicular to rhombic, weakly 3- to 5-lobed, slightly scabrid-setulose. Epicalyx-segments deltate to triangular-lanceolate, ½–⅔ as long as the subacute, triangular sepals. Petals 30–50 mm, contiguous, usually pink but sometimes white or violet. Mericarps 7 mm; dorsal face with deep, narrow furrow; angles rugose, produced into parallel wings; lateral faces appressed-setose. *Of unknown origin; cultivated in gardens throughout Europe and widely naturalized.* [Au Br Bu Cz Ga Ge He Hs Hu It Ju Lu Rm Rs.]

Not known anywhere as an indigenous plant; probably a hybrid of **1** with **3** or with an Asiatic species. The customary ascription of China as the country of origin is certainly false. Cultivated plants show variation in many characters, especially of calyx and epicalyx, but the combination of winged mericarps, wide petals, and absence of tomentum from the mature stem is distinctive. Some garden plants have yellow or yellowish flowers and more deeply divided leaves, but in other respects resemble **2**; they have been given the name of **A. ficifolia** L., *Sp. Pl.* 687 (1753) (*Althaea ficifolia* (L.) Cav.), which does not seem to refer to any wild plant. They are probably hybrids with **6** or a related species.

3. A. pallida (Willd.) Waldst. & Kit., *Pl. Rar. Hung.* **1**: 46 (1801) (*Althaea pallida* Willd.). Stem pubescent-tomentose throughout, and usually also hispid with longer hairs. Leaves cordate-orbicular to rhombic-deltate, crenate, undivided, or divided for *c.* ⅓ of the radius into 3–5 obtuse lobes, greyish-tomentose, especially beneath. Epicalyx-segments triangular, acute, equalling or slightly shorter than the sepals. Petals 30–45 × 25–35 mm, not contiguous. Mericarps 4·5–6 mm; dorsal face with deep, fairly broad furrow; angles rugose, produced into slightly divergent wings. *S.E. & E.C. Europe, northwards to S. Czechoslovakia.* Al Au Bu Cr Cz Gr Hu Ju Rm Rs (W, K) Tu [It].

(a) Subsp. **pallida**: Not very densely tomentose; petals pale pink, usually yellow at the base, deeply emarginate; mericarps 4·5–5 mm, blackish, with lateral faces glabrous. *Throughout the range of the species except Kriti and parts of Greece.*

(b) Subsp. **cretica** (Weinm.) D. A. Webb, *Feddes Repert.* **74**: 27 (1967) (*Althaea cretica* Weinm., *Althaea rosea* auct. plur., non (L.) Cav.): Densely tomentose; petals bright pink or purple, scarcely yellow at the base, not very deeply emarginate; mericarps 5·5–6 mm, pale brown, with lateral faces appressed-setose. *S. & W. parts of Balkan peninsula; Kriti.*

In the S. part of the Balkan peninsula, plants are found with combinations of the differential characters which makes it difficult to assign them to either subspecies.

A. apterocarpa (Fenzl) Boiss., *Fl. Or.* **1**: 830 (1867), which is very like **3(b)** but with unwinged mericarps, is doubtfully recorded from S. Greece; no European material with ripe fruit appears to be available.

4. A. heldreichii (Boiss.) Boiss., *Fl. Or.* **1**: 832 (1867) (*Althaea heldreichii* Boiss.). Like **3(a)** but less robust; flowers rather smaller; epicalyx scarcely half as long as calyx; sepals with conspicuous, raised veins on the back. ● *Bulgaria and Macedonia; S. Ukraine.* Bu Gr Ju Rs (W, ?K).

5. A. lavateriflora (DC.) Boiss., *op. cit.* 828 (1867) (*Althaea pontica* sensu Hayek pro parte, non (Janka) Baker fil.). Stem tomentose or glabrescent, usually setose with long, deflexed hairs. Lower leaves palmatifid, the upper palmatisect, all tomentose with yellowish, stellate hairs, and with prominent veins beneath. Epicalyx-segments narrowly triangular, acute, ⅔–⅘ as long as the sepals. Petals *c.* 35 × 35 mm, violet with yellow base. Mericarps with a rather shallow furrow on dorsal face; angles shortly winged. *Turkey-in-Europe* (*Bosphorus region*). Tu. (*S.W. Asia.*)

6. A. rugosa Alef., *Österr. Bot. Zeitschr.* **12**: 254 (1862) (incl. *A. taurica* Iljin, *A. novopokrovskyi* Iljin). Stem with persistent tomentum and also villous with longer white hairs. Leaves 5-lobed, the upper more deeply (up to ⅔ of the radius), with oblong, obtuse, irregularly crenate lobes, densely tomentose on both surfaces when young, later variably glabrescent; veins very prominent beneath. Epicalyx-segments deltate, as broad as the sepals but only half as long. Petals 35 × 55 mm, pale or clear yellow. Mericarps 5–6 mm, pale brown; dorsal face usually deeply furrowed; angles produced into divergent wings; lateral faces appressed-setose. *Ukraine and S. Russia.* Rs (W, K, E).

9. Abutilon Miller[1]

Herbs, shrubs or small trees. Epicalyx absent. Stigmas terminal, capitate. Fruit a schizocarp; mericarps arranged in a single whorl, each with several seeds, usually dehiscent *in situ* and not separating readily from the central axis.

1. A. theophrasti Medicus, *Künstl. Geschl. Malv.-Fam.* 28 (1787) (*A. avicennae* Gaertner). Erect annual 50–100 cm, the younger parts tomentose, the older pubescent with simple and stellate hairs. Leaves up to 15 cm, long-petiolate, cordate-orbicular, acuminate, slightly crenate. Flowers in small cymes in axils of upper leaves; peduncles shorter than petioles. Sepals

[1] By D. A. Webb.

united in lower half; petals 7–13 mm, yellow. Mericarps c. 13, exceeding the calyx, black, hirsute, with a slender, erecto-patent beak. Seeds finely tuberculate. *Cultivated ground and waste places. S.E. Europe and Mediterranean region, extending to C. Russia and S. Czechoslovakia, but probably introduced in the western and northern parts of its range.* *Al Bu *Co Cr Gr Hu *It Ju Rm Rs (*C, W, K, E) [Cz Ga Hs Lu].

10. Modiola Moench[1]

Herbs. Flowers solitary in the leaf-axils. Epicalyx-segments 3, free. Petals not emarginate. Stigmas terminal, capitate. Fruit a schizocarp; mericarps numerous, in a single whorl, 2-seeded, dehiscent *in situ*, at least in upper half.

1. **M. caroliniana** (L.) G. Don fil., *Gen. Syst.* 1: 466 (1831). More or less setose annual, or sometimes perennial. Stems up to 50 cm, procumbent and rooting at the base, usually ascending towards the apex. Leaves 5–8 cm, deltate to suborbicular-reniform in outline, varying from incise-dentate to deeply palmatisect with 3–7 pinnatifid lobes. Epicalyx-segments lanceolate; sepals broadly triangular-ovate; petals 3–5 mm, slightly exceeding sepals, orange-scarlet. Mericarps c. 20, black; lateral faces rugose in lower half; dorsal face densely setose and bearing 2 stout spines. *Naturalized in damp, grassy places in N. Spain and N.W. Portugal.* [Hs Lu] (*Tropical America and warm-temperate North America; widely naturalized elsewhere.*)

11. Gossypium L.[2]

Shrubs or woody annuals, irregularly dotted with black oil-glands. Flowers solitary in the leaf-axils. Epicalyx-segments 3, free, broadly ovate to deltate, cordate. Calyx cupuliform, with 5 short teeth or lobes. Styles united; stigmas terminal, capitate, more or less united. Fruit a 3- to 5-locular, loculicidal capsule; seeds numerous, hairy.

The cultivated cottons (which include all the European plants) differ from the wild species in having flattened, twisted ('lint') hairs on the seeds. This character seems to have developed under the influence of human selection. The same influence is responsible for the stabilization of the annual habit, which is found only in cultivated plants, and which has enabled the cultivation of cotton to be carried on in regions with a cold winter climate.

No chromosome counts have been made on European plants, but *G. herbaceum*, like all Old World species, is known to be diploid, with $2n = 26$, while *G. hirsutum* is one of the New World tetraploids, with $2n = 52$. Diploids also occur in the New World.

Literature: J. B. Hutchinson, R. A. Silow & S. G. Stephens, *The Evolution of Gossypium*. London, etc. 1947. G. Watt, *The Wild and Cultivated Cotton Plants of the World*. London. 1907.

Teeth of epicalyx-segments usually less than 3 times as long as wide; filaments 1–2 mm, ± equal **1. herbaceum**
Teeth of epicalyx-segments usually more than 3 times as long as wide; filaments 4–6 mm, the upper longer than the lower **2. hirsutum**

1. **G. herbaceum** L., *Sp. Pl.* 693 (1753). Woody annual 1–1·5 m, glabrous or sparsely hairy. Leaves c. 11 × 14 cm, cordate, with 3–7 lobes; lobes c. 5 × 5 cm, ovate-orbicular, usually slightly constricted at the base. Epicalyx-segments 2–2·5 cm, broadly deltate-ovate to semicircular, usually at least as wide as long; margin with 6–8 triangular, acute to shortly acuminate teeth, usually less than 3 times as long as wide. Petals yellow, with

purple claw. Filaments 1–2 mm, more or less equal. Capsule 2–3·5 cm, subglobose, shortly beaked. *Widely cultivated in S. Europe, especially the E. Mediterranean region, and locally naturalized on disturbed ground.* [Al Cr Gr Hs It ?Ju Rm Si.] (*Probably originated in W. Pakistan; cultivated now from India and C. Asia to C. Africa.*)

G. arboreum L., *Sp. Pl.* 693 (1753) (*G. nanking* Meyen), which differs principally in its more conical capsule and epicalyx segments with only 3–4 teeth, is cultivated locally in Kriti.

2. **G. hirsutum** L., *Sp. Pl.* ed. 2, 975 (1763) (*G. mexicanum* Tod.). Like 1 but sometimes hairier; leaf-lobes broadly triangular to lanceolate, seldom constricted at the base and sometimes overlapping; epicalyx-segments c. 4·5 cm, triangular-ovate, with marginal, long-acuminate teeth usually more than 3 times as long as wide; petals entirely yellow; filaments 4–6 mm, the upper longer than the lower; capsule 4–6 cm, ovoid. *Cultivated in S. Europe, and perhaps locally naturalized.* [?Al Cr Gr It Rm.] (*Probably originated in Peru; cultivated in tropical America and southern U.S.A.*)

G. barbadense L., *Sp. Pl.* 693 (1753) (*G. vitifolium* Lam.), like 2 but with longer staminal column and petals spotted with purple, is reported as cultivated in Kriti.

12. Hibiscus L.[1]

Herbs or shrubs. Flowers solitary in leaf-axils. Epicalyx-segments 6–13, linear, free, or united only at the base. Calyx persistent in fruit. Petals not emarginate. Styles free above; stigmas large, capitate, with long papillae. Fruit a 5-locular, loculicidal, subglobose, ovoid or shortly conical capsule; seeds reniform, numerous.

1 Shrub; seeds hairy **1. syriacus**
1 Herb; seeds glabrous
  2 Perennial; leaves softly tomentose beneath **2. palustris**
  2 Annual; leaves glabrous or hispid beneath
    3 Plant not spiny; sepals united for most of their length, membranous in fruit **3. trionum**
    3 Plant ± spiny; sepals united only at base, somewhat woody in fruit **4. cannabinus**

1. **H. syriacus** L., *Sp. Pl.* 695 (1753). Erect, freely-branched shrub 2–3 m. Leaves 4–7 cm, rhombic, dentate-crenate in apical half and usually 3-lobed, sparsely stellate-pubescent beneath; petiole short. Pedicels 1–2 cm. Epicalyx-segments 7–9; sepals united in basal half; petals c. 5 cm, lilac or white, with a dark purple patch at the base. Capsule c. 25 × 13 mm, densely stellate-pubescent with yellow hairs. Seeds with long, white hairs on the margin. *Planted for ornament and for hedges in S. Europe and locally naturalized.* [Ga Gr It Ju Sa Si.] (*S. & E. Asia.*)

2. **H. palustris** L., *Sp. Pl.* 693 (1753) (*H. roseus* Thore ex Loisel.). Perennial herb; stems 80–120 cm, simple, erect. Leaves 10–15 × 4–9 cm, suborbicular to ovate-lanceolate, acuminate, irregularly dentate to crenate, sometimes shortly 3-lobed, whitish-tomentose beneath with soft, stellate hairs; petiole 2–6 cm. Pedicels 5–7 cm. Epicalyx-segments c. 11; sepals united in basal half; petals c. 7 cm, pink (rarely white with red base). Capsule 15–25 mm, subglobose, enclosed in calyx. Seeds glabrous. *Marshes and river-banks. N. & C. Italy; S.W. France; Portugal.* Ga It *Lu. (*E. North America.*)

The European plant (which is found also in Algeria, but doubtfully native) is sometimes separated as **H. roseus** Thore ex Loisel., *Fl. Gall.* 434 (1807), but its differences from American specimens appear to be slight and inconstant.

---

[1] By D. A. Webb.   [2] By D. M. Moore.

3. H. trionum L., *Sp. Pl.* 697 (1753). Somewhat strigose-hispid annual; stem 10–50 cm, erect to decumbent, branched. Leaves 4–7 cm, divided (except the lowest) more or less to the base into 3(–5) oblong-lanceolate, usually deeply pinnatifid lobes. Epicalyx-segments 10–13, bearing long, simple hairs. Sepals united for most of their length, with conspicuous, dark purple veins, strongly accrescent, membranous and vesicular in fruit. Petals *c.* 2 cm, pale yellow with a deep violet patch at the base. Capsule villous, enclosed in calyx. *Cultivated ground and waste places. S.E. & E.C. Europe; naturalized elsewhere in the Mediterranean region and casual further north.* Al Bu Cr *Cz Gr Hu *It Ju Rm Rs (W, K, E) [Au Ga Ge Hs Lu Po *Si].

4. H. cannabinus L., *Syst. Nat.* ed. 10, **2**: 1149 (1759). Annual, with erect, simple stems up to 2 m, glabrous except for calyx and capsule; stem, petioles and sepals usually armed with small prickles. Leaves very variable, the lower usually suborbicular, dentate, scarcely lobed, the upper divided almost to the base into 3–7 linear-oblong lobes. Epicalyx-segments 6–12; sepals united only at the base, triangular, long-acuminate, villous, the free portion accrescent and becoming hard and spiny in fruit. Petals cream or pale yellow, sometimes purple at the base. Capsule subglobose, apiculate, pubescent. *Cultivated for its fibres in parts of S.E. & E.C. Europe.* [Cr Hu Rs (W, E).] (*Tropical Africa and Asia.*)

13. Abelmoschus Medicus[1]

Like *Hibiscus* but calyx tubular almost to the apex, splitting along one side at anthesis and falling before the fruit is ripe; capsule much longer than wide.

1. A. esculentus (L.) Moench, *Meth.* 617 (1794) (*Hibiscus esculentus* L.). Annual, more or less hispid with simple and forked hairs; stems up to 200 cm, erect. Leaves long-petiolate, cordate to orbicular in outline, with 5–7 dentate lobes. Epicalyx-segments 8–10. Petals 2·5–5 cm, bright yellow, with a dark purple patch at the base. Stigmas deep red. Capsule 6–25 cm, linear-oblong, acuminate, somewhat ridged above, strigose. *Cultivated for its edible young fruits in S.E. Europe.* [Al Bu Gr Rm Rs (W, K, E).] (*Tropical Africa.*)

14. Kosteletzkya C. Presl[1]

Herbs. Epicalyx-segments numerous, linear. Petals not emarginate. Stigmas terminal, capitate. Carpels 5; fruit a depressed, 5-angled, loculicidal capsule, with one seed in each loculus.

1. K. pentacarpos (L.) Ledeb., *Fl. Ross.* **1**: 437 (1842). Erect perennial up to 2 m, pubescent with brownish, stellate hairs. Leaves long-petiolate, triangular-ovate, crenate, usually with 3(–5) broadly triangular lobes, but sometimes undivided. Flowers solitary or in small cymes in the leaf-axils. Epicalyx-segments 6–11, much shorter and narrower than the sepals. Petals 20–25 mm, obovate, entire, lilac-pink. Capsule *c.* 5 × 12 mm, black, strigose, dehiscing along the prominent angles. Seeds 4 mm, reniform, striate. *Riversides and marshes. S. Europe, from E. Spain to S.E. Russia; very local.* Bl Hs It Rs (E).

THYMELAEALES

CVII. THYMELAEACEAE[2]

Small shrubs, rarely herbs, with simple, entire, usually alternate, exstipulate leaves. Flowers hermaphrodite or unisexual, regular, 4-merous, usually in small heads or clusters, rarely in racemes or panicles. Sepals often petaloid, arising from the rim of a tubular, campanulate or urceolate hypanthium ('calyx-tube' or 'receptacle' of many authors), usually similar in colour and texture to the sepals. Petals absent in European genera. Stamens 8, inserted in 2 whorls on the wall of the hypanthium; filaments short. Ovary superior, at the base of the hypanthium but free from it, with a single pendent ovule; style terminal or somewhat lateral. Fruit a nut or drupe.

1 Hypanthium articulated near the middle, the lower half persistent in fruit, the upper half, with the sepals, deciduous
      3. Diarthron
1 Hypanthium not articulated, wholly persistent or wholly deciduous
  2 Exocarp succulent, rarely coriaceous; fruit exposed when mature; leaves rarely less than 12 mm; flower usually fragrant   **1. Daphne**
  2 Exocarp thin and dry; mature fruit usually enclosed in the persistent hypanthium; leaves rarely more than 12 mm; flowers scarcely fragrant   **2. Thymelaea**

1. Daphne L.[3]

Dwarf to medium-sized shrubs, usually with tough, flexible branches; leaves often clustered at the ends of the branches. Flowers hermaphrodite, usually fragrant, in terminal heads or axillary spikes or clusters, rarely in terminal panicles. Hypanthium tubular or narrowly campanulate; sepals and hypanthium petaloid; style terminal. Fruit a drupe, exposed at maturity; exocarp succulent, rarely coriaceous.

Literature: K. Keissler, *Bot. Jahrb.* **25**: 29–124 (1898).

1 All flowers terminal, solitary or in ± sessile heads or clusters
  2 Leaves deciduous, not coriaceous
   3 Branches decumbent or ascending; leaves hairy at least when young   **6. alpina**
   3 Branches ± erect; leaves glabrous   **3. sophia**
  2 Leaves evergreen, ± coriaceous
   4 Flowers solitary or in terminal pairs
    5 Leaves 8–11 × 1·5–3 mm, mucronate; flowers usually purple   **15. jasminea**
    5 Leaves 15–18 × 6 mm, not mucronate; flowers white   **16. malyana**
   4 Flowers in terminal heads of 3 or more
    6 Sepals narrowly triangular, acuminate; inflorescence ebracteate   **7. oleoides**
    6 Sepals ovate or broadly triangular, obtuse or acute; flowers subtended by scarious or leaf-like bracts

---

[1] By D. A. Webb.   [2] Edit. D. A. Webb.
[3] By D. A. Webb and I. K. Ferguson.

7 Flowers creamy-white   **8. blagayana**
7 Flowers pink or purplish
  8 Leaves 6–12 mm wide   **9. sericea**
  8 Leaves 2–6 mm wide
    9 Flowers purple, tinged with yellow; leaves ciliate and obscurely denticulate (Islas Baleares)   **14. rodriguezii**
    9 Flowers pink; leaves, entire, not ciliate
      10 Young shoots bright coral-red; leaf-margins strongly revolute (Czechoslovakia)   **12. arbuscula**
      10 Young shoots green to brown; leaf-margins seldom revolute
        11 Leaves strongly keeled beneath, ± trigonous; branches short, tortuous   **13. petraea**
        11 Leaves not keeled; branches fairly long, ± straight
          12 Hypanthium usually hairy; leaves usually 3–4 times as long as wide; fruit brownish-yellow **10. cneorum**
          12 Hypanthium glabrous; leaves usually 5–6 times as long as wide; fruit reddish   **11. striata**
1 Flowers wholly or partly in axillary clusters or racemes, or in terminal panicles
  13 Flowers greenish-yellow, glabrous
    14 Flowers in racemes, arising from the axils of the leaves of the previous year; hypanthium 2–3 times as long as the sepals   **4. laureola**
    14 Flowers in pairs, arising from the axils of reduced leaves of the current year; hypanthium only slightly longer than the sepals   **5. pontica**
  13 Flowers white, cream or pink, often hairy
    15 Mature leaves hairy beneath   **9. sericea**
    15 Mature leaves glabrous beneath
      16 Leaves deciduous, 8–25 mm wide   **1. mezereum**
      16 Leaves evergreen, coriaceous, 3–10 mm wide
        17 Flowers in terminal panicles   **2. gnidium**
        17 Flowers in small, terminal heads, with axillary clusters below them   **17. gnidioides**

1. D. mezereum L., *Sp. Pl.* 356 (1753). Deciduous shrub 25–200 cm, of bushy habit, with erect or ascending, greyish-brown branches; young shoots hirsute. Leaves 30–80 × 8–25 mm, oblong-lanceolate, glabrous or ciliate, thin, narrowed to a short petiole. Flowers pinkish-purple, very fragrant, appearing before or with the leaves, borne in clusters of 2–4 in the axils of the fallen leaves of the previous year, forming intercalary spikes. Hypanthium 5–8 mm, villous; sepals villous beneath, glabrous above, slightly shorter than the hypanthium. Drupe bright red, exposed before maturity. $2n = 18$. *Somewhat calcicole. Most of Europe except the extreme west, south and north.* Al Au Be Bu Cz Fe Ga Ge Gr He Ho Hs Hu It Ju No Po Rm Rs (N, B, C, W, E) ?Si Su [Da].

Both the typical variety and var. *alba* Aiton, with white flowers and yellow drupes and usually a more strict habit, are often cultivated for ornament and are occasionally naturalized by bird-dispersal.

2. D. gnidium L., *Sp. Pl.* 357 (1753). Erect, evergreen shrub up to 200 cm; branches slender, straight, brown, uniformly leafy for some distance below the apex; young shoots appressed-puberulent. Leaves 20–50 × 3–10 mm, glabrous, subcoriaceous, linear to obovate-oblong, acute, glandular beneath. Flowers creamy-white, in small, terminal panicles. Hypanthium 2·5–4 mm, villous, slightly longer than the sepals. Drupe ovoid, red. *S. Europe, mainly in the west, extending northwards to 47° N. in W. France.* Al Bl Co Ga Gr Hs It ?Ju Lu Sa Si.

3. D. sophia Kalenicz., *Bull. Soc. Nat. Moscou* 22(1): 311 (1849). Erect, deciduous shrub up to 150 cm; branches long, slender, strict. Leaves *c.* 45 × 15 mm, oblong-obovate, sessile, glabrous, glaucous beneath, not clustered at the ends of the branches. Flowers white, fragrant, in terminal, ebracteate heads. Hypanthium *c.* 10 mm, narrow, appressed-puberulent; sepals

c. 7 × 3 mm, broadly triangular, acute. Drupe bright red. *Woods and thickets; calcicole.* ● *S.C. Russia and N.E. Ukraine.* Rs (C, W, E).

4. D. laureola L., *Sp. Pl.* 357 (1753). Evergreen shrub with suberect branches; young shoots greenish, glabrous. Leaves 30–120 × 10–35 mm, at least 3 times as long as wide, obovate-oblanceolate, subacute, coriaceous, glabrous, shining. Flowers yellowish-green, glabrous, borne in short, congested, bracteate, axillary racemes on the previous year's growth. Sepals ovate, acute, about ⅓ as long as the hypanthium. Drupe ovoid, black. *S., S.C. & W. Europe, northwards to England and Hungary.* Al Au Az Be Br Bu Co Ga Ge Gr He Hu Hs It Ju Rm Sa Si ?Tu [Da Rs (K)].

(a) Subsp. **laureola**: 50–100 cm or more, erect, with most of the leaves clustered at the ends of the branches. Hypanthium 5–9 mm. $2n = 18$. *Throughout the range of the species.*

(b) Subsp. **philippi** (Gren.) Rouy, *Consp. Fl. Fr.* 225 (1927): 20–40 cm, with spreading, more or less decumbent branches, leafy for some distance below the apex. Hypanthium 3–5 mm. ● *Pyrenees.* Ga Hs.

5. D. pontica L., *Sp. Pl.* 357 (1753). Like 4(a) but of more spreading habit; leaves obovate, 2–2½ times as long as wide; flowers in pairs on a common peduncle, arising from the axils of reduced, bract-like leaves at the base of the current year's growth; hypanthium 8–10 mm, slender; sepals pale yellow, almost as long as the hypanthium. *S.E. Bulgaria and Turkey-in-Europe.* Bu Tu. (*N. Anatolia and Georgia.*)

6. D. alpina L., *Sp. Pl.* 356 (1753). Deciduous dwarf shrub with decumbent, often tortuous branches; young shoots hairy. Leaves 20–40 × 6–10 mm, obovate-cuneate to narrowly oblong, densely sericeous-villous on both surfaces, at least when young. Flowers white, fragrant, appearing after the leaves, subsessile, in terminal, ebracteate heads of 4–10. Hypanthium 4–6 mm, hairy; sepals 2·5–4 mm. Drupe red, pubescent, included in hypanthium till ripe. *Calcicole. Mountains of S. & C. Europe, from E.C. France to the Pyrenees, N. Appennini and Crna Gora.* Au Ga He ?Hs It Ju.

7. D. oleoides Schreber, *Icon. Descr. Pl.* 13 (1766). Evergreen dwarf shrub up to 50 cm (rarely more), with numerous, usually more or less straight branches; young shoots hairy. Leaves 10–45 × 3–12 mm, obovate, oblanceolate, oblong or elliptical, obtuse or acute, coriaceous, usually more or less villous when young, often glabrescent later, at least on upper surface. Flowers white or cream (rarely with hypanthium and lower surface of sepals deep pink), fragrant, subsessile, in terminal, ebracteate heads of 3–6. Hypanthium 6–8 mm, hairy; sepals 5–7 mm, narrowly triangular, acuminate. Drupe red, pubescent, included in hypanthium till ripe. *Usually calcicole. Mountains of S. Europe.* Al Bu Co Cr Gr Hs It Ju Sa Si.

Plants from E. Greece (Evvoia) with large, acute, elliptical leaves have been distinguished as **D. euboica** Rech. fil., *Österr. Bot. Zeitschr.* 104: 176 (1957). They differ strikingly from most specimens of *D. oleoides* from the Balkan peninsula, but can be exactly matched by specimens from E. Spain, where they intergrade with typical plants.

D. kosaninii (Stoj.) Stoj., *Spis. Bălg. Akad. Nauk* 37: 137 (1928), from the mountains of S.W. Bulgaria, requires further investigation. It differs from **7** in taller habit, shining, reddish bark, smaller leaves and deep pink flowers with shorter sepals, but plants can be found intermediate in all such characters between the extreme form and *D. oleoides*. It is possibly the hybrid **7 × 10**.

8. D. blagayana Freyer, *Flora* (*Regensb.*) **21**: 176 (1838). Evergreen dwarf shrub up to 30 cm, with long, decumbent, sparingly branched stems, leafless except at the apex. Leaves 3–6 cm, obovate, obtuse, sessile, glabrous, coriaceous. Flowers creamy-white, fragrant, sessile, in terminal heads of 10–15, subtended by pale, sericeous bracts *c.* 10 mm long. Hypanthium 15–20 mm, narrow, with a few silky hairs. Sepals 6 mm, obtuse, patent. Drupe whitish. $2n = 18$. ● *Balkan peninsula, extending northwards to Slovenija and the S. Carpathians.* Al Bu Gr Ju Rm.

9. D. sericea Vahl, *Symb. Bot.* **1**: 28 (1790) (incl. *D. collina* Sm., *D. vahlii* Keissler). Evergreen shrub up to 70 cm, with erect or decumbent branches; young shoots hairy. Leaves 20–50 × 6–12 mm, oblong-obovate, covered with appressed hairs beneath, glabrous above except for a few hairs on the midrib. Flowers pink, very fragrant, in terminal heads of 5–15, subtended by short, ovate, sericeous bracts about half as long as the hypanthium; some axillary flower-clusters sometimes present as well. Hypanthium 6–8 mm, covered with whitish hairs; sepals 4–6 mm, obtuse. Drupe reddish-brown. *E. & C. Mediterranean region, westwards to c.* 10° 30′ *E. in Italy.* Cr Gr It Si.

10. D. cneorum L., *Sp. Pl.* 357 (1753) (incl. *D. julia* Kos.-Pol.). Evergreen dwarf shrub with usually decumbent or ascending, long, slender, straight, smooth branches; young shoots pubescent, greyish. Leaves 10–18(–25) × (2–)3–5(–6) mm, usually 3–4 times as long as wide, oblong or linear-oblanceolate, obtuse, sometimes mucronate, sessile, glabrous, not clustered at the ends of the branches. Flowers fragrant, pink, subsessile, in heads of 6–10(–20), subtended by bracts similar to the leaves but smaller. Hypanthium 6–10 mm, usually covered with whitish hairs; sepals 4–6 mm, obtuse. Drupe brownish-yellow, included in the hypanthium till ripe. $2n = 18$. *Dry or stony places; usually calcicole.* ● *C. Europe, extending to N.W. Spain, C. Italy, Bulgaria and C. Ukraine.* Al Au Bu Cz Ga Ge He Hs Hu It Ju Po Rm Rs (C, W).

Plants have been recorded from Bulgaria (Pirin Planina) with young shoots and hypanthium glabrous, as in **11**, but with short, wide leaves as in **10**. There is no information available about their fruits.

Plants from W. Hungary, S.E. Austria and N. Jugoslavia, with revolute leaf-margins and a rather erect habit, have been described as f. **arbusculoides** Tuzson, *Bot. Közl.* **10**: 151 (1911). This occurs only on acid soils and may, perhaps, deserve recognition as a subspecies.

11. D. striata Tratt., *Arch. Gewächsk.* **1**: 120 (1814). Like **10** in habit but entirely glabrous, and with somewhat stouter, more freely branched stems; leaves somewhat crowded at the ends of the branches, longer and narrower (usually 5–6 times as long as wide) and less coriaceous; heads with 8–12 flowers; hypanthium 6–12 mm, often longitudinally striped; drupe reddish, exposed before maturity. *Dry and stony places, usually above* 1500 *m; somewhat calcicole.* ● *Alps.* Au Ga Ge He It Ju.

12. D. arbuscula Čelak., *Sitz.-Ber. Böhm. Ges. Wiss.* (*Math.-Nat. Kl.*) **1880**(1): 215 (1890). Like **10** but with shorter branches and young shoots bright coral-red; leaves fleshy, linear or linear-oblong, deeply sulcate above and with revolute margins, crowded at the ends of the branches; flowers fewer but larger (hypanthium 12–20 mm, sepals 6–8 mm); and bracts scarious, much shorter than the flowers. *Calcareous rocks,* 900–1300 *m.* ● *E. Czecho-slovakia* (*Muráň region, E. of Banska Bystrica*). Cz.

The young shoots, leaves, bracts and flowers may be hairy or glabrous.

13. D. petraea Leybold, *Flora* (*Regensb.*) **36**: 81 (1853). Evergreen dwarf shrub with numerous short, stout, tortuous, procumbent, branched stems, forming an intricate mat. Young shoots greenish-brown, sparsely pubescent; older stems covered with raised leaf-scars. Leaves 8–12 × 2–3 mm, linear-oblanceolate, obtuse, glabrous, strongly keeled beneath so as to be triangular in section, clustered at the ends of the branches. Flowers fragrant, bright pink, in heads of 3–5 (rarely more), subtended by scarious bracts much shorter than the flowers. Hypanthium 9–15 mm, villous; sepals 3–5 mm, broadly ovate, obtuse. Drupe sparsely pubescent. *Crevices of calcareous rocks,* 700–2000 *m.* ● *N. Italy* (*in a small region centred on Lago di Idro, N.E. of Brescia*). It.

14. D. rodriguezii Texidor, *Apunt. Fl. Esp.* 64 (1869). Evergreen dwarf shrub up to 50 cm, with numerous short, lateral branches; young shoots pubescent. Leaves 10–20 × 2–6 mm, oblong-oblanceolate, obtuse, obscurely denticulate and sinuate, ciliate, with revolute margins, sessile, not clustered at the ends of the branches. Flowers fragrant, purple tinged with yellow, sessile, in heads of 2–5, subtended by bracts similar to the leaves but smaller. Hypanthium 5–8 mm, covered with whitish hairs; sepals 3–5 mm, obtuse. Drupe greenish-brown, included in hypanthium till ripe. *Littoral scrub.* ● *Menorca.* Bl.

15. D. jasminea Sibth. & Sm., *Fl. Graec. Prodr.* **1**: 260 (1809). Evergreen dwarf shrub up to 30 cm. Stems decumbent or ascending, short, tortuous, freely branched, covered with raised leaf-scars; young shoots glabrous. Leaves 8–11 × 1·5–3 mm, oblong-obovate, mucronate, shortly petiolate, glabrous. Flowers purple (rarely yellow or white) outside, white or pale yellow on upper surface of sepals, in terminal clusters of 2(–3), subtended by very small, hairy, deciduous bracts. Hypanthium 10–12 mm, very slender, glabrous or sparsely pubescent; sepals triangular, acute; ovary glabrous. *Rocky places.* ● *S.E. Greece, from Navplion to Giona and Evvoia.* Gr.

16. D. malyana Blečić, *Bull. Mus. Hist. Nat. Pays Serbe* ser. B, **5–6**: 23 (1953). Like **15** but leaves 15–18 × 6 mm, obovate-spathulate, obtuse, not mucronate, sometimes sparsely hairy beneath; flowers white; ovary more or less sericeous. Drupe greenish, coriaceous. *Limestone rocks.* ● *N.W. Crna Gora* (*gorge of the Piva*). Ju.

17. D. gnidioides Jaub. & Spach, *Ill. Pl. Or.* **4**: 4 (1850). Erect shrub; branches long, strict, stout; young shoots pubescent with brown hairs. Leaves 25–40 × 4–7 mm, oblong-lanceolate, cuspidate-acuminate, pungent, erect, sessile, coriaceous, glaucous, with a few appressed hairs when young, glabrous later. Flowers pink, subsessile, in terminal, ebracteate heads of 5–8 and also in clusters of 2–3 in the upper leaf-axils. Hypanthium broad, covered with silky hairs. Sepals oblong, obtuse, $\frac{1}{2}$–$\frac{2}{3}$ as long as the hypanthium. Drupe with coriaceous, scarcely fleshy exocarp, enclosed in the hypanthium till ripe. *Aegean region* (*Skiathos, Evvoia, Astipalaia*). Gr. (*E. Aegean and S. Anatolia.*)

All European records of this species require confirmation.

2. Thymelaea Miller[1]

Evergreen dwarf shrubs, or rarely perennial or annual herbs. Usually dioecious, but often with some hermaphrodite flowers on male and female plants. Leaves small, sessile. Flowers usually yellow, sometimes tinged with green or purple, solitary or in small clusters in the leaf-axils. Hypanthium urceolate or tubular,

---

[1] By D. A. Webb and I. K. Ferguson.

more or less petaloid, usually yellow or brown; sepals similar in colour and texture. Style short, lateral. Fruit dry, indehiscent, usually enclosed in the persistent hypanthium.

All European species grow in dry places.

Literature: G. Brecher, *Ind. Horti Bot. Univ. Budapest.* **5**: 57–116 (1941).

1 Flowers glabrous
2 Young shoots glabrous   **15. dioica**
2 Young shoots hairy
  3 Leaves with involute margins concealing the tomentose upper surface   **10. broterana**
  3 Leaves flat; upper surface glabrous or hirsute
    4 Leaves sparsely hirsute above when young, with long, straight hairs   **12. ruizii**
    4 Leaves glabrous above, or bearing short, crispate hairs
      5 Leaves *c.* 5 mm wide, elliptical; hypanthium 2·5 mm   **13. subrepens**
      5 Leaves 1·5–3·5 mm wide, oblong; hypanthium 4–5 mm   **14. tinctoria**
1 Flowers hairy (sometimes only sparsely)
6 Annual; flowers greenish or white   **17. passerina**
6 Perennial; flowers yellow or reddish-brown
  7 Leafy stems annual, arising from a short, woody stock
    8 Leaves 15–30 × 4–8 mm   **1. sanamunda**
    8 Leaves 6–14 × 1–4 mm   **2. pubescens**
  7 Leafy stems woody, persistent
    9 Mature leaves glabrous
     10 Flowers in clusters of 2–5; fruit glabrous   **4. tartonraira**
     10 Flowers solitary or in pairs; fruit pubescent
      11 Leaves less than 1 mm wide   **8. coridifolia**
      11 Most of the leaves at least 1 mm wide
       12 Young shoots glabrous   **15. dioica**
       12 Young shoots pubescent   **11. calycina**
    9 Mature leaves hairy, at least on one surface
     13 Leaves imbricate, appressed, white-tomentose on adaxial surface, more or less glabrous on abaxial surface   **3. hirsuta**
     13 Leaves about equally hairy on both surfaces
      14 Leaves and young shoots hirsute, with long, straight, patent hairs   **16. villosa**
      14 Leaves and young shoots sericeous or tomentose, with appressed or intricate hairs
       15 Sepals triangular, acute
        16 Most of the leaves borne on short lateral shoots **5. nitida**
        16 Most of the leaves on long shoots; short lateral shoots few or absent   **4. tartonraira**
       15 Sepals broadly elliptical, obtuse
        17 Stems procumbent; leaves less than 1·5 mm wide; flowers solitary   **9. procumbens**
        17 Stems decumbent to erect; leaves more than 1·5 mm wide; flowers usually in clusters
         18 Leaves sericeous, usually more than 10 mm long   **4. tartonraira**
         18 Leaves villous to tomentose, not more than 10 mm long
          19 Flowers subtended by ovate bracts; hypanthium *c.* 4 mm   **6. myrtifolia**
          19 Flower-clusters without bracts; hypanthium 6·5–8 mm   **7. lanuginosa**

1. T. sanamunda All., *Fl. Pedem.* **1**: 132 (1785) (*Passerina thymelaea* (L.) DC.). Stems 10–30 cm, annual, glabrous, erect, simple, arising from a short, woody stock. Leaves 15–30 × 4–8 mm, elliptical-oblong, acute, glabrous. Flowers unisexual and hermaphrodite, in ebracteate clusters of 2–5. Hypanthium 6–7 mm, tubular, sparsely hairy; sepals 2–2·5 mm, triangular, acute. Fruit glabrous.   ● *C. & E. Spain; S. France.* Ga Hs.

2. T. pubescens (L.) Meissner in DC., *Prodr.* **14**: 558 (1857) (incl. *T. thesioides* (Lam.) Endl., *T. elliptica* (Boiss.) Endl.). Like **1** but smaller in all its parts and sometimes hairy; leaves

6–14 × 1–4 mm, linear to broadly elliptical; flowers in clusters of 2–3; hypanthium *c.* 5 mm.   ● *E. & C. Spain; E. Pyrenees.* ?Ga Hs.

Plants from C. Spain and the region of Valencia are usually less hairy and with smaller and narrower leaves than those from either S.E. Spain or N. Aragon, and have been distinguished as **T. thesioides** (Lam.) Endl., *Gen. Pl., Suppl.* **4**: 66 (1847), but intermediates are numerous and the correlation of these characters with geographical distribution is imperfect.

3. T. hirsuta (L.) Endl., *Gen. Pl., Suppl.* **4**: 65 (1847) (*Passerina hirsuta* L.). Dwarf shrub 40–100 cm, with erect, spreading or decumbent, branched stems densely clothed with imbricate leaves; young shoots white-tomentose. Leaves 3–8 × 1·5–4 mm, ovate to lanceolate, obtuse to acuminate, erect, somewhat fleshy or coriaceous; adaxial surface white-tomentose, abaxial shining, glabrous or sparsely hairy. Flowers unisexual and hermaphrodite, in ebracteate clusters of 2–5. Hypanthium 3–4 mm, densely tomentose; sepals 1 mm, broadly ovate, glabrous above. Fruit glabrous, exposed shortly before maturity. *Mediterranean region, S.E. Portugal.* Bl Co Cr Ga Gr Hs It Ju Lu Sa Si ?Tu.

4. T. tartonraira (L.) All., *Fl. Pedem.* **1**: 133 (1785) (*Passerina tartonraira* (L.) Schrader). Dwarf shrub 20–50 cm, with erect or decumbent stems; young shoots usually sericeous. Leaves 10–18 × 2–7 mm, obovate to narrowly oblong, obtuse to sub-acute, usually sericeous. Flowers unisexual and hermaphrodite, in clusters of 2–5, subtended by numerous small, ovate bracts. Hypanthium 5–6 mm, sericeous or pubescent; sepals 2 mm, broadly triangular, subacute. Fruit glabrous. *Mediterranean region.* Co Cr Ga Gr Hs It Sa Si Tu.

1 Leaves ± glabrous   subsp. (c) **thomasii**
1 Leaves sericeous
  2 Leaves 2–4 times as long as wide   subsp. (a) **tartonraira**
  2 Leaves 4–10 times as long as wide   subsp. (b) **argentea**

(a) Subsp. **tartonraira**: Leaves 2·5–7 mm wide, 2–4 times as long as wide, sparsely to densely sericeous. *Throughout the range of the species, except Kriti.*

(b) Subsp. **argentea** (Sibth. & Sm.) Holmboe, *Stud. Veg. Cyprus* 133 (1914) (*T. argentea* Sibth. & Sm.): Leaves 1·5–3 mm wide, 4–10 times as long as wide, densely silver-sericeous.   ● *Kriti; ?Greece.*
Plants from C. & S. Greece and from S. Spain are often intermediate between subsp. (a) and (b).

(c) Subsp. **thomasii** (Duby) Briq., *Prodr. Fl. Corse* **3**(1): 5 (1938): Leaves and young shoots glabrous, or with a few scattered hairs; leaves 3–4 mm wide, 3–4 times as long as wide. 2*n*=18. ● *Corse (near Ponte Leccia).*

5. T. nitida (Vahl) Endl., *Gen. Pl., Suppl.* **4**: 65 (1847). Like **4** but stems 10–30 cm, slender, erect, with numerous short, lateral branches on which most of the leaves are borne; leaves 6–9 × 1–2 mm; flower-clusters without bracts. *Mountain rocks; calcicole. S. & E. Spain.* Hs.

6. T. myrtifolia (Poiret) D. A. Webb, *Feddes Repert.* **74**: 28 (1967) (*T. velutina* (Pourret ex Camb.) Meissner, *Daphne myrtifolia* Poiret). Dwarf shrub with erect, freely branched stems. Leaves 6–10 × 2·5–4 mm, obovate to elliptic-oblong, densely villous-tomentose. Flowers unisexual and hermaphrodite, solitary or in clusters, subtended by ovate bracts. Hypanthium 4 mm, densely tomentose; sepals 1–1·5 mm, broadly ovate, obtuse. *Maritime sands and calcareous rocks.*   ● *Islas Baleares.* Bl.

Records from N.W. Africa are probably erroneous.

7. T. lanuginosa (Lam.) Ceballos & C. Vicioso, *Estud. Veg. Fl. Forest. Málaga* 235 (1935) (*T. canescens* (Schousboe) Endl., *Daphne lanuginosa* Lam.). Dwarf shrub 60–80 cm, with erect, freely branched stems, young shoots villous. Leaves 3·3–3 × 1·3–2·5 mm, elliptical, grey-villous. Flowers unisexual and hermaphrodite, in ebracteate clusters of 3–9 on short lateral shoots. Hypanthium 6·5–8 mm, villous; sepals 1·5–2 mm, ovate, obtuse. *Maritime sands and calcareous rocks. S. Spain.* Hs.

8. T. coridifolia (Lam.) Endl., *Gen. Pl., Suppl.* **4**: 66 (1847). Dwarf shrub with spreading, freely branched stems 15–35 cm; young shoots subglabrous to pubescent. Leaves 4–7 × 0·5–0·75 mm, linear, patent, crowded, ciliate when young, glabrous later. Dioecious; flowers solitary or in pairs, forming short, terminal spikes, each subtended by 2 small bracts. Hypanthium *c.* 4 mm, covered with short, grey hairs, tubular in male flowers, urceolate in female. Sepals 1·5 mm, ovate, obtuse. Fruit pubescent. *Heaths.* ● *N.W. Spain, eastwards to c. 3° 30' W.* Hs.

9. T. procumbens A. & R. Fernandes, *Bol. Soc. Brot.* ser. 2, **26**: 266 (1952). Like **8** but stems up to 70 cm, procumbent; leaves and young shoots densely covered with silky hairs; leaves 4–10 × 0·5–1·25 mm, linear-lanceolate; flowers solitary, the male with hypanthium 6–7 mm. ● *E. Portugal (Sabugal) and W. Spain (Sierra de Gata).* Hs Lu.

10. T. broterana Coutinho, *Bol. Soc. Brot.* **24**: 145 (1909). Like **8** but stems 15–40 cm, erect; young shoots hairy; leaves with villous-tomentose adaxial surface, but with this surface concealed by strongly involute margins, so that only the glabrous abaxial surface is seen; hypanthium glabrous; sepals triangular. *Mountain heaths.* ● *N. & C. Portugal.* Lu.

11. T. calycina (Lapeyr.) Meissner in DC., *Prodr.* **14**: 555 (1857) (*Passerina calycina* (Lapeyr.) DC.). Dwarf shrub 20–50 cm. with erect or decumbent, branched stems; young shoots pubescent. Leaves 8–15 × 1–3 mm, linear to oblong, glabrous; margins revolute in apical half. Flowers unisexual and hermaphrodite, solitary, subtended by two small, ovate, obtuse, glabrous or ciliate bracts. Hypanthium *c.* 6 mm, tubular or urceolate, sparsely pubescent; sepals 2 mm, broadly ovate. Fruit pubescent. ● *C. & W. Pyrenees and mountains of N. Spain.* Ga Hs.

12. T. ruizii Loscos ex Casav., *Anal. Soc. Esp. Hist. Nat.* **9**: 301 (1880). Like **11** but leaves smaller and sparsely villous on adaxial surface when young; hypanthium *c.* 5 mm, glabrous. *Calcicole.* ● *Mountains of N. Spain.* Hs.

13. T. subrepens Lange, *Overs. Kong. Danske Vid. Selsk. Forh.* **1893**: 193 (1893). Like **11** but stems procumbent, rooting; leaves *c.* 11 × 5 mm, elliptical, acute, flat; hypanthium 2·5 mm, glabrous; sepals 1·5 mm. ● *E. Spain (mountain-ranges between Cuenca and Albarracin).* Hs.

14. T. tinctoria (Pourret) Endl., *Gen. Pl., Suppl.* **4**: 66 (1847) (*Passerina tinctoria* Pourret). Dwarf shrub 20–50 cm, with erect or decumbent, often tortuous, branched stems covered with raised leaf-scars; young shoots crispate-pubescent. Leaves 5–12 × 1·5–3·5 mm, oblong, glabrous or crispate-pubescent on both surfaces. Flowers unisexual and hermaphrodite, solitary, subtended by 2 ovate, obtuse, tomentose bracts, 1·5–3 mm long. Hypanthium 4–5 mm, glabrous; sepals 1–2 mm, broadly ovate. Fruit glabrous. *Rocky woods and scrub; calcicole.* ● *N.E. Spain; Pyrenees; two outlying stations in S. France (Gard).* Ga Hs.

Dwarf plants from the Pyrenees, with small, narrow leaves and sparse indumentum, have been distinguished as **T. nivalis** (Ramond) Meissner in DC., *Prodr.* **14**: 555 (1857). They are easily

confused with **15** but differ in their pubescent young shoots. They may deserve subspecific status.

15. T. dioica (Gouan) All., *Auct. Fl. Pedem.* 9 (1789). Dwarf shrub 20–50 cm, with erect or decumbent, often tortuous, branched stems covered with raised leaf-scars; young shoots glabrous. Leaves 3–12 × 0·75–2·5 mm, linear to cuneate-oblanceolate, glabrous. Flowers unisexual and hermaphrodite, solitary or in pairs, subtended by 2–6 linear bracts, *c.* 1·5 mm long. Hypanthium 4–7 mm, tubular, usually glabrous; sepals 1·5–2 mm, triangular. Fruit pubescent. ● *Mountains of S.W. Europe, from N.W. Italy to the W. Pyrenees and S.E. Spain.* Ga Hs It.

In S.E. Spain (Sierra de Cazorla and adjacent ranges) this species is represented by cushion-like plants with very short, tortuous, much-branched stems, leaves not more than 3·5 × 1·25 mm and often less, and flowers with short, rather obtuse sepals and sometimes a few hairs on the hypanthium. They have been distinguished as **T. granatensis** Pau ex Lacaita, *Cavanillesia* **3**: 40 (1930). Plants from exposed stations in the Pyrenees, however, provide a transition in all these characters to the typical plants of the S.W. Alps.

16. T. villosa (L.) Endl., *Gen. Pl., Suppl.* **4**: 66 (1847). Dwarf shrub 20–40 cm, with erect or decumbent, branched stems; young shoots, flowers and margins and abaxial surface of leaves hirsute with straight, patent, rather stout hairs up to 2 mm long. Leaves 10–13 × 2–4 mm, oblong to narrowly elliptical; adaxial surface glabrous or sparsely hirsute. Dioecious; flowers solitary, without bracts. Hypanthium 7–9 mm, tubular; sepals 3 mm, narrowly oblong. *S.W. Spain, S. Portugal.* Hs Lu.

17. T. passerina (L.) Cosson & Germ., *Fl. Env. Paris* ed. 2, 586 (1861) (*Lygia passerina* (L.) Fasano, *Passerina annua* Wikstr.). Annual; stems 20–50 cm, erect, glabrous or rarely pubescent. Leaves 8–14 × 1–2 mm, linear-lanceolate, acute, glabrous. Flowers hermaphrodite, greenish, solitary or in clusters of 2–3, arising from a tuft of silky hairs and subtended by 2 lanceolate bracts 2–3 mm long. Hypanthium 2–3 mm, pubescent; sepals 1 mm, ovate, obtuse. Fruit pubescent. *S., W., C. & S.E. Europe, northwards to 53° N. in Poland and eastwards to 40° E. in S. Russia.* Al Au Be Bu Co Cz Ga Ge Gr He Hs Hu It Ju Lu Po Rm Rs (C, W, K, E) Sa Si.

Plants from Sicilia, Sardegna and Corse (and also from S.W. Asia), described originally as **Stellera pubescens** Guss., *Fl. Sic. Prodr.* **1**: 466 (1827), appear to differ in their pubescent leaves, whitish flowers in a denser spike, and later time of flowering. Their status is obscure, and they appear to have no valid name, whether as species or variety, under *Thymelaea*.

3. Diarthron Turcz.[1]

Flowers hermaphrodite, in ebracteate, terminal racemes. Hypanthium articulated near the middle, the lower half persistent in fruit, the upper, with the sepals, petaloid, deciduous. Style terminal. Fruit enclosed in the persistent base of the hypanthium; pericarp membranous.

1. D. vesiculosum (Fischer & C. A. Meyer) C. A. Meyer, *Bull. Phys.-Math. Acad. Pétersb.* **1**: 359 (1843). A slender annual, glabrous except for a few hairs on the young leaves and flowers. Stem 20–50 cm, erect, dichotomously branched. Leaves 8–15 × 2–5 mm, alternate, linear-oblong to lanceolate, subacute, shortly petiolate. Flowers greenish-yellow, on very short, clavate pedicels. Hypanthium 2–4 mm, distinctly ribbed in lower half; sepals ¼ as long, linear, obtuse. Fruit 2 × 1 mm, black, shining. *Dry places. W. Kazakhstan*: ?*S.E. Russia.* Rs (E).

---

[1] By D. A. Webb & I. K. Ferguson.

CVIII. ELAEAGNACEAE[1]

Trees or shrubs with peltate or stellate, scale-like hairs. Leaves entire. Flowers perigynous, apetalous. Hypanthium 2- or 4-lobed; lobes valvate; stamens as many as sepals and alternating with them, or twice as many; ovary superior, unilocular; ovule solitary, basal. Fruit drupe-like, the dry fruit being surrounded by the fleshy hypanthium.

Dioecious; calyx 2-lobed 1. **Hippophae**
Polygamous; calyx 4-lobed 2. **Elaeagnus**

1. Hippophae L.[2]

Deciduous. Dioecious. Flowers borne on the previous year's growth. Lower male flowers sessile; hypanthium shorter than the 2 lobes; stamens 4. All female flowers pedicellate; hypanthium longer than lobes.

1. **H. rhamnoides** L., *Sp. Pl.* 1023 (1753). Much-branched, spiny shrub or small tree up to 11 m, suckering freely. Twigs covered with silvery scales. Leaves 1–6 × 0·3–1 cm, linear-lanceolate, covered with silvery or ferruginous scales. Flowers c. 3 mm, appearing before the leaves. Fruit 6–8 mm, subglobose or ovoid, orange. $2n=24$. *On stable dunes and sea-cliffs, and on river-gravel and alluvium in mountain regions. Native throughout a considerable part of Europe, from c. 68° N. in Norway to N. Spain, C. Italy and Bulgaria, and from N.W. France to Finland*

and Moldavia, but local and absent from wide areas. Often planted for ornament, or to stabilize sand or gravel, and naturalized in many places. Au Be Br Bu Cz Da Fe Ga Ge He Ho Hs Hu It Ju No Po Rm Rs (B, W) Su [Hb Rs (N, C, E)].

2. Elaeagnus L.[2]

Flowers shortly pedicellate, all hermaphrodite, or hermaphrodite and male on the same plant, borne on the current year's growth. Hypanthium campanulate or tubular, 4-lobed. Stamens 4.

1. **E. angustifolia** L., *Sp. Pl.* 121 (1753). More or less spiny shrub or small tree up to 7 m. Twigs covered with silvery scales. Leaves 4–8 × 1–2·5 cm, oblong- or linear-lanceolate, green above, covered with silvery scales beneath. Flowers 8–10 mm, appearing with the leaves. Fruit 10–20 mm, ellipsoid, succulent, yellow, covered with silvery scales. *Planted for ornament and widely naturalized in S. Europe, northwards to Czechoslovakia and C. Russia.* [Al Au Bu Cr Cz Ga Gr Hs Hu It Rm Rs (B, C, W, K, E).] (*Temperate Asia.*)

E. commutata Bernh., ex. Rydb., *Fl. Rocky Mount.* 582 (1918) (*E. argentea* Pursh, non Moench) from North America, is often confused with **1**, but differs from it in having brown twigs and dry, mealy fruits. It is cultivated and perhaps locally naturalized.

GUTTIFERALES

CIX. GUTTIFERAE (CLUSIACEAE)[3]

Shrubs or herbs, with translucent glands containing essential oils and sometimes red or black glands containing hypericin. Leaves simple, opposite, or rarely in whorls of 3–4. Flowers actinomorphic. Sepals imbricate in bud. Petals free, contorted in bud. Stamens in fascicles or apparently indefinite. Ovary superior. Placentation axile or parietal. Seeds without endosperm.

1. Hypericum L.[4]

(Incl. *Elodes* Adanson and *Triadenia* Spach)

Flowers hermaphrodite. Sepals (4–)5. Petals (4–)5, yellow, sometimes tinged with red. Stamens in 3 or 5 fascicles of (1–)3 to c. 125, sometimes alternating with sterile fascicles (*fasciclodes*), or in 5 irregular groups; fascicles, if 5, antepetalous; if 3, one antepetalous and two (larger) antesepalous. Ovary (2–)3- to 5-locular or partly or completely 1-locular; ovules numerous. Styles (2–)3–5, free, slender. Fruit a septicidal capsule, rarely fleshy and more or less indehiscent.

Glands are designated *marginal* if they protrude sufficiently to interrupt the line of the margin of a leaf, sepal or petal, *intramarginal* if they abut on the margin but do not interrupt its line, *superficial* if they are quite clear of the margin.

In many species the ovary and capsule have glandular streaks or patches on the wall. These are referred to as *vittae* if flat or slightly swollen, and as *vesicles* if conspicuously swollen. Vittae on or near the midrib of a carpel are described as *dorsal*.

Many European species are cultivated in gardens, as also are several species and hybrids of large-flowered shrubs from Asia belonging to the section *Norysca* (Spach) Endl.

Literature: B. Stefanov, *God. Sof. Univ. (Agron.-Les. Fak.)* **10**: 19–58 (1932); **11**: 139–186 (1933); **12**: 69–100 (1934).

1 Plant without red or black glands on leaves, sepals, petals or anthers
  2 Leaves in whorls of 3–4
    3 Petals and stamens deciduous; leaves smooth   **10. amblycalyx**
    3 Petals and stamens persistent; leaves papillose   **12. ericoides**
  2 Leaves opposite
    4 Broad-leaved shrub; petals and stamens deciduous
      5 Stems and leaves covered with glandular vesicles   **6. balearicum**
      5 Stems and leaves smooth
        6 Styles 5; anthers reddish (Sect. *Eremanthe*)   **1. calycinum**
        6 Styles 3(–4): anthers yellow (Sect. *Androsaemum*)
          7 Petals shorter than sepals; ripe fruit black and succulent   **5. androsaemum**
          7 Petals longer than sepals; ripe fruit red or green, scarcely succulent
            8 Sepals shrivelling and falling before fruit ripens; foliage often goat-scented   **3. hircinum**
            8 Sepals persistent at least until fruit ripens; foliage not goat-scented
              9 Sepals acute or cuspidate; fruit subglobose   **2. foliosum**
              9 Sepals obtuse or subacute; fruit ellipsoid or narrowly ovoid   **4. inodorum**
    4 Herb or microphyllous shrub; stamens and usually petals persistent
      10 Microphyllous shrub; stamens in 3 fascicles (Sect. *Triadenia*)

[1] Edit. T. G. Tutin.   [2] By T. G. Tutin.
[3] Edit. D. A. Webb.   [4] By N. K. B. Robson.

11 Leaves linear-spathulate; inflorescence usually 3-flowered
    7. aciferum

11 Leaves elliptical to narrowly oblong; flowers solitary
    8. aegypticum

10 Herb; stamens 5, or in 5 irregular groups (Sect. *Brathys*)

12 Leaves linear-subulate, closely appressed; branches numerous, fastigiate   **61. gentianoides**

12 Leaves flat, ± patent; branches few or none, not fastigiate

13 Leaves suborbicular to ovate-triangular or broadly oblong

14 Stem simple or nearly so; middle and upper leaves ovate-triangular, ± acute   **57. gymnanthum**

14 Stem usually branched; middle and upper leaves oblong or ovate to suborbicular, obtuse   **58. mutilum**

13 Leaves linear to lanceolate, oblanceolate or narrowly oblong

15 At least the upper leaves lanceolate, with (3–)5–7 veins at the base; sepals 4–7 mm   **59. majus**

15 All leaves linear to linear-lanceolate or oblanceolate, with 1–3(–5) veins at the base; sepals 2–5 mm
    60. canadense

1 Red or black glands present, at least on leaves, sepals or anthers

16 Leaves in whorls of 3–4

17 Petals and stamens deciduous; leaves in whorls of 3
    9. empetrifolium

17 Petals and stamens persistent; leaves usually in whorls of 4

18 Leaves smooth; sepals with sessile marginal glands   **11. coris**

18 Leaves papillose; sepals eglandular or glandular-ciliate
    12. ericoides

16 Leaves opposite

19 Leaves without intramarginal black glands (apical black glands or intramarginal translucent glands sometimes present)

20 Sepals entire, without marginal or intramarginal glands

21 Sepals broadly imbricate; seeds reticulate-pitted

22 Glabrous   **21. olympicum**

22 Pubescent   **22. cerastoides**

21 Sepals not or only slightly imbricate; seeds papillose

23 Leaves 2–6 mm, ovate or ovate-elliptical, glaucous; flowers solitary   **19. taygeteum**

23 Leaves (5–)8–30 mm, linear to narrowly elliptical, not glaucous; inflorescence many-flowered

24 Stems 30–70 cm, scarcely rooting at the base; petals not red-tinged   **20. hyssopifolium**

24 Stems 5–20(–35) cm, rooting at the base; petals usually red-tinged   **15. linarioides**

20 Sepals ciliate, denticulate or fimbriate, with marginal or intramarginal glands

25 Dwarf shrub, with stems conspicuously 4-angled
    13. haplophylloides

25 Herb, with stems not or scarcely 4-angled

26 Lower leaves glandular-denticulate; capsule with orange vesicles   **33. vesiculosum**

26 Leaves entire; capsule with vittae, but without vesicles

27 Glands on sepals red; filaments united for ⅔ of their length   **32. elodes**

27 Glands on sepals black; filaments free almost to the base

28 Leaves pubescent   **14. hirsutum**

28 Leaves glabrous

29 Inflorescence subcorymbose; petioles articulated at base; stems diffuse

30 Leaves glaucous only beneath; petals 10–16 mm
    17. nummularium

30 Leaves glaucous on both sides; petals 6–9 mm
    18. fragile

29 Inflorescence cylindrical or narrowly pyramidal; petioles, if present, not articulated; stems erect or ascending

31 Anthers pink or orange; leaves ovate-cordate to oblong   **16. pulchrum**

31 Anthers yellow; leaves usually linear to linear-lanceolate

32 Stems 30–70 cm, scarcely rooting at the base; petals not red-tinged   **20. hyssopifolium**

32 Stems 5–17(–32) cm, rooting at the base; petals usually red-tinged   **15. linarioides**

19 At least some of the leaves with intramarginal black glands (sometimes very few in **29** and **30**)

33 Sepals without marginal or intramarginal black glands

34 Sepals broadly imbricate; flowerless axillary shoots usually absent

35 Flowers less than 2 cm in diameter; vittae on capsule conspicuous   **50. humifusum**

35 Flowers at least 2 cm in diameter; vittae on capsule faint or absent

36 Glabrous   **21. olympicum**

36 Pubescent   **22. cerastoides**

34 Sepals not, or only slightly imbricate; flowerless axillary shoots present (Sect. *Hypericum*)

37 Stems with 2 raised lines

38 Leaves neither undulate nor amplexicaul; branches of inflorescence ascending   **54. perforatum**

38 Leaves undulate, amplexicaul; branches of inflorescence patent   **55. triquetrifolium**

37 Stems with 4 raised lines or wings

39 Sepals obtuse; stems not winged   **53. maculatum**

39 Sepals acute; stems winged

40 Petals not more than 7·5 mm, pure yellow or rarely red-veined; leaves usually flat   **51. tetrapterum**

40 Petals at least 7·5 mm, usually tinged with red; leaves undulate   **52. undulatum**

33 Sepals with marginal or intramarginal black glands

41 Leaves hairy or papillose on both sides

42 Capsule with vesicles as well as vittae; stems glabrous or finely papillose

43 Stems 1–5(–10) cm, procumbent, glabrous   **44. kelleri**

43 Stems 15–45 cm, erect or decumbent, usually minutely papillose   **45. aviculariifolium**

42 Capsule without vesicles; stem hairy

44 Leaves connate at base   **31. caprifolium**

44 Leaves free

45 Bracts auriculate; sepals with long teeth or cilia

46 Plant puberulent to velutinous; stems ± erect, not rooting at the nodes   **24. annulatum**

46 Plant hirsute; stems decumbent or procumbent, rooting at the nodes

47 Leaves 12–45 mm, sessile   **25. delphicum**

47 Leaves 8–15 mm, shortly petiolate   **26. athoum**

45 Bracts not auriculate; sepals subentire or with short teeth or cilia

48 Sepals glabrous, with superficial black glands

49 Petals at least 8 mm; leaves at least 15 mm, sessile
    27. atomarium

49 Petals not more than 8 mm; leaves not more than 15 mm, usually petiolate   **28. cuisinii**

48 Sepals hairy, without superficial black glands

50 Sepals 5–10 mm, aristate, entire; petals 9–15 mm
    30. pubescens

50 Sepals 3–6 mm, rarely aristate, shortly glandular-ciliate; petals 6–11 mm   **29. tomentosum**

41 Leaves smooth and glabrous, at least above

51 Capsule with longitudinal vittae only; seeds reticulate-pitted

52 Styles 5   **46. thasium**

52 Styles (2–)3(–4)

53 Flowerless axillary shoots present; stems with black glands

54 Sepals entire, with intramarginal black glands; leaves undulate   **52. undulatum**

54 Sepals glandular-denticulate or with sessile marginal glands; leaves flat   **56. elegans**

53 Flowerless axillary shoots absent; stems usually without black glands

55 Sepals equal, regularly dentate or ciliate

56 Stems erect; leaves 20–70 mm   **23. montanum**

56 Stems diffuse or decumbent; leaves 2–15 mm **28. cuisinii**
55 Sepals often unequal, subentire or irregularly ciliate or fimbriate (Sect. *Oligostema*)
  57 Sepals fimbriate or denticulate **47. aucheri**
  57 Sepals entire or ciliate
    58 Sepals unequal, with few or no superficial black glands; petals 4–6(–8) mm **50. humifusum**
    58 Sepals ± equal, with numerous superficial black glands; petals (5–)7–15 mm
      59 Leaves on flowering stems linear to linear-lanceolate; sepals glandular-ciliate **48. linarifolium**
      59 Leaves on flowering stems ovate to lanceolate or oblong; marginal glands of sepals usually sessile **49. australe**
51 Capsule with some vesicles or oblique vittae, or almost smooth; seeds usually longitudinally ribbed (Sect. *Drosocarpium*)
  60 Leaves with dense, conspicuous, reticulate venation
    61 Capsule with black vesicles **38. richeri**
    61 Capsule without black vesicles
      62 Capsule with interrupted vittae; leaves with numerous superficial glands **36. umbellatum**
      62 Capsule with orange vesicles; leaves with few or no superficial glands **37. bithynicum**
  60 Leaves with lax or indistinct reticulate venation
    63 Black glands on petals absent, or confined to margin and apex
      64 Glands on capsule prominent
        65 Capsule narrowly pyramidal; sepals patent or deflexed in fruit **35. montbretii**
        65 Capsule ovoid; sepals erect in fruit
          66 Capsule with dorsal vittae and lateral vesicles; leaves with numerous translucent glands **34. perfoliatum**
          66 Capsule with vesicles only; leaves usually without translucent glands **40. rochelii**
      64 Glands on capsule not prominent
        67 Sepals eglandular-fimbriate **41. barbatum**
        67 Sepals entire or glandular-denticulate to -fimbriate
          68 Leaves 5–14 mm, smooth; seeds longitudinally grooved **43. trichocaulon**
          68 Leaves 2–6 mm, papillose or undulate; seeds reticulate-pitted **44. kelleri**
    63 Black glands dispersed over whole surface of petals
      69 Capsule with prominent vesicles **39. spruneri**
      69 Capsule smooth, or with slightly prominent glands
        70 Sepals eglandular-fimbriate **41. barbatum**
        70 Sepals entire or glandular-denticulate to -fimbriate
          71 Petals 13–20 mm; sepals glandular-ciliate to -fimbriate **42. rumeliacum**
          71 Petals 6–12 mm; sepals entire to glandular-denticulate
            72 Leaves 5–14 mm, smooth; seeds longitudinally grooved **43. trichocaulon**
            72 Leaves 2–6 mm, papillose or undulate; seeds reticulate-pitted **44. kelleri**

Sect. EREMANTHE (Spach) Endl. Glabrous shrubs, without black glands. Stems 4-lined. Leaves opposite. Flowers terminal, solitary or rarely 2–3. Sepals entire. Petals and stamens deciduous. Stamen-fascicles and styles 5. Fruit dry. Seeds reticulate, not or slightly carinate.

1. H. calycinum L., *Mantissa* 106 (1767). Stems 20–60 cm, erect from creeping rhizomes, usually unbranched. Leaves 4·5–8·5 cm, oblong to elliptical or narrowly ovate, subsessile. Sepals 1–2 cm, markedly unequal, elliptical to suborbicular, persistent. Petals 2·5–4 cm, markedly asymmetrical, sometimes shallowly lobed. Anthers reddish. Capsule *c.* 20 mm, ovoid, deflexed. *Shady places. Turkey and S.E. Bulgaria; cultivated elsewhere and locally naturalized.* Bu Tu [Br Ga Hb He It Lu Rs (W, K)]. (*N. Anatolia.*)

Sect. ANDROSAEMUM (Duh.) Godron. Glabrous shrubs, without black glands. Stems 2-lined or 4-angled. Leaves opposite. Flowers in few-flowered, terminal cymes. Sepals entire. Petals and stamens deciduous. Stamen-fascicles 5. Styles 3(–4). Fruit more or less fleshy at first, tardily dehiscent. Seeds reticulate, winged.

2. H. foliosum Aiton, *Hort. Kew.* **3**: 104 (1789). Stems 50–100 cm or longer, erect or spreading, 4-angled. Leaves 3·5–6 cm, narrowly ovate to lanceolate, sessile. Sepals 5–6 mm, markedly unequal, ovate to linear-lanceolate, persistent until fruit ripens. Petals (10–)12–18 mm, narrowly elliptical. Stamens equalling or slightly exceeding petals. Styles 1·5–2·5 times as long as ovary. Fruit 8–13 mm, broadly ovoid to subglobose, thin-walled, soon becoming dry. *Damp, shady places in mountains.* ● *Açores.* Az.

3. H. hircinum L., *Sp. Pl.* 784 (1753) (*Androsaemum hircinum* (L.) Spach). Stems 30–100(–150) cm, erect, 2-lined or 4-angled. Leaves 2–6·5(–7·5) cm, narrowly lanceolate to broadly ovate, sessile or subsessile, often with goat-like smell when crushed. Sepals (2–)3–6(–7) mm, somewhat unequal, lanceolate to ovate-lanceolate, deciduous. Petals 11–18 mm, oblanceolate to narrowly obovate. Stamens exceeding petals. Styles 3–5 times as long as ovary. Fruit 8–13 mm, ellipsoid to subcylindrical, subcoriaceous. *Damp places, often beside rivers. Mediterranean region; naturalized from gardens in W. Europe.* Bl Co Cr Gr It Sa Si [Br Ga Hb He Hs Lu].

4. H. inodorum Miller, *Gard. Dict.* ed. 8, no. 6 (1768) (*H. elatum* Aiton). Intermediate between **3** and **5**. Stems 100–200 cm, erect, 2-lined. Leaves 3·5–9 cm, broadly ovate to oblong-lanceolate, sessile, without goat-like smell when crushed. Sepals 5–8 mm, markedly unequal, narrowly to broadly ovate, persistent at least till fruit ripens. Petals 7–15 mm, obovate. Stamens exceeding petals. Styles 2–2·5 times as long as ovary. Fruit 8–13 mm, ellipsoid to subcylindrical, thin-walled, reddish and succulent at first. *Naturalized from gardens in Britain and France, but perhaps also native in S. France.* [Br *Ga ?It.] (*Madeira.*)

The plants from natural habitats in France may be hybrids between **3** and **5**, but similar and apparently native plants occur in Madeira in the absence of either of these species.

5. H. androsaemum L., *Sp. Pl.* 784 (1753) (*Androsaemum officinale* All.). Stems 30–70 cm, spreading, 2-lined. Leaves (2·5–)4–15(–30) cm, broadly ovate to ovate-oblong, sessile, sometimes amplexicaul, without goat-like smell when crushed. Sepals 8–12(–15) mm, markedly unequal, oblong-ovate to broadly ovate, deflexed and enlarging in fruit, persistent. Petals 6–10(–12) mm, obovate. Stamens shorter than petals, or equalling or slightly exceeding them. Styles shorter than ovary. Fruit 7–10(–12) mm, broadly cylindric-ellipsoid to globose, persistently fleshy, reddish, becoming black, deciduous. *Damp or shady places. W. Europe, and locally in S. Europe eastwards to Turkey.* Be Br Bu Co Ga Hb He Hs It Ju Lu Sa ?Si Tu [Au].

Sect. PSOROPHYTUM (Spach) Endl. Glabrous shrubs, without black glands. Stem and leaves covered with prominent, resinous vesicles. Leaves opposite. Flowers terminal, solitary. Sepals entire. Petals and stamens deciduous. Stamen-fascicles 5. Styles (4–)5. Fruit a capsule. Seeds reticulate, neither winged nor carinate.

6. H. balearicum L., *Sp. Pl.* 783 (1753). Stems 15–120 cm, with ascending branches, 4-angled when young. Leaves (6–)8–10 mm, ovate to narrowly oblong, undulate, coriaceous. Bracteoles appressed to calyx. Flowers 1·5–4 cm in diameter. Sepals suborbicular, patent in fruit. Capsule ovoid-pyramidal. *Dry woods and rocky places.* ● *Islas Baleares.* Bl [It].

Sect. TRIADENIA (Spach) R. Keller. Dwarf, glabrous, microphyllous shrubs, without black glands. Leaves opposite. Flowers terminating long or short shoots, heterostylous. Sepals entire. Petals with entire, ligulate, nectariferous appendage. Stamen-fascicles 3, with filaments united for more than ½ their length, alternating with 3 fleshy fasciclodes. Fruit dry. Seeds slightly carinate, with fleshy caruncle.

7. H. aciferum (W. Greuter) N. K. B. Robson, *Feddes Repert.* **74**: 23 (1967) (*Elodes acifera* W. Greuter). Low, procumbent shrub. Leaves 5–12 mm, narrowly linear-spathulate, coriaceous, somewhat glaucous, not imbricate. Flowers (1–)3, pedicellate. Sepals suberect, elliptical. Petals *c.* 9 mm, deciduous. Stamens persistent. Ovary with 2 ovules in each loculus. *Calcareous rocks near the sea.* ● *S.W. Kriti.* Cr.

8. H. aegypticum L., *Sp. Pl.* 784 (1753) (*Triadenia maritima* (Sieber) Boiss.). Low, spreading shrub. Leaves 3–10 mm, elliptical to narrowly oblong, coriaceous, glaucous, often crowded and imbricate. Flowers solitary, sessile or subsessile. Sepals erect, oblong. Petals 8–14 mm, persistent. Stamens persistent. Ovary with numerous ovules in each loculus. *Cliffs and rocks near the sea. Islands of C. & E. Mediterranean from Sardegna to Kriti.* Cr Gr Sa Si. (*N. Africa.*)

Sect. CORIDIUM Spach. Shrubs or perennial herbs, glabrous or with papillose leaves. Black glands confined to sepals, or rarely absent. Leaves whorled, usually linear; margins revolute. Flowers in cymes, rarely solitary. Sepals usually with glandular margin. Stamen-fascicles 3. Styles 3. Capsule with longitudinal vittae or oblique vesicles. Seeds papillose or rugulose.

9. H. empetrifolium Willd., *Sp. Pl.* 3: 1452 (1802). Shrub. Stems up to 50 cm, erect and caespitose with strict branching, or procumbent, straggling and rooting. Leaves 2–12 mm, in whorls of 3, glabrous. Flowers in elongated panicles or simple cymes, or solitary. Sepals with sessile, marginal black glands. Petals and stamens deciduous. Capsule with oblique vesicles. *Rocky places. Greece and Aegean region; one station in N. Albania.* Al Cr Gr.

10. H. amblycalyx Coust. & Gand., *Bull. Soc. Bot. Fr.* **63**: 14 (1916). Like **9** but with leaves in whorls of 4 and sepals with eglandular margin. *Rocky places. E. Kriti.* Cr.

11. H. coris L., *Sp. Pl.* 787 (1753). Low shrub, or perennial herb with woody base. Stems 10–45 cm, erect or ascending from creeping and rooting base. Leaves 4–18 mm, in whorls of 4 (rarely 3), glabrous. Flowers in elongated or pyramidal panicles, rarely solitary. Sepals glandular-denticulate, or with sessile, marginal black glands. Petals and stamens persistent. Capsule with longitudinal and oblique, swollen vittae. *Sunny, calcareous rocks.* ● *N. & C. Italy, Switzerland, S.E. France.* Ga He It.

12. H. ericoides L., *Sp. Pl.* 785 (1753). Dwarf shrub. Stems 2–12(–25) cm, erect, with strict branching and short internodes. Leaves 1·5–3·5 mm, in whorls of 4, densely papillose. Flowers in corymbs or panicles. Sepals shortly glandular-ciliate or entire. Petals and stamens persistent. Capsule with narrow, longitudinal vittae. *Sunny, calcareous rocks. E. & S.E. Spain.* Hs.

Sect. HAPLOPHYLLOIDES Stefanov. Glabrous dwarf shrubs. Black glands confined to leaf-apex and sepal- and petal-margins. Stem 4-angled. Leaves opposite, subcoriaceous. Flowers few, in cymes. Petals and stamens persistent. Stamen-fascicles and styles 3. Capsule with longitudinal vittae. Seeds papillose.

13. H. haplophylloides Halácsy & Bald., *Verh. Zool.-Bot. Ges. Wien* **42**: 576 (1893). Stems up to 25 cm, straggling, branching and rooting at the base. Leaves 12–28 mm, oblong-linear to elliptic-linear, with apical black gland. Cymes 3- to 5-flowered. Sepals black-glandular-ciliate or -denticulate. Petals with small, sessile, marginal black glands. *Mountain woods and rocks.* ● *S. Albania.* Al.

Sect. TAENIOCARPIUM Jaub. & Spach. Perennial herbs, sometimes woody at the base. Black glands confined to sepal- and petal-margins and sometimes leaf-apex. Stems terete or 2-lined, or very rarely 4-angled. Flowers in cymes or panicles. Petals and stamens persistent. Petals not clawed; pellucid glands elongated (rarely absent). Stamen-fascicles and styles 3. Capsule with longitudinal vittae. Seeds rugulose to papillose.

14. H. hirsutum L., *Sp. Pl.* 786 (1753). Stems 35–110 cm, erect from creeping and rooting base, pubescent. Leaves (20–)25–55(–60) mm, oblong to elliptical or lanceolate, without apical gland, strigose-pubescent. Inflorescence cylindrical. Sepals black-glandular-denticulate or -ciliate. Petals sometimes red-veined. Anthers yellow. $2n = 18$. *Woods, riverbanks and roadsides. Most of Europe except the north-east and the extreme south.* Al Au Be Br Bu Cz Da Fe Ga Ge Gr Hb He Ho Hs Hu It Ju No Po Rm Rs (B, C, W, K, E) Su.

H. confertum Choisy, *Prodr. Monogr. Hypér.* 55 (1821), from S.W. Asia, has once been recorded from Turkey-in-Europe. It is like **14**, but with much smaller leaves and fimbriate sepals.

15. H. linarioides Bosse, *Allgem. Gartenz.* **3**: 99 (1835) (*H. alpestre* Steven, *H. repens* auct., non L.). Stems (5–)8–17(–30) cm, erect from creeping and rooting base, glabrous. Leaves on main stem (5–)8–14 mm, linear to narrowly elliptical, without apical gland, glabrous, usually with deflexed margin; leaves on axillary shoots smaller and crowded. Inflorescence subspicate, or with lowest branches somewhat elongated. Sepals with or without marginal black glands. Petals often red-tinged or red-veined. Anthers yellow. *Mountain rocks and grassland. C. part of Balkan peninsula; Krym.* Bu Gr Ju Rs (K).

16. H. pulchrum L., *Sp. Pl.* 786 (1753). Stems (3–)10–90 cm, erect or ascending, glabrous. Leaves on main stem 6–20 mm, broadly ovate to oblong, cordate-amplexicaul, without apical gland, glabrous, with scarcely deflexed margin; leaves on axillary shoots smaller, oblong, petiolate. Inflorescence narrowly pyramidal or narrowly cylindrical. Sepals with marginal black glands sessile or shortly stalked. Petals red-tinged. Anthers orange to reddish-pink. $2n = 18$. *Woods and heaths, usually on acid soil.* ● *N.W. Europe, extending locally to S.W. Poland, Switzerland and C. Portugal, and possibly to N.W. Jugoslavia and S.E. Italy.* ?Au Be Br Cz Da Fa Ga Ge Hb He Ho Hs It ?Ju Lu No Po Su.

Dwarf, few-flowered plants (f. *procumbens* Rostrup) occur in exposed places in extreme N.W. Europe. This variant is said to breed true, but a complete series of intermediates links it with the typical plant.

17. H. nummularium L., *Sp. Pl.* 787 (1753). Stems 8–30 cm, erect to diffuse, creeping and rooting at the base, glabrous.

Leaves 5–18 mm, broadly ovate to orbicular, green above, glaucous beneath, glabrous, with conspicuous, intramarginal pale glands and 2 apical black glands. Inflorescence 1- to 8-flowered, subcorymbose. Sepals black-glandular-denticulate. Petals sometimes red-veined. Anthers yellow. *Rock-crevices and stony slopes; calcicole.* ● *Pyrenees and N. Spain; S.W. Alps.* Ga Hs ?It.

18. **H. fragile** Heldr. & Sart. ex Boiss., *Diagn. Pl. Or. Nov.* 3(1): 108 (1853). Stems 4–11(–16) cm, slender, suberect or straggling, articulated at the nodes, much branched at the base but not rooting, glaucous. Leaves 2–7 mm, ovate or oblong-orbicular, with conspicuous, intramarginal pale glands, usually without apical black glands, glaucous on both sides. Inflorescence 1- to 8-flowered, subcorymbose. Sepals black-glandular-ciliate. Petals red-tinged. Anthers yellow. *Crevices in calcareous rocks.* ● *E. Greece.* Gr.

Records from Karpathos and Kasos are probably errors for 28. Plants cultivated as *H. fragile* are usually referable to 21.

19. **H. taygeteum** Quézel & Contandr., *Taxon* 16: 240 (1967). Like 18 but stems often procumbent; leaves ovate to elliptical, with superficial as well as intramarginal pale glands; flowers all solitary; sepals entire. *Crevices in calcareous rocks.* ● *S. Greece (Taïyetos).* Gr.

Sect. DROSANTHE (Spach) Endl. Like Sect. *Taeniocarpium* but petals more or less clawed, and with scarcely elongated pellucid glands.

20. **H. hyssopifolium** Chaix in Vill., *Hist. Pl. Dauph.* 1: 329 (1786). Stems 30–70 cm, erect, glabrous. Leaves on main stem 12–28 mm, linear to narrowly elliptical, without apical gland, glabrous, usually with deflexed margin; leaves on axillary shoots smaller, crowded. Inflorescence narrowly cylindrical to subspicate, or rarely narrowly pyramidal. Petals not red-tinged. Anthers yellow. *Mountain regions of S. Europe.* Bu Ga Gr Hs It Ju Rs (K).

(a) Subsp. **hyssopifolium**: Leaves obtuse, not apiculate. Sepals subequal, usually continuously glandular-denticulate. Petals 7–9(–10) mm. Capsule (5–)6–8·5 mm. *Throughout the range of the species, except Greece.*

Plants from Krym, which have been named **H. chrysothyrsum** Woronow in Kusn., N. Busch & Fomin, *Fl. Cauc. Crit.* 3(9): 30 (1906), are very similar to 20(a), but may be better placed in **H. lydium** Boiss., *Diagn. Pl. Or. Nov.* 1(1): 57 (1842).

(b) Subsp. **elongatum** (Ledeb.) Woronow in Kusn., N. Busch & Fomin, *Fl. Cauc. Crit.* 3(9): 32 (1906) (*H. elongatum* Ledeb., *H. hyssopifolium* subsp. *tymphresteum* (Boiss. & Spruner) Hayek): Leaves usually apiculate. Sepals unequal, with margin partly or completely eglandular. Petals 12–15 mm. Capsule 8–13 mm. *C. Greece (Timfristos); Krym.*

Plants from S. Spain (*H. callithyrsum* Cosson) tend to be intermediate between the two subspecies in respect of their sepals.

Sect. OLYMPIA (Spach) Endl. Glabrous, perennial herbs, often woody at the base. Black glands present on anthers, and sometimes elsewhere. Stems 2-lined. Leaves usually with intramarginal black glands. Flowers large, solitary or in few-flowered cymes. Petals and stamens persistent. Stamen-fascicles and styles 3. Capsule smooth or with faint, longitudinal vittae. Seeds reticulate-pitted.

21. **H. olympicum** L., *Sp. Pl.* 784 (1753) (*H. dimoniei* Velen., *H. polyphyllum* sensu Hayek, non Boiss. & Balansa). Stems (8–)10–50(–75) cm, erect to decumbent. Leaves 5–28(–36) mm, narrowly oblong to narrowly elliptical or lanceolate, glaucous. Flowers 20–60 mm in diameter. Sepals unequal, often foliaceous, imbricate, entire, sometimes with superficial black glands. Petals eglandular or with apical black gland, or rarely with a few marginal black glands. *Dry, stony places. Balkan peninsula, mainly in the south and east.* Bu Gr Ju Tu.

Sect. CAMPYLOPUS (Spach) Endl. Perennial herbs, sometimes woody at the base, with pubescent stems, leaves and sepals. Stems 2-lined. Leaves with a few intramarginal or superficial black glands, or with none. Flowers solitary or in small corymbose cymes. Petals and stamens persistent. Stamen-fascicles 3–5, sometimes united at the base. Styles 3(–5). Capsule with faint, longitudinal vittae. Seeds reticulate-pitted.

22. **H. cerastoides** (Spach) N. K. B. Robson, *Feddes Repert.* 74: 22 (1967) (*H. rhodoppeum* Friv., *Campylopus cerastoides* Spach). Stems 7–25 cm, decumbent or ascending, rooting at the base. Leaves 8–30 mm, oblong to elliptical or ovate. Flowers 20–43 mm in diameter. Sepals unequal, foliaceous, broadly imbricate, entire. Petals with intramarginal or sessile, marginal black glands. *Stony places; calcifuge. S.E. part of Balkan peninsula.* Bu Gr Tu.

Sect. ADENOSEPALUM Spach. Perennial herbs, usually pubescent. Black glands present on leaves, sepals, anthers and sometimes petals, usually marginal or intramarginal. Stems usually terete. Leaves with intramarginal and occasionally a few superficial black glands. Flowers in pyramidal to cylindrical or corymbose cymes, rarely solitary. Petals and stamens persistent. Stamen-fascicles and styles 3. Ovary 3-locular. Capsule with longitudinal vittae. Seeds reticulate-pitted or with subscalariform striations.

23. **H. montanum** L., *Fl. Suec.* ed. 2, 266 (1755). Stems 20–80 cm, erect, glabrous. Leaves (20–)25–70 mm, ovate to lanceolate or oblong-elliptical, sessile, glabrous above, usually scabrid beneath. Inflorescence corymbose to shortly cylindrical, usually dense, glabrous. Bracts with glandular-ciliate auricles. Sepals black-glandular-ciliate. Petals without black glands. 2n=16. *Woods and thickets; somewhat calcicole. W. & C. Europe, extending to S.W. Finland, C. Ukraine, C. Jugoslavia and C. Italy.* ?Al Au Be Br Co Cz Da Fe Ga Ge ?Gr He Ho Hs Hu It Ju Lu No Po Rm Rs (B, C, W) Sa Su.

24. **H. annulatum** Moris, *Stirp. Sard.* 1: 9 (1827) (*H. atomarium* subsp. *degenii* (Bornm.) Hayek). Stems 20–65 cm, erect, shortly whitish-pubescent. Leaves 15–55 mm, ovate, sessile, shortly pubescent on both sides. Inflorescence pyramidal to shortly cylindrical or corymbose, rather lax, glabrous. Bracts with densely glandular-ciliate auricles. Sepals black-glandular-ciliate. Petals sometimes with 1–2 superficial black glands. *Scrub and stony places on mountains. Balkan peninsula, from C. Jugoslavia to N. Greece; one locality in Sardegna.* Al Bu Gr Ju Sa.

25. **H. delphicum** Boiss. & Heldr. in Boiss., *Diagn. Pl. Or. Nov.* 3(1): 106 (1853). Stems 11–35(–45) cm, ascending, branching and rooting at the base, strigose-pubescent. Leaves 12–35 mm, ovate to oblong-ovate, sessile, strigose-pubescent. Inflorescence corymbose to shortly cylindrical, glabrous. Bracts with densely glandular auricles. Sepals black-glandular-ciliate. Petals with marginal black glands towards apex. *Woodland and stony places.* ● *Aegean region (Evvoia, Andros).* Gr.

26. **H. athoum** Boiss. & Orph. in Boiss., *Fl. Or.* 1: 794 (1867). Like 25 but smaller in all its parts, with indumentum softer and less dense; stems more slender and diffuse; leaves 8–15 mm,

shortly petiolate; inflorescence with fewer (1–7) flowers. *Stony places in shade.* ● *N. Aegean region (Pangaion to Athos and Samothraki).* Gr.

27. H. atomarium Boiss., *Diagn. Pl. Or. Nov.* **2**(8): 114 (1849). Stems 15–75 cm, erect or ascending, not rooting, shortly pubescent. Leaves 15–45(–55) mm, ovate to oblong or elliptical, sessile, shortly pubescent. Inflorescence cylindrical, rarely corymbose, relatively lax, glabrous. Bracts sometimes with long glandular cilia at the base but not auriculate. Sepals black-glandular-denticulate, with superficial black dots. Petals (8–)9–12 mm, sometimes with a few superficial black glands. *Damp, shady places. S. Greece (Peloponnisos).* Gr. (*E. Aegean region, W. Anatolia.*)

28. H. cuisinii W. Barbey, *Bull. Soc. Vaud. Sci. Nat.* **21**: 220 (1886). Stems 4–15(–28) cm, diffuse or decumbent, rooting at the base, shortly pubescent or glabrous. Leaves 2–15 mm, elliptical to oblong or obovate, usually petiolate, shortly pubescent to glabrous. Inflorescence subcorymbose, lax, glabrous. Bracts not auriculate. Sepals black-glandular-denticulate and with superficial black dots or streaks. Petals 5–7(–8) mm, with superficial black glands. *Fissures of calcareous rocks. Karpathos, Kasos.* Cr.

29. H. tomentosum L., *Sp. Pl.* 786 (1753). Stems 10–90 cm, decumbent, rooting at the base, tomentose. Leaves 5–22 mm, oblong to ovate, hirsute to tomentose or crispate-pubescent. Inflorescence corymbose to cylindrical, becoming elongate-monochasial, tomentose or crispate-pubescent. Bracts not auriculate. Sepals 3–6 mm, ovate or elliptical to broadly lanceolate, acute to acuminate or rarely aristate, hirsute or tomentose, shortly glandular-ciliate and usually with an apical gland. Petals 6–11 mm, with marginal black glands. *Damp places. S.W. Europe.* Bl ?Co Ga Hs It Lu Sa.

30. H. pubescens Boiss., *Elenchus* 26 (1838). Like **29** but with sepals 5–10 mm, lanceolate, aristate, entire, with sessile, marginal black glands and without apical gland; petals 9–15 mm. *S. Portugal, S. Spain; Sicilia, Malta.* Hs Lu ?Sa Si.

Plants intermediate between **29** and **30** are found in S. & E. Spain. Their status is uncertain.

31. H. caprifolium Boiss., *Elenchus* 26 (1838). Stems 20–100 cm, erect or ascending from a creeping, rooting and branching base, crispate-pubescent. Leaves (15–)20–50 mm, ovate to oblong, completely perfoliate, crispate-pubescent. Inflorescence corymbose to broadly pyramidal, usually dense, glabrous. Bracts sometimes auriculate. Sepals lanceolate, aristate, black-glandular-ciliate and with superficial black glands. Petals with marginal black glands. *Damp, shady places.* ● *S.E. Spain.* Hs.

Sect. ELODES (Adanson) Koch. Perennial herbs, tomentose to crispate-pubescent, rarely glabrous. Red glands present on sepals and bracts. Flowers in pseudaxillary cymes. Petals and stamens persistent. Petals with 3-fid, ligulate, nectariferous appendage. Stamen-fascicles 3, with filaments united for more than ½ their length, alternating with 3 scale-like fasciclodes. Styles 3. Ovary with 3 parietal placentae. Capsule with longitudinal vittae. Seeds longitudinally ribbed, with transverse striation.

32. H. elodes L., *Amoen. Acad.* **4**: 105 (1759) (*H. helodes* auct., *H. palustre* Salisb., *Elodes palustris* Spach). Stem terete, erect from a creeping and rooting base, often with swollen internodes. Leaves 5–30 mm, orbicular to broadly ovate or broadly elliptical. Sepals erect, shortly red-glandular-ciliate. Corolla pseudo-tubular; petals eglandular. $2n = 32$. *Damp mud or shallow water.* ● *W. Europe, northwards to 57° 30′ in Scotland, and extending locally eastwards to E. Germany and C. Italy.* ?Au Az Be Br Ga Ge Hb Ho Hs It Lu.

Sect. DROSOCARPIUM Spach. Perennial herbs, usually glabrous and somewhat glaucous. Black glands present on leaves, sepals, petals, anthers, and sometimes ovary. Leaves usually with intramarginal and often superficial black glands. Flowers in subcorymbose or broadly pyramidal panicles, rarely solitary. Petals and stamens persistent. Stamen-fascicles 3(–5). Styles 3(–5). Capsule with interrupted lateral vittae or vesicles, sometimes with dorsal vittae. Seeds usually as in Sect. *Elodes.*

33. H. vesiculosum Griseb., *Spicil. Fl. Rumel.* **1**: 226 (1843). Stems 30–70 cm, erect. Leaves 12–25 mm, ovate, amplexicaul, the lower with margin and auricles orange-glandular-denticulate, all without intramarginal but occasionally with superficial black glands. Sepals obtuse, shortly glandular- or eglandular-ciliate or -denticulate, and with superficial black streaks. Petals with black glandular streaks over the whole surface. Capsule with numerous elongated, orange vesicles. *Mountain woods.* ● *Greece.* Gr.

34. H. perfoliatum L., *Syst. Nat.* ed. 12, **2**: 510 (1767) (*H. ciliatum* Lam.). Stems (15–)25–75 cm, erect or decumbent at the base. Leaves 13–60 mm, ovate to triangular-lanceolate or linear-lanceolate, usually amplexicaul. Sepals obtuse or subacute, densely and irregularly black-glandular-denticulate or -ciliate and with numerous superficial black streaks and dots. Petals sometimes with superficial black dots or streaks towards the apex. Capsule with dorsal vittae and lateral orange vesicles. *Damp meadows or shady places among rocks. Mediterranean region, C. & S. Portugal.* Bl ?Bu Co Cr Ga Gr Hs It Ju Lu Sa Si ?Tu.

Many records of this species from S.E. Europe are errors for **35** or **39**.

35. H. montbretii Spach, *Hist. Vég.* (*Phan.*) **5**: 395 (1836) (*H. cassium* Boiss.). Stems 15–60 cm, erect. Leaves 15–50 mm, ovate to oblong, usually amplexicaul, often black-glandular-ciliate. Sepals acute or acuminate, rather sparsely black-glandular-ciliate and with few or no superficial black dots. Petals sometimes with subapical, superficial black dots. Capsule without dorsal vittae, but with numerous round, orange vesicles. *Damp or shady, stony places. S.E. part of Balkan peninsula.* Bu Gr Ju Tu.

H. setiferum Stefanov, *Bull. Soc. Bot. Bulg.* **3**: 83 (1929), reported from two localities in Bulgaria, differs only in having short, stiff, white hairs on the lower surface of the leaves, and some of the vesicles on the capsule somewhat elongated.

36. H. umbellatum A. Kerner, *Österr. Bot. Zeitschr.* **13**: 141 (1863). Stems 25–50 cm, erect. Leaves 15–40 mm, ovate to ovate-triangular, sessile, entire, with conspicuous reticulate venation and numerous superficial translucent or black glands. Sepals acute, black-glandular-ciliate and with numerous superficial black dots and streaks. Petals with black streaks or dots over the whole surface. Capsule with a few dorsal vittae and interrupted lateral vittae. *Mountain woods.* ● *N. Jugoslavia, C. Romania, Bulgaria; local.* Bu Ju Rm.

37. H. bithynicum Boiss., *Diagn. Pl. Or. Nov.* **2**(8): 112 (1849) (*H. confusum* Vandas). Stems 15–55 cm, erect or decumbent and rooting at the base. Leaves 18–53 mm, ovate, amplexicaul, entire, with conspicuous reticulate venation, sometimes with superficial black glands, usually without translucent glands. Sepals acute, black-glandular-fimbriate and with numerous superficial black dots. Petals with black dots over the whole surface. Capsule without dorsal vittae, but with several round, orange vesicles. *Woods and thickets. Around Instanbul.* Tu. (*N. Anatolia, Georgia.*)

38. H. richeri Vill., *Prosp. Pl. Dauph.* 44 (1779). Stems 10–50 cm, erect, from a creeping and rooting base. Leaves 10–55 mm, ovate to triangular-ovate or elliptical, sessile, sometimes amplexicaul, entire, with conspicuous reticulate venation, without translucent glands or superficial black glands. Sepals acute or acuminate, subentire or variously ciliate or fimbriate, with numerous superficial black streaks and dots. Petals with numerous black dots over the whole surface. Capsule without dorsal vittae, but with numerous round or elongated, black, and sometimes also orange vesicles. *Meadows and woods.* ● *Mountains of S. & S.C. Europe.* Al Bu Ga Gr He Hs It Ju Rm Rs (W, ?K).

1 Sepals acute, denticulate to ciliate; petals (10–)15–25 mm; leaves usually obtuse, amplexicaul **(b) subsp. burseri**
1 Sepals acuminate, subentire to fimbriate; petals 10–17 mm; leaves rarely amplexicaul
2 Sepals fimbriate; leaves usually subacute **(a) subsp. richeri**
2 Sepals ciliate to subentire; leaves usually obtuse **(c) subsp. grisebachii**

(a) Subsp. **richeri**: 2*n* = 14. *S.W. & C. Alps, Jura, Appennini.*

(b) Subsp. **burseri** (DC.) Nyman, *Consp.* 132 (1878) (*H. burseri* (DC.) Spach): *Pyrenees, Cordillera Cantábrica.*

(c) Subsp. **grisebachii** (Boiss.) Nyman, *loc. cit.* (1878) (*H. alpigenum* Kit., *H. alpinum* Waldst. & Kit., non Vill., *H. balcanicum* Velen., *H. grisebachii* Boiss., *H. transsilvanicum* Čelak., *H. richeri* subsp. *alpigenum* (Kit.) E. Schmid): *S.E. Alps, Balkan peninsula; E. & S. Carpathians.*

Subsp. *grisebachii* is variable, especially in habit, form and texture of leaves and form of sepals, but none of the local populations of distinct facies is worthy of taxonomic recognition if the variation of the subspecies throughout its range is considered.

39. H. spruneri Boiss., *Diagn. Pl. Or. Nov.* 2(8): 112 (1849) (*H. perfoliatum* sensu Hayek pro parte, non L.). Stems 30–60 cm, erect, sometimes rooting at the base. Leaves 20–40(–60) mm, triangular-lanceolate or narrowly oblong, sessile, the uppermost sometimes with glandular auricles or with margin black-glandular-ciliate or denticulate. Sepals acute, black-glandular-ciliate and with numerous superficial black dots. Petals with black dots scattered over the whole surface. Capsule without dorsal vittae, but with numerous round, orange vesicles. *Meadows and shady places.* ● *W. part of Balkan peninsula, Istra, S.E. Italy.* Al Gr It Ju.

40. H. rochelii Griseb. & Schenk, *Arch. Naturgesch.* (*Berlin*) 18: 299 (1852) (*H. boissieri* Petrović, *H. pseudotenellum* Vandas). Stems 15–35(–50) cm, erect or decumbent. Leaves (20–)25–50 mm, triangular-lanceolate to linear, usually amplexicaul. Sepals acute to obtuse, black-glandular-ciliate or -fimbriate, usually with superficial black glands. Petals sometimes with a few superficial black dots. Capsule without dorsal vittae, but with numerous round or elongated, orange vesicles. *Rocky pastures.* ● *S.W. & C. Bulgaria, N.E. Jugoslavia, S.W. Romania.* Bu Ju Rm.

41. H. barbatum Jacq., *Fl. Austr.* 3: 33 (1775). Stems 10–45 cm, erect or decumbent. Leaves 6–40 mm, lanceolate to linear-lanceolate or elliptic-oblong, sessile or shortly petiolate, entire. Sepals acute, eglandular-fimbriate, with numerous superficial black dots or streaks. Petals usually with superficial black dots, scattered over the whole surface or near the apex only. Capsule

without or with interrupted dorsal vittae, with round or elongated, orange vesicles or almost smooth. *Meadows and stony places.* ● *Balkan peninsula, extending to Austria and S. Italy.* Al Au Bu Gr Hu It Ju.

42. H. rumeliacum Boiss., *Diagn. Pl. Or. Nov.* 2(8): 113 (1849). Stems 5–40 cm, erect or decumbent, branching but rarely rooting at the base. Leaves 6–35 mm, ovate-lanceolate or oblong to linear, sessile or shortly petiolate, the uppermost ones sometimes black-glandular-ciliate. Sepals acute, black-glandular-fimbriate or -ciliate and with numerous superficial black dots or streaks. Petals with black dots scattered over the whole surface. Capsule without dorsal vittae, with faint, round or elongated vesicles or almost smooth. *Calcareous, stony places.* ● *Balkan peninsula; one station in S. Romania.* Al Bu Gr Ju Rm.

Plants from the south-western part of the range are shorter and more decumbent, and with broader leaves and sepals; they have been distinguished as **H. apollinis** Boiss. & Heldr. in Boiss., *Diagn. Pl. Or. Nov.* 3(1): 105 (1853), but show continuous intergradation with typical plants.

43. H. trichocaulon Boiss. & Heldr. in Boiss., *Diagn. Pl. Or. Nov.* 2(8): 110 (1849). Stems 5–25(–45) cm, procumbent or ascending, slender, sometimes rooting at the base. Leaves 5–11(–14) mm, ovate-oblong to linear, sessile or shortly petiolate, entire. Sepals obtuse, black-glandular-denticulate to entire; sepals and petals sometimes with black dots or streaks scattered over the whole surface. Capsule sometimes with a few dorsal vittae, and with faint, round elongated vesicles or almost smooth. ● *W. & C. Kriti.* Cr.

44. H. kelleri Bald., *Malpighia* 9: 67 (1895). Stems 1–5(–10) cm, procumbent, branching and rooting. Leaves 2–4(–6) mm, oblong or elliptical, petiolate, entire, slightly papillose or almost smooth. Sepals obtuse to subacute, black-glandular-denticulate to subentire, with or without superficial black dots. Petals with a few superficial black dots. Capsule with dorsal vittae and round or elongated, orange vesicles. *W. Kriti.* Cr.

Perhaps not specifically distinct from **43**.

Sect. ORIGANIFOLIUM Stefanov. Perennial herbs. Black glands present on stem, leaves, sepals, petals and anthers. Stem 2-lined. Leaves with intramarginal and often superficial black glands. Flowers in pyramidal to cylindrical panicles. Petals and stamens persistent. Stamen-fascicles 3. Styles 3. Capsule with dorsal vittae and lateral vesicles. Seeds rugulose, or with short, transverse ridges superimposed on faint, longitudinal ribs.

45. H. aviculariifolium Jaub. & Spach, *Ill. Pl. Or.* 1: 59 (1842). Stems 15–35(–45) cm, erect or decumbent at the base, usually finely papillose. Leaves 9–23 mm, oblong or oblanceolate, densely papillose-puberulent. Sepals obtuse, black-glandular-ciliate and with superficial black dots. Petals with black dots scattered over the whole surface. *Stony places. Turkey-in-Europe* (*W. of Istanbul*). Tu. (*Anatolia.*)

The European plants belong to subsp. **byzantinum** (Aznav.) N. K. B. Robson, *Feddes Repert.* 74: 23 (1967) (*H. byzantinum* Aznav.). Other subspecies have glabrous stems and often glabrous leaves.

Sect. THASIA Boiss. Glabrous perennial herbs. Black glands present on stem, leaves, sepals, petals and anthers. Stem 2-lined. Leaves with only intramarginal black glands. Flowers in corymbose cymes. Petals and stamens persistent. Stamen-fascicles 5. Styles 5. Capsule with longitudinal vittae. Seeds reticulate-pitted.

46. H. thasium Griseb., *Spicil. Fl. Rumel.* **1**: 227 (1843). Stems 15–40 cm, erect or decumbent and rooting at the base. Leaves 15–30 mm, lanceolate or narrowly oblong to linear. Sepals broadly imbricate, black-glandular-fimbriate and with intramarginal black dots. Petals sometimes with a few superficial black dots. *Sandy and rocky places.* ● *S.E. part of Balkan peninsula.* Bu Gr Tu.

Sect. OLIGOSTEMA (Boiss.) Stefanov. Glabrous perennial (rarely annual) herbs. Black glands present on leaves, sepals and petals, and sometimes on stem and anthers. Stems terete or 2-lined. Leaves with intramarginal and sometimes a few superficial black glands. Flowers in corymbose or pyramidal cymes or panicles, or rarely solitary. Petals and stamens persistent. Stamen-fascicles 3. Styles (2–)3. Capsule with longitudinal vittae. Seeds reticulate-pitted.

47. H. aucheri Jaub. & Spach, *Ill. Pl. Or.* **1**: 61 (1842) (*H. apterum* Velen., *H. jankae* Nyman). Stems 7–30 cm, suberect or decumbent. Leaves 4–20(–25) mm, oblong to lanceolate or linear, ascending or subappressed, with translucent dots. Sepals lanceolate, equal or unequal, black-glandular-denticulate to -fimbriate and sometimes with a few intramarginal black dots. Petals 2–2·5 times as long as the sepals, without superficial black glands. Capsule scarcely as long as the sepals. *Stony places. S.E. part of Balkan peninsula.* Bu Gr Tu.

48. H. linarifolium Vahl, *Symb. Bot.* **1**: 65 (1790). Stems 5–65(–75) cm, erect or decumbent, sometimes branching and rooting at the base. Leaves 5–35 mm, narrowly oblong to narrowly lanceolate or linear, patent to subappressed, usually without translucent dots. Sepals lanceolate to ovate, subequal, black-glandular-ciliate and with numerous superficial black dots and streaks. Petals 2–4 times as long as the sepals, rarely with superficial black streaks. Capsule about twice as long as the sepals. $2n=16$. *Dry sunny places; calcifuge. W. Europe, northwards to 53° N. in Britain and eastwards to c. 4° E. in S. France.* Br Ga Hs Lu.

49. H. australe Ten., *Fl. Neap. Prodr. App. Quinta* 25 (1826). Stems 8–40 cm, erect to decumbent or ascending, often branching and rooting at the base. Leaves 7–25 mm, ascending or subappressed, without translucent dots, the upper oblong to lanceolate, the lower obovate to oblanceolate. Sepals lanceolate to narrowly oblong, subequal, subentire to shortly black-glandular-ciliate and with numerous superficial black dots and streaks. Petals 1·5–2·5 times as long as the sepals, rarely with a few superficial black glands. Capsule 1·5–2 times as long as the sepals. *W. Mediterranean region.* Bl Co Ga ?Hs It Sa Si.

50. H. humifusum L., *Sp. Pl.* 785 (1753). Stems 3–30(–40) cm, decumbent or procumbent, branching and rooting at the base. Leaves 3–15(–20) mm, patent or ascending, usually with translucent dots, the upper oblong to lanceolate, the lower obovate to oblanceolate. Sepals ovate to lanceolate, unequal, entire or with sessile marginal glands or black-glandular-ciliate or -denticulate, usually with a few superficial black dots. Petals equalling or up to twice as long as the sepals, very rarely with superficial black dots. Capsule equalling or slightly exceeding the sepals. $2n=16$. *Open habitats; usually calcifuge. W. & C. Europe, extending to S. Sweden, White Russia, S. Romania, N. Albania & C. Italy.* Al Au Az Be Br Co Cz Da Ga Ge Hb He Ho Hs Hu It Ju Lu Po Rm Rs (?B, C, W, ?K) Sa Su.

48, 49 and **50** form a group of closely related plants, each with a well-defined geographical area. Where the ranges overlap the species usually remain distinct, but intermediates between **49** and **50** have been recorded from N. Italy, and between **48** and **50** from S.W. England and the Channel Islands.

Sect. HYPERICUM. Glabrous, perennial herbs, usually with axillary shoots developed. Black glands present on stems, leaves, anthers and sometimes sepals and petals. Stems 2- to 4-lined or 4-winged. Leaves with intramarginal and sometimes a few superficial black glands. Flowers in subcorymbose or pyramidal cymes. Petals and stamens persistent. Stamen-fascicles 3(–4). Styles 3(–4). Capsule with longitudinal vittae, or with dorsal vittae and lateral, oblique vesicles. Seeds reticulate-pitted.

51. H. tetrapterum Fries, *Nov. Fl. Suec.* 94 (1823) (*H. quadrangulum* L., nom. ambig., *H. acutum* Moench; incl. *H. corsicum* Steudel). Stems (6–)10–100 cm, erect from a decumbent, rooting base or rarely wholly procumbent, narrowly 4-winged. Leaves (4–)10–35(–40) mm, orbicular, ovate, broadly oblong or broadly elliptical, sessile, with small translucent dots; margin usually plane. Sepals lanceolate or narrowly oblong, acute or acuminate, entire, sometimes with 1–2 superficial black dots. Petals 5–7(–7·5) mm, rarely with 1–2(–4) marginal black dots, not red-tinged (but red-veined in var. *corsicum* (Steudel) Boiss.). $2n=16$. *Damp places. W., S. & C. Europe, extending to S. Sweden and E. Ukraine.* All except Az Bl Fa Fe Is Lu No Rs (N) Sb.

52. H. undulatum Schousboe ex Willd., *Enum. Pl. Hort. Berol.* 810 (1809). Stems 15–100 cm, erect from a decumbent and rooting base, narrowly 4-winged. Leaves 7–40 mm, narrowly ovate to elliptical or narrowly oblong, sessile, with medium or small translucent dots; margin usually markedly undulate. Sepals lanceolate, acute or acuminate, entire, with 3–14 superficial black dots. Petals (7·5–)8–10 mm, without or with a few marginal or superficial black dots, usually red-tinged. *Damp places. S.W. Europe, northwards to Wales.* Az Br Ga Hs Lu.

In Spain and N. Portugal plants are found with characters or combinations of characters intermediate between **51** and **52**. As it is uncertain whether or not such plants are of hybrid origin, it seems best to include them at present in *H. undulatum* as var. *boeticum* (Boiss.) Lange (*H. boeticum* Boiss.).

53. H. maculatum Crantz, *Stirp. Austr.* **2**: 64 (1763) (*H. quadrangulum* auct., non L.). Stems 15–100 cm, erect from a decumbent and rooting base, 4-lined. Leaves (10–)15–40(–50) mm, ovate or ovate-lanceolate to oblong or elliptical, sessile, with densely reticulate venation, without translucent dots or with a few large ones in the upper leaves. Sepals broadly ovate to oblong, with apex rounded or erose-denticulate, without or with a few superficial black dots. Petals usually with numerous superficial black dots or streaks and sometimes with a few marginal black dots. Capsule with longitudinal vittae. *Most of Europe, but rare in the Mediterranean region.* Al Au Be Br Bu Cz Da Fa Fe Ga Ge Gr Hb He Ho Hs Hu It Ju No Po Rm Rs (N, B, C, W, E) Su.

(a) Subsp. **maculatum**: Inflorescence-branches making an angle of *c*. 30° with the stem. Sepals entire, usually wide. Petals with superficial black glands only, mostly in the form of dots or short streaks, or rarely without black glands. $2n=16$. *E., N. & C. Europe, westwards locally to Scotland and the Pyrenees.*

(b) Subsp. **obtusiusculum** (Tourlet) Hayek, *Sched. Fl. Stir. Exsicc.* 23–24: 27 (1912): Inflorescence-branches making an angle of *c*. 50° with the stem. Sepals erose-denticulate, wide to narrow. Petals with a few marginal black glands, and superficial ones mostly in the form of long streaks. 2*n*=32. *Lowlands of N.W. Europe from W. Germany and the Netherlands westwards, and in the Alps at lower altitudes than subsp. (a).*

54. H. perforatum L., *Sp. Pl.* 785 (1753) (*H. noeanum* Boiss.). Stems 10–100 cm, erect from a decumbent, rooting base, 2-lined. Leaves (5–)8–30(–35) mm, ovate to linear, sessile or subsessile, with obscurely reticulate venation and with numerous large translucent dots. Sepals lanceolate or oblong to linear, acute to acuminate or shortly aristate, usually entire, without or with a few superficial black dots. Petals with a few marginal black dots, sometimes also with superficial black dots or streaks. Capsule with dorsal vittae and lateral, oblique vittae or vesicles. 2*n*=32 (?48). *Throughout Europe except the extreme north.* All except Fa Is Sb.

Very variable. Most northern plants have relatively wide leaves; but towards the south of Europe plants with narrow leaves (var. *angustifolium* DC.) or small leaves (var. *microphyllum* DC.) are predominant. Although these southern plants have been found to differ genetically from the wide-leaved ones, there appears to be no morphological discontinuity between them.

Reproduction in *H. perforatum* has been shown to be 97% apomictic (pseudogamous), resulting in the formation of unreduced embryo-sacs with 32 chromosomes. The pollen, however, undergoes normal meiosis. Hybrids with diploid species (2*n*=16) are, therefore, pentaploid if *H. perforatum* is the female parent and triploid if it is the male parent. Triploid hybrids are more or less intermediate between the parents, but pentaploids are usually almost indistinguishable from *H. perforatum*. Two hybrids of this type have been described: **51 × 54** (H. × *medium* Peterm.) which is rather rare, owing to the different habitat-preferences of the parents, and **53(a) × 54** (*H.* × *desetangsii* nm. *carinthiacum* (A. Fröhlich) N. K. B. Robson (*H. carinthiacum* A. Fröhlich)), which is fairly common in E. Europe.
53(b) × 54 (*H.* × *desetangsii* Lamotte nm. *desetangsii*) is, however, tetraploid. It is common and apparently fertile. The primary hybrid is intermediate, with stems usually with 2 strong and 2 weak lines; leaves with laxly reticulate venation and a few pale glandular dots; sepals narrowly oblong or ovate-lanceolate, with the apex apiculate and erose-denticulate. Back-crossing has, however, produced a complete series of intermediates between the parents.

55. H. triquetrifolium Turra, *Farset. Nov. Gen.* 12 (1765) (*H. crispum* L.). Stems 13–55 cm, erect or decumbent, 2-lined, with divaricate branches. Leaves 3–15(–20) mm, lanceolate-triangular or rarely ovate-triangular to linear-oblong, amplexicaul, usually without reticulate venation, sometimes with medium to small translucent dots; margin undulate. Sepals oblong to ovate-oblong, obtuse or apiculate, entire or denticulate, without black dots. Petals without black dots or rarely with one intramarginal dot. Capsule with longitudinal vittae or vesicles. *Dry, stony or sandy places. E. part of Mediterranean region, extending westwards to Sicily; naturalized further west.* Al Cr Gr It Si Tu [Bl Hs].

56. H. elegans Stephan ex Willd., *Sp. Pl.* 3: 1469 (1802). Stems 15–55 cm, 2-lined, erect or decumbent and rooting at the base. Leaves 10–30 mm, lanceolate or ovate-lanceolate to oblong or linear-oblong, sessile, with obscurely reticulate venation, and with several large translucent dots; margin plane. Sepals lanceolate to narrowly oblong, acute to acuminate, black-glandular-denticulate or with sessile marginal glands, occasionally with one

superficial black dot. Petals with only marginal black dots. Capsule with longitudinal vittae. *Dry places; somewhat calcicole. C. & E. Europe, from Turkey to C. Ural and westwards to W. Germany.* Au Bu Cz Ge Hu Ju Rm Rs (C, W, K, E) Tu.

Sect. BRATHYS (Mutis ex L. fil.) Choisy. Glabrous, perennial or annual herbs. Black glands absent. Stem 4-lined. Flowers in pyramidal to corymbose cymes. Petals and stamens persistent. Stamens reduced to 5, or in 5 indefinite fascicles or groups. Styles (2–)3–5. Capsule with longitudinal vittae. Seeds longitudinally ribbed and with transverse striations.
The above description applies to subsect. *Spachium* R. Keller, which alone is represented in Europe.

Five North American species of this subsection have been found in Europe since 1834. All grow in damp places, and some have been recorded in areas previously well known to botanists. Their seeds are small, as in other species of *Hypericum*, and it appears probable that they have been introduced relatively recently from North America by wading birds or among agricultural seeds or fodder. See H. Heine, *Bauhinia* 2: 71–78 (1962).

57. H. gymnanthum Engelm. & A. Gray, *Boston Jour. Nat. Hist.* 5: 212 (1845). Perennial. Stems (20–)25–45(–60) cm, usually branched only in the inflorescence. Leaves 10–25(–30) mm, ovate- to lanceolate-triangular, cordate to rounded at the base, 5- to 7-veined. Sepals 3·5–5 mm. Capsule 3–5 mm, narrowly ovoid. *Drained peat-bog. Poland (near Poznan).* *Po.

First observed in 1884.

58. H. mutilum L., *Sp. Pl.* 787 (1753). Perennial or annual. Stems 10–40 cm, usually branched above the middle. Leaves 7–20(–30) mm, ovate to oblong or lanceolate, rounded to broadly cuneate at the base, 3- to 5-veined. Sepals 1·5–3 mm. Capsule 2·5–4 mm, ellipsoid. *Marshes. Germany, Poland, Italy and France.* *Ga *Ge *It *Po.

First observed in Italy in 1834 (? extinct), in Germany in 1874, in France in 1881, and in Poland in 1885.

59. H. majus (A. Gray) Britton, *Mem. Torrey Bot. Club* 5: 225 (1894). Perennial or annual. Stems 10–35 cm, unbranched or branched above. Leaves 15–40 mm, lanceolate or oblong, rounded to broadly cuneate at the base, (3–)5- to 7-veined. Sepals (4–)5–7 mm. Capsule 5·5–7·5 mm, narrowly ovoid. *Margins of ponds and streams. S. & E. Germany; E. France (Haute-Saône).* *Ga *Ge.

First observed in Germany in 1945 and in France before 1955.

60. H. canadense L., *Sp. Pl.* 785 (1753). Annual or sometimes perennial. Stems 7–25(–40) cm, unbranched or branched above. Leaves 6–20(–30) mm, linear to oblanceolate, narrowly cuneate at the base, 1- to 3(–5)-veined. Sepals 2–4·5(–5) mm. Capsule 4–6 mm, ovoid to cylindrical. 2*n*=16. *Wet heaths. Netherlands (Overijsel); Ireland (Mayo).* *Hb *Ho.

First observed in the Netherlands in 1909, and in Ireland in 1954. Records for Germany are errors for **59**.

61. H. gentianoides (L.) Britton, E. E. Sterns & Poggenb., *Prelim. Cat.* 9 (1888). Annual. Stems 5–20(–40) cm, usually with numerous ascending branches. Leaves 1–3 mm, subulate or scale-like, 1-veined. Sepals 2–2·5 mm. Capsule 5–7 mm, narrowly conical. *Damp, sandy ground. S.W. France (Gironde).* *Ga.

First observed in 1931.

VIOLALES

CX. VIOLACEAE[1]

Small shrubs or herbs. Leaves alternate, stipulate, usually undivided. Flowers hermaphrodite, solitary. Sepals 5, persistent. Petals 5, free. Stamens 5, introrse, connivent round the ovary. Ovary superior, unilocular, with 3 placentas and numerous ovules. Fruit a 3-valved capsule. Seeds endospermic.

1. Viola L.[2]

Leaves petiolate. Sepals prolonged into short appendages below the point of their insertion. Corolla zygomorphic, the lower petal spurred. Connectives of stamens with an apical appendage; the 2 lower stamens spurred. Style thickened above; stigma of various shapes (beaked, bilobed, capitate). Seeds with an elaiosome.

Various species and hybrids of the genus are commonly cultivated, especially the sweet violet, **1**, and the garden pansy, *V. × wittrockiana* Gams (= *V. hortensis* auct.). The origin of the latter is not known with certainty, but it is thought to have arisen from hybrids between **74**, **78** and *V. altaica* Ker-Gawler. Amongst other European species in cultivation are **48**, **63**, **73**, **74** and their hybrids.

Description of flowers in the text refer to open (chasmogamous) flowers. Cleistogamous flowers are produced by species **1–33** and by **88–89**. Flower lengths are measured from apex of spur to apex of lower petal, or from apex of upper to apex of lower petal, whichever is the longer. The term *rosulate* is used for plants with a loose, basal rosette of 3–5 petiolate leaves from the axils of which arise flowering branches, and the term *arosulate* for plants in which this basal rosette is lacking.

Literature: W. Becker, *Violae Europaeae*. Dresden. 1910. (Originally published in *Beih. Bot. Centr.* **26(2)**: 1–44 (1909); 289–390 (1910).)

1 Aerial stems woody, at least below
  2 Spur less than 10 mm
    3 Flowers 2–2·5 cm, purple-violet   **60. allchariensis**
    3 Flowers 1–1·5 cm, whitish, pale violet or yellow
      4 Leaves ovate to linear-lanceolate; petals whitish or pale violet   **88. arborescens**
      4 Leaves obovate; petals yellow   **89. scorpiuroides**
  2 Spur more than 10 mm
    5 Spur *c*. 12 mm   **91. kosaninii**
    5 Spur more than 16 mm
      6 Spur 16–18 mm; lower petal entire   **90. delphinantha**
      6 Spur 20–30 mm; lower petal emarginate   **92. cazorlensis**
1 Aerial stems herbaceous or absent
  7 Lateral petals directed downwards; style usually neither capitate nor bilobed
    8 Style capitate; rhizome thick and fleshy   **32. obliqua**
    8 Style not capitate; rhizome not thick or fleshy
      9 Caulescent, with leafy, aerial flowering stems
        10 Plant lacking basal leaf-rosette; leaves usually longer than wide
          11 Leaves oblong-ovate or ovate, cordate or subcordate
            12 Stipules equalling or exceeding petiole   **25. jordanii**

            12 Stipules usually not more than half as long as petiole   **20. canina**
          11 Leaves ovate-lanceolate or lanceolate, usually not cordate
            13 Spur much exceeding calycine appendages   **21. lactea**
            13 Spur only slightly exceeding calycine appendages
              14 Plant glabrous   **23. pumila**
              14 Plant, or at least leaves, shortly pubescent
                15 Stems up to 50 cm; stipules equalling or exceeding petiole   **24. elatior**
                15 Stems up to 25 cm; stipules not exceeding petiole, often only half as long   **22. persicifolia**
        10 Plant with basal leaf-rosette; leaves usually about as long as wide
          16 Stipules entire; open flowers arising from the base of the rosette
            17 Stems glabrous   **13. willkommii**
            17 Stems with a line of hairs   **12. mirabilis**
          16 Stipules usually dentate or fimbriate; open flowers cauline
            18 Calycine appendages 2–3 mm, conspicuous
              19 Flowers pale-blue or whitish; stigma hairy **19. sieheana**
              19 Flowers deep violet; stigma glabrous or papillose   **18. riviniana**
            18 Calycine appendages not more than 1 mm, inconspicuous
              20 Stipules narrowly lanceolate, long-fimbriate   **15. reichenbachiana**
              20 Stipules broadly lanceolate to ovate, shortly fimbriate
                21 Stems up to 25 cm; spur 4–6 mm   **16. tanaitica**
                21 Stems not more than 10 cm; spur 1·5–4 mm
                  22 Leaves delicate; petiole glabrous or with scattered hairs   **17. mauritii**
                  22 Leaves firm; petiole usually pubescent **14. rupestris**
    9 Acaulescent, lacking leafy, aerial stems, but sometimes with stolons
      23 Stipules semi-adnate to petiole
        24 Leaves deeply palmatifid   **31. pinnata**
        24 Leaves simple, not palmatifid
          25 Leaves cordate with shallow sinus; plant with creeping rhizome   **26. uliginosa**
          25 Leaves cordate with deep sinus; plant without creeping rhizome
            26 Sepals obtuse   **29. jooi**
            26 Sepals acute   **30. selkirkii**
      23 Stipules free
        27 Sepals acute
          28 Leaves acute   **12. mirabilis**
          28 Leaves obtuse   **13. willkommii**
        27 Sepals obtuse
          29 Capsule trigonous, explosive; fruiting peduncles erect
            30 Leaves in pairs; bracts in upper third of peduncle   **28. epipsila**
            30 Leaves 3 or 4 together; bracts at middle of peduncle   **27. palustris**
          29 Capsule spherical, not explosive; fruiting peduncles procumbent
           31 Stolons present
            32 Whole plant glabrous
              33 Stipules broad, triangular; stolons stout **5. jaubertiana**
              33 Stipules narrow
                34 Stipules lanceolate; stolons short, stout   **2. suavis**
                34 Stipules linear-lanceolate; stolons long, slender   **3. alba**
            32 Plant with hairs on stem, leaves or fruit

---

[1] Edit. D. H. Valentine.
[2] By D. H. Valentine, H. Merxmüller and A. Schmidt.

35   Stolons short, stout   **2. suavis**
35   Stolons long, slender
   36   Leaves orbicular, obtuse; stipules broadly ovate
                                1. odorata
   36   Leaves ovate, acute; stipules linear-lanceolate
      37   Leaves sparsely hairy; lateral petals bearded   **3. alba**
      37   Leaves hispid; lateral petals beardless   **4. cretica**
31 Stolons absent
   38   Leaves cordate with deep sinus
      39   Stipules shortly fimbriate; flowers not fragrant **6. hirta**
      39   Stipules long-fimbriate; flowers fragrant
         40   Leaves light green; spur whitish   **7. collina**
         40   Leaves dark green; spur violet   **3. alba**
   38   Leaves cordate with shallow sinus, or truncate
      41   Leaves truncate or subcordate
         42   Flowers pale violet; capsule glabrous   **11. chelmea**
         42   Flowers dark violet; capsule pubescent   **8. ambigua**
      41   Leaves cordate with shallow sinus
         43   Leaves oblong-ovate; sepals and capsule pubescent   **9. thomasiana**
         43   Leaves broadly ovate; sepals and capsule glabrous   **10. pyrenaica**
7 Lateral petals directed upwards; style capitate or bilobed at apex
44   Leaves reniform; style bilobed at apex   **33. biflora**
44   Leaves not reniform; style capitate at apex
   45   Stems entirely subterranean
      46   Leaves ± entire; stipules ovate-orbicular   **45. grisebachiana**
      46   Leaves crenate; stipules oblong or lanceolate   **46. alpina**
   45   Aerial stems present, though sometimes very short
      47   All leaves entire
         48   Annual
            49   Plant densely hairy   **83. parvula**
            49   Plant glabrous or sparsely hairy
               50   Flowers less than 1 cm, lilac-blue   **84. heldreichiana**
               50   Flowers 1 cm, yellow   **85. mercurii**
         48   Perennial
            51   Spur 5–10 mm; flowers 2 cm or more
               52   Leaves linear to oblong-spathulate; flowers yellow (rarely violet)   **40. brachyphylla**
               52   At least the lower leaves broader; flowers pink or violet
                  53   Stipules with 2–7 basal laciniae   **38. valderia**
                  53   Stipules entire or rarely with 1–2 laciniae
                     54   Spur 5–8 mm; petioles long and slender   **36. cenisia**
                     54   Spur 8–10 mm; petioles short and stout **39. magellensis**
            51   Spur less than 5 mm
               55   Plant with a short, dense tomentum
                  56   Upper leaves narrower than lower; stipules with 3 laciniae   **35. diversifolia**
                  56   Upper leaves not narrower than lower; stipules entire or with 1 lacinia
                     57   Spur not more than twice as long as calycine appendages   **42. stojanowii**
                     57   Spur 2–3 times as long as calycine appendages   **43. fragrans**
               55   Plant glabrous or sparsely pubescent
                  58   Spur scarcely exceeding calycine appendages
                     59   Leaves oblong; flowers usually yellow   **41. perinensis**
                     59   Leaves ovate to orbicular; flowers violet to whitish
                        60   Petiole slightly longer than lamina   **34. crassiuscula**
                        60   Petiole at least twice as long as lamina   **45. grisebachiana**
                  58   Spur at least twice as long as calycine appendages
                     61   Flowers up to 2·5 cm; spur slender   **37. comollia**
                     61   Flowers c. 1 cm; spur stout
                       62   Leaves ovate to orbicular, subcordate   **47. nummulariifolia**
                       62   Leaves oblong-linear to oblong-ovate, not subcordate
                          63   Leaves c. 1·5 cm, gradually narrowed to petiole   **43. fragrans**

                          63   Leaves up to 2·5 cm, abruptly contracted into petiole   **44. poetica**
   47   At least the lower leaves crenate or serrulate
      64   Annual or biennial; flowers usually not more than 1·5 cm
         65   Corolla equalling or shorter than calyx
            66   Flowers 1–1·5 cm
               67   Bracts of peduncle concealed by the calycine appendages   **87. occulta**
               67   Bracts of peduncle not concealed   **80. arvensis**
            66   Flowers less than 1 cm
               68   Plant glabrous or sparsely hairy   **84. heldreichiana**
               68   Plant densely hairy
                 69   Hairs short; leaves crenately lobed   **81. kitaibeliana**
                 69   Hairs long; leaves almost entire   **83. parvula**
          65   Corolla distinctly exceeding calyx
            70   Plant hispid   **82. hymettia**
            70   Plant not hispid (sometimes pubescent below)
               71   Leaves almost entire; stipules usually 3-partite   **85. mercurii**
               71   Leaves crenate; stipules divided into more than 3 segments
                 72   Flowers 1–3·5 cm; spur 3–6·5 mm   **78. tricolor**
                 72   Flowers c. 1 cm; spur 2–3 mm   **86. demetria**
      64   Perennial; flowers usually 1·5 cm or more
         73   Spur 7·5 mm or more
            74   Stipules ovate-triangular   **63. cornuta**
            74   Stipules not ovate-triangular
               75   Caespitose   **59. doerfleri**
               75   Not caespitose
                 76   Stipules deeply divided into at least 6 segments
                  77   Plant hairy
                    78   Flowers 3–4 cm   **50. splendida**
                    78   Flowers 2–3 cm
                     79   Stipules 7- to 17-partite; spur 5–8 mm   **66. elegantula**
                     79   Stipules 3- to 8-partite; spur 8–10 mm   **55. pseudogracilis**
                  80   Flowers yellow
                  80   Flowers violet
                    81   Plant puberulent   **67. athois**
                    81   Plant with patent hairs   **75. bubanii**
                 77   Plant glabrous or subglabrous
                  82   Spur not more than 8 mm   **66. elegantula**
                  82   Spur 8–15 mm
                    83   Lower stipules entire or dentate   **48. calcarata**
                    83   Lower stipules 3- to 7-partite   **55. pseudogracilis**
                 76   Stipules entire or dentate, or divided into not more than 5 segments
                  84   Stipules entire or dentate   **48. calcarata**
                  84   Stipules, at least the upper, pinnately divided, with 3–5 segments
                    85   Plant with slender, creeping rhizomes   **57. oreades**
                    85   Plant without slender, creeping rhizomes
                    86   Plant hairy
                     87   Upper leaves linear to lanceolate
                       88   Upper leaves and segments of stipules narrow-linear   **49. bertolonii**
                       88   Upper leaves and segments of stipules linear-lanceolate to lanceolate   **51. aethnensis**
                     87   Upper leaves lanceolate to ovate
                       89   Flowers yellow   **55. pseudogracilis**
                       89   Flowers violet   **67. athois**
                    86   Plant glabrous or subglabrous
                    90   Upper leaves linear to lanceolate, differing greatly from the lower
                     91   Flowers ± square in face view; petals usually contiguous   **49. bertolonii**
                     91   Flowers narrowly rectangular in face view; petals not contiguous   **53. corsica**
                    90   Upper leaves lanceolate to ovate, usually not differing greatly from the lower
                    92   Leaves slightly fleshy; flowers 3–5 cm; spur c. 15 mm   **54. munbyana**

92 Leaves not fleshy; flowers 2–3 cm; spur 8–12 mm

93 Stipules 0·5–1 cm, 3-partite; flowers dark violet   **52. nebrodensis**

93 Stipules 1–2 cm, 3- to 7-partite; flowers yellow or bluish-violet   **55. pseudogracilis**

73 Spur 7 mm or less

94 At least the upper leaves linear or linear-lanceolate

95 Lower leaves orbicular

96 Without leafy stolons; flowers 2–2·5 cm   **72. dubyana**

96 With leafy stolons; flowers 2·5–3·5 cm   **73. declinata**

95 Lower leaves not orbicular

97 Caespitose; plant usually pubescent, rarely glabrous   **60. allchariensis**

97 Not caespitose; plant glabrous

98 Flowers 1·5–2 cm; spur 6 mm   **71. rhodopeia**

98 Flowers 2·5–4 cm; spur 3–4 mm

99 Lower leaves linear-lanceolate   **70. beckiana**

99 Lower leaves ovate   **73. declinata**

94 Leaves not linear or linear-lanceolate

100 Stipules not divided, denticulate   **62. arsenica**

100 Stipules ± deeply divided, or dentate

101 Plant not more than 3 cm

102 Sepals dentate   **61. frondosa**

102 Sepals entire or serrate   **56. eugeniae**

101 Plant more than 3 cm

103 Stipules pinnately divided; terminal segment large, usually crenate

104 Sepals ovate-lanceolate or triangular   **79. aetolica**

104 Sepals lanceolate or linear

105 Flowers 1–1·5 cm, yellow; spur slightly longer than calycine appendages   **77. langeana**

105 Flowers 1·5–4 cm, violet or yellow; spur up to 3 times as long as calycine appendages

106 Petals as wide as long   **56. eugeniae**

106 Petals longer than wide   **78. tricolor**

103 Stipules ± divided but terminal segment not large or crenate

107 Stipules divided to mid-vein or almost to base, the undivided part not wider than the length of the segments

108 Stems hairy

109 Upper leaves narrowly lanceolate   **72. dubyana**

109 Upper leaves ovate or oblong

110 Spur 6–7 mm, much longer than calycine appendages   **68. gracilis**

110 Spur *c.* 4 mm, not much longer than calycine appendages   **76. hispida**

108 Stems glabrous or subglabrous

111 Sepals ovate-lanceolate or triangular   **79. aetolica**

111 Sepals lanceolate or linear

112 Plant with leafy stolons   **73. declinata**

112 Plant without leafy stolons

113 Caespitose   **58. eximia**

113 Not caespitose

114 Stipules 6- to 9-partite; spur curved   **72. dubyana**

114 Stipules 3- to 5-partite; spur straight   **74. lutea**

107 Stipules not very deeply divided, the undivided part wider than the length of the segments

115 Flowers 3–4 cm; petals almost orbicular

116 Spur stout, not much longer than calycine appendages   **57. oreades**

116 Spur slender, twice as long as calycine appendages   **69. speciosa**

115 Flowers usually less than 3 cm; petals ± ovate

117 Spur 3 times as long as calycine appendages   **66. elegantula**

117 Spur not more than twice as long as calycine appendages

118 Plant hairy; spur twice as long as calycine appendages   **64. orphanidis**

118 Plant glabrous or slightly hairy; spur equalling or scarcely exceeding calycine appendages   **65. dacica**

Sect. VIOLA (Sect. *Nomimium* Ging.). Herbaceous. Stipules not leafy. Open flowers blue, violet or white; cleistogamous flowers produced. Style beaked at apex.

Subsect. *Viola.* Perennial; acaulescent, sometimes stoloniferous. Capsules globose, not explosive, on decumbent peduncles. Seeds with conspicuous elaiosome.

1. V. odorata L., *Sp. Pl.* 934 (1753). Perennial 5–15 cm, with leaf-rosette and long, procumbent, rooting stolons. Leaves orbicular-reniform, deeply cordate, widest at about the middle; petiole long. Stipules ovate, glabrous or sparsely ciliate, shortly glandular-fimbriate. Bracts at or above middle of peduncle. Flowers *c.* 1·5 cm, dark violet or white, fragrant. Sepals ovate, obtuse. Spur *c.* 6 mm, exceeding calycine appendages. Stigmatic beak vertical, its length equalling diameter of style. Capsule pubescent. $2n=20$. *Europe, except the extreme north and parts of the Mediterranean region.* All except Bl Fa Fe Is Rs (N) Sb ?Tu.

Widely cultivated in gardens and often naturalized. Its northern limit as a native plant is therefore uncertain; it is certainly native only in S., S.C. and parts of W. Europe.

V. ignobilis Rupr., *Mém. Acad. Sci. Pétersb.* ser. 7, **15**(2): 148 (1869), from the Caucasus and Iran, which is like **1** but has ciliate sepals and a more or less horizontal stigmatic beak, has been recorded from 3 localities in Romania, but further confirmation of identity is needed.

2. V. suavis Bieb., *Fl. Taur.-Cauc.* **3**: 164 (1819) (*V. pontica* W. Becker, *V. sepincola* Jordan). Perennial with short rhizome and leaf-rosette, producing short, stout stolons. Spring leaves 3–8 cm; summer leaves up to 20 cm, glabrous or hairy; lamina ovate-oblong to broadly ovate, broadest below the middle, cordate; petioles eventually very long. Stipules lanceolate, long-fimbriate; fimbriae ciliate. Bracts below middle of peduncle. Flowers 1·5–2 cm, violet with white throat, fragrant. Spur exceeding calycine appendages. Capsule large, glabrous or pubescent. $2n=40$. *S., C. & E. Europe, northwards to 52° N. in C. Russia, and extending to north-west France.* Al Au Bu Co ?Cz Ga ?Gr He Hs Hu It Ju Po Rm Rs (C, W, K, E) [Ge].

A critical and widely distributed species. It was divided by Becker (*loc. cit.*) into a series of geographical subspecies, but his treatment is not satisfactory and further investigation is needed.

V. catalonica W. Becker, *Cavanillesia* **2**: 43 (1929), from N.E. Spain, **V. jagellonica** Zapał., *Bull. Int. Acad. Sci. Cracovie* ser. B, **1914**: 455 (1914), from Poland, and **V. adriatica** Freyn, *Flora (Regensb.)* **67**: 679 (1884), from N.W. Jugoslavia, are somewhat intermediate between **2** and **3**. They may be of hybrid origin, and require further investigation.

3. V. alba Besser, *Prim. Fl. Galic.* **1**: 171 (1809). Perennial 5–15 cm, with short rhizome and leaf-rosette, usually producing long, slender, ascending non-rooting stolons which usually flower in their first year. Leaves persisting during winter, ovate to triangular-ovate, cordate, dark green, hairy or glabrous. Spring leaves *c.* 5 cm, summer leaves 10–15 cm; lamina much shorter than petiole. Stipules linear-lanceolate, long-fimbriate; fimbriae ciliate. Peduncles 4–6 cm; bracts at or above middle. Flowers 1·5–2 cm, fragrant, white or violet; lateral petals bearded. Seeds oblong-ovate. *C. & S. Europe; Öland.* Al Au Bu Co Cr Cz Ga Ge Gr He Hs Hu It Ju ?Lu Po Rm Rs (W, K) Su Tu.

1 Stolons sometimes absent; leaves sparsely hairy to glabrous, with margins usually convex; flowers violet; capsule sparsely hairy to glabrous   **(c) subsp. dehnhardtii**
1 Stolons always present; leaves usually hairy, with straight or somewhat concave margins; flowers usually white; capsule hairy
2 Leaves and capsule light green; spur yellow-green   **(a) subsp. alba**
2 Leaves and capsule dark green; spur violet   **(b) subsp. scotophylla**

(a) Subsp. **alba**: $2n=20$. *North and West of the Alps, from S. France to Poland.*

(b) Subsp. **scotophylla** (Jordan) Nyman, *Consp.* 78 (1878): $2n=20$. *S.E. Europe, extending to N. Italy, Austria and Switzerland.*

(c) Subsp. **dehnhardtii** (Ten.) W. Becker, *Ber. Bayer. Bot. Ges.* 8(2): 257 (1902): $2n=20$. *Mediterranean region.*

V. **alba** subsp. **thessala** (Boiss. & Spruner) Hayek, *Prodr. Fl. Penins. Balcan.* 1: 502 (1925), from Greece, is probably best included in **3(b)**.

V. **cadevallii** Pau, *Mem. Acad. Ci. Artes Barcelona* ser. 3, 2: 62 (1896), from Spain, and V. **pentelica** Vierh., *Verh. Zool.-Bot. Ges. Wien* 64: 266 (1914), from S.E. Greece, are probably best regarded as variants of **3(c)**.

4. V. **cretica** Boiss. & Heldr. in Boiss., *Diagn. Pl. Or. Nov.* 2(8): 51 (1849). Like **3** but stolons very long; leaves hispid; lateral petals beardless; capsule hairy; seeds orbicular to ovate. $2n=20$. ● *Mountains of Kriti.* Cr.

5. V. **jaubertiana** Marès & Vigineix, *Cat. Pl. Baléar.* 37 (1880). Like **3** but completely glabrous; stolons stout; leaves coriaceous, shining, less deeply cordate; stipules broad, triangular, with glandular fimbriae; flowers larger than in **3**, bright violet; capsule glabrous; seeds 3 mm. $2n=20$. *Damp ravines.* ● *Mallorca.* Bl.

6. V. **hirta** L., *Sp. Pl.* 934 (1753). More or less hairy perennial 5–15 cm, with short rhizome and leaf-rosette. Spring leaves cordate, longer than wide, summer leaves oblong-ovate, deeply cordate; petioles more or less glabrous. Stipules broadly triangular to lanceolate, shortly glandular-fimbriate, glabrous or ciliate at apex. Bracts below the middle of peduncle. Flowers *c.* 1·5 cm, violet, not fragrant. Sepals oblong, obtuse. Spur dark violet, exceeding calycine appendages. Capsule pubescent. $2n=20$. *Most of Europe.* Al Au Bc Br Bu ?Co Cr Cz Da Ga Ge Gr Hb He Ho Hs Hu It Ju No Po Rm Rs (N, B, C, W, E) Si Su [Fe].

7. V. **collina** Besser, *Cat. Pl. Horto Cremen.* 151 (1816). Like **6** but leaves with deeper sinus; stipules linear to oblong-lanceolate, longer-fimbriate and more hairy; bracts above middle of peduncle; flowers pale blue, fragrant; spur whitish and shorter. $2n=20$. *Usually calcicole. Scattered over a large part of Europe, but absent from the Balkan peninsula, the islands and much of the north.* Au Be ?Co Cz Fe Ga Ge He Hs Hu It Ju No Po Rm Rs (N, B, C W E) Su.

8. V. **ambigua** Waldst. & Kit., *Pl. Rar. Hung.* 2: 208 (1804). Perennial with leaf-rosette; rhizome stout, more than 2 mm thick. Leaves oblong-ovate with truncate base, usually sparsely hairy, later glabrescent. Stipules 1–1·5 cm, dark green, broadly lanceolate, shortly glandular-fimbriate, ciliate. Peduncles scarcely exceeding the leaves; bracts at or below middle. Flowers 1–1·5 cm, fragrant. Sepals *c.* 3·5 mm, oblong. Petals dark violet. Spur 2–4 mm, curved, stout. Capsule pubescent. $2n=40$. *Calcifuge. From E.C. Russia westwards to Macedonia and E. Austria.* Au Bu Cz ?Ge Hu Ju Rm Rs (C, W, K, E).

9. V. **thomasiana** Song. & Perr. in Billot, *Annot.* 183 (1860). Like **8** but rhizome slender, less than 2 mm thick; leaves shallowly cordate; stipules 0·5–1 cm, linear-lanceolate, densely ciliate; sepals *c.* 2·5 mm; petals lilac or almost white; spur slender. $2n=20$. *Usually above* 1000 *m; calcifuge.* ● *C. & S. Alps.* Au Ga He It.

10. V. **pyrenaica** Ramond ex DC. in Lam. & DC., *Fl. Fr.* ed. 3, 4: 803 (1805). Perennial 8–10 cm, with short, erect rhizome and leaf-rosette. Spring leaves broadly ovate, cordate, glabrescent; summer leaves subacuminate, shining. Stipules lanceolate, shortly glandular-fimbriate, ciliate. Peduncles 3–5 cm; bracts at or above middle. Flowers *c.* 1·5 cm, pale violet with white throat, fragrant. Sepals broadly ovate, obtuse. Spur pale violet. Capsule glabrous. $2n=20$. *Subalpine habitats. Pyrenees, Alps and Jura; one station in C. Appennini; mountains of Balkan peninsula.* Al Au Bu Ga Gr He Hs It Ju.

V. **prenja** G. Beck, *Ann. Naturh. Mus. (Wien)* 2: 81 (1887), described from alpine rocks in W. Jugoslavia (Prenj Planina), is like **10** but has flowers *c.* 1 cm. It is probably best regarded as a variety of **10**.

11. V. **chelmea** Boiss. & Heldr. in Boiss., *Diagn. Pl. Or. Nov.* 3(1): 54 (1853). Perennial 4–8 cm, with short, erect rhizome and leaf-rosette. Lamina of spring leaves 0·5–1·5 cm, much shorter than petiole, triangular, truncate or cuneate at base, rather hairy; summer leaves larger, similar in shape, glabrous. Stipules broadly lanceolate, glandular-fimbriate. Flowers 1 cm, not fragrant. Sepals obtuse or subacute. Petals pale violet, not darkly veined. Spur rather stout, shorter than sepals. Capsule glabrous. *Calcareous rocks,* 500–2200 *m.* ● *W. Jugoslavia and Greece.* Gr Ju.

(a) Subsp. **chelmea**: Stipules fimbriate, glabrous. Bracts of the peduncle narrow, glabrous or sparsely ciliate. $2n=20$. *Mountains of Greece.*

(b) Subsp. **vratnikensis** Gáyer & Degen, *Magyar Bot. Lapok* 13: 309 (1914): Stipules long-fimbriate; fimbriae equalling the breadth of the stipule, ciliate. Bracts of the peduncle wide at the base, ciliate. *W. Jugoslavia.*

V. **vilaensis** Hayek, *Denkschr. Akad. Wiss. Math.-Nat. Kl. (Wien)* 94: 154 (1918), from screes in Crna Gora (near Rikavac, on the Albanian frontier), is a variant of **11** with stipules slightly hairy at the apex, and pale lilac or white petals with dark veins.

Species of Sect. *Viola*, subsect. *Viola* (**1–11**) frequently hybridize with one another, but very rarely with species of other subsections. Of the 30 or more hybrids described the following 5 are probably the most frequent; all but the fifth are stoloniferous: *V. alba × odorata, V. collina × odorata, V. hirta × odorata (V. × permixta* Jordan*), V. alba × hirta, V. collina × hirta.*

Subsect. *Rostratae* Kupffer. Perennial; caulescent; lacking stolons or long creeping rhizomes. Capsules trigonous, explosive, on erect peduncles. Seeds with inconspicuous elaiosome.

12. V. **mirabilis** L., *Sp. Pl.* 936 (1753). Spring leaves in a basal rosette. Stems with a line of hairs, developing in summer to *c.* 20 cm. Mature leaves 4–8 cm, as wide as long, orbicular, cordate, acute. Stipules 1–2 cm, broadly lanceolate, entire, eventually brown. Open flowers 2 cm, arising from the rosette,

fragrant; aerial stems bearing cleistogamous flowers only, on very short peduncles. Sepals acute, with conspicuous appendages. Petals pale violet. Spur 6–8 mm, whitish. Style glabrous. $2n = 20$. *Mainly in woodland, on base-rich soil. Widespread in Europe, but absent from the islands and much of the south and west.* Au Be Bu Cz Da Fe Ga Ge He Hs Hu It Ju No Po Rm Rs (N, B, C, W, K, E) Su.

V. pseudomirabilis Coste, *Bull. Soc. Bot. Fr.* **40**: cxv (1893), from woodland on calcareous soil in S. France (Causse du Larzac, near Millau), is in some ways intermediate between **12** and **18**. The stems, 10–30 cm, are glabrous, the leaves shortly acuminate, the stipules ovate-lanceolate and fimbriate, the flowers large and not fragrant. The open flowers are cauline as in **18**, not basal as in **12**; and the fact that they are fertile and produce capsules (on long peduncles) indicates that the plants are not direct hybrids between **12** and **18**. W. Becker, *Feddes Repert.* **18**: 141 (1922), records *V. pseudomirabilis* from Bulgaria. Further investigation is needed.

13. V. willkommii R. de Roemer, *Linnaea* **25**: 10 (1852). Like **12** but stems glabrous; leaves smaller, ovate-cordate, obtuse; flowers blue, scarcely fragrant; sepals shorter and broader; spur much broader and saccate. *Calcicole.* ● *N.E. Spain.* Hs.

14. V. rupestris F. W. Schmidt, *Abh. Böhm. Ges. Wiss.* ser. 2, **1**: 60 (1791) (*V. arenaria* DC.). Rosulate, with a short, central shoot, a basal rosette of leaves and axillary flowering stems. Plant, including petiole and capsule, finely pubescent (rarely glabrous). Stems up to 10 cm. Leaves 1–3 cm, cordate-reniform, obtuse. Stipules *c.* 8 mm, ovate-lanceolate, entire or dentate. Flowers 1–1·5 cm, reddish-violet, pale blue or white. Calycine appendages inconspicuous. Lower petal usually emarginate. Spur *c.* 3 mm, pale violet. Head of style papillose. *Open habitats on light, base-rich soils. Widespread in Europe, but absent from most of the islands, and much of the south and west.* Au Be Br Bu Cz Fe Ga Ge He Ho Hs Hu It Ju No Po Rm Rs (N, B, C, W, E) Su.

(a) Subsp. **rupestris**: Plant dark green. Petals rather wide, the lower emarginate. Spur of stamen stout. $2n = 20$. *Throughout the range of the species except N.W. Fennoscandia.*

(b) Subsp. **relicta** Jalas, *Ann. Bot. Soc. Zool.-Bot. Fenn. Vanamo* **24**(1): 70 (1950): Plant bright green. Petals narrow, the lower apiculate, scarcely emarginate. Spur of stamen slender. $2n = 20$. *Calcareous screes. N.W. Fennoscandia.*

15. V. reichenbachiana Jordan ex Boreau, *Fl. Centre Fr.* ed. 3, **2**: 78 (1857) (*V. sylvestris* Lam. pro parte). Rosulate, with a short central shoot, a basal rosette of leaves and axillary flowering stems. Stems up to 15 cm. Leaves 2–4 cm, about as wide as long, cordate, subacute. Stipules of cauline leaves 1 cm or more, narrowly lanceolate, fimbriate; fimbriae often equalling width of stipule. Flowers 1·2–1·8 cm. Sepals acute; calycine appendages very short, inconspicuous in fruit. Petals narrow, violet, often darker at the base. Spur 3–6 mm, slender, straight, deep violet. Head of style hairy. $2n = 20$. *Woods and shady places. S., W. & C. Europe, extending northwards to 60° N. in Sweden and eastwards to Estonia and C. Ukraine.* Al Au Be Br Bu Co Cz Da Ga Ge Gr Hb He Ho Hs Hu It Ju Lu Po Rm Rs (B, C, W, K) Sa Si Su Tu.

16. V. tanaitica Grosset, *Feddes Repert.* **26**: 80 (1929). Like **15** in habit but up to 25 cm; stipules of cauline leaves broadly lanceolate or ovate-lanceolate, dentate or shortly fimbriate, with fimbriae less than half the width of the stipule; flowers 10–15 mm, pale violet; calycine appendages short; spur 4–6 mm.

Deciduous woodland and scrub. ● *S.C. Russia (Kurskaja to Kujbyševskaja Obl.).* Rs (C, ?W, E).

17. V. mauritii Tepl., *Bull. Soc. Oural. Sci. Nat.* **7**: 37 (1883). Like **15** in habit. Stems up to 10 cm. Leaves delicate, glabrous or with scattered hairs, deeply cordate, the lower mostly rounded at the apex, the upper ovate, obtuse. Stipules of cauline leaves not more than 1 cm, ovate or ovate-lanceolate, serrate, shortly fimbriate or almost entire. Flowers 1–1·3 cm. Sepals short, acute; appendages inconspicuous. Petals pale violet. Spur 1·5–4 mm, slender, straight or slightly curved. *Coniferous woods. E. Russia (C. Ural and around Perm').* Rs (N, C). (*N. Asia.*)

18. V. riviniana Reichenb., *Pl. Crit.* **1**: 81 (1823). Like **15** in habit, but very variable in size. Leaves from very small to *c.* 4 cm, cordate, subacute. Stipules of cauline leaves more than 1 cm, lanceolate, fimbriate; fimbriae shorter than in **15**. Flowers 1·4–2·5 cm; calycine appendages usually 2–3 mm, more or less conspicuous, the lower accrescent in fruit. Petals rather wide, bluish-violet. Spur 3–5 mm, stout, whitish or light purple, often curved upwards. Stigma glabrous or papillose. $2n = 35, 40, 45, 46, 47$. *Woodland or grassland. Europe except the south-east.* All except Az Bl Cr Rs (W, K, E) Sb ?Tu.

Plants from exposed habitats in W. Europe, with leaves up to 2 cm, flowering branches up to 10 cm and somewhat smaller flowers and fruit than plants from more sheltered habitats, have been distinguished as subsp. **minor** (Murb. ex E. S. Gregory) Valentine, *New Phytol.* **40**: 208 (1941). As, however, so many intermediate plants and habitats exist, it is probably better to relegate subsp. *minor* to the rank of variety.

19. V. sieheana W. Becker, *Bull. Herb. Boiss.* ser. 2, **2**: 751 (1902) (*V. neglecta* sensu Bieb., non F. W. Schmidt). Like **18** but stipules larger, *c.* 4 mm wide, and less fimbriate, or only dentate; flowers larger; petals very wide, pale blue or whitish, the lower deeply cupped; spur very stout, whitish; stigma with a tuft of hairs. *Woodland, or in shade among rocks. S.E. Europe.* Bu Gr Rm Rs (K) Tu.

20. V. canina L., *Sp. Pl.* 935 (1753). Arosulate (without a basal rosette of leaves). Stems 10–40 cm, decumbent, ascending or erect. Leaves ovate to lanceolate, usually cordate or subcordate, usually obtuse or subacute. Stipules dentate or subentire, shorter than petiole. Flowers 1·5–2·5 cm, blue or white. Spur white or greenish-yellow, equalling or up to three times as long as the rather conspicuous calycine appendages. *Most of Europe, but rarer in the south.* All except Az Bl Cr Gr Si.

Very variable. The 3 subspecies described below are sometimes given specific rank.

1 Procumbent or ascending; leaves less than twice as long as wide, cordate; median stipules less than ⅓ as long as petiole; flowers blue   **(a) subsp. canina**
1 Erect; upper leaves twice as long as wide, subcordate; median stipules up to ½ as long as petiole; flowers blue or white
  2 Plant up to 40 cm; petals broadly obovate; spur 1½–2 times as long as calycine appendages, straight   **(b) subsp. montana**
  2 Plant up to 20 cm; petals narrowly elliptical; spur 2–3 times as long as calycine appendages, curved   **(c) subsp. schultzii**

(a) Subsp. **canina**: $2n = 40$. *Throughout the range of the species.*

(b) Subsp. **montana** (L.) Hartman, *Bot. Not.* **1841**: 82 (1841) (*V. montana* L.): $2n = 40$. *Throughout most of the range of the species.*

(c) Subsp. **schultzii** (Billot) Kirschleger, *Fl. Alsace* **1**: 81 (1852): $2n = 40$. *Fens and marshes. C. Europe, extending to N. Italy and S. Romania.*

21. V. lactea Sm. in Sowerby, *Engl. Bot.* **7**: t. 445 (1798). Arosulate. Stems 10–15(–20) cm, ascending. Lamina 1–3(–4) cm, about equalling petiole, lanceolate to ovate-lanceolate, often purplish-tinged, rounded to cuneate at base. Stipules 1–1·5 cm, lanceolate, coarsely dentate, all except the uppermost much shorter than petiole. Flowers 1·5–2 cm, bluish-white. Petals about 3 times as long as wide. Spur 3–4 mm, exceeding calycine appendages. 2n=58. *Dry heaths.* ● *W. Europe, northwards to 53° 30′ in Ireland.* Br Ga Hb Hs Lu.

22. V. persicifolia Schreber, *Spicil. Fl. Lips.* 163 (1771) (*V. stagnina* Kit.). Arosulate. Subglabrous. Stems up to 25 cm, erect; soboliferous. Lamina 2–4 cm, equalling or longer than petiole, triangular-lanceolate, truncate or subcordate at base. Stipules *c.* 1 cm, subentire to fimbriate, shorter than petiole. Flowers 1–1·5 cm, appearing circular in face view, white with violet veins. Petals obovate to orbicular, scarcely longer than wide. Spur 2–3 mm, greenish, as wide as long, only slightly exceeding calycine appendages. 2n=20. *Marshes and fens. Most of Europe except the Mediterranean region, the south-east and the extreme north.* Au Be Br Cz Da Fe Ga Ge He Ho Hs Hu It Ju No Po Rm Rs (N, B, C, W, E) Su.

23. V. pumila Chaix in Vill., *Hist. Pl. Dauph.* **1**: 339 (1786). Arosulate. Glabrous. Stems 5–15(–20) cm, ascending to erect. Lamina 2–3 × 1 cm, usually longer than the petiole, lanceolate, subacute, usually distinctly cuneate or occasionally rounded at base. Stipules large, entire or coarsely dentate, the upper 1–2 cm and longer than the petiole. Flowers *c.* 1·5 cm, pale blue. Spur 2–3 mm, only slightly exceeding calycine appendages. 2n=40. *Grassland. Mainly in C. & E. Europe, northwards to c. 56° in Russia, but extending to N. Italy, W. France and the Baltic islands.* Au Bu Cz Ga Ge He Hu It Ju Po Rm Rs (C, W, K, E) Su.

V. accrescens Klokov in Klokov & Wissjul., *Fl. RSS Ucr.* **7**: 632 (1955), from Ukraine and S. Russia, is like **23** but has larger stems and leaves, papillose stems and petioles, and usually white flowers.

24. V. elatior Fries, *Nov. Fl. Suec.* ed. 2, 277 (1828). Like **23** but up to 50 cm, erect, and larger in all parts. Plants shortly hairy except for sepals and ovary. Lamina 3–9 × 1–2 cm, lanceolate, subcordate, equalling or exceeding petiole. Stipules 2–5 cm, conspicuous, entire or coarsely dentate at base, equalling or exceeding petiole. Flowers 2–2·5 cm, pale blue. Spur 2–4 mm, only slightly exceeding calycine appendages. 2n=40. *Damp grassland and scrub. Mainly in C. & E. Europe, northwards to c. 57° in Eursia, but extending to N. Italy, C. France and the Baltic islands.* Au Bu Cz Ga Ge Gr He Hu It Ju Po Rm Rs (C, W, K, E) Su [Be].

25. V. jordanii Hanry, *Prodr. Hist. Nat. Var* 169 (1853). Stems up to 40 cm, erect. Leaves cordate, dimorphic; the lower about as long as wide, the upper triangular-lanceolate, 2–3 times as long as wide. Stipules up to 4 cm, conspicuous, the lower dentate, the upper almost entire, equalling or exceeding petiole. Flowers *c.* 2·5 cm, pale blue or white; spur *c.* 5 mm, curved upwards. 2n=40. *Marshes and fens. S.E. & E.C. Europe; S.E. France.* Bu Ga Hu Ju Rm Rs (W, K).

Species of Sect. *Viola* subsect. *Rostratae* (**12–25**) frequently hybridize with one another, but very rarely with species of other subsections. Of the 25 or more hybrids described, the following, which are all highly infertile, are probably the most frequent:— *V. mirabilis × riviniana*, *V. riviniana × rupestris*, *V. reichenbachiana × riviniana*, *V. canina × riviniana*, *V. canina × persicifolia*, *V. canina × pumila*, *V. canina × rupestris*.

Subsect. *Repentes* (Kupffer) W. Becker. Acaulescent, with long, creeping rhizome. Stipules broad, semi-adnate to petiole. Capsule and seed as in Subsect. *Rostratae*.

26. V. uliginosa Besser, *Prim. Fl. Galic.* **1**: 169 (1809). Rhizome slender; shoots short, erect, bearing rosettes of leaves. Leaves 4 × 2 cm, cordate, acute; petiole 2–10 cm. Stipules ovate-lanceolate, entire. Flowers 2–3 cm, arising from the base of the rosette, violet. Sepals 5–6 mm, ovate-lanceolate. Spur 3–4 mm, stout, violet. Fruiting peduncles up to 15 cm, erect. 2n=20. *Moorland and marshes. C. & W. parts of U.S.S.R. and S. & E. Fennoscandia, extending very locally westwards to E. Germany and N.W. Jugoslavia.* Da Fe Ge Ju Po ?Rm Rs (N, B, C, W) Su.

Greatly reduced by draining and cultivation, especially in the western part of its range; mainly now in White Russia, W. Ukraine and W.C. Russia. Hybrids with **18** have been described from Russia and Sweden.

Subsect. *Plagiostigma* (Godron) Kupffer. Perennial, acaulescent, with slender, creeping rhizome. Stipules free. Capsule and seed as in Subsect. *Rostratae*.

27. V. palustris L., *Sp. Pl.* 934 (1753). Leaves 3 or 4 in a rosette, 2–6 cm, wider than long, reniform, glabrous; petiole 2–6 cm, slender. Stipules 5–7 mm, ovate-lanceolate, entire or denticulate, free. Peduncles 4–15 cm; bracts at or below the middle. Flowers 1–1·5 cm, pale lilac, not fragrant. Sepals ovate, obtuse. Spur blunt, pale lilac. Capsule glabrous. *Bogs and marshes. Most of Europe, but rare in the south and east.* Au Az Be Br Bu Co Cz Da Fa Fe Ga Ge Hb He Ho Hs †Hu Is It Ju Lu No Po Rm Rs (N, B, C, W) Su.

(a) Subsp. **palustris**: Leaves obtuse; petioles glabrous. Bracts below the middle of peduncle. 2n=48. *Throughout the range of the species, except Portugal and Açores.*

(b) Subsp. **juressi** (Link ex K. Wein) Coutinho, *Not. Fl. Port.* **5**: 12 (1921): Summer leaves subacute; petioles usually with patent hairs. Bracts at about middle of peduncle. *W. Europe, northwards to Ireland.*

28. V. epipsila Ledeb., *Ind. Sem. Horti Dorpat.* 5 (1820). Like **27** but larger in all its parts; leaves always in pairs, cordate-orbicular to reniform, slightly longer than wide, with scattered hairs beneath; bracts in upper third of peduncle; flowers 1·5–2 cm. 2n=24. *Marshes. N. & E.C. Europe.* Cz Da Fe Ge Is No Po Rm Rs (N, B, C, E) Su.

Subsect. *Adnatae* W. Becker. Perennial; acaulescent; without creeping rhizomes or stolons. Stipules semi-adnate to petiole. Capsule and seed as in Subsect. *Rostratae*.

29. V. jooi Janka, *Österr. Bot. Wochenbl.* **7**: 198 (1857). Leaves 10–30 cm, in basal rosette, ovate-triangular, cordate with open sinus, dentate or serrate, glabrous; lamina of mature leaves ⅓–½ as long as petiole. Stipules linear-lanceolate. Flowers 1–2·5 cm, reddish- or violet-purple, fragrant. Sepals obtuse. Spur 4–6 mm, curved, obtuse, rather slender. Capsule glabrous. 2n=24. *Calcareous rocks in mountain regions.* ● *C. Romania.* Rm.

Sometimes regarded as a subspecies of **V. macroceras** Bunge in Ledeb., *Fl. Altaica* **1**: 256 (1829). The other subspecies occur in the Caucasus and the Altai.

30. V. selkirkii Pursh ex Goldie, *Edinb. Philos. Jour.* **6**: 324 (1822). Rhizome slender. Leaves 4–15 cm, in a basal rosette, broadly ovate, cordate with a deep sinus, crenate, subacute, glabrous except for a few scattered hairs; lamina $\frac{1}{3}$–$\frac{1}{2}$ as long as petiole. Stipules 5–8 mm, ovate-lanceolate, remotely fimbriate. Peduncles slightly exceeding leaves. Flowers *c.* 1·5 cm, pale violet. Sepals acute. Spur 5–7 mm, obtuse, stout. *Coniferous woods and damp places. Fennoscandia; N. & C. Russia southwards to c. 54° N.* Fe No Rs (N, C) Su.

31. V. pinnata L., *Sp. Pl.* 934 (1753). Leaves 3–6 cm, in a basal rosette, about as wide as long, deeply palmatifid, almost glabrous; petiole 4–10 cm. Stipules 1 cm, lanceolate, whitish. Flowers 1–2 cm, pale violet, fragrant. Sepals obtuse or subacute. Spur blunt, twice as long as calycine appendages. *Rocks, screes and grassland, mostly between* 1000 *and* 2000 *m; calcicole. Alps.* Au Ga He It Ju.

Subsect. *Borealiamericanae* W. Becker. Perennial; acaulescent; with short, thick, fleshy rhizome and without stolons. Stipules free. Capsule and seed as in Subsect. *Rostratae*.

32. V. obliqua Hill, *Hort. Kew.* 316 (1768) (*V. cucullata* Aiton). Glabrous or very sparsely pubescent. Lamina ovate-cordate with an open sinus, acute, crenate, about $\frac{1}{3}$ as long as petiole. Peduncles equalling or exceeding leaves. Flowers *c.* 2 cm, blue-violet, not fragrant. Sepals acute; calycine appendages 2–6 mm. Style capitate. Cleistogamous fruit 1–1·5 cm, ovoid-cylindrical, only slightly exceeding sepals. *Naturalized from gardens in Switzerland and Italy.* [He It.] (*North America.*)

Sect. DISCHIDIUM Ging. Perennial herbs. Stipules not leaf-like. Cleistogamous flowers produced. Style with 2-lobed stigma.

33. V. biflora L., *Sp. Pl.* 936 (1753). Rhizome slender, creeping. Plant with basal rosette of leaves and ascending, leafy flowering stems up to 20 cm. Leaves reniform, cordate, crenate, with scattered hairs; lamina 3–4 cm; petiole 4–12 cm. Stipules 3–4 mm, ovate to lanceolate, usually ciliate. Flowers 1·5 cm, yellow, not fragrant, the 4 posterior petals of the flower directed upwards as in Sect. *Melanium*. Sepals acute. Spur 2–3 mm. Capsule erect at maturity, glabrous. $2n=12$. *Damp or shady places, mainly in the mountains. Fennoscandia; N. & C. Ural; mountain regions of Europe from C. France and the Carpathians to S. Spain, C. Italy and Bulgaria.* Au Bu Co Cz Fe Ga Ge He Hs Hu It Ju No Po Rm Rs (N, C, W) Su.

Sect. MELANIUM Ging. Herbaceous. Stipules usually large, leaf-like, often divided. Lateral petals directed upwards. Open flowers blue or yellow; cleistogamous flowers not produced. Style geniculate at the base, capitate, with wide stigmatic aperture. Capsule erect at maturity, glabrous.

34. V. crassiuscula Bory, *Ann. Gén. Sci. Phys.* (*Bruxelles*) 3: 16 (1820) (*V. nevadensis* Boiss.). Caespitose perennial up to 15 cm. Leaves *c.* 1·5 cm, ovate-orbicular, entire, glabrous or very shortly pubescent; petiole slightly longer than lamina. Stipules entire, like the leaves but smaller. Peduncles 2–4 cm. Flowers 1–1·5 cm, bright violet, pink or whitish, the lower petal golden-yellow at the base. Spur very short, obtuse, only slightly exceeding calycine appendages. *Open habitats above* 2500 *m.* ● *S. Spain* (*Sierra Nevada*). Hs.

35. V. diversifolia (DC.) W. Becker, *Bull. Herb. Boiss.* ser. 2, 3: 892 (1903). Caespitose perennial 3–8(–15) cm, covered with rather short, stiff hairs. Leaves 1–2 cm, forming a rosette, entire; the lower suborbicular or broadly ovate, the upper ovate-oblong. Stipules oblong, with 3 laciniae. Peduncles 2–4 cm. Flowers up to 2 cm, violet, fragrant. Spur *c.* 5 mm, subulate, curved, greatly exceeding calycine appendages. *Alpine meadows and rocks.* ● *E. and C. Pyrenees.* Ga Hs.

36. V. cenisia L., *Sp. Pl.* ed. 2, 1325 (1763). Glabrous or pubescent, caespitose perennial. Stems 3–5 cm, numerous, procumbent. Leaves *c.* 1 cm, entire, the lower ovate, the upper oblong; lamina shorter than petiole. Stipules like the leaves but smaller, entire or with 1–2 small laciniae at the base. Peduncles 2–4 cm. Flowers 2–2·5 cm, bright violet. Spur 5–8 mm, about as long as the sepals, slender. $2n=20$. *Calcareous screes.* ● *S.W. Alps, and locally in C. Alps eastwards to 9° 20′ E.* Ga He It.

37. V. comollia Massara, *Prodr. Fl. Valtell.* 203 (1834). Caespitose perennial up to 10 cm. Underground stems filiform. Leaves entire, ovate to oblong, glabrous or sparingly pubescent; lamina about $\frac{1}{2}$ as long as petiole. Stipules like the leaves, but smaller, entire, or sometimes with a lacinia on the outer side. Flowers 1·5–2·5 cm. Sepals pubescent. Petals bright violet on inner side, with an orange or deep yellow spot, pale yellow on outer side; lower petal very broad. Spur short, slender, obtuse, curved, twice as long as calycine appendages. *Non-calcareous screes.* ● *S. Alps* (*Alpi Orobie*). It.

38. V. valderia All., *Fl. Pedem.* 2: 98 (1785). More or less pubescent, caespitose perennial 5–10(–20) cm. Leaves 1·5–2(–3) cm, entire; the lower suborbicular, ovate or oblong, the upper oblong or lanceolate. Stipules like the leaves but rather smaller, and with 2–7 laciniae at the base. Peduncles 4–7 cm. Flowers *c.* 2 cm, bright violet; lower petal oblong-obcordate. Spur 7–10 mm, slender, slightly curved, greatly exceeding calycine appendages. $2n=20$. *Open habitats; calcifuge.* ● *Maritime Alps.* Ga It.

39. V. magellensis Porta & Rigo ex Strobl, *Österr. Bot. Zeitschr.* 27: 229 (1877) (incl. *V. albanica* Halácsy). Glabrous or pubescent, caespitose perennial 4–10(–15) cm. Leaves 0·5–1·5 cm, ovate-orbicular, ovate or oblong, entire; lamina equalling petiole. Stipules like the leaves, sometimes smaller, sometimes with 1–2 laciniae. Flowers *c.* 2 cm, violet, pink or with upper petals dark reddish-violet. Spur 8–10 mm, straight or slightly curved, much longer than calycine appendages. $2n=22$. *Alpine pastures and calcareous screes.* ● *C. Appennini; Albania, N. Greece.* Al Gr It.

40. V. brachyphylla W. Becker, *Feddes Repert.* 20: 73 (1924). Glabrous, caespitose perennial, with procumbent stems 5–10 cm. Leaves up to 1 cm, linear- or oblong-spathulate, entire; petiole short. Stipules like leaves but shorter, entire, or rarely with 1 or 2 laciniae. Flowers large, *c.* 2 cm wide, yellow. Spur 6–8 mm, slender, yellow or violet. *Subalpine habitats.* ● *S. Jugoslavia and N. Greece.* Gr Ju.

V. dukadjinica W. Becker & Košanin, *Feddes Repert.* 23: 145 (1926), from N. and C. Albania, on serpentine (? and calcareous) soils at 1400–2000 m, is like **40**, but has leaves 1·5–3 cm, and spur 10–12 mm. Other plants resembling **40**, but with sparse, short hairs, leaves 2 cm and flowers yellow or violet, have been collected from N.W. Greece (Smolikas). Further investigation of all these, based on more material, is needed.

41. V. perinensis W. Becker, *Feddes Repert.* 17: 74 (1921). Glabrous, caespitose perennial 6–8 cm. Leaves 1·5–2·5 cm, entire; lamina oblong, obtuse, rather thick; petiole twice as long as lamina. Stipules like the leaves, with a long petiole. Flowers *c.* 2 cm, yellow, or sometimes violet. Sepals wide, almost parallel-sided, obtuse. Spur very short and rather stout, scarcely exceeding calycine appendages. *Calcareous rocks.* ● *Mountains of S. Bulgaria and N. Greece.* Bu Gr.

42. V. stojanowii W. Becker, *Feddes Repert.* **19**: 332 (1924). Hispid, caespitose perennial 8–10 cm. Leaves 2 cm × 2–3 mm, narrowly oblong-spathulate, obtuse, entire, long-petiolate. Stipules *c.* 1·5 cm, like the leaves, with a single lacinia on the outer side at the base. Flowers 1–1·25 cm, yellow, or the upper petals tinged with violet. Spur 3–5 mm, rather stout, curved upwards, scarcely twice as long as calycine appendages. *Subalpine meadows.* ● *S. Bulgaria & N.E. Greece.* Bu Gr.

43. V. fragrans Sieber, *Reise Kreta* **2**: 320 (1823). Subglabrous to hispid, caespitose perennial 5–10(–15) cm. Leaves *c.* 1·5 cm, oblong to oblong-linear, obtuse, entire, gradually narrowed to a rather long petiole. Stipules like the leaves but shorter. Flowers usually *c.* 1 cm but rather variable in size, 1–2 per stem, yellow or pale violet. Spur *c.* 3 mm, stout, 2–3 times as long as calycine appendages. *Mountain rocks.* ● *Greece, Kriti.* Cr Gr.

44. V. poetica Boiss. & Spruner in Boiss., *Diagn. Pl. Or. Nov.* **1**(6): 21 (1845). Like **43** but glabrous; leaves up to 2·5 cm, oblong-ovate, abruptly contracted into petiole; petiole 3 times as long as lamina; flowers 1–4 per stem, violet or blue. *Rock-crevices at c. 2000 m.* ● *Mountains of S.C. Greece (Fokis, Fthiotis).* Gr.

45. V. grisebachiana Vis., *Mem. Ist. Veneto* **10**: 436 (1861). Acaulescent perennial 3–8 cm. Leaves 1·5–3 cm, ovate-orbicular, obtuse, entire or slightly crenate, subglabrous, abruptly contracted into a relatively long petiole. Stipules like the leaves but smaller. Flowers *c.* 2 cm, basal, violet. Sepals broad, obtuse. Spur 3–4 mm, obtuse, exceeding calycine appendages. 2*n*=22. *Alpine meadows.* ● *C. part of Balkan peninsula.* Al Ju Bu.

46. V. alpina Jacq., *Enum. Stirp. Vindob.* 159 (1762). Acaulescent perennial 4–10 cm, with a stout stock. Leaves 1·5–3·5 cm, in a basal rosette, ovate-orbicular, crenate, truncate or subcordate at base, obtuse, glabrous, long-petiolate. Stipules 5 mm, oblong or lanceolate, semi-adnate to petiole. Peduncles 2–5 cm. Flowers (1·5–)2–3 cm, violet. Spur 3–4 mm, obtuse, only slightly longer than calycine appendages. 2*n*=22. *Alpine meadows and screes; calcicole.* ● *N.E. Alps; Carpathians.* Au Cz Po Rm.

47. V. nummulariifolia Vill., *Prosp. Pl. Dauph.* 26 (1779). Glabrous, delicate perennial 3–5 cm, with numerous, short, procumbent stems. Leaves 1–2 cm, ovate to orbicular, subcordate, entire; petiole equalling or rather longer than lamina. Stipules *c.* 5 mm, oblong-lanceolate, acute, the lower entire, the upper remotely dentate. Flowers 1 cm, bright blue. Spur obtuse, about twice as long as the rather short calycine appendages. *Alpine meadows and rocks; calcifuge.* ● *Maritime Alps, Corse.* Co Ga It.

48. V. calcarata L., *Sp. Pl.* 935 (1753). Low-growing perennial; stems up to 5 cm. Leaves 1–4 cm, rosulate, orbicular, ovate to lanceolate, crenate, glabrous or slightly hairy on the margins; petiole equalling or exceeding lamina. Stipules 0·5–1·5 cm, oblong, entire or dentate or the upper sometimes pinnatifid. Peduncles 3–9 cm, 1–4 per stem. Flowers 2–4 cm long and up to 3 cm wide, violet or yellow. Petals very wide, the lateral and upper often wider than the lower. Spur 8–15 mm, about equalling the petals, straight or slightly curved upwards. *Meadows, pastures and screes, mostly above 1500 m.* ● *Alps and S. Jura; W. part of Balkan peninsula.* Al Au Ga Ge ?Gr He It Ju.

(a) Subsp. **calcarata**: Stems up to 5 cm; leaves medium-sized; stipules subentire or dentate; stems 1- to 2-flowered; flowers usually violet. 2*n*=40. *Alps, S. Jura.*

(b) Subsp. **zoysii** (Wulfen) Merxm., *Feddes Repert.* **74**: 30 (1967) (*V. zoysii* Wulfen): Stems very short; leaves broadly ovate to orbicular; stipules entire or dentate with 1–2 short laciniae; peduncles rather short; stems 1–flowered; flowers usually yellow. 2*n*=40. *Calcareous soils. W. part of Balkan peninsula; S.E. Alps (Karawanken).*

(c) Subsp. **villarsiana** (Roemer & Schultes) Merxm., *Feddes Repert.* **74**: 30 (1967): Stems somewhat elongated; heterophyllous, upper leaves narrower than lower; upper stipules pinnatifid; stems 1- to 4-flowered; flowers yellow, blue or white. 2*n*=40. *S.W. Alps.*

Subsp. (c) is intermediate between **48**(a) and **49**.

49. V. bertolonii Pio, *De Viola* 34 (1813) (*V. heterophylla* Bertol., non Poiret). Glabrous, rarely puberulent or hairy perennial with stems up to 30 cm. Leaves variable in shape and size, but the lower always different in shape from the upper; the lower usually orbicular or ovate, obtuse, crenate, more or less long-petiolate, with stipules much shorter than the petiole and not or only slightly divided; the upper usually narrow, linear, entire, with pinnately or palmately divided stipules which have 3–5 narrow segments, of which the terminal is the longest. Flowers large, violet or yellow, more or less square in face-view; the petals usually contiguous, the lower petal the widest. Spur long, about equalling the petals, acute, straight or slightly curved. *Maritime Alps, N. Appennini; S. Italy, N.E. Sicilia.* Ga It Si.

(a) Subsp. **bertolonii**: Basal leaves usually small, lanceolate to rhombic. Stipules often pinnate. Flowers 2–3 cm in diameter. 2*n*=40. *Maritime Alps and N. Appennini.*
Intermediates occur between this and **48**(c).

(b) Subsp. **messanensis** (W. Becker) A. Schmidt, *Flora (Regensb.)* **154**: 159 (1964): Basal leaves larger than in (a), ovate. Stipules palmate. Flowers usually 3–4 cm in diameter. 2*n*=40. *S. Italy and N.E. Sicilia.*

V. heterophylla Bertol. subsp. **graeca** (W. Becker) W. Becker, *Beih. Bot. Centr.* **26**(2): 326 (1910), from S.E. Italy (Monte Gargano), Albania and Greece, with very narrow upper leaves, very long, pinnate stipules with long, filiform segments, flowers narrowly rectangular in face-view and 2*n*=20, is doubtfully related to **49**, and needs further investigation.

50. V. splendida W. Becker, *Bull. Herb. Boiss.* ser. 2, **2**: 750 (1902). Hairy perennial with ascending stems, 10 cm when young, later up to 50 cm. Lower leaves *c.* 1½ cm, cordate or subcordate, obtuse; upper 2–3 × 1 cm and more acute, all obtusely crenate and with a long, winged petiole. Stipules half as long as leaves, subpinnate, with 6–10 lanceolate laciniae, the terminal larger and entire. Peduncles 1–3, much longer than the leaves. Flowers 3–4 cm; petals yellow or violet, with rounded apex; spur equalling lower petal, straight, subulate. 2*n*=40. *Calcicole.* ● *S. Italy (mountains near Salerno).* It.

51. V. aethnensis Parl., *Fl. Ital.* **9**: 185 (1890). Pubescent perennial *c.* 10 cm, with ascending stems. Leaves 1–1·5 cm, not markedly crenate; the lower orbicular to ovate, the upper lanceolate to linear-lanceolate. Stipules 3-partite, not deeply divided, segments oblong, obtuse, rather broad. Flowers 1·5–2 cm, violet; spur 8–10 mm, usually curved. 2*n*=40. *Volcanic soils, 1500–2500 m.* ● *Sicilia (Etna).* Si.

52. V. nebrodensis C. Presl in J. & C. Presl, *Del. Prag.* 26 (1822). Dwarf, subglabrous perennial. Leaves 1–1·5 cm, crowded, narrowly ovate, obtuse, cuneate; lamina about equalling petiole. Stipules 0·5–1 cm, 3-partite, with a spathulate terminal and 2 filiform lateral segments. Peduncles 3–6 cm. Flowers 2–2·5 cm, dark violet; spur 10–12 mm, slender, straight or slightly curved. $2n = 20$. *Limestone rocks, c.* 1800 *m.* ● *N. Sicilia* (*Le Madonie*). Si.

53. V. corsica Nyman, *Syll.* 228 (1854) (*V. bertolonii* Salis, non Pio). Glabrescent perennial 10–20 cm, with slender stems. Lower leaves small, more or less rhombic, subcrenate; upper 2–3 cm, oblong to linear, subcrenate to entire. Stipules short, linear, foliaceous, with 1–4 laciniae at the base. Peduncles 4–8 cm. Flowers up to 3·5 cm, violet, rarely yellow, narrowly rectangular in face view; petals not contiguous. Spur 10–15 mm, 2–3 times as long as sepals and slightly curved. *Mountain pastures. Corse, Sardegna, Elba.* Co It Sa.

(a) Subsp. **corsica**: Upper leaves narrowly oblong to broadly linear; stipules similar but usually with 1 short, filiform, basal segment. Spur stout. $2n = 52$, *c.* 120. *Corse, Sardegna.*

(b) Subsp. **ilvensis** (W. Becker) Merxm., *Feddes Repert.* **79**: 57 (1968): Upper leaves narrowly linear; stipules similar but usually with 2–4 short, filiform, basal segments. Spur relatively slender. $2n = 52$. *Elba.*

54. V. munbyana Boiss. & Reuter, *Pugillus* 15 (1852). Glabrous, slightly fleshy perennial 10–20 cm. Leaves *c.* 1·5 cm, crowded at the base, forming a cushion, subcordate, obtuse, markedly crenate, the upper somewhat elongated; petiole winged, longer than lamina. Stipules 1·5–2·5 cm, 3- to 5-partite, pinnatifid; terminal segment the largest, lanceolate; lateral segments linear. Flowers 3–5 cm, dark violet or yellow; calycine appendages 4–5 mm; spur *c.* 15 mm. $2n = 52$. *Vertical limestone rocks. N.W. Sicilia.* Si. (*N. Africa.*)

The plants described here have been known previously as *V. nebrodensis* var. *lutea* Guss. and var. *grandiflora* Guss.

55. V. pseudogracilis Strobl, *Österr. Bot. Zeitschr.* **27**: 227 (1877). Hairy or subglabrous perennial 5–15 cm, with ascending stems. Leaves 1–2 cm, the lower orbicular to ovate, truncate at base, obtuse, the upper lanceolate; petiole 1–3 times as long as lamina. Stipules 1–2 cm, irregularly and deeply divided into 3–7 linear to spathulate segments, the terminal the largest. Flowers 2–3 cm, yellow or bluish-violet; spur 8–10 cm. *Open habitats; calcicole.* ● *S.C. Italy.* It.

(a) Subsp. **pseudogracilis**: Always subglabrous. Lower leaves not cordate. Segments of stipules linear. Flowers yellow or bluish-violet. Sepals less than 10 mm. Spur stout. $2n = 34$. *Near Napoli and Salerno.*

(b) Subsp. **cassinensis** (Strobl) Merxm. & A. Schmidt, *Feddes Repert.* **74**: 30 (1967): Hairy or subglabrous. Lower leaves cordate. Segments of stipules linear to spathulate. Flowers yellow. Sepals 10 mm. Spur slender. $2n = 34$. *Prov. Frosinone* (*Monte Cassino*).

56. V. eugeniae Parl., *Nuovo Gior. Bot. Ital.* **7**: 68 (1875). Rather compact perennial usually less than 3 cm. Lower leaves subrosulate, orbicular to ovate, usually markedly crenate; lamina equalling or shorter than petiole. Stipules pinnately divided, with 1–2 linear or oblong segments at each side, usually at the base, and a terminal crenate segment which is usually bigger than the lateral. Flowers 2–4 cm; petals wide, like those of **48**, yellow or violet. Spur 2–6 mm, 2–3 times longer than the calycine appendages. ● *Appennini.* It.

(a) Subsp. **eugeniae**: Dwarf (1–2 cm). All leaves orbicular to ovate. Peduncles 4–10 cm. Spur 2–4 mm, stout, twice as long as calycine appendages. $2n = 34$. *Calcicole; usually above* 1000 *m.*

(b) Subsp. **levieri** (Parl.) A. Schmidt, *Ber. Deutsch. Bot. Ges.* **77**: 96 (1965): Taller than subsp. (a). Upper leaves broadly lanceolate. Peduncles usually more than 10 cm. Spur up to 6 mm. $2n = 34$. *Usually below* 1000 *m.*

57. V. oreades Bieb., *Fl. Taur.-Cauc.* **3**: 167 (1819). Glabrous perennial, with slender, creeping rhizome and ascending stems 4–8 cm. Lower leaves orbicular to ovate, the upper ovate to oblong, remotely crenate, long-petiolate. Stipules 1–2 cm, the lower broadly ovate, the upper oblong, toothed or pinnatifid. Flowers 2–4·5 cm, yellow or violet. Sepals rather large, with conspicuous appendages. Petals, especially the 2 upper, very wide, almost orbicular, twice as long as sepals. Spur 4–8 mm, equalling or somewhat exceeding calycine appendages, stout (2 mm in diameter). *Alpine meadows and rocks at* 1500 *m. Krym.* Rs (K). (*Caucasian region.*)

58. V. eximia Form., *Verh. Naturf. Ver. Brünn* **38**: 221 (1900). Subglabrous, caespitose perennial up to 10 cm. Leaves ovate or oblong-ovate, remotely crenate, lamina equalling petiole. Stipules pinnatipartite, with 1–3 outer and 1–2 inner linear segments, the terminal segment oblong-lanceolate. Flowers 2–2·5 cm, yellow or deep violet. Spur 4–5 mm, stout, curved, about twice as long as calycine appendages. *Alpine meadows.* ● *W. Macedonia.* Gr Ju.

59. V. doerfleri Degen, *Denkschr. Akad. Wiss. Math.-Nat. Kl.* (*Wien*) **64**: 710 (1897). Caespitose perennial 5–12 cm, densely and shortly hairy. Leaves 1–2 cm, slightly crenate; the lower ovate or suborbicular, the upper oblong; petiole equalling or exceeding lamina. Stipules 3- to 5-partite, lateral segments linear, central like the leaf, hairy. Flowers *c.* 2 cm, deep violet. Spur *c.* 7 mm, deflexed, violet, much exceeding the rather large calycine appendages. *Sandy screes,* 1750–2500 *m.* ● *S. Jugoslavia* (*Kaimak-čalan*). Ju.

60. V. allchariensis G. Beck, *Jahres-Kat. Wien. Bot. Tauschver.* **1894**: 6 (1894). Caespitose perennial 8–25 cm, with a cluster of rather stout stems, woody at the base. Stems and leaves grey, with a short, rather dense pubescence. Lower leaves elliptic-oblong, with a cuneate base, remotely crenate-serrate; the upper linear, entire. Stipules 2- to 5-partite; segments linear. Peduncles 6–12 cm. Flowers 2–2·5 cm. Upper petals *c.* 1·5 cm, very wide. Spur *c.* 5 mm, obtuse, curved upwards, about twice as long as calycine appendages. *Rocky hillsides.* ● *Macedonia and E. Albania.* Al Gr Ju.

(a) Subsp. **allchariensis**: 10–25 cm. Lower leaves 4–5 cm, oblong. Flowers violet or yellow. *C. Macedonia, northwards to Jakupica.*

(b) Subsp. **gostivarensis** W. Becker & Bornm., *Feddes Repert.* **17**: 75 (1921): 8–12 cm. Lower leaves 2–3 cm, broadly ovate. Flowers smaller than in (a), yellow. *N.W. Macedonia and E. Albania.*

V. raunsiensis W. Becker & Košanin, *Bull. Inst. Jard. Bot. Univ. Beograd* **1**: 33 (1928), from serpentine at *c.* 1600 m in N. Albania (*c.* 10 km. W. of Kukës), is like **60** but completely glabrous.

61. V. frondosa (Velen.) Hayek, *Prodr. Fl. Penins. Balcan.* **1**: 511 (1925). Compact perennial 1–3 cm. Leaves orbicular, crenate. Stipules pinnately divided; lateral segments 1 on one side, 2–3 on the other; terminal segment subfoliaceous. Flowers medium-sized, yellow and violet. Sepals large, leaf-like, dentate, obtuse. Spur scarcely twice as long as calycine appendages. *Alpine pastures.* ● *S. Makedonija* (*S.E. of Prilep*). Ju.

62. V. arsenica G. Beck, *Jahres-Kat. Wien. Bot. Tauschver.* **1894**: 6 (1894). Glabrous perennial, with ascending stems up to 25 cm. Leaves *c.* 4 cm, broadly ovate or orbicular, truncate or subcordate, crenate, long-petiolate. Stipules *c.* 1 cm, not divided, lanceolate, acuminate, denticulate, shorter than petiole. Peduncles *c.* 10 cm. Flowers 2×2 cm, yellow. Spur rather thick, curved upwards, twice as long as calycine appendages. *Grassy places, on arsenical soil.* ● *S. Makedonija (Alšar, 40 km S.W. of Prilep).* Ju.

63. V. cornuta L., *Sp. Pl.* ed. 2, 1325 (1763). Perennial with slender rhizome. Stems 20–30 cm, ascending. Leaves 2–3(–5) cm, ovate, acute, crenate, hairy beneath. Stipules 0·5–1·5 cm, equalling or exceeding petiole, ovate-triangular, palmately incised. Flowers 2–3(–4) cm, violet or lilac, fragrant. Spur 10–15 mm, slightly curved, greatly exceeding calycine appendages. $2n=22$. *Mountain rocks and pastures.* ● *Pyrenees; naturalized from gardens in several regions.* Ga Hs [Au Br Cz He It Ju Rm].

V. montcaunica Pau, *Anal. Soc. Esp. Hist. Nat.* **23** (Act.): 129 (1895), from N. & C. Spain, has deeply palmatifid stipules and flowers half the size of **63**, with a shorter spur. Further investigation is needed.

64. V. orphanidis Boiss., *Fl. Or.* **1**: 464 (1867). Softly hairy perennial. Stems 20–70 cm, numerous, ascending. Leaves 2–4 cm, ovate or orbicular, sometimes subcordate, crenate or serrate, hairy mainly along the veins and at the margins; lamina equalling or exceeding petiole. Stipules 1–2 cm, obliquely ovate. Flowers 2–3 cm, violet or blue. Spur 4–5 mm, slender, curved, twice as long as calycine appendages. *Mountain grassland and woodland margins.* ● *Mountains of Balkan peninsula from Crna Gora to C. Greece and eastwards to S.W. Bulgaria.* Al Bu Gr Ju.

(a) Subsp. **orphanidis**: Stems up to 50 cm. Stipules deeply dentate or subpinnately divided; terminal segment larger than the rest. Flowers 2 cm. Petals blue, with a yellow spot on the lower petal. *Throughout the range of the species.*
(b) Subsp. **nicolai** (Pant.) Valentine, *Feddes Repert.* **74**: 30 (1967) (*V. nicolai* Pant.): Stems up to 70 cm. Stipules obliquely ovate to ovate-lanceolate, the lower lobed, the upper deeply serrate. Flowers 2–3 cm, violet. *Mountains of N. & E. Crna Gora.*

65. V. dacica Borbás, *Magyar Növ. Lapok* **13**: 79 (1890) (*V. prolixa* (Adamović) Pančić). Perennial 15–35 cm; stems subglabrous or slightly hairy. Leaves 2–4 cm, ovate or elliptical, crenate, glabrous, or margins sometimes ciliate; lamina equalling or exceeding petiole. Stipules *c.* ½ as long as leaves, obliquely ovate-lanceolate, pinnately divided to ½ or ⅓ of their width, or coarsely incise-serrate. Flowers 2–3 cm, usually rather large. Petals violet or yellowish. Spur very short, equalling or scarcely exceeding calycine appendages. *Grassland and forest-margins in mountain districts.* ● *From Albania and Bulgaria northwards to S.E. Poland.* Al Bu ?Cz Ju Po Rm Rs (W).

V. polyodonta W. Becker, *Beih. Bot. Centr.* **26**(2): 332 (1910), from Jugoslavia (E. Bosnia), which differs slightly from **65** in size, indumentum and stipule-shape, is probably best regarded as a variety.

66. V. elegantula Schott, *Österr. Bot. Wochenbl.* **7**: 167 (1857). Glabrous or shortly pubescent perennial 10–30 cm. Leaves 1–2(–3) cm, the lower orbicular, the upper ovate-lanceolate or lanceolate. Stipules ½–¾ as long as leaves, ovate, pinnately divided, the undivided part wider than the length of the segments, with 4–10 segments on the outer and 2–4 on the inner side.

Flowers 2–2·5 cm; petals not very wide, violet or yellow or parti-coloured, rarely white or pink. Spur 5–8 mm, straight or slightly curved, slender, 3 times as long as calycine appendages. *Subalpine and alpine meadows.* ● *W. Jugoslavia and Albania.* Al Ju.

67. V. athois W. Becker, *Bull. Herb. Boiss.* ser. 2, **2**: 854 (1902). Puberulent perennial 10–15 cm. Leaves 1·5–3 cm, the lower orbicular, the upper ovate or oblong, crenulate. Stipules 5- to 7-partite, pinnately divided almost to the base; lateral segments often very short, linear or oblong; terminal segment leaf-like, crenulate. Flowers *c.* 2 cm, violet. Spur *c.* 8 mm, curved. $2n=20$. *Mountain rocks.* ● *N.E. Greece (Athos).* Gr.

68. V. gracilis Sibth. & Sm., *Fl. Graec. Prodr.* **1**: 146 (1806). Puberulent perennial up to 30 cm. Leaves (1–)2–3 cm, orbicular-ovate or oblong, obscurely crenate. Stipules 4- to 8-partite, pinnately divided to the base; segments linear or oblong, the central segment larger, leaf-like, crenate. Flowers 2–3 cm, violet or yellow. Spur 6–7 mm, straight or slightly curved, 2–3 times as long as calycine appendages. $2n=20$. *Rocks and alpine meadows.* ● *Mountains of Balkan peninsula, from Crna Gora to N.E. Greece.* Al Bu Ju Gr.

Taxa which are probably related to **67** and **68** are **V. heterophylla** Bertol. subsp. **euboea** (Halácsy) W. Becker, *Beih. Bot. Centr.* **26**(2): 326 (1910), from E. Greece, and **V. cephalonica** Bornm., *Mitt. Thür. Bot. Ver.* nov. ser., **37**: 50 (1927), from W. Greece (Kefallinia). It is possible that **V. orbelica** Pančić, *Magyar Bot. Lapok* **32**: 9 (1933), from S. Bulgaria, should also be placed here; it has been thought to be related to **78**(c), but an unconfirmed chromosome count of $2n=20$ suggests relationship with **68**.

69. V. speciosa Pant., *Österr. Bot. Zeitschr.* **23**: 79 (1873). Glabrescent or shortly hairy caespitose perennial, with many short stems up to 10 cm. Leaves 1·5–2·5 cm, crenate, the lower ovate, the upper ovate-lanceolate. Stipules ½–¾ as long as leaves, obliquely ovate, deeply pinnatifid; segments linear, entire, about equalling the width of the undivided part. Flowers *c.* 3×3 cm, deep violet; petals almost orbicular. Spur 5–6 mm, straight, twice as long as calycine appendages, slender (1 mm in diameter). *Meadows, pastures and screes.* ● *Mountains of Crna Gora and N. Albania.* Al Ju.

70. V. beckiana Fiala, *Glasn. Muz. Bosni Herceg.* **7**: 423 (1895). Glabrous perennial. Stems 12–20 cm, ascending from procumbent base, thickly clothed with leaves. Leaves 2·5–4·5 cm × 2–5 mm, linear-lanceolate, remotely serrulate. Stipules about half as long as the leaves; the lower linear, subentire, the upper digitately or pinnately divided, with linear segments. Peduncles 5–8 cm. Flowers 2·5–4×2–3 cm, yellow or deep violet. Sepals oblong-lanceolate, obtuse or shortly acuminate. Spur 3–4 mm, curved at the end, slightly longer than calycine appendages. *Serpentine and calcareous rocks, 1000–1800 m.* ● *S. Jugoslavia, Albania.* Al Ju.

V. pascua W. Becker, *Bull. Inst. Jard. Bot. Univ. Beograd* **1**: 34 (1928), from Macedonia, is like **70** but with pubescent stems not more than 12 cm, wider leaves, and spur *c.* 6 mm.

71. V. rhodopeia W. Becker, *Beih. Bot. Centr.* **26**(2): 334 (1910). Like **70** but flowers 1·5–2 cm; sepals lanceolate, long-acuminate; petals yellow; spur 6 mm, straight, slender, pale violet. *Damp mountain pastures.* ● *S. Bulgaria (W. Rodopi).* Bu.

72. V. dubyana Burnat ex Gremli, *Neue Beitr. Fl. Schweiz* **5**: 15 (1890). Glabrous or shortly hairy perennial, without leafy stolons. Stems 10–30 cm; upper internodes short. Lower leaves 2–4 cm, orbicular, the upper narrowly lanceolate or linear. Stipules digitately divided to the base; segments linear, 4–6 on the outer and 2–3 on the inner side, the terminal scarcely wider. Flowers 2–2·5 cm, violet; lower petal with a yellow spot. Spur 5–6 mm, curved downwards, slender, much longer than calycine appendages. $2n=20$. *Dry pastures and rocky places; calcicole.* ● *Italian Alps, between 7° 30' and 11° E.* It.

73. V. declinata Waldst. & Kit., *Pl. Rar. Hung.* **3**: 248 (1807). Glabrous perennial 15–40 cm, with leafy stolons. Leaves 2–3 cm, the lower orbicular or ovate, the upper oblong or linear-lanceolate; petiole short. Stipules $\frac{1}{2}$–$\frac{3}{4}$ as long as leaves, pinnately divided into linear segments, with 3–4 on the outer and 2–3 on the inner side, the terminal scarcely longer. Flowers 2·5–3·5 cm, deep violet. Spur 3–4 mm, curved, slender, scarcely exceeding calycine appendages. *Pastures and meadows,* 800–2000 *m.* ● *E. & S. Carpathians.* ?Bu Cz Rm Rs (W).

74. V. lutea Hudson, *Fl. Angl.* 331 (1762). Perennial, with a slender, branching rhizome and without leafy stolons. Stems 10–20(–40) cm, usually simple, glabrous. Leaves ovate, oblong or lanceolate, glabrous or pubescent. Stipules palmately or pinnately divided into 3–5 segments, sometimes pubescent on the margins and the veins. Peduncles usually *c.* 6 cm. Flowers 1·5–3 cm, yellow, violet or parti-coloured; lower petal with a dense network of veins. Spur 3–6 mm, short, slender, twice as long as calycine appendages, or sometimes scarcely longer. *W. & C. Europe, southwards to the Pyrenees.* Au Be Br Cz Ga Ge Hb He Ho Hs Po.

A very variable species. Typical populations from the east and west of its range can be described as follows:

(a) Subsp. **lutea**: Usually *c.* 10 cm. Stems less than 1 mm in diameter. Leaves and stipules distinctly pubescent. Lower leaves 1–2 cm, ovate; upper 2–4 cm, ovate-lanceolate to lanceolate. Segments of stipules 1–1·5 mm wide. Flowers 1·5–2·5(–3) cm; lower petal 1–1·5 cm wide. $2n=48$. *W. Europe, Switzerland.*

(b) Subsp. **sudetica** (Willd.) W. Becker, *Beih. Bot. Centr.* **18**(2): 388 (1905) (*V. sudetica* Willd.): Usually *c.* 15 cm. Stems more than 1 mm in diameter. Leaves and stipules glabrous or puberulent. Lower leaves *c.* 1·5 cm, ovate to orbicular, median and upper 2–4 cm, narrowly lanceolate. Segments of stipules 1·5–3 mm wide. Flowers 2–3 cm; lower petal up to 2 cm wide. *Mountains of C. Europe, from c. 14° E. eastwards.*

75. V. bubanii Timb.-Lagr., *Congr. Sci. Fr.* 19 *Sess.* (*Toulouse*) **1**: 280 (1852). Perennial, stems 10–25 cm. Leaves and stipules usually with patent hairs. Leaves crenate, the lower orbicular, the upper oblong. Stipules less than 1 cm, deeply palmatifid; segments linear-oblong, 3–5 on the outer and 2–3 on the inner side, terminal slightly wider. Flowers 2–3 cm, violet. Spur *c.* 10 mm, straight or slightly curved. $2n=c.$ 128. ● *Pyrenees and mountains of N. Spain.* Ga Hs.

V. trinitatis Losa, *Contrib. Estud. Fl. Veg. Prov. Zamora* 80 (1949), and **V. palentina** Losa, *Collect. Bot.* (*Barcelona*) **2**: 295 (1950), both from N.W. Spain, have rather large flowers with a long spur, though glabrous or with few hairs. They may be varieties of **75**.

76. V. hispida Lam., *Fl. Fr.* **2**: 679 (1778) (*V. rothomagensis* auct.). Perennial, the whole plant with patent hairs. Stems up to 25 cm. Leaves 1·5–3 cm, crenate, the lower suborbicular, subcordate, the upper ovate or oblong, subcordate or truncate.

Stipules palmatifid; segments linear or oblong, entire, terminal slightly larger than lateral. Flowers *c.* 2 cm, violet, or yellowish. Spur *c.* 4 mm, not much longer than calycine appendages. *Calcareous cliffs.* ● *N.W. France* (*near Rouen*). Ga.

V. cryana Gillot, *Bull. Soc. Bot. Fr.* **25**: 255 (1878), from chalk slopes in N.C. France (S.E. of Tonnerre), is now extinct. It was like **76** but rather smaller and glabrous.

77. V. langeana Valentine, *Feddes Repert.* **79**: 57 (1968). (*V. caespitosa* Lange, non D. Don). Sparsely papillose-hairy perennial, forming a mat up to 30 cm across, from which vertical stems up to 20 cm arise. Lower leaves more or less rosulate, obovate; the upper oblong or narrowly spathulate, all slightly crenate or subentire. Stipules pinnatifid, with linear segments, the terminal slightly the largest. Flowers 1–1·5 cm. Petals yellow, twice as long as calyx. Spur slightly longer than calycine appendages. *Sandy, acid soil.* ● *Mountains of C. Spain & C. Portugal.* Hs Lu.

78. V. tricolor L., *Sp. Pl.* 935 (1753). Glabrous to shortly pubescent annual, biennial or perennial; rhizome absent or short. Stems ascending or erect, usually branched. Lower leaves cordate to ovate, obtuse, crenate; the upper ovate to lanceolate, more or less cuneate at base, crenate. Stipules deeply and pinnately lobed, the terminal segment larger than the others, usually lanceolate, entire or crenate, usually leaf-like. Flowers 1–2·5(–3·5) cm, violet, yellow or parti-coloured. Corolla distinctly exceeding calyx. Spur 3–6·5 mm, variable in length, up to twice as long as the calycine appendages. *Most of Europe, but rare in the south and only on mountains.* All except Az Bl Cr Lu Sa Sb Si.

The following grouping of subspecies is provisional, and further investigation is needed.

1 Usually annual
  2 Stems not more than 40 cm; flowers usually blue-violet
                                        (a) subsp. tricolor
  2 Stems up to 80 cm; flowers usually yellow   **(e) subsp. matutina**
1 Usually perennial
  3 Lateral petals without veins   **(c) subsp. macedonica**
  3 Lateral petals with distinct veins
    4 Low-growing, maritime   **(b) subsp. curtisii**
    4 Ascending or erect, montane   **(d) subsp. subalpina**

(a) Subsp. **tricolor** (incl. *V. luteola* Jordan, *V. nemausensis* Jordan): Annual, rarely perennial. Stems (5–)15–25(–40) cm, usually more or less erect. Median and upper leaves very variable in size (1–5 cm) and shape (ovate to lanceolate). Flowers blue-violet, rarely entirely yellow. Spur 3–5 mm, slightly exceeding calycine appendages. $2n=26$. *Cultivated ground, grassland. Throughout the range of the species.*

(b) Subsp. **curtisii** (E. Forster) Syme in Sowerby, *Engl. Bot.* ed. 3, **2**: 26 (1865) (incl. *V. litoralis* Sprengel): Perennial, rarely annual. Stock vertical, not or scarcely creeping. Stems 3–15 cm. Leaves and stipules narrow, ovate-lanceolate to lanceolate, and fleshy. Flowers variable in colour. Lateral and lower petals distinctly veined. Spur often twice as long as calycine appendages. $2n=26$. *Dunes and dry grassland, usually near the sea. W. Europe and the Baltic region.*

(c) Subsp. **macedonica** (Boiss. & Heldr.) A. Schmidt, *Feddes Repert.* **74**: 30 (1967) (*V. saxatilis* F. W. Schmidt subsp. *macedonica* (Boiss. & Heldr.) Hayek): Usually perennial, but sometimes annual or biennial. Stems *c.* 20 cm, usually erect. Leaves and stipules often narrow, lanceolate. Upper petals bright violet, remainder yellowish; lateral petals without veins, lower petal with few or no veins. $2n=26$. *Meadows, up to more than* 2000 *m. Balkan peninsula.*

(d) Subsp. **subalpina** Gaudin, *Fl. Helv.* **2**: 210 (1828) (incl. *V. elisabethae* Klokov, *V. monticola* Jordan, *V. bielziana* Schur, *V. saxatilis* F. W. Schmidt): Perennial or biennial. Stems 20–30(–40) cm, ascending or erect. Leaves and stipules very variable in size and shape, but wider than in (c). Flowers 2–3·5 cm, yellow, or the upper petals violet. Lateral and lower petals distinctly veined. Spur 5–6 mm, about twice as long as calycine appendages. $2n = 26$. *Subalpine meadows and screes, up to 2700 m. Mountains of S. & C. Europe, from N. Spain to the Carpathians and Krym.*

(e) Subsp. **matutina** (Klokov) Valentine, *Feddes Repert.* **74**: 31 (1967) (*V. matutina* Klokov): Annual or biennial. Stems 10–80 cm. Leaves up to 8 × 3 cm, crenate, with 4–10 teeth on each side. Stipules pinnatipartite; terminal segment up to 4 cm. Flowers 1·5–2·5 cm, yellowish-white to yellow. Spur 4–5 mm. *Wood-margins, thickets and open habitats.* ● *S. & E. Ukraine, N. Moldavia and adjoining parts of S. Russia.*

V. thasia W. Becker, *Bull. Herb. Boiss.* ser. 2, **2**: 855 (1902) may be related to **78**(c).

V. calaminaria (DC.) Lej., *Rev. Fl. Spa* 49 (1824), from soils rich in zinc in Holland, Belgium and Germany, has been placed under both **74** and **78**. Its chromosome number is $2n = 52$, and it is probably best regarded as a variety of **78**(d).

79. V. aetolica Boiss. & Heldr. in Boiss., *Diagn. Pl. Or. Nov.* **3**(6): 24 (1859). Perennial. Stems 15–40 cm. Leaves *c.* 2 cm, ovate to lanceolate, coarsely crenate. Stipules 1–1·5 cm, ciliate, pinnatifid; terminal segment ovate to ovate-lanceolate, sub-crenate or entire. Sepals 2–3 mm wide at base, ovate-lanceolate or triangular, often denticulate, ciliate. Flowers 1·5–2 cm, yellow, the upper petals rarely violet. Spur 5–6 mm, straight, slender, *c.* 1½ times as long as calycine appendages. $2n = 16$. *Montane and subalpine meadows. S. & W. parts of Balkan peninsula.* Al Gr Ju.

80. V. arvensis Murray, *Prodr. Stirp. Götting.* 73 (1770). Annual up to 40 cm, branched, more or less erect, with an indumentum of short, deflexed hairs. Leaves 2–5 cm, oblong-spathulate, acute or obtuse, crenate. Stipules ½–¾ as long as leaves, coarsely pinnatifid; terminal segment lanceolate, leaf-like. Bracts in upper third of peduncle. Flowers 1–1·5 cm; lower petal cream to yellow; others cream to bluish-violet. Sepals lanceolate, equalling or exceeding corolla. Spur equalling calycine appendages. $2n = 34$. *Open and cultivated ground. Almost throughout Europe.* All except Az Bl Cr Fa Is Sb.

The species is very variable, and intermediates between it and **78**, probably of hybrid origin, have frequently been described, e.g. *V. contempta* Jordan.

81. V. kitaibeliana Schultes in Roemer & Schultes, *Syst. Veg.* **5**: 383 (1819). Annual 2–10(–20) cm, with dense indumentum of short, crisped or deflexed hairs. Leaves 1–3 cm, the lowermost orbicular, the others oblong-spathulate, crenately lobed. Stipules 0·5–1 cm, pinnatipartite with an oblong-spathulate, crenately-lobed, shortly stalked terminal segment and smaller lateral segments. Bracts just below the flower. Flowers 0·4–0·8 cm. Sepals lanceolate, exceeding the corolla. Petals cream-white to yellow with a yellow centre. Spur slightly longer than calycine appendages. $2n = 16$, 48. *Dry, open habitats. S. & C. Europe, extending to W.C. France and E. Ukraine.* Al Au Br Bu Co Cr Cz Ga Gr He Hs Hu It Ju Lu ?Po Rm Rs (W, K, E) Si Tu [Ge].

82. V. hymettia Boiss. & Heldr. in Boiss., *Diagn. Pl. Or. Nov.* **3**(1): 57 (1853). Shortly hispid annual 3–10(–20) cm. Leaves 1·5–2 cm, the lower ovate-orbicular, the upper oblong-spathulate,

crenate. Stipules pinnately divided almost to the base; terminal segment stalked and leaf-like, larger than the rest. Peduncles 4–5 cm. Flowers 1–1·5 cm. Petals yellow, the upper often becoming violet, or all violet, about twice as long as sepals. Spur 3–4 mm, stout, slightly longer than calycine appendages. $2n = 16$. *Pastures. Mediterranean region, Portugal; S.W. Romania.* Al Ga Gr It Lu Rm Si Tu.

V. lavrenkoana Klokov, *Ind. Sem. Hort. Bot. Charkov.* 8 (1927) and *V. cretacea* Klokov in Schischkin & Bobrov, *Fl. URSS* **15**: 686 (1949), described from the Ukraine, resemble **82**; their status is uncertain.

83. V. parvula Tineo, *Pl. Rar. Sic. Pug.* 5 (1817). Annual, 2–3(–10) cm, villous with long hairs. Leaves 0·5–1·2 cm, the lowermost oblong-orbicular, almost entire, the others oblong-spathulate. Stipules deeply lobed. Bracts on upper third of peduncles. Sepals ovate-lanceolate; appendages exceeding the spur. Flowers *c.* 0·5 cm, creamy white; corolla scarcely exceeding calyx. $2n = 10$. *Rocks and screes, 1500–2500 m. Mountains of S. Europe, northwards to S. Jugoslavia, Corse and C. Spain.* Co Gr Hs It Ju Si.

84. V. heldreichiana Boiss., *Diagn. Pl. Or. Nov.* **2**(8): 53 (1849). Like **83** but glabrous or sparsely hairy; basal leaves 1–3 cm; stipules like leaves but much shorter, with only 1–2 small teeth at base; flowers lilac-blue; lower petal 0·4–0·6 cm. *Rocks and screes, 1500–2300 m. Kriti.* Cr.

85. V. mercurii Orph. ex Halácsy, *Denkschr. Akad. Wiss. Math.-Nat. Kl.* (*Wien*) **61**: 497 (1894). Sparsely hairy or glabrous annual 2–6(–10) cm. Leaves 0·5–1·5 cm, the lower ovate-orbicular, the upper oblong, almost entire. Stipules usually 3-partite, the terminal segment stalked and leaf-like. Flowers *c.* 1 cm. Petals yellow, about twice as long as sepals. Spur violet, stout, about equalling the calycine appendages. *S. Greece* (*Killini Oros*). Gr.

86. V. demetria Prolongo ex Boiss., *Voy. Bot. Midi Esp.* **2**: 73 (1839). Annual, glabrous or with short hairs. Stems 5–15 cm, slender. Lower leaves 1–4 cm, orbicular, subcordate, long-petiolate; the upper 0·5–1·5 cm, broadly ovate to oblong; margins with 2–3 coarse crenations; petiole short. Stipules often nearly as long as leaves, pinnatifid, with short lateral segments at the base, and a large, leaf-like terminal segment. Flowers *c.* 1 cm. Petals bright yellow, about twice as long as the sepals; upper petals sometimes violet. Spur 2–3 mm, violet, stout, slightly longer than the calycine appendages. *Shady rocks and screes; calcicole. S.W. Spain and W.C. Portugal.* Hs Lu.

87. V. occulta Lehm., *Ind. Sem. Horti Bot. Hamburg.* (1829). Annual. Stems 3–15 cm, erect. Leaves 2–3 cm, narrowly oblong-spathulate, remotely crenate-serrate or almost entire, glabrous to pubescent; the lower petiolate, the upper sessile. Upper stipules deeply and pinnately divided, with the terminal segment the largest, entire. Bracts of peduncle concealed by the calycine appendages. Sepals *c.* 1 cm, broadly lanceolate, as long as or longer than the corolla; appendages large, exceeding the spur. Flowers 1–1·5 cm, white to cream. Petals sometimes with bluish margins. *Ruderal. Krym.* [Rs (K).] (*C. & S.W. Asia.*)

Sect. XYLINOSIUM W. Becker. Perennial, suffruticose. Open flowers violet or yellow; cleistogamous flowers produced. Style neither capitate nor beaked.

88. V. arborescens L., *Sp. Pl.* 935 (1753). Stems 10–20 cm, woody and corky at the base, ascending, greyish, pubescent.

Leaves ovate to linear-lanceolate, acute. Stipules linear-lanceolate, lyrate-pinnatifid, ⅓ as long as leaves. Peduncles with minute bracts. Flowers 1–1·5 cm, whitish or pale violet. Spur *c.* 4 mm, curved, obtuse. Capsule erect at maturity, glabrous. 2*n*=*c.* 140. *Rocky places and thickets, calcicole. W. Mediterranean region, S.W. Portugal.* Bl Ga Hs Lu Sa.

89. V. scorpiuroides Cosson, *Bull. Soc. Bot. Fr.* **19**: 80 (1872) (*V. methodiana* Coust. & Gand.). Like **88** but leaves broadly obovate; stipules linear; flowers yellow. *Rocky places. S. Aegean region* (*Kithira to Kriti.*) Cr Gr. (*N. Africa.*)

Sect. DELPHINIOPSIS W. Becker. Perennial, suffruticose. Open flowers pink, red or violet with very long spur; cleistogamous flowers not produced. Style neither capitate nor beaked.

90. V. delphinantha Boiss., *Diagn. Pl. Or. Nov.* **1**(1): 7 (1843). Stems 5–10 cm, numerous, crowded, erect. Leaves 0·75–1·5 cm,

linear to lanceolate, acute, entire, sessile. Stipules slightly shorter than leaves, the lower bifid, the upper entire. Flowers pinkish- or reddish-purple, on long peduncles. Sepals acute. Petals narrow, the lower entire. Spur 16–18 mm, equalling or exceeding petals. Capsule glabrous. 2*n*=20. *Calcareous rocks.* ● *Mountains of N. Greece, just extending into Bulgaria; one station in S. Greece.* Bu Gr.

91. V. kosaninii (Degen) Hayek, *Denkschr. Akad. Wiss. Math.-Nat. Kl.* (*Wien*) **94**: 155 (1918). Like **90** but flowers lilac-pink; lower petal emarginate; spur 12 mm. *Alpine meadows.* ● *N. Albania, N. Makedonija.* Al Ju.

92. V. cazorlensis Gand., *Bull. Assoc. Fr. Bot.* **5**: 226 (1902). Like **90** but flowers intense pinkish-purple; lower petal emarginate; spur 20–30 mm. *Calcareous rocks,* 1000–2100 m. ● *S.E. Spain* (*prov. Jaén*). Hs.

CXI. PASSIFLORACEAE[1]

Herbaceous or woody climbers with axillary tendrils. Leaves alternate, usually stipulate. Flowers perigynous, usually hermaphrodite, actinomorphic. Hypanthium cup-shaped. Sepals usually 5, free or connate at base, often petaloid. Petals 5, usually free. Corona inserted on the rim of the hypanthium, composed of 1–2 rows of long filaments, 1 row of short filaments and a membrane partially closing the hypanthium. Stamens usually 5, alternating with the petals. Ovary superior, often on a long gynophore, unilocular, with 3(–5) parietal placentae; ovules numerous; stigmas 3–5. Fruit a berry or loculicidal capsule.

1. Passiflora L.[2]

Flowers hermaphrodite. Stamens connate and adnate to the gynophore. Stigmas 3. Fruit a berry.

1. P. caerulea L., *Sp. Pl.* 959 (1753). Glabrous and somewhat glaucous climber up to 10 m; stems terete or obscurely angled. Leaves 10–15 cm, digitately 5(–7)-lobed, the lobes oblong-lanceolate or -ovate; petiole with 2–4 glands. Flowers 7–10 cm in diameter, solitary, axillary on articulated peduncles. Sepals white or pale pink on the inside, with a short horn on the back. Petals white or pale pink. Corona dull purple at the base, white in the middle, purplish-blue at the apex. Stigmas dull purple. Fruit *c.* 5 cm, ovoid or subglobose, orange-yellow. *Naturalized in Açores; frequently cultivated for ornament, particularly in S. Europe.* [Az.] (*C. & W. South America.*)

Some other species are cultivated in S. Europe for ornament and for the edible fruit.

CXII. CISTACEAE[3]

Shrubs or herbs. Leaves simple, usually opposite, mostly with stellate indumentum. Flowers hermaphrodite, actinomorphic, hypogynous, solitary or in cymes. Sepals 5 or 3. Petals 5, rarely 0, usually caducous. Stamens many. Ovary 1-locular or incompletely septate with 3 or 5 (rarely 10) parietal placentae; ovules usually orthotropous, rarely anatropous. Style simple or absent. Fruit a loculicidal capsule. Seeds with a more or less curved embryo; endosperm present.

Literature: W. Grosser in Engler, *Pflanzenreich* **14** (IV. 193) (1903).

1 Capsule with 5, 6 or 10 valves; flowers white to purplish-red
                                           1. Cistus
1 Capsule with 3 valves; flowers white, pink or yellow
  2 Style short, straight or 0; sepals 3 or 5; shrubs   **2. Halimium**
  2 Style long, or if short or 0 then herbs; sepals 5
    3 Annual or perennial, with a basal leaf-rosette (which may be withered at anthesis); stigma sessile or subsessile
                                      3. Tuberaria
    3 Dwarf shrubs or annuals, without a basal leaf-rosette; style present

4 Leaves all opposite and decussate, oblong to linear; stamens all fertile; ovules orthotropous   **4. Helianthemum**
4 Upper leaves usually alternate, ± linear; outer stamens sterile; ovules anatropous   **5. Fumana**

1. Cistus L.[4]

Shrubs. Flowers 2 cm or more in diameter, white to purplish-red. Ovary usually 5-, rarely up to 10-locular; placentation axile; ovules orthotropous, the funicle filiform.

Hybrids occur involving species **1, 2, 5, 7–13,** some being frequent. Hybrids between species **6, 8, 9,** and species of *Halimium* also occur.

All species are plants of dry scrub or open woodland.

Literature: P. Dansereau, *Boissiera* **4**: 1–90 (1939). M. Martín Bolaños & E. Guinea, *Jarales y Jaras* (*Cistografía Hispanica*). Madrid. 1949.

1 Sepals 5
  2 Style filiform, as long as the stamens
    3 Petals white; flowers in unilateral cymes   **6. varius**
    3 Petals purplish-pink or -red; flowers not in unilateral cymes
      4 Leaves sessile, with 3 ± parallel veins

---

¹ Edit. T. G. Tutin.   ² By P. W. Ball.
³ Edit. V. H. Heywood.   ⁴ By E. F. Warburg.

5   Leaves not undulate; pedicels 5–20 mm   **1. albidus**
5   Leaves undulate; pedicels 1–5 mm   **2. crispus**
4   Leaves petiolate, pinnately veined
  6   Leaves usually 20–50 mm, variable in shape, the veins impressed above; sepals ovate-lanceolate, long-acuminate   **3. incanus**
  6   Leaves 5–20 mm, elliptical, the veins not or scarcely impressed above; sepals ovate, acute   **4. heterophyllus**
2   Stigma sessile or subsessile
  7   Petals pink; leaves grey-tomentose, at least beneath   **5. parviflorus**
  7   Petals white; leaves green
    8   Leaves sessile or subsessile
      9   Leaves subsessile, elliptical, narrowed to a cuneate base   **9. albanicus**
      9   Leaves sessile, not or scarcely narrowed at base
        10   Leaves linear-lanceolate or linear; outer sepals broadly cuneate at base   **7. monspeliensis**
        10   Leaves oblong; outer sepals cordate at base   **8. psilosepalus**
    8   Leaves distinctly petiolate
      11   Leaves 10–40 mm, rugose and scabrid above, rounded or cuneate at base   **10. salvifolius**
      11   Leaves 40–100 mm, smooth above, cordate at base   **11. populifolius**
1   Sepals 3
  12   Leaves at least 6 mm wide; linear-lanceolate to ovate or spathulate; flowers at least 5 cm in diameter
    13   Inflorescence usually 4- to 8-flowered; ovary 5-locular   **12. laurifolius**
    13   Flowers solitary; ovary 6- to 10-locular
      14   Leaves linear-lanceolate; ovary 10-locular   **13. ladanifer**
      14   Leaves oblanceolate to spathulate; ovary 6-locular   **14. palhinhae**
  12   Leaves not more than 4 mm wide, linear; flowers not more than 3 cm in diameter
    15   Peduncles, pedicels and calyx clothed with long, white hairs   **15. clusii**
    15   Peduncles, pedicels and calyx subglabrous   **16. libanotis**

1. C. albidus L., *Sp. Pl.* 524 (1753). Compact bush up to 100 cm, erect. Leaves (15–)20–50(–75) × 5–20(–30) mm, oblong to elliptical, flat, 3-veined, densely greyish-white-tomentose, sessile. Cymes 1- to 7-flowered, more or less symmetrical. Pedicels 5–20 mm. Sepals 5, tomentose. Flowers 4–6 cm in diameter, purplish-pink. Styles filiform. $2n = 18$. *S.W. Europe, extending to c. 11° E. in N. Italy.* Bl Co Ga Hs It Lu Sa.

2. C. crispus L., *Sp. Pl.* 524 (1753). Rounded bush up to 50 cm. Leaves 10–40 × 4–15 mm, oblong to elliptical, undulate, 3-veined, greyish-green, villous-tomentose, with mixed stellate and long, simple hairs, sessile. Cymes few-flowered, dense. Pedicels 1–5 mm. Sepals 5, densely villous. Flowers 3–4 cm in diameter, purplish-red. Style filiform. $2n = 18$. *W. Mediterranean region, C. & S. Portugal.* ?Co Ga Hs It Lu Si.

3. C. incanus L., *Sp. Pl.* 524 (1753) (*C. villosus* auct., vix L.; incl. *C. polymorphus* Willk.). Stems up to 100 cm, erect or spreading. Leaves (10–)20–50(–70) × 8–30 mm, ovate, obovate or elliptical, often undulate, pinnately veined, green or greyish, pubescent or tomentose with stellate hairs, with the veins impressed above and prominent beneath. Petioles 3–15 mm. Cymes 1- to 7-flowered, more or less symmetrical. Sepals 5, ovate-lanceolate, long-acuminate, with stellate hairs and long, simple hairs. Flowers 4–6 cm in diameter, purplish-pink. *S. Europe, but rare in the west.* Al Bl Bu Co Cr Gr It Ju Rs (K) Sa Si Tu [Br].

1   Leaves 15–25 × 8–15 mm, distinctly undulate-crispate   **(c) subsp. creticus**
1   Leaves 25–50 × 15–30 mm, flat

2   Sepals with a few long hairs not hiding the dense stellate hairs; stems and pedicels stellate-hairy   **(b) subsp. corsicus**
2   Sepals with many long hairs hiding the stellate hairs; stems and pedicels densely white-villous   **(a) subsp. incanus**

(a) Subsp. **incanus** (*C. tauricus* C. Presl, *C. polymorphus* subsp. *villosus* var. *vulgaris* Willk.): *S. Europe, from Corse & W. Italy eastwards to Krym.*

(b) Subsp. **corsicus** (Loisel.) Heywood, *Feddes Repert.* **79**: 60 (1968) (*C. villosus* subsp. *corsicus* (Loisel.) Rouy & Fouc.): *W. Mediterranean islands, Italy, Jugoslavia.*

(c) Subsp. **creticus** (L.) Heywood, *Feddes Repert.* **79**: 60 (1968) (*C. creticus* L.): *Greece and Aegean region.*

4. C. heterophyllus Desf., *Fl. Atl.* **1**: 411 (1798). Stems up to 100 cm, erect, much-branched. Leaves 5–20 mm, elliptical, pinnately veined, dark green and stellate-pubescent above, with the veins scarcely impressed, paler and tomentose beneath, with prominent veins. Petioles 1–2 mm. Cymes 1- to 5-flowered. Pedicels 15–45 mm. Sepals 5, conspicuously pubescent, with stellate hairs and few long, simple hairs. Flowers 5–6 cm in diameter, purplish-pink. Style filiform. *S.E. Spain (near Cartagena).* Hs. (*N.W. Africa.*)

5. C. parviflorus Lam., *Encycl. Méth. Bot.* **2**: 14 (1786). Stems up to 100 cm, somewhat spreading. Leaves 10–30 mm, ovate, 3-veined in the basal half, grey-tomentose, petiolate. Cymes 1- to 6-flowered, more or less symmetrical. Pedicels 5–10 mm. Sepals 5. Flowers 2–3 cm in diameter, pink. Style absent. *Aegean region; S.E. Italy; Lampedusa.* Cr Gr It Si ?Tu.

6. C. varius Pourret, *Mém. Acad. Toulouse* **3**: 312 (1788) (*C. pouzolzii* Delile). Stems up to 50 cm, somewhat spreading. Leaves oblong, 3-veined, grey-tomentose, sessile. Cymes 2- to 8-flowered, unilateral. Sepals 5. Flowers *c.* 2 cm in diameter, white. Style filiform. *S. France (Gard, Aveyron, Ardèche).* Ga. (*N.W. Africa.*)

Sometimes considered to be a hybrid between **2** and **7**.

7. C. monspeliensis L., *Sp. Pl.* 524 (1753). Compact bush up to 100 cm, erect, viscid. Leaves 15–50 × 4–8 mm, linear-lanceolate or linear, 3-veined, green and sparsely pubescent above, densely stellate-tomentose beneath, sessile. Cymes 2- to 8-flowered, unilateral. Sepals 5, the outer ovate, broadly cuneate at base. Flowers 2–3 cm in diameter, white. Style very short. $2n = 18$. *S. Europe.* Al Bl Co Cr Ga Gr Hs It Ju Lu Sa Si.

8. C. psilosepalus Sweet, *Cistin.* t. 33 (1826) (*C. hirsutus* Lam. (1786), non Lam. (1778)). Stems up to 100 cm, somewhat spreading. Leaves 20–60 × 5–20 mm, ovate-oblong to linear-oblong, green, 3-veined at least in the basal half, more or less hairy, with long, simple and stellate hairs on both surfaces, sessile. Cymes 1- to 5-flowered, more or less symmetrical. Sepals 5, the outer ovate, cordate at base, hirsute and long-ciliate. Flowers 4–6 cm in diameter, white. Style very short. ● *Portugal, W. Spain; N.W. France (doubtfully native).* *Ga Hs Lu.

9. C. albanicus E. F. Warburg ex Heywood, *Feddes Repert.* **79**: 60 (1968) (*C. × florentinus* nm. *adriaticus* Markgraf). Stems *c.* 25 cm. Leaves elliptical, narrowed to a cuneate base, green, hairy with long, simple and stellate hairs above, subsessile. Cymes up to 4-flowered, unilateral. Sepals 5, the outer rounded or cordate at base, with stiff, long, white hairs and short hairs. Flowers 3–4 cm in diameter, white. Style very short. ● *C. Albania.* Al.

Originally described by Markgraf as a hybrid between **7** and **10**, but not growing with either.

10. C. salvifolius L., *Sp. Pl.* 524 (1753). Stems up to 100 cm, spreading or procumbent. Leaves 10–40 × 6–20 mm, ovate or elliptical, rounded or broadly cuneate at base, green, scabrid and rugose above, with stellate hairs on both surfaces, petiolate. Flowers mostly solitary but sometimes up to 4 in a cyme, petiolate. Pedicels 10–100 mm. Sepals 5, the two outer cordate at base. Flowers 3–5 cm in diameter, white. Style very short. 2*n*=18. *S. Europe, extending northwards to c. 47° in W. France.* Al Bl Bu Co Cr Ga Gr He Hs It Ju Lu Sa Si Tu.

11. C. populifolius L., *Sp. Pl.* 523 (1753). Stems up to 150 cm, somewhat spreading. Leaves 40–100 × 30–65 mm, ovate, cordate at base, glabrous, green and smooth above, petiolate. Cymes 2- to 6-flowered, pedunculate. Sepals 5, the two outer cordate at base. Flowers 4–6 cm in diameter, white. Style very short. 2*n*=18. *Spain and Portugal, extending eastwards to c. 3° E in S. France.* Ga Hs Lu.

(a) Subsp. **populifolius**: Leaves usually at least 1½ times as long as wide, not or scarcely undulate. Sepals with few white hairs. ● *France, N. Spain, Portugal.*

(b) Subsp. **major** (Pourret ex Dunal) Heywood, *Feddes Repert.* **79**: 61 (1968) (*C. populifolius* var. *major* Pourret ex Dunal, *C. populifolius* var. *lasiocalyx* Willk.): Leaves usually less than 1½ times as long as wide, undulate. Sepals densely clothed with long white hairs. *C. & S. Spain, Portugal.*

12. C. laurifolius L., *Sp. Pl.* 523 (1753). Stems up to 150 cm, erect. Leaves 30–80(–90) × 10–30 mm, ovate to ovate-lanceolate, 3-veined, dark green and glabrous on upper surface, densely white-tomentose beneath, petiolate. Inflorescence a terminal (1–)4- to 8-flowered, long-pedunculate, umbel-like cyme, often with 1 or 2 opposite pairs of flowers below. Sepals 3. Flowers 5–6 cm in diameter, white. Style very short. *S.W. Europe, extending to W.C. Italy.* 2*n*=18. Co Ga Hs It Lu.

13. C. ladanifer L., *Sp. Pl.* 523 (1753). Stems up to 250 cm, erect, very viscid. Leaves 40–80(–120) × 6–25 mm, linear-lanceolate, 3-veined for ⅓ their length, green and glabrous above, densely white-tomentose beneath, subsessile. Sepals 3. Flowers 7–10 cm in diameter, solitary, white, with or without a crimson spot at the base of each petal. Style very short. Ovary 10-locular. *S.W. Europe.* Ga Hs Lu.

14. C. palhinhae Ingram, *Gard. Chron.* ser. 3, **114**: 34 (1943). Like **13** but stems only up to 50 cm; leaves 20–60 mm, oblanceolate to spathulate; flowers white, unspotted; ovary 6-locular. ● *S.W. Portugal (W. coast of Algarve).* Lu.

15. C. clusii Dunal in DC., *Prodr.* **1**: 266 (1824) (*C. libanotis* auct. mult., non L.). Stems up to 100 cm, erect, much-branched. Leaves 10–25 × 1–2 mm, linear, dark green above, white-tomentose beneath, with revolute margins, subsessile. Inflorescence up to 12-flowered with a terminal, umbel-like cyme and often with opposite pairs of flowers or 3-flowered cymes below. Peduncles, pedicels and calyx clothed with long, white hairs. Sepals 3, 5–8 mm. Flowers 2–3 cm in diameter, white. Style short. *W. & C. Mediterranean region, from S. Spain to S.E. Italy.* Bl Hs It Si.

16. C. libanotis L., *Syst. Nat.* ed. 10, **2**: 1077 (1759) (*C. bourgaeanus* Cosson). Like **15** but leaves (15–)20–35(–40) × 1·5–3 mm; peduncles, pedicels and calyx subglabrous, viscid; sepals 8–10 mm. *S.W. Portugal, S.W. Spain.* Hs Lu.

---

[1] By M. C. F. Proctor and V. H. Heywood.

2. Halimium (Dunal) Spach[1]

Small shrubs. Leaves opposite, decussate, exstipulate. Sepals 3 or 5, the 2 outer much smaller than the 3 inner. Flowers yellow or white. Style short and straight or absent. Capsule 3-valved; ovules orthotropous.

Literature: E. Guinea, *Cistáceas Españolas* (*Bol. Inst. For. Inv. Exper. Madrid*, No. 71) 11–38. Madrid. 1954.

All European species are found in woods or scrub, or on heaths or sandy ground in fairly dry conditions, and usually on siliceous soils.

1 Leaves ovate to lanceolate, mostly more than 4 mm wide, flat or with the margins only slightly revolute
  2 Leaves of flowering shoots subglabrous, sessile; those of sterile shoots smaller, petiolate, tomentose; inflorescence lax, long-pedunculate   **1. ocymoides**
  2 Leaves all alike
    3 Sepals 5, covered with peltate scales   **5. halimifolium**
    3 Sepals 3, villous or hirsute
      4 Flowering branches and pedicels densely tomentose and with patent, purplish, viscid hairs   **4. atriplicifolium**
      4 Flowering branches and pedicels sericeous to villous, without purplish hairs
        5 Inflorescence villous, with short hairs; sepals without purple bristles; petals unspotted   **2. alyssoides**
        5 Inflorescence sericeous, with long hairs; sepals often with long, purple bristles; petals sometimes with a brown spot at the base   **3. lasianthum**
1 Leaves linear, less than 4 mm wide, the margins revolute
  6 Flowers yellow; sepals glabrous   **9. commutatum**
  6 Flowers white; sepals villous
    7 Stems 15–25 cm; branches tortuous; leaves crowded at the ends of branches   **6. umbellatum**
    7 Stems 25–60 cm; branches straight; leaves distributed along the branches
      8 Branches with dense, whitish indumentum; pedicels unequal; sepals hirsute-villous   **7. viscosum**
      8 Branches sparsely stellate-hairy; pedicels equal; sepals pubescent-villous   **8. verticillatum**

1. H. ocymoides (Lam.) Willk. in Willk. & Lange, *Prodr. Fl. Hisp.* **3**: 715 (1878) (*Helianthemum ocymoides* (Lam.) Pers.). Erect, rarely procumbent, up to 100 cm. Leaves of non-flowering shoots 3–15 × 2–5 mm, obovate, grey-tomentose, with 1–3 veins, shortly petiolate, often plicate; leaves of flowering shoots 12–30 × 3·5–7·5 mm, obovate to lanceolate, green, with 3 veins, sessile. Flowers on long pedicels, in lax, terminal panicles. Sepals 3; petals yellow, usually with a dark spot at the base. 2*n*=18. *Iberian peninsula, chiefly in the centre and west.* Hs Lu.

2. H. alyssoides (Lam.) C. Koch, *Hort. Dendrol.* 32 (1853) (*H. occidentale* Willk., *Helianthemum alyssoides* (Lam.) Vent.). Compact, erect or decumbent, 10–100 cm, grey-tomentose. Leaves 5–40 × 3–15 mm, oblong to ovate- or obovate-lanceolate, obtuse, dark green above, white-tomentose beneath, or white-stellate-tomentose on both surfaces; shortly petiolate or sessile. Flowers in short, terminal, shortly villous cymes. Sepals 3, shortly villous; petals yellow, unspotted. ● *W. & C. France, N.W. Spain, N. & C. Portugal.* Ga Hs Lu.

3. H. lasianthum (Lam.) Spach, *Ann. Sci. Nat.* ser. 2 (Bot.), **6**: 366 (1836) (*H. eriocephalum* Willk., *H. occidentale* Willk. pro parte). Like **2** but leaves somewhat obtuse to acute; pedicels and sepals long-sericeous; sepals often with long purple bristles; *S. Portugal, S. Spain.* Hs Lu.

(a) Subsp. **lasianthum**: Corolla 2–3 cm in diameter; petals spotted at the base or unspotted. *S. Portugal, S. Spain.*

(b) Subsp. **formosum** (Curtis) Heywood, *Feddes Repert.* **79**: 59 (1968) (*Cistus formosus* Curtis): Corolla 4–6 cm in diameter; petals spotted well above the claw. *S. Portugal (Algarve).*

4. H. atriplicifolium (Lam.) Spach, *Ann. Sci. Nat.* ser. 2 (Bot.), **6**: 366 (1836). Erect, up to 150 cm. Leaves of non-flowering shoots 20–50 × 10–30 mm, rhombic-elliptical to ovate-lanceolate, obtuse, with 3 prominent veins, shortly petiolate; leaves of flowering shoots 10–40 × 10–20 mm, oblong-cordate, pinnate-veined, sessile; all with a dense covering of peltate scales and stellate hairs. Flowering branches elongate, rigid, leafless, densely tomentose and with patent, purplish, viscid hairs; flowers in lax cymes. Sepals 3; petals yellow, spotted at the base. *C. & S. Spain.* Hs.

5. H. halimifolium (L.) Willk. in Willk. & Lange, *Prodr. Fl. Hisp.* **3**: 717 (1878) (*Helianthemum halimifolium* (L.) Pers.). Much branched, erect, up to 100 cm. Leaves 10–40 × 5–20 mm, elliptical or spathulate-lanceolate, white-tomentose on both surfaces when young, greenish or greyish above when mature with silvery, peltate scales and stellate hairs. Flowers in numerous paniculate cymes. Sepals 5, with a dense covering of peltate scales; petals yellow, spotted or unspotted. $2n=18$. *S.W. Europe, extending to S.E. Italy; mainly near the coast.* Bl Co Hs It Lu Sa ?Si.

(a) Subsp. **halimifolium**: Stems 30–100 cm; flowers in lax panicles; sepals without stellate hairs; petals cuneate. *Almost throughout the range of the species.*
(b) Subsp. **multiflorum** (Salzm. ex Dunal) Maire in Jahandiez & Maire, *Cat. Pl. Maroc* **2**: 494 (1932) (*H. multiflorum* (Salzm. ex Dunal) Willk.): Stems 30–40 cm; flowers in dense panicles; sepals with numerous stellate hairs; petals obcordate. *Portugal, S.W. Spain.*

6. H. umbellatum (L.) Spach, *Ann. Sci. Nat.* ser. 2 (Bot.), **6**: 366 (1836) (*Helianthemum umbellatum* (L.) Miller). Diffuse, 15–25 cm; branches short and tortuous, ascending, with dense, whitish indumentum. Leaves 5–12 × 1·5–2 mm, crowded at the end of branches, linear or linear-lanceolate, with revolute margins, dark green, sparsely pubescent or subglabrous above, densely tomentose beneath. Flowers white, in terminal 3- to 6-flowered, umbellate cymes, sometimes with a secondary and a tertiary 1- to 2-flowered cyme. Pedicels equal, not filiform. Sepals villous. $2n=18$. ● *S.W. & C. France, N. Spain, N. Portugal.* Ga Gr Hs Lu.

7. H. viscosum (Willk.) P. Silva, *Agron. Lusit.* **24**: 165 (1964) (*H. umbellatum* var. *viscosum* Willk.). Like **6** but 25–60 cm; branches more or less straight, suberect; leaves 10–20 mm, distributed along the branches; flowers in 3–5 whorls, each of (4–)5–6(–9) flowers; pedicels more or less unequal; sepals hirsute-villous. *C., S. & E. Spain, E. Portugal.* Hs Lu.

8. H. verticillatum (Brot.) Sennen, *Monde Pl.* **192**: 39 (1931). Like **6** but 25–60 cm; branches short and straight, erecto-patent, sparsely stellate-hairy; leaves distributed along the branches, the margins revolute or not; flowers in 3–5 whorls, each of (3–)4–5(–7) flowers; pedicels filiform; sepals pubescent-villous. ● *S.W. Portugal.* Lu.

9. H. commutatum Pau, *Bol. Soc. Aragon. Ci. Nat.* **3**: 263 (1904) (*H. libanotis* Lange pro parte, *Helianthemum libanotis* Willd. pro parte.). Low-growing, much-branched, up to 50 cm, with annular leaf scars on branches. Leaves 10–35 × 1·5–3 mm, linear, with revolute margins, shining green and glabrous above,

white-tomentose beneath, resembling those of *Rosmarinus*. Flowers solitary or in 2(–5)-flowered, terminal cymes. Sepals glabrous. Petals pale yellow. *Mainly coastal sands. S. Spain, W. & S. Portugal.* Hs Lu.

3. Tuberaria (Dunal) Spach[1]

Annual or perennial, with a basal leaf-rosette; flowering stems erect. Leaves with 3 veins. Flowers yellow, in terminal cymes. Sepals 5, the 2 outer usually smaller than the 3 inner. Stigma more or less sessile. Capsule 3-valved.

1 Perennial, with woody stock and persistent basal rosette; flowering stems with small, bract-like leaves only
  2 Leaves gradually narrowed at the base; petals unspotted
        1. lignosa
  2 Leaves abruptly narrowed into a distinct petiole; petals with a dark spot at the base
    3 Veins of leaf distinctly anastomosing; bracts lanceolate, acute
        2. globularifolia
    3 Veins of leaf not anastomosing; bracts broadly ovate, obtuse
        3. major
1 Annual, with basal rosette often dead at anthesis; flowering stems leafy
  4 Flowers subsessile in dense, ± scorpioid cymes   **10. echioides**
  4 Flowers distinctly pedicellate
    5 Pedicels longer than sepals during anthesis
      6 Outer sepals much smaller than the inner, not accrescent
        7 Leaves obovate to lanceolate or oblong (the uppermost linear), villous, flat (or the uppermost with the margins ± revolute)   **4. guttata**
        7 Leaves lanceolate or linear, at least the upper glabrescent, the margins distinctly revolute   **5. bupleurifolia**
      6 Outer sepals ± equalling the inner, all accrescent in fruit
        8 Stipules ⅓ as long as leaves; plant with appressed hairs
          8. acuminata
        8 Stipules ½–⅔ as long as leaves; plant with patent hairs
          9. macrosepala
    5 Pedicels ± equalling sepals during anthesis
      9 Capsule villous   **6. villosissima**
      9 Capsule glabrous
        10 Pedicels white-puberulent   **7. praecox**
        10 Pedicels glabrous   **4. guttata**

Sect. TUBERARIA. Perennial; all leaves exstipulate; ovary shortly stipitate.

1. T. lignosa (Sweet) Samp., *Bol. Soc. Brot.* ser. 2, **1**: 128 (1922) (*T. vulgaris* Willk., *T. melastomatifolia* Grosser, *Helianthemum tuberaria* (L.) Miller). Perennial up to 40 cm, with branched, woody stock bearing leaf-rosettes resembling those of *Plantago* spp. Leaves 20–55 × 8–25 mm, obovate-lanceolate to elliptical, gradually narrowed at the base into an indistinct petiole, subglabrous (rarely hairy) above, tomentose or sericeous-pubescent beneath. Flowering stems 20–30 cm, unbranched. Bracts lanceolate, acute. Flowers *c.* 30 mm in diameter, 3–7, in bracteate cymes. Sepals 10–15 × 4–7 mm, lanceolate-acuminate. Petals unspotted. $2n=14$. *Scrub and woodland. Iberian peninsula and W. Mediterranean region, extending to S.E. Italy.* Bl Co Ga Hs It Lu Sa Si.

2. T. globularifolia (Lam.) Willk., *Icon. Descr. Pl. Nov.* **2**: 71 (1859). Perennial up to 40 cm, with branched, woody stock. Leaves 25–50 mm, more or less spathulate, abruptly narrowed at the base into a petiole about equalling the lamina, glabrous or hairy above, hairy on veins and margins beneath, the veins distinctly anastomosing. Flowering stems 15–30 cm, unbranched. Bracts lanceolate-ovate. Flowers 30–50 mm in diameter; petals dark-spotted at the base. *Scrub or heath on sandy soil.* ● *N.W. Portugal and N.W. Spain.* Hs Lu.

¹ By M. C. F. Proctor.

3. **T. major** (Willk.) P. Silva & Rozeira, *Agron. Lusit.* **24**: 168 (1964) (*T. globularifolia* var. *major* Willk.) Like **2** but leaves thicker, with the veins not anastomosing; bracts broadly ovate, obtuse. *Coastal scrub.* ● *S. Portugal.* Lu.

Sect. SCORPIOIDES Willk. Annual; uppermost leaves usually stipulate; ovary sessile.

4. **T. guttata** (L.) Fourr., *Ann. Soc. Linn. Lyon* nov. ser., **16**: 340 (1868) (*T. variabilis* Willk.; incl. *T. inconspicua* (Thib.) Willk., *Helianthemum guttatum* (L.) Miller). Villous annual up to 30 cm, with basal rosette often persisting to anthesis. Basal and lower cauline leaves broadly to narrowly elliptical or obovate, exstipulate; upper cauline leaves linear-oblong or linear-lanceolate, more or less revolute, stipulate or not; all leaves with stellate hairs on both surfaces or only simple hairs above. Flowers 10–20 mm in diameter, long-pedicellate, in terminal raceme-like cymes. Outer sepals much smaller than the inner; petals usually dark-spotted at the base. $2n = 36$. *S. & W. Europe, northwards to 54° N. and extending eastwards to 14° E. in E. Germany.* Al Bl Br Bu Co Cr Ga Ge Gr Hb Ho Hs It Ju Lu Sa Si Tu.

Very variable in hairness, shape and size of leaves, size of petals and in the persistence of the basal rosette. In some variants the petals may be inconspicuous or absent.

T. brevipes (Boiss. & Reuter) Willk., *Icon. Descr. Pl. Nov.* **2**: 79 (1859), with very shortly pedicellate flowers, probably represents a variant of this species. It is known only from S. Spain (San Roque).

Species **5–9** are frequently regarded as subspecies or varieties of *T. guttata*, but they appear to be distinct species and do not fit into a subspecific pattern of distribution.

5. **T. bupleurifolia** (Lam.) Willk., *Icon. Descr. Pl. Nov.* **2**: 77 (1859). Like **4** but leaves narrow, the uppermost strongly revolute, bright green, glabrous or sparsely pubescent; inflorescence slender, few-flowered; flowers 10–13 mm in diameter. *Coastal sands. S. Portugal, S.W. Spain.* Hs Lu.

6. **T. villosissima** (Pomel) Grosser in Engler, *Pflanzenreich* **14**(IV, **193**): 59 (1903). Slender, sericeous-tomentose annual. Basal leaves caducous; cauline leaves 25–50 × 8–12 mm, ovate-lanceolate; upper cauline leaves smaller, alternate, with stipules half as long as the leaves; all leaves dark green, villous. Cymes branched at base, distinctly scorpioid when young, dense. Petals unspotted. Capsule villous. *Sicilia (Scoglitti, W. of Ragusa).* Si. (*Algeria.*)

7. **T. praecox** Grosser, *loc. cit.* (1903). Small, unbranched, slender, grey-villous annual. Lower cauline leaves 25–35 × 8–12 mm; stipules $\frac{1}{3}$ as long as upper cauline leaves. Petals scarcely exceeding the sepals, unspotted. Capsule glabrous. *Coastal regions of C. Mediterranean region.* Co It Ju Sa Si.

8. **T. acuminata** (Viv.) Grosser, *loc. cit.* (1903). Robust, erect annual up to 40 cm, grey-pubescent with appressed hairs. Basal and lower cauline leaves 40–60 × 3–5 mm, linear-lanceolate; stipules $\frac{1}{3}$ as long as the upper cauline leaves. Inflorescence dense. Pedicels thick and usually erect in fruit. Outer sepals smaller than the inner at anthesis, all accrescent and subequal in fruit; petals unspotted. *N.W. Italy.* It.

9. **T. macrosepala** (Cosson) Willk., *Icon. Descr. Pl. Nov.* **2**: 80 (1859). Like **8** but often up to 80 cm, with patent hairs; basal leaves 50–80 × 12–30 mm, elliptic-lanceolate, the upper

with stipules $\frac{1}{2}$ to $\frac{2}{3}$ as long as the leaves; inflorescence compact, becoming lax in fruit; pedicels arcuate in fruit; outer sepals equalling or exceeding the inner, rarely shorter. *Open places, often on mountains. S.W. Spain, E.C. Portugal.* Hs Lu.

10. **T. echioides** (Lam.) Willk., *op. cit.* 81 (1859). Robust, erect annual up to 30 cm, variably branched, white-pubescent with patent hairs. Basal leaves caducous and the whole plant often leafless in fruit. Leaves 35–60 × 8–12 mm, oblong-lanceolate. Flowers inconspicuous, subsessile in dense, scorpioid, many-flowered cymes. Outer sepals exceeding the inner, all accrescent in fruit; petals spotted at the base, caducous. *Sandy ground and waste places. S.W. Spain.* Hs. (*N.W. Africa.*)

4. Helianthemum Miller[1]

Dwarf shrubs or annual herbs. Leaves opposite, decussate, usually stipulate. Flowers in raceme-like cymes, often secund. Sepals 5, the 3 inner ovate, the 2 outer smaller, usually linear or oblong. Petals usually yellow, less often white or pink; stamens numerous, all fertile. Style short and straight or filiform and more or less sigmoid. Capsule ovoid, 3-valved; ovules orthotropous.

Literature: E. Guinea, *Cistáceas Españolas* (*Bol. Inst. For. Inv. Exper. Madrid* No. 71) 63–160. Madrid. 1954.

1 Dwarf shrubs
  2 All leaves stipulate (stipules occasionally caducous); style
      slender, filiform, slightly sigmoid at base
    3 Leaves and sepals densely covered with peltate scales
                                   2. squamatum
    3 Leaves and sepals without peltate scales
      4 Inflorescence corymbose   **1. lavandulifolium**
      4 Inflorescence ± simple
        5 Outer sepals broadly ovate, reflexed so as to give the villous
           flower buds the appearance of a cat's head **3. caput-felis**
        5 Outer sepals linear to linear-lanceolate, the flower buds
           not resembling a cat's head
          6 Flowers sessile
            7 Branches slender; sepals 3–4 mm in fruit **14. sessiliflorum**
            7 Branches robust; sepals 5–8 mm in fruit **15. stipulatum**
          6 Flowers pedicellate
            8 Plant glandular-hairy in all its parts **5. viscarium**
            8 Plant not glandular-hairy
              9 Sepals with long, stiff, patent bristles, usually equal-
                 ling or exceeding the intercostal spaces
               10 Intercostal spaces of sepals glabrous and shining
                                     6. asperum
               10 Intercostal spaces of sepals stellate-tomentose
                                     7. hirtum
             9 Sepals without long, stiff bristles
               11 Stipules ± leaf-like, longer than the petioles; stellate
                  hairs (when present) on upper surface of leaf
                  forming a thick felt
                 12 Upper surface of leaves densely stellate-tomentose
                                     8. croceum
                 12 Upper surface of leaves glabrous to pubescent
                               9. nummularium
                 13 Leaves with strongly revolute margins; capsule
                    much shorter than sepals **4. piliferum**
                 13 Leaves nearly or quite flat; capsule about
                    equalling sepals **9. nummularium**
             11 Stipules linear, not leaf-like; stellate hairs (when
                  present) on upper surface of leaf closely appressed,
                  scarcely forming a felt
         14 Petals yellow **10. leptophyllum**
         14 Petals white or pink
           15 Sepals pubescent **11. apenninum**
           15 Sepals glabrous or stellate-tomentose on or
              between the ribs

---

[1] By M. C. F. Proctor and V. H. Heywood.

16 Petals white; sepals in fruit less than 10 mm
12. pilosum
16 Petals pink; sepals in fruit more than 10 mm
13. virgatum
2 Lower leaves exstipulate (the upper leaves sometimes stipulate); style shorter than stamens, strongly sigmoid at base
17 Flowers solitary or few on long pedicels at ends of leafy shoots **31. lunulatum**
17 Flowers in leafless cymes
18 Leaves (at least of non-flowering shoots) densely stellate-tomentose beneath
19 Leaves all exstipulate, or upper leaves of flowering shoots with only small stipules
20 Leaves cuneate at base **23. canum**
20 Leaves rounded or cordate at base
21 Leaves ovate-orbicular to cordate-orbicular
26. pannosum
21 Leaves ovate to lanceolate, acute
22 Inflorescence simple or branched from base; leaves all exstipulate **25. marifolium**
22 Inflorescence usually paniculate; upper leaves of flowering shoots with small stipules **27. cinereum**
19 Upper leaves of flowering shoots with conspicuous, leaf-like stipules
23 Leaves ± revolute **30. viscidulum**
23 Leaves flat
24 Leaves rounded or cordate at base **27. cinereum**
24 Leaves cuneate at base **28. hymettium**
18 Leaves green on both surfaces
25 Upper leaves of flowering shoots with conspicuous stipules **29. rossmaessleri**
25 All leaves exstipulate
26 Leaves elliptical to linear-lanceolate **22. oelandicum**
26 Leaves ovate to cordate-orbicular **24. origanifolium**
1 Annuals
27 Sepals scarious **21. aegyptiacum**
27 Sepals herbaceous
28 Plant viscid; sepals obtuse; pedicels deflexed after anthesis
20. sanguineum
28 Not viscid; sepals acuminate; pedicels erect after anthesis
29 Flowers subsessile; inflorescence dense
30 Bracts shorter than sepals; seeds smooth or foveolate
16. villosum
30 Bracts longer than sepals; seeds covered with crystalline papillae **17. papillare**
29 Pedicels 2–12 mm; inflorescence lax
31 Pedicels erect, shorter than the sepals **18. ledifolium**
31 Pedicels patent, arcuate-erect at apex, exceeding the sepals
19. salicifolium

Subgen. **Helianthemum**. Leaves all stipulate; style straight or slightly sigmoid.

Sect. ARGYROLEPIS Spach. Shrubs or dwarf shrubs. Inflorescence compound; capsule ellipsoid-trigonous.

1. H. lavandulifolium Miller, *Gard. Dict.* ed. 8, no. 13 (1768) (*H. racemosum* Pers.). Small shrub 10–50 cm, densely grey-tomentose. Leaves 10–50 × 3–8 mm, lanceolate to linear-lanceolate, acute, with revolute margins, greyish-green and appressed-tomentose above, white-tomentose beneath. Stipules sometimes caducous. Flowers many, in dense cymes divided from the base into 3–5 branches. Petals 5–10 mm, yellow. Capsule shorter than calyx. $2n=20$. *Calcicole. Mediterranean region.* Cr Ga Gr Hs It Ju Tu.

2. H. squamatum (L.) Pers., *Syn. Pl.* **2**: 78 (1806). Small densely caespitose shrub 10–30 cm, densely covered with sessile, silvery peltate scales. Leaves 10–30 × 3–10 mm, lanceolate to oblanceolate- or linear-spathulate, fleshy, with flat margins.

Stipules about as long as the petiole, subulate to linear-lanceolate, caducous. Inflorescence usually branched from base into usually 3 long-pedunculate, dense, more or less capitate cymes. Petals only slightly longer than the sepals, yellow with a dark spot at the base. $2n=10$. *Gypsum soils. E., C. & S. Spain.* Hs.

3. H. caput-felis Boiss., *Elenchus* 16 (1838). Compact, caespitose dwarf shrub 10–30 cm; branches erect, very leafy, densely tomentose. Leaves 6–15 × 2–10 mm, broadly elliptical to lanceolate, thick, densely tomentose on both surfaces, with revolute margins. Cymes compact, few-flowered. Sepals densely villous, giving the appearance in bud of a cat's head. Petals 9–12 mm, yellow, longer than the sepals. *Coastal limestone cliffs and dry places inland. S.E. Spain (Alicante prov.), Balear* Bl Hs. (*N. Africa.*)

Sect. HELIANTHEMUM. Dwarf shrubs; inflorescence unbranched; capsule ovoid or globose.

4. H. piliferum Boiss., *Elenchus* 17 (1838). Stems 10–20 cm, erect or procumbent, glabrous, leafy below, leafless above. Upper leaves 10–20 × 1–2 mm, the lower smaller, linear-lanceolate, stiff, with strongly revolute margins, glabrous, bright green, often with a white bristle at the apex, very shortly petiolate. Stipules like the leaves, but half as long. Cymes unilateral, few- to many-flowered. Sepals subglabrous. Flowers 25 mm in diameter; petals yellow. Capsule 7 mm, much shorter than the sepals in fruit. *S. Spain.* Hs.

5. H. viscarium Boiss. & Reuter, *Pugillus* 14 (1852). Compact, 15–30 cm, more or less glandular-viscid. Leaves 10–20 × 2–4 mm, linear-lanceolate to linear-elliptical, green on both surfaces or glaucous beneath, with more or less revolute margins. Stipules linear, longer than the petioles. Cymes 4- to 12-flowered, with conspicuous bracts resembling the stipules. Sepals exceeding the capsule, becoming inflated in fruit, with strongly marked ribs. *S.E. Spain (Murcia prov.).* Hs. (*N. Africa.*)

6. H. asperum Lag. ex Dunal in DC., *Prodr.* **1**: 283 (1824). Laxly branched, with erect stems up to 35 cm. Leaves 10–22 × 3–7 mm, ovate-oblong to oblong-linear, roughly hairy with stellate hairs and greenish above, usually grey-tomentose beneath, with more or less revolute margins. Cymes 6- to 9-flowered. Sepals inflated in fruit, prominently ribbed, the ribs with conspicuous, long bristles; intercostal spaces glabrous and shining. Flowers *c.* 20 mm in diameter; petals white. Capsule included in the sepals. *Dry and rocky places on calcareous soil.* ● *E., C. & S. Spain.* Hs.

7. H. hirtum (L.) Miller, *Gard. Dict.* ed. 8, no. 14 (1768). Up to 30 cm, caespitose, with erect branches, rarely procumbent. Leaves 3–20 × 1–6 mm, the lower ones smaller, ovate-orbicular, the upper elliptical to linear-lanceolate, all somewhat fleshy, dark green to grey above, canescent and stellate-tomentose beneath. Cymes 5- to 17-flowered. Sepals prominently ribbed, the ribs and margins with conspicuous, long bristles; intercostal spaces stellate-hairy. Flowers 15 mm in diameter; petals white or yellow. Capsule included in the sepals. *Scrub, woods, dry places, usually on calcareous substrate. S.W. Europe.* Co Ga Hs Lu.

8. H. croceum (Desf.) Pers., *Syn. Pl.* **2**: 79 (1806) (*H. glaucum* Pers.). 5–30 cm, densely or laxly caespitose; branches erect or procumbent. Leaves 5–20 × 2–7 mm, suborbicular to linear-lanceolate, fleshy, usually stellate-tomentose on both surfaces, rarely with stellate, fasciculate and simple hairs above, with flat, or slightly to strongly revolute margins; stipules longer than the petioles. Cymes 3- to 15-flowered. Sepals stellate-tomentose between the ribs, sometimes minutely so, conspicuously stellate-tomentose or hirsute on the ribs. Flowers up to 20 mm in diameter; petals orange-yellow, bright yellow or white. Capsule about equalling the sepals in fruit. $2n=20$. *Scrub, woods and mountain slopes. W. Mediterranean region, Portugal.* Ga Hs It Lu Sa Si.

Extremely variable in habit, indumentum, leaf-dimensions and -shape and flower-colour.

9. H. nummularium (L.) Miller, *Gard. Dict.* ed. 8, no. 12 (1768) (*H. chamaecistus* Miller, *H. vulgare* Gaertner). 5–50 cm; branches procumbent or ascending. Leaves 5–50 × 2–15 mm, oblong or lanceolate to ovate or orbicular, subglabrous to pubescent above, white-tomentose, rarely greenish, beneath, with flat or slightly revolute margins. Stipules lanceolate to linear-lanceolate, longer than the petioles. Cymes 1- to 12-flowered, unilateral. Petals 6–18 mm, golden-yellow, rarely cream, pale yellow, white, orange or pink. Capsule about equalling the sepals. $2n=20$. *Grassy and rocky places, usually on basic soils. Most of Europe except the extreme north.* All except Az Fa Is No Rs (N) Sb.

A complex species variously divided by different authors. A modification of the treatment in Hegi, *Ill. Fl. Mitteleur.* **5(1)**: 565–571 (1925) is followed here.

1 Leaves green on both surfaces
  2 Leaves glabrous, or the margins and midrib ciliate
    3 Petals yellow   **(e) subsp. glabrum**
    3 Petals pink   **(f) subsp. semiglabrum**
  2 Leaves with long hairs on surface and margins
    4 Sepals tomentose or pubescent between the ribs; inner sepals
      5–8 mm; petals 8–12 mm   **(d) subsp. obscurum**
    4 Sepals glabrous between the ribs; inner sepals 7–10 mm;
      petals 10–18 mm   **(h) subsp. grandiflorum**
1 Leaves grey- or white-tomentose, at least beneath
  5 Petals c. 10 mm, pink
    6 Leaves white-tomentose beneath   **(c) subsp. pyrenaicum**
    6 Leaves greyish-pubescent beneath   **(g) subsp. berterianum**
  5 Petals yellow, orange, red-spotted at the base or rarely white
    7 Leaves 2–5(–7) mm wide; petals 6–10(–12) mm
                          (a) subsp. nummularium
    7 Leaves 5–15 mm wide; petals (8–)10–15 mm
                          (b) subsp. tomentosum

(a) Subsp. nummularium (*H. nummularium* subsp. *vulgare* (Gaertner) Hayek; incl. *H. arcticum* (Guss.) Juz.): $2n=20$. *Widespread in Europe.*

(b) Subsp. tomentosum (Scop.) Schinz. & Thell. in Schinz & R. Keller, *Fl. Schweiz* ed. 3, **2**: 249 (1914) (*H. tomentosum* (Scop.) S. F. Gray): *Mountains of S. Europe.*

(c) Subsp. pyrenaicum (Janchen) Schinz & Thell. in Hegi, *Ill. Fl. Mitteleur.* **5(1)**: 570 (1925) (*H. pyrenaicum* Janchen, *H. vulgare* var. *roseum* Willk.): *Pyrenees and neighbouring mountains.*
Widely cultivated in gardens for ornament.

(d) Subsp. obscurum (Čelak.) J. Holub, *Acta Horti Bot. Prag.* **1963**: 53 (1964) (*H. hirsutum* (Thuill.) Mérat, *H. nummularium* subsp. *ovatum* (Viv.) Schinz & Thell., *H. ovatum* subsp. *hirsutum* Hayek, *H. vulgare* var. *genuinum* Willk. pro parte): $2n=20$. *C. Europe and parts of E. & S. Europe, extending northwards to Sweden.*

(e) Subsp. glabrum (Koch) Wilczek, *Annu. Cons. Jard. Bot. Genève* **21**: 453 (1922) (*H. nitidum* G. C. Clementi): *Mountains of C., S. & S.W. Europe.*

(f) Subsp. semiglabrum (Badaro) M. C. F. Proctor, *Feddes Repert.* **79**: 59 (1968) (*H. semiglabrum* Badaro): *Maritime Alps, N. Appennini.*

(g) Subsp. berterianum (Bertol.) Breistr., *Bull. Soc. Sci. Dauph.* **61**: 623 (1947) (*H. berterianum* Bertol.): *Maritime Alps, Appennini.*

(h) Subsp. grandiflorum (Scop.) Schinz & Thell. in Schinz & R. Keller, *Fl. Schweiz* ed. 3, **2**: 249 (1914) (*H. grandiflorum* (Scop.) Lam., *H. ovatum* subsp. *grandiflorum* (Scop.) Hayek): $2n=20$. *Mountains of C., S. & S.W. Europe.*

H. morisianum Bertol., *Fl. Ital.* **5**: 374 (1844), from Sardegna, with pink petals, may deserve recognition as a further subspecies.

10. H. leptophyllum Dunal in DC., *Prodr.* **1**: 279 (1824). Erect, much-branched, slender, 10–20(–30) cm. Leaves 5–10(–20) × 2–5 mm, rather distant and evenly spaced, linear-oblong to elliptic-lanceolate, grey and stellate-tomentose above and greenish beneath, or subglabrous and green on both surfaces. Stipules linear, about as long as the petiole. Cymes (1–)3- to 8-flowered. Flowers c. 20 mm in diameter; petals yellow. Capsule shorter than the sepals in fruit. *Dry hillsides. S. Spain, S. Italy.* Hs It.

Plants from Italy with linear leaves and more pubescent sepals have been recognized as **H. jonium** Lacaita, *Nuovo Gior. Bot. Ital.* nov. ser., **17**: 609 (1910).

11. H. apenninum (L.) Miller, *Gard. Dict.* ed. 8, no. 4 (1768) (*H. polifolium* Miller, *H. pulverulentum* auct.). Lax, somewhat spreading, up to 50 cm, much-branched from the base. Leaves 8–30 × 2–8 mm, linear to linear-oblong, green to grey- or white-tomentose above, densely stellate-hairy to grey- or white-tomentose beneath, with slightly to strongly revolute margins; stipules linear-lanceolate, slightly longer than the lower petioles. Cymes 3- to 10-flowered. Sepals 7–10 mm, pubescent over the whole outer surface. Petals white, with a yellow claw. Capsule about equalling the sepals in fruit. $2n=20$. *S. & W. Europe, northwards to S. England and extending to W. Germany.* Al Be Bl Br Cr Ga Ge Gr He Hs It Lu.

Plants with pink petals occur in the Islas Baleares and N.W. Italy.

12. H. pilosum (L.) Pers., *Syn. Pl.* **2**: 79 (1806). Caespitose or lax, much-branched, up to 30 cm; branches usually white-tomentose. Leaves 10–20 × 1–4 mm, the upper cauline longer than the lower, linear to linear-oblong, greenish-pubescent above, grey-tomentose beneath, or more or less tomentose on both surfaces, with margins strongly or only slightly revolute. Cymes 4- to 10(–15)-flowered. Sepals 5–6 mm, glabrous or stellate-tomentose on the ribs. Petals c. 10 mm, white with a yellow claw. Capsule shorter than the sepals in fruit. *Scrub and forest clearings, on dry, calcareous or clay soils. W. Mediterranean region, Portugal.* Ga Hs Lu It.

H. almeriense Pau, *Mem. Mus. Ci. Nat. Barcelona* (Bot.) **1(3)**: 11 (1925), from S.E. Spain (Almería prov.), resembles some variants of **12** but is usually entirely glabrous and has leaves 5–12 mm, ovate-orbicular to linear-oblong, obtuse. It may deserve recognition as a separate species.

13. H. virgatum (Desf.) Pers., *Syn. Pl.* **2**: 79 (1806). Up to 30 cm; branches slender, white-tomentose. Leaves 10–30 × 2–8 mm, the middle and upper longer than the lower, linear-lanceolate, green above, stellate-tomentose beneath, with revolute margins. Cymes 5- to 10-flowered. Sepals 8–10 mm, stellate-tomentose at least between the ribs. Petals *c.* 10 mm, pink. Capsule shorter than the sepals in fruit. *?Spain.* ?Hs. (*N. Africa.*)

Doubtfully recorded from Spain, where it has been confused with **11** and **12**.

Sect. ERIOCARPUM Dunal. Dwarf shrubs; inflorescence usually unbranched; capsule globose or trigonous; style long, filiform.

14. H. sessiliflorum (Desf.) Pers., *Syn. Pl.* **2**: 78 (1806). Caespitose, much-branched dwarf shrub 30–60 cm; branches and shoots white, stellate-pubescent. Leaves varying according to the season; in rainy periods 15–20 mm, lanceolate, green, with flat margins, caducous; in dry periods 5–12 mm, linear, whitish, strongly revolute, persistent. Cymes 6- to 15-flowered, dense; flowers minute, chasmogamous or cleistogamous. Sepals 1–2 mm, green, with stellate hairs between the ribs and long white hairs on the ribs, white-villous at base. Petals yellow, longer or shorter than the sepals. *Dry hillsides near the sea. S. Italy, Sicilia.* It Si.

Often regarded as a subspecies or variety of **H. lippii** (L.) Pers., *loc. cit.* (1806), from N. Africa and Asia Minor.

15. H. stipulatum (Forskål) C. Chr., *Dansk Bot. Ark.* **4**(3): 20 (1922) (*H. ellipticum* (Desf.) Pers.). Dwarf shrub, usually erect. Leaves 8–15 × 2·5–7·5 mm, linear- to ovate-lanceolate, with slightly or strongly revolute margins according to the season, sparsely stellate-tomentose above, densely so beneath. Cymes 3- to 7-flowered, lax. Sepals 3–4 mm, stellate-hairy between and on the ribs, ciliate; petals as long as the sepals, yellow. *Maritime sands. Greece (N.W. Peloponisos).* Gr. (*S.W. Asia, Egypt.*)

Sect. BRACHYPETALUM Dunal. Annuals; inflorescence mostly unbranched; capsule trigonous; style short, straight.

16. H. villosum Thib. in Pers., *Syn. Pl.* **2**: 78 (1806). Slender, erect annual 8–20 cm. Leaves 5–30 × 3–8 mm, lanceolate to obovate-lanceolate, grey-green with stellate hairs on both surfaces; stipules linear, ⅓–½ as long as the leaves. Cymes dense, unilateral, more or less scorpioid when young, later spike-like; pedicels short. Sepals 5–8 mm, villous; petals inconspicuous, shorter than the sepals, narrow, yellow. Seeds smooth. *Dry or rocky places. S. Spain, S. Portugal.* Hs Lu.

17. H. papillare Boiss., *Voy. Bot. Midi Esp.* **2**: 63 (1839). Somewhat robust annual up to 13 cm; stems erect or ascending, covered with patent hairs. Leaves 10–25 × 5–8 mm, lanceolate to obovate-lanceolate, with patent hairs on both surfaces and with stellate hairs beneath; stipules linear-lanceolate ⅓–½ as long as the leaves. Cymes dense, contracted, 2–5 cm when young, later spike-like; pedicels short. Sepals *c.* 5 mm, accrescent; petals yellow, linear-lanceolate. Seeds covered with shining, crystalline papillae. *Scrub and open habitats on clay or dry soils. S. Spain.* Hs.

18. H. ledifolium (L.) Miller, *Gard. Dict.* ed. 8, no. 20 (1768) (*H. niloticum* (L.) Pers., non Moench; incl. *H. lasiocarpum* Desf. ex Willk.). Villous-tomentose annual 10–60 cm, variable in habit and size of parts. Leaves 10–50 × 3–12 mm, elliptical to lanceolate or obovate, greenish above, greyish beneath, rarely greyish on both surfaces; stipules linear to lanceolate, about ½ as long as the leaves. Cymes 3- to 13-flowered, scorpioid; pedicels erect, thickened. Sepals 6–10 mm, accrescent; petals shorter than the sepals, yellow, cuneate. Seeds foveolate, smooth or papillose. $2n=20$. *Dry places. S. Europe.* Bu Cr Ga Gr Hs It Ju Lu Sa Si.

19. H. salicifolium (L.) Miller, *Gard. Dict.* ed. 8, no. 21 (1768) (*H. intermedium* (Pers.) Thib. ex Dunal). Annual up to 30 cm, variable in habit, usually much-branched. Leaves 5–30 × 3–10 mm, ovate-lanceolate to elliptic-oblong; stipules linear- to ovate-lanceolate. Flowers in simple or branched, 5- to 20-flowered cymes; pedicels long, slender, patent, usually upturned at the apex. Sepals 5–12 mm; petals longer or shorter than the sepals, or absent. $2n=20$. *Dry places. S. Europe.* Al Bl Bu Co Cr Ga Gr He Hs It Ju Lu Rm Rs (K) Sa Si Tu.

20. H. sanguineum (Lag.) Lag. ex Dunal in DC., *Prodr.* **1**: 273 (1824) (*H. retrofractum* Pers.). Much-branched, glandular-hairy annual 2–10 cm; stems often purple-tinged. Lower leaves 10–20 × 8–10 mm, elliptical, obtuse, caducous; upper leaves oblong-lanceolate, persistent; stipules ⅓ as long as the leaves. Cymes lax, 3- to 6-flowered; pedicels stout, curved before anthesis, strongly deflexed in fruit. Sepals 7–8 mm; petals yellow, shorter than the sepals. Capsule glabrous. *Portugal, Spain, Italy, Kriti.* Cr Hs It Lu.

21. H. aegyptiacum (L.) Miller, *Gard. Dict.* ed. 8, no. 23 (1768). Erect, little-branched annual up to 30 cm; stems and branches villous. Leaves 10–30 × 1·5–3 mm, linear-lanceolate or oblong, often with revolute margins, dark green above, grey-tomentose beneath; stipules linear, ⅓–¼ as long as the leaves. Cymes lax, 3- to 9-flowered; pedicels long, filiform, deflexed in fruit. Sepals 6–10 mm; petals yellow, shorter than the sepals. Capsule appressed-pubescent. *Dry, sandy places. Mediterranean region, S. Bulgaria.* Bu Co Cr Gr Hs It Sa Si.

Subgen. **Plectolobum** Willk. Dwarf shrubs; at least the lower leaves exstipulate; style distinctly sigmoid at the base.

Sect. PLECTOLOBUM (Sect. *Chamaecistus* Willk.). Flowers in simple or branched, bracteate cymes.

22. H. oelandicum (L.) DC. in Lam. & DC., *Fl. Fr.* ed. 3, **4**: 817 (1805) (*H. montanum* sensu Willk.). Laxly or densely caespitose dwarf shrub up to 20 cm. Leaves elliptical to linear-lanceolate, acute or obtuse, green on both surfaces, glabrous or with simple or fasciculate but not stellate hairs. Flowers few or many, in simple or branched cymes; petals yellow. *Scattered over much of Europe, but absent from most of the north and most of the islands.* Al Au Cz Ga Ge Gr He Hs It Ju Lu Po Rm Rs (N, C, W, K, E) Su.

Polymorphic, and divisible into at least five fairly clearly delimited subspecies:

1 Lax; cymes 6- to 20-flowered, often produced on lateral branches of the current year's growth; petals 3–6 mm
    (d) subsp. **italicum**
1 Caespitose; cymes 2- to 10-flowered, usually produced from the apex of the previous year's growth; petals 5–10 mm
2 Procumbent and intricately branched; leaves glabrous or subglabrous; sepals subglabrous   (a) subsp. **oelandicum**
2 Compact; leaves pubescent; sepals long-pubescent
3 Hairs on sepals ±patent   (e) subsp. **orientale**
3 Hairs on sepals mainly ±appressed
  4 Leaves oblong-elliptical to oblong-linear, obtuse; petals 7–10 mm   (b) subsp. **alpestre**
  4 Leaves lanceolate, usually acute; petals 5–9 mm
    (c) subsp. **rupifragum**

(a) Subsp. **oelandicum**: $2n=22$. *Öland.*
(b) Subsp. **alpestre** (Jacq.) Breistr., *Bull. Soc. Sci. Dauph.* **61**: 623 (1947) (*H. alpestre* (Jacq.) DC., *H. italicum* subsp. *alpestre* (Jacq.) Beger): $2n=22$. *Mountains of C. & S. Europe.*

(c) Subsp. **rupifragum** (A. Kerner) Breistr., *loc. cit.* (1947) (*H. italicum* subsp. *rupifragum* (A. Kerner) Beger, *H. rupifragum* A. Kerner): *E. & E.C. Europe.*

(d) Subsp. **italicum** (L.) Font Quer & Rothm., *Cavanillesia* **6**: 153 (1934) (*H. italicum* (L.) Pers.): 2n = 20. *Mediterranean region.*

(e) Subsp. **orientale** (Grosser) M. C. F. Proctor, *Feddes Repert.* **79**: 58 (1968) (*H. orientale* (Grosser) Juz. & Pozd.): *Krym.*

23. H. canum (L.) Baumg., *Enum. Stirp. Transs.* **2**: 85 (1816). Stems 4–20(–30) cm, procumbent or ascending. Leaves elliptical, ovate-lanceolate, lanceolate or linear, grey-tomentose beneath, green to grey-tomentose above, with or without stellate hairs. Cymes lax or dense, with 1–5 flowers. Petals 4–8 mm, yellow. *C. & S. Europe; Britain and Ireland; Öland.* Al Au Br Bu Cz Ga Ge Gr Hb He Hs Hu It Ju Po Rm Rs (W, K) Sa Si Su Tu.

Very variable, and divisible with difficulty into a number of subspecies:

1 Shoots arcuate-ascending; inflorescences usually produced on lateral branches of the current year's growth
2 Inflorescence 5- to 15-flowered, eglandular   **(f) subsp. pourretii**
2 Inflorescence 3- to 6-flowered, with purplish-black glandular hairs   **(b) subsp. nebrodense**
1 Shoots forming a more or less procumbent mat; inflorescences usually produced from apex of previous year's growth
3 Robust, very pubescent; sepals densely villous; petals 6–8 mm (Krym)   **(g) subsp. stevenii**
3 Less robust; sepals with more or less appressed pubescence; petals less than 6 mm
4 Leaves lanceolate to linear, sparsely strigose above; inflorescence usually produced on lateral branches of the current year's growth (Öland)   **(c) subsp. canescens**
4 Leaves ovate-lanceolate to lanceolate; indumentum various; inflorescence from apex of previous year's growth
5 Leaves glabrous or subglabrous above; flowers 1–3(–4) (England)   **(d) subsp. levigatum**
5 Leaves ± pubescent or strigose above; flowers usually more than 3
6 Leaves relatively broad, rather persistent on the lower parts of the vegetative shoots   **(e) subsp. piloselloides**
6 Leaves relatively narrow, mostly clustered into rosettes at the apex of the vegetative shoots   **(a) subsp. canum**

(a) Subsp. **canum**: 2n = 22. *Almost throughout the range of the species.*

(b) Subsp. **nebrodense** (Heldr. ex Guss.) Arcangeli, *Comp. Fl. Ital.* 72 (1882) (*H. allionii* Tineo): ● *Sicilia.*

(c) Subsp. **canescens** (Hartman) M. C. F. Proctor, *Feddes Repert.* **79**: 58 (1968): ● *Öland.*

(d) Subsp. **levigatum** M. C. F. Proctor, *Watsonia* **4**: 38 (1957): ● *N. England* (*near Middleton-in-Teesdale*).

(e) Subsp. **piloselloides** (Lapeyr.) M. C. F. Proctor, *Feddes Repert.* **79**: 58 (1968) (*Cistus piloselloides* Lapeyr.): ● *Pyrenees, N. Spain; related forms in W. Ireland.*

(f) Subsp. **pourretii** (Timb.-Lagr.) M. C. F. Proctor, *Feddes Repert.* **79**: 59 (1968) (*H. pourretii* Timb.-Lagr.): ● *S. France, Italy.*

(g) Subsp. **stevenii** (Rupr. ex Juz. & Pozd.) M. C. F. Proctor, *Feddes Repert.* **79**: 59 (1968) (*H. stevenii* Rupr. ex Juz. & Pozd.): ● *Krym.*

24. H. origanifolium (Lam.) Pers., *Syn. Pl.* **2**: 76 (1806). Stems 5–30 cm, procumbent or ascending. Leaves broadly ovate to ovate-lanceolate or orbicular-cordate, variable in size and texture, flat or with revolute margins, subglabrous to densely pubescent, with sparse to dense stellate hairs beneath; stipules

minute and caducous or absent. Cymes 4- to 12-flowered, simple or branched. Petals 3–6 mm, yellow. *Sandy and rocky places. S. Portugal, Spain, Islas Baleares.* Bl Hs Lu.

1 Leaves densely hairy above, stellate-hairy beneath; flowers 5–13 mm in diameter, in simple cymes   **(d) subsp. molle**
1 Leaves sparsely hairy above, with a few long simple and a few stellate hairs beneath; flowers 5–8 mm in diameter, in panicles
2 Stellate hairs on both surfaces of leaf   **(c) subsp. serrae**
2 Stellate hairs absent or on lower surface of leaf only
3 Inflorescence branched; upper leaves stipulate   **(a) subsp. origanifolium**
3 Inflorescence simple; all leaves exstipulate   **(b) subsp. glabratum**

(a) Subsp. **origanifolium**: *S. Portugal; ?E. Spain.* (*N. Africa.*)

(b) Subsp. **glabratum** (Willk.) Guinea & Heywood in Guinea, *Cistác. Esp.* 133 (1954) (*H. origanifolium* var. *glabratum* Willk.): *E. Spain.*

(c) Subsp. **serrae** (Camb.) Guinea & Heywood, *op. cit.* 134 (1954) (*H. serrae* Camb.): *Islas Baleares.*

(d) Subsp. **molle** (Cav.) Font Quer & Rothm., *Cavanillesia* **6**: 162 (1934) (*H. origanifolium* var. *majus* Willk.): *S. & N.E. Spain.*

25. H. marifolium (L.) Miller, *Gard. Dict.* ed. 8, no. 24 (1768) (*H. myrtifolium* Samp.). Laxly caespitose, much-branched; stems 5–30 cm, usually procumbent, rarely erect, grey-tomentose. Leaves 5–25 mm, ovate-lanceolate to broadly ovate, acute, green and glabrous to greyish-tomentose above, grey- or white-tomentose beneath; stipules minute and caducous, or absent. Cymes 4- to 7-flowered, simple or branched. Flowers 10–15 mm in diameter; petals yellow. ● *S. Portugal, S. & E. Spain, S. France.* Ga Hs Lu.

Very variable in leaf-dimensions and indumentum.

26. H. pannosum Boiss., *Elenchus* 15 (1838). Densely caespitose, dwarf shrub, 3–12 cm. Leaves 6–10(–20) × 2–7(–14) mm, ovate-orbicular or cordate-orbicular, covered with a thick, soft, whitish felt on both surfaces or greenish above; stipules minute and caducous, or absent. Cymes few- or many-flowered, simple or branched. Petals yellow. *Limestone rocks and screes.* ● *S. Spain.*

(a) Subsp. **pannosum** (subsp. *boissieri* Font Quer & Rothm.): Inflorescence glandular-hairy; leaves thick and fleshy, with a dense, white covering of simple appressed hairs above. *Sierra Nevada.*

(b) Subsp. **frigidulum** (Cuatrec.) Font Quer & Rothm., *Cavanillesia* **6**: 164 (1934): Inflorescence eglandular; leaves thin, greenish above. *Sierra de Mágina* (*prov. Jaén*).

27. H. cinereum (Cav.) Pers., *Syn. Pl.* **2**: 76 (1806) (incl. *H. rubellum* C. Presl, non Moench, *H. paniculatum* Dunal). Usually laxly caespitose with erect branches, but very variable. Leaves ovate to lanceolate, rounded or cordate at base, green and glabrous to grey-tomentose above, grey-tomentose beneath. Stipules small or large, persistent or caducous. Inflorescence usually paniculate, rarely a simple cyme. Petals yellow. 2n = 20, 22. *Mediterranean region.* Gr Hs It Si.

Very variable in habit, indumentum, leaf-shape and other features. No satisfactory division into infraspecific taxa is possible.

28. H. hymettium Boiss. & Heldr. in Boiss., *Diagn. Pl. Or. Nov.* **3**(1): 52 (1853). Up to 10 cm; branches procumbent or decumbent. Leaves 7–15 × 3–6 mm, oblong-lanceolate to elliptical, white-tomentose on both surfaces or greenish and glabrous above, flat. Stipules of upper leaves 2–3 mm; cymes 7- to 9-flowered, somewhat condensed. Sepals 4 mm. Petals 3–4 mm, yellow. *Stony hillsides.* ● *S. Greece, Kriti.* Cr Gr.

29. H. rossmaessleri Willk., *Linnaea* **30**: 87 (1859). Caespitose dwarf shrub 5–15 cm, often pulvinate, with rigid, projecting shoots. Leaves 5–10 × 3–5 mm, ovate-lanceolate, green, pubescent or glabrous on both surfaces; upper leaves stipulate. Cymes simple or branched, few-flowered, somewhat lax. Petals yellow. ● *S. & S.E. Spain.* Hs.

30. H. viscidulum Boiss., *Elenchus* 15 (1838). Densely caespitose, 7–15 cm. Leaves 5–15 × 5–10 mm, lanceolate to triangular-ovate or ovate-orbicular, long-petiolate, pubescent above, white-tomentose beneath, the margins revolute; upper leaves stipulate. Inflorescence simple or branched. Petals yellow. ● *S. Spain.* Hs.

1 Plant eglandular (b) subsp. **guadiccianum**
1 Plant glandular
  2 Leaves variable, the mid-cauline ovate-triangular, wide at the base; petiole at least as long as the lamina
    (a) subsp. **viscidulum**
  2 Leaves all similar, ovate-elliptical to elliptic-lanceolate, the base as narrow as the apex; petiole shorter than the lamina, or rarely as long (c) subsp. **viscarioides**

(a) Subsp. **viscidulum** (*H. viscidulum* subsp. *texedense* Font Quer & Rothm.): *Sierra Nevada, Sierra Tejeda.*

(b) Subsp. **guadiccianum** Font Quer & Rothm., *Cavanillesia* 6: 170 (1934): *Guadix.*

(c) Subsp. **viscarioides** (Debeaux & Reverchon) Guinea & Heywood in Guinea, *Cistác. Esp.* 160 (1954) (*H. viscarioides* Debeaux & Reverchon): *Sierra de Cazorla, Sierra del Cuarto.*

Sect. MACULARIA Dunal. Flowers long-pedicellate, subsolitary on leafy branches.

31. H. lunulatum (All.) DC. in Lam. & DC., *Fl. Fr.* ed. 3, 4: 816 (1805). Caespitose dwarf shrub 5–20 cm; branches tortuous, often becoming naked and subspinose. Leaves 10 mm, elliptic-lanceolate, green, slightly pubescent. Flowers few, at the apex of leafy branches. Petals yellow, with an orange spot at the base. ● *Alpi Marittime.* It

5. Fumana (Dunal) Spach[1]

Dwarf shrubs. Leaves narrow, ovate-lanceolate to linear, acicular, alternate (rarely opposite), stipulate or exstipulate. The two outer sepals small, the three inner large, scarious, prominently veined. Petals yellow. Outer stamens sterile, moniliform. Style filiform, more or less geniculate at base. Capsule 3-veined, the valves usually patent after dehiscence.

All European species grow on dry, rocky, stony or sandy ground, often in low scrub.

Literature: E. Guinea, *Cistáceas Españolas* (*Bol. Inst. For. Inv. Exper. Madrid* No. 71) 161–181. Madrid. 1954.

1 Leaves ±equally spaced on the stems, not or scarcely reduced above
  2 Leaves ovate- to oblong-lanceolate, stipulate **1. arabica**
  2 Leaves linear, exstipulate
    3 Procumbent; fruiting pedicels as long as or shorter than adjacent leaves, recurved from the base **2. procumbens**
    3 Erect or straggling-ascending; fruiting pedicels much longer than adjacent leaves, patent, with deflexed apex **3. ericoides**
1 Leaves unequally spaced on the stem, ± abruptly reduced above to form small bracts in the inflorescence
  4 Capsule with 3 seeds (Greece) **9. aciphylla**

4 Capsule with 6–12 seeds
  5 Leaves opposite, stipulate **6. thymifolia**
  5 Leaves alternate, usually exstipulate
    6 Leaves linear-setaceous, ±terete, stipulate **8. laevipes**
    6 Leaves linear to linear-lanceolate, trigonous or subtrigonous, exstipulate
      7 Pedicels and calyx eglandular **7. paradoxa**
      7 Pedicels and usually calyx glandular
        8 Leaves glandular-hairy; capsule with (6–)8–12 seeds **4. scoparia**
        8 Leaves glabrous; capsule with not more than 6 seeds **5. bonapartei**

1. F. arabica (L.) Spach, *Ann. Sci. Nat.* ser. 2 (Bot.), **6**: 359 (1836) (incl. *F. viscidula* (Steven) Juz.). Much-branched, laxly caespitose, up to 25 cm. Leaves 5–12 × 0·8–5 mm, alternate, ovate- to oblong-lanceolate, acute, flat, glandular-pubescent to glabrescent (greyish when young in var. *incanescens* Hausskn.); stipules short. Flowers 1–7, forming a distinct, lax, inflorescence. Capsules (6–)8- to 12-seeded; seeds reticulate-foveolate. *S. Europe, from Sardegna to Krym.* Al ?Bu Cr Gr It Ju Rs (K) Sa Si.

F. pinatzii Rech. fil., *Anzeig. Akad. Wiss.* (*Wien*) **93**: 97 (1956), described from Evvoia, Greece, differs only by its 6-seeded capsules; plants of *F. arabica* from the Kikladhes sometimes have only 6 seeds.

2. F. procumbens (Dunal) Gren. & Godron, *Fl. Fr.* **1**: 173 (1847) (*Cistus fumana* L., *Helianthemum procumbens* Dunal, *Fumana nudifolia* Janchen, *F. vulgaris* Spach). Procumbent, with usually spreading branches up to 40 cm. Leaves (4–)10–18 × 0·5–2 mm, alternate, linear, subtrigonous, mucronate, ciliate, exstipulate. Flowers 3–4, solitary in the axils of leaves, not forming an inflorescence; pedicels about as long as the adjacent leaves, recurved from the base. Capsule 8- to 12-seeded; seeds retained for a long time after dehiscence. *W.C. & S. Europe, northwards to N. France; Öland and Gotland.* Al Au Be Bl Bu Cr Cz Ga Ge Gr He Hs It Ju Lu Rm Rs (W, K) Sa Si Su Tu.

3. F. ericoides (Cav.) Gand. in Magnier, *Fl. Select. Exsicc.* no. 201 (1883) (*F. spachii* Gren. & Godron). Erect or straggling-ascending, much branched, up to 20 cm. Leaves (3–)8–12(–15) × 0·5–2 mm, variable in size according to position on plant, alternate, linear, obtuse, glabrous or glandular-puberulent, exstipulate; leaves more or less equally spaced, scarcely reduced in size in inflorescence. Flowers 2–5, axillary or subterminal, scattered among the leaves, not forming a distinct inflorescence; pedicels much longer than the adjacent leaves, patent in fruit and deflexed at the apex. Capsule 8- to 12-seeded, late-dehiscent; seeds not long retained. *Calcicole. Mediterranean region; Portugal.* Bl ?Co ?Cr Ga Gr He Hs It Ju Lu Sa Si.

4. F. scoparia Pomel, *Mat. Fl. Atl.* 10 (1860) (*F. ericoides* auct. pro parte, non (Cav.) Gand.). Like 3 but ascending; leaves crowded below, widely spaced above, abruptly reduced in size in inflorescence; inflorescence terminal, distinct, densely glandular-hairy with long hairs; capsule (6–)8- to 12-seeded; seeds retained for some time after dehiscence. *Mediterranean region.* Al ?Cr Gr Hs It.

5. F. bonapartei Maire & Petitmengin, *Mat. Étude Fl. Géogr. Bot. Or.* **4**: 37 (1908). Procumbent, with spreading stems up to 15 cm. Leaves (3–)4–10(–12) × 0·5–1·5 mm, alternate, linear, subtrigonous, glabrous, exstipulate. Inflorescence up to 7-flowered; pedicels much longer than the subtending bracts, glandular-pubescent, patent in fruit and deflexed at the apex. Capsule 6-seeded. *Balkan peninsula, mainly in the west.* Al Gr Ju.

[1] By V. H. Heywood.

6. F. thymifolia (L.) Spach ex Webb, *Iter Hisp.* 69 (1838) (*F. viscida* Spach, *F. glutinosa* (L.) Boiss., *Helianthemum viride* Ten.). Up to 20 cm; stems erect or ascending. Leaves 5–11 × 0·5–1 mm, opposite at least below, linear, linear-lanceolate or narrowly elliptical, obtuse or mucronate, glabrous, pubescent or glandular-pubescent, with strongly revolute margins, stipulate, with small, leafy, axillary shoots. Inflorescence 3- to 9-flowered; pedicels much longer than the subtending bracts. Capsule (4–)6-seeded. $2n = 32$. *Mediterranean region; Portugal.* Al Bl Co Cr Ga Gr Hs It Ju Lu Sa Si Tu.

Variable in indumentum, more or less glabrous variants having been described as distinct species.

7. F. paradoxa Heywood in Guinea, *Cistác. Esp.* 174 (1954). Laxly or densely caespitose, 3–10(–20) cm. Leaves (3–)4–7(–10) × 0·75–1·25 mm, alternate, linear, trigonous or subtrigonous, mucronate or obtuse, ciliate, exstipulate. Inflorescence 1- to 2(–3)-flowered, the flower terminal when solitary; pedicels much longer than the subtending bracts, arcuate-patent after flowering. Capsule usually 6-seeded; capsule valves widely patent; seeds retained long after dehiscence. *Calcicole.* ● *S.E. Spain* (*Sierra de Cazorla, S. de Segura*). Hs.

A variable and puzzling species showing similarities to **2**, **3** and **4**, possibly the result of former hybridization. Characterized among other features by the fact that from 12 ovules only 6 seeds usually develop which (unlike other species) occupy the space left by the abortive ovules and fill the loculi.

8. F. laevipes (L.) Spach, *Ann. Sci. Nat.* ser. 2 (Bot.), **6**: 359 (1836). Much-branched, laxly caespitose, up to 30 cm; stems slender, ascending. Leaves (3–)4–8 × 0·3–0·4 mm, alternate, linear-setaceous, subterete, bright green or glaucous, glabrous or with scattered glandular hairs, stipulate, with small, axillary, leafy shoots. Inflorescence 3- to 8-flowered, terminal; pedicels much longer than the subtending bracts, patent. Capsule usually 6-seeded; capsule-valves widely patent. $2n = 32$. *Calcicole. Mediterranean region; Portugal.* Bl Cr Ga Gr Hs It Ju Lu Sa Si.

9. F. aciphylla Boiss., *Fl. Or.* **1**: 449 (1867). Up to 30 cm; stems erect, flexuous. Leaves 5–12(–15) × 0·5–1·5 mm, linear-acicular, alternate, glabrous or sparsely pubescent when young, the margins occasionally setose-ciliate, exstipulate. Inflorescence 3- to 5-flowered; pedicels longer than the small subtending bracts, glabrous. Capsule 3-seeded, completely enclosed in the calyx. *Once recorded from C. Greece* (*Thessalia*). ?Gr (*Anatolia*.)

CXIII. TAMARICACEAE[1]

Shrubs or small trees. Leaves simple, alternate, exstipulate, usually ericoid. Flowers solitary or in spike-like racemes, hermaphrodite, actinomorphic. Sepals 4–5, free or slightly united at the base; petals 4–5, free; stamens 4 to numerous, free or partly united. Ovary superior, of 3–5 carpels; placentation parietal. Fruit a septicidal capsule; seeds numerous, with numerous long, unicellular hairs.

1 Flowers solitary; styles 5, filiform   **1. Reaumuria**
1 Flowers in racemes; styles 3, short, or stigma sessile on style-like beak of ovary
   2 Stamens free, or apparently united at the base by a horizontal, fleshy disc; styles 3–4   **2. Tamarix**
   2 Stamens united in a short tube by the expanded, vertical filament-bases; stigma sessile on style-like beak of ovary
                                        3. Myricaria

1. Reaumuria Hasselq. ex L.[2]

Flowers solitary, terminal, 5-merous. Petals with 2 scale-like appendages on inner side at base. Stamens numerous, united at base to form 5 antepetalous bundles. Styles 5, filiform. Seeds covered with hairs except for a slender, glabrous beak.

1. R. vermiculata L., *Syst. Nat.* ed. 10, **2**: 1081 (1759) (*R. mucronata* Jaub. & Spach). Glabrous shrub *c.* 30 cm, with erect branches. Leaves up to 12 mm, semi-cylindrical, mucronate, glaucous, crowded near the base of the branch, widely spaced above, often subtending leafy, axillary short shoots. Flowers subtended by numerous imbricate bracts, which conceal the calyx. Petals *c.* 7 mm, white; appendages fimbriate. Stamens in 5 bundles of *c.* 15. Capsule globose. *Banks near the sea, on gypsaceous clay. S. coast of Sicilia* (*Porto Empedocle*). *Si. (N. Africa.)

2. Tamarix L.[3]

Leaves small, scale-like, amplexicaul or sheathing, with immersed, salt-secreting glands. Flowers small, white or pink, in spike-like racemes, which may be borne on the growth of the current year, or of previous years, or both. Sepals and petals 4–5. Stamens 4–15, of which 4–5 are antesepalous and 0–10 antepetalous; anthers extrorse. In the species without antepetalous stamens there is a more or less fleshy, nectar-secreting, 4- to 5-lobed disc between stamens and ovary. Styles 3–4, short. Seeds with a sessile tuft of hairs.

The racemes may be vernal (produced early in the season, from the woody stems) or aestival (produced later on the growth of the current year). Recent work has shown that this distinction has not the taxonomic value which was formerly attributed to it, as the behaviour of many species varies according to the climate. The information given below refers to the behaviour of plants in their native European habitats. Measurements of racemes refer to the flowering condition.

Literature: A. von Bunge, *Tentamen Generis Tamaricum Species accuratius definiendi*. Dorpat. 1852. F. Niedenzu, *De Genere Tamarice*. Braunsberg. 1895. G. Arendt, *Beiträge zur Kenntnis der Gattung* Tamarix. Berlin. 1926. B. Baum, *Monographic Revision of the Genus* Tamarix. Jerusalem. 1966.

1 Sepals and petals 4
   2 Bracts shorter than or ±equalling the pedicels
     3 Racemes 10–12 mm wide; sepals 2–2·5 mm   **6. hampeana**
     3 Racemes 6–8 mm wide; sepals 1–1·5 mm
       4 Bracts obtuse; racemes mostly vernal   **12. laxa**
       4 Bracts acute; racemes mostly aestival   **13. gracilis**
   2 Bracts distinctly exceeding the pedicels
     5 Racemes 7–10 mm wide; bracts exceeding the calyx
       6 Sepals *c.* 3·5 mm   **7. dalmatica**
       6 Sepals not more than 2·5 mm
         7 Inner sepals denticulate; base of filaments confluent with lobes of disc   **10. boveana**
         7 Sepals all ±entire; filaments inserted in slight notches on lobes of disc   **11. meyeri**
     5 Racemes 3–6(–7) mm wide; bracts not exceeding the calyx
       8 Bark black; bracts herbaceous in proximal half; petals more than 2 mm   **4. tetrandra**

---

[1] Edit. D. A. Webb.   [2] By D. A. Webb.   [3] By B. Baum.

8 Bark brown to purple; bracts entirely scarious; petals less than 2 mm **5. parviflora**

1 Sepals and petals 5

9 Base of filaments confluent with lobes of disc

10 Racemes 8–12 mm wide

11 Bracts exceeding pedicels; petals narrowly obovate, with distinct claw **7. dalmatica**

11 Bracts shorter than pedicels; petals ±elliptical, without distinct claw **6. hampeana**

10 Racemes 3–8 mm wide

12 Petals 2–3 mm, at least some of them persistent **1. africana**

12 Petals 1·25–2 mm, caducous

13 Bracts equalling or exceeding the calyx; petals not more than 1·5 mm **2. canariensis**

13 Bracts not extending beyond middle of calyx; petals 1·5–2 mm

14 Glabrous; sepals entire **3. gallica**

14 Hairy or papillose; sepals finely and closely denticulate **14. hispida**

9 Filaments alternating with lobes of disc, inserted in or below the sinus

15 Petals persistent; racemes 3–5 mm wide

16 Petals obovate, not keeled **8. ramosissima**

16 Petals ovate to suborbicular, keeled **9. smyrnensis**

15 Petals caducous; racemes 6–7 mm wide

17 Pedicel much shorter than calyx; sepals 2–2·5 mm **4. tetrandra**

17 Pedicel longer than calyx; sepals 1 mm **13. gracilis**

1. T. africana Poiret, *Voy. Barb.* **2**: 139 (1789) (*T. hispanica* Boiss.). Tree with black or dark purple bark, glabrous except for papillose bracts and inflorescence-axis. Leaves 1·5–4 mm, acute. Racemes 30–60 × 5–8 mm, usually vernal, but often aestival in Spain and Portugal. Bracts triangular, obtuse to acuminate, usually exceeding the calyx. Flowers subsessile, white or pale pink, 5-merous, without antepetalous stamens. Sepals 1·5 mm, trullate-ovate, acute. Petals 2–3 mm, trullate-ovate, at least some of them persistent. Filaments expanded at the base and confluent with the lobes of the disc. *Coastal marshes and river-banks. S.W. Europe, extending eastwards to S. Italy.* Bl Co Ga Hs It Lu Sa Si [Br].

2. T. canariensis Willd., *Abh. Phys. Kl. Königl. Preuss. Akad. Wiss.* **1812–13**: 79 (1816) (*T. gallica* auct. pro parte). More or less papillose shrub or bushy tree with reddish-brown bark. Leaves 1–3 mm. Racemes 15–45 × 3–5 mm, dense, with papillose axis, usually aestival (but sometimes vernal in Spain), forming panicles. Bracts linear to triangular, acuminate, entire, equalling or somewhat exceeding the calyx. Pedicel equalling calyx. Flowers pink, 5-merous, without antepetalous stamens. Sepals 0·5–0·75 mm, finely and deeply denticulate. Petals 1·25–1·5 mm, obovate, caducous. Filaments as in **1** but disc much smaller. *W. Mediterranean region, Portugal.* Bl Ga Hs Lu Sa Si.

3. T. gallica L., *Sp. Pl.* 270 (1753) (*T. anglica* Webb). Like **2** but entirely glabrous; bark dark brown to dark purple; racemes rather lax; bracts more or less erose-denticulate, not exceeding the calyx; sepals 0·75–1·25 mm, entire; petals 1·5–2 mm, elliptical to elliptic-obovate; disc less fleshy. *S.W. Europe, extending to N.W. France; planted elsewhere for shelter and ornament and sometimes naturalized.* *Az Bl Co Ga Hs It Si [Br].

4. T. tetrandra Pallas ex Bieb., *Fl. Taur.-Cauc.* **1**: 247 (1808). Glabrous or slightly papillose shrub or small tree with black bark. Leaves 3–5 mm, acute, with scarious margin. Racemes 30–60 × 6–7 mm, usually vernal. Bracts oblong, obtuse, herbaceous except for the scarious apex. Flowers white, 4-(5-)merous, with 0–4 antepetalous stamens. Pedicel much shorter than calyx. Sepals 2–2·5 mm, entire, the outer keeled and acute, the inner

obtuse and shorter. Petals 2·5(–3) mm, ovate to ovate-elliptical. Filaments of antesepalous stamens inserted in shallow sinuses between the lobes of the conspicuous, fleshy disc. *Damp places, mainly in the mountains. E. part of Balkan peninsula; Krym.* Bu Gr Ju Rs (K) Tu.

5. T. parviflora DC., *Prodr.* **3**: 97 (1828) (*T. cretica* Bunge). Like **4** but bark brown to purple; racemes narrower (3–5 mm) and usually shorter; bracts almost entirely scarious; flowers smaller, with petals not more than 2 mm; sepals denticulate; filaments of antesepalous stamens confluent with the lobes of the smaller and less fleshy disc. *Hedges and river-banks. Balkan peninsula and Aegean region; widely cultivated for ornament in C. & S. Europe, and perhaps becoming naturalized.* Al Cr Gr Ju Tu [Co Hs It].

6. T. hampeana Boiss. & Heldr. in Boiss., *Diagn. Pl. Or. Nov.* **2(10)**: 8 (1849) (*T. haussknechtii* Niedenzu). Glabrous tree with brown or reddish bark. Leaves 1·75–4 mm. Racemes 20–60 (–130) × 10–12 mm, usually vernal, solitary or aggregated to form a lax panicle. Bracts shorter than or slightly exceeding pedicels. Flowers 4- to 5-merous, with 0–3 antepetalous stamens; sometimes a petal or sepal is 2-fid and can be mistaken for 2. Sepals 2–2·5 mm, the outer keeled, acute, subentire, the inner trullate-ovate, acuminate but with obtuse apex, slightly denticulate. Petals 2·5–4 mm, ovate-elliptical. Filaments of antesepalous stamens confluent with the lobes of the disc. *River-banks and maritime sands. Greece and Aegean region.* Gr Tu.

7. T. dalmatica Baum, *Monogr. Rev. Tamarix* 180 (1966) (*T. africana* auct. balcan., non Poiret). Glabrous tree with blackish bark. Leaves 2·5–4 mm. Racemes 20–60 × 8–10 mm, usually vernal. Bracts broadly triangular, obtuse to acuminate, usually exceeding the calyx. Flowers subsessile, white, 4-(5-)merous. Sepals 3·5 mm, keeled, the outer ovate, acute, the inner trullate, obtuse. Petals 2·5–5 mm, narrowly obovate, with distinct claw. Stamens as in **6**. *Coastal marshes and river-banks. E. Mediterranean region.* Al Cr Gr It Ju *Si.

8. T. ramosissima Ledeb., *Fl. Altaica* **1**: 424 (1829) (*T. eversmannii* C. Presl ex Bunge, *T. pallasii* auct., non Desv.). Glabrous shrub or small tree with reddish-brown bark. Leaves 1·5–3·5 mm, acute. Racemes 15–70 × 3–4 mm, without naked base, usually aestival, in dense panicles. Bracts narrowly triangular, acute, denticulate, exceeding the pedicels. Pedicel shorter than calyx. Flowers pink, 5-merous, without antepetalous stamens. Outer sepals 0·5–1 mm, narrowly trullate, acute, denticulate, the inner wider and more obtuse. Petals 1–1·75 mm, obovate to broadly elliptical. Filaments inserted below the sinuses which separate the emarginate lobes of the disc. *Damp places, especially on saline or alkaline soils. S. part of U.S.S.R.* Rs (W, K, E).

T. chinensis Lour., *Fl. Cochinch.* **1**: 182 (1790), from China, is widely cultivated for ornament (often under the name of *T. gallica*) and perhaps locally naturalized. It is very like **8**, but has smaller, entire sepals, ovate petals and shorter bracts.

9. T. smyrnensis Bunge, *Tent. Tamaric.* 53 (1852) (*T. hohenackeri* Bunge, *T. pallasii* auct., non Desv.). Like **8** but racemes somewhat thicker and naked at the base; sepals 1 mm; petals 2 mm, ovate-orbicular, strongly keeled, especially towards the base; lobes of disc scarcely emarginate. *Coastal marshes and by mountain streams. S.E. Europe, from the Aegean region to S. Ukraine.* ?Al Bu Cr Gr Rm Rs (K, E) Tu.

10. T. boveana Bunge, *Mém. Sav. Étr. Pétersb.* **7**: 291 (1851). Bushy tree with reddish-brown bark; young twigs usually somewhat papillose. Leaves 1·25–4 mm, acuminate. Racemes 50–150 × 8–9 mm, usually vernal. Bracts linear, acute, strongly papillose, much exceeding the flowers. Pedicel shorter than calyx. Flowers 4-merous, without antepetalous stamens. Sepals 1·5–2 mm, the outer broadly ovate-trullate, acute, entire, the inner obtuse, more or less denticulate and somewhat shorter. Petals 3–4 mm, narrowly obovate, with distinct claw, caducous. Filaments confluent with the lobes of the disc. *S.E. Spain (very local).* Hs. (*N.W. Africa.*)

11. T. meyeri Boiss., *Diagn. Pl. Or. Nov.* **2**(10): 9 (1849). Glabrous shrub or small tree with reddish-brown or greyish-brown bark. Leaves 1–4 mm, linear. Racemes 40–100 × 7 mm, usually vernal. Bracts oblong, papillose at least on inner side, equalling or exceeding the flowers. Pedicel shorter than calyx. Flowers 4-merous, rarely with 1 antepetalous stamen. Sepals 2–2·5 mm, subentire, the outer acute and keeled, the inner obtuse and narrower. Petals 3–3·5 mm, elliptical to obovate. Filaments of antesepalous stamens inserted in a slight notch on the lobes of the disc. *Saline soils. S.E. Russia (Volga delta and by Kuma River).* Rs (E). (*Caucasus, S.W. Asia.*)

12. T. laxa Willd., *Abh. Phys. Kl. Königl. Preuss. Akad. Wiss.* 1812–13: 82 (1816) (*T. pallasii* Desv.). Glabrous shrub or small tree with greyish-brown bark. Leaves 1–4 mm, acute, entire. Racemes 10–70 × 6–8 mm, usually vernal but often also some aestival. Bracts spathulate, obtuse, scarious, shorter than the pedicels (except sometimes the uppermost). Pedicel longer than calyx. Flowers 4-merous, without antepetalous stamens. Sepals 1–1·5 mm, trullate-ovate, the outer acute, keeled, subentire, the inner obtuse, erose-denticulate. Petals 2–3 mm, broadly elliptical to ovate, caducous. Filaments inserted in shallow sinuses between the more or less emarginate lobes of the disc. *River-banks and lake-shores. S.E. Russia, W. Kazakhstan.* Rs (E). (*W. & C. Asia.*)

13. T. gracilis Willd., *op. cit.* 81 (1816). Glabrous, bushy tree with brown or blackish bark. Leaves 1–4 mm, much longer than wide. Racemes 20–60 × 6–7 mm, usually aestival, solitary or in lax panicles. Bracts spathulate to narrowly trullate, acute, usually shorter than pedicels. Pedicel longer than calyx. Flowers 4- to 5-merous, the 4-merous sometimes with 1 antepetalous stamen. Sepals 1 mm, ovate, obtuse, subentire to irregularly denticulate. Petals 1·25–2·5 mm, elliptical to obovate, caducous. Filaments of antesepalous stamens inserted in shallow sinuses between the more or less emarginate lobes of the disc. *Sand-dunes, river-banks and salt-marshes. From S.E. Ukraine to W. Kazakhstan.* Rs (W, E). (*W. & C. Asia.*)

European plants are referable to var. *angustifolia* (Ledeb.) Baum (*T. angustifolia* Ledeb.).

14. T. hispida Willd., *op. cit.* 77 (1816). Shrub or small tree with reddish-brown bark, usually densely hispid-pubescent, but sometimes glabrous except for papillose inflorescence-axis. Leaves 1–2·5 mm, acute, cordate-auriculate at base. Racemes 15–70(–150) × 3–5 mm, usually aestival, aggregated in dense panicles. Bracts narrowly triangular, acuminate, longer than pedicels. Pedicel shorter than calyx. Flowers 5-merous, without antepetalous stamens. Sepals 1 mm, trullate-ovate, strongly denticulate, especially in distal half, the outer more or less keeled. Petals 2 mm, obovate to elliptical, caducous. Filaments confluent with the lobes of the disc. *Saline sands. W. Kazakhstan.* Rs (E). (*Dry regions of Asia.*)

3. Myricaria Desv.[1]

Like *Tamarix* but flowers 5-merous, without disc; stamens united in a short tube round the ovary by the expanded filament-bases; anthers introrse; stigma sessile on the beak of the acuminate ovary; tuft of hairs on seed stipitate.

1. M. germanica (L.) Desv., *Ann. Sci. Nat.* **4**: 349 (1825). Glabrous shrub 60–250 cm, with erect branches. Leaves 2–5 mm, linear-lanceolate, obtuse, sessile, somewhat glaucous, closely imbricate on young shoots. Racemes (2–)4–12(–25) × 1 cm, sometimes branched at the base, usually terminal on the main branches, but sometimes terminating short lateral twigs, or sessile on the woody shoots of the previous year. Bracts 5–7(–10) mm, broadly ovate to narrowly oblong, obtuse, acute or long-acuminate, with broad, scarious margin below. Sepals *c.* 4 mm, lanceolate; petals 5–6 mm, pink; stamens 10. Capsule 8–11 mm, narrowly pyramidal. 2n = 24. *River-gravels and other open habitats. Mainly in C. Europe and Fennoscandia, but extending to the Pyrenees and E. Spain, C. Italy and S. Ukraine, with an outlying station on the Lower Volga.* ?Al Au Cz Fe Ga Ge He Hs Hu It Ju No Po Rm Rs (W, K, E) Su.

Var. *bracteosa* Franchet (*M. alopecuroides* Schrenk), which has long-acuminate bracts with patent apex, occurs sporadically throughout the range of the species. **M. squamosa** Desv., *Ann. Sci. Nat.* **4**: 350 (1825), appears simply to designate plants with late-developing, sessile, lateral racemes, which usually have narrow, obtuse bracts and persistent bud-scales at the base.

CXIV. FRANKENIACEAE[2]

Herbs or dwarf shrubs. Leaves opposite, entire, often ericoid; stipules absent. Flowers usually hermaphrodite. Sepals 4–6, connate for more than half their length. Petals 4–6, clawed, with a scale-like appendage on the claw. Stamens usually 6, in 2 whorls. Ovary superior, fertile only in the lower half, 1-locular; placentation parietal; ovules numerous. Capsule loculicidal. Seeds endospermous.

1. Frankenia L.[3]

Annual or perennial. Flowers sessile, solitary or in leafy cymes or spikes. Petals and sepals usually 5; calyx tubular, persistent; petals imbricate. Outer whorl of stamens shorter than the inner. Ovary sessile, of 3(–4) carpels; ripe capsule enclosed in calyx.

Literature: R. Nègre, *Trav. Inst. Sci. Chérif.* ser. bot., **12** (1957).

1 Annual; leaves obovate or oblong-spathulate, usually plane **1. pulverulenta**
1 Perennial, woody at base; leaves with revolute margins, appearing linear or linear-lanceolate
2 Stems procumbent, mat-forming
3 Flowers scattered throughout the upper parts of the stems and branches, not confined to terminal corymbiform clusters **5. laevis**
3 Flowers confined to dense corymbiform clusters which are terminal on the main stems or branches **6. hirsuta**

---

[1] By D. A. Webb.   [2] Edit. D. H. Valentine.
[3] By A. O. Chater.

2 Stems erect or ascending, often caespitose
  4 Flowers in long, terminal, secund spikes; leaves completely covered with a white, calcareous crust   **4. thymifolia**
  4 Flowers mostly clustered in dense terminal cymes; leaves mostly not completely covered by a white crust
    5 Calyx 4–6 mm, with hairs *c.* 0·5 mm   **2. boissieri**
    5 Calyx 2–3 mm, with hairs *c.* 0·1 mm   **3. corymbosa**

1. F. pulverulenta L., *Sp. Pl.* 332 (1753). Annual; stems up to 30 cm, procumbent, often mat-forming, rarely erect, sparsely or densely puberulent. Leaves 1–5(–8) × 0·5–4 mm, obovate to oblong-spathulate, usually plane, glabrous or sparsely puberulent above, densely puberulent beneath, without a white crust. Flowers crowded, secund, in short terminal and axillary spikes. Calyx 2·5–4 mm, puberulent or subglabrous. Petals 3·5–5 mm, oblong to obovate, pale or deep violet. *Maritime sands and shingle, and saline areas inland. S. & S.E. Europe.* Al Az Bl Bu Co Cr Ga Gr Hs It Ju Lu Rm Rs (W, K, E) Sa Si.

2. F. boissieri Reuter ex Boiss., *Voy. Bot. Midi Esp.* **2**: 721 (1845). Perennial; stems up to 30 cm, erect or ascending, branched, woody at base, with long, patent, sparse hairs, or minutely and sparsely puberulent, or subglabrous. Leaves (2·5–)4–7 mm, glabrous above, crystalline-papillose beneath, without a white crust; margins revolute. Flowers in dense terminal cymes. Calyx 4–6 mm, with conspicuous swollen hairs *c.* 0·5 mm on lower half. Petals 5–7 mm, purplish. *Saline places near the sea. S. Portugal, S.W. Spain.* Hs Lu. (*N. Africa.*)

3. F. corymbosa Desf., *Fl. Atl.* **1**: 315 (1798) (*F. webbii* Boiss. & Reuter). Perennial; stems up to 30 cm, erect or ascending, much-branched, woody at base, puberulent. Leaves 2–6 mm, sparsely or densely puberulent on both surfaces, with a white crust or powder which usually does not completely cover the surface; margins revolute. Flowers in dense terminal cymes. Calyx 2–3 mm, sparsely or densely puberulent more or less throughout, with hairs *c.* 0·1 mm. Petals 4–6 mm, pale purplish. *Saline places. S. Spain.* Hs. (*N. Africa.*)

4. F. thymifolia Desf., *op. cit.* 316 (1798) (*F. reuteri* Boiss.). Perennial; stems up to 30 cm, erect or ascending, much-branched, woody at base, densely puberulent. Leaves 2–3·5 mm, completely covered by a white crust; margins revolute. Flowers solitary or in small clusters, secund, arranged in long, terminal spikes. Calyx 2·5–4 mm, sparsely or densely puberulent more or less throughout, with hairs *c.* 0·1 mm. Petals *c.* 7 mm, purplish. *Saline places. C., E. & S. Spain.* Hs.

5. F. laevis L., *Sp. Pl.* 331 (1753) (*F. intermedia* auct., non DC.). Perennial; stems up to 40 cm, procumbent, much-branched, mat-forming, sparsely or densely puberulent with hairs up to 0·1(–0·2) mm. Leaves 2–5 mm, glabrous or subglabrous above, puberulent beneath, sometimes covered by a white crust; margins revolute. Flowers solitary or in small clusters, throughout the upper parts of the main stems and branches, not confined to terminal corymbiform clusters. Calyx 3–4(–5) mm, subglabrous, or puberulent, with hairs up to 0·2 mm in the lower part. Petals 4–6 mm, purplish or whitish. *Maritime sands and shingle. W. Europe, northwards to S. England, and extending eastwards to S.E. Italy.* Az Bl Br Co Ga Hs It Lu Sa Si.

6. F. hirsuta L., *loc. cit.* (1753) (*F. intermedia* DC., *F. hispida* DC.). Perennial; stems up to 40 cm, procumbent, much-branched, mat-forming, densely puberulent or pubescent at least near apex, with hairs 0·1–1 mm. Leaves 2–8 mm, glabrous to puberulent above, puberulent beneath, without a white crust but sometimes with a powdery covering; margins revolute, or rarely some leaves plane. Flowers confined to conspicuous, dense corymbiform clusters terminal on the main stems and branches. Calyx 3·5–5 mm, sparsely puberulent to densely pubescent with hairs up to 0·5 mm. Petals 4–6 mm, pale purplish or white. *Maritime sands and shingle, and saline areas inland. S.E. Europe and Mediterranean region, westwards to c. 3° E.* Al Bl Co Cr Ga Gr It Rm Rs (W, K, E) Sa Si Tu.

CXV. ELATINACEAE[1]

Aquatic or marsh herbs. Leaves simple, opposite or whorled, stipulate. Flowers hermaphrodite, regular, hypogynous, 2- to 5-merous, solitary or in cymes. Sepals free or connate at base, as many as petals. Petals free, imbricate. Stamens as many or twice as many as petals. Ovary superior, 2- to 5-locular; styles free; ovules numerous; placentation axile. Fruit a septicidal capsule.

Sepals 5, with membranous margin, acute; leaves serrate   **1. Bergia**
Sepals 3 or 4, membranous throughout, obtuse; leaves entire   **2. Elatine**

1. Bergia L.[2]

Herbs with serrate leaves. Sepals 5, acute, with wide median vein and membranous margins, free. Petals 5.

1. B. capensis L., *Mantissa Alt.* 241 (1771). Glabrous annual or short-lived perennial. Stems succulent, pink, erect or procumbent and rooting at nodes. Leaves lanceolate to oblanceolate, serrate. Flowers in dense axillary cymes. *Naturalized as a weed in rice-fields in E. Spain.* [Hs.] (*Tropical and subtropical Africa and Asia.*)

2. Elatine L.[2]

Herbs of wet places. Leaves entire. Sepals 3 or 4, membranous, obtuse, connate at base. Petals 3 or 4, membranous, patent in terrestrial plants, closely investing ovary or occasionally absent in aquatic plants. Capsule more or less globose. Seeds straight or curved.

Literature: G. Moesz, *Magyar Bot. Lapok* **7**: 1–34 (1908). H. Glück in A. Pascher, *Die Süsswasser-Flora Mitteleuropas* **15**: 299–313. 1936. H. L. Mason, *Madroño* **13**: 239–240 (1956).

Many authors have regarded the presence or absence of the pedicel as an unreliable character but cultivation experiments (Mason *loc. cit.*) suggest that this character is constant within each species.

All species grow in shallow, usually still water, on wet mud or sand, or in seasonally flooded places.

1 Leaves in whorls of 3–18   **1. alsinastrum**
1 Leaves opposite
  2 Flowers 4-merous
    3 Sepals about 3 times as long as petals at anthesis   **4. hungarica**
    3 Sepals not or little longer than petals at anthesis

---

[1] Edit. S. M. Walters.   [2] By C. D. K. Cook.

4   Sepals shorter than mature capsule   **2. hydropiper**
4   Sepals longer than mature capsule   **3. macropoda**
2 Flowers 3-merous
  5   Stamens 3
    6   Flowers sessile; capsule remaining in leaf-axil at maturity
                                                         5. triandra
    6   Flowers shortly pedicellate; capsule turning away from
           leaf-axil at maturity   **6. ambigua**
  5   Stamens 6
    7   Flowers sessile, in axillary cymes of 2–5   **8. brochonii**
    7   Flowers pedicellate, solitary in leaf-axils   **7. hexandra**

1. E. alsinastrum L., *Sp. Pl.* 368 (1753). Annual or perennial 2–80 cm. Leaves whorled, heterophyllous; in aquatic state linear, up to 18 in a whorl; in terrestrial state lanceolate to ovate, as few as 3 in a whorl. Flowers sessile. Sepals 4, ovate, acute; petals 4, ovate, longer than sepals; stamens 8; carpels 4. Capsule depressed above. Seeds almost straight. *Most of Europe from N. France, S.W. Finland and N. Russia southwards.* Au Bu Co Cr Cz Fe Ga Gr He Hs Hu It Ju Lu Po Rm Rs (N, C, W, E) Sa Si.

2. E. hydropiper L., *Sp. Pl.* 367 (1753) (*E. oederi* Moesz; incl. *E. orthosperma* Düben). Annual 2–16 cm. Leaves opposite. Flowers sessile or subsessile. Sepals 4, ovate, widest at base, shorter than mature capsule; petals 4, ovate, pale red, as long as or longer than sepals; stamens 8; carpels 4. Capsule depressed above, 4-sided. Seeds almost straight or asymmetrically horseshoe-shaped, with long arm 3 times as long as short arm; reticulations on testa rectangular. $2n = c. 40$. *N. & C. Europe, extending locally southwards to C. Spain, N. Italy, Bulgaria and S.E. Russia.* Au Be Br Bu Cz Da Fe Ga Ge Hb He Ho Hs Hu It No Po Rm Rs (N, B, C, W, E) Su.

3. E. macropoda Guss., *Fl. Sic. Prodr.* **1**: 475 (1827) (*E. campylosperma* Seub.). Annual 2–16 cm. Leaves opposite. Flowers with distinct pedicel up to 23 mm. Sepals 4, much longer than mature capsule; petals 4, ovate, pale red, shorter than sepals; stamens 8; carpels 4. Capsule globose or depressed. Seeds almost straight or asymmetrically horseshoe-shaped, with long arm 2–2·5 times as long as short arm; reticulations on testa

usually hexagonal at base of seed. *S.W. Europe and W. Mediterranean region.* Bl Co Ga Hs Lu Sa Si.

4. E. hungarica Moesz, *Magyar Bot. Lapok* **7**: 24 (1908). Annual. Leaves opposite. Flowers with distinct pedicel up to 3 mm. Sepals 4, obovate, about 3 times as long as petals at anthesis; petals 4; stamens 8; carpels 4. Capsule depressed, 4-sided. Seeds symmetrically horseshoe-shaped; reticulations on testa hexagonal at base of seed. *S.E. & E.C. Europe, from S.E. Czechoslovakia to the lower Volga.* Cz Hu Rm Rs (W, K, E) [Lu].

5. E. triandra Schkuhr, *Handb.* **1**: 345 (1791) (incl. *E. callitrichoides* (W. Nyl.) Kaufm.). Annual 1–18 cm. Leaves opposite. Flowers sessile. Sepals (2–)3(–4), obtuse, widest at base; petals 3, white or red, longer than sepals, sometimes absent in submerged plants; stamens 3, opposite sepals; carpels 3. Capsule remaining in leaf-axils at maturity. Seeds slightly curved. $2n = c. 40$. *C., N. & E. Europe, extending to C. France and N. Italy.* Au Az Be Bu Cz Fe Ga Ge Ho Hu It Ju No Po Rm Rs (N, B, C, E) Su.

6. E. ambigua Wight in Hooker, *Bot. Misc.* **2**: 103 (1830). Like **5** but flowers shortly pedicellate; capsules turning to one side away from leaf-axil at maturity. *E. Carpathian region.* [Cz Rm Rs (W).] (*S. & E. Asia.*)

Apparently a recent introduction to Europe.

7. E. hexandra (Lapierre) DC., *Icon. Pl. Gall. Rar.* 14 (1808). Annual or short-lived perennial 2–20 cm. Leaves opposite. Flowers solitary in leaf-axils; pedicels 0·5–10 mm. Sepals 3, obtuse, widest at base; petals 3, longer than sepals; stamens 6; carpels 3. Capsule almost globose, but slightly depressed above. Seeds straight or weakly curved. $2n = 72$. *W. & C. Europe, extending to S. Sweden and N. Italy.* Au Be Br Cz Da Ga Ge Hb He Ho Hs Hu It Ju Lu No Po Rm Su.

8. E. brochonii Clavaud, *Act. Soc. Linn. Bordeaux* **37**: lxiii (1883). Like **7** but flowers sessile, 2–5 in axillary cymes. *S.W. France (Bayonne and S.W. of Bordeaux).* Ga. (*N. Africa.*)

CXVI. DATISCACEAE[1]

Usually dioecious herbs or trees. Leaves alternate, exstipulate. Flowers actinomorphic. Male flowers with 4–9 calyx-lobes; petals usually absent; stamens 4–25. Female and hermaphrodite flowers with 3–8 calyx-lobes; petals usually absent; ovary unilocular, inferior; placentation parietal; ovules numerous, anatropous. Fruit a membranous or coriaceous capsule, dehiscing between the styles.

---

[1] Edit. T. G. Tutin.   [2] By T. G. Tutin.

1. Datisca L.[2]

Herbs with 3-sect or imparipinnate leaves. Petals absent. Stamens with long anthers and short filaments. Styles 3, bifid.

1. D. cannabina L., *Sp. Pl.* 1037 (1753). Robust, glabrous perennial *c.* 1 m, resembling *Cannabis sativa* in general appearance. Leaves imparipinnate; leaflets lanceolate, acuminate, coarsely serrate. Flowers small, interspersed with entire bracts in long racemes subtended by the upper leaves. *Banks of streams. Kriti.* Cr. (*S.W. Asia, Himalaya.*)

CUCURBITALES

CXVII. CUCURBITACEAE[1]

Herbs, often climbing by means of tendrils. Flowers unisexual. Calyx deeply 5(–6)-lobed; corolla deeply 5(–6)-lobed. Stamens 5, sometimes 3 (owing to 4 being connate in pairs); filaments sometimes all connate; anthers with 1 theca, free or coherent. Ovary inferior, unilocular but often more or less divided by the 2–5 placentae; style usually 1, with 3, usually divided stigmas. Ovules 1 to many. Fruit usually fleshy and indehiscent.

The bicollateral vascular bundles are a well-known and apparently constant feature of the family.

Literature: A. Cogniaux in Engler, *Pflanzenreich* **88(IV. 275)I.** (1916); A. Cogniaux & H. Harms in Engler, *op. cit.* II (1924).

1 Tendrils absent; fruit explosive   **2. Ecballium**
1 Tendrils present; fruit not explosive
  2 Tendrils simple
    3 Stamens 5; calyx-tube with a horizontal scale near the base   **1. Thladiantha**
    3 Stamens 3; calyx-tube without a horizontal scale near the base
      4 Flowers greenish-white; male flowers in racemes; fruit 6–10 mm   **3. Bryonia**
      4 Flowers deep yellow; male flowers in axillary clusters, or solitary; fruit at least 20 mm
        5 Connective of the anther prolonged beyond the loculi; disc present   **5. Cucumis**
        5 Connective of the anther not prolonged beyond the loculi; disc absent   **4. Citrullus**
  2 Tendrils branched
    6 Flowers whitish; stamens 5; filaments connate; fruit hispid
      7 Leaves lobed to about the middle; lobes acuminate; female flowers axillary, shortly pedicellate   **7. Echinocystis**
      7 Leaves angled or shallowly lobed; lobes acute; female flowers in dense, long-pedunculate heads   **8. Sicyos**
    6 Flowers deep yellow; stamens 3; filaments free; fruit not hispid
      8 Corolla lobed almost to base   **4. Citrullus**
      8 Corolla lobed to about half way   **6. Cucurbita**

1. Thladiantha Bunge[2]

Dioecious perennial. Tendrils simple. Flowers yellow. Calyx campanulate, 5-fid, with one horizontal scale across the lower part of the tube; corolla deeply 5-fid. Male flowers with 5 free stamens. Female flowers with 5 linear staminodes. Seeds numerous.

1. T. dubia Bunge, *Enum. Pl. Chin. Bor.* 29 (1833). Softly hairy. Stems up to 1·5 m. Leaves 5–10 cm, broadly ovate, acute or acuminate, denticulate. Flowers solitary, or the male in short racemes; calyx-lobes linear-lanceolate; corolla-lobes 25 × 9–12 mm, ovate, subacute, villous beneath, papillose-glandular above. Fruit 4–5 × 2·5 cm, ovoid-oblong, with 10 shallow, longitudinal grooves. *Locally naturalized in C. & S.E. Europe.* [Au Cz Hu Rm Rs.] (*N. China.*)

2. Ecballium A. Richard[2]

Monoecious. Perennial hispid herb with a tuberous root. Stems procumbent, without tendrils. Calyx shortly campanulate, 5-fid; corolla almost rotate, 5-fid. Male flowers in axillary racemes; stamens 3. Female flowers solitary, axillary; staminodes 5, 4 connate in pairs. Seeds numerous.

1. E. elaterium (L.) A. Richard in Bory, *Dict. Class. Hist. Nat.* 6: 19 (1824). Stems 15–60 cm. Leaves long-petiolate; lamina 4–10 cm, cordate to triangular, entire, denticulate or rarely shallowly lobed, undulate, rather fleshy. Corolla of male flowers 18–20 mm, yellowish. Fruit 4–5 × 2·5 cm, green, ovoid, very hispid, dehiscing explosively at the base. *Sandy or stony ground near the sea and often as a ruderal. S. Europe.* Al Bl Bu Co Cr Ga Gr Hs It Ju Lu Rm Rs (W, K, E) Sa Si Tu [Az Br Cz Hu].

3. Bryonia L.[2]

Monoecious or dioecious. Perennial hispid-papillose herbs with tuberous roots. Stems climbing by means of unbranched tendrils. Flowers greenish-white, in axillary, racemose panicles or sub-umbellate fascicles; calyx shortly campanulate, 5-dentate; corolla almost rotate, deeply 5-fid. Stamens 3. Female flowers with 3–5, often almost obsolete staminodes. Seeds few.

Often monoecious; stigma glabrous; berry black   **1. alba**
Always dioecious; stigma papillose-hairy; berry red   **2. cretica**

1. B. alba L., *Sp. Pl.* 1012 (1753). Dioecious in S.E. Europe, elsewhere monoecious. Stems up to 4 m, branched. Leaves 5–10 cm, ovate, cordate, 5-angled or palmately 5-lobed; lobes ovate or triangular, acute or acuminate, sharply dentate, the central much longer than the lateral. Calyx of female flowers usually as long as the corolla. Stigma glabrous. Fruit 7–8 mm in diameter, black. *S., C. & E. Europe; formerly cultivated as a medicinal plant and often naturalized further north and west.* Al Au Bu Ge Gr He Hu It Ju Po Rm Rs (C, W, K, E) Tu [Be Cz Da Fe Ga No Rs (B) Su].

2. B. cretica L., *Sp. Pl.* 1013 (1753). Like **1** but always dioecious; leaf-lobes entire or with few, large, subobtuse teeth, the central usually not markedly longer than the lateral; calyx of female flowers usually about half as long as the corolla; stigma papillose-hairy; fruit 6–10 mm in diameter, red. *S., S.C. & W. Europe, northwards to Britain; formerly cultivated as a medicinal plant and often naturalized.* Al Au Be Br Co Cr Cz Ga Ge Gr He Ho Hs Hu It Ju Lu *Po Sa Si [Da Ge Hb No Su].

1 Male inflorescence eglandular or nearly so; immature fruit white-spotted   **(a) subsp. cretica**
1 Male inflorescence glandular; immature fruit uniformly green
  2 Male inflorescence with few or no long hairs   **(b) subsp. dioica**
  2 Male inflorescence with abundant long hairs   **(c) subsp. acuta**

(a) Subsp. **cretica**: Leaves and young fruit with irregular whitish markings. Male inflorescence eglandular or very nearly so. *Aegean region.*

(b) Subsp. **dioica** (Jacq.) Tutin, *Feddes Repert.* **79**: 61 (1968) (*B. dioica* Jacq., *B. sicula* (Jan) Guss.): Leaves and young fruit uniformly green. Male inflorescence glandular, with few or no long hairs. 2n = 20. *Almost throughout the range of the species.*

(c) Subsp. **acuta** (Desf.) Tutin, *Feddes Repert.* **79**: 61 (1968) (*B. acuta* Desf.): Leaves and young fruit uniformly green. Male inflorescence glandular and with abundant long hairs. ?*Sardegna, Lampedusa.* (*Tunisia, Libya.*)

---

[1] Edit. T. G. Tutin.   [2] By T. G. Tutin.

Plants from Corse with irregular whitish markings on the leaves have been found to have $2n=40$. It is not clear whether this chromosome number is constantly associated with distinctive morphological characters but, if so, an additional subspecies may have to be recognized.

4. Citrullus Schrader[1]

Monoecious. Stems procumbent or climbing; tendrils simple or branched. Flowers solitary, yellow; calyx and corolla deeply 5-fid. Stamens 3; filaments and anthers free; connective not prolonged beyond the loculi. Female flowers with 3 small staminodes. Disc absent. Ovary with 3 placentae and numerous ovules.

Annual; ovary densely lanate 1. lanatus
Perennial; ovary sparsely hispid 2. colocynthis

1. **C. lanatus** (Thunb.) Mansfeld, *Kulturpfl. (Beih.)* **2**: 421 (1959) (*C. vulgaris* Schrader, *Colocynthis citrullus* (L.) O. Kuntze). Annual. Stems *c.* 10 mm in diameter, densely villous. Leaves 8–20 × 5–15 cm, ovate in outline, pinnatisect, with lobed segments. Calyx-lobes narrowly lanceolate; corolla-lobes *c.* 15 mm, ovate-oblong, obtuse. Ovary densely lanate. Fruit *c.* 25 cm in diameter, subglobose or ellipsoid, smooth, greenish; pulp red, succulent. *Widely cultivated in S. Europe for its edible fruits (water-melon).* (*S. Africa.*)

2. **C. colocynthis** (L.) Schrader, *Linnaea* **12**: 414 (1838) (*Colocynthis vulgaris* Schrader). Perennial. Stems *c.* 2 mm in diameter, hispid. Leaves 5–12 × 3–8 cm, 1- to 2-pinnatisect with 3–5 sinuate-lobed segments. Calyx-lobes subulate; corolla-lobes *c.* 5 mm, ovate, acute. Ovary sparsely hispid. Fruit *c.* 4 cm, globose, smooth, yellow; pericarp dry. *S. part of the Mediterranean region; cultivated elsewhere for its purgative fruits and often naturalized.* Gr Hs Si [Hu It Rm]. (*Arid regions of N. Africa and Asia.*)

5. Cucumis L.[1]

Monoecious. Annual or perennial, usually procumbent herbs; tendrils unbranched. Flowers deep yellow, the male usually in clusters, the female solitary; calyx and corolla deeply 5-fid. Stamens 3, free; connective of the anther prolonged beyond the loculi. Female flowers with 3 staminodes. Disc present. Ovary with 3–5 placentae; ovules numerous.

Lagenaria siceraria (Molina) Standley, *Publ. Field Mus. (Chicago) ser. bot.,* **3**: 435 (1930) (*L. vulgaris* Ser.), with white flowers and a rotate corolla, is widely cultivated in S. Europe.

1 Fruit ±cylindrical 3. sativus
1 Fruit globose or ovoid
 2 Flowers 20–30 mm; fruit smooth 1. melo
 2 Flowers 4–5 mm; fruit aculeate 2. myriocarpus

1. **C. melo** L., *Sp. Pl.* 1011 (1753) (*C. dudaim* L., *Melo dudaim* (L.) Sageret, *Melo sativus* auct.). Annual. Stems up to 1 m, *c.* 10 mm in diameter, hispid. Leaves 8–15 × 8–15 cm, suborbicular or reniform, cordate, 5-angled or shallowly 3- to 7-lobed, denticulate, villous, angles or lobes rounded. Calyx-lobes subulate; corolla 20–30 mm; lobes acute. Fruit very variable in size, usually globose or ovoid, pubescent, often becoming glabrous. *Widely cultivated in S. Europe for its edible fruit (melon).* (*Tropical Africa and Asia.*)

2. **C. myriocarpus** Naudin, *Ann. Sci. Nat.* ser. 4 (Bot.), **11**: 22 (1859). Annual. Stems up to 2 m, rather slender, shortly hispid or scabrid. Leaves 4–8 cm, deeply (3–)5(–7)-lobed, deep green, nearly glabrous, somewhat scabrid above, very shortly hirsute and later scabrid beneath. Calyx-lobes subulate; corolla 4–5 mm; lobes acute. Fruit 2–2·5 cm, globose, sparsely and softly aculeate. *Naturalized locally in Spain, C. Portugal and E. Russia; a rather frequent casual in S. Europe.* [Hs Lu Rs (E).] (*S. Africa.*)

3. **C. sativus** L., *Sp. Pl.* 1012 (1753). Annual. Stems stout, hispid. Leaves 7–18 × 7–18 cm, suborbicular in outline, palmately 3- to 5-lobed, dentate, villous and scabrid; lobes acute or acuminate. Calyx-lobes subulate; corolla 20–30 mm; lobes acute. Fruit cylindrical, terete or angled, glabrous, often tuberculate and aculeate. *Widely cultivated for its edible fruit (cucumber) in the southern half of Europe.* (*India.*)

6. Cucurbita L.[1]

Monoecious. Stems procumbent or climbing; tendrils branched (except in some cultivars). Flowers solitary, yellow. Calyx 5-fid; corolla campanulate, lobed to about the middle. Stamens 3; filaments free; anthers coherent. Female flowers with small staminodes. Ovary with 3–5 placentae; ovules numerous.

In addition to the species described below **C. ficifolia** Bouché, *Verh. Ver. Beförd. Gartenb. Preuss.* **12**: 205 (1837) (*C. melanosperma* Gasparr.), a perennial species with fruits 20–30 cm long and with black seeds, and **C. mixta** Pangalo, *Bull. Appl. Bot. Pl.-Breed. (Leningrad)* **23**: 264 (1930), which is like **2** but has the fruiting peduncle hard, inflated and corky, are cultivated for culinary purposes in parts of S.E. Europe.

1 Fruiting peduncle terete, soft, corky, not expanded at attachment to fruit; flowers slightly scented 3. maxima
1 Fruiting peduncle angled, hard, not corky, ± expanded at attachment to fruit; flowers not scented
 2 Plant rather softly hairy; calyx-lobes ligulate but often expanded above; leaves shallowly lobed 2. moschata
 2 Plant harshly hispid-setose; calyx-lobes linear-lanceolate, not expanded above; leaves often deeply lobed 1. pepo

1. **C. pepo** L., *Sp. Pl.* 1010 (1753). Stems up to *c.* 10 mm in diameter, hispid. Leaves 15–30 cm, broadly ovate in outline, cordate, variously lobed, hispid-setose; lobes acute. Peduncles of male and female flowers 5-angled, slightly expanded below the female flowers. Calyx-lobes linear-lanceolate; corolla 7–10 cm in diameter, deep yellow; lobes ovate, acute or acuminate. Fruit 15–40 cm in diameter, globose to cylindrical, green or yellow, smooth or tuberculate; seeds white. *Widely cultivated as a vegetable (vegetable marrow) and for ornament in the southern half of Europe.* (*N. Central America.*)

2. **C. moschata** Duchesne ex Poiret, *Dict. Sci. Nat.* **11**: 234 (1818). Like **1** but leaves shallowly lobed or almost entire, softly hairy; peduncles of male flowers terete, of female flowers strongly expanded at top; calyx-lobes usually expanded and leaf-like at apex; fruit asymmetrical, curved, brown or reddish-yellow, with a musky odour. *Cultivated as a vegetable in S. Europe.* (*Central America.*)

3. **C. maxima** Duchesne in Lam., *Encycl. Méth. Bot.* **2**: 151 (1786). Like **1** but leaves orbicular, not lobed, softly hairy; peduncles terete, not expanded below the female flowers; fruit variously shaped, often glaucous and very large (up to 100 kg.). *Cultivated as a vegetable (pumpkin) in S. Europe.* (*Central America.*)

---

[1] By T. G. Tutin.

7. Echinocystis Torrey & A. Gray[1]

Monoecious, nearly glabrous annual. Stems climbing; tendrils branched. Male flowers in axillary, racemose panicles, the female solitary; both greenish-white; calyx and corolla deeply 5- to 6-fid. Stamens 5; filaments connate. Female flowers without staminodes. Ovary with 2 placentae and 2 ovules on each placenta.

1. **E. lobata** (Michx) Torrey & A. Gray, *Fl. N. Amer.* 1: 542 (1840). Stems 5–8 m. Leaves *c.* 5 cm, 3- to 7-lobed to about the middle, cordate, remotely serrulate; lobes acuminate. Calyx-lobes narrowly triangular, acuminate; corolla-lobes *c.* 5 mm, triangular, acute. Fruit 3–5 cm, ovoid, covered with long slender prickles, splitting irregularly at the apex when mature. *Naturalized in C. and S.E. Europe.* [Au Cz Ge Hu Ju Rm Rs (W).] (*E. North America.*)

8. Sicyos L.[1]

Monoecious annual with climbing stems; tendrils branched. Male flowers in axillary racemes, the female in long-pedunculate heads; both whitish; calyx and corolla deeply 5-fid. Stamens 5; filaments connate; anthers more or less coherent. Female flowers without staminodes. Ovary unilocular, with a solitary ovule.

1. **S. angulatus** L., *Sp. Pl.* 1013 (1753). Stems 5–8 m, more or less viscid-pubescent. Leaves *c.* 7 cm, about as wide as long, deeply cordate, denticulate, 5-angled or -lobed. Calyx-lobes narrowly triangular; corolla-lobes *c.* 5 mm, triangular, subacute. Fruit *c.* 1·5 cm, compressed-ovoid, coriaceous, lanate and covered with long setae. *Naturalized in damp places in S.C. & S.E. Europe.* [Au Cz Hu It Ju Rm Rs (C, W) Si.] (*E. North America.*)

CACTALES

CXVIII. CACTACEAE[2]

Perennial, succulent plants; stems columnar, cylindrical or flattened, often jointed (the individual segments being referred to as joints). Leaves absent or small, subulate, caducous. Branches, spines, flowers and sometimes barbed bristles (glochids) developed from more or less circular, cushion-like structures (areoles) situated in leaf-axils when leaves are present. Flowers hermaphrodite, solitary, sessile. Perianth with tube; sepals numerous and intergrading with petals, all imbricate in several rows. Stamens numerous; filaments inserted on perianth-throat. Style 1; stigmas 2 to many. Ovary inferior, 1-celled; placentae 3 or more, parietal; ovules numerous. Fruit a berry, often spiny or glochidiate, usually many-seeded.

Literature: N. L. Britton & J. N. Rose, *The Cactaceae.* Washington. 1919–23. C. Backeberg, *Die Cactaceae.* Jena. 1958–62.

Stems jointed; areoles with numerous glochids; leaves small, subulate, caducous **1. Opuntia**
Stems not jointed; areoles without glochids; leaves absent **2. Cereus**

1. Opuntia Miller[3]

Somewhat woody; stems with short, cylindrical or flattened, often tuberculate joints. Leaves small, subulate, caducous. Areoles bearing many glochids and usually longer, stouter spines. Ovary spiny or spineless. Seeds hard, pale, more or less discoid or angular.

1 Joints readily detachable **1. tuna**
1 Joints not readily detachable, persistent
  2 Plants less than 1 m, shrub-like, procumbent or ascending
    3 Areoles brown-lanate; spines brownish-white; fruit with flat umbilicus **2. vulgaris**
    3 Areoles not lanate; spines yellow; fruit with depressed umbilicus **4. stricta**
  2 Plants more than 1 m, usually tree-like, often with trunk
    4 Spines absent or rarely 1–2, less than 1 cm; glochids yellow; fruit with depressed umbilicus **5. ficus-indica**
    4 Spines 1–4, more than 1 cm; glochids brownish; fruit with flat umbilicus

5 Joints bright, shining green; areoles lanate; spines 1–2, yellow to dark reddish; fruit reddish-purple **3. monacantha**
5 Joints dull green, slightly glaucous; areoles not lanate; spines 1–4, white; fruit yellowish-red **6. maxima**

1. **O. tuna** (L.) Miller, *Gard. Dict.* ed. 8, no. 3 (1768). Shrub up to 1 m; joints 8–10(–16) cm, obovate to oblong, readily detachable. Leaves 4–5 mm; areoles large, brown; spines in groups of (2–)3–5(–6), up to 5 cm, slightly spreading, light yellow; glochids yellow. Flowers *c.* 5 cm in diameter, yellow, usually tinged with red; filaments short, greenish below; style and stigma-lobes cream or yellowish. Fruit *c.* 3 cm, obovoid, red; seeds 3–4 mm wide. *Locally naturalized in S. Europe.* [Ga It Lu.] (*West Indies.*)

2. **O. vulgaris** Miller, *Gard. Dict.* ed. 8, no. 1 (1768) (*O. humifusa* Rafin.). Procumbent and spreading, or sometimes ascending shrub up to 0·5 m; joints 3–13(–17) cm, orbicular to oblong, thick, dark green. Leaves 4–8 mm, subulate, patent; areoles distant; spines 0–1(–2), up to 5 cm, brownish or whitish; glochids numerous, yellow to dark brown. Flowers 5–9 cm in diameter, bright yellow, sometimes reddish in centre; filaments yellow; stigma-lobes white. Fruit 2·5–5 cm, obovoid to oblong, red, succulent, edible; seeds 4–5 mm wide. *Naturalized on rocks and walls in S. & S.C. Europe.* [Au ?Co Cr Ga Gr He Ho It Ju.] (*E. North America, from Alabama to Ontario.*)

3. **O. monacantha** Haw., *Suppl. Pl. Succ.* 81 (1819) (*O. vulgaris* auct., non Miller). Somewhat tree-like, 2–4(–6) m, often with cylindrical, spiny or smooth trunk 15 cm in diameter, usually much-branched at apex; joints 10–30 cm, ovate to oblong, narrowed at base, bright shining green. Leaves 2–3 mm, subulate; areoles shortly lanate; spines 1–2(–10 on trunk), 1–4 cm, erect, yellowish to dark reddish. Flowers golden-yellow; filaments greenish; style white; stigma-lobes white. Fruit 5–7·5 cm, obovoid, reddish-purple, long-persisting, sometimes proliferous. *Naturalized on the coast of S.E. France and possibly elsewhere.* [?Co Ga ?He ?Hs ?It ?Lu.] (*E. South America.*)

4. **O. stricta** (Haw.) Haw., *Syn. Pl. Succ.* 191 (1812). Low, spreading bush up to 0·8 m; joints 8–15(–30) cm, obovate to oblong, rather thick, green or bluish-green. Leaves 3–4 mm,

[1] By T. G. Tutin.   [2] Edit. N. A. Burges.
[3] By D. M. Moore.

subulate, stout; areoles distant, brownish-lanate; spines 0–2, 1–4 cm, rigid, terete, yellow; glochids short. Flowers 6–7 cm in diameter, yellow; filaments yellow to greenish; style usually white; stigma-lobes white or greenish. Fruit 4–6 cm, obovoid, with slender base and more or less depressed umbilicus, red. *Naturalized on the coast of S.E. France.* [Ga.] (*W. Cuba, S.E. United States.*)

5. O. ficus-indica (L.) Miller, *Gard. Dict.* ed 8, no. 2 (1768). Erect, 3–5 m, with patent branches; joints 20–50 × 10–20 cm, oblong to spathulate-oblong. Leaves 3 mm, subulate; areoles small, whitish; spines usually 0, rarely 1–2, small, pale yellow or white; glochids yellow, numerous, caducous. Flowers 7–10 cm in diameter, bright yellow; filaments pale yellow. Fruit 5–9 cm, ovoid or obovoid, with strongly depressed umbilicus, yellow, red or parti-coloured, edible. *Cultivated for its edible fruit and as a hedge and widely naturalized in the Mediterranean region.* [Bl Co Cr Ga Gr Hs It Ju ?Lu.] (*Tropical America.*)

6. O. maxima Miller, *Gard. Dict.* ed. 8, no. 5 (1768) (*O. amyclaea* Ten.). Tree-like, 1–3 m, much-branched; joints 20–40 cm, broadly elliptical, about twice as long as wide, thick,

dull green, slightly glaucous. Leaves 4 mm, acute, red; areoles small, with 1–2 short bristles from lower part; spines 1–4, up to 3 cm, straight, white or translucent; glochids brown, caducous. Flowers yellow. Fruit yellowish-red. *Occasionally naturalized on rocks and walls in the Mediterranean region.* [It Ju ?Sa ?Si.] (*?Mexico.*)

2. Cereus Miller[1]

Stems elongate, not jointed, with continuous spine-bearing ribs. Leaves absent. Areoles without glochids. Flowers infundibuliform. Ovary scaly.

1. C. peruvianus (L.) Miller, *Gard. Dict.* ed. 8, no. 4 (1768). Tree-like, up to 16 m, much-branched; branches 10–20 cm in diameter, green, sometimes glaucous, with (4–)6–9 ribs; spines 5–10, 1–3 cm, acicular, brown to black. Flowers *c.* 15 cm; tube with thick walls; upper scales and outer perianth-segments red or brownish, inner white. Fruit *c.* 4 cm in diameter, subglobose, orange-yellow, rather glaucous; seeds 2 mm wide, rough, black. *Naturalized on the coast of S.E. France.* [Ga.] (*S.E. South America.*)

MYRTALES

CXIX. LYTHRACEAE[2]

Herbs (non-European genera include trees and shrubs). Leaves simple, entire; stipules minute or absent. Flowers hermaphrodite, regular, perigynous, 4- to 6-merous, solitary or in small cymes or clusters in the leaf-axils, rarely in terminal spikes. Hypanthium pelviform to cylindrical. Epicalyx often present. Petals free, pink or purple, inserted on lip of hypanthium; sometimes 0. Stamens 2–12, inserted on tube of hypanthium. Ovary superior, 2- or 4-locular; style single; stigma capitate. Fruit a capsule; seeds numerous.

In many Floras what is here called the hypanthium is referred to as the calyx, and the sepals and epicalyx-segments are referred to respectively as inner and outer calyx-teeth.

Literature: E. Koehne in Engler, *Pflanzenreich* 17 (IV. 216): 1–326 (1903).

1 Flowers solitary or in pairs in the leaf-axils
  2 Epicalyx present   **1. Lythrum**
  2 Epicalyx absent   **3. Rotala**
1 Flowers, at least in lower part of plant, in cymes or clusters of 3 or more in each leaf-axil
  3 Petals 6–12 mm, conspicuous   **1. Lythrum**
  3 Petals, if present, not more than 2·5 mm
    4 Leaves alternate   **1. Lythrum**
    4 Leaves opposite   **2. Ammannia**

1. Lythrum L.[3]

(Incl. *Peplis* L. and *Middendorfia* Trautv.)

Annual or perennial; leaves alternate, opposite or whorled. Flowers usually 6-merous (sometimes 4- or 5-merous), usually solitary in the leaf-axils, sometimes in pairs or in small, whorl-like cymes, which may form terminal spikes; each flower shortly pedicellate, with 2 bracteoles. Hypanthium usually tubular, sometimes short and wide. Epicalyx present. Petals usually

present. Stamens 2–12. Capsule usually septicidally dehiscent with 2 valves, rarely dehiscing irregularly or by 4 apical teeth.

In several species the flowers are trimorphically heterostylous, with the stamens in two series of different length, and the stigma at a level below the anthers of both series, or above them both, or between the two.

Literature: J. Borja Carbonell, *Anal. Inst. Bot. Cavanilles* 23: 145–62 (1968).

1 Flowers trimorphically heterostylous; stamens 12, at least some of them exceeding the sepals
  2 Flowers mostly in terminal spikes composed of whorl-like cymes of 4 or more; petals at least 7 mm
    3 Leaves truncate or rounded at the base, usually hairy; epicalyx-segments much longer than the sepals   **1. salicaria**
    3 Leaves conspicuously cuneate at the base, glabrous; epicalyx-segments about equalling the sepals   **2. virgatum**
  2 Flowers solitary, rarely in pairs; petals not more than 6 mm
    4 Fruiting hypanthium cylindrical, or somewhat dilated at the obtuse base   **6. flexuosum**
    4 Fruiting hypanthium obconical, tapered gradually to a narrow base
      5 Usually perennial; hypanthium red-spotted near the base, usually shorter than subtending leaf   **3. junceum**
      5 Annual; hypanthium green, or entirely red at the base, but not spotted, often longer than the subtending leaf
        6 Hypanthium with 6 narrow, keel-like ridges decurrent from base of epicalyx-segments, which are about twice as long as sepals   **4. acutangulum**
        6 Hypanthium without conspicuous ridges; epicalyx-segments about half as long as sepals   **5. castellanum**
1 Flowers homostylous; stamens usually less than 12, not exceeding the sepals and usually included in the hypanthial tube
  7 Tube of hypanthium pelviform to broadly campanulate in fruit, scarcely longer than wide; capsule ± globose
    8 Leaves linear-oblanceolate to linear, mostly 4–7 times as long as wide; epicalyx-segments and sepals longer than hypanthial tube   **13. volgense**

---

[1] By D. M. Moore.   [2] Edit. D. A. Webb.   [3] By D. A. Webb.

8 Leaves orbicular to oblong-lanceolate, mostly 1–3 times as
  long as wide; epicalyx-segments and sepals usually shorter
  than hypanthial tube
  9 Tube of hypanthium hemispherical to pelviform in fruit,
  shorter than the capsule; plant glabrous; stems extensively creeping and rooting   **12. portula**
  9 Tube of hypanthium broadly campanulate in fruit, at least as
  long as the capsule; plant often scabrid, at least in
  younger parts; stems ± erect, scarcely rooting
       11. borysthenicum
7 Tube of hypanthium cylindrical or narrowly campanulate in
  fruit, distinctly longer than wide; capsule usually ellipsoid
  or cylindrical
  10 Leaves oblong-lanceolate to broadly obovate, all widest
  above the middle and most of them less than 2·5 times as
  long as wide   **11. borysthenicum**
  10 Leaves linear to oblong, rarely somewhat oblanceolate,
  usually parallel-sided or widest near the middle, most of
  them more than 2·5 times as long as wide
    11 Style *c.* 0·25 mm; hypanthium in fruit less than twice as
  long as wide   **10. thesioides**
    11 Style 1·5–2 mm; hypanthium in fruit 3–6 times as long as
  wide
      12 Epicalyx-segments less than 0·5 mm, triangular, about
  equalling the sepals, involute in fruit   **9. tribracteatum**
      12 Epicalyx-segments at least 1 mm, subulate, longer than the
  sepals, erect or patent in fruit
        13 Flowers (5–)6-merous; stamens (2–)4–6(–12); leaves
  mostly more than 2 mm wide   **7. hyssopifolia**
        13 Flowers 4(–5)-merous; stamens 2–3; leaves mostly less
  than 2 mm wide   **8. thymifolia**

1. L. salicaria L., *Sp. Pl.* 446 (1753) (incl. *L. intermedium*
Colla). Erect perennial 50–150 cm, subglabrous to densely grey-
pubescent. Stem with 4 or more raised lines, sparingly branched.
Leaves mostly opposite or in whorls of 3 (but the upper ones
sometimes alternate), ovate to lanceolate-oblong, acute, sessile,
truncate and semi-amplexicaul at base. Flowers trimorphic, in
whorl-like cymes in the axils of small bracts, forming long,
terminal spikes. Hypanthium 4·5 × 1·5–2 mm, broadly tubular;
epicalyx-segments 2·5–3 mm, subulate, longer and narrower than
the deltate sepals. Petals *c.* 10 mm, reddish-purple. Stamens 12,
some or all exserted. Capsule 3–4 mm, ovoid. 2n = 60. *Fens,
riversides and other damp places. Almost throughout Europe
except the extreme north.* All except Az Bl Cr Fa Is Sb.

2. L. virgatum L., *Sp. Pl.* 447 (1753). Like **1** but less robust
and entirely glabrous; leaves linear-oblong to lanceolate, tapered
to a narrow, cuneate base; epicalyx-segments 1 mm, triangular,
about as long as the sepals; petals *c.* 7 mm. *Marshes and other
wet places. S.E., E. & E.C. Europe, westwards to N.W. Italy and
C. Austria, and northwards to c. 57° N. in E. Russia; sometimes
cultivated for ornament and locally naturalized.* Al Au Bu Cz
Gr Hu It Ju Po Rm Rs (C, W, K, E) [Ga Ge].

3. L. junceum Banks & Solander in A. Russell, *Nat. Hist.
Aleppo* ed. 2, **2**: 253 (1794) (*L. graefferi* Ten., *L. acutangulum* auct.,
non Lag., *L. flexuosum* auct., non Lag.). Glabrous, usually
perennial; stems usually branched from the base with decumbent,
divaricate, straggling branches 20–70 cm. Leaves 8–22 × 2–11
mm, broadly elliptic-oblong to linear-oblong, mostly alternate.
Flowers trimorphic, solitary in the leaf-axils, suberect. Hypanthium 5–6 mm, shorter than the subtending leaf, cylindrical-
obconical, tapered gradually to the base, spotted with red near
the base. Sepals broadly deltate, scarious; epicalyx-segments
c. 1 mm, triangular-subulate, equalling or somewhat exceeding
the sepals. Petals 5–6 mm, purple, sometimes white or cream at
the base. Stamens 12, some or all exserted. Capsule much shorter
than the hypanthium. *Wet places. S.W. Europe and Mediterranean region.* Al Az Bl Co Cr Ga Gr Hs It Lu Sa Si Tu.

4. L. acutangulum Lag., *Gen. Sp. Nov.* 16 (1816) (*L. maculatum*
Boiss. & Reuter). Like **3** but annual, slender, with erect or straggling stems not more than 30 cm; leaves up to 15 × 4 mm, but
usually much less; hypanthium sometimes longer than the subtending leaf, green or uniformly tinged with red, but without
red spots, bearing, especially in fruit, 6 narrow, hyaline, keel-
like ridges decurrent from the base of the epicalyx-segments,
which are about twice as long as the sepals; petals with a sharply
delimited white zone at the base. *Wet places. C. & S. Spain;
S. France.* Ga Hs ?It ?Si.

5. L. castellanum González-Albo ex Borja, *Anal. Inst. Bot.
Cavanilles* **23**: 160 (1968). Like **3** but annual, slender, and smaller
in all its parts; hypanthium 4–5 mm, without red spots and
without prominent ridges; epicalyx-segments *c.* 0·5 mm, ovate-
deltate, inserted below the top of the hypanthium and scarcely
reaching to the sinuses between the sepals. ● *S.C. Spain
(Prov. Albacete and Ciudad Real).* Hs.

6. L. flexuosum Lag., *Gen. Sp. Nov.* 16 (1816). Annual, with
procumbent, flexuous, reddish branches 3–10 cm. Leaves
5–12 × 1–2·5 mm, oblanceolate-elliptical to linear, alternate,
patent, yellowish-green. Flowers trimorphic, patent. Hypanthium
c. 5·5 × 2 mm in fruit, cylindrical, obtuse at both ends, purple-
red. Sepals 1 mm, deltate, white; epicalyx-segments shorter than
the sepals, obtuse, patent. Petals *c.* 4·5 mm, pinkish-purple
distally, creamy-white at the base. Stamens 12, some or all
exserted. Capsule almost as long as the hypanthium. *Seasonally
wet places.* ● *C. Spain.* Hs.

Seldom collected, but very distinct. Most records under this
name are referable to **3**.

7. L. hyssopifolia L., *Sp. Pl.* 447 (1753). Annual, with erect or
ascending branches, more or less glabrous. Leaves up to 25 × 8
mm, but usually much less, alternate, usually suberect, linear to
oblong. Flowers numerous, erect and appressed, (5–)6-merous,
homostylous. Hypanthium 4–6 mm, obconical in flower,
cylindrical in fruit. Epicalyx-segments 1–1·5 mm, subulate,
about twice as long as the deltate sepals. Petals 2–3 mm, pink.
Stamens usually 4–6, included. Capsule about as long as the
hypanthium; style 1·5–2 mm. *Disturbed or seasonally flooded
ground. Europe except the north, where, however, it occurs locally
as a weed or casual.* Al Au Az Be Bl Br Bu Co Cr Cz Ga Ge Gr
He Hs Hu It Ju Lu Po Rm Rs (C, W, K, E) Sa Si Tu [No Su].

8. L. thymifolia L., *Sp. Pl.* 447 (1753). Like **7** but usually
somewhat scabrid; stems 5–12 cm, erect to procumbent; leaves
5–9 × 1–2 mm; hypanthium 3–4(–5) mm; sepals very short,
almost obsolete; flowers 4(–5)-merous; petals 1–2 mm; stamens
usually 2. *Wet places. S.W. Europe, W. & C. Mediterranean
region; S. part of U.S.S.R.* Bl Ga ?Gr Hs It Ju Lu Rs (W, K, E)
Sa.

Although this species usually differs from **7** by a large number
of characters, the correlation between the characters is in some
regions poor, and puzzling intermediate plants exist.

9. L. tribracteatum Salzm. ex Sprengel, *Syst. Veg.* **4**(2): 190
(1827) (*L. bibracteatum* Salzm. ex Guss.). Erect to decumbent,
glabrous or scabrid annual; stem 5–30 cm, with 4 raised lines,
usually with numerous divaricate branches. Leaves 3–12(–18) ×
1–3(–5) mm, linear to oblong-oblanceolate, alternate, those on
the main stem often much larger than those on the branches.
Flowers 4- to 6-merous, solitary in the leaf-axils, numerous,
subsessile; bracteoles usually as long as the hypanthium, but
sometimes minute. Hypanthium 5–6 × 1 mm, narrowly cylindrical,

tapered to the base. Epicalyx-segments and sepals 0·5 mm or less, subequal, inconspicuous and usually involute in fruit. Petals 2–3 mm, oblong, purple; stamens 4–6. Capsule cylindrical, equalling or slightly exceeding the hypanthium; style *c.* 2 mm. *Seasonally wet places. S. Europe, extending northwards to W. France, Hungary and the lower Volga.* Al Bu Ga Gr Hs Hu It Lu Rm Rs (W, K, E) Sa Si Tu.

10. **L. thesioides** Bieb., *Fl. Taur.-Cauc.* **1**: 367 (1808) (*L. geminiflorum* Bertol.). Erect, glabrous, rather glaucous annual; stem 20–40 cm, simple or with a few ascending branches. Leaves 10–20 × 1–2 mm, linear to linear-oblong, acute, alternate. Flowers 4- to 6-merous, in axillary clusters of 1–4. Hypanthium 1·75–2·5 mm, narrowly campanulate in fruit, about twice as long as wide. Epicalyx-segments *c.* 0·5 mm, ovate-lanceolate, green, longer than the deltate, scarious sepals. Petals minute or absent; stamens 4–6, included. Capsule slightly exceeding the hypanthium; style *c.* 0·25 mm. *Lake-margins and other wet places. S. France; Hungary; S.E. Russia; once recorded from N. Italy.* *Ga Hu ?It Rs (E).

A very rare and imperfectly known species. The plants from France have most of the flowers in groups of 2–3; those from Russia have them solitary in the upper part of the plant. Our knowledge is, however, insufficient to justify a division into subspecies.

L. linifolium Kar. & Kir., *Bull. Soc. Nat. Moscou* **14**: 421 (1841), from C. Asia, has been recorded from one locality in Hungary. It seems very doubtful, however, that it is specifically distinct from *L. thesioides*.

11. **L. borysthenicum** (Schrank) Litv. in Majevski, *Fl. Sred. Ross.* ed. 5, 209 (1917) (*Peplis erecta* Req. ex Moris, *P. boraei* (Guépin) Jordan, *Middendorfia borysthenica* (Schrank) Trautv.). Annual, usually more or less scabrid-hispid, at least above; stem 3–10(–18) cm, usually erect, but sometimes creeping and rooting for a short distance at the base. Leaves 5–20 × 2–8(–13) mm, alternate or opposite, oblong-oblanceolate to broadly obovate, sessile. Flowers (5–)6-merous, solitary (rarely in pairs) in the leaf-axils, sessile; bracteoles *c.* 1·5 mm, filiform. Hypanthium 2–3 mm in flower (up to 4 mm in fruit), narrowly to very broadly campanulate, 0·8–1·7 times as long as wide, often vinous red. Epicalyx-segments subulate, equalling or exceeding the deltate sepals. Petals minute, fugacious, purplish, often absent; stamens (5–)6, included. Capsule globular to shortly cylindrical or ovoid-conical, shorter than the hypanthium. $2n = 30$. *Seasonally wet places. S. Europe, extending northwards locally to N.W. France and S.C. Russia, but very rare in the Balkan peninsula.* Co Cr Ga Gr Hs It Lu Rs (C, W, E) Sa Tu.

Very variable, especially in arrangement of leaves, degree of scabridity, and shape of hypanthium.

12. **L. portula** (L.) D. A. Webb, *Feddes Repert.* **74**: 13 (1967) (*Peplis portula* L.). Glabrous annual; stems up to 25 cm, extensively creeping and rooting at the nodes. Leaves 5–15 mm, obovate-spathulate, tapered to a distinct petiole, opposite, rather fleshy. Flowers 6-merous, solitary in the leaf-axils. Hypanthium *c.* 1·5 mm, pelviform to hemispherical, wider than long; epicalyx-segments subulate, equalling or rarely twice as long as the triangular-acuminate sepals. Petals 1 mm, purple, sometimes absent; stamens 6. Capsule subglobose, nearly twice as long as the hypanthium. $2n = 10$. *Wet places and in shallow water; calcifuge. Europe, except the extreme north.* All except Fa Is Rs (K) Sb.

13. **L. volgense** D. A. Webb, *Feddes Repert.* **74**: 13 (1967) (*Peplis alternifolia* Bieb.). Glabrous annual; stem 3–10 cm, erect, branched from the base. Leaves 6–10 × 1–1·5 mm, linear-oblanceolate to linear, shortly petiolate, alternate. Flowers usually 5-merous, solitary in the leaf-axils. Hypanthium 0·75 mm, pelviform, considerably wider than long; epicalyx-segments linear-subulate, longer than the hypanthium and about equalling the triangular-acuminate, red-tipped sepals. Petals absent; stamens 2. Capsule subglobose. *Wet places. U.S.S.R., northwards to c. 58° N.* ?Bu Rs (C, W, E).

2. Ammannia L.[1]

Annuals with opposite, sessile leaves. Flowers 4-merous, in axillary, dichasial cymes, sometimes reduced to congested clusters. Hypanthium broadly campanulate to subglobose, distinctly veined; epicalyx present or absent; petals small, sometimes absent. Stamens 4, rarely more. Capsule irregularly circumscissile, or dehiscing only by decay.

1 Style 2–3 mm
  2 Flowers in ± lax, distinctly pedunculate cymes   **1. auriculata**
  2 Flowers in congested, subsessile clusters   **2. coccinea**
1 Style less than 0·5 mm
  3 Hypanthium and sepals glabrous; epicalyx absent   **3. baccifera**
  3 Hypanthium and sepals pubescent; epicalyx present, its segments exceeding the sepals   **4. verticillata**

1. **A. auriculata** Willd., *Hort. Berol.* **1**: t. 7 (1803). Glabrous; stem up to 1 m, erect, subtetragonal, simple or with ascending branches. Leaves up to 90 × 8 mm, usually longer than the internodes, linear-oblong, acute to acuminate, cordate-auriculate at the base. Cymes 3- to 15-flowered, with peduncles and at least some of the pedicels clearly visible. Hypanthium 2–3 mm, campanulate; sepals reduced to very short teeth; epicalyx-segments even smaller or absent. Petals 2·5 mm, fugacious; stamens 4, rarely more. Style 2–3 mm; capsule globose, slightly exceeding hypanthium. *Naturalized in rice-fields. N.W. Italy (around Novara and Vercelli).* [It.] (*Widespread in tropical and warm-temperate regions.*)

The European plant is referable to var. *arenaria* (Kunth) Koehne, and is probably of American origin.

2. **A. coccinea** Rottb., *Pl. Horti Univ. Rar. Progr.* 7 (1773). Like **1** but cymes very congested, without clearly visible peduncle or pedicels; hypanthium subglobose, up to 5 mm in fruit and slightly exceeding the capsule; epicalyx-segments nearly equalling sepals. *Naturalized in rice-fields. Spain and Portugal.* [Hs Lu.] (*America, from U.S.A. to Brazil.*)

3. **A. baccifera** L., *Sp. Pl.* 120 (1753) (incl. *A. aegyptiaca* Willd., *A. viridis* Willd. ex Hornem.). Glabrous; stem 10–30(–60) cm, erect, simple or branched. Leaves 10–30(–70) × 4–10(–15) mm, linear-oblong to broadly oblanceolate, subacute, with cuneate to subcordate base. Flowers subsessile in crowded, axillary clusters. Hypanthium 1·5–2 mm; sepals deltate, about ⅔ as long as hypanthium; epicalyx and petals absent; stamens 4. Stigma subsessile; capsule slightly exceeding hypanthium. *Rice-fields and other wet places. S.E. Russia.* Rs (E). (*From Africa to Australia and E. Asia.*)

4. **A. verticillata** (Ard.) Lam., *Encycl. Méth. Bot.* **1**: 131 (1783). Like **3** but sepals smaller; hypanthium and sepals pubescent; epicalyx-segments conspicuous, longer than the sepals; reddish petals sometimes present; capsule concealed in hypanthium. *Rice-fields, marshes and shallow water. S.E. Russia and S.E. Ukraine; naturalized in S. & E.C. Europe.* Rs (E) [?Bu Hu It Ju Rm]. (*S.W. Asia.*)

---

[1] By D. A. Webb.

3. Rotala L.[1]

Glabrous annuals with opposite, sessile leaves. Flowers minute, 4-merous, solitary in the leaf-axils, each subtended by 2 linear, whitish bracteoles. Hypanthium broadly campanulate, without visible veins; epicalyx absent; petals minute or absent. Capsule septicidal; walls with very fine horizontal striae (visible only with a microscope).

A mainly tropical genus, known in Europe only as a weed of rice-fields.

Leaves c. 10 mm, mostly longer than the internodes, oblanceolate, with thickened, hyaline margin   1. indica
Leaves 2–5 mm, much shorter than the internodes except near apex of stem, subcordate-ovate, without hyaline margin   2. filiformis

1. **R. indica** (Willd.) Koehne, *Bot. Jahrb.* 1: 172 (1880). Stems 10–25 cm, erect from a creeping base, rather stout; branches few and short. Leaves mostly c. 10 × 3·5 mm, oblanceolate, obtuse, with a narrow, hyaline, cartilaginous margin. Flowers c. 2·5 mm, pinkish. Sepals 1–1·5 mm, triangular, acuminate; petals 0·5 mm, persistent; stamens 4. Style 1·5 mm; capsule 2·5 mm, globose, 4-valved. *Rice-fields. N. Italy, C. Portugal.* [It Lu.] (*S. & E. Asia.*)

The European plant is var. *uliginosa* (Miq.) Koehne, apparently introduced from Japan. Var. *indica* has a narrower hypanthium and is more freely branched.

2. **R. filiformis** (Bellardi) Hiern in Oliver, *Fl. Trop. Afr.* 2: 468 (1871). Like 1 but stems up to 35 cm and more delicate; leaves 2–5 mm, ovate-subcordate (the lowest sometimes linear-oblong), without hyaline margin, much shorter than internodes except near apex of stem; flowers 1·5–2 mm; sepals deltate, not acuminate; petals absent; stamens 2; style 0·5 mm; capsule ellipsoid. *Rice-fields. N. Italy.* [It.] (*C. & S. Africa.*)

CXX. TRAPACEAE[2]

Flowers solitary, perigynous. Sepals, petals and stamens 4. Ovary semi-inferior, 2-locular; ovules 1 in each loculus, pendent. Fruit 1-locular, coriaceous or woody; seed solitary, with unequal cotyledons; endosperm absent. Radicle on germination penetrating the top of the fruit.

1. Trapa L.[3]

Aquatic herbs rooted in the mud. Submerged leaves sessile, linear, entire, the lowest opposite, the others alternate; floating leaves petiolate, alternate, with broad, dentate lamina. Stipules small, caducous. Green, pinnately-branched adventitious roots arise in pairs or whorls from the lower nodes.

1. **T. natans** L., *Sp. Pl.* 120 (1753). Annual, with unbranched stems 0·5–2 m. Submerged leaves caducous; floating leaves 1–4·5 cm long and wide, usually rhombic, dentate, glabrous above, pubescent, at least on the veins, beneath; base broadly cuneate or almost truncate, entire; petiole up to 17 cm, pubescent, often with a fusiform swelling. Flowers in the axils of the floating leaves; pedicels pubescent. Sepals narrowly triangular, keeled, accrescent and indurated in fruit, forming 2, 3 or 4 horns. Petals c. 8 mm, white, caducous. Nut 2–3·5 × 2–5·5 cm. $2n = c.$ 48. *In nutrient-rich but not strongly calcareous water. S. & C. Europe, extending northwards to C. France and to c. 57° N. in C. Russia. Sometimes cultivated elsewhere, and formerly widespread in continental N. Europe, particularly during the Sub-boreal period.* Al Au Bl Bu Cz Ga Ge Gr He Hs Hu It Ju Po Rm Rs (B, C, W, K, E) †Su.

There is great variation in the size of the fruit and in the number and degree of development of the horns. Several of these variants have been described as species or subspecies, and populations in one lake or river-system may show considerable uniformity; there is, however, overlap in characters between populations. In addition some of the variation appears to be due to edaphic factors, such as abnormally high calcium or low potassium and nitrogen concentrations. It therefore seems unjustified to give taxonomic recognition to these populations.

CXXI. MYRTACEAE[4]

Evergreen trees or shrubs. Leaves simple, usually opposite, exstipulate, with aromatic oil-glands. Flowers hermaphrodite, actinomorphic. Calyx and corolla 4- or 5-merous. Stamens numerous. Ovary inferior, syncarpous, with axile placentation; fruit a berry or capsule.

Shrubs; leaves opposite; fruit a berry   1. Myrtus
Trees; leaves on adult shoots alternate; fruit a woody capsule
                                                     2. Eucalyptus

1. Myrtus L.[5]

Shrubs with simple, opposite leaves. Flowers solitary in leaf-axils. Fruit a berry crowned by the persistent calyx-teeth.

1. **M. communis** L., *Sp. Pl.* 471 (1753). Erect, much-branched shrub, up to 5 m. Twigs glandular-hairy when young. Leaves up to 5 cm, ovate-lanceolate, acute, entire, coriaceous, punctate, very aromatic when crushed. Flowers up to 3 cm in diameter, sweet-scented. Pedicels long, slender, with 2 small, caducous bracteoles. Petals suborbicular, white. Berry 7–10 × 6–8 mm, broadly ellipsoid to subglobose, usually blue-black when ripe. *Scrub; usually calcifuge. Mediterranean region and S.W. Europe.* Al Az Bl Co Cr Ga Gr Hs It Ju Lu Sa Si.

Widely cultivated since ancient times; the native range is therefore very uncertain.

In addition to the two subspecies given below, other variation has been described, but seems to show no geographical correlation.

[1] By D. A. Webb. [2] Edit. T. G. Tutin.
[3] By T. G. Tutin. [4] Edit. S. M. Walters.
[5] By M. S. Campbell.

(a) Subsp. **communis**: Up to 5 m; leaves 2–5 cm, not crowded; berry broadly ellipsoid. *Throughout the range of the species.*

(b) Subsp. **tarentina** (L.) Arcangeli, *Comp. Fl. Ital.* 258 (1882): Not more than 2 m; leaves less than 2 cm, crowded; berry sub-globose. *E. Spain to W. Jugoslavia; Kriti; mainly near the coast.* Co Cr Ga Hs It Ju Sa [Lu].

2. Eucalyptus L'Hér.[1]

Trees with persistent or deciduous, smooth or fibrous bark. Leaves dimorphic; juvenile leaves opposite, sessile or shortly petiolate, often glaucous, frequently produced on mature trees in response to wounding; mature leaves alternate, petiolate, pendent, tough, rigid, with a prominent intramarginal vein. Flowers in umbels or solitary, closed in bud by the connate perianth-segments forming a hemispherical or conical operculum which falls off when the flower opens. Fruit a capsule, opening by valves which are described as *exserted* when they project beyond the rim of the capsule and *enclosed* when they remain below the rim.

A large genus centred in Australia. Specific identification is based largely on characteristics of the bark, and on operculum- and capsule-shape, although subdivision of the genus is based principally on anther-morphology. The large number of species examined in Australia all have $2n=22$.

The species described below have been planted in Europe (mainly in S.W. Europe and Italy) on a fairly large scale for timber, for shelter, for soil-stabilization, or as an antimalarial measure. Other species occasionally planted include E. **amygdalinus** Labill., *Nov. Holl. Pl.* **2**: 14 (1806) (*E. salicifolius* auct.), E. **cladocalyx** F. Mueller, *Linnaea* **25**: 388 (1853), E. **gunnii** Hooker fil., *London Jour. Bot. (Hooker)* **3**: 499 (1844), and E. **salignus** Sm., *Trans. Linn. Soc. London* **3**: 285 (1797).

Literature: W. F. Blakely, *A Key to the Eucalypts*, ed. 2. Canberra. 1955. A. R. Penfold & J. L. Willis, *The Eucalypts.* London and New York. 1961.

1 Flowers solitary; fruit more than 10 mm   **9. globulus**
1 Flowers in umbels; fruit not more than 10 mm
  2 Fruit sessile or subsessile
    3 Umbels 3-flowered; peduncles terete   **11. viminalis**
    3 Umbels (3–)5- to 10-flowered; peduncles compressed
      4 Leaves more than 18 cm; bark smooth; fruit glaucous
                              10. maidenii
      4 Leaves less than 18 cm; bark fibrous; fruit not glaucous
        5 Peduncles 7–10 mm; fruit 7–9 × 7–9 mm, cylindrical
                              1. botryoides
        5 Peduncles 25–35 mm; fruit 13–20 × 11–15 mm, campanulate   **5. gomphocephalus**
  2 Fruit distinctly pedicellate
    6 Peduncles compressed, angular or strap-shaped
      7 Fruit 5–8 mm, ovoid to hemispherical; valves strongly exserted   **3. resinifer**
      7 Fruit 12–15 mm, cylindrical to urceolate; valves enclosed or very slightly exserted   **2. robustus**
    6 Peduncles terete or nearly so
      8 Bark smooth, white
        9 Peduncles 10–15 mm; fruit 7–8 × 5–6 mm, hemispherical   **7. camaldulensis**
        9 Peduncles 5–12 mm; fruit 6–9 × 8–10 mm, broadly turbinate   **6. tereticornis**
      8 Bark rough, black or reddish
        10 Leaves with inconspicuous veins   **4. × trabutii**
        10 Leaves with prominent veins   **8. rudis**

---

[1] By N. A. Burges.

1. E. botryoides Sm., *Trans. Linn. Soc. London* **3**: 286 (1797). Tree up to 20 m. Bark subfibrous, persistent. Juvenile leaves 5–8 × 3–4 cm; ovate to broadly lanceolate; mature leaves broadly lanceolate-acuminate. Umbels 6- to 10-flowered; peduncles 7–10 × 4–5 mm, compressed. Operculum hemispherical, obtuse or apiculate. Fruit 7–9 × 7–9 mm, cylindrical, sessile; valves enclosed or slightly exserted. $2n=22$. [It Lu Sa Si.] (*S.E. Australia.*)

2. E. robustus Sm., *Bot. New Holl.* 39 (1795). Tree up to 12 m. Bark rough, subfibrous, persistent. Juvenile leaves up to 11 × 7 cm, broadly lanceolate to elliptic-lanceolate; mature leaves 10–18 × 4–8 cm, broadly lanceolate, long-acuminate. Umbels 5- to 10-flowered; peduncles 20–33 mm, compressed to strap-shaped. Operculum rostrate. Fruit 12–15 × 10–12 mm, cylindrical, pedicellate; valves deeply enclosed. *Planted in swampy ground, often in subsaline areas.* [Ga It Lu Sa.] (*E. Australia.*)

3. E. resinifer Sm. in White, *Jour. Voy. New S. Wales* 231 (1790). Tree up to 40 m. Bark reddish, rough, fibrous, persistent. Juvenile leaves 4–6 × 1·5–2 cm, narrowly lanceolate; mature leaves 10–16 × 2–3 cm, lanceolate, with inconspicuous veins. Umbels 5- to 10- flowered; peduncles 15–20 mm, compressed to angular. Operculum acutely conical to rostrate. Fruit 5–8 × 5–8 mm, ovoid to hemispherical, pedicellate; valves strongly exserted. [Ga Hs It Sa Lu.] (*E. Australia.*)

4. E. × trabutii Vilmorin, ex Trabut, *Bull. Stat. Rech. Forest. Nord Afr.* **1**: 141 (1917). Like **3** but juvenile leaves ovate-lanceolate; peduncles terete; operculum shorter; fruit 6 × 9 mm, more tapering towards the base. $2n=22$. [Ga It Si.]

A hybrid of Algerian origin, probably **1 × 7**.

5. E. gomphocephalus DC., *Prodr.* **3**: 220 (1828). Tree up to 40 m. Bark light grey, closely fibrous, persistent. Juvenile leaves 5–7 × 4–5 mm, broadly lanceolate. Mature leaves up to 17 × 2 cm, narrowly lanceolate, thick. Umbels 3- to 7-flowered; peduncles 25–35 × 10–15 mm, strap-shaped. Operculum hemispherical to conical. Fruit 13–20 × 11–15 mm, campanulate, smooth or with a single rib, sessile; valves shortly exserted. [Hs It Si.] (*Western Australia.*)

6. E. tereticornis Sm., *Bot. New Holl.* 41 (1795) (*E. umbellatus* (Gaertner) Domin, non Dum.-Courset). Tree up to 50 m. Bark smooth, irregularly blotched. Juvenile leaves 6–16 × 5–6 cm, elliptical to broadly lanceolate; mature leaves 10–21 × 1·2–2·5 cm, narrowly lanceolate. Umbels 5- to 12-flowered; peduncles 5–12 mm, almost terete. Operculum long, conical. Fruit 6–9 × 8–10 mm, broadly turbinate, pedicellate; valves strongly exserted. [It Lu.] (*E. Australia and Papua.*)

7. E. camaldulensis Dehnh., *Cat. Pl. Hort. Camald.* ed. 2, 20 (1832) (*E. rostratus* Schlecht., non Cav.). Spreading tree up to 15 m. Bark smooth, dull, white, deciduous. Juvenile leaves 6–9 × 2·5–4 cm, narrowly to broadly lanceolate, slightly glaucous; mature leaves 12–22 × 0·8–1·5 cm, lanceolate, acuminate, thin. Umbels 5- to 10-flowered; peduncles 10–15 mm, terete. Operculum conical to rostrate. Fruit 7–8 × 5–6 mm, hemispherical with a broad raised rim; valves exserted, incurved. $2n=22$. [Al Co Ga Gr Hs It Lu Sa Si.] (*Australia.*)

8. E. rudis Endl. in Endl. et al., *Enum. Pl. Hügel* 49 (1837). Tree up to 15 m with short trunk and spreading branches. Bark black, rough and persistent on trunk, smooth on branches. Juvenile leaves 10 × 7·5 cm, slightly glaucous, ovate to orbicular; mature leaves 10–15 × 1–2 cm, narrowly to broadly lanceolate, with prominent veins. Umbels 4- to 10-flowered; peduncles

10-15 mm, terete, slender. Operculum broadly conical. Fruit 5–9 × 10–12 mm, hemispherical or with a tapering base, pedicellate; valves strongly exserted, incurved. [It Sa Si.] (*Western Australia.*)

9. E. globulus Labill., *Rel. Voy. Rech. La Pérouse* **1**: 153 (1800). Tree up to 40 m. Bark smooth, deciduous. Juvenile leaves 7–16 × 4–9 cm, ovate to broadly lanceolate, cordate, very glaucous; mature leaves 10–30 × 3–4 cm, lanceolate to falcate-lanceolate, acuminate, glossy green. Flowers solitary. Operculum hemispherical, umbonate. Fruit 10–15 × 15–30 mm, depressed-globose, somewhat tapering towards the base, with 4 main ribs, sessile; valves often covered by disc. [Az Bl Ga Hb Hs It Lu Sa Si.] (*Tasmania.*)

10. E. maidenii F. Mueller, *Proc. Linn. Soc. New S. Wales* ser. 2, **4**: 1020 (1890). Tree up to 40 m. Bark smooth, white, decidu-ous. Juvenile leaves 4–16 × 4–12 cm, ovate, cordate, amplexicaul, glaucous; mature leaves *c.* 20 × 2·5 cm, glossy green. Umbels 3- to 7(–10)-flowered; peduncles 10–15 mm, compressed. Operculum hemispherical to broadly conical. Fruit 8–10 × 10–12 mm, turbinate, 1- to 2-ribbed, glaucous, subsessile; valves strongly exserted, partly adnate to disc. [Hs It Lu Sa Si.] (*S.E. Australia.*)

11. E. viminalis Labill., *Nov. Holl. Pl.* **2**: 12 (1806). Tree up to 50 m. Bark smooth, white, deciduous, often hanging from the branches in long ribbons. Juvenile leaves 5–10 × 1·5–3 cm, ovate, pale green, sessile, more or less amplexicaul; mature leaves 11–18 × 1·5–2 cm, lanceolate, acuminate, pale green. Umbels 3-flowered; peduncles 3–6 mm, terete. Operculum hemispherical to conical. Fruit 5–6 × 7–8 mm, spherical or slightly tapered at the base, sessile or shortly pedicellate; valves exserted, patent. [Ga Hs It Lu Si.] (*S. & E. Australia, Tasmania.*)

CXXII. PUNICACEAE[1]

Leaves not gland-dotted. Flowers perigynous. Sepals 5–7, valvate, persistent. Petals 5–7, imbricate and crumpled in bud. Ovary inferior, multi-locular; loculi superposed, the lower ones with axile, the upper ones with parietal placentation; styles simple; ovules numerous. Fruit berry-like, with a coriaceous exocarp and numerous seeds each surrounded by pulp.

1. Punica L.[2]

Shrub or tree, sometimes spiny; twigs 4-angled. Leaves opposite, deciduous; stipules absent. Flowers hermaphrodite, terminal.

1. P. granatum L., *Sp. Pl.* 472 (1753). Spiny shrub or small, unarmed tree. Leaves oblong-lanceolate to obovate, glabrous, entire. Flowers 30–40 mm in diameter. Hypanthium coriaceous, reddish. Petals red, rarely white. Fruit 5–8 cm in diameter, reddish-brown; pulp surrounding seeds translucent, purple, yellowish or white, acid or sweet. *Cultivated for its fruit in most of S. Europe, and widely naturalized in the Mediterranean region and Portugal.* [Al Bl Bu Co Cr Ga Gr He Hs It Ju Lu Rm ?Sa ?Si Tu.] (*S.W. Asia.*)

CXXIII. ONAGRACEAE (OENOTHERACEAE)[3]

Herbs or shrubs. Flowers hermaphrodite, actinomorphic or weakly zygomorphic. Hypanthial tube ('calyx-tube') often present. Sepals 2, 4 or 5; petals 0, 2, 4 or 5. Stamens 2 or 4 in 1 whorl, or 8 or 10 in 2 whorls; pollen connected in masses by fine threads. Style 1; ovary inferior, 1-, 2-, 4- or 5-locular. Fruit a capsule, a berry or dry and indehiscent; seeds without endosperm.

1 Sepals and petals 2; fruit indehiscent, 1- or 2-seeded   **2. Circaea**
1 Sepals 4 or 5; petals 4, 5 or 0; fruit a many-seeded capsule or a berry
  2 Shrubs; fruit a berry   **1. Fuchsia**
  2 Herbs; fruit a capsule
    3 Sepals 4 or 5, somewhat persistent in fruit; petals 0 or 5; stamens 4 or 10   **4. Ludwigia**
    3 Sepals and petals 4, caducous; stamens 8
      4 Seeds with a chalazal tuft of hairs; corolla purplish-pink to white, never yellow   **5. Epilobium**
      4 Seeds without a tuft of hairs; corolla yellow, rarely pink or purplish   **3. Oenothera**

1. Fuchsia L.[4]

Shrubs. Leaves usually opposite or in whorls of 3, sometimes alternate near ends of branches. Flowers actinomorphic, large, axillary, pendent; hypanthial tube long. Sepals 4; petals 4; stamens 8 in 2 whorls; stigma entire; ovary 4-locular. Fruit a berry; seeds numerous, small.

Literature: P. A. Munz, *Proc. California Acad. Sci.* ser. 4, **25**: 1–138 (1943).

1. F. magellanica Lam., *Encycl. Méth. Bot.* **2**: 565 (1788). Up to 3 m. Leaves 1·5–5·5 cm, elliptic-ovate, acuminate. Flowers solitary; pedicels 2–5·5 cm. Hypanthial tube 5–10 mm, red. Sepals 12–20 mm, ovate-lanceolate, red. Petals 6–20 mm, obovate, violet, rarely white. Berry 1–2 cm, black. 2*n*=44. *Planted for hedges in the Açores, Ireland and W. Britain and locally naturalized.* [Az Br Hb.] (*Temperate South America.*)

2. Circaea L.[4]

Stoloniferous perennial herbs. Leaves opposite, ovate, acuminate. Flowers actinomorphic, small, in a terminal, bracteate raceme; hypanthial tube short. Sepals 2; petals 2, white or pinkish; stamens 2; stigma entire or shallowly notched; ovary 1- or 2-locular, with one ovule in each cell. Fruit clavate, indehiscent, densely covered with bristles; seeds 1 or 2.

1 Inflorescence not elongating until petals have dropped, the open flowers clustered at apex; fruit unilocular   **3. alpina**
1 Inflorescence elongating before petals have dropped, the open flowers well-spaced; fruit ± bilocular
  2 Disc 0·2–0·4 mm high; plants fertile   **1. lutetiana**
  2 Disc obscure or rarely up to 0·2 mm high; plants sterile   **2. × intermedia**

---

[1] Edit. T. G. Tutin.   [2] By T. G. Tutin.
[3] Edit. D. M. Moore.   [4] By P. H. Raven.

1. C. lutetiana L., *Sp. Pl.* 9 (1753). 15–60 cm. Leaves truncate or slightly cordate at base, sparsely denticulate and strigulose. Bracteoles usually absent. Hypanthial tube 1–1·2 mm, about as long as ovary. Petals 2–4 × 2–2·5 mm. Filaments 2·5–5·5 mm. Disc 0·2–0·4 mm high. Fruit 3–4 × 2–2·5 mm, bilocular, with hooked bristles 0·7–1·1 mm. 2*n*=22. *Most of Europe except the north-east.* All except Az Bl Cr Fa Fe Is Rs (N) Sb.

2. C. × intermedia Ehrh., *Beitr. Naturk.* **4**: 42 (1789). 10–45 cm. Leaves shallowly cordate at base, dentate, subglabrous. Bracteoles present. Hypanthial tube 0·5–1·2 mm, shorter than ovary. Petals 1·8–4 × 2–2·3 mm. Filaments 2–4 mm. Disc obscure, rarely up to 0·2 mm high. Fruit up to 2 × 1·2 mm, falling in immature state, bilocular but 1 loculus more or less abortive, with hooked bristles 0·5–0·6 mm. 2*n*=22. *N.W. & C. Europe, extending locally to N. Spain, N. Italy, Crna Gora, C. Russia and S. Sweden.* Au Be Br Cz Da Ga Ge Hb He Ho Hs Hu It Ju No Po Rm Rs (B, C, W) Su.

Completely sterile; the hybrid between **1** and **3**, usually growing with one or both parents.

3. C. alpina L., *Sp. Pl.* 9 (1753). 5–30 cm. Leaves cordate, dentate, glabrous. Bracteoles present. Hypanthial tube 0·1–0·2 mm, about as wide as long, much shorter than ovary. Petals 0·6–1·4 × 0·4–0·9 mm. Filaments 1–1·5 mm. Disc obscure. Fruit 2 × 1 mm, unilocular, with often straight bristles 0·1–0·5 mm. 2*n*=22. *N. Europe, extending southwards in the mountains to the Pyrenees, N. Appennini and Crna Gora.* Al Au Be Br ?Co Cz Da Fe Ga Ge He Ho Hs It Ju No Po Rm Rs (N, B, C, W) Su.

3. Oenothera L.[1]

(*Onagra* Miller)

Annual, biennial or perennial herbs. Leaves alternate. Flowers actinomorphic, rather large, in a leafy spike; hypanthial tube conspicuous. Sepals 4, deflexed in flower; petals 4, yellow or rarely pink; stamens 8, in 2 whorls; stigma deeply 4-lobed; ovary 4-locular. Fruit an elongate, loculicidal capsule; seeds small, numerous, without a chalazal plume of hairs.

Plants of American origin cultivated for ornament in Europe and widely naturalized, usually on disturbed ground. Several other species, not described here, have been recorded as casuals.

Literature: O. Renner, *Ber. Deutsch. Bot. Ges.* **60**: 448–466 (1942); **63**: 129–138 (1950); *Planta* **47**: 219–254 (1956). K. Rostański, *Fragm. Fl. Geobot.* **11**: 499–523 (1965).

1 Petals pink to reddish-violet; flowers opening near sunrise
        13. rosea
1 Petals yellow, often becoming reddish or purplish; flowers opening near sunset
  2 Cauline leaves oblong-ovate or broadly oblanceolate, ± truncate at base; plant densely hispid   **10. longiflora**
  2 Cauline leaves lanceolate or elliptical, cuneate at base; plant not densely hispid
    3 Hypanthial tube 70–100 mm   **11. affinis**
    3 Hypanthial tube 15–50 mm
      4 Capsule conspicuously enlarged distally; seeds pendent, not sharply angled   **12. stricta**
      4 Capsule not enlarged distally; seeds horizontal, sharply angled
        5 Sepal-apices divergent in bud; inflorescence ± nodding at anthesis   **7–9. parviflora** group
        5 Sepal-apices appressed to one another in bud; inflorescence erect   **1–6. biennis** group

Subgen. **Oenothera**. Annual or biennial. Cauline leaves up to 20 × 5 cm, lanceolate. Flowers nocturnal; petals yellow, often becoming purplish-red. Capsule elongate, cylindrical; seeds horizontal, sharply angled.

Because of the presence of balanced combinations of lethal genes and of self-pollination in many species of this group, progenies from individual plants breed true; this has led to the establishment of many distinctive variants in Europe and a number of them have been given specific names. Any new combination of chromosomes produces, in effect, a new 'species'. Each is then characterized by a particular complex of chromosomes transmitted only through the pollen and another transmitted only through the egg. Some of the chromosome-complexes are found in more than one of these true-breeding strains, which are thus not comparable with the species recognized in any other group of flowering plants. Only the more widespread species are described here and those which seem to have originated in Europe are treated as native, although their parents were introduced deliberately or accidentally from temperate North America. It is often nearly impossible to determine dried specimens and a number of the distributions are manifestly incomplete.

1–6. O biennis group. Stem up to 300 cm, hairy. Inflorescence erect. Sepal-apices terminal, slender, appressed to one another in bud. Hypanthial tube 18–50 mm. Petals 12–60 mm. Style 3–60 mm. Capsule 10–40 mm, tapering above. Seeds 1·2–2 × 0·7–1 mm.

According to the standards of Munz, *N. Amer. Fl.* ed. 2, **5**: 132–135 (1965), species **1** and probably **2** would be referred to *O. biennis* subsp. *biennis* (subsp. *caeciarum* Munz) and species **3** would be referred to *O. biennis* subsp. *centralis* Munz.

1 Plant greyish-pubescent; ovary and young fruit densely strigose
        4. strigosa
1 Plant green or bluish-green; ovary and young fruit glandular or strigulose
  2 Stem and ovaries without red spots
    3 Veins of mature leaves reddish; petals 24–30 mm   **1. biennis**
    3 Veins of mature leaves not reddish; petals more than 30 mm
        6. suaveolens
  2 Stem and ovaries with red spots
    4 Calyx red-striped or entirely red, at least on later flowers
        5. erythrosepala
    4 Calyx green
      5 Stem up to 200 cm, not easily detached at the base; inflorescence-axis reddish; bracts lanceolate, weakly toothed   **2. rubricaulis**
      5 Stem up to 300 cm, easily detached at the base; inflorescence-axis green; bracts ovate, strongly toothed   **3. chicagoensis**

1. O. biennis L., *Sp. Pl.* 346 (1753) (*Onagra biennis* (L.) Scop.). Stem 10–150 cm, without red spots. Calyx green. Hypanthial tube 18–44 mm. Petals 24–30 mm. Style 3–17 mm. 2*n*=14 (ring of 6 and ring of 8 chromosomes at meiosis). *Waste ground and open habitats.* ● *Europe except the extreme north and parts of the south.* All except Bl Cr Fa Fe Is Rs (N) Sa Sb.

2 contains the same pollen-transmitted chromosome-complex but differs in the complex transmitted through the egg.

2. O. rubricaulis Klebahn, *Jahrb. Hamb. Wiss. Anst.* **31** (Beih. 3): 23 (1914). Stem up to 200 cm, well-branched, with red spots. Inflorescence-axis reddish; bracts lanceolate, weakly toothed. Calyx green. Petals 22–23 mm. 2*n*=14 (ring of 14 chromosomes at meiosis). ● *C. Europe, extending to France and Russia.* ?Br Ga Ge Hu Po Rs (C).

---

[1] By P. H. Raven.

3. O. chicagoensis Renner ex Cleland & Blakeslee, *Proc. Nat. Acad. Sci. U.S.* **16**: 189 (1930). Stem up to 300 cm, well-branched, with red spots. Inflorescence-axis green; bracts ovate, strongly toothed. Calyx green. Petals (12–)20–25 mm. 2n=14 (ring of 12 chromosomes and 1 bivalent at meiosis). *C. Europe.* [Ga Ge It Po.] (*Temperate North America.*)

4. O. strigosa (Rydb.) Mackenzie & Bush, *Man. Fl. Jackson Co. Missouri* 139 (1902) (*O. hungarica* Borbás). Plant greyish-pubescent; stem without red spots. Inflorescence lax. Calyx reddish. Petals *c.* 20 mm. Ovary and young fruit densely strigose. 2n=14 (ring of 14 chromosomes at meiosis). *C. Europe.* [Au Cz Ga Ge Hu Po.] (*Temperate North America.*)

O. renneri H. Scholz, *Wiss. Zeitschr. Pädag. Hochsch. Potsdam* **2**: 206 (1953), from Poland and Germany (near Brandenburg), is like **4** but has a dense inflorescence and green buds. It has 2n=14 (ring of 14 chromosomes at meiosis). Here also may belong plants, having the same chromosomal configuration, which are abundantly naturalized in S. Switzerland (Ticino); they were recently incorrectly reported as *O. elata* Kunth, a species from Mexico and Central America.

5. O. erythrosepala Borbás, *Magyar Bot. Lapok* **2**: 245 (1903) (*O. lamarkiana* auct., non Ser.). Stem 30–150 cm, erect, hairy, with red spots. Leaves broadly lanceolate, with crinkled margins. Calyx red-striped or entirely red, at least on later flowers. Hypanthial tube 30–50 mm. Petals 40–60 mm. Style 20–60 mm; stigma held above anthers. 2n=14 (ring of 12 chromosomes and 1 bivalent at meiosis). *Locally common in W. & C. Europe.* Au Be Br Cz ?Da Ga Ge Ho Hu It Lu Po Rm [He].

Widely cultivated; probably of spontaneous garden origin in Europe from plants introduced from North America, and now naturalized in both continents, where it is generally accepted as a distinct species.

O. coronifera Renner, *Planta* **47**: 239 (1956), from Germany (near Berlin) and W. Czechoslovakia, is like **5** but has plane, narrowly lanceolate leaves and the stigma surrounded by the anthers. It has 2n=14 (ring of 12 chromosomes and 1 bivalent at meiosis).

6. O. suaveolens Pers., *Syn. Pl.* **1**: 408 (1805). Like **5** but leaves not crinkled; stem without red spots; stigma surrounded by anthers. 2n=14 (ring of 12 chromosomes and 1 bivalent at meiosis). *Probably of European origin and sparingly naturalized in scattered localities.* ● Az Cz Ga Ge Hu It Po.

The application of the name is somewhat uncertain. **O. grandiflora** L'Hér. in Aiton, *Hort. Kew.* **2**: 2 (1789), differs in its subglabrous sepals and capsules (pubescent and glandular in **6**). It is probably usually cultivated but may occasionally become naturalized.

7–9. O. parviflora group. Stem 10–200 cm, erect or decumbent, hairy. Inflorescence more as less nodding at anthesis. Sepal-apices subterminal, divergent in bud. Hypanthial tube 15–30 mm. Petals 12–20 mm. Style 4·5–25 mm. Capsule 20–40 mm. Seeds 1·7–2·2 × 1–1·5 mm.

The name *O. muricata* L. (*Onagra muricata* (L.) Moench) refers to this group, but its exact application is uncertain. The distributions of the individual taxa are not at all well-known but the group as a whole occurs in the following territories: Au Be Br Bu Cz Da Ga Ge He Ho Hs Hu It Ju No Po Rm Rs (C).

1 Leaves white-hairy; young capsule red-striped   **9. ammophila**
1 Leaves subglabrous; capsule not red-striped

2 Root thick; inflorescence stout, leafy; sepal-apices short
     7. parviflora
2 Root slender; inflorescence slender, lax; sepal-apices long
     8. silesiaca

7. O. parviflora L., *Syst. Nat.* ed. 10, **2**: 998 (1759). Root thick. Stem slightly nodding, with indistinct red spots at the end of flowering period. Leaves subglabrous. Inflorescence stout, rather dense and leafy. Sepal-apices short, forming U-shaped pairs. Petals *c.* 11 mm. Capsule green. 2n=14. *Disturbed ground and other open habitats. W. & C. Europe.* [Cz Ga Ge Ho Hu It No Po.] (*Temperate North America.*)

O. issleri Renner ex Rostański, *Fragm. Fl. Geobot.* **11**: 514 (1965), differs in its strongly nodding stem which lacks red spots, but which may be red-striped, its sepal-apices forming V-shaped pairs and its longer petals (12–20 mm). It has 2n=14 (ring of 14 chromosomes at meiosis), and is known from E. France (Alsace), where it is the commonest representative of the group, and Poland (near Wrocław).

8. O. silesiaca Renner, *Ber. Deutsch. Bot. Ges.* **60**: 455 (1942). Like **7** but root slender; stem without red spots; inflorescence slender and lax; sepal-apices longer; petals 16–20 mm, cordate. 2n=14 (ring of 14 chromosomes at meiosis). *Disturbed ground and other open habitats.* ● *C. Europe.* Au Cz Ge Po.

O. atrovirens Shull & Bartlett, *Amer. Jour. Bot.* **1**: 239 (1914), which is introduced from E. temperate North America and occurs in E. France (Alsace), is like **8** but has linear, sepal-like petals *c.* 12 × 2–3 mm. It has 2n=14 (ring of 14 chromosomes at meiosis).

9. O. ammophila Focke, *Abh. Nat. Ver. Bremen* **18**: 182 (1905). Stem up to 100 cm, decumbent, often procumbent, without red spots. Leaves white-hairy; cauline leaves dentate, bluish-green. Sepal-apices long, curved. Petals not more than 16 mm. Capsule red-striped when young. 2n=14 (ring of 12 chromosomes and 1 bivalent at meiosis). *Open, sandy habitats, especially on seashores.* ● *Mainly in N. & W. Europe.* ?Br Cz Da Ga Ge Ho.

O. syrticola Bartlett in E. L. Greene, *Cybele Columb.* **1**: 38 (1914), from E. North America, occurs sporadically in W. & C. Europe. It differs from **9** in its erect stems and strongly nodding inflorescence, reddish leaf-margins, short and straight sepal-apices and in its shorter petals (not more than 13 mm). It has 2n=14 (ring of 14 chromosomes at meiosis).

O. rubricuspis Renner ex Rostański, *Fragm. Fl. Geobot.* **11**: 512 (1965), from Germany (Hessen) and Belgium (Limbourg), is like **9** but has erect stems up to 200 cm and entire, dark green, cauline leaves. It has 2n=14 (ring of 14 chromosomes at meiosis).

Subgen. **Raimannia** (Rose) Munz. Usually annual or biennial. Flowers nocturnal; petals yellow, becoming reddish. Capsule elongate, cylindrical; seeds pendent, not sharply angled.

10. O. longiflora L., *Mantissa Alt.* 227 (1771). 10–100 cm, erect, densely hispid. Basal leaves 10–15 × 1·5–2·5 cm, oblanceolate; cauline leaves 2–5 × 0·7–1·6 cm, oblong-ovate or broadly oblanceolate, denticulate; uppermost leaves 1–1·5 cm, ovate. Hypanthial tube 30–80 mm. Petals 20–40 mm. Anthers 8–10 mm; filaments 15–20 mm. Capsule 2–3 × 0·2–0·3 cm, slightly enlarged upwards. Seeds *c.* 1·5 mm, narrowly obovoid, smooth. *Locally naturalized in S. Europe.* [Az Ga Hs.] (*Temperate South America.*)

11. O. affinis Camb. in St-Hil., *Fl. Bras. Mer.* **2**: 269 (1830). 30–100 cm, erect, densely greyish-pubescent. Basal leaves absent at flowering; cauline leaves 5–9 × 0·8–1·2 cm, linear-lanceolate, remotely denticulate, undulate; uppermost leaves 3–4 × 0·4–0·5 cm, otherwise similar. Hypanthial tube 70–100 mm. Petals 32–40 mm. Stamens, capsules and seeds as in **10**. 2n=14. *Coastal areas on sandy soil. Naturalized in C. & S. Portugal.* [Lu.] (*Temperate South America.*)

Perhaps this species should be included in *O. mollissima* L., as suggested by Tandon & Hecht, *Cytologia* 21: 252 (1956).

12. O. stricta Ledeb. ex Link, *Enum. Hort. Berol. Alt.* 1: 377 (1821). 20–100 cm, erect, villous. Basal leaves 5–10 × 0·5–1·8 cm, oblanceolate or linear; cauline leaves 2–5(–9) × 0·5–3 cm, narrowly lanceolate, denticulate, undulate. Hypanthial tube 15–30 mm. Petals 20–45 mm. Anthers 7–8 mm; filaments 10–20 mm. Capsule 2–2·5 × *c.* 0·3 cm, conspicuously enlarged in upper half. Seeds *c.* 1·5 mm, narrowly obovoid, smooth. 2n=14. *Locally naturalized in several districts, mainly in W. & C. Europe.* [Az Co Ga Ge He Hs It Lu Rs (C).] (*Temperate South America.*)

O. laciniata Hill, *Hort. Kew.* 172(4) (1768) (*O. sinuata* L.), with sinuate-pinnatifid leaves and petals 5–18 mm, has been recorded for several countries, but does not seem to be effectively naturalized.

Subgen. **Hartmannia** (Spach) Munz. Petals pink to red-violet. Capsule clavate, attenuate at the base, ribbed or winged. Seeds numerous, rounded.

13. O. rosea L'Hér. ex Aiton, *Hort. Kew.* 2: 3 (1789). Perennial, sometimes flowering in the first year; stems up to 100 cm from a somewhat woody stock, more or less strigulose. Basal leaves 2–5 cm, subentire to coarsely pinnatifid, with petioles 1–2 cm; cauline leaves 1·5–3 cm, subentire or subpinnatifid below. Hypanthial tube 4–8 mm. Petals 4–10 × 3–4 mm. Capsule 8–10 × 3–4 mm, somewhat winged; pedicel 5–20 mm, hollow, ribbed. Seeds *c.* 0·6 mm, oblong-obovoid. *Naturalized in S. Europe; casual elsewhere.* [Az Bl Ga Hs It Lu Si.] (*Warmer regions of North and South America.*)

4. Ludwigia L.[1]

(*Isnardia* L., *Jussiaea* L.)

Perennial herbs of wet places. Leaves opposite or alternate. Flowers actinomorphic, axillary; hypanthial tube absent. Sepals 4 or 5, somewhat persistent in fruit; petals 0 or 5, yellow, showy; stamens 4, or 10 in 2 whorls; stigma entire; ovary 4- or 5-locular. Fruit an irregularly dehiscent capsule; seeds numerous, small, free or embedded in coherent woody blocks of endocarp.

Literature: P. H. Raven, *Reinwardtia* 6: 327–427 (1963).

1 Leaves opposite; petals absent; stamens 4   **3. palustris**
1 Leaves alternate; petals present; stamens 10
  2 Flowering stems coarsely and densely hairy   **1. uruguayensis**
  2 Flowering stems finely hairy to subglabrous   **2. peploides**

1. L. uruguayensis (Camb.) Hara, *Jour. Jap. Bot.* 28: 294 (1953) (*Jussiaea repens* sensu Coste, non L.). Stems up to 1·5 m, floating and subglabrous in vegetative state, erect and densely hairy in flowering state. Leaves 3–13 × 0·3–2·5 cm, lanceolate, alternate. Sepals 5, 6–20 mm, hairy. Petals 12–30 mm, bright yellow. Stamens 10; filaments 2–4 mm. Capsule 13–25 × 3–4 mm, hairy. Fruiting pedicel (0·5–)2·5–6 cm. Seeds 1·5 × 1·5 mm, firmly enclosed in hard endocarp. 2n=80. *Abundantly naturalized in rivers and ditches in parts of S. France and N.E. Spain.* [Ga Hs.] (*North and South America.*)

2. L. peploides (Kunth) P. H. Raven, *Reinwardtia* 6: 393 (1963). Like **1** but leaves 1–6 × 0·4–3 cm, oblong; flowering stems, calyx and capsule finely hairy to subglabrous; fruiting pedicel 1–4 cm. 2n=16. *Naturalized in rivers in S.W. France.* [Ga.] (*North and South America.*)

3. L. palustris (L.) Elliott, *Sketch Bot. South Carol. Georgia* 1: 211 (1817) (*Isnardia palustris* L.). Stems 3–50 cm, glabrous, creeping. Leaves 7–45 mm, narrowly obovate, opposite. Sepals 4, 1·4–2 × 0·8–1·8 mm. Petals absent. Stamens 4; filaments 0·5–0·6 mm. Capsule 2–5 × 2–3 mm, glabrous, with green bands on the angles. Seeds 0·6–0·9 × 0·3 mm, free from endocarp. 2n=16. *Wet places. W., C. & S. Europe; rather local.* Al Au Be Br Bu Co Cz Ga Ge Gr He Ho Hs Hu It Ju Lu Po Rm Rs (W) Sa Tu.

5. Epilobium L.[1]

(*Chamaenerion* Séguier)

Perennial herbs, often flowering in the first year, overwintering by turions or rosettes, which persist about the base of the previous year's stem, or by stolons. Leaves alternate, opposite, or verticillate. Flowers actinomorphic or weakly zygomorphic, small to medium-sized, in a leafy raceme or spike, or axillary; hypanthial tube short or absent. Sepals 4, erect; petals 4, white, pink or purple; stamens 8, in 2 whorls; stigma clavate or capitate and entire, or deeply 4-lobed; ovary 4-locular. Fruit a long and slender, loculicidal capsule; seeds numerous, small, with a chalazal plume of hairs (coma).

Literature: C. Haussknecht, *Monographie der Gattung Epilobium.* Jena. 1884.

1 Leaves alternate; flowers showy, ± zygomorphic, the style deflexed, at least before the anthers have dehisced
  2 Leaves lanceolate or elliptical; seeds smooth
    3 Inflorescence many-flowered; style 10–20 mm, becoming erect after the anthers have dehisced   **1. angustifolium**
    3 Inflorescence few-flowered; style 4–7·5 mm, deflexed throughout anthesis   **2. latifolium**
  2 Leaves linear or linear-lanceolate; seeds finely papillose
    4 Leaves strigulose, at least along veins and on margins; style 7–15 mm, becoming erect after anthers have dehisced   **3. dodonaei**
    4 Leaves glabrous; style 3·5–5 mm, deflexed throughout anthesis   **4. fleischeri**
1 Leaves opposite or verticillate, at least below; flowers actinomorphic, the style very rarely deflexed
  5 Stigma distinctly 4-lobed
    6 Stems patent-pubescent; leaves sessile or subsessile
      7 Leaves semi-amplexicaul; petals usually more than 10 mm   **5. hirsutum**
      7 Leaves not amplexicaul; petals less than 10 mm   **6. parviflorum**
    6 Stems appressed-pubescent or subglabrous; leaves mostly petiolate
      8 Leaves narrowly cuneate at base; petiole 3–10 mm; petals white, later pink   **10. lanceolatum**
      8 Leaves rounded at base; petiole usually less than 3 mm; petals pink to purplish
        9 Overwintering by long, fleshy, hypogeal stolons up to 10 cm; seeds 1·7–2 mm, attenuate, with a prominent beak   **7. duriaei**
        9 Overwintering by turions or short, hypogeal stolons; seeds *c.* 1 mm, without a beak
          10 Leaves 3·5–8 × 1–4 cm; buds acute; petals 6–10 mm   **8. montanum**
          10 Leaves 1–5 × 0·5–1·5 cm; buds obtuse; petals 3–6 mm   **9. collinum**
  5 Stigma clavate or capitate, not lobed
    11 Stolons epigeal, filiform; leaves entire; seeds 1·5–2 mm, with a pellucid appendage   **15. palustre**

---

[1] By P. H. Raven.

11 Stolons hypogeal or epigeal and leafy, or absent; leaves often toothed; seeds usually less than 1·5 mm, with or without appendage
  12 Leaves usually verticillate; seeds 1·8–2 mm, with a pellucid appendage   **11. alpestre**
  12 Leaves usually opposite; seeds 0·7–1·7 mm, with or without appendage
    13 Plant creeping and rooting at the nodes   **27. nerterioides**
    13 Plant not creeping and rooting at the nodes
     14 Inflorescence, including ovary, ±conspicuously pubescent or strigulose
      15 Inflorescence glandular
       16 Seeds not striate-papillose, obovoid and rounded at ends   **14. roseum**
       16 Seeds striate-papillose, acuminate with a pellucid appendage
        17 Overwintering by turions; petals 5–6·5 mm, purplish-pink   **25. glandulosum**
        17 Overwintering by leafy rosettes; petals 2·5–6 mm, purplish-pink or white   **26. adenocaulon**
      15 Inflorescence entirely eglandular or with a few glandular hairs on the calyx only
       18 Seeds with a pellucid appendage at apex (northern or montane plants)
        19 Stigma capitate, held above the anthers at anthesis; petals 5·5–7 mm   **19. atlanticum**
        19 Stigma clavate, surrounded by the anthers at anthesis; petals 3–6 mm
         20 Stolons epigeal, leafy; leaves ovate or elliptical; petals pale violet   **17. nutans**
         20 Stolons absent; leaves linear to narrowly elliptical; petals white, rarely pale pink   **16. davuricum**
       18 Seeds without a pellucid appendage at apex
        21 Stigma capitate; leaves with bulbils or short vegetative shoots in the axils; stems glandular   **24. gemmascens**
        21 Stigma clavate; leaves without bulbils or short vegetative shoots in the axils; stems eglandular
         22 Stolons epigeal, leafy; calyx with a few erect, glandular hairs   **13. obscurum**
         22 Stolons absent; calyx eglandular   **12. tetragonum**
     14 Inflorescence, including ovary, subglabrous (small northern or montane plants)
      23 Stolons epigeal, leafy, sometimes inconspicuous
       24 Stolons short; petals white; inflorescence not or scarcely nodding before anthesis   **22. lactiflorum**
       24 Stolons long; petals purplish-pink, more rarely white; inflorescence nodding before anthesis
        25 Leaves shortly petiolate; stem with weakly elevated lines decurrent from margins of petioles   **20. anagallidifolium**
        25 Leaves sessile; stem with fine, not elevated lines decurrent from margins of petioles   **18. tundrarum**
      23 Stolons hypogeal or absent
       26 Strigulose lines decurrent from margins of petioles not elevated; petals white, rarely pink   **16. davuricum**
       26 Strigulose or smooth lines decurrent from margins of petioles elevated; petals violet or purplish to white
        27 Petals 2·5–4 mm, white; seeds *c.* 1·2 mm   **22. lactiflorum**
        27 Petals 4–11 mm, violet or purplish; seeds *c.* 1 mm
         28 Petals 4–6 mm, violet; seeds finely papillose   **21. hornemannii**
         28 Petals 7–11 mm, bright purplish-pink; seeds smooth   **23. alsinifolium**

Sect. CHAMAENERION Tausch. Roots thick, woody, spreading. Leaves alternate. Flowers more or less zygomorphic, showy; petals entire or shallowly emarginate, violet or purplish. Stigma 4-lobed. Pollen grains shed singly.

1. **E. angustifolium** L., *Sp. Pl.* 347 (1753) (*Chamaenerion angustifolium* (L.) Scop., *E. spicatum* Lam.). Stems up to 250 cm,

erect. Leaves 2·5–20 × 0·4–3·5 cm, lanceolate, with well-developed submarginal vein. Inflorescence a long raceme; flower-buds sharply deflexed. Style 10–20 mm, deflexed, becoming erect after anthers have dehisced and become deflexed. Seeds 1–1·3 mm, smooth. 2*n* = 36. *Almost throughout Europe, but rare in the south.* All except Az Bl Cr Lu Sa Sb Si.

2. **E. latifolium** L., *loc. cit.* (1753) (*Chamaenerion latifolium* (L.) Th. Fries & Lange). Stems 4–55 cm, decumbent. Leaves 1–7·5 × 0·5–2·5 cm, elliptical, veins obscure. Inflorescence 1- to 7(–12)-flowered, leafy; flower-buds nodding just before anthesis. Style 4–7·5 mm, deflexed. Seeds 1·5–2·1 mm, smooth. 2*n* = 72. *Iceland, N.E. Russia.* Is Rs (N). (*N. Asia, North America.*)

3. **E. dodonaei** Vill., *Prosp. Pl. Dauph.* 45 (1779) (*Chamaenerion angustissimum* (Weber) D. Sosn., *C. dodonaei* (Vill.) Schur, *C. palustre* auct. mult., non (L.) Scop., *E. rosmarinifolium* Haenke). Stems 20–110 cm, erect. Leaves 2–2·5 × 0·1–0·35 cm, linear, strigulose, veins obscure. Inflorescence lax; flower-buds slightly nodding just before anthesis. Style 7–15 mm, at first deflexed but later erect after anthers have dehisced. Seeds 1·5–2 mm, papillose. 2*n* = 36. *C. & S. Europe, from C. France to W. Ukraine.* Al Au Bu Cz Ga Ge Gr He Hu It Ju Po Rm Rs (W) Si.

4. **E. fleischeri** Hochst., *Flora (Regensb.)* 9: 85 (1826) (*Chamaenerion rosmarinifolium* sensu Coste pro parte, non (Haenke) Moench). Stems 8–45 cm, decumbent, often many. Leaves 1·5–4 × 0·1–0·6 cm, narrowly lanceolate, glabrous, veins obscure. Inflorescence subcorymbose; flower-buds slightly nodding just before anthesis. Style 3·5–5 mm, deflexed. Seeds 1·2–1·7 mm, papillose. 2*n* = 36. ● *Alps; usually at higher altitudes than 3.* Au Ga Ge He It.

Sect. EPILOBIUM (Sect. *Lysimachion* Tausch). Roots rather slender. Leaves opposite or verticillate, at least below. Flowers actinomorphic or nearly so; petals emarginate, white, pink or purplish. Stigma 4-lobed (spp. **5–10**), or entire. Pollen grains shed in tetrads.

Although self-pollination is normal in most European species of this section, occasional hybrids occur between many of the species, and may best be recognized by their intermediate morphology and (usually) by their high degree of sterility. Most of the species occur on moist or disturbed ground.

5. **E. hirsutum** L., *Sp. Pl.* 347 (1753). Stolons hypogeal, fleshy; stems up to 200 cm, robust, usually villous or tomentose. Leaves 2–12 × 0·5–3·5 cm, oblong to lanceolate, sessile, semi-amplexicaul, sharply serrulate. Petals (6–)10–16 mm, bright purplish-pink. Seeds 1–1·5 mm, obovoid. 2*n* = 36. *Europe except the extreme north.* All except Az Fa Is Sb; only as a casual in Fe No Rs (N).

6. **E. parviflorum** Schreber, *Spicil. Fl. Lips.* 146, 155 (1771). Overwintering by rosettes; stems up to 75 cm, robust, usually villous. Leaves 2·5–10 × 0·7–3 cm, oblong- to linear-lanceolate, subsessile, weakly serrulate. Petals 4–9 mm, purplish-pink. Seeds *c.* 1 mm, obovoid. 2*n* = 36. *Europe except the extreme north.* All except Fa Is Rs (N) Sb.

7. **E. duriaei** Gay ex Godron in Gren. & Godron, *Fl. Fr.* 1: 581 (1849). Stolons hypogeal, long, fleshy; stems 10–40 cm, strigulose. Leaves 1·5–3·5 × 1–2 cm, ovate-acuminate, subsessile, repand with prominent teeth. Petals 6·5–10 mm, pink. Seeds

1·7–2 mm, attenuate with a prominent beak. *Calcifuge.*
● *Mountains of W. Europe from N.W. Spain (Asturias) to the Vosges.* Ga He Hs.

8. E. montanum L., *Sp. Pl.* 348 (1753) (*E. hypericifolium* Tausch). Turions produced in autumn; stems 10–80 cm, strigulose. Leaves 3·5–8 × 1–4 cm, ovate to narrowly ovate, shortly petiolate, repand or with prominent teeth, rarely entire. Buds acute. Petals 6–10 mm, purplish-pink. Seeds *c.* 1 mm, obovoid. 2*n* = 36. *Almost throughout Europe.* All except Az Bl Cr Is Sb Tu.

9. E. collinum C. C. Gmelin, *Fl. Bad.* **4**: 265 (1826). Stolons hypogeal, short, or turions produced; stems 10–40 cm, strigulose. Leaves 1–5 × 0·5–1·5 cm, ovate or narrowly ovate, shortly petiolate, sharply repand-serrulate. Buds obtuse. Petals 3–6 mm; pale purplish-pink. Seeds *c.* 1 mm, obovoid. 2*n* = 36. ● *Most of continental Europe.* Al Au Be Bu Co Cz Fe Ga Ge Gr He Hs Hu Is It Ju No Po Rm Rs (N, B, C, W, E) Sa Su.

10. E. lanceolatum Sebastiani & Mauri, *Fl. Rom.* 138 (1818). Overwintering by rosettes; stems 10–60 cm, strigulose. Leaves 3–12 × 1–3·5 cm, oblong, obtuse, narrowly cuneate at base, serrulate; petiole 3–10 mm. Petals 5–8·5 mm, white, becoming purplish-pink. Seeds *c.* 1 mm, obovoid. 2*n* = 36. *W., C. & S. Europe.* Al Au Be Br Bu Co Cz Ga Ge Gr He Ho Hs Hu It Ju Lu Rm Rs (K) Sa Si Tu.

11. E. alpestre (Jacq.) Krocker, *Fl. Siles.* **1**: 605 (1787) (*E. trigonum* Schrank). Overwintering by turions; stems 20–70 cm, robust, with elevated hirsute lines. Leaves 2·5–8 × 1–2·5 cm, usually verticillate, broadly lanceolate-acuminate, sharply toothed. Petals 5·5–12·5 mm, pinkish-violet. 2*n* = 36. *Mountains of C. & S. Europe, from the Pyrenees to the Carpathians and Bulgaria.* Al Au Bu Cz Ga Ge He Hs It Ju Po Rm Rs (W).

12. E. tetragonum L., *Sp. Pl.* 348 (1753). Overwintering by rosettes; stems 15–110 cm, with elevated strigulose lines. Leaves 2–8 × 0·5–2 cm, oblong or oblong-lanceolate, serrulate. Inflorescence greyish-pubescent. Petals purplish-pink. Capsule 5–8 cm. Seeds 1–1·5 mm, obovoid. *Europe except the extreme north.* All except Az Fa Hb Is Rs (N) Sb; only as casual in No.

1 Petals 7–11·5 mm; stigma usually elevated above the anthers at anthesis **(c) subsp. tournefortii**
1 Petals 2·5–7 mm; stigma surrounded by anthers at anthesis
 2 Leaves oblong, ± decurrent **(a) subsp. tetragonum**
 2 Leaves mostly oblong-lanceolate, shortly petiolate, not decurrent **(b) subsp. lamyi**

(a) Subsp. **tetragonum** (*E. adnatum* Griseb.): 2*n* = 36. *Throughout the range of the species.* Normally self-pollinated.

(b) Subsp. **lamyi** (F. W. Schultz) Nyman, *Consp.* 247 (1879) (*E. lamyi* F. W. Schultz): 2*n* = 36. *Throughout a large part of the range of the species, but only in S.W. part of U.S.S.R. and of doubtful status in Finland.* Normally self-pollinated.

(c) Subsp. **tournefortii** (Michalet) Léveillé, *Monde Pl.* **6**: 22 (1896): *Mediterranean region, apparently local.* Normally cross-pollinated; intergrades with the other subspecies.

13. E. obscurum Schreber, *Spicil. Fl. Lips.* 147, 155 (1771). Stolons epigeal, leafy; stems 20–90 cm, with elevated strigulose lines. Leaves 1·5–8 × 0·5–1·5 cm, narrowly ovate or lanceolate, serrulate. Inflorescence greyish-pubescent, with a few glandular hairs on calyx. Petals 4–7 mm, purplish-pink. Capsule 4–6 cm. Seeds *c.* 1 mm, obovoid. 2*n* = 36. *Wet places. Europe except the extreme north and much of the U.S.S.R.* All except Bl Cr Fa Is Rs (K, E) Sb Si Tu.

14. E. roseum Schreber, *Spicil. Fl. Lips.* 147, 155 (1771). Overwintering by turions; stems 15–80 cm, with elevated strigulose lines. Leaves 3–10 × 1–3·5 cm, lanceolate or elliptical, narrowly cuneate at base, serrulate. Inflorescence canescent and densely glandular-pubescent. Petals 3·5–7 mm, white, becoming pinkish-streaked. Capsule 4–6 cm. Seeds *c.* 1 mm, obovoid. *Most of Europe except the extreme north.* All except Az Bl Co Cr Fa Is Rs (N) Sa Sb Si.

(a) Subsp. **roseum**: Petioles 4–15 mm. 2*n* = 36. *Throughout the range of the species.*
(b) Subsp. **subsessile** (Boiss.) P. H. Raven, *Notes Roy. Bot. Gard. Edinb.* **24**: 194 (1962) (*E. nervosum* Boiss. & Buhse): Petioles 1–3 mm. *S.E. Europe.*

15. E. palustre L., *Sp. Pl.* 348 (1753). Stolons epigeal, filiform; stems 5–70 cm, strigulose above. Leaves 1·5–6 × 0·3–1·5 cm, lanceolate, acuminate, sessile, entire. Inflorescence strigose, rarely glandular-pubescent. Petals 3–7 mm, pale pink or white. Seeds 1·5–2 mm, fusiform, with a conspicuous pellucid appendage. 2*n* = 36. *Wet places. Throughout Europe except for parts of the Mediterranean region.* All except Az Bl ?Co Cr Sa Sb Si Tu.

16. E. davuricum Fischer ex Hornem., *Hort. Hafn.*, *Suppl.* 44 (1819). Overwintering by rosettes; stems 5–30 cm, glandular-pubescent and strigulose above, with slightly elevated lines. Leaves 0·6–5 × 0·1–0·4 cm, linear to elliptical, sessile, entire, remote. Inflorescence subglabrous, rarely greyish-pubescent, nodding until capsules are ripe. Petals 3–5 mm, white, rarely pink. Seeds 1·2–1·7 mm, fusiform, finely papillose, with a conspicuous pellucid appendage. *Fennoscandia and N. Russia, southwards to 60° N. in Norway.* Fe No Rs (N) Su.

(a) Subsp. **davuricum**: Plant not caespitose; leaves linear. 2*n* = 36. *Throughout the range of the species.*
(b) Subsp. **arcticum** (Sam.) P. H. Raven, *Feddes Repert.* **79**: 61 (1968) (*E. arcticum* Sam.): Plant more or less caespitose; leaves rounded; seeds perhaps more finely papillose than in subsp. (a). *Arctic Russia (Vajgač).*

17. E. nutans F. W. Schmidt, *Fl. Boëm.* **4**: 82 (1794). Stolons epigeal, leafy; stems 5–30 cm, strigulose above, with elevated lines. Leaves 1–3 × 0·2–0·8 cm, ovate or elliptical, subentire, rarely as long as internodes. Inflorescence strigulose, nodding before anthesis. Petals 3–6 mm, pale violet; stigma clavate, surrounded by anthers at anthesis. Seeds *c.* 1·5 mm, fusiform, very finely papillose, with a pellucid appendage. ● *Mountains of C. & S. Europe, from the Pyrenees to the Carpathians and Bulgaria.* Au Bu Cz Ga Ge He Hs It Po Rm Rs (W).

18. E. tundrarum Sam., *Bot. Not.* **1922**: 264 (1922). Like 17 but stolons less leafy; stems 4–8 cm, with fine, not elevated, decurrent lines; leaves often longer than internodes; inflorescence with a few fine, appressed hairs. Seeds unknown. *Arctic Russia (Vajgač and adjacent mainland).* Rs (N). (*Arctic Asia.*)

19. E. atlanticum Litard. & Maire, *Mém. Soc. Sci. Nat. Maroc* **26**: 15 (1930) (*E. samuelssonii* P. H. Raven). Like 17 but petals 5·5–7 mm, bright pinkish-purple; stigma capitate, held well above anthers at anthesis. *S. Spain (Sierra Nevada).* Hs. (*N.W. Africa.*)

20. E. anagallidifolium Lam., *Encycl. Méth. Bot.* **2**: 376 (1786) (*E. alpinum* auct. mult., non L.). Stolons epigeal, leafy; stems 2–10 cm, with weakly elevated strigulose lines. Leaves 1–2·5 × 0·5–0·7 cm, ovate or elliptical, shortly petiolate, subentire. Inflorescence glabrescent, nodding. Petals 3–4·5 mm, pale purplish. Seeds *c.* 1·5 mm, fusiform, smooth, with a pellucid appendage. 2*n*=36. *N. Europe, extending southwards in the mountains to Corse and Jugoslavia.* Al Au Br Bu Co Cz Fa Fe Ga Ge Gr He Hs Is It Ju Lu No Po Rm Rs (N, C, W) Sb Su.

21. E. hornemannii Reichenb., *Pl. Crit.* **2**: 73 (1824). Stolons hypogeal, short; stems 6–25 cm, with slightly elevated pubescent lines. Leaves 2–4 × 1–1·5 cm, ovate-elliptical, weakly serrulate; petiole 2–8 mm. Inflorescence glabrescent. Petals 4–6 mm, pale violet. Seeds *c.* 1 mm, fusiform, finely papillose, with a pellucid appendage. 2*n*=36. *Wet places. Arctic and subarctic Europe.* Fe Is No Rs (N) Su.

E. uralense Rupr. in E. Hofmann, *Nördl. Ural* 2 (*Fl. Bor.-Ural.*): 33 (1856) described from N. Ural, is doubtfully distinct. It is said to be more glaucous, with leafy stolons and seeds *c.* 1·2 mm.

22. E. lactiflorum Hausskn., *Österr. Bot. Zeitschr.* **29**: 89 (1879). Like **21** but with short, inconspicuous, leafy, epigeal stolons; leaves very weakly serrulate; petals 2·5–4 mm, white; seeds *c.* 1·2 mm, finely pitted. 2*n*=36. *Arctic and subarctic Europe.* Fa Fe Is No Rs (N) Su.

23. E. alsinifolium Vill., *Prosp. Pl. Dauph.* 45 (1779). Like **21** but stolons long, spreading; leaves shortly petiolate, serrulate; petals 7–11 mm, bright purplish-pink; seeds smooth. 2*n*=36. ● *In the mountains of the greater part of Europe, and at low altitudes in the Arctic.* Au Br Bu Cz Fa Fe Ga Ge Gr Hb He Hs Is It Ju No Po Rm Rs (N, W) Su.

24. E. gemmascens C. A. Meyer, *Verz. Pfl. Cauc.* 173 (1831). Overwintering by turions; stems 12–50 cm, densely glandular-pubescent, with elevated lines. Leaves 2–4 × 0·8–1·5 cm, weakly serrulate, with bulbils or short vegetative shoots in axils; petioles 2–4 mm. Inflorescence canescent. Petals 6–10·5 mm, purplish-pink; stigma capitate. Seeds *c.* 1 mm, attenuate, papillose. *Mountains of the Balkan peninsula.* Al ?Bu Gr Ju.

Plants referred to this species from France and Italy seem close to **14** but further work is needed to clarify their affinities.

25. E. glandulosum Lehm., *Pugillus* **2**: 14 (1830). Overwintering by turions; stems 40–80 cm, with elevated strigulose lines. Leaves 3–8 × 1–3 cm, subsessile, rounded at base, weakly serrulate. Inflorescence glandular-pubescent. Petals 5–6·5 mm, purplish-pink. Seeds *c.* 1 mm, attenuate, with a pellucid appendage. 2*n*=36. *Naturalized in N. Europe.* [Fe No ?Po Rs (N, B) Su.] (*N. North America.*)

26. E. adenocaulon Hausskn., *Österr. Bot. Zeitschr.* **29**: 119 (1879). Overwintering by leafy rosettes; stems 10–140 cm, with elevated strigulose lines. Leaves 3–10 × 1–3 cm, shortly petiolate, serrulate. Inflorescence with glandular, greyish pubescence. Petals 2·5–6 mm, purplish-pink or white. Seeds *c.* 1 mm, attenuate, with a pellucid appendage. 2*n*=36. *Naturalized as a weed and ruderal in a large part of Europe, mainly in the north-west.* [Be Br Cz Da Fe Ga Ge Ho No Po Rs (N, B, C, W) Su.] (*North America.*)

In Scandinavia white-flowered plants have been referred incorrectly to *E. rubescens* Rydb., which is a synonym of *E. saximontanum* Hausskn. Very similar plants, with longer petioles (3–)4–6 mm, from Denmark, Sweden and Finland, may be referable to *E. ciliatum* Rafin. (*E. americanum* Hausskn.), another North American species.

27. E. nerterioides A. Cunn., *Ann. Nat. Hist.* **3**: 32 (1839). Stems procumbent, matted, rooting at the nodes. Leaves 3·5–10 × 3–6 mm, mostly opposite, broadly ovate, weakly toothed, shortly petiolate. Pedicels 2·5–7·5 cm, erect. Petals 2·5–4 mm, white. Seeds *c.* 0·7 mm, papillose. 2*n*=36. *Cultivated in rock-gardens and now widely naturalized in damp, stony places in Britain and Ireland.* [Br Hb.] (*New Zealand.*)

E. inornatum Melville, *Kew Bull.* **14**: 298 (1960), with smooth seeds and more nearly elliptical leaves, is well naturalized in a number of parks and gardens in the Netherlands and Denmark and has been reported as a casual in Ireland. **E. pedunculare** A. Cunn., *Ann. Nat. Hist.* **3**: 31 (1839) (*E. linnaeoides* Hooker fil.), readily distinguishable by its acutely dentate leaves, is well established in two localities in W. Ireland. Both are natives of New Zealand with the same habit as **27** and like it are cultivated as rock-garden plants.

CXXIV. HALORAGACEAE[1]

Herbs with exstipulate leaves. Flowers hermaphrodite or unisexual. Sepals 0, 2 or 4, small. Petals 0, 2 or 4, often caducous. Stamens 2, 4 or 8, epipetalous; anthers basifixed, 2-locular. Ovary inferior, 1- to 4-locular, with 1 pendent, anatropous ovule in each loculus; styles 1–4, usually short; stigmas feathery or coarsely papillose. Fruit a drupe, or a schizocarp separating into 1-seeded nutlets.

Literature: A. K. Schindler in Engler, *Pflanzenreich* 23 (IV. 225): 77–127 (1905).

Terrestrial plants; leaves alternate, ovate to orbicular, palmately lobed; ovary 1-locular **1. Gunnera**
Submerged aquatic plants; leaves whorled, usually pinnate with capillary segments; ovary 4-locular **2. Myriophyllum**

1. Gunnera L.[2]

Stoloniferous or rhizomatous herbs. Leaves alternate, with erect petiole and ovate to orbicular, palmately-lobed lamina.

Sepals 2. Petals 0 or 2. Stamens 2. Ovary 1-locular; styles 2. Fruit a drupe.

1. G. tinctoria (Molina) Mirbel, *Hist. Nat. Pl.* ed. 2, **10**: 140 (1805) (*G. chilensis* Lam.). Gigantic herb with stout, horizontal rhizome. Petiole 20–150 cm, with conical spines; lamina up to 2 m, cordate-suborbicular, palmately 5- to 9-lobed; margin irregularly incise-serrate. Inflorescence a very dense-flowered and much-branched panicle up to 50 × 20 cm; flowers sessile, mostly hermaphrodite. Sepals minute. Petals 2, cucullate. *Cultivated for ornament, and locally naturalized in W. Europe.* [Az Br Ga Hb.] (*W. South America.*)

2. Myriophyllum L.[2]

Glabrous, rhizomatous, aquatic perennial herbs. Leaves (in European species) in whorls of 3–6, usually pinnatisect with capillary segments. Inflorescence an emergent spike; flowers mostly unisexual, male above, female below. Sepals 0 or 4, inconspicuous. Petals 0 or 4, caducous in male, inconspicuous or

[1] Edit. S. M. Walters.　　[2] By C. D. K. Cook.

311

absent in female flowers. Stamens 4 or 8. Ovary 4-locular; stigmas 4, sessile or subsessile. Fruit separating longitudinally into 4 1-seeded nutlets.

1 Upper bracts pinnatisect, with capillary segments
  2 Emergent leaves sparsely glandular; hermaphrodite flowers usually present; fruit smooth **1. verticillatum**
  2 Emergent leaves densely glandular; hermaphrodite flowers absent; fruit finely tuberculate **4. brasiliense**
1 Upper bracts simple, entire or serrate
  3 Stamens 4; bracts longer than flowers **5. heterophyllum**
  3 Stamens 8; bracts shorter than flowers
    4 Flowering spike usually more than 4 cm; all flowers whorled **2. spicatum**
    4 Flowering spike not more than 3 cm; upper flowers solitary or opposite **3. alterniflorum**

1. M. verticillatum L., *Sp. Pl.* 992 (1753). Stems up to 300 cm; perennation by clavate turions. Leaves 25–45 mm, (4–)5(–6) in a whorl, often longer than internodes; segments 24–35. Spike 7–25 cm. Flowers usually in whorls of 5; bracts pinnatisect, 1–15 times as long as the flowers. A few hermaphrodite flowers usually present between male and female flowers. Petals 2·5 mm in male flowers, absent in female. Stamens 8. Fruit *c.* 3 mm, subglobose, smooth. $2n = 28$. *Almost throughout Europe.* All except Az Co Cr Fa Sb.

In the terrestrial state this species may be as small as 3 cm, with leaves 1 cm and with as few as 4 leaf-segments.

2. M. spicatum L., *Sp. Pl.* 992 (1753). Like **1** but turions absent; leaves (3–)4(–5) in a whorl, about equalling internodes; leaf-segments 13–38; flowers in whorls of 4; bracts (except the lowest) entire and shorter than the flowers; female flowers with 4 small petals; fruit finely tuberculate. *Almost throughout Europe.* All except Az Fa Sb.

3. M. alterniflorum DC. in Lam. & DC., *Fl. Fr.* ed 3, **5**: 529 (1815). Stems up to 120 cm; turions absent. Leaves 1–2·5 cm, (3–)4 in a whorl, about equalling the internodes; segments 6–18. Spike up to 3 cm, with apex drooping in bud. Male flowers solitary or in opposite pairs, usually with rudiments of carpels. Female flowers whorled at base, in groups of 2–4 or solitary above. Bracts leaf-like, longer than the flowers at the base of the inflorescence, entire and shorter than the flowers in upper part. Hermaphrodite flowers rare. Petals 2·5 mm, yellow with red streaks. Stamens 8. Fruit 1·5–2 × *c.* 1·5 mm, subcylindrical, finely tuberculate. $2n = 14$. *Mainly in W., N. & C. Europe, but extending south-eastwards to Sicilia & W. Ukraine.* Az Be Br Co Cz Da Fa Fe Ga Ge Hb He Ho Hs Is It Lu No Po Rs (N, C, W) Sa Si Su.

4. M. brasiliense Camb. in St-Hil., *Fl. Bras. Mer.* **2**: 252 (1830). Stems up to 200 cm, often woody at base. Leaves 4–6 in a whorl, usually longer than internodes; emergent leaves light blue-green, covered by minute, transparent, hemispherical glands; segments 8–30. Flowers unisexual, solitary, axillary; bracts pinnatisect, resembling foliage leaves. Petals 5 mm in male flowers, absent in female. Stamens 8. Fruit 1·8 × 1·2 mm, ovoid, finely tuberculate. *Cultivated for ornament and locally naturalized in S.W. France; frequently casual elsewhere in Europe.* [Ga.] (*Tropical and subtropical South America.*)

Male flowers have not been reported from plants grown in Europe or North America.

5. M. heterophyllum Michx, *Fl. Bor. Amer.* **2**: 191 (1803). Stems up to 100 cm. Leaves 4–6 in a whorl, either pinnatisect, with 5–12 capillary segments, or lanceolate to linear and serrate. Spike 3–35 cm, with solitary flowers borne in the axils of lanceolate bracts. Flowers hermaphrodite, or occasionally female at base of inflorescence, male above. Petals 1·5–3 mm. Stamens 4. Fruit 1–1·5 mm, subglobose, each nutlet beaked and with 2 ridges on the outer face, finely tuberculate. *Naturalized in S.E. Austria.* [Au Br.] (*E. North America.*)

The pinnatisect leaves develop when the plant grows in low temperatures (15° C or less), and entire leaves develop when the plant grows at higher temperatures; intermediate leaves are few.

CXXV. THELIGONACEAE[1]

Flowers unisexual, in axillary 1- to 3-flowered clusters. Perianth present. Male flowers with 7–20 stamens. Female flowers with an inferior, unilocular ovary. Fruit a nut-like drupe. Ovule solitary, basal.

1. **Theligonum** L.[2]

(*Cynocrambe* Gagnebin)

Annual, glabrous and somewhat succulent herbs. Flowers green. Perianth of male flowers globose, splitting into 2–5 lobes at anthesis. Female flowers with a tubular, minutely toothed perianth.

1. T. cynocrambe L., *Sp. Pl.* 993 (1753) (*Cynocrambe prostrata* Gaertner). Foetid. Monoecious. Stems 5–30 cm, ascending, swollen at nodes. Leaves ovate, entire, petiolate, the lower opposite, the upper alternate; stipules membranous, sheathing. Male flowers with a 2-partite perianth and usually 7–12 stamens; anthers long, narrowly linear. Fruit *c.* 2 mm, ovoid, adnate to the base of the perianth. $2n = 20$. *Damp or shady rocks and walls. S. Europe.* Al Bl Bu Co Cr Ga Gr Hs It Ju Lu Rs (K) Sa Si Tu.

[1] Edit. T. G. Tutin.   [2] By T. G. Tutin.

CXXVI. HIPPURIDACEAE[1]

Glabrous, aquatic herbs. Leaves simple, exstipulate, whorled. Perianth reduced to a rim around the top of the ovary. Stamen 1, anterior, median. Ovary inferior, 1-locular; ovule solitary, pendent, anatropous; integument single; micropyle closed. Style 1, simple, elongate, with stigmatic papillae throughout its length. Fruit a small nut.

1. Hippuris L.[2]

Flowers male, female, hermaphrodite or sterile, borne in leaf-axils, the upper sessile, the lower usually shortly pedicellate.

Literature: M. E. McCully & H. M. Dale, *Canad. Jour. Bot.* **39**: 611–625, 1099–1116 (1961).

1. **H. vulgaris** L., *Sp. Pl.* 4 (1753). Aquatic perennial with creeping rhizome from which arise erect leafy shoots (4–)30–60(–150) cm. Leaves (0·5–)5–8(–10) cm, obovate to linear-lanceolate, (2–)6–12(–18) in a whorl. Nut 2–3 mm, ovoid, smooth. $2n = 32$. *Almost throughout Europe, but rare in the south-west and extreme south.* All except Az Bl ?Co Gr Sa Si Tu.

A very plastic species. The submerged parts usually have long, linear-lanceolate, flaccid, light green leaves, while the emergent parts have short, linear, rigid, dark green leaves.

H. tetraphylla L. fil., *Suppl.* 81 (1781), is often recognized as a distinct species with leaves very short, obovate and 4 in a whorl. These characters are to some extent under environmental control, and can be induced by conditions of high salinity or low temperature. The status of such variants is therefore doubtful.

UMBELLIFLORAE

CXXVII. CORNACEAE[1]

Trees or shrubs, rarely herbs. Leaves simple, exstipulate. Flowers usually hermaphrodite and 4-merous. Sepals small or almost absent; stamens usually 4, alternating with the petals. Ovary inferior, (1–)2(–4)-locular; ovules pendent, anatropous, solitary in each loculus. Fruit a drupe, rarely a berry.

Literature: W. Wangerin in Engler, *Pflanzenreich* **41** (IV. 229): 1–110 (1910).

Aucuba japonica Thunb., *Nova Gen. Pl.* **3**: 62 (1783), from E. Asia, a dioecious, evergreen shrub, with dark purple flowers and red berries, is often planted for ornament and may be locally naturalized.

1. Cornus L.[3]

(Incl. *Chamaepericlymenum* Hill, *Swida* Opiz, *Thelycrania* (Dumort.) Fourr.)

Deciduous trees, shrubs or rhizomatous herbs. Leaves usually opposite, entire. Petals valvate. Ovary 2-locular. Fruit a drupe with a single, 2-locular pyrene.

Literature: A. I. Pojarkova, *Not. Syst.* (*Leningrad*) **12**: 164–180 (1950).

1 Herb with annual flowering stems; inflorescence an umbel with large, white bracts exceeding the flowers   **5. suecica**
1 Shrubs or small trees; inflorescence ebracteate, or with yellowish-green bracts which do not exceed the flowers
2 Inflorescence an axillary umbel with yellowish-green bracts, appearing before the leaves; petals yellow; fruit red   **4. mas**
2 Inflorescence a terminal, ebracteate corymb, appearing after the leaves; petals dull white or pale yellowish-white; fruit white or purplish-black
3 Petals 4–7 mm; fruit purplish-black; largest leaves with 3–5 pairs of veins   **1. sanguinea**
3 Petals 3–4 mm; fruit white; largest leaves with 5–7 pairs of veins

4 Pyrene ellipsoid, cuneate at base; plant without or with few stolons   **2. alba**
4 Pyrene suborbicular, rounded at base; plant stoloniferous   **3. sericea**

Subgen. **Kraniopsis** Rafin. Inflorescence a terminal, ebracteate, cymose corymb. Fruit more or less globose.

1. **C. sanguinea** L., *Sp. Pl.* 117 (1753) (*Thelycrania sanguinea* (L.) Fourr.). Shrub up to 4 m, with dark red twigs. Leaves (3–)4–10 cm, broadly elliptical or ovate, acuminate, pale green and pubescent; veins 3–5 pairs. Inflorescence 4–5 cm in diameter. Petals 4–7 mm, dull white. Fruit 5–8 mm, purplish-black. $2n = 22$. *Most of Europe except the north-east and extreme north.* All except Az Bl Cr Fa Is Rs (N) Sb; naturalized in Fe.

(a) Subsp. **sanguinea**: Leaves with simple, crispate hairs beneath. *Almost throughout the range of the species.*
(b) Subsp. **australis** (C. A. Meyer) Jáv. in Soó & Jáv., *Magyar Növ. Kéz.* **1**: 398 (1951) (*C. australis* C. A. Meyer, *Thelycrania australis* (C. A. Meyer) Sanadze): Leaves with medifixed hairs beneath. *S.E. Europe, extending northwards to E. Czechoslovakia.*

2. **C. alba** L., *Mantissa* 40 (1767) (*Thelycrania alba* (L.) Pojark.). Shrub up to 3 m, with dark red twigs; usually not stoloniferous. Leaves 4–8 cm, broadly ovate to elliptical, acute, glaucous beneath; veins 5–7 pairs. Inflorescence 3–5 cm in diameter. Petals 3–4 mm, yellowish-white. Fruit *c.* 8 mm, white; pyrene ellipsoid, cuneate at base. *N. Russia, southwards to 56° N; cultivated elsewhere for ornament and locally naturalized.* Rs (N, C) [Br No Su]. (*N. Asia.*)

3. **C. sericea** L., *Mantissa Alt.* 199 (1771) (*C. stolonifera* Michx, *Thelycrania stolonifera* (Michx) Pojark.). Like **2** but with numerous stolons; leaves up to 14 cm, acuminate; pyrene sub-globose, rounded at base. *Cultivated for ornament and locally naturalized.* [Au Br Fe He.] (*North America.*)

---

[1] Edit. S. M. Walters.   [2] By C. D. K. Cook.
[3] By P. W. Ball.

Subgen. **Cornus**. Inflorescence an axillary umbel with yellowish-green bracts. Fruit more or less cylindrical or ellipsoid.

4. **C. mas** L., *Sp. Pl.* 117 (1753). Shrub or small tree up to 8 m, with greenish-yellow twigs. Leaves 4–10 cm, ovate or elliptical, acute or acuminate, dull green beneath; veins 3–5 pairs. Bracts 6–10 × 3–6 mm, deciduous; pedicel about equalling bracts. Petals 2–2·5 mm, yellow. Fruit 12–15 mm, red. *C. & S.E. Europe, extending to C. Italy and C. France; cultivated for ornament and for the edible fruit.* Al Au Be Bu Cz Ga Ge Gr He Hu It Ju Rm Rs (W, K) Tu [Br].

Subgen. **Arctocrania** (Endl.) Reichenb. Inflorescence a terminal umbel with large, whitish bracts. Fruit globose.

5. **C. suecica** L., *Sp. Pl.* 118 (1753) (*Chamaepericlymenum suecicum* (L.) Ascherson & Graebner). Rhizomatous herb with erect, annual flowering stems up to 25 cm. Leaves 1–4 cm, sessile, suborbicular or ovate to elliptical, subacute. Bracts 8–16 × 5–10 mm; pedicels much shorter than bracts. Petals 1–1·5 mm, dark purple. Fruit *c.* 5 mm, red. $2n=22$. Calcifuge. *N. Europe, extending southwards to c. 53° N. in the Netherlands.* Br Da Fa Fe Ge Ho Is No Po Rs (N, B, C) Su.

CXXVIII. ARALIACEAE[1]

Shrubs or woody climbers; leaves alternate. Flowers actinomorphic, small. Calyx small or rudimentary; petals 5, free; stamens 5; ovary inferior. Fruit a berry.

1. Hedera L.[2]

Stems woody, climbing or creeping, with numerous adventitious roots. Leaves simple, exstipulate, evergreen. Indumentum of stellate or peltate hairs. Flowers hermaphrodite, in globose umbels, which may be solitary or grouped in a racemose panicle. Sepals very small, deltate. Ovary 5-celled, surmounted by a conspicuous, domed disc, terminating in a single, short style. Berry globose, with 2–5 rugose, whitish seeds.

There is little agreement among authors on the taxonomic treatment of this genus, the number of European species recognized varying in different accounts from 1 to 6.

Literature: F. Tobler, *Die Gattung* Hedera. Jena. 1912. L. Lammermayr, *Pflanzenareale* 2(7): Karte 65–68 (1930). G. H. M. Lawrence & A. E. Schulze, *Gentes Herb.* 6: 107–173 (1942).

Leaves on non-flowering shoots usually less than 15 cm, distinctly lobed; hairs on young shoots and inflorescence 0·15–0·4 mm in diameter, mostly with 6–16 rays **1. helix**

Leaves on non-flowering shoots often 15–25 cm, scarcely lobed; hairs on young shoots and inflorescence 0·5–0·75 mm in diameter, with 15–25 rays **2. colchica**

1. **H. helix** L., *Sp. Pl.* 202 (1753). Stem up to 30 m, creeping or climbing. Young shoots and inflorescence densely covered with stellate to peltate hairs 0·15–0·4 mm in diameter, with (4–)6–16(–22) rays. Leaves shining, dark green, dimorphic; those of the flowering shoots (often absent in shady or cold situations) 6–10 × 2–12 cm, narrowly elliptical to suborbicular-cordate, entire; those of the non-flowering shoots up to 15 × 15 cm but usually much less, palmately 3- or 5-lobed. Petals 3–5 mm, yellowish-green, patent, later deflexed. Berry with 2–3 seeds. *Climbing on trees, rocks and walls, or covering the ground in woods. W., C. & S. Europe, northwards to 60° 30′ in Norway, and extending eastwards to Latvia and Ukraine.* All except Fa Fe Is Sb Rs (N, E).

Widely cultivated in gardens, where a large number of cultivars differing mainly in size, shape and colour of leaves, are known. Cuttings taken from flowering shoots form erect shrubs with leaves of one type only; these are often cultivated. The wild

populations are also very variable, but beyond the recognition of three subspecies it is difficult to give satisfactory taxonomic expression to this variation.

1 Ripe fruit yellow **(b) subsp. poetarum**
1 Ripe fruit black
　2 Hairs stellate, mostly with 6–10 squarrose rays, united only at the extreme base **(a) subsp. helix**
　2 Hairs stellate-peltate, mostly with 13–16 horizontal rays, united for about ¼ of their length **(c) subsp. canariensis**

(a) Subsp. **helix** (incl. *H. taurica* sensu Pojark., vix Carrière): Leaves dark green, often with paler veins, mostly longer than wide, those of the non-flowering shoots usually conspicuously 5-lobed, those of the flowering shoots elliptical, rhombic or ovate, often undulate. Hairs greyish-white, less often yellowish, stellate, with (4–)6–10(–12) squarrose rays which are united only at the extreme base. Ripe fruit black, 8–10 mm in diameter. $2n=48$. *Throughout the range of the species, but rarer in the extreme south-west and south-east.*

(b) Subsp. **poetarum** Nyman, *Consp.* 319 (1879): Like (a) but leaves rather lighter green, and those of the non-flowering shoots less deeply lobed; ripe fruit yellow, up to 12 mm in diameter. *Greece and Turkey; naturalized in France and Italy.* (*N. Africa, S.W. Asia.*)

(c) Subsp. **canariensis** (Willd.) Coutinho, *Fl. Port.* 428 (1913): Leaves mostly wider than long, those of the non-flowering shoots reniform, rather obscurely 3-lobed, those of the flowering shoots suborbicular-cordate. Hairs usually yellowish-brown, stellate-peltate, with 12–16(–22) rays all in one plane, united at the base for about ¼ of their length. Fruit as in (a). *Açores, Portugal.* (*N.W. Africa, Atlantic Islands.*)

Usually distinct in Europe, but apparently intergrading with (a) and (b) in N. Africa.

H. **hibernica** hort., with very large, deeply cordate leaves on the non-flowering shoots, is somewhat intermediate between (a) and (c). It seems to have originated as a wild plant in S.W. Ireland, but has not been re-discovered. Garden material has $2n=96$.

2. **H. colchica** (C. Koch) C. Koch, *Wochenschr. Gärtn. Pflanzenk.* 2: 74 (1859). Like **1** but leaves very large (up to 25 cm), those of the non-flowering shoots ovate-deltate, only very slightly lobed; hairs yellowish-brown, peltate, with 15–25 horizontal rays united for ⅓–½ of their length. Fruit black. *Cultivated in gardens, especially in S. & W. Europe, and locally naturalized.* [Ga Rs (K).] (*Caucasus, N. Anatolia.*)

---

[1] Edit. D. A. Webb.　　　[2] By D. A. Webb.

CXXIX. UMBELLIFERAE[1]

Herbs, rarely shrubs. Leaves alternate; lamina usually large and much-divided; petiole often inflated and sheathing at base. Stipules absent, except in Subfam. *Hydrocotyloideae*. Inflorescence usually a compound umbel. Flowers epigynous, small, hermaphrodite or unisexual, the plant rarely dioecious. Sepals usually small or absent; petals 5, usually more or less 3-lobed, the middle lobe inflexed; outer petals sometimes much larger than inner (*radiate*); stamens 5; carpels (1–)2, usually attached to a central axis (*carpophore*), from which the mericarps separate at maturity; styles (1–)2, often with a thickened base (*stylopodium*); ovule 1 in each loculus, pendent. Fruit dry; pericarp membranous or exocarp variously indurated; endocarp rarely woody (Subfam. *Hydrocotyloideae*). Mericarps usually joined by a narrow or wide commissure; each mericarp more or less compressed laterally or dorsally, with 5 longitudinal veins, usually with ridges over them, separated by *valleculae* or sometimes with 4 secondary ridges alternating with the primary; resin canals (*vittae*) usually present between the primary ridges and on the commissural face.

In the following account the primary divisions of the leaves are referred to as *segments* and the ultimate divisions, cut nearly or quite to the midrib, as *lobes*. The lobes may themselves sometimes be deeply lobed. The leaves are never truly pinnate, but are described as pinnate, for brevity, when the lamina is divided to the midrib. Descriptions of umbels refer to the terminal, or other well-developed umbel: lateral umbels are often smaller, with fewer rays, and may be entirely male. *Bracts* are the structures which subtend the primary branches (*rays*) of a compound umbel, and *bracteoles* are those which subtend the partial umbel, or the whole of a simple umbel. When the stylopodium is described, the description refers to the stylopodium of a hermaphrodite flower. Descriptions of the ridges of the fruit refer to the primary ridges, unless otherwise specified.

Ripe fruit is essential for the certain identification of some genera, though with a little experience the characters of the ripe fruit can often be deduced from a careful examination of unripe fruit or even the ovary. Some genera which have the ripe fruit strongly compressed and winged (e.g. *Peucedanum*) do not show these characters when young.

It has recently been shown (M. T. Cerceau-Larrival, *Mém. Mus. Nat. Hist. Nat. (Paris)* ser. B, **14**: 1–164 (1962)) that the seedlings and pollen provide valuable generic characters. As information is available so far for little more than half the European genera it has not been practicable to utilize it in the following account.

Literature: H. Wolff in Engler, *Pflanzenreich* **43 (IV. 228)** (1910); **61 (IV. 228)** (1913); **90 (IV. 228)** (1927).

1 Stellate hairs present, at least on under surface of leaves or on fruit; fruit without prickles
2 Leaves simple, lobed; stipules present; petals white **2. Bowlesia**
2 Leaves pinnate; stipules absent; petals yellow
3 Leaf-lobes ovate; fruit strongly compressed **89. Opopanax**
3 Leaf-lobes linear-filiform; fruit not compressed **36. Portenschlagiella**
1 Plant glabrous or with simple hairs, or fruit with prickles
4 Male flowers clustered round a female or hermaphrodite flower, their pedicels ±connate or adnate to the ovary; flowers not in capitula
5 Leaves 3- to 5-lobed **9. Petagnia**
5 Leaves 2- to 3-pinnate **10. Echinophora**
4 Male flowers not clustered round a female or hermaphrodite flower, their pedicels free; flowers sometimes in capitula

[1] Edit. T. G. Tutin.

6 Umbels with slender, flexuous peduncles, arranged in a leafless panicle; pedicels about as long as the rays, filiform **65. Lereschia**
6 Umbels not arranged in a leafless panicle; pedicels shorter than rays, or flowers in capitula or whorls
7 All leaves simple, entire or denticulate
8 Leaves septate, ±fistular; stems prostrate, rooting at the nodes **34. Lilaeopsis**
8 Leaves not septate and fistular; stem erect or 0
9 Flowers in compound umbels; petals yellow **56. Bupleurum**
9 Flowers in sessile or shortly pedunculate capitula; petals whitish **55. Hohenackeria**
7 Leaves deeply and repeatedly divided, or sometimes crenate or dentate
10 Flowers sessile or subsessile in capitula
11 Leaves spinose-dentate or pinnatifid, with spinescent lobes **7. Eryngium**
11 Leaves not spinescent
12 Stems branched; bracteoles small, inconspicuous **4. Sanicula**
12 Stems simple; bracteoles large, leaf-like **5. Hacquetia**
10 Flowers distinctly pedicellate, in umbels or rarely whorls
13 All leaves ±orbicular in outline, crenate or palmately lobed
14 Leaves entire, crenate, or serrate, not deeply lobed
15 All leaves long-petiolate; flowers in simple umbels or whorls **1. Hydrocotyle**
15 Cauline leaves sessile or subsessile; flowers in compound umbels
16 Flowers yellow **20. Smyrnium**
16 Flowers white or pink **92. Heracleum**
14 Leaves deeply lobed; lobes with dentate margins
17 Umbels simple; bracteoles conspicuous **6. Astrantia**
17 Umbels compound; bracteoles small
18 Partial umbels with numerous flowers; fruit strongly compressed **92. Heracleum**
18 Partial umbels with 1–6 flowers; fruit not or scarcely compressed **64. Cryptotaenia**
13 Most or all leaves pinnately or ternately divided, usually distinctly longer than wide, rarely the upper cauline or outer basal entire
19 Cauline leaves in a single whorl; stipules scarious, conspicuous **3. Naufraga**
19 Cauline leaves not in a single whorl; stipules 0
20 Sepals finely pinnatisect; partial umbels with 1 flower; ovary unilocular **8. Lagoecia**
20 Sepals entire or 0; partial umbels with more than 1 flower; ovary bilocular
21 Flowering stem with a flexuous, subterranean part; basal leaves with partly subterranean petioles
22 Stylopodium abruptly contracted into the styles **21. Bunium**
22 Stylopodium gradually narrowed into the styles
23 Fruit linear-oblong; ridges prominent **23. Huetia**
23 Fruit oblong-ovoid; ridges indistinct **22. Conopodium**
21 Flowering stem without flexuous, subterranean part; leaves arising at or above the ground
24 Beak of fruit at least as long as seed-bearing part **14. Scandix**
24 Beak of fruit 0, or much shorter than seed-bearing part
25 Lateral ridges of fruit with a distinct, but sometimes narrow, ±membranous wing, or thickened and rounded at the outer edge; dorsal ridges winged or not
26 Margin of fruit conspicuously thickened, at least at its outer edge
27 Petals yellow

28 Leaves 1-pinnate or 1-pinnatisect; segments of lower leaves ovate **93. Malabaila**
28 Leaves 2- to 3-pinnatisect; lobes of lower leaves linear-lanceolate to setaceous
  29 Bracts 3–5; fruit 5–6·5 mm **82. Palimbia**
  29 Bracts 0; fruit 3 mm **84. Johrenia**
27 Petals white or pinkish
  30 Bracts pinnatisect **110. Artedia**
  30 Bracts simple or 0
    31 Leaves 1-pinnate or simple **94. Tordylium**
    31 Leaves 2- to 3-ternate **95. Laser**
26 Margin of fruit with a wing which is thin, at least at its outer edge
32 Petals yellow
  33 Bracteoles connate, at least at base
     81. Levisticum
  33 Bracteoles free or 0
    34 Leaves ternate **90. Peucedanum**
    34 Leaves pinnate
     35 Leaves 1-pinnate (very rarely some 2-pinnate)
      36 Plant ± hispid, not pruinose; stock without fibres **91. Pastinaca**
      36 Plant nearly glabrous, pruinose; stock with abundant fibres **99. Thapsia**
     35 Leaves at least 2-pinnate
      37 Slender annual **41. Anethum**
      37 Stout perennial
       38 Bracts numerous
        39 Dorsal ridges of fruit with wings as wide as the lateral **52. Cachrys**
        39 Dorsal ridges of fruit not or very narrowly winged
         40 Leaf-lobes ovate-cuneate to ovate-oblong **90. Peucedanum**
         40 Leaf-lobes linear to linear-lanceolate **87. Ferulago**
       38 Bracts 0(–3)
        41 Lateral wings of fruit at least as wide as rest of mericarp, usually shiny
         42 Stock without fibres; dorsal ridges of fruit not prominent, never winged **80. Angelica**
         42 Stock with fibres; dorsal ridges of fruit prominent, often winged
          43 Fruit flat; secondary ridges not winged **99. Thapsia**
          43 Fruit convex on back; secondary dorsal-ridges ± winged **96. Elaeoselinum**
        41 Lateral wings of fruit narrow, not shiny
         44 Fruit pubescent on commissural face **88. Eriosynaphe**
         44 Fruit glabrous on commissural face
          45 Bracteoles 0, rarely 1–2 and caducous; fruit flat **86. Ferula**
          45 Bracteoles (2–)several, persistent; fruit convex on back **90. Peucedanum**
32 Petals white, pink or greenish
46 Dorsal ridges of fruit winged
  47 Bracts 5 or more, persistent
    48 Young fruit densely pubescent all over **97. Guillonea**
    48 Young fruit glabrous, or hispid on the primary ridges only
     49 Stem usually triquetrous; lower leaves simple or 1-pinnate **53. Heptaptera**
     49 Stem terete; lower leaves 2- to 5-pinnate or ternate
      50 Ridges of fruit all equally and narrowly winged
       51 Leaves pubescent on margin and veins beneath **49. Pleurospermum**

51 Leaves glabrous, though sometimes scabrid, on margins and veins beneath
  52 Basal and lower cauline leaves *c.* 100 cm; lateral umbels usually opposite or whorled **16. Molopospermum**
  52 Basal and lower cauline leaves 10–30 cm; lateral umbels absent or alternate
    53 Lateral umbels, if present, not overtopping the terminal umbel **77. Ligusticum**
    53 Lateral umbels overtopping the terminal umbel **50. Aulacospermum**
50 At least the lateral ridges of fruit with wide wings
  54 Dorsal wings of fruit at least ½ as wide as lateral **98. Laserpitium**
  54 Dorsal wings of fruit not more than ¼ as wide as lateral
    55 Plant somewhat hispid; bracts often 2- to 3-fid **100. Rouya**
    55 Plant quite glabrous; bracts always entire **96. Elaeoselinum**
47 Bracts 0, rarely up to 7 and caducous
  56 Wings of fruit all equal and narrow
    57 Primary divisions of leaves erecto-patent **77. Ligusticum**
    57 Primary divisions of leaves patent or deflexed **78. Cenolophium**
  56 Wings of fruit wide, the lateral conspicuously wider than the dorsal
    58 Leaf-lobes broadly ovate to suborbicular **98. Laserpitium**
    58 Leaf-lobes linear to oblong-lanceolate
     59 Rays and top of peduncle glabrous **79. Conioselinum**
     59 Rays or top of peduncle pubescent
      60 Leaf-lobes *c.* 10 mm **76. Selinum**
      60 Leaf-lobes *c.* 30 mm or more **80. Angelica**
46 Dorsal ridges of fruit unwinged
  61 Fruit not obviously compressed
    62 Annual; bracteoles deflexed in flower **35. Aethusa**
    62 Perennial or biennial; bracteoles not deflexed in flower
     63 Petals yellowish **43. Silaum**
     63 Petals white **32. Seseli**
  61 Fruit strongly compressed
    64 Fruit dark brown; wings and strong dorsal ridges sharply denticulate **101. Melanoselinum**
    64 Fruit light brown or greenish; wings and dorsal ridges not denticulate
     65 Fruit narrowly winged, often hairy; vittae conspicuous **92. Heracleum**
     65 Fruit broadly winged, glabrous; vittae inconspicuous
      66 Leaves 1-pinnate or 3-foliolate **90. Peucedanum**
      66 Leaves at least 2-pinnate or ternate
       67 Leaf-lobes less than 30 mm **90. Peucedanum**
       67 Most leaf-lobes 30 mm or more
        68 Leaf-lobes linear to lanceolate **90. Peucedanum**
        68 Leaf-lobes ovate to suborbicular
         69 Bracts 3–6, persistent **90. Peucedanum**
         69 Bracts 0(–3), caducous if present
          70 Rays puberulent on inner side **80. Angelica**
          70 Rays glabrous **90. Peucedanum**
25 Fruit unwinged; ridges without thickened or wing-like edge

71 Flowers yellow
  72 Fruit at least 3 times as long as wide
    73 Lobes of lower leaves broadly ovate
        42. Kundmannia
    73 Leaf-lobes linear to linear-oblong
      74 Most cauline leaves with lamina; fruit with wide, rounded ridges; plant without a ±globose subterranean stock
        12. Chaerophyllum
      74 Most cauline leaves without lamina; fruit with slender, scarcely prominent ridges; plant with a ±globose, subterranean stock
        24. Muretia
  72 Fruit less than 3 times as long as wide
    75 Bracts and bracteoles 0 or few
      76 Leaf-lobes filiform
        77 Perennial or biennial; rays stout, usually c. 20, very unequal; fruit scarcely compressed **40. Foeniculum**
        77 Annual; rays slender, usually c. 40, subequal; fruit strongly compressed **61. Ridolfia**
      76 Leaf-lobes ovate to suborbicular in outline
        78 Stems stout; petioles of cauline leaves strongly inflated **20. Smyrnium**
        78 Stems slender; petioles of cauline leaves not inflated **25. Pimpinella**
    75 Bracts or bracteoles numerous
      79 Leaf-lobes lanceolate to ovate
        80 Leaves pinnate, oblong-lanceolate in outline **83. Bonannia**
        80 Leaves ternate, triangular in outline
          81 Stem freely branched; fruit c. 3 mm **60. Petroselinum**
          81 Stem usually simple; fruit 6–7 mm **39. Xatardia**
      79 Leaf-lobes filiform to linear-obovate
        82 Fruit almost or quite smooth; ridges very wide, corky, ±confluent **52. Cachrys**
        82 Fruit with distinct ridges
          83 Bracts 0(–2) **32. Seseli**
          83 Bracts numerous
            84 Leaf-lobes fleshy, narrowly obovate **29. Crithmum**
            84 Leaf-lobes not fleshy, filiform or linear
              85 Fruit 7–25 mm **52. Cachrys**
              85 Fruit 3–5 mm **32. Seseli**
71 Flowers white, pink, greenish- or yellowish-white
  86 Fruit at least 3 times as long as wide
    87 Fruit without ridges, except in the usually well-developed beak **13. Anthriscus**
    87 Fruit with ridges; beak very short or 0
      88 Bracts 4–15; leaf-margin cartilaginous **70. Falcaria**
      88 Bracts 0(–5); leaf-margin not cartilaginous
        89 Fruit and ovary pubescent
          90 Rays 1–3; nodes swollen **11. Myrrhoides**
          90 Rays 5 or more; nodes not swollen **37. Athamanta**
        89 Fruit and ovary glabrous; rays 4–24
          91 Fruit with very prominent, sharp ridges **15. Myrrhis**
          91 Fruit with low, rounded ridges
            92 Bracteoles numerous; fruit at least 5 times as long as wide **12. Chaerophyllum**
            92 Bracteoles 0 or few; fruit 3–3½ times as long as wide **71. Carum**
  86 Fruit less than 3 times as long as wide
    93 Fruit glabrous, not prickly, though sometimes rugose, muricate or densely papillose
    94 Fruit rugose, muricate or densely papillose
      95 Fruit rugose or muricate

96 Fruit distinctly longer than wide
  97 Annual; rays 2–5 **85. Capnophyllum**
  97 Perennial; rays 6–10 **51. Lecokia**
96 Fruit not or scarcely longer than wide
  98 Stock without fibres; fruit without visible ridges **18. Bifora**
  98 Stock with fibres; fruit with prominent, convolute ridges **57. Trinia**
95 Fruit densely papillose
  99 Bracteoles 0; fruit not or scarcely longer than wide **73. Brachyapium**
  99 Bracteoles present; fruit distinctly longer than wide **32. Seseli**
94 Fruit smooth, except for the longitudinal ridges
  100 Fruit globose; mericarps not separating at maturity **17. Coriandrum**
  100 Fruit usually ovoid; mericarps readily separating at maturity
    101 Lowest leaves 1-pinnate or simple
      102 Stems creeping and often rooting at nodes
        103 Leaf-segments linear **69. Thorella**
        103 Leaf-segments ovate
          104 Bracts 0, or few and entire **59. Apium**
          104 Bracts numerous, large, pinnatisect **28. Berula**
      102 Stems ±erect, not rooting at nodes
        105 Upper leaves with filiform to narrowly oblong, ±parallel-sided lobes, or without lamina
          106 Upper leaves palmately divided into filiform lobes **25. Pimpinella**
          106 Upper leaves pinnately divided into linear to oblong lobes, or without lamina
            107 At least some bracts 3-fid or pinnatisect **66. Ammi**
            107 Bracts entire or 0
              108 Bracteoles 0, rarely 1–3 and caducous
                109 Upper petioles strongly inflated; rays subequal **25. Pimpinella**
                109 Upper petioles scarcely inflated; rays very unequal **71. Carum**
              108 Bracteoles several, persistent
                110 Bracteoles not more than ¼ as long as longest pedicels **62. Sison**
                110 Bracteoles more than ½ as long as longest pedicels
                  111 Stems stout, not branched at base; rays 0·5–1·5 mm in diameter; fruit usually longer than pedicel **33. Oenanthe**
                  111 Stems slender, branched at base; rays not more than 0·25 mm in diameter; fruit usually shorter than pedicel **67. Ptychotis**
        105 Upper leaves with lanceolate to ovate or obovate lobes with distinctly curved sides
          112 Bracts 0, rarely 1–3 and caducous
            113 Stem branched **25. Pimpinella**
            113 Stem simple **74. Endressia**
           112 Bracts numerous, persistent
            114 Stylopodium conical **28. Berula**
            114 Stylopodium nearly flat
              115 Lower leaves with 3–5 segments **48. Hladnikia**
              115 Lower leaves with at least 9 segments
                116 Leaf-segments serrate **27. Sium**
                116 Leaf-segments lobed **32. Seseli**

101 Lowest leaves at least 2-pinnate or 2-ternate
117 Finely divided submerged leaves present at flowering time
118 Rays 1–3   **59. Apium**
118 Rays 5 or more   **33. Oenanthe**
117 No finely divided submerged leaves present at flowering time
119 Larger bracts at least ½ as long as rays, often divided
120 Bracteoles usually 3; rays 1–5   **58. Cuminum**
120 Bracteoles numerous; rays (4–)10–150
121 Sepals conspicuous
122 Rays 4–10; fruit 4–6 mm   **31. Dethawia**
122 Rays (10–)20–60; fruit 2·5–4·5 mm   **32. Seseli**
121 Sepals 0 or very small
123 Fruit 5–10 mm   **49. Pleurospermum**
123 Fruit not more than 2·5 mm
124 Most leaf-lobes 10 mm or more   **66. Ammi**
124 Most leaf-lobes 2–3 mm   **72. Stefanoffia**
119 Bracts much less than ½ as long as rays, sometimes 0
125 Bracteoles strongly dimorphic, some spathulate and often inflated, some subulate   **68. Ammoides**
125 Bracteoles all similar in shape
126 Dioecious   **57. Trinia**
126 Most flowers hermaphrodite, or plant monoecious
127 Cauline leaves 0 or very small
128 Stems junciform, glaucous; rays 2–3   **30. Sclerochorton**
128 Stems not junciform and glaucous; rays more than 3
129 Bracts 0–2
130 Stock with abundant fibres   **32. Seseli**
130 Stock without fibres   **19. Scaligera**
129 Bracts several
131 Basal leaves ternate   **46. Physospermum**
131 Basal leaves pinnate   **71. Carum**
127 Cauline leaves well-developed
132 Roots usually tuberous; most pedicels shorter than fruit and often becoming thickened   **33. Oenanthe**
132 Roots not tuberous; most pedicels longer than fruit and not becoming thickened
133 Bracts 0(–3)
134 Basal leaves with ovate lobes
135 Lateral umbels opposite or whorled; bracteoles numerous   **44. Trochiscanthes**
135 Lateral umbels alternate; bracteoles 0(–2)
136 Rhizome far-creeping; lower leaves 2-ternate; umbels long-pedunculate   **26. Aegopodium**
136 Rhizome 0; lower leaves 2-pinnate; most umbels sessile or subsessile   **59. Apium**
134 Basal leaves with linear to linear-lanceolate lobes
137 Leaf-lobes 5–10 mm wide, incise-serrate   **63. Cicuta**

137 Leaf-lobes 1–2 mm wide, entire or pinnatisect
138 Fruit compressed laterally   **71. Carum**
138 Fruit subterete   **32. Seseli**
133 Bracts several
139 Leaf-lobes filiform
140 Leaves linear-oblong in outline   **71. Carum**
140 Leaves ovate to triangular in outline   **45. Meum**
139 Leaf-lobes linear-lanceolate to ovate
141 Ridges of fruit strongly undulate; stem purple-spotted   **47. Conium**
141 Ridges of fruit smooth; stem not purple-spotted
142 Stem distinctly grooved or angled
143 Leaf-lobes oblong or ovate, ±dentate   **32. Seseli**
143 Leaf-lobes linear-lanceolate, entire or deeply pinnatisect   **75. Cnidium**
142 Stem terete, smooth or faintly striate above
144 Leaf-lobes linear-lanceolate   **32. Seseli**
144 Leaf-lobes ovate
145 Branches numerous, mostly opposite or whorled   **71. Carum**
145 Branches few or 0, alternate   **38. Grafia**
93 Fruit pubescent, hispid, or with prickles
146 Outer mericarp of each fruit with straight prickles; inner mericarp tuberculate or with short, conical projections   **102. Torilis**
146 Both mericarps similar
147 Fruit with broad- or tubercle-based prickles arranged in 1–3 rows on the ridges
148 At least some bracts 3-fid or pinnatisect   **108. Daucus**
148 Bracts simple or 0
149 Rays and upper part of peduncle densely white-pubescent   **109. Pseudorlaya**
149 Rays and upper part of peduncle not densely white-pubescent
150 Bracts 2–5, conspicuous
151 Bracts at least ½ as long as the smooth rays   **107. Orlaya**
151 Bracts not more than ¼ as long as the setose rays   **106. Turgenia**
150 Bracts 0(–3), small and inconspicuous if present
152 Rays more than 6   **103. Astrodaucus**
152 Rays 2–5
153 Outer petals much longer than inner   **105. Caucalis**
153 Outer petals not or little longer than inner   **104. Turgeniopsis**
147 Fruit without broad- or tubercle-based prickles arranged in 1–3 rows on the ridges
154 Most umbels shortly pedunculate and leaf-opposed; fruit with a distinct, glabrous beak   **13. Anthriscus**
154 Umbels long-pedunculate, not leaf-opposed; fruit without a distinct, glabrous beak
155 Fruit densely covered with stiff, rough, minutely glochidiate bristles   **102. Torilis**

155 Fruit ±hairy with smooth, not glochidi-
　　ate hairs
156 Lower leaves simple or 1-pinnate
157 Bracts 0, or few and small **25. Pimpinella**
157 Bracts several, at least 20 mm
158 Lower leaves simple or shallowly
　　lobed **54. Magydaris**
158 Lower leaves pinnate **32. Seseli**
156 Lower leaves at least 2-pinnate
159 Bracteoles 0(–2)
160 Slender annual up to 15 cm
　　　　　　　　　　　73. Brachyapium
160 Perennial (10–)30–100 cm
　　　　　　　　　　　25. Pimpinella
159 Bracteoles 3 or more, sometimes con-
　　nate
161 Bracts as long as or longer than rays
162 Slender annual; rays 1–5 **58. Cuminum**
162 Stout perennial or biennial; rays
　　(10–)20–60 **32. Seseli**
161 Bracts much shorter than rays
163 Bracteoles connate, at least at base
　　　　　　　　　　　32. Seseli
163 Bracteoles free
164 Fruit distinctly narrowed below the
　　stylopodium **37. Athamanta**
164 Fruit ±truncate at apex **32. Seseli**

Subfam. **Hydrocotyloideae**

Leaves usually stipulate. Endocarp woody; vittae absent, at least in the ripe fruit.

1. **Hydrocotyle** L.[1]

Perennial. Stipules present. Inflorescence a simple umbel, or flowers in whorls. Fruit ovoid-ellipsoid to suborbicular, strongly compressed laterally; fruit-wall with a woody inner layer. Carpophore absent. Ridges prominent to obsolete; vittae conspicuous to obsolete.

All species occur in marshy ground or in shallow water.

1 Leaves with a deep basal sinus
2 Petioles 2–3 mm thick, very soft and ±fleshy **1. ranunculoides**
2 Petioles less than 1 mm thick
3 Leaves glabrous or rarely with scattered hairs; umbel 3- to 10-flowered; lateral ridges of fruit prominent
　　　　　　　　　　　4. sibthorpioides
3 Leaves usually hispid to pubescent; umbel 10- to 20-flowered; lateral ridges of fruit conspicuous but not prominent
　　　　　　　　　　　5. moschata
1 Leaves peltate
4 Inflorescence of whorled branches which bear whorls of flowers; fruit distinctly pedicellate **2. bonariensis**
4 Inflorescence simple (rarely branched); flowers in whorls; fruit nearly sessile **3. vulgaris**

1. H. ranunculoides L. fil., *Suppl.* 177 (1781) (*H. natans* Cyr.). Stems floating or creeping, rooting profusely at the nodes. Petioles up to 35 cm; leaves reniform to suborbicular with a deep basal sinus, crenate to lobed. Peduncles much shorter than the leaves. Inflorescence a simple umbel of 5–10 flowers. Fruit 2×3 mm, pedicellate, suborbicular; ridges not prominent. *Perhaps native in Italy, Sardegna and Sicilia.* *It *Sa *Si [Hs]. (*North America.*)

2. H. bonariensis Commerson ex Lam., *Encycl. Méth. Bot.* 3:153 (1789). Stems creeping, not rooting as profusely at the nodes as **1**. Petioles up to 35 cm; leaves peltate, orbicular-ovate, crenate. Inflorescenee of whorled branches which bear whorls of flowers,

equalling or exceeding the leaves. Fruit 1·5×3·5 mm, cordate at the base, pedicellate, with prominent, almost winged ridges. *Brackish dune-slacks. Naturalized in France, Italy, Spain and Portugal.* [Ga Hs It Lu.] (*Temperate South America.*)

3. H. vulgaris L., *Sp. Pl.* 234 (1753). Stems creeping and rooting at the nodes. Petioles up to 25 cm, sometimes with long patent hairs; leaves suborbicular, crenate, with 6–10 veins. Inflorescence ½ as long as subtending petiole; flowers in whorls. Fruit 2 mm wide, very shortly pedicellate, cordate at the base, with evident ridges. 2n=96. *Shallow fresh water and damp places. W., C. & S. Europe, extending to c. 60° N. in Norway and Sweden and to White Russia.* Al Au Az Be Br Co Cz Da Ga Ge Gr Hb He Ho Hs Hu It Ju Lu No Po Rs (B, C) Si Su.

There is some evidence that the closely related **H. verticillata** Thunb., *Diss. Hydrocot.* 2, 5 (1798), which is widespread from Australia, through tropical Africa to America, may occur in S. Europe. It may be distinguished by the leaves with 8–13 veins, glabrous petioles about equalling the inflorescence and fruit with cuneate base.

4. H. sibthorpioides Lam., *Encycl. Méth. Bot.* 3: 153 (1789). Delicate, slender, creeping herb. Leaves glabrous; basal sinus deep. Umbel 3- to 8(–10)-flowered; pedicels very short or absent. Fruit with prominent lateral ridges. *Naturalized near Milano and probably elsewhere.* [It.] (*Widespread in the tropics.*)

5. H. moschata G. Forster, *Fl. Ins. Austral. Prodr.* 22 (1786). Slender creeping herb. Leaves usually more or less densely hispid; basal sinus deep. Umbel 10- to 20-flowered; pedicels very short or absent. Fruit with conspicuous lateral ridges. *Naturalized in S.W. Ireland (Valentia Island).* [Hb.] (*New Zealand.*)

2. **Bowlesia** Ruiz & Pavón[2]

Leaves simple, lobed; stipules present, scarious. Sepals conspicuous, tridentate. Petals white, entire, subobtuse. Fruit ovoid, stellate-pubescent; ridges obsolete; vittae absent.

1. B. incana Ruiz & Pavón, *Fl. Peruv.* 3: 28 (1802). Stellate-pubescent annual. Stems 10–50 cm, slender, procumbent, branched. Leaves 5–30 mm, reniform to cordate in outline, 5- to 7-lobed. Umbels simple, with 1–6 flowers. Bracteoles small. Fruit 1–1·5 mm, subsessile. *Naturalized in S. France (Hérault).* [Ga.] (*Semi-arid parts of temperate South America and southern North America.*)

3. **Naufraga** Constance & Cannon[2]

Leaves 3-foliolate or pinnate, stipulate. Sepals absent. Petals white, ovate, flat or slightly incurved. Carpophore absent. Fruit truncate-globose, didymous. Ridges filiform, inconspicuous; vittae solitary, the commissural absent.

1. N. balearica Constance & Cannon, *Feddes Repert.* **74**: 3 (1967). Glabrous perennial 2·5–4 cm. Basal leaves 5–10 mm, crowded; segments 3–5(–7), 1·5–5×1–3 mm, ovate, entire or the terminal one with 1–2 lobes; petioles with a sheathing, auricled, scarious base. Cauline leaves usually 3-foliolate, in a single whorl; stipules large, whitish, scarious. Umbels simple, 2–4 in a whorl. Bracts and bracteoles absent. Fruit c. 0·8 mm, the truncate apex with a conspicuously scarious-margined disk; stylopodium absent; carpels free from one another, except near the apex, so that they hang down on either side of the pedicel and are attached to it only at the top. *Damp, shady crevices of calcareous sea cliffs.* ● *Mallorca (near Pollensa).* Bl.

---

¹ By J. F. M. Cannon.　　　　² By T. G. Tutin.

Subfam. **Saniculoideae**

Leaves always exstipulate. Endocarp soft; vittae usually present in the ripe fruit. Style surrounded by the annular or pelviform stylopodium. Fruit usually scaly.

4. **Sanicula** L.[1]

Stock short. Leaves palmately 3- to 7-partite. Inflorescence of a number of umbels in a cyme, often forming a false compound umbel. Bracts small. Sepals conspicuous. Fruit covered with hooked bristles. Vittae conspicuous or not.

1 Leaves crenate-serrate, the teeth ending in a short seta; bracts linear, entire or nearly so; fruit 4–5 mm **1. europaea**
1 Leaves sharply toothed, the teeth ending in a long seta; bracts lanceolate, toothed like the leaves; fruit 3 mm **2. azorica**

1. S. europaea L., *Sp. Pl.* 235 (1753). Perennial 20–60 cm, with a stout stock. Basal leaves long-petiolate; segments obovate-cuneate, crenate-serrate and sometimes lobed, the teeth ending in a short seta (*c.* 0·5 mm). False umbels usually with 3 rays. Bracts linear, entire or sometimes with a few teeth; bracteoles entire. Petals pink or white. Fruit 4–5 mm. $2n=16$. *Throughout Europe, except the northern and eastern margins; only on mountains in the south.* All except Az Bl Cr Fa Is Rs (N, E) Sb.

2. S. azorica Guthnick ex Seub., *Fl. Azor.* 41 (1844). Like **1** but usually larger; leaves sharply toothed, the teeth ending in a long seta (*c.* 3 mm); false umbels usually with more than 3 rays; bracts lanceolate, toothed like the leaves; fruit 3 mm. ● *Açores.* Az.

5. **Hacquetia** DC.[1]

Stock shortly creeping. Leaves digitately 3-partite. Inflorescence a simple umbel. Bracts large, leaf-like. Sepals conspicuous. Fruit glabrous, ovoid, slightly compressed laterally; ridges rather stout, prominent; vittae conspicuous, solitary under the ridges.

1. H. epipactis (Scop.) DC., *Prodr.* 4: 85 (1830). Glabrous perennial 10–25 cm, with a creeping rhizome. Basal leaves long-petiolate; segments 2–4 cm, sessile, ovate, cuneate, lobed and toothed in the upper half. Umbels scapose. Bracteoles 1–2 cm, 5–6, like the leaf-segments, longer than the flowers. Petals yellow. Fruit *c.* 4 mm. $2n=16$. *In woods.* ● *E. Alps, mountains of W. Hrvatska, N. Carpathians, S. Poland; local.* Au Cz It Ju Po.

6. **Astrantia** L.[1]

Leaves palmately lobed or palmatipartite. Inflorescence a simple umbel. Bracteoles large, coloured. Sepals conspicuous. Petals whitish, lanceolate; apex long, inflexed. Fruit with inflated ridges, covered with vesicular scales. Carpophore absent.

1 Bracteoles coriaceous, with prominent cross-veins; calyx-teeth long-acuminate **1. major**
1 Bracteoles membranous, with inconspicuous cross-veins; calyx-teeth subobtuse or mucronate
2 At least some basal leaves more than 5-partite
3 Leaf-segments deeply serrate; fruit ellipsoid **2. minor**
3 Leaf-segments obscurely serrate or denticulate; fruit sub-cylindrical **3. pauciflora**
2 All basal leaves 5-partite
4 Middle segment of leaf free nearly or quite to base; bracteoles exceeding the umbel **4. bavarica**
4 Middle segment of leaf united to the lateral ones in the lower part; bracteoles shorter than or equalling the umbel **5. carniolica**

1. A. major L., *Sp. Pl.* 235 (1753). Robust perennial up to 100 cm. Stems usually unbranched, except at the top. Basal leaves long-petiolate, 3- to 5(–7)-partite; segments lanceolate to obovate from a cuneate base, serrate or dentate and often somewhat lobed; middle segment free for at least ⅔ its length. Bracteoles equalling or exceeding the umbel, lanceolate or oblanceolate, acuminate, more or less connate and whitish below, usually pink or purplish towards the apex. Calyx-teeth narrowly lanceolate, long-acuminate. Fruit 6–8 mm, nearly cylindrical. *C. Europe, extending to N. Spain, C. Italy, Bulgaria and White Russia.* Al Au Bu Cz Ga Ge He Hs Hu It Ju Po Rm Rs (C, W) [Br Da Fe].

(a) Subsp. **major**: Bracteoles equalling the umbel. $2n=28$. *Throughout the range of the species, except some of the higher mountains.*

(b) Subsp. **carinthiaca** Arcangeli, *Comp. Fl. Ital.* 265 (1882): Bracteoles twice as long as the umbel, at least in the terminal umbel. *Higher mountains from the S. Alps to N.W. Spain.*

2. A. minor L., *Sp. Pl.* 235 (1753). Slender perennial 20–40 cm. Stems often branched about half-way up. Basal leaves usually 7-partite; segments lanceolate to obovate from a cuneate base, deeply and sharply serrate, mostly free nearly or quite to base. Bracteoles equalling or exceeding the umbel, lanceolate, acute or acuminate, free to base. Calyx-teeth ovate-oblong, mucronulate. Fruit *c.* 3 mm, ellipsoid. $2n=16$. ● *Pyrenees, S.W. Alps, N. Appennini.* Ga He Hs It.

3. A. pauciflora Bertol. in Desv., *Jour. Bot. Appl.* 2: 76 (1813). Like **2** but basal leaves 5- to 7-partite with usually linear-lanceolate, obscurely serrate or denticulate segments; fruit *c.* 4 mm, subcylindrical. *Calcicole.* ● *Italy* (*C. & S. Appennini, Alpi Apuane*). It.

4. A. bavarica F. W. Schultz, *Flora* (*Regensb.*) 41: 161 (1858). Slender perennial up to 60 cm. Stems usually unbranched except at top. Basal leaves long-petiolate, 5-partite; segments lanceolate from a cuneate base, irregularly dentate and often somewhat lobed; middle segment free nearly to base. Bracteoles exceeding the umbel, narrowly lanceolate, acute or acuminate, free to base. Calyx-teeth ovate-oblong, acute. Fruit *c.* 4 mm, oblong or ovoid. $2n=14$. ● *E. Alps.* Au Ge It Ju.

5. A. carniolica Jacq., *Fl. Austr.* 5: 31 (1778). Like **4** but leaf-segments ovate, the middle one united to the lateral ones in the lower part; bracteoles usually shorter than umbel; fruit *c.* 3 mm. $2n=14$. ● *S.E. Alps.* Au It Ju.

7. **Eryngium** L.[2]

Glabrous herbs. Leaves entire to 3-pinnatisect, at least the upper softly to pungently spiny. Inflorescence usually branched; flowers sessile in hemispherical to cylindrical capitula, at the base of which are 3 or more softly to pungently spinescent bracts; entire, 3- or 4-cuspidate bracteoles present at least near the edges of the capitula. Sepals rigid; petals less than 4 mm, erect, emarginate, shorter than sepals. Fruit ovoid to subglobose, nearly always sparsely or densely covered with scales; mericarps plano-convex, slightly ridged; vittae usually slender; carpophore absent.

Descriptions of basal leaves refer to those present at the time of flowering, or to the last ones produced before flowering. Descriptions of capitula, bracts, bracteoles and floral parts refer to the largest, central capitulum of an inflorescence.

---

[1] By T. G. Tutin.   [2] By A. O. Chater.

1 Basal leaves 150–250 cm, ensiform, simple; bracts *c.* 2 mm
　　　　　　　　　　　　　　　　　　　　26. pandanifolium
1 Basal leaves less than 150 cm; bracts more than 3 mm
2 Bracteoles 4-cuspidate　　　　　　　　　　**1. tenue**
2 Bracteoles entire or 3-cuspidate
　3 Most bracts more than 6 cm　　　　　**20. spinalba**
　3 Bracts not more than 6 cm
　　4 Bracts more than 25　　　　　　　　**9. alpinum**
　　4 Bracts fewer than 25
　　　5 Capitula 4–10 cm　　　　　　　　**5. duriaei**
　　　5 Capitula less than 4 cm
　　　　6 Basal leaves 3- to 7-sect, persistent, with linear segments
　　　　　10–30 × 0·2–4 cm
　　　　　7 Basal leaves 3-sect; inflorescence with 3–6 capitula
　　　　　　　　　　　　　　　　　　15. ternatum
　　　　　7 Basal leaves 5- to 7-sect; inflorescence with more than
　　　　　　6 capitula　　　　　　　　　**16. serbicum**
　　　　6 Basal leaves not as above, or decaying early
　　　　　8 Outer bracteoles almost as large as bracts
　　　　　　9 Stems 50–100 cm; basal leaves divided　**22. amorginum**
　　　　　　9 Stems less than 40 cm; basal leaves undivided
　　　　　　　10 Bracts and bracteoles usually more than 12; roots
　　　　　　　　c. 2 mm thick; basal leaves 1 cm or more wide
　　　　　　　　　　　　　　　　　　2. barrelieri
　　　　　　　10 Bracts and bracteoles usually fewer than 12; roots less
　　　　　　　　than 2 mm thick; basal leaves usually less than
　　　　　　　　1 cm wide
　　　　　　　　11 Bracts rigid, pungent; sepals *c.* 2 mm　**3. galioides**
　　　　　　　　11 Bracts soft, with setiform apex; sepals 1–1·5 mm
　　　　　　　　　　　　　　　　　　4. viviparum
　　　　　8 Outer bracteoles much smaller than bracts
　　　　　　12 Bracts entire
　　　　　　　13 Bracts more than 8　　　　　**21. bourgatii**
　　　　　　　13 Bracts fewer than 8
　　　　　　　　14 Inflorescence of fewer than 10 capitula; bracteoles
　　　　　　　　　3-cuspidate　　　　　**13. tricuspidatum**
　　　　　　　　14 Inflorescence of more than 10 capitula; bracteoles
　　　　　　　　　entire
　　　　　　　　　15 Basal leaves undivided, with swollen, segmented
　　　　　　　　　　petiole; axis of capitulum projecting in a bract-
　　　　　　　　　　like appendage　　　　**25. corniculatum**
　　　　　　　　　15 Basal leaves divided; petiole not swollen and seg-
　　　　　　　　　　mented; axis of capitulum not projecting
　　　　　　　　　　16 Bracts 3–4, thickened and folded at base
　　　　　　　　　　　　　　　　　　14. triquetrum
　　　　　　　　　　16 Bracts 5–7, ± flat at base　**24. campestre**
　　　　　　12 Bracts with 1 or more pairs of spines or teeth
　　　　　　　17 Bracts ovate to ovate-lanceolate; sepals 4–5 mm
　　　　　　　　　　　　　　　　　　8. maritimum
　　　　　　　17 Bracts lanceolate to linear; sepals less than 4 mm
　　　　　　　　18 Capitula sessile; stem procumbent, branched from
　　　　　　　　　the base　　　　　　**6. ilicifolium**
　　　　　　　　18 Capitula pedunculate; stem erect, usually not
　　　　　　　　　branched from the base
　　　　　　　　　19 Bracts with more than 6 pairs of teeth or spines
　　　　　　　　　　20 Lamina of leaf decurrent on petiole　**7. aquifolium**
　　　　　　　　　　20 Lamina of leaf not decurrent on petiole
　　　　　　　　　　　　　　　　　　13. tricuspidatum
　　　　　　　　　19 Bracts with 6 or fewer pairs of teeth or spines
　　　　　　　　　　21 Basal leaves undivided (though often deeply
　　　　　　　　　　　dentate)
　　　　　　　　　　　22 Inflorescence strict, with not more than 8 capi-
　　　　　　　　　　　　tula　　　　　**13. tricuspidatum**
　　　　　　　　　　　22 Inflorescence spreading, with more than 10
　　　　　　　　　　　　capitula
　　　　　　　　　　　　23 Sepals mucronate　　**12. creticum**
　　　　　　　　　　　　23 Sepals aristate
　　　　　　　　　　　　　24 Capitula more than half as long as bracts
　　　　　　　　　　　　　　　　　　10. planum
　　　　　　　　　　　　　24 Capitula not more than half as long as bracts
　　　　　　　　　　　　　　　　　　11. dichotomum
　　　　　　　　　　21 Basal leaves divided
　　　　　　　　　　　25 Lamina of basal leaves decurrent on petiole

26 Basal leaves with 2- or 3-pinnatisect segments
　　　　　　　　　　　　　　　　　23. amethystinum
26 Basal leaves with dentate to pinnatisect seg-
　ments
　27 Leaves and bracts very coriaceous, strongly
　　pungent; bracts 3–5 cm, with 1–2(–3) pairs
　　of spines　　　　　　　　　**18. glaciale**
　27 Leaves and bracts slightly coriaceous, scarcely
　　pungent; bracts 1–3 cm, with 3–6 pairs of
　　spines　　　　　　　　　　**19. dilatatum**
25 Lamina of basal leaves not decurrent on petiole
　28 Bracts 10–15　　　　　　　　**21. bourgatii**
　28 Bracts fewer than 10
　　29 Bracteoles entire
　　　30 Basal leaves with dentate to 3-fid segments
　　　　　　　　　　　　　　　11. dichotomum
　　　30 Basal leaves with 2-pinnatisect segments
　　　　　　　　　　　　　　　24. campestre
　　29 Bracteoles 3-cuspidate
　　　31 Inflorescence spreading, usually with 30–100
　　　　capitula　　　　　　　　**12. creticum**
　　　31 Inflorescence strict, with 10 or fewer capitula
　　　　32 Bracts linear-lanceolate, 2·5–4 mm wide in
　　　　　middle; basal leaves palmatisect
　　　　　　　　　　　　　　　17. palmatum
　　　　32 Bracts narrowly linear-lanceolate, 1–2 mm
　　　　　wide in middle; basal leaves 3-fid
　　　　　　　　　　　　　　　13. tricuspidatum

1. E. tenue Lam., *Encycl. Méth. Bot.* **4**: 755 (1798). Annual; stems 2–40 cm, erect. Basal leaves 1–3 cm, obovate, deeply serrate to 3- to 5-fid; lamina decurrent on short petiole. Inflorescence subcorymbose, with usually numerous, pedunculate, globose to ovoid capitula 0·5–1 cm. Bracts 1–2·5 cm, 7–9, narrowly linear-lanceolate, with 6–12 pairs of spines; bracteoles 4-cuspidate. Sepals 1–1·5 mm, ovate, with short awn. Fruit densely covered with scales. *Dry, sandy places on acid soils. Iberian peninsula, except the north-east.* Hs Lu.

2. E. barrelieri Boiss., *Ann. Sci. Nat.* ser. 3 (Bot.), **1**: 125 (1844). Annual or biennial with blackish roots *c.* 2 mm thick; stems 5–30 cm, usually erect. Basal leaves 7–10 × 1–1·5 cm, persistent, linear-oblanceolate or linear-oblong, repand-crenate to crenate-serrate; lamina decurrent on petiole. Inflorescence subcorymbose, with up to 30 usually sessile, depressed-hemispherical capitula 0·5–0·8 cm. Bracts and bracteoles 12–25, not clearly differentiated from each other, the outer 1–2 cm, linear-lanceolate, entire or with 1(–2) pairs of spines. Sepals *c.* 2·5 mm, ovate, aristate. Fruit densely covered with scales. *Places liable to winter flooding. C. & S. Italy, Corse, Sardegna, Sicilia.* Co It Sa Si.

3. E. galioides Lam., *Encycl. Méth. Bot.* **4**: 757 (1798). Annual with brownish roots less than 2 mm thick; stems 3–15(–30) cm, procumbent or erect. Basal leaves 3–4 × 0·7–1 cm, decaying early, lanceolate to linear-lanceolate, deeply incise-serrate; lamina decurrent on petiole. Inflorescence spreading, with usually up to 25, sessile, depressed-hemispherical capitula 0·5–1 cm. Bracts and bracteoles 8–12, not clearly differentiated from each other, the outer 0·5–1 cm, lanceolate-acuminate, with 1–2 pairs of spines. Sepals *c.* 2 mm, ovate, aristate. Fruit sparsely or densely scaly. *Dry, open places.* ● *W. half of Iberian peninsula.* Hs Lu.

A little-known and poorly collected species.

4. E. viviparum Gay, *Ann. Sci. Nat.* ser. 3 (Bot.), **9**: 171 (1848). Biennial; stems 1–8 cm, procumbent. Basal leaves 1–3(–5) × 0·3–0·5 cm, persistent, linear-oblanceolate, closely or remotely serrate; lamina decurrent on petiole; inner rosette-leaves much smaller. Inflorescence spreading, with up to 50 sessile, depressed-hemispherical, 5- to 8-flowered capitula up to 0·5 cm. Bracts and bracteoles 10–12, not clearly differentiated from each other, the outer 0·5–0·8 cm, lanceolate or linear-lanceolate, with 1–2(–3) pairs of spines. Sepals 1–1·5 mm, ovate, aristate. Fruit sparsely scaly. *Places liable to winter flooding.* ● *N. Portugal, N.W. Spain, N.W. France.* Ga Hs Lu.

5. **E. duriaei** Gay ex Boiss., *Voy. Bot. Midi Esp.* **2**: 237 (1839) (*E. duriaeanum* Gay). Monocarpic, living 3–4 years; stems 30–150 cm, erect. Basal leaves 10–45 × 2·5–7 cm, persistent, coriaceous, linear-oblanceolate to linear-spathulate, with large, patent, spinescent teeth; lamina decurrent on petiole. Inflorescence bluish, with up to 10 pedunculate, cylindrical capitula 4–10 × 1·5–2 cm. Bracts 1·5–5 cm, 7–12, linear-lanceolate, with 1(–3) pairs of teeth; bracteoles 3-cuspidate or entire. Sepals *c.* 5 mm, lanceolate, aristate. Fruit densely scaly. *Dry, rocky* Quercus-*woodland.* ● *Mountains of N. half of Iberian peninsula.* Hs Lu.

6. **E. ilicifolium** Lam., *Encycl. Méth. Bot.* **4**: 757 (1798). Annual; stems 2–15 cm, procumbent. Basal leaves 2–6 × 1–3 cm, persistent, coriaceous. obovate, with wide, patent, spinescent teeth; lamina shortly decurrent on petiole. Inflorescence spreading, bluish, with usually numerous, ovoid or globose, sessile capitula 1–1·5 cm. Bracts 1·5–3 cm, 5–7, oblanceolate to elliptical, with 2(–3) pairs of triangular, spinescent teeth; bracteoles tricuspidate, the inner dilated at the base and often entire. Sepals 1–2 mm, ovate, aristate. Fruit densely scaly on the angles only. *Dry places.* S. Spain. Hs. (*N. Africa.*)

7. **E. aquifolium** Cav., *Anal. Ci. Nat.* **3**: 32 (1801) (incl. *E. huteri* Porta & Rigo). Perennial; stems 10–50 cm, erect. Basal leaves 5–10 × 1·5–3 cm, persistent, slightly coriaceous, oblanceolate-spathulate or obovate, coarsely, often doubly dentate, with patent, spinescent teeth; lamina decurrent on petiole. Inflorescence bluish, subcorymbose, with up to 12 pedunculate, subglobose capitula 1–2 cm. Bracts 2·5–5 cm, 5–7, lanceolate, with more than 6 pairs of teeth or spines; at least some bracteoles 3-cuspidate. Sepals *c.* 3 mm, lanceolate, aristate. Fruit densely scaly. *Dry places.* S. Spain. Hs. (*N.W. Africa.*)

8. **E. maritimum** L., *Sp. Pl.* 233 (1753). Perennial, perhaps sometimes monocarpic; stems 15–60 cm, erect, branched above. Lamina of basal leaves 4–10 × 5–15 cm, suborbicular, truncate or cordate at base, 3-(to 5-)lobed, coriaceous, with coarse, patent, spinescent teeth; petiole equalling lamina, unwinged, entire. Inflorescence spreading, bluish, with usually numerous, pedunculate, subglobose capitula 1·5–3 cm. Bracts 2·5–4 cm, 4–7, ovate or ovate-lanceolate, with 1–3 pairs of broad, spinescent teeth; bracteoles 3-cuspidate. Sepals 4–5 mm, ovate-lanceolate, aristate. Fruit densely scaly. 2n = 16. *Maritime sands. Coasts of Europe northwards to 60° N.* Al Be Bl Br Bu Co Cr Da Ga Ge Gr Hb Ho Hs It Ju Lu No Po Rm Rs (B, W, K, E) Sa Si Su Tu.

9. **E. alpinum** L,. *Sp. Pl.* 233 (1753). Perennial; stems 30–70 cm, erect. Basal leaves persistent, soft; lamina 8–15 × 5–13 cm, ovate- or triangular-cordate, irregularly toothed; petiole 2–4 times as long as lamina, unwinged, entire. Inflorescence bluish, with 1–3 pedunculate, ovoid-cylindrical capitula 2–4 × 1·5–2 cm. Bracts 3–6 cm, more than 25, many of them pinnatifid, with numerous pectinately arranged, long, soft spines; bracteoles 3-cuspidate or entire. Sepals *c.* 3·5 mm, ovate-lanceolate, aristate. Fruit densely scaly. *Meadows and grassy places; usually calcicole.* ● *Jura, Alps and mountains of W. & C. Jugoslavia; once recorded from the W. Carpathians.* Au ?Cz Ga He It Ju.

10. **E. planum** L., *Sp. Pl.* 233 (1753). Perennial; stems 25–100 cm, erect. Basal leaves persistent, slightly coriaceous; lamina 5–10 × 3–6 cm, oblong to ovate-oblong, cordate at base, serrate; petiole about as long as lamina, unwinged. Inflorescence usually bluish, subcorymbose, with usually numerous pedunculate, ovoid-globose capitula 1–2 × 1–1·5 cm. Bracts 1·5–2·5 cm, 6–8, sometimes slightly shorter than the capitulum, linear-lanceolate, with 1–4 pairs of spinescent teeth; bracteoles 3-cuspidate or entire. Sepals *c.* 2 mm, ovate-lanceolate, aristate. Fruit densely scaly, the scales overlapping. 2n = 16. *Dry places.* C. & S.E. *Europe.* Au Cz Ge Hu Ju Po Rm Rs (C, W, K, E) Tu.

11. **E. dichotomum** Desf., *Fl. Atl.* **1**: 226 (1798). Perennial; stems 20–100 cm, erect. Basal leaves slightly coriaceous, decaying early; lamina 3·5–6 × 0·7–3 cm, oblong, cordate or truncate at base, simply or doubly serrate (the younger sometimes 3-lobed); petioles equalling or up to twice as long as lamina. Inflorescence usually bluish, with usually numerous pedunculate, subglobose capitula 1–1·5 cm. Bracts 2–4 cm, 4–6(–7), linear-lanceolate, with 1–2 pairs of spinescent teeth; bracteoles entire or rarely 3-cuspidate. Sepals 2–3 mm, ovate-lanceolate, aristate. Fruit densely scaly, the scales overlapping. *Dry places.* W. *Mediterranean region* (*local*); *S.E. Russia.* Hs It Si Rs (E).

The plants from Russia differ from the Mediterranean ones in having the younger basal leaves often 3-lobed and more persistent, and in the larger, bluer inflorescence. They have been called **E. caeruleum** Bieb., *Beschr. Länd. Terek. Casp.* 155 (1800) (*E. biebersteinianum* Nevski), but do not appear to be specifically distinct.

12. **E. creticum** Lam., *Encycl. Méth. Bot.* **4**: 754 (1798). Perennial, or sometimes biennial or annual; stems (12–)25–100 cm, erect, much-branched above. Basal leaves slightly coriaceous, decaying early; lamina 5–15 × 3–15 cm, very variable, oblong-ovate to suborbicular, cordate or truncate at base, undivided and crenate-dentate to 3-sect with 2-pinnatifid segments; petiole 1–2 times as long as lamina, unwinged. Inflorescence bluish, very diffuse, with usually numerous pedunculate, globose capitula 0·5–1 cm. Bracts 1–3 cm, 5–7, linear-lanceolate, pungent, with 1–2 pairs of spines; bracteoles linear-lanceolate, 3-cuspidate. Sepals *c.* 1·5 mm, ovate, mucronate. Fruit sparsely scaly, the scales not overlapping. *Dry places. Balkan peninsula and Aegean region, extending northwards to Slovenija.* Al Bu Cr Gr Ju Tu [It].

13. **E. tricuspidatum** L., *Demonstr. Pl.* 8 (1753). Perennial; stems 15–75 cm, erect. Basal leaves persistent, soft; lamina *c.* 3 × 2–3 cm, oblong-ovate to suborbicular, cordate, crenate-dentate to serrate, undivided or 3-fid for *c.* ½ its length; petiole 1–2 times as long as lamina, unwinged. Inflorescence strict, greenish, with 2–8 pedunculate, hemispherical capitula *c.* 1 cm. Bracts 1·5–5 cm, 5–7, narrowly linear-lanceolate, with (3–)5–8 pairs of spinescent teeth, rarely entire; bracteoles linear-lanceolate, 3-cuspidate. Sepals *c.* 2 mm, ovate, mucronate. Fruit densely scaly, the scales overlapping. *Dry places.* S.W. Spain; Sardegna; Sicilia. Hs Sa Si. (*N. Africa.*)

14. **E. triquetrum** Vahl, *Symb. Bot.* **2**: 46 (1791). Perennial; stems 15–40 cm, erect. Basal leaves decaying early, slightly coriaceous; lamina *c.* 4 × 3 cm, broadly ovate, 3-fid or 3-sect with palmatifid to 2-pinnatifid spinescent-dentate lobes; petiole 1–2 times as long as lamina, unwinged. Inflorescence usually corymbose, bluish, with numerous pedunculate, hemispherical, few-flowered capitula *c.* 5 mm. Bracts 1·5–3 cm, 3–4, linear-lanceolate, entire, pungent, folded and thickened at the base; bracteoles linear-lanceolate, entire. Sepals 2·5–3 mm, ovate-

lanceolate, aristate. Fruit becoming strongly swollen, with numerous small, not overlapping scales. *Dry places. S. Italy and Sicilia.* It Si. (*N. Africa.*)

15. E. ternatum Poiret in Lam., *Encycl. Méth. Bot., Suppl.* **4**: 295 (1816). Perennial; stems 30–60 cm, erect. Basal leaves persistent, slightly coriaceous; lamina 3-sect; segments 10–30 × 0·5–1 cm, linear, undivided, with soft, ascending marginal spines; petiole narrowly winged. Inflorescence with 3–6 pedunculate, hemispherical capitula 1·5–2 cm. Bracts 2–3 cm, 5–9, linear-acuminate, usually with a pair of spinescent teeth near base; bracteoles 3-cuspidate. Sepals 2–3 mm, ovate, shortly aristate. *Dry places.* ● *Kriti.* Cr.

16. E. serbicum Pančić, *Verh. Zool.-Bot. Ges. Wien* **6**: 520 (1856). Perennial; stems 40–75 cm. Basal leaves persistent, slightly coriaceous; lamina palmatisect; segments 10–25 × 0·2–4 cm, 5–7, linear, undivided, with soft ascending marginal spines; petiole winged, sometimes broadly sheathing, often with linear, spinescent-dentate to -pinnatisect appendages in upper part. Inflorescence with usually numerous pedunculate, hemispherical capitula 1–1·5 cm. Bracts 1–4 cm, 5–7, linear-acuminate, often expanded at base, with 1–2 pairs of spinescent teeth; bracteoles 3-cuspidate. Sepals 2–3 mm, ovate-lanceolate, shortly aristate. *Dry places.* ● *Srbija.* Ju.

17. E. palmatum Pančić & Vis., *Mem. Ist. Veneto* **15**: 20 (1870). Perennial; stems 30–75 cm. Basal leaves persistent, slightly coriaceous; lamina 5–9 × 5–11 cm, reniform-orbicular, palmatisect with 5–7 oblanceolate, mostly 3-fid, serrate lobes; petiole unwinged. Inflorescence strict, green, with up to 10 pedunculate, hemispherical capitula 1–1·5 cm. Bracts 2–4 cm, 5–7, linear-oblanceolate, with 1–5 pairs of spinescent teeth; bracteoles 3-cuspidate. Sepals *c.* 3 mm, ovate-lanceolate, aristate. Fruit densely scaly. *Dry places and woods.* ● *C. part of Balkan peninsula.* Al Bu Gr Ju.

Plants from S. Jugoslavia and N.W. Greece have been called E. **wiegandii** Adamović, *Österr. Bot. Zeitschr.* **55**: 178 (1905) (*E. tricuspidatum* auct. balcan., non L.). They differ from **17** only in having less divided, usually 3-fid basal leaves with undivided or weakly 3-fid lobes, and do not merit specific separation.

18. E. glaciale Boiss., *Biblioth. Univ. Genève* ser. 2, **13**: 409 (1838). Perennial; stems 5–20 cm, erect. Basal leaves persistent, coriaceous; lamina 3–5 × 3–6 cm, deeply 3-fid or 3-sect, the lobes with large, lanceolate-acuminate, pungent teeth and the base decurrent as a spiny wing on the petiole; unwinged part of petiole half as long as or as long as lamina. Inflorescence bluish, with 3(–5) pedunculate, globose capitula 1–1·5 cm. Bracts 3–5 cm, 7–8, narrowly linear-lanceolate, pungent, with 1–2(–3) pairs of spines; bracteoles 3-cuspidate. Sepals *c.* 1·5 mm, ovate, shortly aristate. Fruit without scales. *Stony places above 2500 m. S. Spain (Sierra Nevada).* Hs. (*N.W. Africa.*)

19. E. dilatatum Lam., *Encycl. Méth. Bot.* **4**: 755 (1798). Perennial; stems 5–40 cm, erect. Basal leaves 2–10 cm, persistent, slightly coriaceous, obovate, 3-sect above, pinnatisect below, the segments ovate to linear-lanceolate, pinnatisect to coarsely toothed with softly spinescent teeth; petiole not distinct, winged more or less to the base. Inflorescence bluish, with up to 12 pedunculate, globose capitula 0·5–1·5 cm, the lateral often subsessile. Bracts 1–3 cm, 5–10, lanceolate to linear-lanceolate, with 3–6 pairs of spinescent teeth, not pungent; bracteoles entire, or rarely a few 3-cuspidate. Sepals *c.* 2·5 mm, ovate, aristate. *Dry places. Spain and Portugal.* Hs Lu.

20. E. spinalba Vill., *Prosp. Pl. Dauph.* 26 (1779). Perennial; stems 20–35 cm, very stout. Basal leaves coriaceous; lamina 5–8 × 5–10 cm, suborbicular, cordate at base, palmatifid or palmatisect with (3–)4–5 irregularly pinnatifid segments with large, spinescent teeth throughout; petiole about twice as long as lamina, unwinged. Inflorescence usually bluish, with up to 10 pedunculate, ovoid-cylindrical capitula 4–6 × 2–3 cm. Bracts 6–9 cm, 15–30, linear-lanceolate, often 3-fid, with numerous spinescent teeth, pungent; bracteoles mostly 3-cuspidate. Sepals *c.* 3·5 mm, linear-lanceolate, aristate. Fruit densely scaly. *Dry, stony places on mountains.* ● *S.W. Alps.* Ga It.

21. E. bourgatii Gouan, *Obs. Bot.* 7 (1773). Perennial; stems 15–45 cm, erect. Basal leaves slightly coriaceous, persistent; lamina 3–7 cm, suborbicular, 3-sect with pinnatifid or 2-pinnatifid, spinescent-dentate segments; petiole 2–4 times as long as lamina, unwinged. Inflorescence usually bluish, with up to 7 pedunculate, ovoid-globose capitula 1·5–2·5 cm. Bracts 2–5 cm, 10–15, linear-lanceolate, entire or with 1–2(–3) pairs of spinescent teeth; bracteoles entire or 3-cuspidate. Sepals *c.* 3 mm, lanceolate to ovate, aristate. Fruit sparsely scaly. $2n=16$. *Dry, stony places on mountains. Spain, Pyrenees.* Ga Hs.

22. E. amorginum Rech. fil., *Magyar Bot. Lapok* **33**: 9 (1934). Perennial; stems 50–100 cm, erect. Basal leaves slightly coriaceous; lamina *c.* 12 cm, 3-sect, the lobes ovate or ovate-lanceolate, pinnatisect with pinnatifid or irregularly serrate lobes; petiole as long as lamina, unwinged. Inflorescence with *c.* 15 pedunculate, ovoid-globose capitula *c.* 1·5 cm. Bracts up to 1 cm, *c.* 8, linear-oblanceolate, merging into the entire bracteoles. Sepals *c.* 2 mm, ovate, obtuse, aristate. Fruit densely scaly. *Limestone cliffs near the sea.* ● *S. Kikladhes.* Gr.

23. E. amethystinum L., *Sp. Pl.* 233 (1753) (incl. *E. glomeratum* Lam.). Perennial; stems 20–45 cm, erect. Basal leaves usually persistent, coriaceous; lamina 10–15 cm, obovate, palmatisect above and pinnatisect below, the segments 2- or 3-pinnatisect with linear-lanceolate, spinescent-serrate segments; petiole broadly winged. Inflorescence usually bluish, cylindrical to corymbiform with usually numerous pedunculate, globose or ovoid capitula 1–2 cm. Bracts 2–5 cm, 5–9 linear-lanceolate, with 1–4 pairs of spines; bracteoles entire or 3-cuspidate. Sepals 1·5–2·5 mm, ovate-lanceolate, shortly aristate, Fruit sparsely scaly. *Dry places. Balkan peninsula and Aegean region; Italy and Sicilia.* Al ?Bu Cr Gr It Ju Si.

24. E. campestre L., *Sp. Pl.* 233 (1753). Perennial; stems 20–70 cm, erect. Basal leaves usually persistent, coriaceous; lamina 5–20 cm, broadly ovate, 3-sect; the central lobe pinnatisect, with opposite, pinnatisect lobes; the lateral lobes pinnatisect with alternate, often pinnatisect lobes; lobes spinose-serrate; petiole equalling lamina, unwinged. Inflorescence usually corymbiform, pale greenish, with numerous pedunculate, ovoid capitula (0·5–)1–1·5(–2·5) cm. Bracts (1–)1·5–4·5 cm, 5–7, linear-lanceolate, entire or with 1(–2) pairs of spines; bracteoles entire. Sepals *c.* 2·5 mm, ovate-lanceolate, aristate. Fruit densely scaly, the scales overlapping. $2n=14, 28$. *Dry places. C. & S. Europe, extending to S. England.* Al Au Be Bl *Br Bu Co Cr Cz Ga Ge Gr He Ho Hs Hu It Ju Lu Po Rm Rs (C, W, K, E) Sa Si Tu [Da].

25. E. corniculatum Lam., *Encycl. Méth. Bot.* **4**: 758 (1798). Probably biennial; stems 15–60 cm, erect, often branched near base. Basal leaves usually decaying early, soft; lamina 2–5 cm, ovate-oblong, remotely toothed; petiole usually many times as long as lamina, swollen and conspicuously segmented. Inflorescence spreading, bluish, with up to 40 pedunculate, ovoid to subglobose capitula 0·5–1 cm; axis of capitulum projecting in a bract-like appendage. Bracts 1·5–2·5(–5) cm, 3–7, linear-lanceolate, entire; bracteoles entire. Sepals 1–1·5 mm, ovate, shortly aristate. Fruit densely scaly. *Places liable to winter flooding. Portugal and S.W. Spain; Sardegna.* Hs Lu Sa.

26. E. pandanifolium Cham. & Schlecht., *Linnaea* **1**: 336 (1826). Perennial; stems 150–400 cm, erect. Basal leaves 150–250 cm, ensiform, simple, with slender marginal spines. Inflorescence paniculate, of numerous greenish-white, ovoid-globose capitula 6–10 × 4–8 mm; bracts *c.* 2 mm, 6–8, ovate-lanceolate, acute, entire; bracteoles entire, sometimes ciliolate. Sepals *c.* 1 mm, broadly ovate. *Banks of ditches. Naturalized in the Mondego plain, C. Portugal.* [Lu.] (*Subtropical South America.*)

8. Lagoecia L.[1]

Annual. Leaves simply pinnate. Umbels compound. Partial umbels 1-flowered. Bracts and bracteoles leaf-like, pinnatisect. Sepals like the bracteoles, conspicuous. Style 1. Fruit covered with short, brittle, clavate hairs.

1. L. cuminoides L., *Sp. Pl.* 203 (1753). Annual 10–30 cm. Basal leaves with ovate, dentate segments; upper cauline with segments deeply divided into short, lanceolate, aristate lobes. Umbels 0·5–1·5 cm in diameter, dense, globose; rays numerous. Bracts like the leaves; bracteoles 4, 2-pinnatisect, with setaceous lobes. Sepals pinnatisect, the lobes setaceous and sometimes 2- to 3-fid. Fruit *c.* 2 mm. *Mediterranean region, extending to S.E. Portugal and Bulgaria.* Bu Cr Gr Hs It Lu.

9. Petagnia Guss.[1]

Basal leaves usually peltate, deeply lobed; cauline palmately divided. Inflorescence repeatedly branched. Sepals conspicuous. Petals whitish, oblong-cuneate; apex short, inflexed. Ovary 1-locular; ovule 1. Fruit subglobose, hard and nut-like. Ridges prominent above, obsolete below; vittae absent.

1. P. saniculifolia Guss., *Fl. Sic. Prodr.* **1**: 311 (1827). Perennial. Stock stout; stems up to 50 cm. Basal leaves long-petiolate, the lamina 4–6 cm, 5-lobed, the lobes dentate; cauline leaves subsessile, deeply 3- to 5-lobed. Inflorescence cymose, the ultimate branches with a central, sessile female or hermaphrodite flower and 2–4 male flowers, whose pedicels are more or less adnate to the ovary of the central flower. Bracts and bracteoles small. Fruit *c.* 2·5 mm, glabrous. $2n=42$. *Beside woodland streams.* ● *N. Sicilia.* Si.

Subfam. **Apioideae**

Leaves always exstipulate. Endocarp soft; vittae usually present in the ripe fruit. Style terminal on the stylopodium. Fruit never scaly.

10. Echinophora L.[1]

Leaves 2- to 3-pinnate. Sepals pungent, persistent, often unequal in the outer flowers. Petals white or yellow, oblanceolate, emarginate, the outer often larger; apex inflexed. Fruit ovoid-oblong; styles long, persistent, woody. Ridges low, indistinct; vittae solitary.

Leaf-lobes keeled beneath, sulcate above, spine-tipped; petals white, very rarely pink   **1. spinosa**
Leaf-lobes flat, not spine-tipped; petals yellow   **2. tenuifolia**

1. E. spinosa L., *Sp. Pl.* 239 (1753). More or less pubescent perennial up to 50 cm. Leaves 2-pinnate, rigid; lobes thick, keeled beneath, sulcate above, spine-tipped. Rays 4–8, pubescent. Bracts and bracteoles 5–10, oblong-lanceolate to linear, spinose. Each partial umbel with a central hermaphrodite flower and a number of male flowers whose more or less connate pedicels form an involucre round the fruit. Petals white, or very rarely pink, pubescent on the back, the outer larger than the inner. *Maritime sands. Mediterranean region.* Al Bl Co Ga Gr Hs It Ju Sa Si.

2. E. tenuifolia L., *Sp. Pl.* 239 (1753). Greyish-pubescent perennial 20–50 cm. Leaves 2- to 3-pinnate, lanceolate in outline; lobes somewhat fleshy, flat, dentate, not spine-tipped. Rays 2–5, pubescent. Bracts 2–5, lanceolate or linear-lanceolate; bracteoles 5, ovate, deflexed, spinescent in fruit. Partial umbels as in 1. Petals yellow, ciliate, the outer scarcely larger than the inner. *Dry places. S. Europe, from Sicilia eastwards.* Bu Cr Gr It Rm Rs (K) Si Tu.

(a) Subsp. **tenuifolia**: Stem sulcate; lobes of cauline leaves lanceolate. *S. Italy and Sicilia.*

(b) Subsp. **sibthorpiana** (Guss.) Tutin *Feddes Repert.* **74**: 31 (1967) (*E. sibthorpiana* Guss.): Stem striate; lobes of cauline leaves ovate. *From Greece and Kriti to Krym.*

11. Myrrhoides Heister ex Fabr.[2]

(*Physocaulis* (DC.) Tausch)

Leaves 2-pinnate. Sepals absent. Petals white, obovate; apex short, inflexed. Fruit subcylindrical, scarcely beaked; stigmas sessile. Primary ridges 5, obtuse; vittae solitary; endosperm deeply furrowed.

1. M. nodosa (L.) Cannon, *Feddes Repert.* **79**: 65 (1968) (*Physocaulis nodosus* (L.) Koch, *Chaerophyllum nodosum* (L.) Crantz). Annual up to 100 cm. Stem often purplish, hispid, conspicuously swollen below the nodes at maturity. Leaves densely strigose; lobes ovate, irregularly dentate. Bracts absent; bracteoles 5–7. Rays (1–)2–3. Style obsolete; stigmas sessile on the stylopodium. Fruit 4–10 mm, gradually narrowed towards the top, covered with curved, upward-pointing, white bristles, many of which arise from tubercles. $2n=22$. *S. Europe, extending northwards to Hungary.* Al Bu Co Ga Gr Hs Hu It Ju Lu Rm Rs (K) Sa Si Tu.

12. Chaerophyllum L.[2]

Leaves 1- to 3-pinnate or ternate. Sepals obsolete. Petals white, pinkish or yellow, emarginate; apex inflexed. Fruit narrowly oblong to very narrowly ovoid, more or less gradually narrowed towards the scarcely beaked apex, slightly compressed laterally. Ridges wide and rounded; vittae solitary.

1 Petals distinctly ciliate; styles nearly erect, forming a very acute angle   (7–9). **hirsutum** group
1 Petals not ciliate; styles nearly always divergent
  2 Leaf-lobes almost undivided
    3 Leaf-lobes greyish beneath, with a dense, white tomentum over most of the lower surface
      4 Leaf-lobes with deep, rather irregular serrations; petiole with a ± dense covering of patent hairs   **6. azoricum**
      4 Leaf-lobes rather irregularly crenate-serrate; petiole with short, crispate hairs   **3. heldreichii**
    3 Leaf-lobes greenish beneath, glabrous or with hairs ± confined to the veins
      5 Leaf-lobes finely and regularly serrate, obliquely subcordate at base; lobes of upper leaves acute   **2. byzantinum**

---

[1] By T. G. Tutin.   [2] By J. F. M. Cannon.

5 Leaf-lobes coarsely serrate, obliquely cuncatc at basc; lobes of upper leaves acuminate **1. aromaticum**
2 Leaf-lobes deeply divided
6 Bracteoles membranous, very conspicuous in fruit; flowers yellow
7 Bracts 0–1, small, subulate; fruit *c.* 10 mm **4. coloratum**
7 Bracts prominent, membranous, sometimes divided and leaf-like; fruit *c.* 20 mm **5. creticum**
6 Bracteoles herbaceous, with a narrow membranous margin, relatively inconspicuous in fruit; flowers white or pinkish
8 Perennial; fruit 8–12 mm **11. aureum**
8 Biennial; fruit 4–7 mm
9 Plant with an elongate tap-root; stem setose; bracteoles hairy **12. temulentum**
9 Plant with a short, tuberous root; stem glabrous above; bracteoles glabrous **10. bulbosum**

1. C. aromaticum L., *Sp. Pl.* 259 (1753). Robust perennial up to 200 cm, with a long creeping rhizome. Leaves usually 2-ternate or rarely pinnate; lobes 40–100 mm, lanceolate to ovate, acutely and coarsely serrate, obliquely cuneate at base, glabrous to white-puberulent beneath, those of upper leaves usually acuminate. Petals white. Fruit 8–15 mm, narrowly ovoid-oblong; stylopodium depressed-conical, wider than apex of fruit; styles flexuous; pedicels scarcely thickened in fruit. *C. & E. Europe, Balkan peninsula, N. Italy.* Al Au Cz Ge Gr Hu It Ju Po Rm Rs (N, B, C, W).

C. euboeum Halácsy, *Magyar Bot. Lapok* 11: 152 (1912), from E. Greece (C. Evvoia) is an obscure species which appears to have been collected once only. It is probably related to **1**, but is *c.* 30 cm and smaller throughout, and has ternately pinnatisect leaves with wide, undivided segments.

2. C. byzantinum Boiss., *Ann. Sci. Nat.* ser. 3 (Bot.), **2**: 65 (1844). Like **1** but the upper leaf-lobes obliquely cordate at base, with regularly serrate margin and broadly acute apex. *S.E. part of Balkan peninsula.* Bu Tu.

3. C. heldreichii Orph. ex Boiss., *Diagn. Pl. Or. Nov.* 3(2): 104 (1856). Like **1** but lower leaves densely pubescent beneath, the lobes crenate-serrate, broadly acute; fruit 17–25 mm, linear-oblong; stylopodium elongate, passing imperceptibly into the rigid styles; pedicels becoming about as thick as the fruit. ● *Mountains of C. & S. Greece.* Gr.

4. C. coloratum L., *Mantissa* 57 (1767). Annual. Stem up to 100 cm, with long, patent hairs. Leaves 3-pinnate; lobes 0·5–1·5 mm wide, linear. Bracts absent or rarely one, small, subulate; bracteoles lanceolate or ovate-lanceolate, membranous, very conspicuous in fruit. Petals yellow. Fruit 8–12 mm. ● *Albania & W. Jugoslavia.* Al Ju.

5. C. creticum Boiss. & Heldr. in Boiss., *Diagn. Pl. Or. Nov.* 2(10): 51 (1849). Like **4** but stem more or less glabrous; leaf-lobes up to 2 mm wide, linear-lanceolate; bracts present, sometimes divided and leaf-like; bracteoles sometimes divided at apex; fruit 17–25 mm. ● *Kriti (Levka Ori, near Omalos).* Cr.

6. C. azoricum Trelease, *Ann. Rep. Missouri Bot. Gard.* **8**: 116 (1897). Robust herb *c.* 60 cm; stem somewhat purplish. Leaves (1–)2-pinnate; lobes irregularly and jaggedly serrate; petiole, rhachis and lamina with an often dense covering of patent hairs. Bracts 0–3, probably deciduous; bracteoles large and conspicuous. Petals white. Ripe fruit unknown, but probably 15 mm; stylopodium nearly 1 mm, conical; styles patent. ● *Açores.* Az.

(7–9). **C. hirsutum** group. Robust perennials with stems up to 120 cm. Leaves 2- to 3-pinnate. Stems and leaves densely hairy to subglabrous. Bracteoles often unequal. Petals white or sometimes pinkish. Fruit 4–20 mm, narrowly ovoid-oblong, tapering gradually upwards. Styles nearly erect.

A difficult group which requires further investigation. The species here recognized are sometimes treated as subspecies, but as **7** and **9**, at least, occur sympatrically in the Alps without intermediates, specific status seems more appropriate.

1 Carpophore divided to ⅓ of its length or less, somewhat swollen above the base; lowest segments nearly as large as the rest of the leaf **7. hirsutum**
1 Carporphore divided nearly to the base, not noticeably swollen; lowest segments much smaller than the rest of the leaf
2 Lateral umbels mostly opposite or verticillate; stem and leaves with a short, soft, silky pubescence **8. elegans**
2 Lateral umbels alternate; stem and leaves subglabrous or with relatively stiff hairs **9. villarsii**

7. C. hirsutum L., *Sp. Pl.* 258 (1753) (*C. cicutaria* Vill.). Segments of leaves relatively broad and little-divided, tending to overlap one another. Fruit up to 12 mm. $2n=22$. *C. & S. Europe, mainly in mountain regions.* Al Au Bu Cz Ga Ge Gr He Hs Hu It Ju Po Rm Rs (W) [Be Da].

C. magellense Ten., *Fl. Nap.* 3, *Prodr. Suppl.* **4**: 15 (1824), from N. Italy, may deserve specific or subspecific status. It has large, little-divided leaves, and fruits which are 13–20 mm.

8. C. elegans Gaudin, *Fl. Helv.* **2**: 364 (1828). Segments of leaves relatively narrow, not overlapping, long-acuminate. Fruit 8–12 mm. $2n=22$. ● *Alps.* Au Hc It.

9. C. villarsii Koch, *Syn. Fl. Germ.* 317 (1835). Segments of leaves narrow and not overlapping, acute. Fruit 8–20 mm. $2n=22$. ● *C. Europe.* Al Au Ga Ge He It Ju.

10. C. bulbosum L., *Sp. Pl.* 258 (1753) (*C. laevigatum* Vis.). Biennial or monocarpic perennial with a short tuberous root. Stem up to 200 cm, glabrous above, hairy below and often tinged or spotted with purple-brown. Leaves 2- to 3-pinnate; lobes 0·5–2·0 mm wide. Bracts 0–1; bracteoles lanceolate to obovate, glabrous. Petals white. Fruit 5–7 mm; styles about as long as the flattened upper surface of the stylopodium. *E. & C. Europe, extending to N. Italy.* Au Bu Cz Fe Ga Ge He *Ho It Ju Po Rm Rs (N, C, W, K, E) Su [Be].

(a) Subsp. **bulbosum**: Usually 100–200 cm. Leaves finely divided. Bracteoles usually lanceolate; styles patent. *C. & S.E. Europe.*

(b) Subsp. **prescottii** (DC.) Nyman, *Consp.* 300 (1879) (*C. prescottii* DC.): Usually *c.* 50 cm. Leaves less finely divided. Bracteoles usually obovate or ovate-lanceolate; styles at first nearly erect, diverging somewhat later. *Sweden, Finland and USSR.*

11. C. aureum L., *Sp. Pl.* ed. 2, 370 (1762) (*C. maculatum* Willd. ex DC.). Robust perennial up to 150 cm. Stem glabrous or with more or less patent hairs, frequently purple-spotted. Leaves 3-pinnate, yellowish-green, hairy; margins sometimes ciliate; apical lobes long and narrow, serrate. Rays 12–18. Bracts 0–1; bracteoles hairy. Petals white. Fruit 8–12 mm, rather abruptly contracted near the apex; styles much longer than the flattened upper surface of the stylopodium. *C. & S. Europe.* Al Au Bu Cz Ga Ge Gr He Hs Hu It Ju Po Rm Rs (K) Tu [Br].

12. C. temulentum L., *Sp. Pl.* 258 (1753). Erect biennial up to 100 cm. Stem clothed with setose, more or less appressed hairs, purple-spotted to almost entirely purple. Leaves 2- to 3- pinnate; apical lobes with relatively obtuse points, dark green, with appressed hairs on both surfaces. Rays 6–12. Bracts absent; bracteoles hairy. Petals white. Fruit 4–7 mm, gradually narrowed to the apex; styles about as long as the flattened upper surface of the stylopodium. $2n = 14, 22$. *Most of Europe, but absent from or casual in much of the north and rare in the Mediterranean region.* All except Az Bl Cr Fa Fe Is No Sa Sb.

The fruit is always glabrous in W. & N. Europe. In S. & S.E. Europe f. *eriocarpum* Guss., with pubescent fruit, occurs locally.

13. Anthriscus Pers.[1]

(*Cerefolium* Fabr.)

Leaves 2- to 3-pinnate. Sepals minute or absent. Petals white, emarginate; apex inflexed. Fruit narrowly oblong, rarely ovoid, with a usually well-developed beak; commissure constricted. Ridges confined to the beak; vittae solitary.

1 Robust perennial or biennial; fruit with a short, poorly developed beak   **(1–4). sylvestris** group
1 Annual; fruit with a prominent, well-developed beak
   2 Fruit very narrow; beak up to 4 mm, $\frac{1}{3}$–$\frac{1}{2}$ as long as the rest of
     the fruit   **5. cerefolium**
   2 Fruit ovoid to oblong-ovoid; beak up to 2 mm, about $\frac{1}{4}$ as
     long as the rest of the fruit
     3 Fruit usually with stiff, spine-like bristles (rarely glabrous);
       petals minute, inconspicuous   **6. caucalis**
     3 Fruit glabrous; petals ±conspicuous   **7. tenerrima**

(1–4). A. sylvestris group. Robust biennials or perennials up to 150 cm, often perennating by buds in the axils of the basal leaves. Leaves 3-pinnate with pinnatifid lobes. Rays 4–15, glabrous. Bracts 0; bracteoles several, lanceolate-ovate. Pedicels elongating but not becoming noticeably thickened in fruit. Fruit up to 10 mm, narrowly ovoid-oblong, smooth or with bristly tubercles, very shortly beaked.

A group of closely related taxa that have sometimes (e.g. by Thellung in Hegi, *Ill. Fl. Mitteleur.*) been treated as subspecies. Evidence of sympatric occurrence without obvious hybridization suggests that specific status is more appropriate.

1 Fruit black or dark brown when ripe, smooth and shiny, very
   rarely with a few small tubercles
   2 Lowest primary leaf-divisions much smaller than the rest of
     the leaf   **1. sylvestris**
   2 Lowest primary leaf-divisions nearly as large as the rest of the
     leaf   **2. nitida**
1 Fruit greenish to brownish when ripe, rather dull, usually with
   ±prominent tubercles, which often end in a curved bristle
   3 Leaf-lobes not distinctly cuneate   **3. nemorosa**
   3 Leaf-lobes distinctly cuneate   **4. fumarioides**

1. A. sylvestris (L.) Hoffm., *Gen. Umb.* 40 (1814) (*A. torquata* Coste, *Chaerophyllum sylvestre* L., *Cerefolium sylvestre* (L.) Besser). Leaves relatively dull. Partial umbels with 4–8 fruits. Fruit 7–10 mm, with thick, short hairs around the base (sometimes difficult to detect in very ripe material). $2n = 16$. *Almost throughout Europe but rare in the Mediterranean region.* Al Au Be Br Bu Cz Da Fe Ga Ge Gr He Hb Hs Ho Hu It Ju Lu No Po Rm Rs (N, B, C,W, K, E) Su.

2. A. nitida (Wahlenb.) Garcke, *Fl. Nord-Mittel-Deutschl.* ed. 7, 180 (1865) (*A. sylvestris* subsp. *alpestris* (Wimmer & Grab.)

---

[1] By J. F. M. Cannon.

Gremli). Leaves dark green and rather glossy. Partial umbels usually with 3–6 fruits. Fruit 5–7 mm, without a ring of hairs around the base. *C. & E. Europe.* Au Bu Cz Ga Ge He Hu It Ju Po Rm Rs (B, W).

3. A. nemorosa (Bieb.) Sprengel, *Pl. Umb. Prodr.* 27 (1813) (*A. aemula* (Woronow) Schischkin). Leaves glabrous above, hairy on the veins beneath; lobes not distinctly cuneate at base. Fruit 8–10 mm, greenish to blackish-green when ripe, usually tuberculate and hispidulous. *Italy, Balkan peninsula, U.S.S.R.* Al Bu Cr Gr It Ju Rm Rs (N, C, K) Si.

4. A. fumarioides (Waldst. & Kit.) Sprengel, *Pl. Umb. Prodr.* 27 (1813) (? *A. vandasii* Velen.). Like **3** but leaf-lobes cuneate at base; fruit 6–8 mm, yellowish-brown when ripe. ● *Italy and W. part of Balkan peninsula.* Al ?Au Ju It Gr.

5. A. cerefolium (L.) Hoffm., *Gen. Umb.* 41 (1814) (*A. longirostris* Bertol., *Cerefolium cerefolium* (L.) Schinz & Thell.). Wiry annual up to 70 cm. Leaves 3-pinnate with pinnatifid lobes. Rays 2–6, more or less pubescent. Bracteoles linear. Fruit 7–10 mm, almost linear, with a prominent, slender beak up to 4 mm. Styles much longer than the stylopodium, nearly erect. *Probably native in E.C. & S.E. Europe; widespread as an alien elsewhere; var.* cerefolium *cultivated as a herb and often naturalized.* Al Au Bu Cz Gr Hu Ju Po Rm Rs (C, W, E) [Be Br Da Ga Ge He Ho Hs It Rs (B)].

The fruit may be glabrous (var. *cerefolium*, which includes the cultivated plant) or may have numerous hooked hairs (var. *longirostris* (Bertol.) Cannon).

6. A. caucalis Bieb., *Fl. Taur.-Cauc.* 1: 230 (1808) (*A. scandicina* Mansfeld, *A. vulgaris* Pers., non Bernh., *Chaerophyllum anthriscus* (L.) Crantz, *Cerefolium anthriscus* (L.) G. Beck). Wiry annual up to 80(–100) cm, often purplish towards the base. Leaves 2- to 3-pinnate; lobes 1–10 mm, dentate or pinnatisect. Rays 2–6, glabrous. Bracteoles linear-lanceolate to ovate, aristate. Pedicels elongating and becoming thicker than the rays in fruit. Petals minute. Fruit 3 mm, ovoid; beak up to 2 mm, glabrous. $2n = 14$. *Dry places. W., S. & C. Europe, extending to S. Sweden and S. Ukraine.* Al Au Bl Br Bu Co Cz Da Ga Gr Hb He Ho Hs Hu It Ju Lu Po Rm Rs (W, K) Su Tu.

The fruit is usually covered with hooked spines (var. *caucalis*) but is glabrous in var. *neglecta* (Boiss. & Reuter) P. Silva & Franco.

7. A. tenerrima Boiss. & Spruner, *Ann. Sci. Nat.* ser. 3 (Bot.), 2: 60 (1844) (*A. tenella* Hayek). Slender, flexuous annual up to 40 cm, but usually less. Leaves 3-pinnate; lobes of the lower leaves ovate, those of the upper leaves sometimes linear. Rays 2–4. Bracteoles lanceolate-ovate. Pedicels elongating and becoming very strongly thickened in fruit. Petals conspicuous. Fruit 5 mm; beak up to 2 mm, glabrous. *Shady places on mountains.* ● *C. & S. Greece.* Gr.

14. Scandix L.[1]

Leaves (1-)2- to 3-pinnate, with narrow lobes. Umbels with few rays, sometimes reduced to one ray only. Sepals absent. Petals white, oblong, often very unequal in the outer flowers; apex incurved or inflexed. Fruit subcylindrical, slightly compressed laterally; beak up to four times as long as the seed-bearing part. Ridges prominent, slender; vittae very slender.

All species occur in open habitats, often as weeds.

1 Bracteoles obviously pinnate, with narrow, linear, patent lobes
1. stellata
1 Bracteoles entire or with 2 or more coarse, irregular teeth or forward-pointing lobes
2 Beak somewhat compressed laterally, not very clearly differentiated from the seed-bearing part of the fruit; carpophore usually 2-fid at the top
2. australis
2 Beak strongly compressed dorsally and obviously distinct from the seed-bearing part of the fruit; carpophore entire
3. pecten-veneris

1. S. stellata Banks & Solander in A. Russell, *Nat. Hist. Aleppo* ed. 2, **2**: 249 (1794) (*S. pinnatifida* Vent.). Up to 30 cm, slender. Leaves 1- to 3-pinnate, with narrowly linear lobes. Rays 1–3. Bracts absent or 1, like the leaves; bracteoles obviously pinnate, without a distinct membranous margin. Outer petals scarcely radiate. Beak of fruit 1·5–3 times as long as the seed-bearing part, strongly compressed dorsally. *S. Spain; S. Jugoslavia and N.E. Greece; Krym.* Gr Hs Ju Rs (K).

2. S. australis L., *Sp. Pl.* 257 (1753). Leaves 1- to 3-pinnate, with narrowly linear lobes. Rays 1–3. Bracts absent; bracteoles ovate to narrowly oblong-ovate, often with membranous margin, hairy or glabrous. Outer petals sometimes strongly radiate. Fruit 15–40 mm; beak of fruit usually at least twice as long as seed-bearing part, somewhat compressed laterally. *S. Europe.* Al Bu Cr Ga Gr Hs It Ju Lu Rm Rs (K) Sa Si Tu [Ga].

A very complex range of variation has stimulated the description of numerous subspecies. The following are well-marked:

1 Beak not longer than the seed-bearing part of fruit
(b) subsp. brevirostris
1 Beak distinctly longer than the seed-bearing part of fruit
2 Styles about as long as stylopodium
(a) subsp. microcarpa
2 Styles distinctly longer than stylopodium
3 Marginal petals much longer than the others and conspicuously radiate
(c) subsp. grandiflora
3 Marginal petals only slightly longer than others and not conspicuously radiate
(d) subsp. australis

(a) Subsp. **microcarpa** (Lange) Thell. in Hegi, *Ill. Fl. Mitteleur.* **5**(2): 1034 (1926) (*S. microcarpa* Lange): Slender. Bracteoles ovate-lanceolate, usually 2- to 3-fid, about as long as the rays. Styles about equalling the stylopodium, often reddish. Beak longer than the seed-bearing part of fruit. *Portugal & Spain.*

(b) Subsp. **brevirostris** (Boiss. & Reuter) Thell. in Hegi, *loc. cit.* (1926) (*S. brevirostris* Boiss. & Reuter): Bracteoles small and inconspicuous, with long hairs. Styles scarcely longer than the stylopodium. Beak not longer than the seed-bearing part of fruit. *Spain.*

(c) Subsp. **grandiflora** (L.) Thell. in Hegi, *op. cit.* 1035 (1926) (*S. grandiflora* L.): Often relatively robust. Bracteoles broadly elliptical and usually fringed with long hairs. Marginal radiate petals very conspicuous. Styles 4–6 times as long as the stylopodium. Beak longer than the seed-bearing part of fruit. *Balkan peninsula and Aegean region; Italy; naturalized in France.*

(d) Subsp. **australis** (incl. subsp. *balcanica* Vierh., subsp. *gallica* Vierh., subsp. *curvirostris* (Murb.) Vierh., *S. falcata* Loudon): Bracteoles very variable in shape and pubescence. Marginal petals only slightly longer than the rest and not conspicuously radiate. Styles at least twice as long as stylopodium. Beak much longer than the seed-bearing part of fruit. *Mediterranean region; Romania; Krym.*

3. S. pecten-veneris L., *Sp. Pl.* 256 (1753). Up to 50 cm. Leaves 2- to 3-pinnate, with linear lobes. Rays 1–3. Bracts absent or rarely few; bracteoles sometimes with membranous margins,

simple or irregularly divided, with jagged teeth, often with patent hairs. Outer petals often somewhat enlarged and radiate. Fruit 15–80 mm, usually large and robust; beak usually longer than seed-bearing part of fruit, more or less strongly flattened dorsally. *W., C. & S. Europe, extending as a casual northwards to 66° N. in Fennoscandia.* Al Au Be Bl Br Bu Co Cr Cz Da Ga Ge Hb He Ho Hs Hu It Ju Lu Po Rm Rs (K) Si Su Tu.

This species, like **2**, shows a complex range of variation.

1 Beak less than twice as long as the seed-bearing part of fruit; styles less than 0·5 mm; fruit *c.* 15 mm   **(a) subsp. brachycarpa**
1 Beak at least twice as long as the seed-bearing part of fruit; styles 0·4–2·5 mm; fruit 20–80 mm
2 Beak strongly flattened, 3–4 times as long as the seed-bearing part of fruit; styles 1–2·5 mm, 2–4(–6) times as long as the stylopodium; bracteoles conspicuous, longer than the pedicels and persistent   **(b) subsp. pecten-veneris**
2 Beak less strongly flattened, 2–3 times as long as the seed-bearing part of fruit; styles 0·4–0·75 mm, less than twice as long as the stylopodium; bracteoles small, oblong-lanceolate, as long as the pedicels and soon withering
(c) subsp. macrorhyncha

(a) Subsp. **brachycarpa** (Guss.) Thell. in Hegi, *Ill. Fl. Mitteleur.* **5**(2): 1038 (1926) (*S. brachycarpa* Guss.): *Greece, Italy & Sicilia. Naturalized in France & Germany.*

(b) Subsp. **pecten-veneris**: $2n = 26$. *Throughout the range of the species.*

(c) Subsp. **macrorhyncha** (C. A. Meyer) Rouy & Camus, *Fl. Fr.* **7**: 299 (1901) (*S. macrorhyncha* C. A. Meyer, *S. hispanica* Boiss.): *S. Europe.*

S. iberica Bieb., *Fl. Taur.-Cauc.* **1**: 425 (1808) from S.W. Asia, has been recorded from near Istanbul, but probably only as a casual. It is like **3** but with (5–)6–9 rays in the terminal umbel, outer petals strongly radiate, and beak of fruit less strongly flattened.

15. Myrrhis Miller[1]

Leaves 2- to 3-pinnate. Sepals minute. Petals white, cuneate-obovate; apex short, inflexed. Fruit linear-oblong, beaked. Ridges very prominent; vittae obsolete at maturity.

1. M. odorata (L.) Scop., *Fl. Carn.* ed. 2, **1**: 207 (1772). Puberulent, strongly aromatic perennial up to 200 cm. Leaf-lobes oblong-lanceolate, deeply dentate; sheaths conspicuous. Rays 4–20. Bracts usually absent; bracteoles several. Some partial umbels bearing male flowers only, the peduncles of these shorter and more slender than those of the hermaphrodite umbels. Fruit 15–25 mm, scabrid with bristly hairs especially near the top, dark shiny brown when mature. *Alps, Pyrenees, Appennini, mountains of W. part of Balkan peninsula; cultivated for flavouring and for fodder, and widely naturalized elsewhere.* Al Au Ga *Ge He Hs It Ju [Be Br Cz Da Fe Hb Ho Is No Po Rs (N, B, W) Su].

16. Molopospermum Koch[1]

Leaves 2- to 4-pinnate. Sepals present. Petals white, lanceolate; apex more or less inflexed. Fruit ovoid, compressed laterally. Ridges prominent, the dorsal narrowly winged.

1. M. peloponnesiacum (L.) Koch, *Nova Acta Acad. Leop.-Carol.* **12**(1): 108 (1824) (*M. cicutarium* DC.). Robust, glabrous perennial up to 200 cm. Basal and lower cauline leaves *c.* 100 cm; lobes lanceolate, deeply incise-dentate. Lateral umbels often verticillate. Rays 12–21. Bracts and bracteoles 6–9, lanceolate,

---

[1] By J. F. M. Cannon.

acuminate, unequal, sometimes leaf-like, margin broadly membranous. Sepals obtuse, deciduous. Styles longer than the stylopodium. Fruit 12 mm. ● *S. Alps; Pyrenees.* Ga Ge He Hs It Ju.

17. Coriandrum L.[1]

Lowest leaves lobed, others 1- to 3-pinnate. Sepals conspicuous, unequal. Petals white, the outer larger and deeply 2-lobed; apex inflexed. Fruit ovoid or globose, hard; mericarps not separating at maturity. Ridges low; vittae solitary, inconspicuous in fruit.

1. **C. sativum** L., *Sp. Pl.* 256 (1753). Glabrous annual 15–50 cm, foetid when fresh. Segments of lower leaves ovate-cuneate, irregularly toothed; lobes of upper leaves linear. Rays 3–5(–10). Bracts 0 or 1; bracteoles usually 3, linear. Fruit 2–6 × 2–5·5 mm. 2n=22. *Cultivated for its aromatic fruits and widely naturalized in S. Europe, more rarely or casual further north.* [Au Az Cr Cz Ga Ge Gr He Hs Hu It Ju Lu Po Rm Rs (B, C, W, K, E) Si.] (*N. Africa, W. Asia.*)

18. Bifora Hoffm.[1]

Leaves 1- to 2-pinnate; lobes linear or filiform. Sepals small or absent. Petals white, obcordate; apex inflexed. Fruit didymous; mericarps almost spherical, attached by the small commissure but separating when ripe. Ridges scarcely visible.

Lobes of upper leaves linear, flat; rays 1–3(–5); petals of all flowers
  nearly equal; style *c.* 0·2 mm   **1. testiculata**
Lobes of upper leaves filiform; rays 3–8; outer petals of marginal
  flowers much larger than others; style 1–1·5 mm   **2. radians**

1. **B. testiculata** (L.) Roth, *Enum.* **1**(1): 888 (1827). Glabrous annual 20–40 cm. Stem usually freely branched. Leaves 1- to 2-pinnate, oblong in outline; lobes of upper leaves linear, flat, entire or lobed. Rays up to 10 mm, 1–3(–5). Bracts 0 or 1; bracteoles 2–3, subulate. Petals of all flowers nearly equal. Fruit 2·5–3·5 × 4·5–7 mm, rugose, shortly beaked; style not more than 0·2 mm, scarcely as long as stylopodium. *S. Europe.* Al Bl Co Cr Ga Gr Hs It Ju Lu Sa Si Tu.

2. **B. radians** Bieb., *Fl. Taur.-Cauc.* **3**: 233 (1819). Like **1** but lobes of upper leaves filiform; rays up to 25 mm, 3–8; outer petals of marginal flowers much larger than others; fruit rugulose, unbeaked; style 1–1·5 mm, at least twice as long as stylopodium. *S. & S.C. Europe, extending to Ukraine, but absent from most of the Iberian peninsula; naturalized elsewhere in C. Europe.* Al Bl Bu *Cz Ga Ge Hs *Hu It Ju Rm Rs (W, K) Si Tu [Au Ge He Po].

19. Scaligeria DC.[1]

Leaves 2- to 3-pinnate. Sepals absent. Petals white, obcordate, shortly clawed; apex inflexed. Fruit broadly ovoid from a cordate base, compressed laterally, shortly beaked. Ridges slender; vittae 2–3.

1. **S. cretica** (Miller) Boiss., *Diagn Pl. Or. Nov.* **2**(10): 52 (1849). Glabrous biennial 40–50 cm, with a tuberous root. Segments of basal leaves rhombic-ovate, dentate or 3-fid, of cauline leaves pinnatisect, with linear, entire lobes. Rays 6–15(–20), slender. Bracts absent; bracteoles few, linear-lanceolate. Fruit 1·5–2 mm. *Aegean region and S. & W. parts of Balkan peninsula.* Al Cr Gr Ju.

Subsp. **halophila** Rech. fil., *Denkschr. Akad. Wiss. Math.-Nat. Kl. (Wien)* **105**(2): 102 (1943), was described from a single collection from the spray-zone of an islet off Kriti. It has fleshy leaves and larger fruits (2·5 mm) than the typical plant. Until it has been studied further its status remains doubtful.

20. Smyrnium L.[1]

Lower leaves usually 2- to 3-ternate, upper usually simple. Sepals absent. Petals yellow, lanceolate to obcordate; apex inflexed. Fruit ovoid or subglobose, didymous. Ridges slender, the marginal usually inconspicuous; vittae numerous, scattered.

1 Upper cauline leaves not amplexicaul; lamina divided
                                                          1. olusatrum
1 Upper cauline leaves amplexicaul; lamina undivided
  2 Upper leaves and branches opposite; rays 15–20
    3 Lobes of basal leaves cordate-ovate; lower cauline leaves
        cordate-ovate, dentate or shallowly lobed   **2. orphanidis**
    3 Lobes of basal leaves oblong-cuneate; lower cauline leaves
        3-partite   **3. apiifolium**
  2 Upper leaves and branches alternate; rays 5–12
    4 Stems narrowly winged on the angles; upper leaves crenate-
        serrate, sometimes minutely so   **4. perfoliatum**
    4 Stems not winged on the angles; upper leaves entire, rarely
        shallowly denticulate   **5. rotundifolium**

1. **S. olusatrum** L., *Sp. Pl.* 262 (1753). Glabrous biennial 50–150 cm. Stem stout, solid, becoming hollow when old; upper branches often opposite. Leaves dark green and shiny; basal *c.* 30 cm, triangular in outline, ternate; segments 1- to 2-pinnate; lobes 10–60 mm, rhombic-ovate, crenate-dentate and sometimes lobed; cauline leaves smaller and less divided, with short, inflated petioles. Rays (3–)7–15(–18). Bracts and bracteoles few, small, sometimes 0. Fruit 7–8 mm, black. 2n=22. *S. Europe, extending northwards to N.W. France; extensively naturalized in Britain and Ireland.* Al Az Bl Co Cr Ga Gr Hs It Ju Lu Sa Si Tu [Br Hb Ho].

2. **S. orphanidis** Boiss., *Fl. Or.* **2**: 925 (1872). Glabrous biennial up to 130 cm. Stem stout, terete, hollow; upper branches opposite. Basal leaves *c.* 30 cm, 3-ternate; lobes 20–40 mm, cordate-ovate, dentate or shallowly lobed; lower cauline leaves ovate, subcordate, coarsely lobed and obtusely dentate, with short, inflated petioles; upper cauline leaves opposite, ovate-cordate, amplexicaul, crenate-dentate. Rays 15–20. Bracts and bracteoles absent. Fruit 3 mm. *Greece and Aegean region.* Gr Tu.

3. **S. apiifolium** Willd., *Sp. Pl.* **1**: 1468 (1798). Like **2** but lobes of basal leaves smaller, oblong-cuneate, deeply and acutely dentate; lower cauline leaves 3-partite; upper oblong-cordate, sharply dentate. *Aegean region.* Cr Gr.

4. **S. perfoliatum** L., *Sp. Pl.* 262 (1753). Nearly glabrous biennial 50–150 cm. Stem angled and narrowly winged on the angles, solid, pubescent at the nodes; upper branches alternate. Basal leaves 1- to 2-ternate; lobes ovate, dentate or somewhat lobed; upper leaves ovate-cordate, amplexicaul, crenate-serrate, rarely subentire. Rays 5–12. Bracts and bracteoles absent. Styles longer than stylopodium. Fruit 3–3·5 × 5–5·5 mm, brownish-black. *S. Europe, northwards to E. Czechoslovakia.* Al Bu Cr Cz Ga Gr Hs Hu It Ju Lu Rm Rs (K) Sa Si Tu [Au Br Da Ge].

5. **S. rotundifolium** Miller, *Gard. Dict.* ed. 8, no. 2 (1768). Like **4** but stem ridged, not winged; upper leaves entire, rarely shallowly denticulate; styles shorter than stylopodium; fruit 2–2·5 × 3–3·5 mm. *Mediterranean region, from Corse and Sardegna eastwards.* Bu Co Cr Gr It Ju Sa Si.

[1] By T. G. Tutin.

21. Bunium L.[1]

(*Bulbocastanum* Miller)

Stock a more or less globose tuber. Subterranean part of stem flexuous. Leaves 2- to 3-pinnate with narrow lobes. Sepals small or absent. Petals white, obcordate; apex inflexed. Fruit oblong-obovoid or oblong, laterally compressed. Ridges thick and prominent or slender; vittae 1–3. Stylopodium abruptly contracted into the style. Cotyledon solitary through abortion.

1 Fruiting pedicels not more than 0·3 mm in diameter, much thinner than the fruit
  2 Bracts 5–10; pedicels minutely toothed on the inner edge
                                           1. bulbocastanum
  2 Bracts 1–5(–6); pedicels not toothed   **2. alpinum**
1 Fruiting pedicels at least 0·5 mm in diameter, almost as thick as the fruit
  3 Rays 3–6; primary ridges of fruit slender   **2. alpinum**
  3 Rays 6–15; primary ridges of fruit thick
    4 Bracts 0–2; sepals 0 or minute; styles about as long as stylopodium   **3. ferulaceum**
    4 Bracts 6–8; sepals up to 0·4 mm; styles longer than the stylopodium   **4. pachypodum**

1. B. bulbocastanum L., *Sp. Pl.* 243 (1753). Usually erect perennial (5–)30–100 cm. Basal leaves with linear-lanceolate lobes. Rays (5–)10–20, slender; pedicels minutely toothed on inner edge. Bracts 5–10, lanceolate, acuminate; bracteoles like the bracts. Sepals absent or minute. Fruit 3–5 mm, oblong-ellipsoid; primary ridges slender; vittae solitary. *From S. England and Islas Baleares eastwards to C. Germany and N.W. Jugoslavia.* Be Bl Br Co Ga Ge He Ho It Ju Sa Si [*Au Cz Da].

2. B. alpinum Waldst. & Kit., *Pl. Rar. Hung.* 2: 199 (1804). Procumbent or erect perennial up to 50(–75) cm. Basal leaves with linear to elliptical or lanceolate lobes. Rays 3–10(–16), slender or stout; pedicels slender or stout, not toothed on inner edge. Bracts 1–5(–6), linear or lanceolate; bracteoles like the bracts. Sepals absent or minute. Fruit 2–4(–5·5) mm; primary ridges slender; vittae 2–3. *Mountains of S. Europe.* Al Co Ga Hs It Ju Sa Si.

A very variable species, the variation being correlated with altitude and habitat as well as distribution. The following division into subspecies is somewhat tentative.

1 Rays 5–10(–16)
  2 Leaf-lobes not more than 5(–7) mm   **(c) subsp. montanum**
  2 Leaf-lobes 5–10(–20) mm   **(d) subsp. macuca**
1 Rays 3–5(–6)
  3 Fruit 3·5–5·5 mm; leaf-lobes very obtuse   **(e) subsp. petraeum**
  3 Fruit 2·5–3·5 mm; leaf-lobes acute or subobtuse
    4 Fruit oblong or oblong-ellipsoid, widest at the middle   **(a) subsp. alpinum**
    4 Fruit oblong-obovoid, widest above the middle   **(b) subsp. corydalinum**

(a) Subsp. **alpinum**: Stems usually procumbent; leaf-lobes usually 3–5 mm, linear to lanceolate, acute or subobtuse. Rays 3–5(–6), 10–25 mm, slender or somewhat thickened in fruit. Fruit 2·5–3·5 mm, oblong or oblong-ellipsoid. ● *Mountains of N.W. part of Balkan peninsula.*

(b) Subsp. **corydalinum** (DC.) Nyman, *Consp.* 304 (1879) (*B. corydalinum* DC., *B. alpinum* sensu Lange, non Waldst. & Kit.): Like subsp. (a) but rays 5–10(–20) mm; fruit oblong-obovoid. ● *High mountains of S. Spain; Corse, Sardegna.*

---

[1] By P. W. Ball.

(c) Subsp. **montanum** (Koch) P. W. Ball, *Feddes Repert.* **79**: 62 (1968) (*B. montanum* Koch): Stems usually erect; leaf-lobes 3–5(–7) mm, linear to lanceolate, acute. Rays 5–10(–16), 20–40 mm, slender. Fruit 2·5–3·5 mm, oblong. ● *W. part of Balkan peninsula;* ? *Italy.*

(d) Subsp. **macuca** (Boiss.) P. W. Ball, *Feddes Repert.* **79**: 62 (1968) (*B. macuca* Boiss.): Like subsp. (c) but leaf-lobes 5–10(–20) mm. *S. Spain; Sicilia.*

(e) Subsp. **petraeum** (Ten.) Rouy & Camus, *Fl. Fr.* 7: 350 (1901): Stems procumbent; leaf-lobes 4–8 mm, linear-elliptical, very obtuse. Rays 3–5(–6), 4–6 mm, stout. Fruit 3·5–5·5 mm, ellipsoid or oblong-ellipsoid. ● *C. Appennini.*

3. B. ferulaceum Sibth. & Sm., *Fl. Graec. Prodr.* 1: 186 (1806). Erect or ascending perennial 20–60 cm. Basal leaves with linear, acute lobes. Rays 6–15; pedicels almost as thick as the fruit. Bracts 0–2, lanceolate; bracteoles 3–6, lanceolate. Sepals absent or minute. Fruit 4–6 mm, oblong or obovoid-oblong; primary ridges thick and very prominent; vittae solitary; styles about as long as stylopodium. *Balkan peninsula and Aegean region; Krym.* Bu Cr Gr Ju Rs (K) Tu.

4. B. pachypodum P. W. Ball, *Feddes Repert.* **79**: 63 (1968) (*Bulbocastanum incrassatum* Lange pro parte). Like **3** but lobes of leaves subobtuse; bracts 6–8; sepals up to 0·4 mm, distinct; styles longer than the stylopodium. *S.W. Europe.* Bl *Ga Hs Lu.

22. Conopodium Koch[1]

Stock a more or less globose tuber. Subterranean part of stem flexuous. Leaves 2- to 3-pinnate. Sepals absent. Petals white, rarely pink, often with a wide, brown vein on the back, obcordate; apex inflexed. Fruit oblong-ovoid, laterally compressed. Ridges filiform, indistinct; vittae 2–3. Stylopodium more or less attenuate into styles. Cotyledons 2.

A taxonomically difficult genus requiring a thorough revision in the Iberian peninsula. The value and reliability of the characters used to separate the species is not clear, although many difficulties could be resolved by field studies and cultivation.

1 Styles deflexed and appressed to the stylopodium in fruit
  2 Bracts usually solitary; bracteoles 2 or more   **6. thalictrifolium**
  2 Bracts 0; bracteoles 0–1   **7. bunioides**
1 Styles erect or patent, or rarely somewhat deflexed, but not appressed to the stylopodium
  3 Lower part of stem leafless   **1. majus**
  3 Lower part of stem leafy, with ± persistent leaf-bases
    4 Middle and upper cauline leaves with sheath not more than 7 mm and less than ¼ as long as lamina   **4. ramosum**
    4 Middle and upper cauline leaves with sheath usually more than 7 mm and always at least ¼ as long as lamina
      5 Lower cauline leaves with ovate or suborbicular, pinnatifid to dentate lobes   **2. pyrenaeum**
      5 Lower cauline leaves with linear, lanceolate or oblong lobes
        6 Cauline leaves with linear or lanceolate lobes; bracteoles 0–1   **3. bourgaei**
        6 Cauline leaves with long, linear-setaceous to linear lobes; bracteoles usually 5 or more   **5. capillifolium**

1. C. majus (Gouan) Loret in Loret & Barrandon, *Fl. Montpell.* ed. 2, 214 (1886) (*C. denudatum* Koch). Stems 15–50(–90) cm, erect, leafless at the base. Basal leaves with elliptical to ovate, acutely lobed lobes; cauline leaves with linear to filiform lobes; sheaths of middle and upper leaves 1–8 mm, not more than ¼ as long as lamina. Umbels with 6–12 rays; bracts 0–2; bracteoles 2 or more. Fruit 3–4 mm; stylopodium about as long as wide;

styles erect or erecto-patent. ● *W. Europe, extending eastwards to Italy.* Br Co Ga Hb Hs It Lu No ?Si [Fa].

2. C. pyrenaeum (Loisel.) Miégeville, *Bull. Soc. Bot. Fr.* **21**: xxxii (1874). Like **1** but stems 40–70 cm, leafy to the base; cauline leaves with broadly ovate to suborbicular, obtusely lobed lobes; sheaths usually more than 10 mm, at least ¼ as long as the lamina; umbels with up to 16 rays. ● *W. Pyrenees and mountains of N. Spain; C. Portugal.* Ga Hs Lu.

Perhaps only a subspecies of **1**.

3. C. bourgaei Cosson, *Not. Pl. Crit.* 110 (1851) (incl. *Heterotaenia arvensis* Cosson). Stems up to 70 cm, procumbent or erect, leafy to the base. Basal leaves with linear or oblong, acute lobes; cauline leaves with linear or lanceolate lobes; sheaths of middle and upper leaves 7 mm or more, at least ¼ as long as lamina, usually densely pubescent. Umbels with 3–14 rays; bracts 0; bracteoles 0–1. Fruit 2·5–4 mm; stylopodium wider than long; styles erect or patent, rarely somewhat deflexed. ● *Spain and Portugal.* Hs Lu.

4. C. ramosum Costa, *Ind. Sem. Horti Barcin.* **1860** (1860). Like **3** but sheaths of middle and upper cauline leaves 2–7 mm, less than ¼ as long as lamina, usually glabrous or ciliate. ● *C. & E. Spain; C. Portugal.* Hs Lu.

5. C. capillifolium (Guss.) Boiss., *Voy. Bot. Midi Esp.* **2**: 736 (1845) (incl. *C. elatum* Willk., *C. marianum* Lange, *C. subcarneum* (Boiss. & Reuter) Boiss.). Stems 20–80 cm, erect, leafy to the base. Basal leaves with ovate or elliptical, acutely lobed lobes; cauline leaves with very long, linear-setaceous lobes; sheaths of middle and upper leaves 7 mm or more, at least ¼ as long as lamina. Umbels with 6–20(–25) rays; bracts 0–2; bracteoles 5 or more. Fruit 3–5 mm; stylopodium usually wider than long; styles erect or erecto-patent. *S. Europe.* ?Gr Hs Lu It Si.

C. brachycarpum Boiss. ex Lange, *Vid. Meddel. Dansk Naturh. Foren. Kjøbenhaven* **1865**: 44 (1866), from N.W. Spain, is doubtfully distinct from **5**. It is said to differ in the cauline leaves with linear-lanceolate lobes; the bracts with a narrow (not a wide) membranous margin; the petals with a wide, brown vein on the back.

6. C. thalictrifolium (Boiss.) Calestani, *Webbia* **1**: 279 (1905) (*Heterotaenia thalictrifolia* (Boiss.) Boiss.). Stems 30–60 cm, erect, leafy to the base. Basal leaves with ovate to suborbicular, obtusely lobed lobes; cauline leaves with linear-lanceolate lobes. Umbels with 5–8 rays; bracts usually 1; bracteoles 2 or more. Fruit 3–3·5 mm; stylopodium wider than long; styles deflexed and appressed to the stylopodium in fruit. ● *S. Spain.* Hs.

7. C. bunioides (Boiss.) Calestani, *loc. cit.* (1905) (*Butinia bunioides* Boiss.). Stems up to 20 cm, procumbent, leafy to the base. Basal and cauline leaves with lanceolate or elliptical, obtuse, sometimes 3-lobed lobes. Umbels with 3–5 rays; bracts and bracteoles 0. Fruit 2·5–3 mm; stylopodium wider than long; styles deflexed and appressed to the stylopodium in fruit. ● *C. & S. Spain (Sierra Nevada, Sierra de Gredos).* Hs.

23. Huetia Boiss.[1]

(*Biasolettia* Koch, non C. Presl; *Freyera* Reichenb., non *Freyeria* Scop.)

Like *Bunium* but fruit linear-oblong; vittae absent at maturity; stylopodium attenuate into the style.

1 Ridges of fruit minutely toothed; bracteoles pubescent **3. pumila**
1 Ridges of fruit not toothed; bracteoles glabrous, rarely ciliate
2 Bracteoles glabrous **1. cynapioides**
2 Bracteoles ciliate **2. cretica**

1. H. cynapioides (Guss.) P. W. Ball, *Feddes Repert.* **79**: 16 (1968) (*Freyera cynapioides* (Guss.) Griseb.). Procumbent or erect perennial up to 40(–75) cm. Lobes of basal leaves linear or linear-lanceolate, acute, rarely elliptical, obtuse. Rays 3–15. Bracts 0(–3); bracteoles linear or lanceolate, glabrous. Fruit 3–6 mm; ridges not toothed. *Balkan peninsula; C. & S. Italy.* Al Bu Gr It Ju.

A rather variable species in Greece. The three following ill-defined subspecies may be recognized.

1 Leaf-lobes 1·5–4 mm wide, obtuse; rays 3–7 **(c) subsp. divaricata**
1 Leaf-lobes 0·5–1·5 mm wide, acute
2 Pedicels usually with short teeth at apex; rays 6–15
                   (a) subsp. cynapioides
2 Pedicels without teeth at apex; rays 3–8 **(b) subsp. macrocarpa**

(a) Subsp. cynapioides (incl. *Freyera congesta* Boiss. & Heldr., *F. bornmuelleri* (H. Wolff) Hayek, *Conopodium graecum* Freyn & Sint.): Leaf-lobes 0·5–1·5 mm wide, linear or linear-lanceolate, acute; rays 6–15; pedicels usually with short teeth at the apex. *Throughout the range of the species, except for much of C. & S. Greece.*

(b) Subsp. macrocarpa (Boiss. & Spruner) P. W. Ball, *Feddes Repert.* **79**: 16 (1968) (*Butinia macrocarpa* Boiss. & Spruner, *Freyera macrocarpa* (Boiss. & Spruner) Boiss.; incl. *F. parnassica* Boiss. & Heldr.): Leaf-lobes 0·5–1·5 mm wide, linear or linear-lanceolate, acute; rays 3–8; pedicels without short teeth at the apex. *C. & S. Greece, extending locally to S. Albania.*

(c) Subsp. divaricata (Boiss. & Orph.) P. W. Ball, *Feddes Repert.* **79**: 16 (1968) (*Freyera divaricata* Boiss. & Orph.): Leaf-lobes 1·5–4 mm wide, elliptical or linear-elliptical, obtuse; rays 3–6(–7); pedicels without teeth at the apex. *S. & E. Greece.*

2. H. cretica (Boiss. & Heldr.) P. W. Ball, *Feddes Repert.* **79**: 16 (1968) (*Butinia cretica* Boiss. & Heldr., *Freyera cretica* (Boiss. & Heldr.) Boiss. & Heldr.). Procumbent perennial up to 5 cm. Basal leaves with broadly elliptical to oblong, subobtuse lobes. Rays 2–5. Bracts absent; bracteoles oblong or lanceolate, ciliate. Fruit *c.* 4·5 mm; ridges not toothed. ● *Kriti.* Cr.

Possibly only a subspecies of **1**.

3. H. pumila (Sibth. & Sm.) Boiss. & Reuter in Boiss., *Diagn. Pl. Or. Nov.* 3(2): 103 (1856) (*Freyera pumila* (Sibth. & Sm.) Boiss.). Procumbent perennial up to 10 cm. Basal leaves with linear-elliptical, subobtuse lobes. Rays 3–6. Bracts absent; bracteoles lanceolate, pubescent. Fruit 4·5–5 mm; ridges minutely toothed. ● *S.C. Greece (Parnassos).* Gr.

Records from other localities in Greece are erroneous.

24. Muretia Boiss.[2]

Leaves 2-pinnate; lobes linear. Sepals absent. Petals yellow, ovate-lanceolate or lanceolate, emarginate; apex inflexed. Fruit subcylindrical, slightly compressed laterally, not constricted at the commissure. Ridges filiform, scarcely prominent; vittae 2–3.

1. M. lutea (Hoffm.) Boiss., *Fl. Or.* **2**: 858 (1872). Perennial with a tuberous, more or less globose stock. Basal leaves long-petiolate; cauline with remote segments which are cut into few lobes 10–15 mm long. Rays 7–12, slender, very unequal. Bracts

---

[1] By P. W. Ball.   [2] By T. G. Tutin.

lincar, subobtuse, much shorter than the rays; bracteoles like the bracts. Fruit 3·5–4 mm. *Steppes and cultivated ground. S. part of U.S.S.R., northwards to c. 50° N. in the west and c. 53° N. in the east.* Rs (?C, W, E).

25. Pimpinella L.[1]

(Incl. *Reutera* Boiss., *Pancicia* Vis.)

Basal leaves usually entire or trisect, sometimes 1- to 3-pinnate; middle cauline usually 2-pinnate, with narrow lobes. Sepals usually minute. Petals white or yellow, rarely pink or purplish, not or slightly emarginate; apex inflexed. Fruit ovoid-oblong to subglobose, laterally compressed, constricted at the commissure. Ridges filiform, sometimes concealed by hairs, setae or tubercles; vittae (2–)3(–4).

```
 1  Petals yellow
 2    Lower leaves distinctly 2- to 3-pinnate
 3      Plant almost glabrous; fruit 2–2·5 mm, glabrous    4. procumbens
 3      Plant greyish-pubescent; fruit 1·5 mm, ±velutinous    1. rigidula
 2    Lower leaves 1-pinnate; segments occasionally pinnatisect
 4      Leaf-segments 25–60 mm; fruit 3·5–4 mm, ovoid-oblong
                                                               2. lutea
 4      Leaf-segments 8–30 mm; fruit 3 mm, subglobose    3. gracilis
 1  Petals white, pink or purplish
 5    Ovary and fruit quite glabrous
 6      Upper leaves palmately divided to base; lobes numerous,
          long, setaceous                                  12. serbica
 6      Upper leaves pinnately divided; lobes few
 7        Rays 2–7
 8          Plant ±hairy; fruit c. 4 mm              11. anisoides
 8          Plant glabrous; fruit 5–7 mm            15. bicknellii
 7        Rays (5–)10–25
 9          Fruit 5–6 mm; ridges narrowly winged     14. siifolia
 9          Fruit 2–3·5 mm; ridges unwinged
10            Stem usually sharply angled, hollow; ridges of fruit
                prominent, whitish                     13. major
10            Stem usually terete, nearly or quite solid; ridges of fruit
                inconspicuous                          16. saxifraga
 5    Ovary and fruit ±hairy
11      Annual or biennial; stock slender, without fibres or scale-
          like leaf-bases
12        Rays usually more than 15; fruit with long, patent hairs
                                                           7. peregrina
12        Rays 2–15; fruit with short, appressed hairs
13          Fruit 3–5 mm                              5. anisum
13          Fruit 1·5 mm                              6. cretica
11      Perennial; stock stout, with abundant scale-like leaf-bases
14        Lower leaves 1-pinnate
15          Plant densely white-tomentose; fruit c. 3 mm, oblong
                                                           9. pretenderis
15          Plant glabrous to greyish-pubescent; fruit c. 2 mm, ovoid
                                                           8. tragium
14        Lower leaves 2- to 3-pinnate
16          Petals glabrous on the back               11. anisoides
16          Petals hairy on the back
17            Leaf-lobes broadly ovate, crenate in upper half  10. villosa
17            Leaf-lobes oblong, incise-serrate or lobed    8. tragium
```

1. P. rigidula (Boiss. & Orph.) H. Wolff in Engler, *Pflanzenreich* 90 (IV. 228): 227 (1927) (*Reutera rigidula* Boiss. & Orph.). Greyish-pubescent perennial up to 100 cm. Stem terete, pubescent below, glabrous above, freely branched. Lower leaves c. 30 cm, 3-pinnate, triangular in outline; lobes cuneate-ovate, divided into linear-oblong, obtuse lobes; cauline leaves small or reduced to sheathing petioles. Umbels very numerous; peduncles short. Rays 4–8, slender. Bracts and bracteoles absent or few.

Petals yellow. Fruit c. 1·5 mm, ovoid, narrowed at apex, more or less velutinous. *Cultivated fields.* ● *S. Greece.* Gr.

2. P. lutea Desf., *Fl. Atl.* 1: 265 (1798). Somewhat pubescent perennial 80–140 cm. Stem terete, with numerous long branches above. Lower leaves c. 20 cm, pinnate; segments 25–60 mm, 7–11, cordate-ovate to suborbicular, crenate-dentate, sometimes shallowly lobed; upper cauline leaves reduced to petioles. Umbels few; peduncles short. Rays 3–5, filiform. Bracts absent or 1; bracteoles absent. Petals yellow. Fruit 3·5–4 mm, ovoid-oblong, glabrous. *Sunny volcanic rocks. Corse and Pantellaria.* Co Si. (*Algeria, Tunisia.*)

3. P. gracilis (Boiss.) H. Wolff in Engler, *Pflanzenreich* 90 (IV. 228): 228 (1927) (*Reutera gracilis* Boiss.; incl. *R. puberula* Loscos & Pardo). Almost glabrous or puberulent biennial or perennial up to 150 cm. Stock with brown, scale-like leaf-bases. Stem slender, terete, freely branched above. Lower leaves 5–18 cm, pinnate, oblong-lanceolate in outline; segments 8–30 mm, (3–)5–9, ovate, dentate and sometimes pinnatisect; upper cauline leaves reduced to the sheathing petiole. Rays 2–5, very slender. Bracts and bracteoles absent. Pedicels filiform. Petals yellow. Fruit c. 3 mm, subglobose, didymous, glabrous or scabrid-puberulent. ● *Mountains of E. & S. Spain.* Hs.

4. P. procumbens (Boiss.) H. Wolff, *op. cit.* 229 (1927) (*Reutera procumbens* Boiss.). Almost glabrous biennial with slender, terete, decumbent stems 10–30 cm. Stock with scale-like leaf-bases. Lower leaves 5–10 cm, 2- to 3-pinnate, ovate in outline; lobes 2–6 mm, lanceolate, acute, shortly hairy on margins; cauline very small, the lamina pinnatifid or 3-fid. Rays 3–6, filiform. Bracts and bracteoles absent; pedicels filiform. Petals yellow. Fruit 2–2·5 mm, ovoid-globose. *Rock-crevices in mountains.* ● *S. Spain* (*Sierra Nevada*). Hs.

5. P. anisum L., *Sp. Pl.* 264 (1753) (*Anisum vulgare* Gaertner). Finely pubescent, strongly aromatic annual 10–50 cm. Stem terete, striate, branched above. Lowest leaves reniform, incise-dentate or shallowly lobed; next leaves pinnate with 3–5, ovate or obovate, dentate segments; upper cauline leaves 2- to 3-pinnate, with linear-lanceolate lobes and narrow, sheathing petioles. Rays 7–15, sparsely puberulent. Bracts absent or 1; bracteoles usually few, filiform. Petals white. Fruit 3–5 mm, ovoid to oblong, shortly appressed-setose. *Widely cultivated for its aromatic fruits and often naturalized; native distribution unknown, but certainly of Asiatic origin.* [Au Bu Cr Cz Ga Ge Gr Hs Hu It Ju Lu No Po Rm Rs (C, W, E).]

6. P. cretica Poiret in Lam., *Encycl. Méth. Bot.*, *Suppl.* 1: 684 (1811). Puberulent or subglabrous annual 10–30 cm. Lowest leaves suborbicular, simple, cordate at base, crenate-dentate or shallowly lobed; next 3-foliolate, upper 1- to 2-pinnate, with few, linear, entire lobes. Rays (2–)5–10, filiform. Bracts and bracteoles absent. Fruit 1·5 mm, ovoid, densely setulose, grey-green. *Dry places. S. Aegean region.* Cr Gr.

7. P. peregrina L., *Sp. Pl.* 264 (1753) (*P. taurica* (Ledeb.) Steudel). Finely pubescent biennial 50–100 cm. Lowest leaves simple, cordate, serrate; next leaves pinnate, the segments 5–9, suborbicular, more or less cordate at base, crenate; upper cauline 2-pinnate, with linear, often recurved lobes. Rays 8–50, filiform, setulose. Bracts and bracteoles absent. Fruit c. 2 mm, ovoid, with patent hairs. *Dry places. S. Europe, eastwards from E. Spain.* Al Bu Co Cr Ga Gr Hs It Ju Rm Rs (K) Sa Si Tu.

8. P. tragium Vill., *Prosp. Pl. Dauph.* 24 (1779) (incl. *P. tomiophylla* (Woronow) Stankov). Somewhat pubescent to glabrous

---

[1] By T. G. Tutin.

perennial (5–)10–50(–100) cm. Stock with abundant scale-like remains of petioles. Stem striate, solid. Lower leaves pinnate and sometimes further divided, oblong in outline; segments obovate or lanceolate, often obliquely subcordate at base, serrate, crenate, dentate or sometimes lobed, very rarely with an additional pair of segments at the base of the primary ones; cauline leaves few, small. Rays 5–15, glabrous or pubescent. Bracts and bracteoles absent or few. Petals pubescent on back. Fruit *c.* 2 mm, ovoid, shortly tomentose. *S. & E. Europe, northwards to c. 54° N. in E. Russia.* Al Bl Bu Cr Ga Gr Hs It Ju Rm Rs (C, W, K, E) Si Tu.

Very variable in height, dissection of leaves, and hairiness. A number of subspecies have been described and specific rank has been given to some of them. Further investigation is needed, but the following seem to be fairly well-marked subspecies:

1 Leaves 2- to 3-pinnate   (e) subsp. **titanophila**
1 Leaves 1-pinnate
  2 Plant 5–10 cm; stock very stout, with the persistent leaf-bases not or scarcely sheathing it; leaf-lobes up to *c.* 5 mm, ovate, deeply toothed   (d) subsp. **depressa**
  2 Plant (10–)30–100 cm; stock slender to stout, with or without persistent, sheathing leaf-bases; leaf-lobes 5–15 mm, shallowly to deeply toothed
    3 Plant 10–30 cm; stock with long, slender branches; stem slender; cauline leaves few or 0; leaf-lobes suborbicular, not deeply toothed   (a) subsp. **tragium**
    3 Plant 30–100 cm; stock with short, stout branches; leaf-lobes ovate or deeply lobed
      4 Plant slender; leaf-lobes 5–10 mm, incise-serrate   (b) subsp. **lithophila**
      4 Plant stout; leaf-lobes 10–15 mm, not deeply serrate   (c) subsp. **polyclada**

(a) Subsp. **tragium**: *S.E. France; ? Sicilia.*

(b) Subsp. **lithophila** (Schischkin) Tutin, *Feddes Repert.* **79**: 62 (1968) (*P. lithophila* Schischkin; *P. tragium* var. *typica* Halácsy): *S. Europe, from Spain to Krym.*

(c) Subsp. **polyclada** (Boiss. & Heldr.): Tutin, *Feddes Repert.* **79**: 62 (1968) (*P. polyclada* Boiss. & Heldr.): *Balkan peninsula.*

(d) Subsp. **depressa** (DC.) Tutin, *Feddes Repert.* **79**: 62 (1968) (*P. depressa* DC.): *Greece and Kriti.*

(e) Subsp. **titanophila** (Woronow) Tutin, *Feddes Repert.* **79**: 62 (1968) (*P. titanophila* Woronow): *S.E. European U.S.S.R., but absent from Krym.*

9. P. pretenderis (Heldr.) Orph. ex Halácsy, *Consp. Fl. Graec.* **1**: 683 (1901). Like **8** but densely white-tomentose; leaf-lobes obovate to narrowly flabellate, toothed only towards the apex; fruit *c.* 3 mm, oblong. ● *Kikladhes.* Gr.

10. P. villosa Schousboe, *Kong. Danske Vid. Selsk. Skr.* ser. 3, **1**: 139 (1800). Puberulent, somewhat glaucous perennial 30–100 cm. Stem terete, solid, branched from the base. Lower leaves up to 30 cm, 2- to 3-pinnate, ovate-oblong or oblong in outline; segments usually 7–11, all or the lower pair only with an additional pair of segments at the base of the primary ones; lobes 5–30 mm, broadly ovate, crenate in upper half, usually serrulate below, often 3-fid, conspicuously reticulately veined; cauline usually reduced to ovate, sheathing petioles. Rays 3–6, puberulent. Bracts and bracteoles absent or few. Petals villous on back. Fruit *c.* 2 mm, ovoid, densely villous (including the commissural face). *Iberian peninsula, Açores.* Az Hs Lu.

11. P. anisoides Briganti, *Nova Pimp. Spec. Diss.* 11 (1802). Somewhat puberulent to almost tomentose perennial 40–100 cm. Stem terete, finely striate. Lower leaves 2-pinnate, ovate-lanceolate in outline, at least the lower pair of segments usually with an additional pair at base; lobes 5–10 mm, ovate or obovate,

cuneate, dentate or lobed; cauline leaves usually reduced to narrow, sheath-like petioles. Rays 2–6, glabrous. Bracts and bracteoles absent. Petals glabrous on back. Fruit *c.* 4 mm, ovoid or ovoid-oblong, glabrous or appressed-pubescent. ● *C. & S. Italy, Sicilia.* It Si.

12. P. serbica (Vis.) Bentham & Hooker fil. ex Drude in Engler & Prantl, *Natürl. Pflanzenfam.* **3**(8): 195 (1898) (*Pancicia serbica* Vis.). Glabrous perennial *c.* 50 cm. Stem terete, striate, branched above. Lower leaves cordate, serrate, long-petiolate; the next leaves deeply lobed; the upper palmately divided to the base into numerous setaceous lobes. Rays 10–15. Bracts 5–8, linear, scabrid, sometimes caducous; bracteoles 5. Fruit 3–4 mm, ovoid; ridges very narrowly winged. *Mountain meadows.* ● *C. & W. Jugoslavia; N. & E. Albania.* Al Ju.

13. P. major (L.) Hudson, *Fl. Angl.* 110 (1762) (*P. magna* L.). Glabrous or rarely puberulent perennial up to 100 cm. Stock at flowering time with non-flowering rosettes. Stem deeply sulcate (very rarely terete), hollow, branched above. Lower leaves 1(–3)-pinnate with 3–9 segments; segments up to 100 mm, ovate or oblong, dentate, rarely pinnatisect; cauline leaves smaller, with inflated, sheath-like petioles with membranous margins. Rays 10–25, slender. Bracts absent; bracteoles usually absent, rarely few, caducous. Petals white to deep pink. Fruit 2·5–3·5 mm, ovoid-oblong; ridges prominent, whitish. $2n = 20$. *Most of Europe except the extreme north and south and S. & E. Russia.* Au Be Br Bu Cz Da Ga Ge Hb He Ho Hs Hu It Ju No Po Rm Rs (N, B, C, W) Su [Fe].

Very variable in height, degree of dissection of the leaves and colour of flowers. A number of subspecies have been described but the correlation between the characters differentiating these does not appear to be very good. Perhaps the most distinct of the taxa is var. **rubra** (Hoppe) Fiori & Béguinot in Fiori & Paol., *Fl. Anal. Ital.* **2**: 163 (1900), which occurs on the higher mountains of C. Europe. It has short stems and very deep pink flowers, but plants with the same flower colour and up to 100 cm high are found in the W. Carpathians and S.C. France; these seem to intergrade with normal *P. major*. Plants with greatly dissected leaves are common in the eastern part of the range of the species, but occur sporadically elsewhere.

14. P. siifolia Leresche, *Jour. Bot.* (*London*) **17**: 198 (1879). Glabrous perennial up to 60 cm. Stock long. Stem slender, terete, striate, little-branched. Lower leaves 10–30 cm, triangular-lanceolate in outline; segments 3–9, more or less deeply lobed and sharply serrate, somewhat glaucous beneath; petioles inflated, sheathing; upper cauline leaves with long, linear-lanceolate lobes, at least the terminal one entire. Rays 5–12. Bracts 0(–2); bracteoles 1–3(–6), narrowly linear. Petals pink. Fruit 5–6 mm, oblong-ovoid; ridges prominent, very narrowly winged. *Mountain pastures.* ● *W. Pyrenees; N. Spain (Picos de Europa).* Ga Hs.

15. P. bicknellii Briq., *Bull. Herb. Boiss.* **6**: 85 (1898). Perennial up to 50 cm. Stock stout, without fibres. Leaves 2-ternate and ultimately pinnatisect; lobes ovate, incise-dentate. Rays 5–7. Bracts absent; bracteoles usually 5, lanceolate. Pedicels *c.* 0·75 mm in diameter in fruit. Petals white. Fruit 5–7 mm, ovoid; ridges low and rounded. *Rocky slopes.* ● *Mallorca.* Bl.

This species has been placed in a monotypic genus, *Spiroceratium* H. Wolff, mainly on account of the deeply furrowed endosperm. In the other respects it seems to agree well with *Pimpinella*.

332

16. P. saxifraga L., *Sp. Pl.* 263 (1753) (incl. *P. alpestris* (Sprengel) Schultes, *P. dissecta* Retz., *P. laconica* Halácsy). Somewhat pubescent (rarely completely glabrous) perennial up to 60 cm. Stock at flowering time usually without sterile rosettes. Stem usually terete, almost or quite solid, branched, almost leafless above. Lower leaves usually with 3–7 pairs of segments which vary from simple, ovate, to 2-pinnatisect, with linear lobes. Rays 6–25. Bracts absent, rarely 1–4; bracteoles absent, rarely 5–8. Petals white, rarely pinkish or purplish. Fruit 2–2·5 mm, broadly ovoid; ridges not prominent. $2n=40$. *Most of Europe, except the extreme south and most of the islands.* Al Au Be Br Bu Cz Da Fa Ga Ge Gr Hb He Ho Hs Hu It Ju No Po Rm Rs (N, B, C, W, K, E) Sa Su Tu.

Very variable in size, pubescence, leaf-dissection and flower-colour. This variation has received diverse taxonomic treatments, two species, each with several subspecies, being recognized by some authors, while others do not even recognize subspecies. Rather distinct, ecologically specialized, local variants also occur and some resemble **13** so closely that they can be identified with certainty only if ripe fruit is present.

Evidence about the status of the various taxa is at present conflicting and inadequate, and further investigation on a continental scale is necessary before a more detailed account of this polymorphic species can be given.

26. Aegopodium L.[1]

Leaves 1- to 2-ternate, with wide segments. Sepals small. Petals white or sometimes pink, obcordate; apex inflexed. Fruit ovoid, laterally compressed, constricted at the commissure. Ridges slender; vittae absent in ripe fruit.

1. A. podagraria L., *Sp. Pl.* 265 (1753). Perennial. Rhizome long, slender; stems up to 100 cm. Basal leaves deltate in outline; segments 4–8 cm, lanceolate to ovate, serrate, sometimes 3-lobed; petiole longer than lamina, trigonous. Cauline leaves smaller, with a short, inflated petiole; upper usually entire. Rays 10–20. Bracts and bracteoles usually absent. Fruit 3–4 mm. $2n=22$, *c.* 44. *In hedges and open woods; common also as a weed of cultivation. Throughout most of Europe, but rare in the south.* Al Au Be Bu ?Co Cz Da Fe Ga Ge ?Gr He Ho Hu It Ju No Po Rm Rs (N, B, C, W, K, E) Su [Br Fa Hb Is].

27. Sium L.[1]

Submerged leaves 2- to 3-pinnate, with linear lobes; aerial leaves usually pinnate, with broad lobes. Sepals present, sometimes very small. Petals white, obcordate; apex inflexed. Fruit ovoid or ovoid-oblong, slightly compressed laterally. Ridges slender or broad, the lateral marginal on the mericarps; vittae 1–3, superficial; stylopodium nearly flat.

Stem strongly furrowed; sepals conspicuous, lanceolate **1. latifolium**
Stem striate; sepals very small **2. sisarum**

1. S. latifolium L., *Sp. Pl.* 251 (1753). Glabrous perennial up to 150 cm. Stem strongly sulcate, hollow. Submerged leaves (present only in spring) 2- to 3-pinnate, with linear lobes; aerial leaves simply pinnate; lobes up to 10×3 cm, 4–9(–16) pairs, ovate-lanceolate, unequal at base, serrate. Rays usually 20–30. Bracts usually 2–6, often large and leaf-like; bracteoles lanceolate. Sepals *c.* 1 mm. Fruit 3–4 mm, ellipsoid but somewhat compressed; ridges thick. $2n=20$. *In shallow water. Europe, except for much of the Mediterranean region and parts of the west.* Al Au Be

Br Bu Cz Da Fe Ga Ge Hb He Ho Hs Hu It Ju Po Rm Rs (N, B, C, W, E) Su [No].

2. S. sisarum L., *Sp. Pl.* 251 (1753) (incl. *S. sisaroideum* DC.). Like **1** but usually smaller; stem striate; submerged leaves absent; leaf-lobes usually lanceolate; sepals very small; fruit 2–3·5 mm, with slender ridges. *Damp places. From E. Hungary and Bulgaria to C. & S.E. Russia; sometimes cultivated for the edible roots and occasionally naturalized in C. Europe and N. Italy.* Bu Hu Rm Rs (C, W, K, E) [Au Ge It].

The cultivated plant, with tuberous roots, is var. *sisarum* and is of unknown origin. The wild plant is var. *lancifolium* (Bieb.) Thell. (*S. lancifolium* Bieb., non Schrank, *S. sisaroideum* DC.).

28. Berula Koch[1]

Like *Sium* but fruit subdidymous; lateral ridges not marginal on the mericarps; vittae deeply embedded; stylopodium conical.

1. B. erecta (Hudson) Coville, *Contr. U.S. Nat. Herb.* **4**: 115 (1893) (*Sium angustifolium* L., *S. erectum* Hudson). Stoloniferous, glabrous perennial 30–100 cm. Submerged leaves 3- to 4-pinnate, with linear lobes; lower aerial leaves with (5–)7–14(–19) pairs of sessile, oblong-lanceolate to ovate, biserrate segments 2–6 cm; upper cauline leaves small, usually very irregularly serrate. Umbels leaf-opposed; rays 10–20. Bracts and bracteoles numerous, often leaf-like and 3-fid or pinnatisect. Fruit 1·5–2 mm, usually wider than long. $2n=20$. *In shallow water. Almost throughout Europe except the extreme north.* All except Az Cr Fa Fe Is Rs (N) Sb Tu.

29. Crithmum L.[1]

Leaves 1- to 2-pinnate, with subterete, fleshy segments. Sepals minute. Petals yellowish-green, obcordate; apex inflexed. Fruit ovoid-oblong, not compressed; pericarp spongy. Ridges thick and prominent; vittae several.

1. C. maritimum L., *Sp. Pl.* 246 (1753). Glabrous perennial 15–50 cm, woody below. Leaves deltate in outline; lobes 1–5 cm, linear-oblanceolate, subacute. Rays 8–36, rather stout. Bracts and bracteoles triangular-lanceolate to linear-lanceolate, ultimately deflexed. Fruit 5–6 mm, yellowish or purplish. *Maritime rocks, rarely on sand or shingle. Atlantic coast of Europe, northwards to Scotland; Mediterranean and Black sea coasts.* Al Az Bl Br Bu Co Cr Ga Gr Ho Hs It Ju Lu ?Rm Rs (K) Sa Si Tu.

30. Sclerochorton Boiss.[1]

Lower leaves 2- to 3-pinnate, uppermost entire; lobes short, remote, linear. Sepals small. Petals white, obovate, emarginate; apex inflexed. Fruit ovoid-oblong, not compressed. Ridges prominent; vittae 6–7.

1. S. junceum (Sibth. & Sm.) Boiss., *Fl. Or.* **2**: 969 (1872). Glaucous perennial with numerous solid, rigid, striate stems 15–45 cm. Umbels with 2–3 long rays. Bracts and bracteoles 2–3, short, linear. Fruit *c.* 6 mm. *Rocks on high mountains.* ● *S. Greece.* Gr.

31. Dethawia Endl.[1]

Leaves 3-pinnate, with linear, crowded lobes. Sepals conspicuous. Petals white, elliptical; apex not or weakly inflexed. Fruit ovoid-oblong, not compressed. Ridges prominent, obtuse; vittae solitary.

[1] By T. G. Tutin.

1. D. tenuifolia (Ramond ex DC.) Godron in Gren. & Godron, *Fl. Fr.* **1**: 706 (1849). Perennial with flexuous, rather slender stems 10–40 cm. Leaves mostly basal, deltate in outline; lobes mucronate. Rays 4–10, nearly equal. Bracts 1–3, very unequal; bracteoles numerous, linear-lanceolate. Pedicels puberulent. Fruit 4–6 mm. $2n = 22$. *Mountain rocks and screes.* ● Pyrenees, *Cordillera Cantábrica*. Ga Hs.

32. Seseli L.[1]

(Incl. *Libanotis* Hill)

Biennial or perennial herbs; stock usually with fibres. Leaves several times pinnately or ternately divided. Bracts 0–16. Sepals absent or small. Petals white, rarely pink or yellow. Fruit oblong to ellipsoid or ovoid, not strongly compressed. Ridges prominent; vittae usually 1–3.

A taxonomically difficult genus containing a number of rare or very local species. The following are known from very few localities or collections; further investigation is needed to confirm their status, and in a few cases to confirm that they are correctly placed in *Seseli*: 7, 8, 9, 10, 28, 31, 32, and 33.

S. vandasii Hayek, *Feddes Repert.* **21**: 256 (1925), from S. Jugoslavia (near Prilep), is another little-known species of uncertain position. It is a glabrous perennial with the lower leaves 3-sect, the lobes 20–100 × 1 mm, filiform, the rays 3–5, glabrous, the bracteoles free, puberulent, and the petals pink.

1 Rays glabrous and not papillose
  2 Bracteoles less than 1 mm, much shorter than pedicels
          8. intricatum
  2 Bracteoles more than 1 mm, at least ⅓ as long as pedicels
    3 Bracts 5 or more, rarely replaced by 1–3 simple leaves
      4 Petals yellow; rays not more than 20   **4. peucedanoides**
      4 Petals white or pink; rays usually more than 20
        5 Leaf-lobes 20–70 × 1–2(–3) mm, narrowly linear; bracteoles connate at base   **6. lehmannii**
        5 Leaf-lobes either less than 20 mm long or more than 3 mm wide; bracteoles free
          6 Stem ridged; petals white or pink in bud   **1. libanotis**
          6 Stem terete; petals violet in bud   **2. sibiricum**
    3 Bracts 0(–3)
      7 Fruit 4–6 mm; lobes of lower leaves (2–)5–15 mm wide
          14. bocconii
      7 Fruit 2–4(–5) mm; lobes of lower leaves not more than 3 mm wide
        8 Stock stout, branched, with persistent long petioles; fruit tuberculate-verrucose   **21. glabratum**
        8 Stock usually simple, without or with at most some fibrous remains of the leaf-bases; fruit glabrous, pubescent or papillose
          9 Petals yellow; leaf-lobes 20–100 × *c.* 0·5 mm, setaceous
            34. gracile
          9 Petals white, pink or purplish, rarely yellowish-white; leaf-lobes less than 20 mm long or more than 0·5 mm wide, not setaceous
            10 Basal leaves 2-ternate, the lobes 25–100 × 1·5–3 mm, linear to elliptic-lanceolate   **13. cantabricum**
            10 Basal leaves 2- to 4-pinnatisect, the lobes 5–30(–40) × 0·5–2 mm, ± narrowly linear
              11 Petiole of basal leaves not canaliculate; calyx-teeth 0·2–0·5 mm in fruit   **15. elatum**
              11 Petiole of basal leaves canaliculate; calyx-teeth minute or 0   **20. pallasii**
1 Rays pubescent or papillose
  12 Bracteoles connate for at least ⅓ of their length, forming a ± conical sheath around each partial umbel

    13 Lobes of basal leaves 20–120 mm, setaceous or linear-setaceous
      14 Rays 7–10   **28. tomentosum**
      14 Rays 20–30   **29. globiferum**
    13 Lobes of basal leaves not more than 20 mm, linear-setaceous to narrowly obovate
      15 Lower leaves glabrous
        16 Stems puberulent at base; rays 3–6, not more than 6 mm, usually distinctly puberulent all round
            10. peixoteanum
        16 Stems glabrous at base; rays 5–12(–20), at least some more than 6 mm, puberulent on the inner side
            26. hippomarathrum
      15 Lower leaves densely puberulent
        17 Lobes of basal leaves 2–5 mm, narrowly obovate; rays not more than 5 mm   **9. granatense**
        17 Lobes of basal leaves 10–20 mm, linear; rays 12–20 mm
            27. dichotomum
  12 Bracteoles free, or connate only in the lower third, forming a disc at the base of each partial umbel
    18 Lower leaves 1-pinnate; lobes ovate, toothed or pinnatifid
      19 Stem ridged; petals white or pink in bud   **1. libanotis**
      19 Stem terete; petals violet in bud   **2. sibiricum**
    18 Lower leaves 2- to 4-pinnate; lobes various
      20 Bracts (4–)5 or more, rarely replaced by 1–3 simple leaves
        21 Bracteoles connate at base
          22 Lobes of basal leaves (6–)8–30 × 1·5–5(–7) mm, linear to obovate or elliptical; rays densely puberulent all round   **5. gummiferum**
          22 Lobes of basal leaves 20–70 × 1–3 mm, narrowly linear; rays sparsely puberulent on inner side   **6. lehmannii**
        21 Bracteoles free
          23 Sepals 0·5–1 mm, deciduous; rays usually 20–60
            24 Styles not more than ½ as long as fruit   **1. libanotis**
            24 Styles about as long as fruit   **3. condensatum**
          23 Sepals less than 0·5 mm or absent; rays fewer than 20
            25 Rays pubescent on the inner side; fruit glabrous
              16. annuum
            25 Rays puberulent all round; fruit puberulent
              7. vayredanum
    20 Bracts 0(–3)
      26 Lower leaves 1- to 3-ternate; lobes up to 15 mm wide, 2- to 3-fid; fruit 4–6 mm   **14. bocconi**
      26 Lower leaves 2- to 4-pinnatisect; lobes usually less than 5 mm wide; fruit usually less than 4 mm
        27 Rays usually more than (10–)12
          28 Rays puberulent all round
            29 Leaf-lobes long-acuminate, somewhat pungent; petals puberulent; calyx-teeth obscure   **23. rigidum**
            29 Leaf-lobes acute or subobtuse, not pungent; petals glabrous or sparsely puberulent
              30 Calyx-teeth *c.* 0·5 mm; petiole not canaliculate
                5. gummiferum
              30 Calyx-teeth obscure; petiole canaliculate   **12. montanum**
         28 Rays puberulent only on inner side
           31 Fruit 1·5–2·5(–3) mm   **16. annuum**
           31 Fruit 3 mm or more
            32 Petals pubescent on back
              33 Fruit 3–3·5 × 1–2 mm, puberulent   **17. campestre**
              33 Fruit 3·5–5 × 2–2·5 mm, usually glabrous
                18. arenarium
            32 Petals glabrous
              34 Ovary and fruit glabrous   **22. strictum**
              34 Ovary and young fruit puberulent; mature fruit sometimes glabrescent
              35 Lobes of lower leaves 20–40 × *c.* 0·5 mm, linear-setaceous or filiform; longest rays *c.* 30 mm
                36 Bracts 0; bracteoles puberulent   **31. degenii**
                36 Bracts few; bracteoles glabrous except for the ciliate margin   **32. parnassicum**
              35 Lobes of lower leaves usually less than 20 mm long or more than 0·5 mm wide, linear or linear-lanceolate, rarely obovate

37 Longest rays not more than 20(–25) mm; fruit
   2·5–4 × 1·5–2 mm   **12. montanum**
37 Longest rays 30–50 mm; fruit 4–6 × 2–2·5 mm
                                                   24. rhodopeum
27 Rays not more than 10(–12)
38 Rays puberulent all round
39 Lobes of basal leaves up to 60 × 0·5 mm, setaceous
                                                   30. leucospermum
39 Lobes of basal leaves not more than 15 mm, up to
   2 mm wide, linear, elliptical or obovate
40 Bracteoles free   **12. montanum**
40 Bracteoles connate at base
41 Leaf-lobes long-acuminate   **23. rigidum**
41 Leaf-lobes obtuse, sometimes minutely apiculate
42 Stem densely pubescent; bracteoles triangular to
   subulate, densely puberulent   **9. granatense**
42 Stem glabrous, at least in the lower half; bracteoles
   ovate-lanceolate, sparsely puberulent   **25. malyi**
38 Rays puberulent only on inner side
43 Caespitose, the flowering stems not more than
   10(–15) cm; umbels dense, subglobose; rays not
   more than 6 mm   **11. nanum**
43 Not caespitose, the flowering stems more than 10 cm;
   umbels lax; rays usually more than 10 mm
44 Petals glabrous
45 Lobes of lower leaves linear or linear-lanceolate
                                                   12. montanum
45 Lobes of lower leaves setaceous   **33. bulgaricum**
44 Petals puberulent on the back
46 Bracteoles with wide membranous margin, equalling or wider than the herbaceous central part
                                                   19. tortuosum
46 Bracteoles with narrow membranous margin
47 Fruit 3–3·5 × 1–2 mm, puberulent   **17. campestre**
47 Fruit 3·5–5 × 2–2·5 mm, usually glabrous
                                                   18. arenarium

Subgen. **Libanotis** (Hill) Drude. Sepals 0·5–1 mm, deciduous; bracts usually numerous.

1. S. libanotis (L.) Koch, *Nova Acta Acad. Leop.-Carol.*
12(1): 111 (1824) (*Libanotis montana* Crantz). Glabrous or
pubescent biennial or perennial (usually monocarpic) up to
120 cm; stems ridged. Lower leaves 1- to 3-pinnate. Rays
(10–)20–60, up to 50 mm. Bracts usually 5–15 mm, 8 or more,
linear, rarely replaced by 1–3 simple, lobed or toothed leaves;
bracteoles 10–15, free. Sepals linear to ovate-lanceolate; petals
white or pink, pubescent on the back. Fruit 2·5–4 × 1·5–2·5 mm,
ovoid or ellipsoid, subterete, glabrous or pubescent; ridges
obtuse. Styles deflexed, less than ½ as long as the fruit. 2*n*=22.
Most of Europe except the extreme north, west and south. Au Be
Br Bu Cz Da Fe Ga Ge He Hs Hu It Ju No Po Rm Rs (N, B, C,
W, E) Su.

Very variable in leaf-dissection and shape of the lobes, and in
indumentum. Leaf-dissection and shape of the lobes shows an
east-west cline, the plants with the most dissected leaves and
narrowest lobes occurring in the west. Plants with glabrous fruits
occur mainly in the southern half of the range of the species,
sometimes forming discrete populations, sometimes with otherwise almost identical plants with pubescent fruits. Populations
with distinctive combinations of these characters have been described as separate species or as subspecies, but it is generally not
possible to distinguish any of these taxa from the remainder by a
single character. The following, however, seem sufficiently distinct to merit recognition as subspecies.

(a) Subsp. **libanotis** (incl. *S. athamanthoides* Reichenb.,
Libanotis daucifolia (Scop.) Reichenb.): Lower leaves 2- to 3-
pinnate; lobes linear, oblong or lanceolate, often falcate, acute.

● *W., C. and parts of S. Europe, extending northwards to S.
Sweden.*

(b) Subsp. **intermedium** (Rupr.) P. W. Ball, *Feddes Repert.* **79**:
64 (1968) (*Libanotis intermedia* Rupr., *S. libanotis* subsp. *sibiricum*
Thell. pro parte): Lower leaves 1- to 2-pinnate; lobes ovate,
coarsely toothed or pinnatifid, obtuse or subobtuse. *E. & E.C.
Europe.*

2. S. sibiricum (L.) Garcke, *Fl. Nord-Mittel Deutschl.* 139
(1849) (*Libanotis sibirica* (L.) C. A. Meyer). Like **1** but stems
terete, glabrous; lower leaves 1-pinnate; lobes ovate, toothed or
pinnatisect; petals violet in bud; fruit usually glabrous; styles
about ½ as long as fruit. *S. Ural.* Rs (C).

3. S. condensatum (L.) Reichenb. fil. in Reichenb. & Reichenb.
fil., *Icon. Fl. Germ.* **21**: 37 (1867) (*Libanotis condensata* (L.)
Crantz). Like **1** but fruit dorsally compressed, with a narrow
(0·5–1 mm) wing on the margin; styles erect or patent, about as
long as fruit. *N. Russia.* Rs (N). (*N. Asia.*)

Subgen. **Seseli**. Sepals less than 0·5 mm, or absent; bracts
absent or few, rarely numerous.

4. S. peucedanoides (Bieb.) Kos.-Pol., *Bull. Soc. Nat. Moscou*
nov. ser., **29**: 184 (1916) (*Gasparrinia peucedanoides* (Bieb.)
Bertol., *Silaus virescens* (Sprengel) Boiss.; incl. *Seseli elegans*
Schischkin). Glabrous perennial 40–100 cm. Leaves 2- to 3-
pinnate; lobes 3–10 × 0·5–1 mm, linear. Rays (5–)10–20, up to
60 mm, glabrous. Bracts 5–12; bracteoles free. Petals yellow,
glabrous. Fruit 3–5 × 2·5 mm, ovoid or ellipsoid, glabrous.
2*n*=22. *S. & E.C. Europe, from S. France to W. Ukraine and
Bulgaria; one station in S.E. Russia.* Al Bu Cz Ga Gr Hu It Ju
Rm Rs (W, E).

Variable and possibly containing a number of subspecies.

5. S. gummiferum Pallas ex Sm., *Exot. Bot.* **2**: 121 (1807).
Glabrous to densely pubescent perennial up to 150 cm. Leaves
2- to 3-pinnate; lobes (6–)8–30 × 1·5–5(–7) mm, variable in shape.
Rays 20–40(–60), up to 50 mm, pubescent all round. Bracts 8–16,
rarely 0; bracteoles connate at base. Petals white. Fruit 3–4 ×
c. 1·5 mm, oblong, puberulent. *Limestone cliffs below 1000 m.
S. Aegean region; Krym.* Cr Gr Rs (K).

1 Basal leaves glabrous   **(b) subsp. crithmifolium**
1 Basal leaves puberulent or velutinous
2 Lobes of basal leaves 1–5 mm wide, linear or oblong; rays
   20–35   **(a) subsp. gummiferum**
2 Lobes of basal leaves 4–7 mm wide, elliptical or obovate; rays
   30–60   **(c) subsp. aegaeum**

(a) Subsp. **gummiferum**: Glabrous or puberulent. Basal leaves
densely puberulent; lobes 8–30 × 1·5–5 mm, linear or oblong.
Rays 20–35. ● *Krym.*

(b) Subsp. **crithmifolium** (DC.) P. H. Davis, *Notes Roy. Bot.
Gard. Edinb.* **21**: 120 (1953) (*S. crithmifolium* (DC.) Boiss.):
Puberulent. Basal leaves glabrous; lobes (6–)10–30 × 2–5 mm,
oblong. Rays 20–45. ● *Amorgos, Karpathos and neighbouring
islands.*

(c) Subsp. **aegaeum** P. H. Davis, *loc. cit.* (1953) (*S. crithmifolium* auct., pro parte): Velutinous. Basal leaves velutinous;
lobes 10–25 × 4–7 mm, elliptical or obovate. Rays 30–60. ● *E.
Kriti, Folegandros, Sikinos.*

6. S. lehmannii Degen, *Österr. Bot. Zeitschr.* **48**: 121 (1898).
Glabrous or subglabrous perennial 10–75 cm. Leaves 2-pinnate;
lobes 20–70 × 1–2(–3) mm, narrowly linear, acute. Rays 25–35,
up to 60 mm, glabrous or sparsely puberulent on inner side.
Bracts 10–16; bracteoles connate at base. Petals white. Fruit

c. 4 mm, obovoid, puberulent. *Dry, rocky grassland above 1000 m. Calcicole.* ● *S. Krym.* Rs (K).

7. S. vayredanum Font Quer, *Revista Olot.* **1** (**11**): 3 (1926). Glabrous or puberulent perennial 30–60 cm. Leaves 3- to 4-pinnate, glabrous; lobes 5–18 × 1–2 mm, linear-oblong. Rays 5–18, up to 20 mm, puberulent all round. Bracts 6–8; bracteoles free. Petals white, puberulent. Fruit 2·5–5 × 1–2 mm, oblong, puberulent. *Limestone cliffs and screes.* ● *S. Spain.* Hs.

8. S. intricatum Boiss., *Elenchus* 48 (1838). Glabrous, caespitose perennial 20–30 cm. Leaves 2-pinnate; lobes 5–8 × 1–2 mm, linear to elliptical, acute. Rays 2–4, up to 20 mm, glabrous. Bracts 2–3, less than 1 mm; bracteoles less than 1 mm, free. Petals white. Fruit not known. *Limestone cliffs.* ● *S. Spain (Sierra de Gádor).* Hs.

9. S. granatense Willk., *Bot. Zeit.* **5**: 431 (1847). Densely puberulent, caespitose perennial 9–25 cm. Leaves 2-pinnate; lobes 2–5 × 0·5–1 mm, oblong-obovate, obtuse, sometimes mucronate. Umbels dense, subglobose; rays 5–6, not more than 5 mm, densely puberulent all round. Bracts 0–1; bracteoles connate, often up to the middle, the free apex triangular or subulate. Petals white or pink, pubescent on back. Fruit ovoid, densely puberulent. *Limestone cliffs.* ● *S. Spain.* Hs.

10. S. peixoteanum Samp., *Ann. Sci. Nat.* (*Porto*) **10**: 36 (1906). Like **9** but stems 20–75 cm, glabrous except for the puberulent base; leaf-lobes 5–20 × *c.* 0·5 mm, narrowly linear; umbels less dense; rays 3–6, not more than 6 mm, densely puberulent on the inner side, sparsely puberulent to subglabrous on the outer side. *Usually on serpentine.* ● *N.E. Portugal.* Lu.

11. S. nanum Dufour in Bory, *Voy. Souterrain* 363 (1821). Glabrous, caespitose perennial up to 10(–15) cm. Leaves 2-pinnate; lobes 4·5–8 × 0·7–1·5 mm, elliptic-oblong or -obovate, obtuse, mucronate. Umbels subglobose; rays 5–8, not more than 6 mm, puberulent on the inner side. Bracts usually 0; bracteoles free. Petals white or pink, glabrous. Fruit 3–4 × 1·5–2·5 mm, oblong, densely pubescent-scabrid. ● *Pyrenees.* Ga Hs.

12. S. montanum L., *Sp. Pl.* 260 (1753) (*S. glaucum* L.). Glabrous perennial 10–70 cm. Leaves 2- to 3-pinnate; lobes 3–50 × 0·5–2 mm, linear to obovate; petiole canaliculate. Rays 3–25, up to 20 mm, puberulent on inner side, rarely all round. Bracts 0(–3); bracteoles free or connate at base. Petals white or pale pink, glabrous. Fruit 2·5–4·5 × 1·5–2·5 mm, oblong-ovoid, glabrescent or puberulent. *S. Europe, from C. Spain to W. part of Balkan peninsula, and extending northwards to c. 49° N. in France.* Al Ga †Ge He Hs It Ju.

1 Rays 10–20; leaf-lobes 1–2 mm wide   (c) **subsp. polyphyllum**
1 Rays 3–12(–14); leaf-lobes 0·5–1 mm wide
  2 Fruit with thin, acutely angled ridges narrower than valleculae   (a) **subsp. montanum**
  2 Fruit with wide, rounded ridges wider than valleculae   (b) **subsp. tommasinii**

(a) Subsp. **montanum**: Leaf-lobes 5–50 × 0·5–1 mm, linear or linear-lanceolate. Rays 5–12(–14). Fruit puberulent or glabrescent; ridges thin, acutely angled, narrower than valleculae. *C. Spain, S. & E. France, Italy; one station in Switzerland.*
(b) Subsp. **tommasinii** (Reichenb. fil.) Arcangeli, *Comp. Fl. Ital.* 282 (1882) (*S. tommasinii* Reichenb. fil.): Like subsp. (a) but rays 3–9; fruit glabrescent; ridges wide, rounded, wider than valleculae. ● *W. part of Balkan peninsula, E., C. & S. Italy.*
(c) Subsp. **polyphyllum** (Ten.) P. W. Ball, *Feddes Repert.* **79**: 64 (1968): Leaf-lobes 3–14 × 1–2 mm, linear to obovate. Rays

10–20, sometimes puberulent all round. Fruit glabrescent; ridges thin, acutely angled. ● *S. Italy.*

13. S. cantabricum Lange, *Ind. Sem. Horto Haun.* 27 (1855). Glabrous perennial 25–70 cm. Leaves 2-ternate; lobes 25–100 × 1·5–3 mm, linear to elliptic-lanceolate, entire. Rays 5–10, up to 30 mm, glabrous. Bracts 0–1; bracteoles free. Petals very pale yellow or purplish, glabrous. Fruit 2·5–3·5 × 1·5–2 mm, ovoid, glabrous. ● *N. Spain (E. part of Cordillera Cantábrica).* Hs.

14. S. bocconi Guss., *Cat. Pl. Boccad.* 80 (1821). Glabrous perennial 20–50 cm. Leaves 1- to 3-ternate; lobes 10–30 × (2–)5–15 mm, lanceolate or linear, usually 2- to 3-fid at apex, coriaceous. Rays 4–18, subglabrous to puberulent. Bracts 0; bracteoles free. Petals white, glabrous. Fruit 4–6 × *c.* 2·5 mm, oblong, glabrescent. *Rocks and cliffs. Islands of W. Mediterranean region.* Co Sa Si.

15. S. elatum L., *Sp. Pl.* ed. 2, 375 (1762). Glabrous biennial or perennial 15–120 cm. Leaves 2- to 4-pinnate; lobes 10–30 × 0·5–2 mm, linear or linear-lanceolate; petiole terete. Rays 2–20, up to 30 mm, glabrous. Bracts 0(–1); bracteoles free. Sepals *c.* 0·5 mm. Petals white, glabrous or sparsely puberulent. Fruit 2–3·5 × 1·5–2 mm, ovoid or ellipsoid. ● *S. & C. Europe, from C. Spain to S.E. Romania and northwards to Czechoslovakia.* Al Au Cz Ga Hs Hu It Ju Po Rm.

1 Rays usually 2–7
  2 Fruit usually puberulent; sepals inserted at apex of fruit   (a) **subsp. elatum**
  2 Fruit glabrous; sepals inserted *c.* 0·5 mm below apex of fruit   (b) **subsp. gouanii**
1 Rays usually (5–)8–20
  3 Fruit glabrous or sparsely puberulent   (c) **subsp. osseum**
  3 Fruit densely papillose-puberulent   (d) **subsp. austriacum**

(a) Subsp. **elatum**: Rays 3–6. Fruit puberulent, sometimes glabrescent; sepals inserted at apex of fruit. Vittae usually 3. *From C. Spain to S. France and Italy.*
(b) Subsp. **gouanii** (Koch) P. W. Ball, *Feddes Repert.* **79**: 64 (1968) (*S. gouanii* Koch; incl. *S. bosnense* K. Malý, *S. hercegovinium* K. Malý): Rays (2–)4–7(–15). Fruit glabrous; sepals inserted *c.* 0·5 mm below apex of fruit. Vittae (2–)3(–5). *N. Italy, N.W., part of Balkan peninsula.*
(c) Subsp. **osseum** (Crantz) P. W. Ball, *Feddes Repert.* **79**: 64 (1968) (*S. osseum* Crantz, *S. devenyense* Simonkai): Rays (5–)8–15. Fruit glabrous or sparsely puberulent; sepals inserted at apex of fruit. Vittae usually 1. 2*n* = 18. *E.C. Europe.*
(d) Subsp. **austriacum** (G. Beck) P. W. Ball, *Feddes Repert.* **79**: 64 (1968) (*S. austriacum* (G. Beck) Wohlf.): Rays 9–20. Fruit densely papillose-puberulent; sepals inserted at apex of fruit. Vittae 2–3(–4). *From N. Italy and N. Jugoslavia northwards to S. Czechoslovakia.*

16. S. annuum L., *Sp. Pl.* 260 (1753). Puberulent or glabrous biennial or perennial up to 60(–100) cm. Leaves 2- to 3-pinnate; lobes up to 10(–15) × 0·5–2 mm, linear or linear-lanceolate. Rays (8–)12–40, up to 40 mm, puberulent on inner side. Bracteoles free. Petals white or pink, minutely papillose on back. Fruit 1·5–3 × 1–1·5 mm, ovoid, glabrous. 2*n* = 16. *From N. France and C. Russia southwards to N. Spain and Bulgaria.* Au Bu Cz Ga Ge He Hs Hu It Po Rm Rs (B, C, W, E).

(a) Subsp. **annuum**: Leaves ovate in outline; lobes up to 10(–15) × 0·5–2 mm; rays up to 40; bracts 0–1; bracteoles about equalling partial umbel; petals with at least some papillae acute. *Throughout the range of the species.*

(b) Subsp. **carvifolium** (Vill.) P. Fourn., *Quatre Fl. Fr.* 679 (1937) (*S. carvifolium* Vill.): Leaves oblong in outline; lobes up to 6×0·5 mm; rays 12–18; bracts 2–6; bracteoles shorter than partial umbel; petals with all papillae obtuse. ● *S.W. Alps.*

17. S. campestre Besser, *Enum. Pl. Volhyn.* 44 (1822). Glabrous perennial 50–100(–200) cm. Leaves 3- to 4-pinnate; lobes 5–25 × 0·5–1·5 mm, linear or linear-lanceolate. Rays 7–15, up to 40 mm, puberulent on inner side. Bracts 0(–2); bracteoles free. Petals white or yellowish-white, puberulent on back. Fruit 3–3·5 × 1–2 mm, ovoid, puberulent. *S.E. Europe, extending northwards to S.C. Russia.* Bu Rm Rs (C, W, K, E).

18. S. arenarium Bieb., *Fl. Taur.-Cauc.* 3: 242 (1819). Like **17** but leaf-lobes 10–45×0·5–1 mm, linear; fruit 3·5–5×2–2·5 mm, usually glabrous. *S. part of U.S.S.R.* Rs (C, W, K, E).

19. S. tortuosum L., *Sp. Pl.* 260 (1753) (incl. *S. pauciradiatum* Schischkin). Glabrous biennial 10–75 cm. Leaves 3- to 4-pinnate; lobes 5–15×0·5–2 mm, linear-lanceolate to oblong-obovate. Rays 4–11, up to 25 mm, puberulent on inner side. Bracts 0(–3); bracteoles free, with a wide membranous margin. Petals yellowish-white, puberulent on back. Fruit 2–4×(1·5–)2–2·5 mm, ovoid or ellipsoid, puberulent. *S. Europe.* Al Bu Co Ga Gr Hs It Ju Lu Rm Rs (K) Sa Si.

Variable in habit and the shape of the leaf-lobes.

20. S. pallasii Besser, *Cat. Pl. Horto Crem.* 130 (1816) (incl. *S. varium* Trev.). Glabrous biennial or perennial 30–120 cm. Leaves 2- to 4-pinnate; lobes 5–25(–45)×0·5–1(–2) mm, linear to almost filiform; petiole canaliculate on upper side. Rays 7–25(–30), up to 60 mm, glabrous. Bracts 0(–2); bracteoles free. Petals white, glabrous. Fruit 2·5–3·5(–4·5)×1·5–2·5 mm, ellipsoid or oblong, glabrous or slightly tuberculate-verrucose. $2n=16, 20$. *From N. Italy and Czechoslovakia eastwards to C. Ukraine.* Au Bu Cz Gr Hu It Ju Rm Rs (W, K).

21. S. glabratum Willd. ex Schultes in Roemer & Schultes, *Syst. Veg.* 6: 406 (1820). Glabrous perennial 25–40 cm; stock with persistent long petioles. Leaves 2-pinnate; lobes 15–60(–80) ×0·7–2 mm, linear-filiform, rigid. Rays 6–10, up to 20 mm, glabrous. Bracts 0(–1); bracteoles free. Petals white, glabrous. Fruit 2·5–4 mm, ovoid-oblong, tuberculate-verrucose. *S.E. Russia, W. Kazakhstan.* Rs (C, E). (*W. & C. Asia.*)

22. S. strictum Ledeb., *Fl. Altaica* 1: 338 (1829). Glabrous perennial 10–70 cm. Leaves 3-pinnate; lobes 10–50×0·5–1(–2) mm, narrowly linear. Rays 15–35, up to 30 mm, scabrid on the inner side. Bracts 0(–1); bracteoles free. Petals white, glabrous. Fruit 3–4×1–1·5 mm, oblong, glabrous. *S.E. Russia.* Rs (C, E). (*W. Asia*).

23. S. rigidum Waldst. & Kit., *Pl. Rar. Hung.* 2: 156 (1803–4) (incl. *S. serbicum* Degen, *S. gummiferum* var. *resiniferum* Velen.). Perennial up to 100 cm, puberulent, at least at the apex. Leaves 2- to 3(–4)-pinnate; lobes (10–)20–100×0·5–5(–7) mm, linear, long-acuminate, rigid. Rays 5–30, up to 45 mm, puberulent all round. Bracts 0(–3); bracteoles connate at base. Petals white or purplish-pink, puberulent or tomentose. ● *S.E. Europe.* Al Bu Gr Ju Rm Rs (W).

(a) Subsp. **rigidum**: Rays 10–30, up to 45 mm; fruit 4–5 mm. *Cliffs. N. part of Balkan peninsula, W. & C. Romania.*

(b) Subsp. **peucedanifolium** (Besser) Nyman, *Consp.* 295 (1879) (*S. peucedanifolium* Besser, non Mérat): Rays 5–9, up to 25 mm; fruit 2·5–3·5 mm. *From E. Bulgaria to W. Ukraine.*

24. S. rhodopeum Velen., *Sitz.-Ber. Böhm. Ges. Wiss.* (*Math.-Nat. Kl.*) 1890(1): 45 (1890). Like **23 (a)** but stem glabrous; rays puberulent on inner side; petals glabrous; fruit 4–6 mm, puberulent. ● *Bulgaria.* Bu.

25. S. malyi A. Kerner, *Österr. Bot. Zeitschr.* 31: 37 (1881). Biennial 6–15 cm, puberulent above. Leaves 2- to 3-pinnate; lobes 4–15×1 mm, linear, obtuse. Rays 6–12, up to 12 mm, pubescent all round. Bracts 0–1; bracteoles connate at base. Petals pink. Fruit 2·5–3×1·2–1·5 mm, ovoid, sparsely puberulent. *Cliffs.* ● *Mountains of W. Jugoslavia.* Ju.

26. S. hippomarathrum Jacq., *Enum. Stirp. Vindob.* 52, 224 (1762). Glabrous perennial 15–60(–90) cm. Leaves 2- to 3-pinnate; lobes 4–10×0·4–0·7 mm, linear. Rays 5–12(–20), up to 15 mm, pubescent on inner side. Bracts 0; bracteoles connate. Petals white or pale pink, glabrous. Fruit 3·5–6×2–3 mm, oblong-ovoid. *E.C. Europe, W. Ukraine; S. Ural.* Au Cz Ge Hu Ju Rm Rs (?B, C, W).

(a) Subsp. **hippomarathrum**: Bracteoles connate almost to the apex. Fruit puberulent. $2n=20$. ● *E.C. Europe, W. Ukraine.*

(b) Subsp. **hebecarpum** (DC.) Drude in Engler & Prantl, *Natürl. Pflanzenfam.* 3(8): 202 (1898) (*S. ledebourii* G. Don fil.): Bracteoles connate to about the middle. Fruit densely pubescent. *S. Ural.*

27. S. dichotomum Pallas ex Bieb., *Fl. Taur.-Cauc.* 1: 235 (1808). Minutely papillose-velutinous perennial 40–100 cm. Leaves 2-pinnate; lobes 10–20×0·5–1 mm, narrowly linear. Rays 5–9(–16), up to 20 mm, puberulent all round. Bracts 0; bracteoles connate up to the middle. Petals white. Fruit 2·5–3 × 1·5 mm, ovoid, minutely puberulent. *Krym.* Rs (K). (*Caucasus.*)

28. S. tomentosum Vis., *Stirp. Dalm.* 6 (1826). Glabrous perennial 50–100 cm. Leaves 3- to 4-pinnate; lobes 70–120× c. 1 mm, linear-setaceous, rigid. Rays 7–10, up to 7 mm, tomentose. Bracts absent; bracteoles connate almost to the apex. Petals white. Fruit c. 6 mm, tomentose. *Limestone cliffs.* ● *N.W. Jugoslavia.* Ju.

29. S. globiferum Vis., *Flora* (*Regensb.*) 13: 50 (1830). Like **28** but leaf-lobes 20–80×0·4–1 mm; rays 20–30, 5–20 mm; bracteoles connate up to the middle; fruit 4·5–7×2–3·5 mm. *Cliffs.* ● *W. Jugoslavia.* Ju.

Possibly not specifically distinct from **28**.

30. S. leucospermum Waldst. & Kit., *Pl. Rar. Hung.* 1: 92 (1802). Glabrous perennial 50–100 cm. Leaves 4- to 5-pinnate; lobes 20–60× c. 0·5 mm, linear-setaceous. Rays 6–12, up to 15 mm, puberulent all round. Bracts 0–1; bracteoles connate at base, puberulent. Petals white. Fruit 3–4×2–2·5 mm, ovoid-globose, sparsely puberulent. $2n=22$. *Cliffs and screes.* ● *W. Hungary.* Hu.

31. S. degenii Urum., *Magyar Bot. Lapok* 12: 217 (1913). Glabrous perennial up to 100 cm. Leaves 3-pinnate; lobes 30–40×0·5 mm, filiform. Rays 12–20, up to 30 mm, puberulent on inner side. Bracts absent; bracteoles connate at base, puberulent. Petals white. Fruit c. 4×2 mm, oblong, puberulent. *Dry limestone cliffs.* ● *N. Bulgaria.* Bu.

Described as having the fruit 10×2 mm, but this appears to be erroneous.

32. S. parnassicum Boiss. & Heldr. in Boiss., *Diagn. Pl. Or. Nov.* 3(6): 80 (1859). Glabrous perennial up to 40 cm. Leaves 2-

to 3-pinnate; lobes 20–35 × 0·5 mm, linear-setaceous. Rays 15–25, up to 30 mm, puberulent on inner side. Bracts few; bracteoles connate at base, glabrous except for ciliate margin. Petals white. Ovary puberulent. Fruit not known. *Cliffs*, 1000–2000 *m*. ● *C. Greece (Parnassos).* Gr.

33. S. bulgaricum P. W. Ball, *Feddes Repert.* **79**: 64 (1968) (*S. filifolium* Janka, non Thunb.). Glabrous perennial 15–40 cm. Leaves 2-pinnate; lobes of basal leaves 10–15 × 0·5–1 mm, of cauline up to 30 mm, linear-setaceous. Rays 6–11, up to 10 mm, puberulent on inner side. Bracts absent; bracteoles connate at base. Petals white. Fruit densely puberulent. *Cliffs.* ● *C. Bulgaria.* Bu.

34. S. gracile Waldst. & Kit., *Pl. Rar. Hung.* **2**: t. 117 (1802). Glabrous perennial 30–90 cm. Leaves 3-pinnate; lobes 20–100 × *c.* 0·5 mm), setaceous. Rays 5–16, up to 70 mm, glabrous. Bracts 0(–2); bracteoles free. Petals yellow, glabrous. Fruit 2–3 × 1·5–2 mm, ellipsoid or ovoid, glabrous. *Dry stony slopes; calcicole.* ● *N. Jugoslavia, S. & C. Romania.* Ju Rm.

33. Oenanthe L.[1]

Leaves pinnate or pinnatisect. Sepals acute, persistent. Petals white or pale pink, notched, the outer radiating; apex long, inflexed. Fruit ovoid, cylindrical, obconical or globose; commissure wide. Lateral ridges grooved or thickened, sometimes obscure; vittae solitary.

```
1  Fruit globose                                              1. globulosa
1  Fruit not globose
  2  Some umbels leaf-opposed; peduncles shorter than rays
    3  Roots with obovoid tubers                              2. lisae
    3  Roots entirely fibrous
      4  Fruit more than 5 mm                                 12. fluviatilis
      4  Fruit less than 4·5 mm                               13. aquatica
  2  Umbels terminal; peduncles longer than rays
    5  Partial umbels globose in fruit; fruits sessile        3. fistulosa
    5  Partial umbels not globose in fruit; some fruits pedicellate
         (though pedicels sometimes short and thick)
      6  Rays and pedicels not thickened in fruit
        7  Lobes of basal leaves ovate or suborbicular        11. crocata
        7  Lobes of basal leaves spathulate or linear
          8  Root-tubers ovoid, clustered at base of stem
                                                              8. peucedanifolia
          8  Root-tubers not ovoid, not clustered at base of stem
            9  Basal leaves 3-pinnate; styles not more than ¼ as long as
                 fruit                                        10. banatica
            9  Basal leaves 2-pinnate; styles at least ½ as long as fruit
                                                              9. lachenalii
      6  Rays and pedicels thickened in fruit
        10  Lobes of basal leaves cuneate or ovate; stem solid; root-
              tubers ovoid, distant from the stem
                                                              4. pimpinelloides
        10  Lobes of basal leaves linear or linear-lanceolate; stem
              hollow; root-tubers cylindrical or ovoid, near the stem
          11  Upper cauline leaves 2-pinnate; lobes setaceous
                                                              5. millefolia
          11  Upper cauline leaves pinnate; lobes linear or linear-
                lanceolate
            12  Root-tubers obovoid; fruit cylindrical or obconical
                                                              7. silaifolia
            12  Root-tubers cylindrical or fusiform; fruit elliptical
                                                              6. tenuifolia
```

1. O. globulosa L., *Sp. Pl.* 255 (1753). Much-branched perennial rarely more than 50 cm. Roots with ovoid tubers, distant from stem. Stem hollow, grooved. Basal leaves 2-pinnate; lobes

---

[1] By C. D. K. Cook.

ovate to linear. Cauline leaves 1- to 2-pinnate; lobes linear to linear-lanceolate. Umbels terminal or leaf-opposed. Peduncle longer or shorter than rays. Rays 3–16, thickened in fruit. Partial umbels male, or male and hermaphrodite; male flowers pedicellate. Fruit *c.* 5 mm, globose. Styles ⅔ as long as fruit. *Marshes, usually near the sea. W. Mediterranean region, eastwards to S. Italy; C. & S. Portugal.* Bl Co Ga Hs It Lu Sa Si.

(a) Subsp. **globulosa**: Umbels mainly terminal; peduncle usually longer than rays; rays 5–6, of which 2–3 bear fruit and thicken. 2*n* = 22. *Throughout the range of the species, except perhaps Sicilia.*

(b) Subsp. **kunzei** (Willk.) Nyman, *Consp.* 299 (1879): Umbels mainly leaf-opposed; peduncle often shorter than rays; rays 5–16, of which 3–8 bear fruit and thicken. *E. Portugal; S. Spain; Sicilia.*

2. O. lisae Moris, *Mem. Accad. Sci. Torino* 38 (*Sci. Fis. Mat.*): 27 (1835). Much-branched perennial up to 40 cm. Roots with obovoid tubers, narrow at point of attachment to stem. Stem hollow, striate. Basal leaves 1- to 2-pinnate; lobes ovate to linear. Cauline leaves pinnate; lobes linear to linear-lanceolate. Umbels terminal and leaf-opposed; lower umbels sessile. Peduncle absent or shorter than rays, except for terminal umbel. Rays 2–7, not thickened in fruit. Pedicels present only at periphery of partial umbels, not thickened in fruit. Fruit 3–3·5 mm. elliptical. Styles less than ¹⁄₁₀ as long as fruit. *Marshes.* ● *Sardegna.* Sa.

3. O. fistulosa L., *Sp. Pl.* 254 (1753). Erect, slender, stoloniferous perennial up to 80 cm. Roots with fusiform to ovoid tubers. Stem striate, hollow, thin-walled, often constricted at nodes. Basal leaves 1- to 2-pinnate; lobes ovate, lobed. Cauline leaves pinnate; lobes linear-lanceolate to subulate, entire. Submerged or winter leaves 2-pinnate; segments filiform. Umbels terminal. Peduncle longer than rays. Rays 2–4, thickened in fruit. Pedicels present only at periphery of partial umbels, not thickened in fruit. Fruit 3–4 mm, cylindrical or obconical. Styles as long as fruit. 2*n* = 22. *Wet places and in shallow water. W., C. & S. Europe, extending to S. Sweden and White Russia.* Al Au Be Br Bu Co Cz Da Ga Ge Gr Hb He Ho Hs Hu It Ju Lu Po Rm Rs (C,W) Sa Si Su.

4. O. pimpinelloides L., *Sp. Pl.* 255 (1753) (incl. *O. angulosa* Griseb., *O. incrassans* Bory & Chaub., *O. thracica* Griseb.). Erect perennial up to 100 cm. Roots with ovoid tubers, distant from stem. Stem solid, grooved. Basal leaves 2-pinnate; lobes cuneate to ovate, lobed. Cauline leaves 1- to 2-pinnate; lobes linear to linear-lanceolate, entire. Umbels terminal. Peduncle longer than rays. Rays 6–15, thickened in fruit. Pedicels thickened in fruit. Fruit *c.* 3 mm, cylindrical. Styles more than ⅔ as long as fruit. *W. & S. Europe.* Al Be ?Bl Br Bu Co Cr Ga Gr Hb Ho Hs It Ju Lu Rs (K) Sa Si Tu.

5. O. millefolia Janka, *Österr. Bot. Zeitschr.* **22**: 177 (1872). Erect perennial up to 70 cm. Roots with obovoid tubers, narrowed at point of attachment to stem. Stem hollow, grooved. Basal and cauline leaves 2-pinnate; lobes setaceous. Umbels terminal. Peduncle longer than rays. Rays 5–15, slightly thickened in fruit. Pedicels occasionally absent in centre of partial umbel, thickened in fruit. Fruit 2·5–3·5 mm, cylindrical. Styles about as long as fruit. *Dry grassland and thickets.* ● *Bulgaria.* Bu.

6. O. tenuifolia Boiss. & Orph. in Boiss., *Diagn. Pl. Or. Nov.* 3(6): 79 (1859). Erect, slender perennial up to 60 cm. Roots with cylindrical tubers clustered at base of stem. Stem hollow, grooved. Basal leaves 1- to 2-pinnate; lobes linear. Cauline leaves pinnate or simple, linear-lanceolate; lobes linear to linear-lanceolate. Umbels terminal. Peduncle longer than rays. Rays 5–7, slightly thickened in fruit. Pedicels thickened in fruit. Fruit 3·5–4 mm, elliptical. Styles at least as long as fruit. *Marshes.* ● *C. part of Balkan peninsula, from S. Albania and S.E. Bulgaria to C. Greece.* Al Bu Gr.

7. **O. silaifolia** Bieb., *Fl. Taur.-Cauc.* 3: 232 (1819) (*O. media* Griseb.). Erect, sparsely branched perennial up to 100 cm. Roots with obovoid tubers, gradually narrowed to point of attachment to stem. Stem hollow, grooved. Basal leaves 2- to 4-pinnate; lobes linear to linear-lanceolate. Cauline leaves 1- to 2-pinnate; lobes linear-lanceolate. Umbels terminal. Peduncle longer than rays. Rays 4–10, markedly thickened in fruit. Pedicels thickened in fruit, sometimes very short and occasionally absent in centre of partial umbels. Fruit 2·5–4 mm, cylindrical or obconical. Styles almost as long as fruit. *Wet places.* $2n=22$. *C., W. & S. Europe.* Al Au Be Br Bu Co Cz Ga Ge Gr Hu It Ju Rm Rs (W, K) Sa Si Tu.

8. **O. peucedanifolia** Pollich, *Hist. Pl. Palat.* 1: 289 (1776) (incl. *O. stenoloba* Schur). Like 7 but root-tubers ovoid, not narrowed to point of attachment to stem; rays and pedicels not thickened in fruit; fruit ovoid; styles rarely more than ½ as long as fruit. *Wet grassland.* ● *W. & W.C. Europe, from the Netherlands to S. Italy; Romania and Balkan peninsula.* Be Bu ?Cz Ga Ge He Ho Hs It Ju Lu Rm.

Plants from Romania and the Balkan peninsula are usually treated as a separate species, **O. stenoloba** Schur, *Enum. Pl. Transs.* 255 (1866), but do not seem to be distinguishable from **8**.

9. **O. lachenalii** C. C. Gmelin, *Fl. Bad.* 1: 678 (1805) (incl. *O. jordanii* Ten., *O. marginata* Vis.). Erect perennial up to 100 cm. Roots with cylindrical or fusiform tubers. Stem solid, or with a small central cavity when older. Basal leaves 2-pinnate; lobes spathulate to linear. Cauline leaves 1- to 2-pinnate; lobes linear to linear-lanceolate. Umbels terminal. Peduncle longer than rays. Rays 5–15, not thickened in fruit. Pedicels not thickened in fruit. Fruit 2–3 mm, ovoid. Styles ½–⅔ as long as fruit. $2n=22$. *Wet grassland. W. Europe, extending eastwards to N.W. Poland and N.W. Jugoslavia.* Be Br ?Bu Co Da Ga Ge ?Gr Hb He Ho Hs It Ju Lu Po Sa Si Su.

O. foucaudii Tesseron, *Compt. Rend. Soc. Bot. Rochel.* 6: 14 (1884), is a variant of **9** that is up to 175 cm, with a hollow stem and slightly larger and wider leaves; it is confined to W. France.

10. **O. banatica** Heuffel, *Flora (Regensb.)* 37: 291 (1854). Erect, slender perennial up to 100 cm. Roots with cylindrical tubers. Stem hollow, deeply grooved. Basal leaves 1- to 3-pinnate; lobes linear to linear-lanceolate. Umbels terminal. Peduncle longer than rays. Rays 9–16, not thickened in fruit. Pedicels not thickened in fruit. Fruit 3–4 mm, elliptical. Styles ¼ as long as fruit or less. ● *E.C. Europe and Balkan peninsula.* Bu Cz Gr Hu Ju Rm Rs (W).

11. **O. crocata** L., *Sp. Pl.* 254 (1753). Branched, stout perennial up to 150 cm. Roots with obovoid or ellipsoid tubers narrowed at point of attachment to stem. Stem hollow, striate and grooved. Basal leaves 3- to 4-pinnate; lobes ovate to suborbicular, cuneate at base, lobed, crenate. Cauline leaves 2- to 3-pinnate; lobes ovate to linear. Umbels terminal. Peduncle longer than rays. Rays 10–40, not thickened in fruit. Pedicels not thickened in fruit.

Fruit 4–6 mm, cylindrical. Styles ½ as long as fruit. $2n=22$. *Wet places. W. Europe and W. Mediterranean region.* Be Br Co Ga Hb Hs It Lu Sa Si.

O. prolifera L., *Sp. Pl.* 254 (1753), is a teratological variant of **11** with fasciated and proliferating rays. It is confined to the Mediterranean region and S.W. Asia; in Europe it is recorded from Sicilia and Kriti.

12. **O. fluviatilis** (Bab.) Coleman, *Ann. Nat. Hist.* 13: 188 (1844). Submerged, ascending or erect perennial. Roots entirely fibrous. Stem striate, hollow. Basal leaves 2- to 3-pinnate; lobes ovate or suborbicular, cuneate at base, shallowly lobed. Cauline leaves 1- to 2-pinnate; lobes ovate, cuneate at base, shallowly lobed. Submerged or winter leaves 2-pinnate; lobes cuneate, cut at the ends into narrow lobes; lobes linear to filiform. Umbels leaf-opposed. Peduncle usually shorter than rays. Rays 6–15, not thickened in fruit. Pedicels not thickened in fruit. Fruit 5–6·5 mm, elliptical. Styles ⅓ as long as fruit or less. $2n=22$. *In still or slowly flowing water.* ● *W. Europe.* Be Br Da Ga Ge Hb.

O. conioides Lange, *Haandb. Danske Fl.* ed. 2, 199 (1859), is an erect, robust, terrestrial state of **12**, apparently confined to areas liable to flooding in the valley of the lower Elbe and two localities in Belgium.

13. **O. aquatica** (L.) Poiret in Lam., *Encycl. Méth. Bot.* 4: 530 (1798) (*O. phellandrium* Lam.). Much-branched, stout winter annual or biennial up to 150 cm. Roots entirely fibrous. Stem hollow, striate and grooved, often very short. Aerial leaves 3-pinnate; segments deeply lobed; lobes ovate, acute. Submerged or winter leaves 3- to 4-pinnate; lobes linear to filiform. Umbels terminal or leaf-opposed. Peduncle usually shorter than rays. Rays 5–15, not thickened in fruit. Pedicels not thickened in fruit. Fruit 3·5–4·5 mm, oblong-ovoid or elliptical, often curved. Styles less than ¼ as long as fruit. $2n=22$. *In still or slowly flowing water. Most of Europe except the extreme north.* All except Az Bl Cr Is Sb.

34. Lilaeopsis E. L. Greene[1]

Leaves simple. Sepals small. Petals white, ovate; apex involute. Fruit ovoid, slightly compressed laterally. Ridges thick, somewhat corky; vittae solitary.

1. **L. attenuata** (Hooker & Arnott) Fernald, *Rhodora* 26: 94 (1924). Glabrous, with slender, procumbent stems rooting at the nodes. Leaves 1·5–15 cm, erect, fistular, septate. Umbels simple, with 8–15 flowers; peduncles 1–5 cm, slender. Bracteoles few, small, or often absent. Fruit 2–2·25 × 2 mm; ridges acute, the lateral somewhat larger than the dorsal. *Marshes and shallow water. Naturalized in Portugal.* [Lu.] (*S.E. North America, E. temperate South America.*)

35. Aethusa L.[1]

Leaves 2(–3)-pinnate. Sepals absent. Petals white, obcordate, the outer larger; apex inflexed. Fruit ovoid, somewhat compressed dorsally. Ridges thick, the lateral narrowly winged; vittae solitary.

1. **A. cynapium** L., *Sp. Pl.* 256 (1753). Annual or sometimes biennial 5–200 cm. Leaves deltate in outline; lobes lanceolate or ovate, pinnatifid. Rays (4–)10–20. Bracts usually absent; bracteoles usually 3–4, deflexed, subulate, on the outer side of the partial umbels. Fruit 3–4 mm. *A weed of cultivated land, rarely in woods. Most of Europe, but rare in the Mediterranean region.* Au

---

[1] By T. G. Tutin.

Be Br Bu Co Cz Da Fe Ga Ge Hb He Ho Hs Hu It Ju No Po Rm Rs (N, B, C, W. K, E) Su.

A polymorphic species in which the following subspecies can be recognized, though they show considerable phenotypic variation.

1 Stem more than 100 cm; leaf-lobes oblong to linear
                                         (c) subsp. **cynapioides**
1 Stem less than 100 cm; leaf-lobes ovate
2 Bracteoles usually several times as long as the partial umbels; outer pedicels about twice as long as their fruits
                                         (a) subsp. **cynapium**
2 Bracteoles not longer than the partial umbels; outer pedicels usually shorter than their fruits   (b) subsp. **agrestis**

(a) Subsp. **cynapium**: Annual. Stem 30–80 cm, sulcate, green. Leaf-lobes ovate. Bracteoles usually several times as long as the partial umbels. Outer pedicels about twice as long as their fruits. $2n = 20$. *Waste places and arable land. Throughout the range of the species.*

(b) Subsp. **agrestis** (Wallr.) Dostál, *Květena CSR* 1048 (1949) (*A. cynapium* var. *agrestis* Wallr.): Annual. Stem 5–20 cm, angled, green. Leaf-lobes ovate. Bracteoles shorter than to as long as the partial umbels. Outer pedicels usually shorter than their fruits. $2n = 20$. *Arable land. Throughout the range of the species.*

(c) Subsp. **cynapioides** (Bieb) Nyman, *Consp.* 297 (1879): Biennial. Stem 100–200 cm, terete, finely striate, pruinose. Leaf-lobes oblong to linear. Bracteoles about twice as long as their fruits. $2n = 20$. *Woodlands. C. Europe, extending to S. Sweden.*

36. Portenschlagiella Tutin[1]

(*Portenschlagia* Vis., non Tratt.)

Leaves 4- to 5-pinnate. Sepals small. Petals yellow. oblong-spathulate, emarginate, ciliate beneath; apex involute. Fruit ovoid-cylindrical, not compressed, hispid and with stellate hairs. Primary ridges stout, prominent; secondary filiform; vittae 3.

1. **P. ramosissima** (Portenschl.) Tutin, *Feddes Repert.* **74**: 32 (1967) (*Portenschlagia ramosissima* (Portenschl.) Vis.). Stout, somewhat pubescent perennial. Stock subglobose, with abundant coriaceous remains of dead leaves. Leaves with linear-filiform, acuminate lobes. Umbels subglobose, often in a whorl at the top of the main stem; rays 30–50, puberulent. Bracts and bracteoles numerous. *Rocky places.* ● *S. Italy. W. Jugoslavia and N.W. Albania.* Al It Ju.

37. Athamanta L.[1]

Leaves 2- to 5-pinnate, with narrow lobes. Sepals small. Petals white or perhaps rarely yellow, emarginate or 2-lobed, sometimes pubescent beneath or ciliate; apex inflexed, long. Fruit oblong-ovoid to ovoid, scarcely compressed, pubescent, narrowed to a short beak. Ridges low; vittae 1–2.

1 Leaf-lobes ovate to rhombic
2 Each half of stylopodium about twice as long as wide; fruit with conspicuous, pubescent ridges, almost or quite glabrous between them   **1. macedonica**
2 Each half of stylopodium wider than long; fruit with inconspicuous ridges, densely pubescent all over   **2. sicula**
1 Leaf-lobes linear or linear-oblong
3 Hairs on rays straight, shorter than the diameter of the ray
                                        3. turbith

---

3 Hairs on rays flexuous, at least as long as the diameter of the ray
4   Stems simple or with few branches; rays 5–15(–36)   **4. cretensis**
4   Stems freely branched, the umbels forming a dense panicle; rays 15–35
5   Stem 20–40 cm; lateral umbels not overtopping the terminal one; rays 20–35   **5. densa**
5   Stem 3–10 cm; lateral umbels overtopping the terminal one; rays 15–20   **6. cortiana**

1. **A. macedonica** (L.) Sprengel in Roemer & Schultes, *Syst. Veg.* **6**: 491 (1820) (incl. *A. chiliosciadia* Boiss. & Heldr.). Erect perennial up to 200 cm, with numerous short branches in the inflorescence. Lower leaves 2- to 3-pinnate; lobes ovate, dentate; upper leaves simple or pinnatisect. Peduncles and umbels puberulent. Rays 5–18. Bracts 5–8, sometimes pinnatisect; bracteoles linear-lanceolate, acuminate. Each half of the stylopodium about twice as long as wide; styles patent or erecto-patent. Fruit 3–5 mm, ovoid, puberulent on the prominent ridges, nearly or quite glabrous between them. ● *S. part of Balkan peninsula; S. Italy.* Al Gr It Ju.

1 Leaf-lobes 20–50 mm; umbel with arachnoid pubescence
                            (c) subsp. **arachnoidea**
1 Leaf-lobes 2–20 mm; umbels shortly pubescent
2 Leaf-lobes (5–)10–20 mm, puberulent   (a) subsp. **macedonica**
2 Leaf-lobes 2–5 mm. nearly or quite glabrous (b) subsp. **albanica**

(a) Subsp. **macedonica**: Up to 200 cm; lobes of lower leaves (5–)10–20 mm, puberulent; umbels shortly pubescent; fruit 3–4 mm. *Throughout most of the range of the species.*

(b) Subsp. **albanica** (Alston & Sandwith) Tutin, *Feddes Repert.* **79**: 18 (1968) (*A. albanica* Alston & Sandwith): Up to 50 cm; lobes of lower leaves 2–5 mm, somewhat fleshy, nearly or quite glabrous; fruit 4–5 mm. *S. Albania (near Gjinokastër).*

(c) Subsp. **arachnoidea** (Boiss. & Orph.) Tutin, *Feddes Repert.* **79**: 18 (1968) (*A. arachnoidea* Boiss. & Orph.): Leaf-lobes 20–50 mm, puberulent; umbels with arachnoid hairs. *S. Greece (Taïyetos).*

What appears to be a dwarf variant of **1** from N.W. Greece has been described as **Seseli farinosum** Quézel & Contandr., *Taxon* **16**: 240 (1967). It has $2n = 22$.

A. macrosperma H. Wolff, *Feddes Repert.* **18**: 133 (1922) appears to have been collected once only, in S. Greece (near Athinai). It is like **1** and may be an abnormal specimen of it. It has 4–6 rays, obtuse bracteoles and fruit *c*. 8 mm.

2. **A. sicula** L., *Sp. Pl.* 244 (1753). Pubescent perennial up to 100 cm. Leaves 3- to 4-pinnate; lobes 2–5 mm, ovate to rhombic, lobed or dentate. Rays 10–20. Bracts few, linear-lanceolate or rarely pinnatisect; bracteoles linear-lanceolate, acuminate. Each half of the stylopodium wider than long; styles patent. Fruit 6–7 mm, ovoid-oblong, densely pubescent all over; ridges inconspicuous. *C. & S. Italy, Sicilia.* It Si.

3. **A. turbith** (L.) Brot., *Fl. Lusit.* **1**: 435 (1804) (*A. mathioli* Wulfen). Perennial up to 50 cm. Leaves 2- to 4-pinnate, glabrous or sparsely pubescent; lobes up to 35 mm, linear or filiform. Rays (15–)20–35(–50), pubescent, with straight hairs shorter than the diameter of the ray. Bracts 0–8; bracteoles several, lanceolate. Fruit 5–6 mm, ovoid, densely pubescent all over. ● *N.W. part of Balkan peninsula, extending to N.E. Italy; S. Carpathians.* Al It Ju Rm.

1 Styles erect   (a) subsp. **turbith**
1 Styles patent or erecto-patent
2 Rays 15–25   (b) subsp. **haynaldii**
2 Rays 30–50   (c) subsp. **hungarica**

(a) Subsp. **turbith**: Leaves glabrous; lobes 15–35 × 0·3–0·5 mm, filiform; rays (15–)20–35; bracts 0–1(–5); bracteoles abruptly narrowed at apex; each half of stylopodium twice as long as wide; styles erect. *N.E. Italy, N.W. Jugoslavia.*

(b) Subsp. **haynaldii** (Borbás & Uechtr.) Tutin, *Feddes Repert.* **79**: 19 (1968) (*A. haynaldii* Borbás & Uechtr.): Leaves glabrous or pubescent; lobes 8–20 × 0·3–0·6 mm, narrowly linear; rays 15–25; bracts 5–8; bracteoles gradually or rarely abruptly narrowed at apex; each half of stylopodium about as long as wide; styles patent or erecto-patent. *W. Jugoslavia, Albania.*

(c) Subsp. **hungarica** (Borbás) Tutin, *Feddes Repert.* **79**: 19 (1968) (*A. hungarica* Borbás): Leaves glabrous or rarely pubescent; lobes *c.* 25 × 0·8 mm, narrowly linear or filiform; rays 30–50; bracts 5–8; bracteoles gradually or abruptly narrowed at apex; each half of stylopodium 1½–2 times as long as wide; styles patent or erecto-patent. *S. Carpathians.*

4. A. cretensis L., *Sp. Pl.* 245 (1753). Erect, pubescent perennial up to 60 cm. Leaves 3- to 5-pinnate; lobes 3–7 mm, linear or linear-oblong. Rays 5–15(–36), pubescent, with flexuous or crispate hairs at least as long as the diameter of the ray. Bracts 0–5, sometimes pinnatisect; bracteoles several, gradually narrowed at apex. Each half of stylopodium about as long as wide; styles erecto-patent. Fruit 6–8 mm, oblong-ovoid, densely puberulent. 2*n*=22. *Rocky places, mainly in the mountains.* ● *From S.E. Spain and E. France to C. Jugoslavia.* Al Au Ga Ge He Hs It Ju.

Plants which are usually taller and have more numerous rays (15–36) than the typical plant occur in S. Spain, Italy, S.E. Austria, Jugoslavia and Albania. They have been called **A. hispanica** Degen & Hervier, *Bull. Acad. Int. Géogr. Bot.* (*Le Mans*) **17**: 41 (1907) and **A. vestina** A. Kerner, *Sched. Fl. Exsicc. Austro-Hung.* **4**: 37 (1886). They seem to be intermediate between **3(b)** and **4**, differing from the former mainly in having longer hairs on the rays.

5. A. densa Boiss. & Orph. in Boiss., *Diagn. Pl. Or. Nov.* **3**(5): 98 (1856). Densely pubescent perennial 20–40 cm. Stems freely branched from near the base. Leaves mostly basal, crowded, 4- to 5-pinnate; lobes 5–10 mm, oblong. Umbels forming a dense panicle, the lateral not overtopping the terminal one. Rays 20–35. Bracts 3–7, sometimes pinnatisect; bracteoles several, gradually narrowed at apex. Each half of stylopodium about as long as wide; styles erecto-patent. Mature fruit unknown. ● *S. Albania, C. Greece.* Al Gr

6. A. cortiana Ferrarini, *Webbia* **20**: 334 (1965). Like **5** but stem 3–10 cm; lateral umbels overtopping the terminal one; rays 15–20; fruit *c.* 7 mm. *Calcareous rocks.* ● *N. Italy (Alpi Apuane).* It.

A. aurea (Vis.) Neilr. in J. Maly, *Enum. Pl. Austr., Nachtr.* 198 (1848), appears to have been collected once only. The type and the fragment figured by Reichenbach fil. (*Icon. Fl. Germ.* **21**: t. 1935 (1864)) are both missing. It is said to be an annual with yellow flowers and was found in W. Jugoslavia on the east side of the Dinara Planina. It may have been wrongly placed in *Athamanta* but requires further investigation.

38. Grafia Reichenb.[1]

Leaves 3- to 4-pinnate or -ternate. Petals white, obcordate; apex wide and long, inflexed. Fruit ovate-oblong, compressed laterally. Ridges prominent, almost winged; vittae 3–4.

1. G. golaka (Hacq.) Reichenb., *Handb.* 219 (1837) (*Hladnikia golaka* (Hacq.) Reichenb. fil.). Glabrous, glaucous perennial 50–100 cm. Leaf-lobes up to 3·5 cm, nearly rhombic, pinnately lobed or dentate. Rays 12–22. Bracts numerous, ovate-oblong, acute or sometimes bifid at apex; bracteoles usually 3, on the outer side of the partial umbel. Fruit 8–13 mm. *Calcicole.* ● *S.E. Alps, C. Appennini, W. Jugoslavia.* It Ju.

39. Xatardia Meissner[1]

Leaves 2- to 3-pinnate, with narrow lobes. Sepals absent. Petals greenish-yellow, lanceolate, narrowed to the involute apex. Fruit ovoid, slightly compressed dorsally. Ridges stout and prominent; vittae solitary, slender.

1. X. scabra (Lapeyr.) Meissner, *Pl. Vasc. Gen.* **2**: 105 (1838). Glabrous perennial with a long, stout root. Stem 10–25 cm, usually simple, stout, solid. Leaves deltate in outline; lobes mucronate. Rays up to 14 cm, very unequal, scabrid, erect in fruit. Bracts 0–2, bracteoles 4–12, both linear-subulate, caducous. Fruit 6–7 mm. *Alpine, calcareous and schistose screes.* ● *E. Pyrenees.* Ga Hs.

40. Foeniculum Miller[1]

Leaves 3- to 4-pinnate, with long filiform lobes. Sepals absent. Petals yellow, oblong, scarcely narrowed to the involute apex. Fruit ovoid-oblong, scarcely compressed. Ridges stout, prominent, the lateral somewhat wider than the others; vittae solitary.

1. F. vulgare Miller, *Gard. Dict.* ed. 8, no. 1 (1768) (*F. officinale* All.). Glabrous, glaucous perennial or biennial up to 250 cm. Stem striate, shiny, developing a small hollow when old. Leaves more or less triangular in outline; lobes usually 5–50 mm, filiform, acuminate, cartilaginous at apex, usually widely spaced and not all lying in one plane; petioles of upper leaves usually 3–6 cm. Rays 4–30. Bracts and bracteoles usually 0. Fruit 4–10·5 mm, ovoid-oblong; lateral ridges scarcely more prominent than dorsal. *Usually maritime. Most of Europe, except the north, but probably native only in the south and south-west.* Al Az Bl *Br Bu Co Cr Ga Gr *Hb Hs It Ju Lu Sa Si Tu [Au Be Cz Ge He Ho Hu Po Rm Rs (C, W, E, K)].

(a) Subsp. **piperitum** (Ucria) Coutinho, *Fl. Port.* 450 (1913) (*F. piperitum* (Ucria) Sweet): Perennial; leaf-lobes seldom more than 10 mm, rigid and rather fleshy; terminal umbel often overtopped by lateral ones; rays usually 4–10; fruit sharp-tasting. *Dry, rocky places. Mediterranean region.*

(b) Subsp. **vulgare**: Often biennial; leaf-lobes usually more than 10 mm, flaccid; terminal umbel not overtopped by lateral ones; rays usually 12–25; fruit sweet-tasting. *Widely cultivated for flavouring and commonly naturalized.*

A variety of subsp. **(b)** (var. *azoricum* (Miller) Thell.) with a large tuberous stock is cultivated as a vegetable in some Mediterranean countries.

41. Anethum L.[1]

Leaves 3- to 4-pinnate, with long, filiform lobes. Sepals absent. Petals yellow, oblong; apex incurved. Fruit elliptical, strongly compressed dorsally. Dorsal ridges slender, prominent; lateral winged; vittae solitary.

1. A. graveolens L. *Sp. Pl.* 263 (1753). Glaucescent, strongly smelling annual 20–50 cm. Leaves 3- to 4-pinnate with filiform, mucronate lobes. Rays 15–30, unequal. Bracts and bracteoles absent. Fruit 5–6 mm, dark brown with a pale wing. *Widely cultivated as a herb and often more or less naturalized, particularly in the Mediterranean region.* [Au Be Bl Bu Cr Cz Ga Gr He Ho Hs Hu It Ju Lu Rm Rs (B, C, W, K, E) Si.] (*India and S.W. Asia;* ?*N. Africa.*)

42. Kundmannia Scop.[1]

Leaves 1- to 2-pinnate; segments ovate. Sepals small, somewhat accrescent. Petals yellow, broadly ovate; apex involute. Fruit nearly cylindrical. Ridges slender but prominent; vittae numerous, irregularly arranged.

1. **K. sicula** (L.) DC., *Prodr.* 4: 143 (1830) (*Brignolia pastinacifolia* Bertol.). Glabrous perennial 30–70 cm. Lower leaves usually 2-pinnate with a pair of supplementary segments at the base of each pair of primary segments; lobes ovate, crenate-serrate, the lowest sometimes lobed; upper cauline leaves 1-pinnate, the segments incise-serrate or lacerate. Umbels with 5–30 subequal rays. Bracts and bracteoles numerous, linear. Fruit 6–10 mm. *Mediterranean region, extending to S. Portugal.* Bl Co Cr Gr Hs It Lu Sa Si [Ga].

43. Silaum Miller[1]

Leaves 1- to 4-pinnate. Sepals absent. Petals yellowish, ovate; apex short, involute. Fruit ovoid-oblong to nearly cylindrical, scarcely compressed. Ridges slender, prominent, the lateral narrowly winged; vittae numerous, slender, inconspicuous.

1. **S. silaus** (L.) Schinz & Thell., *Viert. Naturf. Ges. Zürich* 60: 359 (1915) (*Silaus pratensis* Besser, incl. *S. besseri* DC., *Silaum alpestre* (L.) Thell., *S. flavescens* (Bernh.) Hayek). Glabrous perennial 3–100 cm. Stem solid, striate. Basal leaves 2- to 4-pinnate, triangular in outline; segments long-stalked; lobes 5–20 mm, lanceolate to linear, finely serrulate, with prominent midrib, acuminate or obtuse and mucronate; apex often reddish; upper cauline leaves 1-pinnate or reduced to an inflated petiole. Rays 5–15, sharply angled. Bracts 0–3; bracteoles several, linear-lanceolate, broadly scarious. Fruit 4–5 mm, ovoid-oblong to subcylindrical. *W., C. & E. Europe.* Al Au Be Br Cz Ga Ge He Ho Hs Hu It Ju Po Rm Rs (B, C, W, K, E) Su.

Very variable in the shape of the leaf-lobes and the size and shape of the fruit, particularly in the eastern part of its range.

Variants with linear leaf-lobes and subcylindrical fruits c. 5 mm have been called **S. alpestre** (L.) Thell. in Hegi, *Ill. Fl. Mitteleur.* 5 (2): 1295 (1926).

44. Trochiscanthes Koch[1]

Leaves 3- to 4-ternate, with large, ovate segments. Sepals conspicuous. Petals greenish-white, clawed; apex inflexed, obtuse. Fruit ovoid, slightly compressed laterally. Ridges slender, prominent; vittae 4.

1. **T. nodiflora** (Vill.) Koch, *Nova Acta Acad. Leop.-Carol.* 12(1): 104 (1824). Glabrous perennial 100–200 cm. Stem with numerous opposite or whorled patent branches. Leaves with irregularly serrate and sometimes shallowly lobed segments 5–11 × 2–4·5 cm. Umbels small, very numerous, terminating the branches; rays 4–8. Bracts 0–1; bracteoles 3–5, subulate. Fruit c. 6 mm. *Mountain woods.* ● *S.E. France, S.W. Switzerland, N. Italy.* Ga He It.

45. Meum Miller[1]

Leaves 3- to 4-pinnate, with crowded, filiform lobes. Sepals absent. Petals white or purplish, ovate; apex more or less inflexed. Fruit ovoid-oblong, scarcely compressed. Ridges very prominent, stout; vittae 3–5.

1. **M. athamanticum** Jacq., *Fl. Austr.* 4: 2 (1776) (incl. *M. nevadense* Boiss.). Glabrous, strongly aromatic perennial 7–60 cm. Stock surrounded by coarse, fibrous remains of petioles. Leaves mostly basal; lobes 2–5 mm. Rays 3–15. Bracts 0–2, setaceous; bracteoles few, setaceous, often small. Fruit 4–10 mm. ● *Mountains of W. & C. Europe, extending locally to Calabria and C. Bulgaria.* Al Au Be Br Bu Cz Ga Ge He Hs It Ju Po ?Rs (W) [No].

46. Physospermum Cusson[1]

Leaves 2- to 3-ternate. Sepals small. Petals white, obovate, emarginate; apex inflexed. Fruit ovoid, didymous, with a cordate base. Ridges filiform; vittae solitary.

Leaf-lobes 10–30 mm, pinnatifid; fruit 3–4 mm   1. cornubiense
Leaf-lobes 50–110 mm, serrate, usually unlobed; fruit 5–9 mm
                                                         2. verticillatum

1. **P. cornubiense** (L.) DC., *Prodr.* 4: 246 (1830) (*P. aquilegifolium* Koch, *Danaa nudicaulis* (Bieb.) Grossh., *D. cornubiensis* (L.) Burnat). Nearly glabrous perennial 30–120 cm. Stem striate, solid. Basal leaves long-petiolate, 2-ternate; lobes 10–30 mm, pinnatifid, cuneate at base, puberulent on margins and veins; cauline leaves small or reduced to petioles. Rays 6–20, glabrous. Bracts and bracteoles lanceolate, acute. Fruit 3–4 mm. *S. Europe, extending northwards to S. England, Hungary and S.C. Russia.* Al Br Bu Co Ga Gr Hs Hu It Ju Lu Rm Rs (C, W, K).

2. **P. verticillatum** (Waldst. & Kit.) Vis., *Fl. Dalm.* 3: 358 (1852) (*Danaa verticillata* (Waldst. & Kit.) Janchen). Like **1** but more robust; stems up to c. 160 cm, sulcate; branches usually whorled; leaves more or less setulose on margins and veins; lobes 50–110 mm on lower leaves, serrate, usually unlobed; fruit 5–9 mm. *N. Jugoslavia, Italy, Sicilia.* It Ju Si.

47. Conium L.[1]

Leaves 2- to 4-pinnate; lobes serrate or pinnatifid. Sepals absent. Petals white, obcordate; apex inflexed. Fruit subglobose, laterally compressed. Ridges prominent, often undulate-crispate; vittae absent.

1. **C. maculatum** L., *Sp. Pl.* 243 (1753). Nearly glabrous winter annual or biennial 50–250 cm. Stem pruinose, usually with reddish-brown spots below, hollow. Lower leaves up to 50 × 40 cm, triangular in outline, 2- to 4-pinnate, soft, entirely glabrous; lobes 10–20 mm, oblong-lanceolate to deltate, pinnatifid, coarsely serrate or crenate-serrate. Rays (6–)10–20, often puberulent. Bracts (0–)5–6, narrowly triangular to ovate-lanceolate, deflexed; margin scarious; bracteoles 3–6, on the outside of the partial umbel, widened at base and often connate. Fruit 2·5–3·5 mm. *Almost throughout Europe, except the extreme north.* All except Fa Is Sb.

48. Hladnikia Reichenb.[1]

Leaves ternate or pinnate, with ovate segments. Sepals conspicuous. Petals whitish. Fruit less than 3 times as long as wide,

---

[1] By T. G. Tutin.

with broad, rounded ridges, forming 3 prominent angles on each mericarp; vittae large.

1. H. pastinacifolia Reichenb., *Pl. Crit.* **9**: 9 (1831). Glabrous biennial or perennial 15–40 cm. Basal leaves 5–16 cm; segments up to 2·5 × 1·5 cm, 3 or 5, ovate, dentate to pinnatifid; petiole with a long sheath-like base; cauline leaves few, small. Rays 9–20, scabrid on their inner angles. Bracts numerous, entire or 3-fid; bracteoles entire; both appressed at first, deflexed later. Fruit 4–5 mm, oblong. 2*n* = 22. *Calcareous rocks*, 1100–1500 *m*. ● *W. Slovenija (mountains north of Ajdovščina).* Ju.

49. Pleurospermum Hoffm.[1]

Leaves 2- to 3-pinnate. Sepals small. Petals white, suborbicular, shortly clawed, papillose within. Fruit ovoid-oblong, slightly compressed laterally; mesocarp spongy, often breaking down and leaving an air-space between exocarp and endocarp as the fruit ripens. Ridges prominent, often with narrow, undulate wings; vittae solitary.

Leaf-lobes sparsely hairy on margins and on veins beneath, abruptly contracted to the acute apex; bracts entire or shallowly lobed **1. austriacum**
Leaf-lobes densely hairy on margins and on veins beneath, gradually narrowed to the acute apex; bracts usually deeply pinnatisect **2. uralense**

1. P. austriacum (L.) Hoffm., *Gen. Umb.* x (1814). Monocarpic perennial or biennial 50–200 cm. Stock with fibres. Stem very stout, hollow, ridged; branches alternate. Lower leaves ternately 2- to 3-pinnate, triangular-ovate in outline; lobes 4–10 cm, ovate, cuneate, pinnatifid, coarsely crenate-dentate, with a wide, obtuse, cartilaginous apex, sparsely hairy on margins and veins beneath. Rays 12–20(–40), weakly angled. Bracts numerous, eventually deflexed, entire or shallowly lobed; bracteoles numerous, lanceolate, entire, with membranous margins. Fruit 9–10 mm; ridges usually unwinged. 2*n* = 22. ● *C. & E. Europe, extending to the S.W. Alps and with an isolated area in Sweden.* Au Bu Cz Ga Ge He Hu It Ju Po Rm Rs (B, C, W) Su.

2. P. uralense Hoffm., *Gen. Umb.* ix (1814). Like **1** but leaf-lobes ovate-lanceolate, densely hairy on margins and veins beneath; rays (15–)20–40, strongly angled; bracts usually deeply pinnatisect; fruit 5–6 mm; ridges narrowly winged. *E. Russia.* Rs (N, C, E). (*N.E. Asia.*)

50. Aulacospermum Ledeb.[1]

Like *Pleurospermum* but sepals absent; mesocarp of fruit not spongy and not breaking down as the fruit ripens.

1. A. isetense (Sprengel) Schischkin, *Fl. URSS* 16: 242 (1950). Glabrous perennial 50–100 cm. Stock surrounded by brown leaf-bases. Stems *c.* 1 cm in diameter, grooved; branches erecto-patent, exceeding the terminal umbel. Basal and lower cauline leaves 8–10 × 4 cm, 3-pinnate, ovate-oblong in outline; lobes 2–6 × 0·3 mm, linear, acute, entire, dentate or 3-fid. Rays 20–30. Bracts 8–10, linear-lanceolate, dentate or pinnatifid; bracteoles entire or rarely dentate. Fruit 5–7 mm, subglobose; lateral ridges with a narrow, erose, crenate-dentate wing. *E. Russia, westwards to 49° E.* Rs (C, E).

51. Lecokia DC.[1]

Leaves 1- to 2-pinnate. Sepals small. Petals white, obovate; apex involute. Fruit ovoid-oblong, somewhat compressed laterally, constricted at the commissure, shortly beaked. Ridges thick, corky, muricate; vittae numerous.

1. L. cretica (Lam.) DC., *Coll. Mém.* **5**: 75 (1829). Glabrous perennial 60–100 cm. Lower leaves broadly triangular in outline; lobes oblong, irregularly toothed or lobed. Rays 6–10. Bracts 0 or 1; bracteoles few, subulate. Fruit 10–12 mm. *Limestone pastures.* Kriti. Cr. (*S.W. Asia.*)

52. Cachrys L.[1]

(Incl. *Hippomarathrum* Link, *Prangos* Lindley)

Leaves 2- to 4-pinnate, with linear lobes. Sepals conspicuous to obsolete. Petals yellow, ovate; apex involute. Fruit subdidymous. Ridges thick, undulately winged and papillose, or wide and smooth; vittae numerous.

1 Sepals absent; ridges of fruit very wide, almost confluent, smooth
2 Plant pubescent **8. odontalgica**
2 Plant glabrous
3 Leaf-lobes 5–30 mm; fruit almost terete **6. trifida**
3 Leaf-lobes 25–60 mm; fruit weakly grooved **7. alpina**
1 Sepals present; ridges of fruit prominent, wedge-shaped, often papillose or dentate
4 Bracts of central umbel (1–)2-pinnatisect **1. sicula**
4 Bracts of central umbel entire or 2- to 3-fid
5 Margins of leaf-lobes scabrid or puberulent **5. ferulacea**
5 Margins of leaf-lobes glabrous, though sometimes remotely denticulate
6 Ridges of fruit smooth or with short, flattened, often appressed papillae **2. libanotis**
6 Ridges of fruit dentate-cristate
7 Leaf-lobes 35–80 mm, filiform **3. pungens**
7 Leaf-lobes not more than 15(–30) mm, flat, 3-fid **4. cristata**

1. C. sicula L., *Sp. Pl.* ed. 2, 355 (1762) (*Hippomarathrum pterochlaenum* Boiss.). Erect, slightly scabrid, glaucescent perennial 30–150 cm. Stem solid, striate; branches opposite or whorled. Leaves 2- to 3-pinnate, broadly rhombic- or triangular-ovate in outline; lobes 15–50 × 1–1·5 mm, linear, often flaccid, mucronate, scabrid on the margin. Rays 20–30. Bracts of central umbel (1–)2-pinnatifid; bracteoles subulate, entire. Fruit 10–15 mm; ridges prominent, wide and rounded, more or less dentate-cristate. *S. part of W. Mediterranean region, extending to C. & S. Portugal.* Hs It Lu Sa Si.

2. C. libanotis L., *Sp. Pl.* 246 (1753) (*Hippomarathrum bocconii* Boiss.). Like **1** but usually smaller and stouter; leaves less divided; lobes 5–10 × 1·5–2·5 mm, rigid, often dentate; rays 8–15; bracts simple or sometimes 2- to 3-fid; fruit not more than 10 mm; ridges smooth or with short, flattened, often appressed papillae. *S. part of W. Mediterranean region extending to S. Portugal.* Hs It Lu Sa Si.

3. C. pungens Jan ex Guss., *Fl. Sic. Prodr.* **1**: add. 7 (1827). Like **1** but leaf-lobes 35–80 mm, filiform; bracts entire or 2- to 3-fid. *S. Italy, Sicilia.* It Si.

4. C. cristata DC., *Prodr.* 4: 238 (1830) (*Hippomarathrum cristatum* (DC.) Boiss.). Erect, glabrous perennial. Stem angled; branches opposite or whorled. Leaves 2- to 4-pinnate, ovate in outline; lobes *c.* 15 × 1 mm, rigid, narrowly linear, flat, 3-fid at apex, divaricate. Rays (4–)10–12. Bracts and bracteoles short, linear. Fruit 7–10 mm; ridges dentate-cristate, almost winged. *S. Italy, S. part of Balkan peninsula, Aegean region.* Bu Cr Gr It Ju.

Plants from N. Greece (Thessaloniki) have leaf-lobes up to 30 mm, but only 0·5 mm wide, which are erecto-patent, not divaricate. In other respects they resemble **4**.

Plants from the Kikladhes with 4–6 rays in the umbels have been called **Hippomarathrum pauciradiatum** (Heldr. & Halácsy) Heldr. & Halácsy in Halácsy, *Consp. Fl. Graec.* **1**: 660 (1901). They do not appear to differ from **4** in any other way.

5. C. ferulacea (L.) Calestani, *Webbia* **1**: 154 (1905) (*Prangos ferulacea* (L.) Lindley). Robust, nearly glabrous perennial up to 180 cm. Stock with numerous coarse fibres. Stem striate, solid. Leaves up to 80 cm, 3- to 5-pinnate; lobes 10–45 mm, linear or filiform, puberulent or minutely scabrid on the margins. Rays 6–18, stout. Bracts and bracteoles several, linear-lanceolate, acuminate. Fruit 10–25 mm; mericarps with 5 equal wings or prominent ridges. *Balkan peninsula, extending to S.W. Romania, C. Italy and Sicilia.* Al Bu Gr It Ju Rm Si.

Variable in length of leaf-lobes and in size, shape and development of wings of the fruit. **Prangos carinata** Griseb. ex Degen, *Term.-Tud. Közl.* **28**: 44 (1896), described from Romania, has suborbicular, strongly ridged but unwinged fruits. Plants with similar fruits occur in S. Italy, where they appear to intergrade with the more widely distributed variant with oblong, winged fruits.

6. C. trifida Miller, *Gard. Dict.* ed. 8, no. 1 (1768) (*C. laevigata* Lam.). Glabrous perennial 60–120 cm. Stem striate, terete, solid; upper branches opposite or verticillate. Lower leaves 30–40 × 30–40 cm, 4- to 7-pinnate, triangular in outline; lobes 5–30 mm, linear, mucronate; petioles of upper cauline leaves narrow. Rays 10–20, glabrous. Bracts few or none, linear; bracteoles several. Fruit 12–15 mm, elliptical, almost terete. ● *W. Mediterranean region, Portugal.* Ga Hs It Lu.

7. C. alpina Bieb., *Fl. Taur.-Cauc.* **1**: 217 (1808). Like **6** but lower leaves 40–50 × 50–60 cm; lobes 25–60 mm, filiform, usually arcuate; bracts and bracteoles very short; fruit weakly sulcate. ● *S.E. Europe, from Macedonia to S.E. Russia; local.* Bu Ju Rm Rs (K, E).

8. C. odontalgica Pallas, *Reise* **3**: 720 (1776). Greyish-puberulent perennial 20–90 cm. Stem terete and striate, or weakly ridged, solid. Lower leaves 10–25 × 8–20 cm, 3- to 4-pinnate, triangular in outline; lobes 3–4 mm, linear, obtuse; petioles of upper cauline leaves inflated. Rays 4–7, glabrous. Bracts and bracteoles 1–3 mm, 5–7, linear-lanceolate, ciliate. Fruit 10–12 mm, subcylindrical. *S. Ukraine, S.E. Russia, W. Kazakhstan.* Rs (W, K, E).

53. **Heptaptera** Margot & Reuter[1]

(*Colladonia* DC., non Sprengel)

Like *Cachrys* but leaf-lobes ovate-lanceolate; fruit oblong or ovate, slightly compressed dorsally.

1 Stem terete; mericarps with lateral wings only   **4. macedonica**
1 Stem triquetrous; mericarps with dorsal and lateral wings
   2 Middle cauline leaves with pinnatifid segments   **3. angustifolia**

---

[1] By T. G. Tutin.

2 Middle cauline leaves with entire or slightly lobed segments
   3 Leaf-segments decurrent on the rhachis; bracts and bracteoles lanceolate or ovate   **1. triquetra**
   3 Leaf-segments not decurrent on the rhachis; bracts and bracteoles linear-lanceolate   **2. colladonioides**

1. H. triquetra (Vent.) Tutin, *Feddes Repert.* **74**: 34 (1967) (*Colladonia triquetra* (Vent.) DC.). Glabrous perennial with a stout, solid, triquetrous stem up to 150 cm. Stock covered with scale-like remains of leaves. Basal leaves simple, the others pinnatifid; segments 7–12 cm, ovate, crenate, sometimes slightly lobed, decurrent on the rhachis. Terminal umbel with 7–14 rays, often overtopped by lateral umbels. Bracts and bracteoles numerous, lanceolate or ovate, membranous. Fruit 10–12 mm, ovate; mericarps with 5 equal wings. *In scrub on hills.* ● *S.E. Bulgaria and Turkey-in-Europe.* Bu Tu.

2. H. colladonioides Margot & Reuter, *Mém. Soc. Phys. Hist. Nat. Genève* **8**: 302 (1839) (*Colladonia colladonioides* (Margot & Reuter) Halácsy). Like **1** but usually up to 60 cm; stems more or less scabrid on the angles; leaf-segments oblique at base, not decurrent, the lateral 1–5 cm, the terminal larger; bracts and bracteoles linear-lanceolate, with scarious margins; fruit 12–18 mm; outer mericarps with 4 wings, inner with 3. *Sunny hillsides.* ● *S. & W. Greece.* Gr.

3. H. angustifolia (Bertol.) Tutin, *Feddes Repert.* **74**: 33 (1967) (*Colladonia angustifolia* Bertol.). Like **1** but segments of middle cauline leaves pinnatifid, not decurrent, the lobes oblong-lanceolate, serrate; bracts and bracteoles linear; fruit 13–15 mm, oblong; lateral wings of mericarps nearly twice as wide as dorsal. *Bushy places.* ● *S. Italy.* It.

4. H. macedonica (Bornm.) Tutin, *Feddes Repert.* **74**: 34 (1967) (*Colladonia anatolica* sensu Hayek, non (Boiss.) Boiss., *C. macedonica* Bornm.). Scabrid perennial with a terete stem up to 100 cm. Basal leaves pinnatifid, with 3–4 pairs of segments; lower segments repeatedly 2- to 3-sect or 3-partite. Rays 9–12. Bracts numerous, linear-lanceolate; bracteoles oblong, without scarious margins. Fruit *c.* 20 mm; mericarps with lateral ridges winged and dorsal ridges obtuse, unwinged. *In scrub on hills.* ● *Makedonija (near Negotino).* Ju.

54. **Magydaris** Koch ex DC.[1]

Leaves simple or pinnate; segments wide. Sepals small. Petals white, obcordate, villous beneath; apex inflexed. Fruit ovoid, hairy, slightly compressed dorsally. Ridges wide and rounded; vittae numerous.

Leaves grey-tomentose beneath; rays 40–50   **1. pastinacea**
Leaves hispid only on the veins beneath; rays 10–20   **2. panacifolia**

1. M. pastinacea (Lam.) Paol. in Fiori & Paol., *Fl. Anal. Ital.* **2**: 205 (1900). Pubescent perennial 100–250 cm, smelling of coumarin. Stem glabrous or shortly hispid, solid, rigid, branched above. Basal leaves simple or shallowly 3- to 5-lobed, ovate-oblong, crenate; cauline pinnate, with 3–5 ovate, crenate-denticulate, obtuse segments; uppermost often reduced to an inflated petiole; all grey-tomentose beneath. Rays 40–50, pubescent. Bracts 50–60 mm, several, deflexed, lanceolate, acuminate or sometimes laciniate or dentate at apex; margins scarious; bracteoles 20–30 mm, linear-lanceolate. Fruit *c.* 5 mm, greyish-brown, densely villous; pericarp spongy. *S. Italy, Sicilia, Sardegna.* It Sa Si.

2. M. panacifolia (Vahl) Lange in Willk. & Lange, *Prodr. Fl. Hisp.* **3**: 62 (1874). Like **1** but leaves hispid only on the veins beneath; rays 10–30, bracts 20–30 mm, linear-lanceolate; bracteoles 10–20 mm. *Portugal, C. & S. Spain, Islas Baleares.* Bl Hs Lu.

55. Hohenackeria Fischer & C. A. Meyer[1]

Leaves simple; margins cartilaginous, minutely denticulate. Inflorescence of sessile capitula. Bracteoles absent. Sepals conspicuous. Petals whitish. Fruit ovoid, glabrous or hispidulous; ridges wide, rounded, corky; vittae inconspicuous. Carpophore absent.

Stems very short, concealed by leaves; fruit glabrous	**1. exscapa**
Stems up to 5 cm, not completely concealed; fruit hispidulous	**2. polyodon**

1. H. exscapa (Steven) Kos.-Pol., *Trudy Bot. Sada Jur'ev.* **15**(2–3): 120 (1914). Small, glabrous annual with very short stems. Leaves up to 10 cm, far exceeding the inflorescence, simple, broadly sheathing at base, with a linear-lanceolate lamina; margin thickened, denticulate. Flowers sessile, in a dense, sessile or subsessile capitulum. Bracteoles absent. Calyx-teeth 3–5, spinescent in fruit. Fruit *c.* 4 mm, glabrous, nearly smooth at top, with wide corky ridges in the lower ¾; mericarps not separating. *S. Spain (Sierra de Baza, Sierra de Gádor).* Hs. (*N. Africa, Caucasus.*)

2. H. polyodon Cosson & Durieu, *Bull. Soc. Bot. Fr.* **2**: 183 (1855). Like **1** but stems sometimes up to 5 cm; calyx-teeth 5, bifid nearly to base; fruit 3 mm, ridged to top, covered with short, stiff, patent hairs. *C. Spain.* Hs. (*N. Africa.*)

56. Bupleurum L.[1]

Leaves simple. Sepals usually absent. Petals yellow, not emarginate; apex inflexed. Fruit usually ovoid or oblong. Ridges usually conspicuous; vittae 1–5.

1 Leaves perfoliate; bracts absent
  2 Rays usually 5–10; bracteoles oblanceolate to ovate; fruit smooth   **1. rotundifolium**
  2 Rays usually 2–3; bracteoles suborbicular; fruit conspicuously tuberculate   **2. lancifolium**
1 Leaves not perfoliate; bracts present, though sometimes deciduous
  3 Perennials with a stout stock and non-flowering stems, or sometimes shrubs
    4 Marginal veins of leaves strongly thickened and at least as prominent as the others   **31. rigidum**
    4 Veins of leaves slender and all similar, or the midrib the thickest
      5 Stems woody, at least at base
        6 Leaves with a well-marked midrib and conspicuous, reticulate lateral veins
          7 Leaves crowded near the top of the woody branches, from the apex of which herbaceous flowering stems arise   **37. foliosum**
          7 Leaves ± evenly spaced along the stems
            8 Primary lateral veins reaching leaf-margin; bracts deciduous   **39. fruticosum**
            8 Primary lateral veins not reaching leaf-margin; bracts persistent   **38. gibraltarium**
        6 Leaves with several well-marked, ± parallel veins; lateral veins few and inconspicuous
          9 Bracts 3(–5)-veined

---

[1] By T. G. Tutin.

10 Leaves not densely crowded, the upper conspicuously longer than lower; bracts and bracteoles linear, not fleshy   **36. acutifolium**
10 Leaves densely crowded, the upper not longer than the lower; bracts and bracteoles oblong-lanceolate, ± fleshy
  11 Leaves 3–6 cm, widest above the middle; pedicels *c.* 1 mm   **34. dianthifolium**
  11 Leaves 7–18 cm, widest below the middle; pedicels 2–4 mm   **35. barceloi**
  9 Bracts 1-veined, or apparently veinless
    12 Flowering stems and rays becoming hard and spinose, persisting for 2–3 years   **32. spinosum**
    12 Flowering stems and rays not becoming hard and spinose, not persistent   **33. fruticescens**
5 Stems herbaceous
  13 Leaves with a prominent midrib and numerous, anastomosing lateral veins
    14 Cauline leaves 1(–2); bracteoles connate for at least ¼ of their length   **4. stellatum**
    14 Cauline leaves 3–5; bracteoles free or very shortly connate
      15 Stem hollow; lower leaves usually ovate; rays 5–12   **3. longifolium**
      15 Stem solid; lower leaves linear to lanceolate; rays 3–6   **5. angulosum**
  13 Leaves ± parallel-veined; lateral veins few and inconspicuous
    16 Basal leaves linear- to oblong-lanceolate, or wider
      17 Bracteoles linear to lanceolate
        18 Stock with few or no remains of leaves; bracts 2–5   **29. falcatum**
        18 Stock covered with persistent remains of leaves; bracts (3–)5–7   **30. elatum**
      17 Bracteoles ovate to suborbicular
        19 Basal leaves usually 3- to 5-veined; rays usually 5–7   **26. ranunculoides**
        19 Basal leaves 7- to 11-veined; rays usually 11–13   **28. multinerve**
    16 Basal leaves linear
      20 Stock with few or no persistent remains of leaves; leaves often falcate   **29. falcatum**
      20 Stock with numerous persistent remains of leaves; leaves not falcate
        21 Cauline leaves not wider than the basal; rays 4–5, filiform (S.E. Spain)   **27. bourgaei**
        21 Cauline leaves usually wider than the basal; rays usually 5–15, stout (not S.E. Spain)
          22 Stock covered with dark brown remains of leaves; fruit 2·5–3 mm   **26. ranunculoides**
          22 Stock covered with light brown leaf-bases; fruit 5–6 mm   **25. petraeum**
3 Annuals with slender root and no non-flowering stems
  23 Bracteoles broadly lanceolate to ovate, often overlapping and ± enclosing the flowers, incurved, aristate or mucronate
    24 Umbels all with 1 ray   **14. capillare**
    24 Most or all umbels with more than 1 ray
      25 Bracteoles with numerous, conspicuous, ascending and then abruptly recurved cross-veins   **15. fontanesii**
      25 Bracteoles without cross-veins or with inconspicuous, ascending, not recurved ones
        26 Bracts more than ½ as long as the longest ray
          27 Bracteoles greenish-brown, with numerous slender cross-veins; margin narrowly scarious   **11. baldense**
          27 Bracteoles yellowish-green, without cross-veins; margin broadly scarious
            28 Bracteoles completely translucent, except for the veins   **6. flavum**
            28 Bracteoles translucent outside the lateral veins, ± opaque inside them   **10. glumaceum**
        26 Bracts less than ½ as long as the longest ray
          29 Bracteoles without cross-veins   **7. gracile**

29 Bracteoles with slender cross-veins
  30 Bracteoles subobtuse, mucronate   **13. karglii**
  30 Bracteoles acute or acuminate, usually aristate
    31 Bracteoles *c.* twice as long as wide   **12. flavicans**
    31 Bracteoles *c.* 3 times as long as wide
      32 Bracteoles translucent outside the lateral veins,
        ± opaque inside them; awn usually less than ½ as
        long as the wide part of bracteole   **10. glumaceum**
      32 Bracteole completely translucent, except for the
        veins; awn at least ½ as long as the wide part of
        bracteole   **9. apiculatum**
23 Bracteoles narrowly lanceolate or narrowly elliptical to
  subulate, not overlapping, ± flat, rarely aristate
  33 At least the lower leaves with the midrib forming a pro-
    minent keel beneath
    34 Middle cauline leaves 5- to 9-veined; petals smooth on
      back   **16. praealtum**
    34 Middle cauline leaves 3-veined; petals papillose on back
             22. asperuloides
  33 Midrib of leaves not forming a prominent keel beneath
    35 Most umbels with 2–3 rays
      36 Fruit conspicuously papillose   **23. tenuissimum**
      36 Fruit not papillose
        37 Most lateral umbels subsessile   **22. asperuloides**
        37 All umbels distinctly pedunculate
          38 Bracteoles exceeding the flowers, linear
               20. trichopodum
          38 Bracteoles about equalling the flowers, lanceolate or
            elliptical   **18. brachiatum**
    35 Most umbels with at least 4 rays
      39 Partial umbels rarely with more than 3 flowers   **8. aira**
      39 Partial umbels with 4 or more flowers
        40 Veins of bracteoles very prominent; fruit almost un-
          ribbed, but with small, white papillae
               24. semicompositum
        40 Veins of bracteoles obscure, at least near apex; fruit
          distinctly ribbed, without white papillae
          41 Branches numerous, short, erecto-patent or appressed
               21. affine
          41 Branches few, long, rarely appressed
            42 Bracteoles lanceolate, distinctly 3-veined throughout
               17. commutatum
            42 Bracteoles subulate to linear-lanceolate, 1-veined, or
              3-veined in the lower half only   **19. gerardi**

Sect. BUPLEURUM (Sect. *Perfoliata* Godron). Annual. Lower leaves sessile or shortly petiolate; upper perfoliate; veins numerous, slender, radiating, anastomosing near the margin and connected by fine cross-veins elsewhere. Bracts absent; bracteoles 4–7, lanceolate to ovate, longer than the partial umbel.

1. B. rotundifolium L., *Sp. Pl.* 236 (1753). Erect, glaucous, often purple-tinged annual 15–75 cm. Leaves elliptic-ovate to suborbicular, obtuse, often mucronate. Rays (3–)5–10, often somewhat thickened at base and apex. Bracteoles 5–6, oblanceolate to ovate or obovate, acuminate, shortly connate at base, yellowish-green and patent in flower, becoming whitish and connivent in fruit; veins conspicuous. Fruit 3–3·25 mm, elliptic-oblong, blackish-brown, smooth; ridges filiform. *Arable land and other dry, open habitats. C. & S. Europe and U.S.S.R. southwards from c. 52° N, but absent from many of the islands.* Al Au Be Bu Co Cz Ga Ge Gr He Hs Hu It Ju Po Rm Rs (?B, C, W, K, E) Sa Tu [Br Ho].

2. B. lancifolium Hornem., *Hort. Hafn.* 267 (1813) (*B. protractum* Hoffmanns. & Link). Like **1** but leaves usually ovate- or oblong-lanceolate; rays 2–3(–5); bracteoles suborbicular, mucronate; fruit 3–5 mm, ovoid-globose, conspicuously tuberculate. *Arable land and other dry, open habitats. S. Europe.* Al Bl Co Cr Ga Gr Hs It Ju Lu Sa Si Tu [Au Be Br].

Sect. DIAPHYLLUM (Hoffm.) Dumort. Perennial. Lower leaves usually long-petiolate; upper cordate-amplexicaul; veins subparallel below, divergent above, anastomozing and becoming inconspicuous near the margin. Bracts and bracteoles leaf-like; bracteoles longer than the partial umbel.

3. B. longifolium L., *Sp. Pl.* 237 (1753). Stout, erect, yellow- or purple-tinged perennial 30–150 cm. Lower leaves elliptic-spathulate, subobtuse, mucronate; petiole longer than or equalling lamina, broadly winged, sheathing at base; cauline leaves ovate- to suborbicular-cordate. Rays 5–12. Bracts 2–4, ovate to suborbicular, obtuse or shortly acuminate; bracteoles 5–8, like the bracts but smaller, usually shortly connate, but sometimes up to half-way; veins conspicuous. Fruit 4–5·5 mm, elliptic-oblong, dark brown or black; ridges very prominent. *C. Europe, extending to C. France, N. Bulgaria and E. Russia.* Al Au Bu Cz Ga Ge He Hu Ju Po Rm Rs (N, C, W, K, E).

(a) Subsp. **longifolium**: Plant rarely yellowish-green; bracts and bracteoles green, sometimes purplish, very rarely yellowish, scarcely translucent. *From C. France to the Carpathians and N. Bulgaria, mainly in mountain regions.*

(b) Subsp. **aureum** (Hoffm.) Soó, *Acta Bot. Acad. Sci. Hung.* **12**: 116 (1966) (*B. aureum* (Hoffm.) Fischer ex Sprengel): Plant usually yellowish-green; bracts and bracteoles yellowish-green, more or less translucent. *C. & E. Russia from c. 51° to 59° N.*

Sect. RETICULATA Godron. Perennial. Lower leaves shortly petiolate; upper dilated at base and more or less amplexicaul; midrib conspicuous; lateral veins slender, divergent, anastomosing, ultimately joining up with a continuous, distinct marginal vein. Bracts leaf-like; bracteoles broadly ovate, connate or free.

4. B. stellatum L., *Sp. Pl.* 236 (1753). Stout, erect perennial 15–40 cm; stock densely covered with remains of dead leaves. Basal leaves numerous, linear to lanceolate; petiole much shorter than lamina, broadly winged; cauline leaves 1(–2), narrowed from an ovate, semi-amplexicaul base. Rays 3–6. Bracts 2–3(–4), like the upper leaves but shorter and relatively wider; bracteoles 8–12, obovate, acute or mucronate, often yellowish, connate for at least ¼ their length; veins conspicuous. Fruit *c.* 5 mm, ovoid-elliptical, dark brown; ridges winged. 2n=16. *Rocky places.* ● *Alps, Corse.* Au Co Ga He It.

5. B. angulosum L., *Sp. Pl.* 236 (1753). Like **4** but cauline leaves 3–5, cordate-amplexicaul at base; bracteoles 4–6, broadly ovate to suborbicular, free or very shortly connate at base; fruit 6–7 mm, oblong. *Rocky places; calcicole.* ● *Pyrenees and mountains of N.E. Spain.* Ga Hs.

Sect. ISOPHYLLUM (Hoffm.) Dumort. Annual or perennial. Lower leaves sessile, linear, rarely wider and petiolate; veins 3-many, more or less parallel, the lateral usually few, short, inconspicuous; marginal vein more or less distinct.

Subsect. *Aristata* (Godron) Briq. Annual. Leaves narrow; veins few, more or less parallel; bracteoles ovate or elliptical, awned or mucronate, 3- to 9-veined.

6. B. flavum Forskål, *Fl. Aegypt.* xxiii, 205 (1775). Slender, erect, divaricately branched annual 20–75 cm. Lower leaves distinctly petiolate, the others sessile, linear-lanceolate or linear, 3- to 5-veined. Rays 3–6(–20). Bracts more than ⅔ as long as longest rays, lanceolate, long-acuminate or aristate, yellowish-green, semitranslucent, broadly scarious-margined; veins 3, usually without cross-veins. Bracteoles semitranslucent, ovate- or oblong-lanceolate, acuminate; apex usually recurved, shortly aristate; margin minutely serrulate; veins prominent, cross-veins absent. Petals 0·7–0·9 × 0·5–0·6 mm, irregularly and remotely dentate. Fruit 1·6–2 mm; ridges filiform. *Dry rocky places, especially near the sea. E. Mediterranean region.* Bu Gr Tu.

7. B. gracile D'Urv., *Mém. Soc. Linn. Paris* **1**: 286 (1822). Like 6 but bracts never more than ½ as long as longest rays; bracteoles quite entire; petals 0·5–0·6 × 0·4–0·6 mm, not dentate; fruit 1·3–1·8 mm. 2n = 14. *S.E. Greece and Aegean region.* Cr Gr.

8. B. aira Snogerup, *Bot. Not.* **115**: 366 (1962). Very slender, erect, much-branched annual up to 50 cm. Lower leaves long-petiolate, linear-lanceolate; upper sessile, linear. Rays 3–6, long, filiform. Bracts ⅓–½ as long as longest ray, 3–4, lanceolate, acuminate, 3-veined. Partial umbels normally with 1–3 flowers. Fruit 1·2–1·8 mm; ridges inconspicuous. 2n = 14. ● *Kikladhes (Naxos).* Gr.

9. B. apiculatum Friv., *Flora (Regensb.)* **18**: 335 (1835). Slender, erect annual 30–60 cm. Lower leaves more or less distinctly petiolate, the others sessile, all narrowly linear-lanceolate or linear, 3- to 5-veined. Rays 6–8(–15). Bracts ⅓–½ as long as rays, linear-lanceolate, aristate, herbaceous, narrowly scarious-margined, minutely serrulate. Bracteoles broadly lanceolate or elliptic- or obovate-lanceolate, whitish, completely semitranslucent; awn at least ½ as long as wide part; margin narrowly scarious, minutely serrulate; veins connected by few (rarely numerous) short cross-veins. Fruit 2·75–3 mm; ridges obscure. *Dry, rocky places.* ● *E. part of the Balkan peninsula, extending to S.E. Romania.* Bu Gr Rm.

10. B. glumaceum Sibth. & Sm., *Fl. Graec. Prodr.* **1**: 177 (1806) (*B. semidiaphanum* Boiss.). Like 9 but rays 3–8(–12); bracts up to as long as the longest ray, with a wide, white, serrulate margin; bracteoles elliptical, completely translucent outside the lateral veins, thicker inside them; awn usually less than ½ as long as broad part. 2n = 16. *Dry places. Albania, Greece and S.E. Jugoslavia.* Al Gr Ju.

11. B. baldense Turra, *Gior. Ital. Sci. Nat. Agric. Arti Commerc.* **1**: 120 (1764). Usually much-branched annual up to 75 cm. Lower leaves more or less distinctly petiolate, the others sessile, all linear-lanceolate or narrowly oblong-spathulate, 3- to 5-veined. Rays (2–)3–8(–10). Bracts more than ½ as long as the longest rays, lanceolate, long-acuminate or aristate, yellowish- or glaucous-green; margin narrowly scarious; veins 3–5, prominent. Bracteoles lanceolate to ovate, slightly concave, aristate, yellowish; margin narrowly scarious, minutely serrulate; veins 3–5, the inner 3 prominent, sometimes with a weaker submarginal vein on each side, connected by numerous slender cross-veins. Fruit c. 2 mm; ridges filiform. *Dry open habitats; calcicole.* ● *S. & W. Europe from Romania to Spain and England.* Al Bl Br Co Ga ?Gr Hs It Ju Rm Sa Si.

(a) Subsp. **baldense** (*B. aristatum* sensu Coste, non Bartl.): Glaucous, usually 5–15 cm and with primary branches only; rays (2–)3–4(–6); bracts glaucous, broadly lanceolate; bracteoles acuminate or shortly aristate. *W. Europe, extending to Sicilia.*

(b) Subsp. **gussonei** (Arcangeli) Tutin, *Feddes Repert.* **74**: 31 (1967) (*B. aristatum* subsp. *gussonei* Arcangeli, *B. veronense* Turra): Bright green, usually 30–75 cm and with primary and secondary branches; rays 5–8(–10); bracts yellowish, linear-lanceolate to lanceolate; bracteoles usually long-aristate. *Balkan peninsula, Italy.*

12. B. flavicans Boiss. & Heldr. in Boiss., *Diagn. Pl. Or. Nov.* 3(6): 74 (1859). Like 11 but bracts not more than ½ as long as the longest rays, yellowish; veins prominent, connected by conspicuous cross-veins; bracteoles broadly ovate, strongly concave, yellowish; awn c. ½ as long as wide part; veins 5, the marginal much weaker than the others, connected by numerous, conspicuous cross-veins. ● *C. & S.W. part of Balkan peninsula.* Al Gr Ju.

Perhaps not specifically distinct from **11**.

13. B. karglii Vis., *Fl. Dalm.* **3**: 35 (1852). Slender annual or rarely biennial up to 50 cm. Lower leaves lanceolate, with a petiole about as long as lamina; upper linear-lanceolate to linear, acute, 5- to 7-veined. Rays 2–5. Bracts ¼ to ½ as long as rays, oblong-lanceolate to elliptical, shortly cuspidate, greenish-yellow, translucent; margin narrowly scarious, minutely serrulate; veins 3–5, the marginal, if present, c. ½ as long as the bract; bracteoles ovate, subobtuse, mucronate, usually larger than the bracts, more or less enclosing the partial umbel, except at anthesis; veins always 5, connected by cross-veins, the marginal disappearing towards the apex. Fruit 1·75–2 mm; ridges slender. *Mountain rocks.* ● *W. part of Balkan peninsula.* Al Ju.

14. B. capillare Boiss. & Heldr. in Boiss., *Diagn. Pl. Or. Nov.* 3(2): 82 (1856). Like 13 but up to 75 cm; bracts longer than the ray, narrowly lanceolate, straw-coloured; margin broadly scarious, entire; veins wide, yellow or brownish-yellow; ray 1. *Mountains.* ● *S. Greece (Delphi-Levadhia region).* Gr.

15. B. fontanesii Guss. ex Caruel in Parl., *Fl. Ital.* **8**: 417 (1889). Divaricately branched annual up to 50 cm. Leaves linear, often somewhat falcate, acuminate; lower more or less petiolate; veins 3–5. Terminal umbels with short, stout peduncles, usually overtopped by lateral ones. Rays 5–7, suberect in fruit. Bracts ½ as long as to about as long as the longer rays, lanceolate, acuminate, whitish and semitranslucent in fruit; veins 3–9, stout, connected by numerous cross-veins which are ascending and then usually abruptly recurved; bracteoles like the bracts but smaller, exceeding the flowers, very shortly connate at base. Fruit 1·5–1·7 mm, oblong-ellipsoid; ridges very slender. *Dry open habitats. Mediterranean region from Sardegna eastwards.* Al Bu Gr It Ju Sa Si Tu [Au Ga].

Subsect. *Juncea* Briq. Annual. Leaves narrow, 3- to 11(–19)-veined; veins more or less parallel; bracteoles herbaceous, flat, acute, 3-veined. Fruit not papillose.

16. B. praealtum L., *Fl. Monsp.* 12 (1756) (*B. junceum* L.). Erect annual up to 150 cm. Leaves linear, acuminate, often subfalcate, abruptly narrowed to a short, sheathing base; veins up to 19, the midrib much thicker than the others and forming a keel beneath. Rays 2–3(–5). Bracts usually much shorter than rays, linear-lanceolate, very rarely wider, long-acuminate; bracteoles like the bracts but smaller, a little longer than the flowers, but exceeded by the fruits. Fruit 4–6 mm; ridges prominent. ● *S. & S.C. Europe.* Al Au Bu Cz Ga Gr Hs Hu It Ju Rm Sa Si Tu [Be].

17. B. commutatum Boiss. & Balansa in Boiss., *Diagn. Pl. Or. Nov.* 3(6): 75 (1859). Erect annual up to 100 cm. Leaves linear to linear-lanceolate, subamplexicaul, the lower 9- to 11-veined at base, the upper 3- to 5-veined. Rays 4–7. Bracts ¼ to ½ as long as the longest rays, 3–6, linear-lanceolate, long-acuminate, conspicuously 3- to 5-veined; bracteoles linear-lanceolate or lanceolate, 3-veined. Fruit 1·25–2·5 mm, ovoid-truncate or subglobose; ridges filiform. *Balkan peninsula, extending northwards to Hungary and N.E. Romania; Krym.* Al Bu ?Gr Hu Ju Rm Rs (K).

1   Rays nearly equal; lateral veins of bracteoles inconspicuous; fruit up to 1·5 mm   **(c) subsp. aequiradiatum**
1   Rays very unequal; lateral veins of bracteoles conspicuous; fruit 2–2·5 mm
2   Green, up to 100 cm; rays of terminal umbels 5–7
                                   (a) subsp. commutatum
2   Glaucous, up to 50 cm; rays of terminal umbels 4–5
                                  (b) subsp. glaucocarpum

(a) Subsp. **commutatum**: Branches patent; rays 5–7, very unequal; bracteoles exceeding the flowers, about equalling the fruit, not hardened in fruit; fruit *c.* 2 mm. *Balkan peninsula; Krym.*

(b) Subsp. **glaucocarpum** (Borbás) Hayek, *Prodr. Fl. Penins. Balcan.* 1: 975 (1927): Branches patent; rays 4–5, very unequal; bracteoles about twice as long as flowers, little longer than fruit, hardened in fruit; fruit 2–2·5 mm. *From Bulgaria to Romania and Hungary.*

(c) Subsp. **aequiradiatum** (H. Wolff) Hayek, *loc. cit.* (1927): Branches strict; rays 5–8, nearly equal; bracteoles scarcely exceeding flowers, shorter than fruit; fruit 1·25–1·5 mm. *Bulgaria; perhaps elsewhere in the Balkan peninsula.*

18. B. brachiatum C. Koch ex Boiss., *Fl. Or.* 2: 844 (1872). Erect annual 80–100 cm. Leaves linear, sessile, the lower with 5–7 veins. Rays 2–3(–5). Bracts ⅓ to ½ as long as the longest rays, 2–5, linear-lanceolate, long-acuminate, conspicuously 3-veined; bracteoles about equalling the flowers, 5, lanceolate or elliptical, cuspidate; margin rather broadly scarious, minutely serrulate; lateral veins inconspicuous. Fruit 1·5–2·5 mm, ovoid-oblong, truncate; ridges filiform. *S. & S.E. Krym.* Rs (K). (*Caucasus, E. Anatolia.*)

19. B. gerardi All., *Mélang. Philos. Math. Soc. Roy. Turin* (*Misc. Taur.*) 5: 81 (1774) (*B. affine* sensu Coste, non Sadler, *B. australe* Jordan). Slender, erect annual up to 75 cm. Leaves linear or linear-lanceolate, often subfalcate, acuminate, semi-amplexicaul at base; veins 5–7. Rays 2–7. Bracts ⅓ to ½ as long as the longer rays, 3–5, linear-lanceolate, acuminate, 3-veined; bracteoles usually distinctly exceeding flowers and fruit, subulate or linear-lanceolate, long-acuminate; veins 1, sometimes with 2 lateral ones in the lower half. Fruit 2–3 mm, ovoid-oblong; ridges filiform. *S. & W. Europe, northwards to N.W. France.* Bu Co Ga Gr Hs It Lu Rs (K) Si [Ge].

20. B. trichopodum Boiss. & Spruner, *Ann. Sci. Nat.* ser. 3 (Bot.), 1: 145 (1884). Slender, erect annual up to 50 cm. Lower leaves long-petiolate, narrowly oblanceolate to spathulate, obtuse or shortly acuminate; upper sessile, linear or oblong from a subcordate, amplexicaul base, acuminate. Rays 2–4(–6), long, slender. Bracts ⅙ to ½ as long as rays, 1–3(–5), linear to ovate-lanceolate, acuminate; veins (1–)3–5; bracteoles exceeding the flowers, about equalling the fruit, linear, acuminate; veins inconspicuous. Fruit 2–3 mm, oblong; ridges slender, prominent. *Dry, open habitats. S. & E. Greece and Aegean region.* Cr Gr.

21. B. affine Sadler, *Fl. Com. Pest.* 1: 204 (1825). Somewhat glaucous, erect annual up to 75 cm, with numerous short, erecto-patent or appressed branches. Leaves linear, somewhat dilated and semi-amplexicaul at base, long-acuminate; veins 3–8, strict, rather stout. Bracts ⅓ to ½ as long as longest rays, 2–5, linear-lanceolate from a rather wide base, long-acuminate, 3-veined; bracteoles somewhat exceeding flowers and fruit, like the bracts, but smaller; lateral veins obscure. Fruit 2–2·5 mm, ellipsoid-

oblong; ridges slender, prominent. *Dry, open habitats. S.C. & E.C. Europe, extending to Bulgaria and S.E. Ukraine.* Au Bu Cz Hu Ju Rm Rs (W, K) ?Si.

22. B. asperuloides Heldr. ex Boiss., *Diagn. Pl. Or. Nov.* 3(6): 76 (1859) (incl. *B. pauciradiatum* Fenzl). Somewhat glaucous, flexuous annual up to 80 cm, much-branched with very stout, appressed secondary branches. Leaves linear, acuminate to sub-obtuse, the lower petiolate, the upper semi-amplexicaul, 3-veined; margin rather narrowly scarious and serrulate. Rays (1–)2–3; most lateral umbels subsessile. Bracts ⅓–½ as long as the longest ray, subulate, acute, 3-veined; bracteoles exceeding the flowers, shorter than the fruit, like the bracts, but smaller. Fruit *c.* 2·25 mm, oblong-ellipsoid; ridges inconspicuous. *Dry, open habitats. S.E. Europe.* Bu Gr Ju Rm Rs (K).

Subsect. *Trachycarpa* (Lange) Briq. Annual. Leaves narrow, 3- to 7-veined; veins more or less parallel; bracteoles herbaceous, flat, acute, 3-veined. Fruit papillose.

23. B. tenuissimum L., *Sp. Pl.* 238 (1753). Usually much-branched, somewhat glaucous annual up to 75 cm. Leaves linear or linear-lanceolate, the lowest shortly petiolate, the others sessile, subobtuse or more or less acuminate; veins 5–7, conspicuous beneath. Rays 1–3. Bracts much shorter than the longer rays, subulate, acuminate, 3-veined; bracteoles acute or sub-obtuse, usually serrulate on the margin and veins. Fruit 1·5–2·25 mm, subglobose, papillose; ridges slender, prominent, crenulate. $2n = 16$. *Usually in more or less saline habitats. S., W. & C. Europe, extending to S.E. Sweden* (*Gotland*) *and S.E. Russia.* Al Au Be Br Bu Co Cz Da Ga Ge Gr Ho Hs Hu It Ju Lu Po Rm Rs (W, K, E) Sa Si Su Tu.

(a) Subsp. **tenuissimum**: Secondary branches short; rays unequal; bracteoles about twice as long as the flowers, subfalcate to linear-lanceolate; vittae obscure. *Throughout the range of the species, except the south part of the Balkan peninsula and S.E. Russia.*

(b) Subsp. **gracile** (Bieb.) H. Wolff in Engler, *Pflanzenreich* 43(IV. 228): 104 (1910) (*B. gracile* (Bieb.) DC., non D'Urv., *B. marschallianum* C. A. Meyer): Secondary branches long; rays nearly equal; bracteoles about equalling the flowers, lanceolate or obovate-lanceolate; vittae conspicuous. *S. part of the Balkan peninsula and S.E. part of U.S.S.R.*

24. B. semicompositum L., *Demonstr. Pl.* 7 (1753) (*B. glaucum* Robill. & Cast. ex DC.). Much-branched, spreading, glaucous annual up to 30 cm. Lower leaves spathulate to linear, petiolate, obtuse to acute; upper linear, sessile, semi-amplexicaul, acuminate; veins 3–5. Rays 3–6, filiform. Bracts ¼ to ½ as long as the longest rays, linear, 3-veined; bracteoles exceeding the flowers, linear-lanceolate or rarely narrowly elliptical, aristate; veins 3, very prominent. Fruit 1·5–2 mm, subglobose or ovoid-oblong, covered with small whitish papillae; ridges slender, inconspicuous. *Dry places, especially on sandy soils. S. Europe.* Bl Co Cr Ga Gr Hs It Ju Lu ?Rs (E) Sa Si.

Subsect. *Nervosa* (Godron) Briq. Usually perennial. Lower leaves often elliptical, usually petiolate, upper sessile and amplexicaul.

25. B. petraeum L., *Sp. Pl.* 236 (1753). Perennial up to 50 cm; stock stout, covered with light brown leaf-bases. Basal leaves numerous, linear; cauline absent or few, linear to ovate-lanceolate, semi-amplexicaul, acuminate. Rays usually 5–15, rather stout. Bracts 3–6, linear or lanceolate; veins up to 9(–13); bracteoles

5–10, free or somewhat connate, very variable in shape; veins 5, conspicuous, with some lateral veins. Fruit 5–6 mm, dark brown, shiny; ridges winged. 2*n* = 14. *Calcicole.* ● *S. & E. Alps.* Au Ga It Ju.

26. B. ranunculoides L., *Sp. Pl.* 237 (1753). Perennial up to 60 cm; stock rather slender, with more or less numerous dark brown remains of dead leaves. Basal leaves linear, linear-lanceolate or spathulate, more or less narrowed into a petiole; veins usually 3–5; cauline leaves wider, the upper usually ovate, acuminate, semi-amplexicaul; veins usually numerous. Rays (3–)5–7(–15), rather stout. Bracts 1–5, like the uppermost leaf; bracteoles 5–(7–9), yellowish, very variable in shape and size; veins usually 3–7, conspicuous, with some lateral veins. Fruit 2·5–3 mm; ridges filiform, prominent or narrowly winged. *Mountains of C. & S. Europe, from the Carpathians to N. Spain, C. Italy and S.W. Jugoslavia.* Au Cz Ga Ge He Hs It Ju Po Rm.

(a) Subsp. **ranunculoides**: Leaves all flat, the lower narrowly lanceolate or oblong-spathulate, distinctly petiolate, the upper ovate, acuminate. 2*n* = 42. *Throughout the range of the species.*

(b) Subsp. **gramineum** (Vill.) Hayek, *Prodr. Fl. Penins. Balcan.* 1: 971 (1927): Lower leaves more or less involute, linear, somewhat narrowed towards the base but scarcely petiolate, the upper linear-lanceolate. 2*n* = 14. *Pyrenees, S. Alps, Appennini, Jugoslavia.*

27. B. bourgaei Boiss. & Reuter in Boiss., *Diagn. Pl. Or. Nov.* 3(2): 84 (1856). Perennial up to 60 cm; stock woody, long, procumbent, covered with remains of dead leaves. Basal leaves crowded, narrowly linear, sessile, amplexicaul, dead at fruiting time; cauline similar in shape, few, remote; veins 5–7. Rays 4–5, filiform. Bracts 1–3, linear to linear-lanceolate, 5-veined; bracteoles equalling the flowers, lanceolate, shortly cuspidate or mucronate, 5-veined. ● *S.E. Spain (Sierra de Alcaraz).* Hs.

Ripe fruit is apparently unknown. Doubtfully distinct from 26(b).

28. B. multinerve DC., *Mém. Soc. Phys. Hist. Nat. Genève* 4: 500 (1828). Perennial *c.* 50 cm; stock rather stout, more or less densely covered with remains of dead leaves. Basal leaves linear-lanceolate, more or less distinctly petiolate; veins 7–11; upper cauline leaves linear-lanceolate to ovate from a cordate-amplexicaul base, often long-acuminate; veins 11–45. Rays (5–)11–13(–20). Bracts 2–7, ovate, acute or acuminate; veins up to 21; bracteoles usually exceeding the flowers, ovate to suborbicular, apiculate, yellowish; veins 5–9(–13). Fruit 3·5–4 mm, dark brown; ridges narrowly winged. *C. & S. Ural, and some outlying stations in C. Russia.* Rs (C, ?W). (*N. Asia.*)

29. B. falcatum L., *Sp. Pl.* 237 (1753) (incl. *B. exaltatum* Bieb., *B. olympicum* Boiss., *B. parnassicum* Halácsy, *B. rossicum* Woronow, *B. sibthorpianum* Sm., *B. woronovii* Manden.). Very variable perennial or rarely annual up to 100 cm; stock almost or quite devoid of remains of dead leaves. Basal leaves obovate, elliptical, oblong or linear, distinctly petiolate or sessile; veins 5–7; cauline leaves sessile, semi-amplexicaul, lanceolate to linear, often falcate. Rays 3–15, filiform. Bracts 2–5, lanceolate to subulate, very unequal; veins 3–5; bracteoles 5, linear-lanceolate, acuminate; veins 3(–5). Fruit 3–6 mm; ridges filiform or more or less winged. *S., C. & E. Europe, extending north-westwards to S. England.* Al Au Be Br Bu Co Cz Ga Ge Gr He Hs Hu It Ju Po Rm Rs (C, W, K, E).

A very variable species which requires investigation before infraspecific taxa can be satisfactorily delimited. The two following are sufficiently distinct morphologically to merit recognition

as subspecies at this stage, though more or less intermediate plants (e.g. *B. parnassicum* Halácsy) occur in some parts of their range.

(a) Subsp. **falcatum**: Leaves elliptical to oblong, distinctly petiolate; fruit *c.* 3 mm; ridges usually unwinged. 2*n* = 16. *Throughout the range of the species.*

(b) Subsp. **cernuum** (Ten.) Arcangeli, *Comp. Fl. Ital.* ed. 2, 590 (1894): Leaves linear, sessile; fruit *c.* 5 mm; ridges more or less winged. *Pyrenees, S. Alps, Appennini, S. Carpathians, mountains of Balkan peninsula, S. Russia.*

Subsp. **dilatatum** Schur, *Enum. Pl. Transs.* 253 (1866), is a tall plant, with usually obovate basal leaves and distinctly but shortly petiolate cauline leaves. It is known from Romania, Hungary and Czechoslovakia and has 2*n* = 32. It is not, however, clear whether there is a constant correlation between morphological characters and chromosome number.

30. B. elatum Guss., *Fl. Sic. Prodr.* 1: 316 (1827). Perennial up to 150 cm; stock stout, woody, covered with remains of dead leaves. Basal leaves oblong-lanceolate to broadly lanceolate, somewhat narrowed to the amplexicaul base; veins 7–9; cauline leaves similar to the basal. Rays 6–14, slender, strict. Bracts (3–)5–7, lanceolate, acuminate; veins 5–7; bracteoles 5–6, much exceeding the flowers, linear-lanceolate, long-acuminate; veins 5(–7). Fruit *c.* 5 mm, oblong; ridges slender. *Shady calcareous rocks.* ● *N. Sicilia (Madonie).* Si.

Subsect. *Marginata* (Godron) H. Wolff. Perennial. Leaves linear to broadly obovate, all petiolate; veins 3–11, parallel, usually with a conspicuous reticulum of small veins between them, the marginal strongly thickened.

31. B. rigidum L., *Sp. Pl.* 238 (1753). Perennial up to 150 cm; stems woody at base, with numerous patent or erecto-patent branches. Leaves coriaceous, very variable in shape; petiole amplexicaul at base. Rays 2–5, slender. Bracts 2–4 mm, 2–5, subulate, appressed to rays; bracteoles like the bracts, shorter than the pedicels in fruit; veins obscure. Fruit *c.* 4 mm, ellipsoid; ridges filiform, prominent. *Dry rocky places. Iberian peninsula, S. France, N. Italy.* Ga Hs It Lu.

(a) Subsp. **rigidum**: Basal leaves oblong to broadly obovate; veins 5–11, with conspicuous small veins between them. *Throughout the range of the species.*

(b) Subsp. **paniculatum** (Brot.) H. Wolff in Engler, *Pflanzenreich* 43 (IV. 228): 154 (1910): Basal leaves linear or linear-spathulate; veins 3(–5), with few, inconspicuous small veins between them. *Spain and Portugal.*

Subsect. *Rigida* (Drude) H. Wolff. Stems woody, at least at base. Leaves evergreen; veins more or less parallel, with inconspicuous small veins between them.

32. B. spinosum Gouan, *Obs. Bot.* 8 (1773). Intricately branched, glaucous perennial up to 30 cm; lower part of stems stout, woody, devoid of remains of dead leaves; upper parts of stems dying after fruiting, becoming hard and spinose and persisting for 2–3 years. Basal leaves linear-subulate, scarcely narrowed at base; veins 3–5, conspicuous beneath; cauline similar but smaller, few, remote. Rays 2–7, becoming rigid and spinose after the fruit has been shed. Bracts 2 mm, usually 5, subulate, 1-veined; bracteoles similar to the bracts, but sometimes lanceolate. Fruit 3–4·5 mm, ovoid-oblong; ridges filiform, prominent. *S. & E. Spain.* Hs.

33. B. fruticescens L., *Cent. Pl.* **1**: 9 (1755). Much-branched, glaucescent, small shrub up to 100 cm. Stem flexuous, the upper part becoming hard and spinose after fruiting. Leaves 2–7 cm, subacute, widest above the middle. Umbels forming a panicle. Rays 2–5(–10), not spinescent; bracts 1–3 mm, recurved at apex, apparently veinless; bracteoles narrowly triangular, recurved at apex; pedicels 1–2 mm. Fruit 3–4 mm, ovoid. *E. & C. Spain.* Hs [Ga].

34. B. dianthifolium Guss., *Fl. Sic. Prodr., Suppl.* 71 (1832). Small shrub with leaves crowded at the apex of the branches and with herbaceous, almost leafless flowering stems up to 40 cm. Leaves 3–6 cm, linear-lanceolate, often somewhat falcate, widest above the middle, cucullate at apex; veins 3–5, without visible small veins between them. Umbels forming a raceme. Rays 3–8, rather stout. Bracts 2–4 mm, 5, oblong-lanceolate, rather fleshy, 3-veined; bracteoles similar but smaller; pedicels *c.* 1 mm. Fruit 4–5 mm, oblong; ridges slender, prominent. *Calcareous rocks.* ● *N. side of island of Marettimo, near Sicilia.* Si.

35. B. barceloi Cosson ex Willk., *Linnaea* **40**: 83 (1876). Like **34** but leaves 7–18 cm, widest below the middle, acute; rays 5–8(–12); bracts 3–6 mm; pedicels 2–4 mm. ● *Islas Baleares.* Bl.

36. B. acutifolium Boiss., *Elenchus* 47 (1838). Perennial up to 100 cm, woody at the base. Leaves linear or linear-lanceolate; lower up to 8 cm, with 7–13 veins, without visible cross-veins; upper up to 18 cm. Rays 3–10, slender. Bracts 4–5 mm, 5, linear, acute, 3- to 5-veined; bracteoles subulate, 3-veined. Fruit *c.* 3 mm. ● *S. Portugal (near Odemira), S. Spain (Sierra de Estepona).* Hs Lu.

Sect. CORIACEA Godron. Shrubs. Leaves evergreen, pinnately veined, with a well-marked midrib.

37. B. foliosum Salzm. ex DC., *Prodr.* **4**: 133 (1830). Small shrub up to 100 cm. Lower leaves linear-lanceolate, shortly petiolate, upper subcordate-ovate, acuminate, sessile. Rays 2–3. Bracts much shorter than rays, 2–3, ovate; bracteoles 5–6, shortly connate at base, similar to the bracts. Fruit *c.* 3 mm. *S.W. Spain (near Algeciras and Gibraltar).* Hs. (*Morocco.*)

38. B. gibraltarium Lam., *Encycl. Méth. Bot.* **1**: 520 (1785) (*B. verticale* Ortega). Small shrub up to 200 cm. Leaves lanceolate, very shortly petiolate, held more or less vertically; primary lateral veins not reaching the margin. Rays 10–30, rather stout. Bracts 5–7, deflexed, persistent, lanceolate to ovate, with up to 12 subparallel veins; bracteoles much shorter than the pedicels, 5, ovate or suborbicular, 5- to 7-veined. Fruit 7–8 mm; ridges wide, narrowly winged. *C. & S. Spain.* Hs.

39. B. fruticosum L., *Sp. Pl.* 238 (1753). Shrub up to 250 cm. Leaves elliptic-oblong to obovate, subsessile, usually erecto-patent; primary lateral veins reaching the margin. Rays 5–25, stout. Bracts 5–6, deflexed, deciduous, elliptical, ovate or obovate, with 5–7 veins and conspicuous cross-veins; bracteoles shorter than the pedicels, 5–6, deciduous, broadly obovate, 4- to 5(–7)-veined. Fruit 7–8 mm; ridges slender, narrowly winged. *S. Europe.* Co Ga Gr Hs It Lu Sa Si [Br Rs (K)].

57. Trinia Hoffm.[1]

Dioecious (rarely monoecious) biennials or perennials, sometimes monocarpic. Leaves usually 2-pinnate, with linear lobes, but sometimes further divided. Bracts and bracteoles present or

---

[1] By J. F. M. Cannon.

absent. Sepals absent or minute. Petals white or yellowish; apex incurved. Fruit ovoid, somewhat compressed laterally; ridges often prominent and rounded; vittae present.

In some species the secondary ridges are conspicuously developed, in others the primary ridges form greatly contorted lobes covering the whole surface of the fruit.

```
1 Secondary ridges not developed on the fruit
  2 Bracteoles 3–6
    3 Fruit scabrid with small spines                    6. muricata
    3 Fruit glabrous
      4 Plant not more than 10 cm                        1. glauca
      4 Plant more than 15 cm
        5 Only some flowers in each umbel of female plant setting
            fruit; fruit c. 3 mm, with well-developed, acute ridges
                                                          5. kitaibelii
        5 Almost all flowers of female plant setting fruit; fruit c.
            2 mm, with rounded ridges with shallow grooves
            between them                                 4. ramosissima
  2 Bracteoles absent
    6 Fruit scabrid-pubescent                            3. hispida
    6 Fruit glabrous
      7 Lobes of leaves 20–70×0·5 mm, very narrowly linear;
          plants often rather erect and fastigiate       2. multicaulis
      7 Lobes of leaves 5–30×c. 1 mm, narrowly linear; plants
          often rather diffuse and spreading             1. glauca
1 Secondary ridges developed, or fruit covered with sinuous lobes
  8 Fruit covered with sinuous lobes, resembling the convolu-
      tions of the brain                                  9. crithmifolia
  8 Fruit with obvious ridges
    9 Primary ridges prominent and smooth; secondary ridges
        rounded                                           7. dalechampii
    9 Primary ridges undulate-verrucose; secondary ridges rugose
                                                          8. guicciardii
```

1. T. glauca (L.) Dumort., *Fl. Belg.* 78 (1827) (*T. stankovii* Schischkin, *T. vulgaris* DC.). Glabrous, glaucous perennial up to 50 cm. Stock with abundant fibres. Stem angled, flexuous, repeatedly branched, the lower branches often nearly as long as the main stem. Lower leaves 2- to 3-pinnate, with 3–5 segments; lobes 5–30 mm, usually longer in the female than in the male plants. Bracts and bracteoles absent or few. Fruit 2–3 mm. *W., C. & S. Europe, northwards to S. England.* Al Au Br Bu ?Co Cz Ga Ge Gr He Hs Hu It Ju Rm Rs (K) Tu.

(a) Subsp. **glauca**: Leaf-lobes often rather long, obscurely veined; bracts and bracteoles absent; pedicels up to five times as long as the ripe fruit. *Throughout the range of the species.*

(b) Subsp. **carniolica** (A. Kerner ex Janchen) H. Wolff in Engler, *Pflanzenreich* **43** (IV. 228): 182 (1910): Leaf-lobes short, with prominent midrib; bracts and bracteoles usually present; pedicels as long as or slightly longer than the ripe fruit. *Appennini, mountains of W. Jugoslavia, Albania and C. Romania.*

It seems likely that **T. dufourii** DC., *Prodr.* **4**: 104 (1830), recorded from E. Spain, is only a minor variant of **1**. The only differential character appears to be that in *T. dufourii* the upper cauline leaves are without a lamina.

2. T. multicaulis (Poiret) Schischkin, *Fl. URSS* **16**: 352 (1950) (*T. henningii* Hoffm.). Erect, glabrous herb, profusely branched especially in the upper part. Stock with abundant fibres. Leaves 1- to 2-pinnate; lobes 20–70 mm, very narrowly linear. Bracteoles absent. Umbels of male plants arranged in verticillate racemes; rays 5–10. Umbels of female plants with 4–8 unequal rays. Fruit 3–4 mm, oblong-ovoid, glabrous; ridges thick and prominent. ● *E. Europe, from E. Romania to E.C. Russia.* Rm Rs (C, W, E).

3. T. hispida Hoffm., *Gen. Umb.* 94 (1814) (*T. hoffmannii* Bieb.). Erect herb up to 35 cm, often much-branched. Stem glabrous to hispid with short stiff hairs. Leaves 2-pinnate, with linear lobes. Bracteoles absent. Umbels of male plants with up to 10 rays; umbels of female plants with up to 9 markedly unequal rays. Fruit ovoid, deeply sulcate, scabrid to sparsely hispid. *S.E. Russia; S. & E. Ukraine.* Rs (W, K, E).

4. **T. ramosissima** (Fischer ex Trev.) Koch, *Nova Acta Acad. Leop.-Carol.* **12**(1): 127 (1824) (*T. ucrainica* Schischkin, *T. kitaibelii* sensu Hayek, non Bieb.). Erect, glabrous herb up to 80 cm. Leaves 2-pinnate, with narrow, linear lobes up to 30 mm. Bracteoles usually 5, conspicuous. Male plants relatively sparsely branched, the branches making a wide angle with the stem. Female plants densely branched. Umbels of male plants with 5–10 rays; umbels of female plants with 4–8(–10) rays. Fruit 2–2·5 mm, oblong-ovoid, much shorter than its pedicel, glabrous; ridges rounded, with shallow grooves between them. 2*n* = 20. ● *S.E. & E.C. Europe, westwards to C. Czechoslovakia.* Au Bu Cz Hu Ju Rm Rs (C, W, E).

It appears possible that this species may include two subspecies. One, occurring in E.C. Europe (*T. kitaibelii* auct., non Bieb.), has smaller fruit (*c.* 2 mm) and relatively robust rays, while the other, from Russia, has rather larger fruit (*c.* 2·5 mm) and finer rays.

5. **T. kitaibelii** Bieb., *Fl. Taur.-Cauc.* **3**: 246 (1819). Glabrous herb up to 40 cm, often branched from near the base, and without a distinct main stem. Leaves 2(–3)-pinnate with linear lobes up to 10 mm; margins with short bristly hairs. Umbels of male plants with 5–7 rays; umbels of female plants with 4–9 rays. Bracteoles 3–5. Pedicels becoming thickened in fruit. Fruit 3–4 mm, ovoid, glabrous, with prominent acute ridges. ● *Krym.* Rs (K).

6. **T. muricata** Godet, *Fl. Jura* **1**: 271 (1852). Like **5** but fruit 4–5 mm, scabrid with small spines. *S.E. Russia.* Rs (C, E).

7. **T. dalechampii** (Ten.) Janchen, *Österr. Bot. Zeitschr.* **58**: 298 (1908). Branched from the base. Branches up to 15 cm, bearing patent umbels. Leaves 2-pinnate; lobes *c.* 10 mm, linear. Umbels of male plants with 5–8 rays; umbels of female plants with 5–7 very unequal rays. Fruit 3 mm, ovoid; primary ridges prominent, smooth; secondary ridges rounded, well developed. *Mountain pastures.* ● *Italy and Balkan peninsula.* Al Gr It Ju.

T. frigida (Boiss. & Heldr.) Drude in Engler & Prantl, *Natürl. Pflanzenfam.* **3**(8): 183 (1898), from Greece, is probably conspecific with **7**. It has a smaller number of rays in the umbels (male 4–6, female 4–5).

8. **T. guicciardii** (Boiss. & Heldr.) Drude in Engler & Prantl, *Natürl. Pflanzenfam.* **3**(8): 183 (1898). Much-branched herb up to 15 cm, distinctly obpyramidal in overall shape. Stems tinged with purple. Leaves 2- to 3-pinnate; lobes up to 10 mm, linear. Umbels of male plants with up to 10 rays; umbels of female plants with 4–6 rays. Fruit 2–3 mm, ovoid, reddish-brown; primary ridges sinuous, verrucose; secondary ridges moderately developed, rugose. ● *C. & S. Greece.* Gr.

9. **T. crithmifolia** (Willd.) H. Wolff in Engler, *Pflanzenreich* **43** (**IV.** 228): 190 (1910) (*Rumia crithmifolia* (Willd.) Kos.-Pol., *R. taurica* Hoffm.). Erect herb up to 50 cm. Stem branched. Leaves 2-pinnate; lobes up to 30 mm, narrowly linear. Umbels of male plants with 7–10 rays; umbels of female plants with 2–4 subequal rays. Mature fruit *c.* 5 mm, whitish or reddish. Primary ridges with greatly developed sinuous lobes, which give the fruit an appearance resembling that of a human brain. ● *Krym.* Rs (K).

58. Cuminum L.¹

Leaves 2-ternate. Sepals subulate, conspicuous. Petals white or pink, emarginate; apex long, inflexed. Fruit ovoid-oblong, dorsally compressed. Ridges filiform, the secondary more conspicuous; vittae solitary.

1. **C. cyminum** L., *Sp. Pl.* 254 (1753). Slender annual 10–50 cm. Leaves with filiform lobes 2–5 cm long. Rays 1–5, rather stout. Bracts 2–4, filiform or 3-fid, usually longer than the rays; bracteoles usually 3, very unequal. Flowers 3–5 in each partial umbel; pedicels stout. Fruit 4–5 mm, ovoid-oblong, setulose or glabrous. *Cultivated in the Mediterranean region for its aromatic fruits and more or less naturalized locally.* [Ga Hs Si.] (*N. Africa, S.W. Asia.*)

59. Apium L.¹

Leaves pinnate, or the upper ternate. Sepals minute or absent. Petals whitish, not emarginate; apex sometimes inflexed. Fruit ovoid, or elliptic-oblong, laterally compressed. Ridges usually stout; vittae solitary.

1 Bracteoles absent   **1. graveolens**
1 Bracteoles 5–7
  2 Leaves all with lanceolate to suborbicular, serrate or shallowly lobed segments; bracteoles with white, membranous margins
    3 Stem procumbent, rooting at lower nodes, then ascending or erect; peduncle usually shorter than rays; bracts 0–2   **2. nodiflorum**
    3 Stem procumbent and rooting at every node; peduncle usually longer than rays; bracts 3–7   **3. repens**
  2 Lower leaves with segments divided into filiform lobes; bracteoles entirely herbaceous
    4 Rays 2(–4); pedicels not thickened in fruit; styles much shorter than stylopodium in fruit   **4. inundatum**
    4 Rays 3–5; pedicels thickened at base in fruit; styles somewhat longer than stylopodium in fruit   **5. crassipes**

1. **A. graveolens** L., *Sp. Pl.* 264 (1753). Stout biennial up to 100 cm, with a strong, characteristic smell. Stem sulcate, solid. Leaves 1- to 2-pinnate; segments 5–50 mm, deltate, rhombic or lanceolate, lobed and serrate or almost crenate. Umbels mostly shortly pedunculate or sessile, often leaf-opposed. Rays 4–12. Bracts and bracteoles absent. Fruit 1·5–2 mm, broadly ovoid. *Damp places, usually near the sea. Coasts of Europe northwards to c. 56° N.* Al Au Az Be Bl Br Bu Co Cr Da Ga Ge Gr Hb Ho Hs It Ju Lu Po Rm Rs (W, K, E) Sa Si [Cz Fe He Hu No Su].

Several varieties are widely cultivated for the edible petioles (celery), leaves or roots, and are locally naturalized.

2. **A. nodiflorum** (L.) Lag., *Amen. Nat.* **1**: 101 (1821) (*Helosciadium nodiflorum* (L.) Koch). Procumbent or ascending perennial up to 100 cm. Stems hollow, rooting at the lower nodes. Leaves 1-pinnate, segments 10–60 mm, 7–13, lanceolate to ovate, serrate and often somewhat lobed. Peduncle usually shorter than rays, often almost absent; umbels leaf-opposed. Rays 3–12. Bracts usually absent, rarely 1 or 2; bracteoles 5–7, ovate or lanceolate, with a white, membranous margin. Fruit 1·5–2 mm, longer than wide, ovoid. 2*n* = 22. *Wet places. Much of Europe, particularly in the west; distribution uncertain owing to confusion with* **3**. Al Az Be Bl Br Bu Co ?Cr Ga Ge ?Gr Hb He Ho Hs It ?Ju Lu ?Rm Sa Si ?Tu.

3. **A. repens** (Jacq.) Lag., *loc. cit.* (1821) (*Helosciadium repens* (Jacq.) Koch). Like **2** but stem creeping throughout its length and rooting at every node; leaf-segments 5–14 mm, 5–11, ovate to sub-

orbicular; peduncle usually 2–3 times as long as rays; rays 3–6; bracts 3–7; fruit usually *c.* 1 mm, wider than long. $2n=22$. *Wet places. Mainly in C. and E. Europe; distribution uncertain owing to confusion with* 2. Au Be ?Bu ?Cr Cz Da Ga Ge ?Gr He Ho Hs Hu It ?Ju Lu Po ?Tu.

4. A. inundatum (L.) Reichenb. fil. in Reichenb. & Reichenb. fil., *Icon. Fl. Germ.* **21**: 9 (1863) (*Helosciadium inundatum* (L.) Koch). Perennial up to 75 cm, usually partly or completely submerged. Leaves pinnate, the lower (whether submerged or not) divided into filiform or linear lobes, the upper with ovate, often 3-lobed segments *c.* 5 mm. Peduncle about as long as rays; umbels leaf-opposed. Rays 2(–4); pedicels not thickened in fruit. Bracts absent; bracteoles 3–6, herbaceous, lanceolate. Styles much shorter than stylopodium in fruit. Fruit 2–3·5 mm, elliptic-oblong. $2n=22$. ● *W. Europe, extending eastwards to Sicilia, Poland and S.E. Sweden.* Be Br Da Ga Ge Hb Ho Hs It Lu Po Si Su.

A. × moorei (Syme) Druce, *Rep. Bot. Exch. Club Brit. Is.* **3**: 20 (1912) (*A. inundatum × nodiflorum*), occurs locally with the parents and is sterile.

5. A. crassipes (Koch ex Reichenb.) Reichenb. fil. in Reichenb. & Reichenb. fil., *loc. cit.* (1863) (*Helosciadium crassipes* Koch ex Reichenb.). Like **4** but rays 3–5; pedicels strongly thickened at base in fruit; bracteoles 5–8; styles distinctly longer than stylopodium in fruit. ● *Corse, Sardegna, Sicilia, S. Italy.* Co It Sa Si.

A. leptophyllum (Pers.) F. Mueller ex Bentham, *Fl. Austral.* **3**: 372 (1867) (*A. tenuifolium* Thell.), from America, with finely divided leaf-segments, umbels with usually 2 rays and no bracts or bracteoles, is recorded as introduced in a number of places; it seems often to be impermanent, but persists in Portugal. It may be incorrectly placed in *Apium.*

60. Petroselinum Hill[1]

Leaves 1- to 3-pinnate. Sepals minute. Petals white or yellowish, emarginate; apex inflexed. Fruit ovoid. Ridges filiform, conspicuous; vittae solitary.

Leaves 3-pinnate; flowers yellowish	**1. crispum**
Leaves simply pinnate; flowers white	**2. segetum**

1. P. crispum (Miller) A. W. Hill, *Hand-list Herb. Pl. Kew* ed. 3, 122 (1925) (*P. hortense* auct., *P. sativum* Hoffm.). Erect, glabrous biennial up to 75 cm. Stem terete, solid; branches ascending. Lower leaves triangular in outline, 3-pinnate; lobes 10–20 mm, cuneate, lobed, often crispate in cultivars. Umbels flat-topped. Rays 8–20. Bracts 1–3, entire or 3-fid; bracteoles 5–8, linear-oblong to ovate-cuspidate. Petals yellowish. Fruit 2·5–3 mm, broadly ovoid. $2n=22$. *Cultivated as a herb and naturalized in much of Europe; origin uncertain, but perhaps S.E. Europe or W. Asia.* [All except Al Fa Fe Ho Is Rs (N) Sb Tu.]

2. P. segetum (L.) Koch, *Nova Acta Acad. Leop.-Carol.* **12**(1): 128 (1824). Slender, more or less glaucous biennial or annual up to 100 cm. Stem terete, solid; branches divaricate. Leaves linear-oblong in outline, simply pinnate; segments 3–10 mm, ovate, serrate or sometimes lobed; margins thickened; teeth cartilaginous, incurved. Rays 2–5, very unequal. Bracts and bracteoles 2–5, subulate. Petals white. Fruit 2–4 mm, ovoid. ● *W. Europe, from the Netherlands and England to Portugal, and extending eastwards to C. Italy.* Be Br ?Co Ga Ho Hs It Lu.

---

[1] By T. G. Tutin.

61. Ridolfia Moris[1]

Leaves 4-pinnate, with filiform lobes. Sepals absent. Petals yellow, ovate; apex inflexed, truncate. Fruit ovoid-cylindrical, compressed laterally. Ridges slender, scarcely prominent; vittae solitary, slender.

1. R. segetum Moris, *Enum. Sem. Hort. Taur.* 43 (1841). Glabrous annual with stems 40–100 cm. Leaves with long, divaricate lobes, the upper often reduced to the inflated petiole. Umbels with 10–60 slender, nearly equal rays. Bracts and bracteoles absent. Fruit 1·5–2·5 mm. *Cultivated land and waste places. Mediterranean region, extending to Portugal.* Bl Co Ga Gr Hs It Ju Lu Sa Si Tu.

62. Sison L.[1]

Leaves pinnate. Flowers hermaphrodite. Sepals absent. Petals white, emarginate; apex inflexed. Fruit subglobose. Ridges filiform; vittae solitary, conspicuous, widest below the middle, much shorter than the fruit.

1. S. amomum L., *Sp. Pl.* 252 (1753). Biennial up to 100 cm, with a nauseous smell when crushed. Leaves 10–20 cm; the lower petiolate, simply pinnate, with 7–9 pairs of pinnae; pinnae 2–7 cm, usually sessile, oblong-ovate, serrate and often lobed; upper cauline leaves usually ternate, with spathulate or linear, dentate or lobed segments. Rays 3–6, slender, unequal. Bracts and bracteoles 2–4, linear, rarely absent. Pedicels very unequal. Fruit 1·5–3 mm. *S. & W. Europe, northwards to 53° 30′ N. in England.* Bl Br Bu Co Ga Gr He Hs It Ju Rm Sa Si Tu.

63. Cicuta L.[1]

Leaves 2- to 3-pinnate. Sepals conspicuous. Petals white or pink, emarginate; apex inflexed. Fruit subglobose, slightly compressed laterally. Ridges wide; vittae solitary, conspicuous.

1. C. virosa L., *Sp. Pl.* 255 (1753). Stout perennial up to 120 cm. Stock ovoid or shortly cylindrical, septate. Leaves up to 30 cm, deltate in outline; lobes 5–10 cm, linear-lanceolate or linear, acutely and deeply serrate, asymmetrical at base; petiole stout. Rays 1–5 cm, 10–20, subequal. Bracts absent; bracteoles 6–8, linear-oblong, about as long as the pedicels in fruit. Flowers 30–50 in each partial umbel. Pedicels slender, divaricate or deflexed in fruit. Fruit 1·75–2 × 1·5–1·75 mm. $2n=22$. *In shallow water or on damp mud. Most of Europe from 45° N. northwards; very rare further south and absent from most of the islands.* Au Be Br Bu Cz Da Fe Ga Ge Gr Hb He Ho *Hs Hu It Ju No Po Rm Rs (N, B, C, W, E) Su.

Extremely poisonous.

64. Cryptotaenia DC.[1]

Leaves ternate. Sepals absent. Petals white, somewhat emarginate; apex inflexed. Fruit subcylindrical. Ridges slender; vittae solitary, inconspicuous in fruit.

1. C. canadensis (L.) DC., *Prodr.* **4**: 119 (1830). Perennial up to 100 cm. Lower leaves long-petiolate with segments 5–10 cm, ovate, biserrate; upper leaves subsessile. Rays 3–10, unequal. Bracts absent; bracteoles absent or small. Flowers 1–6 in each partial umbel. Pedicels slender, strict, very unequal. Fruit 4–6 mm, often curved. *Naturalized in Austria (Steiermark).* [Au.] (*E. North America, Japan.*)

65. Lereschia Boiss.[1]

Like *Cryptotaenia* but fruit more strongly compressed laterally; styles free to base; vittae 2–3.

1. **L. thomasii** (Ten.) Boiss., *Ann. Sci. Nat.* ser. 3 (Bot.), **1**: 128 (1844). Glabrous, rhizomatous perennial 40–60 cm. Leaves mostly basal, ternate; segments rhombic, entire or serrulate in basal half and coarsely dentate to lobed in upper half; teeth and lobes serrulate, aristate; lower leaves long-petiolate. Umbels irregular, arranged in a leafless panicle. Flowers hermaphrodite and male; pedicels slender. Bracts and bracteoles few, small. Base of styles swollen in flower. Fruit *c.* 4 mm, subclavate, often curved. $2n=12$. *Damp places.* ● S. Italy (Calabria). It.

Very similar to *Petagnia* in general appearance and perhaps more closely related to it than *Cryptotaenia*.

66. Ammi L.[1]

Leaves 1- to 3-pinnate or -ternate. Sepals very small or absent. Petals white or yellowish, obcordate, the outer larger; apex inflexed. Fruit ovoid or ovoid-oblong, slightly compressed laterally, constricted at the commissure. Primary ridges filiform, prominent; vittae solitary.

1 Rays patent or erecto-patent in flower, becoming erect, thickened and indurate in fruit **1. visnaga**
1 Rays patent and slender in flower and fruit
2 Plant not more than 10 cm; basal leaves simple **3. majus**
2 Plant *c.* 100 cm; basal leaves 2- to 3-pinnate
3 Lobes of lower leaves narrowly linear or filiform **2. crinitum**
3 Lobes of lower leaves never narrowly linear or filiform
4 Bracts always linear, entire **5. trifoliatum**
4 At least some bracts pinnatisect or 3-fid
5 Lobes of middle and upper cauline leaves lanceolate or linear **3. majus**
5 Lobes of middle and upper cauline leaves obovate or broadly oblanceolate **4. huntii**

1. **A. visnaga** (L.) Lam., *Fl. Fr.* **3**: 462 (1778). Robust annual or biennial up to 100 cm. Lower leaves pinnate; others 2- to 3-pinnate; all with narrowly linear or filiform lobes. Rays up to *c.* 150, slender and patent in flower, becoming erect, thickened and indurate in fruit. Bracts 1- to 2-pinnatisect, equalling or exceeding the rays; bracteoles subulate. Pedicels erect, stout and rigid in fruit. Fruit 2–2·5 mm. *Mediterranean region and Portugal; a frequent weed farther north.* Al *Az Bl Co *Ga Gr Hs It Lu Sa Si Tu.

2. **A. crinitum** Guss., *Pl. Rar.* 128 (1826). Like **1** but leaves very much divided; rays remaining patent and slender in fruit; bracts like the upper leaves. ● S. Italy and Sicilia. It Si.

An imperfectly known species, perhaps not distinct from **1**.

3. **A. majus** L., *Sp. Pl.* 243 (1753). Annual 30–100 cm, very variable in habit and leaf-dissection. Leaves 2- to 3-pinnate; lower usually with elliptical or obovate, obtuse, serrate lobes; middle with lanceolate, acuminate, serrate to dentate lobes; upper with linear, dentate lobes. Rays 15–60, slender and erecto-patent in flower and fruit. Bracts 3-fid or pinnatisect, with filiform lobes, sometimes entire; bracteoles lanceolate, acuminate to linear-lanceolate. Pedicels slender. Fruit 1·5–2 mm. *S. Europe; a frequent weed farther north.* Al *Az Bl Co Cr Ga Gr Hs It Ju Lu Sa Si Tu.

[1] By T. G. Tutin.

A. topalii Beauverd, *Candollea* **7**: 264 (1937) is an extreme dwarf variant of **3** from Evvoia. It is up to 10 cm and has simple basal leaves. Plants intermediate between this and normal *A. majus* occur in rocky places on other Mediterranean islands.

4. **A. huntii** H. C. Watson, *London Jour. Bot.* (*Hooker*) **6**: 382 (1847) (incl. *A. seubertianum* (H. C. Watson) Trelease). Annual or biennial up to 100 cm. Leaves 2- to 3-pinnate or the upper ternate; lobes elliptical to oblong, deeply toothed, with narrow, aristate teeth. Rays 7–20, the longer up to 3 cm. Bracts pinnatisect or 3-fid, with linear lobes; bracteoles linear-lanceolate. Pedicels up to 12 mm, slender. Fruit 1·5–1·75 mm, narrowly ovoid. ● Açores. Az.

5. **A. trifoliatum** (H. C. Watson) Trelease, *Ann. Rep. Missouri Bot. Gard.* **8**: 116 (1897). Like **4** but larger rays 3–5 cm; bracts linear, never pinnatisect, often very unequal; pedicels usually *c.* 5 mm; fruit 1·5–1·75 mm. ● Açores. Az.

4 and **5** have been seldom collected and are not well known. They may, together with *A. seubertianum*, merely represent local populations of one species.

67. Ptychotis Koch[1]

Lower leaves 1-pinnate; segments pinnatisect. Sepals conspicuous. Petals whitish, with a suborbicular lobe projecting on either side of the inflexed, oblong apex. Fruit oblong, constricted at the commissure. Ridges prominent and almost winged; vittae solitary.

1. **P. saxifraga** (L.) Loret & Barrandon, *Fl. Montpell.* 283 (1876) (*P. heterophylla* Koch). Glabrous biennial 30–70 cm. Rosette-leaves simply pinnate, with 3–5(–7) dentate, serrate or somewhat lobed lobes; lower cauline leaves similar but with 5–9 lobes; the others small, 1- to 2-pinnatisect, with linear lobes; petioles of upper leaves expanded and sheathing. Umbels 2–5 cm in diameter; peduncles slender; rays 6–12, slender, unequal. Bracts 2–3, caducous; bracteoles 3–6, setaceous. Pedicels unequal. Fruit 2–3 mm. *Dry places.* *W. Mediterranean region, extending locally to N.E. France and N.E. Italy.* Co Ga He Hs It Sa.

68. Ammoides Adanson[1]

Lower leaves 2-pinnate, with linear segments. Sepals absent. Petals whitish, with a suborbicular lobe projecting on either side of the inflexed, oblong apex. Fruit broadly ovoid, laterally compressed, not constricted at the commissure. Ridges filiform, prominent; vittae solitary.

1. **A. pusilla** (Brot.) Breistr., *Bull. Soc. Sci. Dauph.* **61**: 628 (1947) (*Ptychotis ammoides* Koch). Slender, glabrous annual 10–50 cm. Leaves glaucescent, the lower 2-pinnate, oblong in outline, with 7–11 pairs of very short lobes, the middle with 3–5 pairs of segments, the upper usually with 2–3 long, filiform segments. Rays 5–11, slender, unequal. Bracts absent, or few and caducous; bracteoles 4–6, some linear-lanceolate, others spathulate and inflated distally, with an acuminate apex. Pedicels unequal. Fruit *c.* 1 mm. *Mediterranean region, extending to Portugal.* Al Co Gr Hs It Ju Lu Sa Si.

Ptychotis morisiana Béguinot, *Arch. Bot.* (*Forlì*) **3**: 284 (1927), described from Sardegna (Tavolara), is said to be like **1** but with shorter, linear-lanceolate leaf-lobes, a single bract and 3 linear, aristate bracteoles. It is perhaps a subspecies of *A. pusilla*, but it requires further investigation, especially as ripe fruit is unknown.

69. Thorella Briq.[1]

First leaves usually reduced to the subulate petiole and rhachis; others pinnate, with short, spathulate, sometimes pinnatisect segments. Sepals small. Petals whitish, suborbicular, weakly emarginate; apex inflexed. Fruit ovoid, compressed laterally. Ridges prominent, stout; vittae solitary.

1. **T. verticillatinundata** (Thore) Briq., *Annu. Cons. Jard. Bot. Genève* 17: 275 (1914) (*Ptychotis thorei* Godron & Gren.). Perennial with slender rhizome; stems up to 20 cm, slender. Leaves mostly basal, the first fistular, septate, the others with 7–20 pairs of segments. Rays 3–6, usually unequal. Bracts 3–5, entire or 2- to 3-fid, much shorter than the rays; bracteoles like the bracts. Fruit *c.* 2 mm. *Seasonally flooded places.* ● *S.W. & W.C. France; W. Portugal.* Ga Lu.

70. Falcaria Fabr.[1]

Leaves usually 1- to 2-ternate; margins cartilaginous and serrate. Sepals conspicuous. Petals whitish, broadly obovate, emarginate; apex inflexed. Fruit at least three times as long as wide, oblong, compressed laterally, constricted at the commissure. Ridges low, wider than the grooves; vittae solitary.

1. **F. vulgaris** Bernh., *Syst. Verz. Erfurt* 176 (1800) (*F. rivini* Host, *F. sioides* Ascherson). Glaucous annual, biennial or perennial up to 90 cm. Stems terete, solid, freely branched and often forming a low, tangled mass. Leaves 1- to 2-ternate; segments up to 30 cm, linear-lanceolate or linear, acuminate, somewhat falcate, strongly, sharply and regularly serrate. Rays 12–18. Bracts and bracteoles 4–15, subulate. Fruit 3–4 mm, oblong. $2n = 22$. *Europe from N. France and C. Russia southwards, but absent from the islands.* Au Bu Ga Gr He Hs Hu It Ju Po Rm Rs (*B, C, W, K, E) ?Tu [Be Br Da Ho Su].

71. Carum L.[1]

Leaves 2- to 4-pinnate. Sepals very small or absent. Petals whitish, rarely pink or yellowish, obovate, emarginate; apex inflexed. Fruit obovoid-oblong, laterally compressed. Ridges filiform, prominent or almost winged; vittae solitary and wide, or 2–3 and narrow.

1 Leaf-segments diminishing markedly in size from about the middle of the leaf downwards   **3. verticillatum**
1 Lowest pair of leaf-segments the largest or, rarely, the lower 1–3 pairs smaller than the rest
  2 Lowest leaf-segments at least twice as long as wide
    3 Rays erecto-patent after flowering; bracteoles few or 0, *c.* ¼ as long as the longer pedicels   **1. carvi**
    3 Outer rays almost horizontal after flowering; bracteoles 4–8, *c.* ½ as long as the pedicels   **2. multiflorum**
  2 Lowest leaf-segments about as long as wide
    4 Stems erect; styles longer than stylopodium   **4. rigidulum**
    4 Stems decumbent; styles not longer than stylopodium   **5. heldreichii**

1. **C. carvi** L., *Sp. Pl.* 263 (1753) (incl. *C. velenovskyi* Rohlena). Divaricately branched, glabrous perennial up to 150 cm. Stems striate, leafy. Leaves 2- to 3-pinnate; lobes 3–25 mm, linear-lanceolate or linear. Rays 5–16, very unequal, erecto-patent in fruit. Bracts usually absent, rarely up to 8, and then sometimes 2- to 3-partite; bracteoles absent or few, up to ¼ as long as longest pedicels. Petals whitish or pink. Fruit 3–6 mm, ovoid, strong-smelling when crushed; ridges low, rounded. $2n = 20$. *Most of Europe, except the Mediterranean region; widely cultivated for its*

aromatic fruits (caraway), which are used for flavouring, and frequently naturalized. Al Au Be Bu Cz Da Fe Ga Ge He Ho Hs Hu It Ju No Po Rm Rs (N, B, C, W, K, E) Su [*Br Fa Hb Is Sb.]

2. **C. multiflorum** (Sibth. & Sm.) Boiss., *Fl. Or.* 2: 882 (1872). Biennial or perennial up to 70 cm. Basal leaves up to *c.* 10 cm, triangular in outline, 2- to 3-pinnate; lobes up to 10 mm, ovate to obovate in outline, dentate or pinnatisect with ovate or lanceolate, entire lobes. Rays 5–28, the outer almost horizontal in fruit. Bracts and bracteoles 4–8, oblong to ovate-lanceolate. Petals white. Fruit 2–3 mm, oblong-ellipsoid; ridges very narrowly winged. *S. part of Balkan peninsula; one station in S.E. Italy.* Al Bu Cr Gr It Ju.

(a) Subsp. **multiflorum**: Stems stout, with numerous stout branches; lobes of cauline leaves similar to those of basal leaves; rays usually 15–25. *W. & S. Greece, S. Albania.*

(b) Subsp. **strictum** (Griseb.) Tutin, *Feddes Repert.* 74: 31 (1967) (*Bunium strictum* Griseb., *C. lumpeanum* Dörfler & Hayek): Stems slender, with few, slender branches; lobes of cauline leaves narrower than those of basal leaves; rays 5–12. ● *From N.E. Greece and S.W. Bulgaria to C. & N. Albania and S.E. Italy.*

3. **C. verticillatum** (L.) Koch, *Nova Acta Acad. Leop.-Carol.* 12(1): 122 (1824). Erect, glabrous perennial up to 120 cm. Root of fusiform fibres thickened downwards. Stem striate, little-branched, with few small leaves. Basal leaves 10–25 cm, narrowly oblong in outline, with usually more than 20 pairs of deeply palmatisect segments which are longest in the upper half; lobes up to 10 mm, filiform, appearing as if whorled. Rays up to 12. Bracts up to 10, linear-acuminate; bracteoles numerous, linear-lanceolate, deflexed. Petals white. Fruit 2·5–4 mm, ellipsoid; ridges prominent. *Marshes and damp meadows.* ● *W. Europe, northwards to Scotland and the Netherlands.* Be Br ?Co Ga †Ge Hb Ho Hs Lu.

4. **C. rigidulum** (Viv.) Koch ex DC., *Prodr.* 4: 115 (1830) (incl. *C. graecum* Boiss. & Heldr., *C. adamovicii* Halácsy). Glabrous perennial up to 60 cm. Stems erect, simple or with few long branches, striate, with few small leaves. Basal leaves 10–20 cm, oblong or oblong-lanceolate in outline; segments up to 15 pairs, the largest at or near the base of the lamina; lobes 2–10 × 0·5–2 mm, linear-lanceolate to setaceous. Rays 3–11, erecto-patent. Bracts 0–6, linear; bracteoles 3–8, linear-lanceolate, acuminate, broadly scarious. Petals white or yellowish-white. Styles longer than stylopodium. Fruit 3–4 mm, ellipsoid; ridges prominent. *Mountain rocks.* ● *Balkan peninsula; C. Italy (Alpi Apuane).* Al Bu Gr It Ju.

5. **C. heldreichii** Boiss., *Diagn. Pl. Or. Nov.* 3(2): 78 (1856) (*C. flexuosum* (Ten.) Nyman, non Fries; incl. *C. rupestre* Boiss. & Heldr.). Perennial, up to 40 cm. Stems several, decumbent, then ascending, flexuous. Basal leaves 3–10 cm, oblong-lanceolate in outline; segments up to 8 pairs, the largest at or near the base of the lamina; lobes 2–10 × *c.* 0·5 mm, elliptic-lanceolate, acuminate; bracteoles 3–5, linear to setaceous, acuminate, with narrow scarious margin. Petals white or yellowish-white. Styles not longer than stylopodium. Fruit 3·5–4·5 mm, ellipsoid; ridges prominent. *Mountain rocks.* ● *Greece, Albania and Italy.* Al Gr It.

72. Stefanoffia H. Wolff[1]

Roots napiform. Leaves 2- to 3-pinnate, with narrow lobes. Sepals absent. Petals whitish, obcordate; apex inflexed, broad, reaching nearly to the base of the petal. Fruit subglobose, slightly compressed laterally. Ridges filiform; vittae solitary.

---

[1] By T. G. Tutin.

1. S. daucoides (Boiss.) H. Wolff, *Notizbl. Bot. Gart. Berlin* **9**: 282 (1925). Perennial *c.* 50 cm. Leaves with linear, obtuse, rather rigid lobes 2–3 mm long. Umbels with 10–20 subequal, divaricate rays. Bracts numerous, 2–3 times bifid, $\frac{1}{3}$–$\frac{1}{2}$ as long as rays; bracteoles 6–8, narrowly linear, about equalling the pedicels. Fruit *c.* 1·5 mm. *Dry grassland. N. Greece (Makedhonia), Kikhlades and S.E. Bulgaria.* Bu Gr.

73. Brachyapium (Baillon) Maire[1]

Leaves 2- to 3-pinnate. Sepals absent. Petals white, suborbicular; apex inflexed. Fruit broadly cordate-ovoid, subdidymous, distinctly compressed laterally. Ridges prominent, filiform. Vittae solitary.

1. B. dichotomum (L.) Maire, *Bull. Soc. Hist. Nat. Afr. Nord* **23**: 186 (1932) (*Pimpinella dichotoma* L., *Tragiopsis dichotoma* (L.) Pomel). Slender annual up to 15 cm. Lower leaves 2- to 3-pinnate; lobes 3–10 mm, linear-lanceolate or linear, acute; petioles of upper cauline leaves with wide membranous margins. Rays 4–8, filiform, minutely scabrid. Bracts and bracteoles absent. Fruit *c.* 1 mm, broadly cordate or reniform, densely papillose or shortly hispid. *C. & S.E. Spain.* Hs. (*N. Africa.*)

74. Endressia Gay[1]

Leaf-segments digitately lobed. Sepals subulate, as long as stylopodium. Petals ovate-lanceolate, involute. Fruit ovoid; ridges prominent.

Literature: P. Rey, *Doc. Cartes Vég.*, *Sér. Pyrénées* **3**(3): 1–30 (1945).

Stem glabrous, except for a ring of hairs at base of umbel; rays
  glabrous   **1. pyrenaica**
Stem pubescent below; rays puberulent on inner side   **2. castellana**

1. E. pyrenaica (Gay ex DC.) Gay, *Ann. Sci. Nat.* **26**: 224 (1832). Almost glabrous perennial 5–40 cm. Stock stout, short. Stem terete, strongly striate, hollow, usually simple. Leaves mostly basal; segments 5–9, deeply palmately lobed, the lobes all very narrow, pinnatifid. Umbels small, dense, subglobose in fruit; rays 9–25, glabrous. Bracts usually absent, rarely 1–4, caducous; bracteoles 1–5, subulate. Sepals accrescent. Fruit 2–4 mm, ovoid. *Subalpine grassland.* ● *E. Pyrenees.* Ga Hs.

2. E. castellana Coincy, *Jour. Bot.* (*Paris*) **12**: 3 (1898). Like **1** but stock slender, long; stem pubescent at base and near apex; leaf-segments 7–11, less deeply and more broadly lobed, sometimes broadly ovate and merely dentate; petioles and veins more or less pubescent; rays puberulent on inner side; sepals not accrescent. *Subalpine grassland.* ● *N. Spain (Burgos to Pamplona).* Hs.

75. Cnidium Cusson[1]

Leaves 2- to 4-pinnate. Sepals small. Petals white, obcordate; apex wide, inflexed. Fruit ovoid or subglobose, slightly compressed laterally. Ridges wide and prominent, the lateral a little more prominent than the others; vittae solitary.

Petioles of cauline leaves sheathing the stem throughout their
  length; leaf-segments sessile or subsessile; fruit *c.* 2 mm   **1. dubium**
Petioles of cauline leaves sheathing the stem at their base only;
  leaf-segments long-stalked; fruit *c.* 4 mm   **2. silaifolium**

1. C. dubium (Schkuhr) Thell. in Hegi, *Ill. Fl. Mitteleur.* **5**(2 1305 (1926). Nearly glabrous biennial or perennial 30–100 cm. Stem terete below, furrowed above, hollow. Leaves oblong in outline, 2- to 3-pinnate, the segments sessile or subsessile; lobes 5–20 × 1–2 mm, narrowly oblong, with serrulate, somewhat recurved margins, prominent midrib and whitish, acute apex; cauline leaves with often purplish petioles, sheathing the stem throughout their length. Rays 20–30, narrowly winged and puberulent on the angles. Bracts usually few or absent, subulate; bracteoles numerous, subulate, scabrid. Fruit 2–3 mm, subglobose. $2n = 20$. *C. Europe, extending to S. Sweden and E. Denmark; U.S.S.R.* Au Cz Da Ge Hu Po Rm Rs (N, B, C, W, E) Su.

2. C. silaifolium (Jacq.) Simonkai, *Enum. Fl. Transs.* 259 (1887) (*C. apioides* (Lam.) Sprengel). Glabrous perennial 60–120 cm. Stem striate, solid. Leaves triangular in outline, 2- to 4-pinnate, the segments long-stalked; lobes 4–10 × 0·5–2 mm, linear-lanceolate to obovate or oblanceolate, with scabrid margins; apex acute or obtuse, mucronate or rounded; petioles of cauline leaves stout, sheathing at base only. Rays 20–45, scabrid on the angles. Bracts usually few or absent, subulate; bracteoles numerous, subulate, almost smooth. Fruit 3·5–4 mm, ovoid. *S. Europe, from S.E. France to Romania and Kriti.* Al Bu Cr Ga Gr He It Ju Rm Si [Cz].

(a) Subsp. **silaifolium**: Leaves rather soft, 2- to 4-pinnate; lobes 3–5 times as long as wide, linear-lanceolate to narrowly obovate, acute or narrowed into a mucro. *Throughout the range of the species.*

(b) Subsp. **orientale** (Boiss.) Tutin, *Feddes Repert.* **74**: 31 (1967) (*C. orientale* Boiss.): Leaves rather rigid, 2- to 3-pinnate; lobes twice as long as wide, obovate or oblanceolate, rounded, abruptly mucronate. *Greece, N.W. Bulgaria, Romania.* (*S.W. Asia.*)

C. monnieri (L.) Cusson, *Mém. Soc. Roy. Méd.* (*Paris*) **1782**: 280 (1782), an annual from E. Asia, with leaves resembling those of *Aethusa cynapium* and with ciliate bracteoles, was cultivated in botanic gardens in the 18th century and was formerly naturalized in some of the warmer parts of Europe, but has not been seen for many years.

76. Selinum L.[1]

Leaves 2- to 3-pinnatisect. Sepals absent. Petals white, obcordate; apex inflexed. Fruit ovoid-oblong, compressed dorsally. Ridges winged, the marginal distinctly wider than the others; vittae solitary.

Stem strongly angled, with narrowly winged angles; cauline
  leaves several   **1. carvifolia**
Stem striate, unwinged; cauline leaves 0–2   **2. pyrenaeum**

1. S. carvifolia (L.) L., *Sp. Pl.* ed. 2, 350 (1762). Nearly glabrous perennial 30–100 cm. Stem solid, branched, leafy, strongly angled, the angles narrowly winged. Leaves 2- to 3-pinnate; lobes 3–10 mm, almost linear to ovate, sometimes lobed, finely serrulate, mucronate or aristate. Rays 5–33, puberulent on the angles. Bracts absent, or 1–2, caducous; bracteoles several, linear-lanceolate. Petals white. Fruit 3–4 mm. $2n = 22$. *Most of Europe except for much of the Mediterranean region.* Au Be Br Bu Cz Da Fe Ga Ge He Ho Hs Hu It Ju Lu No Po Rm Rs (N, B, C, W, E) Su.

2. S. pyrenaeum (L.) Gouan, *Obs. Bot.* 11 (1773) (*Angelica pyrenaea* (L.) Sprengel). Glabrous perennial 20–50 cm. Stem simple or slightly branched, striate; cauline leaves 0–2. Basal leaves 2- to 3-pinnatisect; lobes linear-lanceolate to linear, cuspidate. Rays 3–9, stout, very unequal. Bracts absent; bracteoles

---

[1] By T. G. Tutin.

linear-lanceolate, ciliolate. Petals yellowish-white. Fruit 3·5–4 mm. *Mountain pastures.* ● *Mountains of W. Europe, from the Vosges to N.W. Spain.* Ga Hs.

77. Ligusticum L.[1]

Perennial. Leaves 2- to 5-pinnate or 2-ternate. Sepals small. Petals usually white, obcordate; apex inflexed. Fruit not compressed. Ridges very prominent, narrowly winged; vittae numerous.

1 Leaves 2-ternate; lobes 20–50 mm, ovate-cuneate   **7. scoticum**
1 Leaves 2- to 5-pinnate; lobes not more than 15 mm, linear to narrowly ovate-oblong
  2 Bracts 5–10, at least half as long as rays, persistent
    3 Stem simple, usually leafless; stock without fibres; bracts simple, erecto-patent   **1. mutellinoides**
    3 Stem usually branched, leafy; stock with abundant fibres; bracts usually pinnatisect, deflexed   **6. ferulaceum**
  2 Bracts 0 or few, rarely up to 7 and then caducous
    4 Stock without coarse fibres; leaf-lobes narrowly linear   **4. albanicum**
    4 Stock with abundant coarse fibres; leaf-lobes not narrowly linear
      5 Rays 20–50; branches opposite or whorled above   **5. lucidum**
      5 Rays not more than 15; branches alternate
        6 Leaves triangular in outline; fruit with smooth ridges   **2. mutellina**
        6 Leaves oblong in outline; fruit with denticulate ridges   **3. corsicum**

1. L. mutellinoides (Crantz) Vill., *Prosp. Pl. Dauph.* 25 (1779) (*L. simplex* (L.) All., *Gaya simplex* (L.) Gaudin; incl. *Pachypleurum alpinum* Ledeb.). Almost glabrous. Stock stout, with few or no fibres. Stem up to 30 cm, simple, nearly solid, leafless or rarely with 1 leaf. Leaves 3–10 cm, ovate in outline, 2- to 3-pinnate; lobes 2–5 mm, narrowly ovate-oblong to linear-lanceolate, acute. Rays 8–20. Bracts and bracteoles numerous, linear-lanceolate, often 2- to 3-fid. Petals white or pink. Fruit 3–5 mm, ellipsoid; ridges smooth, sometimes with short, stiff hairs between them. $2n=22$. *Alps; Sudety; Carpathians; N. Ural and Arctic Russia.* Au Cz Ga Ge He It Ju Po Rm Rs (N, C).

The Russian populations are sometimes given specific status as **L. alpinum** (Ledeb.) F. Kurtz, *Bot. Jahrb.* **19**: 464 (1894), non Sprengel. Plants having the fruit characters of *L. alpinum* occur sporadically in the Alps; the only reliable distinguishing character of the Russian plants appears to be the somewhat less dissected leaves.

2. L. mutellina (L.) Crantz, *Stirp. Austr.* **3**: 81 (1767) (*Meum mutellina* (L.) Gaertner). Almost glabrous. Stock with abundant coarse fibres. Stem 10–50 cm, hollow, usually with 1–2 branches subtended by small leaves. Leaves 5–10 cm, triangular in outline, 2- to 3-pinnate; lobes 3–5(–15) mm, linear-lanceolate, mucronate. Rays 7–10(–15). Bracts 0–2, small; bracteoles 3 to several, lanceolate, about as long as pedicels. Petals usually red or purple. Fruit 4–6 mm, ovoid-oblong, glabrous; ridges smooth. $2n=22$. ● *Mountains of C. & S. Europe, from S.C. France to the Carpathians and S. Bulgaria.* Al Au Bu Cz Ga Ge He It Ju Po Rm Rs (W).

Two subspecies, based on the division of the leaves, are sometimes recognized, but do not seem to merit this status. 'Subsp. *adonidifolium* (Gay) Beauverd' has long, little-divided leaf-lobes and occurs in the French Alps and Romania; it seems often to be sympatric with the widespread typical subspecies.

3. L. corsicum Gay, *Ann. Sci. Nat.* **26**: 222 (1832). Like **2** but often taller; leaves oblong in outline; fruit with denticulate ridges. *Stony places in the mountains.* ● *Corse.* Co.

4. L. albanicum Jáv., *Bot. Közl.* **19**: 24 (1921). Like **2** but stock without fibres; leaf-lobes narrowly linear, at most 0·5 mm wide; bracts up to 7, caducous; bracteoles often longer than partial umbels. ● *N. Albania.* Al.

5. L. lucidum Miller, *Gard. Dict.* ed. 8, no. 4 (1768) (*L. pyrenaeum* Gouan, *L. seguieri* Vill.). Almost glabrous. Stock with abundant coarse fibres. Stem up to 150 cm, branched, solid, with several cauline leaves; branches opposite or whorled above. Leaves *c.* 30 cm, triangular in outline, 3- to 5-pinnate; lobes (2–)4–15 mm. Rays (11–)20–50. Bracts usually absent; bracteoles 5–8, about ½ as long as the partial umbel. Fruit (4–)5–6(–8) mm, ellipsoid or oblong-ovoid; ridges narrowly winged. ● *Mountains of S. Europe.* Al Bl Ga He Hs It Ju.

(a) Subsp. **lucidum**: Leaf-lobes 4–15 mm, lanceolate, oblanceolate or linear, usually acuminate. Rays 20–50. Fruit usually 5–6 mm. *Throughout the range of the species, except Islas Baleares.*

(b) Subsp. **huteri** (Porta & Rigo) O. Bolós, *Publ. Inst. Biol. Apl. (Barcelona)* **27**: 54 (1958) (*L. huteri* Porta & Rigo): Leaf-lobes 2–5 mm, oblanceolate to almost obovate, obtuse, mucronate. Rays 11–16. Fruit 4–4·5 mm. *Islas Baleares (Mallorca).*

6. L. ferulaceum All., *Mélang. Philos. Math. Soc. Roy. Turin (Misc. Taur.)* **5**: 80 (1774). Nearly glabrous. Stock with abundant coarse fibres. Stem up to 60 cm, solid, usually branched from the base. Leaves 10–30 cm, 3- to 4-pinnate; lobes up to 5 mm, linear-lanceolate to linear, acuminate. Rays 15–25. Bracts numerous, leaf-like, usually pinnatisect; bracteoles as long as the partial umbel. Fruit 3–7 mm, ovoid. *Screes and alpine pastures.* ● *French Jura, S.W. Alps.* Ga It.

7. L. scoticum L., *Sp. Pl.* 250 (1753). Glabrous. Stock without fibres. Stem up to 90 cm, somewhat branched, leafy, hollow. Leaves 10–20 cm, triangular in outline, 2-ternate; lobes 20–50 mm, ovate-cuneate, dentate or shallowly lobed in upper half. Rays 8–20. Bracts and bracteoles 1–7, linear. Fruit 5–8 mm, oblong-ovoid. *Rocky sea-shores. Coasts of N. Europe, from W. Ireland to N.W. Russia; two stations on the Baltic (E. Sweden).* Br Da Fa Hb Is No Rs (N) Su.

78. Cenolophium Koch ex DC.[1]

Leaves 3- to 5-pinnate. Sepals absent. Petals white, broadly ovate, emarginate; apex inflexed. Fruit oblong-ellipsoid, somewhat compressed dorsally. Ridges very prominent, narrowly winged; vittae solitary.

1. C. denudatum (Hornem.) Tutin, *Feddes Repert.* **74**: 31 (1967) (*Athamanta denudata* Hornem., *C. fischeri* (Sprengel) Koch ex DC.). Glabrous perennial 60–150 cm. Stock with fibrous remains of petioles. Stem solid, terete, striate, often purplish. Leaves triangular in outline, 3- to 5-pinnate; primary divisions patent or deflexed; lobes 10–30 mm, oblong-lanceolate, mucronate or acute, often somewhat falcate, scabrid on midrib and margin. Rays 15–20, weakly angled, somewhat papillose. Bracts absent; bracteoles several, subulate. Fruit 3·5–6 mm, 8-angled. *U.S.S.R., except the south.* Rs (N, B, C, W, E).

79. Conioselinum Hoffm.[1]

Leaves 2- to 3-pinnate. Sepals absent. Petals white, broadly ovate, emarginate; apex inflexed. Fruit elliptical, strongly compressed dorsally. Ridges winged, the lateral strongly, the dorsal narrowly; vittae 1–4.

---

[1] By T. G. Tutin.

1. **C. tataricum** Hoffm., *Gen. Umb.* ed. 2, 185 (1816) (*C. vaginatum* Thell.; incl. *C. boreale* Schischkin). Glabrous perennial 50–150 cm. Lower leaves long-petiolate, triangular-rhombic in outline, 2- to 3-pinnate; lobes of lower leaves 3–5(–20) mm, oblong-lanceolate to linear, often pinnatifid; upper leaves with greatly inflated petioles, lamina small. Rays (7–)15–30. Bracts few or absent; bracteoles numerous, narrowly subulate, finely papillose-serrulate. Fruit 5 mm. *E. & E.C. Europe, southwards to the S. Carpathians and westwards to arctic Norway and C. Austria.* Au Cz Fe No Po Rm Rs (N, B, C, W, E).

80. Angelica L.[1]

(Incl. *Archangelica* N. M. Wolf, *Ostericum* Hoffm.)

Leaves 2- to 3-pinnate or -ternate; lobes broad. Calyx-teeth usually inconspicuous, but sometimes well developed. Petals white, pinkish, greenish or rarely yellowish, lanceolate; apex more or less incurved. Fruit ovate or oblong, strongly compressed dorsally. Lateral ridges forming wide wings which are not closely appressed at the margins, and are frequently more or less undulate; dorsal ridges usually well developed; vittae variable in number.

1 Calyx-teeth broadly ovate, whitish and well-developed **1. palustris**
1 Calyx-teeth minute or 0
  2 Petals yellowish **6. laevis**
  2 Petals white, greenish or pinkish
    3 Leaf-lobes (especially the terminal ones) usually strongly decurrent on the rhachis
      4 Fruit with thick, corky wings **3. archangelica**
      4 Fruit with ±membranous wings
        5 Wings of fruit distinctly wider than the mericarps **2. sylvestris**
        5 Wings of fruit not wider than the mericarps **5. razulii**
    3 Leaf-lobes not or only slightly decurrent on the rhachis
      6 Fruit with thick, corky wings; fleshy maritime plant **8. pachycarpa**
      6 Fruit with ± thin, membranous wings
        7 Wings of fruit distinctly wider than the mericarps **2. sylvestris**
        7 Wings of fruit usually narrower than or about as wide as the mericarps
          8 Rays puberulent **4. heterocarpa**
          8 Rays glabrous **7. angelicastrum**

1. **A. palustris** (Besser) Hoffm., *Gen. Umb.* 174 (1814) (*Ostericum palustre* (Besser) Besser). Biennial to perennial up to 120 cm. Leaves 2- to 3-pinnate; sheathing bases well-developed; lobes coarsely serrate; base frequently oblique. Umbels terminal and lateral, with numerous rays. Bracts 0–3, caducous; bracteoles numerous, linear-lanceolate, with a whitish margin. Sepals well-developed, broadly ovate, whitish. Petals white. Fruit 5–6 mm, ovate-elliptical, with prominent dorsal ridges. *Wet places. E. & C. Europe, northwards to Estonia, westwards to C. Germany, and extending southwards to Crna Gora.* ?Au ?Bu Cz Ge Hu Ju Rm Rs (B, C, W, E).

2. **A. sylvestris** L., *Sp. Pl.* 251 (1753) (*A. illyrica* K. Malý, *A. elata* Velen., *A. brachyradia* Freyn). Erect, often robust perennial up to 200 cm or more. Stems usually tinged with purple. Leaves 2- to 3-pinnate, up to 60 cm; lobes obliquely oblong-ovate, acutely serrate, the terminal usually simple. Petioles strongly sheathing at the base. Upper leaves reduced to inflated sheaths, which more or less enclose the developing umbels. Rays numerous, puberulent. Bracts 0 or few, caducous; bracteoles setaceous. Sepals minute. Petals white to pinkish. Fruit 4–5 mm, ovate; dorsal ridges obtuse; wings wider than the mericarps, rather

membranous and somewhat undulate. $2n=22$. *Damp or shady places. Almost throughout Europe.* All except Az Bl Cr Rs (K) ?Sa Sb.

3. **A. archangelica** L., *Sp. Pl.* 250 (1753) (*Archangelica officinalis* Hoffm.). Like **2** but stems less often tinged with purple; leaf-lobes more irregular in outline and more jaggedly cut, often distinctly decurrent on the rhachis; terminal lobe usually 3-lobed; petals greenish-white to cream; fruit with rather thick, corky wings. *Damp places. N. & E. Europe, westwards to the Netherlands and Iceland and southwards to C. Ukraine; often cultivated elsewhere for its aromatic petioles, used in confectionery and for a liqueur; frequently naturalized.* Cz Da Fa Fe Ge Ho Is No Rs (N, B, C, W, E) Su [Au Be Br Bu Ga He Hu It Ju Rm].

(a) Subsp. **archangelica** (incl. *Archangelica decurrens* Ledeb.): Petals greenish to cream. Bracteoles about as long as pedicels. Fruit 6–8×4–5 mm, nearly oblong; dorsal ribs prominent and acute. Odour pleasant and aromatic. $2n=22$. *Almost throughout the range of the species; includes the cultivated plant.*

(b) Subsp. **litoralis** (Fries) Thell. in Hegi, *Ill. Fl. Mitteleur.* **5**(2): 1342 (1926): Petals greenish-white. Bracteoles about half as long as pedicels. Fruit 5–6×3·5–4·5 mm, more elliptical than in (a); dorsal ribs not prominent, obtuse. Odour pungent. $2n=22$. *N. Europe, usually on sea-shores.*

4. **A. heterocarpa** Lloyd, *Bull. Soc. Bot. Fr.* **6**: 709 (1859). Almost glabrous, robust perennial up to 200 cm. Leaves 2-pinnate; lobes up to 10×3 cm, ovate-lanceolate, sometimes oblique at the base and slightly decurrent; margins more or less regularly serrate and slightly cartilaginous. Rays numerous, puberulent. Bracts 0 or few, linear; bracteoles several, about as long as the pedicels. Partial umbels with numerous flowers; pedicels up to 10 mm. Petals white; apex distinctly incurved. Fruit 4–6×2–3 mm, oblong. Wings usually narrower than the mericarp, but apparently rather variable and sometimes wider. *Muddy banks of tidal rivers.* ● *S.W. France.* Ga.

5. **A. razulii** Gouan, *Obs. Bot.* 13 (1773). Like **4** but leaf-lobes narrowly lanceolate, the upper strongly decurrent on the rhachis; petals white to pinkish-white; apex strongly inflexed; fruit *c.* 8 mm, ovate-oblong; wings about as wide as the mericarp. *Meadows and pastures.* ● *Pyrenees, N.W. Spain.* Ga Hs.

6. **A. laevis** Gay ex Avé-Lall., *Ind. Sem. Horti Petrop.* **9**: 58 (1843). Robust, erect perennial. Leaves 2- to 3-pinnate, glabrous above but with short hairs on the veins beneath; lobes narrowly ovate-lanceolate; margin serrate; teeth mucronate. Rays *c.* 10 cm, numerous, very slightly puberulent. Bracts 0; bracteoles several, linear-subulate, unequal and shorter than the pedicels. Petals yellowish, with an inflexed apex. Fruit 7×4·5 mm, oblong; dorsal ridges prominent and acute; wings well developed, about as wide as the mericarps. *Banks of streams.* ● *N.W. Spain; Portugal.* Hs Lu.

7. **A. angelicastrum** (Hoffmanns. & Link) Coutinho, *Fl. Port.* 455 (1913). Robust, erect perennial. Leaves 2(–3)-pinnate; lobes narrowly lanceolate, biserrate, somewhat oblique at the base and sometimes slightly decurrent on the rhachis, with numerous short, rather stiff hairs on the veins beneath. Rays rather unequal, glabrous. Bracts 0(–1 ?); bracteoles linear, numerous. Fruit with prominent, acute dorsal ridges; wings about as wide as the mericarp. ● *C. Portugal (Serra de Estrêla).* Lu.

8. **A. pachycarpa** Lange, *Descr. Icon. Ill.* 7 (1864). Robust, fleshy perennial up to 100 cm. Leaves 2- to 3-pinnate. Lateral lobes broadly elliptical; terminal lobes lanceolate, often 3-lobed;

[1] By J. F. M. Cannon.

margins serrate to rather coarsely dentate. Umbels on short, stout peduncles, with numerous robust, puberulent rays. Bracts 6–8, linear; bracteoles numerous, like the bracts, equalling or shorter than the pedicels. Petals white with green striations. Fruit 10 × 7 mm, broadly oblong; dorsal ridges prominent, acute; wings as wide as the mericarps, rather thick and corky, projecting beyond the stylopodium when mature. Fruit rather glossy, slightly tinged with brown. *Maritime rocks.* ● *N.W. Spain (near La Coruña); W. Portugal (Ilha Berlenga).* Hs Lu.

81. Levisticum Hill[1]

Leaves 2- to 3-pinnate. Sepals absent. Petals greenish-yellow, elliptical, obtuse; apex short, inflexed. Fruit ovoid-oblong, somewhat compressed dorsally. Marginal ridges with rather thick, but distinct, wings; dorsal low, obtuse, or rarely narrowly winged; vittae solitary.

    1. L. officinale Koch, *Nova Acta Acad. Leop.-Carol.* **12**(1): 101 (1824). Stout perennial 100–250 cm, with strong smell; lower branches alternate, upper opposite or whorled; stock with numerous scale-like remains of petioles. Lower leaves up to 70 × 65 cm, triangular-rhombic in outline; lobes long-cuneate, irregularly and deeply dentate and lobed in upper part. Rays 12–20, rather stout, furrowed. Bracts numerous, lanceolate-subulate, deflexed; bracteoles several, connate, at least at base. Fruit 5–7 mm, yellow or brown. *Most of Europe, except the extreme north and south and the islands, particularly in mountainous regions and near habitations; doubtfully native; often cultivated as a herb.* [Al Au Be *Bu Cz Da Fe *Ga Ge He Ho *Hs Hu *It *Ju No Po Rm Rs (N, B, C, W, K, E) Su.] (*Iran.*)

82. Palimbia Besser[1]

Leaves 3-pinnate; lobes linear or setaceous. Sepals absent. Petals yellowish-white, oblong; apex obtuse, inflexed. Fruit ellipsoid-oblong, slightly compressed dorsally. Dorsal ridges not prominent; the lateral thick, prominent but not winged; vittae 3.

    1. P. rediviva (Pallas) Thell. in Hegi, *Ill. Fl. Mitteleur.* **5**(2): 1364 (1926). Perennial. Stem 40–90 cm; stock with fibrous remains of petioles. Lower leaves in a rosette; primary divisions opposite or verticillate; lobes 2–5 × 0·3–0·5 mm, linear or setaceous, mucronate, rigid; cauline leaves reduced to scale-like, amplexicaul petioles. Umbels numerous, forming a panicle; rays 3–20(–30), almost equal. Bracts and bracteoles 3–5, linear-lanceolate. Fruit 5–6·5 mm. *S. part of U.S.S.R., just extending to E. Romania.* Rm Rs (C, W, K, E).

83. Bonannia Guss.[1]

Like *Palimbia* but leaf-lobes oblong-lanceolate; petals yellow; fruit more strongly compressed dorsally; lateral ridges like the dorsal; vittae 3–4.

    1. B. graeca (L.) Halácsy, *Consp. Fl. Graec.* **1**: 641 (1901). Perennial 30–70 cm. Leaves nearly all basal, 2-pinnate; lobes acute, irregularly dentate or lobed, puberulent beneath; cauline leaves usually reduced to sheathing petioles. Rays 6–20. Bracts and bracteoles variable in number, short, linear. Fruit *c.* 6 mm, ovoid, pruinose. ● *Calabria and Sicilia, extending very locally to S. Greece and the Aegean region.* ?Cr Gr It Si.

84. Johrenia DC.[1]

Leaves 1- to 2-pinnate. Sepals absent or small. Petals yellow, oblong; apex involute, obtuse. Fruit oblong-ovoid, compressed dorsally. Ridges not winged, the lateral much thicker than the dorsal; vittae solitary in the ridges.

    1. J. distans (Griseb.) Halácsy, *Consp. Fl. Graec.* **1**: 634 (1901). Perennial up to 100 cm, branched from base. Lower leaves *c.* 10 × 5 cm, 2-pinnate; lobes 1–2 mm wide, linear-lanceolate, pinnatifid; upper leaves reduced to inflated petioles. Rays 3–5. Bracts absent; bracteoles few, setaceous. Fruit 3 mm. ● *Greece.* Gr.

    J. thessala Bornm., *Feddes Repert.* **28**: 37 (1930), was described from C. Greece (Thessalia), but the fruit is unknown, so it is not certain that it belongs to this genus. It has lower leaves *c.* 35 × 40 cm, with lobes *c.* 5 mm wide and entire, and otherwise resembles **1** in general appearance.

    J. selinoides Boiss. & Balansa in Boiss., *Diagn. Pl. Or. Nov.* **3**(5): 99 (1856), known otherwise only from S. Anatolia, was recorded by Velenovsky from Bulgaria, but the record appears to be erroneous.

85. Capnophyllum Gaertner[1]

Leaves 3-pinnate, resembling those of a *Fumaria*. Sepals absent or small. Petals white; apex involute. Fruit ovoid, somewhat compressed dorsally. Ridges very prominent, the lateral thicker than the dorsal and sometimes narrowly winged, all transversely rugose-scabrid; vittae solitary, in the ridges and sometimes between them.

    1. C. peregrinum (L.) Lange in Willk. & Lange, *Prodr. Fl. Hisp.* **3**: 33 (1874). Glabrous annual 10–50 cm. Stem solid, sulcate. Leaf-segments broadly triangular with entire or lobed, lanceolate to linear lobes. Rays 2–5. Bracts absent or few; bracteoles 4–6, shortly triangular. Fruit 4–6 mm. *W. Mediterranean region, southwards from c.* 41° *N., extending to S. Portugal and S.E. Italy.* ?Gr Hs It Lu Sa Si [Ga].

86. Ferula L.[2]

Perennial. Leaves 3- to 4-pinnate or -ternate, with usually linear lobes. Bracts absent; bracteoles absent or few. Sepals absent or minute. Petals yellow. Fruit elliptical or oblong-elliptical, strongly compressed dorsally, with thin lateral wings closely appressed to one another, and filiform or slightly carinate dorsal ridges. Vittae usually numerous.

    Literature: E. Korovin, *Generis* Ferula (*Tourn.*) L. *Monographia illustrata.* Taschkent. 1947.

1	Sheaths of uppermost leaves very large, situated close beneath and enveloping the developing umbels; rays of terminal umbel (15–)20–40	
2	Leaf-lobes up to 50 mm, without distinctly revolute margins	**1. communis**
2	Leaf-lobes not more than 10 mm, with distinctly revolute margins	**2. tingitana**
1	Sheaths of uppermost leaves not very large and conspicuous; rays of terminal umbel not more than 15	
3	Leaf-lobes up to 90 mm	**7. tatarica**
3	Leaf-lobes not more than 30 mm	
4	Leaf-lobes 10–30 mm	**3. sadlerana**
4	Leaf-lobes less than 10 mm	
5	Umbels with 9–15 rays, not proliferating	

       ¹ By T. G. Tutin.   ² By J. F. M. Cannon.

6 Stem up to 40 cm; sheaths of leaves narrow; fruit 6 mm
  5. nuda
6 Stem 80–150 cm; sheaths of leaves broad; fruit 10 mm
  6. orientalis

5 Umbels with 1–7 rays, proliferating
  7 Fruit 8–12 mm; robust plant up to 150 cm **4. heuffelii**
  7 Fruit 4–5(–9) mm; slender plant 30–60 cm **8. caspica**

1. F. communis L., *Sp. Pl.* 246 (1753). Stem up to 200 cm or more, very robust. Leaves with conspicuous sheathing bases; lamina finely divided, with linear lobes up to 50 × 0·5–3 mm; margins not distinctly revolute. Upper leaves of the inflorescence progressively reduced to conspicuous sheathing bases only. Terminal umbel more or less sessile, surrounded by smaller lateral umbels on long peduncles, which may themselves have secondary lateral umbels. Bracts 0; bracteoles few, linear-lanceolate, deciduous. Terminal umbels with 20–40 rays. Fruit *c*. 15 mm. *Mediterranean region.* Al Bl Co Cr Ga Gr Hs It Ju Sa Si Tu.

(a) Subsp. **communis**: Leaf-lobes not more than 1 mm wide, green on both surfaces. *Throughout the range of the species.*
(b) Subsp. **glauca** (L.) Rouy & Camus, *Fl. Fr.* 7: 398 (1901) (*F. glauca* L.): Leaf-lobes up to 3 mm wide, bright green above and distinctly glaucous beneath. *Almost throughout the range of the species.*

2. F. tingitana L., *Sp. Pl.* 247 (1753). Like **1** but leaf-lobes not more than 10 mm, with the margins distinctly revolute, the ultimate lobes partly united to one another. *S. & S.E. Spain; Portugal.* Hs Lu. (*North Africa.*)

3. F. sadlerana Ledeb., *Fl. Ross.* 2: 300 (1844). Stem up to 150(–200) cm, robust. Leaves finely divided, with conspicuous sheathing bases; lobes up to 30 × 3 mm, linear, somewhat scabrid on the margins and veins beneath. Umbels with up to 12 rays; partial umbels with 7–13 flowers. Bracts and bracteoles 0–1. Fruit 7–10 mm. 2*n* = 22. *Dry, rocky places. N. & C. Hungary, S.E. Czechoslovakia, W.C. Romania.* Cz Hu Rm.

4. F. heuffelii Griseb. ex Heuffel, *Flora (Regensb.)* 36: 623 (1853). Stem up to 150 cm, robust. Leaves with smaller sheathing bases than in **3**. Lamina finely divided; lobes 2–7(–9) × 1–1·5 mm, oblong-linear, at first somewhat pubescent, becoming glabrous. Umbels with 4–7 rays, proliferating; partial umbels with 10–15 flowers. Bracts and bracteoles 0. Fruit 8–12 mm. *Dry, rocky slopes; calcicole.* ● *W. Bulgaria, N.E. Jugoslavia, S.W. Romania.* Bu Ju Rm.

5. F. nuda Sprengel, *Sp. Umb.* 81 (1818). Stem up to 40 cm, relatively slender. Leaves glabrous, glaucous, 3- to 4-pinnate; lobes up to 3 mm, remote and deeply cut. Leaf-bases sheathing but narrow. Umbels with 10–15 rays; partial umbels with *c*. 15 flowers. Bracts and bracteoles 0. Fruit *c*. 6 mm. *Saline clays. S.E. Russia (Lower Volga).* Rs (E). (*W.C. Asia.*)

6. F. orientalis L., *Sp. Pl.* 247 (1753). Stem 80–150 cm. Leaves 4-pinnate; lobes *c*. 5 mm, narrowly linear. Leaf-bases well-developed and sheathing. Umbels with *c*. 10 rays; partial umbels with 10–15 flowers. Fruit *c*. 10 mm. *Dry grassland. S. Ukraine; one station in S.E. Bulgaria.* Bu Rs (W, K).

7. F. tatarica Fischer ex Sprengel, *Pugillus* 1: 27 (1813). Stem slender, less than 40 cm. Leaves 3-pinnate or -ternate; lobes up to 90 × 3 mm, narrowly linear. Umbels with 4–7 rays; partial umbels with *c*. 10 flowers. Bracts and bracteoles 0, or the latter

represented by a few minute scales. Fruit 8–9 mm. *Saline soils. E. Ukraine, S.E. Russia.* Rs (C, W, E).

8. F. caspica Bieb., *Fl. Taur.-Cauc.* 1: 220 (1808). Stem 30–60 cm. Leaves 3- to 4-pinnate, with short, closely crowded lobes. Umbels with (1–)3–6(–8) rays, proliferating; partial umbels with 8–10 flowers. Fruit 4–5(–9) mm. *Dry, saline soils. S. & E. Ukraine, S.E. Russia.* Rs (C, W, K, E).

87. Ferulago Koch[1]

Leaves 2- to 3(–4)-pinnate with filiform to linear-lanceolate lobes. Bracts and bracteoles well-developed and usually conspicuous. Sepals minute. Petals yellow. Fruit strongly compressed dorsally, with more or less well-developed lateral wings and filiform to distinctly winged dorsal ridges; vittae numerous.

1 Nodes very conspicuously swollen **1. nodosa**
1 Nodes not conspicuously swollen
  2 Leaves elliptical to almost linear in outline, widest near the middle
    3 Leaf-lobes linear; fruit with acute, filiform dorsal ridges **8. sylvatica**
    3 Leaf-lobes setaceous; fruit with conspicuously winged dorsal ridges **9. sartorii**
  2 Leaves triangular in outline, widest at the base
    4 Leaf-lobes elongate, linear **2. thyrsiflora**
    4 Leaf-lobes setaceous or short and linear
      5 Leaf-lobes finely setaceous; fruit ovate-lanceolate **3. asparagifolia**
      5 Leaf-lobes short, linear; fruit narrowly elliptic-oblong to narrowly obovate **(4–7). campestris** group

1. F. nodosa (L.) Boiss., *Diagn. Pl. Or. Nov.* 2(10): 37 (1849). Erect, glabrous perennial *c*. 60 cm. Stems very conspicuously swollen at the nodes. Leaves triangular in outline; lobes shortly linear. Bracts and bracteoles ovate-lanceolate. Fruit 8–10 mm, with narrow, somewhat undulate lateral wings and narrow dorsal wings. ● *Aegean region and S. part of Balkan peninsula, extending westward to Sicilia.* Al Cr Gr Si.

2. F. thyrsiflora (Sibth. & Sm.) Koch, *Nova Acta Acad. Leop.-Carol.* 12(1): 98 (1824). Robust, glabrous perennial. Leaves triangular in outline; lobes elongate, linear. Umbels closely aggregated into a complex, thyrsoid panicle. Bracts and bracteoles linear-lanceolate. Fruit *c*. 8 mm, narrowly elliptical, with narrow lateral wings and obtuse, filiform dorsal ridges. *Mountain rocks.* ● *Kriti.* Cr.

3. F. asparagifolia Boiss., *Ann. Sci. Nat.* ser. 3 (Bot.), 1: 321 (1844). Erect, glabrous perennial up to 250 cm. Leaves triangular-ovate in outline; lobes setaceous. Bracts and bracteoles lanceolate to linear, often caducous when fully mature. Fruit 9–12 × 5 mm, broadly ovate-lanceolate with narrow lateral wings and filiform dorsal ridges. *Damp, shady places. Karpathos.* Cr. (*S. & W. Anatolia*).

(4–7). F. campestris group. Perennial. Stem up to 200 cm. Leaves up to 60 cm, triangular-ovate in outline; lobes linear or filiform. Bracts and bracteoles usually lanceolate, conspicuous. Fruit 10–12 mm, elliptic-oblong to narrowly obovate; lateral wings very well developed; dorsal ridges slender, obtuse.

A complex of closely related species, all but one of which are confined in Europe to the Iberian peninsula.

1 Stem angled, strongly sulcate; plant dark green
  2 Leaf-lobes short; dorsal ridges of fruit narrowly winged **5. lutea**
  2 Leaf-lobes usually elongate; dorsal ridges of fruit filiform, unwinged **4. campestris**

1 Stem subterete, weakly sulcate; plant light green
  3 Leaf-lobes filiform; bracts setaceous, erect   **6. capillaris**
  3 Leaf-lobes linear; bracts ovate-lanceolate, becoming deflexed
          7. granatensis

4. F. campestris (Besser) Grec., *Consp. Fl. Roman.* 252 (1898) (*F. galbanifera* Koch, *F. nodiflora* sensu Thell., *Ferula ferulago* L.). Plant dark green; stem strongly angled and sulcate; leaf-lobes usually elongate. Dorsal ridges of fruit filiform, unwinged. *S.E. Europe, extending westwards to S. France and northwards to c. 51° N. in W. Russia.* Al Bu Ga It Ju Rm Rs (C, W, K, E) Si.

5. F. lutea (Poiret) Grande, *Bull. Orto Bot. Napoli* **4**: 366 (1914) (*Ferula sulcata* Desf.). Like **4** but leaf-lobes short; dorsal ridges of fruit narrowly winged. *N. Spain, N. Portugal.* Hs Lu.

6. F. capillaris (Link ex Sprengel) Coutinho, *Fl. Port.* 452 (1913). Plant light green; stem subterete, weakly sulcate; leaf-lobes filiform. Bracts setaceous, erect. ● *S. Portugal (near Tavira).* Lu.

7. F. granatensis Boiss., *Elenchus* 48 (1838) (*Ferula granatensis* (Boiss.) Nyman). Like **6** but leaf-lobes linear; bracts ovate-lanceolate, becoming deflexed. ● *Spain, N. Portugal.* Hs Lu.

Plants with shorter and wider leaf-lobes have been distinguished as *F. nodiflora* var. *brachyloba* (Boiss. & Reuter) Thell. (*Ferula brachyloba* (Boiss. & Reuter) Nyman, *Ferulago brachyloba* Boiss. & Reuter).

8. F. sylvatica (Besser) Reichenb., *Pl. Crit.* **4**: 53 (1826) (*F. confusa* Velen., *F. monticola* Boiss. & Heldr.). Glabrous perennial. Stem up to 125 cm. Leaves up to 50 cm, narrowly elliptical to nearly linear in outline, widest at the middle; lobes linear. Bracts and bracteoles conspicuous, ovate-lanceolate to lanceolate. Fruit 6–10 mm, elliptical; lateral wings well-developed; dorsal ridges acute, filiform. *S.E. Europe, extending to Hungary and Italy.* Al Bu Gr Hu It Ju Rm Rs (W) Tu.

Small plants with leaves more or less linear in outline have been wrongly identified as *F. meoides* (L.) Boiss. (*Lophosciadium meoides* (L.) Calestani), a synonym of *Ferula communis* L. They appear to represent one extreme of variation in **8** and to be indistinguishable from it in other respects.

9. F. sartorii Boiss., *Fl. Or.* **2**: 999 (1872). Erect, glabrous perennial *c.* 50 cm. Leaves narrowly elliptical in outline; lobes up to 30 mm, finely setaceous. Bracts and bracteoles ovate-lanceolate to lanceolate. Fruit *c.* 10 mm, elliptical; lateral wings conspicuous and slightly undulate at maturity; dorsal ridges with distinct narrow wings. ● *S.E. Greece, Kiklades.* Gr.

88. Eriosynaphe DC.[1]

Like *Ferula* but mericarps shortly and rather coarsely pubescent on commissural face; vittae solitary between the ridges, none on the commissural face.

1. E. longifolia (Fischer ex Sprengel) DC., *Coll. Mém.* **5**: 51 (1829). Glabrous perennial 50–70 cm. Leaves somewhat glaucous, 3-pinnate; lobes 30–80 × 1–4 mm, narrowly linear. Umbels terminal and lateral, arranged in complex, paniculate inflorescences. Bracts and bracteoles absent. Petals yellow; pedicels up to 25 mm, slender, unequal. Fruit 7–10 mm, elliptical. Dorsal ridges of fruit filiform; lateral wings moderately developed, rather thickened and corky. *Dry, usually calcareous slopes. S.E. Russia, S.E. Ukraine.* Rs (W, E).

---

[1] By J. F. M. Cannon.   [2] By T. G. Tutin.

89. Opopanax Koch[2]

Leaves pinnate, with stellate hairs beneath, segments entire or sometimes deeply lobed. Sepals absent. Petals yellow, ovate-oblong, involute. Fruit obovate to orbicular, strongly compressed dorsally. Lateral ridges united to form a border surrounding the fruit before dehiscence; dorsal ridges slender, low; vittae 2–3.

Fruit 6–7 mm, with a narrow, thickened border   **1. chironium**
Fruit 7–9 mm, with a wide, thin border   **2. hispidus**

1. O. chironium (L.) Koch, *Nova Acta Acad. Leop.-Carol.* **12**(1): 96 (1824). Robust perennial with a stout, solid stem up to 200 cm. Branches verticillate or subopposite, often very close below the terminal umbel. Lower leaves 2-pinnate, with stellate hairs beneath; lobes 4–12 cm, obliquely cordate to cuneate at base, crenate-serrate; rhachis sparsely hispid; upper leaves simple, or reduced to inflated petioles. Rays 9–25. Bracts and bracteoles few, setaceous. Fruit 6–7 mm, elliptical; border narrow, thickened, whitish. $2n = 22$. *W. & C. Mediterranean region, N. & C. parts of Balkan peninsula, just extending to S.E. Romania.* Al Bu Ga Gr Hs It Ju Rm Sa Si.

Plants from the Balkan peninsula tend to have larger leaflets with a more cuneate, decurrent base than those from the western part of the range of the species. Extreme variants have been called **O. bulgaricus** Velen., *Österr. Bot. Zeitschr.* **52**: 51 (1902).

2. O. hispidus (Friv.) Griseb., *Spicil. Fl. Rumel.* **1**: 378 (1843) (*O. orientalis* Boiss.). Like **1** but up to 300 cm; lobes usually 2–4 cm, ovate-lanceolate, hispid (like the rhachis); rays 6–13; fruit 7–9 mm, broadly elliptical; border wide, thin. *S. part of the Balkan peninsula and Aegean region; S. Italy and Sicilia.* Al Bu Cr Gr It Ju Si.

90. Peucedanum L.[2]

Leaves several times pinnate or ternate. Sepals absent or conspicuous. Petals white, yellow or rarely pink or purplish, broadly ovate; apex long, inflexed. Fruit strongly compressed dorsally. Lateral ridges winged; wings closely appressed to one another; dorsal ridges usually prominent; vittae 1–3.

1 Cauline leaves absent   **29. alpinum**
1 Cauline leaves present, the upper sometimes reduced to sheath-like petioles
  2 Lobes of at least the lower leaves linear or narrowly lanceolate, never serrate and usually unlobed; leaves never simply pinnate
    3 Stock without fibres; bracts 4–7, persistent   **23. lancifolium**
    3 Stock with fibres; bracts absent or few, usually caducous
      4 Leaves with at least 5 primary divisions
        5 Lobes of lower leaves narrowly linear; fruit 3–5 mm
          10. oligophyllum
        5 Lobes of lower leaves linear-lanceolate; fruit 6–7·5 mm
          13. schottii
      4 Leaves with 3 primary divisions
        6 Petals white or pink
          7 Lower leaves 3- to 4-ternate; rays 10–20: bracts usually 2–6, caducous   **7. gallicum**
          7 Lower leaves 1- to 2-ternate; rays 5–10; bracts 0(–1)
            8. aragonense
        6 Petals yellow
          8 Fruit $\frac{1}{4}$–$\frac{1}{3}$ as long as pedicel
          9 Leaf-lobes linear, flat, or with slightly recurved margins
            1. officinale
          9 Leaf-lobes filiform, canaliculate   **6. paniculatum**
        8 Fruit about as long as pedicel
          10 Distinct intermediate veins present in the leaf-lobes; leaves 2- to 3-ternate

11 Leaves (2–)3-ternate; intermediate veins conspicuous particularly in the middle of the leaf-lobe; rays very unequal, the longest 3–4 cm   **5. rochelianum**

11 Leaves 2-ternate; intermediate veins usually conspicuous at the base of leaf-lobe; rays nearly equal, not more than 2·5 cm   **9. coriaceum**

10 No distinct intermediate veins in leaf-lobes; leaves 3- to 6-ternate

12 Leaf-lobes 0·5(–1) mm wide, filiform; fruit 7–9 mm   **2. longifolium**

12 Leaf-lobes 1–3 mm wide, linear; fruit 4–7·5 mm

13 Leaves 4- to 6-ternate; lobes 10–45 mm; fruit 4–5(–6) mm   **3. tauricum**

13 Leaves 3- to 4-ternate; lobes 20–90 mm; fruit 6–7·5 mm   **4. ruthenicum**

2 Lobes of at least the lower leaves ovate or oblong in outline, dentate or lobed, or leaves simply pinnate

14 Leaf-lobes ovate, dentate, sometimes with 1–3 broad, dentate lobes

15 Bracts several, persistent; stock usually with fibres

16 Leaves 2-ternate; stock stout, with slender, far-creeping rhizome; bracts erecto-patent   **17. aegopodioides**

16 Leaves 1- to 3-pinnate; stock without slender, far-creeping rhizome; bracts deflexed

17 Leaves 2- to 3-pinnate   **25. cervaria**

17 Leaves 1-pinnate   **18. latifolium**

15 Bracts 0–2, caducous; stock without fibres

18 Stem solid; leaves simply pinnate and often 3-foliolate   **28. hispanicum**

18 Stem hollow; at least some leaves 2-ternate or 2- to 3-pinnate

19 Stem 30–100 cm, simple or with alternate branches; petioles of upper cauline leaves strongly inflated; fruit 4–5 mm   **26. ostruthium**

19 Stem 120–360 cm; upper branches opposite or whorled; petioles of cauline leaves rather narrow; fruit 7–9 mm   **27. verticillare**

14 Leaf-lobes ovate to oblong in outline, but ±pinnatifid; ultimate lobes entire or nearly so

20 Bracts 0, or few and caducous; upper leaves usually simply pinnate, with entire segments

21 Rays scabrid-puberulent on inner side   **12. carvifolia**

21 Rays smooth and glabrous on inner side

22 Stem distinctly sulcate, at least above   **13. schottii**

22 Stem terete and finely striate throughout

23 Stem c. 30 cm; petals white   **11. achaicum**

23 Stem up to 200 cm; petals yellow or yellowish

24 Lower leaves oblong in outline, 2-pinnate   **14. vittijugum**

24 Lower leaves triangular in outline, 3- to 4-pinnate   **16. arenarium**

20 Bracts several, persistent

25 Bracts deflexed

26 Stock with fibres; stalks of primary and secondary divisions of leaves patent or deflexed   **22. oreoselinum**

26 Stock without fibres; stalks of primary and secondary divisons of leaves erecto-patent

27 Stem hollow; wing of ripe fruit less than 1 mm wide   **24. palustre**

27 Stem solid; wing of ripe fruit at least 1·5 mm wide   **21. austriacum**

25 Bracts patent or appressed to rays

28 Rays 5–6, stout   **15. obtusifolium**

28 Rays (6–)10–25, slender

29 Rays glabrous; sepals at anthesis narrow; petals dull yellow; styles c. 1 mm   **19. alsaticum**

29 Rays puberulent; sepals at anthesis almost ovate; petals white; styles at least 1·5 mm   **20. venetum**

1. P. officinale L., *Sp. Pl.* 245 (1753). Glabrous perennial 60–200 cm. Stock often *c.* 5 cm in diameter, with abundant fibres. Stems terete or weakly angled, solid, branched above. Lower leaves 30–60 cm, 2- to 6-ternate; lobes 40–150 × (0·5–)1–3 mm, linear, narrowed at both ends, flat or with margins slightly recurved, with prominent midrib and weaker marginal veins. Rays 10–40. Umbels nodding in bud, later erect. Bracts 0–3, usually caducous; bracteoles several, setaceous. Petals yellow. Fruit 5–10 mm, narrowly elliptical to oblong-obovate, $\frac{1}{4}$–$\frac{1}{3}$ as long as pedicels; wing *c.* $\frac{1}{3}$ as wide as mericarp. $2n = 66$. *Usually in grassland.* ● *C. & S. Europe, extending north-westwards to S.E. England.* Al Au Br Bu Cz Ga Ge ?Gr Hs Hu It Ju Lu Rm.

(a) Subsp. **officinale**: Leaf-lobes not more than 100 mm; styles about as long as stylopodium; fruit elliptical or oblong-obovate. *Throughout the range of the species.*

(b) Subsp. **stenocarpum** (Boiss. & Reuter) Font Quer, *Fl. Cardó* 114 (1950) (*P. stenocarpum* Boiss. & Reuter): Leaf-lobes up to 150 mm; styles shorter than stylopodium; fruit narrowly elliptical or narrowly obovate. *C. & E. Spain.*

2. P. longifolium Waldst. & Kit., *Pl. Rar. Hung.* 3: 279 (1812). Like 1 but leaf-lobes rarely more than 1 mm wide, keeled; ripe fruit about as long as pedicels. *Dry rocky places.* ● *From Bosna and C. Romania to S. Albania and S. Bulgaria.* Al Bu Ju Rm.

3. P. tauricum Bieb., *Fl. Taur.-Cauc.* 1: 215 (1808). Glabrous perennial 40–100 cm. Stock *c.* 1 cm in diameter, with abundant fibres. Stem terete, striate, solid. Lower leaves *c.* 30 cm, (4–)5–6(–7)-ternate; lobes 10–45 × 1–2 mm, linear, acute or mucronate, with 3 prominent veins. Rays 7–28. Bracts 1–3, subulate; bracteoles several, filiform. Petals pale yellow. Fruit 4–5(–6) mm, broadly elliptical, about as long as pedicels; wing *c.* $\frac{1}{3}$ as wide as mericarp. *Dry hillsides and pinewoods.* ● *C. & E. Romania; Krym.* Rm Rs (K).

4. P. ruthenicum Bieb., *Fl. Taur.-Cauc.* 1: 215 (1808). Like 3 but leaves 3(–4)-ternate; lobes 20–90 mm; fruit 6–7·5 mm. *Dry places. S.E. Europe, from N.E. Bulgaria to the middle Volga.* Bu Rm Rs (C, W, E).

5. P. rochelianum Heuffel, *Österr. Bot. Zeitschr.* 8: 27 (1858). Like 3 but lower leaves (2–)3-ternate; lobes 30–70 × 1·5–2·5 mm, with 3 prominent veins and 2 less conspicuous ones between them. ● *C. & S.W. Romania, C. Jugoslavia.* ?Hu Ju Rm.

6. P. paniculatum Loisel., *Fl. Gall.* 722 (1807). Glabrous perennial *c.* 100 cm. Stock stout, with fibres. Stem terete, striate, solid. Lower leaves 3- to 5-ternate; lobes filiform, canaliculate. Umbels numerous; peduncles divaricate. Rays 10–20. Bracts 1–2, caducous; bracteoles several, subulate. Petals pale yellow. Fruit *c.* 10 mm, oblong-elliptical, shorter than pedicels; wing narrow. *Open woodland and scrub.* ● *Corse and Sardegna.* Co Sa.

7. P. gallicum Latourr., *Chlor. Lugd.* 7 (1785). Glabrous perennial 60–100 cm. Stock with abundant fibres. Stem terete, striate, solid. Lower leaves 3- to 4-ternate, the secondary divisions with usually 5 or more lobes; lobes widest below the middle; midrib and marginal veins with a network of much-branched, curved veins between them. Rays (7–)10–20, strongly papillose on inner side, rarely smooth. Bracts usually 2–6, caducous; bracteoles 4–8. Petals white or pink, densely but minutely papillose. Fruit 4–6 mm, elliptical, about as long as the pedicel; wing narrow, rather thin and translucent; dorsal ridges prominent. ● *W. & C. France, N.W. Spain, N. Portugal.* Ga Hs Lu.

8. P. aragonense Rouy & Camus, *Fl. Fr.* 7: 390 (1901). Like 7 but stock with few fibres; stem slender; lower leaves 1- to 2-ternate; rays 5–10, minutely papillose; bracts absent, rarely 1. ● *N. Spain (Oviedo to Teruel).* Hs.

9. P. coriaceum Reichenb., *Fl. Germ. Excurs.* 866 (1832). Like 7 but leaves 2-ternate; midrib and marginal veins of leaf-lobes with ascending simple branches, which unite to form secondary longitudinal veins which are usually conspicuous near the base of the lobe; bracts usually 0–3; rays 4–11, almost smooth; petals pale yellow to white, weakly papillose; wing of fruit thick, not or scarcely translucent; dorsal ridges low. *Grassland.* ● *W. Jugoslavia, just extending into N.E. Italy.* It Ju.

10. P. oligophyllum (Griseb.) Vandas, *Magyar Bot. Lapok* **4**: 110 (1905). Nearly glabrous perennial up to 60 cm. Stem somewhat angled, slender. Lower leaves 1- to 2-pinnate; lobes few, narrowly linear, acute; upper cauline leaves 3-fid or entire. Rays 6–10, glabrous or puberulent. Bracts absent; bracteoles 3–5, filiform. Petals white or pink. Fruit 3–5 mm. *Mountain grassland.* ● *Albania, Jugoslavia, W. Bulgaria.* Al Bu Ju.

(a) Subsp. **oligophyllum**: Stem up to 60 cm, usually branched below the middle. Umbels 2–4; rays 1–5 cm, very unequal, glabrous. Peduncle glabrous. Petals white. *Albania and S. Jugoslavia.*

(b) Subsp. **aequiradium** (Velen.) Tutin, *Feddes Repert.* **79**: 62 (1968) (*P. aequiradium* Velen.): Stem up to 40 cm, simple or with one branch in the upper part. Umbel 1(–2); rays 0·5–2 cm, subequal, puberulent on the upper surface. Peduncle puberulent. Petals pink. *N. Jugoslavia and W. Bulgaria.*

11. P. achaicum Halácsy, *Consp. Fl. Graec., Suppl.* 42 (1908). Glabrous perennial *c.* 30 cm. Stem terete, striate, with few leaves. Lower leaves pinnate, oblong in outline; segments 10–13 mm, sessile, ovate in outline, distant, pinnately lobed; lobes 1- to 3-dentate; upper leaves small, with lanceolate, undivided segments. Rays 8–13, smooth. Bracts absent; bracteoles 1–3. Petals white. Fruit ovate; wing *c.* ½ as wide as mericarp; dorsal ridges somewhat prominent. ● *S. Greece (by the Vouraikos, Akhaia).* Gr.

12. P. carvifolia Vill., *Prosp. Pl. Dauph.* 25 (1779) (*P. chabraei* (Jacq.) Reichenb.; incl. *P. podolicum* (Besser) Eichw.). Almost glabrous perennial 30–100 cm. Stock with fibres. Stem sulcate, at least above, solid, branched. Basal leaves 30–40 cm, oblong in outline, shining on both surfaces, 1-pinnate, the segments pinnatisect; lobes 15–30 mm, usually 2- to 3-fid; ultimate divisions linear-oblong, their margins rough; venation reticulate; upper cauline leaves pinnate; segments linear-lanceolate, often falcate, acuminate, entire; petioles oblong-lanceolate. Rays 6–18, very unequal, papillose-puberulent on the inner side. Bracts absent; bracteoles 1 to few, subulate. Petals yellowish or greenish-white. Styles *c.* 0·5 mm, scarcely as long as stylopodium. Fruit 4–5 mm, broadly elliptical; wing ⅙–¼ the width of the mericarp, translucent; dorsal ridges filiform, rather prominent. *S. & C. Europe, extending north-westwards to the Netherlands and eastwards to S. Russia.* Au Be Bu Cz Ga Ge He Ho Hs Hu It Ju Rm Rs (C, W, E) Si.

13. P. schottii Besser ex DC., *Prodr.* **4**: 178 (1830). Like **12** but leaves dull, somewhat glaucous; rays glabrous and smooth; bracts and bracteoles sometimes several, caducous; petals white or pinkish in bud; styles often up to 2 mm and 2–3 times as long as stylopodium; fruit 6–7·5 mm. $2n = 22$. *Rocky slopes; calcicole.* ● *From S.E. France to C. Jugoslavia; one station in W. Ukraine.* Al Ga It Ju Rs (W).

Perhaps not specifically distinct from **12**.

14. P. vittijugum Boiss., *Fl. Or.* **2**: 1018 (1872). Usually glabrous perennial or biennial up to 140 cm. Stem terete, striate, solid. Lower leaves 2-pinnate, oblong in outline; lobes 7– 10(–15) mm, ovate in outline, pinnately divided, mucronate, with reticulate venation; upper cauline leaves pinnate with long, linear, entire or slightly divided segments. Rays 10–20, very unequal, glabrous. Bracts absent; bracteoles 2–3, linear. Petals yellow. Fruit 5–6 mm, oblong; wing *c.* ⅓ width of mericarp; dorsal ridges filiform. ● *Balkan peninsula.* Al Bu Gr Ju.

P. minutifolium (Janka) Velen., *Fl. Bulg.*, *Suppl.* 122 (1898) is probably conspecific with **14**. It has puberulent stems, nearly equal rays, 6–7 bracteoles and broadly winged fruit.

15. P. obtusifolium Sibth. & Sm., *Fl. Graec. Prodr.* **1**: 189 (1806). Glabrous perennial *c.* 40 cm. Stock with fibres. Stem terete, solid, branched from base. Lower leaves 2-pinnate, narrowly triangular in outline; lobes up to 20 mm, ovate-oblong in outline, deeply dentate or pinnately lobed; cauline leaves similar but smaller. Rays 5–6, stout. Bracts and bracteoles several, short, linear-lanceolate, deflexed. Petals yellowish. Fruit *c.* 12 mm, ovate or obovate; wing more than half as wide as mericarp, rather thick; dorsal ridges filiform. *Maritime sands.* E. *Greece, Turkey.* Gr Tu.

16. P. arenarium Waldst. & Kit., *Pl. Rar. Hung.* **1**: 18 (1800) (incl. *P. borysthenicum* Klokov). Glabrous perennial 90–150 (–200) cm. Stock with fibres. Stem terete, striate, solid, branched above. Lower leaves up to 45 cm, 3- to 4-pinnate; lobes 4–14 (–20) mm, ovate in outline, divided into linear or narrowly oblanceolate, mucronate to shortly acuminate, entire or 2- to 3-fid lobes; upper cauline leaves reduced to the inflated petiole. Rays 2–14, usually very unequal. Bracts 0–2, lanceolate; bracteoles several, linear-lanceolate. Petals yellowish. Fruit 6–9 mm, elliptical; wing *c.* ¼ the width of the mericarp; dorsal ridges conspicuous. *Dry, usually sandy ground. From S. Czechoslovakia to Albania and S.E. Russia.* Al Bu Cz Hu Ju Rm Rs (W, E).

(a) Subsp. **arenarium**: Leaf-lobes narrowly oblanceolate; rays 5–14; pedicel 1–2 times as long as fruit. *Throughout the range of the species, except for Albania and S. & W. Jugoslavia.*

(b) Subsp. **neumayeri** (Vis.) Stoj. & Stefanov, *Fl. Bălg.* ed. 3, 857 (1948) (*P. neumayeri* (Vis.) Reichenb. fil.): Leaf-lobes linear; rays 2–6; pedicel *c.* 4 times as long as fruit. ● *S. & W. Jugoslavia, Albania, S. Bulgaria.*

17. P. aegopodioides (Boiss.) Vandas, *Sitz.-Ber. Böhm. Ges. Wiss.* (*Math.-Nat. Kl.*) **1888**: 449 (1889). Glabrous perennial up to 100 cm. Stock stout; rhizome long, creeping. Stem somewhat angled, hollow, little-branched. Lower leaves 2-ternate; lobes 50–70 × 20–40 mm, ovate or oblong, dentate, truncate or cuneate at base; upper cauline leaves with inflated petioles and 3–5 small segments. Rays 10–25. Bracts (0–)3–6, linear-subulate; bracteoles numerous, as long as pedicels. Petals white or pink. Fruit 5–7 mm, suborbicular; wing 1·5–2 mm wide, somewhat translucent; dorsal ridges filiform. *Beside mountain streams.* ● *Balkan peninsula.* Al Bu Gr Ju.

18. P. latifolium (Bieb.) DC., *Prodr.* **4**: 181 (1830) (incl. *P. macrophyllum* Schischkin). Almost glabrous perennial 40–100 cm. Stock with fibres. Stem terete, striate, somewhat angled above, solid, branched. Leaves simply pinnate; segments 30–60 mm, few, elliptical or ovate, serrate, somewhat fleshy; upper cauline leaves very small, 3-foliolate, with inflated petioles. Rays 10–21, glabrous or puberulent on the inner side. Bracts 4–6, lanceolate or linear-lanceolate, deflexed; bracteoles 5–7, linear. Petals white. Fruit 5–7 mm, elliptical; wing *c.* 0·5 mm wide; dorsal ridges filiform, inconspicuous. *Damp places. From N. Italy to S.E. Russia.* ?Al It Ju Rm Rs (C, W, E).

19. **P. alsaticum** L., *Sp. Pl.* ed. 2, 354 (1762) (*Johrenia pichleri* Boiss.; incl. *P. lubimenkoanum* Kotov). Glabrous, purplish perennial 30–180 cm. Stock stout, branched, with fibres. Stem terete, striate below, somewhat angled above, hollow; branches numerous, short, often whorled. Lower leaves 2- to 4-pinnate, triangular in outline; lobes 10–30 mm, ovate-cuneate in outline and usually lobed, scaberulous on margins and on veins beneath; apex with a small cartilaginous point; petiole and rhachis strongly sulcate on upper surface; upper cauline leaves very small, with broad, auriculate petioles. Rays 6–25, usually 1–2 cm, smooth or slightly rough. Bracts and bracteoles 4–8, persistent, lanceolate to linear-lanceolate, acuminate, not deflexed; margins membranous. Sepals narrow. Petals dull yellow, with few, obtuse, microscopic papillae. Fruit 3–4·5 mm, oblong-elliptical; wing *c.* 0·5 mm wide, rather thin and translucent; dorsal ridges wide. $2n = 22$. ● *C. & S.E. Europe, extending to W. France and C. Russia.* Al Au Bu Cz Ga Ge Hu Ju Po Rm Rs (C, W, K, E).

20. **P. venetum** (Sprengel) Koch, *Syn. Fl. Germ.* 305 (1835). Like **19** but rays denticulate and puberulent along the whole of the inner side; sepals broad, ovate; petals white, with many acute, microscopic papillae; fruit 5·5–6 mm; wing of fruit 0·7–1 mm wide. ● *E. Pyrenees; S. Alps, Appennini.* Ga He Hs It Ju.

21. **P. austriacum** (Jacq.) Koch, *Nova Acta Acad. Leop.-Carol.* **12**(1): 94 (1824). Glabrous or (var. *velutinum* Levier) puberulent perennial 30–120 cm. Stock without fibres. Stem sulcate, solid, branched above. Lower leaves 3- to 4-pinnate; lobes ovate in outline, at least some stalked, deeply pinnately divided into oblong or (var. *rablense* (Wulfen) Koch) linear to linear-lanceolate lobes with cartilaginous apex; upper cauline leaves less divided, with oblong, auricled petioles. Rays 15–40, papillose on the inner side. Bracts and bracteoles numerous, linear-lanceolate to subulate, ciliate, deflexed. Sepals ovate-lanceolate, acute. Petals white, with very sparse, obtuse papillae. Fruit 6–9 mm, elliptical to nearly oblong; wing 1·5–2·5 mm wide, thin; dorsal ridges narrow but prominent. ● *Balkan peninsula and S.C. Europe, extending to E. France.* Al Au Bu Ga Gr He It Ju Rm.

22. **P. oreoselinum** (L.) Moench, *Meth.* 82 (1794). Nearly glabrous, often reddish perennial 30–100 cm. Stock with fibres. Stem terete, striate, solid, glabrous or puberulent. Lower leaves up to 40 cm, 2- to 3-pinnate, triangular in outline, somewhat coriaceous; primary and secondary divisions at right angles to the axis or deflexed, the stalks often flexuous; lobes usually 10–30 mm, ovate in outline, pinnately lobed, their stalks narrowly winged; ultimate lobes obovate-cuneate or rhombic, dentate, apiculate; margins thickened, scabrid; upper cauline leaves less divided, their petioles somewhat inflated. Rays 15–30, puberulent on the inner side. Bracts and bracteoles usually numerous, linear-lanceolate, deflexed. Sepals ovate, subobtuse. Petals white or pinkish, papillose. Fruit 5–8 mm, broadly obovate; wing *c.* 0·75 mm wide, rather thick; dorsal ridges slender, scarcely prominent. $2n = 22$. *Much of Europe, but absent from the islands, the extreme south, and most of the north.* Au Cz †Da Ga Ge He Hs Hu It Ju Lu Po Rm Rs (B, C, W, E) Su.

P. bourgaei Lange in Willk. & Lange, *Prodr. Fl. Hisp.* **3**: 42 (1874), from C. Spain and C. Portugal, is probably a variant of **22**, but ripe fruit appears to be still unknown. It is larger in all its parts, and has broad leaf-lobes which are oblique and unequally lobed.

23. **P. lancifolium** Lange, *Vid. Meddel. Dansk Naturh. Foren. Kjøbenhavn* **1865**: 39 (1866). Glabrous perennial 80–120 cm. Stock without fibres. Stem terete, striate, hollow. Lower leaves 2- to 3-pinnate, triangular in outline; lobes (10–)40–60(–100) mm,

linear, entire; upper cauline leaves pinnate or simple. Rays 5–12, very unequal, puberulent on the inner side. Bracts and bracteoles 4–7, linear, deflexed. Sepals very short, obtuse. Petals white or pale yellowish-white, sometimes suffused with pink. Fruit 5–6 mm, elliptical; wing 0·5–1 mm wide, rather thick; dorsal ridges thick, prominent. *Wet meadows and marshes.* ● *N.W. France, N.W. Spain, W. Portugal.* Ga Hs Lu.

24. **P. palustre** (L.) Moench, *Meth.* 82 (1794). Almost glabrous biennial up to 160 cm. Stock without fibres. Stem sulcate, hollow. often purplish. Lower leaves 2- to 4-pinnate, triangular in outline; petiole strongly canaliculate above, often puberulent beneath; lobes 5–20 mm, ovate in outline, pinnately lobed; ultimate lobes linear or oblong, entire or 2- to 3-fid, with an obtuse cartilaginous apex. Rays 20–40, puberulent on the inner side. Bracts and bracteoles 4 or more, very unequal, lanceolate, sometimes 2- to 3-fid, deflexed. Sepals ovate, obtuse. Petals white, papillose-puberulent above. Fruit 4–5 mm, elliptical; wing 0·5–0·75 mm wide, thick; dorsal ridges wide and prominent; commissural vittae concealed by pericarp. *Wet places. Europe, except the south-west, the extreme south and most of the islands.* Au Be Br Bu Cz Da Fe Ga Ge He Ho Hu It Ju No Po Rm Rs (N, B, C, W, E) Su.

25. **P. cervaria** (L.) Lapeyr., *Hist. Abr. Pyr.* 149 (1813). Nearly glabrous perennial 30–150 cm. Stock with fibres. Stem terete, striate, shallowly sulcate above, with few small cauline leaves. Lower leaves up to 50 cm, (1–)2- to 3-pinnate, triangular in outline; lobes up to 50 mm, ovate to ovate-oblong, obtuse or shortly acuminate, cuneate to cordate and often unequal at base, the lower often more or less deeply 1- to 4-lobed; margins dentate, the teeth and apex with sharp, awn-like points; network of veins conspicuous on both surfaces. Rays 15–30, puberulent on the inner side. Bracts numerous, unequal, the larger often pinnatisect, deflexed; bracteoles numerous, subulate, deflexed. Sepals ovate, acute. Petals white, papillose above. Fruit 4–9 mm, elliptical to suborbicular; wing 0·5–1 mm wide, rather thick; dorsal ridges scarcely prominent. *C. Europe, extending to N. Spain, C. Italy, Albania, E. Ukraine and Lithuania.* Al Au Be Bu Cz Ga Ge He Hs Hu It Ju Po Rm Rs (B, C, W).

26. **P. ostruthium** (L.) Koch, *Nova Acta Acad. Leop.-Carol.* **12**(1): 95 (1824). Almost glabrous perennial 30–100 cm. Rhizome stout, branched. Stem terete, striate, hollow, simple or branched above. Lower leaves up to 30 × 34 cm, usually 2-ternate, triangular in outline; lobes usually 50–100 × 40–70 mm, ovate to lanceolate, acuminate, irregularly dentate; middle lobe sometimes again 3-lobed, lateral ones sometimes 2-lobed; margins and often the veins on the lower surface ciliolate; petioles of cauline leaves strongly inflated. Peduncle puberulent. Rays usually 30–60, puberulent on the inner side. Bracts 0(–1); bracteoles few, setaceous. Petals white or pinkish. Fruit 4–5 mm, suborbicular; wing nearly as wide as mericarp. *Mountain meadows, woods, stream-banks and rocky places.* ● *Mountains of C. & S. Europe, from the Sudety to S. Spain and C. Italy; often naturalized elsewhere from cultivation.* Au Co Cz Ga Ge He Hs It Ju Po ?Rm ?Rs (K) [Be Br Da No Rs (C) Su].

27. **P. verticillare** (L.) Koch ex DC., *Prodr.* **4**: 181 (1830) (*P. altissimum* (Miller) Thell., non Desf., *Tommasinia altissima* (Miller) Thell.). Monocarpic, usually living 4–5 years. Stock without fibres. Stem 120–360 cm, up to 5 cm in diameter, terete, striate, hollow; lower branches alternate, upper opposite or whorled. Lower leaves 30–80 cm, ternately 2- to 3-pinnate, triangular in outline; lobes 25–80 × 20–60 mm, ovate or ovate-oblong, irregularly serrate-dentate or 3-lobed; teeth with short cartilaginous points; petioles of lower leaves puberulent; petioles of upper cauline leaves rather narrow, sheath-like, with small

lamina. Terminal umbel large; lateral umbels smaller, whorled, with male flowers only; rays 10–30, sparsely puberulent at top. Bracts 0–1; bracteoles few, filiform. Sepals triangular. Petals greenish-yellow. Fruit 7–9 mm; wing 2–2·5 mm wide; dorsal ridges prominent. ● *E. & E.C. Alps, extending to Hungary, N. Jugoslavia and C. Italy.* Au He Hu It Ju.

28. P. hispanicum (Boiss.) Endl. in Walpers, *Repert. Bot. Syst.* **2**: 411 (1843). Glabrous perennial 60–100 cm. Stock stout, without fibres. Stem terete, striate, solid, simple or little-branched. Lower leaves simply pinnate, often 3-foliolate; segments 30–100 × 20–50 mm, broadly ovate, dentate, the terminal one sometimes lobed; upper cauline leaves with inflated petioles and often no lamina. Rays 15–40, scabrid-puberulent on the inner side. Bracts 0–2; bracteoles several, setaceous. Petals white. Fruit 3·5–5 mm, elliptical; wing *c.* 1 mm wide; dorsal ridges narrow, prominent. *Damp places. S. & E. Spain.* Hs.

29. P. alpinum (Sieber ex Schultes) B. L. Burtt & P. H. Davis, *Kew Bull.* **1949**: 227 (1949) (*P. creticum* Sprengel). Glabrous perennial up to 15 cm. Stem simple or with one branch. Leaves 2–4 cm, all basal, 3-foliolate or more or less deeply 3-lobed; segments suborbicular-cuneate, dentate, the lateral sessile, the terminal stalked. Rays 3–5, smooth and glabrous, unequal. Bracts and bracteoles 0. Petals greenish-white. Fruit 6–7 mm. *Mountains.* ● *Kriti.* Cr.

91. Pastinaca L.[1]

Leaves simple or pinnate; segments sometimes deeply lobed. Sepals absent. Petals yellow, ovate, incurved. Fruit elliptical, strongly compressed dorsally; lateral ridges winged, the vascular bundles near the outer edge of the wing; vittae 1, rarely 2.

1 Bracts and bracteoles several   **4. hirsuta**
1 Bracts and bracteoles 0, rarely 1–2, caducous
2   Basal leaves simple   **3. lucida**
2   Basal leaves pinnate
3   Petioles of lower leaves slender, not spongy; secondary veins
     of lamina inconspicuous; vittae on the commissural face
     stopping short of the ends of the fruit   **1. sativa**
3   Petioles of lower leaves up to 1 cm in diameter, spongy;
     secondary veins of lamina very conspicuous; vittae on the
     commissural face reaching the ends of the fruit   **2. latifolia**

1. P. sativa L., *Sp. Pl.* 262 (1753). More or less pubescent biennial up to 100 cm. Stem hollow or solid, angled or terete. Basal leaves usually simply pinnate, rarely (subsp. *sativa* var. *fleischmannii* (Hladnik) Burnat) 2-pinnatisect; secondary veins inconspicuous; segments (2–)5–11, acute or obtuse, crenate-dentate, the teeth with a cartilaginous mucro; petioles slender, not spongy. Rays 5–20, more or less angled. Bracts and bracteoles 0–2, caducous. Fruit 5–7 mm, broadly elliptical; wing 0·25–0·5 mm wide; vittae on the commissural face not reaching the ends of the fruit. *Most of Europe except the Arctic, but only as an escape from cultivation in parts of the north.* Al Au Be Br Bu Co Cz Ga Ge Gr He Ho Hs Hu It Ju Po Rm Rs (*B, C, W, K, E) Sa ?Si Tu [Da Fe Hb Lu No Rs (N) Su].

1 Stem, petiole, upper surface of leaves and rays with sparse,
    short, straight hairs; leaf-segments often narrow and acute
    or acuminate, often cuneate at base and ±pinnatisect in
    lower part; stem usually strongly angled   (a) subsp. **sativa**
1 Stem, petiole, upper surface of leaves and rays ±grey-hairy;
    hairs on stem usually long and flexuous; leaf-segments usually

broad, obtuse, subcordate at base, crenate-dentate or shallowly lobed in lower part; stem angled or terete
2 Stem angled   (b) subsp. **sylvestris**
2 Stem terete
3   Terminal umbel not or little larger than the others; rays
    usually 5–7, nearly equal; longest about twice as long as
    partial umbel   (c) subsp. **urens**
3   Terminal umbel much larger than lateral; rays usually
    10–12; longest 3–4 times as long as partial umbel
                   (d) subsp. **divaricata**

(a) Subsp. **sativa** (incl. *P. fleischmannii* Hladnik): Terminal umbel usually with 9–20 very unequal rays; hairs on stem short, straight. $2n=22$. *Widely cultivated for its edible root and often naturalized; almost throughout the range of the species.*

(b) Subsp. **sylvestris** (Miller) Rouy & Camus, *Fl. Fr.* **7**: 372 (1901) (*P. sylvestris* Miller): Terminal umbel usually with 9–20 very unequal rays; hairs on stem soft and flexuous. *C. & W. Europe.*

(c) Subsp. **urens** (Req. ex Godron) Čelak., *Prodr. Fl. Böhm.* 574 (1875) (*P. teretiuscula* Boiss., *P. urens* Req. ex Godron, *P. umbrosa* Steven ex DC.): Hairs usually short and stem often glabrescent. *S., C. & E. Europe.*

(d) Subsp. **divaricata** (Desf.) Rouy & Camus, *Fl. Fr.* **7**: 374 (1901) (*P. divaricata* Desf.): Plant more or less grey-hairy. ● *Corse and Sardegna.*

2. P. latifolia (Duby) DC., *Prodr.* **4**: 189 (1830). Like **1** but commonly up to 200 cm; petioles of lower leaves up to 1 cm in diameter, spongy; lamina with a conspicuous network of secondary veins; segments 3–7; vittae on commissural face running the entire length of the fruit. *River-banks and rocky places.* ● *Corse.* Co.

3. P. lucida L., *Mantissa* 58 (1767). Like **1** but leaves shining above, finely serrate; basal simple, cordate; lower cauline larger, pinnate, with 3–7 entire or shallowly lobed segments, the terminal one 3-lobed and cordate; upper cauline entire, subrhombic. ● *Islas Baleares.* Bl.

4. P. hirsuta Pančić, *Fl. Princ. Serb.* 359 (1874). Perennial with little-branched, angled stem. Leaves pinnate, shortly hispid; segments ovate, lobed and coarsely serrate. Rays 10–16. Bracts and bracteoles several, linear. Fruit 5 mm, elliptical. *Mountain meadows.* ● *E. Jugoslavia, S. & W. Bulgaria.* Bu Ju.

92. Heracleum L.[2]

Leaves simple, rarely subentire, usually apparently digitately lobed, or ternatisect, pinnate or 2-pinnate. Calyx-teeth small. Petals white, greenish-yellow or pink. Fruit elliptical or obovate to suborbicular, strongly compressed dorsally, glabrous to villous. Marginal ridges forming a broad wing; dorsal ridges slender; vittae solitary in the furrows, slender or conspicuously swollen at their lower ends, shorter than the fruit.

Zosima absinthiifolia (Vent.) Link, *Enum. Hort. Berol. Alt.* **1**: 274 (1821), has been recorded from the Lower Volga, but the basis of the record is unknown and it is unlikely that a mountain species such as this would occur there. It is like *Heracleum* but has conspicuous sepals and the fruit has a wide, hyaline wing around which the lateral ridges form a thickened rim.

1 Stems less than 4 mm in diameter, not more than 50(–65) cm
    high; umbels not more than 7(–9) cm in diameter, with
    3–15(–17) rays

---

         [1] By T. G. Tutin.   [2] By R. K. Brummitt.

2 Leaves glabrous   **1. minimum**
2 Leaves pubescent, at least beneath
  3 Leaves pinnatisect, not suborbicular in outline, with 2–3 pairs of segments   **2. austriacum**
  3 Leaves simple or ternatisect, ± suborbicular in outline
    4 Leaves 3-lobed or ternatisect; stem often glandular-pubescent   **3. orphanidis**
    4 Leaves palmately 5- to 7-lobed or almost entire; stem not glandular-pubescent   **4. carpaticum**
1 Stems at least 4 mm in diameter, usually more than 50 cm high; at least the larger umbels more than 10 cm in diameter, with 12–150 rays
  5 Leaves pinnately divided into small segments not more than 3 cm; rhachis and petiolules slender   **6. ligusticifolium**
  5 Leaves simple, or divided into segments more than 3 cm; rhachis and petiolules stout
    6 Vittae not or only slightly expanded at their lower end, not more than 0·4 mm wide   **5. sphondylium**
    6 Vittae conspicuously expanded at their lower end, up to 1 mm wide
      7 Leaves simple, densely white-tomentose beneath   **7. stevenii**
      7 Leaves ternatisect or pinnatisect, not white-tomentose beneath
        8 Plant 60–80 cm; umbels 10–12 cm in diameter; rays 15–20   **8. pubescens**
        8 Plant (170–)200–500 cm; larger umbels 20–50 cm in diameter; rays more than 50   **9. mantegazzianum**

1. H. minimum Lam., *Fl. Fr.* 3: 413 (1778). Rhizomatous perennial. Stem up to 30 cm, 1–2 mm in diameter, glabrous, or sometimes puberulent, above. Leaves resembling those of *Corydalis bulbosa*, glabrous, 2-pinnatisect or 2-ternatisect; segments stipitate, divided into obovate or oblanceolate lobes up to 10 × 3 mm. Umbels up to 6(–9) cm in diameter; rays 3–6, glabrous or puberulent; bracts and bracteoles usually 0. Petals white or pinkish. Fruit 8–10 × 6–8·5 mm, broadly elliptical to suborbicular; vittae absent or very slender and inconspicuous. *Calcareous screes.* ● *Mountains of S.E. France.* Ga.

2. H. austriacum L., *Sp. Pl.* 249 (1753). Rhizomatous perennial. Stem up to 40(–65) cm, 3 mm in diameter, glabrous below, usually pubescent above. Leaves resembling those of *Pastinaca sativa*, pinnate with 2–3 pairs of segments; petiole, rhachis and both surfaces of segments rather sparsely pubescent, rarely glabrous; segments sessile (rarely the lowermost shortly stipitate), ovate, sometimes lobed near the base, serrate or crenate. Umbels up to 9 cm in diameter; rays (5–)6–13(–15), pubescent or puberulent; bracts and bracteoles usually present, linear or triangular. Fruit 7–11 × 6–9 mm, usually broadly obovate; vittae slender. 2*n* = 22. ● *E. Alps; one station in C. Switzerland.* Au Ge He It Ju.

(a) Subsp. **austriacum**: Rays (5–)7–13(–15). Petals white, the outer (3–)5–8(–8·5) mm, with each lobe 0·9–2(–2·3) mm wide. *N.E. Alps, C. Switzerland.*

(b) Subsp. **siifolium** (Scop.) Nyman, *Consp.* 290 (1879): Rays (5–)6–10(–12). Petals pink or red, the outer ones (5–)6–10(–12) mm with each lobe (1·3–)1·8–3·5(–4) mm wide. *S.E. Alps.*

3. H. orphanidis Boiss., *Fl. Or.* 2: 1041 (1872). Stem up to 50 cm, up to 3 mm in diameter, pubescent and sometimes glandular. Leaves simple and 3-lobed, or ternatisect, rather coarsely crenate. Umbels *c.* 5 cm in diameter, with 6–9 rays. Outer petals of each umbel *c.* 4 mm, somewhat larger than the inner. Fruit *c.* 10 × 9 mm, broadly obovate to suborbicular; vittae slender. ● *Macedonia.* ?Gr Ju.

4. H. carpaticum Porc., *Fl. Naseud.* 144 (1881). Rhizomatous perennial. Stem 20–30(–50) cm, up to 3 mm in diameter, with patent hairs. Leaves suborbicular in outline, undivided to deeply palmately 5- to 7-lobed, cordate, crenate to serrate. Umbels 5–7 cm in diameter, with 5–15(–17) rays. Outer petals of each umbel *c.* 4 mm. Fruit 6–7 × 5 mm, broadly obovate to suborbicular; vittae rather slender. ● *E. Carpathians.* Rm Rs (W).

5. H. sphondylium L., *Sp. Pl.* 249 (1753). Biennial or short-lived perennial. Stem up to 250(–350) cm, 4–20 mm in diameter, glabrous to strongly hispid. Leaves varying from simple and shallowly palmately lobed to pinnate with 5(–9) crenate to serrate segments, sparsely pubescent (at least on the veins) to densely hispid or softly white-tomentose beneath, subglabrous to more or less sparsely hispid above. Umbels up to 20 cm in diameter, with 15–45 rays; bracts few or 0; bracteoles usually present. Ovary glabrous to pubescent or hispid; fruit (5–)7–10(–12) × (4–)6–8(–10) mm, elliptical or obovate to suborbicular, glabrous; vittae rather slender, up to 0·4 mm wide. 2*n* = 22. *Throughout Europe except the extreme north and much of the Mediterranean region.* All except Az Bl Co Cr Fa Is Sa Sb.

An extremely variable species, many variants of which have been given specific rank. The main geographical variants are recognized below as subspecies, but there appears to be considerable intergradation between them. Investigation of taxonomy and distribution is hampered by inadequacy of available herbarium material, and good annotated specimens are required. Some variants having short setae on the ovary may also merit subspecific rank, but further investigation is required.

Variants with leaf-lobes very long and narrow occur in several subspecies and are probably best regarded as formae (see Gawłowska, *Fragm. Fl. Geobot.* 7: 3–39 (1961)).

1 Petals white or rarely pinkish, outer radiate
  2 Leaves ternate or pinnate, at least the terminal segment with a distinct petiolule
    3 Larger cauline leaves usually ternate   (d) subsp. **montanum**
    3 Larger cauline leaves nearly always with 5(–9) segments
      4 Upper branches usually not whorled; leaves usually ± coarsely hispid or pubescent beneath   (e) subsp. **sphondylium**
      4 Upper branches whorled; leaves softly villous beneath, with long, soft hairs particularly on the veins   (f) subsp. **verticillatum**
  2 Leaves not divided into separate segments, though often deeply lobed and appearing ± digitate
    5 Lower leaves lobed to ⅓ or less, the lobes usually rounded; commissural vittae very slender or 0   (a) subsp. **alpinum**
    5 Lower leaves lobed to ½ or more, the lobes usually acute or acuminate; commissural vittae conspicuous
      6 Plant usually less than 130 cm; leaves usually less than 30 cm, with 5–7 usually acute lobes, which are simple or with few small secondary lobes   (b) subsp. **pyrenaicum**
      6 Plant more than 130 cm, robust; leaves up to 50 cm with 7–9 ± long, acute lobes, which are usually again conspicuously acutely lobed   (c) subsp. **transsilvanicum**
1 Petals ± greenish or yellowish, outer not or scarcely radiate
  7 Leaves not divided into segments, though often deeply lobed and appearing digitate   (g) subsp. **orsinii**
  7 Leaves ternate or pinnate, at least the terminal segment with a distinct petiolule
    8 Larger cauline leaves usually with 5(–9) segments, each divided into usually acute lobes   (i) subsp. **sibiricum**
    8 Larger cauline leaves usually ternate, each segment divided into usually rounded or acuminate lobes   (h) subsp. **ternatum**

(a) Subsp. **alpinum** (L.) Bonnier & Layens, *Fl. Fr.* 128 (1894) (*H. alpinum* L., *H. sphondylium* subsp. *juranum* (Genty) Thell.): Stem sparsely setose to subglabrous below. Leaves simple, the lower suborbicular with 3–5 shallow, broadly rounded lobes, subglabrous or with sparse setae on veins; rays 12–26. Petals white, outer radiate. Commissural vittae short and very slender or absent. *Open woodland.* ● *Jura*

(b) Subsp. **pyrenaicum** (Lam.) Bonnier & Layens, *loc. cit.* (1894) (*H. pyrenaicum* Lam., *H. pollinianum* Bertol., *H. sphondylium* subsp. *montanum* auct. pro parte, non (Schleicher ex Gaudin) Briq.): Stem pubescent or setose below. Leaves simple, with 5–7 usually acute or acuminate primary lobes, softly pubescent, hispid or densely white-tomentose beneath. Rays 12–45. Petals white, outer radiate. Commissural vittae conspicuous. ● *Pyrenees, Alps, N. Appennini, mountains of the Balkan peninsula.*

Plants with the leaves densely white-tomentose beneath may be referred to var. *pyrenaicum*, while those with leaves pubescent or setose beneath may be referred to var. *pollinianum* (Bertol.) Thell. Both varieties probably occur throughout the range of the subspecies.

(c) Subsp. **transsilvanicum** (Schur) Brummitt, *Feddes Repert.* **79**: 65 (1968) (*H. transsilvanicum* Schur, *H. palmatum* Baumg.): Like subsp. (b) but more robust, up to 250 cm; leaves up to 50 cm, with 7–9 lobes, more conspicuously acute and more divided into secondary lobes. ● *E. & S. Carpathians.*

(d) Subsp. **montanum** (Schleicher ex Gaudin) Briq. in Schinz & R. Keller, *Fl. Schweiz* ed. 2, **1**: 372 (1905) (*H. dissectum* Ledeb., *H. setosum* Lapeyr., *H. montanum* Schleicher ex Gaudin, *H. sphondylium* subsp. *granatense* (Boiss.) Briq.): Larger cauline leaves almost always ternate, sometimes white-tomentose beneath. Rays 12–25(–30). Petals white, outer radiate. *In mountain areas, but probably usually at lower altitudes than* (b). *C. Europe, extending locally southward to Sicilia and S. Spain; E. Russia.*

(e) Subsp. **sphondylium** (subsp. *australe* (Hartman) Ahlfvengren): Larger cauline leaves pinnate, with usually 5(–9) segments, each segment often pinnately lobed; lobes acute, pubescent to densely hispid beneath. Rays 12–25. Petals white, rarely pink, outer radiate. *Usually lowland.* ● *Mainly in N.W. Europe, but extending to Scandinavia, E.C. Europe and the Mediterranean region.*

Plants from the W. Carpathians having short setae on the ovary may be referred to var. *chaetocarpoides* Gawłowska (subsp. *trachycarpum* (Soják) J. Holub).

(f) Subsp. **verticillatum** (Pančić) Brummitt, *Feddes Repert.* **79**: 65 (1968) (*H. verticillatum* Pančić): Like subsp. (e) but leaves softly pubescent, especially on the veins, and the upper branches more or less whorled. ● *Balkan peninsula.*

(g) Subsp. **orsinii** (Guss.) H. Neumayer, *Verh. Zool.-Bot. Ges. Wien* **72**: 169 (1923) (*H. orsinii* Guss.): Leaves simple, with 5–7 acute or acuminate lobes, pubescent to hispid beneath. Rays 12–45. Petals greenish, the outer ones not or only slightly radiate. ● *Mountains of the Balkan peninsula; C. & S. Appennini.*

(h) Subsp. **ternatum** (Velen.) Brummitt, *Feddes Repert.* **79**: 65 (1968) (*H. ternatum* Velen.): Larger cauline leaves usually ternate, the segments usually more or less broadly rounded or acuminate, pubescent or hispid beneath. Rays 12–25. Petals greenish, the outer ones not or only slightly radiate. *Balkan peninsula; N.C. Appennini.*

(i) Subsp. **sibiricum** (L.) Simonkai, *Enum. Fl. Transs.* 266 (1887) (*H. sibiricum* L., *H. lecokii* Gren. & Godron): Larger cauline leaves usually pinnate, with 5(–7) segments, each segment often pinnately lobed, the lobes acute, pubescent or hispid beneath. Rays 12–25. Petals greenish, the outer ones not or only slightly radiate. $2n = 22$. *Mainly in N.E. & E.C. Europe, but also in C. & S.W. France and perhaps in the Appennini.*

Plants of this subspecies with short setae on the ovary may be referred to var. *chaetocarpum* H. Neumayer & Thell.

6. **H. ligusticifolium** Bieb., *Fl. Taur.-Cauc.* **1**: 224 (1808). Stem 40–80 cm, 4–10 mm in diameter, softly and sparsely pubescent. Leaves pinnate to 2-pinnate; segments up to 3 cm, shallowly or deeply divided into more or less cuneate lobes, glabrous to sparsely and softly pubescent; rhachis slender and subglabrous. Rays 12–25. Petals white, outer radiate. Fruit 9–10 × 6–8 mm, broadly elliptical; vittae slender, linear, usually at least ¾ the length of the fruit. *Stony places in mountains.* ● *S. Krym.* Rs (K).

7. **H. stevenii** Manden., *Kavk. Vidy Heracleum* 61 (1950) (*H. laciniatum* auct., ? Hornem., *H. villosum* auct., ? Fischer ex Sprengel). Biennial or perennial up to 100 cm. Leaves simple, 5- to 7-lobed, the lobes obtuse or rounded, serrate, subglabrous above or roughly hairy on veins only, densely white-tomentose beneath. Larger umbels up to 30 cm in diameter; rays up to 70. Petals white, outer radiate. Fruit 10–13 × 7–9 mm, broadly ovate to obovate, with short upwardly-directed hairs; vittae swollen at their lower ends. *Stony slopes. Krym.* Rs (K). (*Caucasus.*)

8. **H. pubescens** (Hoffm.) Bieb., *Fl. Taur.-Cauc.* **3**: 225 (1819) (*H. speciosum* auct., ? Weinm.). Stem 60–80 cm. Leaves ternate to pinnate, the segments pinnately lobed, subglabrous above, shortly pubescent beneath. Umbels 10–12 cm. in diameter; rays 15–20. Petals white, radiate. Fruit 13–14 × 8–10 mm, pubescent or sometimes setose towards the margins; vittae swollen at their lower ends. *Damp, shady places.* ● *S. Krym. Perhaps naturalized in C. & W. Europe.* Rs (K) [?Au ?Cz ?Ga].

9. **H. mantegazzianum** Sommier & Levier, *Nuovo Gior. Bot. Ital.* nov. ser., **2**: 79 (1895). Biennial, monocarpic, or perennial. Stem 200–500 cm, up to 10 cm in diameter, usually with conspicuous purple blotches. Leaves up to 300 cm, ternate, or ternately or pinnately divided in varying degree, with lateral segments up to 130 cm, pinnately lobed, acute, shortly pubescent beneath. Umbels up to 50 cm in diameter, with 50–150 rays. Petals up to 12 mm, white or rarely pinkish, outer radiate. Fruit (7–)9–11 × 6–8(–10) mm, glabrous to villous; vittae strongly swollen, 0·5–1 mm wide or more. *Widely naturalized in Europe.* [Au Br Cz Da Fe Ga Ge Hb He Ho Hu It No Rs (C) Su.] (*S.W. Asia.*)

The taxonomy and nomenclature of the naturalized plants from S.W. Asia require further investigation. They are very variable in duration, height, shape, dissection and pubescence of the leaves, shape and size of fruit, etc., and probably represent more than one species. Perennial plants with leaves pinnately divided into 5 segments may perhaps be referable to **H. lehmannianum** Bunge, *Del. Sem. Horti Dorpat.* 1849: 2 (1850), from C. Asia. **H. persicum** Desf. ex Fischer in Fischer, C. A. Meyer & Avé-Lall., *Ind. Sem. Horti Petrop.* **7**: 50 (1841) (? *H. amplissimum* Wenderoth) is probably distinct and occasionally naturalized; it is like **9** but rarely more than 200 cm high, and usually has 2 pairs of lateral leaf-segments. *H. laciniatum* auct. scand., non Hornem., naturalized in Norway to 70° N and possibly also in Finland, has the leaves pinnately divided into 5(–7) segments and smaller (7–11 mm) fruit than typical *H. mantegazzianum*; it may be identical with *H. persicum*.

For other cultivated species, which may occasionally escape, see Thellung in Hegi, *Ill. Fl. Mitteleur.* **5**(2): 1421–1427 (1926).

93. **Malabaila** Hoffm.[1]

Leaves pinnate. Sepals absent or small. Petals yellow, ovate or obcordate. Fruit orbicular-obcordate. Wing strongly thickened at the margin, with a conspicuous vitta bordering the fruit at the base of the wing.

---

[1] By T. G. Tutin.

1	Bracts 6–8, persistent	**3. involucrata**
1	Bracts 0, rarely few and deciduous	
2	Rays 3–9; styles persistent	**1. aurea**
2	Rays (12–)20–30; styles deciduous	**2. graveolens**

1. M. aurea (Sibth. & Sm.) Boiss., *Fl. Or.* **2**: 1053 (1872). Pubescent and somewhat viscid biennial 30–50 cm. Stem hollow, striate. Leaves pinnate; segments 3–4 pairs, ovate in lower leaves, linear-lanceolate in upper leaves, deeply serrate or sometimes lobed. Rays 3–9. Bracts 0; bracteoles few, linear-lanceolate, deciduous. Fruit 8–10 mm, suborbicular; styles persistent. $2n = 22$. *Balkan peninsula.* Al Bu Gr Ju.

2. M. graveolens (Sprengel) Hoffm., *Gen. Umb.* 126 (1814) (incl. *M. vaginans* (Velen.) Velen.). Shortly pubescent biennial up to 100 cm. Leaves pinnate; segments often 5–7 cm, dentate, sometimes lobed. Rays (12–)20–30. Bracts and bracteoles 0 or few, deciduous. Fruit 6–7 mm, suborbicular; styles deciduous. *Dry places. S.E. Europe, southwards to N.E. Greece and extending to c. 51° N. in E. Russia.* Bu Gr Rm Rs (C, W, K, E).

3. M. involucrata Boiss. & Spruner, *Ann. Sci. Nat.* ser. 3 (Bot.), **1**: 337 (1844). Pubescent biennial 40–50 cm. Leaves pinnate, greyish; segments 3–6 pairs, ovate, deeply dentate, sometimes 3-lobed or -partite. Rays 8–20. Bracts and bracteoles 6–8, persistent, lanceolate, deflexed. Fruit (7–)9–14 mm, glabrous; inner part of wing usually translucent. *S. half of Balkan peninsula; Kikhlades.* Al Gr Ju.

M. psaridiana Heldr., *Österr. Bot. Zeitschr.* **39**: 243 (1889), described from S. Greece (Taïyetos), is said to have the fruit 9 mm, with the inner part of the wing scarcely translucent. In all other respects it appears to be indistinguishable from **3**, and is probably best regarded as a variant of it.

94. Tordylium L.[1]

Leaves simple, pinnatisect or pinnate. Sepals prominent, often unequal. Petals whitish or purplish, cuneate or clawed, outer larger and often 2-lobed; apex inflexed. Fruit orbicular or ovate-elliptical, strongly compressed dorsally; margin strongly thickened, usually corrugated or lobed; dorsal ridges inconspicuous; vittae usually solitary.

1	Rays 20–40	**5. byzantinum**
1	Rays not more than 15	
2	Fruit setose; thickened margin smooth	**1. maximum**
2	Fruit with soft hairs; thickened margin corrugated	
3	Stem stout, sparsely hairy; outer flowers with 1 large, ±equally 2-lobed petal; fruit 5–8 mm	**4. apulum**
3	Stem slender, tomentose; outer flowers with 2 large, unequally 2-lobed petals; fruit 2–5 mm	
4	Stems 20–50 cm, softly hairy; fruit 2–3 mm, with vesicular hairs	**2. officinale**
4	Stems up to c. 15 cm, setose; fruit 3·5–5 mm, with soft flexuous hairs	**3. pestalozzae**

1. T. maximum L., *Sp. Pl.* 240 (1753). Stout, shortly hispid biennial or annual 30–130 cm. Stem with deflexed bristles, ridged, hollow, usually much-branched. Leaves pinnate; basal with ovate or suborbicular, crenate segments which are cordate at base; cauline with ovate-lanceolate to linear-lanceolate, dentate segments which are cuneate at base, sometimes reduced to the terminal segment. Rays 5–15. Bracts and bracteoles numerous, subulate. Outer flowers with 2–3 petals larger than the others (2–3 mm) and unequally 2-lobed. Fruit 5–8 mm, setose; wing

with a thin inner part; thickened margin smooth, not corrugated. *S. & S.C. Europe, though doubtfully native in the northern part of its range.* Al Au *Br Bu Co Cz Ga Gr Hs Hu It Ju Lu Rm Rs (W, K) Sa Si Tu [Be Ge He].

2. T. officinale L., *Sp. Pl.* 239 (1753). Slender, pubescent annual 20–50 cm. Stem with short, rather soft, more or less vesicular, deflexed or patent hairs, ridged, hollow, simple or branched from the base. Lower leaves pinnate with ovate to suborbicular deeply cordate segments, sometimes reduced to the terminal segment; upper leaves simple to pinnatisect, lanceolate or oblong, dentate or crenate-dentate. Rays 8–14. Bracts and bracteoles about as long as the rays, numerous, subulate, stiffly ciliate. Outer flowers with 2 petals much larger than the others (5–8 mm), each very unequally 2-lobed. Fruit 2–3 mm, with soft, vesicular hairs; wing without a thin inner part; thickened margin corrugated. *Italy, Balkan peninsula and Aegean region.* Al Cr Gr It Ju Tu.

3. T. pestalozzae Boiss., *Diagn. Pl. Or. Nov.* **2**(10): 45 (1849). Like **2** but stems up to 15 cm, slender, setose; leaves mostly basal; fruit 3·5–5 mm, with soft, flexuous, not vesicular hairs. *Kikladhes and Karpathos.* Cr Gr. (*S. & W. Anatolia.*)

4. T. apulum L., *Sp. Pl.* 239 (1753). Stout, softly and rather sparsely pubescent annual 20–50 cm. Stem densely hairy at base, with sparse long hairs above, ridged, solid, branched. Leaves pinnate; lower with ovate, deeply crenate segments; uppermost with linear, entire segments. Rays 3–8. Bracts and bracteoles much shorter than the rays, several, subulate, stiffly ciliate. Outer flowers with one petal larger than the others (4–6 mm), more or less equally 2-lobed and appearing like 2 large petals. Fruit 5–8 mm, with soft, vesicular hairs; wing with a thin inner part; thickened margin corrugated, minutely papillose. *Mediterranean region.* Al Bl Co Cr Ga Gr Hs It Ju Sa Si Tu.

5. T. byzantinum (Aznav.) Hayek, *Prodr. Fl. Penins. Balcan.* **1**: 1045 (1927). Annual 30–90 cm. Stem with short stiff bristles and soft hairs below, ridged, branched. Leaves sometimes entire, ovate or cordate, crenate, sometimes pinnate, with 3–5 ovate, crenate-dentate segments. Rays 20–40. Bracts and bracteoles numerous, setaceous. Outer flowers with 2 petals much larger than the others, each very unequally 2-lobed. Fruit c. 3 mm, with soft vesicular hairs; wing without a thin inner part; thickened margin smooth, not corrugated. *Kriti; Turkey-in-Europe (near Istanbul).* Cr Tu.

Ainsworthia trachycarpa Boiss., *Diagn. Pl. Or. Nov.* **2**(10): 43 (1849), has been reported from Makedonija, but probably in error.

95. Laser Borkh.[1]

Leaves 2- to 3-ternate. Sepals conspicuous. Petals white, obovate, long-clawed; apex inflexed. Fruit ovoid or oblong. Primary and secondary ridges very prominent, thickened; vittae solitary under the secondary ridges.

1. L. trilobum (L.) Borkh., *Botaniker (Halle)* 246 (1795) (*Siler trilobum* (L.) Crantz). Glabrous perennial up to 120 cm; stock with fibrous remains of petioles. Leaf-lobes c. 5 cm, ovate or cordate, crenate-dentate and often lobed. Umbels up to 25 cm in diameter; rays 11–20. Bracts few or none; bracteoles few, caducous, lanceolate, long-acuminate. Fruit 5–10 mm. *C. & E. Europe.* Al Au Bu Cz Ga Ge ?Gr Hu It Ju Rm Rs (B, C, W, K, E).

---

[1] By T. G. Tutin.

96. Elaeoselinum Koch ex DC.[1]

Leaves 3- to 5-pinnate. Sepals usually small. Petals yellowish or white, oblong, slightly emarginate to shallowly bifid; apex inflexed. Fruit orbicular, ovoid or oblong, somewhat compressed dorsally, with 4 wide lateral wings and unwinged or narrowly winged dorsal ridges; vittae solitary in the grooves and in the ridges.

1 Petals white   **4. gummiferum**
1 Petals yellow
 2 Lateral wings of mericarps extending much beyond the top of
     the fruit   **3. foetidum**
 2 Lateral wings of mericarps not or scarcely extending beyond
     the top of the fruit
  3 Bracts and bracteoles, if present, few, linear-lanceolate or
      setaceous   **1. asclepium**
  3 Bracts and bracteoles numerous, lanceolate   **2. tenuifolium**

1. E. asclepium (L.) Bertol., *Fl. Ital.* **3**: 383 (1838). Almost glabrous perennial up to 130 cm. Stock with numerous coarse fibres. Stem solid, striate, rather slender, branched above. Basal leaves up to 40 cm, 3- to 5-pinnate; divisions often whorled; lobes 2–3 mm, filiform; petiole and rhachis somewhat pubescent; cauline leaves reduced to inflated petioles. Rays 8–25. Petals yellow. Fruit 8–15 mm, orbicular or ovate-oblong; mericarps with wide, whitish, shiny lateral wings. *Mediterranean region.* Al Bl Gr Hs It Sa Si.

(a) Subsp. asclepium: Basal leaves horizontal; bracts and bracteoles absent or few, setaceous; dorsal ridges of fruit unwinged. *From W. Italy and Sicilia to S.E. Greece.*

(b) Subsp. meoides (Desf.) Fiori, *Nuov. Fl. Anal. Ital.* **2**: 84 (1925) (*E. meoides* (Desf.) Koch ex DC., *E. millefolium* Boiss.): Basal leaves erect; bracts and bracteoles few, linear-lanceolate; dorsal ridges of fruit distinctly but narrowly winged. *From Spain to S. Italy.*

Plants from Spain and Islas Baleares are more or less intermediate between the two subspecies, but are generally assigned to subsp. **(b)**.

2. E. tenuifolium (Lag.) Lange in Willk. & Lange, *Prodr. Fl. Hisp.* **3**: 26 (1874). Like **1**(a) but leaf-lobes lanceolate, rigid; rays 10–27; bracts and bracteoles numerous, lanceolate; fruit 8–10 mm; dorsal ridges of mericarps usually broadly winged. *S. half of Iberian peninsula.* Hs Lu.

3. E. foetidum (L.) Boiss., *Elenchus* 51 (1838). Erect perennial like **1** in general appearance. Leaves 3-pinnate; lobes *c.* 10 mm, ovate in outline, usually 3-fid. Rays 10–20. Bracts 0–1; bracteoles numerous, setaceous. Petals yellow. Fruit 10–12 mm, ovoid-oblong; mericarps with wide lateral wings which extend beyond the top of the fruit, and narrow dorsal wings. *S. Portugal, S.W. Spain.* Hs Lu.

4. E. gummiferum (Desf.) Tutin, *Feddes Repert.* **74**: 33 (1967). (*Margotia gummifera* (Desf.) Lange). Nearly glabrous perennial up to 120 cm. Leaves with inflated petioles and shortly hispid rhachis; basal leaves 3- to 4-pinnate, triangular in outline; lobes ovate in outline, toothed or pinnately lobed; cauline leaves with very small or no lamina. Rays 8–20. Bracts and bracteoles 6–9, linear-lanceolate. Petals white. Fruit 10–12 mm, ovoid-oblong; mericarps with wide lateral and narrow dorsal wings. *C., S. & W. Spain, Portugal.* Hs Lu.

---

[1] By T. G. Tutin.

97. Guillonea Cosson[1]

Like *Laserpitium* but young fruit densely tomentose; dorsal ridges prominent, widened at the top by the development of the pericarp.

1. G. scabra (Cav.) Cosson, *Not. Pl. Crit.* 110 (1851). Pubescent perennial 30–120 cm. Stock with abundant fibres. Stem terete, striate, swollen at the nodes, branched above. Leaves mostly basal, 3-pinnate, triangular in outline, glaucescent and scabrid; lobes 5–20 mm, oblong-lanceolate or cuneate, lobed or dentate. Rays 5–15. Bracts numerous, linear-subulate, ciliate, eventually deflexed; bracteoles villous, about equalling the villous-tomentose pedicels. Petals white, villous beneath. Fruit *c.* 10 mm, densely tomentose between the wings when mature; wings glabrous, wider than the mericarp. ● *S. & E. Spain.* Hs.

98. Laserpitium L.[1]

Leaves several times pinnate or ternate. Sepals conspicuous, ovate to subulate. Petals white, pinkish, pale yellow or greenish-yellow, obcordate; apex inflexed. Fruit ellipsoid or oblong, broadly ovoid or rectangular, terete or slightly compressed dorsally. Primary ridges inconspicuous, secondary winged, the lateral wings usually larger than the dorsal.

1 Leaf-lobes divided into numerous narrow lobes
 2 Rays 15 or more   **9. halleri**
 2 Rays 7–9   **11. pseudomeum**
1 Leaf-lobes entire, dentate or with few broad lobes
 3 Bracts not ciliate, often 0
  4 Leaf-lobes entire
   5 Rays 2–15   **6. peucedanoides**
   5 Rays 20–50   **1. siler**
  4 Leaf-lobes crenate to dentate
   6 Bracts of primary umbel numerous, persistent   **2. latifolium**
   6 Bracts of primary umbel 0–5, caducous
    7 Leaf-lobes suborbicular, shallowly crenate-dentate; fruit
      very narrowly winged   **3. longiradium**
    7 Leaf-lobes ovate or broadly ovate, strongly dentate or
      crenate-dentate; lateral wings of fruit 1–2 mm wide
     8 Wings of fruit equal   **4. nestleri**
     8 Lateral wings of fruit much wider than dorsal   **5. krapfii**
 3 Bracts ciliate, always numerous
  9 Stock without fibres; petals with scattered hairs beneath
   10 Upper petioles narrow; rays rough on the inner side
          12. prutenicum
   10 Upper petioles somewhat inflated; rays hispid   **13. hispidum**
  9 Stock with fibres; petals glabrous beneath
   11 Upper petioles strongly inflated; stem hollow; leaf-lobes
     with a ± regular isodiametric network of veins
          7. archangelica
   11 Upper petioles narrow, appressed; stem solid; leaf-lobes
     with an obscure or irregular and elongated network of
     veins
    12 Membranous margin of bracteoles narrower than her-
      baceous middle part; sepals triangular-ovate
          10. gallicum
    12 Membranous margin of bracteoles as wide as herbaceous
      middle part; sepals lanceolate-subulate   **8. nitidum**

1. L. siler L., *Sp. Pl.* 249 (1753) (*Siler montanum* Crantz). Almost glabrous perennial 30–100(–180) cm. Stock with abundant fibres. Stem terete, striate, simple or branched above. Lower leaves up to 100 cm, 2- to 4-pinnate, triangular in outline; lobes 15–70 × 3–25 mm, linear-lanceolate to oblong-obovate, somewhat glaucous and coriaceous; margin narrowly cartilaginous, smooth or slightly sinuate; midrib prominent, whitish; main lateral veins evenly spaced; network of fine veins elongated parallel to the main lateral veins. Petioles of lower leaves somewhat laterally compressed, of upper cauline strongly inflated. Rays 20–50, papillose on the inner side. Bracts and bracteoles numerous, lanceolate, glabrous, not deflexed. Petals white. Fruit 6–12 mm, oblong; wings 0·5–1 mm wide. ● *Mountains of S. & S.C. Europe.* Al Au Bu ?Cz Ga Ge Gr He Hs It Ju.

1 Leaf-lobes decurrent on rhachis to the next lobe   (c) subsp. **zernyi**
1 Leaf-lobes not or shortly decurrent
   2 Leaf-lobes 5–12 mm wide, acute, shortly stalked   (a) subsp. **siler**
   2 Leaf-lobes 10–25 mm wide, obtuse, narrowed at base
         (b) subsp. **garganicum**

(a) Subsp. **siler**: Lower leaves 4-pinnate; lobes 5–12 mm wide, lanceolate, acute, shortly stalked, the 3 terminal free at base. *Throughout the range of the species except the south-east.*

(b) Subsp. **garganicum** (Ten.) Arcangeli, *Comp. Fl. Ital.* 302 (1882): Lower leaves 4-pinnate; lobes 10–25 mm wide, elliptical or oblong, obtuse, mucronate, narrowed at base, the 3 terminal often confluent and shortly decurrent. *S. part of Balkan peninsula; S. Italy.*

(c) Subsp. **zernyi** (Hayek) Tutin, *Feddes Repert.* **74**: 31 (1967) (*L. zernyi* Hayek): Lower leaves 2- to 3-pinnate; lobes linear-oblong, decurrent on the rhachis to the next lobe. *N.E. Albania, just extending to Jugoslavia.*

2. **L. latifolium** L., *Sp. Pl.* 248 (1753). Nearly glabrous, somewhat pruinose perennial (30–)60–150(–250) cm. Stock with abundant fibres. Stem terete, striate, solid, branched above. Leaves 2-pinnate, triangular in outline; lobes 20–100 mm, ovate, cordate, usually unequal at base, often sparsely hairy beneath; margin narrowly cartilaginous, dentate and serrulate or shortly ciliate; midrib prominent; network of fine veins conspicuous and more or less isodiametric. Petioles of lower leaves somewhat laterally compressed and often sparsely hispid; of upper cauline strongly inflated. Rays (20–)25–40(–50), rough on the inner surface. Bracts numerous, narrow and membranous-margined, sometimes nearly leaf-like, glabrous; bracteoles few, subulate. Petals white. Fruit 5–10 mm, ovoid; wings 2–2·5 mm wide, all equal, usually undulate; primary ridges (between the wings) appressed hispid. $2n = 22$. *Much of Europe, but absent from the margins and most islands; extends from C. Spain to 61° N. in Sweden and to C. Ukraine and Bulgaria.* Au Bu Co Cz Da Fe Ga Ge He Hs Hu It Ju No Po Rm Rs (B, C, W) Su ?Tu [Be].

3. **L. longiradium** Boiss., *Voy. Bot. Midi Esp.* **2**: 734 (1845). Like **2** but leaf-lobes 18–30 mm, usually suborbicular, shallowly crenate-dentate; rays up to 13; bracts absent; bracteoles numerous; fruit narrowly winged. ● *S. Spain (Sierra Nevada).* Hs.

4. **L. nestleri** Soyer-Willemet, *Obs. Pl. Fr.* 87 (1828). Like **2** but leaves 2- to 3-pinnate; lobes dentate or serrate, cuneate or rounded at base; rays 10–30; bracts 1–3, caducous; bracteoles 1–5; fruit oblong; wings 1–2 mm wide, not undulate. *Calcicole.* ● *Mountains of S. France and Iberian peninsula.* Ga Hs Lu.

5. **L. krapfii** Crantz, *Class. Umb.* 67 (1767). Glabrous or somewhat pubescent perennial 60–120 cm. Stock with abundant fibres. Stem terete, striate, slender, branched above. Lower leaves c. 30 cm, 2- to 3-ternate, triangular in outline; lobes 20–80 mm, ovate or suborbicular, acute or obtuse, rounded to cordate at base, often hairy beneath; margin somewhat thickened, dentate or crenate-dentate; midrib prominent; network of fine veins conspicuous and more or less isodiametric. Petioles of upper cauline leaves strongly inflated. Rays (5–)7–15(–24), very unequal. Bracts 0–5, glabrous, caducous; bracteoles several. Petals usually greenish-yellow or pinkish. Fruit 5–11 mm, ellipsoid; lateral wings much wider than dorsal. ● *Mountains of E.C.*

Europe and N. part of Balkan peninsula; S. & E. Alps. Al Au Bu He It Ju Po Rm Rs (W).

(a) Subsp. **krapfii** (*L. marginatum* Waldst. & Kit.; incl. *L. alpinum* Waldst. & Kit.): Stem usually slightly pruinose; upper cauline leaves similar to lower, their lobes ovate, usually dentate; rays nearly always rough or shortly hispid on inner side; primary ridges of fruit with short setae. *N. part of Balkan peninsula, extending to N.E. Italy; Carpathians.*

(b) Subsp. **gaudinii** (Moretti) Thell., *Monde Pl.* **153**: 2 (1925) (*L. gaudinii* Moretti): Stem usually strongly pruinose; upper cauline leaves markedly different from lower, their lobes oblong, usually entire; rays glabrous and smooth on inner side; primary ridges of fruit glabrous. *N. Italy, N.W. Jugoslavia, W. Austria, E. Switzerland.*

6. **L. peucedanoides** L., *Cent. Pl.* **2**: 13 (1756). Slender, glabrous perennial 30–60(–100) cm. Stock with abundant fibres. Stem terete, striate, usually branched. Lower leaves 2- to 3-ternate, triangular in outline; lobes 15–100 × 2–12 mm, linear to ovate-oblong, narrowed at both ends; margin somewhat incurved, entire, lower surface with 3–5 prominent veins arising near the base of lobe, and so nearly parallel; network of fine veins elongated parallel to the midrib. Petioles of upper cauline leaves narrow. Rays 2–15, smooth. Bracts 5–8, linear, glabrous; bracteoles subulate. Petals white. Fruit 4·5–7 mm, broadly ellipsoid; lateral wings 1–1·5 mm wide, somewhat wider than dorsal. $2n = 22$. ● *S.E. Alps; mountains of N.W. Jugoslavia.* Au It Ju.

7. **L. archangelica** Wulfen in Jacq., *Collect. Bot.* **1**: 214 (1787). Pubescent perennial 80–150 cm. Stock with abundant fibres. Stem stout, angled, often reddish-spotted, densely hairy at the nodes and with numerous scattered, long, soft hairs. Lower leaves 3- to 4-ternate, triangular in outline; lobes 20–60 × 10–50 mm, ovate, rounded at base, acute, dentate and sometimes lobed; network of fine veins more or less isodiametric. Petioles of upper cauline leaves very strongly inflated. Rays 14–40, pubescent, rough on the inner side. Bracts numerous, conspicuous, often 3-fid; bracteoles numerous, linear-lanceolate, hairy, soon deflexed. Petals white, or pinkish beneath. Fruit 8–10 mm, ellipsoid; lateral wings 1·5–3 mm wide, wider than dorsal. ● *Mountains of E.C. Europe and N. part of Balkan peninsula.* Bu Cz Ju Po Rm.

8. **L. nitidum** Zanted., *Comment. Ateneo Brescia* **1815**: 89 (1818). Somewhat hispid perennial 30–70 cm. Stock with abundant fibres. Stem terete, striate, solid. Lower leaves 3-pinnate, ovate to triangular in outline; lobes 15–30 × 10–15(–25) mm, ovate, cordate to broadly cuneate at base, glabrous above, hispid beneath, dentate, often lobed, ciliate; network of veins somewhat elongated and irregular. Petioles of upper cauline leaves rather narrow, with wide membranous margin. Rays (10–)15–25(–35), sparsely hispid to glabrous, rough on the inner side. Bracts numerous, lanceolate, often 2- to 3-fid at apex, ciliate; bracteoles numerous, nearly as long as pedicels, largely membranous. Petals white. Fruit 5–6 mm, broadly ovoid; lateral wings c. 1·5 mm wide, dorsal narrower. $2n = 22$. ● *N. Italy (Trentino to Como).* It.

9. **L. halleri** Crantz, *Class. Umb.* 67 (1767) (*L. panax* Gouan). More or less pubescent perennial 15–60 cm. Stock with abundant fibres. Stem terete, striate, simple or little-branched. Lower leaves 4- to 5-pinnate, triangular in outline; lobes 4–7 mm, ovate to linear-lanceolate in outline, usually again divided into 5–7 linear or filiform lobes; midrib prominent, lateral veins obscure. Petioles of upper cauline leaves narrow. Rays 15–30(–45), rough on the inner side. Bracts numerous, conspicuous, linear-lanceolate,

membranous-margined, ciliate, often 3-fid at apex; bracteoles almost entirely membranous. Petals white or pinkish. Fruit 6–9 mm, ellipsoid; lateral wings *c.* 1·5 mm, dorsal often narrower. ● *Alps (local); Corse.* Au Co Ga He It.

(a) Subsp. **halleri**: Pubescent; hairs on leaves and petioles often in small groups; leaf-lobes usually ovate in outline; bracts ciliate; bracteoles lanceolate; sepals usually ciliate. *Alps.*

(b) Subsp. **cynapiifolium** (Viv. ex DC.) P. Fourn., *Quatre Fl. Fr.* 694 (1937) (*L. cynapiifolium* (Viv. ex DC.) Salis): Almost glabrous; leaf-lobes usually linear-lanceolate in outline; bracts sparsely ciliate; bracteoles setaceous or subulate; sepals glabrous. $2n = 22$. *Corse.*

10. L. gallicum L., *Sp. Pl.* 248 (1753). Robust, glabrous or slightly hairy perennial up to 160 cm. Stock with abundant fibres. Stem terete, striate, branched. Lower leaves *c.* 50 cm, 3- to 5-pinnate, triangular in outline; lobes 10–20 mm, linear-lanceolate or oblanceolate, entire or 3-fid; midrib slender, lateral veins obscure. Petioles of upper cauline leaves narrow. Rays 20–50, rough on the inner side. Bracts and bracteoles numerous, linear-lanceolate, ciliate. Petals white or pink. Fruit *c.* 6 mm, oblong; lateral wings *c.* 2 mm wide, the dorsal narrower. *Calcicole.* ● *Mountains of S. Europe from Spain to Italy.* Bl Ga Hs It Sa.

L. paradoxum A. Bolós & Font Quer, *Collect. Bot.* (*Barcelona*) 1: 297 (1947), from the Pyrenees, is a variant of **10** with hairs at the junctions of the branches of the leaves and with considerable development of the pericarp over the primary ridges, rather as in *Guillonea.* Intermediates between this extreme form and the typical plant occur in N.E. Spain.

11. L. pseudomeum Orph., Heldr. & Sart. ex Boiss., *Diagn. Pl. Or. Nov.* 3(2): 95 (1856). Glabrous perennial 10–25 cm. Stem terete, striate, simple or with 1–2 branches. Lower leaves 2- to 3-pinnate, oblong-lanceolate in outline; lobes orbicular in outline, again divided into numerous, crowded, linear-setaceous lobes; cauline leaves few. Petioles rather narrow. Rays 7–9, rough on the inner side. Bracts and bracteoles numerous, lanceolate, broadly membranous, almost glabrous. Petals white. Fruit *c.* 8 mm, oblong; lateral wings *c.* 2 mm, the dorsal narrower. ● *Mountains of C. & S. Greece.* Gr.

12. L. prutenicum L., *Sp. Pl.* 248 (1753). More or less pubescent biennial 30–100 cm, occasionally of longer duration but always monocarpic. Stock without fibres. Stem slender, angled, glabrous or rarely somewhat hispid, branched above. Lower leaves 2- to 3-pinnate, triangular in outline; lobes 10–25 × 2–9 mm, lanceolate or elliptical, sessile or shortly stipitate, usually dentate or lobed, glabrous above, more or less hispid beneath, ciliate; veins obscure. Petioles of upper cauline leaves narrow. Rays (6–)12–20(–30), rough on the inner side. Bracts and bracteoles numerous, conspicuous, linear-lanceolate, broadly membranous, ciliate, soon deflexed. Petals white, or with a yellowish tinge. Fruit 3·5–4·5 mm, broadly ellipsoid; primary ridges hispid; lateral wings *c.* 1 mm wide, the dorsal narrower. ● *C., E. and parts of S. Europe, from N. Germany & C. Russia to N. Portugal, N. Italy and Bulgaria.* Au Bu Cz Ga Ge He Hs Hu It Ju Lu Po Rm Rs (B, C, W, E).

(a) Subsp. **prutenicum**: Stem stout, more or less pubescent; leaf-lobes lanceolate; rays usually more than 12. *Throughout the range of the species, except S.W. France and N.W. Spain.*

(b) Subsp. **dufourianum** (Rouy & Camus)Tutin, *Feddes Repert.* 74: 31 (1967) (*L. prutenicum* forme *dufourianum* Rouy & Camus):

Stem slender, glabrous; leaf-lobes linear-lanceolate; rays 6–12. *S.W. France, N.W. Spain.*

13. L. hispidum Bieb., *Fl. Taur.-Cauc.* 1: 221 (1808). Like **12** but perennial; stem hispid throughout; rays 30–40, hispid; petals yellow; fruit 6 mm, ellipsoid-oblong. *S. & E. Ukraine.* Rs (K, E).

99. Thapsia L.[1]

Leaves (1–)2- to 3-pinnate. Sepals small. Petals yellow, cuneate or long-clawed; apex inflexed. Fruit oblong to ovate, compressed dorsally. Primary ridges slender, inconspicuous; dorsal secondary ridges like the primary or sometimes narrowly winged, the marginal ones broadly winged.

1 Leaves pinnate; lobes 50–120 mm   **2. maxima**
1 Leaves (1–)2- to 4-pinnate; lobes 5–25(–50) mm
2   Ultimate lobes of leaves regularly dentate   **1. villosa**
2   Ultimate lobes of leaves entire or with 1–2 teeth   **3. garganica**

1. T. villosa L., *Sp. Pl.* 261 (1753). Somewhat pubescent, pruinose perennial 30–200 cm. Stock with abundant fibres. Stem terete, striate, solid. Lower leaves 20–35 × 10–30 cm, 3- to 4-pinnate, more or less villous to hispidulous on rhachis and lamina; lobes 5–15 mm, rarely (var. *platyphyllos* P. Silva & Franco) up to 30 mm, ovate-oblong to oblong in outline, dentate or shallowly lobed; teeth with a spinous mucro; petioles of upper cauline leaves inflated, without lamina. Rays 9–24. Bracts and bracteoles few or none. Fruit 8–15 mm, elliptical; lateral wings 2–3 mm wide, deeply emarginate at base and apex; dorsal ridges not or very narrowly winged. *Portugal to S. France.* Ga Hs Lu.

2. T. maxima Miller, *Gard. Dict.* ed. 8, no. 2 (1768). Like **1** but nearly glabrous; lower leaves up to 50 cm, pinnate, ovate in outline; lobes 50–120 mm, ovate-oblong in outline, dentate or shallowly lobed, often decurrent; fruit 7–10 mm. ● *C. & S. Spain, E. & S. Portugal.* Hs Lu.

3. T. garganica L., *Mantissa* 57 (1767) (*T. decussata* Lag.). Glabrous or sparsely hispid, pruinose perennial 30–250 cm. Stock with abundant stiff fibres. Stem terete, striate, solid. Lower leaves 10–40 × 5–20 cm, (1–)2- to 3-pinnate, narrowly triangular in outline, glaucous beneath; lobes 10–50 mm, linear-oblong, often again lobed; ultimate lobes entire or with 1–2 teeth, acute or obtuse. Rays 5–20, glabrous. Bracts and bracteoles absent. Fruit 12–25 mm, oblong or elliptical; lateral wings 3–6 mm wide, deeply emarginate at base and apex; dorsal ridges not winged. *S. part of Mediterranean region, extending to Portugal.* Bl Cr Gr Hs It Lu Sa Si.

100. Rouya Coincy[1]

Like *Thapsia* but bracts and bracteoles numerous; sepals conspicuous, accrescent, exceeding the stylopodium; petals white; fruit strongly compressed; dorsal ridges distinctly winged.

1. R. polygama (Desf.) Coincy, *Naturaliste* (*Paris*) **23**: 213 (1901) (*Thapsia polygama* Desf.). Almost glabrous perennial 10–30 cm. Stock rather slender, flexuous, with whitish, scale-like leaf-bases in upper part. Stem striate, sparsely hairy, solid. Lower leaves 2–8 cm, 2-pinnate, ovate in outline; lobes 5–10 mm, ovate to oblong-lanceolate, sometimes dentate or 3-fid, acute. Rays 10–20. Bracts several, usually 2- to 3-fid, ciliate; bracteoles several, linear-lanceolate. Fruit *c.* 8 mm, oblong-elliptical; lateral wings 1–1·5 mm wide, undulate; dorsal ridges narrowly winged. *Maritime sands. Corse and Sardegna.* Co Sa. (*N. Africa.*)

---

[1] By T. G. Tutin.

101. Melanoselinum Hoffm.[1]

Like *Thapsia* but monocarpic; petals whitish or purplish; fruit pubescent, almost black; lateral wings strongly denticulate.

1. M. decipiens (Schrader & Wendl.) Hoffm., *Gen. Umb.* 177 (1814). Biennial or monocarpic perennial. Stem 120–280 cm, woody and leafless in lower part, smooth, terete. Leaves *c.* 40 × 30 cm, 2- to 3-pinnate, rhachis and midribs pubescent; lobes 20–120 mm, lanceolate or ovate, acute, sharply serrate or dentate; petioles strongly inflated. Inflorescence 60–90 cm, much-branched, leafy, somewhat pubescent. Rays 30–50, pubescent. Bracts 20–30 mm, 10–12, lanceolate or ovate, irregularly cut, pubescent; bracteoles about as long as pedicels, numerous, lanceolate, pubescent. Fruit 12–14 mm, oblong, pubescent, almost black; lateral wings *c.* 1·5 mm wide, dark brown, strongly denticulate. *Açores.* Az. (*Madeira, Canarias.*)

102. Torilis Adanson[2]

Annual, rarely biennial. Leaves 1- to 3-pinnate, the segments jaggedly toothed. Sepals small, rarely conspicuous, persistent. Petals white or pinkish; apex inflexed. Fruit linear to ovoid, narrowed at the commissure. Ridges slender, ciliate, the grooves between the ridges usually filled with spines or tubercles.

```
1  Umbels mostly lateral, leaf-opposed; peduncles up to 5 cm
2    Plant usually procumbent; rays very short, concealed by
       flowers or fruit                              1. nodosa
2    Plant erect; rays evident                 6. leptophylla
1  Umbels mostly terminal on peduncles usually more than 5 cm
3    Bracts 4–12
4      Outer petals only slightly longer than inner; ripe fruit 3–4
         mm; style 2–3 times as long as the stylopodium   3. japonica
4      Outer petals distinctly longer than inner; ripe fruit c. 2 mm;
         style 5–6 times as long as the stylopodium      4. ucranica
3    Bracts 0–1
5      Leaves very finely divided; lobes less than 1 mm wide  5. tenella
5      Leaves coarsely divided; lobes at least 2 mm wide   2. arvensis
```

1. T. nodosa (L.) Gaertner, *Fruct. Sem. Pl.* 1: 82 (1788). Annual up to 50 cm, usually procumbent. Leaves 1- to 2-pinnate, with deeply pinnatifid segments. Umbels sessile to shortly pedunculate, leaf-opposed; rays very short, generally concealed by the flowers or fruit, giving the umbels a capitate appearance. Bracts absent; bracteoles exceeding the subsessile flowers. Petals pinkish-white. Fruit 2–3 mm; outer mericarp with straight, patent spines, the inner with tubercles. 2*n* = 22. *S. & W. Europe; naturalized in C. Europe.* Al Be Bl Br Bu Co Cr Ga Gr Hb Ho Hs It Ju Lu Rm Rs (K) Sa Si Tu [Au Cz Ge He].

2. T. arvensis (Hudson) Link, *Enum. Hort. Berol. Alt.* 1: 265 (1821). Usually erect annual up to 100 cm. Leaves very variable, from 2-pinnate to 3-foliolate; lobes at least 2 mm wide, coarsely toothed, remotely serrate or subentire. Rays 2–12. Bracts 0–1; bracteoles numerous. Fruit 3–6 mm; both mericarps spiny, or the outer spiny and the inner tuberculate; rarely the whole fruit covered with tubercles. 2*n* = 12. *W., S. & C. Europe.* Al Au Az Be Bl *Br Bu Co Cr Cz Ga Ge Gr He Ho Hs Hu It Ju Lu Po Rm Rs (W, K, E) Sa Si Tu.

A difficult complex, here divided into subspecies, all of which have been accorded specific recognition by various authors. It is urgently in need of revision using modern methods.

```
1  Rays 4–12
2    Styles 3–6 times as long as the stylopodium; outer petals
       distinctly radiate                    (a) subsp. neglecta
```

```
2    Styles 2–3 times as long as the stylopodium; outer petals only
       slightly radiate                      (b) subsp. arvensis
1  Rays 2–3(–4)
3    Rays 2(–4), rather robust, widely diverging, usually forming
       an angle of c. 90°; upper leaves similar to lower but reduced;
       fruit 5–6 mm                          (c) subsp. elongata
3    Rays (2–)3(–4), slender, usually forming an angle of 45–60°;
       upper leaves usually with linear, remotely serrate to entire
       segments; fruit 4–5 mm                (d) subsp. purpurea
```

(a) Subsp. **neglecta** (Schultes) Thell. in Hegi, *Ill. Fl. Mitteleur.* **5**(2): 1055 (1926) (*T. radiata* Moench): Stem erect, much-branched. Rays 4–12. Outer petals up to 2 mm or more, distinctly radiate. Styles 3–6 times as long as the stylopodium. *C. & S. Europe.*

(b) Subsp. **arvensis** (subsp. *divaricata* Thell., *T. helvetica* C. C. Gmelin): Stem either ascending and often little-branched, the branches forming a narrow angle with the stem, or low and much-branched, the branches making a wide angle with the stem. Rays 4–12. Outer petals not more than 1·5 mm, very slightly radiate. Styles 2–3 times as long as the stylopodium. *Throughout the range of the species.*

(c) Subsp. **elongata** (Hoffmanns. & Link) Cannon, *Feddes Repert.* **79**: 62 (1968) (*Caucalis elongata* Hoffmanns. & Link): Stem erect, little-branched; upper leaves like the lower but smaller. Rays usually 2. Petals often tinged with violet or purple, the outer not radiate. Styles scarcely longer than the stylopodium. *Mediterranean region.*

(d) Subsp. **purpurea** (Ten.) Hayek, *Prodr. Fl. Penins. Balcan.* **1**: 1057 (1927) (*T. heterophylla* Guss., *T. torgesiana* (Hausskn.) Hayek, *T. arvensis* subsp. *heterophylla* (Guss.) Thell.: Stem erect, with few branches. Basal leaves (often absent at maturity) pinnate, with deeply divided segments; uppermost leaves 3-foliolate, with linear-lanceolate to linear segments, the central segment very long; margin remotely serrate to subentire. More rarely upper and lower leaves similar, but upper smaller and less divided. Rays usually 3. *S. Europe.*

3. T. japonica (Houtt.) DC., *Prodr.* 4: 219 (1830) (*T. anthriscus* (L.) C. C. Gmelin, non Gaertner). Erect annual or rarely biennial up to 125 cm. Leaves 1- to 3-pinnate; apex of lobes ovate to narrowly ovate. Rays 5–12. Bracts 4–6(–12); bracteoles well-developed. Outer petals slightly longer than the inner. Fruit 3–4 mm, with recurved spines. Styles patent, 2–3 times as long as the stylopodium. 2*n* = 16. *Europe, except the extreme north and much of the Mediterranean region.* All except Az Cr Fa Is Rs (N) Sa Sb Si.

4. T. ucranica Sprengel in Roemer & Schultes, *Syst. Veg.* 6: 485 (1820) (*T. microcarpa* Besser). Like **3** but leaf-lobes linear to linear-lanceolate; rays 10–15; outer petals distinctly longer than the inner, radiate; fruit *c.* 2 mm; styles 5–6 times as long as the stylopodium. *S.E. Europe, extending locally to Hungary and S.E. Poland.* Bu Gr Hu Ju Po Rm Rs (C, W, E) Tu [Ge].

Perhaps overlooked in some areas, but some records may be erroneous.

5. T. tenella (Delile) Reichenb. fil. in Reichenb. & Reichenb. fil., *Icon. Fl. Germ.* 21: 84 (1867) (*Caucalis tenella* Delile). Erect annual, with a simple or branched stem up to 60 cm. Leaves 2-pinnate; lobes less than 1 mm wide, narrowly linear. Rays 5–9(–11). Bracts 0–1; bracteoles linear-lanceolate. Sepals conspicuous. Fruit linear in outline. Stigmas sessile on the conical stylopodium. *Rocky hillsides. S.E. Greece.* Gr. (*S.W. Asia, Egypt.*)

6. T. leptophylla (L.) Reichenb. fil. in Reichenb. & Reichenb. fil., *Icon. Fl. Germ.* 21: sub t. 169 (1866) (*T. xanthotricha* (Steven) Schischkin, *Caucalis leptophylla* L.). Erect annual up to 40 cm,

[1] By T. G. Tutin.   [2] By J. F. M. Cannon.

24-2

sometimes branched. Leaves 2-pinnate; lobes linear. Umbels mostly lateral, leaf-opposed; peduncles 2–3(–5) cm. Rays 2–3. Fruit linear-oblong; spines yellowish to straw-coloured. Stigmas sessile on the conical stylopodium. *S. Europe; a frequent casual farther north and perhaps locally naturalized.* Al Bl Bu Co Cr Ga Gr Hs It Ju Lu Rs (K) Tu [Au Cz He Ho Rs (C, W)].

103. Astrodaucus Drude[1]

Leaves several times pinnate. Calyx-teeth short. Petals white or yellowish, the outer ones radiating, unequally 2-lobed; apex inflexed. Fruit prismatic, laterally compressed or subcylindrical; primary ridges 5, filiform, ciliate or with stellate hairs; secondary ridges with triangular or pyramidal spines in 1 or 2 rows, confluent at the base into a wing.

Bracteoles 5; petals 4 mm   **1. orientalis**
Bracteoles 8–11; petals 2·5 mm   **2. littoralis**

1. **A. orientalis** (L.) Drude in Engler & Prantl, *Natürl. Pflanzenfam.* 3(8): 157, 271 (1898). Stem up to 100 cm, erect, branched above. Leaves triangular-ovate in outline, 3- to 4-pinnate, sparsely hairy; lobes minute, oblong, obtuse. Rays 8–15. Bracts 0(–3); bracteoles 5, oblong-lanceolate, ciliate. Petals 4 mm. Fruit 5–6 mm; spines on secondary ridges triangular, 2-seriate, longer than the width of the mericarp. *S. Ukraine.* Rs (W, K) [Cz Rs (C)]. (*S.W. Asia.*)

2. **A. littoralis** (Bieb.) Drude, *loc. cit.* (1898). Stem up to 65 cm, erect, branched. Leaves broadly triangular in outline, 4- to 5-pinnate, sparsely hairy; lobes linear or oblong-linear. Rays 8–20(–25). Bracts 0(–3); bracteoles 8–11, lanceolate to oblong-ovate, ciliate. Petals 2·5 mm. Fruit 6–7 mm; spines on secondary ridges pyramidal, 1-seriate, glochidiate at the apex. *Maritime sands. S.E. Europe, from Bulgaria to S.E. Russia.* Bu Rm Rs (W, K, E).

104. Turgeniopsis Boiss.[2]

Annual. Leaves 3- to 4-pinnate. Sepals small. Petals white, obovate, unequally 2-lobed; apex inflexed. Fruit ellipsoid, obtuse, somewhat compressed laterally. Primary ridges rather indistinct, bearing short bristles. Secondary ridges well-developed and broadly obtuse in section, bearing 2–3 rows of hooked spines, which arise from warty bases.

A monotypic genus distinguished from *Caucalis* by the endosperm, which is not inrolled at the commissural face, and from *Torilis* by the strongly developed secondary ridges.

1. **T. foeniculacea** (Fenzl) Boiss., *Ann. Sci. Nat.* ser. 3 (Bot.), 2: 53 (1844). Stem up to 50 cm, terete, finely striate. Leaves with very fine capillary segments *c.* 0·5 mm wide. Bracts 0–1, linear-subulate; bracteoles 0–3, like the bracts. Rays 2–3. Partial umbels with 2–3 hermaphrodite flowers and a small number of male flowers in the centre. Fruit 8–10 × 4–5 mm. Stylopodium shortly conical; styles short, stiff. *Dry, stony slopes. S. Bulgaria (C. Rodopi).* Bu. (*S.W. Asia.*)

105. Caucalis L.[1]

Leaves 2- to 3-pinnate. Sepals small or obsolete. Petals white or pink, the outer radiating; apex inflexed. Fruit ellipsoid to ovoid, compressed laterally; 3 of the primary ridges with uniseriate cilia; secondary ridges thickened, with aculeate spines.

1. **C. platycarpos** L., *Sp. Pl.* 241 (1753) (*C. daucoides* L. (1767), non L. (1753), *C. lappula* Grande). Annual; stems up to 40 cm, erect, branched, slightly setose or pubescent. Leaf-segments pinnately divided into oblong or lanceolate lobes, almost glabrous. Rays 2–5. Bracts absent, rarely 1–2; bracteoles linear-lanceolate. Petals *c.* 2 mm, white or pink. Fruit 6–13 × 5 mm; secondary ridges of mericarps with 1 row of aculeate spines as long as the width of the mericarp. 2*n* = 20. *Most of Europe except the north.* Al Au Bu Co Cz Ga Ge Gr He Hs Hu It Ju Lu Po Rm Rs (C, W, K, E) Tu.

C. bischoffii Kos.-Pol., *Bull. Soc. Nat. Moscou* nov. ser., **29**: 153 (1916) (*C. muricata* Bischoff, non Crantz) is the name applied to a distinctive variant which occurs mainly in the eastern part of the range of **1** and apparently replaces it in Krym. In this the spines on the secondary ridges of the mericarps are 1 mm (much shorter than the width of the mericarps) and widened at the base. It is variously treated as a species or as a subspecies or variety of **1** and its status is not clear.

106. Turgenia Hoffm.[1]

Like *Caucalis* but the two primary marginal ridges each with a single row of spines or tubercles, and the remaining primary and secondary ridges similar to each other, with spines in 2–3 rows.

1. **T. latifolia** (L.) Hoffm., *Gen. Umb.* 59 (1814) (*Caucalis latifolia* L.). Annual up to 60 cm. Leaves pinnate, the segments lanceolate to oblong, serrate or pinnatifid, pubescent to hispid beneath, the margins often ciliate. Umbels long-pedunculate; rays 2–5. Bracts (2–)3–5; bracteoles 5–7, ovate-lanceolate to oblong, with wide scarious margins. Petals *c.* 5 mm, white, pink or purplish, the marginal 1 or 2 larger, radiating. Fruit 6–10 × 7 mm. *Cultivated and disturbed ground. S. & S.C. Europe; sometimes naturalized or casual farther north.* Al Au Be Bu Cr Cz Ga Ge Gr He Hs Hu It Ju Lu Rm Rs (W, K, E) Si Tu [Br Rs (C)].

107. Orlaya Hoffm.[1]

Leaves 2-pinnate. Sepals small or obsolete. Petals white or pink, the outer larger and radiate. Fruit ovoid, compressed dorsally; primary ridges slender, setulose; secondary ridges with 1 or 2 rows of spines.

1 Rays of umbel 2–4   **1. kochii**
1 Rays of umbel 5–12
   2 Spines on secondary dorsal ridges uniseriate or partly 2-seriate, strongly compressed and confluent at the base; upper cauline leaves 2- to 3-pinnatisect   **3. daucorlaya**
   2 Spines on secondary dorsal ridges 2- to 3-seriate, scarcely compressed and not confluent at the base; upper cauline leaves entire or pinnatisect   **2. grandiflora**

1. **O. kochii** Heywood, *Agron. Lusit.* 22: 13 (1961) (*O. platycarpos* Koch pro parte, *Caucalis platycarpos* auct., non L. (1753)). Annual. Stems up to 40 cm, simple or branched, slightly hairy at the base. Leaves 2- to 3-pinnate, the ultimate segments oblong. Umbels long-pedunculate, with 2–4 rays. Bracts 2–3, lanceolate, usually as long as the rays; bracteoles 2–3, ovate-lanceolate to obovate, with a membranous, glabrous or ciliate margin. Petals white, the outer 2–3 times as long as the others. Fruit 10–15 mm, ellipsoid; secondary ridges with 2–3 rows of spines which are confluent at the base and as long as the width of the fruit. 2*n* = 16. *Dry places. S. Europe.* Al Bu Co Cr Ga Gr Hs It Ju Lu Rs (K) Sa Si Tu.

O. topaliana Beauverd, *Candollea* 7: 262 (1937) described from Greece (Thessalia), has deflexed bracteoles, densely hispid-papillose primary ridges and the spines on the secondary ridges

---

forming conspicuous wings. Other plants from Greece show similar features and they may prove to form a subspecies of **1**.

2. O. grandiflora (L.) Hoffm., *Gen. Umb.* 58 (1814) (*Caucalis grandiflora* L.). Like **1** but rays 5–12; outer petals up to 8 times as long as the others; fruit *c.* 8 mm, ovoid-lanceolate with the spines on the secondary ridges shorter than the width of the fruit, scarcely compressed and not confluent. $2n=20$. *Dry places. S., C. & W. Europe, northwards to Belgium.* Al Au Be Bu Cr Cz Ga Ge Gr He Hs Hu It Ju Rm Rs (K) Tu.

3. O. daucorlaya Murb., *Lunds Univ. Årsskr.* **27**(5): 119 (1892). Annual. Stems up to 80 cm, branched from the middle or the base, glabrous. Leaves 3- to 4-pinnatisect, the ultimate segments linear-lanceolate; upper cauline leaves 2- to 3-pinnate. Umbels long-pedunculate, with 6–10 rays. Bracts lanceolate, half as long as the rays; bracteoles obovate, with a wide, membranous, ciliate margin. Petals white, the outer 8 times as long as the others. Fruit 9–11 mm, ellipsoid; spines on secondary ridges in a single row or partly 2-seriate, strongly compressed and confluent at the base. *Rocky places, scrub, hedgerows. Balkan peninsula, Slovenija; one station in C. Italy.* Al ?Bu Gr It Ju.

108. Daucus L.[1]

Leaves 2- to 3-pinnate. Bracts several, usually pinnatisect. Sepals small or obsolete. Petals white, yellowish or purplish, the outer often radiate; apex inflexed. Fruit ellipsoid to ovoid, cylindrical or somewhat compressed dorsally; primary ridges filiform, ciliate; secondary ridges with a single row of spines.

1 Umbels subsessile, leaf-opposed; styles very short   **1. durieua**
1 Umbels pedunculate; styles medium or long
  2 Lower leaves pinnately divided into sessile, rigid, multifid, apparently verticillate segments
    3 Umbels convex; rays subequal; primary ridges of mericarps densely velutinous; spines on secondary ridges about as long as the width of the mericarp   **9. setifolius**
    3 Umbels ±flat; rays unequal; primary ridges of mericarps with bristles; spines on secondary ridges 1½–2 times as long as the width of the mericarp   **10. crinitus**
  2 Lower leaves 2- to 3-pinnate, with petiolulate lobes which do not appear to be verticillate
    4 Spines on secondary ridges of mericarp not confluent at the base, not longer than the width of the mericarp   **8. carota**
    4 Spines on secondary ridges of mericarp dilated, confluent at the base or winged, (1–)2–3 times as long as the width of the mericarp
      5 Rays 3–4; the petals less than 1 mm   **7. involucratus**
      5 Rays 6 to numerous; at least the outer radiate petals more than 1 mm
        6 Bracts deflexed in flower
          7 Rays markedly unequal; spines on fruits silvery-white, widely confluent at the base; petals remaining white   **2. muricatus**
          7 Rays slightly unequal; spines on fruits yellowish, slightly dilated but not confluent at the base; petals becoming yellowish after anthesis   **5. aureus**
        6 Bracts not deflexed in flower
          8 Bracts usually longer than the rays; spines on secondary ridges of mericarp dilated but not markedly confluent at the base   **6. guttatus**
          8 Bracts as long as or shorter than the rays; spines on secondary ridges of mericarp markedly confluent and winged at the base
            9 Annual; rays 8–14; fruit 4–6 mm   **3. broteri**
            9 Biennial; rays 12–40; fruits 2–3 mm   **4. halophilus**

[1] By V. H. Heywood.

Sect. DURIEUA Batt. Umbels subsessile, leaf-opposed. Styles short, erect.

1. D. durieua Lange in Willk. & Lange, *Prodr. Fl. Hisp.* 3: 23 (1874). Annual; stems 15–30 cm, erect or ascending, flexuous, branched from the base, retrorse-scabrid. Leaves 2- to 3-pinnate; lobes lanceolate, entire or 2- to 3-fid, shortly hispid. Rays 3–5, markedly unequal. Bracts several, unequal, resembling the cauline leaves; bracteoles 3–5, linear-lanceolate, entire or pinnatifid. Petals very small, yellowish-white. Fruit 5–6 mm, oblong-elliptical; primary ridges of mericarps with several rows of whitish spines; secondary ridges with golden-yellow spines. *Dry hillsides and disturbed ground. Spain and E. Portugal.* Hs Lu.

Sect. DAUCUS. Leaf-lobes petiolulate; umbels pedunculate. Styles medium to long, erecto-patent.

2. D. muricatus (L.) L., *Sp. Pl.* ed. 2, 349 (1762). Annual; stems up to 60 cm, branched above, hispid. Leaves 3-pinnate; lobes linear-lanceolate, mucronate, sparsely hairy. Rays numerous, markedly unequal, contracted in fruit. Bracts pinnatisect, the segments linear-setaceous, later deflexed; bracteoles linear-setaceous, entire or 3-fid. Petals white, the outer strongly radiate. Fruit 5–10 mm; spines on secondary ridges silvery-white, 1–2 times as long as the width of the mericarp, strongly dilated and confluent at the base. *Dry, open habitats, especially near the sea. W. & C. Mediterranean region, Portugal.* ?Co Hs It Lu Sa Si.

3. D. broteri Ten., *Fl. Nap.* 4, *Syll. App.* 3: 4 (1830). Annual; stems 15–50 cm, erect or ascending, much-branched from the base, retrorse-scabrid or sometimes hispid near the base. Leaves 2-pinnate; lobes linear-oblong, entire or pinnatifid. Rays 8–14, short. Bracts shorter than or as long as the rays, not deflexed, pinnatifid; bracteoles linear-lanceolate to setaceous, entire or 3-sect. Petals 1–2·5 mm, white or pink, the outer radiate. Fruit 4–6 mm; spines on secondary ridges about as long as the width of the mericarp, strongly dilated and confluent at the base. *Cultivated fields and seashores. Italy, Balkan peninsula, Kriti.* Al Bu Cr Gr It Ju.

4. D. halophilus Brot., *Phyt. Lusit.* ed. 3, **2**: 198 (1827). Biennial; stems 15–40 cm, robust, erect or ascending, little-branched, retrorse-scabrid. Leaves 2-pinnate, the lobes fleshy, ovate, usually entire. Rays 12–40, long. Bracts shorter than the rays, not deflexed, with 5–7 lanceolate lobes; bracteoles cuneate, 3-lobed at apex, the margins wide, scarious and long-ciliate. Petals 1–2 mm, white (sometimes pink in the centre of the umbel), the outer radiate. Fruit 2–3 mm; spines on secondary ridges shorter than or as long as the width of the mericarp, yellowish, confluent at the base and forming a crest. *Rocky seashores. C. & S. Portugal.* Lu.

5. D. aureus Desf., *Fl. Atl.* 1: 242 (1798). Annual or biennial; stems 15–60 cm, branched, scabrid, with sparse, patent hairs or subglabrous. Leaves 3-pinnate, the lobes lanceolate to linear-lanceolate, acute. Rays numerous, slightly unequal. Bracts shorter than rays, deflexed, pinnatisect, the lobes setaceous; bracteoles usually 3-sect, rarely entire, linear. Petals white, becoming yellowish after anthesis, the outer radiate. Fruit 4–6 mm; spines on secondary ridges slightly dilated at base but not usually confluent. *Cultivated fields. E. & S. Spain, Italy, Sicilia.* Hs It Si [Ga].

6. D. guttatus Sibth. & Sm., *Fl. Graec. Prodr.* 1: 184 (1806) (*D. setulosus* Guss. ex DC., *D. bicolor* Sibth. & Sm.). Annual; stems 20–60 cm, usually several, branched, erect or ascending, retrorse-scabrid or sometimes hispid near the base. Lower leaves 2-pinnate, the segments divided into short, linear, acute lobes; upper leaves with linear segments. Rays 8–25. Bracts usually longer than umbels, not deflexed, pinnatifid; bracteoles linear-setaceous. Petals 0·6–2·5 mm, white, those of the central flower of the umbel and those of the central umbel usually dark purple. Fruit 2–4 mm; spines of the secondary ridges 1–2 times as long as the width of the mericarp. *Dry hillsides, especially near the sea. C. & S. Italy, Balkan peninsula, Aegean region, Romania.* Al Bu Cr Gr It Ju Rm Tu.

(a) Subsp. **guttatus**: Fruit 2–3 mm; spines 7–8 on each secondary ridge, twice as long as the width of the mericarp. *C. & S. Italy, Greece, Aegean region.*

(b) Subsp. **zahariadii** Heywood, *Feddes Repert.* **79**: 66 (1968): Fruit 3–4 mm; spines 9–14 on each secondary ridge, about as long as or slightly longer than the width of the mericarp. ● *Romania, Bulgaria, Jugoslavia.*

7. D. involucratus Sibth. & Sm., *Fl. Graec. Prodr.* **1**: 184 (1806). Annual; stems up to 20 cm, usually several, ascending, branched from the base, sparsely hispid or subglabrous. Leaves 1- to 2-pinnate, the segments divided into oblong-lanceolate lobes. Rays 3–4, short. Bracts usually longer than the rays, not deflexed, pinnatisect; bracteoles entire. Petals less than 1 mm, white or purplish. Fruit 2–5 mm; spines of the secondary ridges 3 times as long as the width of the mericarp. *Dry, stony places. S. & E. Greece, Aegean region.* Cr Gr.

8. D. carota L., *Sp. Pl.* 242 (1753) (incl. *D. gingidium* L.). Annual or biennial, variable in habit and branching; stems 10–100(–150) cm, glabrous to hispid. Leaves 2- to 3-pinnate, rarely less divided, the segments linear to lanceolate, glabrous to pubescent, thin or fleshy; upper cauline leaves often bract-like. Umbels concave, flat or convex, with a variable number of rays. Bracts as long as the rays or shorter, 1- to 2-pinnatisect; bracteoles of outer partial umbels 3-sect, those of the inner simple. Petals white to purplish, often with one or several flowers of the central umbel dark purple. Fruit 2–4 mm; spines on the secondary ridges not longer than the width of the mericarps. *Most of Europe.* All except Fa Is Sb, but only as an alien in Fe Rs (N).

Extremely polymorphic and variously divided into a number of subspecies, of which the following appear to deserve recognition. Hybridization between these subspecies is frequent and identification is often difficult. Several authors prefer to recognize subspp. (f) to (l) as a separate species (*D. gingidium* L.).

```
1  Umbels strongly contracted in fruit
2    Stem procumbent or ascending          (j) subsp. gadecaei
2    Stem erect
3      Terminal umbel 3–5 cm across; plant usually glabrous or
         very sparsely retrorse-scabrid      (b) subsp. maritimus
3      Terminal umbels 5–15 cm or more across, plant usually hispid
4        Tap-root swollen, fleshy, orange or whitish   (e) subsp. sativus
4        Tap-root slender, white
5          Terminal umbels (10–)12–20(–30) cm across; spines on the
             secondary ridges usually stellulate  (d) subsp. maximus
5          Terminal umbels 5–10 cm across; spines on the secondary
             ridges simple or 2-pointed, rarely stellulate
6            Segments of lower leaves ovate-lanceolate to lanceolate,
               dentate or pinnatifid with lanceolate lobes; spines
               simple or 2-pointed               (c) subsp. major
6            Segments of lower leaves ovate or cuneate-lanceolate,
               deeply pinnatifid or pinnatisect with linear or linear-
               lanceolate lobes; spines mainly simple  (a) subsp. carota
1  Umbels convex or only slightly contracted in fruit
7    Pinnae forming a right angle with the rhachis
                                                 (g) subsp. commutatus
```

```
7    Pinnae forming an acute angle with the rhachis
8      Leaves not fleshy; stem usually more than 30 cm, erect
9        Stem and leaf-rhachis sparsely pubescent   (h) subsp. hispanicus
9        Stem and leaf-rhachis densely hispid       (i) subsp. hispidus
8      Leaves ± fleshy, usually shiny; stem usually less than 30 cm,
         often procumbent or ascending
10       Stem glabrous or sparsely hairy
11         Spines on secondary ridges straight     (k) subsp. drepanensis
11         Spines on secondary ridges curving upwards
                                                    (f) subsp. gummifer
10       Stem densely hairy
12         Leaf-segments usually 3(–5)-fid, the lobes ovate-
             lanceolate                             (k) subsp. drepanensis
12         Leaf-segments incise-dentate, pinnatifid or pinnatisect
13           Leaves not shiny, densely pubescent; stems c. 10 cm,
               little-branched                      (l) subsp. rupestris
13           Leaves shiny, sparsely pubescent; main stems 10–30
               (–80) cm, much-branched, with erecto-patent
               branches                             (f) subsp. gummifer
```

(a) Subsp. **carota**: $2n = 18$. *Throughout most of the range of the species.*

(b) Subsp. **maritimus** (Lam.) Batt. in Batt. & Trabut, *Fl. Algér.* (*Dicot.*) 382 (1889) (*D. maritimus* Lam.): *Mediterranean region and Portugal, mainly near the coast.*

(c) Subsp. **major** (Vis.) Arcangeli, *Comp. Fl. Ital.* 299 (1882): *S. Europe.*

(d) Subsp. **maximus** (Desf.) Ball, *Jour. Linn. Soc. London* (*Bot.*) **16**: 476 (1878) (*D. maximus* Desf., *D. mauritanicus* L.): *Mediterranean region.*

(e) Subsp. **sativus** (Hoffm.) Arcangeli, *Comp. Fl. Ital.* 299 (1882) (*D. sativus* (Hoffm.) Roehl.): *Cultivated throughout most of Europe for its edible root (carrot).*

(f) Subsp. **gummifer** Hooker fil., *Stud. Fl. Brit. Is.* ed. 3, 185 (1884) (*D. gummifer* Lam., non All., *D. gingidium* subsp. *fontanesii* Onno pro parte): ● *Atlantic coasts of Britain, France and N. Spain.*

(g) Subsp. **commutatus** (Paol.) Thell., *Feddes Repert.* **22**: 312 (1926) (*D. gingidium* subsp. *mauritanicus* Onno): *Mediterranean coasts.*

(h) Subsp. **hispanicus** (Gouan) Thell., *loc. cit.* (1926) (*D. gummifer* All., *D. gingidium* subsp. *gummifer* (All.) Onno, *D. gingidium* subsp. *fontanesii* Onno pro parte.): *Mediterranean coasts.*
Intermediates with subsp. (g) are frequent.

(i) Subsp. **hispidus** (Arcangeli) Heywood, *Feddes Repert.* **79**: 68 (1968) (*D. gummifer* subsp. *hispidus* Arcangeli, *D. carota* subsp. *fontanesii* Thell.): *Coastal sands of S. Portugal and C. Mediterranean region; perhaps elsewhere.*

(j) Subsp. **gadecaei** (Rouy & Camus) Heywood, *Feddes Repert.* **79**: 68 (1968) (*D. communis* prol. *gadecaei* Rouy & Camus): ● *Coast of N.W. France.*

(k) Subsp. **drepanensis** (Arcangeli) Heywood, *Feddes Repert.* **79**: 68 (1968): (*D. gingidium* subsp. *drepanensis* Arcangeli, *D. bocconei* Guss., *D. polygamus* Gouan pro parte): *Mediterranean region.*

(l) Subsp. **rupestris** (Guss.) Heywood, *Feddes Repert.* **79**: 68 (1968) (*D. rupestris* Guss., *D. gingidium* subsp. *rupestris* (Guss.) Onno): ● *Islands of the C. Mediterranean region (Malta, Lampedusa, Lampesina).*

Sect. MEOIDES Lange. Leaf-segments sessile, subverticillate; umbels pedunculate. Styles medium or long, erecto-patent.

9. D. setifolius Desf., *Fl. Atl.* **1**: 244 (1798). Perennial; stems (3–)10–60 cm, erect, branched above, glabrous. Leaves linear to lanceolate in outline, pinnate, the segments pseudo-verticillate, divided into numerous linear, mucronate, rigid, sparsely pubescent lobes; basal leaves numerous, suberect; cauline leaves few. Umbels convex, with 10–20 subequal rays. Bracts 4–6, entire or pinnatifid, shorter than the rays; bracteoles setaceous, deflexed. Petals white, the outer ones scarcely radiate. Fruit 4–6 mm; primary ridges densely velutinous; secondary ridges with numerous setaceous, aculeate spines about as long as the width of the mericarp. *Dry hillsides. C. & S. Spain, C. Portugal.* Hs Lu. (*N. Africa.*)

10. D. crinitus Desf., *op. cit.* 242 (1798). Like **9** but stems sparsely retrorse-scabrid; basal leaves pubescent; male umbels more or less flat, with up to 30 unequal rays; bracts 6–10; bracteoles linear-lanceolate; petals sometimes purplish beneath; primary ridges with biseriate, sericeous bristles; secondary ridges with spines 1½–2 times as long as the width of the mericarp. *Dry hillsides and disturbed ground. C. & S. Spain, Portugal.* Hs Lu.

109. Pseudorlaya (Murb.) Murb.[1]

Leaves 2- to 3-pinnate. Bracts several, linear. Sepals conspicuous. Petals white or purplish, scarcely radiating; apex inflexed. Fruit ellipsoid; primary ridges filiform, ciliate; secondary ridges with 2–3 rows of glochidiate spines.

Fruit 7–10 × 5–6 mm; all secondary ridges with smooth spines 2·5–3·5 mm, those of the lateral ridges widened at the base
1. pumila

Fruit 5–6 × 2·5–3·5 mm; all secondary ridges with scabrid spines 0·5–1·5 mm, not widened at the base
2. minuscula

1. P. pumila (L.) Grande, *Nuovo Gior. Bot. Ital.* nov. ser., **32**: 86 (1925) (*Daucus pumilus* (L.) Hoffmanns. & Link, *Orlaya maritima* (L.) Koch, *Pseudorlaya maritima* (L.) Murb.; incl. *P. bubania* (Philippe) Murb.). Densely hairy annual up to 20 cm, branched from the base. Leaves 2- to 3-pinnate, the segments

divided into ovate lobes. Rays 2–5, unequal. Bracts 2–5, linear, acuminate, green, sometimes 3-fid; bracteoles linear-lanceolate. Petals white or purplish, the outer only slightly larger than the others. Fruit 7–10 × 5–6 mm, ellipsoid; lateral secondary ridges with 8 prominent, smooth spines, widened at the base, the other secondary ridges with *c.* 18 narrow spines 2·5–3·5 mm. 2*n*=16. *Maritime sands; more rarely inland. Mediterranean region, C. & S. Portugal.* Al Bl Co Cr Ga Gr Hs It Lu Sa Si.

A well-marked variant with short spines on the fruits (var. *brevaiculeata* (Boiss.) Heywood) is found in various parts of the range of the species; a similar variant with smaller fruits occurs in S. France and has been confused with the following species.

2. P. minuscula (Pau ex Font Quer) Laínz, *Bol. Inst. Estud. Astur.* (*Supl. Ci.*) ser. C, **5**: 39 (1962) (*P. pycnacantha* H. Lindb.). Like **1** but petals usually white; fruits 5–6 × 2·5–3·5 mm; spines on all secondary ridges 25–40, similar, 0·5–1·5 mm, scabrid, not widened at the base. *Maritime sands.* ● *N. & C. Portugal; Sardegna; probably also in Spain and S. France.* ?Ga ?Hs Lu Sa.

Very like **1** in its vegetative characters and often confused with it, but distinct in its fruits. The distribution is not yet fully known.

110. Artedia L.[2]

Leaves 3-pinnate. Sepals absent. Petals white, the inner ovate, obtuse; the outer very large, obovate, with lobed margin. Fruit ovate, strongly compressed dorsally. Dorsal ridges inconspicuous, lateral secondary ridges forming a thickened border which is deeply divided into ovate-spathulate lobes.

1. A. squamata L., *Sp. Pl.* 242 (1753). Slender, glabrous annual 30–50 cm. Leaf-lobes setaceous. Umbels with numerous rays and in the middle a tuft of violet bristles. Bracts and bracteoles leaf-like, with setaceous lobes. *Turkey-in-Europe (Gelibolu).* Tu. (*S.W. Asia.*)

[1] By V. H. Heywood.   [2] By T. G. Tutin.

APPENDICES

NOTE TO APPENDICES I–III

Considerable variation is found in the orthography of the names of many authors, especially of the earlier ones and of those whose names are transliterated from cyrillic script. Variant spellings are given here only if they are likely to give rise to doubts about identity.

The initials used by some authors vary according to whether the vernacular or latinized form of a Christian name is used (e.g. *Karl* or *Carolus*); the form most frequently used by the author is adopted in these lists.

The dates given for books and periodicals indicate, as far as can be ascertained, the date of effective publication; where this differs from dates on the title-page or elsewhere in the work itself, there is usually a reference to explain the dates given.

Certain publications are of a character intermediate between books and periodicals (e.g. seed-lists, *schedae*). The assignment of these to Appendix II or Appendix III is inevitably somewhat arbitrary.

In Appendix III there is normally no attempt made to indicate whether one periodical is a continuation of another, unless there is some continuity between them in the numbering of the volumes or series.

APPENDIX I

KEY TO THE ABBREVIATIONS OF AUTHORS' NAMES

Abromeit   J. Abromeit (1857–1946)
Acht.   B. Achtarov (1885–1959)
Adamović   L. Adamović (1864–1935)
Adams   M. F. Adams (J. F. Adam) (1780–1838)
Adanson   M. Adanson (1727–1806)
Ade   A. Ade (b. 1876)
Aellen   P. Aellen (b. 1896)
Agardh   C. A. Agardh (1785–1859)
Agardh, J.   J. G. Agardh (1813–1901)
Ahlfvengren   F. E. Ahlfvengren (1862–1921)
Ahti   T. Ahti (b. 1933)
Aichele   D. Aichele (b. 1928)
Airy Shaw   H. K. Airy Shaw (b. 1902)
Aiton   W. Aiton (1731–1793)
Aiton fil.   W. T. Aiton (1766–1849)
Albov   N. M. Albov (Alboff) (1866–1897)
Alechin   V. V. Alechin (1884–1946)
Alef.   F. G. C. Alefeld (1820–1872)
Alexeenko   M. I. Alexeenko (Alexejenko) (b. 1905)
All.   C. Allioni (1728–1804)
Alleiz.   C. d'Alleizette (b. 1884)
Almq.   S. O. I. Almquist (1844–1923)
Alpers   F. Alpers (1841–1912)
Alston   A. H. G. Alston (1902–1958)
Ambrosi   F. Ambrosi (1821–1897)
Amo   Mariano del Amo y Mora (1809–1894)
Anderson, E.   E. S. Anderson (b. 1897)
Anderson, G.   G. Anderson (d. 1817)
Andersson, N. J.   N. J. Andersson (1821–1880)
Andrasovszky   J. Andrasovszky (1889–1943)
Andreas   C. H. Andreas (b. 1898)
Andrews   H. C. Andrews (d. 1830)
Andrz.   A. L. Andrzejowski (1785–1868)
Ångström   J. Ångström (1813–1879)
Antoine   F. Antoine (1815–1886)
Arcangeli   G. Arcangeli (1840–1921)
Ard.   P. Arduino (1728–1805)
Ardoino   H. J. P. Ardoino (1819–1874)
Aresch., F.   F. W. C. Areschoug (1830–1908)
Arnold   (possibly a pseudonym; fl. 1785)
Arnott   G. A. W. Arnott (1799–1868)
Arrh., A.   J. I. A. Arrhenius (1858–1950)
Artemczuk   I. V. Artemczuk (b. 1898)
Arvat   A. Arvat (1890–1950)
Arvet-Touvet   J. M. C. Arvet-Touvet (1841–1913)
Ascherson   P. F. A. Ascherson (1834–1913)
Aspegren   G. C. Aspegren (1791–1828)
Asso   I. J. de Asso y del Rio (1742–1814)
Aublet   J. B. C. F. Aublet (1720–1778)
Aucher   P. M. R. Aucher-Eloy (1792–1838)
Avé-Lall.   J. L. E. Avé-Lallemant (1803–1867)
Aznav.   G. V. Aznavour (1861–1920)
Bab.   C. C. Babington (1808–1895)
Badaro   G. B. Badaro (1793–1831)
Bagnall   J. E. Bagnall (1830–1918)
Bailey, L. H.   L. H. Bailey (1858–1954)
Baillon   H. E. Baillon (1827–1895)

Bailly   E. Bailly (1829–1894)
Baker   J. G. Baker (1834–1920)
Baker fil.   E. G. Baker (1864–1949)
Baksay   L. Baksay (b. 1915)
Balansa   B. Balansa (1825–1891)
Balbis   G. B. Balbis (1765–1831)
Bald.   A. Baldacci (1867–1950)
Balf.   J. H. Balfour (1808–1884)
Balk.   B. E. Balkovsky (b. 1899)
Ball   J. Ball (1818–1889)
Ball, P. W.   P. W. Ball (b. 1932)
Banks   J. Banks (1743–1820)
Barbarich   A. Barbarich (b. 1903)
Barbaz.   F. Barbazita (fl. 1826)
Barbey, W.   W. Barbey-Boissier (1842–1914)
Barc.   F. Barceló y Combis (1820–1889)
Barkley, F. A.   F. A. Barkley (b. 1908)
Barkoudah   Y. I. Barkoudah (b. 1933)
Barn.   F. M. Barnéoud (b. 1821)
Barrandon   A. Barrandon (1814–1897)
Bartal.   B. Bartalini (1746–1822)
Bartl.   F. G. Bartling (1798–1875)
Bartlett   H. H. Bartlett (1886–1960)
Basil.   N. A. Basilevskaja (Bazilevskaja) (b. 1902)
Basiner   T. F. J. Basiner (1817–1862)
Bässler   M. Bässler (b. 1935)
Bast.   T. Bastard (1784–1846)
Batsch   A. J. G. C. Batsch (1761–1802)
Batt.   J. A. Battandier (1848–1922)
Baum   B. R. Baum (b. 1937)
Baumg.   J. C. G. Baumgarten (1765–1843)
Bean   W. J. Bean (1863–1947)
Beauv.   A. M. F. J. Palisot de Beauvois (1752–1820)
Beauverd   G. Beauverd (1867–1942)
Becherer   A. Becherer (b. 1897)
Bechst.   J. M. Bechstein (1757–1822)
Beck, G.   G. Beck von Mannagetta (1856–1931)
Becker, A.   A. Becker (1818–1901)
Becker, W.   W. Becker (1874–1928)
Beger   H. K. E. Beger (b. 1889)
Béguinot   A. Béguinot (1875–1940)
Bellardi   C. A. L. Bellardi (1741–1826)
Belli   S. C. Belli (1852–1919)
Bellot   F. Bellot Rodríguez (b. 1911)
Bell Salter   T. Bell Salter (1814–1858)
Benn., A. W.   A. W. Bennett (1833–1902)
Benn., Ar.   Arthur Bennett (1843–1929)
Benson, L.   L. D. Benson (b. 1909)
Bentham   G. Bentham (1800–1884)
Berchtold   F. von Berchtold (1781–1876)
Berger, A.   A. Berger (1871–1931)
Bergeret, J. P.   J. P. Bergeret (1751–1813)
Berggren, Jakob   Jakob Berggren (1790–1868)
Bergius   P. J. Bergius (1730–1790)
Bergmans   J. Bergmans (b. 1892)
Bernard   P. F. Bernard (1749–1825)
Bernh.   J. J. Bernhardi (1774–1850)

Berth. S. Berthelot (1794–1880)
Bertol. A. Bertoloni (1775–1869)
Bertram F. W. W. Bertram (1835–1899)
Besser W. S. J. G. von Besser (1784–1842)
Beyer R. Beyer (1852–1932)
Biasol. B. Biasoletto (1793–1859)
Bicknell, E. P. E. P. Bicknell (1859–1925)
Bieb. F. A. Marschall von Bieberstein (1768–1826)
Bigelow J. Bigelow (1787–1879)
Bihari J. Bihari (b. 1889)
Billot P. C. Billot (1796–1863)
Binz A. Binz (1870–1963)
Biria J. A. J. Biria (b. 1889)
Bischoff G. W. Bischoff (1797–1854)
Biv. A. de Bivona-Bernardi (1774–1837)
Blakelock R. A. Blakelock (1915–1963)
Blakeslee A. F. Blakeslee (1874–1954)
Blanc — Blanc (fl. 1866)
Blanche E. Blanche (1824–1908)
Blanco F. M. Blanco (1778–1845)
Blečić V. Blečić (b. 1911)
Błocki B. Błocki (1857–1919)
Bloxam A. Bloxam (1801–1878)
Bluff M. J. Bluff (1805–1837)
Blytt M. N. Blytt (1789–1862)
Bobrov E. G. Bobrov (b. 1902)
Boedijn K. B. Boedijn (b. 1893)
Boehmer G. R. Boehmer (1723–1803)
Boenn. C. M. F. von Boenninghausen (1785–1864)
Boguslaw I. A. Boguslaw (fl. 1846)
Boiss. P. E. Boissier (1810–1885)
Bolle, F. F. Bolle (b. 1905)
Bolós, A. A. de Bolós (b. 1899)
Bolós, O. O. de Bolós (b. 1924)
Bolton J. Bolton (c. 1758–1799)
Bolus, L. L. H. M. Bolus (Mrs F. Bolus) (b. 1877)
Bong. H. G. von Bongard (1786–1839)
Bonjean J. L. Bonjean (1780–1846)
Bonnet E. Bonnet (1848–1922)
Bonnier G. E. M. Bonnier (1853–1922)
Bonpl. A. J. A. Bonpland (1773–1858)
Borbás V. von Borbás (1844–1905)
Bord. H. Bordère (1825–1889)
Bordzil. E. I. Bordzilowski (1875–1949)
Boreau A. Boreau (1803–1875)
Borhidi A. Borhidi (b. 1932)
Boriss. A. G. Borissova-Bekrjaševa (b. 1903)
Borja J. Borja Carbonell (b. 1903)
Borkh. M. B. Borkhausen (1760–1806)
Börner C. J. B. Börner (b. 1880)
Bornm. J. F. N. Bornmüller (1862–1948)
Boros Á. Boros (b. 1900)
Borrer W. Borrer (1781–1862)
Bory J. B. G. M. Bory de Saint-Vincent (1778–1846)
Borza A. Borza (b. 1887)
Borzi A. Borzi (1852–1911)
Bosc L. A. G. Bosc (1759–1828)
Bosse J. F. W. Bosse (1788–1864)
Botsch. V. P. Botschantzev (b. 1910)
Bouché C. D. Bouché (1809–1881)
Boulay N. J. Boulay (1837–1905)
Bourgeau E. Bourgeau (1813–1877)
Bout. J. F. D. Boutigny (1820–1884)
Boutelou E. Boutelou (1776–1813)
Bouvet G. Bouvet (1874–1929)
Br., N.E. N. E. Brown (1849–1934)

Br., R. R. Brown (1773–1858)
Brackenr. W. D. Brackenridge (1810–1893)
Bradshaw, M. E. M. E. Bradshaw (b. 1926)
Brand A. Brand (1863–1931)
Brandza D. Brandza (1846–1895)
Braun, A. A. C. H. Braun (1805–1877)
Braun, G. G. Braun (1821–1882)
Braun, H. H. Braun (1851–1920)
Braun, J. J. Braun (later J. Braun-Blanquet) (b. 1884)
Br.-Bl. J. Braun-Blanquet (b. 1884)
Breistr. M. Breistroffer (b. 1906)
Brenan J. P. M. Brenan (b. 1917)
Briganti V. Briganti (1766–1836)
Brign. G. de Brignoli di Brunnhoff (1774–1857)
Briot P. L. Briot (1804–1888)
Briq. J. I. Briquet (1870–1931)
Brittinger C. C. Brittinger (1795–1869)
Britton N. L. Britton (1859–1934)
Brot. F. Avellar Brotero (1744–1828)
Brouss. P. M. A. Broussonet (1761–1807)
Browicz K. Browicz (b. 1925)
Brügger C. G. Brügger (1833–1899)
Brumh. P. Brumhard (b. 1879)
Brummitt R. K. Brummitt (b. 1937)
Bruno — Bruno (fl. 1760)
Bubani P. Bubani (1806–1888)
Buchanan-White F. Buchanan-White (1842–1894)
Buchegger J. Buchegger (b. 1886)
Buchholz J. T. Buchholz (1888–1951)
Buchinger J. D. Buchinger (1803–1888)
Buffon G. L. L. de Buffon (1707–1788)
Buhse F. A. Buhse (1821–1898)
Bunge A. A. von Bunge (1803–1890)
Burgsd. F. A. L. von Burgsdorff (1747–1802)
Burm. fil. N. L. Burman (N. L. Burmannus) (1734–1793)
Burnat E. Burnat (1828–1920)
Burtt, B. L. B. L. Burtt (b. 1913)
Busch, N. N. A. Busch (1869–1941)
Buschm. A. Buschmann (b. 1908)
Buser R. Buser (1857–1931)
Bush B. F. Bush (1858–1937)
Cadevall J. Cadevall i Diars (1846–1910)
Cajander A. K. Cajander (1879–1943)
Caldesi L. Caldesi (1821–1884)
Calestani V. Calestani (b. 1882)
Camb. J. Cambessedes (1799–1863)
Campd. F. Campderá (1793–1862)
Camus, A. A. Camus (1879–1965)
Camus E. G. Camus (1852–1915)
Cañigueral J. Cañigueral Cid (b. 1912)
Cannon J. F. M. Cannon (b. 1930)
Cariot A. Cariot (1820–1883)
Carrière E. A. Carrière (1818–1896)
Caruel T. Caruel (1830–1898)
Casav. J. Ruiz Casaviella (1835–1897)
Cast. J. L. M. Castagne (1785–1858)
Cav. A. J. Cavanilles (1745–1804)
Cavara F. Cavara (1857–1929)
Ceballos L. Ceballos Fernández de Córdoba (1896–1967)
Čelak. L. J. Čelakovsky (1834–1902)
Cesati V. de Cesati (1807–1883)
Chaix D. Chaix (1730–1799)
Cham. L. A. von Chamisso (L.C.A. Chamisseau de Boncourt) (1781–1838)
Charrel L. Charrel ('Abd-ur-Raḥmān-Nadji) (fl. 1888)
Chater A. O. Chater (b. 1933)

Chaub. L. A. Chaubard (1785–1854)
Chenevard P. Chenevard (1839–1919)
Chevall. F. F. Chevallier (1796–1840)
Chiarugi A. Chiarugi (1901–1960)
Ching, R.-C. Ren-Chang Ching (Jên-ch'ang Ch'in) (b. 1899)
Chiov. E. Chiovenda (1871–1940)
Chodat R. H. Chodat (1865–1934)
Choisy J. D. Choisy (1799–1859)
Chowdhuri P. K. Chowdhuri (b. 1923)
Chr., C. C. F. A. Christensen (1872–1942)
Christ H. Christ (1833–1933)
Christm. G. F. Christmann (b. 1752)
Chrshan. V. G. Chrshanovski (b. 1912)
Chrtek J. Chrtek (b. 1931)
Clairv. J. P. de Clairville (1742–1830)
Clapham A. R. Clapham (b. 1904)
Clarke, E. D. E. D. Clarke (1779–1822)
Claus K. Claus (1796–1864)
Clavaud A. Clavaud (1828–1890)
Cleland R. E. Cleland (b. 1892)
Clemente S. de Rojas Clemente y Rubio (1777–1827)
Clementi, G. C. G. C. Clementi (1812–1873)
Clerc O. E. Clerc (1845–1920)
Coincy A. de Coincy (1837–1903)
Coleman W. H. Coleman (?1816–1863)
Colla L. A. Colla (1766–1848)
Collett H. Collett (1836–1901)
Colmeiro M. Colmeiro y Penido (1816–1901)
Commerson P. Commerson (1727–1773)
Comolli G. Comolli (1780–1859)
Conr. P. Conrath (b. 1892)
Constance L. Constance (b. 1909)
Contandr. J. Contandriopoulos (b. 1922)
Conti, P. P. Conti (1874–1898)
Coombe, D. E. D. E. Coombe (b. 1927)
Copel. E. B. Copeland (1873–1964)
Corb. L. Corbière (1850–1941)
Corr. C. F. J. E. Correns (1864–1933)
Cosent. F. Cosentini (1769–1840)
Cosson E. S. C. Cosson (1819–1889)
Costa A. C. Costa y Cuxart (1817–1886)
Coste H. J. Coste (1858–1924)
Coulter J. M. Coulter (1851–1928)
Court. R. J. Courtois (1806–1835)
Coust. P. Cousturier (d. 1921)
Coutinho A. X. Pereira Coutinho (1851–1939)
Covas G. Covas (b. 1915)
Coville F. V. Coville (1867–1937)
Craib W. G. Craib (1882–1933)
Crantz H. J. N. von Crantz (1722–1799)
Crépin F. Crépin (1830–1903)
Crome G. E. W. Crome (1780–1813)
Cuatrec. J. Cuatrecasas (b. 1903)
Cullen J. Cullen (b. 1936)
Cunn., A. A. Cunningham (1791–1839)
Curtis W. Curtis (1746–1799)
Cusson P. Cusson (1727–1783)
Cutanda V. Cutanda (1804–1865)
Cyr. D. Cyrillo (1739–1799)
Czecz. H. Czeczott (fl. 1925–1939)
Czefr. Z. V. Czefranova (fl. 1965)
Czern. V. M. Czernajew (Czernjaew) (1796–1871)
Czetz A. Czetz (1801–1865)
Dahl, O. C. O. C. Dahl (1862–1940)
Dalby D. H. Dalby (b. 1930)
Dalla Torre K. W. von Dalla Torre (1850–1928)

Damanti P. Damanti (b. 1858)
Dandy J. E. Dandy (b. 1903)
Danilov A. D. Danilov (b. 1903)
Danser B. H. Danser (1891–1943)
Darlington, W. W. Darlington (1782–1863)
Daveau J. A. Daveau (1852–1929)
Davidov B. Davidov (1870–1927)
Davies H. Davies (1739–1821)
Davis, P. H. P. H. Davis (b. 1918)
DC. A. P. de Candolle (1778–1841)
DC., A. A. L. P. P. de Candolle (1806–1893)
DC., C. A. C. P. de Candolle (1836–1918)
De Bary H. A. de Bary (1831–1888)
Debeaux J. O. Debeaux (1826–1910)
Decken C. C. von der Decken (1833–1865)
Decker P. Decker (b. 1867)
Decne J. Decaisne (1807–1882)
Degen A. von Degen (1866–1934)
Dehnh. F. Dehnhardt (1787–1870)
Delarbre A. Delarbre (1724–1841)
De la Soie G. A. de la Soie (1818–1877)
De Lens — De Lens (fl. 1828)
Delile A. R. Delile (1778–1850)
Delponte G. B. Delponte (1812–1884)
Dematra Dematra (1742–1824)
Dennst. A. W. Dennstedt (fl. 1800–1820)
De Not. G. de Notaris (1805–1877)
Déséglise P. A. Déséglise (1823–1883)
Desf. R. L. Desfontaines (c. 1751–1833)
Desmoulins C. Desmoulins (1797–1875)
Desportes N. H. F. Desportes (1776–1856)
Desr. L. A. J. Desrousseaux (1753–1838)
Desv. A. N. Desvaux (1784–1856)
Deville L. Deville (fl. 1859)
De Wild. É. de Wildeman (1866–1947)
Dickson J. Dickson (1738–1822)
Diels F. L. E. Diels (1874–1945)
Dietr., A. A. Dietrich (1795–1856)
Dietr., D. D. N. F. Dietrich (1800–1888)
Dietr., F. G. F. G. Dietrich (1768–1850)
Dingler H. Dingler (1846–1935)
Dippel L. Dippel (1827–1914)
Dode L. A. Dode (1875–1943)
Döll J. C. Döll (1808–1885)
Dolliner G. Dolliner (1794–1872)
Domac R. Domac (b. 1918)
Domin K. Domin (1882–1952)
Domokos J. Domokos (b. 1904)
Don, D. D. Don (1799–1841)
Don, G. G. Don (1764–1814)
Don fil., G. G. Don (1798–1856)
Donn J. Donn (1758–1813)
Dörfler I. Dörfler (1866–1950)
Dorthes J. A. Dorthes (1759–1794)
Dostál J. Dostál (b. 1903)
Douglas D. Douglas (1798–1834)
Downar N. V. Downar (fl. 1855–1862)
Drejer S. T. N. Drejer (1813–1842)
Drenowski A. K. Drenowski (Drenovsky) (1879–1967)
Dreves J. F. P. Dreves (1772–1816)
Druce G. C. Druce (1850–1932)
Drude C. G. O. Drude (1852–1933)
Düben M. W. von Düben (1814–1845)
Dubois, F. F. N. A. Dubois (1752–1824)
Duby J. E. Duby (1798–1885)
Duchartre P. E. S. Duchartre (1811–1894)

Duchesne   A. N. Duchesne (1747–1827)
Dudley, T. R.   T. R. Dudley (b. 1936)
Dufour   J.-M. L. Dufour (1780–1865)
Duh.   H. L. Duhamel du Monceau (1700–1781)
Düll   R. Düll (b. 1932)
Dum.-Courset   G. L. M. Dumont de Courset (1746–1824)
Dumort.   B. C. J. Dumortier (1797–1878)
Dunal   M. F. Dunal (1789–1856)
Dupont   — Dupont (fl. 1825)
Durand, B.   B. M. Durand (b. 1928)
Durande   J. F. Durande (1732–1794)
Durieu   M. C. Durieu de Maisonneuve (1796–1878)
Duroi   J. P. Duroi (1741–1785)
D'Urv.   J. S. C. D. D'Urville (1790–1842)
Duthie   J. F. Duthie (1845–1922)
Du Tour   — Du Tour de Salvert (fl. 1803–1815)
Duval-Jouve   J. Duval-Jouve (1810–1883)
Dyer   W. T. Thiselton-Dyer (1843–1928)
Ecklon   C. F. Ecklon (1795–1868)
Edgew.   M. P. Edgeworth (1812–1881)
Edmondston   T. Edmondston (1825–1846)
Ehrenb.   C. G. Ehrenberg (1795–1876)
Ehrh.   J. F. Ehrhart (1742–1795)
Eichw.   K. E. von Eichwald (1794–1876)
Eig   A. Eig (1894–1938)
Ekman, Elis.   H. M. E. A. E. Ekman (1862–1936)
Elias   Frère H. Elias (fl. 1907–1944)
Elkan   L. Elkan (1815–1851)
Elliott   S. Elliott (1771–1830)
Emberger   M. L. Emberger (b. 1897)
Enander   S. J. Enander (1847–1928)
Endl.   S. L. Endlicher (1804–1849)
Engelm.   G. Engelmann (1809–1884)
Engler   H. G. A. Engler (1844–1930)
Engler, V.   V. Engler (1885–1917)
Exell   A. W. Exell (b. 1901)
Fabr.   P. C. Fabricius (1714–1774)
Facch.   F. Facchini (1788–1852)
Farwell   O. A. Farwell (1867–1944)
Fasano   A. Fasano (fl. 1787)
Fauché   M. Fauché (fl. 1832)
Fauconnet   C. I. Fauconnet (1811–1876)
Favrat   L. Favrat (1827–1893)
Fedde   F. K. G. Fedde (1873–1942)
Fedorov   An. A. Fedorov (b. 1908)
Fedtsch., B.   B. A. Fedtschenko (1872–1947)
Fedtsch., O.   O. A. Fedtschenko (1845–1921)
Fée   A. L. A. Fée (1789–1874)
Fenzl   E. Fenzl (1808–1879)
Ferguson, I. K.   I. K. Ferguson (b. 1938)
Fernald   M. L. Fernald (1873–1950)
Fernandes, A.   A. Fernandes (b. 1906)
Fernandes, R.   R. Fernandes (b. 1916)
Ferrarini   E. Ferrarini (b. 1919)
Fiala   F. Fiala (1861–1898)
Fieschi   V. Fieschi (b. c. 1910)
Fil.   N. Filarszky (1858–1941)
Fingerh.   K. A. Fingerhuth (1802–1876)
Fiori   A. Fiori (1865–1950)
Fischer   F. E. L. von Fischer (1782–1854)
Fischer von Wald.   A. A. Fischer von Waldheim (1803–1884)
Fitschen   J. Fitschen (1869–1947)
Fleischm.   A. Fleischmann (1805–1867)
Flerow   A. F. Flerow (1872–1960)
Flod., B.   B. G. O. Floderus (1867–1941)
Flügge   J. Flügge (1775–1816)

Focke   W. O. Focke (1834–1922)
Foggitt   W. Foggitt (1835–1917)
Fomin   A. V. Fomin (1869–1935)
Font Quer   P. Font Quer (1888–1964)
Form.   E. Formánek (1845–1900)
Forskål   P. Forskål (1732–1763)
Forster, E.   E. Forster (1765–1849)
Forster, G.   J. G. A. Forster (1754–1794)
Forster, J. R.   J. R. Forster (1729–1798)
Forster, T. F.   T. F. Forster (1761–1825)
Fouc.   J. Foucaud (1847–1904)
Foug.   A. D. Fougeroux de Bondaroy (1732–1789)
Fourn., E.   E. P. N. Fournier (1834–1884)
Fourn., P.   P.-V. Fournier (1877–1964)
Fourr.   J. P. Fourreau (1844–1871)
Franchet   A. R. Franchet (1834–1900)
Franco   J. do Amaral Franco (b. 1921)
Franklin   J. Franklin (1786–1847)
Fraser, Neill   P. Neill Fraser (1830–1905)
Freyc.   L. C. Desaulses de Freycinet (1779–1842)
Freyer   H. Freyer (1802–1866)
Freyn   J. F. Freyn (1845–1903)
Frid.   K. N. Friderichsen (1853–1932)
Friedrich   H. Friedrich (b. 1925)
Fries   E. M. Fries (1794–1878)
Fries, Th.   T. M. Fries (1832–1913)
Fritsch   K. Fritsch (1864–1934)
Fritze   R. Fritze (fl. 1870)
Friv.   E. Frivaldszky von Frivald (I. Frivaldszky) (1799–1870)
Frodin   D. G. Frodin (b. 1940)
Froelich   J. A. von Froelich (1766–1841)
Fröhlich, A.   A. Fröhlich (b. 1882)
Fröhner   S. E. Fröhner (b. 1941)
Fuchs, H. P.   H. P. Fuchs (b. 1928)
Fuss   M. Fuss (1814–1883)
Gaertner   J. Gaertner (1732–1791)
Gaertner, P.   P. G. Gaertner (1754–1825)
Gagnebin   A. Gagnebin (1707–1800)
Gaill.   C. Gaillardot (1814–1883)
Gams   H. Gams (b. 1893)
Gand.   M. Gandoger (1850–1926)
Garcke   F. A. Garcke (1819–1904)
Gariod   C. H. Gariod (1836–1892)
Gars.   F. A. de Garsault (1691–1776)
Gartner, H.   H. Gartner (fl. 1939)
Gasparr.   G. Gasparrini (1804–1866)
Gaud.-Beaup.   C. Gaudichaud-Beaupré (1789–1854)
Gaudin   J. F. A. T. G. P. Gaudin (1766–1833)
Gaussen   H. Gaussen (b. 1891)
Gavioli   O. Gavioli (1871–1944)
Gawłowska   M. J. Gawłowska (b. 1910)
Gay   J. E. Gay (1786–1864)
Gay, C.   C. Gay (1800–1873)
Gáyer   G. Gáyer (1883–1932)
Gelert   O. C. L. Gelert (1862–1899)
Genev.   L. G. Genevier (1830–1880)
Genn.   P. Gennari (1820–1897)
Genty   P. A. Genty (fl. 1890)
Georgescu   C. C. Georgescu (b. 1898)
Georgi   J. G. Georgi (1729–1802)
Georgiev   T. Georgiev (b. 1883)
Gérard   L. Gérard (1733–1819)
Germ.   J. N. E. Germain de Saint-Pierre (1815–1882)
Gibbs, P.   P. E. Gibbs (b. 1938)
Gibelli   G. Gibelli (1831–1898)
Gilib.   J. E. Gilibert (1741–1814)

Gillet   C. C. Gillet (1806–1896)
Gillies   J. Gillies (1747–1836)
Gillot   F. X. Gillot (1842–1910)
Ging.   F. C. J. Gingins de Lassaraz (1790–1863)
Ginzberger   A. Ginzberger (1873–1940)
Giraud.   L. Giraudias (1848–1922)
Gled.   J. G. Gleditsch (1714–1786)
Glück   C. M. H. Glück (1868–1940)
Gmelin, C. C.   C. C. Gmelin (1762–1837)
Gmelin, J. F.   J. F. Gmelin (1748–1804)
Gmelin, J. G.   J. G. Gmelin (1709–1755)
Gmelin, S. G.   S. G. Gmelin (1744–1774)
Godet   C. H. Godet (1797–1879)
Godman   F. Du Cane Godman (1834–1919)
Godron   D. A. Godron (1807–1880)
Goffart   J. Goffart (1864–1954)
Goiran   A. Goiran (1835–1909)
Goldie   J. Goldie (1793–1886)
Golitsin   S. V. Golitsin (b. 1897)
Gontsch.   N. F. Gontscharov (1900–1942)
González-Albo   J. González-Albo (fl. 1935)
Goodding   L. N. Goodding (b. 1880)
Gordon   G. Gordon (1806–1879)
Gorodkov   B. N. Gorodkov (1890–1953)
Gorschk.   S. G. Gorschkova (b. 1889)
Görz, R.   R. Görz (1879–1935)
Gouan   A. Gouan (1733–1821)
Goulimy   C. N. Goulimy (Goulimis) (1886–1963)
Govoruchin   V. S. Govoruchin (b. 1903)
Grab.   H. E. Grabowski (1792–1842)
Graebner   K. O. P. P. Graebner (1871–1933)
Graells   M. de la P. Graells (1809–1898)
Graham, R. C.   R. C. Graham (1786–1845)
Gram, K.   K. J. A. Gram (1897–1961)
Grande   L. Grande (1878–1965)
Gray, A.   A. Gray (1810–1888)
Gray, S. F.   S. F. Gray (1766–1828)
Grec.   D. Grecescu (1841–1910)
Greene, E. L.   E. L. Greene (1843–1915)
Gregory, E. S.   E. S. Gregory (1840–1932)
Gremli   A. Gremli (1833–1899)
Gren.   J. C. M. Grenier (1808–1875)
Greuter, W.   W. R. Greuter (b. 1938)
Grev.   R. K. Greville (1794–1866)
Grigoriev   J. S. Grigoriev (b. 1905)
Grimm   J. F. K. Grimm (1737–1821)
Grinţ., G.   G. P. Grinţescu (1870–1947)
Griseb.   A. H. R. Grisebach (1814–1879)
Gröntved   J. Gröntved (1882–1956)
Gross, H.   H. Gross (b. 1888)
Grosser   W. C. H. Grosser (b. 1869)
Grosset   H. E. Grosset (b. 1903)
Grossh.   A. A. Grossheim (1888–1948)
Gruner   L. F. Gruner (b. 1838)
Grynj   F. A. Grynj (b. 1902)
Gueldenst.   J. A. von Gueldenstaedt (1745–1781)
Guépin   J. P. Guépin (1779–1858)
Guérin   J. X. B. Guérin (1775–1850)
Guersent   L. B. Guersent (1776–1848)
Guicc.   G. Guicciardi (fl. 1855)
Guimpel   F. Guimpel (1774–1839)
Guinea   E. Guinea (b. 1907)
Guinier   P. Guinier (b. 1876)
Guittonneau   G. Guittonneau (b. 1934)
Gulia   G. Gulia (1835–1889)
Gunnarsson   J. G. Gunnarsson (1866–1944)

Gunnerus   J. E. Gunnerus (1718–1773)
Günther   C. C. Günther (1769–1833)
Gürke   A. R. L. M. Gürke (1854–1911)
Guss.   G. Gussone (1787–1866)
Guşuleac   M. Guşuleac (1887–1960)
Guterm.   W. Gutermann (b. 1935)
Guthnick   H. J. Guthnick (1800–1870)
Habl.   C. von Hablitz (1752–1821)
Hacq.   B. A. Hacquet (1739–1815)
Hadač   E. Hadač (b. 1914)
Haenke   T. Haenke (1761–1817)
Haenseler   F. Haenseler (1766–1841)
Hahne   A. Hahne (1873–1942)
Halácsy   E. von Halácsy (1842–1913)
Hallier   E. Hallier (1831–1904)
Hall, W.   W. Hall (1743–1800)
Haller   A. von Haller (1708–1777)
Haller fil.   A. von Haller (1758–1823)
Halliday   G. Halliday (b. 1933)
Hamet   R. Hamet (fl. 1906–1960)
Hampe   G. E. Hampe (1795–1880)
Hand.-Mazz.   H. von Handel-Mazzetti (1882–1940)
Hanry   H. Hanry (1807–1893)
Hara   H. Hara (b. 1911)
Hartig   H. J. A. R. Hartig (1839–1901)
Hartinger   A. Hartinger (1806–1890)
Hartman   C. J. Hartman (1790–1849)
Hartman fil.   C. Hartman (1824–1884)
Hartmann, F. X.   F. X. von Hartmann (1737–1791)
Hartweg   K. T. Hartweg (1812–1871)
Harz, C. O.   C. O. Harz (1842–1906)
Hasselq.   F. Hasselquist (1722–1752)
Hassk.   J. C. Hasskarl (1811–1894)
Hausskn.   H. K. Haussknecht (1838–1903)
Haw.   A. H. Haworth (1768–1833)
Hayek   A. von Hayek (1871–1928)
Haynald   S. F. L. Haynald (1816–1891)
Hayne   F. G. Hayne (1763–1832)
Häyrén   E. F. Häyrén (1878–1957)
Hazsl.   F. A. Hazslinszky von Hazslin (1818–1896)
Hedberg   K. O. Hedberg (b. 1923)
Hedl.   J. T. Hedlund (1861–1953)
Hedley   G. W. Hedley (1871–1941)
Heer   O. Heer (1809–1883)
Hegelm.   C. F. Hegelmaier (1834–1906)
Hegetschw.   J. J. Hegetschweiler (1789–1839)
Hegi   G. Hegi (1876–1932)
Heimans   J. Heimans (b. 1889)
Heimerl   A. Heimerl (1857–1942)
Heister   L. Heister (1683–1758)
Heldr.   T. von Heldreich (1822–1902)
Helm   G. F. Helm (fl. 1809–1828)
Hendrych   R. Hendrych (b. 1926)
Henry, A.   A. Henry (1857–1930)
Henry, Louis   Louis Henry (1853–1913)
Hepper   F. N. Hepper (b. 1929)
Herbich   F. Herbich (1791–1865)
Hermann, F.   F. Hermann (b. 1873)
Herrmann, J.   J. Herrmann (1738–1800)
Herter   W. G. Herter (1884–1958)
Hervier   J. Hervier-Basson (b. 1846)
Hess, H.   H. Hess (b. 1920)
Heuffel   J. Heuffel (1800–1857)
Heukels   H. Heukels (1854–1936)
Heynh.   G. Heynhold (fl. 1828–1850)
Heywood   V. H. Heywood (b. 1927)

Hicken   C. M. Hicken (1875–1933)
Hiern   W. P. Hiern (1839–1925)
Hieron.   G. H. E. W. Hieronymus (1846–1921)
Hiitonen   H. I. A. Hiitonen (b. 1898)
Hill   J. Hill (1716–1775)
Hill, A. W.   A. W. Hill (1875–1941)
Hitchc., A. S.   A. S. Hitchcock (1865–1935)
Hitchc., E.   E. Hitchcock (1793–1864)
Hladnik   F. Hladnik (1773–1844)
Hochst.   C. F. Hochstetter (1787–1860)
Hoffm.   G. F. Hoffmann (1761–1826)
Hoffm., O.   O. Hoffmann (1853–1909)
Hoffmanns.   J. C. von Hoffmannsegg (1766–1849)
Hofmann, E.   E. Hofmann (fl. 1839–1856)
Hohen.   R. F. Hohenacker (1798–1874)
Holl   F. Holl (fl. 1820–1842)
Holm   T. Holm (1880–1943)
Holmberg   O. R. Holmberg (1874–1930)
Holmboe   J. Holmboe (1880–1943)
Holub, J.   J. Holub (b. 1930)
Holuby   J. L. Holuby (1836–1923)
Holzm.   T. Holzmann (b. 1843)
Hooker   W. J. Hooker (1785–1865)
Hooker fil.   J. D. Hooker (1817–1911)
Hoppe   D. H. Hoppe (1760–1846)
Hormuzaki   K. Hormuzaki (1863–1937)
Hornem.   J. W. Hornemann (1770–1841)
Hornsch.   C. F. Hornschuch (1793–1850)
Hornung   E. G. Hornung (1795–1862)
Horvatić   S. Horvatić (b. 1899)
Horvátovszky   S. Horvátovszky (fl. 1770)
Hose, J. A. C.   J. A. C. Hose (d. 1800)
Hossain   M. Hossain (b. 1928)
Host   N. T. Host (1761–1834)
Houtt.   M. Houttuyn (1720–1798)
Houtzagers   G. Houtzagers (1888–1957)
Howard   H. W. Howard (b. 1913)
Howell   T. J. Howell (1842–1912)
Hruby   J. Hruby (1882–1964)
Huber, J. A.   J. A. Huber (1867–1914)
Hudson   W. Hudson (1730–1793)
Hudziok   G. W. Hudziok (b. 1929)
Huet   A. Huet du Pavillon (1829–1907)
Hülphers   K. A. Hülphers (1882–1948)
Hülsen   R. Hülsen (1837–1912)
Hultén   E. O. G. Hultén (b. 1894)
Humb.   F. H. A. von Humboldt (1769–1859)
Hussenot   L. C. S. L. Hussenot (1809–1845)
Huter   R. Huter (1834–1909)
Huth   E. Huth (1845–1897)
Hy   F. C. Hy (1853–1918)
Hyl.   N. Hylander (b. 1904)
Iljin   M. M. Iljin (Ilyin) (b. 1889)
Iljinsky, A.   A. P. Iljinsky (1885–1945)
Ingram   C. Ingram (b. 1880)
Ionescu   M. A. Ionescu (b. 1900)
Irmisch   J. F. T. Irmisch (1816–1879)
Irmscher   E. Irmscher (b. 1887)
Ivaschin   D. S. Ivaschin (b. 1912)
Jackson, A. B.   A. B. Jackson (1876–1947)
Jackson, B. D.   B. D. Jackson (1846–1927)
Jacq.   N. J. von Jacquin (1727–1817)
Jacq. fil.   J. F. von Jacquin (1766–1839)
Jaeger   H. Jaeger (1815–1890)
Jäggi   J. Jäggi (1829–1894)
Jahandiez   E. Jahandiez (1876–1938)

Jalas   J. Jalas (b. 1920)
Jameson   W. Jameson (1796–1873)
Jan   G. Jan (1791–1866)
Janchen   E. Janchen (b. 1882)
Jancz.   E. Janczewski von Glinka (1846–1918)
Janisch.   D. E. Janischewsky (1875–1944)
Janka   V. Janka von Bulcs (1837–1890)
Jaquet   F. Jaquet (1858–1933)
Jardine, N.   N. Jardine (b. 1943)
Jaub.   H. F. Jaubert (1798–1874)
Jáv.   S. Jávorka (1883–1961)
Jeanb.   E. M. J. Jeanbernat (1835–1888)
Jensen, G.   J. G. K. Jensen (1818–1886)
Jermy   A. C. Jermy (b. 1932)
Joh., K.   K. Johansson (1856–1928)
Jones, B. M. G.   B. M. G. Jones (b. 1933)
Jordan   A. Jordan (1814–1897)
Jordanov   D. Jordanov (b. 1893)
Junge   P. Junge (1881–1919)
Juratzka   J. Juratzka (1821–1878)
Jurišić   Z. J. Jurišić (1863–1921)
Juss.   A. L. de Jussieu (1748–1836)
Juss., A.   A. H. L. de Jussieu (1797–1853)
Juz.   S. V. Juzepczuk (1893–1959)
Kalela   A. Kalela (b. 1908)
Kalenicz.   J. O. Kaleniczenko (1805–1876)
Kalm   P. Kalm (1715–1779)
Kaltenb.   J. H. Kaltenbach (1807–1876)
Kanitz   Á. Kanitz (1843–1896)
Kar.   G. S. Karelin (1801–1872)
Kárpáti   Z. Kárpáti (b. 1909)
Karsten   G. K. W. H. Karsten (1817–1908)
Kaschm., B.   B. F. Kaschmensky (d. 1909)
Kauffm.   N. N. Kauffmann (N. N. Kaufman) (1834–1870)
Kaulfuss   G. F. Kaulfuss (1786–1830)
Kazim.   T. Kazimierski (b. 1924)
Keissler   K. von Keissler (b. 1872)
Keller, J. B.   J. B. von Keller (1841–1897)
Keller, R.   R. Keller (1854–1939)
Kellerer, J.   J. Kellerer (fl. 1905)
Ker-Gawler   J. B. Ker (J. Gawler) (1764–1842)
Kerner, A.   A. J. Kerner von Marilaun (1831–1898)
Kerner, J.   J. Kerner (1829–1906)
Kiffm.   R. Kiffmann (fl. 1952)
Kihlman   A. O. Kihlman (Kairamo) (1858–1938)
Kindb.   N. C. Kindberg (1832–1910)
Kir.   I. P. Kirilow (1821 or 1822–1842)
Kirby   M. Kirby (1817–1893)
Kirchner   G. Kirchner (1837–1885)
Kirschleger   F. R. Kirschleger (1804–1869)
Kit.   P. Kitaibel (1757–1817)
Kitanov   B. Kitanov (b. 1912)
Kittel   M. B. Kittel (1798–1885)
Klásková, A.   A. Klásková (later A. Skalická) (b. 1932)
Klášt.   I. Klášterský (b. 1901)
Klebahn   H. Klebahn (1859–1942)
Kleopow   J. D. Kleopow (1902–1942)
Klika   J. Klika (1888–1957)
Klinggr.   K. J. von Klinggraeff (1809–1879)
Klink.   M. Klinkowski (b. 1904)
Klokov   M. V. Klokov (b. 1896)
Klotzsch   J. F. Klotzsch (1805–1860)
Knaben   G. Knaben (b. 1911)
Knaf   J. Knaf (1801–1865)
Knerr   E. B. Knerr (1861–1942)
Knight   J. Knight (1781–1855)

Knoche H. Knoche (1870–1945)
Koch W. D. J. Koch (G. D. I. Koch) (1771–1849)
Koch, C. C. H. E. Koch (1809–1879)
Koch, Walo Walo Koch (1896–1956)
Koehler J. C. G. Koehler (1759–1833)
Koehne B. A. E. Koehne (1848–1918)
Koelle J. L. C. Koelle (1763–1797)
Koerte F. Koerte (1782–1845)
Komarov V. L. Komarov (1869–1945)
Kondrat. E. N. Kondratjuk (b. 1914)
König, D. D. König (b. 1909)
Korsh. S. I. Korshinsky (1861–1900)
Košanin N. Košanin (1874–1934)
Kos.-Pol. B. M. Koso-Poliansky (1890–1957)
Kossych V. M. Kossych (b. 1931)
Kostel. V. F. Kosteletzky (1801–1887)
Kotov M. I. Kotov (b. 1896)
Kotschy T. Kotschy (1813–1866)
Kotula, A. A. Kotula (1822–1891)
Kovanda M. Kovanda (b. 1936)
Krašan F. Krašan (1840–1907)
Krasch. H. M. Krascheninnikov (1884–1947)
Krause, E. H. L. E. H. L. Krause (1859–1942)
Krause, K. K. Krause (fl. 1958)
Krecz., V. V. I. Kreczetowicz (1901–1942)
Krocker A. J. Krocker (1744–1823)
Krok T. O. B. N. Krok (1834–1921)
Krylov P. N. Krylov (1850–1931)
Krysht. A. M. Kryshtofovicz (1885–1953)
Kühlew. P. E. Kühlewein (1798–1870)
Kuhn M. F. A. Kuhn (1842–1894)
Kulcz. S. Kulczyński (b. 1895)
Kümmerle J. B. Kümmerle (1876–1931)
Kunth C. S. Kunth (1788–1850)
Kuntze, O. K. E. O. Kuntze (1843–1907)
Kunz, H. H. Kunz (fl. 1950)
Kunze, G. G. Kunze (1793–1851)
Kupcsok S. Kupcsok (1850–1914)
Kupffer K. R. Kupffer (1872–1935)
Kuprian. L. A. Kuprianova (b. 1914)
Kurtz, F. F. Kurtz (1854–1920)
Kusn. N. I. Kusnezow (Kuznetzow) (1864–1932)
Kuzen. O. I. Kuzeneva (b. 1887)
Kuzinský P. A. von Kuzinský (fl. 1889)
L. C. von Linné (C. Linnaeus) (1707–1778)
L. fil. C. von Linné (1741–1783)
Labill. J. J. H. de Labillardière (1755–1834)
Lacaita C. C. Lacaita (1853–1933)
Laest. L. L. Laestadius (1800–1861)
Lag. M. Lagasca y Segura (1776–1839)
Lagerh. N. G. von Lagerheim (1860–1926)
Lagger F. Lagger (1799–1870)
Lagrèze-Fossat A. R. A. Lagrèze-Fossat (1818–1874)
Laicharding J. N. von Laicharding (1754–1797)
Laínz M. Laínz (b. 1923)
Laínz, J. M. J. M. Laínz (b. 1900)
Lam. J. B. A. P. Monnet de la Marck (1744–1829)
Lamb. A. B. Lambert (1761–1842)
Lamotte M. Lamotte (1820–1883)
Landolt E. Landolt (b. 1926)
Láng, A. F. A. F. Láng (1795–1863)
Lang, O. F. O. F. Lang (1817–1847)
Lange J. M. C. Lange (1818–1898)
Langsd. G. H. von Langsdorff (1774–1852)
Lapeyr. P. Picot de Lapeyrouse (1744–1818)
Lapierre J. M. Lapierre (1754–1834)

La Pylaie A. J. M. B. de la Pylaie (1786–1856)
Larsen, K. K. Larsen (b. 1926)
Lasch W. G. Lasch (1787–1863)
Lasebna A. M. Lasebna (b. 1922)
Latourr. M. A. L. Claret de Latourrette (1729–1793)
Lauche W. Lauche (1827–1882)
Lauth T. Lauth (1758–1826)
Lavrenko E. M. Lavrenko (b. 1900)
Lawalrée A. Lawalrée (b. 1921)
Lawrance M. Lawrance (fl. 1790–1831)
Lawson, C. C. Lawson (1794–1873)
Lawson, P. P. Lawson (d. 1820)
Laxm. E. Laxmann (1737–1796)
Layens G. de Layens (1834–1897)
Laza M. Laza Palacios (b. 1901)
Láz.-Ibiza Blas Lázaro é Ibiza (1858–1921)
Lebel J. E. Lebel (1801–1878)
Lecoq H. Lecoq (1802–1871)
Lecoyer C.-J. Lecoyer (1835–1899)
Ledeb. C. F. von Ledebour (1785–1851)
Leers J. D. Leers (1727–1774)
Lees E. Lees (1800–1887)
Lefèvre L. V. Lefèvre (b. 1810)
Le Gall N. J. M. le Gall (1787–c. 1860)
Le Grand A. le Grand (1839–1905)
Lehm. J. G. C. Lehmann (1792–1860)
Lehm., C. B. C. B. Lehmann (fl. 1860)
Leins P. Leins (b. 1937)
Lej. A. L. S. Lejeune (1779–1858)
Le Jolis A. F. le Jolis (1823–1904)
Lemaire C. A. Lemaire (1801–1871)
Léman D. S. Léman (1781–1829)
Lemke W. Lemke (b. 1893)
Lengyel G. Lengyel (1884–1965)
Leresche L. Leresche (1808–1885)
Lesp. G. Lespinasse (1807–1876)
Less. C. F. Lessing (1810–1862)
Lester-Garland L. V. Lester-Garland (1860–1944)
Letendre J. B. P. Letendre (1828–1886)
Léveillé A. A. H. Lćvcillć (1863–1918)
Levier E. Levier (1838–1911)
Lewis, P. P. Lewis (b. 1924)
Ley, A. A. Ley (1842–1911)
Leybold F. Leybold (1827–1879)
L'Hér. C. L. L'Héritier de Brutelle (1746–1800)
Libert M. A. Libert (1782–1865)
Lid J. Lid (b. 1886)
Liebl. F. K. Lieblein (1744–1810)
Liebm. F. M. Liebmann (1813–1856)
Liljeblad S. Liljeblad (1761–1815)
Liljefors A. W. Liljefors (b. 1904)
Lindb., H. H. Lindberg (1871–1963)
Lindblad M. A. Lindblad (1821–1899)
Lindblom A. E. Lindblom (1807–1853)
Lindeb. C. J. Lindeberg (1815–1900)
Lindem. E. von Lindemann (1825–1900)
Lindley J. Lindley (1799–1865)
Lindman C. A. M. Lindman (1856–1928)
Lindt. V. H. Lindtner (1904–1965)
Link J. H. F. Link (1767–1851)
Linton, E. F. E. F. Linton (1848–1928)
Lipsky V. I. Lipsky (1863–1937)
Litard. R. V. de Litardière (1888–1957)
Litv. D. I. Litvinov (Litwinow) (1854–1929)
Lloyd J. Lloyd (1810–1896)
Loddiges G. Loddiges (1784–1846)

Loefl.   P. Loefling (1729–1756)
Loesener   L. E. T. Loesener (1865–1941)
Loisel.   J. L. A. Loiseleur-Deslongchamps (1774–1849)
Lojac.   M. Lojacono-Pojero (1853–1919)
Londes   F. W. Londes (1780–1807)
Lonsing   A. Lonsing (fl. 1939)
Lorent   J. A. von Lorent (1812–1884)
Loret   H. Loret (1810–1888)
Losa   M. Losa España (1893–1966)
Loscos   F. Loscos y Bernál (1823–1886)
Loudon   J. C. Loudon (1783–1843)
Lour.   J. de Loureiro (1717–1791)
Löve, Á.   Á. Löve (b. 1916)
Löve, D.   D. Löve (b. 1918)
Lowe   R. T. Lowe (1802–1874)
Lucand   J.-L. Lucand (1821–1896)
Lucé   J. W. L. von Lucé (fl. 1823)
Luerssen   C. Luerssen (1843–1916)
Luizet   D. Luizet (1851–1930)
Lund, N.   N. Lund (1814–1847)
Lundström, E.   E. Lundström (b. 1882)
Lynch   R. I. Lynch (1850–1924)
Lynge   B. A. Lynge (1884–1942)
Maack   R. Maack (1825–1886)
Macbride   J. F. Macbride (b. 1892)
Macfadyen   J. Macfadyen (1798–1850)
Mach.-Laur.   B. Machatschki-Laurich
Mackenzie   K. K. Mackenzie (1877–1934)
Magne   J. H. Magne (1804–1885)
Magnier   C. Magnier (fl. 1883)
Maillefer   A. Maillefer (b. 1880)
Maire   R. C. J. E. Maire (1878–1949)
Majevski   P. F. Majevski (1851–1892)
Major   C. J. F. Major (1843–1923)
Makino   T. Makino (1862–1957)
Malbr.   A. F. Malbranche (1818–1888)
Malinovski   E. Malinovski (b. 1885)
Malladra   A. Malladra (1865–1944)
Malte   M. O. Malte (1880–1933)
Maly, F.   F. de Paula Maly (1823–1891)
Maly, J.   Joseph Karl Maly (1797–1866)
Malý, K.   Karl Malý (1874–1951)
Manden.   I. P. Mandenova (b. 1907)
Mansfeld   R. Mansfeld (1901–1960)
Manton   I. Manton (b. 1904)
Marchesetti   C. de Marchesetti (1850–1926)
Marès   P. Marès (1826–1900)
Margot   H. Margot (fl. 1838)
Mariz   J. de Mariz (1847–1916)
Markgraf   F. Markgraf (b. 1897)
Marsden-Jones   E. M. Marsden-Jones (1887–1960)
Marshall   H. Marshall (1722–1801)
Marshall, E. S.   E. S. Marshall (1858–1919)
Marsson   T. F. Marsson (1816–1892)
Mart., C. F. P.   C. F. P. von Martius (1794–1868)
Mart., H.   H. von Martius (1781–1831)
Martelli, U.   U. Martelli (1860–1934)
Martrin-Donos   J. V. de Martrin-Donos (1801–1870)
Martyn   T. Martyn (1736–1825)
Massara   G. F. Massara (1792–1839)
Masters   M. T. Masters (1833–1907)
Máthé   I. Máthé (b. 1911)
Mattei   G. E. Mattei (1865–1943)
Mattf.   J. Mattfeld (1895–1951)
Mattuschka   H. G. von Mattuschka (1734–1779)
Maurer   W. Maurer (b. 1926)
Mauri   E. Mauri (1791–1836)

Maxim.   K. J. Maximowicz (1827–1891)
Maxon   W. R. Maxon (1877–1948)
Mayer, E.   E. Mayer (b. 1920)
Mayer, J.   J. C. A. Mayer (1747–1801)
McClell.   J. McClelland (1805–1883)
McMillan   C. McMillan (1867–1929)
McNeill   J. McNeill (b. 1933)
Medicus   F. C. Medicus (Medikus) (1736–1808)
Medv.   J. S. Medvedev (1847–1923)
Meerb.   N. Meerburgh (1734–1814)
Meikle   R. D. Meikle (b. 1923)
Meissner   C. F. Meissner (1800–1874)
Mela   A. J. Mela (1846–1904)
Melderis   A. Melderis (b. 1909)
Melville   R. Melville (b. 1903)
Mendes   E. J. S. M. Mendes (b. 1924)
Menéndez Amor   J. Menéndez Amor (b. 1916)
Menyh.   L. Menyhárth (1849–1897)
Mérat   F. V. Mérat (1780–1851)
Merc.   E. Mercier (1802–1863)
Merino   P. B. Merino y Román (1845–1917)
Merr.   E. D. Merrill (1875–1956)
Mert.   F. K. Mertens (1764–1831)
Merxm.   H. Merxmüller (b. 1920)
Metsch   J. C. Metsch (1796–1856)
Mett.   G. H. Mettenius (1823–1866)
Metzger   J. Metzger (1789–1852)
Meyen   F. J. F. Meyen (1804–1840)
Meyer, B.   B. Meyer (1767–1836)
Meyer, C. A.   C. A. von Meyer (1795–1855)
Meyer, D. E.   D. E. Meyer (b. 1926)
Meyer, E. H. F.   E. H. F. Meyer (1791–1858)
Meyer, G. F. W.   G. F. W. Meyer (1782–1856)
Michalet   E. Michalet (1829–1862)
Micheletti   L. Micheletti (1844–1912)
Michx   A. Michaux (1746–1802)
Michx fil.   F. A. Michaux (1770–1855)
Middendorff   A. T. von Middendorff (1815–1894)
Miégeville   Abbé Miégeville (1814–1901)
Milde   C. A. J. Milde (1824–1871)
Miller   P. Miller (1691–1771)
Min.   N. A. Miniaev (b. 1909)
Miq.   F. A. W. Miquel (1811–1871)
Mirbel   C. F. B. Mirbel (1776–1854)
Mitterp.   L. Mitterpacher (1734–1818)
Moench   C. Moench (1744–1805)
Moessler   J. C. Moessler (fl. 1805–1815)
Moesz   G. Moesz (1873–1946)
Mohr   D. M. H. Mohr (1779–1808)
Moldenke   H. N. Moldenke (b. 1909)
Molina   J. I. Molina (1740–1829)
Molinier   R. Molinier (b. 1899)
Monnard   J. P. Monnard (b. 1791)
Monnier, P.   P. C. J. Monnier (b. 1922)
Montandon   P. J. Montandon (fl. 1856)
Montbret   G. Coquebert de Montbret (1805–1837)
Montelucci   G. Montelucci (b. 1899)
Monts., P.   P. Montserrat Recoder (b. 1920)
Moq.   C. H. B. A. Moquin-Tandon (1804–1863)
Moretti   G. Moretti (1782–1853)
Mori   A. Mori (1847–1902)
Moric.   M. E. Moricand (1779–1854)
Moris   G. G. Moris (1796–1869)
Moritzi   A. Moritzi (1806–1850)
Morot   M. L. Morot (fl. 1885)
Morren   C. J. E. Morren (1833–1886)
Morton, C. V.   C. V. Morton (b. 1905)

Möschl   W. Möschl (b. 1906)
Moss   C. E. Moss (1872–1930)
Motelay   L. Motelay (1831–1917)
Mouillefert   P. Mouillefert (1845–1903)
Mueller, F.   F. H. J. von Mueller (1825–1896)
Mueller, P. J.   P. J. Mueller (1832–1889)
Muenchh.   O. Muenchhausen (1716–1774)
Muhl.   G. H. E. Muhlenberg (1753–1815)
Müller Arg.   J. Müller of Aargau (Argoviensis) (1828–1896)
Munby   G. Munby (1812–1876)
Münch   E. Münch (1876–1946)
Munz   P. A. Munz (b. 1892)
Murb.   S. S. Murbeck (1859–1946)
Murr, J.   J. Murr (1864–1932)
Murray   J. A. Murray (1740–1791)
Murray, A.   A. Murray (1812–1878)
Murray, R. P.   R. P. Murray (1842–1908)
Muschler   R. Muschler (b. 1883)
Mutel   A. Mutel (1795–1847)
Mutis   J. C. Mutis (1732–1808)
Mygind   F. Mygind (1710–1789)
Nakai   T. Nakai (1882–1952)
Nasarow   M. I. Nasarow (1882–1942)
Nath.   A. G. Nathorst (1850–1921)
Naudin   C. V. Naudin (1815–1899)
Necker   N. J. de Necker (1730–1793)
Nees   C. G. D. Nees von Esenbeck (1776–1858)
Nees, T.   T. F. L. Nees von Esenbeck (1787–1837)
Neilr.   A. Neilreich (1803–1871)
Nelson, A.   A. Nelson (1859–1952)
Nenukow   F. S. Nenukow (fl. ?1917)
Nestler   C. G. Nestler (1778–1832)
Neuman   L. M. Neuman (1852–1922)
Neumann, A.   A. Neumann (fl. 1960)
Neumayer, H.   H. Neumayer (1887–1945)
Neves, J.   J. de Barros Neves (b. 1914)
Nevski   S. A. Nevski (1908–1938)
Newbould   W. W. Newbould (1819–1886)
Newman   E. Newman (1801–1876)
Neygenf.   F. W. Neygenfind (fl. 1821)
Nicotra   L. Nicotra (b. 1846)
Niedenzu   F. J. Niedenzu (1857–1937)
Nikif.   N. B. Nikiforova (fl. 1947)
Nobre   A. Nobre (b. 1865)
Nolte   E. F. Nolte (1791–1875)
Nordborg   G. Nordborg (b. 1931)
Nordh.   R. Nordhagen (b. 1894)
Nordm.   A. von Nordmann (1803–1866)
Nordstedt   C. F. O. Nordstedt (1838–1924)
Norrlin   J. P. Norrlin (1842–1917)
Norton   J. B. Norton (1877–1938)
Notø   A. Notø (1865–1948)
Noulet   J. B. Noulet (1802–1890)
Novák   F. A. Novák (1892–1964)
Nowacki   E. K. Nowacki (b. 1930)
Nutt.   T. Nuttall (1786–1859)
Nyárády, A.   A. Nyárády (b. 1920)
Nyárády, E. I.   E. I. Nyárády (1881–1966)
Nyl., F.   F. Nylander (1820–1880)
Nyl., W.   W. Nylander (1822–1899)
Nyman   C. F. Nyman (1820–1893)
Ockendon   D. J. Ockendon (b. 1940)
Oeder   G. C. Oeder (1728–1791)
Ohwi   J. Ohwi (b. 1905)
Oken   L. Oken (1779–1851)
Oliver   D. Oliver (1830–1916)
Olivier   G. A. Olivier (1756–1814)

Onno   M. Onno (b. 1903)
Opiz   P. M. Opiz (1787–1858)
Opperman   P. A. Opperman (fl. 1954)
Orlova   N. I. Orlova (b. 1921)
Orph.   T. G. Orphanides (1817–1886)
Örsted   A. S. Örsted (1816–1872)
Ortega   C. Gómez Ortega (1740–1818)
Ortmann   J. Ortmann (b. 1814)
Osbeck   P. Osbeck (1723–1805)
Óskarsson   I. Óskarsson (b. 1892)
Ostenf.   C. E. H. Ostenfeld (1873–1931)
Otth   K. A. Otth (1803–1839)
Otto   C. F. Otto (1783–1856)
Ovcz.   P. N. Ovczinnikov (b. 1903)
Pacher   D. Pacher (1817–1902)
Pacz.   I. K. Paczoski (1864–1942)
Padmore   P. A. Padmore (b. 1929)
Paegle   B. Paegle (fl. 1927)
Palassou   P. B. Palassou (1745–1830)
Palhinha   R. T. Palhinha (1871–1957)
Palitz · R. Palitz (fl. 1935)
Pallas   P. S. Pallas (1741–1811)
Pamp.   R. Pampanini (1875–1949)
Pančić   J. Pančić (1814–1888)
Pangalo   K. I. Pangalo (1883–1965)
Pant.   J. Pantocsek (1846–1916)
Panţu   Z. C. Panţu (1866–1934)
Paol.   G. Paoletti (1865–1941)
Pardo   J. Pardo y Sastrón (1822–1909)
Parl.   F. Parlatore (1816–1877)
Parodi   L. R. Parodi (b. 1895)
Parry   W. E. Parry (1790–1855)
Passer.   G. Passerini (1816–1893)
Patrin   E. L. M. Patrin (1742–1815)
Patze   C. A. Patze (1808–1892)
Pau   C. Pau (1857–1937)
Paulin   A. Paulin (1853–1942)
Paulsen   O. V. Paulsen (1874–1947)
Pavlov   N. V. Pavlov (b. 1893)
Pavón   J. Pavón (1750–1844)
Pawł.   B. Pawłowski (b. 1898)
Pax   F. A. Pax (1858–1942)
Pénzes   A. Pénzes (b. 1895)
Peola   P. Peola (b. 1869)
Pérard   M. Pérard (1835–1887)
Pérez Lara   J. M. Pérez Lara (1841–1918)
Perf.   I. A. Perfiljew (1882–1942)
Perr.   J. O. E. Perrier (1843–1916)
Pers.   C. H. Persoon (c. 1762–1836)
Péterfi   M. Péterfi (1875–1922)
Peterm.   W. L. Petermann (1806–1855)
Petitmengin   M. G. C. Petitmengin (1881–1908)
Petrov   V. A. Petrov (1896–1955)
Petrović   S. Petrović (1839–1889)
Petzold   C. E. A. Petzold (1815–1891)
Philippe   X. Philippe (1802–1866)
Phillips, E. P.   E. P. Phillips (1884–1967)
Phipps, C. J.   C. J. Phipps (1744–1792)
Pierrat   D. Pierrat (1835–1895)
Pignatti   S. Pignatti (b. 1930)
Pilger   R. K. F. Pilger (1876–1953)
Piller   M. Piller (1733–1788)
Pio   G. B. Pio (fl. 1813)
Piré   L. A. H. J. Piré (1827–1887)
Pires de Lima   A. Pires de Lima (b. 1886)
Pirona   G. A. Pirona (1822–1895)
Pissjauk.   V. V. Pissjaukowa (b. 1906)

Pitard   C. J. Pitard (1873–1927)
Planchon   J. E. Planchon (1823–1888)
Planellas   J. Planellas Giralt (1821–1888)
Pobed.   E. G. Pobedimova (b. 1898)
Podl.   D. Podlech (b. 1931)
Podp.   J. Podpěra (1878–1954)
Poech   J. Poech (1816–1846)
Poggenb.   J. F. Poggenburg (1840–1893)
Pohl   J. B. E. Pohl (1782–1834)
Poiret   J. L. M. Poiret (1755–1834)
Poirion   L. P. Poirion (b. 1901)
Poiteau   P. A. Poiteau (1766–1854)
Pojark.   A. I. Pojarkova (b. 1897)
Pollich   J. A. Pollich (1740–1780)
Pollini   C. Pollini (1782–1833)
Pomel   A. Pomel (1821–1898)
Popl.   G. I. Poplavskaja (Poplawska) (1885–1956)
Popov, M.   M. G. Popov (1893–1955)
Porc.   F. Porcius (1816–1907)
Porsild, A. E.   A. E. Porsild (b. 1901)
Porta   P. Porta (1832–1923)
Portenschl.   F. E. von Portenschlag-Ledermayer (1772–1822)
Pospichal   E. Pospichal (1838–1905)
Post   G. E. Post (1838–1909)
Pourret   P. A. Pourret de Figeac (1754–1818)
Pozd.   N. G. Pozdeeva (b. 1913)
Praeger   R. L. Praeger (1865–1953)
Prantl   K. A. E. Prantl (1849–1893)
Presl, C.   C. (K.) B. Presl (1794–1852)
Presl, J.   J. S. Presl (1791–1849)
Price   W. R. Price (b. 1886)
Pritzel, G. A.   G. A. Pritzel (1815–1874)
Proctor, M. C. F.   M. C. F. Proctor (b. 1929)
Prodan   J. Prodan (1875–1959)
Progel   A. Progel (1829–1889)
Prokh.   J. I. Prokhanov (1902–1964)
Prolongo   P. Prolongo y García (1806–1885)
Puget   F. Puget (1829–1880)
Pugsley   H. W. Pugsley (1868–1947)
Pulliat   V. Pulliat (1827–1866)
Purkyně   E. Purkyně (1831–1882)
Pursh   F. T. Pursh (1774–1820)
Putterlick   A. Putterlick (1810–1845)
Quézel   P. Quézel (b. 1926)
Rabenh.   G. L. Rabenhorst (1806–1881)
Racib.   M. Raciborski (1864–1917)
Raddi   G. Raddi (1770–1829)
Rafin.   C. S. Rafinesque-Schmaltz (1783–1840)
Rafn   C. G. Rafn (1769–1808)
Ramond   L. F. E. Ramond de Carbonnières (1753–1827)
Rapin   D. Rapin (1799–1882)
Rau   A. Rau (1784–1830)
Raunk.   C. Raunkiær (1860–1938)
Räuschel   E. A. Räuschel (fl. 1772–1797)
Raven, P. H.   P. H. Raven (b. 1936)
Rayss   T. Rayss (1890–1965)
Rech.   K. Rechinger (1867–1952)
Rech. fil.   K. H. Rechinger (b. 1906)
Rees   A. Rees (1743–1825)
Regel   E. A. von Regel (1815–1892)
Rehder   A. Rehder (1863–1949)
Rehmann   A. Rehmann (1840–1917)
Reichard   J. J. Reichard (1743–1782)
Reichenb.   H. G. L. Reichenbach (1793–1879)
Reichenb. fil.   H. G. Reichenbach (1824–1889)
Rendle   A. B. Rendle (1865–1938)
Renner   O. Renner (1883–1960)

Req.   E. Requien (1788–1851)
Resvoll-Holmsen   H. Resvoll-Holmsen (1873–1943)
Retz.   A. J. Retzius (1742–1821)
Reuss, G.   G. Reuss (1818–1861)
Reuter   G. F. Reuter (1805–1872)
Revel   J. Revel (1811–1887)
Reverchon   E. Reverchon (1835–1914)
Reyn.   A. Reynier (1845–1932)
Ricci   A. M. Ricci (1777–1850)
Richard, A.   A. Richard (1794–1852)
Richard, L. C. M.   L. C. M. Richard (1754–1821)
Richter   H. E. F. Richter (1808–1876)
Richter, K.   K. Richter (1855–1891)
Riddelsd.   H. J. Riddelsdell (1866–1941)
Rigo   G. Rigo (1841–1922)
Rikli   M. A. Rikli (1868–1951)
Rink   H. J. Rink (1819–1893)
Ripart   J. B. M. J. S. E. Ripart (1814–1878)
Risso   J. A. Risso (1777–1845)
Rivas Goday   S. Rivas Goday (b. 1905)
Robert   — Robert (fl. 1838)
Roberts, J.   J. Roberts (1912–1960)
Robill.   L. M. A. Robillard d'Argentelle (d. 1828)
Robson   E. Robson (1763–1813)
Robson, N. K. B.   N. K. B. Robson (b. 1928)
Robyns   W. Robyns (b. 1901)
Rochel   A. Rochel (1770–1847)
Rodr.   J. D. Rodriguez (1780–1846)
Roemer   J. J. Roemer (1763–1819)
Roemer, M. J.   M. J. Roemer (fl. 1835–1846)
Roemer, R. de   R. de Roemer (fl. 1852)
Rogow.   A. S. Rogowicz (1812–1878)
Rohde   M. Rohde (1782–1812)
Rohlena   J. Rohlena (1874–1944)
Röhling   J. C. Röhling (1757–1813)
Rohrb.   P. Rohrbach (1847–1871)
Ronniger   K. Ronniger (1871–1954)
Rose   J. N. Rose (1862–1928)
Ross, J.   J. Ross (1777–1856)
Rossi   M. L. Rossi (1850–1932)
Rössler   W. Rössler (b. 1909)
Rostański   K. Rostański (b. 1930)
Rostock   M. Rostock (fl. 1884)
Rostrup   F. G. E. Rostrup (1831–1907)
Roth   A. W. Roth (1757–1834)
Rothm.   W. Rothmaler (1908–1962)
Rottb.   C. F. Rottboel (Rottbøll) (1727–1797)
Rouleau   E. Rouleau (b. 1916)
Rouy   G. C. C. Rouy (1851–1924)
Roxb.   W. Roxburgh (1751–1815)
Royle   J. F. Royle (1779–1858)
Rozan.   M. A. Rozanova (1885–1957)
Rozeira   A. D. F. Rozeira (b. 1912)
Rudolph, J. H.   J. H. Rudolph (1744–1809)
Ruiz   H. Ruiz López (1754–1815)
Runemark   H. Runemark (b. 1927)
Rupr.   F. J. Ruprecht (1814–1870)
Russell, A.   A. Russell (?1715–1768)
Russell, P.   P. G. Russell (b. 1889)
Rydb.   P. A. Rydberg (1860–1931)
Rylands   T. G. Rylands (1818–1900)
Sabine   J. Sabine (1770–1837)
Sabr.   H. Sabransky (1864–1916)
Sadler   J. Sadler (1791–1849)
Sageret   A. Sageret (1763–1851)
Sagorski   E. Sagorski (1847–1929)
Salis   C. Ulysses von Salis-Marschlins (1760–? 1818)

Salisb.   R. A. Salisbury (1761–1829)
Salmon   C. E. Salmon (1872–1930)
Salzm.   P. Salzmann (1781–1851)
Sam.   G. Samuelsson (1885–1944)
Sambuk   F. V. Sambuk (1900–1942)
Samp.   G. A. da Silva Ferreira Sampaio (1865–1937)
Sanadze   K. S. Sanadze (fl. 1946)
Sándor   I. Sándor (b. 1853)
Sandwith   N. Y. Sandwith (1901–1965)
Sanguinetti   P. Sanguinetti (1802–1868)
Santi, G.   G. Santi (1746–1822)
Sapjegin   A. A. Sapjegin (1883–1946)
Sarato   C. Sarato (1830–1893)
Sarg.   C. S. Sargent (1841–1927)
Sarnth.   L. von Sarntheim (1861–1914)
Sart.   G. B. Sartorelli (1780–1853)
Sauer   F. W. H. Sauer (1803–1873)
Saunders   W. W. Saunders (1809–1879)
Sauter   A. E. Sauter (1800–1881)
Sauzé   C. Sauzé (1815–1889)
Savi   G. Savi (1769–1844)
Savigny   M. J. C. Lelorgne de Savigny (1777–1851)
Săvul.   T. Săvulescu (1889–1963)
Scaling   W. Scaling (fl. 1863–1882)
Schaeffer   J. C. Schaeffer (1718–1790)
Schaeftlein   H. Schaeftlein (b. 1886)
Scheele   G. H. A. Scheele (1808–1864)
Schellm.   C. Schellmann (fl. 1938)
Schenk   J. A. Schenk (1815–1891)
Schenk, E.   E. Schenk (b. 1880)
Scherb.   J. Scherbius (1769–1813)
Scheutz   N. J. W. Scheutz (1836–1889)
Schiffner   V. F. Schiffner (1862–1944)
Schimper, C.   C. F. Schimper (1803–1867)
Schinz   H. Schinz (1858–1941)
Schipcz.   N. V. Schipczinski (1886–1955)
Schischkin   B. K. Schischkin (1886–1963)
Schkuhr   C. Schkuhr (1741–1811)
Schlecht.   D. F. L. von Schlechtendal (1794–1866)
Schleicher   J. C. Schleicher (1768–1834)
Schlosser   J. C. Schlosser (1808–1882)
Schmalh.   I. F. Schmalhausen (1849–1894)
Schmeil   O. Schmeil (1860–1943)
Schmid, E.   E. Schmid (b. 1891)
Schmidel   C. C. Schmidel (1716–1792)
Schmidely   A. I. S. Schmidely (1838–1918)
Schmidt, A.   A. Schmidt (b. 1932)
Schmidt, Franz   Franz Schmidt (1751–1834)
Schmidt, F. W.   Franz Willibald Schmidt (1764–1796)
Schmidt Petrop., Friedrich   Friedrich Schmidt of St Petersburg (1832–1908)
Schneider, C. K.   C. K. Schneider (1876–1951)
Schnittspahn   G. F. Schnittspahn (1810–1865)
Scholz, H.   H. Scholz (b. 1928)
Scholz, J. B.   J. B. Scholz (fl. 1900)
Schönl.   S. Schönland (1860–1940)
Schott   H. W. Schott (1794–1865)
Schousboe   P. K. A. Schousboe (1766–1832)
Schrader   H. A. Schrader (1767–1836)
Schrank   F. von Paula von Schrank (1747–1835)
Schreber   J. C. D. von Schreber (1739–1810)
Schrenk   A. G. von Schrenk (1816–1876)
Schrödinger   R. Schrödinger (1857–1919)
Schroeter   C. Schroeter (1855–1939)
Schultes   J. A. Schultes (1773–1831)
Schultes fil.   J. H. Schultes (1804–1840)
Schultz, C. F.   C. F. Schultz (1765–1837)

Schultz, F. W.   F. W. Schultz (1804–1876)
Schultze, W.   W. Schultze (fl. 1894)
Schulz, O. E.   O. E. Schulz (1874–1936)
Schummel   T. E. Schummel (1785–1848)
Schur   P. J. F. Schur (1799–1878)
Schwantes   G. Schwantes (fl. 1927)
Schwarz, A.   A. Schwarz (1852–1915)
Schwarz, O.   O. Schwarz (b. 1900)
Schwegler   H. W. Schwegler (b. 1929)
Schweigger   A. F. Schweigger (1783–1821)
Schweinf.   G. A. Schweinfurth (1836–1925)
Schwertschl.   J. Schwertschleger (1853–1924)
Scop.   G. A. Scopoli (1723–1788)
Sebastiani   A. Sebastiani (1782–1821)
Sebeók   A. Sebeók de Szent-Miklós (fl. 1780)
Seem.   B. C. Seemann (1825–1871)
Séguier   J. F. Séguier (1703–1784)
Selin   G. Selin (1813–1862)
Sell, P. D.   P. D. Sell (b. 1929)
Semen., N.   N. Z. Semenova-Tjan-Schanskaja (1906–1960)
Sennen   Frère Sennen (E. M. Grenier-Blanc) (1861–1937)
Ser.   N. C. Seringe (1776–1858)
Serg.   L. P. Sergievskaja (b. 1897)
Serg., E.   E. V. Sergievskaja (C. V. Sergievskaja) (fl. 1961)
Serres   J. J. Serres (d. 1858)
Seub.   M. A. Seubert (1818–1878)
Shivas   M. G. Shivas (b. 1926)
Shull   G. H. Shull (1874–1954)
Shuttlew., R. J.   R. J. Shuttleworth (1810–1874)
Sibth.   J. Sibthorp (1758–1796)
Sieber   F. W. Sieber (1789–1844)
Siebert   A. Siebert (1854–1923)
Siebold   P. F. von Siebold (1796–1866)
Siegfr.   H. Siegfried (1837–1903)
Sikura   J. J. Sikura (fl. 1960)
Silliman   B. Silliman (1779–1864)
Silva, P.   A. R. Pinto da Silva (b. 1912)
Sim, R.   R. Sim (1791–1878)
Simkovics   L. Simkovics (later L. von Simonkai) (1851–1910)
Simmler   G. Simmler (b. 1884)
Simmons   H. G. Simmons (1866–1943)
Simon primus, E.   E. Simon (1848–1924)
Simon secundus, E.   E. Simon (fl. 1958)
Simonkai   L. von Simonkai (1851–1910)
Sims   J. Sims (1749–1831)
Sint.   P. E. E. Sintenis (1847–1907)
Širj.   G. I. Širjaev (Schirjaev) (1882–1954)
Skalická   A. Skalická (b. 1932)
Skalický   V. Skalický (b. 1930)
Skeels   H. C. Skeels (1873–1934)
Skvortsov, A.   A. K. Skvortsov (b. 1920)
Slosson   M. Slosson (b. 1873)
Sm.   J. E. Smith (1759–1828)
Sm., A. R.   A. R. Smith (b. 1938)
Sm., K. A. H.   K. A. H. Smith (b. 1889)
Small   J. K. Small (1869–1938)
Smejkal   M. Smejkal (b. 1927)
Smirnov   P. A. Smirnov (b. 1896)
Snogerup   S. E. Snogerup (b. 1929)
Soczava   V. B. Soczava (b. 1905)
Soják   J. Soják (b. 1936)
Solander   D. C. Solander (1733–1782)
Solemacher   J. V. L. A. G. Solemacher-Antweiler (b. 1889)
Solms-Laub.   H. M. C. L. F. Solms-Laubach (1842–1915)
Sommer.   I. Sommerauer (d. 1854)
Sommerf.   S. C. Sommerfelt (1794–1838)
Sommier   C. P. S. Sommier (1848–1922)

Sonder   O. W. Sonder (1812–1881)
Song.   A. Songeon (1826–1905)
Soó   R. de Soó (b. 1903)
Sosn., D.   D. I. Sosnowsky (1885–1952)
Soulié   J. A. Soulié (1868–1930)
Sowerby   J. Sowerby (1757–1822)
Soyer-Willemet   H. F. Soyer-Willemet (1791–1867)
Spach   E. Spach (1801–1879)
Speg.   C. Spegazzini (1858–1926)
Spenner   F. K. L. Spenner (1798–1841)
Sprengel   K. P. J. Sprengel (1766–1833)
Spribille   F. J. Spribille (1841–1921)
Spring   F. A. Spring (1814–1872)
Spruner   W. von Spruner (1805–1874)
Sprygin   I. I. Sprygin (1873–1942)
Standley   P. C. Standley (1884–1963)
Stankov   S. S. Stankov (1892–1962)
Stapf   O. Stapf (1857–1933)
Steele   W. E. Steele (1816–1883)
Stefani   C. de Stefani (1851–1924)
Stefanov   B. Stefanov (b. 1894)
Stefánsson   S. Stefánsson (1863–1921)
Steinh.   A. Steinheil (1810–1839)
Stephan   C. F. Stephan (1757–1814)
Stern, F. C.   F. C. Stern (1884–1967)
Sternb.   C. M. von Sternberg (1761–1838)
Sterns, E. E.   E. E. Sterns (1846–1926)
Steudel   E. G. von Steudel (1783–1856)
Steven   C. Steven (1781–1863)
St-Hil.   A. C. F. P. de Saint-Hilaire (1779–1853)
St-Lager   J. B. Saint-Lager (1825–1912)
Stoj.   N. Stojanov (b. 1885)
Stokes   J. Stokes (1755–1831)
Störk   A. Störk (1741–1803)
Strempel   J. K. F. Strempel (1800–1872)
Strobl   P. G. Strobl (1846–1910)
Stur   D. Stur (1827–1893)
Sturm   J. Sturm (1771–1848)
Suckow, G.   G. A. Suckow (d. 1867)
Sudre   H. Sudre (1862–1918)
Sudworth   G. B. Sudworth (1864–1927)
Suk.   V. N. Sukaczev (Sukatschew) (1880–1967)
Sumnev.   G. P. Sumnevicz (1909–1947)
Sünd.   F. Sündermann (1864–1946)
Suter   J. R. Suter (1766–1827)
Sutulov   A. N. Sutulov (fl. 1914)
Svob.   B. Svoboda
Swartz   O. P. Swartz (1760–1818)
Sweet   R. Sweet (1783–1835)
Swingle   W. T. Swingle (1871–1952)
Syme   J. T. I. Boswell Syme (formerly Boswell) (1822–1888)
Symons   J. Symons (1778–1851)
Szafer   W. Szafer (b. 1886)
Szov.   A. J. Szovits (d. 1830)
Szysz.   I. Szyszylowicz (1857–1910)
Tacik, T.   T. Tacik (b. 1926)
Taliev   V. I. Taliev (1872–1932)
Tanfani   E. Tanfani (1848–1892)
Tardieu-Blot   M. L. Tardieu-Blot (b. 1902)
Taubert   P. H. W. Taubert (1862–1897)
Tausch   I. F. Tausch (1793–1848)
Temesy   E. Temesy (fl. 1957)
Ten.   M. Tenore (1780–1861)
Tepl.   F. A. Teplouchow (1845–1905)
Terechov   A. F. Terechov (fl. 1931)
Terpó   A. Terpó (b. 1925)
Terracc., N.   N. Terracciano (1837–1921)

Tesseron   Y.-A. Tesseron (1831–1925)
Texidor   J. Texidor y Cos (1836–1885)
Teyber   A. Teyber (1846–1913)
Thell.   A. Thellung (1881–1928)
Thév.   A. V. Théveneau (1815–1876)
Thib.   ?E. Thibaud (fl. 1785)
Thomas   E. Thomas (1788–1859)
Thommen   E. Thommen (1880–1961)
Thomson   T. Thomson (1817–1878)
Thore   J. Thore (1762–1823)
Thouars   L. M. A. Aubert du Petit-Thouars (1758–1831)
Thouin   A. Thouin (1747–1824)
Thuill.   J. L. Thuillier (1757–1822)
Thunb.   C. P. Thunberg (1743–1828)
Timb.-Lagr.   P. M. E. Timbal-Lagrave (1819–1888)
Timm   J. C. Timm (1734–1805)
Tineo   V. Tineo (1791–1856)
Tiss.   P. G. Tissière (1828–1868)
Tod.   A. Todaro (1818–1892)
Tolm.   A. I. Tolmatchev (b. 1903)
Top.   S. Topali (fl. 1938)
Topa   E. Topa (b. 1900)
Topitz   A. Topitz (b. 1857)
Torrey   J. Torrey (1796–1873)
Tourlet   E.-H. Tourlet (1843–1907)
Trabut   L. Trabut (1853–1929)
Tratt.   L. Trattinick (1764–1849)
Trautv.   E. R. von Trautvetter (1809–1889)
Travis   W. G. Travis (1877–1958)
Trelease   W. Trelease (1857–1945)
Trev.   L. C. Treviranus (1779–1864)
Trew   C. J. Trew (1695–1769)
Tropea   C. Tropea (fl. 1910)
Trotzky   P. Kornuch-Trotzky (1803–1877)
Truchaleva   N. A. Truchaleva (b. 1927)
Tryon jun., R. M.   R. M. Tryon jun. (b. 1916)
Tubilla   T. Andrés y Tubilla (1859–1882)
Turcz.   N. S. Turczaninow (1796–1864)
Turesson   G. W. Turesson (b. 1892)
Turpin   P. J. F. Turpin (1775–1840)
Turra   A. Turra (1730–1796)
Turrill   W. B. Turrill (1890–1961)
Tutin   T. G. Tutin (b. 1908)
Tuzson   J. Tuzson (1870–1941)
Ucria   Bernadino da Ucria (Michelangelo Aurifici) (1739–1796)
Uechtr.   R. F. C. von Uechtritz (1838–1886)
Ugr.   K. A. Ugrinsky (fl. 1920)
Uhrová   A. Hrabĕtová-Uhrová (b. 1900)
Ujhelyi   J. Ujhelyi (b. 1910)
Ulbr.   E. Ulbrich (1879–1952)
Underw.   J. Underwood (d. 1834)
Unger   F. J. A. N. Unger (1800–1870)
Ung.-Sternb.   F. Ungern-Sternberg (d. 1885)
Urban   I. Urban (1848–1931)
Urum.   I. K. Urumoff (1856–1937)
Vacc.   L. Vaccari (1873–1951)
Vahl   M. H. Vahl (1749–1804)
Vahl, J.   J. L. M. Vahl (1796–1854)
Valck.-Suringar   — Valckenier-Suringar (1865–1932)
Valentine   D. H. Valentine (b. 1912)
Vandas   K. Vandas (1861–1923)
Van den Bosch   R. B. van den Bosch (1810–1862)
Van Hall   H. C. van Hall (1801–1874)
Van Houtte   L. B. van Houtte (1810–1876)
Van Ooststr.   S. J. van Ooststroom (b. 1906)
Vasc.   J. de Carvalho e Vasconcellos (b. 1897)
Vassil., V.   V. N. Vassiliev (b. 1890)

Vassilcz. I. T. Vassilczenko (b. 1903)
Velen. J. Velenovský (1858–1949)
Vendr. X. Vendrely
Vent. E. P. Ventenat (1757–1808)
Vent, W. W. Vent (b. 1920)
Verlot J.-B. Verlot (1825–1891)
Verlot, B. P. B. L. Verlot (1836–1897)
Vest L. C. von Vest (1776–1840)
Vicioso, C. C. Vicioso Martínez (b. 1887)
Vidal L. M. Vidal
Vierh. F. Vierhapper (1876–1932)
Vig. L. G. A. Viguier (1790–1867)
Vigineix G. Vigineix (d. 1877)
Vill. D. Villars (Villar) (1745–1814)
Villar, H. del E. Huguet del Villar (1871–1951)
Vilmorin P. L. F. L. de Vilmorin (1816–1860)
Vindt J. Vindt (b. 1915)
Vis. R. de Visiani (1800–1878)
Viv. D. Viviani (1772–1840)
Vogler J. A. Vogler (1746–1816)
Volk. A. Volkart (1873–1951)
Vollmann F. Vollmann (1858–1917)
Vorosch. V. N. Voroschilov (b. 1908)
Voss A. Voss (1857–1924)
Vuk. L. F. Vukotinović (1813–1893)
Vved. A. I. Vvedensky (b. 1898)
Wagner, H. J. Wagner (H. Wagner) (1870–1955)
Wagner, R. R. Wagner (fl. 1887)
Wahlberg P. F. Wahlberg (1800–1877)
Wahlenb. G. Wahlenberg (1780–1851)
Waisb. A. Waisbecker (1835–1916)
Waldst. F. A. von Waldstein-Wartemberg (1759–1823)
Wale R. S. Wale (d. 1952)
Walker, S. S. Walker (b. 1924)
Wall. N. Wallich (1786–1854)
Wallr. K. F. W. Wallroth (1792–1857)
Walpers W. G. Walpers (1816–1853)
Walsh R. Walsh (1772–1852)
Walter T. Walter (1740–1789)
Walters S. M. Walters (b. 1920)
Wangenh. F. A. J. von Wangenheim (1747–1800)
Warburg, E. F. E. F. Warburg (1908–1966)
Watson, H. C. H. C. Watson (1804–1881)
Watson, S. S. Watson (1826–1892)
Watson, W. C. R. W. C. R. Watson (1885–1954)
Watt D. A. P. Watt (1830–1917)
Webb P. B. Webb (1793–1854)
Webb, D. A. D. A. Webb (b. 1912)
Weber G. H. Weber (1752–1828)
Weber fil. F. Weber (1781–1823)
Weddell H. A. Weddell (1819–1877)
Weevers T. Weevers (1875–1952)
Weigel C. E. von Weigel (1748–1831)
Weihe K. E. A. Weihe (1779–1834)
Weiller M. Weiller (1880–1945)
Wein, K. K. Wein (b. 1883)
Weinm. J. A. Weinmann (1782–1858)
Welden F. L. von Welden (1782–1853)
Welw. F. Welwitsch (1806–1872)
Wendelberger G. Wendelberger (b. 1915)
Wenderoth G. W. F. Wenderoth (1774–1861)
Wendl. J. C. Wendland (1755–1828)
Wendl. fil. H. L. Wendland (1792–1869)
Wenzig T. Wenzig (1824–1892)
Wesmael, A. A. Wesmael (1832–1905)
Weston R. Weston (1733–1806)

Wettst. R. von Wettstein (1863–1931)
White J. White (c. 1750–1832)
Whitehead F. H. Whitehead (b. 1913)
Wibel A. W. E. C. Wibel (1775–1814)
Wibiral E. Wibiral (1878–1950)
Wichura M. E. Wichura (1817–1866)
Widder F. Widder (b. 1892)
Wieg. K. McK. Wiegand (1873–1942)
Wierzb. P. Wierzbicki (1794–1847)
Wiesb. J. Wiesbaur (1836–1906)
Wight R. Wight (1796–1872)
Wikstr. J. E. Wikström (1789–1856)
Wilce J. H. Wilce (b. 1931)
Wilczek E. Wilczek (1867–1948)
Willd. C. L. Willdenow (1765–1812)
Williams, F. N. F. N. Williams (1862–1923)
Willk. H. M. Willkomm (1821–1895)
Wilmott A. J. Wilmott (1888–1950)
Wilson, E. H. E. H. Wilson (1876–1930)
Wimmer C. F. H. Wimmer (1803–1868)
Winge Ö. Winge (1886–1964)
Winter, N. N. A. Winter (1898–1934)
Wirtgen P. W. Wirtgen (1806–1870)
Wissjul. E. D. Wissjulina (b. 1902)
With. W. Withering (1741–1799)
Wittm. M. C. L. Wittmack (1839–1929)
Wohlf. R. Wohlfahrt (1830–1888)
Wolf, N. M. N. M. von Wolf (1724–1784)
Wolf, T. F. T. Wolf (1841–1921)
Wolff, H. H. Wolff (1866–1929)
Wolfner W. Wolfner (fl. 1858)
Wollaston G. B. Wollaston (1814–1899)
Wolley-Dod A. H. Wolley-Dod (1861–1948)
Wolny A. R. Wolny (d. ? 1829)
Wołoszczak E. Wołoszczak (1835–1918)
Wood, W. W. Wood (1745–1808)
Woods, J. J. Woods (1776–1864)
Wormsk. M. Wormskiold (1783–1845)
Woronow J. N. Woronow (Voronov) (1874–1931)
Woynar H. K. Woynar (1865–1917)
Wulf E. V. Wulf (E. W. Wulff, E. V. Vul'f) (1885–1941)
Wulfen F. X. von Wulfen (1728–1805)
Wünsche J. G. Wünsche (fl. 1804)
Zabel H. Zabel (1832–1912)
Zahar. C. Zahariadi (b. 1901)
Zahlbr. J. Zahlbruckner (1782–1850)
Zamels A. Zamels (Zamelis) (1897–1943)
Zanted. G. Zantedeschi (1773–1846)
Zapał. H. Zapałowicz (1852–1917)
Zawadzki A. Zawadzki (1798–1868)
Zenari S. Zenari (b. 1896)
Zerov D. K. Zerov (b. 1895)
Žertová A. Chrtková-Žertová (b. 1930)
Zeyher C. L. P. Zeyher (1799–1858)
Zimm. W. Zimmermann (b. 1892)
Zimmeter A. Zimmeter (1848–1897)
Zinger, N. N. Zinger (1866–1923)
Zinn J. G. Zinn (1727–1759)
Zinserl. Y. D. Zinserling (1894–1938)
Ziz J. B. Ziz (1779–1829)
Zodda G. Zodda (b. 1877)
Zoega J. Zoega (1742–1788)
Zoz I. G. Zoz (b. 1903)
Zsák Z. Zsák (b. 1880)
Zucc. J. G. Zuccarini (1797–1848)
Zuccagni A. Zuccagni (1754–1807)

APPENDIX II

KEY TO THE ABBREVIATIONS OF TITLES OF BOOKS
CITED IN VOLUME 2

Aiton, *Hort. Kew.*
W. Aiton, *Hortus kewensis, or a Catalogue of the Plants cultivated in the Royal Botanic Garden at Kew.* Ed. 1. London. 1789. (**1–3** in 1789.) Ed. 2, by W. T. Aiton. London. 1810–1813. (**1** in 1810; **2** & **3** in 1811; **4** in 1812; **5** in 1813. Cf. C. E. Britton, *Jour. Bot. (London)* **50**, suppl. 3, 1–16 (1912) and F. A. Stafleu, *Taxon* **12**: 53–54 (1963).)

All., *Auct. Fl. Pedem.*
C. Allioni, *Auctuarium ad Floram pedemontanam cum Notis et Emendationibus.* Augustae Taurinorum. 1789.

All., *Fl. Pedem.*
C. Allioni, *Flora pedemontana, sive Enumeratio methodica Stirpium indigenarum Pedemontii.* Augustae Taurinorum. 1785. (**1–3** in 1785.)

Alpers, *Verz. Gefässpfl. Landdr. Stade*
F. Alpers, *Verzeichniss der Gefässpflanzen der Landdrostei Stade.* Stade. 1875.

Arcangeli, *Comp. Fl. Ital.*
G. Arcangeli, *Compendio della Flora italiana, ossia Manuale per la Determinazione delle Piante che trovansi selvatiche od inselvatichite nell'Italia e nelle Isole adiacenti.* Ed. 1. Torino. 1882. Ed. 2. Torino & Roma. 1894.

Arvet-Touvet, *Ess. Pl. Dauph.*
J. M. C. Arvet-Touvet, *Essai sur les Plantes du Dauphiné.* Grenoble. 1871.

Ascherson, *Fl. Brandenb.*
P. F. A. Ascherson, *Flora der Provinz Brandenburg, der Altmark und des Herzogthums Magdeburg.* Berlin. 1859–1864. (**1**: pp. i–xxii, 1–320 in 1860; pp. 321–1034 in 1864; **2** & **3** in 1859.)

Ascherson & Graebner, *Syn. Mitteleur. Fl.*
P. F. A. Ascherson & K. O. P. P. Graebner, *Synopsis der mitteleuropäischen Flora.* Ed. 1. Leipzig. 1896–1938. (**1**: pp. 1–160 in 1896; pp. 161–416 in 1897; **2(1)**: pp. 1–64 in 1898; pp. 65–304 in 1899; pp. 305–544 in 1900; pp. 545–704 in 1901; pp. 705–796 in 1902; **2(2)**: pp. 1–144 in 1902; pp. 145–224 in 1903; pp. 225–530 in 1904; **3**: pp. 1–320 in 1905; pp. 321–560 in 1906; pp. 561–934 in 1907; **4**: pp. 1–80 in 1908; pp. 81–320 in 1909; pp. 321–400 in 1910; pp. 401–640 in 1911; pp. 641–800 in 1912; pp. 801–886 in 1913; **5(1)**: pp. 1–224 in 1913; pp. 225–400 in 1914; pp. 401–480 in 1915; pp. 481–544 in 1916; pp. 545–624 in 1917; pp. 625–784 in 1918; pp. 785–948 in 1919; **5(2)**: pp. 1–160 in 1920; pp. 161–400 in 1921; pp. 401–480 in 1922; pp. 481–560 in 1923; pp. 561–640 in 1926; pp. 641–812 in 1929; **5(3)**: pp. 1–98 in 1935; **5(4)**: pp. 1–160 in 1936; pp. 161–252 in 1938; **6(1)**: pp. 1–64 in 1900; pp. 65–240 in 1901; pp. 241–560 in 1902; pp. 561–640 in 1903; pp. 641–800 in 1904; pp. 801–896 in 1905; **6(2)**: pp. 1–160 in 1906; pp. 161–496 in 1907; pp. 497–688 in 1908; pp. 689–1008 in 1909; pp. 1009–1094 in 1910; **7**: pp. 1–80 in 1913; pp. 81–240 in 1914; pp. 241–320 in 1915; pp. 321–400 in 1916; pp. 401–480 in 1917; **12(1)**: pp. 1–80 in 1922; pp. 81–160 in 1924; pp. 161–400 in 1929; pp. 401–492 in 1930; **12(2)**: pp. 1–160 in 1930; pp. 161–480 in 1931; pp. 481–640 in 1934; pp. 641–790 in 1935; **12(3)**: pp. 1–320 in 1936; pp. 321–480 in 1937; pp. 481–708 in 1938.) Ed. 2.

Leipzig. 1912–1920. (**1**: pp. 1–480 in 1912; pp. 481–630 in 1913; **2(1)**: pp. 1–80 in 1919; pp. 81–160 in 1920.)

Aspegren, *Förs. Blek. Fl.*
G. C. Aspegren, *Försök till en blekingsk Flora.* Carlskrona. 1823.

Asso, *Syn. Stirp. Arag.*
I. J. de Asso y del Rio, *Synopsis Stirpium indigenarum Aragoniae.* Massiliae. 1779.

Avé-Lall., *Ind. Sem. Horti Petrop.*
Cf. Fischer & C. A. Meyer, *Ind. Sem. Horti Petrop.*

Bartal., *Cat. Piante Siena*
B. Bartalini, *Catalogo delle Piante che nascono spontaneamente intorno alla Città di Siena.* Siena. 1876.

Bast., *Essai Fl. Maine Loire*
T. Bastard, *Essai sur la Flore du Département de Maine et Loire.* Angers. 1809.

Batsch, *Beytr. Entw. Pragm. Gesch. Nat.-Reiche*
A. J. G. C. Batsch, *Beyträge und Entwürfe zur pragmatischen Geschichte der drey Natur-Reiche nach ihren Verwandtschaften.* Weimar. 1801. (**1** in 1801.)

Batt. & Trabut, *Fl. Algér.*
J. A. Battandier & L. Trabut, *Flore de l'Algérie. Ancienne Flore d'Alger transformée, contenant la Description de toutes les Plantes signalées jusqu'à ce Jour comme spontanées en Algérie. Dicot., Dicotylédones* by J. A. Battandier. Alger & Paris. 1888–1890. (Pp. 1–184 in 1888; pp. 185–576 in 1889; pp. 577–825 in 1890. Cf. W. T. Stearn, *Jour. Soc. Bibl. Nat. Hist.* **1**: 145 (1938).) *Monocot., Monocotylédones* by J. A. Battandier & L. Trabut. Alger & Paris. 1895.

Baum, *Monogr. Rev. Tamarix*
B. Baum, *Monographic Revision of the Genus* Tamarix. Jerusalem. 1966.

Baumg., *Enum. Stirp. Transs.*
J. C. G. Baumgarten, *Enumeratio Stirpium magno Transsilvaniae Principatui.* Vindobonae. 1816. (**1–3** in 1816.) **4** (*Mantissa*), by M. Fuss. Cibinii. 1846. *Indices*, by M. Fuss. Cibinii. 1860.

Beck, G., *Fl. Bosn. Herceg.*
G. Beck von Mannagetta, *Flora Bosne, Hercegovine i Novipazarskog Sandžaka.* Beograd & Sarajevo. 1903–1950. (*Pteridophyta* in *Glasn. Muz. Bosni Herceg.* **28**: 311–336 (1916). *Embryophyta siphonogama*: **1** as *op. cit.* **15**: 1–48, 185–230 (1903); **2(1)** as *op. cit.* **18**: 69–82 (1906); **2(2)** as *op. cit.* **18**: 137–150 (1906); **2(3)** as *op. cit.* **18**: 469–496 (1906); **2(4)** as *op. cit.* **19**: 15–30 (1907); **2(5)** as *op. cit.* **21**: 135–166 (1909); **2(6)** as *op. cit.* **26**: 451–476 (1914); **2(7)** as *op. cit.* **28**: 41–168 (1916); **2(8)** as *op. cit.* **30**: 177–218 (1918); **2(9)** as *op. cit.* **32**: 83–128 (1920); **2(10)** as *op. cit.* **33**: 1–18 (1921); **2(11)** as *op. cit.* **35**: 49–74 (1923); **3**, titled *Flora Bosnae Hercegovinae et Regionis Novipazar.* / Флора Босне, Херцеговине и Области Новога Пазара [*Flora Bosne, Hercegovine i Oblasti Novoga Pazara*], Beograd & Sarajevo in 1927; **4(1)**, titled *Flora Bosnae et Hercegovinae*, by K. Malý, Sarajevo in 1950). *Pteridophyta* (pp. 1–26), and **1** & **2** (with continuous pagination, pp. 1–484) also reprinted separately with double pagination.

Beck, G., *Fl. Nieder-Österr.*

G. Beck von Mannagetta, *Flora von Nieder-Österreich*. Wien. 1890–1893. (**1**: pp. 1–430 in 1890; **2**(1): pp. 431–[884] in 1892; **2**(2): *Allgem. Th.* pp. 1–74 & 895–1396 in 1893.)

Bentham, *Fl. Austral.*

G. Bentham, *Flora australiensis*. London. 1863–1878. (**1** in 1863; **2** in 1864; **3** in 1867; **4** in 1868; **5** in 1870; **6** in 1873; **7** in 1878. Cf. H. W. Rickett & F. A. Stafleu, *Taxon* **10**: 74 (1961).)

Berchtold & Opiz, *Ökon.-Techn. Fl. Böhm.*

F. von Berchtold & P. M. Opiz, *Ökonomisch-technische Flora Böhmens*. Prag. 1836–1843. (**1** in 1836; **2**(1) in 1838; **2**(2) in 1839; **3**(1) in 1841; **3**(2) in 1843.) **1** by F. von Berchtold & W. B. Seidl.

Berchtold & J. Presl, *Rostlinář*

F. von Berchtold & J. S. Presl, *O přirozenosti Rostlin aneb Rostlinář*. Praze. 1823–1835. (**1** in 1823; **2** in 1825; **3** in 1830–1835.)

Berggren, Jakob, *Res. Eur. Österländ.*

Jakob Berggren, *Resor i Europa och Österländerne* (containing *Växter i Österländerne samlade af J. Berggren; och nogere Bestämda af Göran Wahlenberg*). Stockholm. 1826–1828. (**1** in 1826; **2** in 1827; **3** in 1828.)

Bernh., *Syst. Verz. Erfurt*

J. J. Bernhardi, *Systematisches Verzeichniss der Pflanzen, welche in der Gegend um Erfurt gefunden werden*. Erfurt. 1800.

Bertol., *Amoen.*

A. Bertoloni, *Amoenitates italicae, sistentes Opuscula ad Rem herbariam et Zoologiam Italiae spectantia*. Bononiae. 1819.

Bertol., *Fl. Ital.*

A. Bertoloni, *Flora italica, sistens Plantas in Italia et Insulis circumstantibus sponte nascentes*. Bononiae. 1834–1854. (**1** in 1834; **2** in 1836; **3** in 1838; **4** in 1841; **5** in 1844; **6** in 1847; **7** in 1850; **8** in 1853; **9** & **10** in 1854. Cf. R. V. de Litardière in Briquet, *Prodr. Fl. Corse* **3** (1): vii (1938).)

Bertol., *Hort. Bot. Bon.*

A. Bertoloni, *Horti botanici bononiensis Plantae novae vel minus cognitae*. Bononiae. 1838–1839. (**1** in 1838; **2** in 1839.)

Bertol., *Rar. Lig. Pl.*

A. Bertoloni, *Rariorum Liguriae Plantarum Decas 1 [2, 3]*. Genuae, Pisis. 1803–1810. (**1**, Genuae in 1803; **2**, Pisis in 1806; **3**, Pisis in 1810. **2** & **3** titled *Rariorum Italiae Plantarum....*)

Bertram, *Fl. Braunschw.*

F. W. W. Bertram, *Flora von Braunschweig*. Ed. 1. Braunschweig. 1876. Ed. 2. Braunschweig. 1881.

Besser, *Cat. Pl. Horto Cremen.*

W. S. J. G. von Besser, *Catalogus Plantarum in Horto botanico Gymnasii volhyniensis Cremeneci cultarum*. Cremeneci. 1816.

Besser, *Cat. Pl. Jard. Krzemien.*

W. S. J. G. von Besser, *Catalogue des Plantes du Jardin botanique du Gymnase de Volhynie à Krzemieniec*. Ed. 1. Krzemieniec. 1810. Ed. 2. Krzemieniec. 1811. **Suppl.**, *Supplément* 1. Krzemieniec. 1812. *Supplément* 2. Krzemieniec. 1813. *Supplément* 3. Krzemieniec. 1814. *Supplément* 4. Krzemieniec. 1815.

Besser, *Enum. Pl. Volhyn.*

W. S. J. G. von Besser, *Enumeratio Plantarum hucusque in Volhynia, Podolia, Gub. kiioviensi, Bessarabia cis-tyraica et circa Odessam collectarum, simul cum Observationibus in Primitias Florae Galiciae austriacae*. Vilnae. 1822.

Besser, *Prim. Fl. Galic.*

W. S. J. G. von Besser, *Primitiae Florae Galiciae austriacae utriusque*. Viennae. 1809. (**1** & **2** in 1809.)

Bieb., *Beschr. Länd. Terek Casp.*

F. A. Marschall von Bieberstein, *Beschreibung der Länder zwischen den Flüssen Terek und Kur am caspischen Meere. Mit einem botanischen Anhang*. Frankfurt. 1800.

Bieb., *Cent. Pl.*

F. A. Marschall von Bieberstein, *Centuria Plantarum rariorum Rossiae meridionalis, praesertim Tauriae et Caucasi, Iconibus Descriptionibusque illustrata*. Charkoviae & Petropoli. 1810–1843. (**1**, Charkoviae in 1810; **2**, Petropoli in 1832–1843.)

Bieb., *Fl. Taur.-Cauc.*

F. A. Marschall von Bieberstein, *Flora taurico-caucasica, exhibens Stirpes phaenogamas in Chersoneso taurica et Regionibus caucasicis sponte crescentes*. Charkoviae. 1808–1819. (**1** & **2** in 1808; **3**, *Supplementum*, in 1819.)

Bieb., *Tabl. Prov. Casp.*

F. A. Marschall von Bieberstein, *Tableau des Provinces situées sur la Côte occidentale de la Mer Caspienne entre les Fleuves Terek et Kour*. St.-Pétersbourg. 1798.

Billot, *Annot.*

P. C. Billot, *Annotations à la Flore de France et d'Allemagne*. Haguenau. 1855–1862. (Pp. 1–100 in 1855; pp. 101–116 in 1857 or 1858; pp. 117–140 in 1858; pp. 141–164 in 1859; pp. 165–210 in 1860; pp. 211–244 in 1861; pp. 245–297 in 1862.)

Biv., *Stirp. Rar. Sic. Descr.*

A. de Bivona-Bernardi, *Stirpium rariorum minusque cognitarum in Sicilia sponte provenientium Descriptiones nonnullis Iconibus auctae*. Panormi. 1813–1816. (**1** in 1813; **2** in 1814; **3** in 1815; **4** in 1816.)

Bluff & Fingerh., *Comp. Fl. Germ.*

M. J. Bluff & K. A. Fingerhuth, *Compendium Florae germanicae*. Norimbergae. 1825–1833. (**1** & **2** in 1825; **3** & **4** by K. F. W. Wallroth, in 1831–1833.)

Boenn., *Prodr. Fl. Monast.*

C. M. F. von Boenninghausen, *Prodromus Florae monasteriensis Westphalorum*. Monasterii. 1824.

Boiss., *Cent. Euphorb.*

P. E. Boissier, *Centuria Euphorbiarum*. Lipsiae & Parisiis. 1860.

Boiss., *Diagn. Pl. Or. Nov.*

P. E. Boissier, *Diagnoses Plantarum orientalium novarum*. Lipsiae & Parisiis. 1843–1859. (Ser. 1, **1**(1–3) in 1843; **1**(4–5) in 1844; **1**(6–7) in 1846 or 1847; **2**(8–11) in 1849; **2**(12–13) in 1853; ser. 2, **3**(1) in 1853; **3**(2–3) in 1856; **3**(4) in 1859; **3**(5) in 1856; **3**(6) in 1859. Cf. M. J. van Steenis-Kruseman, *Fl. Males. Bull.* **15**: 733 (1960).)

Boiss., *Elenchus*

P. E. Boissier, *Elenchus Plantarum novarum minusque cognitarum, quas in Itinere hispanico legit....* Genevae. 1838.

Boiss., *Fl. Or.*

P. E. Boissier, *Flora orientalis*. Basileae, Genevae & Lugduni. 1867–1884. (**1** Basileae & Genevae; **2–5** Basileae, Genevae & Lugduni. **1** in 1867; **2** in 1872; **3** in 1875; **4**: pp. 1–280 in 1875; pp. 281–1276 in 1879; **5**: pp. 1–428 in 1882; pp. 429–868 in 1884.) *Suppl. Supplementum*. Basileae, Genevae & Lugduni. 1888.

Boiss., *Voy. Bot. Midi Esp.*

P. E. Boissier, *Voyage botanique dans le Midi de l'Espagne pendant l'Année 1837*. Paris. 1839–1845. (**1**: pp. 1–40, tt. 1–135 in 1839; pp. 41–96, tt. 136–181 in 1840; pp. 97–248, i–x, t. 4a in 1845; **2**: pp. 1–480 in 1839; pp. 481–710 in 1841; pp. 711–757 in 1845.) All new species are described in **2**.

Boiss. & Reuter, *Diagn. Pl. Nov. Hisp.*

P. E. Boissier & G. F. Reuter, *Diagnoses Plantarum novarum hispanicarum, praesertim in Castella nova lectarum*. Genevae. 1842

Boiss. & Reuter, *Pugillus*

P. E. Boissier & G. F. Reuter, *Pugillus Plantarum novarum Africae borealis Hispaniaeque australis*. Genevae. 1852.

Bonnier, *Fl. Compl. Fr.*

G. E. M. Bonnier, *Flore complète illustrée en Couleurs de France, Suisse et Belgique, comprenant la Plupart des Plantes d'Europe*. Paris, Neuchâtel & Bruxelles. 1911–1935. (**1–8** Paris, Neuchâtel & Bruxelles; **9** Paris & Bruxelles; **10–13** Paris.

1: pp. 1–48, tt. 1–24 in 1911; pp. 49–120, tt. 25–60 in 1912; **2** in 1913; **3** in 1914; **4** in 1921; **5** in 1922; **6** in 1923; **7** in 1924; **8** in 1926; **9** in 1927; **10** in 1929; **11**: pp. 1–80, tt. 601–630 in 1930; pp. 81–161, tt. 631–660 in 1931; **12**: pp. 1–24, tt. 661–672 in 1932; pp. 25–80, tt. 673–696 in 1933; pp. 81–133, tt. 697–721 in 1934; **13** in 1935. 7–13 partly by R. Douin. Cf. W. T. Stearn, *Jour. Soc. Bibl. Nat. Hist.* **2**: 212–215 (1950).)

Bonnier & Layens, *Fl. Fr.*
G. E. M. Bonnier & G. de Layens, *Tableaux synoptiques des Plantes vasculaires de la Flore de la France. (La Végétation de la France,* **1**.) Paris. 1894.

Borbás, *Abauj-Torna Vármegye Fl.*
V. von Borbás, *Abauj-Torna Vármegye Flórája* (in *Abauj-Torna Vármegye Monografiája*). 1896.

Boreau, *Fl. Centre Fr.*
A. Boreau, *Flore du Centre de la France.* Ed. 1. Paris. 1840. (**1 & 2** in 1840.) Ed. 2. Paris. 1849. (**1 & 2** in 1849.) Ed. 3. Paris. 1857. (**1 & 2** in 1857.)

Borkh., *Handb. Forstbot.*
M. B. Borkhausen, *Theoretisch-praktisches Handbuch der Forstbotanik und Forsttechnologie.* Giessen & Darmstadt. 1800–1803. (**1**: pp. 1–866 in 1800; **2**(**1**): pp. 867–2070 in 1803.)

Borkh., *Vers. Forstbot. Beschr. Holzart.*
M. B. Borkhausen, *Versuch einer forstbotanischen Beschreibung der in den Hessen-Darmstadtschen Landen, besonders in der Obergrafschaft Catzenellenbogen im Freien wachsenden Holzarten.* Frankfurt am Main. 1790.

Bory, *Dict. Class. Hist. Nat.*
J. B. G. M. Bory de Saint-Vincent (edit.), *Dictionnaire classique d'Histoire naturelle.* Paris. 1822–1831. (**1–17** in 1822–1831.)

Bory, *Expéd. Sci. Morée*
J. B. G. M. Bory de Saint-Vincent, *Expédition scientifique de Morée.* **3**(**2**) *Botanique.* Paris. 1832–1833. (Pp. 1–336 in 1832; pp. 337–367 in 1833.)

Bory, *Voy. Souterrain*
J. B. G. M. Bory de Saint-Vincent, *Voyage souterrain, ou Description du Plateau de Sainte-Pierre de Maestricht et de ses vastes Cryptes...suivi de la Relation de nouveaux Voyages entrepris dans les Montagnes Maudites: par M. Léon Dufour.* Paris. 1821.

Bory & Chaub., *Nouv. Fl. Pélop.*
J. B. G. M. Bory de Saint-Vincent & L. A. Chaubard, *Nouvelle Flore du Péloponnèse et des Cyclades, entièrement revue, corrigée et augmentée par M. Chaubard pour les Phanérogames et M. Bory de Saint-Vincent pour les Cryptogames, les Agames, les Considérations générales, la Distribution des Espèces par Familles naturelles et ce qui a rapport aux Habitats.* Paris & Strasbourg. 1838.

Boulay, *Ronces Vosg.*
N. J. Boulay, *Ronces vosgiennes, Description des Espèces.* Rambervillers. 1864–1866. Saint-Dié. 1867–1869.

Brandza, *Prod. Fl. Romăne*
D. Brandza, *Prodromul Florei romăne.* Bucuresci. 1879–1883. (Pp. vii–lxx, 1–128 in 1879 or 1880; pp. i–vi, lxxi–lxxxiv, 129–568 in 1883. Cf. W. T. Stearn, *Jour. Bot. (London)* **79**: 191 (1941).)

Briganti, *Nova Pimp. Spec. Diss.*
V. Briganti, *De nova Pimpinellae Specie cui Nomen anisoides Dissertatio.* Neapoli. 1805.

Briq., *Cytises Alp. Marit.*
J. I. Briquet, *Études sur les Cytises des Alpes maritimes.* Genève & Bâle. 1894.

Briq., *Prodr. Fl. Corse*
J. I. Briquet, *Prodrome de la Flore corse.* Genève, Bâle, Lyon & Paris. 1910–1955. (**1**, Genève, Bâle & Lyon in 1910; **2**(**1**), Genève, Bâle & Lyon in 1913; **2**(**2**), Paris in 1936; **3**(**1**), Paris in 1938; **3**(**2**), Paris in 1955.) **2**(**2**) & **3**(**1 & 2**) by R. V. de Litardière.

Britton, E. E. Sterns & Poggenb., *Prelim. Cat.*
Preliminary Catalogue of Anthophyta and Pteridophyta reported as growing spontaneously within one hundred Miles of New York City. Compiled by [a] Committee of the Torrey Botanical Club. The Nomenclature revised and corrected by N. L. Britton, E. E. Sterns and Justus F. Poggenburg. New York. 1888.

Brot., *Fl. Lusit.*
F. Avellar Brotero, *Flora lusitanica.* Olisipone. 1804. (**1 & 2** in 1804.)

Brot., *Phyt. Lusit.*
F. Avellar Brotero, *Phytographia Lusitaniae selectior.* Ed. 1. Olisipone. 1800. Ed. 2. Olisipone. 1801. Ed. 3. Olisipone. 1816–1827. (**1** in 1816; **2** in 1827.)

Bunge, *Astrag. Geront.*
A. A. von Bunge, *Generis Astragali Species gerontogeae.* St.-Pétersbourg. 1868.

Bunge, *Del. Sem. Horti Dorpat.*
A. A. von Bunge, *Delectus Seminum Horti botanici dorpatensis.* Dorpat. 1836–1850.

Bunge, *Enum. Pl. Chin. Bor.*
A. A. von Bunge, *Enumeratio Plantarum quas in China boreali collegit....* Petropoli. 1833. (For date cf. W. T. Stearn, *Jour. Bot. (London)* **79**: 63–64 (1941).)

Bunge, *Tent. Tamaric.*
A. A. von Bunge, *...Tentamen Generis Tamaricum Species accuratius definiendi.* Dorpati. 1852.

Burgsd., *Anleit. Erzieh. Holzart.*
F. A. L. von Burgsdorff, *Anleitung zur sichern Erziehung und zweckmässigen Anpflanzung der einheimischen und fremden Holzarten, welche in Deutschland und unter ähnlichem Klima im Freyen fortkommen.* Berlin. Ed. 1. 1787. Ed. 2. Berlin. 1790–1791. (**1** in 1790; **2** in 1791.) Ed. 3. Berlin. 1805. (**1 & 2** in 1805.)

Burm. fil., *Fl. Ind.*
N. L. Burman, *Flora indica.* Lugduni Batavorum & Amstelaedami. 1768.

Burm. fil., *Spec. Bot. Geran.*
N. L. Burman, *Specimen botanicum de Geraniis.* Lugduni Batavorum. 1759. (For date cf. A. Becherer, *Feddes Repert.* **25**: 215–216 (1928).)

Burnat, *Fl. Alp. Marit.*
E. Burnat, *Flore des Alpes maritimes, ou Catalogue raisonnée des Plantes qui croissent spontanément dans la Chaîne des Alpes maritimes, y compris le Département français de ce Nom et une Partie de la Ligurie occidentale.* Genève, Bâle & Lyon. 1892–1931. (**1–6**, Genève, Bâle & Lyon. **7**, Genève. **1** in 1892; **2** in 1896; **3**: pp. 1–172 in 1899; pp. 173–332 in 1901; **4** in 1906; **5** Supplement: pp. 1–96 in 1913; pp. 97–376 in 1915; **6**: pp. 1–170 in 1916; pp. 171–344 in 1917; **7** in 1931.) **5**: pp. 1–96 by F. Cavallier; **5**: pp. 97–376, **6 & 7** by J. Briquet & F. Cavallier.

Buser, *Alchim. Valais.*
R. Buser, *Alchimilles valaisannes.* Zürich. 1894.

Buser, *Not. Alchim.*
R. Buser, *Notes sur quelques Alchimilles critiques ou nouvelles.* Grenoble. 1891.

Cajander, *Suomen Kasvio*
A. K. Cajander, *A. J. Melan Suomen Kasvio.* Helsingissä. 1906. Comprises ed. 5 of A. J. Mela, *Suomen Kasvio.*

Cav., *Icon. Descr.*
A. J. Cavanilles, *Icones et Descriptiones Plantarum, quae aut sponte in Hispania crescunt, aut in Hortis hospitantur.* Matriti. 1791–1801. (**1** in 1791; **2** in 1793; **3** in 1795; **4**: pp. 1–36 in 1797; pp. 37–82 in 1798; **5** in 1799; **6**: pp. 1–40 in 1800; pp. 41–97 in 1801.)

Cav., *Monad. Class. Diss. Dec.*

A. J. Cavanilles, *Monadelphiae Classis Dissertationes decem.* Matriti. 1785–1790. (Pp. 1–48 in 1785; pp. 49–106+4 in *app.* in 1786; pp. 107–266 in 1787; pp. 267–354 in 1788; pp. 355–414 in 1789; pp. 415–464 in 1790.)

Ceballos & C. Vicioso, *Estud. Veg. Fl. Forest. Málaga*

L. Ceballos & C. Vicioso Martínez, *Estudio sobre la Vegetación y Flora forestal de la Provincia de Málaga.* Madrid. 1933.

Čelak., *Prodr. Fl. Böhm.*

L. J. Čelakovsky, *Prodromus der Flora von Böhmen.* (Published as *Arch. Naturw. Landes. Böhm.* **1**, *Bot. Abth.*: 1–112 (1869); **2**(2), *Bot. Abth.*: 113–388 (1873); **3**, *Bot. Abth.*: 389–691 (1875); **4**, *Bot. Abth.*: 693–955 (1881).)

Cesati, *Icon. Stirp.*

V. de Cesati, *Iconographia Stirpium italicarum universa.* Mediolani. 1840–1843. (**1** in 1840; **2** in 1841; **3** in 1843.)

Cesati, Passer. & Gibelli, *Comp. Fl. Ital.*

V. de Cesati, G. Passerini & G. Gibelli, *Compendio della Flora italiana.* Milano. 1868–1886. (Pp. 1–48 in 1868; pp. 49–120 in 1869; pp. 121–168 in 1870; pp. 169–208 in 1871; pp. 209–256 in 1872; pp. 257–328 in 1874; pp. 329–376 in 1875; pp. 377–416 in 1876; pp. 417–440 in 1877; pp. 441–464 in 1878; pp. 465–544 in 1879; pp. 545–592 in 1880; pp. 593–640 in 1881; pp. 641–712 in 1882; pp. 713–760 in 1883; pp. 761–784 in 1884; pp. 785–906 in 1886; index & corrigenda, by Gibelli & Mattirolo, in 1901. Cf. P. A. Saccardo, *Cron. Fl. Ital.* xix (1909).)

Choisy, *Prodr. Monogr. Hypér.*

J. D. Choisy, *Prodromus d'une Monographie de la Famille des Hypéricinées.* Genève & Paris. 1821.

Christ, *Ros. Schweiz*

H. Christ, *Die Rosen der Schweiz.* Basel, Genf & Lyon. 1873.

Clemente, *Ens. Vid*

S. de Rojas Clemente y Rubio, *Ensayo sobre las Variedades de la Vid comun que vegetan en Andalucía.* Madrid. 1807.

Clementi, G. C., *Sert. Or.*

G. C. Clementi, *Sertulum orientale, seu Recensio Plantarum in Olympo bithynico, in Agro byzantino et hellenico nonnullisque aliis Orientis Regionibus Annis 1845–1850.* Taurini. 1855.

Colla, *Herb. Pedem.*

L. A. Colla, *Herbarium pedemontanum.* Augustae Taurinorum. 1833–1837. (Text: **1** in 1833; **2 & 3** in 1834; **4** in 1835; **5 & 6** in 1836; **7 & 8** in 1837. Plates: **1** in 1835; **2** in 1836; **3** in 1837. Cf. R. E. G. Pichi-Sermolli, *Webbia* **8**: 130–140 (1951) and 407–411 (1952).

Corb., *Nouv. Fl. Normand.*

L. Corbière, *Nouvelle Flore de Normandie.* Caen. 1894.

Cosson, *Not. Pl. Crit.*

E. S. C. Cosson, *Notes sur quelques Plantes critiques, rares et nouvelles, et Additions à la Flore des Environs de Paris.* Paris. 1849–1852. (5 fascicles with continuous pagination: pp. 3–24 in 1849; pp. 25–48 in 1849; pp. 49–92 in 1850; pp. 93–140 in 1851; pp. 141–184 in 1852.)

Costa, *Ind. Sem. Horti Barcin.*

A. C. Costa y Cuxart, *Index Seminum Horti botanici barcinonensis.* Barcinone. 1860.

Costa, *Introd. Fl. Cataluña*

A. C. Costa y Cuxart, *Introduccion á la Flora de Cataluña y Catálogo razonado de las Plantas observadas en esta Region.* Barcelona. Ed. 1. 1864. Ed. 2 (containing *Supl.*, Suplemento al Catálogo razonado de Plantas vasculares de Cataluña). Barcelona. 1877.

Coste, *Fl. Fr.*

H. J. Coste, *Flore descriptive et illustrée de la France, de la Corse et des Contrées limitrophes.* Paris. 1900–1906. (**1**:

pp. 1–240 in 1900; pp. 241–416 in 1901; **2**: pp. 1–96 in 1901; pp. 97–224 in 1902; pp. 225–627 in 1903; **3**: pp. 1–208 in 1904; pp. 209–384 in 1905; pp. 385–807 in 1906.) Reprinted in 1937.

Coutinho, *Fl. Port.*

A. X. Pereira Coutinho, *Flora de Portugal (Plantas vasculares).* Ed. 1. Paris, Lisboa, Rio de Janeiro, S. Paulo & Bello Horizonte. 1913. Ed. 2, by R. T. Palhinha. Lisboa. 1939.

Coutinho, *Not. Fl. Port.*

A. X. Pereira Coutinho, *Notas da Flora de Portugal.* Paris etc. 1914–1930. (**1** in 1914; **2** in 1915; **3** in 1916; **4** in 1918; **5** in 1921; **6** in 1926; **7** in 1930.)

Crantz, *Class. Umb.*

H. J. N. von Crantz, *Classis Umbelliferarum emendata cum generali Seminum Tabula et Figuris aeneis in necessarium Instit. Rei herbar. Supplementum.* Lipsiae. 1767.

Crantz, *Stirp. Austr.*

H. J. N. von Crantz, *Stirpes austriacae.* Ed. 1. Viennae & Lipsiae. 1762–1767. (**1** in 1762; **2** in 1763; **3** in 1767.) Ed. 2. Viennae. 1769. (**1 & 2**, with continuous pagination, in 1769.)

Czern., *Consp. Pl. Charc.*

V. M. Czernajew, *Conspectus Plantarum circa Charcoviam et in Ucrania sponte crescentium et vulgo cultarum.* 1859.

Dalla Torre & Sarnth., *Fl. Tirol*

K. W. von Dalla Torre & L. von Sarntheim, *Flora der gefürsteten Grafschaft Tirol, des Landes Vorarlberg und des Fürstenthumes Liechtenstein.* Innsbruck. 1900–1913. (**1** in 1909; **2** in 1901; **3** in 1905; **4** in 1902; **5** in 1904; **6**(1) in 1906; **6**(2) in 1909; **6**(3) in 1912; **6**(4) in 1913.)

Davies, *Welsh Botanol.*

H. Davies, *Welsh Botanology.* / *Llysieuiaith Gymreig.* London. 1813. (**1 & 2** in 1813.)

DC., *Astrag.*

A. P. de Candolle, *Astragalogia, nempe Astragali, Biserrulae et Oxytropidis, nec non Phacae, Colutae et Lessertiae, Historia Iconibus illustrata.* Parisiis. 1802. (2 editions, folio min. and folio max., with different pagination, in same year. Only the former is cited.)

DC., *Cat. Pl. Horti Monsp.*

A. P. de Candolle, *Catalogus Plantarum Horti botanici monspeliensis.* Monspelii. 1813.

DC., *Coll. Mém.*

A. P. de Candolle, *Collection de Mémoires pour servir à l'Histoire du Règne végétal.* Paris. 1828–1838. (**1** (*Mémoire sur la Famille des Mélastomacées*) & **2** (*...Crassulacées*) in 1828; **3** (*...Onagraires*), **4** (*...Paronychiées*) & **5** (*...Ombellifères*) in 1829; **6** (*...Loranthacées*) in 1830; **7** (*...Valérianées*) in 1832; **8** (*Mémoire sur quelques Espèces de Cactées*) in 1834; **9** (*Observations sur la Structure et la Classification de la Famille des Composées*) & **10** (*Statistique de Famille des Composées*) in 1838.)

DC., *Icon. Pl. Gall. Rar.*

A. P. de Candolle, *Icones Plantarum Galliae rariorum.* Parisiis. 1808.

DC., *Mém. Lég.*

A. P. de Candolle, *Mémoires sur la Famille des Légumineuses.* Paris. 1825–1827. (Pp. 1–128 in 1825; pp. 129–440 in 1826; pp. 441–515 in 1826 or 1827. Plates in 1825–1827. Cf. F. A. Stafleu & W. T. Stearn, *Taxon* **9**: 167–171 (1960).)

DC., *Prodr.*

A. P. de Candolle, *Prodromus Systematis naturalis Regni vegetabilis.* Parisiis. 1824–1874. (**1** in 1824; **2** in 1825; **3** in 1828; **4** in 1830; **5** in 1836; **6** in 1838; **7**(1) in 1838; **7**(2) in 1839; **8** in 1844; **9** in 1845; **10** in 1846; **11** in 1847; **12** in 1848; **13**(1) in 1852; **13**(2) in 1849; **14**: pp. 1–492 in 1856; pp. 493–706 in 1857; **15**(1) in 1864; **15**(2): pp. 1–188 in 1862; pp. 189–1286 in 1866; **16**(1) in 1869; **16**(2): pp. 1–160 in

1864; pp. 161–691 in 1868; **17** in 1873. Index **1–4** in 1843; **5–7(1)** in 1840; **7(2)–13** in 1858–1859; **14–17** in 1874.)

DC., A. & C., *Monogr. Phan.*
A. P. de Candolle & A. L. P. P. de Candolle, *Monographiae Phanerogamarum Prodromi nunc Continuatio, nunc Revisio.* Parisiis. 1878–1896. (**1** in 1878; **2** in 1879; **3** in 1881; **4** in 1883; **5(1)**: pp. 1–303 in 1883; **5(2)**: pp. 305–654 in 1887; **6** in 1889; **7** in 1891; **8** in 1893; **9** in 1896.)

De Not., *Ind. Sem. Horti Bot. Genuens.*
G. de Notaris, *Index Seminum Horti botanici genuensis.* Genova. 1840–1871.

Decne, *Jard. Fruit.*
J. Decaisne, *Le Jardin fruitier du Muséum, ou Iconographie de toutes les Espèces et Variétés d'Arbres fruitiers cultivés dans cet Établissement avec leur Description, leur Histoire, leur Synonymie, etc.* Paris. 1858–1871. (Originally published in 123 livraisons; later rearranged to form 9 volumes.)

Dehnh., *Cat. Pl. Horti Camald.*
F. Dehnhardt, *Catalogus Plantarum Horti camaldulensis.* Ed. 1. Neapoli. 1829. Ed. 2. Neapoli. 1832.

Delile, *Descr. Égypte, Hist. Nat.*
A. R. Delile, *Description de l'Égypte. Histoire naturelle.* **2.** Paris. 1813.

Desf., *Fl. Atl.*
R. L. Desfontaines, *Flora atlantica, sive Historia Plantarum, quae in Atlante, Agro tunetano et algeriensi crescunt.* Parisiis. 1798–1799. (**1** in 1798; **2**: pp. 1–160 in 1798; pp. 161–458 in 1799. Cf. M. J. van Steenis-Kruseman & W. T. Stearn in C. G. G. J. van Steenis, *Fl. Males.* **I. 4**: clxxvii (1954).)

Desf., *Hist. Arb.*
R. L. Desfontaines, *Histoire des Arbres et des Arbrisseaux qui peuvent être cultivés en pleine Terre sur le Sol de la France.* Paris. 1809. (**1 & 2** in 1809.)

Desportes, *Ros. Gall.*
N. H. F. Desportes, *Rosetum gallicum, ou Énumération méthodique des Espèces et Variétés du Genre Rosier indigènes en France ou cultivées dans les Jardins, avec la Synonymie française et latine.* Le Mans & Paris. 1828.

Desv., *Jour. Bot. Appl.*
A. N. Desvaux (edit.), *Journal de Botanique, appliquée à l'Agriculture, à la Pharmacie, à la Médecine et aux Arts.* Paris. 1813–1816. (**1 & 2** in 1813; **3 (1 & 2)** in 1814; **3 (3)** in 1816; **3(4)** in 1814; **3(5)** in 1816; **4** in 1814.)

Desv., *Jour. Bot. Rédigé*
A. N. Desvaux *et al.* (edit.), *Journal de Botanique, rédigé par une Société de Botanistes.* Paris. 1808–1809. (**1** in 1808; **2** in 1809.)

Desv., *Obs. Pl. Env. Angers*
A. N. Desvaux, *Observations sur les Plantes des Environs d'Angers, pour servir de Supplément à la Flore de Maine et Loire, et de Suite à l'Histoire naturelle et critique des Plantes de France.* Angers & Paris. 1818.

De Wild., *Pl. Bequaert.*
É. A. J. de Wildeman, *Plantae bequaertianae.* Gand. 1921–1932. (**1**: pp. 1–166 in 1921; pp. 167–593 in 1922; **2**: pp. 1–308 in 1923; pp. 309–570 in 1924; **3**: pp. 1–424 in 1925; pp. 425–576 in 1926; **4**: pp. 1–160 in 1926; pp. 161–308 in 1927; pp. 309–452 in 1928; pp. 453–575 in 1929; **5**: pp. 1–160 in 1929; pp. 161–352 in 1931; pp. 353–496, i–xiii in 1932; **6**: 1–91 in 1932.)

Dietr., F. G., *Vollst. Lexic.*
F. G. Dietrich, *Vollständiges Lexicon der Gärtnerei und Botanik.* Weimar & Berlin. 1802–1811. (**1–10** & *Register* in 1802–1811.) *Nachtrag.* Berlin. 1815–1821. (**1–10** in 1815–1821.) *Neuer Nachtrag.* Berlin & Ulm. 1825–1840. (**1–10** in 1825–1840.)

Don, D., *Prodr. Fl. Nepal.*
D. Don, *Prodromus Florae nepalensis, sive Enumeratio Vegeta-* *bilium, quae in Itinere per Nepaliam proprie dictam et Regiones conterminas, Ann. 1802–1803 detexit atque legit D. D. Franciscus Hamilton (olim Buchanan). . . . accedunt Plantae a D. Wallich nuperius missae.* Londini. 1825. (For date cf. W. T. Stearn, *Jour. Arnold Arb.* **26**: 168 (1945).)

Don fil., G., *Gen. Syst.*
G. Don, *A general System of Gardening and Botany.* Also titled *A general History of dichlamydeous Plants.* London. 1831–1838. (**1** in 1831; **2** in 1832; **3** in 1834; **4** in 1837–1838. Cf. M. J. van Steenis-Kruseman & W. T. Stearn in C. G. G. J. van Steenis, *Fl. Males.* **I. 4**: clxxviii (1954).)

Dörfler, *Herb. Norm.*
I. Dörfler, *Herbarium normale. Conditum a F. Schultz, dein continuatum a K. Keck, nunc editum per I. Dörfler.* Vindobonae. 1894–1915. (**31–56** in 1894–1915.)

Dostál, *Květena ČSR*
J. Dostál, *Květena ČSR.* Praha. 1948–1950. (Preface etc., pp. 1–64 in 1950; pp. 1–800 in 1948; pp. 801–1488 in 1949; pp. 1489–2269 in 1950.)

Duby, *Bot. Gall.*
J. E. Duby, *Aug. Pyrami de Candolle Botanicon gallicum, seu Synopsis Plantarum in Flora gallica descriptarum. Editio secunda. Ex Herbariis et Schedis candollianis propriisque digestum a J. E. Duby.* Paris. 1828–1830. (**1** in 1828; **2** in 1830.) This is Ed. 2 of Lam. & DC., *Syn. Pl. Fl. Gall.*

Duchartre, *Rev. Bot.*
P. E. S. Duchartre (edit.), *Revue botanique. Recueil mensuel renfermant l'Analyse des Travaux publiés en France et à l'Étranger sur la Botanique et sur ses Applications à l'Horticulture, l'Agriculture, la Médecine etc.* Paris. 1845–1847.

Duchesne, *Hist. Nat. Frais.*
A. N. Duchesne, *Histoire naturelle des Fraisiers.* Paris. 1766.

Dumort., *Fl. Belg.*
B. C. J. Dumortier, *Florula belgica, Operis majoris Prodromus. Staminacia.* Tornaci Nerviorum. 1827.

Duroi, *Harbk. Baumz.*
J. P. Duroi, *Die Harbkesche wilde Baumzucht, theils nordamerikanischer und anderer fremder, theils einheimischer Bäume, Sträucher und strauchartigen Pflanzen nach den Kennzeichen, der Anzucht, den Eigenschaften und der Benutzung beschrieben.* Braunschweig. Ed. 1. 1771–1772. (**1** in 1771; **2** in 1772.) Ed. 2 by J. F. Pott. 1795–1800. (**1** in 1795; **2 & 3** in 1800.)

Ehrh., *Beitr. Naturk.*
J. F. Ehrhart, *Beiträge zur Naturkunde.* Hannover & Osnabrück. 1787–1792. (**1** in 1787; **2 & 3** in 1788; **4** in 1789; **5** in 1790; **6** in 1791; **7** in 1792. Cf. F. A. Stafleu, *Taxon* **12**: 59–60 (1963).)

Elliott, *Sketch Bot. South-Carol. Georgia*
S. Elliott, *A Sketch of the Botany of South-Carolina and Georgia.* Charleston. 1816–1824. (**1**: pp. i–ii, 1–200 in 1816; pp. 201–496 in 1817; pp. 497–606 in 1821; **2**: pp. 1–104 in 1821; pp. 105–208 in 1822; pp. 209–312 in ?1822; pp. 313–520 in ?1823; pp. 521–743 in 1824. Cf. M. J. van Steenis-Kruseman & W. T. Stearn in C. G. G. J. van Steenis, *Fl. Males.* **I. 4**: clxxx (1954).)

Endl., *Gen. Pl.*
S. L. Endlicher, *Genera Plantarum secundum Ordines naturales disposita.* Vindobonae. 1836–1841. (Pp. 1–160, i–iv in 1836; pp. 161–400 in 1837; pp. 401–640, v–xii in 1838; pp. 637–960, xiii–xx in 1839; pp. 961–1360, xxi–xl in 1840; pp. 1361–1483, xli–lx in 1841. Pp. 1335–1427 comprise **Suppl. 1**. **Suppl. 2**, titled *Mantissa botanica*, in 1842. **Suppl. 3**, titled *Mantissa botanica altera*, in 1843. **Suppl. 4(2)** in 1847 or 1848. **Suppl. 5** (= 4(3)) in 1850. Cf. M. J. van Steenis-Kruseman & W. T. Stearn in C. G. G. J. van Steenis, *Fl. Males.* **I. 4**: clxxxi (1954).)

Endl. et al., *Enum. Pl. Hügel*

S. L. Endlicher, G. Bentham, E. Fenzl & H. W. Schott, *Enumeratio Plantarum quas in Novae Hollandiae Ora austro-occidentali ad Fluvium Cygnorum et in Sinu Regis Georgii collegit Carolus Liber Baro de Hügel.* Vindobonae. 1837.

Engler, *Pflanzenreich*

H. G. A. Engler (edit.), *Das Pflanzenreich. Regni vegetabilis Conspectus.* (For authors and dates of individual volumes cf. M. T. Davis, *Taxon* 6: 161–182 (1957).) In citation the number of the *Heft* is placed first, followed by the systematic numbers (both Roman and Arabic) in brackets.

Engler & Prantl, *Natürl. Pflanzenfam.*

H. G. A. Engler & K. A. E. Prantl, *Die natürlichen Pflanzenfamilien nebst ihren Gattungen und wichtigeren Arten, insbesondere den Nutzpflanzen.* Ed. 1. Leipzig. 1887–1915. (2(1): pp. 1–172 in 1887; pp. 173–262 in 1889; 2(2): pp. 1–96 in 1887; pp. 96–130 in 1888; 2(3): pp. 1–144 in 1887; pp. 145–168 in 1889; 2(4): pp. 1–48 in 1887; pp. 49–78 in 1888; 2(5): pp. 1–144 in 1887; pp. 145–162 in 1888; 2(6): pp. 1–144 in 1888; pp. 145–244 in 1889; 3(1): pp. 1–48 in 1887; pp. 49–144 in 1888; pp. 145–289 in 1889; 3(1a): pp. 1–48 in 1892; pp. 49–130 in 1893; 3(1b) in 1889; 3(2): pp. 1–96 in 1888; pp. 97–144 in 1889; pp. 145–281 in 1891; 3(2a): pp. 1–48 in 1890; pp. 49–142 in 1891; 3(3): pp. 1–48 in 1888; pp. 49–112 in 1891; pp. 113–208 in 1892; pp. 209–256 in 1893; pp. 257–396 in 1894; 3(4): pp. 1–94 in 1890; pp. 95–362 in 1896; 3(5): pp. 1–96 in 1890; pp. 97–128 in 1891; pp. 129–224 in 1892; pp. 225–272 in 1893; pp. 273–416 in 1895; pp. 417–468 in 1896; 3(6): pp. 1–96 in 1890; pp. 97–240 in 1893; pp. 241–340 in 1895; 3(6a): pp. 1–96 in 1893; pp. 97–254 in 1894; 3(7): pp. 1–48 in 1892; pp. 49–241 in 1893; 3(8): pp. 1–48 in 1894; pp. 49–144 in 1897; pp. 145–274 in 1898; 4(1): pp. 1–96 in 1889; pp. 97–144 in 1890; pp. 145–183 in 1891; 4(2): pp. 1–48 in 1892; pp. 49–310 in 1895; 4(3a): pp. 1–48 in 1891; pp. 49–96 in 1893; pp. 97–224 in 1895; pp. 225–320 in 1896; pp. 321–384 in 1897; 4(3b): pp. 1–96 in 1891; pp. 97–144 in 1893; pp. 145–240 in 1894; pp. 241–378 in 1895; 4(4) in 1891; 4(5): pp. 1–80 in 1889; pp. 81–272 in 1890; pp. 273–304 in 1892; pp. 305–368 in 1893; pp. 369–402 in 1894; *Gesamtregister* pp. 1–320 in 1898; pp. 321–462 in 1899; *Nachtrag* 1 in 1897; *Nachtrag* 2 in 1900; *Nachtrag* 3: pp. 1–192 in 1906; pp. 193–288 in 1907; pp. 289–379 in 1908; *Nachtrag* 4: pp. 1–192 in 1914; pp. 193–381 in 1915. Cf. M. J. van Steenis-Kruseman & W. T. Stearn in C. G. G. J. van Steenis, *Fl. Males.* I. 4: clxxxi–clxxxii (1954).) Ed. 2. Leipzig & Berlin. 1925 →. (13, 14a, 14e, 15a, 16b, 16c, 17b, 18a, 19a, 19b I, 19c, 20b & 21 at Leipzig; 14d, 17a II & 20d at Berlin. 13 in 1926; 14a in 1926; 14d in 1956; 14e in 1940; 15a in 1930; 16b in 1935; 16c in 1934; 17a II in 1959; 17b in 1936; 18a in 1930; 19a in 1931; 19b I in 1940; 19c in 1931; 20b in 1942; 20d in 1953; 21 in 1925.) Photographic reprint. Berlin. 1960.

Engler, V., *Monogr. Gatt. Tilia*

V. Engler, *Monographie der Gattung* Tilia. Breslau. 1909.

Fenzl, *Pugillus*

E. Fenzl, *Pugillus Plantarum novarum Syriae et Tauri occidentalis primus.* Vindobonæ. 1842.

Fiori, *Nuov. Fl. Anal. Ital.*

A. Fiori, *Nuova Flora analitica d'Italia.* Firenze. 1923–1929. (1: pp. 1–480 in 1923; pp. 481–800 in 1924; pp. 801–944 in 1925; 2: pp. 1–160 in 1925; pp. 161–480 in 1926; pp. 481–800 in 1927; pp. 801–944 in 1928; pp. 945–1120 in 1929.)

Fiori & Paol., *Fl. Anal. Ital.*

A. Fiori & G. Paoletti, *Flora analitica d'Italia.* Padova. 1896–1909. (1: pp. i–c in 1908; pp. 1–256 in 1896; pp. 257–610 in 1898; 2: pp. 1–224 in 1900; pp. 225–304 in 1901; pp. 305–493 in 1902; 3: pp. 1–272 in 1903; pp. 273–524 in 1904; 4: pp.

1–217, index pp. 1–16 in 1907; index pp. 17–330 in 1908; 5 in 1909.)

Fischer, *Cat. Jard. Gorenki*

F. E. L. von Fischer, *Catalogue du Jardin des Plantes du Comte Alexis de Razoumoffsky à Gorenki près de Moscou.* Ed. 1. Moscou. 1808. Ed. 2. Moscou. 1812.

Fischer, *Syn. Astrag. Trag.*

F. E. L. von Fischer, *Synopsis Astragalorum tragacantharum.* Mosquae. 1853.

Fischer & C. A. Meyer, *Enum. Pl. Nov.*

F. E. L. von Fischer & C. A. von Meyer, *Enumeratio [altera] Plantarum novarum a clarissimo Schrenk lectarum.* Petropoli. 1841–1842. (1 in 1841; 2 in 1842.)

Fischer & C. A. Meyer, or **Fischer, C. A. Meyer & Avé-Lall.,** *Ind. Sem. Horti Petrop.*

F. E. L. von Fischer & C. A. von Meyer, or F. E. L. von Fischer, C. A. von Meyer & J. L. E. Avé-Lallemant, or J. L. E. Avé-Lallemant, *Index Seminum Horti petropolitani.* Petropoli. 1835–1850. (For dates cf. H. W. Rickett & F. A. Stafleu, *Taxon* 10: 83 (1961).)

Focke, *Syn. Rub. Germ.*

W. O. Focke, *Synopsis Ruborum Germaniae. | Die deutschen Brombeerarten.* Bremen. 1877.

Fomin, *Fl. RSS Ucr.*

A. V. Fomin, *Flora RSS Ucr. | Flora Reipublicae Sovieticae Socialisticae Ucrainicae.* Ed. 1. Kiev. 1936–1965. (1, edit. A. V. Fomin in 1936; 2, edit. E. I. Bordzilowski & E. M. Lavrenko in 1940. Titled Флора УРСР [*Flora URSR*]: 3, edit. M. I. Kotov & A. I. Barbarich in 1950; 4, edit. M. I. Kotov in 1952; 5, edit. M. V. Klokov & E. D. Wissjulina in 1953; 6, edit. D. K. Zerov in 1954; 7, edit. M. V. Klokov & E. D. Wissjulina in 1955; 8, edit. M. I. Kotov & A. I. Barbarich in 1957; 9, edit. M. I. Kotov in 1960; 10, edit. M. I. Kotov in 1961; 11, edit. E. D. Wissjulina in 1962; 12, edit. E. D. Wissjulina in 1965.) Ed. 2. Kiev. 1938. (1, edit. E. I. Bordzilowski in 1938.)

Font Quer, *Fl. Cardó*

P. Font Quer, *Flórula de Cardó.* Barcelona. 1950.

Forskål, *Fl. Aegypt.*

P. Forskål, *Flora aegyptiaco-arabica.* Hauniae. 1775.

Forster, G., *Fl. Ins. Austral. Prodr.*

J. G. A. Forster, *Florulae Insularum australium Prodromus.* Gottingae. 1786.

Fourn., P., *Quatre Fl. Fr.*

P. Fournier, *Les quatre Flores de la France, Corse comprise.* Poinson-les-Grancey. 1934–1940. (Pp. 1–64 in 1934; pp. 65–256 in 1935; pp. 257–576 in 1936; pp. 577–832 in 1937; pp. 833–896 in 1938; pp. 897–992 in 1939; pp. 1065–1092 in 1940.) Photographic reprint with additions and corrections at end. Paris. 1946. Reprint with more additions and corrections. Paris. 1961.

Fries, *Fl. Halland.*

E. M. Fries, *Flora hallandica.* Lund. 1817–1819. (Pp. 1–68 in 1817; pp. 69–100 in 1818; pp. 101–159 in 1819.)

Fries, *Nov. Fl. Suec.*

E. M. Fries, *Novitiae Florae suecicae.* Ed. 1. Lundae. 1814–1823. (Pp. 1–40 in 1814; pp. 43–60 in 1817; pp. 63–80 in 1819; pp. 83–122 in 1823.) Ed. 2. Londini Gothorum. 1828. *Mantissa* 1. Lundae. 1832. *Mantissa* 2. Upsaliae. 1839. *Mantissa* 3. Upsaliae. 1842.

Fries, *Summa Veg. Scand.*

E. M. Fries, *Summa Vegetabilium Scandinaviae.* Holmiae & Lipsiae. 1846–1849. (Pp. 1–258 in 1846; pp. 261–570 in 1849.)

Fritsch, *Exkursionsfl. Österr.*

K. Fritsch, *Excursionsflora für Österreich (mit Ausschluss von Galizien, Bukowina und Dalmatien).* Ed. 1. Wien. 1897. Ed. 2, titled *Exkursionsflora für Österreich (mit Ausschluss von*

Galizien, Bukowina und Dalmatien). Wien. 1909. Ed. 3, titled *Exkursionsflora für Österreich und die ehemals österreichischen Nachbargebiete*. Wien & Leipzig. 1922.

Gaertner, Fruct. Sem. Pl.
J. Gaertner, *De Fructibus et Seminibus Plantarum*. Stuttgardiae, Tuebingae & Lipsiae. 1788–1807. (**1**, Stuttgardiae; **2**, Tuebingae; 3, Lipsiae. **1** in 1788; **2**: pp. 1–184 in 1790; pp. 185–504 in 1791; pp. 505–520 in 1792; **3**: pp. 1–128 in 1805; pp. 129–256 in 1807. Cf. F. A. Stafleu, *Taxon* **12**: 60–62 (1963).) 3 by C. F. Gaertner.

Garcke, Fl. Halle
F. A. Garcke, *Flora von Halle*. Halle. 1848–1856. (**1**(**1** & **2**) in 1848; **2** in 1856.)

Garcke, Fl. Nord-Mittel-Deutschl.
F. A. Garcke, *Flora von Nord- und Mittel-Deutschland*. Ed. 1. Berlin. 1849. Ed. 2. Berlin. 1851. Ed. 3. Berlin. 1854. Ed. 4. Berlin. 1858. Ed. 5. Berlin. 1860. Ed. 6. Berlin. 1863. Ed. 7. Berlin. 1865. Ed. 8. Berlin. 1867. Ed. 9. Berlin. 1869. Ed. 10. Berlin. 1871. Ed. 11. Berlin. 1873. Ed. 12. Berlin. 1875. Later editions with different titles exist.

Gasparr., Not. Piante Lucan.
G. Gasparrini, *Notizie intorno ad alcune Piante della Lucania*. Napoli. 1833. Reprinted as *Progr. Sci. Lett. Arti* nov. ser., **4**: 161–172 (1833).

Gaudin, Fl. Helv.
J. F. A. T. G. P. Gaudin, *Flora helvetica*. Turici. 1828–1833. (**1–3** in 1828; **4** & **5** in 1829; **6** in 1830; **7** in 1833.)

Gay, C., Hist. Chil. Bot.
C. Gay, *Historia física y política de Chile. Botánica*. Paris. 1845–1854. (**1–8** in 1845–1854. For dates cf. I. M. Johnston, *Darwinia* **5**: 154–165 (1941) and M. J. van Steenis-Kruseman & W. T. Stearn in C. G. G. J. van Steenis, *Fl. Males.* I. **4**: clxxxiv (1954).)

Genev., Monogr. Rubus
L. G. Genevier, *Essai monographique sur les* Rubus *du Bassin de la Loire* (as *Mém. Soc. Acad. Angers* **24**: 1–346 (1868)). Reprinted, Angers, 1869. *Premier Supplément* (as *Mém. Soc. Acad. Angers* **28**: 1–96 (1872)). Ed. 2, titled *Monographie des* Rubus *du Bassin de la Loire*. Paris & Nantes. 1880.

Gmelin, C. C., Fl. Bad.
C. C. Gmelin, *Flora badensis alsatica et confinium Regionum cis et transrhenana, Plantas a Lacu Bodamico usque ad Confluentem Mosellae et Rheni sponte nascentes exhibens*. Carlsruhae. 1805–1826. (**1** in 1805; **2** in 1806; **3** in 1808; **4** in 1826.)

Gmelin, S. G., Reise Russl.
S. G. Gmelin, *Reise durch Russland zur Untersuchung der drey Natur-Reiche*. St. Petersburg. 1770–1774. (**1** in 1770; **2** & **3** in 1774.)

Godet, Fl. Jura
C. H. Godet, *Flore du Jura*. Neuchâtel. 1852–1853. (**1**: pp. 1–432 in 1852; **2**: pp. 433–872 in 1853.) *Supplément*. Neuchâtel. 1869.

Godman, Nat. Hist. Azores
F. Du Cane Godman, *Natural History of the Azores or Western Islands*. London. 1870.

Godron, Fl. Lorr.
D. A. Godron, *Flore de Lorraine* (*Meurthe, Moselle, Meuse, Vosges*). Ed. 1. Nancy. 1843–1844. (**1** & **2** in 1843; **3** in 1844). *Supplément*. Nancy. 1845. Ed. 2. Nancy. 1861. (**1** & **2** in 1861.)

Goiran, Spec. Morph. Veg.
A. Goiran, *Specimen Morphographiae vegetabilis*. Veronae. 1875.

Gorodkov, Fl. Murmansk.
B. N. Gorodkov (edit.), Флора Мурманской Области [*Flora Murmanskoj Oblasti*]. Moskva & Leningrad. 1953–1966. (**1** in

1953; **2** in 1954; **3** in 1956; **4** in 1959; **5** in 1966.) **2–5** edit. A. I. Pojarkova.

Gouan, Fl. Monsp.
A. Gouan, *Flora monspeliaca*. Lugduni. 1765.

Gouan, Obs. Bot.
A. Gouan, *Illustrationes et Observationes botanicae, ad Specierum Historiam facientes, seu rariorum Plantarum indigenarum, pyrenaicarum, exoticarum Adumbrationes*. Tiguri. 1773.

Govoruchin, Fl. Urala
V. S. Govoruchin, Флора Урала [*Flora Urala*]. / *Flora medio, boreali el polari uralensis*. Sverdlovsk. 1937.

Gray, S. F., Nat. Arr. Brit. Pl.
S. F. Gray, *A natural Arrangement of British Plants*. London. 1821. (**1** & **2** in 1821.)

Grec., Consp. Fl. Roman.
D. Grecescu, *Conspectul Florei Romaniei*. Bucureşti. 1898. *Supl., Suplement*. Bucureşti. 1909.

Greene, E. L., Cybele Columb.
E. L. Greene, *Cybele columbiana*. Providence, London & Leipzig. 1914. (**1** in 1914.)

Gremli, Beitr. Fl. Schweiz
A. Gremli, *Beiträge zur Flora der Schweiz*. Aarau. 1870.

Gremli, Excurs.-Fl. Schweiz
A. Gremli, *Excursionsflora für die Schweiz*. Ed. 1. Aarau. 1867. Ed. 2. Aarau. 1874. Ed. 3. Aarau. 1878. Ed. 4. Aarau. 1881. Ed. 5. Aarau. 1885. Ed. 6. Aarau. 1889. Ed. 7. Aarau. 1893. Ed. 8. Aarau. 1896. Ed. 9. Aarau. 1901.

Gremli, Neue Beitr. Fl. Schweiz
A. Gremli, *Neue Beiträge zur Flora der Schweiz*. Aarau. 1880–1890. (**1** in 1880; **2** in 1882; **3** in 1883; **4** in 1887; **5** in 1890.)

Gren. & Godron, Fl. Fr.
J. C. M. Grenier & D. A. Godron, *Flore de France, ou Description des Plantes qui croissent naturellement en France et en Corse*. Paris. 1847–1856. (**1**: pp. 1–330 in 1847; pp. 331–762 in 1849; **2**: pp. 1–392 in 1850; pp. 393–760 in 1853; **3**: pp. 1–384 in 1855; pp. 385–779 in 1856. Cf. M. J. van Steenis-Kruseman & W. T. Stearn in C. G. G. J. van Steenis, *Fl. Males.* I. **4**: clxxxv (1954).)

Gren. & Godron, Fl. Fr. Prosp.
J. C. M. Grenier & D. A. Godron, *Flore de France. Prospectus*. Besançon. 1846.

Griseb., Spicil. Fl. Rumel.
A. H. R. Grisebach, *Spicilegium Florae rumelicae et bithynicae*. Brunsvigae. 1843–1845. (**1** in 1843; **2**: pp. 1–160 in 1844; pp. 161–548 in 1845 or 1846. Cf. H. W. Rickett & F. A. Stafleu, *Taxon* **10**: 85 (1961).)

Grossh., Fl. Kavk.
A. A. Grossheim, Флора Кавказа [*Flora Kavkaza*]. Ed. 1. Tiflis & Baku. 1928–1934. (**1** in 1928; **2** in 1930; **3** in 1932; **4** in 1934.) **1** & **2** at Tiflis. **3** & **4** at Baku. Ed. 2. Baku, Moskva & Leningrad. 1939. (**1** in 1939; **2** in 1940; **3** in 1945; **4** in 1950; **5** in 1952; **6** in 1962.) **1–3** at Baku; **4–6** at Moskva & Leningrad.

Guimpel, Otto & Hayne, Abbild. Fremd. Holzart.
F. Guimpel, C. F. Otto & F. G. Hayne, *Abbildung der fremden, in Deutschland ausdauernden Holzarten*. Berlin. 1825.

Guinea, Cistác. Esp.
E. Guinea, *Cistáceas españolas* (*Bol. Inst. Forest. Inv. Exper. Madrid*, No 71). Madrid. 1954.

Günther & Schummel, Sched. Cent. Siles. Exsicc.
C. C. Günther & T. E. Schummel, *Schedae Centuriae Plantarum silesiacarum exsiccatarum*. Wratislaviæ. 1811–1829. (**1** & **2** in 1811; **3** in 1812; **4** & **5** in 1813; **6** & **7** in 1816; **8** in 1818; **9** in 1819; **10** in 1820; **11** in 1821; **12** in 1823; **13** in 1825; **14**(**1**) in 1826; **14**(**2**) in 1827; **15** in 1829.)

Guss., *Cat. Pl. Boccad.*

G. Gussone, *Catalogus Plantarum, quae asservantur in Regio Horto ser. Fr. Borbonii Principis Juventutis in Boccadifalco prope Panormum.* Neapoli. 1821.

Guss., *Fl. Sic. Prodr.*

G. Gussone, *Florae siculae Prodromus.* Neapoli. 1827–1828. (**1** in 1827; **2** in 1828.) *Suppl., Supplementum.* 1832–1834. (Pp. 1–166 in 1832; pp. 167–242 in 1834.)

Guss., *Fl. Sic. Syn.*

G. Gussone, *Florae siculae Synopsis.* Neapoli. 1843–1845. (**1** in 1843; **2**: pp. 1–668 in 1844; pp. 669–903 in 1845. Cf. H. W. Rickett & F. A. Stafleu, *Taxon* **10**: 86 (1961).)

Guss., *Pl. Rar.*

G. Gussone, *Plantae rariores, quas in Itinere per Oras Jonii et Adriatici Maris et per Regiones Samnii et Apruttii collegit....* Neapoli. 1826.

Halácsy, *Consp. Fl. Graec.*

E. von Halácsy, *Conspectus Florae graecae.* Lipsiae. 1900–1904. (**1**: pp. 1–576 in 1900; pp. 577–825 in 1901; **2** in 1902; **3** in 1904. Cf. M. J. van Steenis-Kruseman & W. T. Stearn in C. G. G. J. van Steenis, *Fl. Males.* I. **4**: clxxxv (1954).) *Suppl., Supplementum* 1. Lipsiae. 1908. (*Supplementum* 2 published as *Magyar Bot. Lapok* **11**: 114–202 (1912).)

Hanry, *Prodr. Hist. Nat. Var*

H. Hanry *et al.*, *Prodrome d'Histoire naturelle du Département du Var.* Draguignan. 1853. Pp. 135–397, *Botanique*, by H. Hanry.

Hartinger, *Atlas Alpenfl.* (*Text*)

A. Hartinger, *Atlas der Alpenflora. Text* by K. W. von Dalla Torre. Ed. 1. Wien. 1882. Ed. 2. Graz. 1897.

Hartman, *Handb. Skand. Fl.*

C. J. Hartman, *Handbok i Skandinaviens Flora.* Ed. 1. Stockholm. 1820. Ed. 2. Stockholm. 1832. Ed. 3. Stockholm. 1838. Ed. 4. Stockholm. 1843. Ed. 5. Stockholm. 1849. Ed. 6 by C. Hartman. Stockholm. 1854. Ed. 7 by C. Hartman, Stockholm. 1858. Ed. 8 by C. Hartman. Stockholm. 1861. Ed. 9 by C. Hartman. Stockholm. 1864. Ed. 10 by C. Hartman. Stockholm. 1870. Ed. 11 titled *C. J. Hartman's Handbok i Skandinaviens Flora*, by C. Hartman. Stockholm. 1879. Ed. 12 titled *C. J. och C. Hartman's Handbok i Skandinaviens Flora*, by T. O. B. N. Krok. Stockholm. 1889.

Haw., *Suppl. Pl. Succ.*

A. H. Haworth, *Supplementum Plantarum succulentarum.* Londini. 1819.

Haw., *Syn. Pl. Succ.*

A. H. Haworth, *Synopsis Plantarum succulentarum.* Londini. 1812.

Hayek, *Fl. Steierm.*

A. von Hayek, *Flora von Steiermark.* Berlin & Graz. 1908–1956. (**1**: pp. 1–480 in 1908; pp. 481–960 in 1909; pp. 961–1200 in 1910; pp. 1201–1271 in 1911; **2**(**1**): pp. 1–160 in 1911; pp. 161–480 in 1912; pp. 481–640 in 1913; pp. 641–865 in 1940; pp. 867–870 in 1956; **2**(**2**) in 1956. Cf. A. Lawalrée, *Nat. Mosana* **13**: 50 (1960).)

Hayek, *Prodr. Fl. Penins. Balcan.*

A. von Hayek, *Prodromus Florae Peninsulae balcanicae.* (Published as *Feddes Repert.* (*Beih.*) 30.) (**1**: pp. 1–352 in 1924; pp. 353–672 in 1925; pp. 673–960 in 1926; pp. 961–1193 in 1927; **2**: pp. 1–96 in 1928; pp. 97–336 in 1929; pp. 337–576 in 1930; pp. 577–960 in 1931; pp. 961–1152 in 1932; **3**: pp. 1–208 in 1932; pp. 209–472 in 1933.) **2** & **3** edit. F. Markgraf.

Hayek, *Sched. Fl. Stir. Exsicc.*

A. von Hayek, *Schedae ad Floram stiriacam exsiccatam.* Wien. 1904–1912. (**1** & **2** in 1904; **3** & **4** in 1905; **5** & **6** in 1905; **7** & **8** in 1906; **9** & **10** in 1906; **11** & **12** in 1907; **13** & **14** in 1908; **15** & **16** in 1909; **17** & **18** in 1909; **19** & **20** in 1910; **21** & **22** in 1910; **23** & **24** in 1912; **25** & **26** in 1912.)

Hayne, *Darst. Beschr. Arzn. Gewächse*

F. G. Hayne, *Getreue Darstellung und Beschreibung der in der Arzneykunde gebräuchlichen Gewächse, wie auch solcher, welche mit ihnen verwechselt werden können.* Berlin. **1**–**14**. 1805–1843. (**1** in 1805; **2** in 1809; **3** in 1813; **4** in 1816; **5** in 1817; **6** in 1819; **7** in 1821; **8** in 1822; **9** in 1825; **10** in 1827; **11** in 1830; **12** in 1833; **13** in 1837; **14**: tt. 1–12 in 1843; tt. 13–24 in 1846.) **12** by F. G. Hayne, J. F. Brandt & J. T. C. Ratzeburg. **13** continued by J. F. Brandt & J. T. C. Ratzeburg.

Hegetschw., *Fl. Schweiz*

J. J. Hegetschweiler, *Flora der Schweiz.* Zürich. 1840. *Supplementum* by O. Heer, in same volume with continuous pagination.

Hegi, *Ill. Fl. Mitteleur.*

G. Hegi, *Illustrierte Flora von Mitteleuropa.* Ed. 1. München. 1906–1931. (**1**: pp. 1–72 in 1906; pp. 73–312 in 1907; pp. 313–412 in 1908; **2**: pp. 1–128 in 1908; pp. 129–408 in 1909; **3**: pp. 1–36 in 1909; pp. 37–328 in 1910; pp. 329–472 in 1911; pp. 473–608 in 1912; **4**(**1**): pp. 1–96 in 1913; pp. 97–144 in 1914; pp. 145–192 in 1916; pp. 193–320 in 1918; pp. 321–491 in 1919; **4**(**2**): pp. 497–540 in 1921; pp. 541–908 in 1922; pp. 909–1112a in 1923; **4**(**3**): pp. 1113–1436 in 1923; pp. 1437–1748 in 1924; **5**(**1**): pp. 1–316 in 1924; pp. 317–674 in 1925; **5**(**2**): pp. 679–994 in 1925; pp. 995–1562 in 1926; **5**(**3**): pp. 1567–1722 in 1926; pp. 1723–2250 in 1927; **5**(**4**): pp. 2255–2632 in 1927; **6**(**1**): pp. 1–112 in 1913; pp. 113–304 in 1914; pp. 305–352 in 1915; pp. 353–400 in 1916; pp. 401–496 in 1917; pp. 497–544 in 1918; **6**(**2**): pp. 549–1152 in 1928; pp. 1153–1386 in 1929; **7** in 1931.) Ed. 2. München. 1936→ . (**1** in 1936; **2** in 1939; **3**(**1**): pp. 1–240 in 1957; pp. 241–452 in 1958; **3**(**2**): pp. 453–532 in 1959; pp. 533–692 in 1960; pp. 693–772 in 1961; pp. 773–852 in 1962; **3**(**3**): pp. 1–80 in 1965; **4**(**1**): pp. 1–80 in 1958; pp. 81–160 in 1959; pp. 161–320 in 1960; pp. 321–480 in 1962; pp. 481–548 in 1963; **4**(**2**): pp. 1–80 in 1961; pp. 81–224 in 1963; pp. 225–304 in 1964; pp. 305–384 in 1965; pp. 385–448 in 1966; **6**(**1**): pp. 1–80 in 1965; pp. 81–160 in 1966; **6**(**3**): pp. 1–80 in 1964; pp. 81–160 in 1965; pp. 161–240 in 1966.)

Heldr., *Sched. Herb. Graec. Norm.*

T. von Heldreich, *Schedae Plantarum ad Herbaria graeca normalia.*

Herrmann, J., *Diss. Rosa*

J. Herrmann, *Dissertatio inauguralis botanico-medica de Rosa.* Argentorati. 1762.

Hiitonen, *Suomen Kasvio*

H. I. A. Hiitonen, *Suomen Kasvio.* Helsingissä. 1933.

Hill, *Hort. Kew.*

J. Hill, *Hortus kewensis, sistens Herbas exoticas indigenasque rariores in Area botanica Hortorum aug. Pr. Cambriae dotissae apud Kew in Comitatu surreiano cultas.* Londini. 1768.

Hill, A. W., *Hand-List Herb. Pl. Kew*

Hand-List of herbaceous Plants...cultivated in the Royal Botanic Gardens, Kew. Ed. 3, by A. W. Hill, London. 1925.

Hoffm., *Gen. Umb.*

G. F. Hoffmann, *Genera Plantarum umbelliferarum.* Ed. 1. Mosquae. 1814. Ed. 2 titled *Genera umbelliferarum.* Mosquae. 1816.

Hofmann, E., *Nördl. Ural*

Kaiserliche russische geographische Gesellschaft (edit.), *Der nördliche Ural und das Küstengebirge Pae-Choi.* St. Petersburg. 1853–1856. (**2**, by E. Hofmann, in 1856, containing *Fl. Bor.-Ural., Flora boreali-uralensis* by F. J. Ruprecht.)

Holmboe, *Stud. Veg. Cyprus*

J. Holmboe, *Studies on the Vegetation of Cyprus.* Bergen. 1914. (Also published as *Bergens Mus. Skr.* nov. ser., **1**(**2**) (1914).)

Hooker, *Bot. Misc.*

W. J. Hooker, *Botanical Miscellany*. London. 1829–1833 (**1**: pp. 1–236, tt. 1–50 in 1829; pp. 237–356, tt. 51–75 in 1830; **2**: pp. 1–128, tt. suppl. 1–10 in 1830; pp. 129–416, tt. suppl. 11–19 in 1831; **3**: pp. 1–256, tt. suppl. 21–32, tt. 96–112 in 1832; pp. 257–390, tt. suppl. 33–41 in 1833.)

Hooker, *Exot. Fl.*

W. J. Hooker, *Exotic Flora*. Edinburgh. 1822–1827. (**1**: tt. 1–33 in 1822; tt. 34–79 in 1823; **2**: tt. 80–137 in 1824; tt. 138–150 in 1825; **3**: tt. 151–?170 in 1825; tt. ?171–?190 in 1825 or 1826; tt. ?191–?215 in 1826; tt. ?216–?228 in 1826 or 1827; tt. ?229–232 in 1827. Cf. M. J. van Steenis-Kruseman & W. T. Stearn in C. G. G. J. van Steenis, *Fl. Males.* I. **4**: clxxxviii (1954).)

Hooker, *Fl. Bor.-Amer.*

W. J. Hooker, *Flora boreali-americana, or the Botany of the northern Parts of British America*. London. 1829–1840. (**1**: pp. 1–48 in 1829; pp. 49–96 in 1830; pp. 97–160 in 1831; pp. 161–272 in 1832; pp. 273–328 in 1833; pp. 329–351 in 1834; **2**: pp. 1–48 in 1834; pp. 49–96 in 1837; pp. 97–192 in 1838; pp. 193–240 in 1839; pp. 241–328 in 1840. Cf. H. W. Rickett & F. A. Stafleu, *Taxon* **10**: 88 (1961).)

Hooker fil., *Stud. Fl. Brit. Is.*

J. D. Hooker, *The Student's Flora of the British Islands*. Ed. 1. London. 1870. Ed. 2. London. 1878. Ed. 3. London. 1884. Reprints in 1897, 1930 & 1937.

Hornem., *Enum. Hort. Haun.*

J. W. Hornemann, *Enumeratio Plantarum Horti botanici hauniensis*. Hauniae. 1807.

Hornem., *Hort. Hafn.*

J. W. Hornemann, *Hortus regius botanicus hafniensis, in Usum Tironum et Botanophilorum*. Hauniae. 1813–1815. (**1** in 1813; **2** in 1815.) *Suppl.*, *Supplementum*. 1819.

Host, *Fl. Austr.*

N. T. Host, *Flora austriaca*. Viennae. 1827–1831. (**1** in 1827; **2** in 1831.)

Host, *Syn. Pl. Austr.*

N. T. Host, *Synopsis Plantarum in Austria Provinciisque adjacentibus sponte crescentium*. Vindobonae. 1797.

Hudson, *Fl. Angl.*

W. Hudson, *Flora anglica*. Ed. 1. London. 1762. Ed. 2. London. 1778. Ed. 3. London. 1798.

Humb., Bonpl. & Kunth, *Nov. Gen. Sp.*

F. H. A. von Humboldt, A. J. A. Bonpland & C. S. Kunth, *Nova Genera et Species Plantarum quas in Peregrinatione Orbis novi collegerunt, descripserunt, partim adumbraverunt Amatus Bonpland et Alexander de Humboldt. Ex Schedis autographis Amati Bonpland in Ordinem digessit Carolus Siegesmund Kunth. Accedunt Alexandri de Humboldt Notationes ad Geographiam Plantarum spectantes*. Lutetiae Parisiorum. 1816–1825. (**1**–**7**; for dates cf. M. J. van Steenis-Kruseman & W. T. Stearn in C. G. G. J. van Steenis, *Fl. Males.* I. **4**: cxc (1954) and M. J. van Steenis Kruseman, *Fl. Males. Bull.* **15**: 736 (1960).)

Jacq., *Collect. Bot.*

N. J. von Jacquin, *Collectanea ad Botanicam, Chemiam et Historiam naturalem spectantia*. Vindobonae. 1786–1796. (**1** in 1787; **2** in 1789; **3** in 1790; **4** in 1791; **5** (*Supplementum*) in 1796. Cf. F. A. Stafleu, *Taxon* **12**: 63–64 (1963).)

Jacq., *Enum. Stirp. Vindob.*

N. J. von Jacquin, *Enumeratio Stirpium plerarumque, quae sponte crescunt in Agro vindobonensi, Montibusque confinibus*. Vindobonae. 1762.

Jacq., *Fl. Austr.*

N. J. von Jacquin, *Florae austriacae sive Plantarum selectarum in Austriae Archiducatu sponte crescentium Icones*. Viennae.

1773–1778. (**1** in 1773; **2** in 1774; **3** in 1775; **4** in 1776; **5** in 1778.)

Jacq., *Hort. Vindob.*

N. J. von Jacquin, *Hortus botanicus vindobonensis*. Vindobonae. 1770–1776. (**1** in 1770; **2** in 1772; **3** in 1776.)

Jacq., *Icon. Pl. Rar.*

N. J. von Jacquin, *Icones Plantarum rariorum*. Vindobonae. 1781–1795. (For dates cf. B. Schubert, *Contr. Gray Herb.* **154**: 3–23 (1945) and F. A. Stafleu, *Taxon* **12**: 64–65 (1963).)

Jacq., *Misc. Austr. Bot.*

N. J. von Jacquin, *Miscellanea austriaca ad Botanicam, Chemiam et Historiam naturalem spectantia*. Vindobonae. 1779–1781. (**1** in 1779; **2** in 1781.)

Jacq., *Obs. Bot.*

N. J. von Jacquin, *Observationum botanicarum Iconibus ab auctore delineatis illustratarum Pars* 1[–4]. Vindobonae. 1764–1771. (**1** in 1764; **2** in 1767; **3** in 1768; **4** in 1771.)

Jahandiez & Maire, *Cat. Pl. Maroc*

E. Jahandiez & R. C. J. E. Maire, *Catalogue des Plantes du Maroc (Spermatophytes et Ptéridophytes)*. Alger. 1931–1941. (**1** in 1931; **2** in 1932; **3** in 1934; **4** (*Supplément*) by M. L. Emberger & R. C. J. E. Maire, in 1941.)

Jan, *Cat. Pl. Phaen.*

G. Jan, *Catalogus Plantarum phaenogamarum ad usum Botanophilorum exsiccatarum*. Parmae. 1818.

Jan, *Elench. Hort. Parm.*

G. Jan, *Elenchus Plantarum quae in Horto ducali botanico parmensi Anno MDCCCXXVI coluntur*. Parmae. 1827.

Jaub. & Spach, *Ill. Pl. Or.*

H. F. Jaubert & E. Spach, *Illustrationes Plantarum orientalium, ou Choix de Plantes nouvelles ou peu connues de l'Asie occidentale*. Paris. 1842–1857. (**1**–**5**; for dates cf. M. J. van Steenis-Kruseman & W. T. Stearn in C. G. G. J. van Steenis, *Fl. Males.* I. **4**: cxci (1954) and M. J. van Steenis-Kruseman, *Fl. Males. Bull.* **15**: 736 (1960).)

Jáv., *Magyar Fl.*

S. Jávorka, *Magyar Flóra (Flora hungarica). Magyarország Virágos és Edényes Virágtalan Növenyeinek Meghátarozó Kézikönyve*. Budapest. 1924–1925. (Pp. 1–800 in 1924; pp. 801–1307, i–cii in 1925.)

Jordan, *Obs. Pl. Crit.*

A. Jordan, *Observations sur plusieurs Plantes nouvelles rares ou critiques de la France*. Paris & Leipzig. 1846–1849. (**1**–**4**, Paris & Leipzig in 1846; **5**, Paris in 1847; **6**, Leipzig in 1847; **7**, Paris in 1849.)

Jordan, *Pug. Pl. Nov.*

A. Jordan, *Pugillus Plantarum novarum praesertim gallicarum*. Paris. 1852.

Juss., A., *Euphorb. Tent.*

A. de Jussieu, *De Euphorbiacearum Generibus medicisque earundem Viribus Tentamen*. Parisiis. 1824.

Kaltenb., *Fl. Aachen. Beck.*

J. H. Kaltenbach, *Flora des Aachener Beckens*. Aachen. 1844.

Kauffm., *Mosk. Fl.*

N. N. Kauffmann (N. N. Kaufman), Московская Флора [*Moskovskaja Flora*]. Ed. 1. Moskva. 1866. Ed. 2 by P. F. Maevskij. Moskva. 1889.

Kerner, A., *Sched. Fl. Exsicc. Austro-Hung.*

A. J. Kerner von Marilaun, *Schedae ad Floram exsiccatam austro-hungaricam*. Vindobonae. 1881–1913. (**1** in 1881; **2** in 1882; **3** in 1884; **4** in 1886; **5** in 1888; **6** in 1893; **7** in 1896; **8** in 1899; **9** in 1902; **10** in 1913.) **8** & **9** by K. Fritsch; **10** by R. de Wettstein.

Kirby, *Fl. Leicest.*

M. Kirby, *The Flora of Leicestershire*. Ed. 1. Leicester. 1848. Ed. 2 titled *A Flora of Leicestershire*. London & Leicester. 1850.

Kirschleger, *Fl. Alsace*
F. R. Kirschleger, *Flore d'Alsace et des Contrées limitrophes.* Strasbourg & Paris. 1852–1862. (**1** in 1852; **2** in 1857; **3** in 1862.)

Klokov, *Ind. Sem. Hort. Bot. Charkov.*
M. V. Klokov, *Index Seminum Horti botanici charkoviensis.* Kharkov. 1927.

Klokov & Wissjul., *Fl. RSS Ucr.*
Cf. Fomin, *Fl. RSS Ucr.*

Knoche, *Fl. Balear.*
H. Knoche, *Flora balearica.* Montpellier. 1921–1923. (**1** in 1921; **2** in 1922; **3** & **4** in 1923.)

Koch, *Syn. Deutsch. Fl.*
W. D. J. Koch, *Synopsis der deutschen und schweizer Flora.* Ed. 1. Frankfurt. 1838. Ed. 2. Leipzig. 1847. Ed. 3. Leipzig. 1890–1907. (**1**: pp. 1–320 in 1890; pp. 321–640 in 1891; pp. 641–997 in 1892; **2**: pp. 999–1110 in 1892; pp. 1111–1270 in 1893; pp. 1271–1430 in 1895; pp. 1431–1590 in 1897; pp. 1591–1750 in 1900; pp. 1751–1910 in 1901; pp. 1911–1990 in 1902; **3**: pp. 1991–2070 in 1902; pp. 2071–2390 in 1903; pp. 2391–2710 in 1905; pp. 2711–3094 in 1907.)

Koch, *Syn. Fl. Germ.*
W. D. J. Koch, *Synopsis Florae germanicae et helveticae.* Ed. 1. Francofurti a. M. 1835–1837. (Pp. 1–352 in 1835; pp. 353–844 in 1837.) Ed. 2. Lipsiae. 1843–1845. (Pp. 1–452 in 1843; pp. 451 *bis*–964 in 1844; pp. 965–1164 in 1845.) Ed. 3. Lipsiae. 1857.

Koch, C., *Dendrologie*
C. H. E. Koch, *Dendrologie. Bäume, Sträucher und Halbsträucher, welche in Mittel- und Nordeuropa im Freien kultivirt werden.* Erlangen. 1869–1873. (**1** in 1869; **2**(1) in 1872; **2**(2) in 1873. Cf. W. T. Stearn, *Jour. Soc. Bibl. Nat. Hist.* **3**: 175 (1957).)

Koch, C., *Hort. Dendrol.*
C. H. E. Koch, *Hortus dendrologicus.* Berlin. 1853.

Komarov, *Fl. URSS*
V. L. Komarov *et al.* (edit.), Флора СССР [*Flora SSSR*]. / *Flora URSS.* Leningrad & Mosqua. 1934–1964. (**1–3**, Leningrad; **4–30**, Mosqua & Leningrad. **1** & **2** in 1934; **3** & **4** in 1935; **5** & **6** in 1936; **7** in 1937; **8** & **9** in 1939; **10** & **11** in 1941; **12** in 1946; **13** in 1948; **14** & **15** in 1949; **16** in 1950; **17** in 1951; **18** in 1952; **19** in 1953; **20** & **21** in 1954; **22** in 1955; **23** in 1958; **24** in 1957; **25** in 1959; **26** in 1961; **27** in 1962; **28** in 1963; **29** in 1964; **30** in 1960. **11** reprinted in 1945.) **1–13** edit. V. L. Komarov; **14**, **15**, **18**, **22**, **24**, **26**, **27** & **30** edit. B. K. Schischkin & E. G. Bobrov; **16**, **17**, **19**, **21**, **23** & **25** edit. B. K. Schischkin; **20** edit. B. K. Schischkin & S. V. Juzepczuk; **28** edit. E. G. Bobrov & S. K. Tcherapanov; **29** edit. E. G. Bobrov & N. N. Tzvelev.

Krocker, *Fl. Siles.*
A. J. Krocker, *Flora silesiaca.* Vratislaviae. 1787–1823. (**1** in 1787; **2**(1 & 2) in 1790; **3** in 1814; **4** (*Supplementum*) in 1823. Cf. F. A. Stafleu, *Taxon* **12**: 67 (1963).)

Krylov, *Fl. Zap. Sibir.*
P. N. Krylov, Флора западной Сибири [*Flora zapadnoj Sibiri*]. Tomsk. 1927–1949. (**1** in 1927; **2** in 1928; **3** in 1929; **4** in 1930; **5** & **6** in 1931; **7** in 1933; **8** in 1935; **9** in 1937; **10** in 1939; **11** in 1949.)

Kusn., N. Busch & Fomin, *Fl. Cauc. Crit.*
N. I. Kusnezow, N. A. Busch & A. V. Fomin, *Flora caucasica critica.* / Матеріалы для Флоры Кавказа [*Materialy dlja Flory Kavkaza*]. Jurjev. 1901–1916. (**1**(1): pp. 1–96 in 1911; pp. 97–224 in 1912; pp. 225–248, i–xlvi in 1913; **2**(1): pp. 1–43 in 1911; **2**(4): pp. 1–64 in 1906; pp. 65–144 in 1912; pp. 145–176 in 1913; **2**(5): pp. 1–32 in 1916; **3**(3): pp. 1–32 in 1901; pp. 33–112 in 1902; pp. 113–256, i–xix in 1903; **3**(4): pp. 1–16

in 1904; pp. 17–64 in 1905; pp. 65–144 in 1907; pp. 145–304 in 1908; pp. 305–544 in 1909; pp. 545–820, i–lxxiv in 1910; **3**(5): pp. 1–32 in 1913; pp. 33–48 in 1915; **3**(7): pp. 1–80 in 1908; pp. 81–96 in 1910; pp. 97–112 in 1912; **3**(8): pp. 1–16 in 1911; pp. 17–48 in 1912; **3**(9): pp. 1–64 in 1906; pp. 65–224 in 1909; pp. 225–288 in 1910; pp. 289–320 in 1912; pp. 321–352 in 1913; **4**(1): pp. 1–128 in 1901; pp. 129–208 in 1902; pp. 209–352 in 1903; pp. 353–384 in 1904; pp. 385–464 in 1905; pp. 465–512 in 1906; pp. 513–560 in 1907; pp. 561–590, i–lxii in 1908; **4**(2): pp. 1–32 in 1912; pp. 33–208 in 1913; pp. 209–256 in 1914; pp. 257–320 in 1915; pp. 321–400 in 1916; **4**(3): pp. 1–48 in 1916; **4**(6): pp. 1–16 in 1903; pp. 17–48 in 1904; pp. 49–80 in 1905; pp. 81–144 in 1906; pp. 145–157, i–xviii in 1907.)

L., *Amoen. Acad.*
C. von Linné, *Amoenitates academicae.* Holmiae. 1749–1779. (**3** in 1756; **4** in 1759; **5** in 1760; **6** in 1763; **7** in 1769.)

L., *Cent. Pl.*
C. von Linné, *Centuria Plantarum.* Upsaliae. 1755–1756. (**1** in 1755; **2** in 1756.)

L., *Demonstr Pl.*
C. von Linné, *Demonstrationes Plantarum in Horto upsaliensi 1753.* Upsaliae. 1753.

L., *Fl. Monsp.*
C. von Linné, *Flora monspeliensis.* Upsaliae. 1756.

L., *Fl. Suec.*
C. von Linné, *Flora suecica.* Ed. 2. Stockholmiae. 1755.

L., *Mantissa*
C. von Linné, *Mantissa Plantarum.* Holmiae. 1767.

L., *Mantissa Alt.*
C. von Linné, *Mantissa Plantarum altera.* Holmiae. 1771.

L., *Sp. Pl.*
C. von Linné, *Species Plantarum.* Ed. 1. Holmiae. 1753. Ed. 2. Holmiae. 1762–1763. (Pp. 1–784 in 1762; pp. 785–1684 in 1763.)

L., *Syst. Nat.*
C. von Linné, *Systema Natura.* Ed. 10. Botany in **2**. Holmiae. 1759. Ed. 11. Botany in **1** & **2**. Lipsiae. 1762. Ed. 12. Botany in **2** & **3**. Holmiae. 1767–1768. (**2** in 1767; **3** in 1768.) Ed. 13, by J. F. Gmelin. Botany in **2**. Lipsiae. 1791–1792. (Pp. 1–884 in 1791; pp. 885 1661 in 1792. Cf. H. W. Rickett & F. A. Stafleu, *Taxon* **10**: 85 (1961).)
Titled *Syst. Veg.*, *Systema Vegetabilium.* Ed. 13, edit. J. A. Murray. Gottingae & Gothae. 1774. Ed. nov. (15), by J. J. Roemer & J. A. Schultes. Studtgardtiae. 1817–1830. (**1** & **2** in 1817; **3** in 1818; **4** in 1819; **5** in 1819 or 1820; **6** in 1820; **7** by J. A. & J. H. Schultes: pp. i–xliv, 1–754 in 1829; pp. xlv–cviii, 755–1816 in 1830. Cf. M. J. van Steenis-Kruseman & W. T. Stearn in C. G. G. J. van Steenis, *Fl. Males.* I. **4**: ccix (1954).) Ed. 16 by C. P. J. Sprengel. Gottingae. 1824–1828. (**1** in 1824; **2** in 1825; **3** in 1826; **4** (1 & 2) in 1827; **5** & *Suppl.* in 1828.)

L., *Syst. Pl. Eur.* 1, *Fl. Delph.*
C. von Linné, *Flora delphinalis* in C. von Linné, *Systema Plantarum Europae* **1**. Coloniae-Allobrogum. 1785.

L., *Syst. Veg.*
Cf. L., *Syst. Nat.*

L. fil., *Dec. Prim. Pl. Rar. Hort. Upsal.*
C. von Linné fil., *Decas prima Plantarum rariorum Horti upsaliensis.* Stockholmiae. 1762.

L. fil., *Suppl.*
C. von Linné fil., *Supplementum Plantarum Systematis Vegetabilium Editionis XIII, Generum Plantarum Editionis VI, et Specierum Plantarum Editionis II.* Brunsvigae. 1781.

Labill., *Icon. Pl. Syr.*
J. J. H. de Labillardière, *Icones Plantarum Syriae rariorum.* Parisiis. 1791–1812. (**1** & **2** in 1791; **3** in 1809; **4** & **5** in 1812. Cf. F. A. Stafleu, *Taxon* **12**: 67–68 (1963).)

Labill., *Nov. Holl. Pl.*

J. J. H. de Labillardière, *Novae Hollandiae Plantarum Specimen.* Parisiis. 1804–1807. (**1**: tt. 1–10 in 1804; tt. 11–140 in 1805; **2**: tt. 141–230 in 1806; tt. 231–270 in 1807. Cf. M. J. van Steenis-Kruseman, *Fl. Males. Bull.* **19**: 1141–1142 (1964).)

Labill., *Rel. Voy. Rech. La Pérouse*

J. J. H. de Labillardière, *Relation du Voyage à la Recherche de La Pérouse.* Paris. 1800. (For date cf. H. W. Rickett & F. A. Stafleu, *Taxon* **10**: 111 (1961).)

Lag., *Amen. Nat.*

M. Lagasca y Segura, *Amenidades naturales de las Españas.* Orihuela & Madrid. 1811–1821. (**1**(1): pp. 1–44 in 1811; **1**(2): pp. 47–111 in 1821.) **1**(1) at Orihuela; **1**(2) at Madrid.

Lag., *Gen. Sp. Nov.*

M. Lagasca y Segura, *Genera et Species Plantarum, quae aut novae sunt aut nondum recte cognoscuntur.* Matriti. 1816.

Laicharding, *Veg. Eur.*

J. N. von Laicharding, *Vegetabilia europaea.* Oeniponte. 1790–1791. (**1** in 1790; **2** in 1791.)

Lam., *Encycl. Méth. Bot.*

J. B. A. P. Monnet de la Marck, *Encyclopédie méthodique. Botanique.* Paris. 1783–1817. (**1**: pp. 1–344 in 1783; pp. 345–752 in 1785; **2**: pp. 1–400 in 1786; pp. 401–774 in 1788; **3**: pp. 1–360 in 1789; pp. 361–703 in 1792; **4**: pp. 1–400 in 1797; pp. 401–764 in 1798; **5** in 1804; **6**: pp. 1–384 in 1804; pp. 385–786 in 1805; **7** in 1806; **8** in 1808; *Suppl.* **1** (=**9**): pp. 1–400 in 1810; pp. 401–761 in 1811; *Suppl.* **2**(=**10**): pp. 1–384 in 1811; pp. 385–876 in 1812; *Suppl.* **3**(=**11**): pp. 1–368 in 1813; pp. 369–780 in 1814; *Suppl.* **4**(=**12**): pp. 1–368 in 1816; pp. 369–731 in 1817; *Suppl.* **5**(=**13**) in 1817. Cf. M. Breistroffer, *Proc. Verb. Mens. Soc. Dauph. Ethn. Arch.* **24**: nos. 182–184 (1948) and H. W. Rickett & F. A. Stafleu, *Taxon* **10**: 112 (1961).) **5–13** by J. L. M. Poiret.

Lam., *Fl. Fr.*

J. B. A. P. Monnet de la Marck, *Flore française.* Ed. 1. Paris. 1778. (**1–3** in 1778.) Ed. 2. Paris. 1795. (**1–3** in 1795.) Ed. 3 by J. B. A. P. Monnet de la Marck & A. P. de Candolle. Paris. 1805–1815. (**1–4** in 1805; **5** in 1815. **4** issued in two parts with continuous pagination, the second called '*Tome quatrième. Seconde partie*' and '*Vol. V*'; **5** called '*Tome cinquième, ou sixième volume*' and '*Vol. VI*'. **1–4** also reprinted in 1815.)

Lam., *Tabl. Encycl. Méth. Bot.*

J. B. A. P. Monnet de la Marck, *Tableau encyclopédique et méthodique des trois Règnes de la Nature. Botanique.* **1–3**. Paris. 1791–1823. (**1**: pp. 1–200 in 1791; pp. 201–440 in 1792; pp. 441–496 in 1793; **2**: pp. 1–48 in 1794; pp. 49–72 in 1796; pp. 73–136 in 1797; pp. 137–551 in 1819; **3**: pp. 1–728 in 1823. Cf. H. W. Rickett & F. A. Stafleu, *Taxon* **10**: 112 (1961) and F. A. Stafleu, *Taxon* **12**: 68 (1963).)

Lam. & DC., *Fl. Fr.*

Cf. Lam., *Fl. Fr.*

Lange, *Descr. Icon. Ill.*

J. M. C. Lange, *Descriptio Iconibus illustrata Plantarum novarum vel minus cognitarum praecipue e Flora hispanica.* Hauniae. 1864–66. (Pp. 1–14 in 1864; pp. 15–20 in 1866.)

Lange, *Haandb. Danske Fl.*

J. M. C. Lange, *Haandbog i den Danske Flora.* Ed. 1. Kjøbenhavn. 1851. Ed. 2. Kjøbenhavn. 1856–1859. Ed. 3. Kjøbenhavn. 1864. Ed. 4. Kjøbenhavn. 1886–1888. *Rettelser og Tilfojelser ...* Kjøbenhavn. 1897.

Lange, *Icon. Pl. Fl. Dan.*

G. C. Oeder, *Icones Plantarum sponte nascentium in Regnis Daniae et Norvegiae, in Ducatibus Slesvici et Holsatiae, et in Comitatibus Oldenburgi et Delmenhorstiae: ad illustrandum Opus de iisdem Plantis, Regio jussu exarandum, Florae danicae Nomine inscriptum.* Hauniae. 1761–1883. Fasc. **45–51** by J. M. C. Lange in 1861–1883.

Lange, *Ind. Sem. Horto Haun.*

J. M. C. Lange, *Index Seminum in Horto academico hauniensi collectorum.* Hauniae. 1855, 1857 & 1861.

Lange, *Rev. Crat.*

J. M. C. Lange, *Revisio Specierum Generis Crataegi imprimis earum, quae in Hortis Daniae coluntur.* Kjøbenhavn. 1897.

Lapeyr., *Hist. Abr. Pyr.*

P. Picot de Lapeyrouse, *Histoire abrégée des Plantes des Pyrénées et Itinéraire des Botanistes dans ces Montagnes.* Toulouse. 1813. *Supplément.* 1818.

Latourr., *Chlor. Lugd.*

M. A. L. C. de Latourrette, *Chloris lugdunensis.* Lugdunigallorum. 1785.

Lecoq & Lamotte, *Cat. Pl. Centr. Fr.*

H. Lecoq & M. Lamotte, *Catalogue raisonné des Plantes vasculaires du Plateau central de la France, comprenant l'Auvergne, le Velay, la Lozère, les Cévennes, une Partie du Bourbonnais et la Vivarais.* Paris. 1847.

Ledeb., *Fl. Altaica*

C. F. von Ledebour, *Flora altaica.* Berolini. 1829–1834. (**1** in 1829; **2** in 1830; **3** in 1831; **4** in 1833; Index in 1833 or 1834.)

Ledeb., *Fl. Ross.*

C. F. von Ledebour, *Flora rossica.* Stuttgartiae. 1841–1853. (**1**: pp. 1–240 in 1841; pp. 241–480 in 1842; pp. 481–790 in 1843; **2**: pp. 1–204 in 1843; pp. 205–462 in 1844; pp. 463–718 in 1845; pp. 719–937 in 1846; **3**: pp. 1–256 in 1847; pp. 257–492 in 1849; pp. 493–684 in 1850; pp. 685–853 in 1851; **4**: pp. 1–464 in 1852; pp. 465–741 in 1853.)

Ledeb., *Icon. Pl. Fl. Ross.*

C. F. von Ledebour, *Icones Plantarum novarum vel imperfecte cognitarum Floram rossicam, imprimis altaicam, illustrantes.* Rigae. 1829–1834. (**1** in 1829; **2** in 1830; **3**: tt. 201–250 in 1831; tt. 251–300 in 1832; **4** in 1833; **5** in 1834.)

Ledeb., *Ind. Sem. Horti Dorpat.*

C. F. von Ledebour, *Index Seminum Horti botanici dorpatensis.* Dorpat. 1818–1835.

Lehm., *Ind. Sem. Horti Bot. Hamburg.*

J. G. C. Lehmann, *Index Seminum Horti botanici hamburgensis.* Hamburgi. 1821–1859.

Lehm., *Pugillus*

J. G. C. Lehmann, *Novarum et minus cognitarum Stirpium Pugillus.* Hamburgi. 1828–1857. (**1** in 1828; **2** in 1830; **3** in 1831; **4** in 1832; **5** in 1833; **6** in 1834; **7** in 1838; **8** in 1844; **9** in 1851; **10** in 1857.)

Lej., *Fl. Spa*

A. L. S. Lejeune, *Flore des Environs de Spa.* Liège. 1811–1813. (**1** in 1811; **2** in 1813.)

Lej., *Rev. Fl. Spa*

A. L. S. Lejeune, *Revue de la Flore des Environs de Spa.* Liège. 1824.

Lej. & Court., *Comp. Fl. Belg.*

A. L. S. Lejeune & R. J. Courtois, *Compendium Florae belgicae.* Leodii. 1828–1836. (**1** in 1828; **2** in 1831; **3** in 1836.)

Leresche & Levier, *Deux Excurs. Bot.*

L. Leresche & É. Levier, *Deux Excursions botaniques dans le Nord de l'Espagne et le Portugal en 1878 et 1879.* Lausanne. 1880.

L'Hér., *Geraniologia*

C. L. L'Héritier de Brutelle, *Geraniologia, seu Erodii, Pelargonii, Geranii, Monsoniae et Grieli Historia Iconibus illustrata.* Parisiis. 1792. (For date cf. F. A. Stafleu, *Taxon* **12**: 71–72 (1963).) Reissued in 1813.

Lindeb., *Herb. Rub. Scand.*

C. J. Lindeberg, *Herbarium Ruborum Scandinaviae*. Göteborg. 1882.

Lindeb., *Nov. Fl. Scand.*

C. J. Lindeberg, *Novitiae Florae scandinavicae*. Göteborg. 1858.

Lindley, *Ros. Monogr.*

J. Lindley, *Rosarum Monographia: or, a botanical History of Roses*. London, 1820.

Lindley, *Syn. Brit. Fl.*

J. Lindley, *A Synopsis of the British Flora*. Ed. 1. London. 1829. Ed. 2. London. 1835. Ed. 3. London. 1841.

Lindman, *Svensk Fanerogamfl.*

C. A. M. Lindman, *Svensk Fanerogamflora*. Ed. 1. Stockholm. 1918. Ed. 2. Stockholm. 1926.

Link, *Enum. Hort. Berol. Alt.*

J. H. F. Link, *Enumeratio Plantarum Horti regii botanici berolinensis altera*. Berolini. 1821–1822. (**1** in 1821; **2** in 1822.)

Link, *Handb.*

J. H. F. Link, *Handbuch zur Erkennung der nutzbarsten und am häufigsten vorkommenden Gewächse*. Berlin. 1829–1833. (**1** in 1829; **2** in 1831; **3** in 1833.)

Loisel., *Fl. Gall.*

J. L. A. Loiseleur-Deslongchamps, *Flora gallica*. Ed. 1. Lutetiae. 1806–1807. (Pp. 1–336 in 1806; pp. 337–742 in 1807.) Ed. 2. Paris. 1828.

Loisel., *Not. Pl. Fr.*

J. L. A. Loiseleur-Deslongchamps, *Notice sur les Plantes à ajouter à la Flore de France (Flora gallica) avec quelques Corrections et Observations*. Paris. 1810.

Loret & Barrandon, *Fl. Montpell.*

H. Loret & A. Barrandon, *Flore de Montpellier*. Montpellier & Paris. Ed. 1. 1876. (**1** & **2** in 1876.) Ed. 2. 1886.

Losa, *Contrib. Estud. Fl. Veg. Prov. Zamora*

M. Losa España, *Contribución al Estudio de la Flora y Vegetación de la Provincia de Zamora*. Barcelona. 1949.

Loscos & Pardo, *Ser. Pl. Arag.*

F. Loscos y Bernál & J. Pardo y Sastrón, *Series inconfecta Plantarum indigenarum Aragoniae praecipue meridionalis, e Lingua castellana in Latinam vertit, recensuit, emendavit Observationibus suis auxit atque edendam curavit M. Willkomm*. Dresdae. 1863.

Lour., *Fl. Cochinch.*

J. de Loureiro, *Flora cochinchinensis*. Ulyssipone. 1790. (**1** & **2** in 1790. Cf. F. A. Stafleu, *Taxon* **12**: 72 (1963).) Reprinted, Berolini, 1793.

Mackenzie & Bush, *Man. Fl. Jackson Co. Missouri*

K. K. Mackenzie, B. F. Bush *et al.*, *Manual of the Flora of Jackson County Missouri*. Kansas City. 1902.

Magnier, *Fl. Select. Exsicc.*

C. Magnier, *Flora selecta exsiccata*. Saint-Quentin. 1882–1897. The plants distributed with these labels were again listed in *Scrinia Fl. Select. (Magnier)*.

Majevski, *Fl. Sred. Ross.*

P. F. Majevski, Флора средней России [*Flora srednej Rossii*]. Ed. 1. 1892. Ed. 2 by S. I. Korshinsky. 1895. Ed. 3 by B. A. Fedtschenko. 1902. Ed. 4 by D. I. Litvinov. 1912. Ed. 5 by D. I. Litvinov. 1917. Ed. 6 by B. A. Fedtschenko. 1933. Ed. 7, titled Флора средней Полосы европейской Части СССР [*Flora srednej Polosy evropejskoj Časti SSSR*] by V. L. Komarov. 1940. Ed. 8 by B. K. Schischkin. Moskva & Leningrad. 1954. Ed. 9 by B. K. Schischkin. Leningrad. 1964.

Maly, J., *Enum. Pl. Austr.*

J. K. Maly, *Enumeratio Plantarum phanerogamarum Imperii Austriaci universi*. Vindobonae. 1848.

Manden., *Kavk. Vidy Heracleum*

I. P. Mandenova, Кавказские Виды Рода Heracleum [*Kavkazskie Vidy Roda* Heracleum]. Tbilisi. 1950.

Marès & Vigineix, *Cat. Pl. Baléar.*

P. Marès & G. Vigineix, *Catalogue raisonné des Plantes vasculaires des Îles Baléares*. Paris. 1880.

Marsson, *Fl. Neu-Vorpommern*

T. F. Marsson, *Flora von Neu-Vorpommern und den Inseln Rügen und Usedom*. Leipzig. 1869.

Massara, *Prodr. Fl. Valtell.*

G. F. Massara, *Prodromo della Flora valtellinese ossia Catalogo delle Piante*. Sondrio. 1834.

Medicus, *Gesch. Bot.*

F. C. Medicus, *Geschichte der Botanik unserer Zeiten*. Mannheim. 1793.

Medicus, *Künstl. Geschl. Malv.-Fam.*

F. C. Medicus, *Ueber einige künstliche Geschlechter aus der Malven-Familie, denn der Klasse der Monadelphien*. Mannheim. 1787.

Meerb., *Afbeeld. Zelds. Gewass.*

N. Meerburgh, *Afbeeldingen van zeldzaame Gewassen*. Leyden. 1775–1780. Reissue in 1789, titled *Plantae rariores vivis coloribus depictae*.

Meissner, *Pl. Vasc. Gen.*

C. F. Meissner, *Plantarum vascularium Genera*. Lipsiae. 1836–1843. (**1** (=*Tab. Diagn.*): pp. i–iv, 1–104 in 1837; pp. 105–176 in 1838; pp. 177–256 in 1839; pp. 257–312 in 1840; pp. 313–344 in 1841; pp. 345–408 in 1842; pp. 409–442 in 1843; **2** (=*Comm.*): pp. 1–72 in 1837; pp. 73–120 in 1838; pp. 121–160 in 1839; pp. 161–224 in 1840; pp. 225–252 in 1841; pp. 253–308 in 1842; pp. 309–402 in 1843. Cf. H. W. Rickett & F. A. Stafleu, *Taxon* **10**: 118 (1961).)

Merr., *Interpr. Herb. Amb.*

E. D. Merrill, *An Interpretation of Rumphius's* Herbarium amboinense. Manila. 1917.

Meyer, C. A., *Verz. Pfl. Cauc.*

C. A. von Meyer, *Verzeichniss der Pflanzen, welche während der, auf allerhöchsten Befehl, in den Jahren 1829 und 1830 unternommenen Reise im Caucasus und in den Provinzen am westlichen Ufer des caspischen Meeres gefunden und eingesammelt worden sind*. St Petersburg. 1831.

Michx, *Fl. Bor. Amer.*

A. Michaux, *Flora boreali-americana*. Parisiis & Argentorati. 1803. (**1** & **2** in 1803.)

Miller, *Gard. Dict.*

P. Miller, *The Gardeners Dictionary*. Ed. 8. London. 1768. Ed. 9, by T. Martyn. London. 1797–1804. (**1** in 1797; **2** in 1804.)

Mirbel, *Hist. Nat. Pl.*

C. F. Mirbel, *Histoire naturelle, générale et particulière des Plantes*. Paris. **1–18**. 1802–1806. (**3** & **16–18** by N. Jolyclerc.) Forms one of the supplements of ed. 2 of G. L. L. de Buffon, *Histoire naturelle*.

Moench, *Meth.*

C. Moench, *Methodus Plantas Horti botanici et Agri marburgensis a Staminum Situ describendi*. Marburgi Cattorum 1794. *Suppl.*, *Supplementum*. Marburgi Cattorum. 1802.

Moench, *Verz. Ausl. Bäume Weissenst.*

C. Moench, *Verzeichniss ausländischer Bäume und Stauden des Lustschlosses Weissenstein bei Cassel*. Frankfurt a. M. & Leipzig. 1785.

Moretti, *Bot. Ital.*

G. Moretti, *Il Botanico italiano, ossia Discussioni sulla Flora italica*. Pavia. 1826. Reprinted as *Gior. Fis. Chim. Stor. Nat. (Pavia)* **9**: 65–94, 154–159, 238–250 (1826).

Moris, *Enum. Sem. Hort. Taur.*

G. G. Moris, *Enumeratio Seminum regii Horti botanici taurinensis.* Augustae Taurinorum. 1831–1860.

Moris, *Fl. Sard.*

G. G. Moris, *Flora sardoa.* Taurini. 1837–1859. (**1** in 1837; **2** in 1840–1843; **3** in 1858–1859.)

Moris, *Stirp. Sard.*

G. G. Moris, *Stirpium sardoarum Elenchus.* Carali. 1827–1829. (**1** & **2** in 1827; **3** in 1829.)

Murray, *Prodr. Stirp. Gotting.*

J. A. Murray, *Prodromus Designationis Stirpium gottingensium.* Goettingae. 1770.

Nestler, *Monogr. Potent.*

C. G. Nestler, *Monographia de Potentilla, praemissis nonnullis Observationibus circa Familiam Rosacearum.* Parisiis & Argentorati. 1816.

Neygenf., *Enchirid. Bot. Siles.*

F. W. Neygenfind, *Enchiridium botanicum, continens Plantas Silesiae indigenas. | Botanisches Taschenbuch, welches die in Schlesien einheimischen Pflanzen enthält.* Misenae. 1821.

Nyman, *Consp.*

C. F. Nyman, *Conspectus Florae europaeae.* Örebro. 1878–1882. (Pp. 1–240 in 1878; pp. 241–493 (part) in 1879; pp. 493–677 in 1881; pp. 677–858 in 1882; p. 256 *errata* in 1881; p. 493 (part) and p. 677 published twice. Cf. W. T. Stearn, *Jour. Bot.* (*London*) **76**: 113 (1938).) *Suppl., Supplementum* **1** in 1883–1884; **2**(1): pp. 1–224 in 1889; **2**(2): pp. 225–404 in 1890. *Addit., Additamenta* by E. Roth, in 1886.

Nyman, *Syll.*

C. F. Nyman, *Sylloge Florae europaeae.* Oerebroae. 1854–1855. *Suppl., Supplementum.* Oerebroae. 1865.

Oliver, *Fl. Trop. Afr.*

D. Oliver *et al., Flora of tropical Africa.* London. **1** → . 1868 → .

Ortega, *Hort. Matrit. Descr.*

C. Gómez Ortega, *Novarum, aut rariorum Plantarum Horti reg. botan. matrit. Descriptionum Decades.* Matriti. 1797–1800. (Pp. 1–51 in 1797; pp. 53–108 in 1798; pp. 109–138 in 1800.)

Osbeck, *Dagb. Ostind. Resa*

P. Osbeck, *Dagbok öfwer en ostindsk Resa.* Stockholm. 1757. German edition in 1765.

Palhinha, *Cat. Pl. Vasc. Açores*

R. T. Palhinha, *Catálogo das Plantas vasculares dos Açores.* Lisboa. 1966.

Pallas, *Fl. Ross.*

P. S. Pallas, *Flora rossica.* Petropoli. 1784–1831. (**1**(1) in 1784; **1**(2) in 1788; **2**(1) in 1831. Cf. H. W. Rickett & F. A. Stafleu, *Taxon* **10**: 133 (1961) and F. A. Stafleu, *Taxon* **12**: 75 (1963).)

Pallas, *Reise*

P. S. Pallas, *Reise durch verschiedene Provinzen des russischen Reichs.* Ed. 1. St Petersburg. 1771–1776. (**1** in 1771; **2** in 1773; **3** in 1776.) Ed. 2. St Petersburg. 1801.

Pallas, *Spec. Astrag.*

P. S. Pallas, *Species Astragalorum descriptae et Iconibus coloratis illustratae.* Lipsiae. 1800.

Pančić, *Elem. Fl. Bulg.*

J. Pančić, Грађа за Флору Кнежевине Бугарске [*Graća za Floru Kneževine Bugarske*]. | *Elementa ad Floram Principatus Bulgariae.* Beograd. 1883.

Pančić, *Fl. Princ. Serb.*

J. Pančić, Флора Кнежевине Србије [*Flora Kneževine Srbije*]. | *Flora Principatus Serbiae.* Beograd. 1874. Додатак [*Dodatak*]. | *Additamenta.* Beograd. 1884.

Parl., *Fl. Ital.*

F. Parlatore, *Flora italiana.* Firenze. 1848–1896. (**1**: pp. 1–96 in 1848; pp. 97–568 in 1850; **2**: pp. 1–220 in 1852; pp. 221–638

in 1857; **3**: pp. 1–160 in 1858; pp. 161–690 in 1860; **4**: pp. 1–288 in 1868; pp. 289–623 in 1869; **5**: pp. 1–320 in 1873: pp. 321–671 in 1875; **6**: pp. 1–336 in 1884; pp. 337–656 in 1885; pp. 657–971 in 1886; **7** in 1887; **8**: pp. 1–176 in 1888; pp. 177–773 in 1889; **9**: pp. 1–232 in 1890; pp. 233–624 in 1892; pp. 625–1085 in 1893; **10** in 1894; **11** in 1896.) **6–10** by T. Caruel.

Parry, *Jour. Voy. N.W. Pass.*

W. E. Parry, *Journal of a Voyage for the Discovery of a North West Passage performed in 1819–20 in H.M.S. Hecla and Griper.* London. 1821. *Suppl. App., Botanical Supplement* to the *Appendix* by R. Brown. London. 1824.

Pau, *Not. Bot. Fl. Esp.*

C. Pau, *Notas botánicas á la Flora española.* Madrid & Segorbe. 1887–1895. (**1, 2, 4–6** Madrid; **3** Segorbe. **1** in 1887; **2** & **3** in 1889; **4** in 1891; **5** in 1893; **6** in 1895.)

Pawł., *Zapiski Fl. Tatr*

B. Pawłowski, *Zapiski florystyczne z Tatr. | Notulae ad Floram tatrorum pertinentes.* **4**. Kraków. 1949. (**1–3** published in *Acta Soc. Bot. Polon.* **1, 3** & **7**.)

Pers., *Syn. Pl.*

C. H. Persoon, *Synopsis Plantarum.* Ed. 1. Paris & Tuebingae. 1805–1807. (**1** in 1805; **2**: pp. 1–272 in 1806; pp. 273–657 in 1807.) Ed. 2, titled *Species Plantarum.* Petropoli. 1817–1822. (**1** in 1817; **2** & **3** in 1819; **4** & **5** in 1821; **6** in 1822.)

Peterm., *Deutschl. Fl.*

W. L. Petermann, *Deutschlands Flora.* Leipzig. 1849.

Petrović, *Add. Fl. Agri. Nyss.*

S. Petrović, Додатак Флори околине Ниша [*Dodatak flori okoline Niša*]. | *Additamenta ad Floram Agri nyssani.* Beogradu. 1885.

Petzold & Kirchner, *Arbor. Muscav.*

C. E. A. Petzold & G. Kirchner, *Arboretum muscaviense.* Gotha. 1864.

Pio, *De Viola*

G. B. Pio, *De Viola Specimen botanico-medicum.* Taurini. 1813.

Poiret, *Voy. Barb.*

J. L. M. Poiret, *Voyage en Barbarie, ou Lettres écrites de l'ancienne Numidie, pendant les Années 1785 et 1786; avec un Essai sur l'Histoire naturelle de ce Pays.* Paris. 1789.

Pojark., *Fl. Murmansk.*

Cf. Gorodkov, *Fl. Murmansk.*

Pollich, *Hist. Pl. Palat.*

J. A. Pollich, *Historia Plantarum in Palatinatu electorali sponte nascentium incepta.* Mannhemii. 1776–1777. (**1** in 1776; **2** & **3** in 1777.)

Pomel, *Mat. Fl. Atl.*

A. Pomel, *Matériaux pour la Flore atlantique.* Oran. 1860.

Porc., *Fl. Năseud.*

F. Porcius, *Flora phanerogama din Fostulŭ Districtŭ alu Năseu-duluĭ.* Bucarest. 1881.

Portenschl., *Enum. Pl. Dalmat.*

F. E. von Portenschlag-Ledermayer, *Enumeratio Plantarum in Dalmatia lectarum.* Wien. 1824.

Pospichal, *Fl. Österr. Küstenl.*

E. Pospichal, *Flora des österreichischen Küstenlandes.* Leipzig & Wien. 1879–1899. (**1** in 1879; **2**(1): pp. 1–528 in 1898; **2**(2): pp. 529–946 in 1899.)

Post, *Fl. Syr. Pal. Sin.*

G. E. Post, *Flora of Syria, Palestine, and Sinai, from the Taurus to Ras Muhammad and from the Mediterranean Sea to the Syrian Desert.* Ed. 1. Beirut. 1896. Ed. 2, by J. E. Dinsmore. Beirut. 1932–1933. (**1** in 1932; **2** in 1933.)

Presl, C., *Symb. Bot.*

C. B. Presl, *Symbolae botanicae.* Pragae. 1830–1833. (**1**: tt. 1–10 in 1830; tt. 11–30 in 1831; tt. 31–50 in 1832; **2**: tt. 51–70 in 1833; tt. 71–80 in 1852. Cf. H. W. Rickett & F. A. Stafleu, *Taxon* **10**: 134 (1961).)

Presl, J. & C., *Del. Prag.*
J. S. & C. B. Presl, *Deliciae pragenses, Historiam naturalem spectantes.* Pragae. 1822.

Pursh, *Fl. Amer. Sept.*
F. T. Pursh, *Flora Americae septentrionalis.* Ed. 1. London. 1814. (1 & 2, with continuous pagination, in 1814.) Ed. 2. London. 1816. (1 & 2, with continuous pagination, in 1816.)

Rafin., *Atl. Journ.*
C. S. Rafinesque-Schmaltz, *Atlantic Journal, and Friend of Knowledge.* Philadelphia. 1832–1833. (**1**: pp. 1–154 in 1832; pp. 155–212 in 1833. Cf. H. W. Rickett & F. A. Stafleu, *Taxon* **10**: 135 (1961).)

Rafin., *Précis Découv. Somiol.*
C. S. Rafinesque-Schmalz, *Précis des Découvertes somiologiques ou zoologiques et botaniques.* Palerme. 1814.

Raunk., *Dansk. Ekskurs.-Fl.*
C. Raunnkiær, *Dansk Ekskursions-Flora.* Ed. 1, Kjøbenhavn. 1890. Ed. 2. Kjøbenhavn. 1906. Ed. 3, by C. H. Ostenfeld & C. Raunkiær. København & Kristiania. 1914. Ed. 4, by C. H. Ostenfeld & C. Raunkiær. Kjøbenhavn & Kristiania. 1922. Ed. 5, by K. Wiinstedt & K. Jessen. København. 1934. Ed. 6, by K. Wiinstedt. København. 1942. Ed. 7, by K. Wiinstedt. København. 1950.

Räuschel, *Nomencl. Bot.*
E. A. Räuschel, *Nomenclator botanicus.* Ed. 3. Lipsiae. 1797. Earlier editions published anonymously.

Reichenb., *Fl. Germ. Excurs.*
H. G. L. Reichenbach, *Flora germanica excursoria.* Lipsiae. 1830–1832. (Pp. i–viii, 1–136 in 1830; pp. 137–438 in 1831; pp. ix–xlviii, 435–878 in 1832.)

Reichenb., *Handb.*
H. G. L. Reichenbach, *Handbuch des natürlichen Pflanzensystems.* Dresden & Leipzig. 1837.

Reichenb., *Pl. Crit.*
H. G. L. Reichenbach, *Iconographia botanica, seu Plantae criticae.* Lipsiae. 1823–1832. (**1** in 1823; **2** in 1824; **3** in 1825; **4** in 1826; **5** in 1827; **6** in 1828; **7** in 1829; **8** in 1830; **9** in 1831; **10** in 1832.)

Reichenb. & Reichenb. fil., *Icon. Fl. Germ.*
H. G. L. Reichenbach, *Icones Florae germanicae et helveticae.* Lipsiae & Gerae. 1834–1914. (**1–19**(1), **20 & 21** Lipsiae; **19**(2), **22, 24 & 25**(1 & 2) Lipsiae & Gerae; **23** Gerae. **1**, titled *Agrostographia germanica, sistens Icones Graminearum et Cyperoidearum quas in Flora germanica recensuit Auctor* (=*Pl. Crit.* **11**) in 1834; **2** in 1837–1838; **3** in 1838–1839; **4** in 1840; **5** in 1841; **6** in 1844; **7** in 1845; **8** in 1846; **9** in 1847; **10** in 1848; **11** in 1849; **12** in 1850. By H. G. Reichenbach, **13 & 14** in 1851; **15**: tt. 1–80 in 1852; tt. 81–160 in 1853; **16**: tt. 1–100 in 1853; tt. 101–150 in 1854; **17** in 1855; **18** in 1858; **19**(1): tt. 1–60 in 1858; tt. 61–150 in 1859; tt. 151–260 in 1860. By G. Beck von Mannagetta, **19**(2): pp. 1–10 in 1904; pp. 11–48 in 1905; pp. 49–104 in 1906; pp. 105–152 in 1907; pp. 153–184 in 1908; pp. 185–240 in 1909; pp. 241–288 in 1910; pp. 289–324 in 1911. By H. G. Reichenbach, **20**: pp. 1–48 in 1861; pp. 49–125 in 1862; **21**: tt. 1–70 in 1863; tt. 71–110 in 1864; tt. 111–150 in 1865; tt. 151–190 in 1866; tt. 191–220 in 1867; **22**: pp. 1–96 in ?1869. By G. Beck von Mannagetta, **22**: pp. 97–230 in 1903. By F. G. Kohl, **23**: pp. 1–32 in 1896; pp. 33–44 in 1897; pp. 45–68 in 1898; pp. 69–84 in 1899. By G. Beck von Mannagetta, **24**: pp. 1–24 in 1903; pp. 25–48 in 1904; pp. 49–64 in 1905; pp. 65–88 in 1906; pp. 89–112 in 1907; pp. 113–152 in 1908; pp. 153–160 in 1909; pp. 161–216 in ?1909; **25**(1): pp. 1–12 in 1909;

pp. 13–32 in 1910; pp. 33–48 in 1911; pp. 49–72 in 1912; **25**(2): pp. 1–24 in 1913; pp. 25–40 in 1914.)

Retz., *Obs. Bot.*
A. J. Retzius, *Observationes botanicae.* Lipsiae. 1779–1791. (**1** in 1779; **2** in 1781; **3** in 1783; **4** in 1786 or 1787; **5** in 1788; **6** in 1791. Cf. F. A. Stafleu, *Taxon* **12**: 75 (1963) and H. W. Rickett & F. A. Stafleu, *Taxon* **10**: 137 (1961).)

Risso & Poiteau, *Hist. Nat. Orang.*
J. A. Risso & P. A. Poiteau, *Histoire naturelle des Orangers.* Ed. 1. Paris. 1818. Ed. 2, by M. A. du Breuil. Paris. 1872.

Rochel, *Pl. Banat.*
A. Rochel, *Plantae Banatus rariores.* Pestini. 1828.

Roemer, M. J., *Syn. Monogr.*
M. J. Roemer, *Familiarum naturalium Regni vegetabilis Synopses monographicae.* Vimariae. 1846–1847. (**1 & 2** in 1846; **3 & 4** in 1847.)

Roemer & Schultes, *Syst. Veg.*
Cf. L., *Syst. Nat.*

Röhling, *Deutschl. Fl.*
J. C. Röhling, *Deutschlands Flora.* Ed. 1. Bremen. 1796. Ed. 2. Frankfurt a. M. 1812–1813. (**1 & 2** in 1812; **3** in 1813.) Ed. 3 by F. K. Mertens & W. D. J. Koch. Frankfurt a. M. 1823–1839. (**1** in 1823; **2** in 1826; **3** in 1831; **4** in 1833; **5** in 1839.)

Roth, *Catalecta Bot.*
A. W. Roth, *Catalecta botanica, quibus Plantae novae et minus cognitae describuntur atque illustrantur.* Lipsiae. 1797–1806. (**1** in 1797; **2** in 1800; **3** in 1806.)

Roth, *Enum.*
A. W. Roth, *Enumeratio Plantarum phaenogamarum in Germania sponte nascentium.* Lipsiae. 1827.

Roth, *Tent. Fl. Germ.*
A. W. Roth, *Tentamen Florae germanicae.* Lipsiae. 1788–1800. (**1** in 1788; **2**(1) in 1789; **2**(2) in 1793; **3** in 1800. Cf. F. A. Stafleu, *Taxon* **12**: 76 (1963) and R. E. G. Pichi-Sermolli, *Webbia* **8**: 439 (1952).)

Rottb., *Pl. Hort. Univ. Rar. Progr.*
C. F. Rottboel, *Plantas Horti Universitatis rariores programmate describit....* Hafniae. 1773.

Rouy, *Consp. Fl. Fr.*
G. C. C. Rouy, *Conspectus de la Flore de France.* Paris. 1927.

Rouy & Fouc., or Rouy & Camus or Rouy, *Fl. Fr.*
G. C. C. Rouy, *Flore de France.* Asnières, Paris & Rochefort. 1893–1913. (**1** in 1893; **2** in 1895; **3** in 1896; **4** in 1897; **5** in 1899; **6** in 1900; **7** in 1901; **8** in 1903; **9** in 1905; **10** in 1908; **11** in 1909; **12** in 1910; **13** in 1912; **14** in 1913.) **1–3** in collaboration with J. Foucaud, **6 & 7** with E. G. Camus.

Rouy, *Ill. Pl. Eur. Rar.*
G. C. C. Rouy, *Illustrationes Plantarum Europae rariorum.* Paris. 1895–1905. (**1–4** in 1895; **5–8** in 1896; **9 & 10** in 1898; **11 & 12** in 1899; **13 & 14** in 1900; **15 & 16** in 1901; **17 & 18** in 1902; **19** in 1904; **20** in 1905.)

Royle, *Ill. Bot. Himal. Mount.*
J. F. Royle, *Illustrations of the Botany and other Branches of the natural History of the Himalayan Mountains and of the Flora of Cashmere.* London. 1833–1840. (**1–3**; for dates cf. M. J. van Steenis-Kruseman & W. T. Stearn in C. G. G. J. van Steenis, *Fl. Males.* I. **4**: ccx (1954).)

Ruiz & Pavón, *Fl. Peruv.*
H. Ruiz Lopez & J. Pavón, *Flora peruviana et chilensis.* Matriti. 1798–1802. (**1** in 1798; **2** in 1799; **3** in 1802; **4** in ?1802.)

Russell, A., *Nat. Hist. Aleppo*
A. Russell, *The natural History of Aleppo.* Ed. 1. London. 1756. Ed. 2. London. 1794. (**1 & 2** in 1794.)

Sadler, *Fl. Com. Pest.*
J. Sadler, *Flora Comitatus pestiensis.* Ed. 1. Pestini. 1825–1826. (**1** in 1825; **2** in 1826.) Ed. 2. Pestini. 1840.

Samp., *Lista Esp. Herb. Port.*
G. A. da Silva Ferreira Sampaio, *Lista das Espécies represent-adas no Herbário português.* Porto. 1913. *Ap.* **1**, *Apêndice à Lista....* Porto. 1914. *Ap.* **2**, *Segundo Apêndice à Lista....* Porto. 1914. *Ap.* **3**, *Terceiro Apêndice à Lista....* Porto. 1914.

Sanguinetti, *Fl. Rom. Prodr. Alt.*
P. Sanguinetti, *Florae romanae Prodromus alter.* Romae. 1864.

Saunders, *Refug. Bot.*
W. W. Saunders, *Refugium botanicum; or, Figures and Descriptions from living Specimens, of little known or new Plants of botanical Interest.* London. 1869–1882. (**1** in 1869; **2**: tt. 73–96 in 1869; tt. 97–120 in 1878; tt. 121–144 in 1882; **3** in 1870; **4** in 1875; **5**: tt. 289–312 in 1871; tt. 313–336 in 1872; tt. 337–360 in 1873.)

Savi, *Bot. Etrusc.*
G. Savi, *Botanicon etruscum sistens Plantas in Etruria sponte crescentes.* Pisis. 1808–1825. (**1** in 1808; **2** in 1815; **3** in 1818; **4** in 1825.)

Savi, *Fl. Pis.*
G. Savi, *Flora pisana.* Pisa. 1798. (**1** & **2** in 1798.)

Savi, *Obs. Trif.*
G. Savi, *Observationes in varias Trifoliorum Species.* Florentiae. 1810.

Săvul., *Fl. Rep. Pop. Române*
T. Săvulescu, *Flora Republicii Populare Române.* | *Flora Reipublicae popularis romanicae.* Bucureşti. 1952→. (**1** in 1952; **2** in 1953; **3** in 1955; **4** in 1956; **5** in 1957; **6** in 1958; **7** in 1960; **8** in 1961; **9** in 1964; **10** in 1965; **11** in 1966. **3**–**10** titled *Flora Republicii Populare Romîne.* | *Flora Reipublicae popularis romanicae.* **11** titled *Flora Republicii Socialiste România.* | *Flora Reipublicae socialisticae România.*)

Schinz & R. Keller, *Fl. Schweiz*
H. Schinz & R. Keller, *Flora der Schweiz.* Ed. 1. Zürich. 1900. Ed. 2. Zürich. 1905. (**1** *Exkursionsflora* in 1905; **2** *Kritische Flora* in 1905.) Ed. 3. Zürich. 1909–1914. (**1** *Exkursionsflora* in 1909; **2** *Kritische Flora*, edit. H. Schinz & A. Thellung, in 1914.) Ed. 4. Zürich. 1923. (**1** *Exkursionsflora*, edit. H. Schinz & A. Thellung, in 1923.)

Schischkin, or **Schischkin & Bobrov,** or **Schischkin & Juz.,** *Fl. URSS*
Cf. Komarov, *Fl. URSS.*

Schkuhr, *Handb.*
C. Schkuhr, *Botanisches Handbuch der mehresten theils in Deutschland wildwachsenden, theils ausländischen in Deutschland unter freyem Himmel ausdauernden Gewächse.* Ed. 1. Wittenbergae. 1791–1803. (**1** in 1791; **2** in 1796; **3** in 1803.) Ed. 2. Leipzig. 1808. (**1**–**4** in 1808.)

Schmalh., *Fl. Sred. Juž. Ross.*
I. F. Schmalhausen, Флора средней и южной Россіи, Крыма и сѣвернаго Кавказа [*Flora srednej i južnoj Rossii, Kryma i sěvernago Kavkaza*]. Kiev. 1895–1897. (**1** in 1895; **2** in 1897.)

Schmidt, F. W., *Fl. Boëm.*
F. W. Schmidt, *Flora boëmica inchoata.* Pragae. 1793–1794. (**1** & **2** in 1793; **3** & **4** in 1794.)

Schmidt, Franz. *Östr. Allgem. Baumz.*
Franz Schmidt, *Östreichs allgemeine Baumzucht.* Wien. 1792–1822. (**1** in 1792; **2** in 1794; **3** in 1800; **4** in 1822.)

Schneider, C. K., *Ill. Handb. Laubholzk.*
C. K. Schneider, *Illustriertes Handbuch der Laubholzkunde. Charakteristik der in Mitteleuropa heimischen und im Freien angepflanzten angiospermen Gehölz-Arten und Formen mit Ausschluss der Bambuseen und Kakteen.* Jena. 1904–1921.

(**1**: pp. 1–448 in 1904; pp. 449–592 in 1905; pp. 593–810 in 1906; **2**: pp. 1–240 in 1907; pp. 241–496 in 1909; pp. 497–816 in 1911; pp. 817–1070 in 1912.)

Schott, Nyman & Kotschy, *Analect. Bot.*
H. W. Schott, C. F. Nyman & T. Kotschy, *Analecta botanica.* Vindobonae. 1854.

Schrader, *Neues Jour. Bot.*
H. A. Schrader, *Neues Journal für die Botanik.* Erfurt. 1806–1810. (**1**(1 & 2) in 1805; **1**(3) in 1806; **2**(1) in 1807; **2**(2 & 3) in 1808; **3**(1–4) in 1809; **4**(1 & 2) in 1810. Cf. G. Sayre, *Dates Publ. Musci* 73–75 (1959).)

Schreber, *Icon. Descr. Pl.*
J. C. D. von Schreber, *Icones et Descriptiones Plantarum minus cognitarum.* Halae. 1766.

Schreber, *Spicil. Fl. Lips.*
J. C. D. von Schreber, *Spicilegium Florae lipsicae.* Lipsiae. 1771.

Schultes, *Östreichs Fl.*
J. A. Schultes, *Östreichs Flora.* Ed. 1. 1794. Wien. Ed. 2. Wien. 1814.

Schultz, F. W., *Arch. Fl. Fr. Allem.*
F. W. Schultz, *Archives de la Flore de France et d'Allemagne.* Bitche, Haguenau & Deux-Ponts. 1842–1855. (Pp. 1–48 in 1842; pp. 49–76 in 1844; pp. 77–98 in 1846; pp. 99–154 in 1848; pp. 155–166 in 1850; pp. 167–194 in 1851; pp. 195–258 in 1852; pp. 259–282 in 1853; pp. 283–326 in 1854; pp. 327–350 in 1855.)

Schur, *Enum. Pl. Transs.*
P. J. F. Schur, *Enumeratio Plantarum Transsilvaniae.* Wien. 1866.

Scop., *Fl. Carn.*
J. A. Scopoli, *Flora carniolica.* Ed. 2. Viennae. 1772. (**1** & **2** in 1772.)

Sebastiani, *Rom. Pl.*
A. Sebastiani, *Romanorum Plantarum Fasciculus primus* (=**1**). Romae. 1813. *Romanorum Plantarum Fasciculus alter* (=**2**). Romae. 1815.

Sebastiani & Mauri, *Fl. Rom.*
A. Sebastiani & E. Mauri, *Florae romanae Prodromus.* Romae. 1818.

Seub., *Fl. Azor.*
M. A. Seubert, *Flora azorica, quam ex Collectionibus Schedisque Hochstetteri Patris et Filii elaboravit....* Bonnae. 1844.

Sibth., *Fl. Oxon.*
J. Sibthorp, *Flora oxoniensis.* Oxonii. 1794.

Sibth. & Sm., *Fl. Graec. Prodr.*
J. Sibthorp & J. E. Smith, *Florae graecae Prodromus: sive Plantarum omnium Enumeratio, quas in Provinciis aut Insulis Graeciae invenit Johannes Sibthorp... Characteres et Synonyma omnium cum Annotationibus elaboravit Jacobus Edvardus Smith.* Londini. 1806–1816. (**1**: pp. 1–218 in 1806; pp. 219–442 in 1809; **2**: pp. 1–210 in 1813; pp. 211–422 in 1816. Cf. H. W. Rickett & F. A. Stafleu, *Taxon* **10**: 142 (1961).)

Sieber, *Reise Kreta*
F. W. Sieber, *Reise nach der Insel Kreta im griechischen Archipelagus im Jahre 1817.* Leipzig & Sorau. 1823. (**1** & **2** in 1823.)

Simonkai, *Enum. Fl. Transs.*
L. von Simonkai, *Enumeratio Florae transsilvanicae vasculosae critica.* | *Erdély edényes Flórájának helyesbitett Foglalata.* Budapest. 1887.

Sm., *Bot. New. Holl.*
J. E. Smith, *A Specimen of the Botany of New Holland.* London. 1793–1795. (Pp. 1–24 in 1793; pp. 25–38 in 1794; pp. 39–54 in 1795. Cf. H. W. Rickett & F. A. Stafleu, *Taxon* **10**: 142 (1961).)

Sm., Engl. Fl.
J. E. Smith, *The English Flora*. Ed. 1. London. 1824–1828. (**1 & 2** in 1824; **3** in 1825; **4** in 1828.) Ed. 2. London. 1828–1830. (**1 & 2** in 1828; **3** in 1829; **4** in 1830.) Other volumes on *Musci*, etc.

Sm., Exot. Bot.
J. E. Smith, *Exotic Botany: consisting of coloured Figures and scientific Descriptions of such new, beautiful, or rare Plants as are worthy of Cultivation in the Gardens of Britain*. London. 1804–1808. (**1**: pp. 1–7 in 1804; pp. 9–118 in 1805; **2**: pp. 1–11, 15, 21–22 in 1805; pp. 13–14, 17–20, 23–112, 117–118 in 1806; pp. 121–122 in 1807; pp. 113–116, 119–120 in 1808.)

Sm., Fl. Brit.
J. E. Smith, *Flora britannica*. Londini. 1800–1804. (**1**: pp. 1–436; **2**: pp. 437–914 in 1800; **3** in 1804. Cf. S. W. Greene, *Jour. Soc. Bibl. Nat. Hist.* **3**: 281 (1957).

Soó & Jáv., Magyar Növ. Kéz.
R. de Soó & S. Jávorka, *A Magyar Növényvilág Kézikönyve*. Budapest. 1951. (**1 & 2** in 1951.)

Sowerby, Engl. Bot.
J. Sowerby, *English Botany*. Ed. 1. *Text* by J. E. Smith. London. 1790–1814. (**1–36** in 1790–1814.) *Supplement* by various authors. 1831–1863. (**1–5** in 1831–1863.) Ed. 2. London. 1831–1853. (**1–12** in 1831–1853.) Ed. 3, with new descriptions by J. T. I. Boswell-Syme. London. 1863–1886. (For dates cf. G. Sayre, *Dates Publ. Musci* 77 (1959).)

Soyer-Willemet, Obs. Pl. Fr.
H. F. Soyer-Willemet, *Observations sur quelques Plantes de France, suivies du Catalogue des Plantes vasculaires des Environs de Nancy*. Nancy. 1828.

Spach, Hist. Vég. (Phan.)
E. Spach, *Histoire naturelle des Végétaux. Phanérogames*. Paris. 1834–1848. (**1–3** in 1834; **4** in 1835; **5** in 1836; **6** in 1838; **7 & 8** in 1839; **9** in 1840; **10–13** in 1841; **14** in 1847. Cf. H. W. Rickett & F. A. Stafleu, *Taxon* **10**: 142 (1961).

Sprengel, Neue Entdeck.
K. P. J. Sprengel, *Neue Entdeckungen im ganzen Umfang der Pflanzenkunde*. Leipzig. 1820–1822. (**1 & 2** in 1820; **3** in 1822. Cf. H. W. Rickett & F. A. Stafleu, *Taxon* **10**: 143 (1961).)

Sprengel, Pl. Umb. Prodr.
K. P. J. Sprengel, *Plantarum umbelliferarum denuo disponendarum Prodromus*. Halae. 1813.

Sprengel, Pugillus
K. P. J. Sprengel, *Plantarum minus cognitarum Pugillus*. Halae. 1813–1815. (**1** in 1813; **2** in 1815.)

Sprengel, Sp. Umb.
K. P. J. Sprengel, *Species Umbelliferarum minus cognitae, illustratae*. Halae. 1818.

Sprengel, Syst. Veg.
Cf. L., *Syst. Nat.*

Steele, Handb. Field Bot.
W. E. Steele, *A Handbook of Field Botany, comprising the flowering Plants and Ferns indigenous to the British Isles*. Dublin. 1847.

Stefánsson, Fl. Ísl.
S. Stefánsson, *Flóra Íslands*. Ed. 1. Kaupmannahöfn. 1901. Ed. 2. Kaupmannahöfn. 1924. Ed. 3 by S. Steindórsson. Akureyri. 1948.

Steudel, Nomencl. Bot.
E. G. von Steudel, *Nomenclator botanicus*. Ed. 1. Stuttgardtiae & Tubingae. 1821. (**1 & 2**, with continuous pagination, in 1821.) Ed. 2. Stuttgartiae & Tubingae. 1840–1841. (**1** in 1840; **2**: pp. 1–48 in 1840; pp. 49–176 in 1840 or 1841; pp. 177–810 in 1841. Cf. M. J. van Steenis-Kruseman & W. T. Stearn in C. G. G. J. van Steenis, *Fl. Males.* I. **4**: ccxiii (1954).)

St-Hil., Fl. Bras. Mer.
A. C. F. P. de Saint-Hilaire, *Flora Brasiliae meridionalis*. Parisis. 1825–1834. (Quarto edition. **1**: pp. 1–168 in 1825; pp. 169–240 in 1827; pp. 241–400 in 1828; **2**: pp. 1–116 in 1829; pp. 117–276 in 1830; pp. 277–356 in 1832; pp. 357–384 in 1833; **3**: pp. 1–84, 121–160 in 1833; pp. 86–120 in 1834. Cf. H. W. Rickett & F. W. Stafleu, *Taxon* **10**: 139 (1961).) Folio edition published at same time.

Stoj. & Stefanov, Fl. Bălg.
N. Stojanov & B. Stefanov, Флора на България. [*Flora na Bălgarija*]. Ed. 1. Sofija. 1924–1925. Ed. 2. Sofija. 1933. Ed. 3. Sofija. 1948. Ed. 4, by N. Stojanov, B. Stefanov & B. Kitanov. Sofija. 1966–1967. (**1** in 1966; **2** in 1967.)

Sturm, Deutschl. Fl.
J. Sturm, *Deutschlands Flora*. Nürnberg. 1798–1855.

Sudre, Bat. Eur.
H. Sudre, *Batotheca europaea*. Albi. 1903–1917. (**1–16** in 1903; pp. 17–35 in 1904; pp. 37–48 in 1905; pp. 51–62 in 1906; pp. 63–74 in 1907; pp. 75–84 in 1908; pp. 85–92 in 1909; pp. 93–100 in 1910; pp. 101–107 in 1911; pp. 109–116 in 1912; pp. 117–122 in 1913; pp. 123–129 in 1914; pp. 131–137 in 1915; pp. 139–145 in 1916; pp. 147–153 in 1917.)

Sudre, Rubi Eur.
H. Sudre, *Rubi Europae vel Monographia Iconibus illustrata Ruborum Europae*. Paris. 1908–1913. (Pp. 1–40 in 1908; pp. 41–80 in 1909; pp. 81–120 in 1910; pp. 121–160 in 1911; pp. 161–200 in 1912; pp. 201–294 in 1913.)

Sudre, Fl. Toulous.
H. Sudre, *Florule toulousaine*. Paris, Toulouse & Albi. 1907.

Sweet, Cistin.
R. Sweet, *Cistineae. The natural Order of Cistus or Rock-Rose*. London. 1825–1830. (Tt. 1–12 in 1825; tt. 13–36 in 1826; tt. 37–60 in 1827; tt. 61–84 in 1828; tt. 85–108 in 1829; tt. 109–112 in 1830.)

Sweet, Hort. Brit.
R. Sweet, *Hortus britannicus*. Ed. 1. London. 1826–1827. (Pp. 1–240 in 1826; pp. 241–492 in 1827.) Ed. 2. London. 1830. Ed. 3 by G. Don. London. 1839.

Tausch, Hort. Canal.
I. F. Tausch, *Hortus canalius, seu Plantarum rariorum, quae in Horto illustrissimi, ac excellentissimi Josephi Malabaila Comitis de Canal coluntur, Icones et Descriptiones*. Pragae. 1823–1825. (**1**(1) in 1823; **1**(2) in ?1825. Cf. F. Widder, *Phyton (Austria)* **1**: 258–268 (1949).)

Ten., Cat. Pl. Horti Neap.
M. Tenore, *Catalogus Plantarum Horti regii neapolitani ad Annum 1813*. Neapoli. 1812. *App.*, *Appendix*. Ed. 1. Neapoli. 1815. Ed. 2. Neapoli. 1819.

Ten., Fl. Nap.
M. Tenore, *Flora napolitana*. Napoli. 1811–1838. (**1** in 1811–1815, including *Prodr.*, *Prodromo della Flora napolitana* followed by *Supplimento primo* and *Supplimento secondo* with continuous pagination; **2** in 1820, including *Prodr. Suppl.* 3, *Prodromo della Flora napolitana, Supplimento terzo*; **3** in 1824–1829, including *Prodr. Suppl.* 4, *Prodromo della Flora napolitana, Supplimento quarto*; **4** in 1830, including *Syll.*, *Florae neapolitanae Sylloge* followed by *Addenda et Emendanda* and *Addenda et Emendanda altera* with continuous pagination, and *Syll. App.* 3, *Ad Florae neapolitanae plantarum vascularium Syllogem, Appendix tertia*; **5** in 1835–1838, including *Syll. App.* 4, *Ad Florae neapolitanae Syllogem, Appendix quarta*.) The included works were in most cases reprinted separately later, with different pagination.

Ten., Fl. Neap. Prodr. App. Quinta
M. Tenore, *Ad Florae neapolitanae Prodromum Appendix*

quinta. Neapoli. 1826. This work was never included in the volumes of *Flora napolitana*.

Ten., *Ind. Sem. Horti Neap.*
M. Tenore, or M. Tenore & G. Gussone, *Index Seminum Horti botanici neapolitani*. Neapoli. 1825–1840.

Ten., *Viagg. Calabr.*
M. Tenore, *Viaggio in alcuni Luoghi della Basilicata e della Calabria citeriore effettuato nel 1826*. Napoli. 1827.

Ten. & Guss., *Ind. Sem. Horti Neap.*
Cf. Ten., *Ind. Sem. Horti Neap.*

Texidor, *Apunt. Fl. Esp.*
J. Texidor y Cos, *Apuntes para la Flora de España*. Madrid. 1869.

Thuill., *Fl. Paris*
J. L. Thuillier, *La Flore des Environs de Paris*. Ed. 1. Paris. 1790. Ed. 2. Paris. 1799.

Thunb., *Diss. Hydrocot.*
C. P. Thunberg, *Dissertatio botanica de Hydrocotyle*. Upsaliae. 1798.

Thunb., *Fl. Jap.*
C. P. Thunberg, *Flora japonica*. Lipsiae. 1784.

Thunb., *Nova Gen. Pl.*
C. P. Thunberg, *Nova Genera Plantarum*. Upsaliae. 1781–1801. (**1** in 1781; **2** in 1782; **3** in 1783; **4 & 5** in 1784; **6 & 7** in 1792; **8 & 9** in 1798; **10–12** in 1800; **13–16** in 1801.)

Tineo, *Pl. Rar. Sic. Pug.*
V. Tineo, *Plantarum rariorum Siciliae Pugillus primus*. Panormi. 1817.

Torrey & A. Gray, *Fl. N. Amer.*
J. Torrey & A. Gray, *A Flora of North America*. New York. 1838–1843. (**1**: pp. 1–360 in 1838; pp. 361–712 in 1840; **2**: pp. 1–184 in 1841; pp. 185–392 in 1842; pp. 393–504 in 1843. Cf. H. W. Rickett & F. A. Stafleu, *Taxon* **10**: 145 (1961).)

Tourlet, *Cat. Pl. Indre Loire*
E.-H. Tourlet, *Catalogue raisonné des Plantes vasculaires du Département d'Indre-et-Loire*. Paris & Tours. 1908.

Tratt., *Ros. Monogr.*
L. Trattinnick, *Rosacearum Monographia*. Vindobonae. 1823–1824. (**1–3** in 1823; **4** in 1824.)

Trev., *Ind. Sem. Horto Wratisl.*
L. C. Treviranus, *Index Seminum quae in Horto botanico wratislaviensi prostant*. Wratislaviae. 1818. *Appendix* 1. Wratislaviae. 1819. *Appendix* 2. Wratislaviae. 1820. *Appendix* 3. Wratislaviae. 1821.

Turra, *Farset. Nov. Gen.*
A. Turra, *Farsetia, novum Genus*. Venetiis. 1765.

Unger & Kotschy, *Ins. Cypern*
F. J. A. N. Unger & T. Kotschy, *Die Insel Cypern, ihrer physischen und organischen Natur nach, mit Rücksicht auf ihre frühere Geschichte*. Wien. 1865.

Vahl, *Symb. Bot.*
M. H. Vahl, *Symbolae botanicae*. Hauniac. 1790–1794. (**1** in 1790; **2** in 1791; **3** in 1794. Cf. F. A. Stafleu, *Taxon* **12**: 82 (1963).)

Velen., *Fl. Bulg.*
J. Velenovský, *Flora bulgarica*. Pragae. 1891. *Suppl.*, *Supplementum*. 1898.

Vent., *Descr. Pl. Jard. Cels*
E. P. Ventenat, *Description des Plantes nouvelles ou peu connues, cultivées dans le Jardin de J. M. Cels*. Paris. 1800–1802 (tt. 1–10 in 1800; tt. 11–20 in 1800 or 1801; tt. 21–50 in 1801; tt. 51–100 in 1802. Cf. W. T. Stearn, *Jour. Soc. Bibl. Nat. Hist.* **1**: 199–200 (1939).)

Vill., *Cat. Méth. Jard. Strasb.*
D. Villars, *Catalogue méthodique des Plantes du Jardin de l'École de Médecine de Strasbourg*. Strasbourg. 1807.

Vill., *Hist. Pl. Dauph.*
D. Villars, *Histoire des Plantes de Dauphiné*. Grenoble. 1786–1789. (**1** in 1786; **2** in 1787; **3**(1): pp. 1–580 in 1788; **3**(2): pp. 581–1092 in 1789. Cf. F. A. Stafleu, *Taxon* **12**: 82 (1963).)

Vill., *Prosp. Pl. Dauph.*
D. Villars, *Prospectus de l'Histoire des Plantes de Dauphiné*. Grenoble. 1779.

Vis., *Fl. Dalm.*
R. de Visiani, *Flora dalmatica*. Lipsiae. 1842–1852. (**1** in 1842; **2** in 1847; **3** in 1850–1852.) *Suppl.* 1. Lipsiae. 1872. *Suppl.* 2. Lipsiae. 1877.

Vis., *Stirp. Dalm.*
R. de Visiani, *Stirpium dalmaticarum Specimen*. Patavii. 1826.

Viv., *Elench. Pl. Horti Bot.*
D. Viviani, *Elenchus Plantarum Horti botanici, Observationibus quoad novas vel rariores Species passim interjectis*. Genuae. 1802.

Viv., *Fl. Cors.*
D. Viviani, *Florae corsicae Specierum novarum, vel minus cognitarum Diagnosis*. Genuae. 1824. *App.* 1, *Appendix*. Genuae. 1825. *App.* 2, *Appendix altera*. Genuae. 1830.

Viv., *Fl. Ital. Fragm.*
D. Viviani, *Florae italicae Fragmenta, seu Plantae rariores vel nondum cognitae in variis Italiae Regionibus detectae*. Genuae. 1808.

Viv., *Fl. Lib.*
D. Viviani, *Florae libycae Specimen, sive Plantarum Enumeratio Cyrenaicam, Pentapolim, Magnae Syrteos Desertum et Regionem tripolitanum incolentium*. Genuae. 1824.

Voss & Siebert, *Vilmorin's Blumeng.*
A. Voss & A. Siebert, *Vilmorin's Blumengärtnerei. Beschreibung, Kultur und Verwendung des gesamten Pflanzenmaterials für deutsche Gärten*. Ed. 3. Berlin. 1894–1896. (**1**: pp. 1–592 in 1894; pp. 593–1264 in 1895; pp. i–viii, [1]–[80] in 1896; **2**: pp. 1–96 in 1895; pp. 97–238 in 1896.) An entirely new work with nothing in common with the earlier editions of Vilmorin's *Blumengärtnerei*.

Waldst. & Kit., *Pl. Rar. Hung.*
F. A. von Waldstein-Wartemberg & P. Kitaibel, *Descriptiones et Icones Plantarum rariorum Hungariae*. Viennae. 1799–1812. (**1**: tt. 1–10 in 1799; tt. 11–30 in 1800; tt. 31–50 in 1800–1801; tt. 51–70 in 1801; tt. 71–90 in 1801–1802; tt. 91–100 in 1802; **2**: tt. 101–130 in 1802 or 1803; tt. 131–170 in 1803 or 1804; tt. 171–190 in 1804; tt. 191–200 in 1805; **3**: tt. 201–220 in 1806 or 1807; tt. 221–240 in 1807; tt. 241–250 in 1808 or 1809; tt. 251–260 in 1809; tt. 261–270 in 1810 or 1811; tt. 271–280 in 1812. Cf. M. J. van Steenis-Kruseman & W. T. Stearn in C. G. G. J. van Steenis, *Fl. Males.* **I.** 4: ccxvi (1954).)

Wallr., *Erst. Beitr. Fl. Hercyn.*
K. F. W. Wallroth, *Erster Beitrag zur Flora hercynica*. Halle. 1840.

Walpers, *Repert. Bot. Syst.*
W. G. Walpers, *Repertorium Botanices systematicae*. Lipsiae. 1842–1848. (**1** in 1842; **2** in 1843; **3** in 1844–1845; **4**: pp. 1–192 in 1845; pp. 193–822 in 1848; **5** in 1845–1846; **6** in 1846–1847.)

Webb, *Iter Hisp.*
P. B. Webb, *Iter hispaniense*. Paris & London. 1838.

Webb & Berth., *Phyt. Canar.*
P. B. Webb & S. Berthelot, *Phytographia canariensis*. Vol. 3(2) of *Histoire naturelle des Îles Canaries*. Paris. 1836–1850. (For dates cf. W. T. Stearn, *Jour. Soc. Bibl. Nat. Hist.* **1**: 55 (1937).)

Weber, *Pl. Min. Cogn. Dec.*
G. H. Weber, *Plantarum minus cognitarum Decuria*. Kiloniae. 1784.

Weihe & Nees, *Rubi Germ.*

K. E. A. Weihe & C. G. D. Nees von Esenbeck, *Rubi germanici.* Elberfeldae. 1822–1827. (Pp. 1–28 in 1822; pp. 29–40 in 1824; pp. 41–70 in 1825; pp. 71–78 in 1825 or 1826; pp. 79–86 in 1826; pp. 89–116 in 1827. Cf. W. T. Stearn, *Jour. Soc. Bibl. Nat. Hist.* **4**: 68–69 (1962).) Also German edition, titled *Die deutschen Brombeersträuche.* Elberfeld. 1822–1827. (Pp. 1–28 in 1822; pp. 29–42 in 1824; pp. 43–76 in 1825; pp. 77–86 in 1825 or 1826; pp. 87–96 in 1826; pp. 97–130 in 1827. Cf. W. T. Stearn, *Jour. Soc. Bibl. Nat. Hist.* **4**: 68–69 (1962).) Only the Latin edition is cited.

Weinm., *Bot. Gart. Dorpat*

J. A. Weinmann, *Der botanische Garten der kaiser. Universität zu Dorpat im Jahre 1810.* Dorpat. 1810.

White, *Jour. Voy. New S. Wales*

J. White, *Journal of a Voyage to New South Wales.* London. 1790.

Wibel, *Prim. Fl. Werthem.*

A. W. E. C. Wibel, *Primitiae Florae werthemensis.* Jenae. 1799.

Willd., *Enum. Pl. Hort. Berol.*

C. L. Willdenow, *Enumeratio Plantarum Horti regii botanici berolinensis.* Berolini. 1809. *Suppl.,* *Supplementum* by D. F. von Schlechtendal. Berolini. 1814. For dates, cf. W. T. Stearn, *Jour. Bot.* (*London*) **75**: 234 (1937).

Willd., *Hort. Berol.*

C. L. Willdenow, *Hortus berolinensis, sive Icones et Descriptiones, Plantarum rariorum vel minus cognitarum, quae in Horto regio botanico berolinensi excoluntur.* Berolini. 1803–1816. (**1(1)**: tt. 1–12 in 1803; **1(2–3)**: tt. 13–36 in 1804; **1(4–5)**: tt. 37–60 in 1805; **1(6)** & **2(7)**: tt. 61–84 in 1806; **2(8)**: tt. 85–96 in 1809; **2(9)**: tt. 97–108 in 1812. Cf. W. T. Stearn, *Jour. Bot.* (*London*) **75**: 233 (1937) and M. J. van Steenis-Kruseman & W. T. Stearn in C. G. G. J. van Steenis, *Fl. Males.* I. **4**: ccxviii (1954).)

Willd., *Phytogr.*

C. L. Willdenow, *Phytographia, seu Descriptio rariorum minus cognitarum Plantarum.* Erlangae. 1794.

Willd., *Sp. Pl.*

C. L. Willdenow, ed. 4 of C. von Linné, *Species Plantarum.* Berolini. 1797–1806. (**1(1)**: pp. 1–495 in 1797; **1(2)**: pp. 496–1968 in 1798; **2(1)**: pp. 1–823 in 1799; **2(2)**: pp. 824–1340 in 1800; **3(1)**: pp. 1–847 in 1800; **3(2)**: pp. 848–1474 in 1802; **3(3)**: pp. 1475–2409 in 1803; **4(1)**: pp. 1–629 in 1805; **4(2)**: pp. 630–1157 in 1806. Cf. B. G. Schubert, *Rhodora* **44**: 147 (1942).)

Willk., *Icon. Descr. Pl. Nov.*

H. M. Willkomm, *Icones et Descriptiones Plantarum novarum criticarum et rariorum Europae austro-occidentalis praecipue Hispaniae.* Lipsiae. 1852–1862. (**1(1)**: pp. 1–16, tt. 1–7 in 1852; **1(2)**: pp. 17–24, tt. 8–13 in 1853; **1(3)**: pp. 25–32, tt. 14–20 in 1853; **1(4)**: pp. 33–40, tt. 21–28 in 1853; **1(5)**: pp. 41–48, tt. 29–35 in 1854; **1(6)**: pp. 49–56, tt. 36–41 in 1854; **1(7)**: pp. 57–64, tt. 42–46 in 1854; **1(8)**: pp. 65–80, tt. 47–53 in 1854; **1(9)**: pp. 81–104, tt. 54–63 in 1855; **1(10)**: pp. 105–123, tt. 64–73 in 1856; **2(11)**: pp. 1–24, tt. 74–83 in 1857; **2(12)**: pp. 25–40, tt. 84–93 in 1858; **2(13)**: pp. 41–56, tt. 94–101 in 1858; **2(14)**: pp. 57–68, tt. 102–109 in 1859; **2(15)**: pp. 69–84, tt. 110–118 in 1859; **2(16)**: pp. 85–96, tt. 119–128 in 1859; **2(17)**: pp. 97–108, tt. 129–138 in 1861; **2(18)**: pp. 109–120, tt. 139–148 in 1861; **2(19)**: pp. 121–182, tt. 149–158 in 1862.)

Willk., *Ill. Fl. Hisp.*

H. M. Willkomm, *Illustrationes Florae Hispaniae Insularumque Balearium.* Stuttgart. 1881–1892. (**1(1)**: pp. 1–12, tt. 1–9 in 1881; **1(2)**: pp. 13–28, tt. 10–18 in 1881; **1(3)**: pp. 29–40, tt. 19–28 in 1881; **1(4–6)**: pp. 44–88, tt. 29–56 in 1882; **1(7 & 8)**: pp. 89–120, tt. 57–74 in 1883; **1(9)**: pp. 121–136, tt. 75–83 in 1884; **1(10)**: pp. i–vii, 137–157, tt. 84–92 in 1885; **2(11)**: pp. 1–16, tt. 98–101 in 1886; **2(12)**: pp. 17–32, tt. 102–110 in 1886; **2(13)**: pp. 33–48, tt. 111–119 in 1887; **2(14)**: pp. 49–64, tt. 120–127 in 1888; **2(15 & 16)**: pp. 65–98, tt. 128–146 in 1889; **2(17)**: pp. 99–112, tt. 147–155 in 1890; **2(18)**: pp. 113–126, tt. 156–164 in 1891; **2(19)**: pp. 127–140, tt. 165–173 in 1892; **2(20)**: pp. i–vii, 141–156, tt. 174–183 in 1892. Cf. F. G. Wiltshear, *Jour. Bot.* (*London*) **53**: 372 (1915).)

Willk., *Suppl. Prodr. Fl. Hisp.*

H. M. Willkomm, *Supplementum Prodromi Florae hispanicae.* Stuttgartiae. 1893.

Willk. & Lange, *Prodr. Fl. Hisp.*

H. M. Willkomm & J. Lange, *Prodromus Florae hispanicae.* Stuttgartiae. 1861–1880. (**1**: pp. 1–192 in 1861; pp. i–xxx, 193–316 in 1862; **2**: pp. 1–272 in 1865; pp. 273–480 in 1868; pp. 481–680 in 1870; **3**: pp. 1–240 in 1874; pp. 241–512 in 1877; pp. 513–736 in 1878; pp. 737–1144 in 1880.)

Wimmer, *Fl. Schles.*

C. F. H. Wimmer, *Flora von Schlesien.* Ed. 1. Berlin. 1832. Ed. 2. Breslau. 1841. *Ergänzungsband.* Breslau. 1845. Ed. 3. Breslau. 1857.

Wimmer & Grab., *Fl. Siles.*

C. F. H. Wimmer & H. E. Grabowski, *Flora Silesiae.* Vratislaviae. 1827–1829. (**1** in 1827; **2(1)** & **2(2)** in 1829.)

Wirtgen, *Prodr. Fl. Preuss. Rheinl.*

P. W. Wirtgen, *Prodromus der Flora der preussischen Rheinlande.* Bonn. 1842.

Zimmeter, *Gatt. Potent.*

A. Zimmeter, *Die europäischen Arten der Gattung* Potentilla. Steyr. 1884.

ADDENDA

Cosson & Germ., *Fl. Env. Paris*

E. S. C. Cosson & J. N. E. Germain de Saint-Pierre, *Flore descriptive et analytique des Environs de Paris.* Ed. 1. Paris. 1845. Ed. 2. Paris. 1861.

Rydb., *Fl. Rocky Mount.*

P. A. Rydberg, *Flora of the Rocky Mountains and adjacent Plains.* Ed. 1. New York. 1918. Ed. 2. New York. 1922.

Vis., *Sem. Horto Bot. Patav.*

Semina in Horto botanico Patavius lecta. By R. de Visiani in 1855.

APPENDIX III

KEY TO THE ABBREVIATIONS OF TITLES OF PERIODICALS AND ANONYMOUS WORKS CITED IN VOLUME 2

Abh. Akad. Wiss. (München)
Abhandlungen der mathemat.-physikalischen Classe der königlich-bayerischen Akademie der Wissenschaften. München. Ser. 1, 1–30, 1832–1926. With minor variations of title. Nov. ser., titled Abhandlungen der bayerischen Akademie der Wissenschaften, mathematisch-naturwissenschaftliche Abteilung, 1→ , 1929→ .

Abh. Böhm. Ges. Wiss.
Abhandlungen des böhmischen Gesellschaft der Wissenschaften. Prag. Ser. 1, 1–4, 1786–1789. Ser. 2, titled Neuere Abhandlungen der k. böhmischen Gesellschaft der Wissenschaften, 1–3, 1791–1798. Ser. 3, titled Abhandlungen der königlichen böhmischen Gesellschaft der Wissenschaften, 1–8, 1804–1823. Ser. 4, 1–5, 1817–1837. Ser 5, 1–14, 1838–1867. Ser. 6, 1–12, 1867–1884. Ser. 7, 1–4, 1885–1891. (Ser. 6, 11–12, 1881–1884, also titled Pojedáni královské české Společnosti Nauk; ser. 7 also titled Rozpravy královské české Společnosti Nauk.)

Abh. Nat. Ver. Bremen
Abhandlungen herausgegeben vom naturwissenschaftlichen Verein zu Bremen. Bremen. 1868→ .

Abh. Preuss. Akad. Wiss.
Abhandlungen der physikalischen Klasse der königlich-preussischen Akademie der Wissenschaften. Berlin. 1804→ . With minor changes of title.

Act. Soc. Linn. Bordeaux
Bulletin d'Histoire naturelle de la Société linnéenne de Bordeaux. Bordeaux. 1–3, 1826–1829; titled Actes de la Société linnéenne de Bordeaux, 4→, 1830→ .

Acta Acad. Sci. Nat. Mor.-Sil.
Acta Societatis Scientiarum naturalium Moraviae. / Práce moravské přírodovědecké Společnosti. Brno. 1–18, 1924–1947; titled Acta Academiae Scientiarum naturalium moravosilesiacae. / Práce moravskoslezské Akademie věd přírodních, 19–25, 1948–1953; titled Acta Academiae Scientiarum čechoslovenicae, Basis brunensis. / Práce brněnské Základny československé Akademie Věd, 26→ , 1954→ .

Acta Acad. Theod.-Pal.
Acta Academiae Theodoro-Palatinae. Mannheimii. 1–7, 1766–1794.

Acta Biol. Cracov. (Bot.)
Acta biologica cracoviensia. Series botanica. Cracovie. 1958→ .

Acta Biol. Acad. Sci. Hung.
Acta biologica Academiae Scientiarum hungaricae. Budapest. 1951→ .

Acta Bot. Acad. Sci. Hung.
Acta botanica Academiae Scientiarum hungaricae. Budapest. 1954→ .

Acta Bot. Neerl.
Acta botanica neerlandica. Amsterdam. 1952→ .

Acta Dendrol. Čech.
Acta dendrologica čechoslovaca.

Acta Horti Berg.
Acta Horti bergiani. / Meddelanden från Kongl. Svenska Vetenskaps-Akademiens Trädgård, Bergielund. Stockholm. 1890→ .

Acta Horti Bot. Prag.
Acta Horti botanici pragensis. Praga. 1962→ .

Acta Horti Petrop.
Acta Horti petropolitani. Peterburgi. 1–43, 1871–1931. (With alternative title Труды Имп. С.-Петербургскаго ботаническаго Сада [Trudy Imp. S.-Peterburgskago botaničeskago Sada] 1871–1918, and Труды главнаго ботаническаго Сада [Trudy glavnago botaničeskago Sada] 1918–1931.)

Acta Inst. Bot. Acad. Sci. URSS
Труды ботаническаго Института Академии Наук СССР [Trudy botaničeskago Instituta Akademii Nauk SSSR]. / Acta Instituti botanici Academiae Scientiarum URSS. Leningrad & Mosqua. 1933→ .

Acta Mus. Nat. Pragae
Sborník národního Musea v Praze. / Acta Musei nationalis Pragae. Praha. 1938→ .

Acta Phytogeogr. Suec.
Acta phytogeographica suecica. Uppsala. 1929→ .

Acta Soc. Bot. Polon.
Acta Societatis Botanicorum Poloniae. / Organ Polskiego Towarzystwa botanicznego. / Publication de la Société botanique de Pologne. Warszawa. 1923→ .

Acta Soc. Sci. Fenn.
Acta Societatis Scientiarum fennicae. Helsingforsiae. Ser. 1, 1–50, 1840–1926. Nov. ser., B, 1→ , 1931→ .

Acta Univ. Carol. (Biol.)
Acta Universitatis carolinae. Biologica. Praha. 1954→ .

Agron. Lusit.
Agronomia lusitana. Sacavém. 1939→ . (25→ , 1966→ , at Oeiras.)

Allgem. Gartenz.
Allgemeine Gartenzeitung. Berlin. 1833–1856.

Allgem. Thür. Gartenz.
Allgemeine thüringische Gartenzeitung. 1842→ .

Amer. Jour. Bot.
American Journal of Botany. Lancaster, Pennsylvania. 1914→ . (26–36, 1939–1949, at Burlington, Vermont; 37→ , 1950→ , at Baltimore, Maryland.)

Anais Inst. Vinho Porto
Anais do Instituto do Vinho do Porto. Porto. 1940→ .

Anal. Acad. Române
Analele Academiei Române. Bucureşti. Ser. 1, 1, 1–?, 1867—?. Ser. 2, 1–40, 1878–1922. Ser. 3, titled Mem. Secţ. Şti. (Acad. Română), Memoriile Secţiunii ştiinţifice (Academia Română), 1–20, 1923–1944.

Anal. Ci. Nat.
Anales de Historia natural. Madrid. 1–2, 1799–1800; titled Anales de Ciencias naturales, 3–7, 1801–1804.

Anal. Inst. Bot. Cavanilles
Anales del Jardín botánico de Madrid. Madrid. 1–9, 1941–1950; titled Anales del Instituto botánico A. J. Cavanilles, 10→ , 1950→ .

Anal. R. Acad. Farm. (Madrid)
Anales R. Academia de Farmacía. Madrid. 1942→ .

Anal. Soc. Esp. Hist. Nat.
Anales de la Sociedad española de Historia natural. Madrid. **1–30**, 1872–1902 (**21–30** also called Ser. 2, **1–10**).

Animadv. Syst. Herb. Univ. Tomsk.
Animadversiones systematicae ex Herbario Universitatis tomskiensis. | Систематические Заметки по Материалам Гербария Томского Университета [*Sistematičeskie Zametki po Materialam Gerbarija Tomskogo Universiteta*]. Tomsk. 1927 → .

Ann. Acad. Horti-Viticult. (Budapest)
Cf. *Mitt. Kgl. Ungar. Gartenb.-Lehranst.*

Ann. Bot. Fenn.
Annales botanici fennici. Helsinki. 1964 → .

Ann. Bot. (Roma)
Annali di Botanica. Roma. 1903 → .

Ann. Bot. Soc. Zool.-Bot. Fenn. Vanamo
Soumalaisen eläin- ja kasvitieteellisen Seuran Vanamon kasvitietetellisiä Julkaisuja. | *Annales botanici Societatis zoologicae-botanicae fennicae 'Vanamo'.* Helsinki. **1–35**, 1931–1964.

Ann. Bot. (Usteri)
Annalen der Botanik. Herausgegeben von P. Usteri. Zürich. **1–24**, 1791–1800. (**14–24**, 1795–1800, at Leipzig.)

Ann. Gén. Sci. Phys. (Bruxelles)
Annales générales des Sciences physiques. Bruxelles. **1–8**, 1819–1821.

Ann. Hist.-Nat. Mus. Hung.
Annales historico-naturales Musei nationalis hungarici. | *A Magyar nemzeti Muzeum Természetrajzi Osztályainak Folyóirata.* Budapest. Ser. 1, **1–41**, 1903–1948 (33 → , 1940 → , Hungarian title reads *Az Országos Magyar Természettudományi Muzeum Folyóirata.*) Nov. ser., **1–9**, 1951–1958. Vol. nos. then revert to ser. 1, **51** → , 1959 → . (With several variations of Hungarian title.)

Ann. Mus. Hist. Nat. (Paris)
Annales du Muséum d'Histoire naturelle. Paris. **1–20**, 1802–1813.

Ann. Nat. Hist.
Annals of natural History; or, Magazine of Zoology, Botany, and Geology. London. Ser. 1, **1–5**, 1838–1840; titled *The Annals and Magazine of natural History, including Zoology, Botany and Geology,* **6–20**, 1841–1842. Ser. 2, **1–20**, 1843–1857. Ser. 3, **1–20**, 1858–1867. Many later series.

Ann. Naturh. Mus. (Wien)
Annalen des k.k. naturhistorischen Hofmuseums. Wien. **1–31**, 1886–1917; titled *Annalen des naturhistorischen Hofmuseums,* **32**, 1918; titled *Annalen des naturhistorischen Museums in Wien,* **33** → , 1919 → .

Ann. Rep. Missouri Bot. Gard.
Annual Report. Missouri botanical Garden. St Louis. **1–23**, 1890–1912.

Ann. Sci. Acad. Polyt. Porto
Annaes scientificos da Academia polytechnica do Porto. Coimbra. **1–14**, 1905–1922; titled *Anais da Faculdade de Sciéncias do Porto,* Porto, **15–46**, 1927–1963; titled *Anais da Faculdade de Ciências. Universidade do Porto,* Porto, **47** → , 1964 → .

Ann. Sci. Nat.
Annales des Sciences naturelles. Paris. Ser. 1, **1–30**, 1824–1833. Ser. 2, **1–20**, 1834–1843. Ser. 3, **1–20**, 1844–1853. Ser. 4, **1–20**, 1854–1863. Ser. 5, **1–20**, 1864–1874. Ser. 6, **1–20**, 1875–1885. Ser. 7, **1–20**, 1885–1895. Ser. 8, **1–20**, 1895–1904. Ser. 9, **1–20**, 1905–1917. Ser. 10, **1–20**, 1919–1938. Ser. 11, **1–20**, 1939–1959. Ser. 12, **1** → , 1960 → . (Ser. 2–10, *Botanique.* Ser. 11 → , *Botanique et Biologie végétale.*)

Ann. Sci. Nat. (Porto).
Annaes de Sciencias naturaes. Porto. **1–10**, 1894–1906.

Ann. Sect. Horti-Viticult. Univ. Sci. Agr. (Budapest)
Cf. *Mitt. Kgl. Ungar. Gartenb.-Lehranst.*

Ann. Soc. Bot. Lyon
Annales de la Société botanique de Lyon. Lyon. **1–42**, 1873–1922.

Ann. Soc. Linn. Lyon
Annales de la Société linnéenne de Lyon. Lyon. Ser. 1, **1836–1850/52**, 1836–1852. Nov. ser., **1–80**, 1853–1937.

Annu. Cons. Jard. Bot. Genève
Annuaire du Conservatoire et du Jardin botaniques de Genève. Genève. **1–21**, 1897–1922.

Anzeig. Akad. Wiss. (Wien)
Anzeiger der kaiserlichen Akademie der Wissenschaften. Mathematisch-naturwissenschaftliche Classe. Wien. **1–55**, 1864–1918; titled *Anzeiger. Akademie der Wissenschaften in Wien. Mathematisch-naturwissenschaftliche Klasse,* **56–83**, 1919–1946; titled *Anzeiger. Österreichische Akademie der Wissenschaften. Mathematisch-naturwissenschaftliche Klasse,* **84** → , 1947 → .

Arb. Naturf.-Ver. Riga
Arbeiten des naturforschenden Vereins zu Riga. Rudolstadt. Ser. 1, **1**, 1847–1848. Nov. ser., titled *Arbeiten des Naturforscher-Vereins zu Riga,* Riga, **1–21**, 1865–1937.

Arch. Bot. (Forlì)
Archivio botanico. Forlì. 1925 → .

Arch. Bot. (Roemer)
Archiv für die Botanik. (Herausgegeben von D. Johann Jacob Römer.) Leipzig. **1–3**, 1796–1805.

Arch. Gewächsk.
Archiv der Gewächskunde, von Leopold Trattinick. Wien. 1812–1818.

Arch. Naturgesch. (Berlin)
Archiv für Naturgeschichte. Berlin. 1835 → .

Arch. Sci. Phys. Nat. (Genève)
Archives des Sciences physiques et naturelles. (Bibliothèque universelle de Genève.) Genève. Ser. 1, **1–36**, 1846–1857. Ser. 2, **1–64**, 1858–1878. Ser. 3, **1–34**, 1879–1895. Ser. 4, **1–46**, 1896–1918. Ser. 5, **1–29**, 1919–1947. Ser. 6, titled *Archives des Sciences, éditées par la Société de Physique et d'Histoire naturelle de Genève,* **1** → , 1948 → .

Atti Accad. Ital. (Firenze)
Atti dell'Accademia italiana. Firenze. **1–2**, 1808–1810.

Atti Accad. Sci. Torino
Atti della R. Accademia delle Scienze di Torino. Torino. 1866 → .

Bauhinia
Bauhinia. Zeitschrift der basler botanischen Gesellschaft. Basel. 1955 → .

Beih. Bot. Centr.
Beihefte zum botanischen Centralblatt. Cassel, Jena & Dresden. 1891–1943.

Beitr. Pfl. Russ. Reich.
Матеріалы къ ближайшему Познанію Прозябаемости Россійской Имперіи [*Materialy k bližajšemu Poznaniju Prozjabaemosti Rossijskoj Imperii*]. | *Beiträge zur Pflanzenkunde des russischen Reiches.* St Petersburg. **1–11**, 1844–1859.

Ber. Bayer. Bot. Ges.
Berichte der bayerischen botanischen Gesellschaft zur Erforschung der heimischen Flora. München. 1891 → .

Ber. Deutsch. Bot. Ges.
Berichte der deutschen botanischen Gesellschaft. Berlin. 1883 → . (**59–61**, 1941–1944, at Jena; **62** → , 1949 → , at Stuttgart.)

Ber. Naturw. Ver. Innsbruck
Berichte des naturwissenschaftlich-medizinischen Vereines in Innsbruck. Innsbruck. 1870 → .

Ber. Schweiz. Bot. Ges.
Berichte der schweizerischen botanischen Gesellschaft. | *Bulletin de la Société botanique suisse.* Basel & Genf. 1891 → . (**3–18**, 1893–1909, at Bern; **19–29**, 1910–1920, at Zurich; **30** → , 1922 → at Bern.)

Bibliot. Ital.
Biblioteca italiana, ossia Giornale di Letteratura, Scienze ed Arti. Milano. **1–100**, 1816–1840.

Biblioth. Bot. (Stuttgart)
Bibliotheca botanica. Abhandlungen aus dem Gesammtgebiete der Botanik. Cassel. 1886 → . (**28** → , 1894 →, at Stuttgart.)

Biblioth. Univ. Genève
Bibliothèque universelle des Sciences, Belles-lettres, et Arts, faisant suite à la Bibliothèque britannique. Partie des Sciences. Genève. Ser. 1, **1–60**, 1816–1835. Ser. 2, **1–60**, 1836–1845.

Billotia
Billotia, ou Notes de Botanique. Besançon. **1**, 1864–1869.

Biol. Meddel. Kong. Danske Vid. Selsk.
Biologiske Meddelelser udgivet af det kongelige danske Videnskabernes Selskab. København. 1917 → .

Bjull. Glavn. Bot. Sada
Бюллетень главного ботанического Сада [*Bjulleten' glavnogo botaničeskogo Sada*]. Moskva & Leningrad. 1948 → .

Boissiera
Boissiera. Genève. 1936 → . (Supplement of *Candollea*.)

Bol. Inst. Estud. Astur. (Supl. Ci.)
Boletín del Instituto de Estudios asturianos (Suplemento de Ciencias). Oviedo.

Bol. Soc. Aragon. Ci. Nat.
Boletin de la Sociedad aragonesa de Ciencias naturales. Zaragoza. **1–17**, 1902–1918; titled *Bol. Soc. Ibér. Ci. Nat.*, Boletin de la Sociedad ibérica de Ciencias naturales, **18–33**, 1919–1934.

Bol. Soc. Brot.
Boletim da Sociedade broteriana. Coimbra. Ser. 1, **1–28**, 1880–1920. Ser. 2, **1** → , 1922 → .

Bol. Soc. Esp. Hist. Nat.
Boletin de la real Sociedad española de Historia natural. Madrid. 1901 → .

Bol. Soc. Ibér. Ci. Nat.
Cf. *Bol. Soc. Aragon. Ci. Nat.*

Bol. Soc. Port. Ci. Nat.
Cf. *Bull. Soc. Port. Sci. Nat.*

Bonplandia
Bonplandia. Zeitschrift für die gesammte Botanik. Hannover, London, New York & Paris. **1–10**, 1853–1862.

Boston Jour. Nat. Hist.
Boston Journal of natural History. Boston. **1–7**, 1837–1863.

Bot. Gaz.
Botanical Bulletin. Hanover, Indiana, etc. **1**, 1875–1876; titled Botanical Gazette, **2** → , 1876 → . (**22** → , 1896 → , at Chicago, Illinois.)

Bot. Jahrb.
Botanische Jahrbücher für Systematik, Pflanzengeschichte und Pflanzengeographie. Leipzig. 1880 → . (**69** → , 1938 → , at Stuttgart.)

Bot. Közl.
Cf. *Növ. Közl.*

Bot. Mag.
The botanical Magazine or Curtis's botanical Magazine. London. 1793 → . (For publication dates of **1–6** cf. F. A. Stafleu, *Taxon* **12**: 56 (1963).) Continuous volume numbers are used in citations and series are ignored.

Bot. Mag. Tokyo
The botanical Magazine. Tokyo. **1–19**, 1887–1905; titled The botanical Magazine published by the Tokyo botanical Society, **20–45**, 1906–1931; titled The botanical Magazine published by the botanical Society of Japan, **46** → . 1932 → . (**68** → , 1955 → , also titled The botanical Magazine, Tokyo.)

Bot. Not.
Botaniska Notiser. Lund. 1839 → .

Bot. Reg.
The botanical Register. London. **1–14**, 1815–1829; titled Edwards's botanical Register, **15–33**, 1829–1847. (For dates of publication cf. E. M. Tucker, *Jour. Arnold Arb.* **18**: 183–184 (1937) and M. J. van Steenis-Kruseman & W. T. Stearn in C. G. G. J. van Steenis, *Fl. Males.* I. 4: clxxx (1954).)

Bot. Tidsskr.
Botanisk Tidsskrift. Kjøbenhavn. 1866 → .

Bot. Zeit.
Botanische Zeitung. Berlin. **1–68**, 1843–1910. (**14–68**, 1856–1910, at Leipzig.)

Botaniker (Halle)
Der Botaniker; oder compendiöse Bibliothek alles Wissenswürdigen aus dem Gebiete der Botanik. Halle, Gotha & Eisenach. 1793–1797.

Brit. Flower Gard.
The British Flower Garden. . .by Robert Sweet. London. Ser. 1, **1–3**, 1823–1829. Ser. 2, **1–4**, 1829–1838.

Brotéria (Bot.)
Brotéria. Lisboa. 1902 → . Série botânica, **6–25**, 1907–1931. Série Ciências naturias, **1** → , 1932 → .

Bul. Fac. Şti. Cernăuţi
Buletinul Facultăţii de Ştiinţe din Cernăuţi. Cernăuţi. **1–8**, 1927–1937.

Bul. Grăd. Bot. Cluj
Buletinul de Informaţii al Grădinii botanice şi al Muzeului botanic de la Universitatea din Cluj. Cluj. **1–5**, 1921–1925; titled Buletinul Grădinii botanice şi al Muzeului botanic de la Universitatea din Cluj, **6–28**, 1926–1948, with minor variations of title.

Bull. Acad. Imp. Sci. Pétersb.
Bulletin de l'Académie impériale des Sciences de St.-Pétersbourg. St.-Pétersbourg. Ser. 1, **1–32**, 1860–1888. Ser. 2, **1–2**, 1889–1892. Different series and titles later.

Bull. Acad. Int. Géogr. Bot. (Le Mans)
Cf. *Monde Pl.*

Bull. Appl. Bot. Pl.-Breed. (Leningrad)
Труды по прикладной Ботанике (Генетике) и Селекции [*Trudy po prikladnoj Botanike (Genetike) i Selekcii*]. / Bulletin of applied Botany and Plant Breeding. Leningrad. 1922–1939. Several concurrent series.

Bull. Assoc. Fr. Bot.
Bulletin de l'Association française de Botanique. Le Mans. **1–5**, 1898–1902.

Bull. Assoc. Pyr.
Bulletin de l'Association pyrénéenne pour l'Échange des Plantes. Poitiers. **1–14**, 1891–1904.

Bull. Assoc. Russe Sci. Prague
Записки научно-изслѣдовательскаго Объединения, Русскій свободный Университет в Прагѣ [*Zapiski naučno-issledovatel'skago Ob"edinenija, Russkij svobodnyj Universitet v Pragě*]. / Rozpravy Vědecké společnosti Badatelské při Ruské svobodné Universitě v Praze. / Bulletin de l'Association russe pour les Recherches scientifiques à Prague. Praga. 1935 → .

Bull. Brit. Mus. (Bot.)
Bulletin of the British Museum (Natural History). Botany. London. 1951 → .

Bull. Herb. Boiss.
Bulletin de l'Herbier Boissier. Genève & Bâle. Ser. 1, **1–7**, 1893–1899. Ser. 2, **1–8**, 1900–1909.

Bull. Inst. Jard. Bot. Univ. Beograd
Гласник ботаничког Завода и Баште Универзитета у Београду [*Glasnik botaničkog Zavoda i Bašte Univerziteta u Beogradu*]. / Bulletin de l'Institut et du Jardin botaniques de l'Université de Beograd. Beograd. **1–4**, 1928–1937.

Bull. Int. Acad. Sci. Cracovie
Bulletin international de l'Académie des Sciences de Cracovie. Classe des Sciences mathématiques et naturelles. Série B: Sciences naturelles. | Anzeiger der Akademie der Wissenschaften in Krakau. Mathematisch-naturwissenschaftliche Klasse. Reihe B: Biologische Wissenschaften. Cracovie. 1889–1919. Dates are used as vol. nos. Continued as *Bulletin international de l'Académie polonaise des Sciences et des Lettres,* 1920–1953.

Bull. Jard. Bot. Bruxelles
Bulletin du Jardin botanique de l'État à Bruxelles. Bruxelles. 1–36, 1902–1966. (15–36, 1938–1966, with alternative title *Bulletin van den [de] Rijksplantentuin Brussel);* titled **Bull. Jard. Bot. Nat. Belg.,** *Bulletin du Jardin botanique national de Belgique. Bulletin van de nationale Plantentuin van Belge,* 37→ , 1967→ .

Bull. Jard. Bot. Pétersb.
Bulletin du Jardin impérial botanique de St.-Pétersbourg. | Извѣстія императорскаго С.-Петербургскаго ботаническаго Сада [*Izvěstija imperatorskago S.-Peterburgskago botaničeskago Sada*]. S.-Peterburg. 1–12, 1901–1912; titled *Bulletin du Jardin impérial botanique de Pierre le Grand. |* Извѣстія императорскаго ботаническаго Сада Петра Великаго [*Izvěstija imperatorskago botaničeskago Sada Petra Velikago*], 13–17, 1912–1917; titled **Bull. Jard. Bot. URSS,** *Bulletin du principal Jardin botanique de la République Russe. |* Известия главного ботанического Сада Р.С.Ф.С.Р. [*Izvestija glavnogo botaničeskogo Sada R.S.F.S.R.*], 18–22, 1918–1923; titled *Bulletin du Jardin botanique de la République Russe.* /[Russian title as before], 23–24, 1924–1925; titled *Bulletin du Jardin Botanique principal de l'U.R.S.S. |* Известия главного ботанического Сада С.С.С.Р. [*Izvestija glavnogo botaničeskogo Sada S.S.S.R.*], 25–29, 1926–1930.

Bull. Jard. Bot. URSS
Cf. *Bull. Jard. Bot. Pétersb.*

Bull. Mus. Hist. Nat. Pays Serbe
Гласник Природњачког Музеја Српске Земље [*Glasnik Prirodnjačkog Muzeja Srpske Zemlje.*] | *Bulletin du Muséum d'Histoire naturelle du Pays Serbe.* Beograd. Ser. B, 1–10, 1949–1957.

Bull. Nat. Mus. Can.
Bulletin. Victoria Memorial Museum. Geological Survey. Ottawa. 1, 1913; titled *Museum Bulletin. Geological Survey,* 2–49, 1914–1928; titled *Bulletin. National Museum of Canada.* 50→ , 1928→ .

Bull. Orto Bot. Napoli
Bullettino dell'Orto botanico della R. Università di Napoli. Napoli. 1–17, 1899–1947.

Bull. Phys.-Math. Acad. Petersb.
Bulletin de la Classe physico-mathématique de l'Académie impériale des Sciences. St. Pétersbourg & Leipzig. 1–17, 1843–1859.

Bull. Res. Counc. Israel
Bulletin of the Research Council of Israel. Jerusalem. 1–11 (from 5 onwards *Botany* in Section D), 1951–1963; titled **Israel Jour. Bot.** *Israel Journal of Botany.* 12→ , 1963→ .

Bull. Sci. Acad. Imp. Sci. Pétersb.
Bulletin scientifique publié par l'Académie impériale des Sciences de Saint-Pétersbourg. Saint-Pétersbourg & Leipzig. 1–10, 1836–1842.

Bull. Soc. Bot. Belg.
Bulletin de la Société royale de Botanique de Belgique. Bruxelles. 1862→ .

Bull. Soc. Bot. Bulg.
Извѣстия на българското ботаническо Дружество [*Izvěstija na bălgarskoto botaničesko Družestvo*]. | *Bulletin de la Société botanique de Bulgarie.* Sofija. 1–9, 1926–1943.

Bull. Soc. Bot. Fr.
Bulletin de la Société botanique de France. Paris. 1854→ .

Bull. Soc. Bot. Genève
Bulletin des Travaux de la Société botanique de Genève. Genève. Ser. 1, 1–11, 1879–1905. Ser. 2, titled *Bulletin de la Société botanique de Genève,* 1–43, 1909–1952.

Bull. Soc. Bot. Ital.
Bullettino della Società botanica italiana. Firenze. 1892–1926, 1892–1926.

Bull. Soc. Dauph. Éch. Pl.
Bulletin. Société dauphinoise pour l'Échange des Plantes. Grenoble. Ser. 1, 1–16, 1874–1889 (with continuous pagination). Ser. 2, 1–3, 1890–1892 (with continuous pagination).

Bull. Soc. Étud. Sci. Angers
Bulletin de la Société d'Études scientifiques d'Angers. Angers. 1–89, 1872–1959; titled *Bulletin de la Société d'Études scientifiques de l'Anjou,* 90→ , 1960→ .

Bull. Soc. Hist. Nat. Afr. Nord
Bulletin de la Société d'Histoire naturelle de l'Afrique du Nord. Alger. 1909→ .

Bull. Soc. Nat. Ain
Bulletin de la Société des Naturalistes et des Archéologues de l'Ain. Bourg. 1886→ .

Bull. Soc. Nat. Moscou
Bulletin de la Société impériale des Naturalistes de Moscou. Section biologique. Moscou. Ser. 1, 1–62, 1829–1886. Nov. ser., 1→ , 1887→ . (Nov. ser., 31→ , 1922→ , with alternative title in Russian Бюллетень императорскаго Московскаго Общества испытателей Природы [*Bjulleten' imperatorskago Moskovskago Obščestva ispytatelej Prirody*]. Nov. ser., 52(5)→ , 1947→ , with only Russian title.)

Bull. Soc. Oural. Sci. Nat.
Записки Уральскаго Общества Любителей Естествознанія [*Zapiski Ural'skago Obščestva Ljubitelej Estestvoznanija*]. | *Bulletin de la Société ouralienne d'Amateurs | Amis des Sciences naturelles.* Ekaterinburg. 1–40, 1873–1927.

Bull. Soc. Port. Sci. Nat.
Bulletin de la Société portugaise des Sciences naturelles. Lisbonne. Ser. 1, 1–14, 1907–1943. Ser. 2, with alternative title **Bol. Soc. Port. Ci. Nat.,** *Boletim da Sociedade portuguesa de Ciências naturais,* 1 ›, 1944→ .

Bull. Soc. Sci. Dauph.
Bulletin de la Société de Statistiques des Sciences naturelles et des Arts industriels du Département de l'Isère. Grenoble. Ser. 1, 1–4, 1838 or 1840–1846 or 1848. Ser. 2, 1–7, 1851–1864. Ser. 3, 1–15, 1892–1920. Titled *Bulletin de la Société scientifique de l'Isère, ancienne Société de Statistique des Sciences naturelles et des Arts industriels,* 42(ser. 5, 1)–45(ser. 5, 4), 1921–1924. Titled *Bulletin de la Société scientifique de l'Isère, ancienne Société de Statistique des Sciences naturelles et des Arts industriels du Département de l'Isère,* 46 (ser. 5, 5)–60 (ser. 5, 18) 1925–1944; 61→ (ser. 6, 1–), 1945→ .

Bull. Soc. Vaud. Sci. Nat.
Bulletin de la Société vaudoise des Sciences naturelles. Lausanne. 1842→ .

Bull. Torrey Bot. Club
Bulletin of the Torrey botanical Club. New York. 1870→. (73–78, 1946–1951, titled *Bulletin of the Torrey botanical Club and Torreya.*)

Bull. Trav. Soc. Murith.
Bulletins des Travaux de la Société murithienne. Aigle. 1–5, 1868–1876. (2, 1873, at Genève; 3, 1875, at Sion.)

Bull. Univ. Asie Centr.
Bulletin de l'Université de l'Asie centrale (Tachkent). | Бюллетень Средне-Азиатского Государственного Университета [*Bjulleten' Sredne-Aziatskogo Gosudarstvennogo Universiteta*]. Taškent. 1923–1945.

Butll. Inst. Catalana Hist. Nat.
Butlletí de la Institució catalana d'Història natural. Barcelona. **1–37**, 1901–1949.

Canad. Jour. Bot.
Canadian Journal of Botany. Ottawa. **29** → , 1951 → . (Formerly *Canadian Journal of Research, Sect. C, Botanical Sciences*.)

Canad. Jour. Genet. Cytol.
Canadian Journal of Genetics and Cytology. | Journal canadien de Génétique et de Cytologie. Ottawa. 1959 → .

Candollea
Candollea. Organe du Conservatoire et du Jardin botaniques de la Ville de Genève. Genève. 1922 → .

Cavanillesia
Cavanillesia. Rerum botanicarum Acta. Barcinone. **1–8**, 1928–1938.

Collect. Bot. (Barcelona)
Collectanea botanica a barcinonensi botanico Instituto edita. Barcinone. 1946 → .

Comment. Ateneo Brescia
Commentarj della Accademia di Scienze, Lettere, Agricultura, ed Arti del Dipartimento del Mella. Brescia. **1808–1811**, 1808–1812; titled *Commentarj dell'Ateneo di Brescia*, **1812–1886**, 1814–1886.

Compt. Rend. Acad. (Paris)
Compte rendu hebdomadaire des Séances de l'Académie des Sciences. Paris. 1835 → .

Compt. Rend. Congr. Soc. Sav. (Sci.)
Comptes rendus du Congrès des Sociétés savantes de Paris et des Départements. Section des Sciences. Paris. 1896 → .

Compt. Rend. Soc. Bot. Rochel.
Comptes-rendus des Excursions botaniques. Société botanique rochelaise. La Rochelle. **1–4**, 1879–1882; titled *Comptes-rendus, Descriptions, Notes et Communications*, **5–10**, 1883–1889; titled ***Bull. Soc. Bot. Rochel.***, *Bulletin. Société botanique rochelaise*, **11–24**, 1890–1903.

Compt. Rend. Soc. Hallér.
Compte-rendu des Travaux de la Société hallérienne. Genève. **1–4**, 1852–1856. (**1**: pp. 1–12 in 1852–1853; **2**: pp. 13–76 in 1853–1854; **3**: pp. 77–90 in 1854–1855; **4**: pp. 93–184 in 1854–1856.)

Congr. Sci. Fr.
Sessions des Congrès scientifiques de France. Rouen, Paris, etc. 1833–1878.

Contr. U.S. Nat. Herb.
Contributions from the United States National Herbarium. Washington. 1890 → .

Cytologia
Cytologia. International Journal of Cytology. | Internationale Zeitschrift für Zytologie. | Archives internationales de Cytologie. Tokyo. 1929 → .

Dansk Bot. Ark.
Dansk botanisk Arkiv. København. 1913 → .

Denkschr. Akad. Wiss. Math.-Nat. Kl. (Wien)
Denkschriften der kaiserlichen Akademie der Wissenschaften. Mathematisch-naturwissenschaftlichen Classe. Wien. **1–95**, 1850–1918; titled *Denkschriften. Akademie der Wissenschaften in Wien. Mathematisch-naturwissenschaftlichen Klasse*, **96–107**, 1919–1943; titled *Denkschriften. Österreichische Akademie der Wissenschaften. Mathematisch-naturwissenschaftliche Klasse*, **108** → , 1947 → .

Denkschr. Bayer. Bot. Ges. Regensb.
Denkschriften der königlichen-bayerischen botanischen Gesellschaft in Regensburg. Regensburg. **1–22**, 1815–1946, with minor variations of title. (**7–22**, 1898–1946, also called nov. ser., **1–15**.)

Deutsche Bot. Monatsschr.
Deutsche botanische Monatsschrift. Sondershausen. **1–22**, 1883–1912. (**6–22** at Arnstadt.)

Diana
La Diana. Montbrison. 1797–1816. (?Also a later series, **1–2**, 1863–1865.)

Diar. Med. Flajiani
Diaria medico di Flajiani.

Dict. Sci. Nat.
Dictionnaire des Sciences naturelles. Paris. **1–60**, 1804–1830.

Doc. Cartes Vég.
Documents pour les Cartes des Productions végétales. Paris. (Several series.)

Dokl. Akad. Nauk SSSR
Доклады Академии Наук СССР [*Doklady Akademii Nauk SSSR*]. Moskva & Leningrad.

Edinb. Philos. Jour.
The Edinburgh philosophical Journal. Edinburgh. **1–14**, 1819–1826.

Erdész. Lapok
Erdészeti Lapok. Budapest. 1862–1950.

Feddes Repert.
Repertorium Specierum novarum Regni vegetabilis. Berlin. **1–51**, 1905–1942; titled *Feddes Repertorium Specierum novarum Regni vegetabilis*, **52** → , 1943 . (This work includes *Repertorium europaeum et mediterraneum*, indicated by dual pagination.) ***Beih.***, *Beihefte*. Berlin. **1** → , 1914 → .

Fl. Serres Jard. Eur.
Flore des Serres et des Jardins de l'Europe. Gand. **1–23**, 1845–1880. (**1** in 1845; **2** in 1846; **3** in 1847; **4** in 1848; **5** in 1849; **6** in 1851; **7** in 1851–1852; **8** in 1853; **9** in 1854; **10** in 1855; **11** in 1856; **12** in 1857; **13** in 1858; **14** in 1859; **15** in 1865; **16** in 1865–1867; **17** in 1868–1869; **18** in 1869–1870; **19** in 1873; **20** in 1874; **21** in 1875; **22** in 1879; **23** in 1880.) With minor variations of title. Earlier volumes have no page numbers, but nearly every leaf is numbered, and these are treated as page numbers, sometimes with a & b added.

Fl. Syst. Pl. Vasc.
*Flora et Systematica Plantae vasculares. Acta Instituti botanici Academiae Scientiarum Unionis Rerum publicarum sovieticarum socialisticarum. | *Флора и Систематика высших Растений. Труды ботанического Института Академии Наук СССР [*Flora i Sistematika vysšikh Rastenij. Trudy botaničeskogo Instituta Akademii Nauk CCCP*]. Leningrad. 1933 → . (**2** → , 1936 → , at Mosqua & Leningrad.)

Flora (Regensb.)
Flora oder allgemeine botanische Zeitung. Regensburg. 1818 → . (**72–95** at Marburg; **96** → at Jena.)

Folia Geobot. Phytotax. (Praha)
Folia geobotanica & phytotaxonomica bohemoslovaca. Praha. 1966 → .

Fragm. Fl. Geobot.
Fragmenta floristica et geobotanica. Kraków. 1945 → .

Gard. Chron.
Gardeners' Chronicle and agricultural Gazette. London. Ser. **1**, **1841–1873**, 1841–1873. Ser. **2**, titled *The Gardeners' Chronicle. A weekly illustrated Journal of Horticulture and allied Subjects*, **1–26**, 1874–1886. Ser. 3, **1** → , 1887. (**140–154**, 1956–1963, titled *Gardeners' Chronicle and Gardening illustrated*; **155** → , 1964 → , titled *Gardeners' Chronicle, Gardening illustrated and the Greenhouse*.)

Garten-Zeit. (Wittmack)
Garten-Zeitung. Monatsschrift für Gärtner und Gartenfreunde. Herausgegeben von Dr. L. Wittmack. Berlin. **1–2**, 1882–1883; titled *Garten-Zeitung. Wochenschrift für Gärtner und Garten-*

freunde, **3–4**, 1884–1885; titled *Deutsche Garten-Zeitung. Wochenschrift für Gärtner und Gartenfreunde.* **1886**, 1886.

Gentes Herb.
Gentes Herbarum. Occasional Papers on the Kinds of Plants. Ithaca, N.Y. 1920 → .

Ges. Naturf. Freunde Berlin Mag.
Der Gesellschaft naturforschender Freunde zu Berlin, Magazin für die neuesten Entdeckungen in der gesammten Naturkunde. Berlin. **1–8**, 1807–1818.

Ges. Naturf. Freunde Berlin Neue Schr.
Der Gesellschaft naturforschender Freunde zu Berlin, neue Schriften. Berlin. 1795–1803.

Gior. Bot. Ital.
Giornale botanico italiano. Firenze. **1–2**, 1844–1852.

Gior. Ital. Sci. Nat. Agric. Arti Commerc.
Giornale d'Italia, spettante alla Scienza naturale, e principalmente all' Agricoltura, alle Arti, ed al Commercio. Venezia. **1**, 1764–1765.

Gior. Sci. Sic.
Giornale di Scienze, Lettere ed Arti per la Sicilia. Palermo. **1–79**, 1823–1842.

Glasn. Muz. Bosni Herceg.
Гласник Земаљског Музеја у Босни и Херцеговини [*Glasnik Zemaljskog Muzeja u Bosni i Hercegovini*]. Sarajevo. **1–52**, 1889–1940.

God. Sof. Univ. (Agron.-Les. Fak.)
Годишник на Софийския Университет [*Godišnik na Sofijskija Universitet*]. / *Annuaire de l'Université de Sofia.* Sofija. 1904 → . From 1909 in sections.

Hannover. Mag.
Hannoverisches Magazin. Hannover. 1764–1820. Several series; often with many volumes published in same year.

Hereditas
Hereditas. Genetiskt Arkiv. Lund. 1920 → .

Hung. Acta Biol.
Hungarica Acta biologica. Budapest. **1**, 1948–1950.

Ind. Horti Bot. Univ. Budapest.
Index Horti botanici Universitatis budapestinensis. Budapest. **1–6**, 1932–1943.

Isis
Isis oder encyclopädische Zeitung von Oken. Jena. 1817–1819; titled *Isis von Oken*, 1820–1827. (1828–1848, at Leipzig.)

Israel Jour. Bot.
Cf. *Bull. Res. Counc. Israel.*

Izv. Kavk. Obšč. Ljub. Est.
Извѣстія кавказкаго Общества Любителей Естествознанія [*Izvestija kavkazkago Obščestva Ljubitelej Estestvoznanija*]. Tbilisi. 1879–?.

Jahrb. Hamb. Wiss. Anst.
Jahrbuch der hamburgischen wissenschaftlichen Anstalten. Hamburg. 1884–1921.

Jahresb. Naturf. Ges. Graubündens
Jahresbericht der naturforschenden Gesellschaft Graubündens. Neue Folge. Chur. 1856 → . The first series appeared in the *Bündnerische Volksblatt.*

Jahresb. Schles. Ges. Vaterl. Kult.
Uebersicht der Arbeiten und Veränderungen der schlesischen Gesellschaft für vaterländische Cultur /Kultur. Breslau. **1–28**, 1825–1850; titled *Jahresbericht /Jahres-Bericht der schlesischen Gesellschaft für vaterländische Kultur /Cultur,* **28** → , 1851 → .

Jahresb. Staats-Gymnas. Laibach
Jahresbericht des k.k. i. Staatsgymnasiums zu Laibach. Laibach.

Jahres-Kat. Wien. Bot. Tauschver.
Jahres-Katalog pro ... des wiener botanischen Tauschvereins. Wien. 1894–?.

Jap. Jour. Bot.
Japanese Journal of Botany. Tokyo. 1923 → .

Jour. Arnold Arb.
Journal of the Arnold Arboretum. Cambridge, Massachusetts. 1919 → . (**2(3)–13**, 1921–1932, at Lancaster, Pennsylvania; **14–26(2–3)**, 1933–1955, at Jamaica Plain, Massachusetts.)

Jour. Bot. (London)
The Journal of Botany, British and foreign. London. **1–80**, 1863–1942.

Jour. Bot. (Paris)
Journal de Botanique. Paris. Ser. 1, **1–20**, 1887–1906. Ser. 2, **1–3**, 1907–1925.

Jour. Ecol.
The Journal of Ecology. Cambridge. 1913 → . (**44** → , 1956 → , at Oxford.)

Jour. Jap. Bot.
The Journal of Japanese Botany. Tokyo. 1916 → .

Jour. Linn. Soc. London (Bot.)
The Journal of the Proceedings of the Linnean Society. Botany. London. **1–7**, 1856–1864; titled *The Journal of the Linnean Society. Botany,* **8–46**, 1865–1924; titled *The Journal of the Linnean Society of London. Botany,* **47** → , 1925 → .

Jour. Roy. Hort. Soc.
The Journal of the Royal Horticultural Society. London. 1865 → .

Jour. Washington Acad. Sci.
Journal of the Washington Academy of Sciences. Washington. 1911 → .

Kew Bull.
Bulletin of miscellaneous Information. Royal Gardens, Kew. London. **1887–1941**, 1887–1942; titled *Kew Bulletin,* **1946** → , 1946 → . Volume numbers (**13** →) are given only from 1958.

Kong. Danske Vid. Selsk. Skr.
Cf. *Skr. Kiøbenhavnske Selsk. Laerd. Vid.*

Kulturpfl.
Die Kulturpflanze. Berichte und Mitteilungen aus dem Institut für Kulturpflanzenforschung der deutschen Akademie der Wissenschaften zu Berlin. Berlin. 1953 → . **Beih.**, *Beihefte.* 1956 → .

Kungl. Svenska Vet.-Akad. Handl.
Kongl. svenska Vetenskaps Academiens Handlingar. Stockholm. Ser. 1, **1–40**, 1739–1779 (*svenska* omitted after 1755). Nov. ser., titled *Kongl. Vetenskaps Academiens nya Handlingar,* **1–33**, 1780–1812; titled *Kongl. Vetenskaps Academiens Handlingar,* **1813–1846**, 1813–1846; titled *Kongl. Vetenskaps-Akademiens Handlingar,* **1846–1858**, 1846–1858. Nov. ser., titled *Kongliga svenska Vetenskaps-Akademiens Handlingar. Ny Földj,* **1–35**, 1855–1902; titled *Kungliga svenska Vetenskaps-Akademiens Handlingar,* **36–63**, 1902–1923. Ser. 3, **1–25**, 1924–1948. Ser. 4, **1** → , 1951 → .

Linnaea
Linnaea. Ein Journal für die Botanik in ihrem ganzen Umfange. Berlin. **1–43**, 1826–1882. (**9–34**, 1834–1866, at Halle.) (**1** in 1826; **2** in 1827; **3** in 1828; **4** in 1829; **5** in 1830; **6** in 1831; **7** in 1832; **8** in 1833; **9**: pp. 1–402 in 1834; pp. 403–758 in 1835; **10**: pp. 1–368 in 1835; pp. 369–758 in 1836; **11** in 1837; **12** in 1838; **13** in 1839; **14**: pp. 1–528 in 1840; pp. 529–728 in 1841; **15** in 1841; **16** in 1842; **17**: pp. 1–640 in 1843; pp. 641–764 in 1844; **18**: pp. 1–512 in 1844; pp. 513–774 in 1845; **19**: pp. 1–512 in 1846; pp. 513–765 in 1847; **20** in 1847; **21** in 1848; **22** in 1849; **23** in 1850; **24**: pp. 1–640 in 1851; pp. 641–804 in 1852; **25**: pp. 1–256 in 1852; pp. 257–772 in 1853; **26**: pp. 1–384 in 1854; pp. 385–807 in 1855; **27**: pp. 1–128 in 1855; pp. 129–799 in 1856; **28**: pp. 1–256 in 1856; pp. 257–640 in 1857; pp. 641–767 in 1858; **29**: pp. 1–384 in 1858; pp. 385–764 in 1859; **30**: pp. 1–256 in 1859; pp. 257–640 in 1859–

1860; pp. 641–779 in 1861; **31** in 1861–1862; **32** in 1863; **33** in 1864–1865; **34** in 1865–1866; **35**: pp. 1–512 in 1867–1868; pp. 513–637 in 1868; **36**: pp. 1–256 in 1869; pp. 257–790 in 1870; **37**: pp. 1–544 in 1872; pp. 545–663 in 1873; **38**: pp. 1–144 in 1873; pp. 145–753 in 1874; **39** in 1875; **40** in 1876; **41**: pp. 1–117 in 1876; pp. 118–576 in 1877; pp. 577–655 in 1878; **42**: pp. 1–192 in 1878; pp. 193–667 in 1879; **43**: pp. 1–66 in 1880; pp. 67–252 in 1881; pp. 253–554 in 1882. Cf. R. C. Foster, *Jour. Arnold Arb.* **43**: 400–409 (1962).)

London Jour. Bot. (Hooker)
The London Journal of Botany. London. **1–7**, 1842–1848. (**1** in 1842; **2** in 1843; **3** in 1844; **4** in 1845; **5** in 1846; **6** in 1847; **7** in 1848.)

Lunds Univ. Årsskr.
Lunds Universitets Årsskrift. / Acta Universitatis lundensis. Lund. Ser. 1, **1–40**, 1864–1902. Nov. ser., **1–59**, 1905–1963. Sectio 2, Medica, Mathematica, Scientiae Rerum naturalium, **1→** , 1964→ .

Madroño
Madroño. Journal of the California botanical Society. Berkeley, California. **1–2**, 1916–1934; titled *Madroño. A West American Journal of Botany*, **3→** , 1935→ .

Magyar Biol. Int. Munkái
Archivum balatonicum. Tihany. **1**, 1927; titled *A magyar biologiai Kutató Intézet Munkái.* / *Arbeiten des ungarischen biologischen Forschungs-Institutes*, **2–16**, 1928–1946. Continued under other titles.

Magyar Bot. Lapok
Magyar botanikai Lapok. / Ungarische botanische Blätter. Budapest. **1–33**, 1902–1934.

Magyar Növ. Lapok
Magyar növénytani Lapok. Kolozsvár. **1–15**, 1877–1892.

Malpighia
Malpighia. Rassegna mensuale di Botanica. Messina. **1–34**, 1886–1937. (**3–23**, 1889–1909, at Genova; **24–29**, 1911–1923, at Catania; **30–31**, 1927–1928, at Palermo; **32–34**, 1932–1937, at Bologna.)

Mat. Étude Fl. Géogr. Bot. Or.
Matériaux pour servir à l'Étude de la Flore et de la Géographie botanique de l'Orient (Missions du Ministère de l'Instruction publique en 1904 et en 1906.) Nancy. **1–7**, 1906–1922.

Meddel. Soc. Fauna Fl. Fenn.
Meddelanden af Societas pro Fauna et Flora fennica. Helsingfors. **1–50**, 1876–1925.

Mélang. Philos. Math. Soc. Roy. Turin (Misc. Taur.)
Miscellanea philosophica-mathematica Societatis privatae taurinensis. Turin. **1**, 1759; titled *Mélanges de Philosophie et de Mathématique de la Société royale de Turin. Miscellanea taurinensia.* **2–5**, 1760–1774.

Mem. Acad. Ci. Artes Barcelona
Memorias de la real Academia de Ciencias naturales y Artes de Barcelona. Barcelona. Ser. 1, **1**, 1853–1875. Ser. 2, **1**, 1876–1884; titled *Memorias de la real Academia de Ciencias de Barcelona*, **2**, 1885. Ser. 3, titled *Memorias de la real Academia de Ciencias y Artes*, **1→** , 1892→ .

Mém. Acad. Sci. Pétersb.
Записки имп. Академіи Наукъ (по физико-математическому Отдѣленію) [*Zapiski imp. Akademii Nauk (po fiziko-matematičeskomu Otděleniju)*]. / *Mémoires de l'Académie impériale des Sciences de St. Pétersbourg (Classe des Sciences physiques et mathématiques).* St. Pétersbourg. Ser. 5, **1–11**, 1803–1822. Ser. 6, **1–10**, 1831–1859. Ser. 7, **1–42**, 1859–1897.

Mém. Acad. Sci. (Turin)
Mémoires de l'Académie royale des Sciences. Turin. Ser. 1, **1–22**, 1784–1816; titled **Mem. Accad. Sci. Torino**, Memorie della reale Accademia delle Scienze di Torino, **23–40**, 1818–1838

(each volume contains different classes etc.). Ser. 2, **1→** , 1839→ .

Mém. Acad. Toulouse
Histoire et Mémoires de l'Académie royale des Sciences, Inscriptions et Belles-lettres de Toulouse. Toulouse. Ser. 1, **1–4**, 1782–1792. Ser. 2, **1–6**, 1827–1843. Ser. 3, titled *Mémoires de l'Académie royale des Sciences, Inscriptions et Belles-lettres de Toulouse*, **1–6**, 1844–1850. A further 9 series, with numerous minor changes of title, follow.

Mem. Accad. Ist. Bologna
Memorie della Accademia delle Scienze dell'Istituto di Bologna. Bologna. Ser. 1, **1–12**, 1850–1862. Ser. 2, **1–10**, 1862–1871. Ser. 3, **1–10**, 1871–1880. Ser. 4, **1–10**, 1880–1889. Ser. 5, **1–10**, 1890–1904. Ser. 6, **1–10**, 1905–1914. Ser. 7, **1–10**, 1914–1923. Ser. 8, **1–10**, 1924–1933. Ser. 9, **1–10**, 1934–1943. Ser. 10, **1–10**, 1944–1953. Ser. 11, **1–10**, 1954–1963. Ser. 12, **1→** , 1964→ .

Mem. Accad. Sci. Torino
Cf. *Mém. Acad. Sci. (Turin)*

Mem. Fac. Agric. Kagawa Univ.
Memoirs of the Faculty of Agriculture, Kagawa University. Kagawa-ken. **1→** , 1955→ .

Mem. Ist. Veneto
Memorie dell' i. r. Istituto Veneto di Scienze, Lettere ed Arti. Venezia. **1–13**, 1843–1866; titled *Memorie del reale Istituto Veneto di Scienze, Lettere ed Arti*, **14→** , 1868→ .

Mem. Mat. Fis. Soc. Ital. Sci.
Memorie di Matematica e Fisica della Società italiana. Verona. Ser. 1, **1–25**, 1782–1852. (**9→** , ...*delle Scienze* added to title; **8–13** & **18–25** at Modena.) Ser. 2, Modena **1–2**, 1862–1866. Ser. 3, ?**3–23**, 1879–1930. (**3–9** at Napoli; **10–22** at Roma; **23** at Modena.) Revived, **24→** in 1938→ , with changes of title after **26**.

Mem. Mus. Ci. Nat. Barcelona (Bot.)
Memòries del Museu de Ciències naturals de Barcelona. Sèrie botànica. / Memorias del Museo de Ciencias naturales de Barcelona. Serie botanica. Barcelona. **1**, 1922–1925. (**1**(1) in 1922; **1**(2) in 1924; **1**(3) in 1925.)

Mém. Mus. Hist. Nat. (Paris)
Mémoires du Muséum d'Histoire naturelle. Paris. **1–20**, 1815–1832.

Mém. Mus. Nat. Hist. Nat. (Paris)
Mémoires du Muséum national d'Histoire naturelle. Série B, Botanique. Paris. 1950→ .

Mém. Sav. Étr. Pétersb.
Mémoires des Savants étrangers. Mémoires présentés à l'Académie impériale des Sciences de St.-Pétersbourg pour divers Savants et lus dans ses Assemblées. Saint-Pétersbourg. **1–9**, 1830–1859.

Mem. Sect. Şti. (Acad. Română)
Cf. *Anal. Acad. Române.*

Mém. Soc. Acad. (Angers)
Mémoires de la Société académique de Maine et Loire. Angers. **1–36**, 1857–1881.

Mém. Soc. Émul. Doubs
Mémoires et Comptes rendus de la Société d'Émulation du Doubs (with minor variations of title). Besançon. Ser. 1, **1–3**, 1841–1849. Ser. 2, titled *Mémoires de la Société libre d'Émulation du Doubs*, **1–5**, 1850–1854; titled *Mémoires de la Société d'Émulation du Département du Doubs*, **6–8**, 1855–1857. Ser. 3, titled *Mémoires de la Société d'Émulation du Département du Doubs*, **1–7**, 1856–1864; titled *Mémoires de la Société d'Émulation du Doubs*, **8–10**, 1864–1869. Ser. 4, with same title, **1–10**, 1866–1876. Ser. 5, **1–10**, 1877–1886. Ser. 6, **1–10**, 1887–1896. Ser. 7, **1–10**, 1897–1906. Ser. 8, **1–10**, 1907–1920. Ser. 9, **1–10**, 1922–1931. Ser. 10, **1–8**, 1931–1939).

Mem. Soc. Fauna Fl. Fenn.
Memoranda Societatis pro Fauna et Flora fennica. Helsing-forsiae. 1927→ .

Mém. Soc. Fribourg. Sci. Nat.
Mémoires de la Société fribourgeoise des Sciences naturelles. | Mitteilungen der naturforschenden Gesellschaft in Freiburg (Schweiz). Botanique. Fribourg (Suisse). 1–5, 1901–1947.

Mém. Soc. Linn. Paris
Mémoires de la Société linnéenne de Paris. 1–6, 1822–1828.

Mém. Soc. Nat. Moscou
Mémoires de la Société impériale des Naturalistes de l'Université de Moscou. Moscou. 1–6, 1806–1823.

Mém. Soc. Phys. Hist. Nat. Genève
Mémoires de la Société de Physique et d'Histoire naturelle de Genève. Genève. 1821→ .

Mém. Soc. Roy. Méd. (Paris)
Mémoires de Médecine et de Physique médicale de la Société royale de Médecine. Paris. 1–10, 1779–1798.

Mém. Soc. Sci. Bohême
Cf. *Sitz.-Ber. Böhm. Ges. Wiss.*

Mém. Soc. Sci. Nat. Maroc
Mémoires de la Société des Sciences naturelles du Maroc. Rabat, Paris & Londres. Ser. 1, 1–50, 1921–1952. Nov. ser., titled Mémoires de la Société des Sciences naturelles et physiques du Maroc. Botanique. Rabat, 1→ , 1960→ .

Mém. Soc. Vaud. Sci. Nat.
Mémoires de la Société vaudoise des Sciences naturelles. Lausanne. 1922→ .

Mem. Torrey Bot. Club
Memoirs of the Torrey botanical Club. New York. 1889→ . (18–19, 1931–1941, at Menasha, Wisconsin; 20, 1943–1954, at Lancaster, Pennsylvania; 21→ , 1958→ at Durham, North Carolina.)

Mitt. Bot. Staatssamm. (München)
Mitteilungen aus der botanischen Staatssammlung. München. 1950→ .

Mitt. Deutsch. Dendrol. Ges.
Mitteilungen der deutschen dendrologischen Gesellschaft. Berlin. 1893→ . (1894–1912 at Poppelsdorf-Bonn; 1913–1932 at Thyrow; 1933 at Berlin; 1934 at Berlin & Dortmund; 1935–1942 at Dortmund; 1950→ at Darmstadt.)

Mitt. Kgl. Ungar. Gartenb.-Lehranst.
A m. kir. Kertészeti Tanintézet Közleményei. | Mitteilungen der kgl. ungarischen Gartenbau-Lehranstalt. | Bulletin de l'École royale hongroise d'Horticulture. | Bulletin of the Royal Hungarian Horticultural College. Budapest. Ser. 1, 1–5, 1935–1939. Ser. 2, titled A m. Kertészeti Akadémia Közleményei, with the above alternative titles unaltered, 1–5, 1940–1947 (also called ser. 1, 7–11). Continued as *Ann. Sect. Horti-Viticult. Univ. Sci. Agr. (Budapest)*, Agrártudományi Egyetem Kert- és Szőlő-gazdaságtudományi Karának Évkönyve. | Annales Sectionis Horti- et Viticulturae Universitatis Scientiae agriculturae, 12–16, 1948–1952; titled *Ann. Acad. Horti-Viticult. (Budapest)*, A Kertészeti és Szőlészeti Főiskola Évkönyve. | Annales Academiae Horti- et Viticulturae, 17→ , 1953→ .

Mitt. Naturw. Ver. Steierm.
Mitteilungen des naturwissenschaftlichen Vereines fur Steiermark. Graz. 1863→ .

Mitt. Thür. Bot. Ges.
Mitteilungen der thüringischen botanischen Gesellschaft. Weimar 1–2, 1949–1960.

Mitt. Thür. Bot. Ver.
Mitteilungen des thüringischen botanischen Vereins. Weimar. Ser. 1, 1–9, 1882–1890. Nov. ser., 1–51, 1891–1944.

Monde Pl.
Le Monde des Plantes. Revue mensuelle de Botanique. Organe de l'Académie internationale de Géographie botanique. Le Mans. 1–8: p. 56, 1891–1898. Continued as Le Monde des Plantes, 1→ , 1899→ (later published elsewhere). Also continued as *Bull. Acad. Int. Géogr. Bot. (Le Mans)*, Bulletin de l'Académie internationale de Géographie botanique, 8: p. 47 [57]–19, 1899–1910; titled Bulletin de Géographie botanique. Organe mensuel de l'Académie internationale de Botanique, 21–27, 1911–1919. (16–20 at Paris.)

Monit. Jard. Bot. Tiflis
Вѣстникъ Тифлисскаго ботаническаго Сада [Věstnik Tiflis-skago botaničeskago Sada]. | Moniteur du Jardin botanique de Tiflis. Tiflis. Ser. 1, 1–32, 1905–1914. Nov. ser., 1–5, 1923–1931.

Monogr. Bot. (Warszawa)
Monographiae botanicae. Warszawa. 1953→ .

Naturaliste (Paris)
Le Naturaliste. Journal des Échanges et des Nouvelles. Paris. 1879–1910.

Nederl. Kruidk. Arch.
Nederlandsch kruidkundig Archief. Leiden. Ser. 1, 1–5, 1846–1870. Ser. 2, 1–6, 1871–1895. Ser. 3, 1–2, 1896–1904. Years are then used as vol. nos., 1904–1932, 1904–1932. Vol. nos. then revert to ser. 1, 43–57, 1933–1951. (Ser. 2, ser. 3 & 1904–1913(1) at Nijmegen; 1913(2)–1919 at Groningen; 1920–1921 at Utrecht; 1922–1958 at Amsterdam.)

Neue Denkschr. Schweiz. Ges. Naturw.
Neue Denkschriften der allgemeinen schweizerischen Gesellschaft für die gesammten Naturwissenschaften. | Nouveaux Mémoires de la Société helvétique des Sciences naturelles. Neuchatel. 1–40, 1837–1906; titled Neue Denkschriften der schweizerischen naturforschenden Gesellschaft, same French title, 41–54, 1906–1918; titled Denkschriften der schweizerischen naturforschenden Gessellschaft. | Mémoires de la Société helvétique des Sciences naturelles, 55→ , 1920→ .

New Phytol.
The new Phytologist. London. 1902→ . (29–54, 1930–1955, at Cambridge; 55→ , 1956→ , at Oxford.)

Not. Syst. (Leningrad)
Ботанические Материалы Гербария главного ботаничес-кого Сада Р.С.Ф.С.Р. [Botaničeskie Materialy Gerbarija glavnogo botaničeskogo Sada R.S.F.S.R.]. | Notulae systematicae ex Herbario Horti botanici petropolitani. Mosqua & Leningrad. 1–4, 1919–1923; titled the same in Russian and Notulae systematicae ex Herbario Horti botanici Reipublicae rossicae, 5, 1924; titled Ботанические Материалы Гербария главного ботани-ческого Сада СССР [Botaničeskie Materialy Gerbarija glavnogo botaničeskogo Sada SSSR]. | Notulae systematicae ex Herbario Horti botanici URSS, 6, 1926; titled Ботанические Материалы Гербария ботанического Института Академии Наук СССР [Botaničeskie Materialy Gerbarija botaničeskogo Instituta Akademii Nauk SSSR]. | Notulae systematicae ex Herbario Instituti botanici Akademiae Scientiarum URSS, 7–8(3), 1937–1938; titled Ботанические Материалы Гербария Инсти-тута имени В. Л. Комарова Академии Наук СССР [Botaničeskie Materialy Gerbarija botaničeskogo Instituta imeni V. L. Komarova Akademii Nauk SSSR]. | Notulae systematicae ex Herbario Instituti botanici nomine V. L. Koma-rovii Academiae Scientiarum U.R.S.S., 8(4)–22, 1940–1963.

Notes Roy. Bot. Gard. Edinb.
Notes from the Royal Botanic Garden, Edinburgh. Edinburgh. 1900→ .

Notizbl. Bot. Gart. Berlin
Notizblatt des königl. botanischen Gartens und Museums zu Berlin. Leipzig. 1–15, 1895–1944; with minor changes of title, becoming Notizblatt des botanischen Gartens und Museums zu Berlin-Dahlem from 1920 to 1944. (6–7, 1913–1920, at Leipzig & Berlin; 8–15, 1921–1944, at Berlin-Dahlem.)

Nov. Bot. Horti Bot. Univ. Carol. Prag.
  Novitates botanicae et Delectus Seminum, Fructuum, Spor-
  arumque anno... collectorum, quae Praefectus Horti botanici
  Universitatis carolinae pragensis libentissime pro mutua
  Commutatione offert. Praga. 1960→ . Various minor changes
  of title.

Nov. Syst. Pl. Vasc. (Leningrad)
  Новости систематики высших Растений [*Novosti sistematiki*
  vysšikh Rastenij]. / *Novitates systematicae Plantarum vascu-*
  larium. Moskva & Leningrad. 1964→ .

Növ. Közl.
  Növénytani közlemények. Budapest. **1–7**, 1902–1908; titled
  Bot. Közl., *Botanikai közlemények*, **8**→ , 1909→ .

Nova Acta Acad. Leop.-Carol.
  Nova Acta (Leopoldina) Academiae Caesareae Leopoldino-
  Carolinae germanicae Naturae curiosorum. Norimbergae; later
  volumes elsewhere. Ser. 1, 1757–1928. Nov. ser., 1934→ . Titled
  also in German (in some volumes only in German) *Verhand-*
  lungen der kaiserlichen Leopoldino-Carolinischen deutschen
  Akademie der Naturforscher. The name of the Academy has
  varied greatly; any of the adjectives *Leopoldinisch-Carolinische*,
  kaiserliche and *deutsche* may or may not appear.

Nova Acta Acad. Sci. Petrop.
  Nova Acta Academiae Scientiarum imperialis petropolitana.
  Petropoli. **1–15**, 1787–1806.

Novi Comment. Acad. Sci. Petrop.
  Novi Commentarii Academiae Scientiarum imperialis petro-
  politanae. Petropoli. **1–20**, 1750–1776.

Nuovo Gior. Bot. Ital.
  Nuovo Giornale botanico italiano. Firenze. Ser. 1, **1–25**,
  1869–1893. Nov. ser., **1–68**, 1894–1961; titled *Giornale botanico*
  italiano, **69**→ , 1962→ .

Nuovo Racc. Opusc. Aut. Sic.
  Nuovo Raccolta di Opuscoli di Autori siciliani. Palermo.
  1788–?.

Nyt Mag. Naturvid. (Christiania)
  Nyt Magazin for Naturvidenskaberne. Christiania (Oslo). **1–74**,
  1838–1934; titled *Nytt Magasin for Naturvidenskaperne*,
  75–88, 1936–1951.

Ochr. Przyr.
  Ochrona Przyrody. Kraków. 1920→ .

Op. Bot. (Lund)
  Opera botanica. A Societate botanica lundensi in Supplementum
  Seriei 'Botaniska Notiser' *edita.* Stockholm. 1953→ . (**13**→ ,
  1967→ , at Lund.)

Österr. Bot. Wochenbl.
  Österreichisches botanisches Wochenblatt. Wien. **1–7**, 1851–
  1857; titled *Österr. Bot. Zeitschr.*, *Österreichische botanische*
  Zeitschrift, **8**→ , 1858→ . (**92–93**, 1943–1944, titled *Wiener*
  botanische Zeitschrift.)

Österr. Bot. Zeitschr.
  Cf. *Österr. Bot. Wochenbl.*

Overs. Kong. Danske Vid. Selsk. Forh.
  Oversigt over det kongelige danske Videnskabernes Selskabs
  Forhandlinger. Kiöbenhavn. 1806→ .

Period. Soc. Med. Cádiz
  Periodico de la Sociedad médico-quirúrgica de Cádiz. Cádiz.
  1–4, 1820–1824.

Pflanzenareale
  Die Pflanzenareale. Sammlung kartographischer Darstellungen
  von Verbreitungsbezirken der lebenden und fossilen Pflanzen-
  Familien, -Gattungen und -Arten. Jena. 1926→ .

Phytologist (Newman)
  The Phytologist: a popular botanical Miscellany. Conducted by
  George Luxford. London. Ser. 1, **1** & **2**: pp. 1–372, 1841–1845;
  ...Conducted by Edward Newman, **2**: pp. 373–end, **3** & **4**,
  1846–1853. Ser. 2, **1–6**, 1854–1863.

Phyton (Austria)
  Phyton. Annales Rei botanicae. Horn, Austria. 1948→ .

Planta
  Planta. Archiv für wissenschaftliche Botanik. Berlin. 1925→ .
  (**64**→ , 1965→ , titled simply *Planta.*)

Pollichia
  Jahresbericht der Pollichia, eines naturwissenschaftlichen
  Vereins der bayerischen Pfalz. Landau. **1–32**, 1843–1874.
  (**15–32**, 1857–1874, titled ...*der Rheinpfalz.*)

Predv. Otčet Rabot. Nižegorod. Geobot. Eksped.
  Предварительный Отчёт о Работах Нижегородской геобо-
  танической Экспедиции [*Predvaritel'nyj Otčet o Rabotakh*
  Nižegorodskoj geobotaničeskoj Ekspedicii]. Nižni-Novgorod.
  1925–1928, 1926–1929.

Preslia
  Preslia. Věstník české botanické Společnosti. Praha. **1**, 1914;
  titled *Preslia. Věstník československé botanické Společnosti.* /
  Bulletin de la Société botanique tchécoslovaque à Prague. /
  Reports of the Czechoslovak botanical Society of Prague,
  2–15, 1923–1936; titled *Preslia. Věstník čs. botanické Společ-*
  nosti. / *Bulletin de la Société botanique tchéque à Prague.* /
  Reports of the Czech botanical Society of Prague, **16–17**, 1939;
  titled *Preslia. Věstník české botaničké Společnosti*, **18–21**,
  1940–1942; titled *Preslia. Věstník československé botanické*
  Společnosti v Praze, **22–23**, 1948; titled *Preslia. Časopis*
  československé botanické Společnosti, **24**→ , 1952→ .

Proc. Amer. Acad. Arts Sci.
  Proceedings of the American Academy of Arts and Sciences.
  Boston. **1–85**, 1846–1958.

Proc. Bot. Soc. Brit. Is.
  Proceedings of the botanical Society of the British Isles. London.
  1954→ .

Proc. California Acad. Sci.
  Proceedings of the California Academy of natural Sciences.
  San Francisco. Ser. 1, **1–3**, 1854–1868; titled *Proceedings of*
  the California Academy of Sciences, **4–7**, 1869–1877. Ser. 2,
  1–6, 1888–1896. Ser. 3, *Botany*, **1–2**, 1897–1904. Ser. 4, **1**→ ,
  1907→ .

Proc. Linn. Soc. London
  Proceedings of the Linnean Society of London. London.
  1838→ .

Proc. Linn. Soc. New S. Wales
  The Proceedings of the Linnean Society of New South Wales.
  Sydney. Ser. 1, **1–10**, 1875–1885. Ser. 2, **1–10**, 1886–1895.
  Ser. 1, **21**→ , 1896→ .

Proc. Nat. Acad. Sci. U.S.
  Proceedings. National Academy of Sciences. Washington. **1**,
  1877; titled *Proceedings of the National Academy of Sciences*,
  Baltimore, **1**, 1915; titled *Proceedings of the National Academy*
  of Sciences of the United States of America, **2**→ , 1916→ .
  Later vols. published at Easton, Chicago and Washington.

Prosv. Glasn.
  Просветни Гласник [*Prosvetni Glasnik*]. Beograd. 1880–1928.

Publ. Fac. Sci. Univ. Masaryk
  Publications de la Faculté des Sciences de l'Université Masaryk. /
  Spisy vydávané přírodovědeckou Fakultou Masarykovy Univer-
  sity. Brno. 1921–1958. (1950→ with additional title Труды
  Естественно-исторического Факультета Университета им.
  Т. Г. Масарика [*Trudy Estestvenno-istoričeskogo Fakul'teta*
  Universiteta im. T.G. Masarika].)

Publ. Field Mus. Bot. (Chicago)
  Publications of the Field Columbian Museum. Botanical Series.
  Chicago. **1**, 1895–1902; titled *Publications of the Field Museum*
  of natural History. Botanical Series. **2–8**, 1903–1932; titled
  Publications. Field Museum of natural History. Botanical
  Series, **9–23**, 1941–1947; titled *Fieldiana*, *Fieldiana: Botany*,
  24→ , 1958→ .

Publ. Inst. Biol. Apl. (Barcelona)
Publicaciones del Instituto de Biología aplicada. Barcelona. 1946→ .
Rad Jug. Akad. Znan. Umj.
Rad jugoslavenske Akademije Znanosti i Umjetnosti. Zagreb. 1867→ . (270–272, 1941, titled Rad hrvatske Akademije Znanosti i Umjetnosti).
Redog. Allm. Lärov. Norr.- Söderköping
Redogörelse för Allmänna Läroverken i Norrköping och Söderköping. Norrköping.
Ref. Nauč.-Issled. Rabot. Akad. Nauk SSSR (Biol.)
Рефераты научно-исследовательских Работ. Академия Наук СССР. Биологический Отдел [Referaty naučno-issledovatel'skikh Rabot. Akademija Nauk SSSR. Biologičeskij Otdel]. Leningrad. 1945→ .
Reinwardtia
Reinwardtia. Bogor. 1950→ .
Rendic. Accad. Sci. (Napoli)
Rendiconto delle Adunanze e de' Lavori dell'Accademia delle Scienze. Sezione della Società reale borbonica di Napoli. Napoli. Ser. 1, 1–9, 1842–1850. Ser. 2, with minor changes of title, 1–5, 1852–1856. Ser. 3, with minor changes of title, 1859 & 1(1861), 1860–1861.
Rep. Bot. Exch. Club Brit. Is.
Report of the botanical Society and Exchange Club of the British Isles. Manchester. 1–13, 1867–1947. (2(1903)–3(4), 1904–1913, at Oxford; 3(5)–13, 1913–1947, at Arbroath.)
Result. Wiss. Erforsch. Balaton.
Resultate der wissenschaftlichen Erforschung des Balatonsees. Wien. 1897–1920. (Comprises German edition of A Balaton tudományos Tanulmányozásának Eredményei, Budapest, 1897–1920.)
Rev. Bot. Appl.
Revue de Botanique appliquée et d'Agriculture coloniale. Paris. 1–8, 1921–1928; titled Revue de Botanique appliquée et d'Agriculture tropicale, 9–25, 1929–1945; titled Revue internationale de Botanique appliquée et d'Agriculture tropicale, 26–33, 1946–1953.
Rev. Hort. (Paris)
Revue horticole. Paris 1, 1829–1832. Ser. 1, 1–3, 1832–1841. Ser. 2, 1–5, 1841–1846. Ser. 3, 1–5, 1847–1851. Ser. 4, 1–5, 1852–1856. Then 1857–1865, 1857–1865. Then 37→ , 1866→ .
Revista Fac. Ci. (Lisboa)
Revista da Faculdade de Ciências. 2ª Série. C. Ciências naturais. Lisboa. 1950→ .
Revista Olot
Revista d'Olot. 1926→ .
Rhodora
Rhodora. Journal of the New England botanical Club. Boston. 1899→ .
Sched. Fl. Hung. Exsicc.
Jegyzék Magyarország Növényeinek Gyüjteményéhez, kiadja a Magyar Nemzeti Múzeum Növénytani Osztálya. / Schedae ad Floram hungaricam exsiccatam a Sectione botanica Musei nationalis hungarici editam. Budapest. 1912–1932. (1 in 1912; 2 in 1914; 3 in 1914; 4 in 1916; 5 in 1919; 6 in 1923; 7 in 1925; 8 in 1927; 9 in 1932; 10 in 1932.)
Sched. Herb. Fl. Ross.
Schedae ad Herbarium Florae rossicae, a Sectione botanica Societatis imp. petropolitanae Naturae Curiosorum editum. / Список Растений Гербария Русской Флоры [Spisok Rastenij Gerbarija Russkoj Flory]. Peterburg. 1–8, 1898–1922. (1 in 1898; 2 in 1900; 3 in 1901; 4 in 1902; 5 in 1905; 6 in 1908; 7 in 1911; 8 in 1922.) Titled Sched. Herb. Fl. URSS, Список Растений Гербария Флоры СССР [Spisok Rastenij Gerbarija Flory SSSR]. / Schedae ad Herbarium Florae URSS, 9→ , 1932→ . (9 in 1932; 10 in 1936; 11 in 1949; 12 in 1953; 13 in 1955; 14 in 1957; 15 in 1963; 16 in 1966; 17 in 1967.)

Sched. Herb. Fl. URSS.
Cf. Sched. Herb. Fl. Ross.
Sci. Pap. Appl. Sect. Tiflis Bot. Gard.
Записки научно-прикладныхъ Отдѣловъ Тифлисскаго ботаническаго Сада [Zapiski naučno-prikladnykh Otdělov Tiflisskago botaničeskago Sada]. / Scientific Papers of the applied Sections of the Tiflis botanical Garden. Tiflis. 1–7, 1919–1930.
Scrin. Fl. Select. (Magnier)
Scrinia Florae selectae (Directeur: Charles Magnier.) Saint-Quentin. 1–15, 1882–1896. (1: pp. 1–48 in 1882; 2: pp. 49–56 in 1883; 3: pp. 57–72 in 1884; 4: pp. 73–88 in 1885; 5: pp. 89–104 in 1886; 6: pp. 105–120 in 1887; 7: pp. 121–136 in 1888; 8: pp. 137–156 in 1889; 9: pp. 157–176 in 1890; 10: pp. 177–196 in 1891; pp. 197–228 in ?1891; 11: pp. 229–262 in 1892; 12: pp. 263–298 in 1893; 13: pp. 299–336 in 1894; 14: pp. 337–364 in 1895; 15: pp. 365–383 in 1896.)
Scripta Acad. Hierosol.
Scripta academica hierosolymitana. Jerusalem. 1938→ .
Scripta Bot. Mus. Transs.
Scripta botanica Musei transsilvanici. Kolozsvár. 1–3, 1942–1944.
Sitz.-Ber. Akad. Wiss. Wien
Sitzungsberichte der kaiserlichen Akademie der Wissenschaften. Mathematisch-naturwissenschaftliche Classe. / Sitzungsberichte der mathematisch-naturwissenschaftlichen Classe der kaiserlichen Akademie der Wissenschaften. Wien. 1–123, 1848–1914; titled Sitzungsberichte. Kaiserliche Akademie der Wissenschaften in Wien. Mathematisch-naturwissenschaftliche Klasse, 124–155, 1915–1947 (Kaiserliche dropped from 127, 1918, onwards); titled Sitzungsberichte. Österreichische Akademie der Wissenschaften. Mathematisch-naturwissenschaftliche Klasse, 156→ , 1947→ .
Sitz.-Ber. Böhm. Ges. Wiss.
Sitzungsberichte der königl. böhmischen Gesellschaft der Wissenschaften in Prag. Prag. 1859–1873; titled Zprávy o Zasedání královské české Společnosti Nauk v Praze. / Sitzungsberichte der königl. böhmischen Gesellschaft der Wissenschaften in Prag, 1874–1885; titled Zprávy o Zasedání královské české Společnosti Nauk. Třída mathematicko-přírodovědecká. / Sitzungsberichte der königl. böhmischen Gesellschaft der Wissenschaften. Mathematisch-naturwissenschaftliche Classe, 1886; titled Věstník královské české Společnosti Nauk. Třída mathematicko-přírodovědecká. / Sitzungsberichte der königl. böhmischen Gesellschaft der Wissenschaften. Mathematisch-naturwissenschaftliche Classe, 1887–1915; titled Mém. Soc. Sci. Bohême, Věstník královské české Společnosti Nauk. Třída mathematicko-přírodovědecká. / Mémoires de la Société royale des Sciences de Bohême. Classe des Sciences, 1919–1935; titled Věstník královské české Společnosti Nauk. Třída mathematicko-přírodovědecká. / Mémoires de la Société royale des Lettres et des Sciences de Bohême. Classe des Sciences, 1936–1953. (1940–1944 lacks French title.)
Skr. Kiøbenhavnske Selsk. Laerd. Vid.
Skrifter som udi det Kiøbenhavnske Selskab af Laerdoms og Videnskabers Elskere. København. 1–12, 1745–1779. (11 & 12 titled Skrifter som udi det kongelige Videnskabers Selskab.) Continued as Kong. Danske Vid. Selsk. Skr., Nye Samling af det kongelige danske Videnskabers Selskabs Skrifter. Kiøbenhavn. 1–5, 1781–1799. Continued as Det kongelige danske Videnskabers-Selskabs Skrivter, Ser. 3, 1–7, 1800–1818. Several later series.
Skr. Vid.-Selsk. Kristiania
Skrifter udgivne af Videnskabsselskabet i Christiania. Mathematisk-naturvidenskabelig Klasse. Kristiania. 1894–1925. (With minor changes of title, becoming Skrifter utgit av Videnskapsselskapet i Kristiania, 1911–1925.)

Sovetsk. Bot.
Советская Ботаника [*Sovetskaja Botanika*]. Moskva & Leningrad. 1933–1946.

Specch. Sci.
Specchio delle Science, o Giornale enciclopedico di Sicilia. Palermo. **1–2**, 1814.

Spis. Bălg. Akad. Nauk.
Списание на Българската Академия на Науките [*Spisanie na Bălgarskata Akademija na Naukite*]. Sofija. 1911 → .

Spraw. Kom. Fizyogr. Krakow.
Sprawozdania Komisyi fizyograficznej c. k. Towarsystwa naukowego krakowskiego. Kraków. **1–6**, 1867–1872; titled *Akademia Umiejetnósii w Krakowie. Sprawozdania Komisye fizyografiznej*, **7–52**, 1873–1918; titled *Polska Akademia Umiejetnósii. Sprawozdania Komisji fizjograficznej*, **53–73**, 1920–1939.

Svensk Bot. Tidskr.
Svensk botanisk Tidskrift. Stockholm. 1907 → . (**16** → , 1922 → , at Uppsala.)

Syll. Pl. Nov. Ratisbon. (Königl. Baier. Bot. Ges.)
Sylloge Plantarum novarum itemque minus cognitarum a praestantissimis Botanicis adhuc viventibus collecta et a Societate regia botanica ratisbonensi edita. (Königlich-baierische botanische Gesellschaft in Regensburg.) Ratisbonae. **1–2**, 1824–1828. (**1** in 1824; **2** in 1828.)

Taxon
Taxon. Official News Bulletin of the International Association for Plant Taxonomy. Utrecht. 1951 → .

Term.-Tud. Közl.
Természettudományi Közlöny. Budapest. 1869 → . *Pótfüz., Pótfüzetek a Természettudományi Közlönyhoz.* **1–88**, 1888–1907.

Trans. Camb. Philos. Soc.
Transactions of the Cambridge philosophical Society. Cambridge. **1–23**, 1820–1928.

Trans. Linn. Soc. London
Transactions of the Linnean Society. London. Ser. 1, **1–6**, 1791–1802; titled *Transactions of the Linnean Society of London*, **7–30**, 1804–1875. Ser. 2, *Botany*, **1** → , 1875 → . From 1939 *Botany* and *Zoology* are combined.

Trans. Roy. Soc. Edinb.
Transactions of the Royal Society of Edinburgh. Edinburgh. 1783 → .

Trav. Assoc. Fr. Avanc. Sci.
Travaux de l'Association française pour l'Avancement des Sciences. Paris.

Trav. Inst. Sci. Chérif.
Travaux de l'Institut scientifique chérifien. Série Botanique. Tanger. 1952 → . (1956 → at Rabat.)

Trav. Mus. Bot. Acad. Pétersb.
Travaux du Musée botanique de l'Académie impériale des Sciences de St. Pétersbourg /de Russie). St. Pétersbourg. 1902–1932.

Trudy Bot. Sada Jur'ev.
Труды ботаническаго Сада императорскаго Юрьевскаго Университета [*Trudy botaničeskago Sada imperatorskago Jur'evskago Universiteta*]. / *Acta Horti botanici Universitatis imperialis jurjevensis.* Jur'ev. **1–15**, 1900–1914.

U.S. Dept. Agric. Bur. Pl. Ind. Bull.
United States Department of Agriculture. Bureau of Plant Industry. Bulletin. Washington. 1901–1913.

U.S. Dept. Agric. Techn. Bull.
United States Department of Agriculture. Technical Bulletin. Washington. 1927 → .

Ukr. Bot. Žur.
Українськиї ботанічний Журнал [*Ukrajinskyji botaničnyji Žurnal*]. / *The Ukrainian botanical Review.* Kiev. **1–5**, 1922–1929. Continued as ***Žur. Inst. Bot. URSR***, Журнал Інституту

ботаніки АН УРСР [*Žurnal instytutu botaniky AN URSR*]. / *Journal de l'Institut botanique de l'Académie des Sciences de la RSS d'Ukraine.* **1–23**, 1934–1940. Continued as ***Ukr. Bot. Žur.***, Ботанічний Журнал [*Botaničnyj Žurnal*]. / *Journal botanique de l'Académie des Sciences de la RSS d'Ukraine.* **1–12**, 1940–1955. Titled Український ботанічний Журнал [*Ukrajinskyji botaničnyji Žurnal*], **13** → , 1956 → .

Uppsala Univ. Årsskr.
Uppsala Universitets Årsskrift. Uppsala. 1861 → .

Varied. Ci. Lit. Artes (Madrid)
Variedades de Ciencias, Literatura y Artes. Madrid. **1–2**, 1803–1805.

Verh. Bot. Ver. Brandenb.
Verhandlungen des botanischen Vereins der Provinz Brandenburg. Berlin. 1859 → .

Verh. Naturf. Ver. Brünn
Verhandlungen des naturforschenden Vereines in Brünn. Brünn. **1–75**, 1862–1944.

Verh. Ver. Beförd. Gartenb. Preuss.
Verhandlungen des Vereins zur Beförderung des Gartenbaues in den königlich preussischen Staaten. Berlin. Ser. **1**, **1–20**, 1824–1853. Nov. ser., **1–7**, 1853–1860.

Verh. Zool.-Bot. Ges. Wien
Verhandlungen der k.k. zoologisch-botanischen Verein in Wien. Wien. 1852 → . From 1858, *Verein* is replaced by *Gesellschaft*.

Vid. Meddel. Dansk Naturh. Foren. Kjøbenhavn
Videnskabelige Meddelelser fra den naturhistoriske Forening i Kjøbenhavn. Kjøbenhavn. **1–63**, 1849–1912; titled *Videnskabelige Meddelelser fra dansk naturhistorisk Forening i Kjøbenhavn*, **64** → , 1913 → . Years used as vol. nos. until 1912.

Viert. Naturf. Ges. Zürich
Vierteljahrsschrift der naturforschenden Gesellschaft in Zürich. Zürich. 1856 → .

Vigne Amér. Viticult. Eur.
La Vigne américaine et la Viticulture en Europe. Revue publiée par V. Pulliat et J. E. Robin sous la Direction de J. E. Planchon. Paris. **1–6**, 1877–1882.

Vorl. Churpf. Phys.-Ökon. Ges.
Vorlesungen der kurpfälzischen /churpfalzischen physikalisch-ökonomischen Gesellschaft. Mannheim. **1–5**, 1785–1791.

Watsonia
Watsonia. Journal of the botanical Society of the British Isles. Arbroath. 1949 → . (**3** → , 1953 → , at London.)

Webbia
Webbia. Raccolta di Scritti botanici. Firenze. 1905 → .

Wiss. Zeitschr. Pädag. Hochsch. Potsdam
Wissenschaftliche Zeitschrift der pädogogischen Hochschule Potsdam. Potsdam. 1952 → .

Wiss. Zeitschr. Univ. Halle
Wissenschaftliche Zeitschrift der Martin-Luther-Universität Halle-Wittenberg. Mathematisch-naturwissenschaftliche Reihe. Halle. 1951 → .

Wochenschr. Gärtn. Pflanzenk.
Wochenschrift fur Gärtenerei und Pflanzenkunde. Berlin. **1–3**, 1858–1860; titled *Wochenschrift des Vereines zur Beförderung des Gartenbaues in den königlich preussischen Staaten für Gärtnerei und Pflanzenkunde*, **4–15**, 1861–1872.

Zeitschr. Acker-Pflanzenbau
Zeitschrift für Acker- und Pflanzenbau. Berlin. 1949 → .

Zeitschr. Naturw. Abt. Deutsch. Ges. Wiss. Posen
Zeitschrift der naturwissenschaftlichen Abteilung der deutschen Gesellschaft für Kunst und Wissenschaft. Posen.

Züchter
Der Züchter. Internationale Zeitschrift für theoretische und angewandte Genetik. Berlin, Heidelberg & New York. 1929 → .

Žur. Inst. Bot. URSR
Cf. *Ukr. Bot. Žur.*

APPENDIX IV

GLOSSARY OF TECHNICAL TERMS

The number of technical terms used in *Flora Europaea* has been kept as low as is consistent with a reasonable standard of accuracy and brevity. Most of them are used in well-established traditional senses, and their meanings may be ascertained by reference to glossaries such as H. I. Featherly, *Taxonomic Terminology of the Higher Plants* (Ames, Iowa, U.S.A., 1954). No term is used in a sense inconsistent with that given by Featherly.

Experience has shown, however, that some useful terms are liable to misinterpretation, and others, which can be used in a wider sense, are used in a restricted sense in *Flora Europaea*. This glossary is intended simply to indicate without ambiguity the sense in which these potentially ambiguous terms are employed.

Certain technical terms, which are restricted to descriptions in particular families or genera, are explained under the family or genus concerned.

ABOVE Used to indicate both the upper surface of a normally horizontal organ and the upper part of an organ or of the whole plant.

ALTERNATE Arising singly at a node; includes regularly spiral, as well as distichous arrangements.

ANNUAL Completing its life-cycle from seed to seed in less than 12 months; includes 'overwintering' annuals, which germinate in autumn and flower the following year.

BELOW Used to indicate the basal part of a plant, stem or inflorescence; cf. *beneath*.

BENEATH Used to indicate the lower surface of a normally horizontal organ; cf. *below*.

BIDENTATE With two teeth.

BISERRATE Serrate, with the teeth themselves serrate.

CADUCOUS Falling unusually early.

CILIATE With hairs on the margin.

DECIDUOUS Of leaves: falling in autumn; of other organs: falling before the majority of adjacent or associated organs.

ERECTO-PATENT Diverging at an angle of 15–45° from the axis on which the structure is borne.

FLOCCOSE Clothed with woolly hairs, which are disposed in tufts or tend to rub off and adhere in small masses.

GLABRESCENT Becoming glabrous with increasing age or maturity. For structures very slightly but persistently hairy the term *subglabrous* is used.

HIRSUTE Covered with long, moderately stiff and not interwoven hairs.

HISPID Covered with stiff hairs or bristles.

LANATE Covered with soft, flexuous, intertwined hairs.

PELTATE Denotes an organ on which the stalk is attached to a more or less flat surface, and not to the margin; the attachment is not, however, necessarily central.

PUBERULENT With very short hairs.

PUBESCENT With soft, short hairs.

PYRENE A small stone, consisting of one or two seeds with a hard covering, enclosed in fleshy tissue, e.g. *Crataegus*, *Cornus*.

SEMI-PATENT Between patent and appressed.

SERICEOUS With silky, appressed hairs.

SETOSE Covered with stout, rigid bristles.

SIMPLE HAIR Indicates an unbranched hair; it may or may not bear a gland.

STOCK The persistent, usually somewhat woody base of an otherwise herbaceous perennial.

STOLON A short-lived, horizontal stem, either above or below the surface of the ground, rooting at one or more nodes.

STRIGOSE With stiff, appressed, straight hairs.

TOMENTOSE With hairs compacted into a felty mass.

TUBERCULATE Covered with smooth, knob-like elevations.

VELUTINOUS With a dense indumentum of fine, soft, straight hairs.

VERRUCOSE Covered with rough, wart-like elevations.

VILLOUS Covered with long, soft, straight hairs.

APPENDIX V

VOCABULARIUM ANGLO-LATINUM

IN USUM LECTORUM LINGUAE ANGLICAE MINUS PERITORUM CONFECTUM

N.B. Plurimi termini ad descriptionem botanicam in lingua anglica usurpati aequipollentibus latinis persimiles sunt, e.g. *ovate* (ovatus), *inflorescence* (inflorescentia). Talia verba omnia sunt omissa.

above insuper, supra, super
all omnes
almost fere, paene
always semper
arable fields arva
around circum
arranged dispositus
attached affixus
awn arista
back dorsum
backward(s) retro
bank ripa
barbed pilis hamatis obsitus
bare nudus
bark cortex
basin-shaped pelviformis
beak rostrum
bearded barbatus
become fieri
below infra, sub
beneath infra, subtus
bent inflexus
berry bacca
between inter
bind colligare, firmare
bitter amarus
black niger, ater
blue caeruleus
bloom pruina
boat navicula
border margo
borne prolatus
branch ramus
breadth latitudo
bright laete
bristle seta
broad latus
bronze aeneus
brown fuscus, brunneus
bud gemma
bundle fasiculus
bushy spisse et iteratim ramosus
catkin amentum
chaffy paleaceus
chamber loculus
chequered cancellatus

chestnut castaneus
chief principalis
claw unguis
cliff rupes
climbing scandens
close propinquus, affinis
closed clausus
clothed vestitus
cluster glomerulus
coarse crassus, grossus
coast litus, ora
coat tunica
common vulgaris
completely omnino, ex toto
compound compositus
cone strobilus
corner angulus
cornfield seges
covered obtectus
cream ochroleucus, albido-flavescens
crest crista
crevice fissura
crimson kermesinus, sanguineus; ut flos *Paeoniae officinalis* coloratus
crowded confertus
cultivated cultus, sativus
curled crispus
cushion pulvinus
damp humidus
dark obscure
dead emortuus
decay dissolutio
deep profundus; intense
developed evolutus
die mori
docks navalia
downwards deorsum
downy lanuginosus
dry siccus
dull opace; impolitus
dwarf nanus
early prius, mox, praecoce
eastern orientalis
eastwards orientem versus
edge margo
edible edulis

either...or aut...aut
end pars terminalis
enlarge crescere, augere
entire integer
entirely omnino
equal aequalis, aequans
escape evadere; planta ex horto elapsa
established subspontaneus
evening vesper
evergreen sempervirens
exceeding superans
face facies
fan-shaped flabellatus
female femineus, pistillatus
feebly debiliter, perleviter
few pauci
finely subtiliter
first primus
flap valva, ligula
flat planus
flattened compressus, applanatus
flax *Linum usitatissimum*
flesh-coloured carneus, pallide et opace roseus
fleshy carnosus
floating natans
flooded inundatus
flower flos
fodder bestiarum pabulum
fold plica
following sequens
food cibus
forest silva magna
forwards porro
free liber
fringe fimbriae
fruit fructus
furnished munitus
furrow sulcus
garden hortus
glossy nitidus
golden aureus
grassy graminosus
gravelly glareosus
graze pascere
green viridis

grey cinereus
grooved canaliculatus, sulcatus
ground solum
group grex
grow crescere, habitare
hair pilum
hairy pilis munitus
half dimidium
hard durus
head caput, capitulum
heath ericetum, callunetum
hedge saepes
helmet galea
hill collis
hoary incanus
hollow fistulosus, cavus; cavum, exca-
vatio
hood cucullus
hooked uncinatus
inner interior, internus
inside intus, intra; pagina vel pars interior
introduced inquilinus, allatus
jagged argutus
jointed articulatus
juice succus
keel carina
key clavis
lake lacus
late sero
later postea
leaf folium
leafless foliis carens
leaflet foliolum
length longitudo
less minus
level altitudo, gradus
lid operculum
light clare
limestone calx
lip labium
locally hic inde
low humilis, pusillus
lower inferior
lowland campestris, planitiem incolens
main principalis
male masculus, stamineus
many multi
marbled marmoratus
marsh palus
mat stratum e ramulis procumbentibus
intertextis compositum
meadow pratum
mealy farinosus
medicinal officinalis
middle pars centralis; medius
midrib costa, folii nervus principalis
milky lacteus
mistake error
more plus, magis

most plerique, pars major
mountain mons
mouth os
much multo, multum
naked nudus
narrow angustus
native indigenus
naturalized subspontaneus
near prope
nearly paene, fere
neither...nor nec...nec
net reticulum
never numquam
nodding nutans, cernuus
none nulli
northern borealis
northwards septentrionem versus
notch incisio
nut nux
often saepe
oil oleum
old vetus, antiquus
open apertus
orange aurantiacus
ornament decus
other alius, alter
otherwise aliter
outer exterior, externus
outside extra; pagina vel pars exterior
overlapping imbricatus
pale pallidus
papery chartaceus
pasture pascuum
patch macula
peat-bog turbarium
pink roseus
pitted foveolatus
planted cultus
point acumen
pond stagnum
pool stagnum
poor egens
prickle aculeus
pricklet aculeolus
purple purpurcus
quarter pars quarta
rank ordo
rarely raro
ray radius
red ruber
related affinis
remains reliquiae
rest ceteri
rib costa
rice-field oryzetum
rich abundans
ridge carina
rind fructus cortex
ring anulus
ripe maturus

river flumen
road via
rock saxum, rupes
root radix
rosette rosula
rough asper
rounded rotundatus
rust-coloured ferrugineus
salt-marsh palus salsa
sand arena
scale squama
scanty exiguus
scar cicatrix
scarcely vix
scarlet laete et clare ruber, paullulo auran-
tiaco affectus; ut flos *Salviae splendentis*
coloratus
scattered sparsus
scented fragrans
scree clivus alpestris, saxis deorsum con-
jectis copertus
sea mare
seed semen
seldom raro
several nonnulli, complures
shady umbrosus
shallow haud profundus
shape forma
sharply acute
sheath vagina
shelter tegmen contra ventum
shingle glarea maritima vel fluviatilis
shiny nitidus
shoot caudiculus, surculus
shore litus, ora
short brevis
shoulder angulus obtusus
shrub frutex
side latus, pagina
silky sericcus
silvery argenteus
slender tenuis, gracilis
slightly leviter, paullo
slipper calceolus
slit rima, foramen longum sed angustum
slope clivus, declivitas
small parvus
smell odor
smooth laevis
snow-patch locus in montibus ubi nix sero
perdurat
soft mollis
soil solum
sometimes interdum
southern australis
southwards meridiem versus
spikelet spicula
spot punctum, macula
spreading patens, divaricatus
spring ver

spur calcar
square quadratus
stalk stipes
standard vexillum
stem caulis
stiff rigidus
stock caudex
stony lapidosus
stout crassus, robustus
straight rectus
streak linea
stripe vitta
strong robustus, validus
suddenly abrupte
summer aestas
sunk(en) immersus
surface superficies, pagina
sweet dulcis
swollen tumidus, inflatus
tall altus
taste sapor
tawny fulvus
teeth dentes
thick crassus, densus, spissus
thicket dumetum
thin tenuis
third pars tertia

timber materia; lignum ad usum hominum aptum
tinged suffusus
tip apex
tipped ad apicem munitus vel tinctus
tooth dens
top vertex
tough lentus
tree arbor
true verus
tufted in fasciculos dispositus, caespitosus
twice bis
twig ramulus, virga
twining volubilis
twisted contortus
unarmed inermis
uncertain incertus, dubius
undivided indivisus
unequal inaequalis
united conjunctus, connatus
upper superior
uppermost supremus
upwards sursum
usually plerumque
vegetable olus
veil velum
vein nervus

velvety velutinus
vessel vas
violet violaceus
wart verruca
waste incultus
weak debilis, flaccidus
well bene
western occidentalis
westwards occidentem versus
wet madidus
white albus, candidus
whorled verticillatus
wide latus
widespread late diffusus
width latitudo
wing ala
winter hiems
wiry filo ferreo similis
withered marcidus
without sine
wood silva; lignum
woody lignosus
woolly lanatus
wrinkled rugosus
yellow flavus, luteus
young juvenis

INDEX

This index is intended to serve two purposes: to enable the reader to find the page on which any plant is mentioned, and to cite and explain names relegated to synonymy which occur in 'Standard Floras', but are not in sufficiently wide currency to justify their citation in the text (see p. xvii).

Generic names adopted in *Flora Europaea* are printed in **bold-faced** type; specific and subspecific epithets adopted are printed in ordinary type. (This applies not only to numbered species and genera, but also to those mentioned incidentally in observations, or in the introductory descriptions of families or genera.) All synonyms are printed in *italic* type, and are followed by a page-reference (also in *italics*); for those not cited in the text the page number is followed by a further number or numbers in parentheses to indicate the species (and, where necessary, subspecies, genus and family) on that page to which the synonym is referable. Among these numbers roman numerals denote the family, arabic numerals in ordinary type the genus, arabic numerals in **bold-faced** type the species, and a small letter (also in **bold-faced** type) following the species number the subspecies. Thus,

Spiraea

callosa Thunb., *5* (**6**)

indicates that the name is regarded as a synonym (partial or complete) of the species on p. 5 which is numbered **6**, namely **S. japonica.** Similarly,

Millegrana

radiola (L.) Druce, 211 (LXXXVI, 2, **1**)

indicates that this name is regarded as a synonym of species **1** (**linoides**) in genus 2 (**Radiola**) of family LXXXVI (**LINACEAE**) on p. 211; because more than one family and genus are treated on the page, citation of genus and family is necessary to avoid ambiguity.

Synonyms of taxa mentioned in notes following a numbered species are indexed as being synonyms of that species. In cases where this procedure would be ambiguous or misleading, the synonym in question has been inserted in the text.

Some names of hybrids are similarly indexed with page and number references to their parent species.

All infraspecific taxa are arranged alphabetically, regardless of rank, under the species with which they are combined.

440

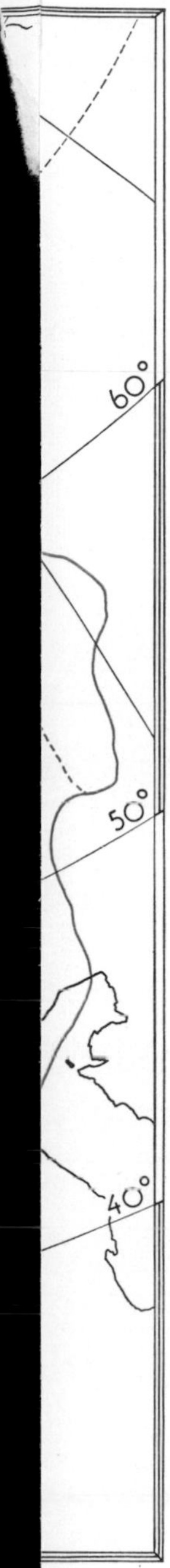

MAP I

To illustrate the boundaries of Europe for the purposes of *Flora Europaea*, and its division into 'territories' which are indicated by two-letter abbreviations after the summary of geographical distribution for each species. These abbreviations are derived from the Latin name of the territory concerned.

Al Albania
Au Austria, with Liechtenstein
Az Açores
Be Belgium, with Luxembourg
Bl Islas Baleares
Br Britain, including Orkneys, Zetland and Isle of Man; excluding Channel Islands and Northern Ireland
Bu Bulgaria
Co Corse
Cr Kriti (*Creta*), with Karpathos, Kasos and Gavdhos
Cz Czechoslovakia
Da Denmark (*Dania*), including Bornholm
Fa Færöer
Fe Finland (*Fennia*), including Ahvenanmaa (Aaland Islands)
Ga France (*Gallia*), with the Channel Islands (Îles Normandes) and Monaco; excluding Corse
Ge Germany (both eastern and western republics)
Gr Greece, excluding those islands included under Kriti (*supra*) and those which are outside Europe as defined for *Flora Europaea*
Hb Ireland (*Hibernia*); both the republic and Northern Ireland
He Switzerland (*Helvetia*)
Hs Spain (*Hispania*), with Gibraltar and Andorra; excluding Islas Baleares
Ho Netherlands (*Hollandia*)
Hu Hungary
Is Iceland (*Islandia*)
It Italy, including the Arcipelago Toscano; excluding Sardegna and Sicilia as defined *infra*
Ju Jugoslavia
Lu Portugal (*Lusitania*)
No Norway
Po Poland
Rm Romania
Rs U.S.S.R. (*Russia*). This has been subdivided as follows, using the floristic divisions of Komarov's *Flora U.R.S.S.*

Rs (N) *Northern division:* Arctic Europe, Karelo-Lapland, Dvina-Pečora
Rs (B) *Baltic division:* Estonia, Latvia, Lithuania, Kaliningradskaja Oblast'
Rs (C) *Central division:* Ladoga-Ilmen, Upper Volga, Volga-Kama, Upper Dnepr, Volga-Don, Ural
Rs (W) *South-western division:* Moldavia, Middle Dnepr, Black Sea, Upper Dnestr
Rs (K) Krym (*Crimea*)
Rs (E) *South-eastern division:* Lower Don, Lower Volga, Transvolga
White Russia falls entirely within Rs (C). Ukraine is largely in Rs (W), but partly in Rs (K), Rs (C) and Rs (E). The European part of Kazakhstan is in Rs (E)

Sa Sardegna
Sb Svalbard, comprising Spitsbergen, Björnöya (Bear Island) and Jan Mayen
Si Sicilia, with Pantelleria, Isole Pelagie, Isole Lipari and Ustica; also the Malta archipelago
Su Sweden (*Suecia*), including Öland and Gotland
Tu Turkey (European part), including Imroz

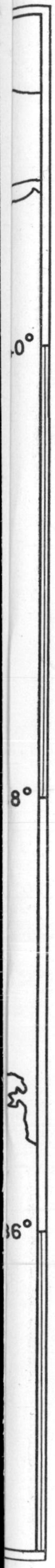

MAP II

To illustrate the boundary between Europe and Asia in the Aegean region.

The boundary is based largely on the proposals of K. H. Rechinger, 'Grundzüge der Pflanzenverbreitung in der Aegäis', *Vegetatio* **2**: 55 (1949). His northern, western and Kikladhes divisions are regarded as entirely in Europe and his eastern division as entirely in Asia; it was, however, necessary to divide his southern and north-eastern divisions.

.0°

8°

36°

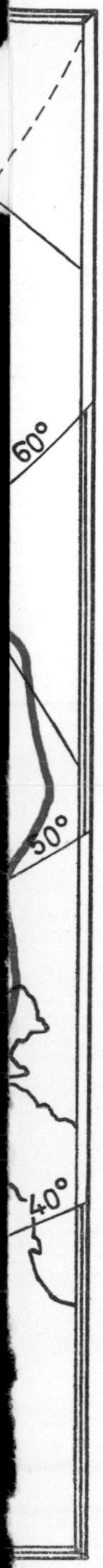

MAP IV

To illustrate the meaning to be attached to certain phrases used in summaries of geographical distribution.

W. Europe: Açores, Portugal, Spain, Islas Baleares, France, Ireland, Britain, Færöer, Iceland, S.W. Norway, Netherlands, Belgium, N.W. Germany, W. Denmark (Jylland), Corse, Sardegna, and small parts of N.W. Italy and W. Switzerland

E. Europe: N.E. Greece and the Aegean islands, Bulgaria, S. & E. Romania, Finland, U.S.S.R.

N. Europe: Svalbard, Iceland, Færöer, Ireland, Britain (excluding S. England), Denmark, Fennoscandia, U.S.S.R. north of a line running through Minsk–Tula–Penza–Orsk

S. Europe: Europe south of a line running through Bordeaux–Chambéry–Aosta–Locarno–Riva–Udine–Zagreb–Beograd–Ploesti–Odessa–Rostov–Astrakhan'.

– – – – – – – eastern boundary of *W. Europe*   – · – · – · – southern boundary of *N. Europe*

○ ○ ○ ○ ○ ○ western boundary of *E. Europe*   × × × × × northern boundary of *S. Europe*

For the definition and illustration of the meaning of S.W., N.W., S.E., N.E. and C. Europe, and of certain other geographical phrases, see map v.